Jane's

MILITARY VEHICLES AND LOGISTICS

Edited by Christopher F Foss and
Terry J Gander

Eighteenth Edition
1997-98

ISBN 0 7106 1554 X
"Jane's" is a registered trade mark

In the USA and its dependencies
Jane's Information Group Inc, 1340 Braddock Place, Suite 300, Alexandria, VA 22314-1651, USA

British Library Cataloguing-in-Publication Data.
A catalogue record for this book is available from the British Library.

Printed and bound in Great Britain by Biddles Ltd, Guildford and King's Lynn

Bv 206 S

Built for Tactical and Strategic Mobility

Bv 206 S, the armoured version of the Bv 206 All Terrain Carrier, is built for the highest level of tactical and strategic mobility in any climate.

It can also be transported by aircraft, helicopter (IAT/EAT), air-dropped, it's fully amphibious and has logistic communality with the Bv 206 All Terrain Carrier.

HÄGGLUNDS VEHICLE
INCENTIVE GROUP

Hägglunds Vehicle AB, S-891 82 Örnsköldsvik, Sweden
Telephone +46 660 805 00, Telefax +46 660 826 12

More and more, environmentally conscious fleet executives are investigating the advantages of alternative fuel vehicles. And no automotive manufacturer offers a wider selection than Ford.
Taurus Flexible-Fuel Vehicle. The same new Taurus that took the industry by storm runs on methanol (M85), unleaded gasoline alone, or any combination of the two. An ethanol (E85) model is also available.
Contour Bi-Fuel NGV. Enjoy all the comfort, sophistication and pure driving fun of Ford Contour in a clean-burning natural gas version that also runs on gasoline.
Crown Victoria NGV. It's the only dedicated natural-gas-powered original equipment manufacturer passenger car available and has the same quality, safety and comfort as the gasoline-powered Crown Victoria.
F-Series NGV. A dedicated natural gas pickup with all the durability, comfort and toughness that have kept F-Series on top of the sales charts year after year.
Econoline/Club Wagon NGV. The full-size van leader for over a decade and a half, Econoline, now available in a dedicated natural gas version, has a reputation for durability, versatility and safety.
New Ranger Electric Pickup. The electric Ranger will be offered to fleet customers in 1997 through the Ford Qualified Vehicle Modifier Program and will be the first electric vehicle with original equipment manufacturer production quality.
Ford Assurance. All Ford AFVs offer standard peace of mind. That's because they're backed by the same limited warranty as regular-production Ford cars and trucks (see your dealer for details). For more information on the quality alternative-fuel vehicle(s) of interest to you, just give us a call at 1-800-ALT-FUEL.
Nobody offers more alternatives in alternative-fuel vehicles.
Ford
"The Ford Fleet Difference"
Always There For You.

Contents

ADMINISTRATION

Publishing Director: Robert Hutchinson

Publisher: Alan Condron

Managing Editor: Peter Howard

Yearbook Database Manager: Ruth Simmance

Editorial Services Manager: Sulann Staniford

Copy Editor: Michael Durkin

EDITORIAL OFFICES

Jane's Information Group Limited, Sentinel House,
163 Brighton Road, Coulsdon, Surrey CR5 2NH, UK

Tel: +44 181 700 3700
Telex: 916907 Janes G
Fax: +44 181 700 3788
email: yearbook@janes.co.uk

SALES OFFICES

Send enquiries to International Sales Manager:
Tony Kingham (Europe, Russian Federation and Associated States (CIS), Africa, Middle East)
Jo Moon (Scandinavia, Far East, UK)
Jane's Information Group Limited, UK address as above

Tel Enquiries: +44 181 700 3759
Fax Enquiries: +44 181 763 1006
Fax Orders: +44 181 763 1005

Send USA enquiries to:
Joe McHale, Senior Vice-President Product Sales,
Jane's Information Group Inc, 1340 Braddock Place, Suite 300,
Alexandria, Virginia 22314-1651

Tel: +1 703 683 3700
Telex: 6819193
Fax: +1 703 836 0029

ADVERTISEMENT SALES OFFICES

Advertisement Sales Manager: Richard West
Jane's Information Group Limited, Sentinel House,
163 Brighton Road, Coulsdon, Surrey CR5 2NH, UK

Tel: +44 181 700 3739
Fax: +44 181 700 3744
email: richard.west@janes.co.uk

Australia: Brendan Gullifer, Havre & Gullifer (Pty) Ltd, Level 7,
82 Collins Street, Melbourne 3000

Tel: +61 3 9650 1100
Fax: +61 3 9650 6611
email: 100017.2676@compuserve.com

Benelux: Jenny Russell, Jane's Information Group (see South Africa)

Brazil: L Bilyk, Brazmedia International S/C Ltda, Alameda Gabriel Monterio da Silva, 366 CEP, 01442-900 São Paulo

Tel: +55 11 853 4133
Telex: 32836 BMED BR
Fax: +55 11 852 6485

France: Patrice Février, Jane's Information Group – France,
BP 418, 35 avenue MacMahon, F-75824 Paris Cedex 17

Tel: +33 (0) 1 45 72 33 11
Fax: +33 (0) 1 45 72 17 95
email: patrice.fevrier@wanadoo.fr

Germany and Austria: Richard West (see Advertisement Sales Manager)

Hong Kong: Jeremy Miller, Major Media Ltd, Room 1402, 14F
Capitol Centre, 5-19 Jardine's Bazaar, Causeway Bay

Tel: +852 2890 3110
Fax: +852 2576 3397

Israel: Oreet Ben-Yaacov, Oreet International Media, 15 Kinneret Street, IL-51201 Bene Berak

Tel: +972 3 570 6527
Fax: +972 3 570 6526

Italy and Switzerland: Ediconsult Internazionale Srl, Piazza Fontane Marose 3, I-16123 Genoa, Italy

Tel: +39 10 583684
Fax: +39 10 566578

Japan: Intermart/EAC Inc, 1-7 Akasaka 9-chome, Minato-Ku, Tokyo 107

Tel: +81 3 5474 7835
Fax: +81 3 5474 7837

Korea, South: Young Seoh Chinn, JES Media International, 6th Floor Donghye Building, 47-16 Myungil-Dong, Kangdong-Gu, Seoul 134-070

Tel: +82 2 481 3411
Fax: +82 2 481 3414

Russian Federation and Associated States (CIS): Richard West (see Advertisement Sales Manager)

Scandinavia: Gillian Thompson, The Falsten Partnership,
11 Chardmore Road, Stamford Hill, London N16 6JA, UK

Tel: +44 181 806 2301
Fax: + 44 181 806 8137

Singapore, Indonesia, Malaysia, Philippines, Taiwan and Thailand: Hoo Siew Sai, Major Media Singapore Pte Ltd, #06-00 Resource Building, 52 Chin Swee Road, Singapore 169875

Tel: +65 738 0122
Fax: +65 738 2108

South Africa: Jenny Russell, Jane's Information Group,
Sentinel House, 163 Brighton Road, Coulsdon,
Surrey CR5 2NH

Tel: +44 181 700 3740
Fax: +44 181 700 3744
email: jenny.russell@janes.co.uk

Spain: Jesus Moran Iglesias, Varex S.L. Modesto Lafuente 4,
28010 Madrid

Tel: +34 1 448 7622
Fax: +34 1 446 0198

UK/Rest of World: Jenny Russell (See South Africa)

USA: Advertising Production Manager/USA and Canada – Maureen Nute, Jane's Information Group Inc, 1340 Braddock Place, Suite 300, Alexandria, Virginia 22314-1651 USA

Tel: +1 703 683 3700
Fax: +1 703 836 0029
email: nute@janes.com

Northeastern USA and Canada: Global Media Services Inc, 299 Herndon Parkway, Suite 308, Herndon,
Virginia 22070 USA

Tel: +1 703 318 5054
Fax: +1 703 318 9728

Southeastern USA: Kristin Schulze, Global Media Services Inc,
5370 East Bay Drive, Suite 104, Clearwater, Florida 34624

Tel: +1 813 524 7741
Fax: +1 813 524 7562
email: kristin@globlmedia.com

Midwest and Western USA and Canada: Kimberley S. Hanson, Global Media Services Inc, 299 Herndon Parkway, Suite 308, Herndon, Virginia 22070 USA

Tel: +1 703 318 5054
Fax: +1 703 318 9728
email: kim@globlmedia.com

Administration: UK/Rest of World: Alexandra Vukic
USA and Canada: Maureen Nute

Alphabetical list of advertisers

The LEGUAN Family

LEGUAN 42 M

Wolverine (LEGUAN M1 A1 Heavy Assault Bridge) in cooperation with General Dynamics Land System/USA with 26 M LEGUAN Bridge

LEGUAN Ferry with 26 M LEGUAN Bridge and Hydraulic Ramps

LEGUAN 26 M

- Minimum of Personnel
- Minimum of Vehicles
- Minimum of Time

Convincing Technology

MAN GHH AG
Department ISB
Ginsheimer Straße 1
D-65462 Ginsheim/Gustavsburg
Phone 49 (0) 6134/55-288
Telefax 49 (0) 6134/55-383

MAN GHH

The Up-Armored HMMWV. Field Proven By The U.S. Army.

Designed to defeat 12 lb. contact-detonated land mines and 7.62 mm AP rounds, the Up-Armored HMMWV has been proven highly effective in Bosnia, Somalia and Haiti. Stealthy, highly mobile and cost-effective, the M1114 is a superior armored light tactical wheeled vehicle for today's land forces. Call us for information on the M1114 Up-Armored HMMWV or about our retrofit HMMWV armor kits for ballistic and blast protection.

For more information, call 513.874.2112 / Fax 513.881.5455 / E-mail at gturner@ogara.com

O'GARA-HESS & EISENHARDT
ARMORING COMPANY

How to use this book

This book has been laid out in a series of main sections grouped under main functional headings, each of which relates to a particular item of equipment or vehicle. The exact headings and groupings can be seen on the Contents page. All sectional entries provide the history of Development and in some cases, contract and associated details relating to the equipment or vehicle. The Description provides the technical details, including main Variants where applicable, while the Specifications table provides the main data for each entry. Other information provided in each entry relates to the current Status of the equipment or vehicle and an indication of the Manufacturer, where known.

If any particular item of equipment or vehicle is needed each heading is laid out in order of nationality. In most cases the latest items are mentioned first, but when relating to vehicles the payload weight order is followed, from the lightest up to the heaviest. If only a designation is available, the Index should be consulted.

In order to indicate where an entry has been updated, Entry tags are placed at the end of each entry as one of the following:

● ***VERIFIED*** The editors have made a detailed examination of the entry's content and checked its relevancy and accuracy for publication in the new edition to the best of their ability.

● ***UPDATED*** During the verification process significant changes have been introduced to the entry to reflect the latest position known to Jane's at the time of publication.

● ***NEW ENTRY*** Information on new equipment or vehicles appearing for the first time in this title.

In future all new pictures will be dated with the year of publication. New pictures this year are dated 1997.

Foreword

One feature of the contents of this years edition of Jane's Military Vehicles and Logistics is the degree of regeneration of existing equipments rather than the introduction of the new.

As ever, there have been innovations and novelties this year, but they have been far outweighed in longer term importance by the expanding practice of either converting existing equipments or somehow refurbishing/upgrading older equipment in order to extend their useful service lives.

This has been emphasised by the many perfectly serviceable older-generation Main Battle Tanks (MBT), for example the German Leopard 1 and US M60 series, made surplus to requirements by the introduction of newer models or their enforced redundancy caused by Treaty limitations.

The users of equipments such as armoured engineer vehicles, armoured recovery vehicles and mechanised bridgelayers can for once, look forward to the prospect of getting some measure of reasonably modern equipments based on reasonably modern chassis. This will finally allow the retirement of at least a fair proportion of the ageing and less capable equipment that they have had to utilise until now.

This process is by no means universal. For some nations it will matter little that large numbers of Leopard 1, M47 and M48 and other tanks become available for conversions to new roles.

While is it commendable that some countries are at last getting new engineer vehicles and bridgelayers it is worth noting that in some cases these will lack the armour and mobility to keep up with its highly mobile mechanised units.

The M1 MBT first entered service with the US Army in 1980 but today the key supporting armoured engineer and armoured vehicle launch bridges are still based on the M60 series MBT chassis which has now been phased out of service with the US Army.

Trials with prototypes of the Heavy Assault Bridge (HAB) and Grizzly counter obstacle vehicle, both based on a much modified M1 MBT chassis have been underway for some time and it is expected that these will enter service with the US Army early in the twenty first century.

For some countries the funds are still not available to modernise their essential supporting elements for what many fund-providers consider luxuries, no matter how hard-pressed their end users might be for what they, often rightly, regard as combat essentials.

The long term results of such a scrimping outlook can be seen in what is fast becoming an extreme example of being penny wise and pound foolish.

That example can be seen in the vehicle fleets of the US armed forces. The US Army fields one of the world's largest military vehicle fleets and expects it to fulfill what are often extended missions in all parts of the world.

Yet every year that passes that requirement gets more and more difficult to fulfill due to the age of the vehicles expected to carry out their widespread tasks. Despite all of the recent initiatives and the introduction of some superb vehicles, the growing age of the US military vehicle fleet has reached the stage where the costs involved in keeping the fleet in running condition now exceeds their replacement costs by new, modern and more cost effective equipments.

The US Family of Medium Tactical Vehicles (FMTV) has yet to live up to expectations and deliveries have been slow and eratic.

The trucks involved appear to be excellent and perform to specifications but such has been their slow introduction into service that alternative methods of keeping the fleet up to the required numerical and capability levels are now being considered.

A second manufacturing source for the FMTV is now being sought while aging 5 ton (6 × 6) trucks, which the MTV of the FMTV family was meant to replace, are now subject to tenders for a rebuild and life extension programme. This could be so extensive, that three old 5 ton trucks will have to be used to replace two extended life examples for the US Army and Marine Corps.

There are also more age-induced problems looming in the not too distant future. The US AM General HMMWV (High Mobility Multi-Purpose Wheeled Vehicle) fleet now numbers over 90 000 and is still growing, yet many examples are already well over seven years old with their planned service life expectancy being only 14 years.

Artist's impression of the Norwegian Hägglunds Moelv pioneer vehicle based on a converted Leopard 1 MBT chassis using a hydraulic arm fitted with a bucket
1997

Already the planners are busy considering life extension programmes and there are mentions of a possible long-term HMMWV replacement, but once again the prospects of funding for what could be a very expensive programme do not look too promising.

The same applies to the relatively low cost CUCV (Commercial Utility Cargo Vehicle) fleet which will be well over its planned retirement age (and economic sustainability age) by the year 2004, only a few years away.

The old problem, common to many similar programmes in many countries, is that military trucks are not attractive items in the manner that weapon platforms such as helicopters, tanks and self-propelled weapon carriers often appear to be.

To the end users all such vehicles are merely tools with which to carry out a task so the military truck holds an important place within that category.

But try telling that to exchequers for whom the provision of military funding is but one demand among many at a time when political considerations require a reduction in government spending combined with restructions on personel and corporate taxation.

Yet to somehow emphasise that it will cost more in the long term by spending less in the short term seems to be a more difficult task than it at first appears.

The defence industries of the world, not only those of the USA, will somehow have to find a method of extending their message in a more positive way.

A number of countries have, or are introducing, new MBTs and infantry fighting vehicles which have a greater cross-country mobility than the vehicles they replace.

It has become apparent that commercial vehicles slightly modified to a military specification will not meet all future military requirements and for this reason, the UK and USA government research and development establishments have each built or funded a wheeled technology demonstrator.

The aim of these is to show their respective armies and other interested users, exactly what can be accomplished using off-the-shelf technology in a new design.

The UK 6 × 6 vehicle is called the High Mobility Load Carrier (HMLC) demonstrator and this integrates many commercially available subsystems such as all wheel drive and steering, electroninc ABS, power steering, inboard air disc brakes and fully independent suspension fitted to a torsionally stiff chassis.

Defence Research Agency High Mobility Load Carrier (HMLC) demonstrator showing cross-country mobility
1997

The forward mounted powerpack consists of a Cummins 350 bhp diesel coupled to a Eaton automatic transmission with full manual override. A central tyre inflation system is fitted as standard and allows the driver to adjust the tyre pressure to one of four settings.

The US vehicle is called the Technology Demonstrator III and is based on the M1074 Palletized Load System (PLS) truck platform which has been integrated by Barnes & Reinecke Inc.

This vehicle features a Detroit Diesel 8V-92T dual turbocharged diesel engine coupled to a Eaton CEEMAT nine-speed automatic transmission.

Other off-the-shelf subsystems include Rockwell fully independent suspension, General Thermodynamics blower cooling system, United air cleaner system, Rockwell WABCO anti-lock air disc brakes, CM automotive central tyre inflation system, Stewart & Stevenson tactical vehicle cab, Ingersoll Rand air starter, Eaton VORAD collision warning system and Truck Life Light Emitting Diode (LED) exterior lighting.

An HMMWV has also been used as a demonstration platform for vehicle safety. This is fitted with side mounted sonar-based Collision Warning System (CWS), front and rear mounted leader-follower radar based collision warning system and a drivers anti-dozing headset.

Military road safety is becoming even more important as far more peacekeeping people have been killed in road traffic accidents in Bosnia, for example, than killed by enemy action, which includes land mines.

With declining threats and reduced defence budgets, companies that are often fighting for the same contract can now be seen to be co-operating in areas where they have a mutual interest.

In the Middle East for example, Giat Industries and Vickers Defence

Systems are competing, along with a number of other companies, for future MBT orders.

The two companies are however co-operating on a bridgelayer based on the Giat Industries Leclerc MBT chassis which would launch bridges designed and built by Vickers Bridging.

There have been many significant developments in the military vehicles and logistics area over the last 12 months as can be seen by the large number of new entries in this edition.

In this forward it is only possible to highlight some of the more recent new developments in this important field and some of the latest orders for equipment placed.

Armoured engineer vehicles

Norway has ordered from the local company of Hägglunds Moelv, 22 pioneer or engineer vehicles, based on a locally modified Leopard 1 MBT chassis which is being supplied to the company as government furnished equipment. Standard equipment will include two winches, front mounted dozer blade and an excavator.

Norway has already standardised its MBT fleet around the Leopard 1 and all of the key support vehicles will now use the same chassis with the obvious logistic and training advantages to the Norwegian Army.

Following trials between prototypes of an armoured engineer tank built by GAMESA and Peugeot Talbot, the latter has recently been selected by the Spanish Army and an order placed for 38 vehicles.

These will be based on a modified M60A1 MBT chassis and be delivered between 1997 and 1999 under the designation of the M60 CZ-10/25E Alacran.

One of the two prototypes of the Grizzly vehicle with hydraulic arm traversed right and showing remote-controlled weapon system to rear **1997**

Standard equipment will include a winch, front-mounted dozer blade, hydraulic arm which can be fitted with various attachments such as shovel/bucket, cutter and hammer. The front-mounted dozer blade can be removed and replaced with mineclearing equipment of the dozer or roller type with a clear lane mineclearing system mounted at the rear.

Although the UK has only recently taken into service the Vickers Defence Systems Chieftain Armoured Vehicle Royal Engineer (AVRE) it is already looking at a replacement, as the baseline Chieftain MBT was phased out of service early in 1996.

The new vehicle will be called the Future Engineer Tank with 51 being required to replace the Chieftain AVRE and a similar number being required to replace the Chieftain Armoured Vehicle Launched Bridge (AVLB) from around 2001.

By this time, all of the Challenger 1 MBTs are expected to have been phased out of service and the new Vickers Defence Systems Challenger 2 will be the only MBT in service with the British Army.

Although some 420 Challenger 1 chassis will become available for sale or conversion to other roles, the UK Ministry of Defence has not specified what chassis the Future Engineer Tank will be based on.

Today the M728 is the standard combat engineer vehicle of the US Army and is based on an M60 series MBT chassis. This will be replaced in the twenty first century by a new vehicle called the Grizzly (or Breacher).

The Grizzly is based on a much modified M1 series MBT with a new superstructure for the crew consisting of commander and driver.

Two prototypes of the Breacher have been built and tested but further modifications are required and recently United Defense LP have been awarded a US$129 million contract for the Engineering and Manufacturing Development (EMD) phase.

Under the terms of this 52 month contract, the two prototypes of the Grizzly will be modified in a number of key areas including the existing architecture and automatic depth control, with main emphasis on cost reduction, performance and production readiness.

The two modified Grizzly vehicles will be delivered back to the US Army for further trials and it is expected that under the Low Rate Initial Production (LRIP) phase a total of 30 vehicles will be built, with the total US Army and Marine Corps requirement being for up to 600 vehicles.

Typical roles of the breacher include clearing minefields, neutralising obstacles, demolishing berms and filling in anti-tank ditches for highly-mobile forces.

Recovery Vehicles

The French Giat Industries Leclerc Armoured Recovery Vehicle (ARV), or Depanneur Nouvelle Generation (DNG) as it is called in France, is now in volume production for the United Arab Emirates (UAE) who have ordered 46 vehicles.

Early in 1997, the French army stated that it was to place a contract with Giat Industries for an initial batch of 15 Leclerc ARVs to support its Leclerc MBTs which cannot always be recovered by the current in-service AMX-30D ARV.

Giat Industries Leclerc ARV with crane in operating position **1997**

It is expected that the French Army Leclerc ARVs will be powered by the German 1500 hp MTU EuroPowerPack which is already installed in the UAE Leclerc MBTs as well as their ARVs.

As this issue of Jane's Military Vehicles and Logistics went to press Giat Industries announced that they were marketing the E-Force of Leclerc ARV based armoured engineer vehicles.

This includes the Leclerc ARV and a engineer vehicle called the EPG which in addition to the dozer blade and winch, has a hydraulic arm which can be fitted with various attachments such as a bucket, auger, grapple, hammer and hook. It is armed with a 12.7 mm machine gun and a 142 mm demolition gun and if required, can be operated by remote control.

Giat Industries have also developed the K2D mineclearing kit which can be quickly fitted to the ARV. This includes the British Pearson Engineering Full Width Mine Plough (FWMP) with an integrated Giat Industries Demeter magnetic signature duplication system.

Mounted on the rear subframe are two Polish Pronit rocket propelled mineclearing systems and two Pearson Engineering Pathfinder marking systems. Although the first application of the K2D kit is in the Leclerc ARV it can be fitted to other armoured vehicles.

The final member of the E-Force family is the PTG armoured bridgelaying vehicle which is mentioned later.

As previously mentioned, a number of countries are now converting obsolete tank chassis into other roles.

Shown for the first time in 1997 was the Jordanian ARV which is based on a stripped down M47 tank chassis. It is fitted with a winch from the US M88 ARV, front mounted dozer/stabiliser blade, crane from the Chieftain ARV, powerpack from the M88 ARV and commanders cupola and 12.7 mm machine gun from the US M113. The side skirts are from the Tariq MBT which is a British Centurion, upgraded in Jordan with many improvements.

So far only a prototype of the Jordanian M47 based ARV has been completed, but it is a good example of what can be accomplished using only limited resources. The vehicle is also fitted with a universal cradle which can carry a complete MBT powerpack.

The US Army has finally received its first United Defense LP M88A2 Hercules Improved ARV which is a further development of the combat proven M88A1.

Jordanian M47 based ARV carrying a complete Khalid powerpack (Rupert Pengelly) **1997**

The first contract covered four preproduction vehicles and nine production vehicles, but since then another 24 have been ordered for the US Army.

The first export customer for the M88A2 Hercules Improved ARV is Kuwait who has ordered 14 under the Foreign Military Sales (FMS) programme to support its 218 General Dynamics Land Systems M1A2 Abrams MBTs.

M88 Hercules ARV vehicle towing an M1 series MBT during extensive trials **1997**

Bridging Systems

Co-operation between defence equipment manufacturers is now becoming much more common and Giat Industries of France has teamed with Vickers Bridging of the UK to offer an Armoured Vehicle Launched Bridge (AVLB) based on the Leclerc MBT chassis.

This is fitted with a Universal Bridge Launching Equipment (UBLE) at the front which can launch the tank bridges used in the Vickers Bridging BR 90 system currently in production for the British Army.

It was announced early in 1997 that Germany and the Netherlands were to co-operate in the development and procurement of a new armoured bridgelayer based on the chassis of the Leopard 2 MBT. Projected in-service date is 2004 and the total requirement is expected to be about 17 systems for the Netherlands and 35 systems for Germany with about double this number of bridges being required.

The British and German Armies have now accepted into service the EWK M3 amphibious bridge system. These are older M2D units stripped down and rebuilt with many improvements, including a reduction in the crew, new powerpack and its pontoons can now be deployed on land or water. The British Army is taking delivery of 38 units while Germany is taking delivery of 30.

Norway has ordered from MAN of Germany 16 additional 26 m long MLC 70 (Military Load Class 70) bridges, ramps and additional equipment for the construction of ferries and floating bridges.

Artist's impression of French Giat Industries E-Force vehicles in action with PTG armoured vehicle launched bridge being deployed, EPG engineer vehicle on far bank and Leclerc ARV fitted with K2D mineclearing kit in the foreground **1997**

Norway already has 14 MAN Leguan (8 × 8) 26 m bridgelayers in service and in 1997 will take delivery of an additional nine bridgelayers (with a total of 13 bridges) based on a German supplied Leopard 1 MBT chassis.

Krauss-Maffei of Germany will integrate the bridgelaying equipment on to the modified Leopard 1 MBT chassis and the bridges will be fitted with special adaptors to allow them to be incorporated into ferries.

Spain has strengthened its bridging capability by placing an order with Eurobridge Mobile Brucken GmbH for the supply of Dornier Foldable Bridges (DfFB).

Under the terms of the contract the company will supply a total of four bridge units having a total length of 188 m and six launching vehicles.

Under an earlier contract the company is supplying the German Army with 10 bridge units. One of these consists of a 40 m and a 14 m bridging system in order to be able to deploy bridges in varying lengths in parallel (14, 20, 27, 33 and 40 m). The bridge segments have a total length of 540 m and final deliveries are due in 1999.

The US Army has down selected two companies to supply prototypes to meet its requirement for a Heavy Dry Support Bridge (HDSB). These are Karlskronavavet AB of Sweden teamed with Systems & Electronics Inc of the USA and Williams Fairey Engineering of the UK. The key requirement is for a bridge with a span of 42 m and with a capacity of 70 tons (US) for tracked vehicles and 96 tons (US) for wheeled vehicles.

Each company will supply the US Army with a complete bridge system for extensive trials, after which one will be selected for production and service to supplement the current British Medium Girder Bridge (MGB).

The MGB has been one of the most successful bridges in recent years and is still suitable for many applications but for some users it is now manpower intensive taking too long to deploy.

For the export market, Systems & Electronics Inc have integrated a variant of their Light Assault Bridge (LAB) on an Alvis Vehicles Stormer flatbed vehicle for Indonesia. This lays a scissors bridge over the rear of the vehicle.

Mine Warfare

Full details of all current in-service anti-tank and anti-personnel mines are given in our companion volume, *Jane's Mines and Mine Clearance*.

Jane's Military Vehicles and Logistics does have complete sections on minefield detection, clearing and marking equipment.

Pearson Engineering Surface Mine Plough fitted to an M113 series APC of the Canadian Army **1997**

Events in Afghanistan, Angola, Bosnia and Cambodia have once again turned the spotlight on the deployment of mines, especially anti-personnel mines and many countries have already banned the manufacture and sale of these.

Considerable effort is now going into the development, testing and perfection of mine detection and clearing systems. While many of these work very well under most European conditions, many of the mechanical clearance devices are of little use in the jungle and in mountainous terrain where mines can be expected to be found in significant numbers.

Following trials with two prototype Keiler flail type mineclearing systems, the German Army placed an order for 24 production units on a modified M48 tank chassis with the prime contractor being MaK System Gesellschaft.

The first vehicles Keiler produced were handed over early in 1997 although before this date the two prototype systems and one of the first production examples were deployed to Bosnia where, according to the manufacturer, they cleared every single mine.

Norway has ordered from the local company of Hägglunds Moelv, seven flail type mineclearing systems based on a modified Leopard 1 MBT chassis and these are expected to be delivered between 1998 and the year 2000.

The UK company of Pearson Engineering is one of the worlds leading designers, developers and manufacturers of dozer blades and mineclearing devices with most current production being for the export market.

Recent orders have included 51 of their latest third generation Surface Mine Plough (SMP) for the Canadian Army for use on M113s. The Canadians call this the Surface Munition Clearance Device. First generation SMPs are used by the British Army. The SMP is used by a number of other countries including the USA and three have been purchased by France for trials purposes.

One of Pearson Engineering's more recent devices is the APC Mine Roller (MR) and 19 of these have been purchased by the US Army for use on Bradley fighting vehicles in Bosnia.

MaK System Gesellschaft Keiler flail type mineclearing system in operation in Bosnia **1997**

Honda Foreman 400 (4 × 4) ATV **1997**

Light vehicles

Because of the increased emphasis being placed on rapid deployment and airborne forces, especially within NATO, some countries are now ordering light vehicles which can be carried internally or slung underneath helicopters such as the Chinook and Puma.

Early in 1997 the Royal Netherlands Army ordered 180 French designed Lohr VLA (4 × 4) vehicles which will be built under licence in the Netherlands with first deliveries due in the second quarter of 1998.

In the UK trials are underway with a number of commercial All Terrain Vehicles (ATV), one of which is the Honda Foreman 400. These are being tested to support foot soldiers, transporting equipment, weapons and in medical evacuation scenarios, casualties. The trials are being carried out by 1st Battalion, The Parachute Regiment.

For some years the UK has used the Supacat (6 × 6) as its All Terrain Mobile Platform (ATMP) but as some additional vehicles were required another competition had to be held.

Early in 1997 Alvis Vehicles were awarded a contract for the supply of an undisclosed quantity of Supacat Mk 3 ATMPs and its associated Self-Loading Lightweight Pallet Trailer (SLLPT) for the British Army. Total requirement is expected to be for 86 ATMP and 84 SLLPT.

A number of armies are now deploying or testing 4 × 4 or 4 × 2 fast attack vehicles or long range strike/reconnaissance vehicles for use by their special forces and a number of new vehicles are covered for the first time in this edition.

Supacat (6 × 6) ATMP fitted with MILAN ATGW launcher and towing an SLLPT, carrying MILAN missile containers **1997**

Transport Vehicles

There are no significant new truck programmes underway in Europe although some are expected to start in the near future, especially in the UK, where a replacement for the Bedford (4 × 4) 8 tonne truck will be required in the near future.

A number of new countries now have the capability to design and manufacture military trucks, although very often some subsystems, such as engines, transmissions and drive lines are brought in.

Malaysia is now building the locally designed Hicom FSS 32G (4 × 4) 3,000 kg truck for its local armed forces and hopes in the future to expand into other areas of vehicle production.

Pakistan is now building the Yasoob Mk 1 (6 × 6) 6000 kg for the Pakistan Army and more recently has built prototypes of the Yasoob Mk II (4 × 4) forward control truck which has a British Leyland Trucks cab.

The Tatra company of Slovakia is well known for their family of 8 × 8 and 6 × 6 vehicles which have exceptional cross-country mobility and are normally fitted with a central tyre pressure regulation system as standard.

Although the truck export market has tended to be dominated by Western truck companies, Tatra is having some success. In 1996 the UAE ordered over 1100 Tatra T815 series 8 × 8 trucks fitted with a Deutz air cooled engine.

As well as the standard Tatra T815 cargo truck the contract also includes a number of specialised types including crane (being supplied by Fassi of Italy) and water tankers (VSS of Slovakia).

Late in 1996, the US Army Tank-automotive Command awarded contracts to AM General and Oshkosh Truck Corporation to cover Phase I of the Medium Tactical Truck Remanufacture (MTTR) programme.

Under the terms of Phase I, prototype design and manufacture contract, each contractor will rebuild five M923 series 5 ton (6 x 6) trucks for the Marine Corps by mid-1997. Within the contract there is also an option for an additional five test vehicles for the US Army.

These trucks will be tested through to 1998 and under Phase II one contractor will be awarded a production contract for the upgrade of about 8,100 US Marine Corps and 3,400 US Army 5 ton trucks with first vehicles expected to be delivered in FY00. Total value of this contract could be as high as US$1.8 billion.

This the second major truck upgrade programme in the US, as the US Army is already upgrading its 2.5 ton (6 × 6) trucks under the Extended Service Programme for which the prime contractor is AM General.

Tank transporters

With the huge investment in the acquisition of brand new MBTs by some countries in the Middle East, for example, Kuwait, Oman, Saudi Arabia and the UAE, a number of countries are now looking to the acquisition of new tank transporter systems, for example tractor truck and semi-trailer.

Early in 1997, Alvis UNIPOWER of the UK unveiled its new MH8875 (8 × 8) rapid deployment tank transporter which is now taking part in trials in the UAE.

The MH8875 has been designed to move MBTs and other armoured vehicles swiftly from rear areas to as far forward as the tactical situation will allow and has a road cruising speed of 80 km/h with a 75 tonne payload. Steering is power-assisted on the front four wheels and standard equipment includes two winches each with a line pull of 25 tonnes.

Its large tyres give it a low ground pressure and it complies with the 12 tonne maximum axle loading legislation applying in the EU and elsewhere.

The prototype of the Alvis Unipower MH8875 is being trialed with a French Nicholas military trailer having six rows of single tyred wheels for maximum manoeuvrability and the rearmost four axles steer. This allows the 25 m long transporter to turn within a 14 m radius.

The British Army has a requirement for up to 80 tank transporters and semi-trailers to replace its current Unipower, previously Scammell, Commander (6 × 4) vehicles of which 125 were delivered between 1983 and 1985.

In addition to the procurement of a brand new system (tractor and semi-trailer) consideration is also being given to the leasing of these vehicles as the UK already does for a number of its roles, for example driver training trucks.

Acknowledgements

The eighteenth edition of *Jane's Military Vehicles and Logistics* would not have been possible without the active assistance of many contributers all over the world.

A special word of thanks are due to Mick Bell, Shaun Conners, Charles Cutsaw, Kensuke Ebata, Scott R Gourley, Lyn Haywood, Arthur Hogben, Andrew Jeacock, Michael Jerchel, Richard Stickland, Pierre Touzin, Steven Zaloga and C R Zwart.

Such a listing of names can never be fully comprehensive and does not even take into account the corps of Jane's Yearbook Editors and staff of the various Jane's magazines.

Once again we do have to acknowledge the valuable work extended to this yearbook by Jane's Production staff headed by Ruth Simmance and Sulann Staniford. Mick Durkin edited many of the entries and Sulann was also the main contact between production and the editors.

Special thanks are also due to the VDU/Typesetters who input and typeset the galleys. They have all made their contributions and our thanks go to all of them.

Peter Howard, Managing Editor, has provided tremendous back-up and support throughout the preparation and production of this yearbook.

Finally we must not forget to express our thanks and appreciation to Butler and Tanner who print and bind this yearbook.

Terry Gander
Christopher F Foss
April 1997

Special Note

The Editors would welcome any additional information, corrections or comments regarding the entries in this yearbook. Please direct all correspondence to the Editors at *Jane's Military Vehicles and Logistics*, Jane's Information Group, Sentinel House, 163 Brighton Road, Coulsdon, Surrey CR5 2NH, United Kingdom.

New entries in this edition

Entry	Country
Hägglunds Moelv Armoured Engineer Vehicle	Norway
IMR-2 Combat Engineer Vehicle	RFAS
GAMESA M60 VZ Engineer Tank	Spain
Future Engineer Tanks	UK
United Defense Combat Engineer Squad Vehicle	USA
GAMESA M47 VR ARV	Spain
Future Beach Recovery Vehicle	UK
Repair Tracked Vehicle	RFAS
EKA Recovery Equipment	UK
MAN GHH Leopard 1 LEGUAN AVLB	Germany
PMCz-94 Armoured Bridgelayer	Poland
MTU-90 Armoured Bridgelayer	RFAS
TMM-6 Truck-mounted Treadway Bridge	RFAS
Aluminium Pontoon Raft	Thailand
NORINCO Type 79 Minelaying Rocket System	People's Republic of China
NORINCO Type 84 Minelaying Rocket System	People's Republic of China
NORINCO Type 1 Anti-personnel Mine Scattering System	People's Republic of China
NORINCO Type 2 Anti-personnel Mine Scattering System	People's Republic of China
Scintrex V-920 Ferrous Ordnance Locator	Canada
Schiebel MIMID miniature mine detector	Austria
FÖRSTER Vehicle Mounted Multi Sensor Systems Germany	
PRS-17K Portable Mine Detector	Korea, South
Reutech VMMD 2000 Vehicle Mounted Mine Detector	South Africa
Metallic Mine Detector	Thailand
NORINCO 253 mm Type 81-II Mineclearing Rocket System	People's Republic of China
NORINCO 305 mm Fuel-air Mineclearing Rocket	People's Republic of China
NORINCO Type 74 Mineclearing System	People's Republic of China
NORINCO Type 84 Tank-mounted Mineclearing Rocket System	People's Republic of China
NORINCO Light Rocket Mineclearing System	People's Republic of China
Božena Remote Control Demining Vehicle	Slovakia
HYDREMA 910 MCV Mineclearing Vehicle	Denmark
SAKR Fateh-3 anti-tank mineclearing system	Egypt
Rolling On Board Anti Magnetic Mine Actuating Device	Israel
RAMTA Combat Vehicle Teleoperation Kit (PELE)	Israel
MKEK Mineclearing System	Turkey
Pearson Mine Roller (MR)	UK
Light Forces Anti-personnel Mine Mini-flail	USA
UAZ-3172 (4 × 4) light vehicle	RFAS
Turbomecanica Special Purpose Off-road Vehicle	Romania
STA Kowari High-Mobility Light Assault Vehicle	Singapore
Mechem BAT Special Forces Utility Vehicle	South Africa
Automotive Technik (4 × 4) IMM Weapons Platform	UK
Alvis Shadow Light Strike Vehicle	UK
Lockheed Martin-Chenowth Advanced Light Vehicle (ALSV)	USA
Lockheed Martin-Chenowth General Purpose (GP) Vehicle	USA
TATRA T 815 VVN 28.400 8 × 8.1 R G3 LIWA Truck	Czech Republic
MAN 45.1000 FFAEG SX 2000 (10 ×10) Demonstrator	Germany
Mercedes-Benz 2638 A (6 × 6) 14,000 kg Truck	Germany
Mercedes-Benz 3550 A (8 × 8) 15,000 kg Truck	Germany
Timoney (8 × 8) High-Mobility Truck	Ireland
IVECO M 110.18 WM (4 × 4) 5,000 kg Truck	Italy
Astra HD 66.45M (6 × 6) Truck	Italy
Astra HD 66.45/5S - C-IMT (6 × 6) Tractor Truck	Italy
IVECO M 320 E 42 W (8 × 8) 17,500 kg Truck	Italy
Hicom FSS 32G (4 × 4) 3,000 kg Truck	Malaysia
South African Aljaba Applications	South Africa
SPA Fox-15D (4 × 4) 1,000 kg airborne vehicle	Spain
Volvo BM A25C (6 × 6) Truck	Sweden
Western Star M6966 (6 × 6) HET	Canada
Astra HD 66.45 TIM (6 × 6) Truck Tractor	Italy
Volvo F16 (6 × 6) Tank Transporter	Sweden
EMPL 3-axle 34 ton Low Bed Semi-trailer	Austria
LOHR SRPB 60 Tank Transporter Semi-trailer	France
LOHR MTS Trailer	France
KOOI-AAP portable forklift truck	Netherlands
Trench Fortification System	Egypt
Marshall DROPS Fuel Rack	UK
Reynolds Boughton Tankers	UK
B.O.I.S. anti-radar camouflage nets	Czech Republic
Croker Rapid Runway Repair System	UK
Marrel ISO 20 ft Container with integrated skid	France
Marrel Amplitainer	France
Kärcher High Mobil Containerised Kitchen HMCK	Germany
Dornier TransHospital mobile field hospital	Germany
SPA NATO shelters	Spain
Marshall Matrix Expandable shelter	UK
MSI-Defence Systems tactical shelters	UK
Marion Composites Army Standard Family (ASF) shelters	USA
Marion Composites General Purpose Rigid ISO shelter	USA
Marion Composites EMI ISO shelter	USA
Marion Composites Knockdown ISO shelter	USA
Marion Composites Relocatable Hospitals	USA
Marion Composites HMMWV shelters	USA
Marion Composites MMSV	USA
Marion Composites cargo bed covers	USA
GEMCO stretcher carrier set (GMS 2058)	Netherlands

Entries deleted from this edition

Entry	Country
Combat Engineer Vehicle based on T-72	RFAS
VCRT Armoured Recovery Vehicle	France
M-47 E2R ARV	Spain
MTP Technical Support Vehicle	RFAS
GKN FV434 ARV	UK
CEFA Gillois Series 2 Ferry System	France
EWK Gillois Bridge and Ferry System	International
MAN Light-Metal Floating Bridge	Germany
MAN GHH Hollow Plate Assembly Bridge	Germany
Class 80 Heavy Ferry	UK
MAN SE Road and Railway Bridge	Germany
MARM Sectional Arch Bridge	RFAS
NZhM-56 Heavy Floating Railway and Road Bridge	RFAS
MAN Short-Span Bridge	Germany
BMK-70 and BMK-90 Bridging Boats	RFAS
GRILLO 90 Manportable Mine Launcher	Italy
Scatter Mine System	South Africa
M131 Modular Pack Mine System	USA
AN/PRS-7 Non-metallic Mine Detector	USA
Scatterable Mine Clearing Device (SMCD)	Israel
Light Forces Countermine Projects	USA
Rapid Operational Minefield Attack and Neutralisation System	UK
PONCIN (6 × 6) 1,000 kg Light Vehicle	France
Terex UAI M1-50 (6 × 6) 5,000 kg Truck	Brazil
Chevrolet 1¼ ton (4 × 4) Series	Canada
BAZ SNA (6 × 6) 4,000 kg Truck	Slovakia
Brimont BRUTT (4 × 4) 2,225 kg Truck	France
Brimont ETR (4 × 4) Trucks	France
IVECO 6607 CM (6 × 6) 6 000 kg Cargo Truck	Italy
Astra BM 309F (6 × 6) Truck	Italy
Bravia Gazela (4 × 4) 1,134 kg Truck	Portugal
SAMIL 20 Mark 1 (4 × 4) 2,000 kg Trucks	South Africa
SAMIL 50 Mark 1 (4 × 4) 4,800 kg Trucks	South Africa
High Mobility Logistic Vehicle (HMLV)	South Africa
Volvo 4151 (4 × 4) Anti-tank Vehicle	Sweden
HUGO (6 × 6) 1,500 kg All Terrain Vehicle	UK
Hotspur One Fifty (6 × 6) Vehicles	UK
M715 (4 × 4) 1¼ ton Truck	USA
M880 (4 × 4) and (4 × 2) 1,133 kg Series	USA
Freightliner All Terrain Tow Vehicle (ATTV)	USA
BMY 3 tonne (4 × 4) Truck, Cargo	USA
ÖAF 34.440 VFA (8 × 8) Tank Transporter	Austria
Nicolas Tractomas Tank Transporters	France
Rampar (4 × 4) 3,000 kg Amphibious Vehicle	India
Amphitruck (4 × 4) 3,000 kg Amphibious Vehicle	UK
Mazur D-350 Medium Tracked Artillery Tractor	Poland
AT-L and AT-LM Light Tracked Artillery Tractors	RFAS
ATS-59 Medium Tracked Artillery Tractor	RFAS
Streaker High-Mobility Load Carrier	UK
GT-SM Tracked Amphibious Over-snow Vehicle	RFAS
Scottorn 125-ration Mobile Kitchen Trailer	UK
Titan T6 R2 3800 kg Trailer	France
Decauville SRPC 48/54 tonne Semi-trailer	France
Titan Tank Transporter Semi-trailers	France
Kärcher NBC Decontamination Trailer	Germany
Trailer, Amphibious: Cargo, ¼ ton, M100	USA
Semi-trailer, Stake: 6 ton, M118 and M118A1	USA
Semi-trailer, Fuel, 5000 gallon, M131 Series	USA
Semi-trailer, MILVAN Container Transporter	USA
Marrel Light Fork-lift Truck	France
BOSS FUG 2.5 tonne All-terrain Lift Truck	UK
High Mobility Materiel Handler	USA
Variable Reach Rough Terrain Forklift Truck	USA
Pumping Assembly POL 180 l/min	Australia
Fruehauf 18,000 litre Fuel Bowser	France
Mobile Fuel Can Filler System	Italy
Mobile Fuel Transfer Station	Italy
SEH 600 Expandable Modular Shelter	France
Marrel Shelters with Ampliroll	France
Relocatable Hospital System	USA
Brunswick Army Standard Family (ASF) Shelters	USA
Brunswick US Marine Corps Shelters	USA
Brunswick Standardised Integrated Command Post Shelter	USA

Glossary

AARADCOM	Army Armament Research and Development Command
AAT	All arms trencher
AAV	Assault Amphibian Vehicle
AAVP	Assault Amphibian Vehicle Personnel
AAVR	Assault Amphibian Vehicle Recovery
ABLE	Automotive Bridge Launching Equipment
ABS	Anti-lock brake system
ABS(T)	Amphibious bridging system (tracked)
ACE	Armored Combat Earthmover
ACEATM	Aimed Controlled Effect Anti-tank Mine
ACP	Airfield clearance plough
ACPM	*Anti-Char à Pose Mecanique*
ACT	Air-portable cargo trailer
ACRV	Armoured Command and Reconnaissance Vehicle
ADAM	Area denial artillery munition
ADS	Ammunition delivery system
AEC	Associated Equipment Company
AEV	Armoured engineer vehicle
AFARV	Armored forward area rearm vehicle
AFDE	Arctic fuels dispensing equipment
AFV	Armoured fighting vehicle
Ah	Ampère hour
AHE	Ammunition handling equipment
AHM	Anti-helicopter mine
ALB	Automatic load-dependent brake
ALSV	Advanced Light Strike Vehicle
AMIDS	Airborne minefield detection system
AMMAD	Anti-magnetic mine actuation device
AMS	Air management system
AMV	Armored Maintenance Vehicle
AMX	*Atelier de Construction d'Issy-les-Moulineaux*
APC	Armoured personnel carrier
APFC	Air-portable fuel container
APU	Auxiliary power unit
ARCE	Amphibious river crossing equipment
ARGES	Automatic Rocket Guardian with Electronic Sensor
ARM	Artillery rearm module
ARMS	Armoured Resupply Maintenance System
ARRV	Armoured recovery and repair vehicle
ARV	Armoured recovery vehicle
ASF	Army standard family (shelters)
ASM	Armored Systems Modernization
ASTAMIDS	Airborne Standoff Minefield Detection System
AT	Anti-tank
ATAC	All-terrain all climate; All-terrain amphibious carrier
ATGW	Anti-tank guided weapon
ATIS	Anti-tank influence sensor
ATLAS	All Terrain Lifter Army System
ATMDS	Anti-tank mine-dispensing system
ATMP	All-terrain mobile platform
ATS	*Atelier de Construction de Tarbes*
ATTD	Advanced technology test demonstrator
ATTV	Armored Treatment and Transport Vehicle
ATV	All-terrain vehicle
AV	American Version
AVE	Armoured vehicle for engineers
AVLB	Armoured vehicle-launched bridge
AVRE	Assault Vehicle Royal Engineers
BAP	Bridge adaptor pallet
BARC	Beach amphibious resupply cargo
BARV	Beach armoured recovery vehicle
BBE	Bridge boat erection
BEML	Bharat Earth Movers Limited
BFTA	Bulk fuel tank assembly
bhp	Brake horse power
BLT	Bridge laying tank
BLU	Bomb live unit
BRE	Battlefield recovery and evacuation
BS	British Standard(s)
CALM	Crane attachment lorry mounted
CAMIS	Computer Aided Munition Identification System
CBU	Cluster bomb unit
CCE	Commercial construction equipment
CCSLEP	Chaparral Chassis Service Life Extension Program
cd	Candela
CEC	*Couverture d'Emplacement de Combat*
CEE	Combat explacement excavator
CEP	Concept evaluation program
CET	Combat engineer tractor
CFM	Crane, field, medium
CFRP	Carbon fibre reinforced plastic
CFV	Cavalry fighting vehicle
CHU	Container handling unit
CIS	Commonwealth of Independent States
CKD	Component knock-down
CLAMS	Clear lane marking system
CLD	*Camion Lourd de Depannage*
CLEWP	Cleared lane explosive widening path charge
CLK	Container lift kit
CMV	Counter mobility vehicle
COCT	Cab-over cargo truck
COE	Cab over engine
COV	Counter obstacle vehicle
CP	Command post
CPP	Cambridge Parallel Processing
CRB	Capsill roller beam
CRDA	Co-operative Research and Development Agreement
CRRC	Combat rubber raiding craft
CSB	Combat support boat
CSF	Combined Service Forces
CST	Combat support trailer
CTIS	Central tyre inflation system
CUCV	Commercial utility cargo vehicle
CVR(T)	Combat vehicle reconaissannce (tracked)
DAP	Distributed Array Processor
DARPA	Defense Advanced Research Projects Agency
DAT	*Direction des Armements Terrestres*
dB	Decibel
DBP	Draw bar pull
DCU	Dispenser control unit
DEF.STAN	Defence Standard
DEU	Deployment and emplacement unit
DEUCE	Deployable Universal Combat Earthmover
DIN	Deutsche Industrie Normen
DROPS	Demountable rack off-loading and pick-up system
EBG	*Engin Blindé de Génie* (combat engineer tractor)
ECV	Expanded capacity vehicle
EFP	Explosively formed projectile
ELCAS	Elevated causeway
EMD	Engineering and manufacturing development
EMI	Electromagnetic interference
EMP	Electromagnetic pulse; Engineering mine plough
EMS	Enhanced mobility system
EOD	Explosive ordnance disposal
EPP	Electric power plant
EPU	Electrical power unit
ERAM	Extended range anti-armor munition
ERV	Electronic repair vehicle
ESAB	Extended Span Assault Bridge
ESBL	Extended Span Bridgelayer
ESMB	Explosive Standoff Minefield Breacher
ESP	Extended service program(me)
EWK	Eisenwerke Kaiserslautern
EWT	Engineer wheeled tractor
FAALS	Forward Area Armored Logistic System
FAASV	Field Artillery Ammunition Support Vehicle
FAC	Forward air controller; fast attack craft
FAE	Fuel-air explosive
FAMV	Forward area multipurpose vehicle
FARE	Forward area refuelling equipment
FARV-A	Fully integrated armored rearm
FASCAM	Family of scatterable mines
FAST	Forward area support team
FAV	Fast attack vehicle
FAWPSS	Forward area water point supply system
FCS	Field-expedient mineclearing system
FCS	Fire control system
FDC	Fire direction centre
FEBA	Forward edge of the battle area
FFLAV	Future Family Light Armoured Vehicles
FFR	Fitted for radio
FLPT	Fork lift pallet trailer
FMS	Foreign military sales
FMTV	Family of Medium Tactical Vehicles
FMU	Folding Medical Unit
FN	Fabrique Nationale
FNNH	Fabrique Nationale Nouvelle Herstal
FOCOS	Foam overhead cover support system
FOM	*Forces d'Outre Mer*
FOPS	Falling object protective structure
FRG	Federal Republic of Germany
FRS-H	Forward Repair System Heavy
FSB	*Faltschwimmbrucke*
FUE	First unit equipped
FV	Fighting vehicle
FVS	Fighting vehicle system
FWAM	Full width attack mine
FWD	Four-wheel drive
FWMP	Full width mine plough
FY	Fiscal year
GCS	Ground control station
GCW	Gross combination weight
GDLS	General Dynamics Land Systems
GEMSS	Ground-emplaced mine-scattering system
GLCM	Ground-launched cruise missile
GMC	General Motors Corporation
GPM	Gallons per minute; *Gepanzerte Pioniermaschine*
GPMG	General purpose machine gun
grp	Glass-reinforced plastic
GV	Giant Viper
GVW	Gross vehicle weight
GW	guided weapon
HAB	Heavy assault bridge
HB	Heavy barrel
HDSB	Heavy dry support bridge
HDRV	Heavy-duty recovery vehicle
HE	High explosive
HEAT	High explosive anti-tank
HELBAT	Human Engineering Laboratory, Battalion Artillery Test
HEMAT	Heavy expanded mobility ammunition trailer
HEMTT	Heavy expanded mobility tactical truck
HESH	High explosive squash head
HET	Heavy equipment transporter
HGV	Heavy goods vehicle
HLVW	Heavy logistic vehicle, wheeled
HMBS	High mobility on board system
HMD	Helicopter mine dispenser
HMMHE	High mobility materiel handling equipment
HMMWV	High mobility multipurpose wheeled vehicle
HMRT	Heavy material recovery team
HMT	High mobility trailer
HMTT	High mobility tactical truck
hp	Horse power
HPD	*Haut pouvoir de destruction*
HRV	Heavy repair vehicle
HTR	Heat transfer reduction
HTT	Heavy tracked tractor
HVSS	Horizontal volute spring suspension
Hz	Hertz
IDA	Improved dogbone assembly
IFP	ISO Compatible Palletised Flatrack
IFV	Infantry fighting vehicle
ILS	Integrated logistic support
IMF	Improved mine fuze
IMMLC	Improved medium mobility load class
IMR	*Inzhenernaia Maschina Razgrazhdeniia*
IOC	Initial operational capability
IOD	Improvised ordnance disposal
IOTE	Initial Operational Test and Evaluation
IPDS	Inland Petroleum Distribution System
IR	Infra-red
IRR	Infrared radiation
IRV	Improved recovery vehicle
ISC	Infantry section carrier
ISO	International Standards Organisation
IVECO	Industrial Vehicle Corporation
JGSDF	Japanese Ground Self-Defence Force
JRA	Jaguar Rover Australia
JSFU	Joint Services flail unit
JSOR	Joint Service operation requirement
KERR	Kinetic energy recovery rope
KHD	Klockner-Humboldt-Deutz
KIFV	Korean Infantry Fighting Vehicle
kT	Kiloton
LAA	Light anti-aircraft
LAAG	Light anti-aircraft gun
LAB	Light assault bridge
LACH	Lightweight amphibious container handler
LAD	Light aid detachment
LAF	Light assault ferry
LAPES	Low altitude parachute extraction system
LAR	Light artillery rocket
LARC	Lighter amphibious resupply cargo
LAV	Light armoured vehicle

LAW	Light anti-armour weapon
LCCV	Light cross-country vehicle
LCD	Liquid crystal display
LCP	Line-carrying projectile
LED	Light-emitting diode
LFV	Light forces vehicle
LHD	Left-hand drive
LIMAS	Lightweight marking system
LIME	Light In-stride Mine Extractor
LMD	Light mobile digger
LMG	Light machine gun
LMMSA	Lightweight modular multipurpose spanning assembly
LMS	Lightweight multi-purpose shelter
LMTV	Light medium tactical vehicle
LMTVT	Light medium tactical vehicle trailer
LOTS	Logistics over the shore
LPC	Launch pod container
LSA	Logistic support analysis
LSV	Light strike vehicle
LSV	Logistic supply vehicle; Logistic support vehicle
LSVW	Light support vehicle wheeled
LTRS	Light tactical recovery system
LVS	Logistic vehicle system
LVT	Landing vehicle tracked
LWB	Long wheelbase
LWCSS	Lightweight camouflage screen system
MAB	Mobile assault bridge; Medium assault bridge
MACH	Mechanically assisted construction by hand
MACI	Military adaption of commercial items; *Mine anti-char indetectable*
MACPED	*Mine Anti-Char Pointable à Effet Dirigé*
MACS	Magnetic countermine system
MARRS	Modular armoured repair and recovery system
MARS	Military amphibious reconnaissance system
MAV	Maintenance assist vehicle
MBB	Messerschmitt-Bölkow-Blohm (now part of Daimler-Benz Aerospace)
MBT	Main battle tank
MCAP	Mineclearing/armor protection
MCBS	Mineclearing blade system
MCESS	Marine Corps Expeditionary Shelter System
MCS	Mobile camouflage screen
MDK	*Mashina dorozhnoy kopatelnoy*
MDR	*Moyen de Deminage Rapide*
MDV	Mine detection vehicle
MEDDS	Mechem Explosive and Drug Detection System
MERADCOM	Mobility Equipment Research and Development Command
MET	Medium equipment transporter
MEV	Medical evacuation vehicle
MEXE	Military Engineering Experimental Establishment
MG	Machine gun
MGB	Medium girder bridge
MHAB	Multi-Hop Assault Bridge
MHC	Materials handling crane
MIACAH	*Mine anti-char d'Action Horizontale*
MICLIC	Mineclearing line charge
MICV	Mechanised infantry combat vehicle
MIDAP	Minefield Detection Algorithm and Processor
MIL	Military
MIMID	Miniature Mine Detector
MIRADOR	Minefield Reconnaissance and Detector System
MiWS	*Minenwerfersystem*
MLAB	Manually launched assault bridge
MLC	Military load class
MLI	Mid-life improvement
MLRS	Multiple launch rocket system
MLVW	Medium logisitic vehicle wheeled
MMLC	Medium mobility load carrier
MMSV	Multipurpose maintenance and shop van
MMW	Medium multipurpose wheeled
MoD	Ministry of Defence
MOPMS	Modular pack mine system
MoT	Ministry of Transport
MR	Mine roller
MRCV	Multi Role Combat Vehicle
MRT	Mobile repair team
MRTFL	Medium rough terrain forklift
MRV	Maintenance-recovery vehicle
MSG	*Minensuchgerat*
MSMW	*Minenstreumittel-Werfer*
MT	Megaton
MTS	Multitrack System
MTT	Medium tactical truck
MTU	*Motoren- und Turbinen-Union*
MTV	Medium tactical vehicle
MTVT	Medium tactical vehicle trailer
MVEE	Military Vehicles and Engineering Establishment
MVP	Motor vehicle plant
MWT	Medium wheeled tractor
NATO	North Atlantic Treaty Organisation
NBC	Nuclear, biological, chemical
NDI	Non-developmental item
OEM	Original equipment manufacturer
OEW	Ordnance and explosive waste
OFC	Overhead foxhole cover
OHC	Overhead cam shaft
OHV	Overhead valve
OP	Observation post
ORATMS	Off-route anti-tank mine system
OVE	Onboard vehicle equipment
Pa	Pascal
PAA	*Pont Automoteur d'Accompagnement*
PADS	Position attack defence system
PARM	*Panzerabwehr-Richtmine*
PBX	Plastic bonded explosive
PDM	Pursuit deterrent mine
PE	Plastic explosive
PETN	Pentaerythritetranitrate
pF	Picofarad
PF	Pocket filter
PFC	Parapet foxhole cover
PFM	*Pont Flottant Motorisé*
PFT	Prefabricated foxhole - twin
PI	Product improvement
PIMD	Pulse induction mine detector
PIP	Product improvement programme
PLS	Palletised load system
POL	Petrol, oil and lubricants
POMINS	Portable mine neutralisation system
POP	Pipeline outfit, petroleum
PSA	Prefabricated surface, aluminium
psi	Pounds per square inch
PTO	Power take-off
pvc	Polyvinyl chloride
RAAC	Royal Australian Armoured Corps
RAAMS	Remote anti-armor mine system
RAC	Royal Armoured Corps
RACE	Rapid Area Clearance Equipment
RAE	Royal Australian Engineers; Royal Aircraft Establishment
RAMBS	Rapid Anti-personnel Minefield Breaching System
RAP	Regimental aid post
RARDE	Royal Armament Research and Development Establishment
RARDE(C)	Royal Armament Research and Development Establishment (Christchurch)
RATT	Rescue All-Terrain Transport
RBEB	Ribbon bridge erection boat
RCATV	Remote control, all-terrain vehicle
RCT	Royal Corps of Transport
R & D	Research and Development
RDB	Rapid deployment bridge
RE	Royal Engineers
REME	Royal Electrical and Mechanical Engineers
RF	Radio frequency
RFI	Radio frequency interference
RFP	Request for proposals
RHD	Right-hand drive
RLC	Royal Logistical Corps
RLT	Rolling liquid transporter
RMID	Road mine detector
ROMANS	Rapid operational minefield attack and neutralisation system
ROPS	Rollover protective structure
ROWPU	Reverse osmosis water purification unit
rpm	Revolutions per minute
RPV	Remotely piloted vehicle
RRR	Rapid runway repair
RSH	*Reservoirs souples heliportable*
RSME	Royal School of Military Engineering
RTE	Rail transfer equipment
RTFL	Rough terrain forklift
SACC	Small area camouflage cover
SAD	Safety and arming device
SAE	Society of Automotive Engineers
SAM	Surface-to-air missile
SAPIR	Sensor Active Passive Infra-Red
SAPLIC	Small arms projected line charge
SAS	*Schnellbrucke auf Stutzen*
SAVA	Standard Army Vectronics Architecture
SEE	Small emplacement excavator
SEP	System enhancement package
SEV	Specially equipped vehicle
SFC	Soldier fighting cover
shp	Shaft horse power
SICPS	Standardised integrated command post shelter
SLEP	Service Life Extension Program
SLUFAE	Surface-launched unit, fuel-air explosive
SMP	Surface mine plough
SMS	Scatterable mine system
SOV	Special operations vehicle
SPB	Section personnel bridge
SPG	Self-propelled gun
SRDE	Signals Research and Development Establishment
SS	Surface-to-surface
SSD	Sensor safety device
SSM	Surface-to-surface missile
STB	Short trackway bridge
STEFVS	Simplified Test Equipment Fighting Vehicle System
STORM	Sensored tactical off-road mine
SUMMADE	System universal modular mine and demolition explosives
SUSV	Small unit support vehicle
SUU	Suspended underwing unit
SWB	Short wheelbase
SWG	Standard wire gauge
SWL	Safe working load
TAB	Towed assault bridge
TACOM	Tank Automotive and Armaments Command
TAD	Trailing arm drive
TARADCOM	Tank Automotive Research and Development Command
TBT	Tank bridge transporter
TCC	Temporary camouflage coating
TD	Tank destroyer
TECOM	Test and Evaluation Command
TEG	Thermo-electric generator
TEXS	Tactical explosive system
TFR	Tank ferry raft
THAAD	Theatre High Altitude Air Defense Missile
TILOS	Tangram Integrated Logistic System
TLB	Trailer-launched bridge
T/MDV	Towing/Mine Detection Vehicle
TNT	Trinitrotoluene
TOC	Tactical operations centre
TPT	Tactical Petroleum Terminal
TQG	Tactical quiet generator
TRM	*Toutes Roues Motrices*
TROSS	Technical and rear operations support system
TRTG	Tactical radar threat generator
TU	Towed unit
TUH	Truck, Utility, Heavy
TUL	Truck, Utility, Light
TUM	Truck, Utility, Medium
TWDS	Tactical water distribution system
TWMP	Track width mine plough
TWVMP	Tactical Wheeled Vehicle Modernization Plan
UA-HHV	Up-Armored HMMWV
UAV	Unmanned aerial vehicle
UBLE	Universal Bridge Launching Equipment
UET	Universal engineer tractor
UK	United Kingdom
ULC	Unit load container
ULCAS	Ultra lightweight camouflage screen
UMIDS	Universal mine-dispensing system
UNPROFOR	United Nations Protection Force
USA	United States of America
UV	Ultraviolet
UXO	Unexploded ordnance/object
VAB	*Véhicule de l'Avant Blindé;* Vickers Armoured Bridgelayer
VAMIDS	Vehicular Array Mine Detection System
VCG	*Véhicule de Combat du Génie*
VCI	*Véhicule de Combat d'Infanterie*
VCR-AT	*Véhicule de Combat a Roues - Atelier Technique*
VEMASID	Vehicle magnetic signature duplicator
VLRA	*Véhicule Léger de Reconnaissance et d'Appui*
VLSMS	Vehicle-launched scatterable mine system
VLTT	*Véhicule de Liaison tout Terrain*
VMB	Vickers Modular Bridge
VMMD	Vehicle mounted mine detector
VMRMDS	Vehicle-mounted road mine detector system
vpm	Vibrations per minute
VRRTFL	Variable reach rough terrain forklift
VRRTFLT	Variable reach rough terrain forklift truck
VSTOL	Vertical short take-off and landing
VTL	Vehicle tactical logistic
VVSS	Vertical volute spring suspension
WAM	Wide area mine
WASPM	Wide area side-penetrating mine
WCP	Water carriage pack
WHO	World Health Organisation
ZF	*Zahnradfabrik Friedrichshafen*

ARMOURED ENGINEER VEHICLES

ARMOURED ENGINEER VEHICLES

AUSTRIA

Steyr-Daimler-Puch Engineer Tank 4KH7FA-AVE

Description

The Steyr-Daimler-Puch Engineer Tank 4KH7FA-AVE, also known as the Steyr Armoured Vehicle for Engineers (AVE) is a further development of the Steyr 4KH7FA-B Greif ARV, described and illustrated in the *Armoured Recovery Vehicles* section. The AVE entered Austrian Army service in 1988 where it is known as the Pionierpanzer 4KH7FA-Pi.

A total of 18 production vehicles were delivered to the Austrian Army but the complete 4KH7FA production line has now closed.

The hull is made of all-welded steel with the fully enclosed crew compartment at the front. Normal means of entry to the crew compartment is by two doors on the left side of the hull but there are additional hatches in the roof. The engine and transmission are at the rear of the hull. The engine compartment is fitted with a fire extinguishing system that can be operated manually or automatically. Over the engine decking is a stowage platform for equipment.

The torsion bar suspension consists of five dual rubber-tyred roadwheels with the drive sprocket at the rear and the idler at the front. There are three track-return rollers and hydraulic shock-absorbers are fitted at the first and fifth roadwheel stations.

Mounted on the right side of the superstructure at the front is a hydraulically operated excavation device (ditcher), the boom of which can be traversed through 234°. Earth-drilling equipment with a 350 mm diameter auger can be mounted in place of the bucket, as can a crane hook.

A winch is situated in the lower part of the hull and leads out through the front part of the hull. It is provided with 60 m of 16 mm diameter cable and has a maximum pulling capacity of 8,000 kg.

A dozer blade is mounted on the front of the hull. This has a straight motion mechanism and no earth-transporting capability. It is 2.5 m wide when travelling (3 m when working) and 0.9 m high.

A full range of engineer equipment is carried, including cutting and welding equipment. The Engineer Tank does not have an NBC system although the crew compartment is provided with a ventilation and heating unit.

Steyr-Daimler-Puch Engineer Tank 4KH7FA-AVE

Specifications

Crew: 4
Combat weight: 21,000 kg
Power-to-weight ratio: 15.23 hp/t
Length: 7.49 m
Width:
(chassis) 2.5 m
(dozer blade in transport position) 2.5 m
(dozer blade in working position) 3 m
Height:
(overall) 3.15 m
(top of hull) 2.3 m
(transported) 2.8 m
Ground clearance: (loaded) 0.4 m
Track: 2.12 m
Track width: 380 mm
Length of track on ground: 3.037 m
Ground pressure: 0.82 kg/cm^2
Max speed: (road) 65.3 km/h
Range: (road) 600 km
Fuel capacity: 500 litres
Fording: 1 m
Gradient: 70%
Side slope: 40%
Trench: 2.1 m
Turning circle: 8 m
Engine: Steyr 7FA 6-cylinder water-cooled 4-stroke turbocharged diesel developing 320 hp at 2,300 rpm
Transmission: ZF type 6HP500 automatic with 6 forward and 1 reverse gears
Electrical system: 24 V
Batteries: 2 × 12 V, 180 Ah
Armament:
1 × 12.7 mm M2 HB MG
4 smoke grenade dischargers

Engineer equipment
Dozer blade
Width, operating: 3 m
Width, travelling: 2.3 m
Max operating depth: 300 mm

Crane winch
Winch pull: (1st gear/2nd gear) 80 kN/6.5 kN
Effective cable length: 57 m
Cable diameter: 16 mm

Ditcher
Digging depth: 2.2 m
Dumping height: 4.5 m
Radius of operation: 6.3 m
Area of swivel: 234°

Ground auger
Max torque: 2.17 kNm
Max speed: 125 rpm
Boring diameter: 350 mm

Status

Production complete. In service with Austria (18).

Manufacturer

Steyr-Daimler-Puch Spezialfahrzeug AG, Postfach 100, A-1111 Vienna, Austria.
Tel: +43 1 764511. Telex: 61 322 1299.
Fax: +43 1 768149.

UPDATED

CANADA

Diesel Division, General Motors of Canada M113 Engineering Specially Equipped Vehicle (SEV)

Description

The M113 armoured personnel carrier can be converted for basic combat engineering tasks by the addition of a bulldozer blade to the front hull. To increase the combat engineering capabilities of the basic M113 and M113A1 the Diesel Division of General Motors of Canada introduced an 'upgrade' kit, known as a Specially Equipped Vehicle (SEV) kit, with five major features. These are: an improved interior layout for personnel, supplies, tools and equipment; external fuel tanks; a hydraulic earth auger; a hydraulic-powered tool system; and a ramp restraining device.

The revised interior layout provides space for eight troops plus engineer section stores, combat supplies, NBC protective equipment and weapons. The communications equipment is repositioned to provide easier access for the driver and commander. There are folding seats for the troops. The external fuel tanks are at the rear and have a combined capacity of 360 litres. The tanks replace the counter-weights on each side of the rear ramp. The Simcoe hydraulic earth auger is on the left-hand side of the vehicle, mounted on the roof. It can be set up, operated and stowed by two men in a few minutes and can be used to drill a 203 mm hole in earth, asphalt and frozen ground. Holes up to 3.048 m can be dug with the auger rotating at a speed of 75 rpm. Hydraulic power at a pressure of 138 bars is provided and the same power source is used to power the hydraulic tool system located on the right-hand rear of the vehicle. This has power tools on the end of a 15.24 m twin pressure hose which is normally stowed on a reel. Tools include a chain saw, jack hammer and an impact wrench that can also be used for wood boring. The ramp restraining system is on the rear ramp and allows the ramp to be used either as a working platform or a load-carrying area for weights up to 500 kg.

Status
In service with the Canadian Forces. At the end of 1988 production was complete with a total of 55 vehicles, plus 15 conversion kits. There are no known exports of the SEV kit.

Manufacturer
Manufacturer (Carrier vehicle only)
Diesel Division, General Motors of Canada Limited, PO Box 5160, London, Ontario, Canada N6A 4N5.
Tel: +1 519 452 5184. Fax: +1 519 452 5688.

UPDATED

Canadian Forces M113 Engineering Specially Equipped Vehicle (R Stickland)

FRANCE

Giat Industries Main Engineer Vehicle

Development
In 1994 Giat Industries completed a feasibility study, as a private venture, for a Main Engineer Vehicle *(Engin Principal du Génie)* based on the chassis of the Leclerc armoured recovery vehicle which is currently in production for the United Arab Emirates who have ordered 46 vehicles.

Description
The Giat Industries Main Engineer Vehicle will share 70 per cent of the components of the Leclerc ARV but will not have a main recovery winch. Armour over the frontal arc will be thicker than that for the Leclerc ARV and it can be fitted with explosive reactive armour.

Standard equipment could include an articulated telescopic arm pivoted at the front right of the chassis which, when not required, can be stowed along the side of the hull. The arm can be fitted with various attachments such as a bucket, a jack hammer or an auger. A front-mounted multipurpose shovel will be fitted for digging, scarifying and the transport of material.

A platform is provided over the rear engine deck to carry specialised engineer equipment such as a mineclearing system, minefield markers, demolition equipment or the Giat Industries Minotaur scatterable minelaying system. The Giat Industries Demeter magnetic signature duplicator can be fitted to the front of the vehicle, as can mineclearing systems such as ploughs or rollers.

Status
Private venture proposal.

Manufacturer
Enquiries to: Giat Industries, 13 route de la Minière, F-78034 Versailles Cedex, France.
Tel: +33 1 30 97 37 37. Fax: +33 1 30 97 39 00.

UPDATED

Artist's impression of the proposed Giat Industries Main Engineer Vehicle

Giat Industries AMX-30 combat engineer tractor (EBG)

Development
The AMX-30 *Engin Blindé du Génie* (EBG) was developed by Giat Industries to meet the requirements of the French Engineers and the prototype was shown for the first time in 1981. The EBG entered production during 1987 with the first of a batch of 20 units being completed by the end of that year. A further 16 vehicles were funded in 1989. The French Army had a total requirement for 126 EBGs but production ceased after 71 had been produced. The last EBG was handed over to the French Army in early 1994. Development and production of the AMX-30 EBG was undertaken at the Giat Industries facilities at Roanne.

AMX-30 combat engineer tractor (EBG) (C R Zwart) ***1996***

Description
The chassis of the EBG is almost identical to that of the AMX-30D ARV described in the following section but uses automotive components of the later AMX-30 B2 MBT including the engine, transmission, torque converter and suspension. The three-man crew consists of the vehicle commander, sapper and driver.

Mounted at the front of the hull is a hydraulically operated dozer blade with a capacity of 250 m³/h for transport and filling, or 120 m³/h for excavating. Mounted at the back of the lower part of the dozer blade are six scarifying teeth, these being used for ripping up the surface of roads to a depth of 200 mm when the vehicle is being driven in reverse. The dozer blade is 3.55 m wide when fully extended and 1.1 m high.

The hydraulic winch has a capacity of 20,000 kg, is provided with 80 m of cable and has a winching speed of 0.2 to 0.35 m/s irrespective of traction force. Automatic winding speed is from 0.2 to 1.4 m/s with traction capability interlocked with the speed of the vehicle. The winch, which leads out through the front of the vehicle, can be used during amphibious operations.

Pivoted at the front of the hull on the right side is a hydraulic arm with a maximum lifting torque of 15,000 kg/m; the double-jointed arm can be extended to 7.5 m and traversed through a full 360°. The arm is provided with a lifting hook and pincer-type grab similar to that fitted to the Russian IMR engineer vehicle described elsewhere in this section. The arm can also be fitted with an auger which can drill 220 mm diameter

Close-up of the two-tier turret of the Giat Industries AMX-30 EBG from the rear (C R Zwart) ***1997***

holes in the ground to a depth of 3 m. A 220 mm cutting saw is carried and a 12.5 kW hydraulic PTO is provided as standard.

Mounted in the centre of the hull, slightly offset to the right, is a two-tier turret. The upper part has a single-piece hatch cover that opens to the rear and is fitted with a 7.62 mm machine gun. To the rear of the turret on either side are two electrically operated smoke grenade dischargers.

Mounted on the forward part of the lower tier is a launching tube for demolition charges and either side of this are two mine-launching tubes, each of which has a launching container of five mines.

The 142 mm calibre demolition charge is 800 mm long, weighs 17 kg and contains 10 kg of explosive. The charge is fin-stabilised and fitted with a nose-mounted point-detonating fuze; range is between 30 and 300 m.

The mines are 139 mm in diameter, weigh 2.34 kg and contain 0.7 kg of explosive. The launcher discharges the mines to a distance of between 60 and 250 m and the mines are then triggered by any vehicle weighing over 1,500 kg. According to Giat, the mines will penetrate a tank floor equivalent to 50 mm of armour plate at 500 mm standoff distance and 60° incidence, or break up a tank's track if the mine is run over. The mines self-destruct after a preset time.

Standard equipment on the EBG includes provision for deep fording with the aid of a snorkel, passive night periscope for the driver, NBC system and a rangefinder telescope for the sapper.

The AMX-30 EBG can carry and launch the CNIM Assault Treadway Bridge which is covered in the *Mechanised Bridges* section.

Specifications

Crew: 3
Weight combat: 41,500 kg
Power-to-weight ratio: 16.86 hp/t
Ground pressure: 0.9 kg/cm^2
Length: (overall) 8.29 m
Width:
(overall) 3.55 m
(without dozer blade extensions) 3.12 m
Height: (overall) 3.02 m
Ground clearance: 0.45 m
Track: 2.52 m
Track width: 570 mm
Length of track on ground: 4.12 m
Max speed: (road) 65 km/h
Fording:
(without preparation) 2.5 m
(with preparation) 4 m
Gradient: 60%
Side slope: 30%
Vertical obstacle: 0.9 m
Trench: 2.9 m
Engine: Hispano-Suiza HS 110-2 12-cylinder, water-cooled supercharged multifuel developing 700 hp at 2,000 rpm, or HS 110-S2 developing 800 hp
Transmission: 5 SD with 5 gears in both directions, or ENC 200 gearbox with lock-up torque converter with 5 forward and 1 reverse gears
Electrical system: 24 V
Armament:
1 × 7.62 mm MG
2 × 2 smoke grenade dischargers
1 × 142 mm demolition charge projector
4 mine projectors
Ammunition:
(MG) 4,000 rounds
(demolition charges) 5
(mines) 40 (8 containers each holding 5)

Status

Production complete. In service with the French Army (71).

Manufacturer

Enquiries to: Giat Industries, 13 route de la Minière, F-78034 Versailles Cedex, France.
Tel: +33 1 30 97 37 37. Fax: +33 1 30 97 39 00.

UPDATED

GERMANY

MaK Pionierpanzer 2 Dachs armoured engineer vehicle

Description

MaK System Gesellschaft mbH, developed and produced three prototypes of the Pionierpanzer 2 Dachs (Badger) armoured engineer vehicle. A contract award for the conversion of a mix of Leopard 1 ARVs and AEVs was awarded by the German MoD during the early part of 1987 for delivery by December 1990. The conversion price was DM1.3 million per unit. A further nine vehicles, known as the Badger, were produced for the Canadian Forces using new chassis and hulls. The Canadian contract was worth DM49 million and was completed in 1990.

For the German conversion programme MaK stripped down the original vehicles, repaired them where necessary, fitted the dozer blade assembly and hydraulic system and completed the final assembly. Jung Jungenthal GmbH was responsible for hull modifications, Eisenwerke Kaiserslautern for the excavator and traversing turntable and Wieger Maschinenbau for the telescopic arm assemblies.

The Pionierpanzer 2 is based on the hull, running gear and suspension of the Leopard 1 AEV (see following entry) and the Leopard 1 ARV (see entry under *Armoured Recovery Vehicles*). The vehicles have an armoured superstructure situated towards the left-hand side of the vehicle with a telescopic excavator arm located on the right with the traversing turntable situated forward. An extended dozer blade is mounted on the front of the vehicle.

The driver is seated at the front of the superstructure under a hatch cover. He is provided with a number of periscopes and night driving aids and he operates the dozer blade. Behind him is seated the commander with his own cupola and all-round vision periscopes. Periscopes provided for the commander and driver can be tilted. The cupola has provision for mounting a 7.62 mm MG3 machine gun. The commander operates the telescopic excavator arm. There is a third crew member seated within the superstructure.

The telescopic excavator arm is hydraulically operated by a 300 bar system and is electrically controlled. It can be used in a variety of ways and can be elevated or depressed to an angle of ±60°. Its horizontal arc of operation is 196.5°. The bucket on the end of the arm has a capacity of 1.1 m^3 and the arm can be extended to a maximum length of 8.3 m. It has an earth working capacity of 140 m^3/h. The arm can also be used as a crane jib. If required the excavator arm can be controlled from a remote dismounted position.

The dozer blade has a maximum width of 3.75 m with side extensions fitted and has scarifiers to rip up road surfaces. Maximum operating speed with the dozer blade in use is 8 km/h and the blade can shift up to 270 m^3/h of spoil.

The Pionierpanzer 2 is equipped with a front-mounted hydraulic winch that can be used for self-recovery or the recovery of other vehicles, in which case the dozer blade can be lowered to improve operating stability. The winch, which is sealed, has a cable tensioning system, is supplied with 90 m of 33 mm cable and has a straight pull capacity of 20,000 kg. This can be increased to 35,000 kg by multireeving the cable. The vehicle can be used to tow disabled vehicles. Canadian Badger vehicles are fitted with a 35,000 kg capstan winch.

German Army Pionierpanzer 2 Dachs armoured engineer vehicle with snorkel tower fitted preparing river crossing point

German Army Pionierpanzer 2 Dachs armoured engineer vehicle in travelling configuration ***1996***

By using a deep wading kit which includes a snorkel tower, the vehicle can be used to prepare underwater approach routes or clear obstacles at depths of up to 4 m. Additional bilge pumps are provided for underwater operations.

Specialist equipment carried on the vehicle includes cutting and welding equipment with an operating voltage of 54 V.

Specifications

Crew: 3
Weight:
(empty) 42,500 kg
(loaded) 43,000 kg
Power-to-weight ratio: 19.30 hp/t
Ground pressure: (combat weight) 0.92 kg/cm^2
Length:
(travelling, roads) 8.925 m
(travelling, cross-country) 9.01 m
Width: 3.25 m
Width, dozer blade extended: 3.75 m
Height: (to top of MG mount) 2.57 m
Ground clearance: 0.44 m
Track: 2.7 m
Track width: 550 mm
Length of track on ground: 4.236 m
Max speed: (road) 62 km/h
Range: approx 650 km
Fuel capacity: 1,410 litres
Fording:
(normal) 1 m
(with preparation) 4 m
Gradient: 50%
Side slope: 30%
Vertical obstacle: 0.9 m
Trench: 2.5 m
Engine: MTU MB 838 Ca M-500 37.4 litres V-10 multifuel developing 830 hp at 2,200 rpm
Transmission: ZF 4 HP 250 planetary-gear shift with hydraulic torque converter, 4 forward and 2 reverse gears
Steering: regenerative double differential
Electrical system: 24 V
Batteries: 6 × 12 V, 300 Ah
Armament:
1 × 7.62 mm MG3 MG
6 smoke grenade dischargers
Dozing capacity: 270 m^3/h
Excavating capacity: 140 m^3/h
Excavator pull force: 125 kN
Excavator push force: 85 kN
Towing force main winch: 350 kN
Scarifying depth: 0.45 m
Dozing and scarifying speed: 8 km/h (max)
Traverse, excavator arm: 195°
Tilt angle, telescopic arm: ±60°

Status

Production complete. In service with Canada (9) and Germany (141).

Manufacturer

MaK System Gesellschaft mbH, PO Box 9333, D-24157 Kiel, Germany.
Tel: +49 431 3995-02. Fax: +49 431 3995 446.

UPDATED

MaK Leopard 1 armoured engineer vehicle

Development

The Leopard 1 armoured engineer vehicle is a direct development of the Leopard 1 ARV and differs from it only in minor details. The first prototype, based on a design by Dr Ing hc P Porsche KG, was completed by MaK of Kiel in 1967, with the first production vehicles completed in 1968. Wherever possible, standard Leopard 1 MBT components were used, for example the engine, transmission and suspension.

Thirty-three Leopard 1 armoured engineer vehicles were converted to Pionierpanzer 2 Dachs armoured engineer vehicles (see previous entry).

Description

The hull is of all-welded construction with the crew compartment at the front and the engine and transmission at the rear. The driver is seated at the front of the vehicle to the left of the crane and is provided with a single-piece hatch cover and three periscopes for observation, one of which can be replaced by a passive night vision periscope. The commander is seated to the rear of the driver and has a single-piece hatch cover and eight periscopes, one of which can be replaced by a passive night vision periscope. There is another hatch to the rear of the commander's position, and three periscopes provide vision to the rear of the vehicle. There is a swivelling periscope in the roof of the superstructure. The crew of four (commander, driver and two mechanics) can enter the vehicle either through the roof hatches or through two doors in the left side of the hull. The suspension is of the torsion bar type and consists of seven roadwheels with the drive sprocket at the rear and the idler at the front. There are four track-return rollers. The first, second, third, sixth and seventh roadwheel stations are fitted with hydraulic shock-absorbers. The tracks are of the double-pin type and have rubber pads.

The Leopard 1 AEV can carry out similar roles to the ARV including recovering disabled vehicles and changing vehicle components (although an auger is carried in place of the spare engine carried on the ARV).

The dozer blade is mounted at the front of the hull and operated by two hydraulic cylinders through two lever arms and is mechanically locked when not in use. The dozer blade has a maximum capacity of 200 m^3/h and the width of the blade can be extended to 3.75 m by additional side attachments It can be fitted with four scarifiers to rip up the surface of roads. The installation of a heat exchanger enables the vehicle to perform unlimited bulldozing activities even at high ambient temperatures. In addition to being used for dozing operations, the dozer blade is also used to stabilise the AEV when the crane is in use or when vehicles are being recovered.

The main winch, with its horizontal cable drum, is in the centre of the crew compartment under the floor. This winch is provided with a total of 90 m of 33 mm cable. The maximum tractive effort of the main winch in the lowest cable position is 35,000 kg, which can be increased to 70,000 kg when using an appropriately

Leopard 1 armoured engineer vehicle with earth auger in action ***1996***

located guide pulley. The exit opening for the main winch is located at the nose of the vehicle. A hydraulically driven cable-tensioning device is located immediately behind the opening and automatically extends or rewinds the cable.

The crane jib is mounted at the front of the hull on the right side and when in the travelling position lies horizontally along the right side of the hull. It has a maximum lifting capacity of 20,000 kg when used with the dozer blade as a stabiliser. A scale on the side plate of the jib shows the allowed load: if it is exceeded the jib is stopped automatically. The two hydraulic cylinders for lifting the jib are arranged so that they are fully splinter-proof in the travelling position and extensively splinter-proof in the operating position. The jib is turned by the traversing gearbox directly underneath the console of the crane. If the hydraulics fail, the traversing gear can be operated by hand. The ladder mounted on the jib assists in the assembly of the earth-drilling equipment and can also be used for other work such as overhead lines. The auger is mounted, when required, at the end of the jib crane and can drill holes 700 mm in diameter to a maximum depth of 2 m. The hydraulic oil is supplied to the radial piston motor of the earth drill through hoses which can be connected to the jib boom by means of quick-disconnect couplings.

The hoisting winch with a vertical cable drum is mounted on the right side of the crew compartment. It is provided with 100 m of 13 mm diameter cable.

Equipment carried on the AEV includes a set of shackles, towbar, electric impact wrench, guide pull and electric welding and cutting equipment. For demolition work a total of 117 kg of explosives is carried.

Armament consists of a 7.62 mm MG3 machine gun mounted in the bow of the tank on the left side; this has an elevation of +15°, a depression of −15°, and a traverse of 15° left and right. A similar machine gun is mounted on the commander's hatch for anti-aircraft defence. Six smoke grenade dischargers are mounted on the left side of the hull.

The Leopard 1 AEV is also provided with an NBC system, crew compartment heater, fire warning and extinguishing system and a hull escape hatch. The basic model can ford to a depth of 2.1 m, but for deep fording operations a snorkel can be quickly installed.

Specifications

Crew: 4
Weight:
(empty) 40,200 kg
(loaded) 40,800 kg
Power-to-weight ratio: 20.70 hp/t
Ground pressure: 0.86 kg/cm^2
Length: (with dozer blade raised) 7.98 m
Width:
(with side elements of blade fitted) 3.75 m
(hull) 3.25 m
Height: (with AA MG) 2.69 m
Ground clearance: 0.44 m
Track: 2.7 m
Track width: 550 mm

Length of track on ground: 4.236 m
Max speed: (road) 62 km/h
Range:
(road) 800 km
(cross-country) 500 km
Fuel capacity: 1,410 litres
Fording: 2.1 m
(with snorkel) 3 m
Gradient: 60%
Side slope: 30%
Vertical obstacle: 1 m
Trench: 3 m
Engine: MTU MB 838 Ca M-500 10-cylinder multifuel developing 830 hp at 2,200 rpm
Transmission: ZF 4 HP 250 with 4 forward and 2 reverse gears
Electrical system: 24 V
Batteries: 6 with total capacity of 300 Ah, charged by 3-phase generator driven from main engine
Armament:
1 × 7.62 mm MG3 MG in bow
1 × 7.62 mm MG3 MG on commander's hatch
4,250 rounds of 7.62 mm ammunition
6 smoke grenade dischargers

Status
Production complete. In service with Belgium (6), Germany (36), Italy (40) and the Netherlands (25). OTO Melara (now OTOBreda) built 28 Leopard 1 AEVs for the Italian Army.

Manufacturer
MaK System Gesellschaft mbH, PO Box 9333, D-24157 Kiel, Germany.
Tel: +49 431 3995-02. Fax: +49 431 3995 446.

UPDATED

Leopard dozer blade unit

Description
The Leopard dozer blade unit can be fitted as an optional extra to Leopard 1 and 2 MBTs. The complete dozer blade unit consists of the dozer blade, two pivot-mounted push arms and a hydraulic unit for raising and lowering the blade. A floating capability is available to enable the dozer blade to follow ground contours. Add-on extension elements allow for the dozing of lanes wider than the width of the tank's running gear.

The complete dozer blade can be attached or removed in approximately 10 minutes without recourse to cranes or other equipment and it can be mechanically locked in the raised position for travelling. All control elements are packaged into a watertight unit so that the dozer blade can be used during fording operations. A control panel is located in the driver's position and the power supply and control cable passes through one of the slots normally occupied by the vision blocks via a cable adaptor.

For using this dozer blade kit with Canadian Leopard C1 MBTs an Engineer Equipment Interface Modification Kit was developed and introduced for the Canadian Leopard C1, as well as into the Australian Leopard 1 MBT fleets. This kit allows the tank to operate the dozer blade kit or one of two mineclearing systems, permitting operation with the driver's hatch closed down and all the driver's vision blocks remaining operable.

Status
Production complete but can be resumed if additional orders are placed. In service with several nations using Leopard MBTs, including Australia (Leopard 1), Canada (Leopard C1) and Denmark.

Manufacturer
Krauss-Maffei-Wehrtechnik GmbH, Krauss-Maffei Strasse 2, D-80997 München, Germany.
Tel: +49 89 8899-0. Telex: 523 3163-31.
Fax: +49 89 812 01 90.

UPDATED

Australian Army Leopard 1 AS MBT equipped with Leopard dozer blade (Australian Army)

INDIA

Armoured Amphibious Dozer

Development
The Armoured Amphibious Dozer was designed for earthmoving operations in a battlefield environment coupled with good cross-country mobility. The vehicle was developed by the Research and Development Establishment (Engineers) at Dighi, Pune and the Vehicle Research and Development Establishment at Ahmednagar; also involved were various Ministry of Defence, government of India, system laboratories. The vehicle is based on the BMP-2 IFV, licence-produced in India as the Sarath. Licence production of the BMP-2 is undertaken in India at the Ordnance Factory Medak. Details of the Indian BMP-2 are given in *Jane's Armour and Artillery 1996-97* pages 290 to 291. As far as it is known, India has not exported any of these vehicles.

A development programme is in progress to produce a similar vehicle on a locally devised chassis.

Description
The Armoured Amphibious Dozer utilises a modified BMP-2 (Sarath) hull and chassis allied to various items of combat engineer equipment. The vehicle has a crew of one or two, with dual control for the driver and dozer bucket operator located back-to-back. The dozer bucket is operated to the rear of the vehicle and has a bucket capacity of 1.5 m^3. Upwards-folding track width mine ploughs, apparently based on the Russian BMP mineclearing ploughs (qv), are located at the front of the hull; the ploughs are lowered by gravity and raised hydraulically. An 8 tonne winch can be combined with a self-recovery rocket-propelled anchor launched from a position over the front hull roof; cable lengths may be 50 or 100 m. Also provided is a 3 t/m jib crane. Hydraulic power for the dozer bucket, winch, mine plough raising mechanism and suspension locks is provided by a 200 bar pressure self-contained hydraulic system.

The armoured and NBC protection, smoke generator and amphibious capabilities of the BMP-2 (Sarath) are retained. Bridge classification is MLC 16.

Specifications
Crew: 1 or 2
Bridge classification: MLC 16
Max speed:
(road) 60 km/h
(water) 7 km/h
Fording: amphibious

Rear view of Armoured Amphibious Dozer in transport configuration **1996**

Armoured Amphibious Dozer in transport configuration **1996**

Gradient: 30°
Vertical obstacle: 0.7 m
Trench: 2 m
Bucket capacity: 1.5 m^3
Winch capacity: 8,000 kg
Jib capacity: 3 t/m

Status
Prototypes.

Manufacturer
Production of the basic BMP-2 is undertaken in India at the Ordnance Factory Medak.

Development Establishment
Research and Development Establishment (Engineers), Dighi, Pune 411 015, India.
Tel: +91 212 667881-7. Telex: 0145 7511 RDEE IN.
Fax: +91 212 666583.

UPDATED

Engineer Reconnaissance Vehicle

Development
The Engineer Reconnaissance Vehicle was developed by the Research and Development Establishment (Engineers) at Dighi, Pune and the Vehicle Research and Development Establishment at Ahmednagar; also involved were various Ministry of Defence, government of India, system laboratories. The vehicle is based on the BMP-2 IFV, licence-produced in India as the Sarath and is intended to be a fully integrated system vehicle capable of acquiring, recording and transmitting combat engineer and military bridging reconnaissance data to a central command post. Data relating to both dry land and water obstacles may be involved, including river bank height and gradient, soil load-bearing capacity, river bed profiles and so on.

About 10 of these Engineer Reconnaissance Vehicles had been produced by mid-1995.

Engineer Reconnaissance Vehicle in operation ***1996***

Description
The Engineer Reconnaissance Vehicle, which is fully amphibious, utilises a modified BMP-2 (Sarath) hull and chassis allied to various items of combat engineer reconnaissance equipment and land navigation instruments. The vehicle has a small turret capable of mounting a machine gun armament and smoke generators are provided.

Inside the hull are a number of wall-mounted navigation systems, control positions and various instruments operating on a 24 V DC power supply with Ni/Cd batteries for back-up. Most of the systems carried in the vehicle are microprocessor based and use modular construction printed circuit boards.

For navigation purposes there is a Gyro Land Navigation System, a Radio Navigation and Guidance System, and a Global Positioning System. Instruments carried include: an echo-sounder, a tripod- or hull-mounted electronic combined rangefinder and theodolite and a portable instrument known as a conpenetrometer, used to test soil conditions and load-bearing capabilities. Other instruments include an electronic distance metre and a water current meter. Reconnaissance data can be gathered from outside or from within the protection afforded by the armoured hull. All data gathered can be integrated in various standard forms at a central control console where it can be logged in a 386 ruggedised IBM-compatible personal computer. Processed data is transmitted to a central control post via a pair of radio modems using FSK modulation techniques and the VHF band.

Located on the hull rear is a 'trail blazing' system using marker rods driven vertically into the ground to mark a path. This electropneumatic equipment contains 50 marker rods in a magazine, each rod being 1.2 m long with a diameter of 10 mm.

Status
About 10 vehicles produced to date.

Manufacturer
Production of the basic BMP-2 is undertaken in India at the Ordnance Factory Medak.

Development Establishment
Research and Development Establishment (Engineers), Dighi, Pune 411 015, India.
Tel: +91 212 667881-7. Telex: 0145 7511 RDEE IN.
Fax: +91 212 666583.

UPDATED

ISRAEL

Urdan RKM bulldozer attachment

Description
The Urdan RKM bulldozer attachment was developed jointly by Urdan RKM Limited and the Israeli Army to provide every MBT with the ability to act as an earthmoving vehicle. In Israel Army service the RKM bulldozer attachment is fitted to Centurion, M48/M60 and locally built Merkava MBTs. It can be fitted to other MBTs and according to the manufacturer no modifications are required to the actual vehicle in order to carry the RKM bulldozer attachment.

The Urdan RKM bulldozer attachment has three main assemblies. The first is a standard bulldozer blade modified slightly to improve its performance when working in rocky terrain and sandy soils. The second part is the electrohydraulic unit which is completely sealed to enable operation underwater. It is connected to the vehicle's electrical system to supply hydraulic power to the main cylinder; the unit can be easily and rapidly replaced for maintenance. The electrohydraulic system and the movements of the blade are controlled from a control box inside the driver's compartment. The third part of the RKM bulldozer attachment is the main structure which includes the adaptor and the arms; it carries the bulldozer blade and the power unit. The structure was designed to withstand vibrations caused by the higher tank driving speeds when crossing difficult terrain. The attachment can be installed and removed by the normal tank crew in about 15 minutes. If required, the blade can be locked in its upper (travel) position manually if the hydraulic system is damaged. The high folding geometry of the attachment enables the carrier tank to negotiate steep banks, terraces and vertical obstacles.

The RKM bulldozer attachment has several operating modes. One is the travel position with the blade in the maximum up position. The bulldozer blade can also be made to 'float' so that the blade rests on the soil, but in the 'blade down' mode the hydraulic cylinder pushes the blade down into the soil for heavy earthmoving. The attachment has automatic hydrostatic locking at any given position and a mechanical safety lock (using the travel lock) at the highest position.

Urdan RKM bulldozer attachment fitted to Merkava Mk 1 tank showing angle of blade when fully lifted

Status
Production as required. In service with the Israeli Army.

Manufacturer
Urdan Industries Limited, Industrial Zone, IL-42378 Netanya, Israel.
Tel: +972 9 338074. Telex: 341 822 uasf il.
Fax: +972 9 610246.

UPDATED

Israel Military Industries ABK-3 add-on bulldozer kit

Description

Israel Military Industries Ltd produces an add-on bulldozer kit known as the ABK-3. This is suitable for use on a range of MBTs including the Centurion, M60 and several other designs. The attachment is joined to the MBT using existing towing and other lugs. Installation takes about 20 minutes without requiring any special tools. When fitted, the dozer blade can dig to 0.25 m below the track level. The blade is 0.99 m high and 3.96 m wide. Weight of the complete kit is approximately 3,400 kg and a 24 V DC power supply is required from the carrier vehicle to drive a 10 hp electric motor. The control box is located in the driver's compartment.

The dozer blade operates in four modes: travelling; 'floating', with the blade resting on the soil surface; blade down for digging; and with blade routers lowered for surface ripping. Blade capacity is 3.1 m³. The dozer blade can also be used for clearing scatterable mines from flat surfaces. Tank speed when digging is approximately 5 km/h.

Status

Production as required. In service with the Israeli Army.

Manufacturer

Israel Military Industries Ltd, Marketing Division, PO Box 1044, IL-47100 Ramat Hasharon, Israel.
Tel: +972 3 542 5222. Telex: 33 719 misbit il.
Fax: +972 3 48 9639.

UPDATED

Israel Military Industries ABK-3 add-on bulldozer kit mounted on Centurion tank

Israel Military Industries bulldozer protection kit

Description

Operational experience gained by the Israeli Defence Forces demonstrated that there is a requirement for some form of protection for combat engineer bulldozers working in combat areas. Protection is required from small arms and artillery fire and also for vehicles operating in areas where chemical attack is possible. Israel Military Industries Ltd designed an add-on kit for such situations which has been fitted to Caterpillar D9 bulldozers, but it can be adapted to suit other models.

The add-on kit requires no modifications to the basic vehicle structure. The armoured cab is mounted on the vehicle's standard safety frame mounts and the added weight has no effect on overall performance characteristics. The armoured cab has space for an operator/driver and a commander/assistant. Reinforced glass block windows for vision forward, to the sides and to the rear, may be removed, leaving an external hardened glass pane to maintain cab isolation and conditioning. Two doors are provided at the front of the cab for entry and exit and an additional door at the rear doubles as a refuelling access point and an emergency exit. The commander has a roof cupola with a full 360° traverse and provision for mounting a machine gun and/or a grenade projector. The cab armour is of a special layered type and armour plating is provided for the engine compartment on three sides complete with integral air vanes for engine compartment ventilation; servicing access is still possible. Protection is also provided for the hydraulic pistons and tubing and for the fuel and hydraulic fluid tanks.

The cab is equipped with an air conditioning unit driven by the vehicle engine and with an internal and external communications system. Internal lighting is provided and an external light projector may also be fitted. Full convoy and blackout lights are provided and other external additions contained in the kit include weapon mountings, brackets for four fuel containers, a camouflage net and three window wipers. Internal stowage includes water tanks, brackets for personal weapons and equipment, stowage for machine gun ammunition and four spare reinforced glass window blocks. The cab is sealed for operations under NBC conditions.

Specifications

Weight of attachment: approx 6,000 kg
Cab dimensions (internal):
(length) 2 m
(width) 2.6 m
(height) 1.7 m
Air conditioner capacity: 24,000 BTU/h

Status

Production as required. In service with the Israeli Army.

Manufacturer

Israel Military Industries Ltd, Marketing Division, PO Box 1044, IL-47100 Ramat Hasharon, Israel.
Tel: +972 3 542 5222. Telex: 33 719 misbit il.
Fax: +972 3 48 9639.

UPDATED

Bulldozer protection kit fitted to Caterpillar D9

ITALY

Astra M113 combat engineer support vehicle

Description

Astra Veicoli Industriali SpA carried out a repowering and updating project on the M113 APC to enable its combat weight to be of the order of 15,000 kg (for details of the Italian M113 series fleet are given in *Jane's Armour and Artillery 1996-97* pages 301 to 302). One application for the uprated vehicle is the SIDAM 25 self-propelled air defence vehicle carrying four 25 mm cannon in a turret. A second application is as a combat engineer support vehicle for the Italian Army Engineer Corps.

The Astra conversion involves changes to the hull, engine, transmission, suspension, fuel tanks and armour. Hull alterations involve the relocation of some engine compartment mounts, modifying the top deck to incorporate an improved reverse-flow cooling system, reinforcement of the idler wheel mounts and repositioning the shock-absorber supports. The existing Detroit Diesel 6V-53 diesel is replaced by a 6V-53T engine developing 265 hp; options include a 300 or 350 hp version of the same engine. The cooling system is fitted with a new fan, a reverse air flow system and a new air cleaner and silencer.

The existing Allison TX-100-1 transmission is replaced by the TX-100-1A. The Allison X-200-4 transmission may be fitted as an option. New roadwheel arms are fitted to the suspension along with new shock-absorbers on four wheel arms each side. New and improved torsion bars are also installed.

Two new fuel tanks are fitted to the hull rear, providing between them a total capacity of 340 litres. Optional equipment includes extra light armour on the hull front and sides.

Basic development vehicle for Astra Veicoli M113-based armoured engineer support vehicle

Specialist equipment to be fitted to production vehicles includes a front-mounted, hydraulically operated dozer blade, a roof-mounted 3 tonne hydraulic crane; maximum jib extension is 3.29 m. A 10 tonne capacity recovery winch is located internally with the cable paying out through the rear hatch. A 12.7 mm M2 machine gun can be fitted on a pintle close to the commander's cupola.

Specifications

Weight: 15,500 kg
Power-to-weight ratio: 17.09 hp/t
Length:
(dozer blade lowered) 6.86 m
(hull) 5.9 m
Width: (hull) 2.94 m
Height:
(cupola) 2.05 m
(top of crane) 3 m
(crane at max elevation) 6 m
Ground clearance: 0.4 m
Max speed: 55 km/h
Range: 500 km
Fuel capacity: 340 litres
Fording: amphibious
Gradient: 60%
Side slope: 30%
Vertical obstacle: 0.61 m
Trench: 1.68 m
Engine: Detroit Diesel 6V-53T diesel developing 265, 300 or 350 bhp
Transmission: Allison TX-100-1A
Armament: 1 × 12.7 mm M2 HB MG

Status

Under development for the Italian Army.

Manufacturer

Astra Veicoli Industriali SpA, Via Caorsana 79, I-29100 Piacenza, Italy.
Tel: +39 523 5431. Telex: 530148.
Fax: +39 523 69293.

UPDATED

JAPAN

Mitsubishi Type 67 armoured engineer vehicle

Description

The Mitsubishi Type 67 armoured engineer vehicle is based on the chassis of the Type 61 MBT and was preceded by a trials vehicle based on the Sherman M4A3E8. This had a dozer blade at the front of the hull and a small crane. The Type 67 AEV was designed to clear obstacles from the battlefield and is provided with a dozer blade and crane.

Specifications

Crew: 4
Weight: 35,000 kg
Power-to-weight ratio: 18.57 hp/t
Length: 7.46 m
Width: 3.2 m
Height: 2.9 m
Ground clearance: 0.4 m
Track: 2.45 m
Track width: 500 mm
Length of track on ground: 3.5 m
Ground pressure: 0.95 kg/cm^2
Max speed: (road) 45 km/h
Range: 200 km
Fording: 1 m
Gradient: 60%
Vertical obstacle: 0.8 m
Trench: 2.7 m
Engine: Mitsubishi Type 12 HM 21 WT V-12 direct injection turbocharged air-cooled diesel developing 650 hp at 2,100 rpm
Transmission: mechanical with 5 forward and 1 reverse gears, with 2-speed auxiliary reduction unit
Electrical system: 24 V
Batteries: 4 × 12 V, 200 Ah
Armament:
1 × 12.7 mm M2 HB MG
1 × 7.62 mm MG
Armour
hull front: 46 mm
hull sides: 25 mm
hull rear: 15 mm

Status

Production complete. In service only with the Japanese Ground Self-Defence Force.

Manufacturer

Production was undertaken at the Maruko, Tokyo, plant of Mitsubishi Heavy Industries, but AFV production is now at the Sagamihara plant near Tokyo.
Mitsubishi Heavy Industries, 5-1, Marunouchi 2-chome, Chiyoda-ku, Tokyo, Japan.

UPDATED

Mitsubishi Type 75 armoured dozer

Development

Development of an armoured dozer began in 1964 and two prototypes were built during Fiscal Year 1972 by Komatsu, one with a straight dozer blade and another with an angled blade. After extensive trials the latter was standardised in 1975 and the dozer is in service with the Japanese Ground Self-Defence Force.

The Type 75 armoured dozer can carry out over 90 per cent of the tasks of the D6 medium dozer but has a much higher road speed to enable it to keep up with the leading elements of a convoy.

Description

When the vehicle is travelling, the dozer blade is to the rear and the fully armoured crew compartment at the front. The two-man crew can enter either by a door on the right side of the hull or through two circular rear-opening roof hatches. The driver is seated on the left with a rectangular shutter hinged at the top immediately

Type 75 armoured dozers (Kensuke Ebata)

1996

Type 75 armoured dozer (K Nogi)

in front of him. This has an integral vision block and can be locked open. The commander is seated to the right of the driver and has a small square shutter hinged at the top, which also has a vision block and can be locked open. Vision blocks are provided in each side of this compartment and there is a shutter with an integral vision block in the rear of the crew compartment for forward vision during dozing operations. There is also a single vision block to the left of the shutter.

The engine and radiator are at the rear of the vehicle. In a combat area they are protected by armoured plates which are removed when the Type 75 is operating in a non-combat area.

The torsion bar suspension consists of five dual rubber-tyred roadwheels with the drive sprocket at the front and the idler at the rear. There are also three track-return rollers. Hydraulic shock-absorbers are fitted at the first, second and fifth roadwheel stations.

The hydraulically operated dozer blade is controlled by the driver from inside the armoured cab. The dozer blade is hinged in the middle and a hydraulic winch can be fitted in front of the cab if required.

Specifications

Crew: 2
Weight: 19,200 kg
Power-to-weight ratio: 17.96 hp/t
Length:
(travelling) 6.84 m
(working) 6.3 m
Width:
(travelling) 2.7 m
(working) 3.45 m
Height: 2.79 m
Max speed: 45 km/h
Fording: 1 m
Gradient: 60%
Engine: water-cooled diesel developing 345 hp at 2,100 rpm on road and 160 hp at 1,850 rpm when working
Armament: none

Status

Production complete. In service with the Japanese Ground Self-Defence Force.

Manufacturer

Komatsu, 3-6 Akasaka 2-chome, Minato-ku, Tokyo 107, Japan.
Tel: +81 3 584 7111. Telex: 22 812 komatsu j.

UPDATED

NORWAY

Hägglunds Moelv armoured engineer vehicle

Development

In 1995 Hägglunds Moelv was awarded a contract by the Norwegian Army for the conversion of 22 Leopard 1 MBT chassis into armoured engineer vehicles.

Main roles of this vehicle can be summarised as:

(1) Repair and re-establish axis of advance
(2) Rapid construction of slit trenches for armoured vehicles and assault personnel
(3) Construction of field locations and fire emplacements for MBTs and armoured infantry fighting vehicles
(4) The preparation of access and egress sites/entry and exit ramps for water crossing and to make river beds trafficable
(5) Load gravel, packaged cargo, timber and building material on to trucks
(6) Construct and remove obstacles and barriers on the battlefield
(7) The destruction of roads.

Description

The conversion includes the removal of the complete turret, the installation of a new all-welded superstructure for the three man crew and new engineer systems.

Mounted at the front of the vehicle is a hydraulically operated dozer blade. This has three hydraulic cylinders; one tilt and two elevation. The primary mission of this is to clear and destroy roads, with the secondary mission being to use the dozer blade as a soil anchor when operating the winch or excavator boom. The dozer blade is 3.75 m wide and 1.2 m high.

Pivoted at the front of the hull is a hydraulically operated arm which is traversed to the rear when not required. This arm can be traversed through 180° and is fitted with a bucket with a capacity of 0.7 m^3 for digging purposes. The arm has a maximum reach of 8 m. If required this can be replaced by a soil drill.

Two hydraulic capstan winches are provided for recovery purposes and these would normally be used in conjunction with the dozer blade. One winch has a towing force of 90 kN with a total theoretical force of 720 kN.

Options include an overhead weapon station (Wegmann Type 2038 or RAFAEL 12.7D), larger capacity (1 m^3) excavator bucket, additional armour protection and a soil drill.

Specifications

Crew: 3
Weight: 45,000 kg
Power-to-weight ratio: 18.44 hp/t
Length: 9.8 m
Width:
(overall) 3.75 m
(rail transport) 3.37 m
Height: 3 m
Ground clearance: 0.44 m
Max road speed: 65 km/h
Max road range: 600 km
Fuel capacity: 985 litres
Fording:
(with preparation) 2.2 m
(with snorkel) 4 m
Vertical obstacle: 2.3 m
Trench: 2.5 m
Gradient: 60%
Side slope: 40%
Engine: MTU MB 838 Ca M-500, 10-cylinder diesel developing 830 hp at 2,200 rpm
Transmission: ZF 4 HP 250 planetary-gear shift with hydraulic torque converter, 4 forward and 1 reverse gears
Steering: double differential
Clutch: torque converter with mechanical interlock
Suspension: torsion bar
Batteries: 8 × 12 V, 100 Ah (each)
Electrical system: 24 V

Status

Under development for the Norwegian Army who have ordered 22 vehicles.

Manufacturer

Hägglunds Moelv AS, PO Box 244, N-2391 Moelv, Norway.
Tel: +47 62 346000. Fax: +47 62 346001

NEW ENTRY

Artist's impression of armoured engineer vehicle fitted with a snorkel for deep fording operations **1997**

Artist's impression of amoured engineer vehicle in operational use **1997**

POLAND

M.I.D. armoured engineering vehicle

Description

During 1992, the Research and Development Centre of Mechanical Appliances at Gliwice developed and produced the prototype of an armoured engineering vehicle based on the hull of the T-72 MBT and incorporating many features of the WZT-3 armoured recovery vehicle (see entry under *Armoured Recovery Vehicles*).

The prototype, known as the M.I.D. Road Engineering Machine (M.I.D. - *maszyna inzyniery jnodrogowa),* uses many engineering features and components from the T-72/WZT-3. The main difference from the WZT-3 is the provision of a hydraulically powered heavy jib crane arm located on the right-hand side of the armoured superstructure. The crane arm is provided either with gripper claws intended for the grasping and removal of battlefield obstacles, or a digger bucket. The jib crane arm has a lift capacity of 7,000 kg and can extend from 5.94 to 7.94 m over a horizontal working angle of 240°; the maximum vertical working angle is +60° and minimum −55°. The gripper claws fitted to the end of the jib arm can open to a maximum width of 1 m, lift 3,000 kg and turn through an angle of 180°. Maximum operating height is 7.5 m

M.I.D. armoured engineering vehicle based on chassis of T-72 MBT ***1996***

and depth 5.5 m. If required, the gripper claws can be replaced by a 120 m^3 scoop bucket with a maximum cutting depth of 5.5 m. When not in use the scoop bucket and other equipment can be carried on the hull rear.

Located on the front of the vehicle is a V-shaped dozer blade with removable side extensions; the maximum width is 4.2 m. Maximum dozing depth is 0.25 m.

The vehicle is provided with two winches for recovery and other purposes. The main winch has a maximum capacity of 900 kN (with a cable pulley) and is provided with 200 m of rope. An auxiliary winch has a capacity of 20 kN and 400 m of cable.

Other special equipment carried includes a 60 to 250 A arc welding system. There are stowage bins on the rear upper decking.

The basic crew is the commander and driver, although there is internal provision for a combat engineer team of three men. They are provided with an NBC air filtration system and detection equipment. Armament is one 12.7 mm NSW (NSV) air defence machine gun, a 7.62 mm RPK light machine gun for local defence and a shoulder-fired RPG-7 rocket launcher. Smoke grenade launchers and a thermal smoke generator are provided. The latter injects diesel fuel into the exhaust outlet on the left side of the hull towards the rear.

If required, the vehicle can be provided with deep wading equipment to enable the vehicle to wade through water obstacles up to 5 m for distances up to 1,000 m. The normal wading depth is 1.2 m.

Specifications

Crew: 2 + 3
Weight: 46,000 kg
Power-to-weight ratio: 16.95 hp/t
Ground pressure: 0.09 MPa
Length: 8.3 m
Width:
(over mudguards) 3.6 m
(dozer blade) 4.2 m
Height: 2.7 m
Ground clearance: 0.43 m
Max speed: (roads) 60 km/h
Range: 650 km
Fording:
(normal) 1.2 m
(with preparation) 5 m
Gradient: 60%
Vertical obstacle: 0.7 m
Engine: V-12 diesel developing 780 hp at 2,000 rpm
Armament:
1 × 12.7 mm NSW MG
1 × 7.62 mm RPK MG
1 × RPG-7 rocket launcher

Jib crane
Max lift capacity: 7,000 kg
Jib length: (max/min) 7.94/5.94 m
Working slope angle: −55 to +60°
Working traverse: 240°

Jib scoop
Capacity: 120 m^3
Cutting depth: 5.5 m

Gripper jaws
Max opening: 1 m
Nominal turn angle: 180°
Lift capacity: 3,000 kg
Max operational height: 7.5 m
Max operational depth: 5.5 m

Dozer blade
Width: (with extensions) 4.2 m
Cutting depth: 0.25 m

Main winch
Max towing capacity: (with pulley) 900 kN
Pay-out speed: 0.3 to 0.71 m/s
Winding speed: 0.31 to 0.74 m/s
Rope length: 200 m

Auxiliary winch
Max towing capacity: 20 kN
Max winding and pay-out speed: 1.5 m/s
Rope length: 400 m

Status

Prototype.

Development Agency

Osrodek Badawczo-Rozwojowy Urzadzen Mechaniczynch (OBRUM), PL-44-101 Gliwice ul. Toszecka 102, Poland.
Tel: +48 32 317241. Telex: 31 61 97.
Fax: +48 32 315887.

UPDATED

MT-LB armoured engineer reconnaissance vehicle

Description

The Polish Army utilises a version of the MT-LB multipurpose tracked vehicle developed specifically for the engineer reconnaissance role. The vehicle is intended for the reconnaissance of roads, bridges and their approaches, structures and potential demolition sites, minefields, battlefield obstacles, structures, water resources and other locations likely to be of interest to combat engineers. The vehicle is also equipped with sensors to detect and plot chemical and nuclear radiation contamination.

The vehicle uses the same general outlines as the conventional MT-LB; for details of the MT-LB refer to *Jane's Armour and Artillery 1996-97* pages 329 to 332 (in addition the vehicle has also been built in Bulgaria and Poland for the home and export markets). The crew of the reconnaissance vehicle is the commander and driver, with provision internally for a further six men. The additional crew members either operate the various items of reconnaissance or radio equipment or dismount for the close inspection of specific objectives. Most of the additional crew members are seated towards the rear of the vehicle where the specialised role equipment is either located or stowed. The reconnaissance equipment includes photographic and other optical instruments, plus specialised instruments for determining the state or characteristics of roads, ground surfaces and levels, river banks and water obstacles. Reconnaissance data is relayed to higher formations via an R-173 radio system which utilises a rail-type antenna mounted over the right-hand side of the hull; the antenna can be dismounted when not required.

Polish armoured engineer reconnaissance vehicle based on Polish-produced MT-LB multipurpose tracked vehicle chassis

The vehicle is amphibious and is provided with NBC air filtration equipment. Under normal operations the vehicle's various NBC sensors are maintained in a permanently on condition. Automatic and hand-operated fire extinguishers are provided.

The commander's position is covered by a small turret provided with vision devices. Mounted over the turret is a 12.7 mm NSW (NSV) machine gun for local and air defence; stowage is provided for 240 rounds. Other armament carried includes six RPG-7 rocket launchers and 20 F-1 hand grenades. Eight NDG-2 smoke grenade launchers are mounted on the hull sides towards the rear, four each side.

Specifications
Crew: 2 + 6
Weight: 12,900 kg ±250 kg
Power-to-weight ratio: 12.9 hp/t
Ground pressure: 0.46 kg/cm²
Length: 6.625 m
Width: (transport condition) 3 m
Height:
(to antenna) 2.7 m
(antenna lowered) 2.35 m
Ground clearance: (max) 0.43 m
Max speed:
(road) 60 km/h
(dirt road) 26-32 km/h
(water) 8 km/h
Range: 500 km
Fording: amphibious
Engine: SW680/167/1 6-cylinder turbocharged diesel developing 245 hp
Armament: 1 × 12.7 mm NSW MG
6 × RPG-7 rocket launchers
8 × smoke grenade launchers

Status
Production as required. In service with the Polish Army.

Manufacturer
Huta Stalowa Wola SA, 1 Kwiatkowskiego Street, PL-37-450 Stalowa Wola, Poland.
Tel: +48 16 435 252. Telex: 06 22 33 hswbh pl.
Fax: +48 16 425 566.

Marketing Agency
Cenzin Co Ltd, 2 Frascati Street, PL-00-489 Warsaw, Poland.
Tel: +48 22 296 738. Telex: 814505 czi pl.
Fax: +48 22 628 6356.

UPDATED

RUSSIAN FEDERATION AND ASSOCIATED STATES (CIS)

IMR-2 combat engineer vehicle

Development
The IMR-2 series of combat engineer vehicles is based on the chassis of the T-72 MBT and is the replacement for the original IMR based on the T-54/T-55 MBT chassis. The current production model is designated the IMR-2M.

Description
The layout of the IMR-2 is the same as the T-72 MBT with driver's compartment at the front, crew compartment in the centre and engine compartment at the rear.

To carry out its combat engineer role, the turret of the T-72 MBT has been removed and a new all-welded steel superstructure has been fitted with bulletproof windows for the commander/operator. On this is the telescopic arm which can be fitted with various attachments.

To enable it to carry out its combat engineer role on the battlefield, the IMR-2M has the following equipment:

(1) Front-mounted hydraulically operated dozer blade which is 1 m high and 3.38 m wide when being used in the V-blade configuration, 3.925 m wide when being used in the straight dozing position and 3.212 m wide when being used in the grading position. Maximum depth of dozer blade is 450 mm. When not required this is folded upwards
(2) Multipurpose operating element which can have a bucket with a maximum capacity of 0.35 m³, pull shovel capacity of 0.17 m³ and a push shovel capacity of 0.17 m³
(3) The above uses a telescopic arm which can also be fitted with a manipulator and can be traversed through a full 360°. With a reach of 8.15 m it can lift 2,000 kg. When travelling this is normally traversed to the rear
(4) Mounted at the front is the KMT-8 mineclearing system with rod fuze and electromagnetic adapter type EMT.

Standard equipment includes night vision equipment for the commander and driver, NBC system of the overpressure type and fire detection and suppression system. The vehicle can lay its own smoke screen by injecting diesel fuel into the exhaust outlet on the left side of the hull.

Specifications
Weight: 44,500 kg
Power-to-weight ratio: 18.87 hp/t
Ground pressure: 0.98 kg/cm²
Crew: 2
Length: 9.55 m
Width: 3.74 m
Height: 3.68 m
Ground clearance: 0.46 m
Max road speed: 60 km
Road range: 500 km
Fording: 1.2 m
Engine: V-12 diesel developing 840 hp
Armament: 1 × 12.7 mm MG

Status
Production as required. In service with the Russian Army and other undisclosed countries.

Manufacturer
Russian state factories.

NEW ENTRY

IMR-2M combat engineer vehicle with dozer blade raised and mineclearing equipment lowered **1997**

IMR-2 in operating configuration using dozer blade

IMR combat engineer vehicle

Description
The combat engineer vehicle IMR (*Inzhenernaya Maschina Razgrazhdeniia* - engineer vehicle for the removal of obstacles) was first seen in 1973 and is based on the T-55 MBT chassis. The turret was removed and replaced by a hydraulically operated crane which can be traversed through 360°. The jib of this crane is telescopic and when in the travelling position rests on a cradle at the rear of the hull. The cradle folds down against the engine deck when the crane is being used.

The crane is provided with a pair of pincer-type grabs which are used to remove trees and other obstacles. The grab can be replaced by a small bucket which is normally carried above the left rear track when not required. The crane operator is provided with an armoured cupola which has observation windows. A searchlight is mounted on the crane for night operations. At the front of the hull is a hydraulically operated dozer blade which can be used in the straight or V-configuration, but cannot angle doze. An unditching beam is carried at the rear of the hull.

Variant
IWT
The IWT (*Inzynieryjny Woz Torujacy*) is the Polish equivalent of the IMR and shares many of the same features. However, the IWT can carry a PW-LWD rocket-launched mineclearing system identical to that carried by Polish Army T-55 and T-72 MBTs (see entry in *Mineclearing equipment* section).

Display example of IMR combat engineer vehicle *1996*

IMR combat engineer vehicle using pincers to remove obstacle *1997*

Specifications (provisional)
Crew: 2
Weight: 34,000 kg
Power-to-weight ratio: 17.05 hp/t
Ground pressure: 0.76 kg/cm²
Length: (dozer blade in operating position and crane stowed) 10.6 m
(hull) 6.45 m
Width:
(hull) 3.28 m
(over dozer blade) 3.48 m
Height:
(crane operator's cupola) 2.48 m
(crane in travelling position) 3.37 m
Ground clearance: 0.425 m
Track: 2.64 m
Track width: 580 mm
Length of track on ground: 3.84 m
Max speed: (road) 48 km/h
Range: 400 km
Fuel capacity: 812 litres
Fording: 1.4 m
Gradient: 60%
Vertical obstacle: 0.8 m
Trench: 2.7 m
Engine: Model V-55 V-12 water-cooled diesel developing 580 hp at 2,000 rpm
Transmission: manual with 5 forward and 1 reverse gears
Electrical system: 24 V
Batteries: 4 × 12 V, 280 Ah
Armament: nil

Armour
hull front: 100 mm at 60°
hull sides: 70 mm
hull rear: 60 mm
hull floor: 20 mm
hull roof: 30 mm

Status
In service with Russia, Yugoslavia and many other countries.

Manufacturer
Russian state factories.

UPDATED

BAT-2 combat engineer vehicle

Description
The BAT-2 (*Bul'dozer na Artilleriyskom Tygache 2*) combat engineer vehicle is intended to supplement the BAT-M high-speed tractor-mounted bulldozers and similar vehicles (see entry in *Field Fortifications and Related Emplacements Equipment* section) but is more readily described as a combat engineer vehicle. In addition to the large dozer blade normally fitted, the BAT-2 has carrying capacity for combat engineer stores behind the forward control cab and also has a crane mounted at the rear of the cab to handle them. The crane can be fitted with pincer-type grabs for clearing obstacles, has a maximum boom outreach of 7.3 m and a load-lifting capacity of 2,000 kg. Mounted on the same platform as the crane is a 25 tonne capacity winch provided with 100 m of cable. The armoured cab has seating for the two-man crew and a compartment for an eight-man combat engineer squad. A NBC system is provided for the cab. The operating temperature range may be from −45 to +45°C.

The BAT-2 is based on the chassis of the MT-T tracked carrier (see entry in *Tracked Prime Movers* section) which uses suspension and running gear components from the T-64 tank and is powered by a V-64-4 V-12 multifuel diesel engine developing 700 hp. Some transmission components are also 'taken from a tank'.

The BAT-2 has the large dozer blade of the earlier BAT-M vehicles mounted at the front but the hydraulic cylinder actuation system is more powerful and uses a different operating layout. There is also a different system to raise the blade vertically when not in use. The blade widths can be varied to suit the task. As a bulldozer the blade is 4.5 m wide, for road or track clearing it is 4.2 m wide, while when grading it may have a width variable between 4.1 and 4.35 m. The blade may also be tilted through 10° for some operations, while a blade-mounted skid allows the blade to follow terrain contours.

The BAT-2 can be employed to drive graded tracks across 'ordinary' terrain at a speed of up to 6.8 km/h; across snow the speed is 8.15 km/h. Scrubby terrain with trees up to 300 mm in diameter can be cleared at a rate of 2.3 km/h. When route-clearing or creating earth barriers the dozer blade has a clearing capacity of from 350 to 450 m³/h. When digging ditches it is 200 to 250 m³/h. Solid or frozen ground can be loosened down to a depth of 500 mm.

Specifications
Crew: 2 + 8
Weight, combat: 39,700 kg
Power-to-weight ratio: 17.63 hp/t
Ground pressure: 0.81 kg/cm²
Length: 9.64 m
Width: 4.2 m
Height: 3.69 m
Ground clearance: 0.425 m
Max speed: (roads) 60 km/h
Range: 500 km
Fording: 1.3 m
Engine: V-64-4 V-12 multifuel diesel developing 700 hp

Status
In service with the Russian Army and probably other countries.

Manufacturer
Kiev Engineer Vehicles Plant, 21 Degtjarevskaga Street, 254119 Kiev, Ukraine.
Tel: +380 044 213 0341. Telex: 131084 Kiev Luna.
Fax: +380 044 213 1037.

UPDATED

BAT-2 combat engineer vehicle

BAT-2 combat engineer vehicle

IRM engineer reconnaissance vehicle

Development

The IRM engineer reconnaissance vehicle (IRM - *Inzhenernaya rezvedivatel 'naya maschina*) was originally known in the West as the IPR amphibious engineer vehicle. It is based on components from the BMP-2 IFV tracked chassis (for details refer to *Jane's Armour and Artillery 1996-97* pages 316 to 318). The IRM is intended for engineer and other reconnaissance over a wide variety of terrain and climatic conditions. It can be carried to its zone of operations by cargo aircraft, assault ships or hovercraft.

Description

The IRM is based on automotive components of the BMP-2 IFV but uses a longer hull with an extra roadwheel, making seven roadwheels each side; the engine is located at the rear. There are also five track-return rollers. The vehicle is fully amphibious and, when travelling, the hull superstructure may carry a folded-down snorkel mast approximately 10 m long which is erected when the vehicle operates fully submerged; this mast is not always carried. When floating, the vehicle is driven and steered by two three-bladed cowled propellers. A trim vane is fitted to the bow.

Internally the vehicle is divided into three main compartments; driver's, fighting and engine. Four further sections include the bow, right- and left-hand sections for the fuel tanks and the stern. The driver's and fighting compartments are protected against nuclear warfare effects with a total radiation penetration reduction ratio of 2.78.

Vehicle services include: a sealed NBC system; an internal air regeneration system, which can switch on automatically to allow the vehicle to remain airtight in contaminated areas for up to 4 hours; a fire extinguishing system; a bilge pump device; and a reusable smoke-generation system which injects diesel fuel into the engine exhausts.

The crew consists of up to six men, the basic crew being the commander, driver and one or two 'reconnaissance engineers'; extra personnel may be carried for dismounted reconnaissance operations. The driver and commander are seated towards the front of the hull and provided with circular hatches; more hatches are provided for the engineers. There is an emergency escape hatch under the hull. The commander is also provided with a small turret at the front of the hull which mounts day and night vision devices and a 7.62 mm PKT machine gun; an infrared searchlight is mounted next to the machine gun. The only other armament consists of the crew's personal weapons, although a smoke-generating apparatus is provided. For closed-down observation the driver is provided with a TNPO-160 observation device for land use; this is changed to a TNP-370 for use when in water.

For the reconnaissance role the IRM is provided with a variety of specialised equipment such as an inertial navigation system; a RShM-2 mine detection system; an auger-type sensor for determining the load-bearing capacities of various types of terrain (described as a PR-1 'penetrometer' and probably coupled to an onboard computer); an echo-sounder providing an accurate determination of water depth from 0.5 to 20 m (this system has an automatic recorder, three hydro-acoustic transducers and can be used to determine the firmness of river bottoms); an AGI-1s artificial horizon mounted on the driver's control panel; a clinometer to measure beach and other terrain angles for the driver; a PAB-2M aiming circle; and seven day plus two night vision devices. Also carried are a PIR retractable engineer reconnaissance periscope 1.5 m long (of which 0.75 m can be raised) and a DSP-30 rangefinder to carry out basic survey operations from within the vehicle. Portable equipment carried for use outside the vehicle includes IMP-2 hand-held mine detectors, a device (*ledobur*) for measuring the thickness of ice and a portable terrain load-bearing measuring instrument. Also carried are two R-147 radio sets with a range of 20 km.

Mounted at the front of the vehicle, in line with the tracks, are two hydraulically operated arms carrying spade-like units which are mine detectors using the induction principle. The detector units are held parallel to the ground surface by a hydraulic terrain-following mechanism mounted on the arms. If the detector senses a mine, the vehicle will stop automatically for the mine to be cleared manually. The vehicle will also stop automatically if the detector strikes an obstacle such as an immovable tree or rock. When not in use, the detector arms are folded back on to the roof and sides of the vehicle. The RShM-2 river-type wide-span mine detector can detect ferromagnetic materials at a depth of up to 300 mm with the vehicle moving at speeds of from 3 to 5 km/h.

Should an IRM engineer reconnaissance vehicle become stranded in soft or difficult terrain under hazardous circumstances, such as under enemy fire, the IRM can make use of a rocket-powered self-recovery system which can be utilised with all crew members remaining under cover. At the top of the hull rear are two banks of up to 12 9M39 solid-propellant rockets which can be ignited to provide thrust to assist the vehicle from difficulties. Each mass-produced 9M39 rocket engine weighs 6.3 kg, is 70 mm in diameter and 800 mm long. Each rocket motor can deliver a tractive force of 312 kg.

Specifications

Crew: up to 6
Weight: 17,200 kg
Power-to-weight ratio: 17.44 hp/t
Ground pressure: 0.69 kg/cm^2
Length: 8.22 m
Width: 3.15 m
Height: 2.4 m
Max speed:
(land) 52 km/h
(water) 12 km/h
Range: 500 km
Fuel capacity: 600 litres
Endurance afloat: 12 h
Max gradient: 36°
Vertical obstacle: 0.65 m
Fording: amphibious
Trench: 2.3 m
Engine: Type UTD-20 6-cylinder in-line water-cooled diesel developing 300 hp at 2,000 rpm
Transmission: manual, 5 forward and 1 reverse gears
Steering: clutch and brake
Suspension: torsion bar
Electrical system: 24 V
Armament:
1 × 7.62 mm PKT MG
smoke generator

Status

In service with the Russian Army and probably other countries.

Manufacturer

Russian state factories.

UPDATED

IRM engineer reconnaissance vehicle with all equipment stowed but without snorkel mast (Christopher F Foss) **1997**

Rear view of IRM engineer reconnaissance vehicle with all equipment stowed but without snorkel mast (Christopher F Foss) **1997**

MT-LB engineer vehicle

Description

The MT-LB engineer vehicle is a variant of the MT-LB multipurpose tracked vehicle (for full details of this vehicle refer to *Jane's Armour and Artillery 1996-97* pages 329 to 332).

The MT-LB engineer vehicle is similar in appearance to the basic MT-LB but modified to carry a plough blade on the hull roof or side. This blade is manually fitted to two hydraulic arm assemblies on the hull rear to allow plough blade operations to the rear only. Other combat engineer equipment is carried internally.

MT-LB engineer vehicle with plough blade in travelling position

The US Army has a number of these vehicles for training purposes. They are based at Fort Irwin, California.

The former East German Army used the MT-LB for combat engineer purposes but their version lacked the plough blade. A rectangular box on the rear upper hull carried engineer equipment.

Variant

Zabot

A prototype of an MT-LB based Czech Army engineer vehicle with a large dozer blade at the rear is known as Zabot. As far as is known this is still at the prototype stage.

Status

Production complete. The standard MT-LB is in service with the following countries, some of whom also operate the engineer variant: Armenia, Azerbaijan, Belarus, Bulgaria, Finland, Hungary, Iraq, Kazakhstan, Moldova, Russia, Sweden, Ukraine and Yugoslavia.

Manufacturer

Bulgarian, Polish and Russian state factories.

UPDATED

Tank-mounted bulldozers BTU and BTU-55

Description

The BTU and BTU-55 tank-mounted bulldozers can both be fitted to either the T-54 or T-55 series of MBTs. The BTU-55 is the more recent version and is an improvement in all aspects. There does not appear to be a bulldozer for the T-62, but the T-64 and T-72 MBTs can incorporate a form of dozer blade for self-emplacement.

A BTU-55M bulldozer unit for attachment to T-55 series tanks is marketed by Kintex of Bulgaria.

Bulgarian BTU-55M bulldozer unit fitted to T-55AM tank ***1996***

Specifications

Model	**BTU**	**BTU-55**
Weight	2,300 kg	1,400 kg
Width	3.4 m	3.8 m
Clearing speed	1.5-5 km/h	5-7 km/h
(in snow)	4-6 km/h	4-6 km/h
Digging performance	100-200 m³/h	130-150 m³/h
Pushing performance	350 m³/h	–
Mounting time	80-90 min	60 min
Dismounting time	30-60 min	45 min

Status

Production complete in Russia but still marketed by Bulgaria. In widespread service.

Manufacturer

Bulgarian and Russian state factories.

UPDATED

SPAIN

GAMESA M60 VZ engineer tank

Development

In 1995, the Spanish Army selected GAMESA to manufacture one prototype of the M60 VZ engineer tank based on the chassis of a modified M60 series MBT.

Prior to this, the company had converted 22 M47E1 tanks into the M47 VR recovery tank configuration and details of this are given later in the *Armoured Recovery Vehicles* section.

The original technical specifications indicated the possibility of different configurations of the M60 VZ engineer tank, including retaining the existing M60 turret or replacing this with a new all-welded superstructure.

In extensive studies, including emphasis on functionality, operability and performance it was decided to adopt a new all-welded superstructure.

If trials with the prototype M60 VZ engineer tank are successful, it is expected that a number of M60 series MBTs will be converted into this configuration for the Spanish Army.

Description

The conversion involves the replacement of the original turret of the M60 with a new armoured superstructure specially designed to house the crew and the controls required to operate the onboard equipment. The normal crew consists of tank commander, driver and radio operator.

The driver station at the front of the vehicle is maintained but has additional controls for the front-mounted bulldozer blade and hydraulic unit.

The tank commander occupies a place in the centre of the armoured superstructure, which retains the original M60 tank cupola, armed with a 12.7 mm machine gun. The commander is also provided with controls of the excavating arm and main winch. The radio operator is located on the left of the hull with his equipment and accessories.

GAMESA M60 VZ engineer tank with dozer blade lowered and excavating arm traversed to the front ***1997***

The armoured superstructure with safety welding, is bolted to the turret ring itself, so allowing for its removal for any necessary maintenance work. The armour provides protection from small arms fire up to 20 mm in calibre.

The front part of the vehicle has been fitted with a bulldozer-type blade which is driven by two hydraulic cylinders with a two-fold mission. First to stabilise the tank when the excavating arm or winch are in operation and second to carry out dozing operations.

The blade has a width of 3.7 m and is 900 mm high and allows the front of the tank to be raised 30 cm. It is also equipped with three ripper-scarifiers which are located on the rear of the blade.

If required, the blade can be disassembled from the vehicle and replaced by plough type minefield clearing equipment (for example Full Width Mine Plough) or mineclearing rollers.

The hydraulic unit is housed inside the hull with a variable flow piston pump. The load sensing is driven through an electromagnetic clutch and cardan joint by an additional power take-off built on to the vehicle's main engine. This is also provided with a constant speed regulator which works when the hydraulic system is operating.

The main 30,000 kg direct-pull winch is housed inside the hull. This is provided with 90 m of 30 mm diameter cable and leaves the vehicle guided by means of a roller box placed on the front of the tank. With the aid of a return pully, which is included in the accessories of the tank, the pulling strength can be doubled to 60,000 kg.

The release speed and rapid collection of the cable are adjustable from 0 to 45 m/min. The working speed or loads of 30 t/m are adjustable from 0 to 5 m/min. For lower loads, the working speed is variable from zero to the values corresponding to consumed power of 50 hp. In this way, hauling work is made easier as well as being more efficient.

The auxiliary winch which is provided for this purpose, can be used to handle 30 mm diameter cable up to a length of 90 m.

Mounted on the right side of the tank is an excavating arm which can be traversed through 360°. This has a digging force of 11,000 kg and a lifting capacity of 10,000 kg by means of a swivel hook fitted to a bucket with an opening of 600 mm.

In addition, the bucket can be replaced with a hammer, shear, graple or 400 mm bucket which are driven hydraulically.

The dual controls of the excavating arm consist of a system installed at the tank commander's station which can work under cover of enemy fire and another remote control system which allows the operator of the excavating arm to get out of the tank (up to a maximum of 10 m) and operate the arm from a convenient position and with better visibility. It also allows quicker and more efficient movements in digging operations.

The arm has a maximum reach of 7.2 m, digging depth of 2.7 m and a capacity of 45 m^3/h in average consistency soil.

Located on the right-hand side of the tank are the external electrical and hydraulic intakes, which allow working with tools equipped with snap-connectors.

The rear of the vehicle is equipped with a minefield opening marking system which can be activated manually or automatically, synchronised with the speed of the tank. This drops fluorescent signals in the form of flares at pre-established intervals in the centre of the breaches cleared of mines.

In cases of an emergency, such as the breakdown of the main engine hydraulic power, the vehicle is fitted with an auxiliary electric pump which allows the implements, excavating arm, bulldozer blade and winch, to be returned to the transport position.

The tank is armed with a 12.7 mm machine gun at the commander's station for which 600 rounds are carried. Two banks of four electrically operated smoke grenade dischargers are mounted forward of the commander's turret and cover 40° to the left and right of the tank.

Protection systems include engine smoke generator, NBC system, fire and explosion detection and suppression in the crew compartment and fire detection and suppression system in the engine compartment.

In order to load the M60 VZ engineer tank on ships, lifting eyes are provided on the hull.

Specifications

Crew: 3
Weight:
(empty) 49,300 kg
(loaded) 51,500 kg
Power-to-weight ratio: 14.75 hp/t
Length:
(chassis) 6.946 m
Width: 3.7 m
Height: 3.2 m
Ground clearance: 0.45 m
Max speed: 48.28 km/h
Max road range: 480 km
Fuel capacity: 1,420 litres
Fording: 1.2 m
Vertical obstacle: 0.9 m
Trench: 2.6 m
Gradient: 60%
Side slope: 40%
Engine: Teledyne (now General Dynamics Land Systems) AVDS-1790-2D diesel developing 760 hp at 2,400 rpm
Transmission: Allison CD-850-6A automatic
Electrical system: 24 V
Batteries: 6 × 12 V, 300 Ah
Armament: 1 × 12.7 mm MG
Ammunition: 600 × 12.7 mm

Status

Prototype. Not yet in production.

Manufacturer

GAMESA, Portal de Gamarra 40, PO Box 628, 01013 Vitoria, Spain.
Tel: +34 45 280700. Fax: +34 45 251642

NEW ENTRY

UNITED KINGDOM

Future Combat Engineer Vehicle (Terrier)

Development

The British Army has a requirement for about 100 Future Combat Engineer Vehicles, to be known as Terrier, to replace the in-service Combat Engineer Tractor (CET - see separate entry). Roles intended for the Terrier include a wide range of battlefield missions such as: route clearance; preparing fire positions; clearing battlefield obstacles; towing trailers carrying combat engineer equipment such as Python mineclearing equipment; trackway and fascines. Mobility compatibility with armoured formations is essential.

In May 1995 the Ministry of Defence selected seven companies to compete for a Feasibility Study (FS) for Terrier. The FS will last 12 months and will entail interim, mid-study and final reports. In addition the Ministry of Defence will also carry out their own studies, including the possibility of meeting the Terrier requirement by an off-the-shelf purchase, or by leasing vehicles.

The seven companies involved at the outset were:
- Royal Ordnance, UK
- Vickers Shipbuilding and Engineering Limited, UK
- GKN Defence, UK
- Pearson Engineering, UK
- Vickers Defence Systems, UK
- Giat Industries, France
- MaK System Gesellschaft mbH, Germany.

It is anticipated that at least some of the five companies receiving FS contracts will arrange among themselves to form teams. Decisions on future procurement strategy will be taken after the evaluation of the FS reports.

Forecast in-service date for Terrier is 2005.

Status

Feasibility studies.

UPDATED

Future Engineer Tanks (FET)

Development/Description

The British Army has a requirement for a total of 102 Future Engineer Tanks (FET) for use by the Royal Engineers in support of armoured formations equipped with Vickers Defence Systems Challenger 2 MBTs with an in-service date of 2001.

It was expected that the MoD would specify that the existing Vickers Defence Systems Challenger 1 or Challenger 2 chassis be used to meet the FET requirement.

Instead, industry is being asked to propose existing vehicles which could be modified to meet the FET requirement. Maximum weight limit for the FET is 62.5 tonnes and it must also be fitted with an NBC system, have a track tensioner operated by the driver and operate 110 V hand tools via a power take-off.

Vickers Defence Systems have confirmed that they will be offering the older Challenger 1 MBT and the current in production Challenger 2 MBT chassis to meet this requirement.

GKN Defence of the UK does not have experience in this area but have teamed with the German company of MaK who have designed and built virtually all the armoured engineer vehicles for the German Army.

With the introduction into service of 386 Challenger 2 MBTs, all of the 420 Challenger 1 MBTs are to be phased out of service. These Challenger 1 MBTs could be sold overseas, or some of the chassis could be used for other applications.

While Vickers Defence Systems is the design authority for the Challenger 1 and 2 MBTs it is possible that other contractors would also be able to build this chassis to meet the FET requirement.

The FET will replace the British Army's current Chieftain Armoured Vehicle Royal Engineers (CHAVRE) and the Chieftain Armoured Vehicle Launched Bridge (CHAVLB). Both of these are based on a Chieftain MBT chassis which entered service with the British Army over 30 years ago.

Vickers Defence Systems were the prime contractor for the CHAVRE with a total of 46 being delivered between 1991 and 1994. The CHAVLB were built by the former Royal Ordnance Factory Leeds, with a total of 51 built. These are now being fitted with elements of the new Vickers Bridging for the 1990s (BR90) system.

The requirements for the FET are such, that it is highly probable that only a full-tracked armoured vehicle would be able to carry out the mission.

The two replacement vehicles are called the Future AVRE Obstacle Breaching (to replace the CHAVRE) and the Future AVLB Gap Crossing (to replace CHAVLB), both will use the same chassis with up to 51 of each system required with an in service date of 2001.

In order to work with the Challenger 2 MBT, the new vehicles are required to have a high level of protection and mobility, with one requirement being that under some circumstances they may be operated by remote control.

The Future AVLB Gap Crossing will mount the new elements of the BR90 system and be fitted with a number of attachments including the Pearson Engineering Universal Dozer Kit, track with mine plough, vehicle signature duplicator and the Mine Individual Detection Neutralisation and Route Marking System (MINDER).

With parts of the BR90 system removed, the Future AVLB Gap Crossing vehicle will be capable of carrying two maxi fascines, tow an AVRE trailer and mark routes.

The Future AVRE Obstacle Breaching vehicle is required to tow and fire the new Royal Ordnance Improved Giant Viper (or Python as it is now called), be fitted with various mineploughs as well as the vehicle signature duplicator and MINDER. It is also required to carry two maxi fascines, carry a roll of Class 60 trackway and tow a Barmine layer.

The current CHAVRE is not fitted with a telescopic arm but this is a key requirement for the Future AVRE Obstacle Breaching Vehicle. Using various attachments the telescopic arm could be used to clear fallen trees and remove other battlefield obstacles.

Status
A future British Army requirement if the first phase, a feasibility study, is completed successfully. Projected in service date is 2001.

Manufacturer
Not selected.

NEW ENTRY

Vickers Defence Systems Chieftain Armoured Vehicle Royal Engineers (AVRE)

Development
Following the introduction into British Army service of locally converted Interim Chieftain AVREs, Vickers Defence Systems was awarded a contract to convert a number of Chieftain MBTs into a more technically advanced AVRE that better met the British Army's requirements. The combined design, development and production contract was placed with Vickers Defence Systems, covering a total of two prototypes and 46 production vehicles. Vickers Defence Systems initially undertook a four-month concept development study, during which, a full-scale wooden mockup was constructed on a Chieftain hull. This was used to demonstrate to the user the exact position of onboard equipment and systems prior to the manufacture of the prototypes.

The two prototypes were completed in mid-1991 and were extensively trialled by both Vickers Defence Systems and the Royal Engineers at the Armoured Trials and Development Unit at Bovington in Dorset. First-off production was late 1992, with production of the full contract quantity proceeding at three units per month until 1994.

The prime role of the Chieftain AVRE is to provide mobility support for armoured battle groups, creating routes through or over natural and manmade obstacles. Typically, breaching of minefields is achieved by using a front-mounted mine plough in conjunction with trailer-mounted Royal Ordnance Giant Viper or Python mineclearing systems. Anti-tank ditches and similar obstacles can be crossed by means of fascines and trackway, both launched from the overhead framework. A dozer blade can be fitted to assist in the clearing of obstructions.

Chieftain AVRE fitted with mineclearing equipment, carrying MAXI pipe fascines and towing Giant Viper mineclearing system (Harry M Steele) ***1997***

Using a hydraulic loader to position MAXI pipe fascines on the front hamper of a Chieftain AVRE

Description
The Chieftain AVRE is built on the chassis of a Chieftain MBT. The running gear remains unchanged, apart from the fitment of a hydraulic track tensioner. This is the same as that fitted to the Chieftain ARRV and AVLB and allows the driver to adjust the track tension from under armour.

The MBT turret and main armament are removed and replaced by a welded steel armoured superstructure referred to as a penthouse. Seating is provided in the crew compartment for three men abreast, with the commander seated in the centre. The driver's station remains unchanged. The commander is provided with a fixed cupola having a single-piece hatch and seven unitary observation periscopes. The forward periscope is a Helio Type 2100 double prism periscope providing close-in vision to the front of the vehicle. The commander's seat can be raised hydraulically to allow both head out and closed down operation. A further five periscopes are fitted to the penthouse roof for crew vision.

In addition to the cupola hatch there are two further hatches in the penthouse for crew access, one in the left-hand roof plate and another in the front plate. The latter is provided with a foul-weather screen with wash/wipe facilities (similar to that provided for the driver on the Warrior APC) to improve crew comfort and forward vision when not in a combat area.

The penthouse is built to be ready to take an NBC pack of Chieftain AVLB dimensions as a Phase Two enhancement. In the interim, the vehicle is provided with a three-speed ventilation fan.

Primary hydraulic power for the Special To Role systems is provided by a Commercial Hydraulics HD2-OC-900 pump driven by the main engine via a power take-off. The hydraulic system is load-sensing to optimise power output against power requirements. An electrically driven Commercial Hydraulics PM125 constant horsepower pump is provided as a back-up.

Mounted on the hull are two overhead frameworks referred to as hampers. These are tilted hydraulically to launch fascines and trackway and, in the case of the rear hamper, to gain access to the engine decks. The hampers can carry combinations of MAXI pipe fascines and Class 60 trackway up to a maximum distributed load of 14.5 tonnes. Loads are released to the front or rear by means of electrically fired blow-out pins activated by the commander or driver from under armour. The hampers may also be used to transport palletised loads and other engineer equipment.

Mounted on the hull rear plate is a Rotzler 10 tonne hydraulic capstan winch. This can be used to the front or rear with up to 60 m of rope available, depending upon reeving configurations.

A Ferrari hydraulic loader is mounted at the rear of the penthouse, stowed along the engine deck beneath the rear hamper. The lifting capacity is 2.9 tonnes at a maximum extension of 5.45 m. The loader can be operated from controls on the base or from a remote control connected by a 10 m umbilical. The loader is capable of lifting both fascines and trackway on to the hampers as an alternative to the winch.

The hull front has the necessary attachments to accept a Pearson mine plough or Combat Dozer UDK1. These can be fitted and removed using equipment carried on the vehicle. When not in use, the plough or dozer can be stowed on the rear hamper using the vehicle's hydraulic loader.

A rotatable tow hook, with the necessary electrical connections, is provided at the rear and is designed to tow either two Giant Viper or Python trailers in tandem or a single AVRE trailer.

The defensive armament consists of electrically fired smoke grenade dischargers mounted on the front hamper and the crew's personal weapons.

Potential Phase Two enhancements include the provision of collective NBC protection for the crew, passive night vision for the commander and the installation of a minefield marking system.

Status
Production complete. In service with the British Army (48).

Manufacturer
Vickers Defence Systems, Scotswood Road,
Newcastle upon Tyne NE99 1BX, UK.
Tel: +44 191 273 8888. Telex: 53104.
Fax: +44 191 273 2324.

UPDATED

Royal Ordnance Combat Engineer Tractor (FV180)

Development

In 1962 a General Staff Target (GST 26) was issued for an engineer equipment which would combine the characteristics of an armoured vehicle and an earthmover. Existing commercial earthmoving equipment could not meet this requirement as it had poor mobility, lacked both armour protection and communications equipment and had no amphibious capability.

The following year, three companies, Caterpillar UK, GKN and Vickers, were invited to put forward their proposals for a vehicle to meet GST 26. GKN and Vickers responded but their proposals were not taken up. In 1965 the Military Engineering Experimental Establishment at Christchurch prepared a design to meet GST 26. At the same time discussions took place between the United Kingdom, France and Germany for the joint development of a Combat Engineer Tractor.

Subsequently, two prototypes were built by the then Royal Ordnance Factory at Leeds, based on a design prepared by the Military Engineering Experimental Establishment. They were powered by a Cummins V-8 diesel which developed 350 hp, were fully amphibious and incorporated some features of the American Universal Engineer Tractor. They were delivered in 1968, but France had meanwhile dropped out of the project as it required a vehicle with a higher water speed and only limited earthmoving capabilities.

In 1970, after trials with the two test rigs in both Germany and the United Kingdom, the Germans withdrew from the programme as they required a heavier vehicle, without an amphibious capability, mainly for clearing river crossing points.

Meanwhile a major redesign of the vehicle had been carried out with two major objectives: to use standard commercial components wherever possible, (engine, transmission, steer unit and winch), and to make the vehicle amphibious with the minimum preparation. In 1969 a GST was issued and the following year Royal Ordnance Factory Leeds was nominated as the prime manufacturer with the Military Vehicles and Engineering Establishment responsible for design work. A contract was awarded to Leeds for seven prototypes, all of which were delivered between February 1973 and January 1974.

Extensive trials were carried out in both the United Kingdom and Germany and in July 1975 the Combat Engineer Tractor (CET) was accepted for service with the British Army. As a result of trials, modifications were incorporated in production vehicles, including improved track tension, replacing the steel-rimmed tyre with a rubber one at the front roadwheel station, and improvements to the buoyancy system.

Production of the CET began at the Royal Ordnance Factory at Nottingham in 1977 and the first production vehicle was accepted in May 1978. Production was completed in March 1981 after 141 had been built for the British Army, but in mid-1984 it was announced that India had ordered the CET. The initial order was for nine vehicles with options (since taken) on a further six. The first vehicle was delivered at the end of 1988 and was first shown publicly during the January 1989 Republic Day parade in New Delhi. The order for 15 was completed by the end of 1990.

Combat Engineer Tractor (CET) towing Giant Viper (Improved) **1997**

In April 1993 it was reported, but not confirmed by Royal Ordnance, that the Singapore armed forces had ordered 18 CETs with an option on a further 18 which was subsequently exercised. Late in 1995 an additional 18 vehicles were ordered bringing the total to 54. In service they will be used for combat engineering duties plus logistics over-the-shore duties during joint operations with the Singapore Navy.

Late production CETs incoporate some automotive assemblies, aluminium welded fabrications and machined components provided by Alvis Vehicles Limited.

Description

The CET was designed to provide integral engineer support for the battle group and typical roles include excavating vehicle and gun pits for defensive purposes, repairing and maintaining roads, preparation of river banks, recovering disabled vehicles and preparing or clearing obstacles.

The hull of the CET is made of all-welded aluminium armour, supplied by Alcan. The vehicle is normally driven with its bucket to the rear and in this position the crew is seated on the left side. The following description is for the vehicle in this condition.

The driver is seated at the front and also operates the winch, with the bucket operator to his rear. Both crew members can reverse their seats and essential controls are duplicated so that either crew member can operate the vehicle. The crew compartment is provided with two hatch covers, which open to the right, and a total of 10 vision blocks. The crew compartment is supplied with cooled air via the NBC system from an air conditioning unit, produced by Gallay Limited. The unit contains a refrigeration package with a compressor driven from the main gearbox.

The engine and transmission are mounted at the right side of the hull with the final drives being mounted at the front of the hull. From the engine, power is passed through two gearboxes to a steering unit and final drives. The first of these gearboxes, the transfer box, provides PTOs for the water propulsion units and hydraulic pumps. Controlled differential steering is used for road and cross-country drive and skid steering is used for bulldozing operations with an independent clutch/brake system provided for this purpose.

The suspension is of the torsion bar type and consists of five roadwheels, with the fifth roadwheel acting as the idler. The drive sprocket is at the front. The tracks are of cast steel with rubber bushes and rubber pads are provided to reduce damage when operating on roads. Hydraulic double-acting ram-type dampers are mounted on the front and rear wheel stations and can be locked from the crew compartment if required.

The CET can ford to a depth of 1.83 m without preparation, but with preparation the vehicle is fully amphibious. It is propelled in the water by two 330 mm Dowty 40/40B water-jets which are mounted one either side of the hull. When in the water, steering is accomplished by deflecting the thrust from the unit on the inside of the turn and deflecting both units gives reverse. The preparation required before entering the water is to unfold the trim board at the front of the hull and inflate two Hycafloat units to the rear of the trim board. The units were developed by FPT Industries of Portsmouth and replaced the original bellows-type units installed on the prototypes. Two plastic-cased polyurethane foam blocks are fitted into the bucket and held in place by retaining straps.

The bucket is of light alloy construction with steel cutting edges and tines and has a maximum capacity of 1.72 m^3. Its maximum lift height is 1.829 m and its minimum lift height is 102 mm below the track line. This can be used for both digging or bulldozing and can also be used as an earth anchor. Its maximum capacity is 200 m^3/h over a 50 m hauling distance.

The two-speed winch has a maximum pull of 8,000 kg and is provided with 113 m of 16 mm diameter rope; maximum winching speed is 1.9 m/s. For self-assistance, the winching speed is matched to the speed of the tractor. A high-speed winching facility is available for ferrying work. The rope can be led to the front or rear of the vehicle by direction-changing blocks.

A self-emplacing earth anchor is mounted on the top of the hull and this can be rocket-propelled to a maximum distance of 91.4 m. The anchor is attached to the vehicle's winch rope and is used to assist the vehicle when leaving a river with a steep bank.

CETs can be provided with an NBC filtration pack installed at the front of the vehicle, providing clean air to the crew compartment. The filters are changed through an external hatch. If required, a passive night vision device can be installed at either crew position. Vehicles produced for India have air conditioning.

The CET can also be used to tow the trailer-mounted Royal Ordnance Giant Viper or Python mine clearance systems. The vehicle is air-portable in C-130 Hercules aircraft.

The following ancillary equipment can be installed on the CET if required:

- Pusher bar for launching bridging
- Class 30 and Class 60 trackway laying equipment
- A jib crane attachment for handling palletised stores up to a maximum weight of 4,000 kg can be installed in the bucket.

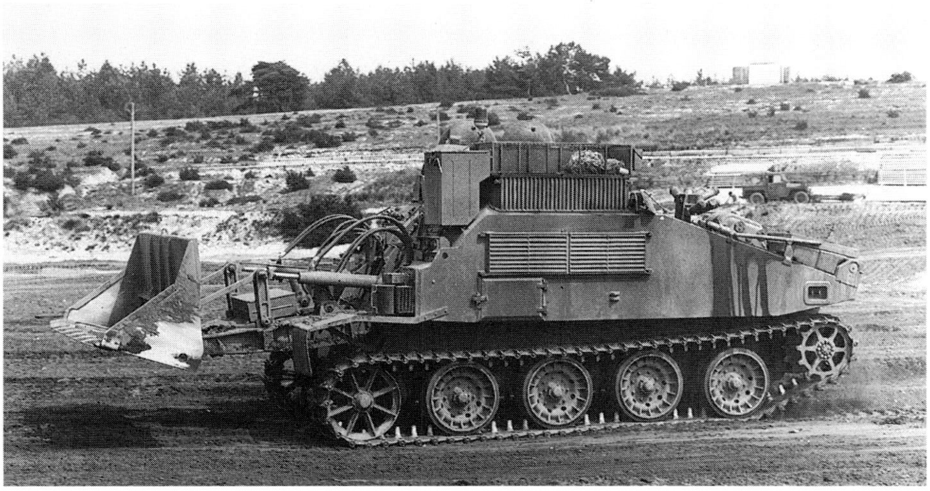

Combat Engineer Tractor (CET) **1996**

Specifications

Crew: 2
Weight combat: 18,350 kg
Power-to-weight ratio: 17.43 hp/t
Length:
(overall) 7.3 m
(hull) 5.334 m
Width:
(bucket) 2.921 m
(hull) 2.87 m
(tracks) 2.769 m
Height:
(without RP anchor) 2.83 m
(with RP anchor) 3.41 m
Ground clearance: 0.425 m
Track width: 508 mm
Length of track on ground: 3.76 m
Ground pressure: 0.435 kg/cm²
Max speed:
(road) 52 km/h
(water) 9 km/h
Range: 320 km
Fuel capacity: 418 litres
Fording: 1.83 m
amphibious with preparation
Gradient: 60%
Vertical obstacle: 0.61 m
Trench: 2.06 m
Engine: Perkins E320TX 12.17 litres 6-cylinder in-line diesel developing 320 bhp at 2,100 rpm
Transmission: TN26 manually controlled power shift with four gears in each direction coupled to Rolls-Royce CGS 312 steering system
Electrical system: 24 V
Batteries: 4 × 6TN rated at 100 Ah connected in series/parallel giving total capacity of 200 Ah
Armament:
1 × 7.62 mm MG (optional)
6 smoke grenade dischargers
Armour: aluminium

Status

Initial production was completed in 1981 but recommenced for an order from India and again for Singapore. In service with the British Army (141), Indian Army (15) and Singapore (54).

Manufacturer

Royal Ordnance Division, Kings Meadow Road, Nottingham NG2 1EQ, UK.
Tel: +44 115 988 2000. Fax: +44 115 968 2139.

UPDATED

Pearson Combat Dozer UDK1

Description

The Pearson Combat Dozer UDK1 was developed as an add-on attachment for METs or armoured engineer vehicles. It can carry out bulldozing tasks such as ground levelling, digging defensive 'tank scrapes', preparing ground for launching bridges, urban obstacle clearance, filling anti-tank ditches and so on.

All Vickers Defence Systems Challenger 1 (UK) and Challenger 2 (Oman and UK) MBTs are fitted to receive the Pearson Engineering Combat Dozer UDK1.

The UDK1 incorporates its hydraulic power pack into the boom arm and is thus a totally self-contained unit. A wedge-block mounting system enables the bulldozer to be removed from one vehicle and fitted to another in under 15 minutes. The bulldozer fits directly to the glacis plate by means of attaching two anchor blocks and lashing eyes pre-welded to the glacis plate. Electrical connection is made between a harness fitted permanently to the bulldozer and a glacis plate harness. The harness passes through armoured ducting to a control box in the driver's compartment. This control box can also be used to control the Pearson Mine Plough (see entry under *Mineclearing Equipment*).

The UDK1 unit comprises a blade and boom fitted to a mounting plate, with the blade position controlled by a hydraulic cylinder and too link. The blade is a fabricated steel unit fitted with replaceable hardened steel cutting edges and corner tips. The blade has an extension in the centre that increases dozing capacity and reduces spillage. An electrohydraulic power pack is fitted within the boom, along with the drive motor and hydraulic oil reservoir.

The control box in the driver's compartment controls blade lift and lowering functions electrically, using a joystick lever. Cutout switches in the power pack provide protection against motor and oil overheating, oil filter blockage and low oil level. These protective switches can be overridden under combat conditions by a 'Battle' switch. The full range of operations are: blade up, blade down, float on, float off, the 'Battle' override and a warning light test.

The Pearson Jacklift Pallet (see entry under Pearson Mine Ploughs in *Mineclearing Equipment* section) can be used with the UDK1.

Vickers Defence Systems Challenger 2 MBT fitted with Pearson combat dozer UDK1 ***1997***

Specifications

Weight: 2,200 kg
Width: 3.69 m
Height: 1.225 m
Cutting depth: 175 mm
Fitting and removal time: 10 min
Power supply: 24-28 V, 250 A max
Operating pressure: 1,250 bar
Power requirements: 250A at 24-28 V DC

Status

In production. In service with British Army for Challenger 1 and 2 MBTs and Chieftain AVRE. Also in service with European and Middle East armies.

Manufacturer

Pearson Engineering, Wincomblee Road, Walker, Newcastle upon Tyne NE6 3QS, UK.
Tel: +44 191 234 0001. Fax: +44 191 262 0402.

UPDATED

Pearson Combat Dozer UDK2

Description

The Pearson Combat Dozer UDK2 is a smaller and lighter version of the UDK1 (see previous entry) and can be fitted to light tanks, self-propelled guns and infantry fighting vehicles. It was developed to meet the requirements of the British Army with the first application being for use with the GKN Defence Warrior mechanised combat vehicle.

The UDK2 weighs 1,100 kg and has virtually no effect on vehicle mobility and does not interfere with any main armament. All working parts are identical to those of the UDK1.

Specifications

Weight: 1,100 kg
Width: 3.27 m
Height: 1.1 m
Cutting depth: 160 mm
Fitting and removal time: 10 min
Power supply: 24-28 V, 250 A max
Operating pressure: 1,250 bar
Power requirements: 250A at 24-28 V DC

Status

In production. In service with the British Army.

Manufacturer

Pearson Engineering, Wincomblee Road, Walker, Newcastle upon Tyne NE6 3QS, UK.
Tel: +44 191 234 0001. Fax: +44 191 262 0402.

UPDATED

Pearson Combat Dozer UDK2 fitted to Warrior mechanised combat vehicle

UNITED STATES OF AMERICA

United Defense LP, Combat Mobility Vehicle - Breacher

Development

Following the termination of the Counter-Obstacle Vehicle (COV) project, the US Army has a requirement for a Combat Mobility Vehicle (CMV). The intention is to provide armoured combat teams with an integrated counter-mine and counter-obstacle capability in a single vehicle. The CMV leads elements of armoured forces to create safe passage lanes for other combat vehicles through complex obstacles such as minefields, craters, anti-tank ditches and so on.

In April 1991, BMY Combat Systems (now part of United Defense LP) was awarded a $10, 945,000 contract for a single CMV Advanced Technology Transition Demonstrator (ATTD). The contract was modified in September 1992 to Breacher, a near-term M1 Abrams-based combat mobility vehicle. Under the terms of the contract, BMY Combat Systems will produce the vehicle based on a modified M1 Abrams MBT chassis provided as Government Furnished Equipment (GFE). The first two prototypes were rolled out in January and February 1995 and were involved in company trials from January to July 1995. After the US Army trials the programme is expected to move into full-scale engineering development and then production. It is anticipated that the full-scale engineering development phase will be completed by 1998, with production deliveries starting in 2000.

Major subcontractors to United Defense LP, Combat Systems Division are GDLS (heavy armour), General Microwave (depth sensor technology), Gradall Corporation (excavating arm) and Pentastor (vectronics).

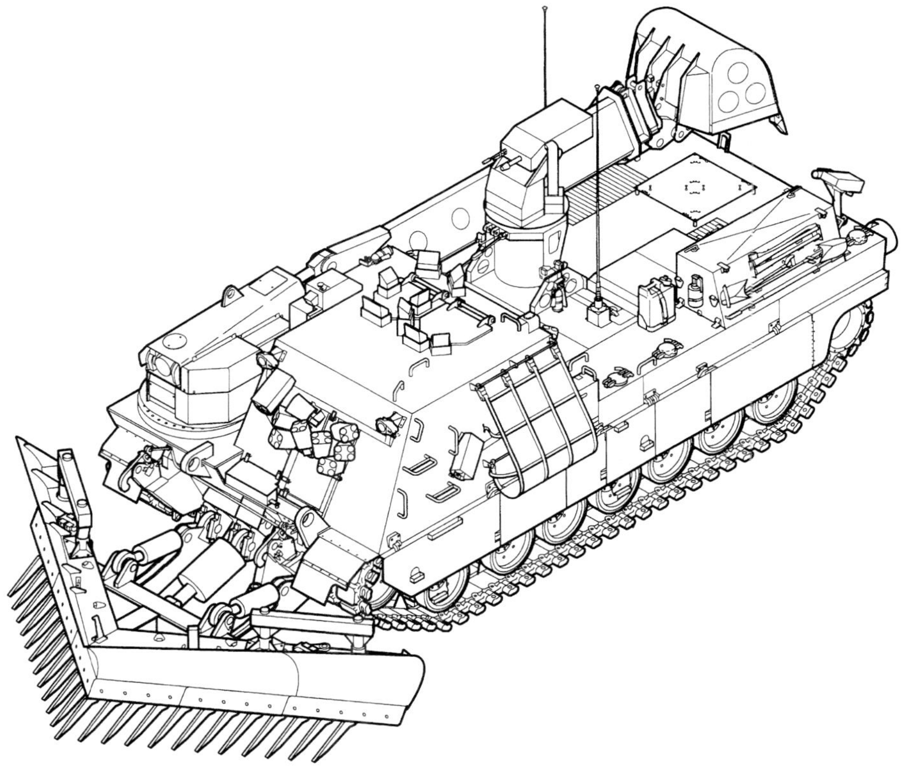

Artist's impression of the Breacher with excavating arm in stowed position and mineclearing blade at front **1997**

Description

The Breacher, also known as the Grizzly, has a crew of two, with the commander and driver seated side by side within an armoured superstructure mounted on the left side of the chassis. The crew compartment has Abrams level armoured protection and survivability. An escape hatch is provided in the rear of the crew compartment. Each crew station can control all vehicle systems. An auxiliary power unit containing the system hydraulics is mounted on the transmission. Ballistic protection is provided for hydraulic components and hoses.

Mobility is stated to be equal to that of the M1A2 Abrams. The M1A2 torsion bar suspension is employed with improved friction dampers and a lockout capability and the M1A1 final drive. The track uses T58 track shoes.

Protection features for the vehicle and crew include a radiation and spall liner, an automatic fire extinguisher system and exhaust generated smoke in addition to smoke grenade launchers. Armament involves a 40 mm MK 19 grenade launcher and a 7.62 mm M240 machine gun on a combination mounting, combined with viewers for night firing operations.

At the front of the vehicle is a full-width mineclearing blade with automatic depth control and rapid attach/detach features. The blade can clear a path 4.02 m wide, is V-shaped and is provided with 23 replaceable tines. Each tine is 305 mm long, retractable and is fitted with a shear pin which breaks when hard obstacles such as rocks are struck; should a shear pin break, the tine will rotate to the rear but will remain attached to the blade. The blade also has reversible and replaceable cutting edges and a conical cross-section to improve soil curl. Folding wing sections on each side are used to reduce the blade width in less than five minutes for transport.

To the right of the superstructure is a hydraulically controlled excavating arm capable of digging, lifting and grappling. The arm is a lightweight single-stage telescopic unit capable of operating in an arc from 46° left and 115° right of the vehicle centreline. The arm has a maximum reach of 9.2 m, a minimum reach of 3.3 m and can dig to a depth of 5.27 m. Maximum lift capacity at full extension is 1,814 kg. The digging bucket is detachable.

The vehicle is provided with seven video cameras to improve vision when operating. A Rapid Observation System (ROS) and Observation Reinforcing System (ORS) are provided to observe the vehicle exterior in a combat scenario.

The Breacher CMV is air-transportable in C-5A/B and C-17 transport aircraft.

Prototype of the Breacher with dozer blade lowered and arm traversed to the right **1997**

Specifications

Crew: 2
Max combat weight: 64,000 kg
Power-to-weight ratio: 23.47 hp/t
Max vehicle width: 3.657 m
Min range: (secondary roads) 483 km
Max speed: 66.6 km/h
Vertical obstacle: 1.067 m
Trench: 2.28 m
Mineclearing speed: 4.8 km/h
Cleared path:
(lane width) 4.02 m
(depth) 380 mm
Arm reach:
(maximum) 9.2 m
(minimum) 3.3 m
Digging depth: 5.27 m
Lift capacity: (full extension) 1,814 kg
Armament:
1 × 40 mm MK 19 grenade launcher
1 × 7.62 mm M240 MG
smoke grenade launchers

Status

Prototypes - see text.

Contractor

United Defense LP, Ground Systems Division, PO Box 15512, York, Pennsylvania 17405-1512, USA.
Tel: +1 717 225 8000. Fax: +1 717 225 8132.

UPDATED

M728 Combat Engineer Vehicle (CEV)

Development

In 1959 the M60 was chosen to become the standard MBT of the US Army. Design work on a new Combat Engineer Vehicle (CEV) based on the M60/M60A1 was soon started and the first prototype was known as the T118E1. After trials this was type classified as the M728 in 1963 and entered production in 1965, entering the US Army inventory in 1968. It is issued to the engineer battalions of armoured, mechanised and infantry divisions. Engineer battalions in armoured and mechanised divisions have eight M728s. Infantry division engineer battalions have only three M728s and separate engineer companies have two M728s each. The US Army has a total of 275 M728 vehicles in service with the only other users being Singapore and Saudi Arabia.

Description

The M728 is based on the M60A1 MBT. The hull is of cast sections welded together, while the turret is cast in one piece. The driver is seated at the front of the hull and is provided with a single-piece hatch cover. He has three M27 periscopes for observation purposes, of which the centre one can be replaced by an M24 infrared periscope on a separate mount for night operations. The other three crew members are in the turret, with the commander and gunner on the right and the loader on the left. The commander's cupola can be traversed through a full 360° and he is provided with a total of eight vision blocks and an M34 periscope, which can be replaced by an M36 infrared periscope for night observation. The gunner has an M105 telescope and an M32C periscope. The loader is seated on the left side of the turret and is provided with a single-piece hatch cover that opens towards the rear and an M37 periscope for observation purposes. The engine and transmission are at the rear of the hull, as are the fuel tanks.

The suspension of the M728 is of the torsion bar type and consists of six roadwheels with the drive sprocket at the rear and the idler at the front; there are three track-support rollers. Hydraulic shock-absorbers are provided at the first, second and sixth roadwheel stations.

The M728 is designed to carry out numerous roles on the battlefield, including the destruction of field fortifications and roadblocks, filling in gaps, craters and ditches, preparing fire positions and roadblocks.

Main armament of the M728 consists of an M135 165 mm demolition gun which has an elevation of +20° and a depression of −10°. Turret traverse is a full 360° at 1.6°/s; traverse and elevation are either powered or manual. A 7.62 mm M219 machine gun is mounted coaxially with the main armament (late production vehicles are equipped with an M240 coaxial machine gun) and a 0.50/12.7 mm M85 machine gun is mounted in the commander's cupola; this has an elevation of +60° and a depression of −15°. Totals of 30 rounds of 165 mm, 3,600 rounds of 7.62 mm and 600 rounds of 0.50/12.7 mm ammunition are carried.

The A-frame is pivoted on the front of the turret and when not required lies back over the rear of the hull. This has a maximum lifting capacity of 15,876 kg; the two-speed winch, which is mounted at the rear of the turret, is provided with 61 m of 19 mm rope and is controlled by the vehicle commander. The dozer blade is hydraulically operated and is mounted at the front of the hull. (Trials were carried out using a two-part mine plough in place of the dozer blade.)

The M728 is provided with infrared night driving equipment and most vehicles have a Xenon infrared searchlight mounted over the top of the main armament. A central air filtration system pipes fresh air to each crew member.

Specifications

Crew: 4
Weight:
(empty) 50,439 kg
(loaded) 53,200 kg
Power-to-weight ratio: 14.87 hp/t
Length:
(with boom erected) 9.3 m
(with blade and boom in travelling position) 8.91 m
Width:
(including blade) 3.7 m
(hull) 3.631 m
Height:
(travelling) 3.2 m
(lowest operable) 3.257 m
Ground clearance: 0.381 m
Track: 2.921 m
Track width: 711 mm
Length of track on ground: 4.235 m
Ground pressure: 0.89 kg/cm²
Max speed: (road) 48.28 km/h
Range: 450 km
Fuel capacity: 1,420 litres
Fording: 1.219 m
(with kit) 2.438 m
Gradient: 60%
Vertical obstacle: 0.76 m
Trench: 2.51 m
Engine: AVDS-1790-2A or -2D, 12-cylinder diesel developing 750 bhp at 2,400 rpm
Transmission: General Motors Corporation (Allison Division) CD-850-6A with 3 ranges (low, high and reverse)
Electrical system: 24 V
Batteries: 6 × 6TN
Armament:
1 × 165 mm demolition gun
1 × 7.62 mm M240 MG coaxial with main armament
1 × 0.50/12.7 mm M85 AA MG
2 banks of smoke grenade dischargers, 6 in each

Armour
turret front: 120 mm
turret sides: 76 mm
turret rear: 50 mm
turret top: 25 mm
hull front: 120 mm
hull sides, front: 76 mm
hull sides, rear: 51 mm
hull top: 57 mm
hull rear: 44 mm
hull floor: 13.63 mm

Status

Production complete. In service with the US Army (275), Saudi Arabia (15) and Singapore (8).

Manufacturer

General Dynamics Land Systems, 38500 Mound Road, Sterling Heights, Michigan 48310-3200, USA.

UPDATED

Side view of US Army M728 combat engineer vehicle with A-frame stowed (Michael Jerchel)

US Army M728 combat engineer vehicle in travelling configuration (Michael Jerchel)

Bulldozer, Earthmoving: Tank Mounting M8, M8A1, M8A3 and M9

Description

Differences in these bulldozer kits occur mainly in the details of their mountings. The M8 kit is used with the M48, M48A1 and M48C MBTs. The M8A1 is used only with the M48A2 MBT while the M8A3 is used with the M48A3. The M9 kit is used on the M60 MBTs.

Each bulldozer has a mouldboard with a reversible blade cutting edge supported on a quadrilateral linkage. The linkage has two push beams connected near the bottom of the mouldboard with two inner and outer tilt arms connected at the mouldboard top. Two double-action hydraulic cylinders are provided for raising and lowering. Pressure for the cylinders is

M60 MBT fitted with M9 bulldozer kit
1997

provided by a hydraulic pump driven by either the tank engine (M8 series) or a power take-off on the transmission (M9). The location of the control valves and hydraulic reservoir varies between the kits. On the M8 they are mounted on the wall of the tank rear hull; on the M8A1 on the right fender and on the M9 on the left-rear fender.

Status

Production complete. No longer used by US Army or Marines but still used by a number of countries.

UPDATED

Specifications

Tank mounting	**M8**	**M8A1**	**M8A3**	**M9**
MBT model	M48, M48A1, M48C	M48A2	M48A3	M60
Weight	3,981.6 kg	3,810 kg	3,810 kg	4,031 kg
Mouldboard length	3.708 m	3.708 m	3.708 m	3.708 m
Mouldboard height	0.914 m	0.914 m	0.914 m	0.914 m
Blade angle (from horizontal)				
(ground level)	58°	58°	–	58°
(float)	59°	59°	65°	60°
(lowest)	65°	65°	65°	60°
Edge position				
(carrying)	0.787 m	0.686 m	0.736 m	0.736 m
(lowest, below ground)	0.254 m	0.254 m	0.254 m	0.254 m
(highest)	0.762 m	0.762 m	0.762 m	0.762 m
Max rate of lift	267 mm/s	267 mm/s	267 mm/s	165.6 mm/s
Max forward speed, dozing	1.6-4.82 km/h	1.6-4.82 km/h	1.6-4.82 km/h	1.6-4.82 km/h

United Defense LP, M9 Armored Combat Earthmover

Development

In 1958, the US Army Mobility Equipment Research and Development Center at Fort Belvoir began work on a vehicle called the All-purpose Ballastable Crawler (ABC). This eventually became known as the Universal Engineer Tractor (UET).

The first prototypes of the Universal Engineer Tractor were built by the Caterpillar Tractor Company and the International Harvester Company. In January 1975 the Pacific Car and Foundry Company completed a further four vehicles, two of which underwent field evaluation at Fort Hood. The result of these trials was considered satisfactory and indicated the vehicle's superiority over available equipment. TECOM testing was completed in August 1976 and type classification (Standard A) was approved in February 1977.

Trials of the M9 were subsequently carried out in both Yuma and Alaska to test modifications made as a result of previous trials and new equipment, such as a winterisation kit. In Fiscal Year 1979, $21.1 million was requested for the production of 75 vehicles, but only $10.6 million was authorised for 29 vehicles. Since production of such a small number was not considered an economic proposition, production was delayed in the hope that the FY80 request, for $40.5 million, plus the 1979 allocation, would allow the production of up to 155 vehicles. No M9 funding was made in either the 1980 or 1981 budgets. In November 1982 a revised contract worth $29 million was awarded for the purchase of 15 equipments of which $19.3 million was for the vehicle and the remainder for product improvements (the original contract was to have been $40.4 million for 36 vehicles). All 15 vehicles had been delivered by the end of 1984. Trials with this first batch of vehicles demonstrated that further improvements were required in some areas. Seven of this first batch were modified by PCF Defense Industries and were delivered to the US Army during January 1985. Tests with this batch of vehicles were carried out at Fort Hood.

RFPs were issued to industry during April 1986 and resulted in offers from ADCOR, BMY, FMC, General Motors of Canada, Ingersoll-Rand and PCF Defense Industries. BMY (now part of United Defense LP) was ultimately awarded a production contract on 25 July 1986. By the end of 1992 434 M9s had been built. The first production examples were delivered to the training base at Fort Leonard Wood during the fourth quarter of FY89. US Army units in Europe started to receive their vehicles during the last quarter of 1990.

During 1991 two overseas orders were placed for M9s. The first, for 18 units, was from a customer in the Far East. The second order was worth $8 million, with deliveries starting in late 1992.

In September 1993 United Defense LP, was awarded a $78 million, 121 vehicle contract consisting of 87 vehicles for the US Marine Corps and 34 vehicles for the National Guard. US Marine Corps deliveries commenced during 1995 with National Guard deliveries starting in 1996.

The M9 ACE has been ordered by the South Korean armed forces. Licence production (the final in-service total will be 197) will be carried out in South Korea by Samsung Aerospace.

Description

The M9 Armored Combat Earthmover (ACE) is intended to operate in forward areas and, due to its high road speed, can be located with the lead tanks in a convoy, closer to where it is needed, rather than at the back, as is the case with vehicles (such as the D7 medium crawler) which the M9 is intended to replace.

The M9 ACE is a general purpose engineer vehicle and can carry out tasks in three critical areas: mobility, counter-mobility and survivability. Mobility tasks include: filling craters and ditches; assisting fighting vehicles (winching or towing); removing roadblocks, trees, rubble and other battlefield obstacles; preparing access/egress for fording sites and river crossings; preparing and maintaining combat routes; and preparing and maintaining assault airfields. Counter-mobility tasks include: the construction of anti-armour obstacles; demolishing fords and bridge bypasses; participating in the digging of tank ditches; destroying landing fields and airfields; participating in the preparation of strong points and hauling obstacle materials. Survivability tasks include: the digging of hull defilade positions for armour; construction of defensive positions for command and control operations; construction of earth berms for protection; hauling material for protective shelters; clearing fields of fire; and digging slots for vehicle-mounted TOWs and other battlefield weapons.

The M9 is air-portable in C-130, C-141B and C-5A transport aircraft. It is unarmed but has a smoke grenade launcher. Other equipment includes a standard NBC system (ventilated facepiece) and a radio and the operator can utilise standard night vision goggles. It has an amphibious capability after minimal preparation and armour protection is provided for the engine, power train and the operator.

The hull of the M9 is made of welded aluminium armour. At the front of the vehicle is the 6.7 m^3 capacity scraper bowl (ballast compartment), hydraulically operated apron and positive load ejector. The driver is seated towards the middle of the vehicle on the left side and is provided with a cupola which provides 360° vision. A lighter hatch cover (weight 81.65 kg) than those fitted to early vehicles has been introduced; the hatch incorporates vision blocks in place of the earlier periscopic vision devices. The engine, a variant of that used in the Bradley IFV, is positioned to the right of the driver's compartment. The transmission is below the engine and the steer unit is to the rear, coupled to the final drives. On-vehicle equipment includes a 15.2 litres/s bilge pump and a Carco P30 planetary-winch with a line pull of 15,900 kg using a 19 mm diameter wire rope 60 m long. The M9 ACE can also be used to tow trailers and other equipment as it has a maximum drawbar pull of 14,059 kg. It is fully amphibious with preparation, being propelled in water by its tracks. The M9 is not suitable for operating in fast-flowing rivers.

The dozer blade is mounted on the apron and dozing and scraping are accomplished by raising and lowering the entire front of the vehicle by means of the hydropneumatic suspension. This consists of eight sets of 711 mm diameter steel roadwheels with the drive sprocket at the rear. The hydropneumatic suspension allows the M9 to be tilted to apply the dozing effort to one corner of the blade. The capability of the vehicle for operations such as dozing can be nearly doubled by self-loading the bowl with approximately 8,000 kg of earth, which is used as ballast.

M9 Armored Combat Earthmover in operating configuration **1997**

Head on view of M9 Armored Combat Earthmover in travelling configuration **1997**

Specifications

Crew: 1
Weight:
(travel mode) 16,327 kg
(gross, ballasted) 24,490 kg
Drawbar pull: 14,059 kg
Power-to-weight ratio: 18.06 hp/t
Length: 6.248 m
Width:
(with dozer wings) 3.2 m
(without dozer wings) 2.79 m
(over tracks) 2.692 m
Height:
(windshield raised) 3 m
(windshield stowed) 2.7 m
Ground clearance:
(sprung, engine running) 0.343 m
(unsprung) variable
Track width: 457 mm
Length of track on ground: 2.67 m
Ground pressure: 0.68 kg/cm² (empty)
Max speed:
(road) 48.3 km/h
(water) 4.8 km/h
Range: (secondary roads) 322 km
Fuel capacity: 507 litres
Fording: 1.83 m
amphibious with min preparation
Freeboard with load of 1,814 kg: 0.28 m
Gradient:
(kerb) 20%
(ballasted) 60%
Side slope: 20%
Vertical obstacle: 0.457 m
Trench: 1.575 m
Engine: Cummins V903C 14.8 litres 8-cylinder diesel developing 295 hp at 2,600 rpm
Transmission: Clark Model 13.5 HR 3610-2 with 6 forward and 2 reverse gears. Geared system for high-speed road and cross-country travel, and clutch and brake system for bulldozing operations
Suspension: hydropneumatic
Steering: hydraulic with gears and clutches
Turning radius:
(geared steer mode) 13.7 m
(clutch brake mode) pivots
Armament: none
Armour: aluminium, Kevlar and steel

Status

In production. In service with the US Army, Marine Corps and National Guard. Ordered by South Korea (197) where it is now being licence-produced by Samsung Aerospace.

Manufacturer

United Defense LP, Ground Systems Division, PO Box 15512, York, Pennsylvania 17405-1512, USA.
Tel: +1 717 225 8000. Fax: +1 717 225 8137.

UPDATED

M113 with dozer blade kit

Description

The M113 armoured personnel carrier can be adapted for general bulldozing work by the addition of a dozer blade kit. Once fitted, the kit does not impair the normal capabilities of the vehicle or its load-carrying ability. The vehicle can remain amphibious and the dozer blade acts as a trim vane in the water once the standard trim vane has been removed.

Status

Production as required. In service with the US Army and other countries. This dozer blade kit is also applicable for other members of the M113 series family including the M113A1 and the design can be adopted to later models of the M113 series of vehicle.

Manufacturer

United Defense LP, Ground Systems Division, 2830 De La Cruz Boulevard, Santa Clara, California 95052, USA.
Tel: +1 408 289 0111.

UPDATED

US Army M113 APC fitted with hydraulically operated dozer blade kit
1997

Sapper Vehicle

Development

In October 1992, it was announced that the US Army Tank Automotive Command (TACOM) was producing the prototype of a 'Sapper Vehicle', based on a stretched version of the M113A3 APC, for use by US Army Engineers. The prototype was manufactured at the Red River Army Depot and utilises experience gained during the TACOM programme that produced the M113A1/A2 Fitter's Vehicle (see entry under *Armoured Recovery Vehicles* for details). Further influence was derived from the Canadian M113 Engineering Specially Equipped Vehicle (see entry in this section).

The mission of the Sapper Vehicle is 'to provide transportation and protection for an eight-man engineer squad and specialised equipment to perform mobility, survivability and counter-mobility tasks for the supported force'.

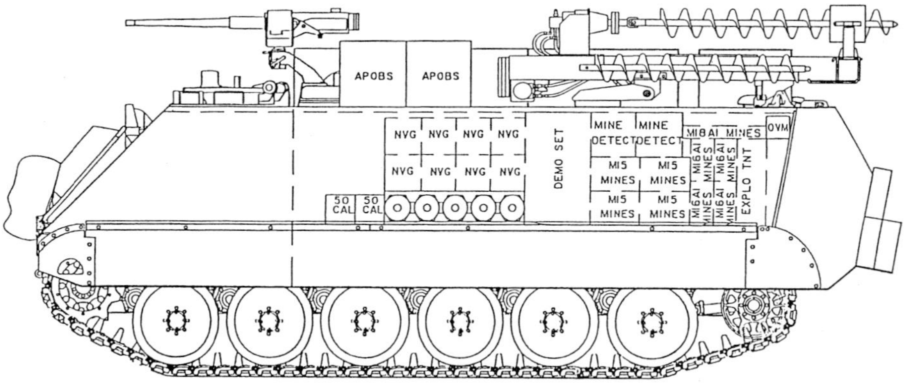

Proposed side layout of the Sapper Vehicle

Description

The Sapper Vehicle is based on a stretched version of the M113A3 APC which is 670 mm longer than the standard M113A3 and features six sets of roadwheels rather than the standard five. The extra internal space thus provided is used for additional combat engineering equipment storage space. Engineer explosives are stored behind spall-suppressant armour panels. Also provided, are built-in blast protection between shaped and cratering charges, bolt-on armour and increased mine blast protection.

Special equipment is expected to include a dozer blade for use against light obstacles, battlefield debris and scatterable mines. A hydraulic auger is stowed on the roof. Extra vision devices are provided for the crew to use during breaching and mineclearing operations.

Status

Prototype.

UPDATED

United Defense LP Combat Engineer Squad Vehicle (CESV)

Development

Further evolution of the Sapper Vehicle, covered in the previous entry, has resulted in the Combat Engineer Squad Vehicle (CESV) which has been proposed to meet the requirements of the US Army Engineers.

The CESV is based on the current in service M113A3 series APC, but would be rebuilt and extended by an additional road wheel either side to extend its internal volume. In many respects the CESV is identical to the Mobile Tactical Vehicle Light (MTVL) (*Jane's Armour and Artillery 1996-97* pages 361 to 362).

According to United Defense, the CESV would allow the eight man combat engineer squad to carry all their equipment to support attack missions without towing a trailer, while providing mobility and survivability equivalent to the manoeuvre units Bradley infantry fighting vehicles.

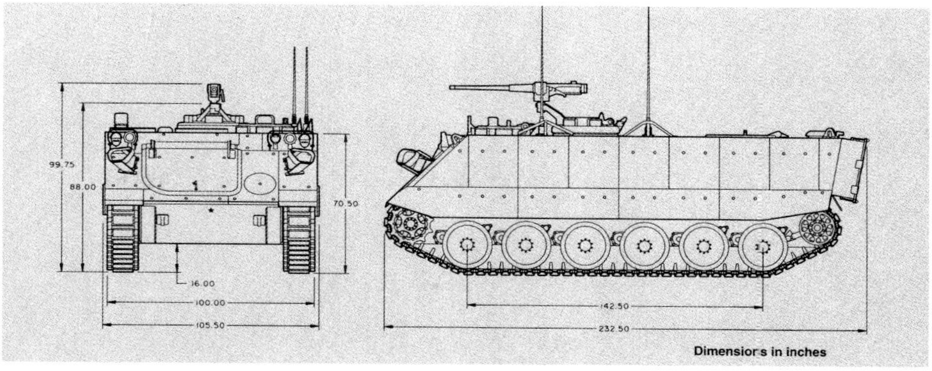

Front and side view of Combat Engineer Squad Vehicle **1997**

Description

The hull of the CESV is of all-welded aluminium armour construction to which an applique armour package has been added for improved battlefield survivability.

The upper front glacis and sides armour package consists of titanium, rear armour package consists of spaced expanded steel while the bottom armour package consists of spaced laminate steel.

The driver is seated at the front left and has a single piece hatch cover that opens to the rear, to his front and left side are four M17 periscopes with provision for an AN/VVS-2 passive night vision device.

The engine compartment is to the right of the driver's position and is fitted with a fire extinguishing system that can be operated by the driver or from outside the vehicle. The air inlet and outlet louvres and the exhaust pipe outlet are in the roof and there is an engine compartment access door in the front of the sloping glacis plate that opens upwards.

The commander is seated to the rear of the engine compartment and has a cupola that can be traversed through a full 360°, five M27 day periscopes and a single piece hatch cover.

The armament is pintle mounted on the forward part of the commander's cupola and can be a 12.7 mm M2 HB machine gun, 40 mm Mk 19 Mod 3 grenade launcher or a 7.62 mm M240 machine gun.

If the 12.7 mm M2 HB machine gun is fitted, then elevation limits are from −21 to +53° with traverse through a full 360°. There are 100 rounds of ready to use ammunition carried for this weapon. Mounted on either side of the hull, firing forwards towards the front, is a bank of four electrically operated smoke grenade dischargers.

The engineers sit on bench seats that run down either side of the hull facing the centre and enter and leave via a large power-operated ramp in the hull rear. There is also a door in the left side of the ramp.

Over the top of the troop compartment is a large roof hatch that opens to the rear, to the rear of this is a domed adjustable ventilator.

The torsion bar suspension either side consists of six dual rubber tyred road wheels with the drive sprocket at the front and the idler at the rear. There are no track return rollers.

Specifications

Crew: 2 + 8
Max combat weight: 16,330 kg
Max payload: 5,942 kg
Power-to-weight ratio: 19.8 hp/t
Ground pressure: 0.6 kg/cm^2
Length: 5.905 m
Width:
(overall) 2.67 m
(over tracks) 2.54 m
Height:
(overall) 2.53 m
(hull top) 1.79 m
Ground clearance: 0.406 m
Track: 2.159 m
Track width: 380 mm
Length of track on ground: 3.62 m
Max road speed: 66 km/h
Acceleration:
(0 to 32.187 km/h) 7.5 s
(0 to 48.28 km/h) 17.7 s
Fuel capacity: 454 litres
Max cruising range: 483 km
Turning radius: pivot to infinity
Vertical obstacle: 0.71 m
Gradient: 60%
Side slope: 40%
Trench: 2.18 m
Engine: Detroit Diesel 6V-53TA 6-cylinder diesel developing 350 hp at 2,800 rpm
Transmission: Allison X200-4A hydrokinetic
Steering: hydrostatic
Brake type: multiple wet plate
Suspension: torsion bar
Electrical system: 28 V
Generator: 200 A (300 A optional)
Batteries: 4 × 6TL, 120 Ah, 12 V
Armament: 1 × 12.7 mm MG (see text)
Ammunition: 1,000 × 12.7 mm
Armour: 12-44 mm (see text)

Status

Prototype. Not yet in production or service.

Manufacturer

United Defense LP, Ground Systems Division, 2830 De La Cruz Boulevard, Box 58123, Santa Clara, California 95052, USA.

NEW ENTRY

RECOVERY VEHICLES AND EQUIPMENT

Armoured recovery vehicles
Armoured repair vehicles
Recovery vehicles
Recovery equipment

ARMOURED RECOVERY VEHICLES

AUSTRIA

Steyr-Daimler-Puch Bergepanzer 4KH7FA-SB 20 Greif armoured recovery vehicle

Description

The Steyr-Daimler-Puch Bergepanzer 4KH7FA-SB 20 Greif armoured recovery vehicle is based on the chassis of the Jagdpanzer SK 105 light tank. The first prototype was completed in 1974 with the first production vehicles following in 1976-77.

The hull is of all-welded steel construction with the winch and crew compartment at the front. Normal means of entry to the crew compartment is via two doors in the left side of the hull. In addition there are hatches in the roof. The engine and transmission are at the rear of the hull and there is a stowage platform over the engine decking, which is used to carry spare components. The torsion bar suspension consists of five roadwheels with the drive sprocket at the rear and the idler at the front. There are three track-return rollers. Hydraulic shock-absorbers are provided at the first and fifth roadwheel stations.

Mounted on the right side of the superstructure, at the front, is a hydraulic crane, which is traversed to the rear for travelling. This crane can be traversed through 234° and its boom elevated from 0 to +60°. The boom can be extended from its normal length of 3 to 3.9 m, and has a maximum lifting capacity of 6,000 kg. The crane is provided with 42 m of cable and is capable of lifting the complete turret of the Jagdpanzer SK 105 light tank.

The main winch is in the lower part of the hull and leads out through the front of the hull. This opening can be sealed for fording operations. The winch is provided with 95 m of 24 mm diameter cable and has a maximum pull of 20,000 kg. This enables the Greif to recover both the Jagdpanzer SK 105 light tank and the Steyr APC. A rake blade is mounted at the front of the hull and is used in conjunction with the main winch. A full range of tools is carried, as is cutting and welding equipment. The Greif does not have an NBC system, but the crew compartment is provided with a ventilation and heating unit.

Steyr developed a pioneer vehicle based on the chassis of the Greif ARV. Details are given in the *Armoured Engineer Vehicles* section. This vehicle is only used by the Austrian Army who took delivery of 18 vehicles.

An upgrading programme involving the installation of a ZF 6HP 500 automatic transmission is in progress for Austrian Army Greif vehicles.

Steyr-Daimler-Puch Bergepanzer 4K H7FA 20 Grief armoured recovery vehicle **1997**

Specifications

Crew: 4
Weight: 19,800 kg
Power-to-weight ratio: 16.66 hp/t
Length: 6.705 m
Width: 2.5 m
Height:
(max) 3.02 m
(top of hull) 2.3 m
Ground clearance: 0.4 m
Track: 2.12 m
Track width: 380 mm
Length of track on ground: 3.037 m
Ground pressure: 0.75 kg/cm^2
Max speed: (road) 67.5 km/h
Range: (road) 625 km
Fuel capacity: 500 litres
Fording: 1 m
Gradient: 70%
Side slope: 40%
Vertical obstacle: 0.8 m
Trench: 2.1 m
Engine: Steyr Type 7FA 6-cylinder turbocharged diesel developing 320 hp at 2,300 rpm
Transmission: ZF manual with 6 forward and 1 reverse gears; being updated to ZF 6HP 500 automatic
Turning circle: 7.25 m
Electrical system: 24 V
Batteries: 2 × 12 V, 180 Ah
Armament:
1 × 12.7 mm M2 HB MG
4 × 80 mm smoke grenade dischargers
Ammunition: 12.7 mm - 1,500 rounds

Status

Production complete. In service with Argentina (10), Austria (50), Bolivia (2), Morocco (11) and Nigeria (15).

Manufacturer

Steyr-Daimler-Puch Spezialfahrzeug AG, PO Box 100, A-1111 Vienna, Austria.
Tel: +43 1 764511. Telex: 61 322 1299.
Fax: +43 1 768149.

UPDATED

BELGIUM

SIBMAS (6 × 6) armoured recovery vehicle

Description

The SIBMAS (6 × 6) armoured recovery vehicle is a conversion of the basic SIBMAS (6 × 6) vehicle range and uses the same basic hull and drive train. No prototype was produced as the first example, produced in mid-1983, retained much of its commonality with the rest of the SIBMAS (6 × 6) range. Full details of the SIBMAS (6 × 6) range can be found in *Jane's Armour and Artillery 1996-97* pages 384 to 385. Production of the SIBMAS range of vehicles was completed many years ago and it is no longer being marketed.

The hull is of all-welded steel and is fully watertight. It provides the crew with full protection against 7.62 mm armour-piercing rounds. The driver is seated in the centre of the hull front behind three bulletproof windscreens. Behind him there is a circular hatch for the commander. Mounted on the hatch rail is a 7.62 mm machine gun which can be elevated 65° and depressed 10°. Between the commander's hatch and crane support there is one large roof hatch and another entry hatch on the right of the hull rear. Internally there is rearward-facing seating for three fitters, each side of the commander's position and next to the side doors.

The engine arrangement, type, suspension and drive train are all identical to the rest of the SIBMAS (6 × 6) range.

Specialist recovery equipment mounted externally consists of two hydraulically operated spades, a hydraulic crane and two winches. The spades are mounted one at the front and the other at the rear and both are fully folding when required. The hydraulic crane is mounted centrally on the hull roof and can be telescopically extended. It has a capacity of 10.5 t/m and can lift loads of up to 3,000 kg at a radius of 3.5 m and 8,000 kg when used with the crane boom support for suspended towing operations. When extended, a crane jib support is used to secure heavy loads for towing. Of the two winches one is used as the main multipurpose winch and is centrally mounted to enable guide pulleys for the cable to be extended front or rear. The hydraulically operated main winch has a direct pull of 20,000 kg. The other winch is used as an auxiliary and is mounted behind the main winch. It has a 1,500 kg capacity and is used to unwind the main winch cable.

Other external equipment includes: towing eyes front and rear; a 360° rotating searchlight on the hull roof; an emergency rotating warning light; a tow rope; a lifting bar; vehicle tool kit and an axe, shovel and pickaxe; sand channels; stowage box; a hydraulic lifting jack with an 8,000 kg capacity; fuel and water jerricans; first aid kit; compressor air outlet for tyre inflation; and two six-barrel 76 mm smoke launchers.

Internally, there is stowage capacity for a wide range of specialist equipment including a portable gas welding kit and a hydraulic lifting jack with a 10,000 kg capacity. There is also a VHF communications radio set

SIBMAS (6 × 6) armoured recovery vehicle

and full internal communications loudspeakers, only the driver and commander have headsets. The driver has a passive periscope for night driving, and air conditioning is provided for the full crew.

Specifications

Crew: 2 + 3
Configuration: 6 × 6
Length overall: (travelling) 7.63 m
Width: 2.54 m
Height: (top of crane) 3.2 m
Ground clearance: 0.4 m
Track: 2.066 m
Wheelbase: 2.8 m + 1.4 m
Angle of approach/departure: 35°/35°
Max road speed: 80 km/h
Fuel capacity: 410 litres
Max range: (road) 800 km
Fording: 1.07 m
Max gradient: 50%
Side slope: 30%
Vertical obstacle: 0.6 m
Trench: 1.8 m
Engine: MAN D 2566 MK 6-cylinder, in-line, water-cooled turbocharged diesel developing 320 hp at 1,900 rpm
Transmission: fully automatic power-shift gearbox with hydrodynamic torque converter Type ZF 6 HP-500
Steering: power-assisted
Turning circle: 18 m
Tyres: 14.00 × 20
Electrical system: 24 V

Status

Production complete and no longer being marketed. In service with Malaysia (24).

Manufacturer

BMF - Belgian Mechanical Fabrication SA, SIBMAS Division, Zoning Industriel, B-4330 Grace-Hollogne, Belgium.
Tel: +32 41 639940. Telex: 42 049.
Fax: +32 41 636562.

UPDATED

BULGARIA

MTP-1 armoured recovery vehicle

Description

The MTP-1 armoured recovery vehicle is a variant of the vehicle known in the West as the Armoured Command and Reconnaissance Vehicle (ACRV - actual designation MT-LBus) as produced in Bulgaria (for details of the ACRV refer to *Jane's Armour and Artillery 1996-97* pages 598 to 602). The MTP-1 is used for the recovery of damaged or disabled vehicles, changing vehicle components, preparing fire positions and vehicle hides, and for general crane lift purposes.

The MTP-1 is virtually identical to the ACRV but mounted on the front right-hand side is a turret identical to that fitted to the MT-LB multipurpose armoured vehicle (pages 264 to 265). This manually operated turret is armed with a 7.62 mm PKT machine gun, although some turrets have been up-gunned to 12.7 mm. Mounted on the roof is a crane with a telescopic jib. When travelling, the crane is traversed to the front and the hook is restrained by an eye on the front of the hull. The crane is mounted in a small turret-like cupola with the crane operator being provided with full armour protection. The crane has a lift capacity of 3,000 kg with the jib extended to 3.4 m, or 2,000 kg extended to the full 5 m.

Mounted at the rear of the vehicle is an entrenching blade similar to that fitted to the engineering variant of the MT-LB (see entry under *Armoured Engineer Vehicles*). The blade is mounted on hydraulic arms and can prepare a vehicle scrape in about 110 minutes. The blade can also act as an anchor and support when the vehicle's onboard winch is in operation. With the blade lowered the winch has a maximum capacity of 30 tonnes. With the blade raised, the winch capacity is lowered to 10 tonnes.

The MTP-1 is fully amphibious, being propelled in the water by its tracks. An NBC system is fitted. Combat weight is 14,000 kg.

Status

Production as required. In service with Bulgarian and other undisclosed countries.

Contractor

Kintex, PO Box 209, 66 Anton Ivanov Boulevard, Sofia, Bulgaria.
Tel: +359 2 662311. Telex: 22471, 23243.
Fax: +359 2 658191.

UPDATED

Rear view of MTP-1 armoured recovery and repair vehicle with entrenching blade lowered

CHINA, PEOPLE'S REPUBLIC

NORINCO Type 653 and 653A armoured recovery vehicles

Description

The Type 653 armoured recovery vehicle is based on the chassis and hull of the Type 69 MBT. The Type 653 bears no relationship to a recovery version of the Type 59 MBT which has a tractor-towing role only. The Type 653A is basically a Type 653 with a more powerful crane.

The Types 653 and 653A are intended for the field recovery and repair of medium battle tanks and also have a limited combat engineering capability. The basic hull and suspension of the Type 69 are retained, but in place of the turret a fixed superstructure is fitted offset to the left and the vehicle also carries a hydraulic crane and an earthmoving dozer blade at the front. The main and auxiliary winches are located within the

NORINCO Type 653 armoured recovery vehicle **1997**

NORINCO Type 653 armoured recovery vehicle (Christoper F Foss) **1996**

superstructure. The vehicle has a crew of five with the driver located forward on the left-hand side in an armoured position raised above the normal height of the Type 69 driver's position. The driver has a single forward vision aperture and a circular overhead access hatch. Infrared lighting and vision devices are provided for night driving. Mounted on the front of the hull and operated by two hydraulic cylinders is the earthmoving dozer blade that can also act as an anchor for heavy winching operations. The dozer blade has an earthmoving capacity of 100 m³/h on dry soil. The main winch access rollers are located between the twin hydraulic cylinders for the dozer blade and the blade has to be in the lower position for recovery. The main winch is hydraulic, with a maximum pulling performance of up to 70,000 kg and the effective length of the cable is 130 m (160 m on Type 653A).

To the right of the driver's position is the hydraulic crane jib with its full 360° turntable in line with the driver's position. The crane jib is telescopic and is normally stowed to the rear for travelling. It can be extended to a maximum lifting height of 6 m and has a rated lifting torque (static) of 40 to 70 t/m on Type 653A. The rated lifting capacity is 10 tonnes on the Type 653 and 20 tonnes on the Type 653A. Recovered vehicles can be towed behind the vehicles using a rigid towing trail which is normally carried on the main hull superstructure side. This superstructure has two circular roof hatches and there is provision for a 12.7 mm machine gun mounting on the roof.

The auxiliary winch is located to the left of the driver on the track fenders, which also support a number of tool and other stowage boxes. The Type 653 is stated to have a capability of carrying out field repairs and the hydraulic crane can lift an entire tank turret complete with gun. The main communications equipment is the Type 889 radio set and the Type 803 is used as the intercom system.

Specifications

Crew: 5
Weight:
(Type 653) 38,000 kg
(Type 653A) 42,000 kg
Power-to-weight ratio:
(Type 653) 15.26 hp/t
(Type 653A) 13.80 hp/t
Length: 7.18 m
Width: (over side skirts) 3.304 m
Height:
(highest point) 3 m
(top of superstructure) 2.34 m
Ground clearance: 0.425 m
Track: 2.64 m
Length of track on ground: 3.845 m
Max speed: 50 km/h
(towing) 18 km/h
Range: (road, normal fuel tanks) 370 km
Fording: 1.4 m
Gradient: 30°
(towing) 15°
Side slope: 30°
Vertical obstacle: 0.7 m
Trench: 2.6 m
Engine power: 580 hp
Armament: 1 × 12.7 mm Type 54 MG

Main winch
Max towing force: 64,000-70,000 kg
Cable delivery and retract speed:
(without load) 28-30 m/min
(with load) 7 m/min

Crane - Type 653
Rated lifting capacity: 10,000 kg
Rated lifting torque: (static) 40,000 kg/m
Max lifting height: 6 m

Dozer blade
Max digging depth: 200 mm
Digging capacity: (dry earth) 100 m³/h
Max supporting force: 70,000 kg

Status

In production. In service with China, Bangladesh, Iraq, Pakistan and Thailand.

Manufacturer

China North Industries Corporation (NORINCO), 7A Yue Tan Nan Jie, PO Box 2137, Beijing, People's Republic of China.
Tel: +86 1 867570. Telex: 22339 CNIC CN.
Fax: +86 1 867840 Beijing.

UPDATED

CZECH REPUBLIC AND SLOVAKIA

ZTS VT-72B armoured recovery vehicle

Development

The VT-72B was developed in Czechoslovakia and is based on the chassis of the T-72M1 MBT, now produced in Slovakia by ZTS Tees Martin with some input from concerns in the Czech Republic. The VT-72B differs from the Russian BREM-1 armoured recovery and repair vehicle in many respects, not the least of which in having the swivelling hydraulic crane on the right-hand side of the superstructure rather than the left.

Of the total of the vehicle's main assemblies, 13 are the same as the T-72 MBTM1, 23 assemblies have been adapted and 12 are of a new design.

In late 1993 India purchased 35 VT-72B at a unit cost of INR28.9 million each. The total order was valued at $31.5 million, including spares and the free transfer of technology. There were 10 vehicles to be provided complete, the first was delivered in May 1994, with the remaining 25 vehicles delivered in semi-knocked-down kit form for assembly at the Bharat Heavy Electricals Limited (BHEL) plant in Trichy, southern India. It is anticipated that production in India will follow, under licence from UNIMPEX of Slovakia. BHEL has signed a contract to supply around 210 ARVs to the Indian Army, with an option to eventually double that number.

The former East German Army received three of these vehicles and had plans to purchase more; they were known as the T-72TK.

A variant known as the VT-72A has been reported with a lighter crane capable of lifting only 7,000 kg, and some other changes. As far as can be determined this variant was not developed past the prototype stage.

Description

The VT-72B has a crew of two, commander and driver, although there is provision to carry a further three

VT-72B armoured recovery vehicle (T J Gander)

VT-72B armoured recovery vehicle (T J Gander)

personnel all housed in an armoured superstructure built onto the chassis of a T-72 MBT. On the vehicle, 13 of the major subassemblies are identical to those of the original T-72 tank, 23 assemblies have been partially adapted and 12 are new.

The VT-72B can be used to recover stranded vehicles and tow them to a place of safety where the 19 tonne crane jib can be used to remove major components such as tank turrets and power packs. Further repair work can be accomplished using the tools and electric welding kit carried on the vehicle, while further spares and equipment can be carried on a load platform capable of bearing up to 4 tonnes and measuring 1.4 × 1.4 m.

An NBC system is provided for the crew compartment.

The hydrostatic main recovery winch is located centrally with a usable 200 m of rope cable on a drum located over the forward part of the engine compartment. The winch has a main traction force of 300 kN with the cable led to the front through a pulley arrangement on the front hull. An auxiliary winch has a 10 kN capability and is provided with 400 m of rope cable.

The main winch is used in conjunction with a front-mounted straight dozer blade that can be used as an earth anchor for heavy recovery tasks. The blade is 3.36 m wide and can be lowered into the ground to a depth of 300 mm. As a secondary function, the blade may also be used to clear battlefield obstacles and debris or create scrapes to assist in the concealment of other vehicles.

The swivelling hydraulic crane is mounted on a small turntable on the right front of the hull next to the superstructure. The crane has a maximum lifting capacity of 19 tonnes and an arm radius of 7.6 m. Arm traverse is 360°; maximum lift angle is 72°.

Racks located over the rear of the vehicle carry extra fuel in both drums and jerricans.

The main armament for air defence is a 12.7 mm NSV machine gun provided with a K 10-T collimator sight.

Specifications
Crew: 2 + 3
Combat weight: 46,500 kg
Power-to-weight ratio: 18.06 hp/t
Length: (with extra tanks) 8.15 m
Width overall: 3.37 m
Height: (incl MG) 2.64 m
Ground clearance: 0.435 m
Max speed: (roads) 60 km/h
Range:
(roads, basic fuel) 485 km
(roads, with extra tanks) 620 km
(dirt roads, basic fuel) 235-470 km
(dirt roads, with extra tanks) 300-600 km
Fording:
(normal) 1.2 m
(with preparation) 1.8 m
(with kit) 5 m
Engine: V-46-6 V-12 diesel developing 840 hp
Armament: 1 × 12.7 mm NSV MG

Main winch
Rated line pull: 300 kN
Usable rope length: 200 m

Auxiliary winch
Rated line pull: 10 kN
Usable rope length: 400 m

Jib crane
Lift capacity: 19,000 kg
Max crane arm radius: 7.59 m
Traverse: 360°

Dozer blade
Width: 3.36 m
Digging depth: 300 mm
Max load capacity: 900 kN

Status
Production. In service with Czech Republic, India and Slovakia. Licenced production in India by BHEL.

Manufacturer
ZTS Tees Martin, Vehicles Division (Vozidlá), ul ČSA 3/1697, 036 57 Martin, Slovakia.
Tel: +42 842 33835. Fax: +42 842 32822.
Marketed by: Kerametal, Jašikova 2, 826 05 Bratislava, Slovakia.
Tel: +42 7 299 1111. Telex: 932 33 KEMET C.
Fax: +42 7 292402.

UPDATED

VPV armoured recovery vehicle

Description
The VPV (*Vyprost'ovaci Pásové Vozidlo*) tracked armoured recovery vehicle is a Czech modification of the Russian BMP-2 infantry fighting vehicle, licence-produced in the Czech Republic and Slovakia as the BVP-2. The VPV is used to support Czech and Slovak armies' motorised infantry units and can recover, tow or push BMP-1, BRDM-2 (4 × 4), BTR-60 (8 × 8) and other vehicles with weights up to 15 tonnes. The VPV is very similar in layout and type of equipment to the Russian BREM-2 (see separate entry in this section).

The VPV has a crew of three: the commander, who also operates the vehicle's crane and winch; the driver; and a recovery fitter. Although the BMP-2 chassis, running gear and engine have been retained, the turret is replaced by a cable drum with a hatch at its centre. An extendable traversing crane is installed on top of the hull rear. It has a maximum lifting capability of 5,200 kg with the boom extended between 2.9 and 3.2 m; maximum boom extension is 4.5 m. The vehicle's fuel capacity has been increased to 480 litres.

The VPV has a recovery winch with a tractive force of a basic 125 kN although this can be increased by the use of return pulleys. The winch has 120 m of 19 mm diameter cable. There is also a spade which is hydraulically lowered when the winch is in use. Also carried are tool kits, a 300 A welding set and cutting devices; the commander and driver are both trained welders.

Other equipment carried on the VPV include a R-123M radio set, a R-124 intercom, fire extinguishers and smoke generating equipment. There are two PR-22 pocket type radio sets provided for dismounted use together with an associated battery charger.

The VPV retains the amphibious characteristics of

Top view of VPV armoured recovery vehicle from rear

VPV armoured recovery vehicle (T J Gander)

the BMP-2 but with some restrictions. It can ford water obstacles if the waves are not higher than 100 mm and in current velocities not more than 1.2 m/s.

A version of this vehicle used by Hungary is known as the BMP-1VPV.

Specifications

Crew: 2 or 3
Combat weight: approx 14,000 kg
Power-to-weight ratio: 20.35 hp/t
Length: 6.47 m
Width overall: 2.94 m
Height: (crane lowered) approx 2.5 m
Max speed:
(road) 70 km/h
(water) 7 km/h
Range: 500 km
Fuel capacity: 480 litres
Engine: V-6 water-cooled diesel developing 285 hp
Armament:
1 × 7.62 mm PKT MG
3 × sub-machine guns
Ammunition: 7.62 mm - 2,000 rounds

Status

In service with the Czech and Slovak armies, Hungary and Russia.

Manufacturer

Podpolianske Strojárne Detva, 962 13 Detva, Slovakia.
Tel: +42 855 955111. Fax: +42 855 955327.
Marketed by: Kerametal, Jašikova 2, 826 05 Bratislava, Slovakia.
Tel: +42 7 299 1111. Telex: 932 33 KEMET C.
Fax: +42 7 292402.

UPDATED

VT-55A Armoured Recovery Vehicle

Description

The VT-55A armoured recovery vehicle is based on the hull and running gear of the T-55A MBT. An earlier production version (in service with the armed forces of Czech Republic and Slovakia) is known as the MT-55 and was based on the T-55: it lacks the forward-mounted dozer blade of the VT-55A. The MT-55 was based on the Russian T-55-T ARV.

Mounted on the right of the hull is a hydraulic crane with a maximum lifting capacity of 1,500 kg. On the roof of the ARV towards the rear, is a 1.6 × 1.6 m platform that can carry a maximum load, such as a power pack of up to 3,000 kg.

The VT-55A has two winches. The main winch is mechanically driven by the main engine. It has a maximum capacity of 25,000 kg and is provided with 28 mm diameter cable. The hydraulically operated auxiliary winch has a capacity of 800 kg and is provided with 400 m of 6.3 mm diameter cable. The seat of the winch operator rotates within a small turret.

At the front of the vehicle is a hydraulically operated dozer blade with a soil-shifting capacity of 150 m³/h.

The vehicle can be fitted with a snorkel tube to allow operations in water obstacles up to 1,000 m wide and 5 m deep, and with a stream velocity of 1.5 m/s. Under normal conditions the vehicle can ford water obstacles up to 1.4 m deep.

All vehicles have several towbars and a tow cable 4.2 m long. Electric welding equipment, a workbench and a vice are mounted above one of the tracks and are pulled out when required. At the rear of the vehicle is a winch anchor spade which can also be used as a dozer blade.

VT-55A armoured recovery vehicle

The machine gun turret can be traversed through 360° and the 7.62 mm machine gun can be used against ground and air targets. The vehicle has a three-man crew and standard equipment includes a crew compartment heater, NBC system and infrared night vision equipment. A snorkel can be fitted for deep fording.

Specifications

Crew: 3
Weight: 36,450 kg
Power-to-weight ratio: 15.91 hp/t
Length overall: 8.3 m
Width:
(overall) 3.4 m
(inside track) 2.04 m
Height: (top of stowed crane) 2.52 m
Ground clearance: 0.425 m
Max speed: 50 km/h
Average speed:
(road) 32-35 km/h
(towing) up to 18 km/h
Range:
(road) 270 km
(cross-country) 100 km
Gradient: 32°
Side slope: 30°
Fording:
(normal) 1.4 m
(with preparation) 5 m
Engine power: 580 hp
Armament: 1 × 7.62 mm MG

Status

Available. The earlier MT-55 is in service with the armed forces of Czech Republic and Slovakia and may have been exported.

Manufacturer

VOP 025 Nový Jičin, s.p., Dukelská 102, 742 42 Šenov u Nového Jičína, Czech Republic.
Tel: +42 656 783316. Telex: 52196.
Fax: +42 656 701734.

UPDATED

FINLAND

Vammas KAM-1 armoured recovery vehicle

Description

The KAM-1 armoured recovery vehicle was produced during 1984 and is a conversion of a T-55 tank chassis. Only one prototype was produced and was handed over to the Finnish Defence Forces in October 1984. The vehicle was developed in answer to a Finnish Defence Forces requirement for a vehicle that would not only act as an armoured recovery vehicle but that could also lift tanks and other vehicles to allow trapped personnel to escape; lift light armoured vehicles onto trailers or railway wagons when disabled for any reason; lift turrets, engines, gearboxes and other heavy vehicle components under field conditions; and tow damaged vehicles.

The KAM-1 conversion involved the removal of the T-55 tank turret, replacing it with a large turntable

KAM-1 armoured recovery vehicle with crane in travelling position

KAM-1 armoured recovery vehicle in travelling configuration and showing rear crane supports raised

mounting a hydraulic crane. The interior of the centre of the vehicle is occupied by the hydraulic packs and winch. The driver is seated at the front of the hull on the left-hand side. To his right is the commander in a separate compartment together with a radio and the intercom equipment.

A BTU-55 dozer blade, complete with routing tines, is fitted to the front of the hull. The dozer blade can be used to clear obstacles or for general earthmoving and also acts as an anchor blade for stability while using the recovery winch. It can also be used as a support when the crane is in use. More support is provided by two hydraulically operated support legs at the rear of the vehicle.

The hydraulic crane has a lift moment of 75 t/m and a maximum boom length of 6.7 m. Traverse is unlimited and the maximum load is 22,000 kg. The crane can be controlled by a remote-control box which also controls the hydraulic support legs at the rear. An axial piston pump is used to drive the hydraulic system for the crane and winch at a maximum pressure of 320 bar.

A hydraulic recovery winch is located towards the front of the hull and has a maximum theoretical line pull of 36,000 kg. Using a snatch block; this can be doubled to 72,000 kg. The winch is provided with 140 m of cable paying out to the front.

Equipment carried on the vehicle includes: a flexible towing device; towing bars; towing ropes; a platform for spare parts and other items; lifting chains; an engine lifting boom; a snatch block; and shackles.

Armament carried is limited to crew weapons and a signal pistol. Smoke generators in the engine exhaust manifolds are carried over from the original T-55 design.

Specifications

Crew: 2
Weight: 44,000 kg
Power-to-weight ratio: 13.8 hp/t
Length: 7.6 m
Width: 3.4 m
Height: 3.6 m
Track: 2.64 m
Track width: 580 mm
Max speed: (road) 50 km/h
Engine: V-55 V-12 water-cooled diesel developing 580 hp at 2,000 rpm
Transmission: manual with 5 forward and 1 reverse gears

Crane
Lift moment: 75 t/m
Max boom length: 6.7 m
Max load: 22,000 kg
Traverse: unlimited

Winch
Max line pull: (theoretical) 36,000 kg
Max line pull: (with snatch block) 72,000 kg
Cable length: 140 m

Status

Development complete. Ready for production when orders are placed.

Manufacturer

Vammas Limited, PO Box 18, FIN-38201 Vammala, Finland.
Tel: +358 32 51971. Fax: +358 32 519 7466.

UPDATED

FRANCE

Giat Industries Leclerc armoured recovery vehicle

Development

The Leclerc armoured recovery vehicle, also known as the DNG (Depannage Nouvelle Generation), is in production by Giat Industries to be the armoured recovery support vehicle for the Leclerc MBT.

The Roanne facility of Giat Industries has built two prototypes of the Leclerc ARV and the UAE has ordered 46 production vehicles.

The French Army has evaluated the Leclerc ARV and is expected to place a production order in the future. French Army Leclerc MBTs have a different power pack to that of the Leclerc MBT and ARV used by the UAE.

Description

The Leclerc armoured recovery vehicle is based on the armoured hull and running gear of the Leclerc MBT, although the chassis is longer to accommodate seven roadwheels and independent suspension units each side instead of six. The SAMM hydropneumatic suspension system is retained, as is the track tensioning system, allowing the driver to adjust the track tension without leaving his protected position. The rear suspension units can be locked by actuators when the crane is in operation.

The vehicle is manned by a crew of three: commander, driver and operator, with one extra foldable seat for a passenger. An air conditioning system is provided for the crew compartment together with an NBC protection system. The Galix close defence system is provided and the commander has a 12.7 mm M2 machine gun mounted over his cupola. Additional armour can be fitted if required.

The vehicle has an armoured superstructure at the front of the hull, with a crane jib with a maximum capacity of 30 tonnes on a turntable on the right of the vehicle. Crane jib traverse is 260° and maximum height below the hook is 7.9 m. A dozer blade with a width of 3.42 m is located at the front to act as an anchor during recovery operations or to clear battlefield obstacles and debris.

The main capstan winch has a 34 tonne capacity and is provided with 160 m of cable; an auxiliary winch has a 15 tonne capacity.

Also carried is a demountable diesel-electric power generator. The actual level of equipments involved, plus items such as tool kits and stowage bins, can vary according to end user requirements. A complete Leclerc power pack can be carried on the hull rear.

Crew and vehicle features include: an NBC system; an automatic fire detection and suppression system in the crew, power pack and hydraulic compartments; and the hydraulic compartment is isolated from the crew. Crew comforts include air conditioning and a chemical toilet.

One optional item of equipment is an auxiliary power unit.

Specifications

Crew: 3 + 1
Weight: approx 60,000 kg
Bridge classification: MLC 65 to 70 according to equipment
Power to weight ratio: 25.4 hp/t
Length: (overall) 9.15 m
Width: (over tracks) 3.38 m
(over dozer blade) 3.42 m
Height: (roof) 2.6 m
Ground clearance: 0.485 m
Track width: 635 mm
Ground pressure: 0.82 kg/cm^2
Max speed: 71 km/h
Fuel capacity: 1,300 litres
Fording: 1 m
Gradient: 60%
Side slope: 30%
Vertical obstacle: 0.7 m
Trench: 3 m
Engine: Europower (MTU 883) turbocharged V-12 diesel power pack developing 1,500 hp
Transmission: Renk HLSW 295 automatic gearbox with integrated torque converter and hydrokinetic retarder; 5 forward and 3 reverse gears
Suspension: hydropneumatic, 2-cylinder
Electrical system: 24 V
Batteries: 8 × 12 V, 125 Ah
Armament:
1 × 12.7 mm M2 HB MG
Galix close defence system
Ammunition: 12.7 mm - 700 rounds

Status

In production. In service with the UAE 46 vehicles ordered.

Manufacturer

Giat Industries, 13 route de la Minière, F-78007 Versailles Cedex, France.
Tel: +33 2 39 49 82 33. Fax: +33 2 39 49 82 10.

UPDATED

Leclerc armoured recovery vehicle **1997**

Leclerc armoured recovery vehicle towing Tropicalised Leclerc MBT **1997**

Giat Industries AMX-30D armoured recovery vehicle

Description

The AMX-30D was designed to carry out three basic tasks: the recovery of disabled and damaged AFVs; major field repairs such as changing engines; and engineer work.

The chassis of the AMX-30D ARV (or *Char AMX-30 Depanneur-Niveleur*) is identical to that of the basic AMX-30 MBT, but the superstructure is new. Its crew of four consists of commander, driver and two mechanics.

The driver is seated at the front of the hull, slightly to the left and is provided with three periscopes for observation. His single-piece hatch cover swings to the left. The commander is seated to his rear and is provided with a TOP 7 cupola. This is similar to that installed on the AMX-30 MBT but does not have the infrared searchlight or the contrarotating equipment. The cupola has 10 periscopes for observation and a ×10 sight for the 7.62 mm machine gun, which is mounted externally but aimed and fired from within the turret. To the rear of the commander's cupola is the entrance hatch for the engineers which opens to the right. There is a single M336 periscope fore and aft of this hatch cover.

The engine and transmission at the rear of the hull are separated from the crew compartment by a fireproof bulkhead. The suspension is of the torsion bar type and consists of five roadwheels with the drive sprocket at the rear and the idler at the front. The first, second, fourth and fifth roadwheels are mounted on bogies which are provided with hydraulic shock-absorbers. Five rollers support the inside of the track.

At the front of the hull is a dozer blade which is hydraulically operated by two cylinders. It is controlled by the driver and is used both for dozing operations and to stabilise the vehicle when the winch crane is being used.

The Griffet crane is mounted at the front right side of the ARV and can lift a load of 12,000 kg through 240°, or 15,000 kg when the crane is towards the front and the dozer blade is in the support position. This crane is normally used for changing major AFV components such as engines and turrets. A spare engine can be carried on the rear of the hull for the rapid replacement of an engine in the field.

The main winch is mounted in the centre of the hull and consists of three subassemblies, frame, transfer gearbox and winch drum with integral reduction gear. The winch cable is led out through the front of the hull. The winch is provided with 100 m of 34 mm diameter cable and is also provided with a safety device which consists of two overload sensing cylinders operated by a pressure switch, which stops the winch automatically when the maximum load of 35,000 kg is reached. The winch has a maximum speed on external layer of 23 m/min and on internal layer of 18.8 m/min.

The auxiliary winch is a Retel TRA 251 mounted at the front of the hull. It is provided with 120 m of 11.2 mm diameter cable and has a maximum capacity of 3,500 kg.

The AMX-30D is provided with an NBC system and a crew heater. A snorkel can be installed over the mechanic's hatch enabling the tank to ford to a depth of 4 m.

Variants

The AMX-30D(S) is a special export model for operations in the Middle East with sand shields over the top half of the tracks and a modified gearbox. The engine develops 620 hp at 2,400 rpm.

The AMX-30DI was developed by the Pinguely Division of Creusot-Loire. The first prototype was completed in July 1974. This model has a boom which can lift a maximum load of 15,000 kg and slew it through 240°.

Specifications

Crew: 4
Weight: (loaded with spare power plant) 40,000 kg
Power-to-weight ratio: 17.5 hp/t
Length: (dozer blade up, jib in travelling position) 7.53 m
Width: 3.15 m
Height: 2.65 m
(with jib at max extension) 6.15 m
(roof) 2.05 m
Ground clearance: 0.45 m
Track: 2.53 m
Track width: 570 mm
Length of track on ground: 4.12 m
Ground pressure: 0.8 kg/cm^2
Max speed: (road) 60 km/h
Range: 650 km
Fuel capacity: 1,100 litres
Fording: 2 m
(with preparation which takes 5 min) 4 m
Gradient: 60%
Vertical obstacle: 0.93 m
Trench: 2.9 m
Engine: Hispano-Suiza HS-110, 12-cylinder, water-cooled, multifuel developing 700 hp at 2,400 rpm
Transmission: automatic with 5 forward gears; reverse gear gives same speeds in reverse
Electrical system: 28 V
Batteries: 8 × 12 V, 100 Ah in 2 groups of 4
Armament:
1 × 7.62 mm MG
3 × smoke grenade dischargers

Status

Production complete. Known users include Cyprus (2), France (134), Greece (14), Qatar (1), Saudi Arabia (57), Spain (10), UAE (4) and Venezuela (4).

Manufacturer

Atelier de Construction Roanne (ARE).
Enquiries to: Giat Industries, 13 route de la Minière, F-78034 Versailles Cedex, France.
Tel: +33 1 30 97 37 37. Fax: +33 1 30 97 39 00.

UPDATED

AMX-30D armoured recovery vehicle (T J Gander) **1997**

Giat Industries VAB Recovery armoured recovery and repair vehicle

Description

The Giat Industries VAB Recovery armoured recovery and repair vehicle is under development as a private venture and will be available in both 4 × 4 and 6 × 6 configurations. The vehicle was developed for what are described as 'assistance missions' and will be equipped with a crane, recovery winch and repair facilities. For details of the VAB ECH repair vehicle refer to the entry in the *Armoured repair vehicles* section, and for full details of the VAB APC refer to *Jane's Armour and Artillery 1996-97* pages 413 to 416.

The VAB Recovery has a crew of three or four; driver and commander in the front compartment and one or two fitters in the rear. The vehicle retains the basic layout of the VAB APC but a hydraulically operated telescopic jib crane is mounted on the left-hand side in a recess over the centre of the hull. The 1,500 kg capacity crane, which has a traverse of 270°, has a maximum height of 4.4 m and a maximum extension of 4 m. When the crane is in use, four stabilisers located at each corner of the hull are lowered hydraulically.

A mechanically powered recovery winch with a

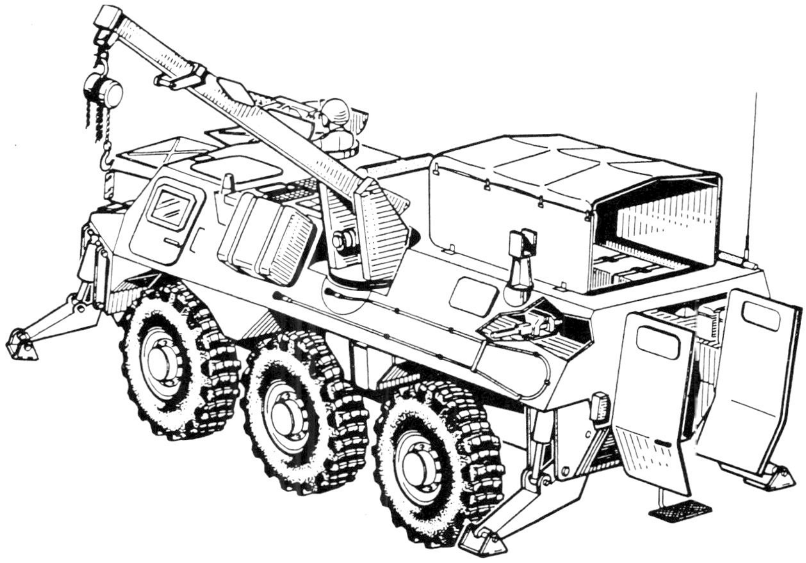

Artist's impression of the 6 × 6 version of the VAB Recovery armoured recovery and repair vehicle

7 tonne capacity is located at the front right-hand side of the vehicle. A block and tackle is carried to increase the winch capacity when required. The winch has a winding speed of 0.27 m/min and is equipped with 60 m of cable. The vehicle is fitted with towing hooks front and rear and a manually operated anchor plate is optional. Among other equipment carried externally are an adjustable floodlight and a ladder.

The rear compartment contains a workbench and, to increase internal working height, the two rear top hatches can be opened to the vertical position and covered by bows and a tarpaulin cover. Equipment carried, other than tool chests, includes a generating set and oxy-acetylene welding equipment. A removable stepladder is also provided.

The vehicle commander is provided with a VHF-FM radio set, with a loudspeaker extension to the rear compartment. Armament may be a 7.62 or 12.7 mm machine gun, as required, with internal stowage for 1,500 rounds of 7.62 mm ammunition and 500 rounds of 12.7 mm.

Total combat weight of the VAB Recovery is 16,000 kg and length overall is 6.54 m.

It is expected that any new production vehicles would be based on the latest VAB New Generation chassis.

Status
Prototype.

Manufacturer (Vehicle)
Renault Véhicules Industriels and Giat Industries.
Enquiries to: Giat Industries, 13 route de la Minière, F-78034 Versailles Cedex, France.
Tel: +33 1 30 97 37 37. Fax: +33 1 30 97 39 00.

UPDATED

GERMANY

MaK Büffel armoured recovery vehicle

Development
Concept studies for a new armoured recovery vehicle capable of recovering up to MLC 60 loads began in 1982, following a series of component studies that commenced in 1977. In 1984 a finalised definition phase contract was placed. This phase was completed in 1986 with the completion of an experimental prototype. A further contract signed in 1987 called for a further two prototypes and the conversion of the experimental prototype.

By 1988 the three prototypes were undergoing technical, tactical and logistic testing. In mid-1990, after an invitation to tender by the Bundesamt für Wehrtechnik und Beschaffung, MaK System Gesellschaft mbH was selected as the main contractor for series production of the new ARV, named Büffel, for the German Bundeswehr and the Dutch armed forces - the Dutch Army contributed to the development costs.

The order was for 100 vehicles. The final allocation was 75 for the Bundeswehr and 25 for the Dutch armed forces. Manufacture was shared by MaK System at Kiel (55) and Krauss-Maffei (45).

The three other Leopard 2 MBT users, Spain, Sweden and Switzerland have yet to place any orders for the Büffel ARV.

Description
The Büffel ARV is based on the chassis of the Leopard 2 MBT. Its main purpose is the support of the Leopard 2 MBT family of vehicles and it can be used for recovery, towing, obstacle removal and crane operations of other vehicles in the same load class.

The general layout of the Büffel ARV follows that of the earlier Leopard 1 ARV (see following entry).

The Büffel features a traversing crane unit with a maximum capacity of 30 tonnes. Jib traverse is 270°. The crane has an electronic load momentum limiter which computes vehicle tilt, crane jib elevation, load mass, and so on, and prevents the crane from being overloaded.

The main winch is a Rotzler GmbH, Treibmatic TR650/3 dual capstan unit with an effective cable length of 180 m and a tractive capacity of 35 tonnes. The winch is installed in the hull front plate with direct connection to the cable roller guide so that there is improved distribution of forces directly into the hull structure. The winch has several advanced features such as a constant maximum pulling force and cable speed over the entire cable length, and an absence of cable friction as the cable runs in grooves on the capstans so that the windings do not touch each other. Other features include reduced cable wear, increased cable life, and safe pay-out and pull-in of the cable under all load conditions and varying loads. No cable tensioner is required. This winch is stated to cover 90 to 95 per cent of all military recovery operations as defined by the German Army.

The vehicle is provided with an enlarged (3.42 m wide and 880 mm high) dozer/support blade at the front to support and assist during recovery and for obstacle-clearing and dozing operations. Other recovery equipment includes: a Rotzler HZ 010/1-8 auxiliary winch with a capacity of 650 kg and 280 m of 7 mm diameter cable; a running gear blocking system; a self-recovery system; various couplings and towbars (including a rapid-recovery bar to be attached to the dozer/support blade); and rapid connect and disconnect couplings for towing. Electrical cutting and welding equipment is carried. An engine pack may be carried on a cradle over the engine compartment decking.

Support systems provided include fire extinguishing and suppression systems, a preheating and heating system, an NBC ventilation system and deep fording equipment with bilge pumps.

Armament carried includes a 7.62 mm MG3 machine gun mounted over the commander's cupola and 16 smoke grenade dischargers mounted on the front hull (eight) and rear (eight). The hull armour is capable of providing protection against 20 mm projectiles.

Specifications
Crew: 3
Combat weight: 54,300 kg
Power-to-weight ratio: 27.62 hp/t
Length: 9.07 m
Width overall: 3.54 m
Height:
(top of MG mount) 2.99 m
(less MG) 2.735 m
Ground clearance: 0.51 m
Track width: 635 mm
Max speed: (road) approx 68 km/h
Range:
(road) approx 650 km
(off-road) approx 325 km
Fuel capacity: 1,620 litres
Fording:
(without preparation) 1.2 m
(with preparation) 4 m
Gradient: 60%
Side slope: 30%
Vertical obstacle: 0.92 m
Trench: 3 m
Engine: MTU MB 873 Ka-501 4-stroke exhaust-gas turbocharged diesel developing 1,500 hp at 2,600 rpm
Transmission: Renk HSWL H354, hydromechanical
Electrical system: 24 V
Batteries: 8 × 12 V/100 Ah, total capacity 400 Ah
Generator: 20 kW
Armament:
1 × 7.62 mm MG3 MG
16 × smoke grenade dischargers

Main winch capacity: 35,000 kg
Cable length: 180 m
Cable diameter: 33 mm
Cable speed: (max) 16 m/min

Crane load:
(lifting) 30,000 kg
(pulling) 70,000 kg
Max hook height: 7.9 m
Max reach:
(front) 4.7 m
(right) 5.9 m
Crane traverse: 270°
Towing capacity: 62,000 kg (MLC 60)

Dozer blade width: 3.42 m
Dozer blade height: 880 mm

Armoured Maintenance System
MaK System GmbH has proposed an Armoured Maintenance System, in which a Büffel ARV is used to tow a tracked field maintenance facility trailer carrying repair and other equipment. The ARV/trailer combination can be used for recovery and the field maintenance and repair support of armoured units. The trailer will use rubber-belted track running over four

Büffel armoured recovery vehicle lifting the rear of Leopard 2 MBT **1997**

Dutch Army Büffel armoured recovery vehicle (Michael Jerchel) **1996**

idler wheels with air over hydraulic disc brakes on all four wheels. Details of the trailer are as follows:

Weight of trailer:
(unloaded) 3 t
(GVW) 10.5 t
Payload: 7.5 t
Length: 4.23 m
Width: 2.4 m
Height: 2.5 m
Ground clearance: 0.419 m
Track width: 686 mm
Towed speed:
(paved road) 80 km/h
(secondary road) 48 km/h
(cross-country) 32 km/h

Status

Production complete. In service with Germany (75) and the Netherlands (25).

Contractors

General Contractor: MaK System GesellschaftmbH, PO Box 9333, D-24157 Kiel, Germany.
Tel: +49 431 3995-02. Fax: +49 431 399 5446.
Co-producer: Krauss-Maffei Wehrtechnik GmbH, Krauss-Maffei Strasse 2, D-8000 Munich 50, Germany.
Tel: +49 89 8899 2393. Telex: 523163-31.
Fax: +49 89 8899 2614.

UPDATED

MaK Leopard 1 armoured recovery vehicle

Development

At an early stage in the development of the Leopard 1 MBT Germany decided to develop an ARV based on its components. Design work was carried out by Porsche with the first prototype being completed by a working group, consisting of Luther and Jordan, Jung Jungenthal and MaK of Kiel, in 1964. The first production ARV (in Germany the Bergepanzer, or BPZ) was completed in September 1966. Production of the ARV and other specialised members of the Leopard 1 family was undertaken by MaK, while production of the MBT was undertaken by Krauss-Maffei of Munich.

The ARV was designed to undertake the following roles:

- recovering vehicles disabled through enemy action or mechanical failure, or that have become bogged down
- towing disabled vehicles
- changing components such as engines and turrets (A spare Leopard engine pack is carried on the rear decking and a complete engine change can be carried out in less than 30 minutes. The fastest engine change to date from the tank's stopping to moving off again is 8 minutes.)
- carrying out dozing operations
- refuelling and defuelling other vehicles.

There were 104 Leopard 1 armoured recovery vehicles converted to Pionierpanzer 2 Dachs armoured engineer vehicles (see entry in *Armoured engineer vehicles* section).

Jung Jungenthal are converting Leopard 1 MBTs to Leopard 1 ARV standard - see following entry.

Description

The Leopard 1 ARV is almost identical to the Leopard 1 AEV, and the reader is referred to this entry for a detailed description of the vehicle. The main differences between the ARV and the AEV are: the ARV can carry a spare power pack and no auger; the ARV carries no explosives for demolition work; no heat exchanger is installed; the support/dozer blade is not provided with scarifiers to rip up the surface of roads.

In 1978 the German Army took delivery of 100 product-improved Leopard 1 ARVs from MaK. The main improvement was the installation of a hydraulically operated rear support on the right side of the hull at the rear. This relieves the suspension on the crane side and enables the crane to lift a maximum load of 16,000 kg and traverse it through 270°. In addition the main winch has a higher cable pay-out speed (74 m/min) than the original vehicle's (22 m/min), which corresponds to that of the creeping speed of a Leopard 1 MBT.

The lifting capability of the product-improved Leopard 1 ARV compared with the original ARV is shown in the Specifications below.

Specifications

(data in square brackets relates to product-improved version where different from basic vehicle)
Crew: 4
Weight: (with spare power pack)
(empty) 39,200 [39,980] kg
(loaded) 39,800 [40,580] kg
Power-to-weight ratio: [20.85] 20.45 hp/t
Length: (support/dozer blade raised) 7.57 [7.68] m
Width: 3.25 m
Height: (incl MG) 2.7 m
Ground clearance: 0.44 m
Track: 2.7 m
Track width: 550 mm
Length of track on ground: 4.236 m
Ground pressure: 0.83 [0.85] kg/cm²
Max speed: (road) 62 km/h
Range:
(road) approx 850 km
(cross-country) 500 km
Fuel capacity: 1,410 litres
Fording: 2.5 m
(with snorkel) 4 m
Gradient: 60%
Side slope: 30%
Vertical obstacle: 1.15 [0.88] m
Trench: 2.5 m
Engine: MTU MB 838 Ca M-500 10-cylinder multifuel developing 830 hp at 2,200 rpm
Transmission: ZF 4 HP 250 with 4 forward and 2 reverse speeds
Electrical system: 24 V
Batteries: 6 with total capacity of 300 Ah, charged by 3-phase generator driven from main engine
Armament:
1 × 7.62 mm MG3 MG in bow
1 × 7.62 mm MG3 MG on commander's hatch
6 × smoke grenade dischargers
Ammunition: 7.62 mm - 4,250 rounds

Status

Production complete, although conversions from Leopard 1 MBTs are available - see following entry. In service with Australia (8), Belgium (36), Canada (8 - called Taurus), Germany (444 plus 100 product-improved versions; 104 converted to Pionierpanzer 2), Greece (4), Italy (137), Netherlands (52), Norway (6) and Turkey (12).

Manufacturer

MaK System GesellschaftmbH, PO Box 9333, D-24157 Kiel, Germany.
Tel: +49 431 3995-02. Fax: +49 431 399 5446.

UPDATED

Netherlands Army Leopard 1 ARV carrying spare power pack on rear decking (C R Zwart) ***1997***

Canadian Forces Taurus (Leopard 1 ARV) with dozer blade deployed (Michael Jerchel)

Jung Jungenthal armoured recovery vehicle conversions

Development

Jung Jungenthal GmbH, is producing modification kits to convert M47, M48 and Leopard 1 MBTs into armoured recovery vehicles with a standard virtually identical to that of the Leopard armoured recovery vehicle (see previous entry).

The M47 conversion was developed mainly to meet a Turkish Army requirement. Following testing with an M47 conversion the concept was approved by the Turkish Army in April 1988. The M47 conversion has carried out trials in Pakistan.

The M48 conversion was originally devised in response to a possible Bundeswehr requirement for up to 100 conversions of M48 MBTs. The conversions will involve the armoured recovery vehicle hulls and equipment made available when 104 existing Leopard 1 armoured recovery vehicles were converted into Pionierpanzer 2 armoured engineer vehicles (see entry in *Armoured engineer vehicles* section). The Bundeswehr has purchased eight units. Four units are in use in Turkey and three units in Greece.

A further application is the possible acquisition of 138 M48 ARV conversions by Turkey, while Greece is understood to have a requirement for 30 M48 conversions. The MBTs used in these conversions will be from German Bundeswehr stocks.

Leopard 1 MBTs are also being used for conversions to ARVs. At least two ex-Bundeswehr MBTs had been converted to ARV standard by mid-1992. It is understood that Austria has a requirement for 18 ARVs, probably using M48s purchased from Norway.

Description

The conversion involves the removal of the turret and most of the front hull superstructure from the MBT involved and replacing them with a new armoured superstructure, a crane, a winch, a pump drive and a dozer blade system. All are virtually identical to the corresponding components used on the Leopard armoured recovery vehicle and the layout is the same. The M48 conversion also involves the replacement of the Continental (now General Dynamics Land Systems) main engine by an MTU diesel unit.

The crane has a jib with a six-stage pulley system that can lift loads up to 20,000 kg. The crane can be traversed over an arc of 200° without the support of the dozer blade and 270° with the dozer blade lowered. The maximum hook height is 6.05 m. The crane system has an overload safety switch and a limit switch.

The main winch, powered by two hydraulic motors, has a cable tensioning device and a basic pull capacity of 35,000 kg. By use of a snatch block this can be increased to 70,000 kg. The 33 mm diameter cable length is approximately 90 m long.

The dozer blade is secured to the front hull by two hydraulically powered lever arms and is lifted and lowered by hydraulic cylinders. The blade can be used for earthmoving and can also be used as a stabiliser when the crane or main winch is operating. When using the dozer blade the maximum vehicle speed is 8 km/h. The earthmoving capacity when working with heavy soil is 200 m³/h.

The hydraulic system that is used to power the crane, dozer blade and main winch operates at a pressure of 140 bar.

Jung Jungenthal armoured recovery vehicle conversion of M48 tank being used to lift Leopard 2 MBT turret

Specifications

(Data for M47 conversion - data in square brackets relates to M48 conversion where applicable)
Crew: 4
Weight:
(empty) 43,500 [48,000] kg
(combat) 44,900 [48,500] kg
(max permissible) 47,800 [49,000] kg
Length: 6.998 [7.88] m
Width: 3.59 [3.666] m
Height: 2.985 [2.965] m
Ground clearance: 0.47 [0.46] m
Track: 2.794 m
Track width: 584 mm
Fording: 1.22 m
Gradient: 60%
Vertical obstacle: 0.91 m
Trench: 2.6 m
Engine:
(M47) Continental (now General Dynamics Land Systems) AV-1790-5B V-12 air-cooled petrol developing 810 bhp at 2,800 rpm or
Continental (now General Dynamics Land Systems) AVDS-1790-2A air-cooled diesel developing 750 hp at 2,400 rpm
(M48) MTU MB 837 Ea-501 liquid-cooled diesel developing 750 hp at 2,300 rpm
Transmission:
(M47) Allison Model CD-850-4, 4A, 4B or 5 cross-drive with 2 forward and 1 reverse ranges
(M48) Allison Model CD-850-5 or 6 cross-drive with 2 forward and 1 reverse ranges
Electrical system: 24 V
Armament: smoke grenade dischargers

Boom winch
Cable speed: (with 130 kN) 1.5-3.5 m/min
Cable diameter: 13 mm
Cable length: (approx) 100 m

Main winch
Traction in lowest position: 350-700 kN
Traction in top position: 200 kN
Cable unwinding speed: approx 22 m/min
Cable winding speed: (1st step) approx 14 m/min
Cable winding speed: (2nd step) approx 44 m/min
Cable length: (approx) 90 m
Cable diameter: 33 mm

Dozer blade
Blade length: 3.25 m
Blade height: 0.591 m
Max dozing speed: 8 km/h
Max dozing capacity: approx 200 m³/h

Status

Series production - see text.

Manufacturer

Jung Jungenthal GmbH, PO Box 20, D-5242 Kirchen, Germany.
Tel: +49 2741 683-0. Telex: 875319.
Fax: +49 2741 683246.

UPDATED

INDIA

Vijayanta armoured recovery vehicle

Description

Little information is available regarding the Vijayanta armoured recovery vehicle other than it is based on the hull, power pack and running gear of the Vijayanta MBT. An armoured superstructure replaces the usual armoured gun turret of the Vijayanta and an A-frame lifting jib is located over the front hull. It is assumed that a recovery winch is located inside the hull.

India is now acquiring the more recent ZTS VT-72B armoured recovery vehicle on the T-72 MBT chassis covered earlier in this section.

Status

Production complete. In service with the Indian Army.

Manufacturer

Government of India, Ministry of Defence, Ordnance Factory Board, Ayudh Bhawan, 10A Auckland Road, Calcutta - 700 001, India.
Tel: +91 33 285077.

UPDATED

Vijayanta armoured recovery vehicle

ITALY

OTOBreda OF-40 armoured recovery vehicle

Description

The OF-40 armoured recovery vehicle is based on the chassis and hull of the OF-40 MBT (for details refer to *Jane's Armour and Artillery 1996-97*, pages 60 to 62. In place of the normal rotating turret the armoured recovery vehicle uses a fixed turret superstructure, and is equipped with a crane, winch and dozer blade.

The driver sits in his normal position offset to the right at the front of the hull. The superstructure houses the commander and two fitters. There is a hydraulic jib crane mounted on the left-hand side of the vehicle with the jib turret providing a 270° traverse. A 6 m boom is fitted with a maximum direct front-lift capacity of 18,000 kg retracted and 5,000 kg when fully extended or when traversing with a load. When stowed, the crane jib lies along the upper hull of the vehicle and the hydraulic actuating cylinders are arranged so that they are protected from damage by artillery splinters and other such hazards. If the hydraulic system fails, the jib can be raised and lowered by hand.

The main winch is installed in the crew compartment and has a direct pull capacity of 36,000 kg, although this can be increased to 72,000 kg by use of a pulley block. The winch is normally used in conjunction with the dozer blade which is fully lowered for maximum pull and partially lowered to provide access to the winch hatch at the front of the hull. Power for the winch is hydraulic and a 540 litre hydraulic fluid tank is located on the left side of the vehicle - this tank also provides the hydraulic fluid for all other systems such as the crane and the dozer blade. The winch is supplied with 80 m of 32 mm cable.

The dozer blade can be used for field combat engineering tasks but is also used as a support stabiliser for winching operations as described above. The blade is controlled by the driver and can be lowered manually.

The OF-40 armoured recovery vehicle carries recovery and repair tools around the hull and inside the superstructure. Some spares can be carried and a spare engine pack can be carried on the hull decks at the rear. The superstructure interior has an automatic fire extinguishing system and there is provision for mounting a 7.62 mm machine gun between the commander's and fitter's hatches on the superstructure roof. A bank of six smoke grenade dischargers is located on the right-hand side of the superstructure. The sides of the superstructure use armour 50 mm thick, the same thickness as the hull sides.

OF-40 armoured recovery vehicle with spare power pack on hull rear

Note: OTOBreda also built 68 Leopard 1 ARVs for the Italian Army.

Specifications

Crew: 4
Weight:
(combat) 45,000 kg
(kerb) 43,000 kg
Ground pressure: 0.92 kg/cm^2
Power-to-weight ratio: 18.5 hp/t
Length: (crane stowed) 7.68 m
Width: 2.47 m
Height: (top of superstructure) 2.35 m
Ground clearance: 0.44 m
Max speed:
(forward) 65 km/h
(reverse) 24 km/h
Range: (road) 600 km
Fuel capacity: approx 1,000 litres
Fording: 1.2 m
Gradient: 60%
Side slope: 30%
Vertical obstacle: 0.7 m
Trench: 3 m
Engine: MTU MB 838 Ca M-500 37.4 litres V-10 diesel or multifuel developing 830 hp at 2,200 rpm
Transmission: ZF 4HP 250 with hydraulic torque converter and lock-up clutch and with 4 forward and 2 reverse gears
Suspension: torsion bar
Electrical system: 24 V
Batteries: 8 × 12 V 100 Ah each, total capacity 400 Ah at 24 V
Armament:
1 × 7.62 mm MG
6 × smoke grenade dischargers

Status

Production complete. In service with the UAE (4).

Manufacturer

OTOBreda, Via Valdilocchi 15, I-19100 La Spezia, Italy.
Tel: +39 187 530111. Telex: 270368.
Fax: +39 187 530669.

UPDATED

JAPAN

Mitsubishi Type 90 armoured recovery vehicle

Description

To support the Type 90 MBT, Mitsubishi Heavy Industries developed a prototype of the Type 90 armoured recovery vehicle based on the same hydropneumatic suspension and power pack but with a new superstructure. On the left-hand front of the upper hull is the pivot for a hydraulically operated crane, with a maximum capacity of 25 tonnes and a remote-control facility. For travelling, the crane is stowed facing the rear along the right-hand side of the hull top. Located alongside the crane pivot is the superstructure with accommodation for the driver, commander and two other crew members. The crew commander's cupola is provided with a mounting for a 12.7 mm M2 HB machine gun; the superstructure also carries smoke grenade dischargers. Mounted on the front of the vehicle is a hydraulically operated dozer blade that can also be used as a stabiliser during recovery operations. The vehicle is provided with a recovery winch capable of recovering the Type 90 MBT which has a combat weight of 50 tonnes. After recovery, the Type 90 armoured recovery vehicle can tow a Type 90 MBT.

The Type 90 armoured recovery vehicle weighs 50 tonnes and has a maximum road speed of 70 km/h. It can pivot turn.

Status

Prototype.

Manufacturer

Mitsubishi Heavy Industries, 5-1 Marunouchi 2-chome, Chiyoda-ku, Tokyo 100, Japan.
Tel: +81 3 212 3111. Telex: 22282.
Fax: +81 3 201 6258.

UPDATED

Prototype of Type 90 armoured recovery vehicle

Mitsubishi Type 78 armoured recovery vehicle

Description

The prototype of an ARV based on the chassis of the Type 74 MBT was completed in 1974 and subsequently standardised as the Type 78 ARV in 1978. The layout of the Type 78 ARV is very similar to that of the German Leopard 1 and French AMX-30D ARVs. Mounted on the right side of the hull at the front is a hydraulically operated crane which is 3.5 m long and has a jib that can be extended a further 1.5 m. The jib can be traversed through 270° and lift a maximum load of 20,000 kg. Mounted at the front of the hull is a hydraulically operated dozer blade which can be used both for dozing operations and as a stabiliser when the winch is being used. The Type 78 retains the hydropneumatic suspension of the Type 74 MBT.

The winch has a maximum capacity of 38,000 kg, is provided with 60 m of 32 mm diameter cable and has two speeds: 6 and 15 m/min. A hydraulic motor is used to extract the cable from the winch at a maximum speed of 30 m/min and there is also a system to apply tension to the cable for rewinding smoothly into the winch when there is no load.

Armament of the Type 78 ARV consists of a pintle-mounted 12.7 mm M2 HB machine gun and six smoke grenade dischargers.

Type 78 ARV in travelling configuration (Kensuke Ebata) ***1996***

Specifications

Crew: 4
Weight: 38,000 kg
Power-to-weight ratio: 18.94 hp/t
Length: 7.95 m
Width: (overall) 3.38 m
Height: (to top of hull) 2.4 m
Ground clearance: 0.4 m
Max speed: 53 km/h
Gradient: 60%
Engine: Mitsubishi 10 ZF 2-cycle 10-cylinder air-cooled diesel, developing 720 hp at 2,200 rpm
Transmission: Mitsubishi power-shift with 6 forward and 1 reverse gears
Suspension: hydropneumatic with suspension lock at kneeled position
Crane capacity: 20,000 kg
Main winch capacity: 38,000 kg
Armament:
1 × 12.7 mm M2 HB MG
6 × smoke grenade dischargers

Status

Production complete. In service with the Japanese Ground Self-Defence Force. Like all Japanese armoured vehicles Type 78 has not been offered on the export market.

Manufacturer

Mitsubishi Heavy Industries, 5-1 Marunouchi 2-chome, Chiyoda-ku, Tokyo 100, Japan.
Tel: +81 3 212 3111. Telex: 22282.
Fax: +81 3 201 6258.

VERIFIED

Mitsubishi Type 70 armoured recovery vehicle

Description

The Type 70 armoured recovery vehicle is based on the chassis of the Type 61 MBT. The turret is replaced by a superstructure which has a winch mounted in a bustle at the rear. The layout of the Type 70 is conventional, with the driver's compartment at the front, crew compartment in the centre and the engine and transmission at the rear. The suspension is of the torsion bar type and consists of six roadwheels with the drive sprocket at the front and the idler at the rear. There are three track support rollers and shock-absorbers are provided for the first, second, fifth and sixth roadwheel stations.

A dozer blade is mounted at the front of the hull for clearing obstacles and for stabilising the vehicle when the A-frame is being used. The latter is pivoted either side of the hull superstructure and is used to change components. A full range of equipment is carried, including towbars, tools and cutting and welding gear.

A 12.7 mm M2 HB machine gun is mounted on top of the superstructure for anti-aircraft defence and an 81 mm mortar can be mounted on the front of the hull.

Type 70 ARV with A-frame to rear (K Nogi)

Specifications

Crew: 4
Weight: 35,000 kg
Power-to-weight ratio: 18.57 hp/t
Length: 8.4 m
Width: 2.95 m
Height: 3.1 m
Ground clearance: 0.4 m
Track: 2.45 m
Track width: 500 mm
Length of track on ground: 3.7 m
Ground pressure: 0.95 kg/cm²
Max speed: (road) 45 km/h
Range: 200 km
Fording: 1 m
Gradient: 60%
Vertical obstacle: 0.5 m
Trench: 2.7 m
Engine: Mitsubishi Type 12 HM 21 WT V-12 diesel injection turbocharged, air-cooled developing 650 hp at 2,100 rpm
Transmission: mechanical with 5 forward and 1 reverse gears, with 2-speed auxiliary reduction unit
Electrical system: 24 V
Batteries: 4 × 12 V, 200 Ah
Armament:
1 × 12.7 mm M2 HB MG
1 × 81 mm mortar

Armour
hull front: 46 mm
hull sides: 25 mm
hull rear: 15 mm

Status

Production complete. In service only with the Japanese Ground Self-Defence Force.

Manufacturer

Production was undertaken at the Maruko, Tokyo, plant of Mitsubishi Heavy Industries, but AFV production is now at the Sagamihara Plant, near Tokyo.
Mitsubishi Heavy Industries, 5-1 Marunouchi 2-chome, Chiyoda-ku, Tokyo 100, Japan.
Tel: +81 3 212 3111. Telex: 22282.
Fax: +81 3 201 6258.

UPDATED

KOREA, SOUTH

Hyundai K-1 armoured recovery vehicle

Development

The Hyundai K-1 armoured recovery vehicle is a derivative of the K-1 MBT, also known as the Type 88 MBT (for details of the Type 88 refer to *Jane's Armour and Artillery 1996-97* pages 68 to 70). The K-1 ARV was developed in co-operation with MaK of Germany and uses an armoured superstructure and recovery system similar to the German Büffel armoured recovery vehicle (see separate entry in this section).

Two separate orders have been placed with Hyundai to manufacture the K-1 for the South Korean armed forces, each for 100 vehicles.

A modified version, the K-1M has been offered to Malaysia.

Description

The K-1 ARV is based on the chassis and running gear of the Type 88/K-1 MBT which was developed by the Land Systems Division of General Dynamics in the USA.

There is a crew of four consisting of commander, driver, operator and a mechanic. An integral crew heating system is provided but NBC protection is restricted to individual systems.

The main winch is of the capstan type and is provided with 150 m of usable cable. The winch capacity is 35 tonnes, or 70 tonnes with a guide pulley. There is also an auxiliary winch with 260 m of usable cable.

The hydraulic lifting crane mounted on a turntable to the right of the main superstructure has a maximum lifting capacity of 25 tonnes and can be elevated to an angle of 70°; the crane traverse is 270°. A straight dozer blade is mounted at the front and can be used for obstacle-clearing or for support when operating the main winch. The blade's soil-moving capacity is 170 m^3/h.

Standard equipment includes fire detection and suppression system, welding and cutting equipment, gas-turbine auxiliary power unit and an integrated test system.

A pintle for a 12.7 mm M2 HB machine gun is mounted over the commander's cupola on the superstructure roof and smoke grenade dischargers are mounted on the hull front.

K-1 armoured recovery vehicle with blade lowered and crane traversed to the front
1997

Specifications

Crew: 4
Weight:
(combat) 51,000 kg
(with spare power pack) 56,000 kg
Length: 8.96 m
Width: 3.60 m
Height: 2.65 m
Ground clearance: 0.46 m
Max speed: (road) 65 km/h
Fording:
(normal) 1.2 m
(with kit) 2.2 m
Gradient: 60%
Side slope: 30%
Vertical obstacle: 1.0 m
Trench: 2.74 m
Engine: MTU MB 871 Ka-501 diesel developing 1,200 hp at 2,600 rpm
Transmission: ZF LSG 3000 automatic with 4 forward and 2 reverse gears
Suspension: hydropneumatic and torsion bar
Electrical system: 24 V
Batteries: 6 × 12 V, 100 Ah
Armament:
1 × 12.7 mm M2 HB MG
2 × 6 smoke grenade launchers

Status

In production. In service with South Korean Army.

Manufacturer

Hyundai Precision & Ind Co Ltd, 140-2, Gye-Dong, Chongro-Ku, Seoul, South Korea.
Tel: +82 2 746 4010. Telex: HDPIC K23238.
Fax: +82 2 741 2248.

UPDATED

Daewoo KIFV armoured recovery vehicle

Description

KIFV stands for Korean Infantry Fighting Vehicle and denotes a family of armoured vehicles. For full details of the KIFV family refer to *Jane's Armour and Artillery 1996-97* pages 305 to 307.

The KIFV family includes an armoured recovery vehicle which is used to support and recover KIFV units in the field. The KIFV ARV is based on the same chassis, running gear and upper aluminium-armoured hull of the basic IFV version but is provided with a recovery winch and crane, together with other equipment and tools to assist in the recovery role. The vehicle can also be used to change components such as power packs. There is space inside the vehicle for a crew of up to four.

The recovery winch is located internally and has a capacity of 20 tonnes. When the winch is in use two stabiliser spades can be lowered from the hull rear.

The hydraulic crane is positioned on the left-hand side of the hull roof, and for transport is stowed with the crane boom facing the front. The crane has a lifting capacity of 3,500 kg.

As with other members of the KIFV family, the KIFV ARV is fully amphibious, being propelled in the water by its tracks.

KIFV armoured recovery vehicle

Specifications

Crew: up to 4
Weight:
(combat) 14,500 kg
(empty) 12,800 kg
Power-to-weight ratio: 24.1 hp/t
Ground pressure: (combat) 0.75 kg/cm²
Length:
(overall) 5.93 m
(hull) 5.345 m
Width:
(overall) 2.91 m
(over tracks) 2.545 m
Height:
(top of crane, stowed) 3.05 m
(hull roof) 1.829 m
Ground clearance: 0.41 m
Track width: 381 mm
Max speed:
(road) 65 km/h
(water) 6.5 km/h
Range: (cruising) 480 km
Fuel capacity: 400 litres
Fording: amphibious
Gradient: 60%
Side slope: 30%
Vertical obstacle: 0.63 m
Trench: 1.68 m
Engine: MAN (Daewoo) D2848T 14.62 litres V-8 turbocharged diesel developing 350 hp or
MAN (Daewoo) D2848M 14.62 litres V-8 diesel developing 280 hp
Transmission: Allison Transmission 200-5K automatic with 4 forward and 1 reverse gears or
Self-Changing Gears (Daewoo) T-300 semi-automatic with 7 forward and 7 reverse gears
Suspension: torsion bar
Electrical system: 28 V
Armament: 1 × 12.7 mm M2 HB MG

Status

In production. In service with the South Korean Army and Malaysia.

Manufacturer

Special Product Division, Daewoo Heavy Industries Limited, 20th Floor, Daewoo Center, 541, 5ga Namdaemun-ro, Jung-ga, Seoul, South Korea.
Tel: +82 2 726 3011. Telex: DHILTD K23301.
Fax: +82 2 756 2679.

UPDATED

POLAND

WZT-3 armoured recovery vehicle

Description

The WZT-3 armoured recovery vehicle is based on the chassis of the T-72 MBT as produced in Poland at the Machinery Industrial Complex 'Bumar-Labedy'. In general layout and appearance it resembles the Russian BREM-1 but there are numerous differences to suit the requirements of the Polish armed forces.

The WZT-3 has a crew of up to four, although only two (the driver and commander) are required during operations. The driver is seated under a hatch cover to the left front of an armoured superstructure with the commander seated on the right under a cupola provided with vision periscopes and a 12.7 mm NSW (NSV) machine gun pintle. Both the commander and driver are provided with night vision devices. A further crew hatch is provided in the centre of the superstructure roof towards the rear of the superstructure. Behind the superstructure is a load-carrying platform for spare power packs or other components. The platform measures 1.91 m long by 2.16 m wide and has sidewalls 0.62 m high; the load capacity is 3,500 kg.

Mounted on a small turntable to the left front of the superstructure is a TD-50 extending jib crane with a lifting capacity of 15 tonnes, a maximum hook height of 8.6 m and a maximum reach of 5.8 m. At the front of the vehicle is a 3.605 m wide dozer blade which can be used for earth-excavating or anchoring and support during recovery operations. The main recovery winch is mechanically driven and has a maximum pulling capacity, with tackle, of 840 kN; the winch has a usable 200 m of 28.5 mm diameter rope cable. Winch speed is 18.6 m/min at full load. There is also an auxiliary hydraulic winch with a pulling capacity of 20 kN and 400 mm of 9 mm cable.

Other recovery equipment carried includes rigid towbars and towcables. Specialised repair equipment includes electrical and gas welding gear, impact wrenches, chain saws, a tool kit and various special tools, spares and mobile work benches.

The WZT-3 can be fitted with a deep wading kit for water obstacles up to 1,000 m wide. Extra fuel can be carried in drums on the hull rear.

Specifications

Crew: up to 4
Weight: (approx) 42,000 kg
Power to weight ratio: 18.6 hp/t
Length (overall) 8.3 m
Width:
(overall) 3.6 m
(over tracks) 3.37 m
Height: (top of hull) 2.17 m
Ground clearance: 0.428 m
Track: 2.79 m
Length of track on ground: 4.27 m
Max speed:
(roads) 60 km/h
(towing 42 t load, road or cross-country) 12 km/h
Range:
(road) 650 km
(dirt road) 420-600 km

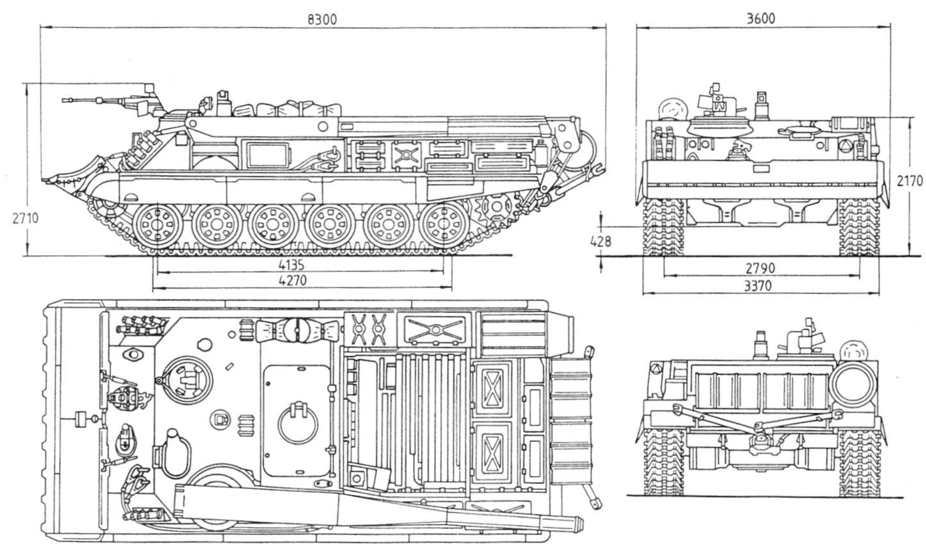

Detailed four-view drawing of Polish WZT-3 armoured recovery vehicle in travelling configuration **1997**

WZT-3 armoured recovery vehicle from the rear **1997**

WZT-3 armoured recovery vehicle

Fuel capacity:
(basic) 1,126 litres
(with extra tanks) 1,526 litres
Fording:
(normal) 1.2 m
(with preparation) 5 m
Vertical obstacle: 0.7 m
Trench: 2.6-2.8 m
Engine: liquid-cooled multifuel supercharged diesel developing 780 hp at 2,000 rpm
Transmission: synchromesh, hydraulically assisted with 7 forward and 1 reverse gears
Steering: clutch and brake
Suspension: torsion bar
Electrical system: 28.5 V
Armament: 1 × 12.7 mm NSW (NSV) MG

Status
In production. In service with the Polish Army.

Manufacturer
Zaklady Mechaniczne 'Bumar-Labedy' SA, ul Mechaników 9, PL-44-109 Gliwice, Poland.
Tel: +48 3 134 5111. Fax: +48 3 134 6966.
Marketed by: Cenzin Limited, ul Frascati 2, PL-00-489 Warsaw, Poland.
Tel: +48 22 296396. Telex: 814505 czi pl.
Fax: +48 2 628 6356.

UPDATED

RUSSIAN FEDERATION AND ASSOCIATED STATES (CIS)

BREM-1 armoured recovery and repair vehicle

Description
The BREM-1 (*Bronirovannaya Remontno-evakuatsionnaya-1*) armoured recovery and repair vehicle is based on the T-72B MBT and was first issued during 1984.

The BREM-1 has a crew of three: commander, driver and a vehicle mechanic. Each of the crew is provided with day and night observation equipment.

The BREM-1 is powered by an 840 hp diesel engine providing a maximum road speed of 60 km/h. The fuel tanks give a range of 700 km on paved roads and 500 km on tracks. When towing another tank, the road range is reduced to 220 km without refuelling. Long-range fuel tanks are mounted on the rear and can be jettisoned if required. An unditching beam is carried under the rear fuel drums. Ditch-crossing performance is between 2.6 and 2.8 m, depending on the firmness of the banks, and the largest vertical step is 0.85 m. For fording, an OPVT snorkel can be fitted over the crew compartment to enable fording of water obstacles up to 5 m deep and 1,000 m wide. When not in use the OPVT snorkel tube is carried at the rear of the vehicle on the right.

On the left side of the vehicle is a 4.4 m hydromechanical crane, used for lifting damaged or ditched vehicles for access or repair. The crane has a lifting capacity of 3,000 kg at the full 4.4 m extension or 19,000 kg at 2 m extension. Hydraulic power for the crane is supplied by a pump driven from the vehicle's main engine via an auxiliary gearbox. If the main engine is not running an electrical pump driven from the vehicle's batteries may be used. The crane is controlled from an elevated position that is equipped with all the necessary controls for the operation. With the maximum load suspended from the crane hook the vehicle may be driven over level ground. If the load exceeds 3,000 kg the vehicle suspension is locked automatically. The crane turntable is also lockable. With the crane outrigger extended to 3 m the crane may be used through a full 360°. When extended 3 to 4 m the crane may only operate between +60 and −10° towards the rear and between +90 and −10° to the front, measured from a line fore and aft. For travelling, the outrigger is stowed along the side of the vehicle and held in a special clamp by a screw fitting.

The BREM-1 is equipped with an electric welding system that comprises an SG-10-1S starter/generator, a switch unit, a working position, a welding socket and a set of welding leads. The maximum welding current is 360 A. Controls for the welding equipment are located on a panel in a hermetically sealed container mounted on a shelf over the left-hand track. Other special tools carried include a universal fitter's tool, spanners and adaptors, all carried in two portable containers on a load platform. In the centre of the roof is a platform 1.706 m long and 1.4 m wide. It is provided with removable side flaps; and can accommodate a maximum load of 1,500 kg.

For recovery operations the BREM-1 is equipped with hoisting and auxiliary winches, add-on plough and bulldozer blades, towing equipment and a full set of accessories. The mechanical main winch has a capacity of 25,000 kg but this may be increased by using snatch blocks to 100,000 kg. Normally this winch is used forward, in conjunction with the lowered dozer blade, but for self-recovery it may be used to the rear. The length of cable used is 200 m. The winch is power-driven from a 'swing frame' gear via a reduction gearbox. The winch is controlled from the elevated operator's position. A 530 kg auxiliary winch and 400 m of cable are used to pull the cable towards the winch hoist and the speed of winding and unwinding is controlled hydraulically. The winch controls are to the right of the operator's position. A detachable device may be used to clean dirt from the cable. In use, this is placed on the fixed element of the cable feeder and comprises two round revolving brushes, a hydraulic motor, the cable guides and a reduction gear to transmit drive to the brushes. The device is switched on once the cable has been extended.

The bulldozer equipment is mounted on the front of the hull and is hydraulically driven, with control via two push-buttons on the driver's main control handles. The blade is 3.1 m wide and can be used to create an MBT firing position in 20 minutes in heavy soil or 12 minutes in light sandy soil.

The towing equipment consists of two towing rods, each 1.68 m long and with internal shock-absorbers, and two 5.5 m long tow lines. Other special equipment carried includes a 30 tonne capacity hydraulic jack.

The BREM-1 has an R-123U radio installation, a TPU tank telephone system and an overpressure NBC system.

The BREM-1 is stated to have the same degree of armour protection as the T-72 MBT. The only armament carried by the BREM-1 is a single 12.7 mm NSV machine gun and 840 rounds of ammunition. Four forward-firing smoke dischargers may be fitted but smoke screens can be produced by injecting diesel fuel into the exhaust outlet which is situated on left side of the hull towards the rear.

Variants
VT-72B
See separate entry under Czech Republic and Slovakia.

WZT-3
This is a Polish armoured recovery vehicle based on the T-72 MBT chassis. Details are provided in a separate entry in this section.

Specifications
Crew: 3
Weight: 41,000 kg
Power-to-weight ratio: 20.5 hp/t
Ground pressure: 0.83 kg/cm^2
Length: 7.98 m
Width: 3.46 m
Height: 2.42 m
Ground clearance: 0.457 m
Max speed: (road) 60 km/h
Range:
(roads) 700 km
(tracks) 500 km
(towing) 220 km
Fording:
(normal) 1.2 m
(with preparation) 5 m
Gradient: 60%
Side slope: 40%
Vertical obstacle: 0.85 m
Trench: 2.6-2.8 m
Engine: V-12 diesel developing 840 hp
Armament: 1 × 12.7 mm NSV MG
Ammunition: 12.7 mm - 840 rounds

Main winch
Rated line pull: 25,000 kg
With snatch block: 100,000 kg
Usable rope length: 200 m

BREM-1 armoured recovery vehicle (Richard Stickland) **1996**

BREM-1 armoured recovery vehicle from rear (Richard Stickland) **1996**

Auxiliary winch
Max line pull: 530 kg
Usable rope length: 400 m

Jib crane
Lift capacity:
(at max 4.4 m extension) 3 000 kg
(2-3 m extension) 12,000 kg
(2 m extension) 19,000 kg
Max hook height: 4.36 m
Traverse: 360°

Status
Production. In service with the Russian Army and other undisclosed countries.

Manufacturer
Uralvagonzavod State Production Association.

UPDATED

BREM-1 armoured recovery vehicle in travelling configuration

BREM-2 armoured repair vehicle

Description
The BREM-2 (*Bronirovannaya Remontno-evakuatsionnaya-2*) armoured recovery vehicle is a conversion based on the hull and chassis of the amphibious BMP-1 IFV. The turret is removed from the BMP-1 and replaced by an armoured plate while the upper hull is modified to accommodate a swivelling jib crane with a 1,500 kg lift capacity capable of lifting a BMP series power unit, plus various items of recovery equipment plus a stowage area over the left rear roof area. Various items of recovery equipment are stowed around the hull roof and sides. The resultant overall layout is very similar to that of the Czech/Slovak VPV armoured recovery vehicle (see separate entry in this section).

The hull interior is rearranged to house a 6,500 kg capacity winch and the four-man crew. The engine remains the usual BMP-1 Type UTD-20 diesel developing 300 hp at 2,000 rpm. Armament is limited to a single 7.62 mm PKT machine gun and six smoke grenade launchers on the hull roof.

The BREM-2 weighs approximately 13,600 kg and has a maximum road speed of 60 km/h.

Status
Conversions of existing BMP-1 vehicles available. The BMP-1 is used by the following countries, some of which also use the BREM-2 armoured recovery vehicle: Afghanistan, Algeria, Angola, Armenia, Azerbaijan, Belarus, Cuba, Czech Republic, Egypt, Ethiopia, Finland, Georgia, Greece, Hungary, India, Iran, Iraq, Kazakhstan, North Korea, Kyrgyzstan, Libya, Mongolia, Mozambique, Poland, Romania, Russia, Slovakia, Sri Lanka, Sweden, Syria, Tajikistan, Ukraine, Vietnam and Yemen.

Manufacturers/Converters
Kharkov Armoured Repair Factory, Kotlava Street 222, 310139 Kharkov, Ukraine.
Tel: +38 0572 225741. Fax: +38 0572 234189.
Zhytomir Armoured Repair Factory, 261109 Zhytomir, Ukraine.
Tel: +380 412 373192. Fax: +380 412 373192.
Marketed by: Progress, 15 Piririchna Street, 252216 Kiev, Ukraine.
Tel: +380 44 411 0589. Fax: +380 44 411 0848.

UPDATED

BREM-2 armoured recovery vehicle roof area showing 1,500 kg capacity crane and other details (Richard Stickland) ***1996***

BREM-2 armoured recovery vehicle (Richard Stickland) ***1996***

BREM-64 armoured repair vehicle

Description
The BREM-64 (*Bronirovannaya Remontno-evakuatsionnaya-64*) armoured recovery vehicle is a conversion based on the hull and chassis of the T-64 MBT. The turret is removed from the T-64 and replaced by an armoured superstructure with a raised cupola for the vehicle commander. A swivelling crane with a folding jib is located on the left-hand side of the superstructure; the crane has a lifting capacity of 2,500 kg. Two winches are provided, the main winch has a basic pulling capacity of 25 tonnes while the secondary winch has a capacity of 2.5 tonnes. When the main winch is in operation vehicle stability is provided by a front-mounted hydraulically operated dozer blade which may also be utilised to clear battlefield obstacles and prepare sites.

The main power unit remains the Type 5DTF 700 hp diesel, as used on the T-64 MBT, but there is also a SMD 21 auxiliary power unit developing 145 hp which may be used to power dismounted welding equipment and other power tools.

The BREM-64 weighs approximately 40,500 kg and has a maximum road speed of 45 km/h. Armament is a 12.7 mm NSV air defence machine gun. The crew comprises of a commander, driver and a fitter.

Status
Production complete. In service with Russia and the Ukraine. It is probable that quantities of these vehicle are held by the Ukraine for export.

Manufacturers/Converters
Kharkov Armoured Repair Factory, Kotlava Street 222, 310139 Kharkov, Ukraine.
Tel: +380 572 225741. Fax: +380 572 234189.
Zhytomir Armoured Repair Factory, 261109 Zhytomir, Ukraine.
Tel: +380 412 373192. Fax: +380 412 373192.
Marketed by: Progress, 15 Piririchna Street, 252216 Kiev, Ukraine.
Tel: +380 44 411 0589. Fax: +380 44 411 0848.

UPDATED

T-54/T-55 armoured recovery vehicles

Description
The first ARV based on the chassis of the T-54 MBT appeared in the 1950s and was designated the T-54-T. Since then at least seven other T-54/T-55 ARVs have been developed. Most of them have very limited capabilities compared to ARVs developed in the West, and few are equipped with a winch, limiting their capabilities to towing damaged vehicles off the battlefield. These vehicles are known as BTS (medium armoured towers) in Russia. The T-55-TK is an improvement over the earlier models as it can lift vehicles weighing up to 20,000 kg.

All these ARVs use a T-54 or T-55 MBT chassis with the turret removed. The driver is seated at the front of the hull on the left side and is provided with two periscopes for observation and a single-piece hatch cover. The vehicle commander is normally seated to his right and is provided with a single-piece hatch cover which opens to the right. The other crew members normally sit in the cargo area, which is often cramped as the snorkel and spare fuel drums are usually carried there. An unditching beam is carried on the right side of the hull and towbars of varying lengths are also carried.

The engine and transmission are at the rear of the hull but unlike the T-54/T-55 MBT there does not appear to be any provision for carrying additional fuel tanks on the rear of the hull. It is assumed that most of these ARVs are provided with an NBC system.

The suspension is of the torsion bar type and consists of five roadwheels with the idler at the front and the drive sprocket at the rear. There are no track-return rollers.

Variants

T-54-T/T-55-T

This was the first model to enter service and performs a similar role to the older T-34-T (B) ARV but is based on a more powerful chassis. A loading platform is mounted in the centre of the vehicle, with sides that can be folded down to facilitate the loading or unloading of spare components such as an engine or a transmission. A large spade is mounted at the rear of the hull and a large-diameter snorkel can be installed to the rear of the driver's position for deep fording operations. A jib crane which can lift a maximum weight of 1,000 kg is provided. There is no winch, so the vehicle is limited to towing operations. The Finnish Army uses a version of the T-54-T called the BTS-2, which has a winch, spade and a small dismountable crane. Another Finnish ARV based on the T-55 is the KAM-1, for which there is a separate entry in this section.

T-54 (A)

This is a German development and can be fitted with a snorkel for deep fording operations. Standard equipment includes a push/pull bar, full range of tools including both welding and cutting equipment, dismountable crane with a lifting capacity of 1,000 kg, radiation warning equipment and a chemical warfare agent detector. This model does not have a winch, nor a spade at the rear. If required, PT-54 or PT-55 roller type mineclearing equipment can be installed at the front.

T-54 (B)

This is also a German development. It is similar to the T-54 (A) but at the rear of the hull are brackets for securing tow ropes and on the glacis plate at the front of the hull is a protective plate. This model is not provided with a winch or a spade.

T-55-TK (previously known as T-54 (C))

This is another German development. It is provided with a stowage platform, snorkel, spade at the rear, dozer blade at the front and a heavy-duty crane which is mounted on the right side of the hull. This has a telescopic jib and can lift a maximum weight of 20,000 kg. When not required the crane is traversed to the rear so that its jib rests along the left side of the hull.

Czechoslovak VT-55A ARV

There is a separate entry for this vehicle in this section under Czech Republic and Slovakia.

Polish T-54/T-55 ARVs

Poland has developed at least two ARVs based on T-54 or T-55 MBT chassis, designated the WZT-1 and WZT-2.

T-55-T armoured recovery vehicle (Michael Jerchel) ***1997***

T-55-TK armoured recovery vehicle (Michael Jerchel) ***1996***

Specifications

(T-54-T; data in square brackets refers to T-54 (B) and T-55-TK where different)
Crew: 3-5
Weight: (empty) 36,000 [32,000], [34,000] kg
Length: 7.12 [7.05], [9.74] m
Width: 3.23 m
Height: 1.89 [2.2], [2.65] m
Ground clearance: 0.425 m
Track: 2.64 m
Track width: 580 mm
Length of track on ground: 3.84 m
Ground pressure: 0.72 [0.72], [0.77] kg/cm^2
Max speed: (road) 48 km/h
Range: 400 km
Fuel capacity: 812 litres
Fording: 1.4 m
Gradient: 60%
Vertical obstacle: 0.8 m
Trench: 2.7 m
Engine: Model V-54, V-12, water-cooled diesel developing 520 hp at 2,000 rpm
Transmission: manual with 5 forward and 1 reverse gears
Electrical system: 24 V
Batteries: 4 with total capacity of 280 Ah
Armament: nil

Armour
glacis plate: 100 mm at 60°
upper hull sides: 70 mm at 0°
hull rear: 60 mm
hull floor: 20 mm
hull roof: 30 mm

Note: Models based on a T-55 chassis are powered by a V-55 engine which develops 580 hp at 2,000 rpm.

Status

Production complete. The T-54 and T-55 are in service with the following countries so it can be assumed that most of these use the T-54/T-55 ARV. The T-54/T-55 MBT is in service with the following countries, many of which also use the various armoured recovery versions: Afghanistan, Albania, Algeria, Angola, Armenia, Azerbaijan, Bangladesh, Belarus, Bosnia-Herzegovina, Bulgaria, Cambodia, Central African Republic, China, Congo, Croatia, Cuba, Czech Republic, Egypt, Finland, Georgia, Guinea, Hungary, India, Iran, Iraq, North Korea, Laos, Lebanon, Libya, Mongolia, Mozambique, Nicaragua, Nigeria, Pakistan, Peru, Poland, Romania, Slovakia, Slovenia, Somalia, Sri Lanka, Sudan, Syria, Tanzania, Togo, Uganda, Ukraine, Vietnam, Yemen, Yugoslavia, Zambia and Zimbabwe.

Manufacturer

Czech/Slovak, Polish and Russian state factories.

UPDATED

BREM-L armoured repair vehicle

Description

The BREM-L (*Bronirovannaya Remontno-evakuatsionnaya-L*) armoured recovery vehicle is based on the hull and chassis of the amphibious BMP-3 IFV. It is manufactured from new and is intended to be a component in a proposed family of vehicles based on the hull and chassis of the BMP-3.

The BREM-L does not have the main turret of the BMP-3 IFV but instead has a small one-man turret offset to the right-hand side of the hull roof. This turret is occupied by the commander who has access to a 7.62 mm PKT machine gun mounted over the turret hatch. A normal crew for the BREM-L is three personnel although there is seating for a further two personnel.

The left-hand side of the hull roof is occupied by a 6 tonne capacity jib crane; the crane capacity can be doubled to 12 tonnes by introducing a pulley block. A 20 tonne capacity winch is located inside the hull; the winch capacity can be doubled to 40 tonnes with a pulley block. When the winch is in operation a hydraulically operated dozer blade can be lowered at the front of the hull. A load area capable of carrying spare engine packs is located over the rear of the vehicle. Welding equipment is carried. Towbars and other recovery equipment and tools are stowed around the hull roof and sides.

Specifications

Crew: 3 + 2
Weight: 18,700 kg
Power-to-weight ratio: 26.73 hp/t
Length: 7.64 m
Width: 3.28 m
Height: 2.71 m
Ground clearance: 0.51 m
Track: 2.76 m
Track width: 380 mm
Length of track on ground: 4.05 m
Max speed:
(roads) 70 km/h
(dirt roads) 52 km/h

(water) 9 km/h
Range: (roads) 600 km
Fording: amphibious
Gradient: 60%
Side slope: 30%
Vertical obstacle: 0.8 m
Trench: 2.50 m
Engine: UTD-29 4-stroke diesel developing 500 hp
Armament: 1 × 7.62 mm PKT MG

Status
Prototypes.

Manufacturers
JSC Kurganmashzavod, 17 Mashinostroitelei Avenue, 640631 Kurgan, Russia.
Tel: +7 35222 74833. Telex: 120112 MICAR.
Fax: +7 35222 33996.

UPDATED

BREM-L armoured recovery vehicle
1996

BREM-K armoured repair vehicle

Description
The BREM-K (*Bronirovannaya Remontno-evakuatsionnaya-K*) armoured recovery vehicle is based on the hull and chassis of the BTR-80 8 × 8 wheeled APC.

The main change from the BTR-80 APC is that on the BREM-L the turret is modified to mount a 7.62 mm PKT machine gun and act as the carrier for two cable drums, one each side. The cable drums are used together with a collapsible jib crane which is erected on the front hull. When not in use the jib components are stowed along the sides of the hull. The crane, which cannot be traversed, has a lifting capacity of 1.2 tonnes and is supplied by two 75 m lengths of cable from the drums behind the turret. There is also an alternative collapsible manually operated jib crane which can be traversed; this has a lift capacity of 800 kg and is provided with two 5 m lengths of cable.

Located inside the hull is the main recovery winch. This has a pull capacity from 4,400 to 6,000 kg although this may be increased to 24,000 kg by the introduction of pulley blocks. When the winch and the crane are in operation, vehicle stability is improved by lowering two hydraulically operated, front-mounted earth spades located under the front of the hull.

Other special-to-role equipment carried by the BREM-K includes a load platform located behind the turret, a GD-4003U2 welding set, towing equipment, and special tools.

Specifications
Crew: 4
Weight: 14,500 kg
Power-to-weight ratio: 17.93 hp/t
Length: 7.65 m
Width: 2.9 m
Height: 3 m
Ground clearance: 0.475 m
Max speed:
(roads) 80 km/h
(water) 9 km/h
Range: (roads) 600 km
Fording: amphibious
Gradient: 60%
Side slope: 40%
Vertical obstacle: 0.5 m
Trench: 2 m
Engine: V-8 water-cooled diesel developing 260 hp
Armament: 1 × 7.62 mm PKT MG

Status
Production. In service with Russian Army.

Manufacturer
Gorky Automobile Plant.

UPDATED

Front view of BREM-K armoured recovery vehicle in travelling configuration (Christopher F Foss)
1996

BREM-K armoured recovery vehicle in travelling configuration (Christopher F Foss) ***1996***

BREhM-D repair and recovery vehicle

Description
The BREhM-D repair and recovery vehicle is based on the chassis of the BTR-D airborne armoured personnel carrier and is equipped with special purpose for the repair and recovery of BMD-I and similar vehicles. Specialised equipment carried includes a hydraulic crane, a recovery winch, a combination spade/dozer blade, towing equipment, a welding kit, plus other tools and stowage points. For details of the BMD-1 and BTR-D see *Jane's Armour and Artillery 1996-97* pages 326 to 329.

When not in use the hydraulic crane boom is carried stowed in a retaining frame on the upper hull of the vehicle. The crane, which has a traverse of 180°, has a capacity of 1,000, 2,000 or 3,000 kg depending on the number of cable runs, with the runs rigged to the top of the vehicle. The crane is operated from the commander's position using power from the vehicle hydraulic system, although a hand pump is provided for when the engine is not running.

For recovery tasks the vehicle uses the hydraulic recovery winch and the spade/dozer blade. Using the main cable the winch has a capacity of 4,000 kg but this can be increased to 12,000 kg using extra cable runs. The winch is controlled from the commander's position. The winch cable is 100 m long and plays out forwards through a framework holding heavy rubber rollers to keep the cable free from mud or snow. When carrying out heavy recovery tasks the spade/dozer blade is lowered manually by two personnel to provide an anchor.

For towing there are two telescopic towbars with internal hydraulic shock-absorbers. When not in use, these are carried on the rear of the hull.

The electrical welding set is powered by a VG-7500 generator driven by the vehicle's main engine. The vehicle also carries special repair and recovery tools, some of which are stowed in special compartments inside the hull.

Armament consists of a 7.62 mm PKT bow machine gun. Also carried are a radio, an intercom and an NBC and ventilation system.

Status
Production as required. In service with Russian Army.

Manufacturer
Volgograd Tractor Plant.

UPDATED

SOUTH AFRICA

Reumech OMC Olifant armoured recovery vehicle

Description
The South Africans have converted a number of Centurion tanks to Olifant (Elephant) armoured recovery vehicles. The conversions involve the replacement of the existing power pack and transmission by the equivalent uprated components fitted into the Olifant tank conversions, also carried out in South Africa. (For details of the Olifant MBT refer to *Jane's Armour and Artillery 1996-97* pages 106 to 107). These conversions were carried out by OMC (Olifant Manufacturing Company) which is now part of Reumech OMC.

The structural conversions include removing the tank turret and fitting a fixed armoured superstructure onto the hull. The superstructure houses a crew of four, with the driver seated in a position raised from his former driving location to behind a protected window on the right-hand side of the superstructure front. An extending boom crane is positioned to the left of the superstructure; it has a possible traverse of 360°. A front-mounted stabilising spade, which can also be used as an obstacle-clearing dozer blade, is brought into use when the internally mounted recovery winch is used in the forward-operating mode. The winch cable can also be deployed through a rear-mounted guide assembly primarily for self-recovery to the rear. Various items of recovery equipment are stowed around the superstructure and there are stowage bins provided over the power pack cover area.

There is provision for machine gun mountings on the cupolas located on the superstructure roof and there are four sets of four-barrelled 81 mm smoke dischargers, one on each corner of the superstructure. No other armament appears to be carried other than the crew's personal weapons.

No further details regarding the Olifant armoured recovery vehicle are available. It has been reported that an armoured repair variant of this vehicle is to be produced. This variant will be able to carry a complete Olifant power pack.

Status
In service with the South African National Defence Forces.

Manufacturer
Reumech OMC, Private Bag X049, Benoni 1500, South Africa.
Tel: +27 11 845 1440. Fax: +27 11 845 1379.

UPDATED

Olifant armoured recovery vehicle

TFM Gemsbok (4 × 4) armoured recovery vehicle

Development
The TFM Gemsbok (4 × 4) armoured recovery vehicle is a member of the Casspir family of vehicles based on a wheeled armoured personnel carrier, the Casspir Mark 2 (for details refer to *Jane's Armour and Artillery 1996-97* pages 470 to 472). These vehicles have been designed to operate as internal security and riot control vehicles but can also operate over long periods in the bush and desert terrain conditions prevailing in South Africa. The Casspir series, including the Gemsbok armoured recovery vehicle, is operated by the South African Police.

Description
The Gemsbok uses the standard Casspir Mark 2 chassis, which in its turn is based on a commercial 15 tonne 4 × 4 truck. The armoured front cab uses a combination of armour plated and high-alloy steels and has a V-shaped monocoque hull to withstand land mine explosions. The hull can withstand three simultaneous TM-57 type anti-tank mine explosions (or an equivalent of between 18 and 20 kg of TNT) without penetration, although mechanical component damage may occur. The armour is protection against NATO 7.62 and 5.56 mm ball ammunition and all windows use bullet-resistant glass. Mounted on the roof next to the driver's position is a one-man steel turret which can mount one 7.62 mm MG4 machine gun. A further MG4 machine gun can be mounted in the front left windscreen.

The armoured cab has seating for three of the crew of a recovered vehicle as well as the recovery vehicle crew of driver and commander. The armoured cab also contains the 185 litre fuel tank. Cab ventilation is provided by two impeller fans.

The Gemsbok recovery unit is mounted on the open chassis behind the armoured cab and consists of a twin boom, two service drum recovery hoist with a maximum capacity of 5,000 kg. Rectangular section-stabilising legs are provided on each side of the chassis when the hoist is in use. Tool box and other stowage is provided on the rear area for extra recovery equipment and special tools. A standard vehicle tool kit is carried inside the main cab.

TFM Gemsbok (4 × 4) armoured recovery vehicle

Specifications
Cab seating: 2 + 3
Configuration: 4 × 4
Weight:
(empty) 9,600 kg
(loaded) 14,800 kg
Max load: 5,000 kg
Towed load: 15,000 kg
Length: 6.74 m
Width: 2.5 m
Height: 3.12 m
Ground clearance: 0.355 m
Wheelbase: 4.3 m
Angle of approach: 46°
Max speed: 87 km/h
Range:
(road) 850 km
(cross-country) 560 km
Fuel capacity: 185 litres

Max gradient: 60%
Side slope: 30%
Vertical obstacle: 0.5 m
Trench: 0.95 m
Fording: 1 m
Engine: ADE 352 T 5.675 litres 6-cylinder vertical in-line turbocharged diesel developing 162.5 hp at 2,800 rpm
Gearbox: Mercedes-Benz DB G 3/60 - 5/7, 5 synchromesh with 5 forward and 1 reverse gears
Clutch: single dry plate
Transfer box: Mercedes-Benz 2-speed
Steering: ball and nut, power-assisted
Turning circle: 18.3 m
Suspension: semi-elliptic leaf springs on needle roller bearings with telescopic shock-absorbers front and rear (double spring pack at rear)
Tyres: 14.00 × 20
Number of tyres: 4 + 1 spare
Brakes:
(main) dual circuit air-assisted hydraulic with engine-mounted exhaust brake
(parking) mechanical
Electrical system: 12 V
Battery: 1 × 12 V
Alternator: 55 A
Armament: 1 or 2 × 7.62 mm MG4 MG

Status
Production complete. In service with South African Police. Production can be resumed if further orders are placed.

Manufacturer
TFM (Pty) Limited, PO Box 46131, Orange Grove 2119, Transvaal, South Africa.
Tel: +27 31 62106. Fax: +27 31 61492.

UPDATED

SPAIN

GAMESA M-47 VR armoured recovery vehicle

Development
Following a competition, in 1994 the Spanish Army selected GAMESA to convert 22 old M-47E1 tanks into armoured recovery vehicles. Final deliveries to the Spanish Army were made late in 1996 under the designation M-47 VR.

Although the original intention was to use these vehicles to support M48 and M60 tanks, experience has shown that they were capable of providing support for the German Leopard 2 MBT which recently entered service with the Spanish Army.

GAMESA M-47 VR recovery tank in travelling configuration ***1997***

Description
The conversion involved replacing the complete turret and part of the hull of the M-47. A new welded armoured superstructure houses the four-man crew consisting of commander, driver, radio operator and the recovery equipment.

The driver's station has been moved with regard to the original vehicle to allow adequate visibility in the converted vehicle. The brakes, steering systems and power pack, of the more recent M-48 tank, have been incorporated in the M-47 VR.

The superstructure is of all-welded steel armour construction and provides protection from small arms fire up 12.7 mm in calibre as well as shell splinters.

The M-47 VR has been fitted with a front-mounted bulldozer-type blade driven by two hydraulic cylinders with two roles. The first is to stabilise the tank when the crane or winch are being used and second is to carry out dozing operations.

The blade is 3.7 m wide and 900 mm high which allows the front of the tank to be raised 30 cm. It is also equipped with three ripper-scarifiers located on the rear of the blade.

The hydraulic unit is housed inside the hull, the variable flow piston pump with load sensing is driven through an electromagnetic clutch and cardan joint, by an additional power-take-off built onto the Teledyne AVDS-1790-2D diesel engine. It is also provided with a constant speed regulator which works when the hydraulic system is operative.

The main 35,000 kg constant pull winch is housed inside the hull and is provided with a cable 120 m long and 33 mm in diameter, this leaves the vehicle guided by means of an air tight roller box placed on the front of the tank. With the return pulley which is included is the accessories of the tank, pulling can be doubled to 70,000 kg.

The release speed and rapid collection of the cable is adjustable from 0 to 40 m/min and working speed under load is adjustable from 0 to 8 m/min.

The auxiliary winch is located on the front of the vehicle and can be used to handle 33 mm diameter cable in lengths of 100 m.

Mounted on the front right side of the vehicle is a telescopic crane which can be traversed through a full 360° and has a torque capacity of 60 m/ton. The crane is provided with 100 m of 16 mm diameter cable which allows it to raise 22,000 kg. An overload prevention system is fitted as standard. The length of the jib is 6.1 m and the telescopic boom adds 2.5 m to its reach. Total length of the jib is 8.6 m and maximum height to hook is 9.75 m.

Armament comprises a 12.7 mm machine gun on the tank commander's cupola and a 7.62 mm machine gun on the front of the vehicle. Two banks of four electrically operated smoke dischargers cover an arc of 40° to the left and right of the vehicle.

Electrical and hydraulic intakes for the use of tools equipped with snap on connectors are located on the right side of the tank.

In case of an emergency, such as the breakdown of the main engine or lack of main hydraulic power, the vehicle is fitted with an auxiliary electric pump which allows the crane, bulldozer blade and winch to be returned to the travelling position.

Standard equipment includes: engine smoke generator; NBC ventilation system; fire detection and suppression system for the engine compartment; fire and explosion suppression system for the crew compartment; and lifting points to enable it to be loaded on board ships.

If required, provision is made for the transport of a complete power pack (engine, transmission and cooling system) for a M48, M60 or Leopard 2 MBT.

Specifications
Crew: 4
Weight:
(empty) 46,450 kg
(combat) 48,500 kg
Power-to-weight ratio: 15.67 hp/t
Length: 8.4 m
Width: 3.7 m
Height: (without armament) 3.07 m
Ground clearance: 0.469 m
Max speed: 48 km/h
Max road range: 500 km
Fuel capacity: 1,420 litres
Fording: 1.2 m
Vertical obstacle: 0.9 m
Trench: 2.6 m
Gradient: 60%
Side slope: 40%
Engine: Teledyne (now General Dynamics Land Systems) AVDS-1790-2D diesel developing 760 hp at 2,400 rpm
Electrical system: 24 V
Batteries: 6 × 12 V, 300 Ah
Armament:
(main) 1 × 12.7 mm MG
(secondary) 1 × 7.62 mm MG
Smoke grenade launchers: 2 × 4
Ammunition:
(main) 600 × 12.7 mm
(secondary) 2,000 × 7.62 mm

Status
Production as required. In service with the Spanish Army (22 units).

Manufacturer
GAMESA, Portal de Gamarra 40, PO Box 628, E-01013 Victoria, Spain.
Tel: +34 45 280 700. Fax: +34 45 251 642.

NEW ENTRY

SBB Blindados SA BMR 3560.55 armoured recovery and repair vehicle

Description

The SBB Blindados SA BMR 3560.55 armoured recovery vehicle, is based on the hull and suspension of the BMR 3560.50 armoured personnel carrier and has many components in common with the basic vehicle. It is intended for the recovery of other vehicles in the BMR range (and similar vehicles) and also has some repair capabilities for immobilised vehicles in the field.

The BMR 3560.55 recovery vehicle retains the basic 6 × 6 configuration and hull of the armoured personnel carrier virtually unchanged, the main alteration being the installation of a hydraulic crane on a roof-mounted turntable. This crane has a full 360° traverse and a lifting capacity of 12,000 kg/m. The horizontal reach is 6 m and the maximum hook elevation is 8 m. To stabilise the crane when deployed, the BMR 3560.55 has four hydraulically operated stabiliser jacks, one on each side between the first and second set of roadwheels and the other two at the rear behind the rear set of wheels.

Access to the interior of the vehicle is gained via a large single-piece rear ramp that opens downwards. The driver's position may be reached from inside the vehicle main cabin or through a roof hatch. There is another roof hatch for the commander and another for the crane operator towards the rear. The commander's position has a cupola with a machine gun mounting (12.7 mm M2 or 7.62 mm MG1A1) that can be operated from within the vehicle. The weapon ports of the basic BMR are retained to allow the crew to use their personal weapons from within and a rack of six Wegmann smoke dischargers is situated on the left-hand side of the hull.

Inside the main vehicle cabin there is provision for various tool cabinets and other special racks but the main space is occupied by a winch with a pulling capacity of 7,000 kg; by using double pulleys this can be increased to 14,000 kg. The cable length is 100 m. The normal front-mounted winch of the BMR is retained. This has a pulling capacity of 4,500 kg. The controls for the main winch are located close to those of the crane, that is, inside the crane operator's turret hatch. The crane and main winch can both be operated from within the main cabin with the rear door closed. When the rear ramp is closed the main winch cable is passed through the rear door emergency hatch.

To enable the vehicle to be used in the repair role the BMR 3560.55 carries a canvas penthouse and supports on the roof and sides. For this purpose, lighting is supplied by a 50 m long string of lamps with power from a 1.5 kW generator. A workshop tool kit, a welding kit, slings, chains and other recovery and repair equipment are also carried. A lockable tool chest is located just inside the left of the rear door. Radio equipment is located against the engine bulkhead.

The BMR 3560.55 is amphibious and can be fitted with water propulsion jets.

SANTA BARBARA BMR 3560.55 armoured recovery vehicle (Pierre Touzin) ***1996***

Specifications

Crew: 5
Configuration: 6 × 6
Combat weight: (APC) 13,750 kg
Power-to-weight ratio: 22.54 hp/t
Length: 6.15 m
Width: 2.5 m
Height: (hull top) 2 m
Ground clearance: (adjustable) 0.45 m
Track: 2.08 m
Wheelbase: 1.65 m + 1.65 m
Angle of approach/departure: 50°/45°
Max speed: 100 km/h
Fuel capacity: 300 litres
Max range: 700 km
Fording: amphibious
Gradient: 60%
Side slope: 30%
Vertical obstacle: 0.8 m
Trench: 1.2 m
Engine: 6-cylinder in-line diesel developing 310 hp at 2,200 rpm
Transmission: ZF 6 HP 500 automatic with 6 forward and 1 reverse gears
Steering: hydraulic
Turning radius: 7.5 m
Suspension: ZF hydropneumatic, independent each wheel
Tyres: 13.00 × 20
Electrical system: 24 V
Batteries: 1 × 12 V, 150 Ah
Armament:
1 × 12.7 mm M2 or 7.62 mm MG1A3 MG
6 × smoke grenade dischargers

Status

Production complete. In service with the Spanish armed forces and the Saudi Arabian and Egyptian armies.

Manufacturer

SBB Blindados SA, Ctra SE-208 Km 2, Alcalá de Guadaira, E-41500 Sevilla, Spain.
Tel: +34 5 561 3780. Fax: +34 5 410 0503.
SBB Blindados SA, Julián Camarillo 32, E-28037 Madrid, Spain.
Tel: +34 1 585 0100. Fax: +34 1 585 0268.

UPDATED

SWEDEN

Hägglunds Vehicle Bärgningsbandvagn 82 armoured recovery vehicle

Development

The Bärgningsbandvagn 82 (or Bgbv 82 as it is usually known) was developed by Hägglunds to meet a Swedish Army requirement for an ARV to recover the S tank and other AFVs. The first prototype was completed in July 1968, and as a result of user trials some modifications were requested by the Swedish Army. The Bgbv 82 was then accepted for service and 24 vehicles were completed between April and December 1973. The vehicle has an almost identical hull to the Brobandvagn 941 (bridgelayer) which was designed at the same time, and to keep costs to a minimum both vehicles share many common components with the Ikv 91 tank destroyer. In addition to being used as a recovery vehicle the Bgbv 82 can be employed as an engineer vehicle, as for example in clearing river exit points.

Description

The hull of the Bgbv 82 is of all-welded steel construction with the crew compartment at the front and the recovery compartment at the rear. The front of the vehicle is immune to attack from projectiles up to and including 20 mm in calibre, while the sides of the

Hägglands Vehicle Bgbv 82 armoured recovery vehicle

hull, above the tracks, are of a double-skinned construction which gives added protection against attack from HEAT projectiles.

The crew consists of four: commander, gunner, driver and winch operator. The gunner is seated in the turret (which is identical to that fitted on the Hägglunds Vehicle Pbv 302 APC) on the left side of the hull. This is armed with a 20 mm Hispano 804 cannon which can be elevated from −10 to +50°. Both elevation and traverse are manual. For engaging ground targets the gunner uses a sight with a magnification of ×8 while a special sight is provided for engaging aerial targets, but to use this the gunner has to open his hatch cover and is therefore exposed to small arms fire. Four M17 periscopes are provided in the turret, three to the front and one to the rear. Eight smoke grenade dischargers are mounted on either side of the turret. The driver is seated in the centre of the hull and is provided with three periscopes to his front for observation, and a single-piece hatch cover which opens to the rear. The vehicle commander is seated to the right of the driver and is provided with six periscopes for all-round observation and a single-piece hatch cover. The winch operator is seated to the rear of the commander facing the rear, with his single-piece hatch cover opening to the rear. All the recovery equipment is hydraulically operated from the crew compartment.

The engine, torque converter, clutch, gearbox and steering gearbox (including the bevel gear) can be removed from under the floor of the load/recovery compartment as a complete unit. The fuel tank is also under the floor.

The six roadwheels on each side are supported by trailing arms individually suspended by transversal torsion bars. Shock-absorbers are mounted on the first and last trailing arms. The idler is at the front and the drive sprocket is at the rear. The track tension is automatically adjusted by a hydraulic system.

The Hägglunds HM 20 winch is mounted to the rear of the crew compartment and is provided with 145 m of wire rope which leads out through the rear of the hull. Two ground anchor spades are mounted at the rear of the hull and are hydraulically positioned. When in position they can absorb a force of 60,000 kg, the maximum traction for a three-part pull.

The HIAB-Foco 9000 lifting crane is mounted on the right side of the hull and can lift 5,000 kg with a jib length of 1.8 m, 3,500 kg with a jib length of 2.5 m and 1,500 kg with a jib length of 5.5 m. At the front of the hull is a hydraulically operated dozer blade which is normally positioned when the winch or lifting crane is being used.

The Bgbv 82 is fully amphibious, being propelled in the water by its tracks. The only preparation required is to erect the trim vane at the front of the hull, switch on the bilge pumps and erect a low flotation screen. If the vehicle is carrying a heavy load, such as a complete power pack for an S tank, a canvas cover should be erected. The canvas cover should also be fitted when crossing rough water.

The vehicle is not provided with an NBC pack, although provision was made in the design for one to be installed at a later date. Infrared driving lights are provided.

Specifications

Crew: 4
Weight:
(empty) 19,800 kg
(loaded) 26,300 kg (with S tank power pack)
Power-to-weight ratio: 15.65 hp/t
Length: 7.2 m
Width: 3.25 m
Height:
(top of turret) 2.45 m
(top of spades) 2.63 m
Ground clearance: (centre of hull) 0.45 m
Track width: 450 mm
Length of track on ground: 3.6 m
Ground pressure: 0.82 kg/cm^2
Max speed:
(road) 56 km/h
(water) 8 km/h
Range: 400 km
Fuel capacity: 550 litres
Fuel consumption: (road) 1 litres/km
Fording: amphibious
Gradient: 60%
Vertical obstacle: 0.6 m
Trench: 2.5 m
Engine: Volvo-Penta Model THD 100C turbocharged 6-cylinder in-line diesel developing 310 bhp at 2,200 rpm
Transmission: manual Volvo-Penta R61 with 8 forward and 2 reverse gears
Electrical system: 24 V
Armament:
1 × 20 mm Hispano 804 cannon
16 × smoke grenade dischargers
Ammunition: 20 mm - 505 rounds

Status

Production complete. In service with the Swedish Army.

Manufacturer

Hägglunds Vehicle AB, S-891 82 Örnsköldsvik, Sweden.
Tel: +46 660 80000. Telex: 6051 HAEGG S 6051.
Fax: +46 660 15190.

UPDATED

Hägglunds Vehicle CV 90 armoured recovery vehicle

Description

The CV 90 armoured recovery vehicle is a member of the Combat Vehicle 90 family of full tracked armoured vehicles, covered in detail in *Jane's Armour and Artillery 1996-97* pages 335 to 337. This is in production and service with the Swedish Army while the Norwegian Army has a slightly different version with a 30 mm cannon called the CV 9030N (pages 333 to 335).

The CV 90 armoured recovery vehicle has been developed by Hägglunds Vehicle with the first prototype completed in 1991. The first production order for this version was placed in 1993 and first production vehicles were completed in 1995. It is now in service with the Swedish Army.

The CV 90 ARV is based on the chassis, running gear, power pack and armoured steel hull of the CV 90, but in place of the usual turret there is a raised cupola for the commander. The rest of the three-man crew comprises the driver and a winch operator. There is also provision for one passenger.

Standard equipment includes a front-mounted stabiliser blade which can also be used for dozing operations. The hull contains two hydraulic capstan winches. Each winch has a constant pulling force and cable speed over the entire cable length. Using a four part pull on each winch the CV 90 ARV has a maximum recovery capability of 72 tonnes to the front. The pulling capacity to the rear is 36 tonnes and 9 tonnes to the side. A light crane with a telescopic jib is fitted on the left side of the hull roof towards the rear. As an option this crane can be replaced by one able to lift a power pack.

The vehicle has an overpressure NBC system and an automatic Halon fire detection and suppression system.

CV 90 ARV in travelling configuration

A 7.62 mm machine gun with 4,000 rounds of ammunition is carried for local defence. A total of 12 smoke dischargers in two groups of six is provided.

Specifications

Crew: 3 + 1
Weight combat: 21,800 kg
Power-to-weight ratio: 25.22 hp/t
Ground pressure: 0.485 kg/cm^2
Length: 7.55 m
Width: 3.1 m
Height:
(over cupola) 2.24 m
(roof) 1.73 m
Ground clearance: 0.45 m
Track: 2.47 m
Track width: 533 mm
Length of track on ground: 3.98 m
Max speed: (road) 70 km/h
Fuel capacity: 800 litres
Gradient: 60%
Side slope: 40%
Vertical obstacle: 1.20 m
Trench: 2.40 m
Engine: Scania DS-14 diesel developing 550 hp at 2,200 rpm
Transmission: Perkins Engines (Shrewsbury) by X-300-5N fully automatic with torque converter and lock-up clutch; 4 forward and 2 reverse gears
Electrical system: 24 V
Armament:
1 × 7.62 mm MG
12 × smoke grenade dischargers
Ammunition: 7.62 mm - 4,000 rounds

Status

Production. In service with the Swedish Army.

Manufacturer

Hägglunds Vehicle AB, S-891 82 Örnsköldsvik, Sweden.
Tel: +46 660 80500. Telex: HAEGG S 6051.
Fax: +46 660 15190.

UPDATED

SWITZERLAND

Entpannungspanzer 65/88 armoured recovery vehicle

Development

In 1961 the Swiss Federal Armament Factory at Thun (now Swiss Ordnance Enterprise) started the development of an armoured recovery vehicle based on the chassis of the Pz 61 MBT. The first prototype was completed in 1967-68 and after trials a modified version, based on the later Pz 68 MBT chassis, was placed in production; series production ceased during the early 1980s. First production Entpannungspanzer 65s (or Entp Pz 65 for short) were completed in 1970 and these are the standard armoured recovery vehicles of the Swiss Army. Modifications to the hydraulic system and some other changes resulted in a change of designation to Entp Pz 65/88. Each battalion of Pz 68 MBTs has two or three Entp Pz 65/88s in its service company and each 155 mm M109A1 self-propelled artillery battalion has one.

Description

The hull of the Entp Pz 65/88 is of cast sections welded together, with the crew and winch compartment at the front and the engine and transmission at the rear. An entry door is provided in the left side of the hull towards the front. In the crew compartment roof on the left side are two periscopes that provide observation to the front of the vehicle. To the rear of this position is a cupola with a two-piece hatch cover which is provided with vision blocks for observation purposes. A 7.5 mm Mg 51 machine gun is mounted externally on this cupola. The driver is seated at the front on the right side and is provided with a one-piece hatch cover which opens vertically to the rear. Three periscopes are mounted to the front of the hatch. To the rear of the driver is another cupola which has vision blocks, a two-piece hatch cover, and four smoke grenade dischargers mounted on its front.

The suspension of the Entp Pz 65/88 is of the Belleville type, which consists of six roadwheels with the drive sprocket at the rear, the idler at the front, and three track-return rollers. Each roadwheel is independently sprung by layers of Belleville washers, or conical springs.

The main winch is provided with 120 m of cable it has a maximum capacity of 12,500 kg at high speed, at a maximum speed of 0.4 m/s, and a maximum capacity of 25,000 kg at low speed, at a maximum speed of 0.2 m/s. The capacity at low speed can be increased to 75,000 kg with snatch blocks. The auxiliary winch, which is used to pull out the main cable, is provided with 240 m of cable. Maximum speed on a full drum is 1.7 m/s and on an empty drum is 1.2 m/s. A hydraulically operated dozer blade mounted at the front of the hull is used either to stabilise the vehicle when the A-frame is being used or to clear away obstacles. Maximum height of the cutting edge above surface level is 0.92 m and below surface level 0.42 m. The A-frame is pivoted at the front of the hull and when not required lies back in the horizontal position on the roof. It has a maximum lifting capacity of 15,000 kg which enables it to lift complete tank turrets. An unusual feature of the frame is that at the lower end of each arm is a hydraulic jack, which allows the operator to make minor adjustments when lifting loads such as tank power packs into position.

A full range of tools, cutting equipment and towbars is carried and storage lockers are provided along each side of the hull. The Entp Pz 65/88 has an NBC system.

Entp Pz 65/88 armoured recovery vehicle in action with A-frame lowered ***1996***

Specifications

Crew: 5
Weight: 39,800 kg
Power-to-weight ratio: ???
Length:
(overall) 9.6 m
(hull) 7.57 m
Width:
(hull) 3.15 m
(blade) 3.15 m
Height: (incl MG) 3.25 m
Ground clearance: 0.45 m
Track: 2.59 m
Track width: 520 mm
Length of track on ground: 4.22 m
Ground pressure: 0.83 kg/cm^2
Max speed: (road) 50 km/h
Range:
(road) 300 km
(cross-country) 110 km
Fuel capacity: 870 litres
Fording: 1.1 m
Gradient: 60%
Vertical obstacle: 0.75 m
Trench: 2.6 m
Engine: MTU MB 837 8-cylinder diesel developing 660 hp at 2,200 rpm
Auxiliary engine: DM OM 836 4-cylinder diesel developing 35 hp at 2,800 rpm
Transmission: Schweizerische Lokomotiv und Maschinenfabrik semi-automatic with 6 forward and 6 reverse gears
Electrical system: 24 V
Batteries: 4 × 6 V, 189 Ah
Armament:
1 × 7.5 mm Mg 51 MG
4 × smoke grenade dischargers
Ammunition: 7.5 mm - 3,200 rounds

Status

Production complete. In service only with the Swiss Army.

Manufacturer

Swiss Ordnance Enterprise, Allmendstrasse 86, CH-3602 Thun, Switzerland.
Tel: +41 33 228 1111. Fax: +41 33 228 2047.

UPDATED

UNITED KINGDOM

Vickers Defence Systems Challenger armoured repair and recovery vehicle

Development

The Challenger Armoured Repair and Recovery Vehicle (CR ARRV - named Rhino by the British Army) was designed and developed by Vickers Defence Systems to support the RVUK Ministry of Defence's Challenger MBT regiments.

Each Challenger armoured regiment has five Challenger ARRVs, one with the REME Light Aid Detachment and one with each armoured squadron.

Production of the CR ARRV was undertaken at Leeds and Newcastle-upon-Tyne with 80 being built for the British Army and four for Oman.

Description

The Challenger ARRV uses the same hydrogas suspension, hull and automotive components as the Challenger 2 MBT (for details of the Challenger 2 MBT refer to *Jane's Armour and Artillery 1996-97* pages 118 to 122). With the David Brown TN54 automatic transmission, and incorporating the Dowty Digital Automotive System Control Unit (DASCU), it carries out the functions of both the previous Main Engine Control Unit and the Gearbox Controller Automatic.

The Challenger ARRV has a basic crew of three (commander, driver and radio operator) but usually carries an additional two fitters. There is seating for a further two passengers in the L-shaped crew compartment. Integral crew heating and cooling facilities are provided, along with full NBC collective protection, automatic explosion detection and suppression and a chemical toilet. The ARRV is equipped with welding and air-powered tools, plus onboard spares holding, allowing the vehicle to be used as a field repair vehicle as well as a recovery vehicle.

The Challenger ARRV can tow vehicles up to 68,000 kg weight on free tracks at speeds up to 30 km/h. Stability is maintained when retarding the load and when towing at low speeds.

The commander's cupola is a modified (no contra-rotation) AV No 32 Mark 1 (known as the No 36 Mark 1) with a demountable quartz spotlight. For day vision there is a main ×1 wide angle day periscope and a ring of six ×1 day periscopes. For night vision a single ×1 image intensifier periscope is fitted in a rotatable mounting. The driver has a main ×1 day periscope and a ×1 auxiliary day periscope; for night vision a ×1 image intensifier periscope is fitted. The commander is also provided with a single 7.62 mm machine gun on a sliding cradle mounted on a fixed rail around his hatch.

The main winch is a Rotzler Treibmatic hydraulically driven double capstan type with a single line pull of 510 kN. It is controlled from the driver's station and is operable in the closed-down position. The auxiliary winch, a Plumett TL 15, has a maximum line pull of 15 kN and it is also operated from the driver's position. It is independent of the main winch.

The front-mounted blade is a multipurpose item that can act as an earth anchor, dozer blade or stabiliser blade for the crane. It is controlled from the driver's compartment and, in its role as earth anchor, is capable of withstanding a double reeved main winch pull of 1,020 kN.

The repair crane is an Atlas AK6000M8. Hydraulically operated, it has a lift capacity of 6,500 kg and is capable of lifting a complete Challenger power pack. It is driven by an electric motor in turn powering a self-contained hydraulic power system.

Challenger ARRV in travelling configuration (R Stickland) ***1997***

Specifications

Crew: 3 + 2 (plus 2 passengers)
Weight: (combat) 62,000 kg
Ground pressure: 0.99 kg/cm²
Power-to-weight ratio: 19.35 hp/t
Length: 9.59 m
Width:
(operating) 3.62 m
(transport, without blade) 3.51 m
Height: (overall) 3.005 m
Ground clearance: 0.5 m
Max speed:
(road) 60 km/h
(cross-country) 40 km/h
Towing speed: (towing 60 t casualty)
(road) 30 km/h
(cross-country, typical) 18 km/h
Range:
(road) 450 km/h
(cross-country) 250 km/h
Fording: 1.07 m
Gradient:
(normal) 58%
(towing 65 t) 25%
Vertical obstacle: 0.85 m
Trench: 2.3 m
Engine: Perkins Engines (Shrewsbury) CV12 TCA 1200 26.1 litre V-12 direct injection 4-stroke diesel developing 1,200 bhp at 2,300 rpm
Transmission: David Brown TN54 automatic with 6 forward and 2 reverse gears
Suspension: hydrogas variable spring rate
Electrical system: 24 V
Main engine generator: 500 A continuous
Auxiliary power unit generator: 350 A continuous
Armament:
1 × 7.62 mm L37A2 MG
20 × smoke grenade dischargers (12 front, 8 rear)

Main winch
Speed: variable up to 60 m/min
Max line pull: 510 kN at 9 m/min
Line pull at 60 m/min: 76 kN
Max pay-out rate: 60 m/min
Rope diameter: 35 mm
Rope min breaking load: 1,000 kN
Rope usable length: (new) 150 m
Winch life: min of 720 full load, full length pulls
Rope life: min of 120 full load, full length pulls

Auxiliary winch
Max line pull and speed: 15 kN at 60 m/min
Rope diameter: 7 mm
Rope min breaking load: 31.4 kN
Rope usable length: 300 m

Crane
Max lift capacity: 6,500 kg at 4.9 m reach
Max reach: 5.15 m with 5,000 kg
Max hook height of lift: 6.82 m above ground level
Max time to elevate boom without max load: 36 s
Max time to elevate boom with max load: 72 s
Lifting speed of hook, max load: 0-4 m/min
Lifting speed of hook without load: 0-8 m/min
Slewing range: (boom up) 360° continuous
Slew speed: (horizontal, no load) 0 to 6°/s
Max working slope: 6°, vehicle at any orientation to slope
Hook travel: 5 m
Rope diameter: 20 mm
Rope breaking load: 357 kN

Status

Production complete. In service with Oman (4) and the UK (80). Production can resume if further orders are placed.

Manufacturer

Vickers Defence Systems, Scotswood Road, Newcastle upon Tyne NE99 1BX, UK.
Tel: +44 191 273 8888. Telex: 53104.
Fax: +44 191 273 2324.

UPDATED

Vickers Defence Systems Chieftain armoured recovery vehicle

Development

The Chieftain armoured recovery vehicle (FV4204) was developed by the Military Vehicles and Engineering Establishment and the Royal Ordnance Factory at Leeds, as the replacement for the Centurion armoured recovery vehicle (FV4006). The final requirement, issued late in 1964, called for a vehicle that could clear obstacles on the battlefield, tow an immobilised Chieftain both on roads and cross country, and be fitted with a winch capacity of 90,000 kg.

The first of two prototypes (designated R1 and R2) was built and delivered to the School of Electrical and Mechanical Engineering at Bordon late in 1971. Trials with the first prototype resulted in a complete redesign of the hydraulic system of the vehicle, undertaken by Lockheed Precision Products (now AP Precision Hydraulics). This redesign with a few more modifications, was considered satisfactory and production was authorised. Production of the Chieftain armoured recovery vehicle was undertaken by Vickers at its Elswick facility and the vehicle entered service with the British Army in 1976.

Iran ordered 71 Chieftain ARVs, the last of which was completed early in 1980. About 21 vehicles had been delivered to Iran by the time of the revolution. The remaining 50 vehicles are believed to have been purchased by Jordan.

Production total for the Chieftain ARV was 257. The Chieftain ARV is no longer used with British MBT regiments as they use the Challenger ARRV covered in the previous entry.

Description

The Chieftain ARV is based on the chassis of the Chieftain Mk 5 MBT and has three main compartments. The winch compartment is at the front of the vehicle on the right side, with the driver seated on the left. To the rear of the driver is a single-piece hatch cover, with a single periscope. The other three crew members are seated in the centre of the vehicle. Provision is made for carrying a fifth passenger. The commander has a No 17 cupola that can be traversed through a full 360°. In the forward part of the cupola is a No 62 sight periscope with magnifications of ×1 and ×10, and seven No 40 periscopes give the commander vision to the sides and rear. An image intensification sight can be fitted if required and a white/infrared spotlight is mounted externally. The 7.62 mm L37A1 machine gun can be aimed and fired from inside the cupola. There is a second hatch in the roof of the vehicle, to the rear of the commander's cupola.

The crew compartment has an NBC system and a heater. The engine and transmission at the rear of the hull are separated from the crew compartment by a fireproof bulkhead. The crew consists of commander, driver/winch operator, radio operator and a recovery mechanic.

The suspension is of the Horstmann type and consists of three bogies per side, each with two sets of roadwheels and a set of three horizontal springs. The first roadwheel station has a hydraulic shock-absorber. The drive sprocket is at the rear and the idler at the front, with three track-return rollers. The top half of the track is covered by armoured skirts, which can be removed for maintenance.

Standard equipment includes a Graviner fire warning, detection and extinguishing system, NBC system, heater and infrared detection equipment.

The main winch is of the double capstan type with electrohydraulic controls. It is provided with 122 m of 28 mm diameter cable and has a maximum speed of 13.73 m/min. The auxiliary winch is also of the double capstan type and is hydraulically operated with 295 m of 11 mm diameter cable and a maximum speed of 137 m/min. Power for both winches is taken from a PTO on the main engine.

The front-mounted dozer blade, which weighs 833 kg, is operated by two hydraulic arms. When lowered it allows the vehicle to exert a pull of up to 90,000 kg.

The vehicle is provided with a full range of recovery equipment including pulleys, cables and towbars. Some of the Chieftain ARVs for Iran have been fitted with an Atlas AK6000M crane which, in this application, can lift a maximum load of 5,803 kg at a reach of 3.62 m. Distance from the hook to the ground at a radius of 3.62 m is 4.67 m and minimum radius with boom extended is 1.445 m. Maximum lifting speed is 3.4 m/min and maximum slewing speed is 9°/s.

Variant

Armoured Repair and Recovery Vehicle

With the introduction of the Challenger 1 MBT into service with the British Army, the FV434 was unable to service the new vehicle as it cannot lift the engine pack. This led to the introduction into service of the Chieftain Armoured Repair and Recovery Vehicle (ARRV), which has a hydraulic crane boom mounted on the left side of the superstructure with the pivot mounted just behind the commander's hatch. A stand for a spare engine pack is located to the rear and over the vehicle's engine compartment.

Specifications

Vehicle with crane
Crew: 4 + 1
Weight: 55,640 kg
Power-to-weight ratio: 13.47 hp/t
Length: 8.57 m
Width:
(incl blade) 3.53 m
(tracks) 3.33 m
Height: 3.43 m

Ground clearance:
(front) 0.5 m
(rear) 0.58 m
Track: 2.718 m
Track width: 610 mm
Length of track on ground: 4.775 m
Ground pressure: 0.9 kg/cm^2
Max speed: (road) 42.2 km/h
Range:
(road) 400-500 km
(cross-country) 200-300 km
Fuel capacity: 955 litres
Fording: 1.067 m
Gradient: 70 %
Vertical obstacle: 0.902 m
Trench: 3.15 m
Engine: Leyland L60 No 4 Mark 8A, 2-stroke, compression ignition, 6-cylinder (12 opposed pistons) vertically opposed multifuel developing 750 hp at 2,250 rpm
Transmission: TN12 with 6 forward and 2 reverse gears plus emergency mechanical selector for second gear forward and low reverse
Electrical system: 28.5 V (24 V nominal) DC
Batteries: 4 × 12 V, 200 Ah
Armament:
1 × 7.62 mm L37A1 MG
2 × 6 smoke grenade dischargers (front)
2 × 4 smoke grenade dischargers (rear)
Ammunition: 7.62 mm - 1,600 rounds

Status
Production complete. In service with Iran, Iraq, Jordan and the UK.

Manufacturer
Vickers Defence Systems, Scotswood Road, Newcastle upon Tyne NE99 1BX, UK.
Tel: +44 191 273 8888. Telex: 53104.
Fax: +44 191 273 2324.

UPDATED

Chieftain armoured repair and recovery vehicle (Richard Stickland) ***1996***

Vickers Defence Systems armoured repair and recovery vehicle

Development
Early in 1977 Kenya placed an order with Vickers Defence Systems for 38 Mk 3 MBTs and three ARVs based on the same chassis. First production MBTs were delivered in 1979 and first production ARVs were delivered in 1981. Two of the ARVs have a crane. In 1978, Kenya placed an additional order for 38 MBTs plus four ARVs, two with a crane and two without. In late 1981 Nigeria placed an order for 36 MBTs, five ARVs and six bridgelayers; since then two repeat orders have been placed.

Description
The all-welded steel armour hull is divided into three main compartments: front, centre and rear. The front compartment is divided into two, with the driver's compartment on the right and the winch compartment on the left. The driver has a single-piece hatch cover which can be locked open. Mounted in the hatch is a single wide-angle periscope.

The centre compartment accommodates the radio operator, commander and mechanic and also contains the radio sets, batteries, machine gun ammunition stowage and recovery kit. The commander's cupola can be traversed manually through 360° and is provided with a single-piece hatch cover that opens to the rear, periscopes for all-round observation and an externally mounted 7.62 mm L37A1 MG, which can be aimed and fired from inside the cupola and has an elevation of +90° and a depression of −10°.

The rear compartment houses the power and transmission units. Panniers above each track contain two bag-type fuel tanks. The vehicle is designed to accommodate either the General Motors 12V-71T or the Perkins Engines (Shrewsbury) Condor V-12 800E engine. Drive is through the TN12 Mark V5 transmission, which combines the Wilson Epicyclic gear change principle with the Merritt steering system.

Suspension is the torsion bar type with each side consisting of six dual rubber-tyred roadwheels with the drive sprocket at the rear, idler at the front and three track-return rollers. All suspension stations mount a secondary torsion bar within the body of the axle and the first, second and sixth roadwheel stations have a hydraulic shock-absorber. The tracks are of manganese steel and, when new, each track comprises 96 links.

The main winch is operated by the driver and is mechanically driven through an input gearbox with hydraulically operated selector box. The engine PTO, mounted on the rear bulkhead of the winch/crew compartment, is engaged by the driver to provide a drive to the input gearbox and raise hydraulic pressure before the winch can be operated. The twin capstan winch is equipped with 122 m of 28 mm diameter cable and has a nominal capacity of 25,000 kg when being used in conjunction with an earth anchor spade fitted to the front of the vehicle. The winch, when achieving 25,000 kg direct line pull, is capable of dealing with all normal recovery operations, but where necessary the line pull can be increased to a nominal figure of 75,000 kg by multireeving the cable using recovery equipment provided with the vehicle. The hydraulically operated earth anchor enables the maximum pull to be achieved without moving the vehicle.

The Vickers ARV hull is manufactured to accept a hydraulic crane with a lifting capacity of 4,000 kg. All vehicles produced to date, other than part of the batch delivered to Kenya, have been fitted with this crane.

Optional equipment for the Vickers ARV includes: passive night vision equipment for both the commander and driver; cradle for transporting a complete power pack and TN12 transmission; auxiliary winch with a capacity of 4,060 kg and 250 m of 11 mm diameter cable; smoke grenade launchers; NBC system; and a heater.

Vickers armoured repair and recovery vehicle ***1997***

Specifications
Crew: 4
Weight: 36,800 kg
Power-to-weight ratio: 19.56 hp/t
Length:
(vehicle) 7.56 m
(vehicle with spade) 8.38 m
Width: 3.16 m
Height: (top of commander's sight) 2.89 m
Ground clearance: 0.432 m
Track: 2.52 m
Track width: 521 mm
Length of track on ground: 4.23 m
Ground pressure: 0.79 kg/cm^2
Max speed: (road) 50 km/h
Range: (road, at 32.2 km/h) 600 km
Fuel capacity: 1,000 litres
Fording: 1.1 m
Gradient: 60%
Side slope: 30%
Vertical obstacle: 0.914 m
Trench: 3 m
Engine: General Motors 12V-71T V-12 diesel developing 720 bhp at 2,500 rpm or
Perkins Engines (Shrewsbury) CV12 800E V-12 diesel developing 800 bhp at 2,300 rpm
Transmission: TN12 Mark V5 automatic
Electrical system: 24 V
Batteries: 4 × 12 V 6TN
Armament: 1 × 7.62 mm L37A1 MG

Status
Production as required. In service with Kenya (7), Nigeria (15) and Tanzania (2).

Manufacturer
Vickers Defence Systems, Scotswood Road, Newcastle upon Tyne NE99 1BX, UK.
Tel: +44 191 273 8888. Telex: 53104.
Fax: +44 191 273 2324.

UPDATED

GKN Warrior recovery and repair vehicle

Development

The Warrior recovery and repair vehicle is a member of the Warrior mechanised combat vehicle family and uses a basically similar hull and superstructure, and a common power train and running gear. Two variants were ordered by the British Army, including 110 FV512 Mechanised Combat Repair Vehicles (MCRVs) with a crane but no winch and 67 FV513 Mechanised Recovery Vehicles (Repair) (MRV(R)s), with both a winch and crane.

For full details of the Warrior family of vehicles refer to *Jane's Armour and Artillery 1996-97* pages 343 to 346. The British Army has taken delivery of 789 Warrior vehicles while Kuwait is now taking delivery of 254, including variants.

Description

The MCRV has a crew of five comprising commander, driver, gunner and two fitters. The main change to the superstructure is the addition of a hydraulically operated crane on the left of the superstructure and with the pivot to the rear. This has a capacity of 6,500 kg and a maximum reach of 4.52 m. The crane has a full 360° traverse and can lift and replace a Challenger MBT or Warrior power pack or similar assemblies from other MBTs, AIFVs or self-propelled artillery. A hydraulically actuated stabiliser leg at the left rear of the vehicle is lowered when the crane is in use.

On the MRV(R) a twin capstan winch is carried internally at the rear and has a 20,000 kg capacity, 38,000 kg when double reeved. There is sufficient cable to enable a single line pull of 100 m. The winch can tow a Warrior Section Vehicle over rough terrain up a gradient of 40 per cent. For the recovery role an earth anchor is lowered from the hull rear. A pilot winch is used to raise and lower the earth anchor and is provided with 200 m of cable to provide a capacity of 1,250 kg maximum pull.

Internally the vehicle is provided with a fully equipped workshop. The vehicle is fitted with a one-man turret armed with a single 7.62 mm L94A1 Chain Gun. There is also a collective NBC unit, along with an air conditioning system.

The Warrior MCRV can tow the GKN High Mobility Trailer (for full details see entry in the *Trailers* section). This trailer was specially designed to operate with the Warrior Mechanised Combat Repair Vehicle and has a gross weight of 10,500 kg. It can carry a Challenger MBT power pack and employs a four-damper jacking system to maintain a safe level platform when not connected for towing.

Specifications

Crew: 5
Weight: (combat, MRV(R)) 30,000 kg
Power-to-weight ratio: 18.33 hp/t
Length:
(overall) 6.675 m
(over tracks) 5.39 m
Width: 3.13 m
Height:
(over cupola) 2.302 m
(roof) 1.935 m
Ground clearance: 0.49 m
Track: 2.54 m
Track width: 460 mm
Length of track on ground: 3.817 m
Max speed: (road) 71 km/h
Range: (road) 500 km
Fuel capacity: 770 litres
Fording: 1.3 m
Gradient: 60%
Side slope: 40%
Vertical obstacle: 0.75 m
Trench: 2.5 m
Angle of approach/departure: 29°/34.5°
Engine: Perkins Engines (Shrewsbury) CV8 TCA V-8 diesel developing 550 hp at 2,300 rpm
Transmission: Perkins Engines (Shrewsbury) X-300-4B fully automatic with torque converter and lock-up clutch, 4 forward and 2 reverse gears
Turning radius: 6 m
Electrical system: 24 V
Armament:
1 × 7.62 mm L94A1 Chain Gun
up to 8 banks of smoke grenade dischargers
Ammunition: 7.62 mm - 4,000 rounds

Armour
hull: aluminium
turret: steel

Status

In production. In service with the British Army and Kuwait.

Manufacturer

GKN Defence, PO Box 106, Hadley Castle Works, Telford, Shropshire TF1 4QW, UK.
Tel: +44 1952 244321. Telex: 35248 sanhad g.
Fax: +44 1952 243910.

UPDATED

Rear view of Warrior Mechanised Recovery Vehicle (Repair) (MRV(R)) with earth anchor raised (Richard Stickland) **1996**

Warrior Mechanised Combat Repair Vehicle (MCRV) (Richard Stickland) **1997**

Alvis Samson armoured recovery vehicle

Development

The Samson (FV106) is a member of the Scorpion range of light tracked vehicles developed by Alvis Vehicles Limited and the Military Vehicles and Engineering Establishment. The first prototype was completed in the early 1970s but, as a result of trials, some redesign work had to be carried out and the final design entered production in 1977-78.

Description

The hull of the Samson is similar to that of the Spartan APC and is of all-welded aluminium construction. The driver is seated at the front of the hull on the left side and is provided with a single-piece hatch cover which opens to the front and a wide-angle periscope for driving in the closed-down position. This can be replaced by a passive night periscope. The engine, mounted to the right of the driver, is common to the Scorpion CVR(T) range.

To the rear of the driver, in the centre of the roof, is a No 27 cupola, which has a single-piece hatch cover that opens to the left and can be traversed manually through 360°. Mounted in the forward part of the cupola is a periscope with a magnification of ×1 which can be replaced by a passive night periscope. There are

Alvis Samson armoured recovery vehicle **1997**

another five periscopes for all-round vision. Mounted externally on the right side of the cupola is a 7.62 mm L7A2 GPMG.

The NBC unit, if carried, is on the right-hand sponson plate, with the radio in a corresponding position on the left side. Alternatively, an air cooling or through-flow air circulation system can be installed.

At the rear of the hull are the winch compartment and a small door with a built-in vision block. Over the winch compartment is a roof hatch.

The suspension is of the torsion bar type and consists of five roadwheels, with the drive sprocket at the front and the idler at the rear. The first and fifth roadwheel stations are provided with a hydraulic shock-absorber.

The Samson can ford to a depth of 1.067 m without preparation, but with the flotation screen erected it is fully amphibious, being propelled in the water by its tracks at a speed of 6.44 km/h. A propeller kit increases its water speed to 9.6 km/h. In British Army service the flotation screen has been removed and the vehicle no longer has an amphibious capability.

The recovery winch is mounted in the rear of the hull and is driven from a PTO on the main engine. This winch is provided with 229 m of rope and has a variable speed of up to 122 m/min. Maximum pull, with a 4:1 snatch block, is 12,000 kg, this enables the Samson to recover vehicles such as the FV432 APC. The cable leads out over the top of the hull at the rear. On prototypes this lead-out was through an opening in the lower part of the hull. Two spades at the rear of the hull are released manually. An additional strengthening piece is then added and the Samson is reversed on to the spades. An A-frame can be mounted at the rear of the vehicle to enable the Samson to change light components in the field. A full range of tools, towbars and blocks is carried.

Stormer ARV

For the export market, Alvis Vehicles have developed an ARV version of the Stormer APC (*Jane's Armour and Artillery 1996-97* pages 349 to 351). The first customer for this vehicle is Indonesia who took delivery of a small quantity of vehicles in 1996.

Specifications

Crew: 3
Weight: (loaded) 8,738 kg
Power-to-weight ratio: 21.74 hp/t
Length: 4.788 m
(incl vice and bench) 5.004 m
Width: 2.43 m
Height:
(top of hull) 1.718 m
(incl MG) 2.254 m
(incl A-frame) 2.83 m
Ground clearance: 0.356 m
Track: 1.7 m
Track width: 432 mm
Length of track on ground: 2.74 m
Ground pressure: 0.358 kg/cm^2
Max speed:
(road) 72.5 km/h
(water) 6.44 km/h
Range: 483 km
Fuel capacity: 404.51 litres
Fording: 1.067 m
Freeboard with screen raised:
(front) 0.965 m
(rear) 0.815 m
Gradient: 60%
Vertical obstacle: 0.5 m
Trench: 2.057 m
Engine: Jaguar J60 No 1 Mark 100B 4.2 litres 6-cylinder in-line petrol developing 190 hp at 4,750 rpm or
Perkins T6-3544 5.8 litres 6-cylinder in-line turbocharged diesel developing 200 hp at 2,600 rpm
Transmission: TN15 cross-drive, 7 speeds in each direction
Electrical system: 28 V
Batteries: 4 × 6TN, 100 Ah; generator output is 140 A at 28 V
Armament:
1 × 7.62 mm L7A2 GPMG
4 × smoke grenade dischargers on each side of hull front
Ammunition: 7.62 mm - 2,000 rounds

Status

Production complete. In service with Belgium, Brunei (2), Oman, UK and Venezuela.

Manufacturer

Alvis Vehicles Limited, The Triangle, Walsgrave, Coventry CV2 2SP, UK.
Tel: +44 1203 535455. Fax: +44 1203 539287.

UPDATED

GKN Saxon armoured recovery vehicle

Description

An armoured recovery vehicle version of the Saxon wheeled armoured personnel carrier, was produced by GKN Defence to support the Saxon infantry battalions allotted to what was then BAOR. The vehicle, known as Saxon Recovery, is basically a converted Saxon APC, with a Williams Fairey Engineering Ltd 5 tonne capstan winch mounted on the left-hand side wall of the hull. The winch can operate to either the front or rear of the vehicle using a swivelling fairlead and fixed roller boxes. The cable can be used over a vertical arc of +40° and −10° front and rear and 15° to either side. The cable is stowed on a reel next to the winch. The winch, which can be controlled by a remote-control unit, has a slow speed of 9.1 m/min and a fast speed of 30.84 m/min. By using a block and tackle, vehicles of up to 16,000 kg weight can be recovered.

The Saxon ARV has a crew of four and in addition to the usual recovery tools and equipment carried, a lightproof tent can be erected over the rear of the vehicle to act as a temporary repair workshop.

For details of the Saxon APC refer to entry in *Jane's Armour and Artillery 1996-97* pages 497 to 499.

Saxon armoured recovery vehicle showing winch mounted on the side (Richard Stickland) ***1996***

Although production has been completed it can be restarted again of additional orders are placed.

Status

Production complete. In service with the British Army (38).

Manufacturer

GKN Defence, PO Box 106, Hadley Castle Works, Telford, Shropshire TF1 4QW, UK.
Tel: +44 1952 244321. Telex: 35248 sanhad g.
Fax: +44 1952 243910.

UPDATED

Future Beach Recovery Vehicle (FBRV)

Development/Description

The Royal Marines currently operate the Centurion Beach Armoured Recovery Vehicle (BARV) during amphibious operations.

This vehicle was designed some 30 years ago and is the last version of the Centurion to remain in operational service with the UK armed forces.

The UK MoD has a requirement to either upgrade these existing BARV vehicles to extend their operational lives into the 21st century, or procure a brand new vehicle under the name Future Beach Recovery Vehicle (FBRV).

The FBRV will be carried by the new Landing Platform Dock (LPD) and operate with Landing Craft Utility (LCU) and Landing Craft Vehicles and Personnel (LCVP).

The main roles of the FBRV will be similar to that of the current BARV, including unbeaching LCVP and the recovery of tracked and wheeled vehicles on the beachead.

Key characteristics of the FBRV include a maximum battle weight of 45 tonnes, the capability of developing 23 tonnes of tractive effort, be able to operate in water to a depth of 2.9 m without preparation and provide protection from small arms fire and shell splinters.

Four FBRVs are required, with an in-service date of 2001. It is expected that the UK MoD will issue tenders for the Feasibility Study (FS) stage in the middle of 1997.

In addition to an upgrade of the existing BARV (which would include the replacement of the current petrol engine with a more fuel efficient diesel), manufacturers could also offer other chassis (for example Chieftain or Challenger) or even a commercial chassis modified for this role.

Potential UK bidders include Alvis Vehicles, GKN Defence and Vickers Defence Systems the design authorities for the Centurion, Chieftain and Challenger MBTs.

The Royal Netherlands Marine Corps also has a requirement for one or two BARV-type vehicles based on a Leopard 1 MBT chassis and this work will be carried out by the Dutch Army depot at Leusden. There is no known intention to combine the Dutch and UK requirements.

Status

Future Royal Marine requirement with an in-service date of 2001. Feasibility study contracts are expected to be awarded in 1997.

Manufacturer

Not selected.

NEW ENTRY

UNITED STATES OF AMERICA

General Dynamics Abrams Recovery Vehicle (ARV)

Development

When the US Army decided to investigate its requirements for recovery support for the M1 Abrams MBT during the early 1980s, one of the options available was the development of an armoured recovery variant of the M1 known initially as the RV90.

The prototype hull was produced on the M1A1 Abrams MBT production line at the Lima Army Tank Plant, Ohio, in only 23 days. The prototype, known as the Abrams Recovery Vehicle or Abrams RV (ARV), was rolled out during February 1988. Contractor trials took place until mid-March 1988, prior to delivery to Aberdeen Proving Ground in April 1988. The US Army eventually decided, on the basis of price, to procure the M88 Improved Recovery Vehicle recovery vehicle (see previous entry). The ARV is available on the international market.

Description

The Abrams RV is automotively identical to the M1 Abrams MBT (over 85 per cent of the Abrams RV chassis is common to the M1A1 Abrams MBT chassis) but uses a new hull armoured superstructure housing a crew of three, with space provided for a further four stranded crewmen. The crew consists of the commander, driver and a fitter/rigger with the driver seated centrally in the upper hull front under a hatch; his field of vision is 167°. The commander is provided with a rotating cupola mounting all-round vision devices and a pintle-mounted 12.7 mm Browning M2 machine gun.

The superstructure and hull provide at least 30 mm of armour protection for the crew throughout a 350° arc. The recovery equipment, crane and winch are provided by MaK GmbH and have a high degree (over 95 per cent) of commonality with the Büffel, Leclerc and Korean K-1 armoured recovery vehicles. Located on the left of the superstructure is a hydraulic jib crane with a fixed boom length and a maximum hook height at 70° elevation of 6.9 m. The crane has a capacity of 35 tonnes and a traverse of 270° — this allows the vehicle to change its own power pack if necessary. Mounted at the front of the hull is a hydraulically operated dozer/stabilising blade. A 45 tonne double capstan winch, the capacity of which can be increased to 90 tonnes by a snatch block carried on the vehicle, is mounted behind the blade. The winch is equipped with 117 m of usable cable. An auxiliary winch has a capacity of 2.2 tonnes and is equipped with 200 m of usable cable. Dual hydraulic circuits are provided throughout.

Other features include an overpressure NBC system, an automatic Halon fire detection and suppression system (plus two portable fire extinguishers), a 300 hp auxiliary power unit and hydraulic lockout for the suspension. Equipment carried includes hydraulic tools, a hydraulic fuel/defuel pump and welding and cutting equipment.

The Abrams RV can carry a complete M1/M1A1/M1A2 Abrams power pack and is fitted with Built-In Test Equipment (BITE), a thermal viewer for the driver, a land navigation system and an intervehicular information system.

Abrams Recovery Vehicle (ARV) prototype with dozer/stabilising blade lowered

Specifications

Provisional
Crew: 3 + 4
Weight: 68,500 kg
Length: (overall, transport) 9.205 m
Width: 3.66 m
Height: (max, to top of MG) 2.987 m
Ground clearance: 0.406 m
Track width: 635 mm
Max speed: (governed) 69.2 km/h
Range: 418 km
Fording: (without kit) 1.22 m
Gradient: (with towed load) 60%
Vertical obstacle: 1.07 m
Trench: 2.74 m
Engine: Textron Lycoming AGT-1500 gas turbine developing 1,500 hp at 3,000 rpm
Transmission: Allison X-1100-3B automatic with 4 forward and 2 reverse gears
Electrical system: 24 V
Armament:
1 × 12.7 mm M2 HB MG
2 × smoke grenade dischargers
Ammunition:
12.7 mm - 1,500 rounds
smoke grenades - 24

Status

Development complete. Ready for production on receipt of orders.

Manufacturer

General Dynamics, Land Systems, PO Box 2074, Warren, Michigan 48090, USA.
Tel: +1 810 825 4000.

UPDATED

United Defense LP, M88, M88A1, M88A1E1 and M88 IRV armoured recovery vehicles

Development

The standard ARV in the US Army immediately after the Second World War was the M74, designed by Bowen-McLaughlin-York (later BMY and now part of United Defense), which produced over 1,000 vehicles. BMY was awarded a contract to build three prototype vehicles under the designation of T88, to use as many components as possible of the M60 tank. These prototypes were followed by 10 preproduction vehicles for troop trials. The production contract was awarded to BMY in 1960 with the first production M88s completed in February 1961. Final vehicles were completed in 1964, by which time 1,075 had been built. The M88 was powered by a Continental (now General Dynamics Land Systems) AVSI-1790-6A, 12-cylinder, air-cooled, supercharged, fuel injection petrol engine.

In April 1972, BMY received a contract from the US Army to design significant improvements for the M88. The basis for these improvements was to change the engine from petrol to diesel fuel, corresponding with the US Army's move towards converting all of its armoured vehicles to diesel engines. In 1973 an M88 was fitted with a variant of the engine fitted to the M60 MBT. The Teledyne Continental Motors (now General Dynamics Land Systems) AVDS-1790-2DR 12-cylinder diesel engine produced 750 bhp at 2,400 rpm and had a power take-off to facilitate operation of the hydraulic system. Trials demonstrated an increase in operating range from 360 to 450 km. This vehicle also had a modified transmission, a diesel-fired personnel heater and auxiliary power unit, and stowage space for a small quantity of LAW anti-tank weapons. The hydraulic system was redesigned to allow the auxiliary power unit to operate the main winch cable as well as stow the boom and spade to prepare the vehicle for recovery, should the hydraulic system fail.

The converted vehicle was designated the M88E1 and BMY built five prototypes for accelerated evaluation. The M88A1 was type classified in March 1975. Due to an increased need for medium recovery vehicles and a desire to improve those already in the field, the US Army contracted in 1975 to reopen the M88 production line at the BMY plant in York, Pennsylvania.

Most of the original M88s (878 of the 1,075 produced) were returned to BMY's production facility for overhaul and conversion. Beginning in 1977 this overhaul/conversion process ran in parallel with new production until early 1982. Production of new M88A1s continued, with a total of 2,167 completed when production ended in early 1989.

When production ended at that stage, and taking into account the overhaul/conversions, the total of M88s was 3,244. Of these, 199 were basic M88s, most in service with overseas armies. Although most of the M88A1s are in service with the US armed forces, some 496 were then in service with NATO and allied countries.

In November 1990 it was announced that orders placed by a Far Eastern nation for the M88A1 would involve the reopening of the production line.

During 1981 an Independent Research and Development (IR&D) programme was initiated to investigate improvements necessary for the recovery support of the M1-series Abrams MBT. An initial result was the development of the M88AX automotive demonstrator (see under Variants in this entry), which led to the acceptance by the US Army of a revised version of the M88A1 known as the M88A1E1 (see also under Variants) following testing completed during 1988. The M88A1E1 was expected to be type classified as the M88A2. The US Army reprogrammed, with Congress approval, $8 million from FY86 Research and Development funds; further Research and Development funding, bringing the total to $25 million, was scheduled for FY87 to FY89. A FY88 request for $24.1 million advance procurement funding was made to cover the procurement of long-lead items such as

Taiwanese Army M88A1 ARV on parade **1996**

M88A1 ARV from the rear in travelling configuration (US Army) **1997**

engines, winches and hull components. The first production funding of $79 million was scheduled for FY89, when 80 vehicles were projected. There was no FY90 funding, leading to further work on the M88A1E1 being terminated from April 1989 until early 1991.

It was expected that between 500 and 1,300 new M88A2s would be produced by BMY with no conversions from existing vehicles being involved. The M88A2 was to have been used to support US Army M60A3, M1 and M1A1 units. Existing US Army M88A1 assets were to have been transferred to supporting the Bradley Fighting System and M109 series self-propelled 155 mm howitzers.

In October 1991 it was announced that the US Army Tank Automotive Command (TACOM) had awarded BMY Combat Systems Division a contract worth $12.8 million to complete the development and testing of the M88A1E1 Improved Recovery Vehicle (IRV). The five existing M88A1E1 prototypes were reconditioned and improved in a number of key areas.

Testing restarted during 1992 and continued into 1993. A contract, eventually worth $30.6 million, was placed in October 1994 and involved 13 vehicles. The award was for Low Rate Initial Production and was completed by January 1996. The US Army plans to reclaim hulls from current M88A1s and build up as new to the IRV.

Description

The hull of the M88 and M88A1 is of cast armour and rolled armour welded together, with the crew compartment at the front and the engine and transmission at the rear. The driver and mechanic are seated at the front of the hull, each with a single-piece hatch cover and periscopes. The commander was originally provided with a cupola with an internally mounted 12.7 mm machine gun but this was subsequently replaced by a simple cupola with a pintle-mounted 12.7 mm machine gun. There is an entry door in each side of the hull.

The suspension is similar to that used on the M60 MBT and consists of six roadwheels, with the drive sprocket at the rear and the idler at the front. There are three track-return rollers. Hydraulic shock-absorbers are provided for the first, second and sixth roadwheel stations.

At the front of the vehicle is a hydraulically operated blade, which stabilises the vehicle when the winch is being used and assists in bulldozing operations. An A-type boom pivoted at the front of the hull can lift a vehicle weighing up to 5,443 kg without using the blade or suspension lockout at the front, or 18,160 kg using lockout without using the blade. The vehicle can lift 22,700 kg when using the blade. The two winches are in the lower part of the hull.

The M88 is not provided with an NBC system (a kit exists for an M88A1 system incorporating a ventilated facepiece) and has no amphibious capability. A fording kit is available for use in water up to 2.6 m deep. Infrared driving lights are provided and the driver can replace one of his day periscopes with an infrared periscope. Standard equipment includes tools, towbars and an auxiliary fuel pump, allowing the vehicle to transfer fuel to other AFVs at 95 litres/min.

The M88A1 is fitted with an Auxiliary Power Unit (APU) powered by an Onan 10.8 hp (8.1 kW) diesel, two-cylinder, four-cycle engine. The APU powers an auxiliary hydraulic system which can be used to power the boom, spade and main winch and hoist cables if the main hydraulic system is inoperative. It also provides power for the refuelling and fuel transfer pump, which allows the M88A1 to act as a mobile filling station. The hydraulic system also provides power to operate a 19 mm drive hydraulic impact wrench, used for track maintenance and other maintenance and recovery tasks.

Variants

Next generation of armoured recovery vehicles M88AX/M88A1E1/M88 IRV/M88A2

In order to determine the growth potential of the basic M88A1 components, BMY developed an M88AX automotive demonstrator in 1984-85. The demonstrator was an up-powered and up-weighted vehicle powered by a 960 bhp Teledyne Continental Motors (now General Dynamics Land Systems) AVDS-1790-8DR diesel engine mated to a slightly modified transmission, the XT-1410-5X. The final drive ratio was changed from 4.63:1 to 4.00:1. During US Army tests held at Aberdeen Proving Ground, Maryland, the 65 tonne M88AX demonstrator recorded a top speed on level ground of 56.8 km/h and towed a 65 tonne load at 46 km/h over the same course. It was also able to tow a 65 tonne load up a 30 per cent slope at 3.38 km/h and to tow a 65 tonne load over the 6.11 km Churchville 'B' unpaved/hilly course at an average speed 51 per cent greater than an M88A1. The primary purpose of the M88AX was to validate engineering predictions in connection with the US Army search for an armoured vehicle to recover the M1A1/M1A2 Abrams MBT.

In late 1985 the US Army decided to establish a Product Improvement Programme (PIP) that capitalised on BMY's initial efforts. This programme was overtaken by the introduction of the M88A1E1.

The US Army awarded BMY a research and development contract in January 1987. The fixed-price contract covered the design, construction and testing of five prototype vehicles plus a separate hull for ballistic testing. The prototype vehicles were designated the M88A1E1.

The M88A1E1 was the follow-on from the M88AX programme and featured a 1,050 hp AVDS-1790-8CR diesel engine coupled to an Allison XT-1410-5A transmission. Improved brakes were fitted along with a brake booster and the sprocket mounting was improved. One of the main changes introduced was a new main winch with a continuous line pull of 63,504 kg, provided with a nominal 100 m of cable. A 2,722 kg lead winch was incorporated to allow a single crewman to deploy the main winch cable. The hoist had a capacity (spade down and with a four-part line) of 31,752 kg and was used in conjunction with an A-frame, lengthened by 0.84 m to increase the lift height. The lengthened A-frame involved the introduction of increased diameter staylines and an increase in length to 8.65 m with the boom stowed.

Combat loaded weight was 63,050 kg, with most of the weight increase introduced by the addition of overlay armour to the existing M88A1 hull armour, providing ballistic protection against up to 30 mm calibre direct fire and 152 mm indirect fire fragmentation. Ballistic skirts were also provided. The suspension torsion bars, shock-absorbers and snubbers were upgraded to cater for the weight increase.

Maximum rated speed was 40.2 km/h and the fuel capacity was increased to 1,628 litres. The draw bar pull was increased to 48,535 kg compared to 40,824 kg for the M88A1.

The M88A1E1 was selected again by the US Army after competitive testing in 1988. Developmental testing was nearly complete and over 11,260 km of durability testing had been logged when further work was cancelled by budget cutbacks. Five of the six prototypes produced were used to support activities at Aberdeen Proving Ground.

In January 1991, at the request of the US Army Tank Automotive Command (TACOM), BMY submitted a Phase I M88 IRV (Improved Recovery Vehicle) proposal. This would involve the tear-down and inspection of the five M88A1E1 prototypes and some limited engineering to further improve selected areas of the design. Phase II would complete the overhaul and upgrade of the prototypes, conduct the required testing and complete a Technical Data Package.

In October 1991 it was announced that the US Army Tank Automotive Command (TACOM) had awarded BMY Combat Systems Division a contract worth $12.8 million to complete the development and testing of the M88A1E1 IRV. The five existing M88A1E1 prototypes were reconditioned and improved in a number of key areas. Approximately 18,000 km of government testing was successfully completed by September 1993, demonstrating that the M88 IRV fully met all M1A2 Abrams MBT support requirements.

Subsequent contract modifications totalling $6.3 million were awarded to United Defense LP, Combat Systems Division (formerly BMY) to define interface drawings with Anniston Army Depot and complete Integrated Logistics Support activities (provisioning, training and technical manuals) that had been deferred due to lack of funding. The US Army plans to reclaim hulls from current M88A1s and build up as new to the IRV.

At a Special In-process Review conducted in January 1994 the US Army decided to move the programme into limited production. A contract award was made in September 1994 with first vehicle deliveries in early 1996. Type classification of the M88A1E1 to the M88A2 was scheduled for late 1996.

Subsequent modifications to this contract were made in 1995 and 1996 which brought the total production quantity of M88 IRVs up from 13 to 63 vehicles of which 14 of these are for the Kuwait land forces.

Israeli M88

Israeli M88 ARVs are fitted with a Blazer reactive armour package and additional anti-personnel weapons for self and local defence.

Specifications (M88A1)

Crew: 3 or 4
Weight (combat) 50,848 kg
Length: (dozer blade raised) 8.267 m
Width: 3.429 m
Height:
(with MG) 3.225 m
(top of commander's hatch) 2.97 m
Ground clearance: 0.776 kg
Track: 2.717 m
Track width: 711 mm
Length of track on ground: 4.61 m
Ground pressure: 0.776 kg/cm^2
Max speed: (road) 42 km/h
Range: 483 km

Fuel capacity: 1,514 litres
Fording: 1.42 m
Gradient: 60%
Vertical obstacle: 1.07 m
Trench: 2.61 m
Engine: AVDS-1790-2DR, 12-cylinder, air-cooled, supercharged fuel injection diesel developing 750 bhp at 2,400 rpm
Transmission: XT-1410-4 cross-drive
Electrical system: 24 V
Armament:
1 × 12.7 mm M2 HB MG
2 × 6 smoke grenade dischargers
Ammunition: 12.7 mm - 1,500 rounds
Armour: (estimate) 12.7-50 mm

Status

In production. The M88A1 is in service with Bahrain (4), Egypt (221), Greece (66), Israel (25), Jordan (32), South Korea (38), Kuwait (10), Morocco (18), Norway (3), Oman (2), Pakistan (52), Portugal (6), Saudi Arabia (139), Spain (1), Sudan (2), Taiwan (33), Thailand (21), Tunisia (6) and the USA (US Army 2,470, Marine Corps 81). The M88 IRV is in production for Kuwait (14) and the US Army (49).

Manufacturer

United Defense LP, Ground Systems Division, PO Box 1512, York, Pennsylvania 17405, USA.
Tel: +1 717 225 8000. Fax: +1 717 225 8132.

UPDATED

M88 IRV with A-frame erected at front of vehicle and showing additional armour
1997

United Defense LP M578 light armoured recovery vehicle

Development

In 1956 the Pacific Car and Foundry Company of Renton, Washington, was awarded a contract to design a new range of self-propelled guns which would all use the same basic chassis. One of the main requirements was to reduce the overall weight of the chassis so that it could be carried by transport aircraft then in service. The following year the programme was expanded to include three armoured recovery vehicles, designated the T119, T120 and T121. The T119 and T121 both had their cranes in an unarmoured mounting, but were not developed beyond the prototype stage. Further development of the T120 (which had a petrol engine) resulted in the T120E1 with a diesel engine, which was accepted for service as the M578. The contract was awarded to the FMC Corporation, which produced the first production vehicle late in 1962. Production was originally completed in the late 1960s.

In 1971 Bowen-McLaughlin-York (now BMY) won a competitive contract to resume production of the M578 and began production the same year. BMY produced a total of 1,018 M578s by the end of 1983 and a total of 1,844 on completion.

Following a Special Mission Adaptation (SMA) a selected number of M578s are used with Combat Electronic Warfare Intelligence (CEWI) battalions. For details see under Variant in this entry.

M578 light armoured recovery vehicle ***1997***

Description

The hull of the M578 is similar to that of the 175 mm M107 and 203 mm/8 in M110 self-propelled guns. The driver is seated at the front of the hull on the left side and is provided with a single-piece hatch cover and three M17 periscopes for driving when closed down. The engine is to his right and the transmission is at the front of the hull.

At the rear of the hull are the turret and crane which can be traversed through 360°. The turret has a door in each side and double doors in the rear. Both the commander and operator are provided with a single-piece hatch cover that opens to the rear and six M17 periscopes for observation.

The torsion bar suspension consists of five roadwheels with the drive sprocket at the front, the fifth roadwheel acting as the idler. Four of the roadwheels have a hydraulic bump-stop and during recovery operations the suspension can be locked to provide a more stable platform.

Armament consists of a single 12.7 mm Browning M2 HB machine gun mounted at the commander's position. The M578 is provided with infrared driving lights and has an NBC filter kit option for the driver and cab crew. The vehicle does not have an amphibious capability.

Equipment carried includes tools, towbars, hydraulic impact wrench, acetylene welding and cutting equipment. A spade is mounted at the rear of the hull. The vehicle has two winches: a tow winch with a maximum capacity of 27,000 kg on a bare drum and a hoisting winch with a maximum capacity of 6,750 kg.

Variant

M578 SMA

Following a Special Mission Adaptation (SMA) a selected number of M578s, unofficially known as the M578 SMA, are used with US Army Combat Electronic Warfare Intelligence (CEWI) battalions. The SMA involves a boom extension of approximately 1.22 m and a lift capacity of 6,800 kg, to enable an M578 to lift tactical shelters containing electronic warfare and intelligence equipment from broken down carrier vehicles and place them on new prime movers. It is believed that the M578 SMA programme is limited to vehicles assigned to US Army CEWI battalions.

Specifications

Crew: 3
Weight: 24,300 kg
(air transport) 20,443 kg
Length:
(overall) 6.426 m
(hull) 5.588 m
Width: 3.149 m
Height:
(top of cupola) 2.921 m
(incl MG) 3.416 m
Ground clearance: 0.44 m
Track: 2.692 m
Track width: 457 mm
Length of track on ground: 3.758 m
Ground pressure: 0.7 kg/cm^2
Max speed: (road) 54.71 km/h
Range: 725 km
Fuel capacity: 1,135.5 litres
Fording: 1.066 m
Gradient: 60%
Vertical obstacle: 1.016 m
Trench: 2.362 m
Engine: Detroit Diesel 8V-71T turbocharged V-8 liquid-cooled diesel developing 425 bhp at 2,300 rpm
Transmission: Allison XTG-411-2A cross-drive with 4 forward and 2 reverse gears (automatic)
Electrical system: 24 V with 300 A generator
Batteries: 4 × 6 TN
Armament:
1 × 12.7 mm M2 HB MG and 500 rounds of ammunition
Dutch vehicles have smoke grenade dischargers installed
Armour: steel

Status

Production complete. In service with Bolivia, Brazil, Canada, Denmark, Egypt, Iran, Jordan, Morocco, Netherlands, Norway, Philippines, Saudi Arabia (88), Spain, UK and USA.

Manufacturer

United Defense LP, Ground Systems Division, PO Box 1512, York, Pennsylvania 17405-1512, USA.
Tel: +1 717 225 8000. Fax: +1 717 225 8132.

UPDATED

United Defense LP ARVs and repair vehicles based on M113 APC chassis

Development

The M113 is the most widely used armoured personnel carrier in the world and in addition to being produced in the USA was also built in Italy by OTO Melara (now OTOBreda). The first M113s were completed in 1960, followed by the M113A1 which has a diesel engine. In the late 1960s a Fitter's Vehicle was developed, fitted with a hydraulic crane and the ability to change AFV components.

United Defense LP (then FMC) developed a recovery vehicle which became known as the XM806 (gasoline), followed by the XM806E1 (diesel), 24 of which were produced for export. As the US Army did not purchase the XM806 type recovery vehicle, FMC called a later version the M113A2 Recovery Vehicle. It has an improved suspension and cooling system and a new dual-air personnel heater compatible with NBC protective systems: it is in service with several countries.

United Defense LP (then FMC) developed a further vehicle combining the features of the Recovery Vehicle with the Fitter's Vehicle. It is called the M113A2 Maintenance Recovery Vehicle (MRV) and uses the same crane and winch as the Fitter's Vehicle. The first prototype was ready early in 1982; since then it has been produced in Belgium and the USA.

The MRV can swim without the need for a water barrier and it has a buoyant trim vane. The crane is hydraulically operated and has its own winch. The Royal Netherlands Army uses the AIFV recovery vehicle which is similar to the M113A2 Maintenance Recovery Vehicle but with torsion bar and tube suspension and a 265 hp turbocharged engine. The AIFV weighs approximately 113 kg more than the M113A2 Recovery Vehicle.

Royal Netherlands Army AIFV recovery vehicle ***1996***

Description

M113A2 Recovery Vehicle

The hull of the M113A1/A2 is made of all-welded aluminium. The driver is seated at the front of the hull on the left side and is provided with a single-piece hatch cover that opens to the rear. For observation purposes he has four M17 periscopes and an M19 infrared periscope in his roof hatch for night driving. The engine is mounted to the right of the driver. The commander's cupola has five periscopes and a single-piece hatch cover that opens to the rear. A 12.7 mm Browning M2 HB machine gun is mounted on the commander's cupola with an elevation of +53° and a depression of −21°. To the immediate rear of the commander is a rectangular hatch which opens to the rear. Normal means of entry and exit are by a power-operated ramp in the rear of the hull.

The torsion bar suspension consists of five roadwheels with the drive sprocket at the front and the idler at the rear. There are no track-return rollers.

The vehicle is fully amphibious, being propelled in water by its tracks. Before entering the water the trim vane is extended at the front of the hull and the bilge pumps are switched on. The basic vehicle does not have an NBC system, but this and a variety of other kits are available as optional extras.

The M113A2 recovery vehicle has a P30 (modified) hydraulic winch which is provided with 91.4 m of 16 mm diameter cable, with a capacity on a full drum of 5,103 kg and on a bare drum of 9,070 kg. A rotating fairlead that guides the cable is at the rear of the vehicle. A cable tensioner, built into the fairlead, allows free cable to be wound tightly. A level winder keeps the cable properly coiled on the winch drum and heavy shrouds over the winch assembly and a swing-up guard set into the ramp opening protect the operator during winching operations.

A single spade is mounted on each side of the hull at the rear and an additional spade unit can be mounted between the two outer ones for recovery in soft soil.

A manually operated hydraulic crane is mounted on the left side of the vehicle with an extensible arm with two positions for flexibility of use. This crane has a maximum lifting capacity of 1,361 kg at a reach of 2.99 m.

Recovery vehicles produced in Belgium incorporate the M113A1 power train and modified (heavy-duty) torsion bars. These vehicles are unique to the Belgian Army.

M113A3 MRV

United Defense LP is currently in production of an M113A3 configured MRV. This latest version is powered by the Detroit Diesel 6V-53T turbocharged RISE unit developing 275 hp.

Specifications

Crew: 3
Weight:
(air transport) 11,991 kg
(combat loaded) 12,896 kg
Length: 5.34 m
Width: 2.69 m
Height: 3.15 m
Track: 2.159 m
Track width: 381 mm
Length of track on ground: 2.67 m
Ground pressure: 0.634 kg/cm^2
Max road speed: 66 km/h
Range: 483 km
Fuel capacity: 363 litres
Fording: amphibious
Gradient: 60%
Side slope: 40%
Vertical obstacle: 0.609 m
Trench: 1.68 m
Engine: Detroit Diesel 6V-53T developing 275 hp
Transmission: Allison X-200-4A hydrokinetic
Electrical system: 24 V
Batteries: 4 × 12 V
Armament: 1 × 12.7 mm M2 HB machine gun
Ammunition: 2,000 × 12.7 mm
Armour: 12 - 38 mm

Fitter's Vehicle

This vehicle has an almost identical hull to the M113A2 but does not have the small crane or heavy winch. Mounted on the left side of the roof is an HIAB hydraulic crane which can lift a maximum of 1,360 kg at a reach of 3.29 m. The commander's cupola and the hatch to his rear are mounted on a large top hatch that opens to the right. This is 2.39 × 1.37 m and enables a spare M113 engine to be carried inside the vehicle and to be lifted out with the aid of the crane. The crane is provided with a hydraulic winch which has a maximum capacity on a full drum of 1,360 kg, or 1,770 kg on a bare drum. For this winch, 15.2 m of 11.1 mm cable is provided.

As of November 1990, 488 Fitter's Vehicles had been built, all of them for export.

Specifications

(M113A2 Recovery Vehicle; data in square brackets relate to Fitter's Vehicle where different)
Crew: 3 [2 + 9 seating]
Weight:
(air transport) 10,834 [10,275] kg
(combat loaded) 11,637 [11,700] kg
Length: 5.34 [4.87] m
Width: 2.69 m
Height: 2.46 [3.15] m
Ground clearance: 0.43 m
Track: 2.159 m
Track width: 381 mm
Length of track on ground: 2.67 m
Ground pressure: 0.57 [0.576] kg/cm^2
Max speed:
(road) 67.6 [67.5] km/h
(water) 5.8 km/h
Range: 483 km
Fuel capacity: 363 [360] litres
Fording: amphibious
Gradient: 70%
Side slope: 30%
Vertical obstacle: 0.61 m
Trench: 1.68 m
Engine: Detroit Diesel Model 6V-53, 6-cylinder, water-cooled developing 215 bhp at 2,800 rpm
Transmission: Allison TX-100-1 with 3 forward and 1 reverse gears
Differential: FMC DS200
Electrical system: 24 V
Batteries: 2 × 12 V
Armament: 1 × 12.7 mm Browning M2 HB MG
Ammunition: 2,000 rounds
Armour: 12-38 mm

Status

Production as required. Both the M113A1/A2 armoured recovery vehicle and the Fitter's Vehicle are known to be in service with the Australian Army. In service with Bahrain (4 M113A2 maintenance recovery vehicles delivered 1991), Belgium, Brazil, Egypt, Israel, Lebanon, Netherlands (based on AIFV), Sudan and Thailand (10).

Manufacturer

United Defense LP, Ground Systems Division, Santa Clara, California 95052, USA.
Tel: +1 408 289 0111. Telex: 6714210.
Fax: +1 408 289 2150.

UPDATED

United Defense LP Assault Amphibian Vehicle, Recovery, Model 7A1

Development

The Assault Amphibian Vehicle, Recovery, Model 7A1 (AAVR7A1) was developed from the Landing Vehicle, Tracked, Recovery, Model 7 (LVTR7) and was renamed during 1985. The original LVTR7 was developed by the FMC Corporation under contract to the US Naval Sea Systems Command to recover other members of the LVT7 family (the LVT7 is now known as the Assault Amphibian Vehicle or AAV). The first prototype, designated the LVTRX2, was completed in 1968 and after trials was standardised as the LVTR7.

Further development by FMC resulted in the LVTR7A1, later known as the AAVR7A1. The AAVR7A1 has improvements to the hull, suspension and power train and changes to the fuel hydraulic and electrical systems. The recovery equipment also underwent major modification. Following the completion of US Marine Corps testing in 1981 a contract was awarded to convert the AAVR7 vehicles to AAVR7A1 configuration.

Description

The hull, which is almost identical to that of the AAV, is made of all-welded aluminium. The engine and transmission are at the front of the vehicle and can be removed as a complete unit. The crew compartment and repair area are at the rear of the vehicle.

The driver is seated at the front of the hull on the left side and is provided with a single-piece hatch cover that opens to the rear. Seven vision blocks are provided for observation and for night driving an AN/VVS-2(V)1A periscope can be mounted in the hatch cover. The commander is seated to the rear of the driver and also has a single-piece hatch cover that opens to the rear and seven direct vision blocks. He has an M27 periscope which extends vertically enabling him to see over the driver's position. The winch/crane operator is seated on the right side and has nine direct vision blocks, with his hatch cover opening to the rear; when operating the crane he uses a seat mounted on the crane structure.

Over the top of the repair area is a large cargo hatch. Normal means of entry and exit are via the large power-operated ramp in the rear of the hull, which is provided with an integral door.

The torsion bar/tube suspension consists of six dual-tyred roadwheels, with the idler at the rear and the drive sprocket at the front. The first, second and sixth roadwheel stations are provided with a hydraulic shock-absorber. The AAVR7A1 is fully amphibious, being propelled in the water by two water-jets, one in each side of the hull at the rear. These water-jets are driven through right-angled gearboxes. Deflectors are used for steering and reverse. If these fail, the vehicle can be propelled in the water by its tracks at a slower speed.

Armament consists of a pintle-mounted 7.62 mm M60D machine gun. The AAVR7A1 is not provided with an NBC system but does have infrared driving lights. Kits for the vehicle include a winterisation kit, visor kit for the driver and a navigation light kit for use when the vehicle is afloat at night.

AAVR7A1 in travelling configuration

On the right side of the hull is a hydraulic crane which can be elevated from 0 to +65°. Its boom is telescopic and can lift 2,722 kg at 6.553 m reach. A two-speed winch, with a maximum capacity of 13,608 kg on a bare drum at low speed and 1,878 kg on a full drum at high speed, is also installed. Equipment carried includes an air compressor, AC generator, workbenches, welding kit and a complete range of tools. If required, a tent can be erected at the rear of the vehicle to enable repairs to be carried out in bad weather or under blackout.

Specifications

Crew: 3
Weight:
(empty) 22,731 kg
(loaded) 12,642 kg
Power-to-weight ratio: 16.88 hp/t
Length: 8.14 m
Width: 3.27 m
Height: 3.28 m
Ground clearance: 0.406 m
Track: 2.609 m
Track width: 533 mm
Length of track on ground: 3.94 m
Ground pressure: 0.55 kg/cm^2
Max speed:
(road) 72.4 km/h
(water) 12.9 km/h
Range:
(land) 482 km at 40 km/h
(water) 7 h at 2,600 rpm
Fuel capacity: 647 litres
Fording: amphibious
Gradient: 60%
Side slope: 40%
Vertical obstacle: 0.914 m
Trench: 2.44 m
Engine:
(LVTR7) Detroit Diesel Model 8V-53T
(AAVR7A1) Cummins VT400
Both are 8-cylinder, water-cooled, turbocharged diesels developing 400 hp at 2,800 rpm
Transmission: FMC HS400-3A1 giving 4 forward and 2 reverse gears, manually operated but with power assistance
Electrical system: 24 V
Batteries: 4 × 12 V 6 TN
Armament: 1 × 7.62 mm M60D MG
Ammunition: 7.62 mm - 880 rounds

Armour
ramp outer: 25.4 mm
ramp inner: 12.7 mm
hull sides: 31-44.5 mm
hull floor and roof: 30.1 mm
hull rear: 35.4 mm

Status
The AAV7RA1 entered production again in 1996 for Brazil who ordered 14 brand new vehicles including one AAV7A1, all will be delivered early in 1997. The basic AAV is in service with Argentina, Brazil, Italy, South Korea, Spain, Thailand, USA and Venezuela.
The AAVR7 and AAVR7A1 are in service with Argentina (1), Brazil (1), South Korea (3), Spain (1), Thailand (1), USA (64) and Venezuela (1).

Late in 1995 South Korea signed a co-production programme for a total of 57 new AAV7A1 vehicles including a quantity of AAVR7A1 vehicles with deliveries to take place from 1996 through to the year 2000.

Manufacturer
United Defense LP, Ground Systems Division, Santa Clara, California 95052, USA.
Tel: +1 408 289 0111. Telex: 6714210.
Fax: +1 408 289 2150.

UPDATED

Cadillac Gage LAV-150S and LAV-300 commando recovery vehicles

Development
This range of 4 × 4 light armoured vehicles was developed as a private venture by Cadillac Gage with the first production models being completed in 1964. These were designated the V-100 and were followed a few years later by the larger V-200. Late production was based on the V-150S, which has many improvements over the earlier V-100 and the similar V-150. By late 1996 over 4,000 vehicles of all types had been built, most for export. More recently the V-150S and V-300 been renamed the LAV-150S and LAV-300. The latest model of the former is the LAV-150ST which is stretched and turbocharged.

Description
The hull of the LAV-150S is of all-welded steel construction which provides the crew with protection from small arms fire. The driver is seated at the front of the hull on the left side with the co-driver to his right. Both are provided with vision blocks and a single-piece hatch cover that opens to the left or right. There is a further vision block on either side of their positions.

The remainder of the crew is seated to the rear of the driver's position. The gunner's hatch is in the centre of the roof and a 7.62 mm machine gun is mounted at this position.

There is a door in each side of the hull and a third door in the rear on the right side, both with a vision block and a firing port. These doors are in two parts, top and bottom; the bottom part opening downwards to form a step and the top half opening left or right. There is a further firing port and vision block in each side of the hull forward of the side doors. There is also a two-part roof hatch to the right of the engine compartment.

The engine and transmission are mounted at the rear of the hull on the left side with access hatches in the roof and side of the hull. The engine compartment is provided with a fire suppression system which is operated by the driver.

The suspension is of the solid axle type with semi-elliptical springs and heavy-duty shock-absorbers at each wheel station. The axles have automatic silent positive locking differentials. The steering is power-assisted and the tyres are of the run-flat type and have a self-cleaning tread. Central tyre inflation is optional. The LAV-150S is fully amphibious, propelled in the water by its wheels; two electric bilge pumps are provided.

The A-frame is supported when in operation by two cables attached to the rear of the vehicle. A hydraulically actuated spade is provided at the front of the hull and is operated through the same controls used for the winch and boom. The winch is mounted in the centre of the hull and leads out through an opening in the forward part of the commander's roof pod. The winch has a maximum capacity of 11,348 kg and has 61 m of 19 mm diameter cable. The boom has a maximum lifting capacity of 4,536 kg with the two jack stands in position. When not in use this rests on the rear of the hull.

Equipment carried on board includes 15.24 m of hose for the compressor, fuel transfer pump, jacks, gunner's platform/workbench, portable spotlight, slave cables, tools and towbars.

LAV-150S (4 × 4) recovery vehicle in travelling order ***1996***

Variant
LAV-300 armoured recovery vehicle

A 6 × 6 armoured recovery vehicle based on the LAV-300 armoured vehicle was produced. This variant has no turret and has a raised and fixed superstructure at the rear. An A-frame jib is mounted on the hull roof front and a folding spade anchor is located under the vehicle front. An 11,348 kg capacity winch is fitted and provided with 61 m of cable. Gross weight of the LAV-300 is 14,969 kg, length is 6.4 m, hull width is 2.54 m and hull height is 1.98 m. Power is provided by a VT-504 turbocharged V-8 diesel developing 270 hp at 2,000 rpm.

For full details of the LAV-300 range of 6 × 6 light armoured vehicles are given in *Jane's Armour and Artillery 1996-97* pages 507 to 508.

Specifications (V-150S)
Crew: 4
Configuration: 4 × 4
Combat weight: 10,886 kg
Power-to-weight ratio: 22.96 hp/t
Length: 6.14 m
Width: 2.26 m
Height: (hull) 1.98 m
Ground clearance: (under hull) 0.647 m
Track:
(front) 1.91 m
(rear) 1.94 m
Wheelbase: 3.12 m
Ground pressure: 1.5 kg/cm^2
Max speed:
(road) 100 km/h
(water) 5 km/h
Range: 644 km
Fuel capacity: 302 litres
Fording: amphibious
Gradient: 60%
Side slope: 30%
Vertical obstacle: 0.91 m
Engine: 6 CTA 8.3 (turbocharged, after-cooled, in-line 6-cylinder diesel), 186 kW (250 hp) at 2,500 rpm (governed), torque 90 kg-M (650 ft-lbs) at 1,584 rpm
Transmission: automatic 6-speed
Transfer box: single speed with spline engagement clutch for front axle drive
Steering: variable ratio power
Turning radius: 8.5 m
Suspension: Solid axles with semi-elliptic multileaf springs with telescopic direct-acting shock-absorbers at each wheel
Tyres: 14.00 R 20
Electrical system: 24 V
Batteries: 2 × 12 V, 100 Ah
Alternator: 60 A
Armament: 1 × 7.62 mm MG
Ammunition: 7.62 mm - 2,200 rounds

Status
Production as required. Users of the LAV-150 include Bolivia, Botswana, Cameroon, Chad, Dominican Republic, Gabon, Guatemala, Haiti (ARV confirmed), Indonesia, Jamaica, Kuwait, Malaysia, Mexico (ARV confirmed), Philippines, Portugal, Saudi Arabia (ARV confirmed), Singapore (ARV confirmed), Somalia, Sudan, Taiwan, Thailand, Turkey, USA, and Venezuela.

Manufacturer
Textron Marine & Land Systems, Combat Vehicle Marketing, 6600 Plaza Drive, New Orleans, Louisiana 70127-2584, USA.
Tel: +1 504 245 6600. Fax: +1 504 245 6960.

UPDATED

ARMOURED REPAIR VEHICLES

BRAZIL

ENGESA EE-11 Urutu (6 × 6) repair and recovery vehicle

Description

The ENGESA EE-11 Urutu repair and recovery vehicle uses the same hull and configuration as the basic EE-11 Urutu armoured personnel carrier, but has some additional equipment to suit it for its repair and recovery role. The main new features are a front-mounted hydraulic winch and a hydraulic crane with a 360° traverse mounted on top of the hull.

The front-mounted hydraulic winch operates through a plate set into the lower front hull and has a capacity of 5,000 kg which can be increased to 10,000 kg by using a pulley block. The winch is supplied with 45 m of 11.1 mm cable. The hydraulic crane has a capacity of 3,000 kg and a horizontal reach of 4 m. When travelling, the crane arms are carried folded down against the hull rear. When the crane is in use, stabilising jacks can be placed under the hull. A wide range of repair and recovery equipment can be carried internally to allow front-line repairs and maintenance to be carried out.

The vehicle has a basic crew of four: driver, commander, crane operator and a fitter. The commander is provided with a manually operated turret, armed with a 12.7 mm Browning (or other) machine gun and two smoke grenades dischargers are located each side of the turret. The welded steel hull uses 8.5 mm plate which provides protection against 7.62 mm armour-piercing ammunition.

The vehicle can be powered by any one of three diesel engine options developing 190, 212 or 260 hp. The driver is provided with a remote tyre pressure control system to cope with various types of terrain.

For full details of the EE-11 Urutu armoured personnel carrier refer to entry in *Jane's Armour and Artillery 1996-97* pages 386 to 387.

ENGESA EE-11 Urutu (6 × 6) armoured repair and recovery vehicle with hydraulic crane deployed but without stabilisers lowered to the ground ***1997***

Specifications

Crew: 4
Configuration: 6 × 6
Weight in action: 15,300 kg
Power-to-weight ratio: 18.57 hp/t
Length overall: 6.26 m
Length of hull: 6.1 m
Width: 2.59 m
Height overall: 2.6 m
Height of hull: 2.125 m
Ground clearance: 340 mm
Track: 2.1 m
Wheelbase: 3.05 m
Angle of approach/departure: 60°/58°
Max road speed: 87 km/h
Fuel capacity: 380 litres
Max cruising range: 850 km
Fording: 1 m
Gradient: 60%
Side slope: 30%
Engine: Mercedes-Benz OM 352-AS 6-cylinder in-line 5.675 litres turbocharged diesel developing 190 hp at 2,800 rpm or
Detroit Diesel 6V-53N 5.22 litres diesel developing 212 hp at 2,800 rpm or
Detroit Diesel 6V-53T 5.22 litres turbocharged diesel developing 260 hp at 2,800 rpm
Transmission: Allison MT-643 automatic 4 forward and 1 reverse gears
Transfer box: ENGESA 2-speed with PTO
Steering: hydraulic
Suspension:
(front) independent with helical springs and double-action telescopic shock-absorbers
(rear) ENGESA 'Boomerang' with walking beams
Tyres: 12.00 × 20 18 PR
Brakes: disc, air over hydraulic
Electrical system: 24 V
Armament:
1 × 12.7 mm M2 HB MG
2 × twin-barrel smoke grenade dischargers
Ammunition:
12.7 mm - 600 rounds
smoke grenades - 12

Status

Production complete. In service with Iraq. The EE-11 is also used by the following countries, some of whom may also operate the repair and recovery version. Angola, Bolivia, Brazil, Chile, Colombia, Cyprus, Ecuador, Gabon, Guyana, Jordan, Libya, Paraguay, Surinam, Tunisia, UAE, Uruguay and Venezuela.

Manufacturer

ENGESA Engenheiros Especializados SA, Avda Tucunaré 125/211, PO Box 152/154, 06400 Barueri, SP, Brazil.
Tel: +55 11 421 4711. Telex: 1171302.
Fax: +55 11 421 4445.
ENGESA has ceased trading.

UPDATED

CANADA

General Motors LAV (8 × 8) armoured repair and recovery vehicle

Development

With the award of a production contract to General Motors to produce the LAV series of vehicles based on the MOWAG (8 × 8) Piranha, General Motors developed a repair and recovery version of the basic vehicle. The LAV recovery was issued to the US Marine Corps which had an initial requirement for 46 of these vehicles. They were delivered during 1986.

In July 1989 the Australian government purchased an LAV repair and recovery vehicle from the US Marine Corps as part of an initial purchase of 15 vehicles. The vehicles entered service with the Australian Army's 2nd Cavalry during 1990.

A maintenance vehicle based on the Infantry Section Carrier (ISC) was developed for the Canadian government. For details of this vehicle, developed from the LAV-L armoured logistics carrier, see next entry.

For details of the LAV-L armoured logistics carrier based on the same chassis see entry under *Tracked Prime Movers, Cargo Carriers and Armoured Logistic Vehicles* section.

Description

The LAV (8 × 8) repair and recovery vehicle uses the basic hull configuration of the original MOWAG Piranha (for details see entry in this section under Switzerland) but it is equipped with American cranes and winches. The vehicle has a crew of four: driver, commander rigger and one additional crew member; with the driver seated forward on the left side and the commander behind him, seated under a cupola. The roof level of the rear hull is raised and there is a hatch for the rigger who controls the hydraulic crane. The HIAB crane has a 4,125 kg lift capacity (1,040 kg at maximum extension) and has a full 360° traverse with the pivoting point just forward and to the left of the rigger's cupola. The crane jib reach radius is from 1.37 to 4.7 m. Normally the crane rests on a support at the rear of the hull. The additional crew member is seated in the rear compartment which also contains extra tools, floodlights and cutting equipment. The vehicle also carries a rear-mounted winch with a 13,608 kg capacity. The vehicle is amphibious and in the water is driven by two propeller units at the rear.

The LAV (8 × 8) repair and recovery vehicle has a fuel transfer subsystem and carries an auxiliary power unit. If required, the crane can be operated using a remote-control box from outside the vehicle. The commander's cupola has a pintle for a 7.62 mm M60 machine gun; 1,000 rounds of 7.62 mm ammunition can be carried. Two M257 smoke grenade dischargers are carried, one on each side, and are controlled from the commander's position. A towing hook is fitted at the rear. Outrigger legs are fitted for use when the crane jib is extended.

Trials were conducted in order to increase the mobility of the LAV series over soft terrain. The trials involved the use of larger tyres producing a 70 per cent wider footprint and lower ground pressure. The larger tyres produce a 30 per cent improvement in soft soil mobility. If the trials result in the adoption of the larger tyres for the LAV it is anticipated that they will also be fitted to the LAV armoured repair and recovery vehicle.

Specifications

Crew: 4
Configuration: 8 × 8
Weight: (combat) 12,860 kg
Power-to-weight ratio: 21.38 hp/t
Length:
(normal) 7.376 m
(reduced) 6.39 m
Width: 2.5 m
Height:
(normal) 2.845 m
(reduced) 2.692 m
Max speed:
(road) 100 km/h
(water) 10 km/h
Range: 660 km
Engine: Detroit Diesel 6V-53T, 6-cylinder diesel developing 275 hp at 2,800 rpm
Transmission: Allison MT-653 DR automatic, 5 forward and 1 reverse gears
Turning diameter: 15.5 m
Wheel travel: 0.33 m
Armament:
1 × 7.62 mm M60 MG
2 × M257 smoke grenade launchers
Ammunition: 7.62 mm - 1,000 rounds

General Motors LAV (8 × 8) armoured repair and recovery vehicle

Status

In production. In service with Australia (1), Saudi Arabia (67) and US Marine Corps (46). These are from the Canadian production line, other users have received these vehicles from Switzerland (MOWAG) or the UK (GKN Defence).

Manufacturer

Diesel Division, General Motors of Canada Limited, PO Box 5160, London, Ontario N6A 4N5, Canada.
Tel: +1 519 452 5184. Fax: +1 519 452 5688.

UPDATED

General Motors Bison (8 × 8) Mobile Repair Team Vehicle

Development

In July 1989 General Motors of Canada received an order from the Canadian government for 199 8 × 8 vehicles based on the US Marine Corps LAV, originally known as the Infantry Section Carrier (ISC) and named Bison. The Bison was designed by General Motors during 1988. Four versions of the vehicle were developed, including 16 maintenance vehicles for use by Mobile Repair Teams (MRT). The other Bison vehicles are the infantry section carrier, a command post and a mortar fire support vehicle. For details of the Bison refer to *Jane's Armour and Artillery 1996-97* pages 394 to 395.

The Bison was developed from the LAV-L armoured logistics carrier by increasing the width of the vehicle at the roof line to create additional internal volume and by adding a hydraulically actuated ramp door at the rear. Other minor changes were made to meet unique Canadian requirements.

The Bison was designed as a common baseline vehicle. By the use of various kits the ISC can be modified for other roles. Each kit is installed in the chassis with a minimum of changes to the hull through the use of a system of integral standard rails to which various components and stowage items can be attached. The vehicle has a flat floor with integral tiedowns and can be used as a cargo carrier.

Bison (8 × 8) Mobile Repair Team Vehicle (R Stickland)

Description

The Bison Mobile Repair Team Vehicle follows the same general lines as the LAV armoured repair and recovery vehicle but differs in many details to meet Canadian Forces requirements. The vehicle can be made amphibious after 5 minutes' preparation, and may be driven through the water by two propeller units and steered using four rudders.

The Bison Mobile Repair Team Vehicle is a basic Bison ISC chassis with the cargo hatch relocated to the rear and with the two troop hatches removed along with two of the four seats. Other additional items include a 265° traverse HIAB crane and stabilisers (the crane is the same as that used on the US Marine Corps repair and recovery vehicle, as are the stabilisers), two detachable floodlights, a utility air line, a machinist's vice, and a quick-release hose for power tools. Provision is made for tool, camouflage net and other equipment stowage.

The vehicle is capable of self-loading and unloading spare wheel assemblies and can carry a complete power pack internally. The vehicle is capable of towing a similar vehicle and is equipped with a standard towbar. Although no heavy recovery winch is installed, all Bison vehicles are equipped with a front-mounted 6,800 kg dynamic pull hydraulic winch.

The vehicle has a crew of two, a driver and a commander, who also acts as a rigger. The driver is seated on the left of the vehicle and when it is fully closed down he is provided with three M17 periscopes; the commander has five for his position behind the driver. Night vision devices are optional. The vehicle is provided with an M8A1 ventilated face mask NBC system and a VHF radio. A Halon 1301 manual fire suppression system is provided for the crew and engine compartments.

There is provision for a 7.62 mm M60 machine gun mounting over the commander's hatch and two Wegmann smoke grenade launchers with eight grenades. An ATGW wire cutter is positioned in front of the driver's position.

Specifications

Crew: 2
Configuration: 8 × 8 or 8 × 4

Weight:
(kerb) 12,055 kg
(max) 13,027 kg
Power-to-weight ratio: 21.[illegible]1 hp/t
Max load: 973 kg
Length: 7.366 m
Width: 2.5 m
Height: 2.692 m
Max speed:
(road) 100 km/h
(water) 9.7 km/h
Range: 665 km
Gradient: 60%
Side slope: 30%
Fording: amphibious after 5 min prep
Trench: 2.06 m
Engine: Detroit Diesel 6V-53T 6-cylinder diesel developing 275 hp at 2,800 rpm
Gearbox: Allison MT-653 DR automatic, 5 forward and 1 reverse gears
Steering: power-assist
Turning radius: 15.5 m
Suspension: independent on all 8 wheels
Electrical system: 24 V
Batteries: 4
Armament: 1 × 7.62 mm M60 MG

Status
Production complete but can be resumed if further orders are placed. In service with the Canadian Forces (16).

Manufacturer
Diesel Division, General Motors of Canada Limited, PO Box 5160, London, Ontario N6A 4N5, Canada.
Tel: +1 519 452 5184. Fax: +1 519 452 5688.

UPDATED

General Motors Husky armoured repair vehicle

Development
In the early 1970s the Swiss MOWAG company developed a range of 4 × 4, 6 × 6 and 8 × 8 vehicles known as the Piranha. These all share many common components such as wheels, suspensions, steering system, hull front and rear, hatches, doors and propellers. In 1977 the Canadian Armed Forces placed an order with the Diesel Division of General Motors of Canada Limited for 350 of the 6 × 6 version. Total value of this order was $52 million and comprised 179 Grizzly APCs, 152 Cougar fire support vehicles (with British-supplied 76 mm Scorpion turrets) and 19 Husky repair and recovery vehicles. The original order for 350 vehicles was increased to 491 vehicles and final deliveries were made in 1982.

The Canadian Forces call this vehicle the Husky wheeled maintenance and recovery vehicle; 27 are in service.

Description
The hull of the vehicle is of all-welded steel armour; the driver is seated at the front of the hull on the left side and is provided with a single-piece hatch cover and three periscopes. The engine is mounted to the right of the driver and the personnel compartment is at the rear. Entry to this is via two doors in the rear of the hull, each with a MOWAG-designed firing port. Hatches are provided in the roof.

The rear suspension is of the torsion bar type, with coil springs and hydraulic shock-absorbers at each wheel station. The front suspension is of the coil spring and wishbone type and each wheel station has a hydraulic shock-absorber. The tyres have Hutchinson run-flat inner tubes. Brakes are hydraulic on all wheels.

The vehicle is fully amphibious, employing two propellers at the rear. Before entering the water a trim vane is erected at the front of the vehicle. Standard equipment on the Husky includes a HIAB 650 crane with a maximum capacity of 3,250 kg.

Details of the complete family of 6 × 6 vehicles are given in *Jane's Armour and Artillery 1996-97* pages 395 to 396.

Specifications
Crew: 4
Configuration: 6 × 6
Weight: (loaded) 10,500 kg
Power-to-weight ratio: 28.57 hp/t
Length: 5.97 m
Width: 2.5 m
Height: (hull top) 1.85 m
Ground clearance:
(max under hull) 0.5 m
(under differential) 0.39 m
Track:
(front) 2.18 m
(rear) 2.2 m
Wheelbase: 2.04 m + 1.04 m
Angle of approach/departure: 40°/45°
Max speed:
(road) 100 km/h
(water) 10 km/h
Range: (road) 600 km
Fuel: 210 litres
Fording: amphibious
Gradient: 70%
Side slope: 30%
Vertical obstacle: 0.5 m
Engine: turbocharged Detroit Diesel Model 6V-53T developing 300 hp at 2,800 rpm
Transmission: Allison MT 650 automatic, 5 forward and 1 reverse gears
Tyres: 11.00 × 16
Electrical system: 24 V
Armament: 1 × 7.62 mm M60 MG

Status
Production complete in Canada. In service with the Canadian Forces (27).

Manufacturer
Canada - Diesel Division, General Motors of Canada Limited, PO Box 5160, London, Ontario N6A 4N5, Canada.
Tel: +1 519 452 5184. Fax: +1 519 452 5688.

UPDATED

Canadian Forces Husky armoured repair vehicle with crane retracted but with stabilisers extended (T J Gander)

FRANCE

Giat Industries AMX-10 ECH repair vehicle

Description
The Giat Industries AMX-10 ECH is the repair vehicle member of the AMX-10P MICV family, it entered service with the French Army in 1973. The vehicle was shown for the first time in 1977. It has no recovery capability and is limited to changing components of other AFVs, for example, the engine of the AMX-10P. Full details of the AMX-10P infantry combat vehicle are given in *Jane's Armour and Artillery 1996-97* pages 277 to 281. Production of the vehicle was undertaken at Roanne.

The hull of the AMX-10 ECH is almost identical to that of the AMX-10P and is of all-welded aluminium construction. The driver is seated at the front of the hull on the left side and is provided with a single-piece hatch cover and three periscopes for observation. The engine, which is manufactured under licence by Renault, is mounted to his right. The crew compartment is at the rear of the hull and entrance to it is by the large power-operated ramp at the rear, which is provided with a door in case the ramp fails to open.

The suspension is of the torsion bar type and consists of five roadwheels with the drive sprocket at the front and the idler at the rear. There are three track-return rollers. Double-acting lever-type shock-absorbers are provided for the first and fifth roadwheel stations.

The crew of five consists of the driver, the commander in the turret and three mechanics. The AMX-10 ECH is fully amphibious, propelled in water by either its tracks at a speed of 6 km/h, or by two water-jets, one in each side of the hull, giving a maximum water speed of 7.92 km/h. Before entering the water a trim board is erected at the front of the hull.

The vehicle has a Toucan I turret, slightly offset to the left side of the hull, fitted with a 20 mm cannon and a coaxial 7.62 mm machine gun. The turret can be traversed through 360° and the armament elevated from −13 to +50°. Elevation and traverse are manual, maximum elevation speed is 16°/s and maximum traverse speed is 12°/s. Optical equipment for the turret includes six periscopes, a sight with a magnification of ×6 for engaging ground targets and a separate sight for anti-aircraft fire. A total of 576 rounds of 20 mm and 2,000 rounds of 7.62 mm ammunition are carried. The 20 mm cannon can fire a variety of ammunition including HE, HE-I and AP, with a maximum muzzle velocity of 1,050 m/s.

On the right side of the roof, at the rear, is the hydraulically operated crane with an extensible jib, which can lift a maximum of 6,000 kg. The operator is provided with a small roof hatch. When using this crane, jacks are placed under the rear of the hull to support the vehicle. Other equipment carried includes tools, two sheer legs and two jacks, which are used to replace suspension components (for example torsion bars) on other AFVs including the AMX-30 MBT.

Optional kits for the AMX-10 ECH include an NBC system and passive night vision equipment.

Specifications
Crew: 5
Weight:
(empty) 11,300 kg
(loaded) 13,800 kg
Power-to-weight ratio: 20.28 hp/t
Length: 5.76 m
Width: 2.78 m
Height:
(top of hull) 1.92 m
(overall) 2.62 m
Ground clearance: 0.45 m
Ground pressure: 0.53 kg/cm^2
Max speed: (road) 65 km/h
Range: (road) 600 km
Fording: amphibious
Gradient: 60%
Vertical obstacle: 0.7 m
Trench: 1.6 m
Engine: Hispano-Suiza HS 115-2, V-8, water-cooled supercharged diesel developing 280 hp at 3,000 rpm
Transmission: preselective with 4 forward and 1 reverse gears
Electrical system: 24 V
Armament:
1 × 20 mm cannon
1 × 7.62 mm MG coaxial with main armament
2 × smoke grenade dischargers either side of turret

Status
Production complete. In service with Saudi Arabia. The basic AMX-10P is also in service with France, Greece, Qatar, Saudi Arabia and the UAE.

Manufacturer
Atelier de Construction Roanne (ARE).
Enquiries to: Giat Industries, 13 route de la Minière, F-78034 Versailles Cedex, France
Tel: +33 1 30 97 37 37. Fax: +33 1 30 97 39 00.

UPDATED

Giat Industries AMX-10 ECH, a member of the AMX-10P family of tracked combat vehicles
1997

Giat Industries VAB ECH repair vehicle

Development
In 1969 the French Army issued a requirement for a new wheeled vehicle called the *Véhicule de l'Avant Blindé* (VAB). It was designed to undertake a wide range of roles including use as an APC, load carrier, ambulance and anti-tank vehicle. Prototypes were built by Panhard and Saviem and, after comparative trials, the Saviem model was adopted by the French Army in 1974. The first production orders were placed the following year and first deliveries of the 4 × 4 version were made to the French Army late in 1976.

Production of the VAB was completed in 1993 after just over 5,000 had been built for the home and export markets. Details of the VAB are given in *Jane's Armour and Artillery 1996-97* pages 413 to 416. Further development has resulted in the VAB New Generation (ibid pages 412 to 413) and it is expected that any new production vehicles would be built to this standard.

See also the entry for the VAB Recovery in the *Armoured Recovery Vehicles* section.

Description
The hull of the VAB ECH is of all-welded steel armour, with the driver seated at the front of the vehicle on the left and the commander, who also operates the machine gun, to his right. Both the driver and commander have a bulletproof windscreen to their front, which can be covered by an armoured plate if required, and a side door with a bulletproof window. The driver has a single-piece hatch cover in the roof which opens to the rear. The armament installation is over the commander/gunner's position.

The engine and transmission are mounted to the rear of the driver's position and are removed as a complete unit through the roof. The engine compartment is provided with a fire extinguishing system.

The crew compartment is at the rear of the vehicle and there is a connecting corridor between the driver and commander's compartment at the front and the rear compartment on the vehicle's right-hand side. At the rear of the vehicle are two doors which open outwards, each with a window with an armoured cover. There is no centre post, therefore bulky equipment can be easily loaded. There are two large roof hatches over the rear compartment and on each side there are three bulletproof windows covered by an armoured shutter.

The VAB ECH can be fitted with at least three different types of armament installation:

Giat Industries VAB ECH repair vehicle with hatches and rear doors open

Giat Industries CB 52 ring mount with a 7.5 mm or 7.62 mm machine gun, which can be traversed by the gunner's shoulder through 360°. Elevation is from −15 to +45°, and 200 rounds of ready-use ammunition are provided.

Giat Industries TLI 52A turret with a 7.5 mm or 7.62 mm machine gun. The turret can be traversed through 360°, and the machine gun elevated from −12 to +45°. The gunner is provided with a tilting-head prism periscope with a magnification of ×1, linked to a sight with a magnification of ×5. For observation purposes there are six episcopes. A white light searchlight is mounted externally on the turret and fumes from firing the MG are extracted by an electric fan system. A total of 200 rounds of ready-use ammunition is provided.

Giat Industries CB127 gun ring shield with 12.7 mm machine gun which can be traversed through a full 360° and elevated from −15 to +65°. 100 rounds of ready-use ammunition are provided.

The steering is power-assisted on the front four wheels (in the case of the 6 × 6 model) and the suspension is of the torsion bar type with telescopic shock-absorbers. The tyres are of the run-flat type.

Power is transmitted from the engine to the wheels through a hydraulic torque converter and the gearbox. Gears are selected by means of a short pneumatically assisted lever which also operates the clutch.

The basic vehicle is fully amphibious, propelled in the water by its wheels, but if required the VAB can be delivered with two water-jets mounted in each side of the hull at the rear. Other optional equipment includes: infrared or passive night vision equipment, NBC system, heater and a front-mounted winch with 60 m of cable and a maximum capacity of 7,000 kg.

The VAB ECH is provided with the following equipment: a welding station with cylinders of oxygen and acetylene; generator; grinding machine; hand drill; hoist; searchlight; storage cabinets and drawers; water tank; workbench; vice; and tool kit. To give the repair crew more headroom a PVC roof extension can be quickly raised over the rear compartment.

Specifications

Crew: 4
Configuration: 6 × 6
Weight:
(empty) 12,000 kg
(loaded) 14,000 kg
Power-to-weight ratio: 15.71 hp/t
Length: 5.98 m
Width: 2.49 m
Height: (top of hull) 2.06 m
Ground clearance: (axles) 0.4 m
Track: 2.035 m
Angle of approach/departure: 45°/45°
Wheelbase: 1.5 m + 1.5 m
Tyres: 14.00 × 20
Max speed:
(road) 92 km/h
(water) 7.2 km/h (with wheels)
Range: (road) 1,000 km
Fuel capacity: 300 litres
Fording: amphibious
Gradient: 60%
Side slope: 30%
Vertical obstacle: 0.5 m
Trench: 1 m (6 × 6 version only)
Engine: Renault MIDS 06.20.45 6-cylinder diesel developing 220 hp at 2,200 rpm
Transmission: semi-automatic with 5 forward and 1 reverse gears
Electrical system: 24 V
Armament: 1 × 7.62 mm MG (see text)

Status

Production complete, in sevice with the French Army and other countries including Brunei, Central African Republic, Cyprus, Ivory Coast, Lebanon, Mauritius, Morocco, Oman, Qatar, UAE and other undisclosed countries. Production can be resumed if additional orders are placed. These would be built to the VAB New Generation standard.

Manufacturers

Renault Véhicules Industriels and Giat Industries, 40 rue Pasteur, BP 302, F-92156 Suresnes Cedex, France.
Enquiries to: Giat Industries, 13 route de la Minière, F-78034 Versailles Cedex, France.
Tel: +33 1 30 97 37 37. Fax: +33 1 30 97 39 00.

UPDATED

Panhard M3 VAT and Buffalo repair vehicles

Development

The first prototype of the Panhard M3 APC was completed in 1969, with the first production vehicles following in 1971. The M3 uses 95 per cent of the mechanical components of the Panhard AML light armoured car which is used by many armies. The Panhard M3 VAT (or *Véhicule Atelier*) is basically the standard M3 APC modified to undertake repairs in the field. It has no recovery capability although it can tow other vehicles.

The M3 is no longer in production and has been replaced by the Buffalo. Details of the M3 are given in *Jane's Armour and Artillery 1996-97* pages 422 to 424.

Description

The hull of the M3 is of all-welded steel construction with the driver seated at the front of the hull. He has a single-piece hatch cover that opens to the right and has three integral periscopes for observation. There are three small hatch covers which lift upwards and a large door in each side of the hull. At the rear of the hull are two doors, each of which has a firing port. There are two circular hatches in the roof and a 7.62 mm machine gun is normally mounted on the forward hatch cover position.

The engine and transmission are to the rear of the driver. The transmission consists of two gearboxes in one, coupled on both sides of the bevel pinion. The low-range box comprises two low gears, top gear and one reverse, which are used in rough country. The high-range box has three low gears and an overdrive. There is no clutch pedal, just a gear selection gate.

The suspension consists of coil springs and hydropneumatic shock-absorbers acting on suspension arms of the wheel mechanism. Brakes are hydraulic, with separate circuits for front and rear wheels. The tyres have punctureproof Hutchinson inner tubes.

The M3 is fully amphibious, propelled in the water by its wheels. Optional equipment includes a ventilation system and night vision equipment of the infrared or passive type.

Equipment installed includes: a generator; tools; workbenches; inspection lamp; towbars; and tow cables. A penthouse can be erected at the rear to enable repairs to be carried out in bad weather. A block and tackle can be erected at the rear to enable the VAT to carry out engine and other component changes.

Variant

Buffalo

In 1985 the basic M3 series, including the VAT, was replaced in production by a basically similar vehicle known as the Buffalo. The Buffalo series includes a light workshop vehicle but no example has yet been manufactured. The basic Buffalo armoured personnel carrier has a combat weight of 6,200 kg and is 4.6 m long, 2 m high and 2.4 m wide. It is powered by a V-6 PRV petrol engine developing 145 hp at 5,500 rpm. As an alternative, it may be powered by a Peugeot XD 3T diesel engine developing 98 hp at 2,250 rpm.

Panhard M3 VAT repair vehicle ***1996***

Specifications

M3
Crew: 5
Configuration: 4 × 4
Weight:
(empty) 5,300 kg
(loaded) 6,100 kg
Power-to-weight ratio: 14.75 hp/t
Length: 4.45 m
Width: 2.4 m
Height: (without armament) 2 m
Ground clearance: 0.35 m
Track: 2.05 m
Wheelbase: 2.7 m
Max speed:
(road) 90 km/h
(water) 4 km/h
Range: (road) 600 km
Fuel: 165 litres
Fording: amphibious
Gradient: 60%
Side slope: 30%
Vertical obstacle: 0.3 m
Trench:
(with 1 channel) 0.8 m
(with 5 channels) 3.1 m
Engine: Panhard Model 4 HD 4-cylinder, horizontally opposed petrol developing 90 hp at 4,700 rpm
Transmission: 6 forward and 1 reverse gears (see text)
Armament:
1 × 7.5 or 7.62 mm MG
4 × smoke grenade dischargers (optional)

Status

The M3 is no longer in production and was replaced by the Buffalo. The M3 is in service with the following countries, some of which also have the VAT: Algeria, Bahrain, Burkina Faso, Burundi, Chad, Gabon, Iraq, Ireland, Ivory Coast, Kenya, Lebanon, Malaysia, Mauritania, Morocco, Niger, Saudi Arabia, Senegal, Sudan (M3 VAT confirmed), Togo, UAE and Zaïre.

Manufacturer

Société de Constructions Mécaniques Panhard et Levassor, 18 avenue d'Ivry, F-75621 Paris, France.
Tel: +33 1 40 77 40 00. Telex: 270 276.

UPDATED

POLAND

MT-LB technical support vehicle

Description

The Polish Army utilises a locally developed version of the MT-LB multipurpose tracked vehicle for the front-line technical support of armoured units. The vehicle is similar to the basic MT-LB (for details of the MT-LB refer to *Jane's Armour and Artillery 1994-95* pages 329 to 332) but is provided with various items of special equipment for its role. In addition to its technical support role the vehicle can also be used as a general or NBC reconnaissance vehicle.

The vehicle has a basic crew of three but further seating is provided internally for up to four passengers. Its main role is the front-line repair and general support of other armoured vehicles, which includes the towing of disabled light vehicles, the evacuation of personnel (including the provision of medical first-aid assistance), and the provision of earth hides or scrapes. For the latter role the vehicle is provided with two hydraulic arms at the rear of the vehicle, onto which a self-entrenching blade can be secured; when not in use, the blade is carried on the hull roof on the right-hand side. Also provided for the support role is a light A-frame crane which can be erected over the rear of the hull to assist when handling vehicle components. For transport the crane is stowed folded flat on the hull roof.

A comprehensive suite of repair and other tools is carried. The vehicle can be used to tow a trailer carrying spare components or supplies.

The vehicle is amphibious and is provided with NBC air filtration equipment. Under normal operations the vehicle's various NBC sensors are maintained in a permanently on condition.

The commander's position is covered by a small turret provided with vision devices. Mounted over the turret is a 12.7 mm NSW (NSV) machine gun for local and air defence.

Specifications

Crew: 3 + 4
Weight: 12,900 kg (±200 kg)
Power-to-weight ratio: 18.99 hp/t
Ground pressure: 0.46 kg/cm²
Towed load:
(trailer) 6,500 kg
(vehicle) 14,000 kg
Length:
(with blade) 7.378 m
(without blade) 7.14 m
(crane extended) 8.72 m
Width: 3.03 m
Height:
(crane stowed) 2.305 m
(crane extended) 4.14 m
Ground clearance: (max) 0.415 m
Max speed:
(road, without trailer) 60 km/h
(road, with trailer) 45 km/h
(when towing vehicle) 6 km/h
(water) 8 km/h
Range: 500 km
Fording: amphibious
Engine: SW680/167/1 6-cylinder turbocharged diesel developing 245 hp
Armament:
1 × 12.7 mm NSW (NSV) MG
smoke grenade launchers

Status

In service with the Polish Army.

Manufacturer

Huta Stalowa Wola SA, 1 Kwiatkowskiego Street, PL-37-450 Stalowa Wola, Poland.
Tel: +48 16 435252. Telex: 062233 hswbh pl.
Fax: +48 16 425566.

Marketing Agency

Cenzin Co Ltd, 2 Frascati Street, PL-00-489 Warsaw, Poland.
Tel: +48 22 296738. Telex: 814505 czi pl.
Fax: +48 22 628 6356.

UPDATED

Polish MT-LB technical support vehicle showing entrenching blade and A-frame crane stowed on hull roof

RUSSIAN FEDERATION AND ASSOCIATED STATES (CIS)

MTP-LB technical support vehicle

Description

Several variations of the MT-LB multipurpose tracked vehicle existed before the MTP-LB technical support vehicle was issued to the Technical Emergency Service. It had the responsibility of vehicle recovery and repair in the front line. For details of the MT-LB refer to *Jane's Armour and Artillery 1996-97* pages 329 to 332.

The basic form of the MT-LB vehicle remains unchanged but several alterations were made to adapt it for recovering armoured fighting vehicles by direct tow or winch, and the recovery of amphibious vehicles from water obstacles. One of the main fitments is an A-frame crane mounted on the front of the vehicle and powered by the winch to provide a 1,500 kg lift capacity. This has two working positions. One is with the crane jib length at 2.15 m, giving a lift height of 3.6 m. The other, with the jib length at 1.35 m, gives a lift height of 4.2 m. With the crane under load the vehicle can be driven at speeds up to 5 km/h and with a list of 5°.

The vehicle is also provided with a cable winch, a jacking device, a towing attachment, chocks and other arresting devices, waterborne salvage equipment, and gas welding and oxy-acetylene cutting equipment for both steel and aluminium. The hydraulic cable winch is driven by the auxiliary gearbox and drives a cable 80 m long; winching capacity is 60 kN. The winch may be operated manually but the maximum pull is then limited to 150 kg. The jacking device on the front of the vehicle acts not only as an anchor when the cable winch is in use, but also as a metalwork workbench. A hand winch is provided for moving the jacking device from its normal transport position on the vehicle front to the cargo platform when the MTP-LB is afloat. The towing attachment is used for towing armoured fighting vehicles with steering that is out of action. An electric welding kit is carried.

Other specialised changes to the MTP-LB are the removal of the MT-LB turret and its replacement by an extra hatch and periscope over the commander's position for use when afloat. Extra frames and containers have been fitted to carry the extra equipment, and apart from the recovery role the MTP-LB carries the equipment required for observation, command, warning, reconnaissance, decontamination and camouflage. Interior lighting is also fitted.

MTP-LB technical support vehicle (Michael Jerchel) ***1996***

Specifications

Crew: 2
Combat weight: 12,300 kg
Power to weight ratio: 19.5 hp/t
Ground pressure: 0.46 kg/cm²
Length: 6.8 m
Width: 2.85 m
Height:
(hull roof) 1.865 m
(recovery array) 2.3 m
Ground clearance: 0.395-0.415 m
Track: 2.3 m
Track width: 350 mm
Max speed:
(land) 55 km/h
(water) 6.5 km/h
Range: (roads) 500 km
Fording: amphibious
Engine: YaMZ 238 V V-8 cylinder diesel developing 240 hp at 2,100 rpm

Status

Production complete. The MT-LB is used by the following countries, some of whom also use the MTP-LB version: Armenia, Azerbaijan, Belarus, Bulgaria, Finland, Hungary, Iraq, Kazakhstan, Russia, Sweden, Ukraine and Yugoslavia (Serbia and Montenegro).

Manufacturer

State factories.

UPDATED

Repair Tracked Vehicle (RTV)

Development

The Repair Tracked Vehicle (RTV) has been developed, by the Rubtsovsk Machine Building Plant, to undertake repairs to tanks and BMP-type armoured vehicles in the forward combat area.

Typical roles of the RTV include:

(1) Towing BMP-type full-tracked vehicles on both land and water
(2) Carrying out repairs and maintenance to MBTs and BMP type armoured vehicles
(3) Carrying out welding repairs
(4) Carrying out technical maintenance
(5) Carrying spare parts up to a maximum weight of 1,000 kg
(6) Charging accumulator batteries
(7) Electric starting of MBT and BMP engines using onboard equipment.

Description

The Repair Tracked Vehicle is based on the chassis of the BMP-2 infantry fighting vehicle and has a similar layout. The driver seated is front left and the vehicle commander to his rear with both being provided with a single-piece hatch cover and periscopes.

The engine compartment is to the right and the crew compartment at the rear. Mounted above the latter is an all-welded steel turret with an electrically operated crane which can lift a maximum load of 3,000 kg. The crane turret can be traversed through 220° and has a maximum lifting height of 4 m.

Mounted on top of the turret is a 7.62 mm PKT machine gun and standard equipment includes an electric welding system, repair set, tools, NBC system, night vision equipment, smoke grenade launchers and an engine-driven smoke generating system.

The suspension either side is the same as that of the BMP with six single rubber tyred roadwheels, drive sprocket at the front, idler at the rear and track-return rollers. The upper part of the track is covered by a steel shroud and an unditching beam can be carried on the left side of the hull.

Scale model of Repair Tracked Vehicle (RTV) on BMP series chassis (T J Gander) ***1997***

Specifications

Crew: 4
Combat weight: 13,800 kg
Power-to-weight ratio: 21.73 hp/t
Ground pressure: 0.65 kg/cm²
Length: 6.736 m
Width: 2.94 m
Height: (with machine gun) 2.4 m
Ground clearance: 0.37 m
Track: 2.55 m
Track width: 300 mm
Length of track on ground: 3.8 m
Max road speed: 65 km/h
Max water speed:
(forward gear) 6 km/h
(reverse gear) 2 km/h
Fuel capacity: 462 litres
Road range: 660 km
Gradient: 60%
Side slope: 30%
Vertical obstacle: 0.7 m
Trench: 2.50 m
Engine: Model UTD-20 6-cylinder, 4-stroke direct injection diesel developing 300 hp at 2,600 rpm
Transmission: manual, 5 forward and 1 reverse gears
Clutch: multiplate, dry, constant engagement
Suspension: torsion bar, hydraulic shock-absorbers on 1st, 2nd and 6th roadwheels
Electrical system: 22-29 V
Armament: 1 × 7.62 mm PKT machine gun
Smoke laying equipment: diesel fuel injected into exhaust 2 × 2 smoke grenade launchers

Status

Production as required. In service with Russian Army.

Manufacturer

Rubtsovsk Machine Building Plant.

NEW ENTRY

SWITZERLAND

MOWAG Piranha (8 × 8) armoured repair vehicle

Development

One of the Piranha series of vehicles developed by the Swiss MOWAG concern during the early 1970s was the 8 × 8 armoured repair vehicle, which may also be used if required as an armoured recovery vehicle. It follows the same general lines as the Husky (6 × 6) and Bison (8 × 8) armoured repair vehicles built in Canada (see entries in this section).

Following an agreement signed in 1990, the MOWAG Piranha series of vehicles is also manufactured in the UK by GKN Defence. In January 1991 it was announced that MOWAG had contracted GKN Defence to produce a 'significant quantity' of Piranha 8 × 8 vehicles for delivery to Saudi Arabia. This order included armoured repair vehicles. Another order was placed by Oman in May 1994, again including armoured repair vehicles. In November 1996, Qatar placed an order with GKN Defence for the supply of 40 8 × 8 Piranha vehicles including two repair and recovery vehicles.

Description

The hull of the vehicle is of all-welded steel armour with the front, sides and rear proof against NATO 7.62 mm ball ammunition. The driver is seated at the front of the hull on the left side and is provided with a single-piece hatch cover and three periscopes. The engine is mounted to the right of the driver and the personnel compartment is at the rear. Entry to this compartment is via two doors at the rear of the hull, each with a firing port. Roof hatches are provided.

The rear suspension is of the torsion bar type, and the front suspension has coil springs and wishbone. All wheels are fitted with hydraulic shock-absorbers. When the crane is in use, the vehicle is stabilised by hydraulic and mechanical ground supports. All tyres have run-flat cores, and the dual circuit brakes are hydraulic, assisted by compressed air.

The vehicle is also available in a fully amphibious form, driven in the water by two propellers at the rear. Before entering the water, a trim vane is erected at the front of the vehicle. The usual crane fitted to the Piranha (8 × 8) is a HIAB unit with a maximum capacity of 3,000 kg, although other cranes may be fitted.

The latest version of the MOWAG Piranha is an 8 × 8 vehicle with a revised superstructure to mount a 7,000 kg hydraulic crane centrally. This version has several layout changes from the original, one of which is the provision of stabiliser legs inset in the sides in a central position. The centrally mounted crane position allows more internal storage space so that more tools and repair equipment can be carried.

MOWAG Piranha (8 × 8) repair vehicle in travelling configuration ***1996***

Specifications

Crew: 4 plus
Configuration: 8 × 8
Weight:
(loaded) 12,300-13,000 kg
(empty) 8,800 kg
Power-to-weight ratio: 23 hp/t
Length: 6.6 m
Width: 2.5 m
Height:
(hull top, front) 1.85 m
(hull top, rear) 2.7 m
Ground clearance: (under hull) 0.5 m

Track:
(front) 2.18 m
(rear) 2.205 m
Tyres: 11.00 × 16
Wheelbase: 1.1 m + 1.335 m + 1.04 m
Angle of approach/departure: 40°/45°
Max speed:
(road) 100 km/h
(water) 10.5 km/h
Range: (road) 780-1,000 km
Fuel: 300 or 500 litres
Gradient: 60%
Side slope: 35%
Vertical obstacle: 0.5 m
Engine: Detroit Diesel 6V-53T developing 300 hp at 2,800 rpm
Transmission: Allison MT-653 automatic, 5 forward and 1 reverse gears
Electrical system: 24 V
Armament: 1 × 7.62 mm MG

Status
In production. Known users of the Piranha, some of whom might use the Piranha repair vehicle, include Chile, Ghana, Liberia, Nigeria, Sierra Leone and Switzerland. Produced by GKN Defence of the UK for delivery to Saudi Arabia, Oman and Qatar.

Manufacturer
MOWAG Motorwagenfabrik AG, CH-8280 Kreuzlingen, Switzerland.
Tel: +41 71 677 5500. Fax: +41 71 672 2886.

UPDATED

UNITED STATES OF AMERICA

United Defense LP, Armored Maintenance Vehicle (AMV)

Development
In 1981 the US Army conducted a Battlefield Recovery and Evacuation (BRE) study to define future guidelines and to analyse the methods of forward area vehicle maintenance and recovery then in use. This BRE study highlighted the fact that the M113 Contact Team repair vehicles then in use were unsuitable for their tasks in forward battle areas. From this the US Army Ordnance Center and School (OC&S), together with the US Army Tank Automotive Command (TACOM) and Training and Doctrine Command (TRADOC), evolved the concept of a forward-area Armored Maintenance Vehicle (AMV). It would operate in a battle area just short of the Forward Line of Troops (FLOT) to a depth of 6 to 10 km to the rear.

In December 1981 TRADOC included a requirement for the AMV in its Operational and Organizational Plan, this in turn led to a Concept Evaluation Program (CEP) study conducted at Fort Hood, Texas, during May 1985. For this study the FMC Corporation (now part of United Defense LP) supplied two prototype vehicles, one based on the M113A3 chassis (the stretched version with an extra roadwheel each side) and the other based on the XM987 Fighting Vehicle System Carrier, used for the Multiple Launch Rocket System (MLRS).

Each vehicle was equipped with a telescopic crane jib capable of lifting a 5 to 6 tonne load such as an M1 Abrams MBT power pack. Control for the crane could be carried out internally or from a remote-control station box. Both vehicles used a rear-mounted crane with hydraulically positioned stabiliser legs emplaced when the crane was in use. The XM987-based AMV can carry 3,266 kg of tools and spare parts and the crane lift capacity is 6,804 kg. There is a crew of four in the XM987-based AMV, with space for a further four, all with NBC protection. Welding and hydraulically operated tools are provided and the vehicle can be used to tow disabled vehicles.

The report on the Fort Hood trials was published by OC&S in October 1985. Since the trials, the Required Operational Capabilities (ROC) document review process required the BRE study to be updated and a Force Mix analysis to be performed. The results of the studies were proven and exceeded in field trials held during the second CEP conducted in Germany during the Reforger '88 exercise with the 8th and 3rd Infantry Divisions. The report on the Reforger '88 CEP was published by OC&S in September 1989. The single XM987-based AMV prototype was leased to the US Army and was deployed to Saudi Arabia during Desert Shield/Storm. The vehicle is currently being utilised at Fort Huachuca, Arizona, in support of government testing of the Ground-Based Common Sensor - Heavy programme. Production of the AMV awaits future Department of Defense budgets.

Armament consists of a 12.7 mm M2 machine gun and banks of electrically operated smoke grenade launchers.

Status
Prototype.

Prototype of Armored Maintenance Vehicle (AMV) based on XM987 Fighting Vehicles System Carrier

Manufacturer
United Defense LP, Ground Systems Division, 2830 De La Cruz Boulevard, Box 58123, Santa Clara, California 95052, USA.
Tel: +1 408 289 0111. Telex: 6714210.
Fax: +1 408 289 2150.

UPDATED

Heavy Repair Vehicle (HRV)

Description
The Heavy Repair Vehicle (HRV) was developed by the US Army Ordnance Center and School to meet a need for a forward area maintenance vehicle capable of supporting armoured, mechanised infantry, combat engineer, field artillery, armoured cavalry and forward support maintenance battalions. The HRV is intended for the repair of battle-damaged vehicles in the forward area and to operate forward of the brigade support areas at equipment breakdown sites and unit maintenance collection points. The HRV will carry tools, selected repair parts, cutting and welding equipment and small quantities of fuel and lubricants, along with the four-man crew, their personal equipment and weapons.

The HRV is based on the chassis of the Oshkosh (10 × 10) M1074 series Palletized Load System (PLS) vehicle - for details of this vehicle refer to the entry in the *Materials Handling Equipment* section. The chassis is modified to accept a repair module containing a work and storage area for a tool set suitable for the type of unit involved. The chassis is further modified to accept a 7.5 tonne lifting crane at the rear of the chassis - the crane is the same as that used on the M984A1 (8 × 8) recovery vehicle and can be used to lift and handle vehicle major components. The main repair/stowage module has a roof hatch to accept components for repair.

Stowage points around the vehicle are used to carry further tools and equipment and provision is made for a minimum of two NATO slave receptacles and a standard diagnostic connector assembly. Other external connectors provide 110 or 220 V power supplies produced by a 10 kW Tactical Quiet Generator (TQG). A spotlight is mounted on the top rear of the repair/stowage module. The HRV is capable of towing other vehicles. To provide a more stable platform when the crane is being used two stabilisers are lowered to the ground at the rear.

Side view of Heavy Repair Vehicle (HRV) with crane jib raised and stabilisers lowered at the rear

The cab and repair/stowage module are armoured against 7.62 mm direct fire projectiles and 152 mm artillery fragments. Both areas also have a collective NBC protection system.

Weapons carried include one 12.7 mm M2 HB machine gun mounted over a hatch in the cab roof, four 5.56 mm rifles, a smoke grenade launcher and light anti-tank weapons.

At the end of 1993 one prototype, based at Fort Hood, had been produced for concept evaluation purposes. It is intended that the HRV will be distributed at the rate of five per armoured battalion and armoured cavalry squadron, six to a mechanised infantry battalion, four to a field artillery or combat engineer battalion and three to a division cavalry squadron. In addition, one HRV will be distributed to each special howitzer, engineer and system support company.

Specifications
Crew: 4
Configuration: 10 × 10
Weight: (kerb, with crane and module) 26,082 kg
Length: 10.973 m
Width: 2.44 m
Height: 3.4 m
Ground clearance: 0.39 m

Wheelbase: 5.71 m
Angle of approach/departure: 42°/62°
Max speed: (laden) 90.1 km/h
Fuel capacity: 378.6 litres
Range: (road) 540 km
Fording: 1.22 m
Engine: Detroit Diesel Model 8V-92TA V-8 diesel developing 500 hp at 2,100 rpm
Transmission: Allison CLT-755 5-speed automatic with torque converter and equipped with Allison Transmission Electronic Controls (ATEC)
Transfer box: Oshkosh 55000 series 2-speed with 30:70 differential with lock
Steering: power-assisted on 1st, 2nd and 5th axle
Suspension: Hendrikson walking beam front and rear
Electrical system: 12/24 V
Armament:
1 × 12.7 mm M2 HB MG
4 × 5.56 mm M16A1 rifles

Status
Prototype - see text.

Development Agency
US Army Ordnance Center and School, Directorate of Combat Development (Proponent Systems), Aberdeen Proving Ground, Maryland 21005-5001, USA.
Tel: +1 410 278 4042/2309.

UPDATED

AV Technology Dragoon Engineer/Maintenance Vehicle

Description
The AV Technology Dragoon Engineer/Maintenance Vehicle, or Dragoon Model ALV, is part of the AV Technology Corporation's range of Dragoon 4 × 4 wheeled armoured vehicles. It is described as a multipurpose logistics support vehicle, equipped with a traversing 4,536 kg capacity hydraulic boom crane capable of lifting an engine pack or turret from other vehicles in the Dragoon family (or similar vehicles). Many of the components used in Dragoon family vehicles are standard components used on other in-service US military vehicles, including the M113A2 APC and the M809/M939 6 × 6 trucks. Details of the complete Dragoon range are given in *Jane's Armour and Artillery 1996-97* pages 513 to 516.

The vehicle has a crew of five, including the driver. Stowage is provided for fitter's and special tools required for field maintenance, while full width doors at the rear allow standard pallets of spares or other supplies to be carried internally. When the boom crane is in use four stabiliser legs can be lowered to the ground. A hydraulic self-recovery winch is provided.

In common with all other vehicles in the Dragoon family, the Dragoon Model ALV uses the Dragoon Power Pack system whereby the engine, transmission, cooling, hydraulic and electrical systems can be removed as a single integral unit for maintenance and repair. A central tyre inflation system is provided as is air conditioning for the interior and a halon fire extinguisher system for the engine compartment. There are five weapon ports for crew weapons provided in the hull along with 12 vision blocks.

The vehicle is amphibious and can be provided with an optional dual propeller kit. This can increase speed when in the water to 8 km/h at gross vehicle weight.

Specifications
Crew: 5
Configuration: 4 × 4
Weights:
(kerb) 11,658 kg
(GVW) 12,700 kg
Power-to-weight ratio: 23.62 hp/t
Max load: 1,044 kg
Length: 5.89 m
Width: 2.49 m
Height: 2.15 m
Ground clearance:
(axles) 0.381 m
(body) 0.658 m
Wheelbase: 3.1 m
Max speed:
(road) 116 km/h
(water) 4.8 km/h (8 km/h with propeller kit)
Fuel capacity: 350 litres
Range: (road) 885 km
Gradient: 60%
Side slope: 30%
Vertical obstacle: 0.609 m
Fording: amphibious
Engine: Detroit Diesel Model 6V-53T V-6 turbocharged diesel developing 300 hp
Transmission: Allison MT-653 5-speed automatic with torque converter
Transfer box: single speed
Turning radius: 8.5 m
Electrical system: 24 V
Batteries: 2 × 12 V
Alternator: 100 A

Status
In production.

Manufacturer
AV Technology, LLC, 50405 Patricia Drive, Chesterfield, Michigan 48051, USA.
Tel: +1 810 949 5850. Fax: +1 810 949 8760.

UPDATED

AV Technology Corporation Dragoon Engineer/Maintenance Vehicle (Dragoon Model ALV) **1996**

AV Technology Corporation Dragoon Engineer/Maintenance Vehicle (Dragoon Model ALV) **1996**

RECOVERY VEHICLES

AUSTRIA

Steyr Type 19S25 (4 × 4) recovery vehicle

Description

The Steyr Type 19S25 is a development of the Steyr 1291 series of 4 × 4 trucks (see entry in *Trucks* section). The first of these vehicles fitted with recovery equipment was produced in 1984 for the Saudi Arabian National Guard.

The vehicle uses a forward control cab-over-engine design and features a hydraulic tilting cab mechanism to provide maintenance access to the main engine and transmission components. The cab has seating for the driver and two passengers. The chassis is constructed using open channel sections, using a parallel frame design for mounting the recovery equipment. There is a heavy-duty suspension to cater for a 13,000 kg rear axle load and 6,500 kg lifted towing.

The main recovery component is a 16 t/m crane manufactured by Penz (Austria). Also provided are two Rotzler winches, one 5 tonne, the other 3.5 tonne. There are four square-section outriggers to provide stability during recovery operations and two crane jib support props for towing loads to the rear, each with a 10 tonne pulling capacity. Tools and recovery equipment are located in a large compartment under the crane boom. A spare wheel is carried on the crane boom. A triangular towbar and lifting tackle is provided for lifted towing.

Specifications

Cab seating: 1 + 2
Configuration: 4 × 4
Weight:
(kerb) 13,000 kg
(GVW) 19,000 kg
Length: 7 m
Width: 2.5 m
Height:
(top of crane) 3.6 m
(top of cab) 3 m
Ground clearance: 0.35 m
Track:
(front) 1.958 m
(rear) 1.8 m
Wheelbase: 3.8 m
Angle of approach/departure: 23°/25°
Max speed: 90 km/h
Fuel capacity: 380 litres
Gradient: 60%
Side slope: 30%
Fording: 0.8 m
Engine: Steyr WD 615.74 10 litre 6-cylinder in-line water-cooled direct injection diesel developing 250 hp at 2,200 rpm or
WD 615.73 diesel developing 290 hp at 2,200 rpm or
WD 615.78 diesel developing 320 hp at 2,200 rpm
Gearbox: ZF 9 S 109 synchromesh or 16 S 130 (for 320 hp engine)
Clutch: single disc dry, hydraulic actuation
Transfer box: Steyr VG 1200 2-speed
Steering: ZF 8046 hydraulic
Suspension: semi-elliptical springs with double-acting hydraulic shock-absorbers front and rear
Tyres: 12.00 R 20
Axle loadings:
(front) 7,500 kg
(rear) 13,000 kg
Brakes: dual circuit air
Electrical system: 24 V
Batteries: 2 × 12 V, 120 Ah
Alternator: 28 V, 30 or 55 A

Status

Production as required. In service with Cyprus, Ghana, Nigeria and the Saudi Arabian National Guard.

Manufacturer

Steyr Nutzfahrzeuge AG, Schönauerstrasse 5, A-4400 Steyr, Austria.
Tel: +43 7252 25351-0. Telex: 28200.
Fax: +43 7252 26746/28650.

UPDATED

Steyr Type 19S25 (4 × 4) recovery vehicle

Steyr Type 19S25 (4 × 4) recovery vehicle

Steyr Type 32S29 (6 × 6) recovery vehicle

Description

The Steyr Type 32S29 recovery vehicle is a development of the Steyr 1491 series of 6 × 6 trucks (see entry in the *Trucks* section). The first of these vehicles fitted with recovery equipment was produced for the Saudi Arabian National Guard.

The vehicle uses a forward control cab-over-engine design and features a hydraulic tilting cab mechanism to provide maintenance access to the main engine and transmission components. The cab has seating for the driver and two passengers. The chassis is constructed using open channel sections, using a parallel frame design for mounting the recovery equipment. There is a bogie suspension to cater for a 2 × 13,000 kg rear load and 7,500 kg lifted towing.

The main recovery component is a 30 t/m crane manufactured by Atlas or Penz (Austria). Also provided are two Rotzler winches, one 10 tonne, the other 6.5 tonne. There are four outriggers to provide stability during recovery operations and two crane jib support props for towing loads to the rear, each with a 20 tonne pulling capacity. Tools and recovery equipment are located in a large compartment under the crane boom. A spare wheel is carried on or behind the crane boom. A triangular towbar and lifting tackle is provided for lifted towing.

Specifications

Cab seating: 1 + 2
Configuration: 6 × 6
Weight:
(kerb) 18,000 kg
(GVW) 32,000 kg
Length: 9 m
Width: 2.5 m
Height:
(top of crane) 3.6 m
(top of cab) 3 m
Ground clearance: 0.35 m
Track:
(front) 1.958 m
(rear) 1.8 m
Wheelbase: 3.5 m + 1.35 m
Angle of approach/departure: 23°/25°
Max speed: 90 km/h
Fuel capacity: 380 litres
Gradient: 60%
Side slope: 30%
Fording: 0.8 m
Engine: Steyr WD 615.73 10 litre 6-cylinder in-line water-cooled direct injection diesel developing 290 hp at 2,200 rpm or
WD 615.78 diesel developing 320 hp at 2,200 rpm
Gearbox: ZF 9 S 109 synchromesh or 16 S 130 (for 320 hp engine)
Clutch: single disc dry, hydraulic actuation

Steyr Type 32S29 (6 × 6) recovery vehicle

Transfer box: Steyr VG 1200 2-speed
Steering: ZF 8046 hydraulic
Suspension: semi-elliptical springs with double-acting hydraulic shock-absorbers front and rear
Tyres: 12.00 R 20
Axle loadings:
(front) 7,500 kg
(rear) 2 × 13,000 kg
Brakes: dual circuit air
Electrical system: 24 V
Batteries: 2 × 12 V, 120 Ah
Alternator: 28 V, 30 or 55 A

Status
Production as required. In service with Cyprus, Ghana, Lebanon and the Saudi Arabian National Guard.

Manufacturer
Steyr Nutzfahrzeuge AG, Schönauerstrasse 5, A-4400 Steyr, Austria.
Tel: +43 7252 25351-0. Telex: 28200.
Fax: +43 7252 26746/28650.

UPDATED

CZECH REPUBLIC AND SLOVAKIA

TATRA AV-15 (8 × 8) heavy recovery truck

Description
The TATRA AV-15 (8 × 8) heavy recovery truck is a variant of the TATRA T 815 8 × 8.1 R truck series (see entry in *Trucks* section) and carries a telescopic boom jib and other equipment for heavy vehicle recovery. The vehicle is also intended to have a secondary role supporting construction and loading operations.

The crane is a traversing jib with one folding adaptor for extended work. It has a maximum lift capacity of 12,000 kg, although with the adaptor extended this is reduced to 4,000 kg.

The main recovery system is a winch supplied with a usable 150 m of cable. Maximum pulling capacity to the front of the vehicle is 270 kN while to the rear it is 290 kN. Stability during these operations is provided by a front-mounted dozer blade which can be lowered 250 mm into the ground surface. The dozer blade can also be used for clearing obstacles and preparing sites. Also provided is an auxiliary winch with a capacity of 6.85 kN and 320 m of cable.

Once recovered, vehicles can be towed by the AV-15 using towbars, towing triangles or other devices. Loads of up to 65 tonnes can be pulled.

Other equipment carried includes a 230 A welding and cutting set and various special recovery tools.

On the road the AV-15 has an overall length of 12.4 m. Angle of approach is 18° and departure 29°.

Other TATRA trucks equipped with recovery or multipurpose cranes include the 6 × 6 AD-20T, a multipurpose crane vehicle with a low-slung forward control cab; the 6 × 6 AD-28 crane truck; and the AVS 6 × 6 medium recovery vehicle.

TATRA AV-15 (8 × 8) heavy recovery truck (T J Gander)

TATRA AV-15 (8 × 8) heavy recovery truck in various configurations ***1997***

Specifications
AV-15
Crew: 3
Configuration: 8 × 8
Weight: 30,000 kg
Max towed load:
(road) 65,000 kg
(cross-country) 15,000 kg
(cross-country, short distance) 30,000 kg
Length: 12.30 m
Width with dozer blade: 2.57 m
Height overall: 3.36 m
Ground clearance: 0.32 m
Angle of approach: 18°
Angle of departure: 29°
Wheelbase: 1.65 + 2.87 + 1.45 m
Max road speed: 70 km/h
Min road speed: 1.7 km/h
Fuel capacity: 480 litres
Vertical obstacle: 0.60 m
Trench: 1.60 m
Fording: 1.30 m
Engines: Tatra 3-930-40 supercharged multifuel, four-stroke, OHV, air cooled with direct injection developing 360 hp at 2,200 rpm
Turning radius: 12.5 m

Status
Production as required. In service with the Czech and Slovak armed forces.

Manufacturer
(Chassis) TATRA, Štefánnikova 1163, 742 21 Kopřivnice, Czech Republic.
Tel: +42 656 40871. Fax: +42 656 42556.
(Recovery hamper) Vychodoslovenske Strojarne as Košice, Južná trieda 82, 042 73 Košice, Slovakia.
Tel: +42 95 50001. Fax: +42 95 20540.
Marketed by: Kerametal, Jašikova 2, 826 05 Bratislava, Slovakia.
Tel: +42 7 299 1111. Telex: 932 33 Kemet C.
Fax: +42 7 522 5896.

UPDATED

FRANCE

Renault TRM 10 000 CLD (6 × 6) heavy wrecker

Development
The French Army has a requirement for a heavy wrecker truck capable of supporting Leclerc MBT operations as well as performing the usual range of heavy-duty recovery operations. Using the Renault TRM 10,000 6 × 6 truck as the basic chassis, trials were carried out with two types of hydraulic crane, both with a 15,000 kg lifting capacity and almost identical in appearance. One crane was a PPM product, the other, the GIC 156, a Pinguely product. The Pinguely GIC 156 was selected by the French Army. In total 300 CLD *(Camion Lourd de Dépannage)* crane and truck combinations were ordered for delivery between 1992 and 1999. This vehicle will replace the old Berliet TBU 15 CLD (6 × 6) wreckers used by the French Army.

Description
For details of the basic Renault TRM 10 000 vehicle refer to the entry in the *Trucks* section.

Designed to take up all the forces supported by the CLD in all its operational configurations, the subframe is fixed onto the side members of the carrier and made up of high yield strength steel plate boxes plus cross members connected to stabilisation beams.

The CLD is equipped with four separately controlled outriggers with sloped telescopic beams. They are actuated by four internal cylinders and are fully protected and equipped with aluminium removable support shoes that can be replaced by anchoring pads for hauling. When the CLD is moved carrying a load, four hydraulic cylinders equipped with resilient limit switch dampers lock the suspension for the rear tandem of the carrier.

The hauling winch is actuated by a hydraulic motor driven from a power take-off on the vehicle's gearbox and is equipped with a rope tensioner with a wheel type fray proof device, a clutching out handle, an anti-drift brake and an unwinding safety device. The winch has a maximum direct pull of 18 tonnes (this can be increased to 36 tonnes with a sheave) a maximum pulling speed of 14.5 m/min and is provided with 90 m of 20 mm diameter cable. The cable is carried in a special trough inside the subframe for unloading/loading using an auxiliary winch with a maximum pulling speed of 40 m/min and 200 m of 5 mm diameter cable.

The CLD is mounted on a constant-traverse turntable with a cab for the operator; maximum traversing speed is 2 rpm. The telescopic box-section jib is raised by a double-acting cylinder equipped with a balancing valve and check valve. The crane can lift 20 tonnes at 2.5 m although it is possible to lift loads such as tank turrets weighing 20 tonnes at 4.2 m. Minimum jib length is 4.4 m and maximum 7 m.

For towing, the CLD can pull a weight of up to 40 tonnes at 25 km/h. It is also possible to lift and tow a weight of 7.5 tonnes by the axle.

The CLD is provided with six toolboxes, a fixed vice and spirit level and working lights. Other equipment provided with the CLD includes four anchoring shoes, an opening sheave, two 36 tonne sheaves for hauling, a safety sleeve used when removing tank turrets, ropes, chains, towing straps, a towing triangle, adjustable connecting rods, a hinged bar, welding and cutting equipment and various other tools. Optional equipment includes a hydraulically operated rear spade-type anchor.

Renault TRM 10 000 CLD heavy wrecker (Pierre Touzin) ***1996***

Specifications
Cab seating: 1 + 3
Configuration: 6 × 6
Weight:
(complete) 23,000 kg
(crane equipment) 11,500 kg
Towed load: up to 40,000 kg
Length: (jib forward) 8.9 m
Width:
(travelling) 2.48 m
(stabilisers extended) 2.903 m
Height: (overall) 3.436 m
Wheelbase: 4.3 m + 1.4 m
Max speed: 90 km/h
Range: 1,000 km
Angle of approach/departure: 45°/33°
Engine: Renault MIDS 06.20-45 6-cylinder diesel developing 275 hp at 2,200 rpm
Transmission: ZF 6 HP 500 automatic with torque converter
Transfer box: 2-speed
Steering: hydraulic power-assisted
Tyres: 14.00 × 20
Electrical system: 24 V

Status
In production. In service with the French Army (300 ordered).

Manufacturers
Vehicle: Renault Véhicules Industriels, 40 rue Pasteur, BP 302, F-92156 Suresnes Cedex, France.
Tel: +33 1 40 99 71 11. Telex: 620 567 sdce f.
Fax: +33 1 40 99 71 08.
Crane: Pinguely Industries, EP 9, F-42152 l'Horme, France.
Tel: +33 4 77 22 82 80. Telex: 330 169 F.
Fax: +33 4 77 22 76 22.

UPDATED

Berliet TBC 8 KT (6 × 6) wrecker

Description
The Berliet TBC 8 KT (6 × 6) CLD (*Camion Lourd de Dépannage*) is a member of the Berliet GBC 8 KT (6 × 6) 4 tonne truck family. The reader is referred to the entry on the latter vehicle in the *Trucks* section for a detailed history of its development and a description of its technical characteristics.

The rear winch is provided with 60 m of cable and has a maximum capacity of 5,000 kg which can be increased to 7,000 kg with the aid of earth anchors. The crane mounted in the rear is hydraulically operated and slews through 270°; two stabilisers on each side are lowered to the ground manually. Equipment carried includes: two telescopic arm brackets to support the jib; two rear radius rods to support the jib on the ground; towbars; two earth anchor spades; tackles; one acetylene bottle and hose; one oxygen bottle and hose; and tools.

Part of the French Army fleet of Berliet TBC 8 KT (6 × 6) trucks is now being upgraded in a number of key areas. This upgrade may include the wrecker.

Variant
A light recovery version of the GBC 8 KT (6 × 6) truck is also in service. This has drop sides and an overhead rail and is provided with a 2,000 kg hoist which can be used to lift light vehicles and change components. A winch with a capacity of 5,000 kg (or 7,000 kg with earth anchors in use) is fitted and can be led out through the front or rear. Loaded weight is 12,600 kg and empty weight is 9,600 kg (for example 4,400 kg on the front axles and 2,600 kg on each of the rear axles).

Berliet GBC 8 KT (6 × 6) wrecker

Specifications
Cab seating: 1 + 2
Configuration: 6 × 6
Weight: (empty) 13,650 kg
Length: 7.693 m
Width: 2.4 m
Height: 2.97 m
Ground clearance:
(max) 0.515 m
(axles) 0.28 m
Track: 1.86 m
Wheelbase: 3.31 m + 1.28 m
Angle of approach/departure: 45°/45°
Max speed: (road) 80 km/h
Range: 800 km
Fuel capacity: 200 litres
Max gradient: 50%
Side slope: 22%
Fording: 1.2 m
Engine: Berliet MK 520 5-cylinder multifuel developing 125 hp at 2,100 rpm
Transmission: Berliet BDSL with 6 forward and 1 reverse gears, 2-speed transfer case
Turning radius: 10.5 m
Steering: worm gear and nut with servo assistance, turning circle indicator provided
Suspension:
(front) longitudinal leaf springs with hydraulic shock-absorbers
(rear) leaf springs
Tyres: 12.00 × 20
Brakes: air with 3 circuits (1 for front, 1 for rear and 1 for trailer) with exhaust retarder
Electrical system: 24 V
Batteries: 4 × 12 V

Status

Production complete. In service with the French Army. The cargo truck version is also used by Algeria, Austria, China, Iraq, Morocco and Portugal, some of which also use the recovery version.

Manufacturer

Automobiles M Berliet, Bourg. Berliet is now part of the Renault Group.
Enquiries to: Renault Véhicules Industriels, 40 rue Pasteur, F-92156 Suresnes Cedex, France.
Tel: +33 1 40 99 71 11. Telex: 620 567 sdce f.
Fax: +33 1 40 99 71 08.

UPDATED

Berliet TBU 15 CLD (6 × 6) wrecker

Description

The TBU 15 CLD (*Camion Lourd de Dépannage*) is the recovery member of the GBU 15 (6 × 6) cargo truck family and uses the same basic chassis and cab. The basic truck model entered production in 1959/60 with the wrecker following in 1962.

Mounted at the rear is a hydraulic crane with a maximum lifting capacity of 10,000 kg, which can be swung through 278°. Its telescopic jib has a maximum length when extended of 5.513 m. Two stabilisers on each side of the hull are lowered manually when heavier loads are being lifted. The front-mounted winch has 60 m of 15 mm diameter cable and a capacity of 5,000 kg, which can be increased to 7,000 kg with the aid of earth anchors. The rear-mounted winch has 60 m of 20 mm diameter cable and a maximum capacity of 8,000 kg which can be increased to 12,000 kg with earth anchors. A full range of tools and other equipment is carried, including oxygen and acetylene cylinders. Its replacement for the French Army is the Renault TRM 10 000 CLD (6 × 6) heavy wrecker which is covered in the previous entry.

Specifications

Cab seating: 1 + 3
Configuration: 6 × 6
Weight: 21,200 kg
Length: 8.88 m
Width: 2.5 m
Height: 3 m
Ground clearance: (max) 0.6 m
Track: 2.04 m
Wheelbase: 3.48 m + 1.45 m
Angle of approach/departure: 45°/45°
Max speed: (road) 68 km/h
Range: 800 km
Fuel capacity: 400 litres (2 × 200 litre tanks)
Max gradient: 60%
Max side slope: 30%
Fording: 1 m
Engine: Berliet MK 640 6-cylinder, multifuel, water-cooled developing 214 hp at 1,800 rpm
Transmission: manual with 5 forward and 1 reverse gears, 2-speed transfer box
Turning radius: 9.2 m
Steering: screw and nut with power assistance
Suspension: longitudinal leaf springs front and rear
Tyres: 14.00 × 20
Brakes: air
Electrical system: 24 V
Batteries: 4 × 6TN

Status

Production complete. In service with the Belgian, French, Irish, UAE and other armed forces.

Manufacturer

Automobiles M Berliet, Bourg. Berliet is now part of the Renault Group.
Enquiries to: Renault Véhicules Industriels, 40 rue Pasteur, F-92156 Suresnes Cedex, France.
Tel: +33 1 40 99 71 11. Telex: 620 567 sdce f.
Fax: +33 1 40 99 71 08.

UPDATED

Berliet TBU 15 CLD (6 × 6) wrecker in service with the USA
1997

GERMANY

MAN Type 27.365 VFAE (8 × 8) recovery vehicle

Description

The MAN Type 27.365 VFAE is a development from the main line of MAN 15 tonne trucks and was produced for use by the US Army in Europe and the Canadian Armed Forces Europe. The first two examples of this recovery vehicle for the US Army were delivered between July and September 1981 under the designation XM1002. They were later approved for service as the M1002. Production ceased in 1987. The US Army vehicles have been disposed of, while the Canadian vehicles are no longer based in Europe.

The main engine is situated just behind the tilt-forward cab which has air-assisted suspension seating for the driver and two passengers. The entire chassis frame is constructed from all-welded steel and is a low torsion box structure. A spare wheel is located to the right of the engine. Behind the engine is an open equipment-carrying box with drop sides 2.3 m long. Behind the box is an Atlas crane with a capacity of 20,000 kg which is provided with a remote-control unit; outriggers for stability can be mounted. Directly over the third and fourth axles is a substructure which can either be used as a 'fifth wheel' load-carrying surface or as the position for a recovery assembly capable of lifting up to 11,340 kg and towing 38,556 kg. The recovery assembly may be fitted with a towing bar or other recovery equipment. Other special equipment includes a heavy-duty recovery winch with a 20,412 kg capacity and a self-recovery winch with a 9,072 kg capacity. Power take-offs for all these equipments can be selected from the main power train. Extra spare wheels can be carried over the main recovery assembly. When originally being marketed a range of optional equipment was availabe including paint that was decontaminant resistant and an artic kit.

Specifications

Cab seating: 1 + 2
Configuration: 8 × 8
Weight:
(kerb) 20,180 kg
(GCW) 45,300 kg
Axle loadings:
(front) 2 × 7,500 kg
(rear) 2 × 10,500 kg
Length: 8.996 m
Width: 2.5 m
Height:
(top of crane) 3 m approx
(top of cab) 2.921 m
Ground clearance: 0.457 m
Track:
(front) 2.066 m
(rear) 2.072 m

MAN recovery vehicle on (8 × 8) Kat 1A1 chassis

Wheelbase: 1.93 m + 3.2 m + 1.5 m or 1.93 m + 2.77 m + 1.5 m
Angle of approach/departure: 42.5°/45°
Max speed: (approx) 90 km/h
Fuel capacity: 600 litres
Max gradient: (without towed load) 60%
Side slope: 40%
Fording: 1.22 m
Engine: MAN D 2840 MFG V-10 water-cooled diesel developing 365 hp at 2,300 rpm
Gearbox: ZF synchromesh 4S-150 8-speed
Clutch: ZF torque converter clutch 400
Transfer box: ZF-GPA
Steering: recirculatory ball hydro-steering
Suspension: coil springs and telescopic shock-absorbers
Tyres: 16.00 × 20
Brakes: two circuit air and air over hydraulic
Electrical system: 24 V
Batteries: 4 × 12 V, 125 Ah
Alternator: 28 V, 95 A

Status
Production complete. In service with the Canadian Forces and other armies. Replaced in production by MAN Kat 1A1.1 - see following entry.

Manufacturer
MAN Nutzfahrzeuge AG, Postfach 500620, D-8000 Munich 50, Germany.
Tel: +49 89 1580-1. Telex: 211023.

UPDATED

MAN Kat 1A1.1 6 × 6 Type 24.360 DFAEG recovery vehicle

Description
This recovery vehicle was developed in close co-operation with the British Army and the subcontractors Atlas Weyhausen and Rotzler. The vehicle has three main components: the chassis supplied by MAN; a self-recovery winch with front pull, provided by Rotzler; and a recovery unit with a crane, heavy towing winch and towing boom, all provided by Atlas Weyhausen.

The MAN Kat 1A1.1 6 × 6 chassis is powered by a 360 hp six-cylinder in-line engine. It has a low torsion frame and provides stability under military conditions. The hydraulically powered recovery unit was designed for military purposes and can be used on roads as well as extreme off-road conditions. The crane can be used to lift tank engines, containers and so on, without difficulty. Maximum payload moment of the crane is 45 t/m, maximum overhang length is 7.7 m and maximum lift height is 9.25 m. Maximum load on the hook (first step) is 15,000 kg. The hydraulically extended recovery arm has a maximum carrying capacity of 8,000 kg. The tractive force on the recovery cable on the first layer of the heavy towing winch is approximately 250 kN.

MAN Kat 1A1.1 6 × 6 Type 24.360 DFAEG recovery vehicle towing MAN (8 × 8) truck

Specifications
Cab seating: 1 + 2
Configuration: 6 × 6
Weight: (kerb, with all recovery equipment) 24,700 kg
Axle loadings, technical permissible:
(front) 7,500 kg
(rear) 2 × 13,000 kg
Length: (less front counterweight and rear pintle) 8.8 m
Width: 2.5 m
Height: (top of crane) 3.3 m
Wheelbase: 4.25 m + 1.5 m
Angle of approach: 38°
Max speed: (without towed load) approx 95 km/h
Fuel capacity: 400 litres
Max gradient: (without towed load) 60%
Side slope: 40%
Engine: MAN D 2866 LXFG 6-cylinder in-line, water-cooled, turbocharged and intercooled, developing 360 hp
Gearbox: ZF synchromesh 16S-160A, 16-speed
Clutch: ZF torque converter WSK 400
Transfer box: ZF GPA flanged onto transmission
Steering: recirculatory ball hydro-steering ZF 8046
Suspension: coil springs and telescopic shock-absorbers
Tyres: 16.00 × 20
Brakes: dual circuit air and air over hydraulic
Electrical system: 24 V
Batteries: 4 × 12 V, 100 Ah
Alternator: 28 V, 120 A

Status
Production as required. In service with undisclosed countries.

Manufacturer
MAN Nutzfahrzeuge AG, Postfach 500620, D-8000 Munich 50, Germany.
Tel: +49 89 1580-1. Telex: 211023.

UPDATED

TADANO FAUN Recovery Crane Type RTF 25

Description
The TADANO FAUN Recovery Crane Type RTF 25 was developed for the recovery and repair of wheeled and tracked vehicles as well as for general crane-lifting purposes. The vehicle has a high degree of cross-country mobility using a 4 × 4 drive configuration and with steering on both axles.

The crane uses a forward control cab constructed from steel and glass fibre and with the crane turntable immediately behind the cab. The crane has a telescopic jib with a maximum extension of 22.34 m. Telescopic stabilisers are provided for use when the crane is operating. The crane is stated to have a lifting capacity of 25,000 kg at 7.54 m.

TADANO FAUN Recovery Crane Type RTF 25

Specifications
Cab seating: 1 + 1
Configuration: 4 × 4
Weight: 21,600 kg
Length: approx 10.45 m
Width: 2.5 m
Height: 3.35 m
Wheelbase: 3.8 m
Max speed: 80 km/h
Fuel capacity: 300 litres
Engine: Mercedes-Benz OM 366 LA water-cooled diesel developing 240 hp at 2,600 rpm
Gearbox: ZF 6 WG 180 power shift with HN 500 converter providing 6 forward and 2 reverse gears
Steering: ZF semi-integral power, dual circuit
Tyres: 14.00 R 25
Brakes: compressed air, dual circuit

Status
Production as required. In service with undisclosed countries.

Manufacturer
TADANO FAUN GmbH, Postfach 100108, D-91205 Lauf ad Pegnitz, Germany.
Tel: +49 9123 185211. Telex: 626093.
Fax: +49 9123 185115.

UPDATED

TADANO FAUN BKF 30.40 recovery vehicle for Patriot/ Roland

Description

The TADANO FAUN BKF 30.40 8 × 8 recovery vehicle for Patriot/Roland was developed in response to a requirement stated in 1985 by the German government. A production contract was placed in 1990 and by 1995 two vehicles were undergoing troop trials. The only known user is the German Air Force which has taken delivery of 10 vehicles for use with its air defence missile units.

The vehicle is based on the MAN 32.422 VFAEG chassis and is modified by fitting a hydropneumatic suspension and larger tyres. The main equipment fitted is a TADANO FAUN recovery crane with a capacity of 15.3 tonnes at a boom length of 4.8 m. Also supplied is: a hydraulic jackplate; hydraulic four-point support plates; a Rotzler Treibmatic winch with a towing capacity of 20 tonnes over the entire cable length of 100 m; auxiliary cable with spooler for a 240 m long; 7 mm cable to tow the main cable; various other items of Rotzler towing and recovery equipment; and a Treibmatic-driven towing attachment with a 13 tonne lifting capacity for towed vehicles. Remote control is standard.

Specifications

Cab seating: 1 + 1
Configuration: 8 × 8
Weight: 29,500 kg
Max axleload, front/rear: 15,000 kg/26,000 kg
Max tow weight: 30,000 kg
Length: approx 10.15 m
Width: 2.75 m
Height: 3.15 m
Wheelbase: 1.93 m + 3.57 m + 1.5 m
Engine: MAN Model D2866 LFG 6-cylinder water-cooled diesel developing 414 hp at 2,000 rpm
Tyres:
(front) 16.00 R 20
(rear) 24 R 20.5

Status

Production as required. In service with German Air Force (10 units).

Manufacturer

TADANO FAUN GmbH, Postfach 100108, D-91205 Lauf ad Pegnitz, Germany.
Tel: +49 9123 185211. Telex: 626093.
Fax: +49 9123 185115.

UPDATED

TADANO FAUN BKF 30.40 recovery vehicle for Patriot/Roland ***1996***

TADANO FAUN BKF 30.40 recovery vehicle for Patriot/Roland ***1996***

ISRAEL

EYAL heavy recovery vehicles

Description

The EYAL Engineering and Industrial Company Limited produces five models of heavy recovery vehicle. They are: the HRV 720 with a lifting moment of 20 t/m; the HRV 930 with a lifting moment of 30 t/m; the HRV 1236 with a lifting moment of 36 t/m; the HRV 1545 with a lifting moment of 45 t/m; and the heaviest model, the HRV 1760 with a lifting capacity of 60 t/m.

All these models are produced using a 6 × 6 chassis with the model and make of chassis according to the customer's choice; most examples used by the Israeli armed forces use locally modified US supplied M809 (6 × 6) series chassis. The recovery hamper is produced as a bolt-on unit and consists of a hydraulic crane, four hydraulically operated outriggers, a main recovery hydraulic winch, a towing hitch, an operating cab and the hydraulic system.

The crane has a full 360° traverse and is equipped with a planetary hydraulic hoisting winch mounted on the main boom. The winch has a built-in automatic brake. Boom operations are carried out by a hydraulic double-acting cylinder with built-in safety devices. The crane block may be assembled with one or two sheaves on a roller bearing and with a swivel hook.

The four outriggers are located with one pair at the front of the upper structure and the other pair at the rear. Each pair is operated separately under hydraulic control and with automatic built-in safety locking valves.

The main recovery winch is mounted on the front of the upper structure and is equipped with 70 m of cable. The winch may have a line pull of 15, 20 or 25 tonnes depending on the model. The cable is drawn under the floor of the recovery unit.

The towing hitch is mounted at the rear and can tow suspended loads of 6, 7, 8 or 10 tonnes, depending on the model involved.

The operating cab is mounted on the right-hand side of the crane boom and levers are used to control the hydraulic system. A rainproof canvas cover is supplied for the operator.

The hydraulic system is driven from the vehicle power take-off.

EYAL also produces a range of hydraulic cranes with varying lift capacities. One of them, the Model 3501, has been mounted on Israeli Army M3 half-tracks.

EYAL heavy recovery vehicle on US supplied 6 × 6 chassis lifting recovery frame complete with crane

Status

Production as required. In service with the Israeli armed forces.

Manufacturer

EYAL Engineering and Industrial Company Limited, PO Box 374, IL-46103 Herzelia, Israel.
Tel: +972 9 541111. Telex: 33597.
Fax: +972 9 541217.

UPDATED

ITALY

IVECO 6605 AG (6 × 6) recovery vehicle

Description

The IVECO 6605 AG is the recovery member of the 6605 range of vehicles for which there is a separate entry in the *Trucks* section. The first prototype of the vehicle was completed in 1972, with first production models following in 1974. The Italian Army designation for the vehicle is the Autogru AG 70-5 t (6 × 6).

The chassis is the ladder type with two longitudinal pressed steel channels, to which are riveted the cross members, brackets and spring supports. The forward control cab is all-steel and has a removable canvas top and side screens and a windscreen which can be folded flat against the bonnet. Standard equipment includes a cab heater and ventilator.

Mounted at the rear of the vehicle is a hydraulically operated crane with an extensible jib which can lift a maximum load of 5,000 kg. Before the crane is used, two stabilisers are extended either side and lowered to the ground. The 6605 AG has two winches, one front-mounted with a capacity of 9,200 kg and one rear-mounted with a capacity of 20,000 kg.

The transmission consists of a set of gears mounted on four shafts (input, primary, layshaft and reverse). Shifting from each gear is controlled by a lever through a pneumatic servo. Shifting from the high to the low range, or vice versa, is by electropneumatic control with a preselector switch, interlocked with the clutch pedal. The transfer box is mechanical and consists of helical constant mesh gears mounted on three shafts (input, intermediate and output) of which the input one carries the dog clutch with a lockable divider differential distributing power to the front and rear axles.

IVECO 6605 AG (6 × 6) recovery vehicle showing front-mounted winch ***1997***

Specifications

Cab seating: 1 + 1
Configuration: 6 × 6
Weight:
(empty) 19,000 kg
(loaded) 25,000 kg
Max load: 5,000 kg
Towed load: 15,000 kg
Length: 8.32 m
Width: 2.5 m
Height:
(cab) 2.74 m
(top of crane in travelling position) 3.01 m
Ground clearance: 0.363 m
Track: 2.072 m
Wheelbase: 3.217 m + 1.365 m
Angle of approach/departure: 45°/40°
Max speed: (road) 80 km/h
Range: 700 km
Fuel capacity: 360 litres
Max gradient: 60%
Max side slope: 20%
Fording: 1.5 m
Engine: IVECO Model 8212.02 6-cylinder in-line water-cooled diesel developing 260 hp at 2,200 rpm
Gearbox: manual with 8 forward and 2 reverse gears
Clutch: dual dry plate
Transfer box: 2-speed
Steering: ZF hydraulic-assisted
Turning radius: 8 m
Suspension:
(front) 2 semi-elliptical constant rate leaf springs with double-acting hydraulic shock-absorbers
(rear) rocker and torque arms with 2 constant rate leaf springs
Tyres: 14.00 × 20
Brakes:
(main) drum air-operated on all wheels, dual circuit with connections for trailer braking
(parking) drum, hand-operated, mounted on transfer rear output shaft
Electrical system: 24 V
Batteries: 4 × 12 V, 90 Ah

Status

Production complete. In service with the Italian Army.

Manufacturer

IVECO SpA, Defence Vehicle Division, Via Volta 6, I-39100 Bolzano, Italy.
Tel: +39 471 905111. Telex: 400541.
Fax: +39 471 905444.

UPDATED

IVECO 90 PM 16 (4 × 4) wrecker with 5 tonne crane

Description

The IVECO 90 PM 16 (4 × 4) truck chassis was developed to mount a 5,000 kg capacity FARID crane boom. The crane is mounted behind the vehicle cab on a sheet steel subframe connected to the chassis by brackets. Onto this subframe are mounted the superstructure, crane and outriggers and when the crane is in use the entire vehicle is raised onto the outriggers for overall stabilisation. The crane is a FARID model F 5 which is hydraulically operated. The maximum boom length is 5.7 m, the maximum boom elevation 75° and the traverse is a full 360°. The crane has a maximum pull to the rear of 5,000 kg but extra pull capacity is provided by a rear-mounted auxiliary winch with a 4,000 kg capacity. A self-recovery winch is mounted at the front and this has a capacity of 3,000 kg. The four outriggers, one at each corner of the subframe, are hydraulically operated from a panel at the side of the vehicle; the crane controls are on a panel by the operator's seat.

Accessories include a drawbar, collapsible snatch blocks, a set of chains with hooks, a spare wheel holder and tool boxes. The deck of the subframe, which is covered with a non-slip steel plate, measures 3.4 m long and 2.3 m wide and may be used for carrying extra equipment or cargo.

IVECO 90 PM 16 (4 × 4) wrecker with 5 tonne crane deployed in operating position

Specifications

Cab seating: 1 + 1
Configuration: 4 × 4
Weight:
(gross) 9,700 kg
(kerb) 8,200 kg
Payload: (load hang to hook) 1,500 kg
Towing capacity: 4,000 kg
Length: (incl crane overhang) 7.042 m
Width: 2.476 m
Height: (cab) 2.627 m
Ground clearance: 0.47 m
Track: 1.851 m
Wheelbase: 3.7 m
Angle of approach/departure: 42°/42°
Max speed: 80 km/h
Range: 700 km
Power-to-weight ratio: 16.5 hp/t
Max gradient: 60%
Side slope: 30%

Fording: 0.7 m
Engine: IVECO Model 8062.24 6-cylinder supercharged direct injection water-cooled diesel developing 160 hp at 3,200 rpm
Gearbox: manual with 5 forward and 1 reverse gears
Clutch: single dry plate
Transfer box: 2-speed
Steering: recirculating ball with hydraulic servo
Turning radius: 7.5 m
Suspension: leaf springs (dual at rear) with hydraulic shock-absorbers (telescopic)
Tyres: 12.5 R20 PR22

Status
Production complete. In service with Venezuela (60).

Manufacturers
Chassis: IVECO SpA, Defence Vehicle Division, Via Volta 6, I-39100 Bolzano, Italy.
Tel: +39 471 905111. Telex: 400541.
Fax: +39 471 905444.
Crane: FARID SpA, Corso Savona 39 bis, I-10024 Moncalieri, Italy.
Tel: +39 11 645454. Telex: 220348.

UPDATED

IVECO 90.17 WM (4 × 4) wrecker

Description
The IVECO 90.17 WM (4 × 4) wrecker replaced the IVECO 90 PM 16 wrecker (see previous entry) in production and is based on the chassis of the IVECO 90.17 WM (4 × 4) truck fitted with a 6,000 kg capacity crane. The hydraulic crane is mounted behind the vehicle on a sheet steel subframe connected to the chassis by brackets. Onto this subframe are mounted the superstructure, crane and outriggers and when the crane is in use, the entire vehicle is raised onto the outriggers for overall stabilisation. The crane maximum boom length is 5.1 m, the maximum boom elevation 60° and traverse is 220°. The crane has a maximum pull to the rear of 6,000 kg but extra pull capacity is provided by a rear-mounted auxiliary winch, also with a 6,000 kg capacity. A self-recovery winch is mounted at the front and this has a capacity of 3,000 kg. The four outriggers, one at each corner of the subframe, are hydraulically operated, while the crane controls are in the crane turret cab.

Accessories include a spare wheel holder and tool boxes. The deck of the subframe is covered with a non-slip steel plate and may be used for carrying extra equipment or stores.

IVECO 90.17 WM (4 × 4) wrecker with 6 tonne crane in travelling configuration

Specifications
Cab seating: 1 + 1
Configuration: 4 × 4
Weight: (gross) 9,080 kg
Payload: (load hang to hook) 2,000 kg
Towing capacity: 4,000 kg
Length: (incl crane overhang) 6.76 m
Width: 2.3 m
Height: (cab) 2.75 m
Ground clearance: 0.47 m
Track: 1.852 m
Wheelbase: 3.7 m
Angle of approach/departure 45°/30°
Max speed: 80 km/h
Range: over 500 km
Max gradient: 60%
Side slope: 30%
Fording: 0.7 m
Engine: IVECO Model 8060.25 6-cylinder supercharged direct injection water-cooled diesel developing 170 hp at 3,000 rpm
Gearbox: manual with 5 forward and 1 reverse gears
Clutch: single dry plate
Transfer box: 2-speed
Steering: recirculating ball with hydraulic servo
Turning radius: 7.5 m
Suspension: leaf springs (dual at rear) with telescopic hydraulic shock-absorbers
Tyres: 14.5 × 20

Status
Production as required. In service with undisclosed countries.

Manufacturer
IVECO SpA, Defence Vehicle Division, Via Volta 6, I-39100 Bolzano, Italy.
Tel: +39 471 905111. Telex: 400541.
Fax: +39 471 905444.

UPDATED

IVECO 230.35 WM (6 × 6) recovery vehicle

Description
The IVECO 230.35 WM (6 × 6) recovery vehicle is a variant of the IVECO 230.35 WM (6 × 6) 10,000 kg truck fitted with a 14,000 kg crane. The recovery equipment is mounted on a welded steel subframe that is linked by brackets to the vehicle chassis, crane, winch outriggers and related control units. The upper part of the subframe carries the 360° rotating platform for the crane which is an Isoli Model M 140 — 49.8 t/m with a maximum lifting capacity of 14,000 kg. The crane hydraulics are powered from the main engine with the crane using a telescopic boom. The crane has a hook-lifting speed of 8 m/min and maximum elevation is 55°. The maximum boom length is 6.35 m and the minimum 3.55 m.

The four outriggers are operated independently of each other and have independent controls for all movements. A hydraulic recovery winch is fitted to the side of the vehicle for front and rear operations. It has a maximum pulling capacity (on the first layer) of 15,000 kg.

The IVECO 230.35 WM (6 × 6) recovery vehicle is fitted with a torque converter to obtain the maximum tractive effort in any gear.

IVECO 230.35 WM (6 × 6) recovery vehicle in travelling configuration

Specifications
Cab seating: 1 + 2
Configuration: 6 × 6
Weight:
(kerb) 19,000 kg
(GVW) 25,000 kg
Load: (load hang to hook) 5,000 kg
Towed load: (off-road) 15,000 kg
Length: (overall) 8.48 m
Width:
(overall) 2.5 m
(cab) 2.494 m
Height:
(overall) 3.06 m
(cab) 3 m
Ground clearance: 0.35 m
Track:
(front) 2.027 m
(rear) 2.032 m
Wheelbase: 3.3 m + 1.38 m
Angle of approach/departure: 45°/30°
Max speed: 83 km/h
Range: 600 km
Fuel capacity: 300 litres
Max gradient: 60%
Side slope: 30%
Fording: 0.85 m
Engine: IVECO Model 8280.02 V-8 17.2 litre water-cooled diesel developing 352 hp at 2,400 rpm
Transmission: Type ZF 4 S 150 GPA with 8 forward and 1 reverse gears, with torque converter
Transfer box: electropneumatically lockable
Steering: power-assist, recirculatory ball
Turning radius: 9.5 m
Suspension:
(front) single flexibility leaf springs with hydraulic telescopic shock-absorbers
(rear) single flexibility leaf springs, reversed, fully articulated
Tyres: 14.00 × 20

Brakes:
(main) drum, air on all wheels
(parking) mechanical on rear wheels
(exhaust) pneumatic, pedal control
Electrical system: 24 V
Batteries: 4 × 12 V, 143 Ah
Generator: 650 W

Status
Production as required. In service with undisclosed countries.

Manufacturer
IVECO SpA, Defence Vehicle Division, Via Volta 6, I-39100 Bolzano, Italy.
Tel: +39 471 905111. Telex: 400541.
Fax: +39 471 905444.

UPDATED

IVECO 260.35 WM (6 × 6) recovery vehicle

Description
The IVECO 260.35 WM (6 × 6) recovery vehicle is a variant of the IVECO 260.35 WM (6 × 6) 10,000 kg truck, fitted with a 12,000 kg hydraulic crane. The recovery equipment is mounted on a welded steel subframe, linked by brackets to the vehicle chassis, crane, winch, outriggers and related control units. The upper part of the subframe carries the 360° rotating platform for the crane, an Isoli Model M 120 – 42.8 t/m with a maximum lifting capacity of 12,000 kg. The hydraulic system uses three pumps driven from the vehicle main engine with the crane having a telescopic extendible boom. The crane has a hook-lifting speed of 10 m/min with the maximum elevation angle of 50°. The maximum boom length is approximately 6.4 m and the minimum about 3.6 m.

The four hydraulic outriggers are operated independently of each other and have independent controls for all movements. An axle lock can lock the rear bogie, raising it off the ground for increased stability during crane operations. A hydraulic recovery winch is fitted to the side of the vehicle cab for front and rear operations. It has a maximum pulling capacity (on the first layer) of 15,000 kg.

The IVECO 260.35 WM (6 × 6) recovery vehicle is fitted with a torque converter to obtain the maximum tractive effort in any gear.

IVECO 260.35 (6 × 6) recovery vehicle with 12 tonne crane in travelling configuration

Specifications
Cab seating: 1 + 1
Configuration: 6 × 6
Weight:
(kerb) 20,600 kg
(GVW) 25,600 kg
Load: (load hang to hook) 5,000 kg
Towed load: (off-road) 15,000 kg
Length: (overall) 8.86 m
Width: (overall) 2.5 m
Height:
(overall) 3.03 m
(exhaust over cab) 3.123 m
Ground clearance: 0.312 m
Track:
(front) 1.927 m
(rear) 1.809 m
Wheelbase: 3.85 m + 1.38 m
Angle of approach/departure: 33°/27°
Max speed: 94 km/h
Range: 600 km
Fuel capacity: 300 litres
Max gradient: 60%
Side slope: 30%
Fording: 0.65 m
Engine: IVECO Model 8280.02 V-8 17.2 litre water-cooled diesel developing 352 hp at 2,400 rpm
Transmission: ZF 4 S 150 GPA with 8 forward and 1 reverse gears, with torque converter
Transfer box: electropneumatically operated
Steering: power-assist, recirculatory ball
Turning radius: 9.5 m
Suspension:
(front) single flexibility leaf springs with hydraulic telescopic shock-absorbers
(rear) single flexibility leaf springs, reversed, fully articulated
Tyres: 12.00 × 20
Brakes:
(main) drum, air on all wheels
(parking) mechanical on rear wheels
(exhaust) pneumatic, pedal control
Electrical system: 24 V
Batteries: 2 × 12 V, 143 Ah
Generator: 650 W

Status
Production as required. In service with undisclosed countries.

Manufacturer
IVECO SpA, Defence Vehicle Division, Via Volta 6, I-39100 Bolzano, Italy.
Tel: +39 471 905111. Telex: 400541.
Fax: +39 471 905444.

UPDATED

JAPAN

Mitsubishi Model FW415M1 (6 × 6) recovery vehicle

Description
This recovery vehicle uses the same chassis as the Type 74 (6 × 6) 10,000 kg cargo truck used by the Japanese Self-Defence Force. The two-door all-steel control cab can be tilted forward to allow access to the engine for maintenance.

Mounted to the rear of the cab is a hydraulically operated crane with a telescopic jib. When the recovery equipment is being used, two stabilisers are lowered each side of the recovery equipment. Winches are provided front and rear and earth anchors can be fitted to the rear of the vehicle if required.

Mitsubishi Model FW415M1 (6 × 6) recovery vehicle **1997**

Specifications
Cab seating: 1 + 2
Configuration: 6 × 6
Weight:
(unladen) 18,100 kg
(laden) 18,340 kg
Length: 9.32 m
Width: 2.49 m
Height: 3.23 m
Wheelbase: 4.15 m + 1.3 m
Track:
(front) 1.915 m
(rear) 1.865 m
Max speed: (road) 101 km/h
Engine: Mitsubishi 8-cylinder diesel developing 300 hp at 2,500 rpm
Turning radius: 10.5 m
Tyres: 10.00 × 20

Status
Production complete. In service with the Japanese Ground Self-Defence Force.

Manufacturer
Mitsubishi Motors Corporation, 33-8, Shiba 5-chome, Minato-ku, Tokyo, Japan.
Tel: +81 3 456 1111. Telex: 26639 bisijiko j.

UPDATED

Isuzu Type 73 (6 × 6) light wrecker

Description

The Type 73 (6 × 6) light wrecker is based on the chassis of the Type 73 (6 × 6) 3,500 kg truck, introduced into Japanese Self-Defence Force service in 1973 - for full details of this vehicle refer to entry in the *Trucks* section. The latest version of the Type 73 light wrecker (manufacturer's designation was SKW440MR but is now SKW463MR) has improvements to the engine, axles, cab and so on. It has a two-man forward control cab with a canvas roof and a windscreen that can be folded forward onto the bonnet. The load-carrying area behind the cab carries a hydraulic crane with a maximum lifting capacity of 4,800 kg. Stabilising jacks are lowered when the crane, which has a full 360° traverse, is in use. There is also a recovery winch with a capacity of 4,500 kg.

Isuzu Type 73 (6 × 6) light wrecker

Specifications

Cab seating: 1 + 1
Configuration: 6 × 6
Weight:
(unladen) 14,200 kg
(laden) 19,160 kg
Towed load: (road) 6,000 kg
Length: 7.81 m
Width: 2.49 m
Height: 3.06 m
Ground clearance: 0.33 m
Wheelbase: 3.295 m + 1.31 m
Track:
(front) 1.97 m
(rear) 1.84 m
Angle of approach/departure: 44°/46°
Max road speed: 95 km/h
Range: 500 km
Fuel capacity: 170 litres
Fording: 0.8 m
Engine: Isuzu 8PD1 V-8 liquid-cooled diesel developing 240 hp at 2,300 rpm
Transmission: manual with 5 forward and 1 reverse gears
Clutch: single dry plate
Transfer box: 2-speed
Steering: power
Turning radius: 9.4 m
Suspension:
(front) elliptical springs and hydraulic shock-absorbers
(rear) semi-elliptical springs
Brakes:
(main) air/hydraulic
(parking) mechanical
Tyres: 11.00 × 20
Electrical system: 24 V
Winch capacity: 4,500 kg

Status

Production complete. In service with the Japanese Ground Self-Defence Force.

Manufacturer

Isuzu Motors Limited, 22-10 Minamioi 6-chome, Shinagawa-Ku, Tokyo, Japan.
Tel: +81 3 762 1111. Telex: 0246 6689.

UPDATED

KOREA, SOUTH

Asia Motors KM502 5 ton (6 × 6) wrecker

Description

The KM502 5 ton (6 × 6) wrecker is part of the Asia Motors KM50 series of 5 ton vehicles. It is fitted with a MAN diesel engine and a locally produced power train. (For details of the other vehicles in the KM50 series refer to entry in the *Trucks* section.) In overall layout and appearance these vehicles are virtually identical to their US counterparts.

The KM502 is equipped with a mechanical 9,070 kg capacity winch fitted with 81 m of cable.

KM502 5 ton (6 × 6) wrecker

Specifications

Cab seating: 1 + 2
Configuration: 6 × 6
Weight:
(kerb, front axle) 4,728 kg
(kerb, rear axle) 11,404 kg
(kerb, total) 16,132 kg
(gross, front axle) 11,329 kg
(gross, rear axle) 20,244 kg
(gross, total) 21,537 kg
Payload: 5,441 kg
Length: (overall) 9.048 m
Width: 2.477 m
Height:
(top of cab) 2.689 m
(top of crane) 2.692 m
Ground clearance: 0.27 m
Track:
(front) 1.88 m
(rear) 1.829 m
Wheelbase: 4.547 m
Angle of approach/departure: 34°/43°
Max speed: 85 km/h
Range: (cruising) 805 km
Fuel capacity: 500 litres
Gradient: 68%
Fording: (without kit) 0.762 m
Engine: MAN D2156HM 10.35 litre 6-cylinder water-cooled diesel developing 236 hp at 2,200 rpm
Gearbox: manual with 5 forward and 1 reverse gears
Clutch: single dry disc
Transfer box: 2-speed
Steering: power-assist
Turning radius: 14.58 m
Suspension:
(front) semi-elliptical leaf springs with shock-absorbers
(rear) semi-elliptical inverted leaf springs
Tyres: 11.00 × 20
Brakes: air over hydraulic
Electrical system: 24 V
Batteries: 4 × 12V, 100 Ah
Alternator: 60 A

Status

Production complete. With the South Korean armed forces.

Manufacturer

Asia Motors Co Inc, 15 Yoido-dong, Yungdeungpo-Ku, Seoul, South Korea.
Tel: +82 2 783 1501/9. Telex: 24374 asiamco k.

UPDATED

NETHERLANDS

DAF YBZ 3300 DKX 500 (6 × 6) heavy recovery vehicle

Description

During early 1990 the DAF YBZ 3300 DKX 500 (6 × 6) heavy recovery vehicle was chosen by the Dutch Ministry of Defence as the replacement for the DAF YB 616/626 recovery vehicles. The new vehicle is equipped with a Norwegian designed Hägglunds Moelv recovery crane which uses hydraulically operated outrigger arms and a large hoist arm at the rear. Three Rotzler winches are provided.

Production started in 1991, with deliveries made during 1992 and 1993. In total, 255 units were ordered, with an option for a further 30 units. The Moelv crane is built under licence by DAF Special Products.

DAF YBZ 3300 DKX 500 (6 × 6) heavy recovery vehicle towing DAF YAZ 2300 DHS 6 × 6 artillery tractor

Specifications

Cab seating: 1 + 1 or 2
Configuration: 6 × 6
Weight:
(empty) approx 11,000 kg
(GVW) 27,500 kg
Towed load: (suspended) 6,000 kg
Length: 8.35 m
Width: 2.5 m
Height: (overall) 3.6 m
Ground clearance:
(front) 0.36 m
(rear) 0.9 m
Track:
(front) 1.97 m
(rear) 1.82 m
Wheelbase: 5 m
Angle of approach/departure: 30°/37°
Max speed: 88 km/h
Range: 500 km
Max gradient: 50%
Side slope: 30%
Fording: 0.9 m
Engine: DAF DKX 1160 ATi 11.6 litre 6-cylinder turbocharged diesel developing 325 hp at 2,200 rpm
Gearbox: ZF 16 S-160 A with 8 forward and 1 reverse gears, with splitter on each gear
Dropbox: ZF A 800
Steering: ZF 8098 hydraulic, power-assist
Turning radius: 12 m
Suspension: DAF semi-elliptic with Koni hydraulic shock-absorbers on front
Tyres: 12.00 R 24 XL
Brakes:
(main) air mechanical, dual line
(parking) spring cylinders on front and 2nd axle
Electrical system: 24 V
Batteries: 2 × 12 V, 125 Ah
Alternator: 55 A

Status

Production as required. In service with Netherlands (255).

Manufacturer

DAF International Military Co-ordination, Hugo van der Goeslaan 1, NL-5643 TW Eindhoven, Netherlands.
Tel: +31 40 142872. Telex: 51980 DAF BP NL.
Fax: +31 40 115456.

UPDATED

RUSSIAN FEDERATION AND ASSOCIATED STATES (CIS)

Ural-375D/Ural-4320 (6 × 6) recovery vehicle

Description

A recovery vehicle based on the Ural-375D/Ural-4320 (6 × 6) 4,000 kg truck was designed to recover vehicles weighing up to 8,500 kg in the half-loaded position (for details of the Ural-4320 refer to entry in the *Trucks* section). The vehicle can tow vehicles weighing up to 5,000 kg on dirt roads and vehicles of up to 10,000 kg on hard-surfaced roads.

The vehicle is fitted with a winch, half-loaded towing gear, a universal rigid towing blade and a trail spade at the rear. Standard equipment includes a repair kit, entrenching tools, fire extinguishers, a searchlight and pivoting signal lamps. The repair kit is carried to enable a recovered vehicle to be made ready for towing or to make a vehicle ready for transporting.

The power-operated winch is mounted at the chassis rear and has a maximum capacity of 7,000 kg. In addition to recovering vehicles it can also load the recovered vehicle onto a transport unit. The transport unit consists of a boom with cross-beam, linking chains and a longeron frame extension with struts and upper frame members. To couple the recovered vehicle to the transport unit, the boom is lowered to the level of the front bumper and locked in place by a special mechanism. The vehicle is chained to the cross-beam and then raised half off the ground by the winch cable.

With the trail spade lowered and in position, the winch capacity is increased from 7,000 kg to 14,000 kg. The boom is also intended to hold the recovery vehicle down while a vehicle is being recovered. The boom is lowered to the ground and attached to the trail spade which in turn is secured to the upper frame members by two cable braces.

The Ural-375D/Ural-4320 recovery vehicle retains the cab heater, engine heater and central tyre pressure regulation system of the standard cargo truck.

A closely allied variant using the Ural-4320 chassis is known as the MTP-A1.1 Technical Assistance Vehicle. This vehicle is used to carry out first and second line repairs and carries special sed repair equipment, tools, spare parts and materials for body repairs. Also carried are lubricants, electrical equipment and hydraulic jacks. A cargo-lifting gantry protrudes over the cab and bonnet and can be used for lifting components. Towing bars and rigs can be used to tow disabled vehicles and a winch is provided.

Status

Production as required. In service with the Russian Army and many other countries.

Manufacturer

Chassis: Ural Motor Vehicle Plant, Miass, Russia.

UPDATED

KET-L (6 × 6) recovery truck

Description

The KET-L (6 × 6) recovery truck is based on the chassis of the Ural-375E (6 × 6) 4,000 kg truck and for details of this vehicle refer to the entry in the *Trucks* section. For recovery operations the KET-L is fitted with two winches which can be used to either the front or rear and it is recommended that vehicles are recovered when in line with the KET-L recovery truck. The rear-mounted winch has a maximum capacity of 15,000 kg while the front winch has a capacity of 5,000 kg. Vehicles up to 1.5 times the weight of the KET-L can be recovered without using supports but for heavier vehicles an anchor plate must be lowered from the rear of the vehicle.

Vehicles weighing up to 5,000 kg can be towed on dirt roads using the Type 3108 rigid towbar carried on

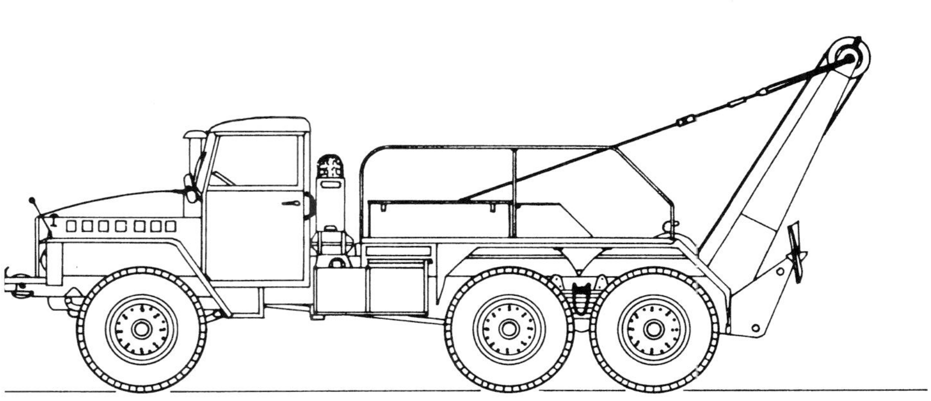

KET-L (6 × 6) recovery truck

the left rear top of the recovery hamper. Vehicles weighing up to 10,000 kg can be towed on paved roads. Mounted on the rear of the chassis is a jib crane with power supplied from the main winch via a power take-off. The crane has a boom 2.4 m long and a hook lift height of 3.5 m. Maximum capacity is 1,500 kg.

Standard equipment carried by the KET-L includes a floodlight at the rear for night recovery operations, timber baulks, a pulley block, two Type OU-2 fire extinguishers (one stowed each side of the cab rear), provision for radio communication systems and metal-cutting equipment.

Status
Production complete. In service with Russia and many other countries.

Manufacturer
Chassis: Ural Motor Vehicle Plant, Miass, Russia.

UPDATED

SOUTH AFRICA

SAMIL recovery vehicles

SAMIL produced two types of recovery vehicle, one based on the chassis of the SAMIL 50 4 × 4 and the other based on the chassis of the 6 × 6 SAMIL 100. Full details of the SAMIL 50 and SAMIL 100 can be found under the SAMIL entry in the *Trucks* section but information relevant to the two recovery vehicles is provided below. Chassis specifications are given under the SAMIL trucks entries. The SAMIL trucks were built in South Africa by Truckmakers but this company is no longer trading.

SAMIL 50 recovery vehicle
The SAMIL 50 recovery vehicle may be provided with a fully protected cab for use against land mine fragments or small arms fire. It has two winches driven by a power take-off from the main gearbox with a combined pull of 20,000 kg and two booms which extend to the rear for a suspended tow. The combined lifting capacity of the booms is 7,250 kg when retracted and 2,500 kg when fully extended. The rear platform is used for recovery equipment stowage and self-recovery rollers are fitted to each corner of the superstructure. The protected cab, when fitted, has seating for four men.

SAMIL 100 recovery vehicle
The SAMIL 100 recovery vehicle may fitted with the standard cab of the SAMIL 100 truck or with a mineproof cab; the standard cab can be enlarged to carry a crew of five men. The recovery hamper is carried to the rear of the vehicle and consists of a twin jib structure on a flatbed heavy-duty body. The main winch is mounted centrally and much of the space at the rear is taken up with recovery equipment stowage.

Status
Production complete. In service with the South African National Defence Force.

Manufacturer
Enquiries to: Armscor, Private Bag X337, Pretoria 0001, South Africa.
Tel: +27 12 428 1911. Telex: 320217.
Fax: +27 12 428 5635.

UPDATED

SAMIL 100 (6 × 6) recovery vehicle with mineproof cab (T J Gander) **1997**

SAMIL 100 (6 × 6) recovery vehicle with mineproof cabs (T J Gander) **1997**

Withings MK 1A (6 × 6) recovery vehicle

Description
The Withings MK 1A (6 × 6) recovery vehicle is based on the chassis of the SAMIL 100 (6 × 6) 10,000 kg truck suitably modified for the recovery role. The vehicle is normally fitted with a mineproof cab.

The main recovery equipment is a centrally located heavy-duty winch which is used to pull disabled vehicles and power a towing arm system which can lift a mass of 8,000 kg from 0.75 m to 1.6 m above ground level and lock it there for towing. The main winch can also be utilised for forward self-recovery and may also be used to provide lifting power for two demountable crane arms. These are normally stowed on top of the operating deck area and in use extend to a maximum of 4.5 m behind the vehicle; maximum lift height is 4 m. Other recovery equipment carried includes four skid pans with chains, tools, sheaves, a cable guard and two adjustable spotlights.

Withings MK 1A (6 × 6) recovery vehicle (T J Gander) **1997**

Specifications
Crew: up to 4
Configuration: 6 × 6
Axle rating:
(front) 7,500 kg
(rear) 10,000 kg/13,000 kg
Length: 9.77 m
Width: 2.495 m
Height: 3.35 m
Ground clearance: 0.355 m
Track:
(front) 2.002 m
(rear) 2.048 m
Wheelbase: 5.94 m
Angle of approach/departure: 30°/35°
Engine: Deutz diesel Type F 10 L 413F
Gearbox: Fuller RT 14609
Clutch: Fichtel & Sachs 42 mm
Transfer box: Getrag Z90
Steering: ZF 8043 ball and nut
Suspension: semi-elliptic leaf springs
Brakes: drum front and rear
Tyres: 16.00 R 20
Electrical system: 24 V
Batteries: 2 × 12 V, 118 Ah
Alternator: 28 V, 55 A

Status
Production complete. In service with the South African National Defence Force.

Manufacturer
TFM (Pty) Limited, PO Box 46131, Orange Grove, Transvaal, South Africa.
Tel: +27 316 2106. Fax: +27 316 1462.

UPDATED

Skimmel 8 × 8 heavy recovery vehicle

Development

The South African National Defence Force has for some time had a requirement for a wheeled heavy recovery vehicle, capable of recovering and towing stranded vehicles over almost any type of local terrain. At one period it seemed as though an uprated 8 × 8 version of the SAMIL series produced by Truckmasters would be selected but with the demise of Truckmasters this option was no longer available. Instead Armscor decided to licence produce a modified version of the Spanish Kynos Aljaba truck tractor (see entry under *Tank Transporters*), already selected for licence production in South Africa as the Cavallo. The hamper and equipment developed for the Truckmasters vehicle have largely been transferred to the Skimmel chassis virtually unmodified although the new vehicle is fitted with a mineproof cab and there have been some other changes to meet local requirements.

Description

The Skimmel 8 × 8 heavy recovery vehicle has a forward control mineproof cab with accommodation for the driver and four passengers on a single bench seat. There is space in the cab for the crew's kit, air conditioning, land navigation and communications equipment; escape hatches are provided in the cab roof. The cab is proof against a single TM-57 anti-tank mine detonating anywhere under the hull or against doubled TM-57s under any wheel. It is also proof against 7.62 mm ball ammunition fired from a range of 30 m. A tyre inflation system can be controlled from within the cab.

The rear body structure has several integrated subassemblies. Centrally mounted is the main capstan winch, a Rotzler Treibmatic TR 200 with a pull capacity of 20 tonnes and provided with 100 m of 24 mm steel rope. Also mounted centrally is a fully traversing low-profile folding crane with a lift capacity of 1,250 kg at the maximum reach of 4 m (minimum reach is 0.7 m). Located at the rear is an auxiliary capstan winch with a minimum pull capacity of 500 kg to assist in paying out the main cable. Also at the rear is a radial tow arm system capable of lifting a mass of 15 tonnes from normal ground level to a height of 1.6 m. The system is operated by the main cable winch running over a pulley and through rollers on the radial arm to the casualty. A

Skimmel Main 8 × 8 heavy recovery vehicle (T J Gander) ***1997***

removable yoke to hook onto the casualty is provided on the end of the main winch cable. This yoke fits into the end of the radial arm when pulled in so that in transit both the radial arm and the yoke are mechanically locked once the main cable is released. Loads weighing up to about 48 tonnes can be towed over almost any terrain.

Hydraulic power for most vehicle services is provided by a system operating at a pressure of 350 bar and 170 litres/min.

Also carried on the rear body is a 1.5 m ISO pattern container carrying a comprehensive array of recovery equipment, spares and tools.

Using the same basic chassis a number of other variants are being developed. So far the prototype of a maintenance and repair vehicle has been completed as has an ammunition resupply vehicle for use with the South African LIW 155 mm G6 (6 × 6) self-propelled artillery system.

Specifications

Cab seating: 1 + 4
Configuration: 8 × 8
Towed load:
(suspended) 15,000 kg
(towed) 48,000 kg
Length overall: 10.132 m
Width: 3.3 m
Height: (overall) 3.58 m
Ground clearance:
(front) 0.36 m
(rear) 0.9 m
Track: 2.6 m
Wheelbase: 1.524 m + 3.476 m + 1.524 m
Angle of approach/departure: 25°/38°
Max speed: approx 80 km/h
Range: up to 1,000 km
Engine: Deutz BF 12 L 513 C V-12 19.144 litre turbocharged intercooled air-cooled diesel developing 525 hp at 2,300 rpm
Gearbox: ZF 16 S-190 A with 16 forward and 2 reverse gears, with splitter on each gear
Torque converter and clutch: ZF WSK 400/ZF-TRANSMATIC
Transfer box: ZF single speed
Suspension: Hendrickson walking beam
Axles: Rockwell, each 13 tonnes
Tyres: 24.00 R 20.5 XS

Status

Ready for production.

Manufacturer

Enquiries to: Armscor, Private Bag X337, Pretoria 0001, South Africa.
Tel: +27 12 428 1911. Telex: 320217.
Fax: +27 12 428 5635.

UPDATED

SWEDEN

Scania P113 HK 6 × 6 recovery vehicle

Description

The Scania P113 HK recovery vehicle is based on the Scania 113 HK 6 × 6Z truck chassis. The vehicle is equipped with a forward control CP19 cab with seating for the driver, a tilting passenger seat and a bench seat for three further passengers at the rear. The cab may be tilted forward hydraulically 60° for maintenance.

The recovery rig is a Norwegian designed Hägglunds Moelv BV 730 which consists of a hydraulic crane with a maximum lifting moment of 28 t/m. The maximum lifting capacity is 8,400 kg. The vehicle is also provided with a towing lift boom with a maximum lift capacity of 7,000 kg and a recovery winch with a maximum single cable pulling force of 20,000 kg. The winch is provided with 85 m of cable and can be slewed through 300°. When the recovery rig is in use, four hydraulic outriggers are deployed for stability. The vehicle is also provided with a self-recovery winch with a maximum pulling force of 10,000 kg, equipped with 50 m of cable.

Ancillary equipment includes a rotating beacon, two working lights, an auxiliary heater for the cab and engine, an electrical engine heater, a 24 V emergency outlet and a spare wheel carrier.

Specifications

Cab seating: 1 + 4
Configuration: 6 × 6
Weight:
(kerb) 21,000 kg
(GVW) 28,000 kg
Length: 8.65 m
Height: (overall) 3.2 m
Wheelbase: 4.25 m + 1.45 m
Max speed: 100 km/h
Fuel capacity: 300 litres

Scania P113 HK 6 × 6 recovery vehicle in action
1997

Engine: Scania DS11 75 11 litre 6-cylinder liquid-cooled turbocharged direct injection diesel developing 310 hp at 2,000 rpm
Gearbox: GRH871 10-speed full synchromesh with hydraulic torque converter acting in low range
Clutch: K432 single dry plate
Transfer box: GT811 2-speed
Steering: hydraulic power-assist
Suspension: semi-elliptic steel leaf springs
Tyres: 14.00 × R20
Brakes:
(main) dual circuit air
(parking) spring
Electrical system: 24 V

Status

Production as required. In service with the Norwegian Army.

Manufacturer

Scania Trucks and Buses, S-151 87 Södertälje, Sweden.
Tel: +46 8 5538 1000.

UPDATED

Volvo F10 (4 × 4) recovery vehicle

Description

The Volvo F10 (4 × 4) recovery vehicle is basically the 4 × 2 model with a slightly lower front axle capacity. The forward control cab is spring-mounted on the chassis and can be hydraulically tilted forward 60° to give access to the engine for maintenance. The cab has a roof hatch and a heater/fresh air system and can be fitted with an air conditioning system if required. The F10 can also be supplied with a sleeper cab with single or twin bunks.

Mounted to the rear of the cab is the Bärgningsbilar EKA D2030 recovery equipment which can be used for three main roles: recovering disabled vehicles; towing disabled vehicles once they have been recovered; and changing components in the field.

The lifting device centres around the main lift boom which is raised by a telescopic ram. The folding boom, which stems from the main boom, houses the hydraulically operated extensible boom which is used for reaching under vehicles. A variety of lifting and towing attachments connect to the extensible boom for direct recovery. The main boom is positioned on the centre line of the vehicle and is extended and retracted by a hydraulic ram and can be mechanically locked in three different positions.

The folding boom houses the extensible boom vertically when not in use and is locked in the stowed position by an automatic mechanical lock. Two extensible legs behind the rear axle provide anchorage and stability when winching or craning. The anchor legs are individually extended and retracted by hydraulic rams and each leg has a wide spade with profiled teeth.

A hydraulic winch on the top of the main boom has a hydraulically released spring-loaded brake, free spool and reversing facilities. The cable runs through the main boom and out via guide rollers fitted in the end of the main boom. The cable tensioner provides tension when winding in without load on the cable. Power is provided by a hydraulic piston pump driven by the vehicle's PTO which has an in-built strainer and return filter fitted. The hydraulic oil reservoir is an integral part of the main boom and the direction control valve with six operating levers is on the offside rear locker. Pressure relief valves are fitted to prevent overloading of the various hydraulic systems and a pressure gauge shows the working pressure and how much lifting or pulling power is being used.

Optional equipment includes: an auxiliary 10,000 kg capacity winch mounted on the main boom; 5,000 or 7,000 kg front winch with ballasted front bumper; second telescopic ram to increase lifting power; extra hydraulic pump for faster winching speed; various towing and lifting equipment; illuminated signs and flashing lights; side leg for extra stability when side winching; second set of operating levers in opposite compartment; and a remote-control unit which allows the operator to operate the controls of the recovery equipment from a distance of up to 100 m.

Specifications

Configuration: 4 × 4
Length: 6.3 m
Width: 2.489 m
Height:
(cab roof) 2.911 m
(air intake) 3.244 m
Track:
(front) 1.945 m
(rear) 1.82 m
Wheelbase: 4.6 m
Fuel capacity: 300 litres
Engine: TD 100A diesel developing 250 hp at 2,200 rpm
Gearbox: manual with 4 forward and 1 reverse gears, 2-speed transfer box
Clutch: twin dry plate
Steering: recirculating ball and nut with built-in servo unit
Turning radius: 7.6 m
Brakes:
(main) dual circuit air
(parking) air-operated spring on rear wheels exhaust brake
Electrical system: 24 V

Boom
Max lifting height: 4.5 m
Max lifting capacity:
(at horizontal level) 22,000 kg
(in crane position) 10,000 kg
Reach: 2.4 m

Winch
Max pulling power: (when stalling) 34,000 kg
Max rope speed:
(at low pulling power) approx 19 m/min
(at high pulling power) approx 1 m/min
Rope diameter: 26 mm
Rope length: 40 m

Status

Production complete. In service with undisclosed countries and Malaysia (30).

Manufacturer

Volvo Truck Corporation, S-405 08 Göteborg, Sweden.
Tel: +46 31 666000. Telex: 27000 volvo s.
Fax: +46 31 510465.

UPDATED

Volvo F10 (4 × 4) recovery vehicle showing HIAB crane to cab rear

Volvo F10 (4 × 4) recovery vehicle from side

UNITED KINGDOM

Unipower M Series (8 × 8) heavy recovery vehicle

Description

The Unipower M Series heavy recovery vehicle is a member of the Unipower M Series 8 × 8 military logistics truck range - see entries under *Trucks*, *Tank Transporters* and *Materials Handling Equipment* for other vehicles in the M Series family.

The M Series of 8 × 8 chassis builds on the experience of Unipower in the design, development and production of the 8 × 8 chassis used in the British Army BR90 system for which Vickers Bridging is the prime contractor. A 6 × 6 chassis has also been built and this has been sold to Oman in a tank transporter semi-trailer configuration.

The M Series 8 × 8 recovery vehicle was designed to meet the most demanding requirements of engineer and other units operating in forward areas. The vehicle has a specified safe towing performance using a lift-and-tow boom, including the ability to suspend-tow a casualty with a 15 tonne axleload at 1.5 m from the recovery vehicle's rear and to tow (at 30 km/h) a tank/heavy recovery vehicle grossing in excess of 115 tonnes.

The vehicle is provided with a three-man forward control all-steel welded cab. The driver's seat is adjustable for height, reach and weight, while a hatch above the central passenger seat provides access to a machine gun mounting, if required.

The chassis frame bolted construction uses heat-treated carbon steel and is of the open channel type with open and tubular cross members. Towing facilities are provided front and rear. The front suspension consists of two slipper-ended semi-elliptic parabolic springs per axle. At the rear, two fully articulating semi-elliptical multi-leaf springs are trunnion-mounted on bearings. The rear axles are located longitudinally by upper and lower radius rods and transversally by eight Panhard rods.

The main power plant can be provided to suit customer requirements. Available is a Perkins Eagle 400 TX turbocharged diesel developing 400 hp or a higher-powered Cummins KTA 600 developing 600 hp. A further option is a Cummins NTA A430 CELECT turbocharged diesel providing 430 bhp, although other similar power plants may be utilised. The power plant is coupled to a ZF Transmatic transmission with a torque converter and a ZF single-speed transfer box.

The vehicle carries three types of recovery equipment: two 25 tonne winches creating a 50 tonne combined line pull; a 30 t/m hydraulic slewing crane with a four-section telescopic boom; and lift and towing gear.

The integrated recovery equipment is mounted on a full-length steel subframe. For the lift and tow gear a two-piece beam, carrying a three-stage hydraulically extending underlift boom section at its rearmost extremity, is hinged at the forward end of the subframe. Luffing is controlled by hydraulic rams. The underlift boom section is stowed vertically in the travelling mode and when hydraulically unfolded is extended horizontally, as required, under the casualty. Casualty lift is achieved by lifting rams. A cross-head pivot locates the load-bearing forks fitted to the third-stage boom for secure location under the casualty. A range of fitting accessories can be supplied. The main boom extremity is fitted with twin swivelling fairleads through which the winch ropes are fed during winch operations.

The vertically inclined section of the main boom is fitted with two independently powered hydraulically extending stiff legs, each capable of operating independently and equipped with ground anchor spades for winch reaction up to 50 tonnes. The two forward-mounted two-speed hydraulic winches are each rated at 25 tonnes maximum line pull. Each drum is provided with 50 m of 24 mm diameter rope.

A 30 t/m folding hydraulic truck crane is located behind the engine, having a three-section hydraulic telescopic extending boom and vertical support legs. Maximum boom reach is 7 m.

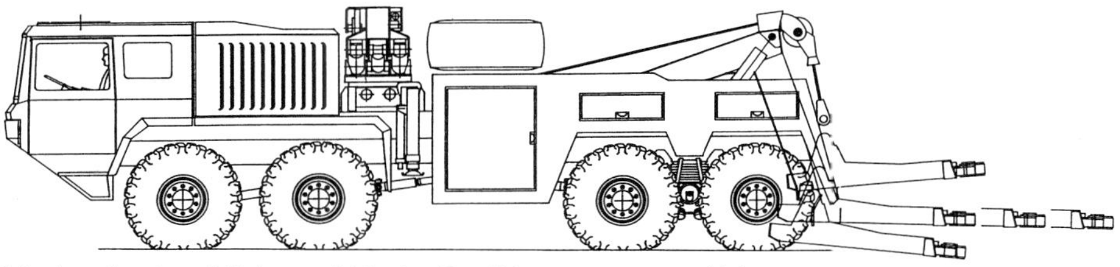

Side-view drawing of Unipower M Series (8 × 8) heavy recovery vehicle

A full range of recovery and winching accessories is carried.

Specifications

Cab seating: 1 + 2
Configuration: 8 × 8
Weight:
(unladen) 27,050 kg
(GCW, design) 140,000 kg
Length: 10.26 m
Width: 3 m
Height: 2.95 m
Ground clearance: 0.33 m
Wheelbase: 6 m
Angle of approach: 38°
Max speed: (laden) 78 km/h
Fuel capacity: 580 litres
Range: 900 km
Gradient: (restart, laden) 54%
Engine: Cummins NTA A430 CELECT 6-cylinder in-line turbocharged and after-cooled diesel developing 430 hp at 1,900 rpm
Transmission: ZF Transmatic 16 S 220 semi-automatic providing 16 forward and 2 reverse gears plus WSK 400 torque converter
Transfer box: ZF single-speed
Steering: ZF 8098 power-assisted
Suspension:
(front) two slipper-ended semi-elliptic parabolic springs per axle
(rear) two fully articulating semi-elliptical multileaf springs trunnion-mounted on bearings
Brakes: air to all axles
Tyres: 24 R 21 XL
Electrical system: 24 V
Batteries: 2 × 12 V, 150 Ah
Alternator: 90 A

Status

Proposal.

Manufacturer

Unipower Vehicles Limited, 34 Greenhill Crescent, Watford Business Park, Watford, Hertfordshire WD1 8QU, UK.
Tel: +44 1923 816555. Telex: 261760 unitrk g.
Fax: +44 1923 228621.

UPDATED

Reynolds Boughton (4 × 4) 6,000 kg recovery vehicle

Description

The Reynolds Boughton (4 × 4) 6,000 kg recovery vehicle was developed as the replacement for the Recovery Vehicle, Wheeled Light, Bedford RL (4 × 4). It uses a Bedford MJP2BMO chassis with a 3.962 m wheelbase and a Bedford TK style cab. The recovery unit is mounted on the rear and is constructed on load-carrying longitudinal main channels with cross-channel bearers gusseted at all intersections. Six stowage lockers are fitted, three to each side. The floor has a non-slip metal covering. Access steps are at the front and rear.

The recovery jib is mounted on a main turret tower. The jib has inner and outer rectangular box sections and a position stay is provided for use when carrying suspended loads. The winch is mounted on the hydraulically operated jib. The cable diameter is 14 mm and the lifting capacity 6,000 kg. Four hydraulically powered stabiliser legs are provided, one at each corner of the recovery unit and each leg is independently operated for levelling. The hydraulic pump is driven from a power take-off on the vehicle gearbox. All crane operations are carried out from the crane turret platform while stabiliser levelling is from the side of the vehicle. Most of the hydraulic piping is carried in flexible two-wire hose.

A recovery winch is mounted in the centre of the vehicle, having a capacity of 6,096 kg. The 14 mm cable for this winch is 65 m long and there is also a heavy-duty ground anchor centrally mounted between the axles. An overload safety device is linked to the crane hydraulics to operate when the hook load exceeds 105 per cent of the safe working load. The crane functions can then only be moved into a safe condition.

Specifications

Cab seating: 1 + 2
Configuration: 4 × 4
Length: 6.579 m

Reynolds Boughton (4 × 4) 6,000 kg recovery vehicle on Bedford chassis and with crane in stowed position ***1997***

Width: 2.438 m
Height: 2.667 m
Wheelbase: 3.962 m
Jib radius retracted: 3.327 m
Jib extension: 1.727 m
Max rear lift:
(0.482 m jib overhang from body rear) 6,000 kg
Crane slewing speed: 270° in 40 s
Jib extension from retract to extension: 16 s

Status

Production as required. In service with the British Army (deliveries complete), Brunei and Oman.

Manufacturer

Reynolds Boughton Limited, Bell Lane, Amersham, Buckinghamshire HP6 6PE, UK.
Tel: +44 1494 764411. Telex: 83132.
Fax: +44 1494 765218.

UPDATED

Foden (6 × 6) recovery vehicle

Description

The Foden (6 × 6) recovery vehicle is a derivative of the Foden FH-70 gun tractor and limber vehicles described in the *Trucks* section, although the recovery vehicle embodies many improvements to enhance overall performance. The British Army has 333 of these vehicles in service. Of these up to 190 are scheduled to undergo a mid-life in-depth repair programme to be undertaken over a period of three years.

The Foden (6 × 6) recovery vehicle (Foden Heavy Recovery Vehicle, Wheeled GS, 6 × 6) retains many of the features of the FH-70 gun tractor and limber medium mobility vehicles with attendant advantages in spares scalings. The Foden non-corrosive, fire-retardant glass-reinforced plastic tilting cab has seating for four men, bunk or stretcher facilities and is fully fitted for radio. A strengthened roof incorporates both observation hatches and machine gun mountings.

The vehicle has a maximum road speed of 97 km/h and is powered by a Perkins Engines Eagle engine, a Fuller main gearbox and GKN hub reduction axles and transfer gearbox. When fully equipped with its full complement of recovery equipment it can climb, hold and restart on 33 per cent gradients.

A front-mounted 10 tonne winch provides a self-recovery capability. Recovery eyes and detachable NATO-pattern towing hooks are provided at both front and rear.

The vehicle is fitted with the Eka Limited hydraulically operated Compact recovery unit Model AK 6500 EA12/1. With its drawbar, support/suspend towing and winching facilities, this unit can recover all of the British Army's in-service logistic support and armoured wheeled vehicles up to a gross vehicle weight of 30,000 kg. A Rotzler type 25000 HS/390 25,000 kg single line pull main winch is fitted and two rear anchor spades provide anchorage for the vehicle.

A slewing crane with 220° of slewing angle and a maximum lift capacity of 12,500 kg provides optimum lifting facilities and for vehicle stability two extensible hydraulically operated outriggers are fitted, one to each side of the vehicle. The boom has a straight lift capacity of 12,500 kg at 2.5 m extension and 5,800 kg at 7.7 m. Detachable load-bearing feet are also provided for the rear ground anchors when these are used as stabilisers.

All recovery and lifting functions are controlled from a locker-mounted console or from a remote-control unit with 30 m of wander lead.

Optional engines and transmissions are available. For use in Bosnia, an appliqué passive armour kit was developed by the Defence Research Agency to provide the cab with protection from small arms fire and shell splinters.

Specifications

Cab seating: 1 + 3
Configuration: 6 × 6
Weight: (loaded) 25,338 kg
Length: 9.055 m
Width: 2.482 m
Height: (cab) 3.35 m
Ground clearance: 0.42 m
Track:
(front) 2.029 m
(rear) 2.06 m
Wheelbase: 3.97 m + 1.516 m
Angle of approach/departure: 28°/29°
Max road speed: 97 km/h
Fuel capacity: 360 litres
Max gradient: 33%
Engine: Perkins Engines Eagle 290 litres 6-cylinder in-line turbocharged 4-stroke diesel developing 290 bhp at 1,950 rpm
Gearbox: Fuller RTX 11609B 9-speed constant mesh
Clutch: Spicer twin plate hydraulically operated
Transfer box: GKN/Kirkstall AGB 7000 Mark 11 2-speed with optional integral PTO
Axles:
(front) Kirkstall SD66-11-1S 10 t with differential lock
(rear) Kirkstall D66-11-1SHF and D66-11-1S with differential locks on both axles, 6.64:1 overall axle ratio. Third differential on foremost axle
Steering: recirculatory ball with integral power-assistance
Turning circle: (kerb to kerb) 25 m
Suspension: semi-elliptical laminated springs front, two-spring fully articulating rear spring
Brakes: air, split-circuit, trailer brake connection to front and rear of chassis
Wheels and tyres: 16.00 × 20 - 28 ply tyres, 11.25 rims
Electrical system: 24 V
Batteries: 2 × 12 V

Status

Production complete. In service with the British Army (333).

Manufacturer

Foden Trucks, a division of Paccar UK Limited, Moss Lane, Sandbach, Cheshire CW11 9YW, UK.
Tel: +44 1270 763244. Fax: +44 1270 762758.

UPDATED

Foden (6 × 6) recovery vehicle (R Stickland)
1997

Scammell S24 (6 × 6) recovery vehicle

Description

The Scammell S24 (6 × 6) recovery vehicle was developed for military use from a range of heavy-duty civilian trucks, ensuring commonality with the large numbers of Scammell trucks in service worldwide. The S24 can operate under a wide range of climatic conditions and, if required, certain appliqué kits can be used to operate under extreme climates.

The recovery equipment used on the S24 recovery vehicle consists of a 15,000 kg capacity hydraulic crane and a 25,000 kg capacity hydraulic winch. The crane is produced by Reynolds Boughton Limited and can traverse through 270°. It is equipped with 16 mm wire rope and when in use the vehicle is stabilised by four outrigger legs that are lowered hydraulically. The hydraulic power for the system is derived from the vehicle gearbox power take-off.

There is space on the vehicle rear deck for a spare wheel, towing bars and other recovery equipment. Optional equipment includes various floodlights and a flashing beacon for the roof.

The S24 is powered by a Cummins NTE 350 turbocharged and after-cooled in-line diesel engine developing 350 hp at 2,100 rpm. Two cooling radiators are used. The transmission is a Fuller RTX 14,614

Scammell S24 (6 × 6) recovery vehicle

15-speed manual gearbox and 16.00 × 20 tyres are used throughout.

The S24 recovery vehicle can maintain a continuous operational speed of 40 km/h climbing a 14 per cent gradient. As a solo unit it can travel at 74 km/h and will climb a 52 per cent gradient.

Status
Production as required. In service with various nations.

Manufacturer
Unipower Vehicles Limited, 34 Greenhill Crescent, Watford Business Park, Watford, Hertfordshire WD1 8QU, UK.
Tel: +44 1923 816555. Telex: 261760 unitrk g.
Fax: +44 1923 228621.

UPDATED

Scammell Crusader (6 × 4) recovery vehicle

Description
In 1977 the British Army ordered 130 Scammell Crusader 6 × 4 recovery vehicles. They have the same chassis, cab, engine and transmission as the Scammell Crusader 35,000 kg tractors used by the Royal Engineers for hauling semi-trailers, which are fully described in the *Trucks* section. The full designation of the vehicle is Recovery Vehicle, Wheeled, CL (Low Mobility Recovery Vehicle).

The vehicle can provide recovery support for wheeled vehicles up to 16 tonnes and can support or suspend tow laden 8 tonne or unladen 16 tonne vehicles and straight tow vehicles up to 30 tonnes gross weight on gradients not exceeding one in five.

The main recovery equipment, of Swedish Eka design, consists of hydraulically operated earth anchors, main boom and a winch, which, together with stowage bins, are mounted on a subframe at the rear of the vehicle. The main boom is lifted by a hydraulic ram and incorporates a folding boom which houses an extending boom that carries a variety of attachments for support or suspend tow. When support or suspend towing, a rear bogie blocking system maintains sufficient front axle loading for safe steering.

A winch with a maximum pulling capacity of 7,000 kg is mounted at the front behind the bumper. The main boom can lift 7,500 kg at a maximum distance of 2.6 m from the rear bogie centreline and the main winch exerts a maximum pull of 20,000 kg. Both winches have fail-safe automatic brakes capable of holding the maximum winch loads. Using the main winch, with suitable attachments to the extending boom, there is a limited craning facility for loads up to 3,000 kg. Apart from main controls for the recovery equipment housed at the right-hand rear side of the vehicle, a hand-held remote-control unit, usable up to 30 m away, can operate the main recovery functions.

Scammell Crusader (6 × 4) recovery vehicle

Specifications
Cab seating: 1 + 3
Configuration: 6 × 4
Weight: 16,700 kg
Length: 8.37 m
Width: 2.98 m
Height: 3.28 m
Track:
(front) 2.05 m
(rear) 1.98 m
Wheelbase: (first axle to centre of rear bogie) 4.59 m
Max speed: 78 km/h
Range: 1,150 km
Fuel capacity: 455 litres
Max gradient: 33%
Engine: Perkins Engines Eagle 305 Mk 111 turbocharged diesel developing 305 bhp at 2,100 rpm
Transmission: manual with 15 forward and 3 reverse gears
Turning radius: 11 m
Steering: ball and nut, power-assisted
Suspension:
(front) longitudinal semi-elliptical springs pivoted front with slipper rear ends and telescopic shock-absorbers
(rear) fully articulated, inverted longitudinal semi-elliptical springs, trunnion-mounted at centre with slipper rear ends
Tyres: 11.00 × 20
Brakes:
(main) air
(parking) hand on all wheels
Electrical system: 24 V
Batteries: 4 × 12 V, 100 Ah

Status
Production complete. In service with the British Army (130) and Royal Air Force (2). Some of these vehicles are being rebuilt to extend their operational lives.

Manufacturer
Unipower Vehicles Limited, 34 Greenhill Crescent, Watford Business Park, Watford, Hertfordshire WD1 8QU, UK.
Tel: +44 1923 816555. Telex: 261760 unitrk g.
Fax: +44 1923 228621.

UPDATED

UNITED STATES OF AMERICA

Stewart & Stevenson M1089 MTV (6 × 6) 5-ton Wrecker

Development
The M1089 MTV (6 × 6) 5-ton Wrecker is the recovery/wrecker component of the Family of Medium Tactical Vehicles (FMTV) being manufactured by Stewart & Stevenson Services Inc, Tactical Vehicle Systems Division, for the US Army. For full details of the FMTV family refer to the entry in the *Trucks* section.

Description
Most of the chassis, forward control cab, automotive components, drive train, central tyre inflation system and other systems used on the M1089 MTV (6 × 6) 5-ton Wrecker are identical with their equivalents on other FMTV variants. The chassis has been provided with some extra stiffening panels but other modifications are few and minor. The main change from the other variants is the provision of a recovery hamper over the rear chassis and axles.

The main recovery component is a recovery winch with a twin 13,608 kg line pull capacity. A lift and tow assembly provides front or rear recovery from the vehicle operator station or remote control. It can handle loads of 4,990 kg, including almost all wheeled military

M1089 MTV (6 × 6) 5-ton Wrecker ***1996***

vehicles. Stabiliser jacks are provided each side for deployment during recovery operations.

Specifications

Cab seating: 1 + 2
Configuration: 6 × 6
Weight: (kerb) 14,892 kg
Length: 9.143 m
Width: 2.438 m
Height:
(operational) 2.845 m
(for air transport) 2.68 m
Ground clearance: 0.559 m
Wheelbase: 5.3 m
Angle of approach/departure: 40°/40°
Max speed: 94 km/h
Range: 483 km plus
Gradient: 60%
Side slope: 30%
Fording:
(without kit) 0.914 m
(with kit) 1.524 m
Engine: Caterpillar 3116 ATAAC 6.6 litre 6-cylinder turbocharged and after-cooled diesel developing 290 hp at 2,600 rpm
Gearbox: Allison MD-D7 automatic, 7-speed
Transfer box: Allison single-speed integral to transmission with planetary inter-axle differential (lockable)
Steering: power
Suspension: parabolic tapered leaf springs
Tyres: 395 R 20 XML
Electrical system: 12/24 V
Alternator: 100 or 200 A

Status

Production. In service with US Army.

Manufacturer

Stewart & Stevenson Services Inc, Tactical Vehicle Systems Division, 5000 I-10 West, Sealy, Texas 77474, USA.
Tel: +1 713 856 7040. Fax: +1 713 885 7910.

UPDATED

M62, Truck, Wrecker: Medium, 5-ton, (6 × 6) M246, Truck, Tractor, Wrecker: 5-ton, (6 × 6) M543, Truck, Wrecker: Medium, 5-ton, (6 × 6)

Description

These vehicles are all members of the M54 series of 5-ton (6 × 6) trucks which were developed after the end of the Second World War, the reader is referred to this entry in the *Trucks* section for the development history and technical description of the basic vehicle.

M62

This has a hydraulically operated crane mounted at the rear with a maximum lifting capacity of 9,072 kg when used with the two stabilisers either side of the vehicle. A winch is mounted at the front, with a maximum capacity of 9,072 kg when being used to the front and 18,144 kg when being used to the rear. Later models of the M62 are the M62A1 and M62A2.

M246

This was designed for use as both a recovery vehicle and for towing semi-trailers. The Austin-Western crane at the rear has a jib which can be extended from 3.504 to 7.924 m. This crane can be slewed through 360° and its jib elevated to a maximum of 45°. Later models of the M246 are the M246A1 and the M246A2, which has a Continental LDS 465-1 or LDS 465-2 multifuel engine. The M246 was replaced in production by the similar M819, for which there is a separate entry.

M543

This has a Gar Wood hydraulically operated crane at the rear which can lift a maximum of 4,536 kg. Two stabilisers are provided each side and a winch is mounted at the front and rear. The M543A1 has a diesel engine. In 1963 the M543A2 was introduced. It has an LDS-465 multifuel engine which develops 180 hp at 2,600 rpm and gives the vehicle a maximum road speed of 84 km/h and a range of 938 km. The M543A2 was replaced in production by the similar M816 for which there is a separate entry.

Status

Production complete. In service with US forces and many other armed forces including Australia (a few M543s and M543A1s are in service) and Spain (all re-engined with Spanish diesels). In US Army service these will slowly be replaced by the new Stewart and Stevenson M1089 MTV (6 × 6) wrecker.

Manufacturers

The basic chassis has been manufactured by various companies since the early 1950s including International Harvester, Kaiser Jeep (now AM General Corporation) and Mack Trucks. Last manufacturer was AM General Corporation, 14250 Plymouth Road, Detroit, Michigan 48232, USA.

Truck, Tractor, Wrecker: 5-ton (6 × 6) M246 with front-mounted winch (US Army)

Truck, Wrecker: Medium, 5-ton (6 × 6) M62 (US Army)

Specifications

Designation	**M62**	**M246**	**M543**
Chassis designation	M40C	M63C	M40C
Configuration	6 × 6	6 × 6	6 × 6
Towed load			
(road)	13,608 kg	20,865 kg	13,608 kg
(cross-country)	9,072 kg	17,010 kg	9,072 kg
Weight (empty)	15,275 kg	14,829 kg	15,603 kg
Length	7.848 m	8.953 m	10.007 m
Width	2.463 m	2.489 m	2.444 m
Height	2.59 m	3.352 m	2.743 m
Ground clearance	0.279 m	0.279 m	0.279 m
Wheelbase	3.86 m + 1.371 m	4.775 m + 1.371 m	3.86 m + 1.371 m
Angle of approach/departure	37°/38°	35°/55°	37°/38°
Max speed (road)	84 km/h	84 km/h	84 km/h
Range	344 km	369 km	360 km
Fuel capacity	295 litres	295 litres	295 litres (some have 504 litres)
Max gradient	36%	47%	61.4%
Fording			
(without preparation)	0.762 m	0.762 m	0.762 m
(with preparation)	1.981 m	–	1.981 m
Engine	Continental 6-cylinder petrol developing 196 hp at 2,800 rpm		
Transmission	5 forward and 1 reverse gears and 2-speed transfer case		
Turning radius	12.648 m	14.325 m	14.5 m
Tyres	11.00 × 20	11.00 × 20	11.00 × 20
Brakes	all have hydraulic brakes, air-actuated		
Electrical system	24 V	24 V	24 V
Batteries	2 × 12 V	2 × 12 V	2 × 12 V

UPDATED

M816, Truck, Wrecker, 5-ton, (6 × 6) M819, Tractor, Wrecker, 5-ton, (6 × 6)

Description

Both these vehicles are members of the M809 series of 5-ton (6 × 6) trucks and the reader is referred to the entry in the *Trucks* section for the development history of the series.

The M816 is provided with a revolving hydraulic crane at the rear which has a self-supported extensible boom and boom to ground supports. Outriggers are provided to stabilise the vehicle when the crane is being used. Winches are supplied. The M819 also has a hydraulic crane mounted on the rear but has a longer wheelbase and can tow a trailer or vehicle weighing up to 13,608 kg on roads, or 9,076 kg across country.

Optional kits for these vehicles include an air brake kit, closure hard top, deep water fording kit, slave receptacle, thermal barrier kit, winterisation personnel heater kit and a power plant kit.

Status

M816 in service with the US Forces, Canada, Australia, South Korea, Philippines (36), Zaïre and many other forces. M819 production complete. In US Army service these will slowly be replaced by the new Stewart and Stevenson M1089 MTV (6 × 6) Wrecker.

Manufacturer

AM General Division, 105 N. Niles Avenue, PO Box 7025, South Bend, Indiana 46617, USA.
Tel: +1 219 284 2942. Fax: +1 219 284 2959.

UPDATED

Specifications

Designation	**M816**	**M819**
Chassis type	M809A1	M811A1
Cab seating	1 + 2	1 + 2
Configuration	6 × 6	6 × 6
Weight		
(unloaded)	16,385 kg	15,392 kg
(on front axle empty)	4,966 kg	6,304 kg
(on rear bogie empty)	11,418 kg	9,251 kg
Length	9.042 m	9.124 m
Width	2.484 m	2.489 m
Height	2.69 m	3.352 m
Ground clearance	0.295 m	0.295 m
Track	1.88 m (front) and 1.829 m (rear)	
Wheelbase	4.547 m + 1.371 m	4.125 m + 1.371 m
Angle of approach/departure	35°/34°	36°/55°
Max speed (road)	84 km/h	83.6 km/h
Range	805 km	563 km
Fuel capacity	295 litres	295 litres
Max gradient	31%	31%
Fording		
(without preparation)	0.762 m	0.762 m
(with preparation)	1.879 m	1.879 m
Engine	Model NHC-250 6-cylinder in-line diesel developing 240 hp (gross) at 2,100 rpm	
Transmission	5 forward and 1 reverse gears, 2-speed transfer box	
Tyres	11.00 × 20	12.00 × 20
Electrical system	24 V	24 V

M816 5-ton (6 × 6) Wrecker in travelling configuration
1997

Oshkosh M984A1 Recovery Vehicle 10-ton (8 × 8)

Description

The M984A1 Recovery Vehicle 10-ton (8 × 8) is the recovery component of the Oshkosh HEMTT family of vehicles. For full details of the HEMTT family refer to the entry in the *Trucks* section. The US Army fleet of HEMTT vehicles is now starting to be upgraded to extend their lives well into the 21st century.

The M984A1 recovery vehicle follows the same general construction lines as the rest of the HEMTT family. The M984A1 recovery system consists of a main recovery winch, a self-recovery winch and ground spades. The main recovery winch has a capacity of 27,216 kg and is an automatic, two-speed, hydraulically operated winch drum with fully proportional controls. It has 56.4 m of usable wire rope. The winch automatically shifts to low speed when recovering heavy loads. The main recovery winch system, which includes a fairlead/tensioner device, provides the capability of recovering stranded vehicles at up to 32° to the left or right, or 23° below the vehicle. The fairlead/tensioner also provides proper winding of the wire rope on the drum winch by maintaining an even rope tension between the rear of the vehicle and the drum. The self-recovery winch is a standard HEMTT winch with forward deployment and a bare drum capacity of 9,072 kg. The ground spades are emplaced manually. The main winch can be operated using a remote-control unit.

The M984A1 uses a lift and tow system that can be deployed and attached to a disabled vehicle in 2 to 3 minutes. The system consists of a crossbar that 'floats' on a 102 mm ball. The ball is supported by three hydraulic cylinders which provide precise positioning of the crossbar during connection to the disabled vehicle. The ball system provides articulation and stability when travelling over uneven terrain and when cornering.

Based on testing by the US Army Combat Systems Test Activity (CSTA), the M984A1 is approved for towing with towbar tracked vehicles up to 30 tons, which includes the M2A2 Bradley Fighting Vehicle.

The M984A1 is equipped with a material handling crane with a lift capacity of 6,350 kg. The crane has the lift and reach capability to support the removal and replacement of power packs from a wide range of

M984A1 (8 × 8) recovery vehicle

wheeled and tracked vehicles including the M1, M1A1, M2, M3, M109A2, M88A1 and M60A3. The crane can be operated using a remote-control unit.

The M984A1 has an equipment body with stowage provisions for support equipment including oxygen/acetylene welding equipment, tow adaptors, chain slings and snatch blocks. The stowage provisions include 10 lockable and watertight individual compartments. The entire body can be removed for air transport weight reduction in less than 10 minutes.

Specifications

Cab seating: 1 + 1
Configuration: 8 × 8
Weight:
(chassis) 14,968 kg
(kerb) 22,226 kg
(GCW) 45,360 kg
Length: 9.957 m
Width: 2.438 m
Height:
(travelling) 2.845 m
(top of cab) 2.565 m
(crane extended) 3.759 m
(loading, flatbed) 1.6 m
Ground clearance: 0.33 m
Wheelbase: 4.85 m
Angle of approach/departure: 43°/45°
Max speed: 88 km/h
Range: 483 km
Fuel capacity: 589 litres
Fording: 1.219 m
Engine: Detroit Diesel 8V-92TA V-8 developing 445 hp at 2,100 rpm
Gearbox: Allison HT740D with torque converter, 4 forward, 1 reverse speeds
Transfer box: Oshkosh 55000 2-speed
Steering: integral hydraulic main and booster gears
Suspension:
(front) Hendrikson RT340
(rear) Hendrikson RT500
Tyres: 16.00R × 20
Electrical system: 24 V
Batteries: 4 × 12 V

Status

In production. In service with the US Army and Saudi Arabia.

Manufacturer

Oshkosh Truck Corporation, PO Box 2566, Oshkosh, Wisconsin 54903, USA.
Tel: +1 414 235 9150. Fax: +1 414 233 9540.

UPDATED

Freightliner (6 × 6) Wrecker

Description

The Freightliner (6 × 6) Wrecker is a member of the Freightliner Military Truck Family; for full details of this family refer to the entry in the *Trucks* section.

The Freightliner Military Truck Family is based on the M915A2 and M916A1 series of long-haul trucks and utilises many components and other items carried over from commercial models. Many well-proven features of the Family are used with the Freightliner (6 × 6) wrecker, including an aluminium cab and an engine of advanced fuel efficiency design.

The main recovery feature is a rotating hydraulic boom crane mounted on a turntable over the rear axle bogies. The boom has a capacity of 18,144 kg and can lift and tow loads of up to 7,031 kg.

Freightliner (6 × 6) Wrecker

Specifications

Cab seating: 1 + 1
Configuration: 6 × 6
Weight:
(kerb) 18,080 kg
(front axle, loaded) 7,258 kg
(rear axles, loaded) 23,587 kg
Length: 9.32 m
Width: 2.49 m
Height:
(overall) 3.24 m
(cab) 2.97 m
Ground clearance: 0.23 m
Wheelbase: 5.79 m
Angle of approach: 26°
Max speed:
(road) 85.3 km/h
(3% gradient) 40.22 km/h
Fuel capacity: 379 litres
Fording: 0.51 m
Engine: Detroit Diesel DDE 12.7 litre 6-cylinder high torque rise diesel developing 400 hp at 2,100 rpm
Transmission: Allison HT-740 AT with torque converter, 4 forward, 1 reverse speeds
Transfer box: Oshkosh Series 55,000, 2-speed
Steering: Ross integral power
Turning radius: 26.33 m
Brakes: air/mechanical; Eaton front, Rockwell rear
Tyres: 315/80 R 22.5
Electrical system: dual 12/24 V
Batteries: 4 × 12 V

Status

Production as required. As far as it is known this vehicle is not in service with the US Army but is being offered for export.

Manufacturer

Freightliner Corporation, 4747 N Channel Avenue, PO Box 3849, Portland, Oregon 97208, USA.
Tel: +1 503 735 7183.

UPDATED

RECOVERY EQUIPMENT

GERMANY

ROTZLER winches

Description
ROTZLER GMBH + CO produces and markets a wide range of hydraulically driven winches for a wide variety of applications with pulling forces from 0.5 to 50 tons. The two main types are: drum winches for pulling, recovery, hoisting and lifting applications; and TREIBMATIC type capstan winches for pulling and recovery.

The ROTZLER TREIBMATIC winches are described as being of a new generation, delivering a constant pulling force and rope speed over the entire (theoretically unlimited) rope length, with reduced rope wear.

Status
Production as required. In service with German and other NATO armed forces.

Manufacturer
ROTZLER GMBH + CC, Postfach 1165, D-79577 Steinen, Germany.
Tel: +49 7627 701-0. Telex: 773716 roste d.
Fax: +49 7627 701-166.

UPDATED

ROTZLER twin winch combination for heavy-duty trailers with a 2 × 25 tons pulling force

ROTZLER TREIBMATIC type TR200 recovery winch with 25 tons pulling force

ITALY

FARID recovery equipment

Description
The Italian company of FARID has developed a range of heavy-duty recovery equipment which can be installed on a variety of tracked and wheeled chassis. The equipment can be divided into three families: slewing cranes, fixed turret cranes and pick-up platforms; all of these are hydraulic but have different applications. Typical chassis include the German Mercedes-Benz and the Italian IVECO.

Slewing Cranes
Model F 25 cross-country
This model can be installed on a 6 × 6 chassis that has a recommended gross vehicle weight of not less than 28,000 kg. The hydraulic crane has a lifting capacity of 25,000 kg and can be slewed through 360°; it is stopped in the desired position by an automatic belt braking device.

The telescopic boom is provided with two hydraulic, plus one mechanical, extensions and can reach a height of 16 m and horizontally 12.7 m from the rear and side boards. Before operations four outriggers are lowered to the ground, two each side, by means of hydraulic and independent control.

Two auxiliary hydraulic winches are fitted, one on the rear having a capacity of 10,000 kg with 75 m of 20 mm diameter steel cable and a self-recovery unit with a capacity of 5,000 kg and 75 m of 12 mm steel cable. Both winches are controlled from the driver's cab.

Power is provided by two hydraulic piston pumps driven by the vehicle's PTO which has an in-built strainer and return filter fitted. The crane is also provided with a pressure relief valve to prevent overloading of the various hydraulic systems, lock valves on the outrigger's cylinder and a pressure gauge, showing the working pressure, in front of the operator's seat.

Standard equipment includes: tow hitch, snatch block chains, tools, yellow flashes and rear spotlight for operations at night.

Fixed Turret Cranes
Model F 20 f
This wrecker crane consists of a fixed turret on which a telescopic boom is hinged. The boom is provided with hydraulic hoisting and two hydraulic extensions. This model can be installed on a chassis and cab that have a gross vehicle weight of not less than 19,000 kg. The two main hydraulic winches are fitted on the turret and are provided with independent control, 50 m of 16 mm diameter steel cable and have a capacity of 10,000 kg each.

The two main steel cables, after passing through their independent rocking heads, end with a steel hook. These two winches can be used both together, having a total capacity of 20,000 kg, or separately. For lateral strength one winch anchors while the other provides pull.

FARID F 25 (6 × 6) cross-country recovery vehicle

FARID F 12 f recovery vehicle

Before the crane is used, two rear stiff legs are lowered to the ground by means of hydraulic and independent controls. A third hydraulic winch, having a capacity of 15,000 kg and 100 m of 20 mm diameter steel cable, is mounted on the rear part of the vehicle so that a total pulling capacity of 35,000 kg can be obtained.

The front-mounted winch has a capacity of 5,000 kg and is provided with 75 m of 16 mm steel cable and is controlled from the driver's cab.

The rear hydraulic lift-and-tow device is made by a foldable boom and has a variety of lifting and towing attachments to permit towing of disabled vehicles once they have been recovered.

All the control levers are fitted on the rear part of the vehicle and the boom and winch controls are duplicated on a remote control. Tow hitch, snatch block, chains, tools, yellow flashes and spotlights for night operations are fitted as standard.

Model F 12 f

This wrecker equipment has the same main features as the Model F 20 f but is a lighter version. It can be installed on a chassis/cab combination that has a gross vehicle weight of not less than 13,000 kg.

The two main hydraulic winches fitted on the turret have a capacity of 6,000 kg each and are provided with 35 m of 16 mm diameter steel cable. The third hydraulic winch also has a capacity of 6,000 kg but is provided with 75 m of 16 mm diameter steel cable. It is mounted on the rear of the vehicle so that a total pulling capacity of 18,000 kg can be obtained.

The front-mounted hydraulic winch has a capacity of 4,000 kg and is provided with 46 m of 11 mm diameter steel cable and is controlled from the driver's cab. Standard equipment includes: remote control, tow hitch, snatch block, chains, standard tools, yellow flashes and rear lights for night operations.

Pick-up Platform

This equipment is to be installed on a chassis/cab combination having a gross vehicle weight of not less than 7,500 kg and has been designed for the recovery of light trucks and Land Rover type vehicles. The pick-up platform truck is made up of three separate parts, each of which has its own task. The boom crane is provided with hydraulic hoisting, hydraulic extension and a 3,000 kg winch; it is for lifting or pulling broken down vehicles and after use it is stowed between the platform rails. The loading platform permits the transport of one vehicle which is loaded on it. It is hydraulically lifted and two rear-loading ramps can be extended by means of a hydraulic and independent control. Before operations are carried out, two rear-mounted hydraulic stiff legs are lowered to the ground. The rear hydraulic 'lift-and-tow' device is for towing a second vehicle, when the first one has already been loaded over the platform. Standard equipment includes: front winch remote control, snatch block, chains, yellow flashes and rear lights for recovery operations at night.

Specifications

Model	**F 25**	**F 20 f**	**F 12 f**	**Pick-up platform**
Turret	360° slewing	fixed	fixed	none
Max lifting capacity	25,000 kg	20,000 kg	12,000 kg	2,000 kg
(fully extended)	3,000 kg	5,000 kg	2,700 kg	800 kg
Max pulling capacity	10,000 kg	35,000 kg	18,000 kg	3,000 kg
Max boom height	17 m	12 m	10 m	5 m
Max horizontal extension				
(from rear end)	9 m	6 m	5 m	2 m
Main cable				
(diameter)	14 mm	16 mm	16 mm	12 mm
(length)	150 m	50 m	35 m	30 m
Max rear auxiliary winch capacity	10,000 kg	15,000 kg	6,000 kg	n/a
Max front-mounted winch capacity	5,000 kg	5,000 kg	4,000 kg	3,000 kg
Rear auxiliary winch cable				
(diameter)	20 mm	20 mm	16 mm	n/a
(length)	70 m	100 m	75 m	n/a
Front auxiliary winch cable				
(diameter)	16 mm	16 mm	11 mm	12 mm
(length)	75 m	75 m	46 m	30 m

Status

In production. In service with many armed forces in Africa and the Middle East.

Manufacturer

FARID SpA, Corso Savona 39 bis, I-10024 Moncalieri (Turin), Italy.
Tel: +39 11 645454. Telex: 220348 farid 1.
Fax: +39 11 641934.

UPDATED

NORWAY

Hägglunds Moelv Recovery Unit RV 730

The Hägglunds Moelv Recovery Unit RV 730 was designed for heavy recovery operations including winching, towing and hoisting. The unit can be mounted on various types of 6 × 6 vehicle chassis including Mercedes-Benz, IVECO, DAF, Scania and other types.

The unit has four hydraulically actuated outriggers that are separately controlled. With the outriggers extended the vehicle is lifted clear of the ground to provide stability in all directions and provide anchoring when using the recovery winch. The outrigger frame is separate from the vehicle chassis so no forces are imparted during operations.

The main recovery winch has a capacity of 20 tonnes single pull and is mounted in the crane slewing frame to provide the capability to pull a load towards the recovery vehicle without extra rigging. The winch speed is dependent on the actual pulling force. The unit can also be equipped with a side-mounted hydraulic self-recovery winch with a capacity of 10 tonnes.

The unit is equipped with a hydraulic crane with arm elevation from horizontal to +70°. The arm has a hydraulic telescopic extension jib that can operate with a full load of 10,000 kg. The crane traverse is symmetrical over the rear of the recovery vehicle; a full 360° traverse is optional. The crane can be operated from a seat to the right of the crane arm or from a remote position.

In addition to the crane, the unit is fitted with a lifting and towing boom mounted at the rear. The boom is hydraulically operated and has a telescopic extension and tilting support point for towing attachments. The boom capacity is variable according to customer requirements. A NATO towing hook is mounted on the boom.

The outriggers are dimensioned so that each can carry the full pulling force of 40 tonnes. This capacity can be used over a rear arc of 30° each side without extra anchoring. Side pulls are limited to 15 tonnes without extra anchoring. Anchor points are provided to utilise the full 40 tonne capacity.

The unit can be provided with a carrying platform for spares and lockable tool cabinets are provided.

Hägglunds Moelv RV 730 recovery unit fitted to Scania P113 HK (6 × 6) recovery vehicle

Specifications

Weight:
(total) approx 20,000 kg
(front axle) 8,000 kg
(bogie axle) 12,000 kg
Length:
(travelling) 8.75 m
(between lowered outriggers) 5.7 m
Width:
(travelling) 2.5 m approx
(between lowered outriggers) 4.2 m
Height: 3.25 m
Crane max lift capacity: 10,000 kg
Max winch capacity, single cable: 20,000 kg
Max anchoring on outriggers: 40,000 kg
Max crane outreach: 6.35 m
Max lift height: 8 m

Status

Production. Known users include Netherlands (DAF chassis, see entry in previous section), Sweden (see entry in previous section) and Norway. Most applications to date have used a Scania (6 × 6) truck chassis as the platform.

Manufacturer

Hägglunds Moelv A S, PO Box 244, N-2391 Moelv, Norway.
Tel: +47 6234 6000. Telex: 76350 hagmo n.
Fax: +47 6234 6001.

UPDATED

SOUTH AFRICA

TFM Model 15 and Model 26 recovery equipments

Description

The TFM Model 15 and Model 26 recovery equipments, of 15 tonne and 26 tonne capacity respectively, are of very rugged construction and have been optimised for military use. They can be mounted on any suitable single- or double-axle rear chassis. The main features include twin lockable and extensible booms with an individual service drum for each boom and a sturdy A-frame tow hitch at the rear.

Each equipment is powered by a slow speed, high torque, radial piston hydraulic motor close-coupled to the transmission. The transmission unit has a single impact, dual output capacity, giving the hydraulic circuit a dual range pressure system to ensure safe operation with either single or twin service cables.

TFM Model 26 recovery equipment fitted to a crew-cabbed SAMIL 100 (6 × 6) recovery vehicle (T J Gander)

Specifications

	Model 15	Model 26
Retracted boom	7,500 kg	12,500 kg
Extended boom	3,700 kg	6,000 kg
Drum rating	7,500 kg	12,500 kg
Single line safe load	3,000 kg	4,500 kg
Cable diameter	13 mm	16 mm

Status

Production as required. In service with undisclosed countries.

Manufacturer

TFM (Pty) Limited, PO Box 46131, Orange Grove, Transvaal, South Africa.
Tel: +27 316 2106. Fax: +27 316 1462.

UPDATED

TFM HW-30 drag winch

Description

The TFM HW-30 is a heavy-duty drag winch for use in conjunction with recovery units or on its own, although it is not designed for use as a personnel hoist. It is powered by a radial piston hydraulic motor via a hydraulic dog clutch.

The clutch can be manually disengaged for manual cranking operation and a manually engaged winch brake is standard. The drum winch, hydraulic motor and all controls are coupled together as a compact unit and mounted on a transarm. Design features include high-grade precision castings and a precision-machined brass main worm wheel and gear.

TFM HW-30 drag winch

Specifications

Unit weight: 630 kg
Rope length: 102 m
Rope diameter: 26 mm
Max rope pull:
(1st layer) 25,000 kg
(2nd layer) 21,500 kg
(3rd layer) 18,700 kg
(4th layer) 16,600 kg

Status

Production as required. In service with undisclosed countries.

Manufacturer

TFM (Pty) Limited, PO Box 46131, Orange Grove, Transvaal, South Africa.
Tel: +27 316 2106. Fax: +27 316 1462.

UPDATED

TFM 25 tonne recovery winch

Description

This is a self-spooling, constant pull, 25 tonne winch with a nominal rope speed of 9 m/min at rated load. The technology utilised is centred in a three-stage planetary reduction drive and hydraulic circuitry. All other components are fabricated in-house. The winch utilises a proprietary self-aligning platform which inherently assures correct spooling. The winch is intended mainly for the recovery role but could have other applications.

Specifications

Weight: (fully installed) 2,080 kg
Drum diameter: 495 mm (grooved)
Drum width: 600 mm
Rope length: 100 m (120 m optional)
Rope diameter: 26 mm
Max rope pull:
(1st layer) 25 t
(2nd layer) 23 t
(3rd layer) 21 t

Status

Production as required. In service with undisclosed countries.

Manufacturer

TFM (Pty) Limited, PO Box 46131, Orange Grove, Transvaal, South Africa.
Tel: +27 316 2106. Fax: +27 316 1462.

UPDATED

TFM 25 tonne recovery winch

UNITED KINGDOM

Reynolds Boughton recovery cranes

Description

Reynolds Boughton Limited has developed a range of hydraulic slewing recovery cranes with lifting capacities from 6 to 15 tonnes. These equipments can be installed on a variety of vehicles. The range comprises the following:

15/50 SHR

This is a 15 tonne lift capacity 270° slewing hydraulic recovery crane with a telescopic jib. It is available with a 25/50 tonne line pull winch. The crane is suitable for recovering heavy-duty vehicles and MBTs from most situations by using the main winch and a ground anchor. Recovery is completed by using the suspended tow equipment provided. The main winch can be used for self-recovery.

10/50 SHR

This is a 10 tonne lift capacity 270° slewing hydraulic recovery crane with a telescopic jib. It is available with a 25/50 tonne line pull winch. Using the main winch and a ground anchor, this equipment is suitable for recovering all wheeled vehicles and can be adapted for general workshop and stores handling, using four independent hydraulic stabilising legs. The main winch can be used for self-recovery.

6/16 SHR

This is a 6 tonne lift capacity 270° slewing hydraulic recovery crane with a telescopic jib that is available with a 6/8 tonne line pull winch. It is suitable for recovering smaller vehicles in the 2/4 tonne range using the main winch and a ground anchor. A special A-frame at the chassis rear is fitted for suspended towing. Stores and equipment can also be handled. The main winch can be used for self-recovery.

Reynolds Boughton 15/50 SHR recovery equipment on Scammell S24 (6 × 6) recovery vehicle

10/50 UAL

This is a 15/10 tonne lift 270° slewing hydraulic recovery crane with a telescopic jib and an 8 tonne lift under-axle equipment. Using one-man operation it is suitable for all types of wheeled vehicles. It is available with a 25/50 tonne line pull recovery winch for heavy duty, MBT and self-recovery.

Note: There is a separate entry for the Reynolds Boughton (4 × 4) 6,000 kg recovery vehicle in the previous section as this is a complete system.

Status

In production. In service with many armed forces in Africa, the Middle East and the Far East.

Manufacturer

Reynolds Boughton Limited, Engineering Division, Bell Lane, Amersham, Buckinghamshire HP6 6PE, UK.
Tel: +44 1494 764411. Telex: 83132.
Fax: +44 1494 765218.

UPDATED

Boughton winches

Description

Reynolds Boughton Limited produce a range of mechanical and hydraulic freespool winches suitable for a wide range of applications. Many are used in conjunction with recovery equipments. Details of the range are provided in the following table and it should be noted that in each case the rated performance is based on bare drum performance.

Recent winch contracts included the following:

H5000 to the US Marine Corps for Ribbon Bridge launching equipment

H7500 for the UK Ministry of Defence's winch variant of the Leyland 4 tonne truck

H10000 for the Case 721 Medium Wheeled Tractors for the British Army

H25000 for Royal Navy aircraft tugs

VH8 for the UK Ministry of Defence's 2 tonne TUH.

Status

Production as required. In service with a number of countries including the UK and the USA.

Manufacturer

Reynolds Boughton Limited, Engineering Division, Bell Lane, Amersham, Buckinghamshire HP6 6PE, UK.
Tel: +44 1494 764411. Telex: 83132.
Fax: +44 1494 765218.

Boughton H15000 hydraulic winches on Foden transporter ***1996***

Mechanical drive winches

Type	Rated capacity	Transmission	Brake	Fairlead	Anchor
TTB100	6.8 t	forward/reverse	auto	option	–
1N	7 t	single speed	manual	option	option
VM16	8.2 t	forward/reverse	manual	option	–
20	10 t	two speed	manual	standard	standard
30	10 t	forward/reverse	manual	option	option
40	15 t	two speed or forward/reverse	manual	option	option
5	25 t	forward/reverse	manual[1]	option	option
80	35 t	forward/reverse	auto	option	–
120	55 t	forward/reverse	auto	option	–

Hydraulic drive winches

Type	Rated capacity	Transmission	Brake	Fairlead	Anchor
VH6	3 t	forward/reverse	auto[1]	option	–
VH8	3.8 t	forward/reverse	auto[1]	option	–
VH10	5 t	forward/reverse	auto[1]	option	option
H5000	5 t	forward/reverse	auto	option	option
H7500	7.5 t	forward/reverse	auto	option	option
H10000	10 t	forward/reverse	auto	option	option
H12500	12.5 t	forward/reverse	auto	option	option
H15000	15 t	forward/reverse	auto	option	option
H20000	20 t	forward/reverse	auto	option	option
H25000	25 t	forward/reverse	auto	option	option
1NH	7 t	forward/reverse	manual	–	option
2H	12 t	forward/reverse	manual[1]	option	option
5H B200	19 t	forward/reverse	manual[1]	option	–
5H B270	26 t	forward/reverse	manual[1]	option	–
VH45	20 t	forward/reverse	auto	option	–
VH55	25 t	forward/reverse	auto	option	–

[1] Auto is optional

UPDATED

Boughton recovery equipment

Description

A complete range of recovery equipment has been developed by Reynolds Boughton Limited. This equipment is suited for use with the Reynolds Boughton range of recovery cranes but is applicable to other vehicles. The range includes: recovery chains, ground anchors, A-frames, tow poles, sand wedges, gun planks, pioneer kits and so on.

Status

Production as required. In service with many armed forces.

Manufacturer

Reynolds Boughton Limited, Engineering Division, Bell Lane, Amersham, Buckinghamshire HP6 6PE, UK.
Tel: +44 1494 764411. Telex: 83132.
Fax: +44 1494 765218.

UPDATED

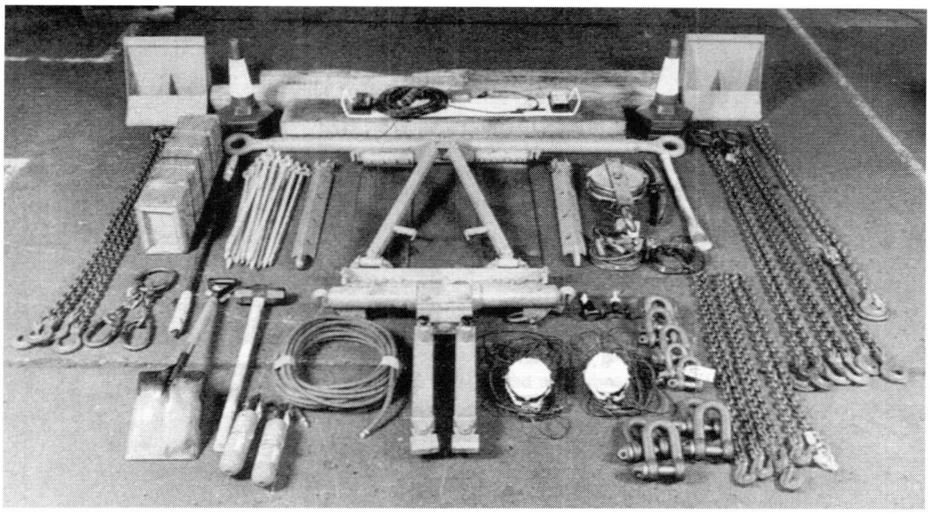

A selection of Boughton recovery equipment

Marlow Kinetic Energy Recovery Rope (KERR)

Description

The Marlow Kinetic Energy Recovery Rope (KERR) system was developed in conjunction with the British Army to provide a fast and effective recovery method and is particularly suited to armoured vehicles in tactical situations where wheeled or tracked recovery vehicles are not available. Using KERR it is possible for lighter vehicles to recover heavier vehicles; the rope sized for the recovering vehicles should be used (see under Specifications).

In use, the towing vehicle reverses as close as possible to the vehicle to be recovered. The rope is connected and snaked to allow tangle-free deployment. For situations where it is not possible to get close to the bogged vehicle, extension cables may be used. The KERR must be connected directly to the recovery or towing vehicle.

The towing vehicle then accelerates to the maximum speed possible and snatches the rope with its total energy at that maximum speed. The towing vehicle is slowed or halted and its kinetic energy is converted into the potential energy of a stretched rope. If the bogged vehicle has power to assist the transfers of energy by the KERR, a faster recovery can be accomplished. After a slight pause the bogged vehicle rises free. If the vehicle is not freed by the first attempt the process is repeated. Once free, continued towing by KERR is possible.

Marlow Ropes Limited also manufactures lightweight bridles which can be used in conjunction with the KERR. These bridles are described as ideal replacements for the heavy wire ropes used on many armoured vehicles and they can be made to suit any size of vehicle. They are available in Polyester and the high modulus polyethylene fibres such as Spectra/ Dyneema.

Status

In production. In service with the British Army and more than 20 other armed forces, including Canada, South Africa and the US Army.

Manufacturer

Marlow Ropes Limited, Diplocks Way, Hailsham, East Sussex BN27 3JS, UK.
Tel: +44 1323 847234. Telex: 87676.
Fax: +44 1323 440093.

Typical recovery operation involving a Marlow Kinetic Energy Recovery Rope

Specifications

Gross weight of vehicle	Rope size	Rope strength	Overall length of rope	Rope weight
Under 7,000 kg	32 mm	20,000 kg	12.2 m	9.5 kg
7,000 to 7,990 kg	36 mm	24,800 kg	12.2 m	12.5 kg
8,000 to 11,990 kg	40 mm	30,000 kg	13.7 m	17 kg
12,000 to 15,490 kg	48 mm	42,000 kg	13.7 m	25 kg
15,500 to 18,490 kg	52 mm	48,800 kg	13.7 m	29 kg
18,500 to 21,990 kg	56 mm	56,000 kg	13.7 m	34 kg
22,000 to 23,990 kg	60 mm	63,800 kg	13.7 m	39 kg
24,000 to 30,990 kg	64 mm	72,000 kg	15 m	49 kg
31,000 to 43,990 kg	68 mm	79,500 kg	19 m	67 kg
44,000 to 54,990 kg	72 mm	90,000 kg	21 m	84 kg
55,000 to 65,000 kg	80 mm	110,000 kg	21 m	103 kg

UPDATED

EKA recovery equipment

Development/Description

The EKA company designs, develops and manufactures heavy recovery equipment for use in civil and military applications.

Typical of these is the EKALIFT 2500 which is a heavy duty military recovery body specifically developed to provide field support for off-road eight-wheel military load carriers with high front-end pick-up weights.

Designed primarily for mounting on the British Army DROPS IMMLC chassis, the EKALIFT provides all of the facilities required for rapid, labour saving, recovery and associated equipment handling whilst minimising axle weights to achieve good mobility and route availability.

The hydraulic capstan winch has a line pull of 25 tonnes over the entire rope length, so there is no need to draw all the rope out to make a maximum capacity pull.

The winch powers the rope out and a 1 tonne auxiliary winch assists in taking the rope to the casualty. Two hydraulic anchor spades are rated at 50 tonnes to enable two-part pulls to be made.

The electronic remote control allows full variable rope speed control from 0 to 6 m/min at full load and up to 40 m/min for rapid retrieval of a light rope.

A 10 tonne line pull hydraulic drum winch is mounted in front of the cab for self-recovery and secondary winching tasks.

The low weight recovery boom is raised by the main winch rope. It functions with a range of specially designed articulating recovery beams and adaptors which connect to the casualty recovery eyes via quick release pins. This provides a simple and robust method of locating the casualty. The boom has a lift capacity of 12 tonnes.

Whilst the EKALIFT 2500 is designed primarily for towing fully laden large eight-wheeled off-road vehicles, other logistic vehicles down to Land Rover (4 × 4) can be successfully suspended by means of auxiliary adaptors.

To the rear of the cab is fitted a 12.5 tonne hydraulic folding crane which fulfils both equipment handling and recovery roles.

In addition to lifting larger items of recovery equipment the crane is used to transfer the quick release stowage compartment to the ground for ease of removal of tools and fittings. The crane is also essential for spare wheel handling and the EKALIFT 2500 is equipped to carry spare wheels for this purpose.

In its recovery role the crane is able to unload casualties if required and to provide restraint whilst righting an overturned vehicle.

Other recovery units include the EKALIFT 2000, Compact and the D series. Over 400 Compact systems have been built with known users including Ireland, Singapore and the UK. The latter application is for the

Foden (6 × 6) recovery vehicle covered in an earlier section.

The company also designs and manufactures container transfer units including the SRTE on a Leyland 8 × 6 chassis for the UK MoD and the civilian Stevedore system. They are also the UK agent for Rotzler winches (qv).

Status

Production as required. In service with many countries including Angola, Brunei, Ireland, Kenya, Malaysia, Nigeria, Singapore, Tanzania, United Kingdom and Zambia.

Manufacturer

EKA Limited, Valkyrie House, Packhorse Road, Gerrards Cross, Buckinghamshire SL9 8EB, UK.
Tel: +44 1753 889818. Fax: +44 1753 880004.

NEW ENTRY

Foden 6 × 6 recovery vehicle of the British Army is fitted with the EKALIFT 200 system shown here in the deployed position
1997

Tirfor recovery equipment

Description

The Tirfor models TU-8, TU-16 and T-35 are all portable hand-operated machines which can be used for a wide range of loading, lifting, hauling and lowering operations. They are reversible and are available with a range of accessories which includes slings, shackles, snatch blocks, ground anchors, automatic wire rope reeling devices, wire rope grips and girder clamps. The latest model is the TU-16H, which is a hydraulically operated strengthened TU-16. Hydraulic pump units allow the operation of one, two or four TU-16Hs and the system can work both forwards (pulling or lifting) or reverse (lowering). The hydraulic power system gives the TU-16H a nominal capacity of 1,600 kg. This model can be operated from a unit with the following capability:

Flow: 8 -13 litres/min maximum
Pressure: 120-140 bars
Hydraulic oil: 2° to 5° Engler viscosity at 50° C
Viscosity index: over 100
Filter: 40 µm

For light recovery operations the company offers the Jockey, which weighs less than 1.8 kg and has a SWL of 300 kg. The handle has been designed to bend when the rated capacity is exceeded by 50 per cent.

Tirfor T-35 being used with Williams Fairey Medium Girder Bridge

Other hand-operated Tirfor machines:

Model	**J5 Junior**	**T508**	**T-30**
SWL for lifting operations	500 kg	800 kg	3,000 kg
Weight	4 kg	6.6 kg	27 kg
Breaking strength	3,000 kg	4,800 kg	16,800 kg
Diameter of wire rope	6.5 mm	8.2 mm	16.3 mm

Status

All this equipment is in production. The TU-8, TU-16 and T-35 are used by many armed forces around the world.

Manufacturer

Tirfor Limited, Old Lane, Halfway, Sheffield S19 5GZ, UK.
Tel: +44 114 248 2266. Fax: +44 114 247 5649.

Specifications

Model	**TU-8**	**TU-16**	**T-35**
Safe working load for lifting operations	800 kg	1,600 kg	3,000 kg
Effort	31 kg	46 kg	43.2 kg (slow speed) 68.1 kg (fast speed)
Distance of rope travel for one complete cycle of operating lever	65 mm	70 mm	38 mm (slow speed) 48 mm (fast speed)
Mechanical advantage	28:1	38:1	70:1 (slow speed) 45:1 (fast speed)
Weight of machine only	8.4 kg	18 kg	26.8 kg
Overall dimensions (without handle)	528 × 284 × 113 mm	660 × 356 × 152 mm	718 × 140 × 318 mm
Length of operating handle (fully extended)	0.73 m	1.13 m	1.157 m
Standard length of wire rope	20 m	20 m	20 m
Rope diameter	8.2 mm	11.5 mm	16.3 mm
Rope end fitting	safety hook	safety hook	shackle

UPDATED

Superwinch recovery winches

Description

Superwinch is a major manufacturer in the design, manufacture and worldwide provision of electrical, mechanical and hydraulic winches. With single-line-pull capacities from 453.6 to 20,412 kg, including both capstan and drum variants.

Superwinch produces are regularly procured by the UK MoD, military vehicle manufacturers and armed forces worldwide. Typical applications include vehicle and equipment recovery, logistics handling and target launch systems.

Ongoing commitment to quality and continuous design improvements in winch efficiency have enabled Superwinch to maintain the availability of a cost-effective and durable product range demanded by both the civilian and military markets.

1997-98 will see additional new products within the Superwinch range both as replacements to existing models and further extensions to the range in response to specific market requirements.

Status

Production as required. In service with many countries including UK.

Manufacturer

Superwinch Limited, Abbey Rise, Whitchurch Road, Tavistock, Devon PL19 9DR, UK.
Tel: +44 1822 614101/7. Fax: +44 1822 615204.

UPDATED

Superwinch H14 lightweight hydraulic winch mounted on Leyland T244 (4 × 4) truck ***1996***

Superwinch electric worm drive winch mounted on Land Rover Defender 110 ***1996***

Rubery Owen recovery equipment

Description

A special range of lightweight recovery equipment has been designed by Rubery Owen, using safety factors as low as 1.5 times the working load limit. Meeting the needs of skilled recovery personnel, all the equipment has been subjected to rigorous field and destruction test programmes.

Rubery Owen Group Services Limited is also a design authority to MoD(PE) Log Veh and a Post Design Services contractor to Quarter Master General for general and special purpose trailers.

Status

Manufactured as required. In service with the British Army.

Manufacturer

Defence Equipment Division, Rubery Owen Group Services Limited, Darlaston, PO Box 10, Wednesbury, West Midlands WS10 8JD, UK.
Tel: +44 121 526 3131. Fax: +44 121 526 2869.

UPDATED

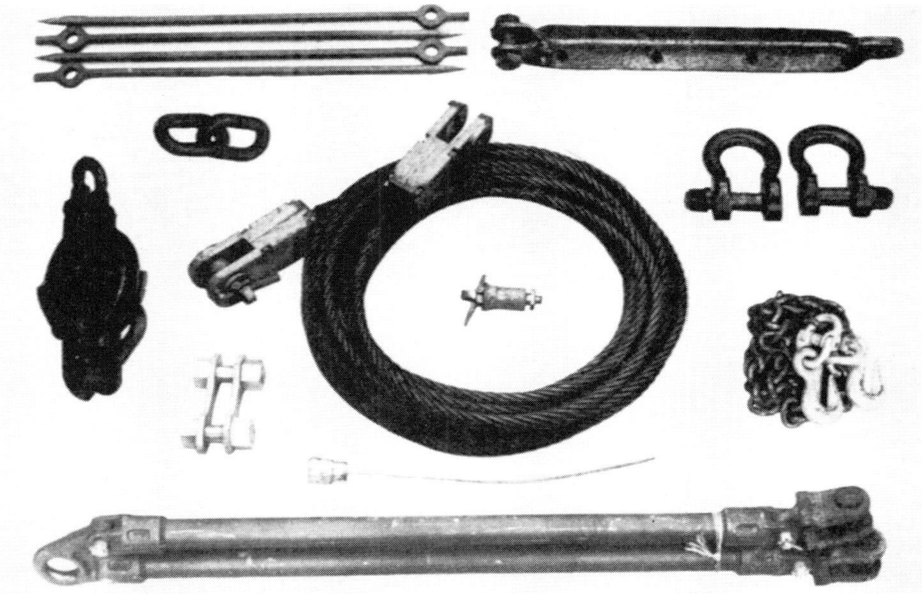

Selection illustrated is typical of recovery equipment supplied by Rubery Owen

UNITED STATES OF AMERICA

Holmes International recovery equipment

Development

From 1973 to 1987 Holmes International was a division of the Dover Corporation. In 1987 Holmes International Inc became a privately owned company. Since 1914 it has developed a wide range of twin-boom equipment which can be mounted on a variety of wheeled and tracked chassis, including chassis available from Kenworth, Navistar, Chevrolet, GMC, Ford, Dodge, Bedford, Foden and Mercedes-Benz. A complete line of single-boom hydraulic wreckers from 2 to 45 tonnes is also available. Steel and aluminium deck tilt-bed carriers are also manufactured.

The main advantage of the twin-boom recovery equipment is that the vehicle can recover disabled vehicles from the side as well as the rear, because the booms can be swung to the left and right of the vehicle. By securing the cable from one of the booms to a suitable anchorage point, the second boom can handle vehicles equalling the weight of the recovery vehicle itself.

Descriptions

Holmes FW-35 towing unit

The Holmes FW-35 towing unit is designed to be fitted over the 'fifth wheel' turntable unit of heavy tractors for extreme heavy-duty lift and towing of tactical vehicles, such as tank recovery vehicles and missile carriers. When fitted, the FW-35 is capable of towing vehicles with a gross cargo weight of up to 38,555 kg and it can lift up to 15,875 kg. Once lifted, the towed vehicle travels on its rear axle(s). These units are in use by US armed forces in Europe.

Holmes ProStar wheel lift

This is an all-hydraulic add-on wheel lift with 1,360 kg lifting capacity and 3,400 kg towing capacity. It can be mounted on most light chassis wreckers and has combined lift and tilt functions. Power is derived from an electric/hydraulic power pack or directly from a hydraulic system.

Holmes Protech PT1102

This equipment includes a hydraulic wheel lift (1,360 kg lifting capacity and 3,400 kg towing capacity) with independently controlled lift, tilt and extension functions plus a special boom wrecker. The wrecker has a fixed boom angle and a manual extension. This equipment can be installed on a two-axle chassis with dual rear wheels and a gross vehicle weight of not less than 4,536 kg.

Holmes Protech PT3105 and PT3210

These equipments can be installed on a two-axle chassis with dual rear wheels and a gross vehicle weight of not less than 4,536 kg. The equipment includes a hydraulic wheel lift (1,360 kg lifting capacity and 3,400 kg towing capacity) with independently controlled lift, tilt and extension functions and a single-boom hydraulic-powered wrecker. There are dual hydraulic cylinders to raise and lower the boom. The main difference between the two models is that the PT3105 has a single winch while the PT3210 has two.

Holmes 750 Wrecker

This equipment can be installed on a two- or three-axle chassis that has a gross vehicle weight of not less than 14,514 kg. Equipment that can be fitted (including optional equipment) includes: two stiff legs and a bolster, chain and pinion guards, extensible or non-extensible booms, controls recessed into the body, towing chains, rear jacks to increase lifting stability, dual power unit with a gear reduction of 214 to 1, 14 anchor points, backlash brakes, telescoping outboard legs, cable drums with guards and two snatch blocks. This version is known to be used by the Australian Army based on a Mack (6 × 6) truck chassis.

Holmes 1601 Python hydraulic wrecker

The Holmes 1601 is a single-boom wrecker with a hydraulic extension of 1.64 m. Its structural capacity is 19,954 kg retracted and 6,350 kg at full extension. The structure incorporates the Holmes mast which allows for greater recovery capability during side pulls. Optional outboard legs increase the performance capacity during side pulls. A minimum front axle rating of 3,175 kg and 10,433 kg for the rear axle are required.

Holmes 1625 Hydraulic Wrecker

The Holmes 1625 incorporates many well-proven Holmes features plus some new technological features. It has a long or short boom configuration to suit the chassis and to suit single or tandem axles. The single boom extends 1.6 m with a retracted capacity of 22,680 kg while the fully extended capacity is 9,072 kg. Height above the truck frame is 1.4 m. Optional side legs can increase the side pull of the unit. Minimum gross vehicle weight required is 12,777 kg.

Holmes Model 1701 Lancer Wrecker

The Holmes Model 1701 uses a single, non-traversable boom. The type is described as a 35 tonne power wrecker and uses a system of stabilising jacks and stays for lifting heavy loads. The single boom has a single-stage maximum extension of 2.438 m and has a maximum lift of 31,750 kg retracted and 10,886 kg extended. The boom is located on a steel mast structure which also carries a single adjustable extending stabiliser leg on each side. The mast can also be used as the main load-carrying structure when side pulls are used, as the Model 1701 uses a twin winch system. The twin winches can be used either over twin cable pulleys at the end of the boom or over the sides of the central mast. Twin hydraulic cylinders raise and lower the boom. With side pulls it is possible to use one winch cable as an anchor for improved stability. The winches, which are similar to those used on the Model 750, are specially produced for wrecker use and have such features as manual rewind and nylon bearings. The winch gear reduction is 459 to 1.

Holmes Model 1801 Victor Wrecker

The Holmes Model 1801 Victor wrecker follows the same general lines as the Model 1701 but the lifting capacity is much greater. The single non-traversable boom of the Model 1801 has a maximum extension of 2.54 m. Retracted it has a maximum lift capacity of 40,823 kg and fully extended it can lift 13,608 kg. This lift capacity enables the vehicle to assist disabled AFVs using other vehicles as anchors to assist in side pulls. The Model 1801 boom has the lifting capacity to enable it to be used as a heavy repair vehicle, lifting tank turrets and engine packs when necessary. The Model 1801 has a three-stage hydraulic motor. One 321.7 litre/min stage drives one winch and the boom extension cylinder. Another 321.7 litre/min stage drives the other winch and the boom lift cylinders. A third 56.8 litre/min stage drives the outboard legs and rear jacks. Each stage can be operated independently.

Holmes TH395 Super Heavy-Duty Truck Hitch

The TH395 super heavy-duty truck hitch is rated at 11,340 kg lift capacity and 38,102 kg towing capacity and uses 15.8 mm alloy chains. A full contact anchor bar reduces towed vehicle bumper damage and 10 grease fittings provide adequate lubrication to maintain operational life.

Specifications

Model	**1701 Lancer**	**1801 Victor**
Basic wrecker rating	31,750 kg	40,823 kg
Boom rating		
(retracted)	31,750 kg	40,823 kg
(fully extended)	10,886 kg	13,608 kg
Winch characteristics (each drum)		
(load rating, first cable layer)	18,144 kg	23,587 kg
(drum diameter)	381 mm	508 mm
(line speed)	5.79 m/min	6.096 m/min
(cable length, max)	86.87 m	94.5 m
(gearbox/cable drum gear reduction)	214 to 1	485 to 1
Winch type	worm drive	worm drive
Winch cable diameter	15.875 mm	19 mm
Cable length (each drum)	76.2 m	91.44 m
Working load limit (single line)	5,883 kg	8,373 kg
Breaking strength (single line)	20,593 kg	29,302 kg
Wrecker height above truck frame	1.963 m	2.083 m
Wrecker body width	2.438 m	2.438 m
Weight		
(wrecker with body approx)	5,693 kg	7,824.5 kg
(wrecker without body approx)	4,241 kg	6,373 kg

Specifications

Model	**PT1102**	**PT3105**	**PT3210**	**750**	**1601**	**1625**	**1701**	**1801**
Type	mechanical	hydraulic	hydraulic	mechanical	hydraulic	hydraulic	hydraulic	hydraulic
Rating								
(both booms)	n/a	n/a	n/a	22,680 kg	n/a	n/a	n/a	n/a
(each boom)	3,630 kg	9,090 kg	9,090 kg	11,340 kg	19,958 kg	22,680 kg	31,750 kg	40,823 kg
(boom fully extended)	1,815 kg	2,720 kg	2,720 kg	5,443 kg	6,450 kg	9,072 kg	10,886 kg	13,608 kg
Drum rating								
(number of winches)	1	1	2	2	2	2	2	2
(each drum)	2,630 kg	4,536 kg	4,536 kg	11,340 kg	9,979 kg	15,876 kg	23,452 kg	30,535 kg
(cable diameter)	10 mm	10 mm	10 mm	14.3 mm	14.3 mm	16 mm	16 mm	19 mm
(cable length)	23 m	30 m	30 m	61 m	61 m	61 m	76.2 m	94.5 m
Recommended working limit of each cable	1,585 kg	1,585 kg	1,585 kg	4,309 kg	3,493 kg	4,990 kg	5,883 kg	8,373 kg
Boom								
(max extension)	0.75 m	1.53 m	1.53 m	1.93 m[1] 2.44 m*	1.64 m	1.64 m[1] 2.44 m*	2.44 m	2.54 m
(horizontal swing)	n/a	n/a	n/a	90°	n/a	n/a	n/a	n/a
(vertical swing)	fixed	60°	60°	60°	60°	60°	60°	60°
Height of equipment above truck chassis	varies with chassis	varies with chassis	varies with chassis	1.99 m	1.52 m	1.4 m	1.96 m	2.08 m

[1] single axle
* tandem axle

Status

In production. In service with many armed forces throughout the world including Australia and the USA.

Manufacturer

Holmes International Inc, 2505 East 43rd Street, Chattanooga, Tennessee 37407, USA.
Tel: +1 615 867 2142. Telex: 558409.

UPDATED

Holmes 750 equipment installed on Mack R (6 × 6) truck chassis for the Australian Army

Paratech MAXIFORCE air lifting bags

Description

The Paratech MAXIFORCE air lifting bag is a thin, strong, moulded envelope made from neoprene embedded with three full layers of Kevlar aramid reinforcement on each side. Each lifting bag has a single inlet/outlet used to inflate and deflate the bag; the usual inflation agent is compressed air. The bags have a wide range of applications and can be used on soft ground, under water or when it is not possible to use jacks or other lifting devices. They can also be used for odd-shaped or cylindrical loads and for many other load-lifting applications.

A MAXIFORCE system can be quickly set up as even the largest bag can be fully inflated in less than 60 seconds. It is also completely silent in use. All that is required to operate the system is positioning a MAXIFORCE bag under the load involved and inflating the bag. Using the energy of compressed air a load is moved within seconds. The bags require only 25 mm of insertion space and operate on air pressure up to 8 bar (118 psi), so air sources such as air cylinders, truck airbrake systems, compressors and foot pumps can be used for inflation. Inflation and deflation is operator-controlled by use of a safety control valve which can control two bags independently.

All MAXIFORCE bags have a positive grip surface designed for maximum friction and hold. As an additional safety precaution during lifting operations a bright yellow 'X' is moulded into both sides of each bag. This ensures high visibility and exact centering of the bag under a load for maximum lift heights and stability.

The weight of a MAXIFORCE air lifting bag is 4.2 to 20 kg depending on size and type.

Status

Production as required. In service with undisclosed countries.

Manufacturer

Paratech Incorporated, 1025 Lambrecht Road, Frankfort, Illinois 60423, USA.
Tel: +1 815 469 3911. Fax +1 815 469 7748.

UPDATED

A Paratech MAXIFORCE air lifting bag being used to lift a tracked construction vehicle

A selection of Paratech MAXIFORCE air lifting bags and accessories

Vulcan Model LTRS-1 light tactical recovery system

Description

The Vulcan Model LTRS-1 light tactical recovery system is a military adaptation of the Vulcan Model 837 Wheel Snatcher commercial underlift recovery system. The LTRS-1 has a lift rating of 1,588 kg and a tow rating of 3,402 kg. Strapless wheel retainers are employed which pick up high on the wheel tyre and free automatically when the tyre is lowered to the ground so no straps are required during normal use; safety chains are recommended for severe cross-country conditions. The LTRS-1 wheel retainers utilise 15 adjustments to accommodate a full range of tyre sizes and can secure a wheel with a flat tyre.

The LTRS-1 provides underlift extensions of 762 mm to 1.625 m and can power tilt under load.

The company also supplied the recovery equipment for the Steyr/UTDC Percheron (6 × 6) series of vehicles manufactured in Canada to meet the requirements of the Canadian Forces. Details of these are given in the *Trucks* section under International with a total of 124 recovery vehicles being supplied by 1991.

Status

Production as required. In service with undisclosed countries.

Manufacturer

Vulcan Equipment Company, Inc, PO Box 826, 8530 Sandidge Road, Olive Branch, Mississippi 38654, USA.
Tel: +1 601 895 2400. Fax: +1 601 895 3465.

UPDATED

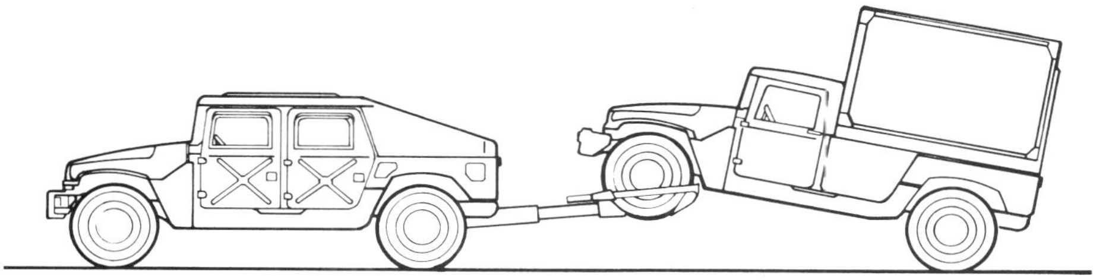

Outline drawing of Vulcan Model LTRS-1 light tactical recovery system in use with HMMWVs

Percheron (6 × 6) 10,000 kg recovery vehicle with Vulcan recovery equipment
1997

BRIDGING SYSTEMS

Mechanised bridges
Tactical floating bridges and ferries
Tactical (non-floating) and line of communication bridges
Bridging boats

MECHANISED BRIDGES

CHINA, PEOPLE'S REPUBLIC

NORINCO Type 84 armoured bridgelayer

Description

The NORINCO Type 84 armoured bridgelayer uses the hull and suspension of the Type 69 MBT. The vehicle has a crew of three with the driver in the normal position at the hull front. The other two are seated inside the hull with a small observation hatch provided for the commander. This is fitted with a periscope and an infrared night vision device. The vehicle is equipped with a Type 899 radio set and a Type 803 set is used for crew intercommunications. Also carried is a single 7.62 mm machine gun.

A hydraulically operated stabiliser blade, used during the final stages of bridgelaying, is mounted under the front of the hull and may also be used for obstacle clearing and dozing operations.

The bridge is of the Biber pattern with some alterations to the laying system. It is constructed of light steel and is in two halves, one half carried on top of the other. Total length is 18 m of which 16 m can normally be used and the maximum weight carried on the main trackways is 40,000 kg. A narrower internal auxiliary trackway is limited to 8,000 kg. The bridgelaying mechanism is driven using hydraulic actuators and rams carried on forward and rear operating assemblies and a rail framework is used to carry and move the bridge into position.

The laying sequence commences with the positioning of the carrier vehicle close to the obstacle to be bridged. Operations commence with both halves of the bridge being moved backwards along the carrier frame. As this is carried out the rear of the upper bridge half is raised slightly, followed by the front. This upper bridge half lifts clear of the lower half, allowing the hydraulic system to move the latter forwards along the carrier frame as far as it will go on the forward positioning assembly. At this point the hydraulic ram system at the rear of the vehicle is swung downwards to the rear, after which the lower half is moved forward yet again by a ram on the forward assembly. This allows both halves to be lowered to the same level and they can then be joined. The ram system at the rear is then disconnected to allow the forward assembly to push the entire bridge forward to its maximum extent. Before this operation is completed the stabiliser blade at the front is lowered to the ground to provide support for the vehicle as the bridge is extended. At the full extension point the carrier frame is tilted forward by the forward hydraulic assembly until the far end of the bridge touches the ground. The carrier vehicle is then reversed slowly as the near end of the bridge is lowered, after which the carrier frame is disconnected to complete the operation. Recovery is carried out in the opposite sequence.

The bridge can be assembled and carried to the laying point positioned centrally on the carrier frame in order to save time at the actual bridging position. The time required for a complete bridge launch operation is between 4 and 5 minutes and between 4 and 6 minutes for recovery. A height difference of 4 m between banks is possible, with the slope limit for the carrier vehicle being 10° under such conditions. The bridge can be launched with the carrier vehicle positioned on a side slope of between 8 and 10°.

NORINCO Type 84 armoured bridgelayer in travelling configuration

Specifications

Crew: 3
Weight: (combat) 38,500 kg
Length: (with bridge) 9.879 m
Width: (overall) 3.27 m
Height: (with bridge) 2.978 m
Ground clearance: 0.35 m
Track: 2.64 m
Length of track on ground: 3.845 m
Ground pressure: 0.852 kg/cm^2
Max speed: (road) 50 km/h
Range: 365 km
Fuel capacity: 930 litres
Fording: 1.1 m
Gradient: 25°
Side slope:
(travelling) 15°
(launching bridge) 8-10°
Engine: Model 12150 L-7BW V-12 water-cooled diesel developing 580 hp at 2,000 rpm
Transmission: manual, 5 forward and 1 reverse gears
Electrical system: 24 V
Batteries: 4 × Type 65
Armament: 1 × 7.62 mm MG
Ammunition carried: 1,500 rds

Bridge
Weight: 8,000 kg
Load capacity:
(main bridge) 40,000 kg
(auxiliary bridge) 8,000 kg
Length:
(complete) 18 m
(usable) 16 m
Width: 3.2 m
Track width:
(main bridge) 0.6 m
(auxiliary bridge) 0.55 m

Status

In production. In service with the Chinese Army and Thailand.

Manufacturer

China North Industries Corporation (NORINCO), 7A Yue Tan Nan Jie, PO Box 2137 Beijing, Beijing, China.
Tel: +86 1 867570. Telex: 22339 cnic cn.
Fax: +86 1 867840.

VERIFIED

Heavy mechanised bridge Type 84A

Description

The Heavy Mechanised Bridge Type 84A is a close derivative of the TMM truck-mounted treadway bridge (see separate entry in this section) and can be used by tracked vehicles up to a maximum weight of 50 tonnes and by wheeled vehicles with a maximum axle load of 13 tonnes. The bridge sections are carried and placed

Heavy mechanised bridge Type 84A under construction
1996

into position by an 8 × 8 truck and the general laying sequence and operation are much the same as those for the TMM (qv).

Specifications
Weight of bridgelayer: (total) approx 20 t
Length of single span: 10.5 m
Roadway width: 3.8 m
Total length of bridge: 50 m
Bridge height:
(max) 3.8 m
(min) 2.2 m
Capacity: (tracks) 50,000 kg

Status
In production.

Manufacturer
China Shipbuilding Trading Company Limited, 10 Yuetan Beixiaojie, Beijing, China.
Tel: +86 10 6858 2562.
Fax: +86 10 6858 3380.

UPDATED

CZECH REPUBLIC AND SLOVAKIA

ZTS MT-72 armoured bridgelayer

Development/Description
The MT-72 armoured bridgelayer was developed using the chassis of the T-72 MBT as a basis but otherwise follows the same general lines as the MT-55A armoured bridgelayer covered by the following entry. However, the MT-72 uses a 20 m long bridge in place of the 18 m bridge used by the MT-55A. The MT-55A can carry and launch the MT-72 bridge if required.

The MT-72 bridge is of the scissors type and is constructed using high-strength steel. The bridge has a carrying capacity of 50,000 kg and has a total length of 20 m, allowing gaps of up to 18 m to be bridged. When gaps longer than 20 m have to be bridged it is possible to lay a second bridge on the first and second bridges. Gradients are ±50° and side slopes ±6°.

Width of the bridge is 3.3 m while each bridge track is 1.15 m wide. Weight of the complete bridge is 6,000 kg. When folded the bridge is 10.6 m long.

The hydraulic bridgelaying mechanism used on the MT-72 is the same as that used on the MT-55A; working pressure is 22 MPa and maximum pressure 28 MPa. An automatic laying sequence lasts 3 minutes, with bridge retrieval from 3 to 8 minutes.

The MT-72 is provided with a fording device which allows water obstacles up to 4.2 m deep to be negotiated.

Reference has been found to a Czech/Slovak armoured bridgelaying project known as the JTTP. This will be a 50,000 kg capacity bridge with an overall length of 25 to 30 m and a width of 4 m. The bridgelayer and bridge combination will weigh a maximum of 45,500 kg and be 12 m long overall.

Rear view of ZTS MT-72 armoured bridgelayer in travelling configuration **1996**

Specifications
Crew: 2
Weight: (with bridge) approx 41,500 kg
Length: (with bridge) 10.6 m
Width: (with bridge) 3.59 m
Height: (max, with bridge) 3.8 m
Fording: 4.2 m
Engine: V-64-4 V-12 diesel developing 840 hp

Bridge
Weight: 6,000 kg
Length:
(extended) 20 m
(folded) 10.6 m
Width: 3.3 m
Track width: (each) 1.15 m
Carrying capacity: 50,000 kg

Status
Available for production.

Manufacturer
ZTS Tees Martin, ul ČSA 3/ 1697, 036 57 Martin, Slovakia.
Tel: +42 842 33835.
Fax: +42 842 32822.
Marketed by: Kerametal, Jašikova 2, 826 05 Bratislava, Slovakia.
Tel: +42 7 299 1111. Telex: 932 33 Kemet C.
Fax: +42 7 522 5896.

VERIFIED

MT-55A armoured bridgelayer

Development/Description
The MT-55A armoured bridgelayer, a joint Czech/Soviet project, was the replacement for the earlier MT-34 and was adopted by Czechoslovakia and the Soviet Union, before the break, and other countries. It is based on a T-55A tank chassis with the turret removed. The basic chassis is similar to that used for the T-54/T-55 ARVs and, as many vehicles are converted MBTs rather than 'new' vehicles, the reader is referred to this entry in the *Armoured Recovery Vehicles* section for a description.

There are two types of scissors bridge. The first model has circular holes in the sides of the bridge, similar to those on the bridge carried by the earlier MT-34 bridgelayer, while the more recent model has solid panels. It is possible to mistake the latter for the German BLG-60 armoured bridgelayer. The main feature that distinguishes between the two is that the bridge surface of the MT-55A has a pattern whereas the BLG-60's is smooth with a plastic covering.

The steel bridge is launched over the front of the vehicle hydraulically by an automatically controlled electrohydraulic system. In an emergency the hydraulic system can be operated with the aid of three mechanical levers and, if the engine fails, the tank's batteries can provide sufficient power to launch the bridge.

The launching procedure is as follows: the MT-55A

MT-55A armoured bridgelayer in travelling configuration (T J Gander)

MT-55A armoured bridgelayer on rail transport and without its bridge **1996**

VICKERS **DEFENCE SYSTEMS**

Bridging & Engineering Equipment
Scotswood Road
Newcastle-upon-Tyne
NE99 1BV
England
Tel: 0191 273 8888
Fax: 0191 273 2324

stops short of the gap and raises the bridge slightly from its horizontal travelling position, the launching girder foot is lowered to the ground and the span is raised to the vertical. The bridge is then unfolded and lowered across the gap. The bridge takes 3 minutes to lay in position and from 3 to 8 minutes to retrieve. The bridge itself weighs 6,500 kg and when opened out is 18 m long, 3.34 m wide and 0.9 m high. It has a maximum clear span of 16 m and a maximum capacity of 50,000 kg.

The MT-55A can also be used to carry and launch the 20 m bridge normally used with the MT-72 (see previous entry).

Standard equipment on the MT-55A includes an NBC system, snorkel, inclinometer and other equipment for determining the width of the gap before the bridge is laid in position.

Specifications

Crew: 2
Weight: (with bridge) 36,000 kg
Length: (with bridge) 9.88 m
Width: (with bridge) 3.3 m
Height: (with bridge) 3.35 m
Ground clearance: 0.425 m
Track: 2.64 m
Track width: 580 mm
Length of track on ground: 3.84 m
Ground pressure: 0.81 kg/cm^2
Max speed:
(road) 32-35 km/h
(cross-country) 16-20 km/h
Range: 485-500 km
Fuel capacity: 900 litres
Fording:
(normal) 1.4 m
(with preparation) 4 m
Gradient: 30°
Vertical obstacle: 0.7 m
Trench: 2.7 m
Engine: V-55, V-12, water-cooled diesel developing 580 hp at 2,000 rpm
Transmission: manual with 5 forward and 1 reverse gears
Electrical system: 24 V
Batteries: 4, 280 Ah

Bridge
Weight: 6,500 kg
Length:
(extended) 18 m
(folded) 9.6 m
Width: 3.3 m
Track width: (each) 1.15 m
Carrying capacity: 50,000 kg

Status

In service with the Czech Republic, India, Iraq, Russian Federation and Associated States (CIS) (568 declared), Slovakia, Yugoslavia (Serbia and Montenegro) and some countries in the Middle East.

Manufacturer

VOP 025 N. Jičín, sp, Dukelská 102, 742 42 šenov u Hoveho Jicina, Czech Republic.
Tel: +42 656 793316. Telex: 52196.
Fax: +42 656 701734.

UPDATED

AM-50 and AM-50B truck-mounted scissors bridges

Description

In 1972 Czechoslovakia introduced a scissors bridge launched over the rear of a TATRA 813 (8 × 8) truck and known as the AM-50 *(Automobilni Most 50)* or AM-50A. It is launched in a similar fashion to the TMM treadway bridge but the Czech/Slovak system has the added advantage that the trestle columns have hydraulic rather than manual adjustment. Another feature is that the bridge has a full-width roadway rather than the two single tracks of the TMM system.

The AM-50B is an essentially similar bridge carried on the TATRA T 815 VP 19.28 265 8 × 8.1 R truck with a central tyre pressure control system; this was introduced in 1989. A complete AM-50B bridge system on the move weighs 25,900 kg when a dozer blade is included. Late production AM-50 and all AM-50B bridges are interchangeable.

A single AM-50B bridge weighs 5,700 kg and has a load capacity of 50 tonnes for tracked vehicles, or 70 tonnes for multi-axle wheeled vehicles. Using a crew of three, each bridge can span a 12.5 m gap (the actual bridge length when opened is 13.5 m), the width with the bridge rims unfolded being 4 m. As the bridge can be laid onto hydraulic trestle columns, extra bridges can be added to span large gaps. For instance, up to eight AM-50B bridges can be used to cross a water obstacle 106 m wide and up to 6 m deep. The time to lay a single AM-50B bridge is up to 7 minutes and up to four can be laid in line in about 30 minutes. At night or in unfavourable conditions the erection time is multiplied by a factor of 1.5. Each telescopic trestle leg can be varied in height from 1.988 to 6.052 m; each leg weighs 1,760 kg.

To prevent the front axles of the AM-50B carrier vehicle rising during the final stages of the laying operation the front of the vehicle is anchored to the ground or to an existing bridge section by a rope or cable.

Once the bridge is in position, wheeled vehicles can cross at a maximum speed of 30 to 40 km/h; the maximum speed for tracked vehicles is from 15 to 20 km/h.

To prepare bridging sites and to remove obstacles the TATRA truck carrier can be provided with a BZ T 815 dozer blade at the front of the vehicle. NBC protection or air conditioning systems for the cab are optional.

Specifications

AM-50B
Crew: 2 or 3
Weight:
(AM-50B, vehicle and bridge) 25,900 kg
(AM-50, vehicle and bridge) 24,450 kg
(bridge) 5,700 kg
(trestle leg) 1,760 kg
(ramp section) 70 kg
(vehicle only) 17,400 kg
Bridge capacity:
(tracked vehicles) 50,000 kg
(multi-axle wheeled vehicles) 70,000 kg
Bridge span: 13.5 m
Bridge width: (rims unfolded) 4 m
Length of bridge travelling: (folded) 8.05 m
Width of bridge travelling: 3.15 m
Height of bridge travelling: 3.8 m
Angle of approach/departure:
(with dozer blade) 18°/28°
(without dozer blade) 32°/28°
Fording: 1.4 m
Length of trestle legs: 1.988-6.052 m
Time to lay one bridge: up to 7 min
Vehicle crossing speeds:
(wheeled vehicles) 30-40 km/h
(tracked vehicles) 15-20 km/h

Vehicle length with dozer: 12 m
Vehicle width with dozer: 3.15 m
Vehicle height, loaded: 3.8 m
Max vehicle speed, loaded:
(roads) 80 km/h
(cross-country) 30 km/h
Range, roads: 900 km

Status

Available for production. In service with the armies of the Czech Republic, India and Slovakia.

Manufacturer

Vychodoslovenské Strojárne a s Košice, Južná trieda, 82, 042 73 Košice, Slovakia.
Tel: +42 95 50001.
Fax: +42 95 20540.
Marketed by: Kerametal, Jašikova 2, 826 05 Bratislava, Slovakia.
Tel: +42 7 299 111. Telex: 932 33 Kemet C.
Fax: +42 7 522 5896.

UPDATED

AM-50B scissors bridges in position **1996**

AM-50B truck-mounted scissors bridge in travelling configuration **1996**

FRANCE

Giat Industries Leclerc armoured bridgelayer

Development/Description

A proposal has been made to develop and produce an armoured bridgelayer based on the chassis of the Leclerc MBT. No details of this proposal have yet been released other than an artist's impression.

In June 1996 Giat Industries and Vickers Defence Systems signed a cooperation agreement for the development and manufacture of Universal Bridge Launching Equipment (UBLE) to be placed on Leclerc or Challenger tank chassis. The bridge involved is the 26 m No 10 MLC 70 tank bridge, already part of the British Army's BR90 programme (qv). Under the agreement the UBLE will be first proposed to the French and British armies within the framework of their current bridging systems modernisation and/or replacement. Giat Industries and Vickers Defence Systems will also market the UBLE worldwide.

Status

Proposal.

Manufacturers

Enquiries to: Giat Industries, 13 route de la Minière, F-78034 Versailles Cedex, France.
Tel: +33 1 30 97 36 48.
Fax: +33 1 30 97 39 58.

UPDATED

Artist's impression of the proposed Leclerc armoured bridgelayer

Giat Industries AMX-30 armoured bridgelayer

Development/Description

The prototype of the AMX-30 bridgelayer was built in 1966-67, but there were initial technical problems and CODER-Industries was formed to take over the manufacturing activities for the bridge.

This bridgelayer was produced only for Saudi Arabia.

The Class 50 bridge is composed of two hinged elements with removable widening panels and wheel guides. When opened out it is 22 m long and will span a gap of up to 20 m. The bridge is 3.1 m wide without the widening panels and 3.92 m wide with them. The bridge is launched hydraulically over the rear of the vehicle and takes about 5 minutes to lay into position. It can be recovered from either end. The bridge can be launched on relative slopes of 30 per cent with a relative slant reaching 15 per cent.

The basic chassis is almost identical to that of the standard AMX-30 MBT. The hull provides the crew with protection against small arms fire and artillery splinters, and is fitted with the same NBC system as the MBT.

The driver is seated at the front of the vehicle on the left side and is provided with a single-piece hatch cover and three periscopes for observation. One of these can be replaced by a night periscope of the infrared or image intensification type. The commander and bridge operator are seated to the rear of the driver. The engine and transmission are at the rear of the tank and are separated from the crew compartment by a fireproof bulkhead.

The suspension is of the torsion bar type and consists of five roadwheels with the drive sprocket at the rear and the idler at the front. There are five track support rollers which support the inside of the track only. The first two and last two roadwheels are mounted on bogies and are provided with hydraulic shock-absorbers.

AMX-30 bridgelayer with bridge folded for travelling (ECP Armées)

Specifications

Crew: 3
Weight:
(with bridge) 42,500 kg
(without bridge) 34,000 kg
Length:
(with bridge) 11.4 m
(hull only) 6.7 m
Width:
(with bridge) 3.95 m
(without bridge) 3.15 m
Height: (with bridge) 4.29 m
Ground clearance: 0.4 m
Track: 2.53 m
Track width: 570 mm
Length of track on ground: 4.12 m
Ground pressure: (with bridge) 0.93 kg/cm^2
Max road speed:
(with bridge) 50 km/h
(without bridge) 60 km/h
Range: 600 km
Fuel capacity: 1,100 litres
Fording: 1 m
Gradient: 50%
Side slope: 25%
Vertical obstacle: 0.93 m
Trench: 2.9 m
Engine: Hispano-Suiza HS-110, 12-cylinder, water-cooled, multifuel diesel developing 700 hp at 2,400 rpm
Transmission: automatic with 5 forward gears; a reverse gear gives the same speeds in reverse
Electrical system: 28 V
Batteries: 8 × 12 V, 100 Ah in 2 groups of 4
Armament: nil

Status

Production complete. In service with Saudi Arabia.

Manufacturers

Bridge: CODER-Industries, BP 4, F-13367 Marseilles Cedex 11, France.
Chassis: Centre de Roanne.
Enquiries to: Giat Industries, 13 route de la Minière, F-78034 Versailles Cedex, France.
Tel: +33 1 30 97 37 37.
Fax: +33 1 30 97 39 00.

VERIFIED

CNIM Assault Treadway Bridge MLC 65

Development

The Assault Treadway Bridge (*Travure Courte Portée* - TCP) was developed by CNIM, under contract to the Direction Technique des Armements Terrestres, to support crossings of 12 m gaps (that is, 80 per cent of gaps in the European theatre) by armoured units. A feature of this bridge is that it does not require a dedicated armoured vehicle for launching. The first prototype was shown in June 1990.

Description

The CNIM Assault Treadway Bridge is of the scissors type, with a Military Load Classification (MLC) of 65; this could be increased to MLC 70. It is transported on a single-axle trailer towed by a light truck. The bridge is towed to near the launching site and, after unfolding and arranging the trackways, is lifted from the trailer by an armoured vehicle equipped with a dozer blade, a winch and a crane jib. The French Army intends to use the AMX-30 combat engineer vehicle (EBG) for the launching role but other suitably equipped armoured vehicles weighing around 40 tonnes could be used. An accessory kit is used to adapt the launching vehicle. Once lifted from the trailer the bridge is carried to the launch point and the entire launching operation is carried out with the vehicle crew under their armoured cover. During the launch operation the crane jib and winch are used to lift the folded bridge to an upright position, after which the bridge opens and is deployed without any external power.

The time taken to prepare and take the bridge from its trailer is about 10 minutes while the actual launch operation takes approximately 3 minutes.

CNIM Assault Treadway Bridge being lifted from its transport trailer by an AMX-30 combat engineer vehicle (EBG) (T J Gander) ***1996***

Specifications

Bridge
Load classification: MLC 65
Weight: 4,800 kg
Length:
(travelling) 9.6 m
(deployed) 14 m
Width:
(folded) 2.14 m
(roadway) 4 m
Height: (folded) 2.9 m

Trailer
Weight:
(loaded) 6,000 kg
(unloaded) 1,200 kg
Length: 9.45 m
Width: 2.5 m
Height: (loaded) 3.7 m

Status

Prototype.

Manufacturer

Constructions Navales et Industrielles de la Méditerranée (CNIM), Zone Industrielle de Brégaillon, BP 208, F-83507 La-Seyne-sur-Mer Cedex, France.
Enquiries to: CNIM Head Office, 35 rue de Bassano, F-75008 Paris, France.
Tel: +33 1 44 31 11 00. Telex: 642 869 f.
Fax: +33 1 44 31 11 30.

UPDATED

CNIM PAR 70 Towed Support Bridge MLC 70

Development

The CNIM PAR 70 Towed Support Bridge MLC 70 is a development which involves many existing in-service and fully developed components. The intention is to produce a fast deployment close support bridge, 19.5 m long, with a bridging classification of MLC 70.

The scissors bridge and launch/retrieve systems involved are already employed on the PAA self-propelled bridge system (see following entry), while the semi-trailer axle assembly is taken from the PFM floating bridge system which has been in French Army service since 1985. The non-specific truck tractor involved may be a 6 × 6 or 6 × 4 unit. The only completely new elements in the system are a hydraulic power generating set and the semi-trailer frame.

Description

The CNIM PAR 70 Towed Support Bridge MLC 70 has a twin trackway bridge formed from two folding symmetrical sections. When fully opened the bridge is 19.5 m long and 4 m wide, enabling gaps up to 19 m wide to be bridged. Construction involves light alloys and the load classification of the PAA scissors bridge is MLC 70.

As the non-specific truck tractor prime mover involved provides a measure of stability during bridge-laying and recovery operations it has to weigh a minimum of 12 tonnes; for towing 330 hp is required.

For laying the semi-trailer carrying the system is reversed to the launch point. A hydraulic power generating set located over the semi-trailer gooseneck provides the power to the launch/retrieve system which involves a stabiliser leg under the point of launch, a bridge-raising arm cylinder, a winch, and two deployment jacks within the bridge to provide the power to open the bridge once it has been raised past the vertical. As the bridge unfolds it is further stabilised by the cable winch which also provides the motive power during the bridge retrieval stage. Further stabilisation is provided by the weight of the towing vehicle which remains connected to the launch semi-trailer throughout launch and recovery sequences. Once the bridge is fully opened it is held horizontally prior to being lowered to the far bank under winch cable control. The near side is then lowered to the ground on the launch side by a further cable. Once the bridge is emplaced two members of the three-man crew (the third member is the driver) can add extra side plates and guidance lamps on posts; these increase the overall emplaced bridge width to 4.5 m. The side plates and posts are carried on the semi-trailer. Each trackway is 1.523 m wide. Maximum difference in the level between banks is ±3 m.

The time required to deploy the bridge is 6 minutes. Recovery can be from either side of the gap involved.

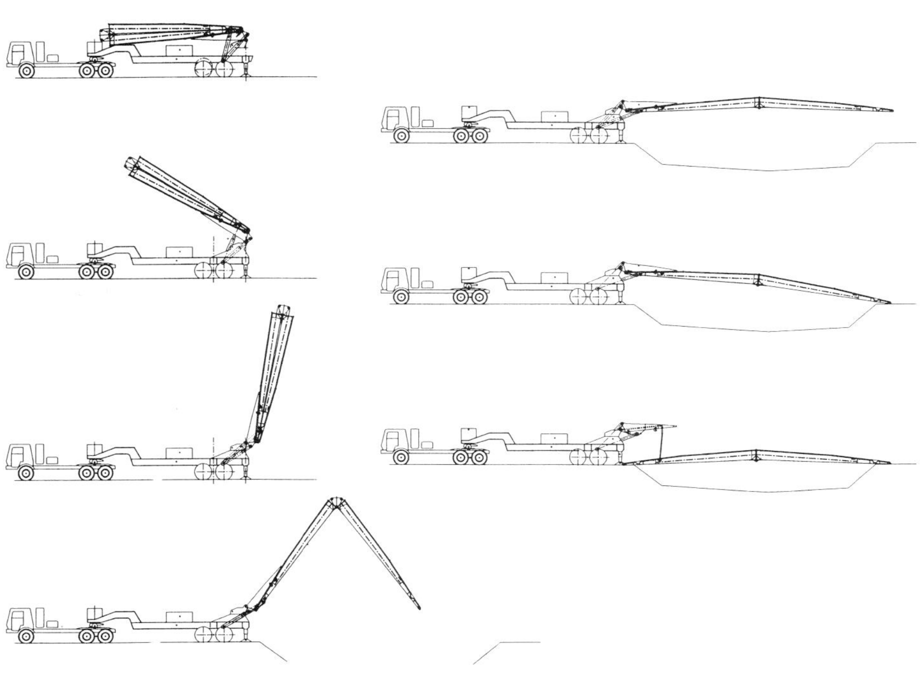

Laying sequence for the CNIM PAR 70 Towed Support Bridge MLC 70, from top left downwards and then from top right ***1996***

Specifications

Provisional
Bridge
Weight: 9,000 kg
Length: (unfolded) 19.5 m
Span: <19 m

Width:
(overall, emplaced) 4.5 m
(roadway) 4 m
(trackways, each) 1.523 m
Height: (emplaced) 1.15 m
Max difference in bank levels: ±3 m
Load classification: MLC 70

Bridge and vehicle, loaded
Crew: 3
Weight:
(complete) approx 35,000 kg
(semi-trailer) approx 23,000 kg
Length: <18 m
Width: 4 m
Height: approx 3.9 m
Max speed:
(road) 80 km/h
(cross-country) 25 km/h

Status
Development.

Manufacturer
Constructions Navales et Industrielles de la Méditerranée (CNIM), Zone Industrielle de Brégaillon, BP 208, F-83507 La-Seyne-sur-Mer Cedex, France.
Enquiries to: CNIM Head Office, 35 rue de Bassano, F-75008 Paris, France.
Tel: +33 1 44 31 11 00. Telex: 642 869 f.
Fax: +33 1 44 31 11 30.

VERIFIED

DCAN PAA self-propelled bridge system

Development
The PAA (*Pont Automoteur d'Accompagnement*) was developed as a bridging system to support assault units, pending the arrival of engineer units with permanent bridges.

Development can be traced back to the late 1950s when the German company Eisenwerke, under contract to the Service du Matériel du Genie, built three prototype vehicles. Between 1963 and 1968 the Direction des Constructions et Armes Navales (DCAN) at Lorient and the Etablissement Technique d'Angers completely redesigned the vehicle. There were two prototypes built, the first of which was completed in 1968 and the second in 1970. After extensive trials, the PAA was adopted by the French Army in 1972 and production began at Lorient in 1973. First production vehicles were delivered to the French Army in 1974, and by early 1978, 51 had been delivered. Production was completed the same year.

The scissors bridge system involved with the PAA is also employed, in an upgraded load classification form, on the CNIM PAR 70 Towed Support Bridge MLC 70 (see previous entry).

Description
The PAA consists of a rigid and watertight hull of light alloy (AZ5G) construction with a removable span in two sections on the front and an access ramp at the rear. The hull is divided into three compartments, front, centre and rear. The front section is the crew compartment with the driver seated on the left and the operator on the right. This compartment is provided with an NBC and air conditioning system. The engine compartment is in the centre and is provided with an electric pump which automatically operates as soon as water reaches a certain level. The rear compartment contains the fuel tanks.

Power is transmitted by the engine to the four roadwheels through a Guinard-Clark hydrokinetic torque converter, a power-assisted Clark-Genemat gearbox and a sliding joint transmission shaft. Steering is power-assisted.

The hydraulic system operates the following: retraction and extension, and locking and unlocking, of the forward and rear sets of wheels; tilting of the prismatic boom supporting the span; opening and closing of the two access ramp hinged links; locking and unlocking of the span on the prismatic boom; and the hoisting and lowering of the span.

The PAA can be used in two basic roles: first by laying its bridge and then leaving the bridge in place, and second by laying the span and remaining in position, with the vehicle acting as the ramp of the bridge.

When deployed, the four wheels of the PAA are raised off the ground and the hull is supported on shoes. The bridge and ramp have a normal width of 3.05 m but this can be increased to 3.55 m with widening panels. These are normally carried in a Berliet GBC 8 KT (6 × 6) truck which also carries the wheel guide sections, hand-rope posts and ropes. The widening panels can be left on the vehicle for travelling, but are normally removed to reduce the overall width of the vehicle.

The removable span has a total length of 21.72 m and consists of two symmetrical sections coupled by a hinge. The access ramp also consists of two symmetrical hinge-coupled sections which permit the access ramp to be folded on the body of the bridge when not in use. The prismatic boom is hinged at the front part of the body by a rugged hinge, and houses the hydraulically operated hoisting, gripping and locking devices.

With the vehicle remaining in position the PAA can be used to span a gap, with soft banks, up to 17.4 m wide, or a gap of up to 22.4 m with hard banks. The bridge itself can span a gap, with soft banks, of up to 15.63 m, or, with hard banks, up to 20.63 m. This bridge can be used by vehicles up to Class 40 or, with precautions, Class 45. By using two PAA vehicles and a special jointing section, it is possible to bridge a gap of up to 40 m. A triangular support boom has also been developed which has an adjustable leg and is used to span 40 m gaps.

Specifications
Weight:
(loaded) 34,500 kg
(without span but incl 18 widening panels) 25,620 kg
Length:
(travelling) 13.15 m
(bridge unfolded) 38.25 m
Width:
(without widening panels) 3.05 m
(with widening panels) 3.55 m
Height: 3.99 m
Max speed: (road) 60 km/h
Range: (road) 800 km
Fording: 1.5 m
Gradient: 50%
Side slope: 20%
Turning radius: 20.5 m
Engine: Deutz V-12 diesel developing 300 hp
Brakes: air/oil, 2 independent front and rear systems, disc brakes on all axles
Electrical system: 24 V
Batteries: 4

Status
Production complete. In service with the French Army.

Manufacturer
DCAN Lorient.
Enquiries to: Direction Techniques des Constructions Navales, 2 rue Royale, BP 1, F-75200 Paris Naval, France.
Tel: +33 1 42 86 04 36. Telex: 650 421.
Fax: +33 1 42 86 05 20.

VERIFIED

PAA bridge in travelling configuration

PAA bridge in operation

CEFA Engin de Franchissement de l'Avant (EFA)

Development
The *Engin de Franchissement de l'Avant* (EFA) was originally known as the *Matériel Amphibie de Franchissement* (MAF). It was under competitive development as the replacement for the Gillois system then used by the French Army (it has now been withdrawn from service). The Etablissement Technique d'Angers was in charge of the study and evaluation of two prototypes and overall project management was under the Direction Technique des Armements Terrestres. There were two prototypes built and evaluated: the petrol-engined MAF 1 under the direction of the Lorient Board for the Construction of Naval Weapons (Direction des Constructions et Armes Navales de Lorient) and the MAF 2 under the direction of the Alsace Metals and Steel Company (Chaudronnerie et Forges d'Alsace - CEFA).

The French Army decided to provide funds for

further testing of the MAF 2 after the first prototype had been completed. These trials were completed in 1987 and the French Army ordered 80 units (later reduced to 70) to replace the Gillois bridge and ferry system.

Description

The hull of the EFA is made of welded light alloy. On the top of the vehicle, at each end, is a 12 m jointed ramp also of welded light alloy. Before the unit enters the water, airbags are positioned under hinged flaps on either side of the hull and inflated. The unit has four large wheels with low-pressure tyres and is driven in the water by two Schottel SPJ 55 swivelling propellers via a Renk transmission, one each end of the hull, which can be rotated through 360°.

The EFA can be used both as a bridge and a ferry. As a bridge it has a 23.5 m span, can take loads of up to MLC 70, and be used in rivers flowing at up to 3 m/s. As a ferry two EFAs can carry loads of up to 150 tonnes provided the load is distributed over the loading surface. The total load area of each EFA ferry unit is 96 m^2.

There can be four EFAs connected to provide a MLC 70 bridge 100 m long. Such a bridge could be formed within 15 minutes and would require a crew of 16.

Each EFA vehicle unit cab has NBC and fire protection.

Due to the overall 3.6 m width of the EFA vehicle a training simulator for driver training, complete with a motion system and a wide angle visual system, has been produced by Thomson Training & Simulation. The simulator was delivered during early 1995.

Engin de Franchissement de l'Avant (EFA)

Specifications

Crew: 4
Weight: (total) 42,000 kg
Length:
(travelling) 12.35 m
(as bridge) 23.54 m
(as ferry) 34.7 m
Width:
(travelling) 3.6 m
(airbags in position) 6.8 m
Height: (travelling) 4 m
Track: 2.17 m
Wheelbase: 6.6 m
Max speed:
(road) 70 km/h
(water, with MLC 70 load) 3.5 m/s
(water, empty) 14 km/h
Range:
(road) approx 650 km
(in water) approx 12 h
Fording: (without preparation) 1.2 m
Gradient: 50%
Side slope: 30%
Engine: Badouin 12 F 120 SR water-cooled diesel developing 730 hp at 2,700 rpm
Suspension: hydropneumatic with level correction
Tyres: 26.5 × 25 XL
Electrical system: 24 V

Status

Ordered for the French Army (70). First deliveries during 1992.

Manufacturer

Chaudronnerie et Forges d'Alsace (CEFA), BP 11, route de Woerth, F-67250 Soultz-sous-Forêts, France.
Tel: +33 3 88 80 49 47. Telex: 870 030 f.
Fax: +33 3 88 80 50 05.

UPDATED

GERMANY

BLG-60, BLG-67 and BLG-67 M2 armoured bridgelayers

Development/Description

The BLG-60 was a joint development between East Germany and Poland. The bridge is launched in a similar fashion to the Czech/Slovak MT-55A and the reader is referred to this entry for the method of launching the bridge.

In position the bridge has a total length of 21.6 m, a width of 3.2 m and a height of 0.8 m. It can span a maximum gap of 20 m compared with the maximum span of 16 m of the MT-55A, and has a maximum capacity of 50,000 kg.

The main external differences between the MT-55A and the BLG-60 are that the surface of the BLG-60's bridge is smooth and covered in plastic (the MT-55A has a patterned surface), the bridge girder ends are closed (on the MT-55A they are open), and the pulley wheels for the scissors-action operating gears are solid. The BLG-60 is provided with an NBC system and a snorkel.

There is also a BLG-67 bridgelayer, an updated version of the basic BLG-60, the main users being Bulgaria and Poland. The BLG-67 M2 is a special variant designed to lay two- and three-span bridges. All its main assemblies (chassis, track bridge, span-laying gear, span-fastening system, hydraulics, electrical control system and so on) have been modified, as have the bridges. The base of the extendible arm of the span-laying gear has been lengthened and the surface on which this arm rests is lined with rubber. The edges of the roadway have been widened to provide a width of 3.47 m. Anchorages are provided on the left- and right-hand side of the base span while the bridge is being laid. The BLG-67 M2 system has been carried over to the Polish PMCz-90 armoured bridgelayer based on the T-72M1 MBT.

When two- or three-span bridges are laid in water with a current speed above 0.2 m/s, a bridge anchorage system is employed. This uses a tie, reel and hoist with the reel holding 40 m of steel cable. One end of the cable is attached to the front of a bridge while the other

BLG-60 armoured bridgelayer (Michael Jerchel)

is passed through the hoist affixed to a tie driven into the ground. The hoist draws and grips the cable with two pairs of jaws operated by a hand lever. Each BLG-67 M2 carries two anti-current anchors.

Multispan bridges are formed by laying the first span with one end on a bank and the other at the bottom of the gap to be bridged. A second span is then laid from the first and, if necessary, a third span is then emplaced either to extend the span or to accommodate a difference in height on the far bank. The span support points should be situated not more than 2.5 m from the span ends on which the next span rests. Once emplaced, the spans are connected and anchored by three grab links, two further links, a block and the hoist and line from the anti-current anchor system. A suitable track bridge or pontoons can be used as intermediate supports.

Multispan bridges have to be monitored for alignment after 10 vehicles have crossed (two spans) or, in the case of three spans, every three vehicles. Vehicles have to cross the multispan bridges in low gear at a constant speed.

A special training version of the BLG-67 M2 was introduced for trainee drivers. The vehicle is used for training in laying bridges, vehicle maintenance and repair.

Wojskowe Zakłady Inzynieryjne (Army Engineer Workshops) of Deblin offer a service to bring earlier BLG-60 and BLG-67 models up to full BLG-67 M2 standards.

Specifications
BLG-60
Crew: 2-3
Weight:
(with bridge) 37,000 kg
(without bridge) 31,000 kg
Length: (with bridge) 10.57 m
Width: (with bridge) 3.48 m
Height: (with bridge) 3.4 m
Ground clearance: 0.425 m
Track: 2.64 m
Track width: 580 mm
Length of track on ground: 3.84 m
Ground pressure: 0.83 kg/cm²
Max speed: (road) 50 km/h
Range: 500 km
Fuel capacity: 960 litres
Fording: 1.4 m
Gradient: 58%
Vertical obstacle: 0.8 m
Trench: 2.7 m
Engine: V-55 V-12 water-cooled diesel developing 580 hp at 2,000 rpm
Transmission: manual with 5 forward and 1 reverse gears
Electrical system: 24 V
Batteries: 4, 280 Ah
Armament: nil

Armour
glacis plate: 100 mm at 60°
upper hull side: 70 mm at 0°
hull rear: 60 mm
hull floor: 20 mm
hull roof: 30 mm

BLG-60 armoured bridgelayer (C R Zwart)

Status
Production complete. In service with the Indian, Iraqi (at least 24 BLG-60s) and Polish armies (including BLG-67 M2). Bulgaria uses the BLG-67.

Manufacturers
Chassis: Polish state factories.
Bridge and conversion work: Germany.

UPDATED

LEGUAN bridgelayer

Description
The LEGUAN mobile bridgelayer with its 26 m MLC 70 bridge was jointly developed by KRUPP and MAN GHH. The technical concept aims at a higher efficiency, reliability and compatibility with different chassis types. The standard chassis is a modified MAN-ÖAF 36.422 VFAE 8 × 8 all-terrain truck chassis, with a low set driver's cab. Other wheeled and tracked chassis can be used as carrier vehicles.

The LEGUAN bridge and bridgelayer are attached to the vehicle via an auxiliary frame on which the bridge can be pushed out over the rear of the vehicle.

The bridge consists of four interchangeable wheel tread girders with a flat ramp design, each 13 m long, 2 m wide and weighing 2,500 kg. Bolt connections join the bridge girders and when travelling the bridge halves lie one on top of the other. The laying equipment consists of the laying arm at the rear of the shifting frame which moves the bridge longitudinally by means of a roller guide and a geared drive.

The top bridge half is taken up by an arm at the front end and is supported by an auxiliary arm at the rear. The bridge can be laid with a traverse and longitudinal inclination of up to 10 per cent. Laying is carried out automatically with the two-man crew monitoring the sequence. No preparation of the crossing site is normally required.

The laying sequence commences with the vehicle reversing to the bank until the end of the shifting frame is approximately 7.5 m from the edge. After raising the top bridge half, which is suspended at the front in the slewing support arm, and with the roller at its rear resting on the top chord of the lower wheel girder, the lower bridge half is advanced and then lowered by 10 per cent. It is held at the front by the slewing support arm and at the rear by the coupling support arm. The bridge halves are then coupled automatically by lowering the coupling supporting arm.

The bridge is then pushed back until its centre of gravity is over the two rear axles of the carrying vehicle, where it is held until the shifting frame has been moved under the bridge and the hydraulic supports have been lowered. The next step is for the bridge to be fully advanced, at which point it is clamped in the laying arm by the roller carriage. At this stage the ramp tip is placed on the far bank. This removes the load from the rollers and the laying arm can then be removed to lower the bridge. For the final stage the supports are retracted and the shifting frame is moved back. The bridge is then ready for use. The entire laying process takes approximately 8 minutes. Reloading the bridge is carried out in the reverse order and can be carried out from either bank.

The bridge is constructed of standard commercial aluminium alloy, grade Al Zn 4.5 Mg 1 and is designed to meet US Standards for the bridge life.

LEGUAN bridgelayer in travelling configuration **1996**

Variants
The LEGUAN bridgelayer can also be used to create ferries. For details refer to entry in the *Tactical Floating Bridges and Ferries* section.

A LEGUAN 42 m bridgelayer system capable of laying bridges up to 42 m long has been developed. For details see following entry.

LEGUAN bridges have been adapted to be carried and launched by Leopard 1 and M47/M48 MBT chassis. For details see separate entries in this section.

In association with General Dynamics of the USA a MAN GHH LEGUAN bridge is to be the US Army's Wolverine Heavy Assault Bridge (HAB). For details see under USA in this section.

LEGUAN bridge principles are employed with the KRUPP-LOHR Modular Bridge System (MBS). See entry under International in this section.

Specifications
Crew: 2
Weight:
(vehicle) 25,200 kg
(bridge) approx 10,000 kg
(total) approx 35,600 kg
Length:
(overall, travelling with bridge) 15.3 m
(overall, with bridge) 13.4 m
Width:
(overall, with bridge) 4.01 m
(without bridge) 3.15 m
Height:
(overall with bridge) 4 m
(overall without bridge) 2.73 m
(top of cab) 2.285 m
Ground clearance: 0.45 m
Wheelbase: 1.6 m + 3.25 m + 1.55 m
Angle of approach/departure: 20°/25°
Max road speed: 71.5 km/h
Range: 600 km
Fuel capacity: 400 litres
Max gradient: 60%
Engine: MAN D 2866LD/422 water-cooled diesel developing 412 hp at 2,000 rpm
Transmission: WSK 400 + 16 S 151
Transfer box: 2-speed 1700-1

Steering: ball and nut, ZF Model 8046
Turning circle: 29 m
Suspension: Koni adjustable shock-absorbers with stabilisers on front and rear axles
Brakes:
(main) dual circuit air
(parking) mechanical on rear wheels
(engine) pneumatically actuated on exhaust
Tyres: Michelin 16.00 R 20
Electrical system: 24 V
Batteries: 2 × 12 V, 173 Ah each

Bridge
Classification: Class 70
Weight: approx 10,000 kg
Length:
(bridge) 26 m
(effective span) 25 m
Width:
(bridge) 4.01 m
(wheel tread girder) 1.555 m
Height:
(bridge centre) 1.1 m
(ramp tip) 0.075 m

Status
In production. In service with Norway and Singapore and other countries.

Manufacturers
KRUPP Fördertechnik GmbH, Altendorfer Strasse 120, D-45143 Essen, Germany.
Tel: +49 201 82801.
Fax: +49 201 828 2566.
MAN Gutehoffnungshütte AG, Ginsheimer Strasse 1, D-65463 Ginsheim-Gustavsburg, Germany.
Tel: +49 6134 55292.
Fax: +49 6134 55383.

UPDATED

LEGUAN 42 m bridgelaying system

Description
The LEGUAN 42 m bridgelaying system can be used to lay bridges up to 42 m long and closes the gap between the short LEGUAN bridges, up to 26 m long, and floating bridges which are normally used above a length of approximately 40 m.

The LEGUAN 42 m system is based on the standard LEGUAN all-terrain 8 × 8 wheeled carrier and launch vehicle (see previous entry) which is modified so that it can lay a 42 m bridge as well as the original 26 m version. With LEGUAN 42 m the following bridge lengths can be built:

26 m; two ramps
34 m; two ramps, one centre-section
42 m; two ramps, two centre-sections.

The bridge is designed for a normal loading of MLC 70 tracked vehicles and MLC 96 wheeled vehicles 'with caution'.

With the basic LEGUAN 42 m set the LEGUAN vehicle, a MAN-ÖAF 42.422 VFAE 8 × 8 with a 412 hp engine, carries two 13 m ramp sections ready to be launched as a normal 26 m LEGUAN bridge if required (for details see previous entry). Carried on a separate all-terrain support vehicle are two extra centre-sections, each 8 m long. The support vehicle (typically a MAN-ÖAF 34.370 8 × 8) carries handling equipment to allow the centre-sections to be inserted between the two ramp halves to form a 34 or 42 m bridge as required.

Once the launch vehicle has arrived at the launch site it is aligned and a shifting frame is extended towards the gap to be crossed. Supporting arms are extended and the two ramp halves are separated. An extra bridge section, or sections, can then be inserted and connected between the ramp halves by the support vehicle using its crane-type handling equipment. A launching girder is then extended over the gap and once the bridge has come to rest on the far bank the bridge is lowered. The launch vehicle separates from the bridge and the bridge is ready for use.

The advantages of this laying method are that the bridge sections remain coupled to the launch girder and that the bridge sections are connected to each other at exactly defined positions on the launch vehicle. Thus, using the handling equipment, the centre-sections need only be roughly aligned on the launch vehicle. Guides on the vehicle allow the bridge sections to be aligned exactly so that they can be coupled. All coupling procedures are fully automatic and require no manual intervention. Laying and loading the bridge are controlled step-by-step by an electronic control system that eliminates operating errors and allows the bridge to be laid at night.

The laying operation for a LEGUAN 42 m bridge requires a space measuring approximately 25 × 10 m and involves only four men (the crews of the laying and support vehicles). The laying time for a 42 m bridge is approximately 25 minutes.

The LEGUAN 42 m bridge was one submission for the US Army's Heavy Dry Support Bridge (HDSB) requirement. For this programme MAN GHH teamed with the AAI Corporation of Hunt Valley, Maryland. For this application it was proposed that the two ramp sections would be carried on a suitably modified Oshkosh M1074 10 × 10 PLS truck, with the centre-sections carried on two Oshkosh M977 8 × 8 HEMTT trucks. The AAI/MAN GHH submission was not successful.

Specifications
Weight:
(laying vehicle with laying equipment) 29,600 kg
(2 ramp sections) 13,800 kg
(total) 43,400 kg
Length: (laying vehicle with bridge, overall) 15.3 m
Width:
(with bridge) 4.01 m
(without bridge) 3.15 m
Height: (overall) 4 m
Bridge lengths: 26, 34 or 42 m
Bridge width: 4.01 m
Bridge height: 1.135 m
Weight:
(ramp section, complete) 6,900 kg
(centre section, complete) 6,500 kg

Status
Prototype completed.

Manufacturers
KRUPP Fördertechnik GmbH, Altendorfer Strasse 120, D-45143 Essen, Germany.
Tel: +49 201 82801.
Fax: +49 201 828 2566.
MAN Gutehoffnungshütte AG, Ginsheimer Strasse 1, D-65463 Ginsheim-Gustavsburg, Germany.
Tel: +49 6134 55292.
Fax: +49 6134 55383.

UPDATED

LEGUAN 42 m bridge in fully extended form

LEGUAN 42 m bridge launch vehicles preparing to assemble bridge

Loaded tank transporter crossing a completed LEGUAN 42 m bridge **1996**

MAN GHH Leopard 1 LEGUAN AVLB

Development

In December 1995 the Norwegian Army Materiel Command awarded MAN GHH, as general contractor in association with Krauss-Maffei of Munich, a contract for the conversion of nine Leopard 1 tank chassis to bridgelayers. MAN GHH will supply the bridgelaying equipment plus 13 LEGUAN bridges, each 26 m long with load classification MLC 70. The bridges are fitted with special adaptors to allow them to be incorporated in ferries. Krauss-Maffei undertook the chassis conversions and the integration of the bridgelaying equipment.

Description

The MAN GHH Leopard 1 LEGUAN AVLB is based on the hull and chassis of the Leopard 1 MBT with all main and subcomponents unchanged. Various items from the removed turret and other components are modernised and integrated into the AVLB.

For the conversion the driver's hatch remains unchanged, while the commander's hatch is adopted from the MBT turret. The driver's seat also remains unchanged, while the commander's seat is adopted from the MBT turret. A hatch providing access to the hydraulic and stowage compartment is of a new form. The fording depth is 1.8 m while sealing, NBC protection, bilge pumps and heating remain the same as the MBT, as does most of the electrical system, apart from extra lighting.

Other unchanged systems carried over include the engine and transmission, although the dual-circuit braking system is reinforced to accommodate the increased overall weight of approximately 49.5 tonnes. Some adjustments are necessary to the torsion bar suspension while the tracks are to Diehl 640 A standard.

Some extra ballistic protection is added externally and an anti-spall liner is provided for the crew compartment.

The 26 m LEGUAN aluminium alloy bridge follows standard LEGUAN construction and laying procedures. The main change from the standard procedures is that before laying a hydraulically operated tiltable support blade is lowered under driver control from the front of the hull. The support blade can be deployed as a light dozer blade for light obstacle clearing.

For laying, a laying arm supports and guides the LEGUAN bridge during transport and launching. The arm is hinged to the hull and positioned by a hydraulic cylinder. The arm is a welded box-type structure carrying the roller carriages for guiding the bridge, the bridge advance unit and the mechanism for bridge unlocking. A rear arm holds the bridge at the rear end during transport and guides it for coupling and uncoupling. This arm is connected to the hull by four bolts and is actuated by a hydraulic cylinder. An electronic control system monitors the position of all drive units, operates the programmed laying/retrieval sequences and indicates any faults.

For laying, once the AVLB is in position with the support blade lowered, the rear arm swings into the vertical position. It raises the rear of the upper bridge half to allow the lower half to move forward until it reaches the coupling position. An auxiliary arm then rises to support the forward part of the upper bridge, as the laying arm moves down to move the lower half to the coupling position. The rear arm then swings down for the upper half to assume the coupling position so that when the auxiliary arm is lowered the two bridge halves can be automatically coupled.

At this point the rear arm is unlatched to allow the entire bridge to be moved forward. Once fully extended the far end is lowered onto the opposite bank. Tilting the laying arm lowers the nearest end of the bridge onto the near bank. Once the support blade is raised the bridge is ready for use. Launching and retrieval times can be less than 5 minutes.

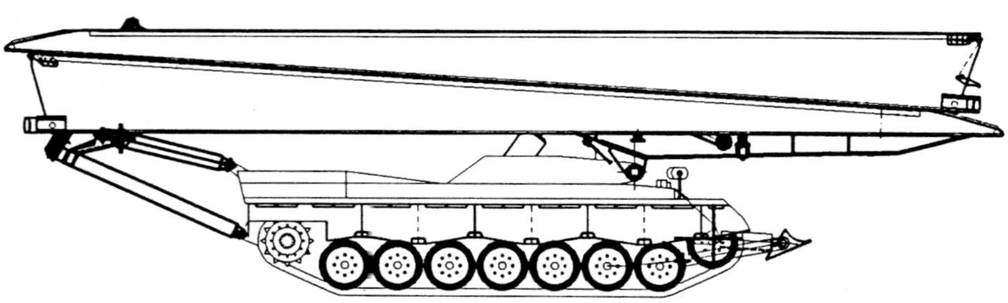

Outline drawing of MAN GHH Leopard 1 LEGUAN AVLB in travelling position
1997

Specifications

Crew: 2
Weight: (with bridge) approx 49,500 kg
Length: (with bridge) 13.349 m
(without bridge, rear arm in transport position) 12.705 m
Width: (with bridge) 4.01 m
(without bridge) 3.37 m
(dozer blade) 3.25 m
Height: (with bridge) 3.815 m
Fording: approx 1.8 m
Engine: MTU MB 838 CaM multifuel developing 830 hp at 2,200 rpm
Transmission: ZF 4 HP 250 with hydraulic torque converter
Suspension: torsion bar
Electrical system: 24 V

Bridge weight: 11,000 kg
Load classification: MLC 70
Bridge length: (extended) 26 m
Gap spanned: 24 m
Bridge width: 4.01 m
Roadway width: (each) 1.55 m

Status

Conversions in production for Norway.

Manufacturers

MAN Gutehoffnungshütte AG, Department ISB, Ginsheimer Strasse 1, D-65463 Ginsheim-Gustavsburg, Germany.
Tel: +49 6134 55288.
Fax: +49 6134 55383.
Krauss-Maffei Wehrtechnik GmbH, Krauss-Maffei Strasse 2, D-80997 München, Germany.
Tel: +49 8988 992060.
Fax: +49 8988 992461.

NEW ENTRY

MAN GHH M47/M48 LEGUAN armoured bridgelayer

Description

The MAN GHH M47/M48 LEGUAN armoured bridgelayer is an adaptation of an M47 or M48 MBT chassis, designed to carry and launch the standard MLC 70 26 m LEGUAN bridge (see entry in this section). One trial M47 conversion for tests has been successfully conducted in Spain in co-operation with Peugeot Talbot España SA. It is based on an M47 MBT chassis and known as the VLPD-26/70 E 'Lanzador'. The conversion can be carried out on MBTs such as the Centurion, Leopard 1/2, M1 Abrams and so on.

The M47/M48 LEGUAN conversion utilises as many standard LEGUAN components as possible, however, the overall bridge launch position is forward instead of to the rear as on the wheeled LEGUAN. This entails displacing the normal launch components through 180° but the basic LEGUAN launch procedures are maintained.

The bridgelaying M47/M48 tank chassis has to undergo some modifications in addition to the removal of the turret. The laying components include a traversing laying arm on the hull roof, roller carriages, a bridge advance unit, auxiliary arms and a hydraulic cylinder supported by the hull. A rear arm is used for coupling the bridge halves and there are hydraulic and electrical switching and control components inside the hull.

Modification to the tank hull includes removing the top frontal armour plate and front roofplate together with the turret ring. New armour and roofplates with hatches are welded into place and extra plates are added along the sidewalls and engine cover. A new forward support blade, described as an outrigger system, is added to the front hull. Once lowered, by two hydraulic cylinders, the support blade levels the laying vehicle on adverse slopes and provides support during the bridgelaying operations.

MAN GHH M47 LEGUAN armoured bridgelayer in its Spanish VLPD-26/70 E 'Lanzador' form

The LEGUAN laying armrests in a bearing welded to the hull roof, with its traversing cylinder pivoting on girders welded inside the hull. A line and cable bushing close to the bearing provides protection for the electrical cables and hydraulic lines, connecting the control unit inside the hull and the switches and actuators in the laying arm. Extra welded stiffening webs transmit loads from the bearing to the rear roofplate and the area around the traversing cylinder is kept watertight by a plastic cover.

A rear arm assembly is secured to baseplates welded to the top rear corners of the sidewalls. The laying arm is supported by the hull roof and controlled by a hydraulic cylinder inside the hull. The laying arm accommodates the roller carriages for the bridge as

COMPETENCE IN ALUMINIUM

EWK

Development – Production – Support for Combat Engineer Equipment – world-wide

M3 Amphibian

Mainhull M3

Folding Float Bridge

M3 as Bridge MLC 70/100 ▼

Fueltank 4,600 ltr

Infantrybridge, light

Tugboat

EWK Eisenwerke Kaiserslautern GmbH, Barbarossastr. 30, D-67655 Kaiserslautern, phone + 49 631 8 57-0, fax + 49 631 8 57-3 00

well as the bridge advance unit and auxiliary arms, for use during the coupling of the bridge halves. A two-element outrigger arm is provided at the hull rear to assist the coupling of the bridge sections; this arm can be lowered to a travelling position when a bridge is not being carried. The rear arm base also supports the bridge.

On arrival at a bridge launch site, the bridgelayer is driven to the point to be bridged and the support blade is lowered and extended to ensure the bridge is in a horizontal position. Forward slopes up to 20 per cent up and down can be accommodated, as can transverse slopes of up to 10 per cent. The maximum downwards laying step is 0.8 m.

To commence the laying operation the rear end outrigger arm is swung vertically to lift the bridge upper section. The bridge advance unit moves the lower bridge section forward towards the coupling position as the auxiliary arms move upwards to lift the upper section. At that point the rear end outrigger arm is lowered and the laying arm swings the front section ready for coupling. Claw couplings then slide into each other and the rear end outrigger is disengaged. The laying arm swings upwards and coupling is complete. The bridge advance unit can then move the complete bridge forward into position. Once fully extended the laying arm lowers first the far end and then the near end, ready for the support blade to be raised and the laying arm disengaged from the bridge. The bridgelayer can then withdraw. Only two men, the driver and commander, are involved in the laying operation, which takes between 3.5 and 4 minutes.

Specifications

VLPD-26/70 E 'Lanzador'
Crew: 2
Weight:
(with bridge) 50,370 kg
(without bridge) 43,000 kg
(bridge) 10,000 kg
Ground pressure:
(with bridge) 1.15 kg/cm^2
(without bridge) 0.93 kg/cm^2
Length:
(with bridge) 13.5 m
(without bridge) 11.73 m
Width:
(with bridge) 4.1 m
(without bridge) 3.56 m
Height:
(with bridge) 3.9 m
(without bridge) 2.81 m
Ground clearance: 0.47 m
Angle of approach/departure:
(with bridge) 24°/30°
(without bridge) 26°/30°
Max speed: (with bridge) 50 km/h
Range: 600 km
Gradient: (with bridge) 45%
Side slope: 30%
Vertical obstacle: 0.48 m
Trench: 2.45 m
Engine: Teledyne Continental AVDS-1790-2D turbo-charged diesel developing 760 hp at 2,400 rpm
Transmission: GMC-Allison CD-850-6A
Bridge weight: 10,000 kg
Load classification: MLC 70
Bridge length:
(extended) 26 m
(folded) 13.5 m
Gap spanned: 24 m
Bridge width: 4 01 m
Roadway width: (each) 1.555 m
Bridge height: (max) 0.9 m

Status

Prototype conversion and tests successfully completed in Spain.

Manufacturers

MAN Gutehoffnungshütte AG, Ginsheimer Strasse 1, D-65463 Ginsheim-Gustavsburg, Germany.
Tel: +49 6134 55292.
Fax: +49 6134 55383.
Peugeot Talbot España SA, Productos Especiales, Departmento Ventas/Marketing, Apartado no 140, Madrid, Spain.
Tel: +34 1 347 2370. Telex: 27590.
Fax: +34 1 347 2243.

VERIFIED

MaK Brückenlegepanzer Biber armoured bridgelayer

Development

Development of an armoured bridgelayer based on the Leopard 1 MBT commenced during 1965. In 1969 prototypes of two different types of bridgelayer based on the chassis of the Leopard 1 MBT were built, known as Type A and Type B. The Type A had a telescopic boom which was extended to the far bank; the bridge was then slid across and the telescopic boom removed. The Type B was of the cantilever type. The bridge was designed by Klöckner-Humboldt-Deutz, with Porsche in charge of overall development. After comparative trials, the Type B was selected for production by MaK of Kiel and the first production bridgelayers, known as the *Biber*, were completed in 1975. The official German Army designation is *Brückenlegepanzer Biber* or BRP-1.

Description

The hull of the *Biber* (Beaver) is almost identical to that of the Leopard 1 MBT. The driver is seated at the front of the hull on the right side, the commander in the centre and the engine and transmission is at the rear. The torsion bar suspension consists of seven roadwheels, with the drive sprocket at the rear and the idler at the front. There are four track-return rollers. The first, second, third, sixth and seventh roadwheel stations are provided with a hydraulic shock-absorber. The tracks are of the double-pin type and have rubber pads. Standard equipment includes an NBC system.

The bridge is of aluminium construction and has a total span of 22 m, which allows a gap of up to 20 m to be breached. The main advantage of the *Biber* is that its bridge is extended horizontally rather than vertically as with the majority of bridgelayers. When travelling, the bridge is carried in two 11 m halves, one above the other. The bridge may be taken up from either end and, according to MaK, allows some 60 per cent of all watercourses and most of the soil cuttings in Germany to be crossed.

The bridge was designed to take AFVs and other vehicles up to MLC 60. The bridge can be laid on longitudinal and lateral slopes of 10 per cent; the opposite bank can be 2 m higher or lower, or have a difference in inclination of 10 per cent towards the bank on the laying vehicle's side.

When the *Biber* arrives at a gap or river, the vehicle first lowers the support blade (which can also be used for dozing operations such as preparing a river bank) at the front of the hull. The lower half of the bridge slides forward until its end is lined up with the end of the upper half and the two sections are then locked together and extended over the gap. The bridge is lowered into position and the cantilever arm withdrawn. The *Biber* then raises the support blade and pulls away. The bridge can be retrieved from either end by performing the laying operations in the reverse order. Power for the bridgelaying operation comes from a central hydraulic system consisting of hydraulic pumps, an oil tank, valve blocks, control valves and hydraulic cylinders. All operations are controlled by electrical non-contact limit switches, using a sequence control system.

A Biber armoured bridgelayer in the process of laying a bridge ***1996***

Specifications

Crew: 2
Weight:
(with bridge) 45,300 kg
(without bridge) 35,100 kg
Length:
(with bridge) 11.82 m
(without bridge) 10.56 m
Width:
(with bridge) 4 m
(without bridge) 3.25 m
Height:
(with bridge) 3.55 m
(without bridge) 2.56 m
Ground clearance: 0.42 m
Track: 2.7 m
Track width: 550 mm
Length of track on ground: 4.236 m
Ground pressure: 0.97 kg/cm^2
Max speed: (road) 62 km/h
Range:
(road) approx 450 km
(cross-country) 300 km
Fuel capacity: 995 litres
Fording:
(without preparation) 1.2 m
(with preparation) 1.65 m
Gradient: 60%
Side slope: 30%
Vertical obstacle: 0.7 m
Trench: 2.5 m
Engine: MTU MB 838 Ca M-500 10-cylinder multifuel developing 830 hp at 2,300 rpm
Transmission: ZF HP 250 with 4 forward and 2 reverse gears
Electrical system: 24 V
Batteries: 6, 300 Ah, charged by 3-phase generator driven from main engine
Armament: 8 smoke grenade dischargers

Armour
nose: 70 mm at 55°
glacis: 70 mm at 60°
glacis top: 25 mm at 83°
sides upper: 35 mm at 50°
sides lower: 25 mm at 90°
hull rear: 25 mm at 88°
hull roof: 10 mm
hull floor: 15 mm

Bridge
Weight: 9,940 kg
Length:
(total) 22 m
(effective) 20 m
Width: 4 m
Width of one track: 1.55 m
Classification: MLC 60

Status
Production complete. In service with Australia (5), Canada (9), Denmark (10), Germany (104) and the Netherlands (17). Production in Italy by OTOBreda (formerly OTO Melara) for the Italian Army (64) is complete.

Manufacturer
MaK System Gesellschaft mbH, PO Box 9333, D-24157 Kiel, Germany.
Tel: +49 431 3995-02.
Fax: +49 431 399 5446.

VERIFIED

Jung-Jungenthal M48 and Leopard 1 armoured bridgelayer conversions

Description
In addition to its involvement in M48 and Leopard 1 armoured recovery vehicle conversions (see entry under *Armoured Recovery Vehicles*), Jung-Jungenthal GmbH also carries out armoured bridgelayer conversions based on the same vehicles.

The bridgelaying system involved was developed by the Swiss Eidgenössische Konstruktionswerkstätte (now Swiss Ordnance Enterprise) in Thun in co-operation with several Swiss companies. The system was originally developed in Switzerland for the Swiss BrüPz 68/88 (see entry in this section).

As fitted to the M48 and Leopard 1 chassis the bridge consists of two four-part rails connected by two girders. The laying process is hydraulically operated as the bridge is advanced and laid by a guide frame and tracer. The bridge can be laid within 2 minutes and recovered in 3 minutes.

The bridge has a normal load capacity of 50 tonnes that can be increased to 60 tonnes for exceptional loads. The bridge itself is 18.2 m long and 3.79 m wide. It is possible to use the bridge in multiples of two or three to cross wide gaps.

Jung-Jungenthal Leopard 1 armoured bridgelayer conversion ***1996***

Specifications
Bridge weight: 6,800 kg
Bridge length: 18.2 m
Bridge width: 3.79 m
Width of guide rails: 1.48 m
Load capacity:
(normal load) 50,000 kg
(exceptional load) 60,000 kg

Status
Prototypes.

Manufacturer
Jung-Jungenthal GmbH, PO Box 20, D-5242 Kirchen, Germany.
Tel: +49 2741 683-0. Telex: 875319.
Fax: +49 2741 683246.

VERIFIED

Jung-Jungenthal Leopard 1 armoured bridgelayer conversion ***1996***

EWK M3 amphibious bridging and ferry system

Development/Description
The M3 amphibious bridging and ferry system is under development to be the replacement for the earlier M2 system (see following entry). There have been four German Army M3 prototypes produced plus three for the British Army. These prototypes underwent trials with the German and British armies and successfully completed reliability and other trials, prior to the production contract being put out to tender.

Total requirements were understood to be 64 units to be jointly procured for the German and British armies, with the British Army taking 38. It was anticipated that deliveries would take place between 1996 and 1999; British Army in-service date is scheduled for October 1997.

The M3 incorporates several improvements over the existing M2 versions, including the ability to be driven on land and water from the same end (when the M2 is in the water the cab is at the rear). Three bridging ramps are carried in place of the M2's four and each bridge module has a length increase of 2.3 m. Extra flotation is provided by flotation bags inside the wheel arches. In general terms the M3 operates in much the same manner as the M2 (see following entry). All-wheel steering is possible and a tyre pressure adjusting system is fitted. The M3 has a load classification of MLC 70.

For standard configurations with an M3 bridge, the bridge length (L) is calculated based on the number (N) of units by using the formula: L = N × 11.5 + 8.35 m.

Specifications
Crew: 2 or 3
Configuration: 4 × 4
Weight: (total) approx 25,800 kg
Front axle load: 12,800 kg
Rear axle load: 13,000 kg
Length: 12.882 m
Width:
(side floats folded) 3.35 m
(side floats unfolded) 6.57 m
Height: (wheels in normal position) 3.93 m
Ground clearance: (wheels in normal position) 0.7 m
Track: (front and rear) 2.4 m
Wheelbase: 6.5 m
Max speed: (road) approx 80 km/h
Range:
(land) 725 km
(as ferry) 6.25 h
Max gradient: 60%
Fording: amphibious
Engine: Deutz (KHD) BF 8 L 513 LC diesel developing 360 hp at 2,100 rpm
Turning circle diameter: (all-wheel steering) 25.4 m

Performance (with/without wheel arch inflation)
Immersion:
(one-bay ferry, without payload) 0.53/0.49 m
(two-bay ferry, with MLC 70 payload) 0.99/0.92 m
(bridge with MLC 70 payload) 0.84/0.77 m
Useful bridge roadway width: 4.76 m
Water travelling speed:
(one-bay ferry, without payload) approx 14 km/h

(one-bay ferry with 12 t payload) approx 13 km/h
(two-bay ferry with MLC 70 payload) approx 11 km/h
Average fuel consumption:
(land) 55 litres/100 km
(water) 64 litres/h

Status
In production by converting existing M2 systems.

Manufacturer
Eisenwerke Kaiserslautern GmbH, Barbarossastrasse 30, D-67655 Kaiserslautern, Germany.
Tel: +49 631 8570.
Fax: +49 631 857300.

UPDATED

M3 amphibious bridging and ferry system unit in travelling configuration

EWK M2 amphibious bridging and ferry system

Development
The M2 amphibious bridging and ferry system was developed by Eisenwerke Kaiserslautern (EWK) and Klöckner-Humboldt-Deutz (KHD). First production units were delivered to German Army Pioneer battalions early in 1968.

There have been five versions of the M2, as follows:
M2A; used for trials only
M2B; production version for British and German armies, now modified to M2D (see below)
M2C; version for Singapore with Deutz Model F 8 L 413 F diesel engines
M2D; conversion of M2B to accommodate Class 70 (tracked) and Class 92 wheeled vehicles following the introduction of the Challenger 1 MBT into British Army service. The extra load is accommodated by the use of engine-inflatable bags at the centre, front and rear along with some other modifications
M2E; proposed interim updated version with a hydraulic crane and a more powerful engine; although one vehicle was fitted with a new engine this version was not proceeded with.

Production of the M2 is complete; 385 units were produced.

For details of the M3, the replacement for the M2 series, see previous entry.

Description
The M2 has a chassis of high-alloy construction, with the four-man crew seated in the cab at the front of the vehicle. All four wheels can be steered and the suspension is adjustable.

Before entering the water the hydraulically operated hinged buoyancy tanks, which when travelling are on the top of the vehicle to reduce the overall width, are swung through 180° into position. The decking is positioned in a few minutes by a light alloy crane which when travelling is on the centreline of the vehicle. When assembled the roadway is 8.7 m long and 5.6 m wide. When the roadway has been positioned, the crane is traversed 90° to the centre of the unit. Once in the water, the units are coupled together to form a Class 50 bridge or ferry. For the latter, three M2s are required.

The M2 is fully amphibious: one of the main engines drives two 600 mm propellers for sideways propulsion while the second engine powers the 650 mm diameter steering propeller. One of the two side propellers can also be used for steering. When swimming, the cab of the M2 is to the rear and the wheels are raised to reduce drag.

British Army M2D in travelling configuration (Michael Jerchel) ***1997***

Specifications
Crew: 4
Configuration: 4 × 4
Weight: 22,000 kg
Length: (travelling) 11.315 m
Width: (travelling) 3.579 m
Overall width: (with ramps and buoyancy tanks in position) 14.68 m
Height: (travelling) 3.579 m
Ground clearance: 0.6-0.84 m
Track:
(front) 2.13 m
(rear) 2.161 m
Wheelbase: 5.35 m
Max speed:
(road) 60 km/h
(water, single unit) 14 km/h
Range:
(road) 1,000 km
(water) 6 h
Max gradient: 60%
Engines: Deutz Model F 8 L 714a V-8 diesels developing 178 hp at 2,300 rpm each (Singapore vehicles have Deutz model F 8 L 413 F diesels developing 180 hp at 2,300 rpm)
Turning circle: 25.4 m
Tyres: 16.00 × 20

Status
Production complete. In service with Germany, Singapore (36) and the UK.

Manufacturer
Eisenwerke Kaiserslautern GmbH, Barbarossastrasse 30, D-67655 Kaiserslautern, Germany.
Tel: +49 631 8570.
Fax: +49 631 857300.

VERIFIED

INDIA

Kartik armoured bridgelayer

Description

During the 1989 Republic Day parade held in New Delhi, the Indian Army displayed the prototype of an armoured bridgelayer known as the Kartik. The Kartik is based on the chassis of the Catapult self-propelled 130 mm gun (SP-130), an Indian-produced version of the Vijayanta (Vickers Mk 1 MBT) lengthened by the addition of an extra roadwheel each side. The vehicle has a crew of three.

The Kartik uses a hydraulically operated scissors-type bridge apparently similar to that used on the German BLG-60 AVLB, also in service with India. The bridge has a total length of 20 m and a maximum load capacity of MLC 60; bridge weight is 8,500 kg.

Status

In service with the Indian Army.

Manufacturer

Government of India, Ministry of Defence, Ordnance Factory Board, Ayudh Bhawan, 10A Auckland Road, Calcutta 700 001, India.
Tel: +91 33 285077.

VERIFIED

The Kartik armoured bridgelayer

Bridge Layer Tank T-72 (BLT T-72)

Description

The Bridge Layer Tank T-72 (BLT T-72) is based on the turretless chassis of the T-72M1 MBT licence-produced in India as the Ajeya. It is one component in an indigenous military bridging system encompassing all aspects of military bridging from bridge site reconnaissance to site preparation, various designs of assault bridging and line of communication bridges.

The BLT T-72 can lay two types of bridge. One is the same 20 m MLC 60 scissors bridge as that already in service with the Kartik armoured bridgelayer (see previous entry). The second bridge is 22 m long and is also a scissors bridge, with an MLC 70 classification. The gap between the trackways is capable of being filled in with decking and end panels once the bridge is emplaced.

Both types of bridge are laid using a modular high-pressure hydraulic system and cylinders. The 22 m bridge is constructed using a high-strength weldable self-ageing aluminium zinc-magnesium alloy (RDE-40); weight of the bridge alone is 8,800 kg. Once emplaced the full roadway is 4 m wide, with each trackway 1.64 m wide. A bridge takes between 3 and 5 minutes to launch using either manual or electronic control. The vehicle crew is three, including the driver, who are provided with full NBC protection.

Armament is limited to the crew's personal weapons and a machine gun for air defence, although a multibarrel smoke discharger is provided. KMT-6 mine ploughs may be added at the front of the vehicle.

Bridge Laying Tank T-72 (BLT T-72) in travelling configuration with mine ploughs fitted ***1996***

Under development is a BLT T-72 carrying a variable span bridge capable of bridging gaps from 10 to 22 m. No further details are available.

Status

Development complete.

Development Agency

Government of India, Ministry of Defence, Defence Research and Development Establishment (Engineers), Dighi, Pune 411 015, India.
Tel: +91 212 667881. Telex: 0145 7511 REDE IND.
Fax: +91 212 666583.

VERIFIED

Bridge Laying Tank T-72 (BLT T-72) in the process of launching a 22 m assault bridge ***1996***

Bridge Laying Tank T-72 (BLT T-72) minus its bridge ***1996***

Tank-mounted Multi-Hop Assault Bridge (MHAB)

Description

For development and initial production purposes the Multi-Hop Assault Bridge (MHAB) is based on the chassis of the Kartik armoured bridgelayer (see separate entry in this section) although it may eventually also be carried on the turretless chassis of the T-72M1 MBT licence-produced in India as the Ajeya. (Some test systems employed Centurion MBT chassis.) It is one component in an indigenous military bridging system encompassing all aspects of military bridging from bridge site reconnaissance to site preparation, various designs of assault bridging and line of communication bridges.

The bridge used with the MHAB is constructed using a high-strength weldable self-ageing aluminium zinc-magnesium alloy (RDE-40). The bridge, which is flat-topped when emplaced, is 20 m long and 4 m wide with side kerbs, and is capable of carrying MLC 60 loads; with caution the bridge may carry MLC 70 loads. A four-legged trestle is provided at the far end to support the bridge when it is employed as part of a multispan bridge. As the bridge is launched it is lowered through a cradle and boom and scissored open using a 22 tonne winch and spreading cylinders. In the final phases of the launch the trestle is released from its nesting position and telescoped to the required height using hydraulic motors. Power for the modular high-pressure hydraulic controls and circuits is provided by the main carrier vehicle engine via a PTO. The trestle height can be varied from 1.7 to 2.7 m. This allows the bridge to be used over canals and river beds having channels up to 2.5 m deep.

The final phase of the launch of a Multi-Hop Assault Bridge (MHAB) with the trestle about to be lowered from its nesting position **1996**

A complete MHAB launch sequence can be completed in 7 to 8 minutes. Side folding decks on both sides of the bridge can be laid manually in 10 to 15 minutes after launch. The launch of each additional span, with docking, takes about 20 minutes by day or 30 minutes by night.

The MHAB can be employed as a ramp to negotiate bank heights up to 2.5 m.

Status

Development.

Development Agency

Government of India, Ministry of Defence, Defence Research and Development Establishment (Engineers), Dighi, Pune 411 015, India.
Tel: +91 212 667831. Telex: 0145 7511 REDE IND.
Fax: +91 212 666583.

UPDATED

A Kartik armoured bridgelayer chassis carrying a Multi-Hop Assault Bridge (MHAB) in its travelling configuration **1996**

A TATRA Mat-laying vehicle crossing an emplaced Multi-Hop Assault Bridge (MHAB) **1996**

Extended Span Assault Bridge (ESAB)

Description

The Extended Span Assault Bridge (ESAB) is carried on and launched from the chassis of the Kartik armoured bridgelayer (see separate entry in this section) which is then known as the Extended Span Bridgelayer (ESBL). (Some test systems employed Centurion MBT chassis.) It is one component in an indigenous military bridging system encompassing all aspects of military bridging from bridge site reconnaissance to site preparation, various designs of assault bridging and line of communication bridges.

The ESAB is constructed using a high-strength weldable self-ageing aluminium zinc-magnesium alloy (RDE-40). The bridge, which is flat-topped when emplaced, has three segments, two with lengths of 10 m and one 8 m, so that when extended it is 28 m long and 4 m wide with side kerbs, and is capable of carrying MLC 24 loads. When fully folded the bridge is 10.5 m long and 3.5 m wide; height when travelling is 3.96 m. The bridge could be redesigned for higher load classifications if a heavier carrier vehicle is available.

The initial phase of the launch of an Extended Span Assault Bridge (ESAB) **1996**

Extending the third segment of an Extended Span Assault Bridge (ESAB) **1996**

The launch of the ESAB follows the same lines as those of the Multi-Hop Assault Bridge (MHAB - see previous entry). As the bridge is launched it is lowered through a cradle and boom and scissored open using a 22 tonne winch and spreading cylinders. Scissoring of the third segment is carried out with the assistance of the winch and a hydraulic cylinder mounted on the second bridge segment, after which the entire bridge is lowered. Power for the modular high-pressure hydraulic controls and circuits is provided by the main carrier vehicle engine via a PTO. The entire launch sequence can be carried out under manual or electronic control.

A complete ESAB launch sequence can be completed in 7 to 8 minutes. Side folding decks on both sides of the bridge can be laid manually in 10 to 15 minutes after launch.

If required, the ESAB could be launched from a MHAB and docked with the MHAB. The ESAB can be employed as a ramp to negotiate bank heights up to 3 m.

Status
Development.

Development Agency
Government of India, Ministry of Defence, Defence Research and Development Establishment (Engineers), Dighi, Pune 411 015, India.
Tel: +91 212 667881. Telex: 0145 7511 REDE IND.
Fax: +91 212 666583.

UPDATED

The initial phase of the launch of an Extended Span Assault Bridge (ESAB) **1996**

Sarvatra truck-mounted bridging system

Development
The Sarvatra truck-mounted bridge is one component in an indigenous military bridging system. It encompasses all aspects of military bridging from bridge site reconnaissance to site preparation, various designs of assault bridging and line of communication bridges.

The Sarvatra has been developed under the auspices of the Defence Research and Development Establishment (Engineers), Dighi, Pune, using what is described as a concurrent engineering approach. This entails numerous advanced design and evaluation techniques involving several teams, including the co-operation of several commercial concerns involved in all design and development stages and that will eventually manufacture the fully developed system. Commercial concerns involved to date have been:

BHEL
Binny Limited
L & T McNeil
Oscar Equipments
Badawe Engineers
Tata Electric Companies
Oilgear Towler Polyhydron Limited.

Sarvatra is the motto of the Indian Corps of Engineers. Translated it means the same as the motto of the British Royal Engineers, namely Ubique, or Everywhere. It may be termed a universal bridging system capable of overcoming gaps up to 100 m (following further development).

A Sarvatra truck-mounted bridging system laying one half of a Sarvatra bridge during trials **1996**

Description
Sarvatra has many concept features in common with the RFAS TMM series of bridging (see separate entry in this section) in that folded bridge sections are launched from the back of a carrier truck. However, the two differ in design detail, the Sarvatra having, to name but one difference, five sections as opposed to the four of the TMM.

Each Sarvatra bridge section is 15 m long, 4 m wide and fully decked. A trestle is located at the far end of each section so that, as each bridge section is hydraulically launched from a re-engineered TATRA T 815 VVN 8 × 8 carrier vehicle, the telescopic trestle can be adjusted in height before the next section is added (the fifth section rests on the far bank). To assist the laying operation an auxiliary cab is added at the rear of the TATRA chassis, not only for controlling the laying process but to allow the vehicle to be driven safely in reverse along newly laid bridge sections.

Up to five 15 m sections may be laid at present, making an overall Sarvatra bridge length of 75 m. Following further development, each bridge section will eventually be extended to 20 m with only a marginal bridge weight increase overall, so 100 m Sarvatra bridges will be possible.

Status
Development.

Development Agency
Government of India, Ministry of Defence, Defence Research and Development Establishment (Engineers), Dighi, Pune 411 015, India.
Tel: +91 212 667881. Telex: 0145 7511 REDE IND.
Fax: +91 212 666583.

VERIFIED

INTERNATIONAL

Future French/German assault bridge system PTB1/PSB2

Development
In 1993 the French *Délégation Générale pour l'Armement* (DGA) and the German *Bundesamt fur Wehrtechnik und Beschaffung* (BWB), placed an order with Eurobridge Mobile Brücken GmbH (a joint subsidiary of the Dornier-led Command and Information Systems of DASA and the French CNIM) to define an assault bridge system suited to both French and German military requirements.

The French/German Assault Bridge System is to be known as System PTB1/PSB2. The French requirement calls for a specific wheeled launcher to be known as the Poseur de Travures Automoteur, also known as Design B1 or PTB1. The German approach is to use a tracked launcher, the Panzer Schnell Brücke 2, also known as Design B2, or PSB2. The two systems will be compatible.

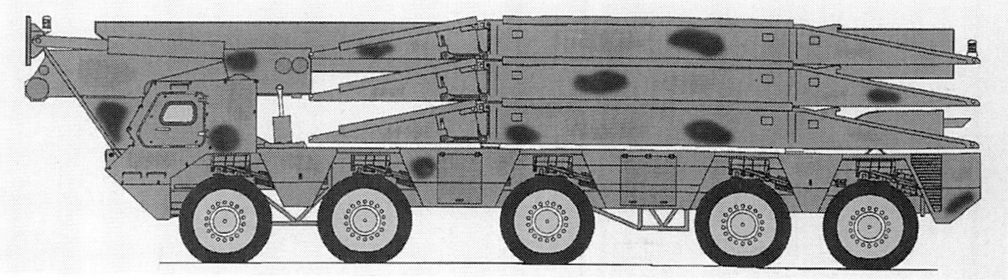

Drawing of the planned PTB1 assault bridge system, also known as the PTA **1997**

Procurement of the system by France is expected to be completed by 2000. Procurement by the German armed forces is expected to follow.

Description
Both the wheeled PTB1 and the tracked PSB2 will carry a bridge-launching device and a set of three modular bridges. The three modular bridges will allow assault bridges with lengths of 10, 18 or 27 m to be built, all with a bridge-crossing capacity of MLC 70; bridge width is 3.95 m. The launch vehicles will allow forward launching and will be controlled by two operators from inside the armoured driver's cab. Launch time for a 27 m bridge will be between 5 and 10 minutes.

The PTB1 bridges (also known as the PTA), will be launched from a 10 × 10 wheeled vehicle weighing around 53 tonnes and with a 1,000 hp engine providing a road speed of 100 km/h. Steering will be provided on four axles. Length of the vehicle will be about 15 m with a height and width of 4 m. A long-stroke suspension and a central tyre pressure control system will be provided for cross-country travel. NBC and ballistic protection (proof against 14.5 mm projectiles) will be provided for the crew, smoke dischargers will give further protection. Each PTB1 launch vehicle will be supported by a truck and semi-trailer carrying extra modular bridges.

The PSB2 bridges will be carried on and launched from a converted Leopard 2 tank chassis. Weight of this system will be about 59 tonnes.

Status
Development.

Manufacturers
Eurobridge Mobile Brücken GmbH, D-88039 Friedrichshafen, Germany.
Tel: +49 7545 83611.
Fax: +49 7545 84634.

Enquiries for PTB1 (PTA) to: CNIM, 35 rue de Bassano, F-75008 Paris, France.
Tel: +33 1 44 31 11 00. Telex: 642 869 f.
Fax: +33 1 44 31 11 30.

UPDATED

KRUPP-LOHR LEGUAN/LOGUAN Modular Bridge System (MBS)

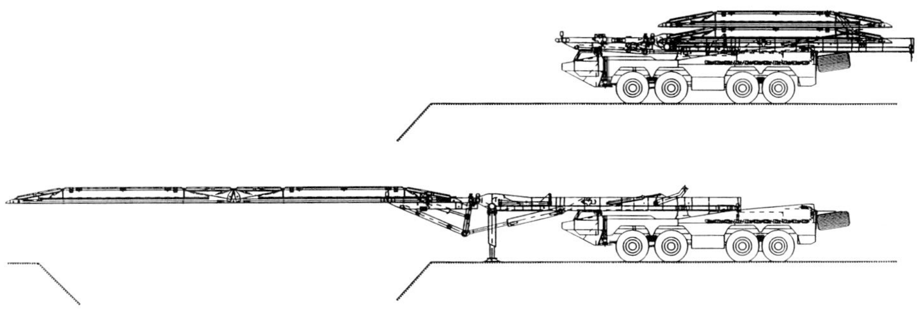

Outline drawing of wheeled 8 × 8 prototype of LEGUAN Modular Bridge System (MBS) before (above) and during two-module bridgelaying operation (below)

Development
Following an analysis of future requirements the French and German armed forces have a requirement for a new mechanised bridgelaying system. For the French armed forces the new system will be carried on a heavy wheeled vehicle and will be known as the *Porteur Travure Automoteur,* or PTA. The German approach is that the bridgelaying system should be carried on an armoured tracked chassis to be known as the *Panzerschnellbrücke 2,* or PSB2.

The Franco-German joint development between KRUPP and LOHR has been undertaken to address this requirement. The joint development is known as the LEGUAN Modular Bridge System (MBS) and combines state-of-the-art technology with the tried and tested LEGUAN bridgelaying family (see entries under Germany in this section and in the *Tactical Floating Bridges and Ferries* section), known in France as the LOGUAN. Interoperability is therefore possible between the various versions of LEGUAN, enabling the combination of different bridgelayers and bridge types.

A version of the MBS system with extended bridge modules 13 m long has been designed. For details see under Variants.

Description
KRUPP-LOHR LEGUAN Modular Bridge System (MBS) features include: three bridge modules each 10 m long and carried on the bridgelayer; laying the bridge modules either individually or coupled together; front-laying of bridges; automatic laying of the bridge from the driver's cab where the two-man crew is provided with ballistic and NBC protection; and the possibility of modifying the laying system and bridges to fit different tracked and wheeled carriers.

The LEGUAN MBS has three weldable aluminium alloy bridges used individually or coupled together, with a carrying capacity of MLC 70/100 (normal load/caution crossing) and a bridge length of up to 28 m. A bridge is approximately 4 m wide, having two track girders each with a width of approximately 1.5 m. Each bridge section weighs approximately 4.9 tonnes. There are no electrical or hydraulic components in the bridges. Bridges are laid by a wheeled (PTA) or tracked (PFB) carrier.

There are three main sections of the LEGUAN MBS laying equipment: shifting frame; outrigger system; and laying arm. Laying operations may be either automatic or semi-automatic.

At the commencement of the laying procedure, for emplacing a 28 m LEGUAN MBS bridge, the three 10 m bridge modules are transported on top of each other. After reaching the bridging site the middle bridge is pushed forward on the laying arm and the top bridge is lowered into the coupling position with the auxiliary arm. Both bridges are then coupled automatically when the ramp tips are lifted and the two bridges are pushed together. Both bridges are then fastened tightly together to create a level trackway.

The two coupled bridges are pushed forward on the laying arm and the bottom bridge is lifted into the coupling position. After being pushed backwards the two already-coupled bridges couple with the third bridge.

The fully coupled bridge is then pushed over the rear of the bridgelayer towards the land (that is, away from the gap to be crossed) so that a counterweight is created for the advancement of the shifting frame. The shifting frame is then advanced and supported with hydraulically operated support legs. The bottom truss of the laying arm is then extended and the bridge is advanced over the obstacle on the laying arm in the free launching position. The bridge is lowered and, after being separated from the bridgelayer, is ready for crossing.

The laying operation for a 28 m LEGUAN MBS lasts less than 8 minutes. Retrieval is carried out in the reverse order, is possible from either bank and takes approximately 12 minutes.

A KRUPP-LOHR LEGUAN MBS prototype first exhibited during mid-1994 was based on an 8 × 8 wheeled chassis which could carry only two LEGUAN MBS bridge modules due to load-carrying restrictions. It is planned that future PTA wheeled carriers will have a 10 × 10 configuration capable of carrying three bridge modules. Such a vehicle would weigh approximately 50 to 52 tonnes and be in the MLC 60 class. A tracked PFB carrier using a modern main battle tank chassis would weigh from 62 to 63 tonnes and be in the MLC 70 class.

Variant
A variant of the KRUPP-LOHR LEGUAN MBS has been specifically designed to meet the latest requirements for an enhanced strategic mobility and air-transportability. A model of this variant is shown in the accompanying photograph and has the following features: two bridge modules each 13 m long; a bridge classification of MLC 30; the bridges may be laid individually or coupled together; front and automatic laying; a bridge module weight of 5 tonnes; an 8 × 8 wheeled carrier; and a crew of two under ballistic and NBC protection. For air transport, typically in a C-130, the bridges and laying vehicle are carried separately and transported in two loads. To prepare the system for air transport and reassembly afterwards takes approximately 1 hour in each case.

Specifications
Bridges
Weight of each module: 4.9 t
Bridge lengths: 10, 19, 28 m
Max obstacle width: 8, 17, 26 m
Load class: MLC 70/100
Laying time for 28 m bridge: 8 min

Status
Prototypes.

Manufacturers
KRUPP Fördertechnik GmbH, Altendorfer Strasse 120, D-45143 Essen, Germany.
Tel: +49 201 82801.
Fax: +49 201 828 2566.
LOHR Industrie, F-67980 Hangenbieten, France.
Tel: +33 3 88 38 98 00. Telex: 870 082 F
Fax: +33 3 88 96 06 36.

UPDATED

Model of LEGUAN Modular Bridge System (MBS) carrying two 13 m bridge modules

Wheeled 8 × 8 prototype of LEGUAN Modular Bridge System (MBS) carrying two 10 m bridge modules
1997

ISRAEL

IMI Alligator Medium Assault Bridge (MAB)

Description

The Alligator Medium Assault Bridge (MAB) consists of three main components. One is the bridge set of two bridges, one 22 m long and the other 15 m long. The second component is the launching mechanism which is mounted on an AMX-13 light tank chassis. The system is completed by the control system.

The lightweight bridges are made of high-strength aluminium alloy with a load capacity of MLC 30. The bridges are constructed with two symmetric treadways transversely connected by beams. There is also a folding and locking mechanism. Each trackway has a centre-section and two ramps. There is no hydraulic system in the bridge.

The launch mechanism is carried on a modified AMX-13 light tank chassis with the gun turret removed. An armoured superstructure protects the launch mechanism and has a position for the commander. Launching is carried out using a launcher and outriggers which are hydraulically operated to move and position the bridge using a flip cantilever mode. Launch and retrieval can be carried out from either end. Launching takes 3 to 4 minutes and retrieval takes 4.5 to 5.5 minutes.

The control system controls the hydraulics in either an automatic or manual mode. The system can be operated by either member of the two-man crew (driver and commander) or from outside the system vehicle.

If required, two identical bridges can be used in a tandem combination to increase the gap crossing capability. Two 15 m bridges can then be used to cross a 21 m gap while two 22 m bridges can cross a 32 m gap.

A bridge can be crossed even when one main structural chord is damaged. Vehicles crossing a bridge can do so at speeds of up to 25 km/h in crosswinds of less than 20 km/h.

Specifications

System		
Bridge length	**22 m**	**15 m**
Weight		
(complete system)	20,300 kg	18,500 kg
(vehicle with system)	16,000 kg	15,000 kg
(bridge)	4,300 kg	3,500 kg
Length		
(overall)	9 m	9 m
(vehicle with system)	7.6 m	7.6 m
Width		
(complete system)	3.2 m	3.2 m
(vehicle with system)	2.6 m	2.6 m
(bridge)	3.2 m	3.2 m
Height		
(overall)	3.6 m	3.4 m
(vehicle with system)	3 m	3 m
Ground clearance	0.4 m	0.4 m
Max speed		
(road)	51 km/h	51 km/h
(cross-country)	30 km/h	30 km/h
Range (road)	350 km	350 km
Fording	0.8 m	0.8 m
Gradient	50%	50%
Side slope	30%	30%
	22 m	**15 m**
Vertical obstacle		
(forward)	0.65 m	0.65 m
(to rear)	0.45 m	0.45 m
Trench	1.6 m	1.6 m

Bridge		
Bridge length	**22 m**	**15 m**
Load capacity	MLC 30	MLC 30
Max gap capability	20 m	14 m
Folded length	9 m	9 m
Width		
(overall)	3.2 m	3.2 m
(roadway)	3 m	3 m
(trackway)	1.2 m	1.2 m
Height		
(folded)	1.47 m	1.24 m
(deployed)	0.735 m	0.735 m
Centre gap	0.6 m	0.6 m
Ramp slope	8%	12%
Max twist angle	10%	10%
Transverse slope	15%	15%
Uphill slope	25%	25%
Downhill slope	15%	15%

Status

In production.

Manufacturer

Israel Military Industries Limited (IMI), PO Box 1044, IL-47100 Ramat Hasharon, Israel.
Tel: +972 3 542 5222. Telex: 33 719 misbit il.
Fax: +972 3 489639.

UPDATED

Deploying an Alligator Medium Assault Bridge (MAB)

JAPAN

Mitsubishi Type 91 Armoured Vehicle-Launched Bridge (AVLB)

Development/Description

The Type 91 AVLB is understood to be a variant of the Type 90 MBT - for details of this vehicle refer to *Jane's Armour and Artillery 1995-96* pages 90 and 91. Few details of the Type 91 AVLB have yet been released, and it is not yet certain that a prototype has been manufactured. It is reported that the bridge carried is 20 m wide when emplaced and has a weight classification of MLC 60. Other available information follows.

Specifications

Crew: 2
Weight: 41,800 kg
Length: 10.9 m
Width: 4 m
Height: 3.8 m
Ground clearance: 0.4 m (variable)
Max speed: (road) 50 km/h
Engine power: 720 hp at 2,200 rpm
Transmission: preselect
Suspension: hydropneumatic

Status

Development.

Manufacturer

Mitsubishi Heavy Industries, Sagamihara Limited, 5-1, Marunouchi 2-chome, Chiyoda-ku, Tokyo, Japan.
Tel: +81 3 212 3111.

VERIFIED

Mitsubishi Type 67 Armoured Vehicle-Launched Bridge (AVLB)

Development/Description

The Type 67 AVLB is basically a turretless Type 61 MBT carrying a scissors bridge and launching mechanism. The bridge is similar in design and construction to the US M48 AVLB bridge but is much shorter. When opened out it is 12 m long, compared with the M48's bridge which is 19.202 m long; it will span a gap of up to 10 m. The bridge is opened out over the front of the Type 67 and can be picked up from either end. Launching time is 3 to 5 minutes. The bridge has a maximum loading capacity of 40,000 kg.

The chassis is almost identical to that of the Type 61 MBT, with the driver seated at the front of the hull on the right side, three periscopes for observation and a single-piece hatch cover. The other two crew members are positioned in the centre of the hull, with the engine and transmission at the rear. The suspension is of the torsion bar type and consists of six roadwheels, with the drive sprocket at the front and the idler at the rear, and three track-return rollers per side. Hydraulic shock-absorbers are provided for the first, second, fifth and sixth roadwheel stations.

Type 67 armoured vehicle launched bridge in travelling configuration (Kensuke Ebata) ***1996***

Specifications

Crew: 3
Weight: 35,000 kg
Length:
(with bridge) 7.27 m
(hull) 6.32 m
Width:
(with bridge) 3.5 m
(without bridge) 2.95 m
Height: (with bridge) 3.5 m
Ground clearance: 0.4 m
Track: 2.45 m
Track width: 500 mm
Length of track on ground: 3.7 m
Ground pressure: 0.95 kg/cm^2
Max speed: (road) 45 km/h
Range: 200 km
Fording: 1 m
Gradient: 60%
Vertical obstacle: 0.8 m
Trench: 2.7 m
Engine: Mitsubishi Type 12 HM 21 WT V-12 turbocharged air-cooled diesel developing 650 hp at 2,100 rpm
Transmission: mechanical with 5 forward and 1 reverse gears, with 2-speed auxiliary reduction unit
Electrical system: 24 V
Batteries: 4 × 12 V, 200 Ah
Armament: 1 × 7.62 mm MG
Armour
hull front: 46 mm
hull sides: 25 mm
hull rear: 15 mm

Status

Production complete. In service only with the Japanese Ground Self-Defence Force.

Manufacturer

Production was undertaken at the Maruko, Tokyo, plant of Mitsubishi Heavy Industries, but AFV production is undertaken at the Sagamihara Plant, near Tokyo.
Mitsubishi Heavy Industries, 5-1, Marunouchi 2-chome, Chiyoda-ku, Tokyo, Japan.
Tel: +81 3 212 3311

UPDATED

Type 70 self-propelled pontoon bridge

Development/Description

During the early 1960s, the Japanese Ground Self-Defence Force issued a requirement for a self-propelled pontoon bridge system which could also be used as a ferry. Development by the Hitachi Manufacturing Company began in 1965 and the first prototype bridge was completed the following year. After further trials it was standardised as the Type 70 self-propelled pontoon bridge.

In concept the Type 70 is similar to the German M2. Before entering the water, a hydraulic mechanism rotates the floats from the deck through 180° so that they lie along the sides of the vehicle and provide extra flotation. When afloat, the wheels, which have large low-pressure tyres, are retracted into the hull to reduce water resistance. To provide additional buoyancy the wheel wells are pressurised. The vehicle is fitted with a central tyre pressure regulation system, allowing the tyre pressure to be adjusted to suit the type of ground being crossed.

Once afloat, a built-in crane emplaces the three treadways and the units are joined to form a ferry bridge. The crane is also used to emplace drive-on ramps. Three Type 70 units coupled together have a capacity of 40,000 kg and a roadway width of 3.9 m.

Type 70 self-propelled pontoon bridge unit

Specifications

Configuration: 4 × 4
Weight: 26,000 kg
Length: 11.4 m
Width:
(without floats) 2.8 m
(floats extended) 5.4 m
Height: 3.4 m
Superstructure length: 8.5 m
Superstructure width: 3.8 m
Engine: Nissan V-8 diesel developing 330 hp at 2,200 rpm
Max speed:
(road) 56 km/h
(water)12 km/h
Gradient: 47%

Status

In service with Japanese Ground Self-Defence Force.

VERIFIED

Type 81 bridgelayer

Description

The Type 81 bridgelayer is a Type 74 (6 × 6) 10,000 kg truck chassis with a hydraulically launched bridge, mounted to the rear and laid from the rear of the cab.

To lay a bridge, which can take armoured vehicles weighing up to 42,000 kg, two vehicles are required. Each vehicle carries a bridge span in two halves with one also carrying two telescopic pier legs. The latter vehicle approaches the gap to be bridged tail first and the telescopic pier legs are attached to the end of the bottom pier half. The bottom pier half is then hydraulically driven by rams out across the gap. As the bridge half is driven across the gap the two pier legs fall downwards and, when the full length of the first bridge half is reached, the top half is attached and the hydraulic rams then push the second half out to the full extent. When the full extent of the bridge half is reached, the two are lowered with the telescopic pier legs resting on the ground. The legs can accommodate slight variations in level and the full weight of the two connected bridge halves is taken up by lowering the bridge to the ground at the truck end. The weight of the span in the centre is then taken onto foot pads, each about 1 m^2, and a ramp is fitted to the truck end, ready for the second truck to approach backwards onto the bridge to lower the second half of the bridge span. This is carried out in exactly the same way as the first except that there are no pier legs. Once the second bridge section has been laid, a further drive-on ramp can be fitted to the far end and the bridge is ready for use. Some additional levelling may be required before heavy vehicles use the bridge.

The above relates to a two-section bridge but extra sections, along with extra pier legs, may be employed to build longer bridges when necessary. The maximum bridging length is 60 m which requires 10 bridging sets.

Type 81 mechanised bridge in travelling configuration (K Nogi)

Specifications

Crew: 2
Configuration: 6 × 6
Weight: 21,800 kg
Length: 9.6 m
Width: 2.85 m
Height: 3.4 m
Max road speed: 85 km/h
Vertical obstacle: 0.8 m

Bridge
Capacity: MLC 42
Length: 10 m
Width: 3.75 m
Max bridging height: 4 m
Bridging height variation: 2 m
Max bridging length: 60 m (10 sets)

Status

In service with Japanese Ground Self-Defence Force.

VERIFIED

KOREA, SOUTH

Hyundai K1-M Armoured Vehicle-Launched Bridge (AVLB)

Development

The K1-M Armoured Vehicle Launched Bridge (K1-M AVLB) is a derivative of the K-1 MBT, also known as the Type 88 MBT (for details of the Type 88 refer to *Jane's Armour and Artillery 1995-96* pages 95 and 96). The K1-M AVLB was developed in co-operation with Vickers Defence Systems, UK, which was awarded a contract to design and build the bridge and launch system in early 1989. The bridge and launch system was built in the UK and in 1990 was shipped to South Korea where it was integrated with the K-1 (Type 88) chassis.

In September 1993 the Hyundai Precision & Ind Co Ltd awarded Vickers Defence Systems, UK a $34.47 million contract to manufacture eight bridges and 41 launching systems for installation on production K1-M AVLBs by Hyundai. The total South Korean requirement is for 56 AVLBs so the balance of bridges and launching systems will be built in South Korea by Hyundai under a technology transfer agreement.

Description

The K1-M AVLB is based on the chassis and running gear of the Type 88/K-1 MBT, developed by General Dynamics Land Systems, USA. It is powered by a 1,200 hp water-cooled, multifuel MTU MB871 ka-501 diesel coupled to a LSG-3000 electrohydraulic transmission.

The crew is two, commander and driver. The commander is located under a cupola in the hull roof. An integral crew heating system is provided but NBC protection is restricted to individual systems. Automatic fire detection and suppression systems are provided in the engine and crew compartments.

The bridge is of the scissors type and can bridge a gap with a clear span of 18.5 m; the load classification is MLC 60 and the bridge alone weighs 1 ,910 kg. The anticipated service life is over 8,000 crossings. Launch time is 3 to 5 minutes and retrieval time 10 minutes. The bridge can be launched with a side slope of 3° and a longitudinal slope of ±10°; bank height difference is 2.4 m.

A pintle for a 7.62 mm M60 machine gun is mounted over the commander's cupola on the hull roof and two six-barrel smoke grenade dischargers are mounted on the hull front.

K1-M armoured vehicle-launched bridge ***1997***

Specifications

Crew: 2
Weight:
(vehicle and bridge) 49,900 kg
(bridge) 11,910 kg
Ground pressure: 0.85 kg/cm^2
Length overall: 12.5 m
Width overall: 4 m
Height overall: 4 m
Ground clearance: 0.46 m
Max speed:
(road) 65 km/h
(cross-country) 40 km/h
Fording: 1.2 m
Gradient: 60%
Side slope: 30%
Vertical obstacle: 0.9 m
Trench: 2.7 m
Engine: MTU MB 871 ka-501 diesel developing 1,200 hp at 2,600 rpm
Transmission: LSG 3000 automatic with 4 forward and 2 reverse gears
Suspension: hydropneumatic and torsion bar
Electrical system: 24 V
Armament: 1 × 7.62 mm M60 MG

Status

In production. In service with the South Korean Army.

Manufacturer

Hyundai Precision & Ind Co Ltd, 140-2, Gye-Dong, Chongro-Ku, Seoul, South Korea.
Tel: +82 2 746 4010. Telex: HDPIC K23238.
Fax: +82 2 741 2248.

UPDATED

PAKISTAN

M47M Armoured Vehicle-Launched Bridge (AVLB)

Description

The Military Vehicle Research and Development Establishment at Rawalpindi, has developed an Armoured Vehicle-Launched Bridge (AVLB) based on the turretless chassis of an M47M tank for the Pakistan Ministry of Defence. The project has been described as a step towards indigenous development and self-reliance. For details of the original M47M tank refer to *Jane's Armour and Artillery 1994-95* page 140.

The bridge-launching mechanism is installed on the hull of the turretless M47M and is used to launch an aluminium alloy folding scissors bridge with a maximum span of 21.4 m. The bridge has a maximum clear span of 20.3 m. The vehicle has a crew of two, a driver and a commander/operator who uses 13 hydraulic cylinders to launch the bridge in about 3 minutes. Power for the hydraulic system is provided by the M47M launch vehicle's main engine.

The bridge-launching sequence commences with the main launching cylinder lifting the main launching beam from its transport position. When the bridge reaches an angle of 45° the bridge unloading cylinder is operated and starts to retract. As it retracts it pulls a set of two steel ropes, passing over a system of pulleys and joined to the further part of the bridge. This initiates the bridge unfolding sequence. The travel of the main launching cylinder and the bridge-operating cylinder continues until the nearest part of the bridge rests on the ground in front of the launcher. In this position the launching cylinders block the movement of the launching frame and the bridge is pushed further by auxiliary cylinders. The fulcrum of operation shifts and the operation of the bridge-opening cylinders continues until the bridge is fully open. The bridge is then lowered into position by the main operating cylinder. At this point the operator actuates three unlocking cylinders disengaging the launching beam from the bridge. A mechanical lock is opened at the same time allowing the launching vehicle to move away from the bridge. The bridge is recovered in the reverse sequence. If the launch vehicle engine or the hydraulic pumps fail, there is provision for manual launching of the bridge.

The M47M launcher can also be used to lay a short single-span bridge. This bridge is one half of the folding 21.4 m bridge fitted with an extra set of folding ramps. No details are available regarding this variant but it is assumed that it is launched in the same manner and sequence as the full-size folding bridge.

M47M armoured vehicle-launched bridge with 21.4 m bridge in transport position

Specifications

AVLB
Crew: 2
Weight: 54,000 kg
Width: 4 m
Height: 3.83 m

Bridge
Weight: 9,800 kg
Max span of bridge: 21.4 m
Max clear span: 20.3 m
Bridge width: 4 m
Distance between treadways: 0.96 m

Status

Prototype.

Manufacturer

Military Vehicle Research and Development Establishment, PO Box 240, Abid Majid Road, Rawalpindi, Pakistan.
Tel: +92 51 679 1115.

VERIFIED

POLAND

PMCz-90 armoured bridgelayer

Description

The PMCz-90 armoured bridgelayer, also known as the PMCz-94, is based on the turretless hull of a modernised T-72M1 main battle tank. It closely follows the same lines as other similar T-72 based bridgelayers such as the Czech/Slovak ZTS MT-72 (qv). The German BLG-67 M2 series of MLC 50 load capacity steel-constructed scissors bridge and bridgelayer system are carried over to the PMCz-90. It is, therefore, possible for up to three bridges to be placed one on the other to span extended gaps, although the load capacity is then reduced to MLC 40.

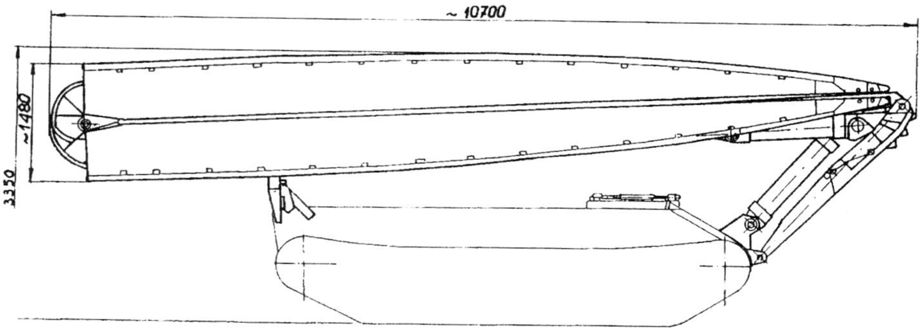

Schematic drawing of PMCz-90 armoured bridgelayer bridgelaying arrangements ***1997***

Specifications

Crew: 2
Weight:
(combat ready) approx 40,000 kg
(chassis) 32,900 kg
Power to weight ratio: 21/25 hp/t
Length:
(overall) 10.7 m
(chassis) 6.86 m
Width: 3.59 m
Height (overall): 3.35 m
Ground clearance: 0.42 m
Max speed:
(road) 60 km/h
(cross-country) 40 km/h
Range: (cruising, road) 700 km
Gradient: 30°
Side slope: 25°
Vertical obstacle: 0.85 m
Trench: 2.8 m
Fording:
(without preparation) 1.2 m
(with 5 min preparation) 1.8 m
Engine: S12-U V-12 diesel developing 850 hp at 2,300 rpm
Transmission: synchromesh, hydraulically assisted with 7 forward and 1 reverse gears
Suspension: torsion bar
Electrical system: 24 V

Bridge
Load capacity: 50,000 kg
Length: 20 m
Width: 3.47 m

Status

Available.

Manufacturer

Zaklady Mechaniczne 'Bumar-Labedy' SA, ul Mechaników 9, PL-44-109 Gliwice, Poland.
Tel: +48 3 134 5111.
Fax: +48 3 134 6966.
Marketed by: Cenzin Limited, ul Frascati 2, PL-00-489 Warsaw, Poland.
Tel: +48 22 296396. Telex: 814505 czi pl.
Fax: +48 2 628 6356.

NEW ENTRY

SMT-1 Truck-mounted treadway bridge

Description

The SMT-1 consists of four 11 m spans, each of which is carried on the rear of a Star 660 M1 or Star 660 M2 2,500 kg (6 × 6) truck. They are launched over the front of the vehicle, unlike the TMM and KMM treadway bridges which are launched over the rear.

Unlike the TMM and KMM treadways, the SMT-1 treadways do not have to be spread before launching, as they are fixed. Each treadway consists of tubular steel trusses welded together by struts, crosspieces and diagonal stiffeners, with steel mesh panels mounted on the top of the trusses to form the treadway. The SMT-1 is much lighter than the TMM treadway: the TMM weighs 666 kg/m and the SMT-1 210 kg/m.

Each span is launched in 3 to 5 minutes with the launching controlled from the cab. Individual spans of the SMT are used for other purposes, for example as ramps on pontoon bridges. The SMT-1 is often used as a single-span bridge without the trestle legs. Additional spans are carried on a single-axle trailer which has dual tyres.

SMT-1 bridges being operated during a training exercise

The SMT-1 truck-mounted treadway bridge is used in conjunction with the PSMT-1 intermediate support, which, is also used with other bridges such as the German BLG-60 AVLB. The PSMT-1 is positioned by crane or floated into position. It consists of a platform and four trestles adjustable in height. The trestles are 5 m long and can be folded for travel. The PSMT-1 is capable of floating, primarily because of the plastic foam material used in its construction. It is transported on a single-axle trailer and can be floated directly by reversing the prime mover into the water.

Specifications

Bridge
Weight: 2,300 kg
Length: 11 m
Width: 3 m
Trestle leg height: adjustable to 3.5 m
Capacity: 40,000 kg

Status

Production complete. In service with the Polish Army.

Manufacturer

Polish state factories.

UPDATED

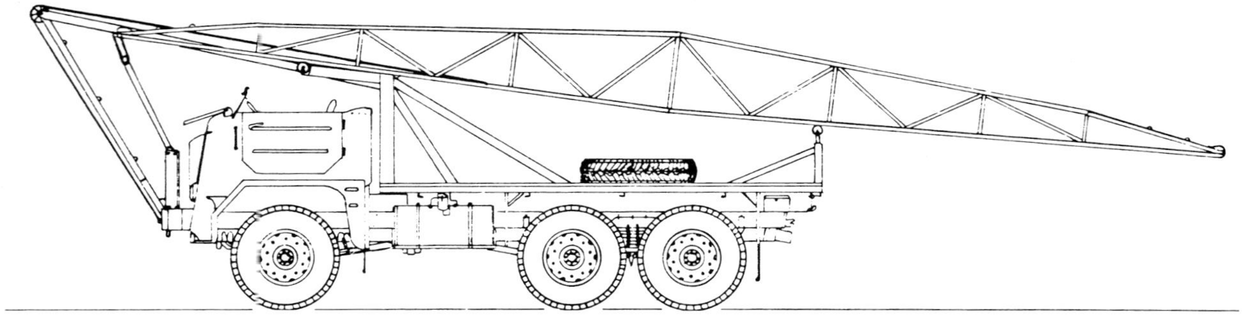

SMT-1 bridgelayer carried on Star 66 (6 × 6) truck chassis

RUSSIAN FEDERATION AND ASSOCIATED STATES (CIS)

PMM-2 amphibious bridging and ferry system

Development

The PMM-2, or *Paromno-Mostovaya Maschina-2*, floating bridging and ferry system was originally known in the West as the Amphibious Bridging System (Tracked), or ABS(T). The vehicle has a tracked suspension based on the use of T-64 MBT track and roadwheel components and uses a chassis similar to that employed with the PTS-2 amphibious carrier, the MDK-3 trench-digging machine and, what is probably the base vehicle for the series, the MT-T heavy tracked transporter. It is thus extremely likely that the engine used to power the PMM-2 is the V-64-4 diesel.

An updated PMM-2M is known to exist; the available data provided in the Specifications table refers to this version.

Description

The PMM-2 is the replacement vehicle for the GSP heavy amphibious ferry (see entry in *Floating Bridges and Ferries* section). It uses a similar chassis and float layout but with the PMM-2 the float units unfold for use and can be folded back for road transport. The PMM-2 is able to ferry loads of up to around 42.5 tonnes.

In use the PMM-2 follows the same general lines as the GSP but the float units are mechanically unfolded using hydraulic mechanisms just before or as the vehicle enters the water. The top float unfolds to the right and the second section, which is slightly shorter than the top section as it is located behind the cab, to the left. Also unfolded mechanically are two two-section loading ramps each side. Once in the water, propulsion is provided by two propeller units at the rear. These are lowered once the vehicle is in the water and the drive is taken from the main engine. The propellers are concealed within prominent cowls to prevent damage; the cowls also support small steering rudders.

PMM-2 units can be connected together to form floating bridges by using latching mechanisms on the outer edges of the float units. Bridges up to 10 units wide have been reported and no bridging boats are apparently involved in the formation of such bridges. The PMM-2 system can operate in rivers with a flow velocity of up to 2 m/s.

Individual PMM-2 units can also be used to bridge small water gaps. The overall width of the PMM-2 on the road is 3.36 m, so once the side floats are lowered the overall width is 10 m. The loading ramps each have a usable length of at least 5 m so a single PMM-2 can probably be used to bridge water gaps almost 20 m wide. In practice this width is probably reduced to around 17 m.

It has been reported that the PMM-2 has a crew of five but the cab appears to be too small to accommodate this number and the number of personnel required to actually operate the vehicle as a ferry would probably be limited to two or three. The PMM-2 probably has a collective NBC protection system for the cab, in common with other vehicles using the same or similar chassis.

PMM-2 amphibious bridging and ferry unit

Specifications

PMM-2M
Length, overall: 13.35 m
Width, overall: 3.36 m
Height, overall: 3.65 m
Max speed:
(road) 55 km/h
(water) 10 km/h
Engine power: 710 hp
Max load: 42,500 kg

Status

In service with RFAS armed forces.

Manufacturer

Kryukovskiy Vegonostroitel'nny Zavod, 315307 Kremenchug, Ukraine.
Tel: +38 20467/69505/69785.

UPDATED

MTU-90 armoured bridgelayer

Description

Very little information is yet forthcoming regarding the MTU-90 armoured bridgelayer, first announced mid-1996. The hull and chassis are based on the T-72 MBT while the bridging system is of a completely novel type. The bridge appears to have a total gap crossing length of 24 m. There are three bridge sections involved, two identical ramps and a centre-section, to scissor open in an up-and-over sequence, during which a forward support frame is lowered to the ground to assume most of the bridge weight as it unfolds. Total laying time is of the order of 2.5 to 3 minutes.

The MTU-90 bridge is operationally compatible with the TMM-6 system (qv).

Status

Prototypes.

Manufacturer

State factories.

NEW ENTRY

MTU-90 armoured bridgelayer in travelling configuration ***1997***

MTU-72 armoured bridgelayer

Description

A number of T-72B MBTs were converted to the MTU-72 armoured bridgelayer configuration by removing the turret and replacing it with a bridge-launching system and bridge. One reference mentions that a small number (nine) of these bridgelayers were manufactured from new, utilising the T-72S export chassis. The single-span bridge involved has an overall length, when laid, of 20 m. The bridge, which is constructed using aluminium alloys, has a maximum capacity of 50,000 kg, is 3.3 m wide, and can span a gap of 18 m. By itself the bridge weighs 6,400 kg. For transport the overall length of the bridge can be reduced by folding both ends of the bridge through 180°.

The overall layout and the operating method of the resultant MTU-72 are similar to those of the MTU-20 and MTU bridgelayers (see following entries). The time required to lay a bridge is 3 minutes and 8 minutes for retrieval.

The vehicle is fitted with a hydraulically operated front-mounted stabiliser blade. This can also double as a dozer blade to clear obstacles or prepare bridging sites.

Armament is limited to a light machine gun and crew weapons.

A bridging vehicle based on the chassis of the T-72 MBT, but using a scissors bridge and known as the BLP-72, was developed to the prototype stage by the former East German Army. Few details are available but it is believed to have been generally similar to the MTU-72.

ZTS Tees Martin of Slovakia has developed a bridging vehicle based on the T-72 chassis and known as the MT-72. See under *Czech Republic and Slovakia* for details.

MTU-72 armoured bridgelayer in travelling configuration

Armoured bridgelayers based on the T-72M1 MBT chassis have been developed in India and Poland (qv).

Specifications

Crew: 2
Weight: approx 40,000 kg
Power-to-weight ratio: 21 hp/t
Ground pressure: 0.808 kg/cm²
Length with bridge: 11.64 m
Width: 3.46 m
Height: (with bridge) 3.38 m
Ground clearance: 0.49 m
Track width: 580 mm
Length of track on ground: 4.278 m
Max speed: (road) 60 km/h
Range: (road) 500 km
Fuel capacity: 1,000 litres
Fording:
(normal) 1.2 m
(with preparation) 4.5 m
Gradient: 60%
Side slope: 40%
Vertical obstacle: 0.85 m
Trench: 2.8 m
Engine: V-46-6 multifuel diesel developing 840 hp
Suspension: torsion bar
Electrical system: 24 V
Armament: 1 × 7.62 mm MG

Bridge
Bridge capacity: 50,000 kg
Bridge weight: 6,400 kg
Bridge length:
(extended) 20 m
(folded) 9.42 m
Bridge width: 3.3 m
Gap spanned: 18 m

Status

In service with the FFAS.

Manufacturer

State factories.

UPDATED

MTU-20 armoured bridgelayer

Development/Description

The MTU-20 was introduced into the Soviet Army in 1967 as the replacement for the older MTU bridgelayer. It was based on the T-55 tank chassis rather than the T-54 tank chassis used for the MTU.

The method of laying the bridge is almost identical to that of the MTU bridgelayer and the reader is referred to this entry for the method of launching and for a description of the chassis (see following entry). Stabilisers at the nose of the tank are lowered into position before the bridge is launched.

The bridge weighs 7,000 kg and is of the box construction type. It has a maximum length of 20 m and can span a gap up to 18 m. It is 3.3 m wide, 1 m high and has a maximum capacity of 60,000 kg. When travelling, the ends of the bridge are folded back through 180° so that they lie on top of the bridge. They are lowered and locked in position before the bridge is launched. The bridge takes 5 minutes to lay in position and between 5 and 7 minutes to recover.

MTU-20 armoured bridgelayer

As the MTU-20 is based on a T-55 chassis it is thought likely that an NBC system is installed.

Specifications

Crew: 2
Weight: (with bridge) 37,000 kg
Length: (with bridge) 11.64 m
Width: (with bridge) 3.3 m
Height: (with bridge) 3.4 m
Ground clearance: 0.425 m
Track: 2.64 m
Track width: 580 mm
Length of track on ground: 3.84 m
Ground pressure: 0.83 kg/cm^2
Max speed: (road) 50 km/h
Range: 500 km
Fuel capacity: 960 litres
Fording: 1.4 m
Gradient: 40%
Vertical obstacle: 0.8 m
Trench: 2.7 m
Engine: V-55 V-12 water-cooled diesel developing 580 hp at 2,000 rpm
Transmission: manual with 5 forward and 1 reverse gears
Electrical system: 24 V
Batteries: 4 × 12 V, 280 Ah
Armament: nil

Armour
glacis plate: 100 mm at 60°
upper hull sides: 70 mm at 0°
hull rear: 60 mm
hull floor: 20 mm
hull roof: 30 mm

Status

Production complete. The MTU-20 is known to be in service with Afghanistan, Egypt, Finland, India, Israel, Nigeria, RFAS (563 declared) and Syria.

Manufacturer

State factories.

VERIFIED

MTU armoured bridgelayer

Development

The MTU bridgelayer (MTU - *Mostoykladchik tankoviy ustroystvo*), entered service in 1958 as the replacement for an older bridgelayer based on a T-34 MET chassis. Since 1967 the MTU has been replaced in many units by the MTU-20 (see previous entry), which has the ability to span a gap of up to 18 m compared with the 11 m of the MTU. The MTU is also sometimes referred to as the MTU-1.

MTU armoured bridgelayer in travelling configuration

Description

The MTU is based on the chassis of the T-54 MBT with its turret removed. (Late production models used a T-55 chassis rather than the T-54 chassis.) The driver is seated at the front of the hull on the left side and is provided with two periscopes for observation and a single-piece hatch cover. The commander is seated in the crew compartment to the rear of the driver with the engine and transmission at the rear of the hull. The suspension is of the torsion bar type and consists offive roadwheels with the idler at the front and the drive sprocket at the rear. There are no track-return rollers.

A 12.7 mm DShKM machine gun, mounted in the centre of the hull between the two treadways, has to be removed before the bridge can be laid in position.

The MTU does not have an NBC system and has no deep fording capability. Auxiliary fuel drums and an unditching beam are often carried at the rear of the hull.

The bridge itself is 12.3 m long, 3.27 m wide and 1 m high, and will span a gap of 11 m. It has a maximum loading capacity of 50,000 kg.

The bridge consists of four box-truss panels. The outer treadways are used by tracked vehicles and the inner treadways by smaller vehicles. When travelling, the ramp sections of the inner treadways are folded on top of the main treadways.

The bridge is positioned as follows: a chain drive mechanism moves the bridge over the cantilever launching girder until the far bank is reached, the cantilever launching girder is depressed and lowers the span onto the near bank. The bridge takes between 3 and 5 minutes to launch and can be retrieved from either end.

Specifications

Crew: 2
Weight: (with bridge) 34,000 kg
Length: (with bridge) 12.3 m
Width: (with bridge) 3.27 m
Height: (with bridge) 2.87 m
Ground clearance: 0.425 m
Track: 2.64 m
Track width: 580 mm
Length of track on ground: 3.84 m
Ground pressure: 0.76 kg/cm^2
Max speed: (road) 48 km/h
Range: 400 km
Fuel capacity: 812 litres
Fording: 1.4 m
Gradient: 60%
Vertical obstacle: 0.8 m
Trench: 2.7 m
Engine: V-54 V-12 water-cooled diesel developing 520 hp at 2,000 rpm
Transmission: manual with 5 forward and 1 reverse gears
Electrical system: 24 V
Batteries: 4,280 Ah
Armament: 1 × 12.7 mm DShKM MG

Armour
glacis plate: 100 mm at 60°
upper hull sides: 70 mm at 0°
hull rear: 60 mm
hull floor: 20 mm
hull roof: 30 mm

Status

Production complete. The T-54 is used by the following countries, some of which also use the MTU bridgelayer: Afghanistan (confirmed), Albania, Algeria, Angola, Bangladesh, Bulgaria, Cambodia, China, Congo, Cuba, Cyprus, Czech Republic (do not use the MTU), Egypt (confirmed), Equatorial Guinea, Ethiopia, Finland (confirmed), Guinea, Guinea-Bissau, Hungary, India, Iraq, Israel (confirmed), North Korea, Libya, Mali, Mongolia, Morocco, Mozambique, Nigeria, Pakistan, Peru, Poland, Romania, RFAS (confirmed), Slovakia, Somalia, Sudan, Syria (confirmed), Vietnam, Yemen, Yugoslavia (Serbia and Montenegro) and Zimbabwe.

Manufacturer

State factories.

VERIFIED

TMM truck-mounted treadway bridge

Description

The TMM (heavy mechanised bridge) consists of four 10.5 m spans, each of which is carried and launched from the rear of a modified KrAZ-214 (6 × 6) 7,000 kg truck or the more recent KrAZ-255B (6 × 6) 7,500 kg truck. The latter model has improvements in the bridgelaying mechanism and is recognisable by the spare tyre, which is carried on the roof of the cab, rather than at the rear of the cab as on the KrAZ-214. The model carried on the KrAZ-255B (6 × 6) truck is designated the TMM-3. Three of the spans have integral mounted adjustable trestle legs; the fourth (or far-shore) span does not, as it is the link between the third span and the far bank.

The system operates as follows: before launching the treadway the trestle legs must be adjusted to the correct height so the roadway is level when the bridge has been positioned. During transit the legs are folded and stored beneath the folded scissors span. The treadways are then spread to the full roadway width of 3.8 m. The truck backs up to the river and the hydraulic launching girder raises the folded span to the vertical position, the span is straightened by a cable and winch system and then lowered. As it is lowered into position, the integral trestle legs swing into place. Once it is in position the cables are disconnected, the launching girder is brought back into the travelling position and the truck moves off. This procedure is repeated until the bridge is complete. If required, the bridge can be extended past the basic four spans by further additions. The launched spans can be recovered from either end and recovery takes about the same time as launching.

A complete TMM with four spans can cross a gap of up to 40 m in 45 to 60 minutes in daylight, or 60 to 80 minutes at night. These times are for an average crew and can be halved by a well-trained crew. To reduce the possibility of detection, the TMM can also be laid under the surface of the water, which takes 50 per cent longer than the normal method. Log and metal plate supports can be used to support the trestle legs when obstacles up to 5 m deep are encountered or when the slope is too great.

A Chinese-produced version of the TMM is known as the Heavy Mechanised Bridge Type 84A. See separate entry in this section for known details. The Indian Sarvatra bridging system employs many principles similar to those used on the TMM series (qv).

Specifications

Vehicle with bridge
(full technical characteristics of the KrAZ-214 and KrAZ-255B will be found in the *Trucks* section)
Cab seating: 1 + 2
Configuration: 6 × 6
Weight: 19,500 kg
Length: 9.3 m
Width: 3.2 m
Height: 3.15 m
Ground clearance: 0.36 m
Track: 2.03 m
Wheelbase: 4.6 m + 1.4 m
Max speed: (road) 55 km/h
Range: 530 km
Fuel capacity: 450 litres

TMM-3 bridge carried on KrAZ-255B 7,500 kg truck (Michael Jerchel)

TMM-3 bridge carried on KrAZ-255B 7,500 kg truck (Michael Jerchel)

TMM-3 truck-mounted scissors bridge being lowered into position, showing front wheels of KrAZ-255B 7,500 kg (6 × 6) truck off ground as centre of gravity is well to rear when bridge is being lowered into position

Fording: 1 m
Engine: YaMZ M206B 6-cylinder water-cooled diesel developing 205 bhp at 2,000 rpm
Transmission: manual with 5 forward and 1 reverse gears and 2-speed transfer box
Tyres: 15.00 × 20

Bridge
Weight: 7,000 kg
Length: 10.5 m
Width: 3.8 m
Trestle leg length: adjustable between 1.7-3.2 m
Capacity: 60,000 kg

Status
Production complete. In service with Bulgaria, Hungary and other armed forces including those of China, Nicaragua (20) and Yugoslavia (Serbia and Montenegro). The Bulgarian Army developed a modified bridge using tubular section metal parts. It is not known if it was adopted or produced in quantity.

Manufacturer
State factories.

VERIFIED

TMM-6 truck-mounted treadway bridge

Description
Little information is yet available regarding the TMM-6 truck-mounted treadway bridge, other than it is an enlarged version of the standard TMM system described in the previous entry, presumably operating along similar lines. The carrier vehicle is an MAZ-543 8 × 8 truck with a lengthened wheelbase. From this it can be estimated that each folded bridge section is approximately 10 m long, forming a 20 m bridge section when unfolded. Each has a trestle assembly that can unfold as each section is opened or can be kept folded if the far bank is reached. Five TMM-6 sections can form a 100 m bridge may then be further extended to 125 m by the use of an MTU-90 armoured bridgelayer system.

Status
Prototypes.

Manufacturer
State factories.

NEW ENTRY

The only available illustration of a TMM-6 truck-mounted treadway bridge, seen here in the travelling configuration ***1997***

KMM truck-mounted treadway bridge

Description
The KMM (mechanised treadway bridge) consists of five 7 m spans, each of which is carried and launched from the rear of a modified ZIL-157 (6 × 6) 2,500 kg truck or a similar vehicle. Four of these spans have integral-mounted adjustable trestles while the fifth (or far-shore) span does not, as it is the link between the fourth span and the far bank.

Each trestle has a pair of octagonal shoes with a tip protruding about 0.6 m below the shoe. When buried in the soil, the tip provides additional stabilising strength to the bridge support. When travelling, the trestle shoes are detached from the trestle columns.

The system works as follows: before launching the treadway, the trestle legs must be adjusted to the correct height so the treadway is level when the bridge has been positioned, the treadway is spread to the full roadway width of 2.95 m. Each clear span can bridge a gap of 6 m and a set of five spans can bridge 34 m. Bridge capacity is reported to be 12,000 kg. The truck then backs up to the river and the span is raised hydraulically to the vertical with the launching girder. Once in this position, the launching girder serves as a brace for the cables which support the downward movement of the span. Once the launch has been completed, the launching girder is lowered back to the travel position. The procedure is repeated with more spans until the far bank has been reached. The launched spans can be recovered from either end, with recovery taking about the same time as launching.

A single KMM can span a gap of up to 9.5 m and can be launched in about 15 minutes. A complete five-span KMM bridge can span a gap of up to 34 m in 45 to 60 minutes in daylight, or 60 to 80 minutes at night. These times are for an average crew and can be halved by a well-trained crew. If required, the KMM can also be laid under water to reduce the possibility of detection, although this takes approximately 50 per cent longer than the normal method.

Specifications

Vehicle with bridge
Cab seating: 1 + 2
Configuration: 6 × 6
Weight: 8,800 kg
Length: 8.3 m
Width: 3.15 m
Height: 3.36 m
Ground clearance: 0.31 m
Track:
(front) 1.755 m
(rear) 1.75 m
Wheelbase: 3.665 m + 1.12 m
Max speed: (road) 40 km/h
Range: 430 km
Fuel capacity: 215 litres
Max gradient: 28%
Fording: 0.8 m
Engine: ZIL-157K 6-cylinder water-cooled petrol developing 109 hp at 2,800 rpm
Transmission: manual with 5 forward and 1 reverse gears and 2-speed transfer box
Tyres: 12.00 × 18

KMM truck-mounted treadway bridge on ZIL-157 (6 × 6) 2,500 kg truck with TMM truck-mounted treadway bridge on KrAZ-255B (6 × 6) 7,500 kg truck behind (Egyptian Ministry of Defence)

Bridge
Weight: 1,420 kg
Length: 7 m
Width:
(without trestle leg) 2.95 m
(with trestle leg) 3.95 m
Trestle leg height: (adjustable) 1-3 m
Capacity: 15,000 kg
Crossing speed:
(tracked vehicles) 1 km/h
(wheeled vehicles) 15-20 km/h
Min distance between crossing vehicles: 15 m

Status

Production complete; replaced in many front-line units by the TMM truck-mounted treadway bridge. In service with members of the former Warsaw Pact and other armed forces including those of China and Egypt.

Manufacturer

State factories.

VERIFIED

SINGAPORE

Light Assault Bridge LAB 30

Description

The Light Assault Bridge LAB 30 is very similar to the French AMX-13 armoured bridgelayer (now withdrawn from service) and is based on an AMX-13 light tank chassis. The bridgelayer is produced by removing the AMX-13 tank turret and replacing it with the mechanism to lay a folding or sliding assault bridge. The hydraulically operated mechanism is incorporated on the front and rear of the tank chassis and operates in a manner very similar to that of the French AMX-13 system.

The folding bridge employed is manufactured entirely from aluminium alloy and weighs 3,200 kg. Length is 14 m and width 2.9 m. The bridge can be launched within 3 to 4 minutes and will take loads up to Class 30. The bridge can cross a gap of 12 m (on soft ground) or 13 m (on hard ground). With two bridges placed side by side the load class can be increased to Class 60.

When travelling complete with its bridge the LAB 30 is 7.38 m long and 3.2 m high.

Status

Available.

Manufacturer

Singapore Technologies and Engineering Limited, 7 Benoi Road, Singapore 629832.
Tel: +65 861 2244. Telex: 21206 singa rs.
Fax: +65 861 3028.

VERIFIED

SWEDEN

Hägglunds Brobandvagn 941 armoured bridgelayer

Development

The first prototype of the *Brobandvagn* 941 (or Brobv 941) was completed by Hägglunds in 1968. It entered production late in 1972 and 17 vehicles were delivered to the Swedish Army the following year.

Description

The chassis and hull of the Brobv 941 are almost identical to those of the Bgbv 82 armoured recovery vehicle (see *Armoured Recovery Vehicles* section for a full description of the chassis). The main difference between the two vehicles, apart from the bridge, is that the Brobv 941 does not have a 20 mm gun turret.

The gunner can fire the 7.62 mm machine gun from either his open hatch or from an alternative position on top of the bridgelaying mechanism when engaging aircraft. In addition, provision was made for mounting a 7.62 mm machine gun in the loading platform at the rear, and on the commander's hatch. The crew of four consists of the commander, driver, gunner and operator, all seated in the front compartment.

At the front of the vehicle is a hydraulically operated bulldozer blade, which can be used for levelling operations as well as stabilising the vehicle when the bridge is being laid. On arriving at the river or ditch, the dozer blade is first lowered into position; a telescopic beam is extended until the point of the beam reaches the other bridge abutment; the aluminium bridge is slid across the beam which is then withdrawn so that the bridge is horizontal. The laying mechanism then returns to the travelling position, the dozer blade is raised, and the vehicle can cross the bridge. This complete operation takes less than 5 minutes with all the crew under armour protection. If required, one of the crew can be outside the vehicle directing the laying of the bridge via a communications link with the vehicle. The bridge itself is 15 m long and has a maximum load capacity of 50,000 kg.

The Brobv 941 is fully amphibious, being propelled in the water by its tracks. Before entering the water a trim vane is erected at the front of the vehicle and the electric bilge pumps switched on. When the Brobv 941 is afloat, the bridge, which is buoyant, is towed behind the vehicle.

Infrared night vision equipment is provided but the vehicle does not have an NBC system, although provision was made for one in the design of the vehicle.

Specifications

Crew: 4
Weight:
(with bridge) 28,400 kg
(without bridge) 21,400 kg
Length:
(with bridge) 17 m
(without bridge) 6.7 m
Width:
(with bridge) 4 m
(without bridge) 3.25 m
Height:
(with bridge) 3.24 m
(without bridge) 2.9 m
Ground clearance: 0.41 m
Track width: 450 mm
Length of track on ground: 3.6 m
Ground pressure: 0.84 kg/cm²
Max speed:
(road) 56 km/h
(water) 8 km/h
Range: 400 km
Fuel capacity: 550 litres
Fording: amphibious
Gradient: 60%

Brobv 941 in travelling order

Side slope: 30%
Vertical obstacle: 0.6 m
Trench: 2.5 m
Engine: Volvo-Penta THD 100C 6-cylinder in-line turbocharged diesel developing 310 bhp at 2,200 rpm
Transmission: Volvo-Penta R61 with 8 forward and 2 reverse gears
Electrical system: 24 V
Batteries: 2 × 12 V, 152 Ah
Armament:
1 × 7.62 mm MG
12 smoke grenade dischargers

Status
Production complete. In service with the Swedish Army.

Manufacturer
Hägglunds Vehicle AB, S-891 82 Örnsköldsvik, Sweden.
Tel: +46 660 80000. Telex: 6051 haegg s.
Fax: +46 660 15190.

VERIFIED

SWITZERLAND

Brückenlegepanzer 68/88

Development
The first prototype of the *Brückenlegepanzer* 68 (BrüPz 68) was based on the chassis of the Pz 61 MBT but production vehicles were based on the chassis of the later Pz 68 MBT. The prototype vehicles carried a steel bridge which was replaced on production vehicles by an aluminium bridge. Production was completed in June 1977. Since then several modifications have been introduced, including a fully automatic hydraulic bridgelaying system, resulting in a change of designation to BrüPz 68/88.

The bridgelaying system was developed by the Eidgenössische Konstruktionswerkstätte (now Swiss Ordnance Enterprise) in Thun in co-operation with several other Swiss companies.

The bridge and bridgelaying system used with this vehicle have been utilised by Jung-Jungenthal GmbH of Germany for conversions of M48 and Leopard 1 MBTs to the armoured bridgelayer role. See entry under Germany in this section for details.

Description
The hull of the *Brückenlegepanzer* 68/88 (BrüPz 68/88) is almost identical to that of the Pz 68 MBT. The driver is seated at the front of the hull and is provided with three periscopes for observation and a single-piece hatch cover. The commander and the bridge operator are seated in the centre of the vehicle with the engine and transmission at the rear.

The suspension of the BrüPz 68 is the Belleville type and consists of six roadwheels with the drive sprocket at the rear and the idler at the front. There are three track-return rollers. Each roadwheel is independently sprung by layers of Belleville washers.

The bridge itself has an overall length of 18.23 m and a width of 3.79 m, maximum trackway width being 3.55 m. Its maximum capacity is 60,000 kg but it is normally limited to a maximum load of 50,000 kg. The bridge is launched as follows: the bridgelayer approaches the obstacle and halts; the bridge is tilted forwards and a beam slid across to the far bank; the bridge is slid across the beam until it reaches the far bank; and the beam is then withdrawn back into the horizontal position. The laying sequence is fully automatic once initiated. A bridge takes about 2 minutes to lay and under 5 minutes to recover.

BrüPz 68/88 armoured bridgelayer crossing a laid bridge

Specifications
Crew: 3
Weight:
(with bridge) 47,000 kg
(without bridge) 37,200 kg
Length: (with bridge) 20.1 m
Width: (with bridge) 4 m
Height: (with bridge) 3.3 m
Ground clearance: 0.4 m
Track: 2.59 m
Track width: 520 mm
Length of track on ground: 4.22 m
Ground pressure: 0.98 kg/cm^2
Max speed: (road) 55 km/h
Range:
(road) 300 km
(cross-country) 160 km
Fuel capacity: 855 litres
Fording: 1.1 m
Gradient: 70%
Vertical obstacle: 0.7 m
Trench: 2 m
Engine: MTU MB 837 8-cylinder diesel developing 660 hp at 2,200 rpm
Transmission: semi-automatic with 6 forward and 6 reverse gears
Electrical system: 24 V
Batteries: 4 × 6 V, 360 Ah
Armament: 8 × 80 mm smoke grenade dischargers
Armour: 20-60 mm

Status
Production complete. In service with the Swiss Army.

Manufacturer
Swiss Ordnance Enterprise, Allmendstrasse 86, CH-3602 Thun, Switzerland.
Tel: +41 33 228 1111.
Fax: +41 33 228 2047.

UPDATED

UNITED KINGDOM

Vickers Bridging Modular Bridge - tank-launched bridges

Development
The Vickers Modular Bridge (VMB) was originally developed by Thompson Defence Projects (part of the Industrial Power Group of Rolls-Royce plc) for the British Army's Bridging for the 1990s (BR90) programme. It underwent an eight year development programme and is a comprehensive bridging system covering tank-launched or close support bridges (mechanised bridges) and line of communication or dry support bridges, together with dedicated logistic support vehicles. (See also entry under *Tactical (Non-floating) and Line of Communication Bridges*.)

In November 1993, the UK Ministry of Defence awarded a contract worth £140 million for the supply of production quantities of the Bridging for the 1990s system for the Royal Engineers.

BR90 is divided into four types of bridge: General Support Bridge; Close Support Bridge; Long Span Bridge; and Two Span Bridge.

The Close Support Bridges are launched by the Chieftain AVLB and supported by the Unipower 8 × 8 Tank Bridge Transporter. The latter will also be used for transporting, handling and launching dry support bridges.

In June 1996, Giat Industries and Vickers Defence Systems signed a co-operation agreement for the development and manufacture of Universal Bridge-Launching Equipment (UBLE) to be placed on Leclerc or Challenger tank chassis. The bridge involved is the 26 m No 10 tank bridge, already part of the British Army's BR90 programme.

Description
All the BR90 bridges are built from a range of seven

Chieftain AVLB in process of laying the first of a pair of No 12 13.5 m close support bridges ***1997***

A Unipower 8 × 8 Tank Bridge Transporter (TBT) loading a No 10 26 m close support bridge ***1997***

modular panels of advanced aluminium alloy fabrication. These are interchangeable through the various bridge types, to form two interconnecting trackways with a 4 m overall bridge width and a 1 m girder depth.

There are three basic Tank-Launched Bridges (also known as Close Support or Assault Bridges) the No 10, No 11 and No 12.

The No 10 bridge is a single-fold scissoring bridge 26 m long and capable of spanning 24.5 m. A shorter bridge length can be achieved by the removal of two 4 m panels, reducing to a 22 m bridge. It can also be fitted with an Assault Trestle to provide a 21 m trestle bridge. The No 10 bridge is constructed from 8 m ramp panels, 4 m panels and 2 m hinge panels. The folding mechanism uses scissoring parts which eliminate the need for hydraulics on the bridge, thus allowing the bridge to be launched on sites where longitudinal slopes would defeat horizontally launched bridges. The No 26 bridge is part of the Giat/Vickers Universal Bridge Launching Equipment (UBLE).

The No 11 bridge is built from four 8 m ramp panels and is a 16 m long, 14.5 m span up-and-over launched bridge.

The No 12, shortest of the Tank-Launched Bridges, is 13.5 m long and capable of spanning armour 12 m. Its girder depth of 0.71 m enables two bridges to be conveyed and launched by the Chieftain AVLB. The two No 12 bridges with a unique launching system allow the AVLB to construct two bridges without having to be resupplied, for the crossing of two short gaps, or in combination to cross a wider single gap, all without crew exposure.

Full decks and kerbs can be added to all BR90 bridges to enable the full range of military in-service vehicles to cross.

A small modification kit enables the existing Chieftain AVLB to convey and launch the new range of bridges from under armour, while still allowing existing in-service bridges to be carried and launched. Alternatively, any suitable armoured tracked or wheeled vehicle could be modified as a suitable launch platform.

All the bridges are designed and tested in accordance with the UK/US/GE Trilateral Design and Test Code and are classified MLC 70 for both wheeled and tracked vehicles. They can all be crossed by fully laden tank transporters at MLC 105. Launching without crew exposure is possible with all bridges in 3 minutes and they can all be recovered in under 5 minutes from either end (except those fitted with a trestle).

To increase the length capability, combination bridges may be built using two or three of the above bridges, the first bridge being launched from the home bank with the far end supported on the base of the gap, a floating pier, or a trestle. Gaps up to a nominal 60 m span and 5 m depth can be bridged in this way. Combination bridges are classified as MLC 70 for tracked vehicles.

A wheeled Tank Bridge Transporter (TBT) a suitably modified Unipower M Series 8 × 8 improved medium mobility vehicle with an extended wheelbase, forms part of the VMB system, its primary use being the rapid resupply of the AVLB. The TBT can carry all variants of the tank-launched bridges and is equipped with mechanisms to load and offload them. An excellent cross-country performance and flexibility allow it to travel virtually anywhere tracked vehicles can go. The TBT has a high degree of commonality with the vehicles used to launch and transport Dry Support Vehicles (see also entry under *Tactical (Non-floating) and Line of Communication Bridges*).

Chieftain AVLB in process of laying a No 10 26 m close support scissors bridge over an already emplaced bridge ***1996***

Specifications

No 10 Tank Bridge
Length overall: 26 m
Ramp length: 6 m
Width overall: 4 m
Depth of section: 1 m
Max span: 24.5 m
Weight: 13,000 kg
Panels: 4 × 8 m ramp panels, 4 × 4 m panels, 2 × 2 m hinge panels

No 10 Trestle Bridge
Length overall: 21 m
Ramp length: 6 m
Width overall: 4 m
Depth of section: 1 m
Max span: 18 m usable for combination bridge
Weight: 13,000 kg with trestles
Panels: 2 × 8 m ramp panels, 4 × 4 m panels, 2 × 2 m hinge panels, 2 trestle adaptor panels, 2 assault trestles

No 11 Tank Bridge
Length overall: 16 m
Ramp length: 6 m
Width overall: 4 m
Depth of section: 1 m
Max span: 14.5 m
Weight: 7,000 kg
Panels: 4 × 8 m ramp panels

No 12 Tank Bridge
Length overall: 13.5 m
Ramp length: 5.25 m
Width overall: 4 m
Depth of section: 0.71 m
Max span: 12 m
Weight: 5,700 kg
Panels: 2 × 13.5 m trackways

Combination Bridge
Bridges: No 10 and trestle bridge, No 11 and No 12 bridges

BR90 No 10 close support bridge being emplaced across a water gap

Quantity: 2 or 3 bridges
Max span: 60 m
Max gap depth: 5 m

Status

In production. In service with the British Army.

Manufacturer

Vickers Bridging, PO Box 37, Spring Road, Ettingshall, Wolverhampton WV4 6YN, UK.
Tel: +44 1902 405050.
Fax: +44 1902 491459.

UPDATED

Vickers Defence Systems Chieftain AVLB

Development

The Chieftain AVLB (FV4205) was developed as the replacement for the Centurion AVLB (FV4002) and the Centurion ARK (FV4016). Development of the Chieftain bridgelayer began in 1962 but, owing to some necessary redesign, the first production vehicles were not delivered to the Army until 1974. Design work on the AVLB was carried out by the then Military Vehicles and Engineering Establishment, AP Precision Hydraulics and the Hydraulic Controls Department of Tubes (Birmingham).

The Chieftain AVLB is used by only one regiment of the British Army: 32 Armoured Engineer Regiment, Royal Engineers. Within the British Army each Chieftain AVLB is part of a system that includes the bridgelayer carrying a No 8 tank bridge and another No 9 tank bridge carried on a Crusader tractor and semi-trailer. Under certain circumstances the No 9 bridge can be laid directly from the semi-trailer.

The production total of Chieftain AVLBs for the British Army was 37, with a further 14 produced for Iran.

In order to increase the number of Chieftain AVLBs available to the British Army, 11 Mark 1/4 Chieftain MBTs were converted by Vickers Defence Systems of Newcastle to the bridgelayer role. These are known as the Mark 6 and are some 3,000 kg heavier than the earlier AVLBs. They have a more powerful hydraulic pump and system plus other modifications such as a toe plate for fitting mine ploughs and an intercom system built into the bridge boom. The first Mark 6 AVLB was completed in 1985 with first deliveries later the same year. Deliveries were completed by the end of 1986.

It has been proposed that the Chieftain AVLB could be replaced in British Army service by Challenger 1 MBT chassis converted to the armoured bridgelayer role. If this happens the eventual Challenger AVLBs will have full BR90 capabilities (see previous entry).

Description

The hull of the AVLB is similar to that of the basic Chieftain MBT. The driver is seated at the front of the hull and is provided with a single-piece hatch cover and a periscope. The commander and radio operator are behind the driver. The engine and transmission are at the rear of the hull. The suspension is of the Horstmann type and consists of three bogies per side, each bogie with two sets of roadwheels and a set of three horizontal springs. The first and last roadwheel stations have a hydraulic shock-absorber. The drive sprocket is at the rear and the idler at the front, with three track-return rollers. The top half of the track is covered by armoured skirts which are removable for maintenance. The Chieftain AVLB has an NBC system.

The Chieftain AVLB can utilise either a No 8 or a No 9 tank bridge, the latter manufactured by Laird.

The No 8 bridge is carried folded and is launched over the front of the vehicle. The hull-mounted hydraulic pump, which is driven from a PTO on the main engine, operates five cylinders arranged to launch the bridge in three manually sequenced stages. Once the clamps have been removed the launching can begin. The Stage 1 cylinders pivot the folded bridge about the forward part of the glacis casting. This operation is continued until the launching pad meets the ground or Stage 1 cylinders reach the end of their stroke, when the folded bridge will be at an angle of about 30° to the horizontal. The Stage 2 cylinders then tilt the launching pad on the ground until approximately level. At this point the bridge is partly scissored by the rods connecting the central quadrant to the launching frame and is approximately vertical. Once the Stage 2 cylinders have reached the end of their strokes, the single Stage 3 cylinder is actuated, causing the bridge to scissor further until the far end reaches the required bank height. At this point the rods slacken and on further lowering the bridge becomes a rigid connection at its centre point. The rods are then free and can be detached using the remotely operated release mechanism, thus separating the rods from the launching structure. The tank then reverses and the launching structure is removed from the bridge. The tank retracts all its cylinders and then either moves to the rear to pick up another bridge or crosses the bridge and picks up the bridge from the other end. The launch normally takes between 3 to 5 minutes, with recovery of the bridge taking about 10 minutes.

The bridge girders and launching structure are made of high-strength nickel-alloy maraging steel developed by the International Nickel Company. The deck and kerbing are of weldable aluminium alloy. The bridge is made up of two tracks, each 1.62 m wide, with a 0.76 m centre gap between. Each track is divided into four parts, two toe pieces each 4.55 m long and two centre pieces each 7.6 m long. A hinge is provided at each joint. The bridge has a maximum capacity of Class 60; Land Rovers and similar small vehicles can cross on a single track, allowing two-way traffic with vehicles of this size. By hinging the 4.55 m long toes of the bridge and dropping them at will, it is possible to cater for a range of bank conditions. A bank-sensing device is incorporated in the ramp sections.

In order to permit crossing gaps that are wider than the standard No 8 bridge, an extension trestle has been introduced. This trestle is fitted in place of the far bank end ramps on a standard No 8 bridge in such a fashion that the trestle is allowed to unfold and swing downwards as the bridge is fully extended. The trestle then supports the No 8 bridge to allow another AVLB to cross, in order to lay a further bridge.

The No 9 bridge consists of two interconnected trackways 13.411 m long and 1.62 m wide, with a gap between them of 0.76 m. Total effective roadway width is 4.01 m. The bridge takes between 3 and 5 minutes to launch and can be recovered from either end. It is launched in three stages: Stage 1 raises the bridge from the horizontal to an angle of 45°; Stage 2 raises it to the vertical; and Stage 3 lowers it to the horizontal. The launching arm is then disengaged, the vehicle is reversed away and the launching structure retracted.

Each Chieftain bridgelayer normally has one No 8 and one No 9 tank bridge, one carried on the vehicle and the other on a specially adapted Scammell Crusader prime mover towing a semi-trailer.

Following a small modification, Chieftain AVLBs can also carry No 10, No 11 and No 12 tank-launched bridges. These are part of the BR90 modular bridging system produced by Vickers Bridging (see previous entry).

Some Chieftain bridgelayers, in addition to the Mark 6, have been converted to carry mine ploughs. Vehicles operating in the Gulf region during Desert Shield/Storm were fitted with extra armour and other protection measures.

Specifications

With No 8 bridge
Crew: 3
Weight: 53,300 kg
Length: 13.741 m
Width: 4.165 m
Height: 3.923 m
Ground clearance: 0.5 m
Track width: 610 mm
Length of track on ground: 4.8 m
Ground pressure: 0.9 kg/cm^2
Max speed: (road) 48 km/h
Range: 400 km
Fuel capacity: 950 litres
Fording: 1.066 m
Gradient: 60%
Vertical obstacle: 0.9 m
Trench: 3 m
Engine: Leyland L60, 2-stroke, compression ignition, 12-cylinder vertically opposed multifuel developing 730 hp at 2,100 rpm
Transmission: TN12 with 6 forward and 2 reverse gears, plus emergency mechanical selection for second gear forward and low reverse
Electrical system: 28.5 V (24 V nominal)
Batteries: 4 × 12 V, 200 Ah
Armament: 1 or 2 × 7.62 mm L7A2 GPMGs

No 8 Tank Bridge
Overall length: 24.384 m
Centre-section length: 7.62 m
Ramp end length: 4.572 m
Overall width: 4.165 m
Roadway width: 4.012 m
Centre gap: 0.762 m
Depth of centre-section: 0.914 m
Max span:
(firm banks) 22.86 m
(soft banks) 22.25 m
Weight: 12,200 kg

No 9 Tank Bridge
Overall length: 13.411 m
Overall depth: 0.914 m
Overall width: 4.165 m
Max clear span: (firm bank) 12.192 m
Weight: 9,144 kg

Status

In service with the British Army and the Iranian Army.

Manufacturer

Chassis: Royal Ordnance Leeds built all production AVLB chassis but in late 1986 this facility was taken over by Vickers Defence Systems.
Enquiries to: Vickers Defence Systems, Scotswood Road, Newcastle upon Tyne NE99 1BX, UK.
Tel: +44 191 273 8888. Telex: 53104.
Fax: +44 191 273 2324.

UPDATED

Chieftain AVLB carrying a No 9 bridge (T J Gander) **1996**

Chieftain AVLB with bridgelaying arm lowered but without a bridge (T J Gander) **1996**

Vickers Defence Systems Armoured Bridgelayer

Development

Preliminary design work on the Vickers Defence Systems Armoured Bridgelayer (VAB) started during the late 1970s, to meet the requirement for a family of support vehicles for the Vickers MBT which was being ordered in increasing numbers. The first order for six VABs was placed by Nigeria in 1981. A further six were ordered in 1985.

Description

The hull, automotive and running gear components of the VAB are based on the Vickers MBT with the hull structure divided into three main compartments. There is accommodation for a crew of three: the driver, radio operator and commander. The forward compartment is divided into two sections by a panel on the right of the driver who is seated on the centreline of the compartment. The panel provides a mounting face for some vehicle control units and behind it are the batteries and a stowage space. The vehicle controls and the bridge control units are located within the driver's compartment. Three wide-angle periscopes are provided for driving with the hatch closed. The driver's seat has provision for vertical and fore and aft adjustments. The maximum possible opening is provided between the centre and forward compartments. The driver, by dropping his seat backrest, can enter or leave his position through the centre compartment, thus making a change of crew possible without opening up the vehicle.

The centre or crew compartment has seating for the commander and radio operator. The commander's seat is positioned directly under a rotating cupola fitted to the centre of the roof plate. Behind the commander and to his right is the radio operator and radio sets. Beneath the radio equipment are two batteries and space for personnel kit in the left-hand pannier space. The crew ventilation system is in the right-hand pannier space. To the centre and left rear of the commander's seat are hydraulic pumps, filters, tank, valves and other components for operating the bridge launch equipment.

The rear compartment accommodates the 800E power pack and transmission unit. In common with the other Vickers MBT series the VAB may be fitted with a Perkins CV12 800E or a General Motors 12V-71T diesel. Panniers above each track contain bag-type fuel tanks. Lifting and towing eyes are welded to the hull front and rear.

Both batches of VABs for Nigeria are fitted with a conventional tank bridge. This bridge is 13.41 m long with a military load classification of 60/70. The bridge launching equipment is hydraulically operated with the power provided by a pump driven from a power take-off from the main engine. The launching operation is in four stages:

1) bridge securing and clamping devices released
2) bridge launched hydraulically
3) release of bridge from launching structure
4) vehicle reverses to withdraw clear of bridge.

To recover the bridge after use, the bridgelayer vehicle is aligned with the bridge so that the launching arm engages with the bridge lifting brackets. The hydraulic lifting mechanism then recovers the complete bridge onto the roof of the vehicle for reuse.

Vickers Armoured Bridgelayer (VAB)

Specifications

Crew: 3
Weight: (with bridge) 43,910 kg
Length: (with bridge) 13.7 m
Width: (with bridge) 4.16 m
Height: (with bridge) 3.25 m
Ground clearance: 0.432 m
Max range: 600 km
Fuel capacity: 1,000 litres
Fording: 1.1 m
Gradient: 60%
Side slope: 30%
Vertical obstacle: 0.914 m
Trench: 3 m
Engine: Perkins CV12 800E V-12 diesel developing 800 bhp at 2,300 rpm
or Detroit Diesel 12V-71T V-12 diesel developing 720 bhp at 2,500 rpm
Transmission: TN12 automatic
Clutch: centrifugal
Steering: Merritt regenerative
Electrical system: 24 V
Armament: 1 × 7.62 mm MG

Bridge
Length: (overall) 13.4 m
Width: 4.16 m
Classification: MLC 60/70

Status

In service with the Nigerian Army (17).

Manufacturer

Vickers Defence Systems, Scotswood Road, Newcastle upon Tyne NE99 1BX, UK.
Tel: +44 191 273 8888. Telex: 53104.
Fax: +44 191 273 2324.

VERIFIED

UNITED STATES OF AMERICA

General Dynamics/MAN GHH Wolverine Heavy Assault Bridge (HAB)

Development

The General Dynamics/MAN GHH Wolverine Heavy Assault Bridge (HAB), is the outcome of concept evaluation work by the US Army following on from the cancellation of the 'Bridging in the 1980s' programme in 1981. The result of the evaluation was that a new heavy assault bridge with a Class 70 load classification would be required. A minimum gap-crossing capability of 30 m was initially specified. A request for quotations was issued to industry and from the resultant designs the US Army selected the submission from BMY (now part of United Defense LP) with Israel Military Industries (IMI), Haifa Division, as a major subcontractor. BMY was awarded a contract for the design and manufacture of one prototype HAB system in April 1983 and design approval was given in November 1984. Additional systems were added to the contract later; three were produced and system testing was completed on one of them.

General Dynamics/MAN Wolverine Heavy Assault Bridge (HAB) on converted M1 Abrams chassis ***1996***

Testing by the US government commenced in late 1988 and was scheduled to continue until July 1990. However, during 1989 the US Army requested that all work on the project be stopped.

During late 1989 a US Army requirement for a new version of the HAB was issued. This HAB was to have a total length of 26 m and be able to span a gap of up to 24 m. The bridge would include aluminium components in place of the original composite materials. The bridge was to be so constructed that infantry would be able to walk across each deck.

BMY Combat Systems (now part of United Defense LP) was awarded a contract to design and construct a new 26 m MLC 70 HAB, with IMI as a major subcontractor. The HAB, known as the HAB-26, was carried and launched from a M1 Abrams MBT chassis. Two launcher systems and bridges were completed and tested.

In March 1990, General Dynamics, Land Systems Division, was also awarded a contract for a complete

M1-based HAB. This HAB version used the German MAN GHH LEGUAN bridge system (see separate entry in this section) with its launching system modified for the M1 chassis. MAN GHH was a subcontractor to General Dynamics, Land Systems Division.

The General Dynamics/MAN GHH HAB contract involved two prototype demonstrators for the Fort Belvoir Research, Development and Engineering Center and was worth $12.459 million. The contract called for the complete assembly and testing sequence to be completed by the end of October 1992.

In August 1992 it was announced that a third system was to be evaluated in connection with the HAB programme. Southwest Mobile Systems of St Louis, Missouri, was awarded a $5,481,383 firm fixed-price contract, to deliver and support two Chieftain AVLBs and two No 10 tank bridges (see entry under United Kingdom in this section for details of these equipments).

The US Army conducted a comparative testing of the three systems from late 1992 and throughout 1993. In February 1994 it was announced that the US Army Tank Automotive Command had awarded General Dynamics Land Systems a $26.05 million, 39 month, engineering, manufacturing and development Phase II contract for the HAB. Up to six pilot vehicles plus two prototypes were planned to take part in these two phases. Low rate initial production will begin in 1998. The US Army plans to procure 599 HAB systems with the First Unit Equipped (FUE) scheduled for the year 2000.

Description

The General Dynamics/MAN GHH Wolverine HAB uses the M1 Abrams chassis retaining over 95 per cent compatibility between the HAB chassis and the M1 Abrams MBT chassis. It is planned that production vehicles will incorporate some aspects of the Abrams System Enhancement Package (SEP).

The two-man crew is provided with NBC protection, while fire protection is provided by an automatic three-zone extinguisher system.

The LEGUAN bridge involved has a length of 26 m and a load classification of MLC 70/96, allowing gaps up to 24 m wide to be crossed. Designed for a 5,000 MLC 70 crossings requirement, the bridge can be launched in less than 5 minutes and recovered in under 10 minutes.

The LEGUAN bridge consists of four interchangeable quarter sections with a flat ramp design, each 13 m long, 2 m wide and weighing 2,500 kg. Bolt-on crossmembers join the bridge sections and when travelling the bridge halves lie one on top of the other. A laying arm moves the bridge longitudinally by means of a roller guide and a geared drive.

The top bridge half is taken up by an arm at the ront end and is supported by an auxiliary arm at the rear. The bridge can be laid with a traverse and longitudinal inclination of up to 10 per cent. Laying is carried out automatically with the crew monitoring the sequence from within the protection of the vehicle. No preparation of the crossing site is normally required.

The laying sequence commences with the Wolverine vehicle approaching the bank until the spade stabilisation system is close to the edge. The spade system is supplied by Caterpillar of Peoria, Illinois. The bridge halves are then coupled automatically by lowering the coupling supporting arm.

The next step is for the bridge to be fully advanced, at which point it is clamped in the laying arm by the roller carriage. At this stage the ramp tip is placed on the far bank. This removes the load from the rollers and the laying arm can then be removed to lower the bridge. After the final stage the stabilising blade is retracted. The bridge is then ready for use. The entire laying process takes less than five minutes. Recovering the bridge is carried out in the reverse order and can be carried out from either bank. Recovery time is less than 10 minutes.

The bridge is constructed of standard commercial aluminium alloy, grade Al Zn 4.5 Mg 1 and is designed to meet US Standards for the bridge life.

Power for the LEGUAN hydraulics is provided by a 65 hp auxiliary power unit designed by Stewart and Stevenson of Houston, Texas.

Specifications

Crew: 2
Bridge length: 26 m
Load capacity: MLC 70/96
Weight: approx 68,700 kg
Length:
(overall, with bridge) 13.4 m
(hull) 7.636 m
Width: (with bridge) 4 m
Height:
(with bridge) 3.95 m
(without bridge) 2.906 m
Ground clearance: 0.482 m
Length of track on ground: 2.845 m
Track width: 635 mm
Max speed: (governed) 72.4 km/h
Range: 418 km
Fording: (without kit) 1.219 m
Gradient: 60%
Vertical obstacle: 1.07 m
Trench: 2.69 m
Engine: Lycoming Textron AGT 1500 gas turbine developing 1,500 hp at 30,000 rpm
Transmission: Allison Transmission Division of General Motors X-1100-3B automatic with 4 forward and 2 reverse gears
Electrical system: 24 V

Status

Prototypes under test by US Army.

Manufacturers

General Dynamics Land Systems Division, 38500 Mound Road, Sterling Heights, Michigan 48310-3200, USA.
Tel: +1 810 825 4000.
MAN Gutehoffnungshütte AG, Ginsheimer Strasse 1, D-65462 Ginsheim-Gustavsburg, Germany.
Tel: +49 6134 55292.
Fax: +49 6134 55383.

UPDATED

M48 and M60 Armoured Vehicle-Launched Bridges (AVLB)

Development

In the 1950s the standard AVLB of the US Army was the M48A2 AVLB. This was an M48 MBT chassis, fitted with a scissors bridge, designed by the US Army Engineer Research and Development Laboratories (now the Belvoir Research, Development and Engineering Center) at Fort Belvoir, Virginia. Production of the M48 was completed in 1959 and from 1963 the chassis of the M60 MBT was used. The chassis of these two vehicles is almost identical, the major difference being the type of engine. The M48, M48A1 and M48A2 were all powered by a 12-cylinder petrol engine which developed between 810 and 825 hp at 2,800 rpm, while the M48A3 was powered by a 12-cylinder diesel (AVDS-1790-2A) engine which developed 750 hp at 2,400 rpm, giving the vehicle an increased operational range. In FY78 the US Army requested $20.9 million to convert 136 M48A1 and M48A2 tanks into M48A5 AVLBs. (These vehicles now have M60 AVLB launcher components and the AVDS-1790-2D engine.)

NAPCO Inc of Hopkins, Minnesota, can provide complete AVLB system support including training, technical assistance, special tools packages, spare parts and technical manuals.

Description

The basic chassis of the M60 AVLB is almost identical to that of the M60 MBT, except that the driver is located aft of his MBT location and accommodation is made in the same area for the commander, since the MBT turret is removed for the AVLB configuration. The engine and transmission are at the rear. Early M48 AVLBs had two turrets, each with a 0.50/12.7 mm Browning M2 HB machine gun, but they were later removed and replaced by two conventional hatch covers. The crew of two consists of the driver and commander.

M48 AVLB with bridge about to reach fully laid position

The suspension is of the torsion bar type and consists of six roadwheels with the idler at the front and the drive sprocket at the rear, with three track-return rollers. Hydraulic shock-absorbers are fitted at the first, second and sixth roadwheel stations.

The bridge weighs 13,380 kg and is made primarily of aluminium. It is carried folded and launched over the front of the vehicle hydraulically as follows: the AVLB is driven up to the obstacle and halted, the bridge is raised hydraulically into the vertical, unfolded and lowered into place and the launcher detached. The bridge, which weighs 13,290 kg complete, takes 3 minutes to launch and can be recovered from either end. Recovery time is between 10 and 60 minutes depending on the ground conditions. The bridge has an overall length of 19.202 m and can span a gap of up to 18.288 m. Overall bridge width is 3.96 m, 3.81 m of which is usable roadway, with each trackway 1.486 m wide. Its maximum capacity is 54,431 kg.

Variant

M60 Razorback
See following entry.

Specifications

M60 AVLB
Crew: 2
Weight:
(with bridge) 55,205 kg
(without bridge) 41,730 kg
Length:
(with bridge) 11.28 m
(chassis) 8.648 m
Width:
(with bridge) 4.002 m
(chassis) 3.64 m

Height:
(with bridge) 3.9 m
(without bridge) 3.04 m
Ground clearance: 0.36 m
Track: 2.921 m
Track width: 711 mm
Length of track on ground: 4.235 m
Ground pressure: 0.92 kg/cm²
Max speed: (road) 48.28 km/h
Range: 500 km
Fuel capacity: 1,420 litres
Fording: 1.219 m
Gradient: 30%
Vertical obstacle: 0.914 m
Trench: 2.59 m
Engine: Continental AVDS-1790-2A or AVDS-1790-2D 12-cylinder diesel developing 750 bhp at 2,400 rpm
Transmission: Allison CD-850-6 with 2 forward and 1 reverse ranges
Electrical system: 24 V
Batteries: 6 × 12 V, 100 Ah
Armament: nil
Armour
hull front: 101-120 mm
hull sides front: 76 mm
hull sides and rear: 51 mm
hull top: 57 mm
hull floor: 12.7-63 mm
hull rear: 44 mm

Status
Current users are known to include Germany, Israel, Pakistan (M48), Singapore (12 M60), Spain, Taiwan, and the US Army and Marine Corps.

Manufacturers
General Dynamics Land Systems Division, PO Box 2074, Warren, Michigan 48090-2074, USA.
Tel: +1 313 825 4000.
Systems and Electronics Inc (SEI), 201 Evans Lane, PO Box 5639, St Louis, Missouri 63121-1126, USA.
Tel: +1 314 553 4911.
Fax: +1 314 553 4950.
Marvin Land Systems Inc, 12637 Beatrice Street, PO Box 66999, Los Angeles, California 90066-0999, USA.
Tel: +1 310 306 2800.
Fax: +1 310 306 5108.

UPDATED

M60 Razorback Armoured Vehicle-Launched Bridge (AVLB)

Development
The M60 Razorback programme was begun under a Co-operative Research and Development Agreement (CRDA) between Tank Automotive and Armaments Command (TACOM) and a team of contractors. The objective was to provide M60 series vehicles with a speed and agility performance comparable to that of the M1 Abrams MBT series. The programme emerged as the result of an examination of US Marine Corps armoured vehicle requirements when it emerged that they would be unlikely to obtain the Heavy Assault Bridge (HAB - see separate entry). It was then decided to utilise a number of newly available vehicle components, to provide an upgrade package for existing M60 AVLB systems originally known as the AVLB-70 but later called M60 Razorback.

Razorback team participants include General Dynamics Land Systems (GDLS) and Allison Transmission. Conversion work on the prototype vehicle commenced during late 1993 and was carried out at the Anniston Army Depot. The prototype was rolled out during the Spring of 1994.

GDLS provided an enhanced power AVDS-1790-9A diesel engine delivering 1,200 hp and an externally mounted bolt-on hydropneumatic suspension system. Allison Transmission installed its X-1100-5 transmission, M1 final drives and the upgrade package with the integrating hardware in the driver's compartment.

The MLC 70 bridge involved with Razorback will lack the length of the HAB bridge. It is expected to feature modifications to the standard M48/M60 AVLB bridge and will result from a separately funded bridge development programme being conducted by the US Army Tank Automotive Command. Any Razorback package is also likely to incorporate a number of additional subsystem upgrades that have been developed for M60-based AVLBs, including modified hardware designed to enhance maintenance on the bridge hydraulics and electronic subsystems.

Prototype of M60 Razorback ***1996***

Trials with the Razorback prototype have demonstrated the forecast increases in speed and agility to the extent that it has been proposed that the automotive features could form part of an overall M60 series tank and other variant upgrade package.

Specifications
Provisional
Max speed:
(cross-country) 62.7 km/h
(3% gradient) 35.4 km/h
(10% gradient) 19.3 km/h
Acceleration: 0-32 km/h in 8 s

Status
Prototype.

Contractors
General Dynamics Land Systems, 38500 Mound Road. Sterling Heights, Michigan 48310-3200, USA.
Tel: +1 810 825 4000.
Allison Transmission, PO Box 894-L7, Indianapolis, Indiana 46206-0894, USA.
Tel: +1 317 242 2715.

UPDATED

Mobile Assault Bridge/Ferry (MAB)

Development
The Mobile Assault Bridge (MAB) was developed from 1959 by the US Army Engineer Research and Development Laboratories at Fort Belvoir, Virginia (now the Belvoir Research, Development and Engineering Center). The first production batch had hulls of riveted construction. Totals of 32 end bays and 66 interior bays were delivered to the US Army between April 1963 and December 1967. The basic vehicle was manufactured by the FMC Corporation of San Jose, California, with the interior and end bridge bays being built by the Consolidated Diesel Electric Corporation of Schenectady, New York.

Mobile Assault Bridge in travelling configuration

After the initial batch of MABs was built, a product improvement programme was initiated which was completed in September 1970. This version has an all-welded hull and an improved electrical and hydraulic system. In total 220 MABs of the improved design were built for the US Army between 1973 and 1976.

The MAB is issued on the scale of 24 units for each division and non-divisional bridge units. The total requirement, for 298 units to equip NATO earmarked units, was completed with FY74 funding. It was for 44 MAB end bays at a cost of $1.5 million, 88 interior bays at $1.6 million and 132 MAB transporters at $18.4 million.

Description
The MAB consists of a 4 × 4 transporter which is fitted with either an interior bay or an end bay superstructure. These two types of superstructure can be interchanged, with the aid of a crane, in approximately 15 minutes. The hull is of all-aluminium construction which is either welded or riveted together. The sides and deck are 3.175 mm thick while the bottom, bow and stern are 4.7 mm thick. Reinforcing ribs provide additional strength to the sides and bottom of the transporter.

The crew consists of three: a driver who is also the bridge crew chief, an assistant driver who is also the bridge pilot, and a crewman. The three-man watertight cab is mounted at the front of the hull and can be removed for transportation if required. Entrance to the cab is by two hatches in the roof and rear of the cab.

The air compressor provides power for the windscreen wiper, tyre inflation and brake system. The MAB has two hydraulic systems. The high-pressure system provides power for positioning the superstructure, raising and lowering the wheels and operating the capstans. The low-pressure system operates the power steering (all four wheels can be steered), marine drive functions and the air blower for the wheel-well pressurisation. Each MAB has four electrically operated bilge pumps, which are located at the four corners of the engine compartment and have a combined capacity of 509 litres/min.

Two hydraulically operated capstans are provided, one on the forward and one on the aft deck, to pull units together for pinning during superstructure connection. The capstans have a capacity of 2,268 kg at 15.24 m/min. An anchor with 60.96 m of line is stowed on the top of the cab.

The MAB is propelled in water by a single 711 mm propeller which is mounted at the rear of the hull. This has a maximum thrust of 2,272 kg and when in the travelling position is swung up horizontally. The propeller system can be rotated 360° for steering and reversing. Once in the water, the wheels are raised and the propeller lowered into position. Draught when loaded is 1.066 m.

The interior bay superstructure is constructed of welded steel girders and an extruded aluminium decking. It is carried parallel to the axis of the hull for road travel and is rotated 90° for use during water operations. When extended it has an effective length of 7.924 m and is 3.657 m wide. After the 228 mm kerbs are raised, the bridge deck has an overall width of 4.224 m.

The end bay superstructure is constructed of aluminium girders with aluminium decking and consists of two sections hinged together. The end bay superstructure section is 7.315 m long and 3.657 m wide with the kerbs down. The combined ramp and deck section is 10.972 m long. The superstructure can be rigged before entering the water and can be rotated either to the left or the right. The ramp can be articulated to reach different bank heights down to and including 0.914 m below water level.

The MAB can be used for two basic roles, either as a ferry unit, or as a floating bridge. Ferries and bridges are assembled by joining successive MAB units. In ferry assembly an end bay and an interior bay are jointed to form a double bay, two of which are then joined to form a four-unit ferry which can be used to carry a 60 tonne load at 12.875 km/h.

Bridges are constructed by connecting units and then lowering the ramps from the end bays to the shore. The combined crews can assemble a four-unit ferry in about 6 minutes, and dismantle it in about 4 minutes.

In a stream, bridges can be assembled at the rate of 4.572 to 6.096 m/min and dismantled at the rate of 4.572 to 7.62 m/min. When assembled, the bridge has a maximum capacity of 60 tonnes with a gap of at least 30.48 m between MBTs and other heavy vehicles crossing.

Specifications

Crew: 3
Configuration: 4 × 4
Weight:
(basic vehicle) 15,386 kg
(vehicle with interior bay) 21,850 kg
(vehicle with end bay) 24,599 kg
Length: 13.03 m
Width:
(travelling) 3.657 m
(over wheels) 2.844 m
Height:
(travelling, interior bay) 3.3 m
(travelling, end bay) 3.568 m
Ground clearance: 0.457 m
Track: 2.336 m
Wheelbase: 5.943 m
Angle of approach/departure: 22°/20°
Max speed:
(road) 64 km/h
(water) 16.126 km/h
Range:
(land) 596 km
(water) 185 km
Fuel capacity: 380 litres
Gradient: 60%
Side slope: 20%
Fording: amphibious
Engine: 336 hp diesel
Electrical system: 24 V
Batteries: 4 × 12 V

Status

Production complete. In service with Belgium and Israel.

Manufacturer

ConDiesel Mobile Equipment Division, CONDEC Corporation, 84 Progress Lane, Waterbury, Connecticut 06705, USA.

VERIFIED

Light Assault Bridge (LAB)

Development

In early 1983 the US Army Mobility Equipment Research and Development Command (MERADCOM, now the Belvoir Research, Development and Engineering Center, Fort Belvoir, Virginia) awarded Foster-Miller Associates of Waltham, Massachusetts, a contract for the design of a new bridge for the US Army's Light Infantry Divisions. The MLC 30 bridge is to be used during combat assaults by and for logistics support of light combat forces. Three prototypes were constructed in 1987 by Southwest Mobile Systems Corporation (now SEI) for US Army testing.

Alvis Vehicles of the UK have entered an agreement with the ESCO Electronics Group (of which SEI is a subsidiary), whereby Alvis will offer the LAB on three possible carriers; the Unipower M-series 8 × 8 truck, a standard trailer, and/or a flatbed version of the Alvis Stormer APC. A small number of the Stormer-carried bridges are being produced for Indonesia. The MLC 30 bridges involved with Stormer have only two ramp sections, each 7.62 m long.

Description

The Light Assault Bridge (LAB) is a self-contained three-section bridge (one centre-section and two ramp sections) mounted on its own launcher. It is capable of spanning a 23 m gap with a military load classification of MLC 30. The launcher/trailer is an integral part of the bridge system and is used during the launch, recovery and transport of the bridge.

The LAB was designed and constructed using aircraft construction techniques to combine lightness with strength. It can be towed by a variety of vehicles including most tracked vehicles and tactical trucks over 5 tonnes capacity. It is normally air-transportable in a C-141 or larger transport aircraft, although it is possible for the LAB to be carried in a C-130.

The LAB uses its own diesel/hydraulic power unit to raise and lower the bridge. Launching takes up to 5 minutes and recovery up to 15 minutes, using a crew of two. Control of launch and recovery can be carried out from a position on the launch trailer or using a remote-control unit. Launch and recovery can be carried out using either end of the bridge. The trailer is fitted with run-flat tyres.

Launching a Light Assault Bridge (LAB)

Specifications

Load capacity: MLC 70
Weight: 8,160 kg
Lengths:
(total, overall) 24.08 m
(crossing) 23 m
(folded) 9.8 m
(ramps) 7.72 m
(centre-section) 8.84 m
Width of bridge:
(operational) 3.37 m
(transport) 2.8 m
Track width: 1.2 m

Status

US Army technical evaluation completed. Further development or production is not currently funded. 15 m variant in production for Indonesia.

Manufacturer

Systems & Electronics Inc (SEI), 201 Evans Lane, PO Box 5639, St Louis, Missouri 63121-1126, USA.
Tel: +1 314 553 4950.
Fax: +1 314 553 4909.

UPDATED

YUGOSLAVIA (SERBIA AND MONTENEGRO)

Truck-mounted scissors treadway bridge

Description

This truck-mounted scissors bridge consists of three 13 m spans carried and launched from the rear of a modified FAP 2220BDS (6 × 4) truck.

The system is laid in a similar manner to the TMM truck-mounted scissors treadway bridge and has similar capabilities. The trestle legs can be extended to a maximum of 4 m, whereas the TMM's trestle legs can be extended to only 3 m, enabling the bridge to be used for deeper crossings.

The treadway bridge is laid as follows: before launching the treadway, the trestle legs are adjusted to the correct height so that the roadway is level when the bridge is in position; the truck then backs up to the gap; and the treadway is raised to the vertical position. The span is then straightened and lowered into position and at the same time the trestle legs swing into place. Once this is complete, the truck drives off and the procedure is repeated until the bridge is complete. Each individual span takes 5 to 6 minutes to lay in position and can be recovered from either end. Recovery takes about the same time as launching. A complete three-span bridge will span a gap up to 36 m wide.

Truck-mounted scissors treadway bridge in travelling position on FAP 2220BDS (6 ×4) truck

Specifications

Vehicle with bridge
(full technical characteristics of the FAP 2220BDS will be found in the *Trucks* section)
Cab seating: 1 + 1
Configuration: 6 × 4
Weight: 20,000 kg
Length: 8.2 m
Width: 2.5 m
Height: 2.56 m
Track:
(front) 1.97 m
(rear) 1.745 m
Wheelbase: 3.12 m + 1.2 m
Max road speed: 60 km/h
Range: 300 km (estimate)
Fuel capacity: 200 litres
Engine: Famos 2F/002A 6-cylinder water-cooled diesel developing 200 hp
Tyres: 11.00 × 20

Bridge
Length: 13 m
Width: 2.5 m
Trestle leg length: 4 m

Status

In service with the former Yugoslav Army.

Contractor

FDSP - Federal Directorate of Supply and Procurement, PO Box 89, Bulevar umetnosti 2, YU-11070 Novi Beograd, Yugoslavia.
Tel: +381 11 222 2743. Telex: 71000 JU SPDR.
Fax: +381 11 324 8738.

VERIFIED

TACTICAL FLOATING BRIDGES AND FERRIES

AUSTRIA

4 tonne ferry boat

Description

Designed to replace existing ferry systems, the 4 tonne ferry boat was produced by a Linz shipyard and was first used for trials by Austrian Army engineer units in early 1985. The basis of the ferry is an aluminium boat designed specifically for combat engineer purposes that can be joined together in units of three by a ferry bridge equipped with hydraulic ramps.

The stackable ferry boat uses an aluminium welded structure of cross-rib design and fitted with a seal-welded double bottom. Each ferry boat has a payload of 4,000 kg and is unsinkable. The ferry superstructure connects three of the ferry boats side by side and is formed of two aluminium trackways that are themselves buoyant. Four spacers between the hulls of the boats serve as an assembly aid and as treadboards when the ferry is in use.

The ferry is driven by three 40 hp outboard motors. The full crew consists of a ferry commander, three outboard motor controllers and eight sappers. The same crew is sufficient to construct the ferry.

Specifications

Boat
Weight: approx 500 kg
Length: 7.37 m
Width: 1.97 m
Height: (front/rear) 0.82 m/0.67 m

Ferry
Weight: (fully equipped) approx 3,000 kg
Length of trackways: 7.5 m
Overall width: 14.5 m
Draught: between 0.2-0.4 m
Min navigable depth: approx 0.7 m

Status

In service with the Austrian Army.

VERIFIED

Aluminium footbridge

Description

This consists of a string of aluminium hollow-plate floats connected to form a flush deck walkway without any additional superstructure. Each float is reinforced internally with truss stiffeners and can support a load of 136 kg in water. The deck surface is covered with a plastic material.

The footbridge is assembled by joining the floats end to end with pot-shaped connecting devices attached to the adjoining corners, which also float. In strong currents an additional float is added either side of the bridge at intervals. Guy ropes extending from eyelets on the connectors to the shore anchor the bridge.

Specifications

Float
Weight: 81 kg
Length: 5 m
Width: 1.2 m
Depth: 0.13 m

Status

Believed to be in service with the Austrian Army.

VERIFIED

CHINA, PEOPLE'S REPUBLIC

Type 79 and Type 79-A Ribbon Bridge

Description

The Type 79 Ribbon Bridge was developed from the RFAS PMP Heavy Folding Pontoon Bridge and the US Ribbon Bridge. The construction, operating and other details are similar to those for the PMP (reference can be made to the PMP entry in this section).

Figures supplied for the Type 79 state that when a pontoon bridge is constructed across a river flowing at 2.5 m/s, a 50 tonne capacity bridge 6.5 m wide and 312 m long can be erected. A 20 tonne capacity bridge under similar circumstances can be 3.2 m wide and 527 m long. Construction of a floating bridge with a capacity of up to 110 tonnes takes only 15 minutes.

A later version is known as the Type 79-A Ribbon Bridge. This appears to differ only in dimensions from the original version and is launched from a 6 × 6 truck. The details of a single Type 79-A pontoon unit follow:

Specifications

Pontoon
Weight: 7,200 kg
Length: 6.9 m
Width: 8.092 m
Height: 1.1 m

Status

In production. In service with the Chinese armed forces and offered for export.

Manufacturer

China Shipbuilding Trading Company Limited, 10 Yuetan Beixiaojie, Beijing, China.
Tel: +86 10 6858 2562.
Fax: +86 10 6858 3380.

UPDATED

Assembling a Type 79-A Ribbon Bridge
1997

CZECH REPUBLIC AND SLOVAKIA

LMS light pontoon bridge

Description

The LMS light pontoon bridge is similar in design to the larger Czech SMS medium/heavy pontoon bridge. Both the centre and bow pontoons are aluminium. The centre pontoon is fully enclosed while the bow pontoon has an open deck. Once the pontoons are in the water they are assembled, balks added and bolted together. The trackway is then added, which can be of the full-width roadway type, or a dual trackway, the latter being more common. The LMS can also be used as a raft with the following types being constructed: 10, 15, 20 and 24 tonne. The rafts are propelled by a 22 hp outboard motor and can attain a maximum speed in water of 11 km/h.

A complete LMS light pontoon bridge set consists of 48 bow and 24 centre sections, carried on Praga V3S (6 × 6) 3,000 kg trucks.

Specifications

	Bow pontoon	Centre pontoon
Weight	470 kg	450 kg
Length	5.43 m	4 m
Width	2.02 m	2.02 m
Depth	0.8 m	0.8 m

Bridge construction

Bridge type	**15 t**	**20 t**
Roadway width	3 m	3 m
Pontoon sections per support	3	3
Assembly time	40 min	n/a

Status
In service with the Czech and Slovak armies.

Manufacturer
Czech and Slovak state factories.

VERIFIED

FRANCE

Giat Industries MLF light river crossing equipment

Description
The MLF light river crossing equipment (*Matériel Léger de Franchissement*), also known as the CASTOR light spanning equipment Type F2, was developed by the Etablissement Technique d'Angers and was adopted by the French Army in 1976.

The MLF can be used as a floating bridge a raft or an individual flotation unit can be used as an assault or river boat. When being used as the latter it can carry 32 men, and is propelled in the water by paddles or an outboard motor.

The flotation units, that are of light alloy construction, are made up of two flat-bottom boats assembled stern to stern by pins, and have an assembled length of 9.8 m. The gunwales are capped by stiffeners that are attached to the boat by four locking pins. Each unit is also provided with handrails, outboard motor attachment and a floor.

The bridge deck comprises two tracks made up of deck components, steel assemblies, articulation units and ramp units. The central deck units fill the space between the tracks and are of reinforced plastic. The bridge decks are secured to the flotation units by the sets of tiltable pegs on the gunwale stiffeners. The bridge deck components are made of light alloy sections welded together and are provided with anti-skid strips. The deck bridge components are arranged in pairs, end to end, and spaced 0.7 m apart.

The articulated joints provide the connection between the bridge deck components located on the floating support and those acting on the ramp. Each articulation joint weighs 165 kg and is adjusted manually. The ramp units are of welded light alloy construction and assembled by means of pins. Each ramp weighs 50 kg and is provided with a steel contact edge. Each bridge deck unit can be fitted with a track guide in tubular light alloy, which is articulated on bosses welded to the outside of the bridge deck. The track guide can be retracted into a free space on the side of the bridge deck.

The French Army version of this system consists of four flotation units and three deck sections forming a loading length of 10.5 m with two ramp sections on each side. The resultant raft has a load capacity of Class 16 in a river flow rate of 1.5 m/s. Propulsion is provided by two 40 hp outboard motors.

Rafts

Various types of raft of two, three, four, five or six flotation units can be assembled. A 4/3 type basic raft can be assembled by three teams of eight men in about 40 minutes. This raft can be carried on a Renault 4000 (4 × 4) 4,000 kg truck, towing an SKD 3536 trailer carrying the eight flotation unit halves stacked one inside the other. The raft is either propelled in the water by outboard motors or pushed by a bridging boat. For example, a 4/3 raft would be propelled by two 40 hp outboard motors on the first and fourth flotation units.

Bridges

Two types of bridge can be built using the MLF system: the medium bridge composed of bridge deck units with flotation units under them, and the heavy-duty bridge with the flotation units positioned one alongside the other. The medium bridge (1/1) is 8.07 m long and will span a river 7 m wide, while the heavy-duty bridge has an overall length of 16.13 m and will span a river of 15 m.

Specifications
Individual flotation unit
Length: 4.9 m
Width: 1.75 m
Height: 0.75 m
Weight:
(without equipment) 160 kg
(with floor and superstructure) 235 kg
(with floor, superstructure and gunwale stiffeners) 250 kg

Bridge decks
Length: 3.44 m
Width: 1 m
Depth: 0.26 m
Weight: 205 kg

Raft and bridge load capacities in tonnes

Type of raft or bridge	**Current speed** 1.5 m/s	2 m/s	2.5 m/s
2/2 raft	6	–	–
3/3 raft	10	8	6
4/3 (basic) raft	16	13	10
6/4 raft	22	20	15
Medium-duty bridge	15	13	8
Heavy-duty bridge	22	20	16

Status
Available for production. In service with the French Army and one overseas army.

Manufacturer
Atelier de Construction de Tarbes, 2 rue Alsace-Lorraine, BP 313, F-65013 Tarbes, France.
Tel: +33 5 62 37 62 62. Telex: ARSEN 531 581 F.
Enquiries to: Giat Industries, 13 route de la Minière, F-78034 Versailles Cedex, France.
Tel: +33 1 30 97 36 48.
Fax: +33 1 30 97 39 58.

VERIFIED

AMX-10P IFV crossing bridge built from MLF light river crossing equipment

A Renault TRM 2000 (4 × 4) 2,000 kg truck crossing a bridge built from MLF light river crossing equipment

CNIM Pont Flottant Motorisé Mle F1

Development
The *Pont Flottant Motorisé Modèle* F1 (PFM Mle F1) was developed by Constructions Navales et Industrielles de la Méditerranée (CNIM), under contract to the Direction Technique des Armements Terrestres. It was tested in competition with a similar system developed by Creusot-Loire and compared to other equipment available in Western nations during 1979-80. It was adopted by the French Army in early 1981.

Production of French Army PFM units commenced during February 1984 with the first deliveries to operational units starting at the end of 1985. In service the PFM is used in forward areas for tactical river crossings, while in rear areas it can be used to form line of communications bridges.

Description
The PFM Mle F1 may be employed as either a bridge or a raft, and can be used in rivers with a current of up to 3 m/s as an MLC 70 raft, and in currents of up to 3.5 m/s as an MLC 70 bridge (MLC 80 with restrictions). The equipment can be launched from banks up to 4 m high, used as a raft or bridge from banks 3 m high and recovered from banks 2 m high. The main advantages claimed for the PFM Mle F1 are that floating bridges and rafts can be assembled rapidly with minimum manpower, and the main elements can be launched and recovered without cranes, bridging boats or other ancillary equipment. Section assembly is eased by the use of prelocking devices. The PFM Mle F1 is practically unsinkable as the various elements are filled with polyurethane foam or are partitioned. The equipment is claimed to be easy to use and maintain.

The PFM Mle F1 consists of two major components, the centre-section and the approach ramp. The centre-section has locking elements at both ends, two wing tanks and two ballast tanks. The centre and wing units are constructed from light alloy, the ballast tanks from laminated GRP. Included in each ballast tank is a light alloy framework which contains the outboard motor (55 kW or 75 hp). The approach ramps are also light alloy and their ends can be set hydraulically into various positions to suit the river bank height. The centre-section and ramp are transported folded on a special semi-trailer towed by a 6 × 6 tractor. The resultant articulated unit normally has a 10 × 6 configuration but the trailer wheels can be hydraulically driven to create a 10 × 10 unit when crossing rough terrain. The semi-trailer is provided with power for unfolding, launching, recovery and refolding the section. The components are launched as follows: the tractor reverses to the water's edge and the section is unfolded, the sliding frame is set back, hydraulic rams push up the forward end of the tilting frame and the section slides down into the water. A number of 10 m long sections are assembled in the water and the bridge is completed by adding the approach ramps. Each section is propelled in the water by two outboard motors.

A team of 21 men can assemble a Class 70 raft in 20 minutes. Such a raft would consist of three centre-sections and two approach ramps. A team of 45 men can build a 100 m bridge within 45 minutes although this time can be reduced to 15 minutes if there are no obstacles on the river side. This time reduction is due to the efficiency of the launch system which makes it possible to carry out simultaneous launchings from unprepared banks.

Berliet GBC 8 KT (6 × 6) truck crossing a PFM Mle F1 bridge

Specifications

Main section
Weight: 10,500 kg
Length: 10 m
Width:
(folded) 3.6 m
(unfolded) 9.8 m
Track width: 4 m
Height:
(folded) 2.1 m
(unfolded) 0.73 m
Draught: (unloaded) 0.18 m
Positive buoyancy: 42 t

Approach ramp
Weight: 7,500 kg
Length: 12 m
Track width: 4 m

Trailer and section with TRM 10 000 tractor
Weight: 31,000 kg
Length: 18 m
Width: 3.6 m
Height: 4 m
Ground clearance: 0.3 m
Max road speed: 80 km/h
Max off road speed: 10 km/h
Fording: 1.2 m

Status

In production. In service with the French, Italian and Swiss armies.

Manufacturer

Constructions Navales et Industrielles de la Méditerranée (CNIM), Zone Industrielle de Brégaillon, BP 208, F-83507 La-Seyne-sur-Mer Cedex, France.
Tel: +33 4 94 30 30 00. Telex: 401 235 f cnim f.
Fax: +33 4 94 30 31 00.
Enquiries to: CNIM Head Office, 35 rue de Bassano, F-75008 Paris, France.
Tel: +33 1 44 31 11 00. Telex: 642 869 f.
Fax: +33 1 44 31 11 30.

VERIFIED

French Army TRM 10 000 (6 × 6) tractor trucks towing PFM Mle F1 centre-section bridging unit (P Touzin)

French Army TRM 10 000 (6 × 6) tractor truck carrying PFM Mle F1 ramp unit on semi-trailer (P Touzin)

CNIM BAC 20 and BAC 60 autonomous ferries

Description

The BAC 20 and BAC 60 autonomous ferries are intended for use in assault crossings by engineer units in infantry and armoured divisions. Both can carry wheeled or tracked vehicles, the BAC 20 up to MLC 20 or 30, and the BAC 60 up to MLC 50 or 60.

The BAC 20 ferry is derived directly from the *Pont Flottant Motorisé* (PFM - see previous entry) and consists of a single PFM centre-section with folding ramps fitted. The ferry is launched in the same way as the normal PFM sections and is ready to operate immediately after launching. Ready-to-use time from arrival on site is thus less than 5 minutes. When in the water, propulsion is provided by two 75 hp outboard engines.

The BAC 60 consists of two PFM centre-sections each permanently fitted at one end with a folding ramp. The two 'half BAC 60' units can be assembled when on the water and ready-to-use time can be around 7 minutes.

BAC 20 and the half BAC 60 units are carried, launched and retrieved by the PFM transport vehicle. Both equipments can carry their own transport vehicle.

Specifications

Model	**BAC 20**	**BAC 60**
Crew	4	8
Weight	11,000 kg	22,500 kg
Length		
(ramps folded)	11.6 m	21.6 m
(ramps unfolded)	17.4 m	27.4 m
Width		
(folded)	3.6 m	3.6 m
(unfolded)	9.8 m	9.8 m
Max useful length	15 m	25 m
Height		
(folded)	2.1 m	2.1 m
(unfolded)	0.7 m	0.7 m
Assembly time	5 min	7 min

Model	**BAC 20**	**BAC 60**
Special semi-trailer		
Weight loaded	20,500 kg	20,500 kg
Length with TRM 10000 tractor	18 m	18 m
Width		
(loaded)	3.6 m	3.6 m
(unloaded)	2.5 m	2.5 m
Height	4 m	4 m
Speed		
(road)	80 km/h	80 km/h
(off road)	10 km/h	10 km/h

BAC 20 autonomous ferry carrying engineer equipment

BAC 60 autonomous ferry carrying PFM bridging unit

Status
Prototype being evaluated by the French Army and one other European army.

Manufacturer
Constructions Navales et Industrielles de la Méditerranée (CNIM), Zone Industrielle de Brégaillon, BP 208, F-83507 La-Seyne-sur-Mer Cedex, France.
Tel: +33 4 94 30 30 00. Telex: 401 235 f cnim f.
Fax: +33 4 94 30 31 00.

Enquiries to: CNIM Head Office, 35 rue de Bassano, F-75008 Paris, France.
Tel: +33 1 44 31 11 00. Telex: 642 869 f.
Fax: +33 1 44 31 11 30.

VERIFIED

GERMANY

Infantry bridge

Description
This light bridge consists of small rubber dinghies, footboards, handrails and ropes, and retaining ropes. It is used by infantry to cross rivers and streams.

Status
In service with the German Army.

VERIFIED

German infantry crossing lightweight bridge of rubber dinghies and footboards (Federal German Ministry of Defence)

KRUPP Festbrücke floating bridges

Description
The KRUPP *Festbrücke* system, also known as the Light-Metal S bridge, can be adapted to become a floating bridge or ferry by adding extra modular components. The main components are still the twin-triangular cross-section, wheel tread girders but to this are added a hydraulic ramp adjusting section, pontoons with load-bearing saddles and anchoring and drive units. These floating bridges and ferries can carry loads up to MLC 60.

The floating bridge can be adapted to almost all shore conditions by using the ramp adjusting section. One side of this section is joined to the floating bridge and the height and angle of the outer section can be raised or lowered 20° by means of a hand-operated hydraulic pump acting on a pressure cylinder. This angle corresponds to a difference in height of approximately 7.6 m with a ramp length of 19.3 m. The time required for raising and lowering the ramp from the highest to the lowest point is 5 minutes. A special pressure compensating unit between pressure cylinders positioned close to each other in the bridge cross-section prevents tilting or jamming during raising or lowering. These adjusting sections can also be used as part of the floating bridge itself so that the bridge can be assembled from one shore and can assume angles to suit the required conditions.

KRUPP Festbrücke in use as a floating bridge

The *Festbrücke* pontoons are light metal components each 8.8 m long, 2.25 m wide and 1.23 m high. Each has a carrying capacity of 11,400 kg at which weight each will have a freeboard of 300 mm. For supporting a bridge two single pontoons are joined stern to stern using twin pins inserted vertically over the full height. Each pontoon has a foam-filled decking covered by metal plating and each weighs 830 kg. The pontoons can be powered by 115 hp outboard engines and various accessories are available including mooring posts, trimming tanks, an anchor, a winch with 100 m of cable and so on. Normally these pontoons are carried on special trailers 9.4 m long, onto which the pontoons can be stacked five high to a height above the trailer carrying area of 2.43 m.

The bridge and pontoon are connected by longitudinal saddle girders set on each pontoon wall and joined by crossbeams. The lower chords of the *Festbrücke* are engaged by clamping jaws. For launching from a shore each pontoon is set afloat using at least three inflatable load rollers each with a diameter of 0.5 m and a length of 3 m.

The floating *Festbrücke* is anchored with winches on each pontoon. If this system cannot be used, each pontoon uses its outboard engine. When used as a floating bridge the load capability can be altered by changing the pontoon spacing.

Most floating bridges are between 60 and 200 m long. These bridges can be used in water velocities of up to 3 m/s. Normal crossing speeds should not exceed 10 km/h.

When used as a ferry, the *Festbrücke* is powered by the pontoon outboard engines alone, and can reach speeds of up to 20 km/h on smooth water.

Status

Production complete. In service with European and other armies.

Manufacturers

KRUPP Fördertechnik GmbH, Altendorfer Strasse 120, D-45143 Essen, Germany.
Tel: +49 201 82801.
Fax: +49 201 828 2566.
MAN Gutehoffnungshütte AG, Ginsheimer Strasse 1, D-65462 Ginsheim-Gustavsburg, Germany.
Tel: +49 6134 55292.
Fax: +49 6134 55383.

UPDATED

KRUPP-MAN FFB 2000

Development

The *Faltschwimmbrücke* 2000 (FFB 2000 or Folding Float Bridge 2000) was developed jointly by KRUPP Fördertechnik GmbH and MAN Gutehoffnungshütte AG ,as a result of experience gained in the manufacture and use of the German version of the US Ribbon Bridge. The construction of the FFB 2000 followed extensive testing with models in the Research Institute for Inland Waterway Shipbuilding at Duisburg.

Description

The FFB 2000 can be used in water with stream velocities of up to 3.5 m/s with a load of MLC 70, or up to MLC 80 in exceptional cases. It can be used in a similar fashion to the existing Ribbon Bridge and can be used for the construction of ferries and floating bridges. It is completely compatible with existing US and German Ribbon Bridges.

The FFB 2000 consists of inner and ramp sections with a new streamlined cross-section with 30° bows. Each inner and ramp section is folded into a 'W' shape for transport on an all-terrain 7,000 kg capacity truck. The sections unfold automatically as they are launched from the carrier truck. Once in the water, the sections can be assembled by a bridging boat into ferries or bridges. A ferry can be assembled in a modular fashion from two ramp sections and the necessary number of inner sections. The effective roadway width is 4.1 m (5.2 m for two-lane traffic up to MLC 16) with footpaths 2.2 m wide on either side of the road. The assembly time for a 100 m bridge can be as low as approximately 60 minutes, and up to 200 MLC 80 vehicles can cross the bridge every hour. A ferry consisting of two inner sections and two ramp sections can carry vehicles up to MLC 70 in stream velocities of up to 3.5 m/s. To accommodate various bank conditions and heights the ramp sections can be adjusted hydraulically to a height of approximately 2.2 m above water level. Minimum water depth for operations is 1.2 m. Each section has four crane lifting points for handling.

FFB 2000 ramp section being launched from its carrier vehicle

Advantages claimed for the FFB 2000 over similar equipments include: improved floating stability due to the new cross-sectional shape; a higher ferry capacity; the ability to cope with greater bank heights; a flatter approach slope onto the bridge; an improved bridge-crossing performance; a continuous road panel; no heavy ramp plates; a stronger overall construction; and improved handling in the water and at launch.

Specifications

Inner and ramp sections
Weight:
(inner section) 4,800 kg
(ramp section) 5,200 kg
Effective length: 6.7 m
Width:
(folded) 3.03 m
(unfolded) 8.85 m
(roadway) 4.1 m
(walkways) 2 × 2.22 m
Height:
(unfolded, bow) 1.27 m
(unfolded, deck) 0.737 m
(folded) 2.35 m
(road transport) 3.95 m
Material: AlMgSil

Status

In production.

Manufacturers

KRUPP Fördertechnik GmbH, Altendorfer Strasse 120, D-45143 Essen, Germany.
Tel: +49 2065 783816. Fax: +49 2065 782340.
MAN Gutehoffnungshütte AG, Ginsheimer Strasse 1, D-65462 Ginsheim-Gustavsburg, Germany.
Tel: +49 6134 55292.
Fax: +49 6134 55383.

UPDATED

EWK Bundeswehr Faltschwimmbrücke (FSB)

Development

In 1976 the West German Army decided to participate in the US Army's test and evaluation programme of the first production Ribbon Bridge, which is fully described in the USA section. The trials were carried out in Germany between September 1976 and March 1977. Early in 1977 the West German Army decided to adopt the Ribbon Bridge System and subsequently undertook a redesign to meet European requirements allowing production in Germany. Production by EWK began in 1978 with first deliveries to the German Army in December of that year. Each of the German Army's 12 Division Combat Engineer Companies has one complete set of 26 segments: 18 bridge segments and eight ramp elements. In Germany the system is known as the *Faltschwimmbrücke*, or FSB.

An order for FSB for the Australian Army was completed in 1989. The order was worth approximately AUS$15 million, with the bridges being manufactured in Australia by Transfield Construction.

Australian Army FSB pontoon unit loaded on a Mack 6 × 6 truck (Paul Handel) ***1997***

Description

The German version of the Ribbon Bridge, unlike the US model, is built to metric standards and incorporates a number of improvements to suit German requirements, none of which affect its compatibility with the original US version.

The hydraulic system for raising the ramp bay to adjust to varying bank conditions has been modified and non-polluting hydraulic fluid introduced; on the original model, oil sometimes leaked out and water got in. Other improvements include: winch-operated approach ramps; changes interior bays to prevent the cables from getting crushed as the section folds; non-skid coating on roadways and walkways; and stops on the bridge sections to keep the bridging boats in place when the bridge is being used as a raft.

The *Faltschwimmbrücke* used by the German, Portuguese and Belgian armies is carried on the rear of a MAN (6 × 6) 7,000 kg truck, designated the Lkw 7t gl *Brückentransporter*. The following 'next generation' is known as the *Faltschwimmbrücke* Export, or FSB-E, used by the Australian, Belgian, Brazilian, Canadian, Nigerian, Portuguese, Singapore, Swedish and Turkish armies and carried on different trucks, usually the IVECO 6 × 6. Full specifications of these trucks are given in the *Trucks* section.

EWK also produces an all-steel version which is compatible with the original PMP system. This version was originally developed for, and is now in service with, Egypt. The main differences between the steel and the aluminium versions follow.

Specifications

Vehicle with folded unit
Cab seating: 1 + 2
Configuration: 6 × 6
Weight: (loaded) 18,800 kg
Length: (overall) 10.4 m
Width: (overall) 3.31 m
Height: (overall) 3.93 m
Ground clearance: 0.415 m
Wheelbase: 5.4 m
Angle of approach/departure: 45°/31°
Max speed: 90 km/h
Range: 700 km
Max gradient: 60%
Fording: 1.2 m

Bridge section	**Aluminium**	**Steel**
Weight	5,400 kg	7,700 kg
Length	6.7 m	6.7 m
Width unfolded	8.12 m	8.02 m
Road width	4.1 m	6.6 m

Ramp section	**Aluminium**	**Steel**
Weight	5,500 kg	7,500 kg
Length	5.6 m	5.5 m
(with unfolded plate)	7.6 m	7.5 m
Road width	4.1 m	6.6 m

Status

In production. In service with the Australian, Belgian, Brazilian, Canadian, Egyptian, German, Nigerian, Portuguese, Singapore, Swedish and Turkish armies.

Manufacturer

Eisenwerke Kaiserslautern GmbH, Barbarossastrasse 30, D-67655 Kaiserslautern, Germany.
Tel: +49 631 857339. Telex: 45810.
Fax: +49 631 857300.

UPDATED

LEGUAN Ferry

Description

The LEGUAN bridgelayer can be used to launch floating ferries up to MLC 70. For this the LEGUAN requires additional pontoons and hydraulically operated ramp aprons fixed to the bridge tips. For details of the LEGUAN bridgelayer and its normal operating sequence refer to the entry in the *Mechanised Bridges* section.

The complete LEGUAN floating ferry consists of a modified LEGUAN bridge with ramps, a hydraulic system and pontoons with bridge saddles. To connect the ramp girder to a bridge, use is made of two openings in each LEGUAN bridge wheel tread girder. Normally these openings are closed by adaptors held in place by plug-in strips and an interlocking system. Each adaptor weighs 23 kg. When the bridge is used as a ferry the openings are closed by the ramp connection systems that form the connection between the bridge on one side and the ramp girder and hydraulic system on the other. The units are 1.4 m wide and the ramp connection systems weigh 137 kg for each tread girder. All have the same anti-slip coating as the LEGUAN bridge.

Each ramp girder is 4.8 m long and can accommodate a difference in levels of approximately 1.32 m; this is a maximum gradient of 29 per cent (16°). One ramp girder weighs 620 kg. The hydraulic system is used as an adjusting element. The total weight of the hydraulic system for one ramp is 600 kg.

The pontoons used with the ferry are unsinkable aluminium alloy boats coupled at their bluff sterns to form a pontoon. The pontoons are connected to the LEGUAN bridge and are fixed to the lower chord sections of the outer sides of the wheel tread girders. Two sets of three pontoons are required for MLC 60 loads and three pontoons for loads up to MLC 27.

To lay the LEGUAN bridge across the pontoons the laying procedure is much the same as for a normal LEGUAN bridge.

It is claimed that the LEGUAN ferry has good manoeuvrability and is well suited for use in shallow water.

MLC 27 LEGUAN ferry in use

Two or more individual LEGUAN ferries can be joined together to form a floating bridge. The only additional equipment required for this capability is a coupling element for joining two ferry units together. In operation this coupling element is installed on one end of a ferry before the other ferry is moved in and attached to the coupling element. Coupling elements have been supplied to Norway.

A typical drive unit used with the LEGUAN Ferry is a Schottel Pump Jet SPJ 20 driven by a KHD F 61 912 diesel engine which can be swivelled through 360°. When pontoons are stacked for storage or transport the drive unit is always in the top pontoon.

Specifications

Pontoons
Weights:
(pontoon with drive) 2,055 kg
(pontoon without drive) 1,185 kg
Lengths:
(single pontoon) 8.8 m
(double pontoon) 17.6 m
Width: 2.25 m
Height: 1.225 m

Ramps
Load capacity: MLC 60 (MLC 70 with caution)
Weight: (ramp with hydraulic system) 2,200 kg
Length: 4.8 m
Width: 1.97 m
Height: (overall) 0.645 m

Status

In production. In service with Norway.

Manufacturer

KRUPP Fördertechnik GmbH, Altendorfer Strasse 120, D-45143 Essen, Germany.
Tel: +49 201 82801.
Fax: +49 201 828 2566.
MAN Gutehoffnungshütte AG, Ginsheimer Strasse 1, D-65462 Ginsheim-Gustavsburg, Germany.
Tel: +49 6134 55292.
Fax: +49 6134 55383.

UPDATED

Heavy ferry

Description

The *Pionier Fahre*, or heavy ferry, was designed and built by the Boden-Werft shipyards at Kressbronn. It consists of 12 pontoons, each of which weighs 10,300 kg and can be transported on a railway flatcar with a loading space of at least 8.8 m, or on truck trailers. On arrival at the launch site the pontoons are joined side by side in three rows, with four pontoons in each row. When assembled, the ferry is 36.6 m long and 7.6 m wide, with the deck of the pontoons serving as the roadway, which has a usable width of 5.2 m. The 120 tonne heavy ferry is powered by four 125 hp diesel engines, one at each corner pontoon, each of which powers a propeller that can be raised or lowered hydraulically and can be traversed through 360°. Maximum water speed is 16 km/h.

A feature of the heavy ferry is that the bridge, on the right side, can be raised or lowered to permit the ferry to pass under bridges and other river obstacles. There are ramps fore and aft for loading and unloading and a 20 mm cannon can be mounted on each side for air defence.

Heavy ferry carrying British soldiers and equipment across the Rhine (UK Ministry of Defence)

Specifications

Pontoons
Length: 9.1 m
Width: 2.59 m
Depth: 1.9 m

Assembled ferry
Length: 36.6 m
Width: 7.6 m
Capacity: 120,000 kg

Status

Production complete. In service with the German Army.

VERIFIED

GREECE

TEMAK aluminium floating footbridge

Description
The aluminium floating footbridge, as produced by TEMAK SA, is a licence-produced version of the US aluminium floating footbridge; for details and specifications see the relevant entry in the United States of America section.

Status
In production. In service with the Greek Army.

Manufacturer
TEMAK SA, 21 K Varnali Street, GR-12134 Peristeri, Athens, Greece.
Tel: +30 1 513 5886. Telex: 224268 tema gr.

VERIFIED

TEMAK aluminium floating footbridge

TEMAK light tactical bridge and raft

Description
This equipment is a licence-built version of the US Bridge Floating: Raft Section, Light Tactical. For a full description and specifications see the relevant entry in the United States of America section.

Status
In production. In service with the Greek Army.

Manufacturer
TEMAK SA, 21 K Varnali Street, GR-12134 Peristeri, Athens, Greece.
Tel: +30 1 513 5886. Telex: 224268 tema gr.

VERIFIED

TEMAK light tactical bridge and raft

ISRAEL

IMI Two-Tank Ferry Raft (2TFR)

Description
The Two-Tank Ferry Raft (2TFR) is intended for use as a ferry raft carrying one or two MBTs, or for connecting with one or more 2TFRs to form a floating bridge. On land the 2TFR is sufficiently mobile to allow it to be towed into position by the MBTs that will use it to cross water obstacles, while on roads the 2TFR may be towed by heavy trucks. Each 2TFR is an independent unit handled by a crew of three. A 33-man force with 11 2TFRs can bridge 300 m of water. Linking time for two 2TFR units is about 5 minutes.

The 2TFR consists of a floating loading platform, 21 m long and 5 m wide, with two side floats, each 21 m long and 1.6 m wide. The side floats are folded onto the loading platform during transit and are usually opened before reaching the operational area. At each end of the platform is a hinged loading ramp, 7.5 m long and 5 m wide, which is folded up onto the platform until the unit is launched. Both side floats and platform are filled with rigid polyurethane foam to ensure buoyancy in the event of battle damage.

2TFR carrying Centurion and M48 MBTs

The 2TFR is propelled by two identical power units that may be produced by Schottel (Germany) or Stewart and Stevenson (USA). The two power units may be controlled from either of two control units, or from a remote-control unit that may be mounted anywhere on the 2TFR. The power units propel the raft in the water and raise and lower the side floats and loading ramps. Each power unit is connected to the propulsion unit by a 270° pivot for steering. The propeller shaft can be raised and lowered to suit varying water depths and in transit the whole propulsion unit can be angled upward and over for road clearance. On land the raft travels on a wheeled undercarriage with four pairs of wheels, each pair with its own suspension. The wheels are low pressure and specially designed. Once waterborne the entire undercarriage can be uncoupled and dropped. A hydraulically operated tow boom, powered by either of the power units, is located on the stern of the raft, and is self-uncoupling.

In use, the 2TFR is towed by a heavy truck, semi-trailer or wheeled tractor to the nearest possible point by road. When the road is left, the 2TFR is hitched to a tank by the hydraulic boom on the stern of the raft, and towed to the edge of the water obstacle. The MBT then pulls the raft round and pushes it into the water. The tow boom is uncoupled from the raft and drops off the MBT. The 2TFR can then proceed under its own power and the undercarriage can be separated from the raft. The loading ramps are extended and, once lowered, the raft can take on its load of two MBTs. The entire sequence, once the raft is in the water, takes under 5 minutes.

The wheels on the undercarriage may be fitted with high-speed brakes, and for road towing the width of the 2TFR may be reduced to 4.25 m. For water use, the undercarriage can be supplied in a retractable form, in which state the wheels will extend only 0.5 m. Propulsion hydrojets may be used in place of the normal propulsion units, and another option is that the 2TFR can be fitted with only one loading ramp, allowing the raft to be used as a one-direction raft with both propulsion units at the rear. A smaller version of the 2TFR can be produced which will carry only one tank. This unit will have a transit length of 17 m and a width of 4.45 m.

For maintenance, the power units can be removed and stored separately from the rest of the raft. Tank drivers need no special training to tow the 2TFR.

Specifications

Crew: 3
Length:
(transit) 21 m
(water) 31 m
(bridging formation) 27.5 m
Width:
(road) 5.25 m
(water, side floats down) 8.25 m
Height:
(road) 5.1 m
(water) 3.5 m
Loading platform width: 5 m
Weight:
(transit, less power units) 46,000 kg
(ready for launch) 54,000 kg
(waterborne) 48,000 kg
Towing speed:
(road) 60 km/h
(tank-towed, off road) 25 km/h
Uphill inclination angle: (tank-towed) 11°
Lateral inclination angle: (tank-towed) 15°
Turn radius: 20 m
Speed in water:
(unloaded) 14 km/h
(one tank) 11 km/h
(two tanks) 9 km/h
Draught:
(unloaded) 0.35 m
(one tank) 0.74 m
(two tanks) 1.2 m
Manoeuvrability: 360°
Operational time before refuelling: 12 h
Sea-going capability: to Sea State 3
Bridging capability:
(resistance to current) velocities up to 2.5 m/s
(length) unlimited

Status

In service with the Israeli Army.

Manufacturer

Israel Military Industries Limited (IMI), PO Box 1044, IL-47100 Ramat Hasharon, Israel.
Tel: +972 3 542 5222. Telex: 33719 misbit il.
Fax: +972 3 48 9639.

VERIFIED

JAPAN

Aluminium floating footbridge

Description

The Japanese aluminium floating footbridge is similar to the old US M1938 footbridge. Major differences are that aluminium alloy is used instead of white pine for the float body and the duckboard stringer, and cedar wood is substituted for pine in the transverse slats of the duckboard. The floats are filled with foam rubber to give increased buoyancy.

A bridge set, which is carried on two 2,500 kg trucks, consists of 21 duckboards and 42 floats, this is sufficient to provide 76 m of walkway 0.55 m wide and has a carrying capacity of 75 personnel/min in single column. It can be assembled in 11 minutes by a working party of 45 personnel. Alternatively, a reinforced footbridge 25.2 m long and 1.65 m wide with a capacity of 150 personnel/min in two columns can be assembled by 45 personnel in 18 minutes.

Specifications

	Floats	Duckboards
Weight	43 kg	37 kg
Length	3 m	3.6 m
Width	0.25 m	0.55 m
Depth	0.4 m	0.15 m

Status

Production complete. In service with the Japanese Ground Self-Defence Force.

VERIFIED

Light tactical raft

Description

This is similar to the US Bridge Floating: Raft Section, Light Tactical and is transported on 2,500 kg (6 × 6) trucks. It consists of Japanese folding assault boats which are joined stern to stern to form a pontoon. Either two bow sections or one bow and one stern section may be joined. Attached to the gunwales of the pontoons are dual plywood treadways forming the raft deck. Siderails are then placed along the inside edge of the treadways to stop vehicles being driven off.

Rafts of 9, 10 and 13 tonnes capacity can be assembled with three, four or five pontoons respectively. A three-pontoon raft is normally propelled by one 25 hp outboard motor and the four- and five-pontoon rafts are propelled by two outboard motors. Maximum speed of a loaded three-pontoon raft is 6.5 km/h.

Specifications

	Pontoon	Treadway
Weight	269 kg	100 kg
Length	7.7 m	2.5 m
Width	1.5 m	0.9 m
Depth	0.65 m	0.18 m

Assembled raft

Raft Class	**9 t**	**10 t**	**13 t**
Number of pontoons	3	4	5
Number of treadways	8	10	12
Overall length of deck	9 m	12 m	15 m
Distance between pontoon centres	2.5 m	2.5 m	2.5 m

Status

Production complete. In service with the Japanese Ground Self-Defence Force.

VERIFIED

NETHERLANDS

Damen FAC 540 floating bridges and ferries

Description

The Damen FAC 540 fast assault craft can be used in the construction of a floating footbridge or light ferries, for loads up to MLC 8. All the bridges and ferries are based on the use of the aluminium-hulled FAC 540, designed to carry a fully equipped infantry section of 12 personnel and two crew.

A tactical ferry raft for loads up to MLC 8 can be assembled using four FAC 540 boats joined stern to stern in a catamaran configuration. The same configuration can be used as a light tactical bridge. Each set of four FAC 540 boats is supplemented by a set comprising eight aluminium track units with a width of 762 mm for vehicles with single or dual rear tyres. The set also contains four aluminium track couplings and accessories, eight aluminium gunwale saddles to support the tracks, 12 nylon track fasteners to secure the tracks to the boats, six stainless steel pins to couple

A tactical raft using FAC 540 fast assault craft carrying a truck during United Nations operations in Cambodia

Loading a Volvo 4140 (6 × 6) truck onto a tactical raft based on four FAC 540 fast assault craft

the tracks and four nylon rigging lines to raise or lower the hinged tracks. This type of raft can carry a 3 tonne truck at speeds of 7 to 10 kt.

The same set can be used to assemble a trimaran ferry raft with a total of six FAC 540 boats joined stern to stern. This type of raft has a load capacity of MLC 8 'plus' and can carry vehicles up to 7.3 m long. A raft created using racks laid directly across three FAC 540 boats can have a capability of MLC 2. Both types of raft can travel at 10 to 15 kt.

Light tactical footbridges can be assembled using the same components as the ferry rafts but with special end ramps, anchor winches, heavy anchors and steel cables and shore anchors. The FAC 540 boats may be joined stern to stern for MLC 8 bridges or used as single units to carry the single trackway.

When the FAC 540 boats are used as ferry rafts they can be powered by either a Johnson Workhorse 25 hp longshaft outboard motor with 22.7 litres of fuel or a Yamaha 40 hp longshaft model with 24 litres of fuel. The number of engines depends on the raft capacity; a MLC 8 raft requires two motors.

For rapid deployment forces and similar applications a Triple Raft Pack is available. This contains three FAC 540 boats together with wheel tracks, couplings, saddles and other components to form a light ferry raft. The pack can be towed on a light trailer or carried slung under a helicopter.

Specifications

FAC 540
Weight: 235 kg
Payload: 2,200 kg
Length: 5.4 m
Beam: 1.83 m
Draught: 0.62 m
Max speed:
(with 4 men) 25-30 kt
(1,200 kg payload) 10-12 kt

Status

In production. In service with the Corps of Engineers in Brunei, Malaysia, the Royal Netherlands Marine Corps and the International Red Cross in Nicaragua.

Manufacturer

Damen Shipyards, Industrieterrein Avelingen West 20, PO Box 1, NL-4200 AA Gorinchem, Netherlands.
Tel: +31 1830 39911. Telex: 25335 dame nl.
Fax: +31 1830 32189.

VERIFIED

POLAND

PP-64 heavy folding pontoon bridge

Description

In 1964, following the success of the PMP heavy folding pontoon bridge, Poland started design work on a folding pontoon bridge. The initial prototype was finished in 1965, with first production units being completed the following year.

In design the PP-64 is different from the PMP and has a much faster construction rate. According to some sources this is over 25 m/min. The basic bridge has a capacity of 40 tonnes whereas the PMP has a capacity of 60 tonnes, which is not a great tactical disadvantage as the PP-64 can still handle T-54/T-55 and T-62 MBTs. The PP-64 can probably accommodate tanks such as the T-64 and T-72.

A PP-64 set consists of: 48 river pontoons carried on trucks; six shore pontoons carried on trucks; 12 ramps for ferries carried on trucks which also tow the KH-200 bridging boats; six KH-200 bridging boats; and one special connecting piece for use in joining PP-64 with PMP bridges.

Three types of bridge can be built: 40 tonne (A), 40 tonne (B) and 80 tonne. The 40 tonne (A) is single roadway type and is 186 m long, has a 4.35 m wide roadway, can take a 12 tonne axle load and can be constructed in rivers with a maximum velocity of 1.2 m/s. A working party consists of 60 pontoon workers, 60 drivers, six powerboat operators and 10 NCOs. The 40 tonne (B) is also a single roadway type and is 145 m long, 4.35 m wide, can take a 12 tonne axle load and can be constructed in rivers with a maximum velocity of 2 m/s. A working party consists of 54 pontoon workers, 60 drivers, six powerboat operators and 10 NCOs. The 80 tonne bridge is double width, 97 m long, has a roadway width of 8.7 m, can take a 12 tonne axle load and be constructed in rivers with a maximum velocity of 3 m/s. The working party consists of 54 pontoon workers, 60 drivers, six powerboat operators and 10 NCOs.

PP-64 heavy folding pontoon bridge in travelling position on rear of Star 660 (6 × 6) 2,500 kg truck

The following ferries can be constructed with the basic PP-64 set: six 40 tonne ferries 14.8 m long and 12.8 m wide, each with eight river pontoons, one shore pontoon and two ramps, a working party of 20 pontoon workers and three NCOs, and two powerboats per ferry; or two large ferries 37 m long and 12.5 m wide, each with 20 river pontoons, two shore pontoons and four ramps, a working party of 30 pontoon workers and four NCOs and three powerboats per ferry.

The bridge is launched as follows: the truck backs up to the river and the pontoon, which is in two parts hinged in the middle, is unfolded and then launched. The pontoons are carried on the rear of a Star 660 M2 (6 × 6) 2,500 kg truck. PMP pontoons are much heavier and therefore have to be carried on a 7,000 or 7,500 kg (6 × 6) truck chassis. Bridging boats used with the PP-64 are the RFAS BMK-130 and the Polish KH-200.

Wojskowe Zaklady Inzynieryjne (Army Engineer Workshops) of Deblin, Poland, offers a complete GSP repair service as well as anti-corrosion treatments using up-to-date materials.

Specifications

Pontoon
Weight: 1,000 kg
Length: 3.7 m
Width: 6.2 m
Depth: 0.85 m

Status

In service with the Polish Army.

Manufacturer

Wojskowe Zaklady Inzynieryjne, ul 15 Pulku Piechoty Wilków, PL-08-522 Deblin, Poland.
Tel: +48 831 52996.
Fax: +48 831 52996.

VERIFIED

ROMANIA

PR-60 heavy pontoon bridge

Description

The PR-60 was developed by Romania as a replacement for the Soviet-supplied TMP heavy pontoon bridge. It can also be used as a raft. The PR-60 is not of the folding type like the RFAS PMP and Polish PP-64 and is therefore slower to construct.

The PR-60 consists of enclosed shore and river pontoons, which are launched from the rear of the Bucegi SR-114 (4 × 4) 4,000 kg truck by gravity and are recovered by being lifted out of the water by a crane as the truck has no recovery capability. Once launched the pontoons are connected to form a continuous roadway. Both single- and double-lane bridges can be constructed, the latter with double pontoons.

A complete PR-60 pontoon set consists of 56 river pontoons and four shore pontoons. Each Bucegi SR-114 (4 × 4) truck carries either one shore pontoon or two river pontoons resting on top of each other.

Specifications

	River pontoon	Shore pontoon
Weight	1,200 kg	1,350 kg
Length	6 m	6.2 m
Width	2.4 m	4.2 m
Height/depth	0.7 m	0.7 m

Bridge construction		
Bridge type	**40 t**	**60 t**
Roadway width	3.8 m	7.6 m
Bridge length	142.8 m	80.4 m
Assembly time	1 h	1.5 h
Number of pontoons*	50	60

* Including shore pontoons

Status

In service with the Romanian Army. Offered for export.

Manufacturer

Romanian state factories.

VERIFIED

RUSSIAN FEDERATION AND ASSOCIATED STATES (CIS)

DPP-40 floating bridge equipment

Description
Reports have been made of a new floating bridge equipment known as the DPP-40. The DPP-40 uses inflatable pontoon supports and is capable of carrying light armoured vehicles. From what little information is available it would appear that the DPP-40 is used mainly by airborne formations.

Status
In service with the RFAS armed forces.

VERIFIED

GSP heavy amphibious ferry

Description
The GSP (*Gusenichniy Samokhodniy Parom*), introduced in 1959, was the standard heavy amphibious ferry of the former Warsaw Pact armed forces. Numbers were used by the Egyptian Army during the crossing of the Suez Canal in October 1973.

A complete GSP ferry consists of two units, left and right, they are not interchangeable, enter the water separately and are then linked up. Once linked together the pontoon, which when travelling on land is in the inverted position, is swung through 180° into the floating position. Each unit carries retractable trackways which enable vehicles to be loaded or unloaded either side. The GSP can carry a maximum load of 52,000 kg. According to US reports, under favourable circumstances a tank is able to fire its main armament when loaded on the GSP.

The track suspension of the GSP is similar to that used on members of the PT-76 light amphibious tank family and consists of seven roadwheels with the idler at the front and the drive sprocket at the rear. There are no track-return rollers. The hull of the GSP is of lightweight welded steel filled with plastic foam, both to increase its buoyancy and to reduce its vulnerability to damage by enemy fire. Before entering the water a trim vane is erected at the front of the hull, although this was not fitted to early units. Each unit is propelled in the water by two propellers mounted in separate tunnels under the hull. As a direct result of user experience the GSP has been modified in recent years. These modifications include an improved suspension which has given the unit a higher ground clearance, a stronger hull and an improved cab for the crew at the front of the hull.

The GSP has two major disadvantages: GSPs cannot be joined together to form a floating bridge and the bank where the GSP unloads must be no higher than 0.5 m with a minimum water depth of 1.2 m, otherwise the GSP will be damaged.

Wojskowe Zaklady Inzynieryjne (Army Engineer Workshops) of Deblin, Poland, offers a complete GSP overhaul service as well as undertaking major repairs and spare parts supply. It can also replace some GSP components with parts of local manufacture.

GSP heavy amphibious ferry (Michael Jerchel) ***1996***

Specifications
Half ferry unit
Crew: 3
Weight: 17,000 kg
Length: 12 m
Width: 3.24 m
Height: 3.2 m
Ground clearance: 0.35 m
Track: 2.62 m
Track width: 360 mm
Length of track on ground: 4.83 m
Ground pressure: 0.52 kg/cm^2
Max speed: (road) 40 km/h
Range: (road) 300 km (estimate)
Fuel capacity: 370 litres
Fording: amphibious
Gradient: 45%
Vertical obstacle: 0.8 m (estimate)
Trench: 3 m (estimate)
Engine: Model 8D6 6-cylinder water-cooled diesel developing 240 hp at 1,800 rpm

Full ferry
Weight: 34,000 kg
Length: 12 m
Width: 12.63 m
Draught:
(unloaded) 0.97 m
(loaded) 1.5 m
Max water speed:
(unloaded) 10.8 km/h
(loaded) 7.7 km/h
Roadway width: 3.54 m
Roadway track: 1.66 m
Max payload: 52,000 kg

Status
Production complete. In service with Afghanistan (6 - present status uncertain), Egypt, India (18), Iraq, Israel, Nicaragua, Uganda, Yugoslavia (Serbia and Montenegro) and members of the former Warsaw Pact, including Poland. It was known as the GSP-55 in East Germany.

Manufacturer
State factories.

UPDATED

PMP heavy folding pontoon bridge

Description
The PMP (*Pomtommo Mostovoj Park*, or pontoon bridge set) was considered to be a major breakthrough in the design of floating bridges. It was used successfully by the Egyptian forces during the crossing of the Suez Canal in October 1973. Since the introduction of the PMP (or Ribbon Bridge as it is also known) the US Army has developed and produced a similar bridge system called the Ribbon Bridge. The pontoons on the US system are made of aluminium rather than steel and are therefore much lighter: the US river pontoon weighs 5,440 kg while the RFAS unit weighs 6,676 kg, the US shore pontoon weighs 5,310 kg while the RFAS equivalent weighs 7,252 kg.

Each pontoon of the PMP is constructed of SKhL-4 steel and is in four major sections which are hinged together. These are carried in the folded position on the rear of a 6 × 6 or 8 × 8 truck. Initially the truck used the KrAZ-214 (6 × 6) 7,000 kg truck chassis but this was later supplemented by the more powerful KrAZ-255B (6 × 6) 7,500 kg truck.

The pontoon is launched as follows: the travel locks on the pontoon are disengaged and the truck backs to the edge of the water; it brakes sharply and the pontoon slides over a roller system into the water where it almost immediately unfolds; six locking devices are activated stiffening the pontoon; the pontoons are then normally connected together on the near shore to form a continuous roadway which is swung into position by bridging boats. Once in position the bridge is ready for immediate use. The surface of the roadway is ribbed to prevent vehicles from skidding when crossing.

The pontoon is recovered as follows: the pontoon truck backs up to the water's edge and an integral jib is unfolded from the truck bed; two cables are strung from the winch, which is to the rear of the cab, through

PMP pontoon bridge units on their transporter trucks with centre-section on left and ramp section on right

PMP pontoon bridge unit carried by KrAZ-214 (6 × 6) 7,000 kg truck (Michael Jerchel)

Tanks crossing PMP pontoon bridge during exercise

Raft construction						
Type	**40 t**	**60 t**	**80 t**	**110 t**	**150 t**	**170 t**
Length	13.5 m	20.25 m	27 m	39.25 m	52.75 m	59.6 m
Rafts per set	16	10	8	4	4	4
Pontoons per raft	2	3	4	5 + 1 shore	7 + 1 shore	8 + 1 shore
Assembly time	8 min	10 min	12 min	15 min	20 min	20 min

the jib pulley, around the pontoon retrieval guides and secured to the pontoon retrieval studs. The winch then simultaneously folds and lifts the pontoon on to the truck bed. The jib is folded back into the truck bed and the pontoon winched over the roller and secured.

The basic bridge has a capacity of 60 tonnes but it is also possible to build a half-width bridge of 20 tonne capacity and of greater length. This is achieved by splitting the pontoons lengthwise once they are launched. Full-length pontoons are placed at intervals to give greater stability.

The PMP can also be used to construct rafts of varying sizes, the maximum having a capacity of 170 tonnes.

The Czech and Slovak armies use a TATRA 813 or 815 (8 × 8) 8,000 kg roadway truck in conjunction with the PMP; details will be found in the *Portable Roadways* section. They also have some of their TATRA 813 and 815 (8 × 8) 8,000 kg trucks fitted with a dozer blade for preparation of river banks before launching the pontoons. The PMP was used by East German forces to assist in the unloading of landing ships.

A complete PMP pontoon set consists of 32 river pontoons, four shore pontoons and 12 BMK-T (or BMK-130/BMK-130M or BMK-150/BMK-150M) bridging boats. A half set has 16 river and two shore pontoons.

A variant of the PMP is produced in China as the Type 79 Ribbon Bridge (qv).

The Serbian Floating Bridge Set M71 (KPM-M71) is based on the PMP but differs in some details and dimensions.

PMP bridges and components remain in production in the Czech Republic, where the system is known as the PMS. Czech and Slovak Army PMS components are carried on TATRA T 815 VPR9 28.265 8 × 8.1R trucks.

In Germany, EWK produces an all-steel version of the US Ribbon Bridge which is very similar to the original PMP (qv).

PP-91

Reference has been found to a PP-91 pontoon bridge set which is intended to replace the PMP. No details have yet been discovered but the chief designer was V Trefilov of the Scientific Research Institute, together with specialists from the 'Vympel' Central Design Bureau.

Specifications

Bridge construction		
Type	**20 t**	**60 t**
Roadway width	3.27 m	6.5 m
Length of bridge		
(whole set)	389 m	227 m
(half set)	281 m	119 m
Working party	82 men*	82 men*
Assembly time	50 min*	30 min*
Number of whole pontoons incl shore		
(whole set)	34	34
(half set)	18	18

* Data is for whole set
Two additional shore pontoons are held in bridge set

The bridge can be constructed in water with a maximum velocity of 2 m/s

	River pontoon (open)	Shore pontoon (open)
Weight	6,676 kg	7,252 kg
Length	6.75 m	5.58 m
Width	7.1 (3.21 m)*	7.02-7.32 m (3.3 m)*
Depth	0.915 m	0.73 m

* Folded dimensions are in brackets

Status

In production. In service with members of the former Warsaw Pact (except Poland which uses the Polish-designed PP-64 and Romania which uses the Romanian-designed PR-60). It is also used by Afghanistan (15), China (Type 79), Egypt (some carried on German Magirus-Deutz 6 × 6 chassis), India (156), Iraq, Israel and Serbia. The PMP is known as the PMS in the Czech and Slovak armies.

Manufacturer

State factories.

UPDATED

PPS heavy girder floating bridge

Description

The PPS heavy girder floating bridge, first observed in 1962, consists of large sectional pontoons and Warren-type truss girders with a full flushdeck roadway mounted on top. Each pontoon consists of a bow section, two centre-sections and a stern power unit. Each section is carried on the rear of a ZIL-151 (6 × 6) 2,500 kg or similar truck. The pontoon is approximately 23 m long when assembled. The deck is about 380 mm below the gunwales, which support six rows of truss girders, with the top chords carrying I-beam stringers. Decking is laid on these I-beams to form a roadway about 6 m wide. The bridge has a maximum capacity of at least 60 tonnes.

Status

In service with the RFAS.

Manufacturer

State factories.

VERIFIED

SINGAPORE

Floating Bridge System FBS 60

Description

The Floating Bridge System FBS 60 is intended for the construction of floating bridges and ferries and for ease of erection and speed of recovery. The complete FBS 60 consists of bridge sections, a bridge transporter, a bridging boat (described as a push-pull tug boat) and a boat trailer or boat carriage. For details of the associated bridging boat and its trailer or carriage refer to the entry in the *Bridging Boats* section.

An FBS 60 floating bridge consists of two types of section, inner and ramp. Both can be folded upwards for transport and will unfold automatically when launched into the water. The inner and ramp sections are made of aluminium and can be coupled together using pins to form floating bridges or rafts capable of carrying Class 60 loads. Each section has four main assemblies, two outer pontoons and two inner pontoons. The inner pontoons form the roadway.

A typical FBS 60 floating raft, consisting of two inner sections with a ramp section at each end, is 27.2 m long and can be constructed by a team of 16 personnel. A normal FBS 60 kit consists of nine inner sections and four ramp sections. Using the nine inner sections and two of the ramp sections a floating bridge 74.1 m long can be constructed. If all the inner sections of two FBS 60 sets (that is, 18 inner sections) are combined with two ramp sections, a floating bridge 134.4 m long is formed.

The FBS 60 sections are normally carried on 6 × 6 7,000 kg trucks. These are carried on bridge transporters that are also used for the launching and retrieval of the sections. The complete length of a typical loaded bridge transporter is of the order of 10.8 m and height is around 4 m. Width overall is 3.4 m.

Specifications

Inner and ramp sections
Weight: 5,500 kg
Width:
(folded) 3.2 m
(unfolded) 8.1 m
Height: (unfolded) 1.1 m
Roadway width: 4.1 m

Status
Available.

Manufacturer
Singapore Technologies Shipbuilding and Engineering Limited, 7 Benoi Road, Singapore 629882.
Tel: +65 861 2244. Telex: singa rs 21206.
Fax: +65 861 3028.

UPDATED

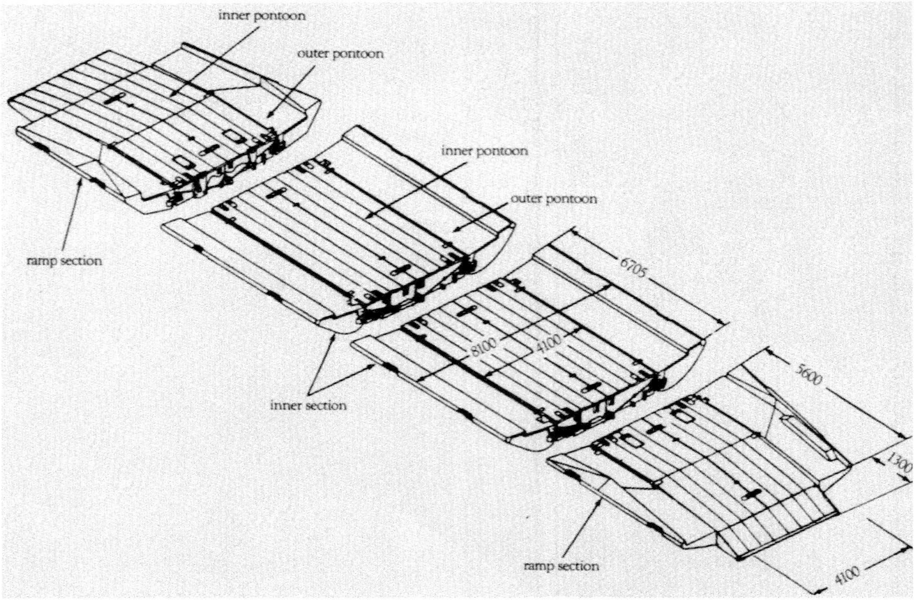

Inner and ramp bridge sections of the Floating Bridge System FBS 60

THAILAND

Aluminium pontoon raft

Description
This aluminium pontoon raft, described in sales brochures as a metal alloy flotation for raft building, is essentially similar to the US M4T6 floating bridge (see entry under USA for details). By combining 16 pontoons, each of which is basically an aluminium box structure supported by pneumatic floats, it is possible to carry MLC 55 loads, including tracked vehicles. Other loads can be accommodated, a typical example being a 155 mm M198 gun and its tractor vehicle with a combined weight of 25 tons. By combining pontoons in pairs it is possible to carry MLC 70 loads.

Specifications
Pontoon
Load capacity: (used in pairs) MLC 70
Weight: approx 500 kg
Length: 5.5 m
Width: 2.03 m
Height: 1.1 m

Status
In service with the Royal Thai Army.

Manufacturer
Office of the Army Research Development and Co-ordination, Royal Thai Army, Rajsima Road, Bangkok 10300, Thailand.
Tel: +66 2824161.

NEW ENTRY

Aluminium pontoon raft in use by Royal Thai Army
1997

UNITED KINGDOM

Class 16 air-portable bridge

Description
The Class 16 air-portable bridge was designed by the Military Engineering Establishment at Christchurch (now the Royal Armament and Research Development Centre, Christchurch). Production and marketing of the bridge were undertaken by Laird (Anglesey) Limited (now FAUN Municipal Vehicles Limited).

The bridge is constructed of a high-strength aluminium zinc-magnesium alloy and can be used as a clear span bridge up to 15.2 m in length, as a floating bridge of any length or as a powered raft with a maximum speed of 6 kt. All have a non-skid roadway 3.3 m wide when assembled.

The basic components of the Class 16 bridge are the deck boxes, ramps, articulator boxes, floats and the sponsons. These components are normally carried by Land Rovers, which also tow a single-axle trailer. A Puma helicopter can carry any component of the Class 16 bridge and a C-130 transport aircraft can carry all the components required to build a ferry.

The deck box combines the functions of bridge girders and cross girders, and contributes buoyancy in floating applications. The top of the box forms the deck of the bridge. At either end of the bridge tapered ramps provide access. When being used as a floating bridge or raft, hydraulically operated articulator boxes are fitted between the deck boxes and the ramps to allow adjustment for varying heights. For floating bridges and rafts, pneumatic floats are fitted to give additional buoyancy and stability. When being used as a floating raft for powered raft operations, sponsons with integral turntable brackets are fitted at each corner of the raft, each sponson having a 40 hp outboard motor.

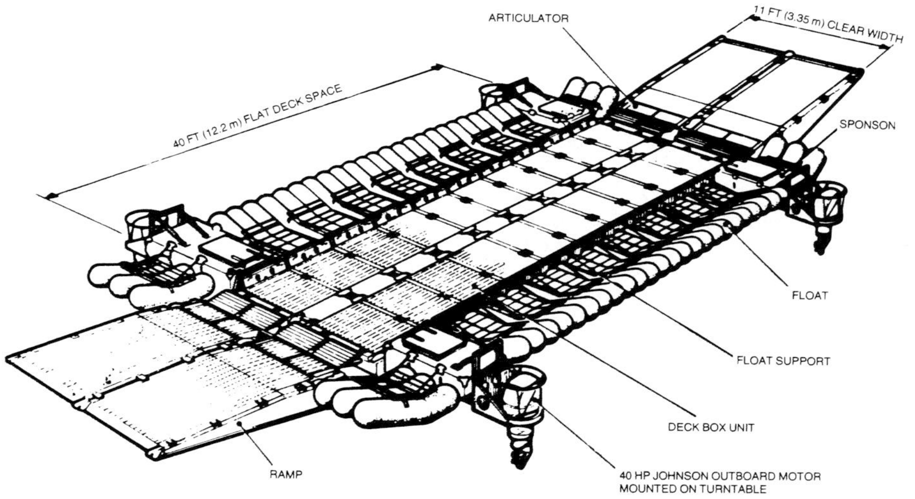

Class 16 air-portable bridge configured as a raft

Clear span bridge
A 15.2 m clear span bridge is assembled from seven deck boxes and two pairs of ramps. Three Land Rovers and trailers carry all the components for one 15.2 m clear span bridge. The bridge can be constructed and positioned by 16 personnel in approximately 20 minutes, using a special launching nose and rollers. A launching nose and three pairs of rollers are used for building and launching the bridge. One pair of rollers is subsequently used as a landing roller on the far bank. The angle of the launching nose can be altered by means of its built-in jack to allow for varying heights of the far bank. The far end of the bridge is lowered to the ground by operating a jack.

Floating bridge
A floating bridge 58 m long can be assembled from 40 deck boxes, two pairs of articulator boxes and two pairs of ramps and floats using a standard bridge set. Five Land Rovers and trailers carry all the necessary

components for a floating bridge. A team of 24 men takes 45 minutes to build the 58 m bridge and there is no limit to the length of floating bridge that can be built. The floating bridge is assembled in the same manner as the raft but the sponsons are omitted and floats fitted in their place.

Powered raft

The standard raft has a 12.2 m level deck and is 22 m overall. It is assembled from 10 deck boxes, two pairs of articulator boxes, two pairs of ramps, floats and four sponsons with outboard motors. There are 20 Land Rovers and trailers carrying all the components necessary to build one standard powered raft. Building time is approximately 40 minutes with 24 personnel. The raft is constructed as follows: the ramps and articulators are assembled and launched in the water, deck boxes added, then the rear articulator and finally the rear ramps, the sponsons and outboard motors are fitted to the raft and the floats are fitted to the boxes. The floats are inflated from the Land Rovers' exhaust gases. Longer rafts may be built up to a maximum length of 28 m overall, with an 18.3 m level deck. These are powered by six outboard motors and have a total distributed load of 24 tonnes.

Specifications

	Deck box	Ramp	Articulator box	Float and support frame	Sponson (with motor and accessories)
Length	3.6 m	3.6 m	1.8 m	2 m	2 m
Width	1.2 m	1.8 m	1.2 m	1.2 m	1.2 m
Height	0.38 m	0.38 m	0.38 m	0.75 m	0.75 m
Weight	305 kg	346 kg	279 kg	23 kg	281 kg

Status

Production as required. In service with Australia, Canada, Nigeria and the UK.

Manufacturer

FAUN Minicipal Vehicles Limited (formerly Laird (Anglesey) Limited), Beaumaris, Anglesey LL58 8HY, UK.
Tel: +44 1248 810431.
Fax: +44 1248 810300.

UPDATED

Williams Fairey Engineering Floating Medium Girder Bridge

Description

Using the Medium Girder Bridge (MGB) pontoon (powered or unpowered), floating bridges can be constructed using MGB system components (for full details of the MGB refer to entry under *Tactical (Non-floating) and Line of Communication Bridges*). Single- or double-storey configurations can be constructed.

For bank heights of 2 m or less, single-storey construction can be adopted. This is the fastest time of construction where 100 m of bridge can be constructed on a normal site in under 1 hour. This configuration requires more pontoons than double-storey bridges.

Double-storey floating construction allows long landing bays of up to 26.5 m and is suitable for higher bank heights, difficult shore lines or where there is a considerable rise and fall in the water level.

The aluminium pontoons used for floating MGBs are the same as those used for the MGB ferry—see following entry.

Status

In production and widespread service.

Manufacturer

Williams Fairey Engineering Limited, PO Box 41, Crossley Road, Heaton Chapel, Stockport, Cheshire SK4 5BD, UK.
Tel: +44 161 432 0281. Telex: 667866.
Fax: +44 161 431 3575.

VERIFIED

A double-storey floating MGB under construction

Williams Fairey Engineering Medium Girder Bridge Powered Pontoon and Ferry

Description

The Williams Fairey Engineering Medium Girder Bridge (MGB) pontoon can be powered by means of a Schottel water jet driven by a Deutz air-cooled diesel engine. The engine assembly is mounted in a lightweight frame, secured to the floor by four clamps and can be rapidly removed for either maintenance or complete in-field replacement. The Schottel pump jet is mounted by clamps in the floor of the pontoon and can also be rapidly removed for servicing.

There are no protrusions on the underside of the pontoons, enabling the unit to operate in shallow or heavily obstructed water without problems of fouling. Steering is effected by rotating the jet outlet which, with its 360° movement, provides the pontoon with a high level of manoeuvrability.

The powered pontoon is primarily intended for use on MGB ferries, with four units on a Class 70 and two each on Class 27 and Class 20 ferries. However, they can also be effectively used during the construction of floating bridges, providing a rapid means of placing piers.

Each pontoon is of open-top construction which allows stacking for transport and storage. The aluminium pontoons have sealed in-built buoyancy which alone provides adequate flotation for the ferry or bridge structure. They are self-draining and anchor winches are fitted as standard. Each pontoon has a (net) buoyancy of 12,000 kg at 300 mm freeboard. They can be stacked with four pontoons one inside the other to form a load 2.7 m high. Such a stack can be carried on a non-dedicated truck of the DROPS type which can be used to launch and recover the pontoons.

The MGB ferry is essentially a single-storey bridge connected to MGB pontoons to provide a range of ferries from MLC 20 upwards. The pontoons may be connected to form rafts by the use of raft saddles which are constructed in three sections: a centre-section and two arms. They pivot to allow the assembly to be stowed within the pontoon during transportation. The construction sequence is similar to that of a long single-

MGB Class 60 ferry carrying Chieftain MBT

Unloading stack of four MGB pontoons using Boughton DROPS/PLS system fitted to Foden 8 × 4 Low-Mobility truck

storey MGB. The end of the bridge is supported on a roller beam and the first landing bay is completed by adding top panel bays followed by hinge panels. Further top panels are fitted and the girder boomed out to allow connection between the piers or rafts at the saddle or hinge panel position. The construction continues with further top panels up to the second pair of hinge panels and then completion of the second landing bay.

Williams Fairey Engineering Limited markets a standard range of three ferries: MLC 20, 27 and 70. Other options up to MLC 100 are possible.

Specifications
MGB pontoon
Weight: (less power pack) approx 1,100 kg
Length: 7.96 m
Width: 2.6 m
Height: 1.17 m
Max current speed: (laden) in excess of 3 m/s
Available buoyancy: (net) 12,000 kg at 300 mm freeboard
Power pack: Deutz air-cooled diesel
Water-jet: Schottel pump jet providing thrust through 360°

Status
In production. In service with three countries.

Manufacturer
Williams Fairey Engineering Limited, PO Box 41, Crossley Road, Heaton Chapel, Stockport, Cheshire SK4 5BD, UK.
Tel: +44 161 432 0281. Telex: 667866.
Fax: +44 161 431 3575.

VERIFIED

MEXEFLOTE - multipurpose pontoon and harbour equipment

Description
MEXEFLOTE, introduced into British Army service during the 1960s, is a multipurpose pontoon equipment designed specifically for marine applications. It is of MLC 60 and can be rapidly constructed as lighterage rafts, for use at sea and in harbours, and can also be assembled as causeways, jetties and other floating structures. The equipment can be used in the following wave conditions pontoons connected into rafts, causeways and jetties in 0.61 m waves; operation of rafts, causeways and jetties in 1.22 to 1.52 m waves; survival of causeway, raft or jetty (unladen) at moorings in 2.74 to 3.05 m waves; survival of causeways or rafts (unladen) in tow in 3.65 m waves.

The system is based on the use of three steel pontoons: bow, centre and stern. These can be connected end to end and side to side to form rafts, causeways, jetties and floating platforms of any shape. The pontoons are of welded steel construction with flush sides. Built into the sides and ends of the pontoons are recessed slots into which the connectors are fitted. There is a 50 mm gap between pontoons.

The bow pontoon consists of a forward section, an aft section and a ramp. The forward section is hinged to the bottom edge of the box-shaped aft section and can articulate vertically to a maximum of 457 mm above the deck level and be lowered to a maximum of 380 mm below the surface of the aft section. The manually operated, demountable articulator is mounted in a recess in the aft section and is connected to the forward section by an articulator ram. The articulator has a safe working load of 81,280 kg. The pontoon ramp is hinged to the forward section and slides over the forward end of the aft section to bridge the gap between the sections.

The centre pontoon is a box-shaped unit with an internal lateral bulkhead dividing the interior into two watertight compartments. Each compartment has a hatch cover, air-line connector and a bilge discharge outlet fitted with nylon plugs set flush with the deck surface.

The stern pontoon is also a box-shaped unit with an internal lateral bulkhead dividing the interior into two watertight compartments. Each compartment has a hatch cover, air-line connector and a bilge discharge outlet fitted with nylon plugs set flush with the deck surface. The bottom edge of the stern pontoon is chamfered to allow the propeller and skeg of the propelling unit to be rotated through 360° when the unit is mounted at the stern of the pontoon.

The pontoon connector is a rectangular unit weighing 73.94 kg, fits into any full-length connector slot of a pontoon and is used to join together pontoons side by side and end to end. Each connector has one fixed pin and one movable pin at the bottom that can be raised or lowered by a handle in the top of the connector to make the bottom connection of adjoining pontoons. The top connection is made by passing a short bolt on each adjoining pontoon through a hole in the connector.

The pontoon link is a triangular box-shaped unit weighing 20.87 kg. This has short bolt holes through the top and a fixed pin at the bottom which fits through a hole in a jaw at the bottom of the short slot in the side of the bow pontoon forward section and in the end of the stern section.

Specifications
MEXEFLOTE

	Bow pontoon	Centre pontoon	Stern pontoon
Weight	5,909 kg	4,654 kg	4,418 kg
Length	7.92 m	6.1 m	6.1 m
Width	2.44 m	2.44 m	2.44 m
Depth	1.45 m	1.45 m	1.45 m

Rafts
The most common types of raft are:
20.22 × 7.42 m which can carry one Class 60 tank or three 4,000 kg trucks.
38.4 × 7.42 m which can carry two Class 60 tanks or six 4,000 kg trucks.
38.4 × 12.9 m (Maxi-MEXEFLOTE) which has a maximum capacity of 198,000 kg and can carry three Class 60 tanks or equivalent vehicles.

A Landing Ship Logistic, or similar ship, is capable of carrying one 38.4 × 7.42 m raft on each side or two 20.22 × 7.42 m rafts on each side. When approaching the beach these are released and can be used as rafts (with the addition of propulsion units) or connected end to end to form a causeway to the beach.

The MEXEFLOTE pontoons can easily be handled as they are compatible with ISO container storage and handling systems: the centre and bow pontoons conform to the 6.1 × 2.4 m container dimensions.

During late 1994 the British Army issued a requirement for the production of 50 MEXEFLOTE centre pontoons.

90 tonne MEXEFLOTE being used as general lighterage raft at Marchwood Military Port ***1996***

Status
In service with the British Army and other armed forces.

Manufacturer (Original)
GEC Engineering (Accrington) Limited, Blackburn Road, Clayton-le-Moors, Accrington, Lancashire BB5 5JW, UK.
Tel: +44 1254 382151. Telex: 63124.

VERIFIED

Storey Uniflote system

Description
The Storey Uniflote was conceived in the late 1950s as a flotation system based on unit construction principles, in which identical flotation units can be assembled together to form rafts of various load-carrying capacities, as well as pontoon bridges.

Storey Uniflote is used by military forces throughout the world for line of communication bridges, floating bridges, ship-to-shore causeways and as vehicle and personnel rafts. It is also widely used for civil purposes such as landing stages, ro-ro terminals, and temporary applications to carry land-based plant such as cranes, excavators and pile-driving equipment for marine works.

All Storey Uniflote equipment can be carried on standard commercial vehicles and if necessary can be skidded into the water. Assembly in the water is accomplished by a maximum of four personnel, and the units are held together by locking pins inserted in the couplers.

The standard Uniflote is 5.283 m long, 2.438 m wide and 1.219 m deep. It is a structural steel-framed unit of all-welded construction with 4 mm skin-plates welded to the frame. Two internal watertight bulkheads are incorporated to provide three watertight compartments, each with a watertight hatch. Individual compartments in the flotation units can be flooded or emptied by compressed air. Under a load of 10,000 kg, each Uniflote maintains a freeboard in the region of 0.23 m. Couplers are placed so that Uniflotes can be joined end to end, side to side and end to side. Steel gunwales are provided along the side of the Uniflote and are drilled to allow simple saddles to be fitted. The couplers allow for the transmission of loads throughout a Uniflote raft. Concentrated loads can be applied to the gunwales through saddle attachments. Runners are attached to the bottom to assist in skidding operations on shore, and four lifting shackles are fitted. The Uniflote is also available in a 1.828 m deep version.

The standard Uniflote has the deck set approximately 80 mm below gunwale level so that a replaceable timber deck may be fitted if required. This is fitted in the form of three preassembled mats (one centre and two outer sections) to each Uniflote. Each mat is designed so that it is retained laterally by the gunwales and is of such a thickness that it stands 50 mm above the gunwale level. Side rings enable Uniflotes to be craned from the shore to the water. Cross junction and side junction mats are also available. The special military version has an integral steel deck set level with the top of the gunwales over which traffic may drive direct.

Scow ends are designed to be attached to either end or side of the Uniflote and have identical end sections. They also have short lengths of gunwale to correspond with the gunwales in the Uniflote. The bottom plate slopes up at an angle of 30° and terminates in a reinforced nosing plate on which a bollard is mounted.

The ramp unit can be connected to the end or sides of the basic Uniflote by ramp connectors which allows the ramp to articulate and gives access for shore loading over a wide variation in bank heights.

The various Interflote connectors available allow Uniflotes to be spaced apart, giving greater stability to floating platforms where required without the need for additional buoyancy.

A range of saddles is available to allow the secure fixing of winches and propulsion units. These saddles can be fitted over a bow or stern unit, over the junction between a bow or stern unit and a Uniflote, over a Uniflote, over the junction between two Uniflotes, or an outrigger (fitted to the side of a Uniflote). Other saddles are available for fixing Acrow panel or Bailey bridging, and other ancillary equipment.

Details of the Storey Motorflote are given in the *Bridging Boats* section.

Specifications

Standard Uniflote (U4/1A)
Weight: 2,895 kg
Length: 5.283 m
(coupler to coupler) 5.41 m
Width: 2.438 m
Height: (without gunwale) 1.219 m
Gunwale size: 76.2 × 76.2 mm

Bow Unit (female U4/3A, male U4/2A)
Weight: 843 kg
Length: 1.82 m
Width: 2.438 m
Height: 1.219 m

Ramp Unit (U4/6A)
Weight: 1,879 kg
Length: 3.658 m
Height: 1.219 m

Status

In production. In service with many armed forces all over the world, including Belgium and Brazil.

Manufacturer

Thos Storey (Engineers) Limited, 52 Queens Road, Weybridge, Surrey KT13 0AN, UK.
Tel: +44 1932 858133. Telex: 21868.
Fax: +44 1932 855588.

50 tonne Storey Uniflote ferry carrying Grove Coles 315M crane truck

VERIFIED

FBM Marine Mexecell modular logistics system

Development

The Mexecell modular logistics system was developed by FBM Marine Limited, using experience gained during the production of a similar pontoon system for the British Army's Military Port at Marchwood, near Southampton. FBM Marine Limited won a contract to design and project manage the construction of a Diving Platform for the US Navy's Underwater Construction Teams (UCTs). FBM's US partner for the Diving Platform is the Marinette Marine Corporation, Marinette, Wisconsin.

The contract was to design and construct a modular floating platform for use in diving operations. The UCTs are responsible for the repair and maintenance of various types of US Navy fleet systems such as moorings and cables; and to form harbours during a contingency or active service. The current platform is being phased out and this contract covers the prototype platform, the first of six to be delivered. The 60 × 32 ft platform is constructed from the FBM Mexecell ISO-compatible steel pontoon units.

The Mexecell modular concept provides natural building blocks, together with a simple but effective connector system offering total versatility, rugged construction, and allowing the cell units to be assembled on site without the need for specially trained personnel. The ISO-compatible cells based on the 20 × 8 × 4.5 ft half height container module enables the component parts of the platform, or any other Mexecell-based structure, to be transported to the operational area using conventional containerised shipping, road or rail systems.

Description

The Mexecell is an ISO-compatible steel cell capable of being used in multiple configurations for logistical support and maritime related applications. The system consists of four basic units: bow, centre, stern and jet cell modules. They can be connected together end to end and side to side to form any desired configuration. To increase flexibility the Mexecell bow module can be raised above deck level to give a raked end, or lowered below the bottom of the module to provide a loading ramp.

Applications for the Mexecell system include floating causeways, beach landing ferries, ro-ro discharge platforms, underwater construction platforms, drilling platforms, offshore elevated platforms, barge trains, diving support platforms, elevated structures and offshore wave breaks.

Artist's impression of offshore offloading facility and ferry raft constructed using FBM Marine Mexecell modular logistics system

Mexecell is designed for total integration for transport by container ships, container road vehicles and ISO container handling equipment. The deck loading capability can cope with wheeled vehicles up to MLC 60 and tracked vehicles up to MLC 70. The Mexecell design features an open cell structure free of any internal bracing, which allows the interior to be used for the carrying of cargo such as fuel or water in flexible bags for bulk liquid transfer, general stores and system components. The cell has flush sides and decks to minimise the risk of damage during assembly.

Mexecell can be clipped and unclipped afloat in conditions up to Sea State 3 and assemblies can survive in conditions up to Sea State 6.

The Mexecell uses high-precision jaw units that form the key to the fabrication concept. The jaws can be replaced in the field with the minimum of support equipment. Assemblies are not compromised when individual modules are removed for repair or replacement.

Mexecell modules are available in two lengths, 6.096 and 12.19 m. Accessories for the system include a range of 'quick fit' items including bitts, cleats, fender attachments, navigation lighting, beach ramps, spud wells (for external pile fixing) and spud cells (for internal pile fixing). The last two are for fixing up to 0.609 m piles.

Also available is a self-contained propulsion unit which consists of a 360° steerable water-jet. The unit has a flush bottom profile for shallow water operations with a 210 hp marine diesel powering the water-jet. The engine, pump jet, fuel tank and all ancillary equipment are contained in a Mexecell stern module. All manoeuvring controls are remotely led via an umbilical cord to a driver's harness pack to allow complete control and give the driver mobility across the deck. For very large Mexecell structures propulsion can be provided by propulsion units connected as bow thrusters, with all controls grouped together in one pack. A fixed console can also be fitted.

For use in the elevated causeway (ELCAS) application Mexecell is connected using a cantilevered construction technique above the surf zone. An offloading shore facility can be fully operational within seven days of the arrival of its transporting vessel at the erection site. ELCAS represents an entire system of cranes, pile drivers, pile extractors, lighting, generators, turntables, vehicle handling equipment and Mexecell modules. All the equipment is transported from a seaward discharging location by Mexecell floating causeways.

Specifications

Individual Mexecell
Length:
(bow pontoon) 7.92 m
(centre and stern pontoons) 6.1 m
Width: (all pontoons) 2.44 m
Height: (all pontoons) 1.45 m

Status

In service with US Department of Defense.

Manufacturer

FBM Marine Limited, Cowes Shipyard, Cowes, Isle of Wight PO31 7DL, UK.
Tel: +44 1983 297111. Telex: 86466 fambro g.
Fax: +44 1983 299642.

VERIFIED

UNITED STATES OF AMERICA

Aluminium floating footbridge

Description

The aluminium floating footbridge consists of treadways, pontoons, handrail posts, ropes, holdfasts and approach posts. One bridge set contains 144 m of bridge in normal mode of assembly and is allocated on the scale of one bridge set per corps/army engineer float bridge company. Half the bridge set is carried on a 2½ tonne (6 × 6) truck which tows a 2½ tonne utility pole-type trailer. The complete bridge can be carried in a C-130 Hercules transport aircraft and one half set of the footbridge can be delivered by parachute from a C-130 using two 2,721 kg bearing platforms.

The treadways are made of aluminium and consist of two I-beams carrying traverse channels supporting a corrugated aluminium sheet tread. The ends of the I-beam are fitted with spring-loaded connectors, male at one end and female at the other to provide connection between the treadways.

The pontoons are of sheet aluminium reinforced with light aluminium members. The pontoon has a false bottom 171 mm above its true bottom which provides a compartment filled with a light cellular plastic material. Each gunwale is fitted with two hooks which grip the outer bottom flanges of a treadway I-beam. Each pontoon has a hole 25 × 51 mm in the bow and stern just above the false bottom, to make the pontoon self-bailing.

The aluminium handrail post is 32 mm in diameter and is mounted on a base of aluminium bar. This is installed by inserting the base in a socket in the treadway and rotating the post 180° to lock it in position. Each bridge set contains 12.7 mm diameter manila rope which is cut as required for handrail line, guy lines and bridle ties. There are two 182 88 m reels of 9.5 mm galvanised wire rope for use as anchor cables and guys for improvised cable towers provided, together with 20 wire rope clips, four holdfasts (each complete with nine pickets) and 16 approach posts.

The bridge is normally assembled using the successive bay method but if the water is too deep for this method the bridge is assembled in sections, which takes twice as long. The bridge can also be assembled on the shore, which requires a straight shore assembly area as long as the bridge, cleared to a width of 6.096 m. Finally the bridge can be assembled in the water or on steep banks. In this case two skids are set in the water and inclined against the bank. This allows the bays or bridge sections to be launched into the water. In currents up to 2.438 m/s the number of troops that can cross single file at two-pace intervals is as follows:

Daylight: 75 personnel/min, single file, double time
Moonlight: 40 personnel/min, single file, quick time
Blackout: 25 personnel/min, single file, quick time

If the current velocity is between 2.743 and 3.352 m/s, the crossing rates should be reduced by 20 per cent.

If required, the pontoons can also be used as rafts. An expedient two-pontoon raft is formed by lashing two pontoons side by side with one treadway across them. This will carry one wounded man and two paddlers, one in the bow and one in the stern. A three-pontoon expedient raft is formed by lashing three pontoons side by side with one treadway across them and an additional treadway along each side of the centre treadway. This can carry four wounded men and four paddlers, two in the bow and two in the stern.

An expedient bridge 30.48 m long, which will take a Jeep-type vehicle and trailer, can be constructed from the components of the aluminium footbridge set. This may not be used where the current exceeds 1.524 m/s.

Colombian Army bridging truck loaded with aluminium floating footbridge pontoons

Specifications

Assembly time
Day: 15 min plus 1 min/4.572 m of bridge
Night: (with illumination or moonlight) 20 min plus 1¼ min/4.572 m of bridge
Blackout: 30 min plus 2 min/4.572 m of bridge

Maximum safe allowable deflection

Number of bays	Effective length of bridge	Max allowable deflection
4	13.716 m	0.05 m
6	20.57 m	0.1 m
8	27.43 m	0.152 m
10	34.29 m	0.254 m
12	41.148 m	0.381 m
18	61.72 m	0.863 m
24	82.29 m	1.549 m
36	123.44 m	3.479 m

Treadway
Weight: 38.1 kg
Length:
(overall) 3.555 m
(effective) 3.428 m
Width:
(overall) 0.711 m
(walkway) 0.527 m
Depth: 0.133 m

Pontoon
Weight: 45.36 kg
Length: 4.267 m
Width: 0.609 m
Depth: 0.367 m (gunwale)

Post
Weight: 1.02 kg
Length: 1.092 m

Weight of complete bridge set: 4,105 kg

Status

In service with the US Army and other countries including Australia, Colombia and South Korea. Licence-produced in Greece (qv).

VERIFIED

Bridge Floating: Raft Section, Light Tactical

Description

This bridge set has sufficient components for one four-pontoon reinforced raft, or 13.41 m of normal bridge. The same set is used in assembling a combination of floating bridges or rafts of various classes. One bridge set is carried by two 2½ ton (6 × 6) trucks, one of which tows a 2½ ton pole-type trailer that carries the eight half-pontoons stacked one on top of the other.

Two aluminium half-pontoons are joined stern to stern to form a complete pontoon. The deck panel is positioned and retained laterally by four retainer lugs on the pontoons. One end of each deck panel is male and the other female. Two deck filler panels are used to fill the space between one set of deck panels. These are normally retained in position by pintles, but provision is made for bolting the filler panels to the deck panels. Two types of ramp panel are provided, one to mate with the male end of a deck section and one with the female end. The articulating assembly is provided to permit variations in the end span abutment elevations in bridges and rafts from 1.041 m above the horizontal position of the deck to 0.482 m below. It consists of male and female sections. A connecting pin and adjusting bar are used to join the two sections.

There are two sizes of kerb. The deck panel kerb is used on the normal bay of the superstructure and the short ramp kerb is used on the ramp and articulator panels. These are held in position by holding lugs that extend from the bottom of the kerb and bear directly on the underside of the top flange of the deck panel.

A raft normally consists of three bays of decking with ramps at each end on four complete pontoons. The loading space from kerb to kerb is 2.743 m, and loading length is 8.992 m.

The bridge is rated as Class 11 in a current of 2.438 m/s and the four-pontoon three-bay raft as Class 12 in a current of up to 2.438 m/s. The light tactical bridge has loading space kerb to kerb of 2.743 m with the pontoons normally being spaced at 3.352 m centres. Reinforced bridges have the pontoons spaced at 2.438 m centres.

Each pontoon is provided with a fluked marine anchor and each raft set is issued with four outboard motor brackets. The outboard motors normally used are rated at 25 hp.

A six-pontoon, four-bay reinforced raft has a length of 21.336 m, a loading length of 12.192 m and a roadway width (kerb to kerb) of 3.352 m.

Ferry conversion set
Trail ferry method
A bicycle traveller allows the raft to move smoothly along a ferry cable. The ferry cable is stretched across the river and made fast. The bicycle traveller is attached to the cable so that its sheaves roll smoothly. The hauling line is attached to the grommet in the bicycle traveller and the manoeuvre lines to the snatch blocks which are attached to the sheave housings of the traveller. On the raft, the hauling line is attached to the upstream end of the centre pontoon. The manoeuvre lines are attached to the gunwales of the outside pontoons. The ferry is operated by adjusting the manoeuvre lines so that the raft is at an angle to the stream current. The upstream end of the pontoon inclines towards the opposite shore. The current pushes against the upstream side of the pontoon and forces the raft across the river. As the pontoons are pushed into the current, the speed of the raft increases. This method may be used when the current exceeds 0.914 m/s.

The flying ferry works on a similar principle to the trail ferry except that the raft is held in the stream by an anchor well upstream from the crossing site. As the raft moves from shore to shore, it swings in an arc of a circle centred on the anchor. This ferry requires a maximum stream or river velocity of 1.219 m/s.

Specifications
Half pontoon
Weight: 295 kg
Length: 5.638 m
Width: 2.044 m
Depth: 0.863 m

Deck Panel
Weight: 256.28 kg
Length: 3.701 m
(effective) 3.352 m
Width: 1.066 m
Depth: 0.323 m

Deck filler panel
Weight: 43 kg
Length: 1.647 m
Width: 0.787 m
Height: 0.161 m

Ramp panel
Weight:
(male) 149.9 kg
(female) 181 kg
Length:
(male) 2.438 m
(female) 2.136 m
Width: 1.066 m
Height: 0.323 m

Articulating assembly
Weight: 290 kg
Length: 2.2 m
Width: 1.066 m
Height: 0.514 m

Kerb
Weight:
(normal) 49.9 kg
(short) 9.07 kg
Length:
(normal) 3.295 m
(short) 0.865 m

Construction time
4-pontoon, 3-bay raft: 30 min
5-pontoon, 5-bay raft: 35 min
6-pontoon, 4-bay raft: 45 min

Status
In service with the US Army and other armed forces including South Korea. Licence-produced in Greece (qv).

VERIFIED

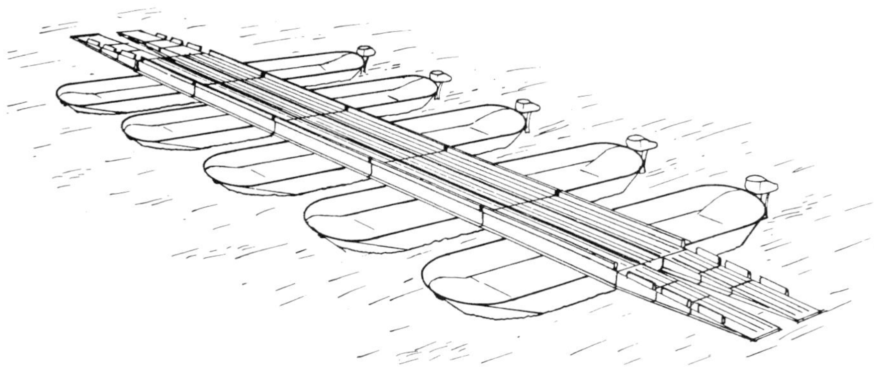

Bridge Floating: Raft Section, Light Tactical (five-bay normal raft)

Ribbon Bridge

Development
The Ribbon Bridge was developed by the Pacific Car and Foundry Company of Renton, Washington, and the US Army Mobility Equipment Research and Development (MERADCOM) Center (now the Belvoir Research, Development and Engineering Center) at Fort Belvoir, Virginia and is based on the design of the Soviet PMP Ribbon Bridge. The Ribbon Bridge has two main differences from the PMP: it is aluminium rather than steel and so weighs about a third less, and the PMP uses torsion bars to help unfold the pontoons, which is effective in launching but means that the torsion force must be overcome during retrieval, which results in longer retrieval times. Development of the US bridge began in 1969 and the bridge was type classified as standard A in June 1972, less than three years after development started. The first production contract, worth $10 million, was awarded to the Consolidated Diesel Electric Division of the CONDEC Corporation at Old Greenwich, Connecticut (later ConDiesel Mobile Equipment of Waterbury, Connecticut), and was for 250 interior bays, 50 ramp bays and 300 transporter trucks.

The Ribbon Bridge augments and replaces the mobile assault bridging emplaced by the forward elements. It can be emplaced 10 times faster with a fifth of the personnel required to emplace standard floating bridges. The bridge can support 80 tonne loads in currents of up to 2.438 m/s and up to 110 tonnes when conditions are ideal.

The Ribbon Bridge system was first fielded by the US Army in Germany and South Korea in 1976. A US Army bridge company has 18 interior bays and 12 ramps.

A FY84 request was for 585 interior bays and 257 ramp bays. The resultant contract was awarded to ConDiesel Mobile Equipment.

Southwest Mobile Systems (now Systems and Electronics Inc (SEI)) was the only producer of the

US Army Ribbon Bridge assembled during exercise in Holland (C R Zwart)

M812 5 ton (6 × 6) truck carrying Ribbon Bridge interior bay unit

Ribbon Bridge Transporter between 1984 and 1993, during which period it completed two multiyear contracts with the US Army for over 500 systems.

In July 1993 Precision Machining Inc, of Pensacola, Florida, was awarded a $5 million firm fixed price contract for 67 Ribbon Bridge interior bays and 27 ramp bays. Work was completed by May 1995.

Description

The Ribbon Bridge consists of integral float-deck elements connected longitudinally to form a continuous floating roadway. There are two basic bridge elements, the interior bay and the ramp bay. The interior bay consists of the roadway pontoons and two bay pontoons joined by hinges and pins. During deployment, the bow and roadway pontoons automatically unfold in the water. The ramp bay unfolds in the same manner as the interior bay. Ramp bays also have the capability to adjust to river bank slopes of up to 20° through the use of hydraulic cylinders.

The bays are carried folded on the rear of specially modified M812 5 ton (6 × 6) trucks (or other 6 × 6 trucks having the same capacity). Launching of bridge bays may be accomplished by one of three modes: free, controlled or high bank. In the free launch mode the truck is backed into the water. A minimum depth of 0.914 m is required to free launch an interior bay (1.125 m for ramp bay) with a 10 per cent river bank slope. Water pressure and the pontoon hinging arrangement permit the bay to automatically unfold. The bridge assembly crew then lock the bay sections using a series of latches. Bays are joined together using a lock pin mechanism and a special T-wrench.

The bridge is normally assembled along the shore and the complete bridge is then swung across the river using bridge erection boats. Bridging boats are used to hold the bridge in position and in a current with a velocity of 2.5 m/s are normally positioned every 20 m. If the bridge is to remain in place for extended periods, overhead anchor cables may be employed. It can also be constructed using the method of successive bays or rafts. If required, individual bays can be emplaced by helicopter. The Ribbon Bridge can be constructed at the rate of 6.705 m/min over rivers with a current of up to 2.438 m/s. The roadway itself is 4.089 m wide and there is a 1.219 m walkway on either side.

To retrieve a bridge bay the transporter is backed into the water and the boom is raised to the vertical. The lifting cable runs from the winch over a sheave at the top of the boom and down to the bridge bay. The cable hook is attached to the roadway hinge pin and, when the cable is reeled in, one end of the bay is lifted out of the water until water pressure acting on the other end causes the bay to fold up. At this time, a series of latches automatically lock the bay in the folded position. The bay is then lowered into the aft tiedown locks at the back of the transporter, the boom is lowered, and the bay is winched onto the transporter. The front locking pin is engaged and the vehicle is ready to drive away. Bridge sections can be launched or retrieved by one man but each transporter carries an assistant to speed up operations, especially during retrieval, and to assist in assembling the bridge.

The bridge transporter was originally a dedicated vehicle of limited utility when not carrying bridge equipment. In order to overcome this shortcoming, a special pallet was developed which permits the transporter to haul up to 9,000 kg of cargo (4,500 kg across country). The pallet can be self-loaded or offloaded while carrying up to 4,500 kg, a feature which lends itself to pre-palletised loads.

US Army Ribbon Bridge in use during an exercise in Germany

Specifications

Interior bay
Weight: 5,443 kg
Length:
(folded) 6.93 m
(unfolded) 6.93 m
Width:
(folded) 3.22 m
(unfolded) 8.13 m
Height:
(folded) 2.31 m
(at bow, unfolded) 1.12 m
(at roadway, unfolded) 1.09 m

Ramp bay
Weight: 5,307 kg
Length:
(folded) 5.79 m
(unfolded) 7.79 m
Width:
(folded) 3.2 m
(unfolded) 8.13 m
Height:
(folded) 2.31 m
(of roadway, unfolded) 0.736 m
(of shore end, unfolded) 0.381 m

Transporter loaded
Weight: 17,931 kg
Length: 9.6 m
Width: 3.43 m
Height: 3.91 m

Status

In service with the US Army and some other armed forces. Adopted by the German Army (see separate entry), South Korea and the Netherlands.

Manufacturer

Ribbon Bridge Transporter
Systems and Electronics Inc (SEI), 201 Evans Lane, PO Box 5639, St Louis, Missouri 63121-1126, USA.
Tel: +1 314 553 4950.
Fax: +1 314 553 4909.

UPDATED

Bridge, Floating Aluminium, Highway Type, Deck Balk Superstructure on pneumatic floats (M4T6)

Description

The M4T6 floating bridge consists of the substructure of the Class 60 bridge (24 tonne pneumatic float with saddle assembly) adapted to the superstructure (aluminium balk) of the M4 bridge. It is hand-erectable and air-portable and can carry all vehicles in the armoured and infantry division. The bridge is issued on the scale of division (armour, infantry and mechanised) engineer battalion: four sets, and corps/army float bridge company: five sets.

The M4T6 floating bridges, rafts and ferries consist of a deck built of square, hollow aluminium sections (balk), supported on pneumatic floats. The pneumatic float consists of two half-floats assembled stern to stern. The saddle panels rest on the float and are attached to it with straps through the D-rings. The saddle beams rest on the saddle panels with the carrying handles of the saddle beams in line with the D-rings on the saddle panels. The straps on the floats are run through the D-rings on the saddle panels, and then through the carrying handles of the beam. The spring-actuated catches of the saddle panels are then placed over the flanges of the saddle beams. The saddle adaptors rest on the saddle beams and are connected using sliding retainer lugs. The balk-connecting stiffeners rest on the saddle adaptors and are connected to them by four connecting pins. The balk-connecting stiffeners support the balk. Each of the latter has lugs which fit into recesses in the balk-connecting stiffener and are secured by balk-connecting pins. Kerb adaptors inserted between the balk and the balk-connecting stiffener are used to raise the balk to form a kerb.

The M4T6 bridge is normally assembled using the successive bay or successive raft method. Reinforced floating sections are constructed by placing the floats closer together than normal. Minimum float spacing is 3.048 m centre to centre. The bridge set can also be used to assemble bridges with a reduced capacity. A full-width bridge with floats spaced at intervals of 9.144, 13.716 or 18.288 m can be constructed for light loads, and when heavier loads are required, the capacity of the bridge can be increased by adding floats and/or deck balk without the necessity of breaking the bridge. A half-width bridge using the 12 tonne half-float spaced on 4.572 m centres can also be constructed, but its capacity cannot be increased without dismantling the bridge.

An essentially similar floating bridge/raft to the M4T6 is produced in Thailand. See separate entry for basic details.

Raft constructed from several M4T6 float sections ferrying M728 combat engineer vehicle (US Army)

Specifications

Pneumatic half-float
Weight: 340 kg
Length: 6.705 m
Width: 2.743 m

Float connections
Length of bar: 2.2 m

Saddle
Weight:
(interior saddle panel) 74.84 kg
(end panel) 81.64 kg
(centre beam) 211 kg
(end beam) 125.74 kg

(outriggers) 3 kg
(saddle adaptor, normal) 118 kg
(saddle adaptor, offset) 125 kg

Normal deck balk (22 per floating bay)
Weight: 102 kg
Length: 4.572 m
Width and height: 222 × 234 mm

Short deck balk
Weight: 55.33 kg
Length: 2.54 m
Width and height: 222 × 234 mm

Tapered deck balk
Weight: 453 kg

Kerb adaptor
Weight: 6.804 kg

Abutment plates
Bearing plate
Weight: 74.84 kg
Length: 1.771 m
Width: 0.304 m
Height: 95 mm
Cover plate
Weight:
(short cover plate) 12.7 kg
(long cover plate) 43.99 kg
Length:
(short cover plate) 0.457 m
(long cover plate) 1.625 m

Ramps (4 required each end)
Weight: 107 kg

Status
In service with the US Army and other armed forces.

UPDATED

YUGOSLAVIA (SERBIA AND MONTENEGRO)

Floating bridge set M71 (KPM-M71)

Description
The design of the Floating Bridge Set M71, or KPM-M71, is based on the RFAS PMP heavy folding pontoon bridge. Each set is composed of 32 floating sections and four shore sections loaded onto 36 FAP 2026 BDS/AV-PMP (6 × 4) trucks equipped with special carrying platforms. A further two FAP 2026 BDS/AV-PMP trucks carry a set of bridge surfacing and anchor posts. The set is completed by up to 12 R-M88 bridging boats towed by FAP 2026 BDS/AVG trucks, and 12 C(H)A 70 aluminium boats with 40 hp outboard engines.

The KPM-M71 can be used to bridge water with a minimum depth of 0.4 m at the shore and 0.4 to 0.8 m under the floating sections if the bottom is flat. The maximum water velocity possible during use is 2 m/s. Vehicles driving over the bridge can attain a maximum speed of 20 km/h.

Launching a floating section from a floating bridge set M71 (KPM-M71)

The KPM-M71 bridging set can be assembled to build a pontoon bridge with a 20 or 60 tonne capacity. Ferries with payloads of 40, 60, 80, 110 or 150 tonnes can also be assembled. Bridge assembly time is given as 50 minutes while a ferry can be assembled in 8 to 20 minutes, depending on the size.

When being driven on good roads, a complete KPM-M71 set can reach speeds of up to 60 km/h.

All bridge unit maintenance and minor repairs can be carried out using a set of tools, spare parts and materials supplied with each KPM-M71 set.

Specifications

Section	**Floating**	**Shore**
Length	6.62 m	5.5 m
Width		
(transport)	3.15 m	3.192 m
(floating)	8.03 m	7.186 m
(roadway)	6.5 m	6.5 m
Height		
(transport)	2.22 m	2.2 m
(floating)	1.1 m	1.019 m
Weight of unit	8,380 kg	8,730 kg
Weight with truck	20,540 kg	20,890 kg

Status
In service with the former Yugoslav Army.

Contractor
FDSP - Federal Directorate of Supply and Procurement, PO Box 89, Bulevar umetnosti 2, YU-11070 Novi Beograd, Yugoslavia.
Tel: +381 11 222 2743. Telex: 71000 JU SPDR.
Fax: +381 11 324 8738.

VERIFIED

M-70 light pontoon bridge

Description
The M-70 is of local design and construction and can be used both as a light pontoon bridge and as a raft. Each pontoon consists of two aluminium half-pontoons joined together. The balk and decking are also of aluminium construction. The load of the bridge depends on the velocity of the stream. For example, a 12 tonne bridge can be constructed in a stream with a maximum velocity of 2.5 m/s, and a 16 tonne bridge can be constructed in a stream with a maximum velocity of 2.1 m/s.

A raft normally consists of four complete pontoons (eight half-pontoons) and the balk from three bridge bays, with or without articulators. Again, the load depends on the velocity of the stream. Rafts of 12 tonnes can be constructed for use in streams with a maximum velocity of 2.5 m/s and 16 tonne rafts can be constructed for use in streams with a maximum velocity of 2.1 m/s.

A complete M-70 light pontoon bridge set has a total of 60 half-pontoons plus deck balk and other ancillary equipment; the deck balk is carried in the rear of a TAM 4500 (4 × 4) 4,500 kg truck, which also tows a single-axle pole-type trailer carrying six half-pontoons inverted on top of each other.

Specifications
Pontoon
Weight: 295 kg
Length: 5.64 m
Width: 2.04 m
Depth: 0.86 m

Bridge construction	
Type	**12 t and 16 t**
Roadway width	2.8 m
Length	120 m
Half-pontoons per support	2
Assembly time	45-60 min

Raft construction	
Type	**12 t and 16 t**
Length	9 m
Rafts per set	7
Half-pontoons per raft	8
Working party	30 men
Assembly time	15 min

Status
In service with the former Yugoslav Army.

Contractor
FDSP - Federal Directorate of Supply and Procurement, PO Box 89, Bulevar umetnosti 2, YU-11070 Novi Beograd, Yugoslavia.
Tel: +381 11 222 2743. Telex: 71000 JU SPDR.
Fax: +381 11 324 8738.

VERIFIED

TACTICAL (NON-FLOATING) AND LINE OF COMMUNICATION BRIDGES

AUSTRALIA

Transfield-MBK line of communication bridging system

Development

The Australian Company Transfield Construction, in association with design consultants McMillan, Britton and Kell of Sydney, New South Wales, designed a new line of communication bridging system for the Australian Army. The bridge, built at Transfield Construction's manufacturing plant at Seven Hills, Sydney, New South Wales, is constructed using steel. Bridge construction and launching principles are similar to those of the Bailey Bridge. By using material handling equipment, component size and weight can be increased without jeopardising the speed and ease with which construction, launch and recovery can be accomplished. Thus individual sections can be heavier and are simpler and faster to put together than the equivalent Bailey-type bridge.

Testing was conducted by the School of Civil and Mining Engineering at the University of Sydney and full-scale user trials were conducted by the Australian Army School of Military Engineering. The Australian Army issued an initial contract for 32 bridging sets worth A$7.5 million. The contract value was later increased to A$11 million; production was completed during 1989.

Transfield-MBK line of communication bridging system in use

Pier sets and floating pontoons compatible with the system are under development.

Description

The bridge is two lanes wide with steel decking. The main load-carrying components comprise through-trusses each side which are assembled from standard 5 m long steel panels and chord reinforcements connected by pins. The arrangement of panels and reinforcement depends on the span and loading. Universal jackable end posts enable construction time to be shortened and launching rollers are adjustable for level and tilt. Each Universal end post assembly incorporates a 1.5 m range jacking system. The deck units incorporate huck bolts to avoid distortion problems. The design loading is MLC 70 (single lane) or MLC 35 (twin lane).

The steel used in the bridge construction is Grade 350 to AS1250, with Grade 450 for panel diagonals. Pins are Grade 630 stainless steel and pin plates are Bisalloy 80 (Grade 690). All components are hot dip galvanised to AS1650, ensuring optimum life in all operating environments during storage and deployment.

Status

In service with the Australian Army.

Manufacturer

Transfield Construction Pty Limited, 100 Arthur Street, North Sydney, New South Wales 2060, Australia.
Tel: +61 2 929 8600. Telex: AA21396.
Fax: +61 2 929 7187.

UPDATED

BULGARIA

Prefabricated bridge modules

Description

These prefabricated bridge modules, used by the Bulgarian Army, are intended to be utilised with other rapid bridge-building equipment or else assembled on the remains of existing bridge structures. Each bridge module is 5 m long and can be used to assemble bridges up to 30 m long. For transport the modules are carried two to a standard 5 tonne or similar load capacity truck.

Once assembled a complete bridge has a weight classification of MLC 60 for wheeled vehicles, MLC 80 for tracked vehicles, and MLC 104 for vehicles such as loaded tank transporters. Once assembled the modules can be arranged so that a track width of 4.2 m can carry single lane traffic, and a trackway 6.3 m wide can carry two-way traffic.

The linear weight per metre of bridge is 340 kg and the building rate is between 15 and 20 m/h.

Status

Understood to be in service with the Bulgarian Army. Offered for export sales.

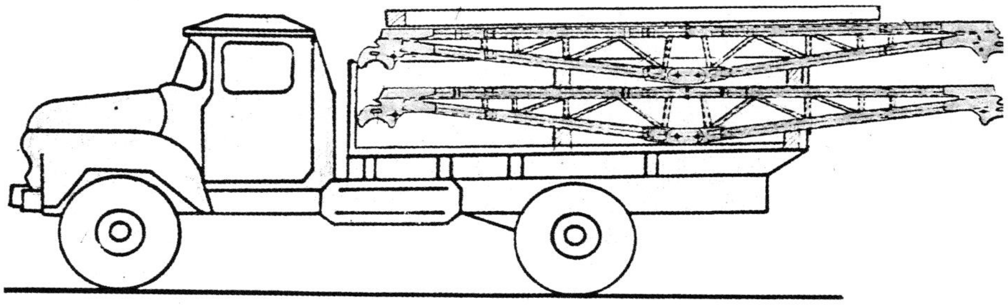

Outline drawing of Bulgarian prefabricated bridge modules being carried on a truck

Marketing agency

KINTEX, PO Box 209, 66 James Baucher Street, 1407 Sofia, Bulgaria.
Tel: +359 2 662311. Telex: 23 243, 22 471.
Fax: +359 2 658191/658101.

VERIFIED

CZECH REPUBLIC AND SLOVAKIA

MS-1 (SM-60) heavy panel bridge

Description

The MS-1 (SM-60) is a single-storey through-truss type bridge which in some respects resembles the Bailey Bridge. The MS-1, however, has truss members made of triangles placed back-to-back to form verticals while the Bailey bridges are of the diamond-shaped pattern.

The bridge has a maximum capacity of 60 tonnes and a maximum clear span of 21 m. Its roadway is 4 m wide. Multispan bridges can be constructed with the aid of trestles. These have circular baseplates and can be adjusted for heights between 1.5 and 7 m. The bridge is assembled with the aid of cranes and is launched by the cantilever method. The heaviest part of the bridge weighs 3,150 kg. The bridge components are carried on standard TATRA 111 (6 × 6) 10,000 kg trucks or similar vehicles.

Status

Production apparently complete. In service with the armies of the Czech Republic, Pakistan and Slovakia.

Manufacturer

Czech and Slovak state factories.

VERIFIED

MS-1 (SM-60) heavy panel bridge showing trestles

GERMANY

Eurobridge Foldable Bridge

Development

During 1984 the Federal Procurement Office of the German armed forces awarded a contract to Dornier System GmbH, for the final development and production of an operational prototype of a foldable dry support bridge with a span of 40 m. The first Foldable Bridge set was handed over to the German Ministry of Defence proving ground at Koblenz in May 1987 for intensive tests. The German Army certified the bridge in early 1991 and approved a modified version in July that year.

A 45.9 m preseries type bridge was built for company use and was demonstrated following a short Ministry of Defence test starting in August 1990.

In June 1992, it was announced that the Dornier-led DASA and CNIM of France had founded Eurobridge Mobile Brücken GmbH for the further development, production and marketing of the Dornier Foldable Bridge and for the joint development of future bridges. Eurobridge Mobile Brücken GmbH was formally founded in October 1992 as a subsidiary of Dornier GmbH, a corporate unit of Deutsche Aerospace AG (DASA/Munich), and the French company Constructions Industrielles de la Méditerranée (CNIM/Paris).

In November 1993, it was announced that Eurobridge Mobile Brücken GmbH had concluded an agreement to co-operate with the US company Teledyne Continental Motors, General Products (now General Dynamics Land Systems (GDLS)). This was nominally in the field of dry support bridges but was specifically for the US Army's Heavy Dry Support Bridge (HDSB) programme.

Description

The main advantages claimed for the Eurobridge Foldable Bridge (*Faltfestbrücke*) are: high mobility; easy transport due to compact folded bridge units during transport; limited bank space requirements (15 to 20 m); assembly without preparation and physical strain; long automatically foldable bridge units; a closed road surface over the entire width of 4.4 m; a high crossing capacity; ease of driving even small vehicles and motor cycles when crossing the bridge; high transverse stiffness; and flexible spans from 14 up to 39.5 m. (A company-owned bridge has a span of 45.9 m.)

A 40 m Dornier Foldable Bridge train on the road ***1997***

The Eurobridge Foldable Bridge is a fixed aluminium structure, transported on trucks with suitable rigs so that during transport the bridge bays do not exceed 2.75 m in width and 4 m in height. The bridge deck consists of sections folded parallel to the bridge axis and a supporting traversing beam with bank beams. The bridge has been designed for a nominal load of MLC 70 and a crossing speed of 25 km/h. It has been exceptional load tested with a 110 tonne load of two Leopard 2 MBTs in a tractor-trailer combination.

The 39.5 m bridge can be assembled by a five-man crew in 60 minutes without physical strain. The launching vehicle is equipped with a crane, a fulcrum system and launching equipment. The hydraulic crane has a load-carrying capacity of 4,800 kg for a working radius of 7.7 m. It is equipped with a hoisting device for carrying the bridge bays and traversing beam. The bridge bays are so arranged that using the hoisting device the bays will automatically unfold to full 4.4 m road width as they are lifted.

On arrival at the bridging site the fulcrum system of the launching vehicle is extended, the launching beam is placed into the launching position and, after placing a traversing beam ramp section into position on the launching beam, a bank beam is attached. The traversing beam ramp with the bank beam attached is moved outwards. The first traversing beam inner section is positioned on the launching vehicle and coupled. The traversing beam sections are gradually moved outwards as more beam sections are coupled until the second traversing beam ramp is placed onto the launching vehicle and coupled. The traversing beam is then completely extended and the first bank beam is lowered onto the opposite bank. After fitting a second bank beam at the launch end the traversing beam is ready for the bridge assembly.

Bridge assembly commences with the first bridge ramp bay being placed onto the traversing beam. A set of access (climbing) ramps is placed on the ramp bay and the launching beam is used to move the ramp bay along the traversing beam and outwards over the gap. The hydraulic drive for the bridge is then connected to the traversing beam via a power take-off on the launching vehicle. This is used to drive the ramp bay further along the traversing beam to allow the first of the bridge interior bays to be placed into position and coupled using a snap ring mechanism on the top chord and a pin-joint mechanism on the lower chord (the latter can be operated manually or using a power tool). The coupled bridge bays are then driven along the traversing beam and the process is repeated until the last interior bay is in position and the second ramp bay is coupled. The traversing beam on both sides of the gap can then be lowered into the bank beams, until the bridge ramp bays rest on the bank beams and the access ramps are attached. The bridge is then ready for use.

The bridge is recovered in the reverse order. As the bridge bays are lifted by the crane hoisting device they fold automatically to transport width of 2.75 m.

An operational bridge set for a 39.5 m Foldable Bridge consists of the launching vehicle and five trucks. The Bundeswehr uses 8 × 8 trucks for launch and logistic purposes but 6 × 6 trucks can be used to carry the bridge components. A typical 39.5 m bridge set could be transported as follows:

8 × 8 launching vehicle carrying one traversing beam ramp

one truck carrying one traversing beam ramp and two bank beams

one truck carrying four traversing beam sections and 40 access ramps

one truck carrying one bridge ramp bay and one bridge interior bay

one truck carrying two bridge interior bays

one truck carrying one bridge interior bay and one bridge ramp bay.

Bridge lengths that can be built include 13.9 m (two bays), 20.3 m (three bays), 26.7 m (four bays), 33.1 m (five bays), and 39.5 m (six bays). All bridges are constructed by the truck crews only.

A company-owned bridge set has a span of 45.9 m (MLC 60) and an overall length of 54 m.

Specifications

Max bridge span: 39.5 m (or 45.9 m)
Max gap: 38.8 m (or 45.2 m)
Length overall: 47.6 m (or 54 m)
Road width: 4.4 m
Height: 1.26 m
Ramp length: 9.7 m
Weight:
(bridge interior bay) 4,000 kg
(bridge ramp bay) 4,000 kg
(bridge access ramps) 70 kg
(traversing beam inner section) 525 kg
(traversing beam ramp section) 680 kg
(bank beam) 600 kg

Status

Production for the German Army started in 1994 (six-bay 39.5 m version).

Manufacturer

Eurobridge Mobile Brücken GmbH, subsidiary of Dornier and CNIM, Postfach 1420, D-88039 Friedrichshafen, Germany.
Tel: +49 7545 83611. Telex: 0734209-0.
Fax: +49 7545 84634.

UPDATED

KRUPP Festbrücke system

Description

The KRUPP *Festbrücke* is a single-lane road bridge of wheel-tread girder design which can be used by tracked and wheeled vehicles of all types up to MLC 60 for normal loading and MLC 70 in special load cases.

The main load-bearing elements are the two triangular, torsion-resistant, wheel tread girders in between which are suspended intermediate plates. The roadway width is 4.026 m and the structural height 1.5 m with a ramp inclination of approximately 13 per cent. The commercially available construction material is a weldable, cold-hardened aluminium alloy, Al Zn 4.5 Mg 1. Weight of the bridge is 435 kg/m and the heaviest single component is the 260 kg wheel tread girder plate.

The bridge is assembled by hand and is pushed over an obstacle using a special erection platform. It is possible to assemble the bridge from a helicopter.

After training, a 34.4 m bridge can be assembled in 25 minutes, a 47 m bridge in 100 minutes and a 60 m bridge in 120 minutes (all times approximate).

For obstacles of up to 60 m the *Festbrücke* can be constructed as a single-field girder with or without a reinforcement kit or as a two-field girder with pendulum support. Obstacles over 60 m in width can be crossed using the *Festbrücke* as a continuous girder bridge supported on trestles. Use of the *Festbrücke* as a floating bridge or ferry is covered in the appropriate section.

Specifications

Bridge length, single-span girder: 17.6-59.6 m
Roadway width: 4.026 m
Height: 1.5 m
Ramp inclination: 13%
Bridge weight: 435 kg/m
Heaviest component: 260 kg

Status

Production complete. In service with European and other armies, including Argentina.

Manufacturers

KRUPP Fördertechnik GmbH, Altendorfer Strasse 120, D-45143 Essen, Germany.
Tel: +49 201 82801.
Fax: +49 201 828566.
MAN Gutehoffnungshütte AG, Ginsheimer Strasse 1, D-65462 Ginsheim-Gustavsburg, Germany.
Tel: +49 6134 55292.
Fax: +49 6134 55383.

UPDATED

KRUPP Festbrücke bridge with pendulum support

KRUPP Festbrücke bridge with reinforcement kit

MAN GHH F bridge

Description

The MAN GHH F bridge (F - framework), is a bridge system consisting of standardised components for building single- or two-lane bridges with a maximum span of 70 m in 3 m graduations. The bridges can be installed temporarily or permanently.

The F bridge components are constructed of St 52 grade steel and are joined together by high-strength bolts. Steel or wood can be used for the deck materials. Constructed bridges can be dismantled at any time to be set up elsewhere in similar or different combinations. The usual method of installation is to roll the bridge into position using a launching nose and a short cantilever section at the rear to carry ballast. The launching nose and the cantilever section use the same type of components as the rest of the bridge. All elements of the F bridge system are of such a size that they can be handled by standard lifting tackle and carried on standard trucks.

A reinforced version, known as the FT bridge, is available for use as a railway bridge.

MAN GHH F bridge erected in Ghana

Specifications

Bridge type	**Single-lane**	**Two-lane**
Deck width	4 m	6.5 m
Free span		
single-storey	44 m	30 m
double-storey	61 m	47 m
three-storey	70 m	58 m

Status
In production. In service in Burma (Myanmar) and Ghana.

Manufacturer
MAN Gutehoffnungshütte AG, Ginsheimer Strasse 1, D-65462 Ginsheim-Gustavsburg, Germany.
Tel: +49 6134 55292.
Fax: +49 6134 55383.

VERIFIED

MAN GHH Short Trackway Bridge (STB)

Description
The MAN GHH Short Trackway Bridge (STB) was designed to cross small gaps, ditches and gullies. It consists of identical trackways which are laid separately and then connected by detachable braces. Each trackway consists of one trackway girder with a fixed ramp, one hinged ramp and one connection bolt. The complete structure is constructed using extruded sections and sheets of a high-tensile aluminium alloy.

The Standard Bridge consists of one pair of trackways. It is 5.21 m long, can span gaps of up to 4 m and can carry MLC 24 loads.

The Extended Bridge is 9.21 m long and can span gaps of up to 8 m. This bridge is formed from two pairs of trackways joined by a connecting bolt and without any further reinforcement. This bridge can carry loads up to MLC 18 with a maximum crossing speed of 25 km/h.

On both bridges the distances between the trackways can be adjusted to intervals of 1, 1.2 or 1.4 m.

Standard and Extended Bridges can be vehicle launched and retrieved using a launching frame and a vehicle winch. The Standard Bridge can also be launched manually if bank conditions are suitable. Retrieval can be from either end of the bridge.

A single trackway weighs 180 kg and can be carried by six men using integral carrying handles. Bridges can be transported on truck cargo areas or laterally attached to armoured vehicles such as M113 APCs.

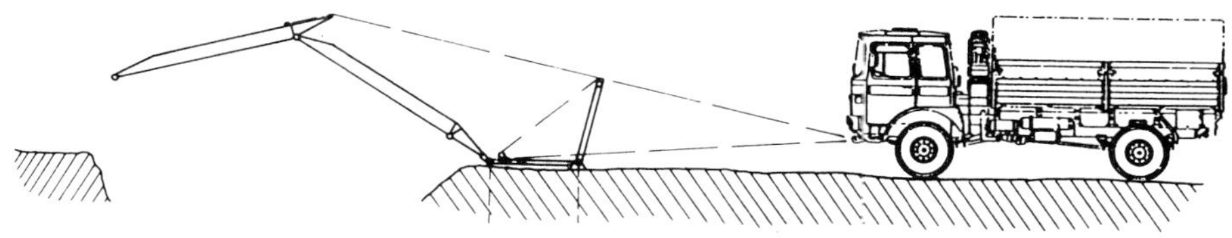

Truck using a launching frame to lay a lengthened MAN GHH Short Trackway Bridge

AMX-13 light tank crossing emplaced MAN GHH Short Trackway Bridge

Specifications

Unit	**Trackway**	**Standard Bridge**	**Extended Bridge**
Weight	180 kg	365 kg	730 kg
Length	5.21 m	5.21 m	9.21 m
Max span	4 m	4 m	8 m
Track width	0.6 m	0.6 m	0.6 m
Width overall	0.7 m	0.7 m	0.7 m
Trackway height	0.28 m	0.28 m	0.28 m
Ramp slope	1:4.5	1:4.5	1:4.5
Laying slope	10%	10%	10%
Load class	–	MLC 24	MLC 18

Status
In production. In service with Singapore.

Manufacturer
MAN Gutehoffnungshütte AG, Ginsheimer Strasse 1, D-65462 Ginsheim-Gustavsburg, Germany.
Tel: +49 6134 55292.
Fax: +49 6134 55383.

VERIFIED

EWK infantry bridge

Description
The EWK infantry bridge is issued as a bridge set comprising seven 4.43 m long bridge modules, a roller block, a swimmer float, a stretcher adaptor, an anchoring rope and a transport frame. The main components are constructed of high-tensile aluminium. Total weight is 340 kg, or 550 kg when packed in the transport frame.

A bridge set can be carried by four groups of two personnel with each carrying 43.5 kg. The main elements can be coupled in any order with a complete 30 m span assembled by eight personnel in about 6 minutes. The bridge will support loads of 135 kg with a distance between loads of 10 m. Crossing capacity is 34 soldiers every 5 minutes.

For free launch operations the entire bridge is assembled and pushed across a gap using the roller block. When bridging dry gaps which exceed a clear span of 14.5 m a further bridge set is used as a counterbalance. When bridging wet gaps wider than 16 m, the swimmer float is placed at the end of the assembled bridge and the bridge is pushed across. The bridge can be launched in currents up to 2.5 m/s. Multiple bridges can be assembled.

Status
In service with the British Army (20 sets).

The EWK infantry bridge at a British Army demonstration ***1997***

Manufacturer
Eisenwerke Kaiserslautern GmbH, Barbarossastrasse 30, D-6750 Kaiserslautern, Germany.
Tel: +49 631 8570.
Fax: +49 631 857300.

VERIFIED

HUNGARY

Hungarian barge bridge

Description

Hungary uses a standard nose-to-stern barge bridge using standard inland waterway barges. Each barge is 80.4 m long and 10 m wide and can be used for road or rail traffic, but not both together. Modified barges can be used normally in peacetime and then quickly converted when needed.

Each barge has a payload of 1,600 tonnes, and the draught, depending on the weight carried, may be from 0.4 to 2.5 m. Normal deck loading for an evenly distributed load is 3,000 kg/m^2, but an individual load can produce a deck loading as high as 10,000 kg/m^2.

Status

In service with Hungarian forces.

Manufacturer

Hungarian state factories.

VERIFIED

Hungarian barge bridge for railway traffic being assembled

Hungarian barge bridge in position being used for railway traffic

INDIA

Manually Launched Assault Bridge (MLAB)

Development

The Manually Launched Assault Bridge (MLAB) was designed and developed by the Research and Development Establishment (Engineers) at Dighi, Pune. It is in service with the Indian Defence Forces as an assault bridge as well as a line of communication bridge and can be delivered in several bridging kit configurations to meet requirements.

Development continues to provide longer bridging spans and to introduce mechanical assembly techniques.

Description

The MLAB is constructed of a lightweight, high strength weldable, self-ageing aluminium zinc-magnesium alloy capable of withstanding extremes of temperature and humidity. The heaviest component can be carried by a team of six personnel while palletised loads can be configured for truck transport or for slinging under helicopters such as the Mil Mi-17. For vehicle transport a standard 3 ton pallet is available although 7 ton pallets or any other configuration can be accommodated.

The MLAB is a deck-type two girder bridge system involving a fully modular, pin-jointed structure with interlocking components, providing a 4 m wide roadway. A standard double-storey bridge can bridge gaps of up to 31 m (MLC 60). If longer gaps are encountered this can be extended to 49 m using ancillary equipment. If MLC 70 is required slightly shorter spans will be involved.

A single span MLAB is constructed using two parallel girders consisting of top panels, each 1.83 m long and pinned together. With the addition of deck panels and bank-seat beams, bridges can be formed providing MLC 60 at 9.8 m or MLC 16 at 22 m.

By adding bottom panels to single-storey girders during construction the girder depth is increased and strengthened to form a double-storey bridge capable of carry MLC 60 loads over a 31 m span or MLC 16 vehicles over a 49 m single span. A link reinforcement kit can enhance the load carrying capacity to MLC 60 for a 49 m span or MLC 70 with caution (or on a slightly shorter span).

The integration of a span-junction set with portable piers enables the construction of MLAB multispan bridges up to 75 m long for MLC 60 vehicles.

Assembled bridges are launched over building platform roller beams using a centrally located nose girder. Launch site preparation is minimal. Deck and ramp units are added once the bridge girder is jacked down on both banks.

Approximate build times

Length	9.8 m	31 m	49 m	49 m
Men	10	30	30	40
Build time	15 min	90 min	120 min	150 min

Status

In service with the Indian Army.

Development Agency

Research and Development Establishment (Engineers), Defence Research and Development Organisation, Ministry of Defence, Government of India, Dighi, Pune 411015, India.
Telex: 0145 7511 RDEE IN.
Fax: +91 212 666583.

VERIFIED

A Vijayanta MBT crossing a completed 31 m MLC 60 double-storey Manually Launched Assault Bridge (MLAB) **1996**

A Shaktiman (4 × 4) 4,000 kg truck crossing a 22 m MLC 16 single-storey Manually Launched Assault Bridge (MLAB) **1996**

ISRAEL

IMI RDB 11 to 62 m Rapid Deployment Bridge (RDB)

Description

Israel Military Industries Limited (IMI) has developed a lightweight, aluminium alloy 62 m modular Rapid Deployment Bridge (RDB). It can be transported by a fleet of 10 trucks and mechanically launched by a team of 12 personnel within 100 minutes. The RDB is capable of MLC 70 loads for tracked vehicles and MLC 96 loads for wheeled vehicles. The launching configuration uses the auxiliary bridge method. All launching operations are performed from the near side to the far bank. The RDB can bridge gaps of up to 60 m and has two trackways.

Status

A 34 m version has been tested successfully by the Israeli Defence Forces.

Manufacturer

Israel Military Industries Limited (IMI), PO Box 1044, IL-47100 Ramat Hasharon, Israel.
Tel: +972 3 542 5222. Telex: 33 719 misbit il.
Fax: +972 3 489639.

VERIFIED

Merkava MBT crossing an early example of an RDB

IMI SPB36 sectional personnel bridge

Description

Intended for infantry use, the SPB36 sectional personnel bridge is transported in nine 4 m sections and assembled on site by three personnel in 10 minutes. An earlier version was known as the SPB24 and used six 4 m sections. The bridge can be built to any length and as a floating bridge, for which floats and anchoring devices are required.

Each SPB36 section weighs 40 kg and is carried by two personnel using its retractable handles. When assembled the bridge is 36 m long and can be launched across a gap by one person using a roller mounted on a small frame. The bridge has a handrail on each side and is only 0.98 m wide. Stretchers can be carried across the bridge by placing them on a two-wheel bogie mounted on the handrails; the stretcher can then be pushed across the bridge by one man. When launching the bridge across water obstacles a small pontoon can be fitted to the leading-edge. The SPB36 has been tested by the US Army.

It is possible to pack up to six bridge sections together with a launching pontoon for transport by truck or other vehicle, including a Jeep-towed two-wheel trailer platform.

Specifications

Weight: (each section) 40 kg
Length:
(complete bridge) 36 m
(section) 3.986 m
Width:
(overall) 0.98 m
(walkway) 0.378 m
Height: 0.75 m
Dimensions packed
Length: 3.986 m
Width: 2.1 m
Height: 1.56 m
Crossing capacity:

Span	Personnel	Interval
36 m	3	17 m
32 m	3	17 m
28 m	4	11 m
24 m	6	4 m
20 m	9	2 m
16 m	10	1.5 m
8 m	close order formation	

Status

Available.

Manufacturer

Israel Military Industries Limited (IMI), PO Box 1044, IL-47100 Ramat Hasharon, Israel.
Tel: +972 3 542 5222. Telex: 33 719 misbit il.
Fax: +972 3 489639.

VERIFIED

Crossing an IMI SPB36 sectional personnel bridge
1997

IMI TAB 12 AT 12 m, air-transportable Towed Assault Bridge (TAB)

Description

The IMI TAB 12 AT 12 m, air-transportable Towed Assault Bridge (TAB) is intended for bridging anti-tank ditches or other obstacles up to 10 m wide and can carry vehicles weighing up to 70 tonnes. The bridge was designed to be separated easily into two sections along its longitudinal axis for transport in a C-130 transport aircraft, and it was also designed to absorb the loss of up to 50 per cent of its structural beams as well as its complete wheel and suspension system from mines or direct fire and still fulfil its MLC 70 mission.

The main body is constructed of four prestressed

IMI TAB 12 AT 12 m, air-transportable towed assault bridge being pushed into position over dry obstacle with front horns erected and pusher bar still connected to MBT

beams made of 2.2 mm steel plates. The upper surface consists of formed steel plates with anti-sliding elements. The front section is manufactured from steel bars which are extended to enable the bridge to reach the opposite bank of the obstacle when deployed for use.

The front section can be folded by operating a simple mechanism which is activated by the tank crew. Normally the bridge is towed into action by an M1, M60 or Centurion tank but many other types of tank can be used. The bridge is towed on four wheels on a single axle in the centre of the bridge. A coupling hook is also fitted for either towing the bridge or deploying the bridge when it is pushed across an obstacle.

In use the bridge is towed to the obstacle with the bridge coupled to the rear of the tank. At the obstacle the tank and bridge make a 180° turn ready for the bridge to be deployed in one of two possible ways. One is for the tank to reverse and push the bridge across the obstacle. For the other, the bridge can be uncoupled for the tank to make a further 180° turn and recouple the bridge to the front of the tank. This coupling operation is normally carried out some distance before the obstacle is reached. The bridge is then pushed across the obstacle. The bridge hitch is then disconnected from the tank using a mechanical device; early versions used a pyrotechnic disconnect.

The bridge is supplied with an adaptor which is bolted onto the tank's forward section, permiting connection with the bridge by means of the coupling hook.

Specifications

Weight: (total) 9,840 kg
Length:
(incl coupling hook and extensible front section) 19 m
(bridge) 12 m
Width:
(with siderails) 5 m
(crossing) 4.2 m
Effective max height of obstacle: 2.5 m
Bridge classification: MLC 70

Status

In production. In service with the Israeli Defence Forces and the US Marine Corps.

Manufacturer

Israel Military Industries Limited (IMI), PO Box 1044, IL-47100 Ramat Hasharon, Israel.
Tel: +972 3 542 5222. Telex: 33 719 misbit il.
Fax: +972 3 489639.

VERIFIED

RADOM Lightweight Bridging System

Description

The RADOM Lightweight Bridging System is an inexpensive portable system, consisting of two to eight bridging units that can carry loads from 650 to 4,000 kg. It is intended for a variety of light bridging tasks and has been deployed by special forces.

A RADOM bridge is made up of segments of trackway with raised edges to form guide walls for vehicle wheels up to 225 mm wide. There is normally no interconnection between trackways. Bridges can be assembled in less than 3 minutes.

Light vehicle about to cross a RADOM lightweight bridge

There are three basic bridging systems.

The RL100 is a 4 m folding bridge consisting of four segments, each 2 m long, and designed for loads up to 4,000 kg over rugged terrain. It is possible to add a central 1 m long plug to form the RL110 bridge with the same load capacity. Each segment of an RL100 weighs 32 kg, with each side of the RL110 weighing 40 kg.

The single length RL200 has four segments, each 2 m long, and is intended for use as a vehicle unloading ramp. Maximum load is 4,000 kg and each side weighs 36 kg.

The RL300 is intended for use in mountainous terrain, being a 4 m folding bridge designed for loads up to 650 kg. Each side has four 1 m long segments and weighs 17 kg. By reducing the number of segments to three each side the span is reduced to 3 m but the load capacity is increased to 850 kg. Weight each side is then 12.75 kg.

Status

In production. In service with special forces in many parts of the world.

Manufacturer

RADOM Aviation Technology Limited, PO Box 3034, Kiryat Arie, IL-49180 Petach Tikva, Israel.
Tel: +972 3 924 4434.
Fax: +972 3 924004.

VERIFIED

ITALY

Drago modular mountain bridge

Description

Drago Engineering SpA has been involved in the construction of light structures for over 40 years and is currently producing a range of products including light modular bridges for use in mountain areas. The materials used for its modular mountain bridges are mainly light magnesium and aluminium alloys.

The modular mountain bridge can be constructed to accommodate three weight classes, MLC 5, 8 and 12. The bridges are built up from a number of 2.5 m span bays using tubular alloy components as the main structural elements. Deck baulks are added when the bridge structure is complete. The bridges can be constructed without recourse to special tools or equipment and it is possible to bring all the bridge components to the site using pack transport by animals or light vehicles. It is also possible to use many of the bridge components for other purposes such as the construction of aerial ropeway towers or observation towers, and for the internal bracing of underground shelters and observation posts.

Troops crossing a Drago modular mountain bridge
1997

Specifications

Bridge class	**5**	**8**	**12**
Span of one bay	2.5 m	2.5 m	2.5 m
Max span	37.5 m	35 m	30 m
Useful width	3.05 m	3.05 m	3.05 m
Useful height	4.15 m	4.15 m	4.15 m
Length of longest element	3.25 m	3.25 m	3.25 m
Weight of heaviest element	125 kg	125 kg	125 kg
Average weight/m	470 kg	470 kg	470 kg
Bridge total weight/m	19,700 kg	16,920 kg	14,000 kg
Useful deck surface	114.37 m²	106.75 m²	91.5 m²

Status
In production.

Manufacturer
Drago Engineering SpA, Viale Trieste 24, I-39100 Bolzano, Italy.
Tel: +39 471 916339. Telex: 500 198.
Fax: +39 471 935604.

UPDATED

JAPAN

Bridge, Panel, Class 30

Description
The Bridge, Panel, Class 30, a modified version of the US M2 Bailey Bridge, was introduced in 1960. The main difference between the US and Japanese versions is that the Japanese components are smaller and lighter, facilitating handling and reducing transport requirements.

Typical examples are the truss of the Japanese model which is 2.5 m long, 1.3 m high and weighs 222 kg compared with the US model which is 3.3 m long, 1.5 m high and weighs 262 kg. The Japanese bridge does, however, have a narrower roadway (3.4 m) and a smaller carrying capacity than the US M2 bridge. An MLC 30 bridge, which is assembled with a 40-man team, is 25 m long and has a roadway width of 3.4 m.

Specifications
Truss panel
Weight: 222 kg
Length: 2.5 m
Width: 0.2 m
Height: 1.3 m

Transoms
Weight: 241 kg
Length: 5 m
Width: 0.2 m
Height: 0.25 m

Stringers
Weight: 88 kg
Length: 2.5 m
Width: 0.6 m
Height: 0.1 m

Status
Production complete. In service with the Japanese Ground Self-Defence Force.

Manufacturer
Kisha Seizo Kaisha Limited, Tokyo, Japan.

VERIFIED

POLAND

DMS-65 heavy girder bridge

Description
The DMS-65 is basically a development of the Bailey panel bridge. It is normally built as a single-storey, through-truss road bridge, but can also be constructed as a deck-truss railway bridge. There are only five basic elements of the bridge which are carried in Star 660 (6 × 6) 2,500 kg cargo trucks. The DMS-65 can be assembled manually or with the aid of cranes, and bridges of different lengths, with either single or multiple spans, can be built. The roadway consists of metal plates that can be covered with crushed stone.

Status
In service with the Polish Army.

Manufacturer
Polish state factories.

VERIFIED

Star 660 (6 × 6) 2,500 kg truck carrying DMS-65 components

Bailey type panel bridges

Description
The British-designed Bailey Bridge was adopted by the US Army during the Second World War, and the Soviet Union obtained some of these under the Lend-Lease programme. Some of these bridges may still be in service with former Warsaw Pact countries, together with local copies of the bridge. Poland is one of the countries that copied the Bailey Bridge with some minor modifications. This is distinguishable from the standard Bailey Bridge by the increased number of holes in the transom.

There are separate entries in this section for the Czech/Slovak MS-1 (SM-60) and the Polish DMS-65 panel bridges, both of which resemble the Bailey Bridge in some respects.

Status
In service with the Polish Army.

Manufacturer
Polish state factories.

VERIFIED

RUSSIAN FEDERATION AND ASSOCIATED STATES (CIS)

PVM foot suspension bridge
LVM light suspension bridge
TVM heavy suspension bridge

Description

These suspension bridges were designed primarily for use in mountainous country and are normally transported by pack animals or, if the terrain is suitable, by motor transport.

The PVM (pedestrian suspension footbridge) set contains sufficient components to construct either two bridges 60 m long and 0.7 m wide, or a single bridge 120 m long and 0.7 m wide. A team of 18 personnel can erect the former in 2 hours or the latter in 3 hours. The complete PVM bridge set weighs 4,360 kg and can be transported by 46 pack animals.

The LVM (light suspension bridge) set contains sufficient components to construct either two bridges 40 m long and 2 m wide or a single bridge 80 m long and 2 m wide. With a team of 27 personnel, the former can be erected in 4 hours and the latter in 2 hours. The complete LVM bridge set weighs 13,500 kg and can be transported by 160 pack animals, the maximum weight of each individual component not exceeding 50 kg. The LVM has a maximum rated capacity of 2,000 kg but the maximum axle load of any vehicle must not exceed 635 kg.

The TVM (heavy suspension bridge) is 60 m long and has a maximum rated capacity of 10,000 kg.

Status

In service with the RFAS armed forces.

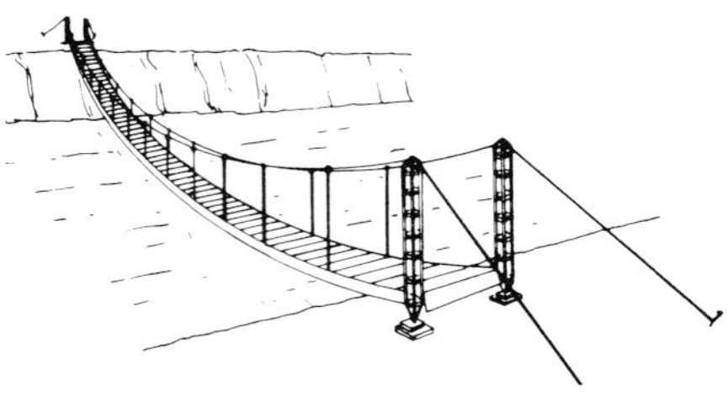

PVM foot suspension bridge

Manufacturer

State factories.

VERIFIED

Underwater bridge set

Description

This set consists of hollow pontoons carried on specially modified 6 × 6 3,500 kg trucks. The pontoons are estimated to be 8 to 8.5 m long and 3 to 3.5 m wide, and are offloaded from the trucks by gravity (similar to the TPP and PMP pontoons). They are then assembled on the shoreline to form a ribbon bridge and brought into position by bridging boats. Bridge piers or pilings are then placed, after which the bridge is sunk into position by flooding the pontoons.

Status

In service with the RFAS armed forces.

Manufacturer

State factories.

VERIFIED

Underwater bridge set in use with ZIL-131 (6 × 6) 3,500 kg truck crossing

SARM sectional deck truss bridge

Description

The SARM is a sectional deck truss bridge made up of triangular and rectangular sections bolted together to support a roadway of steel deck panels. A single roadway is 4.2 m wide and has a capacity of 40 tonnes and a dual roadway has a capacity of 60 tonnes and is 7.2 m wide. The bridge is carried on single-axle semi-trailers which are towed by MAZ-504 or ZIL-130V (4 × 2) 5,000 kg tractor trucks.

Deck truss is made of welded steel sections, roadway panels with kerbing sections welded together. Clear spans of 18.6, 25.6 and 32.6 m are feasible.

Where possible, existing piers are used in the construction of the SARM bridge but where none are available and the span required is greater than 32.6 m, piers have to be provided. For low-level bridges, either pile bents or trestle bents are used, while for high-level bridges panel-type crib piers are used. Steel towers can be used on existing bridge supports, piles or floating supports. Tower supports are adjustable at 2 m intervals, the maximum height being 8.8 m.

The SARM takes a relatively long time to build compared with more recent tactical bridges. For example a 158 m bridge takes about 24 hours to construct. As the maximum weight of any individual component is only 4,400 kg, a 5,000 kg capacity crane is sufficient for construction of the SARM. Other elements in the SARM bridge system include pile drivers, crane trucks, gang saws and powerboats.

Status

In service with the RFAS armed forces.

Manufacturer

State factories.

VERIFIED

REM-500 railway and road section bridge

Description

The REM-500 is a sectional railway bridge that can also be used as a road bridge with the addition of a wooden floor. It is employed over wide and shallow water barriers, closing breaches in railway lines and as approaches for floating bridges.

The bridge is composed of individual steel spans, 12.51 m long and weighing 10,700 kg, which rest on trestles that can be adjusted in height from 3 to 12.7 m and weigh between 5,000 and 7,000 kg. The trestles' feet are provided with large circular baseplates so that they will not sink into soft ground. The footers have an area of 7 m^2 and a ground pressure of 1.2 kg/cm^2. Both longitudinal and transverse bracing is provided where necessary. The bridge can be built across rivers with a maximum water depth of 7 m and a maximum water velocity of 1.2 m/s.

REM-500 section bridge being built

The bridge can have a single track of either European (1.435 m) or RFAS (1.524 m) standard gauge. The speed of a train crossing the bridge must not exceed 30 km/h, with a maximum axle load of 20 tonnes. The REM-500 can also be used in combination with floating railway bridges which use river barges as their floating supports. The bridge is constructed a span at a time with each span and trestle positioned at once by an overhead gantry (called the SRK-2D) travelling along the completed spans.

Status

In service with the Polish and RFAS armed forces.

Manufacturer

State factories.

VERIFIED

Heavy barge bridges

Description

Former Warsaw Pact armies make wide use of heavy floating road and railway bridges, constructed from standard inland waterway barges of various capacities. RFAS forces normally use standard 600 and 1,000 tonne barges suitably modified. These can be used to bridge a river in three different ways. The most common method is by placing the barges end to end with the vehicles crossing from one barge to another over the bows of the barge. A river 400 m wide would require seven 600 tonne standard barges that would take about 24 hours to position. Equipment required to position this bridge would include truck-mounted cranes, bulldozers, bracing equipment and three river tugboats, one with a 300 hp engine and two with 475 hp engines.

The second method is connecting the barges side by side with the barges resting in the direction of the current. Some of the barges used for this method have been provided with special supports which can carry an integral roadway.

The third method is to connect the barges side by side, but aligning them on a slant towards the current, with the vehicles crossing from one barge to another via the sides of the barges. The main disadvantages of these bridges are the length of time required to position the barges and the delay to other river traffic when the barges are in position.

These barges are also used for railway bridges, where the barges are used as pontoons. Finally, river barges have been used to construct heavy ferries which are pulled by tugboats.

Status

Produced in most former Warsaw Pact countries using locally manufactured components. Known to be used in Hungary and the RFAS.

Manufacturer

State factories.

VERIFIED

SINGAPORE

Portable Assault Bridge PAB 15

Description

The Portable Assault Bridge PAB 15 was designed to be man-portable, carried by wheeled or tracked vehicles to bridge small gaps or other obstacles. It can carry loads of up to 15 tonnes in any configuration.

The PAB 15 consists of two standard 4 m panels and one 2 m panel connector which can be coupled together to bridge gaps up to 10 m wide. For a complete PAB 15 the 2 m panel connector is placed in the centre of the two 4 m panels and a supporter unit is placed underneath. A 4 m panel can be used by itself to bridge small gaps. Two PAB 15 bridges are placed side by side to carry vehicles.

A complete PAB 15 bridge can be constructed by four personnel within 10 minutes.

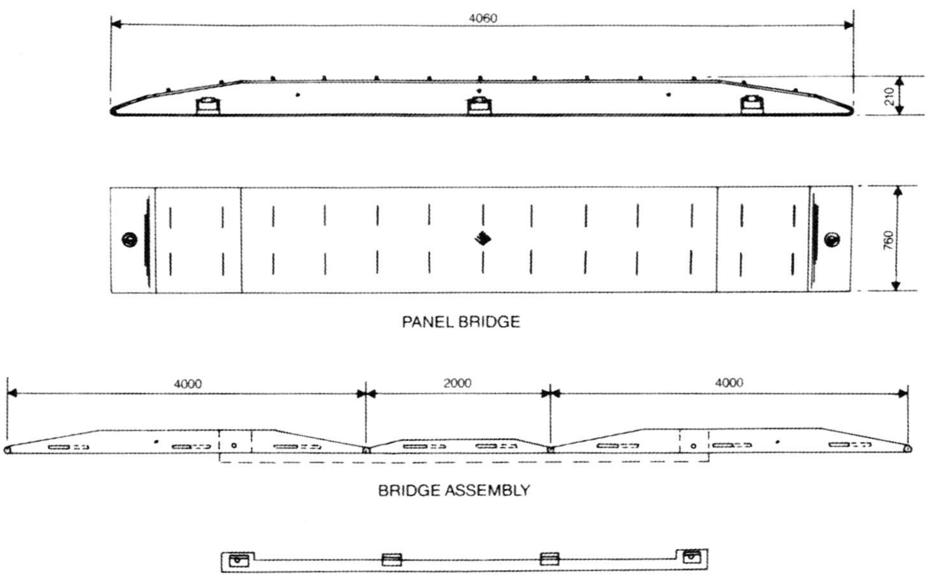

Main components of Portable Assault Bridge PAB 15 with supporter unit below

Specifications

Panel	4 m	2 m
Weight	135 kg	62 kg
Length	4.06 m	2.07 m
Width	0.76 m	0.76 m
Height	0.21 m	0.21 m

Status

In production.

Manufacturer

Singapore Technologies Shipbuilding and Engineering Limited, 7 Benoi Road, Singapore 629882.
Tel: +65 861 2244. Telex: singa rs 21206.
Fax: +65 861 3028.

UPDATED

SWEDEN

Fast Bridge 48 mobile bridge system

Development

The Fast Bridge 48 (FB48) is the result of co-operative research and development between the Swedish Defence Materiel Administration (FMV) and Karlskronavarvet AB. This began in 1985 so a modern and efficient mobile bridge system for the Swedish Army could be designed. The bridge is constructed according to NATO standards and the 'Tri-lateral Design and Test Code for Military Bridging and Gap-crossing Equipment'. The design is patented. FB48 has been thoroughly tested and approved by FMV throughout development and final design.

The first serial order for 10 FB48 (Kb 5) systems was placed with Karlskronavarvet AB in October 1993; the order included an option for a further 10 to 20 systems.

The Fast Bridge family is being developed to include the following variants:

FB56—Emergency (and/or military) single-span 56 m MLC 50 bridge, now available

FB88—Military double-span 88 mm MLC 70 bridge

FB Multi—Military multispan MLC 100 bridge of unlimited length.

In September 1996 Karlskronavarvet AB, as one of two competing prime contractors, received an order from the US Army Tank-Automotive and Armaments Command (TACOM) for the development of a Heavy Dry Support Bridge (HDSB). The order included the development, design, manufacturing, testing and evaluation of one bridge prototype with launching equipment and continues until the year 2000 when

The remotely controlled FB48 mobile bridge system crane handling a bridge section

A Leopard 2 (Improved) MBT crossing a Fast Bridge 48 (FB48) mobile bridge system

The nose assembly of a 48 m FB48 resting on the opposite bank of the gap about to be bridged

The FB48 mobile bridge system crane about to lift a bridge section from a standard 6 × 6 truck

TACOM plans to order 35 to 50 bridge systems from one of the two prime contractors. The work will be carried out in close co-operation with TACOM and US industry team partners of Systems and Electronics Inc (SEI), XMCO and Barnes & Reinecke. The bridge involved is the Fast Bridge 48 AV (American Version). The US Army requires a 48 m crossing capability for MLC 70 tracked or MLC 96 wheeled vehicles. This could be increased to MLC 80 tracked or MLC 100 wheeled at 42 m. It is anticipated that the carrier vehicles will be based on the Oshkosh M1075 10 × 10 PLS carrier. Six trucks and three trailers will carry one 42 m bridge or two 24 m bridges.

Description

The Fast Bridge 48 (FB48) is a combined tactical and line of communication bridge claimed to represent a new generation of dry support bridging and is primarily intended for the support of armoured and mechanised units. With a clear span of 46 m FB48 can take the load of all types of modern battle tank up to MLC 70.

The FB48 mobile bridge system consists of a 48 m long bridge with two additional 15 m ramps; the construction material is weight-optimised extra high tensile steel (1,100 Mpa). Launching is carried out from one side and the bridge can be retrieved from either side of a gap.

A specially designed hydraulic launch unit, with a remotely controlled crane, both powered by an integral air-cooled diesel engine with two separate hydraulic systems, is used to connect launching beam segments which are gradually extended over the gap to be bridged. When the beam reaches the other side it is supported on a nose support assembly ready for one launch crew member to cross to the opposite bank on a special trolley travelling along the beam. Bridge sections, ramps and other components are thereafter connected sequentially, using manually operated push-pull rods, and pushed out along the launching beam to complete the bridge. The total launch or retrieval time, day or night, is less than 75 minutes using a crew of seven personnel. The working area required for the operation measures 6 × 30 m (180 m²).

Maximum clear span, that is, between the innermost points of the supporting areas, is 46 m. A bridge length can vary from 24 to 48 m, made up of 8 m sections, so completed bridges may have lengths of 24, 32, 40 or 48 m. The usable trackway width is 4 m.

A Fast Bridge 48 mobile bridge system emplaced **1997**

Bridge length	56 m	48 m	40 m	32 m	24 m
Clear span	54 m	46 m	38 m	30 m	22 m
Max load	MLC 60/70	MLC 70	MLC 100	MLC 100	MLC 100

A single FB48 bridge section weighs 5,300 kg, measures 8 × 4 × 1.5 m and can be delivered fixed (4 m) or foldable (3 m).

The FB48 is designed to be transported on six standard non-specific trucks with a 12 to 14 tonne capacity. Bridge length and load combinations are as above:

Specifications

Launch crew: 7
Max bridge length: 48 m
Max clear span: 46 m
Bridge section weight: 5,300 kg
Bridge section length: 6-8 m
Bridge section width: 3-4 m
Bridge section height: 1.5 m
Trackway width: 4 m
Max launch time: 75 min

Status

FB48/MLC 70 version in serial production for the Swedish Army. Other variants under development. Heavy Dry Support Bridge prototype ordered for test by US Army.

Manufacturer

Karlskronavarvet AB, a member of the Celsius Group, S-371 82 Karlskrona, Sweden.
Tel: +46 455 334100.
Fax: +46 455 17934.

UPDATED

UNITED KINGDOM

Vickers Modular Bridge (VMB)

Development

As the British Army's Bridging for the 1990s (BR90) the Vickers Modular Bridge (VMB) underwent an eight year development programme. It is a comprehensive bridging system covering line of communication bridges (also known as dry support bridges) and tank-launched bridges (see separate entry in the *Mechanised Bridges* section for that aspect of the system). All bridges are built from a range of seven modular panels which are interchangeable through the various bridge types. General support bridges are used for longer spans and where AVLBs are not available.

In November 1993 the UK Ministry of Defence awarded a contract worth £140 million for the supply of production quantities of the Bridging for the 1990s (BR90) system for the Royal Engineers.

The four types of Thompson Modular Bridge involved will enter service as follows:

General Support Bridge - January 1996
Close Support Bridge - November 1997
Two Span Bridge - May 1997
Long Span Bridge - June 1997.

Description

Vickers Modular Bridges (VMB) are launched by the Automotive Bridge Launching Equipment (ABLE), with components transported to the bridging site on bridging vehicles. The ABLE consists of a truck-mounted crane and a launch mechanism. The launch is achieved by deploying a launch rail across the gap and lowering it onto the far bank. The bridge is assembled beneath the launch rail and winched across the gap. Due to the launch rails not being integral with the bridge structure they are recovered, thus allowing the ABLE to proceed to other bridge sites to carry out further

Assembling a 32 m General Support Bridge **1997**

A TMB Automotive Bridge Launching Equipment (ABLE) vehicle (T J Gander)

launches. Recovery by ABLE can be carried out from either end of the bridge and ABLE can be used to recover any of the tank-launched bridges in the VMB system. The bridging vehicles (BVs) carry all the bridge components to the bridging site. They are fitted with truck-mounted cranes which facilitate loading/unloading and assist in the launch of the bridge. The launch can be completed with only one crane operable while the bridging vehicles can be used for carrying general cargo when not required for bridging purposes.

A British Army Bridge Set consists of 8 m ramp panels, 8 m panels, 4 m panels and 2 m panels to enable a bridge to be constructed from a minimum length of 16 m in 2 m increments up to 32 m. (Bridge sets can be produced to meet specific requirements.) The bridge is 4 m wide and, with in-fill deck units fitted, provides a continuous trackway over the width. A Bridge Set is carried on two Bridging Vehicles (BV).

To enable longer bridges to be built, Long Span Equipment is added to the 32 m bridge set. This allows bridges up to 56 m to be built in 2 m increments, using reinforcement links and a king post system.

Two Span Equipment provides the flexibility for the bridge to be supported mid-span by an intermediate floating or fixed pier. An articulator panel and a rocking roller set provide both the support and change of angle at that support. The Long Span Equipment can also be used to reinforce one span of a two span bridge. The maximum Two Span Bridge length is 60 m with a fixed pier, and 62 m with a floating pier. Two Span Equipment, including the necessary trackway to extend the bridge to 62 m, is carried on two extra BVs.

A 32 m General Support Bridge can be launched in under 30 minutes and dressed with in-fill decks and kerbs, and so on, in an additional 10 minutes by a crew of 10.

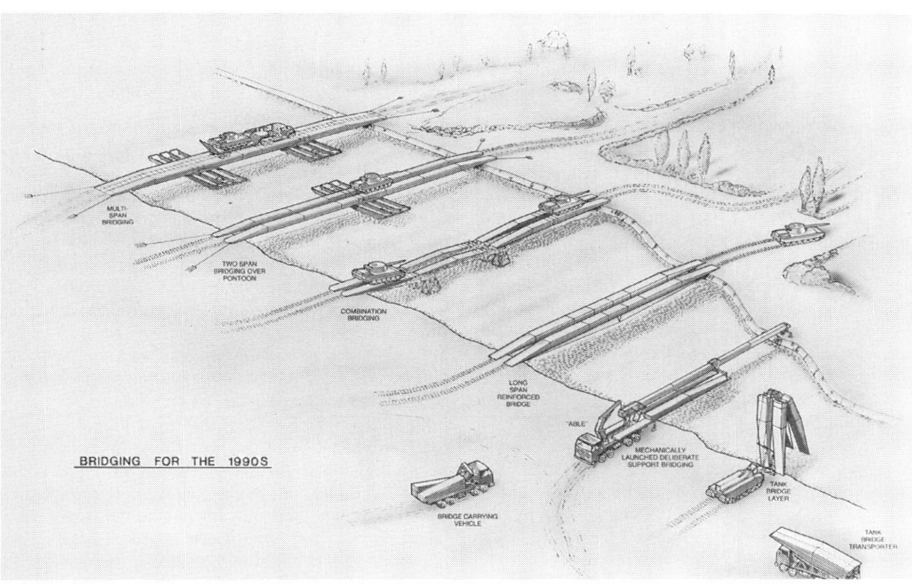

The complete range of the British Army's Bridging for the 1990s (BR90) programme of which the Vickers Modular Bridge forms a major part

The initial stages of emplacing a 32 m General Support Bridge as the ABLE launch rail is extended across the gap **1996**

Classification of the bridges is MLC 70 tracked and MLC 105 wheeled for all spans, including the 56 m Long Span Bridges. The bridges were designed using the US/UK/GE Trilateral Design and Test Code.

The ABLE and BV are Improved Medium-Mobility Load Carriers (IMMLC) giving a good cross-country capability with 8 × 8 configuration, enabling them to reach most bridging sites accessed by AFVs. The vehicles have a high degree of commonality with each other and with the TBT used as part of the Tank Launched Bridge system (see separate entry under *Mechanised Bridges*), although other vehicles could be used.

Specifications

32 m Bridge Set
Length overall: 32 m
Panels: 4 × 8 m ramp panels; 2 × 8 m panels; 3 × 4 m panels; 2 × 2 m panels
Ramp length: 6 m
Width overall: 4 m
Depth of section: 1 m
Max span: 30 m
Weight: 18,900 kg
Launch/recovery times: 30 min/40 min
Dressing time: 10 min
Crew: 10

Long Span Equipment (current system capability)
Length overall: 44 m (with 32 m Bridge Set)
Extra panels: 16 m of panels; 2 × 4 m reinforcement panels; reinforcement anchorages; 2 additional launch rail panels
Width overall: 4 m
Depth of section: 1 m
Max span: 42 m
Weight: (total, 44 m bridge) 31,093 kg
Launch/recovery times: 45 min/60 min
Dressing time: 15 min
Crew: 11

Two Span Equipment
Length overall:
(fixed pier) 60 m
(floating pier) 62 m
Extra panels: extra 28 m of panels; 2 × 2 m articulator panels; 2 rocking rollers
Width overall: 4 m
Depth of section: 1 m
Max span:
(fixed pier) 58 m
(floating pier) 60 m
Launch/recovery times: 60 min/90 min
Dressing time: 20 min
Crew: to be defined

Status

Entering production for the British Army.

Manufacturer

Vickers Bridging, PO Box 37, Spring Road, Ettingshall, Wolverhampton WV4 6YN, UK.
Tel: +44 1902 405050.
Fax: +44 1902 491459.

UPDATED

Thos Storey Bailey Bridge

Development/Description

The Bailey Bridge was originally developed by Sir Donald Bailey at the (then) Military Engineering Experimental Establishment at Christchurch and was widely used during the Second World War. In 1948, Thos Storey (Engineers) gained a licence from the National Research Development Corporation to manufacture and sell Bailey bridging components and equipment. In 1950 this licence was extended, granting exclusive rights to the patent, production and selling of the Bailey Bridge system. Since then the company has been continually improving the system, which remains in production to meet a continuing demand.

The Bailey Bridge was designed as a universal unit-construction military bridging system, with the Bailey panel as its basic component. The great advantage of the system lies in its use of these standard interchangeable components, which, combined with the simplicity of design, enables it to be erected in a short time by unskilled labour under limited specialist supervision.

The Bailey panel, the basic component, is made of high-tensile steel. Panels can be connected together to form beams or columns. They are connected by panel pins and chord bolts to give a series of composite girders with varying strengths to meet loading conditions.

Using basic equipment, the maximum span is 61 m and the maximum military load is MLC 80. The bridge can be constructed in three widths, 3.28 3.81 and 4.19 m. The bridge is normally constructed on rollers on one side of the gap and then launched into position using a skeleton, cantilever nose, which is detached after the bridge has crossed the gap.

Over the years, Thos Storey has improved the original design. These improvements include a new steel decking which is quick and easy to erect, provides an anti-skid surface and, unlike the wooden deck, has a long life. This can be used with the standard, standard widened, extra wide or double width Bailey Bridges.

Through-type Bailey Bridge with decking system carried between two side girders formed from Bailey panels

Another development is the Thos Storey Bailey panel, which provides an extra 40 per cent safe working shear load plus increased bending capacity and is completely interchangeable with the standard Bailey panel. Finally, a double width Bailey 7.23 m wide was developed to permit two-way traffic.

In 1991 Thos Storey (Engineers) Limited introduced a new decking system comprising a main deck element, kerb units, transoms and fixings. The decking was designed in three weights: light, medium and heavy. Road widths are: standard, 3.285 m; single carriageway, 3.685 m; extra wide, 4.422 m; double wide, 7.37 m; and triple wide 11.055 m. New fixing systems make the decks easier and quicker to handle and install. The clamp design utilises a 'Halfen'-type channel system which allows deck elements to be placed in position with the hold-down bolts introduced afterwards. Normal installations require one bolt and clamp to hold down four deck units at the corners with outer edge corners needing only a single bolt and clamp. The units weigh between 309 and 339 kg and are also compatible with Thos Storey Acrow panel bridging systems.

In addition to being used as a road bridge, the Bailey Bridge has been widely used for other applications including rail and foot bridges, retractable lift bridges, derrick supports and mobile gantries. It can also be used in conjunction with the Storey Uniflote for floating bridges.

Status

In production. In service with armed forces all over the world.

Manufacturer

Thos Storey (Engineers) Limited, 52 Queens Road, Weybridge, Surrey KT13 0AN, UK.
Tel: +44 1932 858133. Telex: 21868.
Fax: +44 1932 855588.

VERIFIED

Acrow Panel Bridge

Description

The Bailey Bridge was designed to take military traffic up to Class 80 standard and to cope with maximum single-axle loadings of 20 tonnes. Today a bridge may have to be Class 150 and cope with single-axle loadings of 60 tonnes or more, particularly where heavy earthmoving equipment is using the bridge. Such capacities could be achieved only by making extensive modifications to the standard Bailey design.

Thos Storey decided that the new design should be based on the Bailey system and principles, but take advantage of modern developments in bridge design and steel technology. The result is the Acrow panel bridge which uses the Bailey unit construction system with its ease of assembly, but employs higher tensile steels and has an advanced design to give enhanced load-carrying capacity. In particular the Acrow bridge has a fatigue life many times that of the Bailey system at less cost.

Like the Bailey Bridge, the decking can be either steel or timber. Varying strengths of decking can be provided depending on the axle load the bridge is required to carry. The Acrow panel bridge can have roadway widths of 3.43, 4.13, 4.84 and 7.23 m to give two-lane traffic using standard equipment. Again, as with the Bailey, the Acrow panel bridge can be launched on rollers and constructed using unskilled labour without a crane if necessary.

Acrow panel equipment can be used even more successfully than Bailey equipment to produce structures other than road bridges, for example rail bridges and gantries. All components of the bridge have been designed to be carried in standard military transport vehicles. Bridges can be installed either for permanent use, or subsequently dismantled and stored for reuse when required.

The major advantages of the Acrow panel bridge can be summarised as follows: improved bending-moment capacity; improved shear capacity; improved stability against buckling under load; increased efficiency of stress transfer; greater rigidity and stability overall; and a fatigue life approximately four times that of the standard Bailey Bridge.

The Acrow panel bridge is constructed on rollers on one side of the gap to be bridged and an additional temporary structure, the launching nose, utilising standard components, is built on the leading end. The length of the nose is so judged that when the construction is moved across the gap, the nose tip engages on rollers on the opposite bank before the point of balance is reached. With the bridge positioned across the gap, the launching nose is removed and dismantled. The bridge is then jacked up off the rollers and lowered onto permanent bearings on the abutments on either bank.

Basic Components

The Acrow panel is the basic truss component. This has a shear capacity in a single storey of 25 tonnes, 67 per cent more than the standard Bailey panel, and a shear capacity in a double storey of 41 tonnes, again 67 per cent more than the standard Bailey Bridge.

Panel pins with circlips are used to connect panels end to end. These provide a locking feature which is also a useful precaution against tampering.

Chord reinforcements can be bolted to top and bottom panels for additional bending movement capacity and are staggered over panel joints in the compression chord to provide greater stability under load. Four bolts are used to connect panels to chord reinforcements to increase the efficiency of stress transfer.

Chord bolts are used to connect panels one on top of another in the case of two or more panel heights or to connect chord reinforcements to panels.

Transoms of different lengths give four roadway widths; there are two basic profiles and light, heavy and super heavy versions. Transoms are located by dowels and bolted down to the transom seatings at intersections between diagonals and the bottom chord of panels at exactly 1.52 m centres.

The steel decking is available in different versions for different load requirements. Deck units, consisting of chequer plate decking welded to the stringers, are clamped at each transom. Decking can terminate 0.762 m inside the ends of the bridge or may be extended 0.762 m onto abutments. Sloping ramps can be provided.

In the timber decking, deck chesses are held in position by steel kerb units which are bolted down to the outer stringers.

An Acrow four-panel type pier can be constructed for use as a supporting member for the bridge. This form of construction is very rigid and will carry an axial load of 400 tonnes on a pier as high as 41 m. Two panel adjustable towers, capable of carrying axial loads of up to 200 tonnes (dependent on height) can also be provided.

The Acrow panel bridge can also be used in conjunction with the Acrow Uniflote system, for floating bridges, jetties and ro-ro terminals.

In 1978 Thos Storey (Engineers) introduced the AB1X hybrid panel for use with either Bailey or Acrow panel bridging. It is manufactured from the same materials as the AB1 Acrow panel, which has better shear capacity, bending-movement capacity and fatigue life than the Bailey. Used one way up the hybrid panel accepts the Acrow panel decking system at standard 5 ft centres in the normal way and the other way up reverts to a Bailey panel and can be used with standard Bailey decking and other components to form Bailey bridging, with the added advantage that the shear capacity is increased by the full 67 per cent of an Acrow panel.

In 1991 Thos Storey (Engineers) Limited introduced a new decking system comprising a main deck element, kerb units, transoms and fixings. The decking was designed in three weights: light, medium and heavy. Road widths are: standard, 3.285 m; single carriageway, 3.685 m; extra wide, 4.422 m; double wide, 7.37 m; and triple wide, 11.055 m. New fixing systems make the decks easier and quicker to handle and install. The clamp design utilises a 'Halfen'-type channel system which allows deck elements to be placed in position with the hold-down bolts introduced afterwards. Normal installations require one bolt and clamp to hold down four deck units at the corners with outer edge corners needing only a single bolt and clamp. The units weigh between 309 and 339 kg and are also compatible with Thos Storey Bailey bridging systems.

Bridge widths

Roadway width	Type of Bridge	Clearance between inner trusses
	Single Traffic Lane Bridges	
3.43 m	Standard Acrow Panel Bridge	3.76 m
4.13 m	Extra Wide Acrow Panel Bridge	4.78 m
4.85 m	Ultra Wide Acrow Panel Bridge	5.48 m
	Two Traffic Lane Bridges	
7.23 m	Double Width Acrow Panel Bridge	7.6 m

Status
In production. In service with armed forces around the world.

Manufacturer
Thos Storey (Engineers) Limited, 52 Queens Road, Weybridge, Surrey KT13 0AN, UK.
Tel: +44 1932 858133. Telex: 21868.
Fax: +44 1932 855588.

VERIFIED

Heavy earthmoving equipment crossing Ultra Wide Acrow Panel Bridge

Acrow Heavy Bridge

Description
Acrow panel bridges, like the Bailey, are normally limited to a clear span of 61 m. There are, however, occasions when a longer span is required, to meet this requirement Thos Storey developed the Acrow heavy bridge. The bridge is of the Warren truss type and has all the advantages of unit construction based on the uniformity of similar components produced by a strict jig control during the manufacturing process. The bridge has been designed for two-lane traffic on single spans up to 100 m in length, or for single-line railway bridges. The maximum span depends on the loading required.

The Acrow heavy bridge was designed to be launched in the same way as the Bailey, with a cantilever nose on the front. Although it has been designed primarily for use as a permanent bridge, it is quick and easy to erect and can, if necessary, be dismantled and reused elsewhere. A crane is required to assist erection of the bridge but because of the simplicity of the design and the method of assembly no skilled labour is required.

Alternative *in situ* reinforced concrete, with or without asphalt wearing surface, steel, or timber decking systems are available. In each case the decking is rigidly connected to the bridge structure, forming an integral part of the bridge. Footwalks can be cantilevered on either side of the bridge structure outside the main trusses.

Each component of the bridge is an all-welded fabrication designed not only for its structural purpose but also for ease of handling, transportability and for rapid and simple construction on site.

82.3 m Acrow Heavy Bridge for two-lane traffic

Status
In production. In service in many parts of the world.

Manufacturer
Thos Storey (Engineers) Limited, 52 Queens Road, Weybridge, Surrey KT13 0AN, UK.
Tel: +44 1932 858133. Telex: 21868.
Fax: +44 1932 855588.

VERIFIED

Mabey and Johnson Military Bridging System

Description
The Mabey and Johnson Military Bridging System has its roots in the Bailey design, but the application of high-grade steels and advanced computer design and analysis techniques led to more effective structures. The Military Bridge is based on the proven Mabey Compact 200 system upgraded to meet military requirements.

The design of the system is modular, allowing full spans of up to 61 m to be built in 3 m increments. Floating bridges or multipier systems can be used to bridge any width of wet gap. Loads up to MLC 100 have been proven during extensive testing. A typical 30.5 m MLC 70 bridge can be built by 30 personnel within 2 hours.

Status
Ready for production.

Manufacturer
Mabey and Johnson Limited, Floral Mile, Twyford, Reading, Berkshire RG10 9SQ, UK.
Tel: +44 1734 403921. Telex: 848649 mabeyt g.
Fax: +44 1734 403941.

VERIFIED

Mabey and Johnson Military Bridging System MLC 70 bridge

Mabey and Johnson Compact Bailey Bridging

Development

Mabey and Johnson is one of Europe's foremost organisations involved in the supply of steel bridging, with almost 150 years of experience. Continuous development has led to a range of products capable of meeting a wide variety of needs. These products feature clear spans up to 100 m; one- two- or three-lane road widths; ease of transportation; quick and simple erection; assembly by hand; modular pre-engineered designs; and ex-stock availability. Systems are suitable for permanent, floating and emergency bridging, jetties and support schemes and can be adapted to a wide range of special applications. Components are fully galvanised and require minimal maintenance.

Mabey and Johnson Compact Bailey Bridging in position

Status

In production. In service.

Manufacturer

Mabey and Johnson Limited, Floral Mile, Twyford, Reading, Berkshire RG10 9SQ, UK.
Tel: +44 1734 403921. Telex: 848649 mabeyt g.
Fax: +44 1734 403941.

VERIFIED

Mabey Universal Bridging

Description

Mabey and Johnson used its wide experience of the Bailey Bridge to develop the Universal Bridge, which entered production and service in 1975.

The equipment extends the capability of speedily erected panel bridging as it is available in one-, two- or three-lane carriageway widths and for spans of up to 81 m.

As with Bailey, the construction strength can be varied to suit the loading, and bridge construction is chosen by reference to load tables.

The bridges can be launched by the standard cantilever method, thus combining most of the benefits of Bailey, including ease of erection, multiple use, and guaranteed load-carrying capacities with permanent bridge capability.

Status

In production and service.

Mabey Universal Bridging in position

Manufacturer

Mabey and Johnson Limited, Floral Mile, Twyford, Reading, Berkshire RG10 9SQ, UK.
Tel: +44 1734 403921. Telex: 848649 mabeyt g.
Fax: +44 1734 403941.

VERIFIED

Williams Fairey Engineering Medium Girder Bridge

Development

The Medium Girder Bridge (MGB) was designed and developed by the Military Vehicles and Engineering Establishment at Christchurch (MVEE(C)) (now the Defence Research Agency) with production, marketing and further development undertaken by Williams Fairey Engineering Limited. The MGB entered service with the British Army in 1971 and since then has been adopted by 36 countries.

The main advantages of the MGB can be summarised as follows: it is quickly and easily built by hand; little training is required; no site preparation or grillages are required; it is easily transported by road or air as palletised loads; it is of light and sturdy construction; it has a multispan capability with span junction set; a portable pier set is available; it can also be used as a ferry or floating bridge of unlimited length; it has a load capacity of up to MLC 70 (tracked) or MLC 100 (wheeled); and little or no maintenance is required.

49 m link-reinforced MGB being crossed by a Chieftain MBT

Description

The MGB is a lightweight, easily transported bridging system which can be quickly erected by hand to provide a flexible and manoeuvrable bridging system covering the full range of military bridging requirements. The MGB has also been adopted by a number of governments for emergency bridging operations.

Much of the success of the MGB lies in the material from which the components are manufactured. The RARDE(C) developed a weldable alloy of aluminium, zinc and magnesium (DGFVE 232B) necessary to ensure both the strength and the lightness of the bridge. Seven major components are used in the basic bridge construction. All except two weigh under 200 kg and they can be easily handled by a team of four. The two heavier components are handled by teams of six.

The MGB is a two-girder deck-type bridge in which the longitudinal girders with the deck units between produce a 4 m wide roadway. Girders of top panels, joined at each end by a bankseat, form a shallow single-storey construction for heavy loads at short spans and lighter loads at longer spans up to 22.4 m. A double-storey configuration using deeper girders of top and bottom panels, together with additional end-of-bridge components, can be constructed for heavier loads over longer spans.

The MGB can be supported on unprepared and uneven ground without foundations. It is constructed on one roller beam for single-storey construction and on two roller beams 4.6 m apart in a building frame for double-storey construction. The ends of the roller beam can be adjusted in height within the building frame so that no levelling or other preparation of the ground is required.

All components of the MGB are transported in standard loads on a special pallet. The British Army uses the Bedford MK (4 × 4) or Leyland 4 × 4 which also tow a two-wheeled trailer carrying another pallet. There is a separate entry for the MGB trailer (FV2842) in the *Trailers* section. When the vehicle and trailer arrive at the bridge-building area the pallet is attached to a stationary vehicle, or an alternative anchorage point, and the towing vehicle is driven away pulling off the pallet, which falls onto the ground. Rubber buffers cushion the fall. A single 9.8 m bridge is transported on two pallet loads while a 31 m double-storey bridge is transported on 10 pallet loads. Any vehicle of suitable capacity may be used and special pallets for specific vehicles have been introduced.

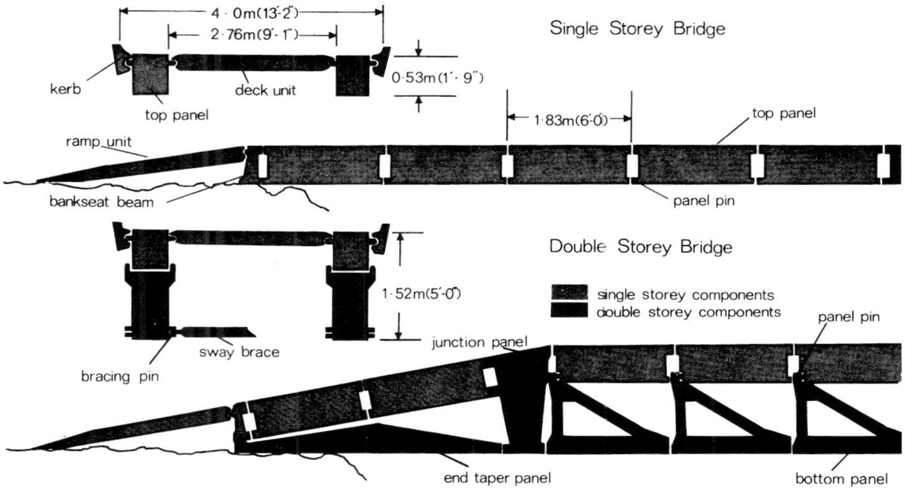

Main components of MGB

Link-reinforced 49 m MGB

Single-span bridging

The MGB can be built either in the single- or double-storey configuration. Single-span single-storey bridges can be used by Class 70 vehicles at a length of 9.8 m and by lighter vehicles up to a maximum length of 22 m at Class 16. Single-span double-storey bridges can be used by Class 70 vehicles at lengths of up to 31 m and at increasing lengths for lower load classes, to a maximum length of 49 m at Military Load Class 16.

The single-storey bridge is constructed using top panels pinned together forming two girders and joined at each end by a bankseat beam. In the double-storey configuration, the girders consist of top and bottom panels with junction panels, and end taper panels forming the sloping end of the bridge. In both cases, ramp, deck and kerb units complete the construction.

Single-span bridges are launched using a centrally mounted launching nose made up of 3 m sections. During launching, the bridge is supported on roller beams at the home bank and a roller at the far bank. Single-storey bridges are launched over one roller beam, supported on baseplates. Double-storey bridges are launched over two such roller beams supported 4.6 m apart in a building frame. To adjust the height and levels of the roller beams, a hydraulic jack (with a lift of 300 mm) can be mounted at each corner of the building frame. These jacks are also used to lower the bridge on both banks after construction.

Reinforcement set

This set extends the single-span capability of the MGB to MLC 60 at 49.4 m and MLC 70 at 45.8 m. It complements the capability of the MGB pier and span junction equipment as it can be used in terrain where deep valleys make it impractical to use piers. The kit contains all the additional components required for use with normal double-storey bridge sets to construct a length of reinforced Class 60 MGB from 32.9 to 49.4 m. More recent reinforcement sets provide a capability of MLC 70 at a span of 49.4 m. The kit consists of reinforcing links 3.66 m long (with one pair of links 1.83 m long to provide the greatest range of spans), these are connected to form a pair of chains, one beneath each bridge girder. Links are connected to bottom panels at the ends of the bridge by an anchorage assembly. The links are positioned 2 m below the bridge by reinforcing posts which enable the system to be tensioned towards a vertical position by Tirfor cable jacks (details of these will be found in the *Recovery Equipment* section). The reinforcing is added by an extra party of eight personnel while the bridge is being constructed by the normal 24-man team.

The reinforcement set is in service with the British Army, US Army and Marine Corps and 16 other countries.

Reduced slope End of Bridge

The standard slope of a double-storey MGB End of Bridge is 1:5. A reduced slope can be desirable when a bridge is being used in a line of communication role and thus by second echelon vehicles with longer wheelbases and lower ground clearances. For this purpose an End of Bridge with a slope of 1:10 is available with an overall length of 4.267 m, as opposed to the normal 3.086 m.

Span junction set

The basic single-span MGB can be given a multispan capability by the use of the MGB span junction set. This equipment enables bridges of any length up to Class 70 load capacity to be hand-built over any fixed or floating supports, existing or improvised.

The span junction set consists of span junction posts pinned to top and bottom panels. The span junction posts are themselves pinned together at the top and hinged by an articulator connected to the lower halfof the span junction posts. A span junction link is fitted to join the bottom chords of the junction post topermit movement over rollers during launching. After launching the junction link is removed to allow articulation of the junction posts. The other main component is the capsill. This is a beam to which rocking rollers incorporating bridge bearings can be pinned. The capsil fitted with rocking rollers can be used on top of a pier to carry the bridgeduring launching or in the MGB adjustable roller beam support to form the heavy-duty Capsill Roller Beam (CRB) required when launching most multispan bridges.

Portable pier set

The portable pier set provides the MGB with its own two-legged pier which can be assembled during the building of the bridge. The legs pass through housings at each end of a pier beam. The legs are constructed in 3 m sections and piers of up to 12 m in height can be used in dry gaps and wet gaps with current speeds of up to 5.5 m/s.

The pier beam with housing, leg base section and one standard leg section at each end is preassembled and then pinned to the span junction post on the home bank and launched with the bridge. Once the bridge is in position, the adjustable braces are used to set the pier beam vertical before the legs are lowered and any additional leg sections are added. The pier components are also transportable on the standard MGB pallet, the heaviest component weighing 408 kg.

A MACH MGB under construction

A MACH MGB with reduced slope End of Bridge being crossed by engineer equipment

Close-up of Universal Light Pier in position

MGB configurations

Type	Length	Class	Building party	Building time	Transport in 4,000 kg pallet loads
Single-storey	9 m 22 m	70 16	9 17	15 min 30 min	2 5
Double-storey	31 m 49 m	70 16	25 25	45 min 1 h 20 min	10 16
2-span + pier	51 m	60	40	3 h	20
3-span + piers	76 m	60	40	6 h	27
Reinforcing kit					
Long Post	45.8 m 49.4 m	70 60	32 32	2 h 2 h	18 18
Short Post	49.4 m	70	32	2 h	18
Single-storey floating bridge	any	60	32+	dependent on length	dependent on length
Double-storey floating bridge	any	70	40+	dependent on length	dependent on length

Multispan

Using the MGB portable pier set for multispan bridges, the piers can be launched with the bridge by temporarily placing the first pier in position as the next is brought up, then moving the piers successively by booming the bridge back and forward.

The span junction components allow double-storey floating bridges of any length to be constructed on virtually any type of floating support which has sufficient buoyancy.

The articulation provided by the span junction bay and the use of long landing bays means that the MGB is the only modern military bridge equipment that can be used in a situation where bank heights range from 0 to 5 m. Even this range can be extended to 6 m if the MGB portable pier is used to form trestle bays.

Floating MGB/MGB Ferry/MGB Pontoon

See entry in the *Tactical Floating Bridges and Ferries* section.

MACH MGB

MACH (Mechanically Aided Construction by Hand) was introduced to meet the reduced manning requirements of modern armed forces. With a minimal purchase of additional components the normal MGB set can be held as part-assembled modules and constructed using a suitable crane or Crane Attachment Lorry-Mounted (CALM). The normal MGB team on an operational site can be reduced from 1 + 24 to 1 + 14 (optimum) or a minimum of 1 + 8.

Universal Light Pier

The Universal Light Pier was developed to meet the needs of the Swiss Army for a rapidly built pier for use with the MGB in mountain terrain. It is of modular construction and was designed to be independently erected from the bottom of a gap by a team of 12 personnel in 1 hour. The pier enables long two-span bridges to be built, for example 60 m at MLC 60 and 90 m at MLC 20. The pier legs can be up to 20 m high and can take loads of up to 120 tonnes.

Ski jump for V/STOL aircraft

The components of the MGB readily adapt to the Ski jump profile when raised at one end. The ramp surface comprises three MGB single-storey girders supporting MGB deck units between them. Outrigger units provide a total ramp width of 10 m.

Container ship flight deck

The MGB will provide a rapid easily erected flight deck on a wide range of commercial container ships. An MGB deck structure is supported on beams mounted on ISO containers which creates a hangar space underneath and/or side dispersal.

Specifications

NATO stock number	Service designation	Unit weight
8714	Top panel	175 kg
8710	Bottom panel	197 kg
8713	Junction panel	182 kg
8611	End taper panel	272 kg
8687	Bankseat beam	258 kg
8723	Ramp unit	120 kg
8698	Deck unit	74 kg

Bridge lengths and load classification

	Single-storey		Double-storey	
Load Class	Bays	Span	2 ends +bays	Span
100	5	9.8 m	10	27.4 m
70	5	9.8 m	12	31.1 m
60	5	9.8 m	13	33.0 m
50	5	9.8 m	14	34.8 m
40	6	11.6 m	16	38.5 m
30	8	15.2 m	18	42.0 m
24	9	17.1 m	20	45.8 m
20	10	19.0 m	21	47.6 m
16	12	22.5 m	22	49.4 m

The load class equates to a multi-axled wheeled vehicle of about the same weight in tonnes as the load class, and to a tracked vehicle of weight less than the load class (about 6 tonnes less for MLC 60). The exact load class of a vehicle depends on axle spacing, wheel loadings, track length and so on.

Span junction set

NATO stock number	Service designation	Unit weight
8159	Post span junction	260 kg
8157	Articulator, hydraulic	172 kg
8153	Capsill bridging	160 kg
8154	Roller assembly, rocker bearing	111 kg
8125	Beam, pier, half-section	445 kg
8123	Housing, pier leg	272 kg
8122	Pier leg, base section	162 kg
8121	Pier leg, standard section	208 kg
8119	Beam, grillage, pier	82 kg
8118	Sleeper pier	64 kg

Approximate weight of portable pier in bridge with length 12.2 m is 4,026 kg.

Status

In production. In service with Australia, Brunei, Canada, Denmark, Germany, Ghana, India, Iraq, Ireland, Italy, Jordan, Kenya, South Korea, Malaysia, Morocco, Netherlands, New Zealand, Nigeria, Pakistan, Peru, Philippines, Singapore, South Africa, Sweden, Switzerland (Festbrücke 69), Tanzania, Thailand, UK, USA (over 100 sets delivered to the US Army and Marine Corps), Venezuela and Zambia.

Manufacturer

Williams Fairey Engineering Limited, PO Box 41, Crossley Road, Heaton Chapel, Stockport, Cheshire SK4 5BD, UK.
Tel: +44 161 432 0281. Telex: 667866.
Fax: +44 161 431 3575.

UPDATED

Williams Fairey Engineering Axial Folding Bridge

Development

The Williams Fairey Engineering Axial Folding Bridge was developed using experience gained from the Medium Girder Bridge and is a dry support bridge capable of forming a span up to 60.96 m long. The Axial Folding Bridge was designed for ease of transportation, rapid assembly with minimum labour and operational simplicity.

An Axial Folding Bridge was supplied to the US Navy for use as a Lightweight Modular Multipurpose Spanning Assembly (LMMSA). This is used in association with the ELCAS (Elevated Causeway) programme carried out at the National Civil Engineering Laboratory at Port Hueneme in the USA. The LMMSA will be used as part of the ELCAS RESTORE programme to cross damaged sections of ELCAS, and make rapid connections between ships and ELCAS and ELCAS and shore.

Description

The Axial Folding Bridge is formed from three main modules: parallel (forming the main part of the bridge), ramp and reinforcement. Using only the parallel and ramp modules, bridges from 43 m at MLC 70 to 61 m at MLC 20 can be constructed. Adding the reinforcement set increases the load classification of the longest bridge to MLC 70.

The design of the end modules enables the bridge to be built with a ramp slope of 16 per cent or, where site conditions allow, as a flat-decked rampless bridge.

The modules are transported in pairs on ISO, DROPS or PLS flatracks or, alternatively, transport can be by standard 10 tonne capacity trucks. The flatracks can be handled by standard ISO material handling systems.

The Axial Folding Bridge is constructed and launched by conventional mobile cranes, using a roller frame and a launching nose. The crane is used to lift the modules from their packs, with the crane attachment arranged such that as the modules are lifted they automatically open outwards, that is, the sidebeams rotate about the centre decking with its integral roadway. Once opened, the module is 4.038 m wide (each sidebeam is 0.99 m wide and 1.03 m deep). The crane then lifts each module onto a building frame where it is connected to the previous module and pushed towards and across the gap to be bridged. The first module to be used is a ramp module which also carries a forward pylon and rear attachment to hold the launching nose in place. Extra parallel modules are added until the launching nose reaches the far bank. By that time the full bridge length should have been assembled and the entire bridge can be launched across the gap. The reinforcement system consists of a centre module containing a king post and tensioner and reinforcement links. The reinforcement links are connected to the bridge by support chains.

Retrieval is the reverse of the build procedure and can be carried out from either bank.

Timings have indicated that a 36.57 m bridge can be trestle-launched by eight personnel in 40 minutes.

Status

In service with the US Navy.

Manufacturer

Williams Fairey Engineering Limited, PO Box 41, Crossley Road, Heaton Chapel, Stockport, Cheshire SK4 5BD, UK.
Tel: +44 161 432 0281. Telex: 667866.
Fax: +44 161 431 3575.

VERIFIED

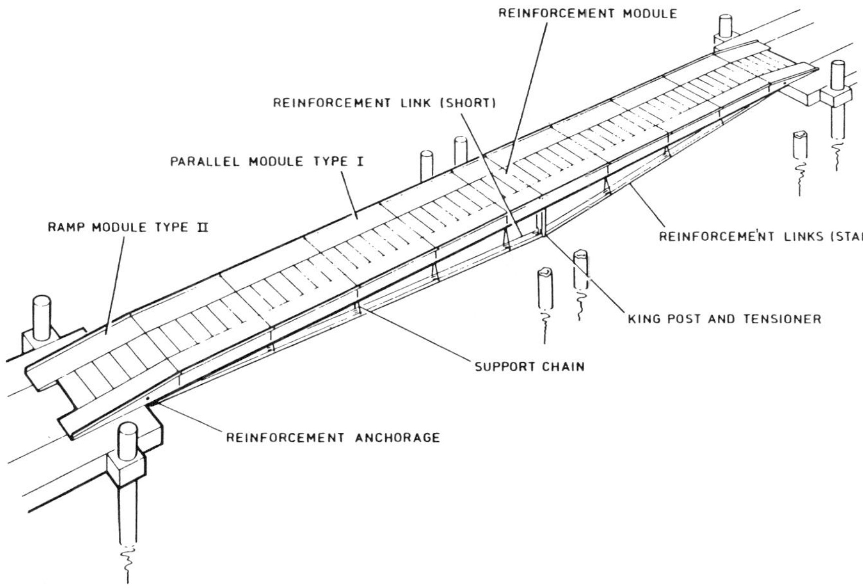

Outline drawing showing the main components of a 60.96 m Axial Folding Bridge

US Army truck crossing a 41 m Axial Folding Bridge

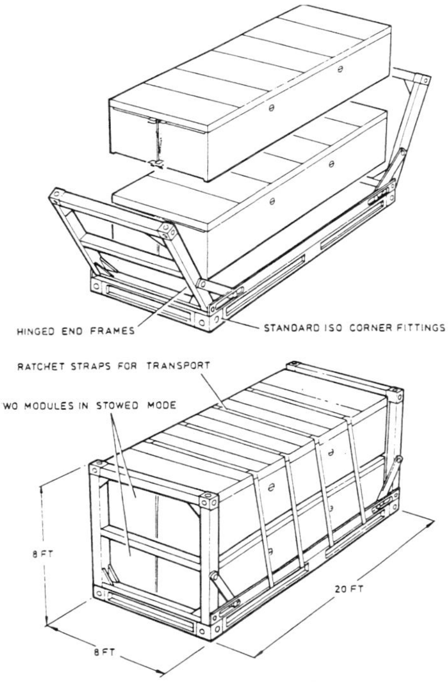

ISO modular packing for Axial Folding Bridge modules

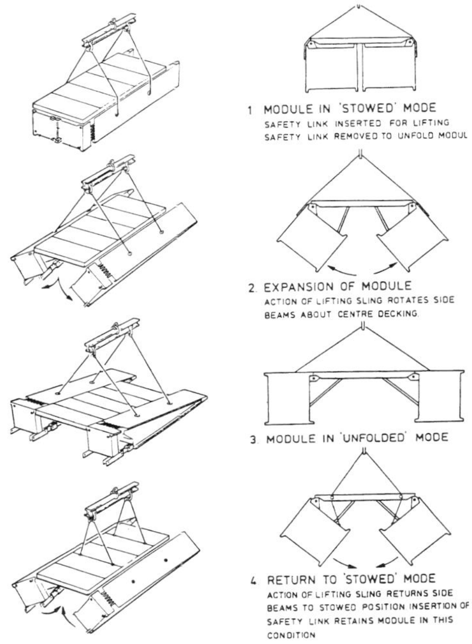

Sequence of folding and unfolding used for Axial Folding Bridge

UNITED STATES OF AMERICA

M2 Bailey Bridge

Description

The M2 Bailey Bridge was developed by the US Army Research and Development Laboratories. It has been well proven both as a tactical and line of communication bridge and is capable of carrying heavy traffic loads. It is used for both temporary and permanent applications. In emergencies it can be open to traffic in one to three days.

The M2 Bailey Bridge is an all-purpose prefabricated steel panel bridge designed for portability and speed of erection under adverse conditions. Optimum spans are 12.2 to 61 m. Width is 3.809 m between steel kerbs and 4.343 m between trusses. The components are manufactured in fixtures to ensure accuracy and interchangeability. The heaviest component weighs 281 kg.

The Bailey roadway is supported between two trusses or multiple-truss girders. It consists of longitudinal runner planks over transverse planks

M2 Bailey Bridge being launched (top) and completed (bottom)

(chess) laid over steel stringers supported by floorbeams (transoms) which rest on and are clamped to truss bottom chords. Steel kerbs secure the chess to the stringers. The basic truss element is the panel which is 3.048 m long, 1.448 m deep and 165.1 mm wide. Pin-connected end to end, trusses of any length are formed. Where strength exceeding that of single trusses is needed, multiple-truss girders can be assembled with either two or three panels side by side in single-, double- or triple-storey heights. End ramps extend the deck 3.048 m onto the approach runway.

Bailey truss panels, end posts, transoms and ramps are of low-alloy high-tensile steel having a yield point of 3,515 kg/cm^2 and an ultimate strength of 4,921.7 kg/cm^2.

The cantilever method of erection is accomplished without falsework. The bridge is assembled on stationary rollers and then pushed or pulled across the gap. A skeleton launching nose is assembled from standard bridge components and fixed to the leading end of the bridge. The nose precedes the leading end of the bridge while the bridge proper, acting as a counterweight, enables the nose to reach and land on rollers on the far bank. The bridge is then rolled into position, the nose is removed and the span is lowered onto its bearings. A 24.4 m double-truss bridge has been completed in less than 40 minutes in competitive trials.

There are approximately 15 major components in an average Bailey Bridge and about 50 components, fittings, accessories, special items and tools are available.

Specifications

Length of single span:
(min) 9.144 m
(max) 61 m
Width of roadway: 3.809 m

Status

Production complete. In service with the US Army and other armed forces.

Supplier

Bailey Bridges Inc, PO Box 1186, San Luis Obispo, California 93406, USA.
Tel: +1 805 543 8083.
Fax: +1 805 543 8983.

VERIFIED

BRIDGING BOATS

BELGIUM

Advanced Bridge Erection Boat

Description

In mid-1986 it was announced that a contract to supply FSB erection boats for the Belgian Army had been won by Meuse et Sambre SA, a member of the Belgian Shipbuilders Corporation NV. The boat involved was the Advanced Bridge Erection Boat (ABEB), which may also be used as a light tug, pusher, personnel or cargo transport, or as an assault craft.

The ABEB hull form has a moderate deep vee, a hard chine and spray deflection rails. The 7.76 m long hull form offers a high planing speed coupled with stability. The craft is configured such that flooding any one compartment will not affect the boat's ability to stay afloat and upright. A large working deck is provided aft and projects over the propulsion nozzles to provide them with protection. Draught is 0.53 m.

The propulsion systems are placed amidships in two separate compartments. Should either one of these compartments flood, forcing an engine shutdown, the craft will still be able to operate at reduced power. The main engines are two Deutz BF 6 L 913 C air-cooled diesels with each engine coupled to a resiliently mounted Twin Disc MG 506 reverse/reduction gearbox unit. Two KaWeMa 32 S62/6 water-jet units are driven by a shaft from each gearbox unit via an SKF gear type coupling. The water-jets were specially developed for the ABEB and are fitted with a six-bladed NAB impeller.

The twin water-jet units are widely spaced and are controlled by a single crew member using single levers. A hand hydraulic steering system is also incorporated for the nozzle direction control.

The ABEB can be carried on a special trailer but was designed to be carried on a MAN FSB carrier truck or a similarly equipped vehicle.

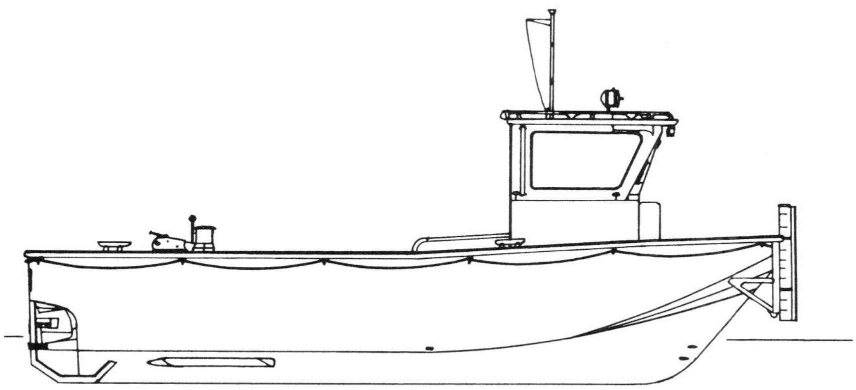

Advanced Bridge Erection Boat (ABEB)

Specifications

Length: 7.76 m
Beam: 3.23 m
Draught: 0.53 m
Max speed: 52 km/h
Engines: 2 × Deutz BF 6 L 913 C air-cooled diesels, each developing 191 hp at 2,500 rpm
Fuel capacity: 400 litres
Endurance at full power: 5 h

Status

In service with the Belgian Army.

Manufacturer

Meuse et Sambre SA, 16 Rue de Namur, B-5000 Beez, Belgium.
Tel: +32 81 220233.
Fax: +32 81 225920.

UPDATED

CANADA

Versatech Model 27/4 Combat Support and Bridge Erection Boat

Development

Versatech Products Inc of Burnaby, British Columbia, had been producing its Sealander work boats since 1972. Thus, when the Canadian Forces issued a requirement for a Bridge Boat Erection (BBE), Versatech was able to meet the Army's specification and the prototype was used in joint Canadian/US Army exercises. As a result of these trials the Canadian Forces ordered five Model 27/4 BBEs and a further 15 units were delivered during 1984.

Description

The Model 27/4 may be regarded as a combat support boat in addition to its role as a bridge erection boat. The vessel has enough foredeck space to carry a Jeep-sized vehicle and/or personnel and cargo. A vehicle loading ramp is fitted to the bluff bow which mounts two bridge erection push bars which can also be used to nudge pontoons. A towing bollard is also fitted and there is provision for a removable davit and winch. The Model 27/4 has a glass fibre hull with extra thickness in all stress areas. A balsa core stiffening is applied to the bow area and bottom. Buoyancy is increased by the use of polyurethane foam, making the vessel virtually unsinkable. Full-length Kevlar sponson protection shoes are fitted and the jet pan is a patented design made from aluminium. Marine grade aluminium is used for the superstructure and there is an open-rear cabin located amidships. Power comes from the twin Mercury M1E-340s driving twin Jacuzzi 20 YS water-jets providing a top speed in excess of 32 km/h when fully loaded.

Additional features include: a vessel lifting rig, removable push bars, a sternguard/swim grid, removable full-length safety rails, a demountable cabin top, an exhaust muffler to reduce heat signature radiation, a Halon fire suppression system, four automatic and two manual bilge pumps, screen wipers, and vehicle loading tracks. The instrument panel is a plug in/plug out component used to disable the vessel if required. The Model 27/4 can be operated by a single crew member. In routine bridging operations the vessel has an endurance of 8 hours.

Versatech Model 27/4 loading Jeep

Specifications

Weight:
(bare hull) 1,820 kg
(full fuel and crew) 7,315 kg
Payload:
(sheltered waters) 3,360 kg
(sea conditions) 2,270 kg
Length:
(overall) 9.34 m
(waterline) 8.33 m
Beam: 3.02 m
Draught:
(unladen) 0.229 m
(laden) 0.406 m
Max speed:
(full fuel and crew) 40 km/h+
(laden) 32 km/h+
Fuel capacity: 946 litres
Endurance: 8 h
Engine: twin Mercury M1E-340 with 2.5:1 transmissions
Max static thrust:
(forward) 2,497 kg
(reverse) 1,135 kg

Status

In service with the Canadian Forces. No longer marketed.

Manufacturer

Versatech Products Inc, 2479 Bellevue Avenue, Burnaby, West Vancouver, British Columbia V7V 1E1, Canada.
(This concern is no longer trading.)

UPDATED

CZECH REPUBLIC AND SLOVAKIA

Mo-108, Mo-111 and Mo-930 bridging boats

Mo-108

This is identical to the Mo-111 but is powered by a T-108, V-8 air-cooled diesel which develops 105 hp, giving the boat a lower performance than the Mo-111.

Mo-111

This is basically a modification of a Second World War German bridging boat and retains the German design of three rudders and a Kort nozzle, a metal ring-guard which houses the screw. The Mo-111 is transported on a large two-wheeled trailer called the MP-4. It is also known as the M-111 and the TATRA 111. The Mo-111 was introduced in the 1950s and built at the CXD plant at Decin.

Mo-930

This is the latest design and uses the same engine as the TATRA 813 (8 × 8) truck to power a single screw. The Mo-930 is transported on a single-axle trailer weighing 2,500 kg.

Specifications

Type	Mo-111	Mo-930
Crew	2	2
Weight	3,200 kg	4,000 kg
Length	7.5 m	7.68 m
Beam	2.2 m	2.2 m
Depth	1.2 m	n/a
Draught	0.85 m	0.85 m
Max speed	24 km/h	20 km/h
Fuel capacity	178 litres	n/a
Endurance	n/a	7 h
Engine model	T-111A-4	T-930-53
Engine type	diesel	petrol
Cylinders	V-12	V-12
Hp	170	200
Cooling	air	air
Towing power		
(forward)	n/a	2,275 kg
(reverse)	n/a	1,200 kg

Status

In service with the Czech and Slovak armies.

Manufacturer

State factories.

VERIFIED

FRANCE

Giat Industries F1 pontoon boat

Development

The F1 pontoon boat was designed to meet the requirements of the French Army Corps of Engineers by the Etablissement Technique d'Angers (ETAS), with production being undertaken by the Atelier de Construction de Tarbes (ATS). The first prototype was built by the ETAS with the second being built with the assistance of the ATS. The two prototypes were followed by five preproduction boats, that were completed in 1971. A total of 135 production model F1 pontoon boats was built for the French Army between 1972 and 1975.

The boat was designed to undertake a wide range of roles including the pushing and pulling of pontoons, installation of protective nets, transport of personnel and cargo, use as a platform for divers, and other bridging and mooring roles.

Description

The hull of the F1 pontoon boat is made of all-welded alloy (AG4MC) with 4.1 m³ of the boat filled with polyurethane foam to make it virtually unsinkable. The boat consists of two sections, fore and aft. The fore section contains the engine and crew compartment while the aft section contains the two propellers, their protective frames and steering motor. This is hinged to the main hull and can be swung upwards by two hydraulic joints to enable the boat to be operated in very shallow water very close inshore.

The engine is mounted forward of the crew compartment, and transmits power to a reduction gearbox with twin output shafts coupled via electromagnetic clutches to two variable displacement hydraulic pumps. Each pump feeds oil under pressure to a hydraulic motor mounted in a submerged nacelle directly behind the propeller. Each propeller has four blades and is 0.9 m in diameter. The boat has no rudder as the propellers, which are pod-mounted, can be traversed through 360°. The boat has exceptional acceleration and deceleration.

The boat is provided with three bilge pumps with a capacity of 1,200 litres/h each. Hand-operated capstan winches with a 3,500 kg capacity are mounted on either side of the hull and there is a towing post at the front of the hull and a second towing post at the rear of the crew compartment. The height of the rubber-clad bow fender post can be adjusted. There are two white light searchlights (one movable and one fixed) with a range of 100 m provided, together with two infrared lights and a range of 50 m (one fixed and one movable).

F1 pontoon boat

The F1 pontoon boat is normally carried in two sections, the bow section is carried on the rear of a Berliet GBC 8 KT (6 × 6) 4,000 kg truck, which also tows a single-pole 2,500 kg trailer carrying the rear half of the boat. These are offloaded by a crane and assembled in the water. It takes only 3 minutes to assemble the two units.

Specifications

Crew: 2-3
Weight:
(forward section) 3,000 kg
(aft section) 1,800 kg
Length: 8.3 m
Beam: 2.49 m
Height above waterline: 1.3 m excl lights
Draught:
(max) 1.15 m
(min) 0.45 m
Max speed: 25 km/h
(carrying 3 men plus 400 kg of cargo) 24.4 km/h
(pushing 4 pontoon floating bridge with a 40 t load) 9.36 km/h
Engine: Deutz F 10 L 413 V-10 air-cooled diesel developing 237 hp at 2,400 rpm
Electrical system: 24 V
Batteries: 4 × 6 V

Status

In service with the French Army.

Manufacturer

Atelier de Construction de Tarbes, 2 rue Alsace-Lorraine, BP 313, F-65013 Tarbes, France.
Enquiries to: Giat Industries, 13 route de la Minière, F-78034 Versailles Cedex, France.
Tel: +33 1 30 97 36 48.
Fax: +33 1 30 97 39 58.

VERIFIED

GERMANY

EWK bridging boat

Description

The EWK bridging boat, known as the *Bugsierboot* or *M-boot*, is used in the construction of floating bridges (including the German version of the Ribbon Bridge) and ferries. It has two adjustable pushing knees at the front, a windscreen that can be fitted with a removable canvas cover to give some degree of protection to the crew, and is propelled in water by two propellers. The

German Bugsierboot *bridging boat* (Michael Jerchel)

hull is aluminium and the built-in air compartments make it very difficult to sink. The boat is carried on and launched from a two-wheeled single-axle trailer.

Specifications

Crew: 2
Weight: 4,600 kg
Length: 7.45 m
Beam: 2.48 m
Depth: (overall) 1.85 m
Max speed: 22.5 km/h
Engine: air-cooled diesel developing 250 hp

Trailer plus boat
Weight: 8,100 kg
Length: 9.75 m
Width: 2.48 m
Height: 3.36 m

Status

In production. In service with the Belgian and German armies.

Manufacturer

Eisenwerke Kaiserslautern GmbH, Barbarossastrasse 30, D-67655 Kaiserslautern, Germany.
Tel: +49 631 8570.
Fax: +49 631 857300.

UPDATED

ITALY

SAI Ambrosini bridging boat

Description

This bridging boat was designed by SAI Ambrosini to meet the requirements of the Italian Army and entered service in 1974. The stainless steel hull is divided into six watertight compartments to guarantee buoyancy. The inboard-mounted engine is towards the front and the semi-enclosed cabin towards the rear.

Specifications

Crew: 2-3
Weight: 3,800 kg
Length: 7.5 m
Beam: 2.45 m
Height:
(without cabin) 1.8 m
(with cabin) 2.3 m
Draught: 0.7 m
Max speed:
(pushing or towing half MLC 60 raft in 1.75 m/s current) 3 km/h
(in calm water without load) 25 km/h
Endurance: 6 h
Turning radius: 8.5 m
Engine: Deutz model SF 12 L 413 12-cylinder 16.96 litres air-cooled diesel developing 260 hp

Status

Production complete. In service with the Italian Army.

Manufacturer

SAI Ambrosini - Societa Aeronautica Italiana SpA, Viale Roma 25, I-06065 Passignano sul Trasimeno (PG), Italy.
Tel: +39 75 827592. Telex: 660140 sai i.
Fax: +39 75 827650.

VERIFIED

SAI Ambrosini bridging boat being lowered into water by Astra BM 20MB1 crane truck. Boat has cockpit/cabin housing removed (T J Gander)

SAI Ambrosini 121 Thruster bridging boat

Description

The SAI Ambrosini 121 Thruster bridging boat was designed to cope with loads involved in the assembly of Class 60 floating bridges on rivers and inland water obstacles. It can be used with several types of floating bridge and can also be used in the assembly of non-floating bridging such as the Bailey Bridge. The hull is constructed from stainless steel with light alloy components. The layout is conventional with the engine compartment at the rear, the main crew compartment amidships and a bluff bow. On the bow is an angled assembly for nudging bridging components into place and behind this there is a small working area for assemblers. The crew compartment can be covered with a canvas tilt for extra shelter. The cockpit area can accommodate four persons or a load of 500 kg.

SAI Ambrosini 121 Thruster bridging boat

Specifications

Weight: (fully equipped) 4,450 kg
Length: 7.45 m
Beam: 2.44 m
Max draught: 0.78 m
Max speed: 20 km/h
Endurance: 9 h
Turning radius: 4 m
Fuel capacity: 274 litres
Engines: 2 × AIFO 8061 SM 5.5 litres 6-cylinder water-cooled (closed-circuit) turbo diesels developing 173 hp each at 2,950 rpm
Propulsion: Schottel SRP 50/50 steering traction screws traversing through 360°
Tow and thrust values:
(forward) 2,400 kg
(reverse) 1,200 kg

Status

In production. In service with the Italian Army.

Manufacturer

SAI Ambrosini - Societa Aeronautica Italiana SpA, Viale Roma 25, I-06065 Passignano sul Trasimeno (PG), Italy.
Tel: +39 75 827592. Telex: 660140 sai i.
Fax: +39 75 827650.

VERIFIED

SAI Ambrosini 126 multipurpose boat

Description

The SAI Ambrosini 126 multipurpose boat is intended for a variety of military roles but is equipped with a bluff bow to enable it to push heavy floating structures, such as bridging components, in shallow waters. It is also suitable for towing similar equipment. The SAI Ambrosini 126 is powered by two inboard FIAT diesel engines each driving a Castoldi TD 318 hydrojet that can be traversed for a high level of manoeuvrability.

The hull of the SAI Ambrosini 126 can be manufactured either in steel or in 5083 light alloy and is designed to render the vessel unsinkable. The cockpit is located well forward although other configurations are possible and it is possible to fit either a diving platform over the stern or carry stores or pallets behind the cockpit.

SAI Ambrosini 126 multipurpose boat

Specifications

Weight: 2,250 kg
Length: 7.5 m
Total width: 2.58 m
Draught: 0.45 m
Max speed: 35 kt
Endurance: 7 h at 20 kt
Engines: 2 × FIAT AIFO 8061 SM06 diesels developing 170 hp each
Hydrojets: 2 × Castoldi TD 318

Status

In production.

Manufacturer

SAI Ambrosini - Societa Aeronautica Italiana SpA, Viale Roma 25, I-06065 Passignano sul Trasimeno (PG), Italy.
Tel: +39 75 827592. Telex: 660140 sai i.
Fax: +39 75 827650.

VERIFIED

NETHERLANDS

Damen Bridge Support Boat 700

Description

The Damen Bridge Support Boat 700 (BSB 700) was designed specifically to operate with the Ribbon Bridge system. It has a high-tensile steel hull and one partly elevated deck with the engine located forward. The hull is divided into four watertight compartments accessible via flush deck-mounted hatches. All hatches and the mast are made of aluminium. Just aft of midships is an open cockpit with seating for a helmsman and all the controls.

Power is derived from a single 246 hp air-cooled diesel engine which provides power for two hydraulic pumps driving two four-bladed propellers. The swivelling hydraulic propellers are also used for steering through 360°. The engine provides a static bollard pull of 2.6 tonnes and for positioning bridge components in a water flow of around 5 knots the boat develops a thrust of 2.1 tonnes.

Standard equipment includes a helmsman seat in the cockpit, an aluminium sun awning over the cockpit, a polyethylene fender around the vessel (except the stern), two hand winches, two hoisting eyes, a dismountable mast, a hand bilge pump and an anchor and line.

For transport and launching the Damen Bridge Support Boat 700 is carried on the DAF YGZ 2300 (6 × 6) 10,000 kg Ribbon Bridge launcher truck.

Two Damen BSB 700s in operation with Ribbon Bridge ferry

Specifications

Weight: 6,000 kg
Length overall: 7 m
Beam: 2.9 m
Depth: 1.35 m
Draught: 0.75 m
Max speed: 16 km/h
Engine: Deutz BF 8L 513 air-cooled diesel developing 246 hp at 2,300 rpm

Status

In service with the Royal Netherlands Army (58).

Manufacturer

Damen Shipyards, Industrieterrein Avelingen West 20, PO Box 1, NL-4200 AA Gorinchem, Netherlands.
Tel: +31 1830 39911. Telex: 25335 dame nl.
Fax: +31 1830 32189.

VERIFIED

POLAND

KH-200 bridging boat

Description

The KH-200 bridging boat was developed in the late 1960s and after four years of trials was approved for production in 1971. It has a hull of all-steel construction and a pusher knee mounted at the bow. The crew is seated in a cabin towards the bow which has an open back. The engine is mounted to the rear of the crew and drives a single propeller at the rear. The KH-200 is used with the PP-64 heavy folding pontoon bridge and can also be used as a transport carrying up to 15 troops. The boat is transported on a large two-axle trailer which weighs 2,800 kg unladen.

Polish KH-200 bridging boat

Specifications
Crew: 2-3
Weight: 3,865 kg
Length: 8.14 m
Beam: 2.3 m
Draught: 0.72 m
Max speed: 25 km/h
Endurance: 12.1 h
Towing power:
(forward) 2,500 kg
(reverse) 1,200 kg
Engine: Leyland UE 680 6-cylinder water-cooled diesel developing 169 hp

Status
In service with the Polish Army.

Manufacturer
Polish state factories.

VERIFIED

RUSSIAN FEDERATION AND ASSOCIATED STATES (CIS)

BMK-T bridging boat

Description
The BMK-T is the most powerful bridging boat in the RFAS inventories and is used with pontoon bridges such as the PMP. The boat is carried on the rear of a KrAZ-214 (6 × 6) 7,000 kg or KrAZ-255B (6 × 6) 7,500 kg truck chassis and is launched by gravity, often with its engine running ready for immediate use. The recovery technique is as follows: the truck is reversed into the water and a cable from the truck's winch is connected to the stern of the BMK-T. The boat is then pulled out of the water over runners at the rear of the truck until it is back in the travelling position.

The hull of the BMK-T is of the 'sled' design for greater stability and has four enclosed compartments. Even if two non-adjacent compartments are flooded, the boat will not sink. The crew cabin is towards the front of the boat and is fully enclosed. The engine compartment is to the rear of the crew cabin and is also fully enclosed, allowing the boat to operate in very rough water. It is moved through the water by two propellers at the stern which have a maximum speed of 945 rpm. On encountering an obstacle during forward travel the two propellers are lifted out of the water automatically by hinges at the stern. When in the travelling position, the propellers are swung through 180° so that they are on top of the boat. The BMK-T is highly manoeuvrable and can be steered equally well in reverse. All controls are electrical and if required the boat can be remote-controlled using a cable from a maximum distance of 30 m. A pusher knee is mounted at the bow, although the boat is also used for towing pontoons.

BMK-T bridging boat being carried on KrAZ-214 (6 × 6) 7,000 kg truck

Specifications
Crew: 2
Weight: 6,000 kg
Length: 8.6 m
Beam: 2.7 m
Depth: 2.2 m
Draught: 0.75 m
Max speed: 17 km/h
Fuel capacity: 300 litres
Fuel consumption: 20 litres/h
Endurance: 15-17 h
Towing power:
(forward) 2,000 kg
(reverse) 750 kg
Engine: YaMZ-236, SP4 V-6, water-cooled diesel developing 180 hp at 2,100 rpm

KrAZ-255B with BMK-T Boat
Weight: 19,100 kg
Length: 10.6 m
Width: 2.8 m
Height: 3.76 m

Status
In service with some former Warsaw Pact nations.

Manufacturer
State factories.

VERIFIED

BMK-130 and BMK-150 bridging boats

Description
BMK-130/BMK-130M
The hull of the BMK-130 is of steel construction and is divided into individual watertight compartments. Mounted each side of the hull is a strut and a wheel which assist in the launching and recovery of the boat. Once afloat, the wheels are folded up alongside the hull. The BMK-130M, introduced in the mid-1960s, has a modified hull and when afloat the wheels are swung forward and stowed in wells on either side of the hull. This reduces not only drag but also the chances of damage to the wheels when afloat. The launching sequence is as follows: the bow attachment is connected to the front bumper of the truck, the boat is pushed into the water stern first and once the boat is afloat it is disconnected from the truck and the wheels of the boat are swung forward and stowed in the wells.

Yugoslavia produced a bridging boat based on the BMK-130M known as the RPR M68. For details see entry under Yugoslavia (Serbia and Montenegro) in this section.

BMK-150/BMK-150M
Until the introduction of the BMK-T this was the most powerful bridging boat used by the Warsaw Pact. It has a hull of aluminium construction is much lighter than the BMK-130 and has two engines with separately controlled twin screws. Unlike the BMK-130, this boat is provided with a windscreen and a cover that can be erected in bad weather. The basic BMK-150 has wheels which fold up on the outside of the hull, while the later BMK-150M has wheel wells.

BMK-130M bridging boat (Michael Jerchel)

Specifications

Type	BMK-130M	BMK-150	BMK-150M
Crew	2	2	2
Weight	3,450 kg	2,500 kg	3,800 kg
Length	7.85 m	8.2 m	7.4 m
Beam	2.1 m	2.55 m	2.55 m
Depth	1.5 m	2 m	n/a
Draught	0.622 m	0.66 m	0.75 m
Max speed	21 km/h	22 km/h	22 km/h
Fuel capacity	n/a	300 litres	150 litres
Endurance	12 h	7 h	6 h
Engine model	YaAZ-M204VKr-2, 5	M51-SPE-3, 5(2)	M70 SPA 3.5(2)
Engine type	diesel	petrol	diesel
Number of cylinders	4	6	4
Hp	120	62 × 2	119 × (2)
Cooling	water	water	water
Towing power			
(forward)	1,450 kg	1,500 kg	n/a
(reverse)	800 kg	n/a	n/a

Status
Production complete. In service with some members of the former Warsaw Pact.

Manufacturer
State factories.

VERIFIED

SINGAPORE

FBS 60 bridging boat

Description
The FBS 60 bridging boat is an integral part of the Floating Bridge System FBS 60 (for details of this system refer to entry in the *Tactical Floating Bridges and Ferries* section) and is intended for manoeuvring FBS 60 bridge sections.

The craft is referred to as a push-pull tug boat and uses a single conventional screw propulsion system powered by a 250 hp engine. The boat is constructed of steel and is conventional in layout with the control position well forward and the engine located amidships. A pushing knee is fitted over the squared-off bows.

For road transport the awning over the control position can be lowered or removed. The boat can be carried on a special trailer towed behind a 6 × 6 7,000 kg bridge transporter vehicle. The trailer is 9.7 m long overall. As an alternative the boat may be carried on a special boat carriage carried by a bridge transporter vehicle. Using the boat carriage, launching and retrieval times are quicker than using the trailer. The boat carriage is 7 m long, 2.9 m wide and 2 m high. Weight is 1,400 kg.

Specifications
Length overall: 7.45 m
Length at waterline: 7 m
Beam: (moulded) 2.4 m
Depth: (moulded) 1.3 m
Max draught: (aft) 0.6 m
Weight: 4,600 kg
Engine output: 250 hp
Bollard pull: 1,900 kg
Free running speed: 21.3 km/h (11.5 kt)

Status
Available.

Manufacturer
Singapore Technologies Shipbuilding and Engineering Limited, 7 Benoi Road, Singapore 629882.
Tel: +65 861 2244. Telex: singa rs 21206.
Fax: +65 861 3028.

VERIFIED

UNITED KINGDOM

FBM Marine CSB Bridge Erection Boat

Development
Based on a requirement issued by the UK Military Vehicles and Engineering Establishment at Christchurch (now the Defence Research Agency, Christchurch), the Fairey company (now FBM Marine Limited) started development of the 8 m CSB in the Autumn of 1975, with the first prototype being delivered to the British Army for trials early in 1977. As a result of extensive trials by the British Army both in the UK and on the Rhine in Germany, a preproduction boat was ordered by the British Army for further trials. It was delivered in May 1978. The boat was accepted for service with the British Army in February 1979, and an order placed for 56 boats the following year. In 1983 the UK Ministry of Defence ordered a further 12 CSBs for the British Army to replace those lost during the Falkland Islands campaign. After extensive evaluation by the US Army MERADCOM (now the Belvoir Research, Development and Engineering Center, Fort Belvoir, Virginia) the CSB was accepted under NATO Standardisation and Rationalisation arrangements, and 280 are in service with the US Army.

The Stealth version of the CSB (see under Variant for details) is in full licensed production in Southeast Asia.

FBM Marine Combat Support Boat (CSB) (T J Gander) ***1997***

Description
Based on the Allday 8 m hull, the boat features a pusher bow and an aft-mounted capstan/tow hook. The mast and cabin top are both removable for low-profile operations, and transport/stowage. The two self-draining cockpits are each designed to accommodate a standard 1,000 kg NATO pallet.

The CSB bridge erection boat is made of welded marine-grade aluminium alloy to give a lightweight yet extremely strong boat. Based on a well-tried hull, the craft was designed to meet the requirements of the British Army for a boat to assist with bridging operations and other river and estuary support and assault duties. Ultra Hydraulics' water-jets enable the boat to be used in shallow water with twin 212 shp turbocharged marine diesels providing the power. Transport, launching and recovery is by special cradle on a Ribbon Bridge Transporter Truck or by a towed off-road trailer.

Standard equipment includes: navigation and towing lights and a searchlight; fire extinguishers; windscreen wipers; bilge pumping system; heavy-duty fendering; Explosafe filled fuel tank; and built-in buoyancy to float craft in an upright position.

The batch of 12 ordered by the Ministry of Defence during 1983 all have 'winterisation' kits for operations at low temperatures. This includes extra cable insulation, more powerful screen wipers, pump and engine heaters, defrosting equipment and personnel heaters. Current production craft incorporate 'keel' or 'remote' engine cooling depending on the operating environment.

The US Army refers to the CSB as the Bridge Erection Boat (BEB).

In September 1992 FBM Support Services completed a two year $10 million modification programme on the 248 CSBs in service with the US Army at locations in Europe and the USA. The modifications updated the cooling system and associated electrical circuitry and enable the engines to be started prior to launch.

Variant

In response to a requirement from a Far Eastern customer (South Korea) FBM Marine developed a version of its CSB bridge erection boat offering a drastic reduction in noise levels. The noise level in the cab is reduced from 101 dBA to 84 dBA and the optimum speed for this reduced noise level is 7 kt. At this speed the boat is inaudible at approximately 285 m with a background noise of 42 dBA approaching the listener and at 530 m when departing.

The twin Sabre Marine 212 diesel engine installation is enclosed by adding a glass fibre hood and transverse bulkheads, all lined with 32 mm black PVC-faced foam with a heavy polymeric barrier. Entry points for services through the bulkheads are sealed. Hull side panels and hatch covers are lined with 3 mm bitumen foil vibration-damping material.

Burgess 100 mm ADS through-adsorption type exhausts have replaced the side-shell water-injected units. The engines are installed on Metzeler Fluidastic rubber hydraulic resilient mounts. Extra cooling for the enclosed engine compartment is provided by two Jabsco 24 V electric fans.

Other changes introduced with the Stealth CSB involve the replacing of the original goal-post mast by a simple pole mast in front of the wheelhouse. The Stealth CSB prototype had rotatable nozzles fitted to the Dowty water-jets. These improve reversing and obviate icing problems.

The Stealth CSB has been in licensed production in South Korea with subcontractors Korea Tacoma Marine Industries (KTMI) since 1985. During 1992 FBM Support Services Limited was awarded a contract worth more than $7 million for the supply of materials and technology transfer to KTMI. The initial contract involved the materials for 55 craft. It was anticipated that further contracts would result in the same number of sets being supplied each year until 1997—material for a further 56 units was delivered during 1993, a further 55 during 1994 plus the same again in 1995. South Korea has thus become the largest single user of the CSB.

Specifications

Length:
(with pushing knee) 8.38 m
(waterline) 6.98 m
Beam over fenders: 2.49 m
Height:
(without cab) 1.98 m
(with cab) 2.79 m
Weight:
(fully fitted out and fuelled) 4,080 kg
Maximum payload: 2,000 kg
Draught:
(fully fitted out and fuelled) 0.56 m
(fully laden) 0.66 m
Engines: twin Sabre 212 turbocharged marine diesels each continuously rated at 212 shp at 2,450 rpm
Propulsion units: twin Ultra 300 mm 2-stage water-jets coupled to engines through Borg Warner 72C hydraulic direct drive reverse gearbox
Max static forward thrust: 2,000 kg
Max static reverse thrust: 1,000 kg
Max speed:
(unladen) 44 km/h
(fully laden) 31 km/h
Fuel tank capacity: 227 litres

Status

In production. In service with the engineer units of five armies including Greece (18), Turkey, UK and the USA (280). The Stealth CSB is in licensed production in South Korea.

Manufacturer

FBM Marine Limited, Cowes Shipyard, Cowes, Isle of Wight PO31 7DL, UK.
Tel: +44 1983 297111. Telex: 86466 fambro g.
Fax: +44 1983 299642.

VERIFIED

RTK Combat Support Boat Type CSB 508

RTK Combat Support Boat Type CSB 508

Description

The CSB 508 was designed to undertake a wide range of duties including bridging, tug boat, supply and diving. The boat is capable of static thrusts of up to 1,700 kg for tugging purposes and can transport 3,050 kg of cargo. The wide deck and bow ramp permit the loading of men and equipment from either beach or quay and with the jet propulsion system the boat has a draught of 0.42 m, allowing operation in very shallow water.

The hull of the CSB 508 is made of glass fibre-reinforced plastic and has a built-in foam buoyancy in excess of the craft's gross displacement, making the boat unsinkable. The hull requires the minimum of maintenance and can easily be repaired in operation.

The deck has a non-slip bonded grit surface applied. The gunwales are of stainless steel rolled section and full peripheral fendering is provided at the gunwale with a solid nylon rubbing strake installed at the waterline. The winch-operated ramp is reinforced and has a non-slip surface. Scuppers for the removal of water from the deck are located at the transom.

There are two pushing knees mounted at the bow and a towing post mounted aft of the helmsman's seat.

Standard equipment includes six warping bollards and four lifting points mounted on the gunwale rail, navigation lights, floodlight, klaxon and a rechargeable fire extinguisher.

Optional equipment includes an illuminated helmsman's compass, deck gear, cargo/vehicle lashing kit, demountable jib crane with a capacity of 250 kg and a four-wheel close-coupled boat trailer.

Specifications

Length: (overall) 8.23 m
Beam: (overall) 2.9 m
Height:
(overall) 1.88 m
(of gunwales above deck) 0.75 m
Deck area: 15.1 m^2
Fuel capacity: 2 × 135 litres pannier tanks

Status

In production. In service with several countries.

Manufacturer

RTK Marine Limited, 446 Blandford Road, Hamworthy, Poole, Dorset BH16 5BL, UK.
Tel: +44 1202 685531.
Fax: +44 1202 683347.

VERIFIED

Thos Storey Motorflote

Description

The Thos Storey Motorflote was designed for use with the Thos Storey Uniflote System. It consists of a modified Uniflote box-section pontoon powered by an internal marine diesel engine. The engine drives a propeller situated in a protected area underneath the rear of the Motorflote. Steering is carried out by a rudder operated from a wheel in the control cabin in the centre of the Motorflote pontoon. Having a simple box-section the Motorflote can be readily secured to any Uniflote pontoon combination such as a raft or ferry. The advantages of using the Motorflote in place of orthodox outboard engines on a Uniflote raft or float are

Uniflote raft being propelled by Motorflote

Motorflote with auxiliary bow fitted for use when not pushing Uniflote ferry

said to be that the Motorflote can be easily connected and disconnected to any Uniflote System construction, the propeller is less vulnerable to damage as it is contained within a protected area, and the Motorflote is quieter and cheaper.

The control cabin may be fully open or enclosed, and handrails can be fitted if required. Two Motorflotes have operated in the Falkland Islands propelling a 12-Uniflote ferry with a capacity of over 100 tonnes.

Specifications

Length: 3.05 m
Width: 2.44 m
Depth:
(deck to skids) 1.71 m
(deck to cabin top) 2.14 m
(total) 3.85 m
Draught: 1.1 m
Freeboard: 0.61 m
Weight: approx 4,500 kg
Fuel capacity: 226 litres
Engine: Mermaid Mariner 6-cylinder 6.227 litre diesel developing 115 bhp at 2,600 rpm continuous
Propeller diameter: (4 blades) 0.7 m
Static bollard pull: 13.9 kN
Speed in still water:
(unladen pushing) 2.4 m/s
(unladen towing) 1.8 m/s
(loaded pushing 150 t) 1.7 m/s

Status

Available.

Manufacturer

Thos Storey (Engineers) Limited, 52 Queens Road, Weybridge, Surrey KT13 0AN, UK.
Tel: +44 1932 858133. Telex: 21868.
Fax: +44 1932 855588.

VERIFIED

YUGOSLAVIA (SERBIA AND MONTENEGRO)

RPR M68 bridging boat

Description

Described as a tug vessel, the RPR M68 bridging boat is based on the RFAS BMK-130M (see separate entry in this section) and is intended for pushing and towing floating bridges and ferries. It can also be used to carry personnel and cargo.

The overall design of the RPR M68 closely follows that of the BMK-130M. It has an all-steel hull and is equipped with travelling wheels that fold upwards into wheel wells when the boat is in the water; when the wheels are lowered the boat can be towed by a truck. Power is supplied by a diesel engine located amidships and developing approximately 140 hp.

A small cabin can be erected for the helmsman and a small mast carrying a floodlamp can also be provided.

Specifications

Crew: 2
Weight:
(standard) 3,940 kg
(full load) 4,220 kg
(transport) 4,300 kg
Length: 7.88 m
Beam: 2.1 m
Draught: (mean) 0.64 m
Max speed: (full load) 15.55 km/h
Tractive force:
(forward) 16.5 kN
(astern) 8.25 kN

Status

In service with the Yugoslav Army.

Contractor

FDSP - Federal Directorate of Supply and Procurement, PO Box 89, Bulevar umetnosti 2, YU-11070 Novi Beograd, Yugoslavia.
Tel: +381 11 222 2743. Telex: 71000 JU SPDR.
Fax: +381 11 324 8738.

VERIFIED

RPR M68 bridging boat

RPR M68 bridging boat

MINE WARFARE EQUIPMENT

Minelaying equipment
Mine detection equipment
Mineclearing equipment
Minefield marking equipment

MINELAYING EQUIPMENT

AUSTRIA

AID 2000 minelaying system

Description

The AID 2000 (Austrian Intermittent Digger 2000) is a trailer-mounted minelaying system intended for use with the Intertechnik ATM 2000E anti-tank mine, although it can be adapted to handle other mines. The AID 2000 can lay up to 250 mines/h (at a laying distance of 2 m in normal soil) at depths of 50 to 300 mm in 50 mm increments and at intervals of 1.5 to 12 m, as required.

The self-contained AID 2000 is operated by one man and utilises a 54 hp air-cooled diesel engine to power the system. The trailer carrying the system can be towed by virtually any in-service truck, with mines being fed into the system via rails protruding over the truck cargo area. Once the system has been started the laying process is virtually automatic and soil disturbance after laying is minimal.

AID 2000 minelaying system in operation

Specifications

Crew: 1
Weight: approx 6,500 kg
Length: approx 6 m
Width: approx 2.5 m
Height: approx 2.7 m
Towing speed: (road) up to 100 km/h
Fording: 0.8 m
Engine: air-cooled diesel developing 54 hp

Laying capacity: up to 250 mines/h
Laying interval: 1.5-12 m
Laying depth: 50, 100, 150, 200, 250 or 300 mm

Status

In production.

Manufacturer

Intertechnik Techn. Produktionen-GesmbH,
Industriezeile 56, Postfach 100, A-4040 Linz, Austria.
Tel: +43 70 732 7892-0. Telex: 221522.
Fax: +43 70 732 7892-13.

VERIFIED

CHINA, PEOPLE'S REPUBLIC

NORINCO scatterable mine system

Description

In mid-1993, it was disclosed that China North Industries Corporation (NORINCO) had developed a prototype of a scatterable mine system, it had been trialled on various types of truck and tracked armoured personnel carrier chassis.

The largest of the truck versions has eight launchers located on the cargo area, with four launchers on each side. Each launcher is mounted on a turntable which can be traversed through 180° and elevated 40°. Each launcher carries 36 magazines, which double as launch tubes, in six rows of six. A typical carrier for the eight launcher version would be the Mercedes-Benz 2026A (6 × 6) truck or a similar vehicle. Lighter capacity vehicles, such as the Dong Feng 240 (6 × 6) truck, carry only four launchers. There is also a version carrying six launchers on a tracked WZ 534 armoured personnel carrier chassis.

Each magazine has a diameter of 122 mm and can carry either five SATM (Scatterable Anti-Tank Mines), 15 SAPM tripwire-activated anti-personnel fragmentation mines, or 45 SAPEM pressure-activated anti-personnel explosive mines. Magazines containing a mix of SAPM and SAPEM anti-personnel mines are available. All three mines have a self-destruct mechanism preprogrammed before launch for a time of 1 to 272 hours in 1 hour increments. If required, individual magazines can be deployed by hand.

An operator in the vehicle cab can control the minelaying sequence and select the number and types of mine to be laid, as well as setting the launching angle. Using an eight launcher system, an anti-tank minefield can be laid with a density of 0.2 to 1 mine/m. Anti-personnel minefields can have a density of 0.5 to 5 mines/m. A typical minefield would have a width of 20 m at a launch distance of no less than 20 m and no more than 150 m. A single launcher can be reloaded in 2 minutes.

Specifications

Mercedes-Benz 2026A chassis
Launchers: 8
Magazines: 288
SATM mines: 1,440
SAPM anti-personnel mines: 4,320
SAPEM anti-personnel mines: 12,960
Weight, empty assembly: 150 kg
Weight, loaded assembly: 720 kg
Dimensions of launcher platform:
(8 launchers) 5.5 × 2.35 m
(6 launchers) 4.1 × 2.35 m
(4 launchers) 3.445 × 2.18 m
Max elevation angle: 40°
Traverse: 180°
Reload time: 2 min
Typical minefield width: 20 m
Min launch distance: 20 m
Max launch distance: 150 m
Operating temperature range: –32 to +50°C

SATM anti-tank mine
Weight: 2.2 kg
Diameter: 114 mm
Height: 125 mm
Penetration: (steel) 140 mm

SAPM anti-personnel mine
Weight: approx 900 g
Diameter: 52 mm
Height: 97 mm
Charge: 52 g desensitised RDX
Lethal radius: 6 m

NORINCO scatterable mine system mounted on a Mercedes-Benz 2026A (6 × 6) truck **1997**

NORINCO scatterable mine system mounted on a WZ 534 tracked armoured personnel carrier chassis **1997**

SAPEM anti-personnel mine
Weight: 85 g
Length: 71 mm
Width: 54 mm
Height: 31 mm
Charge: 14 g desensitised PETN

Status
Ready for production.

Manufacturer
China North Industries Corporation (NORINCO), 7A Yue Tan Nan Jie, PO Box 2137 Beijing, Beijing, People's Republic of China.
Tel: +86 1 853 3695. Telex: 22339 cnin cn.
Fax: +86 1 853 3236.

UPDATED

NORINCO Type 74 minelaying rocket system

Description
The NORINCO Type 74 minelaying rocket system is used by combat engineer units of the People's Liberation Army for the rapid laying of anti-tank minefields in forward areas. The Type 74 units are organised into batteries of four launchers carried on Jiefang CA-30A (6 × 6) trucks. Each battery can lay a 400 × 400 m minefield to a maximum range of 1,500 or 3,500 m, depending on the type of rocket launched.

Each CA-30A truck carries on the rear a set of 10 launcher rails 4 m long. On the road each truck can carry some of the four-man launcher crew seated behind the cab on an open bench seat. Before firing, the crew has to take cover and the cab windows are covered by steel shutters. The rockets are loaded on to the rails manually, with five rockets above and five below the launcher rails. The rails may be elevated between 7 and 48° and traverse is 90° left and 45° right; when on the move the rails point forward. A salvo of 10 rockets can be fired in 15 seconds.

The Type 74 minelaying rocket system can launch either the Type 74 or Type 85 minelaying rockets.

Type 74 minelaying rockets each weigh 127 kg and contain 10 Type 69 or Type 70 plastic-cased anti-tank mines. The diameter of the rocket head is 284 mm. The rocket motor body has an estimated diameter of 160 mm and four tailfins are used for stabilisation. Each Type 74 rocket is 2.47 m long and has a maximum range of 1,500 m.

Type 85 minelaying rockets each weigh 156 kg and carry 10 anti-tank mines to a maximum range of 3,500 m. The warhead has a diameter of 284 mm and each rocket is 2.6 m long.

Specifications
Launcher weight:
(complete with rockets and crew) approx 8,780 kg
(empty) 6,954 kg
Length: (complete, travelling) 6.44 m
Width: 2.27 m
Height: 2.93 m
Max speed: 40 km/h
Length of launcher rail: 4 m
Traverse: 90° left/45° right
Elevation: 7 to 48°

Type 74 rocket
Head diameter: 284 mm
Length: 2.47 m
Weight: 127 kg
Number of mines carried: 10
Max range: 1,500 m

Type 85 rocket
Head diameter: 284 mm
Length: 2.6 m
Weight: 156 kg
Number of mines carried: 10
Max range: 3,500 m

Status
Type 74 and Type 85 in service with the Chinese armed forces.

Manufacturer
China North Industries Corporation (NORINCO), 7A Yue Tan Nan Jie, PO Box 2137 Beijing, Beijing, People's Republic of China.
Tel: +86 1 853 3695. Telex: 22339 cnin cn.
Fax: +86 1 853 3236.

UPDATED

Type 74 305 mm minelaying rocket system mounted on CA-30A (6 × 6) trucks

Type 74 minelaying rockets on Type 74 launcher

NORINCO Type 79 minelaying rocket system

Description
The NORINCO Type 79 minelaying rocket system is in production for the Chinese armed forces. This system is carried by the same CA-30A truck as the Type 74 system (see previous entry) but the 305 mm diameter warhead rockets are larger overall and have ring stabilisers at the tail. The launchers use frames instead of rails and are constructed from tubular steel welded into two banks of four cells. Each launcher has a crew of four and weighs 8,730 kg loaded.

The Type 79 rocket system has a maximum scatterable range of from 2,100 to 2,500 m. Each Type 79 rocket carries 10 Type 69 plastic cased anti-tank mines inside the 305 mm diameter warhead and is 2.927 m long; weight at launch is 156 kg. Rockets can be launched at 1.75 second intervals.

The Type 79 rocket system can also be used to launch 305 mm fuel-air mineclearing rockets to breach minefields. For details refer to the entry under *Mineclearing equipment*.

Specifications
Type 79 launcher
Crew: 4
Launcher weight: (complete with rockets and crew) approx 8,730 kg
Length: (complete, travelling) 6.918 m
Width: 2.2 m
Height: 3 m
Max speed: 65 km/h
Fording: 0.85 m
Traverse: 95° left/45° right
Elevation: +4° to +50°

Type 79 rocket
Head diameter: 305 mm
Length: 2.927 m
Weight: 156 kg
Number of mines carried: 10
Max range: 2,500 m
Launch velocity: 45 m/s
Max velocity: 185 m/s
Operating temperature range: −40 to +40°C

Status
In service with the Chinese armed forces. Offered for export.

Manufacturer
China North Industries Corporation (NORINCO), 7A Yue Tan Nan Jie, PO Box 2137 Beijing, Beijing, People's Republic of China.
Tel: +86 1 853 3695. Telex: 22339 cnin cn.
Fax: +86 1 853 3236.

NEW ENTRY

NORINCO Type 84 minelaying rocket system

Description

The NORINCO Type 84 minelaying rocket system is a variant of the Type 83 122 mm multiple artillery rocket system, itself based on the Soviet 122 mm Grad series. The Type 84 has 24 launch tubes firing Type 84 minelaying rockets over an operational range of 3,500 to 6,000 m; the maximum range of a Type 84 rocket is 7,000 m. The launch system is carried on a 6 × 6 truck chassis with a four-door, fully enclosed cab.

The launch tubes have a maximum elevation of 55° (although the normal maximum launch angle is 49°) and a traverse of 95 ° to the left and 75° to the right. All 24 rockets can be fired in 13 to 16 seconds. The rocket body breaks open in flight over the target area to dispense the mines which then fall to the ground steadied by a parachute. Once on the ground three hinged legs open to orientate the mine in the vertical position and provide support. A single 24-rocket salvo can create a minefield 400 m long and 650 m wide.

Fin-stabilised Type 84 minelaying rockets each weigh 62 kg and carry eight contact- or sensor-detonated anti-tank mines to a maximum possible range of 7,000 m. The warhead has a diameter of 122 mm and each rocket is 2.77 m long. Initial launch velocity is 41 m/s rising to 350 m/s.

The mines used with the Type 84 rocket are both 114 mm in diameter and weigh 3.12 kg. Once dispensed and standing on their three support legs they are 160 mm high.

NORINCO Type 84 122 mm minelaying rocket system ***1997***

Specifications

Launch tube diameter: 122.4 mm
Launcher weight: (complete with rockets and crew) approx 9,033 kg
Length: (complete, travelling) 6.51 m
Width: 2.2 m
Height: 2.77 m
Max speed: 60 km/h
Traverse: 95° left/75° right
Elevation: 0 to +55°

Type 84 rocket
Head diameter: 122 mm
Length: 2.77 m
Weight: 62 kg
Number of mines carried: 8
Max range: 7,000 m
Operational range: 3,500-6,000 m

Status

In service with the Chinese armed forces.

Manufacturer

China North Industries Corporation (NORINCO), 7A Yue Tan Nan Jie PO Box 2137 Beijing, Beijing, People's Republic of China.
Tel: +86 1 853 3695. Telex: 22339 cnin cn.
Fax: +86 1 853 3236.

NEW ENTRY

NORINCO Type 1 anti-personnel mine scattering system

Description

The NORINCO Type 1 anti-personnel mine scattering system is a man-portable mine dispensing system, intended to produce rapid anti-personnel barriers to protect infantry and special forces units. It can be carried and operated by one man.

The Type 1 dispenser consists of a cylindrical container set at a fixed angle of 44° by a base frame and two folding legs. To operate, the dispenser is placed on the ground, the support legs are unfolded and the unit is then ready. A mechanical friction pull igniter is employed to launch the dispenser contents after a delay of 8 seconds.

The Type 1 can create a 25 × 25 m minefield over an area of 220 to 300 m from the dispenser. The dispenser scatters 30 diminutive Type 1 anti-personnel mines to form the minefield barrier.

The mines involved with the Type 1 system are only 43 mm in diameter and 24 mm high, containing a pyrotechnic chain which delays arming until after the mines have reached the ground. The explosive content of each mine is estimated to be 22 g. Pressure on either side of the mine will cause it to function.

A single Type 1 anti-personnel mine scattering system weighs 8 kg complete.

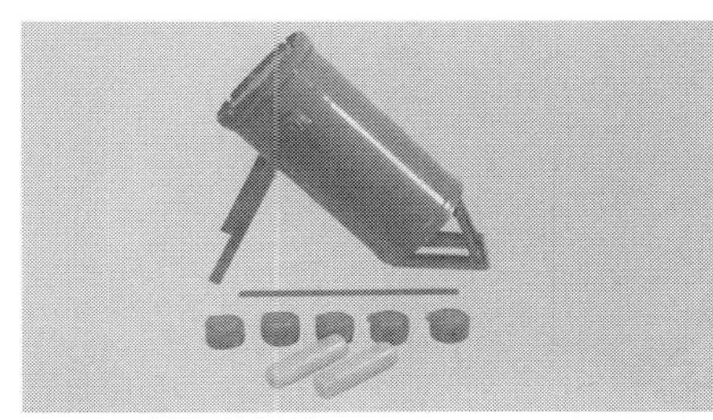

NORINCO Type 1 anti-personnel mine scattering system ***1997***

Status

In production. Offered for export.

Manufacturer

China North Industries Corporation (NORINCO), 7A Yue Tan Nan Jie, PO Box 2137 Beijing, Beijing, People's Republic of China.
Tel: +86 1 853 3695. Telex: 22339 cnin cn.
Fax: +86 1 853 3236.

NEW ENTRY

NORINCO Type 2 anti-personnel mine scattering system

Description

The NORINCO Type 2 anti-personnel mine scattering system is a man-portable mine dispensing system, intended to produce rapid anti-personnel barriers to protect infantry and special forces units. It can be carried and operated by one man.

The Type 2 dispenser resembles a suitcase, inside which are five cylindrical mine dispensers held in foam packing. Each dispenser contains 16 anti-personnel mines, making 80 mines in all. To operate, the case is placed on the ground at an angle between 35 and 40° and the case lid is removed. A friction pull igniter is used to launch the mines to a range between 120 and 170 m where they will cover an area measuring approximately 80 m long and 60 m wide.

The type of anti-personnel mine used with the Type 2 system is not known but it is believed that it is similar to the mine used with the Type 1 system (see previous entry).

A single Type 1 anti-personnel mine scattering system weighs 19 kg complete.

Status

In production. Offered for export sales.

Manufacturer

China North Industries Corporation (NORINCO), 7A Yue Tan Nan Jie, PO Box 2137 Beijing, Beijing, People's Republic of China.
Tel: +86 1 853 3695. Telex: 22339 cnin cn.
Fax: +86 1 853 3236.

NEW ENTRY

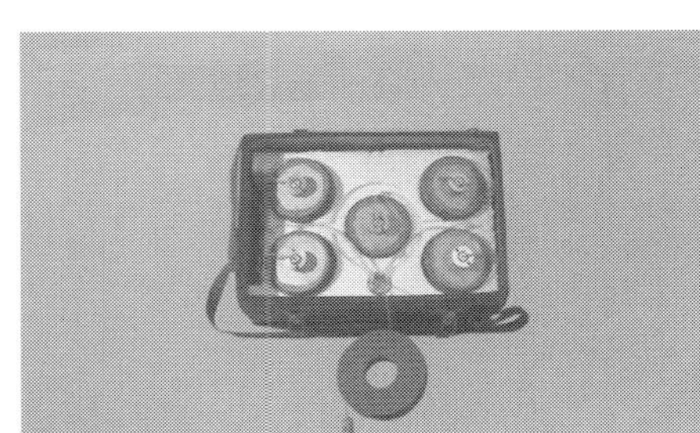

NORINCO Type 2 anti-personnel mine scattering system ***1997***

CZECH REPUBLIC AND SLOVAKIA

MV-90 tracked minelaying vehicle

Description

The MV-90 tracked minelaying vehicle (*minový ukladač*) is a turretless variant of the BVP-1, the Czech/Slovak version of the Soviet BMP-1 infantry fighting vehicle, (for full details of the BVP-1/BMP-1 refer to *Jane's Armour and Artillery 1995-96* pages 294 to 299).

The MV-90 is used to dispense surface-laid anti-tank mines via a chute, which is lowered from within the vehicle hull once the right-hand rear door has been opened. The hull contains racks for up to 100 PT Mi-U and/or PT Mi-Ba-III anti-tank mines. Laying speed is between 3 and 7 km/h.

The MV-90 has a crew of three: commander, driver and mine handler.

Status

Prototype.

Manufacturer

Czech and Slovak state factories.

VERIFIED

Rear view of MV-90 tracked minelaying vehicle with mine-dispensing chute lowered (T J Gander)

MV-90 tracked minelaying vehicle (T J Gander)

VZ 92 Križan mechanical and rocket minelaying system

Description

The VZ 92 Križan mechanical and rocket minelaying system is a modular minelaying system based on the chassis of the TATRA 815 VP 14 36.265 8 × 8 truck. There are two major components in the system, the combined rocket and mechanical minelayer vehicle, and an ammunition resupply vehicle, both of which have armoured cabs.

The combined rocket and mechanical minelayer has an armoured forward control cab for the crew of two, in front of which is a dozer blade. Behind the cab is a spare wheel rack over which are 12 smoke grenade projectors. Behind this is a platform mounting a 40 barrel 122 mm rocket launcher which can be elevated and traversed as required. The area under the launcher platform is used for the stowage of a further 40 rockets.

The mines used with the rocket-dispensed system can be anti-tank, anti-personnel, or a mixture of both.

The anti-tank mines are known as the PT Mi-D1, a variant of the PT Mi-D1 anti-tank mines fired to a maximum range of 17,000 m by the RM-70/85 multiple rocket launcher (for details of this system refer to *Jane's Armour and Artillery 1995-96* pages 717 to 719). The PT Mi-D1 is a cylindrical mine held erect by spring-loaded legs once on the ground. The mine is 160 mm high and 116 mm in diameter, containing 800 g of A-IX-1 explosive; total weight is 2.5 kg. The mine is detonated by means of an electronic detonator, against tracks via a contact sensor or against the bottom of the hull via a sensor-mast. The electronic detonator can be preset for a self-destruct time of 3, 12 or 48 hours. The mine can penetrate up to 130 mm of armour.

VZ 92 combined rocket and mechanical minelayer vehicle **1996**

The anti-personnel mine used with the VZ 92 system is the PP Mi-S1. This cylindrical mine has a diameter of 116 mm and weighs 1.75 kg, containing 170 g of PL-U-EP14 explosive. It is a bounding type mine which creates about 1,900 fragments when initiated via triplines 10 m in length.

With the VZ 92 system, these mines are fired using 122 mm rockets with cargo warheads. The 122 mm Krizhna rocket contains four anti-tank mines or five anti-personnel mines. The 122 mm Kuš rocket carries five anti-personnel mines. Both rockets have a maximum range of 3,000 m when used with the VZ 92 system.

On the vehicle, to the rear of the rocket launcher and its platform, are four rotary anti-tank mine magazines which dispense their mines downwards in a spiral fashion so they can be guided by a chute with an option of laying on the surface or burying to a depth of 200 mm. The anti-tank mines involved may be the PT Mi-U or PT Mi-Ba-III, with the vehicle carrying a total of 292 ready for use. Maximum vehicle speed when laying anti-tank mines is 10 km/h.

At the rear of the vehicle, behind the four anti-tank mine magazines are two crate magazines carrying a total of 4,032 PP-Mi-Na 1 anti-personnel mines which are dispensed to each side of the vehicle.

In the rocket/mechanical dispensing form the total

VZ 92 combined rocket and mechanical minelayer vehicle **1996**

VZ 92 combined rocket and mechanical minelayer vehicle **1996**

vehicle weight is 32,800 kg, but as the overall design of the system is modular it can be readily converted into a mechanical minelayer or a rocket minelayer.

As a mechanical anti-tank minelayer the number of anti-tank mine magazines can be increased to eight, carrying a total of 584 anti-tank mines. The dispensers for 4,032 anti-personnel mines are retained. In this form the vehicle weighs 32,000 kg.

As an all-rocket minelayer the mine magazines are replaced by an extra 40-barrel rocket launcher and a magazine for 40 rockets. Total vehicle weight is then 31,900 kg.

The ammunition resupply vehicle may be configured as required, up to a safe working load of 33,000 kg. The crew of this vehicle is two.

For both vehicles the maximum road speed is 70 km/h and 20 km/h cross-country.

Status
Prototypes.

Manufacturer
Konštrukta Defence, AS, PO Box 62, 912 50 Trenčín, Slovakia.
Tel: +42 831 535850. Fax: +42 831 534451.

UPDATED

MV-3 mine launcher

Description
The three-barrel MV-3 mine launcher is an adjunct of the VZ 92 Križan mechanical and rocket minelaying system (see previous entry) as it fires the same mine-carrying 122 mm rockets to a maximum range of 3,000 m.

The MV-3 (*Minový vrhač 3*) is towed by hand to the launch location on a folding two-wheeled carriage and then installed on a folding tripod carriage. If required, the launcher may be fired from a vehicle platform using a control unit which also acts as a system check for the launcher and rockets, as well as for setting the mines' self-destruct times. After aiming, three rockets can be fired in 4 seconds over a range of 500 to 3,000 m. One three-rocket salvo will create a minefield approximately 100 m wide and 100 m deep.

The mines used with the MV-3 launcher can be either anti-tank, anti-personnel, or a mixture of both.

The anti-tank mines are known as the PT Mi-D1, a variant of the PT Mi-D anti-tank mines fired to a maximum range of 17,000 m by the RM-70/85 multiple rocket launcher (for details of this system refer to *Jane's Armour and Artillery 1995-96* pages 717 to 719). The PT Mi-D1 is a cylindrical mine held erect by spring-loaded legs once on the ground. The mine is 125 mm high (after laying) and 116 mm in diameter, containing 800 g of A-IX-1 explosive; total weight is 2.5 kg. The mine is detonated by means of an electronic detonator, against tracks via a contact sensor or against the bottom of the hull via a sensor-mast. The electronic detonator can be preset for a self-destruct time of 3, 12 or 48 hours. The mine can penetrate up to 130 mm of armour.

The anti-personnel mine involved is the PP Mi-S1. This cylindrical mine is 116 mm in diameter and weighs 1.75 kg, containing 170 g of PL-U-EP14 explosive. It is a bounding type mine which creates about 1,900 fragments. An electronic block provides initiation via 10 m long triplines or a self-destruct device. A self-destruct time of 3, 12 or 48 hours can be selected.

With the MV-3 launcher, these mines are fired using 122 mm rockets with cargo warheads. The 122 mm Krizhna rocket contains four anti-tank mines or five anti-personnel mines. The 122 mm Kuš rocket carries five anti-personnel mines.

Specifications
MV-3
Crew: 2
Weight: 106 kg
Number of barrels: 3
Length of each tube: 1.35 m
Elevation: +15 to +40°
Traverse: 13° left; 13° right
Barrel dispersion angle: 8°
Range: 500-3,000 m

Status
In production.

Manufacturer
Konštrukta Defence, AS, PO Box 62, 912 50 Trenčín, Slovakia.
Tel: +42 831 535850. Fax: +42 831 534451.

UPDATED

Cutaway example of 122 mm Krizhna rocket showing PT Mi-D1 anti-tank mines ***1996***

MV-3 mine launcher ***1996***

EGYPT

Anti-tank mine dispensing system

Description
The Egyptian anti-tank mine dispensing system consists of a number of mine scattering modules mounted on either a Fahd (4 × 4) armoured personnel carrier (four modules) or a CJ-7 Jeep (one module). The system could also be carried by other vehicles but with the number of modules varying according to the vehicle size and load capacity.

Each of the box-shaped modules contains 20 or 30 tubes from which mines are fired, creating an anti-armour minefield with a density and coverage dependent on the speed of the carrier vehicle and other factors such as the rate of firing and the number of tubes involved.

Each launch tube contains five T-93 anti-tank mines, each with a diameter of 112 mm, a height of 140 mm and weighing 2 kg. Each mine is fitted with an electromagnetic sensor unit that can detect

Anti-tank mine dispensing system mounted on Fahd APC ***1997***

approaching armour and detonate the mine to penetrate 80 mm of armour at a distance of 500 mm. The mines also have anti-clearing and self-destruct features, the latter functioning after a predetermined time of up to 240 hours.

The Fahd minelaying vehicle is a flatbed variant of the Kader Fahd (4 × 4) armoured personnel carrier (for full details refer to *Jane's Armour and Artillery 1995-96* pages 326 and 327). It can carry 600 T-93 mines ready for launch from 30-tube launcher modules. The CJ-7 version carries a single 20-tube launcher module containing 100 T-93 anti-tank mines. This system can also be mounted on tanks.

Status

In production. In service with the Egyptian Army.

Manufacturer

SAKR Factory for Developed Industries, PO Box 33, Heliopolis, Cairo, Egypt.
Tel: +202 2 902155. Telex: 92175-92770 CERVA UN.
Fax: +202 2 901978.

UPDATED

FRANCE

Giat Minotaur scatterable minelaying system

Description

The Minotaur scatterable minelaying system consists of a series of modules carried on a tracked or other vehicle. The system may also be mounted on a light towed carriage or carried by infantry in a man-transportable pack form (see under Variants). The system is designed to create rapid minefields to act as a barrier to enemy movement and as a counter to a surprise attack.

The system can be used to create minefields on either side of the moving carrier vehicle, to one side only, or to the rear. A carrier vehicle could carry up to six or eight modules. Each module contains 20 LU 980 launching tubes and mounted on a traversing base that can be traversed 40° right and left of centre. Mines can be dispensed to a maximum range of 300 m to each side and 90 m to the rear.

Each LU 980 glass fibre launch tube contains five anti-tank mines or 10 anti-personnel mines. The AC DISP anti-tank mines used with the system are the same as those used with the mine launching tubes mounted on the AMX-30 combat engineer tractor, or EBG (see entry under *Armoured Engineer Vehicles* for details) and the Giat 155 mm OMI 155 H1 artillery projectile. A LU 980 weighs 15.5 kg and is 760 mm long; diameter is 142 mm.

The LU 981, also known as AS DISP, anti-tank mines are 139 mm in diameter, weigh 2.25 kg and contain 0.7 kg of explosive. The launcher discharges the mines to a distance of up to 300 m and the mines are then triggered via a magnetic sensor by any vehicle weighing over 1,500 kg. According to Giat Industries, the mines will penetrate a tank floor equivalent to 50 mm of armour plate at 500 mm standoff distance and 60° incidence, or break up a tank's track if the mine is run over. Self-destruct of the mines is automatic after a preset time, which can vary from one to 96 hours in 1 hour steps. Inert biodegradable practice mines have been developed.

The overall weight of a LU 980 container of five LU 981 anti-tank mines is 15.5 kg and a loaded module weighs 690 kg. The space requirement for a single 20-container module measures 1 × 1 m.

From May 1995 Giat Industries, Euro Impact, started to deliver to the French Army the first of a contracted 2,500 LU 980 mine-launch containers and 12,500 LU 981 anti-tank mines. These are being delivered to engineer units fielding the EBG armoured engineer vehicle (71 in service); units fielding the Matenin mine distributor 4 × 4 vehicle will be equipped later.

To date the Minotaur system has been mounted on

Matenin mine launch vehicle for the Minotaur minelaying system (T J Gander) ***1997***

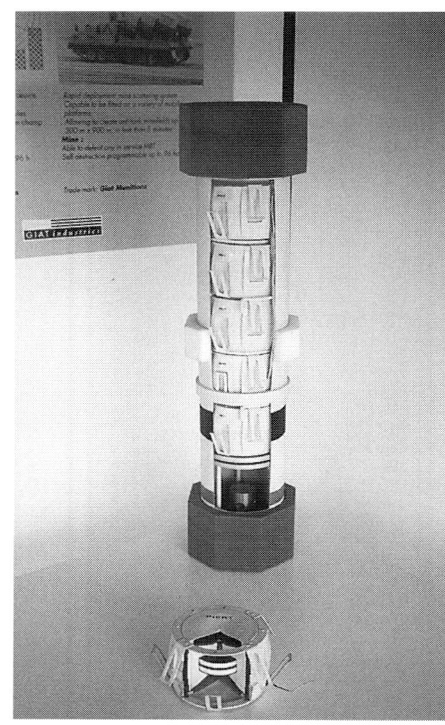

Cutaway launch container and AC DISP mine used with the Minotaur scatterable minelaying system (Christopher F Foss) ***1996***

Minotaur scatterable minelaying system mounted on Matenin mine distributor 4 × 4 vehicle ***1997***

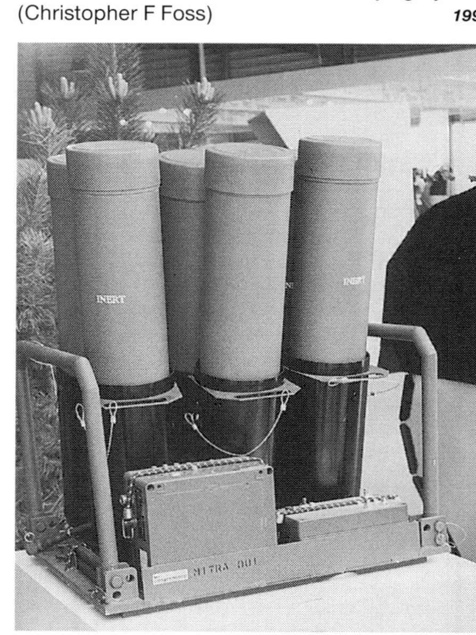

The Mitra portable remote-controlled mine scattering system ***1996***

tracked Alvis Sapper and Stormer flatbed system carriers, ACMAT (6 × 6) trucks and Matenin mine distributor (4 × 4) vehicles. The French Army qualified the Matenin mine distributor vehicle for service.

The Minotaur system was one of the contenders for the UK Vehicle-Launched Scatterable Mine System (VLSMS) and a number of Minotaur systems (four or six) mounted on Stormer tracked carriers were fielded by the British Army during Desert Shield/Storm. The Stormer carried six 20-container Minotaur modules.

Variant

Mitra

Mitra *(Module de Minage Portable Telecommandе)* is still at the prototype stage and is intended for use by infantry and light forces. Utilising the same basic components as Minotaur, it can be carried by two men and contains six mine tubes carrying a total of 30 AC DISP mines, sufficient to lay a road or bridge blocking minefield measuring 30 × 40 m out to a range of 70 m. The launcher weighs 40 kg empty.

8060 Pod Dispenser

Giat Industries, in association with Alkan, is developing the 8060 pod dispenser for the AC DISP mine. The pod will contain 54 launch tubes, each carrying five mines, giving a total of 270 mines. A Puma or Super Puma helicopter will carry one pod each side, each programmed to dispense the mines depending on the height and speed of the helicopter.

Specifications

Minotaur launcher vehicle
Crew: 2
Configuration: 4 × 4
Weight: (loaded) 14,000 kg
Length: 7.15 m
Width: 2.5 m
Height: 2.9 m
Wheelbase: 3.3 m
Ground clearance: 0.51 m
Max speed: (road) 80 km/h
Gradient: 50%
Fording: 1.2 m
Engine: 6-cylinder supercharged diesel developing 226 hp
Transmission: Powershift with torque converter and two 4 gear ranges
Steering: hydraulic assisted
Tyres: 16.00 × 25 low pressure

Mine range: up to 300 m
Mine container capacity: 120
Mine capacity: (anti-tank) 600

Status

Mass production commenced during 1995. In production for the French and British armies.

Manufacturer

Giat Industries, 13 route de la Minière, F-78034 Versailles Cedex, France.
Tel: +33 1 30 97 36 48. Fax: +33 1 30 97 39 58.

UPDATED

Giat ARE Type SOC plough-type minelayer

Description

The ARE Type SOC plough-type minelayer was developed specifically for use with the HPD anti-tank mine in production for the French Army by TRT Défense, although it can be adapted for use with other mines. The minelayer is towed behind a 4 tonne truck or tracked vehicle, which also carries the crew and mines.

A crew of four is required to operate the complete system: commander, two loaders and the driver. A canvas top can be fitted for weather protection.

The minelayer opens up a furrow in the ground, lays the mines and then replaces the top soil. If required, the mines can be laid on the surface. The HPD mines can be buried to a maximum depth of 250 mm and can be spaced 2.5, 3.3 or 5 m apart. Maximum laying rate is between 900 and 1,500 mines/h.

ARE Type SOC plough-type minelayer

Specifications

Crew: 4
Weight: 2,500 kg
Length:
(travelling) 4.86 m
(operating) 5.685 m
Width: 2.335 m
Height:
(travelling) 2.245 m
(working, canvas hood up) 2.07 m
(working, canvas hood down) 2 m
Max towing speed:
(on roads) 80 km/h
(operating) 4.5-10 km/h
Towing effort: 40 kN
Laying rate: 900-1,500 mines/h
Burying depth: 0-250 mm

Status

In service with one non-French army.

Manufacturer

Giat Industries, 13 route de la Minière, F-78034 Versailles Cedex, France.
Tel: +33 1 30 97 36 48. Fax: +33 1 30 97 39 58.

VERIFIED

Matenin mine distributor

Description

To meet the French Army's requirement for at least 150 systems of this type and in competition with Creusot-Loire, Etablissements Matenin (now Matenin SA) designed and built a mine distributor for the HPD anti-tank mine. Following trials held in 1982 the Matenin system was chosen as the model for the French Army.

Towed by a truck carrying the mines, the mines are manually placed into the layer to slide down the chute under gravity. They are automatically armed in the process. The laying distance between mines is adjustable. Weight of the minelayer is 150 kg and it can be broken down into three parts for transport purposes. The maximum laying speed is up to 1,000 mines/h.

Status

In service with the French Army.

Manufacturer

Matenin SA, 34 avenue des Champs Elysées, F-75008 Paris, France.
Tel: +33 1 43 59 22 00. Telex: 801 835 F.
Fax: +33 1 42 25 44 19.

UPDATED

Matenin mine distributor (T J Gander)

Matenin mine burier

Description

The Matenin mine burier is a self-propelled 4 × 4 vehicle based on components used in the Matenin trench digger. Development began in 1971. It is capable of cross-country operations and can lay or bury HPD or ACPR anti-tank mines on any ground which can be negotiated by armoured fighting vehicles. It can also lay mines in river and stream beds not more than 1.2 m deep.

The mines are buried by a hydromechanical push tool that enters and leaves the soil by the same route. There is no exterior change in vegetation as the entry point is automatically closed by a roller which stamps down the vegetation after the mine is buried. The stopping, deceleration and acceleration of the vehicle after each burial are fully automatic due to its hydrostatic transmission.

The vehicle carries four containers each loaded with 112 mines, making 448 mines in all. The mines may be laid in straight lines, curves (with diameters under 20 m) and zigzags. The distance between mines can be varied between 2.5 and 10 m during laying. The average burying speed is 400 mines/h but with a 3 m interval between mines this can be increased to 500 mines/h. The vehicle is used by the French Army to lay HPD anti-tank mines.

Variants

Mine Burier on MAN Truck

The Matenin mine burier system has been adapted to be carried on a MAN (6 × 6) truck. In this mode, handling the mines is carried out manually and involves a crew of four (including the vehicle driver) but the operating system is otherwise similar to the standard mine burier. The mine capacity of the truck system is increased to 560 anti-tank mines; eight containers with 70 mines each.

This variant was produced for the Belgian Army. The first batch of 25 units out of a total of 274 was ordered in early 1988. In Belgian service the truck-mounted system is used to bury HPD F2 anti-tank mines. The MAN truck variant has been demonstrated in Abu Dhabi.

Minotaur

The Matenin mine burier vehicle has been selected as a carrier for the Minotaur scatterable minelaying system – see separate entry in this section for details of the Minotaur system. The vehicle underwent qualification trials for the Minotaur system during 1994 and 1995.

Specifications

Matenin vehicle
Crew: 1 + 2
Configuration: 4 × 4 or 4 × 2
Weight:
(empty) 12,700 kg
(loaded) 16,000 kg
Length: 7.68 m
Width: 2.5 m
Height: 3 m
Ground clearance: 0.51 m
Wheelbase: 3.3 m
Max speed: (road) 80 km/h
Average speed: (road) 50 km/h
Speed with hydrostatic transmission: 3.6 km/h
Range: (road) 600 km
Endurance while laying: 11 h
Max gradient: 50%
Max side slope: 25%
Vertical obstacle: 0.45 m
Trench: 0.65 m
Fording: 1.2 m
Engine: MAN Type D2156 HM 6-cylinder water-cooled diesel developing 225 hp at 2,400 rpm
Transmission: power-shift hydromechanical unit with torque converter and two 4 gear ranges
Steering: power-assisted
Turning radius: 11 m
Tyres: 16.00 × 25 low pressure
Brakes: air
Winch capacity: 5,000 kg (10,000 kg when tackle is used)
Electrical system: 24 V

Status

In production. In service with the French Army. Truck-mounted system in production for Belgium (total requirement 274) and has been demonstrated in Abu Dhabi.

Manufacturer

Matenin SA, 34 avenue des Champs Elysées, F-75008 Paris, France.
Tel: +33 1 43 59 22 00. Telex: 801 853 F.
Fax: +33 1 42 25 44 19.

UPDATED

Matenin mine burying system on MAN (6 × 6) truck during demonstrations in Abu Dhabi (T J Gander)

Matenin mine burier

GERMANY

Mine Launching System (MiWS) Skorpion

Development

The Dynamit Nobel Mine Launching System (*Minenwurf* system) has been under active development since 1978 and the first deliveries of the system (including training ammunition) were made to the German Army in 1986. The system was originally known as the *Minenstreumittel* (MSM) system and encompassed the MSM-Fz vehicle launching system (later named as Skorpion – see below) and the MSM-Hs helicopter launching system which was flight tested. The vehicle-mounted version is marketed as the Vehicle-Launched Scatterable Mines System (VLSMS) Skorpion.

Contracts called for the delivery of 301 M548GA1 tracked carriers (conversions carried out by Krauss-Maffei with last deliveries during 1989), more than 2,400 training magazines and 32,000 service mine magazines (until 1992).

Mine production is continuing with the advanced DM 1399 MLRS-AT2 mine, which can be adapted to operate with the Skorpion system.

Description

The MiWS (Skorpion) consists of a modified M584GA1 tracked carrier, on the rear of which is mounted the

MiWS Skorpion-AB with Bv 206 all-terrain carrier being loaded into CH-53 helicopter

mine launching assembly with six mine launcher units, each mounted on a turntable. Each mine launcher unit has five preloaded magazines, each of which holds 20 AT2 anti-tank mines. The magazines have a metal body with foam filling and contain four ejection units with glass-reinforced plastic tubes. Each unit holds five AT2 anti-tank mines, all of which are ejected from the unit at once. Each magazine unit has an external cable which is connected to the launcher system after loading. Magazine units are delivered packed 10 to a standard NATO pallet and are discarded after use.

Carried inside the vehicle cab is the EPAG timing, testing and firing device, which acts as the main control unit for the system. The unit can be used to preselect the resultant mine barrier data and can be placed into several laying modes which can be changed while the vehicle is moving. Data that can be keyed into the unit prior to operations include: six mine self-destruct times; six mine densities of 0.1 to 0.6 mines/m; automatic or single launching; and alternate launching on both sides or from one side only. The EPAG unit also has a self-test mode with module fault display, testing modes for the launcher electrical system, a mode for monitoring operational conditions and a display showing ammunition ready to launch.

Depending on the settings entered into the EPAG the Skorpion MiWS will eject the groups of five AT2 anti-tank mines as the vehicle moves forward independent of its speed. Mines are ejected obliquely to the rear. The resultant mine barrier density will depend on the EPAG settings. A typical example is that a single combat load of 600 mines can form a mine barrier strip 1,500 m long with a density of 0.4 mines/m of front in approximately 5 minutes, travelling at a speed of 20 km/h. A typical mine barrier strip is 50 m wide. A typical Skorpion mine barrier has two or more parallel strips.

Various training aids, including reusable training ammunition, are available for the system.

A helicopter-carried system based on the Bell UH-1D (other helicopter types can be used following appropriate modifications) was developed and underwent flight tests. This system uses two magazine frames and the EPAG. The frames are held on a transverse support with damping and suspension facilities. Helicopter installation time for the system is about 15 minutes. The mines are dispensed from an altitude of between 5 and 15 m. A typical combat load of 2 × 5 mine magazines (200 AT2 anti-tank mines) can produce a 500 m long mine barrier strip (typical example) with a density of 0.4 mines/m of front in 20 seconds (flying at a speed of 50 kt).

A third member of the Skorpion family involves a continuing programme for the Swedish Army to mount the system on the Bv 206 tracked all-terrain carrier. The launching assembly with four magazine frames (400 AT2 DM 1274 anti-tank mines) is mounted on the rear unit of the Bv 206. In March 1991 the system was tested by Swedish Army engineers in 1 m deep soft snow with the mines fitted with parachutes. A light armoured version, the Bv 206S, is available. The Skorpion-AB, an airborne version, can be carried inside a CH-53 helicopter together with the front section of the Bv 206S. The German Army plans to test and order this system for use with the German Rapid Reaction Forces.

Specifications

Mine launcher
Crew: 2
Combat weight: 12,000 kg
Length: 5.87 m
Width:
(loading and transport) 2.68 m
(laying position, loaded) 2.87 m
Height: (top of machine gun) 3.52 m
Max speed: 40 km/h (FRG limit)
Range: (cruising) 500 km
Engine: diesel developing 210 hp
Armament: 1 × 7.62 mm MG

Mine launching assembly
Weight:
(combat, loaded) approx 4,000 kg
(empty) approx 1,900 kg
Length: 3.347 m
Width: (platform) 2.285 m

Mine magazine
Weight: 70 kg
Length: 728 mm
Width: 130 mm
Height: 520 mm
Contents: 20 mines

Status

In service with the German Army and under consideration by several other countries.

Prime Contractor

Dynamit Nobel AG, Defence and Industrial Products Division, D-53839 Troisdorf, Germany.
Tel: +49 22 41 89-0. Telex: 885 666 dn d.
Fax: +49 22 41 89 1669.

UPDATED

MiWS Skorpion with M548GA1 (Michael Jerchel)

Loading MiWS Skorpion magazines into launcher units carried on an M548GA1 (Michael Jerchel)

INDIA

Mechanical Mine Layer

Development

The towed Mechanical Mine Layer is an indigenous Indian development carried out by the Research and Development Establishment (Engineers), Defence Research and Development Organisation, Dighi, Pune, for the Indian Army. The Mechanical Mine Layer is designed to be towed by a wide variety of vehicles and can be used to lay the Indian Barmine, the Adrushy and other Indian anti-tank mines. It can also be adapted to lay anti-tank mines from many other countries.

Description

The Mechanical Mine Layer is towed by a suitable vehicle which also carries the mines to be fed manually, directly into the conveyor system for laying and/or burial. Mines can be laid either on the surface or buried in furrows up to 250 mm deep; spacing between the mines can be predetermined. A crew of four can lay several hundred mines per hour for as long as mine supplies are maintained. Indian-produced Barmines (the Mine Anti-tank Non-detectable 3A) are armed as they pass through the layer. This is also possible for other mines following minor modifications. No additional power pack is required for operation and no hydraulics, pneumatics or electronics are involved.

The Indian Mechanical Mine Layer (T J Gander)
1996

The Indian Mechanical Mine Layer seen from the rear with its mine dispenser section raised (T J Gander)
1996

For transport, the Mechanical Mine Layer can be towed on roads at speeds up to 45 km/h; for long journeys it may be carried on a vehicle. Cross-country towing speeds are up to 25 km/h. Typical towing vehicles include heavy 6 × 6 or 8 × 8 cross-country trucks or armoured personnel carriers. The height of the minelayer towing arm and conveyor height can be adjusted to suit a wide variety of vehicles.

When minelaying, if the plough/ripper blade encounters buried obstacles the towing vehicle becomes detached from the towing vehicle due to the breaking of a shear pin in the towing attachment, leaving the Mechanical Mine Layer undamaged.

Nearly all components on the Mechanical Mine Layer are mild steel so repairs or part replacements require no special techniques. No specialised maintenance is required and user training requirements are minimal.

Status
In service with the Indian Army.

Development Agency
Research and Development Establishment (Engineers), Defence Research and Development Organisation, Ministry of Defence, Government of India, Dighi, Pune - 411 015, India.
Telex: 0145 7511 RDEE IN. Fax: +91 212 666583.

VERIFIED

ITALY

BPD ST-AT/V anti-tank minelayer

Development
The BPD ST-AT/V anti-tank minelayer was designed for use with the BPD SB-MV/1, SB-81 and SB-81/AR anti-tank mines. The system is of the towed type and can be hitched to any standard vehicle towing hook.

The ST-AT/V minelayer is also produced in Spain by Explosivos Alaveses SA (EXPAL). An enhanced version is known as the EXPAL SEM-1A towed minelayer (qv).

Description
The ST-AT/V is constructed on to a welded steel frame fitted with a single axle having large diameter low-pressure tyres for cross-country travel. The system is provided with a hydraulic circuit to carry out all the mechanical operations under the control of a single crew member operating an electronic switching control unit; another loads the mine conveyor from the towing vehicle. As an alternative the system can be connected to the towing vehicle electrical supply. The hydraulic power is supplied by a pump driven from the wheels via a gearbox on the axle's self-locking differential unit.

The electronic control unit is based on a low-energy consumption microprocessor and controls all the laying operations from a remote switch panel. The panel is located close to the mine conveyor to the front of the towed unit. The position is provided with an intercom for communication with the towing vehicle driver.

Mines are placed on the mine conveyor by a loader in the rear of the towing vehicle. The mine safety devices are removed as the mine is loaded and the mine is conveyed to the unit plough share under the control of the electronic control unit. The spacing intervals can be varied from 2 to 99.9 m, and the laying depth is from 150 to 200 mm; the mine arming safety device is removed as the mine is emplaced in the furrow. Various types of plough can be used and there is a safety device that raises the share if an underground obstacle is encountered. Scraper devices behind the plough smooth off the ground surface after laying for added mine concealment.

For road transport the plough is raised from its operational position and locked. A canvas cover is fitted when the unit is not in operation and lifting eyes are provided for handling and for slinging under helicopters.

ST-AT/V anti-tank minelayer as produced in Spain by EXPAL

Specifications
Crew: (on unit) 1
Weight: approx 2,400 kg
Length:
(operational) 5.4 m
(on tow) 4.4 m
Width: 2.2 m
Height: 2.2 m
Max towing speed: 80 km/h
Laying speed: 2-8 km/h
Laying depth: 150-200 mm
Mine interval programming: 2-99.9 m
Interval accuracy: 0.1 m
Electrical system: 24 V
Batteries: 2 × 12 V, 80 Ah

Status
In production. In service with the Australian Army (8).

Manufacturers
BPD Difesa e Spazio srl, Corso Garibaldi 20-22, I-00034 Colleferro (Rome), Italy.
Tel: +39 6 97891. Telex: 611434 bpd cf 1.
EXPAL, Avenida del Partenón, 16, 5a, Campo de las Naciones, E-28042 Madrid, Spain.
Tel: +34 1 722 0235. Fax: +34 1 722 0295.

VERIFIED

Valsella GRILLO 128 man-portable mine launcher

Description
The GRILLO 128 is a portable, disposable, single-tube mine launcher intended to create small rapidly emplaced anti-tank minefields. It launches the mines without generating noise or flash by using a gas generator. A single launcher can launch a set of five anti-tank mines of the VS-SATM1 type.

The GRILLO 128 is a mortar-type launcher using mainly reinforced plastic components and is operated by a firing lever. It is provided with a carrying sling on which are two metal labels to be used as references for setting the angle of elevation. Having positioned the GRILLO 128 on the ground and setting the elevation angle to either +45 or +60° by placing one foot on the relevant metal label and elevating the barrel until the sling is tensioned, the launcher is fired by removing the safety and actuating the firing lever. The GRILLO 128 is stated to be completely safe in functioning, handling and storage, even in the case of a misfire.

Practice versions with reusable barrels and percussion devices are available for training purposes.

Specifications
Weight: (ready to fire) 14.5 kg
Length, transport: 920 mm
Base plate dimensions: 140 × 140 mm
Launch tube diameter:
(external) 134 mm
(internal) 128 mm
Range:
(45°) 60 m
(60°) 50 m
Standard deviation: approx 4 m
Noise at 30 m: 120 dBA
Temperature range:
(operating) −32 to +60°C
(storage) −40 to +70°C

Status
In production.

Manufacturer
Valsella Meccanotecnica SpA, I-25014 Castenedolo, Brescia, Italy.
Tel: +39 30 21371. Telex: 300495 emmeti i.
Fax: +39 30 273 1687.

VERIFIED

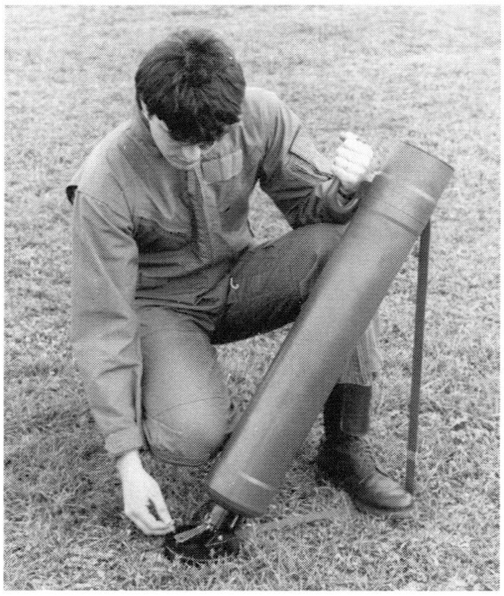
GRILLO 128 man-portable mine launcher in operation

Valsella Istrice land mine scattering system

Description

The Istrice land mine scattering system was designed to deploy minefields with fast and safe operation, high flexibility in the pattern and density of the resultant minefield and a short preparation time.

The Istrice system consists of the following components:

a launcher which can be installed on several types of military vehicle
a fire-control unit
canisters and magazines
the carrier vehicle
the mines.

The mines are stored in launch tubes and are ejected by an electrically fired gas generator. Arrays of tubes are assembled in modular canisters.

The launcher consists of up to eight launch units arranged in pairs, with each pair on a single frame. The units, mounted on a trunnion bearing, are adjustable in traverse by hand while the angle of elevation is fixed.

The system is controlled by a fire-control unit installed in the carrier vehicle cab. Using this unit the status of the launcher can be monitored, launch parameters can be set and data relevant to the launch operation are displayed in real time. A log of the deployed minefield is automatically provided.

The mines are stored in sealed launch tubes which also act as magazines for storage and transport. The magazines are assembled into canisters fitted with mechanical and electrical connections to the magazines and to the launch units. The launcher can be loaded using either magazines or full canisters.

The launcher is installed on the carrier vehicle in minutes using an interface frame. No mechanical or electrical modification of the vehicle is required. The minimum width of the vehicle platform required for installation is 2.2 m. Depending on the platform length, two, three or four pairs of launch units can be installed. The minimum loading capacity required for the vehicle is 3.9, 5.9 or 7.9 tonnes respectively. Istrice has been field tested on the IVECO 90 PM 16 (4 × 4) 4,000 kg truck (two pairs of launch units) and the IVECO 230 PM 35 (6 × 6) 10,000 kg truck (four pairs of launch units). The launcher can also be installed on several models of tracked cargo vehicle such as the M548 and the ARIS-converted M113 APC.

Istrice can scatter several types of anti-personnel and anti-tank mine, although the only mines now available are anti-tank types with self-neutralising devices.

The Istrice is available in two versions: Short Range (50 m) and Long Range (250 m).

Specifications

4-unit launcher
Mine payload:
(anti-tank only) 720 mines
(anti-personnel only) 3,840 mines
Weight:
(launcher, loaded with anti-tank mines) 3,900 or 4,100 kg
(launcher, loaded with anti-personnel mines) 3,100 or 3,400 kg
(launcher without canisters) 1,000 kg
Width: 2.2 m
Height over platform: 1.4 m
Traverse: 90°
Elevation:
(short range) +60°
(long range) +45°
Scatter rate: (anti-tank mines) max 500 mines/min
Launch distance: 50 or 250 m
Power requirements: 24 V DC, 4 A

Status

In production.

Manufacturer

Valsella Meccanotecnica SpA, I-25014 Castenedolo, Brescia, Italy.
Tel: +39 30 21371. Telex: 300495 emmeti i.
Fax: +39 30 273 1687.

UPDATED

Four-unit Istrice land mine scattering system on an ARIS-converted M113 APC

Four-unit Istrice land mine scattering system on an IVECO 90 PM 16 (4 × 4) truck

Valsella VS-MDH scatter-dropping mine system

Development

The VS-MDH scatter-dropping mine system (at one time known as the VS/MD system) was designed to allow the scatter-dropping of 2,080 anti-personnel or 200 anti-tank mines in mixed loads and varying ratios. It should be noted that the only mines currently available are of the anti-tank type with self-neutralising devices.

Description

The VS-MDH system consists of two main components: the control panel in the helicopter and the mine dispenser. This consists of a metal rack designed to accept 40 modular magazines. On the bottom of each magazine is a stop which keeps the mines in place during storage, shipment and while inserting the magazines into the rack of the dispenser. Once in the dispenser the stop is removed.

Anti-personnel mines are dropped in groups of 52 and anti-tank mines in groups of five. They are dispensed through doors in the bottom of each magazine. The door opening sequence was designed to prevent any change in the centre of gravity of the dispenser. At the end of the scatter-dropping programme, a safety device actuated by a push-button closes all the doors.

The dispenser is attached to the centre of gravity hook of the helicopter by a special ring connector. The system's remote controls are connected to the control panel in the helicopter via a cable fitted with a multipole slip ring connector. This has been designed for quick release, both on the ground and in the air, in case of an emergency.

The compact control panel can be mounted on the console of the helicopter, or in any other suitable position. The system has been designed for both manual and automatic scatter-dropping and consists of an electronic control unit which can open the doors at intervals of 0.1 to 0.6 second. This, together with the speed of the helicopter, allows the mines to be dispersed according to the tactical requirement.

A readout on the control console indicates when each door of the dispenser is opened. The operator can also monitor, at a glance, how many mines have been dispensed and how many are left. When operated manually, one door opens every time the button is depressed. There is also an emergency device which, when operated, causes the complete dispenser to be emptied at once. The system requires a 28 V DC electrical supply.

UH-1 helicopter carrying Valsella VS-MDH scatter-dropping mine system laying minefield

Specifications
Weight:
(dispenser) 310 kg
(loaded with anti-tank mines) 1,160 kg
(control unit) 1.2 kg
Dispenser:
(length) 1.75 m
(width) 1.55 m
(height) 1.6 m
Control unit:
(length) 180 mm
(width) 120 mm
(height) 100 mm
Number of mines carried:
(anti-personnel) 2,080
(anti-tank) 200
Power supply: 28 V DC
Max power load: 8 A

Status
In service with various undisclosed countries.

Manufacturer
Valsella Meccanotecnica SpA, I-25014 Castenedolo, Brescia, Italy.
Tel: +39 30 21371. Telex: 300495 emmeti i.
Fax: +39 30 273 1687.

UPDATED

BPD SY-AT helicopter anti-tank and anti-personnel mine-scattering system

Development
The BPD SY-AT helicopter-carried scattering system, was designed for use with the BPD SB-81 anti-tank and BPD SB-33 anti-personnel mines. The first production SY-AT system was completed in April 1978.

The SY-AT system is also produced in Spain by Explosivos Alaveses SA (EXPAL). A variant is known as the EXPAL DMH-1 helicopter mine dispenser (qv).

Description
The SY-AT system is composed of a basic distributor module and two ancillary modules. The former is connected to the baricentric hook of the helicopter by a special suspension sling and connector hook which also contains the electrical connection. The unit can be carried slung under standard helicopter types.

The mines are dispensed through hatches in the bottom of the distributors. The hatches can be opened at a programmed frequency, or manually by pushing a button on the control console mounted in the helicopter. Safety blocks ensure that the hatches do not open until the helicopter reaches the target area. The scattering ratio of anti-tank to anti-personnel mines is variable according to mission requirements and is controlled from the control panel. Indicators on the panel advise the operator when the magazines are empty as well as how many full magazines remain. Safety pins are removed from the mines as they leave the plastic magazines. In an emergency it is possible to discharge all the mines in the distributor simultaneously.

The basic module contains 32 magazines which contain 2,496 anti-personnel or 160 anti-tank mines. The ancillary distributors are hooked on each side of the basic module, each with eight magazines for 624 anti-personnel or 40 anti-tank mines, or a combination. The complete system, basic module and two ancillary modules, contains a total of 3,744 anti-personnel mines, 240 anti-tank mines or a combination of both types.

Specifications
Magazine of anti-tank mines
Weight:
(empty) 3.5 kg
(loaded) 19.5 kg
Number of SB-81 mines: 5
Dimensions: 246 × 110 × 1,142 mm

Magazine of anti-personnel mines
Weight:
(empty) 3.6 kg
(loaded) 15 kg
Number of SB-33 mines: 78
Dimensions: 246 × 110 × 1,142 mm

Basic distributor module
Empty weight: 150 kg
Loaded weight:
(with anti-tank mines) 774 kg
(with anti-personnel mines) 630 kg
Capacity: 32 magazines containing 5 anti-tank or 78 anti-personnel mines each
Dimensions: 1,545 × 1,320 × 1,380 mm

Ancillary distributor module
Weight empty: 35 kg
Loaded weight:
(with anti-tank mines) 191 kg
(with anti-personnel mines) 155 kg
Capacity: 8 magazines containing 5 anti-tank or 78 anti-personnel mines each
Dimensions: 1,545 × 399 mm

Complete unit
Weight: (empty) 220 kg
Weight loaded:
(with anti-tank mines) 1,156 kg
(with anti-personnel mines) 940 kg
Capacity: 48 magazines each holding 5 anti-tank or 78 anti-personnel mines each
Dimensions: 1,545 × 2,119 × 1,380 mm
Voltage requirements: 24-28 V DC

BPD SY-AT helicopter anti-tank and anti-personnel mine-scattering system slung under Lama helicopter

Status
In production. In service with various countries.

Manufacturers
BPD Difesa e Spazio srl, Corso Garibaldi 20-22, I-00034 Colleferro, Rome, Italy.
Tel: +39 6 97891. Telex: 611434 bpd cf 1.
EXPAL, Avenida del Partenón, 16, 5a, Campo de las Naciones, E-28042 Madrid, Spain.
Tel: +34 1 722 0235. Fax: +34 1 722 0295.

VERIFIED

Tecnovar DAT scatter-dropping mine system

Description
The DAT scatter-dropping mine system provides for automatic or manual dropping of loads of either TS-50 anti-personnel or MATS anti-tank mines, or a combination of both. The complete system consists of two main components, the electronic programmer fitted to the instrument panel of the carrier helicopter and the automatic dispenser system. There are two versions of the dispenser: model A, which can carry 128 anti-tank or 1,280 anti-personnel mines; and model B, which can carry 64 anti-tank, 640 anti-personnel or 32 anti-tank and 320 anti-personnel. The mines are carried in preloaded magazines slotted into the dispenser. The anti-tank magazine carries eight mines and the anti-personnel magazine 40 mines. The two models can be coupled together to form combinations, such as A + A, A + B and A + A + B. The type of combination used depends on the underslung cargo capacity of the helicopter and the type of minefield to be laid.

The dispenser system is attached to the helicopter on the end of a sling fitted with a rotary hook. The distributor doors are opened only after the correct signal is received from the electronic programmer unit, which controls the interval between door openings. The interval may be set between 0.1 and 9.9 seconds in increments of 0.1 second.

For an anti-personnel minefield the mines are dropped in groups of 80, for an anti-tank minefield in groups of eight and for a mixed minefield in groups of four anti-tank mines and 40 anti-personnel mines. The typical flight altitude for laying is up to 100 m at speeds of up to 200 km/h.

Specifications
Anti-personnel mine magazine
Construction: steel
Length: 280 mm
Width: 145 mm
Height: 910 mm
Weight: (empty) 8 kg
Capacity: 40 TS-50 anti-personnel mines

Anti-tank mine magazine
Construction: steel
Length: 290 mm
Width: 200 mm
Height: 910 mm
Weight: (empty) 11.5 kg
Capacity: 8 MATS anti-tank mines

Model A dispenser unit
Length: 1,214 mm
Width: 1,224 mm
Height: 1,244 mm
Weight:
(empty, including sling, cables and hook) 170 kg
(filled with 128 anti-tank mines) 800 kg
(filled with 1,280 anti-personnel mines) 740 kg

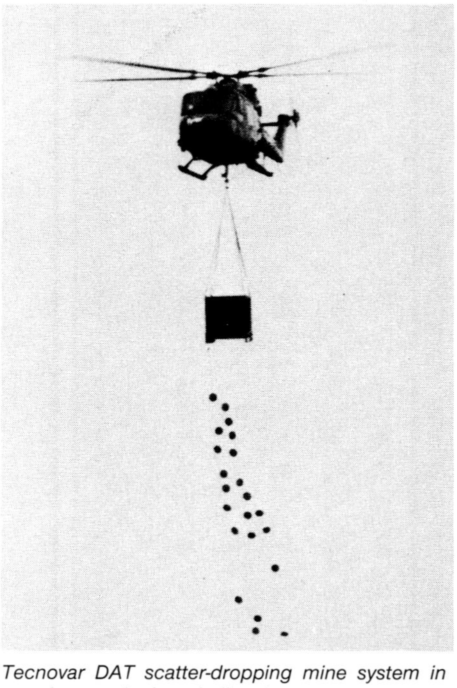
Tecnovar DAT scatter-dropping mine system in use slung under Lynx helicopter

Model B dispenser unit
Length: 622 mm
Width: 1,224 mm
Height: 1,244 mm
Weight:
(empty, including sling, cables and hook) 100 kg
(filled with 64 anti-tank mines) 425 kg
(filled with 640 anti-personnel mines) 355 kg
(filled with 32 anti-tank and 320 anti-personnel mines) 385 kg

Typical combinations of Models A and B dispenser units

	A + A	A + B	A + A + B
Length	2.428 m	1.836 m	3.050 m
Height	1.244 m	1.244 m	1.244 m
Width	1.224 m	1.224 m	1.224 m

Power requirement for electronic programmer: 28 V DC

Status
Available.

Manufacturer
Tecnovar Italiana SpA, 95 Via Argiro, I-70121 Bari, Italy.
Tel: +39 80 521 1744. Telex: 810345 tecvar i.
Fax: +39 80 370400.

VERIFIED

JAPAN

Japanese minelayer being towed by Type 60 armoured personnel carrier (Kensuke Ebata)

Japanese minelayer being towed by Type 60 armoured personnel carrier and showing onboard operator position (Kensuke Ebata)

POLAND

SUM tracked minelayer

Description
The SUM tracked minelayer was designed specifically for the rapid laying of surface and buried anti-tank mines. The minelayer has a crew of two and is carried on a special tracked chassis with six roadwheels and a torsion bar suspension.

The minelaying system is carried at the rear of the vehicle with mines being automatically fed into the system from two magazines carried within the armoured superstructure; each magazine can hold up to 125 mines of the TM-62M type. The mine magazines are replenished or replaced using a K10 crane mounted at the rear of the vehicle which has a lifting capacity of 2,800 kg. The mine magazines can also be reloaded manually while the magazines are inside the vehicle.

Supervision of the laying process can be carried out using a rear-mounted video camera connected to a monitoring screen inside the vehicle. The monitoring screen is located over a control panel used to control the minelaying operations. Provision is made for operations under NBC conditions by an air filtration system and a GO-27 detection device.

When burying mines the vehicle can travel at a speed between 6 and 10 km/h. When laying surface mines the speed range is from 6 to 20 km/h. It is possible to lay mines in water up to 0.9 m deep at a speed of 6 km/h.

The vehicle is armed with a 12.7 mm NSW machine gun; 902A smoke grenade launchers are also provided. An R173 radio communication system is provided and crew intercommunications are carried out using an R124 system.

SUM tracked minelayer in travelling position and with handling crane partially raised

Variant
A special fire-fighting vehicle carrying a wheeled pumping station on the hull rear is based on the same chassis as the SUM and is known as the Kalina.

Specifications
Crew: 2
Weight: (combat) approx 32,000 kg
Length: (travelling) 10.41 m
Width: (travelling) 3.354 m
Height: (travelling) 2.87 m
Max speed: (road) 65 km/h
Range: 500 km
Gradient:
(travelling) 25°
(operating) 19°
Fording:
(normal) 0.9 m
(with preparation) 1.5 m
Engine: diesel developing 700 hp
Suspension: torsion bar
Armament: 1 × 12 7 mm NSW (NSV) MG

Laying speeds:
(buried mines) 6-10 km/h
(surface mines) 6-20 km/h
(under water) 6 km/h

Status
Available. Offered for export.

Manufacturer
Research and Development Centre of Mechanical Appliances (OBRUM), PL-44-101 Gliwice, ul. Toszecka 102, Poland.
Tel: +48 32 31724-. Telex: 0136 197.
Fax: +48 32 315887.

VERIFIED

RUSSIAN FEDERATION AND ASSOCIATED STATES (CIS)

Remote mine dispensing systems

Description

The term remote mine dispensing system covers three mine dispensing systems employed by helicopter and land-based carriers. Information regarding these systems is sparse and the information provided here should be regarded as provisional.

All three systems are based around cylindrical mine canisters of which there are three externally similar types, each with different mine contents. One type of canister contains a single PTM-3 scatterable anti-tank mine. Each mine has four dished sectors so arranged along the body length that whichever way the mine lands, one shaped charge face will be facing upwards. A second canister contains four POM-2S scatterable anti-materiel mines which are orientated after landing by six spring-loaded legs before ejecting sensor tripwires over the adjacent area. The third type of canister contains 64 PFM-1S scatterable anti-personnel mines. All three types of mine have a self-destruct feature.

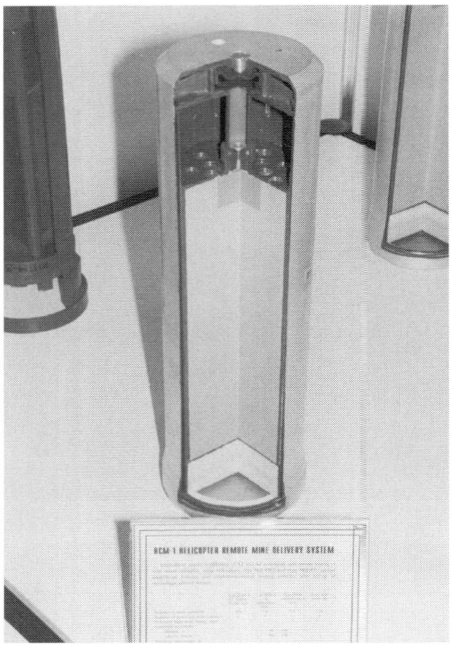

Cross-section of a mine canister containing a single PTM-3 scatterable anti-tank mine (T J Gander)

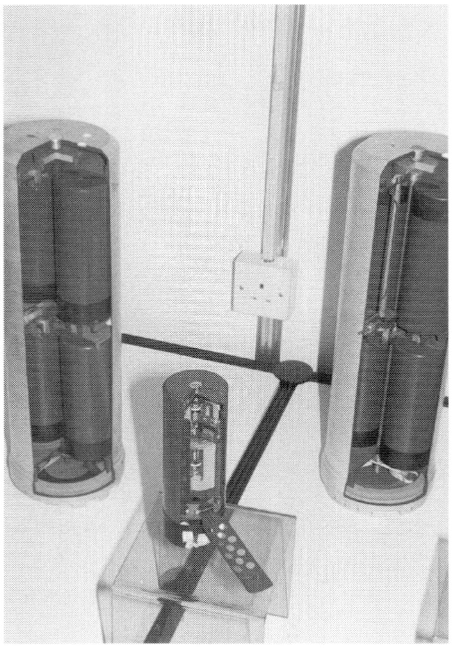

Cross-sectioned POM-2S scatterable anti-materiel mines with a single cross-sectioned POM-2S mine in the foreground (T J Gander)

The helicopter-carried mine dispensing system is the Type VSM-1 which has up to four K-29 boom-mounted mine containers with a total of 116 downward-ejected mine canisters which may be all of one type or a mix of types. This system can rapidly emplace a minefield some 20 to 30 m wide and varying in length according to the type of mine involved. A minefield involving only PTM-3 mines will be 380 m long, while a POM-2S minefield will be 2,300 m long. PFM-1S anti-personnel minefields will be up to 1,800 m long.

The UM3 system is carried on a 6 × 6 truck or a tracked vehicle and has up to six clusters of 30 launch tubes containing 180 canisters. These are deemed sufficient to produce up to three barrier minefields, each of which will be 20 to 30 m wide. The UM3 system can produce a PTM-3 anti-tank minefield approximately 600 m long. The POM-2S equivalent will be 3,600 m long, while a PFM-1S anti-personnel minefield will be 2,800 m long. The maximum range of the UM3 scattering system depends on the type of mine and is 60 to 80 m for the PTM-3, 40 to 110 m for the POM-2S, and 40 to 50 m for the PFM-1S.

The third type of mine dispensing system is the man-portable PKM which is a single ground-mounted launcher tube projecting a single canister. Ranges of the various types of canister are the same as for the UM3 system.

Status

Offered for export.

Manufacturer

Scientific Research Engineering Institute, Entuziastov Highway, Balashikha-1, Moscow Region, Russia.
Tel: +7 095 521 7171. Telex: 346435 Epokha.

VERIFIED

GMZ-3 tracked minelayer

Description

The GMZ-3 (*Gusenichniy Minniy Zagraditel 3*) tracked minelayer is based on the chassis of the SA-4 (Ganef) surface-to-air missile system. The driver is seated at the front of the vehicle on the left side with the engine to his right. This leaves the rear of the vehicle clear for the mounting of the minelaying equipment and the stowage of mines. The suspension of the GMZ-3 consists of seven roadwheels with the drive sprocket mounted at the front and the idler at the rear. There are four track-return rollers. The minelayer is of the plough type. Mines are fed into the minelayer via two chutes, one either side.

The GMZ-3 can carry 208 anti-tank mines which are loaded into the rear armoured compartment through two large roof hatches. The laying system spaces the mines at intervals of between 4 and 5.5 m at a speed of 2 to 3 km/h and a rate of 4 mines/min. When laying mines on the surface the rate is 8 mines/min at a road speed of between 6 and 16 km/h. Mines may be laid under either manual or automatic control and can be buried in up to 120 mm of soil or up to 500 mm of snow. Mines of up to 12 kg can be handled by the system. During minelaying operations the system operator is seated at the right rear, facing to the rear from under a small raised superstructure on the right of the hull.

A land navigation system is carried by the GMZ-3. Infrared vision equipment enables the GMZ-3 to carry out minelaying operations during darkness. A platoon of three GMZ-3s is attached to the engineer battalion of each motorised rifle or tank division, replacing the earlier PMR-3.

The armament of the GMZ-3 is a single 14.5 mm KPVT machine gun mounted over the commander's position. Smoke grenade launchers are provided and there is a smoke production system incorporated into the vehicle's exhaust system.

Specifications

Crew: 3
Weight: 28,500 kg
Power-to-weight ratio: 18.42 hp/t
Length: (travelling) 8.62 m
Width: 3.25 m
Height: (travelling) 2.7 m
Ground clearance: 430-470 mm
Max speed:
(road) 60 km/h
(road, cruising) 40-45 km/h
(dirt roads) 25-30 km/h
Range: (road) 500 km
Fording: 1 m

GMZ-3 tracked minelayer from right rear showing mine plough in raised position

GMZ-3 tracked minelayer burying anti-tank mines with a foreground indication of how deep the mines are buried

Gradient: 57.7%
Side slope: 46.7%
Vertical obstacle: 0.7 m
Trench: 2.5 m
Engine: KaMAZ-7482 water-cooled diesel developing 525 hp
Armament: 14.5 mm KPVT MG
Mine capacity: 208
Work speed:
(surface laying) 6-16 km/h
(burying) 2-3 km/h
Minelaying rate:
(surface laying) 8 mines/min
(burying) 4 mines/min
Mine spacing: 4-5.5 m apart
Reload time: 12-15 min

Status
In service with the RFAS. Offered for export.

Manufacturer
State factories.

VERIFIED

Close-up of GMZ-3 tracked minelayer mine plough in travelling configuration (T J Gander)

UMZ tracked mine dispensing vehicle

Description
The UMZ tracked mine dispensing vehicle is based on the chassis and armoured hull of the MT-LBu general purpose tracked carrier and command vehicle and features two arrays of mine dispensers tubes carried over the rear half of the hull. The arrangement of launch tubes involved may vary according to the type of mine involved, but is normally around 90 in each array. Mine types dispensed from the system include the PFM-1, PFM-1S and KSO-1 anti-personnel mines, and the POM-1, POM-2, PTM-1 and PTM-3 anti-tank mines, including various mixes of all those types. The weight of a complete loaded launcher is approximately 1,600 kg.

The vehicle can dispense mines while travelling at speeds of 10 to 40 km/h, creating one or two belts of minefield in a single pass. The banks of launchers can be traversed through 360° and elevated to an angle of +50°. To prepare a fully loaded vehicle for mine dispensing from the travelling state takes less than 5 minutes. The two-man crew can reload the launchers in 120 minutes, although a sapper squad will take only 20 minutes.

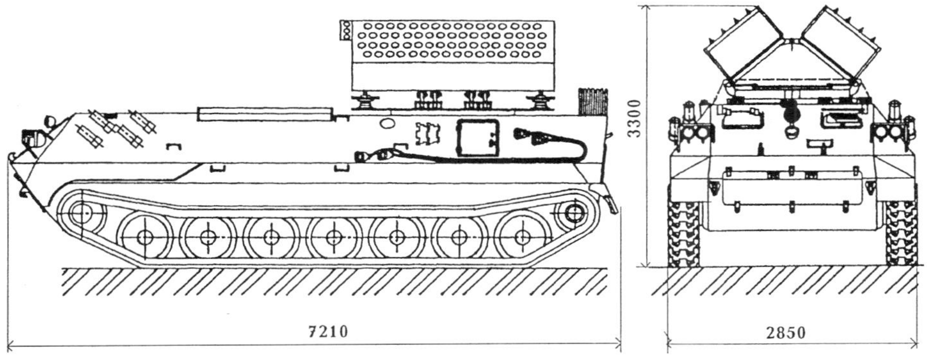

Outline drawing of UMZ tracked mine dispensing vehicle ***1996***

Specifications
Crew: 2
Length: 7.21 m
Width:
(travelling) 2.85 m
(firing) 3.45 m
Height:
(travelling) 2.2 m
(firing) 3.3 m
Max speed: (road, loaded) 61.5 km/h
Speed, mine dispensing: 10-40 km/h
Range: (road) 500 km
Max dispensing gradient: 15°
Operational temperature range: −45 to +45°

Status
Offered for export

Manufacturer
Kruikov Railway Car Building Works, 139 Karl Liebknecht Street, 315307 Kremenchug, Ukraine.
Tel: +7 05366 61101. Telex: 064 131330 OSA.
Fax: +7 05366 295 7203.

VERIFIED

PMR-2, PMR-3 and PMZ-4 towed minelayers

Description
The PMR-2 (*Pritsepniy Minniy Zagraditel-2*) is a two-wheeled trailer with two chutes. The upper part of the chute has a wide mouth into which the anti-tank mines are loaded. These then slide down a double roller conveyor into the distributing mechanism, which spaces them at intervals of 2 or 4 m. The chain drive distribution mechanism is controlled by a three-position lever mounted on the control box. The mines are laid on the surface and buried by a follow-up team if required. The trailer is normally towed by a 6 × 6 truck or armoured personnel carrier. The work speed of the PMR-2 is 3 to 5 km/h and mine load is 120 mines.

The PMR-3 (*Pritsepniy Minniy Zagraditel-3*) has a single chute and the operator, seated on the two-wheeled trailer, can select either surface or buried laying, to 300 or 400 mm depth, and also controls the spacing of the mines. The PMR-3 is normally towed by a specially modified BTR-152 (6 × 6) armoured personnel carrier which carries 120 TM-46 or similar anti-tank mines. A fully loaded BTR-152, carrying 120 mines, can lay a minefield 500 m long (when the mines are spaced at 4 m intervals) in 5 minutes.

For both systems the capacity varies if the tow vehicle is other than the BTR-152, as follows: ZIL-157, 200; BTR-60, 100 to 130; Ural-375, 350. The PMZ-4 towed minelayer is the same as the PMR-3 but with a 200 mine capacity.

The Polish Army Engineering Works at Deblin offers manufacture, repair and overhaul services for the PMR-3 towed minelayer and its associated spare parts.

PMR-2 minelayer in travelling position

Specifications
PMR-3
Crew: 4 or 5
Length: 3 m
Width: 2 m
Height: 2.5 m
Tyres: 7.50 × 20
Mine spacing: 4-5.5 m apart
Burial depth: (soft soil) 300-400 mm
Work speed:
(surface laying) 4-10 km/h
(burying) 2-3 km/h
Minelaying rate: 10-12 mines/min
Reload time: 10-12 min

Status
In service with former Warsaw Pact countries and other countries that have received former Soviet aid, including Afghanistan (PMR-3) and Iraq.

PMR-3 towed minelayer with towing eye nearest camera (Christopher F Foss)

Manufacturer
State factories.

VERIFIED

SPAIN

EXPAL SEM-11 self-propelled minelayer

Development

The EXPAL SEM-11 self-propelled minelayer was developed for the fast automatic laying or burying of SB-81 series anti-tank mines (also produced by EXPAL). The system can be adapted to many types of tracked vehicle although the prototype was mounted on a flatbed derivative of the FMC (now United Defense LP) M548 tracked carrier, the FMC-548-6.

EXPAL have also produced, under licence, the ST-AT towed minelayer and the SY-AT helicopter mine-scattering system, both originally produced by BDP of Italy (qv).

Description

The EXPAL SEM-11 is a device which, once fitted to a tracked vehicle, buries SB-81 series anti-tank mines and conceals them automatically. The mining depth is adjustable and the distance between mines is programmable - both details can be recorded. Safety pin removal and electronic mine fuze activation are also carried out automatically. Mining operations are automatically cancelled if excessive ground resistance is encountered, avoiding damage to the system. Mining operations can continue once an obstacle has been circumvented.

The SEM-11 assembly consists of: a mine burying unit, an electronic control unit, a manual and automatic control panel, mine containers which transport the mines and feed the system and a crane to handle the mine containers. The system can be installed by EXPAL on any vehicle with suitable traction, carrying capacity and dimensions.

In operation, once the mine burying depth and separation distance have been set, the operator takes mines from the containers and introduces them to the system feed ramp. As the vehicle advances the system opens a furrow, safety pins are automatically removed, and the mine is placed within the furrow at a preset rate. The furrow is automatically covered by the system. If electronic mines are involved, their fuzes will be automatically activated and will be armed after their normal safety interval. If required, minefield data can be electronically recorded.

Mines can be buried at depths of up to 220 mm and at intervals of between 2 and 12 m. The maximum mine dispensing rate is one mine every 2 seconds, with optimum mine burying speed being 4 to 5 km/h. Mines can also be laid on the surface, if required.

When mounted on the FMC-548-6 the system can carry 540 mines. An optional trailer can be towed to transport further quantities of mines.

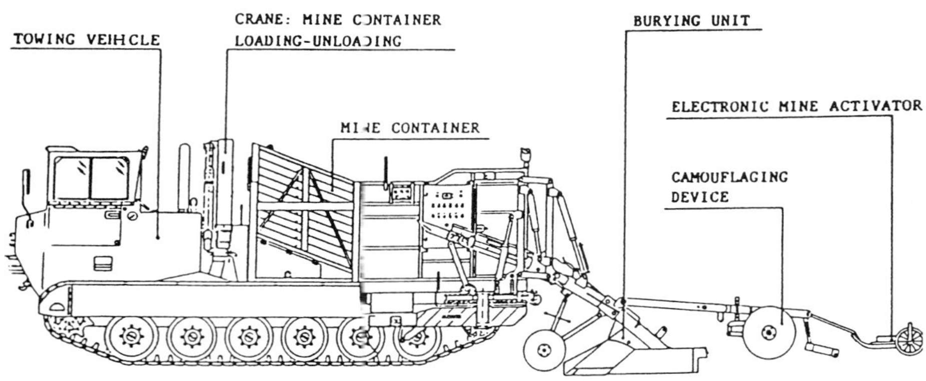

SEM-11 self-propelled minelayer in operational mode on M548 tracked carrier chassis with main assemblies indicated

SEM-11 self-propelled minelayer on modified M548 tracked carrier chassis

Specifications

SEM-11 on M548
Crew: 2 + driver
Weights:
(total, loaded) 12,957 kg
(vehicle only) 8,252 kg
(loaded mine containers) 3,000 kg
(minelayer only) 1,000 kg
(crane) 705 kg
Lengths:
(operating) 11.3 m
(transport) 6.9 m
Width: 2.79 m
Height: 3.13 m

Mine burying depth: up to 220 mm
Mine intervals: 2-12 m
Max minelaying rate: 1 mine/2 s
Optimum mine burying speed: 4-5 km/h
Max mines carried: 540

Status

Prototypes.

Manufacturer

EXPAL, Avenida del Partenón, 16, 5a, Campo de las Naciones, E-28042 Madrid, Spain.
Tel: +34 1 722 0235. Fax: +34 1 722 0295.

VERIFIED

EXPAL SEM-1A towed minelayer

Development

The EXPAL SEM-1A towed minelayer may be regarded as a revised version of the BPD ST-AT/V anti-tank minelayer, licence-produced in Spain by EXPAL, with enhancements to allow the system to accommodate electronically fuzed versions of the SB-81 anti-tank mine, also produced by EXPAL.

Description

The EXPAL SEM-1A towed minelayer buries SB-81 series anti-tank mines and conceals them automatically. The mining depth is adjustable and the distance between mines is programmable. Safety pin removal and electronic mine fuze activation are also carried out automatically.

The SEM-1A consists of a ploughshare, scraper device to close the furrow, activator for electronic mines, electronic control unit and a manual and automatic control panel. The system can be towed by any vehicle with suitable traction capacity.

Mines can be buried or laid on the surface, with the mine intervals set automatically or manually. As the vehicle advances, the system opens a furrow, safety pins are automatically removed and the mine is placed within the furrow at a preset rate. The furrow is then automatically covered by the system. If electronic mines are involved, their fuzes will be automatically activated and armed after their normal safety interval.

Mines can be buried at depths of up to 200 mm and at intervals of between 2 and 9.9 m. The maximum mine dispensing rate is one mine every 2 seconds, with optimum mine burying speed being 4 to 5 km/h. Mines can also be laid on the surface if required.

EXPAL SEM-1A towed minelayer **1996**

Specifications

Crew: 2
Weight: 3,200 kg
Lengths:
(operating) 7.7 m
(transport) 5.85 m

Width: 2.25 m
Height: 2.7 m
Max towing speed: 80 km/h

Mine burying depth: up to 200 mm
Optimum mine burying depth: approx 120 mm
Mine intervals: 2-9.9 m
Max minelaying rate: 1 mine/2 s
Optimum mine burying speed: 4-5 km/h
Min distance between rows: approx 2.25 m

Status
Prototypes.

Manufacturer
EXPAL, Avenida del Partenón, 16, 5a, Campo de las Naciones, E-28042 Madrid, Spain.
Tel: +34 1 722 0235. Fax: +34 1 722 0295.

VERIFIED

EXPAL DHM-1 helicopter mine dispenser

Development
The EXPAL DHM-1 helicopter mine dispenser may be regarded as a revised version of the BPD SY-AT mine scattering system, licence-produced by EXPAL, with enhancements to allow the system to accommodate electronically fuzed versions of the SB-33 anti-personnel and SB-81 anti-tank mines, also produced in Spain by EXPAL.

Description
The EXPAL DHM-1 helicopter mine dispenser consists of a dispenser frame which can be loaded with 160 SB-81 series anti-tank mines loaded into 32 CAT mine magazines, or 2,496 SB-33 anti-personnel mines loaded into 32 CAP mine magazines. If required, CAT and CAP mine magazines can be loaded into the DHM-1 dispenser frame in predetermined proportions.

Mine scattering is carried out in a programmed mode so that differing minefield densities can be produced. The DHM-1 is slung under the helicopter and connected with it mechanically and electronically. A control panel is provided in the helicopter cabin.

The DHM-1 system is provided with a device which automatically removes mine safety pins, while an electronic activator automatically arms electronic mine fuzes at the instant they are scattered. Prior to loading, detonators can be installed in the mines without having to remove the mines from their magazines. The mine magazines can then be loaded into the DHM-1 dispenser. Minefield densities and mine type proportions are programmed using the system control unit in the helicopter cabin. When the area to be mined is reached, the flight speed and altitude are adjusted and the mines are scattered according to the programmed data. In an emergency, all the loaded mines can be released instantaneously.

The DHM-1 system may be rapidly installed on any current combat helicopter.

Specifications
Weights:
(loaded with 160 SB-81 mines) approx 830 kg
(loaded with 2,496 SB-33 mines) approx 655 kg
(empty dispenser) approx 158 kg
Length: 1.495 m
Width: 1.321 m
Height: 1.545 m
Max release speed: 30-80 kt
Max release altitude: 100 m

Status
In production.

Manufacturer
EXPAL, Avenida del Partenón, 16, 5a, Campo de las Naciones, E-28042 Madrid, Spain.
Tel: +34 1 722 0235. Fax: +34 1 722 0295.

VERIFIED

SWEDEN

Bofors Minelayer 5821

Development
This mechanical minelayer was developed by FFV (now Bofors AB) and was designed specifically for use with the 028 anti-tank mine. Development began in 1976.

During 1985, the German Army placed an order worth DM43 million for 300 of these minelayers. The Netherlands placed an order in late 1985, followed by a further order in September 1989.

Description
The Minelayer 5821 can lay mines either on the surface or buried to a depth of 200 mm. The distance between the mines can be varied between 1 and 99 m and maximum laying capacity is 20 mines/min at a speed of 7 km/h. All the crew has to do is remove the ransport safety tab and place the mine on the chute. The minelayer then buries the mine and replaces the soil. The system can be towed behind most types of vehicle.

Specifications
Crew: 1 + 3
Weight: 2,800 kg
Length: 5.3 m
Width: 2.5 m

Status
Available. In service with the Dutch and German armies.

Manufacturer
Bofors AB, S-691 80 Karlskoga, Sweden.
Tel: +46 586 81000. Fax: +46 586 581 45.

VERIFIED

Minelayer 5821 in action

Minelayer 5821 in action

UNITED STATES OF AMERICA

M128 vehicle-towed mine dispenser

Development
The Ground Emplaced Mine Scattering System (GEMSS) M128 (or Frisbee Flinger as it is more popularly known), was developed by Picatinny Arsenal and the Ordnance Division of the FMC Corporation of San Jose, California (now part of United Defense LP). The M128 was type classified for US Army service in 1980 but introduction into service was delayed. During Fiscal Year 1982 and FY83 59 and 52 units respectively were purchased. None were purchased in FY84 but in FY85 81 units at a cost of $30.1 million were scheduled. The first unit was equipped in the first quarter of FY86. The final equipment buy was completed during the first quarter of FY88. The M128 is issued on the basis of one to selected engineer companies, although it will be eventually replaced by the M139 Volcano system (see following entry).

Description
The M128 is mounted on a modified M794 flatbed trailer designated the M979, and can be towed behind a truck or tracked vehicle such as an M113 APC, both on roads and cross-country. Anti-tank and anti-personnel mines of the same dimensions are carried in the drum-type magazine and can be dispensed concurrently or separately in preselected patterns and minefield densities. Weight of the M128 empty is 4,773 kg and loaded 6,364 kg.

The M128 feeds mines out of a hydraulic drum-type magazine, flips them through a directional tunnel and sends them spinning to the ground. The mines are left on the surface and a percentage of mines in each batch are fitted with an anti-disturbance mechanism.

The system uses the M74 anti-personnel and M75 anti-tank mines. It can carry up to 800 mines in its storage drums and dispense 2 mines/s. The mines have a factory-set long (more than a day) self-destruct time, which can be changed to a still longer time by the M128 while the mines are being dispensed. The anti-personnel mine is tripwire-activated and has a blast kill mechanism. The anti-tank mine is very similar to the mine used in the Remote Anti-Armor Mine System (RAAMS). Both types of mine have anti-disturbance features fitted.

Inside their crates the mines are packed in tubular sleeves each carrying five mines. Loading the M128 dispenser takes from three to five men up to 20 minutes with the normal mine mix being five anti-tank to one anti-personnel.

Status

In service with US Army. Being replaced by the M139 Volcano system (see following entry).

UPDATED

M128 vehicle-towed mine dispenser ***1996***

M128 vehicle-towed mine dispenser ***1996***

M139 Volcano multiple delivery mine system

Development

Following earlier development, Alliant Techsystems Inc. was awarded an initial $10.5 million contract for the advanced development of the Volcano multiple delivery mine system. The materiel developing agency is the Project Manager, Mine Countermine Demolition at Picatinny Arsenal. The US Army Engineer School and Center at Fort Leonard Wood is in charge of combat development.

In late September 1987 a firm fixed-price contract was awarded to Alliant Techsystems Inc. The contract was worth $19,590,592 and was for six units. The US Army's Fiscal Year 1988 request was for 23,000 M87 canisters at a cost of $34.5 million. Day and Zimmerman Inc of Texarkana, Texas, was also involved in the programme.

By October 1990 the 5 ton truck system was type classified and in production by Brunswick Defense; it was awarded a $4 million contract to produce 24 launch systems for delivery during 1991. In August 1993 Brunswick Corporation was awarded a further $11,000,136 firm fixed-price contract for the production of 98 mine dispensers, 98 air system mounting kits, 20 maintenance self-testers and 20 post-jettison rebuild kits. The contract was completed by the end of June 1995.

In August 1995 Technical Products Group, Intellitec Division, were awarded a $28,619,650 contract for 315 M139 Mine Dispensers, 290 mounting kits for the M548A1 tracked vehicle, 60 maintenance self-testers, and 45 mounting kits for ground systems. Work on the contract is expected to be completed by the end of June 1998.

The UH-60A and M548A1 systems were type classified during June 1991.

In late 1995 Volcano was selected to meet the British Army requirement for a Vehicle-Launched Scatterable Mine System (VLSMS). A contract worth over $100 million was placed with Alliant Techsystems in June 1996, initially covering the testing and delivery of 29 Volcano systems to be mounted on an Alvis Stormer tracked carrier. Production of the initial units should commence by mid-2000 although the contract includes various options which can be exercised until March 1999. The complete system, to be deployed by the Royal Engineers, is known as Shielder and will be used to dispense anti-tank mines only.

In June 1996 Alliant Techsystems announced a collaborative agreement with Kvaerner Eureka of Norway to pursue a contract for an anti-tank system for Norwegian Army infantry and armoured forces. It is understood that the projected system will be based on the British Army's Shielder.

Volcano Vehicle Launched Scatterable Mine System (VLSMS) mounted on M548A1 tracked flatbed carrier ***1996***

Description

The M139 Volcano multiple delivery mine system can rapidly dispense up to 960 GATOR anti-tank mines and anti-personnel mines from expendable canisters mounted on as many as four launcher racks. The system is designed to be carried by either the M817 engineer dump-truck, M814 5 ton cargo truck, M548A1 tracked carrier, or the UH-60 Black Hawk helicopter. Other ground vehicles have been tested and authorised for use.

The main system components are: the expendable M87 Mine Canister which holds six mines; five GATOR BLU-91/B anti-tank mines; and one BLU-92/B anti-personnel mine. The other component is the M139 Dispenser, a reusable item consisting of the launcher rack, the Dispenser Control Unit (DCU) and mounting hardware. Each launcher rack can hold up to 40 canisters and is connected to the mounting hardware (an angled mounting bracket holding up to four launcher racks). Only one DCU is used for each system to provide fire signals, both for testing the system and for setting the arming and one of three self-destruct times.

After the Volcano system has been mounted in the carrier vehicle or helicopter and the DCU has been programmed for the carrier vehicle speed and the minefield length, the firing sequence can be initiated by the operator. Upon receiving the initial firing command, the DCU begins to fire canisters on alternating sides of the carrier vehicle in the sequence prescribed for uniform minefield density. The propellant charge at the base of the canister expels the mine stack and, as it leaves the end of the canister, the mines receive electrical impulses through a strap interlaced between them. These electrical impulses arm the mine and set its self-destruct time. As the mine stack reaches the slack limit of the strap the mines are dispersed by the increasing tautness of the interlaced strap.

A Volcano minefield can have a nominal density of 0.9 mines/m along a 1 km front, although this can be varied. The mine delivery speed is from 8 to 88 km/h for ground vehicles and from 20 to 120 kt for helicopters. At a ground speed of 88 km/h 960 mines can be dispensed in 43 seconds. A helicopter flying at 120 kt can dispense 960 mines in 17 seconds. A typical Volcano-dispensed minefield strip is 125 m wide and 1,150 m long.

The 5 ton truck version of Volcano weighs 3,255 kg. The UH-60A version weighs 2,940 kg and the M548A1 version 3,325 kg.

For training there are two Volcano training canisters: the M88 practice canister holding six dummy mines and the M89 canister with an electronic dummy load.

Variant

Volcano Light

As a private venture Alliant Techsystems Inc have developed a lightweight Volcano system known as Volcano Light. This consists of a 20 canister dispenser assembly carried on the rear of a M998 HMMWV. The assembly can carry 120 anti-armour or 300 anti-personnel mines and can be fixed on a swivel mount in one of five positions for dispensing.

Alliant Techsystems has developed a man-portable variant which allows the firing of single canisters.

Status

In production for the US Army. Ordered by the British Army (29 units).

System Contractors

Alliant Techsystems Inc, 600 Second Street Northeast, Hopkins, Minnesota 55343, USA.
Tel: +1 612 931 6815. Fax: +1 612 931 6654.
Intellitec, 2000 Brunswick Lane, DeLand, Florida 32724-2099, USA.
Tel: +1 904 736 1700. Fax: +1 904 736 2250.

UPDATED

M57 Anti-Tank Mine Dispensing System (ATMDS)

Description

The M57 Anti-Tank Mine Dispensing System (ATMDS) is a side-lift plough with a side-insertion mine chute mounted on a two-wheeled trailer towed by a 2½ ton (6 × 6) truck which carries the crew and a supply of mines. The M57 can lay M15 anti-tank mines on the surface, or buried at a sustained rate of 375 mines/h. The mines can be buried to a maximum depth of 152 mm. The system is not suitable for use in very hard soil or on side slopes. It is issued on the scale of one per combat engineer company in Europe.

Status

In service with the US Army.

VERIFIED

Sidelift plough of M57 ATMDS with mine dispenser chute above (US Army)

Complete M57 ATMDS from mine pallets on left to dispenser on right (US Army)

Remote Anti-Armor Mine System (RAAMS)

Development

The Remote Anti-Armor Mine System (RAAMS) consists of a 155 mm projectile carrying anti-armour mines which are fired from M109 series self-propelled howitzers. It is a member of the Family of Scatterable Mines (FASCAM). Initial production commenced during 1978 and is now complete.

Description

There are two types of RAAMS projectile: the M741 projectile carries nine mines with a factory-set, short self-destruct time of under 24 hours; the M718 carries nine mines with a factory-set, long self-destruct time of over 24 hours. M741A1 and M718A1 mines with revised electronic arming systems and fuzes were also produced.

Each circular M75 anti-tank mine weighs about 2.26 kg. Arming is automatic on landing. Actuation is by a magnetic influence fuze that operates against the width of the tank. The detonation then occurs in two stages. The first stage is when a clearing charge blows off the mine cover and removes any ground debris. The second stage occurs 30 ms later when the main charge, which consists of 572 g of PBX 0280 (95:5 RDX:Estane) and two plate charges so as to make the ground orientation of the mine non-critical, is detonated.

The mine is capable of destroying any tank passing over it and is said to cause a spalling effect when it penetrates tank belly armour. The projectile is fired to a maximum range of 17,000 m and a single battery of six 155 mm self-propelled howitzers can lay a minefield 300 m long and 250 m deep with two salvos. The 12 projectiles fired, each weighing 46.7 kg, provide a density of 0.001 mine/m^2 for the minefield. The density can be changed to meet the tactical situation by varying the lay of the guns and the number of rounds fired.

Cutaway of Remote Anti-Armor Mine System (RAAMS) projectile

Status

In service with the US Army and Marine Corps.

Manufacturer

Alliant Techsystems Inc, 600 Second Street Northeast, Hopkins, Minnesota 55343, USA.
Tel: +1 612 931 6815. Fax: +1 612 931 6654.

UPDATED

M138 Flipper portable minelaying device

Description

The M138 portable minelaying device Flipper was developed by the US Army Armament, Research, Development and Engineering Center and the Project Manager for Mines, Countermine and Demolitions at Picatinny Arsenal, New Jersey, as an in-house programme. Production was carried out by the Peerless Instrument Company of Queens, New York, and the first units were produced during 1990.

The M138 Flipper is intended to augment the M128 vehicle-towed mine dispenser in scattering M74 anti-personnel and M75 anti-tank mines. The intention is to provide field commanders with the ability to emplace minefields at short notice and under varying field conditions.

The unit is transported in a packing box which can be air-dropped. Two men can install a Flipper on the rear of a vehicle such as a 2½ ton or larger truck in about 10 minutes using a clamp assembly and a power lead to connect the unit to the vehicle's electrical system. As the truck is driven along, a soldier manually feeds mines into the launcher which uses a spinning tyre encased in a metal frame. After it is fed, the mine is armed, spun and launched at the rate of one every 10 seconds.

Status

In service with the US Army.

Development Agency

US Army Armament Research, Development and Engineering Center and the Project Manager for Mines, Countermine and Demolitions, Picatinny Arsenal, Dover, New Jersey 07806, USA.
Tel: +1 201 724 6460.

VERIFIED

Soldier loading the M138 Flipper portable minelaying system

YUGOSLAVIA (SERBIA AND MONTENEGRO)

Towed minelayer

Description

This towed minelayer resembles the PMR-3 in layout and operation but is used to emplace the TMRP-6 anti-tank mine. Both were produced and designed in the former Yugoslavia but neither appear to be currently in production.

The minelayer is towed behind a suitable 4 × 4 or 6 × 6 wheeled vehicle and mines are fed into the dispenser chute direct from the vehicle's rear cargo area. At the end of the dispenser chute two sets of ploughs open up the soil to bury the mine and then push back the soil to disguise the mine's position.

No details regarding performance or dimensions are available.

Status

Probably no longer in production. In service with the Yugoslav Army.

VERIFIED

Towed minelayer laying TMRP-6 anti-tank mines

MINE DETECTION EQUIPMENT

AUSTRALIA

TM-4 image processing magnetometer system

Description

The TM-4 is a self-contained digital magnetometer system for the computer-aided location of ferrous ordnance. Fully positioned, magnetic field measurements to 0.01 nT resolution may be digitally recorded at up to 200 per second. An in-built odometer eliminates the premarking of search lanes. A dual sensor configuration allows two lanes to be traversed simultaneously. On-site image processing and computer-aided unexploded object anomaly recognition and identification, enable assurance confidence levels to be specified for each search environment and provide a verifiable record of an Explosive Ordnance Disposal (EOD) operation. Typically, the detection assurance depth equals or exceeds the maximum expected penetration depth for all ferrous ordnance. Explosive Ordnance Disposal insurance underwriting has been obtained by users of the system.

In the dual sensor mode up to 4 hectares/day may be searched on foot, providing 100,000 positioned magnetic field measurements. The system may also be vehicle-mounted on request, with multiple, optically pumped, atomic resonance type magnetic sensors. A real-time digital filter suppresses interference from mains electricity.

Power is supplied by 30 V rechargeable batteries with a life of 7 hours per pack.

Specifications

Resolution: 0.01 nT
Position accuracy: 0.1% of distance between controls
Survey rate: up to 200 positioned measurements/s
Power supplies: 30 V rechargeable batteries
Battery pack life: 7 h

Status

In service with the Australian Department of Defence and others.

Manufacturer

Geophysical Technology Pty Ltd, PO Box U9, Armidale, New South Wales 2351, Australia.
Tel: +61 67 732617. Fax: +61 67 711661.

VERIFIED

TM-4 magnetometer in action in dual sensor mode

AUSTRIA

Schiebel AN-19/2 mine detecting set

Development

The AN-19/2 mine detecting set was originally developed through co-operation between the Dipl Ing Hans Schiebel Elektronische Geräte GesmbH of Vienna and NobelTech Electronics AB (now CelsiusTech) of Sweden. It is a design derivative of the AN-17/1 originally developed by Hans Schiebel. NobelTech Electronics delivered a large quantity of AN-19/2s to the Swedish Army during 1986 and 1987.

Since then, this instrument has been marketed by Schiebel Elektronische Geräte GesmbH of Vienna, one result being the sale of approximately 30,000 units to the US Army where the equipment is known as the AN/PPS-12. The AN-19/2 is also used by the United Nations peace-keeping forces operating in Afghanistan, Angola, Cambodia, Lebanon, Mozambique, Somalia and in the former Yugoslavia.

The AN-19/2 has been approved for procurement by NATO and some other countries. The German armed forces procured quantities of this instrument with deliveries extending from late 1989 to 1994, to be followed by a further order. The United Kingdom ordered 700 units for use in the Gulf region during 1991. Italy ordered a quantity in February 1993. Other NATO users include Canada, Greece and the Netherlands.

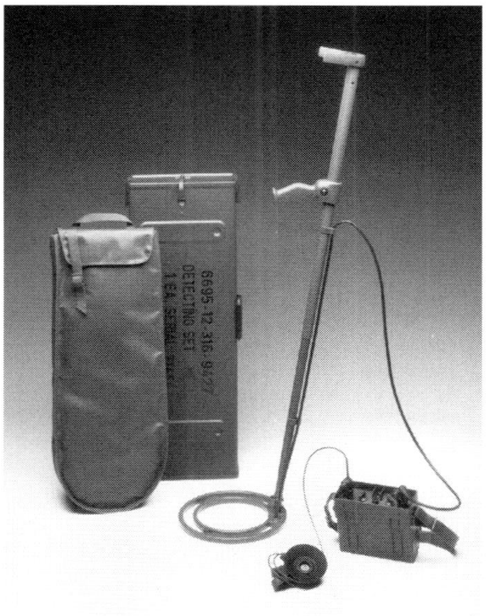

AN-19/2 mine detecting set

AN-19/2 mine detecting set in use **1996**

Description

The AN-19/2 mine detecting set is designed to detect very small metallic objects such as plastic mines with a small metal content and uses a detector system with minimum mutual interference between adjacent detectors. The system produces a high degree of discrimination independent of ambient temperature and battery condition. Also inherent in the system is the ability to maintain search sensitivity independent of search head speed, from standstill to normal speed. The lightweight search head can detect metallic objects in a limited depth of fresh or seawater.

The main components of the AN-19/2 are the electronics unit, the search head with its telescopic pole and cable, the headphone and the carrying bag. An aluminium transport case is available.

The electronics unit contains the control panel, the internal printed circuit board and the battery container. There are sockets on the control panel for the headphone cable plug and the search head cable plug. When the plugs are removed the sockets are protected by rubber caps. The control panel contains the power switch, headphone volume control, sensitivity control knob and the failure warning lamp. The printed circuit board is sealed within the case and the waterproof and leakproof battery container holds four 1.5 V R20 batteries. There is a built-in test circuit for function and battery condition checks.

The search head consists of concentric transmitter and receiver coils embedded in a plastic head. This lightweight head permits visual inspection of the ground when searching. The coils are connected to the electronics unit via a cable and plug. The cable is made of special rubber material that remains flexible down to −40°C. The cable is secured to the telescopic pole with cable clamps allowing the search head to be changed in the field. The telescopic pole consists of an inner and outer tube, an arm support cup and adjustable handle. The inner tube is made of Glass-Reinforced Plastic (GRP) and the joint between the search head and the pole has a plastic shear bolt that will break if excessive force is applied to the pole. A spare bolt is supplied in the carrying bag.

The headphone has a single side speaker so the operator can also hear surrounding noises. It can be worn under a helmet.

The carrying bag can be adjusted as a backpack and contains all the parts of the AN-19/2 plus spare batteries.

In operation the transmitting coil in the search head is energised by electrical pulses to build up a magnetic field. The field induces eddy currents in metal objects in the vicinity which give rise to a secondary field to be detected by the receiving coil. The signal from this coil is processed in the electronics unit to create a sound in the headphone. Approximately 3 minutes are required to prepare the equipment for operation. An operator can be fully qualified to use and maintain the equipment after 8 hours of training.

Specifications

Weight:
(set with bag and batteries) 6.2 kg
(search head with telescopic pole) 1.3 kg
(electronics unit) 1.1 kg
(headphone) 0.15 kg

(carry bag) 1.2 kg
(transport case) 2.35 kg
Dimensions:
(transport case) 800 × 300 × 130 mm
(electronics unit) 185 × 80 × 150 mm
(search head diameter) 270 mm
(telescopic pole length) 770 to 1,620 mm
Battery voltage: 6 V
Battery life: 70 h
Operating temperature: −40 to +55°C
Detection range:
(mine with 0.15 g metal content) up to 100 mm
(typical anti-tank mine) up to 500 mm

Status
In production. In service with Austria, Canada, Germany, some Gulf states, India, Italy, Netherlands, Sweden, UK (700), US Army (over 30,000 units) and United Nations forces operating in Afghanistan, Angola, Cambodia, Lebanon, Mozambique, Somalia and the former Yugoslavia.

Manufacturer
Dipl Ing Hans Schiebel Elektronische Geräte GesmbH, Margaretenstrasse 112, A-1050 Wien, Austria.
Tel: +43 1 546260. Fax: +43 1 545 2339.
American office: Schiebel Instruments Inc, 2127 California Street NW, Washington DC 20008, USA.
Tel: +1 202 483 8311. Fax: +1 202 483 8316.
Canadian office: Schiebel Systems Limited, BDO Center, Suite 306, 301 Moodie Drive, Nepean, Ontario, Canada K2H 9C4.
Tel: +1 613 596 1546. Fax: +1 613 596 1551.

UPDATED

Schiebel MIMID miniature mine detector

Development
The Schiebel MIMID miniature mine detector is based on proven Schiebel AN-19/2 mine detecting set technology (see previous entry) and was jointly developed by Schiebel and SAIC to meet a specific operational requirement. It was designed to be easily operated by all personnel working in countries where mines are a constant threat to personal safety. It is not intended to replace the AN-19/2 but to supplement in-theatre mine detection capability by providing all military and civilian personnel with a practical and easy means of enhancing safety in mined areas.

Description
The MIMID can detect mines with minimum metal content and is ideally suited for special operations where soldiers have to carry heavy equipment The complete equipment can be carried folded into a compact unit in a rucksack or on a belt. For use, the rectangular search head is unfolded from the main battery and electronics unit and a two-section arm is extended to the required length. Controls are minimal and easy to operate.

Power is provided by four 1.5 V cells. Weight is approximately 1 kg.

Status
Advanced development.

Manufacturer
Dipl Ing Hans Schiebel Elektronische Geräte GesmbH, Margaretenstrasse 112, A-1050 Wien, Austria.
Tel: +43 1 546260. Fax: +43 1 545 2339.
American office: Schiebel Instruments Inc, 2127 California Street NW, Washington DC 20008, USA.
Tel: +1 202 483 8311. Fax: +1 (202) 483 8316.
Canadian office: Schiebel Systems Limited, BDO Center, Suite 306, 301 Moodie Drive, Nepean, Ontario, Canada K2H 9C4.
Tel: +1 613 596 1546. Fax: +1 613 596 1551.

NEW ENTRY

Schiebel MIMID miniature mine detector
1997

Schiebel AN-20/1 underwater mine detecting set

Development
The Schiebel AN-20/1 underwater mine detecting set was developed specifically for underwater operation and also for use on the battlefield. The technology involved is based on that employed with the AN-19/2 mine detecting set. The AN-20/1 can detect small metallic objects in fresh or saltwater up to 100 m below sea level.

Description
Due to the compact design and low mutual interference between two detectors, the AN-20/1 underwater mine detecting set can detect small metal objects such as minimal metal content mines in a wide variety of conditions. The search coils are essentially similar to those used on the AN-19/2 mine detecting set. For accurate object location, the search head can be moved crosswise over a target area with aural signals transmitted via an earphone.

A mine with very small metal content (0.15 g) can be detected at a distance of 100 mm. A typical anti-tank mine can be detected at a distance of 500 mm.

The AN-20/1 can be operated to full effect by unqualified personnel following about 8 hours of training. Approximately 3 minutes are required to prepare the set for use. No preventive maintenance is required apart from a simple functioning check and cleaning. All main components are interchangeable with the need for recalibration. For storage and transport the AN-20/1 is packed in a carry bag which also contains some parts such as a test piece for checking sensitivity, spare cable clamps and a spare plastic shear bolt for the telescopic pole.

Power for the AN-20/1 underwater mine detecting set is provided by four 1.5 V AA dry battery cells.

Specifications
Weight:
(set with bag and batteries) approx 4.6 kg
(set in operation) 2.62 kg
(headphone) approx 0.2 kg
(carry bag) approx 1.7 kg
Dimensions:
(carry bag) 775 × 375 × 120 mm
(search head diameter) 270 mm
(telescopic pole length) 700 to 1,190 mm
Battery voltage: 6 V
Operating temperature: −35 to +55°C
Detection range:
(mine with 0.15 g metal content) up to 100 mm
(typical anti-tank mine) up to 500 mm
Max operating depth: 100 m

Status
In production.

Manufacturer
Dipl Ing Hans Schiebel Elektronische Geräte GesmbH, Margaretenstrasse 112, A-1050 Wien, Austria.
Tel: +43 1 546260. Fax: +43 1 545 2339.
American office: Schiebel Instruments Inc, 2127 California Street NW, Washington DC 20008, USA.
Tel: +1 202 483 8311. Fax: +1 202 483 8316.

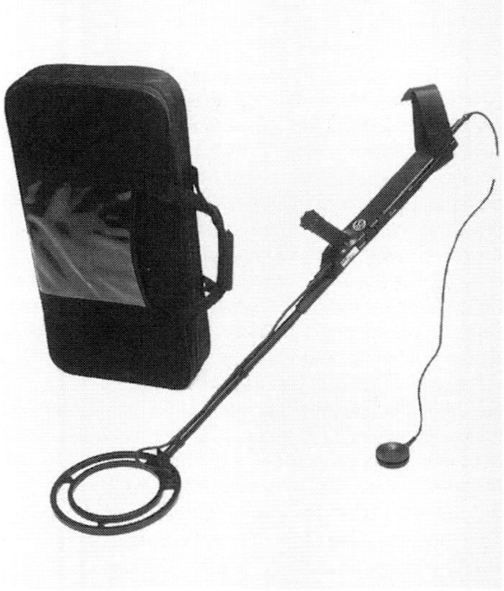

Schiebel AN-20/1 underwater mine detecting set
1996

Canadian office: Schiebel Systems Limited, BDO Center, Suite 306, 301 Moodie Drive, Nepean, Ontario, Canada K2H 9C4.
Tel: +1 613 596 1546. Fax: +1 613 596 1551.

VERIFIED

Schiebel AN-23/2 bomb locating set

Description
The Schiebel AN-23/2 bomb locating set is designed to locate ferro-magnetic objects underground or underwater. Two three-dimensional sensor arrays are used so the absolute and differential values of the earth's magnetic field can be measured in spheric components which can be utilised to calculate two vectors, so a high detection performance is possible.

The maximum search depth which can be obtained depends on several factors, including object type and mass, the depth of the object below the ground surface and the time the object has been in the ground.

The sensor detection head is fully microprocessor-controlled with corrected data transferred via an RS-485 interface. On the electronics unit, an easy-to-use graphic user interface based on the extensive use of icons, enables operators to operate and fully utilise the device. The basic version of the AN-23/2 includes all features to facilitate computer-aided search techniques, such as the manually or automatically triggered storage of search data. Stored data can be

Schiebel AN-32/2 bomb locating set ***1997***

transferred via a serial interface to a personal computer for advanced analysis and visualisation. The unit can be combined with positioning systems such as the Global Positioning System (GPS) or a precision beacon system. In addition a VHF data transmission link can be connected to the unit. All the necessary interfaces are provided on the standard version.

The detection head of the AN-23/2 includes two highly sensitive three-dimensional sensor arrays placed 400 mm apart and based on the magneto-resistive effect. Each sensor array consists of three single-axis sensors. In this way six absolute values and three differential values are provided for processing by the measuring electronics, creating in effect vector magnetometer and gradiometer data, together with total field magnetometer and gradiometer data.

The complete AN-23/2 comprises an electronics unit, the detection head, a handle including a battery compartment and a stowage/transport case.

The electronics unit contains a printed circuit board, a control panel, an LCD graphic display, sockets for the detection head and the power supply and additional sockets for peripheral equipment.

The back-lit graphic display presents data to the user using standard icons such as those used with Windows; there is an on-line help feature. Also provided is a headphone which can be worn under a helmet.

The detection head contains the sensor arrays and their associated electronics. Although normally attached to the carrying handle the head may be used at a distance of more than 1,000 m from the electronics unit.

The handle contains two sections with the battery compartment holding the size D batteries at the rear. All operations are menu-led once a self-test routine is completed after switching on; thereafter self-test routines are continuous. Manual or automatic search programmes in the electronics unit can then enable the user to build up a graphic display formed from the data obtained. Transferring the stored data to a personal computer, enables various types of software to be utilised for area and multi-operative analysis and to create visual displays of the area covered; the data can also be transferred to GIS software. Other software available includes two- and three-dimensional graphics, a vector plot, a guided search and target identification, all of which are incorporated in the electronics unit.

Specifications

Weight:
(set with batteries) 6 kg
(transport case) 6.5 kg

Dimensions:
(electronic unit) 200 × 125 × 60 mm
(detection head) 50 × 500 mm
(handle, length) 1.6 m
(transport case) 820 × 320 × 170 mm

Status

In production.

Manufacturer

Dipl Ing Hans Schiebel Elektronische Geräte GesmbH, Margaretenstrasse 112, A-1050 Wien, Austria.
Tel: +43 1 546260. Fax: +43 1 545 2339.
American office: Schiebel Instruments Inc, 2127 California Street NW, Washington DC 20008, USA.
Tel: +1 202 483 8311. Fax: +1 202 483 8316.
Canadian office: Schiebel Systems Limited, BDO Center, Suite 306, 301 Moodie Drive, Nepean, Ontario, Canada K2H 9C4.
Tel: +1 613 596 1546. Fax: +1 613 596 1551.

UPDATED

Schiebel Vehicular Array Mine Detection System (VAMIDS)

Development

The Schiebel Vehicular Array Mine Detection System (VAMIDS) is designed to detect metallic land mines, including those with minimal metal content and unexploded ordnance (UXO), from a vehicle platform. With the use of special array segments, VAMIDS can detect buried UXO to depths of several metres. Although the system is designed for use along established routes, it has a cross-country capability, depending on the terrain and the mobility of the vehicle platform.

Description

The Schiebel VAMIDS is a modular integrated system comprising a control unit, a segmented array and a system console.

The control unit is housed in a watertight and ruggedised case including an anti-vibration mounted standard military 19 in/483 mm rack. It is suitable for adverse environmental conditions and vehicle applications.

The control unit contains: a power card; up to 12 sensor cards, each interfacing with a group of four detection coils; and one processor card for converting and processing the data received from the sensor cards and for transmitting information from the system console via a system data port. The VAMIDS main interface is a RS-232 selectable system data port. The control unit also provides an RS-232 interface on the processor card face plate for local system access; a synchronous serial data interface capable of indicating the exact location of a target to a marking system; and an output switched to the supply voltage under control of the processor to flag a global event warning. The control unit circuit cards provide face plates with control and display components for maintenance purposes.

Schiebel Vehicular Array Mine Detection System (VAMIDS) mounted on Mamba APC ***1997***

The segmented arrays are available as rigid or flexible panels which can be either mounted on a vehicle structure or pulled over the ground surface. Individual segments provide overlap areas on either side and may be arrayed up to a maximum width of 6 m. On the rigid version of VAMIDS array segments the individual detection coils are embedded in one panel. Flexible arrays are designed to be pulled over the ground to provide optimum coverage; the individual detection heads are mounted on a flexible drawbed structure which serves as a wearsheet both supporting and protecting the assembled detection coils.

A marking system can accurately mark the location of targets by spraying a marking fluid through up to 8 nozzles.

The system console serves as the operator interface and can be an industry standard personal computer such as a notebook computer. Alternatively, the software functions assigned to the system console may be implemented into a higher level data processing system such as a vehicle control computer. The system console provides a user-friendly real-time graphical display of up to 48 input detection channels from a 6 m wide segmented array. The system configuration, set-up, and diagnostics reporting are tasked by the system console.

Specifications

Control unit
Weight: (max) 14 kg
Width: 534 mm
Height: 217 mm
Depth: 450 mm
Voltage: 12-24 V
Current consumption: (max) 30 A

Pulse frequency (PRF): (max) 19,200 Hz
Data transfer rate: (max) 38,400 Baud

One metre rigid array segment
Weight: 17 kg
Length: 390 mm
Width: 1.18 m
Height: 17 mm
Detection width: 1 m
Number of detection coils: 8

One metre rigid UXO array segment
Weight: 57 kg
Length: 700 mm
Width: 1.18 m
Height: 35 mm
Detection width: 1 m
Number of detection coils: 12

One metre flexible array segment
Weight: 20 kg
Length: 613 mm
Width: 1.168 m
Depth: 150 mm
Effective width: 1 m
Number of detection coils: 8
Max bend angle: ±14°

System console requirements
Processor: 32-bit x86 based microprocessor (such as Intel 80486/25 or higher) or Intel Pentium
Display: standard VGA, 16-colour, 640 × 480 or higher resolution, video display adapter
Hard disk: 10 MB min free
Floppy disk drive: high density 3.5 in
Interfaces: one RS-232 or RS-485 serial data port
Memory: 4 MB min

Status
Advanced development. A flexible array version is under development for the Canadian Forces under the codename JUMBO.

Manufacturer
Dipl Ing Hans Schiebel Elektronische Geräte GesmbH, Margaretenstrasse 112, A-1050 Wien, Austria.
Tel: +43 1 546260. Fax: +43 1 545 23 39.
American office: Schiebel Instruments Inc, 2127 California Street NW, Washington DC 20008, USA.
Tel: +1 202 483 8311. Fax: +1 202 483 8316.
Canadian office: Schiebel Systems Limited, BDO Center, Suite 306, 301 Moodie Drive, Nepean, Ontario, Canada K2H 9C4.
Tel: +1 613 596 1546. Fax: +1 613 596 1551.

VERIFIED

BULGARIA

MTA-86 mine detector

Description
The MTA-86 mine detector can detect metallic mines or plastic mines with some metallic components such as metallic fuzes. The detector head assembly is circular and mounted on a telescopic search handle. The power source and operating controls are carried in a separate shoulder-slung, moisture-proof pack.

Power for the detector is provided by a 12 to 16 V lithium battery which will keep the detector operational for periods of up to 100 hours.

The MTA-86 can detect anti-tank mines buried at depths of up to 130 mm. Detection signals are provided aurally through a pair of operator headphones.

Weight of the complete equipment is 4 kg. The operational temperature range is from −40 to +50°C.

Status
Offered for export.

Marketing Agency
Kintex, PO Box 209, 66 James Baucher Street, 1407 Sofia, Bulgaria.
Tel: +359 2 662311. Telex: 23 243.
Fax: +359 2 658101.

VERIFIED

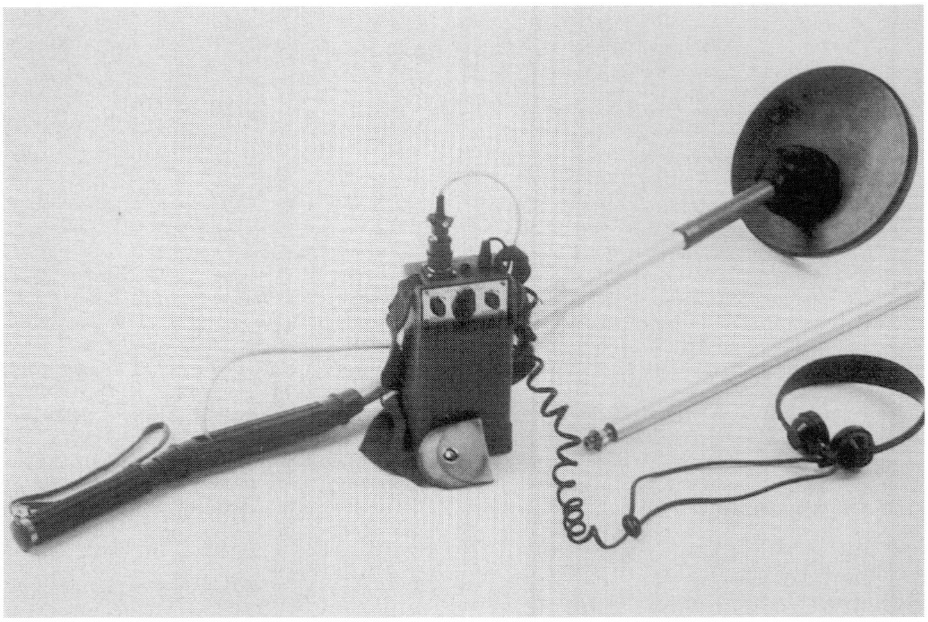

MTA-86 mine detector

CANADA

Scintrex V-920 ferrous ordnance locator

Description
The Scintrex V-920 ferrous ordnance locator is particularly suitable for the sweeping of large areas of terrain and for the detection of magnetic anomalies produced by buried ferrous objects, including mines and other ordnance. The V-920 employs a 64 × 240 LCD graphic screen, visible under all lighting conditions, which is used to show a variety of graphic displays to allow an operator to rapidly analyse the subsequent survey data. In addition, the V-920 is provided with user friendly, menu-driven software for IBM compatible personal computers to which survey data can be transferred providing data processing, three dimensional mapping and anomaly analysis functions.

The system may be deployed as a hand-held equipment or carried on light vehicles with the sensor probe suspended from a mast in front of the vehicle. Sampling rates as high as 10 times a second are possible and a large memory capacity is expandable to 200,000 readings. A sweep mode provides aural feedback for large areas.

Status
Available.

Manufacturer
Scintrex Security Systems, 222 Snidercroft Road, Concord, Ontario, Canada L4K 1B5.
Tel: +1 905 669 2280.
Fax: +1 905 669 5132.

NEW ENTRY

Scintrex V-920 ferrous ordnance locator
1997

CHINA, PEOPLE'S REPUBLIC

Model 82 portable mine detector set

Description

The Model 82 portable mine detector set contains two separate detectors, the main set being used for detecting non-metallic mines and the auxiliary set being used for detecting metallic mines. Both are contained in a sealed metal carrying case that is water and vibration-proof. The entire equipment is claimed to be waterproof to the extent that the main set can be immersed in water for 24 hours and still remain operational.

Both types of detector can be carried on a telescopic aluminium pole. The detector head is shrouded in rubber and is capable of absorbing impacts such as being dropped 1 m. The entire set is also corrosion-proofed. All cables used with the set can be used in temperatures as low as −40°C (upper temperature limit +50°C).

Power for the set is derived from a single non-rechargeable 12 V zinc-mercury battery and 1.5 V R20 and R14 batteries.

It is claimed that an operator can use the Model 82 after only one hour of training using the instruction manual supplied with each set.

The main set weighs 3.5 kg and the auxiliary set 1.5 kg.

Status

In production. In service with the Chinese armed forces.

Manufacturer

China National Electronics Import and Export Corporation, Electronics Building A23, PO Box 140, Fuxing Road, Beijing, People's Republic of China.
Tel: +86 1 821 9534. Telex: 22475 CEIEC CN.
Fax: +86 1 821 2361.

VERIFIED

Shanghai Research Institute mine detectors

Description

The Shanghai Research Institute of Microwave Technology has designed five models of portable mine detector, the Models 82-G, 120, GTL-112B, 85-S and 90.

Model 82-G mine detector

The Model 82-G is designed to detect metallic and non-metallic mines or packed explosives buried in the earth's subsurface. It is powered by an R6 battery and can remain operational for up to 20 hours continuously. The detector has a rectangular search head and has its electronics unit mounted over the carrying handle. The headphones are worn inside a flying-type helmet.

Specifications

Model 82-G
Weight: less than 3 kg
Detecting head dimensions: 225 × 225 mm
Length of handle: 750 to 1,524 mm
Range of detecting head above ground: 20 to 80 mm
Maximum detection depth with head 50 mm over ground:
(anti-tank mines with 275 mm diameter) not less than 100 mm
(anti-personnel mines with 78 mm diameter) not less than 30 mm
Operating temperature range: −40 to +55°C

Model 120 mine detector

The Model 120 can detect mines with small metal components and has a ground rejection function for use when operating over soil with a magnetic content. The detector is powered by an R6 battery and can be used continuously for up to 30 hours. The search head is circular and the electronic unit is mounted to the side of the carrying handle. Headphones are worn inside a flying-type helmet.

This model has been in service with the People's Liberation Army since 1986.

Specifications

Model 120
Weight: < 3.7 kg
Detecting head diameter: 220 mm

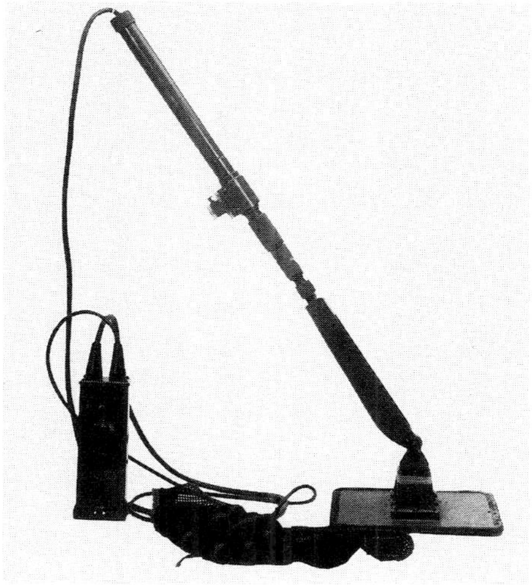

Model 82-G mine detector

Length of handle: 700 to 1,524 mm
Detection sensitivity: max detection distance for M25 anti-personnel mines (distance of detection head above mine) not less than 140 mm
Operating temperature range: −40 to +55°C

Model GTL-112B metallic mine detector

The Model GTL-112B is an improved version of the Model 120 mine detector and employs single-coil electromagnetic induction techniques. It is designed to eliminate false alarm signals resulting from background soil, grass and water. Other design features include low weight, ease of operation, modular components for maintenance and repair, and ease of adjustment. The detector can operate in rain or shallow water.

The main components are the electronics unit, search head with telescopic handle and cable, earphone, and a carrying case. The electronics unit contains the control panel, internal printed circuit board and the battery container with 10 R6 batteries which provide an operational life of not less than 20 hours. There are two sockets on the control panel for the earphone, together with an adjustable switch knob for controlling the power and sensitivity adjustment.

Model GTL-112B metallic mine detector in operation

The Model GTL-112B can detect a plastic anti-personnel mine buried over 120 mm deep.

The Model GTL-112B is in production and in service with the People's Liberation Army.

Specifications

Model GTL-112B
Weight: less than 2.5 kg
Weight of empty carrying case: 4 kg
Search head diameter: 210 mm
Telescopic handle length: 0.7-1.5 m
Detection sensitivity: (M25 anti-personnel mines) not less than 120 mm
Operating temperature range: −40 to +55°C

Model 85-S special purpose mine detector

The Model 85-S is a special purpose mine detector with a small narrow search head and is suitable for detecting

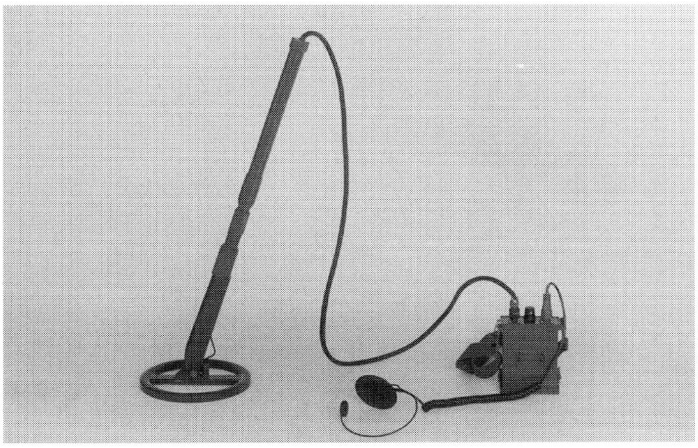

Model GTL-112B metallic mine detector

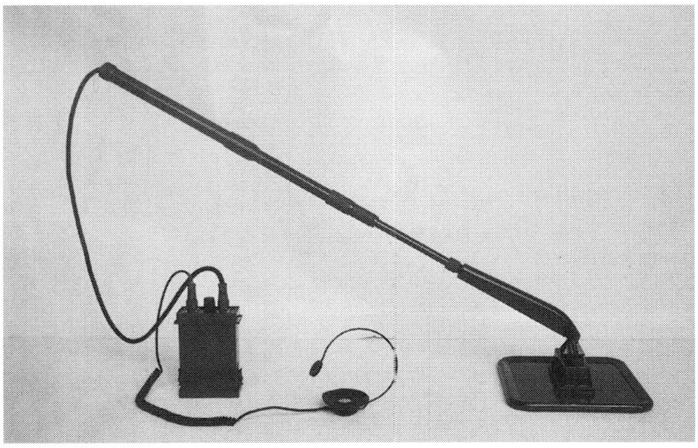

Model 90 dual-purpose mine detector

mines with minimal metal content. It is characterised by its small size, low weight and high reliability. It has a ground rejection function for use under conditions of differing types of soils.

Specifications

Model 85-S
Weight: less than 2 kg
Detecting head dimensions: 54 × 220 mm
Detection sensitivity: (M25 anti-personnel mines) not less than 80 mm
Operating temperature range: −40 to +55°C

Model 90 dual-purpose mine detector

The Model dual-purpose mine detector can be used to detect both metallic and non-metallic mines. Design features claimed include a low false alarm rate, high reliability and speed of operation. The Model 90 can detect mines with minimal metal content such as firing pins.

The main components are the electronics unit, search head with telescopic handle and cable, earphone and a carrying case.

The Model 90 can detect a 275 mm diameter anti-tank mine buried 80 mm deep with the search head 50 mm above the surface. For 78 mm diameter anti-personnel mines the figure is not less than 20 mm.

The Model 90 is in production and in service with the People's Liberation Army.

Specifications

Model 90
Weight: 3.5 kg
Search head dimensions: 255 × 255 mm
Telescopic handle length: 0.7-1.524 m
Battery life: not less than 8 h
Operating temperature range: −40 to +55°C

Status

In production. In service with the Chinese armed forces.

Manufacturer

Shanghai Research Institute of Microwave Technology, 423 Wu Ning Road, Shanghai, People's Republic of China.
Tel: +86 21 254 7600. Fax: +86 21 257 8506.

VERIFIED

CZECH REPUBLIC AND SLOVAKIA

M-10 and M-11 metallic mine detectors

Description

Both the M-10 and M-11 metallic mine detectors operate on the beat-frequency oscillation principle, the only difference between the two models being the construction of their search heads. The M-10 search head consists of two detachable plates 300 mm in diameter which overlap by 100 mm, while the M-11 has a single plate of the same effective area.

The equipment consists of: the search head assembly; search handle consisting of four jointed aluminium sections each 500 mm long; tuning box with tone regulator; power supply of three dry cells (25 V); and the headphones. The tuning-box tone regulator and the batteries are carried in a canvas pack on the operator's back. The batteries, when not in use, are kept in a small pouch on the side of the pack.

The complete equipment weighs 12 kg.

Status

In service with the Czech and Slovak armies and Indonesia (50).

Manufacturer

Czech and Slovak state factories.

VERIFIED

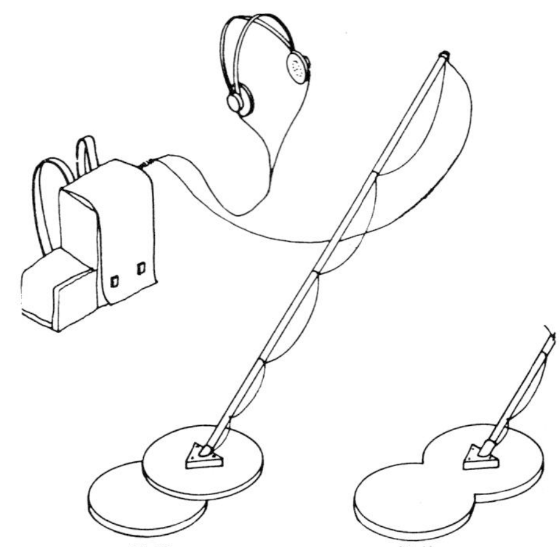

M-10 and M-11 mine detectors showing different heads

EGYPT

Metallic mine detector MMD 140 non-metallic mine detector PMD 156

Description

These two detectors are similar in appearance but vary in performance. The MMD 140 metallic mine detector uses a square detection head and operates at a frequency of 2,500 Hz. The PMD 156 uses a visually similar search head but operates at a frequency in the RF range, that is, from 300 to 600 MHz. Both detectors are used together with headphones which provide aural indications of mines under the search head and also provide a low battery warning signal on a differing frequency. Each detector has a telescopic pole with the control unit and battery holder attached.

Power is derived from 6 V batteries inside the control unit. The batteries can be of various types, including rechargeable cells. Both detectors weigh 3.5 kg when in use and 10 kg with their shipping cases.

Status

Believed to be in service with the Egyptian Army.

Manufacturer

Benha Company for Electronic Industries, PO Box 865, Cairo, Egypt.
Tel: +20 2 322891. Telex: 92360 katel un cairo.

UPDATED

Metallic mine detector MMD 140
1997

FRANCE

Giat metallic mine detector Model F1

Description

The electromagnetic metallic mine detector Model F1 (*Le Détecteur Electromagnétique de Mines*) was designed for detecting anti-tank and anti-personnel mines that contain at least one metal component.

The complete detector is packed into a case 655 mm long, 320 mm wide and 160 mm high, which has two foam packing pieces to protect the equipment from vibration. The complete equipment can be assembled for use in less than 5 minutes.

The mine detector consists of five major components: the detecting head; telescopic handle; electronics box; connecting cable; and a set of standard dry batteries. The batteries provide sufficient power for up to 30 hours of continuous use. Accessories include the flexible cover for the detection head and headset with cable and connector. The former is fitted to the detection head when the terrain has fluctuating temperatures, for example shaded and sunlit areas.

There are two methods of using this mine detector, upright and prone. For the upright method, the telescopic handle is extended and the angle of the search head on the handle is adjusted to enable the operator to sweep the head in front of him from side to side as he moves forward. For the prone position, the sweep motion is similar but the operator lies on the ground with the handle and the head parallel to the ground.

Specifications

Weight ready for use: 3.5 kg
Dimensions: 650 × 300 × 155 mm
Operational life of batteries: 30 h (continuous)
Operating temperature range: −31.5 to +51°C

Detection capability	**Detection range**
1 g of steel	100 mm
20 g of aluminium	300 mm
100 g of aluminium	450 mm

Status
In service with the French Army, designated DHPM-1A.

Manufacturer
Giat Industries, 13 route de la Minière, F-78034 Versailles Cedex, France.
Tel: +33 1 30 97 36 48. Fax: +33 1 30 97 39 58.

VERIFIED

Metallic mine detector Model F1

GERMANY

Vallon MC1 computer-aided detection system

Description
The Vallon MC1 computer-aided detection system is specifically designed for the operating conditions encountered by Explosive Ordnance Disposal (EOD) personnel. It consists of a microcomputer and the corresponding self-contained software to support detection work. For location of objects, the course of magnetic field change is graphically indicated on a display, indicating the course of the previously passed 5 m of the measuring track. Object depth and position can then be calculated on site. Maximum track length is 110 m, with the maximum number of tracks being 40.

The MC1 microcomputer is connected to the ferrous locator only and is thus independent of any other power supplies. The integral software leads the operator using a simple menu technique.

Operation is possible after a short training period. Software modules for land and borehole detection are included. To relate the exact X-Y co-ordinates to the measuring data, the Sensor Position System (SEPOS) can be added as can an alternative Global Positioning System (GPS) unit.

Status
In production. In service with various undisclosed countries.

Manufacturer
Vallon GmbH, Im G-und 3, D-72800 Eningen, Germany.
Tel: +49 7121 9855-0. Telex: 729924 valon d.
Fax: +49 7121 83643.

VERIFIED

Vallon MC1 Computer-aided Detection System

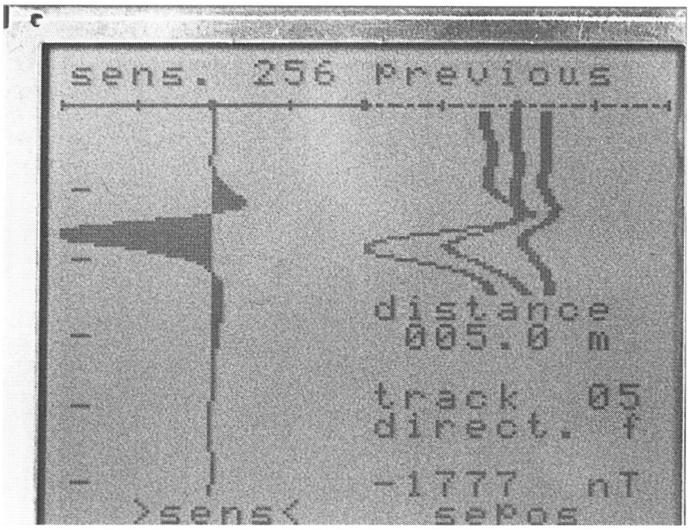

A typical Vallon MC1 computer-aided detection system display **1996**

Vallon software EVA

Description
Development of the MC1 system for large-scale detection for unexploded ordnance (UXOs) work led to the Vallon EVA (Evaluation and Analysis) software package. This is based on survey data transferred from the MC1 microcomputer to a personal computer. These data are not computed but are displayed directly in graph form.

With EVA, magnetometer data is analysed and the exact position of an object is calculated by means of ideal graphs. The result, that is X-position, Y-position, depth, inclination, declination, signal length, is printed out as a true-to-scale object map.

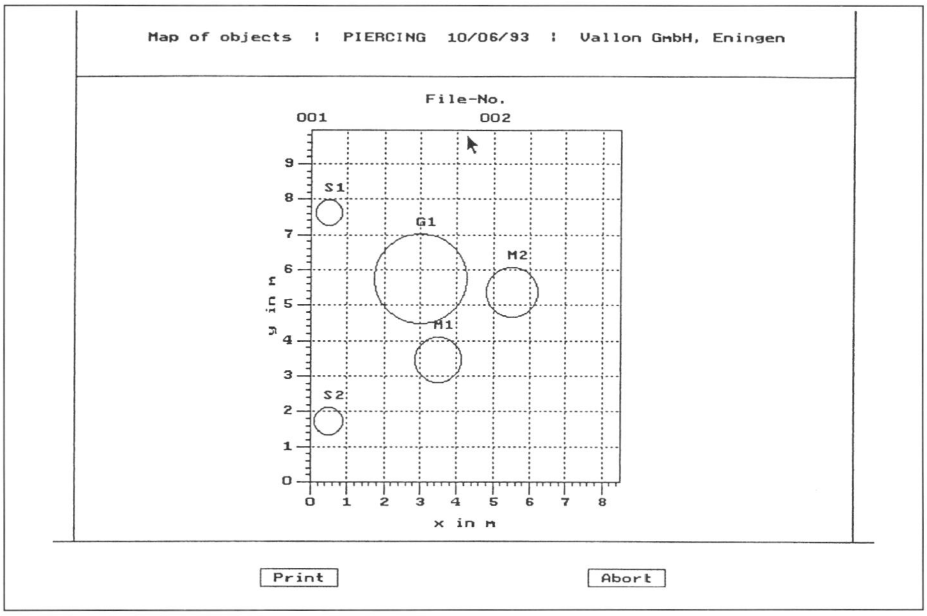

Typical example of true-to-scale object map produced using EVA software

Typical example of object list produced using EVA software

Obj No	Tr No	Latitude	Longitude	Depth in m	Max field in nT	Signal length in m	Phi	Th	Fil No
100	0	48°28.94008'N	9°15.75765'E	1.13	438	1.55	335	28	014
101	1	48°28.94745'N	9°15.75802'E	1.01	543	2.29	133	353	014
102	3	48°28.94826'N	9°15.75879'E	0.36	823	0.57	179	2	014
103	7	48°28.94736'N	9°15.76045'E	0.35	654	0.80	43	3	014
104	7	48°28.94878'N	9°15.76044'E	0.46	401	1.45	92	6	014
105	9	48°28.94843'N	9°15.76137'E	1.03	435	1.44	33	14	014
106	10	48°28.94274'N	9°15.76158'E	1.02	557	2.88	88	2	015
107	12	48°28.94967'N	9°15.76273'E	0.22	463	0.37	180	355	015
108	8	48°28.94107'N	9°15.76088'E	0.36	547	0.56	1	47	014
109	4	48°28.93979'N	9°15.75936'E	0.64	557	1.89	264	0	014
110	2	48°28.94242'N	9°15.75861'E	0.21	547	0.35	337	59	014
111	2	48°28.94427'N	9°15.75863'E	0.25	547	0.41	355	42	014
112	9	48°28.94334'N	9°15.76138'E	0.35	547	0.84	547	45	014
113	11	48°28.94508'N	9°15.76193'E	0.87	364	1.76	25	0	015
114	3	48°28.94114'N	9°15.75878'E	1.12	531	1.55	325	325	014
115	6	48°28.94333'N	9°15.75989'E	2.13	532	2.81	335	45	014
116	6	48°28.94786'N	9°15.75993'E	1.87	547	3.85	225	15	014
117	7	48°28.94674'N	9°15.76045'E	1.09	547	3.09	270	15	014
118	8	48°28.93944'N	9°15.76087'E	0.97	547	1.40	180	340	014
119	0	48°28.93952'N	9°15.75765'E	1.60	547	2.17	180	0	014
120	11	48°28.94857'N	9°15.76194'E	1.28	158	2.59	140	0	015
121	1	48°28.93769'N	9°15.75807'E	0.51	252	0.79	0	0	014
122	2	48°28.93887'N	9°15.75858'E	0.29	252	0.71	135	0	014
123	7	48°28.93904'N	9°15.76049'E	1.49	252	3.15	225	0	014
124	11	48°28.93915'N	9°15.76190'E	0.21	244	0.35	0	0	015

Status
In production.

Manufacturer
Vallon GmbH, Im Grund 3, D-72800 Eningen, Germany.
Tel: +49 7121 9855-0. Telex: 729924 valon d.
Fax: +49 7121 83643.

VERIFIED

Vallon ferrous locator EL 1302A1

Description
The Vallon EL 1302A1 ferrous locator is a highly sensitive difference magnetometer used for the detection of bombs, shells, mortar projectiles and other unexploded ordnance. It senses the earth's magnetic field and indicates any disturbance caused in the field by buried objects such as mines, bombs and artillery projectiles. Indications are provided both acoustically and visually. This equipment is used for land-based detection and the locator has an interface for the MC1 microcomputer.

The EL 1302A1 is designed for military use, using a construction involving fibre glass and carbon fibre materials making the detector very light.

During comparison tests held at the Jefferson Proving Ground in the USA the EL 1302A1 was first in overall target detection ratio tests (EL 1302A1 79 per cent, next best 56 per cent).

Specifications
Weight of instrument: approx 4 kg (with batteries)
Weight of instrument in case: approx 11 kg
Dimensions of case: 785 × 285 × 140 mm
Power supply: 6 × 1.5 V monocells type IEC R 14
Detection depth: up to 6 m

Status
In production. In service with various undisclosed countries.

Manufacturer
Vallon GmbH, Im Grund 3, D-72800 Eningen, Germany.
Tel: +49 7121 9855-0. Telex: 729924 valon d.
Fax: +49 7121 83643.

VERIFIED

Vallon ferrous locator EL 1302A1 in use

Vallon ferrous locator EL 1302A1 with GPS

Description
This man-portable system consists of: the Vallon EL 1302A1 ferrous locator, microcomputer MC1, a backpack containing a Global Positioning System (GPS) receiver with antennae and a GPS reference station receiver. A laptop computer with the Vallon EVA software program is also utilised with the system. Operation of the system is simple and only a short training period is required prior to use.

The MC1 microcomputer (carried on the locator) displays in real time the measured nano-Tesla values as a curve. In addition the GPS position is displayed (corrected once every second), along with an integrated GPS window that guides the operator using simple symbols for the walking direction and starting and stopping positions; the operator is thus guided by the MC1. It is not necessary to presurvey a search area.

After a survey the recorded data from the MC1 is downloaded to the laptop computer with the EVA software program. The operator can then evaluate any suspected unexploded ordnance targets. A complete target list and true-to-scale map can be printed out, scaled in metres or feet, for follow-up operations. Both the lists and maps indicate the field and each target by its GPS position in latitude and longitude.

Status
In production.

Manufacturer
Vallon GmbH, Im Grund 3, D-72800 Eningen, Germany.
Tel: +49 7121 9855-0. Telex: 729924 valon d.
Fax: +49 7121 83643.

VERIFIED

Vallon ferrous locator EL 1302A1 with GPS in use

1996

Vallon ferrous locator ET 1340

Description

The Vallon ferrous locator ET 1340 is used for defined area examination of various forms of ordnance as well as buried pipes, cables and covers to a depth up to 15 m. The instrument monitors the earth's magnetic field and signals any disturbance of the field by a buried object by visual or acoustic means.

The ET 1340 locator has an interface for the MC1 microcomputer and for the Sensor Position System (SEPOS).

Specifications

Weight of set: approx 15 kg
Weight in case complete: approx 37 kg
Dimensions of case: 2,040 × 340 × 230 mm
Power supply: 5 × 1.5 V monocells type IEC R 20
Detection depth: up to 15 m

Status

In production. In service with various undisclosed countries.

Manufacturer

Vallon GmbH, Im Grund 3, D-72800 Eningen, Germany.
Tel: +49 7121 9855-0. Telex: 729924 valon d.
Fax: +49 7121 83643.

VERIFIED

Vallon ferrous locator ET 1340 in use

Vallon mine detectors ML 1612B and ML 1620B

Description

The Vallon mine detectors ML 1612B and ML 1620B are capable of detecting all kinds of ferrous and non-ferrous metals buried underground, down to the metallic striker pins of plastic mines. The ML 1620B is recommended for detection tasks involving saltwater or mineralised soils. The detectors are waterproofed up to a depth of 2 m. The computer-optimised design of the search head allows the precise location of small metallic objects without loss of detection speed compared to large ring-shaped coils. An optional stick probe with a solid tip can be used for the detection of objects in undergrowth, hay and other loose material.

The electronics unit has an automatic battery test facility, a volume control and continuously adjustable sensitivity. It adapts itself automatically to the ambient ground conditions so that adjustment and readjustment procedures are not necessary.

Specifications

Weight of set: 3 to 3.5 kg
Shipping weight complete: approx 12 kg
Dimensions of case: 782 × 300 × 142 mm
Carrying bar length: adjustable 0.96 to 1.9 m
Power supply: 4 × 1.5 V monocell type IEC R 20 or rechargeable Ni/Cd accumulators KR 35/62

Status

In production. In service with various undisclosed countries.

Manufacturer

Vallon GmbH, Im Grund 3, D-72800 Eningen, Germany.
Tel: +49 7121 9855-0. Telex: 729924 valon d.
Fax: +49 7121 83643.

VERIFIED

Vallon mine detector ML 1612B in use

Vallon Multisensor Vehicle (MSV 5)

Description

The Vallon MSV 5 was developed for the detection of unexploded objects (UXO) over large areas and can be used in lakes and rivers. The vehicle offers a combined arrangement of ferrous locators operating with an MC1 microcomputer network.

The vehicle carries standard Vallon ferrous detectors which are moved on the MSV 5 special trailer; other special vehicles carry the locators on or underwater. Each locator can be removed from the array and used for hand-carried operations. Each ferrous locator is connected to one MC1 microcomputer. Up to 14 MC1s can be connected in a network and operate in a master-slave operation in which only the master is served. During detection operations the operator can examine each MC1 display to monitor the data. Positional data is determined using SEPOS or GPS in the MC1. After a search operation the complete memory of all the MC1s is transferred to a personal computer and evaluated using the EVA software package.

Status

In production.

Manufacturer

Vallon GmbH, Im Grund 3, D-72800 Eningen, Germany.
Tel: +49 7121 9855-0. Telex: 729924 valon d.
Fax: +49 7121 83643.

VERIFIED

A Vallon MSV 5 in use

Vallon underwater mine detector MW 1630/MW 1630B

Description

The Vallon underwater mine detector MW 1630 was designed for use by divers, is watertight up to a depth of 60 m and is balanced for use underwater. The electronics were designed specifically for underwater use and provide a high level of detection sensitivity under all environmental conditions (for example, fresh or saltwater, mineralised terrain and so on). Standard equipment includes accessories required for detection work on land, along river banks and in shallow water.

For detection tasks in fast-flowing rivers the MW 1630B is available. The MW 1630B has been in service with the German Army since 1994.

Status

In production. In service with the German Army and various undisclosed countries.

Manufacturer

Vallon GmbH, Im Grund 3, D-72800 Eningen, Germany.
Tel: +49 7121 9855-0. Telex: 729924 valon d.
Fax: +49 7121 83643.

VERIFIED

Vallon underwater mine detector MW 1630 in use

FÖRSTER search instrument FEREX 4.021

Description

The FÖRSTER FEREX 4.021 search instrument was developed to replace the earlier 4.015 and 4.016 detectors. It may be regarded as a search instrument for ferro-magnetic objects, for the detection and exact location of unexploded bombs, pipes, cables and similar objects. Depending on the size and position of these objects, they can be detected to a depth of 6 m.

The FEREX 4.021 meets military standards of construction and operation and uses lightweight and splashproof components. The probe is watertight to a depth of 100 m. The instrument uses 1.5 V standard cells for the power supply and it also has built-in functional test and auto-zero compensation.

The FEREX 4.021 has three basic components: control unit, power supply and probe. Depending on the application, three versions of the FEREX 4.021 can be supplied by using the basic components with additional accessories. These are:

FEREX K. This is the complete equipment for land, underwater and borehole applications

FEREX L. For searching, locating and measuring on land

FEREX W. For underwater and borehole use.

The complete FEREX K can be supplied packed in a shockproof aluminium transit case. The FEREX L can be packed into a special rucksack, as can the FEREX W.

Specifications

Weight of L version complete: 5.5 kg
Weight of basic components: 3.8 kg
Dimensions of transit case: 1.12 × 0.175 × 0.31 m
Operating voltage: 6-12 V
Batteries: 6 monocells each 1.5 V (IEC R 20)
Sensitivity range: 0.3-10,000 nT (8 steps)
Operating temperature range: −30 to +65°C
Probe underwater limit: 100 m

Status

In production. In service with the US Navy, British Army, Canadian, Danish and other NATO armed forces, the Australian and Singapore armies and many other undisclosed countries.

Manufacturer

Institut Dr Förster GmbH and Co KG, In Laisen 70, PO Box 1564, D-72705 Reutlingen, Germany.
Tel: +49 7121 140489. Telex: 729 781 ifr d.
Fax: +49 7121 140404.

UPDATED

FEREX 4.021 Model L in use

FÖRSTER detection gradient magnetometer FEREX 2000 SL

Description

Based on FEREX (MK26) sensor technology, it is approved by the US Army, British and German authorities for the environmental clean-up of ordnance and explosive waste (OEW). The FÖRSTER detection gradient magnetometer FEREX 2000 SL can be used for all conventional search operations on land, in boreholes and underwater. The instrument works with the MAGNETO microprocessor-aided magnetic measurement data collection and personal computer evaluation system (see following entry). Various types of FEREX probe and probe arrangements are available for both hand-held and vehicle-mounted search systems.

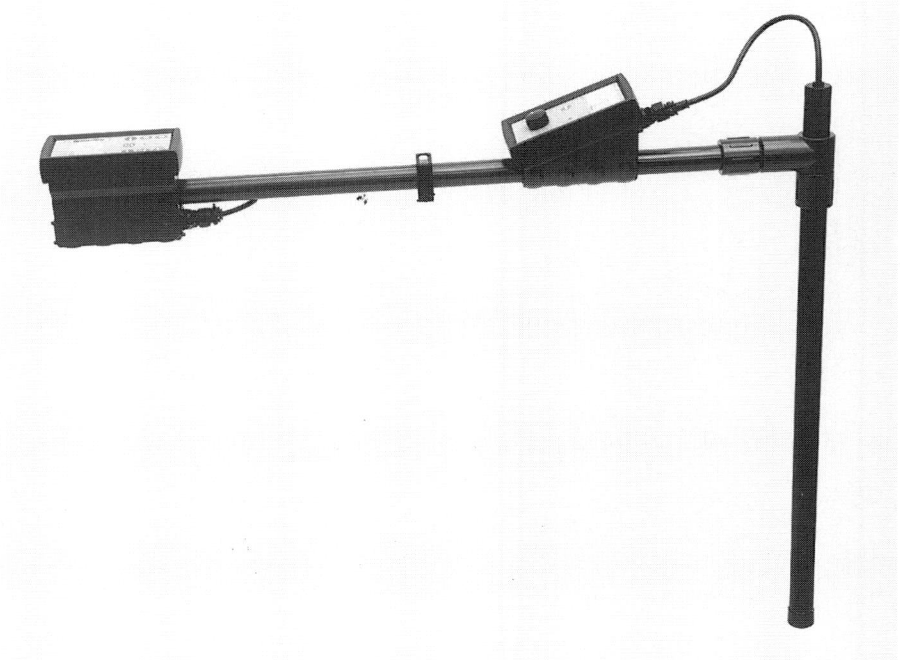

Detection gradient magnetometer FEREX 2000 SL

1996

The FEREX 2000 SL is light and easily assembled and has a conventional and proven meter display. It is claimed to have a higher search sensitivity by using larger sensor spacing. There is an additional digital/analogue multifunction display for borehole and underwater searches; the instrument is fully waterproofed. The power supply is provided by four standard cells. A light case is provided for transport and storage.

Status
In production.

Manufacturer
Institut Dr Förster GmbH and Co KG, In Laisen 70, PO Box 1564, D-72705 Reutlingen, Germany.
Tel: +49 7121 140489. Telex: 729 781 ifr d.
Fax: +49 7121 140404.

VERIFIED

FÖRSTER FEREX MAGNETO data acquisition system

Description
The FÖRSTER FEREX MAGNETO is a data collection and evaluation system for magnetometer scanning with the FEREX 4.021 search instrument. The system consists of a portable data logger with high resolution and storage capacity for track scanning of areas up to four hectares in one measurement. After the transfer of collected data to a personal computer via a standard interface, interactive Windows 3. X software is used for a true-to-scale presentation of the data either by colour-coded magnetic maps, ISO lines, a three-dimensional display, or by a display of individual tracks. The final results are an object list and true-to-scale object maps.

Status
In production.

Manufacturer
Institut Dr Förster GmbH and Co KG, In Laisen 70, PO Box 1564, D-72705 Reutlingen, Germany.
Tel: +49 7121 140489. Telex: 729 781 ifr d.
Fax: +49 7121 140404.

VERIFIED

FEREX MAGNETO data acquisition system in use

FÖRSTER mine detector METEX 4.125

Description
The FÖRSTER mine detector METEX 4.125 has replaced the earlier METEX 4.122; the latest model is the METEX 4.125.04. It may be used to detect ferrous and non-ferrous metal objects but its primary use is in the detection of mines underground and in shallow water. Metallic anti-tank mines can be detected to a depth of approximately 700 mm and it is also possible to detect plastic mines with a small metal content.

The METEX 4.125 is constructed to military standards of design and operation using light and splashproof components; the search coil may be used completely underwater. Standard 1.5 V batteries are used for the power supply and the instrument has built-in functional and battery test features.

The METEX 4.125 has three basic components, a search coil with telescopic handle; a control unit with power supply; and headphones. The complete system can be supplied in an aluminium transit case which also contains a carrying bag with strap, test piece, handgrip, armrest and rucksack.

For searches under difficult environmental conditions the METEX 4.125 may be equipped with a truncheon probe approximately 400 to 1,100 mm in length.

METEX 4.125.04 in use ***1997***

Specifications
Weight of components: 3.5 kg
Weight with transit case: 15 kg
Dimensions of transit case: 0.74 × 0.4 × 0.16 m
Length of search coil with handle:
(minimum) 0.94 m
(max) 2 m
Search coil diameter: 0.26 m
Truncheon probe diameter: approx 30 mm
Truncheon probe length: 400 to 1,100 mm
Batteries: 6 × 1.5 V (IEC R14)
Battery life: 80 h approx

Status
In production. In service with undisclosed countries.

Manufacturer
Institut Dr Förster GmbH and Co KG, In Laisen 70, PO Box 1564, D-72705 Reutlingen, Germany.
Tel: +49 7121 140489. Telex: 729 781 ifr d.
Fax: +49 7121 140404.

UPDATED

FÖRSTER mine detector MINEX 2FD 4.400

Description
The FÖRSTER MINEX 2FD 4.400 mine detector is a product-improved version of the METEX 4.125 (see previous entry) and includes the latest technology in electronics, material and design. It is claimed to be the first metal detector to incorporate two frequencies operating simultaneously. It is also claimed that the 2FD 4.400 is the first instrument of its kind to use a multilayer printed circuit board 'coil system'.

Together with its microprocessor-controlled phase selection, the 2FD 4.400 can achieve maximum sensitivity to all metals and optimal ground suppression at the same time. The instrument is also microprocessor-controlled in operation for ground exclusion balance, sensitivity adjustment and for displayed detection and fault signals; the instrument has functional and battery test features.

The instrument is constructed to full military standards and uses strong waterproof components; it may be used under any climatic conditions and under both fresh and seawater.

Power is provided by three 1.5 V IEC R20 alkaline batteries. The complete equipment can be supplied in an aluminium transit case which also contains a carrying harness, test piece, handgrip and armrest and a rucksack.

Typical anti-personnel mine detection distances possible with the 2FD 4.400 are as follows:

Type 72 - 80 mm
M14 - 110 mm
SB-33 - 120 mm
PMN-6 - 220 mm
VS-50 - 280 mm.

Specifications

Search head diameter: approx 257 mm
Probe length:
(maximum) approx 1.85 m
(minimum) approx 0.8 m
Operating time at ambient 20°C: 30 h
Operating temperature: −32 to +60°C
Storage temperature: −55 to +75°C

Status

In production.

Manufacturer

Institut Dr Förster GmbH and Co KG, In Laisen 70, PO Box 1564, D-72705 Reutlingen, Germany.
Tel: +49 7121 140489. Telex: 729 781 ifr d.
Fax: +49 7121 140404.

UPDATED

MINEX 2FD 4.400 in use

FÖRSTER metal detector MINEX 2000 SL

Description

The FÖRSTER metal detector MINEX 2000 SL is the successor to the earlier MINEX 2FD and is based on FEREX 4.021 sensor technology. It involves a significant weight reduction due to a redesign of the probe and the introduction of a carbon fibre telescopic handle. The instrument employs a microprocessor-controlled two-frequency search system which provides good search sensitivity, selectivity and pinpointing. The search system is not affected by magnetic soils, fresh or saltwater and can be used in any climate or environment. The MINEX 2000 SL may be employed in computer-controlled multiprobe and vehicle-mounted systems.

Probes available for use with the MINEX 2000 SL include the standard probe, a three-point ander array, borehole and underwater probe and a deep search probe.

Weight of the MINEX 2000 SL is 4.5 kg.

Status

In production.

Manufacturer

Institut Dr Förster GmbH and Co KG, In Laisen 70, PO Box 1564, D-72705 Reutlingen, Germany.
Tel: +49 7121 140489. Telex: 729 781 ifr d.
Fax: +49 7121 140404.

UPDATED

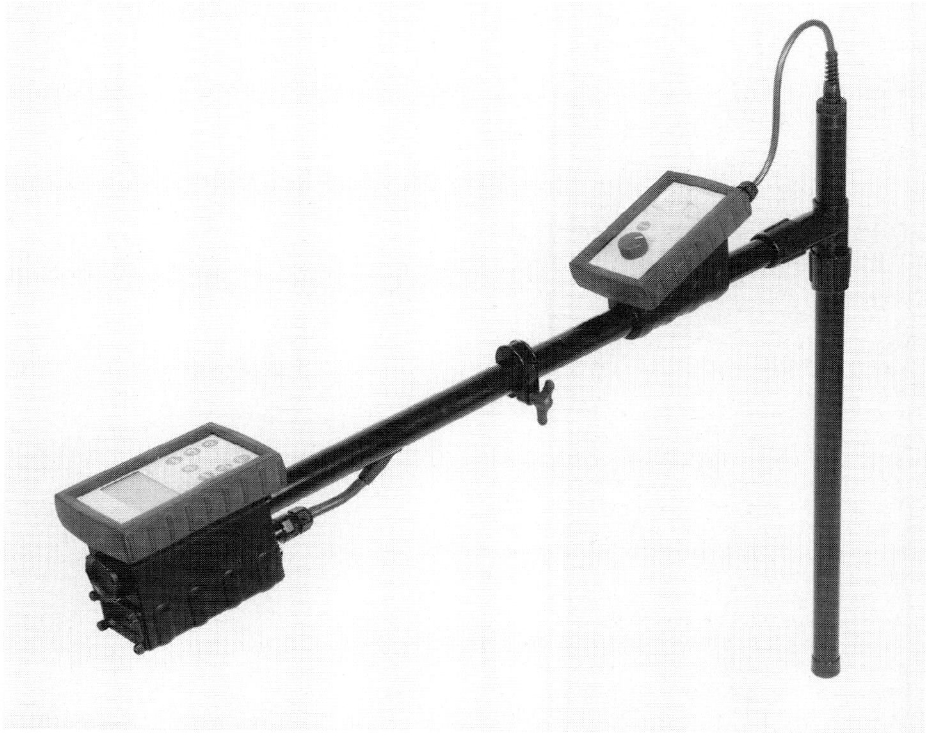

FÖRSTER metal detector MINEX 2000 SL **1997**

FÖRSTER metal detector MINEX 2000 P

Description

The FÖRSTER MINEX 2000 P metal detector is the first pulse induction metal detector produced by FÖRSTER and was designed as an economic instrument intended to fulfil a wide range of search tasks in all climates and environments, including fresh and saltwater.

The MINEX 2000 P has a lightweight probe and a carbon fibre telescopic handle with an embedded probe cable. It can employ state-of-the-art pulse detection techniques during both static and dynamic search operations; there is no interference between two or more detectors operating in close proximity. The instrument is easy to adjust and there is automatic monitoring of battery condition, including an indication of the need for battery replacement.

Status

In production.

Manufacturer

Institut Dr Förster GmbH and Co KG, In Laisen 70, PO Box 1564, D-72705 Reutlingen, Germany.
Tel: +49 7121 140489. Telex: 729 781 ifr d.
Fax: +49 7121 140404.

VERIFIED

FÖRSTER Vehicle Mounted Multisensor Systems

Description

FÖRSTER have developed several types of vehicle mounted multisensor systems suitable for surveying large areas of militarily contaminated terrain where hazardous ordnance might be present. Land and underwater systems are under development, all employing multiple FEREX standard, deep search and precise location gradiometer sensors as well as MINEX sensor arrays. Involved are the following systems, all of which can be equipped and built according to customer requirements.

MULTI-CAT land system

Constructed using totally non-metallic materials, the MULTI-CAT land system is a towed FEREX sensor platform carried on special wheels capable of operations over terrain which a wheeled towing vehicle can traverse. Sensors covering a width of 3 m are totally protected inside the body on the standard version. An integrated GPS system serves for navigation and mapping at speeds up to 2.5 m/s. A similar model based on tracks is projected.

MULTI-CAT underwater system

The MULTI-CAT underwater system is generally similar to the land system but it can be towed by a boat to either float, or ballasted to travel on the sub-surface bottom.

Artist's impression of a projected tracked version of the FÖRSTER MULTI-CAT Land System vehicle mounted multisensor system ***1997***

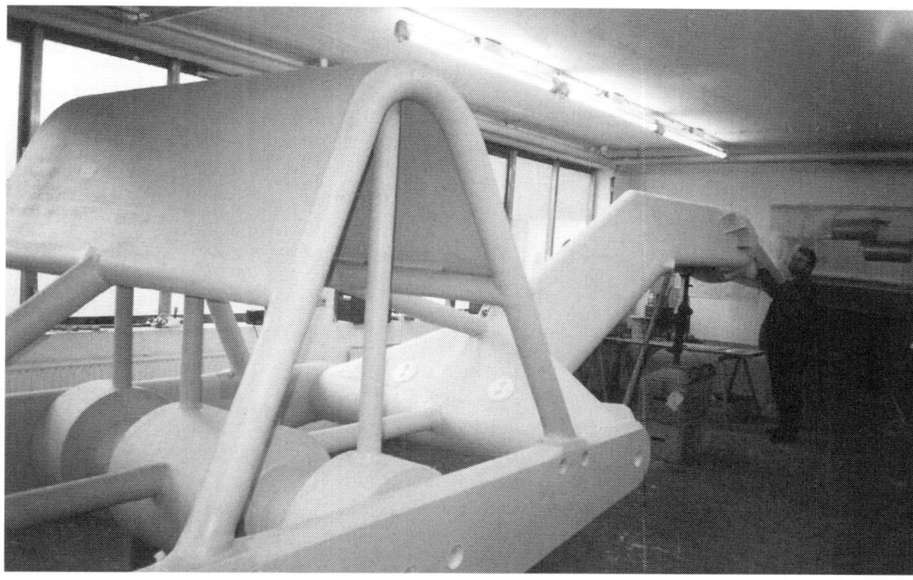

A FÖRSTER FLEX-CAT vehicle mounted multisensor system under construction ***1997***

FLEX-CAT

FLEX-CAT is a totally non-metallic sensor array carried on tracks capable of smooth progress over uneven and difficult terrain, travelling with a ground pressure of less than 10 g/cm^2.

Status

Developments.

Manufacturer

Institut Dr Förster GmbH and Co KG, In Laisen 70, PO Box 1564, D-72705 Reutlingen, Germany.
Tel: +49 7121 140489. Telex: 729 781 ifr d.
Fax: +49 7121 140434.

NEW ENTRY

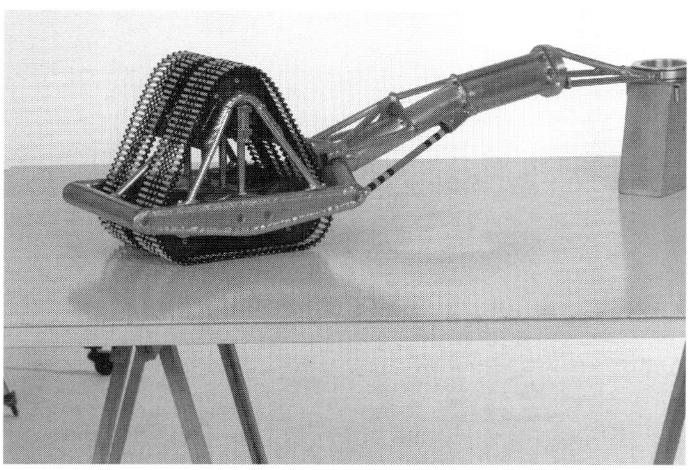

A 1:10 model of a FÖRSTER FLEX-CAT vehicle mounted multisensor system ***1997***

A FÖRSTER MULTI-CAT Land System vehicle mounted multisensor system ***1997***

IRAQ

Mansour-7 mine detector

Description

The Mansour-7 mine detector was designed to detect buried metallic mines and the metal components of plastic-bodied mines at depths between 50 and 400 mm below the ground surface. No further details are available.

Status

In service with the Iraqi armed forces.

Manufacturer

State factories.

VERIFIED

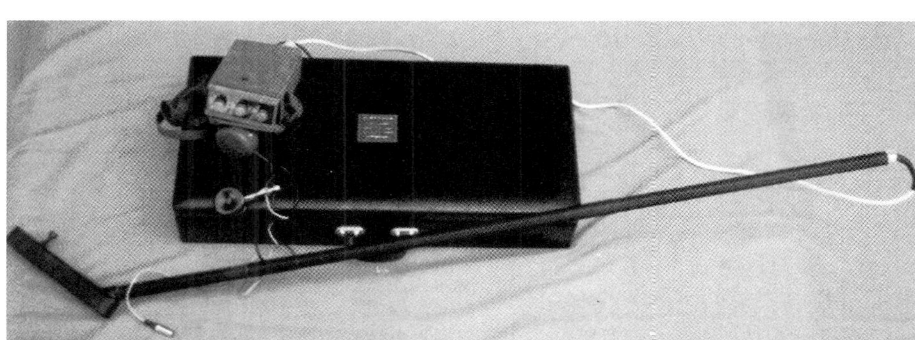

Mansour-7 mine detector showing main components

ISRAEL

BETA BMD-34 metallic mine detector

Description

The BETA BMD-34 metallic solid-state mine detector is a sensitive, sophisticated, miniature detector which can be used to detect metallic anti-tank or anti-personnel mines in both magnetic or non-magnetic soils. The instrument detects any conductive material, ferrous or non-ferrous, in any type of soil.

The instrument comprises a search head mounted on the end of a telescopic handle, a belt-carried electronic control unit, and a headset for the audible output signal. As an optional extra, a visual display can be clamped on the telescopic handle. The visual display consists of a 40 mm diameter meter with a central zero and a self-illuminated scale.

The controls consist of an on/off control, search head balance knob and the battery check toggle switch. The mine detector is human-engineered for the minimum of adjustments during operation and features

an automatic compensation for variables such as search head characteristics, amount of magnetic material in the soil and temperature variations.

The mine detector has been designed for military use under severe environmental conditions. It is immersion-proof in 2 m of water, shock- and vibration-proof and has electromagnetic interference protection.

Specifications

Weight of equipment ready for use: 2.9 kg
Detection capability: metallic anti-tank mine 300 mm deep

BETA BMD-34 metallic mine detector in use

Operating temperature range: −15 to +72°C

Status

In production since 1976. In service with undisclosed countries.

Manufacturer

BETA Engineering and Development Limited, PO Box 98, IL-84100 Beer Sheva, Israel.
Tel: +972 7 077 34341. Telex: 5388 beta il.

VERIFIED

KOREA, SOUTH

NMD-9 and GDS-17 mine detection sets

Description

Developed by the Agency for Defence Development of Korea, the NMD-9 is a rugged, solid-state portable device capable of detecting metallic and non-metallic mines. The GDS-17 is used for the detection of metal objects only.

The search head of the NMD-9 generates and radiates a continuous wave RF signal and receives and demodulates the resultant signal for aural presentation in a headset. The GDS-17 uses a pulse induction detection system.

The search head is encased in a rubber cover and mounted on the end of a long telescopic carrying handle which also mounts the control box. The control box contains the system electronics on a single board and the battery which has an operating life of 15 hours (NMD-9) in continuous use (GDS-17, 28 hours). The control box also has a 'search' switch used to locate the general presence of a mine when a 'point' mode with three varying degrees of sensitivity may be selected.

Specifications

NMD-9
Weight:
(in use) 4 kg
(in carrying case) 10.8 kg

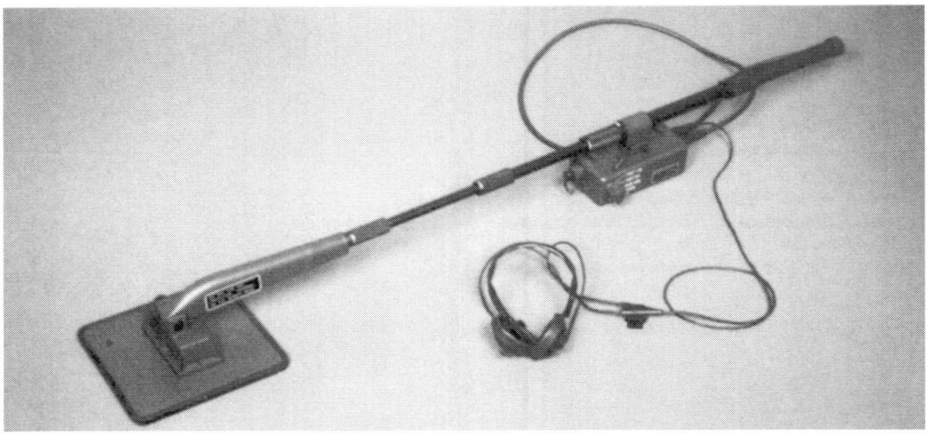

NMD-9 mine detection set

Length of telescopic handle: 540 to 1,460 mm
Battery voltage:
(NMD-9) +15 V/−15 V
(GDS-17) +12 V
Dimensions of carrying case: 600 × 380 × 200 mm

Status

In production.

Manufacturer

Gold Star Electric Company Limited, 27-2, Yeoeuido-dong, Youngdungpo-ku, Seoul 150, South Korea.
Tel: +82 2 783 9311. Telex: 28295 gswave k.
Fax: +82 2 784 3549.

UPDATED

PRS-17K portable mine detector

Description

The PRS-17K portable mine detector utilises pulse induction techniques with an oscillation frequency of 100 Hertz. Using a 220 mm diameter search head assembly on the end of long or short carrying handles, the PRS-17K can detect metallic anti-personnel mines, such as the M16, at ranges of 500 to 600 mm. Metallic anti-tank mines such as the M19 can be detected at distances from 700 to 800 mm.

A complete PRS-17K detector consists of: a search head assembly, telescopic long handle and a fixed length (300 mm) short handle, control box, signal processing module and the batteries. The complete detector can be carried in a specially fitted carrying case together with an arctic pouch for cold conditions. Eight 1.5 V LR6A-3 alkali batteries are contained within the control box and provide power for up to 40 hours of continuous operation. The control box, which also contains the signal processing module, features an automatic signal control function and has a connector for external batteries.

Specifications

Weight: (total, complete) 6 kg
Length of telescopic handle: 560 to 1,460 mm
Length of short handle: 300 mm
Battery voltage: (total) 12 V
Dimensions of carrying case: 560 × 350 × 150 mm
Operating temperature range: −32 to +52°C

Status

In production.

Manufacturer

LG Precision Company Limited, Kookmin Life Insurance B/D 168, Kongdok-dong, Mapo-gu, Seoul 121-020, South Korea.
Tel: +82 2 3271-5456-9. Telex: GSRADAR K22838.
Fax: +82 2 3271-5553.

NEW ENTRY

SOUTH AFRICA

Reutech MIDAS Pulse Induction Mine Detector (PIMD)

Description

The MIDAS Pulse Induction Mine Detector (PIMD) is a self-contained, lightweight equipment with self-nulling and confidence-tone features. It uses a three-part modular construction to minimise maintenance and has three selective sensitivity levels (high, medium and low) to ensure early and accurate target acquisition. MIDAS can work continuously for up to 28 hours powered by a 16.5 V alkaline battery (up to 11 hours with standard cells); an alarm is given once the battery has discharged.

The foam-filled Kevlar search head is rectangular or round. The dimensions of the rectangular head are 633 × 288 mm and a diameter of 305 mm for the round head; these provide instantaneous search areas of 182,304 mm^2 and 73,061 mm^2 respectively. The heads can be made to a size or shape to suit customer requirements.

The head is mounted on a telescopic handle, made from wound phenolic impregnated glass fibre cloth, that can be extended to a maximum length of 2.11 m; fully closed the length is 1.64 m. The most commonly used operating length is 2 m. The system's electronic unit and battery holder is at the top end of the handle, secured in such a way that it counterbalances the weight of the search head and handle.

Weight of the complete equipment, with large head and battery, is 4.3 kg. Operating temperature range is from −25 to +65°C, at an operational humidity of up to 95 per cent.

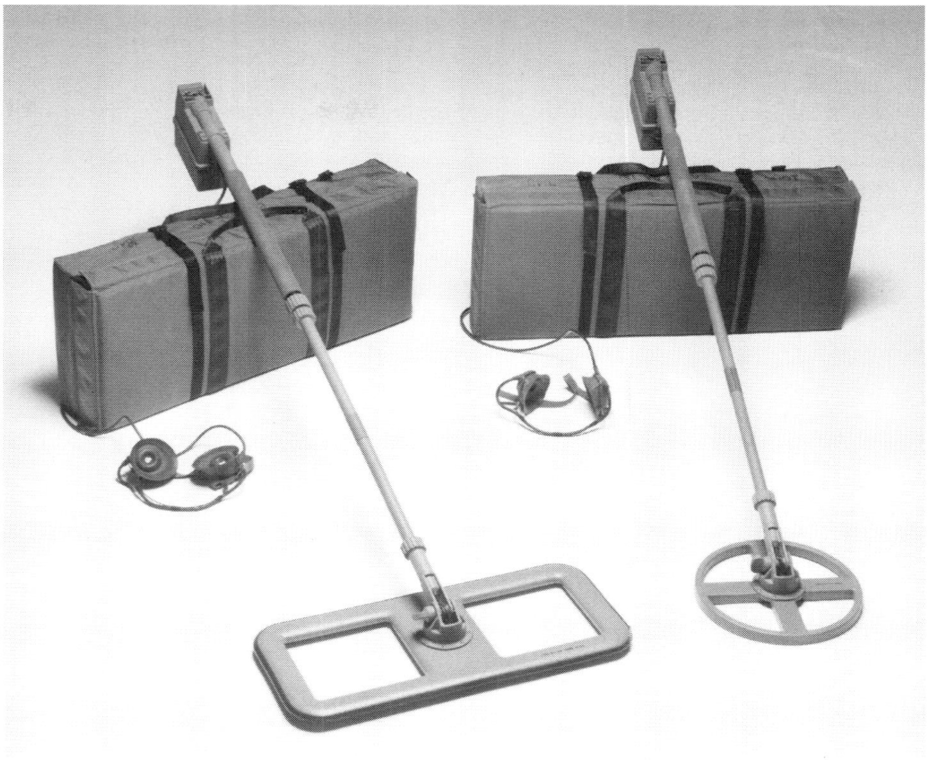

MIDAS pulse induction mine detectors

Specifications

Weight complete: (large head) 4.3 kg
Large head dimensions: 633 × 288 mm
Round head diameter: 305 mm
Length overall, extended: 2.11 m
Length overall, closed: 1.64 m
Max reach from centre of gravity: 1.76 m
Min reach from centre of gravity: 1.22 m
Carry case dimensions: 840 × 360 × 180 mm
Operational temperature range: −25 to +55°C

Status

In production. In service with the SANDF.

Manufacturer

Reutech Defence Industries, 10 Pineside Road, New Germany, Kwa Zulu Natal, 3620, South Africa.
Tel:+27 31 719 5711. Fax: +27 31 723569.

UPDATED

Reutech VMMD 2000 Vehicle Mounted Mine Detector

Description

The VMMD 2000 vehicle mounted mine detector was developed from the earlier VMMD 1000. It is typically deployed with the system detector heads located one each side of the Meerkat mine detection vehicle, part of the mine-protected Dorbyl Chubby mine detection and clearing system (qv), although other vehicles could be involved.

The VMMD 2000 employs similar pulse induction techniques to those used with the MIDAS pulse induction mine detector (see previous entry). The system has two search heads, each measuring 1,500 × 600 mm, thus producing a total sweep width of 3 m. In addition to a control unit and its mounting rack, the Meerkat driver/operator is provided with a headset providing variable volume audible warning signals.

Using the control unit the operator can select several operating modes, including a dual coil operation with variable sensitivity (high, medium or low) according to the selected target size discrimination required. As a typical example a TM-57 anti-tank mine can be detected at a distance of 600 mm. There is also a single coil operation mode with no size discrimination which permits small target recognition. The latter mode also has variable sensitivities, although the audible tone differs from that used with dual coil operation to indicate the mode selected. A pinpoint mode for object detection at low speeds is another mode option, with simultaneous left and right operation, again with variable sensitivities.

The VMMD 2000 has an automatic self-nulling feature which cancels out the metallic content of the vehicle body, thereby allowing the system to respond only to changes in the metallic environment beneath the detector heads. This allows the system to be deployed on a variety of vehicle types. A single test button confirms the system's operational state.

The VMMD 2000 operates from the host vehicle's 12 or 24 V DC electrical supply. The operational temperature range is from −20 to +55°C.

VMMD 2000 vehicle mounted mine detector heads deployed between the axles and each side of a Meerkat mine-protected mine detection vehicle (T J Gander) ***1997***

Status

In production. In service with the SANDF.

Manufacturer

Reutech Defence ndustries, 10 Pineside Road, New Germany, Kwa Zulu Natal, 3620, South Africa.
Tel:+27 31 7195711. Fax: +27 31 723569.

NEW ENTRY

Mechem Explosive and Drug Detection System (MEDDS)

Description

The Mechem Explosive and Drug Detection System (MEDDS) is a system for the detection of explosive and drug vapours which can also be applied to mine detection. It combines the chemical concentration of vapours with the acute sense of smell of specially trained dogs, offering a means of detection more sensitive than any known laboratory method. The system has been in use in Southern Africa for over 10 years and has been employed operationally to detect unexploded ordnance in Mozambique and Croatia.

MEDDS is employed in three stages. The first is in the collection of vapours. For this stage air is drawn through a specially prepared sampling tube to concentrate the volatile vapours of explosives. When clearing mined areas the vapour concentrating tubes are suspended over the front of a mine protected vehicle, such as a Casspir, or carried by an operator on foot with the suction equipment in a backpack. By changing the tubes at fixed intervals, determined at times by the Global Positioning System (GPS) or a similar location system, each sector of a minefield or route may be checked. Each sampling tube is labelled and recorded.

In the second stage of MEDDS the sampling tubes are checked by specially trained dogs at a location remote from the sampling site(s). Groups of tubes are placed on a row of stands and the dogs then sniff along the rows of samples. The dogs indicate positive samples by lying down. This operation can be carried out in ideal locations well away from the point of source and free of distractions for the dogs. Once a tray has been indicated the individual tubes are then separated on to a fresh row of stands for the same, or another dog to repeat the sniffing process. This enables a single location, or locations, to be isolated for the third stage. If a dog ignores a sample tube the corresponding area or section of route is deemed to be clean.

The third stage is the actual physical searching of the area indicated by the detected sample tube. This reduces the subsequent search task to a fraction of the work which would be otherwise necessary. The actual search may be conducted by dogs or sappers.

There are numerous refinements of the MEDDS system. One is that more than one sample tube may be taken for an area or length of route for independent testing by dogs at different locations, both for system qualification and assurance or training.

MEDDS has proved to be virtually 100 per cent successful and safe during repeated operations. All items containing explosives will be indicated, including small arms ammunition and unexploded ordnance, as well as mines. During operations in Mozambique during 1994, between 15 and 20 km of overgrown road were swept per day for explosive debris with a quality assurance level better than 99.6 per cent obtained via an independent survey.

As the full title of MEDDS indicates, the system can also be employed to detect drugs and other specific substances.

Status

In widespread use throughout South Africa, some neighbouring countries, and Croatia.

System developer

Mechem Consultants, PO Box 912454, Silverton 0127, South Africa.
Tel:+27 12 803 7290. Telex: 321053 SA.
Fax: +27 12 803 7189.

UPDATED

A Casspir mine protected vehicle showing the sniffer sampling heads used with the initial stage of MEDDS hanging over the front of the vehicle; the device at the rear of the vehicle is a remotely controlled mineclearing rake and is not part of MEDDS ***1996***

Close-up of a pair of MEDDS sniffer sampling heads ***1996***

RUSSIAN FEDERATION AND ASSOCIATED STATES (CIS)

IMP and IMP-2 mine detectors

Description

The IMP and IMP-2 can detect metallic mines, or plastic mines which contain some metallic components.

The IMP mine detector consists of: the detector head assembly; a four-piece search handle; tuning box which is combined with the battery pack; power supply which consists of four 1.6 V batteries; and the headphones.

When dismantled, all the components are carried in a lightweight rectangular metal box. The detector head assembly consists of two transmitting antennas and a receiving antenna, all encased in a bakelite cylinder (417 mm long × 38 mm diameter). The handle consists of four aluminium sections which, when joined together, are 1.58 m long. One or more of these sections can be removed to reduce the length of the equipment for use in confined areas. The tuning box, or amplifier, consists of five transistors and associated components which are mounted in a lightweight metal box, with the tone regulator on top of the tuning box.

An advantage of the IMP over earlier RFAS mine detectors is that the equipment can be used underwater to a maximum depth of 1 m.

The IMP-2 is similar overall to the IMP but is much lighter and may be powered by several types of battery or accumulator. It also has a longer telescopic probe handle, a rectangular head, and is generally more sensitive.

A commercial model based on the IMP-2 and known as the PR-507 is available, with the ability to utilise various types of power source and batteries.

Specifications

IMP-2
Weight of detector head: 2 kg
Probe handle length: 830-1,500 mm
Detector head dimensions: 200 × 200 × 10 mm
Processor unit dimensions: 195 × 130 × 45 mm

IMP mine detector in use on construction site

Detection ranges
Copper coin (1 g): 80-100 mm
Gold wedding ring: 200 mm
Steel nail (70 × 3 mm): 120 mm
Steel needle: 90-100 mm
7.62 mm pistol: 400 mm
Steel disc (300 mm diameter): 400-600 mm

Status
In service with the RFAS and some other nations, including Bangladesh (60 IMP).

Manufacturer
KB Proekt, Tomsk 51, RFAS.
Tel: +7 44 40 65.

Marketing Agency
V/O Mashproborintorg 6, 2nd Spasonalivkovky Per, 11909 Moscow, Russia.
Tel: +7 095 230 1652/238 0687. Telex: 411236.
Fax: +7 095 230 2126.

VERIFIED

PR-507 metal detector

Description
The portable PR-507 metal detector operates on the induction principle and was designed to detect objects with a minimal metal content. The detector head is square in outline and located on the end of a telescopic carrying handle which can be varied in length from 0.83 to 1.5 m. Other components include a processing unit, headphones and a power supply unit. Power is supplied by a 9 V battery although various other lower voltage battery combinations can be utilised. The detector can be used over snow and water (both fresh or salt) up to 1 m deep.

Specifications
Detector head dimensions: 200 × 200 ×10 mm
Probe handle length: 830-1,500 mm
Processor unit dimensions: 195 × 130 × 45 mm
Operating temperature range: −50 to +50°C

Detection ranges
Copper coin (1 g): 80-100 mm
Gold wedding ring: 200 mm
Steel nail (70 × 3 mm): 120 mm
Steel needle: 90-100 mm
7.62 mm pistol: 400 mm
Steel disc (300 mm diameter): 400-600 mm

Status
In production.

Manufacturer
Tomsk Measuring Apparatus Factory, 51 Voikova Street, 634042 Tomsk, Russian Federation.
Telex: +7 3822 128159 MOST.

VERIFIED

DIM Vehicle-mounted mine detector

Description
The DIM vehicle-mounted mine detector consists of a UAZ-69 or UAZ-469 (4 × 4) light vehicle modified to carry a frame on which is mounted a non-magnetic sensing head. This is supported by a frame attached to the front bumper. When travelling, the equipment is swung upwards through 180° and rests on top of the cab. When required for use, the equipment is swung forward and rests on two rubber-tyred wheels which are in contact with the surface of the road. The sensing head is mounted in front of the two wheels and is provided with three sets of dual wheels.

The DIM system will detect mines to a maximum depth of 250 mm, or when fording, to a depth of 700 mm with a sweep width of 2.2 m. The vehicle is normally driven at a maximum speed of 10 km/h during mine detection operations. Once a mine or metallic object has been detected an audio alarm sounds and the vehicle stops automatically. Once the vehicle comes to a halt, its operator can adjust the search coils to pinpoint the exact location of the metal object. No provision is made for remote control of the vehicle.

As the equipment lacks mobility, it is normally used only for clearing roads and airfields. The system was mounted on T-62 MBTs in Afghanistan.

Status
In service with former Warsaw Pact members and some countries in the Middle East.

Manufacturer
State factories.

VERIFIED

DIM mine detector mounted on front of UAZ-69 (4 × 4) vehicle (Christopher F Foss) ***1996***

DIM mine detector sensor array ***1996***

THAILAND

Metallic mine detector

Description
Few details are available regarding the metallic mine detector developed for the Royal Thai Army. The detector was designed to detect mines with some metallic content up to a maximum detection depth of 200 mm. Designed for one-man operation, the detector weighs 4.5 kg without its storage and carrying case. Power is provided by 10 AA batteries which provide an operating life of 20 hours. It is assumed that the detector operates along conventional lines.

Status
In service with the Royal Thai Army.

Manufacturer
Office of the Army Research Development and Coordination, Royal Thai Army, Rajsima Road, Bangkok 10300, Thailand.
Tel: +66 2 824161.

NEW ENTRY

Royal Thai Army metallic mine detector ***1997***

UNITED KINGDOM

Mine detector No 4C

Description

The United Scientific Instruments mine detector No 4C has been the standard mine detector of the British Army since 1968. Earlier models were the No 4 and No 4A. It comprises a search head mounted on a telescopic handle, an amplifier unit, headphones and connecting cables. For transport, the equipment is disassembled and carried in a wooden transport case which also contains the spare battery, test boxes, extension cables and other accessories.

The search head contains two mutually reacting inductance coils. Inductance is adjustable by means of a dust core trimmer assembly and when adjusted the mutual inductance of the two coils is in balance. When any metal component is brought within the field of the coils, the balance is distorted, resulting in an oscillating difference potential. This is then amplified by the amplifier unit and fed to the earphone as a distinctly audible signal.

There are two modes of operation, normal and in ferrite-bearing soil (pavé). The former is used for detecting metallic mines in normal soil while the latter is used for detecting metallic mines in ground (pavé), or if there are likely to be any metallic splinters on the surface. The preferred mode of operation is selected by a selector switch on the amplifier unit.

For use, the telescopic pole is extended, one end strapped to the forearm and the other attached to the search head. The amplifier unit, which also contains the batteries, is clipped to the operator's belt and the cable from the search head is plugged into the unit. The lead of the earphones is connected to the amplifier unit and the equipment is then ready for immediate use.

There are two methods of searching for mines, upright and prone. For the upright method, the telescopic handle is extended and the angle of the search head on the handle is adjusted to enable the operator to sweep the head in front of him from side to side as he moves forward. For the prone position the sweep motion is similar but the operator lies on the ground, the handle is not extended and the head is parallel to the ground.

Specifications

Weight of detector in transit case: 14.4 kg

Transit case
Length: 533 mm
Width: 254 mm
Height: 203 mm

Search head
Weight: 1.8 kg
Length: 285 mm
Width: 185 mm
Height: 108 mm

Telescopic handle
Weight: 1.15 kg
Length:
(collapsed) 380 mm
(fully extended) 1.28 m

Amplifier unit
Weight with battery: 1.8 kg
Length: 216 mm
Width: 108 mm
Height: 108 mm

Battery type: 9 V dry cell PP6 or equivalent
Current consumption: 3.3-3.6 mA
Operational life of battery: 300 h of intermittent use

Detection capacity	Detection depth
British Mk 7 anti-tank mine normal ground	510 mm
British Mk 7 anti-tank mine pavé ground	320 mm

Status

In service with the British Army and several overseas nations.

Manufacturer

United Scientific Instruments plc, Unit 5, Quinn Close, Manor Park, Whitley, Coventry CV3 4LH, UK.
Tel: +44 1203 539299. Telex: 31332.
Fax: +44 1203 539281.

VERIFIED

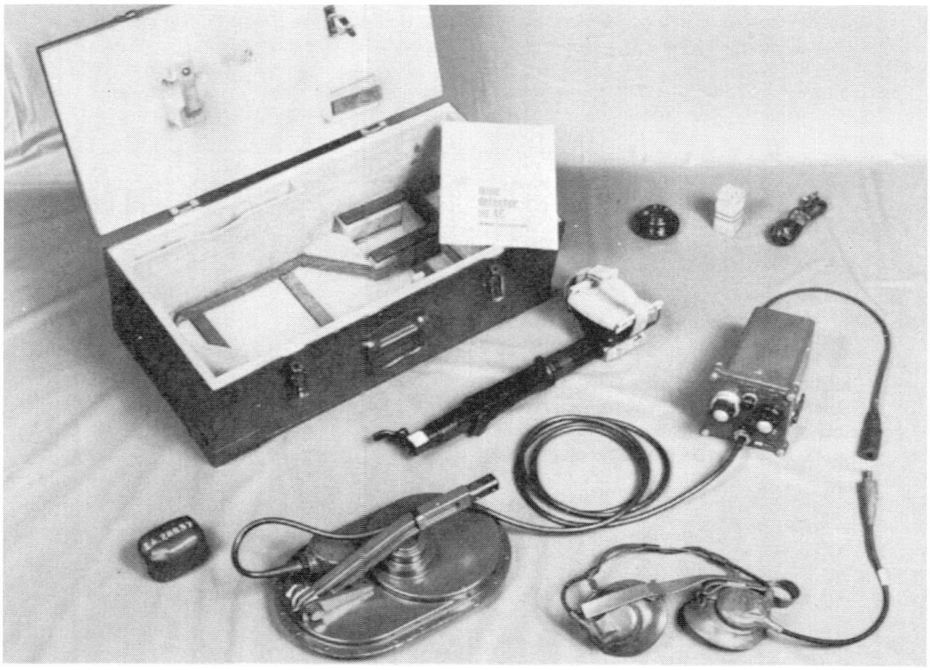

Component parts of mine detector No 4C

MD2000 metallic mine detector with 300 mm loop in use

Guartel MD2000 metallic mine detector

Description

The MD2000 metallic mine detector is a multi-role, manportable instrument for the detection of ferrous and non-ferrous objects. The equipment operates on the pulse induction principle and may be used with probe heads in addition to the normal halo search heads. Upgrades have improved sensitivity, particularly to minimum metal targets. The MD2000 is easy to operate, being completely automatic in operation and requiring no tuning for different ground conditions. The search heads may be operated immersed in water with no adjustments.

MD2000 in use with search probe

The full field pack is supplied equipped with two search probes, a search halo head, loudspeaker and earpiece audio devices and the control unit which houses the batteries packed into a weather-resistant soft carry case. A short pack may be specified for mine detection operations only but lacks the search probes.

The MD2000 may be operated using dry cells or rechargeable batteries. It is NATO codified.

Detection ranges

Item	**Search loop**	**Probe**
7.62 mm cartridge case	400 mm	190 mm
Parkray timer	410 mm	200 mm
Walther pistol	590 mm	300 mm
0.22 rifle	700 mm	430 mm
Pressure pad (large)	1,100 mm	710 mm
Barmine fuze	330 mm	150 mm
Mark 7 a/t mine	1,200 mm	580 mm
PMN a/p mine	570 mm	260 mm

Specifications

	Dimensions	Weight
Field pack complete	1,030 × 400 × 80 mm	6 kg
Electronics less batteries	248 × 172 × 42 mm	1.42 kg
300 mm search loop (handle extended)	1,300 × 318 × 59 mm	0.9 kg
1 m search probe	980 × 42 mm	0.695 kg
400 mm search probe	457 × 42 mm	0.485 kg

Status
In production. In service with the British Army and police forces and with similar organisations worldwide.

Manufacturer
Guartel Limited, 32 St Mary's Road, London W5 5EU, UK.
Tel: +44 181 567 0702. Fax: +44 181 579 2158.

VERIFIED

Guartel MD4 metal detector

Description

The MD4 metal detector was developed in collaboration with the UK Ministry of Defence and was specifically designed for Improvised Ordnance Disposal (IOD) and Explosive Ordnance Disposal (EOD) operations, although it may be used for mine detection and clearance. It is in service with the British Army for the above mentioned applications.

The construction of the MD4 is such that all electronics, controls, detection head and batteries are contained within a hand-held probe. There are only two operating controls, ON/ZERO and OFF. Sensitivity is stated to be high. The operating principle is multisampling pulse induction and detection indication is by a built-in speaker or earphone. The equipment is fully ruggedised, waterproof and ECM compatible.

The MD4 is NATO codified and supplied packed into a shockproof aluminium transit case with a soft carry case, an earpiece attachment and a spare battery cassette.

Specifications
Weight: 2 kg
Length: 1.231 m
Diameter: 39 mm
Battery life: 12-14 h

Status
In production and service with the British Army and other undisclosed organisations.

Manufacturer
Guartel Limited, 32 St Mary's Road, London W5 5EU, UK.
Tel: +44 181 567 0702. Fax: +44 181 579 2158.

VERIFIED

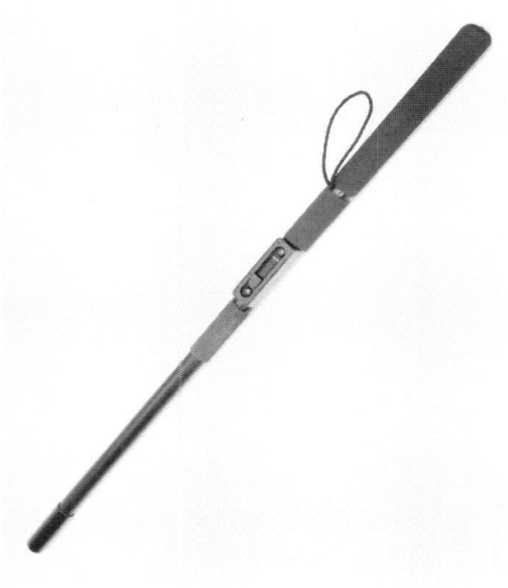

MD4 metal detector

MD8 minimum metal mine detector

Description

MD8 was designed from the outset to address the problem of mines with minimum metal content by employing a high frequency version of pulse induction. The use of two microprocessors in the system permits very precise signal processing. Coupled with a unique head design, it allows the machine to recognise very small targets even in the close proximity of other very much larger objects and virtually eliminates interference from ground effects.

There are two self-diagnostic systems. One is running at all times and monitors the principal activities of the machine to ensure optimum performance. The second may be user-initiated at any time and examines every aspect of the instrument's functions.

A sighting bar incorporated within the search head facilitates target pinpointing to within millimetres.

The physical configuration of the unit is unusual. All elements of the instrument, power supply, electronics search head and so on, have been incorporated into a single unit. The nuisance of interconnecting cables and straps is avoided in this layout. The MD8 can be unpacked and in use in seconds.

The equipment is automatic in use and requires very little training and familiarisation. It is fully ruggedised, waterproof and ECM compatible.

The detector is supplied in composite materials transit case complete with a soft haversack and lightweight earphone.

Specifications
Total operating weight: 2 kg
Max operating length: 1.5 m
Min operating length: 1.05 m
Power source: 2 × 'D' cell
Battery life: 18 h continuous with alkaline cells
Audio: built-in or earpiece

Detection performance
Min metal mines: Type 72 APM 100 mm min at 500 mm from metal ATM; TMA 3/5/4 200 mm min at 500 mm from metal ATM
Large metal target grenade: 500 mm

Status
In production.

Manufacturer
Guartel Ltd, 32 St Mary's Road, London W5 5EU UK.
Tel: +44 181 567 0702. Fax: +44 181 579 2158

VERIFIED

MD8 minimum metal mine detector

White's Electronics military mine detector

Description

White's Electronics (UK) Limited is a subsidiary of a US electronics company and, with the co-operation of several governments and military advisers, has produced a modern concept military mine and bomb detector. The unit is extremely sensitive to both very small objects such as firing pins and large objects such as bombs. This sensitivity remains constant until the batteries are discharged. Mineralised soil and saltwater have little or no effect on sensitivity levels as the unit is designed to automatically compensate for these conditions. The equipment produces virtually no DC magnetic field. Visual and aural indications of metal are provided; the visual element can be on the search head if required.

The aluminium control box is waterproof, resilient and weighs 1 kg. There are only three controls: on; push auto-tune; and a lockable audio threshold control. Three LEDs are used to indicate unit faults when batteries need replacing and are approaching the end of their operational life. There is also an audio alert system which operates in conjunction with the LEDs. The box is carried in a weatherproof canvas bag on a shoulder strap or belt loops.

An aluminium battery box contains four 1.5 V type D alkaline manganese batteries with an operational life at 20°C of approximately 20 hours. Normally the battery box is carried next to the control box but an extension lead allows the box to be kept under clothing in cold conditions. Battery box weight when empty is 0.53 kg.

The three-piece rod set carries a 280 mm search coil which can operate when submerged. The rod length can be varied from 1.6 m down to 596 mm and has a handgrip and armrest, both of which are adjustable.

The equipment is completed by a waterproof headset which can be worn over or under a helmet. All the components can be carried in a special shoulder bag which also carries an extra battery box and spares.

Two probes are available as accessories to this equipment. One is 1 m long and the other 400 mm. Both may be used underwater.

Total weight of the equipment when in use is 4.61 kg. Total weight complete with spares, tools and eight batteries packed into an optional aluminium carry case is 11.42 kg. Operating temperature range is from −40° to +85°C.

Detection sensitivity
Steel strip: (50 × 12.5 × 1 mm) 350 mm
Aluminium sheet: (20 × 40 mm) 320 mm
20 mm anti-personnel projectile: 450 mm
Metal anti-tank mine: (diameter 200 mm) 800 mm
Coil spring: (23 mm long, diameter 7.1 mm) 90 mm
Ball bearing: (diameter 4.7 mm) 80 mm

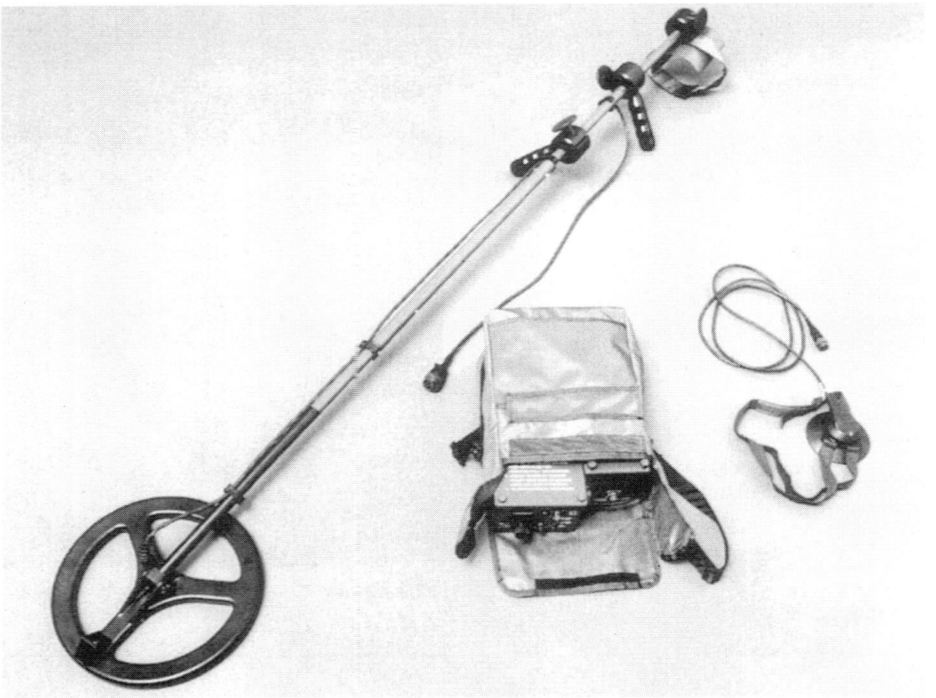

White's Electronics military mine detector

Aluminium cap: (6.5 mm long, diameter 5.9 mm) 120 mm
Steel pin: (10 mm long, diameter 1.5 mm) 140 mm

Status

In production. In service with Chile, Malawi, Norway, Tunisia, US Marine Corps, United Nations forces and some other nations.

Manufacturer

White's Electronics (UK) Limited, 13 Harbour Road, Inverness IV1 1SY, UK.
Tel: +44 1463 223456. Fax: +44 1463 224048.

UPDATED

Computer Aided Munition Identification System (CAMIS)

Description

The Computer Aided Munition Identification System (CAMIS) was developed by Negretti Aviation Limited as a purpose-designed computer-based system providing Explosive Ordnance Disposal (EOD) departments and teams with rapid access to detailed munitions data. The data provides the operator with neutralisation, render safe and counter-mining procedures. The system is supplied complete and ready to use with all software installed. A basic set of data is available for the database while further data may be readily added by the user or by the supplier.

The system may be used on-site and includes a video camera which may be used to provide direct on-screen comparisons for identification purposes or for the insertion of a new record into the database. If required, the system can be security protected and it may also be used for simulation and training.

Status

Available. In service with an undisclosed organisation.

Manufacturer

Negretti Aviation Limited, 6 Manor Court, Barnes Wallis Road, Fareham, Hampshire PO15 5TH, UK.
Tel: +44 1489 579090. Telex: 47671.
Fax: +44 1489 579013.

VERIFIED

UNITED STATES OF AMERICA

Airborne Standoff Minefield Detection System (ASTAMIDS)

Development

The Airborne Standoff Minefield Detection System (ASTAMIDS) will provide a highly reliable airborne minefield detection system which will provide a commander with the accurate location of minefields. The system will be required to detect and identify buried and surface patterned mines plus scatterable mines and buried nuisance mines on unpaved roads. The minefield location, combined with other intelligence, surveillance and target acquisition operations, will provide alternatives for command decisions.

Following a development phase during which Westinghouse Electronic Systems and Raytheon were contenders, the then Raytheon, Missile Systems Division, combined with Cambridge Parallel Processing (CPP) of Irvine, California in early 1994. They were awarded a $1.52 million increment as part of a $22,261,448 cost plus incentive fee contract to further develop ASTAMIDS over a 32 month demonstration and validation period, during which ASTAMIDS will be demonstrated with a surrogate air vehicle and a mobile display station. At the end of that period a single Engineering and Manufacturing Development contract will follow until the end of 1999, with production scheduled for the beginning of 2000.

The contracting activity was the US Army Aviation and Troop Command, Fort Belvoir, Virginia. Project Manager is the Office of the Project Manager for Mines, Countermine and Demolitions, Picatinnery Arsenal, New Jersey.

Description

ASTAMIDS is an airborne mine detection system consisting of an imaging sensor mounted on an Unmanned Aerial Vehicle (UAV) and a processor/algorithm integrated into the UAV Ground Control Station (GCS). The system allows the user to obtain information for formulating an operational plan and gather further information to further modify that plan.

The ASTAMIDS sensor, weighing less than 45 kg, will be mounted on the UAV and controlled by the UAV GCS. Minefield imagery is transmitted via a datalink to the GCS for processing in near real time by a very high-speed parallel processor from CPP. The latter involves a Distributed Array Processor (DAP) with a growth capability of up to 450 per cent. The resultant information will be displayed and disseminated to users in a manner similar to other reconnaissance, intelligence, surveillance and target acquisition data.

The Raytheon-developed sensor system involves laser diode technology, including a laser receiver and a UAV-mounted passive farband infrared and active laser sensor detection system. The ground-based portion of the ASTAMIDS includes the Minefield Detection Algorithm and Processor (MIDAP).

Status

Development - see text.

Manufacturer

Raytheon, Missile Systems Division, Bedford, Massachusetts 01730, USA.
Tel: +1 617 274 2222.

VERIFIED

GDE non-metallic mine detectors

Description

The US Army has a requirement for a system capable of detecting non-metallic mines with a minimal or non-existent metal content. General Dynamics, Electronics Division, (now GDE Systems Inc) was awarded a contract by the Defense Advanced Research Projects Agency (DARPA) to develop such a system. Following investigation into the various technologies possible, they decided to concentrate on the use of Radio Frequencies (RF) to provide aural and visual methods of buried object display. The use of RF means that any buried objects can be detected, while nearly all existing systems rely upon the presence of metals and thence magnetic anomaly or similar detection methods.

The first GDE unit involving the use of RF is a hand-held detector with a search head containing a four-element antenna. Using the antenna, a low-powered RF source operating on three sequenced frequencies electronically illuminates the ground and any targets, while backscattered radiation is measured for target presence and location. A preprogrammed microprocessor switches the transmit and receive ports of the antenna to collect both phase and amplitude reflected data from a buried target. After data acquisition, special algorithms operate on the data for the extraction of information which signifies the presence of a target. With the first hand-held detector model, evidence of a target is provided aurally as a 300 to 3,000 Hz tone in a headset and visually on a small LCD display mounted on the detector handle. System electronics are carried in a backpack.

The first hand-held detector using RF principles was produced by late 1991. A second and more advanced model employs a search head with 40 micro antennas. The dimensions of the search head and the antennas are optimised for search and detection sensitivity while the operator's backpack is used to contain and carry

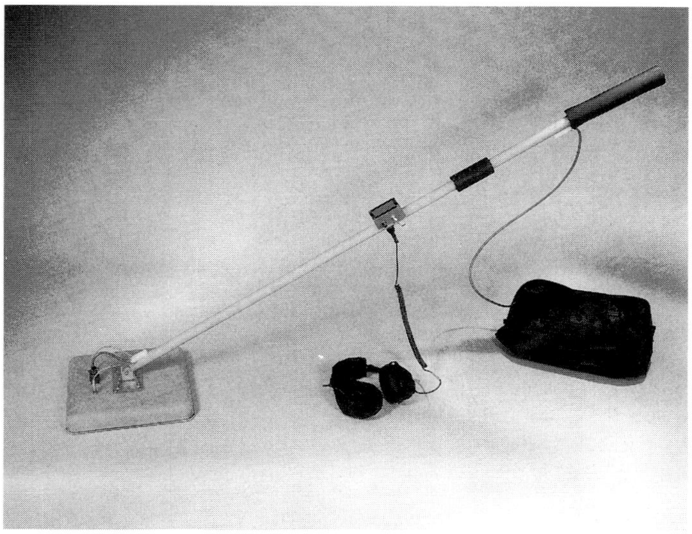

Prototype of GDE Systems hand-held non-metallic mine detector showing electronics unit backpack and LCD display on handle

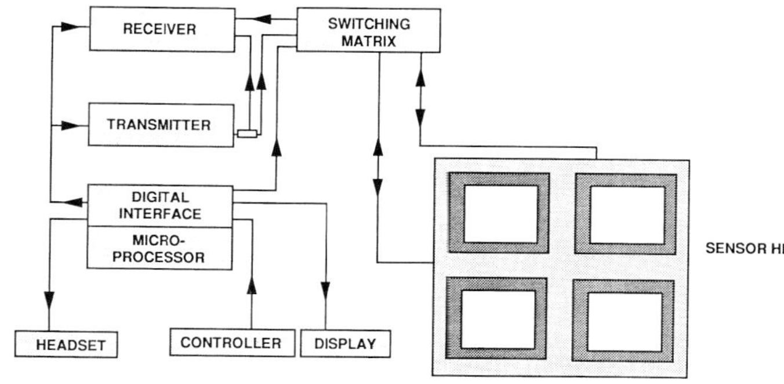

Schematic layout of GDE Systems non-metallic mine detection system

the circuitry necessary for the system. This provides a graphic display of targets on a colour screen carried on the detector handle. The antennas' sweeps and responses provide sufficient data for the system to provide a graphic indication of the shape and location of buried objects, whatever their material composition.

The hand-held detector makes use of lightweight materials and has a length (with handle) of about 1.524 m. On the first model the search head measures 356 × 356 mm and total weight is 5.4 kg. The batteries used are of the BA-3090/U and BA-3030/U type, providing a life of 8 hours in continuous use at 22°C.

Detection capabilities for plastic objects are as follows:

Small - 0 to 51 mm
Medium - 0 to 152 mm
Large - 0 to 305 mm

It is anticipated that, using the system, there will be an 80 per cent detection probability during development tests and a 90 to 95 per cent detection probability with production versions.

A vehicle-mounted detection system is also under development, using similar RF principles to the hand-held detectors.

Status
Development.

Manufacturer
GDE Systems Inc, PO Box 85468, San Diego, California 92186-5468, USA.
Tel: +1 619 592 5567.

VERIFIED

AN/PRS-8 metallic and non-metallic mine detector

Description
In Fiscal Year 1976 the US Army Mobility Equipment Research and Development Command (MERADCOM) initiated an AN/PRS-7 Product Improvement Program (PIP) with the Cubic Corporation Defense Systems Division. Under this, the detector was improved using new solid-state electronics and signal processing equipment for better mine detection in any type of soil, the designation AN/PRS-8 being assigned to the detector. The new solid-state transmitter and receiver sweeps over a wide radio frequency band instead of the single frequency of the earlier AN/PRS-7 providing an order of magnitude increase in the search head data gathered. This is then processed by a single board processor module fitted with a 16 bit microprocessor and memory on a single chip. The algorithm used is developed from that used on the AN/VRS-5 vehicle-mounted mine detector system (now cancelled) and results in a vastly increased performance over the AN/PRS-7, specifically in a higher detection rate and fewer false alarms.

The detector adapts its threshold sensitivity to various soils automatically. Handling characteristics have been improved with end-point blanking to eliminate false alarms produced by the operator's tendency to tilt the search head at the ends of the swing, a metronome pacing signal to set the optimum search sweep rate and more distinctive audio signals. An automatic switch from 'search' to 'point' is also included.

In late 1981 a $7.3 million contract was placed for 1,461 AN/PRS-8 detectors. The contract included retrofit kits to update 9,500 AN/PRS-7 detectors in service to AN/PRS-8 standard.

Specifications
Search head assembly
Weight: 0.77 kg
Length: 221 mm
Width: 221 mm
Height: 56 mm

Control box (without battery)
Weight: 0.857 kg
Length: 180 mm
Width: 64 mm
Height: 78 mm

Battery
Weight: 0.25 kg
Length: 95 mm
Width: 38 mm
Height: 65 mm

Long handle
Weight: 0.62 kg
Length: 0.48 m
Height:
(nested) 35 mm
(extended) 1.46 m

Short handle
Weight: 0.11 kg
Length: 0.3 m
Width: 30 mm
Height: 56 mm

Miscellaneous (headset, cables, straps and so on)
Weight: 1.2 kg

Carrying case
Weight:
(without unit) 6.2 kg
(with unit) 10 kg
Length: 0.6 m
Width: 0.4 m
Height: 0.19 m

RF frequency range: 300-600 MHz
Audio frequency:
(search mode) 1 click/2.5 s
(point mode) 3-150 clicks/s
Temperature range:
(operating) −32 to +52°C
(storage) −55 to +70°C
Battery complement: Type BA-5847 (Lithium)/U
Battery voltage: +5.6 V nominal
Battery life: 20 h at ambient room temperatures

Status
In service with the US Army. A total of 9,500 AN/PRS-7 detectors underwent product improvement.

VERIFIED

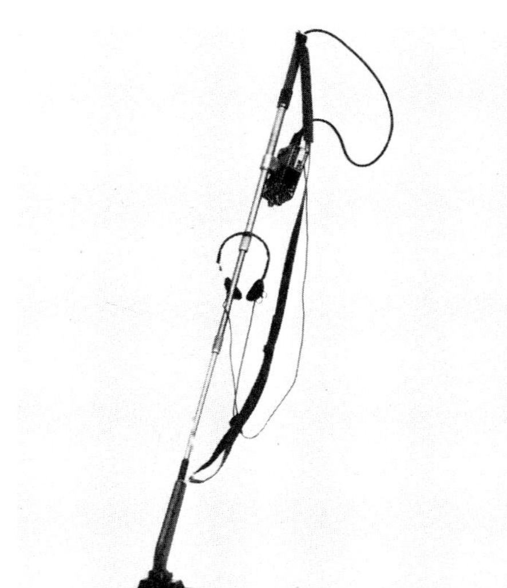

AN/PRS-8 hand-held mine detector

AN/PSS-11 metallic mine detector

Development
The AN/PSS-11 was developed by the US Army with production being undertaken by Cubic Corporation's Defense Systems Division under contract to the US Army Mobility Equipment Research and Development Command (MERADCOM), now the Belvoir Research, Development and Engineering Center.

The AN/PSS-11 is licence-produced in Egypt by the Benha Company for Electronic Industries, Cairo, under the designation MMD 2000.

Description
The AN/PSS-11 is a simple, rugged, solid-state equipment which can detect metallic anti-tank and anti-personnel mines of ferrous and non-ferrous materials.

The equipment consists of five major components: detection head, telescopic handle, control box, receiver/transmitter and headset.

The detection head is square and contains four coils which are electrically balanced. These transmitter coils send out electromagnetic waves which set up a magnetic field with a radius of 0.914 to 1.219 m. When a metallic object is encountered, the magnetic field is distorted, resulting in an imbalance in the mutual inductance bridge.

The receiver/transmitter, that is carried in a pouch on the operator's belt, amplifies the imbalance and causes a signal increase in the headset and indicates the presence of the metallic object. A phase discriminating circuit cancels false signals caused by saltwater or magnetic soils and only signals caused by metallic objects are transmitted to the headset.

The control box is on the handle and nulls the signal to the headset to establish initial balance condition and has low, intermediate and high sensitivity selections to locate metallic objects and pinpoint their exact positions.

The telescopic handle, which is detachable from the detector head, can be extended from 0.482 to 1.752 m. It is identical to the handle on the AN/PRS-8. The handle is normally used in the walking position with the search head about 1.219 m from the body. The detector head can also swivel to allow the operator to use the equipment while prone.

The headset consists of two headphones which produce a 625 Hz audible sound.

A complete AN/PSS-11 set consists of a carrying case; mine detector; spare modules; spare batteries; arctic kit; and operations and maintenance handbook.

Specifications

Weight: 10 kg
Case:
(length) 600 mm
(width) 400 mm
(height) 180 mm
Audio frequency of headset: 625 Hz
Battery voltage: 10.7 to 14 V
Battery type: BA-1568/U
Battery life: 70 h continuous operation
Detection capability:
(small metal object) 304-355 mm
(large metal object) 1.219-1.828 m
Operating temperature range: −32 to +52°C
Storage temperature range: −55 to +70°C
Relative humidity operational: 100%
Relative humidity storage: 100%

Status

In production. US$10 million contract placed for US Army use, April 1983. Also ordered by Egypt (US$4.5 million contract 1983, now licence-production). Known to be in service in Brazil.

UPDATED

AN/PSS-11 hand-held mine detector in use

MINECLEARING EQUIPMENT

CANADA

MREL LEXFOAM Liquid Explosive Foam mineclearing systems

Description

MREL (Mining Resource Engineering Limited) LEXFOAM liquid explosive foam is a patented explosive based on nitromethane which may be used for clearing minefields of unexploded ordnance. Other applications include the breaching of walls and vehicles for access and the explosive welding of metals.

LEXFOAM stock solution is transported in 45 gallon (204.5 litre) drums as a flammable liquid and its explosive properties can be tailored to suit the application. The detonation velocity and resultant pressure vary with the foam density and layer thickness. The suggested foam density and foam layer thickness is capable of neutralising greater than 95 per cent of hardened fuzed anti-tank and anti-personnel mines buried to depths of 100-150 mm are 0.5 g/cm^3 and 50 mm respectively. Detonation pressures and impulses can range between 5 and 40 kbar and 100 and 1,500 bar/ms for foam densities between 0.1 and 0.6 g/cm^3.

In use, drums containing the LEXFOAM concentrate are taken close to the operational area and thoroughly mixed with liquid propane in a dispersal tank prior to being over-pressurised with nitrogen. The mixture is prepared in 10 kg capacity backpack units or 200 kg capacity palletised or truck-mounted units. For large-scale area coverage LEXFOAM can be dispensed from 2,400 kg capacity towable units. For small-scale use LEXFOAM can be dispensed from hand-held aerosol cans. If required, LEXFOAM can be prepared at a central depot.

In a typical mineclearing operation using the 10 kg capacity backpack unit, the LEXFOAM is sprayed using a 3 m long wand. The foam may be sprayed generally in large patches or in more defined patches over detected mines which may then be explosively linked to each other by thin strips of LEXFOAM or detonating cord to allow all the patches to be detonated from a single detonator source. Once deployed, LEXFOAM can be detonated by standard military or industrial blasting caps. The foam is cap sensitive above a density of 0.2 g/cm^3.

Other MREL explosive systems include the following: BREACHCASE, a wall-breaching device; R-SEF rigid solid explosive foam; F-SEF flexible solid explosive foam; and L-PWG, a liquid plane wave generator.

Status

In production.

Manufacturer

Mining Resource Engineering Limited, 1555 Sydenham Road, R.R.8 Kingston, Ontario, Canada K7L 4V4.
Tel: +1 613 545 0466. Fax: +1 613 542 8029.

VERIFIED

Spraying LEXFOAM from a 10 kg capacity backpack

CHILE

FAMAE Bangalore torpedo

Description

The FAMAE Bangalore torpedo is delivered packed in wooden cases each holding six torpedoes ready for individual use or to be assembled into lengths to suit any mineclearing or demolition task. Each section of the torpedo is 1.518 m long and consists of a tubular body with an ogive and coupling device. Diameter is 60.3 mm and the weight of each unit is 9 kg. Standard No 8 detonators are used to fire the assembled torpedoes. A complete wooden box weighs approximately 65 kg.

Status

In production. In service with the Chilean armed forces.

Manufacturer

FAMAE Fabricaciones Militares, Avenida Pedro Montt 1568/1606, Santiago, Chile.
Tel: +56 2 556 1011. Telex: 242 346 famae cl.
Fax: +56 2 556 9044.

VERIFIED

CHINA, PEOPLE'S REPUBLIC

NORINCO 425 mm mineclearing rocket system

Description

The NORINCO 425 mm mineclearing rocket, also known as the Type 762, is launched from the 425 mm mineclearing rocket launcher based on the hull and chassis of the Type 83 152 mm self-propelled gun-howitzer (for details refer to *Jane's Armour and Artillery 1995-96* pages 539 to 541). This launcher vehicle follows the same general lines as the Type 83 and carries two 425 mm rockets on two rails mounted on a low turret at the hull rear which can be traversed 90° right and left. The launcher rails can be elevated hydraulically from +5 to +45° and when fired the rockets have a maximum range of between 800 and 1,000 m. Each rocket can clear a path through a minefield 80 to 100 m long and 10 to 20 m wide.

NORINCO 425 mm mineclearing rocket system with launcher rails elevated ***1997***

The launcher turret mounts a 7.62 mm machine gun provided with 500 rounds for anti-aircraft, ground target or dismounted use. Firing ports for crew weapons are provided on each side and to the rear of the turret for self-defence. The hull armour is stated to be proof against 7.62 mm projectiles and shell splinters. A door is provided at the rear of the hull. Equipment provided includes a Type WG002 panoramic telescope, a 0.5 m stereo-optical rangefinder, a Type TYG-1 night vision device for the driver, a Type 889 radio set and a Type 803 intercom set.

Specifications

Launch vehicle
Crew: 4
Weight: (combat) 27 000 kg
Power to weight ratio: 19.3 hp/t
Length:
(overall) 7.162 m
(hull) 6.405 m
Width: 3.236 m
Height: 3.2 m
Ground clearance: 0.45 m
Track: 2.62 m
Track width: 480 mm
Length of track on ground: 4.601 m
Ground pressure: 0.32 kg/cm^2

Max speed: (road) 55 km/h
Average speed:
(on highway) 40-42 km/h
(on country road) 30-32 km/h
Range: (highway) 450 km
Fording: 1.3 m
Gradient: 31°
Side slope: 25°
Vertical obstacle: 0.7 m
Trench: 2.7 m
Engine: Type 12150L diesel developing 520 hp
Armament: 1 × 7.62 mm MG
Ammunition carried: (7.62 mm) 500 rds

Rocket
Rocket diameter: 425 mm
Total length: 4.7 m
Total weight: 750 kg
Warhead weight: 600 kg
Rocket section weight: (with fins) 129 kg
Temperature range: −40 to +50°C
Launch velocity under standard conditions: 31 m/s
Max velocity under standard conditions: 98 m/s
Max range: 800-1,000 m
Cleared path: (1 rocket)
(length) 80 to 100 m
(width) 10 to 20 m

Status
In production. In service with the People's Liberation Army.

Manufacturer
China North Industries Corporation, 7A Yue Tan Nan Jie, PO Box 2137 Beijing, Beijing, People's Republic of China.
Tel: +86 1 853 3695. Telex: 22339 cnin cn.
Fax: +86 1 853 3236.

UPDATED

NORINCO 253 mm Type 81-II mineclearing rocket system

Description
There are two versions of the NORINCO 253 mm Type 81-II mineclearing rocket system, one with ten launcher tubes carried on a Dong Feng 6 × 6 cross-country truck and the other with 24 launcher tubes mounted on a turretless Type 69 tank chassis. Both versions utilise the same 253 mm Type 81-II minecleariing rocket. The blast created when a single warhead is detonated via a fuze activated by a long stand-off spike on the rocket nose, will clear mines over a 35 m radius. As well as clearing mines, Type 81-II rockets can also be used to clear battlefield obstacles and demolish strongpoints.

The launcher tubes on the truck version can be traversed through 135° and elevated from +20 to +50°. The rockets are launched at intervals of from 1.2 to 1.4 seconds, with a complete salvo creating a minefield breach 60 m wide and 10 m deep at an operational range of from 1,200 to 1,400 m.

The fire control for the 24 launcher tube tracked carrier version is enhanced by the introduction of a laser rangefinder. On this version the rockets have the option of removing an efflux cone from the rocket exhaust venturi, thereby increasing the maximum possible range of each rocket from 2,000 to 3,000 m.

In addition to firing mineclearing rockets, the truck-mounted system can also launch Type 62 depth-charge rockets for coastal defence.

NORINCO 253 mm Type 81-II minecleariing rocket system mounted on Dong Feng 6 × 6 truck **1997**

Specifications
Wheeled launch vehicle
Weight: (combat, approx) 8,300 kg
Length: (travelling) 6.8 m
Width: 2.2 m
Height: 2.45 m
Max speed: (road) 80 km/h
Average speed: (on highway) 12-40 km/h
Fording: 0.85 m
Elevation: +20 to +50°
Traverse: (total) 135°

Type 81-II rocket
Rocket diameter: 253 mm
Length:
(with efflux cone) 1.961 m
(without efflux cone) 1.921 m
Weight:
(with efflux cone) 74.5 kg
(without efflux cone) 74 kg

Status
In production. In service with the People's Liberation Army.

Manufacturer
China North Industries Corporation, 7A Yue Tan Nan Jie, PO Box 2137 Beijing, Beijing, People's Republic of China.
Tel: +86 1 853 3695. Telex: 22339 cnin cn.
Fax: +86 1 853 3236.

NEW ENTRY

NORINCO 305 mm fuel-air minecleariing rocket

Description
The NORINCO 305 mm fuel-air minecleariing rocket is fired from the same launcher as used for the NORINCO Type 79 minelaying system (qv). The high pressures produced by the fuel-air warheads of these rockets are used to clear anti-personnel minefields at ranges of up to 2,700 m from the launch vehicle. The intense blast from each rocket can clear anti-personnel mines from an impact area 48 m in diameter and will be lethal against exposed personnel over an area of from 24 to 60 m in diameter. The rockets can also be used to clear foliage and grass.

For safety purposes the rocket warheads and motors are transported in separate wooden boxes and assembled only before loading.

Specifications
Rocket diameter: 305 mm
Length: (total) 2.927 m
Weight: (at launch) 156 kg
Max velocity: 185 m/s
Max range: 2,700 m
Operating temperature range: −15 to +50°C

Status
In production. In service with the People's Liberation Army.

Manufacturer
China North Industries Corporation, 7A Yue Tan Nan Jie, PO Box 2137 Beijing, Beijing, People's Republic of China.
Tel: +86 1 853 3695. Telex: 22339 cnin cn.
Fax: +86 1 853 3236.

NEW ENTRY

NORINCO Type 84 tank-mounted minecleariing rocket system

Description
The NORINCO Type 84 tank-mounted minecleariing rocket system is carried on the rear decking of a main battle tank and is prepared for use before the tank arrives at the point where a minefield is to be breached. This involves removing the lid of the wooden crate in which the system is transported, opening up the internal canvas packing to reveal the line charges to be projected and placing the two rockets used to carry the line charge on their short launch rails.

Launching is carried out with the tank crew protected inside the vehicle. Each 153 mm calibre rocket carries a

NORINCO Type 84 tank-mounted minecleariing rocket system on Type 69 MBT **1997**

line charge to a maximum range of 340 m. Once the line charges are in position they are detonated from within the carrier tank. Anti-tank mines will be cleared from a path 6 to 8 m wide and from 80 to 100 m long, along with battlefield obstacles such as wire barriers. Once the system has been used the packing crate can be discarded.

Specifications

Rocket calibre: 153 mm
Weight of rocket: (each) 35 kg
Length of rocket: 1.13 m
Max range: 340 m
Width of cleared path: 6-8 m
Length of cleared path: 80-100 m
Operational temperature range: −40 to +50°C

Status

In production. In service with the People's Liberation Army.

Manufacturer

China North Industries Corporation, 7A Yue Tan Nan Jie, PO Box 2137 Beijing, Beijing, People's Republic of China.
Tel: +86 1 853 3695. Telex: 22339 cnin cn.
Fax: +86 1 853 3236.

NEW ENTRY

NORINCO Type 74 mineclearing system

Description

The NORINCO Type 74 mineclearing system involves a rocket-carried explosive line charge which is transported in man-portable crates. The line charge is carried in a total of 14 boxes which are opened up before deployment and arranged in two parallel lines of seven boxes. The charges contained in the boxes are connected together before use with one end yoked to two 153 mm calibre rocket motors operating in tandem, the front motor firing first and igniting the second once its propellant has been consumed. The rockets are launched from a short rail held at an angle by two supporting legs.

Launching is carried out with the operators behind cover. The rockets will carry the line charge to a maximum distance of 400 m. When the charge is detonated it will clear anti-personnel mines from a path 3 to 6 m wide and from 80 to 100 m long.

Status

In production. In service with the People's Liberation Army.

Manufacturer

China North Industries Corporation, 7A Yue Tan Nan Jie, PO Box 2137 Beijing, Beijing, People's Republic of China.
Tel: +86 1 853 3695. Telex: 22339 cnin cn.
Fax: +86 1 853 3236.

NEW ENTRY

NORINCO Type 74 mineclearing system ready for use ***1997***

NORINCO light rocket mineclearing system

Description

The NORINCO light rocket mineclearing system is transported with all its components inside a portable steel case which also acts as the launch platform. At a point in front of the minefield to be breached, the case is opened up and the launch tube is erected inside the open lid. A 50 mm calibre light rocket, attached to a line charge by a length of wire cable, is inserted in the launch tube. The system is then ready for use.

Launching is carried out with the operator behind cover. The rocket carries the line charge to a maximum distance of 66 m. When the charge is detonated, it will clear anti-personnel mines from a path 0.5 m wide and 20 m long.

It is claimed that up to three of these equipments can be launched in series, a combination of three equipments producing a maximum range of 126 m and clearing a path 54 m long. Exactly how this is achieved is uncertain.

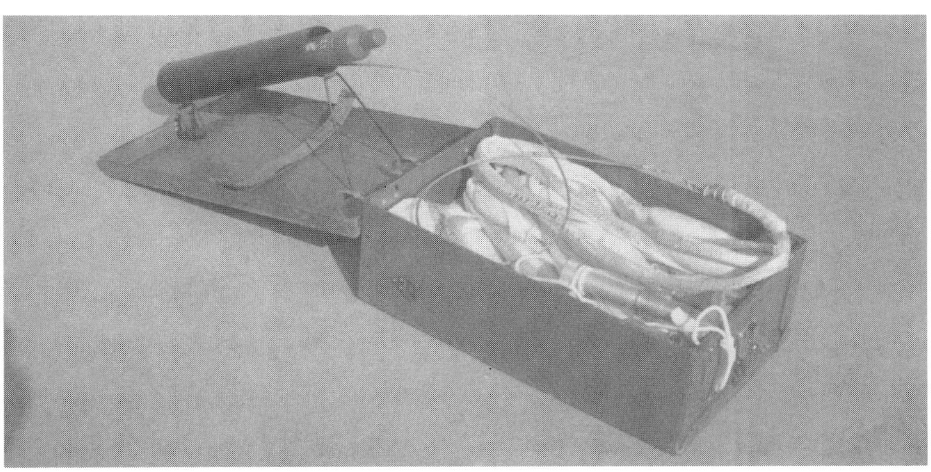

NORINCO light rocket mineclearing system ready for use ***1997***

Specifications

Rocket calibre: 50 mm
Weight: (complete) 15.5 kg
Length of case: 375 mm
Max range: 66 m
Width of cleared path: 0.5 m
Length of cleared path: 20 m

Status

In production. In service with the People's Liberation Army.

Manufacturer

China North Industries Corporation, 7A Yue Tan Nan Jie, PO Box 2137 Beijing, Beijing, People's Republic of China.
Tel: +86 1 853 3695. Telex: 22339 cnin cn.
Fax: +86 1 853 3236

NEW ENTRY

Type 84 portable mineclearing equipment

Description

The Type 84 portable mineclearing equipment is a rocket-based system normally carried into action as a backpack, although handles are provided on the case. The complete system is contained in a light steel case fitted with shoulder straps and, when in position, the case is opened so that the lid acts as a firing support. The rocket is then fired from a safe distance and, as it travels forward, the rocket pulls an explosive-filled hose from the case to a maximum range of from 90 to 100 m. The explosive hose has a length of about 20 m and once it is on the ground the explosive is detonated to clear mines from a path approximately 20 m long and up to 0.5 m wide.

Total weight of the Type 84 is 17 kg.

Status

In production. In service with the Chinese armed forces and offered for export.

Manufacturer

China North Industries Corporation, 7A Yue Tan Nan Jie, PO Box 2137 Beijing, Beijing, People's Republic of China.
Tel: +86 1 853 3695. Telex: 22339 cnin cn.
Fax: +86 1 853 3236.

UPDATED

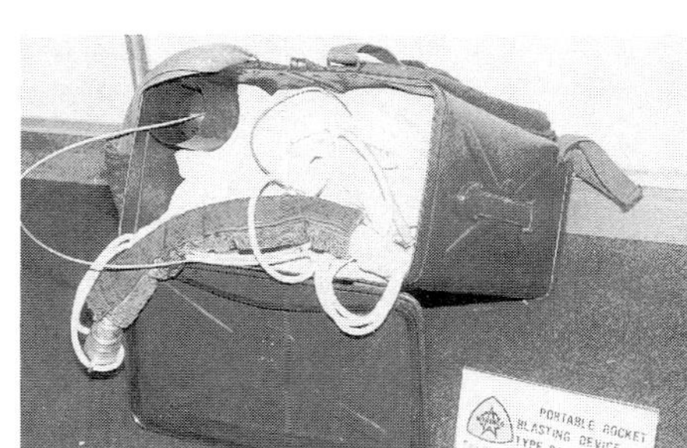

Type 84 portable mineclearing equipment (T J Gander)

Bangalore torpedoes

Description

The armed forces of the People's Republic of China employ at least four types of Bangalore torpedo for clearing minefields and other obstacles. Three types use a heavy steel body and the other a lightweight metal body. All four employ pull friction igniters that allow them to be used as extemporised land mines with tripwires if required. The torpedoes have conventional rounded noses, although simple plugs may be fitted at both ends. Booster charges may be used, but are not always fitted. As with other such devices, lengths of torpedo can be joined to make up any length as required.

Also available are sets of Bangalore torpedoes consisting of five rigid units and four flexible units which can be combined to suit virtually any combat engineering requirement. It is claimed that a single rigid unit, 41 mm in diameter, 110 mm long and weighing 4.5 kg, can be used as a hand grenade.

Specifications

Type	**Type 1**	**Type 2**	**Type 3**	**Lightweight**
Length	0.998 m	0.99 m	1.187 m	0.99 m
Diameter	33 mm	45.7 mm	35.5 mm	54 mm

Status

In service with the People's Liberation Army and exported to some other countries.

Manufacturer

China North Industries Corporation, 7A Yue Tan Nan Jie, PO Box 2137 Beijing, Beijing, People's Republic of China.
Tel: +86 1 853 3695. Telex: 22339 cnin cn.
Fax: +86 1 853 3236.

UPDATED

CZECH REPUBLIC AND SLOVAKIA

SVO mineclearing system

Description

SVO stands for *Samohybný Výbuš ný Odmínovač*. It is an explosive mineclearing system based on the chassis of the BVP-1, the Czech/Slovak produced version of the BMP-1 IFV (for details of the BVP-1/BMP-1 refer to *Jane's Armour and Artillery 1995-96* pages 294 to 299). On the SVO the turret is removed and the roof of the rear compartment is left open to accommodate 24 projectiles with high explosive warheads. Although the SVO can have a crew of three, it may be operated by a crew of two, driver and system operator.

On the SVO the driver and commander occupy their normal positions as they would on a BVP-1/BMP-1. Behind them is an open compartment with mounting racks for the 24 projectiles. Traverse and elevation of the racks is carried out solely by the BVP-1 chassis. At the rear of the open compartment, which is normally protected when not in action by a canvas cover, is a bulkhead. To the right rear of the bulkhead is a compartment for the system operator, entered through a door in the hull rear. The compartment has seating for the system operator plus a raised superstructure with vision ports and contains the projectile aiming devices and controls.

Each projectile has a 245 mm diameter warhead filled with 26.5 kg of THAF explosive and a 73 mm diameter tail boom supporting six fixed tail fins. Total weight of each projectile is 41.5 kg and length overall, including the stand-off fuze spike, is 1.457 m. On impact each projectile warhead will create an above-surface over-pressure over a diameter of 9 m.

The projectile is fired using a modified 30 mm PLdVK cartridge case 110 mm long, filled with 92 g of Nctp 8/1 powder propellant; total weight of each cartridge is 420 g. Base diameter of the cartridge is 48 mm and neck diameter 43 mm. The electrical primer designation is EZS-1.

The projectile's piezoelectric point detonating fuze is known as the PEZ-1, actuated by the stand-off spike. The fuze will arm 1.15 seconds after firing. If the fuze does not function on impact it will self-destruct after a minimum of 20 seconds.

SVO mineclearing vehicle showing the 24 projectiles in their firing racks ***1996***

In action the SVO is driven to the edge of the minefield to be cleared. After preparation, which takes two minutes, all 24 projectiles can be fired in 64 seconds in a decreasing range pattern. The first projectile has the longest range, a maximum of 530 m and a minimum of 350 m. The last projectile, number 24, has a maximum range of 430 m and a minimum range of 250 m. The projectiles are arranged to fall in an overlapping pattern, clearing a path 100 m long and 5 m wide. The system may be used against underwater mines up to depths of 5 m and can be used to clear non-explosive battlefield obstacles.

The SVO vehicle weighs 13,830 kg and has a radius of action of 450 km.

Status

Understood to be in service with the Czech and Slovak Armies.

Manufacturer

Konštrukta Defence, PO Box 62, 912 50 Trenčin, Slovak Republic.
Tel: +42 8313 5850. Fax: +42 8313 4451.

UPDATED

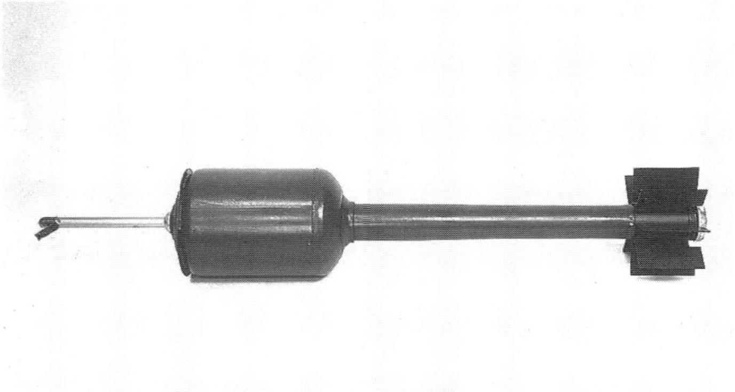

Projectile used with SVO mineclearing system ***1996***

Side view of SVO mineclearing vehicle showing superstructure over operator's compartment at rear (T J Gander)

Tank-mounted mineclearing roller
Tank-mounted mineclearing plough

Description

The Czech and Slovak Armies use tank-mounted mineclearing rollers and ploughs of their own design. The former consists of three to five rollers of varying thickness mounted on an arm in front of each track in a manner similar to the RFAS equipment. The Czech/Slovak rollers are bigger than their RFAS counterparts and have a different type of serrated edge. Between and in front of the two sets of rollers is a frame for detonating the tilt-rod fuzes of any mines before the tank passes over them.

The plough type equipment is similar in concept to the RFAS equipment but uses a larger plough and also has a frame for detonating mines with tilt-rod fuzes.

Status

In service with the Czech and Slovak Armies.

Manufacturer

Czech and Slovak state factories.

T-55 MBT fitted with Czech/Slovak mineclearing rollers

VERIFIED

Božena remote control demining vehicle

Description

Relatively few details regarding the Božena remote control demining vehicle have been released other than the limited data provided here. It is intended for clearing both anti-personnel and anti-tank mines from a safe distance by employing a rotating chain flail. Control is by remote line-of-sight from an air-conditioned armoured shelter carried on a truck or a similar vehicle. The same truck also tows a single axle trailer on which the Božena is transported.

A complete Božena system includes two flail units and a special pallet on which system components are carried. Up to 38 separate attachments, most of them intended for other uses than demining, are available.

The Božena flail unit is 2.18 m wide and it can clear a path 1.985 m wide to a depth of 150 mm. The host vehicle unit is 5.282 m long, 2.002 m high and 1.708 m wide.

Status

Prototypes.

Marketing Agency

Technopol, Military and Police Group, Kutlikova 17, 852 50 Bratislava, Slovak Republic.
Tel: +42 7 829 995. Fax: +42 582 2180.

Božena remote control demining vehicle ***1997***

NEW ENTRY

Mineclearing charges

Description

The Czech and Slovak armies use a two-wheeled trailer towed behind a T-54/T-55 MBT fitted with mineclearing equipment of the roller or plough type. The trailer lays a series of explosive cords behind the tank on those parts of the ground which have not been cleared by the mineclearing equipment installed on the front of the tank. The cords are detonated from inside the tank and clear any remaining mines.

The Czech and Slovak armies also have a trailer-mounted mine clearance system. This is an armoured four-wheeled trailer towed behind an OT-64 (8 × 8) APC or a similar vehicle, and contains rockets to which a number of flexible mineclearing explosive lines are attached. This is fired over a minefield and then detonated.

Status

In service with the Czech and Slovak armies.

Manufacturer

Czech and Slovak state factories.

VERIFIED

DENMARK

HYDREMA 910 MCV mineclearing vehicle

Description

The HYDREMA 910 MCV (mine clearing vehicle) was developed for the clearing of anti-tank and anti-personnel mines by means of a rotating flail. The vehicle can drive over and clear mines from very rough terrain as well as on roads.

The vehicle is based on the 10 tonne HYDREMA 910 off-road dump truck with an articulated chassis, constant four wheel drive and four equal sized wheels to provide good off road performance together with a continuous road speed at a minimum of 30 km/h. Power is provided by a 94 hp turbocharged diesel engine connected to a six-speed powershift transmission with an electronic shift system. Steering is fully hydrostatic with an emergency steering capability. Brakes are dual-circuit servo-assisted with maintenance free oil immersed discs on all four wheels. Heavy duty rigid axles with hub reduction are employed.

For mineclearing, the vehicle has a fully independent hydrostatic drive system which gives a clearing speed variable from 0 to 3 km/h. The two drive systems can be used independently, providing the possibility to drive out of a minefield following an emergency caused by engine failure. During mineclearing the vehicle is driven by seat-mounted joysticks similar to those used on HYDREMA loader excavators.

The flail has its own power provided from a separate powerpack mounted on the rear of the chassis. Flail width is 3.4 m with drive coming from two hydraulic motors providing continuously variable speed rotation against or away from the vehicle. Behind the flail is an armoured blast deflector plate. The flail height from the

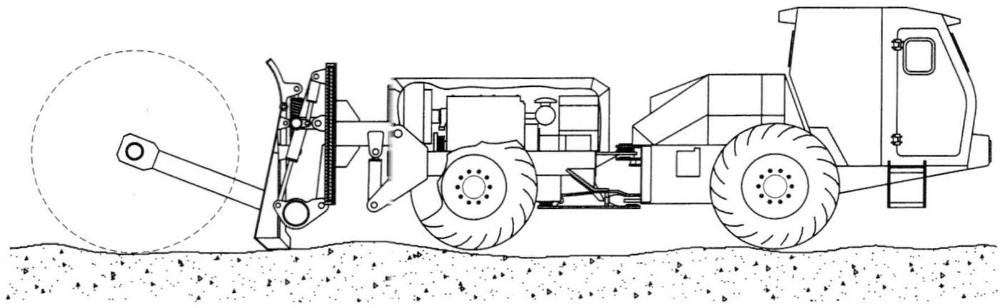

Outline drawing of HYDREMA 910 MCV mineclearing vehicle **1997**

ground can be adjusted either manually or automatically.

For transport the complete mineclearing unit, complete with the flail and deflector plate, can be stowed over the powerpack. This operation is carried out using hydraulic cylinders and does not involve tools. In the stowed configuration the vehicle can be transported legally on roads and may be carried in a C-130 cargo aircraft or on railways. Time for preparation after transport is less than 10 minutes.

The HYDREMA 910 MCV has a spacious cabin with two suspended and damped seats for the driver and one passenger. Heating and ventilation systems are provided, with air conditioning as an option. The cabin is armoured to protect against 7.62 mm small arms fire.

Specifications

Cab seating: 1 + 1
Weight: 11,000 kg
Length, travelling: 8 m
Width, travelling: 2.37 m
Height, travelling: 2.67 m
Ground clearance: 0.4 m
Speed:
(road) up to 30 km/h
(mineclearing, off road) 0-3 km/h
(mineclearing, on roads) 0-10 km/h
Engine:
(main) 4-cylinder turbocharged diesel developing 94 hp at 2,200 rpm
(powerpack) 6-cylinder turbocharged diesel
Transmission: 6-speed powershift with torque converter; hydrostatic drive unit for mineclearing
Brakes: dual circuit servo-assisted with oil immersed discs on all wheels
Tyres: (standard) 17.5 R - 25 non-directional with optional foam filling or run flat
(option) 600/55 - 26 low ground pressure
Electrical system: 24 V

Status

Prototype.

Manufacturer

A/S Hydrema, Gl Kirkevej 16, DK-9530 Stoøvring, Denmark.
Tel: +45 9837 1333. Telex: 938 1805.
Fax: +45 9837 1996.

NEW ENTRY

EGYPT

SAKR Fateh-1 anti-personnel mineclearing system

Description

The SAKR Fateh-1 anti-personnel mineclearing system is man-portable by two men and is transported in a carrying box containing a line charge, a rocket motor and a rocket launching rail frame. In use, the system is carried in its camouflaged box to the launching point and the launching frame is erected on one side of the box. The rocket is then placed on the launch rail and launched electrically to pull the line charge across the anti-personnel minefield to be breached. The line charge is then detonated to clear a path up to 110 m long and up to 0.6 m wide. The furthest point of clearance from the launching point is 140 m.

Inert drill and practice versions are available.

Specifications

Weight: 72 kg
Dimensions: 1 × 0.5 × 0.3 m
Weight of line charge: 50 kg
Length of line charge: 120 m

Status

In production.

Manufacturer

SAKR Factory for Developed Industries, PO Box 33, Heliopolis, Cairo, Egypt.
Tel: +20 2 418 3314. Telex: 92175 cerva un.
Fax: +20 2 290 1978, +20 2 290 1210.

VERIFIED

SAKR Fateh-1 anti-personnel mineclearing system ready for launch

SAKR Fateh-2 anti-tank mineclearing system

Description

The SAKR Fateh-2 anti-tank mineclearing system is towed into action on a single-axle trailer by a tank or other armoured vehicle. Once at the launch position the 3 m long launch rail is raised to an angle of 35° via a hydraulic system and the 122 mm two-stage rocket is fired, which pulls the line charge from its box. The rocket, which weighs 65 kg, has a range of 200 m and tows a line charge 120 m long weighing 720 kg. When the rocket trajectory is completed, the mechanical fuze is armed to detonate the plastic explosive line charge, which has a 5 kg/m plastic explosive content, after a 10-second delay. The maximum length of the cleared path is 110 m and width is 8 m.

After detonation of the line charge the trailer can be separated from the towing vehicle by a pyrotechnic charge. All control procedures are carried out from inside the towing vehicle by a firing box.

An inert charge for training purposes is available.

SAKR Fateh-2 anti-tank mineclearing system being towed by a T-62 MBT

Specifications

Trailer capacity: 2,000 kg
Trailer flat bed dimensions: 3.4 × 1.4 m
Explosive weight of line charge: 5 kg/m
Length of line charge: 120 m
Diameter of rocket: 122 mm
Length of rocket: 2 m
Weight of rocket: 65 kg

Status

In production.

Manufacturer

SAKR Factory for Developed Industries, PO Box 33, Heliopolis, Cairo, Egypt.
Tel: +20 2 418 3314. Telex: 92175 cerva un.
Fax: +20 2 290 1978.

UPDATED

SAKR Fateh-3 anti-tank mineclearing system

Description

The SAKR Fateh-3 anti-tank mineclearing system represents the third generation of anti-tank mineclearing systems developed by the SAKR Factory for Developed Industries. Based on the Fateh-2 system, the Fateh-3 is carried on tank chassis, typically a T-54 or T-55 chassis. The launcher is operated electrically to increase system reliability and two rockets can be fired in series to increase the cleared path length to 220 m; cleared path width is 8 m. All control procedures are carried out from within the tank chassis using a firing control box. The rocket and line charge are the same as for the Fateh-2 system.

SAKR Fateh-3 anti-tank mineclearing system
1997

Specifications

Explosive weight of line charge: 5 kg/m
Length of line charge: 120 m
Diameter of rocket: 122 mm
Length of rocket: 2 m
Weight of rocket: 65 kg

Status

In production. In service with the Egyptian Army.

Manufacturer

SAKR Factory for Developed Industries, PO Box 33, Heliopolis, Cairo, Egypt.
Tel: +20 2 418 3314. Telex: 92175 cerva un.
Fax: +20 2 290 1976.

NEW ENTRY

FINLAND

Sisu RA-140 DS mineclearing vehicle

Description

The Sisu RA-140 DS, introduced in 1994, is a 4 × 4 minecleaning vehicle developed by Sisu Defence in collaboration with the Finnish Army. The basic requirements included the capability of making breaches for troop movement in barrages of scattered mines laid out on roads and runways, and of clearing anti-tank and anti-personnel mines laid in the ground. Adequate mobility, high operational safety and good serviceability were among other features specified.

The mines are cleared by means of a horizontally rotating assembly of iron chains located in the rear end of the vehicle, which moves in the reverse direction when operating, driven by a hydraulic motor. By forcefully beating the ground, the rotating chains detonate the mines over a 3.4 m wide path, proceeding at a maximum clearing speed of 12 km/h. The automatic clearing depth adjustment ensures that no mines can remain active. Tests have been carried out with mines in different sizes of up to 10 kg in explosive weight. The results obtained have demonstrated that both the operator and the vehicle can operate in complete safety and that the Sisu RA-140 DS is well suited for use on uneven terrain.

The RA-140 DS vehicle, with a fully armoured (proof up to 7.62 mm small arms fire) NBC-protected two-man cab and a reinforced overall structure, features an air-cooled 189 hp (141 kW) Deutz diesel, together with an automatic transmission and stepless hydraulic driving speed control. There are separate driving instruments for forward and reverse directions. The vehicle's road speed is 70 km/h. For transport over long distances, the minecleaning assembly can be folded to a travelling position.

If required, a 12.7 mm machine gun can be mounted over the cab roof.

Specifications

Cab seating: 1 + 1
Configuration: 4 × 4
Weight: 15,000 kg
Length: 7.33 m
Width: 2.92 m
Height: 2.86 m
Ground clearance: 370 mm
Max speed:
(road) 70 km/h
(clearing) 0-0.6 km/h, stepless speed variation
Fording: 0.8 m
Engine: Deutz BF 6L 913 C air-cooled 6-cylinder diesel developing 189 hp at 2,500 rpm
Transmission: Renk Doromat 874 AM/PTO 4-speed automatic; hydraulic motor for minecleaning operations
Transfer box: Sisu single-speed, lockable differential gear
Axles: Sisu BEB at front, Sisu BTE at rear, both axles with differential locks
Suspension: coil springs and shock absorbers (front); leaf springs and shock absorbers (rear)
Brakes: dual-circuit air-controlled fluid disc brakes (front); drum brakes (rear)
Steering: powered
Tyres: 14.00 R 20 with Hutchinson VPPV run-flat devices
Electrical system: 24 V
Batteries: 2 × 12 V

Clearing unit: chain flail; hydraulic drive; manual/automatic clearing depth control
Clearing path width: 3.4 m
Clearing speed: 0-0.6 km/h, stepless speed variation

Status

In service with the Finnish Army and contracted to ATS in Denmark.

Manufacturer

Sisu Defence Oy, PO Box 189, FIN-13101 Hämeenlinna, Finland.
Tel: +358 17 6451. Fax: +358 17 619 6710.

VERIFIED

Sisu RA-140 DS minecleaning vehicle in action

Sisu RA-140 DS minecleaning vehicle ***1996***

FRANCE

Giat Demeter Vehicle Signature Duplicator System

Description

Few details have been released regarding the Demeter Vehicle Signature Duplicator System under development by Giat Industries, other than that it is a vehicle magnetic signature duplication device radiating a false magnetic field in front of the carrier vehicle to cause magnetically fuzed mines to detonate before the carrier vehicle reaches them. The system is carried inside two units carried on the front of the host vehicle and is intended for use in conjunction with mine ploughs or rollers to provide the maximum level of mine clearance. The Demeter system is claimed to be capable of triggering magnetically fuzed mines at a distance of 5 m in front of the carrier vehicle and 2.5 m to either side.

Status

Development.

Manufacturer

Giat Industries, 13 route de la Minière, F-78034 Versailles Cedex, France.
Tel: +33 1 30 97 36 48. Fax: +33 1 30 97 39 58.

VERIFIED

Demeter units located on Challenger 1 MBT to be used in conjunction with a Pearson Track Width Mine Plough (TWMP)

GERMANY

MIPAG scatterable mineclearing system

Description

Krauss-Maffei has developed a scatterable mineclearing system known as MIPAG which can be attached to armoured tracked and wheeled vehicles.

MIPAG is mechanical in operation and operates on a pulsating principle. It is pushed in front of the carrier and can clear scatterable mines from roads, paths and flat terrain. The attachment can be fitted to any suitably equipped armoured vehicle in about 10 minutes and weighs 1,380 kg. It extends 6.4 m in front of the carrier vehicle and clears a lane 4.8 m wide. Maximum clearing speed on paved roads is 30 km/h.

Status

In production.

Manufacturer

Krauss-Maffei Wehrtechnik GmbH, Krauss-Maffei Strasse 2, D-80997 Munich 50, Germany.
Tel: +49 89 88990. Telex: 523163 31.
Fax: +49 89 812 0190.

VERIFIED

MIPAG scatterable mineclearing system fitted to Leopard 2 MBT

MaK Keiler armoured mineclearing vehicle

Development

In 1973 the Federal German Defence Ministry and the Procurement Office issued a request for proposals for a rapid land mine clearance system *Landminen-Schnellräummittel*). A number of companies, including Krupp MaK (now MaK System Gesellschaft mbH), AEG/Telefunken, Dynamit Nobel, Industriewerke Karlsruhe and Rheinstahl (now Thyssen Henschel) submitted proposals for the system.

In 1983 MaK became the general contractor for further development of the resultant vehicle, named Keiler and was responsible, as general contractor, for the manufacture of two prototypes in 1985.

Preliminary and final company testing was carried out with the two prototypes and troop tests by the Federal German Army commenced during the second half of 1990. A contract award for the manufacture of 24 units was awarded during the second half of 1993. Production units have been deployed in Bosnia.

Description

The Keiler (Wild Boar) armoured mineclearing vehicle is based on a modified M48A2 MBT hull and unmodified M48 running gear and carries a flail-type mineclearing system at the front. The vehicle has a crew of two (commander and driver). Power is provided by a power pack consisting of an improved MTU MB 871 KA-501 diesel engine and a Renk HSWL 284 M transmission.

The mineclearing system, a Kaelble Gmeinder concept, is powered by hydraulic motors on a pump distribution gearbox. In order to suppress low frequency vibrations which would make the clearing process impossible, the mineclearing vehicle was supplied with a front support. With the front support in position, the vehicle can reach speeds of up to 25 km/h.

The mineclearing system uses a cantilever arm which operates the system from the travelling position to the 20° clearing position. There is a tilt and elevation system for terrain adjustment and a carrier arm and clearing shaft frame carrying hydraulic motors. The system is completed by two clearing shafts carrying 24 flails and a height measuring device.

There are three operating modes: driving; clearing standby; and clearing. During the clearing mode it is possible to select either automatic or manual modes; the latter is also used as an emergency mode when the automatic mode is inoperable. In either mode, paths up to 4.7 m wide can be cleared and a clearing depth of +50, −50 or −250 mm can be preselected. The height measuring device located on one side operates the clearing depth mechanically. Variations in terrain level are catered for by a self-regulating device.

The rotating flails can either destroy mines in their path, throw them to one side or detonate them before they pass under the clearing system. It is claimed that the system will clear 98 per cent of mines in its path. A typical clearing speed in medium-heavy soil is 120 m in 10 minutes at a 250 mm clearing depth. During clearing operations it is not normally necessary to replace flail heads until a path 3,000 m long has been cleared.

An automatic cleared path lane-marking system is

Keiler armoured mineclearing vehicle in clearing mode with front support lowered

Keiler armoured mineclearing vehicle (Michael Jerchel) ***1996***

carried on the rear of the hull. When the mineclearing system is not in use it is carried on racks over the hull.

Smoke grenade dischargers are the only self-defence equipment carried.

Specifications

Crew: 2
Combat weight: 52,000 kg
Power-to-weight ratio:
(driving) 19 hp/t
(cruising) 21 hp/t
Length:
(driving) 7.9 m
(clearing) 10.7 m
Width:
(driving) 3.76 m
(clearing) 6.35 m
Height:
(driving) 3.75 m
(clearing) 2.76 m
(hull deck) 1.99 m
Ground clearance:
(overall) approx 390 mm
(front support) 260 mm
Max cruising speed: 50 km/h
Fording: (without preparation) 1.2 m
Gradient: 60%
Range: (cruising, driving only) approx 600 km
Trench: 2.6 m
Engine: MTU MB 871 KA-501 diesel developing 986 hp in driving mode and 1,092 hp in clearing mode
Transmission: Renk HSWL 284 M
Armament: smoke grenade dischargers

Clearing speed: 0.2-4 km/h
Clearing width: 4.7 m
Max clearing depth: 250 mm

Status

24 units ordered during 1993. First units handed over during 1996 and deployed to Bosnia.

Manufacturer

General contractor: MaK System GmbH, PO Box 9333, D-24157 Kiel, Germany.
Tel: +49 431 399502. Fax: +49 431 399 5446.

UPDATED

Ahlmann Area Clearance System

Development

The Ahlmann Area Clearance System is intended for the rapid clearance of surface mines and submunitions from airfield areas such as runways, aprons, taxiways and other hard surfaces. The device and the carrier vehicle, a modified Ahlmann Swing Shovel Loader are both armoured.

The German Luftwaffe has 100 of these clearance systems, the last of which was delivered in 1993.

Description

The Ahlmann Area Clearance System is mounted on the front of a modified Ahlmann Swing Shovel Loader, a heavy wheeled construction vehicle which is fitted with an armoured protection kit for the driver/operator position and some vehicle components. The armour provided is proof against 5.56 and 7.62 mm ball and armour-piercing rounds, anti-personnel mines and 155 mm artillery air bursts. All four tyres are foam-filled.

Mounted on a quick-change device on the vehicle operating arms, the system can clear an area of approximately 15,000 m^2 in one hour. Cleared area width is 3.5 m and operating speed up to 14 km/h. In operation the device rests on rollers when operating on hard surfaces and skids when operating on grass. The system weighs a total of 18 tonnes and sweeps unexploded munitions into a clearance bucket for later disposal. The driver/operator can also use a rifle from a protected turret over the driving position for direct projectile attack against unexploded munitions when appropriate.

When not required, the clearance device can be removed from the carrier vehicle within minutes, allowing the vehicle to assume other duties.

Ahlmann Area Clearance System in operation

Specifications

Crew: 2
Weight: 18,000 kg
Length: 6.99 m
Width:
(overall) 3.98 m
(clearing) 3.5 m
Height: 3.96 m
Speed:
(travelling) 35 km/h
(clearing) 0-14 km/h
Engine: Deutz BF6L 913 air-cooled diesel limited to developing 152 hp at 2,500 rpm
Transmission: hydrostatic
Steering: hydrostatic
Turning radius: (outside) 6.9 m
Tyres: 20-25 R 12 PR
Electrical system: 24 V
Batteries: 2 × 12 V, 110 Ah

Status

Available. 100 systems delivered to German Luftwaffe by 1993.

Manufacturer

Ahlmann-Baumaschinen GmbH, PO Box 964, D-24758 Rendsburg, Germany.
Tel: +49 4331 351-01. Telex: (17) 433 111 amb d.
Fax: +49 4331 32046.

VERIFIED

Comet No 3001 one-man minesweeping line device

Description

The basic system consists of an 18 kg box containing an 82 m long detonating cord with 75 × 100 g plastic explosive charges placed on the cord at 1 m intervals. A smokeless solid propellant rocket with a launching device is fitted with a footplate and ground spike. The rocket is equipped with a propellant fuze which has a six-second delay and the detonating cord has a fuze, also with a six-second delay. The rocket's towing gear consists of two 2.6 m long steel wire ropes fitted with thimbles.

In operation the elements of the system are assembled by one man approximately 20 m in front of the minefield to be breached. The rocket is fired by the fuze and the minesweeping line is extended to its full length. The line is then ignited either automatically by the fuze or electrically by hand. As detonation takes place, all the mines lying under the explosives will be detonated. The consecutive craters created result in a narrow mine-free path approximately 74 m long. If a longer path is required then the procedure is repeated with a second unit. A training system (No 3002) is available and can be used for repeated firings.

Status

Development complete.

Manufacturer

Comet GmbH Pyrotechnik-Apparatebau, PO Box 10 02 67, D-27502 Bremerhaven-1, Germany.
Tel: +49 471 3930. Fax: +49 471 39394.

VERIFIED

Comet No 3001 one-man minesweeping line device ready to fire

Comet No 3030 minesweeping system MRL-80

Description

The Comet No 3030 mine sweeping system MRL-80 consists of two transport units, each with two carrying bars. Each of the units can be either carried by two men or dragged along on integral skids.

One transport unit contains the rear blasting unit, the ignition system with the blasting cap fuzes and the retardation system. The second unit contains the front blasting unit, the rocket, a shock absorber and one rocket igniter with a 40 second delay. The ladder-type blasting units consist of 281 blasting tubes, each with 285 g of explosive, connected at both ends by a steel rope. Parallel to the steel ropes is a detonating cord to initiate the blasting tubes.

Before use, both transport units and their blasting units are linked mechanically and pyrotechnically in front of the minefield to be breached. As the rocket is fired it pulls out the blasting units to their maximum length and after six seconds the system is ignited automatically. This leaves a clearly visible cleared path approximately 70 m long and 0.6 m wide.

A training version which can be reused is available.

Status

Development complete.

Manufacturer

Comet GmbH Pyrotechnik-Apparatebau, PO Box 10 02 67, D-27502 Bremerhaven-1, Germany.
Tel: +49 471 3930. Fax: +49 471 39394.

VERIFIED

ISRAEL

RAMTA On Board Anti Magnetic Mine Actuating Device (OB AMMAD)

Description

The On Board Anti Magnetic Mine Actuating Device (OB AMMAD) detonates magnetically fuzed mines in front of the host vehicle outside their effective range. It is an add-on kit easily adaptable to a variety of wheeled and tracked armoured vehicles.

The OB AMMAD system consists of two magnetic field emitters which are mounted on the front of a vehicle and a control panel mounted in the driver's compartment. The system generates a magnetic field simulating the magnetic pattern of the host vehicle. Magnetically fuzed mines detect the magnetic field change and sense it as a target vehicle.

On Board Anti Magnetic Mine Actuating Device (OB AMMAD) mounted on a M60 MBT **1996**

Exhibit showing an On Board Anti Magnetic Mine Actuating Device (OB AMMAD) with the control unit at bottom right (T J Gander)

The system weight is approximately 113 kg. OB AMMAD requires an electrical supply of less than 20 A at 24 V, produces no electro-magnetic interference for other vehicle systems, does not impede vehicle mobility, is reliable under all weather conditions and is designed to survive in a combat environment.

Status
Delivered to the Israeli, Italian, Swiss and Swedish armies as well as the US Army and Marine Corps and others.

Manufacturer
RAMTA Division, Israel Aircraft Industries Limited, PO Box 323, IL-84102 Beer Sheva, Israel.
Tel: +972 7 272231. Fax: +972 7 276770.

UPDATED

RAMTA rolling onboard Anti Magnetic Mine Actuating Device (AMMAD)

Description
The rolling Anti Magnetic Mine Actuating Device (AMMAD) is intended to actuate magnetically fuzed mines before the host vehicle reaches them. Also known as the Improved Dogbone Assembly (IDA), the rolling AMMAD can be deployed as a stand alone unit or as an add-on kit to be combined with mineclearing ploughs or rollers.

The rolling AMMAD operates by projecting a magnetic field ahead of the host vehicle. The magnetic field is generated as a result of the rolling motion of the AMMAD. As it rolls the AMMAD projects a fluctuating field which duplicates the magnetic pattern of the host vehicle and detonates magnetically fuzed mines about 3 to 5 m in front of the ploughs or rollers.

Mounted between mine ploughs or between the wheels of a mine roller, the rolling AMMAD can be quickly and easily installed without special tools. When fitted it does not affect the mobility and firepower of the host vehicle.

In extended service and use by the Israel Defence Forces, the US Army and US Marine Corps, the rolling AMMAD has proven to be 100 per cent efficient in neutralising magnetically fuzed mines.

Status
In service with the Israel Defence Forces. More than 1,000 AMMAD systems were deployed by the US Army and Marine Corps during Operations Desert Shield and Desert Storm.

Manufacturer
RAMTA Division, Israel Aircraft Industries Limited, PO Box 323, IL-84102 Beer Sheva, Israel.
Tel: +972 7 272231. Fax: +972 7 276770.

NEW ENTRY

A RAMTA rolling Anti Magnetic Mine Actuating Device (AMMAD) mounted between the mine ploughs carried by a M60 MBT
1997

RAMTA Lightweight In-stride Mine Extractor (LIME)

Description
The Lightweight In-stride Mine Extractor (LIME) is a lightweight full width countermine system which provides the capability to breach areas threatened with rapidly deployed scatterable mines or with surface laid mines of various sizes and types. The LIME is designed for mounting on the front of light to medium weight armoured vehicles such as the M113 APC, the Bradley IFV, the AAV7A1, the LAV, self-propelled artillery vehicles and others.

Suitable for use over any type of terrain, the LIME can be easily installed by using a special design adaptor for each specific vehicle. Using this adaptor, the LIME is attached to the host vehicle by two pins inserted in the front tow lugs. In addition, two secured adjusting bolts are used to ensure a stable connection between the LIME and the vehicle hull. These attachment and securing methods do not require any modification to the host vehicle.

The LIME is comprised of the following major assemblies:

Scraper sections (usually three) which clear mines from the full width path of the vehicle. These scrapers span approximately 1.2 times the outer dimensions of the track width of the host vehicle and are arranged in the form of a V for paved roads and cross-country configurations (weight 1,300 kg, width 4.3 m), or as a 'one side push' for airfield runways and aprons (weight 1,000 kg, width 4.3 m). In the mineclearing position, each scraper section can move independently of the other units. The lower part of each scraper has skid shoes that follow ground contours and prevent any ploughing action by the scraper. A set of heat-treated spring steel leaf springs are attached to the upper part of each scraper with a further spring forcing each leaf spring downwards. In this way the lower front part of each individual leaf spring is always in contact with the ground or surface being cleared.

Pushbeams (usually three) are connecting arms that connect the scrapers to the mainframe.

A single mainframe interfaces between the host vehicle and other parts of the LIME.

An electrical lifting system raises and lowers the elements of the LIME.

An electrical system contains the control elements necessary to operate the lifting system.

A control box is located in the host vehicle driver's compartment to raise and lower the LIME.

The LIME system can be combined with the OB AMMAD (qv) to enhance its survivability and to provide a solution to the threat of scatterable magnetic mines.

Status
Available. Systems have been delivered to the Swiss Army, the US Air Force, and others.

Manufacturer
RAMTA Division, Israel Aircraft Industries Limited, PO Box 323, IL-84102 Beer Sheva, Israel.
Tel: +972 7 272231. Fax: +972 7 276770.

UPDATED

Lightweight In-stride Mine Extractor (LIME) in a 'one side push' configuration mounted on a M113 APC
1997

Lightweight In-stride Mine Extractor (LIME) in a V configuration mounted on a M113 APC
1997

RAFAEL Soil Mill

Description

The Soil Mill is offered as a safe, reliable, quick and cost-effective method of mineclearing. It is propelled by a modified bulldozer of the Caterpillar D9 or similar type with the front end of the machine having a rotating shaft on which chain-hung hammers are mounted. While the shaft rotates at high speed and advances, the hammers mill the soil at a depth of 250 mm.

The Soil Mill system has three main units. The driving vehicle can be an armoured tractor, bulldozer or similar vehicle. There is a 900 hp power unit installed and driven on a protected wheeled platform linked to the driving vehicle front. The third unit is the front end system containing the Soil Mill.

The Soil Mill's front end system is a rotating shaft on which chain-hung hammers are attached. As the shaft rotates at a speed of around 500 rpm the direction of rotation is such that when the hammer hits the ground the speed vector is in a forward direction. As the system moves forwards the hammers mill the ground at high speed and force, pounding any object in their way and flinging forward any mass encountered. A mine hit by a hammer will be mechanically destroyed and flung forward or detonated.

As the Soil Mill operates, a layer of topsoil is removed and pulverised to a depth of 250 mm. The cleared strip is about 3 m wide. The advance velocity in hard soils containing a high percentage of stone is 0.2 m/s, while on light soils the velocity is about 0.8 m/s. More than 95 per cent of mines in the cleared strip will be destroyed.

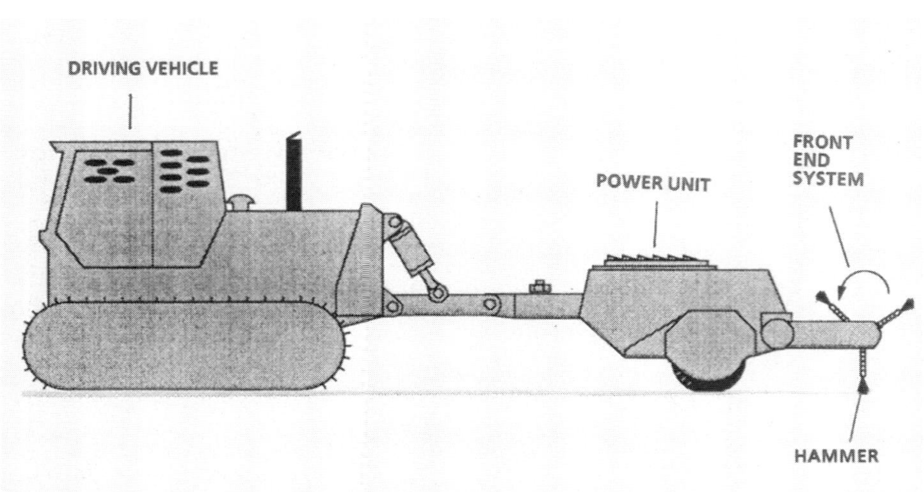

Diagrammatic outline of the main units constituting the RAFAEL Soil Mill

Status

Available.

Manufacturer

RAFAEL, PO Box 2082, IL-31021 Haifa, Israel.
Tel: +972 4 706965. Telex: 471508 vered il.
Fax: +972 4 794657.
Marketed by: Hermes Tech. Corp., 1 Yonge Street, Suite 1801, Toronto, Ontario, Canada M5E 1W7.
Tel: +1 416 868 1539. Fax: +1 416 512 0198.

VERIFIED

IMI Full Width Mine Plough

Description

The IMI Full Width Mine Plough is used for the mechanical clearance of a 5 m wide path through anti-tank minefields to open safety lanes for tanks, APCs and other vehicles. The plough can be used in most soil types other than rocky ground and can also be used to clear scatterable mines.

The plough, designed as an 'add-on' kit for Caterpillar D9H bulldozers, has a V-shaped blade with hardened teeth. It is mounted on three points of the bulldozer blade and clamped in place. Two hinged feeler arms with electric sensors travel in front of the plough to assess the terrain level and transmit information to an electrohydraulic control system which raises or lowers the plough accordingly. The tractor's hydraulic actuators are used for elevation and tilt. This automatic depth control enables operations at night and in poor visibility.

The Full Width Mine Plough can be adapted to fit on to other models of bulldozer apart from the D9H.

Full Width Mine Plough fitted to D9H bulldozer

Specifications

Weight of plough: 5,500 kg
Width:
(plough) 6 m
(cleared path) 5 m
Depth of clearing: up to 400 mm
Diameter of cleared mines: 200 mm and more
Clearing speed: up to 8 km/h depending on soil conditions

Status

Development complete.

Manufacturer

Israel Military Industries Limited (IMI), 64 Bialik Boulevard, PO Box 1044, IL-47100 Ramat Hasharon, Israel.
Tel: +972 3 542 5222. Telex: 33 719 misbit il.
Fax: +972 3 489639.

UPDATED

RAMTA Track Width Mine Plough (TWMP)

Development

The RAMTA Track Width Mine Plough (TWMP) can be fitted to all types of modern MBT in use. The TWMP has been fitted to Centurion, M48, M60, M1 and M1A1 Abrams, Leopard 1, Leopard 2 and Merkava MBTs.

In January 1989, it was announced that the US Army would receive more than 400 RAMTA TWMPs, known to the US Army as the Mine Clearance Blade System (MCBS). The MCBS was operated by the Allied forces during Operation Desert Storm in the Persian Gulf.

Description

The RAMTA TWMP consists of two plough units with separate lifting mechanisms and depth control systems. No alterations to the tank are needed as the plough is attached to the towing lugs on the tank front hull. Transferring the TWMP from one tank to another takes less than one hour using the tool kit supplied.

The TWMP is operated by the tank driver. When travelling, the ploughs are raised and secured by a locking device. In operation, the driver releases the locking devices and the ploughs drop free for the teeth

RAMTA Track Width Mine Plough (TWMP) fitted to Leopard 2 MBT

1996

to dig into the ground and dislodge buried mines which are then pushed aside. Any tilt-rod mines in the central unploughed area between the ploughs, will be detonated by a chain suspended between the ploughs. Each plough unit operates independently and follows the natural terrain contours, clearing a path wider than the tank tracks.

After use, the TWMP can be lifted back to the travelling position by a mechanical device connected to the track tension wheel or by 24 V DC electrical motors.

Specifications

As fitted on Leopard 2 MBT
Ploughing depth: 300 mm
Ploughed width each side: 1.312 m
Unploughed width between tracks: 1.069 m
Skidshoe track each side: 0.35 m
Chain track in centre: 0.197 m
Clearing speed:
(loess) 6.5 km/h
(sand and clay) 9.5 km/h
(stony) 6.5 km/h

Status

In production. In service with Israeli armed forces, Australia and the US Army (more than 400).

Manufacturer

RAMTA Division, Israel Aircraft Industries Limited, PO Box 323, IL-84102 Beer Sheva, Israel.
Tel: +972 7 272231. Fax: +972 7 276770.

UPDATED

Urdan RKM mineclearing rollers

Description

The RKM mineclearing rollers can be used on M47, M48, M60, M60A1, M1, Centurion, AMX-30 and all types of Leopard MBTs.

The RKM mineclearing rollers are track width rollers mounted in two banks in front of the carrier tank, pushed and suspended from two pusher arms. The weight and suspension of the rollers is used to detonate any mine over which they travel and a weighted chain between the two roller banks is used to detonate any tilt-rod actuated mines that might explode under the tank's belly. As the rollers detonate a mine, the force of the explosion causes the roller to be lifted upwards on its suspension arm thus reducing the force of the blast which is distributed in the mass of the roller and arm and in the mass of the carrier vehicle attachment point.

The two sets of rollers can articulate independently, with each set being able to move upwards 152 mm and downwards 254 mm to take account of ground surface irregularities. The entire roller banks can castor up to 30° either side when the carrier tank makes a turn. No modification is necessary to couple the mineclearing roller system to a tank, as special adaptors shaped to the front glacis of each MBT are provided. The rollers are connected to the tank by a standard 5-tonne lifting device which is available in any tank unit for maintenance purposes. Using such a device, two soldiers can fit the system in about 15 minutes. In an emergency the system can be disconnected from the tank from within the carrier tank without the crew having to leave the interior. This manual release facility can be completed in 30 seconds.

RKM mineclearing rollers fitted to Australian Army Leopard AS1 MBT (Paul Handel)

Status

In production. In service with the Israeli armed forces and exported to several countries, including Australia.

Manufacturer

Urdan Industries Limited, Industrial Zone, IL-42378 Netanya, Israel.
Tel: +972 9 338074. Telex: 341822 uasf il.
Fax: +972 9 610246.

VERIFIED

RAMTA combat vehicle teleoperation kit (PELE)

Description

The RAMTA combat vehicle teleoperation kit, known as PELE, is a modular add-on kit which can be adapted to any combat vehicle to convert it into a remote-controlled vehicle at a fraction of the cost of developing special purpose robotic vehicles. The conversion does not interfere with the vehicle's normal functions.

PELE is designed for multi-role applications and enables the performing of hazardous missions such as breaching minefields, leading convoys, performing unmanned surveillance missions and attracting enemy fire. It consists of electrical/mechanical actuators, a vehicle control assembly, a navigation system, cameras, RF, UHF, data and video communication links and an operator control unit. These components can be adapted to suit user requirements.

Initially developed for an M60A3 tank in a minefield breaching application, PELE can control not only the hull and turret functions (including the weapons) but also control a mine plough, an OB AMMAD (qv), a Clear Lane Marking System (CLAMS - qv), the vehicle smoke generation system and a line charge (if one is being pulled by the tank). The system has been successfully tested, evaluated and demonstrated in both manned and unmanned modes of operation.

The PELE teleoperated control system has been integrated as a humanitarian demining system used as a solution for clearing mine infected areas by humanitarian aid organisations without risking civilian lives. PELE equipped vehicles could also be used for detection, mapping, risk assessment and emergency response at disaster or contaminated sites with minimum risk to human health. Other possible roles for PELE could include border patrol, firefighting, outdoor security, policing firing ranges and as a mobile radio relay link.

A M60A3 MBT equipped with PELE and controlled from a RAMTA Ram light AFV in a typical operation setting, with the control unit shown in inset **1997**

Status

Delivered to the Swedish Army and US Army, and the Swedish Humanitarian Aid Organisation (SIDA).

Manufacturer

RAMTA Division, Israel Aircraft Industries Limited, PO Box 323, IL-84102 Beer Sheva, Israel.
Tel: +972 7 272231. Fax: +972 7 276770.

NEW ENTRY

IMI POMINS II Portable Mine Neutralisation System

Description

Developed and proved by the Israel Defence Force, POMINS II is described as a second-generation portable mine neutralisation system that can be carried by two men. Designed for infantry use, it can be set up and used in less than one minute to clear a path through anti-personnel minefields, barbed wire or both. The cleared path is 50 m long and 500 mm wide.

POMINS II consists of two subsystems each carried by one man using a special backpack harness that is an integral part of the system. The two subsystems are the launcher, comprising a rocket motor and the forward part of a line charge and the container with the rear part of the line charge. In use, the two subsystems are set approximately 1 m from the starting point of the path to be cleared. The covers are removed from both subsystems and the line charges are connected by a quick-connector. The system is then ready for use and when the launch pin is withdrawn a delay unit is activated to allow time (20 seconds) for nearby troops to take cover. A canister, propelled by the rocket motor, then leaves the launcher, which draws out the forward and rear interconnected line charges. When the entire line charge is extended it falls to the ground and after five seconds the fuze ignites the line charge to clear a path 50 m long and 500 mm wide. At night, a marker indicates the beginning of the cleared path.

Safety features incorporated into POMINS II are such that the launch pin cannot be removed until the covers of the sections have been removed. Even then two actions are required to withdraw the launch pin. These are that the arms of a safety catch have to be pressed together to release a pull ring and only then can the launch pin be pulled. Another safety feature is the 20-second delay period that allows troops to take cover before the launch. The system is trajectory-safe as only deployment of the front line charge completes the arming of the fuze. The fuze meets the safety requirements of MIL-STD-1316B.

POMINS II can withstand submersion in 1 m of water, a 5 m drop on to concrete (simulating a parachute drop) and loose cargo transportation. Both subsystems are watertight and dustproof. For transport, both subsystems are wrapped in clear, heat-sealed polyethylene bags and packed in a foamed polyurethane case. The packed case weighs approximately 70 kg.

See also the entry on APOBS in the United States of America section.

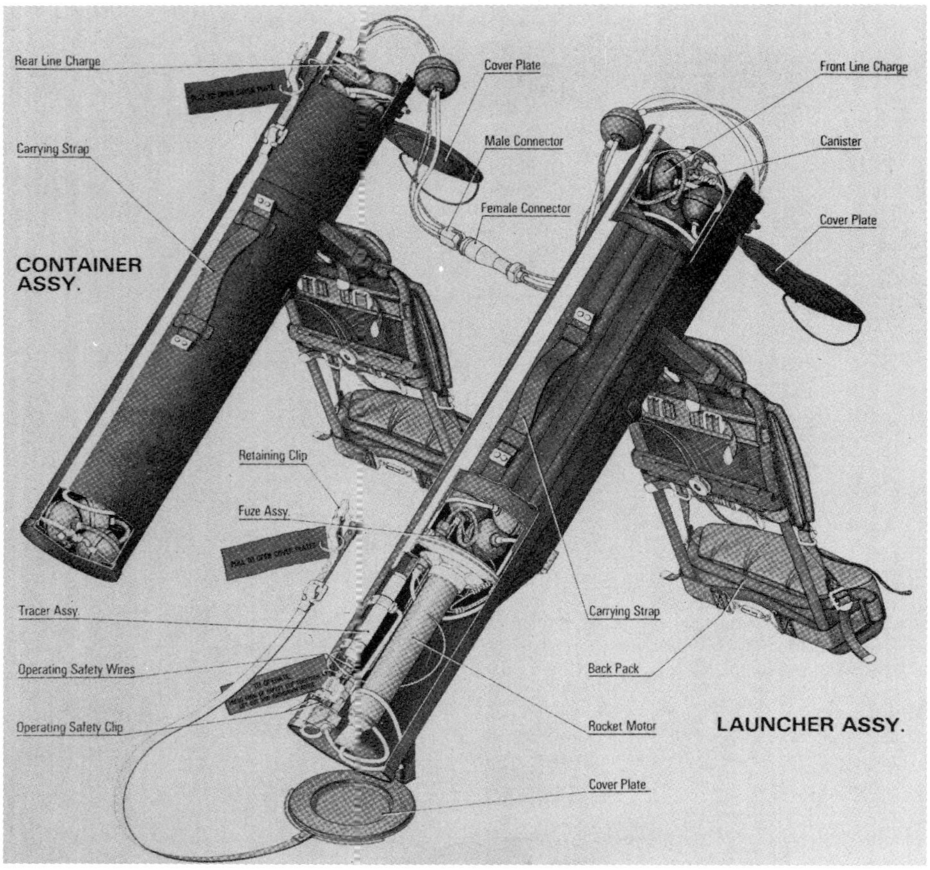

Main components of POMINS II portable mine neutralisation system

Specifications

Subsystem	**Launcher**	**Container**
Material	aluminium	aluminium
Length	1.1 m	0.84 m
Diameter	170 mm	155 mm
Weight with backpack harness	25 kg	22 kg

Status

In production. In service with the Israeli armed forces.

Manufacturer

Israel Military Industries Limited (IMI), 64 Bialik Boulevard, PO Box 1044, IL-47100 Ramat Hasharon, Israel.
Tel: +972 3 542 5222. Telex: 33 719 misbit il.
Fax: +972 3 489639.

VERIFIED

Cleared Lane Explosive Widening and Proofing charge (CLEWP)

Description

The Cleared Lane Explosive Widening and Proofing (CLEWP) charge, an explosive line charge, is used to clear a 150 m long path through a minefield thereby allowing the safe passage of tanks and other vehicles. It is commonly used in conjunction with a tank carrying a mine plough or rollers to ensure that the cleared path is free of anti-personnel mines and other mines that may have escaped the mechanical clearing method.

CLEWP is carried in an armoured container on the back of the clearing tank. As the carrier tank moves into the minefield a pyrotechnic device blows the spring-loaded lid off the container and on to the ground where it acts as an anchor for the charge. As the tank moves forward the charge is laid along the path behind it. The charge has 6 kg of explosive per metre and the charge string is 150 m long. When the end is reached, the armoured container is dropped from the vehicle which then moves forward at least 20 m. The charge is then fired by remote-control from within the carrier tank. The resultant explosion removes any mines or other explosive devices from the cleared path. A fail-safe protection circuit is used to prevent the possibility of an explosion within the container. It is possible to activate only part of the charge.

Lid of Cleared Lane Explosive Widening and Proofing charge (CLEWP) resting on ground as charge is pulled from container

Status

Development.

Manufacturer

Israel Military Industries Limited (IMI), 64 Bialik Boulevard, PO Box 1044, IL-47100 Ramat Hasharon, Israel.
Tel: +972 3 542 5222. Telex: 33 719 misbit il.
Fax: +972 3 489639.

VERIFIED

No 21 demolition Bangalore torpedo charge

Description

Although this charge is used primarily to breach barbed wire fences it may also be employed against minefields. The assembly consists of four sections of aluminium pipe charge, four sleeve couplings and two head units. The charge is actuated by a No 41 firing device with an 18 second delay.

The 1.1 m long charge sections are assembled into the charge unit and pushed into the minefield. Detonation is then accomplished using a No 41 firing unit and the resultant explosion clears a narrow path by detonating the mines underneath it.

Specifications

Total charge assembly weight: 19.35 kg
Single charge section
Length: 1.1 m
Diameter: 57 mm
Weight: 4 kg
Charge type: cast TNT
Charge weight: 3.3 kg

Status

In service with the Israeli Army.

Manufacturer
Israel Military Industries Limited (IMI), 64 Bialik Boulevard, PO Box 1044, IL-47100 Ramat Hasharon, Israel.
Tel: +972 3 542 5222. Telex: 33 719 misbit il.
Fax: +972 3 489639.

VERIFIED

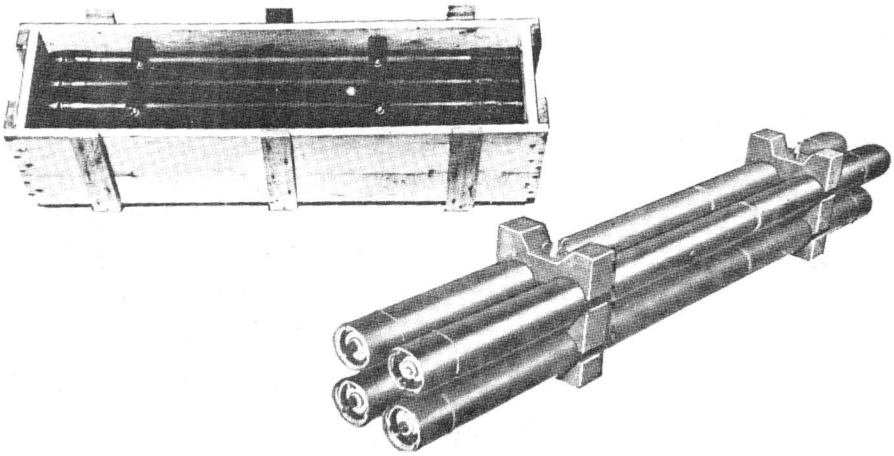

No 21 demolition Bangalore torpedo charge

No 30 demolition pipe charge

Description

The No 30 pipe charge assembly consists of 26 steel pipe charge sections, two front skids fitted with obstacle clearing wheels and four detonating units fitted with No 41 firing devices that have 18 second delays.

The pipe charge is used for breaching barbed wire fences and minefields. The 2.162 m long sections are assembled into a unit of the required length and an obstacle-clearing skid unit is fixed to the front to facilitate pushing the charge into position. Detonation is accomplished by the 380 mm long detonating unit fitted with a No 41 firing device. The wide path cleared by the resultant explosion is some 50 m long.

Specifications

Pipe charge section
Length: 2.162 m
Diameter: 55 mm
Weight: 8.5 kg
Charge type: Composition B
Charge weight: 3 kg

Pipe skid unit
Length: 2.7 m (with pipe section)
Weight: 17 kg

Detonating unit
Length: 380 mm (with No 41 firing device)
Weight: 1.5 kg (with No 41 firing device)
Charge type: Composition B
Charge weight: 0.5 kg

Status

In service with the Israeli Army.

Manufacturer

Israel Military Industries Limited (IMI), 64 Bialik Boulevard, PO Box 1044, IL-47100 Ramat Hasharon, Israel.
Tel: +972 3 542 5222. Telex: 33 719 misbit il.
Fax: +972 3 489639.

VERIFIED

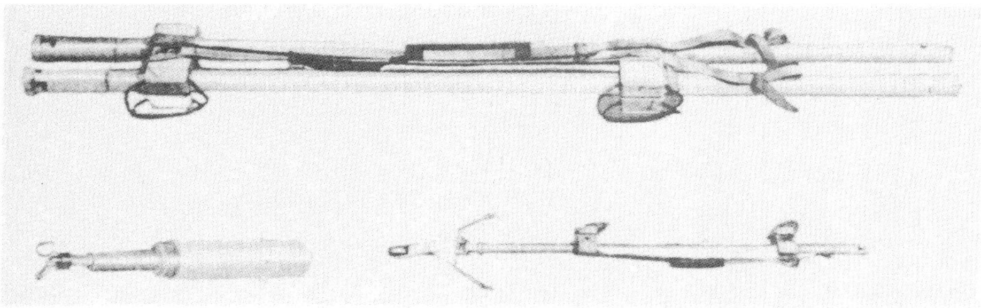

No 30 demolition pipe charge

Israeli mineclearing systems

Description

In addition to the Israeli mineclearing systems mentioned elsewhere in this section the Israel Defence Force also operates two further items.

One is a special turretless Centurion mineclearing tank. This vehicle has its turret replaced by an armoured enclosed superstructure provided with roof hatches and mountings for at least three machine guns. The tank pushes a set of RKM mineclearing rollers and more combat engineering and mineclearing equipment is stowed along each side of the vehicle and superstructure. Some of these vehicles have been observed fitted with Blazer reactive armour.

The second item of equipment is a towed minefield clearance system based on a single-axle trailer. The system uses a rocket and an explosive-filled hose which are fired electronically from a towing vehicle.

Status

In service with the Israel Defence Force.

VERIFIED

Israeli towed mineclearing system

Centurion tank converted to mineclearing role and fitted with RKM mineclearing rollers

JAPAN

Japanese mineclearing equipment

Description

Japan is known to have developed at least two items of mineclearing equipment.

One system was first demonstrated in mid-1992 and is a special mineclearing vehicle built on to a variant of the Type 73 APC, elongated by an extra roadwheel each side. The vehicle carries a large elevating launcher box over the hull, containing two rockets and line charges to be fired to a range of 800 m. The system was developed by Nissan Motors, with the first prototype completed in September 1989.

The second system is a rocket-propelled, explosive-filled mineclearing charge, known as the Type 70 Mine Clearing System, designed for clearing anti-personnel minefields. The Type 70 can project a 70 m line charge to a range of 550 m where it can clear a path 100 m long and 0.5 m wide. The towing rocket has a diameter of 150 mm, is 750 mm long and weighs about 24 kg. The launcher weighs 18 kg. The Type 70 has been observed mounted on the roof of a Type 73 APC.

Status

In service with the Japanese Ground Self-Defence Force.

Manufacturer

Type 70 Mine Clearing System: Aeronautical and Space Division, Nissan Motor Company Limited, 5-1 3-chome, Momoi, Suginami-ku, Tokyo, Japan.
Tel: +81 3 390 1111. Fax: +81 3 399 9100.

VERIFIED

Type 70 Mine Clearing System on display (K Nogi) ***1996***

PAKISTAN

POF Bangalore torpedo

Description

The Bangalore torpedo produced by the Pakistan Ordnance Factories (POF) is orthodox in design and uses a steel tube filled with Amatol 50:50. It can be fired from a distance using a variety of standard detonators and may be used for demolition purposes as well as mineclearing. Each unit is fitted with connecting lugs to allow units to be connected in line to create longer obstacle breaches.

Specifications

Weight: 6.6 kg
Length: 1.83 m
Diameter: 41.3 mm
Filling: Amatol 50:50

Status

In production.

Manufacturer

Pakistan Ordnance Factories, Wah Cantt, Pakistan.
Tel: +92 51 83219. Telex: 5840 pofac pk.
Fax: +92 51 584175.

VERIFIED

POLAND

PW-LWD tank-mounted rocket-propelled mineclearing equipment

Description

This mineclearing equipment has been observed mounted on Polish Army T-55A and T-72 MBTs and also on modified S-60 57 mm anti-aircraft gun carriages mounted on PTS tracked amphibians. It has been marketed with the name 'Giant Viper'.

The complete equipment consists of two PW-LWD launcher-containers which may be carried on the rear decking of an armoured vehicle. The two containers each hold a coiled LWD 110/5000 line charge weighing 920 kg and 110 m long. Also included in the full kit are a set of shields, assembly jigs and other components for the launching system, four PS illuminating ground markers and a PSK launcher for the markers.

The two line charges are towed by rockets across the minefield to be breached and then detonated. If the line charge is being used at night, the PS illuminating ground markers provide an indication of the fall of the charge. Deflection of the line charge in flight is 10 m either side and range is 276 m (±14 m). The resultant cleared path is 110 m long and 4 m wide.

Inert firing and non-firing training versions are available.

Shelf life of the complete equipment is five years.

A single PW-LWD rocket-propelled mineclearing unit ***1997***

Specifications
Weights:
(line charge) 920 kg
(launcher without shields) 230 kg
Length of line charge: 110 m
Launcher length: 2.816 m
Launcher width: 1 m
Launcher height: 680 mm

Range: 276 m (±10 m)
Length of cleared path: 110 m
Width of cleared path: 4 m

Status
In production. In service with the Polish Army.

Manufacturer
Zaklady Tworzyw Sztucznych 'PRONIT', ul. Zakladowa 7, P-26-940 Pionki, Poland.
Tel: +48 48 124717. Telex: 0672232.
Fax: +48 48 125534.
Marketed by: Cenzin Limited, ul Frascati 2, P-00-489 Warsaw, Poland.
Tel: +48 22 296396. Telex: 814505 czi pl.
Fax: +48 2 6286356.

UPDATED

TW-92 track width mine ploughs

Description
The TW-92 track width mine ploughs are intended for installation on T-55 and T-72 MBTs and are claimed to be more effective than the KMT-4 or KMT-8 mine ploughs. Mounted on attachment points on the MBT hull front the TW-92 has two plough units, each with three tines, which are raised and lowered pneumatically from within the tank. Mines lifted by the plough units from a depth of up to 440 mm are removed from the tank track path and moved outwards away from the tank.

For travelling, the plough units are locked in the raised position and do not affect the normal operational characteristics of the vehicle involved. In an emergency the plough units may be raised or lowered mechanically. When operating, the two plough units, each of which can clear a path 500 mm wide, are joined by a folding metal rod to actuate tilt-rod actuated mines.

Specifications
Weight: 1,200 kg
Length, plough front to front of MBT: 1.165 m ±10 mm
Width of cleared path: 2 × 500 mm
Width between plough units:
(T-55) 2.15 m
(T-72) 2.3 m
Ploughing depth: 440 mm
Ploughing speed: 6-15 km/h

Status
In production. Offered for export.

Manufacturer
Wojskowe Zaklady Inzynieryjne, ul. 15 Pulku Piechoty Wilków, PL-08-522 Dęblin, Poland.
Tel: +48 831 52996, ext 123. Telex: 0643306 wzn pl.
Fax: +48 831 52996.
Marketed by PHZ CENZIN Sp. z o o, PL-00-489 Warszawa, Poland.
Tel: +48 2 628 6874. Telex: 814505 czi pl.
Fax: +48 2 628 6874.

VERIFIED

TEM-7 mineclearing device

Description
The TEM-7 mineclearing device is an electromagnetic unit secured over the front hull of a MBT to detonate any magnetically fuzed mines before the carrier vehicle can reach them. It is normally used in association with mine ploughs but can be used as a stand-alone unit.

The TEM-7 coil unit with its casing is located over the towing attachments on the front of the carrier MBT hull and powered from a source located inside the hull via a special control unit in the driver's compartment. Installation time is from 25 to 30 minutes. If necessary, the carrier tank can be towed using the towing attachments as normal.

The TEM-7 will clear magnetically fuzed mines from a distance of between 2 to 8 m in front of the carrier MBT. The normal clearing speed is 8 km/h. Weight of the unit is 320 kg.

Status
In production. Offered for export.

Manufacturer
Wojskowe Zaklady Inzynieryjne, ul. 15 Pulka Piechoty Wilków, PL-08-522 Dęblin, Poland.
Tel: +48 831 52996 x123. Telex: 0643306 wzn pl.
Fax: +48 831 52996.
Marketed by PHZ CENZIN Sp. z o o, PL-00-489 Warszawa, Poland.
Tel: +48 2 628 6874. Telex: 814505 czi pl.
Fax: +48 2 628 6874.

VERIFIED

ROMANIA

D-5M mineclearing system

Description
The Romanian D-5M mineclearing system is essentially a copy of the RFAS KMT-5M mineclearing plough and roller system (see separate entry in this section) and is normally observed mounted on a T-55 series tank.

The system can be mounted on the tank by a team of four in 45 minutes and can be dismounted in an emergency in one second, without any crew member having to leave the vehicle. The driver can select to use either the roller or plough mode; as with the KMT-5 and KMT-5M they cannot be used together.

The mineclearing speed will vary according to a number of factors but is typically between 6 and 12 km/h. Each block of rollers clears a path 730 to 810 mm wide.

Specifications
Weight: between 7,300 and 7,500 kg
Width: 3.8 m
Length: 3.16 m
Length with carrier tank: 10.05 m

Status
In service with the Romanian Army.

Manufacturer
Romanian state factories.

VERIFIED

Romanian Army T-55 MBT fitted with D-5M mineclearing system

RUSSIAN FEDERATION AND ASSOCIATED STATES (CIS)

Tank-mounted mineclearing rollers and ploughs

Description

PT-54

The PT-54 was introduced into service in the 1950s as the replacement for the earlier PT-3 system. The equipment consists of two independent roller sets which are mounted on arms in front of each track. Each roller set has six rollers and provision is made for clearing mines fitted with tilt-rod fuzes before they can detonate under the hull. To enable following tanks to see which ground has been cleared, the equipment leaves a furrow in the ground 80 mm deep and 100 mm wide. As the tank moves forward the rollers should detonate any mines in its path. The area between the tracks remains uncleared. To widen the lane, additional tanks fitted with mineclearing rollers are used. Normally three tanks work in a wedge pattern to clear a minefield. The PT-54 system weighs 8,800 kg and clears two paths, each 1.3 m wide, with a gap 1.2 m wide between them.

KMT-6 tank-mounted mineclearing plough mounted on Finnish Army T-55M MBT **1996**

PT-54M

This is a modified PT-54 system with five rollers in each section in place of the six of the original. Weight is 7,000 kg and the cleared path is reduced to 0.89 m each side.

PT-55

This is similar to the PT-54 but has only four rollers in each section and clears a narrower path 0.83 m wide each side with a central 1.7 m gap), but at a higher speed than the PT-54. The PT-54/55 series takes three to five minutes to detach and can survive the explosions of 10 swept anti-tank mines. Weight of the PT-55 is 6,700 kg.

KMT-4

This was introduced into service in the 1960s and consists of a 600 mm wide cutting device with five cutting tines mounted at an angle in front of each track. Each cutting device is lowered to the ground by a hydraulic ram. This equipment has a number of advantages over the roller-type mineclearing equipment which can be summarised as follows: reduced weight of equipment; retention of tank's cross-country mobility; and the removal of mines from a path instead of their detonation. The KMT-4 is normally allotted at the rate of three sets to each tank company.

KMT-4M

The KMT-4M was introduced during the late 1960s and is basically the KMT-4 with improvements to the blade attachment system.

KMT-5

The KMT-5 was introduced into service to combine the plough and roller mineclearing equipment. The plough system used is the KMT-4 but the roller design is new, lighter, and has only three rollers per section. Each KMT-5 roller sweeps approximately 800 mm. When not in use the system is carried on a KrAZ-214 (6 × 6) truck fitted with a special KM-61 auxiliary crane.

The roller and plough cannot be used simultaneously other than on good, flat ground. The choice of using the rollers or ploughs depends on the type of terrain, soil or minefield to be breached. The system is fitted with a quick-release disconnect unit to allow the tank driver to release both systems rapidly. The KMT-5 system can survive eight to ten 5 to 6 kg explosions, but cannot be used in snow thicker than 200 mm or over loose, ploughed land. Weight is approximately 7,300 kg.

KMT-5M

The KMT-5M is an improved version of the KMT-5 that incorporates a lane marking plough and the PSK equipment which marks the clear lane at night with a luminous substance. For details of the PSK refer to entry in the *Minefield Marking Equipment* section.

The Romanian D-5M mineclearing system is a virtual copy of the KMT-5M - see separate entry in this section.

KMT-6

The KMT-6 mine plough was introduced during the late 1960s and is an improved KMT-4M developed for the T-64 MBT; it has largely replaced the earlier model in front-line service. It has also been fitted to T-72 series MBTs. The KMT-6 can be fitted with a hard surface attachment for clearing scatterable mines. Each plough can clear a path 0.75 m wide, leaving an intermediate space of 1.9 m.

KMT-6 M2

The KMT-6 M2 is a modified KMT-6. One of the modifications involves three holes in the equipment drawbar in place of the original one to simplify procedures when moving the plough to the travelling position. Some components on the blade sections have been reinforced and the device for triggering tilt-rod fuzes has been modified to allow it to remain attached when travelling; leather straps secure the device in the travelling position. For handling, a BRANO Type 11-10 0.8-tonne capacity ratchet lift is used in place of the usual hand winch. Some extra tools are also provided.

KMT-7

The KMT-7 roller is similar to the KMT-5 but utilises a box channel frame and the rollers are less 'cleated'. The ploughs used with the KMT-7 are similar to those used with the KMT-6. The supporting frames for the rollers are L-shaped to allow the rollers to swing past the frame when a mine is detonated. Chains are fitted between the rollers to detonate tilt-mast fuzes. As with previous models, the rollers are carried on a separate vehicle and are fitted using a crane. The KMT-7 can clear tracks 1.65 m wide with an intermediate space 2.16 m wide.

KMT-1990

This mine plough may be known as the KMT-10 and is a modified KMT-6 with a modified operating mechanism which may allow the simultaneous fitting of a roller mechanism while also enhancing the clearance depth. The KMT-1990 can be used as part of the KMT-7 combination.

Associated equipment

The KM-61 crane on the KrAZ-214 (6 × 6) chassis was designed specifically for use with the plough and roller type mineclearing equipment and is used to position

KMT-6 mine plough

KMT-5 roller mineclearing equipment **1996**

Single KMT-5 mine roller unit

and remove the equipment. The crane has a maximum capacity of 3,200 kg with boom at a reach of 2 m. It can be rotated a maximum of 80° from the forward axis of the truck and can be used on slopes up to a maximum of 4° in any direction.

Specifications

Type	**PT-54**	**PT-54M**	**PT-55**	**KMT-5**
Weight (equipment)	8,800 kg	7,000 kg	6,700 kg	7,300 kg
(section)	3,752 kg	2,700 kg	n/a	2,265 kg
(roller)	500 kg	500 kg	500 kg	n/a
(plough)	—	—	—	420 kg
Assembly (length)	n/a	2.64 m	n/a	3.18 m
(width)	3.9 m	3.8 m	n/a	4 m
Lane swept (each)	1.3 m	0.89 m	0.83 m	0.73-0.81 m
Width unswept	1.2 m	n/a	1.7 m	2.1 m
Operating speed	6-10 km/h	n/a	8-12 km/h	n/a
Safe turning radius	40 m	n/a	65 m	65 m
Ditch crossing ability	3 m	n/a	n/a	2.5 m
Attachment time	10-15 min	10-25 min	10-15 min	n/a

Status
In service with former Warsaw Pact countries. Tank-mounted mineclearing rollers are also known to be in service with India (mounted on T-72), the former Yugoslavia and a number of countries in the Middle East including Egypt, Iraq and Syria. The equipment is usually mounted on T-54/T-55 series MBTs but some T-62s and the more recent T-72, T-74 and T-80 MBTs have provision for installing mineclearing equipment of the roller or plough type.

Manufacturer
State factories.

VERIFIED

KMT-8 track width mine ploughs

Description
The KMT-8 track width mine ploughs are intended for installation on all types of RFAS MBTs provided with the necessary universal attachment points. The two ploughshares, each fitted with three tines, are raised and lowered by pneumatic cylinders. Once lowered they are kept at the required digging depth by sensor shoes which travel on the ground surface and follow changes in terrain contours.

Each plough can clear a path up to 600 mm wide in front of each tank track. In use the two plough units are joined by a metal rod which trips tilt-rod actuated mines.

Installing the plough units takes approximately 1.5 hours. Once installed, they can operate without repair for up to 30 km of normal digging operations. Digging speed can be up to 15 km/h, depending on the type of terrain involved. Once installed, the ploughs do not degrade the normal operating characteristics of the MBT involved.

The vehicle and mine plough combination illustrated is also fitted with an electro magnetic counter-mine system intended to detonate magnetically fuzed mines before the carrier vehicle can reach them.

Specifications
Weight: 1,200 kg
Length, plough front to front of MBT: 1.225 m
Width of cleared path: 2 × 600 mm
Ploughing speed: up to 15 km/h

Status
In production. Offered for export.

Manufacturer
State factories.

VERIFIED

Front view of a T-54/55 MBT carrying KMT-8 track width mine ploughs in the operational position; the MBT also carries an electromagnetic counter-mine system to detonate magnetically fuzed mines

Mineclearing plough for BMP

Description
The mineclearing plough used on the BMP-1 and BMP-2 has also been referred to as the KMT-8 (designation not confirmed) and is much smaller than the rest of the RFAS mineclearing plough equipments. The BMP mineclearing ploughs are narrow and have only two digging tines which probably limits them to clearing surface-laid or shallow-buried mines; the intention is probably to clear scatterable mines. To reduce the travelling width, the outer section of each plough can be folded inwards and held in place by a spring lock. Despite the small size of the ploughs, they apparently affect the carrier vehicle's amphibious abilities, so an extra flotation device is secured to the lower front of the hull. It is possible that this device could be a layer of extra composite or other armour to provide extra protection when clearing mines.

Status
In production. In service with the RFAS and probably some other former Warsaw Pact armies.

Manufacturer
State factories.

VERIFIED

Mineclearing plough fitted to BMP IFV

M1977 armoured mineclearing vehicle

Description

In 1986 reports from Afghanistan mentioned a new armoured mineclearing vehicle. This emerged as a variant of the M1977 armoured recovery vehicle (for details refer to the entry in the *Armoured Recovery Vehicles* section) converted to use the tank-mounted mineclearing plough and roller KMT-5 (see earlier entry).

On the M1977 mineclearing vehicle, the driver is seated at the front right with a circular turret to his left which appears to be armed with a 14.5 mm KPVT heavy machine gun. The KMT-5 system is mounted on the front hull. Conditions in Afghanistan mainly limited use of the KMT-5 system to the mineclearing rollers only, to avoid the mine ploughs damaging tracks and roads to the extent that wheeled vehicles could not follow the cleared path. M1977 mineclearing vehicles observed elsewhere carry the full KMT-5 array.

As the M1977 ARV was based on the T-55 MBT many of the hull and running gear features of the MBT are carried over to the mineclearing vehicle, including the ability to produce smoke screens by injecting diesel fuel into the exhaust outlet above the fourth and fifth left roadwheel stations.

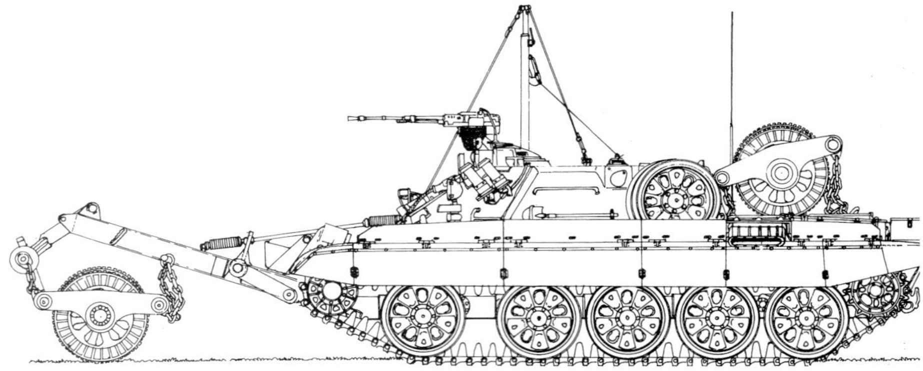

Side view drawing of M1977 armoured mineclearing vehicle fitted with mineclearing rollers from KMT-5 set

Status

In service with the RFAS.

Manufacturer

State factories.

VERIFIED

MTK armoured mineclearing vehicle

Description

The MTK armoured mineclearing vehicle is a derivative of the BTR-50PK tracked armoured personnel carrier carrying a UR-67 rocket launcher system using a length of UZR-3 high explosive line charge 170 m long. The UZR-3 charge is carried in a fabric tube container carried inside the hull of the vehicle. In use, the MTK vehicle is driven to the edge of a minefield and aligned before the rocket is fired from its elevated launcher at the rear of the hull. As the rocket travels it tows the line charge, which is secured by a line to the launch vehicle, across the minefield. The line charge is then positioned by the vehicle crew using the securing line and detonated to clear any mines in its vicinity. The MTK carries three UR-67 rocket and line charge systems.

The MTK weighs just over 14 tonnes and is 7.1 m long.

For details of the BTR-50PK APC refer to *Jane's Armour and Artillery 1995-96* pages 305 to 307.

Status

In service with the RFAS.

Manufacturer

State arsenals.

UPDATED

MTK armoured mineclearing vehicle with UR-67 launcher elevated for firing

MTK-2 armoured mineclearing vehicle

Description

The armoured mineclearing vehicle based on the chassis of the amphibious 122 mm 2S1 self-propelled howitzer was known in the West as the M-1979 but is now known to be the MTK-2. It has a low turret-like superstructure that contains three UR-77 rockets on launch ramps. These, together with the upper part of the superstructure, are hydraulically elevated for firing. Range of the rockets is up to 500 m with each rocket connected, via a towing line, to two 93 m lengths of UZR-77 mine clearance hose stowed folded in the uncovered base of the turret on the vehicle roof. Total weight of the mineclearing equipment is 2,400 kg. Each metre of line clearing charge contains 8 kg of a plastic explosive known as PVV-7.

MTK-2 armoured mineclearing vehicle

The explosive-filled hoses are connected by a cable to the vehicle which, by using guide rails and reversing the vehicle, allows the two-man vehicle crew to position the hoses in the optimum breaching position once the launching has been carried out. The hoses are then command detonated electrically to clear a path up to 90 m long and not less than 6 m wide through minefields provided with pressure fuzes. When tilt-mast fuzes are involved the cleared path can be as much as 8 m. An entire breaching operation can be carried out in from three to five minutes.

The UR-77 launching equipment may be used at least 75 times before repair measures are required.

The MTK-2 is capable of operating in an NBC environment, is amphibious, and has a good cross-country performance; maximum speed is 61.5 km/h and road range 500 km. It has a crew of two, weighs approximately 15,500 kg and is 8.435 m long; the width is 2.86 m and height overall 3.91 m.

Details of the 122 mm 2S1 self-propelled howitzer are given in *Jane's Armour and Artillery 1995-96* pages 550 to 553.

Status
Thought to have entered RFAS service in 1981/82.

Manufacturer (UR-77)
Scientific Research Engineering Institute, Entuziastov Highway, Balashikha-1, Moscow Region, Russia.
Tel: +7 095 521 7171/24 9112.
Telex: 346435 Epokha.

Marketing Agency
VO GED, General Export for Defence, 113324 Moscow, Russia.
Tel: +7 095 231 0049. Telex: 411428.
Fax: +7 095 233 1813.

UPDATED

UR-83P mineclearing system

Description
The UR-83P mineclearing system is carried in sections and assembled at a point close to the minefield to be breached. The system is based around the use of two flexible line charges containing 22 separate plastic explosive charge elements. These two line charges are carried across the minefield by a single rocket weighing 230 kg (360 kg packed), fired from a simple bipod launcher.

The combined line charges weigh 1,380 kg or 1,810 kg when packed. Line charge length is 114 m with maximum delivery range 440 m (+20 m/−40 m). When the line charge is detonated, a breach up to 114 m long and 6 m wide can be cleared.

When deployed, the UR-83P mineclearing system is 1.5 m long, 1.5 m wide and 2.6 m high.

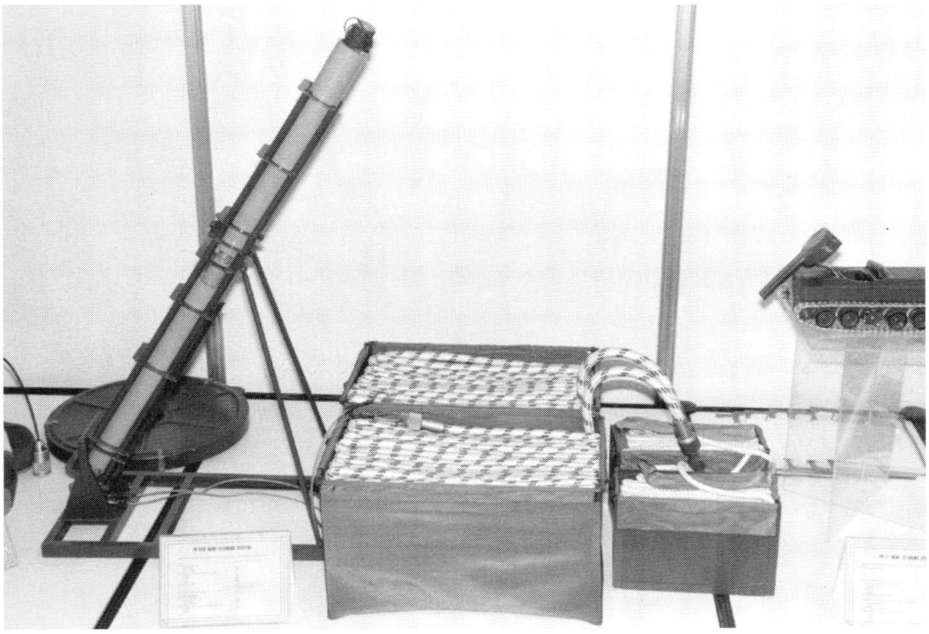

UR-83P mineclearing system (T J Gander)

Status
In production. Offered for export.

Manufacturer
Scientific Research Engineering Institute, Entuziastov Highway, Balashikha-1, Moscow Region, Russia.
Tel: +7 095 521 7171/24 9112. Telex: 346435 Epokha.

Marketing Agency
VO GED, General Export for Defence, 113324 Moscow, Russia.
Tel: +7 095 231 0049. Telex: 411428.
Fax: +7 095 233 1813.

VERIFIED

ZRP-2 mineclearing line charge

Description
The ZRP-2 mineclearing line charge is carried packed in a webbing satchel container and unpacked and deployed close to the edge of the anti-personnel minefield to be breached. The main line charge is a flexible hose 60 m long carried by a small rocket to a range of between 140 and 160 m. When detonated, the charge will clear anti-personnel mines from a path 0.4 m wide and 55 m long.

The ZRP-2 mineclearing line charge weighs 49 kg complete and 33 kg when emplaced ready to fire.

Status
In production. Offered for export.

Marketing Agency
VO GED, General Export for Defence, 113324 Moscow, Russia.
Tel: +7 095 231 0049. Telex: 411428.
Fax: +7 095 233 1813.

VERIFIED

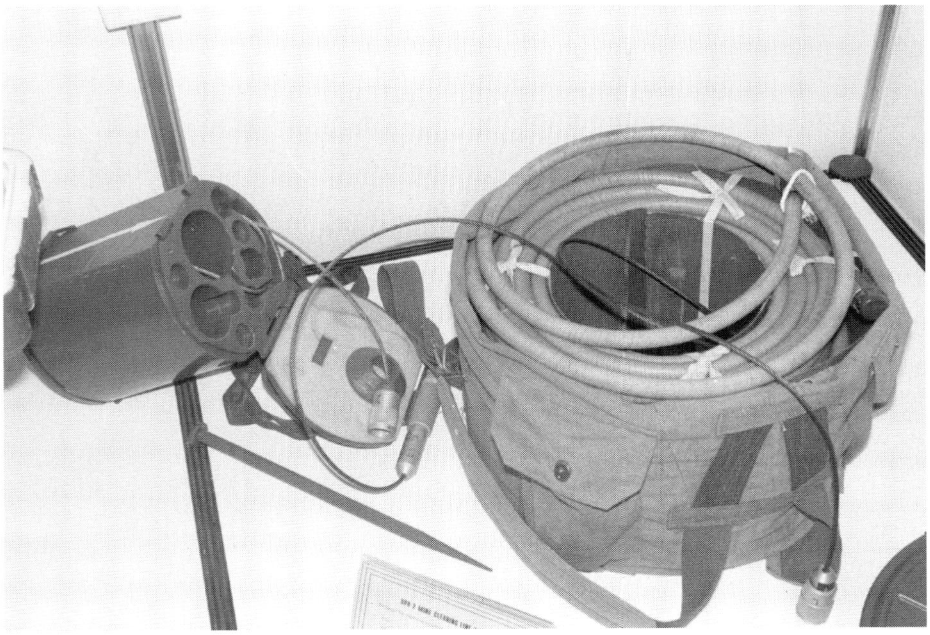

ZRP-2 mineclearing line charge (T J Gander)

ITB-2, SPZ-2 and SPZ-4 mine clearance charges

Description

ITB-2
A rocket-launched anchor and cable is propelled across the minefield and then a winch, or other source of motive force, is used to draw more cable with explosive charges toward the anchor and on to the minefield. Once in position the charge is detonated.

SPZ-2
Placed at the forward edge of a minefield, the SPZ-2 uses a metal-framed anchor guide to winch a cable with explosive charges attached across the area to be cleared at 200 m/h. The charges may be either single, double or triple and, depending on type, may clear a path 300 to 500 m long and up to 6 m wide.

SPZ-4
This consists of a double or triple charge which can either be pushed into the minefield by a tank at a rate of 100 m/h or, if the tank is equipped with roller or plough type mineclearing equipment, laid behind the tank in the gap between the cleared paths left across the minefield by the mineclearing equipment. In the former case, the charge is detonated after it has been positioned by the tank, while in the latter case the charge is detonated by the tank crew.

Status
In service with the RFAS.

Manufacturer
State factories.

VERIFIED

BDT mineclearing charge

Description

The BDT mineclearing charge consists of three separate linear charges which are connected in parallel to form a triple charge. The BDT can also be disassembled to form a single or double charge. Each charge consists of a light metal tube 50 mm in diameter filled with 8.62 kg of cast TNT per linear 305 mm. Lengths of charges can be coupled together to form any desired length. Maximum practical length is 500 m and a squad can assemble a 500 m triple charge in 60 to 90 minutes.

Once the charge has been assembled to the desired length, the nose and detonator are added, and a roller is fitted to the forward end of the charge with a shield mounted above to ensure that there is no premature detonation by enemy small arms fire. The BDT is normally assembled to the rear and towed by a tracked vehicle to the minefield at a maximum speed of 10 km/h. The tank then pushes the BDT into the minefield where it is detonated by either an electric blasting cap, which is initiated by a firing cable connected to the batteries of the pushing tank, or by the detonation box. This contains a number of percussion detonators, connected by a booster charge, which are detonated by machine-gun fire from the pushing tank. When detonated, a triple BDT line charge will clear a path 6 m wide.

Status

In service with the RFAS armed forces and others.

Manufacturer

State factories.

VERIFIED

Bangalore torpedoes

Description

Bangalore torpedoes, also known as extended charges, are used for clearing paths through minefields and barbed wire. Three types are known to be available in the RFAS, the rigid UZ-1 (now largely replaced by later types), the similar UZ-2 and the flexible UShZ-6.

Each section of a UZ-1 Bangalore torpedo consists of a metal tube 1 m long and 53 mm in diameter containing 5.3 kg of explosive. Each section of a UZ-2 Bangalore torpedo is 2 m long, 53 mm in diameter, weighs 10.2 kg and contains 2.65 kg of explosive for every metre length, making a total explosive content for each length of 5.3 kg. Collars are provided for connecting sections together as required.

The flexible UShZ-6 is issued in one or two metre lengths which can be connected in series or parallel as required. Each metre length contains up to 6 kg of explosive. Section diameters can vary from 83 to 88 mm.

When assembled, a Bangalore torpedo is pushed across the minefield and detonated, clearing a path between 2.5 and 3 m wide. If a wider path is required, double or triple charges may be assembled with special collars. These charges may also be placed on carts, sleds or rollers, not only for ease of movement and deployment but also to attain a more favourable detonation height. A metal shield is sometimes positioned on the forward end of the torpedo to ensure that there is no premature detonation by enemy small arms fire. Fuzes used with the UZ-2 include the MD-5M and the EDP-R.

The UZ-2 and UShZ-6 are produced in Bulgaria and are probably manufactured by other former Warsaw Pact countries.

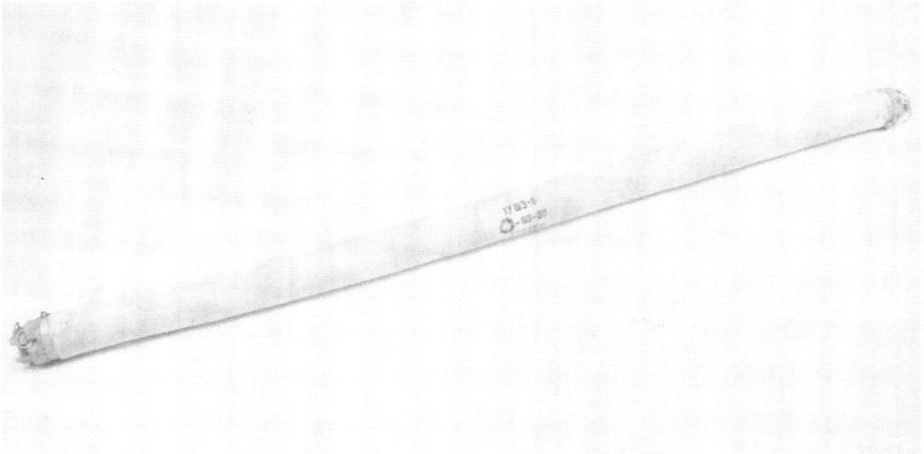

A length of UShZ-6 flexible Bangalore torpedo **1997**

Status

In service with former Warsaw Pact countries.

Manufacturer

State factories.

UPDATED

Electromagnetic counter-mine system

Description

Reference has been made of an electromagnetic counter-mine system fitted to some RFAS MBTs. The system, which may be known as the EMT-7, involves the fitting of a flat assembly on the front glacis of a tank hull, an installation which involves the relocation of some equipment such as the driving lights. Cables connect the assembly to internal junction boxes so it is assumed that the system is powered by the host vehicle's electrical system.

It is also assumed that this system operates in much the same manner as the US VEMASID (qv) in that the assembly houses an electrical coil which creates and projects an electromagnetic field forward of the host vehicle. The field will then detonate any mines with electromagnetic fuzes before the host vehicle reaches them.

Status

In service with the RFAS armed forces.

Manufacturer

State factories.

VERIFIED

Electromagnetic counter-mine system fitted to front glacis of T-72M1 MBT (Michael Jerchel) **1996**

SINGAPORE

Portable Mine Clearing Charge

Description

The Portable Mine Clearing Charge, which has been referred to as Cobra, is a two-man portable rocket-propelled minefield clearing system. It consists of two units, with each unit carried by a soldier by means of an integrated backpack harness.

The two units are the launcher or propulsion unit containing the fuze, front line charge and a separately packed rocket motor, and the container or dispenser unit containing the rear line charge. Each unit has protective cushions to protect it against accidental drops, and light alloy adjustable legs which allow the user to set up the unit at different launch angles.

The system can clear a path approximately 45 m long and 500 mm wide. It is able to cut a two-strand cattle fence or cut a gap through triple concertina wire fences. Anti-tank mines buried up to 70 mm below the explosive points on a line charge will be destroyed or exposed.

At a safe location close to the launch point the legs of the two units are set at the desired launch angle and the rocket motor is connected to the launcher. The prepared units are then set up about 5 m from the starting point of the path to be cleared. The top cushions are removed from both units and the line charges are linked by a connector. When the firing device is triggered, a delay unit is activated to provide a time delay before the system is launched. The rocket-propelled canister then leaves the launcher, drawing out the interconnected line charges, to fall to the ground when the line charges are fully extended. The line charges then detonate after about 7 seconds.

The system is designed not to detonate unless the following sequential events have taken place: the front and rear line charges are connected; the safety pin is removed; and the rocket motor is ignited. The system will not launch when there is accidental rocket motor ignition as it is not connected to the launcher during transport. The rocket motor cannot be fired unless the firing device is connected to it and is activated; its ignition system is non-electric. From the activation of the firing device to the ignition of the rocket motor there is sufficient delay time for the operators to take cover.

The launcher and container units are both 1 m long and 340 m wide. The launcher unit weighs 21.5 kg of which 6.9 kg is the main Composition B charge. The container unit weighs 21 kg of which 9.1 kg is the main Composition B charge.

Status

In production.

Manufacturer

Chartered Ammunition Industries, A subsidiary of Chartered Industries of Singapore, 249 Jalan Boon Lay, Singapore 2261.
Tel: +65 265 1066. Telex: RS 38951 CIS.
Fax: +65 268 7579.

UPDATED

Portable Mine Clearing Charge ready for launch
1996

Bangalore torpedo

Description

The Bangalore torpedo produced by Chartered Industries of Singapore is virtually identical to the US Torpedo, Bangalore M1A1. It is issued in four-section tubes made from aluminium. Each section weighs 4.5 kg and is loaded with 3.25 kg of Grade 1 TNT. Each section is 1.1 m long and has a diameter of 55 mm. The sections are joined by connecting sleeves. Four compressed PETN boosters are used. Standard non-electric detonators can be employed to detonate the charges.

Status

In production.

Manufacturer

Chartered Ammunition Industries, A subsidiary of Chartered Industries of Singapore, 249 Jalan Boon Lay, Singapore 2261.
Tel: +65 265 1066. Telex: RS 38951 CIS.
Fax: +65 268 7579

VERIFIED

Bangalore torpedo

Charge, Mine Clearing, 50 m

Description

This mineclearing pipe charge is issued as a kit consisting of 24 main sections, two front skid units and four detonating units. The 24 main sections can be joined together to clear a path through a minefield 50 m long and 0.5 m wide, but the set may be used to form shorter charge lengths as required.

The front skid unit contains no explosive, is 2.6 m long and weighs 12 kg. The main function of this unit is to carry the remainder of the charge across the gap to be breached. It has on its nose a hinged carrier assembly with three small plastic wheels which enable the charge to be pushed manually across the ground from the rear. The charge may be pushed into place fully assembled or assembly can be gradually carried out as the charge is emplaced.

Each main section is 2.14 m long and has a diameter of 51 mm. Weight is 10 kg and the explosive filling is 3 kg of Composition B. Each unit has push-on connectors to join it to other sections, the front skid unit or a detonator unit.

A detonator unit is 210 mm long, 51 mm in diameter and weighs 1 kg, of which 500 g is the Composition B filling. The detonator unit, and thus the charge, may be detonated electrically or by a blasting cap.

Status

In production.

Manufacturer

Chartered Ammunition Industries, A subsidiary of Chartered Industries of Singapore, 249 Jalan Boon Lay, Singapore 2261.
Tel: +65 265 1066. Telex: RS 38951 CIS.
Fax: +65 268 7579.

VERIFIED

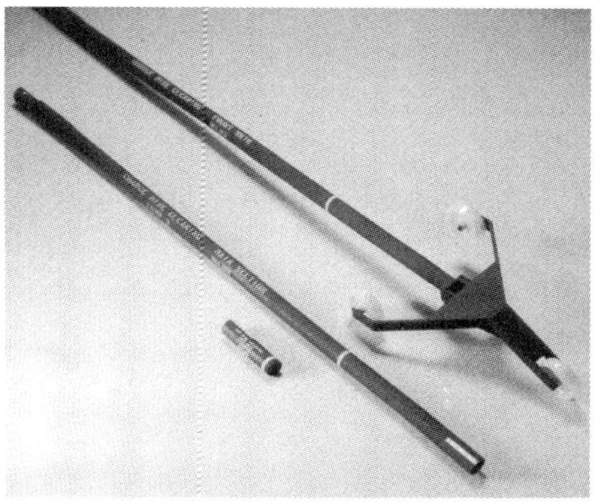

Charge, Mine Clearing, 50 m

SOUTH AFRICA

160AT minefield breaching system

Development

The 160AT anti-tank minefield breaching system, at one time known as the Plofadder 160 (Puff Adder), was developed to provide fast-moving armoured attack groups with the capability to breach enemy anti-tank minefields with the minimum interruption to forward movement. The system was developed and is produced by Somchem, a division of Denel. The system was design and process qualified in 1989, although prequalification units were used in Angola during 1987 and 1988. The system was modified as a result of lessons learnt during that conflict and is now regarded as a fully battle-proven system.

Description

The 160AT anti-tank minefield breaching system can be fitted to any suitably customised armoured vehicle and will create a 170 m long, 9 m wide breach through minefields within seconds of being deployed. The system is armed and made ready for deployment in a rear area before approaching the edge of the targeted minefield; this takes approximately three minutes. The system then moves forward with an attack group and 50 m from the edge of the minefield the system is fired. All operations are carried out from within the vehicle's armoured cab.

The 160AT system consists of a 170 m long explosive rope deployed from a steel container by a double-base propellant rocket. The explosive rope detonates automatically 20 seconds after full deployment and is protected against premature detonation by a Safety and Arming Device (SAD) at the tail end of the rope. The SAD will initiate detonation only when it is 20 m from the vehicle.

The 160AT system is supplied as a complete self-contained system ready for mounting on to the carrier vehicle. The system consists of a steel container measuring 2.5 × 1 × 1.3 m in which are contained the explosive rope, the rocket and the SAD. The mass of the unit is 1,550 kg.

The 160AT system can be used operationally over a temperature range of −15 to +65°C, while the shelf-life is more than five years over a temperature range of +5 to +35°C.

Status
In production. In service with the South African National Defence Forces.

Manufacturer
Somchem, a division of Denel Pty Ltd, PO Box 187, Somerset West 7129, South Africa.
Tel: +27 21 850 3058. Fax: +27 21 850 2992.

VERIFIED

Loading a 160AT minefield breaching system launch pack on to a vehicle

A 160AT anti-tank minefield breaching system loaded on to a Casspir APC

150AP and 38AP minefield breaching systems

Development
The 150AP and 38AP anti-personnel minefield breaching systems, at one time known as the Plofadder (Puff Adder) 150 and Plofadder 38, were developed to provide infantry with the capability to breach enemy anti-personnel minefields quickly and safely. The systems were developed and are produced by Somchem, a division of Denel.

Description
The 150AP and 38AP anti-personnel minefield breaching systems were designed to be fully self-contained units that can be easily transported by personnel on foot. The 150AP system is packed in a wooden container measuring 0.9 × 0.65 × 0.65 m and weighing 135 kg. It is equipped with handles and can be easily carried across most types of terrain by four men. The 38AP one-man system is packed into a standard issue backpack and weighs 35 kg.

In use, both systems are positioned 50 m from the edge of a targeted minefield and can be set up and fired by one man within five minutes. Both systems work on the same basic principle; an explosive rope carried by a rocket. The 150AP system will create a breach 150 m long and 0.9 m wide. The 38AP will create a breach 38 m long and 0.9 m wide.

Both systems are protected against premature detonation by a Safety and Arming Device (SAD) that ensures the system will not detonate unless the rope is fully deployed.

The 150AP and 38AP systems can be used operationally over a temperature range of −15 to +65°C, while the shelf life is more than five years over a temperature range of +5 to +35°C.

Status
In production. In service with the South African National Defence Forces.

Manufacturer
Somchem, a division of Denel Pty Ltd, PO Box 187, Somerset West 7129, South Africa.
Tel: +27 21 850 3058. Fax: +27 21 850 2992.

UPDATED

A 150AP anti-personnel minefield breaching system ready to fire (T J Gander)

A 38AP anti-personnel minefield breaching system ready to fire (T J Gander) **1997**

Dorbyl Chubby mine detection and clearing system

Description
The Dorbyl Chubby mine detection and clearing system is intended for the clearance of mines from roads and tracks and is a system in that several components are involved.

The Chubby has been described as a road train, comprising a Mine Detection Vehicle (MDV, named Meerkat), followed by a second vehicle known as a Towing/Mine Detection Vehicle (T/MDV, named Husky). If required, the latter may tow one or two Mine Detonation Trailers (MDT, named Duisendpoot) which are basically ballasted roadwheels intended to set off mines by pressure alone. A recent innovation is a third three-wheel Duisendpoot towed at the rear. These are followed by a command vehicle and a support vehicle towing spare wheel modules (named Redpacks) and other spares and tools. The convoy can proceed at a speed of 35 km/h and can cover 200 km in one day.

The MDV is a wheeled vehicle with a central mine-protected driver's cab with a V-shaped heavy gauge steel underside to deflect blast. The four wheel stations are frameworks which are so designed that if they are damaged by a mine the entire assembly can be rapidly replaced. This feature is continued throughout the

Drawing of the Meerkat Mine Detection Vehicle (MDV)

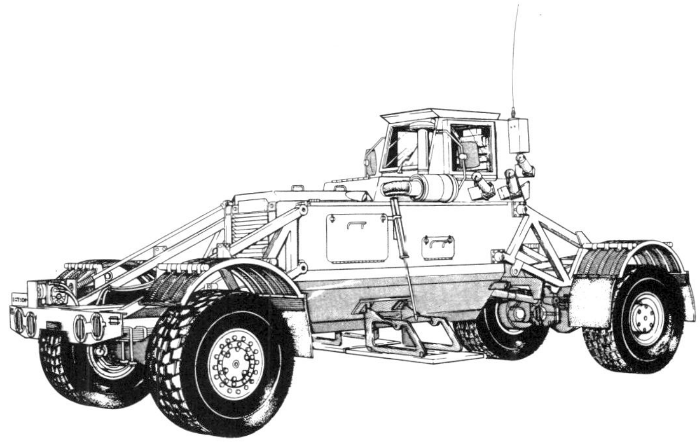

Drawing of the Husky Towing/Mine Detection Vehicle (T/MDV)

The lead vehicles of a Chubby mine detection and clearing system prior to a demonstration (T J Gander) ***1997***

system. Each side of the cab are mine detector coil panels which are lowered to the horizontal for use. The engine is a four-cylinder, water-cooled, direct injection diesel providing drive to the rear two wheels.

The Vehicle Mounted Mine Detector (VMMD) is produced by Reutech and can detect metallic anti-tank mines at speeds of up to 35 km/h. Maximum detection depth is 500 mm below the bottom surface of the array.

The T/MDV is basically a more powerful version of the MDV, as it is intended to tow up to two MDTs. The general construction and layout of the T/MDV resembles the MDV but the engine is a more powerful six-cylinder diesel.

The First and Second Mine Detonation Trailers (F-MDT and S-MDT) are weighted assemblies with their axle lengths so arranged that when two trailers are towed one behind the other the full width of a vehicle is presented to detonate any pressure-fuzed mine which may have escaped detection by the previous vehicles. As with all the other components in the system, if a wheel station is damaged it can be rapidly replaced by spare assemblies following in the convoy.

The Chubby mine detection and detonation vehicles are followed by a command vehicle, usually an armoured personnel carrier and a 10 tonne truck, both towing spare wheel stations and with the truck carrying tools and other spares.

Status
In production. In service in South Africa. Also procured by the French, British and US Armies for operations in Bosnia.

Manufacturer
RSD, a Division of Dorbyl Limited, Victor Street Industrial Sites, Boksburg East, PO Box 229, Boksburg 1460, South Africa.
Tel: +27 914 1400. Telex: 7-46007.
Fax: +27 914 1884.

UPDATED

SPAIN

EXPAL Bangalore torpedo

Description
The EXPAL Bangalore torpedo follows conventional lines and is issued in six units. The first unit is a sheet steel ogival head and the other five are steel tubes each filled with 2.7 kg of Amatol 80:20. The units are joined together by sleeves and the weight of the complete torpedo is 28 kg. Detonation can be carried out by electrical means or by a blasting cap.

Status
In production.

Manufacturer
EXPAL, Avenida Partenón, 5a Planta, E-28042 Madrid, Spain.
Tel: +34 1 722 0235. Fax: +34 1 722 0295.

VERIFIED

SWEDEN

Bofors Demining Vehicle

Development
The Bofors Demining Vehicle is a joint venture between Bofors Weapon Systems, Swedish DeMining and Tonstad Maskinfabrik in Norway; Bofors Weapon Systems has overall system responsibility. An order has been placed by the Norwegian Norsk Folkehjelp which is engaged in anti-personnel mineclearing in Kampuchea, Northern Iraq, Mozambique and Angola. The first prototype was completed by the end of 1995.

Bofors Demining Vehicle showing demining roller ***1996***

Description
The Bofors Demining Vehicle is based on a modified, turretless Leopard 1 MBT chassis. A demining roller is carried on hydraulic support arms at the front of the chassis and powered by a 900 hp engine mounted over the centre of the hull. The demining roller, which is tiltable to follow ground undulations and adjustable for depth, is made up of a series of plates fitted with tungsten carbide teeth which either chew anti-personnel mines into small pieces or cause them to detonate. Damaged plates can be easily replaced by oxyacetylene cutting and welding in the field. A complete roller change can be effected in less than 10 minutes using the hydraulic support arms to lift the roller for access or to load/unload the roller on to or off a transport vehicle. The roller can clear anti-personnel mines to a depth of 500 mm over a width of 4 m. If required, the demining roller can be fitted with a special tool for clearing flat surfaces such as roads.

The vehicle, which weighs approximately 45,000 kg, may be operated either by remote control, using on-board television cameras, or from the protection of the usual driving position which is further protected against fragments by a raised armoured superstructure.

It is estimated that the Bofors Demining Vehicle can clear up to 20,000 m²/hour and can remain operational for periods of up to 12 hours.

Status
In production for Norwegian Norsk Folkehjelp.

Manufacturer
Bofors AB, S-691 80 Karlskoga, Sweden.
Tel: +46 586 81000. Telex: 73210.
Fax: +46 586 58142.

VERIFIED

Minvält 1

Development
Development of a mine roller system to equip the Swedish Army's tank divisions began at the Tank Warfare School at Skövde in 1979-80. Production of the *Minvält 1* (Mine Roller 1 - also known as the NMH 1K) system began during late 1987 and deliveries began to all combat vehicle divisions in February 1989.

Description
The *Minvält 1* system uses two sets of three heavy steel discs suspended from a coupling pushed in front of the carrier vehicle's tank tracks. Two types of coupling have been produced, Coupling 1 (*kopplingsfäste 1*) for use with variants of the Centurion tank (*stridsvagn 101, 102, 104*) and Coupling 2 for use with the S-tank (*stridsvagn 103*). The mine rollers are intended to detonate pressure-activated anti-tank mines. Tilt-rod triggered anti-tank mines located between the rollers are activated by a chain slung between each set of rollers. Each set of rollers is able to withstand the detonation of 15 to 20 heavy anti-tank mines and still continue to detonate mines across the full roller width.

Swedish Minvält 1 mine rollers with Coupling 1 for attachment to Centurion-type MBTs

The overall weight of the mine rollers is 6,000 kg with the coupling weighing a further 1,000 kg. The mine roller, complete with the couplings, can be transported in sections on trucks fitted with cranes. A tank crew can fix the mine roller to a vehicle in 30 to 40 minutes and minefield clearing can then commence immediately. The minefield breaching speed is between 8 and 15 km/h, depending on the terrain. Short distances can be cleared at a maximum speed of 25 km/h.

The mine rollers can be removed by the crew in 20 minutes, but in emergencies the mine roller is removed in five seconds by releasing four explosive spring bolts in the coupling assembly. The emergency release system is controlled by the driver using a TA 11 igniter system (*tändapparat TA 11*) connected to a 24 V supply.

The mine roller system is intended to be used in conjunction with a cleared lane marking system which is still under development. One system under examination involves a central spray device mounted on the rear of the mine roller carrier tank.

Status
In service with the Swedish Army.

Manufacturer
Norbergs Maskin & Hydraulik AB, Box 69, S-778 01 Norberg, Sweden.
Tel: +46 223 23150. Telex: 7524 norberg s.
Fax: +46 223 20432.

VERIFIED

Bofors mineclearing device m/60

Description
The mineclearing device m/60 (referred to as a mineclearing snake) consists of a rocket which is fired from a simple launching device and to which a hose filled with explosive is connected. At the rear end of the hose a 3.6 m diameter parachute is fastened to stretch the hose so that, when it falls to the ground, the hose is practically straight. There are igniters in the front and rear ends of the hose which ignite delay compositions when the rocket is fired and which, after 15 seconds, initiate the explosive in the hose via primers.

Depending on the sensitivity of the mines, all within 0.15 to 1 m are exploded on detonation. Anti-tank mines with normal sensitivity are also exploded while more shockproof mines are thrown aside or uncovered. The device produces a clearly defined path with a width of 0.3 to 2 m and a length of up to 150 m. The rocket has a range of 200 to 225 m.

The device can be prepared for firing in less than one minute. It is carried by a section of seven men supervised by a section commander. One man carries the rocket with the launching device (which also serves as a carrying rack) and the detonator. The other six men each carry 25 m of hose. One of the six also carries the parachute. The rocket pack weighs 18 kg, the hose pack with parachute 15.5 kg and the other hose packs 14 kg.

Specifications
Rocket
Calibre: 103 mm
Weight:
(complete) 9.5 kg
(connection extras) 1.7 kg
(propellant) 4.3 kg
Velocity: 50 m/s
Range: 200 to 225 m

Explosive hose
Material: terylene
Total length: 150 m (6 × 25 m sections)
Diameter: 32 mm
Weight per 25 m length: 12.5 kg
Weight of filling: 0.4 kg/m
Type of filling: desensitised PETN

Status
Production complete. In service with the Swedish Army.

Manufacturer
Bofors AB, S-691 80 Karlskoga, Sweden.
Tel: +46 586 81000. Telex: 73210.
Fax: +46 586 58142.

VERIFIED

TURKEY

MKEK mineclearing system

Description
The MKEK mineclearing system was originally developed by Elroksan of Ankara and is marketed by Makina ve Kimya Endüstrisi Kurumu (MKEK). It consists of a trailer bearing an explosive line charge to be carried across a minefield by a cluster of eight 2.75 inch G1 rockets. The single axle trailer can be towed by a tank or any other suitable vehicle.

The cluster of eight 2.75 inch G1 rockets is contained in a streamlined housing. It is launched from an adjustable elevation angle launching spigot on a raised mast at the rear of the trailer. The flexible line charge weighs 400 kg and is filled with C-4 plastic explosive.

In use the trailer is towed to a point 80 m from the edge of the minefield to be breached and the angle of the launcher spigot is adjusted as required. The trailer is pointed in the required direction and personnel in the vicinity take cover. The rockets are fired to a maximum range of 200 m by one operator using a firing and control unit powered by a 24 V 90 A DC electrical supply. The firing and control unit features several multi-step safeties.

Once the line charge is in position across the minefield it can be detonated by the operator using the firing and control unit. The resultant explosion will clear anti-tank mines from a path 100 m long, up to 10 m wide and to a depth of 1 m, with a claimed reliability of

MKEK mineclearing system showing line charge ***1997***

99 per cent. Minefields more than 100 m deep can be cleared using further systems.

A training version is available with the explosive content of the line charge replaced by a synthetic polymer. The procedure for using the training version is exactly the same as for the operational equipment except that the detonator produces coloured smoke on activation.

Storage life of the MKEK mineclearing system is 10 years.

Specifications

Rocket calibre: 69.85 mm (2.75 in)
Number of rockets: 8
Weight of line charge: 400 kg
Explosive type: C-4 plastic
Length of cleared path: 100 m
Width of cleared path: 10 m
Electrical requirements: 24 V 90 A DC
Operating temperature range: −30 to +52°C

Trailer
Weight:
(loaded) 2,530 kg
(empty) 1,170 kg
Height: 785 mm
Width: 2.2 m
Launcher height: 1.13 m
Launcher length: 1.5 m
Launcher elevation angle: 30°(±3°)
Towing speed:
(road) 80 km/h
(cross country) 40 km/h

MKEK mineclearing system ***1997***

Status

In production.

Manufacturer

Makina ve Kimya Endüstrisi Kurumu (MKEK), T-06330 Tandoğan, Ankara, Turkey.
Tel: +90 312 223 2011. Telex: 42 223 MKGA TR.
Fax: +90 312 222 2241.

NEW ENTRY

UNITED KINGDOM

Royal Ordnance Giant Viper anti-tank mineclearing equipment

Description

The Giant Viper (L5) mine clearance system consists of a 230 m long, 67 mm diameter hose filled with PE6/Al aluminised plastic explosive. This is packed coiled in a wooden box which is mounted on a single-axle L7 trailer, which can be towed by a variety of tracked vehicles (for a photograph of this trailer refer to entry in the *Trailers* section). The British Army uses the Chieftain AVRE, FV432 APC, Combat Engineer Tractor (CET) or Challenger MBT to tow the trailer. The hose is fired across the minefield by a cluster of eight rocket motors. The tail end of the hose is fitted with arrester gear in the form of three parachutes which straighten the hose during flight and operate the striker mechanism, which detonates the charge after the hose has landed.

The equipment is used as follows: the Giant Viper is towed to the firing point between 115 and 140 m from the edge of the minefield and the trailer, which remains attached to the towing vehicle during the operation, is lined up on the proposed line of flight. Taking into account any factors such as wind which will alter the trajectory, the Giant Viper is then fired electrically from the towing vehicle. The Giant Viper was designed to blast a passage for vehicles through a minefield from 183 to 230 m long and 7.3 m wide. During trials, up to 90 per cent of anti-tank mines were cleared, provided that they were not blast-proofed or multi-impulse fuzed.

An extension of the Giant Viper system known as the Tandem Firing Concept was developed for use with various towing vehicles, including MBTs. The vehicle tows two Giant Viper trailers, one behind the other. The rear Giant Viper is fired outside the minefield as normal and the trailer is jettisoned. The vehicle and the remaining Giant Viper trailer then proceed into the minefield to the second firing point where the process is repeated. This allows minefields up to 400 m deep to be breached.

Each Giant Viper system requires the support of a 4 tonne truck. The expendable stores are packed in containers which can be manhandled, except the charge box containing the hose, which weighs 2,136 kg and is handled by a forklift truck or crane. The charge box is 3.15 m long, 1.74 m wide and 1.18 m high.

In addition to the special L7 trailer used to carry Giant Viper, the Royal Engineers also make use of L8 trailers, which were formerly employed to carry now obsolete Heavy Girder Bridge sections. This trailer can be recognised by the two coupled wheels on each side. Other 3 tonne capacity single- and double-axle trailers could be modified to carry the Giant Viper.

For training purposes, practice equipment is available which is identical to the live system in all respects except that the line charge contains an inert filling. The training version is supplied with sufficient components to enable a maximum of six projections to be made.

Variant

Improved Giant Viper

In mid-1992 the Ammunition Division of Royal Ordnance was awarded a contract to carry out a Mid-Life Improvement (MLI) programme to the Giant Viper mineclearing system.

The main improvements involve the replacement of the eight-rocket cluster by a single new rocket using a new launcher and having a slightly increased impulse to provide an improved lay of the hose with an overall directional accuracy of 10° either side of aim. With the revised system the rocket can be elevated on command to the firing angle and automatically tilts to the true horizontal. Other changes include: a new safety, arming and firing unit; an advanced electrical system to allow all prefiring checks to be carried out from within the towing vehicle; replacing the drogue chutes with tear webbing retardation to ensure maximum hose straightness; and replacing the practice hose with a new ballistically matched 6 kg/m woven nylon rope for training purposes over a life of at least 100 firings.

The Improved Giant Viper, known to the British Army as Python, will provide a user with reduced life cycle costs, quicker into-action times, improved reliability and safety and a reduction in training times.

Specifications

Hose box
Weight: 2,136 kg
Length: 3.15 m
Width: 1.74 m
Height: 1.18 m

Hose
Length: 230 m
Diameter: 67 mm
Filling: aluminised plastic explosive PE6/Al

Head range: (average) 300-350 m
Tail range: (average) 110-120 m
Accuracy: 10° either side of aim
Cleared zone: 183 m (minimum) × 7.3 m wide

Status

In service with the British Army and Jordan (100 units). Undergoing a mid-life improvement programme (see text).

Manufacturer

British Aerospace Defence Limited, Royal Ordnance Division, Euxton Lane, Chorley, Lancashire PR7 6AD, UK.
Tel: +44 1257 265511. Fax: +44 1257 242199.

UPDATED

Improved Giant Viper being fired from trailer towed by Chieftain AVRE ***1997***

Schermuly Rapid Anti-personnel Minefield Breaching System Mark 3 (RAMBS 3)

Description

The Rapid Anti-personnel Minefield Breaching System Mark 3 (RAMBS 3) is a self-contained, lightweight (7 kg), one-shot method of quickly clearing safe lanes through surface-laid anti-personnel minefields. RAMBS 3 uses a rocket-assisted Line Carrying Projectile (LCP) fired from a standard assault rifle to propel an explosive line 60 m across a minefield. Detonation of the line charge enables breaching of the minefield (including tripwires) by secondary detonation, disruption or dispersal of the mines.

The breach is based on the use of 60 m of explosive line which is coiled ready for use in the bottom tray of a camouflaged haversack. Before launching, the kit is secured by a restraining strap and the LCP, which is attached to the explosive line by a bridle, is fitted over the rifle muzzle. Using a standard rifle round and bullet, the LCP is propelled across the minefield, towing the explosive line in the direction in which the breach is to be made. The explosive line is then command detonated using a simple percussion firing device. The resultant explosion clears a path approximately 60 m long and 0.6 m wide. With training, the entire operation can take less than 60 seconds. Lightsticks are provided as breach markers.

RAMBS 3 uses an RDX plastic bonded explosive (PBX) line which provides an improved blast overpressure in the mine killing zone. It is insensitive to bullet attack and provides an improved discernible pathway in day, night or low visibility.

Status

In production for an undisclosed NATO customer.

Manufacturer

Pains-Wessex (Schermuly) Limited, High Post, Salisbury, Wiltshire SP4 6AS, UK.
Tel: +44 1722 411611. Telex: 47486 pw sch g.
Fax: +44 1722 412121.

VERIFIED

A Rapid Anti-personnel Minefield Breaching System Mark 3 (RAMBS 3) about to be fired

Pearson Mine Ploughs

Development

The Pearson Mine Plough is available in three versions: the Track Width Mine Plough (TWMP); the Engineer Mine Plough (EMP); and the Full Width Mine Plough (FWMP). The TWMP has five tines per side, the EMP has seven, while the FWMP clears the whole width of the vehicle. The only difference between the TWMP and the EMP is the number of tines; they are interchangeable.

The EMP comprises two independent ploughs, left and right, pivoted from a common mounting plate. Unlike ploughs which have a single pivot, the Pearson plough is mounted on a sophisticated linkage which provides maximum ploughing stability throughout the range of movement. The top link is a hydraulic cylinder which lowers and raises the plough from the control box operated by the driver.

The FWMP is the EMP with the addition of a third plough in the centre and two arms which locate the depth control skids in front of the ploughs. An EMP can be converted into an FWMP in about 20 minutes.

The only modifications to the vehicles required to fit the plough are two wedge blocks welded to the toe plate, an entry point for the single electrical harness and a mounting bracket next to the driver for the control box. On most tanks the existing eyes at the top of the toe plate are used to pin the plough in place, enabling the plough to be fitted or removed within minutes. In an emergency the plough can be jettisoned using explosive pins.

Specifications

Type	FWMP	EMP	TWMP
Weight	2,950 kg	2,400 kg	2,300 kg
Cleared lane width	3.8 to 4.25 m	2 × 1.4 m	2 × 1.1 m
Cleared lane depth	200-400 mm	200-400 mm	200-400 mm
Max operating current	240 A	240 A	240 A
Operating voltage	24/28 V	24/28 V	24/28 V
Ploughing speed range	1-10 km/h	1-10 km/h	1-10 km/h
Min safe ploughing radius	50 m	50 m	50 m

The mounting system, harnesses and control box are identical to those used by the Pearson Combat Dozer (see entry under *Armoured Engineer Vehicles*).

The British Army has a requirement for a new track width mine plough.

Jacklift Pallet

The Pearson Jacklift Pallet enables front end attachments to be fitted and removed from fighting vehicles without the need for cranes or other lifting equipment. The hydraulic pallet enables two men to install a mine plough or bulldozer unit on any MBT in 10 to 15 minutes.

The Jacklift consists of a pallet with three independently operated hydraulic legs mounted on a demounting sledge. Unloading the pallet and sledge from a vehicle or container is achieved by towing the entire configuration down two ramps until it sits on the ground. The pallet, sledge and equipment may then be towed any distance into the required position.

The mine plough or bulldozer can then be raised to the required height for fitting to the vehicle by manually jacking the three hydraulic legs. This takes about two minutes. Each hydraulic leg incorporates an individual pump and ram unit which allows for precise control of height and pitch of the equipment to be fitted.

The Jacklift Pallet weighs 480 kg, is 4.1 m wide and 2.27 m long. Maximum lift is 555 mm and load capacity 5,500 kg. Deck height is 190 mm.

Status

Production as required. In service with the British Army and in the Middle East and Far East (including Singapore). Sold to and under evaluation by several European armies, including France, which ordered an undisclosed number in mid-1994.

Manufacturer

Pearson Engineering, Wincomblee Road, Walker, Newcastle upon Tyne NE6 3QS, UK.
Tel: +44 191 234 0001. Fax: +44 191 262 0402.

UPDATED

Pearson Full Width Mine Plough (FWMP) fitted to a Chieftain AEV

Pearson Full Width Mine Plough (FWMP) fitted to a Leopard 2 MBT; note the deployed Pathfinder marker system

1996

Pearson Surface Mine Plough (SMP)

Description

The Pearson Surface Mine Plough (SMP) was designed to meet the requirements of United Kingdom Ministry of Defence Staff Target 3916, and was selected after a programme of competitive evaluation. Two models are produced; the SMP-1 for vehicles such as Warrior MCV and the SMP-2 for lighter vehicles such as trucks.

The SMP-1 is a multi-section, multi-bladed V-plough which physically removes mines from the path of a vehicle. The combination of the articulated blades and the inbuilt hydraulic system enables the plough to conform to the local topography, maintaining blade contact with the ground across its width. The blades are articulated in such a way that they are able to ride up and over small obstacles without damage, even at speed. The SMP-1 has been demonstrated working with 100 per cent efficiency on forest tracks at speeds of up to 25 km/h.

Pearson Surface Mine Plough (SMP-1) fitted to Warrior MCV

The SMP-1 and SMP-2 have been extensively tested on all types of terrain including hard surfaces, rutted tracks, grassland, moorland, sand and stony desert, all under wet and dry conditions. Tests were carried out using both a tracked APC and a 4 × 4 truck.

The SMP-2 is light enough to be carried by vehicles from a 4 × 4 truck (with a load capacity heavier than 4 tonnes) to an MBT and is robust enough to pass blast trials and remain operational. It uses the Pearson Wedge Block Mounting System which enables it to be fitted to or removed from vehicles in minutes. The only vehicle modifications required are two wedge blocks welded to the toe plate (or a small adaptor bracket fitted), an entry point for the control/power harness and an internal mounting bracket for the control box.

Specifications

Model	**SMP-1**	**SMP-2**
Weight	1,250 kg	800 kg
Cleared lane width, blades	3.56 m	3.16 m
Cleared lane width, springs	4.65 m	4.45 m
Number of tines	18	16
Max operating current	140 A	145 A
Operating voltage	24/28 V DC	24/28 V DC

Status

In service with the British Army. Sold to some undisclosed European armies (including France) and Canada.

Manufacturer

Pearson Engineering, Wincomblee Road, Walker, Newcastle upon Tyne NE6 3QS, UK.
Tel: +44 191 234 0001. Fax: +44 191 262 0402.

VERIFIED

Pearson Area Clearance Plough (ACP)

Description

The Area Clearance Plough (ACP) is a development of the SMP (see previous entry) with the major difference being that the ACP uses a straight-angled 18 segment blade preceded by a heavy chain curtain suspended from two booms. It is also heavier and stronger. Mines and other surface debris are cleared to one side so that by repeated passages, a large area can be rapidly cleared. As with the SMP, the ACP requires no power while operating and can be fitted to a wide range of vehicles with little modification.

In an emergency the ACP can be jettisoned from under armour by explosive pins. Blades can be fitted or removed by the crew in 10 minutes. Raising and lowering time for the ACP is 15 seconds.

Specifications

Weight: 1,500 kg
Width overall: 4 m
Width of cleared path: 3.8 m
Ploughing speed: 3-30 km/h
Max operating current: 140 A (intermittent)
Operating voltage: 24/28 V

Status

Under development with several European armies and the United States.

Manufacturer

Pearson Engineering, Wincomblee Road, Walker, Newcastle upon Tyne NE6 3QS, UK.
Tel: +44 191 234 0001. Fax: +44 191 262 0402.

VERIFIED

Pearson Area Clearance Plough (ACP) fitted to Warrior MCV
1996

Pearson Crawler-mounted Full Width Mine Plough (C/FWMP)

Description

The Pearson Crawler-mounted Full Width Mine Plough (C/FWMP) was developed as part of a minefield breaching and clearance system capable of being fitted to most heavy commercial tracked tractors such as the Caterpillar D9. The overall design is a development of the Pearson Full Width Mine Plough (FWMP - see under Pearson Mine Ploughs in this section) for use with MBTs and their derivatives, although the C/FWMP is much heavier and capable of clearing a mine-free path up to 4.6 m wide. Unlike the Pearson FWMP, which can be fitted to, or removed from, any MBT in 10 to 15 minutes, the C/FWMP is part of a dedicated minefield breaching and clearing machine.

The system is designed for driverless operations, using a video link at distances of up to 3,000 m and incorporates dual Pearson Pathfinder Markers mounted at the rear and a Vehicle Signature Duplicator (VSD), manufactured by Giat Industries of France, built into the plough structure. The VSD activates magnetically influenced mines up to 5 to 6 m ahead of the vehicle.

Ploughing depth control is similar to that on the FWMP. The C/FWMP is mounted on a purpose-built 'U' frame which replaces the normal bulldozer frame on the track frame fittings and is attached to the bulldozer lift cylinders. Each plough unit may have two or three tines on replaceable adaptors. The C/FWMP is deployed and stowed by operating the crawler hydraulics.

Specifications

Weight: 7,500 kg
Width of cleared path: 3.5-4.6 m
Cleared lane depth: 250-350 mm

Status

Production as required. In service with the French Army.

Manufacturer

Pearson Engineering, Wincomblee Road, Walker, Newcastle upon Tyne NE6 3QS, UK.
Tel: +44 191 234 0001. Fax: +44 191 262 0402.

VERIFIED

Pearson Crawler-mounted Full Width Mine Plough (C/FWMP) mounted on Caterpillar D9 tractor
1996

Pearson Mine Roller (MR)

Description

The Pearson Mine Roller (MR) is a track width countermine vehicle protection device for armoured vehicles such as armoured personnel carriers. The MR is designed to detonate mines which would otherwise detonate under the vehicle's track, thus reducing the risk of damage to the vehicle and its crew. Protection against tilt rod mines is built into the MR system.

The downward force required to activate mines is created by a hydraulically controlled mechanism. The castered wheel sets include four wheels each side, each wheel providing an average downwards force of approximately 400 kg. Wheel sets can be readily replaced or repaired as required as the equipment is designed to be sacrificial.

The hydraulic installation required to operate the MR is identical to that used with the Pearson Surface Mine Plough, Area Clearance Plough and UKD-2 dozer blade and all are interchangeable.

Pearson Mine Roller (MR) attached to a Canadian M113A2 APC ***1997***

Specifications

Weight: 1,150 kg
Operating speed: up to 50 km/h
Number of rollers: 2 × 4
Average force on each roller: 400 kg
Roller track width: 2 × 700 mm
Raise cycle: 6 s
Lower cycle: 8 s
Voltage: 24-28 V DC
Current: (max) 140 A

Status

In production. In service with the British Army.

Manufacturer

Pearson Engineering, Wincomblee Road, Walker, Newcastle upon Tyne NE6 3QS, UK.
Tel: +44 191 234 0001. Fax: +44 191 262 0402.

NEW ENTRY

Aardvark Rapid Area Clearance Equipment (RACE)

Description

The Aardvark Rapid Area Clearance Equipment (RACE) was developed to meet the British Army's General Staff Target 3916 for a scatterable mine clearance device.

RACE is similar in concept to the Aardvark Joint Services Flail Unit (see entry in this section) and uses the same type of striker heads. The system has two 1.524 m rotors that operate independently but in unison. The two rotors have their own individual motors and together clear a path approximately 3.048 m wide. Chain heads can be varied to suit the operational conditions and types of mine being cleared. The flail chains are 0.609 m long and were designed to disrupt, lift and throw mines to one side. The system was designed to clear mines with an explosive content of up to 2 kg. The direction of rotation of the flails can be changed in three seconds. Mines can be cleared from most surfaces including cultivated fields, front banks and some types of ditch. Typical clearance speeds are 15 km/h on hard surfaces and from 8 to 15 km/h cross-country, depending on the operating conditions.

The RACE system can be installed on cross-country trucks and light armoured vehicles such as M113 and FV432 APCs; the only modification to the latter is the provision of four bolts to which RACE can be quickly attached. The size of the system is such that it can be left on the vehicle without degrading its operational capabilities. Normally the system would run off the prime mover's hydraulic power take-off, although the prototype uses an auxiliary power unit.

Aardvark Rapid Area Clearance Equipment (RACE) undergoing trials

RACE(C) is a variant intended for installation on medium-sized (and upward) tracked or wheeled engineer plant, such as the Volvo L90 wheeled loader with an armoured cab provided by Glover Webb of Southampton. The RACE(C) version is essentially similar to the standard RACE, although the minimum clearance width is 3.2 m; the clearance width of scatterable munitions from hard surfaces is 4 m.

Status

Prototypes.

Manufacturer

Aardvark Clear Mine Limited, Shevock Farm, Insch, Aberdeenshire AB5 6XQ, UK.
Tel: +44 1464 20122. Telex: 73509 aard g.
Fax: +44 1464 20985.

VERIFIED

Aardvark Joint Services Flail Unit (JSFU)

Description

The Aardvark Mark 3 Joint Services Flail Unit (JSFU) is a half-track vehicle designed to carry the Aardvark flail device in operational conditions. The flail assembly is attached to the rear chassis and controlled from a two- or three-man fully armoured crew cab. The power train, incorporating a 120 hp in-line diesel engine with an integrally mounted 16 speed transmission and lockable differential, is located between the crew cab and the flail assembly and is fully protected by an armour-plated compartment hood and a full length belly plate.

The flail assembly consists of a 3.05 m wide, 175 mm diameter double-sleeved rotor carrying 72 chains along a triple helix design (seven chains per 305 mm). Each chain can be fitted with a choice of different striker heads (for example, disc, ball or T-head hammer).

Power to the rotor is supplied from a power train power take-off via the assembly support arms. A resistant blast deflector plate is attached between the flail and the prime mover. The boom arms are raised and lowered hydraulically and incorporate automatic contouring and depth control features. The air force/hard surface version can accommodate an angle dozer plate and support wheel attachments.

Clearance speeds vary depending on the type and quantity of devices encountered. An optimum beat or sweep pattern should be maintained throughout active operations. Trials have shown that rough terrain/cross-country speeds average 3 km/h while the unit has cleared surface-laid objects on hard surfaces (for example, airfields) at 15 km/h.

Optional extras include a 7.62 mm machine-gun turret installation on the cab roof and side-mounted smoke dischargers.

A development is the JSFU MKIIID with an interchangeable drive arrangment which can be either tracked or wheeled. The large tractor-type wheels involved provide the JSFU with a much higher road speed and manoeuvrability.

Aardvark Joint Services Flail Unit (JSFU) MKIIIC in action in Kuwait

Specifications

Crew: 1 + 1 or 2
Weight: approx 12,000 kg
Length: 8.08 m
Width: 3.55 m
Height: 2.67 m
Max road speed: 20 km/h
Max operating gradient: 33%
Beaten path: 3.05 m

Status

In production. Over 100 units in service with 12 countries, including Egypt, Finland, France (MKIIID), Italy, Jordan, Kuwait, Pakistan, Saudi Arabia, Sweden, UK and USA. Also used by UNIFIL.

Manufacturer

Aardvark Clear Mine Limited, Shevock Farm, Insch, Aberdeenshire AB5 6XQ, UK.
Tel: +44 1464 20122. Telex: 73509 aard g.
Fax: +44 1464 20985.

UPDATED

Aardvark Combi Valliant Mineclearing Vehicle

Description

The Aardvark Combi Valliant Mineclearing Vehicle is a heavy-duty development of earlier Aardvark mineclearing flail systems. The system is a 'bolt-on' design on a 350 hp dozer vehicle of the D9 type, with the flails driven by a hydraulic system designed by Commercial Hydraulics Liverpool Limited. The hydraulic system was designed so that it can be fitted on to a wide variety of compatible dozers and is powered by an auxiliary power unit mounted behind the sound-proofed vehicle cab. The driver controls the variable speed flail drive by electrical signals to achieve optimum performance.

The flails can clear a path 3 m wide of all types of mine.

Status

Prototypes.

Manufacturer

Aardvark Clear Mine Limited, Shevock Farm, Insch, Aberdeenshire AB5 6XQ, UK.
Tel: +44 1464 20122. Telex: 73509 aard g.
Fax: +44 1464 20985.

VERIFIED

Aardvark Combi Valliant Mineclearing Vehicle

Mine Mole

Description

The Mine Mole is the result of two years of extensive research and development into a mechanical mine-clearing device which is self-propelled, requires no support transportation, can be remotely driven and controlled. The vehicle is armoured to withstand blast and fragments and is capable of operating in most terrains and environments where mines and minefields exist.

The design of the Mine Mole is based on existing earth-scrapers used in the construction industry but highly modified for the mineclearing role. It is a 35 tonne wheeled armoured vehicle driven by tandem turbocharged diesel engines, 330 hp at the front and 225 hp at the rear; the engines have shields to protect them from damage. An eight-speed transmission is provided with individual clutch modulation. If required, the Mine Mole can be driven using only one of the engines.

The front unit contains an armoured compartment for the driver/operator who is further protected by mine ploughs mounted in front of the unit and a boat-shaped armoured hull underside. The compartment is sound suppressed and has a Roll-Over Protection System (ROPS).

The rear unit is basically a scraper which digs up to 333 mm into the soil to scoop up mine-contaminated topsoil and place it into a central compartment known as the magazine. Inside the magazine the soil is sifted and subjected to vortices created by low-pressure air lines run by separate compressors and routed along each side of the magazine. The vortices cause the soil to be sifted through vents in the sides of the magazine so that mines and other debris are left inside. Once the magazine is full the vehicle can be driven to a suitable location where the magazine can be hydraulically emptied in safety.

The magazine can receive up to 15.3 m^3 of spoil weighing approximately 21,773 kg at one time. A path up to 3 m wide can be cleared.

One advantage of the Mine Mole is that, having a digging depth of up to 333 mm, it will dig and lift under a mine, while most mines have their fuze systems on the top. The chances of fuzes functioning while being lifted by the Mine Mole are thus considerably reduced.

Status

Design stage completed. Ready for production.

Designer

Michael G Fellows, PO Box 1227, Arundel, West Sussex BN18 0AU, UK.
Tel: +44 1243 551025. Fax: +44 1243 551025.

VERIFIED

UNITED STATES OF AMERICA

Project Remote

In response to an urgent request from the US Army and Marine Corps operating in Saudi Arabia during the run-up to Operation Desert Storm, the US Army Tank-Automotive Command (TACOM) Research, Development and Engineering Center assembled and outfitted 20 remote-control modification kits for M60A1/A3 tanks to allow them to be operated with Track Width Mine Ploughs (TWMPs) for remotely controlled mineclearing in forward areas. The programme was named Project Remote.

The Project Remote kits were based on target tank systems modified and supplemented with TACOM-supplied equipment, including an interface for the TWMP. The kits contained various safety override systems with duplication of some critical systems and mechanisms, including the communications systems.

The remotely controlled M60s with TWMPs were intended to operate in two-vehicle teams, the M60 breaching tank and a controlling M60 or M1 Abrams tank. The operator in the control tank guides the breaching tank using the control tank's main gun optics or thermal imaging equipment.

The short duration of Operation Desert Storm did not allow any operational use of the Project Remote kits but they have since been fitted to tanks at several US Army locations in the United States. The system was demonstrated in Germany during late 1992, fitted to a M1 Abrams tank fitted with mine ploughs.

Status

In service with the US Army.

VERIFIED

Vehicle Magnetic Signature Duplicator (VEMASID)

Development

The Vehicle Magnetic Signature Duplicator (VEMASID) is a system designed to detonate magnetically fuzed mines before the carrier vehicle reaches them, by projecting an electromagnetic field in front of the host vehicle. Any magnetically fuzed mines in the path of the carrier will then be detonated before the host vehicle reaches them.

During 1987 Alliant Techsystems Inc was awarded a $4,000,402 increment as part of a $20,537,900 contract for full-scale engineering development, fabrication, test and demonstration of VEMASID kits. The contract covered a total of 30 testbed units, six test sensors and 25 training mines. The work was completed by September 1992. The system was developed under contract to the US Army's Belvoir Research, Development and Engineering Center and Project Manager Mines, Countermine and Demolitions (PM-MCD) at Fort Belvoir, Virginia.

Vehicle Magnetic Signature Duplicator (VEMASID) coil mounted on front hull of M109A3 self-propelled howitzer ***1996***

Description

VEMASID operates by using an electromagnetic coil mounted on the front of the carrier vehicle. Radiation from the coil induces the magnetically fuzed mines to detonate. It has been fitted on to the M1 and M60 tank series and M109 series self-propelled 155 mm howitzers and can be adapted to a number of combat vehicles.

On all vehicles the coil is carried over the front hull of the vehicle. The coil is usually held in place by tie-down straps and can be fitted by four men within 60 minutes. VEMASID can be used in conjunction with other mineclearing devices such as mine rollers or ploughs for breaching minefields with both pressure and magnetically fuzed mines.

The coil for the majority of vehicles weighs 129.7 kg. A typical assembly measures 1.625 × 0.86 × 0.305 m. The system operates off the carrier vehicle's 28 V power supply and the weight of the electronics package carried internally is 12.7 kg. The full VEMASID kit consists of the coil, amplifier, control panel (placed near the tank commander or driver) and cables.

Status

Type classified for the M109 and M992 series of vehicles.

Manufacturer

Alliant Techsystems Inc, 600 Second Street Northeast, Hopkins, Minnesota 55343, USA.
Tel: +1 612 931 6723. Fax: +1 612 931 6654.

VERIFIED

Field-expedient Countermine System (FCS)

Development

The Field-expedient Countermine System (FCS) was developed by Alliant Techsystems Inc in under 100 days in order to meet a US Marine Corps 'Desert Storm' requirement for a method of neutralising magnetically fuzed mines. The FCS was developed in co-operation with the Program Manager Engineer Systems at the US Marine Corps Research, Development and Acquisition Command in Quantico, Virginia and was tested under desert conditions at the Marine Corps Air-Ground Combat Center in Twenty Nine Palms, California.

US Marine Corps Assault Amphibian Vehicle (AAV) displaying the Field-expedient Countermine System (FCS) on the front hull

Description

The Field-expedient Countermine System (FCS) projects an electromagnetic field ahead of the host vehicle to allow a magnetic mine fuze to sense the field and detonate prematurely. The system is self-contained and consists of an electronic controller which generates a magnetic signature through a flexible coil that is field-wound from two spools of wire and secured to the front of the host vehicle. A support kit contains tools and minimal mounting material. A remote status indicator detects faults in the system and allows the vehicle driver or commander to monitor system operation.

The system weighs less than 43 kg and has no effect on host vehicle performance. Externally mounted components and electrical connections are waterproofed and operable when submerged in saltwater. No through-hull modifications are required and the system can be installed or removed by two men within 60 minutes without auxiliary support equipment. The coil is flexible and allows the normal operation of components such as an Assault Amphibian Vehicle (AAV) bowplane or a Light Armoured Vehicle (LAV) trim vane. The kit can be adapted to a variety of wheeled and tracked vehicles. No auxiliary power supply is required.

Status

In service with the US Marine Corps.

Manufacturer

Alliant Techsystems Inc, 600 Second Street Northeast, Hopkins, Minnesota 55343, USA.
Tel: +1 612 931 6723. Fax: +1 612 931 6654.

UPDATED

Caterpillar Track Width Mine Plough

Description

In response to a US Marine Corps requirement, Caterpillar Inc developed a Track Width Mine Plough (Plow) for use with the AAV7 series of amphibious assault vehicles. The plough is low in weight (2,041 kg) for amphibious operations and uses six tines per blade; the tine depth is operator adjustable and they can be replaced. For transport, the plough is carried stowed within the vehicle width and there are automatically positioned fold-out blade wings. The plough follows ground contours independently of vehicle attitude and can be released should a blade hit an immovable object. A skid plate is located behind each blade.

Specifications

Weight: 2,041 kg
Overall plough width:
(travelling) 3.276 m
(deployed without wing) 3.632 m
(deployed with wing) 4.75 m
Blade width: (each) 1.117 m
Tines per blade: 6

Status

Prototype.

Manufacturer

Caterpillar Inc, Defense & Federal Products,
Mossville, Illinois 61552-0470, USA.
Tel: +1 309 578 4852
Fax: +1 309 578 3739

VERIFIED

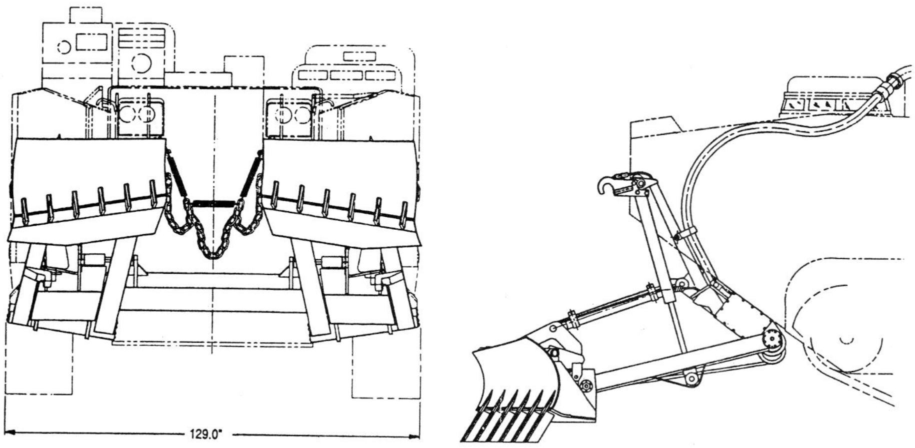

Drawings showing outline details of Caterpillar Track Width Mine Plough

Mine Clearing/Armor Protection (MCAP) kit

Development

Under an Urgent Procurement contract from the US Army Tank Automotive Command, Caterpillar Inc manufactured Mine Clearing/Armor Protection (MCAP) kits for installation on Caterpillar T-9 Track Type Tractors to be used by US Army Engineers during Operation Desert Storm. MCAP kits are retained by US Army engineer units for countermine training.

Description

The MCAP kit consists mainly of an angled rake assembly fitted to the front of the T-9 tractor and an armoured enclosure for the operator. The angled rake assembly side-casts to the right all known varieties of surface-laid or buried land mines down to a depth of 305 mm (12 in) and can clear a path 3.576 m (12 ft) wide; clearing efficiency is stated to exceed 95 per cent. The assembly is provided with replaceable tines and is fitted with an automatic flotation device. It is used to widen existing breaches established in minefields by mineclearing line charges, track width mine ploughs and other similar devices. Subsequent passes by a T-9 fitted with the MCAP can further widen a path and over a period of time entire minefields can be cleared. The rake assembly can be installed by two men in 10 minutes.

The armoured enclosure for the operator consists of a number of rolled homogeneous armour panels proof against 7.62 mm armour-piercing projectiles and land mine or artillery fragments. Panels are provided around the operator's compartment, engine, belly, fuel tank, and lift cylinders. The operator's station has two access doors and two fans for ventilation. Eight transparent armoured vision ports are provided and there are service access panels for maintenance. The armour can be installed by two men.

Mine Clearing/Armor Protection (MCAP) kit fitted to T-9 Track Type Tractor ***1997***

Specifications

Weight:
(rake assembly) 1,769 kg
(armour) 3,084 kg
Length:
(rake assembly) 3.9 m
(T-9 tractor with rake fitted) 9.187 m
Width: (overall) 3.874 m
Height: (to top of operator's cab) 3.246 m
Cleared path width: 3.657 m
Raking depth: 305 mm

Status

In service with the US Army.

Manufacturer

Caterpillar Inc, Defense & Federal Products,
Mossville, Illinois 61552-0470, USA.
Tel: +1 309 578 4852
Fax: +1 309 578 3739

VERIFIED

M173 mineclearing equipment

Description

The M173 mineclearing equipment was developed by Picatinny Arsenal and the Martin Company of Baltimore. The equipment consists of a boat-shaped glass fibre sled which is divided up into three compartments. The front compartment contains an M95 JATO rocket attached to a 91.4 m length of M96 linear demolition charge coiled in the centre compartment. The rear compartment contains the accessories and towing cable when not being used.

The sled, which will float, can be towed behind a tracked vehicle such as an M113 APC. The sled is normally brought up to the edge of the minefield and the rocket is launched across the minefield trailing the M96 linear demolition charge. The rocket is halted by an arresting wire attached to the sled. The line then falls onto the minefield and detonates, exploding any mines in its path.

Specifications

Weight: 1,406 kg
Length: 3.683 m
Width: 1.435 m
Height: 0.609 m
Type of explosive: Composition C4
Length of M96 charge: 91.4 m
Charge weight:
(total) 780 kg
(explosive) 680 kg
Number of charges on line: 1,200 (600 pairs)
Electrical supply: 24 V

Status

In service with the US Army.

VERIFIED

Explosive Standoff Minefield Breacher (ESMB)

Development

In May 1995 it was announced that Tracor Aerospace Inc had been awarded a competitive $19.9 million cost plus incentive fee contract for a joint US Army and Marine Corps programme for the design and development of a Demonstration and Validation Explosive Standoff Minefield Breacher (ESMB). By July 1997 Tracer Aerospace Inc will deliver three ESMB systems plus 15 reloads, five flight test reloads, two trainers, one test trailer, plus technical field support and training. The ESMB will undergo extensive trials before a decision on production is taken.

The contract was placed by the US Army Communications-Electronics Command, C41EW Acquisition Center – Ground Support Branch, Fort

Belvoir, Virginia. The programme is managed by the US Army Project Manager for Mines, Countermine and Demolitions, at Picatinny Arsenal, New Jersey.

Description

The ESMB will be trailer mounted and will be required to rapidly and safely neutralise mines and submunitions from a position outside their lethal distance during both amphibious and assault operations. The system is required to be capable of neutralising all types of known surface laid and buried mines down to a depth of 200 mm in the breached lane which will be 145 m long and 5 m wide.

Status

Under development for the US Army and Marine Corps.

Manufacturer

Tracor Aerospace Inc, 6500 Tracor Lane, Austin, Texas 76725-2996, USA.
Tel: +1 612 926 2800.

VERIFIED

Mineclearing Line Charge M58/M59 (MICLIC)

Development

The Mineclearing Line Charge (MICLIC) M58A1 was originally a US Marine Corps system. It was type classified by the US Army in August 1985 and the first units were issued for training purposes at the beginning of October 1986. The equipment was modified to M58A3 standard and the M58A4 version is the main production model.

With the US Army, the M58A1 MICLIC is issued to US Army engineer units and towed by M113 APCs, M2/3 Bradley IFVs and the M9 ACE.

The M59 MICLIC is used by the US Marine Corps which also operates the slightly revised M59A1, both units being capable of operating through the open roof hatches of AAVP7A1 amphibious assault vehicles.

As of April 1993 there were 3,512 MICLIC sets in the US armed forces inventory. It is planned that the M58A1 MICLIC will be replaced in service by the Explosive Standoff Minefield Breacher (ESMB — see previous entry).

Description

The M58A4 linear demolition charge is used to clear a path for tanks, vehicles and personnel through minefields.

The linear charge is approximately 106.7 m long and consists of approximately 816 kg of Composition C-4 explosive distributed in unit charges assembled around a core of nylon rope and detonating cord. A steel pallet serves as a storage and transport container for the linear charge, an arresting cable and the M1134A1E1 electric fuze. For ground launch, the pallet containing the linear charge is secured at a set location by driven steel posts. The launch may also be made from a pallet mounted on an M353 (or similar) trailer equipped with an M155 launching system and towed behind an armoured vehicle. At the target area the pallet is aligned to the direction of intended flight. For ground launch, the ground launcher assembly is attached to the pallet and an MK 22 MOD 4.5-inch/127 mm rocket motor is positioned on the launcher. The towed launcher has an integral launcher assembly. After fuzing the linear charge, the charge rocket motor harness is secured to the rocket motor. An M34 firing device is used to fire the rocket from a remote or protected position.

Training version of Mineclearing Line Charge (MICLIC) M58A4 mounted on a Caterpillar Combat Support Trailer (T J Gander) ***1997***

When fired, the rocket pulls the linear charge out of the pallet and over the target area. Forward travel of the charge is limited by the burn time of the rocket motor and the 62.48 m long arresting cable. After the charge is deployed, an electrical pulse from the firing device detonates the charge via an M1134 fuze to clear a path approximately 12 m wide and 100 m long.

The M68A2 Inert Linear Demolition Charge is an inert version of the M58A3 used for practice and training. This inert charge uses moulded synthetic rubber pellets in place of high explosive pellets and an inert fuze, the M1147.

Specifications

Weight complete: (loaded) 1,406 kg
Pallet:
(length) 2.362 m
(width) 1.346 m
(height) 0.711 m
Charge weight: 926.25 kg
Length of linear charge: approx 106.7 m
Number of pellets: 1,400
Pellet weight: 0.567 kg
Type of explosive: Composition C4
Weight of explosive: approx 826 kg
Arresting cable length: 62.48 m

Rocket motor
Weight:
(ready to fire) 52.16 kg
(motor propellant) 20.865 kg
(packed) 84.37 kg
Length: 1.93 m
Diameter: 127 mm
Length of motor cable: 3.05 m

Status

In service with the US Army and Marine Corps.

Manufacturer

Thiokol Corporation, Ordnance Operations, 401 Edwards Street, Suite 1111, Shreveport, Louisiana 71101, USA.
Tel: +1 318 222 7725. Telex: 22101.
Fax: +1 318 222 8151.

UPDATED

Anti-personnel mine clearance device M1 and M1A1

Description

This kit, officially designated Cable, Detonating, Anti-Personnel, Mineclearing, M1, consists of a circular aluminium storage and carrying case, a 52 m length of detonating cable, propulsion unit, launcher unit and the firing equipment.

The detonating cable consists of 19 strands of nylon-covered detonating cord, each of which contains 100 g of oil-soaked PETN per 305 mm. The launcher unit consists of a folding stand constructed of aluminium angles with connected legs.

The propulsion unit consists of a rocket unit with a length of wire rope attached to the front of the detonating cable, length of time fuze and two M2 fuze lighters. The cable is detonated after a 15 second delay.

The equipment is used as follows: the complete equipment is positioned just outside the minefield and one end of the detonating cable is staked to the ground, the cord is projected by the launcher across the minefield by the jet propulsion system, falls to the ground and is detonated.

The M1 is supplemented by the M1A1 (formerly the M1E1) which follows the same general lines. It is slightly larger and heavier (weight crated 64.4 kg with the explosive line weighing 28.5 kg). When detonated it can clear a path about 2.44 m wide of anti-personnel mines.

Specifications

Weight of complete equipment: 42 kg
Diameter of storage case: 420 mm
Type of explosive: PETN
Weight of explosive: 21 kg

Status

In service with the US Army.

VERIFIED

Projected Charge M3A1

Description

The Projected Charge M3A1 consists of two parallel, linear, corrugated aluminium or steel plates bolted together to form a rigid assembly. This is flexible in the vertical plane to allow it to pass over rough ground and rigid in the horizontal plane to maintain as straight a course as possible when being pushed.

When assembled it is 360 mm wide, 130 mm high and 122 m long. Aluminium models weigh 4,082 kg and steel models between 5,670 and 6,800 kg. Both contain up to 2,042 kg of high explosive. Once assembled, the charge is pushed by a tracked vehicle such as an M48 or M60 MBT across the minefield and then detonated by rifle or machine-gun fire. The complete charge then explodes along its entire length and the pressure waves created should detonate any mines in the immediate area.

Status

In service with the US Army.

VERIFIED

Projected Charge M157

Description

The Projected Charge M157 is an updated version of the Projected Charge M3A1 and follows the same general lines. The M157 is issued in kit form and takes from six to eight personnel hours to assemble. It is drawn from stocks only as and when required. Each M157 section is made up from: 62 centre loading sections, each 1.524 m long and weighing 64.4 kg; 13 body sections which are inert, 1.524 m long and weighing 35.8 kg; two impact fuze sections each 1.524 m long and weighing 69.4 kg with an M603 fuze; a single fuze; a single tail section weighing 70.3 kg. Each centre loading section and impact fuze section contains 20.4 kg of Composition B and 2.268 kg of Composition B4.

The M157 is assembled in a rear area and towed to the edge of the minefield where it is pushed across the area to be cleared. The M157 is detonated by firing at the impact plate on the impact fuze sections with either 7.62 mm or 0.50/12.7 mm ball ammunition. Once a hit is registered the charge explodes leaving a path 90 m long, 4 to 5 m wide and 1 to 1.5 m deep.

Specifications

Length: 122.225 m
Width: 305 mm
Height: 178 mm
Weight: 4,989 kg

Status

In service with the US Army.

VERIFIED

Torpedo, Bangalore M1A1 and M1A2

Description

The M1A1 Bangalore torpedo is used for clearing minefields and other battlefield obstacles such as barbed wire. It consists of 1.524 m lengths of high-explosive filled steel tubes with connecting sleeves. A complete item of issue consists of 10 loaded assemblies, 10 connection sleeves and one nose sleeve. Each sleeve, which is 54 mm in diameter, is filled with Amatol, with 102 mm of TNT at each end. Total weight of explosive in each assembly is 4.09 kg. Each end of the tube contains a threaded well to accommodate a firing device with a crimped-on non-electric blasting cap or a delay detonator. The firing device can consist of six turns of detonating cord wrapped around one end of a loading assembly, to which is attached an 8 to 15 second delay detonator. When one loading assembly is detonated, the entire series will be detonated. The connecting sleeve consists of a short tube into which the ends of the loading assemblies fit, held in position by spring clips. The nose sleeve has a round point to assist in pushing the torpedo through obstacles.

The lengths of tube are connected together up to the required length and then pushed across the minefield and detonated. The shock waves created by the explosive should detonate any mines in its path, although many modern mines have been designed to resist detonation by this method.

The M1A1 is supplemented by the M1A2 which follows the same general lines but each loading charge weighs 6.8 kg and total weight is 89.8 kg.

Status

In service with the US Army and Marine Corps.

VERIFIED

Anti-Personnel Obstacle Breaching System (APOBS)

Description

The US Marine Corps is developing a countermine system known as the Anti-Personnel Obstacle Breaching System (APOBS). Development is being conducted at the Naval Coastal Systems Center in Panama City, Florida.

APOBS is based on the Israeli POMINS II portable mine neutralisation system (see entry in this section for details) but with some operational and safety modifications. These include a command firing mode to allow the user to fire at will from a set back location.

It was anticipated that APOBS would be in service with US Marine Corps engineer and infantry units by the end of Fiscal Year 1993 or early FY94 but has been delayed. It is intended to be the replacement for the Bangalore torpedo currently in use.

Status

Development.

VERIFIED

Light Forces Anti-personnel Mine Mini-flail

Development

The Light Forces Anti-personnel Mine Mini-flail was developed from a series of light forces countermine projects originally investigated by the US Army Belvoir Research, Development and Engineering Center at Fort Belvoir, Virginia, during the early 1990s. The programme is now a co-operative effort with the US Army Engineering School together with the Office of Special Technology, Fort Washington, Maryland, the US Army Research Laboratory, Human Research and Engineering Directorate, at Aberdeen Proving Ground, Maryland and the Unmanned Ground Vehicles/Systems Joint Project Office at Redstone Arsenal, Alabama.

During 1996 two prototypes were deployed to Bosnia with the 130th Engineer Brigade, 16th Engineer Battalion.

Light Forces Anti-personnel Mine Mini-flail (T J Gander) ***1997***

Description

The Light Forces Anti-personnel Mine Mini-flail is an expendable, line-of-sight remotely controlled mini-flail unit intended to neutralise anti-personnel mines by detonating or destroying them from a safe distance. The unit is based on a John Deere 375 skid-steer chassis powered by a 17 hp, three-cylinder diesel engine. A series of mine flail chains are held in front of the vehicle on a spring-supported frame and rotated using power supplied from the main engine.

The original requirement called for a UH-1H helicopter external lift capability. Single units can be carried on a light trailer towed by a HMMWV or similar vehicle.

Specifications

Weight: 1,111 kg
Length: 3.05 m
Width: 1.27 m
Height: 1.092 m
Speed: (flailing) 4 to 8 km/h
Engine: 3-cylinder diesel developing 17 hp
Electrical system: 24 V

Status

Prototypes.

Manufacturer

Lockheed Martin Advanced Environmental Systems, 103 Chesapeake Park Plaza, Baltimore, Maryland 21220, USA.
Tel: +1 410 682 0892. Fax: +1 410 682 3648.

NEW ENTRY

MINEFIELD MARKING EQUIPMENT

FRANCE

Giat minefield lane marking lamp F1

Description

This flashing lamp (French Army designation is *Lampe Clignotante pour lot léger de balisage de champ de mines modèle D1F1*) is used for marking cleared lanes through minefields and can be seen from a distance of up to 100 m, although this can be reduced by a screen if necessary.

The lamp can be mounted on a pole or left on the ground and a green filter can be fitted over the lamp if required.

Specifications

Weight: (without battery) 185 g
Height: 170 mm
Diameter: 50 mm
Power supply: 1 × BA 30 battery
Operating period: 100 h
Operating temperature range: −31.5 to +51°C

Status

In service with the French Army.

Manufacturer

Giat Industries, 13 route de la Minière, F-78034 Versailles Cedex, France.
Tel: +33 1 30 97 36 48. Fax: +33 1 30 97 39 58.

VERIFIED

GERMANY

German Army minefield marking set

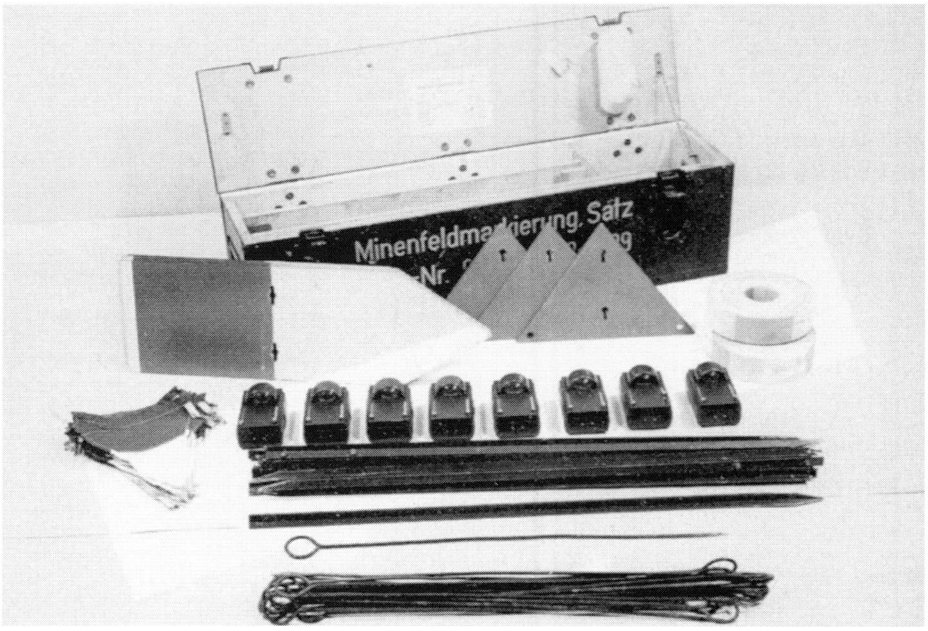

German Army minefield marking set

ISRAEL

IMI Clear Lane Marking System (CLAMS)

Description

The Clear Lane Marking System (CLAMS) is a self-contained add-on system which can be mounted on the rear of an armoured vehicle (other types of vehicle can also use the system) to mark a cleared path through a minefield. Normally only one CLAMS dispenser is centrally mounted on a vehicle to indicate the centre of a cleared path but if two CLAMS units are mounted one each side of a vehicle they can indicate the boundaries of a cleared path.

CLAMS has two main components, a dispenser and the markers. The dispenser is a sheet metal box with a cover, into which 150 markers can be loaded. It also has a release mechanism and a light stick activator. There is an accessory compartment for stowage of the control box, ground travel meter, electrical harness and the loading accessories when not in use. In use the dispenser may be fitted to the user vehicle by two men in about 10 minutes. An adaptor is mounted on the vehicle, which may incorporate a heat shield if required and a cable harness connects the dispenser to the control box mounted in the driver's compartment. The system may be operated manually or automatically. In the automatic mode the interval for release of the markers may be set at 6, 12, 18, 24, 36 or 48 m with the intervals being measured by a ground travel meter attached to the vehicle's speedometer. Manual release is carried out by pressing a release button on the control unit.

The marker consists of a sheet metal base disc on which is mounted a light stick and a flag. The light stick is automatically activated as it is released from the dispenser and glows for approximately 12 hours. For marking by day the flag alone may be used. The base is so designed that it will ensure the light stick is erect once dropped and is base-heavy to remain upright even in high winds. After use the markers may be retrieved and reused. Light sticks are available in green, blue and orange, the flags are red and the normal colour for the base is white, although other colours are available.

In addition to indicating cleared paths through minefields, CLAMS may be used to indicate passages through difficult terrain, to indicate the periphery of chemical or nuclear fallout contamination, or to mark emergency runways for aircraft.

In April 1987 a contract worth $2,000,185 was awarded to the BMY Corporation of York, Pennsylvania, for the production of CLAMS units for the US Army. With the US Army, CLAMS is used by minefield-breaching tanks such as M1 Abrams and M60 series MBTs fitted with mine ploughs.

CLAMS markers in position
1997

Specifications

Markers
Light stick height: 127 mm
Base disc diameter: 155 mm
Weight: 300 g

Dispenser
Length: 860 mm
Width: 280 mm
Height: 970 mm
Weight:
(empty) 165 kg
(loaded) 210 kg
Operating voltage: 24 V DC (12 V optional)

Status

In service with Israel Defence Forces and the US Army.

Manufacturer

Israel Military Industries Limited (IMI), 64 Bialik Boulevard, PO Box 1044, IL-47100 Ramat Hasharon, Israel.
Tel: +972 3 542 5222. Telex: 33719 misbit il.
Fax: +972 3 489639.

UPDATED

RUSSIAN FEDERATION AND ASSOCIATED STATES (CIS)

PSK clear lane marking system

Description

The PSK clear lane marking system is a flare dispenser for marking a cleared lane through minefields. The PSK can be fitted to virtually all T-54, T-55, T-62 and T-72 MBTs except the command tank variants. Suitable vehicles have a distinctive square panel with four bolts on the upper hull rear on to which the device is fitted. The PSK holds a number of cassettes, each of which can dispense six phosphorous-based flares. The flares are dispensed downwards at 15 m intervals and each flare burns on the ground for about 20 minutes.

The PSK is usually used in conjunction with the KMT-5M mineclearing plough and roller equipment (see entry in the *Mineclearing equipment* section).

Status

In service with former Warsaw Pact armed forces and other armed forces that have received RFAS military aid.

Manufacturer

State factories.

VERIFIED

PSK clear lane marking system mounted on rear of Iraqi modified T-55 MBT

Close-up of top of PSK clear lane marking system dispenser showing four flare cassettes

UNITED KINGDOM

Pearson Pathfinder marker system

Description

Pathfinder is a vehicle-mounted electropneumatic ground marking system that can be operated from under armour. It was designed to fire marker poles into the ground at preset intervals to delineate hazardous or dangerous areas by day or night. The Pathfinder system can also be used for NBC contaminated area marking, identifying unfriendly terrain and reconnaissance waypoints. The variable spacing of the marker poles is controlled by a unit mounted in the vehicle.

Each Pathfinder unit has four main elements: a Deployment and Emplacement Unit (DEU); a compressor unit; a control unit; and 100 marker poles. The system can be installed as twin units either side of a vehicle, as a single unit for boundary marking, with a deployment and stowage capability, or in a fixed operating position.

The Deployment and Emplacement Unit (DEU) fires the marker poles into the ground and is mounted on one or both sides of a vehicle. Each DEU magazine is 1.1 m long and contains 100 marker poles in four concentric rings. Mounted directly above the magazine is the firing and indexing mechanism operated by compressed air. After each firing the magazine automatically advances the next marker pole to the firing position. When a fire signal is received from the control unit a marker pole is fired from the barrel and driven into the ground.

The DEU is common to all vehicle types and the standard option includes a deployment assembly. This allows the DEU to be folded away and stowed when not in use. The DEU is fully protected against vibration, heat and dust and it is watertight. Magazine reloading takes about 5 minutes.

The compressor unit provides the clean, dry air source to operate the DEU. Driven by a 24 V DC motor drawing up to 140 A from the vehicle supply, the unit supplies compressed air on demand with monitoring via a pressure switch.

The control unit is fitted in the vehicle crew compartment. There are three operating modes: manual, timed and via an external command source.

The reusable marker poles are constructed using lightweight flexible tubing with stainless steel tips. The standard colour is white with fluorescent dayglo bands. A night vision marker pole is available. Each marker pole is 996 mm long with a maximum diameter of 12 mm; weight is 110 g.

Pathfinder marking device deployed for use on AMX-30B D2 MBT

Pathfinder marking device deployed in travelling position on AAV7A1 amphibious assault vehicle

1997

Specifications

Compressor unit
Weight: 129 kg
Length: 890 mm
Width: 746 mm
Height: 510 mm
Power supply: 24 V DC

Deployment/Emplacement Unit (DEU)
Weight: 200 kg
Length: 1.57 m
Width: 585 mm
Height: 370 mm

Magazine
Weight full: 58 kg
Weight empty: 47 kg
Height: 2.055 m
Diameter: 286 mm
Capacity: 100 rods

Control unit
Weight: 3.4 kg
Length: 145 mm
Height: 126 mm
Width: 320 mm

Status

In production. In service with the British Army, France, Saudi Arabia and the US Marine Corps. Under evaluation by European armies.

Manufacturer

Pearson Engineering, Wincomblee Road, Walker, Newcastle upon Tyne NE6 3QS, UK.
Tel: +44 191 234 0001. Fax: +44 191 262 0402.

UPDATED

Hunting Engineering Lightweight Marking System (LIMAS)

Description

The Hunting Engineering Lightweight Marking System (LIMAS) delineates minefields and other hazards, route and lane markings and similar operations. It utilises lightweight picket posts with one end slotted to hold tape, wire or light rope; flanged interface units secured by steel pins which are club-hammered into the ground or tarmac surface; yellow and red tape on rolls 125 m long and 16 mm wide; and a simple hand-held tape dispenser. Minefield and other marking triangles and signs can be provided. Using this equipment, a two-man team can mark perimeters faster than 1 km/h and therefore it is compatible with mechanical minelaying systems.

Packaging can be arranged to suit customer requirements. Satchels are available to enable one man to carry a standard LIMAS set sufficient for 250 m of fencing.

Status

In production. In service with the British Army.

Manufacturer

Hunting Engineering Limited, Reddings Wood, Ampthill, Bedford MK45 2HD, UK.
Tel: +44 1525 841000. Fax: +44 1525 405861.

VERIFIED

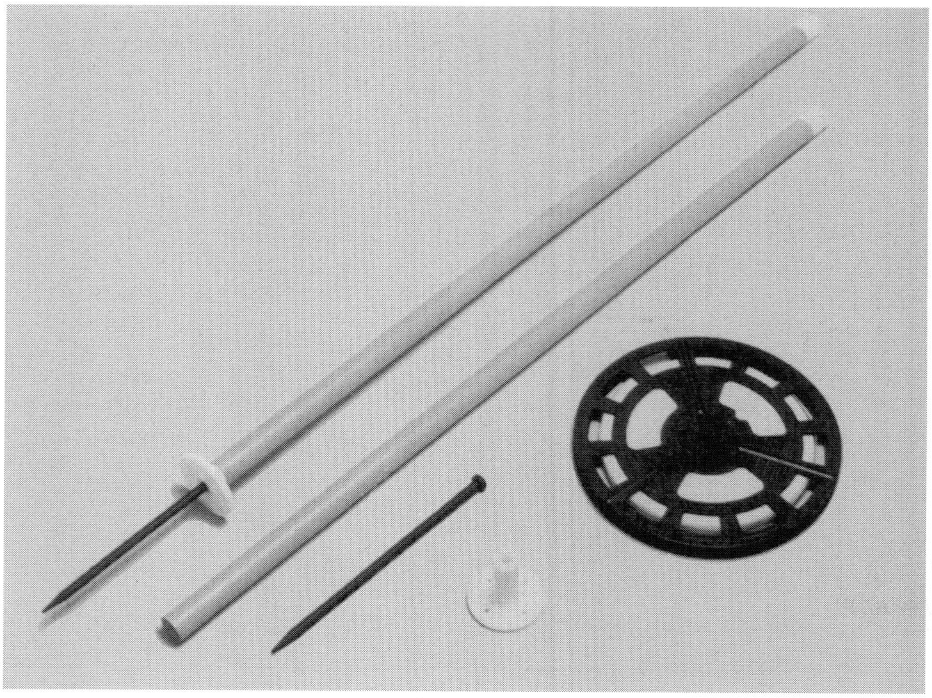

Components of Hunting Engineering Lightweight Marking System (LIMAS)

Betalight illuminated defile marker

Description

The Saunders-Roe Developments Limited illuminated defile marker was originally designed for use with the Medium Girder Bridge but has since undergone extensive trials by the British Army for a wide range of other applications, including marking routes through minefields.

When being used for bridge marking, two inward-facing defile markers show the commencement of the defile, upward-facing ones guide through the defile and two outward-facing ones indicate the end of the defile.

The defile marker consists of a white arrow on a black background with a Betalight illuminated head, in a tough plastic housing. The arrow can be rotated to one of eight fixed positions, allowing positional or other information to be indicated.

Illumination is provided by SRDL Betalight self-powered light sources which are borosilicate glass capsules internally coated with phosphor and filled with tritium gas which activates the phosphor to emit light. The arrow is clearly visible in both day and night conditions. They are perfectly safe for the user and have a long maintenance-free life of between 15 and 20 years. The marker is small, compact and soldier-proof. It will withstand normal army mechanical environmental requirements and can be used in all climatic conditions.

The marker can be attached to military police mounting poles, screwed or attached to mortar baseplates or to the entry flag of a bridging layout, or even to a bayonet. The ring allows the marker to be attached to a nail, tree or other available support. A total of 12 defile markers can be carried in a standard 7.62 mm ammunition box.

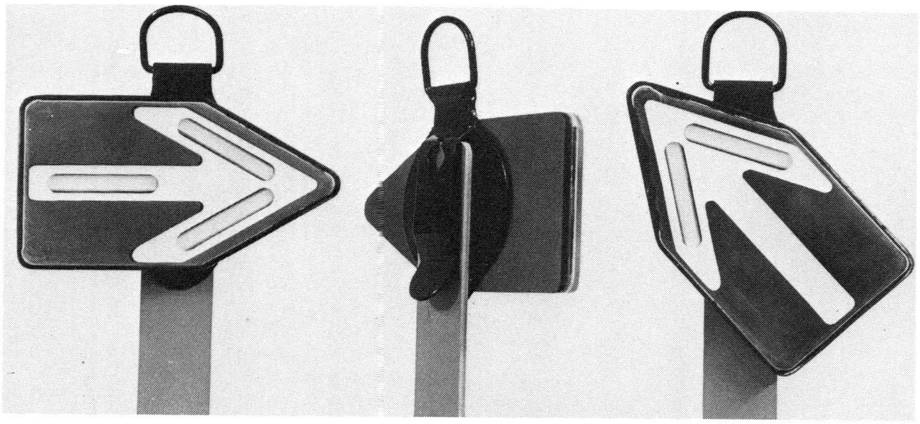

Saunders-Roe Developments Betalight illuminated defile marker in different positions

Specifications

Weight: 170 g
Length: 130 mm
Width: 70 mm
Height: (including backplate) 27 mm

Status

In service with the British Army and other armed forces.

Manufacturer

Saunders-Roe Developments Limited, Millington Road, Hayes, Middlesex UB3 4NB, UK.
Tel: +44 181 573 3800. Fax: +44 181 561 3436.

VERIFIED

UNITED STATES OF AMERICA

Hand-emplaceable minefield marker set M133

Description

The hand-emplaceable minefield marker set M133 was developed by the US Army Mobility Equipment Research and Development Command (MERADCOM) (now the Belvoir Research, Development and Engineering Center at Fort Belvoir, Virginia) and is used for the temporary rapid marking of indefinite-life minefields by troops on foot. The major components of the set are a pole, orange fluorescent marking tape and an orange flashing light (152.4 mm long and 50.8 mm in diameter) which is mounted on top of the pole. The lamp is neon and initially flashes at about 82 times/minute. A shield inside the dome of the flasher prevents the light being seen from directly overhead. A reflector behind the neon lamp directs the light so that it can be seen from only one direction. The outside of the plastic case has a reflective surface to make it visible in light from vehicle headlights. Additional components are provided in the kit for emplacing and retrieving the poles. The poles are normally spaced at intervals of between 10 and 15 m, except for those adjacent to corner poles which are then spaced 4 m from the corner pole; the marking tape is threaded through a holder on the side of the pole. The equipment was designed to mark a 700 to 1,000 m perimeter minefield, if the average pole spacing is not less than 10 m, for 15 days under intermediate cold temperature conditions and to be reused between three and ten times.

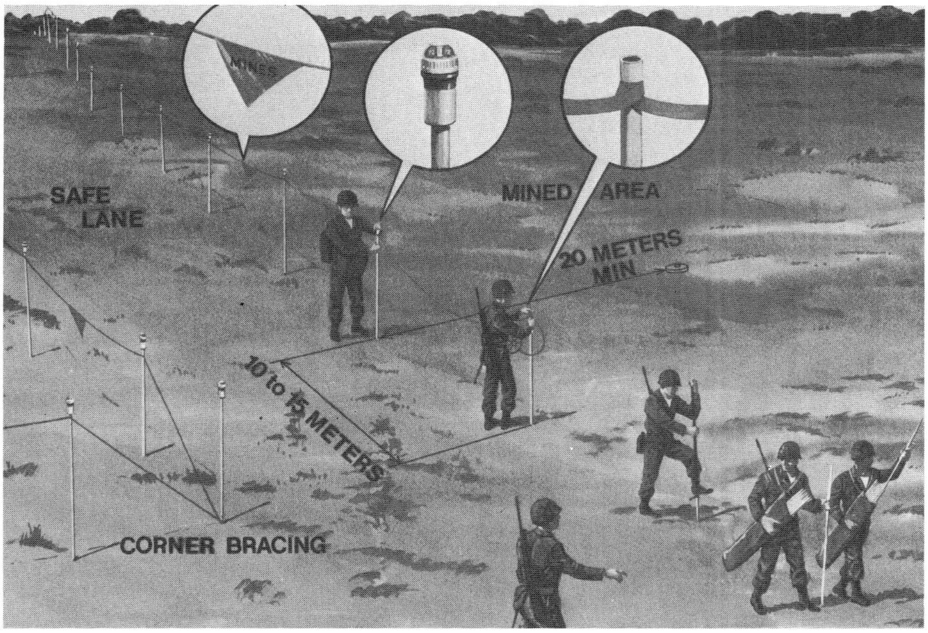

Method of employing hand-emplaceable minefield marker set M133 (US Army)

Status

Entered US Army service late 1980.

VERIFIED

Cyalume lightsticks

Description

Cyalume lightsticks may be used for a variety of military purposes ranging from emergency lighting to marking the perimeters of minefields by simply placing them end-on into the ground. The lightsticks are activated by snapping the plastic body to allow two liquids to mix, thereby producing a phosphor-based chemical reaction which emits light. The lightsticks may be red, yellow, blue or green and are used in the CLAMS system (qv). Normal illumination life is up to 12 hours.

Cyalume produces a holder for these lightsticks known as the Combat Light. This plastic container may be worn on a uniform and is so arranged that a simple shutter allows the light to be blocked off if required. The Combat Light holder may be used on a stake or pole to act as a route guide or denote a route through a minefield. For this purpose a keyhole clip is provided. The Combat Light holder is 70 mm long and can be reused indefinitely.

The Combat Light and the lightsticks are produced under licence in Israel by Rabintex.

Status

In production.

Manufacturer

Omninglow Corporation, 231 West Parkway, Pompton Plains, New Jersey 07444, USA.
Tel: +1 201 831 0770. Fax: +1 201 831 1877.

VERIFIED

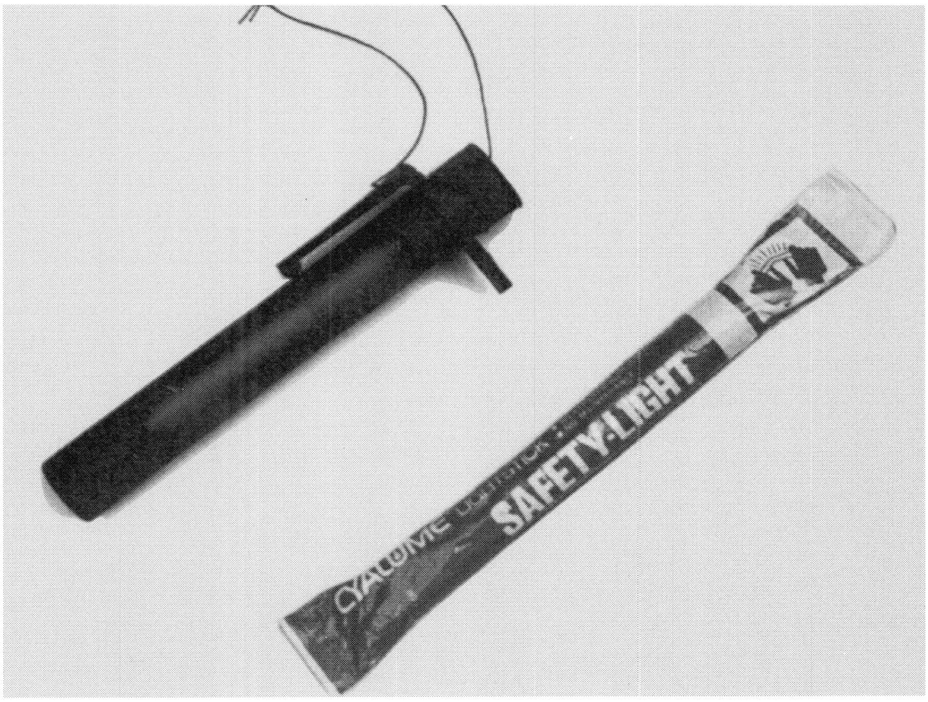

Cyanamid Combat Light on left with lightstick foil wrapper on right

Infantry and Sapper Minefield Breaching and Marking Set

Description

Intended to form part of the Light Forces Countermine programme, the Infantry and Sapper Minefield Breaching and Marking Set, also known as the Squad Sapper Set, is intended to provide dismounted combat engineer or infantry squads with the ability to conduct a clandestine minefield breach and mark the resultant clear path. The expendable set consists of an AN/PRS-12 mine detector carrier assembly, plus a set of items including the following: a box of chemical lights plus 10 holders with stakes; two rolls of non-adhesive white tape; nine non-metallic mine probes; a set of bolt cutters; 14 mine markers; a roll of minefield markers; a carrying bag; and a grapnel with a 30.5 m line.

Status

Project.

Development Agency

Belvoir Research, Development and Engineering Center, Fort Belvoir, Virginia 22060-5606, USA.
Tel: +1 703 704 1047.

VERIFIED

TRANSPORT EQUIPMENT

LIGHTWEIGHT VEHICLES

BRAZIL

ENGESA EE-12 (4 × 4) 500 kg light vehicle

ENGESA EE-12 (4 × 4) 500 kg light vehicle

Description

The EE-12 is a small four-seat vehicle of distinctive appearance. It has a conventional layout with the four-cylinder petrol or diesel engine at the front under a sloped bonnet and with separate seats for the driver and a passenger; a bench seat at the rear can accommodate a further two passengers. Normally the EE-12 is driven open with no weather protection for the occupants other than a forward-folding windscreen, but it is possible to fit a canvas tilt with roll-down blinds. There are no side doors other than canvas covers. A spare wheel is carried at the rear along with a jerrican.

The overall construction of the EE-12 is very robust. The front and rear bumpers are equipped with towing shackles and a towing pintle is provided at the rear.

The EE-12 can be provided in a number of variations. One is as a carrier for a 7.62 or 12.7 mm machine gun or a 106 mm recoilless rifle. This carrier has an open rear with the jerrican and spare wheel moved to the vehicle sides and the windscreen is split to allow the gun barrel to protrude over the bonnet. A hardtop version is available and was exported to Angola. Other variants include an anti-tank missile carrier, a radio communications vehicle and a front-line stretcher carrier.

Specifications

Seating: 1 + 3
Configuration: 4 × 4
Weight: (loaded) 2,160 kg
Max load: 500 kg
Towed load: 250 kg
Length: 3.57 m
Width: 1.77 m
Height:
(overall) 1.9 m
(windscreen down) 1.5 m
Ground clearance: 0.23 m
Track: 1.454 m
Wheelbase: 2.16 m
Angle of approach/departure: 70°/50°
Max speed: (road) 110 km/h
Fuel capacity: 92 litres
Max gradient: 70%
Side slope: 30%
Fording: 0.6 m
Engine: GM 151 2.47 litre, 4-cylinder petrol developing 85 hp at 4,400 rpm or
Perkins 2.8 litre, 4-cylinder diesel developing 70 hp
Transmission: Clark 240 V manual with 5 forward and 1 reverse gears
Transfer box: single-speed for front axle engagement
Steering: mechanical
Turning radius: 6 m
Suspension: coil springs and telescopic shock-absorbers, location by longitudinal and transverse control links; 250 mm wheel travel
Tyres: 7.50 × 16
Brakes: servo-assist dual circuit; disc front, drum rear
Electrical system: 24 V
Batteries: 2 × 12 V, 63 Ah
Alternator: 40 A

Status

Production complete. In service with Angola.

Manufacturer

ENGESA Engenheiros Especializados SA, Avenida Tucunare 125/211, Caixa Postal 152/154, 06400 Barueri, SP, Brazil.
Tel: +55 11 421 4711. Telex: 1171 302 enes br.
This company is no longer trading.

VERIFIED

Bernardini Xingu BT25 and BT50 (4 × 4) light vehicles

Description

The Bernardini Xingu BT25 and BT50 (4 × 4) light vehicles are basically Jeep-type vehicles. Both are described as ½ tonne vehicles. The main difference is that the BT50 has a longer wheelbase but in most other respects the two vehicles are similar.

Both vehicles are conventional in layout with the engine forward under a steel bonnet and seating for the driver and three passengers under an optional soft top. Canvas screen doors are normally fitted. The BT50 could be used as a light pick-up truck with the rear passenger bench seat removed. A protective grill for the radiator is fitted across the front of the vehicle and a spare wheel is usually carried at the rear on a small tailgate. A towing hitch is fitted under the tailgate. Floating axles are used front and rear.

In July 1990 the previously fitted OM-314 diesel engine was replaced on the production lines by an OM-364 unit delivering the same horsepower.

Variants of these vehicles include a ¾ tonne pick-up truck version of the BT50 with a canvas tilt carried over the bows and the spare wheel on the left-hand side in front of the cab door. A variant of the BT25 is used as a carrier for an M40A2 106 mm recoilless rifle — this version has a split front windscreen to accommodate the recoilless rifle barrel when travelling. Other variants of both models include communications vehicles with various radio fits.

Optional equipment on both models includes a front-mounted 2,500 kg capacity winch, a folding aerial bracket and an extra spare wheel carrier. A light cargo trailer is available.

Specifications

Cab seating: 1 + 3
Configuration: 4 × 4
Weight:
(BT25) 1,600 kg
(BT50) 1,700 kg
Length:
(BT25) 3.93 m
(BT50) 4.4 m
Width: 1.8 m
Height: 1.965 m
Ground clearance: 0.23 m
Track:
(front) 1.415 m
(rear) 1.4 m
Wheelbase:
(BT25) 2.285 m
(BT50) 2.755 m
Max speed: 115 km/h
Range: 950 km
Max gradient: 60%

Bernardini Xingu BT25 (4 × 4) light vehicle carrying 106 mm recoilless rifle

Bernardini Xingu BT50 (4 × 4) ¾ tonne pick-up truck

Side slope: 30%
Fording: 0.7 m
Engine: OM-364 4-cylinder water-cooled diesel developing 94 hp at 2,800 rpm
Gearbox: manual, 4 forward and 1 reverse gears; from 1993 all models have 5 forward and 1 reverse gears
Clutch: single dry plate
Transfer box: 2-speed
Steering: worm and roller; from 1993 all models have power steering
Turning radius:
(BT25) 5.3 m
(BT50) 6.3 m
Suspension: semi-elliptical leaf springs with double-action hydraulic shock-absorbers
Tyres: 7.50 × 16
Brakes: dual circuit hydraulic with vacuum auxiliary
Electrical system: 12 or 24 V
Batteries: 2 × 12 V, 60 Ah
Alternator: 35 A

Status
In production. In service with the Brazilian armed forces.

Manufacturer
Bernardini SA Indústria e Comércio, Rua Hipólito Soares 79, São Paulo - SP - 04201, Brazil.
Tel: +55 11 273 8996. Telex: 011 21605 bsaibr.

VERIFIED

ENGESA EE-34 (4 × 4) 750 kg light vehicle

Description
The prototype ENGESA EE-34 (4 × 4) 750 kg light vehicle was completed in 1980 and production commenced in 1982.

The body of the EE-34 is made of pressed steel and the layout is conventional, with a cargo area at the rear. The standard form is a pick-up truck for general purpose load carrying with seating for the driver plus a bench seat for two passengers. Standard equipment includes a windscreen that folds downwards onto the bonnet when not in use, vinyl doors, a removable vinyl top and a drop tailgate for loading stores. The maximum payload across country is 750 kg and 1,000 kg on roads.

Variants include a light truck version and an ambulance with a pressed steel body. This has space for four stretchers or two stretchers plus three seated casualties or six seated casualties. The ambulance body is entered by two large rear doors.

Specifications
Cab seating: 1 + 2
Configuration: 4 × 4
Weight:
(unladen) 2,500 kg
(laden, road) 3,580 kg
(laden, cross-country) 3,330 kg
Max load:
(road) 1,000 kg
(cross-country) 750 kg
Length: 5.32 m
Width: 2.04 m
Height:
(tarpaulin) 2.2 m
(cab top) 2 m
(windscreen down) 1.493 m
Ground clearance: 0.23 m
Track: 1.55 m
Wheelbase: 2.95 m
Angle of approach/departure: 42°/52°
Max speed: (road) 100 km/h
Range: 600 km
Max gradient: 60%
Max side slope: 30%
Engine: Perkins 4.236 4-cylinder diesel developing 82 hp at 2,800 rpm
Gearbox: manual, 4 forward and 1 reverse gears
Clutch: single dry plate
Transfer box: mechanical, 2-speed
Steering: mechanical
Suspension: leaf springs and hydraulic shock-absorbers
Tyres: 7.50 × 16
Brakes:
(main) hydraulic, disc front and drum rear
(parking) drum, operating on transmission output shaft
Electrical system: 24 V

Status
Production complete. In service with the Brazilian armed forces.

Manufacturer
ENGESA Engenheiros Especializados SA, Avenida Tucunaré 125/211, Caixa Postal 152/154, 06400 Barueri, SP, Brazil.
Tel: +55 11 421 4711. Telex: 1171 302 enes br.
This company is no longer trading.

VERIFIED

ENGESA EE-34 (4 × 4) 750 kg pick-up truck

ENGESA EE-34 (4 × 4) ambulance

UAI M1-34 (4 × 4) 750 kg light vehicle

Description
The UAI M1-34 (4 × 4) 750 kg light vehicle is a conventional design powered by an 87 hp diesel engine through a five-speed mechanical transmission and dual-reduction transfer box. The cargo capacity for cross-country use is 750 kg plus a 1,000 kg trailer. On roads this can be increased to 1,000 kg plus a towed load of 1,800 kg.

Status
In production.

Manufacturer
Terex do Brasil SA, Rodovia BR-252, Anel Rodoviário -31250 Belo Horizonte, MG - Brazil.
Tel: +55 31 441 3444.

VERIFIED

UAI M1-34 (4 × 4) 750 kg light vehicle

CANADA

Bombardier Iltis (4 × 4) 500 kg light vehicle

Development
In October 1981 Volkswagen AG of Germany agreed to cede and transfer to Bombardier Inc of Canada the design and manufacturing technology of its Iltis (4 × 4) 500 kg light vehicle as used by the German Army. The agreement also granted to Bombardier the worldwide rights for the marketing, sale and distribution of the Iltis, as well as the rights to manufacture and market the civilian version. Bombardier also acquired the tooling equipment used in the manufacture of the Iltis.

Before this announcement the Canadian Department of National Defense asked the company to submit a proposal, based on the Iltis, for the replacement of the existing utility 4 × 4 vehicles of the Canadian Forces. In November 1983 it was announced that the Canadian government had awarded a contract for 1,900 (later increased to 2,500) Iltis vehicles to the Logistic Equipment Division of Bombardier Inc. The contract included the supply of spare parts, manuals and training programmes for operators and maintainers and was valued at C$68 million. Production for this contract began at the end of Summer 1984 and was completed by the end of 1985. During 1984 an order for over 2,500 was placed for the Belgian armed forces (2,673 were delivered).

Production ceased in 1988.

Description
Originally a Volkswagen design, the Iltis uses components from the Volkswagen Polo and Audi

Bombardier Iltis (4 × 4) light vehicle in service with Canadian Forces in Bosnia (R Stickland) ***1997***

Quattro vehicles. The all-steel body has a conventional layout with the engine at the front and the cargo area to the rear. The driver and front passenger have separate seats with seating for two more passengers to the rear. The rear seats can be folded down to increase the load-carrying area and the vehicle is provided with a folding hood, removable side flaps and a folding windscreen. The driver can select either a full 4 × 4 drive for cross-country or 4 × 2 for road drive, in which case the front axle is disengaged. An integral roll bar is provided.

The Iltis can be lifted by the UH-60 Black Hawk helicopter and has been tested for low altitude parachute extraction from transport aircraft and for parachute dropping.

Optional extras for the Iltis include a turbocharged diesel engine, electric winch kit, rear pintle hook, three-door glass fibre hardtop and up to three radio equipment outlets. Also provided are weapon mounting systems for 7.62 mm and 0.50/12.7 mm machine guns, the MK19 40 mm grenade launcher and anti-tank missile systems such as MILAN and TOW.

Variants of the basic Iltis include an ambulance, command vehicle, cable layer, communications vehicle and a version carrying artillery sound ranging equipment.

Specifications

Cab seating: 1 + 3
Configuration: 4 × 4
Weight:
(empty) 1,550 kg
(loaded) 2,050 kg
Max load: 500 kg
Max towed load:
(with brakes) 2,000 kg
(without brakes) 750 kg
Length: 3.887 m
Width: 1.52 m
Height: 1.837 m
Ground clearance: 0.225 m
Track:
(front) 1.23 m
(rear) 1.26 m
Wheelbase: 2.017 m
Angle of approach/departure: 41°/32°
Max speed: 130 km/h
Fuel capacity: 85 litres
Range: 700 km
Max gradient: 77%
Fording: 0.6 m
Engine: VW 183 4-cylinder OHC petrol developing 75 hp at 5,000 rpm
Gearbox: manual with 4 forward and 1 reverse gears plus a cross-country gear
Clutch: single dry plate
Steering: rack and pinion
Turning radius: 5.5 m
Suspension: overhead semi-elliptical leaf springs and double-acting shock-absorbers
Tyres: 6.50 × R-16
Brakes:
(main) hydraulic, dual circuit, drum
(parking) mechanical
Electrical system: 24 V
Batteries: 2 × 12 V, 45 Ah

Status

Production complete. In service with Belgium (2,673), Cameroon armed forces, Canadian Forces (2,500) and the Oman Royal Guard Brigade.

Manufacturer

Bombardier Inc, Logistic Equipment Division, Valcourt, Québec, Canada J0E 2L0.
Tel: +1 514 532 2211. Telex: 5832 575.

VERIFIED

CHINA, PEOPLE'S REPUBLIC

Beijing BJ212, BJ212A, BJ2020S and BJ2020SA light vehicles

Description

Production of the basic BJ212 began in 1965. These light vehicles are similar in appearance to the UAZ-469B, the BJ212 seating five people and the BJ212A seating up to eight. The latter also has a slightly increased payload. The BJ212A has no rear side door and has a conventional tilt (if fitted) as opposed to the raised 'convertible' cover of the BJ212. The BJ2020S and BJ2020SA are much improved versions of the earlier models.

All versions have an identical layout, with the engine at the front and the passenger or cargo compartment to the rear. Next to the driver is a seat for one passenger. On the BJ2020S a single bench seat accommodates three more passengers, while on the BJ2020SA, there are two bench seats along the sides for six passengers. The windscreen can be folded forward over the bonnet on both versions. The BJ2020SA has a rear tailgate that can be folded down for access to allow it to be used as a light utility vehicle.

A variant of the Beijing BJ212 is used to carry the NORINCO Type 75 105 mm recoilless rifle. This model can be identified by the split windscreen, provided to accommodate the length of the barrel.

Beijing BJ2020S (4 × 4) light vehicle

Model	**BJ2020S**	**BJ2020SA**
Seating	1 + 4	1 + 7
Configuration	4 × 4	4 × 4
Weight		
(unladen)	1,580 kg	1,550 kg
(laden)	2,005 kg	2,150 kg
Max load	425 kg	600 kg
Towed load	800 kg	800 kg
Length	4.028 m	3.86 m
Width	1.768 m	1.75 m
Height		
(unladen)	1.99 m	2.1 m
(laden)	1.81 m	1.97 m
Wheelbase	2.3 m	2.3 m
Track		
(front and rear)	1.44 m	1.44 m
Ground clearance	0.21 m	0.21 m
Max speed	100 km/h	100 km/h
Gradient		
(less trailer)	30°	30°
(with trailer)	22.5°	22.5°
Fording	0.5 m	0.5 m
Engine	2.445 litre water-cooled 4-cylinder in-line 4-stroke petrol developing 75 hp at 3,800 rpm	
Gearbox	manual, 3 forward and 1 reverse, synchromesh on 2nd, 3rd	

Clutch	single dry plate	
Transfer box	2-speed	2-speed
Suspension	longitudinal semi-elliptical leaf springs with hydraulic double-acting telescopic shock-absorbers	
Steering	recirculating ball	
Turning radius	6 m	6 m
Brakes	hydraulic drum	
Tyres	6.50 R 16	6.50 R 16
Electrical system	12 V	12 V
Battery	12 V, 54 Ah	12 V, 54 Ah

Status
BJ2020S and BJ2020SA in production. In service with People's Liberation Army, Chad, Pakistan and some other nations. Over 460,000 units have been produced.

Manufacturer
Beijing Jeep Corporation Limited, Chuiyangliu Chaoyang District, Beijing, People's Republic of China.
Tel: +86 1 771 2233. Telex: 20039 bjc cn.

VERIFIED

Beijing Jeep Corporation XJ Cherokee

Development/Description
In May 1983 it was announced that a production agreement between the American Motors Corporation and the Chinese Beijing Automotive Works had been signed. The agreement established the Beijing Jeep Corporation to produce Jeeps on a new production line in China, initially using prepacked kits delivered from an AMC plant in Ontario, Canada.

The model selected for initial production was the Jeep XJ Cherokee and the first models were produced in Beijing during September 1985. Details of the XJ Cherokee can be found under the Jeep entry in this section.

Initial production plans for 1986 envisaged a rate of 4,000 vehicles a year but this was not achieved because of foreign exchange problems and the expected rate became 2,100 vehicles a year. The XJ Cherokee was evaluated by the People's Liberation Army. Over 66,000 units have been produced. The Chrysler Motor Corporation took over from AMC in 1986.

Status
In production.

Manufacturer
Beijing Jeep Corporation Limited, Chuiyangliu Chaoyang District, Beijing, People's Republic of China.
Tel: +86 1 771 2233. Telex: 20039 bjc cn.

VERIFIED

Beijing Jeep Corporation-produced XJ Cherokee

EGYPT

Egyptian Jeep production

Development
Egypt is manufacturing under licence from the Chrysler Corporation the YJ (4 × 4) Jeep. This agreement is part of the Arab Organisation of Industrialisation Plan passed to the Chrysler Corporation from the American Motors Corporation during their 1987 merger.

The first production Egyptian Jeeps (CJ-7) were completed in 1978 and 2,500 vehicles were built in 1979, all of which were supplied to the Egyptian Army. Since then well over 14,000 vehicles have been built, including the CJ-7, CJ-8 (both with an average of 30 per cent local content), J-20, AM-720 and Jeep Wagoneer.

The YJ-L, sold commercially as the Wrangler, entered production in 1989 with 800 units for the Egyptian Army and 350 for the Kuwait Military Forces. The YJ-L differs from its predecessor by having an improved heavy duty air flow system, larger 7.50 × 16 tyres, heavy-duty shock-absorbers and springs and a payload of 750 kg. Optional extras include a Ramsey 3,630 kg winch and blackout lights.

Variants
Several variants of Egyptian-produced Jeeps are in Egyptian Army use. They include versions of the CJ-7 used as Swingfire anti-tank missile launch vehicles (carrying four missiles) or Swingfire missile resupply vehicles (carrying 13 missiles). Other missile-carrying CJ-7s include versions for the SA-7 SAM and its Egyptian-produced version, the Sakr Eye.

Variants of the CJ-8 in Egyptian Army service include: 500 TOW anti-tank missile carriers delivered direct from the USA, plus a number of TOW missile resupply vehicles; a refrigerated mobile blood bank with demountable electrical generators; an 80 mm multiple rocket launcher carrier (for details refer to *Jane's Armour and Artillery 1995-96* pages 722 and 723); a mobile workshop; an NBC reconnaissance vehicle; an ambulance; a communications vehicle; a scatterable mine dispenser vehicle; and a command vehicle.

A border patrol unit supplied to Kuwait was a YJ-L. Fittings include a 0.5 0/12.7 mm machine gun mount, 24 V system, electronic land navigation system and an auxiliary fuel tank providing a total fuel capacity of 100 litres.

Specifications
YJ-L
Configuration: 4 × 4
Weight:
(kerb) 1,450 kg
(GVW) 2,200 kg
Max load: 750 kg
Length: 4.489 m
Width: 1.676 m
Height:
(soft top) 1.828 m
(hard top) 1.85 m
Ground clearance: 0.23 m
Track: 1.473 m
Wheelbase: 2.627 m

Jeep Corporation YJ-L (4 × 4) light vehicle, long wheelbase version of YJ-S (Wrangler)

Egyptian-built Jeep YJ-L border patrol vehicle

Angle of approach/departure: 37°/30°
Fuel capacity: 76 litres
Engine: 4 litre 6-cylinder MPI petrol developing 180 hp at 4,500 rpm or
2.5 litre 4-cylinder MPI petrol developing 117 hp at 5.250 rpm
Gearbox: manual with 5 forward and 1 reverse gears
Transfer box: NP 231 2-speed
Suspension:
(front) Dana Model 30 semi-floating axle with multileaf XHD springs mounted below plus track bar and stabiliser bar
(rear) Dana Model 44 w/Trac-Loc semi-floating with multileaf XHD springs mounted below; a track bar is utilised
Tyres: 16 × 7.5 sand
Brakes: disc front, drum rear
Electrical system: 12 V (24 V optional)

Status

In production. In service with Egypt (CJ and YJ), Kuwait (YJ) and Oman (CJ).

Manufacturer

Arab American Vehicles Co (AAV), PO Box 2419, Horreia, Heliopolis, Cairo, Egypt.
Tel: +20 2 290 3554. Telex: 93791 aav un.
Fax: +20 2 290 3562.

UPDATED

FRANCE

LOHR Fardier FL 500 and FL 501 (4 × 4) 500 kg light vehicles

Description

These light air-portable vehicles were developed by SOFRAMAG (now LOHR) for use by airborne troops and 300 were delivered to the French Army. A C-130 or C-160 Transall transport aircraft can carry six FL 500s ready for air-dropping or 12 FL 500s for delivery as cargo. The SA 330 Puma helicopter can carry one FL 500 and one 120 mm mortar. The vehicle can carry a maximum load of 500 kg, including the driver and tow a trailer or weapon such as a 120 mm Thomson Brandt mortar, weighing a maximum of 500 kg.

The FL 500 has a chassis of tubular steel welded construction with the load area covered by aluminium sheeting. The driver is seated at the front of the vehicle on the left side and, if required, an inverted U-shaped safety bar can be fitted to the rear of the driver's position.

The engine is mounted crosswise in the centre of the vehicle and connected to the front and rear axles by two Cardan-driven central transmission units with shock-absorbing guides. The light alloy reinforced axles are mounted on cushioned rubber and the suspension arms consist of helicoidal springs placed horizontally on each side of the chassis, in two suspension boxes.

Brakes are hydraulic on all wheels, with an independent parking brake for each rear wheel. A complete lighting system is installed, enabling the vehicle to be driven on roads.

The final production version was the FL 501 with an optional 36 hp engine fitted as standard. This allows loads of up to 600 kg to be towed.

The FL 500 was used for trials with the MILAN ATGW system and has been equipped with light machine guns. A light field ambulance version carrying two stacked stretchers has been observed. Alternative loads could include radios. The FL 500 and FL 501 can be used to tow 120 mm mortars.

Although numerous FL500/FL 501 vehicles remain in service they are scheduled to be replaced when the LOHR VLA (see following entry) enters production.

LOHR Fardier FL 501 (4 × 4) 500 kg light vehicle configured as a field ambulance

Specifications

Cab seating: 1
Configuration: 4 × 4
Weight:
(empty) 680 kg
(loaded) 1,180 kg
Max load: 500 kg
Towed load:
(FL 500) 500 kg
(FL 501) 600 kg
Load area: 1.93 m^2
Length:
(laden) 2.41 m
(unladen) 2.375 m
Width: 1.5 m
Height:
(steering wheel) 1.18 m
(load area) 0.92 m
Ground clearance:
(unloaded) 0.25 m
(loaded) 0.2 m
Track: 1.26 m
Wheelbase: 1.735 m
Angle of approach/departure: 90°/90°
Max speed: (road) 80 km/h
Range: (road) 200 km
Fuel capacity: 25 litres
Max gradient: 60%
Max side slope: 30%
Vertical obstacle: 0.2 m
Fording: 0.4 m
Engine: (FL 500) Citroën AK 2 flat twin petrol developing 29 hp (DIN) at 6,750 rpm
(FL 501) Citroën V06/630 flat twin, air-cooled petrol developing 36 hp at 5,500 rpm
Gearbox: Citroën manual with 4 forward and 1 reverse gears
Steering: rack and pinion
Turning radius: 4.3 m
Suspension: independent, with helical spring and hydraulic shock-absorber at each wheel station
Brakes:
(main) hydraulic disc
(parking) handbrake on each rear wheel
Electrical system: 12 V
Battery: 1 × 12 V, 30 Ah
Alternator: 390 W

Status

Production complete. In service with the French Army (300), Argentina, Spain (parachute brigade) and Tunisia.

Manufacturer

LOHR SA, 29 Rue du 14 Juillet, F-67980 Hangenbieten, France.
Tel: +33 3 88 38 98 00. Telex: 870082 f.
Fax: +33 3 88 96 06 36.

UPDATED

LOHR VLA Light Airmobile Vehicle

Description

The LOHR VLA *(Véhicule Léger Aéromobile)* was developed in response to a requirement from the French Army for a light vehicle capable of being transported in or under an NH 90 helicopter. Three prototypes have been completed, with the second prototype undergoing French Army trials.

The VLA is a 4 × 4 multipurpose light vehicle with a welded tube chassis and aluminium and polyester panel body; one or two roll bars can be incorporated. The engine is located in the centre of the chassis and is allied to an automatic transmission with an integral differential. The driver is seated far forward in a well with seating for one passenger on the right. Behind the well is a small load platform over the engine compartment with a further cargo/passenger well at the rear; this can accommodate three passengers. The cargo well may be used to mount various weapons, such as a MILAN or TOW ATGW and a machine gun may be mounted in front of the passenger.

For air transport, a C-130 or C-160 transport aircraft can carry up to 10 VLAs stacked in twos one above each other, or four vehicles on paradrop pallets. For transport slung under helicopters the VLA is provided with four lashing rings and stowage space for slings.

Optional equipment includes a winch, tarpaulin to cover the load and rear areas and removable light modular armour. Proposed versions include an air defence variant carrying a light surface-to-air missile system, fire support vehicle with a 12.7 mm heavy machine gun, field ambulance, light recovery version, light reconnaissance vehicle with radios, 120 mm mortar tractor, light load carrier towing a small trailer and a command version with an extendable canvas penthouse.

Specifications

Seating: 2 + 3
Configuration: 4 × 4
Weight empty: 1,350 kg
Combat weight: 2,200 kg
Max load: 850 kg

LOHR Véhicule Léger Aéromobile (VLA) (T J Gander) **1997**

LOHR Véhicule Léger Aéromobile (VLA) (T J Gander) **1997**

Towed load: 1,200 kg
Length: (body) 3.36 m
Width: 1.72 m
Height:
(body) 1.03 m
(roll bar) 1.66 m
Ground clearance: 0.29 m
Wheelbase: 1.965 m
Max speed: 80 km/h
Range: 700 km
Max gradient: 60%
Max side slope: 50%
Fording: 0.55 m
Engine: Peugeot XUD 9A diesel developing 70 hp
Transmission: automatic with 4 forward and 1 reverse gears
Steering: rack bar
Suspension: MacPherson front; trailing link and transverse leaf spring at rear
Tyres: 7.00 R 16 Michelin XC type L

Status
Prototypes.

Manufacturer
LOHR SA, 29 Rue du 14 Juillet, F-67980 Hangenbieten, France.
Tel: +33 3 88 38 98 00. Telex: 870082 f.
Fax: +33 3 88 96 06 36.

UPDATED

Auverland Type A3 (4 × 4) 480 kg light vehicle

Description
First produced in 1988, the Auverland Type A3 (4 × 4) light vehicle is manufactured in a range of types to meet various user requirements. The base vehicle, the Type A3, is available in one of two chassis lengths and various diesel or petrol engines (including a 105 hp unit) can be fitted.

The base model of the Type A3 is powered by a diesel engine, usually the Peugeot XUD 9, although from October 1995 the 95 hp XUD 9TF turbo diesel became available. The Type A3MH is intended for transport by helicopter, while the Type A3L has a longer wheelbase (2.65 m) and can carry up to nine passengers or a payload of 800 kg. There is also a 4 × 2 version of the basic Type A3.

The Type A3 is a small Jeep-type vehicle available in various forms. The layout is conventional, with the engine at the front and seating for the driver and two passengers. A small area at the rear can be used to carry 500 kg of payload or two extra passengers. The vehicle is constructed so that the body is isolated from the rectangular tube chassis.

There are two chassis lengths available (3.65 m and 3.85 m) and the vehicle can be supplied in soft or hardtop versions with standard or comfort pick-up variants. Cargo platform lengths may be 0.8, 1.4 or 1.9 m. A small winch can be fitted at the front if required.

The standard engine is a Peugeot XUD 9 diesel. Engine options include the Peugeot XUD 11 diesel developing 84.5 hp (for the Type A3 SL) and the Peugeot XY 52C petrol developing 94 hp. From October 1995 onwards, the 95 hp XUD 9TF turbo diesel became available. A 24 V electrical system can be fitted.

Auverland Type A3L (4 × 4) 480 kg light vehicle armed with 106 mm recoilless rifle **1996**

Variants include the Type A3 MHCB light armoured vehicle, Type A3 SL 4 × 2 armed with a 12.7 mm machine gun, Type A3 'Gendarmerie' with a long wheelbase and hardtop and Type A3 produced for the French Air Force armed with a 7.62 mm AA 52 machine gun. There is also a Type A3L armed with a 106 mm recoilless rifle or configured to carry MILAN or other anti-tank missiles.

A special Attack Vehicle variant, the Type A3F, has been developed - see entry in *Special attack vehicles* section.

Specifications
Type A3 4 × 4
Cab seating: 1 + 2
Configuration: 4 × 4
Weight:
(empty) 1,230 kg
(loaded) 1,710 kg
Max load: 480 kg
Towed load:
(without brake) 605 kg
(with brake) 1,250 kg
Length: 3.85 m
Width: 1.54 m
Height: 1.7 m
Ground clearance: 0.25 m
Track: 1.342 m
Wheelbase: 2.25 m or 2.65 m
Angle of approach/departure: 50°/45°
Max speed: 115 km/h
Range: 800-900 km
Fuel capacity: 80 litres
Gradient: 100%
Side slope: 40°
Fording: 0.58 m
Engine: Peugeot XUD 9 1.905 litre diesel developing 64 hp (DIN) at 4,600 rpm

Auverland Type A3 (4 × 4) 480 kg light vehicle as delivered to the French Air Force **1996**

Auverland Type A3 (4 × 4) 480 kg light vehicle with hardtop body **1996**

Gearbox: Peugeot BA 7/5 with 5 forward and 1 reverse gears
Clutch: single dry plate
Transfer box: Auverland A80 2-speed
Turning radius: 5 m
Suspension: coil springs with telescopic shock-absorbers
Brakes: disc front, drum rear, dual circuit
Tyres: 6.50 × 16 or 20.5 R 16
Electrical system: 12 V (24 V optional)
Battery: 1 × 12 V, 60 Ah

Status
In production. In service with the French Army (1st Regiment of Infantry, 210 vehicles), Gendarmerie Nationale (350), Gendarmerie Mobile (380) and the French Air Force (200).

Manufacturer
Auverland SA, BP 12, F-42260 Saint-Germain-Laval, France.
Tel: +33 4 77 65 54 44. Telex: 310663 f.
Fax: +33 4 77 65 49 18.

VERIFIED

Auverland Type A3 (4 × 4) 480 kg light vehicle prior to delivery to the French Army's 1st Regiment of Infantry **1996**

Citroën Méhari Armée light vehicle

Description
The Citroën Méhari Armée is a military version of the standard civilian Méhari vehicle. Its low weight enables it to be used for a variety of military applications such as traffic control, command/radio vehicle and light supplies carrier. Over 8,000 of these vehicles were delivered to the French armed forces and police.

The vehicle has an all-steel chassis with a plastic body. The engine is mounted at the front with the driver's and passenger's seats in the centre and the cargo area at the rear.

There are two basic models of the Méhari, two plus two and two-seat utility. The first has two seats at the front, removable or foldaway back seat, fold-down windscreen, black cotton canopy, safety door chains and an anti-theft device. Optional equipment includes a complete hood with transparent side panels and doors. The two-seat utility model has the two seats at the front, a flat load area at the rear, fold-down windscreen, black cotton canopy, safety door chains and an anti-theft device. Optional equipment includes a hood without transparent side panels and doors.

Trials confirmed that the Méhari Armée is suitable for dropping by parachute when strapped to a landing platform. The basic vehicle can be used for a variety of roles including ambulance, traffic control and radio vehicle.

French Army Citroën Méhari Armée light vehicle (Pierre Touzin) **1996**

Specifications
Cab seating: 1 + 1
Configuration: 4 × 2 (front wheel drive)
Weight:
(empty) 570 kg
(loaded) 955 kg
Max load: 385 kg
Load area: 1.34 × 1 m
Length: 3.52 m
Width: 1.53 m
Height:
(overall) 1.635 m
(load area) 0.65 m
Wheelbase: 2.37 m
Max speed: (road) 100 km/h
Range: 300 km
Fuel capacity: 25 litres
Max gradient: 40%
Engine: AK 2 2-cylinder petrol, air-cooled developing 29 hp at 5,500 rpm
Gearbox: manual with 4 forward and 1 reverse gears
Clutch: single dry plate
Steering: rack and bar
Turning radius: 5.35 m
Suspension: both axles suspended by arms with lateral interplay on spiral springs. Hydraulic shock-absorbers front and rear
Brakes:
(main) hydraulic
(parking) mechanical
Electrical system: 12 V or 24 V
Batteries: 2 × 12 V

Status
Production complete. In service with Spain and the French Army, Air Force, Navy and Police.

Manufacturer
SA Automobiles C troën, 62 boulevard Victor Hugo, F-92200 Neuilly-sur-Seine, France.
Tel: +33 2 47 48 41 41. Telex: citr 614830 f.

VERIFIED

Citroën A FAF 400 kg (4 × 2) and 400 kg (4 × 4) light vehicles

Description
The Citroën A FAF is basically a standard civilian vehicle with the minimum of modifications for military use. Two models are currently in service, a 4 × 4 version and a 4 × 2 version, both with a 400 kg payload. For ease of maintenance and obtaining spare parts, all the automotive components are from Citroën A-type vehicles, of which, well over seven million were built.

The chassis and body are all-steel and the windscreen can be folded forward on to the bonnet. The layout is conventional, with the engine at the front, passenger and driver seats in the centre and the load area at the rear. A model is also available with an additional two-man bench seat behind the driver's and front passenger's seats, which folds forward to increase the cargo area. In wet weather a cover can be erected over the top of the vehicle.

The Citroën A FAF is suitable for a number of roles, including cargo carrier, command/radio and reconnaissance and can be dropped by parachute. Optional equipment includes a 24 V electrical system and hydraulic shock-absorbers.

Specifications
(4 × 2 version; data in square brackets relate to 4 × 4 model where different)
Configuration: 4 × 2 [4 × 4]

Portuguese Army Citroën A FAF (4 × 4) 400 kg light vehicle (T J Gander)

Weight:
(empty) 690 [850] kg
(loaded) 1,090 [1,250] kg
Max payload: 400 kg
Length: 3.62 m
Width: 1.55 m
Height: 1.68 m
Wheelbase: 2.4 m
Track:
(front) 1.26 m
(rear) 1.28 m
Ground clearance: 0.238 m
Max speed: 100 [110] km/h
Gradient: 50%
Fuel capacity: 66 litres
Fording: 0.35 m
Engine: air-cooled Citroën 4-stroke petrol developing 28.5 hp (DIN) at 6,750 rpm [same engine or 4-cylinder petrol developing 34 hp (DIN) at 5,500 rpm]
Gearbox: manual with 4 forward and 1 reverse gears (road, 4 forward and 1 reverse gears; cross-country, 3 forward and 1 reverse gears, rear differential)
Clutch: single dry plate
Steering: rack and pinion
Turning radius: 6 m
Brakes:
(main) (front) disc on gearbox
(main) (rear) disc on differential
(parking) acts on front wheels
Suspension: arms with lateral interaction, working with helicoidal springs and inertia stabilisers
Tyres: 155.14 × CM
Electrical system: 12 V, 48 Ah (24 V optional)

Status
Production complete. The 4 × 4 model is in service with Burundi and Portugal. Produced in Greece at one time as the Namco Pony.

Manufacturer
SA Automobiles Citroën, 82 boulevard Victor Hugo, F-92200 Neuilly-sur-Seine, France.
Tel: +33 2 47 48 41 41. Telex: citr 614830 f.

VERIFIED

Peugeot P4 (4 × 4) 750 kg light vehicle

Development
Until the mid-1980s, the standard light vehicle of the French Army was the Hotchkiss M 201 (4 × 4) 400 kg 'Jeep', production of which was completed in 1969. To replace this fleet, three manufacturers each submitted 12 vehicles for extensive trials; these were the Renault TRM 500 (4 × 4) 500 kg based on the Italian FIAT 1107 AD, the Citroën C 44 based on the German Iltis and the Peugeot P4 based on the Mercedes-Benz vehicle. Early in 1981 the French Army selected the Peugeot P4 and placed an order for 15,000 vehicles. Mass production, by Panhard, began in early 1983 with the first 2,400 vehicles being fitted with petrol engines. Production ceased in 1992.

The P4 is often referred to as the VLTT (*Véhicule de Liaison Tout Terrain).*

Description
The layout is conventional, with the engine at the front, driver and passenger immediately behind the engine and the cargo area at the rear, with a two-man bench seat down either side and an opening tailgate on which the spare wheel is mounted.

The front seats are adjustable and hinge forward to give access to the rear seats and the tool boxes beneath each of them. The driver and passengers are provided with seat belts with automatic rollers. The bench seats in the rear can be folded down to enlarge the load-carrying area.

The chassis is formed from two parallel longitudinal

Model	**SWB**	**LWB**
Configuration	4 × 4	4 × 4
Weight (empty) (petrol/diesel)	1,815/1,895 kg	1,985/2,065 kg
(loaded) (petrol/diesel)	2,565/2,645 kg	2,985/3,065 kg
Max load	750 kg	750 kg
Length	4.2 m	4.65 m
Width (with antenna mount)	1.83 m	1.83 m
(less antenna mount)	1.7 m	1.7 m
Height	1.9 m	1.9 m
Ground clearance	0.24 m	0.24 m
Track	1.4 m	1.4 m
Wheelbase	2.4 m	2.85 m
Angle of approach/departure	42°/37°	42°/37°
Fording	0.5 m	0.5 m
Gradient (loaded) (petrol/diesel)	73.5/55%	73.5/55%
Side slope (loaded)	30%	30%
Max speed (petrol/diesel)	118/108 km/h	118/108 km/h
Fuel capacity	75 litres	75 litres
Engine	Peugeot 1.971 litre 4-cylinder petrol developing 79 hp at 4,750 rpm or Peugeot 2.498 litre 4-cylinder diesel developing 70.5 hp at 4,500 rpm	
Gearbox	manual, 4 forward and 1 reverse gears	
Clutch	single dry disc with ball-bearing thrust mechanism	
Transfer box	2-speed	2-speed
Steering	rack and pinion with damper	
Turning radius	5.5 m	6.45 m
Suspension (front)	coil springs, anti-sway bar and double-acting telescopic hydraulic shock-absorbers	
(rear)	coil springs and double-acting telescopic hydraulic shock-absorbers	
Tyres	700 RC 16 x C type L	700 RC 16 x C type L
Brakes	hydraulic dual circuit, discs front and drum rear	
Electrical system	24 V	24 V
Generator	1,200 W	1,200 W

French Army Peugeot P4 (4 × 4) 750 kg light vehicle serving with IFOR in Bosnia (R Stickland) ***1997***

Peugeot P4 (4 × 4) 750 kg light vehicle armed with 7.62 mm machine gun (Pierre Touzin) ***1996***

Armoured Peugeot P4 (4 × 4) 750 kg light vehicle (R Stickland) ***1997***

beams with rectangular sections connected by five round transverse members. The rear axle is fitted with a hydraulically controlled differential lock.

The sheet metal body is connected to the chassis with eight flexible mounts and has a removable roll-over bar and a folding windscreen of bonded safety glass. Fixtures, bumpers, inner bumper liners and the front portion are mounted to the body structure. For ease of replacement these are bolted on. The two cloth doors have translucent window panels and are removable. The rear troop/cargo area is covered by a tarpaulin of plastic-coated cloth, with translucent sides and rear windows, which folds to the rear and is easily removable.

The fuel tank is below the floor of the vehicle between the chassis beams. A fuel can is carried externally at the rear of the vehicle on the left side. The P4 is fitted with a towing eye at the front and a towbar at the rear with an electric trailer plug The electrical system includes a park plug, lantern p ug and a connection box for a two-way radio.

Optional equipment includes: power-assisted steering, front locking differential, PTO front and rear, front-mounted elec:ric winch, a 15 litre fuel can and various adaptors for mounting machine guns and other weapons.

The basic vehicle is used by the French Army for a variety of roles, including one armed with twin F1 machine guns for use by scout teams and traffic control squads, while another is equipped with a MILAN ATGW launcher and four missiles and used by motorised infantry. The P4 has also been demonstrated carrying the Mygale air defence system and various forms of radar. An LWB mo:lel was also available with a drop tailgate, bows, tarpaulin cover and back-to-back seating for eight fully equipped soldiers at the rear. The P4 has been demonstrated carrying a 106 mm recoilless rifle. Both the SWB and LWB versions are fitted with radios for use in the command role and fully enclosed van-type versions of both the LWB and SWB models are available. There is also a fully equipped workshop vehicle version and an ambulance, both LWB. The P4 has been demonstrated carrying NBC decontamination equipment.

An armoured protection kit for the P4 is produced by Panhard et Lavassor. It is proof against 7.62 mm ball rounds at 100 m.

Status
Production comple:e. In service with the French Army, Gabon and Togo.

Manufacturer
Automobiles Peugeot, 75 Avenue de la Grande-Armée, F-75116 Paris Cedex 16, France.
Tel: +33 1 40 66 55 11. Telex: 610700 f.

VERIFIED

GERMANY

VW Iltis (4 × 4) 500 kg light vehicle

Development
In 1976, the Federal German Army issued a requirement for a 4 × 4 vehicle which could carry 500 kg of cargo both on roads and cross-country. Both Daimler-Benz and Volkswagen built prototypes for trials and in 1977 the Volkswagen Iltis was selected for production as the Lkw 0.5t tmil gl.

Production of the Iltis continued until December 1981 when the last of 8,800 vehicles was handed over to the German armed forces. Of this total, 8,470 went to the Army, 310 to the Luftwaffe and the remaining 20 to the Navy.

Production in Germany ceased in 1982 but in 1983 production of the Iltis was transferred to Bombardier of Canada. For a description and the specifications of this vehicle refer to the entry under Canada in this section.

Status
Production complete. In service with the German Army (approx 8,800). In October 1981 the Canadian company Bombardier Inc obtained a licence to produce the Iltis. Further details are given in this section under Canada.

Manufacturer
Volkswagenwerk AG, D-3180 Wolfsburg 1, Germany.
Tel: +49 5361 90. Telex: 95 860.
Fax: +49 5361 928282.

VERIFIED

Volkswagen Iltis (4 × 4) 500 kg light vehicle configured as field ambulance (Bob Morrison - Military Scene)

Mercedes-Benz (4 × 4) 700 kg light vehicle

Development
Mercedes-Benz produces its 4 × 4 range for both civil and military use. Vehicles produced for military purposes may have a standard wheelbase of 2.4 m, longer wheelbases of 2.85 m and 3.12 m and, as a late introduction, a long wheelbase of 3.4 m. The main components of the vehicle, with the exception of the specially designed transfer case, are from Mercedes-Benz passenger car and light truck series production.

Gelaendewagenfahrzeug Gesellschaft (GFG) began production in 1979 with an initial target of 9,000 vehicles a year. The Austrian-based GFG, owned jointly by Mercedes-Benz and Steyr-Daimler-Puch, ceased to exist in the early 1980s. From then on production was continued with Steyr-Daimler-Puch as a contract partner of Mercedes-Benz. Steyr-Daimler-Puch remains responsible for sales in Austria, Switzerland and the former Yugoslavia under the name of Puch, whereas Mercedes-Benz sells to all other countries under the name of Mercedes-Benz.

In 1980 the Mercedes-Benz (4 × 4) light vehicle was entered by Peugeot in the French Army competition for a new light veh cle to replace the then current Hotchkiss M 201. After comparative trials the P4 was selected by the French Army and entered service in 1982. There is a separate entry in this section for the Peugeot P4 (4 × 4) light vehicle and its variants.

The Mercedes-Benz vehicle is licence-produced in Greece (5,000 units) by Steyr Hellas SA and called the 240 GD (see entry under Greece).

Description

The layout of the base Mercedes-Benz (4 × 4) light vehicle is conventional, with the engine at the front. Driver and passenger sit immediately behind the engine, two passenger seats and a small cargo area are at the back. The floor of the load area is made of sheet steel and is fitted with C-type rails, which allows communications equipment to be installed quickly. The vehicle has a folding hood, removable side flaps and a folding windscreen. A self-recovery front winch may be fitted.

The same frame design is used for both short and long wheelbases. They have closed box-shaped side members in conjunction with tubular cross members, providing high bending and torsional stiffness in a ladder-type frame to which the noise-insulated bodywork is attached by rubber mountings. The frame is corrosion protected and is also waxed and sealed.

At one time as many as four engine options were available. This has been reduced to a basic two, both of which meet current exhaust emission requirements. The 230 GE has the M102/230 GE 2.298 litre petrol engine while the 290 GD has the OM 602/D29 2.874 litre five-cylinder diesel. Engine output is transmitted to the wheels using an all-synchromesh five-speed manual gearbox and a two-shaft transfer case. The four-wheel drive can be engaged when the vehicle is either stationary or moving. The cross-country ratio can be engaged and disengaged when driving, to double the power to the wheels and reduce the speed by half. The 100 per cent locking differentials on the front and rear axles can be engaged and disengaged individually during travel.

Payloads vary from a minimum 700 kg for the base 2.4 m wheelbase models up to 1,600 kg (depending on body) for the 3.4 m wheelbase models - the latter can include flatbed/shelter carriers, ambulances and radio or workshop vehicles.

There are station wagon and van variants of the basic vehicle and various companies produce special bodies such as forward-area ambulances. There is also a light cable-laying version with an open rear and a side-mounted spare wheel on the right. Various seating arrangements are possible, with the short wheelbase version having up to eight seats and the long wheelbase version up to 10. Radio/command variants, a cable-laying vehicle, an NBC decontamination variant and variants carrying anti-tank missiles have been demonstrated. The vehicles can also be used as tugs for light aircraft or helicopters.

Optional military equipment includes a radiator and engine guard, brush guards for the headlights, weapons rack, swivelling spare wheel carrier and a carrier in the rear for a 20 litre fuel container.

An armoured protection kit for this range of

Long wheelbase (3.4 m) flatbed-bodied version of Mercedes-Benz (4 × 4) light cross-country vehicle

Mercedes-Benz vehicles is produced in France by Panhard et Lavassor. It is proof against 7.62 mm ball rounds at 100 m.

Specifications

Standard 2.4 m wheelbase, open soft top version
Configuration: 4 × 4
Weights:
(permissible front axle load) 1,200 kg
(permissible rear axle load) 1,600 kg
Max load: 700 kg
Max towed load:
(braked) 2,800 kg
(unbraked) 750 kg
Length: 4.165 m
Width: 1.7 m
Height: (overall) 2.005 m
Ground clearance: 0.24 m
Track: 1.425 m
Wheelbase: 2.4 m
Angle of approach/departure: 39°/34°
Max road speed:
(230 GE) 143 km/h
(290 GD) 127 km/h
Fuel capacity: approx 96 litres
Max gradient: 80%
Side slope: (static) 25°
Fording: 0.6 m
Engine:
(230 GE) M 102 /230 GE 2.298 litre 4-cylinder petrol developing 126 hp at 5,000 rpm
(290 GD) OM 602/D29 2.874 litre 5-cylinder diesel developing 95 hp at 4,000 rpm
Transmission: GL76/27-P5 5-speed manual synchromesh
Clutch: single dry plate
Transfer box: 2-speed VG 080 for mechanical front-axle engagement and synchronised cross-country gear
Steering: LS 2 B power-assisted
Turning radius: 5.7 m
Suspension: front and rear, coil springs and telescopic shock-absorbers. Both front and rear axles are located by 1 transverse and 2 longitudinal control links
Tyres: 20.5 R 16 (other sizes optional)
Brakes: (main) hydraulic, dual circuit with vacuum booster, disc front, drum rear
Electrical system: 24 V
Batteries: 2 × 12 V, 60 Ah
Alternator: 27 or 55 A

Status

In production. In service with Abu Dhabi (25), Algeria (90), Argentina (1,200), Austria (1,200), Denmark (1,300), France (15,000, local production - see separate entry), Germany (12,000 plus 870 for Federal German Border Guard and 200 for Federal German foreign aid), Greece (over 5,000, local production), Guatemala (50), Ireland, Kuwait (400), Luxembourg (38), Netherlands (3,195 ordered in mid-1990), Norway (3,500), Oman, SHAPE (350), Singapore (311), Switzerland (4,100), UNO (100), UK (captured Argentine vehicles used in Falkland Islands), US Army Europe (380), Yugoslavia (500), Zambia (50) and other undisclosed African countries (50).

Manufacturer

Mercedes-Benz AG, Dept SGW R.102, D-70322 Stuttgart-Untertürkheim, Germany.
Tel: +49 711 17 83662. Telex: 725420 db d.
Fax: +49 711 17 83432.

VERIFIED

Danish Army Mercedes-Benz (4 × 4) light vehicle serving with IFOR in Bosnia (R Stickland) ***1997***

Mercedes-Benz (4 × 4) light vehicle configured as ambulance for the Abu Dhabi armed forces (T J Gander)

GREECE

Hellenic Vehicle Industry Type 240 GD (4 × 4) light vehicle

Description

This is the 2.4 m wheelbase model of the Mercedes-Benz (4 × 4) 700 kg light vehicle (see previous entry) licence-produced in Greece by Hellenic Vehicle Industry SA. The size of the production batch involved has been reported as over 5,000.

The Type 240 GD is so named as it is powered by the Mercedes-Benz 240 GD 2.4 litre four-cylinder diesel, developing 72 hp at 4,400 rpm coupled to a four-speed synchromesh gearbox. Other details are as provided in the previous entry.

The Greek-produced vehicles are produced in several versions, the base vehicle being a general purpose vehicle with a soft top. Open-topped weapons carrier variants include a machine gun carrier with the machine gun mounted on a roll-over bar and anti-tank versions. These include a 106 mm M40 recoilless rifle carrier with a split front windscreen and a blast screen over the bonnet and TOW or MILAN anti-tank missile carriers. A front-line ambulance version with a soft top is also available.

Status

In production. In service with the Greek armed forces.

Manufacturer

Hellenic Vehicle Industry SA, PO Box 10239, GR-54110 Thessaloniki, Greece.
Tel: +30 31 798502. Telex: 412586.
Fax: +30 31 798426.

UPDATED

Type 240 GD (4 × 4) light vehicle (T J Gander)
1997

Type 240 GD (4 × 4) light vehicle configured as TOW anti-tank missile carrier

Type 240 GD (4 × 4) light vehicle configured as 106 mm recoilless rifle carrier; note blast shield over bonnet

INDIA

VV501 Toofan (4 × 4) 500 kg light vehicle

Description

The VV501 Toofan (4 × 4) light vehicle was developed at the Indian Vehicle Research and Development Establishment (VRDE). Seven prototypes were completed and used for Indian Army trials.

The VV501 is a Jeep-type vehicle of conventional layout with the water-cooled petrol engine mounted forward under a prominent square bonnet. The radiator is large for a vehicle of its size and the headlamps are set into large square mudguards. The cab has seating for the driver and one passenger and the cargo area, to the rear, may be adapted for up to four passengers. The prototypes were fitted with a canvas top and sides but it is expected that more specialised body types will be fitted to production examples.

Specifications

Cab seating: 1 + 1
Configuration: 4 × 4
Weight: (gross) 2,500 kg
Max load: 500 kg
Length: 4 m
Width: 1.72 m
Height: 1.95 m
Max speed: 120 km/h
Range: 500 km
Engine: 6-cylinder water-cooled 4 litre petrol developing 110 bhp at 3,200 rpm
Gearbox: manual, 3 forward and 1 reverse gears
Transfer box: 2-speed

Status

Prototypes. User trials completed. The current status of this project is uncertain.

VERIFIED

Mahindra (4 × 4) army model vehicles

Development

Mahindra and Mahindra Limited first assembled imported Jeep kits in 1947 in technical association with the Willys Overland Corporation of the USA. A phased local manufacture of the vehicle was undertaken from 1954 onwards and was completed by 1968. A wide range of petrol and diesel utility vehicles is manufactured and well over 500,000 vehicles have been sold within India; Mahindra and Mahindra Limited is the chief supplier of light 4 × 4 vehicles to the Indian Defence Forces.

Exports commenced in 1968 and since then nearly 40,000 vehicles have been exported. From 1980, 25,000 units were delivered in knock-down form for assembly in Iran.

In 1979, the company entered into an agreement with Automobiles Peugeot of France to manufacture the XPD 4.90 diesel engine at its Igatpuri plant. Peugeot diesel engines were manufactured in the same plant from 1991 onwards.

Mahindra and Mahindra Limited also produce single and double cab models with open rear cargo areas and fully enclosed models with varying standards of body fit.

Mahindra and Mahindra Limited has a joint venture company in Greece, Mahindra Hellas SA, engaged in the assembly of CL series vehicles to European specifications and fitments.

Description

Two series of 4 × 4 vehicles are in production, the CL and MM series. The CL series is available in 2.032 m and 2.286 m wheelbase versions and the MM series with a 2.426 m wheelbase. Both series come with a variety of options for seating, body configurations and with diesel or petrol engines.

Standard military fittings include: blackout lighting, a starting crank assembly, radiator chaff screen, rifle bracket, shovel, pick and helve stowage, starting

Mahindra CL340 Army Model with 106 mm recoilless rifle

Mahindra MM540 (4 × 4) light vehicle

Specifications

Model	CL340	CL540	MM540
Seating	1 + 2	1 + 2 (or 4)	1 + 2 (or 4)
Configuration	4 × 4	4 × 4	4 × 4
Gross vehicle weight	1,760 kg	1,810 kg	1,810 kg
Towing capacity	1,800 kg	1,800 kg	1,800 kg
Length	3.39 m	3.75 m	3.81 m
Width	1.6 m	1.6 m	1.78 m
Wheelbase	2.032 m	2.286 m	2.426 m
Track (with 7.00 × 15 tyres)	1.23 m	1.23 m	1.29 m
Max speed (2.1 litre diesel)	105 km/h	105 km/h	105 km/h
Fuel capacity	40 litres	40 litres	40 litres
Engine (diesel)	2.1 or 2.5 litres		
(petrol)	2.2, 2.3 EFI or 2.5 litres		
Transmission	3-speed (1st gear non-synchromesh) or 4-speed (1st gear non-synchromesh) or 5-speed (all synchromesh)		
Transfer case	2-speed	2-speed	2-speed
Turning radius	5.3 m	5.86 m	6.34 m
Suspension	semi-elliptical underslung leaf springs with double-acting telescopic shock-absorbers		
Tyres	6.00 × 16 or 7.00 × 15 or P235/75 R 15		
Brakes (main)	hydraulic power-assist		
(parking)	mechanical or transmission drum		
Electrical system	12 V	12 V	12 V
Battery (diesel)	90 Ah	90 Ah	90 Ah
(petrol)	60 Ah	60 Ah	60 Ah
Alternator	35 A	35 A	35 A

handle stowage, four grip handles, dashlight switch, fuel tank dipstick, sign plates, speedometer seal, fuel tank strainer and headlight brush guards.

The CL combat vehicle version can carry a 106 mm recoilless gun. This version has a split windscreen through which the barrel of the recoilless gun protrudes when travelling and there are two squab seats for the gunners at the rear.

Status

In production. In service with the Indian armed forces and some other nations including Bangladesh (100 CL) and Sri Lanka (175 CL).

Manufacturer

Mahindra and Mahindra Limited, Overseas Operations, Gateway Building, Apollo Bunder, Bombay 400 039, India.
Tel: +91 22 202 1031. Telex: 2361/5944.

UPDATED

Jonga (4 × 4) 400 kg light vehicle

Description

The Jonga is basically the Japanese Nissan Patrol (4 × 4) light vehicle manufactured under licence at the Vehicle Factory at Jabalpur. It is in widespread service with the Indian armed forces.

The Jonga has undergone numerous changes during its production life. It has been produced with a 145 hp 4 litre petrol engine in place of an earlier 125 bhp unit and has been available with a Perkins P4(V) diesel. The current standard engine is a 4 litre Hino WO4D diesel developing 108 hp.

The Jonga is produced in both 4 × 4 and 4 × 2 versions, with the latter being mainly for commercial sales. Two wheelbase lengths are produced, 2.35 m and 2.8 m and there are three basic bodies. The 2.35 m wheelbase model may have either a three-door soft top or five-door hard top, with seating for the driver and nine or ten passengers respectively. The 2.8 m wheelbase model has a five door hard top with seating for the driver and up to 11 passengers. Other body variations include a pick-up and an ambulance. At one time the Indian Army used an open-backed version carrying SS-11 ATGWs.

Optional equipment includes a 2 tonne self-recovery winch and an extra fuel tank.

Jonga (4 × 4) light vehicle ***1996***

Specifications

2.35 m WB, soft top
Cab seating: 1 + 9
Configuration: 4 × 4
Weight: (GVW) 2,700 kg
Max load: 500 to 700 kg
Towed load: 2,000 kg
Length: 4.045 m
Width: 1.715 m
Height: 2.015 m
Ground clearance: 0.22 m
Track:
(front) 1.382 m
(rear) 1.4 m
Wheelbase: 2.35 m
Max speed: 120 km/h
Fuel capacity: 80 litres
Max gradient: 60%
Engine: Hino WO4D 4.009 litre 4-cylinder in-line direct injection diesel developing 108 hp at 3,200 rpm
Gearbox: synchromesh with 3 forward and 1 reverse gears
Clutch: single dry plate

Transfer box: 2-speed
Steering: worm and roller
Turning radius: 5.3 m
Suspension: semi-elliptical leaf springs (4 front, 5 rear) with double-acting telescopic shock-absorbers
Tyres: 7.00 × 16
Brakes: hydraulic, dual circuit
Electrical system: 12 V
Alternator: 35 A

Status
In service with the Indian Army and Mozambique.

Manufacturer
Vehicle Factory, Indian Ordnance Factories, Jabalpur 482009, Madhya Pradesh, India.
Tel: +91 761 330520. Fax: +91 761 330436.

VERIFIED

Carrier (4 × 4) 750 kg light truck

Description
The Carrier (4 × 4) 750 kg light truck is a Japanese Nissan D4W73 vehicle produced under licence in India. The first production units were completed at the Defence Unit (Ordnance Factory) at Jabalpur during 1961 and production continues.

The basic version of the Carrier is a general purpose light utility truck with a two-person hardtopped cab and a cargo area covered by a canvas tilt over removable bows. The cab windscreen can be folded forwards on to the bonnet and some vehicles feature a square roof hatch over the passenger seat. The cargo area has a drop tailgate and folding seats may be provided along the sides.

Some vehicles are fitted with a front-mounted winch. Variants include a desert operations vehicle with oversize sand tyres, field ambulance and a light fire tender.

Specifications
Cab seating: 1 + 1
Configuration: 4 × 4
Weight:
(empty) 2,690 kg
(loaded) 4,190 kg
Max load:
(road) 1,500 kg
(cross-country) 750 kg
Length: 4.73 m
Width: 2.045 m
Height: (overall) 2.355 m
Ground clearance: (axles) 0.26 m
Track: 1.6 m
Wheelbase: 2.8 m
Max speed: 95 km/h
Fuel capacity: 110 litres
Engine: 3.956 litre 6-cylinder water-cooled petrol developing 145 hp at 3,800 rpm
Gearbox: manual, 4 forward and 1 reverse gears
Clutch: single dry plate
Transfer box: 2-speed
Suspension: semi-elliptic leaf springs
Brakes: hydraulic
Electrical system: 12 V

Status
In production. In widespread service with the Indian Army.

Manufacturer
Vehicle Factory, Indian Ordnance Factories, Jabalpur 482009, Madhya Pradesh, India.
Tel: +91 761 330520. Fax: +91 761 330436.

VERIFIED

Carrier (4 × 4) 750 kg light truck (T J Gander)

Carrier (4 × 4) 750 kg light truck (T J Gander)

ISRAEL

M-240 Storm (4 × 4) 800 kg MultiMission Vehicle (MMV)

Development
The M-240 Storm MultiMission Vehicle (MMV) was designed to fulfil several military functions including general utility, patrol and reconnaissance, command, and anti-tank weapons vehicle.

After decades of experience with 4 × 4 vehicles by the Israeli Defence Forces, the M-240 introduced innovations such as a reinforced body and frame, an updated drive train and a 180 hp petrol engine with a fuel injection system. An automatic transmission is an option.

Description
The M-240 Storm is produced with two frame lengths, 4.5 m and 4.15 m. The layout of the vehicle is conventional, with the engine at the front, the driver and one passenger seated behind the engine and space for further passengers or other loads at the rear. The front axle is fully floating while the rear axle is semi-floating.

Apart from the main general purpose utility vehicle version, which can be fitted with a soft top, there are several variants of the M-240. The short frame version may be used for the command vehicle variant which can be fitted with machine gun mountings and other special equipment; it can be fitted with a soft top to protect the equipment carried. The long frame version is available with a hard top and an air conditioning system. The anti-tank weapon vehicle can carry a 106 mm recoilless rifle or a missile system. A reconnaissance and patrol variant, armed with machine

M-240 Storm (4 × 4) MultiMission Vehicle (MMV)

M-240 Storm (4 × 4) 106 mm recoilless gun carrier

guns and with extra stowage for fuel and water, is also available.

One further vehicle is a police vehicle intended to handle riots in crowded urban areas or off-roads. This version is provided with transparent polycarbonate sheets which protect the occupants against stones, incendiary bombs and splinters. The side panels and doors have firing ports for riot control weapons.

Optional equipment for all models includes blackout lamps, radio antenna base, container mounting and a 3,629 kg capacity winch. The electrical system may be 12 or 24 V.

Specifications

MMV
Cab seating: 1 + 3 to 5
Configuration: 4 × 4
Gross vehicle weight: 2,350 kg
Max load: (off-road) 750 kg
Length:
(long frame) 4.492 m
(short frame) 4.15 m
Width: 1.676 m
Height: (overall) 2.02 m
Ground clearance: 0.24 m
Wheelbase: 2.63 m
Angle of approach/departure:
(SWB) 40°/37°
(LWB) 40°/26.5°
Max speed: 150 km/h
Fuel capacity: 75 litres
Max gradient: 75%
Vertical obstacle: 0.35 m
Engine: 4 litre 6-cylinder in-line petrol with fuel injection developing 180 hp at 4,700 rpm, fitted with Vortox 2-stage air cleaner
Transmission: Aisin AX15 with 5 forward and 1 reverse gears
Clutch: 266.7 mm hydraulic
Transfer box: NP 231 2-speed
Steering: recirculating ball
Turning radius: 5.73 m
Suspension: front and rear leaf springs with double-acting shock-absorbers
Tyres: 7.00 × 15 8PR
Brakes: dual hydraulic with rear drum self-aligning and disc front
Electrical system: 12 V (24 V optional)
Batteries: 1 or 2 × 12 V, 75 Ah
Alternator: 90 A

Status

In production. In service with the Israeli Defence Forces.

Manufacturer

Automotive Industries Limited, PO Box 535, IL-17105 Nazareth Illit, Israel.
Tel: +972 6 558111. Telex: 46217 ail il.
Fax: +972 6 558103.

VERIFIED

ITALY

Fresia F18 (4 × 4) 530 kg mountain power truck

Development

The Fresia F18 (4 × 4) 530 kg mountain power truck was developed to meet the requirements of Italian mountain troops for a vehicle which could be used to transport equipment and stores over mountain roads and tracks. The prototype was completed in 1974 and, following extensive trials, a preproduction batch was delivered for operational trials with Italian mountain units in June 1978. The original vehicles were limited in load capacity to 200 kg but development increased the capacity to 400 kg and later 530 kg.

Description

The Fresia F18 is a small all-wheel drive vehicle with an open platform for carrying up to 530 kg of cargo. The vehicle can be steered by the operator when seated on the platform or when walking behind. The engine is mounted in the centre of the vehicle under the load area which has a folding handrail on the sides and rear. The transmission is hydrostatic and steering is power-assisted on both axles. Each wheel has helical spring independent suspension.

Fresia F18 (4 × 4) 400 kg mountain power truck

Specifications

Configuration: 4 × 4
Weight:
(empty) 530 kg
(loaded) 1,060 kg
Max load: 530 kg
Length: (over load platform) 1.915 m
Width: 1.06 m
Height:
(over steering wheel) 1 m
(load platform) 0.75 m
Ground clearance: 0.22 m
Track: 0.71 m
Wheelbase: 1.2 m
Max speed: 20 km/h
Fuel capacity: 40 litres
Max gradient: 70%
Max side slope: 30%
Engine: Briggs and Stratton Type 422400 694 cc 2-cylinder air-cooled petrol developing 18 hp at 3,600 rpm
Gearbox: hydrostatic closed circuit with variable displacement radial piston primary pump and 2 fixed displacement orbital hydraulic motors
Steering: hydraulic, power-assisted on all 4 wheels
Turning radius: 2.5 m
Suspension: independent with helical springs for each axle and hydraulic shock-absorbers
Brakes:
(main) operating on hydrostatic transmission
(parking) disc brakes on drive line, hand-operated
Tyres: 26 × 12.00
Electrical system: 12 V
Battery: 1 × 12 V, 45 Ah

Status

In service with Italian Army mountain units.

Manufacturer

Fresia SpA, Via Trento Trieste 30, I-17017 Millesimo (Savona), Italy.
Tel: +39 19 565065. Telex: 571254 fresof i.
Fax: +39 19 565586.

VERIFIED

FIAT Campagnola 1107 AD (4 × 4) 750 kg light vehicle

Description

The FIAT Campagnola 1107 AD (4 × 4) light vehicle entered production in 1974 as the replacement for the FIAT AR-59. The main improvements included a more powerful engine, increased load-carrying capability and the option of a hardtop.

The vehicle was produced in hard and soft top versions and there was also a longer model, with an overall length of 4.025 m and with the same wheelbase, also in hard and soft top models. The longer model has three-man bench seats down either side behind the front seat rather than two as in the basic model.

Special equipment for operation in dusty and tropical zones includes: an air cleaner with centrifugal pre-cleaner; fuel filter between feed pump and carburettor; sealed type declutch thrust bearing; engine compartment front protection; and a low octane petrol engine with 7.5 compression ratio, developing 75 hp.

The body is of all-steel construction and has two side doors and a tailgate. Seating is provided for three, including the driver, at the front and four at the rear, two each side on bench-type seats. The canvas top and side curtains can be removed and the windscreen folded forward flat against the bonnet.

Standard equipment includes a spare wheel, heater, ventilating and defrosting system, pintle towing hook at the rear, towing eyes at the front, fire extinguisher, pick and shovel and additional fuel cans.

The Italian Army uses the FIAT 1107 AD as a TOW anti-tank missile launcher carrier. The vehicle can easily be adapted for a wide variety of other roles such as an ambulance and a command or radio vehicle.

Specifications

Soft top model; data in square brackets relates to hardtop where different
Cab seating: 3 + 4 [3 + 6, LWB version]
Configuration: 4 × 4
Weight:
(empty) 1,670 [1,740] kg
(loaded) 2,420 [2,490] kg
Max load: 750 kg
Towed load:
(road) 1,300 kg
(off-road) 900 kg
Length: 3.775 [4.025] m
Width: 1.58 m
Height: 1.901 m
Ground clearance: 0.275 m
Track:
(front) 1.365 m
(rear) 1.404 m
Wheelbase: 2.3 m
Angle of approach/departure: 44°/45° [44°/28°]
Max speed: (road) 120 km/h
Range: 400 km
Fuel capacity: 57 litres
Max gradient: (SWB) 100%
Max side slope: 40%
Fording: 0.7 m
Engine: FIAT 4-cylinder in-line petrol developing 80 hp (DIN) at 4,600 rpm
Gearbox: manual with 5 forward and 1 reverse gears

Clutch: single dry plate
Transfer box: 2-speed
Steering: hour-glass and roller
Turning radius: 5.4 m
Suspension: independent McPherson type with longitudinal torsion bar, single telescopic shock-absorbers at front and twin at rear
Tyres: 7.00 × 16 C
Brakes:
(main) drum, hydraulically operated on all 4 wheels, dual circuit
(parking) operates on rear drums
Electrical system: 24 V
Batteries: 2 × 12 V, 45 Ah
Generator: 26 A

Status
Production complete. In service with Italy and Tunisia (400). Produced at one time in the former Yugoslavia.

Manufacturer
FIAT, Direzione Mezzi Speciali, Corso G Marconi 10/20, Turin, Italy.

UPDATED

FIAT Campagnola 1107 AD 750 kg light vehicle (R Stickland) ***1997***

JAPAN

Mitsubishi (4 × 4) light vehicles

Description
In 1953, Mitsubishi obtained a licence from the American Willys (later Kaiser Jeep) company to manufacture the Jeep in Japan for both civilian and military use. The first model was the CJ3B-JB and was followed by many other models, all available with petrol or diesel engine, left- or right-hand drive, long or short wheelbase and with an open or enclosed passenger/cargo area.

For many years the standard ¼ tonne vehicle of the Japanese Self-Defence Force was the Mitsubishi J54A. Powered by a four-cylinder petrol engine, it developed 75 hp at 3,800 rpm, was coupled to a manual gearbox with three forward and one reverse gears and had a two-speed transfer case. Basic details of the J54A are: 3.33 m length, 1.595 m width, 1.85 m height, 2.03 m wheelbase, 92 km/h maximum road speed and 6.00 × 16 tyres. Variants in service included an ambulance, anti-tank armed with a 106 mm Type 60 recoilless rifle and anti-tank armed with two KAM-3D (Type 64) wire-guided anti-tank missiles.

The J54A was replaced by the J24A, designated the Type 73 by the Japanese Ground Self-Defence Force (Army).

Specifications
J24A
Seating: 1 + 5
Configuration: 4 × 4
Weight:
(empty) 1,470 kg
(loaded) 1,970 kg
Max load: 480 kg
Length: 3.75 m
Width: 1.665 m
Height: 1.95 m
Ground clearance: 0.21 m
Track: 1.296 m
Wheelbase: 2.225 m
Fuel capacity: 48 litres
Max speed: 100 km/h
Engine: 4-cylinder diesel developing 80 bhp at 3,500 rpm
Gearbox: manual with 4 forward and 1 reverse gears
Clutch: single dry plate
Transfer box: 2-speed
Steering: recirculating ball
Turning radius: 5.8 m
Suspension: semi-elliptical leaf springs with hydraulic shock-absorbers
Tyres: 7.00 × 15
Brakes:
(main) hydraulic
(parking) mechanical
Electrical system: 24 V

Status
In service with the Japanese Self-Defence Force, Burma (Myanmar), Indonesia and Thailand.

Manufacturer
Mitsubishi Motors Corporation, 33-8, Shiba 5-chome, Minato-ku, Tokyo, Japan.
Tel: +81 3 212 3311. Telex: 26639 bisijiko j.

VERIFIED

Mitsubishi (4 × 4) Jeep used by the Japanese Ground Self-Defence Force (Kensuke Ebata)

Mitsubishi (4 × 4) Jeep used by the Japanese Ground Self-Defence Force and fitted with Type 64 (KAM-3D) ATGWs (Kensuke Ebata)

Toyota Land Cruiser (4 × 4) series light vehicle

Description
The Toyota Land Cruiser was developed in the 1950s primarily for the civilian market but has since been adopted by a number of armies in the Middle East and South America.

The basic model has a conventional layout, with the engine at the front and the passenger and cargo areas at the rear. The windscreen can fold flat against the bonnet and the side doors and vinyl top are removable. Six basic models of the Toyota Land Cruiser are in production; these are: hard top (regular wheelbase of 2.285 m and super long wheelbase of 2.95 m); vinyl top series (regular wheelbase of 2.285 m, long wheelbase of 2.43 m and super long wheelbase of 2.95 m); and pick-up (super long wheelbase of 2.95 m). In addition

Toyota Land Cruiser (4 × 4) light vehicle with hard top

Toyota Land Cruiser (4 × 4) light vehicle in long wheelbase pick-up configuration

Specifications

Model	**Regular wheelbase**	**Long wheelbase**	**Super long wheelbase**
Configuration	4 × 4	4 × 4	4 × 4
Weight (empty)	1 680 kg	1,740 kg	1,865 kg
(loaded)	2 295 kg	2,445 kg	3,035 kg
Max load	615 kg	705 kg	1,170 kg
Length	3 915 m	4.275 m	4.955 m
Width	1 665 m	1.665 m	1.665 m
Height	1 955 m	1.96 m	2.03 m
Ground clearance	0 21 m	0.21 m	0.225 m
Track (front)	1 415 m	1.415 m	1.415 m
(rear)	1 4 m	1.4 m	1.41 m
Wheelbase	2 285 m	2.43 m	2.95 m
Fuel capacity	85 litres	85 litres	85 litres
Engine	6-cylinder in-line OHV petrol developing 135 hp (DIN) at 3,600 rpm 6-cylinder in-line OHV diesel developing 103 hp (DIN) at 3,500 rpm 4-cylinder in-line OHV petrol developing 90 hp (DIN) at 3,500 rpm 4-cylinder in-line OHV diesel developing 80 hp (DIN) at 3,600 rpm		
Gearbox	all have manual, 4 forward and 1 reverse gears		
Clutch	single dry plate	single dry plate	single dry plate
Transfer box	2-speed	2-speed	2-speed
Steering	recirculating ball	recirculating ball	recirculating ball
Turning radius	5.3 m	5.5 m	6.5 m
Suspension (front and rear)	semi-elliptical springs with hydraulic double-acting shock-absorbers at each wheel station		
Tyres	7.00-15-6 PR	7.00-15-6 PR	7.50-16-8 PR
Brakes (main)	hydraulic, all wheels	hydraulic, all wheels	hydraulic, all wheels
(parking)	mechanical	mechanical	mechanical
Electrical system	12 V	12 V	12 V

there is a four-door station wagon version of the Land Cruiser.

Status
In production. In service with armed forces in the Middle East (including Egypt), South Africa and South America.

Manufacturer
Toyota Motor Corporation, 1 Toyota-cho, Toyota, Aichi Prefecture 471, Japan.
Tel: +81 565 282121. Telex: 4528371 toyota j.
The vehicle is also built under licence in Australia and Brazil.

VERIFIED

KOREA, SOUTH

Asia Motors KM41 series of light vehicles

Development
The Asia Motors KM41 series of light vehicles is the standard light vehicle of the South Korean armed forces. An order for 9,000 examples was placed during mid 1996, most of those ordered being the KM410L long wheelbase utility version. An initial 200 were to be delivered during 1997 and 300 during 1998.

The KM41 series was developed in co-operation with the ADD (Agency of Defence Development) in Korea and, although similar in appearance to the Jeep CJ series, there are many locally introduced changes to the original design.

Description
The layout of the KM41 base vehicle is conventional, with the engine at the front and the driver and passenger/load-carrying area to the rear. The same engine is used throughout the series, a KG2000 1.985 litre four-cylinder petrol engine developing 73 hp at 4,000 rpm.

The vehicles in the KM41 series are as follows:
KM410 utility vehicle
KM410L long wheelbase version of KM410
KM411 ambulance
KM412 TOW anti-tank guided missile launcher carrier
KM413 TOW anti-tank missile carrier
KM414 106 mm recoilless rifle carrier
KM415 searchlight carrier

Specifications

Model	**KM410**	**KM410L**	**KM411**	**KM412**	**KM413**	**KM414**	**KM415**
Purpose	utility	long wheelbase utility	ambulance	TOW launcher carrier	TOW missile carrier	106 mm RR carrier	searchlight carrier
Cab seating	1 + 3	1 + 5	1 + 1	1 + 3	1 + 1	1 + 3	1 + 1
Configuration	4 × 4	4 × 4	4 × 4	4 × 4	4 × 4	4 × 4	4 × 4
Weight (kerb)	1,180 kg	1,286 kg	1,380 kg	1,293 kg	1,290 kg	1,320 kg	1,290 kg
(GVW)	1,720 kg	1,786 kg	1,860 kg	1,880 kg	1,660 kg	1,969 kg	1,660 kg
Max load	540 kg	620 kg	480 kg	587 kg	370 kg	649 kg	370 kg
Length	3.35 m	3.925 m	3.841 m	3.405 m	3.276 m	3.301 m	3.276 m
Width	1.475 m	1.675 m	1.68 m	1.85 m	1.85 m	1.734 m	1.85 m
Height	1.705 m	1.946 m	1.946 m	1.705 m	1.73 m	1.705 m	1.73 m
Ground clearance	0.21 m	0.21 m	0.233 m	0.21 m	0.21 m	0.21 m	0.21 m
Track	1.23 m	1.23 m	1.23 m	1.23 m	1.23 m	1.23 m	1.23 m
Wheelbase	2.032 m	2.41 m	2.032 m	2.032 m	2.032 m	2.032 m	2.032 m
Angle of approach	45°	45°	51°	45°	45°	45°	45°
Angle of departure	30°	23°	19°	26°	30°	30°	30°
Max speed	105 km/h	105 km/h	105 km/h	105 km/h	105 km/h	105 km/h	105 km/h
Range	340 km	340 km	340 km	340 km	340 km	340 km	340 km
Fuel capacity	45.5 litres	45.5 litres	45.5 litres	45.5 litres	45.5 litres	45.5 litres	45.5 litres
Gradient	75%	75%	75%	75%	75%	75%	75%

KM410L (4 × 4) light vehicle, the long wheelbase (2.41 m) variant of the KM410

KM410 (4 × 4) light vehicle with open top

	KM410	KM410L	KM411	KM412	KM413	KM414	KM415
Fording	0.51 m	0.51 m	0.51 m	0.51 m	0.51 m	0.51 m	0.51 m
Engine	KG2000 1.985 litre 4-cylinder OHV petrol developing 73 hp at 4,000 rpm						
Gearbox	manual with 4 forward and 1 reverse gears with synchromesh on all gears						
Clutch	all single dry plate						
Transfer box	all 2-speed						
Steering	all ball and nut						
Turning radius	5.06 m	6 m	5.06 m	5.06 m	5.06 m	5.06 m	5.06 m
Tyres	all 6.00 × 16						
Brakes	all dual-circuit drum						
Suspension (front)	semi-elliptical leaf springs with hydraulic double-acting shock-absorbers						
(rear)	progressive semi-elliptical leaf springs with hydraulic double-acting shock-absorbers						
Electrical system	24 V	24 V	24 V	24 V	24 V	24 V	24 V
Batteries	2 × 12 V, 45 Ah	2 × 12 V, 45 Ah	2 × 12 V, 45 Ah	2 × 12 V, 45 Ah	2 × 12 V, 45 Ah	2 × 12 V, 45 Ah	2 × 12 V, 45 Ah
Alternator	25 A	25 A	25 A	25 A	25 A	25 A	180 A

Status
In production. In service with the armed forces of South Korea.

Manufacturer
Asia Motors Company Inc, 15 Yoido-dong, Yungdeungpo-gu, Seoul, South Korea.
Tel: +82 2 783 1501/9. Telex: 24374 asiamco k.

UPDATED

PAKISTAN

Nispak (4 × 4) 400 kg light vehicle

Description
The Nispak was designed under the direction of the Inspector of Vehicles and Engineering Equipment, Pakistan Army and is patterned on a Japanese Nissan (4 × 4) light vehicle. At least 90 per cent of the components of the Nispak, including the engine, transmission, chassis and body are manufactured in Pakistan.

The layout of the Nispak is conventional with the engine at the front, driver and passenger in the centre and the cargo area at the rear. The single-piece windscreen folds forward on to the bonnet and the canvas cover folds down to the rear.

Specifications
Configuration: 4 × 4
Weight:
(empty) 1,580 kg
(loaded, road) 2,180 kg
(loaded, cross-country) 1,980 kg
Weight on front axle: (empty) 853 kg
Weight on rear axle: (empty) 727 kg
Max load:
(road) 600 kg
(cross-country) 400 kg
Towed load:
(road) 907 kg
(cross-country) 680 kg
Length: 3.772 m
Width: 1.689 m
Height: 1.981 m
Ground clearance: 0.235 m
Track:
(front) 1.382 m
(rear) 1.4 m
Wheelbase: 2.2 m
Max speed: 125 km/h
Range: 290 km
Fuel capacity: 50 litres
Gradient: 65%
Fording: 1.88 m
Engine: 6-cylinder petrol developing 145 hp at 3,800 rpm
Gearbox: manual 3 forward and 1 reverse gears
Clutch: single dry plate
Transfer box: 2-speed
Turning radius: 5.69 m
Suspension: semi-elliptical springs and hydraulic shock-absorbers
Tyres: 7.00 × 16
Brakes:
(main) hydraulic
(parking) mechanical
Electrical system: 24 V

Status
In service with the Pakistan armed forces.

Manufacturer
Facility at Chaklala, Pakistan.

VERIFIED

PHILIPPINES

Delta Explorer (4 × 4) light vehicle

Description
The Delta Explorer (4 × 4) light vehicle, originally known as the Delta Mini-Cruiser, was developed from 1972 by the Delta Motor Corporation with the assistance of the Research and Development Centre of the Philippines armed forces. It was the country's first major project called for by the Self-Reliance Defence Posture Program initiated by the then President Marcos.

Following trials with 40 preproduction models, the Philippine armed forces placed an order for an initial batch of 500 vehicles under the designation RJ-2.

The left-hand drive model is designated the RJ-2BL and the right-hand drive version the RJ-2BR. The more recent diesel model, the DJ-2, is powered by an Isuzu C-190 4-stroke vertical in-line OHV diesel developing 55 bhp at 4,000 rpm.

The vehicle entered service with the Philippine Army in 1975 and all foreign-made light utility vehicles were planned to be phased out of service. The export of the vehicle to any friendly country was approved in principle by the Ministry of National Defence in February 1980.

The Explorer is powered by the same 12RM OHV petrol engine as in the Toyota Corona car, also manufactured in the Philippines, where practically all the components of the vehicle are manufactured. The remaining imported components, about 10 per cent, are limited to relatively unimportant parts which will all eventually be made by Delta Motor Corporation. The company has the capacity to produce about 300 Explorer vehicles a month for export.

The Explorer chassis is of the channel section girder type with five cross members. The main panels are made of 1.6 mm thick steel and other panels of steel 1.2 mm thick. The layout is conventional, with the engine at the front, driver and passenger in the centre and two two-man seats down each side at the rear. The

Military version of Delta Explorer (4 × 4) light vehicle with soft top

Delta Explorer (4 × 4) light ambulance

Specifications

Model	**RJ-2B**	**DJ-2B**
Configuration	4 × 4	4 × 4
Cab seating	1 + 5	1 + 5
Weight (empty)	1,070 kg	1,220 kg
(loaded)	1,740 kg	1,740 kg
Weight on front axle (empty)	600 kg	670 kg
Weight on rear axle (empty)	470 kg	550 kg
Length	3.575 m	3.575 m
Width	1.595 m	1.595 m
Height (overall)	1.925 m	1.925 m
(reduced)	1.505 m	1.505 m
Ground clearance	0.215 m	0.215 m
Track (front)	1.3155 m	1.3155 m
(rear)	1.2693 m	1.2693 m
Wheelbase	2.185 m	2.185 m
Angle of approach/departure	52°/39°	52°/39°
Fuel capacity	45 litres	45 litres
Engine	Toyota 12RM, 1.587 litre 4-cylinder OHV petrol developing 90 bhp at 5,400 rpm	Isuzu C-190 4-stroke vertical in-line OHV diesel developing 55 bhp at 4,000 rpm
Gearbox	manual, 4 forward and 1 reverse gears	
Clutch	single dry plate	single dry plate
Transfer box	2-speed	2-speed
Steering	2-shaft with worm and recirculating ball sector	
Turning radius	4.75 m	4.75 m
Suspension	front and rear, semi-elliptical leaf springs front and rear with hydraulic telescopic shock-absorbers	
Axles (front)	full-floating with steering knuckles and hub locks	
(rear)	semi-floating hypoid type	
Tyres	6.00 × 16	6.00 × 16
Brakes (main)	dual system hydraulic drum type	
(parking)	mechanical	mechanical
Electrical system	12 V	12 V
Battery	1 × 12 V	1 × 12 V

crew compartment is covered by a tarpaulin cover which folds down at the back and the windscreen can be folded forward on to the bonnet. Apart from the basic military model there are two other canvas-topped versions, one with canvas doors and the other with steel doors.

The basic vehicle can also be used as a command vehicle and weapons carrier. More specialised variants are the ambulance and police versions.

Status

In service with Philippines, Qatar and UAE. Sold to Australia, Colombia, Italy, Pakistan and Thailand.

Manufacturer

Delta Motor Corporation, PO Box 305, MCC, Makati, Metro Manila 3117, Philippines.
Tel: +63 2 865061. Telex: 5009 delta pm.

VERIFIED

POLAND

Tarpan (4 × 4) light vehicle

Description

The Tarpan (4 × 4) light vehicle, also known as the Tarpan Honker, was originally produced as an agricultural light vehicle but also has military and paramilitary applications. It is available in two wheelbase lengths; the short wheelbase version is the 4012 and the long wheelbase the 4022. The 4012 can have accommodation for up to 10 personnel (including the driver) or carry up to 1,000 kg of cargo on roads. A 4022 ambulance version is available.

The layout of the Tarpan light vehicles is conventional, with the engine forward, the driver with seating for one passenger next to him and a load area to the rear with a number of optional soft or hardtops and possible seating arrangements. The normal personnel carrier version has a soft top and bench seats for up to eight personnel in the rear. Other rear area arrangements include seating for six or two personnel or an open 'pick-up' light utility cargo load platform.

Two IVECO 2.5 litre diesel engines are available. The 4012 is normally powered by a turbocharged IVECO 8141.27 while the 4022 has the IVECO 8141.07

Long wheelbase version of Tarpan (4 × 4) light vehicle

1996

Long wheelbase version of Tarpan (4 × 4) light vehicle configured as police vehicle ***1996***

Tarpan (4 × 4) light vehicle ***1996***

normally aspirated diesel. Both are coupled to a five-speed manual transmission.

Specifications

Data in square brackets refer to long wheelbase (ambulance) version

Cab seating: 2 + 2 to 8
Configuration: 4 × 4
Weight empty: 2,050 [1,675] kg
Max load: (off road) 800 [1,175] kg
Length: 4.55 [4.95] m
Width: 1.85 m
Height: (overall) 2.15 [2.04] m
Wheelbase: 2.827 [3.027] m
Track: (front/rear) 1.575 m/1.535 m
Angle of approach/departure:
(SWB) 40°/42°
(LWB) 42°/31°
Max speed: 125 [110] km/h
Fuel capacity: 110 litres
Max gradient: 68%
Side slope: 42%
Fording: 0.8 m
Engine: (4012) IVECO 8141.27 2.499 litre turbocharged diesel developing 102 hp or
IVECO 8141.07 2.499 litre diesel developing 74 hp
Transmission: manual with 5 forward and 1 reverse gears
Transfer box: Borg Warner 2-speed
Steering: ZF power-assisted
Tyres: 840-15 or 235/75 R 15

Status

In production.

Manufacturer

Tarpan Sp z o o, ul. Warszawska 349, PL-61-057 Poznan, Poland.
Tel: +48 61 761200. Telex: 041 32 16.
Fax: +48 61 768280.

VERIFIED

PORTUGAL

UMM (4 × 4) light vehicle

Description

The UMM (4 × 4) light vehicle is available with Peugeot 2.5 litre diesel engines, either naturally aspirated (76 hp) or turbo intercooled (110 hp). The vehicle has a chassis formed from 120 × 60 mm steel channel section 4 mm thick, which is reinforced by three cross members. The body is formed from 2 mm steel plate welded to the chassis. The body layout is conventional, with the engine forward and the driver's position central. The rear of the body is an open platform on to which various types of body can be built. These range from a hardtop to various forms of soft top. The driver's cab portion can be fitted with a canvas roof but may also accommodate a hardtop. The rear area may be configured for personnel seating in various layouts or left open for cargo carrying or for mounting various forms of light weapon or missiles.

The standard military version is powered by a 2.5 litre diesel engine but a 110 bhp turbocharged version of the same engine is available.

Variants include an ambulance with a box body, a mortar carrier, a 106 mm M40 recoilless rifle carrier and 7.62 and 12.7 mm machine gun carriers. Special missile-carrying variants include a version with a hardtop cab and a three-missile launcher for SS-11 anti-tank missiles, a MILAN or TOW anti-tank guided weapon launcher version and a special version with racks in the rear for ATLAS or Mistral surface-to-air missiles. Also produced are police versions and one variant carrying a hydraulic inspection platform.

There is also a long wheelbase (3.078 m) version of the UMM. This version has a maximum load capacity of 1,700 kg or up to 10 men in the rear.

The chassis cab model can carry standard NATO shelters.

The UMM can be paradropped.

Specifications

Data in square brackets refer to long wheelbase version

Cab seating: 2 + 6 [2 + 10 or 3 + 10]
Configuration: 4 × 4
Weight:
(empty) 1,610 [1,770] kg
(loaded) 2,720 [3,470] kg
Max load: (road) 1 110 [1,700] kg
Towed load:
(trailer with brakes) 3,500 kg
(trailer without brakes) 1,000 kg
Length: 4.14 [4.77] m
Width: 1.57 m
Height: 1.955 m
Ground clearance: 0.26 m
Track: 1.342 m
Wheelbase: 2.56 [3.078] m
Angle of approach/departure: 48°/44° [48°/35°]
Max speed: 120 km/h (turbocharged version 140 km/h)
Fuel capacity: 60 litres (120 litres optional)
Max gradient: 45°
Fording: 0.6 m
Engine: 2.498 litre 4-cylinder in-line water-cooled diesel developing 76 hp at 4,500 rpm or
turbocharged version of same engine developing 110 hp at 4,150 rpm
Gearbox: manual with 5 forward and 1 reverse gears
Clutch: single dry plate
Transfer box: 2-speed

UMM (4 × 4) light vehicle armed with 106 mm M40 recoilless rifle ***1996***

UMM (4 × 4) light vehicle armed with 12.7 mm M2 heavy machine gun ***1997***

UMM (4 × 4) light vehicle configured as personnel carrier ***1996***

UMM (4 × 4) light vehicle equipped with ATLAS air defence missile launcher ***1996***

Steering: worm and roller, assisted
Turning radius: 5.9 [6.5] m
Suspension: semi-elliptical leaf springs with double-acting telescopic shock-absorbers front and rear
Tyres: 7.00 × 16
Brakes: ventilated discs front and drums rear, servo-assisted, dual circuit
Electrical system: 12 or 24 V (FFR)
Battery: 1 × 12 V 63 Ah
Alternator: 750 W (turbocharged version 1,000 W)

Status
In production. In service with the Portuguese, Angolan, Cape Verde, Congo, French and Netherlands armed forces.

Manufacturer
UMM - União Metalo Mecânica Limitada, Estrada do Cemitério Novo, Cerro da Murta, P-2745 Queluz, Portugal.
Tel: +351 1 435 2518. Fax: +351 1 436 3322.

VERIFIED

RUSSIAN FEDERATION AND ASSOCIATED STATES (CIS)

LuAZ-967M amphibious battlefield support vehicle

Description
This vehicle was developed in the 1960s and was first seen during trials with Russian troops based in the former East Germany during the latter part of that decade. It was generally deployed during 1976 and was originally used as a battlefield medical evacuation vehicle but is now also used as a general light support vehicle. The LuAZ-967M uses some components of the LuAZ-969 (4 × 4) light vehicle described in the following entry.

The body of the LuAZ-967M is made of all-welded steel with the engine at the front. The driver sits immediately behind the windscreen which folds forward to reduce the overall height. His seat is on the centreline of the vehicle and can be folded down so that he can drive lying flat. The steering column and steering wheel can also be lowered.

As a battlefield medical evacuation vehicle the LuAZ-967M has two folding seats for walking wounded as well as two stretchers, but its normal load is two patients plus the driver. It is fully amphibious, being propelled in the water by its wheels. A treadway is carried on each side of the vehicle for crossing trenches and other obstacles. There is a winch with a capacity of 200 kg and 100 m of cable mounted at the front of the vehicle; this winch may be used to winch in wounded personnel on a canvas mat.

The LuAZ-967M is also used by airborne forces as a light support vehicle and as a weapon carrier; in the latter role it has been observed carrying an AGS-17 grenade launcher, the AT-4 'Spigot' anti-tank guided missile and an 82 mm B-10 recoilless rifle. An ammunition carrier version can carry up to 320 kg of small arms ammunition.

Specifications
Seating: 1 + 2 or 3
Configuration: 4 × 4
Weight:
(empty) 930 kg
(loaded) 1,350 kg
Max load: 420 kg
Max towed load: 300 kg
Length: 3.682 m
Width: 1.74 m
Height: (incl windscreen) 1.625 m
Ground clearance: 0.285 m
Wheelbase: 1.8 m
Angle of approach/departure: 34°/36°
Max road speed: 75 km/h
Max water speed: 5-6 km/h
Range: 285 km
Fuel capacity: 34 litres
Fuel consumption: 9 litres/100 km at 40 km/h
Gradient: 58%
Fording: amphibious
Engine: MeMZ-967A 4-cylinder air-cooled petrol developing 37 hp at 4,300 rpm
Gearbox: manual, 4 forward and 1 reverse gears, plus cross-country gear
Turning radius: 5.8 m
Suspension: individual torsion bar
Electrical system: 12 V

Status
Production complete. In service with the RFAS.

Manufacturer
Lutsk Motor Vehicle Plant, Ukraine.

VERIFIED

LuAZ-967M amphibious battlefield medical evacuation vehicle (Michael Jerchel)

LuAZ-967M amphibious battlefield medical evacuation vehicle

LuAZ-969 (4 × 4) light vehicle

Description

In 1965, the Zaporozhe Motor Vehicle Plant developed a 250 kg vehicle, based on the ZAZ-966 passenger car, called the ZAZ-969, but it did not enter production. In the late 1960s the project was transferred to the Lutsk Machine Building Plant and the designation was changed from the ZAZ-969 to the LuMZ-969. Production finally began in 1972 under the designation LuAZ-969 as the plant name was changed from the Lutsk Machine Building Plant to the Lutsk Motor Vehicle Plant. The vehicle is intended primarily for civilian use, although it has military and paramilitary applications. The chassis of the LuAZ-969 light vehicle was also used as the basis for the LuAZ-967M amphibious battlefield medical evacuation vehicle described in the previous entry.

LuAZ-969M (4 × 4) light vehicle

The vehicle has an all-steel body with the engine at the front and the crew compartment at the rear, two side doors, a windscreen which can be folded forward on to the bonnet and a removable canvas top. The driver can select either 4 × 4 or 4 × 2 drive, in which case the rear instead of the front axle is disengaged. A winch with a capacity of 200 kg and 100 m of cable can be fitted.

The latest model is the LuAZ-969M. This is powered by an engine developing 40 hp at 4,200 rpm, giving the vehicle a maximum road speed of 85 km/h. Empty weight is 970 kg, overall length is 3.37 m and width is 1.64 m.

Specifications

LuAZ-969
Cab seating: 1 + 1
Configuration: 4 × 4
Weight:
(empty) 820 kg
(loaded) 1,200 kg
Weight on front axle: (empty) 510 kg
Weight on rear axle: (empty) 310 kg
Towed load: 300 kg
Length: 3.2 m
Width: 1.6 m
Height: 1.77 m
Ground clearance: 0.3 m
Track: 1.32 m
Wheelbase: 1.8 m
Max speed: 75 km/h
Range: 400 km
Fuel capacity: 32 litres
Gradient: 58%
Fording: 0.45 m
Engine: MeMZ-946 V-4 air-cooled petrol developing 27 hp at 4,000 rpm
Gearbox: manual with 4 forward and 1 reverse gears
Transfer box: 2-speed
Turning radius: 3 m
Suspension: torsion bar
Tyres: 5.90 × 13
Brakes:
(main) hydraulic
(parking) mechanical
Electrical system: 12 V

Status

In production.

Manufacturer

Lutsk Motor Vehicle Plant, Ukraine.

VERIFIED

UAZ-469B (4 × 4) 695 kg light vehicle

Development

For many years the standard Jeep-type vehicle of the Russian and other Eastern Bloc armies was the UAZ-69 (or GAZ-69), but for a number of reasons it was never considered a satisfactory design. In 1960 a new 4 × 4 light vehicle called the UAZ-460B was tested. This used many components of the UAZ-450 truck, but did not enter production. Further development of this vehicle resulted in the UAZ-469B, which was first produced late in 1972 and entered service the following year. The UAZ-469B (also known as the UAZ-31512) uses many components, including the axles, brakes, engine, transmission and parts of the chassis of the UAZ-452 series of 4 × 4 light vehicles.

The latest model of the basic UAZ-469 is known as the UAZ-315136 (see under Variants in this entry). This model replaced all others on the production lines from 1990 onwards.

The UAZ-469 is licence-produced in the Czech Republic by Auto Závod (see following).

Description

The basic vehicle has an all-steel body with the engine at the front and the four-door crew compartment towards the rear, with a removable canvas top and windscreen that can be folded down flat against the bonnet. The tops of the doors can also be removed. There are two individual seats at the front, a three-person seat in the centre and two personnel can sit facing each other at the rear. Normal load is two personnel plus 600 kg of cargo or seven personnel and 100 kg of cargo. A hardtop can be fitted if required.

In 1973 the UAZ-469 was shown. This has a number of improvements including a Portal axle which gives the vehicle both a higher ground clearance (300 mm) and improved cross-country capability, but the engine is identical to that in the standard UAZ-469B.

The UAZ-469 has been sold in the West as the Tundra.

Variants

Variants include: the UAZ-469BG ambulance (latest designation UAZ-3152) which can carry a driver and four patients; van-type vehicles; and a special version for dispensing lane marking pennants into the ground.

The UAZ-315136 is an improved version of the basic UAZ-469. The overall appearance of the new model is similar to that of the original vehicle but the independent front suspension uses coil springs and a more powerful UAZ four-cylinder engine, developing 85 hp; petrol and diesel engines are available. Other changes include improved seating and heating. Extra shock-absorbent panels have been fitted to the body. Weight empty is 1,350 kg and loaded 2,150 kg. The maximum towed load is increased to 800 kg.

Iraqi Army UAZ-469s have been observed carrying four RPG-7 anti-tank rocket launchers on a post-type mounting in the rear area.

Czech Production

Auto Závod of the Czech Republic produces civilian versions of the UAZ-315136 powered by a choice of three 2.5 litre diesel engines, including the UMZ 41796.10 or a Peugeot XD3P. The turbocharged VM HR 4924 HT from Italy is also available. Models available include the Maraton, Tuareg and UAZ (4 × 4) 2.5 TD; a long wheelbase chassis and station wagon body are available for the latter. Production of these models commenced during 1992.

Specifications

Cab seating: 1 + 6
Configuration: 4 × 4
Weight:
(empty) 1,490 kg
(loaded) 2,290 kg
Weight on front axle: (loaded) 960 kg
Weight on rear axle: (loaded) 1,330 kg
Max load: 695 kg
Max towed load:
(unbraked) 600 kg
(braked) 2,000 kg
Length: 4.025 m
Width: 1.785 m
Height: 2.015 m
Ground clearance: 0.22 m
Track: 1.422 m
Wheelbase: 2.38 m
Angle of approach/departure: 52°/42°
Max speed: 100 km/h
Range: 620 km
Fuel capacity: 78 litres
Max gradient: 62%
(towing trailer) 40%
Vertical obstacle: 0.45 m
Fording: 0.7 m

Czech Army UAZ-469B (4 × 4) light vehicle (R Stickland) ***1997***

UAZ-469B (4 × 4) 695 kg light vehicle in police markings in Russia (Stefan Marx) ***1996***

Engine: ZMZ-451 MI 4-cylinder water-cooled petrol developing 75 hp at 4,000 rpm
Gearbox: manual with 4 forward and 1 reverse gears
Transfer box: 2-speed
Turning radius: 6.5 m
Tyres: 8.40 × 15
Brakes:
(main) hydraulic
(parking) mechanical
Electrical system: 12 V

Status
Replaced in production from 1990 onwards by the UAZ-315136. The UAZ-469B is in service with former members of the Warsaw Pact including the Czech Republic and Slovakia, Hungary, Poland and the RFAS. Also exported to other countries including Afghanistan, Cuba, Egypt, Iran, Iraq and Syria.

Manufacturer
Ulyanovsk Motor Vehicle Plant JSC, Moskovkoye Shosse, Ulyanovsk 322008, Russia.
Tel: +7 8422 315161 Fax: +7 8422 323212

UPDATED

UAZ-3172 (4 × 4) light vehicle

Description
The UAZ-3172 (4 × 4) light vehicle may be regarded as a progressive update of the earlier UAZ-469B series (see previous entry). However, the UAZ-3172 is larger overall, being able to carry nine or ten personnel (including the driver) or cargo loads weighing up to 1,000 kg. The overall layout and appearance remains much the same as the earlier vehicle and it appears that many components are shared, although the UAZ-3172 has a revised front end and grille.

The overall layout remains conventional, with the engine under a forward bonnet, the driving position central and the load or passenger carrying area at the rear. The vehicle body is all metal with a removable canvas top over two safety bows. Five body doors are provided, one at the rear.

Power is provided by a UMZ-421.10 petrol engine developing 105 hp and coupled to a four-speed manual gearbox. There is an inter-axle locking differential. The suspension is of the rigid axle, coil spring type with hydraulic shock-absorbers.

The vehicle is intended to operate under extreme climatic conditions varying in temperature from −50 to +50°C, with relative humidities up to 98 per cent and at altitudes up to 4,500 m above sea level. Loads weighing up to 1,200 kg can be towed over all types of road and terrain.

Various types of light weapon, such as machine guns, can be mounted on the vehicle.

UAZ-3172 (4 × 4) light vehicle ***1997***

Specifications
Cab seating: 1 + 8 or 9
Configuration: 4 × 4
Weight: (empty) 1,950 kg
(loaded) 2,950 kg
Max load: up to 1,000 kg
Max towed load: 1,200 kg
Length: 4.6 m
Width: 1.88 m
Height: 2.1 m
Ground clearance: 0.315 m
Track: 1.46 m
Wheelbase: 2.7 m
Max speed: 115 km/h
Range: 650 km
Fuel capacity: 105 litres
Max gradient: 31°
Side slope: 20°
Fording: 1 m
Engine: UMZ-421.10 petrol developing 105 hp
Gearbox: manual with 4 forward and 1 reverse gears
Clutch: single dry plate
Transfer box: 2-speed
Turning radius: 7 m
Suspension: rigid axle, coil spring with hydraulic shock-absorbers
Tyres: I-288 275/8 R 16
Electrical system: 12 V

Status
In production. Offered for export sales.

Manufacturer
Ulyanovsk Motor Vehicle Plant JSC, Moskovkoye Shosse, Ulyanovsk 322008, Russia.
Tel: +7 8422 315161. Fax: +7 8422 323212.

NEW ENTRY

GAZ-69AM and GAZ-69M (4 × 4) light vehicles

Description
The GAZ-69 series of light 4 × 4 vehicles entered production at the Gor'kiy Plant in 1952 and stayed in production there until 1956 when production was transferred to the Ul'yanovsk Plant. They were then also known as the UAZ-69 and UAZ-69A and production continued until the UAZ-469B was introduced. These vehicles used many components of the UAZ-450 range of civilian vehicles.

There were two basic models in service, the GAZ-69AM and the GAZ-69M. The former has two doors and has been designed to carry 500 kg of cargo. Bench seats down each side in the rear hold two personnel on each side facing each other. The spare wheel is mounted externally on the left side of the body. This model is widely used as a command/radio vehicle and for towing recoilless rifles and light anti-aircraft guns such as the 23 mm ZU-23. The second model is the GAZ-69M, which has four doors and can carry five personnel plus 100 kg of cargo. The spare wheel is carried under the rear of the vehicle. Both models have a removable top, a windscreen that can be folded forward on to the bonnet and removable door tops.

Late production models of the UAZ-69 were the UAZ-69M which has a 65 hp M-21 engine and the UAZ-69-68. Late production models of the UAZ-69A were the UAZ-69AM, which also has a 65 hp engine and the UAZ-69A-68.

Anti-tank (recoilless rifle)
Some countries in the Middle East have mounted a recoilless rifle on the rear of the vehicle.

Aircraft starter
This is the GAZ-69AM with a modified rear on which is mounted an aircraft starting unit.

Mine detector vehicle
The GAZ-69AM can be fitted with the DIM mine detection system; there is a separate entry for this in the *Mine detection equipment* section.

North Korea
North Korea built a version of the GAZ-69M with a revised front.

Romania
Romania produced the GAZ-69AM under the designation M-461. This was almost identical to the original but had a four-speed gearbox.

UAZ-456
This is a tractor version for towing semi-trailers.

Specifications

Model	**GAZ-69AM**	**GAZ-69M**
Configuration	4 × 4	4 × 4
Weight (empty)	1,535 kg	1,525 kg
(loaded)	1,960 kg	2,175 kg
Weight on front axle		
(loaded)	925 kg	940 kg
Weight on rear axle		
(loaded)	1,035 kg	1,235 kg
Max load	425 kg	650 kg
Towed load	750 kg	850 kg
Length	3.85 m	3.85 m
Width	1.75 m	1.85 m
Height	1.92 m	2.03 m
Ground clearance	0.21 m	0.21 m
Track	1.44 m	1.44 m

Czech Army GAZ-69AM (4 × 4) 500 kg light vehicle

Czech Army GAZ-69AM (4 × 4) 500 kg light vehicle

Wheelbase	2.3 m	2.3 m
Angle of approach/ departure	45°/35°	45°/35°
Max speed	96 km/h	96 km/h
Range	525 km	600 km
Fuel capacity	60 litres	75 litres
Fuel consumption	19 litres/100 km at 40 km/h	
Max gradient	57%	55%
Fording	0.55 m	0.55 m
Vertical obstacle	0.3 m	0.3 m
Trench	0.35 m	0.35 m
Engine	UAZ-69B 4-cylinder water-cooled petrol developing 65 hp at 3,800 rpm	
Gearbox	manual with 3 forward and 1 reverse gears	
Clutch	single disc, dry	single disc, dry
Transfer box	2-speed	2-speed
Steering	globoid worm with double collared cone	
Turning radius	6 m	6.5 m
Suspension	longitudinal semi-elliptical springs with double-acting hydraulic shock-absorbers front and rear	
Tyres	6.50 × 16	6.50 × 16
Brakes (main)	hydraulic on all wheels	
(parking)	on transmission	
Electrical system	12 V	12 V
Batteries	2 × 6 V ST 54	2 × 6 V ST 54
Generator	250 W	250 W

Status

Production complete. Increasingly being replaced by later vehicles, but may still be encountered in service with several members of the former Warsaw Pact and most countries that have received Soviet/RFAS aid, for example Cambodia, Cuba, Egypt, Finland, Syria and Vietnam.

Manufacturers

Gor'kiy Motor Vehicle Plant, Gor'kiy, Russia (1952-56) and Ul'yanovsk Motor Vehicle Plant, Ul'yanovsk (from 1956 onwards).

UPDATED

UAZ-452 (4 × 4) 800 kg light vehicles

Description

The UAZ-452 (4 × 4) light vehicle entered production at the Ul'yanovsk Motor Vehicle Plant in 1966. Although used primarily for civilian applications, the vehicle and its variants are used in some numbers by the military, especially as ambulances and command vehicles.

The UAZ-452D has a two-door all-steel forward control type cab. The rear cargo area has drop sides and a drop tailgate.

Variants of the vehicle are the UAZ-452 (van), UAZ-452A (ambulance carrying three seated patients plus three stretcher patients), UAZ-452P (tractor truck), UAZ-452E (with shielded electrical system), UAZ-452DE (with shielded electrical system) and the UAZ-452V (10-seat bus).

The vehicle was preceded by the UAZ-450 series of 4 × 4 vehicles which were in production from 1958 to 1966 and included the UAZ-450 (van), UAZ-450A (ambulance), UAZ-450B (bus) and the UAZ-450D (cargo truck). These were all powered by a 65 hp four-cylinder petrol engine and had a gearbox with three forward and one reverse gears and a two-speed transfer case.

In 1962 a series of similar 4 × 2 vehicles, the UAZ-451 (van) and the UAZ-451D (cargo), entered production. These were replaced by the improved UAZ-451M in 1966. The UAZ-451DM is also a 4 × 2 vehicle which can carry 1,000 kg of cargo.

The UAZ-452 series is being replaced in production by the UAZ-3251. This is a generally improved version of the earlier models and is produced with UAZ diesel or petrol engines developing 85 hp.

Cambodian Army UAZ-452 (4 × 4) 800 kg light vehicle with general purpose body (T J Gander) **1996**

Specifications

UAZ-452 van
Cab seating: 1 + 1
Weight:
(empty) 1,720 kg
(loaded) 2,520 kg
Weight on front axle: (loaded) 1,260 kg
Weight on rear axle: (loaded) 1,410 kg
Max load: 800 kg
Towed load: 850 kg
Load area: 2.6 × 1.87 m
Length: 4.36 m
Width: 1.94 m
Height: 2.09 m
Ground clearance: 0.22 m
Track: 1.442 m
Wheelbase: 2.3 m
Angle of approach/departure: 36°/30°
Max speed: 95 km/h
Range: 500 km
Fuel capacity: 56 + 30 litres
Fuel consumption: 13 litres/100 km
Gradient: 57%
Fording: 0.55 m
Engine: ZMZ-451M1 4-cylinder water-cooled petrol developing 72 hp at 4,000 rpm
Gearbox: manual with 4 forward and 1 reverse gears
Clutch: single dry disc
Transfer box: 2-speed
Turning radius: 6.6 m
Suspension: longitudinal semi-elliptical springs with hydraulic double-acting shock-absorbers on both axles
Tyres: 8.40 × 15
Brakes:
(main) hydraulic
(parking) mechanical
Electrical system: 12 V
Battery: 1 × STE-54EM
Generator: 250 W

Status

Replaced in production by the UAZ-3251. In service with former Warsaw Pact countries, Cambodia and Egypt.

Manufacturer

Ul'yanovsk Motor Vehicle Plant, Ul'yanovsk, CIS.

VERIFIED

SOUTH AFRICA

Jakkals (4 × 4) lightweight airborne vehicle

Description

The Jakkals (Jackal) lightweight airborne vehicle is a 4 × 4 utility vehicle designed for use by airborne forces. It resembles a small jeep and is conventional in layout, with the engine at the front and the driver and single passenger in the centre; space behind the seats is very limited. It is assembled using standard, commonly available vehicle components, including the engine. Production is carried out by Associated Automotive Distributors (Pty) Limited of Cape Town.

The Jakkals has an impact-resistant glass fibre body that can withstand an unlimited number of parachute drops. The vehicle is normally used together with a small single-axle trailer. Weight of the vehicle and trailer together is 1,120 kg. Power is derived from a 1.6 litre petrol engine and a high- and low-range transmission with selectable 4 × 4 or 4 × 2 drive.

The vehicle carries a front-mounted hand-operated self-recovery winch and a collapsible casualty evacuation stretcher, with securing straps, can be mounted over the passenger seat in less than one minute.

The dimensions of the Jakkals and its trailer are such, that two vehicles with trailers will fit on to a standard 8 × 9 foot (2.438 × 2.74 m) air supply pallet. Thus a cargo aircraft such as a C-130 can carry 10 units. For paradropping, two 30.5 m or three 18.3 m parachutes are used for each pallet. The descent speed is typically 7 m/s. Once on the ground, the vehicle can be unpacked from its pallet and prepared for use by two men in 10 minutes.

The Jakkals is produced in four basic versions: a general purpose section vehicle carrying a 7.62 mm machine gun; an attack vehicle with dual 0.50/12.7 mm machine guns; a mortar vehicle carrying an 81 or 120 mm mortar with ancillary equipment and first-line ammunition; and a stretcher vehicle for casualty evacuation. The Jakkals can also be used as a stores carrier. Various adaptations of these basic variants are possible and any vehicle (other than the attack version) can be fitted with communications equipment and used as a command vehicle. A complete front line mobile medical centre has been produced using the Jakkals.

Jakkals (4 x4) lightweight airborne vehicle (T J Gander)

Specifications

Seating: 2
Configuration: 4 × 4
Weight:
(tractor) 940 kg
(trailer) 180 kg
Max load: (tractor) 350 kg
Towed load: 350 kg
Length: 2.408 m
Width:
(tractor) 1.211 m
(trailer) 1.191 m
Height:
(packed, tractor) 0.963 m
(packed, trailer) 0.91 m
(operational) 1.23 m
Ground clearance: 0.19 m
Track: 1.027 m
Wheelbase: 1.539 m
Angle of approach/departure: 45°/44°
Range: 200 km
Fuel capacity: 60 litres
Engine: 1.6 litre 4-cylinder petrol developing 63 hp at 5,200 rpm
Gearbox: manual, synchromesh, with 4 forward and 1 reverse gears
Clutch: single dry plate
Transfer box: 2-speed
Turning circle: 9.74 m
Suspension: leaf springs front and rear with double-action shock-absorbers
Tyres: 670 × 14 6-ply
Brakes: discs front, self-adjusting drums rear

Status

In service with the South African Army.

Manufacturer

Enquiries to: Armscor, Private Bag X337, Pretoria 0001, South Africa.
Tel: +27 12 428 1911. Telex: 30217.
Fax: +27 12 428 5635.

UPDATED

SPAIN

Santana Model 88 Militar (4 × 4) 500 kg light vehicle

Description

The first of these vehicles appeared during 1956 following an arrangement between Land Rover and Metalurgica de Santa Ana SA. In the years that followed, the Land Rover holding in the concern was reduced to about 47 per cent and in early 1990 the association with Land Rover was terminated. The first Santana Model 88 was produced in 1970 in a variety of military and civilian forms.

The Santana Model 88 Militar closely resembles the British air-portable 500 kg Land Rover. The chassis uses a stair-type construction with two parallel side members with welded cross members. The chassis is painted and may be finished with a special galvanised anti-corrosion finish if the vehicle is used in a salt water

Santana Model 88 Militar (Pierre Touzin) ***1996***

Santana Model 88 Militar carrying MILAN anti-tank guided weapon launcher and three more missiles in containers at rear

environment. A 2.286 litre petrol engine is fitted, although a diesel engine is available if required. The front and rear axles both use the same differential, while the front axle half shafts have universal joints; the rear axle is fully floating. Front and rear suspension uses semi-elliptical springs combined with double-action shock-absorbers.

The driver's position is provided with a single seat for the driver and another for a passenger. Up to four personnel can be carried in the rear and a soft or hardtop may be fitted. A special deep wading version, capable of coping with water obstacles up to 1.9 m deep is available; this uses a depressurising system for the engine, gearbox and axles and all components are waterproofed.

Apart from the basic personnel and load (up to 500 kg) carried, the Model 88 Militar can be readily converted to a number of special purpose configurations and weapon carriers. The Model 88 Militar may also be paradropped on a platform which, together with the parachute and fittings, weighs 2,200 kg.

Variants

Communications version

This has a hardtop and the rear interior is fitted with a folding table for radio equipment or for use in the command role. The vehicle may be fitted with a 42 or 90 A alternator.

106 mm recoilless rifle

The 106 mm recoilless rifle is loaded into the rear of the vehicle via short ramps and, once in position, the barrel may be pointed through the centre of a split windscreen that can be folded forwards over the bonnet. This version has a crew of four, the driver and three personnel, two of whom sit on squab seats at the rear on either side of the gun. Stowage is provided for eight rounds.

7.62 mm machine gun carrier

The 7.62 mm Browning machine gun is mounted on a pivot secured to the rear area floor. This mounting allows a full 360° traverse and a maximum elevation of 36°. A similar arrangement which allows the mounting of a 0.50/12.7 mm Browning M2 HB machine gun limits maximum elevation to 30°.

MILAN

The Model 88 Militar can be configured to carry a MILAN ATGW launcher on a special rear-mounted pivot. The pivot allows a full 360° traverse and elevation and depression of 10°. Up to four missiles can be carried, one on the launcher itself and three in rear-mounted racks.

60 mm mortar

A 60 mm mortar is secured to the cargo area floor by a robust support plate. This allows firing through a full 360° arc.

Specifications

Cab seating: 1 + 1 (up to 4 in rear)
Configuration: 4 × 4
Weight:
(max) 2,160 kg
(empty) 1,660 kg
Load: 500 kg
Length: 3.725 m
Width: 1.574 m
Height: (with hood) 1.905 m
Track: 1.309 m
Wheelbase: 2.235 m
Fuel capacity: 97 litres
Angle of approach/departure: 48°/32°
Engine: 2.286 litre 4-cylinder in-line water-cooled petrol developing 61 hp at 4,000 rpm or
2.286 litre 4-cylinder in-line water-cooled diesel developing 59 hp at 4,000 rpm
Gearbox: manual with 4 forward and 1 reverse gears
Clutch: hydraulic diaphragm and disc
Transfer box: 2-speed
Steering: worm and recirculating ball
Turning radius: 6.25 m
Suspension: semi-elliptical springs front and rear with double-action telescopic shock-absorbers
Tyres: 6.00 × 16
Brakes:
(main) hydraulically operated drums front and rear
(parking) mechanical
Electrical system: 24 V
Batteries:
(petrol) 2 × 12 V, 57 Ah
(diesel) 2 × 12 V, 70 Ah
Alternator: 35 A (50 or 90 A optional)

Status

Production complete. In service with Spanish armed forces and some other nations including Egypt and Morocco.

Manufacturer

Land Rover Santana SA, Carretora de Vadollamas, S/N, E-23700 Linares, Spain.
Tel: +34 53 693051. Fax: +34 53 653201.

VERIFIED

Santana Model 109 Militar (4 × 4) light vehicle

Description

Metalurgica de Santa Ana SA was established in 1955 and in 1956 commenced production of the Land Rover in an agreement with Land Rover Limited. From that time Santana produced both military and civil versions of the Model 109 and introduced its own in-house modifications to production models. The association with Land Rover ceased in 1990.

The Santana Model 109 Militar bears a close resemblance to its British counterpart and is constructed along similar lines. Main differences occur with the engine, which may be either a locally produced 2.286 litre petrol or diesel engine or a 3.429 litre six-cylinder petrol or diesel engine. The basic model is produced in several versions (see following) and can be configured to a wide variety of customer requirements.

A deep wading version of the Model 109 Militar was produced and is in service with the Spanish Marines. The Model 109 Militar can also be prepared for parachute drops by lashing it to a special platform in a stripped-down state. The complete load of prepared vehicle, fixtures and platform weighs 2,800 kg.

Basic troop-carrying version of Santana Model 109 Militar (left) with ambulance (right) (Bob Morrison - Military Scene)

Variants

Basic troop carrier

This is fitted with a soft top but a hardtop version is available. The vehicle carries the driver and nine personnel and the rear area holds cargo or supplies. It is also possible to mount various weapons in the cargo area including machine guns, a 60 mm mortar and a MILAN ATGW launcher.

Communications

This version usually has a hardtop and is specially produced for the communications role with extra screening of components and wiring looms, a 90 A alternator, extra internal racking and stowage and an engine oil cooler. Of the five versions that have been produced, one has three long-range radios, another two ground-to-ground radio facilities. A further version acts as a 30-line telephone exchange, while a fourth version has three telex equipments. The fifth version is a special ground-to-ground communications station.

Light recovery vehicle

This variant carries a light electrical crane in the cargo area. The crane winch can lift loads of up to 1,000 kg while the jib can tow loads of up to 750 kg if the towed vehicle uses only one axle and 2,040 kg if both axles are used.

Ambulance

This variant has an insulated ambulance body, capable of carrying either four stretcher cases or two stretcher cases and four seated casualties. Alternatively eight seated casualties can be accommodated. Two large access doors are provided at the rear and the interior is equipped with medical equipment stowage.

Specifications

Cab seating: 1 + 1 (plus 6 in rear)
Configuration: 4 × 4
Weight:
(max) 3,150 kg
(empty) 1,890 kg
Max load: 1,000 kg
Length: 4.546 m
Width: 1.574 m
Height: (with hood) 2.008 m
Track: 1.309 m
Wheelbase: 2.768 m
Fuel capacity: 114 litres
Angle of approach/departure: 52°/31°
Engine: 2.286 litre 4-cylinder in-line water-cooled petrol developing 61 hp at 4,000 rpm or
2.286 litre 4-cylinder in-line water-cooled diesel developing 59 hp at 4,000 rpm or
3.429 litre 6-cylinder in-line water-cooled petrol developing 95 hp at 4,000 rpm or
3.429 litre 6-cylinder in-line water-cooled diesel developing 91.7 hp at 4,000 rpm
Gearbox: manual with 4 forward and 1 reverse gears
Clutch: hydraulic diaphragm and disc
Transfer box: 2-speed

Steering: worm and recirculating ball
Turning radius: 6.25 m
Suspension: semi-elliptical springs front and rear with double-action telescopic shock-absorbers
Tyres: 7.50 × 16
Brakes:
(main) hydraulically operated drums front and rear
(parking) mechanical
Electrical system: 24 V
Batteries:
(petrol) 1 × 12 V, 57 Ah
(diesel) 1 × 12 V, 70 Ah
Alternator: 35 A (50 or 90 A optional)

Status
Production complete. In service with the Spanish armed forces and some other nations including Egypt and Morocco.

Manufacturer
Land Rover Santana SA, Carretora de Vadollamas, S/N, E-23700 Linares, Spain.
Tel: +34 53 693051. Fax: +34 53 653201.

VERIFIED

SWEDEN

Volvo L3314 (4 × 4) Laplander and Volvo C202 (4 × 4) cross-country vehicles

Development
The first prototype of the Volvo Laplander L3314 vehicle was completed in 1961 with production beginning the following year. Over 10,000 Laplanders were built for delivery to some 40 countries before production was completed. Two basic versions of the Laplander were built, a pick-up (designated PU) and a hardtop (designated HT). Both are used by the Swedish Army, the PU being known as the Pltgbil 903 and the hardtop as the Pltgbil 903B. Further development of the Laplander resulted in the C202.

Description
Volvo L3314 (4 × 4) Laplander
The chassis of the Laplander consists of an all-welded box-section frame of pressed and welded 3 mm sheet steel, two tubular cross members, one box-section cross member and one U-section cross member.

Both models have an all-welded steel body with a forward control two-door fully enclosed cab with a heater and defroster. The hardtop model has the same body as the pick-up but in addition has a fully enclosed hardtop steel body built integrally with the cab, containing six removable folding seats and a heater.

The vehicle is powered by a Volvo B18A four-cylinder petrol engine which develops 75 bhp (SAE) at 4,500 rpm and is coupled to an M40 four-speed gearbox with four forward and one reverse gears. The transfer box has four gear positions: four-wheel drive in low ratio; neutral; four-wheel drive in high ratio; and reverse drive in high ratio.

The suspension consists of semi-elliptical springs which have progressive springing with rubber bush-type helper springs and double-acting telescopic hydraulic shock-absorbers at each wheel station. Brakes are hydraulic, with the parking brake operating on the propeller shaft.

Optional fittings include Rockinger type towing hook, hitch plate, PTO (centre and rear), speed governor for PTO and side-mounted winch with a capacity of 2,100 kg which can be used to the front or rear of the vehicle. Variants in use with the Swedish Army include an ambulance, firefighting and radio vehicles.

The L3315 is a 24 V version intended mainly for radio-equipped roles.

Volvo C202 (4 × 4)
The C202 was also manufactured in Hungary by Csepel. The vehicle was offered in three basic versions: a pick-up, a fully enclosed hardtop model and a canvas top model. The pick-up model has a fully enclosed two-door all-steel cab and a single tailgate at the rear. The hardtop model has five doors, two each side and one at the rear. At the rear are two foldable seats and two foldable bench seats. All versions have a heater and defroster for the front cab and the hardtop model also has a heater for the rear compartment. A wide range of specialised variants has been developed by Volvo, including firefighting, ambulance, repair and rescue versions.

The C202 is powered by a Volvo B20A four-cylinder petrol engine which develops 82 bhp at 4,700 rpm and is coupled to a manual Volvo M45 gearbox with four forward and one reverse gears. The transfer gearbox is a Volvo FD51. All-wheel drive is engaged by a press button in the high range and automatically when low range is engaged.

Brakes are of the vacuum-hydraulic drum type. The parking brake is mechanical and operates off the propeller shaft. Suspension consists of semi-elliptical leaf springs, telescopic double-acting shock-absorbers and hollow-rubber springs to soften bottoming from extreme spring deflection. Optional equipment included an air conditioning system, canvas top, electric engine heater, fire extinguisher, PTO, sand tyres, snow tyres, spotlight, tyre inflation pump, tropical radiator and a winch with a capacity of 2,000 kg.

Volvo L3315 (4 × 4) Laplander with fully enclosed hardtop body (T J Gander)

Specifications

Model	**Laplander PU**	**C202 HT**	**C202 PU**
Cab seating	1 + 1	1 + 7	1 + 1
Configuration	4 × 4	4 × 4	4 × 4
Weight (empty)	1,520 kg	1,795 kg	1,600 kg
(loaded)	2,450 kg	2,525 kg	2,525 kg
Weight on front axle (empty)	935 kg	1,020 kg	990 kg
Weight on rear axle (empty)	585 kg	775 kg	610 kg
Max load	900 kg	730 kg	925 kg
Towed load	1,000 kg	1,500 kg	1,500 kg
Load area	2.3 × 1.535 m	2.27 × 1.4 m	2.27 × 1.53 m
Length	3.985 m	4.015 m	4.015 m
Width	1.66 m	1.68 m	1.68 m
Height	2.09 m	2.09 m	2.09 m
(load area)	0.665 m	0.665 m	0.665 m
Ground clearance	0.285 m	0.285 m	0.285 m
Track (front/rear)	1.338 m/1.338 m	1.348 m/1.338 m	1.348 m/1.338 m
Wheelbase	2.1 m	2.1 m	2.1 m
Angle of approach/departure	40°/40°	39°/48°	39°/48°
Max speed (road)	90 km/h	115 km/h	115 km/h
Range	330 km	330 km	330 km
Fuel capacity	46 litres	51 litres	51 litres
Max gradient	60%	65%	65%
Max side slope	40%	40%	40%
Fording	0.8 m	0.45 m	0.45 m
Steering	cam and roller	cam and roller	cam and roller
Turning radius	5.4 m	5.4 m	5.4 m
Tyres	8.90 × 16	8.90 × 16	8.90 × 16
Brakes (main)	hydraulic	hydraulic	hydraulic
(parking)	mechanical	mechanical	mechanical
Electrical system	12 V	12 V	12 V

Status
The Volvo Laplander is no longer in production but remains in service with the Swedish Army and Civil Defence Corps and with Norway (small numbers only). The Volvo C202 was manufactured in Hungary by Csepel.

Manufacturers
Volvo Truck Corporation, S-405 08 Göteborg, Sweden.
Tel: +46 31 666000. Telex: 27000 volvo s.
Fax: +46 31 664660.
Csepel, Hungary.

VERIFIED

UNITED KINGDOM

Supacat (6 × 6) light vehicle

Development

The Supacat (6 × 6) light vehicle, known as the ATMP (All Terrain Mobile Platform) by the British Army, was developed primarily as a marginal terrain vehicle which retained a good hard surface performance. It is used in both military and civilian applications and has been in service with the British Army and RAF since 1984. Prime users are the airborne and airmobile brigades with others, such as Marines, making occasional training and more specific operational use. Both Mark 1s and Mark 2s are used as range service vehicles.

The Supacat was first produced in 1982 and since then has been continuously developed to cater for increasing demands on its abilities. From skid steering and a limited capacity petrol engine it now has a dual steering system, a 1.9 litre turbo diesel engine (introduced in May 1994) and increased load capacity.

In 1995 an agreement was established between Supacat Limited and Alvis Vehicles Limited for Alvis to undertake the marketing and production of the Supacat 6 × 6 vehicle for military markets; Supacat Limited retains the production and marketing rights for civilian applications.

In August 1996 the United Kingdom Ministry of Defence announced a requirement for a further 86 ATMPs and 84 of the associated SLLPT trailers for delivery between 1997 and 1999. A total of 10 companies were invited to tender.

Supacat Mark 2 (6 × 6) light vehicle configured for the MILAN ATGW carrier role ***1996***

Description

The Supacat is a 6 × 6 low ground pressure vehicle running on 31 × 15.50 × 15 wide section low-pressure tyres. The vehicle is powered by a Volkswagen ADE 1900 turbocharged engine matched to an automatic gearbox. Drive to the wheels is mechanical with a single differential; traction control is via the handlebar brakes. Steering is by handlebars which steer the front four wheels conventionally and also operate the brake steering as required. A conventional foot brake is also provided. Both foot and hand throttles are fitted.

The Supacat is capable of a speed of 64 km/h and has a limited amphibious capability. It can be fitted with tracks for extremely soft conditions and for deep snow. The Supacat is primarily an open vehicle but can be fitted with a variety of hard, soft or lightweight removable canopies. Load capacity is 1,000 kg but it is acceptable to carry 1,400 kg with reduced performance. It is possible to self-load a pallet using its winch and ramps; the same ramps can be used for bridging and recovery.

The Supacat is designed to be a towing vehicle for trailers, light artillery pieces such as the 105 mm Light Gun and for vehicle recovery. Special trailers have been developed for use with the Supacat, including the Fork Lift Pallet Trailer (FLPT - see entry under *Materials handling equipment*). Cranes can also be fitted.

The Supacat is fully air-portable. Schemes are approved for single, dual and quadruple vehicle underslung loads under suitable helicopters such as the Sea King (1), Black Hawk (2), Puma (1) and Chinook (4). Two Supacats can be driven into the Chinook and the vehicle is compatible with the EH 101. Supacat can be air-dropped and a number of schemes have been proved. These include two vehicles on one platform or one vehicle plus trailer or mixed stores. Supacats have been airlifted in a variety of aircraft such as the C-130 Hercules and trials have been carried out for the CASA 235. Vehicles can be stacked one on the other for air transport, using the vehicle's own equipment.

Supacats can be used for troop transport and as weapon platforms and carriers. They are used in the anti-tank and air defence roles with, for example, MILAN, Javelin and Rapier. Close support weapons can be fired from the vehicle, including 60 and 81 mm mortars. British Army ATMPs are fitted with 7.62 mm GPMG mounts as standard.

Specifications

1.9 litre turbo diesel engine version
Seating: 1 + 5
Configuration: 6 × 6
Weight:
(maximum) 2,650 kg (3,050 kg limited application)
(empty) 1,650 kg
Payload: 1,000 kg (up to 1,400 kg with reduced mobility)
Load area: 1.445 × 1.87 m
Towed load: up to approx 2,000 kg
Length: 3.335 m
Width: 2 m
Height:
(roll bar) 1.895 m
(cab) 2.01 m
(folded down) 1.21 m
(load platform) 0.94 m
Ground clearance: 0.215 m
Track:
(front) 1.62 m
(rear) 1.61 m
Wheelbase: 0.923 m + 0.923 m
Angle of approach/departure: 57°/58°
Max speed: (road) 64 km/h
Fuel capacity: 50 litres
Max gradient: 100%
Side slope: 40°
Vertical obstacle: 0.5 m
Fording: amphibious
Engine: VW ADE 1900 1.896 litre 4-cylinder turbo diesel developing 78 bhp at 4,000 rpm
Gearbox: VW fully automatic 3-speed with park and reverse
Drive: twin helical reduction gearboxes to central axles and mechanical coupling to front and rear axles
Steering: combined power-assisted Ackermann (1st and 2nd axles) and skid steer (all axles) — lock-up facility for use with tracks
Brakes: hydraulic inboard disc
Tyres: low pressure 31 × 15.5 × 15
Electrical system: 12 V (24 V optional)
Battery: 1 × 12 V, 65 Ah
Alternator: 65 A

Status

In service with the British Army (50) and Royal Air Force (5) and Mexico.

Manufacturers

Alvis Vehicles Limited, The Triangle, Walsgrave, Coventry, West Midlands CV2 2SP, UK.
Tel: +44 1203 535455. Fax: +44 1203 539287.
Supacat Limited, The Airfield, Dunkeswell, Honiton, Devon EX14 0RA, UK.
Tel: +44 1404 891777. Fax: +44 1404 891776.

UPDATED

Supacat Mark 2 (6 × 6) light vehicle in an ammunition supply configuration with a hydraulic crane and towing a ½ tonne trailer ***1996***

Supacat Mark 2 (6 × 6) light vehicle towing a flexible fuel container ***1996***

Octad T8 × 8 Mk3 all-terrain vehicle

Description

The Octad T8 × 8 Mk3 uses hydraulic transmission, with the option of adding tracks and a variety of ancillary functions, including hydraulic power take-off for winches and power tools.

The main body is steel and carries the engine and twin hydraulic swash-plate pumps on a subframe. It also houses the hydraulic transmission, oil reservoir, axles and axle drives. One pump drives each side. From the pumps the axle drive is by a closed loop system via fixed displacement hydraulic motors to sprockets and chains giving a full 8 × 8 drive. The hydraulic drive provides an inherent braking facility in addition to multidisc brakes within the motor frames. The independent joystick control of each set of four wheels allows neutral turns and fine control. The hydraulic transmission provides high torque at low speeds. The vehicle can be supplied with customer-specified superstructures and in its standard form carries a payload of 1,500 kg in addition to three crew. It is readily adaptable to remote control.

Octad T8 × 8 1,000 kg all-terrain vehicle fitted with optional tracks over wheels **1996**

Specifications

Cab seating: 1 + 2
Configuration: 8 × 8
Weight: (empty) 1,670 kg
Max load: 1,500 kg
Length: 3.4 m
Width: 2 m
Height: 2.005 m
Ground clearance: 0.314 m
Max speed: (over rough terrain) 30 km/h
Fuel range: 6 h
Fuel consumption: (nominal) 5.68 l/h
Max gradient: 100%
Engine: Ford XLD 1.8 litre diesel or Peugeot XUD 1.9 litre diesel or to customer requirement
Transmission: hydraulic closed loop
Steering: hydraulic
Suspension: tyres
Tyres: 26 × 12 × 12
Tracks: multisegment steel/rubber composite
Brakes: hydraulic with disc brakes inside hydraulic motors
Electrical system: 12 V

Status

In production.

Manufacturer

TRAMMIC Limited, 20b Tatbank Road, Oldbury, Warley, West Midlands B69 4NU, UK.
Tel: +44 121 544 5100. Fax: +44 121 544 5498

VERIFIED

Land Rover (4 × 4) vehicles

Development

Land Rover, the world's largest dedicated manufacturer of 4 × 4 vehicles, is part of the Rover Group within BMW. The company has been manufacturing 4 × 4 vehicles since 1948 and of the two million Land Rover vehicles produced it is estimated that 70 per cent are still in regular use. Some 100,000 vehicles a year are produced, of which 70 per cent are exported. Land Rover vehicles are currently in service with more than 100 military and paramilitary forces worldwide in every role from ambulances to weapon carriers.

At its factory in Solihull, West Midlands, the company currently produces a comprehensive range of permanent four-wheel drive vehicles covering the payload range of 0.5 to 1.5 tonnes. All current military variants have a high degree of engine and transmission commonality and share the majority of suspension, body, electrical and braking components.

After the end of the Second World War the British government was compelled to ration steel to the motor industry in proportion to the value of its exports. This created serious difficulties for the Rover company since its luxury cars were not proving to be exportable. It became obvious that the company would have to produce something that would have a world appeal and be outside the luxury class, a working vehicle that would attract buyers from agricultural and industrial markets of the world.

Early in 1947 a decision was made to build a new all-purpose cross-country vehicle and by late 1947 prototypes of a 4 × 4 vehicle suitable for both agricultural and industrial applications had been built. Trials proved the concept and Rover introduced the Land Rover at the Amsterdam Motor Show in April 1948. Quantity production began at Solihull in July 1948. The first model had an 80 in (2.032 m) wheelbase and was powered by a 1.6 litre petrol engine, which used many components, including the engine of the Rover P3 '60' car. The basic model was fitted with a canvas hood and was followed late in 1948 by a fully enclosed estate model which could seat six plus the driver. The British government placed its first order for the Land Rover in 1949.

By 1950 over 24,000 Land Rovers had been built and in 1952 the 1.6 litre petrol engine was replaced by a 2 litre petrol engine. In 1954 the original 80 in (2.032 m) wheelbase model was replaced in production by an 86 in (2.184 m) model. In the same year the first long wheelbase model (with a wheelbase of 107 in (2.717 m) and which could carry 750 kg of cargo) was introduced.

In 1956 the 86 in (2.184 m) model gave way to the 88 in (2.23 m) and the 107 in (2.717 m) was replaced by the 109 in (2.768 m) model. The same year the British Army adopted the Land Rover as its standard ¼ tonne (4 × 4) vehicle. In 1957 Land Rover offered a diesel engine in place of the standard petrol engine.

In February 1958 the Series II Land Rover was introduced in both 88 in (2.23 m) and 107 in (2.717 m) configurations, with a 2.25 litre petrol engine. Late in 1958 production of the 107 in (2.717 m) was discontinued.

In September 1961 the Series IIA was introduced and the 2.25 litre engine replaced the 2 litre engine. In 1962 the 109 in (2.768 m) 12-seater station wagon and a new 1,000 kg forward control Land Rover were introduced.

By April 1966 half a million Land Rovers had been completed and late in 1966 a 110 in (2.794 m) model of the forward control Land Rover was introduced, powered by a six-cylinder petrol engine and capable of carrying 1,500 kg of cargo.

From early 1967 a 2.6 litre six-cylinder petrol engine was offered for the 109 in (2.768 m) Land Rover. This was deleted in 1980 with the introduction of the 3.5 litre V-8 version.

Early in 1968, to comply with new legal requirements governing vehicle lighting in the Netherlands, Belgium and Luxembourg, a headlight modification was introduced and for the first time the headlights were incorporated in the wings instead of the grille panel.

In September 1968 two new Land Rovers specifically for military use were introduced, the ½ tonne and 1,000 kg, for which there are separate entries in this section. In 1970 the Range Rover, for which there is a separate entry in this section, was introduced.

By July 1971 half a million Land Rovers for export had been completed. The following year production of the forward control model ceased.

In October 1971 the Series III Land Rover was introduced with a restyled grille, redesigned safety facia, improved gearbox and other detailed modifications.

The millionth Land Rover was completed in June 1976 and in early 1979 production of the standard Land Rover was running at 1,600 units a week, increased to 2,700 units a week by 1981. About 70 per cent of Land Rovers are exported and, according to the company, Vietnam is the only country not to have ordered the vehicle.

In 1979 Land Rover Limited, a new and autonomous company, was formed to manufacture, develop and market Land Rovers and Range Rovers.

In March 1982, a 'High Capacity' version of the 109 in long wheelbase vehicle was introduced, which was available in both full length hood and pick-up forms and in both cases provided greater passenger or cargo carrying ability than the standard body. This was available in both 12 V and 24 V configurations. In its standard form the High Capacity 109 in had a gross vehicle weight of 3,020 kg and a payload of up to 1,450 kg, or seating for up to 13 including the driver.

Production of the 109 in and 88 in models was phased out during 1985.

A new range of models was introduced in 1983, the long wheelbase version with a 110 in (2.794 m) wheelbase and a short wheelbase version (introduced in 1984) with a nominal wheelbase length of 90 in (actually 2.36 m). An extended wheelbase version with a 127 in (3.226 m) wheelbase was also made available in crew cab and chassis cab form. In 1990, to distinguish the vehicles from other models manufactured by Land Rover, the range was given the name 'Defender'. The Defender family consists of the short wheelbase Defender 90, the long wheelbase Defender 110 and the extended wheelbase Defender 130.

Current models in production are the Core Military Defender 90, 110 and 130 and the company has developed the Defender XD (eXtra Duty) range of 90 and 110 military vehicles and 130 field ambulance to meet the arduous operational requirements specified by the UK Ministry of Defence and selected in 1996 as its replacement TUL/TUM and 4 × 4 field ambulance into the next century.

Additionally, in 1989, the company introduced the Discovery, which is used extensively by armed forces in command and support roles, and increasingly by police and peacekeeping organisations, including the UN. Also used in these roles is the revised Range Rover introduced in 1994, building on the success of the Classic Range Rover, which ceased production in 1995. These vehicles have their own entries in this section.

Description

88 in (2.23 m)

No 88 in military model is now offered as it was initially replaced by the ½ tonne air-portable vehicle and later the Land Rover Defender 90, but quantities of the civilian version were sold for military and paramilitary applications.

The Land Rover has an all-welded box-section ladder-type chassis. The body is of aluminium panels with the steel bumpers, cappings and other vital components galvanised. The layout is conventional, with the engine at the front, driver and two passengers in the centre and the cargo area at the rear with a drop tailgate. The basic model has a galvanised steel hood frame and a full length hood.

Optional equipment for the standard and the long wheelbase models includes a front-mounted 2,270 kg capacity mechanical drum winch controlled from the driver's seat, capstan winch with a 1,360 kg capacity, overdrive, rubber helper springs, lamp guards and a fire extinguisher.

Land Rover developed special models of the 88 in Land Rover to meet special overseas requirements. In 1977 the Danish Army ordered a version based on the commercial model with a 2.25 litre diesel engine and a 12 V and supplementary 24 V electrical system which entered service in 1978. Many armies adapted the vehicle to their own specific requirements, for example the Australians had some armed with a 106 mm M40 series recoilless rifle.

109 in (2.768 m)

This has a similar layout to the 88 in (2.23 m) model and was the standard vehicle in its load class in the British Army and many other armed forces, but it is being progressively replaced by the Defender 110 series. Standard equipment on the military version includes an FV design towing hook, twin fuel tanks, vehicle lashing eyes at front and rear, freight lashing points in the rear, oil cooler, FV pattern lights, stowage for shovel, pick and axe on tailboard, water jerrican at the bulkhead and provision for stowing a rifle on the dashboard. The Fitted-For-Radio (FFR) version has a 90 A rectified electrical system with provision for charging radio batteries, full suppression of electrical equipment, radio table battery carrier, two 100 A batteries, coaxial leads and operator's seat.

There were numerous variants of the long wheelbase Land Rover in service, including an anti-tank version armed with a US 106 mm M40 recoilless rifle, ambulance, command, fire control and numerous other roles. It is also used for towing artillery such as the Italian 105 mm Pack Howitzer, carrying or towing missiles such as the Bofors RBS 70 or Rapier and towing radars such as Cymbeline. The chassis was also used as a basis for the Shorland armoured patrol vehicle and the SB.401 armoured personnel carrier.

Land Rover built modified versions of the long wheelbase model to meet specific overseas countries' requirements. In 1973 the Norwegian Army ordered the vehicle with a 24 V electrical system and fitted with lifting and towing facilities, petrol-burning heater, radio and aerial mountings and inter-vehicle starting capability. In 1976 the Royal Netherlands Army ordered a version based on the commercial long wheelbase Land Rover with a four-cylinder OHV diesel engine, 24 V electrical system, fresh air heater, sun visors, shovel and axe stowage, radio and aerial mountings and radio battery stowage.

In February 1979 Land Rover launched the 3.5 litre V-8 Land Rover which served with the Iraqi Army. The vehicle was developed for FFR and as a half-track (by Laird, but development was discontinued) and a 6 × 6 version was developed in conjunction with Hotspur Limited (details of which are given in the *Trucks* section). This model was available in 3.175 m (standard body and armoured versions) and 3.529 m (standard) wheelbases.

The Range Rover's 3.5 litre engine was modified to give maximum torque for the Land Rover V-8 at a lower engine speed. Maximum 187 lb/ft at 2,500 rpm compares with the Range Rover figure of 207 lb/ft at 3,200 rpm and the 122 lb/ft at 2,000 rpm produced by the six-cylinder engined Land Rover, which considerably improved the vehicle's off-road performance.

The Land Rover V-8 has permanent four-wheel drive, with high and low ratio gears and a differential lock, as on the Range Rover. Modification of the bodywork in the engine compartment provided plenty of room for the larger engine and allows easy access for servicing.

Military equipment for the V-8 Land Rover included an FV design towing hook, twin fuel tanks, front and rear lifting and towing rings, stowage for shovel, pick and axe on rear tailboard, rifle clips, bumperettes, wire lamp guard, oil cooler, split charge facility, hand throttle, bonnet-mounted spare wheel and raised air intake. To meet military requirements, 24 V electrics were developed and are in service with Middle East defence forces. The 2.23 m (109 in) V-8 Land Rover is no longer in production.

Specifications

88 in (2.23 m)
Cab seating: 1 + 2 in front (4 in rear)
Configuration: 4 × 4
Weight:
(empty) 1,430 kg
(loaded) 2,120 kg
Weight on front axle: (loaded) 930 kg
Weight on rear axle: (loaded) 1,190 kg
Max load: 690 kg
Length: 3.65 m
Width: 1.68 m
Height: 1.97 m
Ground clearance: 0.2 m
Track: 1.33 m
Wheelbase: 2.23 m
Angle of approach/departure: 46°/30°
Max speed: (road) 105 km/h
Range: 560 km
Fuel capacity: 45 litres
Max gradient:
(high) 39%
(low) 115%
Fording: 0.5 m
Engine: 4-cylinder OHV diesel developing 51 bhp at 4,000 rpm or 4-cylinder OHV petrol engine developing 69 bhp at 4,000 rpm
Gearbox: manual with 4 forward and 1 reverse gears
Clutch: single dry plate
Transfer box: 2-speed
Steering: recirculating ball
Turning radius: 5.79 m
Suspension: semi-elliptical springs front and rear with double-acting telescopic shock-absorbers
Tyres: 7.50 × 16
Brakes:
(main) drum, hydraulic, tandem braking system
(parking) drum, mechanical on transmission
Electrical system: 12 V (24 V optional)

109 in (2.768 m) LWB
Cab seating: 1 + 2 in front (optional seating for 8 in rear)
Configuration: 4 × 4
Weight:
(empty) 1,750 kg
(loaded) 2,600 kg
Max load: 850 kg
Load area: 1.85 × 1.619 m
Length: 4.56 m
Width: 1.68 m
Height:
(overall) 1.98 m
(reduced) 1.52 m
Ground clearance: 0.2 m
Track: 1.33 m
Wheelbase: 2.768 m
Angle of approach/departure: 45°/29°
Max speed: 90 km/h
Range: 600 km
Fuel capacity: 90 litres
Gradient:
(high) 25%
(low) 58%
Fording: 0.7 m
Engine: 4-cylinder in-line OHV petrol developing 69 bhp at 4,000 rpm
Gearbox: manual with 4 forward and 1 reverse gears
Clutch: single dry plate
Transfer box: 2-speed
Steering: recirculating ball
Turning radius: 7.5 m
Suspension: semi-elliptical springs front and rear with hydraulic double-acting shock-absorbers
Tyres: 7.50 × 16
Brakes:
(main) hydraulic, tandem braking system, servo-assisted
(parking) drum, mechanical on transmission
Electrical system: 12 V (24 V optional)

Status

Production complete. In service with over 140 armed forces.

Manufacturer

Land Rover, Lode Lane, Solihull, West Midlands B92 8NW, UK.
Tel: +44 121 722 2424. Telex: 338641 lanrov g.
Fax: +44 121 742 1927.

Land Rovers have also been assembled or manufactured in Australia, Kenya, Malaysia, Morocco, New Zealand, Nigeria, Trinidad, Turkey, Zaïre, Zambia and Zimbabwe.

UPDATED

British Army Series III Land Rovers (Bob Morrison - Military Scene)

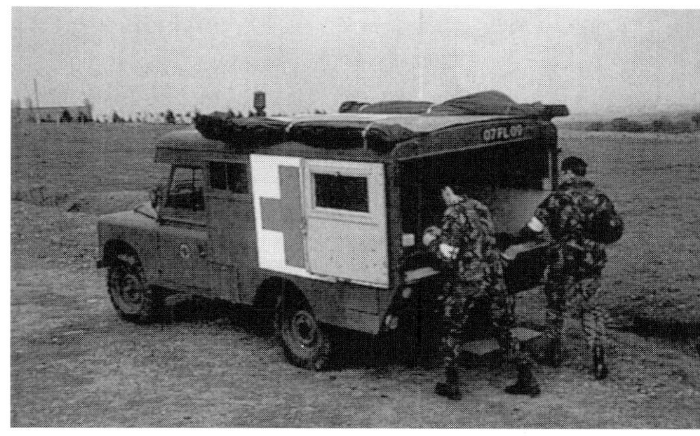

British Army Land Rover ambulance (Bob Morrison - Military Scene)

Land Rover Defender 90, Core Military Defender 90 and Defender XD (4 × 4) light vehicles

Development

In 1990, the Land Rover range of vehicles was given the name 'Defender' so the Land Rover Ninety was renamed the Land Rover Defender 90. The original Land Rover Ninety was introduced in early 1984 with a 2.36 m wheelbase. Like the Land Rover One Ten (now the Defender 110), the Ninety was fitted with long travel coil spring suspension and front disc brakes. Both vehicles share a number of common components, in particular engines, transmissions and front end body panels.

At the same time as the adoption of the Defender name a significant package of product improvements was introduced, including the introduction of the 200 Tdi 2.5 litre direct injection turbocharged intercooled diesel engine, with improvements in power, torque and economy over the previous direct injection unit. Other improvements concerned the cab ergonomics and electrical system.

The Land Rover Defender 90 is part of the Land Rover family of military vehicles which includes the Land Rover Defender 110, Land Rover Defender 130 and the Land Rover 6 × 6. The Defender 90 is in service with the British armed forces as the Truck, Utility, Light (TUL). Production ceased during 1994.

The current production model is the Core Military Defender 90. The company has also developed the Defender XD 90, part of the Defender XD range, to meet the operational requirements of the UK Ministry of Defence into the 21st century. These vehicles were unveiled at the Royal Navy and British Army Equipment Exhibition in September 1995. These vehicles were selected by the UK Ministry of Defence for the British Armed Forces requirement during 1996.

Description

The Land Rover Defender 90 is based on a boxed section steel chassis with the centre section approximately 190 mm deep. A cross member is bolted in place to facilitate removal of the gearbox and transfer box and wiring looms are routed along the insides of the box sections for maximum protection. The Defender 90 uses long travel coil springs all round to provide outstanding traction, ride and handling. Suspension movement is controlled at the front and rear by long stroke hydraulic dampers. The Land Rover front beam axle is located by radius arms with Panhard rods providing lateral location. The rear axle is located by tubular trailing links with a centrally mounted 'A' frame.

Four engines were available for the Defender 90: a 2.5 litre petrol, 3.5 litre V-8 petrol, 2.5 litre naturally aspirated diesel and a 2.5 litre direct injection turbocharged intercooled diesel. The Defender 90 uses the same gearbox and transfer box as the Defender 110 and has permanent four-wheel drive with a lockable centre differential.

Options included: a truck cab; open top with folding windscreen and detachable door tops; a roll-over bar; NATO standard towing jaw and 12-pin trailer socket; blackout lighting systems; air conditioning; small arms clips; hand throttle; wire mesh lamp guards; disruptive pattern camouflage paint; raised air intake; waterproof screw-lens lamps; various front and rear seating options; external pick and shovel stowage and power take-offs.

The Land Rover Defender 90 was developed to accommodate a variety of weapon system mounts, including an M40 106 mm recoilless rifle, machine guns, and the MILAN ATGW. For details of the Defender 90 Multi Role Combat Vehicle (MRCV) see entry in *Special attack vehicles* section.

Core Military Defender 90

The Core Military Defender 90 is the result of years of development of the basic vehicle's capabilities to meet the increasing operational requirements of military customers into the 21st century. In appearance, the Core Military Defender 90 retains the heritage of earlier Land Rover members. The large number of major improvements over earlier models has resulted in higher levels of interior comfort and visibility, combined with better manoeuvrability and performance, both on and off road. There are also chassis, steering and suspension improvements.

Land Rover Defenders can tow loads up to 4 tonnes and, depending on vehicle specifications, have payload capabilities up to nearly 1 tonne on the Defender 90. This strength is based on a box-section steel chassis similar to that used on the Range Rover and Discovery, but considerably strengthened. The electrical harness is routed inside the chassis for maximum protection in off-road use. Long travel coil spring suspension is utilised.

Maximum braking is provided, even after fording rivers or driving through deep mud, by dual-line servo-assisted front disc brakes. The provision of a transmission parking brake means that braking effort is applied to all four wheels for maximum holding power on the steepest gradients.

The five-speed synchromesh transmission combined with the permanent four-wheel drive, lockable centre differential and two-speed transfer box, gives the Defender 90 vehicles outstanding mobility and excellent engine braking through all four wheels. Core Military Defenders are available with either the Tdi, V-8 or four-cylinder petrol engine, with 12 volt electrics and right- or left-hand steering.

All vehicles have six-way lighting systems and convoy lights, military bumpers, towing pintles and military chassis with NATO rotating tow hook and 12-pin trailer socket. The two front vinyl seats have head restraints and lap and diagonal seat belts. The front doors are fitted with removable window tops, full length soft tops and roll bars. Options available include Fitted For Radio (FFR) fitments and rear seating.

Cutaway drawing of Land Rover Defender XD 90 **1996**

Defender XD 90

The Defender XD 90 was developed to meet the arduous operational requirements specified by the UK Ministry of Defence in considering its replacement TUL fleet into the next century. It successfully completed UK Ministry of Defence exhaustive battlefield simulating trials, was selected for the British armed forces and now forms part of Land Rover's military derivative range available worldwide. A total of 8,000 Defender XD 90 and Defender XD 110 vehicles was ordered by the UK Ministry of Defence in January 1996.

The eXtra Duty (hence XD) vehicles resulted from consultations and trials with in-depth engineering analysis to establish and satisfy vehicle needs for enhanced operational requirements well into the 21st century. They were developed using state-of-the-art computer-aided engineering technology, also used for the new Range Rover. They built on the capabilities of the Core Military Defender and offer even more performance, payload carrying, durability, improved ergonomics providing greater comfort, extra safety and reduced maintenance. They also provide suppression specifications to high levels.

The extra duty off-road capability of the Defender XD is provided by the 300 Tdi engine in combination with the increased robustness of the chassis frame, cab, body structure and major assemblies. It can carry up to 20 per cent more equipment or payload than the earlier equivalent models.

Specifications

Defender 90
Cab seating: 1 + 2 (4 in rear)
Configuration: 4 × 4
Weight (GVW):
(standard suspension) 2,400 kg
(high-load suspension) 2,550 kg
Max load: between 920 and 1,036 kg, depending on model
Towed load:
(off road) 500 kg
(on road) 4,000 kg
Length: 3.722 m
Width: 1.79 m
Height: 1.993 m
Ground clearance: (7.50 × 16 tyres) 0.229 m
Track: 1.486 m
Wheelbase: 2.36 m
Angle of approach/departure: 51°/52°
Fuel capacity: 54.5 litres
Engine: 2.495 litre 4-cylinder in-line water-cooled petrol developing 83 bhp at 4,000 rpm or
3.528 litre V-8 water-cooled petrol developing 134 bhp at 5,000 rpm or
2.495 litre in-line water-cooled indirect injection naturally aspirated diesel developing 68 bhp at 4,000 rpm or
2.495 litre in-line water-cooled direct injection turbocharged intercooled diesel developing 107 hp at 3,800 rpm
Gearbox: manual with 5 forward and 1 reverse gears
Clutch: diaphragm spring
Transfer box: 2-speed
Steering: manual, recirculating ball (power-assisted optional)
Turning radius: 5.75 m
Suspension: beam axles front and rear with coil springs controlled by telescopic dampers
Tyres: 7.50 × 16, 6.00 × 16 or 20.5 × 16
Brakes: dual servo-assisted, discs front and rear
Electrical system: 12, 24 or 12/24 V

Specifications

Core Military Defender 90
Cab seating: 1 + 2 (4 in rear)
Configuration: 4 × 4
Weight (GVW): 2,550 kg
Towed load: up to 4,000 kg
Length: 3.88 m
Width: 1.79 m
Height: 2.04 m
Ground clearance: 0.191 m
Track: 1.49 m
Wheelbase: 2.36 m
Angle of approach/departure: 48°/49°
Fuel capacity: 54 litres
Engine: 2.5 litre Tdi direct injection turbocharged water-cooled diesel developing 111 hp at 4,000 rpm
Gearbox: manual with 5 forward and 1 reverse gears
Clutch: diaphragm spring
Transfer box: 2-speed
Steering: manual, recirculating ball (power-assist optional)
Suspension: beam axles front and rear with coil springs controlled by telescopic dampers
Brakes: dual servo-assisted, discs front and rear
Electrical system: 12, 24 or 12/24 V

Land Rover Defender XD 90 Truck Utility Light (TUL) **1997**

Land Rover Defender 90 with soft top **1997**

Specifications
Defender XD 90
Cab seating: 1 + 1 (6 in rear)
Configuration: 4 × 4
Weight (GVW): 2,600 kg
Maximum load: 600 kg
Towed load: up to 4,000 kg
Length: 3.84 m
Width: 1.79 m
Height: 2.03 m
Ground clearance: 0.24 m
Track:
(front) 1.52 m
(rear) 1.51 m
Wheelbase: 2.36 m
Fuel capacity: 55 litres
Engine: 2.5 litres in-line water cooled direct injection turbocharged inter-cooled diesel developing 111 hp at 4,000 rpm
Gearbox: manual with 5 forward and 1 reverse gears
Clutch: diaphragm spring
Transfer box: 2-speed
Steering: manual, recirculating ball (power-assist optional)
Suspension: beam axles front and rear with coil springs controlled by telescopic dampers
Brakes: dual servo-assisted, discs front and rear
Electrical system: 12, 24 or 12/24 V

Status
Defender 90, production complete. In service with the British and other armed forces.
Core Military Defender 90, in production.
Defender XD 90, in production and entered service with the British armed forces in late 1996.

Manufacturer
Land Rover, Lode Lane, Solihull, West Midlands B92 8NW, UK.
Tel: +44 121 722 2424. Telex: 338641 lanrov g.
Fax: +44 121 742 1927.
Land Rovers are also assembled or manufactured in Australia, Kenya, Malaysia, Morocco, Turkey, Zaïre, Zambia and Zimbabwe.

UPDATED

Land Rover Defender 110, Core Military Defender 110 and Defender XD (4 × 4) light vehicles

Development
In 1990, the Land Rover range of vehicles was given the name 'Defender' so the Land Rover One Ten was redesignated the Land Rover Defender 110. The original Land Rover One Ten was first shown publicly in early 1983 with a 110 in (2.794 m) wheelbase and combining a revised body style with a new chassis and coil sprung suspension. Early production in 1983 was mainly for the commercial market but since 1984 a large number of vehicles has been supplied for military contracts. In June 1984 it was announced that the first military order had been placed by an unspecified customer in the Far East. That order was for 900 vehicles and many similar orders followed.

The UK Ministry of Defence has taken into service a large number of Defender 110s for use by the British armed forces, in General Service Cargo (GSC) and Fitted For Radio (FFR) variants. It is known to the British armed forces as the Truck, Utility, Medium (TUM).

At the same time as the adoption of the Defender name a significant package of product improvements was introduced, including the introduction of the 200 Tdi 2.5 litre direct injection turbocharged intercooled diesel engine with improvements in power, torque and economy over the previous direct injection unit. Other improvements concerned the cab ergonomics and electrical system.

The Land Rover Defender 110 is part of the Land Rover family of military vehicles which includes the Land Rover Defender 90, Land Rover Defender 130 and the Land Rover 6 × 6.

The current production model is the Core Military Defender 110. The company has also developed the Defender XD 110, part of the Defender XD range, to meet the operational requirements of the UK Ministry of Defence into the 21st century. These vehicles were unveiled at the Royal Navy and British Army Equipment Exhibition in September 1995 and selected for the British Armed Forces requirement in 1996. A total of 8,000 Defender XD 90 and Defender XD 110 vehicles was ordered by the UK Ministry of Defence in January 1996.

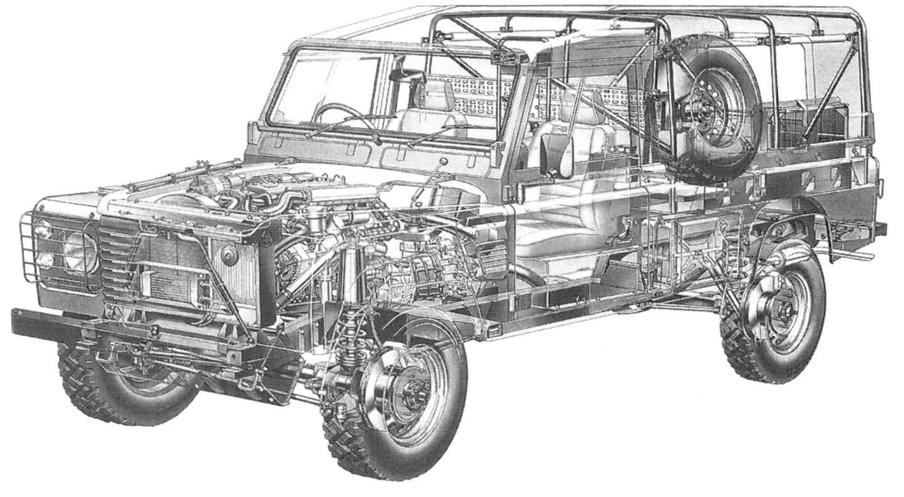
Cutaway drawing of Land Rover Defender **1996**

Description
In appearance, the Land Rover Defender 110 is similar to earlier Land Rover models, but can be identified by the new moulded injection grille and one-piece windscreen. The Defender 110 is available with either the integral pick-up body in soft or hardtop versions, the high capacity pick-up body with soft top, or as a station wagon. The increased track involved the introduction of GRP deformable 'eyebrows' over the wheels which are held in place by frangible pins.

The Defender 110 is based on the box-section steel chassis used on the Range Rover but is considerably stronger. The chassis is robot welded and the centre section is over 190 mm deep. The overall strength is said to withstand the most demanding off-road conditions. Detail design of the chassis is such, that the fuel tank has added protection. A bolt-in cross member facilitates removal of the gearbox and transfer box and the wiring looms are routed inside the chassis members for maximum protection.

The Defender 110 suspension uses coil springs in place of the earlier leaf springs. The coil springs are reinforced and provide 180 mm of wheel travel at the front and 210 mm at the rear. Minimum ground clearance is 0.21 m. Suspension movement at the front and rear is controlled by long-stroke hydraulic dampers. The Land Rover beam front axle is located by radius arms with a Panhard rod providing lateral location. A Salisbury axle is used on the rear, with 300 mm disc brakes being fitted all round.

The increased track of the Defender 110 led to a redesigned steering linkage that provides a turning radius of 6.4 m. Factory-fitted power-assisted steering is optional.

Four engines are available for the Defender 110: a 2.5 litre petrol, 3.5 litre V-8 petrol, 2.5 litre naturally aspirated diesel and a 2.5 litre direct injection turbocharged intercooled diesel.

The transmission is permanent four-wheel drive with a five-speed gearbox matched to a two-speed transfer box. This arrangement provides 10 forward and two reverse gears and the transfer box incorporates a lockable differential.

Factory-fitted options include: truck cab; open top with folding windscreen and detachable door tops; roll-over bar; NATO standard towing jaw and 12-pin trailer socket; blackout lighting systems; additional 68 litre fuel tank; air conditioning; small arms clips; hand

Land Rover Defender XD 110 **1997**

Australian Army Land Rover 110 (Barry Marriot) **1997**

throttle; wire mesh lamp guards; raised air intake; disruptive pattern camouflage paint; waterproof screw-lens lamps; various front and rear seating options; cabinets in the body sides for stowing up to four jerricans; external pick and shovel stowage; and power take-offs.

A variety of weapon system mounts for the Defender 110 was developed with weapon system manufacturers, including the MILAN ATGW and the LAU-97 70 mm multiple launch rocket system.

A wide range of conversions was developed for the Defender 110, including a special forces patrol vehicle, armoured patrol vehicle, remotely piloted vehicle launcher system, mobile workshops, shelter vehicles, ambulances, back hoe, fire tenders, armoured and discreetly armoured personnel carriers, towlift recovery vehicles and a hydraulically operated multiloader.

Core Military Defender 110

The Core Military Defender 110 is the result of years of development of the basic vehicle's capabilities to meet the increasing operational requirements of military customers into the 21st century. In appearance, the Core Military Defender 110 retains the heritage of earlier Land Rover members, but the large number of major improvements over earlier models have resulted in higher levels of interior comfort and visibility, combined with better manoeuvrability and performance, both on and off road. There are also chassis, steering and suspension improvements.

Land Rover Defenders can tow loads up to 4 tonnes and, depending on vehicle specifications, have payload capabilities up to nearly 1.5 tonnes on the Defender 110 (HD). This strength is based on a box-section steel chassis similar to that used on the Range Rover and Discovery, but considerably strengthened. The electrical harness is routed inside the chassis for maximum protection in off-road use. Long travel coil spring suspension is utilised.

Maximum braking is provided, even after fording rivers or driving through deep mud, by dual-line servo-assisted front disc brakes. The provision of a transmission parking brake means that braking effort is applied to all four wheels for maximum holding power on the steepest gradients.

The five-speed synchromesh transmission combined with the permanent four-wheel drive, lockable centre differential and two-speed transfer box, gives the Defender 90 vehicles outstanding mobility and excellent engine braking through all four wheels. Core Military Defenders are available with either the Tdi, V-8 or four-cylinder petrol engine, with 12 volt electrics and right- or left-hand steering.

All vehicles have six-way lighting systems and convoy lights, military bumpers, towing pintles and military chassis with NATO rotating tow hook and 12 pin trailer socket. The two front vinyl seats have head restraints and lap and diagonal seat belts. The front doors are fitted with removable window tops, full length soft tops and roll bars. Options available include FFF fitments and rear seating.

Defender XD 110

The Defender XD 110 was developed to meet the arduous operational requirements specified by the UK Ministry of Defence in considering its replacement TUM fleet into the next century. It successfully completed UK Ministry of Defence exhaustive battlefield simulating trials, was selected for the British Armed Forces requirement in 1996 and now forms part of Land Rover's military derivative range available world-wide.

The eXtra Duty (hence XD) vehicles result from consultations and trials with in-depth engineering analysis to establish and satisfy vehicle needs for enhanced operational requirements well into the 21st century. They have been developed using state-of-the-art computer-aided engineering technology also used for the new Range Rover. They built on the capabilities of the Core Military Defender and offer even more performance, payload carrying, durability, improved ergonomics providing greater comfort, extra safety, and reduced maintenance. They also provide suppression specifications to high levels.

The extra duty off-road capability of the Defender XD is provided by the 300 Tdi engine in combination with the increased robustness of the chassis frame, cab, body structure and major assemblies. It can carry up to 20 per cent additional equipment or payload than the earlier equivalent models.

Variants

Australian Land Rover 110

Land Rover 110s are produced in Australia by Land Rover Australia at their facility in Moorebank, New South Wales. An initial order for 2,500 vehicles for the Australian Army was placed and the first was handed over in August 1987. The 1,000th vehicle was handed over in December 1988.

Australian 110s are powered by an Isuzu 4BD1T 3.9 litre turbocharged diesel engine developing 121 hp at 3,000 rpm. The same engine is used to power the Australian Land Rover 110 Heavy Duty (6 × 6) truck - see separate entry in the *Trucks* section for details of this vehicle.

Special Operations Vehicle (SOV)

See entry in *Special attack vehicles* section.

Specifications

Defender 110
Cab seating: 1 + 2 in front (4-10 in rear)
Configuration: 4 × 4
Weight:
(standard suspension) 3,050 kg
(levelled suspension) 2,950 kg
Max load: 1,486 kg
Towed load:
(on road) up to 4,000 kg
(off road) 500 kg
Length: 4.631 m
Width: 1.79 m
Height: 2.035 m
Cargo bed length: 2.01 m
Ground clearance: 0.216 m
Track: 1.486 m
Wheelbase: 2.794 m
Angle of approach/departure: 50°/35°
Fuel capacity: 79.5 litres
Engine: 3.528 litre V-8 water-cooled petrol developing 134 bhp at 5,000 rpm or
2.495 litre 4-cylinder water-cooled in-line petrol developing 83 bhp at 4,000 rpm or
2.495 litre in-line water-cooled indirect injection naturally aspirated diesel developing 68 bhp at 4,000 rpm or
2.495 litre in-line water-cooled direct injection turbocharged intercooled diesel developing 107 hp at 3,800 rpm
Gearbox: manual with 5 forward and 1 reverse gears
Clutch: diaphragm spring
Transfer box: LT230T, 2-speed
Steering: manual, recirculating ball (power-assist optional)
Turning radius: 6.4 m
Suspension: beam axles front and rear with coil springs controlled by hydraulic telescopic dampers
Tyres: 7.50 × 16
Brakes: dual servo-assisted, discs front and rear
Electrical system: 12, 24 or 12/24 V

Specifications

Core Military Defender 110
Seating: 10
Configuration: 4 × 4
Weight:
(GVW) 3,050 kg
(heavy duty) 3,500 kg
Towed load: up to 4,000 kg
Length: 4.44 m
Width: 1.79 m
Height: 2.04 m
Ground clearance: 0.215 m
Track: 1.49 m
Wheelbase: 2.79 m
Angle of approach/departure: 48°/35°
Fuel capacity: 80 litres
Engine: 2.5 litre direct injection turbocharged water-cooled diesel developing 111 hp at 4,000 rpm
(export only) 3.5 litre water-cooled petrol developing 134 hp at 5,000 rpm
Gearbox: manual with 5 forward and 1 reverse gears
Clutch: diaphragm spring
Transfer box: 2-speed
Steering: manual, recirculating ball (power-assist optional)
Suspension: beam axles front and rear with coil springs controlled by hydraulic telescopic dampers
Brakes: dual servo-assisted, discs front and rear
Electrical system: 12, 24 or 12/24 V

Specifications

Defender XD 110
Seating: 10
Configuration: 4 × 4
Weight: (GVW) 3,344 kg
Max load: 1,200 kg
Towed load: up to 4,000 kg
Length: 4.55 m
Width: 1.79 m
Height: 2.08 m
Ground clearance: 0.23 m
Track:
(front) 1.52 m
(rear) 1.51 m
Wheelbase: 2.79 m
Angle of approach/departure: 50°/35°
Fuel capacity: 82 litres
Engine: 2.5 litre direct injection turbocharged water-cooled diesel developing 111 hp at 4,000 rpm
Gearbox: manual with 5 forward and 1 reverse gears
Clutch: diaphragm spring
Transfer box: 2-speed
Steering: manual, recirculating ball (power-assist optional)

Suspension: beam axles front and rear with coil springs controlled by hydraulic telescopic dampers
Brakes: dual servo-assisted, discs front and rear
Electrical system: 12, 24 or 12/24 V

Status

Defender 110, production complete. In service with the British armed forces and some other nations including Australia, Belgium, Ireland, Netherlands and Turkey.

Core Military Defender 110, in production.
Defender XD 110, in production and entered service with the British armed forces in late 1996.

Manufacturers

Land Rover, Lode Lane, Solihull, West Midlands B92 8NW, UK.
Tel: +44 121 722 2424. Telex: 338641 lanrov g.
Fax: +44 121 742 1927.

Land Rover Australia, PO Box 59, Liverpool, New South Wales 2170, Australia.
Tel: +61 2 600 1333. Telex: AA 120222.
Fax: +61 2 600 8935.
Land Rovers are also assembled or manufactured in Kenya, Malaysia, Morocco, Turkey (by Otokar), Zaïre, Zambia and Zimbabwe.

UPDATED

Land Rover Defender 130 (4 × 4) light vehicle

Development

In 1990, the Land Rover range of vehicles was given the name 'Defender'. Thus the Land Rover 127 was designated the Land Rover Defender 130. As the Land Rover 127 (4 × 4) light vehicle, the Defender 130 was first shown early in 1983 when it was displayed to potential customers as a concept vehicle in crew cab form. Production started during 1984 to supply vehicles to the commercial market but since 1987 the Defender 130 has entered service with a number of military customers. In August 1989 it was announced that 214 Land Rover 127s (now Defender 130s) were to be supplied to the Royal Air Force to support RAF Regiment Rapier surface-to-air missile squadrons.

As with the Land Rover Defender 90 and Defender 110 models (see entries in this section) the Defender 130 is fitted with long travel coil spring suspension and front disc brakes. It has a high degree of commonality with other Land Rover models, including engines, transmissions, brakes, steering, electrical systems and front end body panels.

At the same time as the adoption of the Defender name, a significant package of product improvements was introduced. This included the introduction of the 200 Tdi 2.5 litre direct injection turbocharged intercooled diesel engine with improvements in power, torque and economy over the previous direct injection unit. Other improvements concerned the cab ergonomics and electrical system.

Although designated the Defender 130, this vehicle retains the 127 in (3.226 m) wheelbase.

Description

The Land Rover Defender 130 is an extended chassis variant of the Land Rover Defender 110, with a payload capacity of up to 1,600 kg dependent upon the type of body fitted. It is available either with a crew cab and pick-up rear body or with a chassis cab to which any one of a number of different body styles can be fitted.

The suspension uses long travel coil springs front and rear with suspension movement controlled by telescopic hydraulic dampers. The front and rear beam axles are located by front radius arms and rear trailing links and lateral location is provided by a Panhard rod at the front of the vehicle and a centrally mounted 'A' bracket at the rear.

Two engine options are available for the Defender 130; the 3.5 litre V-8 petrol engine and 2.5 litre direct injection turbocharged intercooled diesel. The Defender 130 uses the same transfer box, five-speed gearbox and permanent four-wheel drive system as the Land Rover Defender 90 and 110.

Among the factory-fitted options available on military versions of the Defender 130 are: air conditioning; raised air intake; open top with folding windscreen and detachable door tops; power steering; range of winches and power take-offs; NATO towing jaw with 12-pin trailer socket; blackout lighting systems; 12, 24 or 12/24 V electrical systems; additional fuel tank; wire mesh lamp guards; disruptive pattern camouflage paint finish; various rear seating options; small arms clips; and external pick and shovel stowage.

The crew cab configuration is suitable for use as a light artillery tractor, housing both crew and equipment. The extra body length enables bulky cargo or missiles to be carried.

A number of variants of the Defender 130 have been developed including mobile workshops, shelter-carrying vehicles and ambulances.

Core Military Defender 130

The Core Military Defender 130 builds on the capabilities of its predecessors to meet the increasing operational requirements of military forces into the 21st century.

Defender XD 130 field ambulance

The Defender XD 130 field ambulance, developed in association with Marshall SPV of Cambridge, was unveiled at the Royal Navy and British Army Equipment Exhibition in September 1995, and was selected as the new field ambulance for the British Armed Forces in 1996. It is intended for world-wide deployment and provides exceptional levels of performance, ride and stability on or off road.

The cab accommodates two crew with easy internal access to the ambulance compartment. This compartment seats a medical attendant and can easily be adapted for up to four stretcher casualties, two stretcher and three seated casualties, or six seated patients.

Comprehensive stowage space is provided for ambulance and medical equipment and supplies, blankets and personal kit. An independent ambulance compartment, heating and forced air ventilation provide a comfortable patient environment. Winterised and tropical variants are available.

In January 1996 Marshall Specialist Vehicles of Cambridge were awarded a £15 million contract to supply bodies for the Defender XD 130 field ambulances.

Specifications

Defender 130
Cab seating: 1 + 2 (1 + 5 in crew cab)
Configuration: 4 × 4
Weight: (GVW) 3,400 kg
Max load: up to 1,600 kg (dependent upon type of body fitted)
Towed load:
(off road) 750 kg
(on road) 4,000 kg
Length: 5.132 m
Width: 1.79 m
Height: (cab roof) 2.073 m
Ground clearance: 0.215 m
Track: 1.486 m
Wheelbase: 3.226 m
Angle of approach/departure: 50°/31°
Engine: 3.528 litre V-8 water-cooled developing 134 hp at 5,000 rpm or
2.495 litre in-line water-cooled direct injection turbocharged intercooled diesel developing 107 hp at 3,800 rpm
Gearbox: manual with 5 forward and 1 reverse gears
Clutch: diaphragm spring
Transfer box: LT230T, 2-speed
Steering: manual, recirculating ball (power-assisted optional)
Turning radius: 7.54 m
Suspension: beam axles front and rear with coil springs controlled by hydraulic telescopic dampers
Tyres: 7.50 × 16
Brakes: dual line servo-assisted, disc front, drums rear
Electrical systems: 12 V, 45/65 A; 24 V with outputs of up to 90 A; or 12/24 V

Specifications

Core Military Defender 130
Seating: 14
Configuration: 4 × 4
Weight: (GVW) 3,500 kg
Towed load: up to 4,000 kg
Length: 5.13 m
Width: 1.79 m
Height: 2.04 m
Ground clearance: 0.215 m
Track: 1.49 m
Wheelbase: 3.23 m
Angle of approach/departure: 48°/35°
Engine: 2.5 litre water-cooled direct injection turbocharged diesel developing 111 hp at 4,000 rpm
(export only) 3.5 litre water-cooled petrol developing 134 hp at 2,500 rpm
Gearbox: manual with 5 forward and 1 reverse gears
Clutch: diaphragm spring
Transfer box: 2-speed
Steering: manual, recirculating ball (power-assisted optional)
Suspension: beam axles front and rear with coil springs controlled by hydraulic telescopic dampers
Brakes: dual line servo-assisted, disc front, drums rear

Specifications

Defender XD 130 Field Ambulance
Cab seating: 2 + patients
Configuration: 4 × 4
Weight: (GVW) 3,741 kg

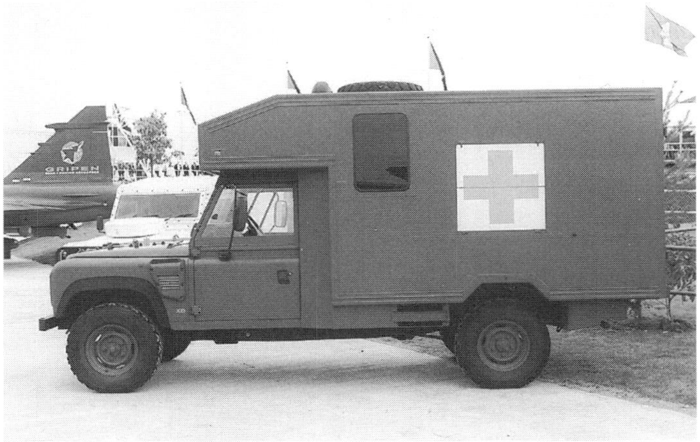

Land Rover Defender XD 130 ambulance (T J Gander) ***1997***

Land Rover Defender 130 (4 × 4) Rapier missile support vehicle ***1996***

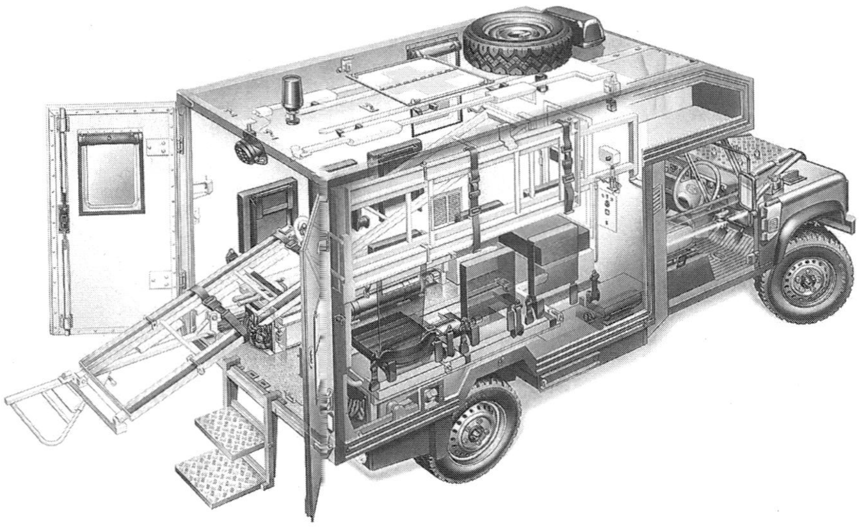

Land Rover Defender XD 130 ambulance **1996**

Max load: 1,200 kg
Length: 5.19 m
Width: 2.09 m
Height: 2.57 m
Ground clearance: 0.23 m
Track:
(front) 1.52 m
(rear) 1.51 m
Wheelbase: 3.23 m
Angle of approach/departure: 50°/33°
Engine: 2.5 litre water-cooled direct injection turbocharged diesel developing 111 hp at 4,000 rpm
Gearbox: manual with 5 forward and 1 reverse gears
Clutch: diaphragm spring
Transfer box: 2-speed
Steering: manual, recirculating ball (power-assisted optional)
Suspension: beam axles front and rear with coil springs controlled by hydraulic telescopic dampers
Brakes: dual line servo-assisted, disc front, drums rear

Status

Defender 130 in service with the Royal Air Force and other armed forces. Licence-produced in Turkey by Otokar.
Core Military Defender 130, in production.
Defender XD 130, in production and began entering service with the British armed forces in late 1996.

Manufacturer

Land Rover, Lode Lane, Solihull, West Midlands B92 8NW, UK.
Tel: +44 121 722 2424. Telex: 338641 lanrov g.
Fax: +44 121 742 1927.

UPDATED

Range Rover (4 × 4) light vehicle

Development

The Range Rover, first announced in June 1970, combines a formidable cross-country capability with the comfort and road performance of a high-speed saloon car. It is ideally suited to a variety of military applications including use as an ambulance, command/radio vehicle, fire tender, personnel carrier and border patrol vehicle. A number of factory approved conversions are available including 6 × 4 configurations equipped for use as fast response emergency units. The four-door variant extended the vehicle's versatility still further, particularly for personnel carrier or security duties where ease of access for all passengers is important.

With the introduction of the new Range Rover in 1994, the existing version became known as the Classic Range Rover. It ceased production in 1995 but remains in service with military, paramilitary and aid organisations worldwide.

The all-new Range Rover resulted from a £300 million investment programme, building on the success of the Classic Range Rover. Range Rovers served as staff and command cars during the 1990-1991 Gulf War and since then with the UN in Bosnia and elsewhere. Among the military and paramilitary applications is a complete police specification variant.

Description

The Range Rover is based on a rigid box-section steel chassis. The rubber-mounted steel body frame is clad in separate body panels, most of which are formed in lightweight, corrosion-resistant aluminium alloy.

From 1990 power was provided by a 3.9 litre V-8 petrol engine with fuel injection developing up to 185 bhp in high compression form. Power is transmitted via a five-speed manual or four-speed automatic gearbox. A turbocharged 2.5 litre Tdi diesel developing 111 bhp is also available. The chain-driven two-speed transfer box features a viscous control unit for the permanent four-wheel drive system to automatically transfer torque away from a slipping axle to the one with more grip. This provides better off-road traction without the need for driver intervention.

The suspension features long travel coil springs to give maximum on-road comfort and good off-road performance. A ride-levelling device provides a good towing capability. A new suspension system was introduced on Vogue and Vogue SE models in 1991. This features front and rear anti-roll bars and revised damper and spring rates to minimise body sway when cornering and preserve off-road abilities and good on-road ride. This suspension is available as an option on other models in the range.

The 1991 Range Rovers also featured a fuel system with an increase in capacity up to 81.8 litres. A modified filler system gives faster fill with reduced splashback.

To reflect the increased emphasis on long distance cruising, a cruise control system was introduced in 1991. This was standard on Vogue SE automatic vehicles and optional on Vogue automatic.

Disc brakes (ventilated at the front) with asbestos-free pads are fitted all round and an anti-lock braking system (ABS) with brake power booster is available as an option (standard on Vogue SE models).

Luxury-standard accommodation is provided for up to five persons, entry being gained through two or four doors. Central door locking is provided. The rear load area is accessed through a two-piece tailgate and may be extended by folding the rear seat squab which is asymmetrically split on four-door models.

Late models featured an improved heating and ventilation system with a heated front window being available. From 1991, a glass sunroof became available to replace the steel version fitted to earlier vehicles; a blind provides screening when necessary. This feature was standard on the Vogue SE and optional on other models. Other improvements introduced in 1991 included an automatic dipping rear mirror on Vogue SE models, heated door locks, improved sound insulation and an uprated stereo system. A range of optional equipment includes a heavy-duty suspension, air conditioning (standard on some models), winches and towing equipment.

In 1993, the extended wheelbase Vogue LSE was introduced, powered by a 4.2 litre catalyst-equipped V-8 engine and offering additional legroom and ride/handling refinements. This model features Electronic Air Suspension (EAS) which provides five alternative ride heights for easy access, reduced road noise within the vehicle and improved off-road capability. EAS is standard on the Vogue SE and Vogue LSE (it is optional on the petrol-engined Vogue). Other models have long travel coil springs and all vehicles are fitted with anti-roll bars front and rear. For long distance cruising, a cruise control system is standard on the Vogue LSE and SE and optional on Vogue petrol variants with both manual and automatic transmission.

Disc brakes (ventilated at the front) are fitted all round and an anti-lock brake system (ABS) with power brake booster is available as an option (it is standard on Vogue LSE and SE models). Electronic traction control (ETC), which offers enhanced traction in severe and off-road conditions, is available on ABS-equipped vehicles

British Army Range Rover serving with IFOR in Bosnia (R Stickland) **1997**

Range Rover Vogue SE

(standard on Vogue LSE and SE and optional on others).

Luxury-standard accommodation is provided for up to five persons. Central door locking is provided and the latest models are available with a security alarm system and remote locking. The rear load area is accessed through a two-piece tailgate and may be extended by folding the rear seat squab which is asymmetrically split.

The new (1995) Range Rover has more headroom, more legroom, a larger rear seat compartment and 50 per cent more luggage space. It is available with three new engines. Diesel versions are powered by a derivative of BMW's 2.5 litre turbocharged and intercooled six-cylinder power unit, specially developed for the Range Rover. Land Rover's V-8 has been totally redesigned and rebuilt and is available in 4 litre and 4.6 litre versions. Standard features include side intrusion rails, electronic air suspension, advanced anti-lock braking, a remotely controlled anti-theft system and, dependent on the market, twin airbags and knee bolsters.

There are three models in the range, starting with the 4 litre V-8 or 2.5 litre diesel version, followed by SE derivatives in petrol or diesel and then the HSE model with the 4.6 litre V-8 engine. The 4 litre petrol and 2.5 litre diesel derivatives are available with a choice of manual transmission or an H-gate electronic automatic gearbox.

Land Rover's Special Vehicles Division produces a number of Range Rover conversions including police and rapid assault vehicles and discreetly armoured specifications for the transport of VIPs and personnel. The discreetly armoured specification features Level 1 ballistic protection, a specially developed heavy-duty suspension and a full range of security options.

Specifications

Cab seating: 1 + 4
Configuration: 4 × 4
Weight: (EEC kerb)
(2.5 litre diesel, manual) 2,115 kg; (automatic) 2,130 kg
(4 litre petrol, manual) 2,090 kg; (automatic) 2,100 kg
(4.6 litre petrol, auto) 2,220 kg
(GVW, all models) 2,780 kg
Weight on front axle: (max) 1,320 kg
Weight on rear axle: (max) 1,840 kg
Length: 4.72 m
Width: 1.89 m
Height: 1.82 m
Ground clearance: 0.21 m
Track:
(front) 1.54 m
(rear) 1.53 m
Wheelbase: 2.75 m
Angle of approach/departure: 38°/26°
Max speed:
(4 litre manual) 190 km/h
(4 litre automatic) 187 km/h
(2.5 litre manual) 170 km/h
(2.5 litre automatic) 162 km/h
(4.6 litre automatic) 200 km/h
Fuel capacity:
(petrol) 100 litres
(diesel) 90 litres
Engine: Land Rover V-8 4 litre petrol with fuel injection developing 190 bhp at 4,750 rpm with catalyst or Land Rover V-8 4.6 litre petrol with fuel injection developing 225 bhp at 4,750 rpm with catalyst or BMW 6-cylinder diesel with turbocharged intercooled indirect injection developing 136 bhp at 4,400 rpm
Gearbox:
(manual) 5 forward and 1 reverse gears
(automatic) 4 forward and 1 reverse gears
Clutch: (manual) single dry plate
Transfer box: 2-speed chain-driven with viscous control unit
Steering: power assisted recirculating ball with safety column and steering damper
Turning radius: 11.39 m
Suspension: Electronic Air Suspension with variable ride height
Air springs, composite radius arms, panhard rod, anti-roll bar and dual action telescopic hydraulic dampers
Brakes:
(main) Electronic 4-channel hydraulic power anti-lock braking system with outboard reverse ventilated discs on front and outboard solid discs at rear
(parking) mechanical on transmission
Electrical system: 12 V

Status

The latest model of the Range Rover has been in production since September 1994.

Manufacturer

Land Rover, Lode Lane, Solihull, West Midlands B92 8NW, UK.
Tel: +44 121 722 2424. Telex: 338641 lanrov g.
Fax: +44 121 742 1927.

UPDATED

Land Rover Discovery (4 × 4) light vehicle

Development

The Land Rover Discovery (4 × 4) light vehicle was first announced in November 1989 and is an all-terrain four-wheel drive vehicle. It was designed to meet the requirements of customers requiring the ability to traverse the most difficult off-road conditions yet retain a degree of style for driving on roads. The vehicle retains all the strength of the other models in the Land Rover range, including the ability to tow fully braked trailers weighing up to 4,000 kg.

The Discovery was launched as a three-door estate; a five-door version was introduced during 1990. The vehicle features the long travel coil spring suspension used on the Range Rover and Defender vehicles and has an all-disc braking system, beam axles and power steering as standard.

The Discovery is suitable for a variety of military and security tasks such as escort vehicle, police patrol vehicle and staff car. It was used extensively by armed forces in command and support roles during the 1990-1991 Gulf War and is being increasingly used in UN aid and peacekeeping roles, notably in Bosnia.

In 1993, the Discovery Commercial (IEDD) was introduced. Specifically developed to meet the needs of explosive device detonation teams, this variant has a spacious van-type body containing stowage racks for tools and equipment, including the remote-control Wheelbarrow Mark 8 and remote monitoring equipment.

Description

The Land Rover Discovery (4 × 4) light vehicle is based on a box-section chassis designed to minimise mud traps and treated with an electrophoretic anti-corrosion process. Exterior body panels are of aluminium alloy to provide a corrosion-resistant finish combined with low weight, enhancing the vehicle's low centre of gravity.

Power is provided by an all-alloy 3.9 litre V-8i fuel injected petrol engine or the 2.5 litre 300 Tdi direct injection turbocharged intercooled diesel engine. Transmission is via a five-speed manual or a four-speed automatic gearbox and a two-speed transfer box with permanent four-wheel drive and a lockable centre differential. Beam axles are coupled to a long travel coil spring suspension all round and the steering has power assistance as standard. Brakes are disc all round with dual circuit and servo-assistance.

The interior, with seating for up to seven, was designed to combine durability, practicality and style. There is internal stowage for maps and documents while entry is by either three or five doors with central locking standard on the five-door version (optional on the three-door). The rear compartment has a large capacity for kit or stores and this may be augmented by lowering the asymmetrically split rear seat. Folding seats in the rear compartment are available on all models.

A range of conversions is available from Land Rover's Special Vehicles Division. Conversions include vehicles for police and paramilitary applications. A range of Discovery paramedic ambulances is also available in standard and extended wheelbase versions with medical and rescue equipment options. The Discovery Commercial features a van-type body.

Specifications

Cab seating: 1 + 4 (2 extra seats available)
Configuration: 4 × 4
Weight: (3-door/5-door)
(unladen, V-8i petrol) 1,919/1,986 kg
(unladen, 300 Tdi petrol) 2,008/2,053 kg
(laden, all models) 2,720 kg
Towed load: (braked) up to 4,000 kg
Length: 4.521 m
Width: 1.793 m
Height: 1.928 m
Ground clearance: 0.241 m
Track: 1.486 m
Wheelbase: 2.54 m
Angle of approach/departure: 41°/32°
Max speed:
(V-8i) 165.7 km/h
(300 Tdi) 147.4 km/h
Fuel capacity: 88.6 litres
Engine: 3.947 litre V-8 diesel with fuel injection developing 180 bhp at 4,750 rpm with catalyst or 300 Tdi 2.495 litre direct injection turbocharged

Land Rover Discovery operating in Saudi Arabia during Operation Granby (Bob Morrison - Military Scene)

Land Rover Discovery Commercial IEDD (4 × 4) light vehicle

diesel with intercooler developing 111 bhp at 4,000 rpm
Gearbox: manual, 5 forward and 1 reverse gears or automatic, 4 forward and 1 reverse gears
Transfer box: 2-speed giving permanent 4-wheel drive with lockable centre differential
Steering: recirculating ball with steering damper and power-assistance
Turning radius: 5.95 m
Suspension: coil springs, hydraulic dampers; axle location by radius arms
Brakes:
(main) hydraulic, disc all round, dual circuit with servo-assistance
(parking) mechanical drum on transmission
Electrical system: 12 V

Status
In production. Ordered for supply to UK Ministry of Defence and UN organisations.

Manufacturer
Land Rover, Lode Lane, Solihull, West Midlands B92 8NW, UK.
Tel: +44 121 722 2424. Telex: 338641 lanrov g.
Fax: +44 121 742 1927.

UPDATED

Land Rover (4 × 4) 564 kg air-portable light vehicle

Development
This vehicle was designed and developed by Rover in conjunction with the Military Vehicles and Engineering Establishment to meet the special air-portable requirements of the British Army, Royal Air Force and Royal Marines. It was shown for the first time at the Commercial Motor Show at Earls Court in September 1968. The vehicle is in service as a light 4 × 4 vehicle with the British forces but is progressively being replaced by the Land Rover Defender 90 (see previous entry). The engine, transmission, axles, suspension and brakes were from the commercial 88 in (2.23 m) Land Rover modified for military use.

Production was phased out during 1985.

The current Core Military Defender 90 and Defender XD 90 are also air-portable (see previous entry).

Description
The layout of the vehicle is similar to other Land Rovers, with the engine at the front, driver and two passengers in the centre and cargo area at the rear, with a drop tailgate and a bench seat down either side.

The main feature of this vehicle is that the hood, body sides, doors, windscreen, bumpers and spare wheel can easily be removed to facilitate transport by aircraft or helicopter.

The basic model has a 12 V electrical system but a 24 V Fitted-For-Radio (FFR) version is available with the two batteries installed between the front seats in place of the third seat in the 12 V model. This model features a unitary radio kit containing radio, batteries, operator's seat and radio mounting equipment installed in the rear of the vehicle.

Land Rover developed special models of the ½ tonne Land Rover to meet specific overseas requirements, a vehicle developed for the Netherlands Army, with a diesel engine (56 bhp at 4,000 rpm) and 24 V 60 A electrical system, being typical.

Ambulance
This version can carry two stretchers in an emergency.

Anti-tank
To meet the requirements of a foreign army, Marshall of Cambridge (Engineering) Ltd designed a 106 mm Land Rover gun vehicle which entered service in 1977. This is the standard vehicle modified to mount a 106 mm M40 recoilless rifle in the rear of the vehicle. The design permits forward arc of fire of 180°. Seats for the crew of the weapon and stowage of 106 mm and 12.7 mm/0.50 in ranging machine gun ammunition are provided. A blast shield is fitted to the bonnet and a barrel clamp is fitted to the dashboard. Known users of this model include Saudi Arabia and Sudan.

Specifications
Cab seating: 1 + 2
Configuration: 4 × 4
Weight:
(empty) 1,386 kg
(stripped down) 1,206 kg
(loaded) 2,020 kg
Max load: 564 kg
Towed load: 1,130 kg
Load area: 1.14 × 1.4 m
Length: 3.65 m
Width: 1.52 m
Height:
(overall) 1.95 m
(reduced) 1.47 m
(load area) 0.71 m
Ground clearance: 0.21 m
Track: 1.31 m
Wheelbase: 2.23 m
Angle of approach/departure: 49°/36° (58°/38° in stripped down form)
Max speed: (road) 105 km/h
Range: 600 km
Fuel capacity: 90 litres
Max gradient:
(high) 39%
(low) 115%
Fording: 0.5 m
Engine: 4-cylinder 2.286 litre in-line OHV petrol developing 70 bhp at 4,000 rpm or 4-cylinder 2.286 litre in-line diesel developing 60 bhp at 4,000 rpm
Gearbox: manual with 4 forward and 1 reverse gears
Clutch: single dry plate
Transfer box: 2-speed
Steering: recirculating ball
Turning radius: 6.4 m
Suspension: semi-elliptical springs with hydraulic shock-absorbers
Tyres: 6.50 × 16 or 7.50 × 16
Brakes:
(main) drum, hydraulic oval line, servo-assisted
(parking) drum, mechanical on transmission
Electrical system: 12 V (24 V optional)
Battery: 1 × 12 V

Status
Production complete. In service with Belgium, Brunei, Guyana, Hong Kong, Indonesia (812), Jamaica, Libya, Netherlands (24 V and a diesel engine), Saudi Arabia, Sudan and the UK.

Manufacturer
Land Rover, Lode Lane, Solihull, West Midlands B92 8NW, UK.
Tel: +44 121 722 2424. Telex: 338641 lanrov g.
Fax: +44 121 742 1927.

VERIFIED

Netherlands Army air-portable (4 × 4) 564 kg Land Rover

Royal Air Force air-portable (4 × 4) 564 kg Land Rover with hard top body (Bob Morrison - Military Scene)

Land Rover (4 × 4) 1,000 kg truck

Development
The 101 in (2.57 m) wheelbase forward control Land Rover was developed by the Rover Company in co-operation with the Military Vehicles and Engineering Establishment to meet a British Army requirement for a 1,000 kg payload 4 × 4 vehicle with the added capacity of towing a 1,500 kg powered axle trailer. The vehicle was first shown to the public in September 1968. A pre-production run of vehicles was completed in 1972 and the vehicle entered service with the British Army in 1975. British Army vehicles are being phased out of service.

Description
The vehicle is of the forward control type with the engine at the front and the cargo area at the rear. The chassis is of steel frame construction with two U-sections overlapped and welded on the vertical faces, full length flat-topped side members and intermediate support bearing for rear PTO on bolt-in

RAF Land Rover (4 × 4) 1,000 kg truck (Michael Jerchel) ***1996***

cross members between the third and fourth cross members.

The 1,000 kg Land Rover is powered by a military version of the successful lightweight aluminium Rover 3.5 litre V-8 engine, which is basically the same as the car engine but with a reduced compression ratio of 8.5:1 to enable it to operate on low octane fuels.

The four-speed all-synchromesh Range Rover type transmission with a high and low transfer box provides eight forward and two reverse gears. A third differential between the front and rear axles obviates transmission wind-up associated with four-wheel drive transmission. This differential can be locked by a vacuum-actuated control to provide maximum traction through both axles. The front axle drive provides a 35° steering lock through constant velocity joints.

The complete hood, body sides, windscreen, bumpers and spare wheel can be removed from the vehicle for air transport. The vehicle can be lifted by medium lift helicopters such as the Puma.

Tool lockers are provided in the rear body behind the wheel arches. As a result of trials with the preproduction run, production vehicles have an anti-roll bar at the front, mud flaps and a folding rear entry step.

Standard equipment includes rear bumperettes, helicopter lifting rings front and rear, rotating towing hooks front and rear and trailer brakes, operated from the towing vehicle using single line vacuum system on foot and handbrake.

The British Army uses this vehicle for a wide range of roles including towing the 105 mm Light Gun and carrying MILAN ATGW firing posts and 81 mm mortars. The Australian and British armies have used the vehicle to tow the Rapier SAM system.

In 1975 the Royal Luxembourg Army ordered a number of vehicles with a 24 V electrical system (FFR), a cable drum winch with front or rear pull and a fresh air heater to cab and rear body.

Variants

Ambulance

In July 1976 Marshall of Cambridge (Engineering) Limited received a contract from the UK Ministry of Defence for a two- to four-stretcher ambulance version of the Land Rover. This was built and designed to meet an MoD requirement for an improved pattern lightweight ambulance for casualty evacuation from forward areas, able to carry four stretcher or six sitting patients. The aluminium alloy thermally insulated body provided a high standard of patient comfort while a new pattern stretcher gear simplified loading and unloading in the upper position. Additional equipment which may be fitted includes an Eberspacher heater, an air conditioner and automatic resuscitation equipment.

This ambulance served with the UK United Nations contingent in Bosnia, although from 1993 onwards it was replaced by the Defender 130 ambulance.

MILAN carrier

The British Army used this vehicle to carry MILAN ATGW teams. Each vehicle carried six personnel, two MILAN firing posts and 14 missiles.

Communications vehicle

The British Army had a number of these vehicles fitted with a hardtop and carrying various forms of communications or electronic warfare equipment.

Anti-tank

The Egyptian Army uses this vehicle fitted with the British Aerospace Swingfire ATGW system. Four missiles carried in the rear of the vehicle can be launched from the vehicle with the aid of a separation sight and controller.

Specifications

Cab seating: 1 + 1 (up to 8 in the rear)
Configuration: 4 × 4
Weight:
(empty) 1,924 kg
(loaded) 3,120 kg
(stripped) 1,580 kg
Weight on front axle: (loaded) 1,608 kg
Weight on rear axle: (loaded) 1,512 kg
Max load: 1,000 kg
Load area: 2.49 × 1.727 m
Length: 4.127 m
Width: 1.842 m
Height: 2.138 m
Ground clearance: 0.254 m
Track:
(front) 1.524 m
(rear) 1.549 m
Wheelbase: 2.565 m
Angle of approach/departure: 50°/46°
Max speed: 120 km/h
Range: 560 km
Fuel capacity: 109 litres
Max gradient: 60%
Engine: Rover 3.5 litre V-8 petrol developing 128 bhp at 5,000 rpm
Gearbox: manual with 4 forward and 1 reverse gears
Clutch: single dry plate
Transfer box: 2-speed with permanent 4-wheel drive
Steering: recirculating ball with steering damper
Turning radius: 5.14 m
Suspension: semi-elliptical taper leaf springs with telescopic hydraulic dampers, anti-roll bar at front
Tyres: 9.00 × 16
Brakes:
(main) hydraulic, split system, servo-assisted
(parking) drum, mechanical on transmission
Electrical system: 12 V (24 V optional)

Status

Production complete. In service with Australia, Royal Air Force (being replaced by Land Rover Defender 130), Egypt, Iran and Luxembourg (57).

Manufacturer

Land Rover, Lode Lane, Solihull, West Midlands B92 8NW, UK.
Tel: +44 121 722 2424. Telex: 338641 lanrov g.
Fax: +44 121 742 1927.

VERIFIED

UNITED STATES OF AMERICA

M151 (4 × 4) 362 kg light vehicle and variants

Development

In 1950, the Continental Army Command placed a requirement with the Ordnance Corps for a new ¼ ton vehicle. Research and development began that year at the Ordnance Tank Automotive Command. In 1951, the Ford Motor Company was awarded a development contract for the new vehicle and the first prototypes were completed in 1952. A further batch was completed in 1954 under the designation XM151 and in 1956 the XM151E1 (of conventional steel construction) and the XM151E2 (of aluminium construction) were built and tested. The former was selected for production in 1959; Ford was awarded the first production contract and first production vehicles were completed in 1960 at Ford's Highland Park Plant. The M151 replaced the M38 (4 × 4) ¼ ton vehicle in service with the US armed forces. Later production contracts were awarded to the AM General Corporation. The

M718A1 (4 × 4) ambulance

Portuguese Army M151A1 (4 × 4) ¼ ton light vehicle (Bob Morrison - Military Scene)

Fiscal Year 1978 request was for only 3,880 vehicles at a cost of $29.1 million. It was then stated that the engine of the vehicle no longer met emission standards and that this procurement would be the last for several years. Production ceased in 1985 only to start again in 1988 to supply vehicles for overseas sales to Pakistan and other nations.

Production of the M151 series has ceased in the USA. Nearly all US armed forces M151 models were replaced by the HMMWV series.

From 1994 onwards TRACE SA of Cascais, Portugal and Sao Paulo, Brazil, commenced the marketing of a M151A2 clone powered by a Renault J89-240 2-litre diesel engine and known as the TRACE Comando Mk 1.

Description

The body and chassis of the M151 are integral and are of all-welded construction. The layout of the vehicle is conventional, with the engine at the front and the crew area at the rear with a removable canvas top and side curtains and a windscreen that can be folded forward flat on the bonnet.

In 1964 the M151 was replaced in production by the M151A1 which had improved suspension as the earlier model had a tendency to oversteer. The M151A2 followed in 1970 with modifications to the rear suspension, dual brake system, collapsible steering wheel, two-speed wipers and modified lighting. The M151A2LC has a gearbox with three forward and one reverse gears (other vehicles have four forward and one reverse gears) and two-speed transfer case (other vehicles have a single-speed transfer case). The vehicle also has a different suspension which consists of semi-elliptical springs and shock-absorbers instead of the coil springs and telescopic shock-absorbers fitted to other versions.

The basic vehicle can also be armed with a pintle-mounted 7.62 mm M60 or a 0.50/12.7 mm M2 HB Browning machine gun and many have been fitted with the Hughes TOW ATGW system which replaced the 106 mm recoilless rifle.

Many kits are available including a front-mounted winch, heater, hardtop, 100 A alternator and a Xenon searchlight. The vehicle can ford to a depth of 0.533 m without preparation and 1.524 m with the aid of a kit.

From October 1986 a programme was initiated to retrofit rollbars to about 11,000 M151 series vehicles. Most of the vehicles involved belong to National Guard and Army Reserve units. A contract to provide kits for 6,000 vehicles was awarded to the Wright Automotive Corporation of Colorado Springs, Colorado in August 1987. The contract was worth just over $4 million.

M151A2 (4 × 4) ¼ ton light vehicle (T J Gander)

Specifications

Model	M151	M151A1	M151A2	M718A1	M825
Cab seating	1 + 3	1 + 3	1 + 3	1 + 1	1 + 3
Configuration	4 × 4	4 × 4	4 × 4	4 × 4	4 × 4
Weight (empty)	1,012 kg	1,088 kg	1,107 kg	1,247 kg	1,174 kg
(loaded)	1,575 kg	1,633 kg	1,652 kg	1,656 kg	1,959 kg
Weight on front axle (empty)	574 kg	607 kg	618 kg	620 kg	655 kg
Weight on rear axle (loaded)	448 kg	481 kg	490 kg	626 kg	519 kg
Max load (road)	554 kg	544 kg	545 kg	409 kg	785 kg
(cross-country)	362 kg	362 kg	363 kg	409 kg	785 kg
Towed load (road)	970 kg	970 kg	970 kg	not authorised	not authorised
(cross-country)	680 kg	680 kg	680 kg	not authorised	not authorised
Length	3.352 m	3.371 m	3.371 m	3.631 m	3.645 m
Width	1.58 m	1.634 m	1.633 m	1.819 m	1.943 m
Height (overall)	1.803 m	1.803 m	1.803 m	1.94 m	1.621 m
(reduced)	1.332 m	1.332 m	1.333 m	1.313 m	1.346 m
Ground clearance	0.26 m	0.24 m	0.24 m	0.23 m	0.21 m
Track	1.346 m	1.346 m	1.346 m	1.346 m	1.346 m
Wheelbase	2.159 m	2.159 m	2.159 m	2.159 m	2.159 m
Angle of approach/departure	66°/37°	65°/37°	66°/37°	67°/36°	62°/33°
Max speed	106 km/h	104 km/h	90 km/h	90 km/h	80 km/h
Range	482 km	482 km	482 km	442 km	483 km
Fuel capacity	56 litres	56 litres	59.8 litres	59.8 litres	59.8 litres
Max gradient	60%	60%	60%	60%	50%
Max side slope	40%	40%	40%	40%	30%
Fording (without preparation)	0.533 m	0.533 m	0.533 m	0.533 m	0.533 m
(with preparation)	1.524 m	1.524 m	1.524 m	1.524 m	1.524 m
Engine	L-142 4-cylinder liquid-cooled OHV petrol developing 72 hp at 4,000 rpm				
Gearbox	manual with 4 forward and 1 reverse gears				
Clutch	single dry disc				
Transfer box	single-speed, integral with transmission. Driver can select either 4 × 4 or 4 × 2 drive				
Steering	worm and double roller				
Turning radius	5.486 m	5.638 m	5.638 m	5.638 m	5.638 m
Suspension	coil springs with hydraulic shock-absorbers				
Tyres	7.00 × 16	7.00 × 16	7.00 × 16	7.00 × 16	7.00 × 16
Brakes (main)	hydraulic	hydraulic	hydraulic	hydraulic	hydraulic
(parking)	mechanical	mechanical	mechanical	mechanical	mechanical
Electrical system	24 V	24 V	24 V	24 V	24 V
Battery	1 × 2 HN	1 × 2 HN	1 × 2 HN	1 × 2 HN	1 × 2 HN

Variants

The M107 and M108 are communications vehicles with the radios installed in the rear of the vehicle. The passenger seat is rear facing enabling the radio operator to work his equipment.

The M718 and M718A1 are ambulance versions and have a crew of two, a driver and a medical attendant. This version can carry a stretcher and three seated patients, two stretcher and two seated patients or three stretcher patients.

The M825 is fitted with the M40 106 mm recoilless rifle mounted in the rear. This model was replaced in US Army service by the Hughes TOW ATGW, with kits available to convert the M825 to the M151A2 standard. However, during mid-1987 25 M151A2s were converted to M825 standard by TACOM in order to meet a requirement from Somalia.

The UK company, A F Budge (Sales) Limited of Retford, Nottinghamshire, proposed a diesel engine retrofit package for the M151A1/A2 series in which the original engine is replaced by a Perkins Prima 80T turbocharged 2 litre diesel unit developing 80 hp at 4,500 rpm. This would result in an increase in range, reduced fuel consumption and improved reliability.

In the US, Cummins Military Systems proposed a retrofit programme for M151 series vehicles in which the engine would be replaced by a Cummins B3.9-105 diesel engine developing 105 hp at 2,500 rpm. This would be coupled to a Chrysler A 727 automatic transmission to provide improved fuel economy, increased range and ease of maintenance.

Status

Production complete. In service with the US Army, Air Force and Marines (although scheduled for replacement by the HMMWV series) and over 100 other countries including Ecuador (including M825), Egypt, Gambia, Greece, Indonesia, South Korea (141), Netherlands, Pakistan, Peru, Philippines, Portugal, Saudi Arabia, Senegal, Singapore (including M825), Somalia (M151 and M825), Spain (marines only), Thailand, Turkey, Venezuela (including M825) and Zaïre.

By early 1980 AMG had built 95,000 M151 series. The company held the production contract to supply M151A2 series vehicles on a 'requirement' basis.

Manufacturer

AM General Corporation, 14250 Plymouth Road, Detroit, Michigan 48232, USA.
Tel: +1 313 493 3300. Telex: 023 5652.

UPDATED

Jeep (4 × 4) light vehicle

Development

The original Jeep (general purpose) vehicle was developed and manufactured for the US Army by Willys Overland. In 1963 Willys Overland became Kaiser Jeep which, in 1970, became the Jeep Corporation, a subsidiary of American Motors Corporation. The Jeep Corporation manufactures the XJ range of compact 4 × 4 utility vehicles (see next entry) and the YJ range of small 4 × 4 utility vehicles, replacing the CJ range.

The CJ-5, CJ-6, CJ-7 and CJ-8 are no longer produced by Jeep Corporation but are in widespread use around the world. The latest model is the YJ, commercial name Wrangler, a successor to the CJ series. Militarised versions of the long wheelbase version (YJ-L) commenced production in Egypt during 1988.

Jeep Corporation has manufacturing or assembly licensees in Australia, China (Beijing Jeep Corporation), Egypt (see separate entry), India (by Mahindra and Mahindra, see separate entry), Indonesia, Mexico, Pakistan and Venezuela.

Description

The layout of the YJ series is conventional, with the engine at the front, individual seats for the driver and one passenger and folding transverse benches or two longitudinal seats in the rear. A vertically hinged tailgate provides access to the cargo area.

Automatic and manual four-speed and five-speed transmissions are available for the YJ series with a four-wheel drive system. A four-cylinder petrol engine is standard, with a six-cylinder petrol engine as an available option.

A wide range of optional and heavy-duty equipment is available including air conditioning, soft or hardtops, heavy-duty cooling, battery, tyres and suspension and power steering. The military versions of the YJ are available with additional equipment including military wrap springs, lifting eyes, tow hooks and pintle hook, a blackout lighting system, 24 V auxiliary power and severe-use oil, air and fuel filtration systems.

In the US, Cummins Military Systems has proposed a retrofit programme for Jeep series vehicles in which the engine would be replaced by a Cummins B3.9-105 diesel engine developing 105 hp at 2,500 rpm. This would be coupled to a Chrysler A 727 automatic transmission to provide improved fuel economy, increased range and ease of maintenance.

Specifications

Model	**YJ regular wheelbase**	**YJ long wheelbase**
Cab seating	1 + 5	1 + 5
Configuration	4 × 4	4 × 4
Weight (empty)	1,316 kg	1,471 kg
(loaded)	1,882 kg	2,200 kg
Max load	566 kg	729 kg
Towed load	907 kg	907 kg
Length	3.862 m	4.49 m
Width	1.676 m	1.676 m
Height (overall)		
(open body)	1.744 m	1.744 m
(soft top)	1.828 m	1.828 m
(hard top)	1.76 m	1.785 m
Ground clearance	0.207 m	0.207 m
Track	1.473 m	1.473 m
Wheelbase	2.373 m	2.627 m
Angle of approach/departure	31°/32°	32°/22°
Fuel capacity	76 litres	76 litres
Engine	2.5 litre 4-cylinder petrol developing 104 hp at 5,000 rpm or 4.2 litre 6-cylinder petrol developing 112 hp at 3,000 rpm	
Gearbox	manual with 4 forward and 1 reverse gears	
Transfer box	2-speed	2-speed
Clutch	single dry plate	single dry plate
Steering	recirculating ball (power steering optional)	
Turning radius	10.3 m	10.9 m
Suspension	semi-elliptical leaf springs with hydraulic shock-absorbers	
Brakes	hydraulic dual-circuit disc, front; drum rear (both power-assisted)	
Electrical system	12 V (24 V optional)	

YJ-L long wheelbase version of YJ (4 × 4) series

Status

Production of commercial YJ-S (Wrangler) began in February 1986. The YJ series is under evaluation by China and Pakistan.

Manufacturer

Jeep Corporation, International Operations, 27777 Franklin Road, Southfield, Michigan 48034, USA. Telex: 023 1255.

VERIFIED

Jeep XJ (4 × 4) light truck/wagon series

Description

The XJ utility/truck series (Cherokee, Wagoneer and Comanche pick-up truck) was introduced by Jeep Corporation in 1984 (wagon) and 1986 (truck). The Cherokee and Wagoneer are five-passenger wagons, capable of towing up to 2,268 kg. The Comanche is a 1,000 kg payload pick-up truck with a seating capacity for up to 13 when equipped with longitudinal rear seats in a rear cabin.

Automatic and manual four-speed and five-speed transmissions are available with a choice of 4 × 4 or 4 × 2 drive or permanent 4 × 4 (Cherokee only). Engines include a 2.5 or 4 litre petrol or a 2.1 litre turbocharged diesel.

The Cherokee is produced in China by the Beijing Jeep Corporation. For details see separate entry under China in this section.

Specifications

Model	**XJ Wagon (Cherokee)**	**XJ Truck (Comanche)**
Cab seating	1 + 4	1 + 2
Configuration	4 × 4	4 × 4
Weight (empty)	1,363 kg	1,464 kg
(loaded)	2,213 kg	2,464 kg
Max load	750 kg	1,000 kg
Towed load	2,268 kg	2,268 kg
Length	4.2 m	4.928 m
Width	1.79 m	1.822 m
Height (hardtop)	1.63 m	1.666 m
Ground clearance	0.21 m	0.213 m
Track	1.473 m	1.473 m
Wheelbase	2.576 m	3.033 m
Angle of approach/departure	42°/32°	44°/27°
Fuel capacity	76 litres	76 litres
Engine	2.5 litre 4-cylinder petrol developing 104 hp at 5,000 rpm or 4 litre 6-cylinder petrol developing 173 hp at 4,500 rpm or 2.1 litre 4-cylinder turbocharged diesel developing 85 hp at 3,750 rpm	

Jeep Corporation XJ (Cherokee) utility truck

Jeep Corporation XJ (Comanche) pick-up truck

Model	**XJ Wagon (Cherokee)**	**XJ Truck (Comanche)**
Gearbox	manual with 4 forward and 1 reverse gears	
Transfer box	2-speed	2-speed
Clutch	single dry plate	single dry plate
Steering	recirculating ball (power steering optional)	
Turning radius	10.9 m	12 m

Model	**XJ Wagon (Cherokee)**	**XJ Truck (Comanche)**
Suspension	coil springs, front, semi-elliptical leaf springs rear, all with hydraulic shock-absorbers	
Brakes	hydraulic dual circuit disc, front; drum rear (both power-assisted)	
Electrical system	12 V	12 V

Status
In production. Military versions under evaluation by People's Liberation Army (China).

Manufacturer
Jeep Corporation, International Operations, 27777 Franklin Road, Southfield, Michigan 48034, USA.
Telex: 023 1255.

VERIFIED

R-1 Rescue All-Terrain Transport (RATT)

Description
The R-1 Rescue All-Terrain Transport (RATT) was developed by RACECO as an emergency medical evacuation vehicle. The prototype contract was issued by the US Air Force Special Tactics Squadron in early 1991 with the prototype being delivered some six months later. After tests, a contract for a further 14 vehicles was placed in January 1992. All vehicles have been delivered, replacing the M151 vehicles then in use.

The R-1 chassis is built of aircraft-quality high-strength steel tubing welded into a 'trussed bridge' structure. Secondary structures are made of high-strength aluminium alloys or aircraft-quality plastics. A thick aluminium alloy plate provides a smooth bottom skid plate for the entire length of the vehicle. The suspension employs multiple long-stroke shock-absorbers coupled with torsion and coil springs. Each wheel station has a tie-down strap to reduce the overall vehicle height by approximately 100 mm for transport in some types of aircraft.

Various engines can be installed, although the in-service R-1s are powered by a rear-mounted 110 hp Type IV Porsche 914 petrol unit coupled to an automatic transmission with 4 × 4 drive and power steering. The wheels use run-flat tyres on the rear and smaller tyres at the front. A spare is carried under the rear platform to be used on any wheel station in an emergency. In addition the R-1 has an HMMWV bolt pattern to allow the use of HMMWV wheels as emergency spares.

The R-1 can carry a crew of three (driver and two attendants) and six medical litters, with options for a further two litters. The litters are carried on folding racks on the side and rear of the vehicle with the central area available for the two medical attendants to treat casualties while on the move. An oxygen generator is provided to produce and distribute oxygen to all six patients and each patient can be treated with plasma and medical equipment during transit. Lighting, navigation and communications equipment is provided for night operations.

Other equipment carried includes a removable 2,268 kg winch capable of operating to the front, sides or rear.

Specifications
Crew: 3 + 6 patients
Configuration: 4 × 4
Weight:
(kerb) 1,270 kg
(loaded) 2,132 kg
(max, gross) 2,631 kg
Length: 3.86 m
Width: 1.8 m
Height:
(max) 1.93 m
(compressed) 1.8 m
Ground clearance: 0.4 m
Wheelbase: 2.667 m
Angle of approach/departure: 75°/60°
Max speed: 112 km/h
Range: 402 km
Fuel capacity: 75.7 litres
Fording: 0.76 m
Gradient: 60%
Side slope: 60%
Vertical obstacle: 0.406 m
Engine: Type IV Porsche 914 2.3 litre petrol developing 110 hp
Transmission: automatic, 3 forward and 1 reverse gears
Steering: Saginaw, power-assisted
Suspension:
(front) A-arm with or without stage shock and spring
(rear) swing arm torsion bar with or without stage secondary spring
Electrical system: 12 and 24 V

Status
In service with the US Air Force.

Manufacturer
RACECO/HSMV LP, 11611 San Vicente Blvd, Suite 615, Los Angeles, California 90049, USA.
Tel: +1 818 710 9340. Fax: +1 310 821 8261

VERIFIED

R-1 Rescue All-Terrain Transport (RATT) (Scott Gourley)

R-1 Rescue All-Terrain Transport (RATT) (Scott Gourley)

SPECIAL ATTACK VEHICLES

FRANCE

Giat Buggy VRA (4 × 2) fast attack vehicle

Description

The Buggy VRA (4 × 2) fast attack vehicle is produced by Giat Industries and is a development of a US Aureco design for which Giat Industries acquired a licence in 1992. The Buggy VRA is claimed to offer exceptional all-terrain high-speed performance coupled with a low profile and the ability to mount a powerful armament. Intended roles include reconnaissance, air defence, use by special forces, deep attack missions, raiding and other such missions.

The Buggy VRA is based on a chromium-molybdenum tubular chassis frame powered by a 2.6 litre multifuel or petrol engine coupled to a semi-automatic gearbox. Maximum possible speed is 140 km/h. The front suspension has two coil springs and two shock-absorbers on each wheel. At the rear each wheel suspension involves a torsion bar, a secondary torsion bar and four shock-absorbers; suspension travel is 580 mm.

The vehicle seats two or three personnel and has a rear platform to carry mission equipment and supplies. Baskets along each side of the vehicle can carry further stores. The tubular frame has lifting or lashing rings which may be used for helicopter transport.

Equipment carried on the vehicle includes blackout lights, aerial and radio brackets, a ground positioning system, an electric winch, a fire extinguisher, a toolkit and fuel and water cans.

Armament carried can vary but usually involves a 7.62 mm machine mounted at the front. Above the centrally located rear seat is a pivot mounting for a 7.62 or 12.7 mm machine gun or a 12.7 mm machine gun on a rotary support mounting, a 40 mm grenade launcher, and air defence or anti-tank missiles.

A 4 × 4 version of this vehicle is available, as well as a diesel-engined version.

Specifications

Seating: 1 + 1 or 2
Configuration: 4 × 2
Weight:
(empty) 1,200 kg
(combat) 2,000 kg
Payload: 800 kg
Length: 4.45 m
Width: 2.1 m
Height: 1.55 m
Ground clearance: 0.43 m
Track:
(front) 1.88 m
(rear) 1.84 m
Wheelbase: 3.05 m
Angle of approach/departure: 75°/60°
Max speed: 140 km/h
Fuel capacity: 80 litres
Max gradient: 70%
Side slope: 60%
Fording: 0.91 m
Engine: 2.6 litre air-cooled multifuel or petrol unit developing 140 hp at 6,000 rpm
Gearbox: semi-automatic with 3 forward and 1 reverse gears
Clutch: single dry plate
Steering: power-assisted
Turning radius: 6.5 m
Suspension:
(front, each wheel) 2 coil springs and 2 shock-absorbers
(rear, each wheel) 1 torsion bar, 1 secondary torsion bar, with 4 shock-absorbers
Tyres:
(front) 7.50 × 16
(rear) 31 × 10.5 × 15
Brakes: front and rear independent systems; discs front and rear
Electrical system: 12 V (24 V optional)
Battery: 1 × 12 V

Status

Available.

Manufacturer

Giat Industries, 13 route de la Minière, F-78034 Versailles Cedex, France.
Tel: +33 1 30 97 36 48. Fax: +33 1 30 97 39 58.

VERIFIED

Buggy VRA (4 × 2) fast attack vehicle

Buggy VRA (4 × 2) fast attack vehicle

Auverland Type A3F (4 × 4) fast attack vehicle

Description

The Auverland Type A3F (4 × 4) fast attack vehicle was first shown publicly in June 1994. The vehicle was developed in response to a French Army requirement for an airborne light strike vehicle (VEHA) and a similar logistic/combat variant (VLA/VCA).

The Type A3F is a development of the Auverland Type A3 (4 × 4) light vehicle (see entry under France in the *Lightweight Vehicles* section) modified with a helical spring suspension, revised axles and drive shaft arrangement, and a frame superstructure. The basic A3F has a short rear platform while the A3F SL has a longer rear platform. Various weapon mountings can be carried, including MILAN and similar anti-tank missiles. The usual seating arrangement is two forward seats for the driver (on the left or right, as required) and one passenger with a further central seat behind them for one further passenger or weapon operator.

Specifications

Seating: 1 + 2
Configuration: 4 × 2 or 4 × 4
Weight:
(A3F) 2,100 kg
(A3F SL) 2,300 kg

Auverland Type A3F (4 × 4) fast attack vehicle (T J Gander)

Payload:
(A3F) 760 kg
(A3F SL) 960 kg
Length: 3.4 m
Width: 1.58 m
Height: approx 1.7 m
Ground clearance: 0.235 m
Wheelbase: 2.25 m
Max speed: 140 km/h
Range: 800 km
Engine: Peugeot XUD 11 ATE turbo diesel developing 110 hp
Gearbox: Peugeot with 5 forward and 1 reverse gears
Transfer box: Auverland A-80 2-speed
Suspension: coil spring
Electrical system: 12 V

Status
Ready for production.

Manufacturer
Auverland SA, BP 12, F-42260 Saint-Germain-Laval, France.
Tel: +33 4 77 65 54 44. Telex: 310663 f.
Fax: +33 4 77 65 49 18.

UPDATED

HUNGARY

Szöcske light strike vehicle

Description
The Szöcske Light Strike Vehicle may be produced in either a 4 × 2 or 4 × 4 drive configuration and with a choice of air- or water-cooled engines. The general layout of the vehicle appears to have been influenced by design trends elsewhere, consisting of a low open platform construction built onto a tubular frame well provided with roll-over bars. There are lockable differentials on the drive shafts and on the 4 × 4 version there is a central differential gear with a hydraulic clutch between the two axles. The front suspension has double cross swinging arms while the rear suspension has a single longitudinal swinging arm.

Seating is two at the front and two at the rear. The rear portion of the vehicle has a small open load-carrying area and there is provision for further stowage along the sides.

The Szöcske roll-over frame can be used to mount several types of weapon, including missile launchers or recoilless guns of various calibres. The vehicle can also be used to tow a trailer or light weapons such as automatic mortars. An electrically driven winch is standard.

Szöcske light strike vehicle ***1997***

Specifications
Seating: 1 + 3
Configuration: 4 × 2 or 4 × 4
Weight: 1,000 or 1,100 kg
Payload: 500 kg
Length: 3.8 m
Width: 1.8 m
Height: 1.5 m
Ground clearance: 0.3 m
Wheelbase: 2.48 m
Angle of approach/departure: 80°/35°
Max speed: 130 km/h
Range: (cruising) 600 km/h
Max gradient: 80%
Fording: 0.4 m
Engine: 1.9 litre air-cooled petrol developing 87 hp or 1.9 litre water-cooled petrol developing 77 hp
Gearbox: (4 × 4) manual with 4 forward plus 1 climbing gear
Clutch: hydraulic
Steering: rack and pinion
Suspension:
(front) double cross swinging arms
(rear) single longitudinal swinging arm
Tyres: 215/75 R 16
Brakes: dual circuit; discs front, drums rear
Electrical system: 12 V

Status
Available.

Manufacturer
GÉPFET Kft, Hofherr A u 3-15, H-1195 Budapest, Hungary.
Tel: +36 1 177 7266. Fax: +36 1 282 9795.

UPDATED

ROMANIA

Turbomecanica special purpose off-road vehicle

Description
The Turbomecanica special purpose off-road vehicle is intended to fulfil a wide range of military and other roles where high mobility is required. The vehicle is based around a welded steel tube structure with the 1.3 litre air-cooled, four-cylinder petrol engine at the rear. Seating is provided for the driver and up to three passengers although the rear two seats can be removed to increase cargo-carrying space. The main cargo carrying area is behind the rear passenger seats, while more space could be provided in the area normally occupied by a spare wheel.

The information provided in the Specifications table refers to a basic model. Various other configurations and equipment levels can be provided according to customer specifications.

Turbomecanica special purpose off-road vehicle ***1997***

Specifications
Seating: 1 + 3
Configuration: 4 × 2
Weight:
(kerb) 750 kg
(loaded) 1,500 kg
Payload: 750 kg
Towed load: 1,000 kg
Length: 3.555 m
Width: 1.535 m
Height: 1.05 m
Ground clearance: 0.2 m
Track: (front and rear) 1.356 m
Wheelbase: 2.6 m
Max speed: (road) 80 km/h
Max gradient: 40°
Side slope: 30°
Engine: 1.299 litre air-cooled 4-cylinder petrol developing 61.5 hp at 5,500 rpm
Gearbox: manual, 5 forward and 1 reverse gears
Clutch: single dry plate

Suspension:
(front) torsion rods and hydraulic shock-absorbers
(rear) coil springs and hydraulic shock-absorbers
Tyres: 155 SR 14
Brakes: dual circuit, disc
Electrical system: 12 V
Battery: 1 × 12 V, 45 Ah
Alternator: 40 A

Status
Available.

Manufacturer
Societatea Comercială Turbomecanica SA, Boulevard Pacii 244, Sector 6, Bucharest 77826, Romania.
Tel: +401 312 92 06. Fax: +401 312 91 65.

NEW ENTRY

SINGAPORE

STA Kowari high mobility light assault vehicle

Description
The STA Kowari light assault vehicle is based on a modified version of the British Wessex Saker (see separate entry in this section) and is intended for highly mobile reconnaissance and strike roles. Power is provided by a rear-mounted 2 litre turbocharged diesel engine while the tubular space frame structure and 70:30 weight distribution enables the Kowari to literally 'jump' over low obstacles. A ground clearance of 350 mm and a flat undertray ensures that all systems are well protected. Agility is enhanced by a turning radius of 6.15 m.

The crew comprises the driver, the commander seated next to the driver and a gunner seated to the rear. The primary weapon station manned by the gunner mounts heavy calibre weapons such as the CIS 50MG heavy machine gun, the CIS 40AGL automatic grenade launcher, or MILAN or TOW anti-tank missiles. The commander operates the front weapon mounting which accommodates 7.62 mm weapons. Ample cargo space is provided by side panniers and a rear load bed behind the gunner station. The kerb weight of 1,250 kg allows the Kowari to be air-transported by helicopters and cargo aircraft. Payload is a rated 650 kg.

Although the standard version mounts a manual gearshift with rear-wheel drive only, an automatic four-wheel drive version is under development. A combination of leaf torsion and coil springs provides a long travel independent suspension system.

STA Kowari high-mobility light assault vehicle ***1997***

Specifications
Seating: 1 + 2
Configuration: 4 × 2 or 4 × 4
Weight: (kerb) 1,250 kg
Payload: 650 kg
Length: 3.94 m
Width: 1.89 m
Height: 1.97 m
Ground clearance: 0.35 m
Wheelbase: 2.82 m
Angle of approach/departure: 70°/50°
Max speed: 112 km/h
Range: 500 km
Side slope: 40%
Fording: 0.45 m
Engine: 1.993 litre Perkins 4-cylinder turbocharged diesel developing 80 bhp at 4,500 rpm
Turning radius: 6.15 m
Suspension: long travel independent with leaf torsion and coil springs

Status
Prototype completed. Awaiting production.

Manufacturer
Singapore Technologies Automotive Limited, 5 Portsdown Road, Singapore 139296.
Tel: +65 473 6311. Telex: RS 25755 SINENG.
Fax: +45 471 0662.

NEW ENTRY

SOUTH AFRICA

Mechem BAT special forces utility vehicle

Description
The Mechem BAT lightweight, air-droppable special forces utility vehicle is based on a civilian lightweight overland vehicle design and embodies numerous standard commercial components. It is based on a galvanised steel, ladder-type chassis with an all-welded tubular steel frame body faced with armoured plate. The split windscreens are armoured glass.

The BAT is supplied with a choice of either a 3-litre V-6 petrol engine or a 2 litre four-cylinder diesel, both coupled to a five-speed manual transmission with a cable-operated differential lock on the rear axle. Radial or cross-ply tyres may be fitted, depending on terrain. Disc brakes are fitted on the front axle and self-adjusting drum brakes on the rear.

Mechem BAT special forces utility vehicle armed with a single 7.62 mm MG4 machine gun (T J Gander) ***1997***

Mechem BAT special forces utility vehicle armed with a single 7.62 mm MG4 machine gun (T J Gander) ***1997***

The BAT can carry a variety of palletised loads. A six-seater personnel pallet may be replaced by three seats and palletised infantry weapons such as a multiple 68 mm rocket launcher or a 107 mm rocket launcher. The pallets were designed for ease of mounting and can be rapidly interchanged without a hoist. The weapon pallets provide storage space underneath for ammunition. A mounting point for single or twin machine guns is standard between the bullet protected twin windscreens.

Optional extras are available, including a removable cabin top, a towbar, a front axle differential lock, and an electric winch.

Specifications

Seating: 1 + up to 5
Configuration: 4 × 4
Axle loadings:
(front) 1,200 kg
(rear) 1,600 kg
Length: 3.1 m
Width: 1.8 m
Height: 1.8 m
Ground clearance: 0.35 m
Max speed: 120 km/h
Range: 500 km
Engine: 3 litre V-6 petrol developing 138 bhp at 5,000 rpm or
2-litre 4-cylinder turbocharged diesel
Gearbox: manual, 5 forward and 1 reverse gears
Transfer box: 2-speed
Turning radius: 5.25 m
Suspension: leaf springs and double-acting shock-absorbers on all wheels
Tyres: 15 × 7J R
Brakes: vacuum assist, discs front, drums rear

Status

In production. In service with the South African National Defence Force.

Manufacturer

Mechem, a division of Denel (Pty) Limited, PO Box 912454, Silverton 0127, South Africa.
Tel: +27 12 803 7290. Fax: +27 12 803 7189.

NEW ENTRY

UNITED KINGDOM

Automotive Technik (4 × 4) IMM weapons platform

Description

The Automotive Technik (4 × 4) Improved Medium Mobility (IMM) weapons platform is based on the short wheelbase (2.2 m) version of the Steyr-Daimler-Puch Pinzgauer Turbo D (4 ×4) 1,500 kg truck. As such, it retains the reliability and durability of the original vehicle, combined with increased agility and the ability to be configured for a number of special forces and other roles.

The Automotive Technik IMM weapons platform has an open tubular steel top hamper which doubles both as a roll-over safety structure and as the base for mounting a heavy machine gun, a 20 mm cannon or a similar weapon; it is also possible to incorporate a full 360° ring mounting. The body platform rides on rubber insulators held by the torsion resistant centre tube chassis. All axle movements are controlled by the centre tube so there is no body twist, providing a stable platform; this is enhanced by independent coil spring suspension with a pneumatic level control system. The centre tube chassis and independent suspension also combine with half axles to improve mobility by ensuring the wheels are kept in constant contact with the ground.

Normal drive is via the rear wheels, with full 4 × 4 drive and cross-axle differential engagement available at any speed by the flick of a switch on the dashboard.

Various equipment suites and stowage arrangements are possible and can be carried according to user requirements.

Automotive Technik (4 × 4) IMM weapons platform ***1997***

Specifications

Seating: 2, 3 or 4
Configuration: 4 × 2 or 4 ×4
Weight: (GVW) 3,500 kg
Towed load: 2,000 kg
Length: 4.252 m
Width: 1.8 m
Height: 2.045 m
Ground clearance: 0.36 m
Track: 1.52 m
Wheelbase: 2.2 m
Angle of approach/departure: 40°/45°
Max speed: 120 km/h
Fuel capacity: up to 140 litres
Max gradient: 100%
Fording: 0.7 m
Trench: 1 m
Engine: 2.383 litre 6-cylinder water-cooled turbocharged diesel developing 115 hp at 4,500 rpm
Gearbox: manual ZF S 5 18/3 with 5 forward and 1 reverse gears or
automatic ZF 4 HP22 with 4 forward and 1 reverse gears
Transfer box: 2-speed with synchromesh
Steering: ZF Gemmer, power-assisted
Turning circle: 10.5 m
Suspension: independent, with swing half axles in Portal axle design; coil springs and shock-absorbers on all wheels
Tyres: B F Goodrich 285/75 R 16
Brakes: discs on all wheels
Electrical system: 24 V
Batteries: 2 × 12 V, 75 Ah
Alternator: 28 V/55 A

Status

Available.

Marketed by

Automotive Technik Limited, Unit 7, Lawson Hunt Industrial Park, Guildford Road, Broadbridge Heath, Horsham, West Sussex RH12 3JR, UK.
Tel: +44 1403 211976. Fax: +44 1403 260001.

NEW ENTRY

Wessex Saker (4 × 2) Light Strike Vehicle (LSV)

Description

The Wessex Saker (4 × 2) Light Strike Vehicle (LSV) is a high-mobility vehicle intended for a number of tactical roles such as long-range incursion, reconnaissance, airfield security and installation defence. It combines a low profile with good ground clearance, and has a low thermal signature.

The Saker is based on a monocoque space frame chassis that can be readily adapted to a number of forms to suit role and customer requirements. The basic form is a two-seat vehicle with a rear-mounted diesel engine (a 2 litre air-cooled VW petrol engine is optional) with the two-man crew seated in the centre of the vehicle; three- and four-seat versions are available. The area over the engine can be used for stowage and the crew position is protected by a rollbar. Further stowage panniers or racks can be located along both sides of the vehicle.

The vehicle can be used to carry a number of types of weapon including 7.62 and 12.7 mm machine guns, the 30 mm ASP-30 cannon, anti-tank missiles (including Hellfire) and grenade launchers. Total payload is 700 kg and the Saker can tow trailers or weapons such as 120 mm mortars.

Numerous options are available including bush guards, extra fuel tanks, communications and navigation equipment, run-flat tyres, and so on. Kevlar armour can be used to provide extra protection for the crew.

The Saker is built by DML at Devonport Royal Dockyard.

In July 1991 Wessex (UK) plc concluded a licensing agreement with Manufacturing Assembly Inc (MAI) of Hunt Valley, Maryland, whereby MAI has the right to manufacture and market the Saker in the USA.

The Kowari LSV produced by Singapore Automotive Technologies (qv) is a development of the Wessex Saker.

Specifications

Seating: 2, 3 or 4
Configuration: 4 × 2
Weight: (kerb) 1,100 kg
Max load: 700 kg
Length: 3.9 m
Width: 1.86 m
Height: 1.73 m
Ground clearance: 0.35 m
Angle of approach/departure: 80°/65°
Side slope: 48°
Engine: Perkins Prima 80T 1.993 litre 4-cylinder water-

Wessex Saker (4 × 2) 400 kg light strike vehicle

Wessex Saker (4 × 2) 400 kg light strike vehicle

cooled turbocharged diesel developing 80 hp at 4,500 rpm
Gearbox: 4 forward and 1 reverse gears incorporating limited slip differential
Clutch: single dry plate operated by hydraulic sealed split system
Steering: rack and pinion, power-assisted
Turning circle: 12.3 m
Suspension:
(front) torsion bar and single shock-absorber with spring coil over
(rear) torsion bar and double shock-absorber, one with single coil over
Tyres: 30 9.5 R 15LT
Brakes: discs front and rear operated by hydraulic sealed split system
Electrical system: 12 V (24 V optional)
Batteries: 2 × 12 V, 60 Ah

Status
In production.

Manufacturer
Wessex (UK) plc, Woolhampton Court, Woolhampton, nr Reading, Berkshire RG7 5ST, UK.
Tel: +44 1734 712663. Fax: +44 1734 713110.

UPDATED

Cobra Light Strike Vehicle (LSV)

Description
The Cobra (originally Longline) Light Strike Vehicle (LSV) was developed to be a highly mobile vehicle for military use in hostile environments. It was designed as a weapon platform and for military missions such as reconnaissance and airborne operations. Commercially available major components and assemblies are used in its construction where possible.

The LSV is based on a high tensile tubular steel space frame with a 1.9 litre water-cooled diesel engine located at the rear together with the gearbox. Engine cooling is through twin radiators with independently operated thermostatically controlled electric pans. A separate oil cooler is provided for the engine lubricant. The drive configuration may be 4 × 2 or 4 × 4 with lockable differentials or viscous couplings as required. The frame incorporates a full roll-over cage and an integral weapon-mounting baseplate. A Makralon skidplate extends the full length of the underside and side panniers are provided for equipment or additional fuel tanks. The low sloping bonnet plate is aluminium and may be used for equipment stowage.

The driver is seated on the left with a separate seat for the navigator/gunner on the right; both seats are provided with safety harnesses. Two fuel tanks are located just behind the seats and may be provided with an optional ignition retardant system. A dual 12 V electrical system is provided. The exhaust system was designed to have a low aural and thermal signature.

Weapons that can be mounted on the LSV's UMI (Universal Mount Interface) include: 7.62 and 12.7 mm machine guns (including the 12.7 mm GECAL Gatling gun), a 40 mm grenade launcher, a 30 mm ASP-30 cannon, or six MILAN anti-tank missiles and their launcher; a 51 or 60 mm mortar may be carried for dismounted use.

The LSV can be transported inside a standard ISO container and may be slung under CH-47D, Puma and Sea King helicopters. It has been paradropped from a C-130 transport aircraft.

Cobra Light Strike Vehicle (LSV) (Michael Jerchel) ***1996***

Specifications
Seating: 1 + 1
Configuration: 4 × 4
Weight:
(unladen) 1,220 kg
(fully laden) 1,810 kg
Load capacity: 590 kg
Length: 4.166 m
Width: 1.728 m
Height: (top of roll cage) 1.6 m
Ground clearance: 0.3 m
Wheelbase: 2.54 m
Max speed: 130 km/h
Range: over 610 km
Fuel capacity: 75 litres
Gradient: 50%
Side slope: (with weapons fitted) 50°
Fording: (unprepared) 0.5 m
Trench: 1 m
Engine: VAG 1.9 litre turbo diesel developing 95 bhp
Transmission: manual with 4 forward (plus crawler) and 1 reverse gears
Steering: rack and pinion
Turning radius: approx 6 m
Suspension: double wishbone front and trailing arm rear, both with coil springs on heavy-duty shock-absorbers, rubber bump stops front and rear
Tyres: 30 × 9.50 R 15LT
Brakes: disc front, drum rear
Electrical system: 12 and 24 V
Batteries: 3 × 12 V

Status
In production. In service with the British Army.

Manufacturer
Longline Limited, 9 The Grange Industrial Estate, Albion Street, Southwick, Brighton, West Sussex BN42 4EN, UK.
Tel: +44 1273 870888. Fax: +44 1273 870808.
Marketed by: Cobra International Marketing Limited, Cobra House, Southwick Street, Southwick, Brighton, West Sussex BN42 4TE, UK.
Tel: +44 1273 871408. Fax: +44 1273 871420.

VERIFIED

Land Rover Defender Special Operations Vehicle (SOV)

Description

The Land Rover Defender Special Operations Vehicle (SOV) was introduced during 1992 and was specifically designed to meet a requirement for a rapid reaction, air-portable, all-terrain weapons platform. An initial order for 60 units was placed by the US Army for use by their Special Forces (Rangers) during late 1992

The SOV is based on the Defender 110 chassis and retains a high level of commonality with other Defender models in terms of permanent four-wheel drive, long travel coil spring suspension, transfer box and the latest generation of gearbox, steering and braking systems.

The SOV is configured as a long-wheelbase all-terrain weapons platform. The suspension uses long travel coil springs front and rear combined with telescopic dampers, rear trailing links, a Panhard rod at the front and a centrally mounted 'A' bracket at the rear. The SOV is available with either the 300 Tdi direct injection turbocharged intercooled diesel or the 3.5 litre V-8 petrol engine.

The SOV can carry a wide range of weapons. The frame rollbar mounted over the rear can carry one or two machine guns, including the 30 mm ASP-30 cannon, or a 40 mm grenade launcher, plus a further machine gun next to the driver. Racks and stowage bins located around the vehicle can be used to carry a 50 or 81 mm mortar and mounting and/or an anti-tank missile launcher. Further stowage bins can carry mines or demolition equipment in addition to the crew's personal kit.

The SOV may be carried inside a C-130 transport aircraft or the CH-47 and EH 101 helicopters and may be carried slung under medium- and heavy-lift helicopters. The SOV may be paradropped.

Land Rover Defender Special Operations Vehicle (SOV)

Specifications

Crew: up to 6
Configuration: 4 × 4
Weight: (GVW) 3,400 kg
Towed load:
(off road) 1,000 kg
(on road, full trailer brakes) 4,000 kg
Length: 4.445 m
Width: 1.89 m
Height: (gun ring) 1.93 m
Ground clearance: 0.216 m
Track: 1.486 m
Wheelbase: 2.794 m
Engine: 3.528 litre V-8 water-cooled petrol developing 134 bhp at 5,000 rpm or
2.5 litre water-cooled direct injection turbocharged intercooled diesel developing 111 bhp at 4,000 rpm
Gearbox: manual, 5 forward and 1 reverse gears
Clutch: diaphragm spring
Transfer box: LT 230T, 2-speed
Steering: manual, recirculating ball
Turning radius: 6.4 m
Suspension: beam axles front and rear controlled by hydraulic telescopic dampers
Brakes: dual servo-assisted; discs front and rear
Tyres: 7.50 × 16
Electrical system: 12 V
Alternator: 65 A

Status

In production. In service with the US Army (60).

Manufacturer

Land Rover, Lode Lane, Solihull, West Midlands B92 8NW, UK.
Tel: +44 121 722 2424. Telex: 338641 lanrov g.
Fax: +44 121 742 5012.

VERIFIED

Land Rover Defender 90/110/130 Multi Role Combat Vehicle (MRCV)

Description

The Land Rover Defender 90 Multi Role Combat Vehicle (MRCV) was introduced during 1993. It is fully air-portable and was designed to be a versatile and highly mobile platform with powerful firepower capabilities. Although the MRCV was initially demonstrated based on the Defender 90, variants based on the Land Rover Defender 110 and Defender 130 are now available, with a high commonality of equipment.

The MRCV is based on the Defender 90 chassis and retains a high level of commonality with other Defender models in terms of permanent four-wheel drive, long travel coil spring suspension, transfer box and the latest generation of gearbox, steering and braking systems.

The MRCV is configured as an all-terrain weapons platform. The suspension uses long travel coil springs front and rear combined with telescopic dampers, rear trailing links, a Panhard rod at the front and a centrally mounted 'A' bracket at the rear. The MRCV is available powered by the 300 Tdi diesel engine.

The MRCV uses a modular construction and can be readily adapted to respond to the requirements of rapid reaction forces, reconnaissance and patrol units. A mounting with a rising system is provided for a MILAN ATGW missile launcher and six missiles. Three mounting locations are also provided for 7.62 or 12.7 mm machine guns. The vehicle can be adapted to accommodate a 360° ring mounting for further weapons, including a 40 mm MK 19 grenade launcher or a 20 mm cannon. Roll-over bars are provided and, if required, a soft-top cover can be fitted.

Seating can be provided for a crew of up to four, with two seated in the rear. There are side and rear mountings for ammunition panniers with further space for communications and navigation equipment, personal weapons and kit. Numerous ratchet tie-down straps are provided for further equipment. Sand channels can be carried. Under-vehicle protection against mines is provided.

Specifications

Crew: from 2 up to 8
Configuration: 4 × 4
Weight: (GVW) 2,550 kg
Length: 3.722 m
Width: 1.79 m

Land Rover Defender 90 Multi Role Combat Vehicle (MRCV) equipped with ring mounting for 40 mm MK 19 grenade launcher **1997**

Land Rover Defender 90 Multi Role Combat Vehicle (MRCV) equipped with MILAN ATGW

Ground clearance: 0.23 m
Track: 1.486 m
Wheelbase: 2.36 m
Engine: 300 Tdi 2.5 litre diesel developing 111 bhp at 4,000 rpm
Gearbox: manual, 5 forward and 1 reverse gears
Clutch: diaphragm spring
Transfer box: 2-speed
Steering: worm and roller
Turning radius: 5.85 m
Suspension: beam axles front and rear with coil springs controlled by hydraulic telescopic dampers
Brakes: dual servo-assisted; discs front and rear
Tyres: 7.50 × 16
Electrical system: 12 V
Alternator: 65 A

Status
In production.

Manufacturer
Land Rover, Lode Lane, Solihull, West Midlands B92 8NW, UK.
Tel: +44 121 722 2424. Telex: 338641 lanrov g.
Fax: +44 121 742 5012.

UPDATED

BIRST MWP Fast Strike Vehicle

Description
The BIRST MWP (Mobile Weapons Platform) Fast Strike Vehicle was introduced during 1995. It is based on the Land Rover chassis and drive train and can be produced as a fully furnished vehicle built to customer specifications or supplied as a modular package for local assembly.

The chassis uses spaceframe construction with an integral bulkhead. The front body wings, bonnet and body panels are constructed of fire retardant fibreglass with underbonnet steel tubular reinforcement; the bonnet opens forward for engine access. Limited mine protection plates are fitted under the passenger seats, both front and rear. A heavy-duty rollbar is an integral part of the chassis and frame and provides the basis for various weapon and equipment mountings. The complete floor area is aluminium with non-slip projections.

There are two front seats provided with seat belts and another two seats in the rear area. Lightweight side and roof covers can be added for weather protection. Ammunition and equipment stowage is provided front and rear. Ancillary equipment carried includes fuel, oil and water containers, a radio, a map tray, a wind deflector for the driver and various tools. An electrically operated 4.5-tonne winch is mounted on the front bumper bar. Other equipment can include a Global Positioning System, snow chains, a first aid kit, and a searchlight. Recovery eyes are provided front and rear and there is a rear-mounted NATO tow hook.

The front weapon mounting, in front of the passenger, is for a 7.62 mm machine gun. At the rear an adjustable height pillar can be used for a MILAN ATGW (or a similar system), a 40 mm grenade launcher, or a 12.7 mm heavy machine gun. The rollbar may also be used to mount a heavy machine gun or grenade launcher. The crew's personal weapons are stowed in various stowage points and there is provision for smoke grenade launchers front and rear.

Possible vehicle configurations include a command version with a hard or soft top, and a mortar carrier.

BIRST MWP Fast Strike Vehicle armed with MILAN ATGW ***1996***

Specifications
Provisional, variable according to user specifications
Crew: 1 + 3
Configuration: 4 × 4
Length: 4.45 m
Width: 1.89 m
Height: 1.93 m
Fuel capacity: 79.5 litres
Engine: 300 Tdi 2.495 litre diesel developing 107 bhp at 3,800 rpm
Gearbox: LT77 manual, 5 forward and 1 reverse gears
Transfer box: 230T 2-speed
Steering: worm and roller, power-assisted
Turning radius: 6.2 m
Suspension: beam axles front and rear with coil springs controlled by hydraulic telescopic dampers
Brakes: dual servo-assisted; discs front and rear
Tyres: 7.50 R 16 LT
Electrical system: 12 or 24 V

Status
Available.

Manufacturer
British Independent Reconnaissance & Strike Technology (UK) Limited, Deon House, Ilton Business Park, nr Ilminster, Somerset TA19 0DU, UK.
Tel: +44 1460 55955. Fax: +44 1460 55955.

VERIFIED

Alvis Shadow Light Strike Vehicle (LSV)

Development
The initial development of what was to become the Alvis Shadow Light Strike Vehicle (LSV) started in the aftermath of the 1990-1991 Gulf War when a requirement for a new generation of LSVs was raised. A bid request was issued in November 1994 and a contract to develop the new vehicle was awarded to Alvis Vehicles in 1995. Two preproduction examples were ready for trials by mid-1996 and were handed over to the UK Ministry of Defence in October 1996.

The Shadow rolling chassis is provided by Reumech OMC of South Africa and is based on their A-Wagon (the Cheko). Supacat Limited produces the top hamper frame while Alvis acts as prime contractor and integrator. All development was carried out in conjunction with special forces end users.

Description
The Alvis Shadow LSV is based on a Reumech A-Wagon all-welded steel ladder construction chassis with the wheelbase shortened to 3 m. Power is provided by a General Motors 6.5-litre V-8 turbocharged diesel engine coupled to a General Motors Hydramatic four-speed automatic transmission and a Borg Warner two-speed transfer box. Both the

Alvis Shadow light strike vehicle ***1997***

engine and gearbox have a high degree of component commonality with the AM General M998 HMMWV series. There are Reumech hypoid differentials with selectable locks on both axles and ZF independent coil spring suspension with Reumech shock-absorbers.

The chassis underside is reinforced with renewable polymer plates and a 9 mm armour-steel skidplate at the front of the chassis. There are two double-skinned fuel tanks, both provided with fire suppression inserts, located low in the vehicle and provided with a cross feed system.

The tubular steel top hamper was designed for air-dropping with the underslinging plates engineered into the rollcage and bullbar at the front of the vehicle. Numerous tiedown points are provided for securing equipment and for securing the vehicle inside aircraft. The top hamper can also be employed for various automatic and other weapon mountings. Two Shadows can be carried inside a CH-47 Chinook helicopter and the vehicle can be carried on a DROPS flatrack or container.

An electric winch with a 3.7 tonne capacity is located within the front bullbar. Ground anchor pins for the winch form part of the grill guard. At the rear the full width tailgate can serve as a workbench or medical platform and can be detached to act as a stretcher or a lifting platform. The driver's and commander's seat backs can be folded to provide an elevating standing surface while the reinforced seat backs can be raised to permit the wearing of webbing equipment in relative comfort. All flat aluminium panel surfaces are reinforced. The protective light covers are wired so that they can accept an interchangeable light cluster or a smoke discharger. Multiple pressurised air points are provided for tyre inflation or power tools.

Specifications

Seating: 1 + 2 or 3
Configuration: 4 × 2 and 4 × 4
Weight: (GVW) 6,000 kg
Payload: up to 2,250 kg
Length: 4.95 m
Width: 2.08 m
Height: loaded, 1.8 m
Ground clearance: 0.24 m
Wheelbase: 3 m
Angle of approach/departure: 40°/40°
Engine: General Motors 6.5 litre V-8 turbocharged diesel developing 190 hp
Gearbox: General Motors Hydramtic automatic with 4 forward and 1 reverse gears
Transfer box: Borg Warner 2-speed
Suspension: ZF all round independent, double wishbone
Steering: worm gearbox
Turning radius: 6.8 m
Tyres: Michelin XZL 275/70-R20
Brakes: dual circuit, power-assisted outboard discs on all wheels
Electrical system: 12/24 V

Status

Preproduction examples.

Manufacturer

Alvis Vehicles Limited, The Triangle, Walsgrave, Coventry CV2 2SP, UK.
Tel: +44 1203 535455. Fax: +44 1203 539287.

NEW ENTRY

UNITED STATES OF AMERICA

Lockheed Martin Defense Systems-Chenowth Advanced Light Vehicle (ALSV)

Development

In December 1995 Lockheed Martin Defense Systems and the Chenowth Corporation signed a manufacturing licence agreement that gives Lockheed Martin Defense Systems global rights to manufacture and sell Chenowth high-performance lightweight military vehicles. The manufacturing licence agreement makes Lockheed Martin Defense Systems the prime contractor and Chenowth its major subcontractor for design and manufacturing services on all new orders. The agreement also provides a basis for joint development work.

In October 1996 the Lockheed Martin Defense Systems-Chenowth partnership unveiled a development model of their Advanced Light Strike Vehicle (ALSV), described as a third-generation high-performance surveillance, light strike and reconnaissance vehicle.

Description

The ALSV has seating for a crew of two, three or four protected by a tubular rollbar upper frame hamper. It features a main weapon station with a full 360° traverse at the rear. The station can mount either a 12.7 mm M2 heavy machine gun or a 40 mm MK 19 grenade launcher. Light machine guns and man-portable anti-tank weapons can also be carried. Remote-control and stabilised weapon platforms are possible, as are integral and configurable light armour and various stowage arrangements.

The ALSV has full 4 × 4 drive and is powered by a 140 hp water-cooled diesel engine. Extensive use is made of commercial off-the-shelf components.

The ALSV can be carried internally by CH-47 and CH-53 helicopters. Two may be carried by a C-130 Hercules for paradropping, or three as cargo.

Lockheed Martin Defense Systems-Chenowth Advanced Light Vehicle (ALSV) ***1997***

Specifications

Seating: 1 + up to 3
Configuration: 4 × 4
Weight:
(kerb) 1,250 kg
(loaded) 2,350 kg
Payload: 1,100 kg
Length: 4.25 m
Width: 2.11 m
Height: 1.9 m
Ground clearance: 0.38 m
Wheelbase: 2.85 m
Max speed:
(road) 130 km/h
(off road) 110 km/h
Range: 500 km
Gradient: >60%
Side slope: >40%
Engine: 2.5 litre 5-cylinder water-cooled diesel developing 140 hp
Gearbox: manual, 5 forward and 1 reverse gears
Suspension: independent
Steering: rack and pinion
Electrical system: 12 V (24 V optional)

Status

Development.

Manufacturers

Chenowth Military Products Inc, 1101 30th Street NW, Suite 302, Washington DC 20007, USA.
Tel: +1 202 625 2424. Fax: +1 202 338 3368.
Lockheed Martin Defense Systems, 100 Plastics Avenue, Pittsfield, Massachusetts 01201, USA.
Tel: +1 413 494 3066. Fax: +1 413 494 5643.

NEW ENTRY

Lockheed Martin Defense Systems-Chenowth Light Strike Vehicle (LSV)

Description

Lockheed Martin Defense Systems and Chenowth have teamed to promote the Light Strike Vehicle (LSV), a development of Chenowth's original off-road racing vehicle which combines low weight, high speed and good cross-country mobility and manoeuvrability. The original military model was the two-seat Fast Attack Vehicle (FAV) produced during the mid-1980s from which the three-seat LSV was further developed.

The LSV chassis is an open high-strength 4130 chrome-moly tubular frame with an integral chrome-moly steel roll cage. Other safety features include a racing-type fuel cell, which is explosion-resistant, and run-flat capable rims on all four tyres. The standard engine, a 125 hp petrol unit, is air-cooled and rear-mounted. A diesel engine is available.

Just forward of and above the engine, the gunner/vehicle commander has the capability to fire forward and to the rear on a swivel seat. The assistant driver, the lower gunner, rides in the right-hand seat. All positions have suspension-type bucket seats fitted with retaining straps.

Weapons can include two 7.62 mm machine guns, mounted fore and aft, together with one 0.50/12.7 mm M2 machine gun or a 40 mm MK 19 automatic grenade launcher. The LSV can also carry a TOW 2 anti-tank missile launcher, a 30 mm ASP-30 cannon, an AT-4 anti-tank rocket launcher or the Stinger SAM. Light armour may be added as required, as can special stowage configurations.

The LSV can be carried internally by CH-47 and CH-53 helicopters. Two may be carried by a C-130 Hercules for para-dropping, or three as cargo.

During Operation Desert Shield/Storm the FAV and LSV were operated by US and UK Special Forces and by the US Marines.

The LSV was one entry for the Australian Army's Project Mulgara in cooperation with an Australian company, MMA.

Specifications

Seating: 1 + 2
Configuration: 4 × 2
Weight:
(kerb) 960 kg
(loaded) 1,660 kg
Payload: 700 kg
Length: 4.08 m
Width: 2.11 m
Height: 2.01 m
Ground clearance: 0.41 m
Wheelbase: 2.84 m
Angle of approach/departure: 81°/48°
Max speed:
(road) 130 km/h
(off road) 110 km/h
Range: 500 km
Gradient: 75%
Side slope: >45%
Engine: 2.2 litre air-cooled petrol delivering 125 hp

Lockheed Martin Defense Systems-Chenowth Light Strike Vehicle (LSV) armed with TOW anti-tank missile launcher

Gearbox: manual, 4 forward and 1 reverse gears
Suspension: independent
Steering: rack and pinion
Electrical system: 12 V (24 V optional)

Status

In production. In the inventory of US Navy, Marine Corps and Army (two-man FAV) and in Greece, Israel, Portugal, Oman, Mexico, Spain and the UAE.

Manufacturers

Chenowth Military Products Inc, 1101 30th Street NW, Suite 302, Washington DC 20007, USA.
Tel: +1 202 625 2424. Fax: +1 202 338 3368.
Lockheed Martin Defense Systems, 100 Plastics Avenue, Pittsfield, Massachusetts 01201, USA.
Tel: +1 413 494 3066. Fax: +1 413 494 5643.

UPDATED

Lockheed Martin Defense Systems-Chenowth General Purpose (GP) Vehicle

Description

The Lockheed Martin Defense Systems-Chenowth General Purpose (GP) vehicle was originally marketed by TPC Logistics Services Inc of Farmingdale, New York, as the RAMP-V Rapid Multipurpose Vehicle. It now forms the utility/logistics variant of the Lockheed Martin-Chenowth family of light high-mobility vehicles.

The 4 × 4 GP vehicle may be equipped for various support roles, from seating for up to eight passengers, to cargo carrier, casualty evacuation ambulance, and command and control. The main chassis frame consists of 4130 chromoly steel tubing. Total rollover protection is provided by a tubular frame upper hamper while light armour can be added if required. Various stowage configurations can be adopted as can mountings for weapons such as light machine guns.

Power is provided by a 2.2-litre petrol engine mounted at the rear and using a cross-drive transaxle transmission. A diesel engine is optional.

The GP vehicle can be carried internally by CH-47 and CH-53 helicopters. Two may be carried by a C-130 Hercules for paradropping, or three as cargo.

Specifications

Seating: 1 + 1 (up to 6 in rear)
Configuration: 4 × 4
Weight:
(kerb) 1,100 kg
(loaded) 2,100 kg
Payload: 1,000 kg
Length: 4.27 m
Width: 2.11 m
Height: 1.93 m
Ground clearance: 0.38 m
Wheelbase: 2.84 m
Max speed:
(road) 130 km/h
(off road) 110 km/h
Range: 500 km
Gradient: >60%
Side slope: >40%
Engine: 2.2 litre 4-cylinder petrol
Gearbox: manual, 4 forward and 1 reverse gears
Suspension: independent
Steering: rack and pinion
Electrical system: 12 V (24 V optional)

Status

In production.

Service

US armed forces.

Manufacturers

Chenowth Military Products Inc, 1101 30th Street NW, Suite 302, Washington DC 20007, USA.
Tel: +1 202 625 2424. Fax: +1 202 338 3368.
Lockheed Martin Defense Systems, 100 Plastics Avenue, Pittsfield, Massachusetts 01201, USA.
Tel: +1 413 494 3066. Fax: +1 413 494 5643.

NEW ENTRY

Lockheed Martin Defense Systems-Chenowth General Purpose (GP) vehicle ***1997***

Warrior NMC-40 Long-range Fast Attack Vehicle (FAV)

Description

The Warrior NMC-40 fast attack vehicle was designed and developed as part of the Fast Attack Vehicle (FAV) programme initiated by the US Army Tank Automotive Command (TACOM). It was first shown publicly in July 1983 and is offered for sale and export.

The Warrior NMC-40 is a four-wheeled light vehicle with a high-strength tubular steel roll-cage frame. Seating is provided for three men, two facing forward and one facing the rear. The air-cooled engine is a German-manufactured rear-mounted unit capable of generating land speeds in excess of 161 km/h. Provision has been made for mounting two weapons, usually two M60 machine guns or one M60 and an automatic 40 mm grenade launcher, one at the front and the other at the rear.

The Warrior NMC-40 is supplied complete with equipment ranging from personal body armour to radios and special tools. Also included are map cases, grenades in racks and a neoprene vehicle cover.

Specifications

Cab seating: 1 + 2
Configuration: 4 × 2
Weight: (less crew and armament) 748 kg
Ground clearance: 0.355 m
Range: (cross-country) 483-644 km

Gradient: 75%
Side slope: 40%
Max speed: 153-161 km/h
Engine: air-cooled 4-cylinder petrol developing 80 hp
Gearbox: manual with 4 forward and 1 reverse gears
Suspension: torsion bars and shock-absorbers providing 0.355 m wheel travel front and rear
Steering: rack and pinion with dampers
Brakes: drums front and discs rear with hydraulic steering brake
Electrical system: dual 12 V, 100 A

Status
Ready for production.

Manufacturer
Nordac Manufacturing Corporation, Route 12, Box 124, Fredericksburg, Virginia 22405, USA.
Tel: +1 703 752 2552. Telex: 89 7440.

VERIFIED

Warrior NMC-40 long-range Fast Attack Vehicle (FAV)

HSMV R-12 (4 × 4) High-Speed Mobility Vehicle

Description
Known as The Flyer™ R-12, this vehicle is a 4 × 4 high-agility lightweight military vehicle with a 1,200 kg payload. HSMV designed the world's first 4 × 4 vehicle in this class (now in service with the US armed forces) and has designed a family of lightweight, agile tactical vehicles using the latest engine, transmission, steering suspension and frame technology, all proven during off-road sport racing.

The chassis is constructed using welded chrome-moly seamless steel tubing and is powered by a rear-mounted 110 hp water-cooled diesel engine coupled to a semi-automatic transmission. A long travel progressive suspension is provided for virtually unimpeded travel over most terrains.

Many options are available to allow this vehicle to be tailored to satisfy a wide range of customer requirements.

Specifications
Crew: 4
Configuration: 4 × 4 or 4 × 2
Weight:
(kerb, 4 × 2) 1,100 kg
(kerb, 4 × 4) 1,200 kg
Payload:
(4 × 2) 1,100 kg
(4 × 4) 1,250 kg
Length: 4.5 m
Width: 2 m
Height: 1.75 m
Ground clearance: 0.4 m
Wheelbase: 2.87 m
Angle of approach/departure: 70°/45°
Max speed: 110 km/h
Range: 500 km (750 km optional)
Fuel capacity: 110 litres (160 litres optional)
Fording: 0.6 m
Gradient: 80%
Side slope: 65%
Vertical obstacle: 0.45 m
Engine: water-cooled turbocharged diesel developing 110 hp
Transmission: torque converter, semi-automatic
Steering: power-assisted
Suspension: long travel progressive
Electrical system: 24 V
Batteries: 2 × 12 V, 60 Ah

Status
Available.

Manufacturer
RACECO-HSMV LP, 11611 San Vicente Blvd, Suite 615, Los Angeles, California 90049, USA.
Tel: +1 818 710 9340. Fax: +1 310 821 8261.

VERIFIED

HSMV R-12 (4 × 4) High-Speed Mobility Vehicle **1996**

HSMV R-12 (4 × 4) High-Speed Mobility Vehicle **1996**

TRUCKS (over 1,000 kg)

AUSTRALIA

OKA (4 × 4) 2,000 kg truck

Description

The OKA (4 × 4) 2,000 kg truck is a military derivative of a series of trucks designed for a long service life when used under extreme conditions. The overall construction is based on readily available proprietary components and a modular construction which can be readily adapted to accommodate a variety of utility and other bodies.

The chassis is a ladder frame construction with box section straight parallel rails to provide a flatbed onto which a wide variety of bodies can be placed, including non-standard loads such as weapon systems. The welded Duragel and sheet steel panel forward control cab has a Roll-Over Protection Structure (ROPS) plus falling object protection. Provision is made in the cab for a driver and one passenger. If a bus-type personnel carrier body is provided, up to 14 passengers can be carried although personnel/utility versions normally carry nine. Air conditioning for the cab is an option.

The military version of the OKA (4 × 4) is powered by a Perkins Phaser 110 MT 4 litre turbocharged diesel coupled to a Dana Spicer manual five-speed transmission. The New Process transfer case has a power take-off on the left-hand side.

The front suspension has a single-stage semi-elliptic spring assembly with a rubber second stage and axle stops. The rear suspension is similar but has a double-stage spring assembly. Both axles are from Dana Spicer, both fully floating.

Winches may be fitted at the front, rear or centrally. Optional equipment includes long-range fuel and water tanks, extra spare wheel carriers, convoy or blackout lighting, various tyre sizes, and various electrical systems, among other options. A towing hook can be fitted at the front and/or rear.

Specifications

Cab seating: 1 + 1
Configuration: 4 × 4
Weight: (cab and chassis) 2,734 kg
Max load: (nominal) 2,000 kg
Towed load: 3,500 kg
Length overall: 5.455 m
Width: (overall) 2 m
Height: (cab, unladen) 2.51 m
Ground clearance: 0.24 m
Track:
(front) 1.651 m
(rear) 1.6 m
Wheelbase: (unladen) 2.995 m
Angle of approach/departure: (laden) 40.5°/32°
Fuel capacity: (per tank) 105 litres
Max gradient: 60%
Fording: 0.91 m
Engine: Perkins Phaser 110 MT 3.99 litre turbocharged direct injection water-cooled diesel developing 108 hp at 2,800 rpm
Transmission: Dana Spicer T5-290 manual with 5 forward and 1 reverse gears
Transfer box: NP 205 2-speed
Steering: Kirby Bishop integral power
Suspension:
(front) single-stage semi-elliptic leaf spring assembly with rubber Aeon second stage and deceleration axle stops
(rear) dual-stage semi-elliptic leaf spring assembly with rubber Aeon second stage and deceleration axle stops
Tyres: 8.25 R 16LT, 14 ply
Brakes: power-boosted dual hydraulic
Electrical system: 12 or 24 V

Status

In production. Production in United Arab Emirates planned. In service with the Australian Army.

Manufacturer

OKA Motor Company Limited, PO Box 1238, Fremantle, Western Australia 6160.
Tel: +61 9 434 1300. Fax: +61 9 434 1659.

VERIFIED

OKA (4 × 4) 2,000 kg truck (T J Gander)

OKA (4 × 4) 2,000 kg truck (T J Gander)

Land Rover 110 heavy-duty (6 × 6) truck

Development

During 1981 Jaguar Rover Australia (JRA) (later Land Rover Australia) carried out market research that indicated a requirement for a heavy-duty (6 × 6) derivative of the Land Rover 110 (now the Defender 110) with a payload of up to 3,000 kg and a rear chassis length of up to 3.3 m. Several alternatives were investigated before a 6 × 6 version of the 110 was selected based on a trials vehicle produced by SMC Engineering (Bristol) Limited in the United Kingdom. Once in Australia this vehicle was fitted with an Isuzu 3.9 litre diesel engine and the original coil springs at the rear were replaced by semi-elliptical leaf springs. The resultant vehicles were involved in the Australian Army's Project Perentie. The first of 400 production vehicles was handed over to the Australian Army in March 1989.

Some of these vehicles have been fitted with Rover 3.5 litre V-8 petrol engines and are marketed as part of the Land Rover range in the UK.

Description

The Land Rover 110 heavy-duty (6 × 6) truck maintains a great deal of commonality with the basic 4 × 4 Land

Australian Army Land Rover 110 heavy-duty (6 × 6) truck with workshop body (Barry Marriott) ***1997***

Land Rover 110 heavy-duty (6 × 6) truck with workshop body (Barry Marriott) ***1997***

Land Rover 110 heavy-duty (6 × 6) truck (Barry Marriott) ***1997***

Rover Defender 110, although the body of the Australian vehicle is wider. The basic engine is an Isuzu 3.9 litre diesel but the standard Rover 3.5 litre V-8 petrol engine may be fitted if required. The cab and front end of the JRA vehicle are the same as the Defender 110 but the spiral bevel front axle is uprated. The chassis is a heavy-duty rigid chassis frame fabricated from steel tubing and steel pressings with special anti-corrosion treatment on all surfaces. The twin rear axles are Salisbury 8HA fully floating hypoid bevel gear axles with the centre axle differential offset to the right of the centre line and the rear axle differential offset to the left.

The rear tray on the basic vehicle carries a cargo/personnel carrier 3.2 m long and 2.082 m wide. It is normally covered by a heavy-duty canvas canopy and inward- or outward-facing seats can be provided for up to six men each side. Various options are on offer including a front-mounted Thomas T9000M 4,000 kg recovery winch, a brush guard, a towing pintle, various stowage racks and tie-down points.

The standard electrical system is 12 V. Fitted-For-Radio (FFR) vehicles have a 24 V system with two 12 V batteries in a slide-out carrier under the left-hand front corner of the rear body.

The initial contract covered seven derivatives: cargo; cargo with winch; air defence FFR with winch; ambulance FFR with winch; general repair with winch; electronics repair; and an armed long-range patrol vehicle for the Australian SAS which carries on the rear a frame for a motor cycle. Other derivatives planned include a communications centre, a command post and a firefighting vehicle. A crew cab version has been developed.

Modular rear bodies, such as those used for the electronics and general repair vehicles, can be unloaded and used as free-standing units if required. The sides of some shelters can expand to increase working space and shelters can be clipped together when off the vehicle. The bodies were produced by, and are still available from, JAKAB Industries Pty Ltd of Tamworth, New South Wales.

Specifications

Cab seating: 1 + 1 (up to 12 in rear)
Configuration: 6 × 6
Weight: (cargo/personnel FFR with winch)
(empty) 3,600 kg
(loaded) 5,600 kg
(gross combination mass) 7,100 kg
Load area: 3.2 × 2.082 m
Length overall: 6.001 m
Width:
(overall) 2.2 m
(over mirrors) 2.43 m
Height:
(cab, unladen) 2.08 m
(cab, laden) 2.05 m
Ground clearance:
(front) 0.235 m
(rear) 0.215 m
Track: 1.698 m
Wheelbase: 3.04 m + 0.9 m
Angle of approach/departure:
(unladen) 45°/33°
(laden) 41°/30°
Max speed: 100 km/h
Range: (roads) 600 km
Fuel capacity: 2 × 65 litres
Max gradient: 60%
Engine: Isuzu 4BD1 T 4-cylinder 3.856 litre turbocharged diesel developing 121 hp at 3,000 rpm
Gearbox: Land Rover LR95A with 4 forward and 1 reverse gears
Transfer box: 2-speed
Steering: Adwest power-assisted variable ratio worm and peg with power-assistance
Suspension:
(front) long travel coil springs with telescopic dampers
(rear) dual rate semi-elliptical leaf springs linked via shackles to rubber-bushed load-sharing rocker beam
Tyres: 7.50 × 16, 10 ply
Brakes: dual-vacuum/hydraulic
Electrical system: 12 V (24 V optional)
Battery:
(standard) 1 × 12 V, 98 Ah
(FFR) 2 × 12 V, 95 Ah
Alternator: 70 A

Status

In service with the Australian Army (400).

Manufacturer

Rover Australia Pty Limited, 12 Riverside Centre, 148-309 James Ruse Drive, Paramatta, New South Wales 2150, Australia.
Tel: +61 2 685 5111. Fax: +61 2 687 2766.

UPDATED

Mack Model RM6866RS (6 × 6) truck

Development

The Mack Model RM6866RS (6 × 6) truck was produced in Australia by Mack Trucks Australia Pty Limited as the Truck, Cargo, Heavy, MC3. It is a version of the basic US Mack 'R' series. Three prototypes were produced (in Australia) in 1979. After evaluation an order was placed in 1981 for 906 units. These were all delivered by the end of 1986.

During 1993 a contract to upgrade 541 of these trucks was awarded to Mack Trucks Australia Pty Limited. The contract, worth A$8.77 million, involves the replacement of mechanical suspension units with a SA44W air suspension system and was scheduled to take two years. In 1994 an extension to the contract was issued to cover the remainder of the Australian Army's Mack fleet, together with modifications on Cargo/Cargo with Winch variants to fit container twistlocks. This work was completed in April 1996.

Description

The basic vehicle is the Truck, Cargo, Heavy, MC3 which may be fitted with a winch. It is classified as an MLC 22 8 tonne capacity vehicle and many components and subassemblies are produced in Australia. The cargo body is a Mack design built by Walsh Engineering of Toowoomba, Queensland, and the EDE alternator is produced by the Ordnance Factory at Maribyrnong, Victoria.

The layout of the vehicle is conventional, with the engine located at the front under a prominent bonneted

Truck, Cargo, Heavy, MC3 (Barry Marriott) ***1997***

Truck, Tank, Water, Heavy, MC3 (Barry Marriott) ***1997***

Mixer, Concrete, Truck-Mounted, Heavy, MC3 (Barry Marriott) ***1997***

Truck, Wrecker, Heavy, MC3 (Barry Marriott) ***1997***

cab and the cargo area to the rear. The front axle is a Kelsey Hayes SDA 18B (Fabco) rated at 8,165 kg. The tandem rear axle unit is a Mack SA44W bogie rated at 20,000 kg.

The cab has seating for the driver and one or two passengers while the cargo area can have seating for 26 fully equipped troops.

Power is provided by an 11 litre Mack EM6-285 Maxidyne turbocharged diesel and a Maxitorque gearbox.

The standard winch carried is an Ateco 24L with a maximum pull of 10,575 kg and 60 m of cable.

The data in the Specifications table refers to the basic cargo vehicle.

Variants

Truck, Cargo, Heavy, With Crane, MC3

On this variant, an Abbey CTM 3000-1 hydraulic crane is mounted directly behind the cab. The crane has a maximum lift capacity of 3,400 kg at 1.8 m. Power for the crane's hydraulic system is provided by a Powauto AH23BR11 power take-off on the transfer case. The cargo area is 5.016 m long and 2.39 m wide with seating reduced to 22 fully equipped troops. Twistlocks for containers have been retrofitted. This variant weighs 12,810 kg unladen.

Truck, Wrecker, Heavy, MC3

On this variant the cargo area is replaced by a Holmes A750 recovery rig with a maximum towing lift of 7,800 kg (for more details refer to entry under USA in the *Recovery equipment* section). The vehicle also has a 10,500 kg recovery winch provided with 100 m of cable. This variant weighs 23,100 kg unladen.

Truck, Cargo, Army Tractor/Medium Gun, Heavy, MC3 Ammunition

This variant is used to tow the Australian Army's 155 mm M198 howitzers and may also be used to carry ammunition. A collapsible crew shelter for the gun crew is provided behind the driving cab; it may be removed to provide more ammunition-carrying capacity. The cargo area is replaced by a flatbed platform divided by an Abbey CTM 3000-2 ammunition-handling hydraulic crane. The crane has a maximum lift capacity of 2,000 kg at 3.053 m and a full 360° traverse. Power for the hydraulic system is provided by a Powauto AH23BR11 on the transfer case. A winch is fitted. This variant weighs 13,180 kg unladen.

Mixer, Concrete, Truck-Mounted, Heavy, MC3

This variant carries a Fowlerex Rheem Model 5 KH produced at Bulimba in Queensland. The hydraulically driven mixer has a capacity of 5.5 m³, and is used in association with a 204 litre water tank. This variant uses a ZF Model PK1400 gearbox and has an unladen weight of 13,700 kg.

Truck, Dump, Heavy, MC3

On this variant the cargo area is replaced by an 8 m³ capacity dump body produced by Hockney Alcan at Rocklea, Queensland. It is possible to arrange troop seating inside the dump body and up to 18 fully equipped troops can be carried. A winch may be fitted. This variant weighs 11,520 kg unladen.

Truck, Tank, Fuel, Heavy, MC3

For the refuelling role this variant carries a three-compartment tank produced by Holmwood Industries of Rochedale, Queensland. The tank has a capacity of 11,380 litres and the vehicle also carries the associated electrical and refuelling equipment such as four hose reels. This variant is 9.475 m long and weighs 12,160 kg unladen.

Truck, Tank, Water, Heavy, MC3

Carrying an aluminium three-compartment water tank with a highway operation capacity of 9,240 litres, this variant also carries a spray bar, two hose reels, a 60 m hose and three dispensing cocks. This variant is 9.474 m long and weighs 11,800 kg unladen.

Distributor, Bituminous Substance, Tank Type, Truck-Mounted, Heavy, MC3

This variant differs from the others in the Australian Mack range in having a 6 × 4 drive configuration. The steel tank has aluminium lagging and a highway operations capacity of 7,740 litres. A spray bar is provided along with two tar burner tubes. This variant is 10.194 m long and weighs 13,960 kg unladen.

Truck, Bridge Recovery, Heavy, MC3

This variant is equipped to launch and recover Ribbon Bridge sections and carries a Techlaunch ARK221 roll-on hook loader that can also be used to load a cargo tray when the Ribbon Bridge is not required. The loader is produced by Techmark Australia and manufactured in Melbourne, Victoria. It has a capacity of 6,000 kg. The cargo tray is 5.8 m long and 2.5 m wide.

When carrying a Ribbon Bridge section this variant is 10.3 m long. When carrying the cargo tray the length is 10.5 m. The unladen weight is 14,800 kg.

Specifications

Truck, Cargo, Heavy, MC3
Cab seating: 1 + 1 or 2
Configuration: 6 × 6
Weight: (unladen)
(front axle) 5,070 kg
(rear axle) 6,510 kg
(total) 11,570 kg
Load area: 5.714 × 2.39 m
Length: 9.6 m
Width: 2.5 m
Height:
(unladen) 3.155 m
(laden) 3.125 m
Ground clearance:
(unladen) 0.318 m
(laden) 0.28 m
Track:
(front) 1.911 m
(rear) 1.81 m
Wheelbase: 4.765 m + 1.4 m
Angle of approach/departure: (laden) 31.5°/30.5°
Range: (first-class roads) 550 to 590 km
Fuel capacity: 265 litres
Max gradient: 50%
Side slope: 22.5°
Fording: 0.8 m
Engine: Mack EM6-285 Maxidyne 11 litre intercooled 6-cylinder in-line turbocharged diesel developing 283 hp at 2,100 rpm
Gearbox: Maxitorque with 5 forward and 1 reverse gears
Clutch: Spicer twin dry plate
Transfer box: 2-speed
Steering: Sheppard Model 592 integrated power-assisted
Turning radius: 12.1 m
Suspension: multileaf springs (541 trucks upgraded to air suspension from 1993 onwards)
Tyres: 12.00 × 20
Brakes:
(main) dual-circuit, air
(parking) spring on rear axles
(trailer) dual-circuit, air
Electrical system: 24 V
Batteries: 2 × 12 V, 61 Ah
Alternator: 140 A

Status

In service with the Australian Army (906).

Manufacturer

Mack Trucks Australia Pty Limited, PO Box 364, Darra, Queensland 4076, Australia.
Tel: +61 7 3853 3333. Fax: +61 7 3375 4936.

UPDATED

AUSTRIA

Steyr-Daimler-Puch Pinzgauer Turbo D (4 × 4) 1,500 kg and (6 × 6) 2,000 kg vehicles

Development

The Pinzgauer Turbo D (4 × 4) and (6 × 6) diesel-engined versions of the Pinzgauer range were developed from 1983 onwards and were first shown during 1985. Although the Turbo D resembles many of the earlier models it has many innovations apart from the six-cylinder diesel engine.

Commencing in early 1995 the British Army took delivery of of an initial 394 Pinzgauer Turbo D 4 × 4 vehicles based on the 716 M model but revised slightly to meet the British Army's requirement for a Truck Utility Medium (Heavy Duty) (TUM(HD)) for service with Airmobile and Commando forces, some of them as artillery tractors for 105 mm Light Guns; others are employed as Javelin air defence missile carriers while others have been configured as signals or other specialised vehicles. The data in the Specifications table outlines the details for this model. The British Army's TUM(HD) fleet is made up of the following variants:

- GS Soft Top
- GS Soft Top with winch
- Waterproofed GS Soft Top
- Waterproofed Winterised GS
- Hard Top FFR.

Automotive Technik Limited of the United Kingdom market a IMM Weapons Platform based on the Pinzgauer Turbo d 4 × 4. See entry under *Special attack vehicles* section.

Description

In appearance and general layout the 4 × 4 and 6 × 6 versions of the Turbo D resemble the earlier Pinzgauer models. The main change is to a water-cooled and turbocharged six-cylinder diesel engine. The wheelbases are lengthened on both the 4 × 4 and 6 × 6 and the track is wider. Changes were made to the driver's position instrument layout, disc brakes introduced and a larger fuel tank fitted to improve range. An automatic locking system is incorporated.

The 4 × 4 version has an automatic level control system acting on the rear axle. A pneumatic system is controlled by sensors that raise and lower the vehicle superstructure to suit the load being carried. This enables the centre of gravity to remain constant.

The Pinzgauer Turbo D is produced in a number of versions as follows:

Pinzgauer 716 M 4 × 4; personnel carrier with canvas top carrying the driver and up to nine passengers

Pinzgauer 716 MK 4 × 4; cargo or personnel carrier with closed body carrying the driver and one front passenger plus approximately 1,250 kg of payload or eight passengers in the rear compartment

Pinzgauer 716 K 4 × 4; command vehicle carrying the driver and one front passenger plus three seats in the rear compartment (seating for a further five passengers is an option)

Pinzgauer 716 T 4 × 4; carrier for shelters and weapon systems carrying the driver, one passenger and approximately 1,300 kg of payload

Pinzgauer 718 M 6 × 6; personnel carrier with canvas top carrying the driver and up to 13 passengers

Pinzgauer 718 MK 6 × 6; cargo or personnel carrier with closed body carrying the driver and one front passenger plus approximately 1,700 kg of payload or 12 passengers in the rear compartment

Pinzgauer 718 K 6 × 6; command vehicle carrying the driver, one front passenger and three seats in the rear compartment (seating for seven more passengers is an option)

Pinzgauer 718 T 6 × 6; carrier vehicle for shelters and weapon systems carrying the driver, one passenger and approximately 1,800 kg of payload.

The data in the Specifications table refers to the Pinzgauer 716 M 4 × 4 and the Pinzgauer 718 M 6 × 6.

Status

In production. In service with Austria, Cyprus, Malaysia, Nigeria, Oman, Pakistan, Saudi Arabia, UK (over 500) and USA.

Manufacturer

Steyr-Daimler-Puch Fahrzeugtechnik GesmbH, Postfach 823, Liebenauer Hauptstrasse 317, A-8011 Graz-Thondorf, Austria.
Tel: +43 316 404-0. Fax: +43 316 404 2986.

UPDATED

Steyr-Daimler-Puch Pinzgauer (4 × 4) Turbo D in service with the Royal Marines ***1997***

Specifications

Model	**Pinzgauer 716 M**	**Pinzgauer 718 M**
Configuration	4 × 4	6 × 6
Weight empty	approx 2,350 kg	2,650 kg
Weight loaded	3,850 kg	4,500 kg
Max load	1,400 kg	1,850 kg
Towed load	1,500 kg	1,800 kg
Length	4.528 m	5.26 m
Width	1.8 m	1.8 m
Height	2.078 m	2.045 m
Ground clearance	0.345 m	0.335 m
Track	1.52 m	1.52 m
Wheelbase	2.4 m	2.2 m + 0.98 m
Angle of approach/departure	40°/45°	40°/45°
Max speed	122 km/h	112 km/h
Fuel capacity	145 litres	145 litres
Gradient	100%	84%
Side slope	45%	45%
Fording	0.7 m	0.7 m
Engine	2.383 litre 6-cylinder in-line water-cooled turbocharged diesel developing 115 hp at 4,350 rpm	
Transmission	ZF S 5-18/3 manual with 5 forward and 1 reverse gears or ZF automatic with 4 forward and 1 reverse gears	
Transfer box	2-speed	2-speed
Turning circle	11.5 m	13 m
Suspension	helical springs front and rear	helical springs front and rear with leaf springs at rear
Brakes	dual-circuit power-assisted discs	
Electrical system	24 V	24 V

Steyr-Daimler-Puch Pinzgauer (4 × 4) Turbo D in United Nations markings ***1996***

Steyr-Daimler-Puch Pinzgauer (4 × 4) Turbo D configured as front-line ambulance

Steyr-Daimler-Puch Pinzgauer (4 × 4) 1,000 kg and (6 × 6) 1,500 kg vehicles

Development

The Pinzgauer range of all-terrain vehicles was developed by Steyr-Daimler-Puch as the successor to the 700 AP Haflinger range of 4 × 4 vehicles. The first prototype of the 4 × 4 Pinzgauer was completed in 1965 with first production models being completed in 1971. The vehicle was adopted by the Austrian Army in 1973. The 6 × 6 model was first shown in 1968 and entered production in 1971-72. Production was completed in 1990.

Description

Both the 4 × 4 and 6 × 6 models are available with two basic types of body, fully enclosed or with a military-type cargo body. The former has an all-steel fully enclosed body with two doors in each side and a single door at the rear, and the military-type body has a single door each side for the driver and one passenger. The tops of the doors are removable and the windscreen can be folded forward onto the bonnet. The rear cargo area has removable bows and a tarpaulin cover and there are bench seats down each side of the rear body for eight men in the 4 × 4 model and 12 in the 6 × 6 model.

The chassis consists of a torsion-resistant central tube with independent swing axles incorporating the transfer case and axle drive. The engine is mounted towards the front of the vehicle and is coupled to a ZF manual gearbox which transmits power via a propeller shaft to the transfer case, which also supports the hydraulically operated shifting clutch for the power transmission to the front axle. The drive shaft is to the front, and rear differentials are within the central tube chassis. The differentials have a hydraulically operated mechanical lock.

Optional equipment for the Pinzgauer includes: antenna holder (middle and side), camouflage net holders, convoy lights, blackout blinds, divided windscreen, foldable table, jerrican holders, rear mounting trays, rifle racks, mounting points for shovels, split battery system, rear towhook and a 125 litre petrol tank.

Variants

Pinzgauer Turbo D

See previous entry.

Ambulance

Versions of both the 4 × 4 and 6 × 6 Pinzgauer were available with a fully enclosed rear body which can be fitted with air conditioning.

The Swiss Army ordered 310 Pinzgauer 6 × 6 ambulances in addition to the 1,000-plus 4 × 4 and 6 × 6 cargo versions in service. As well as the driver and one medical assistant, the vehicle can carry four stretcher patients, or two stretcher and four seated patients, or six seated patients.

Anti-aircraft

The Austrian Army fitted a number of 6 × 6 Pinzgauers with 20 mm Oerlikon anti-aircraft cannon at the rear, with spare drum magazines stowed to the immediate rear of the driver's position.

Others

The basic military-type vehicles with cargo-type body are designated the 710 M (4 × 4) and 712 M (6 × 6), the radio vehicles the 710 K (4 × 4) and 712 K (6 × 6), the 6 × 6 firefighting vehicle the 712 FW, the 6 × 6 workshop vehicle the 712 W, and the flatbed version, used to carry the 20 mm cannon, is designated the 712 T.

In 1981 the 712 DK (6 × 6) was introduced; this had an enlarged four-door cab which extends to just over the second axle but still has a payload of 1,500 kg.

Steyr-Daimler-Puch Pinzgauer (6 × 6) 1,500 kg vehicle with open cargo body

Specifications

Configuration	**4 × 4**	**6 × 6**
Weight (empty, soft top/hard top)	1,950/2,100 kg	2,350/2,600 kg
(loaded, soft top/hard top)	3,050/3,100 kg	3,900/4,100 kg
Max weight on front axle (loaded)	1,500 kg	1,450 kg
Max weight on rear axle (loaded)	1,550 kg	2,600 kg (axles)
Max load	1,000 kg	1,500 kg
Towed load (road)	5,000 kg	5,000 kg
(cross-country)	1,500 kg	1,800 kg
Load area	2.25 × 1.592 m	3.03 × 1.592 m
Length	4.175 m	4.955 m
Width	1.76 m	1.76 m
Height	2.045 m	2.045 m
(load area)	0.94 m	0.94 m
Ground clearance	0.335 m	0.335 m
Track	1.44 m	1.44 m
Wheelbase	2.2 m	2 m + 0.98 m
Angle of approach/departure	45°/45°	45°/45°
Max speed	110 km/h	100 km/h
Fuel capacity	75 litres (125 litres optional)	75 litres (125 litres optional)
Fuel consumption (road)	17 litres/100 km	19 litres/100 km
(cross-country)	6-10 litres/h	7-11 litres/h
Gradient	100%	100%
Side slope	40%	40%
Fording	0.7 m	0.7 m
Engine	Steyr 4-cylinder in-line air-cooled petrol developing 87 hp at 4,000 rpm	
Gearbox	manual with 5 forward and 1 reverse gears	
Clutch	single dry plate	single dry plate
Transfer box	2-speed	2-speed
Steering	worm with roller	worm with roller
Turning circle	10.37 m	12.28 m
Suspension	independent, 4 × 4: coil springs at all wheel stations, 6 × 6: coil springs at front wheel stations; leaf springs and hydraulic shock-absorbers rear suspension	
Tyres	7.50 × 16	7.50 × 16
Brakes	hydraulic drum	hydraulic drum
Electrical system	24 V	24 V
Batteries	2 × 12 V, 66 Ah	2 × 12 V, 66 Ah

Status

Production completed during 1990. In service with Austria, Ghana, Nigeria, Oman, Sudan, Switzerland, Tunisia and Yugoslavia. Also in service with various nations in the Middle East and South America.

Manufacturer

Steyr-Daimler-Puch Fahrzeugtechnik GmbH, Postfach 823, Liebenauer Hauptstrasse 317, A-8011 Graz-Thondorf, Austria.
Tel: +43 316 404-0. Fax: +43 316 404 2986.

UPDATED

Steyr 680 M (4 × 4) 4,500 kg truck

Description

The Steyr 680 M is a 4 × 4 cargo truck designed to carry a maximum of 4,170 kg of cargo (when fitted with a winch) across country, although for practical purposes it is normally 4,570 kg, or 6,570 kg on roads.

The all-steel cab is of the forward type and has two doors, observation hatch in the roof, heater and fresh air ventilator. The rear cargo area is provided with removable bows, tarpaulin cover, collapsible seats for up to 20 men and a drop tailgate. A winch with 90 m of 13 mm diameter cable is fitted as standard.

The Steyr 680 M is being supplemented in Austrian Army service by the Steyr 12 M 18 and 12 M 21 (see entry in this section). An overhaul and upgrading programme for the 680 M is being undertaken by Steyr for the Austrian Army. The programme will involve 2,000 Steyr 680 M 4 × 4 and 6 × 6 trucks and will extend their service lives by some 15 years. Part of the programme will involve the fitting of disc brakes.

The Steyr 680 M was produced by Steyr Hellas (now ELBO SA) in Greece as the 680 MH. It was produced for use by the Greek armed forces and for export.

A fuel tanker version is in service. The Steyr 680 M3 (6 × 6) 3,500 kg truck is a development of the Steyr 680 M (4 × 4) truck (see following entry).

Steyr 680 M (4 × 4) 4,500 kg truck

Specifications
Cab seating: 1 + 1
Configuration: 4 × 4
Weight:
(empty, with winch) 5,830 kg
(road, loaded) 11,000 kg
(cross-country, loaded) 8,500 kg
Weight on front axle: (empty) 3,260 kg
Weight on rear axle: (empty) 2,570 kg
Max load:
(road) 6,500 kg
(cross-country) 4,500 kg
Towed load:
(road) 8,000 kg
(cross-country) 4,000 kg
Load area: 4.06 × 2.2 m
Length: 6.57 m
Width: 2.4 m
Height:
(cab) 2.63 m
(tarpaulin) 2.85 m
(load area) 1.16 m
Ground clearance: 0.3 m
Track:
(front) 1.81 m
(rear) 1.67 m
Wheelbase: 3.7 m
Angle of approach/departure: 28°/28°
Max speed: (road) 80 km/h
Range: 450 km
Fuel capacity: 160 litres
Max gradient: (in low range at weight of 10,000 kg) 62%
Fording: 0.8 m
Engine: Steyr WD 610.23 6-cylinder direct injection water-cooled diesel developing 132 hp (SAE) at 2,800 rpm
Gearbox: manual with 5 forward and 1 reverse gears
Clutch: single dry plate
Transfer box: 2-speed
Steering: ZF hydraulic
Turning radius: 7.25 m
Suspension: semi-elliptical springs front and rear
Tyres: 9.00 × 20
Brakes:
(main) hydraulic disc, air-assisted
(parking) mechanical
Electrical system: 24 V
Batteries: 2 × 12 V, 110 Ah

Status
Production complete. In service with Austria, Greece, Indonesia (750), Nigeria and Switzerland.

Manufacturers
Steyr Nutzfahrzeuge AG, Schönauerstrasse 5, A-4400 Steyr, Austria.
Tel: +43 7252 585-0. Fax: +43 7252 46746/48650.
Steyr Hellas SA (now ELBO SA), Industrial Area Sindos, PO Box 10239, GR-541 10 Thessaloniki, Greece.
Tel: +30 31 798502. Telex: 412586 elbo gr.
Fax: +30 31 798426.

VERIFIED

Steyr 680 M3 (6 × 6) 3,500 kg truck

Description
The Steyr 680 M3 is a 6 × 6 cargo truck designed to carry a maximum of 3,500 kg of cargo cross-country or 6,500 kg of cargo on roads. It is a direct development of the Steyr 680 M (4 × 4) 4,500 kg truck (see previous entry) sharing many common components.

An overhaul and upgrading programme for the 680 M is being undertaken by Steyr for the Austrian Army. The programme will involve 2,000 Steyr 680 M 4 × 4 and 6 × 6 trucks and will extend their service lives by some 15 years. Part of the programme involves the fitting of disc brakes at the front and rear.

The all-steel cab is of the forward control type and has two doors, observation hatch in the roof, heater and fresh air ventilator. The rear cargo area is provided with removable bows, tarpaulin cover, drop tailgate and collapsible seats down the centre. Most models have a 4,500 kg capacity winch with 90 m of 13 mm diameter cable. Some vehicles have a hydraulic crane for unloading mounted to the rear of the cab.

Steyr 680 M3 (6 × 6) 3,500 kg truck

Variants
Dump Truck
This is known as the *Dreiseitenkipper* and can tip to either side or the rear. Its loaded weight for road use is 13,000 kg and for cross-country use is 10,000 kg.

Tanker
This is known as the *Einheitstankwagen* (or ETW for short) and carries 8,000 litres of fuel.

680 MH 3
The 680 MH 3 is a licence-built version at one time produced by Steyr Hellas in Greece for the Greek Army. It was also offered for export. Some 55 per cent of this vehicle was produced in Greece.

Specifications
Cab seating: 1 + 1
Configuration: 6 × 6
Weight:
(empty) 6,500 kg
(loaded, road) 12,000 kg
(loaded, cross-country) 10,000 kg
Weight on front axle: (empty) 3,200 kg
Weight on rear bogie: (empty) 3,300 kg
Towed load:
(road) 8,000 kg
(cross-country) 4,000 kg
Load area: 4.06 × 2.198 m
Length: 6.73 m
Width: 2.4 m
Height:
(cab) 2.63 m
(tarpaulin) 2.85 m
(load area) 1.21 m
Ground clearance: 0.3 m
Track:
(front) 1.81 m
(rear) 1.72 m
Wheelbase: 2.76 m + 1.2 m
Angle of approach/departure: 28°/32°
Max speed: (road) 79.7 km/h
Range: 500 km
Fuel capacity: 180 litres
Max gradient: (low range at weight of 10,500 kg) 99%
Fording: 0.8 m
Engine: Steyr model WD 610.74 6-cylinder direct injection supercharged diesel developing 165 hp (SAE) at 2,800 rpm
Gearbox: manual with 5 forward and 1 reverse gears
Clutch: single dry plate
Transfer box: 2-speed
Steering: ZF hydraulic
Turning radius: 7.25 m
Suspension:
(front) semi-elliptical springs with telescopic shock-absorbers
(rear) semi-elliptical springs, reverse mounted with reversible spring hangers positioned in middle of rear axles
Tyres: 9.00 × 20
Brakes:
(main) hydraulic disc, air-assisted
(parking) mechanical
Electrical system: 24 V
Batteries: 2 × 12 V, 110 Ah

Status
In service with the Austrian and Greek armies. It was built by Steyr Hellas in Greece for the Greek Army.

Manufacturers
Steyr Nutzfahrzeuge AG, Schönauerstrasse 5, A-4400 Steyr, Austria.
Tel: +43 7252 585-0. Fax: +43 7252 46746/48650.
Steyr Hellas SA (now ELBO SA), Industrial Area Sindos, PO Box 10239, GR-54110 Thessaloniki, Greece.
Tel: +30 31 798502. Telex: 412586 elbo gr.
Fax: +30 31 798426.

VERIFIED

Steyr 10 M 14 and 10 M 18 (4 × 4) 4,000 kg trucks

Description

The Steyr 10 M 14 (4 × 4) 4,000 kg truck was introduced in December 1990 and is a derivative of the Steyr 14 M 14 (4 × 4) 8,000 kg truck but is fitted with single-tyred rear wheels and with a lockable front axle differential for increased mobility. The Steyr 10 M 18 is essentially the same vehicle but fitted with a more powerful (177 hp) engine.

The layout uses a cab-over-engine forward control cab with seating for the driver and two passengers. The cab is the Steyr 152 (short version) with a hydraulic tilting mechanism. Left- and right-hand drive can be provided. The chassis has a low-torsion ladder-type frame with riveted tubular cross members. The 140 hp diesel engine used on the 10 M 14 is turbocharged and powers a synchromesh five-speed gearbox (six-speed on 10 M 18). Using a Steyr VG 450 two-speed transfer box, front-wheel drive can be engaged. There are differential locks on both axles.

The cargo body has drop sides and is provided with a spare wheel at the front. The body can be covered by a tarpaulin over a frame. Seating for passengers in the rear can be along mid- or side-positioned bench seats. An optional 5,000 kg capacity winch can be used to the front or rear.

These vehicles are also available as water or fuel bowsers, ambulances, fire tenders, shelter carriers and in other configurations.

Specifications

Cab seating: 1 + 2 (up to 16 in rear)
Configuration: 4 × 4
Weight:
(empty) approx 6,000 kg
(loaded) approx 10,000 kg
Max load: 4,000 kg
Towed load: 6,000 kg
Load area: 4.18 × 2.42 m
Length: 6.54 m
Width: 2.5 m
Height:
(top of tilt) 3.28 m
(cab roof) 3.11 m
Ground clearance: 0.295 m
Track:
(front) 1.907 m
(rear) 1.9 m
Wheelbase: 3.7 m
Angle of approach/departure: 41°/40°
Max speed: approx 100 km/h
Range: approx 900 km
Fuel capacity: 200 litres
Gradient:
(without trailer) 60%
(with trailer) 40%
Side slope: 30%
Fording: 0.8 m
Engine: (10 M 14) Steyr WD 612.91 6.6 litre 6-cylinder direct injection water-cooled diesel with turbocharger developing 140 hp (ISO) at 1,300 rpm
(10 M 18) Steyr WD 612.73 6.6 litre 6-cylinder water-cooled diesel with turbocharger developing 177 hp at 2,200 rpm
Gearbox: ZF S6-36 synchromesh with 6 forward and 1 reverse gears
Clutch: single dry disc, hydraulic actuation
Transfer box: Steyr VG 450, 2-speed
Steering: ZF 8043, power-assisted
Turning radius:
(10 M 14) 7.6 m
(10 M 18) 7.8 m
Suspension: semi-elliptical springs with telescopic shock-absorbers at front
Tyres: 12.5 R 20 or 10.00 R 20
Brakes:
(main) dual circuit hydropneumatic
(parking) spring actuated, air
Electrical system: 24 V
Batteries: 2 × 12 V, 135 Ah
Alternator: 28 V/840 W

Status

In production.

Manufacturer

Steyr Nutzfahrzeuge AG, Schönauerstrasse 5, A-4400 Steyr, Austria.
Tel: +43 7252 585-0. Fax: +43 7252 46746/48650.

UPDATED

Steyr 10 M 14 (4 × 4) truck

Steyr 12 M 18 and 12 M 21 (4 × 4) 5,000 kg trucks

Development

In December 1985 the Austrian Ministry of Defence placed an order for 1,000 12 M 18 and 12 M 21 (4 × 4) trucks to replace existing Steyr 680 trucks. The 12 M 18 was used as the basis for the US Stewart and Stevenson FMTV (Family of Medium Tactical Vehicles - see entry under USA in this section).

The Steyr 12 M 18 and 12 M 21 (4 × 4) 5,000 kg trucks are described as a new generation of medium military trucks and are designed to carry a payload of 5,350 kg both on and off roads. The main differences between the two models involve the type and power of the engine fitted and the gearboxes involved.

Description

Both models use a chassis consisting of a low-torsion ladder-type frame with bolted tubular cross members. The forward control cab is of all-steel construction and can be tilted forward up to an angle of 70° for access to the engine and transmission. The two-door cab has seats for the driver and up to three passengers. It is equipped with a roof observation hatch, a heater and a fresh air blower.

The front and rear axle differentials, as well as the inter-axle differential in the transfer box, are all lockable.

The drop-side body has a tarpaulin frame which can be lowered for rail transport and mid-positioned bench seats for up to 18 men can be fitted. Alternative bodies could include fuel and water tankers, fire tenders, ambulances and other options. A driver training version is available.

Optional equipment includes a blackout lighting system, a 5,000 kg capacity hydro-winch with 50 m of cable, a machine gun mounting over the cab, air conditioning, an electrohydraulic tailgate (the Steyr LBW 1500 electrohydraulic tailgate – see entry under *Materials handling equipment*) and a loading crane fitted behind the driver's cab.

Variants include ambulances, recovery vehicles, fire-fighting vehicles, communication shelter carriers, mobile workshops and generators, tankers, light artillery tractors and weapons carriers.

Steyr 12 M 21 (4 × 4) 5,000 kg truck with winch

Specifications

Cab seating: 1 + 1 to 3, with up to 18 in rear
Configuration: 4 × 4
Weight:
(unladen) approx 6,500 kg
(laden) 11,500 kg
Weight on front axle: 5,000 kg
Weight on rear axle: 6,500 kg

Max load: 5,000 kg
Towed load:
(road) 10,000 kg
(cross-country) 8,500 kg
Platform dimensions: 4.1 × 2.42 m
Length overall: 6.36 m
Width: 2.5 m
Height:
(cab) 2.93 m
(top of tilt) 3.24 m
(load area) 1.46 m
Ground clearance: 0.3 m
Track:
(front) 2.061 m
(rear) 2.005 m
Wheelbase: 3.5 m
Angle of approach/departure: 41°/44°
Max speed: (road) approx 100 km/h
Range: approx 700 km
Fuel capacity: 180 litres
Gradient: approx 80%
Side slope: 30%
Fording: 0.8 m
Engine: (12 M 18) Steyr Model WD 612.74 6.6 litre 6-cylinder direct injection turbocharged water-cooled diesel developing 177 hp at 2,400 rpm
(12 M 21) Steyr Model WD 612.76 6.6 litre 6-cylinder direct injection turbocharged (with boost air cooler) water-cooled diesel developing 210 hp at 2,400 rpm
Clutch: single dry plate, hydraulically actuated
Gearbox: ZF 9 S 109 with 9 forward (8 + crawler) and 1 reverse gears
Transfer box: Steyr VG 1200, 2-stage
Steering: ZF Model 8043 power
Turning radius: 7.3 m
Suspension:
(front) parabolic leaf springs with telescopic shock-absorbers
(rear) parabolic leaf springs with telescopic shock-absorbers and stabiliser bar
Tyres: 14.5 R 20 MPT
Brakes:
(main) drum, dual-circuit hydropneumatic
(parking) spring-actuated, air-operated
Electrical system: 24 V
Batteries: 2 × 12 V, 135 Ah
Alternator: 980 W

Status

In production. In service with the Austrian Army (over 1,000), Kuwait, Pakistan, Thailand and United Nations forces. Used by Canadian Forces for rapid airfield repair.

Manufacturer

Steyr Nutzfahrzeuge AG, Schönauerstrasse 5, A-4400 Steyr, Austria.
Tel: +43 7252 585-0. Fax: +43 7252 46746/48650.

UPDATED

Steyr 17 M 29 (4 × 4) 7,000 kg truck

Description

The Steyr 17 M 29 (formerly the 1291.280.4 × 4 M) is a 4 × 4 cargo truck designed to carry a maximum load of 7,000 kg both on roads and cross-country. It shares many common components with the Steyr 24 M (6 × 6) 12,000 kg truck series.

The chassis consists of a low-distortion parallel ladder frame with bolted tubular cross members. The forward control cab is of all-steel construction and can be tilted forwards at an angle of 70° to allow access to the engine and transmission for maintenance purposes. The driver's suspension seat is adjustable and the cab can be fitted with a single or twin passenger seat. The cab has two doors, observation/ machine gun hatch in the roof, heater and a fresh air ventilator.

The rear cargo area is provided with removable bows, tarpaulin cover, drop tailgate and, down the centre, collapsible seats for 20 fully equipped troops. Seat benches, bows and the tarpaulin cover can be stowed behind the headboard.

Optional equipment includes a winch with a capacity of 10,000 kg and 60 m of cable; alternative engines and transmissions are also available.

The Steyr 17 M 29 has been produced in airfield crash tender form.

Various versions in 4 × 4 and 4 × 2 are licence-produced in China as part of the Hongyan heavy-duty truck range manufactured at the Sichuan Truck Plant, Chongqing.

Specifications

Cab seating: 1 + 1 or 2
Configuration: 4 × 4
Weight:
(empty) approx 10,000 kg
(loaded) 17,000 kg
Weight on front axle: (loaded) 7,000 kg
Weight on rear axle: (loaded) 10,000 kg
Max load: (road and cross-country) 7,000 kg
Towed load:
(cross-country) 10,000 kg
(road) up to 11,500 kg
Load area: 5 × 2.43 m
Length: 7.73 m
Width: 2.5 m
Height: (cab platform) 3.246 m
Ground clearance: 0.37 m
Track: 2.072 m
Wheelbase: 4.2 m
Angle of approach/departure: 31°/36°
Max speed: (road) 95 km/h
Range: 700 km
Fuel capacity: 320 litres (400 litres optional)
Gradient: 100% (low range)
Fording: 0.8 m (1 m optional)
Engine: Steyr model WD 615.73, 9.726 litre 6-cylinder, direct injection turbocharged intercooled water-cooled diesel developing 287 hp (SAE) at 2,200 rpm
Gearbox: ZF 9 S 109 GP manual with 9 (8 + crawler) forward and 1 reverse gears
Clutch: single dry plate (or optional ZF hydraulic torque converter in combination with synchromesh gearbox or Allison automatic)
Transfer box: Steyr VG 1200 2-speed
Steering: ZF hydraulic
Turning radius: 9.8 m
Suspension:
(front) semi-elliptical springs with telescopic shock-absorbers
(rear) semi-elliptical springs with telescopic shock-absorbers
Tyres: 14.00 × 20
Brakes:
(main) two circuit compressed air
(parking) spring energy, air-operated
Electrical system: 24 V
Batteries: 2 × 12 V, 135 Ah

Status

In production. In service with Egypt, Ghana and Indonesia (200 plus). In licensed production in China.

Manufacturer

Steyr Nutzfahrzeuge AG, Schönauerstrasse 5, A-4400 Steyr, Austria.
Tel: +43 7252 585-0. Fax: +43 7252 46746/48650.

VERIFIED

Steyr 17 M 29 7,000 kg truck with brushguard

Steyr 17 M 29 7,000 kg truck with bows and tarpaulin cover

Steyr 14 M 14 and 14 M 18 (4 × 4) 8,000 kg trucks

Description

The Steyr 14 M 14 (4 × 4) 8,000 kg truck was originally designed as a 4 × 2 truck for Nigeria. It was converted to 4 × 4 by installing the transfer case from the Steyr 680 M and the front axle of the 12 M 18/21 (see entries in this section for details of both vehicles). This vehicle is licence-produced in Greece by the Hellenic Vehicle Industry SA (ELBO). The 14 M 18 is an essentially similar vehicle with a more powerful (177 hp) engine.

The layout of the 14 M 14 is conventional, with a forward control cab and a cargo area to the rear. The cab has seating for the driver and two passengers and the cargo area has steel drop sides and a tailgate all covered by a tilt over bows. The cargo area has folding back-to-back benches for carrying troops.

Options include a right-hand drive version, a 5,000 kg capacity rear winch, a driver training version, an electrohydraulic tailgate, a blackout lighting system, a machine gun mount over the cab, air conditioning, and a 24 V electrical system.

Specifications

Cab seating: 1 + 2
Configuration: 4 × 4
Weight:
(empty) approx 6,000 kg
(loaded) 14,000 kg

Max front axle load: 5,000 kg
Max rear axle load: 9,500 kg
Max load: 8,000 kg
Towed load: 6,000 kg
Load area: 4.065 × 2.42 m
Length: 6.62 m
Width: 2.5 m
Height:
(top of tilt) 3.205 m
(cab) 3.025 m
(load area) 1.43 m
Ground clearance: 0.275 m
Track:
(front) 1.872 m
(rear) 1.721 m
Wheelbase: 3.7 m
Angle of approach/departure: 39°/53°
Max speed: approx 80 km/h
Range: approx 700 km
Fuel capacity: 160 litres
Max gradient:
(without trailer) 100%
(with trailer) 40%
Side slope: 30%
Fording: 0.8 m
Engine: (14 M 14) Steyr WD 612.22 6.6 litre 6-cylinder direct injection water-cooled diesel developing 136 hp (DIN) at 2,400 rpm
(14 M 18) Steyr WD 612.73 6.6 litre 6-cylinder water-cooled turbocharged diesel developing 177 hp at 2,200 rpm
Gearbox: ZF S6-36 synchromesh with 6 forward and 1 reverse gears
Clutch: single dry plate
Transfer box: Steyr VG 450, 2-speed
Steering: ZF 8043, power-assisted
Turning radius: 7.6 m

Steyr 14 M 14 (4 × 4) 8,000 kg truck

Suspension: semi-elliptic leaf springs with telescopic shock-absorbers front and rear
Tyres: 9.00 × 20
Brakes:
(main) single-circuit hydropneumatic (dual-circuit optional)
(parking) spring-actuated, air-operated
Electrical system: 12 V (24 V optional)
Batteries: 1 × 12 V, 165 Ah (2 × 12 V, 88 Ah optional)
Alternator: 396 W

Status
In production. In service with Canadian Forces, Cyprus, Ghana, Greece, Nigeria, Thailand and Uganda. Licence-produced in Greece (see separate entry).

Manufacturers
Steyr Nutzfahrzeuge AG, Schönauerstrasse 5, A-4400 Steyr, Austria.
Tel: +43 7252 585-0. Fax: +43 7252 46746/48650.
ELBO SA, Industrial Area Sindos, PO Box 10239, GR-541 10 Thessaloniki, Greece.
Tel: +30 31 798502. Telex: 412586 elbo gr.
Fax: +30 31 798426.

UPDATED

ÖAF Type 20.320 (6 × 6) 10,000 kg truck

Development
The ÖAF Type 20.320 (6 × 6) 10,000 kg truck is essentially a modified version of the German Army MAN 7,000 kg truck built in Austria. Modifications have included the replacement of the Klockner-Humbolt-Deutz engine by a MAN unit, installation of a hydraulic crane and provision for mounting a machine gun on the roof of the cab. The initial order was for 350 trucks, the first of which was handed over in 1975. Some are employed as light air defence gun and associated radar tractors.

Description
The chassis is constructed from welded box-section longitudinal girders strengthened by tubular cross members. An auxiliary frame to the rear of the cab carries the loading crane.

The cab is made of welded sheet steel and is mounted on rubber supports on the chassis. It is divided into two compartments, crew at the front and engine at the rear. Forward-opening doors are fitted left and right and there are seats for three people which can be arranged to provide two couchettes. The centre seat has a 180° vertical movement and serves as a step for the machine gun slewing ring mounted on the roof. The roof has a sliding hatch. The spare wheel is housed in the upper part of the engine compartment and there is a small hoist provided to raise and lower it. The battery box is on the right side of the engine compartment.

The base for the body comprises seven sheet steel cross members bolted firmly to the chassis, the floor of which is 40 mm tongue-and-groove pine. The side and tailboards consist of 25 mm hollow-section aluminium. The front is 1.5 m high and the sides and rear boards 700 mm high. The side boards are divided and removable seats for eight men are fitted to the floor either to the front or rear. If required, a second similar seat-group can be fitted. With only one eight-seater in place there is a load area for 72 ammunition boxes on the floor. The three adjustable bows support the canvas tarpaulin which has windows towards the top.

The vehicle is radio-suppressed and a junction for an external current supply is provided. An Eberspeicher independent diesel-driven warm air heating unit is fitted as standard. Mounted to the rear of the cab is a Palfinger PK 7500 loading crane with a capacity of 7,000 kg.

ÖAF Type 20.320 (6 × 6) 10,000 kg truck

The last production version was the Type 20.320/G2 which has a centrally mounted winch with a 7,500 kg capacity front and rear.

Specifications
Cab seating: 1 + 2
Configuration: 6 × 6
Weight:
(empty with crane and winch) 12,500 kg
(loaded, cross-country) 22,000 kg
Max load: (cross-country) 10,000 kg
Length: 8.85 m
Width: 2.5 m
Height:
(tarpaulin cover) 3.34 m
(cab) 3 m
Ground clearance: 0.415 m
Track: 2.07 m
Wheelbase: 3.8 m + 1.4 m
Angle of approach/departure: 42°/38°
Max road speed: 84 km/h
Max range: 700 km
Fuel capacity: 500 litres
Gradient: 60%
Fording: 0.8 m
Engine: MAN D 2538 MTX V-8 diesel developing 320 bhp at 2,500 rpm
Gearbox: hydraulic torque converter with lock-up clutch (foot-operated), 6-speed gearbox (6 forward and 1 reverse gears) and 3-axle transfer case with differential lock between front and rear axle drive
Steering: recirculating ball with hydraulic assistance
Turning circle: 19.5 m
Suspension: progressively acting helical springs and hydraulic shock-absorbers
Tyres: 14.00 × 20
Brakes:
(main) air/hydraulic
(parking) mechanical
Electrical system: 24 V

Status
Production complete. In service with the Austrian Army.

Manufacturer
Österreichische Automobilfabrik ÖAF-Gräf und Stift AG, Brünner Strasse 72, A-1211 Vienna, Austria.

VERIFIED

Steyr 24 M (6 × 6) 12,000 kg truck series

Description

The base vehicle in the Steyr 24 M series (formerly the 1491.6 × 6 M) is a 6 × 6 cargo truck designed to carry a maximum load of 12,000 kg both on roads and cross-country, and to share many common components with the Steyr 17 M (4 × 4) 7,000 kg trucks.

The chassis consists of a low-distortion parallel ladder frame with bolted tubular cross members. The forward control cab is of all-steel construction and can be tilted forwards at an angle of 70° to allow access to the engine and transmission for maintenance purposes. The driver's suspension seat is adjustable and the cab can be fitted with a single or twin passenger seat. The fully enclosed cab has two doors, observation/machine gun hatch in the roof, heater and a fresh air ventilator.

There is a lockable differential lock between the two rear axles and a lockable differential on each of the rear two axles.

The rear cargo area is provided with removable bows, tarpaulin cover, drop tailgate and collapsible seats for troops down the centre. Seat benches, bows and the tarpaulin cover can be stowed behind the headboard.

Optional equipment includes a winch with a capacity of 10,000 kg and 60 m of cable.

Various versions in 6 × 4 and 6 × 6 are licence-produced in China as part of the Hongyan heavy-duty truck range manufactured at the Sichuan Truck Plant, Chongqing.

Variants

Steyr 1491.260. K34/6 × 6 recovery vehicle

This variant mounts a lifting crane in place of the cargo area. Four telescopic stabiliser legs are provided for use when the crane is in operation and extra stowage is provided for tools and equipment. The crane has a lifting capacity of 13,000 kg.

Steyr 24 M 32 040/6 × 6 M

This vehicle was produced to participate in the Canadian Heavy Logistic Vehicle Wheeled (HLVW) contest. During 1988 the Canadian Department of National Defence ordered 1,200 units to be assembled in Canada by UTDC Inc of Kingston, Ontario. For further details of these vehicles refer to separate entry under International in this section.

A military firefighting version of this vehicle is produced by Hellenic Vehicle Industry SA (ELBO) in Greece. It carries a 5,500 litre water tank and a 700 litre foam tank.

Steyr 24 M 34 (6 × 6)

This version is used as a heavy artillery tractor and is equipped with an ammunition handling crane. Part of the cargo area is covered by a tarpaulin and is used as a crew compartment for the gun crew.

Steyr 24 M 34 S40/6 × 6 truck tractor

This variant is used as the prime mover for a tank transporter. For further details refer to the entry in the *Tank Transporters* section.

Specifications

Steyr 24 M 34
Cab seating: 1 + 1 (or 1 + 2)
Configuration: 6 × 6
Weight:
(empty) 12,000 kg
(loaded) 22,500 kg
Weight on front axle: (loaded) 7,500 kg
Weight on each rear axle: (loaded) 10,000 kg
Max load: 12,000 kg
Towed load:
(cross-country) 15,000 kg
(road, up to) 40,000 kg
Load area: 6.2 × 2.43 m
Length: 8.92 m
Width: 2.5 m
Height: (cab platform) 3.246 m
Ground clearance: 0.37 m
Track: 2.072 m
Wheelbase: 4 m + 1.4 m
Angle of approach/departure: 31°/36°
Max speed: (road) 95 km/h
Range: (approx) 700 km
Fuel capacity: 400 litres
Max gradient: 100% (low range at weight of 22,500 kg)
Fording: 0.8 m (1 m optional)
Engine: Steyr model WD 815.74, 11.97 litre V-8 direct injection, turbocharged, water-cooled diesel, developing 342 hp at 2,200 rpm
Gearbox: ZF 5 S 111 GP manual with 9 forward (8 + 1 crawler) and 1 reverse gears or optional ZF hydraulic torque converter in combination with synchromesh gearbox or Allison automatic
Clutch: single dry plate (mechanically operated with assistance of compressed air)
Transfer box: Steyr VG 1200 2-speed
Steering: ZF 8046 hydraulic
Turning radius: 10.45 m
Suspension: semi-elliptical springs with telescopic shock-absorbers front and rear
Tyres: 14.00 × 20
Brakes:
(main) dual-circuit compressed air
(parking) spring energy, air-operated
Electrical system: 24 V
Batteries: 2 × 12 V, 135 Ah

Status

In production. In service with Canada, Cyprus and Egypt. Licence-produced in China. Was manufactured in Canada for the Canadian Department of Defence.

Manufacturer

Steyr Nutzfahrzeuge AG, Schönauerstrasse 5, A-4400 Steyr, Austria.
Tel: +43 7252 585-0. Fax: +43 7252 46746/48650.

VERIFIED

Steyr 24 M 32 12,000 kg truck

Steyr 26 M 39 (6 × 6) 12,000 kg truck

Description

The Steyr 26 M 39 (6 × 6) 12,000 kg truck is an improved version of the 24 M series cargo truck (see previous entry) developed to carry a 12,000 kg payload and tow a similar or heavier load. It is fitted with a 10 t/m material handling crane and a 10 tonne self-recovery winch. A central tyre inflation system is available.

Specifications

Cab seating: 1 + 1 or 1 + 2
Configuration: 6 × 6
Weight:
(empty) 19,000 kg
(loaded) 26,000 kg
Weight on front axle: 7,500 kg
Weight on rear axles: 2 × 9,500 kg
Max load: 12,000 kg
Towed load: (cross-country) 15,000 kg
Load area: 5.2 × 2.4 m
Length: 8.4 m
Width: 2.55 m
Height: 3.1 m
Ground clearance: 0.42 m

Steyr 26 M 39 (6 × 6) 12,000 kg truck during desert trials

Track: 2.1 m
Wheelbase: 4 m + 1.4 m
Angle of approach/departure: 35°/40°
Max speed: 95 km/h
Range: 700 km
Fuel capacity: 400 litres
Gradient: 100%
Fording: 1 m
Engine: Steyr WD 815.76 12 litre V-8 direct injection water-cooled turbocharged intercooled diesel developing 390 hp at 2,200 rpm
Gearbox: ZF 4 S 150 GP 8-speed synchromesh
Transfer box: Steyr VG 2000, 2-speed
Steering: ZF 8046 power-assist
Turning radius: 10.5 m
Suspension: semi-elliptical springs with shock-absorbers front and rear
Tyres: 16.00 R 20
Brakes: dual-circuit air
Electrical system: 24 V
Batteries: 2 × 12 V, 165 Ah each
Alternator: 28 V, 55 A

Status
In production.

Manufacturer
Steyr Nutzfahrzeuge AG, Schönauerstrasse 5, A-4400 Steyr, Austria.
Tel: +43 7252 585-0. Telex: 28200.
Fax: +43 7252 26746/28650.

VERIFIED

Steyr 26 M 42 P40 (6 × 6) 14,000 kg truck series

Description
The Steyr 26 M 42 P40 (6 × 6) 14,000 kg truck series is a further developed and enhanced capacity version of the 24 M series cargo truck (see separate entry) developed to carry a 14,000 kg payload and tow a similar or heavier load. Numerous body versions can be adapted to the basic chassis, including a Multilift Mark 4 load-handling system (or a similar system), tanker bodies, recovery equipment, shelters, firefighting equipment, material-handling cranes, troop carriers, and many others.

Specifications
Cab seating: 1 + 2
Configuration: 6 × 6
Weight:
(empty) approx 12,000 kg
(loaded) 26,000 kg
Max load: approx 14,000 kg
Towed load: (cross-country) 15,000 kg
Length: (with LHS) 8.469 m
Width: 2.5 m
Height: 3.276 m
Ground clearance: 0.4 m
Track: 2.081 m
Wheelbase: 4 m + 1.4 m or 4 m + 1.47 m
Angle of approach/departure: 31°/36°
Max speed: approx 100 km/h
Range: 700 km
Fuel capacity: 400 litres
Gradient: approx 80%
Fording: 0.8 m
Engine: Steyr WD 815.73 12 litre V-8 turbocharged intercooled diesel developing 416 hp at 2,200 rpm
Gearbox: ZF 4 S 150 GP 8-speed synchromesh
Transfer box: Steyr VG 2000, 2-speed
Steering: ZF power-assisted
Turning radius: 10.45 m
Suspension: semi-elliptical springs with shock-absorbers front and rear, rear axles suspended independently
Tyres: 16.00 R 20
Brakes: dual-circuit air
Electrical system: 24 V
Batteries: 2 × 12 V, 135 Ah each
Generator: 28 V, 840 W

Status
In production.

Manufacturer
Steyr Nutzfahrzeuge AG, Schönauerstrasse 5, A-4400 Steyr, Austria.
Tel: +43 7252 585-0. Fax: +43 7252 46746/48650.

VERIFIED

Steyr 26 M 42 P40 (6 × 6) 14,000 kg truck with Multilift Mark 4 load-handling system

Steyr 26 M 42 P40 (6 × 6) 14,000 kg truck with cargo body and material-handling crane

BRAZIL

ENGESA EE-15 (4 × 4) 1,500 kg truck

Description
The EE-15 (4 × 4) 1,500 kg truck was designed by ENGESA for both civil and military use. Its chassis is of high-tensile steel with cross members riveted into position. A rear towing hook and four lifting eyes are fitted. The cab has a vinyl top and doors, fold-down windscreen with wipers, adjustable seat for the driver and a seat for the two passengers. The standard all-steel cargo body has a tailgate, bows and a tarpaulin cover.

A wide range of optional equipment is available including a body with tilting troop seats, a power take-off and engine starting pilot for cold areas.

ENGESA can supply the EE-15 with other types of body including ambulance, crash tender and van.

Specifications
Cab seating: 1 + 2

ENGESA EE-15 (4 × 4) 1,500 kg truck fitted with van-type body

ENGESA EE-15 (4 × 4) 1,500 kg truck

Configuration: 4 × 4
Weight:
(empty) 4,050 kg
(loaded, road) 7,050 kg
(loaded, cross-country) 5,550 kg
Max load:
(road) 3,000 kg
(cross-country) 1,500 kg
Max towed load: 1,500 kg
Length: 5.65 m
Width: 2.25 m
Height:
(cab) 2.4 m
(tarpaulin) 2.5 m
Ground clearance: 0.27 m
Track: 1.71 m
Wheelbase: 3.3 m
Angle of approach/departure: 42°/35°
Max speed: (road) 80 km/h
Range: 600 km
Fuel capacity: 136 litres
Max gradient: (loaded, cross-country) 60%
Max side slope: 30%
Fording: 0.9 m
Vertical obstacle: 0.4 m
Engine: Mercedes-Benz OM-352 6-cylinder in-line water-cooled diesel developing 130 hp at 2,800 rpm
Gearbox: Mercedes-Benz G-3-60 manual with 5 forward and 1 reverse gears
Clutch: single dry disc
Transfer box: mechanical ENGESA, 2-speed with constant mesh helical gears, optional front or rear left side PTO
Suspension:
(front) semi-elliptical springs and double-acting hydraulic shock-absorbers
(rear) 2 semi-elliptical springs and double-acting hydraulic shock-absorbers
Steering: mechanical
Turning radius: 8 m
Tyres: 9.00 × 20-12 PR
Brakes:
(main) drum type, air/hydraulic
(parking) mechanical acting on transfer case
Electrical system: 24 V
Batteries: 2 × 12 V, 95 Ah

Status
Production complete. In service with Angola, Brazil, Chile, Colombia, Gabon and other undisclosed countries.

Manufacturer
ENGESA Engenheiros Especializados SA, Avenida Tucunaré 125/211, Caixa Postal 152/154, 06400 Barueri, SP, Brazil.
Tel: +55 11 421 4711. Telex: 1171 302 enes br.
This company is no longer trading.

VERIFIED

ENGESA EE-25 (6 × 6) 2,500 kg truck

Description
The EE-25 (6 × 6) truck was developed by ENGESA for both civil and military applications. Its chassis is of high-tensile steel with cross members riveted into position. A rear towhook is provided. The layout of the vehicle is conventional, with the engine at the front, cab in the centre and the cargo area at the rear. The cab has a vinyl top and doors, fold-down windscreen with wipers, adjustable seat for the driver, and a single seat for the two passengers. The all-steel body has a tailgate, bows and a tarpaulin cover.

With the award of a contract to supply EE-25 trucks to Angola, some modifications to the vehicle were introduced. The main changes were a strengthened chassis and a new body and cab. The engine and transmission remain the same but an air cleaner is introduced to the right-hand side of the bonnet. A 24 V electrical system was incorporated. Vehicles delivered to Angola have a brushguard over the front.

Optional equipment includes: a 6 t/m capacity hydraulic crane; a centre or rear winch with a 7,500 kg capacity and two drums with 80 m of 16 mm diameter cable; a power take-off; and an engine start pilot for cold areas. A 4 × 4 model of the EE-25 is also available.

The basic chassis can be used for a variety of roles including ambulance, crash tender, dump truck (4 × 4), recovery, tanker (fuel or water), lubrication vehicle, van (for example, command or workshop) and NBC decontamination truck (4 × 4).

ENGESA EE-25 (6 × 6) 2,500 kg truck

Specifications
6 × 6 model
Cab seating: 1 + 2
Configuration: 6 × 6
Weight:
(empty) 7,100 kg
(loaded, road) 12,100 kg
(loaded, cross-country) 9,600 kg
Max load:
(road) 5,000 kg
(cross-country) 2,500 kg
Length: 6.99 m
Width: 2.25 m
Height:
(cab) 2.5 m
(tarpaulin) 3.1 m
Ground clearance: (axles) 0.27 m
Track: 1.78 m
Wheelbase: (front axle to centre of rear axle) 4.2 m
Angle of approach/departure: 46°/50°
Max speed: (road) 80 km/h
Range: 700 km
Fuel capacity: 200 litres
Max gradient: 60%
Max side slope: 30%
Fording: 0.9 m
Vertical obstacle: 0.6 m
Engine: Mercedes-Benz OM-352A 6-cylinder in-line diesel developing 156 hp at 2,800 rpm
Gearbox: manual with 5 forward and 1 reverse gears
Clutch: single dry disc
Transfer box: ENGESA 2-speed
Steering: hydraulic
Turning radius: 9.5 m
Suspension:
(front) semi-elliptical springs with double-acting hydraulic shock-absorbers
(rear) semi-elliptical springs; rear axle is ENGESA Boomerang type with walking beams
Tyres: 11.00 × 20-14 PR
Brakes:
(main) air/hydraulic drum
(parking) mechanical acting on transfer case
Electrical system: 24 V

Status
Production complete. In service with Angola, Bolivia, Brazil, Chile, Colombia, Gabon, Libya and other undisclosed countries.

Manufacturer
ENGESA Engenheiros Especializados SA, Avenida Tucunaré 125/211, Caixa Postal 152/154, 06400 Barueri, SP, Brazil.
Tel: +55 11 421 4711. Telex: 1171 302 enes br.
This company is no longer trading.

VERIFIED

ENGESA EE-50 (6 × 6) 5,000 kg truck

Description
The ENGESA EE-50 (6 × 6) 5,000 kg truck was designed to meet the requirements of the Brazilian Army. Following trials with prototypes it was selected to become the standard 5,000 kg truck of the Brazilian Army in the 1980s. First production vehicles were completed during late 1980.

The EE-50 chassis is made of riveted steel rails and cross members. The all-steel forward control cab can be tilted forward to allow access to the engine. The windscreen is in two parts, left and right, both hinged at the top and with a windscreen wiper.

The all-steel rear cargo area is provided with lateral tilting troop seats, drop tailgate, removable bows and a tarpaulin cover. Optional equipment includes an engine-starting pilot for cold weather and a fire detection and extinguishing system.

The EE-50 could be supplied with a number of types of body including an ambulance, van, fuel tanker, light recovery, water tanker, fire tender and mobile workshop.

Specifications
Cab seating: 1 + 1
Configuration: 6 × 6
Weight:
(empty) 12,000 kg
(loaded, road) 22,000 kg
(loaded, cross-country) 17,000 kg
Max load:
(road) 10,000 kg
(cross-country) 5,000 kg
Max towed load: 6,000 kg
Length: 7.85 m

ENGESA EE-50 (6 × 6) 5,000 kg truck

Width: 2.6 m
Height:
(cab hatch) 3.025 m
(tarpaulin cover) 3.5 m
Ground clearance: 0.35 m
Track: 2.1 m
Wheelbase: (first axle to centre of rear bogie) 4.5 m
Angle of approach/departure: 40°/50°
Max speed: 80 km/h
Range: 700 km
Fuel capacity: 300 litres
Max gradient: 60%
Max side slope: 30%
Fording: 1.4 m
Vertical obstacle: 0.6 m
Engine: Scania D11 6-cylinder water-cooled diesel developing 202 hp (SAE) at 2,100 rpm
Gearbox: manual, 5 forward and 1 reverse gears
Clutch: single dry plate
Transfer box: mechanical, 2-speed
Suspension:
(front) leaf springs, front wheels driven by hypoid angular transmission and bevel differential gears
(rear) leaf springs, ENGESA Boomerang articulated axle with hypoid angular transmission, bevel differential gears and side walking beams with helical gears
Steering: integral hydraulic
Tyres: 14.00 × 20-18 PR
Brakes:
(main) air
(parking) air, on transfer case output shaft
Electrical system: 24 V

Status
Production complete. In service with Angola, the Brazilian Army and at least one other country.

Manufacturer
ENGESA Engenheiros Especializados SA, Avenida Tucunaré 125/211, Caixa Postal 152/154, 06400 Barueri, SP, Brazil
Tel: +55 11 421 4711. Telex: 1171 302 enes br.
This company is no longer trading.

VERIFIED

TECTRAN military trucks

Description
TECTRAN is producing a range of military trucks with payloads of 2.5, 5 and 10 tonnes. The trucks share a common cab and many other components, the 2.5 tonne model having a 4 × 4 configuration while the others are 6 × 6.

The cab is of the normal forward-control type and may have a hard or canvas top. An armoured cabin weighing 2,100 kg is optional for the heavier models. Internal equipment can include air conditioning and a radio installation.

A family of body types is also produced and it is possible for all chassis types to have an auxiliary frame for the installation of a torsion-free platform. Standard body types include a general cargo platform, a cargo/personnel body, a liquid load transporter, a dumper body, mechanical and electrical mobile workshops, command, control and communication posts, and a mobile satellite communications post body. By the addition of a fifth wheel over the rear axle(s) the vehicles can tow various semi-trailers.

The data in the Specifications table refers to the basic chassis only.

Status
Ready for production.

Manufacturer
TECTRAN Engenharia, Indústria e Comércio SA, Rodovia Presidente Dutra, Km 155/156, Caixa Postal 165, 12200 São José dos Campos, São Paulo, Brazil.
Tel: +55 123 318200. Telex: 123 3519.
Fax: +55 123 318775.

VERIFIED

Specifications

Model	**2.5 tonne**	**5 tonne**	**10 tonne**
Cab seating	1 + 1	1 + 1	1 + 1
Configuration	4 × 4	6 × 6	6 × 6
Weight (chassis)	4,800 kg	6,700 kg	8,500 kg
(payload and body)	3,200 kg	6,300 kg	11,500 kg
(GVW)	8,000 kg	13,000 kg	20,000 kg
Length	6.7 m	7 m	7.85 m
Width	2.36 m	2.36 m	2.36 m
Height (cab)	3 m	3.05 m	3.05 m
Wheelbase	4.2 m	3.6 m + 1.3 m	3.8 m + 1.45 m
Track	2.053 m	1.943 m	2.07 m
Max speed (road)	80 km/h	85 km/h	90 km/h
Fuel capacity	210 litres	300 litres	300 litres
Max gradient	60%	60%	60%
Fording	0.9 m	1 m	1 m
Engine	diesel, water-cooled, turbocharged		
Power	180 hp	200 hp	280 hp
Rpm	2,600	2,200	2,300
Transmission	mechanical, fully synchronised		
Gears	5 f, 1 r	6 f, 1 r	6 f, 1 r
Clutch	single dry disc	single dry disc	single dry disc
Transfer box	2-speed	2-speed	2-speed
Steering	worm and gear, power-assisted		
Turning radius	8 m	8.7 m	10.2 m
Tyres	11.00 × 20	12.00 × 20	14.00 × 20
Brakes	drum, air	drum, air	drum, air
Electrical system	24 V	24 V	24 V

CANADA

Western Star M 1500 1,500 kg (4 × 4) Light Support Vehicle Wheeled (LSVW)

Development
In March 1992 Western Star Trucks Inc of Kelowna, British Columbia, was awarded the contract (valued in 1993 at C$250 million) for a new Light Support Vehicle Wheeled (LSVW) to replace Chevrolet 1¼ ton (4 × 4) vehicles in service. Following Canadian trials covering 250,000 km, vehicle fielding started in March 1994. By October 1995 2,815 Western Star M 1500 (4 × 4) LSVW vehicles in 46 different configurations had been delivered, along with a full Logistics Support Analysis (LSA) and Integrated Logistic Support (ILS) deliverables.

Description
The Western Star M 1500 LSVW is based on the IVECO 40.10 WM modified to incorporate a higher power engine, a longer wheelbase and an automatic transmission. The LSVW is a high-mobility 1,500 kg payload vehicle powered by a 2.5 litre turbocharged and intercooled IVECO FIAT diesel engine producing 116 hp, coupled to a four-speed automatic transmission and a Borg Warner transfer case delivering power to the front and rear axles. A front independent suspension coupled with leaf springs at the rear provide good ground clearance and mobility with a full payload.

The semi-forward control hardtop cab can have two- or three-man configurations and is provided with three seatbelt anchor points and a centralised electrical system.

The M 1500 LSVW meets all Canadian Motor Vehicle Safety Standards. As the standard operating temperature range is −40 to +49°C, all vehicles are equipped with a fuel-fired coolant heater mounted in the On-Vehicle Equipment (OVE) module, located behind the cab. All bodies are attached to the frame through six flexible mounts and can be heated via a heat exchanger mounted in the on-vehicle equipment

Western Star M 1500 1,500 kg (4 × 4) Light Support Vehicle Wheeled (LSVW)

Western Star M 1500 1,500 kg (4 × 4) Light Support Vehicle Wheeled (LSVW) with armoured cab (R Stickland) ***1997***

which facilitates rapid body interchange. A soft-top cab and ballistic protection are also available and are being provided. The three basic body variations are cargo, a multipurpose van body, and utility, upon which are based troop carriers, an ambulance, a communications vehicle, a cable layer, shelter carriers, mobile workshops, and other bodies. The LSVW can be fitted with a mid-mounted front-pull winch.

Specifications

Cab seating: 1 + 1 or 2
Configuration: 4 × 4
Weight: (GVW) 5,250 kg
Max load: 1,500 kg
Towed load: 1,650 kg
Length overall: 5.61 m
Width: (overall) 2.01 m
Height:
(overall) 2.579 m
(cab) 2.2 m
Ground clearance: 0.28 m
Track: 1.67 m
Wheelbase: (unladen) 3.2 m
Angle of approach/departure: 40°/30°
Fuel capacity: (per tank) 90 litres
Max speed: (road) 100 km/h
Gradient: 60%
Fording: (without preparation) 0.7 m
Engine: IVECO FIAT 8142-47-1811 2.5 litre diesel developing 116 hp at 2,200 rpm
Transmission: ZF 4HP 220 automatic with torque converter giving 4 forward and 1 reverse gears
Transfer box: 2-speed
Suspension: (front) independent; (rear) leaf springs
Tyres: 9.00 × 16 XL
Electrical system: 24 V
Alternator: 110 A

Status

Production complete. In service with the Canadian Forces.

Manufacturer

Western Star Trucks Inc, Defense Products Group, 2076 Enterprise Way, Kelowna, British Columbia, Canada V1Y 6H8.
Tel: +1 604 860 3319. Fax: +1 604 860 1252.

VERIFIED

Western Star M4800 Series Medium Logistic Vehicle (MLV)

Description

The Western Star M4800 Series Medium Logistic Vehicle (MLV) is based around a high-mobility 4 × 4 truck with a payload of either 5,000 or 7,000 kg. The M4800 series MLV has a conventional configuration, with the engine in front under a prominent steel hood behind which is an enclosed three-man cab. The rear is occupied by the load area covered by a tilt over bows.

The Western Star galvanised steel cab has provision for the driver and two passengers. A Western Star 'Glacier' radiator, capable of cooling the engine at temperatures of over 60°C, is located in front of the 6.6 litre Caterpillar M3116TA JWAC engine, which is equipped with a three-stage air cleaner and an after-cooling system. A Cummins C-series engine can be installed as an option. The main driveline is an Eaton/Fuller CEEMAT nine-speed automatic transmission with a torque converter. A manual transmission with from 9 to 15 forward speeds can be provided as an option.

The chassis frame and tubular cross members are constructed using high-tensile alloy steel. The two standard wheelbase lengths are 4.32 m (5,000 kg payload) and 5.08 m (7,000 kg payload) but other wheelbase lengths to suit special requirements can be provided in 50 mm increments. Front and rear towing devices are provided.

A computer-controlled central tyre-inflation system is provided as standard.

Options include a front- or mid-mounted winch, a two-speed transfer case, air conditioning for the cab interior, an enlarged five-passenger crew cab, larger single or dual fuel tanks, fording and electrical waterproofing, various tyre sizes, and a 12 V electrical system. Interior and exterior blackout lights are provided as standard.

M4800 series variants include a range fire tender, which has been sold to both the Canadian Department of National Defence and the British Army for use at the Suffield training area in Canada. An air-transportable 6 × 4 variant capable of being carried in a C-130 aircraft has been produced and can accommodate various military shelter bodies. Various types of purpose-built truck and other bodies can be produced. A Multi-Modal truck capable of operating using either conventional roadwheels or hydraulically-actuated guide wheels for use over rail tracks has been produced for public utility and other services.

A M4800 series 6 × 6 truck with an optional double channel frame is available.

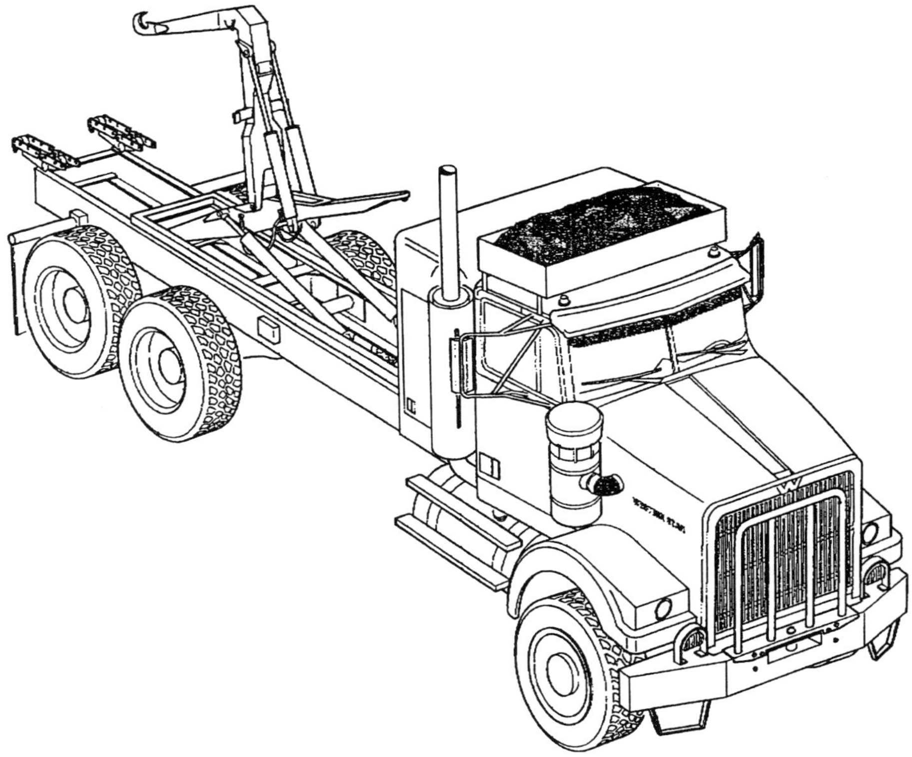

Drawing of Western Star M4800 Series 6 × 6 PLS vehicle ***1997***

Variant

In April 1996 the Canadian Department of National Defence awarded Western Star a contract for the design and manufacture of 44 6 × 6 variants in three-way dump truck and PLS configurations. Features required in this contract include central tyre inflation, run-flat tyres, a mid-mounted winch (front and rear pull), and hydraulic power tool take-offs.

The PLS variant is capable of loading and transporting all current flatracks used by the Canadian Department of Defense, 20 ft/6.096 m ISO containers and floating bridge modules. It can also load and unload these units on drawbar trailers. A new dump module is under development as part of this contract which will allow thse vehicles to operate as rear-tipping dump trucks. These vehicles commenced fielding with the Engineering Corps in the third quarter of 1996.

Specifications
Cab seating: 1 + 2
Configuration: 4 × 4
Front axle capacity: 7,256 kg
Rear axle capacity: 10,400 kg
Payload: 5,000 or 7,000 kg
Towed load: 10,000 kg
Ground clearance: 0.584 m
Track: (front and rear) 2.045 m
Wheelbase:
(5,000 kg payload) 4.32 m
(7,000 kg payload) 5.08 m
Max speed: 100 km/h
Range: 750 km
Fuel capacity: 227 litres
Gradient: 60%
Engine: Caterpillar M3116TA JWAC 6.6 litre diesel developing 220 hp at 2,200 rpm
Transmission: Eaton/Fuller CEEMAT automatic with 9 forward speeds
Transfer case: single-speed with torque proportioning differential
Electrical system: 24 V (12 V optional)
Alternator: 70 Ah

Status
In production.

Manufacturer
Western Star Trucks Inc, 2076 Enterprise Way, Kelowna, British Columbia, Canada V1Y 6H8.
Tel: +1 604 860 3319. Fax: +1 604 860 1252.

UPDATED

Western Star M4800 Series Medium Logistic Vehicle (MLV) (T J Gander) ***1996***

CHINA, PEOPLE'S REPUBLIC

Dong Feng EQ2080E4DY (6 × 6) 2,500 kg truck

Description
The Dong Feng (export name Aeolus) EQ2080E4DY (6 × 6) 2,500 kg truck is the latest version of the EQ2080E range (formerly the EQ240 range) and is fitted with a Cummins 6B5.9 5.88 litre diesel engine; earlier models were powered by an EQ6100 5.417 litre petrol engine. It is a right-hand drive vehicle intended for operations in mountainous areas and over rough roads and tracks.

Layout is conventional, with the engine forward, the steel cab having seating for the driver and two passengers, and the load area to the rear. The cargo body has a steel and wood floor, steel side racks and a tailgate. Optional extras include a power take-off and a 4,500 kg winch mounted behind the front bumper. A left-hand drive version (the EQ2080E4D) is available.

Specifications
Cab seating: 1 + 2
Configuration: 6 × 6
Weight:
(unladen) 5,090 kg
(laden, road) 9,300 kg
(laden, cross-country) 7,800 kg
Max load:
(road) 4,000 kg
(cross-country) 2,500 kg
Towed load: 2,500 kg
Load area: 3.545 × 2.18 m
Length: 6.34 m
Width: 2.4 m
Height: (cab, unladen) 2.4 m
Track: 1.774 m
Wheelbase:
(front axle to centre of rear bogie) 3.74 m
(between rear axles) 1.1 m
Angle of approach/departure: 45°/44°

Dong Feng EQ2080E4D (6 × 6) 2,500 kg truck

Max speed: 77 km/h
Gradient: 57%
Fording: 0.85 m
Engine: Cummins 6B5.9 5.88 litre 6-cylinder 4-stroke in-line water-cooled diesel developing 130 hp at 2,800 rpm
Gearbox: manual with 5 forward and 1 reverse gears
Clutch: single dry disc
Transfer box: 2-speed
Turning radius: 8.2 m
Tyres: 11.00 × 18
Brakes: air

Status
In production. In service with Thailand (235).

Manufacturer
Dong Feng Motor Corporation, Shiyan, Hubei Province, People's Republic of China.
Tel: +86 719 224700. Telex: 40104 HBDFM CN.
Fax: +86 719 226845.

VERIFIED

CA-30 (6 × 6) 2,500 kg truck

Description
The CA-30 (6 × 6) 2,500 kg truck entered production at the First Automobile Works, Changchun, in 1959 and was the standard vehicle in its class in the People's Liberation Army. Production ceased in 1986.

The CA-30 is very similar to the ZIL-157 (6 × 6) 2,500 kg truck but has a different cab, with the headlamps sometimes in the mudguards rather than mounted externally as on the ZIL-157.

The layout of the vehicle is conventional, with the engine at the front, two-door fully enclosed cab in the centre and at the rear the cargo area, which consists of a wooden platform with sides, bench seats down each side which can be folded up when the vehicle is carrying cargo, and a drop tailgate. If required, the vehicle can be fitted with bows and a tarpaulin cover.

The CA-30 has been used for a number of military roles including that of artillery tractor and for carrying multiple 122 mm rocket launchers, including the Type

CA-30 (6 × 6) 2,500 kg truck with winch and with box body for use in command post role (Christopher F Foss)

74 Minelaying Rocket System (see entry in *Minelaying equipment* section).

Specifications

Cab seating: 1 + 2
Configuration: 6 × 6
Weight:
(empty) 5,450 kg
(loaded, road) 9,950 kg
(loaded, cross-country) 7,950 kg
(on front axle, empty) 2,360 kg
(on rear bogie, empty) 3,090 kg
Max load:
(road) 4,500 kg
(cross-country) 2,500 kg
Towed load: (cross-country) 3,600 kg
Load area: 3.57 × 1.8 m
Length: 6.684 m
Width: 2.315 m
Height: (cab) 2.36 m
Ground clearance: 0.305 m
Track: (front and rear) 1.752 m
Wheelbase: (first axle to centre of rear bogie) 4.785 m
Max speed: (road) 65 km/h
Range: 680 km
Fuel capacity: 150 litres
Max gradient: 30%
Fording: 0.85 m
Engine: Chieh'Fang 120 6-cylinder liquid-cooled petrol developing 95 hp at 2,800 rpm
Gearbox: manual, 5 forward and 1 reverse gears
Clutch: single dry plate
Transfer box: 2-speed
Turning radius: 12.1 m
Suspension: (front and rear) semi-elliptical leaf springs
Tyres: 12.00 × 18
Brakes:
(main) air
(parking) mechanical
Electrical system: 12 V

Status

Production on request. In service with the People's Liberation Army and has been exported.

Manufacturer

First Automobile Works, Dong Feng Street, Changchun, Jilin, People's Republic of China.
Tel: +86 431 504051. Telex: 83009 jffaw cn.
Fax: +86 431 867780.

VERIFIED

Jiefang CA-10 (4 × 2) Liberation 3,540 kg truck

Development/Description

The CA-10 (4 × 2) Liberation 3,540 kg truck entered production at the First Automobile Works at Changchun in July 1956. It is a modified ZIL-150 (4 × 2) 3,500 kg truck. Although it has a very limited cross-country capability and was primarily produced for commercial purposes, it is used in some numbers as a cargo carrier and as a prime mover for light artillery. Compared with the original it has a less efficient engine and a lack of climbing power.

Over 1,000,000 CA-10s of all types had been produced by the beginning of 1983.

The initial CA-10 was followed by the CA-10B in 1960. The CA-10BX, with a higher compression ratio, maximum speed and power output, appeared in September 1980. It also had changes to the camshaft, carburettor, fuel pump, oil sump, radiator, clutch, starter motor and electrical system, all of which went some way to overcoming earlier technical drawbacks.

The CA-10C appeared in January 1982 and is rated at 4,500 kg. The CA-10CJ is fitted with a 100 hp engine and the CA-10CT with a 110 hp engine. The CA-15 has a 5.55 litre 115 hp engine.

The layout of the vehicle is conventional, with the engine at the front, two-door fully enclosed cab in the centre and the cargo area at the rear with drop sides and a drop tailgate. Both left- and right-hand drive models are in service.

Variants are known to include a tractor truck, fire-fighting vehicles, tankers, crane carriers and dump trucks. Various wheelbase lengths have been produced.

The QH-140 truck, produced at the Qinghai plant, is a special variant for use at high altitudes. It was produced in petrol- and diesel-engined forms.

Jiefang CA-10B (4 × 2) Liberation 3,540 kg truck

Specifications

Cab seating: 1 + 1
Configuration: 4 × 2
Weight:
(empty) 3,840 kg
(loaded, road) 7,375 kg
(on front axle, empty) 1,735 kg
(on rear axle, empty) 2,100 kg
Max load: (road) 3,540 kg
Load area: 3.54 × 2.26 m
Length: 6.502 m
Width: 2.29 m
Height: 2.33 m
Ground clearance: 0.27 m
Track:
(front) 1.69 m
(rear) 1.74 m
Wheelbase: 3.98 m
Angle of approach/departure: 40°/24°
Max speed: 80 km/h
Range: 415 km
Fuel capacity: 158 litres
Max gradient: 37%
Engine: CA-10 6-cylinder in-line liquid-cooled petrol developing 95 hp at 2,800 rpm
Gearbox: manual, 5 forward and 1 reverse gears
Clutch: single dry plate
Turning radius: 9 m
Suspension:
(front) semi-elliptical leaf springs and hydraulic shock-absorbers
(rear) semi-elliptical springs with overload springs
Tyres: 9.00 × 20
Brakes:
(main) air
(parking) mechanical
Electrical system: 12 V
Batteries: 2 × 6 V

Status

Production complete. In service with the People's Liberation Army. Replaced from 1987 onwards by the CA-1091 and CA-141 series of trucks (qv).

Manufacturer

First Automobile Works, Dong Feng Street, Changchun, Jilin, People's Republic of China.
Tel: +86 431 504051. Telex: 83009 jffaw cn.
Fax: +86 431 867780.

VERIFIED

Jiefang CA-1091 (4 × 4) 5,000 kg trucks

Description

The Jiefang CA-1091 series of 4 × 4 trucks are the military purpose successors to the Jiefang CA-10 truck series (see previous entry) and share many components with the Jiefang CA-141 series of 4 × 2 trucks (see next entry).

There are three basic models in the series. The CA-1091E2 uses a petrol engine and has a 12 V electrical system. The CA-1091K2E2 and CA-1091K3E2 have diesel engines and a 24 V system. All three models share the same drive train, chassis and cargo body.

The chassis uses a side girder-type structure with riveted pressed longitudinal beams and eight cross members. Towing hooks are provided at the front and rear.

The cab has seating for the driver and two passengers. The one-piece hood can be tilted forward 50° for engine access. Optional equipment for the cab includes a heater and defrosting equipment. The cargo body uses mixed steel and timber construction and is fitted with side racks. A tailboard is provided along with a ladder for access and a rainproof tarpaulin.

Jiefang CA-1091 series (4 × 4) 5,000 kg truck

Specifications

Cab seating: 1 + 2
Configuration: 4 × 4
Weight: (kerb)
(CA-1091E2) 4,700 kg
(CA-1091K2E2) 4,350 kg
(CA-1091K3E2) 4,195 kg
Max load: 5,000 kg
Load area: 4.2 × 2.3 m
Length: 7.205 m
Width: 2.484 m
Height: 3.09 m
Ground clearance: (rear axle) 0.262 m
Track:
(front) 1.8 m
(rear) 1.74 m
Wheelbase: 4.05 m
Angle of approach/departure: (loaded) 28°/19°
Max speed: 90 km/h
Gradient: 28%
Engine:
(CA-1091E2) Model CA-6102 5.56 litre 6-cylinder water-cooled petrol developing 132 hp at 3,000 rpm
(CA-1091K2E2) Model CA-6110A 6.842 litre 6-cylinder water-cooled direct injection diesel developing 157 hp at 2,900 rpm
(CA-1091K3E2) Model CY-6102BQ-6 5.785 litre 6-cylinder water-cooled direct injection diesel developing 138 hp at 3,000 rpm
Gearbox: Model LFO6S-CB with 6 forward and 1 reverse gears
Transfer box: 2-speed
Steering: circulating ball
Turning radius: 8 m
Suspension: leaf springs with double-acting hydraulic shock-absorbers
Tyres: 9.00 × 20 or 9.00 R 20
Brakes:
(main) air, drum type
(parking) mechanical
Electrical system:
(petrol) 12 V
(diesel) 24 V

Status

In production.

Manufacturer

First Automobile Works, Dong Feng Street, Changchun, Jilin, People's Republic of China.
Tel: +86 431 504051. Telex: 83009 jffaw cn.
Fax: +86 431 867780.

VERIFIED

Jiefang CA-141 (4 × 2) 5,000 kg trucks

Description

The Jiefang CA-141 (4 × 2) 5,000 kg truck entered trial production during 1983 in its original forward control cab form. A more orthodox model with a conventional bonnet entered production during 1987 and replaced the earlier CA-10 (qv). The CA-141 shares many components with and has the same general layout as the CA-1091 series of 4 × 4 trucks (see previous entry).

The CA-141 may be regarded as a fully updated version of the earlier CA-10 series and retains the same general layout of a bonneted cab with the cargo area open to the rear. The bonnet can be tipped forward to an angle of 50° for maintenance access and the cab has an adjustable seat for the driver and a dual seat for two passengers. The driver is provided with a comprehensive instrument panel, cabin and windscreen ventilation and large wide-vision windows.

The CA-141 is produced in a range of nine models, primarily for commercial purposes, but some are used by the Chinese armed forces for a variety of roles. There is a choice of a petrol or diesel engine.

The CA-141K series of trucks have Model 6110 6.842 litre diesel engines. Included in this range are the CA-141K2Y and longer wheelbase CA-141K2YL trucks, both with a 6,000 kg load capacity, and the CA-341K2Y dump truck with a 4.3 m³ capacity.

Jiefang CA-141 (4 × 2) 5,000 kg truck

Specifications

CA-141K2Y
Cab seating: 1 + 2
Configuration: 4 × 2
Weight:
(kerb) 4,350 kg
(loaded) 10,350 kg
Max load: 6,000 kg
Towed load: 6,000 kg
Length: 7.205 m
Width: 2.476 m
Height: 2.425 m
Load area: 4.2 × 2.3 m
Ground clearance: 0.262 m
Track:
(front) 1.8 m
(rear) 1.74 m
Wheelbase: 4.05 m
Max speed: 110 km/h
Max gradient: 28%
Engine: Model 6110 6.842 litre 6-cylinder in-line diesel developing 157 hp at 2,900 rpm
Gearbox: manual with 6 forward and 1 reverse gears
Clutch: single dry plate
Steering: recirculating ball
Turning radius: 8.2 m
Suspension:
(front) leaf springs with shock-absorbers
(rear) torsion bar
Brakes: dual-circuit
Electrical system: 24 V
Battery: 6-QA-100
Alternator: 700 W

Status

In production at a rate of up to 100,000 units per year.

Manufacturer

First Automobile Works, Dong Feng Street, Changchun, Jilin, People's Republic of China.
Tel: +86 431 504051. Telex: 83009 jffaw cn.
Fax: +86 431 867780.

VERIFIED

Dong Feng EQ2100E (6 × 6) 3,500 kg truck

Description

The Dong Feng (export name Aeolus) EQ2100E (6 × 6) 3,500 kg truck (formerly the EQ245) is a high-mobility vehicle powered by an EQ6105-1 petrol engine. Layout is conventional, with the engine forward, the steel cab having seating for the driver and two passengers, and the load area to the rear. The cargo body has a steel and wood floor and steel side racks. There is provision for towing a trailer weighing up to 4,500 kg. A self-recovery winch may be fitted.

This vehicle is also used for commercial purposes such as geological and oil exploration, and may also carry various forms of technical service vehicle bodies.

Dong Feng EQ2100E (6 × 6) 3,500 kg truck

Specifications

Cab seating: 1 + 2
Configuration: 6 × 6
Weight: (without winch)
(laden, road) 11,290 kg
(laden, cross-country) 9,790 kg
Max load:
(road) 5,000 kg
(cross-country) 3,500 kg
Towed load: 4,500 kg
Length: 6.844 m
Width: 2.4 m
Height: (cab, unladen) 2.484 m
Track:
(front) 1.876 m
(rear) 1.87 m
Wheelbase:
(front axle to centre of rear bogies) 4.025 m
(between rear axles) 1.25 m
Angle of approach/departure: 44.5°/45°
Max speed: 82 km/h
Fording: 0.9 m
Engine: EQ6105-1 petrol
Gearbox: manual with 5 forward and 1 reverse gears
Clutch: double dry disc, 300 mm
Transfer box: 2-speed
Turning radius: 9.3 m
Tyres: 12.00 × 20
Brakes: air

Status

In production. In service with the Chinese armed forces and offered for export.

Manufacturer

Dong Feng Motor Corporation, Shiyan, Hubei Province, People's Republic of China.
Tel: +86 719 224700. Telex: 40104 HBDFM CN.
Fax: +86 719 226845.

VERIFIED

Dong Feng EQ1112F8D (4 × 2) truck

Development

The Dong Feng Motor Corporation at Shiyan in Hubei Province manufactures 200,000 medium-weight trucks annually and is the largest truck manufacturer in China. One of its main products is the Dong Feng EQ1092F, fitted with an EQ6100 5.4 litre six-cylinder 135 hp petrol engine. Under the export name Aeolus they have been exported to more than 10 nations in Asia, Africa (including Cameroon) and South America.

Description

The EQ1112F8D (formerly the EQ8D142) is only one vehicle in the EQ1110 range and is a conventional design in the 11 tonne gross weight class. It is powered by a diesel unit but other models have been powered by a Renault 133 hp petrol engine. The drive configuration is 4 × 2, with the engine at the front and a conventional steel cab. The usual cargo body is steel with a raised 'ladder' grill to protect the cab rear. Twin tyres are fitted on the rear axle.

A similar vehicle with right-hand drive, the EQ1112F8DY, was also produced. The EQ1092 series is available for both military and commercial users.

Dong Feng EQ1112F8D (4 × 2) truck ***1996***

Specifications

Cab seating: 1 + 2
Configuration: 4 × 2
Weight:
(kerb) 4,480 kg
(GVW) 11,000 kg
Max load: 6,400 kg
Length: 8.145 m
Width: 2.47 m
Height: 2.5 m
Track:
(front) 1.81 m
(rear) 1.8 m
Wheelbase: 3.95 m
Angle of approach/departure: 39.5°/21°
Max speed: 85 km/h
Gradient: 25%
Engine: Chao Yang 6102 4-stroke in-line water-cooled diesel developing 140 bhp at 3,000 rpm
Gearbox: manual with 5 forward and 1 reverse gears
Clutch: single dry disc
Turning radius: 8.9 m
Tyres: 9.00 × 20 14PR
Brakes: air

Status

Production of various models in this series continues.

Manufacturer

Dong Feng Motor Corporation, Shiyan, Hubei Province, People's Republic of China.
Tel: +86 719 224700. Telex: 40104 HBDFM CN.
Fax: +86 719 236845.

VERIFIED

Dong Feng EQ1141G (4 × 2) 8,000 kg truck

Description

The Dong Feng (export name Aeolus) EQ1141G (4 × 2) 8,000 kg truck is described as a high-performance logistics vehicle. It has a forward control cab that can be tilted forward 43° to gain access to the Cummins 6BT118-01 5.88 litre diesel engine and features a conventional load area.

Specifications

Cab seating: 1 + 2
Configuration: 4 × 2
Weight:
(empty) 5,900 kg
(GCW) 14,100 kg
Max load: 8,000 kg
Load area: 5.3 × 2.294 m
Length: 7.73 m
Width: 2.47 m
Height: 2.71 m
Track:
(front) 1.94 m
(rear) 1.86 m
Wheelbase: 4.5 m
Angle of approach/departure: 34°/18°
Max speed: 85 km/h
Gradient: 25%
Engine: Cummins 6BT118-01 5.88 litre in-line water-cooled diesel developing 160 hp at 2,600 rpm

Gearbox: manual with 5 forward and 1 reverse gears
Clutch: single dry disc
Turning radius: 8 m
Suspension: semi-elliptical springs with double-acting hydraulic shock-absorbers
Tyres: 10.00 R 20, 16PR
Brakes: air

Status
In production.

Manufacturer
Dong Feng Motor Corporation, Shiyan, Hubei Province, People's Republic of China.
Tel: +86 719 224700. Telex: 40104 HBDFM CN.
Fax: +86 719 226845.

VERIFIED

Dong Feng EQ1141G (4 × 2) 8,000 kg truck
1996

Hongyan CQ 261 (6 × 6) 8,250 kg truck

Description
The CQ 261 (6 × 6) 8,250 kg truck can trace its design origins back to the early 1960s when Berliet GBU and GCH trucks were imported into China. Licence production in China followed and in 1977 the CQ 261 appeared.

The CQ 261 bears an overall resemblance to the Berliet original but has many changes to suit local production methods. The large forward control cab has a distinctive radiator grille and seating for at least two passengers in addition to the driver, and some cabs are fitted with a roof hatch. At least two forms of passenger door have been seen, with later models having 'straight line' doors in place of the earlier sloping rear edge. The cargo area to the rear has steel drop sides and a tailgate and towing hook are provided. Bench seats can be fitted along each side of the cargo area. A canvas tilt can be fitted.

A tractor version of this vehicle is known as the CQ 261Q25 and is fitted with a Model 614OZ engine developing 250 hp. There is also a long wheelbase tractor truck known as the CQ 261C25 and an 8 × 4 crane truck is known as the CQ 40D.

An 8 × 8 version is used to carry the components of the Type 79 Ribbon Bridge.

Similar vehicles are the Yan'An SX 250 (6 × 6) truck and the SX 161 (6 × 4) truck. The Yan'An QD 360 is a 6 × 4 dump truck variant of the SX 161.

Specifications
Cab seating: 1 + 2 or 3
Configuration: 6 × 6
Weight loaded:
(roads) 26,460 kg
(cross-country) 22,460 kg
Max load:
(roads) 12,260 kg
(cross-country) 8,266 kg
Length: 7.947 m
Width: 2.5 m
Height: (top of cab) 3 m
Wheelbase: 4.93 m
Max speed: 61 km/h
Engine: Chongfa Model 6150 14.78 litre 6-cylinder diesel developing 200 hp
Gearbox: 5 forward and 1 reverse gears
Clutch: single dry plate
Transfer box: 2-speed
Brakes: dual-circuit air

Status
In production. In service with the People's Liberation Army.

Manufacturer
Sichuan Automobile Factory, Dazu, Sichuan Province, People's Republic of China.

VERIFIED

CQ 261 (6 × 6) 8,250 kg truck towing Type 66 152 mm gun-howitzer

Other Chinese military vehicles

The People's Liberation Army still uses large numbers of vehicles supplied by the former USSR in the 1950s including the GAZ-63, GAZ-51, ZIL-150 (Chinese model CA-10), ZIL-164, ZIL-151 and the ZIL-157 (Chinese model the CA-30, with workshop versions known as the QD402, X01000 and X8000). Chinese-designed vehicles include the 7,000 kg capacity JN-150 Huang Ho dump truck and the CN-130 truck for towing semi-trailers.

In the 1950s vehicles were also obtained from other Eastern Bloc countries including the former Czechoslovakia and East Germany, plus Romania and Hungary. China also obtained vehicles from a number of Western countries including Austria (Steyr), France (Berliet, now Renault Véhicules Industriels) and Sweden (Volvo). The Berliet trucks were the 6 × 6 GBC 8MT and the heavier GBU (the Chinese model was known as the GCH).

VERIFIED

CZECH REPUBLIC AND SLOVAKIA

Avia (4 × 4) TNK 2,500 kg truck

Description
The Avia (4 × 4) TNK 2,500 kg truck was developed in co-operation with the Czech Army and the Avia factory. It was designed as a general utility load carrier and also as a carrier for shelter/containers, hence terrain container carrier – *Terénní Nosič Kontejnerových* (TNK). It was first shown during 1993.

The vehicle has a forward control cab with seating for the driver and one passenger. The Avia engine is located over the front axle with space behind the cab for a spare wheel on the left and a stowage locker on the right. As a utility load carrier the vehicle is equipped with steel drop sides and a tailgate.

The vehicle may also be used as a chassis frame carrier for a special shelter/container produced by the Orličan Works at Choceň, with a loaded weight of 1,500 kg and base dimensions of 2.991 × 2.438 m. The shelter/container is secured using normal ISO 1D fittings and it may be loaded onto the vehicle either by a crane or manually, using side-mounted stand legs on the container base. To assist loading, the vehicle has air suspension which enables the entire vehicle, and hence the load area, to be lowered by 200 mm and then raised again for travelling. It is possible to carry NATO standard containers.

Equipment supplied with each vehicle includes: a NBC decontamination kit, provision for night driving devices, a power generator set location for a 3 kW generator driven from the vehicle's engine, night driving headlights, a cab roof-mounted spotlamp, and a removable internal light for map-reading. Optional equipment includes: snow chains, fittings for container and shelter stands, and a three-phase generator to power shelter systems.

The vehicle may be used to carry an elevating platform for a variety of applications. Commercial variants are available.

Specifications
Cab seating: 1 + 1
Configuration: 4 × 4
Weight: (empty) 3,500 kg
Max load: 2,500 kg
Towed load: 2,000 kg
Length: 5.32 m
Width: 2.44 m
Height: 2.63 m
(cab) 2.51 m
**Angle of approach/departu

Max speed: 80 km/h
Engine: Avia D407.01.00 diesel developing 87 hp at 2,600 rpm
Gearbox: 5 P 20.20 manual
Suspension: air
Tyres: 8.25 × 20 HD Special or NB 60
Electrical system: 24 V

Status
In production.

Manufacturer
Vojenský Opravárenský Podnik 081 sp, 535 39 Přelouč, Czech Republic.
Tel: +42 0457 2261. Fax: +42 0457 2092.

VERIFIED

Avia (4 × 4) TNK 2,500 kg truck (T J Gander)

Praga V3S (6 × 6) 3,000 kg truck

Description
The V3S was developed in the early 1950s. Initial production was undertaken by Praga but in 1964 it was transferred to Avia as Praga began to concentrate on automotive components. It has been replaced in many Czech and Slovak units by the RFAS ZIL-131 (6 × 6) 3,500 kg truck and other similar vehicles, but large numbers are still retained for second-line and rear area duties.

The layout of the Praga V3S is conventional, with the engine under a distinctive attenuated bonnet at the front, the two-man cab in the centre and the cargo area at the rear. There is provision for stowing fuel containers behind the front bumper. The two-door cab is of all-steel construction. There is a circular observation hatch in the right side of the roof. The windscreen is of the split type and both parts can be opened horizontally for increased vision. The rear cargo area has a tarpaulin cover and removable bows which can be stowed to the rear of the cab when not required. Some V3S trucks have a winch with a 3,500 kg capacity.

Variants include: a crane truck, hopper-type dump truck (V3S-K), various shop/van models (including some with an A-frame mounted at the front of the vehicle), tanker (special), 3,000 litre capacity tanker (V3S-C), ARS-12M NBC decontamination apparatus carrier, and tractor truck (V3S-A). The chassis is also used to mount the 130 mm (32-round) M51 multiple rocket system (*Jane's Armour and Artillery 1995-96*, page 717).

A 4 × 2 model of the Praga V3S was built under the designation S5T.

Specifications
Cab seating: 1 + 1
Configuration: 6 × 6
Weight:
(empty) 5,350 kg
(loaded, road) 10,650 kg
(loaded, dirt road) 8,650 kg
(loaded, cross-country) 8,350 kg
Max load:
(road) 5,300 kg
(dirt road) 3,300 kg
(cross-country) 3,000 kg
Towed load:
(road) 5,500 kg
(cross-country) 3,100 kg
Length: 6.91 m
Width: 2.31 m
Height:
(cab) 2.51 m
(tarpaulin) 2.92 m
Ground clearance: 0.4 m
Track:
(front) 1.87 m
(rear) 1.755 m
Wheelbase: 3.58 m + 1.12 m
Angle of approach/departure: 72°/32°
Max speed: 62 km/h
Range: 500 km
Fuel capacity: 120 litres
Fuel consumption: 27 litres/100 km
Max gradient: 60%
Fording: 0.8 m
Engine: T-912 in-line air-cooled diesel developing 98 hp at 2,100 rpm (late production vehicles have T-912-2 developing 110 hp at 2,200 rpm, which is also installed in the Praga S5T2 truck)
Gearbox: manual with 4 forward and 2 reverse gears
Clutch: single dry plate
Transfer box: 2-speed
Turning radius: 10.5 m
Suspension: (front and rear) semi-elliptical springs
Tyres: 8.25 × 20
Brakes:
(main) hydraulic, air-assisted
(parking) mechanical
Electrical system: 12 V

Status
Production complete. In service with the Czech and Slovak armed forces, other former Warsaw Pact countries, and some countries in the Middle East.

Manufacturer
Avia Závody NP, Letnany, Czech Republic.

UPDATED

Praga V3S (6 × 6) 3,000 kg truck with office body (R Stickland)

1997

Praga V3S (6 × 6) 3,000 kg truck (T J Gander)

16°/33°

TATRA 813 (6 × 6 and 8 × 8) trucks

Description

The TATRA 813 series of trucks, also known as the Kolos series, was developed in the former Czechoslovakia in the early 1960s with the first 8 × 8 production vehicles being completed in 1968. The vehicle is related to the OT-64 (8 × 8) amphibious armoured personnel carrier.

The standard cargo truck version has an all-steel forward control cab which usually has one or two circular observation hatches in the roof. The cab is pressurised and is provided with an effective ventilation system and individual masks for each member of the crew. The engine is in the centre of the cab at the front, with the transmission to the rear of the cab, under the cargo area. The rear cargo area has all-steel drop sides and a drop tailgate, removable seats, and can be equipped with bows and a tarpaulin cover if required. All military versions are fitted with a central tyre pressure regulation system and a 22,000 kg capacity winch. The vehicle is sometimes fitted with the BZ-T hydraulically operated dozer blade at the front; a snow plough can also be fitted.

The TATRA 813 (8 × 8) is still used by the Czech Republic and Slovakia as a cargo/personnel carrier, for towing heavy artillery up to 152 mm in calibre, and for towing trailers carrying engineer equipment and MBTs. It is being replaced by the later TATRA 815 series (see following entry).

A more recent model was the TATRA 813-12 powered by a 310 hp multifuel engine.

Variants

The basic chassis (8 × 8) is used for a variety of roles including carrying and launching PMP heavy floating pontoon bridge units (these vehicles are often fitted with the BZ-T hydraulically operated dozer blade) and for carrying and laying the truck-mounted AM-50 scissors bridge, described in the *Mechanised Bridges* section. There is also a special roadway-laying version which is used with the PMP system, details of which will be found in the *Portable Roadways* section.

The TATRA 813 (road prime mover) is the basic truck with the rear cargo area replaced with ballast. There is also a 6 × 6 cargo truck version with a multifuel engine, designated the TATRA 813-8.

The TATRA 813 (8 × 8) chassis was used as the basis for the RM-70 122 mm (40-round) multiple rocket system with a fully armoured cab with 40 rockets in the ready to launch position and another 40 rockets ready to reload the launcher. Details of this model are given in *Jane's Armour and Artillery 1995-96,* pages 717 to 719. These are often fitted with the BZ-T hydraulically operated dozer blade for clearing obstacles and preparing fire positions.

There are two prime movers for civil use, the TATRA 4-813 T-3 (6 × 6) and the TATRA 4-813 T-2 (4 × 4), which can tow a trailer weighing up to 65,000 kg. These do not have the central tyre pressure regulation system as fitted to the military versions. There are at least three dump truck models, the TATRA 813 S1 (6 × 6), TATRA 813 S3 (6 × 6) and the TATRA 813 S1 (8 × 8), which also do not have the central tyre-pressure regulation system. Crane versions are also in service for both civil and military use.

TATRA 813 (8 × 8) truck

VOP 028 of Prešov in Slovakia offers a complete overhaul and repair service for TATRA 813 trucks and offers refurbished vehicles in a virtually as-new condition.

Specifications

Cab seating: 1 + 6
Configuration: 8 × 8
Weight:
(empty) 13,800 kg
(loaded) 22,000 kg
Weight on 1st and 2nd axles, loaded: 5,150 kg
Weight on 3rd and 4th axles, loaded: 5,850 kg
Max load: 8,200 kg
Towed load: (road) 100,000 kg
Length: 8.8 m
Width: 2.5 m
Height:
(cab) 2.69 m
(tarpaulin) 3.355 m
Ground clearance: 0.425 m
Track: (front/rear) 2.03 m/2.03 m
Wheelbase: 1.65 m + 2.2 m + 1.45 m
Max speed: (road) 80 km/h
Range: (road) 1,250 km
Fuel capacity: 520 litres
Max gradient: 65%
Trench: 1.4 m
Vertical obstacle: 0.6 m
Fording: 1.4 m
Engine: TATRA T-930-3 12-cylinder air-cooled diesel developing 250 hp at 2,000 rpm
Gearbox: dual-range 5-speed plus overdrive providing a total of 20 forward and reverse gears
Steering: power-assisted
Tyres: 15.00 × 21
Brakes:
(main) air
(parking) mechanical
Electrical system: 24 V

Status

Production complete. In service with the Czech Republic, India and Slovakia.

Manufacturer

TATRA, Štefánnikova 1163, 742 21 Kopřivnice, Czech Republic.
Tel: +42 656 40871. Fax: +42 656 42556.

UPDATED

TATRA T 815 (8 × 8) truck series

Development

The TATRA T 815 series of 8 × 8 trucks replaced the earlier TATRA 813 series in production in the former Czechoslovakia in 1982 and for a while the name of Kolos was applied to both. The TATRA T 815 series is produced in 8 × 8, 6 × 6 and 4 × 4 configurations for both military and civil use (see following entries for details of the 6 × 6 and 4 × 4 models) in a variety of forms including platform trucks, ballast tractors and special versions.

By 1993 the original 8 × 8 and 6 × 6 versions of the TATRA T 815 had been joined by a 4 × 4 version (see separate entry).

The TATRA T 815 VVN 26.265 8 × 8.1R and TATRA T 815 VT 26.265 8 × 8.1R are produced under licence in India by Bharat Earth Movers Limited (BEML).

In April 1995 it was announced that Twin Disc Inc of the USA and TATRA had signed a letter of intent to allow Twin Disc to supply TATRA chassis and supension systems for military and commercial truck applications in North America.

VOP 028 of Prešov, Slovakia, offers a complete repair and overhaul service for time-expired TATRA trucks, including the TATRA T 815 (8 × 8) series and associated vehicles.

TATRA T 815 VVN 20.235 8 × 8.1R truck

1997

TATRA T 815 VVN 26.265 8 × 8.1R truck **1996**

Basic chassis of TATRA 815 VT 26.265 8 × 8.1R truck **1997**

Description

The TATRA T 815 (8 × 8) series followed the same general lines as the earlier TATRA 813 but there were many overall improvements. There are two basic military versions of the T 815 (8 × 8). One is the T 815 VVN 26.265 8 × 8.1R, described as an off-road truck, having a two-door cab. The TATRA 815 VT 26.265 8 × 8.1R is described as a special heavy road tractor and has a four-door cab and a correspondingly shorter (5.35 m as opposed to 6.25 m) load area. The load capacity for the T 815 VVN is 12,000 kg off-road, or a towed load of 20,000 kg off-road (65,000 kg on hard roads). For the T 815 VT the figures are 10,000 kg and towed load 25,000 kg (100,000 kg on hard roads).

Both types of cab are all-metal, have a roof hatch, and can be tilted forward for maintenance purposes. The VVN two-door cab has seating for a driver and three passengers, with provision for an occasional berth. The four-door VT cab has seating for the driver and five passengers, with provision for an occasional berth slung above the front seats. Both types of cab have a multifuel heater. They are also equipped with a collective NBC protection system that connects to individual masks, in addition to an air filtering and ventilation system.

The chassis is of frameless construction with the load-bearing area consisting of the individual main assembly unit cases interconnected by load-bearing tubes. Mounted on the cross members is a through-frame to which the cab, engine and clutch, steering box, winch, truck platform and front and rear bumpers (together with the trailer coupling) are all fitted. The load-carrying platform is mounted on the frame behind the cab and is fitted with a fixed head, drop sides and a hinged tailgate. Side board extensions can be converted into benches for carrying troops. A tarpaulin over sliding bows can be used to cover the load-carrying area.

The swinging half-axles, with independent wheel suspension on the four-axle vehicles, are constantly driven. Front-wheel drive on the three-axle vehicle can be selected by the driver. Reduction gears in the wheels reduce power train stress and increase the vehicle's tractive effort. For uniform driving torque division, between the individual axle pairs a torque divider is inserted in the power train. There is also a system of inter-axle and axle differential locks.

Suspension on the four-axle vehicles is by longitudinally located leaf springs. The front axle of the three-axle vehicle uses torsion bars and is equipped with hydraulic telescopic dampers.

The 10-speed gearbox employs constant mesh helical gears. With the exception of the first and reverse, all gears are provided with block synchromesh. Engaging five of the gears is in direct with a pneumatic servo; two optional direct gears are engaged electropneumatically using a preselector on the gear lever. There is also an auxiliary two-speed gearbox directly coupled to the gearbox for winch drive.

The winch has a capacity of 12,000 kg on all models and is supplied with 85 m of cable.

All 8 × 8 models are equipped to carry the ŠSP 1000 arrow-type snow plough and can accommodate the BZ-T 815 dozer blade.

Variants

TATRA T 815 VVN 26.265 8 × 8.1R. Platform off-road truck – see text above; carrier for ALS-M airfield lighting set.

TATRA T 815 VT 26.265 8 × 8.1R. Heavy off-road tractor – see text above.

TATRA T 815 VVN 28.400 8 × 8.1R G3 LIWA. See following entry.

TATRA T 815 VTV 26.265 8 × 8.1R. Carrier for PMS bridge roadway components.

TATRA T 815 VPR9 28.265 8 × 8.1R. Bridging equipment carrier for PMS (Ribbon) bridge ramp and pontoons; chassis for 122 mm RM-70 and 70/85 multiple rocket system, with armoured cab; carrier for SB-4H pile-driver system.

TATRA T 815 VP13 32.265 8 × 8.1R. Truck chassis adaptable to carry specialised bodies; carrier for ST-T 815 NBC decontamination system (for details refer to *Jane's NBC Protection Equipment 1996-97* page 106).

TATRA T 815 VP14 32.265 8 × 8.1R. UP-82 command shelter carrier with armoured cab and permanent dozer blade; chassis for VZ-92 minelaying system and support/reload vehicle, both with armoured cabs.

TATRA T 815 VP19 32.265 8 × 8.1R. Carrier for AM-50B bridge components; carrier for TAMARA passive radar reconnaissance system aerial.

TATRA T 815 VP31 29.265 8 × 8.1R. Chassis for 152 mm self-propelled gun-howitzer DANA; chassis for 152 mm self-propelled gun-howitzer ONDAVA; chassis for 155 mm self-propelled gun-howitzer ZUZANA; chassis for STROP air defence system.

AV-15. Heavy recovery vehicle.

AV-20. Heavy recovery vehicle.

Indian-produced TATRA T 815 (8 × 8) trucks are used for many of the above purposes and also act as the carrier for part of the INDRA radar system and the carrier and launch vehicle for various Indian missile and bridging systems. BEML produce a heavy recovery vehicle based on the TATRA T 815 (8 × 8) chassis; this vehicle has a 10 tonne recovery crane and a 20 tonne winch.

Numerous commercial models based on the TATRA T 815 8 × 8 chassis are available. Many of these are basic chassis suitable for conversion to construction industry vehicles, some of which are used by the Czech and Slovak armed forces.

Specifications

T 815 VT 26.265 8 × 8.1R

Cab seating: 1 + 3

Configuration: 8 × 8

Weight:

(kerb) 15,700 kg

(loaded) 25,700 kg

Weight on axles, loaded:

(front) 2 × 5,775 kg

(rear) 2 × 7,075 kg

Max load: 10,000 kg

Towed load:

(off road) 15,000 kg

(on road) 70,000 kg

Load area: 5.36 × 2.41 m

Length: 9.34 m

Width: 2.5 m

Height:

(overall) 3.65 m

(cab roof) 3.03 m

Ground clearance: 0.41 m

Track: (front/rear) 2.044/1.988 m

Wheelbase: 1.65 m + 2.97 m + 1.45 m

Angle of approach/departure: 32°/35°

Max speed: (road) 80 km/h

Fuel capacity: 2 × 230 litres

Gradient: 100%

Side slope: 30°

Trench: 2 m

Vertical obstacle: 0.6 m

Fording: 1.4 m

Engine: TATRA 3-930-51 19 litre V-12 multifuel diesel developing 265 hp at 2,200 rpm

Gearbox: 10 forward, 2 reverse

Clutch: single plate

Steering: worm and roller, power-assisted

Suspension:

(front) torsion bars, shock-absorbers

(rear) leaf springs

Tyres: 15.00-21 TO-3

Brakes: dual-circuit

Electrical system: 24 V

Batteries: 2 × 12 V, 175 Ah

Alternator: 28 V, 60 A

Status

In production. In service with the armed forces of the Czech Republic, India, Saudi Arabia, Slovakia and the UN.

Manufacturers

TATRA, Štefánnikova 1163, 742 21 Kopřivnice, Czech Republic.

Tel: +42 656 40871. Fax: +42 656 42556.

Bharat Earth Movers Limited, BEML Soudha, Mission Road, Sampangirama Nagar, Bangalore 560 027, India.

Tel: +91 80 222 4141. Telex: 0845 2278.

Fax: +91 90 222 6883.

UPDATED

TATRA T 815 VVN 28.400 8 × 8.1 R G3 LIWA truck

Development
Although the TATRA T 815 VVN 28.400 8 × 8.1 R G3 LIWA truck outwardly resembles other 8 × 8 TATRA series trucks it differs in many automotive respects, having been developed to meet the specific requirements of the United Arab Emirates (UAE). Following extensive trials held in July/August 1995, in February 1996 it was announced that the UAE had awarded TATRA a $180 million contract for 1,100 of these trucks in various forms, including cargo/personnel carriers, and fuel and water tankers, the tanks for the latter being provided by VSS of Slovakia. Deliveries will be carried out between 1996 and 1997.

Description
The overall layout of the TATRA T 815 VVN 28.400 8 × 8.1 R G3 LIWA truck follows closely that of the other 8 × 8 models of the TATRA family (see previous entry). The cross-country load capacity is 10,500 kg, rising to a maximum possible 18,500 kg. A two-door all-steel forward cab with a four-element windscreen is provided, with the cab tilting forward hydraulically for maintenance access to the engine. Air-sprung seats are provided for the driver and three passengers.

Power is provided by a German KHD Deutz 544 hp diesel providing a maximum speed of 115 km/h. The engine is coupled to a Twin Disc automatic gearbox and torque converter. Twin Disc also produces the SM-231 Cardan shafts and the differential gears on each axle. To permit operations under desert conditions a central tyre inflation system produced by Syegon of France is provided for all wheels while cooling systems for the transmission oil, water, charge air and fuel are provided from Langerer and Reich of Germany.

The standard 15 m³ cargo/personnel body is produced by TATRA, with fuel and water tanks produced by VSS of Slovakia. Rear-mounted loading cranes are provided by Fassi of Italy. A winch can be carried.

TATRA T 815 VVN 28.400 8 × 8.1 R G3 LIWA truck **1997**

Specifications
Cab seating: 1 + 3
Configuration: 8 × 8
Weight:
(kerb) 17,500 kg
(GVW) 28,000 kg
Max load: 10,500 kg
Load area: 5.06 × 2.45 m
Length: 9.33 m
Width: 2.5 m
Height: (overall) 3.115 m
Ground clearance: 0.41 m
Track:
(front) 2.046 m
(rear) 1.99 m
Wheelbase: 1.65 m + 2.97 m + 1.45 m
Angle of approach/departure: 36°/44°
Max speed: (road) 115 km/h
Fuel capacity: 2 × 230 litres
Gradient: 100%
Side slope: 30%
Trench: 2 m
Vertical obstacle: 0.6 m
Fording: 1.4 m
Engine: KHD Deutz BF8M 1015C 15.87 litre watercooled diesel developing 544 hp at 2,100 rpm
Gearbox: Twin Disc TD-61-1175 automatic with 6 forward and 1 reverse gears plus torque converter
Steering: HR-420-7-LR-17 hydrosteer
Turning circle: 25 m
Suspension: independent, semi-axle, leaf spring
Tyres: Michelin 16.00 R 20 XZL
Brakes: pneumatic
Electrical system: 24 V
Alternator: 80 A

Status
In production. In service with the United Arab Emirates Land Forces.

Manufacturer
TATRA, Štefánnikova 1163, 742 21 Kopřivnice, Czech Republic.
Tel: +42 656 40871. Fax: +42 656 42556.

NEW ENTRY

TATRA T 815 (6 × 6) truck series

Development
The TATRA T 815 (6 × 6) truck series entered produced in 1982 alongside the 8 × 8 series (see previous entry); many of the major assemblies and components are interchangeable.

The TATRA T 815 VVNC 20.235 6 × 6.1R is produced under licence in India by Bharat Earth Movers Limited (BEML).

VOP 028 of Prešov, Slovakia, offers a complete repair and overhaul service for time-expired TATRA trucks, including the TATRA T 815 6 × 6 series and associated vehicles.

Description
The overall layout and construction of the TATRA T 815 6 × 6 variants follow the same general lines as the 8 × 8 versions and many of the major assemblies and components are interchangeable. The main military 6 × 6 version is the TATRA T 815 VVN 20.235 6 × 6.1R, which uses the same two-door cab as the 8 × 8 VVN model. The load capacity of the T 815 VVN 6 × 6 is 8,000 kg off-road and it can tow up to 16,000 kg off-road (65,000 kg on hard roads).

Variants
TATRA T 815 VVN 20.235 6 × 6.1R. Platform off-road truck with 10,000 kg off-road load capacity.

TATRA T 815 VP11 20.235 6 × 6.1R. Tanker using truck chassis.

TATRA T 815 VP12 28.235 6 × 6.1R. Special body carrier with disengageable front axle and central tyre inflation system; carrier for V-24 universal drilling machine; carrier for UDS-114a hydraulic shovel.

TATRA T 815 VP33 24.235 6 × 6.1R. Special body carrier; carrier for ACHR-90 NBC decontamination system (for details refer to *Jane's NBC Protection Equipment 1996-97* pages 106 and 107).

TATRA T 815 VPR8 20.235 6 × 6.1R. Special body carrier with weight up to 10,650 kg; derivative of T 815 VVN 20.235 6 × 6.1R.

TATRA T 815 VNT 28.235 6 × 6.1R. Semi-trailer tractor.

Czech Army TATRA T 815 VVN 20.235 6 × 6.1R truck (R Stickland) **1997**

TATRA T 815 VVN 20.235 6 × 6.1R truck **1997**

TATRA T 815 PR2 22.235 6 × 6.1. ISO container carrier with self-loading equipment.
TATRA T 815 PR4 28.208 6 × 6.1. Bridge component carrier.
TATRA T 815 VPR8 20.235 6 × 6.1R. Carrier for VOSK antenna; carrier for ALS-M airfield lighting set.
TATRA T 815 VP21 21.265 6 × 6.1R. Chassis for TATRApan armoured personnel carrier; chassis for VESPRA mobile command post.
TATRA T 815 S1 6 × 6. 10,700 kg rear-tipping dump truck.
TATRA T 815 S3 6 × 6. 10,700 kg three-way tipping dump truck.
SONDA. Multipurpose environmental monitoring vehicle.
AD-20T. Crane truck.
AD-28. Crane truck with low-slung forward control cab and longer wheelbase (3.95 m + 1.32 m).
AV-14. Multipurpose crane truck.
AVS. Medium recovery vehicle.
CAP-6. Forward area refuelling tanker.
CAP-16. Aircraft refuelling vehicle.
CA-18. Fuel tanker.
CNPL-45. Tractor truck for towing fuel semi-trailers.
CAHD-15. Ground equipment refuelling tanker.
CKV-7. Drinking water tanker.
PHA-32. Foam firefighting truck.
TATRA T 815-2. Commercial model; models include a tractor truck and a 37-seater bus.

In India, BEML produces an airfield crash tender and a medium recovery vehicle with a 7.5 tonne crane and a 12 tonne winch. Both are based on the TATRA T 815 6 × 6 VVN chassis.

Numerous commercial models based on the TATRA T 815 6 × 6 are available. Many of these are basic chassis suitable for conversion to construction industry vehicles, some of which are used by the Czech and Slovak armed forces.

Specifications
T 815 VVN 20.235 6 × 6.1R
Cab seating: 1 + 3
Configuration: 6 × 6
Weight:
(kerb) 12,700 kg
(loaded) 20,700 kg
Weight on axles:
(front) 6,800 kg
(rear) 2 × 6,950 kg
Max load: 8,000 kg
Towed load:
(off road) 10,000 kg
(on road) 50,000 kg
Load area: 5.36 × 2.41 m
Length: 8.35 m
Width: 2.5 m
Height:
(overall) 3.69 m
(cab roof) 3.03 m
Ground clearance: 0.41 m
Track:
(front) 2.044 m
(rear) 1.988 m
Wheelbase: 3.27 m +1.45 m
Angle of approach/departure: 32°/35°
Max speed: (road) 80 km/h
Fuel capacity: 2 × 230 litres
Gradient: 100%
Side slope: 30%
Trench: 0.9 m
Vertical obstacle: 0.6 m
Fording: 1.4 m
Engine: TATRA 3-930-31 19 litre V-12 multifuel diesel developing 235 hp at 2,200 rpm
Gearbox: 10 forward, 2 reverse
Clutch: single plate
Steering: worm and roller, power-assisted
Suspension:
(front) torsion bars, shock-absorbers
(rear) leaf springs
Tyres: 15.00-21 TO3
Brakes: dual-circuit
Electrical system: 24 V
Batteries: 2 × 12 V, 175 Ah
Alternator: 28 V, 60 A

Status
In production. In service with the armed forces of the Czech Republic, India and Slovakia.

Manufacturers
TATRA, Štefánnikova 1163, 742 21 Kopřivnice, Czech Republic.
Tel: +42 656 40871. Fax: +42 656 42556.
Bharat Earth Movers Limited, BEML Soudha, Mission Road, Sampangirama Nagar, Bangalore 560 027, India.
Tel: +91 80 222 4141. Telex: 0845 2278.
Fax: +91 90 222 6883.

UPDATED

TATRA T 815 (4 × 4) truck series

Development
The TATRA T 815 (4 × 4) truck series was developed during 1990-91 and placed in production in 1992. It has a nominal load capacity of 5,600 kg and can tow trailers weighing up to 16,000 kg on hard roads. This vehicle follows the same general layout and construction lines as the larger TATRA T815 military vehicles and again shares many assemblies and components. Some design details have been adapted from the TATRA T 815 4 × 4 VE commercial truck series.

The TATRA T 815 (4 × 4) series is produced under licence in India by Bharat Earth Movers Limited (BEML).

Description
The basic model of the TATRA T 815 4 × 4 is the TATRA T 815 VV 15.170 4 × 4.1. The general overall construction and many components are shared with the larger TATRA T 815 military trucks, including the cab.

The cargo area can be configured to carry 20 passengers but may also be arranged to carry ZSK shop/van bodies to be occupied by a number of control and communications, electronic warfare, battery servicing, workshop and office systems – see listing under variants.

TATRA T 815 VV 15.170 4 × 4.1 truck ***1997***

Variants
TATRA T 815 VV 15.170 4 × 4.1. Platform off-road truck.
TATRA T 815 VPS 15.170 4 × 4.1. Special body carrier; carrier for PDZ-2/TA tank and automotive equipment mobile workshop; carrier for SOJKA drone reconnaissance system launch and control system.
TATRA T 815 VPK 15.170 4 × 4.1. Special body carrier; carrier for PD-OA mobile workshop for accumulators; carrier for PD-2ES series of mobile electrical workshops; carrier for PDZ-2T series of mobile armourers and light engineering workshops.
TA-4. Firefighting technical support truck.
TATRA T 815 4 × 4 VE. Commercial model

Specifications
T 815 VV 15.170 4 × 4.1
Cab seating: 1 + 3
Configuration: 4 × 4
Weight:
(kerb) 9,200 kg
(loaded) 15,000 kg
Max load: 5,600 kg
Towed load:
(off road) 16,000 kg
(on road) 16,000 kg
Load area: 5.055 × 2.42 m
Length: 7.8 m
Width: 2.5 m
Height:
(overall) 3.62 m
(cab roof) 3.03 m
Ground clearance: 0.38 m
Track:
(front) 2.03 m
(rear) 2.036 m
Wheelbase: 4.09 m
Angle of approach/departure: 30°/30°
Max speed: (road) 86 km/h
Fuel capacity: 230 litres
Gradient: 100%
Side slope: 30°
Trench: 0.9 m
Vertical obstacle: 0.6 m
Fording: 1.4 m
Engine: TATRA 3-928-33 12.7 litre 8-cylinder diesel developing 228 hp at 2,200 rpm
Gearbox: 10 forward, 2 reverse
Clutch: single plate
Steering: worm and roller, power-assisted
Suspension:
(front) torsion bars, shock-absorbers
(rear) leaf springs
Tyres: 15.00-21 TO-4
Brakes: dual-circuit
Electrical system: 24 V
Batteries: 2 × 12 V, 175 Ah
Alternator: 28 V, 60 A

Status
In production. In service with the Czech and Slovak armed forces, and the Indian armed forces.

Manufacturer
TATRA, Štefánnikova 1163, 742 21 Kopřivnice, Czech Republic.
Tel: +42 656 40871. Fax: +42 656 42556.
Bharat Earth Movers Limited, BEML Soudha, Mission Road, Sampangirama Nagar, Bangalore 560 027, India.
Tel: +91 80 222 4141. Telex: 0845 2278.
Fax: +91 90 222 6883.

UPDATED

TATRA 148 (4 × 4 and 6 × 6) trucks

Description

The TATRA 148 range of trucks was introduced in 1972 as the replacement for the TATRA 138 series. They are similar in appearance to the earlier vehicles but have more powerful engines and an increased payload. A further development of the TATRA 148, the TATRA 157, was built.

The basic cargo model is designated the TATRA 148. This has a conventional layout, with the engine at the front, all-steel two-door cab and the cargo area at the rear with a drop tailgate, bows and a tarpaulin cover. Vehicles fitted with a winch have the suffix N, for example TATRA 148 N.

Variants

The TZ-74 decontamination vehicle is based on the chassis of the TATRA 148 (6 × 6) truck chassis. For details refer to *Jane's NBC Protection Equipment 1996-97* pages 105 and 106.

The CAS 32 is a fire engine with a 6,000 litre water tank and a 600 litre foam tank.

A hydraulic servicing/inspection platform carrier was produced between 1975 and 1979.

Dump trucks included the TATRA 148 S1 (6 × 6), TATRA 148 S3 (6 × 6), TATRA 148 S3 CH-HMH (6 × 6), TATRA JMH (6 × 6, single-seater cab on left side), TATRA TMCH (forward control type cab, 6 × 6) and the TATRA 148 S3 CH-JMH (4 × 4).

Tractor trucks included the TATRA 148 NTt (6 × 6), TATRA 148 NTPt (6 × 6), TATRA NTPst (6 × 6), TATRA NTt (4 × 4) and TATRA TTt (4 × 4).

VOP 028 of Prešov in Slovakia has instituted a complete repair and overhaul service to produce the following vehicles virtually as new:

T 148 VNM: 13,000 kg capacity cargo truck capable of towing trailers up to 20,000 kg

T 148 VVN: 12,000 kg capacity truck (8,000 kg off-road) capable of towing trailers up to 30,000 kg on roads (10,000 kg off-road); has five-speed gearbox

T 148 CL: uses T 148 PP 36 VN 6 × 6 chassis to carry 11 m^3 fuel tank

T 148 CAPL 15: as T 148 CL but with 18 m^3 tank

T 148 AM-369: concrete carrier/mixer with chassis load capacity of 16,740 kg

T 148 S1: 15,340 kg capacity tipper truck

T 148 S3: 15,200 kg capacity three-sided tipper truck

T 148 HARA 60 Klanice: flatbed truck with hydraulic grab crane

T 148 HARA 60 VNM: cargo truck with hydraulic grab crane

Specifications

TATRA 148

Cab seating: 1 + 2

Configuration: 6 × 6

Weight:

(empty) 11,060 kg

(loaded) 25,640 kg

Max load: (road) 14,580 kg

Towed load: (road) 13,580 kg

Length: 9 m

Width: 2.5 m

Height: 2.44 m

Ground clearance: 0.29 m

Track: (front/rear) 1.966 m/1.77 m

Wheelbase: 4.8 m + 1.32 m

Max speed: 71 km/h

Fuel capacity: 200 litres

Fording: 1.4 m

Engine: TATRA 2-928-litre V-8 air-cooled diesel developing 212 hp at 2,000 rpm

Tyres: 11.00 × 20

Status

Production complete. In service with the armed forces of the Czech Republic and Slovakia and the German Army (former East German Army vehicles).

Manufacturer

TATRA, Štefánnikova 1163, 742 21 Kopřivnice, Czech Republic.

Tel: +42 656 40871. Fax: +42 656 42556.

UPDATED

TATRA 148 (6 × 6) truck (Michael Jerchel)
1996

TATRA 148 CL (6 × 6) 11 m^3 tanker truck (Michael Jerchel) ***1996***

EGYPT

Egyptian truck production

Description

Truck production in Egypt is concentrated on three main types. The first is a licence-manufactured version of the Jeep AM 720 1,135 kg light vehicle (see entry under USA in this section for details) which is produced by the Arab American Vehicles Company at Horreia, Heliopolis. About 40 to 45 per cent of the vehicle is local content and the AM 720 is produced in various forms such as troop carrier, ambulance, mobile workshop (with a ¼ tonne trailer) and communications shelter carrier. Another locally produced Jeep product is the J-20, which is produced in pick-up form for a variety of civil and military roles. About 20 per cent of the J-20 is local content. The company also produces CJ-7 and CJ-8 Jeeps (see entry under *Lightweight Vehicles*).

The third vehicle is a locally produced version of the Magirus-Deutz (now IVECO Magirus) Mercur, known as the NASR. This is a 4 × 4 vehicle, the German origins of which date back to the 1950s. It is produced in several forms including a basic troop and cargo carrier with a 3,350 kg payload, a mobile surgery and at least two types of NBC decontamination vehicle. The mobile surgery has an air conditioned office-type body with its own 4 kW electrical generator, a clean water supply system and enough medical equipment and supplies (including a refrigerated blood bank) to remain in the field for 3.3 days. When in position, two penthouse tents can be erected each side of the vehicle or separately to provide extra working space. Each tent measures 4 × 4 m on each side and is 2.4 m high. The NASR is produced under the auspices of the Egyptian Ministry of Defence Vehicles Department.

Specifications

NASR

Cab seating: 1 + 1 or 2

Configuration: 4 × 4

Weight:

(empty) 3,800 kg

(loaded) 7,150 kg

Max load: 3,350 kg

Length: 7.2 m

Width: 2.4 m

Ground clearance: 0.315 m

Wheelbase: 4.2 m

Fuel capacity: 150 litres

Engine: NASR 112 H6 air-cooled diesel developing 112 hp at 2,300 rpm (licence-produced version of Deutz F 6 L 614)

Gearbox: ZF 35 AK5 manual with 5 forward and 1 reverse gears

Clutch: single dry plate

Transfer box: 2-speed

Turning radius: 9 m

Suspension: semi-elliptic leaf springs with auxiliary springs at rear and shock-absorbers in front

Tyres: 20.00 × 10

Status

All the above are in production and in service with the Egyptian armed forces.

VERIFIED

FINLAND

Sisu A-45 (4 × 4) 3,000 kg truck

Development

Introduced in 1965, the Sisu A-45 (4 × 4) truck was the first all-terrain truck to be developed specifically to meet the requirements of the Finnish Defence Forces. As such it can be regarded as the predecessor of the current Sisu SA series of all-terrain trucks. The vehicle is used by the Finnish Army both in Finland and overseas in the course of United Nations duties. The vehicle is also used for a variety of civil applications. Production ceased during the late 1970s.

Description

The cab of the Sisu A-45 (4 × 4) 3,000 kg truck has a steel floor and walls with a detachable glass fibre roof. The driver's seat is adjustable and the cab has a heater as a standard feature. The engine is mounted to the immediate rear of the cab and projects into the load

Sisu A-45 (4 × 4) 3,000 kg truck

Sisu A-45 (4 × 4) 3,000 kg truck

area of the vehicle. The rear platform is of steel construction and attached to the chassis by rubber mountings. The side boards and tailgate can be folded down. The tailgate has access steps. The load area can be fitted with bows and a tarpaulin cover if required. A towhook is mounted at the rear of the vehicle and a pulling jaw is mounted in the centre of the front bumper.

Optional equipment included: a hydraulic mechanism for the operation of Sisu hydraulic motors in the trailer (for example, when a special two-wheel trailer is being towed the wheels of the trailer are powered, giving the truck/trailer combination full 6 × 6 drive); trailer air pressure brake system; cold-starting equipment; blowlamp-operated coolant and oil sump heater; and a Sisu 6,500 kg capacity hydraulic winch, with 60 m of cable, which can be used to the front or rear of the truck.

There are ambulance, fire, radio, command and workshop variants.

Specifications

Cab seating: 1 + 2
Configuration: 4 × 4
Weight:
(empty) 5,950 kg
(loaded, road) 10,000 kg
(loaded, cross-country) 9,000 kg
Max load:
(road) 4,050 kg
(cross-country) 3,050 kg
Towed load: 9,000 kg
Load area: 4.28 × 2.16 m
Length: 6 m
Width: 2.3 m
Height:
(cab) 2.6 m
(tarpaulin) 2.8 m
(load area) 1.29 m
Ground clearance: (axles) 0.4 m
Track: 1.89 m
Wheelbase: 3.7 m
Angle of approach/departure: 38°/38°
Max speed: (road) 100 km/h
Range: 700 km
Fuel capacity: 210 litres
Max gradient: 60%
Fording: 1 m
Engine: 6-cylinder direct injection diesel developing 130 hp (DIN) at 2,600 rpm or turbocharged diesel developing 160 hp (DIN) at 2,600 rpm
Gearbox: manual with 5 forward and 1 reverse gears
Transfer box: 2-speed (front and rear axles have differential lock, planetary gears in wheel hubs)
Steering: powered
Turning radius: 8.2 m
Suspension: leaf springs front and rear with shock-absorbers on front axle only
Tyres: 14.00-20 MPT
Brakes:
(main) air
(parking) mechanical
Electrical system: 24 V
Batteries: 2 × 12 V, 145 Ah

Status

Production complete. In service with the Finnish Army.

Manufacturer

Sisu Defence Oy, PO Box 189, FIN-13101 Hämeenlinna, Finland.
Tel: +358 17 6451. Fax: +358 17 619 6710.

VERIFIED

Sisu SA-130 VKH (4 × 4) 6,500 kg all-terrain truck

Description

The Sisu SA-130 VKH (4 × 4) 6,500 kg all-terrain truck is the medium-heavy model in the Sisu SA series of all-terrain trucks developed by Sisu Defence for the Finnish Army during the late 1970s and early 1980s. The SA-130 VKH is an all-terrain truck primarily designed to tow medium and heavy artillery while having a good cargo-carrying and hauling capacity. The vehicle has an off-road carrying capacity of 6,500 kg and can tow a 5,000 kg trailer at the same time.

The SA-130 VKH has a well-balanced load distribution while a twisting frame and balanced parabolic springs ensure that all wheels are in contact with the ground at all times. The cab and cargo body are attached to the chassis, maintaining balance when traversing rough terrain yet insulating them from all frame movements. The planetary drive axles are Sisu's own design.

The all-steel forward control cab has seating for the driver and two passengers. Standard equipment includes a roof hatch and a camouflage net support frame. A heavy machine gun can be mounted on the cab roof. The rear cargo area has a steel frame construction with removable sides. Standard equipment includes an 8,000 kg winch with 50 m of cable that can be used to the front or rear of the vehicle.

The Sisu XA-180 armoured personnel carrier uses many components of the SA-130. For details refer to *Jane's Armour and Artillery 1995-96*, pages 331 to 333.

Sisu SA-130 VKH (4 × 4) 6,500 kg truck

Specifications

Cab seating: 1 + 2
Configuration: 4 × 4
Weight:
(empty) 7,500 kg
(loaded) 14,000 kg
Weight on front axle: (loaded) 7,500 kg
Weight on rear axle: (loaded) 8,000 kg
Max load: 6,500 kg
Towed load: 10,000 kg
Length: 6.75 m
Width: 2.58 m
Height:
(cab) 2.9 m
(tarpaulin) 3.1 m
(load area) 1.5 m
Ground clearance: (axles) 0.4 m
Wheelbase: 3.85 m
Track: 2.18 m
Angle of approach/departure: 43°/40°
Max speed: 100 km/h
Range: 800 km
Fuel capacity: 225 litres
Fording: 1 m
Engine: Valmet 620 DSI diesel developing 175 hp (until 1994) Valmet 620 DS 6-cylinder (in-line), water-cooled turbocharged diesel developing 180 hp at 2,500 rpm
Gearbox: ZF S6-66 manual with 6 forward and 1 reverse gears

Transfer box: 2-speed
Steering: powered
Turning radius: 8.2 m
Suspension: parabolic leaf springs and hydraulic shock-absorbers front and rear
Tyres: 14.00 R 20
Brakes:
(main) dual-circuit air-operated
(parking) spring-operated, controlled by compressed air operating on rear wheels
Electrical system: 24 V
Batteries: 2 × 12 V, 150 Ah

Status
In production. In service with the Finnish Army.

Manufacturer
Sisu Defence Oy, PO Box 189, FIN-13101 Hämeenlinna, Finland.
Tel: +358 17 6451 Fax: +358 17 619 6710.

UPDATED

Sisu SA-241 CKH (6 × 6) 12,000 kg all-terrain truck

Description
The Sisu SA-241 CKH (6 × 6) 12,000 kg all-terrain truck is a straightforward derivation of the Sisu SA-130 (see previous entry) and many components and the overall construction are the same as for the latter, including the frame and cab. The SA-241 CKH has a more powerful Cummins engine and a stronger drive train providing a permanent 6 × 6 drive configuration, plus reinforced structures overall. A torque converter is provided.

Along with the lighter models of the Sisu SA series of all-terrain trucks, several hundreds of these vehicles have been in service with the Finnish Army since the early 1980s as ammunition carriers and heavy artillery tractors, among other roles. Variants include a longer wheelbase version, firefighting vehicles, recovery/rescue vehicles and missile system platforms, such as for the RBS-15AK anti-ship missile. One highly specialised civil application is an advanced rescue vehicle family delivered to the Finnish State Railways.

Sisu SA-241 CKH (6 × 6) 12,000 kg truck

Specifications
Cab seating: 1 + 2
Configuration: 6 × 6
Weight:
(empty) 10,000 kg
(loaded) 22,000 kg
Weight on front axle:
(empty) 4,500 kg
(loaded) 7,500 kg
Weight on rear bogie:
(empty) 5,500 kg
(loaded) 8,000 kg + 8,000 kg
Max load: 12,000 kg
Towed load: 15,000 kg
Length: 7.8 m
Width: 2.6 m
Height:
(overall) 3.13 m
(top of cab) 2.13 m
Ground clearance: (axles) 0.4 m
Wheelbase: 3.74 m + 1.44 m
Track: 2.01 m
Angle of approach/departure: 38°/54°
Max speed: 95 km/h
Range: 600 km
Fuel capacity: 250 litres
Gradient: 45°
Side slope: 40°
Fording: 1 m
Engine: Cummins LTA 10-330 turbocharged and intercooled diesel developing 340 hp at 2,200 rpm
Gearbox: ZF 4S-150 GPA fully synchronised manual, 8 forward and 1 reverse gears plus torque converter
Clutch: single dry plate
Transfer box: single-speed
Steering: powered
Suspension: parabolic twin-leaf springs and hydraulic shock-absorbers
Tyres: 14.00 R 20
Brakes: dual-circuit pneumatic
Electrical system: 24 V
Batteries: 2 × 12 V 150 Ah

Status
In production. In service with the Finnish Army.

Manufacturer
Sisu Defence Oy, PO Box 189, FIN-13101 Hämeenlinna, Finland.
Tel: +358 17 6451. Fax: +358 17 619 6710.

VERIFIED

FRANCE

Renault B110 Turbo (4 × 4) light vehicle

Development
The Renault B110 Turbo (4 × 4) light vehicle is the military version of the Renault 'Messenger' commercial vehicle that can be used for a variety of tactical roles, ranging from a 2,285 kg light truck to a military ambulance. Also included in the range is an enclosed van-bodied Fourgon light vehicle.

There are two chassis versions, the B110-35D and B110-45D which differ in their gross vehicle weight capacities, with the B110-45D having the greater capacity.

In April 1994 300 units were ordered for the French Gendarmerie. In 1993 170 examples of a 4 × 2 version were ordered for the French Gendarmerie.

Description
The B110 Turbo uses a flat frame chassis with parallel channel side members treated with anti-corrosion protection. The vehicle uses a steel monocoque semi-forward control cab with a rearwards-sloping laminated windscreen. Inside the cab there is seating for the driver and a bench seat for two passengers. The bonnet can be raised for engine maintenance and repair and a rear engine hood can be removed for further engine access.

Renault B110 (4 × 4) platform truck fitted with drop sides and tarpaulin **1997**

As a light tactical truck the cargo area can be covered by a fabric tilt on removable bows. There is also a tailgate. The van-bodied version can be supplied with or without side-opening doors. For the ambulance version the bodywork alterations are carried out by sub-contractors. The ambulance can carry six stretchers or six seated patients. Entrance to the ambulance section is normally through two large doors at the rear but there is a side door on the right-hand side of the body.

Optional equipment includes: an airlift kit, making the vehicle suitable for transport in the C-130, C-160 or similar aircraft; right-hand drive; a removable towing bracket for mounting front or rear; and a removable electric winch.

Specifications

Fourgon van model
Cab seating: 1 + 2
Configuration: 4 × 4
Weight:
(chassis and cab) 2,775 kg
(GVW, B110-35D) 3,500 kg
(GVW, B110-45D) 4,500 kg
Max load:
(B110-35D) 725 kg
(B110-45D) 1,725 kg
Length: 5.652 m
Width: (over cab) 2 m
Height: (van body) 2.8 m
Ground clearance: (loaded) 0.287 m
Track:
(front) 1.714 m
(rear) 1.703 m
Wheelbase: 3.26 m
Angle of approach/departure: 55°/26°
Max speed: 108 km/h
Fuel capacity: 67 litres
Engine: Type 8140.27 IDS 2.5 litre 4-cylinder in-line turbocharged diesel developing 106 hp at 3,800 rpm
Gearbox: Type S 5-24 manual synchromesh with 5 forward and 1 reverse gears
Clutch: single dry plate
Transfer box: Type 100 2-speed
Steering: worm and screw with hydraulic power-assist
Suspension: leaf springs front and rear with hydraulic shock-absorbers and front and rear anti-roll bars
Tyres: 9.00R × 16 XZL
Brakes:
(main) discs front, drums rear, dual-circuit, hydraulic vacuum-assisted
(parking) mechanical on rear axle
Electrical system: 12 V
Batteries: 1 × 12 V, 90 Ah
Alternator: 70 A

Status

In production.

Manufacturer

Enquiries to: Renault Véhicules Industriels, 40 rue Pasteur, BP 302, F-92156 Suresnes Cedex, France.
Tel: +33 1 40 99 71 11. Telex: 620 567 f.
Fax: +33 1 40 99 71 08.

UPDATED

Peugeot 504 Dangel (4 × 4) 1,110 kg light vehicle

Description

The Peugeot 504 is basically a standard civilian 4 × 2 vehicle modified to 4 × 4 configuration and fitted with seats, bows and a tarpaulin cover over the rear cargo area. It is also known as the Pick Up Armée.

The layout of the Peugeot 504 is conventional, with the engine at the front, three-man fully enclosed cab in the centre and the load area at the rear. The load area has a drop tailgate, removable bows, tarpaulin cover and bench seats running down the centre facing outwards. In addition to the driver and two passengers in the cab, 10 fully equipped troops can be carried in the rear.

The front and rear axles have Peugeot limited-slip differentials. The Dangel transfer case is mounted in the tunnel and is independent of the transmission due to a short universal joint shaft turning in a reaction tube. It gives four road and four cross-country gears. Shifting from road gears to all-terrain gears is fully synchronised and can be carried out while the vehicle is travelling.

Mounted at the front of the vehicle is a winch. Special protection is provided for the underside of the vehicle.

An ambulance version can carry three stretcher patients in the fully enclosed rear body which is also provided with a dual seat. A 4 × 2 version is also available.

Specifications

Cab seating: 1 + 2
Configuration: 4 × 4
Max load: 1,110 kg
Load area: 2 × 1.5 m
Length: 4.8 m
Width: 1.73 m
Height: (cab) 1.74 m
Max road speed:
(petrol model, no tarpaulin) 135 km/h
(petrol model, with tarpaulin) 125 km/h
(diesel model, no tarpaulin) 120 km/h
(diesel model, with tarpaulin) 115 km/h
Engine: XN-1 4-cylinder petrol developing 96 hp at 5,200 rpm or XN-2 4-cylinder diesel developing 70 hp at 4,500 rpm
Gearbox: manual with 4 forward and 1 reverse gears
Clutch: single dry plate
Transfer box: 2-speed

Status

In service with the French Navy and Marines.

Manufacturer

Automobiles Peugeot, 75 avenue de la Grande-Armée, F-75016 Paris, France.
Tel: +33 1 40 66 59 76. Telex: 649 237 f.

VERIFIED

Peugeot 504 Pick Up Armée (4 × 4) truck (T J Gander)

Peugeot 504 (4 × 4) ambulance

SOVAMAG TC10 DT (4 × 4) 1,100 kg multipurpose vehicle

Development

The SOVAMAG concern was taken over by Auverland in 1990 with production of the two companies integrated by the end of that same year. Since then 1,300 TC10 DT series vehicles have been produced with 300 vehicles going to the French Air Force and 840 to the French Ministry of Cooperation. The Mauritania Ministry of Defence procured a further 160 vehicles.

Description

The SOVAMAG TC10 DT (4 × 4) 1,100 kg multipurpose vehicle has the general appearance of the US Jeep but is a larger vehicle and can carry a larger (1,100 kg) payload. It has square military lines and is ruggedly constructed, having a tubular chassis construction and galvanised body panels. Layout is conventional, with the IVECO diesel engine and radiator protected behind a sturdy front bumper and a grill guard. This grill guard can be hinged forward to provide extra front stowage capacity. Prior to April 1994 the vehicle was powered by a Peugeot XD3P 72 hp diesel.

There are seats for the driver and a passenger, while at the rear is the main cargo-carrying area. This has a reinforced floor and bench seats for four men along each side. These seats can be folded up for access to three stowage compartments each side. Access to the cargo area is via a small hatch that opens to the right and carries the spare wheel. In addition to carrying cargo or passengers, the area can be used to mount light weapons such as machine guns or a 106 mm M40 recoilless rifle. A canvas roof carried on a removable light tubular steel frame provides overhead and side weather protection.

Standard equipment includes two 40 litre drinking water tanks and a jerrican rack at the rear. A standard towing hook is also provided. There is provision for a radio aerial and the normal electrical system is 24 V. The forward-opening bonnet allows easy access to the engine and all maintenance and repairs can be carried out using normal tools.

Variants include a right-hand drive model, gun and missile carriers, van bodies, workshop vehicles, a 1,000 litre water tanker, and firefighting vehicles. Right-hand drive and dual-control driver training versions are available.

In 1996 the SOVAMAG TC10 SL appeared. This is basically similar to the TC10 DT and is powered by the same engine but has a longer 3.07 m wheelbase, increasing the overall length to 4.85 m. This enables the vehicle to carry up to 12 personnel and their equipment or a payload of up to 1,560 kg. Hard- and soft-top versions are available.

Specifications

TC10 DT
Cab seating: 1 + 1 (8 in rear)
Configuration: 4 × 4

SOVAMAG TC10 DT (4 × 4) 1,100 kg multipurpose vehicle (T J Gander) **1997**

SOVAMAG TC10 SL (4 × 4) 1,500 kg multipurpose vehicle with 3.07 m wheelbase and hardtop body (T J Gander) **1997**

Weights:
(unladen) 1,900 kg
(GVW) 3,000 kg
(front axle, max) 1,300 kg
(rear axle, max) 2,400 kg
Payload: 1,100 kg
Towed load:
(with brake) 3,500 kg
(without brake) 750 kg
Length: 4.45 m
Width: 1.65 m
Height: (unladen) 2.12 m
Ground clearance: 0.23 m
Wheelbase: 2.77 m
Track: (front and rear) 1.4 m
Angle of approach/departure: 53°/44°
Max speed: 130 km/h
Fuel capacity: 105 litres
Gradient: 100%
Side slope: 40°
Fording: 0.6 m
Engine: SOFIM 2.5 litre 4-cylinder in-line water-cooled turbo diesel developing 103 hp at 3,800 rpm
Gearbox: 2826-5 manual, 5 forward and 1 reverse gears
Transfer box: Auverland A80 2-speed
Steering: mechanical
Turning radius: 6.5 m
Suspension: semi-elliptical leaf springs with double-acting shock-absorbers front and rear
Brakes: dual-circuit, servo assist hydraulic, discs front, drums rear
Tyres: 7.50 × 16
Electrical system: 24 V
Batteries: 2 × 12 V, 55 A

Status
In production. In service with the French Air Force (300), French Ministry of Cooperation (840) and Mauritania Ministry of Defence (160).

Manufacturer
Auverland SA, BP 12, F-42260 Saint-Germain-Laval, France.
Tel: +33 4 77 65 54 44. Telex: 310 663 f.
Fax: +33 4 77 65 49 18.

UPDATED

Renault TRM 1200 (4 × 4) 1,200 kg truck

Description
The Renault TRM 1200 was previously known as the Saviem TP3 and was in production at Renault's Blainville factory from 1969; by mid-1981 some 9,200 had been produced, of which 4,000 were for the French Army and 4,000 for overseas military sales.

The Renault TRM 1200 is based on standard commercial components and can carry 12 fully equipped men or 1,200 kg of cargo.

The chassis consists of cold-drawn steel side members with the cross members welded into position. The vehicle can be delivered with a two-door all-steel fully enclosed cab, or a cab with a tarpaulin roof, removable door tops and a windscreen which folds forward onto the bonnet when not required. The rear cargo area has removable drop sides, a drop tailgate, removable bows and a tarpaulin cover. Removable seats can be installed down the centre of the vehicle (for example back to back) or along each side of the vehicle.

Optional equipment includes: a 72 hp petrol engine; a winch with a capacity of 2,000 kg; a 120 litre fuel tank in place of the standard 70 litre fuel tank; 7- and 12-pin outlet sockets; 10.50 × 20 tyres and a fire extinguisher.

Variants include: an ambulance capable of carrying six stretcher patients or 12 seated patients, which is in service with the French Army; command post (in service with the French police); dump truck; firefighting vehicle; and a light recovery vehicle.

Specifications
Cab seating: 1 + 1
Configuration: 4 × 4
Weight:
(empty) 2,620 kg
(loaded) 3,950 kg
Max load: (cross-country) 1,330 kg
Load area: 3.025 × 1.85 m
Max towed load: 1,000 kg
Length: 5.005 m
Width: 1.996 m
Height:
(cab) 2.4 m
(tarpaulin) 2.594 m
(load area) 1.094 m
Ground clearance: 0.27 m
Track: 1.641 m
Wheelbase: 2.64 m
Angle of approach/departure: 31°/40°
Max speed: 95.5 km/h
Range: 600 km
Fuel capacity: 70 litres
Max gradient: 58.9%
Fording: 0.5 m
Engine: Renault 712 4-cylinder diesel developing 72 hp at 3,200 rpm
Gearbox: Renault 321-4 manual with 4 forward and 1 reverse gears
Clutch: single dry disc

Saviem TP3/Renault TRM 1200 (4 × 4) 1,200 kg truck with command post body

Renault TRM 1200 (4 × 4) 1,200 kg truck with soft top cab, bows and tarpaulin cover

Transfer box: Renault 433 2-speed
Steering: cam and roller
Turning radius: 8.25 m
Suspension: leaf springs with Evidgom pads and hydraulic shock-absorbers
Tyres: 9.00 × 16
Brakes:
(main) hydraulic
(parking) mechanical
Electrical system: 24 V
Batteries: 2 × 12 V, 96 Ah

Status
Production complete. In service with the French Army, Algeria, Morocco and other undisclosed countries.

Manufacturer
Enquiries to: Renault Véhicules Industriels, 40 rue Pasteur, BP 302, F-92156 Suresnes Cedex, France.
Tel: +33 1 40 99 71 11. Telex: 620 567 f.
Fax: +33 1 40 99 71 08.

VERIFIED

SUMB (4 × 4) 1,500 kg MH 600 BS and SUMB (4 × 4) 3,000 kg trucks

Description
The SUMB (Simca-Unic Marmon-Bocquet) 1,500 kg model MH 600 BS truck was developed in the late 1950s by M Bocquet, who was president of Marmon-Herrington which subsequently became Marmon-Bocquet. Series production was undertaken from 1964 and, although production has ceased, it remains one of the standard 1,500 kg trucks of the French Army.

The engine and cab are at the front with the cargo area at the rear. The cab has two doors and a canvas top mounted on an articulated frame, which enables the top to be folded to the rear to reduce the vehicle's overall height. The windscreen can be folded forward onto the bonnet. The rear cargo area has a drop tailgate, bows and a removable tarpaulin cover. A bench seat can be fitted down the centre of the rear cargo area when the truck is carrying passengers. A locking differential is fitted to the rear axle. Some versions have a front-mounted winch with 60 m of cable.

A long wheelbase model of this vehicle, known as the SUMB (4 × 4) 3,000 kg truck, was produced from 1971 with the same engine, transmission and cab as the 1,500 kg version.

Variants include a fuel servicing vehicle and a light digger with a Poclain light shovel mounted to the rear of the cab. The SUMB is also used to carry communication shelters.

SUMB Upgrade
Renault Véhicules Industriels (RVI) has upgraded over 500 SUMB (4 × 4) 1,500 kg trucks out of a planned programme involving some 2,300. The upgrade involves the replacement of the existing petrol engine with the same Type 720 SPC diesel developing 115 hp as used in the RVI TRM 2000 truck, along with the transmission from the TRM 2000. Other upgrade changes include electrical components, also from the TRM 2000, and the braking system is updated to meet current EC directives.

SUMB (4 × 4) 1,500 kg truck (Michael Jerchel)

Specifications
(Data in square brackets applies to the 3,000 kg truck where it differs from the 1,500 kg model)
Cab seating: 1 + 1
Configuration: 4 × 4
Weight:
(empty) 3,670 [4,220] kg
(loaded) 5,300 [7,420] kg
Max load: 1,500 [3,000] kg
Towed load: 2,000 kg
Load area: 2.95 × 1.97 [4.4 × 2.29] m
Length: 5.195 [6.55] m
Width: 2.305 [2.41] m
Height:
(cab) 2.27 m
(tarpaulin) 2.88 [2.97] m
Ground clearance: 0.33 [0.38] m
Track: 1.704 m
Wheelbase: 2.9 [4.1] m
Angle of approach/departure: 43°/42°
Max speed: (road) 85 [82] km/h
Range: 550 km
Fuel capacity: 130 litres
Fuel consumption: 30-32 l/100 km
Max gradient: 60 [50]%
Max side slope: 30%
Fording:
(without preparation) 0.8 m
(with preparation) 1.2 m
Engine: 8-cylinder petrol developing 100 bhp at 3,000 rpm or 6-cylinder diesel developing 100 bhp at 3,000 rpm [6-cylinder diesel developing 100 bhp at 3,000 rpm]
Gearbox: manual with 4 forward and 1 reverse gears
Clutch: single dry plate
Transfer box: 2-speed
Turning radius: 7.5 [11.5] m
Suspension: coil springs and hydraulic shock-absorbers
Tyres: 10.00 × 20 [12.50 × 20]
Brakes:
(main) hydraulic
(parking) mechanical
Electrical system: 24 V
Batteries: 2 × 12 V

Status
In service with the French Army.

Manufacturer
FFSA, 3 bis rue Salomon de Rothschild 92, Suresnes, France.

VERIFIED

Renault TRM 2000 (4 × 4) 2,000 kg truck

Development
Following a competition for a new 2,000 kg truck in which five vehicles were tested by the French Army, it was announced in March 1981 that the High Mobility version of the Renault TRM 2000 (4 × 4) 2,000 kg truck had been selected. The French Army then had a requirement for some 12,000 vehicles, with first production vehicles completed at Renault's Blainville facility during 1983. The first 2,000 examples were delivered to signal units in the three French Army corps to carry RITA communications equipment. Production is complete.

Description
The Renault TRM 2000 is produced in two versions, Standard and High Mobility. The main difference between the two is that the Standard has a ground clearance under the axle of 0.302 m as opposed to the High Mobility's 0.425 m.

The chassis consists of U-beam side members (180 mm × 70 mm × 4 mm) with the cross members riveted and bolted together. A towing bracket and a protective skid pad are provided at the front with a 5 tonne towing hook at the rear.

The forward control cab is of all-steel construction but may have a tarpaulin cover, and is mounted onto the chassis at three points with elastic pads and two shock-absorbers at the rear. The cab tilts forward and is provided with a torsion bar tilting mechanism with double lock and warning indicator lamp, toughened glass windscreen, fully wind-down door windows, rear quarter lights, two-man passenger seat and an adjustable driver's seat. To the rear of the cab is a spare wheel rack. In the cab roof a Creusot-Loire Industrie STR TA rail-mounted rotary support is fitted to a hatch. This mounting can accommodate a 7.5 or 7.62 mm machine gun and is a standard fitting.

The cargo area is provided with drop sides and rear, bows and a tarpaulin cover. If required, seats can be fitted down the centre of the vehicle to carry 12 fully equipped troops (for example six down each side facing outwards).

Optional equipment includes a front-mounted winch with a capacity of 2,500 kg and a pioneer tool holder.

Variants include: a fully enclosed ambulance; command post with shelter-mounted RITA system; forward air control; 2,000 litre fuel tanker; missile carrier; prime mover for heavy mortars and the 20 mm Tarasque 53 T2 anti-aircraft cannon; recovery vehicle; water tanker; a field ambulance carrying six casualties; and a workshop vehicle. In 1990 LOHR SA announced its Mo.120 RA, a TRM 2000 carrying a 120 mm mortar which can be fired from a flat bed area in place of the normal cargo body. A hydraulic system at the rear can lower a recoil spade assembly to counter recoil forces or the mortar can be lowered to the ground for firing.

Specifications
Cab seating: 1 + 2 (up to 12 in rear)
Configuration: 4 × 4
Weight:
(empty) 3,980 kg
(loaded) 6,300 kg
Max load: 2,320 kg
Towed load: 2,000 kg

Length: 5.02 m
Width: 2.2 m
Height: (cab) 2.713 m
Ground clearance:
(axles) 0.425 m
(centre of vehicle) 0.604 m
Track: (front and rear) 1.8 m
Wheelbase: 2.7 m
Angle of approach/departure: 42°/48°
Maximum road speed: 89 km/h
Range: 1,000 km
Fuel capacity: 130 litres
Max gradient:
(1st gear, high range) 35%
(1st gear, low range) 50%
Max side slope: 30%
Fording: 0.9 m
Engine: Type 720S 4-stroke 4-cylinder in-line turbo-compressor supercharged diesel developing 115 hp at 3,000 rpm

Renault TRM 2000 (4 × 4) 2,000 kg truck **1997**

Gearbox: S 5-24/3, 5 forward and 1 reverse gears
Clutch: single dry plate
Transfer box: 2-speed
Axles:
(front) axle nose with single, non-locking differential
(rear) axle nose with single reduction differential lock, total reduction 6.37. Electropneumatic rear axle differential lock control
Steering: ball race, hydraulic
Turning radius: 7 m
Suspension: semi-elliptical leaf springs front and rear, with Evidgom pads, hydraulic telescopic shock-absorbers, anti-roll bar on rear axle
Tyres: 12.50 × 20
Brakes:
(main) air, dual-circuit with load equaliser on rear circuit
(parking) mechanical
Electrical system: 24 V
Batteries: 2 × 12 V, 6TN, 125 Ah

Status
Production complete. In service with the French Army and other armed forces including Morocco (190), Namibia (50), Qatar (10) and FINUL (85).

Manufacturer
Enquiries to: Renault Véhicules Industriels, 40 rue Pasteur, BP 302, F-92156 Suresnes Cedex, France.
Tel: +33 1 40 99 71 11. Telex: 620 567 f.
Fax: +33 1 40 99 71 08.

UPDATED

SOVAMAG TC24 (4 × 4) 2,570 kg Multipurpose Vehicle

Development
The SOVAMAG TC24 (4 × 4) 2,570 kg multipurpose vehicle may be regarded as an enlarged version of the SOVAMAG TC10 DT (see separate entry for details) with an increased payload capacity that allows the carrying of shelter/containers. Various applications for the basic vehicle include a firefighting vehicle, tanker, and an armoured version, the Armoured TC24.

Description
The SOVAMAG TC24 (4 × 4) 2,570 kg multipurpose vehicle is based on a ladder type chassis with a galvanised sheet steel body. The Jeep-type cab has seating for a driver and one or two passengers while the cargo body can be provided with seating for up to 12 passengers on folding bench seats.

Variants include gun and missile carriers, van bodies, workshop vehicles, a 1,000 litre water tanker, and fire-fighting vehicles. A lightened variant (total weight 3,500 kg) has a nominal payload of 1,070 kg. Right hand drive and driver trainign models are available.

SOVAMAG TC24 (4 × 4) 2,570 kg multipurpose vehicle (T J Gander) **1997**

Specifications
Cab seating: 1 + 1 (12 in rear)
Configuration: 4 × 4
Weights:
(unladen) 2,430 kg
(GVW) 5,000 kg
(front axle, max) 2,200 kg
(rear axle, max) 3,200 kg
Payload: 2,570 kg
Towed load:
(with brake) 5,000 kg
(without brake) 750 kg
Length: 4.96 m
Width: 1.97 m
Height: 2.27 m
Ground clearance: 0.28 m
Wheelbase: 3.6 m
Track: (front and rear) 1.7 m
Angle of attack/departure: 53°/44°
Max speed: 110 km/h
Fuel capacity: 2 × 80 litres
Range: (cruising) 1,000 km
Gradient: 100%
Side slope: 40°
Fording: 0.6 m
Engine: 2.5 litre 4-cylinder water-cooled turbo diesel developing 115 hp at 4,000 rpm
Gearbox: 2826-5 manual, 5 forward and 1 reverse gears
Transfer box: Type 241 2-speed
Steering: power
Turning radius: 7 m
Suspension: semi-elliptical leaf springs plus helper pneumatics, with double-acting shock-absorbers front and rear
Brakes: dual circuit, servo assist hydraulic, discs front, drums rear
Tyres: 9.00 × 16
Electrical system: 24 V

Status
Available for production.

Manufacturer
Auverland SA, BP 12, F-42260 Saint-Germain-Laval, France.
Tel: +33 4 77 65 54 44. Telex: 310 663 f.
Fax: +33 4 77 65 49 18.

UPDATED

ACMAT/ALM VLRA 1,500 kg and 2,500 kg (4 × 4) and 4,500 kg and 5,500 kg (6 × 6) vehicles

Description

The VLRA *(Véhicule de Liaison, de Reconnaissance et d'Appui)* was developed in the late 1950s specifically for operations in the Sahara. Since then it has been adopted by the French Army and many other countries throughout the world and is also used for a variety of civil applications, for example by oil companies.

The layout of the vehicle is conventional, with the engine at the front, cab in the centre and cargo area at the rear. The cab has a windscreen which can be folded forward onto the bonnet and a removable nylon canvas roof and side doors. A European cab (metal door and fabric top) and an all-metal cab are available. The spare wheel is mounted to the left of the driver's position, or in the rear cargo space. The cargo area at the rear has a drop tailgate, removable bows and nylon canvas covers. Seats can be fitted down the sides or centre of the cargo area for carrying passengers.

Special equipment fitted as standard includes long-range fuel tanks, which allow 1,600 km of travel or 34 hours of cross-country operation, and a 200 litre water tank; these figures are reduced to 900 km and 100 litres for the TPK 4.15 SM3. Standard equipment also includes four jerricans, two sand channels, twin dry and oil-based air filtration equipment and a six-bladed 500 mm fan driven by twin fanbelts, as well as a specially designed water radiator. A heavy-duty chassis, gearbox and suspension are also employed. The reduction box and axles are of ACMAT design.

All vehicles are suppressed to the second degree (NATO standard) and a 3,000 kg winch can be fitted to the front of the vehicle if required. A 6,000 kg towing hook is standard and a multiposition version is also available for artillery-towing duties. Full EC lighting equipment is available if required, as is power-assisted steering. Finally, all VLRA vehicles can be produced in left- or right-hand drive forms.

The basic vehicle can be armed with a 5.56 or 7.62 mm machine gun, which is pedestal-mounted, and a 12.7 mm machine gun ring-mounted above the passenger's seat if required. TDA 60 or 81 mm mortars can be mounted in the cargo space over the rear axle. For anti-tank use the VLRA can be armed with a launcher for the MILAN ATGW system.

It is important to note the high degree of commonality of wearing parts throughout the entire range of ACMAT vehicles. Not only is the same engine used in every ACMAT vehicle, but a commonality level of over 80 per cent exists across the entire range of chassis and cabs. While detailed items are the subject of continued research and development, the overall dimensions of the major components do not change, thus ACMAT is able to guarantee the availability of exchange components for the life of the vehicle.

Variants

Troop/Cargo Carrier - TPK 4.15 SM3

This parachutable, air-portable (4 × 4) troop/cargo carrier has a carrying capacity of 15 fully equipped soldiers or 1,500 kg of cargo. An operating range of 900 km is available, and a 100 litre drinking water tank is standard. Virtually all wearing parts are identical to the 4.20 SM3. It can be delivered by the Low Altitude Parachute Extraction System (LAPES).

Dimensions are:

Length: (overall) 5.28 m
Width:
(overall) 2.07 m
(internal) 1.94 m
Height:
(overall) 2.38 m
(internal) 1.44 m
Wheelbase: 3.3 m
Track: 1.66 m

Long-range Patrol - TPK 4.15 FSP

This vehicle is a long-range patrol version of the existing TPK 4.15 SM3 and 4.15 STL (4 × 4) 1,500 kg troop/cargo carrier. The mechanical details are the same and the TPK 4.15 FSP can be readily adapted as a weapons carrier for machine guns and anti-tank missile systems such as MILAN. Weight of a fully loaded vehicle is 5,650 kg.

Command and Transmission Vehicle - TPK 4.20 VCT

This parachutable, air-portable command vehicle has

VLRA (4 × 4) TPK 4.20 SM3 troop carrier, the base vehicle of the ACMAT range, carrying the ASPIC short range surface-to-air air defence missile system

VLRA (6 × 6) 6.35 TSR tractor truck towing Roland surface-to-air missile system

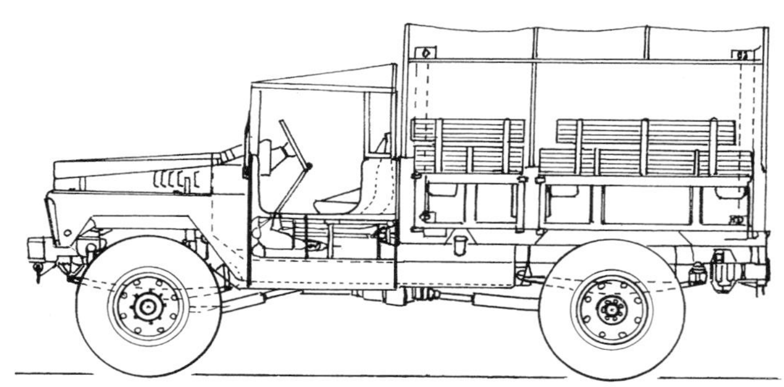

VLRA (4 × 4) ALM Type TPK 4.15 SM3 troop/cargo carrier

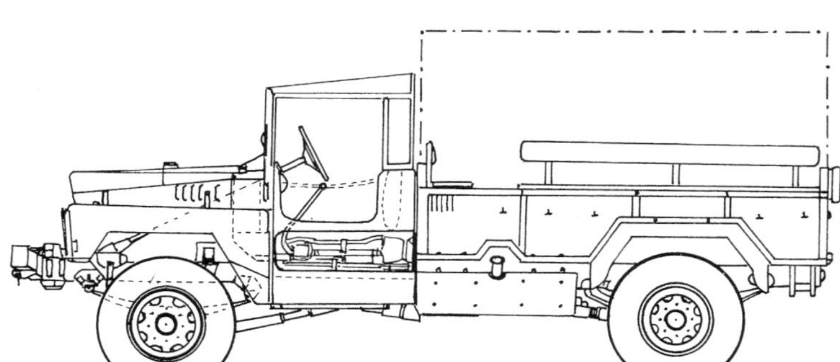

VLRA (4 × 4) ALM Type TPK 4.20 STL troop/cargo carrier

nine seats of which two are folding and used by the radio operators. Other equipment fitted as standard includes: five stowage compartments; folding table; antenna mounting; two supports for identification flag on front wings; handrail on commander's position for use during parades; and a fold-down hood. Basic specifications are: overall length 5.035 m, overall width 2.07 m, height with windscreen and hood lowered 1.331 m, wheelbase 3.3 m and track 1.66 m.

Multiple Rocket Launching Vehicle - TPK 4.15 LRM
This vehicle mounts a 40-tube launcher for the Belgian 70 mm LAU97 multiple rocket system. The launcher is mounted on the cargo-carrying area and is aimed using a hydraulic system. All 40 rockets can be fired in 6 seconds to a maximum range of 8,000 m. Weight of the vehicle and loaded launcher is 5,650 kg.

Troop Carrier - TPK 4.20 SM3
This parachutable, air-portable (4 × 4) troop-carrying vehicle has the capacity for 17 fully equipped soldiers. It can be adapted to carry a wide range of weapons, as noted in the Description. It is the base vehicle of the ACMAT range.

Other adaptations on the identical chassis include:
Split Dropside Truck - TPK 4.25 STL
Dropside Truck - TPK 4.36 STL
These two vehicles have hinged removable sides and tailgate for carrying pallets or shelters. The rear suspension is modified to give a floor height of 1.04 m above ground level when unladen.

Ambulance Vehicle - TPK 4.20 SAM
An air-portable vehicle with a fully enclosed rear body with lateral windows, two large rear access doors, four stretchers with swivelling frames for ease of access, a folding seat, washbasin, hand pump, 200 litre water tank, first aid kit and cabinet.
Dimensions are:
Length: (overall) 5.8 m
Width:
(overall) 2.04 m
(internal) 1.95 m
Height:
(overall) 2.52 m
(internal) 1.58 m
Wheelbase: 3.6 m
Track: 1.66 m

Radio Command Post - TPK 4.20 PCR
An air-portable vehicle designed for command and listening-in duties. The internal design, which includes wiring for wireless transmitters, two desks to carry equipment, two cupboards, chairs, and so on, can be completed to customer's individual requirements. Leading dimensions are identical to those given above, including the internal dimensions.

Also available are:
TPK 4.20 SL7 - Light Recovery Truck
TPK 4.20 SC - 2,500 litre Bowser Vehicle (water, oil, petrol, and so on)
TPK 4.20 VPL2 - Scout Car
TPK 4.20 VBL - 10-man Light Armoured Personnel Carrier
TPK 4.20 SM 2,000 kg Shelter Carrier (2.96 × 2.07 m platform)
TPK 4.20 FFM Firefighting Vehicle
TPK 4.20 SPM Foam Firefighting Vehicle

Double Cabin Patrol Vehicle - TPK 4.21 SM3
This patrol vehicle introduced the ACMAT double cabin which has four doors and cab seating for the driver and five men. There is seating for a further 12 men in the rear which is identical to the rear cargo area of the ACMAT TPK 4.15 SM3. The all-metal cab has roll-over bars incorporated and a heavy-duty air conditioner is optional.
Dimensions are:
Length: 6.445 m
Width: 2.07 m
Height: (cab) 2.29 m
Wheelbase: 4.2 m
Weight:
(unladen) 5,300 kg
(laden) 6,900 kg

Air-portable four-man Ambulance - TPK 4.25 SAM
This ambulance has a steel body with glass fibre

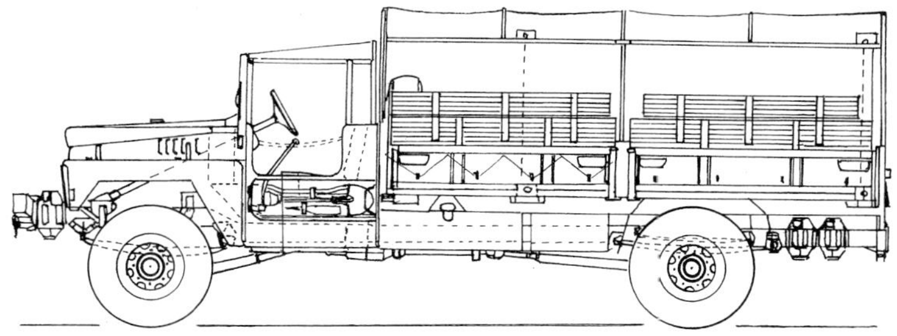

VLRA (4 × 4) ALM Type TPK 4.36 STL 21-seat troop carrier

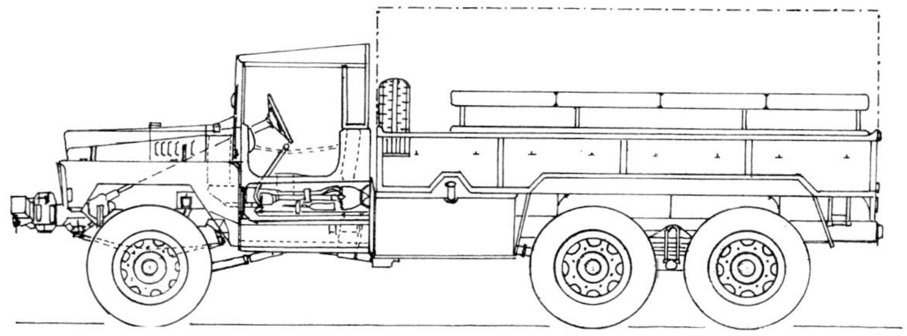

VLRA (6 × 6) 6.40 SM3 heavy reconnaissance and support vehicle

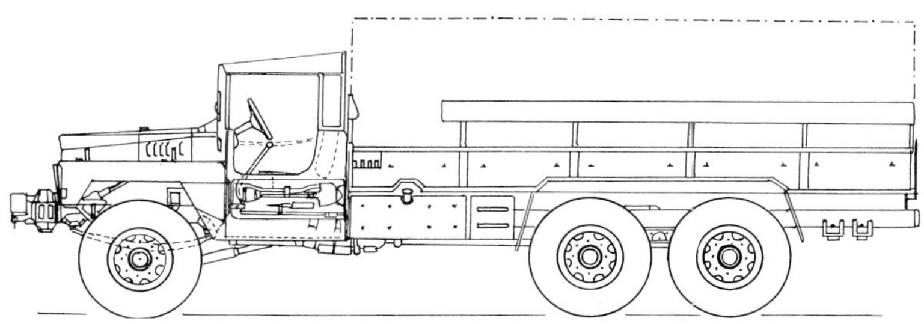

VLRA (6 × 6) 6.50 SM3 heavy reconnaissance and support vehicle

insulation with internal racks to accommodate four stretchers. The internal dimensions of the ambulance body are 3.307 m long, 1.956 m wide and 1.675 m high. The driver's cab, which can be fitted with air conditioning, may be all-metal. This ambulance is in service with the French Army.
Dimensions are:
Length: 6.33 m
Width: 2.17 m
Height: (overall, laden) 2.625 m
Wheelbase: 3.9 m
Weight:
(unladen) 5,500 kg
(total, all up) 7,000 kg

RITA Shelter Carrier - TPK 4.25 SH/STL
This communications shelter carrier has a carrier platform 3.4 m long and 2.07 m wide — when loaded the platform is 1.025 m above the ground. The platform has various securing points and a spare wheel is carried between the cab and the carrier platform. A toolbox is carried on the left-hand side.
Dimensions are:
Length: 6.33 m
Width: 2.07 m
Height: (platform, laden) 1.025 m
Wheelbase: 3.9 m
Weight:
(unladen) 4,600 kg
(laden) 7,000 kg

Troop Carrier - TPK 4.35 SM3
This air-portable (4 × 4) vehicle has a troop-carrying capacity of 21 fully equipped soldiers or 3,000 kg of cargo.

Other adaptations on the same chassis, a 'stretched' version of the TPK 4.20 SM3 above, include:
Workshop Vehicle - TPK 4.30 FAM
This vehicle is supplied as a fully equipped workshop vehicle and the following items are included: 28 kVA generating set, engine and transmission oil tanks, large working surfaces, lubricating and so on, oil lines, air compressor, battery charging equipment, portable welder, seven power points for handtools, column drill, two tool cabinets (one with 14 and one with 28 drawers) and tyre galvanising equipment.
Dimensions are:
Length: (overall) 6.9 m
Width:
(overall) 2.22 m
(internal) 2.14 m
Height:
(overall) 2.79 m
(internal) 1.8 m
Wheelbase: 4.3 m
Track: 1.66 m

Off-Highway Bus - TPK 4.30 SB
This air-portable, rough-terrain bus can carry 28 passengers with their luggage in comfort. Pneumatic doors are fitted front and rear. Air conditioning can be supplied if required.
Dimensions are:
Length:
(overall) 7.12 m
(internal) 4.3 m (driver excluded)
Width:
(overall) 2.2 m
(internal) 2.12 m
Height: 2.6 m
Wheelbase: 4.3 m
Track: 1.66 m

Other adaptations on the identical chassis include:
TPK 4.35 SCM - Mechanical Handling Vehicle, fitted with flat cargo area and telescopic hydraulic crane.
TPK 4.35 VPC - Convoy Protection Vehicle, armed with Giat 20 mm M693 cannon.
TPK 4.30 FAE/FME, FMA1, FMA2 and FRA range of electrical and armament workshop vehicles.

Troop Carrier - TPK 6.40 SM3
This air-portable 6 × 6 vehicle has a troop-carrying capacity of 21 fully equipped soldiers or 4,300 kg of cargo.

VLRA (4 × 4) TPK 4.15 FSP long-range patrol vehicle with HOT ATGW

VLRA (4 × 4) TPK 4.36 STL split drop side truck

Reconnaissance and Support Vehicle - TPK 6.50 SM3
This 6 × 6 vehicle has an effective load of 5,550 kg or 25 troops. Wheelbase from the front axle to the centre of the rear bogie is 4.5 m, and weight loaded is 11,500 kg.

Drop Side Truck - TPK 6.50 CTL
The ACMAT TPK 6.50 CTL is a drop-sided version of the TPK 6.50 SM3 (6 × 6) truck. The vehicle has a maximum effective load of 5,550 kg and a cargo floor height of 1.195 m to facilitate pallet loading. Standard ISO fixing points are incorporated in the body and shelters may be carried. Weight of the vehicle fully loaded is 11,800 kg.

Other adaptations on the same chassis, a 6 × 6 version of the TPK 4.20 SM3 above, include:
Shelter Carriers - TPK 6.40/50/60 SH
This vehicle is capable of carrying standard width and height shelters, up to a maximum length of 5.2 m. Corresponding sizes of refrigerating or plain shelters are available.
Dimensions are:
Length: (overall) 7.23-8.13 m
Width: (overall) 2.4 m
Height: (overall) 3.46 m
Wheelbase: 4.1-4.8 m
Track: 1.8 m

Bowser Vehicle - TPK 6.40 SC
An air-portable 4,000 litre bowser available for carrying water, oil or petrol.
Dimensions are:
Length: (overall) 6.94 m
Width: (overall) 2.25 m
Height: (overall) 2.13 m
Wheelbase: 4.1 m
Track: 1.8 m

NBC Decontamination Vehicle - UMTH 1000
This vehicle carries various NBC decontamination systems and is in service with the French Army. For full details refer to *Jane's NBC Protection Equipment 1996-97* page 109.

Medium Recovery Vehicle - TPK 6.35 SL7
Uses a strengthened rear platform and a 3,000 kg gantry crane plus a 9,000 kg capacity winch at the front.

Heavy Duty Fire Tender - TPK 6.35 FFL
Carries a 3,500 litre tank.

Tractor - TPK 6.35 TSR
This is based on a shortened chassis version of the TPK 6.40 SM3. It can tow a maximum load of 13,300 kg and is powered by a Perkins T6.354.4 engine rated at 180 hp. It is used in association with an ACMAT semi-trailer (see entry in *Trailers* section).

Also available are the TPK 6.40 SG - Tar Spreader, TPK 6.40 SB - 6,000 kg Tipper and the TPK 6.40 WRT - 5,000 kg Recovery Vehicle.

Specifications

Model	**TPK 4.15 SM3/STL truck**	**TPK 4.20 SM3 truck**	**TPK 4.35 SM3/4.36 CTL LWB truck**	**TPK 6.40 SM3/CTL truck**	**TPK 6.50 SM3/CTL truck**
Cab seating	1 + 2	1 + 2	1 + 2	1 + 2	1 + 2
Configuration	4 × 4	4 × 4	4 × 4	6 × 6	6 × 6
Weight (empty)	4,150 kg	4,300 kg	5,300 kg	5,700 kg	5,950 kg
(loaded)	5,650 kg	6,800 kg	8,800 kg	10,000 kg	11,500 kg
Max load	1,500 kg	2,500 kg	3,500 kg	4,500 kg	5,550 kg
Max towed load	3,000 kg	3,000 kg	3,000 kg	6,000 kg	6,000 kg
Load area	2.6 × 1.94 m	2.8 × 1.94 m	4 × 1.94 m	3.9 × 2.15 m	4.8 × 2.15 m
Length	5.28 m	6 m	7.25 m	6.94 m	7.83 m
Width	2.07 m	2.07 m	2.07 m	2.25 m	2.25 m
Height (steering wheel)	1.83 m	1.83 m	1.83 m	1.9 m	1.9 m
(tarpaulin)	2.38 m	2.62 m	2.62 m	2.64 m	2.64 m
(load area)	0.92 m	1.096 m	0.99 m	1.12 m	1.12 m
Ground clearance	0.287 m	0.287 m	0.287 m	0.3 m	0.3 m
Track (front)	1.76 m	1.76 m	1.76 m	1.76 m	1.76 m
(rear)	1.66 m	1.66 m	1.66 m	1.8 m	1.8 m
Wheelbase	3.3 m	3.6 m	4.3 m	4.1 m	4.5 m
Angle of approach/departure	51°/41°	43°/41°	43°/26°	43°/41°	43°/30°
Max road speed	100 km/h	100 km/h	100 km/h	85 km/h	85 km/h
Range	900 km	1,600 km	1,600 km	1,600 km	1,600 km
Fuel capacity	2 × 120 litres	2 × 180 litres	2 × 180 litres	2 × 210 litres	2 × 210 litres
Max gradient	65%	65%	65%	55%	55%
Fording	0.9 m	0.9 m	0.9 m	0.9 m	0.9 m
Engine (Perkins 6.354.4)	120 hp at 2,800 rpm	120 hp at 2,800 rpm	120 hp at 2,800 rpm	138 hp at 2,800 rpm	138 hp at 2,800 rpm
Gearbox (manual)	4 + 1 speed	4 + 1 speed	4 + 1 speed	5 + 1 speed	5 + 1 speed
Clutch	single dry plate	single dry plate	single dry plate	single dry plate	single dry plate
Transfer box	2-speed	2-speed	2-speed	2-speed	2-speed
Steering	worm and nut	worm and nut	worm and nut	worm and nut	worm and nut
Turning radius	8.4 m	8.5 m	11 m	10.35 m	10.35 m
Suspension	all have leaf springs and double-action hydraulic shock-absorbers				
Tyres	12.50 × 20 XL	12.5 × 20 XL	12.5 × 20 XL	12.5 × 20 XL	12.5 × 20 XL
Brakes (main)	air/hydraulic	air/hydraulic	air/hydraulic	air/hydraulic	air/hydraulic
(parking)	mechanical	mechanical	mechanical	mechanical	mechanical
Electrical system	24 V	24 V	24 V	24 V	24 V
Batteries	2 × 12 V, 100 Ah	2 × 12 V, 100 Ah	2 × 12 V, 100 Ah	2 × 12 V, 100 Ah	2 × 12 V, 100 Ah

Status
In production. In service with French forces and many other countries including Benin, Burkina Faso, Cameroon, Chad, Gabon, Gambia, Ireland, Ivory Coast, Morocco, Senegal, Somalia, Togo and Zaïre. ACMAT vehicles are in service in 35 countries throughout the world.

Manufacturer
ACMAT (Ateliers de Construction Mécanique de l'Atlantique), Le Point du Jour, F-44600 Saint-Nazaire, France.
Tel: +33 2 40 22 33 71. Telex: 700 913 f.
Fax: +33 2 40 66 30 96.

ACMAT MTV (4 × 4) ALM WPK 4.40 CTL 4,000 kg logistics vehicle

Description

A prototype of the ACMAT MTV (4 × 4) ALM WPK 4.40 CTL 4,000 kg logistics vehicle was first shown in 1992. It may be regarded as a 4 × 4 version of the WPK 8.70 CTL (8 × 8) mentioned in the following entry and shares many components. The vehicle was produced as a private venture.

The WPK 4.40 CTL has a semi-forward control steel cab (the same as that used on the WPK 8.70 CTL) with a hard top and seats for the driver and one or two passengers. A 12.7 mm machine gun can be mounted on the roof over the passenger's seat and the one-piece windscreen can be folded forward or dismounted. To the rear of the cab is the multipurpose cargo body with drop sides, drop tailgate, removable bows and a tarpaulin cover. The rear area can be configured to carry 20 troops, ammunition pallets or other stores. It can also be configured to carry a shelter or container. Standard equipment includes a 24 V electrical system and two fuel tanks to provide a range of 1,300 km. A 150 litre water tank is mounted between the body and the chassis.

Variant

MTV (6 × 6) ALM WPK 6.55 CTL

First shown in mid-1994, this 6 × 6 model completes the ACMAT Medium Tactical Vehicle (MTV) range with a 5,700 kg capacity vehicle powered by a 185 hp turbo diesel engine. Overall length is 7.455 m and wheelbase 4.2 m. Weight fully loaded is 12,500 kg.

ACMAT MTV (4 × 4) ALM WPK 4.40 CTL 4,000 kg logistics vehicle

Specifications

ALM WPK 4.40 CTL
Cab seating: 1 + 1 or 2
Configuration: 4 × 4
Weight:
(unloaded) 5,800 kg
(loaded) 9,800 kg
(front axle, loaded) 4,200 kg
(rear axle, loaded) 5,600 kg
Max load: 4,000 kg
Length: 6.645 m
Width: 2.4 m
Height: (overall, unloaded) 2.895 m
Ground clearance: 0.3 m
Wheelbase: 3.9 m
Track: 1.95 m
Angle of approach/departure: 35°/30°
Max speed: 100 km/h
Fuel capacity: 2 × 210 litres
Range: 1,300 km
Fording: 1 m
Gradient: (loaded) 65%
Side slope: 30%
Engine: Perkins 6-cylinder turbocharged diesel developing 165 hp at 2,900 rpm
Transmission: Type CR 650 with 6 forward and 1 reverse gears
Transfer box: ALM ACMAT 2-speed
Steering: ACMAT Burman recirculating ball
Suspension: semi-elliptic springs with shock-absorbers
Brakes: air/hydraulic
Tyres: 12.5 × 20 XS or 14.5 × 20 XS G 139
Electrical system: 24 V
Batteries: 2 × 12 V, 100 Ah

Status

Prototype.

Manufacturer

ACMAT, (Ateliers de Construction Mécanique de l'Atlantique), Le Point de Jour, F-44600 Saint-Nazaire, France.
Tel: +33 2 40 22 33 71. Telex: 700 913 f.
Fax: +33 2 40 66 30 96.

VERIFIED

ACMAT MTV (8 × 8) ALM WPK 8.70 CTL 6,000 kg logistic vehicles

Development

The ACMAT MTV (8 × 8) ALM WPK 8.70 CTL 6,000 kg logistic vehicle was first shown in 1987 and was primarily designed to be a front-line carrier for various forms of communications and radar shelters and containers. The ALM WPK 8.70 SH has the shelter carrier platform equipped to carry 15 ft/4.572 m ISO containers while the ALM WPK 8.75 is equipped to carry 20 ft/6.096 m ISO containers. The vehicle was designed to a French Army requirement and is air-transportable. The WPK 8.70 CTL 6,000 kg logistic vehicle is a general purpose carrier but may be configured as a shelter carrier if required.

This vehicle acted as the base model for ACMAT's Medium Tactical Vehicle (MTV) family.

Description

The ACMAT MTV (8 × 8) ALM WPK 8.70 CTL has a full 8 × 8 drive configuration with both front axles being steerable. As with all other vehicles in the ACMAT range a high degree of component commonality has been maintained, with many components of the 8 × 8 vehicle being the same as those used with other ACMAT 4 × 4 and 6 × 6 models. The main visible change on the 8 × 8, apart from the two twin-axle groupings, is to the cab, which is a semi-forward control design. The cab may be either soft- or metal-topped with the windscreen folding forward on both types. The metal-topped version has provision for a 12.7 mm machine gun ring mounting over a roof hatch.

Under the rear carrier platform is provision for a spare wheel, side storage chests and a chest at the rear for a shelter access ladder housing. There is a side-mounted 210 litre fuel tank with another 210 litre tank under the platform that can be used for either fuel or water. Towing hooks are provided front and rear. A toolkit is provided.

It has been proposed that this vehicle could be used as an artillery rocket carrier and launch vehicle. A vehicle carrying a Bennes Marrel load handling system has been produced.

Specifications

WPK 8.70 CTL
Cab seating: 1 + 2
Configuration: 8 × 8
Weight:
(empty) 8,000 kg
(loaded) 14,000 kg
(front bogie, loaded) 6,050 kg
(rear bogie, loaded) 7,950 kg
Max load: 6,000 kg
Length: 7.5 m
Width: 2.4 m

ACMAT MTV (8 × 8) ALM WPK 8.75 SH logistic vehicle

ACMAT MTV (8 × 8) ALM WPK 8.75 SH logistic vehicle in bare chassis form

Height: (overall, unloaded) 2.725 m
Ground clearance: 0.3 m
Track: 1.95 m
Wheelbase: 1.22 m + 3.75 m + 1.22 m
Angle of approach/departure: 35°/33°
Max speed: 85 km/h
Range: 1,000 km
Fuel capacity: 2 × 210 litres
Max gradient: 55%
Side slope: 30%
Fording: 1 m
Engine: Perkins 6-cylinder turbocharged diesel developing 185 hp at 2,900 rpm
Gearbox: Type CR 650 manual with 6 forward and 1 reverse gears
Clutch: single dry plate
Transfer box: ALM ACMAT type AL 450, 2-speed
Steering: ACMAT Burman recirculating ball
Suspension: slide-mounted inverted springs front and rear with radius arms mounted on silent blocks; telescopic shock-absorbers on front
Tyres: 12.5 × 20
Brakes: air/hydraulic
Electrical system: 24 V
Batteries: 2 × 12 V, 100 Ah

Status
In production. In service with the French Army and Morocco.

Manufacturer
ACMAT, (Ateliers de Construction Mécanique de l'Atlantique), Le Point du Jour, F-44600 Saint-Nazaire, France.
Tel: +33 2 40 22 33 71. Telex: 700 913 f.
Fax: +33 2 40 66 30 96.

VERIFIED

Renault TRM 4000 (4 × 4) 4,000 kg truck

Development
The Renault TRM 4000 (4 × 4) 4,000 kg truck was selected to be the standard vehicle in its class in the French Army which, in the early 1970s, stated that it then had a requirement for 15,000 vehicles of this type. Production commenced at Renault's Blainville facility in 1973. At the end of 1987 the French Army's adjusted total requirement was 7,500. Production is complete.

Renault TRM 4000 (4 × 4) 4,000 kg truck carrying an RAC 3-D radar system ***1997***

Description
The Renault TRM 4000 (4 × 4) 4,000 kg truck is essentially a commercial vehicle (the Saviem SM8) modified to meet the requirements of the French Army and uses proven commercial components, including cab, engine, transmission, chassis and axles.

The basic model has single rear wheels and a payload of 4,000 kg but there is also a model with dual rear wheels and a payload of 5,800 kg. The chassis consists of longitudinal side members with the cross members welded into position. The two-door cab (type 812) is of all-steel construction and is of the forward control type. It can be tilted forward to an angle of 50° to allow access to the engine and can also be delivered with an observation hatch in the roof. The rear cargo area has drop sides, drop tailgate, removable bows and a tarpaulin cover, and removable seats can be installed down the centre or sides of the vehicle.

Optional equipment includes long-range fuel tanks, different tyres, 4,500 kg capacity winch, twin rear wheels, differential lock, additional seat in the cab, exhaust retarder, 24 V electrical system, blackout lights, 12-pin current outlet inspection socket on dashboard and a cab which can be split at door level to reduce overall height for air transport.

Variants include: a Crotale missile carrier vehicle complete with hydraulic crane, a crane truck, a dump truck, firefighting vehicles, a light recovery vehicle, a 5,000 litre tanker/refueller (Decauville), a radar carrier, a van/command vehicle and a water tanker.

Specifications
Cab seating: 1 + 1
Configuration: 4 × 4
Weight:
(empty) 5,680 kg
(loaded, road) 11,200 kg
(loaded, cross-country) 10,000 kg
Weight on front axle: (loaded, cross-country) 3,190 kg
Weight on rear axle: (loaded, cross-country) 2,490 kg
Max load:
(road) 5,340 kg
(cross-country) 4,320 kg
Towed load: 6,000 kg
Load area: 4.48 × 2.296 m
Length: 6.538 m
Width: 2.47 m
Height:
(cab) 2.75 m
(tarpaulin) 3.26 m
(load area) 1.26 m
Ground clearance: 0.28 m
Track:
(front) 1.836 m
(rear) 2.018 m
Wheelbase: 3.85 m
Angle of approach/departure: 37°/39°
Max speed: (road) 87 km/h
Range: 700 km
Fuel capacity: 150 litres
Max gradient: 50%
Max side slope: 30%
Fording: 0.9 m
Engine: Renault model 797 6-cylinder diesel developing 133 hp at 2,900 rpm
Gearbox: ET 301 manual with 5 forward and 1 reverse gears
Clutch: single dry plate
Transfer box: G 300 2-speed
Steering: cam and roller with hydraulic power-assistance
Turning radius: 10 m
Suspension: leaf springs with Evidgom pads, hydraulic shock-absorbers
Tyres: 12.00 × 20
Brakes:
(main) air, dual-circuit
(parking) mechanical
Electrical system: 24 V
Batteries: 2 × 12 V, 95 Ah
Alternator: 25/30 A

Status
Production complete. In service with the French Army and undisclosed countries.

Manufacturer
Enquiries to: Renault Véhicules Industriels, 40 rue Pasteur, BP 302, F-92156 Suresnes Cedex, France.
Tel: +33 1 40 99 71 11. Telex: 620 567 f.
Fax: +33 1 40 99 71 08.

UPDATED

Berliet GBC 8 KT (6 × 6) 4,000 kg truck

Development
In the 1950s Berliet developed a 6 × 6 truck for use in North Africa called the Gazelle. With modifications it was subsequently adopted by the French Army in the late 1950s and, by the time production had been completed, over 18,000 trucks had been built.

It is planned to update some 6,000 GBC 8 KT trucks to the Renault TRM GBC 180 standard (see following entry for details).

Description
The layout of the vehicle is conventional, with the engine at the front, cab in the centre and cargo area at the rear. The two-door cab has a removable top, there are removable door tops and the windscreen can be folded forward onto the bonnet. The rear cargo area is provided with a drop tailgate with an integral step, removable side boards, bows and a tarpaulin cover. The height of the bows can be adjusted for road or rail transport and bench-type seats can be fitted down the centre of the vehicle if required. Optional equipment included a Pan-Bonnier winch with a capacity of between 5,000 and 7,000 kg and drop sides for the rear cargo area. The KT series is powered by a multifuel engine while the MT series is powered by a diesel which develops 125 hp at 2,100 rpm.

Variants
Renault TRM GBC 180 (6 × 6) 4,000 kg truck
See following entry.

Long wheelbase
This is 8.32 m long, 2.4 m wide and 3.23 m high unladen; the wheelbase is 3.71 m + 1.28 m and empty weight of the vehicle is 9,400 kg.

4 × 4 version
Small numbers of a 4 × 4 version were built for the Portuguese Army under the designation of GBC 8 (4 × 4).

Recovery vehicles
Details of these will be found in the entry for the Berliet TBC 8 KT (6 × 6) wrecker in the *Recovery Vehicles* section.

Berliet GBC 8 KT (6 × 6) 4,000 kg tipper truck (Pierre Touzin) *1996*

Berliet GBC 8 KT (6 × 6) 4,000 kg truck (Pierre Touzin) *1996*

Tipper

Sometimes known as the GBC 8 KT Genie, this can carry 4 m³ of soil. Unladen weight is 12,500 kg and principal dimensions are: length 7.175 m, width 2.4 m, height 2.845 m, length inside tipper body 3.8 m and width inside tipper body 2.3 m.

Tractor truck

This has an overall length of 6.521 m, width of 2.4 m and a height (reduced) of 2 m.

Other variants include: a firefighting vehicle, a 5,000 litre fuel tanker, a mobile compressor, and a number of communications and command post vehicles carrying container/shelters.

Specifications

Cab seating: 1 + 2
Configuration: 6 × 6
Weight:
(empty) 8,370 kg
(loaded) 12,370 kg
Weight on front axle: (empty) 4,180 kg
Weight on rear bogie: (empty) 4,190 kg
Max load: (cross-country) 4,000 kg
Load area: 4.36 × 2.35 m
Length: 7.28 m
Width: 2.4 m
Height:
(cab) 2.7 m
(tarpaulin) 3.3 m
Ground clearance: (axles) 0.28 m
Track: 1.86 m
Wheelbase: 3.31 m + 1.28 m
Angle of approach/departure: 45°/45°
Max speed: 80 km/h
Range: 800 km
Fuel capacity: 200 litres
Max gradient: 50%
Fording: 1.2 m
Engine: Berliet MK 520 5-cylinder OHV water-cooled multifuel developing 125 hp at 2,100 rpm
Gearbox: Berliet BDSL 13 manual with 6 forward and 1 reverse gears
Clutch: single dry plate
Transfer box: 2-speed
Steering: worm gear and nut, servo-assisted
Turning radius: 10.5 m
Suspension:
(front) semi-elliptical springs and hydraulic shock-absorbers
(rear) leaf springs on oscillating pivot
Tyres: 12.00 × 20
Brakes:
(main) air
(parking) mechanical
Electrical system: 24 V
Batteries: 4 × 12 V, 100 Ah

Status

Production complete. In service with Algeria, Austria, China, France, Iraq, Morocco and Portugal.

Manufacturer

Berliet is now part of Renault Véhicules Industriels, 40 rue Pasteur, BP 302, F-92156 Suresnes Cedex, France.
Tel: +33 1 40 99 71 11. Telex: 620 567 f.
Fax: +33 1 40 99 71 08.

UPDATED

Renault TRM GBC 180 (6 × 6) 4,000 kg truck

Description

Renault Véhicules Industriels (RVI) has undertaken a major upgrade of the Berliet GBC 8 KT (6 × 6) 4,000 kg truck (see previous entry) to produce a virtually new vehicle with a much extended life. The upgraded vehicle is known as the Renault TRM GBC 180 (6 × 6) 4,000 kg truck and the overall programme is expected to involve the upgrading of some 6,000 trucks. The programme will be supervised by the French Army's Direction des Systemes Terrestres et d'Information (DSTI) and will involve a joint testing and trials programme.

One of the main upgrade measures is the replacement of the existing Berliet engine with a MIDR 06.02.26 diesel rated at 174 hp at 2,500 rpm. This engine is the same as that fitted to the RVI TRM 180 and 200 series of trucks. The new engine is coupled to an Eaton 4106 OD gearbox providing six forward and one reverse gears and further coupled to a new HWT 2030 two-speed transfer box. TRW power steering is provided and the entire electrical system is updated to full 24 V standard — components from the RVI TRM 2000 and TRM 10 000 are involved. The braking system is also revised to meet the latest EU directives.

The old cab assembly is replaced by a new steel and glass fibre cab based on the GBC 180 cab design. Enclosed or open desert-type cabs will be involved.

Renault TRM GBC 180 (6 × 6) 4,000 kg truck (left) with uncoverted Berliet GBC 8 KT (6 × 6) 4,000 kg on right for comparison *1997*

Status

Upgrading programme planned — see text.

Manufacturer

Renault Véhicules Industriels, 40 rue Pasteur, BP 302, F-92156 Suresnes Cedex, France.
Tel: +33 1 40 99 71 11. Telex: 620 567 f.
Fax: +33 1 40 99 71 08.

UPDATED

Renault TRM 180 and TRM 200 (4 × 4) truck series

Description

The Renault TRM 180 and TRM 200 (4 × 4) truck series, referred to as the militarised civilian range, was developed from the earlier Renault TRM 150.11 (4 × 4)

Various configurations of Renault TRM 180 and TRM 200 trucks

Renault TRM 180.11 (4 × 4) truck

Renault TRM 200.13 (4 × 4) truck **1997**

truck to meet the military requirements of less technically advanced nations for a logistic vehicle with a degree of cross-country mobility. Many of the components used on the vehicles are of commercial origin or have been taken from models already in production.

The vehicles are equipped with a multipurpose platform suitable for a variety of military roles including cargo or troop carrier, shelter carrier (3.05 or 4.57 m), recovery vehicle, water or fuel carrier, or weapon carrier.

The forward control cab is similar to that used on the Renault TRM 2000 (see separate entry in this section). The all-steel cab can be tilted forward for engine and transmission access. The standard hard-topped cab may be fitted with a roof hatch for access to a light machine gun mounting. As an alternative, the cab may be produced in 'torpedo' form with a canvas top or a removable hard top. On the latter the windscreen can be folded forward onto the bonnet. With all cab types the driver has his own adjustable seat and a bench seat is provided for two passengers. Steering can be right- or left-handed.

The chassis is formed from beams measuring 234 × 70 × 7 mm.

Various options are available on request. These include: a spare wheel holder behind the cab; a power take-off from the gearbox; a fuel preheater; a removable electric winch; and an automatic gearbox.

Specifications

Model	**TRM 180.11**	**TRM 180.13**	**TRM 200.11**	**TRM 200.13**
Cab seating	1 + 2	1 + 2	1 + 2	1 + 2
Configuration	4 × 4	4 × 4	4 × 4	4 × 4
Weight (cab and chassis)	5,260 kg	5,490 kg	5,370 kg	5,490 kg
(loaded)	11,500 kg	13,500 kg	11,500 kg	13,500 kg
Load capacity	6,240 kg	8,060 kg	6,130 kg	8,030 kg
Towed load (road)	6,000 kg	6,000 kg	6,000 kg	6,000 kg
(off road)	4,000 kg	4,000 kg	4,000 kg	4,000 kg
Length	6.88 m	6.88 m	6.88 m	6.88 m
Width (cab)	2.365 m	2.365 m	2.365 m	2.365 m
(rear tyres)	2.37 m	2.37 m	2.37 m	2.37 m
Height, top of cab, empty	2.89 m	2.897 m	2.897 m	2.897 m
Ground clearance, rear axle, loaded	0.311 m	0.311 m	0.311 m	0.311 m
Track (front)	1.896 m	1.96 m	1.96 m	1.96 m
(rear)	2 m	2 m	2 m	2 m
Wheelbase	3.85 m	3.85 m	3.85 m	3.85 m
Angle of approach/departure	42°/41°	42°/41°	42°/41°	42°/41°
Max speed	95 km/h	95 km/h	95 km/h	95 km/h
Range	950 km	950 km	900 km	900 km
Fuel capacity	275 litres	275 litres	275 litres	275 litres
Gradient	50%	50%	50%	50%
Max side slope	30%	30%	30%	30%
Engine (type)	MIDR 06-02-26 V	MIDR 06-02-26 V	MIDR 06-02-26 W	MIDR 06-02-26 W
(power)	182 hp/2,800 rpm	182 hp/2,600 rpm	196 hp/2,500 rpm	196 hp/2,500 rpm
Gearbox	S 6-36 manual	S 6-36 manual	S 6-36 manual	S 6-36 manual
gears	6f/1r	6f/1r	6f/1r	6f/1r
Transfer box	2-speed	2-speed	2-speed	2-speed
Electrical system	24 V	24 V	24 V	24 V
Alternator	50 A	50 A	50 A	50 A

Status

In production. TRM 200-13 ordered by the French armed forces (approx 150, including a tanker version). Bangladesh ordered 275 TRM 180.11 trucks in July 1996 with a follow-on order for a further 200 anticipated.

Manufacturer

Enquiries to: Renault Véhicules Industriels, 40 rue Pasteur, BP 302, F-92156 Suresnes Cedex, France.
Tel: +33 1 40 99 71 11. Telex: 620 567 f.
Fax: +33 1 40 99 71 08.

UPDATED

Renault TRM 9000 (6 × 6) 9,000 kg truck

Development

The Renault TRM 9000 (6 × 6) (formerly known as the Berliet GBD) truck uses many components, such as the cab, gearbox, transfer case and steering, of the TRM 6000 (4 × 4) 6,000 kg truck.

Production was undertaken at Vénissieux. First production vehicles were completed in 1975. The largest orders came from Algeria (500) and Morocco (1,500). Production is complete.

Further development of the TRM 9000 resulted in the TRM 10 000 6 × 6 truck adopted by the French Army, for which there is a separate entry in this section.

Description

The two-door forward control cab has a removable canvas roof, removable side screens and a windscreen that can be folded forward against the bonnet if required. The rear cargo area is fitted with drop sides and a drop tailgate, removable bows and a canvas cover. If required, seats can be fitted in the rear. Optional equipment includes: a four-door six- or seven-seater cab; fully enclosed two- or four-door cab; a winch with a capacity of 3,500/4,500 kg and 60 m of cable; different tyres; dual wheels on the rear axles; an additional 200 litre fuel tank; and an exhaust brake.

Renault TRM 9000 (6 × 6) 9,000 kg truck

Variants include: an artillery tractor able to tow 155 mm artillery weapons as well as carrying a gun crew and ammunition; a command/radio vehicle fitted with a container on the rear which could be fitted with communications equipment; a 6,000 litre water or fuel tanker; and a tipper with a two-man cab. The chassis has been used to mount Exocet or Otomat (for Egypt) anti-shipping missiles, as an RPV launching vehicle, and for supporting the Shahine anti-aircraft system supplied to Saudi Arabia; 38 vehicles were supplied to Saudi Arabia for this role. A recovery version known as the TRM 9000 CLD is fitted with a four-door cab. A tractor variant can tow a semi-trailer weighing a maximum of 27,000 kg; unladen weight is 15,000 kg.

Specifications

Cab seating: 1 + 1
Configuration: 6 × 6
Weight:
(empty) 11,000 kg
(loaded) 20,000 kg
Max load: (cross-country) 9,000 kg
Max towed load: 10,000 kg
Length: 9.87 m
Width: 2.48 m
Height: (cab) 3.066 m
Ground clearance: 0.382 m
Track:
(front) 1.971 m
(rear) 1.886 m
Wheelbase: 3.8 m + 1.4 m
Angle of approach/departure: 45°/50°
Max speed: 82 km/h
Range: 800 km
Fuel capacity: 200 litres
Max gradient: 45%
Max side slope: 30%
Fording: 1 m
Engine: Renault MIDS 06-20-30 6-cylinder turbocharged diesel developing 228 hp at 2,200 rpm
Gearbox: BDS manual with 6 forward and 1 reverse gears
Clutch: air power-assistance
Transfer box: BT 20 2-speed
Steering: power-assisted
Turning radius: 11 m
Suspension:
(front) leaf springs with auxiliary springs and 2 mechanical buffer stops, telescopic shock-absorbers
(rear) balanced by leaf springs and 4 mechanical buffer stops
Tyres: 14.00 × 20
Brakes:
(main) air
(parking) mechanical
Electrical system: 24 V
Batteries: 4 × 6TN, 190 Ah

Status

Production complete. In service with armed forces including Algeria (500), Egypt (50), Morocco (1,500) and Saudi Arabia (38). This vehicle is not used by the French Army.

Manufacturer

Enquiries: to Renault Véhicules Industriels, 40 rue Pasteur, BP 302, F-92156 Suresnes Cedex, France.
Tel: +33 1 40 99 71 11. Telex: 620 567 f.
Fax: +33 1 40 99 71 08.

VERIFIED

Berliet GBU 15 (6 × 6) 6,000 kg truck

Description

In the 1950s the Rochet-Schneider company developed a 6 × 6 truck called the T-6. This concern was subsequently taken over by the Berliet Company and production of the vehicle, called the GBU 15, began in 1959. Production is complete and its replacement in the French Army is the Renault TRM 10 000 (6 × 6) 10,000 kg truck.

The forward control cab has four doors, removable canvas top and side screens, and its windscreen can be folded down onto the front of the cab if not required. The rear cargo area is provided with a drop tailgate, removable sides, removable bows and a canvas cover.

Mounted at the rear of the vehicle is a winch with a capacity of 8,000 kg. The engine is the multifuel type and will run on a variety of fuels including petrol, paraffin, JP4, gas-oil, light fuel and mineral or vegetable oils in the lower power ranges. The rear wheel train consists of two tandem axles operating with a flexible progressive air-driven differential device. This ensures power even when two wheels on one side of the vehicle are not in contact with the ground.

Variants include an artillery tractor used to tow French Army 155 mm Model 1950 howitzers, and a tractor, designated the TBU 15, with an unladen weight of 13,500 kg; it can tow a semi-trailer carrying a maximum weight of 22,000 kg. A wrecker variant is called the TBU 15 CLD; full details are given in the *Recovery Vehicles* section. There are also tanker and tipper variants.

Berliet GBU 15 (6 × 6) 6,000 kg truck (C R Zwart)

Specifications

Cab seating: 1 + 3
Configuration: 6 × 6
Weight:
(empty) 14,500 kg
(loaded, road) 24,500 kg
(loaded, cross-country) 20,500 kg
Max load:
(road) 10,000 kg
(cross-country) 6,000 kg
Max towed load: 15,000 kg
Length: 7.974 m
Width: 2.5 m
Height:
(cab) 3 m
(tarpaulin) 3.25 m
Track: 2.04 m
Wheelbase: 3.48 m + 1.45 m
Angle of approach/departure: 45°/45°
Max speed: 75 km/h
Range: 800 km
Fuel capacity: 400 litres
Max gradient: 60%
Fording: 1 m
Engine: Berliet 6-cylinder multifuel, water-cooled developing 214 hp at 1,800 rpm
Gearbox: manual with 5 forward and 1 reverse gears
Clutch: dry
Transfer box: 2-speed
Steering: screw and nut, power-assisted
Turning radius: 9.2 m
Suspension: longitudinal springs
Tyres: 14.00 × 20
Brakes:
(main) air, 3 circuits; front, rear and trailer
(parking) operates on rear wheels only
(emergency) pneumatic handbrake operates on rear wheels
Electrical system: 24 V

Status

Production complete. In service with Belgium, China, France and United Arab Emirates.

Manufacturer

Berliet is now part of Renault Véhicules Industriels, 40 rue Pasteur, BP 302, F-92156 Suresnes Cedex, France.
Tel: +33 1 40 99 71 11. Telex: 620 567 f.
Fax: +33 1 40 99 71 08.

VERIFIED

Renault TRM 10 000 (6 × 6) 10,000 kg truck

Development

The Renault TRM 10 000 (6 × 6) 10,000 kg truck was developed from the Renault TRM 9000 (6 × 6) 9,000 kg truck. The main differences between the TRM 9000 and the TRM 10 000 are that the latter has a longer wheelbase, a more powerful engine and a different transmission.

The TRM 10 000 (6 × 6) 10,000 kg truck was selected by the French Army to be its standard truck in this class; the initial requirement was 5,000 vehicles. Preproduction manufacture commenced at Renault's Blainville factory in 1985. Production of the first full production batch of 178 commenced during mid-1987 and later the same year an order for a further 759 was placed.

By 1994 an increased GVW weight variant had been developed, resulting in a new range of versions — see under Variants below.

Description

The chassis of the Renault TRM 10 000 consists of two U-shaped side members (302 × 85 × 8 mm) with cross members bolted and riveted into position. The front bumper is provided with a front towing shackle, and impact buffers and shackles are provided front and rear.

The front axle is a Type PA 721 double reduction with bevel gear and reducers in the hubs. The rear tandem is a Type PMR 2021 with double reduction (taper gear and reducers in hubs) and inter-wheel and inter-differential locking.

The basic version has a forward control cab which can be tilted forwards for maintenance purposes and is provided with two doors, tarpaulin cover over removable roof bows, suspended adjustable driver's seat, adjustable passenger's seat, fold-down windscreen, heating, defrosting and ventilation equipment as standard.

The rear cargo area is provided with removable sides and tailgate, bows and a tarpaulin cover, and can be fitted with removable seats for 24 fully equipped troops. With the sides and rear removed the TRM 10 000 can carry standard 20 ft/6.096 m ISO containers.

Optional equipment includes: an 8,000 kg hydraulic winch which can be used to the front or rear; 10 to 40 tonne pivoting hook; rear impact buffers; fully enclosed four-door cab; hardtop on two-door cab; four seats in the two-door cab; one bunk in two-door cab; 12.7 mm anti-aircraft machine gun mount on top of two- or four-door cab; PTO on gearbox; automatic gearbox model 6 HP 500 with two-gear VG 500 transfer box and torque distributor; trailer braking device; ISO coupling plugs (two circuits) and SO front coupling plugs; class A anti-interference with 50 A alternator; 7- or 12-pin

Renault TRM 10 000 (6 × 6) truck carrying field refuelling equipment ***1997***

trailer plug; additional 250 litre fuel tank on left-hand side of chassis; toolbox; and boxes for wheel chains.

Variants include: an artillery prime mover for 155 mm TR gun with four-door cab (adopted by the French Army); a dump truck; fuel or water tankers; hydraulic cranes of various types; a mine-carrying vehicle; a Tropomil aerial carrier; vehicles carrying multiple rocket launcher or various types of surface-to-surface missile; a missile resupply vehicle (for example Crotale or Shahine); and a lubrication vehicle. Other variants are described below.

Uprated variants

In 1990 Renault demonstrated a TRM 10 000 equipped with a Bennes Marrel load-handling system capable of handling a 13,000 kg capacity flatrack. This version is powered by a 326 hp diesel engine and the gross vehicle weight is 29,000 kg. This variant was adopted by the French Army with a total order for 792 vehicles to be delivered over a four-year period. The eventual French Army requirement may be over 3,000 units.

Further development of the uprated model resulted in a fuel tank carrier intended for field refuelling of Leclerc MBT and combat helicopter units – it is anticipated that there is a requirement for 200 of these. Another truck version acts as the carrier for the ASTER launching unit together with its firing command and control systems – it is anticipated that 150 units will be involved for this requirement. The uprated model is also used as the carrier/layer for the Krauss-Maffei Folding Roadway System (qv).

For these models the basic cab and chassis weight is 11,010 kg, with a total payload (including body) of 17,990 kg.

TRM 10 000 T

Although the wheelbase of this variant remains the same as the standard models, the overall length is reduced to 7.336 m. A fifth wheel is provided to tow semi-trailers to carry bridging and pontoon units or engineer equipment semi-trailers.

TRM 10 000 CLD heavy wrecker

See entry in *Recovery vehicles* section.

Specifications

Flatrack carrier version
Cab seating: 1 + 2
Configuration: 6 × 6
Weight:
(cab and chassis) 13,540 kg
(loaded) 29,000 kg
(front axle, loaded) 6,950 kg
(rear axles, loaded) 6,590 kg
Max load: 13,000 kg
Length: 8.023 m
Width: 2.48 m
Height: (cab roof) 3.11 m
Ground clearance: (loaded) 0.36 m
Track: (front) 2.015 m
Wheelbase: 4.3 m + 1.4 m
Angle of approach/departure: 42.5°/65°
Max speed: (road) 88 km/h
Fuel capacity: 250 + 250 litres
Max gradient: 50%
Side slope: 30%
Fording: 1.2 m
Engine: Renault MIDR 06-20-45 9.839 litres 6-cylinder supercharged exhaust diesel developing 326 hp at 2,000 rpm
Gearbox: Model B.9.150, 9 forward and 1 reverse gears
Clutch: hydropneumatic control
Transfer box: A 800 3D
Steering: Type 8046, hydraulic
Turning radius: 11.25 m
Suspension:
(front) semi-elliptical leaf springs (auxiliary and main springs), mechanical stops and telescopic shock-absorbers
(rear) semi-elliptical leaf springs, mechanical stops
Tyres:
(front) 14.00 × 20
(rear) 475/80 R 20
Brakes:
(main) air, dual-circuit
(parking) locking on rear wheels
Electrical system: 24 V
Batteries: 4 × 12 V, 6TNB, 125 Ah
Alternator: 50 A

Status

In production. In service with the French Army and Saudi Arabia (67). Ordered by some European armies.

Manufacturer

Enquiries to: Renault Véhicules Industriels, 40 rue Pasteur, BP 302, F-92156 Suresnes Cedex, France.
Tel: +33 1 40 99 71 11. Telex: 620 567 f.
Fax: +33 1 40 99 71 08.

UPDATED

Renault TRM 10 000 (6 × 6) 10,000 kg truck with Bennes Marrel load-handling system handling a 13,000 kg capacity flatrack (T J Gander) ***1997***

Renault TRM 10 000 (6 × 6) artillery tractor (T J Gander)
1997

Renault G 290.26 (6 × 4) VTL transport vehicle

Development

In January 1988 the French Army awarded a contract worth FFr 3,500 million for 3,500 Renault G 290.26 (6 × 4) trucks, 3,500 Bennes Marrel Ampliroll 155NB load-handling systems, 2,500 LOHR RM19 trailers, 12,000 20 ft/6.096 m platforms, and 100 tilt-type semi-trailers. The order was completed by December 1993. The equipments will be used by about 40 transport regiments and various other army units. The combined logistic package is known as the *Véhicule de Transport Logistique* (VTL).

A total of 2,050 examples of a revised version with a 290 hp engine, the G 340, have been delivered to the French Army. These have a 26,000 kg payload and some are employed to carry MLRS reload rocket pods.

Description

The Renault G 290.26 is a logistic transport vehicle fitted with a hydraulic loading system produced by Bennes Marrel. The system is capable of loading and unloading platforms carrying loads of up to 16,000 kg. Loading time is 38 seconds and unloading 34 seconds. The system can withstand misaligned loads of up to 10° either side. Apart from the usual cargo-carrying platforms the system can be used with tanker and tipper bodies.

The Renault G 290.26 vehicle can carry a maximum load of 18,000 kg. It is powered by a Type MIDR 06-20-45 turbocharged diesel developing 291 hp at 2,100 rpm and has a Type B9 gearbox with nine forward and one reverse gears. The cab has seating for the driver and two passengers. Gross vehicle weight is 26,000 kg.

Status

Production complete. French Army Service Corps.

Manufacturers

Renault Véhicules Industriels, 40 rue Pasteur, BP 302, F-92156 Suresnes Cedex, France.
Tel: +33 1 47 72 33 33. Telex: 620567 sdce f.
Fax: +33 1 40 99 71 08.
Bennes Marrel SA, ZI St Etienne Bouthéon, BP 56, F-42160 Andrezieux-Bouthéon, France.
Tel: +33 4 77 36 55 50. Telex: 330657 f.
Fax: +33 4 77 55 35 77.

VERIFIED

Renault G 290.26 (6 × 4) VTL transport vehicle (Pierre Touzin)
1996

Renault TRM 340.34 (6 × 6) 16,000 kg truck

Development

The Renault TRM 340.34 (6 × 6) truck replaced the earlier TRM 12 000 (6 × 6) 12,000 kg truck in production. It was developed to meet a number of heavy military engineer requirements and was produced in a number of body forms. Production is complete.

Description

The Renault TRM 340.34 has a semi-forward control cab with an adjustable seat for the driver and a bench seat for two passengers. The cab is constructed of high-stress steel and is provided with air conditioning and standard instruments. For maintenance the cab can be tilted forward to an angle of 70°.

The chassis is constructed from high-strength special steel U-members with section dimensions of 356 × 100 × 8 mm. Chassis cross members are bolted to the webs of the side members while the integral front bumper is fitted with a towing block; various types of towing hook can be fitted. A Type PA 721 double reduction front axle is fitted while the rear axles are Type PMR 2032, also with double reduction. The rear axles are fitted with an inter-axle and inter-wheel differential locking device.

Optional equipment includes various spare wheel carriers, reinforced rear cross members for towing trailers weighing up to 75,000 kg, enlarged or extra fuel tanks, power take-offs from the gearbox, and accessories such as a fully sprung driving seat.

The TRM 340.34 was produced in tipper form with an 11 m³ tipping body. Other body forms were available. Details of the TRM 340.34 T tractor truck can be found in the *Tank Transporters* section.

Renault TRM 340.34 (6 × 6) tipper trucks (Pierre Touzin)

Specifications

Cab seating: 1 + 2
Configuration: 6 × 6
Weight:
(empty) 9,580 kg
(loaded) 26,000 kg
Max load: 16,420 kg
Towed load: 12,000 kg
Length: 7.78 m
Width: 2.49 m
Height: (top of cab, empty) 3.003 m
Ground clearance: 0.312 m
Track:
(front) 1.91 m
(rear) 1.825 m
Wheelbase: 3.9 m + 1.35 m
Angle of approach/departure: 28°/48°
Max speed: (road) 87.8 km/h
Fuel capacity: 250 litres
Max gradient: 50%
Side slope: 30%
Fording: 0.85 m
Engine: Type MIDR 06-35-40 12 litre 6-cylinder water-cooled diesel supercharged by turbocompressor on exhaust and developing 335 hp at 1,900 rpm
Gearbox: Type B9 with 8 synchronised forward and 1 reverse gears
Clutch: Type 430DTP2400 single dry plate
Transfer box: Type VG 1200 2-speed
Steering: worm and nut, power-assisted
Turning radius: 12 m
Suspension:
(front) semi-elliptical slide-mounted leaf springs with telescopic shock-absorbers
(rear) leaf springs flanged to oscillating pivots with compensator rods
Tyres: 12.00 × 20
Brakes:
(main) air, independent front and rear circuits
(parking) mechanical
Electrical system: 24 V
Batteries: 2 × 12 V, 143 Ah
Alternator: 55 A

Status

Production complete. In service with French Army.

Manufacturer

Enquiries to: Renault Véhicules Industriels, 40 rue Pasteur, BP 302, F-92156 Suresnes Cedex, France.
Tel: +33 1 40 99 71 11. Telex: 620 567 f.
Fax: +33 1 40 99 71 08.

UPDATED

GERMANY

IFA L60 LA/PVB (4 × 4) 5,000 kg trucks

Description

IFA L60 LA/PVB (4 × 4) trucks are of East German origin placed in production to supersede the earlier W50 series. The vehicle was produced for commercial as well as military purposes with the first examples appearing during 1986.

The L60 follows the same overall design layout as the earlier W50 (see following entry) but has a new design of forward control cab (which can be tilted forward for engine access), a revised cab interior, a more powerful engine with improved fuel consumption, a new transmission and a revised chassis. The chassis uses a torsionally soft ladder frame and can accommodate a variety of types of body. The standard cargo body uses an all-steel platform covered by a tarpaulin stretched over bows with side benches along the side boards to accommodate passengers.

Most variants carry a 50 kN winch with 60 m of cable.

Variants

L60 AP-N: standard 4 × 4 cargo truck
L60 APB-N: short wheelbase (3.26 m) 4 × 4 cargo truck with 5,500 kg payload; no winch
L60 AC-N: LAK and LAP shelter carrier (4 × 4)
L60 APB: standard 4 × 4 cargo truck with double wheels on rear axle to increase payload to 5,600 kg; no winch
L60 AP: 4 × 2 cargo truck
L60: 4 × 2 tractor truck
L60 N: 4 × 4 maintenance truck
L60 DSK-N: 4 × 4 tipper truck
L60 AIKB: 4 × 2 insulated shelter carrier
L60 ANKB: 4 × 4 or 4 × 2 shelter carrier

L60 AKKB: 4 × 2 shelter carrier
L60 ADK100-N: 4 × 4 crane truck

Specifications

L60 AP-N
Cab seating: 1 + 2
Configuration: 4 × 4
Weight:
(empty) 7,000 kg
(loaded) 12,000 kg
Weight on front axle, loaded: 4,700 kg
Weight on rear axle, loaded: 7,500 kg
Max load: (road) 5,000 kg
Towed load: (road) 12,000 kg
Length: 7.13 m
Width: 2.95 m
Height: 3.375 m
Track:
(front) 2 m
(rear) 1.97 m
Wheelbase: 3.816 m
Angle of approach/departure: 33°/32°
Max speed: 100 km/h

L60 APB (4 × 4) 5,600 kg truck

Fuel consumption: 28 litres/100 km
Fuel capacity: 180 litres
Max gradient: 55%
Fording: 1.2 m
Vertical obstacle: 0.4 m
Trench: 0.75 m
Engine: Model 6-VD-13 13.5/12 SRF 6-cylinder in-line 4-stroke diesel developing 173 hp at 2,300 rpm
Gearbox: manual, synchronised, with 8 forward, 1 reverse and 1 crawling gears or manual with 5 forward and 1 reverse gears
Clutch: with hydropneumatic assist
Transfer box: 2-speed, lockable
Steering: hydraulic power-assisted
Turning radius: 8.75 m
Brakes: combined air/hydraulic, dual-circuit
Tyres: 18.00 × 20
Electrical system: 24 V
Alternator: 840 W

Status

Production complete. In service with the armed forces of Germany and some other nations, including Cambodia, Hungary and Iraq.

Manufacturer

VEB IFA-Automobilwerke, Ludwigsfelde, Bezirk Potsdam, Germany.

VERIFIED

IFA W50 LA (4 × 4) trucks

Description

IFA W50 L trucks entered production in 1965 as the replacement for the earlier S 4000-1 4 × 2 truck series and were built in large numbers for military and civilian markets.

Military versions of the W50 L series have larger tyres than civilian models and may also feature a central tyre pressure regulation system. The basic 3,000 kg payload military vehicle has an all-steel two-man forward control cab with a circular observation hatch in the roof and a rear cargo area with removable bows and a drop tailgate.

Most variants carry a 4,500 kg capacity winch with 60 m of cable.

Variants

The letters LA after the W50 designation denote a 4 × 4 model; L denotes a 4 × 2 model
Not all the following variants may remain in service.
W50 L/A: standard 4 × 2 3,000 kg cargo truck
W50 LA/A: standard 4 × 4 3,000 kg cargo truck
W50 LA/PV: 4 × 4 cargo truck
W50 L/SP/A: 4 × 2 truck for bulky loads
W50 L/L/A: 4 × 2 cargo truck with loading crane
W50 L/LB/A: 4 × 2 cargo truck with loading crane
W50 L/S: 4 × 2 tractor truck
W50 L/S/A: 4 × 2 tractor truck; no winch
W50 LA/WT80: 4 × 4 maintenance truck
W50 LA/TLR16: 4 × 4 fire engine
W50 LA/K 3SK5: 4 × 4 tipper truck; no winch
W50 LA/K/A 3SK5: 4 × 4 tipper truck; no winch
W50 LA/Z 3SK5: 4 × 4 tipper truck; no winch
W50 LA/MK 5-5: 4 × 4 special tipper truck
W50 LA/A ADK70: 4 × 4 crane truck
W50 LA/A ADK80: 4 × 4 crane truck
W50 LA/A-1 ADK80: 4 × 4 crane truck
W50 L/H/FA1.1: 4 × 2 waste water truck
W50 L/H/FA1.2: 4 × 2 waste water truck
W50 LA/AB/A: 4 × 4 wrecker
W50 L/RK: 4 × 2 street cleaning truck
W50 L/KM2301: 4 × 2 street cleaning truck
W50 L/U/A: 4 × 2 elevating platform carrier
W50 L/IKB/A: 4 × 2 insulated shelter carrier
W50 LA/A/C: 4 × 4 LAK shelter carrier
W50 LA/A-1/C: 4 × 4 LAK shelter carrier
W50 LA/WK/CA: 4 × 4 gas cylinder carrier

Specifications

W50 LA/A
Cab seating: 1 + 1
Configuration: 4 × 4
Weight:
(empty) 6,400 kg
(loaded) 9,400 kg
Weight on front axle, loaded: 3,550 kg
Weight on rear axle, loaded: 5,950 kg
Max load: 3,000 kg
Towed load: (road) 8,000 kg
Length: 6.16 m
Width: 2.5 m
Height: 3.2 m
Track:
(front) 1.866 m
(rear) 1.952 m
Wheelbase: 3.2 m
Angle of approach/departure: 40°/36°
Max speed: 85 km/h
Fuel consumption: 26 litres/100 km
Fuel capacity: 150 litres
Max gradient: 67%
Fording: 1 m
Vertical obstacle: 0.4 m
Trench: 1 m
Engine: Model 4-VD14.5/12-1 6.56 litre 4-cylinder in-line 4-stroke diesel developing 125 hp at 2,300 rpm
Gearbox: manual, 5 forward and 2 reverse gears
Transfer box: 2-speed
Steering: hydraulic power-assisted
Turning radius: 8 m
Tyres: 16.00 × 20
Electrical system: 24 V

Status

Production complete. In service with the German armed forces and some other nations, including Cambodia, Egypt, Hungary and Iraq.

Manufacturer

VEB IFA-Automobilwerke, Ludwigsfelde, Bezirk Potsdam, Germany.

VERIFIED

IFA W50 LA/A (4 × 4) 3,000 kg truck fitted with snow plough (Michael Jerchel)

IFA W50 LA/A (4 × 4) 3,000 kg truck fitted with office body (Michael Jerchel)

Mercedes-Benz Unimog (4 × 4) series

Development

The Unimog (*Universal Motor Gerät*, or universal power-plant) was originally designed in 1946, primarily for industrial and agricultural use. Since then large numbers of Unimogs have been manufactured in many different models for both civilian and military use.

The number of Unimog models has grown considerably over the years. In 1993, the lightweight range of vehicles (U 650 L to U 1150 L) were replaced by completely redesigned models, the U 100 L and the U 140 L. Within the new Unimog range the largest 4 × 4 model is the U 2150 L with a maximum payload of 5,000 kg. The 6 × 6 U 2450 L with a payload of 7,000 kg offers the maximum payload capability of the complete Unimog range (and is also used as the chassis for the Giat Industries/Lohr Industrie CAESAR 155 mm self-propelled gun).

Only the later ranges of models are described here. For the earlier models refer to *Jane's Military Logistics 1988* pages 410 to 413 and 418.

Unimogs have been licence-produced in Argentina, Australia, Indonesia, South Africa and Turkey.

Description

The layout of all vehicles in the Unimog series is basically the same, with the engine and cab at the front and the cargo area at the rear. Most military models have a two-door cab with a hard or soft top and a windscreen which folds forward onto the bonnet, and a rear cargo area with a drop tailgate, drop sides, removable bows and a tarpaulin cover.

The Unimog vehicle is well-known for its excellent cross-country capabilities and all vehicles are fitted with a differential lock on both the front and rear axles. For normal road use the vehicle is driven with only the rear wheels engaged, while for cross-country use the front wheels are also engaged, with the differential locks being used in very rough terrain.

A wide range of optional equipment is available for the vehicle including a fully enclosed two- or four-door cab, generator, front-mounted pump, snow ploughs, cranes and winches.

In 1993 Mercedes-Benz modified its range of production models and devised new designations. The Specifications table indicates all models of the current production range of Unimogs for military use.

Mercedes-Benz Unimog U 100 L (4 × 4) 1,800 kg light truck

Specifications

Model	**U 100 L**	**U 140 L**	**U 1350 L**	**U 1550 L**	**U 1550 L/38**	**U 2150 L**	**U 2450 L**
Configuration	4 × 4	4 × 4	4 × 4	4 × 4	4 × 4	4 × 4	6 × 6
Cab seating	1 + 1	1 + 1	1 + 2	1 + 2	1 + 2	1 + 2	1 + 2
Weight (empty)	3,000 kg	4,000 kg	5,000 kg	5,200 kg	6,500 kg	7,000 kg	8,100 kg
(loaded)	4,800 kg	7,500 kg	7,500 kg	8,500 kg	10,600 kg	12,500 kg	17,000 kg
(axle load, front/rear)	3,000/3,500 kg	4,500/4,500 kg	4,000/4,000 kg	4,000/4,600 kg	5,300/4,600 kg	5,300/7,200 kg	5,300/2 × 6,000 kg
Max load	1,800 kg	2,500 kg	3,000 kg	3,000 kg	4,000 kg	5,000 kg	7,000 kg
Load area	2.6 × 1.7 m	3 × 2 m	3.15 × 2.2 m	4.05 × 2.2 m	4.05 × 2.2 m	4.25 × 2.38 m	5.2 × 2.43 m
Length	5.07 m	5.53 m	5.54 m	6.4 m	6.47 m	6.67 m	7.58 m
Width	1.912 m	2.05 m	2.3 m	2.3 m	2.343 m	2.5 m	2.5 m
Height (cab)[1]	2.46 m	2.48 m	2.63 m	2.63 m	2.73 m	2.83 m	2.8 m
(canvas cover)[1]	2.65 m	2.67 m	2.8 m	2.8 m	3.04 m	3.095 m	3.3 m
Ground clearance[1]	0.4 m	0.43 m	0.44 m	0.44 m	0.485 m	0.5 m	0.48 m
Track	1.54 m	1.69 m	1.86 m	1.86 m	1.92 m	1.92 m	1.87 m
Wheelbase	3.22 m	3.47 m	3.25 m	3.7 m	3.85 m	3.85 m	3.9 m + 1.4 m
Angle of approach/departure	47°/52°	50°/50°	46°/51°	40°/47°	43°/42°	48°/36°	47°/49°
Max speed	103 km/h	89 km/h	85 km/h	85 km/h	87 km/h	85 km/h	95 km/h
Fuel capacity	90 litres	110 litres	160 litres	160 litres	160 litres	160 litres	165 litres
Gradient	80%	80%	80%	80%	80%	80%	80%
Max side slope	40°	40°	40°	40°	40°	40°	40°
Fording	0.8 m	1 m	1.2 m	1.2 m	1.2 m	1.3 m	1.3 m
Engine type	OM 602	OM 364 LA	OM 366 A	OM 366 A	OM 366 A	OM 366 LA	OM 366 LA
Number of cylinders	5	4	6	6	6	6	6
Capacity	2.874 litres	3.972 litres	5.958 litres	5.958 litres	5.958 litres	5.958 litres	5.958 litres
Hp/rpm	122/3,800	102/2,600	136/2,800	155/2,400	155/2,400	214/2,600	240/2,600
Gears[2]	5 forward, 1 reverse	6 forward, 1 reverse	8 forward, 8 reverse	8 forward, 8 reverse	8 forward, 8 reverse	8 forward, 8 reverse	8 forward, 8 reverse
Clutch	single dry disc	single dry disc	single dry disc	single dry disc	single dry disc	single dry disc	single dry disc
Turning radius	6.6 m	6.6 m	7.2 m	7.6 m	8.1 m	8.1 m	10 m
Tyres	10.5 R 20	12.5 R 20	12.5 R 20	12.5 R 20	12 R 22.5	13 R 22	14.5 R 20
Electrical system	12 or 24 V	12 or 24 V	12 or 24 V	12 or 24 V	12 or 24 V	12 or 24 V	12 or 24 V

[1]Unloaded [2]Transfer box fitted

Mercedes-Benz Unimog U 1350 L (4 × 4) 3,000 kg truck **1997**

Mercedes-Benz Unimog U 1350 L (4 × 4) ambulance (Michael Jerchel) **1996**

Mercedes-Benz Unimog U 140 L (4 × 4) 2,500 kg truck

Mercedes-Benz Unimog U 2450 L (6 × 6) 7,000 kg truck

Status

In service with Algeria, Angola, Australia, Belgium (1,512 U 1350 L ordered), Chile, Denmark, Egypt, France, Germany, Indonesia, Iraq, Kenya, Kuwait, Libya, Morocco, Mozambique, New Zealand, Nigeria, Pakistan, Peru, Portugal, Saudi Arabia, Singapore, South Africa, Syria, Tanzania, Tunisia, Turkey, United Arab Emirates and United Kingdom (Army and Royal Air Force). The U 140 L is out of production.

Unimogs have been licence-produced in Argentina, Australia, Indonesia, South Africa and Turkey.

Manufacturer

Mercedes-Benz AG, Department VMC/R Z503, D-70322 Stuttgart, Germany.
Tel: +49 711 179 0326. Fax: +49 711 179 0876.

UPDATED

CAESAR 155 mm self-propelled gun mounted on a Mercedes-Benz Unimog U 2450 L (6 × 6) 7,000 kg truck chassis
1997

Mercedes-Benz L 508 DG MA (4 × 2) 2,105 kg truck

Description

The Mercedes-Benz L 508 DG MA (4 × 2) 2,105 kg truck is used for transporting men and cargo in rear areas and has little cross-country mobility. It is a standard commercial vehicle with a minimum of modifications to suit it for military use. Final deliveries were made to the German Army during 1990.

The chassis is a bend-resistant parallel ladder frame with tubular cross members, except for the one at the rear. A coupling member is incorporated in the front cross member suitable for a towbar as used by the German armed forces.

The all-steel forward control cab is noise and temperature insulated. A heater is fitted as standard on all models and the L 508 D/35 van-type model, used as a command vehicle, has a heat exchanger and a special vent in the front part of the cab roof

The rear cargo platform has a plywood floor with hinged steel side and tailboards, stowage boxes and support points for the centre bench and a tarpaulin frame as well as locating and lashing points for fitting a shelter or transporting pallets.

Swedish Army Mercedes-Benz L 508 DG MA (4 × 2) 2,105 kg truck (T J Gander)

Variants

L 508 D/35

This is the standard pick-up model with a wheelbase of 3.5 m, cargo area 4.06 m long and 2.26 m wide, and maximum cargo capacity of 2,460 kg.

L 608 D/41

This is a pick-up with a wheelbase of 4.1 m, cargo area 5.06 m long and 2.26 m wide, and maximum cargo capacity of 3,500 kg.

Specifications

L 508 DG MA
Cab seating: 1 + 2
Configuration: 4 × 2
Weight:
(empty) 3,495 kg
(loaded) 5,600 kg
Max load: 2,105 kg
Load area: 3.15 × 2.2 m
Max towed load: 1,150 kg
Length: 5.325 m
Width: 2.32 m
Height:
(cab, unloaded) 2.345 m
(cab, loaded) 2.3 m
(tarpaulin, unloaded) 2.68 m
(tarpaulin, loaded) 2.535 m
Track:
(front) 1.76 m
(rear) 1.54 m
Wheelbase: 2.95 m
Max speed: 98 km/h
Range: 750 km
Fuel consumption: 11.2 litres/100 km
Gradient: 27%
Engine: OM 314 4-cylinder diesel developing 85 hp at 1,800 rpm
Transmission: manual, 5 forward and 1 reverse gears
Clutch: single dry plate
Steering: recirculating ball

Turning radius: 5.7 m
Suspension: semi-elliptical leaf springs, hydraulic shock-absorbers with both axles fitted with torsion bar stabilisers
Tyres: 7.00 R 16 C 10 PR
Brakes:
(main) air/hydraulic
(parking) mechanical
Electrical system: 24 V
Batteries: 2 × 12 V

Status
In production. In service with the German Army and other armed forces, including Sweden.

Manufacturer
Mercedes-Benz AG, Department VN/RM - 2503, D-70322 Stuttgart, Germany.
Tel: +49 711 179 0326. Fax: +49 711 179 0934.

VERIFIED

IVECO Magirus 75-13 AWM (4 × 4) 2,500 kg truck

Description
There are two versions of the IVECO Magirus 75-13 AWM (4 × 4) 2,500 kg truck, one using banjo axles and the other with a longer wheelbase and Portal axles. Both have forward control cabs and a cargo area that can be used to carry 12 personnel, a 2,500 kg load or a shelter/container. These trucks were formerly known as the 130M7FAL. Production ceased during 1994.

The forward control cab is all steel and has seating for the driver and two passengers. The cab may be tilted forward for access to the engine and transmission. At the rear the cargo body is also all steel and has tiltable and removable side boards and a tailgate. Folding benches are provided for 12 personnel. The load area is covered by a tarpaulin resting on bows. There is a spare wheel holder between the cab and the cargo body. Both versions have a standard 130 litre fuel tank but the version with Portal axles has an optional 200 litre tank.

Other options include a 3,000 kg mechanical or hydraulic winch, an anti-roll bar at the rear, an exhaust brake and a roof hatch for the cab. It is also possible to provide a wading kit for depths up to 1.2 m.

Specifications
Cab seating: 1 + 2
Configuration: 4 × 4
Weight:
(empty) 4,900 kg
(loaded) 7,500 kg
Max load: 2,600 kg
Towed load: 7,500 kg
Length: 5.51 m
Width: 2.4 m
Height:
(cab, SWB) 2.745 m
(cab, LWB) 2.835 m
(tarpaulin, SWB) 3.18 m
(tarpaulin, LWB) 3.24 m
(load area, SWB) 1.38 m
(load area, LWB) 1.44 m
Ground clearance:
(SWB) 0.26 m
(LWB) 0.44 m
Track:
(SWB) 1.852 m
(LWB) 1.943 m
Wheelbase: 2.85 m or 3.1 m
Angle of approach/departure:
(SWB) 35°/38°
(LWB) 39°/50°
Max speed: 83 km/h
Range: 900 km
Fuel capacity: 130 litres (200 litres optional on LWB)
Gradient: 60%
Side slope: 30%
Fording: 0.8 m (optional 1.2 m)
Engine: Deutz Diesel F 6 L 913 6.128 litre 6-cylinder air-cooled developing 130 hp at 2,500 rpm
Gearbox: manual, 5 forward and 1 reverse gears
Clutch: single dry disc
Transfer box: 2-speed
Steering: ball and nut, assisted
Turning radius:
(SWB) 6.7 m
(LWB) 7.1 m
Suspension: semi-elliptical springs with hydraulic shock-absorbers front and rear
Tyres: 12.50 × 20
Brakes:
(main) drum type, dual-circuit air over hydraulic
(parking) mechanical on rear wheels
Electrical system: 24 V
Batteries: 2 × 12 V, 100 Ah
Alternator: 55 A

Status
Production complete. In widespread service.

Manufacturer
IVECO Magirus AG, Defence Vehicles Division, D-89070 Ulm, Germany.
Tel: +49 731 729 6826. Fax: +49 731 76317.

VERIFIED

IVECO Magirus 75-13 AWM (4 × 4) 2,500 kg truck with banjo axles

IVECO Magirus 75-13 AWM (4 × 4) 2,500 kg truck with Portal axles

IVECO Magirus 90-13 ANWM (4 × 4) 4,000 kg truck

Description
The IVECO Magirus 90-13 ANWM (4 × 4) 4,000 kg truck was formerly known as the 130 D9AL.

The layout of the 90-13 ANWM is orthodox, with the engine set forward under a square-cut bonnet. The radiator is usually protected by a grille guard and the bonnet top can be raised to the rear. The all-steel cab has seating for the driver and two passengers and a roof hatch is optional. The cab is soundproofed with cavity corrosion protection. Laminated glass is used for the one-piece windscreen. The cargo body has folding bench seating for 16 personnel and is covered by a tarpaulin resting on bows. The tailgate and side boards are both tiltable and removable.

Optional equipment includes: a 5,000 kg mechanical or hydraulic winch; additional or enlarged fuel tanks; jerrican or extra equipment stowage; and a power take-off on the gearbox or transfer box. A wading kit for depths up to 1.2 m can be fitted.

IVECO Magirus 90-13 ANWM (4 × 4) 4,000 kg truck

Specifications
Cab seating: 1 + 2
Configuration: 4 × 4
Weight:
(empty) 6,100 kg
(loaded) 10,100 kg
Max load: 4,000 kg
Towed load: 5,000 kg
Length: 7.195 m
Width: 2.5 m
Height:
(air intake) 2.87 m
(tarpaulin) 3.2 m
(load area) 1.4 m

Ground clearance: 0.31 m
Track:
(front) 1.97 m
(rear) 1.902 m
Wheelbase: 4.2 m
Angle of approach/departure: 42°/31°
Max speed: 82 km/h
Range: 900 km
Fuel capacity: 130 litres (optional 200 litres or 200 + 130 litres)
Gradient: 43%
Side slope: 30%
Fording: 0.8 m (1.2 m optional)
Engine: Deutz Diesel F 6 L 913 6.128 litre 6-cylinder air-cooled developing 130 bhp at 2,500 rpm
Gearbox: manual with 5 forward and 1 reverse gears
Clutch: single dry disc
Transfer box: 2-speed
Steering: ball and nut, assisted
Turning radius: 9.9 m
Suspension: leaf springs and hydraulic shock-absorbers front and rear
Tyres: 13.00 × 20
Brakes:
(main) drum, dual-circuit air over hydraulic
(parking) mechanical on rear wheels
Electrical system: 24 V
Batteries: 2 × 12 V, 100 Ah
Alternator: 35 A

Status
In production. In widespread service.

Manufacturer
IVECO Magirus AG, Defence Vehicles Division, D-89070 Ulm, Germany.
Tel: +49 731 729 6826. Fax: +49 731 76317.

VERIFIED

IVECO Magirus 110-16 AWM (4 × 4) 5,000 kg truck

Description
The IVECO Magirus 110-16 AWM (4 × 4) 5,000 kg truck was formerly known as the 168M11FAL and is available in two forms, one with single tyres on the rear axle and the other with dual tyres and a shorter wheelbase. Both can be used as cargo trucks to carry a 5,400 or 5,500 kg load, a shelter/container or 18 personnel.

The 110-16 AWM has a forward control cab with an adjustable seat for the driver plus seating for two passengers. The cab is all steel and can tilt forward for engine access. A ventilation flap is provided on the roof and a roof hatch is available as an optional extra. A spare wheel holder is located between the cab and the cargo body which is also all steel with removable or tiltable sides and tailgate. A tarpaulin resting on bows is used to cover the cargo area and folding benches are provided for up to 18 personnel.

Optional equipment includes a 5,000 kg mechanical or hydraulic winch, stowage for five jerricans, an enlarged fuel tank and an anti-roll bar at the rear.

Variant
IVECO Magirus 110-17 AM (4 × 2) 5,000 kg Truck
This is a 4 × 2 version of the 110-16 AWM and was formerly known as the 168M11FL. It is similar to the 4 × 4 models, having dual tyres on the rear axle and a spare wheel stowed under the cargo body.

Status
Production being phased out; replaced during 1994. In widespread service.

Manufacturer
IVECO Magirus AG, Defence Vehicles Division, D-89070 Ulm, Germany.
Tel: +49 731 729 6826. Fax: +49 731 76317

VERIFIED

Specifications

Model	110-16	110-17
Cab seating	1 + 2	1 + 2
Configuration	4 × 4	4 × 2 or 4 × 4
Weight		
(empty)	6,300 kg	6,400 kg
(loaded)	11,800 kg	11,800 kg
Max load	5,500 kg	5,400 kg
Towed load	11,800 kg	11,800 kg
Length	7.56 m	7.1 m
Width	2.37 m	2.37 m
Height		
(cab)	2.905 m	2.835 m
(tarpaulin)	3.39 m	3.29 m
(load area)	1.59 m	1.59 m
Ground clearance	0.31 m	0.264 m
Track (front/rear)	1.988/1.988 m	1.92/1.724 m
Wheelbase	4.2 m	3.6 m
Angle of approach/departure	40°/30°	35°/24°
Max speed	85 km/h	80 km/h
Range	600 km	600 km
Fuel capacity	130 litres	130 litres
Max gradient	60%	60%
Side slope	30%	30%
Fording	0.8 m	0.8 m
Engine	Deutz Diesel BF 6 L 913 6.128 litre 6-cylinder air-cooled turbocharged developing 168 bhp at 2,650 rpm	
Gearbox	manual with 6 forward and 1 reverse gears	
Clutch	single dry disc	single dry disc
Transfer box	2-speed	2-speed
Steering	ball and nut, power-assisted	
Turning radius	8.8 m	7.9 m
Suspension	leaf springs with hydraulic telescopic shock-absorbers front and rear	
Tyres	13.00 × 20 PR-18	10 × 22.5
Brakes		
(main)	drum, independent, dual circuit pneumatic	
(parking)	spring energy brake cylinder	
Electrical system	24 V	24 V
Batteries	2 × 12 V, 100 Ah	2 × 12 V, 125 Ah
Alternator	35 A	35 A

Danish Army IVECO Magirus 110-16 AWM (4 × 4) 5,000 kg truck with single tyres on rear axle

IVECO Magirus 120-19 ANWM (4 × 4) 5,000 kg truck

Description

The IVECO Magirus 120-19 ANWM (4 × 4) 5,000 kg truck was formerly known as the 192D12AL and may be regarded as an enlarged version of the 4,000 kg 90-13 ANWM. It has a more powerful engine and a longer wheelbase, and a number of different types of body can be fitted.

The standard cargo body version is covered by a tarpaulin resting on bows and has folding bench seating for 18 personnel. The cab has an adjustable driver's seat and seating for two passengers. The one-piece windscreen is of laminated glass and the cab is soundproofed with cavity corrosion protection. A ventilation flap is provided on the roof and a roof hatch is optional. A spare wheel is carried between the cab and cargo body. The cargo body tailgate and side boards are both tiltable and removable.

Optional equipment includes a 5,000 or 8,000 kg mechanical or hydraulic winch, an enlarged fuel tank or extra fuel tanks, jerrican or extra equipment stowage and a power take-off on the gearbox or transfer box. A wading kit for depths up to 1.2 m is available.

IVECO Magirus 120-19 ANWM (4 × 4) 5,000 kg truck

Specifications

Cab seating: 1 + 2 (up to 18 in rear)
Configuration: 4 × 4
Weight:
(empty) 7,300 kg
(loaded) 12,300 kg
Max load: 5,000 kg
Towed load: 6,000 kg
Length: 8.405 m
Width: 2.5 m
Height:
(air intake) 2.93 m
(tarpaulin) 3.38 m
(load area) 1.58 m
Ground clearance: 0.37 m
Track:
(front) 1.985 m
(rear) 2.03 m
Wheelbase: 4.9 m
Angle of approach/departure: 37°/33°
Max speed: 85 km/h
Fuel capacity: 130 litres (optional 200 litres or 200 + 130 litres)
Max gradient: 60%
Side slope: 30%
Fording: 0.8 m (1.2 m optional)
Engine: Deutz Diesel F6L 413 F 9.572 litre 6-cylinder air-cooled developing 192 hp at 2,500 rpm
Gearbox: manual with 6 forward and 1 reverse gears
Clutch: single dry disc
Transfer box: 2-speed
Steering: ball and nut, assisted
Turning radius: 12 m
Suspension: leaf springs with hydraulic shock-absorbers front and rear
Tyres: 14.00 × 20
Brakes:
(main) drum, dual-circuit plus exhaust brake
(parking) fail-safe acting on rear wheels via air-controlled spring energy cylinder
Electrical system: 24 V
Batteries: 2 × 12 V, 100 Ah
Alternator: 35 A

Status

Production complete. In widespread service.

Manufacturer

IVECO Magirus AG, Defence Vehicles Division, D-89070 Ulm, Germany.
Tel: +49 731 729 6826. Fax: +49 731 76317.

VERIFIED

IVECO Magirus 160-23 ANWM (6 × 6) 7,000 kg truck

Description

This truck may be regarded as a lengthened, three-axled version of the lighter 90-13 ANWM and 120-19 ANWM versions. It can either be used in the basic cargo body form or to carry a wide variety of other types of body such as tankers, office or workshop bodies and so on. The 160-23 ANWM and the similar 160-25 ANWM, with a payload capacity of 8,000 kg (formerly known as the 256D18AL), are no longer in production having been replaced by the 240-25 ANWM (6 × 6) 12,000 kg truck.

The layout is orthodox, with the engine forward under a bonnet top that can be raised to the rear. The radiator grille can be folded forward. The all-steel cab has an adjustable seat for the driver and seating is provided for two other passengers. The roof has a ventilation flap and a roof hatch is optional. Laminated glass is used for the one-piece windscreen and the cab is soundproofed with cavity corrosion protection. Two air intake pipes are located each side of the cab windscreen. A spare wheel holder is located between the cab and the cargo body. The cargo area is covered by a tarpaulin resting on bows, has folding benches for 18 personnel and is all steel with tiltable and removable side boards and tailgate.

Optional equipment includes an 8,000 kg or 10,000 kg mechanical or hydraulic winch, an enlarged or extra fuel tank, stowage for five 20 litre jerricans and a power take-off on the gearbox or transfer box. A wading kit for the electrical system is available for depths of up to 1.2 m.

IVECO Magirus 160-23 ANWM (6 × 6) 7,000 kg truck

Specifications

Cab seating: 1 + 2 (up to 18 in rear)
Configuration: 6 × 6
Weight:
(empty) 9,400 kg
(loaded) 16,400 kg
Max load: 7,000 kg
Towed load: 12,000 kg
Length: 8.55 m
Width: 2.5 m
Height:
(air intake) 2.99 m
(tarpaulin) 3.365 m
(load area) 1.565 m
Ground clearance: 0.37 m
Track:
(front) 2.005 m
(rear) 2.048 m
Wheelbase: 4.2 m + 1.38 m
Angle of approach/departure: 38°/38°
Max speed: 80 km/h
Range: 700 km
Fuel capacity: 200 litres (optional 300 litres or 300 litres plus 200 litres)
Max gradient: 60%
Side slope: 30%
Fording: 0.8 m (optional 1.2 m)
Engine: Deutz Diesel F 8 L 413 11.31 litre 8-cylinder air-cooled diesel developing 232 hp at 2,650 rpm
Gearbox: manual with 6 forward and 1 reverse gears
Clutch: single dry disc
Transfer box: 2-speed
Steering: ball and nut, assisted
Turning radius: 10.4 m
Suspension:
(front) single flexibility leaf springs with hydraulic telescopic shock-absorbers
(rear) single flexibility leaf springs, reversed, fully articulated
Tyres: 14.00 × 20

Brakes:
(main) drum, dual-circuit plus exhaust brake
(parking) mechanical on rear wheels
Electrical system: 24 V
Batteries: 2 × 12 V, 100 Ah
Alternator: 35 A

Status
Production complete. In widespread service.

Manufacturer
IVECO Magirus AG, Defence Vehicles Division, D-89070 Ulm, of Germany.
Tel: +49 731 729 6826. Fax: +49 731 76317.

VERIFIED

IVECO Magirus 200-23 ANWM and 210-32 ANWM (6 × 6) 10,000 kg trucks

Description
There are two basic types of truck under this designation with the main difference between the two being the type and power of engine fitted — the 200-23 has an eight-cylinder 232 hp engine and the 210-32 a 10-cylinder 320 hp engine. Short and long wheelbase models of each version were produced.

The IVECO Magirus 200-23 and 210-32 follow the same basic lines as the IVECO Magirus 160-23 ANWM and 160-25 ANWM only differing in some weights and dimensions. They can be produced in a number of body types in addition to the basic cargo body version. The cargo version has seating for up to 18 personnel in the rear.

This model was replaced in production by the 240-25 ANWM — see separate entry in this section.

Status
Production complete. In widespread service.

Manufacturer
IVECO Magirus AG, Defence Vehicles Division, D-89070 Ulm, of Germany.
Tel: +49 731 729 6826. Fax: +49 731 76317

VERIFIED

Specifications
(Data for short wheelbase versions; data for long wheelbase in square brackets where different.)

Model	**200-23**	**210-32**
Cab seating	1 + 2	1 + 2
Configuration	6 × 6	6 × 6
Weight		
(empty)	10,400 kg	10,700 kg
(loaded)	20,400 kg	21,000 kg
Max load	10,000 kg	10,300 kg
Towed load	18,000 kg	18,000 kg
Length	7.495* [9.6] m	7.495 [9.6] m
Width	2.5 m	2.5 m
Height		
(air intake)	2.925 m	2.925 m
(tarpaulin)	3.365 m	3.365 m
(load area)	1.565 m	1.565 m
Ground clearance	0.335 m	0.305 m
Track		
(front)	2.005 m	2.005 m
(rear)	2.048 m	2.048 m
Wheelbase	3.9 m + 1.38 m [5.25 m + 1.38 m]	3.9 m + 1.38 m [5.25 m + 1.38 m]
Angle of approach/departure	32°/60°* [35°/30°]*	32°/60°* [35°/30°]*
Max speed	80 [81] km/h	80 [81] km/h
Range	600 km	600 km
Fuel capacity	200 litres	200 litres
Max gradient	60%	60%
Side slope	30%	30%
Fording	0.8 m	0.8 m
Engine	Deutz Diesel F 8 L 413F 11.310-litre 8-cylinder air-cooled	Deutz Diesel F 10 L 413F 15.953-litre 10-cylinder air-cooled
Engine output	232 hp at 2,650 rpm	320 hp at 2,500 rpm
Gearbox	manual with 6 forward and 1 reverse gears	
Clutch	single dry disc	single dry disc
Transfer box	2-speed	2-speed
Steering	ball and nut, power-assisted	
Turning radius	11.8 [12.5] m	11.8 [12.5] m
Suspension		
(front)	single flexibility leaf springs with hydraulic shock-absorbers	
(rear)	single flexibility leaf springs, reversed, fully articulated	
Tyres	14.00 × 20	14.00 × 20
Brakes		
(main)	drum, dual-circuit, exhaust brake	
(parking)	mechanical on rear wheels	
Electrical system	24 V	24 V
Batteries	2 × 12 V, 100 Ah	2 × 12 V, 100 Ah
Alternator	35 A	35 A

* chassis only

IVECO Magirus 200-23 ANWM (6 × 6) 10,000 kg truck

IVECO Magirus 240-25 ANWM (6 × 6) 12,000 kg truck

Description

The IVECO Magirus 240-25 ANWM (6 × 6) 12,000 kg high-mobility truck replaced the earlier 160-23 ANWM and 200-23 ANWM (6 × 6) trucks in production. Short and long wheelbase models are available.

The IVECO Magirus 240-25 follows the same basic lines as the earlier models. Both wheelbase length models can be produced in a number of body types, in addition to the basic cargo body version.

Status

In production.

Manufacturer

IVECO Magirus AG, Defence Vehicles Division, D-89070 Ulm, Germany.
Tel: +49 731 729 6826. Fax: +49 731 76317.

VERIFIED

Specifications

Model	**SWB**	**LWB**
Cab seating	1 + 2	1 – 2
Configuration	6 × 6	6 × 6
Weight		
(kerb, chassis/cab)	8,800 kg	8,800 kg
(GVW)	24,500 kg	24,500 kg
Max load	12,000 kg	12,000 kg
Towed load	11,500 kg	11,500 kg
Length	8.55 m	9.2 m
Width	2.5 m	2.5 m
Height		
(air intake)	3.088 m	3.088 m
(tarpaulin)	3.365 m	3.365 m
(load area)	1.565 m	1.565 m
Ground clearance	0.37 m	0.37 m
Track		
(front)	2.005 m	2.005 m
(rear)	2.048 m	2.048 m
Wheelbase	4.2 m + 1.38 m	5.25 m + 1.38 m
Angle of approach/departure	38°/38°	35°/30°
Max speed	84 km/h	84 km/h
Range	600 km	600 km
Fuel capacity	300 litres	300 litres
Max gradient	56%	56%
Side slope	30%	30%
Fording	0.8 m	0.8 m
Engine	Deutz Diesel F 8 L 513 13.382 litre 8-cylinder air-cooled diesel developing 264 hp at 2,300 rpm	
Gearbox	manual with 6 forward and 1 reverse gears	
Clutch	single dry disc	single dry disc
Transfer box	2-speed	2-speed
Steering	ball and nut, assisted	
Turning radius	11.4 m	13.5 m
Suspension		
(front)	semi-elliptic leaf springs with hydraulic shock-absorbers	
(rear)	cantilever leaf springs	
Tyres	14.00 R 20	14.00 R 20
Brakes		
(main)	drum, dual-circuit, exhaust brake	
(parking)	mechanical on rear wheels	
Electrical system	24 V	24 V

Short wheelbase version of IVECO Magirus 240-25 ANWM (6 × 6) 12,000 kg truck

Mercedes-Benz LA 911B (4 × 4) 4,500 kg truck

Description

The Mercedes-Benz LA 911B (4 × 4) 4,500 kg truck is also known as the LA 1113 and before 1963 was known as the LA 328. It has been in production for many years and is in widespread military use, particularly in Africa and Asia.

The LA 911B is produced in two wheelbase lengths, 3.6 m and 4.2 m. The layout is orthodox, with the engine forward under a distinctive curved bonnet, and the cab is all steel. The cab has seating for the driver and two passengers with the passenger's bench seat folding down to provide access to the roof hatch. The cargo area can be used to seat up to 24 personnel and can be covered by a canvas tilt. A spare wheel is carried under the load area. The electrical system is splashproof and to NATO standard, with an external plug socket.

Various types of bodies can be fitted to the LA 911B. In service are ambulance, workshop, light recovery, firefighting and other bodies. The LA 911B can be used as a light artillery tractor.

Specifications

Cab seating: 1 + 2
Configuration: 4 × 4
Weight:
(empty) 5,500 kg
(loaded) 10,000 kg
(GCW) 21,600 kg
Weight on front axle: 4,000 kg
Weight on rear axle: 6,400 kg
Max load: 4,500 kg
Max towed load: 11,000 kg

Length:
(SWB) 5.98 m
(LWB) 6.9 m
Width: 2.46 m
Height:
(cab) 2.71 m
(tarpaulin) 3.13 m
Ground clearance: 0.34 m
Track: 1.97 m
Wheelbase: 3.6 m or 4.2 m
Angle of approach/departure: 37°/46°
Max speed: 84 km/h
Fuel capacity: 135 litres
Gradient: 68%
Fording: 0.9 m
Engine: Mercedes-Benz OM 352 5.675 litre 6-cylinder in-line water-cooled diesel developing 130 hp at 2,000 rpm or Mercedes-Benz OM 352A 5.675 litre 6-cylinder in-line water-cooled turbocharged diesel developing 168 hp at 2,000 rpm
Gearbox: Mercedes-Benz G 3/50-5/8.5 with 5 forward and 1 reverse gears
Clutch: single dry disc
Transfer box: Mercedes-Benz VG 500-3W/1.6 2-speed
Steering: recirculating ball (power steering optional)
Turning radius: 8.1 m
Tyres: 12.00 × 20
Brakes: single-circuit hydraulic, compressed air assisted
Electrical system: 24 V
Batteries: 2 × 12 V, 100 Ah
Alternator: 55 A

Status
In production. In military service worldwide.

Manufacturer
Mercedes-Benz AG, Department VN/RM - 2503, D-70322 Stuttgart, Germany.
Tel: +49 711 179 0326. Fax: +49 711 179 0934.

VERIFIED

Mercedes-Benz LA 911B (4 × 4) 4,500 kg truck

Mercedes-Benz Model 1017 (4 × 2) and Model 1017 A (4 × 4) 5,000 kg trucks

Development
These two vehicles were developed by Mercedes-Benz to meet the requirements of the German Army and are essentially standard commercial vehicles with the minimum of modifications to suit them for military use. Adaptations include a modified electrical system, use of low-torsion platform with stowage boxes and adequate structure plus mounting points for van bodies, additional stowage space on the rear wall of the cab for military equipment and installation of an observation hatch in the roof of the cab. The major differences between the 1017 and the 1017 A are that the latter has 4 × 4 drive and larger tyres. Final deliveries were made in 1987. It was replaced in production by the Model 1217 CA (see entry in this section).

Description
The layouts of both vehicles are identical, with the forward control cabs at the front and the cargo areas at the rear.

The parallel ladder chassis is bend-resistant but flexible; the side members are the fish belly type in that the height of the web is tailored to the load acting on the chassis at the respective point. Open hat section cross members, cold-riveted to the frame, provide the desired torsional elasticity. With this design the chassis adapts to the surface of the road or track and imposes little stress on the material.

The all-steel forward control cab is noise and temperature insulated and the front rests on two rubber-bushed pivot bearings and its rear end on two vibration-damped spring struts.

The heating system keeps the inside temperature between −4 and +25°C and the cab has a ventilation system incorporating a blower. Adjustable air inlets

Specifications

Model	**1017**	**1017 A**
Cab seating	1 + 1	1 + 1
Configuration	4 × 2	4 × 4
Weight		
(empty)	6,250 kg	6,800 kg
(loaded)	11,700 kg	12,200 kg
Max load	5,450 kg	5,400 kg
Max towed load	12,800 kg	12,300 kg
Length	7.19 m	7.19 m
Width	2.47 m	2.47 m
Height (cab, unloaded)	2.7 m	2.88 m
(cab, loaded)	2.665 m	2.843 m
(load area, unloaded)	1.425 m	1.525 m
Load area	5 × 2.38 m	5 × 2.38 m
Ground clearance	0.263 m	0.288 m
Track (front)	1.954 m	2.067 m
Wheelbase	3.6 m	3.6 m
Angle of approach/departure	23°/20°	30°/23°
Max speed	87 km/h	81 km/h
Range	730 km	695 km
Fuel capacity	135 litres	135 litres
Fuel consumption	18.5 litres/100 km	19.4 litres/100 km
Gradient (road)	45%	46%
(cross-country)	n/a	80%
Fording	0.5 m	0.5 m
Engine	OM 352 A 6-cylinder turbocharged diesel developing 172 hp at 2,800 rpm	
Clutch	hydraulically operated	hydraulically operated
Transfer box	none	2-speed
Steering	hydraulic	hydraulic
Turning radius	7.4 m	8.75 m
Suspension (front and rear)	leaf springs, telescopic shock-absorbers and torsion bar stabilisers, rear axle has secondary leaf spring	
Tyres	22.50/6.00 × 25	22.50/7.50 × 22.50
Brakes (main)	hydraulic, dual-circuit	hydraulic, dual-circuit
(parking)	mechanical	mechanical
(exhaust brake)	standard	standard
Electrical system	24 V	24 V
Batteries	2 × 12 V	2 × 12 V

Mercedes-Benz 1017 A (4 × 4) 5,000 kg truck with bows stowed and tarpaulin cover removed

Mercedes-Benz 1017 A (4 × 4) 5,000 kg truck complete with bows and tarpaulin cover (Michael Jerchel)

ensure well-balanced air distribution and window defrosting.

The platform has a wooden bed with sectional steel seam and is provided with lashing eyelets for pallets and supplies, plus countersunk anchors for the centre seat benches for the transport of troops. The top edges of the drop type wooden side boards are covered by U-sections and side racks are installed for troop transport. The tarpaulin can be removed and the bows, which are adjustable for height, can be removed and stowed away. The truck can be loaded from the sides, even with the bows installed.

Status

Production complete. In service with Germany, Belgium, Portugal and NATO units.

Manufacturer

Mercedes-Benz AG, Department VN/RM - 2503, D-70322 Stuttgart, Germany.
Tel: +49 711 179 0326. Fax: +49 711 179 0934.

VERIFIED

MAN-Volkswagen Type 8.150 FAE and 9.150 FAE (4 × 4) 3,000 kg trucks

British Army MAN-Volkswagen Type 8.136 FAE 3,000 kg truck in use as shelter carrier; these vehicles are being withdrawn

Development

These trucks, originally the Type 8.136 FAE and 9.136 FAE, were developed jointly by MAN and Volkswagen and first shown during Spring 1983. The two models vary in chassis weights and capacities but use the same engine.

In late 1984 the Danish Army ordered more than 725 of the Type 8.136 FAE version, at a reported cost of DM45 million, for delivery between 1985 and 1989; by late 1990, 989 had been delivered and a further 60 were on order. These trucks were fitted with engines developing 136 hp at 3,000 rpm.

In 1987 MAN received an order from the British Army for 42 Type 9.136 FAE vehicles. In early 1988 the engine rating was increased to 150 hp and the British Army ordered 15 units of the new model Type 9.150 FAE. By the end of September 1990 100 units had been delivered.

A 3.5 m wheelbase version of these trucks is licence-produced in Hungary by the RÁBA Works Truck and Tractor Factory at Gyor. This model is known as the RÁBA H9.114-4.4, and is powered by a MAN D 0828 GF 01 diesel developing 155 hp at 2,700 rpm.

Description

The Type 8.150 FAE and Type 9.150 FAE (4 × 4) trucks are of conventional layout, with all-steel forward control cabs. Seating is provided inside the cab for the driver and one or two passengers and there is space for a spare wheel and tool stowage behind the cab. The load area may be covered by a canvas tilt.

Single tyres are provided front and rear and the 4 × 4 drive configuration is permanent. The suspension is based on the use of semi-elliptical springs front and rear allowing a considerable degree of axle articulation.

A winch is located behind the rear axle with the cable leading through rollers mounted on the front bumper. Other items that may be fitted include a material-handling crane behind the cab, and the cab may have a roof hatch. A NATO towing hitch is usually fitted. The vehicles for the Danish order are fitted with external engine starting, stowage over the cab, movable running boards and chassis shackles for use when carried on rail cars; and the alternator and fan have been moved to allow fording through water obstacles 1 m deep. Other optional extras include a special oil sump with two pumps to ensure optimum oil flow on up to 60 per cent gradients.

Specifications

Type 8.150 FAE; data for 9.150 FAE in square brackets where different
Cab seating: 1 + 1 or 2
Configuration: 4 × 4
Weight:
(empty, 3.1 m WB) 3,580 [3,670] kg
(empty, 3.5 m WB) 3,620 [3,710] kg
(loaded) 7,490 [8,990] kg
Max load: 3,330 [3,240] kg
Length: 5.92 m
Width: 2.245 m
Height: 2.7 m
Ground clearance:
(8.150 FAE) 0.325 m
(9.150 FAE) 0.34 m
Wheelbase: 3.1 m or 3.5 m
Track: 1.8 m
Angle of approach/departure: 35°/36°
Max speed: 95 km/h
Fuel capacity: 100 litres
Max gradient: (with special sump) 60%
Fording: (prepared) 1 m
Engine: D 0826 6.87 litre 6-cylinder in-line water-cooled direct injection diesel developing 150 hp at 2,700 rpm
Gearbox: VW7 with 5 forward and 1 reverse gears
Transfer box: MAN G 300, 2-speed
Clutch: single dry disc
Steering: ZF Type 8036 hydraulic-assist
Turning circle: 15.5 m
Suspension: (front and rear) semi-elliptic leaf springs with double-action hydraulic shock-absorbers
Tyres: 9.00 × 20 S + G, 10.00 × 20, 12.5 R 20 (22PR)
Brakes:
(main) dual-circuit air
(parking) mechanical on wheels
Electrical system: 24 V

Status

Production complete. In service with Danish Army (more than 725). Licence-produced in Hungary.

Manufacturer

MAN Nutzfahrzeuge AG, Postfach 500620, D-80976 Munich, Germany.
Tel: +49 89 15801. Fax: +49 89 1580 4716.

VERIFIED

Mercedes-Benz 1114 A and 1117 A (4 × 4) 4,500 kg trucks

Development

The Mercedes-Benz 1114 A and 1117 A (4 × 4) 4,500 kg trucks are two-axle, cross-country trucks designed for the transport of personnel or supplies and as a tractor for light artillery. Both were developed from the Mercedes-Benz commercial LN2 series and are outwardly identical, differing mainly in the engine installed. The Mercedes-Benz 1114 A has a 136 hp diesel while the 1117A has a 170 hp turbocharged diesel.

Description

The Mercedes-Benz 1114 A and 1117 A have forward control cabs with seating for the driver and two passengers. The LN2 standard cab can be tilted forward to an angle of 55° for engine and other maintenance. Day-to-day engine checks can be carried out via a flap in the cab front. Cab heating and ventilation systems are provided.

A permanent 4 × 4 drive configuration is provided and differential locks are fitted for cross-country travel. All major drive components are protected against splash water and mud, allowing a fording depth of 0.5 m.

The cargo body has drop sides and a drop tailgate and may be protected by a cover and tilt. Purpose-built bodies are available for a number of applications including fuel and water tankers, recovery vehicles, ISO container carriers and shelter-type bodies for mobile offices and workshops.

Optional equipment includes a wading kit allowing wading depths of up to 1 m.

Specifications

Cab seating: 1 + 2
Configuration: 4 × 4
Weight:
(kerb) 6,500 to 7,500 kg
(GVW) 11,500 kg
(GCW) 21,000 kg
Front axle load: 4,700 kg
Rear axle load: 7,000 kg
Max load: 4,000 to 5,000 kg
Towed load: 10,000 kg
Load area: 4.5 × 2.35 m
Length: 6.98 m
Width: 2.48 m
Height:
(top of cab hatch) 2.768 m
(tilt) 3.225 m
(load area) 1.37 m
Ground clearance: 0.314 m
Track: 2.066 m
Wheelbase: 3.62 m
Angle of approach/departure: 26°/30°
Max speed:
(1114 A) 96 km/h
(1117 A) 98 km/h
Fuel capacity: 200 litres
Max gradient:
(road) 40%
(cross-country) 80%
(road, GCW with trailer) 19%
(cross-country, GCW with trailer) 35%
Fording: 0.5 m (1 m optional)
Engine: (1114 A) Mercedes-Benz OM 366 5.959 litre 6-cylinder in-line water-cooled diesel developing 136 hp at 2,600 rpm
(1117 A) Mercedes-Benz OM 366 A 5.959 litre

Mercedes-Benz 1117 A (4 × 4) 4,500 kg truck

Mercedes-Benz 1117 A (4 × 4) 4,500 kg truck

6-cylinder in-line turbocharged water-cooled diesel developing 170 hp at 2,600 rpm
Transmission: Mercedes-Benz G 3/55-6/8.5 synchromesh with 6 forward and 1 reverse gears
Clutch: single dry disc
Transfer box: Mercedes-Benz VG 500-3W with differential lock, 2-speed
Steering: Mercedes-Benz LS 5 F, power-assisted
Turning circle: 15.9 m
Tyres: 14.75/80 R 20
Brakes: dual-circuit compressed air with automatic load-dependent brake (ALB) pressure control, plus exhaust brake
Electrical system: 24 V
Batteries: 2 × 12 V, 88 Ah

Status
In production.

Manufacturer
Mercedes-Benz AG, Department VN/RM - 2503, D-70322 Stuttgart, Germany.
Tel: +49 711 179 0326. Fax: +49 711 179 0934.

UPDATED

MAN 630 Series (4 × 4) 5,000 kg trucks

Description
The MAN 630, together with the MB LG 315/46, was the standard 5,000 kg truck of the German Army until the introduction of the new generation of tactical trucks. It is in service in two basic models, the MAN 630 L2A with dual rear wheels and the MAN 630 L2AE with single rear wheels.

The layout of the vehicle is conventional, with the engine and cab at the front and the cargo area at the rear. The cab has a removable canvas top, windscreen which can be folded forward onto the bonnet and a single door on each side, the tops of which can be removed. The all-steel rear cargo area has a drop tailgate, removable bows and a tarpaulin cover. Some vehicles have been fitted with a winch at the front.

An earlier MAN truck was the 3,000 kg MAN 400 L1AE which had an empty weight of 4,480 kg. This was followed by the MAN 415 series of (4 × 4) 4,000 kg trucks powered by six-cylinder multifuel engines developing 100 bhp and with 4.2 m wheelbases. The MAN 415 is produced in India (with a number of modifications) as the Shaktiman, details of which appear in the Indian section.

Variants include (or have included) an ambulance which can carry up to four stretcher patients, a decontamination vehicle called the TER-Kz, a drone carrier for the Canadian AN/USD-501 reconnaissance system, a field kitchen, a radar vehicle for AN/TPS-1E system, a tanker with two fuel tanks in the rear and dispensing equipment, a tipper with a short wheelbase with either single or dual wheels at the rear, a tractor with a wheelbase of 4.1 m and dual rear wheels, and a van.

Specifications
Data in square brackets relates to L2AE where different from the L2A
Cab seating: 1 + 1
Configuration: 4 × 4
Weight:
(empty) 7,515 [7,980] kg
(loaded) 13,000 [13,200] kg
Load area: 5 × 2.35 m
Length: 7.9 [7.73] m
Width: 2.5 m
Height: (tarpaulin) 2.845 [2.98] m
Ground clearance: 0.35 [0.4] m
Track: (front/rear) 1.922 m/1.763 m [2.506 m/2.506 m]
Wheelbase: 4.6 m
Angle of approach/departure: 32°/45°
Max speed: (road) 66 km/h
Range: 440 km
Fuel capacity: 110 litres
Fuel consumption: 25 litres/100 km
Max gradient: 60%
Fording: 0.85 m
Engine: MAN Model D 1246 MV3A/W 6-cylinder multifuel developing 130 hp at 2,000 rpm
Gearbox: manual with 6 forward and 1 reverse gears
Transfer box: 2-speed
Turning radius: 9.75 m
Suspension: semi-elliptical springs with hydraulic shock-absorbers
Tyres: 11.00 × 20 [14.00 × 20]
Brakes:
(main) air
(parking) mechanical
Electrical system: 24 V
Batteries: 2 × 12 V, 100 Ah

Status
Production complete. In service in Belgium (assembled in Belgium by Ets Hocké and known as the L2AE-B), Germany, India, and several nations in Central and South America, including El Salvador.

Manufacturer
MAN Nutzfahrzeuge AG, Postfach 500620, D-80976 Munich, Germany.
Tel: +49 89 15801. Fax: +49 89 1580 4716.

VERIFIED

Belgian Army MAN 630 L2AE 5,000 kg trucks (Michael Jerchel)

Belgian Army MAN 630 L2AE 5,000 kg truck

MAN 11.136 HA (4 × 4) 5,000 kg truck

Description

The MAN 11.136 HA (4 × 4) 5,000 kg truck is based on standard commercial components with suitable modifications for military use. MAN delivered 3,000 of these vehicles to the Belgian Army which also uses the older MAN 630 L2AE-B series of 5,000 kg (4 × 4) trucks.

The chassis consists of U-frame longitudinal members with riveted and bolted cross members with four reinforced couplings at the front and trailer coupling, type RU, size K3D, at the rear.

The layout of the vehicle is conventional, with the engine at the front, two-door fully enclosed cab in the centre and cargo area at the rear. The cab has a reinforced bonnet, hydraulically cushioned seat for the driver and a buzzer connected to the differential lock that informs the driver when the differential is engaged. The rear cargo body has a steel frame, sheet steel drop sides and tailgate, pinewood floor, removable bows and a tarpaulin cover.

A 4 × 2 model was also available under the designation 11.136 H.

Specifications

Cab seating: 1 + 2
Configuration: 4 × 4
Weight:
(empty) 6,000 kg
(loaded) 11,000 kg
Max load: 5,000 kg
Load area: 4.6 × 2.35 m
Length: 7.305 m
Width: 2.3 m
Height:
(cab) 2.605 m
(load area) 1 m
Ground clearance:
(front) 0.33 m
(rear) 0.334 m
Track:
(front) 1.82 m
(rear) 1.664 m
Wheelbase: 4.4 m
Max speed: 83.9 km/h
Range: 720 km
Fuel capacity: 200 litres
Max gradient: 60%
Fording: 0.75 m
Engine: MAN/Renault model 797/06 6-cylinder water-cooled diesel developing 150 hp (SAE) at 3,000 rpm
Gearbox: ZF synchromesh model S 5-35 with 5 forward and 1 reverse gears
Clutch: single dry plate
Transfer box: G 300, 2-speed
Steering: ZF Gemmer worm and roller
Turning radius: 9.65 m
Suspension:
(front) semi-elliptical leaf springs with progressively working hollow rubber springs
(rear) leaf springs with progressively working stepped springs
Tyres: 9.00 × 20
Brakes:
(front) air/hydraulic
(rear) air
(parking) spring loaded on rear wheels
(engine brake) exhaust
Electrical system: 24 V
Batteries: 2 × 12 V, 100 Ah

Status

Production complete. In service with the Belgian Army (3,000).

Manufacturer

MAN Nutzfahrzeuge AG, Postfach 500620, D-80976 Munich, Germany.
Tel: +49 89 15801. Fax: +49 89 1580 4716.

VERIFIED

Belgian Army MAN 11.136 HA (4 × 4) 5,000 kg truck (P Touzin)

Belgian Army MAN 11.136 HA (4 × 4) 5,000 kg truck with van body (C R Zwart)

MAN 14.240 HAE and 19.240 HAE (4 × 4) 5,000 kg trucks

Description

The MAN 14.240 HAE (4 × 4) 5,000 kg truck was based on standard commercial components with suitable modifications for military use. It may be regarded as a logical progression from the MAN 11.136 HA series (see previous entry). MAN delivered more than 250 of these vehicles to various North African countries. Production ceased in June 1994.

The chassis consists of U-frame longitudinal members with riveted and bolted cross members with one reinforced coupling at the front and a trailer coupling, type RU, size K4D, at the rear.

The layout of the vehicle is conventional with the engine at the front, a two-door fully enclosed cab in the centre and the cargo area at the rear. The cab has a reinforced bonnet, a hydraulically cushioned seat for the driver, and a buzzer connected to the differential lock to inform the driver when the differential is engaged. The rear cargo body has a steel frame, sheet steel drop sides and tailgate, a pinewood floor, removable bows and a tarpaulin cover.

An upgraded version, the 19.240 HAE, was also produced. A 4 × 2 model with the designation 19.240 H was available. Also available were 6 × 6 and 6 × 4 models with payloads up to 10,000 kg with rear axles single-tyred, or 23,000 kg with twin tyres on the rear axles.

MAN 19.240 HAE (4 × 4) truck

Specifications

14.240 HAE
Cab seating: 1 + 2
Configuration: 4 × 4
Weight:
(empty) 6,600 kg
(loaded) 14,000 kg
Max load: 5,000 kg
Load area: 5 × 2.44 m
Length: 8.2 m
Width: 2.5 m
Height: (cab, empty) 2.863 m
Ground clearance:
(front) 0.44 m
(rear) 0.37 m
Track:
(front) 2.035 m
(rear) 2.078 m
Wheelbase: 5.2 m
Max speed: 80 km/h
Range: approx 800 km
Fuel capacity: 300 litres
Max gradient: 60%
Fording: 0.75 m
Engine: MAN model D 2566 6-cylinder water-cooled diesel developing 240 hp (DIN) at 2,200 rpm
Gearbox: ZF synchromesh model S 6-90 with 6 forward and 1 reverse gears

Clutch: single dry plate
Transfer box: G 801, 2-speed
Steering: ZF hydraulic
Turning radius: 9.65 m
Suspension:
(front) semi-elliptic leaf springs with progressively working hollow rubber springs
(rear) leaf springs with progressively working stepped springs
Tyres:
(4 × 4) 14.00 R 20
(4 × 2) 12.00 R 20
Brakes:
(main) air
(parking) spring loaded on rear wheels
(engine) exhaust
Electrical system: 24 V
Batteries: 2 × 12 V, 140 Ah

Status
Production complete. In service with various North African and South American countries.

Manufacturer
MAN Nutzfahrzeuge AG, Postfach 500620, D-80976 Munich, Germany.
Tel: +49 89 15801. Fax: +49 89 1580 4716.

VERIFIED

Mercedes-Benz 1222 A (4 × 4) 5,500 kg truck

Description
The Mercedes-Benz 1222 A (4 × 4) 5,500 kg truck is produced in two versions, one with a wheelbase of 3.55 m and the other with a wheelbase of 4 15 m, both powered by the same engine and using the same transmission.

Two cab versions are available, both providing seating for the driver and two passengers with the centre seat folding down to provide access to the roof hatch. One is the standard cab and the other the medium-length cab with more internal space for equipment stowage. Both cabs can be tilted forward for engine and transmission access. The normal body is a cargo platform with drop sides and tailgate that can be protected by a cover over the bows. Other body options are possible, including vehicle-mounted cranes. Mechanical or hydraulic cable winches with tractive capacities up to 5,000 kg can be fitted, with cable control possible to front or rear. A spare wheel is carried between the cargo body and the cab.

Specifications
Data for short wheelbase model; data for long wheelbase model in square brackets where different
Cab seating: 1 + 2
Configuration: 4 × 4
Weight:
(kerb) approx 8,500 kg
(GVW) 14,000 kg
(GCW) 24,500 kg
Front axle load: 6,000 kg
Rear axle load: 8,500 kg
Max payload: approx 5,500 kg
Towed load: 10,500 kg
Load area: 4.35 × 2.5 [4.85 × 2.5] m
Length: 6.87 [7.4] m
Width: 2.5 m
Height:
(cab) 2.93 m
(top of tilt) 3.4 m
(loading platform) 1.537 m
Ground clearance: 0.33 m
Track: 2.091 m
Wheelbase: 3.55 [4.15] m
Angle of approach/departure: 30°/30°

Mercedes-Benz 1222 A (4 × 4) 5,500 kg truck

Max speed:
(road) 93 km/h
(cross-country) 55 km/h
Fuel capacity: 200 litres
Max gradient:
(road) 50%
(cross-country) 80%
(road, GCW with trailer) 26%
(cross-country, GCW with trailer) 48%
Fording: 1 m
Engine: Mercedes-Benz OM 421 10.964 litre V-6 water-cooled diesel developing 216 hp (DIN) at 2,300 rpm
Gearbox: Mercedes-Benz G4/95-6/9.0 synchromesh with 6 forward and 1 reverse gears
Clutch: GF 380 single dry disc
Transfer box: Mercedes-Benz VG 900-3W/1.6, 2-speed
Steering: Mercedes-Benz LS 5F, power
Turning circle: 16.8 [19] m
Tyres: 13.00 × 20 XL (option 14.00 × 20 XL)
Brakes: dual-circuit compressed air with automatic load-dependent brake (ALB) pressure control
Electrical system: 24 V
Batteries: 2 × 12 V, 110 Ah

Status
In production.

Manufacturer
Mercedes-Benz AG, Department VN/RM - 2503, D-70322 Stuttgart, Germany.
Tel: +49 711 179 0326. Fax: +49 711 179 0934.

VERIFIED

Mercedes-Benz 1628 A (4 × 4) 7,000 kg truck and 2028 A (6 × 6) 10,000 kg truck

Development
These two Mercedes-Benz cross-country trucks are produced for the export market and have many common components. Some civilian model design experience is incorporated but the design is basically new. See also entry on Mercedes-Benz 2628 A in this section.

The 1628 A is a 4 × 4 model which is virtually identical to the 1627 A but it has a slightly reduced load area and is powered by an OM 422 V-8 diesel limited to 265 hp.

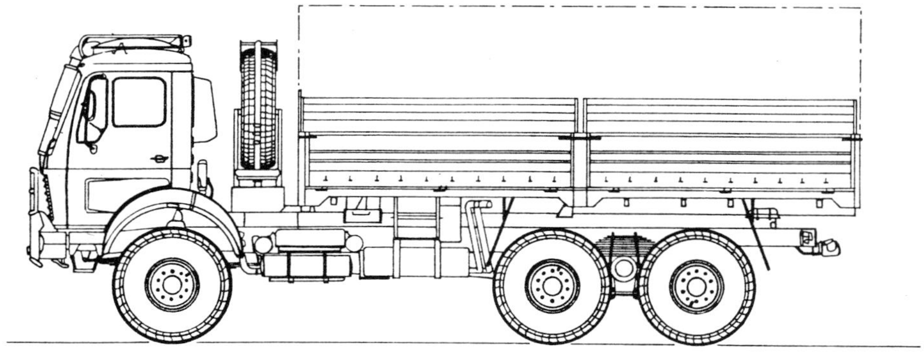

Mercedes-Benz 2028 A (6 × 6) 10,000 kg truck

Description
Both vehicles use the same design of cab with seating for the driver and two passengers. The centre seat has a folding backrest to allow clear access to the roof-mounted hatch which is provided with a rail grip. There is a medium length cab available which has extra space for crew equipment. Both cab types can be tilted forward for engine access to an angle of 65°. For normal day-to-day maintenance access flaps are provided. Behind the cab there is spare wheel stowage, and on the front of the cab the headlights and radiator grille are protected by heavy-duty tubular guards.

All the undersides and major drive components are fully waterproofed for a fording depth of 1 m without preparation and the air intake pipe is routed upwards over the cab roof side. The load area has 500 mm high drop sides and tailgate, and is provided with lash-down points and a canvas tilt. A towing hook for trailers or artillery is provided at the rear. Optional equipment includes a hydraulic 10,000 kg winch with cable control to the front or rear. Various box and cab bodies may be provided, and both models may be produced as tankers, fire tenders or command and communication vehicles.

Single-tyred versions of both trucks are available with tyre pressure regulation systems as an option.

Status

In production. 2028 A in service with Pakistan and the United Arab Emirates.

Manufacturer

Mercedes-Benz AG, Department VN/RM - 2503, D-70322 Stuttgart, Germany.
Tel: +49 711 179 0326. Fax: +49 711 179 0934.

VERIFIED

Specifications

Model	**1628 A**	**2028 A**
Cab seating	1 + 2	1 + 2
Configuration	4 × 4	6 × 6
Weight		
(empty)	9,000 kg	11,200 kg
(loaded)	16,000 kg	22,000 kg
(front axle load)	7,300 kg	7,500 kg
(rear axle load)	9,000 kg	2 × 8,000 kg
Max load	7,000 kg	10,000 kg
Length	7.43 m	8.3 m
Width	2.5 m	2.5 m
Height (cab)	3.08 m	3.04 m
(canvas cover)	3.485 m	3.5 m
Load area	4.6 × 2.35 m	5.5 × 2.35 m
Ground clearance	0.435 m	0.435 m
Track	2.07 m	2.07 m
Wheelbase	4.5 m	3.8 m + 1.45 m
Angle of approach/departure	32°/42°	32°/46°
Max speed	91 km/h	91 km/h
Fuel capacity	300 litres	300 litres
Gradient	80%	80%
Fording	1.2 m	1.2 m
Engine	Mercedes-Benz OM 422 V-8 water-cooled diesel, 14.618 litre, developing 280 bhp at 2,300 rpm	
Transmission	ZF 5 S-111 GP fully synchronised with 9 forward and 1 reverse gears or optional ZF torque converter system	
Turning radius	10.15 m	10.15 m
Tyres	14.00 × 20 XL	14.00 × 20 XL
Brakes	dual-circuit compressed air	dual-circuit compressed air
Electrical system	24 V	24 V
Batteries	2 × 12 V	2 × 12 V

MAN 14.240 FAEG (4 × 4) 6,000 kg and 20.280 DFAEG (6 × 6) 10,000 kg trucks

Description

MAN developed the Category III vehicles basing them on the Category I vehicles developed for the German Army. They use the same suspension and drive train technology as the Category I vehicles, torsionally stiff chassis, coil springs, axles with planetary gear hub reduction, but use MAN engines, commercial tilt cabs and allow a high payload; 6,000 kg for the 14.240 FAEG (4 × 4) and 10,000 kg for the 20.280 DFAEG (6 × 6).

The chassis consists of hollow section longitudinal members welded with tubular cross members. The cab is a MAN two-door all-steel forward control type which can be tilted forward to allow access to the engine for maintenance purposes. The rear cargo area has a steel subframe, aluminium drop sides with two side walls and one tailgate, fixed corner stakes in the front, insertable stakes in the middle and rear, pinewood floor, removable bows and a tarpaulin cover.

Production of these models ceased during 1989.

Status

Production complete. In service with Algeria, Ireland, Oman, Peru, Singapore and Venezuela.

Manufacturer

MAN Nutzfahrzeuge AG, Postfach 500620, D-80976 Munich, Germany.
Tel: +49 89 15801. Fax: +49 89 1580 4716.

VERIFIED

Specifications

Model	**14.240 FAEG**	**20.280 DFAEG**
Configuration	4 × 4	6 × 6
Weight (empty)	8,350 kg	10,300 kg
(loaded)	14,350 kg	20,300 kg
Max load	6,000 kg	10,000 kg
Load area	5 × 2.44 m	6.1 × 2.44 m
Length	7.85 m	8.95 m
Width	2.49 m	2.49 m
Height (cab)	3.01 m	3.01 m
Wheelbase	4.5 m	4 m + 1.4 m
Angle of approach/departure	40°/43°	40°/40°
Max road speed	80 km/h	80 km/h
Fuel capacity	310 litres	310 litres
Max gradient	limit of adhesion	limit of adhesion
Fording	1 m	1 m
Engine	MAN D 2566 6-cylinder in-line water-cooled diesel	MAN D 2566 6-cylinder in-line water-cooled diesel
Hp/rpm	240/2,200	280/2,200
Gearbox	manual, 6 forward and 1 reverse gears	manual, 6 forward and 1 reverse gears
Clutch	torque converter	torque converter
Transfer box	MAN G801 with differential lock	MAN G801 with differential lock
Tyres	14.00 × 20	14.00 × 20
Brakes	air over hydraulic dual-circuit with spring-loaded auxiliary brake and starting auxiliary brake (climbing brake), exhaust brake	
Steering	ZF recirculating ball hydro-steering	ZF recirculating ball hydro-steering
Suspension	progressive coil springs, hydraulic telescopic shock-absorbers	
Electrical system	24 V	24 V
Batteries	2 × 12 V, 110 Ah	2 × 12 V, 110 Ah

Irish Army MAN 14.240 FAEG (4 × 4) 6,000 kg truck (T J Gander)

MAN 4 × 4, 6 × 6 and 8 × 8 high-mobility tactical trucks

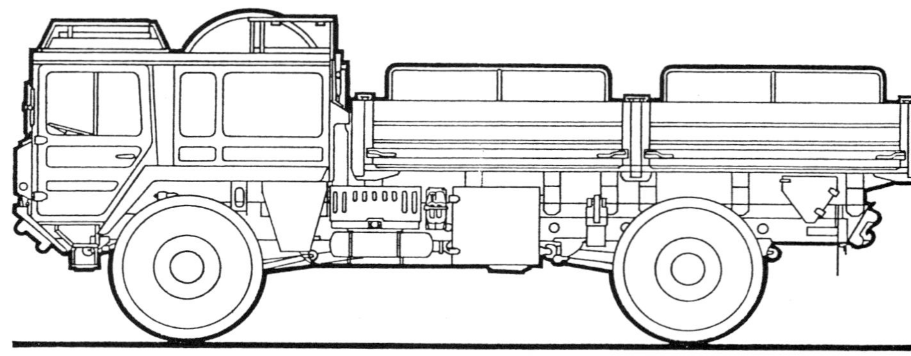

MAN Cat I A1.1 (4 × 4) truck (not to 1/76th scale)

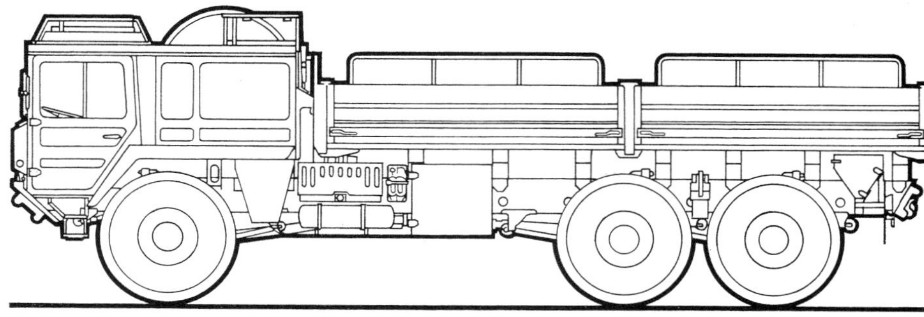

MAN Cat I A1.1 (6 × 6) truck (not to 1/76th scale)

Development/Description

In the late 1950s and early 1960s the West German Technical Office for Armament and Military Purchases drew up requirements for a new range or second generation) of tactical trucks for the Army. This range was to have consisted of 4 tonne (4 × 4), 7 tonne (6 × 6) and 10 tonne (8 × 8) vehicles, 4 × 4 and 6 × 6 armoured amphibious load carriers and an 8 × 8 amphibious reconnaissance vehicle.

Since development of these trucks began in the 1960s there were some changes in the requirements of the army and in 1972 the amphibious specification and the requirement that the vehicles should be powered by a multifuel engine were dropped. In 1975 the 4 tonne was uprated to 5 tonne and at the same time its rear cargo platform was lengthened and the wheelbase was increased from 4.3 to 4.5 m.

MAN 6 × 6 trucks are licence manufactured in Hungary by RÁBA.

Category I

In December 1975 MAN was awarded a contract to build 8,385 4 × 4, 6 × 6 and 8 × 8 vehicles at a cost of DM1,400 million. The 8 × 8 version was the first model to enter production and first deliveries were made in 1976. Deliveries of the 6 × 6 tipper and the 4 × 4 cargo truck began in 1977. Deliveries of the 6 × 6 cargo truck began in January 1979 and final deliveries were made in 1981. By 1983, 8,617 vehicles of all configurations had been delivered to the West German armed forces. By the end of 1986, 411 had been delivered to the Austrian Army and 82 to the Belgian Army.

This range of vehicles, which was produced at MAN's Watenstedt plant, was known as the Category I (or Kat I) it was designed specifically for cross-country operations and to keep up with mechanised forces operating across country.

Many components of this range, such as axles, engines and gearboxes, are standard commercial components and spare parts can be easily obtained.

Category II

The next generation was the Category II (or Kat II) version which was developed at the request of and in close co-operation with the US armed forces. By the end of 1986, 534 of these trucks had been delivered to the US armed forces, 157 to France and 42 to Canada.

Category III

In a parallel development the so-called Category III (or Kat III) vehicles were developed. These are similar to the Category I vehicles but are fitted with a cab from the MAN civilian truck range, modified for military use. A total of 1,594 Category III trucks was delivered to Algeria (280), Ireland, Oman (95), Peru (165), Singapore (104) and Venezuela (405).

Category I A1 extra wide versions

In April 1985 MAN presented the first of a new vehicle, the Category I A1, to the German Federal Office for Military Technology and Procurement (BWB) and to the Test Centre 41 facility. It was planned that there would be 35 test vehicles of all configurations (including the extra-wide 8 × 8 version) and versions for Patriot, Roland, MLRS, a folding bridge and a folding roadway, and for the LUER air surveillance radar. Production of the new category vehicles began in 1987.

The Category I A1 generation was a development of the Category I with improved performance on and off roads due to higher powered engines and an improved design of frame and cab.

Planned changes introduced on the Category I A1 vehicles included the following:

A hand-tiltable cab, for better access to the engine which can then be removed or installed in about 30 minutes as opposed to the previous 10 hours

A rearrangement of components so that the cab no longer has to be tilted for routine checks

A fully integrated anti-corrosion treatment package for extreme climate conditions and operational use for more than 20 years

A paint finish that fulfils all military requirements including camouflage pattern, infrared reflection and NBC decontamination needs

Extended use of more commercial components to reduce costs

A range of water-cooled MAN engines with performances of up to 550 hp

All vehicles able to use 16.00 × 20 radial tyres

A removable cab roof to assist rail and air transport

Logistic commonality with earlier vehicles would be maintained.

The layout of all the basic vehicles is almost identical. All vehicles are fitted with a hydraulically tiltable uniform three-man cab specifically designed for military purposes. The plastic roof can support the weight of a man and is fitted with an observation hatch; it can be removed for air transport. A low-torsion frame is used together with coil springs and telescopic shock-absorbers. Most vehicles are fitted with a Rotzler winch. All attachments, such as stowage compartments and canister brackets, are fastened to the frame so that different bodies and platforms can be fitted.

Variants

MAN Cat I A1 (4 × 4) truck

This was designed for a chassis carrying capacity up to 7,500 kg. It is used as a cargo truck and is able to carry the Dornier military container, in production for the German Army.

MAN Cat I A1 (6 × 6) truck

This truck was designed for a chassis carrying capacity of up to 13,000 kg. The cargo version is also used to tow 155 mm FH-70 howitzers used by the German Army. Specialised versions include a tipper, a bridging truck which transports and launches the German version of the Ribbon Bridge (this has a longer wheelbase) and a CL-289 drone carrier. The 110 mm Light Artillery Rocket Systems (LARS), previously mounted on the older Magirus-Deutz chassis, have

MAN Cat I A1.1 (8 × 8) truck (not to 1/76th scale)

been installed on this MAN (6 × 6) truck chassis. (Full details of the 110 mm LARS are given on pages 691 and 692 of *Jane's Armour and Artillery 1994-95.)* The Austrian company Gräf und Stift built 350 6 × 6 versions rated at 10 tonnes for the Austrian Army, with a slightly different cab and a hydraulic crane mounted to the rear of the cab. Details of this version are given in this section under Austria.

Variants introduced during 1988 included a minelaying version for the Belgian Army and a carrier for a Dornier mobile telescopic mast.

MAN Cat I A1 (8 × 8) truck

This truck was designed for a chassis carrying capacity of up to 16,000 kg, and by using 13 tonne axles, up to 25,000 kg. Many of the cargo trucks are fitted with a 1,000 kg Atlas hydraulic crane for loading and unloading. The 8 × 8 chassis is also used to carry the German AEG-Telefunken TRMS radar and a shelter-mounted version system has been proposed to carry and launch the Dornier Keibitz RPV. The range has been increased by the addition of a truck with a hydraulic crane mounted at the rear. This version has the cargo area reduced to 5.5 × 2.5 m. Other specialised versions include a palletised loading system, and a wrecker equipped with a recovery winch, towing device and a crane.

MAN has produced a number of special container-body versions. For details of these vehicles see entries in the *Shelters and Containers* section.

MAN Cat I A1 (8 × 8) Width 2.9 m

MAN offers an extra-wide version of the 8 × 8 with an increased width of 2.9 m for the transport of sensitive loads with high centres of gravity. This extra-wide version uses a low-torsioning frame equipped with coil springs, mounted as wide apart as possible, and with telescopic shock-absorbers to provide good lateral stability.

MAN has proposed that this chassis, for missile and other tactical applications, could be fitted with four (two each side) hydraulic extended outriggers, combat tyres with an emergency run-flat capability, a central tyre inflation system, a coil spring suspension system with lockout capability, EMP hardening and an onboard camouflage system.

FlaRakRad Roland II

The MAN Cat I A1 (8 × 8) Width 2.9 m chassis is used to carry the shelter body that acts as the command and launching post for the Roland II mobile anti-aircraft weapon system. The vehicle/weapon combination is known as the FlaRakRad Roland II. The German Air Force ordered 68 of these systems, the Navy 20 and the Air Force ordered 27 to defend US Air Force bases in Germany.

Patriot

Germany adopted the Patriot (MIM-104) tactical air defence system and mounts the various mobile components on various MAN chassis. The 4 × 4 truck is used to carry a Dornier mobile telescopic mast associated with the Patriot system; the mast is 34 m

MAN Cat I A1.1 (8 × 8) high-mobility truck (Michael Jerchel)

Specifications

Model	**Cat I A1 4 × 4**	**Cat I A1 6 × 6**	**Cat I A1 8 × 8**	**Cat I A1 8 × 8 Width 2.9 m**
Cab seating	1 + 2	1 + 2	1 + 2	1 + 2
Configuration	4 × 4	6 × 6	8 × 8	8 × 8
Weight (empty)	9,500 kg	11,300 kg	13,400 kg	12,800 kg
(loaded)	16,000 kg	26,000 kg	32,000 kg	32,000 kg
Chassis carrying capacity	7,500 kg	13,500 kg	20,000 kg	20,000 kg
Load area	5.1 × 2.44 m	6.2 × 2.44 m	7.2 × 2.44 m	7.2 × 2.44 m
Length	8.17 m	9.27 m	10.27 m	10.27 m
Width	2.5 m	2.5 m	2.5 m	2.9 m
Height (cab)	2.64 + 0.29 m	2.64 + 0.29 m	2.64 + 0.29 m	2.64 + 0.29 m
(load area)	1.54 m	1.54 m	1.54 m	1.54 m
Ground clearance	0.41 m	0.41 m	0.41 m	0.41 m
Track	2.07 m	2.07 m	2.07 m	2.47 m
Wheelbase	4.5 m	4.5 m + 1.5 m	1.93 m + 3.57 m + 1.5 m	
Angle of approach/departure	34°/39°	34°/39°	34°/39°	34°/39°
Max cruising speed (road)	90 km/h	90 km/h	90 km/h	90 km/h
Fuel capacity	400 litres	400 litres	400 litres	400 litres
Max gradient	limit of adhesion			
Fording	1.2 m	1.2 m	1.2 m	1.2 m
Engine	range of MAN 6-cylinder in-line diesel, water-cooled with turbocharging or combined intercooling turbocharge system (optional MAN V-10 diesel, water-cooled with intercooling and turbocharging)			
Engine hp (DIN)	performance range from 252 hp up to 550 hp			
Gearbox	range of ZF synchronised gearboxes between 9 and 16 gears with or without integrated transfer case			
Clutch	mechanical or hydraulic converter with lock-up clutch			
Steering	recirculating ball hydro-steering type:			
	ZF 8046	ZF 8046	ZF dual circuit 8096	
Turning radius	9.65 m	10.6 m	13.7 m	13.8 m
Suspension	progressively acting helical springs			
Tyres	14.00 R 20 (optional 16.00 R 20)			
Brakes	air/2 circuit/2 lines			
Electrical system	24 V	24 V	24 V	24 V
Batteries	2 × 12 V, 170 Ah (optional 4 × 12 V, 100 Ah)			

MAN Cat I A1 (6 × 6) cargo truck

MAN Cat I A1 (6 × 6) truck carrying Dornier mobile telescopic mast

MAN Cat I A1.1 (8 × 8) truck carrying COBRA battlefield radar **1996**

high and carries four dish antennas. The 6 × 6 chassis is used to carry a generator and a hydraulic mast system similar to that carried by the 4 × 4 truck. The (8 × 8) Width 2.9 m chassis is used to carry a radar system with a folding antenna and a four-round Patriot missile launcher.

Elevating combat platform

Using a MAN Cat I A1 chassis as a basis, a prototype of an elevating combat platform was developed in close co-operation with Krauss Maffei and MBB for possible use in anti-tank and anti-helicopter operations. For this project an armoured cab variant is under development to provide a higher degree of protection for the vehicle crew; this will eventually be suitable for use with the entire MAN range.

Status

In service with Belgium and the German armed forces. See also text for users of other Category vehicles.

Manufacturer

MAN Nutzfahrzeuge AG, Postfach 500620, D-80976 Munich, Germany.
Tel: +49 89 15801. Fax: +49 89 1580 4716.

UPDATED

MAN SX 2000, LX 2000 and FX 2000 high-mobility off-road trucks

Development

During 1989 MAN introduced its LX 2000 range of high mobility trucks incorporating all the experience gained during the extended production of its family of High Mobility Tactical Vehicles. Since then the new range has been extended with the introduction of the SX 2000 and FX 2000 series of vehicles.

The full range of MAN high mobility off-road vehicles are available with drive configurations of 4 × 4, 6 × 6 and 8 × 8 and with a choice of engines with ratings from 370 hp to 1,000 hp.

Description

SX 2000

The MAN SX 2000 heavy vehicles may be regarded as the logical successors to the MAN High Mobility Tactical Vehicles (see previous entry). They employ a torsion-free vehicle frame consisting of box section main members and welded-in tubular cross members. The axles are precisely located by control arms, while coil springs and shock-absorbers on all wheels combine to ensure that they remain in contact with the ground on difficult terrain and at all speeds. A torque converter and permanent all-wheel drive are standard. The vehicles can be equipped with central tyre inflation systems. The cooling system is located above the engine to reduce the effects of dust and mud.

A 2.9 m wide version of the SX 2000 chassis is available for special bodies such as airport crash tenders or Patriot and Roland missile systems.

The SX 2000 series has a choice of engines from 370 hp to 1,000 hp. Chassis carrying capacities cover the 7,000 to 28,000 kg range.

MAN 11.233 FAEG LX2000 (4 × 4) truck powered by a 230 hp engine **1997**

MAN SX 2000 8 × 8 32.422 VFAEG with 2.9 m wide chassis to accommodate Patriot radar system **1996**

MAN 32.403 VFAEG SX2000 (8 × 8) truck equipped with a load handling system and powered by a 400 hp engine **1997**

MAN 27,403 VFAEG SX2000 (6 × 6) truck, powered by a 400 hp engine, deploying a 43 m hydraulically-supported aerial mast **1997**

MAN FX 2000 25.372 DFAX 6 × 6 platform vehicle powered by 370 hp diesel; GVW is 25 tonnes ***1996***

MAN FX 2000 19.372 FALSX tractor truck powered by 370 hp diesel. Front axle has leaf spring suspension, with air suspension at the rear ***1996***

LX 2000

The LX 2000 series are lighter than the SX 2000 series. They retain the MAN SX 2000 modular cab but the chassis is built around a low-torsion ladder frame with bolted-in tubular cross members, and steel leaf suspension for the front and rear axles. Produced in 4 × 4 and 6 × 6 versions with engine ratings ranging from 190 hp to 270 hp, the chassis carrying capacity varies from 3,500 to 10,000 kg. Large tyres and light construction combine to produce a low specific ground pressure.

FX 2000

The FX 2000 series are a relatively low-priced range of 4 × 4, 6 × 6 and 8 × 8 military vehicles available with or without all-wheel drive. They are based on the F90 civilian heavy truck range combined with the MAN modular military cab.

Specifications

Data for 32.403 VFAEG SX2000 with 14.00 R 20 tyres
Cab seating: 1 + 2
Configuration: 8 × 8
Weight:
(empty) approx 12,140 kg
(GCW) 48,000 kg
Weight on front axles: 2 × 7,500 kg
Weight on rear axles: 2 × 9,000 kg
Max load: approx 20,000 kg
Max towed load: 16,000 kg
Length: 10.04 m
Width: 2.5 m
Height: 2.967 m
Ground clearance: 0.43 m
Wheelbase: 1.93 m + 3.3 m + 1.5 m
Angle of approach/departure: 40°/36°
Max speed: 88 km/h
Fuel capacity: 400 litres
Range: more than 800 km
Gradient: 60%
Fording: 1.2 m
Vertical obstacle: 0.6 m
Trench: 1.9 m
Engine: MAN D2866 11.967 litre 6-cylinder water-cooled direct injection diesel with exhaust turbocharging developing 400 hp at 2,000 rpm
Transmission: ZF 16 S-151 semi-automatic providing 8 forward and 2 reverse gears
Clutch: ZF WSK 400
Transfer box: MAN VG 1700-2 2-speed
Steering: ZF power-assisted
Brakes: dual-circuit, compressed air on all wheels
Tyres: 14.00 R 20
Electrical system: 24 V
Batteries: 2 × 12 V, 140 Ah
Alternator: 55 A

Status

In production.

Manufacturer

MAN Nutzfahrzeuge AG, Postfach 500620, D-80976 Munich, Germany.
Tel: +49 89 15801. Fax: +49 89 1580 4716.

UPDATED

MAN 45.1000 FFAEG SX 2000 (10 × 10) Demonstrator high-mobility truck

Development

During 1996 MAN introduced its 45.1000 FFAEG SX 2000 (10 ×10) Demonstrator high mobility five-axle truck. Developed according to specifications issued by the procurement office of the French Army (ETAS), the Demonstrator utilises standard MAN modular components and is intended to be the forerunner of a series of vehicles intended to carry bridge components, heavy cranes and similar loads. To date the Demonstrator has been shown with a test superstructure.

Description

The MAN 45.1000 FFAEG SX 2000 (10 ×10) Demonstrator high mobility truck utilises a resilient torsion-resistant frame constructed from channel-section chassis rails with welded-in tubular cross members. Each axle is located by one wishbone and two link arms combined with a hydropneumatic suspension system with load-sensitive shock-absorbers, axle load equalisation, ride height control and axle locking. This arrangement permits stability under extreme conditions.

The three front axles have mechanical, power-assisted steering. The fifth axle is hydraulically steered at speeds up to 40 km/h.

The electronically controlled tyre pressure system has four pressure settings which the driver can select while the vehicle is in motion. Once set the pressure is checked automatically.

Power is provided by a 1,000 hp MAN V-12 diesel coupled to a Renk five-speed automatic gearbox with an integrated retarder brake. The engine can be installed behind the cab to permit the carrying of special bodies. If required the cab can be armoured.

MAN 45.1000 FFAEG SX 2000 (10 ×10) Demonstrator high-mobility truck ***1997***

Specifications

Cab seating: 1 + 2
Configuration: 10 × 10
Weight:
(kerb) approx 23,000 kg
(GCW) 50,000 kg
Max load: up to 27,000 kg
Length: 13.18 m
Width: 3.18 m
Height: (roof) 2.85 m
Ground clearance: 0.6 m
Wheelbase: 1.85 m + 1.85 m + 3 m + 1.85 m
Max speed: 95 km/h
Range: approx 800 km
Gradient: >60%
Side slope: >40%
Fording: 1.8 m
Vertical obstacle: 0.8 m

Trench: 2.74 m
Engine: MAN 21.93 litre V-12 water-cooled turbocharged and intercooled diesel developing 1,000 hp
Transmission: Renk automatic providing 5 forward and 1 reverse gears
Transfer box: MAN 2-speed
Steering: power-assisted on front 3 axles
Turning circle diameter: approx 27 m
Brakes: gearbox retarder; also dual-circuit, compressed air on all wheels

Status
Demonstrator.

Manufacturer
MAN Nutzfahrzeuge AG, Postfach 500620, D-80976 Munich, Germany.
Tel: +49 89 15801. Fax: +49 89 1580 4716.

NEW ENTRY

Mercedes-Benz 2628 A (6 × 6) 12,000 kg truck

Development
The Mercedes-Benz 2628 A (6 × 6) 12,000 kg truck was developed along the same lines as the Mercedes-Benz 1628 A and 2028 A trucks (see separate entry) and shares many features and components. The Mercedes-Benz 2628 A does, however, have a much heavier payload and may be used as a tractor for artillery with calibres up to 155 mm. It may also have a materials-handling crane located behind the cab.

This truck is licence produced by TECTRAN in Brazil, where it is employed with an armoured chassis in several forms for the AVIBRAS ASTROS II multiple rocket system.

Description
The Mercedes-Benz 2628 A has an enlarged cab to accommodate the driver and two passengers plus space for other personnel or equipment. A roof hatch is provided. The standard cargo area may be fitted with a canvas tilt and has dimensions of 5.5 × 2.35 m.

Various purpose-built bodies may be fitted to the 2628 A, one being the Marrel torsion-free military platform. The 2628 A can also carry the Marrel Ampliroll AL 160 demountable platform system capable of carrying vehicles as large as the M113 armoured personnel carrier. Other versions can carry tanker or box and cabin bodies for a variety of roles. One option is a side-mounted 10,000 kg hydraulic winch with the cable paying out to front or rear. An automatic tyre pressure adjusting system is also available as an optional extra.

The Mercedes-Benz 2628 A has been produced as a heavy recovery vehicle carrying a Marrel MH 45 or Hägglunds Moelv recovery hamper.

Variants
The Mercedes-Benz 2634 A is externally identical to the 2628 A but is powered by a OM 422 A turbocharged V-8 diesel developing 340 hp at 2,100 rpm. Maximum speed is 89 km/h.

The 2628 AS was a 6 × 6 tractor truck intended to tow engineer equipment on a semi-trailer with a gross combination weight of 38,000 kg.

Mercedes-Benz 2628 A (6 × 6) 12,000 kg truck (T J Gander)

Specifications
Cab seating: 1 + 2 (space for more)
Configuration: 6 × 6
Weight:
(empty) approx 11,300 kg
(loaded) 25,000 kg
(GCW) 38,000 kg
Weight on front axle: 7,500 kg
Weight on rear axle: 2 × 9,000 kg
Max load: approx 10,000 kg
Length: 8.3 m
Width: 2.5 m
Height:
(cab) 3.04 m
(tarpaulin) 3.5 m
Load area: 5.5 × 2.35 m
Ground clearance: 0.435 m
Track: 2.07 m
Wheelbase: 3.8 m + 1.45 m
Angle of approach/departure: 32°/46°
Max speed: 93 km/h
Fuel capacity: 300 litres
Gradient:
(max) more than 80%
(with trailer, GCW 38,000 kg, on-road) 34%
(with trailer, GCW 38,000 kg, off-road) 66%
Fording: 1.2 m
Engine: Mercedes-Benz OM 422 14.618 litre V-8 liquid-cooled direct injection diesel developing 280 hp at 2,300 rpm
Transmission: ZF 5 S - 111 G fully synchronised with crawler providing 9 forward and 1 reverse gears or ZF S 6-90 6-speed with torque converter/mechanical clutch (WSK)
Clutch: GF 420 single dry disc
Transfer box: VG 1400 2-speed
Turning radius: 10.15 m
Brakes: dual-circuit, compressed air
Tyres: 14.00 × 20 XL
Electrical system: 24 V
Batteries: 2 × 12 V, 110 Ah

Status
In production.

Manufacturer
Mercedes-Benz AG, Department VN/RM - Z503, D-70322 Stuttgart, Germany.
Tel: +49 711 179 0326. Fax: +49 711 179 0934.

VERIFIED

Mercedes-Benz 2638 A (6 × 6) 14,000 kg truck

Development
The Mercedes-Benz 2638 A (6 × 6) 14,000 kg truck is only one member of a series of similar heavy-duty trucks, all based on commercial designs but with suitable modifications to suit them for military purposes. The other members of the family are as follows:

2638, 6 × 4, wheelbase 4.5 m + 1.35 m
2638 S, 6 × 4, wheelbase 3.2 m + 1.35 m
2638 AS, 6 × 6, wheelbase 3.5 m + 1.35 m
2638 K, 6 × 4, wheelbase 3.2 or 3.5 m + 1.35 m
2638 AK, 6 × 6, wheelbase 3.5 or 3.85 m + 1.35 m

The data in this description will refer only to the 2638 A, a 6 × 6 truck with a 4.1 m + 1.35 m wheelbase.

Description
The Mercedes-Benz 2638 A forward control cab seats the driver and two passengers plus space for

Mercedes-Benz 2638 A (6 × 6) 14,000 kg truck with load handling system
1997

equipment. A roof hatch is provided. The vehicle is based on a double ladder frame with a high load-bearing capacity, bending resistance and torsional flexibility. The standard cargo area, if used, may be fitted with a canvas tilt.

Numerous types of body can be used with the 2638 A, including load handling systems. A Krauss-Maffei armouring system is available for this vehicle.

Specifications

Cab seating: 1 + 2
Configuration: 6 × 6
Weight on front axle: 7,500 kg
Weight on rear axles: 2 × 13,000 kg
Max load: approx 14,000 kg
Length: (chassis) 8.84 m
Width: 2.49 m
Height: (cab roof) 3.32 m
Ground clearance: 0.347 m
Track:
(front) 1.99 m
(rear) 1.804 m
Wheelbase: 4.1 m + 1.35 m
Max speed: 94 km/h
Fuel capacity: 300 litres
Engine: Mercedes-Benz OM 402 LA Euro 1 12.763 litre V-8 liquid-cooled direct injection intercooled diesel developing 381 hp at 1,900 rpm
Transmission: Mercedes-Benz G180-16/11.9 providing 16 forward and 2 reverse gears
Clutch: single dry disc
Transfer box: VG 2400 2-speed
Turning radius: 9.5 m
Brakes: dual-circuit, compressed air
Tyres: 12.00 R 22.5
Electrical system: 24 V
Batteries: 2 × 12 V, 165 Ah
Alternator: 55 A

Status

In production.

Manufacturer

Mercedes-Benz AG, Department VN/RM - Z503, D-70322 Stuttgart, Germany.
Tel: +49 711 179 0326. Fax: +49 711 179 0934.

NEW ENTRY

Mercedes-Benz 3550 A (8 × 8) 15,000 kg truck

Description

The Mercedes-Benz 3550 A (8 × 8) 15,000 kg truck is a four-axle, cross-country vehicle based on a commercial design but modified to meet military requirements. The forward control cab has seating for the driver and two passengers, the centre seat tilting forward to act as a platform for access to a roof hatch. The cab can be tilted forward 70° under hydraulic control to permit engine access. Routine checking access is via flaps.

Power is provided by a 503 hp V-8 turbocharged diesel coupled to a ZF 16-speed transmission. There are differential locks on all axles plus three inter-axle locks. Cross-country mobility can be enhanced by the provision of a tyre pressure regulation system which can be operated when travelling.

Numerous types of body can be utilised, including flatrack load-handling systems. The 3550 A has been configured as a bridge component carrier. Special versions for desert operations can be provided.

Specifications

Cab seating: 1 + 2
Configuration: 8 × 8
Weights:
(kerb) approx 18,000 kg
(GVW) 35,000 kg
(front axles) 2 × 8,000 kg
(rear axles) 2 × 10,000 kg
Max load: approx 15,000 kg
Load area: 6.1 × 2.5 m
Length: 9.635 m
Width: 2.5 m
Height: (overall) 3.425 m
Ground clearance: 0.462 m
Track: (front) 2.08 m
Wheelbase: 1.77 m + 3.4 m + 1.45 m
Max speed: 93 km/h
Fuel capacity: 400 litres
Gradient: 70%
Fording: 1.2 m
Engine: Mercedes-Benz OM 442 LA 14.618 litre V-8 water-cooled intercooled diesel developing 503 hp at 2,100 rpm
Transmission: ZF 16 S 220 WSK providing 16 forward and 2 reverse gears
Transfer box: VG 2400 2-speed
Turning circle: approx 27.3 m
Brakes: dual-circuit, compressed air
Tyres: 16.00 R 20
Electrical system: 24 V
Batteries: 2 × 12 V 115 or 165 Ah
Alternator: 55 or 100 A

Status

In production.

Manufacturer

Mercedes-Benz AG, Department VN/RM - Z503, D-70322 Stuttgart, Germany.
Tel: +49 711 179 0326. Fax: +49 711 179 0934.

NEW ENTRY

Mercedes-Benz 3550 A (8 × 8) 15,000 kg truck with load handling system **1997**

GREECE

ELBO 14 ME 14 (4 × 4) 8,000 kg truck

Description

The ELBO 14 ME 14 (4 × 4) 8,000 kg truck is produced in Greece under licence from Steyr Nutzfahrzeuge AG of Austria.

The overall layout closely follows that of the Steyr 14 M 14 original and thus involves a forward control tiltable cab and the superstructure area to the rear. There are some changes from the original to suit Greek requirements. The cab has seating for the driver and one passenger.

Body options produced in Greece include an 8 tonne general cargo version with steel sides and tailgate with folding benches, all covered by with removable bows and a tarpaulin cover. Other bodies include a 5,000 litre fuel or water tanker, a 5,000 litre water-foam firefighting body, an 8 m^3 refuse body, a 6 tonne sewage treatment body, an 8 tonne container carrier, mobile workshops and so on.

Options include a right-hand drive version, a 5,000 kg capacity hydraulic winch with front or rear pull, a driver training cab, and a two-passenger cab seat.

Specifications

Cab seating: 1 + 1 or 1 + 2
Configuration: 4 × 4
Weight:
(empty, general cargo body) 6,000 kg
(loaded) 14,000 kg
Max front axle load: 5,000 kg
Max rear axle load: 9,200 kg
Max load: 8,000 kg
Towed load:
(on-road) 10,000 kg
(off-road) 5,000 kg
Load area: 3.99 × 2.28 m
Length: (with general cargo body) 6.56 m
Width: 2.4 m
Height: 3.15 m
Ground clearance: 0.28 m
Track:
(front) 1.879 m
(rear) 1.721 m
Wheelbase: 3.7 m
Angle of approach/departure: 38°/35°
Max speed: 80 km/h
Range: 800 km
Fuel capacity: 200 litres
Max gradient:
(without trailer) 55%
(with trailer) 40%
Side slope: 30°
Fording: 0.8 m
Engine: ELBO/Steyr WD 612.91 6.6 litre 6-cylinder direct injection water-cooled diesel developing 136 hp at 2,400 rpm
Gearbox: ZF S6-36 with 6 forward and 1 reverse gears
Clutch: single dry plate
Transfer box: Steyr VG 450, 2-speed
Steering: ZF 8043, power-assisted

ELBO 14 ME 14 (4 × 4) 8,000 kg general cargo truck

ELBO 14 ME 14 (4 × 4) 8 m³ refuse truck

Turning radius: 7.6 m
Suspension: semi-elliptic leaf springs front and rear with telescopic shock-absorbers at front
Tyres: 9.00 × 20
Brakes:
(main) dual-circuit hydropneumatic
(parking) spring-actuated, air-operated
Electrical system: 24 V
Batteries: 2 × 12 V, 110 Ah

Status
In production. In service with the Greek armed forces.

Manufacturers
ELBO SA, Industrial Area Sindos, PO Box 10239, GR-570 22 Thessaloniki, Greece.
Tel: +30 31 798502. Telex: 412586 elbo gr.
Fax: +30 31 798426.

UPDATED

ELBO 26 S 28 P38 (6 × 6) 20,000 kg general cargo train

Description
The chassis for the ELBO 26 S 28 P38 (6 × 6) truck is produced in Greece under licence from Steyr Nutzfahrzeuge AG of Austria. The trailer and the general cargo bodies were developed and produced in Greece.

Layout is conventional with a forward control tiltable cab and the general cargo body to the rear. The general cargo body is all steel with three hinged side doors along each side, a tailgate, and removable bows and a tarpaulin cover. The cab has seating for the driver and a co-driver. Various bodies other than the general cargo can be fitted.

Options include a 9,000 kg capacity hydraulic winch with a front or rear pull, and a second co-driver's seat.

Variants
ELBO 26 S 28 K35 (6 × 6) 8 m³ Dump Truck
This vehicle chassis is closely allied to the general cargo truck described above but is fitted with a rear-tipping 8 m³ dump body. Other bodies, such as a 10 tonne general cargo body, can be fitted to this chassis.

ELBO 26 S 31 S38 (6 × 6) Tractor Truck
See entry under Greece in *Tank Transporters* section.

Specifications
Cab seating: 1 + 1 or 1 + 2
Configuration: 6 × 6
Weight:
(empty, general cargo body) 13,000 kg
(loaded) 32,000 kg
Max front axle load: 6,500 kg
Max rear axle load: 2 × 13,000 kg
Max train weight: 38,000 kg
Towed load:
(on-road) 12,000 kg
(off-road) 6,000 kg
Load area: 4.6 × 2.3 m
Length: (general cargo train) 17 m
Width: 2.5 m
Height: 3.13 m
Ground clearance: 0.3 m
Track:
(front) 1.958 m
(rear) 1.8 m
Wheelbase: 3.8 m + 1.35 m
Angle of approach/departure: 30°/40°
Max speed: 80 km/h
Range: 500 km
Fuel capacity: 400 litres
Max gradient:
(without trailer) 70%
(with trailer) 60%
Side slope: 30°
Fording: 0.8 m
Engine: Steyr WD 615.73 9.73 litre 6-cylinder direct injection turbocharged with intercooler water-cooled diesel developing 278 hp at 2,200 rpm
Gearbox: ZF 5S111GP with 8 forward plus 1 crawler and 1 reverse gears
Clutch: single dry plate
Transfer box: Steyr VG 1200, 2-speed
Steering: ZF 8043, power-assisted
Turning radius: 9.3 m
Suspension:
(front) semi-elliptic reinforced leaf springs with telescopic shock-absorbers
(rear) semi-elliptic reinforced reversed leaf springs with torsion bars
Tyres: 12.00 R 20
Brakes:
(main) dual-circuit air
(parking) spring-actuated, air-operated
Electrical system: 24 V
Batteries: 2 × 12 V, 143 Ah

Status
In production. In service with the Greek armed forces.

Manufacturers
ELBO SA, Industrial Area Sindos, PO Box 10239, GR-570 22 Thessaloniki, Greece.
Tel: +30 31 798502. Telex: 412586 elbo gr.
Fax: +30 31 798426.

VERIFIED

ELBO 26 S 28 P38 (6 × 6) 20,000 kg general cargo train

ELBO 26 S 28 K35 (6 × 6) 8 m³ dump truck

HUNGARY

Csepel D-594.00 (4 × 4) 3,000 kg truck

Description

The Csepel D-594.00 (4 × 4) 3,000 kg truck is the 4 × 4 component in the Csepel Toldi series.

The D-594.00 has a 3,000 kg payload across country and, in its standard form, has a forward control cab with single-row seating for the driver and two passengers. The cab is of the tilting type while the standard cargo body is constructed using light steel, with folding sides, a tailgate, and a tarpaulin over bows. The chassis is of the ladder type with riveted construction.

The engine is a Cummins B 5,9-180 delivering 180 hp and coupled to a Csepel gearbox. Power steering is provided.

Variant

The Csepel D-592.00 is a 4 × 2 variant with the same 3.5 m wheelbase but a longer overall length at 6.455 m and dual rear tyres each side. On-road payload is 6,000 kg.

Csepel D-594.00 (4 × 4) 3,000 kg truck

Specifications

Cab seating: 1 + 2
Configuration: 4 × 4
Weight:
(kerb) 6,000 kg
(GVW) 11,000 kg
Weight on front axle, loaded: 4,000 kg
Weight on rear axle, loaded: 7,000 kg
Max load:
(road) 5,000 kg
(cross-country) 3,000 kg
Length: 6.05 m
Width: 2.446 m
Height: (overall) 2.802 m
Track: 1.86 m
Wheelbase: 3.5 m
Max speed: 98.2 km/h
Fuel capacity: 200 litres
Range: 750 km
Max gradient: 51.7%
Engine: Cummins B 5,9-180 6-cylinder in-line diesel developing 180 hp
Gearbox: Csepel S5-70U with 5 forward and 1 reverse gears
Clutch: Csepel 082.43
Transfer box: TAM 5560784 2-speed
Steering: hydraulic power-assisted
Turning circle: 17.28 m
Brakes: combined air/hydraulic, dual-circuit
Electrical system: 24 V

Status

In production.

Manufacturer

Csepel Automotive Works, PO Box 38, H-2311 Szigetszentmiklós, Hungary.
Tel: +36 24 277 5303. Telex: 22 40 47.
Fax: +36 24 367311.

VERIFIED

Csepel D-562 (4 × 2), D-564 (4 × 4), D-566 (6 × 6), D-566.02 (6 × 6) and D-588 (8 × 8) trucks

Description

The D-566 (6 × 6) cargo truck appeared in 1970 and entered production the following year. It has a two-door all-steel forward control cab, with the cargo area at the rear with a drop tailgate, bows and a tarpaulin cover. A central tyre pressure regulation system is fitted as standard. The D-566.02 is identical to the D-566 but has a 7,000 kg capacity winch with 60 m of cable.

The D-562 (4 × 2) and D-564 (4 × 4) vehicles were introduced in 1972, with first production vehicles being completed in 1974. These also have a two-door all-steel forward control cab, with the cargo area at the rear with a drop tailgate, drop sides, a tarpaulin cover and bows. Both are provided with a central tyre pressure regulation system. A dump truck model was produced.

The D-588 was an 8 × 8 truck with a forward control cab and a central tyre pressure regulation system.

Variants of the D-566 include a recovery vehicle and a shop/van.

Status

The D-566 is in service with the Hungarian Army.

Manufacturer

Csepel Automotive Works, PO Box 38, H-2311 Szigetszentmiklós, Hungary.
Tel: +36 24 277 5303. Telex: 22 40 47.
Fax: +36 24 367311.

VERIFIED

Specifications

Model	D-562	D-564	D-566
Configuration	4 × 2	4 × 4	6 × 6
Weight			
(empty)	5,200 kg	5,400 kg	9,000 kg
(loaded, cross-country)	9,200 kg	9,400 kg	14,000 kg
Load			
(road)	5,000 kg	5,000 kg	8,000 kg
(cross-country)	4,000 kg	4,000 kg	5,000 kg
Towed load			
(road)	n/a	n/a	10,000 kg
(cross-country)	n/a	n/a	5,000 kg
Length	6.43 m	6.43 m	7.18 m
Width	2.5 m	2.5 m	2.5 m
Height	2.67 m	2.67 m	2.74 m
Ground clearance	n/a	n/a	0.575 m
Track	2 m	2 m	2.05 m
Wheelbase	3.7 m	3.7 m	2.9 m + 1.4 m
Max speed	87.9 km/h	87.9 km/h	80 km/h
Range	n/a	n/a	700 km
Fording	n/a	n/a	1.2 m
Engine model	D-614.33	D-614.33	RÁBA-MAN
Engine type	6-cylinder water-cooled diesel	6-cylinder water-cooled diesel	6-cylinder water-cooled diesel
Hp/rpm	145/2,300	145/2,300	200/2,200
Gearbox	n/a	n/a	manual, 5 forward and 1 reverse gears
Transfer box	none	n/a	2-speed
Tyres	16.5 × 19.5	16.5 × 19.5	14.00 × 20
Electrical system	24 V	24 V	24 V
Batteries	n/a	n/a	2 × 120 Ah
Generator	n/a	n/a	750 W

Csepel D-566 (6 × 6) recovery vehicle

Csepel D-566 (6 × 6) shop/van

INDIA

Azad (4 × 4) 2,500 kg truck

Description

At least four prototypes of the Azad (4 × 4) 2,500 kg truck were produced. The truck has a forward-mounted steel cab with seating for a driver and at least one passenger and it would appear that the cab can be tilted forward for access to the engine and gearbox. The cab roof appears to have hatches for access to a machine gun roof mounting and the radiator grill is protected by a steel guard mounted on the front bumper; this guard also protects the headlights. The cargo area to the rear can be covered by a canvas tilt and the side and rear flaps either fold downwards or can be removed to enable the Azad to carry a variety of loads, including container bodies. Apart from use as a cargo truck the Azad can be used as a personnel carrier, a tractor for light artillery field pieces, a tanker (fuel or water) or a missile launcher vehicle.

Specifications

Cab seating: 1 + 1 or 2
Configuration: 4 × 4
Weight: (loaded, max) 7,500 kg
Max load: 2,500 kg
Length: 5.4 m
Width: 2.25 m
Height: (cab roof) 2.45 m
Ground clearance: 0.32 m
Max speed: 84 km/h
Range: 800 km
Gradient: 60°
Fording: 0.8 m
Engine: Perkins P6.354 water-cooled diesel developing 120 hp at 2,800 rpm
Gearbox: ZF AK 5-35 with 5 forward and 1 reverse gears
Transfer box: 2-speed
Steering: power-assisted

Status

Prototypes.

VERIFIED

Shaktiman (4 × 4) 4,000 kg truck

Description

The Shaktiman 4,000 kg (4 × 4) truck is basically the German MAN 415 L1 AR truck built under licence in India. The main visual difference between the German and Indian vehicle is that the latter has dual rather than single rear wheels. The vehicle entered production at Jabalpur in 1958 and was initially assembled from components supplied by Germany, but as production built up an increasing number of the components were supplied by Indian companies.

The Shaktiman is available in both 4 × 4 and 4 × 2 versions and they can be produced in either hard or soft top form. Also available is a version with a deeper chassis frame, a heavy-duty suspension and larger tyres for a GVW of 11,000 kg. An artillery tractor version with a separate crew shelter behind the driver's cab is used by the Indian Army to tow 105 mm field guns.

Shaktiman (4 × 4) 4,000 kg truck (T J Gander)

Specifications

Cab seating: 1 + 2
Configuration: 4 × 4 or 4 × 2
Weight:
(empty) 5,075 kg
(GVW) 9,000 kg
Max load:
(front axle) 3,300 kg
(rear axle) 6,600 kg
Load area: 4.45 × 2.25 m
Length: 7.16 m
Width: 2.35 m
Height:
(top of cab, loaded) 2.435 m
(load area) 1.13 m
Ground clearance: 0.308 m
Track:
(front) 1.824 m
(rear) 1.632 m
Wheelbase: 4.2 m
Fuel capacity: 300 litres
Engine: D 0026 M 8 A 5.88 litre 6-cylinder in-line diesel developing 110 hp at 2,500 rpm
Gearbox: AK5-35 manual with 5 forward and 1 reverse gears
Clutch: single dry plate
Transfer box: 2-speed
Steering: heavy-duty roller
Turning radius: 8.075 m
Suspension: semi-elliptical springs front and rear
Tyres: 8.25 × 20
Brakes:
(main) hydropneumatic dual-circuit
(parking) mechanical
Electrical system: 12 V
Batteries: 2 × 12 V, 92 Ah
Alternator: 40 A

Status

In production. In service with India and Bangladesh.

Manufacturer

Ordnance Factory Board, 10-A Auckland Road, Calcutta 700 001, India.

UPDATED

INTERNATIONAL

Percheron (6 × 6) 10,000 kg Truck Heavy Logistic Vehicle Wheeled (HLVW)

Description

In March 1988 the Canadian Department of Defence signed a contract with UTDC Inc of Kingston, Ontario to produce the Heavy Logistic Vehicle Wheeled (HLVW) for the Canadian Forces.

The UTDC submission was an Austrian vehicle, the Steyr 1491.6 × 6 M. (For full details of this vehicle refer to the entry under Austria in this section.) Other submissions came from MAN, Oshkosh (which entered a 6 × 6 version of its 8 × 8 HEMTT) and Scania (the SBAT 111).

The HLVW is based on the Steyr 1491.310/040/6 × 6 M and in Canadian service is known as the Percheron. The Percheron differs from the original vehicle in several respects. These include a camouflage net storage space behind the cab, a cold weather start capability, a brush guard, a cargo bay walkway and a radio installation kit.

There are eight variants of the basic Percheron, produced in the following numbers:

- Basic cargo - 276
- Basic cargo with winch - 147
- Basic cargo with winch and crane - 388
- Recovery - 124
- Tractor - 46
- Dumper - 25
- Bridge transport/flatrack system - 176
- Heavy Mobile Repair Team - 18.

The three cargo variants are essentially similar and have dropping sides and a rear tailgate. The version with the material-handling crane has its cargo area reduced in length from 6 to 5 m.

The recovery variant is a combination of the Vulcan recovery unit and the Pitman PK 19000 material-handling crane, forming an integrated system. The Vulcan recovery unit consists of a heavy-duty hydraulic lift and carry system at the rear together with two stabilising and anchoring legs. There is also a recovery winch with a capacity of 25,000 kg. A special bogie-type suspension is used with this variant.

The dumper has a three-way tipper body with a capacity of 8,500 kg.

The tractor can tow loads of up to 25,000 kg with a 12,000 kg fifth wheel capacity. The rear suspension is of the bogie type.

The bridge transporter is used to carry Ribbon Bridge components and is equipped with a Multilift Mark 4 load handling system. The system can also be used to carry cargo flatracks.

The Heavy Mobile Repair Team (HMRT) vehicle is used to provide forward maintenance support for the Canadian Forces' Leopard C1 MBT and M109 155 mm self-propelled howitzer units. A Pitman PK 19000 material-handling crane is mounted at the rear. This

Percheron Heavy Logistic Vehicle Wheeled (HLVW) with load handling system (R Stickland) *1997*

Basic cargo version of the Heavy Logistic Vehicle Wheeled (HLVW), the Percheron

crane has a lifting capacity of 19,000 kg and can lift a Leopard C1 engine pack. The HMRT can carry major assemblies and also carries a workshop area.

Specifications

Cab seating: 1 + 2
Configuration: 6 × 6
Weight:
(kerb) 9,700 kg
(GVW) 22,000 kg
Max load:
(without winch) 10,000 kg
(with winch) 9,600 kg
Length: 9.12 m
Width: 2.5 m
Height: (cab) 3.025 m
Ground clearance: (axles) 0.37 m
Track: 2.072 m
Wheelbase: 4 m + 1.4 m
Angle of approach/departure: 31°/35°
Max speed: 95 km/h
Range: 1,000 km
Fuel capacity: 400 litres
Max gradient: 100%
Fording: 0.8 m
Engine: Steyr model WD 615.98m 9.7 litre 6-cylinder in-line direct injection water-cooled turbocharged diesel developing 306 bhp at 2,300 rpm
Transmission: ZF WSK 400 +4 S 150 GP hydrodynamic torque converter with single dry disc, synchromesh gearbox with 8 forward and 1 reverse gears
Transfer box: Steyr VG1200 2-speed
Steering: ZF 80 46 hydraulic power-assisted
Turning radius: 10.5 m
Suspension: semi-elliptic leaf springs with double-acting hydraulic shock-absorbers
Tyres: 14 × 20
Brakes:
(main) air, dual-circuit
(parking) spring-loaded on both rear axles
Electrical system: 24 V
Batteries: 4 × 12 V, 100 Ah

Status

Production for the Canadian Forces completed November 1991 (1,200 units).

Manufacturer

UTDC Inc, PO Box 70, Station A, Kingston, Ontario, Canada K7M 6P9.
Tel: +1 613 384 3100. Fax: +1 613 389 6382.
This concern is no longer trading.

VERIFIED

IRELAND

Timoney (8 × 8) high-mobility truck

Development

The Timoney (8 × 8) high-mobility vehicle is a front-engined version of the Timoney airport rescue and firefighting chassis in use with the US Air Force. The 8 × 8 truck is a demonstrator for a range of high-mobility trucks employing the Timoney Independent Suspension System, also manufactured under licence by Rockwell. Trucks with configurations from 4 × 2 up to 10 × 10 are possible with axle capacities of from 7,000 to 15,000 kg or even higher, and with wheel movements of 350 to 500 mm, depending on application.

Description

The Timoney (8 × 8) high-mobility vehicle is based on a torsionally rigid chassis frame with straight members bolted to the structural axle housings. The Timoney Independent Suspension System is fitted front and rear with twin variable rate coil springs and shock-absorbers. Four Rockwell Timoney Series 5000 independent suspension axles are provided with electropneumatically operated cross differential locks front and rear. Axle capacity on the demonstrator is 10,000 kg.

The cab provided on the demonstrator model is of the Volvo FL12 forward control type. A Volvo 12 litre diesel provides 420 hp and is coupled to a Volvo VT1906 automatic six-speed transmission. All-wheel steering is an option.

Various body options are possible, including load-handling systems. Options include 24.00 R 20 tyres, a central tyre inflation system, an antilocking brake system, a power take-off, a manual gearbox, and hot weather cooling.

Timoney (8 × 8) high-mobility truck chassis *1997*

Specifications

Chassis only
Cab seating: 1 + 1 or 1 + 2
Configuration: 8 × 8
Weight:
(kerb) 13,360 kg
(GVW) 32,000 kg
Length: 9.155 m
Width: (over wheels) 2.5 m
Height: (cab roof) 3.2 m
Track: (front and rear) 2.052 m
Wheelbase: 1.524 m + 3.326 m + 1.524 m
Angle of approach/departure: 36°/40°
Max speed: (governed) 90 km/h
Fuel capacity: 205 litres
Engine: Volvo D12A4420 12.1 litre 6-cylinder diesel developing 420 hp at 1,900 rpm
Transmission: Volvo VT1906 automatic with 6 forward and 2 reverse gears
Transfer box: Steyr VG 2001/396, 2-speed
Steering: ZF power-assisted
Turning radius: 13 m
Suspension: Timoney Independent System front and rear with twin variable rate coil springs and shock-absorbers
Tyres: 16.00 R 20 XZL
Brakes: Rockwell RADB 2000 reaction beam type disc
Electrical system: 24 V
Alternator: 80 A

Status

Available.

Manufacturers

Timoney High Mobility Vehicles, Gibbstown, Navan, Co. Meath, Ireland.
Tel: +353 46 54183. Fax: +353 46 54424.

NEW ENTRY

ISRAEL

M-462 ABIR (4 × 4) 1,800 kg multipurpose tactical truck

Description

The M-462 ABIR (4 × 4) 1,800 kg multipurpose tactical truck was designed and developed using the considerable experience gained by Automotive Industries Limited during its involvement with the M-325 Commandcar series of 4 × 4 trucks (see following entry).

The layout of the ABIR is conventional but many new features have been incorporated to improve crew comfort, the smoothness of the ride and enhance all-round visibility. The cab is a modern design and can comfortably accommodate the driver and two passengers. The standard cargo body can seat up to 12 passengers under a canvas tilt.

Fully automatic three-speed transmission and integral power steering are provided. Maintenance costs and time are stated to be low.

The standard version of the ABIR can have a canvas top to the cab and a canvas tilt over the cargo area. The cab doors are also canvas and can be readily removed. Metal doors are available as an option, together with a hardtop cabin cover and air conditioning.

Other versions of the ABIR are: cargo version with roll-bars; version with a shortened cargo area; reconnaissance vehicle armed with machine guns at front and sides; TOW ATGW missile carrier; mine-shielded troop carrier; communication centre shelter carrier; command, communications, control and intelligence (C³I) vehicle; police vehicle with the rear area enclosed with transparent polycarbonate screens; and maxi and mini ambulances. There are also fire tender versions. The Model W-1000 is a special forest firefighting vehicle.

Specifications

Cab seating: 1 + 2
Configuration: 4 × 4
Weight:
(empty) 2,610 kg
(loaded) 4,500 kg
Max load: 1,800 kg
Towed load: 1,800 kg
Length: 5.12 m
Width: 2.04 m
Height: 2.01 m
Ground clearance: 0.26 m
Wheelbase: 3.2 m
Track: 1.715 m
Angle of approach/departure: 52°/30°
Max speed: 110 km/h
Range: 600 km
Fuel capacity: 144 litres
Max gradient: 78%
Side slope: 60%
Vertical obstacle: 0.45 m
Fording: 0.76 m
Engine: GM 6.2 litre water-cooled diesel developing 145 hp at 3,600 rpm or optional Chrysler 318 5.2 litre V-8 petrol
Transmission: automatic with 3 forward and 1 reverse speeds or manual with 4 forward and 1 reverse gears or NVG 4500 manual with 5 forward and 1 reverse gears
Transfer box: NP 241 2-speed
Steering: integral power steering
Turning radius: 7.1 m
Suspension:
(front) semi-elliptic multileaf (11 leaves) plus shock-absorbers
(rear) semi-elliptic multileaf plus shock-absorbers
Tyres: 9.00 × 16 - 10 ply NATO tread
Brakes: hydro/booster power (vacuum booster for petrol engines); rear drum self-aligning, front disc
Electrical system: 24 V
Batteries: 2 × 12 V, 100 Ah
Alternator: 60 A (100 A optional)

Status

In production. In service with the Israel Defence Forces and some other nations.

Manufacturer

Automotive Industries Limited, PO Box 535, IL-17105 Nazareth Illit, Israel.
Tel: 972 6 558111. Telex: 46217 ail il.
Fax: 972 6 558103.

VERIFIED

M-462 ABIR (4 × 4) 1,800 kg multipurpose tactical vehicle fitted with cargo body

Mine-shielded version of M-462 ABIR (4 × 4) 1,800 kg multipurpose tactical vehicle

1996

M-325 Commandcar (4 × 4) 1,800 kg truck

Description

The first M-325 Commandcars were constructed during 1966 and have remained in production ever since. Although the basic design remains unchanged some innovations have been introduced and several variants now exist. Commandcars are produced for export as well as for use by the Israel Defence Forces.

The M-325 Commandcar is produced using mainly commercial components allied to a locally developed chassis design. The basis is a chassis using a ladder-type frame with a reinforced drop centre section channel side rail. Cross members are bolted or riveted to the frame. Carbon steel with a minimum yield strength of 2,200 kg/cm is used for the side rails. Dana Spicer axles are used front and rear and a dual-hydraulic brake system using Bendix disc brakes is employed. The front-mounted engine is a Chrysler in-line petrol OHV 225-2 LC. A Borg & Beck clutch system connects power to a four-speed Model 435D transmission system and NP-200-D transfer case. This has four forward gears and one reverse. There is a New Process two-speed transfer case. A Chelsea 350 series power take-off is located on the left side. The propeller shaft is a Chrysler 5380. Steering is carried out by a Ross 378 worm and roller steering gearbox.

The bonnet and cab are standard for nearly all versions but the cab may be of the open type with a simple canvas hood for weather protection or of the enclosed type for the ambulance and some export versions. Recent changes to the cab involve a more comfortable seat for the driver and a revised 'through the steering wheel' layout for the instrument panel. Most M-325s have a single-piece windscreen that can be folded forward over the bonnet but some variants use a two-part split windscreen so that a machine gun can be mounted for forward use by the passenger. Night driving lights can be fitted and forward of the front bumper a front-mounted winch may be fitted. This is a Braden AMSU3-10F with a capacity of 4,500 kg and 76 m of 11 mm cable. A Donaldson Cyclopac air cleaner is a virtual standard for Middle East operations.

The cargo body is a multipurpose cargo and personnel carrier with an all-steel, one-piece body with two longitudinal benches facing inwards and containing lockable equipment compartments. For patrol and reconnaissance versions the benches are located centrally and face outwards. Machine gun and other weapon pintles can be mounted on the sides along with searchlights for night use. A towing hook at the rear allows the M-325 to tow loads up to 1,500 kg, including weapons such as 120 mm mortars.

There is a 24 V electrical system and a distribution panel powerful enough for up to four radios is fitted. Two batteries are located in compartments under the cargo area and may be pulled out on rails for maintenance, whereas earlier models had the batteries behind the cab.

Variants

Basic cargo and personnel carrier

Bench seats are provided for up to 12 passengers in the cargo area. The tilt is light PVC-coated tarpaulin. Weapons may be carried and the towing hook can be used to tow weapons such as a 120 mm mortar or a special trailer known as the MT-100 (see entry in the *Trailers* section for details). This version can be used to carry a communications shelter; for this role a 100 A alternator is used in place of the normal 60 A item.

Patrol and reconnaissance version

This normally uses the Type B body with weapon and searchlight pintles along the sides of the rear area. This version may be used to carry a 1,300 litre tank that can be fitted onto the cargo area for carrying fuel or water.

Mine-shielded version

This is a variation of the Type B-bodied patrol and reconnaissance vehicle with the cab and body constructed of heavy-duty metal plates and with the cargo body floor strengthened by honeycomb partitions between the floor plates. There are fitted seats for eight personnel in the rear. The cab and sides are left open and the spare wheel is relocated to the rear. Large roll-bars are fitted over the cab and at the rear. The seats can be removed easily for carrying cargo.

Ambulance

This uses a hard cab and body with access from the cab into the ambulance body. Folding racks for up to four stretchers are provided and up to 12 seated casualties may be carried. A folding seat is provided for an attendant. Ventilation is either a three-speed blower system or a full air conditioning unit.

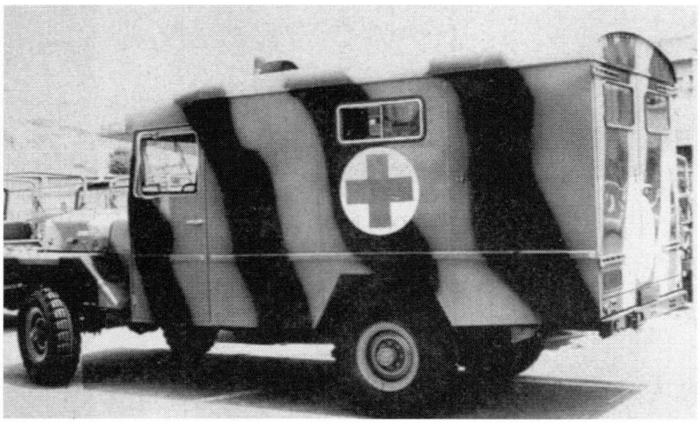

Ambulance version of M-325 Commandcar (T J Gander)

M-325 Commandcar with Type B body and machine gun mounted in split front windscreen (T J Gander)

Box body
A solid body similar to that of the ambulance may be used as a mobile workshop, a mobile office or for other similar roles.

Specifications
Cargo version, Type A body
Cab seating: 1 + 1 (up to 12 in rear)
Configuration: 4 × 4
Weight: (loaded) 4,500 kg
Max load: 1,950 kg
Towed load: 1,500 kg
Length: (without winch) 5.073 m
Width: 2.08 m
Height: (top of tilt) 2.37 m
Ground clearance: 0.4 m
Track: 1.7145 m
Wheelbase: 3.2 m
Angle of approach/departure:
(without winch) 62°/33.5°
(with winch) 45°/33.5°
Max speed: 100 km/h
Fuel capacity: 144 litres
Max gradient: 73%
Side slope: 70%
Vertical obstacle: 0.45 m
Fording: 0.76 m
Engine: Chrysler 3.687 litre 6-cylinder in-line OHV 225-2 LC petrol developing 100 hp at 3,600 rpm
Gearbox: manual with 4 forward and 1 reverse gears
Transfer box: NP-200-D 2-speed
Steering: worm and roller, manual
Turning radius: 7.1 m
Suspension: front and rear leaf springs with double-acting shock-absorbers
Tyres: 9.00 × 16 - 10 ply
Brakes: hydraulic disc, dual-circuit
Electrical system: 24 V
Batteries: 2 × 12 V 100 Ah
Alternator: 28 V, 60 or 100 Ah

Status
In production. In service with the Israel Defence Forces and some other nations.

Manufacturer
Automotive Industries Limited, PO Box 535, IL-17105 Nazareth Illit, Israel.
Tel: +972 6 558111. Telex: 46217 ail il.
Fax: +972 6 558103.

VERIFIED

ITALY

IVECO 40.10 WM (4 × 4) 1,500 kg light truck

Development
The IVECO 40.10 WM (4 × 4) light truck is a military derivative of the commercial Daily model and was developed via the interim 40 PM. The prototypes were produced at IVECO's Brescia factory with the intention of providing a light truck for airborne and other similar formations. Production vehicles are manufactured at Bolzano. In its current form the vehicle has a payload of 1,500 kg but a gross vehicle weight of 4,300 kg, making it ideally suited for many tactical roles as a troop and cargo carrier. It is suitable for mounting various forms of weapon including machine guns, anti-tank weapons and rocket launchers, and can be used as a tractor for light artillery.

In October 1991 it was announced that the Canadian Forces had selected the IVECO 40.10 WM (4 × 4) light vehicle as the winner of a competition for the Light Support Vehicle Wheeled (LSVW) project. For details refer to separate entry under Canada in this section.

IVECO 40.10 WM (4 × 4) 1,500 kg light truck towing 120 mm mortar (T J Gander) ***1997***

Description
The IVECO 40.10 WM has a conventional layout, with the engine at the front and the cargo area to the rear. The semi-forward cab has seating for the driver and two passengers and is normally provided with a soft canvas top; a hardtop is optional. The all-steel cargo area has folding seats along each side facing inwards for 10 personnel. The tailgate folds downwards and when down the loading height is 0.98 m.

The bonnet may be removed for engine access and maintenance with routine maintenance operations designed to be kept to a minimum. The Model 8142 diesel engine is supercharged and is a military version of a well-established commercial model. The transfer box is of the two ratio chain-type and is controlled by a single lever. The axles are of the single reduction type front and rear, both with locking differentials. Hydraulic power steering is standard. Left- and right-hand steering versions are available.

The standard truck can accommodate an S-250 container shelter at the rear and various other body configurations are possible. A van body is offered as is a four stretcher ambulance body. The same body may also be used as a command and communications centre, and various weapons, including a 106 mm recoilless rifle, may be carried in the cargo area. The standard truck can be used to tow a 105 mm light artillery piece or 120 mm mortar. Optional extras include a 2,000 kg capacity front winch. Two vehicles can be carried in a C-130 Hercules and a vehicle, stripped down, can be carried with an OTOBreda 105 mm Pack Howitzer in an Alenia G222 transport aircraft.

Specifications
Cab seating: 1 + 2 (10 in rear)
Configuration: 4 × 4
Weight:
(empty) 2,900 kg
(max) 4,400 kg
Max load: 1,500 kg
Towed load: (off-road) 1,500 kg
Load area: 2.075 x 1.8 m
Length: 4.645 m
Width: 2 m
Height:
(top of tilt) 2.38 m
(top of cab) 2.24 m
(load area) 0.98 m
Track: 1.67 m
Wheelbase: 2.8 m
Angle of approach/departure: 45°/40°
Max speed: over 100 km/h
Max gradient: over 60%
Side slope: over 30%
Range: over 500 km
Fuel capacity: 70 litres
Fording: 0.7 m
Engine: 8142 2.5 litre turbocharged 4-cylinder water-cooled diesel developing 103 hp at 3,800 rpm
Gearbox: manual with 5 forward and 1 reverse gears
Clutch: single dry plate
Transfer box: 2-speed

IVECO 40.10 WM (4 × 4) ambulance

IVECO 40.10 WM (4 × 4) with mobile workshop body

Steering: hydraulic power-assisted
Turning radius: 6 m
Suspension: independent longitudinal torsion bars at front, leaf spring at rear, telescopic shock-absorbers front and rear
Tyres: 9.00 × 16 14PR
Brakes:
(main) disc front, drum rear, split circuit
(parking) mechanical on rear wheels
Electrical system: 24 V
Batteries: 2 × 12 V, 55 Ah
Alternator: 28 V, 30A

Status
In production. In service with Italy (1,300), Pakistan (2,200) and the Belgian Gendarmerie, Canada (approx 2,750), Netherlands, Portugal and several other countries. Also serving with EC organisations operating in the former Yugoslavia.

Manufacturer
IVECO SpA, Defence Vehicles Division, Via Volta 6, I-39100 Bolzano, Italy.
Tel: +39 471 905111. Telex: 400541.
Fax: +39 471 905444.

UPDATED

IVECO 75.14 WM (4 × 4) 2,500 - 3,000 kg cargo truck

Description
This vehicle is an improvement of the 75 PM 13 (4 × 4) 2,000 kg truck and 90 PM 16 (4 × 4) 4,000 kg truck, now in service with the Italian Army with the designation ACL/75 and ACM-80. The two vehicles have a high degree of standardisation.

The chassis consists of longitudinal members to which the cross members are cold-riveted. The cab is all steel and is of the forward control type. It can be tilted forward 55° to allow access to the engine. A circular observation hatch is provided in the roof of the cab on the passenger side. The rear cargo area has drop sides and a drop tailgate and, if required, bows and a tarpaulin cover can be installed.

Optional equipment includes a front- or rear-mounted winch with a capacity of 4,000 kg and an air-operated lockable front differential. Left- and right-hand drive versions are available.

The vehicle can be fitted with a wide range of different bodies and various other types of specialised equipment including aerial cage, bus body, crane, digger, fire appliance, mobile lubrication and greasing unit and a three-way tipper body.

Specifications
Cab seating: 1 + 1 (up to 12 in rear)
Configuration: 4 × 4
Weight:
(kerb) 5,250 kg
(loaded) 7,750-8,250 kg
Max load: 2,500-3,000 kg
Towed load: 4,000 kg
Load area: 3.01 × 2.195 m
Length: 5.178 m
Width: 2.3 m
Height:
(cab) 2.646 m
(tarpaulin) 2.96 m
(load area) 1.4 m
Ground clearance: 0.445 m
Track: 1.852 m
Wheelbase: 2.75 m
Angle of approach/departure: 45°/45°
Max speed: over 80 km/h
Range: over 500 km
Fuel capacity: 155 litres
Max gradient: over 60%
Max side slope: 30%
Fording: 0.7 m
Engine: Model 8060.05 5.861 litre 6-cylinder direct injection water-cooled diesel developing 139 bhp (DIN) at 3,000 rpm
Gearbox: manual with 5 forward and 1 reverse gears
Clutch: single dry plate
Transfer box: 2-speed
Steering: recirculating ball with hydraulic servo
Turning radius: 6 m
Suspension: leaf springs (dual at rear) with hydraulic telescopic shock-absorbers
Tyres: 12.50 × 20
Brakes:
(main) air-hydraulic drum on all wheels with separate circuits
(parking) on rear wheels
Electrical system: 24 V
Batteries: 2 × 12 V, 110 Ah
Alternator: 28 V, 30 A

IVECO 75-14 WM 14 (4 × 4) 2,500 kg truck

Status
In service with the Italian Army as the ACL/90, Somalia and several other countries.

Manufacturer
IVECO SpA, Defence Vehicles Division, Via Volta 6, I-39100 Bolzano, Italy.
Tel: +39 471 905111. Telex: 400541.
Fax: +39 471 905444.

VERIFIED

IVECO 90.17 WM (4 × 4) 4,000 kg cargo truck

Description
This vehicle is an improvement of the 90 PM 16 (4 × 4) 4,000 kg truck and the 75 PM 13 vehicle family in service with the Italian Army as the ACM-80 and the ACL/75. The two vehicles have a high degree of standardisation.

The chassis consists of longitudinal members to which the cross members are cold-riveted. The forward control cab is all steel and can be tilted forwards to an angle of 55° to allow access to the engine. There is a circular observation hatch in the roof of the cab on the right side. The rear cargo area has drop sides and a drop tailgate and if required bows and a tarpaulin cover can be installed.

Optional equipment includes a front- or rear-mounted winch with a capacity of 4,000 kg, an air-operated lockable front differential and 14.5 R 20 tyres.

Variants produced include a light recovery vehicle (see entry in the *Recovery Vehicles* section), a tanker, a mobile workshop, a light artillery tractor and a firefighting vehicle.

The IVECO M 110.18 WM (4 × 4) 5,000 kg truck is an essentially similar uprated model (see following entry).

Specifications
Cab seating: 1 + 1 (up to 18 in rear)
Configuration: 4 × 4

IVECO 90-17 WM (4 × 4) fuel tanker

IVECO 90-17 WM (4 × 4) firefighting vehicle

Weight:
(kerb) 5,740 kg
(loaded) 9,740 kg
Max load: (off-road) 4,000 kg
Towed load: (off-road) 4,000 kg
Load area: 4.19 × 2.21 m
Length: 6.358 m
Width: 2.3 m
Height:
(cab) 2.646 m
(tarpaulin) 2.96 m
(load area) 1.4 m
Ground clearance: 0.445 m
Track: 1.852 m
Wheelbase: 3.7 m
Angle of approach/departure: 45°/35°
Max speed: over 80 km/h
Range: over 500 km
Fuel capacity: 155 litres
Max gradient: over 60%
Side slope: over 30%
Fording: 0.7 m

Engine: Model 8060.25 5.861 litre 6-cylinder supercharged direct injection water-cooled diesel developing 170 hp (DIN) at 3,000 rpm
Gearbox: manual with 5 forward and 1 reverse gears
Clutch: single dry plate
Transfer box: 2-speed
Steering: recirculating ball with hydraulic servo
Turning radius: 7.5 m
Suspension: leaf springs (dual at rear) with hydraulic telescopic shock-absorbers
Tyres: 12.5 R 20 PR22
Brakes:
(main) air-hydraulic on all wheels with separate circuits
(parking) on rear wheels
Electrical system: 24 V
Batteries: 2 × 12 V, 110 Ah

Status
In production. In service with Italy (as ACM/90), Portugal, Singapore, Somalia (100) and several other countries.

Manufacturer
IVECO SpA, Defence Vehicle Division, Via Volta 6, I-39100 Bolzano, Italy.
Tel: +39 471 905111. Telex: 400541.
Fax: +39 471 905444.

UPDATED

IVECO 90-17 WM (4 × 4) 4,000 kg truck carrying workshop body

IVECO M 110.18 WM (4 × 4) 5,000 kg truck

Description
The IVECO M 110.18 WM (4 × 4) 5,000 kg truck may be regarded as an uprated and updated version of the IVECO 90.17 WM (see previous entry) with which it share many components and features.

The chassis consists of longitudinal members to which the cross members are cold-riveted. The forward control cab is all steel and can be tilted forwards to an angle of 55° to allow access to the engine. There is a circular observation hatch in the roof of the cab on the right side. The rear cargo area has drop sides and a drop tailgate and, if required, bows and a tarpaulin cover can be installed. Two wheelbases are available, 3.7 m and 4 m, with corresponding cargo area lengths.

Optional equipment includes a front- or rear-mounted winch. Variants include a tanker, a mobile workshop, a light artillery tractor and a firefighting vehicle.

Specifications
Data refers to 4 m wheelbase version
Cab seating: 1 + 1
Configuration: 4 × 4
Weight:
(kerb) 6,500 kg
(GVW) 11,500 kg
Max load: 5,000 kg
Towed load: 6,400 kg

IVECO M 110.18 WM (4 × 4) 5,000 kg truck

1997

Load area: 4.5 × 2.25 m
Length: 6.718 m
Width: 2.395 m
Height:
(cab) 2.7 m
(tarpaulin) 2.995 m

Ground clearance: 0.48 m
Wheelbase: 4 m
Angle of approach/departure: 45°/35°
Max speed: over 90 km/h
Range: over 500 km
Fuel capacity: 155 litres
Gradient: >60%
Engine: IVECO diesel developing 178 hp (DIN)
Gearbox: manual with 6 forward and 1 reverse gears
Clutch: single dry plate
Transfer box: 2-speed
Steering: recirculating ball with hydraulic servo
Suspension: leaf springs (dual at rear) with hydraulic telescopic shock-absorbers
Tyres: 14.5 R 20
Brakes:
(main) air-hydraulic on all wheels with separate circuits
(parking) on rear wheels
Electrical system: 24 V

Status
In production.

Manufacturer
IVECO SpA, Defence Vehicle Division, Via Volta 6, I-39100 Bolzano, Italy.
Tel: +39 471 905111. Telex: 400541.
Fax: +39 471 905444.

NEW ENTRY

Astra SM 44.30 (4 × 4) truck

Description
Developed to meet very exacting military requirements, the Astra SM 44.30 (4 × 4) truck is part of the new IVECO DVD range which also encompasses 6 × 6 and 8 × 8 designs. The SM 44.30 is designed to have a gross vehicle weight of 15 tonnes and is available with three possible wheelbase lengths; 3.5 m, 4.2 m and 4.8 m. Cross-country mobility is enhanced by the use of extra-large 'single track' tyres to provide high flotation over soft terrain and deserts; the provision of a central tyre inflation system is an option to further improve mobility over soft ground.

The forward control three-man cab is constructed of glass fibre to resist corrosion while the chassis frame is constructed using a special steel. Due to the cab outline the vehicle is rail-portable to STANAG 2832 requirements, while a tight turning circle is assured due to the front axle's wide joint spacing coupled with the advanced steering and suspension system design. All wheels have disc brakes and an Anti-lock Braking System (ABS) is provided as standard. Drum brakes are available as an option.

Specifications
Cab seating: 1 + 2
Configuration: 4 × 4
Weights:
(empty, 3.5 m WB) 7,880 kg
(empty, 4.2 m WB) 7,950 kg
(empty, 4.8 m WB) 8,050 kg
(GVW) 15,000 kg
Lengths:
(3.5 m WB) 6.325 m
(4.2 m WB) 7.425 m
(4.8 m WB) 8.025 m
Width: 2.5 m
Height: (cab) 2.87 m
Ground clearance: 0.43 m

Astra SM 44.30 (4 × 4) truck chassis ***1996***

Track:
(front) 2.05 m
(rear) 2.06 m
Wheelbase: 3.5 m, 4.2 m or 4.8 m
Angle of approach/departure: 45°/45°
Max road speed: 88 km/h
Range: 800 km
Fuel capacity: 300 litres
Max gradient: 100%
Side slope: 40%
Fording: 1.2 m
Engine: IVECO 8460 TCA 9.5 litre 8360 6-cylinder turbo intercooled diesel developing 300 hp at 2,200 rpm
Gearbox: ZF 5 HP 500 automatic with 6 forward and 1 reverse gears or
ZF 16 S 109 manual with 16 forward and 2 reverse gears
Transfer box: 2-speed
Steering: ZF Servocom Type 8098
Turning radius:
(3.5 m WB) 14.5 m
(4.2 m WB) 16.94 m
(4.8 m WB) 19.04 m
Suspension:
(front) semi-elliptic springs with 2 double-acting hydraulic shock-absorbers and bumpers
(rear) reversed semi-elliptical springs with centre pin and slider connection to axles, 4 lower reaction rods and 2 upper reaction triangles
Tyres: 14.00 R 20 (run-flat optional)
Brakes:
(main) hydraulic disc
(parking) on 2nd axle
Electrical system: 24 V
Batteries: 4 × 12 V, 110 Ah
Alternator: 80 A

Status
Ready for production.

Manufacturer
Astra Veicoli Industriali SpA, Via Caorsana 79, I-29100 Piacenza, Italy.
Tel: +39 523 5431. Telex: 530148 astra i.
Fax: +39 523 593293.
Astra Veicoli is part of the IVECO organisation.

UPDATED

IVECO 6602 CM (4 × 4) 6,000 kg cargo truck

Description
The IVECO 6602 CM (4 × 4) truck was designed to carry 6,000 kg of cargo both on roads and across country and is designated the Autocarro Pesante CP.70 by the Italian Army. Many of the components of the 6602 CM are also used in the 6607 CM (6 × 6) 6,000 kg cargo truck (see following entry).

The chassis is of the ladder type with two longitudinal steel channels to which the cross members, brackets and spring supports are riveted. The cab is of the forward control type and is of all-steel construction with removable canvas top and side curtains. The windscreen can be folded flat against the front of the cab if not required. The cab is equipped with a heater.

The rear cargo area is of all-steel construction with a wood-lined floor. The drop tailgate has swing-out steps and is removable. Along each side of the cargo area are seats which are folded flat against the sides when not required. The cargo area is covered by removable bows and a tarpaulin cover.

IVECO 6602 CM (4 × 4) 6,000 kg cargo truck complete with bows and tarpaulin cover

The 9,200 kg capacity winch is mounted at the rear and is provided with 60 m of 16 mm diameter cable. The electrical system is waterproof and radio suppressed.

A fuel tanker version is in service.

Specifications

Cab seating: 1 + 1
Configuration: 4 × 4
Weight:
(empty) 7,400 kg
(loaded) 13,540 kg
Weight on front axle: (loaded) 4,790 kg
Weight on rear axle: (loaded) 8,750 kg
Max load: 6,140 kg
Towed load:
(road) 10,000 kg
(cross-country) 4,000 kg
Load area: 4.28 × 2.23 m
Length: 6.55 m
Width: 2.46 m
Height:
(cab) 2.7 m
(tarpaulin) 2.974 m
(load area) 1.417 m
Ground clearance: 0.271 m
Track:
(front) 1.873 m
(rear) 1.785 m
Wheelbase: 3.57 m
Angle of approach/departure: 45°/35°
Max speed: (road) 74 km/h
Range: 700 km
Fuel capacity: 230 litres
Max gradient: 60%
Max side slope: 30%
Fording: 0.85 m
Engine: Model 8202.02 6-cylinder in-line liquid-cooled diesel developing 193 hp (DIN) at 2,500 rpm
Gearbox: manual with 5 forward and 1 reverse gears
Clutch: single dry plate
Transfer box: 2-speed
Steering: worm and sector, power-assisted
Turning radius: 8.1 m
Suspension: semi-elliptical springs front and rear with double-acting hydraulic shock-absorbers on the front axle only
Tyres: 11.00 × 20
Brakes:
(main) air
(parking) mechanical
Electrical system: 24 V
Batteries: 2 × 12 V, 90 Ah

Status

Production complete. In service with the Italian Army.

Manufacturer

IVECO SpA, Defence Vehicle Division, Via Volta 6, I-39100 Bolzano, Italy.
Tel: +39 471 905111. Telex: 400541.
Fax: +39 471 905444.

VERIFIED

IVECO 6605 (6 × 6) series

Development

In the 1960s IVECO built a light artillery tractor (6 × 6) called the Model 6606 (TL65) and a medium artillery tractor (6 × 6) called the Model 6605 (TM65). The TM65 (TM standing for *Trattore Medio*) was powered by a petrol engine; further development resulted in the diesel-engined TM69 which was adopted by the Italian Army. The range consists of the 6605 TM used for towing artillery such as the 155 mm FH-70, the 6605 FH used for carrying ammunition, the 6605 AG recovery vehicle and the 6605 A cargo truck.

Description (6605 TM)

The chassis is of the ladder type with two longitudinal pressed steel channels, to which are riveted the cross members, brackets and spring supports. The forward control cab is all steel and has a removable canvas top and side screens and a windscreen which can be folded flat against the bonnet. A cab heater and ventilator are standard on all versions. The gap to the rear of the cab and body houses the spare wheel, two water and two fuel containers, tyre chain locker and the exhaust pipe.

The rear cargo area is all steel with a wood-lined floor. The sides consist of two boards, the front a drop type and the rear fixed. The drop tailgate is fitted with integral steps and is removable. The rear cargo area is covered by removable bows and a tarpaulin cover. When not required the bows can be stowed under the central area of the body. The cargo space is divided into three compartments by removable partitions: the first is used for stowing the charges, the second for the projectiles and the rear for stores and the tarpaulin cover.

The transmission consists of a set of gears mounted on four shafts (input, primary, layshaft and reverse). Shifting from each gear is controlled by a lever through a pneumatic servo. Shifting from the high to the low range, or vice versa, is by an electropneumatic control with a preselector switch, interlocked with the clutch pedal. The transfer box is mechanical and consists of helical constant mesh gears mounted on three shafts (input, intermediate and output) of which the input shaft carries the dog clutch with a lockable divider differential distributing power to front and rear axles.

Specifications

Model	6605 TM	6605 FH	6605 A
Cab seating	1 + 11	1 + 11	1 + 1
Configuration	6 × 6	6 × 6	6 × 6
Weight (empty)	11,800 kg	12,600 kg	12,000 kg
(loaded)	17,000 kg	19,220 kg	20,500 kg
Max load	5,000 kg	6,620 kg	8,500 kg
Towed load	15,000 kg	15,000 kg	15,000 kg
Length	7.33 m	7.33 m	7.33 m
Width	2.5 m	2.5 m	2.5 m
Height (cab)	2.92 m	2.92 m	2.78 m
(tarpaulin)	2.87 m	2.78 m	3.08 m
(load area)	1.533 m	1.533 m	1.522 m
Ground clearance	0.363 m	0.363 m	0.363 m
Track	2.072 m	2.072 m	2.072 m
Wheelbase	3.217 m + 1.365 m	3.217 m + 1.365 m	3.217 m + 1.365 m
Angle of approach/departure	45°/40°	45°/40°	45°/40°
Max speed (road)	80 km/h	80 km/h	80 km/h
Range	700 km	700 km	700 km
Fuel capacity	360 litres	360 litres	360 litres
Max gradient	60%	60%	60%
Max side slope	20%	20%	20%
Fording	1.5 m	1.5 m	1.5 m
Engine	Model 8212.02.500 6-cylinder in-line water-cooled diesel developing 260 hp at 2,200 rpm		
Gearbox	manual with 8 forward and 2 reverse gears		
Clutch	dry, dual-plate	dry, dual-plate	dry, dual-plate
Transfer box	2-speed	2-speed	2-speed
Steering	ZF hydraulic assisted	ZF hydraulic assisted	ZF hydraulic assisted
Turning radius	8 m	8 m	8 m
Suspension (front)	2 semi-elliptical constant rate leaf springs with double-acting hydraulic shock-absorbers		
(rear)	rocker and torque arms with 2 constant-rate leaf springs		
Tyres	14.00 × 20	14.00 × 20	14.00 × 20
Brakes (main)	drum air-operated on all wheels, dual-circuit with connections for trailer braking		
(parking)	drum, hand-operated, mounted on transfer rear output shaft		
Electrical system	24 V	24 V	24 V
Batteries	4 × 12 V, 90 Ah	4 × 12 V, 90 Ah	4 × 12 V, 90 Ah

Mounted at the rear of the vehicle is a worm type winch with a capacity of 10,000 kg which can be used to the front or rear of the truck and has 60 m of 18 mm diameter cable.

Variants

6605 FH

This is almost identical to the 6605 TM but has a shorter rear cargo area as a hydraulic crane is mounted

IVECO 6605 TM69 (6 × 6) artillery tractor fitted with removable cab roof, bows and tarpaulin cover

IVECO 6605 FH (6 × 6) with hydraulically operated crane mounted to rear of cab

between the cab and the cargo area for unloading pallets of ammunition. The prototype was completed in 1974 with first production vehicles completed in 1976.

6605 AG

Full details of this recovery vehicle are given in the *Recovery Vehicles* section.

6605 A

This is the truck version and has a two-man cab with the cargo area at the rear with tiltable troop seats down each side, drop tailgate with integral steps, removable bows and a tarpaulin cover. The prototype was completed in 1974 and first production vehicles in 1976.

Status

In service with the Italian Army, Libya and Somalia. Somalia received 10 6605 TM, 200 6605 A, 12 10,500 litre fuel tankers, 10 recovery vehicles, four command posts, five aircraft refuellers and 50 8,000 litre water tankers.

Manufacturer

IVECO SpA, Defence Vehicle Division, Via Volta 6, I-39100 Bolzano, Italy.
Tel: +39 471 905111. Telex: 400541.
Fax: +39 471 905444.

VERIFIED

IVECO 230.35 WM (6 × 6) 10,000 kg artillery truck/tractor

Description

The IVECO 230.35 WM (6 × 6) 10,000 kg artillery truck/tractor is the only vehicle of its type produced by IVECO and is an updated version of the former 230 PM 35. An essentially similar vehicle, the 230 PM 26 is no longer available.

The vehicle has a forward control cab for the driver and two passengers. The cab has a canvas roof that can be easily removed and can tilt forward for engine and transmission access. The cargo area is covered with a tarpaulin resting on bows. The tailgate and side boards are all tiltable and removable and there are folding benches for 22 personnel. A spare wheel holder is located between the cab and the cargo body.

Options include a 10,000 kg hydraulic winch, sand tyres, a hardtop cab with observation hatch, a power take-off on the gearbox, brake couplings for towing, a spare wheel holder with a tyre-handling crane, a towing hook, a 900 kg material-handling crane, and a jib for handling artillery trail legs.

The IVECO 230.35 WM can be produced in an artillery tractor version with the cargo body divided into two sections. One section is used to carry a gun crew of 10 personnel, while the other section is used to carry ammunition and other supplies up to a weight of 8,500 kg. The crew compartment can be made removable if required. The tractor can tow artillery pieces of up to 15,000 kg.

Other variants of this vehicle include a recovery vehicle (refer to *Recovery Vehicles* section for details), a tanker, a mobile workshop, a mobile hospital, a logistics vehicle carrying a pallet-loading system, and a missile system carrier.

IVECO 230.35 WM (6 × 6) 10,000 kg artillery truck/tractor

Specifications

Cab seating: 1 + 2 (up to 22 in rear)
Configuration: 6 × 6
Weight:
(kerb) 13,000 kg
(GVW) 23,000 kg
Max load: 10,000 kg
Towed load: 15,000 kg
Load area: 5.05 × 2.35 m
Length: 7.762 m
Width: 2.5 m
Height:
(cab) 3.05 m
(tarpaulin) 3.43 m
(load area) 1.58 m
Ground clearance: 0.36 m
Track:
(front) 2.027 m
(rear) 2.032 m
Wheelbase: 3.3 m + 1.38 m
Angle of approach/departure: 42°/40°
Max speed: over 80 km/h
Range: over 600 km
Fuel capacity: 300 litres
Max gradient: 60%
Side slope: 30%
Fording: 1.2 m
Engine: 8280.02 17.174 litre V-8 water-cooled diesel developing 352 hp at 2,400 rpm
Gearbox: ZF Type 4 S 150 GPA, 8 forward and 1 reverse gears
Transmission: torque converter
Transfer box: electropneumatically lockable
Steering: recirculating ball, power-assisted
Turning radius: 9.5 m
Suspension:
(front) single flexibility leaf springs with hydraulic shock-absorbers
(rear) single flexibility leaf springs, reversed, fully articulated
Tyres: 14.00 × 20
Brakes: drum, air-operated
Electrical system: 24 V
Batteries: 2 × 12 V, 143 Ah
Generator: 650 W

Status

Available. In service with Abu Dhabi.

Manufacturer

IVECO SpA, Defence Vehicle Division, Via Volta 6, I-39100 Bolzano, Italy.
Tel: +39 471 905111. Telex: 400541.
Fax: +39 471 905444.

VERIFIED

IVECO 260.35 WM (6 × 4) and (6 × 6) 10,000 kg trucks

Description

The IVECO 260.35 WM series of 6 × 4 and 6 × 6 10,000 kg trucks was developed by the company from standard commercial vehicles. The chassis consists of section side members connected by cross members with a sheet steel bumper at the front and a towing hook and bumper at the rear.

The all-steel two-door fully enclosed cab is of the forward control type and can be tilted forward by one man to give access to the engine for maintenance. The steering wheel and both seats are adjustable and a two-speed electrical ventilating system, combined with a water heater and defroster, is fitted as standard.

The rear cargo body is of sheet steel with a wood-lined platform, fixed headboard, tiltable side boards with folding benches (22 troop seats), and a tiltable and removable tailgate with tiltable ladders for access to the platform. The body is attached to the chassis by means of elastic brackets with the spare wheel between the cab and the cargo body. Standard equipment includes

IVECO 260.35 WM (6 × 6) 10,000 kg truck

removable bows and tarpaulin cover, blackout lights, transceiver power input radio suppressed, auxiliary receptacle and master switch, NATO standard 12-point socket for trailer and tyre inflation system.

The 6 × 6 model has a combined gearbox and transfer box split from the engine and converter. The transfer box allows the torque distribution between front axle and rear bogie through a central differential which is pneumatically lockable.

The 6 × 4 model has a steering dead front axle with the 6 × 6 model having steer drive axle with epicyclic hub reduction. Both models have the same high articulation bogie for on/off highway use composed of two drive axles with hub reductions and pneumatically lockable differentials.

Optional equipment includes an OM DF 0.9 torque converter and a 1,000 kg jib crane to the cab rear.

The chassis can be used for a variety of other applications including a mobile workshop, water or fuel tanker, artillery tractor and a recovery vehicle. The last can be fitted with a hydraulically operated crane with a capacity of 12,000 kg and an auxiliary recovery winch.

Status
In production.

Manufacturer
IVECO SpA, Defence Vehicle Division, Via Volta 6, I-39100 Bolzano, Italy.
Tel: +39 471 905111. Telex: 400541.
Fax: +39 471 905444.

VERIFIED

Specifications

Configuration	**6 × 4**	**6 × 6**
Cab seating	1 + 1	1 + 1
Weight (loaded)	24,000 kg	24,000 kg
Max load	12,000 kg	14,000 kg
Towed load	20,000 kg	20,000 kg
Length (excluding body)	8 m	8 m
Width	2.5 m	2.5 m
Height (cab, unloaded)	3.123 m	3.123 m
Ground clearance	0.316 m	0.312 m
Track		
(front)	2.069 m	1.927 m
(rear)	1.809 m	1.809 m
Wheelbase	3.85 m + 1.38 m	3.85 m + 1.38 m
Angle of approach/departure	28°/22°	33°/27°
Max speed	97 km/h	95 km/h
Range	600 km	600 km
Fuel capacity	300 litres	300 litres
Max gradient	60%	60%
Side slope	30%	30%
Fording	0.65 m	0.65 m
Engine	8280.02 V-8 liquid-cooled diesel developing 352 hp at 2,400 rpm	
Gearbox	12802, 8 forward and 2 reverse gears	ZF 4S-150 GPA combined gearbox and transfer box, 8 forward and 1 reverse gears
Steering	ZF recirculating ball with integral power-assistance	
Turning radius	9.75 m	9.75 m
Suspension (front)	single flexible leaf springs and hydraulic shock-absorbers	
(rear)	single flexible leaf springs reversed and hydraulic telescopic shock-absorbers	
Tyres	12.00 R 20 PR 18	12.00 R 20 PR 18
Brakes (main)	air	air
(parking)	spring-loaded acting on rear axle	
(exhaust)	pneumatic, controlled by separate foot control	
Electrical system	24 V	24 V
Batteries	2 × 12 V, 143 Ah	2 × 12 V, 143 Ah

Astra SM 66.37 (6 × 6) truck

Description
Developed to meet very exacting military requirements, the Astra SM 66.37 (6 × 6) truck is part of the IVECO DVD range which also encompasses 4 × 4 and 8 × 8 designs. The SM 66.37 is designed to have a gross vehicle weight of 25 tonnes and is available with three possible wheelbase lengths; 3.3 m + 1.45 m, 3.75 m + 1.45 m and 4.2 m + 1.45 m. Cross-country mobility is enhanced by the use of extra-large 'single track' tyres to provide high flotation over soft terrain and deserts; the provision of a central tyre inflation system is an option to further improve mobility over soft ground.

The forward control three-man cab is constructed using glass fibre to resist corrosion while the chassis frame is constructed using a special steel. Due to the cab outline the vehicle is rail-portable to STANAG 2832 requirements while a tight turning circle is assured due to the front axle's wide joint spacing coupled with the advanced steering and suspension system design. All wheels have disc brakes and an Anti-lock Braking System (ABS) is provided as standard.

Drum brakes are available as an option. A 10,000 kg capacity winch with front and rear operation is available.

Astra SM 66.37 (6 × 6) truck chassis

Specifications
Cab seating: 1 + 2
Configuration: 6 × 6
Weights:
(empty, 3.3 m + 1.45 m WB) 10,470 kg
(empty, 3.75 m + 1.45 m WB) 10,500 kg
(empty, 4.2 m + 1.45 m WB) 10,560 kg
(GVW) 25,000 kg
Lengths:
(3.3 m + 1.45 m WB) 7.7 m
(3.75 m + 1.45 m WB) 8.15 m
(4.2 m + 1.45 m WB) 8.6 m
Width: 2.5 m
Height: (cab) 2.87 m
Ground clearance: 0.43 m
Track:
(front) 2.05 m
(rear) 2.06 m
Wheelbase: 3.3 m + 1.45 m, 3.75 m + 1.45 m or 4.2 m + 1.45 m
Angle of approach/departure: 45°/45°
Max road speed: 92 km/h
Range: 800 km
Fuel capacity: 300 litres
Max gradient: 100%
Side slope: 40%
Fording: 1.2 m
Engine: IVECO 8210 TCA 13.798 litre 6-cylinder turbo intercooled diesel developing 370 hp at 1,900 rpm
Gearbox: ZF 6 HP 900 automatic with 6 forward and 1 reverse gears or
ZF 16 S 151 + WSK 400 with torque converter, with 16 forward and 2 reverse gears
Transfer box: 2-speed
Steering: ZF Servocom Type 8098
Turning radius:
(3.3 m + 1.45 m WB) 16.32 m
(3.75 m + 1.45 m WB) 17.9 m
(4.2 m + 1.45 m WB) 19.47 m
Suspension:
(front) semi-elliptic springs with 2 double-acting hydraulic shock-absorbers and bumpers
(rear) reversed semi-elliptical springs with centre pin and slider connection to axles, 4 lower reaction rods and 2 upper reaction triangles
Tyres: 14.00 × 20 (run-flat optional)
Brakes:
(main) hydraulic disc
(parking) on 2nd axle
Electrical system: 24 V
Batteries: 4 × 12 V, 110 Ah
Alternator: 80 A

Status
Ready for production.

Manufacturer
Astra Veicoli Industriali SpA, Via Caorsana 79, I-29100 Piacenza, Italy.
Tel: +30 523 5431. Telex: 530148 astra i.
Fax: +39 523 593293.
Astra Veicoli is part of the IVECO organisation.

UPDATED

Astra HD 66.45M (6 × 6) truck

Description

In 1994 Astra Veicoli Industriali introduced a range of 6 × 6 trucks known as the HD series, produced in a variety of wheelbase lengths, engine fits cabs and configurations. The Astra HD series replaced the earlier Astra 300 Series.

Models are produced with three wheelbases: 3.5 m + 1.4 m; 3.8 m + 1.4 m; and 4.1 m + 1.4 m. Five different engines are available with outputs of 345, 370, 420, 440 and 520 hp. Single and double cabs are available as are numerous other options such as tyre sizes, power take-offs and other options.

Variants include the Astra HD 66.45/5S — C-IMT tractor truck (see following entry) and the Astra 66.45 TIM tractor truck (see entry in *Tank transporters* section).

The military version, the HD 66.45M is derived from the HD 66.45 model, details of which are provided in the Specifications table.

Specifications

HD 66.45
Cab seating: 1 + 2
Configuration: 6 × 6
Weights:
(empty) 10,435 kg
(GVW) 33,000 kg
Length: 7.735 m
Width: 2.5 m
Height: (overall) 3.405 m
Ground clearance: 0.31 m
Track:
(front) 2.02 m
(rear) 1.825 m
Wheelbase: 3.5 m + 1.4 m, 3.8 m + 1.4 m or 4.1 m + 1.4 m
Max road speed: 80 km/h
Fuel capacity: 300 litres
Max gradient: 100%
Engine: IVECO 8280 TCA 17.17 litre V-8 turbo intercooled diesel developing 440 hp at 1,900 rpm
Gearbox: ZF 16 S 221 with 16 forward and 2 reverse gears
Clutch: single dry disc
Transfer box: 2-speed
Steering: ZF Servocom Type 8098
Turning radius: 17.3 m
Suspension:
(front) semi-elliptic springs with shock-absorbers, torsion bar and bumpers
(rear) reinforced oscillating springs with pivot system
Tyres: 12.00 × 20
Brakes:
(main) pneumatic
(parking) mechanical on rear axles
Electrical system: 24 V
Batteries: 4 × 12 V, 110 Ah
Alternator: 80 A

Astra SM 66.45M (6 × 6) dump truck ***1997***

Status

In production.

Manufacturer

Astra Veicoli Industriali SpA, Via Caorsana 79, I-29100 Piacenza, Italy.
Tel: +30 523 5431. Telex: 530148 astra i.
Fax: +39 523 593293.
Astra Veicoli is part of the IVECO organisation.

NEW ENTRY

Astra HD 66.45/5S - C-IMT (6 × 6) tractor truck

Description

The Astra HD 66.45/5S - C-IMT (6 × 6) tractor truck is derived from the Astra HD series and modified to render it suitable for towing a special semi-trailer used to carry and launch CNIM PFM pontoon bridge components. The modifications included placing a torque converter between the engine and gearbox, a four-door five-passenger cab with a hydraulic tilting system, a hydraulic system to launch and recover the bridge and to power the motors on the semi-trailer wheels, a winch for self-recovery and a 5th wheel.

Specifications

Cab seating: 1 + 4
Configuration: 6 × 6
Weight: (empty, with hydraulic system and winch) 13,500 kg
Load on 5th wheel: 7,000 kg
Length: 8.035 m
Width: 2.5 m
Height: 3.31 m
Ground clearance: 0.37 m
Track:
(front) 2.053 m
(rear) 2.117 m
Wheelbase: 4.1 m + 1.4 m
Max road speed:
(possible) 103 km/h
(with limiter) 85 km/h
Fuel capacity: 300 litres
Engine: IVECO 8280 TCA 17.17 litre V-8 turbo intercooled diesel developing 440 hp at 1,900 rpm
Gearbox: ZF WSK 400.90 torque converter plus ZF 16 S 221 with 16 forward and 2 reverse gears
Transfer box: 2-speed
Steering: ZF Servocom Type 8098
Turning radius: 11.3 m
Suspension:
(front) semi-elliptic springs with shock-absorbers, torsion bar and bumpers
(rear) reinforced oscillating springs with pivot system and torsion bar
Tyres: 14.00 × 20
Brakes:
(main) pneumatic
(parking) mechanical on rear axles
Electrical system: 24 V
Batteries: 4 × 12 V, 110 Ah
Alternator: 80 A

Astra HD 66.45/5S - C-IMT (6 × 6) tractor truck ***1997***

Status

In production for the Italian Army.

Manufacturer

Astra Veicoli Industriali SpA, Via Caorsana 79, I-29100 Piacenza, Italy.
Tel: +30 523 5431. Telex: 530148 astra i.
Fax: +39 523 593293.
Astra Veicoli is part of the IVECO organisation.

NEW ENTRY

Astra SM 88.42 (8 × 8) 15,000 kg high-mobility truck

Description

Developed in response to an Italian Army requirement for a heavy high-mobility vehicle, the Astra SM 88.42 (8 × 8) 15,000 kg truck was presented to the Italian Ministry of Defence in May 1989. The vehicle can be fitted with load-handling systems and can carry payloads up to 15,000 kg.

The Astra SM 88.42 (8 × 8) truck is part of the IVECO DVD range which also encompasses 4 × 4 and 6 × 6 designs. Gross vehicle weight is approximately 32 tonnes. Cross-country mobility is enhanced by the use of extra-large 'single track' tyres to provide high flotation over soft terrain and deserts; the provision of a central tyre inflation system is an option to further improve mobility over soft ground.

The forward control three-man cab is constructed using glass fibre to resist corrosion while the chassis frame is constructed using a special steel. Due to the cab outline the vehicle is rail-portable to STANAG 2832 requirements while a tight turning circle is assured due to the front axle's wide joint spacing coupled with the advanced steering and suspension system design. All wheels have disc brakes and an Anti-lock Braking System (ABS) is provided as standard.

Drum brakes are available as an option. Also available is a 10,000 kg capacity winch for front and rear traction.

Specifications
Cab seating: 1 + 2
Configuration: 8 × 8
Weights:
(empty) 12,200 kg
(GVW) 32,000 kg
Load area: (LWB) 7.5 × 2.4 m
Length:
(SWB) 9.34 m
(LWB) 10.06 m
Width: 2.5 m
Height: (cab) 2.87 m
Ground clearance: 0.43 m
Track:
(1st and 2nd axles) 2.05 m
(3rd and 4th axles) 2.06 m
Wheelbase:
(SWB) 1.955 m + 2.995 m + 1.45 m
(LWB) 1.955 m + 3.595 m + 1.45 m
Angle of approach/departure: 45°/40°
Max speed: 92 km/h
Range: 800 km
Fuel capacity: 300 litres
Max gradient: 100%
Side slope: 40%
Fording: 1.2 m
Engine: IVECO 8210 TCA 13.798 litre 6-cylinder turbo-intercooler diesel developing 420 hp at 1,900 rpm
Transmission: ZF 6 HP 600 automatic with 6 forward and 1 reverse gears or

Astra SM 88.42 (8 × 8) 15,000 kg high-mobility truck

ZF 16 S 221 900 + WSK 400 with torque converter with 16 forward and 2 reverse gears
Transfer box: 2-speed
Steering: ZF Servocom type 8099
Turning radius:
(SWB) 10.3 m
(LWB) 11.3 m
Suspension:
(front) semi-elliptic leaf springs with 2 double-acting hydraulic shock-absorbers and bumpers
(rear) reversed semi-elliptic leaf springs with centre pin and slider connection to axles, 4 lower reaction rods and 2 upper reaction triangles
Tyres: 14.00 R 20 (run-flat optional)
Brakes:
(main) hydraulic disc
(parking) on 2nd and 3rd axle
Electrical system: 24 V
Batteries: 4 × 12 V, 110 A
Alternator: 80 A

Status
In production for Italian Army.

Manufacturer
Astra Veicoli Industriali SpA, Via Caorsana 79, I-29100 Piacenza, Italy.
Tel: +39 523 5431. Telex: 530148 astra i.
Fax: +39 523 69293.
Astra Veicoli is part of the IVECO organisation.

UPDATED

Astra HD6 84.45 (8 × 4 × 4) logistic transport truck

Description
The Astra HD6 84.45 (8 × 4 × 4) logistic transport truck was derived from the Astra 'Heavy Duty' civil vehicle range to meet Italian Army specifications for the new STL logistic transport system. The vehicle carries a Multilift Mark 4 load-handling system provided by Multilift Isoli – for details of this system refer to the entry in the *Materials handling equipment* section.

The vehicle can carry loads up to 15,000 kg that can be lowered to the ground in 33 seconds or retrieved in 25 seconds. The system is cross-operational with other load-handling systems already in use with NATO and other nations and can handle 20 ft/6.096 m ISO standard platforms or ISO 1C standard containers.

The vehicle is powered by a 440 hp diesel coupled to a 16-speed gearbox. The front two axles can be steered.

Astra HD6 84.45 (8 × 4 × 4) logistic transport truck ***1997***

Specifications
Cab seating: 1 + 1
Configuration: 8 × 4 × 4
Weights:
(empty) 16,200 kg
(loaded) 31,200 kg
(GVW) 32,000 kg
Payload: (incl flatrack) 15,000 kg
Length: 8.97 m
Width: 2.5 m
Height: (cab) 3.36 m
Ground clearance: 0.31 m
Track:
(front) 2.03 m
(rear) 1.825 m
Wheelbase: 1.76 m + 3.24 m + 1.4 m
Max road speed: 82 km/h
Range: 600 km
Fuel capacity: 300 litres
Max gradient: 50%
Engine: IVECO 8280 TCA 17.17 litre V-8 turbo intercooled diesel developing 440 hp at 1,900 rpm
Gearbox: ZF 16 S 221 with 16 forward and 2 reverse gears
Clutch: single dry disc
Steering: ZF Servocom 8099
Turning radius: 10.3 m
Suspension:
(front) semi-elliptic springs with shock-absorbers, torsion bar and bumpers
(rear) reinforced oscillating springs with pivot system
Tyres: 12.00 R 20
Brakes:
(main) pneumatic
(parking) mechanical on rear axles
Electrical system: 24 V
Batteries: 4 × 12 V, 110 Ah
Alternator: 80 A

Status
In production for the Italian Army.

Manufacturer
Astra Veicoli Industriali SpA, Via Caorsana 79, I-29100 Piacenza, Italy.
Tel: +39 523 5431. Telex: 530148 astra i.
Fax: +39 523 69293.
Astra Veicoli is part of the IVECO organisation.

UPDATED

IVECO M 320 E 42 W (8 × 8) 17,500 kg Truck

Description

The IVECO IVECO M 320 E 42 W (8 × 8) 17,500 kg truck is the top load carrier in IVECO's tactical vehicle range. It has a forward control cab with seating for the driver and at least two passengers while the standard cargo body is an open platform with drop sides and a tailgate. Power is provided by an IVECO diesel developing 420 hp and coupled to a six-speed automatic transmission.

Specifications

Cab seating: 1 + 2 (space for more)
Configuration: 8 × 8
Weights:
(kerb) 14,500 kg
(GVW) 32,000 kg
Max load: 17,500 kg
Towed load: 12,000 kg
Length: 10.4 m
Width: 2.5 m
Height: (top of cab) 2.864 m
Track: (front) 2.054 m
Ground clearance: 0.462 m
Wheelbase: 1.955 m + 3.595 m + 1.45 m
Angle of approach/departure: 45°/40°
Max speed: over 90 km/h
Max gradient: over 60%
Engine: IVECO diesel developing 420 hp
Transmission: automatic with 6 forward and 1 reverse gears
Transfer box: 2-speed
Steering: hydraulic power-assisted
Tyres: 14.00 R 20
Electrical system: 24 V

Status

Available.

Manufacturer

IVECO SpA, Defence Vehicles Division, Via Volta 6, I-39100 Bolzano, Italy.
Tel: +39 471 905111. Telex: 400541.
Fax: +39 471 905444.

NEW ENTRY

IVECO M 320 E 42 W (8 × 8) 17,500 kg truck
1997

JAPAN

Kohkidohsha 1,500 kg (4 × 4) high-mobility vehicle

Description

The term *Kohkidohsha* translates as high-mobility vehicle and it is a 4 × 4 1,500 kg vehicle developed as a possible replacement for many existing Type 73 2,000 kg and Isuzu 2,500 kg trucks currently in Japanese Ground Self-Defence Force service. A request for 85 vehicles was made during 1992, with at least nine configured as Kin-Sam close-range air defence missile system carriers.

The Kohkidohsha visually resembles the US M998 HMMWV series (qv) but differs in many respects. It can carry at least four people including the driver. Power is derived from a 4 litre liquid-cooled 150 hp diesel coupled to an automatic transmission and there is permanent four-wheel drive. Steering is power-assisted on all four wheels and independent suspension is provided, also on all four wheels.

Kohkidohsha 1,500 kg (4 × 4) high-mobility vehicle

Specifications

Cab seating: 1 + 3
Configuration: 4 × 4
Weight: approx 2,440 kg
Max load: approx 1,500 kg
Length: 4.91 m
Width: 2.15 m
Height: 2.09 m
Ground clearance: 0.4 m
Wheelbase: 3.4 m
Angle of approach/departure: 73°/51°
Max speed: 100 km/h
Engine: 4 litre liquid-cooled diesel developing 150 hp
Transmission: automatic
Transfer box: permanent 4-wheel drive
Steering: power-assisted on all 4 wheels
Suspension: independent
Tyres: run-flat

Status

In production.

VERIFIED

Type 73 (4 × 4) 2,000 kg truck

Description

The Type 73 (4 × 4) 2,000 kg truck was standardised in 1973 as the successor to Nissan and Toyota (4 × 4) 750 kg trucks in service since the 1950s.

The Type 73 has an over-engine cab with a steel back and sides up to seat-top height and steel doors with glass windows. The single-piece windscreen can be folded forwards. The roof is a single-piece waterproof tarpaulin that can be fixed in place by removable braces. With the tarpaulin in place a high-capacity heater-blower-defroster system can be used. Seats for the driver and one passenger are provided each with space for kit stowage behind.

The cargo body is all steel but has wooden side stakes and rails that are held in place by latches and unfold to serve as benches. The tailgate has a built-in step that will unfold when a ring on each side is pulled. A heavy canvas tilt and braces can be used to

Type 73 (4 × 4) 2,000 kg truck (Kensuke Ebata)

cover the cargo area which is 3 m long and 1.95 m wide.

Standard equipment includes a rear fender, tool box, canvas tilt and spare tyre and carrier. Optional equipment includes a winch (front-mounted), a power take-off, and a tropical cooling system.

An ambulance version with a fully enclosed cab and rear body has been produced. The truck version has been used to carry mortars.

Specifications

Cab seating: 1 + 1
Configuration: 4 × 4
Weight:
(empty) 3,195 kg
(gross weight, on-road) 5,355 kg
(gross weight, off-road) 4,855 kg
Max load:
(on-road) 2,000 kg
(off-road) 1,500 kg
Front axle load: (on-road, loaded) 2,325 kg
Rear axle load: (on-road, loaded) 3,030 kg
Length: 5.36 m
Width: 2.09 m
Height: 2.49 m
Ground clearance: 0.28 m
Wheelbase: 2.9 m
Track:
(front) 1.61 m
(rear) 1.635 m
Angle of approach/departure: 37°/32°
Max speed: 87 km/h
Fuel capacity: 115 litres
Max gradient: 60%
Fording: 0.8 m
Engine: Hino DQ100 diesel developing 95 hp at 3,000 rpm
Gearbox: 5 forward and 1 reverse gears
Transfer box: 2-speed
Clutch: single dry disc
Steering: recirculating ball
Turning radius: 6.4 m
Suspension: semi-elliptic leaf springs with shock-absorbers
Tyres: 8.25 × 20
Brakes: hydraulic with vacuum servo
Electrical system: 24 V
Batteries: 2 × 12 V, 100 Ah
Generator: 400 W

Status

In production. In service with the Japanese Ground Self-Defence Force

Manufacturer

Toyota Motor Corporation, 4-18, Koraku 1-chome, Bunkyo-ku, Tokyo 112, Japan.
Tel: +81 3 5391 5407. Fax: +81 3 5391 5016.

VERIFIED

Type 73 (6 × 6) 3,500 kg truck

Description

The Type 73 (6 × 6) 3,500 kg truck was developed by Isuzu from the late 1960s to replace the range of 2,500 kg (6 × 6) trucks then used by the Japanese Self-Defence Force. After trials and modifications it was standardised in 1973 as the Type 73 large truck series and since then it has been produced in large numbers for the JASDF.

The first model was the SKW440 but since then many improvements have been made to the engine, axles, cab and so on. The latest model is the SKW464.

The forward control cab has a windscreen that can be folded forward onto the bonnet and a removable canvas top. The basic cargo body is all steel and has a drop tailgate, bench troop seats down either side, removable bows and a tarpaulin cover. This basic model can carry 3,500 kg of stores across country or 6,000 kg on roads.

There are two basic models, one with single-tyred

Specifications

Model	**SKW464M**	**SKW464M**	**SKW464MV**	**SKW464MV**	**SKW464MV**	**SKW464MD**	**SKW464MR**	**SKW464**
Type	Cargo truck	Cargo truck (W/W)	Tanker (utility)	Water tanker	Crane	Dump truck	Light wrecker	Cargo (long)
Cab seating	1 + 1	1 + 1	1 + 1	1 + 1	1 + 1	1 + 1	1 + 1	1 + 1
Configuration	6 × 6	6 × 6	6 × 6	6 × 6	6 × 6	6 × 6	6 × 6	6 × 6
Weight								
(empty)	8,120 kg	8,340 kg	8,960 kg	9,100 kg	9,080 kg	9,600 kg	14,220 kg	8,410 kg
(loaded, road)	14,280 kg	14,500 kg	14,220 kg	14,260 kg	14,240 kg	14,760 kg	14,380 kg	13,570 kg
(loaded, cross-country)	11,780 kg	12,000 kg	–	–	12,740 kg	13,260 kg	–	12,070 kg
Max load								
(road)	6,000 kg	6,000 kg	5,100 kg	5,000 kg	5,000 kg	5,000 kg	–	5,000 kg
(cross-country)	3,500 kg	3,500 kg	–	–	3,500 kg	3,500 kg	–	3,500 kg
Towed load								
(road)	6,000 kg	6,000 kg	6,000 kg	6,000 kg	6,000 kg	6,000 kg	6,000 kg	6,000 kg
(cross-country)	4,000 kg	4,000 kg	4,000 kg	4,000 kg	4,000 kg	4,000 kg	4,000 kg	4,000 kg
Length	7.030 m	7.230 m	7.190 m	6.810 m	7.355 m	7.580 m	7.810 m	8.130 m
Width	2.485 m	2.485 m	2.485 m	2.485 m	2.485 m	2.485 m	2.490 m	2.485 m
Height	3.080 m	3.080 m	2.865 m	2.890 m	3.250 m	3.355 m	3.060 m	3.080 m
Ground clearance	0.330 m	0.330 m	0.330 m	0.330 m	0.330 m	0.330 m	0.330 m	0.330 m
Track (front)	1.970 m	1.970 m	1.970 m	1.970 m	1.970 m	1.840 m	1.840 m	1.970 m
Wheelbase	3.295 m + 1.310 m	3.295 m + 1.310 m	3.295 m + 1.310 m	3.295 m + 1.310 m	3.295 m + 1.310 m	3.295 m + 1.310 m	3.295 m + 1.310 m	3.795 m + 1.310 m
Max road speed	95 km/h	95 km/h	95 km/h	95 km/h	95 km/h	95 km/h	95 km/h	95 km/h
Range	500 km	500 km	500 km	500 km	500 km	500 km	500 km	500 km
Fording	0.8 m	0.8 m	0.8 m	0.8 m	0.8 m	0.8 m	0.8 m	0.8 m
Turning radius	9.2 m	9.2 m	9.2 m	9.2 m	9.2 m	9.4 m	9.4 m	10.5 m
Fuel capacity	170 litres	170 litres	170 litres	170 litres	170 litres	170 litres	170 litres	170 litres
Engine	Isuzu 8PE1 V-8 liquid-cooled diesel developing 250 hp at 2,200 rpm							
Gearbox	all have a manual gearbox with 5 forward and 1 reverse gears							
Clutch	single dry plate	single dry plate	single dry plate	single dry plate	single dry plate	single dry plate	single dry plate	single dry plate
Transfer box	2-speed	2-speed	2-speed	2-speed	2-speed	2-speed	2-speed	2-speed
Steering	power	power	power	power	power	power	power	power
Brakes	(main) air/hydraulic (parking) mechanical							
Suspension	(front) semi-elliptical spring and hydraulic shock-absorber, (rear) semi-elliptical spring							
Electrical system	24 V	24 V	24 V	24 V	24 V	24 V	24 V	24 V
Tyres	11.00 R20 -14PR	11.00 R20 -14PR	11.00 R20 -14PR	11.00 R20 -14PR	11.00 R20 -14PR	11.00 R20 -14PR	11.00 R20 -14PR	11.00 R20 -14PR

Type 73 (6 × 6) 3,500 kg dump truck

Type 73 (6 × 6) water tanker

rear wheels (cargo, tanker and so on) and the other with dual-tyred rear wheels (dumper and light wrecker).

A 4,500 kg capacity winch can be mounted on the front of the vehicle.

Variants

Models with single rear wheels

Truck, Cargo with winch (SKW464M 6 × 6)

This is the base model of the cargo series and can carry 3,500 kg of cargo across country and 6,000 kg on roads. The base version can be used to carry shelters and has been used to carry radar systems and surface-to-air missile systems (such as the Tan SAM). There is a version with a longer wheelbase.

Tanker, Utility (SKW464MV 6 × 6)

This tanker has a maximum payload of 5,100 kg of fuel.

Tanker, Water (SKW464MV 6 × 6)

Carries a tank for 5,000 litres of water.

Models with dual rear wheels

Truck, Dump (SKW464MD 6 × 6)

This has an all-steel rear tipping body with a payload of 5,000 kg. It can also be used to carry cargo.

Truck, Wrecker, Light (SKW464MR 6 × 6)

See entry in the *Recovery Vehicles* section.

Status

In production. In service with the Japanese Ground Self-Defence Force.

Manufacturer

Isuzu Motors Limited, 22-10, Minami-oi 6-chome, Shinagawa-ku, Tokyo 140, Japan.
Tel: +81 3 762 1111. Telex: 0246 6689.

UPDATED

Mitsubishi W121P (6 × 6) 6,000 kg truck

Description

In the 1950s Mitsubishi introduced the Fuso range of 6 × 6 trucks which were based on US Second World War type trucks. In the early 1960s these were replaced in production by the improved 6W series which had a redesigned cab as well as numerous other improvements. Production was completed in 1973.

The layout of all models is similar, with the engine at the front and cab in the centre. The cab has a removable canvas top and a windscreen that folds forward onto the bonnet. The rear cargo area has troop seats, drop tailgate, removable bows and a tarpaulin cover. Behind the cab, on the right side, is a winch that can be used to the front or rear.

Variants

Tractor truck

The 6,000 kg model is used to tow semi-trailers weighing up to 20 tonnes and the 10,000 kg model for semi-trailers up to 25,000 kg. The 6,000 kg model is used by JGSDF engineer units and the 10,000 kg by construction battalions and engineer equipment companies. Basic specifications are length 7.57 m, width 2.48 m, height 3 m, weight unloaded 9,860 kg, maximum road speed 70 km/h and engine six-cylinder water-cooled diesel developing 200 hp.

Wrecker

Two models are in service, one with a mechanical and the other with a hydraulic crane and both with front- and rear-mounted winches. Basic specifications are as follows (those in square brackets relate to the model with a hydraulic crane): length overall 9.9 [9.32] m, width 2.48 [2.48] m, height 3.14 [3.01] m, empty weight 13,400 [16,570] kg and maximum road speed 72 km/h.

Specifications

Cab seating: 1 + 1
Configuration: 6 × 6
Weight:
(empty) 9,700 kg
(loaded, road) 21,700 kg
(loaded, cross-country) 15,700 kg
Max load:
(road) 12,000 kg
(cross-country) 6,000 kg
Length: 7.42 m
Width: 2.48 m
Height: 3 m
Max road speed: 72 km/h
Engine: 6-cylinder water-cooled diesel developing 200 hp

Status

Production complete. In service with the Japanese Ground Self-Defence Force.

Manufacturer

Mitsubishi Motors Corporation, 33-8 Shiba 5-chome, Minato-ku, Tokyo, Japan.
Tel: +81 3 456 1111. Telex: 26639 bisijiko j.

VERIFIED

Late production model (6 × 6) 6,000 kg truck towing twin 35 mm Oerlikon AA gun (Kensuke Ebata)

Late production model (6 × 6) tractor truck towing dozer on Type 73 special semi-trailer (Kensuke Ebata)

Mitsubishi FW415 (6 × 6) trucks

Description

The Mitsubishi FW415 series of 6 × 6 trucks was introduced in 1984 when it replaced the FW115 series (see next entry) in production.

The basic FW415 (6 × 6) truck has an all-steel forward control cab with seating provided for the driver and two passengers. The doors have extra vision panels for the driver and warning lights are set into the upper cab roof. The space immediately behind the cab is used for tool and spare wheel stowage on most models, and a heavy-duty fender extends behind the front wheel. The front headlights are set into heavy-duty bumpers.

Variants

FW415L heavy-duty truck

This is the base model of the FW415 series and has a conventional heavy-duty rear cargo deck with provision for a canvas tilt. When not in use the folded tilt supports are stowed forward. The cargo area has a tailgate at the rear and a spare wheel is stowed in the space between the cab rear and the front of the cargo area. The FW415L can be used as an artillery tractor for weapons with calibres of up to 35 mm.

Specifications

Model	FW415L	FW415L3	FW415LD	FW415LD1	FW415M	FW415M1
Role	truck	truck w/winch	truck	truck w/winch	drop side	recovery
Cab seating	1 + 2	1 + 2	1 + 2	1 + 2	1 + 2	1 + 2
Configuration	6 × 6	6 × 6	6 × 6	6 × 6	6 × 6	6 × 6
Weight						
(empty)	9,580 kg	12,950 kg	10,430 kg	11,070 kg	9,860 kg	18,100 kg
(loaded)	19,820 kg	19,670 kg	19,670 kg	19,810 kg	19,850 kg	18,340 kg
Max load (off road)	7,000 kg	6,000 kg	7,000 kg	7,000 kg	7,000 kg	—
Length	8.18 m	8.59 m	8.135 m	8.135 m	9.215 m	9.32 m
Width	2.49 m	2.49 m	2.49 m	2.49 m	2.49 m	2.49 m
Height	3.05 m	3.05 m	3.1 m	3.1 m	3.05 m	3.23 m
Wheelbase	3.88 m + 1.3 m	3.88 m + 1.3 m	3.88 m + 1.3 m	3.88 m + 1.3 m	4.15 m + 1.3 m	4.15 m + 1.3 m
Max road speed	101 km/h	101 km/h	101 km/h	101 km/h	101 km/h	101 km/h
Engine	Mitsubishi Model 8DC9-1A diesel developing 320 hp at 2,200 rpm					

FW415L3 heavy-duty truck with winch

This is a version of the FW415L with a 10,000 kg capacity rear-mounted winch and a 2,000 kg crane. It can be used as an artillery tractor for light weapons.

FW415LD heavy-duty truck

The FW415LD heavy-duty truck has an all-steel heavy-duty dump body with a cab protector. It is provided with a centrally mounted winch with the controls on each side of the cargo area between the body and the cab rear.

FW415LD1 heavy-duty truck

This is a version of the FW415LD with a 6,000 kg rear winch.

Mitsubishi FW415L3 (6 × 6) heavy-duty truck with winch

FW415 variant known as the Prime Mover Medium used to tow the 155 mm FH-70 howitzer (Kensuke Ebata)

FW415M drop side truck
This is a lengthened wheelbase version of the FW415L. The length of the cargo area is 6.71 m.

FW415M1 recovery vehicle
See separate entry in the *Recovery Vehicles* section.

FW415LR2 truck tractor
See separate entry in the *Tank Transporters* section.

Prime Mover Medium
This is a tractor vehicle used to tow 155 mm FH-70 howitzers and is equipped with a 10 tonne crane to handle ammunition stowed on a shortened cargo area.

Status
In service with the Japanese Ground Self-Defence Force.

Manufacturer
Mitsubishi Motors Corporation, 33-8 Shiba 5-chome, Minato-ku, Tokyo, Japan.
Tel: +81 3 456 1111. Telex: 26639 bisijiko j.

VERIFIED

Mitsubishi FW115 (6 × 6) trucks

Description
The Mitsubishi FW115 series of trucks was introduced in 1979 when it replaced the FW105 series in production. The FW105 series, in its turn, had replaced the earlier FW103 trucks which went out of production in 1979.

The basic FW115 (6 × 6) truck is a forward control design, having an all-steel cab with seats for the driver and two passengers. The doors have extra vision panels for the driver and warning lights are set forward on the roof. The area immediately behind the cab is used for tool and spare wheel stowage on most models, and a heavy-duty fender extends behind the front wheel to just forward of the second wheel to protect the fuel tanks, transmission and other engine extensions. A towing hook is usually provided at the rear. The front headlights are set into the heavy-duty bumpers.

Variants
FW115 L1 heavy-duty truck
This is the base model of the FW115 series and has a conventional heavy-duty rear cargo area with provision for a canvas tilt. When not in use the folded tilt supports are stowed forward. The cargo area has a tailgate at the rear and a spare wheel is stowed in the space between the cab rear and the front of the cargo area. The FW115 L1 can be used as an artillery tractor for calibres up to 155 mm.

FW115 LD1 heavy-duty truck
The FW115 LD1 heavy-duty truck has an all-steel heavy-duty cargo body with a steel bulkhead protecting the rear of the cab. It is provided with a centrally mounted winch with the controls on each side of the cargo area between the body and the cab rear.

Specifications

Model	**FW115 L1**	**FW115 LD1**	**FW115 M1**	**FW115 M2**	**FW115 M4**
Configuration	6 × 6	6 × 6	6 × 6	6 × 6	6 × 6
Weight					
(empty)	9,660 kg	11,300 kg	18,290 kg	11,480 kg	10,130 kg
(loaded)	19,160 kg	19,800 kg	18,530 kg	18,970 kg	19,870 kg
Length	8.39 m	8.135 m	9.32 m	9.205 m	9.285 m
Width	2.49 m	2.49 m	2.49 m	2.49 m	2.49 m
Height	3.05 m	3.1 m	3.23 m	3.05 m	3.09 m
Max road speed	95 km/h	95 km/h	95 km/h	95 km/h	95 km/h

FM115 M1 recovery vehicle
This variant has the rear occupied by a large telescopic crane with limited traverse. The crane has a large cable-actuated pulley and is secured when travelling. The crane is controlled from the left-hand side and a spare wheel is stowed on the right. A high-capacity winch is located centrally and the cable pays out to the rear. A smaller self-recovery winch is situated under the front bumper on the right-hand side. There are four stabiliser legs for the crane which are located one leg fore and aft of each of the four rear wheels. Their stabilising feet can be lowered directly to the ground in use. Various tool and equipment lockers are provided.

FW115 M2
The FW115 M2 has a lengthened wheelbase and cargo body to carry floating bridge components. The cargo body extends forward to just behind the cab.

FW115 M4
The FW115 M4 has the same lengthened wheelbase and body as the FW115 M2 but is used as a conventional heavy-duty cargo truck and is equipped with drop sides. There is provision for a canvas tilt.

Status
Production complete. In service with the Japanese Ground Self-Defence Force.

Manufacturer
Mitsubishi Motors Corporation, 33-8 Shiba 5-chome, Minato-ku, Tokyo, Japan.
Tel: +81 3 456 1111. Telex: 26639 bisijiko j.

VERIFIED

Mitsubishi FW115 M2 (6 × 6) long-bodied floating bridge carrier

Mitsubishi FW115 M4 (6 × 6) drop side truck

Type 74 (6 × 6) 10,000 kg truck

Description

This is a standard commercial truck modified to meet the requirements of the Japanese Ground Self-Defence Force for a 6 × 6 vehicle with a payload of 10,000 kg. It was standardised in 1974 as the Type 74 special large truck.

The all-steel cab is the forward control type and the rear cargo area has drop sides, drop tailgate, removable bows and a tarpaulin cover. The short model of the Type 74 does not have drop sides. A model of the Type 74 is also used to carry bridging equipment for which special unloading equipment is fitted.

Type 74 (6 × 6) 10,000 kg truck (Kensuke Ebata)

Specifications

Data in square brackets relate to short model where different

Configuration: 6 × 6
Weight:
(empty) 9,850 [9,600] kg
(loaded) 19,850 [19,100] kg
Max load: 10,000 [9,500] kg
Length: 9.22 [8.3] m
Width: 2.49 m
Height: 3.055 [3] m
Max road speed: 100 km/h
Engine: 300 hp diesel

Status

In service with the Japanese Ground Self-Defence Force.

VERIFIED

KOREA, SOUTH

KM45 (4 × 4) 1¼ ton truck series

Description

The Asia Motors KM45 series of 4 × 4 trucks are similar in appearance to the US M715/AM 715 1¼ ton trucks but are fitted with MZBA1 diesel engines and a new power train.

The base model of the series is the KM450 light cargo truck. The driver's cab has a canvas top and seats are provided for the driver and one passenger. The cargo area at the rear can be covered by a canvas tilt and side boards are usually fitted. Bench seats can be fitted to carry four men along each side There is a drop tailgate at the rear.

The KM451 is an ambulance with a fully enclosed van body and two outward-opening doors at the rear. The ambulance body has space for up to eight seated casualties or five stretchers and a medical orderly. Weight loaded is 3,235 kg, payload 1,080 kg and length overall is 5.446 m. Overall height is 2.443 m.

The KM452 also uses a van-type box body but is fitted with windows at the side to allow it to be used as a communications vehicle. Other similar vehicles are the KM453 to KM458, all of them equipped with various office or workshop interiors; for instance the KM458 is a mobile workshop van.

Specifications

KM450 cargo truck
Cab seating: 1 + 1
Configuration: 4 × 4
Weight:
(kerb) 2,550 kg
(GVW) 4,080 kg
Max load: 1,530 kg
Length: 5.328 m
Width: 2.008 m
Height: 2.37 m
Ground clearance: 0.254 m
Track:
(front) 1.675 m
(rear) 1.71 m
Wheelbase: 3.2 m
Angle of approach/departure: 45°/25°
Max speed: 105 km/h
Range: (cruising) 450 km
Fuel capacity: 106 litres
Max gradient: 75%
Fording: (without kit) 0.76 m
Engine: MZBA1 4.052 litre 6-cylinder water-cooled diesel developing 115 hp at 3,600 rpm
Gearbox: manual, with 5 forward and 1 reverse gears
Clutch: single dry plate
Transfer box: 2-speed
Steering: ball and nut
Turning radius: 7.1 m
Suspension: (front and rear) semi-elliptical leaf springs with hydraulic double-acting shock-absorbers
Tyres: 9.00 × 16
Brakes:
(main) vacuum-assisted hydraulic
(parking) mechanical
Electrical system: 24 V
Batteries: 2 × 12 V, 100 Ah
Alternator: 45 A

Status

In production. In service with the armed forces of South Korea.

Manufacturer

Asia Motors Company Inc, 15 Yoido-dong, Yongdungpo-gu, PO Box 5022, Seoul, South Korea.
Tel: +82 2 783 1501. Telex: 24374 asiamco k.

VERIFIED

KM450 (4 × 4) 1¼ ton cargo truck

KM25 (6 × 6) 2,500 kg truck series

Description

The KM25 series of 2,500 kg (6 × 6) trucks closely follow the appearance of the US M44A2 truck and are fitted with MAN diesel engines and locally produced power trains. They are produced in five basic forms: KM250 cargo truck; KM254 decontamination truck; KM255 1,000 gallon fuel tanker; KM256 1,200 gallon water tanker; and KM258 shop van.

A 4,500 kg capacity winch equipped with 61 m of cable is optional for all models.

Specifications

KM250 cargo truck
Cab seating: 1 + 2
Configuration: 6 × 6
Weight:
(kerb) 5,790 kg
(GVW) 10,330 kg
Max load: 4,540 kg
Length:
(without winch) 6.712 m
(with winch) 7.068 m
Width: 2.438 m
Height:
(top of tilt) 2.845 m
(top of cab) 2.467 m
Ground clearance: 0.278 m
Track:
(front) 1.721 m
(rear) 1.778 m
Wheelbase: 3.911 m
Angle of approach/departure:
(without winch) 48°/40°
(with winch) 40°/40°
Max speed: 96 km/h

Range: 560 km
Fuel capacity: 200 litres
Max gradient: 75%
Fording: 0.762 m
Engine: MAN DO846HM 7.225 litre 6-cylinder water-cooled diesel developing 160 hp at 2,500 rpm

KM250 (6 × 6) 2,500 kg truck

Gearbox: manual with 5 forward and 1 reverse gears
Clutch: single dry plate
Transfer box: 2-speed
Steering: cam and twin lever
Turning radius: 10.6 m
Suspension:
(front) semi-elliptical leaf springs with shock-absorbers
(rear) semi-elliptical inverted leaf springs
Tyres: 9.00 × 20
Brakes:
(main) air over hydraulic, internal expanding
(parking) dual grip mounted on transfer case
Electrical system: 24 V
Batteries: 2 × 12 V 100 Ah
Alternator: 60 A

Status
In production. In service with the armed forces of South Korea.

Manufacturer
Asia Motors Company Inc, 15 Yoido-dong, Yongdungpo-gu, PO Box 5022, Seoul, South Korea.
Tel: +82 2 783 1501. Telex: 24374 asiamco k.

VERIFIED

KM50 (6 × 6) 5 ton truck series

Description
The KM50 series of 5 ton (6 × 6) trucks are similar in appearance to the US M809 truck chassis but are fitted with MAN diesel engines and locally produced power trains. The KM50 series is produced in at least five forms: KM500 cargo truck; KM501 dump truck; KM502 wrecker (see entry in the *Recovery Vehicles* section); KM503 tractor truck; and KM507 expansible van.

A 9,070 kg capacity winch with 61 m of cable is optional on all models.

Specifications
KM500 cargo truck
Cab seating: 1 + 2
Configuration: 6 × 6
Weight:
(kerb) 9,440 kg
(GVW) 18,690 kg
Max load: 9,250 kg
Length:
(without winch) 7.652 m
(with winch) 8.045 m
Width: 2.477 m
Height:
(top of tilt) 2.946 m
(top of cab) 2.68 m
Ground clearance: 0.27 m
Track:
(front) 1.88 m
(rear) 1.829 m
Wheelbase: 4.547 m
Angle of approach/departure:
(without winch) 46°/38°
(with winch) 34°/38°
Max speed: 90 km/h
Range: 563 km
Fuel capacity: 295 litres
Max gradient: 75%
Fording: 0.762 m
Engine: MAN D2156 HM 10.35 litre 6-cylinder water-cooled diesel developing 236 hp at 2,200 rpm
Gearbox: manual with 5 forward and 1 reverse gears
Clutch: single dry plate
Transfer box: 2-speed
Steering: integral power-assisted
Turning radius: 12.75 m
Suspension:
(front) semi-elliptical leaf springs with shock-absorbers
(rear) semi-elliptical inverted leaf springs
Tyres: 11.00 × 20
Brakes:
(main) air over hydraulic, internal expanding
(parking) dual grip mounted on transfer case
Electrical system: 24 V
Batteries: 2 × 12 V, 100 Ah
Alternator: 60 A

Status
In production. In service with the armed forces of South Korea.

Manufacturer
Asia Motors Company Inc, 15 Yoido-dong, Yongdungpo-gu, PO Box 5022, Seoul, South Korea.
Tel: +82 2 783 1501. Telex: 24374 asiamco k.

VERIFIED

KM500 (6 × 6) 5 ton cargo truck

KM501 (6 × 6) 5 ton dump truck

MALAYSIA

Hicom FSS 32G (4 × 4) 3,000 kg truck

Development
First unveiled in April 1996, the Hicom FSS 32G (4 × 4) 3,000 kg truck was developed in close collaboration with the Malaysian Army and is manufactured by Diversified Resources Berhad (DRB). Following trials lasting six months a contract worth more than RM10 million ($4 million) was awarded for the initial supply of 60 units to the Malaysian Army. It is anticipated that the Malaysian Army will procure a further 3,000 to 4,000 units over a period lasting until 2001.

Several trucks have carried out trials in the Philippines and interest has been shown by Pakistan.

Description
The Hicom FSS 32G (4 × 4) 3,000 kg truck is based on a ladder chassis frame with channel sections with fully floating axles front and rear (FD 039 split at front, RO 65 banjo at rear). The suspension employs semi-elliptic steel alloy leaf springs with hydraulic double-acting telescopic shock-absorbers front and rear. Power-assisted steering is provided.

The forward control cab has seating for the driver and two passengers. Special attention is provided to supplying cooling air to the cab interior. A circular roof hatch is provided. The standard body is a cargo platform which may be covered by a canvas tilt over bows. A full 24 V electrical system is provided with a trailer socket provided close to the towing hook at the rear.

Standard equipment includes a rotary pintle hook at

Hicom FSS 32G (4 × 4) 3,000 kg truck ***1997***

the rear, two tie-down hooks at the front, a bull bar, a tyre inflator and hose, tool kits, a fire extinguisher and two rifle racks. Optional equipment includes an air conditioning system, an automatic lubrication system, and a vehicle monitoring system.

Specifications

Cab seating: 1 + 2
Configuration: 4 × 4
Weight:
(empty) approx 5,800 kg
(GVW) 8,980 kg
Max load: 3,000 kg
Towed load: 3,000 kg
Length: 6.935 m
Width: 2.4 m
Height:
(cab roof) 2.72 m
(top of tarpaulin) 3.23 m
Ground clearance: 0.295 m
Track: (front and rear) 1.945 m
Wheelbase: 3.8 m
Angle of approach/departure: 33°/32°
Max speed: (road) 105 km/h
Fuel capacity: 140 litres
Max gradient: (at GVW) 93%
Side slope: 30%
Fording: 0.8 m
Engine: DRB 6HE1 7.127 litre 6-cylinder in line water-cooled diesel developing 180 hp at 3,000 rpm
Gearbox: MGB 5A manual with 5 forward and 1 reverse gears
Clutch: single dry plate
Transfer box: 2-speed
Steering: recirculating ball, power-assisted
Turning radius: 17 m
Suspension: semi-elliptic steel leaf springs with telescopic shock-absorbers, front and rear
Tyres: 11.00 R 20-14PR
Brakes: air over hydraulic, dual-circuit
Electrical system: 24 V
Batteries: 2 × 12 V, 65 Ah
Alternator: 24 V/40 A

Status

In production for the Malaysian Army (60).

Manufacturers

Diversified Resources Berhad (DRB), Lot 3, Jalan Perusahaan Dua, Kawasan Perindustrian Batu Caves, 68100 Batu Caves, Selangor Darul Ehsan, Malaysia.
Tel: +603 6880225. Fax: +603 6894433.

NEW ENTRY

NETHERLANDS

DAF YA 4442 DNT (4 × 4) 4,000 kg truck

Description

In December 1985 the Dutch Ministry of Defence ordered 5,125 DAF YA 4442 DNT (4 × 4) 4,000 kg trucks. After verification trials, production commenced in mid-1988 and continued until 1994.

The DAF YA 4442 DNT is virtually identical to the YA 4440 (see following entry) but has a new radiator grill, the F 218 tilt cab front has revised protection, and there are dual or single tyres on the rear wheels. The main change was the fitting of the DAF DNT 620 six-cylinder turbocharged diesel engine matched to a ZF six-speed fully synchronised gearbox. Changes were also made to the electrical and brake systems.

The DAF YA 4442 DNT is produced in three main forms: the YA 4442 DNT is a general-purpose cargo vehicle, the YAL 4442 DNT is a general-purpose cargo vehicle with a driver training cab, and the YAK 4442 DNT is a cargo vehicle with a HIAB loading crane. The YAM 4442 is a sub-variant of the YA 4442 produced for the Dutch Marines.

The YF 4442 DNT 'Refueller' was developed by DAF for the Dutch Army and carries a 4,000 litre fuel tank and pump unit. A total of 395 was ordered with the last to be delivered in December 1994.

Specifications

YA 4442 DNT
Cab seating: 1 + 1 (up to 18 in rear)
Configuration: 4 × 4
Weight:
(empty) 7,620 kg
(loaded) 11,750 kg
Front axle, permissible: 6,000 kg
Rear axle, permissible: 6,750 kg
Max load: 4,000 kg
Towed load: 6,000 kg
Length: 7.3 m
Width: 2.47 m
Height: 3.42 m
Ground clearance: 0.29 m
Track:
(front) 1.9 m
(rear) 1.8 m
Wheelbase: 4.05 m
Angle of approach/departure: 36°/30°
Max speed: 93 km/h
Range: approx 500 km
Fuel capacity: 200 litres
Max gradient: 50%
Side slope: 30%
Fording: 0.9 m

DAF YF 4442 DNT 'Refueller' carrying Signaal Skyspy surveillance radar

DAF YA 4442 DNT (4 × 4) 4,000 kg truck (Michael Jerchel)

Engine: DAF DNT 6.242 litre 6-cylinder in-line turbocharged diesel developing 172 hp at 2,600 rpm
Gearbox: ZF S6-36 synchromesh with 6 forward and 1 reverse gears
Clutch: single dry plate (350 mm)
Transfer box: Steyr VG 450 2-speed
Steering: ZF 8043 hydraulic
Turning radius: approx 9.8 m
Suspension: semi-elliptical springs (8 front, 10 rear) with double-acting shock-absorbers
Tyres: 13 × 22.5
Brakes: air, dual-circuit plus exhaust brake
Electrical system: 24 V
Batteries: 2 × 12 V, 125 Ah
Alternator: 95 A

Status
Production of 5,125 units for the Dutch Ministry of Defence commenced in mid-1988.

Manufacturer
DAF Military Sales, Hugo van der Goeslaan 1, NL-5643 TW Eindhoven, Netherlands.
Tel: +31 40 143440. Telex: 51085 daf nl.
Fax: +31 40 144318.

VERIFIED

DAF YA 4440 (4 × 4) 4,000 kg truck

Development
The YA 4440 (4 × 4) 4,000 kg medium-mobility vehicle was developed by DAF to meet the requirements of the Dutch Army. The first five prototypes were handed over to the Army for trials late in 1974, and late in 1976 the Dutch Army placed an order for 4,000 for delivery between 1977 and 1980. Late in 1977 a further order for 2,500 was placed, with final deliveries made in mid-1983. The truck is based on proven commercial components and shared many common components with the DAF YA 2442 (4 × 4) 2 tonne truck (this was developed to prototype stage but not placed in production), including the cab, engine, transfer case and transmission. It was designed to carry 4,000 kg of cargo both on roads and across country and to tow a trailer with a maximum weight of 4,000 kg.

In February 1984 it was announced that Portugal had ordered 300 YA 4440 trucks following tests with four vehicles. The order was stated to be worth 26 million guilders and 35 per cent of the value of the vehicles was offset by assembling the vehicles in Portugal and using some Portuguese components. Starting in July 1984 the first vehicles were assembled in Portugal by EVICAR of Setubal. The order was completed by the end of January 1985.

By the end of 1985 DAF had supplied 7,250 YA 4440 trucks to the Dutch Ministry of Defence.

Description
The all-steel cab is of the forward control type and can be tilted forward to allow access to the engine for maintenance. The engine is mounted at the front of the chassis and power is transmitted to the two-speed transfer box via the five-speed synchromesh transmission. From the transfer case power is taken to the front and rear axles.

The reinforced cab roof allows the installation of a ring mounting for a light machine gun. The frame which supports this ring is bolted directly to the cab roof over the manhole cover. The rear cargo platform has bows, tarpaulin cover, side boards and a tailgate which can be quickly removed to enable the truck to carry containers or pallets.

The truck is fitted with air brakes, with the parking brake acting on the front axle, an exhaust brake and a trailer brake. The electrical system is waterproof and radio suppressed.

Optional equipment includes: a hydraulic crane to the rear of the cab with a capacity of 7,000 kg; stabilisers on each side of the chassis for when the hydraulic crane is being used; automatic transmission; two-man passenger seat in place of the standard one-man seat; and a manually operated crane with a 1,000 kg capacity.

The YAL 4440 is used for training drivers and has a four-man tilt cab with seats for the driver under instruction, instructor and two student drivers. In total, 375 of these vehicles were built for the Dutch Army.

DAF delivered 17 examples of the YAK 4440 for trials fitted with a hydraulic crane for rapid unloading of ammunition and cargo. These have the same chassis as the YA 4440 but a shorter cargo body with no bows or tarpaulin cover. Immediately behind the cab is a hydraulic crane and stabilisers, either side of the cab rear, which are lowered to the ground before the crane is used. Of the 17 units, eight had a PESCI (P445G) crane and nine a HIAB (850S) crane. Following trials with these prototype vehicles, the Dutch Army placed an order for 200 units fitted with the HIAB hydraulic crane.

Specifications
Cab seating: 1 + 1
Configuration: 4 × 4
Weight:
(empty) 7,000 kg
(loaded) 11,000 kg
Max load: 4,000 kg
Towed load: 4,000 kg
Length: 7.19 m
Width: 2.44 m
Height:
(tarpaulin) 3.42 m
(load area) 1.43 m
Ground clearance: 0.3 m
Track: 1.91 m
Wheelbase: 4.05 m
Angle of approach/departure: 36°/30°
Max speed: (road) 80 km/h
Range: 500 km
Max gradient:
(without trailer) 50%
(with loaded trailer) 20%
Side slope: 30%
Fording: 0.9 m
Engine: DAF DT 615 liquid-cooled 6-cylinder in-line turbocharged diesel developing 153 hp (DIN) at 2,400 rpm
Gearbox: ZF S5-35/2 with 5 forward and 1 reverse gears
Clutch: hydraulic, single dry plate
Transfer box: ZF VG 250/2 2-speed
Steering: ZF 8042 hydraulic power-assisted
Turning radius: 9 m
Suspension: semi-elliptical springs with double-acting hydraulic shock-absorbers on both axles
Tyres: 12.00 × 20
Brakes:
(main) air, dual-circuit (plus exhaust brake)
(parking) mechanical, also holding brake
Electrical system: 24 V
Batteries: 2 × 12 V, 100 Ah
Alternator: 85 Ah

Status
Production complete. In service with the Dutch Army and Navy and Portuguese Army (300). 200 passed to Botswana.

Manufacturer
DAF Military Sales, Hugo van der Goeslaan 1, NL-5643 TW Eindhoven, Netherlands.
Tel: +31 40 143440. Telex: 51085 daf nl.
Fax: +31 40 144318.

VERIFIED

DAF YA 4440 (4 × 4) 4,000 kg truck carrying specialised container (T Neate)

DAF YAK 4440 (4 × 4) 4,000 kg truck fitted with HIAB hydraulic crane and stabilisers (T Neate)

DAF YA 5444 DNT (4 × 4) 5,000 kg truck

Description
The DAF YA 5444 DNT (4 × 4) 5,000 kg truck is a further development of the DAF YA 4442 series designed for logistic roles such as a carrier for generator sets, fuel tanks, munitions, shelters and so on. The three-seater cab has a purpose-designed reinforced roof with a hatch and provision for a machine gun mounting. Production is in progress with first deliveries going to the Royal Netherlands Air Force. The YA 5444 will eventually replace the YA 5441 and YA 5442 models in service.

Specifications
Cab seating: 1 + 2
Configuration: 4 × 4
Weight:
(empty) 8,400 kg
(loaded) 13,600 kg
Front axle, permissible: 6,000 kg
Rear axle, permissible: 8,000 kg
Max load: 5,000 kg
Towed load: 6,000 kg
Length: 8.2 m
Width: 2.5 m

DAF YA 5444 DNT (4 × 4) 5,000 kg truck

Height: (cab) 3 m
Ground clearance: 0.26 m
Track:
(front) 1.9 m
(rear) 1.8 m
Wheelbase: 4.05 m
Angle of approach/departure: 36°/20°
Max speed: 95 km/h
Range: approx 500 km
Fuel capacity: 200 litres
Max gradient: 50%
Side slope: (static) 30%
Fording: 0.9 m
Engine: DAF DNT 6.242 litre 6-cylinder in-line turbocharged diesel developing 172 hp at 2,600 rpm
Gearbox: ZF S6-36 with 6 forward and 1 reverse gears
Clutch: single dry plate (350 mm)
Transfer box: Steyr VG 450 2-speed
Steering: ZF 8043 power-assist
Turning radius: approx 9.8 m
Suspension: semi-elliptic leaf spring (8 front, 10 rear) plus telescopic, hydraulic, double-acting shock-absorbers front and rear
Tyres: 10.00 × 20
Brakes:
(main) dual-circuit, 2-line air
(parking) spring cylinders on rear axle
Electrical system: 24 V
Batteries: 2 × 12 V, 125 Ah
Alternator: 95 A

Status
In service with the Dutch armed forces.

Manufacturer
DAF Military Sales, Hugo van der Goeslaan 1, NL-5643 TW Eindhoven, Netherlands.
Tel: +31 40 143440. Telex: 51085 daf nl.
Fax: +31 40 144318.

VERIFIED

DAF YA 5441 and YA 5442 (4 × 4) 5,000 kg trucks

Description
The DAF YA 5441 (4 × 4) 5,000 kg truck was designed to carry 5,000 kg of cargo both on and off roads. It was developed from a standard commercial design and uses many components of the YA 4440 (4 × 4) 4,000 kg truck in service with the Dutch Army and Navy. A second production series with slight modifications is known as the YA 5442.

The cab is all steel and has a circular observation hatch in the roof. The cab can be tilted forward to allow access to the engine for maintenance. The rear cargo area is of all-steel construction and has drop sides and a drop tailgate with integral steps. Removable bows and a tarpaulin cover are fitted as standard.

The engine is mounted at the front of the vehicle, with power being transmitted to the two-speed transfer box via the five-speed (plus one reverse) synchromesh gearbox. From the transfer case power is transmitted to the front and rear axles.

Standard equipment includes a front bumper with a built-in push/pull pin, rear towing hook, exhaust brake retarder and connections for the brakes on the trailer.

Variants
YAK 5442
The YAK 5442 was produced during 1984 for the Dutch Air Force (16) and is a crane truck mounting a HIAB Type 2027 AVL 20 t/m hydraulic crane over the rear axle. Telescopic outrigger legs on the ends of sliding jibs are provided at each corner of the cargo body area. Some vehicles have 8,000 kg recovery winches.

Specifications

Model	YA 5441	YA 5442
Cab seating	1 + 2	1 + 2
Configuration	4 × 4	4 × 4
Weight		
(empty)	7,290 kg	7,300 kg
(loaded)	12,290 kg	12,600 kg
Weight on front axle (max load)	4,900 kg	4,800 kg
Weight on rear axle (max load)	8,100 kg	8,100 kg
Max load	5,000 kg	5,000 kg
Towed load	4,000 kg	4,000 kg
Length	7.54 m	7.54 m
Width	2.47 m	2.44 m
Height		
(overall)	2.96 m	2.96 m
(load area)	1.3 m	1.28 m
Ground clearance (axle)	0.265 m	0.265 m
(chassis)	0.45 m	0.45 m
Track (front)	1.938 m	1.93 m
(rear)	1.729 m	1.8 m
Wheelbase	3.85 m	3.85 m
Angle of approach/departure	32°/23°	32°/24°
Max speed	80 km/h	80 km/h
Range	500 km	500 km
Fuel capacity	200 litres	200 litres
Max gradient (without trailer)	59%	60%
(with trailer)	38%	40%
Side slope	30%	30%
Fording	0.6 m	0.6 m
Engine	DAF model DT 615 6-cylinder in-line 6.17 litre liquid-cooled turbocharged diesel developing 153 hp at 2,400 rpm	
Gearbox	ZF S5-35/2 manual with 5 forward and 1 reverse gears	
Clutch	single dry plate	single dry plate
Transfer box	ZF VG250/2 2-speed	ZF VG250/2 2-speed

DAF YAK 5442 with HIAB crane for Dutch Air Force

DAF YA 5442 DT (4 × 4) 5,000 kg truck

Specifications

Model	**YA 5441**	**YA 5442**
Steering	ZF 8065 power-assisted	ZF 8065 power-assisted
Suspension	semi-elliptical springs front and rear with double-acting telescopic shock-absorbers front and rear	
Tyres	10.00 × 20	10.00 × 20
Brakes (main)	air, 2-line, dual-circuit	air, 2-line, dual-circuit
(parking)	spring-brake cylinders on rear axle	
Electrical system	24 V	24 V
Batteries	2 × 12 V, 100 Ah	2 × 12 V, 100 Ah

YA 5442 DH
This version has a naturally aspirating diesel engine.

YA 5442 DT
This variant uses a slightly different cargo body and is used by Dutch Army Lance missile batteries for logistic support. This vehicle was also produced for issue to the NATO NORTHAG Transport Company.

The basic chassis can be adapted for other specific roles such as water tanker, fuel carrier and radio vehicle. DAF also delivered to the Dutch Air Force a number of 4 × 2 tractor trucks based on the commercial FT 1600 series chassis.

Status
Replaced in production by YA 5444. In service with the Dutch Army, Air Force and Navy and some NATO forces.

Manufacturer
DAF Military Sales, Hugo van der Goeslaan 1, NL-5643 TW Eindhoven, Netherlands.
Tel: +31 40 143440. Telex: 51085 daf nl.
Fax: +31 40 144318.

VERIFIED

DAF YAV 2300 DHTD (4 × 4) 7,000 kg truck

Description
The DAF YAV 2300 DHTD (4 × 4) 7,000 kg truck was derived from the YAZ 2300 (6 × 6) 10,000 kg series (see next entry). It was designed to meet military specifications using components from DAF's commercial vehicle range.

In 1984 the Dutch Ministry of Defence placed an order, on behalf of NATO, for 85 of these trucks. Deliveries were made in 1985. Some of these are used as prime movers for the 20 tonne trailers used with NATO's AFCENT Mobile War Headquarters. Some of the vehicles involved are used to carry shelters and electronic equipment.

The basic DAF YAV 2300 DHTD (4 × 4) truck is fitted with a cargo body equipped with a load platform and side boards; these can be removed to provide a platform for a container/shelter or a fuel distribution unit. The forward control cab is derived from the commercial short F 218 model and has a reinforced roof to allow the installation of a machine gun over the roof hatch. It can be tilted forward for engine and transmission access. The frame uses a ladder-type construction that is partially welded and partially bolted with two U-type reinforced longitudinal members and Omega-type cross members.

A spare wheel with a handling winch is provided behind the cab. At the front of the vehicle is located a NATO push/pull pin while the rear is equipped with a 24 tonne towing hook. Two towing and two lifting eyes are provided on the front bumper. A trailer brake system is provided.

YTV 2300
The DAF YTV 2300 is a 4 × 4 tractor intended for towing semi-trailers. Loads towed can include water or fuel tank semi-trailers or light armoured vehicle semi-trailers. An ambulance semi-trailer is another possibility. The YTV 2300 has a 7 tonne capacity fifth wheel (2 in/51 mm) and a kerb weight of 8,000 kg. Length overall is 6.1 m, width 2.49 m and height (cab) 3.02 m. Wheelbase is 3.6 m.

DAF YAV 2300 DHTD (4 × 4) 7,000 kg truck carrying electronics container

Specifications
Cab seating: 1 + 1 or 2
Configuration: 4 × 4
Weight:
(kerb) 10,200 kg
(GVW) 15,500 kg
Max permissible weight on front axle: 6,500 kg
Max permissible weight on rear axle: 9,000 kg
Max load:
(road) 7,000 kg
(cross-country) 5,300 kg
Length: 7.82 m
Width: 2.46 m
Height: (overall) 3.1 m
Load area: 5 × 2.35 m
Ground clearance: 0.32 m
Track:
(front) 1.98 m
(rear) 1.82 m
Wheelbase: 4.5 m
Angle of approach/departure: 30°/30°
Max speed: 88 km/h
Range: (with 300 litre fuel tank) 1,000 km
Fuel capacity: 300 litres
Max gradient: 50%
Side slope: 30%
Fording: 0.9 m
Engine: DAF DHTD 825 Mil. 8.25 litre 6-cylinder in-line direct injection turbocharged diesel developing 213 hp (ISO) at 2,400 rpm
Torque converter: ZF WSK 400/25
Gearbox: ZF 5S-111 GPA with 8 forward and 1 reverse gears
Steering: ZF 8046 hydraulic, power-assisted
Turning radius: min 11 m
Suspension:
(front) DAF trapezium with adjustable KONI hydraulic telescopic double-acting shock-absorbers
(rear) DAF trapezium with auxiliary spring and adjustable KONI hydraulic telescopic double-acting shock-absorbers
Tyres: 13 R 22.5K (14.75/80 R 20 optional)
Brakes:
(main) dual-circuit air-mechanical plus exhaust retarder
(parking) spring-brake cylinders on rear axle
Electrical system: 24 V
Batteries: 2 × 12 V, 125 Ah
Alternator: 55 A

Status
In service with NATO AFCENT HQ.

Manufacturer
DAF Military Sales, Hugo van der Goeslaan 1, NL-5643 TW Eindhoven, Netherlands.
Tel: +31 40 143440. Telex: 51085 daf nl.
Fax: +31 40 144313.

UPDATED

DAF YAZ 2300 (6 × 6) 10,000 kg truck

Development
In mid-1981, following trials with a number of prototypes submitted by several manufacturers, the Dutch Ministry of Defence placed an order with DAF Trucks worth Dfl236.5 million for a family of new 10,000 kg (6 × 6) trucks to replace the then current DAF YA 616 6,000 kg (6 × 6) trucks. The YA 2300 series is based on the commercial 2300 range and is powered by a DAF DHS 825 250 hp diesel.

Delivery of the first batch commenced in mid-1983 and was completed during 1985. In 1988 the Dutch Ministry of Defence ordered over 1,400 YA 2300 series vehicles.

Description
The YAZ 2300 (6 × 6) general cargo truck may be taken as being typical of the other vehicles in the series. It has a chassis designed as a common ladder-type platform and comprises two U-type and Omega-type cross members. The longitudinal members are strengthened over their full length.

The fully enclosed two-door all-steel cab was developed from the civilian F 218 type and can be tilted forward to an angle of 60° allowing access to the engine for maintenance. The cab has a large capacity heating and ventilation system and thermal and sound insulation. The driver's and co-driver's seats are adjustable, with the third seat with foot rest being mounted against the cab rear wall. The cab roof, in the cargo model only, is fitted with an observation hatch on which can be mounted a light machine gun.

The rear cargo area is of aluminium construction and is provided with removable drop sides and a drop tailgate. To the rear of the cab is a hydraulic crane for unloading cargo but, before this is used, stabilisers are lowered to the ground either side of the vehicle to the rear of the cab.

The front axle is a Kirkstall DS 65 and is fitted with stabilisers. The rear bogie is a DAF 2699 T with

DAF YTZ 2301 (6 × 6) truck tractor with handling crane in use with Patriot SAM system

DAF YTZ 2301 (6 × 6) truck tractor towing Patriot tactical air defence system missile launcher

lockable differentials and a lockable inter-axle differential.

Standard equipment includes: a 7 and 12 pole trailer connector, slave connector, trouble light connector at front and rear, standard commercial traffic lighting and blackout lighting to NATO specifications. At the front is a push/pull NATO pin and at the rear is a 24 tonne towing hook. Optional equipment includes a ZF automatic transmission and a larger four-man fully enclosed cab; when the latter is fitted, however, the crane is not mounted to the rear of the cab.

Variants

YAZ 2300 Snow Fighting Vehicle
This is a special snow clearing vehicle used by the Dutch Air Force for clearing airfield runways. It is equipped with a de-icing fluid tank, pumping gear and swing-out spray bars. It is possible to fit a snow plough on the front.

YKZ 2300 Tipper
This was developed as the replacement for the DAF YK 616 three-way tipper. It has a maximum payload of 14,000 kg and a kerb weight of 13,500 kg.

YHZ 2300 Artillery Tractor
This variant was ordered by the Dutch Ministry of Defence and production of 78 commenced in 1987. It is used to tow 155 mm M114 and M114/39 howitzers and has seating for 12 personnel and two 12 round racks for ammunition.

YGZ 2300 Bridging Vehicle
This is in production to carry and launch Ribbon Bridge units using a DAF Special Products system in place of the normal cargo body. The system can also be used to carry and launch rolls of trackway.

YBZ 2300 Recovery Vehicle
This is a special recovery vehicle equipped with a side-mounted hydraulic crane for the recovery of Patriot SAM vehicles and units.

YAZ 2301 Truck
Although described as a general cargo truck, the YAZ 2301 (6 × 6) is used to carry container/shelters associated with the Patriot tactical air defence system.

Specifications

Model	**YAZ 2300 Truck**	**YKZ 2300 Tipper**	**YHZ 2300 Tractor**	**YGZ 2300 Bridging**
Cab seating	1 + 2	1 + 2	1 + 2	1 + 2
Configuration	6 × 6	6 × 6	6 × 6	6 × 6
Weight (kerb)	13,500 kg	13,500 kg	12,500 kg	13,700 kg
(max permissible)	27,500 kg	27,500 kg	25,500 kg	20,150 kg
Max load	10,000 kg	14,000 kg	10,000 kg	6,500 kg
Max gross combination weight	40,000 kg	40,000 kg	40,000 kg	40,000 kg
Length	9.55 m	8.03 m	7.74 m	8.75 m
Width	2.49 m	2.49 m	2.49 m	3.45 m
Height (overall)	3.59 m	3.54 m	3.59 m	4 m
(load area)	1.5 m	1.5 m	1.63 m	1.55 m
Ground clearance	0.32 m	0.32 m	0.32 m	0.32 m
Track (front)	1.98 m	1.98 m	2.05 m	2.05 m
(rear)	1.82 m	1.82 m	2.01 m	2.1 m
Wheelbase	4.85 m	4.45 m	4.45 m	5.35 m
Angle of approach/departure	30°/20°	30°/30°	30°/32°	30°/20°
Max speed	88 km/h	88 km/h	88 km/h	88 km/h
Range (minimum)	600 km	600 km	600 km	600 km
Fuel capacity	300 litres	300 litres	300 litres	300 litres
Max gradient	50%	50%	50%	50%
Side slope	30%	30%	30%	30%
Fording	0.75 m	0.75 m	0.75 m	0.75 m
Engine	DAF DHS 825 8.25 litre 4-stroke, 6-cylinder in-line direct injection, turbocharged liquid-cooled diesel developing 250 hp at 2,400 rpm			
Torque converter	ZF WSK 400/1	ZF WSK 400/1	ZF WSK 400/1	ZF WSK 400/1
Gearbox	ZF 5S-110 GPA with 8 forward and 1 reverse gears. Transfer box is a ZF A 600/3D with lockable differential and is an integral part of the gearbox			
Steering	ZF 8046, hydraulic power-assisted			
Turning radius	11.2 m	11 m	11 m	12 m
Suspension (front)	semi-elliptical springs and hydraulic telescopic double-action shock-absorbers			
(rear)	semi-elliptical springs			
Tyres	13 R 22.5	13 R 22.5	14.75 R 20	14.75/80 R 20
Brakes (main)	air, mechanical, 2-line, dual-circuit			
(parking)	spring brake	spring brake	spring brake	spring brake
(retarder)	exhaust brake	exhaust brake	exhaust brake	exhaust brake
Electrical system	24 V	24 V	24 V	24 V
Batteries	2 × 12 V, 125 Ah	2 × 12 V, 125 Ah	2 × 12 V, 125 Ah	2 × 12 V, 125 Ah
Alternator	55 A	55 A	55 A	55 A

YTZ 2301 Truck Tractor
This is used to tow the Patriot tactical air defence system missile launcher.

Status

In service with the Dutch Army, Marines and Air Force (Patriot).

Manufacturer

DAF Military Sales, Hugo van der Goeslaan 1, NL-5643 TW Eindhoven, Netherlands.
Tel: +31 40 143440. Telex: 51085 daf nl.
Fax: +31 40 144318.

UPDATED

PAKISTAN

Yasoob (6 × 6) 6,000 kg truck

Development

The decision to proceed with the development of an indigenous family of military and commercial trucks was taken by the Pakistan government in July 1989. The first two prototypes of the resultant military truck, named Yasoob after the fastest horse of the Prophet, were handed over to the Pakistan Army for trials during 1990. These vehicles had 40 per cent local content, which is expected to increase when full production commences. Components for all major vehicle units, such as the engine, transmission, axles and transfer, will be manufactured under licence (with provision for buy-back arrangements). Production was scheduled to commence about eight months after the completion of the initial test and trial programmes.

Description

The Yasoob (6 × 6) 6,000 kg truck is an all-purpose vehicle intended for general logistic purposes and can be used to tow artillery or trailers. The standard model is a cargo truck with a maximum highway load capacity of 12,000 kg (6,000 kg cross-country). Layout is conventional, with the 8.3 litre diesel engine at the front, a two-door, three-seat, fully enclosed cab in the centre and the cargo area with high drop sides and a tailgate at the rear. The cargo area can be covered by bows and a tarpaulin, with the bows stowed behind the cab when not in use. Bench seats can be fitted for troop carrying.

The frame uses rolled channel construction with six reinforced cross members. The front and rear axles have spiral bevel hub reduction.

Three wheelbase lengths (front axle to rear bogie) are available: 4.3 m, 4.6 m and 5.5 m. The standard cargo version has the 4.6 m wheelbase chassis but there is also a lengthened cargo bed variant which uses the 5.5 m chassis. Van-bodied variants (for command post or mobile stores functions) can be produced for all three wheelbase lengths. Other variants in the Yasoob range include a dump truck, a wrecker and a tractor truck for towing loads up to 25,000 kg; all three use the 4.3 m wheelbase chassis. There will also be fuel or water tanker versions (4.6 m chassis) and a bridging equipment carrier (5.5 m chassis).

Commercial versions of the Yasoob will have gross vehicle weights of from 14,000 to 24,000 kg.

Yasoob (6 × 6) 6,000 kg truck with hydraulic load handling crane ***1997***

Specifications

standard cargo version
Cab seating: 1 + 2
Configuration: 6 × 6
Weight:
(loaded, front axle) 6,500 kg
(loaded, rear bogie) 15,000 kg
(total) 21,500 kg
Max load:
(road) 12,000 kg
(cross-country) 6,000 kg
Towed load: 10,000 kg
Length: 7.73 m
Width: 2.5 m
Height:
(unloaded, overall) 3.1 m
(reduced) 2.75 m
Ground clearance: 0.335 m
Track:
(front) 2.063 m
(rear) 2.112 m
Wheelbase: 4.6 m + 1.4 m
Angle of approach/departure: 40°/35°
Max speed with towed load:
(road) 80 km/h
(cross-country) 50 km/h
Range: (cruising) 550 km
Fuel capacity: 270 litres
Gradient: (with towed load) 60%
Fording: 0.75 m
Engine: 8.3 litre 6-cylinder direct injection diesel developing 240 hp at 2,400 rpm
Gearbox: manual with 9 forward and 2 reverse gears
Clutch: twin dry plate
Transfer box: 2-speed
Steering: ZF 8097 power-assisted
Turning radius: 12.75 m
Suspension:
(front) semi-elliptic 8-leaf springs
(rear) inverted semi-elliptic 13-leaf springs
Tyres: 13.00 R 20
Brakes:
(main) air
(parking) spring/air
Electrical system: 24 V
Batteries: 2 × 12 V, 150 Ah
Alternator: 150 A

Status

In production.

Manufacturer

Pakistan state factories.

VERIFIED

POLAND

Star 266 (6 × 6) 3,500 kg truck

Description

The Star 266 3,500 kg truck was the replacement for the older Star 66 2,500 kg range of 6 × 6 trucks. The first prototype was completed in 1971 and production began in 1976. Major improvements over the earlier vehicle are its increased load-carrying capability, more powerful engine, all-steel fully enclosed cab and larger tyres. The cab is of the forward control type and the all-steel rear cargo area has removable bows, tarpaulin cover and drop tailgate. A winch with a capacity of 6,000 kg is fitted.

A tropical version of the Star 266 is produced. This is similar to the standard model but is 7.07 m long and weighs 7,350 kg unladen.

The following special bodies are produced for the Star 266.

SARNA II: van body configured as workshop, command post, ambulance, mobile dressing station, laundry, bakery or canteen.
786: sealed container body for electronic equipment.
Waggon bodies for special laboratories.
DSRT-0081: 8,000 kg lift capacity mobile crane.
DSR-0101: 10,000 kg lift capacity mobile crane.
WUS-3: mobile cleaning station for heavy equipment.
A3-578: fuel tanker.
A3-579.2: fuel tanker.
A3-522: fuel tanker.
A3-531: fuel dispenser station.
A3-518: NBC decontamination station for personnel and equipment; also used for firefighting.
Platform truck for bridging pontoons.

Specifications

Cab seating: 1 + 2
Configuration: 6 × 6
Weight:
(empty) 7,350 kg
(loaded) 12,200 kg
Max load: 3,500 kg
Towed load: 4,000 kg
Load area: 3.9 × 2.2 m
Length: 6.9 m
Width: 2.5 m
Height:
(cab) 2.69 m
(tarpaulin) 3.22 m
Ground clearance: 0.325 m
Track: 1.97 m
Wheelbase: 2.99 m + 1.25 m
Angle of approach/departure: 37°/42.5°
Max speed: (road) 90 km/h
Fuel capacity: 300 litres
Range: 800 km
Max gradient: 76%
Fording: 1.8 m
Engine: S-359M 6.842 litre 6-cylinder water-cooled diesel developing 147 hp at 2,800 rpm
Gearbox: ZF S5-45 manual, 5 forward and 1 reverse gears
Clutch: single dry plate
Transfer box: 2-speed
Steering: ZF 8060 with hydraulic servo
Turning radius: 8.5 m
Suspension: semi-elliptic leaf springs with double-acting hydraulic shock-absorbers
Tyres: 12.00 × 20

Brakes:
(main) hydraulic
(parking) mechanical
Electrical system: 24 V

Status
In service with the Polish Army.

Manufacturer
Fabryka Samochodów Ciężarowych, PL-27-202 Starachowice, Poland.
Telex: 0612571.

UPDATED

Star 266 (6 × 6) 3,500 kg truck

Star 28 and Star 200 series (4 × 2) trucks

Description
In 1968 the Star 28 and Star 29 entered production as the replacements for the Star 25 and Star 27. The Star 28 is powered by an S-530A1 six-cylinder diesel which develops 100 hp at 2,600 rpm and gives the vehicle a maximum road speed of 81 km/h. Payload on roads is 5,000 kg. The Star 29 is powered by an S-474 six-cylinder water-cooled petrol engine, which develops 105 hp at 3,000 rpm and gives the vehicle a maximum road speed of 81 km/h. Payload on roads is 5,000 kg and towed load is 5,250 kg.

The Star 28 was replaced by the Star 200 and Star 244 (see following entry). The Star 200 is powered by an S-533 (90 hp) or an S-359 (150 hp) engine and can carry a maximum payload of up to 6,000 kg on roads; towed load is 8,500 kg. The basic Star 200 is a 4 × 2 general-purpose platform truck with a 3.4 m wheelbase; the Star 200L has a 3.9 m wheelbase. Variants include the C200 tractor truck, the A200 produced to carry box bodies, the A200K produced for various utility bodies, the A200A (wheelbase 4.1 m) for bus bodies, and the W200 for tipper bodies.

The Star 200 is powered by an S-359 engine which develops 150 hp at 2,800 rpm and is coupled to a ZF S5-45 manual gearbox with five forward and one reverse gears. Brief specifications are: length 6.51 m, width 2.33 m, payload 6,000 kg, towed load 8,500 kg and maximum road speed 90 km/h.

There are many variants of the Star 28 and Star 200 series including tankers (for example Star 20C, 21C (3,500 litres), A3-573 and C-28), tippers (for example W-28) and crane trucks (for example Z-28).

Status
In service with Polish armed forces.

Manufacturer
Fabryka Samochodów Ciężarowych, PL-27-202 Starachowice, Poland.
Telex: 0612571.

VERIFIED

Star 244 (4 × 4) 5,000 kg truck

Description
The Star 244 (4 × 4) 5,000 kg truck series was developed from the 4 × 2 Star 200 series and differs from the earlier model mainly by the addition of front-axle drive and a reduction gear. The Type 359 6.842 litre diesel engine is retained. The forward control all-steel cab was developed in association with Chausson of France.

Variants of the basic Star 244 are: the Star 244RS, produced primarily for agricultural purposes the A244 adapted to carry various forms of cross-country bodies; the P244L (wheelbase 3.9 m) for mounting fire truck bodies; and the 3W244 (wheelbase 3 m) for tipper bodies.

Specifications
Configuration: 4 × 4
Weight: (empty) 5,500 kg
Max load: 5,000 kg
Towed load: 8,500 kg
Length: 6.34 m
Width: 2.46 m
Height: (cab) 2.59 m
Ground clearance: 0.28 m
Track:
(front) 1.9 m
(rear) 1.8 m
Wheelbase: 3.4 m
Angle of approach/departure: 40°/27°
Engine: S-359 6.842 litre 6-cylinder water-cooled diesel developing 150 hp at 2,800 rpm
Gearbox: ZF S5-45 manual with 5 forward and 1 reverse gears
Transfer box: 2-speed
Clutch: single dry plate
Steering: ZF 8060 with hydraulic servo
Tyres: 8.25 × 20
Electrical system: 24 V
Batteries: 2 × 12 V, 135 Ah
Alternator: 30 A

Status
In production.

Manufacturer
Fabryka Samochodów Ciężarowych, PL-27-202 Starachowice, Poland.
Telex: 0612571.

VERIFIED

Star 244 (4 × 4) 5,000 kg truck (Henry Dodds)

PORTUGAL

Bravia Leopardo Mk III (6 × 6) 3,000 kg truck

Description
The Leopardo Mk III, which is based on the US M44 series, was designed to carry 3,000 kg of cargo across country or up to 6,000 kg on roads. Its layout is conventional, with the engine at the front, cab in the centre with a windscreen which folds forward onto the bonnet and a removable canvas top, and the cargo area at the rear with hinged seats, drop tailgate, removable bows and a tarpaulin cover. It is available with single or dual rear wheels and tracks can be fitted to the dual rear wheel version to increase cross-country mobility. The vehicle can be encountered with one of four types of engine: Model V-8 318-3 petrol developing 202 hp at 2,400 rpm; a V-8 361-3 petrol developing 210 hp at 4,000 rpm; or a British Perkins diesel type 6-354-2D; the later models are powered by an 8 litre turbo diesel developing 180 hp at 3,600 rpm. Optional equipment includes: a fully enclosed cab; Braden Model MU2-3 winch with a capacity of 4,536 kg; 24 V electrical system; power steering; and an additional 250 litre fuel tank.

The Leopardo Mk I is a 4 × 2 truck with a 4.179 m wheelbase, gross vehicle weight of 8,845 kg and a 10,206 kg gross combination weight. The Leopardo Mk II is also a 4 × 2 truck, with a 4.444 m wheelbase, gross vehicle weight of 10,206 kg and a gross combination weight of 16,330 kg.

Variants include a dump truck, fuel and water tankers, van bodies, a light wrecker and firefighting vehicles.

Specifications
Cargo version
Cab seating: 1 + 2
Configuration: 6 × 6
Weight:
(empty) 5,900 kg
(loaded, road) 10,400 kg
Max load:
(road) 6,000 kg
(cross-country) 3,000 kg
Length: 6.7 m

Bravia Leopardo III (6 × 6) 3,000 kg truck with single rear wheels

Bravia Leopardo III (6 × 6) 3,000 kg truck with tracks fitted to rear wheels to increase traction in soft soil

Width: 2.4 m
Height:
(overall) 3.2 m
(reduced) 2.1 m
Ground clearance: 0.28 m
Track:
(front) 1.721 m
(rear) 1.778 m
Wheelbase: 2.9 m
Angle of approach/departure: 47°/40°
Fuel capacity: 189 litres
Max speed: 100 km/h
Range: 480 km
Max gradient: 60%
Side slope: 45°
Fording: 0.78 m
Engine: see text
Transmission: Allison automatic, 6 forward and 1 reverse gears
Transfer box: T-136 2-speed
Steering: manual
Turning radius: 13 m
Suspension:
(front) semi-elliptical springs and hydraulic double-acting shock-absorbers
(rear) Hendrickson type with semi-elliptical springs
Tyres:
(double) 9.00 × 20
(single) 11.00 × 20
Brakes:
(main) air over hydraulic
(parking) mechanical
Electrical system: 24 V
Batteries: 2 × 12 V 6TN, 100 A

Status
In service with the Portuguese armed forces.

Manufacturer
Bravia SA International, Sociedade Luso-Brasileira de Viaturas e Equipamentos, Estrada Nacional No 10, KM 107 Porto Alto, P-2135 Samora Correia, Portugal.
Tel: +351 63 93256/7. Telex: 12715 bravia p.

UPDATED

Bravia Pantera (6 × 6) 6,000 kg truck

Description
The Pantera, which is based on the US M809/M939 trucks, was designed to carry 12,000 kg of cargo on roads or 6,000 kg of cargo across country. Its layout and appearance are almost identical to the Leopardo Mk III and it has the same wheelbase. It can be delivered with one of four types of engine: Perkins V-8-510 diesel developing 170 hp at 2,800 rpm; Perkins V-8-540 diesel developing 180 hp at 2,600 rpm; Cummins 378C diesel; or a 361-4 petrol engine. Optional equipment is the same as the Leopardo except that the winch installed is an MU12 rated at 9,072 kg. Tracks may be fitted to the dual rear wheels to increase mobility on soft ground.

Variants include a long wheelbase (4.775 m + 1.371 m) cargo truck, dump trucks, a tractor truck, a wrecker, a van-body truck and firefighting vehicles.

Bravia Pantera (6 × 6) 6,000 kg truck with front-mounted winch, bows and tarpaulin cover

Specifications
Cab seating: 1 + 2
Configuration: 6 × 6
Weight:
(empty) 9,733 kg
(loaded, road) 18,985 kg
Max load:
(road) 12,000 kg
(cross-country) 6,000 kg
Towed load:
(road) 13,620 kg
(cross-country) 6,802 kg
Length: (with winch) 7.52 m
Width: 2.464 m
Height:
(overall) 2.946 m
(reduced) 2.172 m
Ground clearance: (axles) 0.295 m
Track:
(front) 1.88 m
(rear) 1.829 m
Wheelbase: 3.86 m + 1.371 m
Angle of approach/departure: 46°/35°
Max speed: 100 km/h
Fuel capacity: 500 litres
Range: 746 km
Max gradient:
(without towed load) 67°
(with towed load) 42°
Side slope: 30%
Fording: 0.76 m
Engine: see text
Gearbox: manual, with 5 forward and 1 reverse gears
Clutch: single dry plate
Transfer box: 2-speed
Turning radius: 12.75 m
Tyres:
(double) 11.00 × 20
(single) 14.00 × 20
Brakes:
(main) air over hydraulic
(parking) mechanical
Electrical system: 24 V
Batteries: 4 × 12 V, 100 Ah

Status
In service with the Portuguese armed forces.

Manufacturer
Bravia SA International, Sociedade Luso-Brasileira de Viaturas e Equipamentos, Estrada Nacional No 10, KM 107 Porto Alto, P-2135 Samora Correia, Portugal.
Tel: +351 63 93256/7. Telex: 12715 bravia p.

UPDATED

ROMANIA

SR-131 trucks and variants

Description
The SR-131 Bucegi (4 × 2) truck replaced the SR-101 vehicle in production. It is powered by an SR-211 eight-cylinder diesel engine which develops 140 hp and gives the vehicle a maximum speed of 90 km/h. Payload on roads is 3,000 kg. The 4 × 4 model, designated the SR-132 Bucegi, is almost identical and is powered by the same engine. The layout of both vehicles is identical, with the engine at the front, two-door all-steel cab in the centre and the cargo area at the rear with drop sides and a drop tailgate.

Further development of the SR-131/SR-132 vehicles resulted in the SR-113 (4 × 2) and SR-114 (4 × 4)

SR-132 Bucegi (4 × 4) truck

Carpati trucks powered by the same engine as the earlier vehicles but with a wheelbase of 4 m compared with the earlier vehicle's wheelbase of 3.4 m. The SR-113's payload on roads is 5,000 kg and the SR-114's is 4,000 kg. Both vehicles are fitted with 9.00 × 20 tyres. The SR-114 is used as a carrier for Romanian Army 21 round BM-21 multiple rocket systems (for details of the BM-21 see *Jane's Armour and Artillery 1995-96,* page 743 or pages 715 and 716.

Tractor truck models (4 × 2) are designated the SR-116, while dump trucks (4 × 2) include the ABS-116 and AB45-116. Ambulance and van-type versions are also in service. Other Bucegi vehicles include the 7-BA-1 L (4 × 2) 7,000 kg truck, 7-BA-1 N (4 × 2) 6,000 kg truck and the 18-BTA-1 (4 × 2) tractor truck which can pull a semi-trailer weighing up to 18,000 kg. The AC-302 is a 4,000 litre tanker based on the SR-113 chassis. As far as it is known these are not used by the Romanian Army.

Specifications

SR-132 (4 × 4) truck
Cab seating: 1 + 2
Configuration: 4 × 4
Weight:
(empty) 3,750 kg
(loaded, road) 6,750 kg
(loaded, dirt road) 5,750 kg
Max load:
(road) 3,000 kg
(dirt road) 2,000 kg
Towed load:
(road) 3,000 kg
(dirt road) 2,000 kg
Length: 5.78 m
Width: 2.263 m
Height: (cab) 2.1 m
Ground clearance: 0.27 m
Track: 1.75 m
Wheelbase: 3.4 m
Max speed: (road) 95 km/h
Engine: SR-211 V-8 water-cooled petrol developing 140 hp at 3,600 rpm
Gearbox: manual with 4 forward and 1 reverse gears
Transfer box: 2-speed
Tyres: 9.75 × 18

Status

Production ceased in 1979. In service with the Romanian Army.

Manufacturer

ROMAN SA, Poienelor Str. No 5, R-2200 Braşov, Romania.
Tel: +40 68 311 212. Telex: 61324.
Fax: +40 68 311 222.

UPDATED

DAC 11.154 FAEG (4 × 4) 2,500 kg truck

Description

The DAC 11.154 FAEG (4 × 4) 2,500 kg truck is closely related to the DAC 15.215 DFAEG (6 × 6) 5,000 kg truck (see following entry) and shares many components.

The forward control all-steel cab has seating for the driver and up to three passengers; the driver's seat is adjustable. The cab hinges forward to allow access to the engine for maintenance and there is an observation hatch in the roof on the right side. The spare wheel is mounted to the rear of the cab on the right-hand side.

The rear cargo area has a wood platform, a drop tailgate, removable bows and a tarpaulin cover. Folding bench seats are normally fitted along each side. Optional equipment includes a 6 tonne capacity hydrostatic winch with 60 m of cable, and a central tyre pressure regulation system which can be operated while the vehicle is moving. Other optional equipment includes an engine preheater.

Outline drawing of DAC 11.154 FAEG (4 × 4) 2,500 kg truck

Specifications

Cab seating: 1 + 4
Configuration: 4 × 4
Weights:
(chassis with cab) approx 5,700 kg
(GVW) 10,700 kg
(front axle load) 5,500 kg
(rear axle load) 6,200 kg
Max load:
(road) 4,000 kg
(cross-country) 2,500 kg
Towed load:
(road) 5,500 kg
(cross-country) 2,500 kg
Load area: 3.55 × 2.245 m
Length: 6.46 m
Width: 2.5 m
Height:
(cab, loaded) 2.785 m
(load area, loaded) 1.4 m
Ground clearance: 0.395 m
Track: (front/rear) 2 m/1.99 m
Wheelbase: 3.1 m
Fuel capacity: 310 litres
Max speed: 88.6 km/h
Max gradient: 50%
Fording: 0.65 m
Engine: Type 798-05 5.49 litre 6-cylinder in-line water-cooled turbocharged diesel developing 154 hp at 2,800 rpm
Gearbox: S5-45/AK5-35 with 5 forward and 1 reverse gears
Clutch: single dry disc
Transfer box: G-450, 2-speed
Steering: Type 8065 hydraulic
Turning circle: 15 m
Suspension:
(front) semi-elliptical leaf springs with progressively acting hollow rubber springs and telescopic shock-absorbers
(rear) semi-elliptical leaf springs
Brakes: dual-circuit, hydropneumatic front, pneumatic rear
Tyres: 14.00 × 20
Electrical system: 24 V
Batteries: 2 × 12 V, 110 Ah
Alternator: 28 V/40 A

Status

In service with the Romanian Army.

Manufacturer/Agency

ROMAN SA, Poienelor Str. No 5, R-2200 Braşov, Romania.
Tel: +40 68 311 212. Telex: 61324.
Fax: +40 68 311 222.

UPDATED

DAC 15.215 DFAEG (6 × 6) 5,000 kg truck

Description

The DAC 15.215 DFAEG (6 × 6) 5,000 kg truck, at one time known as the DAC 665T, is the standard truck in its class in the Romanian Army and is used by several other former Warsaw Pact nations, including Hungary. This vehicle is a larger version of the DAC 11.154 FAEG – see previous entry.

The forward control all-steel cab has seating for the driver and up to five passengers; seat belts are an optional extra. The cab hinges forward to allow access to the engine for maintenance and there is an observation hatch in the roof on the right side. A spare wheel is mounted to the rear of the cab on the right-hand side. The chassis is of the ladder type based on U-section longitudinal members with riveted cross members.

The all-metal rear cargo area has a drop tailgate, removable bows and a tarpaulin cover. Folding bench seats are normally fitted along each side. Optional equipment includes a 10 tonne capacity hydrostatic winch with 60 m of cable, and a central tyre pressure regulation system which can be operated while the vehicle is moving. Other optional equipment includes a brake power regulator on the rear axle, a trailer brake system, a hydraulic cab tilting device, and a hydraulic spare wheel handling device.

One variant of the DAC 15.215 DFAEG is used to

carry a locally produced 122 mm (40-round) multiple rocket launcher. This vehicle caries a crew of up to five and weighs 17,250 kg loaded. A further variant is used as a resupply vehicle for this system and carries two launcher reload modules and a rear-mounted MH 90 crane. The crane has a maximum lifting capacity of 6,000 kg and a maximum lifting height of 6 m. The resupply vehicle weighs 15,000 kg loaded and can tow a four-wheeled two-axle trailer carrying more rockets to be loaded into empty modules by hand. For full details of this system refer to *Jane's Armour and Artillery 1995-96* page 741.

DAC 15.215 DFAEG trucks are used to carry the ADTT-1 jet decontamination truck-mounted installation. For details refer to *Jane's NBC Protection Equipment 1996-97* page 131.

DAC 15.215 DFAEG trucks are also employed for many other roles, typical being the ARTM mobile workshop for the testing and repair of infantry weapons and ammunition. For this role box bodies with an integral 38kVA generator are utilised and the vehicles involved tow either a twin-axle workshop trailer or a 2R96M single-axle trailer.

DAC 15.215 DFAEG (6 × 6) 5,000 kg truck **1997**

Specifications

Cab seating: 1 + 5
Configuration: 6 × 6
Weights:
(chassis with cab) approx 9,600 kg
(GVW) 15,600 kg
(front axle load) 5,200 kg
(rear axle load) 2 × 5,200 kg
Max load: 5,000 kg
Towed load:
(road) 10,000 kg
(cross-country) 5,000 kg
Load area: 4.66 × 2.245 m
Length: 7.975 m
Width: 2.49 m
Height:
(overall) 3.265 m
(cab) 2.86 m
(load area) 1.45 m
Ground clearance: 0.426 m
Track: 2.403 m
Wheelbase: 3.075 m + 1.35 m
Fuel capacity: 2 × 300 litres
Max speed: 84.9 km/h
Max gradient: 62.5%
Fording: 0.65 m
Engine: D 2156 HMN8 10.344 litre 6-cylinder in-line water-cooled diesel developing 215 hp at 2,200 rpm
Gearbox: AK6-80 with 6 forward and 1 reverse gears
Clutch: single dry disc
Transfer box: G-450, 2-speed
Steering: Type 8065 hydraulic
Turning circle: 21.25 m
Suspension:
(front) semi-elliptical leaf springs with progressively acting hollow rubber springs and telescopic shock-absorbers
(rear) semi-elliptical leaf springs fitted in an oscillating system
Brakes: dual-circuit, hydropneumatic front, pneumatic rear
Tyres: 14.00 × 20
Electrical system: 24 V
Batteries: 2 × 12 V, 110 Ah
Alternator: 28 V/40 A

Status

In service with the Romanian Army, Hungary and Iraq.

Manufacturer/Agency

ROMAN SA, Poienelor Str. No 5, R-2200 Braşov, Romania.
Tel: +40 68 311 212. Telex: 61324.
Fax: +40 68 311 222.

UPDATED

DAC 13.215 FAEG (4 × 4) 5,000 kg truck

Description

The DAC 13.215 FAEG is used by the Romanian Army. This has the same two-door all-steel cab of the forward control type as the DAC 11.154 FAEG (4 × 4) 2,500 kg truck (see separate entry); the overall appearance of these two trucks is very similar although the DAC 13.215 FAEG has a longer wheelbase and is longer overall. The cargo area has a drop tailgate and can be fitted with seats. Optional equipment includes a 6 tonne capacity winch with 60 m of cable.

Specifications

Cab seating: 1 + 4
Configuration: 4 × 4
Weights:
(chassis with cab) approx 7,070 kg
(GVW) 13,500 kg
(front axle load) 5,500 kg
(rear axle load) 8,000 kg
Max load: 5,000 kg
Towed load:
(road) 8,500 kg
(cross-country) 5,000 kg
Load area: 4.6 × 2.38 m
Length: 7.64 m
Width: 2.49 m
Height:
(overall) 7.38 m
(cab, loaded) 2.8 m
(load area, loaded) 1.5 m
Ground clearance: 0.39 m
Track: (front/rear) 2 m/2 m
Wheelbase: 3.8 m
Fuel capacity: 220 litres
Max speed: 97.9 km/h
Gradient: 59%
Engine: D 2156 HMN8 10.344 litre 6-cylinder in-line water-cooled diesel developing 215 hp at 2,200 rpm
Gearbox: AK6-80 M with 6 forward and 1 reverse gears
Clutch: single dry disc
Transfer box: G-550, 2-speed
Steering: Type 8065 hydraulic
Turning circle: 18 m
Suspension:
(front) semi-elliptical leaf springs and telescopic shock-absorbers
(rear) semi-ellipt cal leaf springs
Brakes: dual-circuit, hydropneumatic front, pneumatic rear
Tyres: 14.00 × 20
Electrical system: 24 V
Batteries: 2 × 12 V, 110 Ah
Alternator: 28 V/40 A

Status

In production. In service with the Romanian Army.

Manufacturer/Agency

ROMAN SA, Poienelor Str. No 5, R-2200 Braşov, Romania.
Tel: +40 68 311 212. Telex: 61324.
Fax: +40 68 311 222.

UPDATED

ROMAN SA Trucks

Description

The latest generation of trucks produced by the ROMAN SA Truck Company Braşov, bearing the DAC trademark, have been revised to meet more diverse customer requirements. Since 1993, vehicles produced by ROMAN SA can be provided with imported engines (Renault, Steyr, Caterpillar, and International, etc) which meet Euro 1 and Euro 2 requirements.

ROMAN SA trucks manufactured at Brasov and suitable for military applications include the following models:

DAC 11.154 FAEG: see separate entry
DAC 12.133 FA: 4 × 4 chassis truck
DAC 15.215 DFAEG: 6 × 6 truck — see separate entry
DAC 15.215 DFAEG: 6 × 6 chassis truck
DAC 15.215 DFAEGS: 6 × 6 prime mover
DAC 15.230 DFAEGS: 6 × 6 prime mover
DAC 15.230 DFAEG: 6 × 6 chassis truck
DAC 15.230 DFAEG: 6 × 6 truck
DAC 16.230 FA: 4 × 4 chassis truck
DAC 16.230 FA: 4 × 4 truck
DAC 21.320 VFAEG; 8 × 8 cross country truck with 7,300 kg payload

Status

All of the above are available.

Manufacturer/Agency

ROMAN SA, Poienelor Str. No 5, R-2200 Braşov, Romania.
Tel: +40 68 311 212. Telex: 61324.
Fax: +40 68 311 222.

UPDATED

RUSSIAN FEDERATION AND ASSOCIATED STATES (CIS)

GAZ-66 (4 × 4) 2,000 kg truck

Description

The GAZ-66 (4 × 4) 2,000 kg truck was the replacement for the GAZ-63 and entered production at the Gorky Automobile Plant in 1964. All vehicles built since 1968 have a central tyre pressure regulation system fitted as standard. The GAZ-66 is widely used by the RFAS armed forces and is also used for a variety of civilian roles.

The two-door all-steel forward control cab hinges forward to allow access to the engine for maintenance. The all-steel rear cargo body has fixed sides and a drop tailgate. The vehicle can be fitted with five bows and a tarpaulin cover if required. Standard equipment includes a cab heater and an engine preheater and many vehicles have a winch.

Variants

GAZ-66-02: with crane
GAZ-66-04: with shielded electrical system
GAZ-66-05: with shielded electrical system and crane
GAZ-66-01: with tyre pressure regulation system (also known as GAZ-66-51)
GAZ-66-02: with tyre pressure regulation system and winch
GAZ-66-03: with shielded electrical system
GAZ-66-04: with shielded electrical system and tyre pressure regulation system
GAZ-66-05: with shielded electrical system, tyre pressure regulation system and winch
GAZ-66-51: with tyre pressure regulation system and tropical equipment
GAZ-66-52: with winch and tropical equipment
GAZ-66-54: with shielded electrical system and tropical equipment
GAZ-66-55: with shielded electrical system, winch and tropical equipment
GAZ-66A: with tyre pressure regulation system and winch
GAZ-66E: with shielded electrical system
GAZ-66P: tractor truck, not placed in production
GAZ-66B: for airborne forces, with collapsible canvas cab, removable doors and windscreen, telescopic steering wheel and tie-down points for parachute dropping
DDA-53C: NBC decontamination vehicle
AVTs-1.7: 1,700 litre water tanker, built at Dalmatovo Molmashstroy plant
MZ-66: motor oil supply vehicle (820 litres)
Shop/van for command, communications and other roles (weight loaded 5,800 kg)
Multiple Rocket Launcher: the GAZ-66B is used as a carrier for a lightweight 12-tube version of the BM-21 known as the BM-21V Grad-P. For details of the BM-21 system refer to *Jane's Armour and Artillery 1995-96* pages 715 to 717.

Specifications

GAZ-66
Cab seating: 1 + 1
Configuration: 4 × 4
Weight:
(empty) 3,640 kg
(loaded) 5,770 kg
Weight on front axle, loaded: 2,930 kg
Weight on rear axle, loaded: 3,040 kg
Max load: 2,130 kg
Towed load: 2,000 kg
Load area: 3.33 × 2.05 m
Length: 5.655 m
Width: 2.342 m
Height:
(cab) 2.44 m
(load area) 1.11 m
Ground clearance: (axles) 0.31 m
Track:
(front) 1.8 m
(rear) 1.75 m
Wheelbase: 3.3 m
Angle of approach/departure: 42°/32°
Max speed: (road) 90 km/h
Range: 800 km
Fuel capacity: 210 litres (2 × 105 litres tanks)
Max gradient: 67%
Fording: 1 m
Engine: ZMZ-66 V-8 4.254 litre water-cooled petrol developing 115 hp at 3,200 rpm
Gearbox: manual with 4 forward and 1 reverse gears
Clutch: single dry plate
Transfer box: 2-speed
Steering: globoid worm with 3-collar roller and hydraulic booster
Turning radius: 10 m
Suspension: longitudinal semi-elliptical springs and hydraulic double-acting shock-absorbers
Tyres: 12.00 × 18
Brakes:
(main) hydraulic
(parking) mechanical operating on transmission
Electrical system: 12 V
Battery: 1 × 6-ST-68

Status

Production of a replacement design was anticipated but appears to have been delayed by economic factors. In service with former Warsaw Pact countries. Also exported to many other countries, including Angola, Cambodia, Cuba, Egypt, Finland, Iran, Iraq and Syria.

Manufacturer

Gorky Automobile Plant, RF-603046 Nizhny Novgorod, Russia.
Tel: +7 8312 564 206. Telex: 151127 OKA SU.
Fax: +7 8312 539 842.

UPDATED

Estonian armed forces GAZ-66 (4 × 4) 2,000 kg truck (Stefan Marx) **1996**

Estonian Army GAZ-66 (4 × 4) 2,000 kg truck with office body used to transport EOD teams and their equipment (Stefan Marx) **1996**

GAZ-3937 (4 × 4) 2,500 kg high-mobility vehicle

Development

When the GAZ-3937, known as the Vodnik, was first displayed in mid-1995 it was thought to be a lightly armoured personnel carrier. Subsequent information indicated that it is an unarmoured high-mobility carrier with a load capacity of 2,500 kg, or up to nine personnel in cramped conditions in the rear. Some light armour could be applied to this vehicle, especially when it is deployed as a weapon platform, although this was not shown on the example first displayed. Production is anticipated during 1997.

GAZ-3937 (4 × 4) 2,500 kg high-mobility vehicle (C F Foss) **1996**

GAZ-3937 (4 × 4) 2,500 kg high-mobility vehicle (C F Foss) **1996**

Description

When the GAZ-3937 is deployed as a personnel carrier the driver is seated on the left in a forward-located cab with the engine compartment on his right. Personnel are carried in an open compartment to the rear. This has been observed covered with canvas over low bows, with integral windows in the sides of the cover. There are no access doors or flaps so personnel have to enter over the sides.

The hull of the GAZ-3937 is welded steel and components common to other GAZ vehicles are used wherever possible. The vehicle is amphibious, being propelled in the water by its wheels at a maximum speed of 5 km/h. A central tyre inflation system is provided.

The GAZ-3937 shares many components with the GAZ-3934 (SIAM) bullion transporter already in production.

Specifications

Provisional
Cab seating: 1 + up to 9 in rear
Configuration: 4 × 4
Weight:
(combat) 6,500 kg
(unladen) 4,000 kg
Max load: 2,500 kg
Power to weight ratio: 26.15 hp/t
Length: 4.5 m
Width: 2.8 m
Height: 1.9 m
Ground clearance: 0.47 m
Track: 2.4 m
Wheelbase: 3.05 m
Max speed:
(road) 120 km/h
(water) 5 km/h
Range: 800 km
Fording: amphibious
Engine: diesel developing 170 hp

Status

Prototypes.

Manufacturer

Gorky Automobile Plant, RF-603046 Nizhny Novgorod, Russia.
Tel: +7 8312 564 206. Telex: 151127 OKA SU.
Fax: +7 8312 539 842.

VERIFIED

ZIL-157 (6 × 6) 2,500 kg truck

Description

The ZIL-157 (6 × 6) truck replaced the ZIL-151 (6 × 6) truck in production from 1958, and in 1961 the improved ZIL-157K entered production, to be replaced in 1966 by the more powerful ZIL-131 (6 × 6) 3,500 kg truck. In appearance the ZIL-157 is very similar to the ZIL-151 but has a slightly different cab (which was also fitted to late production ZIL-151s), and single instead of dual rear wheels.

The layout of the vehicle is conventional, with the engine at the front, two-door fully enclosed cab in the centre, and the cargo area at the rear, which consists of a wooden platform with sides, bench seats down each side which can be folded up when the vehicle is carrying cargo, and a drop tailgate. If required, the vehicle can be fitted with bows and a tarpaulin cover. Standard equipment includes a cab heater and an engine preheater and many vehicles also have a winch.

Variants

ZIL-157V and ZIL-157KV: tractor trucks for towing semi-trailers (for example carrying SA-2, FROG-3, FROG-4 or FROG-5 missiles)
ZIL-157KG: with shielded electrical system
ZIL-157KE: temperate climate export model of ZIL-157K
ZIL-157KYu: tropical climate export model of ZIL-157K
ZIL-157GT: tropical climate export model of ZIL-157K with shielded electrical system
ZIL-157KYe: ZIL-157K chassis for special bodies
ZIL-157KYeI: ZIL-157KYe chassis with high-output generator
ZIL-157KYeG: ZIL-157KG chassis with shielded electrical system
ZIL-157YeGT: ZIL-157KYeG chassis, tropical climate export model
ZIL-157KYeGT: ZIL-157KYeG chassis, tropical climate export model (chassis has features of the ZIL-157KYu)
ZIL-157YeT: ZIL-157KYe chassis, tropical climate export model (chassis contains features of ZIL-157KYu)
ARS-12U: decontamination vehicle
AGV-3M and AGW-3M: decontamination stations
ATsM-4-157K: 4,000 litre fuel truck (also on ZIL-157 chassis)
ATZ-3-157K and ATZ-3, 8-157K: fuel service trucks
ATsMM-4-157K: 4,000 litre oil tank truck
AVTs 28-157: 2,800 litre water tanker
CAS 16: Czech fire engine
Carrying pontoons and other bridging equipment
Crane truck
Carrying KMM treadway bridge system (details of this version are given in the *Mechanised Bridges* section)
Multiple rocket system: 200 mm BMD-20 (4-round) (*Jane's Armour and Artillery 1987-88* page 785), 240 mm BM-24 (12-round) (*Jane's Armour and Artillery 1990-91* pages 738 and 739) and 140 mm (16-round) BM-14-16 (*Jane's Armour and Artillery 1995-96* page 714)
VMZ-ZIL-157K: water and oil service truck which carries 1,400 litres of water and 700 litres of oil, equipped with a heater system which maintains the temperature of the water at between +15 and +90°C and the oil at +80°C.

Specifications

Cab seating: 1 + 1
Configuration: 6 × 6
Weight:
(empty, with winch) 5,800 kg
(loaded, with winch) 8,450 kg
Max load:
(road) 4,500 kg
(cross-country) 2,500 kg
Towed load:
(dirt road) 3,600 kg
(cross-country) 2,500 kg
(roads) 3,600 kg
Load area: 3.57 × 2.09 m
Length: 6.922 m (with winch)
Width: 2.315 m
Height:
(cab) 2.36 m
(tarpaulin) 2.915 m
(load area) 1.388 m
Ground clearance: 0.31 m
Track:
(front) 1.755 m
(rear) 1.75 m
Wheelbase: 3.655 m + 1.12 m
Angle of approach/departure: 33°/43°
Max speed: (road) 65 km/h
Range: 460 km
Fuel capacity: 150 litres
Max gradient: 53%
Fording: 0.85 m
Engine: ZIL-157K 5.55 litre 6-cylinder water-cooled petrol developing 109 hp at 2,800 rpm
Gearbox: manual with 5 forward and 1 reverse gears
Clutch: single dry plate
Transfer box: 2-speed
Steering: cone worm with 3-ridge roller
Turning radius: 11.2 m
Suspension:
(front) longitudinal semi-elliptical springs with hydraulic double-acting shock-absorbers
(rear) bogie with semi-elliptical springs
Tyres: 12.00 × 18
Brakes:
(main) air
(parking) mechanical
Electrical system: 12 V
Batteries: 2 × ST-84

Status

Production complete. In service in declining numbers with former Warsaw Pact countries and other armed forces.

Manufacturer

Likhachev Motor Vehicle Plant, Moscow, Russia.

VERIFIED

ZIL-157K (6 × 6) 2,500 kg truck with shielded box body containing R-140 communications equipment (Stefan Marx) ***1996***

ZIL-157K (6 × 6) 2,500 kg truck with shielded box body containing R-138 ground-to-air communications equipment (Stefan Marx) ***1996***

ZIL-131 (6 × 6) 3,500 kg truck

Description

The ZIL-131 (6 × 6) truck entered production in December 1966 as the replacement for the earlier ZIL-157 (6 × 6) 2,500 kg truck and uses many components of the ZIL-133 (6 × 4) truck. Main improvements over the ZIL-157 can be summarised as increased load-carrying capacity, more powerful engine, power steering, shorter wheelbase, waterproof ignition and the central tyre pressure regulation system. The front axle is engaged automatically when the driver selects first gear and the driver can also engage the front axle manually when in second gear. The vehicle is used for transporting cargo or personne and as a prime mover for towing artillery such as the 122 mm D-30 howitzer.

The layout of the vehicle is conventional, with the engine at the front, fully enclosed two-door all-steel cab in the centre and the cargo area at the rear, consisting of a wooden platform with metal fittings and a hinged tailgate. The platform has recesses for the bows and hinged bench seats are provided down either side of the platform.

The ZIL-131 has a central tyre pressure regulation system and a 4,500 kg capacity winch. Standard equipment includes a cab heater and an engine preheater.

Variants

ZIL-131A: this is a truck with a standard ignition system rather than the shielded ignition system as fitted on the standard ZIL-131
ZIL-131D: dump truck
ZIL-131V: for towing semi-trailers
ZIL-131PM: fire engine
ZIL-137: tractor truck for towing a two-axle powered trailer (the complete unit then becomes a 10 × 10); also used to tow SA-6 'Gainful' SAMs on a single-axle trailer
ARS-14: decontamination vehicle
ATs 4, 2-131: 4,200 litre fuel tank truck
ATs 4, 2-131: 4,100 litre fuel tank truck
ATs 4, 3-131: 4,400 litre fuel tank truck
ATZ 3, 8-131: 4,300 litre fuel service truck
ATZ 4, 3-131: 4,300 litre fuel service truck
AVTs 28-131: 2,800 litre water tanker
PBU 50M: German well drilling plant

Shop/van
Various models are in service including bakery, maintenance and a field kitchen.

MA-41
This is a combined fuel, lubricant and water service vehicle. It carries 1,700 litres of diesel, 340 litres of petrol, two 170 litre tanks of oil and 700 litres of water. Both the diesel and water tanks are heated as this unit is used in very cold climates.

SA-6 'Gainful' SAM Resupply Vehicle
Three missiles are carried in a triangular formation with the warhead sections projecting over the roof of the cab. A foldable crane is mounted at the rear of the truck to facilitate reloading of the SA-6 tracked launcher vehicle.

SA-3 'Goa' SAM Resupply Vehicle
Two missiles are carried adjacent to each other on a rail system with the warhead sections projecting over the roof of the cab. The SA-3 launcher is reloaded by backing the vehicle in line with the twin launcher rails, connecting up the resupply rails and then winching the missiles onto the launcher rails.

Multiple Rocket Launcher
A number of ZIL-131 trucks have been fitted with the 140 mm BM-14-16 multiple rocket system normally carried on the rear of a ZIL-151 (6 × 6) truck chassis (*Jane's Armour and Artillery 1995-96*, page 714) and the 122 mm BM-21 multiple rocket system (*ibid* pages 715 to 717).

Specifications

Cab seating: 1 + 2
Configuration: 6 × 6
Weight:
(empty, with winch) 6,700 kg
(loaded, cross-country) 10,425 kg
Weight on front axle, loaded: 3,360 kg
Weight on rear axles, loaded: 7,065 kg
Max load:
(road) 5,000 kg
(cross-country) 3,500 kg
Towed load:
(road) 6,500 kg
(dirt road) 4,000 kg
Load area: 3.6 × 2.32 m
Length:
(without winch) 6.9 m
(with winch) 7.04 m
Width: 2.5 m
Height:
(cab) 2.48 m
(tarpaulin) 2.975 m
(load area) 1.43 m
Ground clearance: 0.33 m
Track: (front and rear) 1.82 m
Wheelbase: 3.35 m + 1.25 m
Angle of approach/departure: 36°/40°
Max speed: (road) 80 km/h
Range: 645 km
Fuel capacity: 340 litres
Max gradient: 58%
Fording: 1.4 m
Engine: ZIL-131 6 litre V-8 water-cooled petrol developing 150 hp at 3,200 rpm
Gearbox: manual with 5 forward and 1 reverse gears
Clutch: single dry plate
Transfer box: 2-speed
Steering: screw and nut with hydraulic booster
Turning radius: 10.1 m
Suspension:
(front) longitudinal semi-elliptical springs with double-acting hydraulic shock-absorbers
(rear) equaliser arm on longitudinal semi-elliptical springs
Tyres: 12.00 × 20 ND
Brakes:
(main) air
(parking) mechanical
Electrical system: 12 V
Battery: 1 × 6 ST 78

Status

In service with Finland, RFAS and other armed forces.

Manufacturer

Likhachev Motor Vehicle Plant, Moscow, Russia.

UPDATED

ZIL-131 (6 × 6) truck with box body containing R-145 radio relay station (Michael Jerchel) ***1996***

ZIL-131 (6 × 6) 3,500 kg truck (Richard Stickland) ***1996***

Ural-375D (6 × 6) 4,000 kg truck

Description

The first model of the Ural-375 entered production in 1961. It had an open cab with a canvas top and a stake-type rear cargo body. The second model was the Ural-375A, which had a number of automotive improvements as well as a fully enclosed all-steel cab. The Ural-375A was standardised as the Ural-375D; those with a winch (capacity of 7,000 kg) are designated Ural-375T. The vehicle is fitted with a central tyre pressure regulation system. The Ural-377 (6 × 4) truck uses many components of the Ural-375.

The layout of the vehicle is conventional, with the engine at the front, fully enclosed two-door all-steel cab in the centre and the cargo area at the rear with hinged bench-type seats, removable bows and a tarpaulin cover, and a drop tailgate. Standard equipment includes a cab heater and an engine preheater.

In 1973 a Ural-375D was tested with a YaMZ-740 V-8 water-cooled diesel engine, which develops 210 hp and is used in the KAMAZ range of 6 × 4 trucks. It is now in production as the Ural-4320 (6 × 6) 4,500 kg truck (see latter entry).

Variants

Ural-375DK-1: Ural-375 adapted for operations in northern RFAS
Ural-375K: Ural-375 adapted for operations in northern RFAS
Ural-375L: timber truck
Ural-375N: cargo truck with drop sides
Ural-375S: tractor truck for towing semi-trailers
Ural-375Yu: designed for use in the tropics
ATs 5-375: 5,000 litre fuel tank truck
ATsM 4-375: fuel tank truck
TZ 5-375: 5,000 litre fuel service truck which also tows a PTs 4-754 trailer with another 4,200 litres of fuel. This unit is provided with three 3 m hoses, a 15 m dispensing hose and a 9 m pressure hose
TMS-65: truck-mounted decontamination apparatus used for the rapid decontamination of vehicles and equipment; for details refer to *Jane's NBC Protection Equipment 1996-97* page 103
ATsG-5-375: 5,000 litre fuel tanker, equipped with four SRGS-70 3 m and four SRGS-32 3 m metal hoses
ATMZ-4.5-375: fuel lubricant truck (4,500 litres), equipped with six dispensing hoses
ATO-40: workshop body truck
APA-5: aircraft starting and refuelling truck
AC-40: fire engine

Cambodian Air Force APA-5 aircraft starter and refuelling vehicle based on chassis of Ural-375D (6 × 6) 4,000 kg truck (T J Gander) ***1996***

Ural-375D (6 × 6) truck with office body for SDR-2 ECM system (Stefan Marx) ***1996***

Crane truck: the crane model is fitted with an 8 T 210 crane on the rear
Shop/van: the basic chassis is used to mount a wide range of van type bodies for command, communications and other roles
Recovery vehicle: designated the KET-L (see entry in *Recovery Vehicles* section)
Multiple Rocket Launcher: the Ural-375D is used to mount the BM-21 (40-round) 122 mm multiple rocket system (*Jane's Armour and Artillery 1995-96*, pages 715 to 717)
RMT-35M electrical testing laboratory
Pontoon-carrying vehicle
SA-4 'Ganef' transporter-loader (TZM).

Specifications

Cab seating: 1 + 2
Configuration: 6 × 6
Weight:
(empty, with winch) 8,400 kg
(loaded, road) 13,200 kg
Weight on front axle, loaded: 3,900 kg
Weight on rear axles, loaded: (each) 4,650 kg
Max load:
(road) 4,800 kg
(cross-country) 4,000 kg
Towed load:
(road) 10,000 kg
(dirt road) 5,000 kg
Load area: 3.9 × 2.43 m
Length: 7.35 m
Width: 2.69 m
Height:
(cab) 2.68 m
(tarpaulin) 2.98 m
(load area) 1.42 m
Ground clearance: 0.4 m
Track: 2 m
Wheelbase: 3.5 m + 1.4 m
Angle of approach/departure: 44°/40°
Max speed: (road) 75 km/h
Range: 570 km
Fuel capacity: 360 litres
Max gradient: 65%
Vertical obstacle: 0.8 m
Trench: 0.7 m
Fording:
(without preparation) 1 m
(with preparation) 1.5 m
Engine: ZIL-375 7 litre V-8 water-cooled petrol developing 180 hp at 3,200 rpm
Gearbox: manual with 5 forward and 1 reverse gears
Clutch: twin dry discs
Transfer box: 2-speed
Steering: double-thread worm, hydraulic booster
Turning radius: 10.5 m
Suspension:
(front) longitudinal semi-elliptical springs with hydraulic shock-absorbers (all interchangeable with same units on the MAZ-500)
(rear) bogie with longitudinal semi-elliptical springs
Tyres: 14.00 × 20
Brakes:
(main) air/hydraulic
(parking) mechanical
Electrical system: 12 V
Battery: 1 × 6-STEN-140M
Generator: G51

Status

In production. In service with the RFAS and other former Warsaw Pact countries. Also used by some Middle East nations.

Manufacturer

Ural Automobile Plant, Miass, RF-456300 Chelyabinsk Region, Russia.
Tel: +7 35135 33570. Telex: 124849 URAL SU.
Fax: +7 35135 34189.

UPDATED

KAMAZ-4350 (4 × 4) 4,000 kg truck

Description

The KAMAZ-4350 (4 × 4) 4,000 kg truck is one of a family of three trucks, the others being the KAMAZ-5350 (6 × 6) and KAMAZ-6350 (8 × 8) — see separate entries for these two vehicles. All three trucks share many design features and components, not the least being the forward control cab which is very similar to that utilised on the KAMAZ-5320 series (qv). The intention appears to be to provide a family of similar vehicle capable of meeting a wide range of military logistic requirements. Various enhancements, such as an improved efficiency diesel engine coupled to an automatic gearbox, have been introduced, although full details have yet to be determined. Many components are also shared with the extensive range of KAMAZ commercial trucks.

The cargo body fitted to the example observed is entirely conventional, being an all-steel flatbed with drop sides, a tailgate and a canvas cover over removable bows.

Specifications

Provisional
Cab seating: 1 + 2
Configuration: 4 × 4
Weight: 7,500 kg
Max load: 4,000 kg
Towed load: 5,000 kg
Load area: 4.8 × 2.42 m
Length: 7.87 m
Width: 2.5 m
Height:
(cab) 2.98 m
(tarpaulin) 3.35 m
(load area) 1.57 m
Ground clearance: 0.385 m
Track: 2.01 m
Wheelbase: 4.2 m
Angle of approach/departure: 32°/33°
Max speed: (road) 100 km/h
Range: 1,390 km
Fording: 1.75 m

KAMAZ-4350 (4 × 4) 4,000 kg truck (C F Foss) ***1996***

KAMAZ-4350 (4 × 4) 4,000 kg truck (C F Foss) ***1996***

Engine: KAMAZ-7405 10.85 litre V-8 diesel developing 240 hp at 2,200 rpm
Gearbox: automatic
Tyres: 16.75/85 R 21
Electrical system: 24 V

Status
Available for production.

Manufacturer
KAMAZ, Naberezhnye Chelny, RFAS.

UPDATED

Ural-4325 (4 × 4) 4,000 kg truck

Description
The Ural-4325 (4 × 4) 4,000 kg truck was developed from the Ural-4320-10 (6 × 6) 4,500 kg truck (see following entry). Apparently it was developed for potential export sales, the single rear axle and reduction in unladen weight being a more attractive and less expensive option than the 6 × 6 model in some market areas.

The Ural-4325 carries over the same cab together with most of the mechanical attributes and design details of the Ural-4320-10, although the diesel engine is derated to 180 hp. The standard cargo body is all-steel with a tailgate and removable bows and tilt; there is provision for seating up to 24 personnel. Maximum payload on hard roads is 5,000 kg (4,000 kg on dirt roads or cross-country). Maximum towed weight is 7,000 kg. The central tyre inflation of the 6 × 6 model is also provided.

One option is a 7,000 to 9,000 kg winch with 60 m of cable. Power is taken from a power take-off (also optional).

Ural-4325 (4 × 4) 4,000 kg truck ***1996***

Specifications
Provisional
Cab seating: 1 + 2 (up to 24 in rear)
Configuration: 4 × 4
GCW: 11,630 kg
Max load: 4,000 kg
Towed load: 7,000 kg
(loaded) 10,000 kg
Max speed: 85 km/h
Max gradient: 60%
Fording: 1.2 m
Engine: KAMAZ-740 10.85 litre liquid-cooled diesel developing 180 hp
Gearbox: manual with 5 forward and 1 reverse gears
Transfer box: 2-speed
Tyres: 12.00 × 500-508

Status
Available for production.

Manufacturer
Ural Automobile Plant, Miass, RF-456300 Chelyabinsk Region, Russia.
Tel: +7 35135 33570. Telex: 124849 URAL SU.
Fax: +7 35135 34189.

VERIFIED

Ural-4320-10 (6 × 6) 4,500 kg truck

Description
The Ural-4320-10 (6 × 6) 4,500 kg truck is a diesel-engined development of the Ural-375D truck produced by the Ural Motor Vehicle Plant, Miass. Development of the Ural-375D with a new diesel engine started in 1972 and the first production examples were produced in 1978.

The diesel engine is the YaMZ-740, produced at the Kama Motor Vehicle Plant, Naberezhnye Chelny. It is a 10.85 litre, eight-cylinder engine developing 210 hp at 2,600 rpm. Fitting the new engine enabled the payload to be increased by 500 kg to 4,500 kg, and the maximum speed is increased to 85 km/h. Fuel consumption is decreased by 30 per cent and maintenance by 10.8 per cent. Using typical RFAS figures the engine produces 16 per cent more power at 19 per cent fewer rpm. To use this extra power the transmission and rear-axle gear ratios were reduced from 8:9 to 7:32 and numerous other small changes were made.

The main outward appearance is that the radiator shell has been lengthened in front to accommodate the longer diesel engine, which is also some 280 kg heavier than the earlier ZIL-375 petrol engine, leading to some revision of the axle loading. The front axle now assumes 4,350 kg of the total weight instead of 3,865 kg; the rear axle assumes 9,075 kg instead of 9,335 kg. This revision improves the traction of the Ural-4320-10 over soft ground.

Fuel tanker version of Ural-4320 (6 × 6) 4,500 kg truck (Stefan Marx) ***1996***

Other improvements on the Ural-4320-10 relate to the fuel, cooling, exhaust and electrical systems, which were all redesigned. Modifications were made to the frame, clutch, front suspension and other assemblies. Safety alterations include individual brake control links, an auxiliary motor brake and some warning devices. Inside the cab, alterations were made to the instrument layout and to the driver's seat, which is now fully adjustable. A more powerful heater was installed and changes were made to the layout of the steering wheel and other controls. The batteries were moved from the driving cab.

There are four main versions of the Ural-4320-10: the base model, the Ural-4420-10 and Ural-44202-10 tractors capable of towing trailers up to 15 tonnes and 18.5 tonnes respectively, and the Ural-43202-10 truck. Differences occur in tyre size and the fact that the Ural-4320-10 and Ural-44202-10 have tyre pressure regulation systems while the other two do not. In addition, chassis without loading platforms, stretched chassis for box bodies and some other models are produced.

A winch with 65 m of 17.5 mm cable is an option on all versions.

A recovery version has been produced (see entry in the *Recovery Vehicles* section).

One variant of the Ural-4320 is known as the MTP-A1.1. This is a special Technical Assistance Vehicle intended to provide field repairs to broken-down vehicles. It is equipped with tools, repair kits and a 700 kg capacity traversable telescopic hoist jib for use when conducting repairs and has limited recovery and towing facilities.

The APA-5D is an aircraft starting and refuelling truck.

The Ural-4325 is a 4 × 4 export model with a payload of 4,000 kg. See previous entry for details.

Specifications

Model	Ural-4320	Ural-4420	Ural-43202	Ural-44202
Use	truck	tractor	truck	tractor
Cab seating	1 + 2	1 + 2	1 + 2	1 + 2
Configuration	6 × 6	6 × 6	6 × 6	6 × 6
Weight (empty)	8,020 kg	7,900 kg	7,800 kg	7,390 kg
Front axle loading (empty)	4,560 kg	3,830 kg	4,040 kg	3,950 kg
Rear axle loading (empty)	4,500 kg	4,070 kg	3,760 kg	3,440 kg
Payload/semi-trailer capacity				
(on-road)	5,225 kg	7,000 kg	5,500 kg	7,500 kg
(off-road)	–	5,000 kg	–	5,500 kg
Towed load/semi-trailer weight				
(on-road)	7,000 kg	11,500 kg	15,000 kg	18,500 kg
(off-road)	–	7,000 kg	–	12,500 kg
Length	7.366 m	7.611 m	7.1 m	6.836 m
Width	2.674 m	2.674 m	2.674 m	2.475 m
Height	2.87 m	2.6 m	2.68 m	2.6 m
Ground clearance	0.4 m	0.345 m	0.4 m	0.345 m
Track (front)	2 m	2.02 m	2 m	2.02 m
(rear)	2 m	2.02 m	2 m	2.02 m
Wheelbase	3.525 m + 1.4 m	3.525 m + 1.4 m	3.525 m + 1.4 m	3.525 m + 1.4 m
Max speed	75 km/h	80 km/h	72 km/h	72 km/h
Fuel capacity (main tank)	210 litres	210 litres	300 litres	300 litres
(auxiliary)	60 litres	none	60 litres	none
Max gradient	58%	50%	28%	28%
Angle of approach/departure	45°/40°	42°/38°	44°/45°	42°/70°
Fording	0.7 m	0.8 m	1.5 m	0.8 m
Engine	KAMAZ-740 V-8 10.85 litre, 4-stroke diesel developing 210 hp at 2,600 rpm			
Gearbox	manual with 5 forward and 1 reverse gears			
Clutch	single dry disc			
Steering	worm with gear quadrant			
Turning circle	22.8 m	22.8 m	22.8 m	22.8 m
Suspension				
(front)	longitudinal semi-elliptical springs with hydraulic shock-absorbers			
(rear)	longitudinal semi-elliptical springs with check springs			
Tyres	14.00 × 20	1.110 × 400.533	14.00 × 20	1.110 × 400.533
Brakes	dual-circuit air plus auxiliary engine brake			
Electrical system	24 V	24 V	24 V	24 V
Trailer/semi-trailer coupling				
(height)	0.75 m	0.81 m	1.67 m	1.39 m

Status
In service with the RFAS armed forces.

Manufacturer
Ural Automobile Plant, Miass, RF-456300 Chelyabinsk Region, Russia.
Tel: +7 35135 33570. Telex: 124849 URAL SU.
Fax: +7 35135 34189.

UPDATED

ZIL-130 (4 × 2) 4,500 kg truck

Description
The ZIL-130 (4 × 2) truck entered production in late 1964 as the replacement for the older ZIL-164 (4 × 2) truck. Despite its age the ZIL-130 continues to serve on as a general utility vehicle and in a wide range of military and commercial variants (see below), some of which may no longer be in service. The ZIL-130 formed the basis for the ZIL-131 (6 × 6) and ZIL-133 (6 × 4) trucks.

The layout of the ZIL-130 is conventional, with the engine at the front, a two-door fully enclosed sheet-steel cab in the centre and the cargo area at the rear with drop sides and tailgate.

ZIL-130 (4 × 2) 4,500 kg truck (Michael Jerchel)

Variants
ZIL-130A1: cargo truck capable of pulling trailers
ZIL-130D1: chassis for ZIL-MMZ-555 dump truck with 3.3 m wheelbase
ZIL-130D2: dump truck chassis
ZIL-130D1E: ZIL-MMZ-555 dump truck chassis, temperate climate export model
ZIL-130D1T: export model of ZIL-130D1E for tropical climates
ZIL-130G: long cargo bed (length 4.686 m) with 4.5 m wheelbase
ZIL-130GE: temperate climate export model of ZIL-130G
ZIL-130GT: tropical climate export model of ZIL-130G
ZIL-130S: for use in northern RFAS
ZIL-130T: temperate climate export model
ZIL-1361: export model with Perkins diesel engine
ZIL-1361G: long wheelbase export model with Perkins diesel engine
ZIL-1361D1: dump truck chassis with Perkins diesel engine
ZIL-130V1: tractor truck with 3.3 m wheelbase
ZIL-130V1T: tropical climate export model of ZIL-130V1
ZIL-130Ye: ZIL-130 with shielded electrical system
ZIL-130YeT: tropical climate export model of ZIL-130Ye
ZIL-130YeE: temperate climate export model of ZIL-130Ye
ZIL-MMZ-554: agricultural dump truck
ZIL-MMZ-555: agricultural dump truck
ATs 4,2-130: 4,200 litre fuel tanker
ATs 3,8-130: 3,800 litre fuel servicing truck
AVTs 2,8-130: 2,800 litre water tanker
DDA-53B: NBC decontamination vehicle
PAZS-3152: mobile fuelling unit
PSG-160: POL transfer pump unit
Shop-van: for various applications
Crane truck

Specifications
ZIL-130 with radio body
Cab seating: 1 + 2
Configuration: 4 × 2
Weight:
(empty) 4,500 kg
(loaded) 9,000 kg
Weight on front axle, loaded: 2,575 kg
Weight on rear axle, loaded: 6,950 kg
Max load: 4,500 kg
Towed load: 6,400 kg
Width: 2.5 m
Height:
(cab) 2.35 m
(tarpaulin) 3.5 m
(load area) 1.43 m

Ground clearance: 0.275 m
Track:
(front) 1.8 m
(rear) 1.79 m
Wheelbase: 3.8 m
Angle of approach/departure: 38°/27°
Max speed: (road) 85 km/h
Range: 500 km
Fuel capacity: 150 litres
Max gradient: 36%
Fording: 0.7 m
Engine: ZIL-130 6 litre V-8 water-cooled petrol developing 170 hp at 3,600 rpm
Gearbox: manual with 5 forward and 1 reverse gears
Clutch: single dry plate
Steering: screw and nut with hydraulic booster
Turning radius: 8 m
Suspension:
(front) longitudinal semi-elliptical springs with hydraulic telescopic 2-way shock-absorbers
(rear) semi-elliptical springs and check springs
Tyres: 9.00 × 20
Brakes:
(main) air
(parking) mechanical
Electrical system: 12 V
Battery: 1 × 6-ST 78EMSZ
Generator: 350 W

Status
In service with the RFAS and other armed forces.

Manufacturer
Likhachev Motor Vehicle Plant, Moscow, Russia.

VERIFIED

KAMAZ-5350 (6 × 6) 6,000 kg truck

Description
The KAMAZ-5350 (6 × 6) 6,000 kg truck is one of a family of three trucks, the others being the KAMAZ-4350 (4 × 4) and KAMAZ-6350 (8 × 8) – see separate entries for these two vehicles. All three trucks share many design features and components, not the least being the forward control cab which is very similar to that utilised on the KAMAZ-5320 series (qv).

The intention appears to be to provide a family of similar vehicle capable of meeting a wide range of military logistic requirements. Various enhancements, such as an improved efficiency diesel engine coupled to an automatic gearbox, have been introduced, although full details have yet to be determined. Many components are also shared with the extensive range of KAMAZ commercial trucks.

The cargo body fitted to the example observed is entirely conventional, being an all-steel flatbed with drop sides, a tailgate and a canvas cover over removable bows. The KAMAZ-5350 shares the same cargo body as the KAMAZ-4350.

KAMAZ-5350 (6 × 6) 6,000 kg truck (Christopher F Foss) ***1996***

Specifications
Provisional
Cab seating: 1 + 2
Configuration: 6 × 6
Weight: 8,890 kg
Max load: 6,000 kg
Towed load: 8,000 kg
Load area: 4.8 × 2.42 m
Length: 7.94 m
Width: 2.5 m
Height:
(cab) 2.98 m
(tarpaulin) 3.25 m
(load area) 1.57 m
Ground clearance: 0.385 m
Track: 2.01 m
Wheelbase: 3.34 m + 1.32 m
Angle of approach/departure: 37°/35°
Max speed: (road) 100 km/h
Range: 1,140 km
Fording: 1.75 m
Engine: KAMAZ-7406 10.85 litre V-8 diesel developing 260 hp at 2,200 rpm
Gearbox: automatic
Tyres: 16.75/8 R 21
Electrical system: 24 V

Status
Available for production.

Manufacturer
KAMAZ, Naberezhnye Chelny, RFAS.

VERIFIED

Ural-377 (6 × 4) 7,500 kg truck

Description
The Ural-377 is a 6 × 4 version of the Ural-375 (6 × 6) truck and entered production in 1965. The layout of the vehicle is conventional, with the engine at the front, fully enclosed all-steel two-door cab in the centre and the cargo area at the rear. It has a longer cargo area than the Ural-375D, as the spare wheel is stowed on the right side of the chassis instead of to the rear of the cab as on the Ural-375D. The rear cargo area has a wooden platform with drop sides and a drop tailgate. Standard equipment includes a cab heater and an engine preheater.

Variants
Ural-377D: adapted for use in northern regions with an insulated cab, cold weather tyres, central tyre pressure regulation system and windows with double glass
Ural-377M: the military version of the basic vehicle with a central tyre pressure regulation system not found on the basic vehicle
Ural-377S: the tractor truck version for towing semi-trailers
Ural-377V: side dump truck version

Specifications

Model	**Ural-377**	**Ural-377M**	**Ural-377S**
Cab seating	1 + 2	1 + 2	1 + 2
Configuration	6 × 4	6 × 4	6 × 4
Weight (empty)	7,275 kg	6,635 kg	6,970 kg
(dirt road)	15,000 kg	14,635 kg	n/app
Weight on front axle (loaded)	4,000 kg	n/app	n/app
Weight on rear bogie (loaded)	11,000 kg	n/app	n/app
Towed load (road)	10,500 kg	n/app	19,000 kg
(dirt road)	5,600 kg	n/app	n/app
Load area	4.5 × 2.326 m	4.5 × 2.326 m	n/app
Length	7.6 m	7.86 m	6.9 m
Width	2.5 m	2.5 m	2.5 m
Height (cab)	2.62 m	2.535 m	2.62 m
(load area)	1.6 m	1.6 m	n/app
Ground clearance	0.4 m	0.4 m	0.32 m
Track	2 m	2 m	2 m
Wheelbase	3.5 m + 1.4 m	3.5 m + 1.4 m	3.5 m + 1.4 m
Angle of approach/departure	44°/42°	44°/42°	44°/65°
Max speed (road)	75 km/h	70 km/h (est)	65 km/h
Range	550 km	500 km	740 km
Fuel capacity	300 litres	300 litres	300 litres
Fording	0.8 m	0.8 m	0.8 m
Engine	ZIL-375Ya V-8 petrol developing 175 hp at 3,000 rpm		
Gearbox	manual with 5 forward and 1 reverse gears		

Ural-377 (6 × 4) 7,500 kg truck

Ural-377S (6 × 4) tractor truck (A McKrill)

Specifications

Model	**Ural-377**	**Ural-377M**	**Ural-377S**
Clutch	twin dry disc	twin dry disc	twin dry disc
Transfer box	2-speed	2-speed	2-speed
Steering	double-thread worm with hydraulic booster		
Turning radius	10.5 m	10.5 m	10.5 m
Suspension (front)	semi-elliptical springs with hydraulic shock-absorbers (interchangeable with MAZ-500)		
(rear)	bogie with semi-elliptical springs		
Tyres	14.00 × 20	14.00 × 20	14.00 × 20
Brakes	air/hydraulic	air/hydraulic	air/hydraulic
Electrical system	12 V	12 V	12 V
Battery	1 × 6-STEN-140M	1 × 6-STEN-140M	1 × 6-STEN-140M
Generator	G130	G130	G130

Status
In service with former Warsaw Pact countries and also exported for military use.

Manufacturer
Ural Automobile Plant, Miass, RF-456300 Chelyabinsk Region, Russia.
Tel: +7 35135 33570. Telex: 124849 URAL SU.
Fax: +7 35135 34189.

VERIFIED

KrAZ-255B (6 × 6) 7,500 kg truck

Description
The KrAZ-255B replaced the KrAZ-214 (6 × 6) truck (in production from 1967 until 1979). It is very similar to the earlier vehicle but has a more powerful engine and a central tyre pressure regulation system.

The layout of the vehicle is conventional, with the engine at the front, a fully enclosed two-door cab in the centre and the cargo area at the rear with a hinged tailgate, removable bows and a tarpaulin cover. Standard equipment on all vehicles includes a cab heater, engine preheater, winch with 12,000 kg capacity and suspension locking mechanism.

Variants
There are at least three variants of the KrAZ-255B, including the KrAZ-255L timber truck and the KrAZ-255V tractor truck. The latter weighs 10,600 kg and can tow a semi-trailer weighing 26,000 kg on roads or 18,000 kg across country. Principal dimensions are length 7.685 m, width 2.75 m and height 2.94 m. The fuel service truck is designated the TZ 8-255B and carries 8,000 litres of fuel. The KrAZ-256 is a 6 × 4 tipper truck with a load capacity of 12,000 kg.

Many roles previously undertaken by the KrAZ-214 are now being carried out by the KrAZ-255B including carrying and laying the TMM treadway bridge, carrying and launching the PMP heavy floating pontoon bridge, carrying and launching the BMK-T bridging boat, crane truck, E-305 BV crane shovel, excavator EDV-4421, ATsM 7-255B 7,000 litre fuel tanker towing the PTsM 8925 5,800 litre trailer, and USM pile driving set.

Specifications
KrAZ-255B
Cab seating: 1 + 2
Configuration: 6 × 6
Weight:
(empty) 11,950 kg
(loaded, cross-country) 19,950 kg
Weight on front axle, loaded: 5,600 kg
Weight on rear axles, loaded: 7,000 kg
Max load: 8,000 kg
Towed load:
(road) 30,000 kg
(cross-country) 10,000 kg
Load area: 4.56 × 2.5 m
Length: 8.645 m
Width: 2.75 m
Height:
(cab) 2.94 m
(tarpaulin) 3.17 m
(load area) 1.65 m
Ground clearance: 0.36 m
Track: (front and rear) 2.16 m
Wheelbase: 4.6 m + 1.4 m
Angle of approach/departure: 48°/32°
Max speed: 70 km/h
Range: 490 km
Fuel capacity: 330 litres
Max gradient: 57%
Vertical obstacle: 0.6 m
Trench: 0.68 m
Fording: 1 m
Engine: YaMZ-238 14.86 litre V-8 water-cooled diesel developing 240 hp at 2,100 rpm
Gearbox: manual with 5 forward and 1 reverse gears
Clutch: dual dry disc
Transfer box: 2-speed with differential between axles
Steering: screw and nut with rolling ball, hydraulic booster
Turning radius: 13 m
Suspension:
(front) longitudinal semi-elliptical springs with telescopic double-acting shock-absorbers
(rear) equaliser type on 2 longitudinal semi-elliptical springs
Tyres: 13.00 × 530-533
Brakes:
(main) air
(parking) mechanical
Electrical system: 24 V
Batteries: 2 × 6-TST-165EMS
Generator: 500 W

Status
Production completed in 1979. In service with the RFAS and other former Warsaw Pact countries.

Manufacturer
Kremenchug Motor Vehicle Plant, Kremenchug, Ukraine.

VERIFIED

KrAZ-255B (6 × 6) 7,500 kg truck (Michael Jerchel)

E-305 BV crane shovel mounted on chassis of KrAZ-255B (6 × 6) 7,500 kg truck (Stefan Marx)
1996

KAMAZ-5320 (6 × 4) 8,000 kg trucks

Description

The KAMAZ-5320 (6 × 4) 8,000 kg truck is the base vehicle in a complete family of trucks available in both 6 × 4 and 6 × 6 configurations. Originally intended for commercial use, they are widely used by the RFAS armed forces for long-haul logistics. All are powered by a V-type water-cooled diesel engine with the same bore and stroke (120 × 120 mm). Brief specifications of the various engines are:

Model	**YaMZ-7401**	**YaMZ-740***	**YaMZ-741**
Type	V-8	V-8	V-10
Capacity	10.85 litres	10.85 litres	13.58 litres
Output	180 hp	210 hp	260 hp

* The YaMZ-740 engine is also used in the Ural-4320, described earlier in this section.

The basic cargo model is the KAMAZ-5320 (6 × 4) that has a fully enclosed all-steel forward control cab that can be tilted forward for maintenance purposes. The rear cargo area has drop sides, removable bows and a tarpaulin cover. Standard features include: an air filter working from the exhaust gases; air intake above the cab roof on the left side; centrifugal oil cleaner; chemical- and heat-treated crankshaft; hydraulic power steering; pneumatic clutch control; pneumatic preselector (a two-stage gearbox fitted between the clutch and the conventional gearbox); heater; driver's seat with suspension; and a heat- and soundproofed cab. In long-distance versions a bunk is provided.

For travelling on rough or dirt roads the 5320 and 5410 have their rear axles rated at 5,500 kg each. Models travelling on improved roads, such as the 53212 and 54112 have axle ratings of up to 10,000 kg. The KAMAZ-5320 is powered by a YaMZ-740 V-8 diesel engine that develops 210 hp. This engine is also installed in the latest version of the Ural-375 (6 × 6) truck and the Ural-4320 (4 × 4) 4,500 kg truck.

Variants include the KAMAZ-5511 (6 × 4 rear tipper truck. This has a wheelbase of 2.84 m – 1.32 m, a length of 7.125 m, width of 2.5 m and height of 2.708 m. Front track is 2.01 m and rear 1.85 m.

The KAMAZ-4310 is a 6 × 6 version with a cross-country payload of 6,000 kg and a towed load of 7,000 kg. It is fitted with a central tyre pressure regulation system.

Variants

(6 × 4)

KAMAZ-53201: cargo truck chassis based on KAMAZ-5320, 180 hp engine
KAMAZ-53202: long wheelbase cargo truck
KAMAZ-53203: longer flatbed KAMAZ-53202, 180 hp engine
KAMAZ-53212: long flatbed cargo truck, sleeper cab, long-range fuel tanks, strengthened leaf springs at rear
KAMAZ-5410: tractor truck, sleeper cab
KAMAZ-5411: rear tipper truck
KAMAZ-54112: tractor truck, long-range fuel tanks, strengthened leaf springs at rear
KAMAZ-54101: tractor truck for dump semi-trailers
KAMAZ-54102: tractor truck for heavier loads, 260 hp engine
KAMAZ-5510: 7,000 kg dump truck, 180 hp engine
KAMAZ-5511: rear tipper truck
KAMAZ-55102: agricultural dump truck with trailer
KAMAZ-55103: tipper truck

(6 × 6)

KAMAZ-4310: cargo truck with fixed side boards
KAMAZ-43101: cargo truck with drop sides
KAMAZ-43102: same as KAMAZ-43101, bunk in cab
KAMAZ-43103: same as KAMAZ-4310 with bunk in cab
KAMAZ-43105: empty weight 8,200 kg, load capacity 7,000 kg
AA-40 (43105)-189: airfield fire tender based on KAMAZ-43105
KAMAZ-4410: tractor truck

7,000 litre tanker based on KAMAZ-4310 (6 × 6) truck series (Stefan Marx) ***1996***

Specifications

Model	**KAMAZ-5320**	**KAMAZ-53202**	**KAMAZ-5410**	**KAMAZ-54102**	**KAMAZ-4310**
Type	cargo truck	cargo truck	tractor truck	tractor truck	cargo truck
Configuration	6 × 4	6 × 4	6 × 4	6 × 4	6 × 6
Weight (empty)	7,080 kg	7,240 kg	6,445 kg	n/a	9,000 kg
Max load	8,000 kg	8,000 kg	n/a	n/a	6,000 kg*
Max towed load	11,500 kg	11,500 kg	19,000 kg	n/a	7,000 kg*
Length	7.435 m	8.295 m	6.14 m	6.43 m	7.895 m
Width	2.506 m	2.496 m	2.48 m	n/a	2.5 m
Height (cab)	2.83 m	2.63 m	2.63 m	2.63 m	3.09 m
Ground clearance	0.28 m	0.285 m	0.285 m	n/a	0.365 m
Track (front)	2.026 m	2 m	2.01 m	n/a	2.01 m
(rear)	1.856 m	1.85 m	1.85 m	n/a	2.01 m
Wheelbase	3.19 m + 1.32 m	3.69 m + 1.32 m	2.84 m + 1.32 m	3.14 m + 1.32 m**	3.34 m + 1.32 m***
Max speed (road)	85 km/h	85 km/h	85 km/h	85 km/h	85 km/h
Engine	YaMZ-740	YaMZ-740	YaMZ-740	YaMZ-741	YaMZ-740

* cross-country ** also quoted as 3.04 m + 1.32 m *** also quoted as 3.39 m + 1.32 m

Status

In production.

Manufacturer

KAMAZ, Naberezhnye Chelny, RFAS.

UPDATED

KrAZ-260 (6 × 6) 9,000 kg trucks

Description

The KrAZ-260 (6 × 6) 9,000 kg truck replaced the earlier KrAZ-255B and has been in production since 1979. It was first seen in public in 1985 towing the 2A36 152 mm nuclear-capable gun. That version had an open body with forward-facing bench seats for the gun crew, although it is possible that this was purely a 'parade' arrangement. The normal cargo body of the KrAZ-260 has a hinged tailgate with the cargo area covered by a tilt over removable bows. The KrAZ-260 has an increased payload (9,000 kg as opposed to 7,500 kg) compared to the KrAZ-255B and uses the YaMZ-238L 14.68 litre V-8 turbocharged diesel engine. The gearbox has eight forward speeds and one reverse, and the two-speed transfer box has a lockable differential.

The KrAZ-260 is fitted with a 118 kN capacity winch provided with 53 m of 22 mm cable.

Variants

KrAZ-260V. A 6 × 6 tractor truck version with a fifth wheel for towing semi-trailers is known as the KrAZ-260V. It replaced the earlier KrAZ-255V1 and KrAZ-258B1 tractors in production. Maximum fifth wheel load is 9,500 kg and the full weight of a towed semi-trailer is 27,500 kg.

One application for this vehicle is as a tractor for various components, including the erector/launcher and missile resupply trailer, of the SA-10a 'Grumble' surface-to-air missile system (for details of this system refer to *Jane's Land-based Air Defence 1995-96* pages 126 to 128).

KrAZ-250. This is a 6 × 4 variant with twin-tyred rear wheels and an on-road load capacity of 16,575 kg. Wheelbase is 4.88 m + 1.4 m. As well as being used for cargo bodies the chassis may also be used to carry cranes.

KrAZ-5444. A 4 × 2 truck tractor with a maximum fifth wheel load of 9,400 kg and a towed semi-trailer maximum weight of 32,000 kg. Wheelbase is 4.7 m. Twin rear wheels are fitted.

KrAZ-6437. A 6 × 6 tractor truck with twin-tyred rear wheels and configured for carrying stake, pipe-carrying and logging semi-trailers weighing up to 30,500 kg, particularly the GBK-9632 pole trailer manufactured by the Tavda Mechanical Works. Apparently mainly for commercial use. GCW is 47,000 kg.

KrAZ-6443. A 6 × 6 tractor truck intended for towing fuel tank semi-trailers weighing up to 48,000 kg.

KrAZ-64431. A 6 × 4 tractor truck for commercial use. Maximum towed weight is 32,700 kg.

KrAZ-6444. A 6 × 4 tractor truck for commercial use. Maximum towed weight is 32,000 kg.

KrAZ-6510. A 6 × 4 dump truck with an 8 m³ tipper body carrying loads up to 15,000 kg. GVW is 26,380 kg.

KrAZ-65034. A 6 × 6 dump truck with a 12 m³ tipper body carrying loads up to 18,000 kg. GVW is 31,250 kg.

A KrAZ-260 chassis is used as the carrier for the Protector-N truck-mounted decontamination apparatus. Refer to *Jane's NBC Protection Equipment 1996-97* pages 103 and 104 for available details.

Specifications

KrAZ-260
Cab seating: 1 + 2
Configuration: 6 × 6
Weight:
(empty) 12,250 kg

KrAZ-260 (6 × 6) 9,000 kg truck (T J Gander) ***1996***

(loaded) 21,475 kg
Weight on front axle, loaded: 6,440 kg
Weight on rear axles, loaded: 15,035 kg
Max load: 9,000 kg
Towed load:
(empty) 30,000 kg
(loaded) 10,000 kg
Length: 9.006 m
Width: 2.724 m
Height: (top of cab) 2.985 m
Ground clearance: 0.37 m
Track: (front and rear) 2.16 m
Wheelbase: 4.6 m + 1.4 m
Angle of approach/departure: 35°/26°
Max speed: 80 km/h
Range: 700 km
Fuel capacity: 2 × 165 litres + 1 × 45 litres
Max gradient: 58%
Vertical obstacle: 0.5 m
Trench: 0.68 m
Fording: 1.2 m
Engine: YaMZ-238L 14.86 litre V-8 water-cooled turbocharged diesel developing 288 hp at 2,100 rpm
Gearbox: manual with 8 forward and 1 reverse gears
Clutch: twin dry discs
Transfer box: 2-speed
Turning radius: 13.5 m
Tyres: 13.00 × 530-533
Electrical system: 24 V

Status

In production. In service with the RFAS armed forces and possibly some other former Warsaw Pact armed forces.

Manufacturer

KrAZ, 62 Kievskaya Street, 315331 Kremenchug, Ukraine.
Tel: +53 66 5 95 48. Telex: 131476 GROZA.
Fax: +380 44 224 2097.

UPDATED

Ural-5323 (8 × 8) 9,000 kg truck

Description

The Ural-5323 (8 × 8) 9,000 kg truck is a high-mobility vehicle produced using numerous components and systems taken from other Ural and KAMAZ trucks. The vehicle can be used for transporting supplies or personnel and may be used as the basis for mounting weapons or various combat systems.

The Ural-5323 weighs 10,550 kg empty and can carry 9,000 kg of payload. It has a forward control all-metal cab placed directly over the engine to provide the maximum cargo-carrying area length and to spread the load weight more evenly over the axles. The cab can be tilted forward for engine and transmission access and has seating for the driver and one passenger.

The metal cargo platform is an open, low-sided area with hinged side benches that can be removed. The front and side boards are corrugated for extra strength and a tarpaulin can be erected over bows.

The standard engine is a KAMAZ-7403 10.85 litre V-8 water-cooled turbocharged diesel developing 260 hp, although an air-cooled Ural-745 diesel developing 290 hp may be used. A third engine option is the Ural-746 air-cooled diesel developing 320 hp. To simplify starting at low temperatures the engine is equipped with a PZhD-30A preheater and an electric glow plug.

The transmission is operated via a dual dry plate KAMAZ-142 clutch with mechanical linkage provided with pneumatic assist. The gearbox is a three-range, five-speed KAMAZ-141 unit. The two-speed transfer case has a cylindrical power splitter that symmetrically supplies power to the differentials. Differential and shifting speed locks are remotely controlled using pneumatics on the transfer case. The main transmission supplies power to the axles via a two-stage reduction mechanism with a pair of cylindrical helical gears matched to a pair of spiral gears.

The forward and rear axle suspensions are mounted on rocker units. Each pair of axles has two longitudinal semi-elliptic leaf springs with the front (steering) axles provided with four hydraulic shock-absorbers.

The service brakes are of the drum type with a twin-chamber hydropneumatic assist mechanism. One of the chambers acts as a reserve mechanism while the parking brake is mounted on the transfer case and operates on the output shafts. Power steering is provided.

The vehicle is equipped with a central tyre pressure regulation system. To enable the vehicle to drive over various types of terrain the truck is provided with OI-25 14.00 × 20 tyres having a directional tread pattern.

The vehicle uses a 24 V electrical system with two 6ST-190TR or 6ST-190TRN batteries and a 1 kW generator. For operations in extremely cold conditions the batteries can be provided with an automatic heating system.

A standard component is a mechanical drum winch with a 7,000 kg capacity and 60 m of cable. Power for the winch is taken from an auxiliary power take-off mounted on the transfer case.

Status

In production.

Manufacturer

Ural Automobile Plant, Miass, RF-456300 Chelyabinsk Region, Russia.
Tel: +7 35135 33570. Telex: 124849 URAL SU.
Fax: +7 35135 34189.

VERIFIED

KAMAZ-6350 (8 × 8) 10,000 kg truck

Description

The KAMAZ-6350 (8 × 8) 10,000 kg truck is one of a family of three trucks, the others being the KAMAZ-4350 (4 × 4) and KAMAZ-5350 (6 × 6) – see separate entries for these two vehicles. All three trucks share many design features and components, not the least being the forward control cab which is very similar to that utilised on the KAMAZ-5320 series (qv).

The intention appears to be to provide a family of similar vehicle capable of meeting a wide range of military logistic requirements. Various enhancements, such as an improved efficiency diesel engine coupled to an automatic gearbox, have been introduced, although full details have yet to be determined. Many components are also shared with the extensive range of KAMAZ commercial trucks

The cargo body fitted to the example observed is entirely conventional, being an all-steel flatbed with drop sides, a tailgate and a canvas cover over removable bows.

KAMAZ-6350 (8 × 8) 10,000 kg truck (C F Foss) ***1996***

KAMAZ-6350 (8 × 8) 10,000 kg truck (C F Foss) ***1996***

Specifications

Provisional
Cab seating: 1 + 2
Configuration: 8 × 8
Weight: 11,400 kg
Max load: 10,000 kg
Towed load: 12,000 kg
Load area: 6.1 × 2.42 m
Length: 9.235 m
Width: 2.5 m
Height:
(cab) 3.23 m
(tarpaulin) 3.25 m
(load area) 1.57 m
Ground clearance: 0.385 m
Track: 2.01 m
Wheelbase: 1.941 m + 3.34 m + 1.32 m
Angle of approach/departure: 36°/50°
Max speed: (road) 95 km/h
Range: 1,250 km
Fording: 1.75 m
Engine: KAMAZ-7482 11.75 litre V-8 diesel developing 360 hp at 2,200 rpm
Gearbox: automatic
Tyres: 16.75/8 R 21
Electrical system: 24 V

Status

Available for production.

Manufacturer

KAMAZ, Naberezhnye Chelny, RFAS.

VERIFIED

ZIL-135 Series (8 × 8) trucks

Description

The ZIL-135 series of 8 × 8 trucks was introduced during the 1960s. The original design and development was carried out at the Likhachev Motor Vehicle Plant near Moscow, but production was carried out at the Bryansk Automobile Works, for which reason the range is sometimes known as the BAZ-135L4 (the cargo truck version is also known as the ZIL-135L4). These trucks are widely used by members of the former Warsaw Pact for both military and civil applications.

The vehicle is powered by two petrol engines, each of which drives the four wheels on one side of the vehicle. Steering is by the front and rear wheels only and these wheels are carried on axles with a torsion bar suspension. The two central axles are fixed and a central tyre pressure regulation system is fitted. The fully enclosed forward control cab has two doors, one on each side. When being used for missile launching roles the windscreen is fitted with covers to protect it from blast when the missile is launched.

Variants

9M21 Transporter/Launcher Vehicle
This carries and launches the FROG-7 (9M21 Luna M R-75) surface-to-surface tactical rocket system. Mounted on the right side of the vehicle is a hydraulic crane for reloading purposes. Stabiliser jacks are lowered to the ground before the missile is launched to provide a more stable firing platform.

9T29 Resupply Vehicle
This carries three FROG-7 rockets.

Sepal Transporter/Launcher Vehicle
This has a large cylindrical container which acts as the launcher for the Sepal cruise missile. This version has a different cab from the other members of this series and two stabiliser jacks are lowered to the ground either side before the missile is launched.

BM-22 Multiple Rocket System
The ZIL-135 series chassis is also used for the BM 9P140 220 mm Uragan 16-round multiple rocket launcher and its associated 9T452 transloader. For details see *Jane's Armour and Artillery 1995-96*, pages 711 to 713.

ZIL-135 Convoy Escort Vehicle
Reports from Afghanistan mentioned the use of an improvised convoy escort vehicle formed by mounting a ZU-23 twin 23 mm cannon on the rear of a ZIL-135 truck. Some form of armour was provided for the gun crew.

BAZ-135L4 Cargo Truck
This is produced at the Bryansk Plant and is used for carrying cargo. The rear cargo area is provided with a drop tailgate. Also referred to as ZIL-135L4.

ZIL-135 Tractor Truck
As far as is known, this is only used for civilian roles, primarily for carrying long lengths of pipe used in constructing pipelines.

Specifications

BAZ-135L4 cargo truck
Cab seating: 1 + 2
Configuration: 8 × 8
Weight:
(loaded) 19,000 kg
(empty) 9,000 kg
Max load: 10,000 kg
Towed load:
(road) 20,000 kg
(cross-country) 18,000 kg
Load area: (estimated) 4.57 × 2.65 m
Length: 9.27 m
Width: 2.8 m
Height: 2.53 m
Ground clearance: (axles) 0.58 m
Track: 2.3 m
Wheelbase: 2.415 m + 1.5 m + 2.415 m
Max speed: 70 km/h
Range: 500 km
Fuel capacity: 768 litres
Fuel consumption: 160 litres/100 km
Max gradient: 57%
Vertical obstacle: 0.685 m
Trench: 2.63 m
Fording: 0.58 m
Engines: 2 × ZIL-375 7 litre V-8 water-cooled petrol developing 180 hp (each) at 3,200 rpm
Gearbox: hydromechanical
Steering: power-assisted, 1st and 4th axles
Turning radius: 12.5 m
Suspension: 1st and 4th axles have torsion bars, 2nd and 3rd axles are fixed to chassis and have no suspension
Tyres: 16.00 × 20
Brakes: (main) hydraulic, air-assisted
Electrical system: 24 V

Status

In service with Afghanistan (Uragan), Algeria (FROG-7), Bulgaria (FROG-7), Cuba (FROG-7), Czech Republic and Slovakia, Egypt (FROG-7), Hungary (FROG-7), Iraq (FROG-7), North Korea (FROG-7), Kuwait (FROG-7), Libya (FROG-7), Poland (FROG-7), RFAS, Romania (FROG-7), Syria (Uragan and FROG-7), Yemen (FROG-7) and the former Yugoslavia.

Manufacturer

Bryansk Motor Vehicle Plant, Bryansk, Russia.

UPDATED

ZIL-135 (8 × 8) truck configured as transloader vehicle for BM-22 220 mm multiple rocket system

FROG-7 transporter/loader vehicle based on ZIL-135 (8 × 8) truck chassis

---

MAZ-543 (8 × 8) trucks

Development

Military vehicles produced at Minsk are based on the 8 × 8 MAZ-543 truck and MAZ-537 tractor trucks. For details of the latter refer to the separate entry in the *Tank Transporters* section.

Description

The MAZ-543 (8 × 8) truck uses many automotive components of the MAZ-537 (8 × 8) tractor truck, including a similar engine. It was seen in public for the first time in November 1965.

The MAZ-543 is a four-axle 8 × 8 truck chassis designed for off-road use, with steerable wheels on the front two axles. The basic chassis has a nominal load capacity of 20,000 kg and can be used for many special applications apart from as a cargo or load carrier.

The MAZ-543 series is powered by a 525 hp D12A-525A 38.88 litre V-12 diesel engine located between two separate two-man cabs. There is a hydromechanical transmission with a torque converter and three-speed planetary gearbox. The drive axles have central reduction gears with self-locking differentials, semi-axle Cardan joints and constant velocity universal joints in the steerable wheel linkages, plus planetary wheel reduction gears. The wide-profile tyres have a pressure control system. The suspension features independent torsion bar stations with double-acting telescopic type shock-absorbers.

Variants

Military 8 × 8 variants of the MAZ-543 produced at Minsk include the following models.

MAZ-543A: On this model the nominal load capacity is increased to 22,000 kg. The engine is located further forward, resulting in one of the two cabs having seating for only one person. GVW is 43,300 kg and axle loading is increased to 11,300 kg.

MAZ-543M: With this model the engine is moved to the front of the vehicle, resulting in a single two-man cab on the left-hand side. This allows a further increase in load capacity to 22,600 kg, although the GVW and axles loadings remain the same as those for the MAZ-543A.

MAZ-7313: This model is a cargo carrier with a load

MAZ-543M (8 × 8) chassis carrying accommodation body for personnel operating SA-10 SAM systems (T J Gander)

MAZ-543M (8 × 8) chassis carrying 54K6 SA-10 command post body (T J Gander)

capacity of 21,000 kg in a conventional flatbed cargo body.
MAZ-73132: On this tractor truck model the chassis is shortened to 10.815 m and the wheelbase to 2.2 m + 2.65 m + 1.7 m. GVW is 45,150 kg and axle load 12,500 kg.

Variants of the MAZ-543 are used to carry and launch the 'Scud A' (SS-1b), 9P117M 'Scud B' (SS-1c) and 'Scaleboard' (SS-12) surface-to-surface missiles. The Scud launchers are known as the MAZ-543LTM or MAZ-7310LTM Uragan (Hurricane), or Kashalot (Sperm Whale); the vehicle is 12 m long, and, with the missile, weighs 29,000 kg. A version with single forward cab is used to carry the SSC-3 coastal defence missile system. Other missile-related variants include the fire-control vehicle, trailer-erector-launcher and other units for the SA-10b 'Grumble' surface-to-air system. The 300 mm BM 9A52 Smerch 12-round multiple launch rocket system is based on the MAZ-543 and there is also an associated reload vehicle, the 9T234-2. Many vehicles carry control centre, workshop, accommodation and office bodies for various weapon systems in addition to launch vehicles.

MAZ-543 variants are used to carry components of a mobile 130 mm A-222 BEREG coastal artillery system. Another mobile coastal defence system involving the MAZ-543 is the long-range multiple rocket-based 3K-60; this eight-rocket launcher system is not yet in full production.

Other MAZ-543 variants include an aircraft crash tender, a heavy crane carrier, and a tanker with a full-width forward control cab. A trials vehicle fitted with a gas turbine developing the equivalent of 1,100 hp has been produced.

Specifications

MAZ-543
Cab seating: 2 + 2
Configuration: 8 × 8
Weight:
(kerb) 20,500 kg
(GVW) 40,500 kg
Max load: (nominal) 20,000 kg
Axle load: 10,450 kg
Length: 11.46 m
Width: 3.07 m
Height: (top of cab) 2.9 m
Ground clearance: (min) 0.4 m
Track: (front and rear) 2.375 m
Wheelbase: 2.2 m − 3.3 m + 2.2 m
Angle of approach/departure: 28°/35°
Max speed: 63 km/h
Range: 1,525 km
Max gradient: 57%
Vertical obstacle: 0.78 m
Trench: 3.38 m
Fording: 1.1 m
Engine: D12A-525A 38.88 litre V-12 water-cooled diesel developing 525 hp at 2,100 rpm
Transmission: hydromechanical with lockable torque converter and 3-speed planetary gearbox with gearshift; 1 reverse gear
Transfer box: 2-speed
Turning radius: 13.5 m
Suspension: independent torsion bar with double-acting telescopic shock-absorbers
Brakes: air-hydraulic drum on all wheels
Tyres: 15.00 × 600-635 with pressure control system
Electrical system: 24 V
Batteries: 4 × 6 V

Status

In production. In service with former Warsaw Pact countries, Iraq and the former Yugoslavia.

Manufacturer

Minsk Wheel Tractor Plant, 150 pr-t Partizanski, 220021 Minsk, Belarus.
Tel: +7 0172 469433. Telex: 252519 VIR BY.
Fax: +7 0172 429861.

The 130 mm Bereg coast defence gun system is based on the MAZ-543M truck chassis ***1997***

UPDATED

MAZ heavy trucks

Development

The Minsk Wheel Tractor Plant was established in 1991, based on the Minsk Automobile Plant which for many years had been specialising in manufacturing specialised heavy vehicles for mobile missile systems. The new concern has diversified into heavy construction plant and multiwheel drive configurations for tasks such as oil drilling and exploration or construction applications. This has led to the 'Volat' series of commercial 8 × 8 vehicles which utilise chassis and many components originally intended for military purposes but having mechanical transmissions in place of the hydromechanical equivalents used by military vehicles; they also have lower powered engines. They nearly all have commercial pattern forward control cabs.

Among the vehicles produced at the Minsk Wheel Tractor Plant are 10 × 10 and 12 × 12 chassis for various purposes, the ultimate being a 16 × 16 monster apparently intended for the oil industry. A summary of these specialised chassis, some of which were originally developed for mobile missile applications, is provided below.

Description

The commercial 'Volat' family includes the following 8 × 8 models, all with a 470 hp YaMZ-8424.10-03 diesel and 9-speed mechanical gearbox.
MAZ-79092. Load carrier with a 22,000 kg load capacity.
MAZ-75165. Dump truck with a 24,500 kg load capacity.
MAZ-79093. Special truck body for carrying pipes, timber or stakes.
MAZ-79094 and MAZ-79096. Truck tractors, with the MAZ-79096 having uprated axles to accommodate a slightly heavier kerb weight.
MAZ-7429. Truck tractor with lowered fifth wheel height and GVW of 44,050 kg.
MAZ-74131. On this heavy-haul truck tractor model the engine is a 650 hp YaMZ-8401.10 diesel connected to an automatic 4-speed gearbox. Axle load is 15,500 kg and GVW 54,000 kg.

In addition to the commercial carriers a series of specialised load carriers are produced, primarily for commercial purposes but many based on mobile missile and other system carriers. A summary of the main models, most of which have commercial pattern forward control cabs similar to those of the 'Volat' series, is provided here for reference.
MAZ-6923. An 8 × 4 model with a load capacity of 22,800 kg. Powered by a 330 hp YaMZ-238D diesel via a manual 8-speed gearbox. No central tyre pressure control system.
MAZ-69232. Essentially similar to the MAZ-6923 but powered by a 360 hp MAN D-2866 via a 16-speed gearbox. No central tyre pressure control system.
MAZ-79091. Full 8 × 8 drive configuration. Powered by a 470 hp YaMZ-8424.10 diesel. Load capacity 24,000 kg. Cab has two sleeper positions. A central tyre pressure control system is provided as standard.
MAZ-79092. As MAZ-79091 but with shorter chassis produced by reduced distance between two rear axles.
MAZ-79301. 8 × 8 with cab located in front of engine to reduce overall height to 2.93 m.
MAZ-7916. 12 × 12 drive configuration with 'military' style three-seat cab. Powered by 38.88 litre V-12

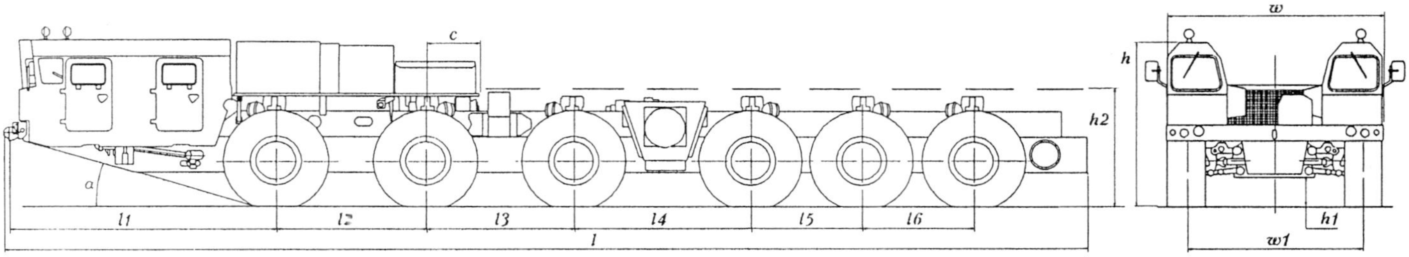

Outline drawing of MAZ-7916 (12 × 12, 50,000 kg heavy vehicle chassis **1996**

turbocharged diesel developing 710 hp Automatic four-speed transmission with single-speed transfer case. Load capacity is 50,000 kg with kerb weight 32,000 kg. Top speed limited to 45 km/h. The data in the Specifications table relates to this model.

MAZ-79191. Another 12 × 12 but with 'commercial' style forward control two-position cab. Load capacity is 56,000 kg and GVW 90,000 kg. Powered by 650 hp YaMZ-8410.10 turbocharged diesel via a four-speed automatic transmission and single-speed transfer case.

Specifications

MAZ-7916
Cab seating: 1 + 2
Configuration: 12 × 12
Weight:
(kerb) 32,000 kg
(GVW) 82,000 kg
Max load: (nominal) 50,000 kg
Axle load: 15,000 kg
Length: 19.6 m
Width: 2.78 m
Height: (top of cab) 3.13 m
Ground clearance: 0.465 m
Track: (front and rear) 2.78 m
Wheelbase: 2.3 m + 2.3 m + 2.8 m + 1.75 m +1.75 m
Angle of approach: 16°
Max speed: 45 km/h
Engine: 38.88 litre V-12 turbocharged diesel developing 550 hp
Transmission: hydromechanical with lockable torque converter and 4-speed planetary gearbox with gearshift; 2 reverse gears
Transfer box: single-speed
Turning radius: 24 m
Suspension: independent torsion bar with double-acting telescopic shock-absorbers
Brakes: air-hydraulic drum on all wheels
Tyres: 16.00 × 600-685 with pressure control system
Electrical system: 24 V

Status

In production.

Manufacturer

Minsk Wheel Tractor Plant, 150 pr-t Partizanski, 220021 Minsk, Belarus.
Tel: +7 0172 46 94 33. Telex: 252519 VIR BY.
Fax: +7 0172 42 98 61.

VERIFIED

BAZ-5937 and BAZ-5939 (6 × 6) special purpose vehicles

Description

The BAZ-5937 and BAZ-5939 are basically similar 6 × 6 special purpose vehicles that are amphibious and highly mobile. The vehicles have many uses and are employed as the carriers for components of the SA-8a and SA-8b 'Gecko' surface-to-air missile systems and the SS-21 'Scarab' surface-to-surface missile system. The BAZ-5937 is used for the SA-8a and SA-8b 'Gecko' transporter/erector/launcher and radar systems and for the SA-8 transport and reloading vehicle; the BAZ-5939 is used as the basis for the SS-21 'Scarab' launcher and reloading vehicles. The BAZ-5937 has its engine located just in front of the rear wheels, while the SS-21 versions of the BAZ-5939 have the engine located between the first and second set of wheels. Many components are interchangeable between the two vehicles.

Being essentially similar, the two vehicles are often referred to as the BAZ-5937/39. Both use a long boat-shaped hull with the driver/operators' cab set forward in front of the load-bearing area. The cab has room for three occupants with one occupant seated in a semi-prone position behind the driver. The cab is equipped with an NBC protection system with the air filtration unit inside the cab over the left front wheel. A diesel heater is provided, with the heater blower coupled to the air filtration system.

The load area is constructed from welded steel plating with supports for strength and stabilisation. A door is used for access to the engine and generator area. Both vehicles use a type 5020B six-cylinder, four-stroke, water-cooled diesel engine rated at 290 hp. The air intake system uses a two-stage filter; the air can be heated in winter. The primary starter system uses air pressure with a back-up electrical starter. Fuel is carried in a 320 litre tank mounted in front of the left rear wheel.

The five-speed automatic transmission can be operated manually during amphibious operations. There is a two-stage hydraulic system. Final drive is transmitted to all six roadwheels independently. For lubrication there is a dry sump oil injection system with pumps, a centrifugal filter and a cooling system. The engine cooling system is similar to that used on the BMP ICV, using injection cooling. The pump involved is combined with the oil pump. A diesel prewarmer for the engine is provided by routeing the exhaust pipe through the oil and hydraulic fluid reservoirs.

Both vehicles are provided with power steering and brakes. The brakes are a two-part system with the first part controlling the right front, left middle and right rear wheels; the second part controls the left front, right middle and left rear wheels. A two-drum emergency brake system is provided.

A 24 V electrical system is provided and the lighting system is similar to that used on the Ural-4320 (6 × 6) 4,500 kg truck. The BAZ-5937 carries a 50 kW generator driven by a gas turbine along with a back-up 35 kW generator powered by the transmission. Both generators can be used to provide power for the various weapon systems carried. An R-123 radio and an R-124 intercom are carried.

For amphibious operations the BAZ-5937/39 is powered by two water-jets located in the rear. Maximum speed in water is 8 km/h. In the water, steering is effected by a rudder and the cab is protected by a raised baffle plate; the driver and co-driver use special vision devices.

Specifications

Provisional
Cab seating: 1 + 2
Configuration: 6 × 6
Weight: (basic configuration) 10,650 kg
Length: 9.16 m
Width: 2.78 m
Height: 1.95 m
Ground clearance: 0.43 m
Max speed:
(land) 70 km/h
(water) 8 km/h
Fuel capacity: 320 litres
Fuel consumption: 40 litres/100 km
Max gradient: 60%
Engine: 5D20B V-6 four-stroke diesel developing 295 hp
Gearbox: 5 forward and 1 reverse gears
Clutch: hydraulically operated double disc dry
Suspension: independent torsion bar with vibration damping

BAZ-5939 special purpose vehicle (T J Gander)

BAZ-5937 special purpose vehicle (T J Gander)

Turning radius: 7 m
Electrical system: 24 V

Status
In production. BAZ-5937 (SA-8) in service with Algeria, Angola, Czech Republic and Slovakia, Guinea-Bissau, Hungary, India, Iraq, Jordan, Kuwait, Libya, Poland, RFAS, Syria and the former Yugoslavia.

BAZ-5939 (SA-21) in service with the Czech Republic and Slovakia, RFAS and Syria.

Manufacturer
Likhachev Motor Vehicle Plant, Moscow, Russia.

VERIFIED

SLOVENIA

IMR TARA (4 × 4) 1,200 kg truck

Description
The IMR TARA (*Terenski Automobili Rakovica*) (4 × 4) truck appears to be based on the Austrian Steyr-Daimler-Puch Pinzgauer light vehicle but the IMR vehicle has a longer wheelbase and numerous body changes.

The IMR TARA uses a forward control cab with the engine located between the driver's and front passenger's positions. An open area to the rear can accommodate two rows of lateral seats for a further six passengers or it can be used to carry cargo. The all-steel cargo body can be covered by a tarpaulin supported on bows. If required, the bows can be removed, as can the upper parts of the front doors; the windscreen can be folded forwards. The cab area can be heated either by using a system employing the engine coolant or by a Webasco heater.

The vehicle is supplied with a spare wheel, a spare fuel container, a maintenance toolkit, spare parts, engineer tools, NBC equipment and a camouflage net. The vehicle can be screened against radio interference and there is a connection for an infrared driving or vision system.

Optional equipment includes power steering, a heavy-duty air cleaner or air precleaner, 7.50 R 16 XS tyres, a towing pintle, an engine preheating system, snow chains, air conditioning and a light electric winch.

The IMR TARA can be configured as a personnel carrier with seating for the driver and seven passengers, a minibus, a general utility pick-up (with or without an enlarged crew cab), a van, or as a shelter/container carrier. Other specialised bodies are available according to requirements.

Specifications
Cab seating: 1 + 7
Configuration: 4 × 4
Weight: (GVW) 3,500 kg
Max load: 1,200 kg
Towed load:
(without brakes) 900 kg
(with brakes) 1,500 kg
Load area: 2.4 × 1.7 m
Length: 4.2 m
Width: 1.85 m
Height: 2.1 m
Ground clearance: 0.295 m
Track: 1.54 m
Wheelbase: 2.4 m
Angle of approach/departure: 45°/50°
Max speed: 105 km/h
Fuel capacity: 70 litres
Range: approx 500 km
Max gradient:
(without trailer) 70%
(with trailer) 45%
Side slope: 35%
Fording: 0.6 m
Engine: IMR S-54 V 2.45 litre 4-cylinder water-cooled in-line diesel developing 75 hp at 4,200 rpm
Gearbox: manual with 5 forward and 1 reverse gears
Clutch: single dry plate
Transfer box: 2-speed
Steering: Gemmer mechanical
Turning radius: 6.25 m
Suspension: coil springs with rubber elements and double-action hydraulic shock-absorbers
Tyres: 6.50K × 16
Brakes:
(main) dual-circuit hydraulic
(parking) mechanical on rear axle
Electrical system: 24 V
Batteries: 2 × 12 V, 143 Ah

Status
In production.

Manufacturer
Tovarna Avtomobilov in Motorjev Maribor dd, Ptujska cesta 184, 62000 Maribor, Slovenia.
Tel: +386 62 411320. Fax: +386 62 414945.

VERIFIED

IMR TARA (4 × 4) 1,200 kg truck

IMR TARA (4 × 4) 1,200 kg truck

TAM 110 T7 BV (4 × 4) 1,500/2,500 kg truck

Description
The TAM 110 T7 BV has been referred to as the TAM 1500 and was developed specifically to meet the requirements of the former Yugoslav Army, using the design of the German Magirus-Deutz 130T7FAL (which did not enter production) as a basis. It is the 4 × 4 component of two basically similar vehicles, the other being the TAM 150 T11 BV 6 × 6 vehicle (see following entry). Both vehicles use versions of the TAM 413 engine which was produced in what was then Yugoslavia under licence from Klockner-Humboldt-Deutz.

The two-door all-steel forward control cab has a reinforced PVC material roof which can be removed, as can the side windows. The cab has seating for two and can be tipped forward to an angle of 55° for engine access. The windscreen can be tipped forward over the bonnet if required. Two forms of cab heating are available. The cargo body is of all-steel construction and has two collapsible benches fitted to each side to seat 12 personnel. The cargo area has a capacity of 3.31 m^3 and is 3.02 m long, 2.12 m wide and the sides are 0.64 m high. A canvas tilt may be fitted.

The engine is mounted forward under the cab, as is a 2,500 kg capacity winch that can be used to the front or rear. A towing hook is provided at the rear under the downward-opening tailgate. Special equipment carried on the vehicle includes pioneer tools, equipment for the winch and an NBC kit. The vehicle is fitted with radio interference protection and space is provided for the fitment of an infrared device for the driver. There is provision for changing the tyre pressures in the range 0.7 to 3.5 atm while the vehicle is being driven.

A signals vehicle variant houses its signals equipment in a removable container, weighing 6,700 kg, that can be lifted off the vehicle using telescopic support legs. The container is 5.17 m long, 2.34 m wide and 3 m high. A generator may be fitted inside the container.

An ambulance variant has a box-body and can carry up to four stretchers and associated medical equipment. Total weight is 6,400 kg.

Specifications
Cab seating: 1 + 1 (up to 12 in rear)
Configuration: 4 × 4
Weight:
(empty) 4,500 kg
(loaded, road) 7,000 kg
(loaded, off road) 6,000 kg
Max load:
(road) 2,500 kg
(off road) 1,500 kg
Towed load: 1,800 kg
Load area: 3.02 × 2.12 m
Length: 4.85 m
Width: 2.275 m
Height: (top of cab) 2.47 m
Ground clearance: 0.3 m
Wheelbase: 2.85 m
Angle of approach/departure: 49°/45°
Max speed:
(road) 90.8 km/h
(off road) 45 km/h
Fuel capacity: 100 litres

TAM 110 T7 BV (4 × 4) signal truck

TAM 110 T7 BV (4 × 4) 1,500/2,500 kg truck

Max gradient:
(without trailer) 67%
(with trailer) 44.5%
Fording: 1 m
Engine: TAM (KHD) F 4 L 413 FR 4-cylinder 6.381 litre air-cooled diesel developing 115 hp at 2,650 rpm
Gearbox: Z5-35S with 5 forward and 1 reverse gears
Clutch: GF 310 KR hydraulic with single dry plate
Transfer box: R 28 NP 2-speed
Steering: PPT (ZF) 8038 hydraulic
Turning radius: 6.5 m
Suspension: leaf-type springs with rubber buffers and telescopic shock-absorbers
Tyres: 12.00 × 18 PR 8
Brakes:
(main) dual-circuit air-hydraulic on all wheels
(parking) mechanical with air-servo assist on rear wheels
Electrical system: 24 V
Batteries: 2 × 12 V, 110 Ah

Status

In service with the former Yugoslav Army and Saudi Arabia.

Manufacturer

Tovarna Avtomobilov in Motorjev Maribot dd, Ptujska cesta 184, 62000 Maribor, Slovenia.
Tel: +386 62 411320. Fax: +386 62 414945.

VERIFIED

TAM 150 T11 BV (6 × 6) 3,000/5,000 kg truck

Description

The TAM 150 T11 BV (6 × 6) truck may be regarded as an enlarged version of the TAM 110 T7 BV (4 × 4). It uses the same cab and general layout but the TAM 150 is larger and uses a version of the F 6 L 413F in a V-6 configuration. This engine was built in the former Yugoslavia under licence from Klockner-Humboldt-Deutz.

The TAM 150 T11 BV has a reinforced PVC cab roof which can be removed, as can the side windows. The cab has seating for two and, if required, the windscreen may be folded forward over the bonnet. The cargo body is all-steel and has collapsible benches along each side to seat 18 personnel. A downward-opening tailgate is provided. A canvas tilt may be fitted. There is a 5,000 kg capacity winch fitted under the cargo body rear which may be used forwards for self-recovery or to the rear. Power for the winch is taken from a reduction gear and a power take-off for tools may be fitted.

Tyre pressures may be altered while the TAM 150 T11 BV is being driven and a rear-mounted hook may be used to tow a trailer or light artillery such as anti-tank guns.

Specifications

Cab seating: 1 + 1 (up to 18 in rear)
Configuration: 6 × 6
Weight:
(empty) 6,200 kg
(loaded, road) 11,200 kg
(loaded, off road) 9,260 kg
Max load:
(road) 5,000 kg
(off road) 3,600 kg
Towed load: 3,600 kg
Load area: 4.17 × 2.12 m
Length: 6.55 m
Width: 2.275 m
Height:
(max, loaded) 2.82 m
(cab) 2.42 m
Ground clearance: 0.305 m
Wheelbase: 3.1 + 1.2 m
Track: 1.86 m
Angle of approach/departure: 48°/40°
Max speed:
(road) 94.7 km/h
(off road) 55.2 km/h
Fuel capacity: 150 litres
Max gradient:
(without trailer) 71%
(with trailer) 42.8%
Fording: 1 m
Engine: TAM F 6 L 413 F 9.572 litre V-6 air-cooled diesel developing 154 hp at 2,650 rpm
Gearbox: Z5-35S with 5 forward and 1 reverse gears
Clutch: G 350 KR single dry plate
Transfer box: R 28 NP 2-speed
Steering: PPT (ZF) 8038 hydraulic
Turning radius: 7.6 m
Suspension: leaf-type springs with rubber buffers and telescopic shock-absorbers, additional front rubber springs
Tyres: 12.00 × 18 PR 10
Brakes:
(main) dual-circuit air-hydraulic on all wheels
(parking) mechanical with air-servo assist on rear wheels
Electrical system: 24 V
Batteries: 2 × 12 V, 143 Ah

Status

In service with the former Yugoslav Army and Saudi Arabia.

Manufacturer

Tovarna Avtomobilov in Motorjev Maribot dd, Ptujska cesta 184, 62000 Maribor, Slovenia.
Tel: +386 62 411320. Fax: +386 62 414945.

VERIFIED

TAM 150 T11 BV (6 × 6) 3,000/5,000 kg truck

TAM 150 T11 BV (6 × 6) 3,000/5,000 kg trucks used to tow anti-tank guns

TAM 4500/5000/5500/6500 trucks

Description

The TAM 4500 series of trucks was manufactured in the former Yugoslavia under licence from Magirus-Deutz. The TAM 4500 is powered by an F 4 L 514 four-cylinder air-cooled diesel developing 85 hp, giving the vehicle a maximum road speed of 75 km/h. Payload on roads is 4,500 kg. The layout of the vehicle is conventional, with the engine at the front, fully enclosed two-door all-steel cab in the centre and the cargo area at the rear with drop sides and a drop tailgate. Suffixes are used to designate different versions: B for forward control, D for 4 × 4 drive, and K for dump truck. The TAM 4500D is the 4 × 4 model used by the former Yugoslav Army and is similar in layout to the civil version but has a different rear cargo area with a drop tailgate. Seats can be installed if required. The forward control version is the TAM 4500B and the dump truck version is the 4500K.

The TAM 5000 is powered by the same engine as the TAM 4500 and was available in two versions: 4 × 2 (TAM 5000) and 4 × 4 (TAM 5000DV). The DV version was developed specifically for the former Yugoslav Army and has a central tyre pressure regulation system. A dump truck model is designated the TAM 5000K (4 × 2).

TAM 4500D truck towing 128 mm M-63 Plamen multiple rocket system launcher

The TAM 5500 is powered by a V-6 F 4 L 614 air-cooled diesel engine, developing 85 hp and giving the vehicle a maximum road speed of 85 km/h. Payload is 5,000 kg. The dump truck model is the TAM 5500DK (4 × 4) and there is also a tractor truck model.

The largest model in the range is the TAM 6500, which can carry 6,500 kg, and there is also a dump truck model.

Specifications

TAM 5000DV
Configuration: 4 × 4
Weight:
(empty) 4,500 kg
(loaded road) 9,500 kg
Max load: (road) 5,000 kg
Length: 6.75 m
Width: 2.19 m
Height: (cab) 2.75 m
Wheelbase: 4.2 m
Engine: F 4 4-cylinder air-cooled diesel developing 85 hp at 2,300 rpm
Tyres: 12.00 × 18

Status

Production complete. In service with the former Yugoslav Army.

Manufacturer

Tovarna Avtomobilov in Motorjev Maribot dd, Ptujska cesta 184, 62000 Maribor, Slovenia.
Tel: +386 62 411320. Fax: +386 62 414945.

VERIFIED

SOUTH AFRICA

SAMIL 20 Mark 2 (4 × 4) 2,000 kg truck

Description

Development of the SAMIL 20 Mark 2 (4 × 4) 2,000 kg truck began in 1981, with the vehicle being unveiled in 1985. The SAMIL 20 Mark 2 retains the overall configuration and general appearance of the earlier SAMIL 20 Mark 1 from which it was developed, but the basic specification was altered. The most significant change was that the engine became a South African manufactured ADE 352N (ADE - Atlantis Diesel Engines) water-cooled diesel weighing 420 kg and corresponding alterations to the gearbox and transfer box ratios were introduced. The forward control cab became a fully enclosed hard cab. The fuel tank is constructed using high density polyethylene and the associated piping systems are made of a nylon-based material. These changes, along with other new locally produced components such as propshafts, mean that over 90 per cent of the vehicle is of local content.

SAMIL 20 Mark 2 (4 × 4) 2,000 kg truck

Variants

Protected transport

This version has the driver's cab protected against mine blast and small arms fire by an angled arrangement of armoured plates around the cab and engine. The cargo area is slightly reduced in size from the normal version.

Bulldog

This is a completely armoured personnel and load carrier originally based on the SAMIL 20 Mark 1. There is a centrally placed cab for the driver. The rear carrying area is enclosed by folding armour plates round the sides. The open top may be covered by a canvas cover.

Protected troop carrier (Rhino)

This version has a fully armoured body with a two-person front cab and protection for 10 personnel in the rear. Rear access is by two outward-opening armoured doors and 13 firing ports are provided on the sides and rear. Armoured glass windows and vision ports are provided. The rear passengers normally face inwards.

Valkiri rocket launcher

The SAMIL 20 Mark 1 was originally used for the vehicle component of the production version of the Valkiri multiple artillery rocket system. The SAMIL 20 can be used as the launcher vehicle as well as the meteorological vehicle associated with the system (early versions of the Valkiri system used Unimog light trucks). When in use as the launch vehicle the SAMIL 20 carries 24 launching tubes and when covered with a canvas tilt it is virtually identical to the standard SAMIL 20 truck. Full details of the Valkiri system can be found in *Jane's Armour and Artillery 1995-96,* pages 745 and 764.

Light repair workshop

On this version of the SAMIL 20 the rear area is occupied by a cabin constructed from glass fibre panels reinforced by steel strips and brackets. The rear and sides are flaps that fold upwards for access to the body interior and when raised act as partial weather protection. The interior contains racks for repair equipment.

Container body carrier

When the SAMIL 20 is used as a container body carrier, the container is fixed directly on to the chassis frame and can then carry a payload of up to 2,000 kg. The containers may also be used as mobile offices, control or command posts, radio and other equipment.

Specifications

Cab seating: 1 + 1
Configuration: 4 × 4
Weight:
(chassis/cab) 4,730 kg
(loaded) 7,700 kg
Max load: 2,000 kg
Towed load: 1,360 kg
Length: 5.692 m
Width: 2.438 m
Height: 2.82 m
Ground clearance: 0.47 m
Track: (front and rear) 1.852 m
Wheelbase: 2.9 m
Angle of approach/departure: 40°/45°
Max speed: (road) 85.9 km/h
Range: 800 km
Fuel capacity: 200 litres
Max gradient: 70%
Side slope: 18°
Fording: 1.2 m at 5 km/h
Engine: ADE 352N 5.675 litre 6-cylinder in-line water-cooled direct injection diesel developing 106 hp at 2,800 rpm
Gearbox: manual, 5 forward and 1 reverse gears
Transfer box: 2-speed, pneumatic
Clutch: single sintered metal with automatic adjustment
Steering: ball and nut, power-assisted
Turning radius: 6 m
Suspension: semi-elliptic leaf springs with telescopic shock-absorbers; stabiliser fitted to rear axle

Tyres: 14.5 × 20 PR 12
Brakes: dual circuit air-assisted hydraulic with spring brakes on rear axle and engine-mounted exhaust retarder
Electrical system: 24 V
Batteries: 2 × 12 V, 118 Ah

Status
In service with South African National Defence Forces and Mozambique.

Manufacturer
TruckMakers (Pty) Ltd, PO Box 911-255, Rosslyn 0200, South Africa.
Tel: +27 12 582361. Telex: 3 20297 SA.
Fax: +27 12 541 2802.
This concern is no longer trading.

UPDATED

SAMIL 50 Mark 2 (4 × 4) 5,000 kg truck

Description
Development of the SAMIL 50 Mark 2 (4 × 4) 5,000 kg truck began in 1981 with first examples being shown in 1985. The SAMIL 50 Mark 2 was developed from the earlier SAMIL 50 Mark 1 and retains its general configuration. However, its basic specification was altered significantly, with the most important alteration being the inclusion of a South African produced ADE 409N diesel engine weighing 675 kg and the transmission was modified accordingly. The new engine is housed under an enlarged bonnet in front of a locally produced cab. The front axle was upgraded for increased durability. Other changes included the two 200 litre fuel tanks manufactured using high density polyethylene and the associated piping made from a nylon-based material. These changes mean the local content of the vehicle is over 90 per cent.

The SAMIL Mark 2 retains the personnel and general cargo body of the SAMIL Mark 1, with removable bench seating for 32 personnel, and can be readily converted to carry containers/shelters.

Variants

Recovery Vehicle
See entry in the *Recovery vehicles* section.

Telecommunications workshop
This version has a glass fibre superstructure to house work tables and drawers for telecommunications repair requirements. Small windows are provided in the sides with burglar proofing bonded to the superstructure. An air conditioner is standard. Access to the rear door is via a ladder which extends to the roof.

Battery charging vehicle
This has a box-type body with side and rear panels hinging upwards if required. Inside there are two sets of battery charging equipment each capable of charging 18 batteries arranged on a bench.

Mobile welding workshop
This follows the same general lines as the battery charging vehicle but encloses a fully equipped welding workshop.

Water tanker
The superstructure of this variant consists of a 4,500 litre water tank fitted with a tropical roof. There are five taps each side of the tank from which water can be drawn. At the rear is a single draw-off point and a spray bar for sprinkling. Two sections of armoured hose 75 mm in diameter and 4 m long are stored in boxes alongside the tank. These hoses can be quickly coupled to the rear draw-off point. A water pump capable of passing 500 litres/min for filling or disposal is driven by a power take-off from the gearbox.

Fuel tanker
This variant has an oval-section mild steel fuel tanker body with a capacity of 5,000 litres. The tank has a tropical roof cover and three manholes on the top. A 400 litres/min fuel pump is provided and a meter and control platform is situated at the tank rear. Four delivery lines for can filling are provided.

Mobile pantry
On this SAMIL 50 variant the superstructure box body is formed from rigid sandwich sections of foamed polyurethane contained in seamless glass fibre skins with reinforcing members. Internally there are two compartments, with the forward section refrigerated and the rear for the stowage of canned and non-perishable goods. Each section has shelves. The front section is equipped with stainless steel trays while the rear section is equipped with bins and baskets for handling and stowage purposes. The temperature can be controlled between −20 and +20°C in the front compartment.

Container shower
The SAMIL 50 can carry a specially developed 4.5 m container shower for use by personnel in field locations. Showers can be provided for eight personnel at one time with each of the shower heads being individually operated. The container is mounted on twist-locks and equipped with three liquid gas draw-off water heaters, a petrol engine driven self-priming pumping plant and liquid draw-off gas cylinders mounted outside the front panel. The water-heating compartment is provided with fire protection using an automatic roof-mounted fire extinguisher. Internal lighting is provided by a 12 V battery charged by the pump engine. The shower container is non-stackable and can only be lifted by a fork-lift.

Specifications
Cab seating: 1 + 1
Configuration: 4 × 4
Weight:
(chassis/cab) 6,060 kg
(loaded) 12,400 kg
Body and payload allowance: 5,800 kg
Towed load: 6,000 kg
Length: 7.78 m
Width: 2.5 m
Height: 2.955 m
Ground clearance: 0.355 m
Track:
(front) 1.985 m
(rear) 2.03 m
Wheelbase: 4.9 m
Angle of approach/departure: 32°/37°
Max speed: 85 km/h
Range: 1,000 km
Fuel capacity: 400 litres (200 + 200 litres)
Max gradient: 70%
Side slope: 18°
Fording: 1.2 m
Engine: ADE 409N 9.51 litre 5-cylinder in-line water-cooled direct injection diesel developing 165 hp at 2,200 rpm
Gearbox: manual, 6 forward and 1 reverse gears
Transfer box: 2-speed
Clutch: single sintered metal with automatic adjustment
Steering: ball and nut, power-assisted
Turning radius: 11.5 m
Suspension: semi-elliptic leaf springs and hydraulic telescopic shock-absorbers
Tyres: 14.00 × 20 PR 18
Brakes: dual circuit air with spring brakes on rear axle plus exhaust retarder
Electrical system: 24 V
Batteries: 2 × 12 V, 118 Ah

Status
In service with South African National Defence Forces and Mozambique (30).

Manufacturer
TruckMakers (Pty) Ltd, PO Box 911-255, Rosslyn 0200, South Africa.
Tel: +27 12 582361. Telex: 3 20297 SA.
Fax: +27 12 541 2802.
This concern is no longer trading.

UPDATED

SAMIL 50 Mark 2 (4 × 4) carrying ISO container (T J Gander) ***1997***

SAMIL 100 (6 × 6) 10,000 kg trucks

Description
The SAMIL 100 (6 × 6) 10,000 kg truck shares many components with the SAMIL 50, including the cab and, as with the other vehicles in the SAMIL range, is used as the basis for a wide range of bodies. The basic SAMIL 100 is a cargo truck. The rear body is constructed of pressed steel to form a sturdy unit and fixed to the chassis by spring-loaded mountings for flexibility in off-road conditions. The floor is of 5 mm flat sheet metal with supporting cross members fitted with drop-out twist locks to transport containers (one 6 m, two 3 m or four 1.5 m). The drop sides and tailgate are bottom-hinged to fold down and are detachable. On the mine-protected versions expanded metal extensions are side-hinged to the support pillar and swing open by removing two pins either side or can be detached by removing all four pins. All support pillars can fold down and are detachable. A 6 t/m elephant trunk-type crane

is mounted between the driver's cab and rear body for cargo handling. This crane is fully power-controlled through an arc of 360° and is capable of lifting 1,200 kg at a jib length of 5 m.

Variants

Protected transport
This is the same vehicle as the basic SAMIL 100 but is fitted with an armoured cab and engine cover.

Armoured personnel carrier
This version of the SAMIL 100 has a fully protected cab and engine cover and the rear is fully enclosed in an armoured hull. Seating is provided internally and there are five bullet-resistant vision windows on each side.

Tipper
This is a conventional tipper truck with a 7.5 m^3 capacity body and dual-tyred rear wheels.

Water tanker
This follows the same general lines as the SAMIL 50 water tanker but the tank capacity on the SAMIL 100 is 9,100 litres. A water pump with a rate of 910 litres/min is provided. The oval-section stainless steel tank has a manhole on the top and internal baffles. Spray devices are fitted at the rear and all controls are situated between the cab and the tank.

Fuel tanker
The tank for this variant is subdivided into two baffled compartments and has a capacity of 13,000 litres. Each compartment has its own sump drain and manhole on the top. A 64 mm diameter hose is used to fill containers directly from the tank and four smaller hoses can be used to fill cans. All controls are on the left-hand side of the tank.

Artillery tractor
Developed to tow the 155 mm G5 howitzer, the SAMIL 100 artillery tractor has a superstructure consisting of a crew compartment, cargo drop side body, elephant trunk-type crane and a 10,000 kg winch. The crew compartment provides seating for seven crew members and storage space for their personal equipment. A 150 litre hydraulic oil tank is situated under the rearward-facing seats and a 200 litre water tank is located under the forward-facing seats. An observation hatch is fitted centrally in the roof with provision for mounting a machine gun. The cargo drop side body, immediately behind the crew compartment, consists of built-in storage for 60 propellant charges and winching equipment. On top of these compartments storage is provided for 15 projectile pallets, each weighing 189 kg. A canvas cover is provided to cover the cargo body and ground pegs are supplied to allow the cover to be used as a ground shelter. A 360° traverse crane is mounted behind the rear cargo area with an 800 kg lift at a jib length of 3.5 m. It is used mainly to handle the projectile pallets. The hydraulic drag winch located at the rear between the longitudinal chassis members has a 10,000 kg capacity on the first layer with the cable routed front or rear.

Production of this variant is complete.

Canteen
The canteen box body has a load capability of 8,500 kg and consists of front and rear compartments manufactured from glass fibre and polyurethane sandwich-constructed panels. A cooling unit is provided on the front panel of the front compartment which is used for storing cooled products. The rear compartment is fitted with a food warmer, microwave oven, till, mini-safe, counter, shelves, potato chip dispenser, bulk storage compartment, roof lights, ventilation fans and a trickle charger. A service platform and a power plant are fitted and stored under the body.

SAMIL 100 (6 × 6) 10,000 kg with mine-protected cab and standard cargo body (T J Gander) **1997**

Cargo panel body
Designed mainly for the transport of perishable products, the panel body has a maximum load of 9,700 kg. The body is manufactured from glass fibre and polyurethane sandwich-constructed panels. The interior is equipped with lashing eyes and roof lights. There are two rear doors and one side door.

Mobile beer cooler
This box-bodied variant of the basic SAMIL 100 can carry 500 cases (4,140 litres) of cooled beer; each case holds 24 cans. The body is manufactured from glass fibre and polyurethane constructed panels. A cooling unit is fitted to the front panel. Roof lights and two rows of four racks each are fitted to the interior to hold the beer cases. A ladder and a platform are provided on the outside to assist in servicing the cooling unit.

Medical provisions carrier
A number of SAMIL 100 trucks have been converted to carry medical and associated provisions in a compartmented body with sides that fold up to provide shelter when the vehicle is at a medical location. More shelter can be provided by erecting wide screens and the mineproof bonnet can accommodate a folding tent for the crew.

Withings
For details of this recovery vehicle refer to entry in *Recovery vehicles* section.

Recovery vehicle
For details refer to entry in *Recovery Vehicles* section.

Valkiri Mk II
The SAMIL 100 chassis was adapted to act as a carrier for a 40-round version of the Valkiri Mk II 127 mm multiple rocket launcher system, also known as Bateleur. The chassis used for this version has its front axle capacity increased from 6,500 to 7,500 kg and extra fuel tanks are provided. Combat weight of the system is 21,500 kg and length travelling is 8.53 m. A mineproof cab is provided for the five-personnel crew. For full details of this system refer to *Jane's Armour and Artillery 1995-96*, pages 744 and 745.

Zumlac
The SAMIL 100 chassis with an armoured mineproof cab has been adapted to carry the Zumlac 23 mm air defence gun system.

Specifications

Basic SAMIL 100 truck
Cab seating: 1 + 1
Configuration: 6 × 6
Weight:
(chassis/cab) 9,135 kg
(loaded) 24,000 kg
Payload: 10,000 kg
Towed load: 14,000 kg
Length: 10.27 m
Width: 2.5 m
Height: 3.35 m
Ground clearance:
(front axle) 0.355 m
(rear axles) 0.359 m
Track:
(front) 2.002 m
(rear) 2.048 m
Wheelbase: 5.25 m + 1.38 m
Angle of approach/departure: 30°/30°
Max speed: (road) 90 km/h

Rear view of SAMIL 100 (6 × 6) 10,000 kg truck configured as artillery vehicle for towing 155 mm G5 howitzer (T J Gander) **1997**

Zumlac 23 mm air defence gun system based on SAMIL 100 (6 × 6) 10,000 kg truck with mine-protected cab (T J Gander) **1997**

Range: 800 km
Fuel capacity: 400 litres
Max gradient: 70%
Side slope: 18°
Fording: 1.2 m at 5 km/h
Engine: V-10 air-cooled diesel developing 268 hp at 2,650 rpm
Gearbox: manual, 6 forward and 1 reverse gears
Transfer box: 2-speed
Clutch: single sintered metal, self-adjusting, hydraulic
Steering: power-assisted, ball and nut
Turning radius: 11.9 m
Suspension: telescopic hydraulic shock-absorbers, leaf springs
Tyres: 14.00 × 20 PR 18
Brakes: duplex, compressed air
Electrical system: 24 V
Batteries: 2 × 12 V, 120 Ah

Status

In service with the South African National Defence Forces.

Manufacturer

TruckMakers (Pty) Ltd, PO Box 911-255, Rosslyn 0200, South Africa.
Tel: 27 12 582361. Telex: 3 20297 SA.
Fax: 27 12 541 2802.
This concern is no longer trading.

UPDATED

South African Aljaba applications

Description

Following the adoption by the South African National Defence Force (SANDF) of the Kynos Aljaba (8 × 8) truck tractor as the Cavallo, the Kynos Aljaba chassis has been adapted for several other applications with the SANDF. A check list of these is provided below. For full details of the Kynos Aljaba refer to the entry in the *Tank transporter* section. The South African vehicles can have mine-protected cabs.

Cavallo truck tractor

See separate entry in the *Tank transporter* section.

Kameelperd gunnery control vehicle

The Kameelperd (Giraffe) gunnery control vehicle acts as a battery command post for self-propelled air defence gun batteries. It carries a 6 m ISO standard sub-frame on which is mounted a 60 km range surveillance radar and its 13 m extending antenna arm. The Kameelperd is 10.64 m long and has a wheelbase of 1.524 m + 5 m + 1.524 m.

Zebra maintenance and repair vehicle

The Zebra is a front line maintenance and repair vehicle for mobile forces and carries three ISO modules and a 28,000 kg/m folding crane with a reach of 3.2 m. One of the modules is a 3 m ISO container containing workshop tools. One of the two 1.5 m ISO modules containers a compressor, generator, welding set, steam washer and other equipment. The third module carries special tools and spares. The Zebra is 11.84 m long and has a wheelbase of 1.524 m + 5.8 m + 1.524 m.

Kameelperd gunnery control vehicle with antenna raised (T J Gander) ***1997***

Bridgelayer

This variant is as yet unnamed and in late 1996 had yet to appear in prototype form. The bridge involved is the LEGUAN MLC 70 26 or 42 m bridge. The vehicle has a wheelbase of 1.524 m + 5 m + 1.524 m and is 13.297 m long when carrying a bridge.

Skimmel heavy recovery vehicle

Intended to recover heavy armoured vehicles, the Skimmel is provided with a radial tow arm system with a lift capacity of 12,000 kg from 750 mm to 1.6 m above normal ground level. Maximum casualty vehicle mass is 48,000 kg. Also carried is a 20,000 kg capacity main winch, a 500 kg auxiliary winch, and a 7,600 kg/m folding crane with a reach of 4 m. The Skimmel is 10.132 m long and has a wheelbase of 1.524 m + 5 m + 1.524 m.

Ammunition and cargo carrier

Produced in prototype form in late 1996 this ammunition and cargo carrier was developed for the front line support of 155 mm G5 and G6 artillery batteries, carrying ammunition and other supplies in up to four 1.5 m ISO standard containers on a flat bed cargo area. Also carried is a 21,315 kg/m folding crane with a maximum reach of 9.25 m. The vehicle is 11.84 m long and has a wheelbase of 1.524 m + 5.8 m + 1.524 m.

Status

In production. In service with South African National Defence Forces.

Manufacturer

FAF Engineering, PO Box 123872, Alrode 1450, South Africa.
Tel: +27 11 864 5033. Fax: +27 11 864 7510.

NEW ENTRY

Skimmel heavy recovery vehicle with mine-protected cab (T J Gander) ***1997***

Ammunition and cargo carrier version of South African Kynos Aljaba (T J Gander) ***1997***

SPAIN

SPA Fox-15D (4 × 4) 1,000 kg airborne vehicle

Description

The SPA Fox-15D (4 × 4) 1,000 kg light vehicle was developed to meet the requirements of the Spanish Army's airborne units and the parachute brigade in particular. It was designed for high mobility, while retaining a useful personnel or cargo payload The high quality steel chassis and aluminium body provide mechanical strength, ensuring a long service life under extreme environmental conditions, while the independent suspension on all wheels provide stability.

All-wheel drive is permanent with self-locking differentials front and rear, plus a central distributor. The driver's seat is located forward and centrally with seating for a further six passengers, plus space for kit and other loads.

The SPA Fox-15D can be transported by aircraft, either internally or underslung and para-dropped. Up to eight units can be stacked inside a C-130 or C-160 transport aircraft.

Optional equipment includes a winch and cargo loading rails. Also available is a 1.5 ton trailer. A soft top can be supplied to protect the driver's area or the entire vehicle.

The SPA Fox-15D is considered suitable for various applications including ambulance, communications, reconnaissance, firefighting, search and rescue and water desalination equipment carrier roles. It may also be used as a platform for 12.7 mm machine guns, recoilless rifles and anti-tank missiles, as well as for towing 120 mm mortars and light artillery.

SPA Fox-15D (4 × 4) 1,000 kg light vehicle **1997**

SPA Fox-15D (4 × 4) 1,000 kg light vehicle **1997**

Specifications
Seating: 1 + 6
Configuration: 4 × 4
Weight: (empty) 1,680 kg
Max load: 1,000 kg (up to 1,500 kg with reduced performance)
Towed load: 2,000 kg
Length: 3.395 m
Width: 1.93 m
Height:
(body) 1.05 m
(roll bar) 1.98 m
Ground clearance: 0.3 m
Angle of approach/departure: 50°/58°
Max speed: 75 km/h
Fuel capacity: 55 litres
Max gradient: 100%
Side slope: 30%
Vertical obstacle: 0.35 m
Fording: 0.8 m
Engine: inter-cooled turbo diesel developing 92 hp
Gearbox: automatic with 4 forward and 1 reverse gears
Steering: rack bar
Suspension: independent with variable-action springs on all wheels plus hydraulic shock-absorbers
Tyres: 7.50 R 16 XZL
Electrical system: 12 V
Batteries: 1 × 12 V, 105 Ah
Alternator: 90 A

Status
In production. In service with the Spanish Army's parachute brigade and other airborne units (48).

Manufacturer
Servicios y Protectos Avanzados, SA, C/ Torres Quevedo, 24 - Naves 2 y 3, E-28022 Madrid, Spain.
Tel: +34 1 329 4455. Fax: +34 1 329 4805.

NEW ENTRY

Santana S-2000 Militar (4 × 4) light truck

Description
The first forward control vehicle developed by Land Rover Santana was the Model 1300, first produced in 1967 in both petrol- and diesel-engined versions. From this evolved the S-2000, produced in both military and civil versions. The Model Militar has a payload of 2,000 kg and was produced in several forms for military use.

In its basic form as a troop carrier, the S-2000 has a forward control cab for the driver and passenger and seating for up to 12 personnel in the rear. In the rear area the troops are seated along two outward-facing bench seats down the centre line of the body. The body sides and tailgate can be quickly folded down for easy access and the rear area can be covered by a non-flammable canvas tilt carried on bows. The cab roof may have a hatch and provision for a light machine gun mounting.

The chassis has a stair-type construction with two parallel side members and seven cross members welded into position. All the chassis members are rectangular in cross-section and are painted once finished; a special galvanised finish is applied if the vehicle is to be used in a saline environment. There is a six-cylinder 3.5 litre petrol engine but a diesel engine may be fitted.

Several versions of the S-2000 exist and the model can be converted for a number of roles.

Variants

Command post
This version uses a van body with sliding rear windows and wide rear doors with a concealed rear step. The body is made from duraluminium sheet with insulating panels and fibre panels for the interior. The normal command interior includes a table, two swivelling seats, document cabinet, lamps, two portable fans and a fire extinguisher. Electrical equipment can be powered by a 42 or 90 A alternator or two auxiliary batteries. An office version can be produced.

Communications version
This version also uses a van body but can contain various radio communication installations.

Mobile workshop
On this version the items of workshop equipment are contained in a series of lockers situated each side of the vehicle rear. The centre area of the rear body is used for equipment stowage and there are lighting masts carried over the lockers. Equipment provided includes a generating set powered from a power take-off and capable of delivering 15 kW, compressor, welding equipment and electrical, pneumatic and manual tools. One side of the body rear folds down to form a workbench, complete with a vice. A 500 kg crane is also supplied.

Tanker
This version has a rear-mounted tank body that can be used to carry 1,600 litres of fuel or water. The tank is 2.2 m long and 0.8 m high and the equipment will vary according to the liquid carried.

Generating set
This carries a 15 kW generator powered from a power take-off on the vehicle. This equipment is used mainly for lighting purposes and carries up to four strings of lamps or 1,000 W halogen lamps together with all the necessary leads and components.

Container carrier
This version is used for the carriage of S-250 and other container/shelters.

A fire tender was also produced.

Specifications
Cab seating: 1 + 1 (up to 12 in rear)
Configuration: 4 × 4
Weight:
(empty) 2,360 kg
(loaded) 4,360 kg
Max load: 2,000 kg

Basic troop-carrying version of S-2000 (4 × 4) light truck

Flatbed version of S-2000 (4 × 4) light truck carrying S-250 container/shelter

Length: 4.949 m
Width: 1.96 m
Height: 2.235 m
Ground clearance: 0.27 m
Track:
(front) 1.461 m
(rear) 1.486 m
Wheelbase: 2.565 m
Angle of approach/departure: 37°/34°
Fuel capacity: 157 litres
Engine: 3.429 litre 6-cylinder in-line water-cooled petrol developing 100 hp at 4,000 rpm
Gearbox: manual with 4 forward and 1 reverse gears
Clutch: diaphragm and disc, hydraulic
Transfer box: 2-speed
Steering: worm and recirculating ball with power-assist
Turning radius: 6.75 m
Suspension: semi-elliptical underslung springs with double-acting shock-absorbers
Tyres: 9.00 × 16
Brakes: hydraulic drum, front and rear, dual circuit
Electrical system: 24 V
Batteries:
(petrol) 2 × 12 V, 57 Ah
(diesel) 2 × 12 V, 120 Ah
Alternator: 35 A (50 or 90 A optional)

Status
Production complete. In service with the Spanish armed forces and some other nations.

Manufacturer
Land Rover Santana SA, Carretora de Vadallanos, S/N, E-23700 Linares, Spain.
Tel: +34 53 693051. Fax: +34 53 653201.

VERIFIED

URO MT 2,000 kg (4 × 4) trucks

Development
As a result of trials held between April and late June 1983, a new 2,000 kg (4 × 4) tactical truck was introduced to the Spanish armed forces. The only company to submit a vehicle for the competition was URO Vehiculos Especiales SA, formed in 1981. This company produced a truck prototype and a 10 vehicle preproduction series. The first order for the Spanish Army was placed in October 1984 and a further order was placed in 1990. Initial models utilised a 115 hp Perkins Military diesel engine although production models now have Fiat-IVECO engines producing 143 or 177 hp.

The following models are available:
MT-15.12.13, 143 hp, GVW 12,000 kg
MT-15.12.13A, 143 hp, GVW 12,000 kg
MT-18.12.13A, 177 hp, GVW 12,000 kg
MT-18.12.13S, 177 hp, GVW 12,000 kg
MT-18.14.10, 177 hp, GVW 14,000 kg
MT-18.14.13, 177 hp, GVW 12,000 kg.

All models are available in a choice of four wheelbases, 3.2 m, 3.5 m, 3.8 m and 4 m.

URO MT 2,000 kg (4 × 4) truck (T J Gander) ***1997***

Description
URO MT series trucks are built on a welded steel C-section ladder rail chassis with four cross members designed to combine strength and rotational flexibility. The chassis is arranged to allow fixtures such as spring brackets to be bolted on and removed as required using modular components. Towing hooks and other equipment can be mounted front and rear. The chassis is attached to the engine and cab at only three points to allow maximum chassis torsion. The suspension uses leaf springs allied to telescopic shock-absorbers.

The driver is seated in an all-metal forward control cab that can be tipped forward to an angle of 48 or 56° for engine access. The cab structure uses steel tube for the frame and GRP panels are fitted for insulation.

The engine is mounted forward over the front axle which is a URO U-ED with double reduction gears in the wheel hubs and URO differentials; the rear axle is a URO U-ET.

Three types of gearbox and transmission are available. The standard type has a manual gearbox with six or nine forward speeds and one reverse. An Allison automatic gearbox and torque converter is one alternative, while a Clark powershift gearbox and torque converter is another. A power take-off can be taken from the transfer box. Crawler gears are available and a variety of tyre sizes can be selected.

All models can be configured for a variety of special bodies including general load carriers, wireless vehicle, four-stretcher ambulance, water tanker, enlarged cabs, fuel tankers, cranes, workshop vans, fire tenders, tippers, wreckers, snow ploughs, anti-aircraft gun or other weapon platform and a minibus for 16 passengers plus the driver.

Specifications
MT-18.14.13
Cab seating: 1 + 1
Configuration: 4 × 4
Weight: (GVW) 12,000 kg
Payload: 2,000 kg
Towed load: 3,500 kg
Length: (chassis only)
(3.2 m wheelbase) 4.42 m
(4 m wheelbase) 6.15 m
Width: 1.996 m
Height: (cab) 2.67 m
Ground clearance: 0.384 m
Track:
(front) 1.553 m
(rear) 1.552 m
Wheelbase 3.2 m, 3.5 m, 3.8 m or 4 m
Max speed: 94 km/h
Range: (road) 500 km
Fuel capacity: 180 litres
Angle of approach/departure: 45°/60°
Gradient: 100%
Side slope: 40%
Fording: 0.8 m
Engine: Fiat-IVECO 806025V 5.861 litre 6-cylinder in-line water-cooled diesel developing 177 hp
Gearbox: manual with 6 or 9 forward and 1 reverse gears
Transfer box: URO U-40 2-speed
Clutch: single dry disc
Steering: hydraulic-assist
Suspension: leaf springs plus telescopic shock-absorbers
Tyres: 13.80 R 20 (other options)
Brakes: drums, dual circuit air on all wheels
Electrical system: 24 V
Batteries: 2 × 12 V, 110 Ah
Alternator: 35 A

Status
In production. In service with the Spanish armed forces.

Manufacturer
URO, Vehiculos Especiales SA, Via Edison 17, Poligono del Tambre, Santiago de Compostela, Spain.
Tel: +34 81 580322. Fax:+34 81 565370.

UPDATED

Pegaso 3045 (4 × 4) 3,000 kg cargo truck

Description
The Pegaso 3045 (4 × 4) truck was designed by DAF of the Netherlands to meet the requirements of the Spanish Army. While under development it had the DAF designation YA 414, and is basically an improved model of the DAF YA 314 (4 × 4) 3,000 kg truck once used by the Netherlands armed forces.

The Pegaso 3045 entered service in 1970 and initial production vehicles used many DAF components, but as production increased more components of Spanish manufacture were used and eventually almost all of the truck was built in Spain.

First production models were powered by a petrol engine coupled to a manual gearbox with five forward and one reverse gears and had single rear wheels. This was replaced in production by the Pegaso 3045D powered by a Pegaso diesel engine built under licence in Spain from British Leyland, having a gearbox with six forward and one reverse gears and dual rear wheels. To meet the requirements of the Spanish Marines, a special model was developed known as the Pegaso 3045DV which can ford to a maximum depth of 2 m.

The two-door forward control cab is made of steel having a tarpaulin roof that can be folded backward and a windscreen that can be folded forward on to the radiator. The rear cargo body is all steel and has a drop tailgate, five removable bows and a tarpaulin cover. The wheel arches extend all the way along the cargo body on either side and provide seats when the vehicle is carrying passengers. Wooden panels can be inserted between the wheel arches to enable the vehicle to carry wider loads. A winch with a 4,000 kg capacity can be installed.

Variants include crane trucks (Marines use one version and Army use at least two different versions), dump truck, refueller, water carrier and workshop.

Specifications
Pegaso 3045D
Cab seating: 1 + 1

Pegaso 3045D (4 × 4) 3,000 kg cargo truck

Pegaso 3045DV (4 × 4) fuel tanker fitted with deep wading equipment

Configuration: 4 × 4
Weight:
(empty) 6,750 kg
(loaded, cross-country) 9,750 kg
Max load:
(road) 6,000 kg
(cross-country) 3,000 kg
Towed load: 4,500 kg
Load area: 4.2 × 2.15 m
Length: 6.47 m
Width: 2.48 m
Height: (cab) 2.617 m
Ground clearance: 0.32 m
Track: 1.9 m
Wheelbase: 3.7 m
Angle of approach/departure: 48°/34°
Max speed: 72 km/h (3045, 90 km/h)
Range: 650 km
Fuel capacity: 260 litres
Max gradient: 60%
Fording: 1 m
(3045) 0.7 m
(3045DV) 2 m
Engine: Pegaso Model 9026/13 6-cylinder diesel developing 125 hp at 2,400 rpm (3045, DAF 6-cylinder petrol developing 134 hp at 3,500 rpm)
Gearbox: manual with 6 forward and 1 reverse gears (3045, manual with 5 forward and 1 reverse gears)
Transfer box: 2-speed
Turning radius: 6.96 m
Suspension: semi-elliptical springs with double-acting hydraulic shock-absorbers
Tyres: 11.00 × 20 (3045, 12.00 × 20)
Brakes:
(main) air/hydraulic
(parking) mechanical
Electrical system: 24 V

Status
Production complete. In service with the Spanish armed forces. Also used by Burkina Faso (33), Chile and Nicaragua.

Manufacturer
Pegaso, Empresa Nacional de Autocamiones SA, Defence Division, Avenida de Aragón 402, E-28022 Madrid, Spain.
Tel: +34 1 750 1000. Telex: 46 972 maba e.
Fax: +34 1 747 1307.

VERIFIED

Pegaso 3046 (4 × 4) cargo trucks

Development
The Pegaso 3046 trucks were developed primarily as export/commercial models. The first widely available model was the Pegaso 3046/50 which had a nominal payload of 3,000 kg. This was replaced by the Pegaso 3046/10 with a nominal payload of 5,000 kg; this model was sold to Egypt. The first contract to export 2,650 Pegaso 3046/10 trucks to Egypt was signed in November 1980. In July 1981 a contract was signed for a further 5,000 trucks and a contract for another 4,000 trucks was awarded. More than 4,000 units were supplied to the Spanish Army and a further 1,000 units were delivered to Somalia, Peru and Morocco.

A variant is the Model 7222A with a 220 hp engine and a longer wheelbase. The 7222 designation is a Pegaso numbering system in which 7 denotes a military vehicle, 2 refers to the number of axles and 22 to the 220 hp of the engine. A workshop-bodied version of the Model 7222A was produced.

A later model was the Model 7217 with a Pegaso 170 hp diesel engine. Production of this model commenced in 1987 with 200 units delivered to the Spanish armed forces. The similar Model 7223 has a 225 hp turbocharged engine. The Pegaso BRP 2223 (4 × 4) forest firefighting vehicle was derived from the Model 7223.

Description
The Pegaso 3046/50 and 3046/10 are fundamentally identical. The two-man cab is situated well forward over the engine and has a soft top which can be removed if required. The two-piece windscreen can fold forward over the short bonnet to reduce the vehicle height. In front of the bonnet a large steel frame provides protection for the radiator grille and headlamps which are set into the bodywork. At the rear, the normal load-carrying body is an open cargo/personnel area that can be used to carry 5,000 kg of stores or 23 personnel on bench seats situated along each side. Off-road the maximum load is 3,000 kg. A canvas tilt can be fitted. A spare wheel is carried under the load area on the left-hand side; the corresponding area on the right is occupied by the fuel tank. On some versions the spare wheel is carried behind the cab.

Optional equipment for the 3046/10 includes: a metal roof for the cab, fording kit, power take-off on the gearbox and a 4,500 kg capacity front winch. Versions other than the cargo/personnel type include: a tanker, recovery vehicle with a swivelling crane, ambulance, communications post vehicle, mobile workshop, refrigerated cold storage vehicle and a command post. Various weapons can be fitted.

Specifications
Model 3046; data for Model 7217 in square brackets where different
Cab seating: 1 + 1 (22 in rear)
Configuration: 4 × 4
Weight:
(empty) 7,200 [7,250] kg
(loaded, on road) 12,200 [13,250] kg
(loaded, off road) 10,200 [11,250] kg
Max payload:
(on road) 5,000 kg
(off road) 3,000 [4,000] kg
Towed load:
(on road) 7,500 kg
(off road) 4,500 kg
Length: 6.085 [7.05] m
Width:
(cab) 2.4 m
(over rear axle) 2.406 m
Height: (cab) 2.765 [2.755] m
Ground clearance: 0.34 [0.32] m
Track: 1.96 m
Wheelbase: 3.7 [4.14] m
Angle of approach/departure: 49°/49° [49°/36°]
Max speed: 89 [90] km/h
Range: 900 km
Fuel capacity: 350 litres
Max gradient: 70%
Side slope: 30%
Engine: Pegaso Model 9100/42, 6-cylinder, in-line, 10.17 litre 4-stroke diesel developing 170 hp at 2,100 rpm
Gearbox: Pegaso Model 8256.10 manual with 6 forward and 1 reverse gears
[ZF synchronised with 6 forward and 1 reverse gears]
Clutch: Model 8500.01 single dry disc
Transfer box: Pegaso pneumatic 2-speed
Steering: block, servo-hydraulic

Pegaso Model 7217 (4 × 4) 4,000 kg cargo truck

Pegaso Model 7217 (4 × 4) 4,000 kg cargo truck

Turning radius: 9.5 [11.3] m
Suspension: semi-elliptical leaf springs and hydraulic shock-absorbers, front and rear
Tyres: 13.00 × 20 or 14.00 × 20
Brakes: dual circuit air
Electrical system: 24 V
Batteries: 2 × 12 V, 110 Ah
Alternator: 840 W

Status
In service with Egypt (more than 9,000), Morocco, Peru, Somalia and the Spanish armed forces.

Manufacturer
Pegaso, Empresa Nacional de Autocamiones SA, Defence Division, Avenida de Aragón 402, E-28022 Madrid, Spain.
Tel: +34 1 750 1000. Telex: 46 972 maba e.
Fax: +34 1 747 1307.

VERIFIED

Pegaso 3050 (6 × 6) 6,000 kg cargo truck

Description
The Pegaso 3050 (6 × 6) 6,000 kg cargo truck was developed from the earlier Pegaso 3040 (4 × 4) 4,000 kg truck and uses the same cab as the Pegaso 3045 (4 × 4) 3,000 kg truck. The two-door forward control cab is of steel construction, with a tarpaulin roof that can fold back and a windscreen which folds forward on to the radiator. The rear cargo area is all steel and has a drop tailgate, seven removable bows and a tarpaulin cover. Most vehicles have a 6,000 kg capacity winch.

Variants
These include a bridging vehicle (for carrying components of the MAN bridge), dump truck, two recovery vehicles (one with a Bazán Onara crane, one with an IASA crane), shop/van and tractor truck. A Pegaso 3050 was used as a prototype launcher for the Teruel artillery rocket. It carried 40 rockets and had an enlarged armoured cab with two rows of seats for the crew.

Basic specifications of the Pegaso 3040 are: weight loaded, road 12,000 kg; weight loaded, cross-country 9,000 kg; unladen weight 5,000 kg; length 6.78 m; width 2.45 m; height overall 3.2 m; and powered by a six-cylinder diesel engine developing 125 bhp. The Pegaso 3040 is used in small numbers by the Spanish Army and some were sold to the Nigerian Army. The 3040 has a hard top cab and the earlier 3020 is similar but with a soft top cab.

Specifications
Cab seating: 1 + 1
Configuration: 6 × 6
Weight:
(empty) 8,500 kg
(loaded, cross-country) 14,500 kg
Max load:
(road) 10,000 kg
(cross-country) 6,000 kg
Towed load:
(road) 14,500 kg
(cross-country) 7,500 kg
Length: 7.2 m
Width: 2.5 m
Height: (cab) 2.64 m
Ground clearance: 0.32 m
Track: 1.9 m
Wheelbase: 3.987 m + 1.426 m
Angle of approach/departure: 48°/44°
Max speed: (road) 68 km/h
Range: 500 km
Fuel capacity: 250 litres
Max gradient: 89% (low-range, bottom gear)
Fording: 1 m
Engine: Pegaso Model 910/40 6-cylinder in-line diesel developing 170 hp at 2,000 rpm
Gearbox: manual with 6 forward and 1 reverse gears
Transfer box: 2-speed
Steering: power-assisted
Suspension:
(front) semi-elliptical springs and hydraulic shock-absorbers
(rear) oscillating arms and spring common to both axles. Rear axles joined to chassis frame by torsion bars
Tyres: 13.00 × 20
Brakes:
(main) air/hydraulic
(parking) mechanical
Electrical system: 24 V
Batteries: 2 × 12 V, 175 Ah

Status
In service with Spanish forces and Nigeria.

Manufacturer
Pegaso, Empresa Nacional de Autocamiones SA, Defence Division, Avenida de Aragón 402, E-28022 Madrid, Spain.
Tel: +34 1 750 1000. Telex: 46 972 maba e.
Fax: +34 1 747 1307.

VERIFIED

Pegaso 3050 (6 × 6) 6,000 kg cargo truck

Pegaso 3050 (6 × 6) recovery vehicle

Pegaso 3055 and Model 7323 (6 × 6) 6,000 kg trucks

Description
The Pegaso 3055 is a 6 × 6 derivative of the Pegaso 3046 (4 × 4) truck. In March 1982 it was announced that the Pegaso 3055 was to be the standard 6 000 kg off-road truck chassis for all three Spanish armed services.

The cab of the Pegaso 3055 is the same as that used on the Pegaso 3046. It has a canvas hood as standard with an optional metal top. The cab has seating for the driver and one passenger and, if required, the hood can be removed and the windscreen folded forward over the short bonnet. The main load-carrying area is at the rear and on the standard cargo/personnel body there is bench seating for 30. The engine fitted could have either a 200 or 220 hp output, with the latter selected for use with the Spanish armed forces.

The Pegaso 3055 was produced in the following versions: medium truck; medium fuel tanker; medium water tanker; heavy crane; medium tipper truck; firefighting vehicle; tractor; van body for workshops, stores, refrigerated bodies and so on. A tractor truck for use as a tank transporter is in service.

The Pegaso 3055 has been used as an artillery tractor or Teruel artillery rocket launcher (for details refer to *Jane's Armour and Artillery 1996-97* page 774) and may be used to carry engineering stores such as bridging components. A 6,000 kg winch may be fitted.

A late production version of the Pegaso 3055 was the longer wheelbase Model 7323 powered by a Pegaso 225 hp naturally aspirated diesel engine. The Model 7323 replaced the Model 3055 in production in 1987 (the prototype was completed in 1986) and during 1987 and 1988 more than 1,000 units were delivered to Morocco. The Model 7323 was produced as an artillery tractor, fuel or water tanker, recovery vehicle with a crane, mobile workshop, ambulance, firefighting vehicle, tipper and as a multiple rocket launcher carrier. Optional extras include: a hardtop cab, power take-off, run-flat tyres and a 4,500 kg winch.

Specifications
Data in square brackets apply to the Model 7323 where it differs from the Pegaso 3055
Cab seating: 1 + 1 (30 in rear)
Configuration: 6 × 6
Weight:
(empty) 9,000 [9,700] kg
(loaded, on road) 19,000 [19,700] kg
(loaded, off road) 15,000 [15,700] kg
Max payload:
(on road) 10,000 kg
(off road) 6,000 kg
Towed load:
(on road) 14,500 kg
(off road) 7,500 kg
Length: (cab and chassis) 6.956 [7.483] m
Width:
(cab) 2.4 m
(rear wheels) 2.406 m
Height: (cab) 2.71 [2.8] m
Ground clearance: 0.34 [0.32] m
Track: 1.96 [1.93] m
Wheelbase: 3.245 m + 1.484 m [3.7 m + 1.484 m]
Angle of approach/departure: 47°/50°
Max speed: 80 [90] km/h
Range: 550 km
Fuel capacity: 350 litres [360 litres - 200, 400 and 560 litres optional]
Max gradient: 51% [60%]
Side slope: 30%
Fording: 1.1 m [1.9 m with preparation]
Engine: Pegaso model 9220/10 6-cylinder, in-line, 10.518 litre 4-stroke, turbocharged diesel developing 200 or 220 hp at 2,000 rpm
[Pegaso 10.52 litre 6-cylinder naturally aspirated diesel developing 225 hp at 2,000 rpm]
Gearbox: Pegaso Model 8256.10.09 manual with 6 forward and 1 reverse gears [ZF with 6 forward and 1 reverse gears]
Transfer box: Pegaso [ZF] pneumatic 2-speed
Clutch: Model 8500.01 single dry disc
Steering: recirculating ball, power-assisted
Turning radius: 10.2 [11.1] m
Suspension:
(front) semi-elliptical leaf springs and telescopic shock-absorbers
(rear) semi-elliptical leaf springs
Tyres: 13.00 × 20 or 14.00 × 20

Brakes: dual circuit air
Electrical system: 24 V
Batteries: 2 × 12 V, 105 Ah
Alternator: 840 [1,300] W

Status

Production complete. In service with the Spanish armed forces. Model 7323 in service with the Spanish and Moroccan armed forces.

Manufacturer

Pegaso, Empresa Nacional de Autocamiones SA, Defence Division, Avenida de Aragón 402, E-28022 Madrid, Spain.
Tel: +34 1 750 1000. Telex: 46 972 maba e.
Fax: +34 1 747 1307.

UPDATED

Pegaso Model 7323 (6 × 6) 6,000 kg truck

SWEDEN

Volvo 4140 Series of 4 × 4 and 6 × 6 cross-country vehicles

Development

In the early 1960s, the Swedish Army Materiel Department drew up its requirements for a new generation of tactical vehicles for the 1970s and 1980s. In 1966, Volvo was awarded the development contract for the Class 1 and 2 vehicles in the 1,000 to 2,500 kg range and Saab-Scania the contract for the heavier Class 3 and 4 vehicles, subsequently known as the Scania SBA (4 × 4) and SBAT (6 × 6).

Primary requirements were: a high power-to-weight ratio; forward control cab; good angle of approach and departure; high ground clearance; tough suspension; chassis which could be adapted to accept a wide variety of bodies; commercial components to be used wherever possible; ease of repair and maintenance; low training requirement; and a minimum total service life cost.

During the development stage it was decided to increase the payload of the 4 × 4 version to 2,000 kg and the 6 × 6 model to 2,500 kg. An 8 × 8 version was developed to the prototype stage but was not placed in production.

Description

(4 × 4) 4140/4141 (or C303)

The first prototypes completed in 1966 were powered by a B-20 (94 hp) engine, which was subsequently replaced by the more powerful B-30 (145 hp) engine. First deliveries were made to the Swedish Army in 1974.

Two basic models were built, the 4140 cargo and the 4141 fully enclosed, or hardtop. Both have a two-door fully enclosed cab which can be split above the waist line. The cargo model has an all-steel rear cargo area with a drop tailgate, removable bows and a tarpaulin cover. The hardtop model has a fully enclosed steel rear body with an aluminium roof and a large door at the rear and a door in each side.

All-wheel drive is engaged by a press-button in the high range and automatically when low range is

Specifications

Model	**C303 hard top**	**C304 chassis and cab**	**C306 chassis and cab**
Cab seating	1 + 6	1 + 1	1 + 1
Configuration	4 × 4	4 × 4	6 × 6
Weight			
(empty)	2,250 kg	1,940 kg	2,400 kg
(loaded)	3,450 kg	3,900 kg	5,500 kg
Weight on front axle			
(loaded)	1,650 kg	1,800 kg	1,000 kg
Weight on rear axles			
(loaded)	1,800 kg	2,100 kg	1,800 kg
Max load	1,200 kg	1,960 kg (incl body)	3,100 kg (incl body)
Towed load	2,500 kg	2,500 kg	2,500 kg
Load area	2.25 × 1.8 m	n/a	n/a
Length	4.25 m	4.28 m	5.735 m
Width	1.9 m	1.87 m	1.88 m
Height			
(cab)	2.13 m	2.13 m	2.13 m
(load area)	0.83 m	n/app	n/app
Ground clearance	0.386 m	0.386 m	0.386 m
Track	1.54 m	1.54 m	1.54 m
Wheelbase	2.3 m	2.53 m	2.72 m + 1.05 m
Angle of approach/ departure	45°/45°	45°/45°	45°/40°
Max speed			
(road)	120 km/h	100 km/h	90 km/h
Fuel capacity	83 litres	125 litres	150 litres
Max gradient	100%	100%	100%
Max side slope	40%	40%	40%
Fording			
(without preparation)	0.7 m	0.7 m	0.7 m
Engine	Volvo B-30 in-line 6-cylinder OHV petrol developing 125 hp at 4,250 rpm		
Gearbox	all have manual gearbox with 4 forward and 1 reverse gears		
Clutch	single dry plate	single dry plate	single dry plate
Transfer box	2-speed	2-speed	2-speed
Steering	all cam and roller type		
Turning radius	5.77 m	5.73 m	8.25 m
Tyres	8.90 × 16	8.90 × 16	8.90 × 16
Electrical system	12 V	12 V	12 V
Battery	1 × 12 V, 60 Ah	1 × 12 V, 60 Ah	1 × 12 V, 60 Ah

Volvo 4143 (6 × 6) in Tgb 211A configuration (S Bengtson)

Volvo 4141 (4 × 4) fully enclosed version (T J Gander)

engaged. Both front and rear axles have vacuum-operated mechanical differential locks which can be engaged separately or together. The chassis consists of box side members with tubular cross members welded into position. The chassis is torsionally stiff to avoid stressing the superstructure.

The front suspension consists of underslung semi-elliptical leaf springs carried in rubber mountings, hollow-rubber springs and double-acting telescopic shock-absorbers. The rear suspension consists of overslung semi-elliptical leaf springs carried in rubber mountings, hollow-rubber springs and double-acting telescopic shock-absorbers. The main brakes are vacuum-hydraulic drum-type, dual circuit, with one vacuum cylinder per circuit.

The handbrake is mechanical and acts on the propeller shaft. Optional equipment includes: an air conditioning system; electric engine heater; electric compressor with 10 m of hose for pumping tyres; PTO; protective wooden floor ribs; roof ventilator; tow hook; trailer electrical socket; Webasto engine and passenger area heater; and a 2,200/3,000 kg capacity winch. The basic model is fitted with a thermostatically controlled heater, defroster system and a two-speed blower in the cab, with a heater and two-speed blower for the rear compartment.

The C304 (4 × 4) version is almost identical to the C303 but has a wheelbase of 2.53 m.

(6 × 6) 4143 (or C306)

The 6 × 6 model is based on the 4 × 4 model and has the same engine, gearbox, transfer box and cab. The vehicle can be fitted with a variety of bodies including cargo, fully enclosed and ambulance. The cargo model has drop sides, drop tailgate, removable bows and a tarpaulin cover. The fully enclosed body version has a large door in the rear and a door in each side and can be used for a variety of roles including radio/command.

The front suspension consists of underslung semi-elliptical leaf springs, hollow-rubber springs and double-acting telescopic shock-absorbers. The rear suspension is of the double cantilever type with parabolic springs with progressively acting hollow-rubber springs and double-acting telescopic shock-absorbers. The main brakes are of the vacuum-hydraulic drum-type, with the mechanical parking brake operating on the transfer box output shaft. Optional equipment is similar to that available for the 4 × 4 version.

Status

Production complete. In service with the Swedish armed forces and Malaysia (4 × 4 and 6 × 6).

Manufacturer

Volvo Truck Corporation, S-405 08 Göteberg, Sweden.
Tel: +46 31 666000. Telex: 27000 volvo s.
Fax: +46 31 664660.

VERIFIED

Scania SBA 111 4,500 kg (4 × 4) and SBAT 111S 6,000 kg (6 × 6) trucks

Development

In the early 1960s, the Swedish Army Materiel Department drew up its requirements for a new generation of tactical vehicles for the 1970s and 1980s. Volvo was awarded the contract for the lighter Class 1 and 2 vehicles and Scania the contracts for the heavier Class 3 (4 × 4) and Class 4 (6 × 6) vehicles.

Scania started design work in 1966 and received its first development contract in 1968. Primary requirements of the Swedish Army were for a vehicle which would be easy to handle and maintain, use proved and standardised commercial components wherever possible, be reliable, have a low repair cost and finally have a low total service life cost. The first prototypes were completed early in 1971 with the second series of prototypes being completed late in 1972. The Swedish Army placed its first production order in 1974 for 2,000 vehicles (both 4 × 4 and 6 × 6) at a total cost of SEK225 million for delivery between 1976 and 1979. In 1977 a further 258 vehicles were ordered for the Swedish Air Force and Navy. A total of 2,500 vehicles was delivered to the Swedish Army. Production was completed in 1989.

A modified version of the SBAT was produced for the Indian Army to tow Bofors 155 mm FH-77B howitzers.

Description

The layout of both vehicles is almost identical, the only major differences being in their engines and configurations. The components of both vehicles are 90 per cent interchangeable.

The chassis consists of two longitudinal U-shaped members with the cross members riveted into position. The two-door forward control cab is all steel and has a hatch in the right side of the roof. The driver's windscreen is hinged at the top and can be opened upward for improved visibility. The cab can be tilted forward to an angle of 55° with the aid of a double-acting hydraulic pump. The grille on the front of the cab opens upwards to allow access to the oil dipstick, oil filter and oil tank for daily checks.

The engines of the two vehicles differ only in that the SBAT (6 × 6) has a turbocharger. The engine is fitted with a special fuel pump which allows the engine to continue running when inclined at an angle of 35°.

The gearbox is of the automatic split type in which two-thirds of the power is always transmitted mechanically. The gearbox consists of the automatic gearbox, distributor gearbox and a torque converter. The main gearbox has six speeds, three hydraulic and three mechanical. The transfer box has one ratio for cross-country operations and another for road operations.

All axles on both trucks are identical and each axle has a central bevel gear, hub reduction gears and a differential lock.

The rear platform is torsionally rigid and the basic cargo models have removable drop sides and a drop tailgate.

Both vehicles have an 8,000 kg capacity winch mounted on the right side of the chassis, driven by a PTO on the transfer box and operated by a switch on the dashboard and a winch brake control in front of the

Indian Army version of SBAT 111S 6,000 kg (6 × 6) truck

Specifications

Model	**SBA 111**	**SBAT 111S**
Cab seating	1 + 2	1 + 2
Configuration	4 × 4	6 × 6
Weight		
(empty)	9,150 kg	11,650 kg
(loaded, cross-country)	13,700 kg	20,650 kg
Max load		
(cross-country)	4,500 kg	6,000 kg
(road)	6,000 kg	9,000 kg
Towed load	6,000 kg	12,000 kg
Load area	4.2 × 2.35 m	4.75 × 2.35 m
Length	6.75 m	7.78 m
Width	2.48 m	2.48 m
Height		
(cab)	2.9 m	2.9 m
(load area)	1.472 m	1.472 m
Ground clearance	0.4 m	0.42 m
Track	2.02 m	2.02 m
Wheelbase	4 m	3.55 m + 1.48 m
Angle of approach/departure	45°/40°	45°/40°
Max speed	85 km/h	85 km/h
Range (road)	600 km	550 km
Fuel capacity	167 litres	167 litres
Max gradient	60%	60%
Max side slope	40%	40%
Fording	0.8 m	0.8 m
Engine	D11 6-cylinder diesel developing 202 hp at 2,200 rpm	DS11 (supercharged) 6-cylinder diesel developing 296 hp at 2,200 rpm
Steering	hydraulic	hydraulic
Turning radius	9 m	10.3 m
Suspension	semi-elliptical springs with double-acting hydraulic shock-absorbers	
Tyres	14.00 × 20	14.00 × 20
Brakes	air, dual circuit. Hand, spring type operating on front and rear wheels	
Electrical system	24 V	24 V

steering wheel. The winch can be used to the front or rear of the vehicle.

Cold weather equipment includes: an engine heater, battery heater, fuel preheater, starting pilot and connections for starting cables.

Variants

SBA (4 × 4)

This basic model was designed for carrying cargo although some were delivered to the Swedish Army with a 1,500 kg hydraulic crane at the rear for unloading. Some have been fitted with bows, canvas cover and bench seats down each side for carrying passengers; others have a small hardtop crew shelter.

SBAT (6 × 6)

The basic model was designed for carrying cargo but other variants include a version with a 5,500 kg hydraulic crane mounted to the rear of the cab, which is used for ammunition resupply for the Bofors 155 mm FH-77 howitzer and the 120 mm KARIN towed coast defence gun. The FH-77 and KARIN are towed by an SBAT truck fitted with a fully enclosed cabin for the 10 man crew at the forward end of the platform. Mounted at the rear of the vehicle is a 1,500 kg hydraulic crane. A 6 × 6 recovery vehicle was built to the prototype stage but was not ordered by the Swedish Army. The 6 × 6 version is used to carry the Giraffe radar system used in conjunction with the Bofors RBS 70 and RBS 90 SAM systems. In March 1989, the Swedish Defence Materiel Administration ordered 83 of these vehicles.

Crash rescue vehicles

Following trials with two prototype bodies, the Swedish armed forces ordered 47 crash rescue vehicles for delivery in 1979. They have a loaded weight of 16,000 kg and are fitted with both firefighting and crash rescue equipment.

Snow-clearing vehicle

This version is based on the chassis of the SBA (4 × 4) truck and used by the Swedish Air Force for clearing snow from airfield runways; 45 of these vehicles were ordered at a cost of SEK12 million. When being driven forward on the road the vehicle is driven from the normal driver's position, but when clearing snow it is driven in reverse from the second cab which faces the rear. Maximum speed when clearing snow is 30 km/h. Steering during snow-clearing operations is by a duplicated hydrostatic steering system which acts on the ordinary steering mechanism of the vehicle. The snow-clearing equipment fitted is a Rolba 1500S, which is of the cut and sling type and has a capacity of 30 to 35,000 kg of snow/min.

Status

Production complete. In service with Finland (33 SBAs), India (660 SBATs ordered in 1986) and Sweden.

Manufacturer

Scania, S-151 87 Södertälje, Sweden.
Tel: +46 8 5538 1000. Telex: 10200 scania s.

UPDATED

Volvo L 4854 and N86 (4 × 4) trucks

Description

The Volvo L 4854 (4 × 4) truck is basically a standard production Viking truck modified to meet the requirements of the Swedish Army. The first prototype was completed in 1960 and the vehicle was in production for the army between 1961 and 1963. The L 4854 is known as the *Lastterrängbil* 939 (cross-country truck) by the army and is in service in three basic models, the 939 BF with a 5.2 m wheelbase, the 939 AF with a 4.4 m wheelbase and the 939 E with a 4.4 m wheelbase and a longer cab.

The layout of the vehicle is conventional, with the engine at the front, cab in the centre and the cargo area at the rear. The two-door all-steel cab has an observation hatch in the roof, over which anti-aircraft machine guns can be mounted. The rear cargo area has drop sides and a drop tailgate. A winch with a capacity of 5,000 kg is mounted under the right side of the chassis and can be used to the front or rear. Both front and rear axles have a hand-operated differential lock. Some L 4854s have been fitted with a HIAB crane to the rear of the cab for unloading.

The Volvo N86 is a later variant of the L 4854 produced only for the Swedish coast artillery and air force. It has a more powerful engine and a revised transmission.

The Swedish armed forces use over 20 variants of the basic L4854 and more are added as existing vehicles are converted for various roles.

Volvo N86 (4 × 4) truck fitted with personnel shelter as used by coast artillery units (T J Gander)

Specifications

L 4854
Cab seating: 1 + 2
Configuration: 4 × 4
Weight:
(empty) 6,720 kg
(loaded, cross-country) 9,720 kg
Max load:
(road) 4,500 kg
(cross-country) 3,000 kg
Length: 7.5 m
Width: 2.28 m
Height: 2.82 m
Ground clearance: 0.25 m
Track:
(front) 1.83 m
(rear) 1.74 m
Wheelbase: 4.4 m
Angle of approach/departure: 38°/25°
Max speed: (road) 77 km/h
Range: 300 km
Fuel capacity: 120 litres
Engine: Volvo D67C 6-cylinder diesel developing 125 hp at 2,400 rpm
Gearbox: Volvo K17 manual with 5 forward and 1 reverse gears
Clutch: single dry plate
Transfer box: 2-speed
Steering: ZF power-assisted
Turning radius: 10 m
Tyres: 10.00 × 20

Status

Production complete. In service with the Swedish armed forces.

Manufacturer

Volvo Truck Corporation, S-405 08 Göteborg, Sweden.
Tel: +46 31 666000. Telex: 27000 volvo s.
Fax: +46 31 664660.

VERIFIED

Scania-Vabis (4 × 2) L-36A and L-50 5,920 kg trucks

Description

The Scania-Vabis L-36A is the standard civilian model L3642 (4 × 2) truck with a minimum of modifications to suit it for military use. In total 800 were built for the Swedish Army between 1964 and 1967.

The layout of the L-36A is conventional, with the engine at the front, all-steel fully enclosed cab in the centre and the cargo area at the rear with drop sides and a drop tailgate. Some vehicles have been fitted with a HIAB crane mounted to the rear of the cab for loading.

A fully enclosed mobile workshop version is in service under the designation of the Materielvardsbil 111. Scania-Vabis also delivered 1,500 L50 series 4 × 2 vehicles to the Swedish Army with various bodies. The L-50 is based on the civilian L5042 (4 × 2) truck and has a wheelbase of 4.2 m. Basic specifications are: length 6.5 m; width 2.206 m; height (cab) 2.538 m; powered by a four-cylinder diesel developing 110 hp at 2,400 rpm; and tyres 9.00 × 20.

Specifications

Cab seating: 1 + 2
Configuration: 4 × 2
Weight:
(empty) 4,580 kg
(loaded) 10,500 kg
Max load: 5,920 kg
Length: (chassis) 6.59 m
Width: 2.19 m
Height: (cab) 2.76 m
Ground clearance: 0.355 m
Track: 1.85 m
Wheelbase: 4.2 m
Angle of approach/departure: 36°/30°
Max speed: (road) 77 km/h
Range: 450 km
Fuel capacity: 100 litres
Engine: Model D5 4-cylinder diesel developing 102 bhp at 2,400 rpm
Gearbox: manual Model S-5-35 with 5 forward and 1 reverse gears
Clutch: single dry plate
Steering: cam and triple roller
Turning radius: 7.3 m
Suspension:
(front) semi-elliptical springs and double-acting

Scania-Vabis (4 × 2) L-36A 5,920 kg truck (T J Gander)

Scania-Vabis (4 × 2) L-50 5,920 kg truck (Christopher F Foss)

hydraulic shock-absorbers
(rear) semi-elliptical springs
Tyres: 8.25 × 20
Brakes:
(main) air
(parking) mechanical

Electrical system: 24 V
Batteries: 2 × 12 V

Status
Production complete. In service with the Swedish Army.

Manufacturer
Scania-Vabis, now Scania, S-151 87 Södertälje, Sweden.
Tel: +46 8 553 81000. Telex: 10200 scania s.

VERIFIED

Scania trucks

Description
Scania builds an extensive range of heavy trucks with a gross vehicle weight of 16 tonnes and upwards. By drawing on a range of standardised modules a large number of truck models and variants can be built using a strictly limited selection of standard components.

The Scania truck range is powered by four basic diesel engine models with capacities of 9, 11, 12 and 14.2 litres. These power plants are available in turbocharged and intercooled variants. The Scania truck engine range features nine engine types covering a power range from 220 to 530 bhp in closely stepped increments.

There are three different chassis frame strength classes available (M, H and E) with several axle configurations in each class. There is also a range of transmission options such as a manual gearbox with torque converter, a fully automatic transmission, hub reduction and all-wheel drive.

There is also a wide range of cab options of two major types, forward control (P and R) and bonneted (T).

There are about 250 Scania chassis variants including several all-wheel drive off-road models. Gross vehicle weights range from 17,000 up to 48,000 kg. Many of the models are all-wheel driven for cross-country operation.

Scania has delivered a large number of special vehicles both to military and civilian customers. Among the former are the Norwegian armed forces, which ordered 1,700 Scania vehicles of different types in 1986 at a cost of SEK750 million. These included the P93 M (4 × 4), P113 H (6 × 6) and the R143EK (6 × 6) tank transporter. Vehicles are delivered as chassis from Scania with bodies fitted by Norsk Scania-Vabis A/S in Norway. Deliveries continued until August 1995.

In total 255 Scania T113 HK (6 × 6) trucks plus spare parts were ordered by the Brazilian ENGESA concern. It received locally produced vehicles from Scania's

A selection of the 1,700 Scania trucks ordered by the Norwegian Army ***1997***

Specifications

Model	**P93 MK 4 × 4-5T**	**P93 MK 4 × 4-8T**	**P93 HK 4 × 4 + 2/S42**	**P113 HK (6 × 6)**
Purpose	truck	truck	bridge carrier	truck, desert version
Cab seating	1 + 2	1 + 2	1 + 2	1 + 2
Configuration	4 × 4	4 × 4	6 × 4	6 × 6
Weight (empty)	9,300 kg	9,500 kg	13,360 kg	13,400 kg
Max load	5,000 kg	8,000 kg	11,140 kg	11,540 kg
Length	7 m	8.56 m	8.41 m	7.6 m
Width (loaded)	2.48 m	2.48 m	3.4 m	2.66 m
Height	3.3 m	3.28 m	3.34 m	3.4 m
Ground clearance				
(front)	0.41 m	0.4 m	0.4 m	0.4 m
(rear)	0.41 m	0.335 m	0.285 m	0.335 m
Track				
(front)	2.02 m	2.03 m	2.03 m	2.03 m
(rear)	2.02 m	2.03 m	1.83 m	2.03 m
Wheelbase	4 m	4.6 m	4.6 m + 1.315 m	3.55 m + 1.35 m
Angle of approach/ departure	34°/41°	33°/23°	n/a	n/a
Max speed	90 km/h	105 km/h	90 km/h	90 km/h
Engine type	DS9 diesel	DS9 diesel	DS9 diesel	DS11 diesel
Engine power/rpm	252 hp/2,200 rpm	252 hp/2,200 rpm	252 hp/2,200 rpm	310 hp/2,000 rpm
Gearbox	5-speed	5-speed	5-speed	10-speed
Tyres	14.00 × 20	14.00 × 20	13 × 22.5 or 12.00 × 20	20.00 × 20 rib

Swedish Army Scania P113 HK 6 × 6 truck with Mulitilift HL26 load handling system ***1997***

Scania P113 HK 8 × 6/4 truck with Multilift Mk 4 load handling system as produced for Norwegian Army ***1997***

Brazilian subsidiary, provided them with new coachwork and delivered them to Angola. The order was worth SEK120 million.

A further model is the Scania P113 HK 6 × 6, ordered for the Swedish Army carrying a Multilift HL26 DROPS load handling system. The Norwegian Army has the Scania P113 HK 8 × 6/4 with a Multilift Mk 1V load handling system. This model has a GVW of 34,000 kg and is powered by a DS11 75 six-cylinder turbocharged diesel developing 310 hp at 2,000 rpm.

The data provided in the tables may be taken as typical for only a small portion of the full Scania truck range.

Status

In production. 3-series production was completed during 1996 and superseded by the Scania 4-series. The first 4-series trucks were expected to be fielded by the end of 1996.

Manufacturer

Scania, S-151 87 Södertälje, Sweden.
Tel: +46 8 5528 1000. Telex: 10200 scania s.
Fax: +46 8 5538 1037.

UPDATED

Volvo FL6 (4 × 4) trucks

Description

Volvo FL6 (4 × 4) trucks were designed and developed in Sweden but are produced at the Volvo Europa Truck plant at Ghent in Belgium. The FL6 is best described as a militarised version of a commercial design which is versatile and easy to handle, having a forward control, steel, crash-proof cab with various optional layouts and equipment fits. Other options include various wheelbase lengths, engines (with outputs from 180 to 250 hp), gearboxes, transfer boxes, suspensions, wheel and tyre sizes. Payloads can vary from 3,000 to 5,000 kg.

Special versions include a version for airborne operations, a desert version and a mobile communications shelter carrier. A crew cab with seating for five or six is also produced.

Specifications

Cab seating: 1 + 2
Configuration: 4 × 4
Weight: (basic 4.1 m wheelbase chassis) approx 5,000 kg
Max payload: 4,960-6,920 kg incl superstructure
Length: 6.96 m
Width: (overall) 2.368 m
Height: (cab) 2.86 m
Wheelbase: 4.1 m (other lengths available)
Fuel capacity: 200 to 300 litres
Engine: Volvo TD61GC 6-cylinder 4-stroke direct injection turbocharged diesel developing 180 hp at 2,800 rpm or
Volvo TD61F 6-cylinder 4-stroke direct injection turbocharged intercooled diesel developing 207 hp at 2,800 rpm or
Volvo TD61FUG 6-cylinder 4-stroke direct injection turbocharged intercooled diesel developing 230 hp at 2,800 rpm
Gearbox: T600B or T700B 6-speed fully synchronised or R1000 8-speed fully synchronised with crawler
Clutch: single dry plate
Transfer box: VG450, VG750 or VG1200, all 2-speed
Steering: ball and nut, power-assisted
Suspension: tapered leaf springs with shock-absorbers front and rear
Tyres: 11.00 R 22.5, 365/85 R 20 or 12.00 R 20
Brakes: dual circuit air
Electrical system: 24 V

Volvo FL6 (4 × 4) truck (T J Gander) ***1997***

Status

In service with the Swedish armed forces and other nations, including Kuwait.

Manufacturers

Volvo Truck Corporation, S-405 08 Göteborg, Sweden.
Tel: +46 31 666000. Telex: 27000 volvo s.
Fax: +46 31 664660.
Volvo Europa Trucks, Smalleheerweg 31, PO Box 10, B-9041 Ghent, Belgium.
Tel: +32 91 504211. Fax: +32 91 511994.

UPDATED

Volvo F10, F12 and F16 trucks

Description

The Volvo F10, F12 and F16 trucks are available in a wide range of drive configurations and types which are produced in both military and commercial forms. All models utilise an all-welded, forward control, steel crash-proof cab available with various seating arrangements, including a five/six-passenger crew cab and optional roof hatches or other openings.

As an example, the F10 is available in basic 4 × 4 and 6 × 6 configurations as well as 4 × 2, 6 × 2, 6 × 4, 8 × 2 and 8 × 4. Engine power can vary from 285 to 320 hp and there is a choice of manual or automatic gearboxes, driven axles with or without hub reduction and differential locks between the wheels or axles.

The Volvo F10 may be fitted with fixed cargo or container-carrying bodies and may also be used as a fuel or water tanker or as a tipper truck. A tractor truck version is available.

The Volvo F12 tractor truck is powered by a Volvo TD122FL 12 litre turbocharged intercooled diesel developing up to 400 hp. A typical front axle capacity is 8,000 kg and rear axles 26,000 kg. There are 4 × 4, 6 × 6 and 8 × 4 versions available, with GCWs from 44,000 kg up to 150,000 kg, enabling F10 tractor trucks to be employed in the tank transporter role if required.

The Volvo F16 is a tractor truck for particularly heavy haulage work. It is powered by a Volvo TD163ES 16 litre turbocharged and intercooled diesel developing 500 hp, coupled to a Volvo five- or six-speed Powertronic automatic gearbox. GCW is up to 150,000 kg.

Swedish Army Volvo F10 (4 × 4) cargo truck

Swedish Army Volvo F10 (6 × 6) cargo truck with loading crane

Swedish Army Volvo F12 (6 × 6) tractor truck towing VM 70-5 RKSZHML semi-trailer with a load capacity of approximately 75,000 kg

Specifications
F10 6 × 6
Cab seating: 1 + 2
Configuration: 6 × 6
Weight on front axle: 8,000 kg
Weight on rear axles:
(12.00 R 20 tyres) 26,000 kg
(14.00 R 20 tyres) 18,000 kg
Payload incl superstructure: typically 25,200 kg
Length: 7.63 m
Width: (overall) 2.5 m
Height: (cab) 3.1 m
Wheelbase: 4 m + 1.37 m
Fuel capacity: 300 litres
Engine: Volvo TD102F 10 litre turbocharged intercooled diesel developing 320 hp at 2,050 rpm
Gearbox: manual 8-speed with crawler
Clutch: KFD117C single dry plate
Transfer box: FD7 2-speed
Steering: recirculating ball and nut, power-assisted
Suspension: tapered leaf springs with helpers
Tyres: 12.00 R 20 or 14.00 R 20
Brakes: dual circuit air
Electrical system: 24 V

Status
In production in Sweden, Belgium, Morocco and the UK (Scotland). In service with the Swedish Army (approx 200) and with the UN in the former Yugoslavia.

Manufacturer
Volvo Truck Corporation, S-405 08 Göteborg, Sweden.
Tel: +46 31 666000. Telex: 27000 volvo s.
Fax: +46 31 664660.

UPDATED

Volvo N10/N12 (6 × 6) trucks

Description
The Volvo N10 and N12 6 × 6 trucks are heavy-duty commercial vehicles that can be used for military purposes. They are produced in both 4 × 4 and 6 × 6 forms but to date the 6 × 6 versions are the only ones known to have been adopted for military use. There is also a choice of rigid rear axles or a Volvo T-ride bogie on 6 × 6 models. The main difference between the two models is that the N10 has a turbocharged and intercooled engine while that on the N12 lacks the intercooling.

The layout of the N10/N12 is conventional, with the engine located forward under a squared-off bonnet, the cab having seating for the driver and at least one passenger and with the cargo area to the rear. Some vehicles are fitted with a handling crane just behind the cab and the all-steel cargo body has drop sides and a tailgate. Three wheelbase lengths are available, 4.835 m, 5.235 m, and 5.635 m, and there are various chassis lengths.

Gross vehicle weights vary between 16,500 to 19,500 kg for the 4 × 4 models and 26,500 to 33,500 kg for the 6 × 6 models.

Belgium adopted the N10, with local production being carried out in Belgium by Hocke at Alsemberg. A total of 1,228 units will be procured. Of these 683 will be cargo trucks, 78 will have enclosed bodies for offices, maintenance and so on, 133 will have tipper bodies, 40 units will have the Bennes Amplirol load handling system, 145 will be equipped as recovery vehicles and 20 will have bodies for the medical services. Manumat cranes fitted to Belgian vehicles have a capacity of 1,100 kg and a reach of 5.6 m.

Specifications
N10 6 × 6
Cab seating: 1 + 1 or 2
Configuration: 6 × 6
Weight:
(basic chassis, 5.235 m wheelbase) 9,155 kg
Max payload: 17,345 kg incl superstructure
Length: (chassis) 7.992 m
Width: (overall) 2.4 m
Height: (cab) 3.02 m
Wheelbase: 5.235 m
Fuel capacity: 300 litres
Engine: 9.6 litre 6-cylinder turbocharged and intercooled diesel developing 275 hp at 2,200 rpm
Gearbox: R1400, 8 forward (plus crawler) and 2 reverse gears
Clutch: KFD214A twin dry plate
Transfer box: FD6 2-speed
Steering: ball and nut, power-assisted
Suspension: leaf springs and hydraulic shock-absorbers front and rear
Tyres: 12.00 R 20, 14.00 R 20 or 12.00 R 24
Brakes: dual circuit air
Electrical system: 24 V

Status
In production. In service with the Belgian Army (1,228 ordered). Also used by UN.

Manufacturer
Volvo Truck Corporation, S-405 08 Göteborg, Sweden.
Tel: +46 31 666000. Telex: 27000 volvo s.
Fax: +46 31 664660.

VERIFIED

Belgian Army Volvo N10 (6 × 6) truck with cargo body (C R Zwart)

Belgian Army Volvo N10 (6 × 6) truck fitted with crane for use as ammunition resupply vehicle (C R Zwart)

Volvo NL10 (4 × 4) and NL12 (6 × 6) trucks

Description
The Volvo NL10 (4 × 4) and NL12 (6 × 6) rigid axle trucks are the successors to the Volvo N10 and N12 trucks (see previous entry) and follow the same general layout. As with other Volvo trucks, components such as engines, gearboxes, axles, cabs and most chassis parts were designed and produced in-house by Volvo.

NL trucks are available with a wide range of optional equipment for different military roles. Also available are NL trucks with 4 × 2, 6 × 2 and 6 × 4 drive configurations, various types of suspension, differing rear axle ratios, differential locks, single or hub reductions, crew cabs and various wheelbase lengths.

A 10 tonne desert transporter version has been developed. This, in common with other NL10/12 6 × 6 trucks, can be equipped with a DROPS type load handling system. This version has an on-road payload of 16 tonnes.

Specifications

Cab seating: 1 + 1 or 2
Configuration:
(NL10) 4 × 4
(NL12) 6 × 6
Weight:
(NL10) 6,880-6,920 kg
(NL12) 9,000-9,150 kg
Max payload:
(NL10) 12,770-12,820 kg incl superstructure
(NL12) 23,700 kg incl superstructure
Length: (chassis only)
(NL10) 6.952-8.292 m
(NL12) 7.962-9.662 m
Width: (overall) 2.4 m
Height: (cab) 3.02 m
Wheelbase: 4.3-5.1 m
Fuel capacity: 300 to 600 litres
Engine:
(NL10) Volvo 9.6 litre 6-cylinder turbocharged and intercooled diesel developing 275-320 hp
(NL12) Volvo 12 litre 6-cylinder turbocharged and intercooled diesel developing 329-405 hp
Gearbox:
(NL10) R1400 range and SR1400 range, 8 forward (plus crawler) and 2 reverse gears

Volvo N10/12 (6 × 6) truck equipped with DROPS type load handling system ***1997***

(NL12) R1700 range and SR1700 range, 8 forward (plus crawler) and 2 reverse gears
Clutch: KFD 117C single dry plate
Transfer box: Volvo FD7 2-speed
Steering: ball and nut, power-assisted
Suspension: leaf springs and hydraulic shock-absorbers front and rear
Tyres: 12.00 R 20, 14.00 R 20 or 12.00 R 24
Brakes: dual circuit air
Electrical system: 24 V

Status

In production in Sweden, Australia, Brazil, Iran, Morocco and Peru.

Manufacturer

Volvo Truck Corporation, S-405 08 Göteborg, Sweden.
Tel: +46 31 666000. Telex: 27000 volvo s.
Fax: +46 31 664660.

UPDATED

Volvo BM A25C (6 × 6) truck

Description

The Volvo BM A25C (6 × 6) truck is one of a series of similar Volvo trucks built on an articulated chassis to provide a payload of about 19,000 kg while operating over rough terrain. The BM A25C is described as a multipurpose carrier having an HIAB 8801/3 hydraulic crane to support Swedish Army 155 mm self-propelled gun batteries, although it could be readily adapted to other similar heavy duty roles.

Due to the articulated chassis, the front section and cab can turn to an angle of 45° either side and independently of the cargo area. The Volvo BM two-seat cab is roll-over protected and mounted on rubber pads. A 7 litre Volvo diesel with turbocharging and intercooler provides 255 hp and is connected to a five-speed fully automatic gearbox with a torque converter and integrated hydraulic retarder. A Volvo drop-box has a power take-off and differential lock providing full six-wheel drive with the possibility of 100 per cent differential locking on all axles.

The front axle is suspended on a three-point suspension while the Volvo rear bogie permits individual oscillation between the axles.

If required, the Volvo BM A25C can be equipped with a Cargotec Multilift load handling system.

An essentially similar Volvo 6 × 6 chassis is employed for the Bofors TRIDON 40 mm L/70 self-propelled air defence gun system with the cab provided with add-on armour. A similar arrangement is undertaken for the Bofors 155 mm FH-77 howitzer and the Bofors 120 mm KARIN coast defence gun.

Specifications

Cab seating: 1 + 1
Configuration: 6 × 6
Weight:
(front axle loaded) 11,500 kg
(rear axles loaded) 28,770 kg
(GVW) 40,270 kg
Max payload: (off road) 19,000 kg
Load area: 6.42 × 3 m

Volvo BM A25C (6 × 6) truck (T J Gander) ***1997***

Length: 11.7 m
Width: (overall) 3.1 m
Height:
(cab) 3.24 m
(crane) 3.91 m
Ground clearance: 0.465 m
Track: 2.15 m
Wheelbase: 5.165 m + 1.8 m
Max speed: 52 or 63 km/h
Fuel capacity: 280 or 360 litres
Fording: 1 m
Engine: Volvo 7 litre 6-cylinder turbocharged and intercooled diesel developing 255 hp
Transmission: ZF 5 HP 500 automatic with torque converter with 5 forward and 1 reverse gears
Transfer box: Volvo FL652 2-speed
Steering: hydromechanical, power-assisted
Suspension:
(front) 3 point axle suspension
(rear) Volvo bogie with individual axle movement
Tyres: 23.5 R 25
Brakes: dual circuit air
Electrical system: 24 V

Status

In production. In service with Sweden.

Manufacturer

Volvo Construction Equipment, S-405 08 Göteborg, Sweden.
Tel: +46 31 666000. Telex: 27000 volvo s.
Fax: +46 31 664660.

NEW ENTRY

SWITZERLAND

Bucher DURO (4 × 4) and (6 × 6) trucks

Development

Development of the Bucher DURO (4 × 4) and (6 × 6) trucks began in 1979 and (4 × 4) prototypes were produced during 1985. The (6 × 6) version was announced in October 1986. Prototypes were tested by the Defence Technology and Procurement Agency of the Swiss Military Department. In 1993, 2,000 DURO (4 × 4) were ordered for the Swiss Army, 1,800 as troop transports and 200 with command post bodies. Delivery commenced during 1994 and will continue

Specifications

Model	**4 × 4 S**	**4 × 4 L**	**6 × 6**
Cab seating	1 + 2	1 + 2	1 + 2
Configuration	4 × 4	6 × 6	4 × 4
Weight			
(cab and chassis)	2,750 kg	2,800 kg	3,250 kg
(total)	5,800 kg	6,000 kg	7,000 kg
Max load	2,450 kg	2,600 kg	3,150 kg
Towed load	2,000 kg	2,000 kg	2,000 kg
Load area inside length	2.61 m	3.74 m	3.74 m
Length	5.2 m	6 m	6.35 m
Width	1.96 m	1.96 m	1.96 m
Height (unloaded)	2.7 m	2.7 m	2.7 m
Ground clearance	0.36 m	0.36 m	0.36 m
Track	1.68 m	1.68 m	1.68 m
Wheelbase	3.23 m	3.88 m	3.23 m + 1.15 m
Angle of approach/departure	46°/36°	46°/32°	46°/36°
Max speed	110 km/h	110 km/h	110 km/h
Fuel capacity	120 litres	120 litres	120 litres
Max gradient	80%	80%	80%
Fording	0.8 m	0.8 m	0.8 m
Engine	VM EP 638 LI 3.749 litre 6 cylinder turbocharged diesel developing 147 hp at 4,000 rpm		
Transmission	Allison AT 542 automatic with torque converter with 4 forward speeds and 1 reverse		
Transfer box	2-speed	2-speed	2-speed
Steering	ZF power-assist	ZF power-assist	ZF power-assist
Turning circle	13.5 m	15.5 m	15 m
Suspension	coil spring with gas pressure shock absorber and roll stabiliser		
Tyres	275/80 R 20/28K	275/80 R 20/28K	275/80 R 20/28K
Brakes	hydraulic disc	hydraulic disc	hydraulic disc
Electrical system	24 V	24 V	24 V

until 1999. By the end of 1996 over 800 units were in operation.

The vehicle was originally known as the Puma. DURO stands for *Dauerhaft, Unabhängig, Robust, Oekonomisch* — Durable, Self-reliant, Robust, Economic.

Description

Bucher DURO (4 × 4) and (6 × 6) trucks are constructed using a low-slung, torsion-free chassis with an integral roll stabiliser. The frame is of the tubular ladder type. Both vehicles use a forward control cab of aluminium construction with reinforced plastic panels. There is seating for the driver and two passengers. Wide-vision windows are provided for all cab occupants and the entire cab can be tilted forward for maintenance access. The rear platform is of aluminium construction with provision for bench seats along each side and there is provision for the body to be covered by a tarpaulin carried on sliding bows. There is space at the rear to carry four or six standard pallets.

The engine for all three models is a 147 hp six-cylinder diesel. An automatic transmission with a torque converter is used. There is permanent all-wheel drive with Torson self-locking differentials. Steering is power-assisted. De Dion rigid axles with patented roll stabilisers are fitted and the hydraulic disc brakes operate with Y-type braking effort distribution. The wheel geometry remains stable under braking conditions and there is no brake fading and cleaning of brakes after fording.

Two 4 × 4 models are available. The 4 × 4 S has a 3.23 m wheelbase while the wheelbase for the 4 × 4 L is 3.88 m.

DURO trucks are produced in cargo-carrying, personnel or shelter carrier and ambulance forms. Other versions proposed include a command and communications vehicle, mobile workshops, firefighting vehicle and an NBC decontamination vehicle.

Status

In production. 2,000 ordered for the Swiss Army.

Manufacturer

Bucher-Guyer AG, Vehicles, CH-8166, Niederweningen/Zürich, Switzerland.
Tel: +41 1 857 2211. Fax: +41 1 857 2249.

UPDATED

Bucher DURO (4 × 4) truck with communications body
1997

Bucher DURO (4 × 4) S truck ***1997***

Bucher DURO (6 × 6) truck ***1997***

Saurer 6 DM (4 × 4) 6,000 kg and 10 DM (6 × 6) 10,000 kg trucks

Development

During the late 1970s Adolph Saurer Limited started the development of a new family of 6,000 and 10,000 kg trucks to meet the requirements of the Swiss Army. The first prototypes, the D250MF (4 × 4) and D3000MF (6 × 6) were completed in 1978. They were subsequently redesignated the 6 DM (4 × 4) and 10 DM (6 × 6). By late 1981 the vehicles had been tested by the Swiss Army and production commenced soon after. Production ceased in February 1986.

Description

Both vehicles share many common components, including the engine. Both have a two-door fully enclosed all-steel forward control cab with a circular observation hatch in the roof. The rear cargo area is provided with removable drop sides and a drop tailgate, removable bows and a tarpaulin cover. A 10,000 kg capacity hydraulic winch can be fitted.

The chassis is semi-flexible having independent leaf spring suspension with telescopic dampers, all axles are fitted with differential locks. On both vehicles the front and rear axles are interchangeable.

In-service variants include a 6 × 6 Ribbon Bridge carrier and a crane version.

Status

In service with the Swiss armed forces.

Manufacturer

Adolph Saurer Limited, CH-9320 Arbon, Switzerland. (This facility has closed).

VERIFIED

Specifications

Model	**6 DM**	**10 DM**
Configuration	4 × 4	6 × 6
Weight (empty)	10,000 kg	12,000 kg
(loaded)	16,000 kg	22,000 kg
Max load	6,000 kg	10,000 kg
Length	7.705 m	8.905 m
Width	2.5 m	2.5 m
Height (cab)	3.335 m	3.46 m
Ground clearance	0.38 m	0.38 m
Wheelbase	4.35 m	4 m + 1.4 m
Track	2.1 m	2.1 m
Angle of approach/departure	40°/40°	40°/40°
Fuel capacity	300 litres	300 litres
Fording	1.15 m	1.15 m
Engine	Saurer D4KT 12 L 6-cylinder in-line 4-stroke turbocharged diesel developing 250 hp at 2,200 rpm	
Gearbox	ZF S-6 90 with torque converter, retarder and splitter group	
Transfer box	ZF A 800 3D electropneumatically lockable	
Turning radius	8.75 m	9.75 m
Axles	Saurer with lockable bevel gear differential	
Tyres	14.00 × 20	14.00 × 20
Brakes (main)	air, dual circuit	air, dual circuit
Electrical system	24 V	24 V

Saurer 10 DM (6 × 6) 10,000 kg truck
1996

TURKEY

MANAS 16.230 FAEG (4 × 4) 7,500 kg and 26.372 DFAEG (6 × 6) 14,500 kg trucks

Description

These heavy-duty trucks were originally developed by the German firm MAN Nutzfahrzeuge AG to meet requirements for front-line logistic supply vehicles capable of crossing rough terrain. The two vehicles share many components, with the main difference between the two, apart from the differing drive configurations, being the engines. The 16.230 FAEG (4 × 4) vehicle is fitted with a MAN D 0826 6.87 litre diesel developing 230 hp (ISO) at 2,400 rpm. The 26.372 DFAEG (6 × 6) has an 11.97 litre supercharged engine, the MAN D 2866 LFG 04, developing 370 hp (SAE) at 2,000 rpm.

The two vehicles share the same all-steel FX type cab with seating for the driver and two passengers. A roof hatch is provided. The cab can be tilted forward for maintenance. The chassis has longitudinal members of pressed U-profile made of high quality steel. Cross members are fixed to the main members by rivets or bolts. There is a tow pin at the front and a NATO-type tow hook at the rear; electrical and brake circuits are provided for trailers. A cable winch is located in the centre of the vehicle between the frame members. It has a 5 tonne pull capacity to the front and 10 tonnes to the rear. The winch has 65 m of cable.

The body is steel and covered with a tarpaulin. Swing-open side and rear walls are provided. A tool box and steps are provided at the rear and there is stowage for extra fuel canisters. A swing-out spare wheel carrier is located behind the cab.

Specifications

Model	**16.220 FAEG**	**26.372 DFAEG**
Cab seating	1 + 2	1 + 2
Configuration	4 × 4	6 × 6
Weight (empty)	7,350 kg	10,050 kg
(GVW)	16,500 kg	25,000 kg
Max load	8,150 kg	14,500 kg
Length	7.45 m	8.5 m
Width	2.5 m	2.5 m
Height (top of cab)	approx 2.85 m	3.104 m
Ground clearance	0.433 m	0.428 m
Wheelbase	4.2 m	3.9 m + 1.4 m
Angle of approach/departure	38°/35°	38°/40°
Max speed	97 km/h	106 km/h
Fuel capacity	300 litres	300 litres
Max gradient	60%	60%
Fording	1 m	1.2 m
Engine	see text	see text
Gearbox	9 forward, 1 reverse	16 forward, 2 reverse
Clutch	single dry plate	single dry plate
Transfer box	G1000, 2-speed	G1700, 1- or 2-speed
Steering	ZF 8098	ZF Servocom
Tyres	14.00 R 20	14.00 R 20
Brakes	dual circuit air	dual circuit air
Electrical system	24 V	24 V
Batteries	2 × 12 V, 120 Ah	2 × 12 V, 140 Ah
Alternator	35 A	55 A

Tipper truck, tanker, snow plough and firefighting versions of both vehicles are available.

Status

In production.

Manufacturer

MAN Kamyon Ve Otobüs San, Esenboğa Havaliman, Yolu 06105, Akyurt Ankara, Turkey.
Tel: +90 312 398 0220. Fax: +90 312 398 0340.

VERIFIED

UNITED KINGDOM

HUGO (8 × 8) 2,000 kg all terrain vehicle

Description

The HUGO (High Utility, Global Operation) 8 × 8 all terrain vehicle was extensively developed from the original ESARCO design (for details refer to *Jane's Military Logistics 1989* page 508). While retaining a substantial use of Land Rover components in the suspension, axles and transfer boxes, the engine is a Perkins Phaser turbocharged diesel coupled to a specially developed Chrysler automatic gearbox.

The chassis was substantially modified to create a robust four-box section, with stronger cross members, carrying the axle suspension, engine and body components. The hull is mild steel and both body and hull are extensively corrosion treated, with a further layer of wear-resistant paint.

This forward control vehicle has a robust steering layout and utilises a strong built-in roll cage in the four-door, four-seat cab. An occasional seat is available between the rear seats and the cab. Full safety harness is fitted all round.

The front axle is heavily reinforced and the front and rear axles both steer, giving the vehicle great agility off road and directional stability on road. All axles can be locked together using differential locks in the two transfer boxes. The central non-steering pair of axles can also engage a lock on the differential which is selectable and deselectable, both on the move or stationary.

There is a fully redesigned electrical system allowing easy access to fuses, relays, and so on, from within the cab. The vehicle can fast wade unprepared in over 1 m of water. Disc brakes are fitted all round and are operated by twin servos. Heating/demisting is by use of a fuel-burning heater, allowing the heater to be used with the engine switched off. Air conditioning is an option.

The vehicle will carry 2,000 kg both on and off roads and can tow an additional 2,000 kg at the same time. The vehicle can be self-loading with an optional rear-mounted crane. The load deck, which is completely flat, can have a variety of body types, including ambulance, canvas tilt and frame, mobile workshop, hydraulic lift bucket for servicing overhead cables and other specialised equipment.

The HUGO was granted full UK type approval during 1991 and is thus certified for unrestricted general use on public highways.

HUGO (8 × 8) 2,000 kg all terrain vehicle **1997**

Specifications

Cab seating: 1 + 3 or 4 (up to 10 in rear)
Configuration: 8 × 8
Weight:
(empty) 3,550 kg
(loaded) 5,550 kg
Max load: 2,000 kg
Towed load: 2,000 kg
Load area: 2.77 × 2.2 m
Length: 5.37 m
Width: 2.2 m
Height:
(cab/roll bar) 2.36 m
(cargo bed, unladen) 1.17 m
Ground clearance: (under differentials) 0.268 m
Track: 1.779 m
Angle of approach/departure: 45°/45°
Max speed: (road, empty) 113 km/h
Range: (road) 960 km
Fuel capacity: 136 litres
Max gradient: 45°
Side slope: 30°
Fording: 1 m
Engine: Perkins Phaser 110T 4 litre 4-cylinder turbocharged diesel developing 110 bhp at 2,600 rpm
Gearbox: Chrysler 3-speed automatic
Transfer box: 2 × Land Rover LT230, 2-speed
Steering: Land Rover, power-assisted
Turning circle: (kerbs) 11.1 m
Suspension: coil springs all round with double-acting shock-absorbers
Tyres: Trelleborg 305/75-16
Brakes: disc all round
Electrical system: 12 V
Battery: 12 V
Alternator: 60 A

Status

Ready for production.

Manufacturer

HUGO All Terrain Vehicles Limited, PO Box 2, Guildford, Surrey GU3 3EH, UK.
Tel: +44 1483 60516. Fax: +44 1483 502234.

UPDATED

JCB 150M (4 × 4) general support tractor

Description

The JCB 150M (4 × 4) general support tractor is based on the JCB Fastrac all terrain tractor already in service with the Royal Air Force (26). The JCB 150M is an on/off road high traction improved medium mobility vehicle designed for a maximum gross train weight of 24,000 kg.

Designed for military use, the 150M is equipped with a low profile cab for C-130 air transport; maximum road speed is 75 km/h. The two-person operator cab is mid-mounted affording all-round visibility for the driver. Ancillary equipment can be front, rear deck or rear mounted.

The JCB 150M is suitable for winch, crane, snow plough, snow blower, salt spreading, de-icing, armament towing and similar applications as well as aircraft deployment and support operations. Some Royal Air Force Fastracs are equipped for operations at temperatures as low as −35°C.

JCB 150M (4 × 4) general support tractor **1996**

Specifications

Seating: 1 + 1
Configuration: 4 × 4
Weight: 6,200 kg
Max GVW: 8,850 or 12,000 kg
Max GTW: 24,000 kg
Length: 4.8, 5.525 and 6.3 m
Width: 2.279 m
Height: (with 445/70 R 24 tyres) 2.633 m
Ground clearance: 0.385 m
Track: 1.875 m
Angle of approach/departure: 39°/74°
Max speed: (road) 75 km/h
Fuel capacity: 280 litres
Fording: 0.75 m
Engine: Perkins Phaser 160TW 6-cylinder water intercooled diesel developing 160 bhp at 2,400 rpm
Transmission: JCB/Eaton with 36 forward speeds and 12 reverse speeds fully synchronised; 2-speed change under load (no need to declutch)
Steering: power-assisted

Turning circle: (kerbs) 14.1 m
Tyres: 445/70 R 24 XM 47 standard
Brakes: 4 wheel disc, dual split system with air/hydraulic actuation
Electrical system: 12 V
Battery: 2 × 12 V, 83 Ah
Alternator: 85 Ah

Status
In service with Royal Air Force (26) and some overseas armed forces.

Manufacturer
JCB Military Products Division, Rocester, Staffordshire ST14 5JP, UK.
Tel: +44 1889 590312. Fax: +44 1889 590588.

VERIFIED

Reynolds Boughton RB-44 (4 × 4) truck

Development
The RB-44 was developed as a private venture by the Boughton Group and was originally called the RB-510. In addition to its use as a cargo troop carrier, it can be used for a variety of other roles such as towing a 105 mm artillery piece, the British Aerospace Dynamics Rapier SAM system or the Marconi Command and Control Systems Blindfire radar.

The RB-44 Mark 2 was one of the two final contenders for the British Army's 2 tonne Truck Universal Heavy (TUH) requirement. During mid-1988 it was announced that the RB-44 had been selected as the successful contender. The initial order was for 846 vehicles. The first examples entered service with the British Army during 1992 and production was completed in 1993. In-service problems were experienced resulting in a lengthy correction programme but it is now understood to be in full service.

Reynolds Boughton RB-44 (4 × 4) trucks **1997**

Description
The Reynolds Boughton RB-44 (4 × 4) truck has a ladder-type bolted chassis which can be fitted with various types of body. A three-seat semi-forward control pressed steel cab is standard, but an extended crew cab conversion is available for carrying extra personnel and equipment. Three wheelbase lengths are available: 3.226, 3.68 and 4.06 m.

The standard production engine is the Perkins Phaser 110MT diesel but other engines may be fitted if required. A Boughton designed and manufactured two-speed transfer box incorporating six- and eight-stud SAE power take-offs capable of between 30 and 90 hp is fitted and the transfer box provides a permanent 4 × 4 drive configuration incorporating a differential lock.

Conventional semi-elliptical springs are used for the suspension, fitted with double-acting telescopic shock-absorbers. Vacuum-assisted brakes, with a dual-servo split system are fitted front and rear. The front and rear axles have hypo gearing with a ratio of 4.1:1. The front axle's plated capacity is 2,500 kg with 2,800 kg for the rear.

RB-44 variants include a general service cargo body, soft and hardtop bodies FFR (including a hardtopped version for the VIXEN tactical electronic warfare system), transportable container body FFR, ambulance and a lubrication body. A front-mounted winch (the Boughton VH8) can be added to any version.

RB-44 TUH variants for the British Army include general service cargo FFR, soft and hardtop and truck container body variants. Reynolds Boughton is undertaking the private venture development of ambulance, command and administration and recovery vehicle variants.

Specifications
Cab seating: 1 + 2 (up to 12 in rear)
Configuration: 4 × 4
Weight:
(kerb, basic wheelbase chassis) 3,000 kg
(loaded) 5,300 kg
Max load: 2,250 kg
Towed load: 2,200 kg
Length:
(3.226 m wheelbase) 5.06 m
(3.68 m wheelbase) 5.65 m
(4.06 m wheelbase) 6.03 m
Width: 2.1 m
Height: (cab top, loaded) 2.348 m
Ground clearance: 0.255 m
Wheelbase: 3.226 m, 3.68 m or 4.06 m
Track: (front and rear) 1.789 m
Angle of approach/departure: 45°/40°
Max speed: 109 km/h
Gradient: 60%
Side slope: 33°
Fording: 0.75 m
Engine: Perkins Phaser 110MT 4-cylinder direct injection diesel developing 109 bhp at 2,800 rpm
Gearbox: Spicer T5-250 manual with 5 forward and 1 reverse gears or
Chrysler A727 Torqueflite automatic with 3 forward and 1 reverse gears
Clutch: single plate, hydraulic
Transfer case: Boughton 2-speed
Steering: ZF hydraulic power-assisted
Turning radius: (3 226 m wheelbase) 7.4 m
Suspension: (front and rear) heavy-duty leaf springs with AEON rubber assisters and double-action shock-absorbers
Tyres: 9.00 × 16 XL or 11.00 × 16 XL or XS
Brakes: hydraulic, servo-assisted, dual-line system
Electrical system: 12 V or 12 V with 24 V FFR
Alternator: (12 V) 65 A

Status
Production complete. In service with the British Army (846).

Manufacturer
Reynolds Boughton Limited, Bell Lane, Amersham, Buckinghamshire HP6 6PE, UK.
Tel: +44 1494 764411. Telex: 83132.
Fax: +44 1494 765218.

UPDATED

Volvo Trucks (GB) M6 (4 × 4) truck

Description
The Volvo Trucks (GB) M6 (4 × 4) truck is a derivative of the commercial Volvo FL6 range and is called the Highlander. It uses a high level of British components, including the gearbox and axles, to the point where there is over 80 per cent local content. The prototypes used a Volvo FL6 forward control cab with an optional low roof design and seating for the driver and two passengers. For maintenance, the steel cab can be tilted forward 55°. The chassis is a conventional ladder frame and carries the load area. A fuel tank is located centrally on the left-hand side and towing points are provided front and rear.

The Highlander can be fitted with a winch or a crane. It is air-portable in a C-130 transport aircraft.

Volvo Trucks (GB) M6 (4 × 4) 4,000 kg truck

Specifications
Cab seating: 1 + 1 or 2
Configuration: 4 × 4
Weight:
(kerb) 4,890 kg
(front axle) 3,480 kg
(rear axle) 1,500 kg

Max load: (body and payload) 9,020 kg
Length: 7.66 m
Width: 2.368 m
Height:
(top of cab) 2.826 m
(frame height, empty) 1.176 m
(frame height, loaded) 1.07 m
Ground clearance: 0.325 m
Track: 2 m
Wheelbase: 3.7 m
Fuel capacity: 200 litres
Engine: Volvo D6A180 Euro II 5.48 litre 6-cylinder in-line direct injection diesel developing 180 hp at 2,400 rpm
Gearbox: Eaton T600A fully synchronised 6-speed
Clutch: KB114C single plate hydraulic
Transfer box: Steyr VG 450 2-speed, provision for PTO
Steering: power-assisted ball and nut
Turning circle: 14.71 m
Suspension:
(front) twin-leaf parabolic springs with rubber bump stops
(rear) 3-leaf parabolic springs with telescopic dampers and rubber bump stops with 2-leaf helper spring
Tyres: 12.00 R.20 18 PR D002 all-terrain (other options available)
Brakes: dual-line air-actuated dual-circuit wedge-operated drum-type
Electrical system: 24 V
Batteries: 160 Ah
Alternator: 55 A

Status
Available as special build. In service with the Royal Swedish Air Force (63).

Manufacturer
Volvo Trucks (GB) Limited, Wedgnock Lane, Warwick CV34 5YH, UK.
Tel: +44 1926 401777. Fax: +44 1926 490991.

UPDATED

Leyland Trucks (4 × 4) 5,000 kg truck

Development
This (4 × 4) 4,000 kg truck was developed by Leyland Trucks under the project number T 244 to participate in the British Army's 4 tonne truck replacement contest along with Volvo (GB) Limited and Bedford. In December 1987 eight prototypes were handed over to the Ministry of Defence to take part in a series of technical trials. The result of the contest was, that in June 1989, Leyland Trucks was awarded a contract worth £155 million for 5,350 vehicles over a five-year period. That contract was completed in April 1995 after some 4,500 had been delivered.

The vehicle is produced at the Leyland Trucks Assembly Plant in Lancashire using largely British-sourced components. The first six production vehicles were handed over for acceptance in July 1990, with deliveries to the Ministry of Defence commencing in early August 1990. During 1995 the vehicle's payload was increased from 4,000 to 5,000 kg.

Description
The Leyland Trucks (4 × 4) 5,000 kg truck is conventional in layout and design. Some key components used on the vehicle are shared with the commercial 45 Series light truck. The T 244 uses the sleeper cab version of the C44 forward control cab which has room for the driver, two passengers and stowage for their full kit. As an alternative, the space provided may be used for driver training (the driver plus four personnel) or radio communications equipment. The cab roof is reinforced to take the weight of two personnel and has a roof hatch and provision for a machine gun installation over an observer's platform inside the cab. For maintenance the cab can be hydraulically tilted forward 50°.

The vehicle is powered by a Leyland 313 turbocharged diesel engine coupled to a transmission using a five-speed, all-synchromesh gearbox. There is permanent 4 × 4 drive. The front axle has a rating of 5,000 kg, with an offset bowl to reduce overall cab height. The rear axle has a rating of 7,500 kg.

Fitments include: front and rear end rotating tow hooks, lugs for suspended or supported recovery, Helicopter lift points and an infrared reflective paint finish. The vehicle can be carried in a C-130 Hercules transport aircraft.

The body is produced and fitted by Edbro Limited and has a fixed flat platform and bulkhead with provision for interchangeable drop sides, tailboard, superstructure and canopy. The flat platform can be used to carry various forms of military equipment including containers, seats, fuel pods, NATO pallets, Class 30 trackway and Medium Girder Bridge (MGB) sections. A vehicle carrying a 'mini DROPS' load handling system has been developed as a concept exploration vehicle.

Options include: left- or right-hand drive, winch, hydraulic crane, tipping body, or a chassis and cab-only arrangement. The crane has a capacity of 6.5 t/m while the winch (produced by Reynolds Boughton) has a capacity of 5,500 kg front and rear and is provided with 75 m of cable.

The vehicle shares many similarities and components with Leyland Trucks's civilian 45 series and Comet trucks.

Specifications
Cab seating: 1 + 2
Configuration: 4 × 4
Weight:
(empty) 6,010 kg
(loaded) 11,210 kg
GVW: 11,600 kg
GCW: 16,260 kg
Max load: 5,000 kg
Towed load: (on/off road) 8,000 kg
Length: 6.65 m
Width: 2.49 m
Height: 3.43 m
Platform height: (empty) 1.41 m
Ground clearance: 0.32 m
Track: (front and rear) 2.1 m
Wheelbase: 3.95 m
Angle of approach/departure: 41°/38°
Max speed: 89 km/h
Range: approx 500 km
Fuel capacity: 135 litres
Max gradient: (fully loaded, with trailer) 33%
Side slope: (static) 33°
Fording: 0.75 m
Engine: Leyland 313 5.9 litre 6-cylinder direct injection turbocharged diesel developing 145 hp at 2,600 rpm
Gearbox: Turner T5-350 synchromesh with 5 forward and 1 reverse gears
Clutch: servo-assisted self-adjusting single dry plate (330 mm)
Transfer box: Getrag 304 2-speed + neutral
Steering: ZF 8045 power-assisted
Turning radius: 9 m
Suspension:
(front) 2 taper leaf with telescopic hydraulic double-acting shock-absorbers
(rear) dual-rate taper leaf with telescopic hydraulic double-acting shock-absorbers
Tyres: 12.00 R 20
Brakes:
(main) Girling air-actuated 2-line dual system
(parking) spring actuators on all wheels
Electrical system: 24 V
Batteries: 2 × 12 V, 95 Ah
Alternator: 40 A

Status
In production. In service with the British Army (ca 4,500), Irish Defence Force, Malaysian and Indonesian forces operating in Bosnia, and Royal Brunei armed forces (41).

Manufacturer
Leyland Trucks Limited, Croston Road, Leyland, Preston PR5 3LZ, UK.
Tel: +44 1722 621400. Fax: +44 1722 625800.

VERIFIED

British Army Leyland Trucks (4 × 4) 5,000 kg truck (T J Gander) ***1997***

British Army Leyland Trucks (4 × 4) 5,000 kg truck carrying an equipment shelter (T J Gander) ***1997***

Bedford MT series 4 × 4 and 6 × 6 trucks

Development

Bedford Trucks was acquired by Marshall SPV Limited in October 1992 when production moved to Cambridge where the Marshall Group's aerospace and vehicle facilities are based. In January 1994, Marshall SPV Limited acquired the Bedford name and parts business from General Motors. Bedford trucks and genuine Bedford parts are therefore available through distribution centres worldwide.

The family of Bedford 4 × 4 and 6 × 6 trucks is available in the range from 12 to 30 tonnes gross vehicle weight (GVW), with payloads up to 21 tonnes.

Description

Bedford MT 4 × 4 Range

The Bedford MT 4 × 4 range are the next generation of the current in-service M-Type 4 × 4 trucks (see following entry). All 16 standard models in the MT 4 × 4 range are of conventional layout and design with a forward control cab on a flat-topped ladder-type chassis frame capable of accommodating a wide range of bodies.

The cab is all-steel with hydraulic tilting for easy engine access having hinged panels for day-to-day checks. A reinforced cab roof, able to take the weight of two personnel, has a hatch with hip ring and cover for access to a mounting for a light machine gun. There is

Specifications

Type	**MT12-14/16/18**	**MT15-16/18**	**MT17-21**	**MT19-21**	**MT24-21**	**MT26-30**
Cab seating	1 + 1 or 1 + 2	1 + 1 or 1 + 2	1 + 1 or 1 + 2	1 + 1 or 1 + 2	1 + 1 or 1 + 2	
Configuration	4 × 4	4 × 4	4 × 4	4 × 4	6 × 6	6 × 6
Kerb weight nominal	4,760 kg	4,900 kg	5,026 kg	5,026 kg	7,286 kg	7,586 kg
Front axle design weight	4,750 kg	4,750 kg	7,000 kg	7,000 kg	7,000 kg	7,000 kg
Rear axle design weight	7,620 kg	10,000 kg	10,500 kg	13,000 kg	20,000 kg	20,000 kg
GVW	12,000 kg	14,750 kg	17,000 kg	19,000 kg	24,000 kg	26,000 kg
GTW	17,100 kg	19,750 kg	24,000 kg	24,000 kg	n/avail	32,000 kg
Max load nominal	6,000 kg	8,000 kg	10,000 kg	12,000 kg	14,500 kg	17,000 kg
Length (chassis)	6.53/6.87 m	6.53/6.87 m	6.60/6.95 m	6.6 m	9.36 m	8.87 m
Width	2.48 m	2.48 m	2.48 m	2.49 m	2.47 m	2.47 m
Height (top of cab)	2.63 m	2.7 m	2.7 m	2.59 m	2.72 m	2.72 m
Track front	2.09 m	2.09 m	2.07 m	1.93 m	2.02 m	2.02 m
Track rear	2.05 m	2.05 m	2.09 m	1.82 m	2.08 m	2.08 m
Wheelbase	3.94/4.24 m	3.94/4.24 m	4.01/4.24 m	4.01 m	5.01/5.4 m	5.01 m
Angle of approach/departure	44°/39°	44°/39°	46°/41°	46°/41°	44°/38°	44°/38°
Max speed	87 km/h (12-14) 104 km/h (12-16/18)	104 km/h	105 km/h	105 km/h	86 km/h	104 km/h
Fuel capacity	109 litres	209 litres	209 litres	209 litres	209 litres	209 litres
Restart gradability	49% (12-14) 46% (12-16) 52% (12-18)	37% (15-16) 42% (15-18)	39%	35%	45%	72%
Engine-diesel	Perkins Phaser 6-cylinder 135 bhp, 97 kW (12-14) 160 bhp, 119 kW turbocharged (12-16) 180 bhp, 134 kW turbocharged and intercooled (12-18)	Perkins Phaser 6-cylinder 160 bhp, 119 kW turbocharged (15-16) 180 bhp, 134 kW turbocharged and intercooled (15-18)	Perkins Phaser 6-cylinder 210 bhp, 156 kW wastegated turbocharged and intercooled	Perkins Phaser 6-cylinder 210 bhp, 156 kW wastegated turbocharged and intercooled	Perkins Phaser 6-cylinder 210 bhp, 156 kW wastegated turbocharged	Perkins Peregrine 6-cylinder 300 bhp, 224 kW turbocharged and air-to-air intercooled
Gearbox	Spicer T5-350/4 5-speed (12-14) Spicer T5-400 5-speed (12-16/18)	Spicer T5-4292 5-speed	Eaton 4106B 6-speed	Eaton 5206B 6-speed	Eaton 6109 9-speed	Fuller RT-11609A 9-speed
Clutch - air assisted	Borg & Beck Organic (12-14) Lipe 13-330 Cerametallic (12-16/18)	Lipe 13-330 Cerametallic	Lipe 14-1DLB Cerametallic	Lipe 14-1DLB Cerametallic	Lipe 14-1DLB Cerametallic	Lipe 14 in twin plate
Transfer box	Bedford 2-speed	Bedford 2-speed	Rockwell SVI 2-speed	Rockwell SVI 2-speed	Steyr VG1600/300 2-speed	Steyr VG1600/300 2-speed
Steering	Power-assisted - recirculating ball type					
Turning circle (kerbs)	18 m	18 m	19.3 m	19.3 m	21.5 m	21.5 m
Front suspension	8-leaf semi-elliptic	8-leaf semi-elliptic	8-leaf semi-elliptic	8-leaf semi-elliptic	2-leaf tapered semi-elliptic	2-leaf tapered semi-elliptic
Rear suspension	9-leaf semi-elliptic taper leaf helper	9-leaf semi-elliptic taper leaf helper	9-leaf semi-elliptic taper leaf helper	9-leaf semi-elliptic taper leaf helper	Hendrickson Norde HN460 rubber	Hendrickson Norde HN460 rubber
Tyres	12.00 R 20 (single rears)	14.75/80 R 20 (single rears)	14.75/80 R 20 (single rears)	12.00 R 20 (dual rears)	12.00 R 20 (dual rears)	12.00 R 20 (dual rears)
Brakes	full air	air/hydraulic	full air	full air	full air	full air
Electrical system	24 V	24 V	24 V	24 V	24 V	24 V
Batteries	2 × 12 V 90 Ah	2 × 12 V 90 Ah	2 × 12 V 90 Ah	2 × 12 V 90 Ah	2 × 12 V 90 Ah	2 × 12 V 90 Ah
Alternator	55 A	55 A	55 A	55 A	55 A	55 A

Bedford MT 12-18 (4 × 4) truck

Bedford MT (6 × 6) truck configured as recovery vehicle

1997

internal stowage for personal kit, rifles and other equipment. Power steering and engines from the Perkins range give enhanced off-road mobility with increased payloads.

Numerous equipment options are available including different axle options (Rockwell all round), power take-off (front and rear), extreme climate modifications for the engine, different wheel and tyre sizes plus many others. During 1997 Cummins engines will be offered as power options at each model size.

With specialised bodies and equipment the Bedford MT range meets a multitude of military requirements including the transport of ammunition, fuel or water, NATO pallets, containers, shelters and troops, as well as general cargo.

Bedford MT 6 × 6

The Bedford MT 6 × 6 trucks are a development of the MT 4 × 4 and extend the payload capabilities up to 21 tonnes. Power is provided by either Perkins Phaser 210 bhp engines or the 300 bhp Peregrine engines. Cummins engines will be offered during 1997.

With specialised bodies and equipment options available the Bedford MT 6 × 6 range can be used in a wide variety of applications, including recovery units, PLS/DROPS load handling systems, ammunition carrier, water/fuel tankers, refuellers, firefighting vehicles and tipper/dumpers.

Status

In production. In service with armed forces in Africa, the Middle East and South-East Asia and on behalf of NATO.

Manufacturer

Marshall SPV Limited, Bedford Truck Division, Airport Works, Cambridge CB5 8RX, UK.
Tel: +44 1223 373068. Fax: +44 1223 373073.

UPDATED

Bedford MK (4 × 4) 4,000 kg truck

Development

In the early 1960s, the British Army issued a requirement for a 4 × 4 4,000 kg (nominal payload) truck to replace the then current Bedford RL. To meet this requirement Austin submitted the FJ (FV 13701), Commer the CB (FV 13901) and Vauxhall the RK (FV 13801). After comparative trials the Vauxhall model, based on its civilian TK (4 × 2) truck, was selected and standardised as the Truck Cargo (Bedford MK 4 tonne 4 × 4).

As from April 1981 Bedford changed the designation of the MK to MJ as the K multi-fuel engine was superseded by the J diesel engine. The designation MK is an abbreviated form of the alpha designation MKP2BMO, MJ being MJP2BMO. B becomes W for winch variants.

In mid-1982, a £46 million order was announced for new MJP trucks fitted with a new 5.42 litre turbocharged diesel engine designated the 5.4/105TD. This series of trucks incorporated many minor modifications to take advantage of the new engine's power and production of the new model commenced in September 1982.

Production totals for M-series vehicles, including both military and civilian models, stood at 48,318 in December 1988 with a further 200 being built between 1989 and 1992, when production ceased in favour of the MT series (see previous entry).

In October 1992 Marshall SPV Limited acquired the product designs and some of the assets of AWD Limited.

Description

The chassis is of the ladder type with six cross members, two of 'alligator jaw' design. The all-steel two-door forward control cab has a circular observation hatch in the roof. Access to the engine for maintenance is via the top-hinged panels on the rear quarters of the cab at either side. The cab rear panel, between the driver's and passenger's seats, is easily removed for engine access.

The all-steel rear cargo area has drop sides and a drop tailgate, which can be removed quickly to provide a platform for the carriage of containers, pallets, or the Simon Saro demountable fuel pod. Removable bows and a tarpaulin cover are fitted as standard. A hydraulic crane can be fitted for unloading. Detachable outward-facing seats can be fitted in the centre of the cargo area for carrying passengers. The body of the MJP is manufactured by Marshall of Cambridge (Engineering) Limited.

The MJP2 has a 3.962 m wheelbase but a model with a wheelbase of 3.505 m is also available (MOR1). Power is provided by a 5.42 litre diesel engine in turbocharged or naturally aspirated form.

The basic models of the M-type are designated the FV 13801/FV 13802 or, when fitted with a winch, the FV 13803/FV 13804. The winch has a capacity of 5,080 kg and 76 m of cable. Provision is made for a PTO to be driven from either the front or rear of the transfer box casing. In addition, a standard six-stud SAE PTO facing is fitted to the side of the gearbox. The MJP has single rear wheels with either 12.00 × 20 or 13.00 × 20 tyres, but dual rear wheels with 9.00 × 20 tyres can be fitted. Standard equipment includes a heater and defroster and stowage racks for small arms.

Variants

Drone carrier, dump truck, refueller, mineproof cab (conversion work carried out by Reynolds Boughton), portable roadway laying vehicle and with special bodies. The Royal Air Force uses the M-type for carrying bombs and other munitions to support Harrier V/STOL aircraft.

Bedford TK trucks (4 × 2) are used for a wide variety of roles by all three British services including driver training vehicles, stores carriers and aircraft refuellers.

Specifications

Cab seating: 1 + 1
Configuration: 4 × 4
Weight:
(empty) 5,129 kg
(loaded, on road) 11,180 kg
(loaded, off road) 9,650 kg
Weight on front axle: (max load) 4,060 kg
Weight on rear axle: (max load) 7,620 kg
Max load: (single rear wheels) 6,060 kg
Towed load: 4,570 kg
Load area: 4.28 × 2.01 m
Length: 6.579 m
Width: 2.489 m
Height:
(cab) 2.501 m
(tarpaulin) 3.404 m
Ground clearance: 0.343 m
Track:
(front) 2.05 m
(rear) 2.03 m
Wheelbase: 3.962 m
Angle of approach/departure: 41°/38°
Max speed: (road) 77 km/h
Range: 560 km
Fuel capacity: 155 litres
Max gradient: 49%
Fording: 0.762 m
Engine: Bedford 5.42 litre 6-cylinder naturally aspirated diesel developing 97.9 bhp (gross) at 2,600 rpm
Gearbox: manual with 4 forward and 1 reverse gears
Clutch: single dry plate
Transfer box: 2-speed
Steering: semi-irreversible worm and sector
Turning radius: 9 m
Suspension: semi-elliptical springs with telescopic hydraulic double-action shock-absorbers
Tyres: 12.00 × 20, 13.00 × 20, or 9.00 × 20
Brakes:
(main) air/hydraulic
(parking) mechanical
Electrical system: 24 V
Batteries: 2 × 12 V, 128 Ah

Status

Production complete. In service with Bangladesh, Belgium, Brunei, Indonesia, Ireland, Kenya, Netherlands, Turkey (300), Uganda, United Arab Emirates, UK and other countries.

Manufacturer

Marshall SPV Limited, Bedford Truck Division, Airport Works, Cambridge CB5 8RX, UK.
Tel: +44 1223 373068. Fax: +44 1223 373073.
Main production was carried out by Bedford Trucks at Dunstable.

UPDATED

Late version of MJP (4 × 4) 4,000 kg truck with mobile stores body (T J Gander)
1997

Bedford MK (4 × 4) 4,000 kg truck carrying MGB components (T J Gander)
1997

Bedford TM 4-4 (4 × 4) 8,000 kg truck

Development

In the early 1970s, the British Army issued a requirement for a new 4 × 4 8,000 kg (nominal payload) cargo truck as a part of its Medium Mobility Vehicle Programme. To meet this requirement prototypes were built by Foden, Leyland and Bedford. After comparative trials, in September 1977, Bedford was awarded a contract worth almost £40 million for 2,099 of its model TM 4-4. Production began in September 1980 and first vehicles were delivered to the British Army in April 1981.

In October 1992 Marshall SPV Limited acquired the product designs and some of the assets of AWD Limited.

Description

The basic cargo model is designated the WNV3NPO by Bedford and has a GVW of 17,000 kg and a GTW of 25,000 kg.

The TM 4-4 has a ladder-type chassis with 450 N/mm^2 yield structural steel channel-section side members and constant depth and section throughout, with special rear cross members for drawbar trailer operations. The full width heavy-duty front bumper is bolted to the first chassis cross member. Front towing pintle and brake pipeline couplings for trailer brake operations and vehicle recovery are incorporated.

The front axle is a Kirkstall fully floating, single speed, spiral bevel with 3.857:1 hub reduction with overall axle ratio of 5.887:1. The rear axle is also a Kirkstall fully floating, single speed, spiral bevel with 3.857:1 hub reduction and air-actuated differential lock. Overall axle ratio is 5.887:1.

The all-steel two-door forward control cab has an observation hatch in the roof. The cab can be tilted forwards hydraulically through 60° locking in a number of intermediate positions. The hinged front grille provides access to the engine for checking oil and other services such as heater and steering systems. The water level can be checked in the expansion tank at the rear of the cab and fuses can be changed from inside. Standard equipment includes an inter-vehicle starting system, heater and ventilation system and a spare wheel carrier and spare wheel.

The cargo body was built and mounted on to the chassis by Marshall of Cambridge (Engineering) Limited. It is steel with a wooden floor, removable drop side boards and tailboard, lashing shackles for NATO pallets and longitudinal folding bench seats for personnel. An alternative platform body was available with headboard only. Twistlock attachments or lashing hooks for securing containers are also fitted.

The central roof-mounted hip ring has a glass fibre-reinforced cover which is stowed on the rear of the cab when removed. The gunner's platform is centrally mounted in the cab, for use with the hip ring. The roof is reinforced to withstand the load of a light machine gun and two personnel.

The standard vehicle is equipped with a 24 V electrical system, an exhaust system and fire screening which comply with UK petroleum-carrying regulations.

Power is transmitted from the engine to the gearbox and then to the transfer case which provides drive to the front and rear hub reduction axles.

The following optional equipment is available for the Bedford TM 4-4: exhaust brake, cab painted in NATO IRR green, ringfeeder drawbar coupling, flitch plates, NATO batteries, cab roof front marker lights, dry charged batteries (for CKD purposes), reverse lamps, rear fog guard lamps, tachograph, steering column lock and electric stop control, mid-mounted winch, Atlas hydraulic crane, inertia reel seat belts for driver and passenger, multi-leaf springs, transfer box differential, transfer box with PTO adaptor, transfer box with differential and PTO adaptor and military-type cargo body.

Bedford TM 4-4 (4 × 4) 8,000 kg truck (T J Gander) ***1997***

Variants

TM 4-4 with tipper body

These have the Bedford designation of WNV6NPO + RPO 585, a wheelbase of 3.883 m, GVW of 16,300 kg and GTW of 24,300 kg. The Edbro military tipper body has a capacity of 6.5 m^3. The body can be tipped to an angle of 54° and is of all-steel construction with a three-stage front end ram. The tailgate is hinged at the top and bottom with hydraulic drive via the main gearbox PTO.

TM 4-4 with Atlas self-loading crane

Over 700 of these were delivered to the British armed forces and have the Bedford designation of WNV3NPO + RPO 355. Mounted to the cab rear is an Atlas self-loading crane controlled by the operator standing at the hip ring position. The hydraulic crane is driven via the main gearbox PTO and has a slew angle of 193° and the following capabilities: 3,650 kg lift at 2.02 m reach, 2,140 kg lift at 3.46 m reach and 1,700 kg lift at 4.32 m reach.

TM 4-4 with winch

These are fitted with a mid-mounted winch with a capacity of 8,000 kg powered from a transfer box from a PTO. The cable is 75 m long and has a minimum speed of 4.5 m/min and a maximum speed of 23 m/min. The controls are air-operated with fairleads and pulleys for front and rear winching capability, band-type winch brake and safety overload cutout. The Bedford designation for this model is WNV3NPO + RPO 414.

Specifications

Cab seating: 1 + 1
Configuration: 4 × 4
Weight:
(empty) 8,300 kg
(loaded) 17,000 kg
Max load: 8,000 kg
Towed load: 10,000 kg
Length: 6.623 m
Width: 2.476 m
Height:
(cab) 2.997 m
(tarpaulin) 3.454 m
Ground clearance:
(axles) 0.352 m
(mid-wheelbase) 0.47 m
Track:
(front) 2.02 m
(rear) 2.08 m
Wheelbase: 4.325 m
Angle of approach/departure: 41°/38°
Max road speed: 93 km/h
Max range: 500 km
Fuel capacity: 155 litres
Gradient: 55.55%
Side slope: (empty) 43°
Fording: 0.75 m
Engine: Bedford 8.2/205 TD 8.2 litre water-cooled direct injection turbocharged diesel developing 206 bhp at 2,500 rpm
Gearbox: Spicer T6-47026 manual with 6 forward and 1 reverse gears
Clutch: twin dry plate
Transfer box: Rockwell T226-133, 2-speed
Steering: recirculating ball with integral power assistance
Suspension: (front and rear) semi-elliptical taper leaf springs with hydraulic double-acting telescopic shock-absorbers
Tyres: 15.50/80 × 20 radial
Brakes:
(main) air
(parking) air released spring brakes
Electrical system: 24 V

Status

Production complete. In service with Abu Dhabi, Bahrain, Oman and the UK.

Manufacturer

Marshall SPV Limited, Bedford Truck Division, Airport Works, Cambridge CB5 8RX, UK.
Tel: +44 1223 301133. Fax: +44 1223 373083.
Main production was carried out by Bedford Trucks at Dunstable.

VERIFIED

Foden Medium Mobility (6 × 6) range of vehicles

Development

The Foden Medium Mobility range of vehicles was developed for use with the FH-70 155 mm howitzer in service with the British, German and Italian armies. The British Army placed a production order for 116 tractors and limbers, with the bodies built by Marshall of Cambridge (Engineering) Limited. Final deliveries to the British Army were made late in 1979. These vehicles are now deployed by the Territorial Army.

The Medium Mobility range was designed for both on- and off-road conditions and shares many common components with the Foden Low Mobility range.

Description

The chassis is of bolted construction with a flitch plate and the full width front bumper has a towing jaw. Mounted at the rear is a rigid rear-pulling jaw and there are lift and recovery lugs front and rear.

The forward control cab is identical to that used for the Low Mobility range and the reader is referred to that entry for a description.

The Rolls-Royce engine is coupled to a Fuller gearbox with eight forward and one reverse gears. This has a single control stick and range change with overdrive.

The front axle is a Kirkstall SD.65-11-1 rated at 10,000 kg and fitted with a differential lock. The foremost tandem axle is a Kirkstall D65-11-1AF with an air-operated differential and the rearmost tandem axle is a Kirkstall D65-11-1 with a differential driving head. A cross-axle differential lock is fitted to each of the rear axles and is air-operated with an electrical warning switch which indicates when the lock is engaged. The

Foden (6 × 6) medium mobility vehicle used as gun tractor for FH-70 (Christopher F Foss) ***1996***

Foden (6 × 6) medium mobility vehicle used as FH-70 limber vehicle ***1996***

rear bogie has high articulation for cross-country operations.

The transfer (or auxiliary) gearbox is a Kirkstall AGB7000 which comprises a two-speed three-shaft, constant mesh helical gearbox with disengageable front-wheel drive. Engagement is pneumatically actuated via spur dogs designed to prevent any possibility of clutch throw-out. Provision for optional high or low ratio live PTO is provided by utilising a duplicate front drive cartridge mounted at the rear of the input shaft.

Various optional engines and transmissions are available.

Variants

Gun tractor

This is used to tow the 155 mm FH-70 howitzer. The platform comprises two sections of underframe with hardwood boarded floors mounted fore and aft of the Atlas hydraulic crane. Mounted to the rear of the cab is a removable cabin which provides heated accommodation for eight personnel. Two NATO ammunition pallets are carried both fore and aft of the crane. Stowage for the spare wheel and other gun equipment is provided at the rear of the vehicle. Spare wheels for the gun are stowed on the roof of the cabin.

Gun limber

This is used to carry eight NATO pallets containing 155 mm ammunition for the FH-70 howitzer. The platform comprises two sections of underframe with hardwood boarded floor mounted fore and aft of the crane.

Recovery vehicle

This uses the 6 × 6 chassis and cab. Full details of this vehicle are given in the *Recovery vehicles* section.

Specifications

Cab seating: 1 + 2
Configuration: 6 × 6
Weight:
(loaded) 27,440 kg
(chassis and cab) 10,990 kg
Max towed load: 9,300 kg
Length: 9.16 m
Width: 2.5 m
Height: 3.75 m
Track:
(front) 2.029 m
(rear) 2.032 m
Wheelbase: 3.97 m + 1.516 m
Angle of approach/departure: 40°/29°
Max speed: (road) 104 km/h (tyre limitation 80 km/h)
Fuel capacity: 409 litres
Max gradient: 33%
Engine: turbocharged Rolls-Royce 305 Mk III 6-cylinder liquid-cooled diesel developing 305 bhp at 2,100 rpm
Gearbox: Fuller with 8 forward and 1 reverse gears
Transfer box: 2-speed
Steering: recirculatory ball with integral power-assistance
Turning radius: 12 m
Suspension:
(front) semi-elliptical springs
(rear) 2 springs fully articulated
Tyres: 16.00 × 20
Brakes:
(main) air
(parking) mechanical
Electrical system: 24 V
Batteries: 2 × 12 V

Status

Production complete. In service with the British Army.

Manufacturer

Foden Trucks, A Division of Paccar UK Ltd, Moss Lane, Sandbach, Cheshire CW11 9YW, UK.
Tel: +44 1270 763244. Fax: +44 1270 762758.

VERIFIED

Bedford TM 6-6 (6 × 6) 14,000 kg truck

Development

The Bedford TM 6-6 (6 × 6) 14,000 kg truck was developed from the Bedford TM 4-4 (4 × 4) 8,000 kg truck (see separate entry in this section). It was in response to a British Army General Staff Requirement for a cost-effective 6 × 6 vehicle with the same 14 tonne payload as the TM 4-4 and trailer combination (such as 10 standard NATO pallets). The TM 6-6 uses many of the same components as the TM 4-4 and the first pretest prototype was completed in November 1981. A further four prototypes were built during 1983 and another eight Ministry of Defence validation vehicles were built during 1984 for trials at the RARDE Chertsey and user trials that extended into 1985. Volume production commenced in September 1986 with initial Ministry of Defence contracts for 1,045 vehicles.

During 1988 the payload allowance of the TM 6-6 was uprated to 16,000 kg.

In October 1992, Marshall SPV Limited acquired the product designs and some of the assets of AWD Limited.

Bedford TM 6-6 (6 × 6) 14,000 kg truck with platform body (T J Gander) ***1997***

Description

The cabs of the TM 4-4 and TM 6-6 are identical, with the roof having a central hip ring with a detachable cover. The interior layout is also the same as the TM 4-4 except for modifications to controls associated with permanent six-wheel drive and the revised transmission. The roof is reinforced to take the weight of a machine gun and two personnel. Provision is made for stowage of a rifle at each end of the instrument panel. By using a hydraulic pump the cab can be tilted forward 62° allowing engine maintenance. The engine is the same as that used with the TM 4-4 but the TM 6-6 has an exhaust brake operated from a foot pedal. Features common to both vehicles include the cooling system, exhaust, front axle and suspension, steering wheel and tyres. The electrical and braking systems are modified to suit the 6 × 6 function.

Features of the TM 6-6 designed for ease of operation include the synchroniser engagement of the transfer box for on-the-move ratio changes. This provides more effective use of the transfer box combining the advantages of the wider gear ratio spread with the advantages of a range change.

A six-link rear bogie suspension provides equalised axle loading on wide wheel centres for optimum traction and large axle articulation through maintenance-free rubber bushed links for good axle control. Tapered-leaf chevron-profile springs provide high durability with substantial weight savings.

Three versions of the TM 6-6 were produced. The load-carrier with a standard cargo body was manufactured by Edbro Limited. This is a basic flatbed with body sides, tailboard, tilt and canvas covers. An alternative cargo-bodied version has a centrally mounted hydraulic winch produced by FW Engineering Limited. This can be used for the recovery of loads up to 10,000 kg front and rear. A third variant is a platform body version equipped with an Atlas crane capable of lifting 1,400 kg at a radius of 6.55 m. The crane, with controls immediately behind the cab, is only operable when the stabiliser legs are extended. All versions are fitted with a towing pintle, lashing and lifting eyes.

Specifications

Cab seating: 1 + 1
Configuration: 6 × 6

Weight:
(kerb) 8,682 kg
(loaded) 24,390 kg
Max load: 14,000 kg (max 16,000 kg)
Length: 8.59 m
Width: 2.47 m
Height: (cab) 3.04 m
Ground clearance: (axle) 0.35 m
Track:
(front) 2.02 m
(rear) 2.08 m
Wheelbase: (mean) 5 m
Max speed: (road) 89.7 km/h
Fuel capacity: 227 litres
Range: 500 km
Side slope: 28°
Fording: 0.75 m
Engine: Bedford 8.2/205TD 8.195 litre water-cooled direct-injection turbocharged diesel developing 205 bhp at 2,500 rpm
Gearbox: ZF S6-80 manual with 6 forward and 1 reverse gears
Transfer box: Kirkstall AGB 42 2-speed with synchronised shift
Clutch: twin dry plate
Steering: recirculating ball with power-assistance
Turning circle: 21.5 m
Suspension:
(front) semi-elliptical taper leaf springs with telescopic shock-absorbers
(rear) six-leaf bogie-type with chevron-profile springs and telescopic shock-absorbers
Tyres: 15.50/80 × 20 radial
Electrical system: 24 V

Status
Production complete. In service with the British Army and Abu Dhabi.

Manufacturer
Marshall SPV Limited, Bedford Truck Division, Airport Works, Cambridge CB5 8RX, UK.
Tel: +44 1223 373068. Fax: +44 1223 373073.
Main production was carried out by Bedford Trucks at Dunstable.

VERIFIED

Bedford TM 30-30 and TM 30-43 (6 × 6) trucks

Description
The Bedford TM 30-30 and TM 30-43 (6 × 6) trucks were derived from the TM 6-6 truck (see previous entry) and were developed specifically for military off-road operations.

The layout of the TM 30-30 and TM 30-43 is conventional, with a forward control cab. Power from a 300 hp Caterpillar 10.5 litre (TM 30-30) or 14.6 litre (TM 30-43) turbocharged and after-cooled diesel is transmitted to the front and rear Kirkstall hub reduction axles via a synchromesh gearbox and single-speed transfer box on the TM 30-30 and an Allison automatic gearbox and two-speed transfer box on the TM 30-43. This drive line, plus cross- and inter-axle differential locks on the rear bogie, provides tractive effort for outstanding mobility under demanding conditions, with the capacity to tow guns or trailers weighing up to 20 tonnes.

The vehicles meet fuel and explosives carrying requirements and with specialised bodies can be used for the carriage of ammunition, fuel or water and the loading and unloading of palletised equipment.

Specifications
Cab seating: 1 + 1
Configuration: 6 × 6
Weights:
(design weight, front axle) 8,000 kg
(design weight, rear axle) 22,000 kg
(GVW) 30,000 kg
Body and payload capacity: approx 20,700 kg
Length: (chassis)
(TM 30-30) 8.91 m
(TM 30-43) 10.27 m
Width: (over tyres)
(TM 30-30) 2.47 m
(TM 30-43) 2.6 m
Height: (cab)
(TM 30-30) 3.1 m
(TM 30-43) 3.23 m
Ground clearance: 0.42 m
Track:
(front) 2 m
(rear) 2.1 m
Wheelbase: (TM 30-30/TM 30-43)
(centre of front axle to centre of rear bogie) 5 m/5.5 m
(between centres of rear axles) 1.61 m/1.42 m
Angle of approach/departure: 38°/30°
Max speed:
(TM 30-30) 90 km/h
(TM 30-43) 96 km/h
Fuel capacity: 2 × 340 litres
Gradient:
(TM 30-30) 50%
(TM 30-43 low ratio) 42%
(TM 30-43 high ratio) 86%
Engine: (TM 30-30) Caterpillar 3306B ATAAC 10.487 litre 6-cylinder turbocharged diesel developing 300 bhp at 2,100 rpm
(TM 30-43) Caterpillar 3406B ATAAC 14.6 litre turbocharged diesel
Gearbox: (TM 30-30) ZF synchromesh range change splitter giving 16 forward and 2 reverse gears
(TM 30-43) Allison CL(B) T754 automatic with TC 496 torque converter
Transfer box:
(TM 30-30) ZF A800 permanent 6 × 6 drive with lockable inter-axle differential
(TM 30-43) 6KN Kirkstall AGB 7000 Mark II pneumatically controlled from cab
Steering: recirculating ball with power-assistance
Suspension: leaf springs and hydraulic shock-absorbers front and rear
Tyres: 16 R 20
Brakes: air
Electrical system: 24 V
Batteries: 2 × 12 V, 120 Ah

Status
Production complete. In service with Abu Dhabi.

Manufacturer
Marshall SPV Limited, Bedford Truck Division, Airport Works, Cambridge CB5 8RX, UK.
Tel: +44 1223 373068. Fax: +44 1223 373073.
Main production was carried out by Bedford Trucks at Dunstable.

VERIFIED

Bedford TM 30-30 (6 × 6) truck in service with Abu Dhabi armed forces (T J Gander) ***1997***

Foden Low Mobility (8 × 4) and (6 × 4) Range of Vehicles

Development
The Foden Low Mobility range of vehicles was developed to meet a British Army requirement for a vehicle that would operate satisfactorily when loaded with its gross vehicle weight on both roads and unsurfaced tracks. The range was based on standard commercial components with emphasis on ease of maintenance. Deliveries to the British Army amounted to 1,275 units, of which over 60 per cent were delivered with left-hand drive for use by the British Army in Germany. A total of 40 more 8 × 4 tankers and 95 6 × 4 tankers were supplied fitted with GRP cabs and the Foden FF20 rubber rear suspension to update the overall specification.

Description
The vehicles are provided with a circular observation hatch in the centre of the cab roof. Standard cab equipment includes suspension seats, twin heaters/demisters, sun visors and a visible warning system for air pressure, water temperature level and oil pressure.

The cab has two stages of access for maintenance. The first, for daily and weekly servicing, involves lifting the front grille panel to gain access to the power steering and clutch reservoirs, engine oil filter and dipstick, windscreen washer bottle, wiper motor and linkage, cab heater system, air auxiliary feed connections and wiring harness and so on. The second stage involves using a hand-operated pump and ram assembly to tilt the cab forward to an angle of 68° to enable the engine to be removed without removing the cab. The cab can be removed as a complete unit if required, as quick-detachable multi-point electrical connections are provided.

The front axles on the 8 × 4 versions are rated at 6,604 kg; the rear axles have a capacity of 10,160 kg each, with single reduction hubs. A third differential is fitted to the foremost differential housing, with provision for a differential lock on both rear axles.

The rear bogie is designed to a rating of 19,500 kg and is fitted to Foden FF20 rubber suspension units with a roll control device and telescopic shock-absorbers.

Standard equipment on all models includes full width front bumper with central towing eye, rigid rear pulling jaw and lift and recovery lugs front and rear.

Variants
8 × 4 Cargo
A total of 800 was delivered. These can carry a maximum of 20,000 kg of cargo. The rear cargo body has drop sides, drop tailgate, removable bows and a canvas cover.

8 × 4 Tanker
A total of 256 was delivered. The chassis is fitted with a 22,500 litre tank body which has five compartments,

Foden (8 × 4) 16,000 kg Low Mobility cargo truck (Ministry of Defence)

Foden (8 × 4) Low Mobility tanker (T J Gander)
1997

each of which holds 4,500 litres of fuel. The tank was manufactured by Thompson Tankers of Bilston, Staffordshire.

8 × 4 Tipper
A total of 70 was delivered. These have a shorter wheelbase than the other two 8 × 4 vehicles and are fitted with an Edbro 11 m^3 all-steel end tipping body and an Edbro 6 NC single ram tip gear.

6 × 4 Tanker
A total of 183 was delivered; the tank has a capacity of 12,000 litres. The tank body was manufactured by Charles Roberts (Engineering) Limited of Wakefield, West Yorkshire.

Specifications

Type	**Cargo**	**Tanker**	**Tipper**	**Tanker**
Cab seating	1 + 2	1 + 2	1 + 2	1 + 2
Configuration	8 × 4	8 × 4	8 × 4	6 × 4
Weight (loaded)	29,553 kg	28,888 kg	29,705 kg	22,786 kg
(chassis and cab)	9,553 kg	9,553 kg	9,770 kg	9,030 kg
Weight on front axle (loaded)	10,000 kg	11,570 kg	10,659 kg	5,532 kg
Weight on rear axles (loaded)	19,553 kg	17,318 kg	19,046 kg	17,254 kg
Length	10.278 m	10.27 m	8.69 m	8.75 m
Width	2.497 m	2.502 m	2.497 m	2.497 m
Height	3.317 m	3.25 m	3.319 m	3.214 m
Wheelbase	1.372 m + 3.614 m + 1.516 m	1.372 m + 3.614 m + 1.516 m	1.372 m + 2.522 m + 1.516 m	3.97 m + 1.516 m
Angle of approach/departure	23°/23°	23°/23°	23°/23°	36°/28°
Max speed (road)	76 km/h	76 km/h	76 km/h	87 km/h
Fuel capacity	227 litres	227 litres	227 litres	227 litres
Max gradient	31.64%*	31.64%*	31.3%	28.6%
Engine	Rolls-Royce 220 Mk III 6-cylinder liquid-cooled diesel developing 220 bhp at 2,100 rpm			
Gearbox	Fuller with 9 forward and 1 reverse gears			
Steering	recirculatory ball with integral power assistance			
Turning radius	12.25 m	12.25 m	10.35 m	11.4 m
Suspension (front)	semi-elliptical springs with telescopic hydraulic dampers			
(rear)	Foden FF20 rubber			
Tyres	11.00 × 20	11.00 × 20	11.00 × 20	12.00 × 20 (front) 14.00 × 20 (rear)
Brakes (main)	2-line air, dual circuit			
(parking)	spring brake chambers on rear axles only. Air brake couplings front and rear for recovery			
Electrical system	24 V	24 V	24 V	24 V
Batteries	2 × 12 V	2 × 12 V	2 × 12 V	2 × 12 V

* At GVW of 24,000 kg

Status
Production complete. In service with the British Army.

Manufacturer
Foden Trucks, A Division of Paccar UK Ltd, Moss Lane, Sandbach, Cheshire CW11 9YW, UK.
Tel: +44 1270 763244. Fax: +44 1270 762758.

VERIFIED

Scammell Crusader (6 × 4) tractors

Description
The Scammell Crusader (6 × 4) tractor is basically a standard civilian vehicle adapted to meet military requirements. Two basic military models were produced, both of which are in service with the British Army. The first, known as the 20 tonne payload tractor, has a two-man cab and the second, the 35 tonne tractor, has a four-man cab with provision for two bunks.

20,000 kg (6 × 4) Tractor
The two-door forward control pressed steel cab is mounted on the chassis by two rubber-bushed trunnion mountings at the front and two coil springs with integral telescopic dampers at the rear. The military version is powered by a Rolls-Royce Eagle 305 Mk III

Scammell Crusader (6 × 4) 35,000 kg tractor towing RE semi-trailer

Scammell Crusader (6 × 4) 35,000 kg tractor (T J Gander)
1996

turbocharged diesel which develops 305 bhp at 2,100 rpm, coupled to a manual transmission with nine forward and two reverse gears.

35,000 kg (6 × 4) Tractor

The two-door forward control pressed steel cab has two individual seats at the front (one for the driver and one for a passenger) and two seats at the rear. The rear seats convert to bunks. The cab is mounted on the chassis by two rubber-bushed trunnion mountings at the front and two coil springs with integral telescopic dampers at the rear. The engine is the same as in the 20,000 kg model but is coupled to an RTO 915 manual gearbox with 15 forward and three reverse gears. Mounted to the rear of the cab is a Plummett capstan Model CA80 winch with 120 m of 16 mm cable which has a maximum capacity of 8,000 kg at a speed of 27.5 m/min. The winch can be used either to the front or rear and is fitted with an overload warning bell.

Late in 1977 the British Army ordered 130 Scammell Crusader (6 × 4) recovery vehicles. For details see the *Recovery vehicles* section.

Status

Production complete. In service with the British Army.

Manufacturer

Alvis Unipower Limited, 34 Greenhill Crescent, Watford Business Park, Watford, Hertfordshire WD1 8QU, UK.
Tel: +44 1923 816555.
Fax: +44 1923 228621.

VERIFIED

Specifications

Type	**20,000 kg**	**35,000 kg**
Cab seating	1 + 1	1 + 3
Configuration	6 × 4	6 × 4
Weight (empty)	9,200 kg	11,095 kg
Towed load	20,000 kg	35,000 kg
Length	6.66 m	6.66 m
Width	2.502 m	2.502 m
Height (overall)	3.3 m	3.3 m
(5th wheel)	1.412 m	1.549 m
Track (front)	2.05 m	2.05 m
(rear)	1.845 m	1.845 m
Wheelbase (1st axle to centre of rear bogie)	3.962 m	3.962 m
Max speed (road)	85 km/h	65 km/h
Range	500 km	500 km
Fuel capacity	318 litres	455 litres
Max gradient (stop and restart)	24%	20.9%
Engine	Rolls-Royce Eagle 305 Mk III turbocharged diesel developing 305 bhp at 2,100 rpm	
Gearbox	manual with 9 forward and 1 reverse gears	manual with 15 forward and 3 reverse gears
Clutch	twin dry plate	twin dry plate
Steering	ball and nut, power-assisted	
Turning radius	9.5 m	9.5 m
Suspension (front)	longitudinal semi-elliptical springs pivoted front with slipper rear ends, and telescopic shock-absorbers	
(rear)	fully articulated, inverted longitudinal semi-elliptical springs, trunnion-mounted at centre with slipper rear ends	
Tyres	11.00 × 20	11.00 × 20
Brakes	air on all axles incorporating 3-line brake system	
Electrical system	24 V	24 V
Batteries	4 × 12 V, 100 Ah	4 × 12 V, 100 Ah

Alvis Unipower M Series (8 × 8) high-mobility load carrier

Development

The Alvis Unipower M Series is a new generation of heavy logistic vehicles developed to meet the needs of armed forces into the 21st Century. The M Series ML8840 8 × 8 high-mobility load carrier is designed to carry up to 20 tons of combat supplies over terrain not normally accessible to wheeled vehicles and its stealth features reduce the risk of detection. It is suited to following tracked armoured vehicles at speed and undetected into forward areas for the rapid replenishment of combat supplies and, when fitted with 475/80 R 20 tyres, it is C-130 air-portable. The ML8840 can be equipped either as a conventional load carrier, or with a PLS/DROPS load handling system. There is also a 15,000 litre tactical refueller suited for helicopter refuelling; it is C-130 air-portable.

The ML8840 high-mobility load carrier shares a number of important features with the 8 × 8 BR90 bridging vehicle in service with the British Army. There were 139 chassis built for both the Close Support and General Support bridging roles.

Alvis Unipower M Series (8 × 8) high mobility load carrier carrying PLS/DROPS load handling system
1997

Description

The Alvis Unipower M Series (8 × 8) high-mobility load carrier is powered by an 11 litre 400 hp Cummins diesel engine coupled to a ZF six-speed fully automated transmission with an integral transfer box and hydrodynamic torque converter. The four GKN drive axles have both cross- and inter-axle differential locks. The rear suspension is a centrally pivoted spring pack providing equal axle loading over uneven ground and the front taper leaf springs provide a superior ride.

The engine-behind-cab configuration ensures that there is no engine intrusion into the cab, maximising cab space for personnel and their equipment. The cab is low, facilitating rapid entry and exit and ease of camouflage. The ergonomically designed interior minimises driver fatigue.

Standard tyres are 24 R 21. When fitted with 465/80 R 20 tyres the ML8840 is air-portable.

Alvis Unipower M Series (8 × 8) high mobility load carrier configured as 15,000 litre combat refueller
1997

Specifications

Cab seating: 1 + 2
Configuration: 8 × 8
Weights:
(1st and 2nd front axles) 9,500 kg
(1st and 2nd rear axles) 10,000 kg
(kerb) 15,600 kg
(GVW) 39,000 kg

Body and payload: 23,000 kg
Length: 10.5 m
Width: 3 m
Height: 2.78 m
Ground clearance: 0.33 m
Track: 2.8 m
Wheelbase: 6.4 m
Max speed: (laden) 84 km/h
Fuel capacity: 450 litres
Range: 1,000 km
Gradient: (restart, loaded) 40%
Engine: Cummins M11-T405E 11 litre 6-cylinder in-line turbocharged and after-cooled diesel developing 400 hp at 1,900 rpm
Transmission: ZF 6HP 900 fully automatic providing 6 forward and 1 reverse gears plus torque converter
Transfer box: ZF single-speed
Steering: ZF 8098 power-assisted
Suspension:
(front) two slipper-ended semi-elliptic parabolic springs per axle
(rear) two fully articulating semi-elliptical multileaf springs trunnion-mounted on bearings
Brakes: air to all axles
Tyres: 24 R 21 or 475/80 R 20
Electrical system: 24 V
Batteries: 2 × 12 V, 150 Ah
Alternator: 90 A

Status
In production. Tank Bridge Transporter (TBT) and Automatic Bridge Laying Equipment (ABLE) versions ordered for British Army.

Manufacturer
Alvis Unipower Limited, 34 Greenhill Crescent, Watford Business Park, Watford, Hertfordshire WD1 8QU, UK.
Tel: +44 1923 816555.
Fax: +44 1923 228621.

UPDATED

Leyland Trucks (8 × 6) DROPS logistic support vehicle

Development
Two Leyland Trucks logistic support vehicles, a 6 × 6 and an 8 × 6, were developed as part of the British Army's DROPS programme. The 8 × 6 version was selected in late 1986 as the DROPS medium mobility load carrier. Under a contract worth £150 million, 1,522 units were ordered in April 1989. The first vehicle came off the production line in November 1989 and the first examples entered British Army service in early 1990. The contract was completed in March 1994. Regular repeat orders have been placed by the British Army. Some 1,650 vehicles are in service throughout the world.

Description
The Leyland Trucks (8 × 6) logistic support vehicle can carry a 15,000 kg payload with a mobility equivalent to Medium Mobility Load Class (MMLC). It uses a forward control S26 cab with seating for the driver and a second crew member, with space for their kit. A hatch with a machine gun mounting is located in the roof and the cab roof can take the weight of two personnel.

The engine is a Perkins Eagle 350LM six-cylinder turbocharged four-stroke diesel developing 350 bhp. The transmission uses a five-speed fully automatic ZF 6 HP 600 gearbox with a lockable torque converter. Drive to the axles is via a single-speed auxiliary gearbox. Kirkstall axles are fitted.

The vehicle is fitted with the Multilift Mark 4 load handling system (see separate entry in *Materials handling equipment* section). To meet the requirements for United Nations Relief Operations in Bosnia, developments include armour protection on the cab, containerised troop support facilities and container handling equipment.

A total of 24 vehicles were delivered fitted with Rail Transfer Equipment (RTE).

Leyland Trucks (8 × 6) DROPS logistic support vehicle supplying water from a flatrack carrying a water tank ***1997***

Specifications
Cab seating: 1 + 1 or 2
Configuration: 8 × 6
Weight:
(kerb) 14,038 kg
(loaded) 32,000 kg
Max load: 15,000 kg (plus flatrack)
Towed load: 20,000 kg
Length: 9.11 m
Width: 2.5 m
Height: 3.18 m
Ground clearance: 0.29 m
Track: 1.99 m
Wheelbase: 5.54 m
Angle of approach/departure: 34°/38°
Max speed: 75 km/h
Range: 500 km
Fuel capacity: 272 litres
Max gradient: 61%
Fording: 0.75 m
Engine: Perkins Eagle 350LM 12.17 litre 6-cylinder turbocharged diesel developing 350 hp at 2,100 rpm
Gearbox: ZF 6 HP 600 with 6 forward and 1 reverse gears
Clutch: lockable torque converter
Transfer box: Leyland single-speed
Steering: ZF 8098 power-assisted
Turning radius: 12.5 m
Suspension: semi-elliptic 3-leaf springs with telescopic hydraulic shock-absorbers, double-acting on both front axles only
Tyres: 18 R 22.5 Michelin XL
Brakes:
(main) air, dual line, dual circuit
(parking) spring on 3 axles
Electrical system: 24 V
Batteries: 4 × 6TN
Alternator: 100 A

Status
Production for British Army complete. In service with the British Army and Malaysian Army units operating with the UN in Bosnia.

Manufacturer
Leyland Trucks, Croston Road, Leyland, Preston PR5 3LX, UK.
Tel: +44 1722 621400. Fax: +44 1772 625800.

VERIFIED

Foden (8 × 6) DROPS IMMLC logistic support truck

Development
Foden Trucks developed two logistic support trucks for evaluation by the Ministry of Defence in connection with the British Army's DROPS programme. Termed the Medium Mobility Load Carrier (MMLC) and Improved Medium Mobility Load Carrier (IMMLC) they completed concept field trials and the IMMLC was selected for production commencing in January 1994. Over 400 vehicles are involved in the £75 million four-year contract. In service the main role for the Foden DROPS vehicles is the supply of ammunition flatracks to the Royal Artillery's 155 mm AS90 batteries.

Description
The Foden (8 × 6) DROPS IMMLC logistic support truck uses a 10 tonne capacity forward steering drive axle, a second steer axle also of 10 tonne capacity and 20 tonne double drive rear bogies, all manufactured by

Foden (8 × 6) DROPS IMMLC logistic support truck ***1997***

GKN. They are powered by a Perkins (Shrewsbury) Eagle 350 MX diesel engine using a ZF 'Ecomat' HP 600 fully automatic six-speed transmission. The load handling system is the Multilift Mark 4 (see separate entry in *Materials handling equipment* section) using flatracks manufactured by Marshall of Cambridge that can carry up to 15,000 kg of palletised military stores.

The vehicle can tow, load and offload a custom-built drawbar trailer incorporating ISO twistlocks and designed to carry standard 20 ft containers or flatracks.

The IMMLC is built to a width of 2.9 m and is fitted with 20.5 × 25 R tyres. Wheelbase is 5.06 m and turning circle 22.4 m. Operational gross vehicle weight is 32,960 kg.

Status

In production for the British Army's DROPS programme (over 400).

Manufacturer

Foden Trucks, A Division of Paccar UK Limited, Moss Lane, Sandbach, Cheshire CW11 9YW, UK.
Tel: +44 1270 763244. Fax: +44 1270 762758.

UPDATED

UNITED STATES OF AMERICA

General Motors Commercial Utility Cargo Vehicles (CUCV)

Development

Starting in 1981 a 14 month programme known as the 'Special Analysis of Wheeled Vehicle Requirements' (usually abbreviated to just 'WHEELS') was carried out by the US Army Tank Automotive Command. Its objective was to find a commercial vehicle that could be easily procured for US Army use in areas where extreme environmental conditions would not be met and where expensive vehicles were not necessary. The objective was to replace 20 per cent of the M151 Jeeps in use and the bulk of the M880 series of pick-up trucks. The vehicle types chosen would also be used by the US Air Force and Marine Corps.

During the programme the US Army purchased 26 commercial vehicles and subjected them to exhaustive tests at the Aberdeen Proving Grounds, Maryland. The final choice was a General Motors design, the Model K which is one of the GM C/K series, over nine million of which were produced.

In July 1982, a contract was awarded to General Motors worth US$689 million for a total of 53,248 vehicles known as the Commercial Utility Cargo Vehicles (CUCV). First deliveries were made in August 1983 and the initial production programme lasted three years.

Projected procurement was for 57,349 CUCVs for all the armed services with 54,087 for the US Army. Total production at the end of 1986 was 70,889, including FMS. Production ceased in 1987.

The US Marine Corps received 3,159 CUCVs of the following types: M1008 - 2,194; M1009 - 166; M1010 - 290; M1031 - 95; M1028 - 414.

Description

The CUCV fleet comprises the following five types of vehicle:

Type A, Truck, Utility, Tactical ¾ ton, (4 × 4), M1009
Type B, Truck, Cargo, Tactical 1¼ ton, (4 × 4), M1008
Type C, Truck, Ambulance, Tactical 1¼ ton, (4 × 4), M1010
Type D, Truck, Chassis, Tactical 1¼ ton, (4 × 4), M1031
Type E, Truck, Cargo, Shelter Carrier, Tactical 1¼ ton, (4 × 4), M1028

All five types are based on a common power train, but utilise different body and/or chassis variations to fulfil a variety of purposes, and have varying payloads to suit their role. All five types have the 6.2 litre diesel engine, automatic transmission and two-speed transfer box. The rear and front axles are the same in all types apart from the Type A. The same 28 V electrical system is used in all types apart from the Type C ambulance which has a system with doubled power.

The five types are all commercial models with the following alterations incorporated: blackout lights; camouflage paint (three colour, NATO); engine diagnostic connector assembly; military markings, including a removable red cross for the Type C ambulance; NBC warfare protection and kit provisions; rear pintle hook; slave-start capability; towing capability; weapon holders; and a winterisation kit to allow operations down to −46°C.

The Type C ambulance version has provision for four stretcher patients or eight seated casualties. An air filter system is also provided along with air conditioning and extra lighting, including spotlights. The Type E cargo, shelter carrier is equipped with shelter tiedown brackets.

All five types are air-transportable.

Specifications

Type	A	B	C	D	E
Function	utility	cargo	ambulance	truck	cargo, shelter carrier
Model	M1009	M1008	M1010	M1031	M1028
GM model number	K10516 Blazer	K30903 Pick-up	K30903 Chassis/cab	K30903 Chassis/cab	K30903 Pick-up
Cab seating	1 + 1 + 3	3	1 + 1	3	3
Configuration	4 × 4	4 × 4	4 × 4	4 × 4	4 × 4
Weight (loaded)	2,903 kg	3,992 kg	4,287 kg	4,334 kg	4,264 kg
Max payload	544 kg	1,315 kg	943 kg	1,792 kg	1,633 kg
Length	4.873 m	5.607 m	5.784 m	5.408 m	5.607 m
Width	2.022 m	2.062 m	2.062 m	2.062 m	2.062 m
Height (overall)	1.905 m	1.915 m	2.581 m	1.938 m	1.92 m
Track (front)	1.485 m	1.488 m	1.488 m	1.488 m	1.488 m
(rear)	1.382 m	1.438 m	1.438 m	1.438 m	1.438 m
Wheelbase	2.705 m	3.34 m	3.34 m	3.34 m	3.34 m
Fording	0.508 m	0.508 m	0.508 m	0.508 m	0.508 m
Engine	General Motors 6.2 litre V-8 diesel developing 135 hp at 3,600 rpm				
Gearbox	General Motors THM-400 3-speed automatic with 1 reverse				
Transfer box	2-speed	2-speed	2-speed	2-speed	2-speed
Suspension (front)	tapered leaf springs with 1,021 kg capacity				
(rear)	semi-elliptical multileaf springs with 850, 1,588, 1,588, 1,701, 1,701 kg capacity respectively				
Tyres	10.00 × 15	9.50 × 16.5	9.50 × 16.5	9.50 × 16.5	9.50 × 16.5
Electrical system	28 V, 100 Ah	28 V, 100 Ah	28 V, 200 Ah	28 V, 100 Ah	28 V, 100 Ah

Status

Production complete. There were 70,889 units delivered to all branches of the US armed forces. The total includes FMS to nations such as Columbia, Grenada, Honduras, Israel, Jamaica, Liberia (M1009), Panama and Taiwan.

Manufacturer

General Motors Truck and Bus Group, Flint, Michigan, USA.
General Motors Corporation, Military Vehicles Operation, 2619 Product Lane (Suite 130), Rochester Hills, Michigan 48309-3807, USA.
Tel: +1 810 853 3873. Fax: +1 810 853 3876.

VERIFIED

Truck, Utility, Tactical ¾ ton, (4 × 4), M1009 (Michael Jerchel)

1996

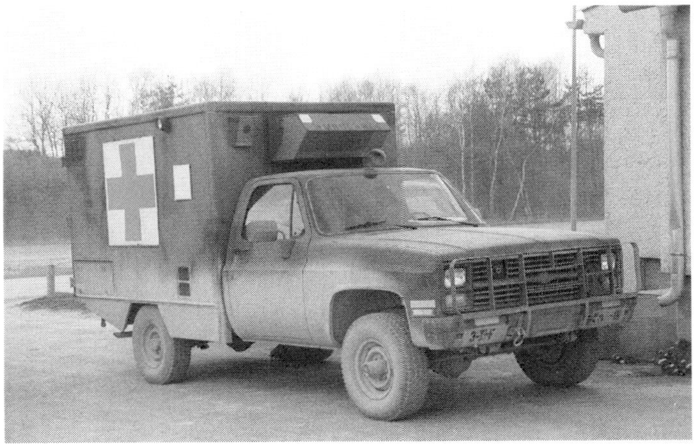

Truck, Ambulance, Tactical 1¼ ton, (4 × 4), M1010 (Michael Jerchel)

General Motors Commercial Utility Cargo Vehicles II (CUCV II)

Description

Following the completion of production of the original CUCV range (see previous entry), General Motors Corporation, Military Vehicles Operation, introduced a new range of economical tactical support vehicles known as the CUCV II. The CUCV II range follows the same general lines as the original CUCV range and similar basic General Motors commercial models are involved, but updated and revised to the latest commercial and military specifications. The CUCV II range includes petrol as well as diesel engined options.

There are four basic models in the CUCV II range, they are:

The Type A Utility based on the K10516 Yukon and generally similar to the M1009

The Type B Truck Cargo based on the K20903 Pick-up and generally similar to the M1008

The Type C Truck Ambulance based on the K30903 Chassis/cab and generally similar to the M1010

The Tactical Truck, 1¼ ton based on the K20903 Chassis/cab intended for special operations, security patrols and similar duties. Also known as the Special Operations Vehicle (SOV).

As yet the US Army has not fielded CUCV II vehicles but Types B and C have been purchased by TACOM for Foreign Military Sales (FMS).

The military standard options available for CUCV II vary from model to model but include a 24 V electrical system, blackout lighting, a battery booster connector, a radiator brush guard, front and rear tiedowns, a trailer pintle hook and electrical connector, weapon racks and similar specialised items.

The base engine for the CUCV II range is a GM 6.5 litre V-8 petrol developing 190 bhp. Also available is a 5.7 litre petrol developing 245 bhp or a 7.4 litre V-8 petrol developing 290 bhp. A further option is provided by a 6.5 litre diesel developing 150 bhp. All engines can be coupled to General Motors automatic and manual transmissions.

One closely allied vehicle to the CUCV II range is the General Motors Pursuit Vehicle. Based on the K20906 Suburban chassis this hard topped (4 × 4) vehicle can carry a 12.7 mm machine gun over a round hatch in the roof. A number of these vehicles have reportedly been sold to countries in the Middle East. Also based on the K20906 Suburban chassis is the Personnel Carrier ¾ ton (4 × 4), with seating for a driver and five passengers, and a Truck Utility, 1 ton, (4 × 4) Security Vehicle for the command, control and personnel transport roles.

Specifications

Type	**A**	**B**	**C**	**SOV**
Function	utility	carrier	ambulance	tactical truck
GM model number	K10516	K20903	K30903	K20903
Cab seating	2 + 3	3 + 8	3 + 4	1 + 2
Configuration	4 × 4	4 × 4	4 × 4	4 × 4
Weight (loaded)	2,835 kg	4,172 kg	4,173 kg	4,172 kg
Payload	475 kg	1,701 kg	907 kg	1,663 kg
Length	4.902 m	5.669 m	5.9 m	n/avail
Width	1.957 m	1.95 m	1.961 m	n/avail
Height (overall)	1.869 m	1.93 m	2.713 m	n/avail
Ground clearance	0.229 m	0.206 m	0.206 m	0.206 m
Wheelbase	2.832 m	3.34 m	3.34 m	3.34 m
Max speed	88 km/h	88 km/h	88 km/h	88 km/h
Range	400 km	400 km	400 km	400 km
Fording	0.508 m	0.508 m	0.508 m	0.508 m
Engine	GM 6.5 litre V-8 petrol developing 190 hp at 3,400 rpm			
Transmission	GM THM 4L60E automatic with 4 forward and 1 reverse gears			
Transfer box	2-speed	2-speed	2-speed	2-speed
Suspension (front)	independent heavy-duty torsion bar with shock-absorbers			
(rear)	two-stage semi-elliptical leaf springs with shock-absorbers			
Tyres	16 × 6.5	16 × 6.5	16 × 6.5	16 × 6.5
Electrical system	24 V	24 V	24 V	24 V

Status

In production.

Manufacturer

General Motors Corporation, Military Vehicles Operation, 2619 Product Lane (Suite 130), Rochester Hills, Michigan 48309-3807, USA.
Tel: +1 810 853 3873. Fax: +1 810 853 3876.

UPDATED

CUCV II Type A, Truck Utility **1997**

CUCV II Type B, Truck Cargo (Scott Gourley) **1997**

CUCV II Type C, Truck Ambulance **1997**

General Motors Suburban Personnel Carrier **1997**

AM 715 Series (4 × 4) 1¼ ton trucks

Description

When production of the M715 (4 × 4) 1¼ ton truck was completed for the US Army in 1969, AM General developed another version for export called the AM 715. In appearance this was almost identical to the earlier M715 but with a more powerful engine and other small modifications. The chassis was designated the AM 724.

The layout of the AM 715 is conventional, with the engine at the front, two-door cab with a removable canvas top, windscreen which folds forward onto the bonnet and side windows which can be wound down in the centre; and the cargo area at the rear with a drop tailgate, stake-type sides, removable bows and a tarpaulin cover. A fold-up bench seat, capable of seating five fully-equipped personnel, is provided down each side of the cargo area. The AM 715 can also carry a standard S-250 military shelter and can be used as a command or communications vehicle.

The AM 725 is an ambulance that can carry five stretcher patients plus an attendant. The AM 715C is a weapon carrier (for example the TOW ATGW), while the AM 715S is a field service vehicle.

Specifications

AM 715; data in square brackets relates to AM 725 where different

Cab seating: 1 + 2
Configuration: 4 × 4
Weight:
(empty) 2,132 [2,676] kg
(loaded) 3,537 [3,583] kg
Weight on front axle: (loaded) 1,361 [1,383] kg
Weight on rear axle: (loaded) 2,177 [2,200] kg
Max load: 1,134 [726] kg
Towed load: 1,288 kg [n/a]
Load area: 2.355 × 1.626 m [n/a]
Length: 5.321 [5.359] m
Width: 2.032 [2.134] m
Height:
(reduced) 1.36 m [n/a]
(overall) 2.286 [2.451] m
(load area) 0.838 m [n/a]
Ground clearance: (axles) 0.22 [0.21] m
Track:
(front) 1.641 [1.626] m
(rear) 1.674 [1.661] m
Wheelbase: 3.327 [3.319] m
Angle of approach/departure: 41°[32°]/25°
Max speed: (road) 97 km/h
Range: 322 km
Fuel capacity: 72 litres
Max gradient: 60%
Fording: 0.457 m
Engine: Model 6-258 6-cylinder in-line OHV petrol developing 150 hp at 3,800 rpm
Gearbox: manual, with 4 forward and 1 reverse gears
Clutch: single plate
Transfer box: 2-speed
Steering: recirculating ball
Turning radius: 7.11 [7.32] m
Suspension: horizontal semi-elliptical springs with hydraulic shock-absorbers
Tyres: 7.50 × 16
Brakes: front, power; rear, drum (vacuum-assisted)
Electrical system: 12 V
Battery: 1 × 12 V, 63 Ah

Status

Production complete.

Manufacturer

AM General Corporation, 105 North Niles Avenue, South Bend, Indiana 46617, USA.
Tel: +1 219 284 2942. Fax: +1 219 284 2959.

VERIFIED

AM General AM 715 (4 × 4) 1¼ ton cargo truck

Jeep AM 720 (4 × 4) 1,135 kg light truck

Description

The Jeep AM 720 (4 × 4) 1,135 kg light truck is an updated version of the AM 715 (see previous entry), in that it resembles the general layout and appearance. The AM 720 is based on the commercial J-20 truck and is available with two main types of cargo body known as the Military Cargo Box or the Townside Cargo Box, with the latter intended mainly for the civilian market. Either can carry up to 13 personnel seated on folding benches and can be covered by a vinyl top resting on bows. The cab also has a vinyl top.

Numerous options are available for the AM 720, including various wheelbase lengths and left- or right-hand drive. An electric winch with a capacity of 2,832 kg can be fitted and a 24 V conversion kit is available.

This vehicle is licence-produced in Egypt - see separate entry in this section.

Specifications

Cab seating: 1 + 2 (up to 13 in rear)
Configuration: 4 × 4
Weight: (off road)
(empty) 2,175 kg
(loaded) 3,814 kg
Max load:
(off road) 1,135 kg
(on road) 1,639 kg
Towed load: 3,629 kg
Length: 5.36 m
Width: 2.004 m
Height:
(with cargo top) 2.286 m
(cargo bed) 0.84 m
Ground clearance: 0.208 m
Track:
(front) 1.626 m
(rear) 1.661 m
Wheelbase: (standard) 3.322 m
Angle of approach/departure: 35°/21°
Fuel capacity: 69 litres
Fording: 0.457 m
Engine: Model 258 CID 4.2 litre 6-cylinder OHV developing 112 hp at 3,000 rpm or 5.9 litre 8-cylinder developing 144 hp at 3,200 rpm
Gearbox: manual, 4 forward and 1 reverse gears
Transfer box: 2-speed
Clutch: single dry disc
Steering: variable ratio power-assisted
Turning radius: 6.78 m
Suspension: longitudinal leaf springs with severe-use shock-absorbers
Tyres: 9.50 × 16.5 or 9.00 × 16
Brakes: power, disc front; drum rear
Electrical system: 12 V (24 V optional)

Status

In production. In widespread service.

Manufacturer

AM General Corporation, 105 North Niles Avenue, South Bend, Indiana 46617, USA.
Tel: +1 219 284 2942. Fax: +1 219 284 2959.

VERIFIED

Jeep AM 720 (4 × 4) 1,135 kg light truck

Jeep AM 720 (4 × 4) 1,135 kg light truck showing Military Cargo Box body

AM General HMMWV M998 series multipurpose wheeled vehicles

Development

Based on the draft specification for the High-Mobility Multipurpose Wheeled Vehicle (HMMWV) issued by the US Army in mid-1979, AM General Corporation designed and built a prototype in the weapons carrier configuration.

The first prototype was completed in August 1980 and was sent to the Nevada Automotive Test Center for extensive trials and by February 1981 the prototype had accumulated 21,000 km of instrumented and dynamic testing.

AM General became one of three contenders awarded a US Army contract for the design and construction of 11 prototype HMMWVs (six weapons carriers and five utility), these were delivered in May 1982. In March 1983 AM General was awarded a US$59.8 million contract by the US Army Tank Automotive Command (TACOM) for 2,334 HMMWVs, which were then designated the M998 Series (unofficially HUMMER). This was the first increment in a five-year contract for 54,973 vehicles worth approximately US$1.2 billion. Of these, approximately 39,000 were for the US Army and the remainder divided between the US Air Force, Navy and Marine Corps (initially 3,123). Production commenced at Mishawaka, Indiana, early in 1985. Contract options for a further 15,000 were exercised to bring the total production in mid-1991 to over 72,000 vehicles, including overseas sales.

In August 1989, the US Army awarded AM General a further multiyear contract worth approximately US$1 billion. The contract called for a further 33,331 vehicles until 1993, with two further option years. Production under the new contract began in January 1990 and production for the US armed forces continued until March 1995, by which time they had manufactured approximately 100,000 HMMWVs. A further 20,000 units have been ordered by 30 foreign governments since 1991, including 2,300 units purchased by Saudi Arabia in mid-1991 at a cost of US$123 million. Total production to date is well over 140,000.

In service the M998 Series HMMWVs replaced some M151 Jeeps, the M274 Mule (830 in service), the M561/M792 Gama Goat (11,000 in service) and the M880 series (40,000 in service), with 20 per cent of the fleet of M151s and many of the M880s being replaced by the Commercial Utility Cargo Vehicle (CUCV).

Description

The HMMWV has either 2 × 2 or 4 × 4 seating (depending on the model) on each side of the drive train which is in a midship position allowing the front differential to be raised. This, together with the geared hubs, provides a ground clearance of 0.41 m. The location of the crew on each side of the drive train also allows a low centre of gravity. The windshield frame is strong enough to serve as a rollbar and support for various equipment kits. Other pillars also make the ballistically protected weapon station inherently strong and a steady location on which to mount a variety of weapons such as TOW, 7.62 and 12.7 mm machine guns and the MK19 MOD3 40 mm grenade launcher. At the rear the cargo bed is large enough to accommodate a S-250 or similar shelter without overhang; it also accommodates the Standard Integrated Command Post Systems (SICPS) shelter, with a 15 in (381 mm) overhang, and all Mobile Subscriber Equipment (MSE) shelters.

To reduce life cycle and initial procurement costs, standard automotive components are used wherever possible, as in the engine, transmission, transfer case, brakes and steering.

The independent suspension, front and rear, gives good manoeuvrability, ease of handling and part commonality. The geared hubs give 0.41 m ground clearance (0.38 m on the later, heavier models) incorporating raised axles for a 1.92:1 torque output multiplication at the ground.

The suspended carrier front and rear axles are identical, have differentials and are mounted high-directly in the chassis frame. The front propeller shaft has double cardan joints and the rear propeller shaft has single cardan joints which, according to AM General, give minimal motion, improved torque characteristics and higher reliability with resultant lower support costs. Vehicle handling is enhanced by the front stabiliser bar being attached to the lower control arms and pivot bracket reducing shock from the lower A-frame member to the chassis.

Acceleration of the HMMWV is such that it can move from a standstill to 48 km/h in 8 seconds and from a standstill to 80 km/h in 24 seconds.

Three HMMWVs can be carried in a C-130 Hercules transport aircraft, six in a C-141B and 15 in a C-5A Galaxy.

The basic HMMWV can be converted into numerous variants by changing the body configuration. These configurations are:

M998 Cargo/troop Carrier without winch
M1038 Cargo/troop Carrier with winch
M966 TOW Missile Carrier, basic armour, without winch
M1036 TOW Missile Carrier, basic armour, with winch
M1045 TOW Missile Carrier, supplemental armour, without winch
M1046 TOW Missile Carrier, supplemental armour, with winch
M1025 Armament Carrier, basic armour, without winch
M1026 Armament Carrier, basic armour, with winch
M1043 Armament Carrier, supplemental armour, without winch
M1044 Armament Carrier, supplemental armour, with winch
M997 Maxi-ambulance, four-litter, basic armour
M1035 Soft top Ambulance, two-litter
M1037 Shelter Carrier, without winch
M1042 Shelter Carrier, with winch.

In addition to the above, selected application kits are produced as follows:

Cargo/troop Carrier, soft top enclosure (two-door cab) for M998
Cargo/troop Carrier, soft top enclosure (four-door cab) for M998
Cargo/troop Carrier, soft top enclosure (two-door cab, troop seats) for M1038
Cargo/troop Carrier, soft top enclosure (two-door cab, troop/cargo) for M1038
Cargo/troop Carrier, soft top enclosure (four-door cab, cargo) for M1038
Armament Carrier, basic armour with M60 7.62 mm machine gun for M1026
Armament Carrier, supplemental armour with M2 0.50/12.7 mm machine gun for M1044
Armament Carrier, supplemental armour with MK19 MOD3 grenade launcher for M1043
Retrofitable weapon station kit for the M998/M1038 (weight 114 kg).

The HMMWV has been adapted for use as a missile launch vehicle. Launch vehicles produced to date include the following:

Boeing Avenger Pedestal-Mounted Stinger (PMS) carrying eight Stinger surface-to-air missiles and one 12.7 mm machine gun. This became operational with the US Army in 1989-90 and the total requirement is for 273 over a five-year period. For full details refer to *Jane's Land-based Air Defence 1995-96* pages 173 to 176.

Boeing/Matra Guardian air defence system, a variant of the Avenger (see above) carrying Mistral missiles. For full details refer to *Jane's Land-based Air Defence 1995-96* pages 161 and 162.

Shorts/Boeing Starstreak Avenger, a further variant of the Avenger (see above) carrying Starsteak missiles. For full details refer to *Jane's Land-based Air Defence 1995-96* page 162.

An HMMWV was used for firing trials involving the Hellfire anti-tank guided weapon in a surface-to-surface role.

HMMWVs have been adapted to carry the M40A2 106 mm recoilless rifle for overseas sales.

An HMMWV chassis is used as the basis for the Swiss MOWAG Eagle armoured reconnaissance vehicle and the Turkish Otokar Cobra light armoured vehicle.

In addition to the above-mentioned variants, AM General has also produced the M1097 'Heavy Hummer' Variant (HHV) to increase the GVW to 4,536 kg - all existing HMMWV versions can be produced to this level, known as the Heavy Weapons Carrier or Heavy Armament Carrier. The 'heavy' chassis permits a payload increase to 1,996 kg in the case of the M1097 - GVW is 4,536 kg. The M1097 chassis incorporates improved front and rear differentials, a new transfer case, new front and rear propshafts, an improved frame mounting for the steering gear, variable rate rear springs and new lower ball joints. The

Common ground station (CGS) shelter, the ground component of Joint STARS (surveillance target attack radar system) mounted on a M1097A2 HMMWV
1997

HMMWV M1026A1 Armament Carrier

heavier chassis allows the use of an up-armoured ballistic panel protection kit with 5.56 and 7.62 mm NATO ball protection and permits the carriage of heavy weapons such as the McDonnell Douglas Helicopters 30 mm ASP-30 cannon and Giat Industries 20 and 30 mm cannon systems. Production of the HHV commenced in September 1992.

Modification kits developed by and available directly from AM General include: a selective up-armour kit; a brushguard; spare tyre and jerrican carriers; a driveline skid protection kit; a CTIS and a special desert operations package including secondary oil and fuel filtration; sealed dipsticks; a constant drive fan and enhanced oil filtration. A retrofitable weapon station kit is an adaptation of that used on standard HMMWV weapon carriers (M998 and M1038 only) and weighs 114 kg.

The M1097 HHV was used as the basis for the Hummer Cab-Over-Engine (COE) (4 × 4) 2,495 kg cargo truck - see following entry.

Civilian and public utility versions of the HMMWV are available. Production of the civilian HUMMER® began in 1992 and an extensive dealer network has been established in the USA as well as distributors in Canada, the Middle East, Asia and Europe. Applications for the civilian HUMMER include a backhoe, manlift platform, stake bed, utility box body or it can be equipped as a fire/rescue vehicle. The 1995 civilian HUMMER can be equipped with a General Motors 6.5 litre naturally aspirated diesel engine or a 5.7 litre petol engine. Both are coupled to a General Motors 4L80-E four-speed automatic transmission.

M998A1 series

In late 1993 AM General began producing the M998A1 series of HMMWV for the US Army. All A1 models incorporate the M1097 'Heavy Hummer' chassis components plus new front seats, an improved parking brake lever with safety release, a metal hood grill, improved slave receptacle, solid-state glow plug controller, modified rifle mounts, and upgraded rear half-shafts. Using these common chassis components enhances HMMWV standardisation across all models, leading to improvements in logistic support, training and fleet durability.

With the introduction of the A1 series the number of models was reduced to the following:

M998A1 Cargo/troop Carrier without winch
M1038A1 Cargo/troop Carrier without winch
M966A1 TOW Missile Carrier, basic armour, without winch
M1045A1 TOW Missile Carrier, supplemental armour, without winch
M1046A1 TOW Missile Carrier, supplemental armour, with winch
M1025A1 Armament Carrier, basic armour, without winch
M1026A1 Armament Carrier, basic armour, with winch
M1043A1 Armament Carrier, supplemental armour, without winch
M1044A1 Armament Carrier, supplemental armour, with winch
M997A1 Maxi-ambulance, four-litter, basic armour
M1035A1 Soft top ambulance, two-litter
M1097A1 Heavy HUMMER Variant (HHV).

GVW for all A1 models, except the M1097A1, increased by approximately 81 kg although vehicles payloads and loads remained the same.

AM General offered a new front seat retrofit kit for the existing HMMWV fleet as part of the series production.

Production of the M998A1 series ceased in 1995.

Specifications

M1038A1 Cargo/troop Carrier without winch
Cab seating: 1 + 1 (other models 1 + 3)
Configuration: 4 × 4
Weight:
(kerb) 2,544 kg
(GVW) 3,574 kg
Max load: 1,077 kg
Max towed load: 1,542 kg
Length: 4.72 m
Width: (mirrors folded) 2.18 m
Height: 1.83 m
Ground clearance: 0.41 m
Track: 1.82 m
Wheelbase: 3.3 m
Angle of approach/departure: 47°/45°
Max speed: 113 km/h
Range: 482 km
Fuel capacity: 94.6 litres
Max gradient: 60%
Side slope: 40%
Fording: 0.76 m
(with preparation) 1.52 m
Engine: GM V-8 6.2 litre diesel developing 150 hp at 3,600 rpm
Transmission: automatic with 3 forward and 1 reverse gears
Transfer box: 2-speed, full-time 4-wheel drive
Suspension: (front and rear) independent, double A-arm, coil spring
Steering: power-assisted
Turning radius: 7.62 m
Brakes: hydraulic disc front and rear
Tyres: 37 × 12.50 R 16.5 LT load range D
Electrical system: 24 V
Batteries: 2 × 12 V

M998A2 series

AM General started developing the M998A2 series of the HMMWV for the US Army in early 1993. Using the M1097A1 HHV as a baseline, the new series incorporates a EPA approved General Motors 6.5 litre naturally aspirated diesel engine generating 160 hp, with an electronically controlled four-speed transmission. Other enhancements designed to improve performance, vehicle maintenance and user comfort include increased payload, increased cargo bed tie-down capacity, improved heating and ventilation, new rear high-back seats, self-cancelling turn signals, an improved steering wheel and column, LED side marker lights, and revised transport tie-down provisions. All M998A2 vehicles are ready to accept AM General's Central Tyre Inflation System (CTIS) as a field installed item.

M1114 Heavy HMMWV armoured by the O'Gara, Hess and Eisenhardt Armoring Company ***1997***

HMMWV M1038 cargo/troop carrier (Michael Jerchel) ***1996***

The A2 Series covers six models, all without winch, including:
M1097A2 HHV (replacing the M998A1 Cargo/troop Carrier)
M1025A2 Armament Carrier, basic armour
M1043A2 Armament Carrier, supplemental armour
M1045A2 TOW Missile Carrier, supplemental armour
M997A2 Maxi-ambulance, four-litter, basic armour
M1035A2 Soft top Ambulance, two-litter.

The M1097A2 is available either as a base platform, as a shelter carrier, as a cargo/troop carrier with optional hard top, or as a carrier/prime mover with optional hard top and troop seats. Also available are a light wrecker and a special operations model available in several configurations.

In February 1996 a right-hand drive version of the M998A2 was introduced.

All models are rated at 4,672 kg GVW, except the M997A2 which retains its current 4,345 kg GVW.

US Army pilot vehicles were produced in mid-1995 and series production began in September 1995. By September 1996 the US Army Tank-automative and Armaments Command (TACOM) had ordered 3,093 M998A2 series HMMWVs worth US$161 million.

Specifications

M1097A2 Cargo/Troop/Shelter Carrier
Cab seating: 1 + 1 (other models 1 + 3)
Configuration: 4 × 4
Weight:
(kerb) 2,676 kg
(GVW) 4,672 kg
Max load: 1,996 kg
Max towed load: 1,905 kg
Length: (without winch) 4.84 m
Width: (mirrors folded) 2.18 m
Height: 1.83 m
Ground clearance: 0.38 m
Track: 1.82 m
Wheelbase: 3.3 m
Angle of approach/departure:
(without winch) 63°/33°
(with winch) 46°/33°
Max speed: 113 km/h
Range: 640 km
Fuel capacity: 94.6 litres
Max gradient: 60%
Side slope: 40%
Fording: 0.76 m (1.52 m with kit)
(with preparation) 1.52 m
Engine: GM V-8 6.5 litre diesel developing 160 hp at 3,400 rpm
Transmission: automatic with 4 forward and 1 reverse gears
Transfer box: 2-speed, full-time 4-wheel drive
Suspension: (front and rear) independent, double A-arm, coil spring
Steering: power-assisted
Turning radius: 7.62 m
Brakes: hydraulic disc front and rear
Tyres: 37 × 12.50 R 16.5 LT load range D
Electrical system: 12/24 V waterproof
Batteries: 2 × 12 V
Alternator: 100 or 200 A

M1113 Expanded Capacity Vehicle (ECV)

AM General, under contract to the US Army, developed a HMMWV with a greater payload capacity known as the M1113 Expanded Capacity Vehicle, or ECV. The ECV provides the payload capacity allowing for larger and heavier communications shelters, and improved armour protection level for scouts, military police, security police, and explosive ordnance disposal platforms.

To meet the desired performance and mobility characteristics the ECV is provided with a General Motors 6.5 litre 190 hp turbo diesel engine coupled to a General Motors 4L80E four-speed automatic transmission. The ECV utilises many of the improved components developed for the Cab-Over-Engine (HMMWV) truck (see following entry for details). These include modified differentials, improved half-shafts, a new exhaust system, improved brake and cooling systems, an improved suspension system, a modified steering system, upgraded wheels and a reinforced frame.

To meet an urgent requirement, the US Army began procurement of the XM1109 Up-Armored Heavy HMMWV (UA-HHV) in mid-1993. Based on the M1097/M1079A1 HHV, the XM1109 provides the crew compartment with 360° 7.62 mm armour-piercing perimeter protection, as well as artillery airburst and underbody blast protection. The payload of the XM1109 is approximately 590 kg.

During 1995 production of the M1114 based on the improved ECV chassis began. The M1114 meets US Army and Air Force requirements for a scout, military police, security police, and explosive ordnance disposal vehicle with improved ballistic protection levels. The M1114 provides full 360° protection against 7.62 mm armour-piercing projectiles as well as protection against 155 mm artillery airbursts and 5.5 kg anti-tank mine blast. It provides the same 1,016 kg payload as the M1025 models which it will replace for many applications. The M1114 is equipped with a Central Tyre Inflation System (CTIS) and air-conditioning. It accepts all weapons currently used with the M1025/M1025A1. Kerb weight is 4,447 kg and payload 1,043 kg.

US Army pilot vehicles were produced in mid-1995. Production began in late 1995. The first overseas country to purchase the M1114 was Luxembourg. In June 1996 the US Army purchased 390 M1114s for operations in Bosnia with follow-on orders anticipated. In 1996 40 were ordered for the Qatari Police.

Armour for the M1114 is manufactured by O'Gara-Hess & Eisenhardt Armoring Company of Fairfield, Ohio. They also produce two and four door armour and blast kits for retrofitting to M998 or M1097 models and a four door armour and blast kit for the M1025.

AM General have attained ISO 9002 registration for the manufacture of all HMMWV variants.

Specifications

M1113
Cab seating: 1 + 1 (other models 1 + 3)
Configuration: 4 × 4
Weight:
(kerb) 2,858 kg
(GVW) 5,216 kg
Max load: 2,359 kg
Max towed load: 2,041 kg
Length: (with winch) 4.99 m
Width: (mirrors folded) 2.18 m
Height: 1.83 m
Ground clearance: 0.38 m
Track: 1.82 m
Wheelbase: 3.3 m
Angle of approach/departure: 46°/33°
Max speed: 113 km/h
Range: 380 km
Fuel capacity: 94.6 litres
Max gradient: 40%
Side slope: 40%
Fording: 0.76 m
(with preparation) 1.52 m
Engine: GM V-8 6.5 litre turbo diesel developing 190 hp at 3,400 rpm
Transmission: GM 4L80E automatic with 4 forward and 1 reverse gears
Transfer box: 2-speed, full-time 4-wheel drive
Suspension: (front and rear) independent, double A-arm, coil spring
Steering: power-assisted
Turning radius: 7.62 m
Brakes: hydraulic disc front and rear
Tyres: 37 × 12.50 R 16.5 LT load range D
Electrical system: 12/24 V waterproof
Batteries: 2 × 12 V

Status

M998A2 series and ECV in production. Earlier models are no longer in production.

In service with US Army, Air Force, Navy and Marine Corps (at the end of FY93 the US Army had 89,486 HMMWVs and the US Marine Corps 19,598). Also in service with Abu Dhabi (4), Algeria (1), Bahrain (277), Bolivia (30), Chad (20), China (5), Columbia (86), Djibouti (18), Ecuador (95), Egypt (681), Honduras (125), Israel (16), Japan (5), Jordan (129), Kuwait (770), Luxembourg (69 plus M1114), Mexico (1,838), Philippines (440), Qatar (40), Saudi Arabia (3,829), Switzerland (5), Taiwan (5,429), Thailand (275), Tunisia (341), Turkey (2) and Venezuela (5). (Figures relate to totals as at August 1993.)

Civilianised versions have been sold to the US Border Patrol and other civilian agencies.

In 1991 AM General announced it would sell a civilian HUMMER® to the general public. Production began in 1992.

Manufacturer

AM General Corporation, 105 North Niles Avenue, South Bend, Indiana 46617, USA.
Tel: +1 219 284 2942. Fax: +1 219 284 2959.

UPDATED

AM General Hummer Cab-Over-Engine (4 × 4) 2,495 kg cargo truck

Development

Using many of the components common to the HMMWV A2 series (see previous entry), the Hummer Cab-Over-Engine (COE) (4 × 4) 2,495 kg cargo truck was developed to meet possible US Military and friendly foreign requirements for a lightweight highly mobile and heavy capacity truck. The first two prototypes were produced and tested by AM General during 1990. Of the nine subsequent prototypes, six underwent full US government testing and two were tested by the Belgian Army as part of their VTMTT programme.

Description

The Hummer COE utilises the basic steel box frame chassis of the M1097 HHV allied to a new forward control three-man cab. The full 4 × 4 drive configuration and front and rear independent suspension of the HMMWV are also retained. The main change is a new transfer case with power take-off, and a General Motors 6.5 litre 190 hp turbo diesel with catalyst, common to the ECV.

The forward control cab is high and distinctive in appearance and utilises a soft top and canvas doors. The location of the cab is such that a load area measuring (externally) 3.18 × 2.16 m is available at the rear. This load area may be used for a cargo body with drop sides, a flatbed capable of carrying shelters and/or containers, or up to 10 personnel on outward-looking (back-to-back) troop seats. A low bed version with the same cargo bed as the HHV can carry Standard Integrated Command Post (SICP) or other shelters, plus a generator.

Optional equipment includes a central tyre inflation system and a 3,629 kg capacity hydraulic winch weighing, in kit form, approximately 136 kg.

The vehicle may be air-transported in C-130, C-141B and C-5A aircraft.

Hummer Cab-Over-Engine (COE) (4 × 4) 2,495 kg truck

Specifications

Cab seating: 1 + 2
Configuration: 4 × 4
Weight:
(kerb) 3,175 kg
(GVW) 5,670 kg
Max load: 2,495 kg
Towed load: 1,905 kg
Load area: (high bed version) 3.06 × 2.02 m
Length:
(low bed) 5.06 m
(high bed) 5.22 m
Width: 2.16 m
Height: (top of exhaust) 2.6 m
Ground clearance: 0.39 m
Wheelbase: 3.3 m
Angle of approach/departure: 47°/33.5°
Max speed: 125 km/h
Range: 483 km
Fuel capacity: 182 litres
Gradient: 60%
Fording:
(standard) 0.76 m
(optional) 1.5 m
Engine: GM 6.5 litre V-6 turbo diesel developing 190 hp
Gearbox: automatic with 4 forward and 1 reverse gears
Transfer box: 2-speed
Steering: power-assisted
Suspension: (front and rear) independent, double A-arm, coil spring
Tyres: 37 × 12.50 R 16.5 LT load range D
Brakes: hydraulic, disc front and rear
Electrical system: 12/24 V
Batteries: 2 × 12 V
Alternator: 100 or 200 A

Status

Prototypes.

Manufacturer

AM General Corporation, 105 North Niles Avenue, South Bend, Indiana 46617, USA.
Tel: +1 219 284 2942. Fax: +1 219 284 2959.

UPDATED

Stewart & Stevenson Family of Medium Tactical Vehicles (FMTV)

Development

In October 1991, the US Army Tank-Automotove Command announced that Stewart & Stevenson Services Inc had been awarded a five year US$1.2 billion multiyear contract to produce 10,843 Family of Medium Tactical Vehicles (FMTV). These Stewart & Stevenson vehicles met or surpassed all US Army requirements for its 21st Century Medium Tactical Fleet.

The Austrian Steyr 12 M 18 truck (for details see entry under Austria in this section) was used as the basis for the Stewart & Stevenson FMTV.

Produced in Stewart & Stevenson's dedicated Tactical Vehicles Systems Division plant in Sealy, Texas, the FMTV is in service with the US Army and is in full rate production. During acceptance trials the FMTV achieved the highest acceptance test scoring of any wheeled vehicle at the US Army's proving grounds, scoring 337 per cent (fleet average) above requirements for reliability, availability and maintainability.

Complementing the FMTV vehicles are 2.5 and 5 ton tactical trailers. Details of these can be found in the *Trailers* section of this book.

Stewart & Stevenson have associated with Perry Engineering of Adelaide, South Australia, for the Australian Army's Bushmaster armoured infantry mobility vehicle. The Stewart & Stevenson LMTV chassis is used as the basis for this project.

A MTV (6 × 6) chassis is used as the launcher/carrier for the proposed High Mobility Artillery Rocket System (HIMARS). For details refer to *Jane's Armour and Artillery 1995-96* pages 729 and 730.

Description

The FMTV is produced in two versions, the Light Medium Tactical Vehicle (LMTV) 2½ ton (4 × 4) and the Medium Tactical Vehicle (MTV) 5 ton (6 × 6). There is over 85 per cent commonality of components and assemblies between the two vehicles, including the same Caterpillar 3116 ATAAC diesel engine, rated at 225 hp for the LMTV and 290 hp for the MTV. The

Stewart & Stevenson 2½ ton (4 × 4) LMTV in M1078 standard cargo truck form ***1996***

Stewart & Stevenson 5 ton (6 × 6) MTV in M1084 standard cargo truck with crane form (T J Gander) ***1996***

Stewart & Stevenson 5 ton (6 × 6) MTV in M1086 long cargo with crane form

Stewart & Stevenson 5 ton (6 × 6) MTV in M1087 expansible van form (T J Gander) **1996**

power pack was designed for easy and rapid removal and replacement. Other systems shared by the two vehicles include: the transmission, fan system, air system and air compressor, wheels and tyres, starter motor, alternator, front suspension and axle, cab, steering system, exhaust and intake systems, fuel system and cooling.

The basic overall layout of the LMTV and MTV is similar for both vehicles, with a forward control cab over the engine. The stamped panel steel cab can be tilted forward for power train maintenance access and has space for the driver and two passengers, with enough internal space to allow all three occupants to wear and exchange arctic or NBC clothing. Internal space is also provided for chemical detection equipment and crew equipment. The cab has been tested for roll-over and crash resistance and has a high-strength roof structure to support a ring-type machine gun mount. All FMTV vehicles share the same waterproof instrument and control panel. A spare tyre is mounted behind the cab.

The Allison MD-D7 automatic seven-speed transmission is coupled to an Allison full-time all-wheel drive transfer case with a torque bias inter-axle differential. There are traction differentials on front and rear axles. The 'smooth ride' suspension system allows over 254 mm of spring travel. On the MTV the rear suspension uses a combination of bogie and walking beam elements. Radial steel-belted tyres are used with a Central Tyre Inflation System (CTIS). This system has settings for roads, cross-country, air transport, sand/

Specifications - LMTV Series

Model	**M1078**	**M1079**	**M1080**	**M1081**
Type	standard cargo	van	chassis	cargo-airdrop (LVAD)
Cab seating	1 + 2	1 + 2	1 + 2	1 + 2
Configuration	4 × 4	4 × 4	4 × 4	4 × 4
Weight (kerb)	7,484 kg	8,156 kg	6,150 kg	7,803 kg
Max load	2,268 kg	2,268 kg	n/a	2,268 kg
Towed load	4,700 kg	4,700 kg	4,700 kg	4,700 kg
Length	6.429 m	6.701 m	6.429 m	6.429 m
Width	2.438 m	2.438 m	2.438 m	2.438 m
Height (overall)	2.845 m	2.845 m	2.845 m	2.845 m
(air transport)	2.68 m	2.68 m	2.68 m	2.68 m
Ground clearance (loaded)	0.559 m	0.559 m	0.559 m	0.559 m
Wheelbase	3.9 m	3.9 m	3.9 m	3.9 m
Angle of approach/departure	40°/40°	40°/40°	40°/40°	40°/40°
Max speed	94 km/h	94 km/h	94 km/h	94 km/h
Range (max, loaded)	645 km	645 km	645 km	645 km
Max gradient	60%	60%	60%	60%
Fording (with preparation)	1.524 m	1.524 m	1.524 m	1.524 m
(without preparation)	0.914 m	0.914 m	0.914 m	0.914 m
Engine	Caterpillar 3116 ATAAC 6.6 litre 6-cylinder turbocharged and after-cooled diesel developing 225 hp at 2,600 rpm			
Transmission	Allison MD-D7 automatic 7-speed			
Transfer case	Allison single-speed integral to transmission with planetary inter-axle differential (lockable)			
Steering	power	power	power	power
Suspension	parabolic tapered leaf springs			
Tyres	395 R 20 XML	395 R 20 XML	395 R 20 XML	395 R 20 XML
Electrical system	12/24 V	12/24 V	12/24 V	12/24 V
Alternator	100 A	100 A	100 A	100 A

Specifications - 5 ton MTV

Model	**M1083**	**M1084**	**M1085**	**M1086**	**M1087**	**M1088**	**M1089**
Type	standard cargo	cargo/crane	LWB cargo	LWB cargo/crane	expansible van	tractor	wrecker
Cab seating	1 + 2	1 + 2	1 + 2	1 + 2	1 + 2	1 + 2	1 + 2
Configuration	6 × 6	6 × 6	6 × 6	6 × 6	6 × 6	6 × 6	6 × 6
Weight (kerb)	8,889 kg	10,740 kg	9,451 kg	11,198 kg	12,994 kg	8,455 kg	15,682 kg
Payload	4,536 kg	4,536 kg	4,536 kg	4,536 kg	2,268 kg	11,340 kg[1]	4,990 kg[2]
Towed load	9,526 kg	9,526 kg	9,526 kg	9,526 kg	9,526 kg	27,538 kg[3]	16,709 kg
Length	6.935 m	7.768 m	8.863 m	9.674 m	9.927 m	7.145 m	9.143 m
Width	2.438 m	2.438 m	2.438 m	2.438 m	2.438 m	2.438 m	2.438 m
Height (overall)	2.845 m	2.845 m	2.845 m	2.845 m	3.684 m	2.845 m	2.845 m
(air transport)	2.680 m	2.680 m	2.680 m	2.680 m	2.680 m	2.680 m	2.680 m
Ground clearance (loaded)	0.559 m	0.559 m	0.559 m	0.559 m	0.559 m	0.559 m	0.559 m
Wheelbase	4.1 m	4.5 m	4.5 m	5.5 m	5.5 m	4.1 m	5.3 m
Angle of approach/departure	40°/60°	40°/36°	40°/20°	40°/21°	40°/22°	40°/44°	40°/40°
Max speed	94 km/h	94 km/h	94 km/h	94 km/h	94 km/h	94 km/h	94 km/h
Range (max, loaded)	483+ km	483+ km	483+ km	483+ km	483+ km	483+ km	483+ km
Max gradient	60%	60%	60%	60%	60%	60%	60%
Fording (with kit)	1.524 m	1.524 m	1.524 m	1.524 m	n/a	1.524 m	1.524 m
(without kit)	0.914 m	0.914 m	0.914 m	0.914 m	0.914 m	0.914 m	0.914 m
Engine	Caterpillar 3116 ATAAC 6.6 litre 6-cylinder turbocharged and after-cooled diesel developing 290 hp at 2,600 rpm						
Transmission	Alison MD-D7 automatic 7-speed						
Transfer case	Allison single-speed intergral to transmission with planetary inter-axle differential (lockable)						
Steering	power	power	power	power	power	power	power
Suspension	parabolic tapered leaf springs						
Tyres	395 R 20 XML	395 R 20 XML	395 R 20 XML	395 R 20 XML	395 R 20 XML	395 R 20 XML	395 R 20 XML
Electrical system	12/24 V	12/24 V	12/24 V	12/24 V	12/24 V	12/24 V	12/24 V
Alternator	100 A	100 A	100 A	100 A	100 A	100 A	100 A

[1] 5th wheel load
[2] max load on lift and tow stinger
[3] 27,538 kg on-road/20,867 kg off-road

Stewart & Stevenson 5 ton (6 × 6) MTV in M1088 tractor truck form

Stewart & Stevenson 5 ton (6 × 6) MTV in M1091 fuel tanker form

Specifications - 5 ton MTV Series

Model	**M1090**	**M1091**	**M1092**	**M1093**	**M1094**	**M1096**
Type	dump	1,500 gallon (US) fuel tanker	chassis	cargo-airdrop (LVAD)	dump-airdrop (LVAD)	chassis 4,500 mm
Cab seating	1 + 2	1 + 2	1 + 2	1 + 2	1 + 2	1 + 2
Configuration	6 × 6	6 × 6	6 × 6	6 × 6	6 × 6	6 × 6
Weight (kerb)	9,969 kg	10,974 kg	7,515 kg	9,257 kg	10,203 kg	7,754 kg
Payload	4,536 kg	4,797 kg (fuel)	n/a	4,536 kg	4,536 kg	n/a
Towed load	9,526 kg	n/a	9,526 kg	9,526 kg	9,526 kg	9,526 kg
Length	7.182 m	8.054 m	6.935 m	6.935 m	7.182 m	8.840 m
Width	2.438 m	2.438 m	2.438 m	2.438 m	2.438 m	2.438 m
Height (overall)	2.845 m	2.845 m	2.845 m	2.845 m	2.845 m	2.845 m
(air transport)	2.680 m	2.680 m	2.680 m	2.680 m	2.680 m	2.680 m
Ground clearance (loaded)	0.559 m	0.559 m	0.559 m	0.559 m	0.559 m	0.559 m
Wheelbase	4.1 m	4.5 m	4.1 m	4.1 m	4.1 m	4.5 m
Angle of approach/departure	40°/44°	40°/40°	40°/60°	40°/60°	40°/44°	40°/20°
Max speed	94 km/h	94 km/h	94 km/h	94 km/h	94 km/h	94 km/h
Range (max, loaded)	483+ km	483+ km	483+ km	483+ km	483+ km	483+ km
Max gradient	60%	60%	60%	60%	60%	60%
Fording (with kit)	1.524 m	1.524 m	1.524 m	1.524 m	1.524 m	1.524 m
(without kit)	0.914 m	0.914 m	0.914 m	0.914 m	0.914 m	0.914 m
Engine	Caterpillar 3118 ATAAC 6.6 litres 6-cylinder turbocharged and after-cooled diesel developing 290 hp at 2,600 rpm					
Transmission	Allison MD-D7 automatic 7-speed					
Transfer case	Allison single-speed intergral to transmission with planetary inter axle differential (lockable)					
Steering	power	power	power	power	power	power
Suspension	parabolic tapered lead springs					
Tyres	395 R 20 XML	395 R 20 XML	395 R 20 XML	395 R 20 XML	395 R 20 XML	395 R 20 XML
Electrical system	12/24 V	12/24 V	12/24 V	12/24 V	12/24 V	12/24 V
Alternator	100 A	100 A	100 A	100 A	100 A	100 A

mud/snow and emergency. The controls for the CTIS are mounted in the cab.

On both versions the front axle is a Rockwell Model R-611 Type 1. The LMTV rear axle is a Rockwell Model R-611 Type 2; those for the MTV are a Type 3 and a Type 4. The wrecker and tractor truck versions of the MTV have the Type 5 and Type 6 rear axles.

For helicopter or sling loading the FMTV vehicles make use of a special sliding outrigger system that allows the vehicle to retain the full use of available cargo space and eliminates sling contact with the cargo drop sides. For air transport two vehicles, LMTV or MTV, can be carried without preparation inside a C-130 and four in a C-141B.

FMTV production models are as follows:

M1078 - LMTV Standard Cargo
M1079 - LMTV Van
M1080 - LMTV Chassis
M1081 - LMTV Cargo-airdrop LVAD
M1082 - LMTV Trailer
M1083 - MTV Standard Cargo
M1084 - MTV Cargo with Crane
M1085 - MTV Long Wheelbase Cargo

Prototype of Stewart & Stevenson MTV/PLS with a Bennes Marrel Ampliroll palletised load system (T J Gander)
1997

Stewart & Stevenson 5 ton (6 × 6) MTV in M1090 dump truck form

Stewart & Stevenson 2½ ton (4 × 4) LMTV in M1079 van form

M1086 - MTV Long Wheelbase Cargo with Crane
M1087 - MTV Expansible Van
M1088 - MTV Tractor Truck
M1089 - MTV Wrecker
M1090 - MTV Dump
M1091 - MTV 1,500 gallon Fuel Tanker
M1092 - MTV Chassis
M1093 - MTV Cargo-airdrop LVAD
M1094 - MTV Dump-airdrop LVAD
M1095 - MTV Trailer
M1096 - MTV Long Wheelbase (4.5 m) Chassis.

Cargo versions without a dedicated crane may be fitted with an optional electrical loading/unloading crane capable of lifting 680 kg at 1.8 m.

In addition to the above models, in 1996 Stewart & Stevenson introduced the MTV/PLS as a private venture. The MTV/PLS is a standard MTV 6 × 6 chassis modified to accept a Bennes-Marrel Ampliroll 8-ton palletised load system (PLS). The MTV/PLS has a gross vehicle weight of 17,710 kg and a payload of 7,530 kg with a flatrack.

LMTV variants can be fitted with a P-10J winch with a 4,990 kg line pull. MTV variants use a DP-515 winch with a 7,031 kg line pull.

Status

In series production for US Army.

Manufacturer

Stewart & Stevenson Services Inc, Tactical Vehicle Systems Division, 5000 I-10 West Sealy, Texas 77474, USA.
Tel: +1 713 856 7040. Fax: +1 409 885 7910.

UPDATED

M35/M44A2 (6 × 6) 2½ ton cargo truck series

Development

In the late 1940s Reo and the Truck and Bus Division of General Motors Corporation each developed a new 2½ ton (6 × 6) truck for the US Army to replace wartime vehicles. Reo was awarded the initial production contract for 5,000 vehicles and delivered the first vehicle in 1950. Originally it was to have been only an interim solution pending large-scale production of the General Motors design, but as soon as the Korean War broke out it was apparent that Reo alone could not meet the requirements of the Army so the General Motors models were placed in immediate production. They were the M135 with single rear wheels and the M211 with dual rear wheels, but they were phased out of production after the end of the Korean War in favour of the Reo design which was also built by Studebaker and was commonly known as the Eager Beaver.

The first vehicles were powered by a Reo (Model OA-331) or Continental (COA-331) petrol engine which developed 146 bhp at 3,400 rpm, but later models with the suffix A1 (for example M35A1) were powered by a Continental LDS 427-2 multifuel engine. Current models (for example M35A2) have the LD 465-1 multifuel engine which develops 140 bhp (gross) at 2,600 rpm; this engine is manufactured by the Hercules Engine Company.

In 1964 the Kaiser Jeep Corporation bought the Studebaker facilities in South Bend, Indiana, and was awarded contracts to build both 2½ ton (6 × 6) and 5 ton (6 × 6) trucks for the US Army.

In 1967 Kaiser Jeep formed the Defense and Government Products Division to handle its government contracts, but in 1970 Kaiser Jeep was acquired by American Motors and the Defense and Government Products Division was named the General Products Division of the Jeep Corporation. In 1971 it was renamed the AM General Corporation, then a wholly owned subsidiary of American Motors Corporation.

By early 1980 AM General had produced over 150,000 M35/M44 series 6 × 6 trucks. A product-improved prototype designated the M963 series was developed by the company under contract to the US Army but it did not enter production.

The M963 series was powered by a Caterpillar Model 3208 V-8 diesel developing 210 hp coupled to an Allison MT643 four-speed automatic transmission. New axles gave the vehicle a wider track and larger tyres improved soft soil mobility, allowing single instead of the usual dual rear wheels to be fitted. Other improvements included redesigned suspension, brakes and steering, a forward-tilting bonnet for easier maintenance and a wider three-man cab with a spring-mounted seat for the driver.

By 1988 M35/M44 production was being carried out by AM General Corporation. An order for a further 399 M44A2s was placed for delivery by October 1988. In September 1988 it was announced that AM General would discontinue medium and heavy truck production to concentrate on production of the HMMWV series. The South Bend plant closed in late 1989.

In October 1990 it was announced that the Southeastern Equipment Company Inc of Augusta, Georgia, had been awarded a US$10,908,791 contract to rebuild 285 M35A2C trucks.

The latest 'rebuild' configuration of this series is the M35A3 Extended Service Program - see below for details.

Description

In the 1950s there were two basic models in this series: the M34 with single rear wheels and the M35 with dual rear wheels. The former still remains in service, especially in Europe, Asia and South America.

The layout of the basic cargo model is conventional, with the engine at the front; two-door cab in the centre with a windscreen that can be folded forward onto the bonnet and a removable canvas top; and the cargo area at the rear with a drop tailgate, removable bows, tarpaulin cover and troop seats down either side.

A wide range of kits was available including A-frame, alcohol evaporator, arctic, cargo body closure (arctic), central troop seats, electric brakes, fording, hard top for cab, hoist and rail for installation in rear of vehicle for carrying out minor repair work, ring mount for machine gun over cab, hot water personnel heater, fuel-burning personnel heater, power plant heater, slave cable and a thermal barrier. Many vehicles are fitted with a 4,536 kg capacity winch that can be used to the front or rear of the vehicle, and has 61 m of 13 mm diameter cable with two speeds forward and one speed in reverse.

Variants

Models with single rear wheels

M34: cargo truck with an empty weight of 5,332 kg, chassis designated M44
M47: dump truck with an empty weight of 6,100 kg, chassis designated M57
Truck, maintenance: earth-boring machine and pole setter V18A/MTQ, chassis M44
Truck, maintenance: telephone construction and maintenance V17A/MTQ, chassis M44.

Models with dual rear wheels

M35: this is the basic member of the series and is based on the M45 chassis; the final production model was designated the M35A2. The M35A2C (chassis M46A2C) is almost identical but has drop sides. Both drop sides and tailgate of this model are interchangeable with those of the M54A2C (5 ton) truck.
M36: this cargo truck is the long wheelbase version of the M35 and its chassis is designated the M36. Final production model was the M36A2
M48: tractor truck for towing semi-trailers, chassis designation M45
M49: fuel tanker, final model the M49A2C (chassis M45A2), has an all-steel tank which holds 4,542 litres of fuel for road travel or 2,271 litres of fuel for cross-country travel. Fuel can be dispensed by gravity or pumped under pressure at a maximum rate of 303 litres/min and the pump can also be used to refill the tank. The tanker can be fitted with bows and a tarpaulin cover to make detection more difficult
M50: water tanker, final model the M50A3, has an aluminium tank which holds 3,785 litres of water for road travel or 1,893 litres of water for cross-country travel. Water can be dispensed by gravity or pumped under pressure at a maximum rate of 303 litres/min and the pump can also be used to refill the tank. The tanker can be fitted with bows and a tarpaulin cover to make detection more difficult
M59: dump truck, chassis designation M58
M60: light wrecker, details of which will be found in the *Recovery vehicles* section
M108: wrecker, details of which will be found in the *Recovery vehicles* section
M109: shop van, chassis designated M45, final production model the M109A3, has a fully enclosed rear body and is used for a variety of roles such as workshop and maintenance
M132: medical van
M185: repair van, M45 chassis; late models were M185A3
M275: tractor truck for towing semi-trailers weighing up to 16,329 kg on roads and up to 7,712 kg across country. Late production model was M275A2
M292: shop van with extensible sides
M342: dump truck, final production model was the M342A2 (chassis M45A2G) which can carry 1.9 m³ of soil
M756A2: pipeline maintenance truck (chassis M45A2), fitted with rear winch, PTO and rear- or side-mounted A-frame, removable cargo rack and sides
M763: telephone maintenance truck
M764: truck, maintenance, earth-boring machine and pole setter, equipped with rear winch, PTO, boring machine. Pole setting is accomplished using derrick tube of boring machine and rear winch.

M50A3 water tanker version of M35A2 2½ ton (6 × 6) truck

M36A2 long wheelbase version of M35A2 2½ ton (6 × 6) truck

Canadian Forces M35 CDN 2½ ton (6 × 6) cargo/personnel truck with winch (Michael Jerchel) **1996**

M35A3 2½ ton (6 × 6) cargo/personnel truck following the Extended Service Program (ESP) incorporation (T J Gander) **1997**

M35A3 2½ ton Truck Extended Service Program (ESP)

In December 1990 the US Army Tank Automotive Command (TACOM) issued a requirement for an Extended Service Program (ESP) involving M44A2 series vehicles. The intention was to overhaul a portion of the US Army's 2½ ton fleet and equip them with new EPA certified diesel engines and matching automatic transmissions, Super Single radial tyres and a central tyre inflation system, upgraded steering, FMVSS braking, seat belts, plus other major improvements. In addition to extending vehicle service life by a further 16 years, the ESP will improve fuel economy, enhance off-road performance and reduce servicing costs and exhaust emissions. The resulting savings in operating and support costs will pay for the vehicle in less than half of the vehicle's useful life.

Prototype ESP contracts were awarded to AM General Corporation and Cummins Military Systems Company on 8 May 1992. Each company provided eight vehicles: four with new engines, manual transmissions and other improvements (these acted as baseline vehicles); and four enhanced vehicles with new engines, automatic transmissions and Central Tyre Inflation Systems (CTIS). These vehicles were delivered to the US Army for preproduction qualification test in October 1992. The tests were completed in June 1993, during which time the US Army decided not to test the baseline vehicles and settled on the enhanced versions. The production solicitation was issued in February 1993 with the production contract awarded in September 1993 for the remanufacture of 2,483 to 8,407 A2 standard trucks to A3 standard over five years. AM General was awarded the ESP contract worth US$150 million for an initial 2,483 vehicles (M35A3 cargo, M35A3C drop side and M36A3 long bed cargo) in September 1993. By mid 1996 520 options had been exercised so that by September 1996 the US Army had ordered 3,488 remanufactured M35A3 trucks worth US$201.6 million. The ESP covers Continental US Army Reserve and National Guard vehicles only.

Government Initial Production Testing (IPT) was completed in June 1995. Full rate production was achieved in late 1995, and fielding began in early 1996.

The ESP includes the installation of a Caterpillar 3116 diesel engine, an Allison automatic transmission, a hydraulic winch and a central tyre inflation system. Other changes incoude rebuilt brakes, a power-assisted steering system, a new exhaust system, revised electrics, a new fuel system and tank, a new radiator and a new battery. Three-point seat belts are also included.

If required, the components involved in the ESP can be provided by AM General in the form of kits, with full instructions, for in-country installation.

Specifications

M35A3 Cargo ESP
Cab seating: 1 + 2
Configuration: 6 × 6
Weight:
(kerb) 6,504 kg
(GVW) 8,774 kg
Max load: 2,270 kg
Towed load: 2,724 kg
Load area: 3.7 × 2.2 m
Length: (without winch) 6.713 m
Width: 2.438 m
Height:
(overall) 2.868 m
(reduced) 2.088 m

Specifications

Model	**M35A2**	**M36A2**	**M49A2C**	**M50A3**	**M109A3**	**M342A2**
Type	cargo	cargo	fuel	water	van	dump
Cab seating	1 + 2	1 + 2	1 + 2	1 + 2	1 + 2	1 + 2
Configuration	6 × 6	6 × 6	6 × 6	6 × 6	6 × 6	6 × 6
Weight (empty)	5,900 kg	6,900 kg	6,500 kg	6,644 kg	6,800 kg	6,800 kg
(loaded, road)	10,400 kg	11,500 kg	10,100 kg	10,408 kg	10,200 kg	11,300 kg
Weight on front axle (loaded)	2,700 kg	3,200 kg	2,700 kg	2,880 kg	3,000 kg	3,100 kg
Weight on rear axle (loaded)	7,700 kg	8,300 kg	7,400 kg	7,528 kg	7,100 kg	8,300 kg
Max load (road)	4,535 kg	4,535 kg	4,542 litres	3,785 litres	3,401 kg	4,535 kg
(cross-country)	2,268 kg	2,268 kg	2,271 litres	1,893 litres	2,268 kg	2,268 kg
Towed load (road)	4,535 kg	4,535 kg	4,535 kg	4,535 kg	3,628 kg	4,535 kg
(cross-country)	2,721 kg	2,721 kg	2,721 kg	2,721 kg	2,721 kg	2,721 kg
Load area	3.7 × 2.2 m	5.3 × 2.2 m	n/a	n/a	n/a	3.3 × 1.9 m
Length (without winch)	6.7 m	8.4 m	6.7 m	6.7 m	6.8 m	6.6 m
Width	2.4 m	2.4 m	2.4 m	2.4 m	2.4 m	2.4 m
Height (reduced)	2.1 m	2.1 m	2.3 m	2.4 m	3.3 m	2.1 m
(overall)	2.9 m	3.2 m	2.6 m	2.6 m	3.3 m	2.7 m
(load area)	1.32 m	1.4 m	n/a	n/a	1.3 m	1.3 m
Ground clearance (axles)	0.28 m	0.28 m	0.28 m	0.28 m	0.28 m	0.28 m
Track (front/rear)	1.721/1.778 m	1.721/1.778 m	1.721/1.778 m	1.721/1.778 m	1.721/1.778 m	1.721/1.778 m
Wheelbase	3.912 m	4.826 m	3.912 m	3.912 m	3.912 m	3.912 m
Angle of approach/departure	47°/40°	47°/24°	47°/40°	47°/40°	47°/40°	47°/70°
Max speed (road)	90 km/h	90 km/h	90 km/h	90 km/h	90 km/h	90 km/h
Range	480 km	480 km	480 km	480 km	480 km	480 km
Fuel capacity	189 litres	189 litres	189 litres	189 litres	189 litres	189 litres
Max gradient	60%	60%	60%	60%	60%	60%
Fording (without preparation)	0.76 m	0.76 m	0.76 m	0.76 m	0.76 m	0.76 m
(with preparation)	1.98 m	1.98 m	1.98 m	1.98 m	1.52 m	1.83 m
Engine	LDT-465-1C 6-cylinder in-line multifuel diesel developing 140 net hp at 2,600 rpm					
Gearbox	all have a manual gearbox with 5 forward and 1 reverse gears					
Clutch	single dry plate is fitted to all models					
Transfer box	2-speed	2-speed	2-speed	2-speed	2-speed	2-speed
Steering	all have cam and twin lever type					
Turning radius	11 m	13.7 m	11 m	11 m	11 m	11 m
Suspension (front/rear)	semi-elliptical springs/semi-elliptical springs inverted					
Tyres	9.00 × 20	9.00 × 20	9.00 × 20	9.00 × 20	9.00 × 20	9.00 × 20
Brakes (main)	air over hydraulic	air over hydraulic	air over hydraulic	air over hydraulic	air over hydraulic	air over hydraulic
(parking)	internal/external	internal/external	internal/external	internal/external	interna /external	internal/external
Electrical system	24 V	24 V	24 V	24 V	24 V	24 V
Batteries	2	2	2	2	2	2

Ground clearance: 0.3 m
Wheelbase: 3.911 m
Track:
(front) 1.722 m
(rear) 1.778 m
Angle of approach/departure: 51°/41°
Max speed: 82 km/h
Range: 644 km
Fuel capacity: 189 litres
Gradient: 60%
Fording:
(without kit) 0.762 m
(with kit) 1.828 m
Engine: Caterpillar 3116 ATAAC diesel developing 170 hp at 2,700 rpm
Transmission: Allison AT-1545P automatic with 4 forward and 1 reverse gears
Transfer box: Rockwell T-136-27 2-speed
Steering: Sycon, cam and lever type with air assist
Turning radius: 11 m
Suspension: semi-elliptical springs
Tyres: 14.5 R 20
Brakes: dual circuit air/hydraulic
Electrical system: 24 V
Batteries: 2 × 12 V 6TL
Alternator: 60 A

Status
Production complete, with ESP in progress. In service with the armed forces of the USA and many other countries including Bolivia, Brazil, Canada, Chad, El Salvador, Guatemala, Haiti, Honduras, Israel, Djibouti, South Korea, Lebanon, Liberia, Morocco, Pakistan, Panama, Philippines, Saudi Arabia, Senegal, Somalia, Spain (over 2,000 of which were re-engined), Sudan, Taiwan, Thailand, Turkey and Zaïre. An undisclosed number were supplied to the People's Republic of China.

Manufacturer
AM General Corporation, 105 North Niles Avenue, South Bend, Indiana 46617, USA.
Tel: +1 219 284 2942. Fax: +1 219 284 2959.

UPDATED

GM Commercial Enhanced-Mobility Medium-Duty Truck (CEMMV)

Description
The GM Commercial Enhanced-Mobility Medium-Duty Truck (CEMMV) has the General Motors Corporation truck model number GMC C7H042. It is a 3 ton capacity (4 × 4) vehicle using various components of commercial origin. The cab and bonnet are adaptations of commercial components, as is much of the other bodywork. Seating for the two-door cab is a three-passenger bench seat, although bucket seats, with suspension under the driver's seat, are available as options. The tapered leaf and multileaf suspension have been strengthened and heavy-duty shock-absorbers have been added.

The engine is a 6.6 litre Caterpillar 3116 diesel, similar to the unit fitted to the FMTV and ESP 2½ ton trucks. It is coupled to an Allison four-speed automatic transmission and a two-speed transfer case. All terrain tyres are provided for off-road traction.

The standard military pattern cargo body is provided with troop seats and there is also a cargo cover. Tie-down shackles are provided front and rear. Other standard equipment includes an Eaton tyre pressure system with four inflation settings selected from within the cab, and a 4,990 kg capacity self-recovery winch mounted centrally. A fully swivelling towing pintle is provided along with NATO 7- and 12-pin sockets. A spare wheel is located behind the cab together with a crane handling system. A brush guard is positioned in front of the radiator grill.

Optional equipment includes cab air conditioning, an adjustable tilt steering wheel, and dual circuit power steering.

Variant
6 × 6 CEMMV
The 6 × 6 CEMMV is essentially a lengthened version of the 4 × 4 CEMMV and has the GM truck model number GMC C7H064. It has a payload, with crew, of 5,442 kg and a GVW of 15,873 kg. The load area measures 4.3 × 2.4 m. Wheelbase is 4.978 m.

4 × 4 version of GM Commercial Enhanced-Mobility Medium-Duty Truck with 3 tonne payload **1996**

Specifications
Cab seating: 1 + 2
Configuration: 4 × 4
Weight: (GVW) 10,431 kg
Max load: 3,000 kg
Load area: 3.6 × 2.4 m
Wheelbase: 4.978 m
Fuel capacity: 189 litres
Engine: Caterpillar 3116 6.6 litre 6-cylinder diesel developing 230 hp at 2,600 rpm
Transmission: Allison MT643D automatic with 4 forward and 1 reverse gears
Transfer box: Rockwell T-1138 2-speed
Steering: power-assist
Suspension: multileaf (front), tapered spring (rear), with heavy-duty shock-absorbers
Tyres: 395/85 R 20G SBR
Brakes: air
Electrical system: 12/24 V
Batteries: 3 × 12 V

Status
Available.

Manufacturer
General Motors Corporation, Military Vehicles Operation, 2619 Product Lane (Suite 130), Rochester Hills, Michigan 48309-3807, USA.
Tel: +1 810 853 3873.

VERIFIED

M54 (6 × 6) 5 ton cargo truck series

Development
The M54 series of (6 × 6) 5 ton trucks was developed in the immediate post-Second World War period to replace vehicles in the 4, 5, 6 and 7½ ton class. First production models were completed by Diamond T and International Harvester in 1950 with later production undertaken by Kaiser Jeep (now AM General) and Mack. The first vehicles were powered by a six-cylinder petrol engine, which was replaced by a six-cylinder diesel engine in 1962, in turn replaced by a six-cylinder multifuel engine the following year. The multifuel engine was a supercharged version of the LD465 as used in the 2½ ton truck, it developed 230 hp but was not successful as it became overheated. The AM General Corporation replaced this with a commercial engine and the series then became known as the M809 range, for which there is a separate entry. However, the Hercules Engine Company had a LDS465 turbocharged engine available as a one-for-one replacement for the LD465.

There was a wide range of kits developed for the M54 series including A-frame, airbrake, electric brake, engine heater, deep fording, hard top, machine gun mount, personnel heater, thermal barrier and a front-mounted winch with a capacity of 9,072 kg.

Description
The layout of the basic cargo version is conventional with the engine at the front; two-door cab in the centre with a windscreen that can be folded forward onto the bonnet and a removable canvas top; and the welded steel cargo area at the rear with fixed sides, removable side racks, troop seats, removable bows and a tarpaulin cover.

Variants
AM General retrofit package
AM General designed a retrofit kit to upgrade M54 Series vehicles to an M809 Series configuration. These kits, produced for the US Marine Corps, include a power pack, frame, steering system, brake system, fuel and exhaust system, and electrical and sheet metal components to achieve the M809 configuration.

The power pack includes a Cummins NHC-250 diesel engine, radiator and cooling lines, a 24 V alternator and starter, clutch, flywheel, five-speed transmission and an air cleaner assembly. The frame includes rails and cross members for the new power pack while the steering includes steering gear, a power steering pump, power-assist cylinder and hoses. The brake system includes air reservoir tanks, air and hydraulic lines and couplings, adaptors and mounting hardware. Included in the fuel system are a 295-litre tank, a cold start primer, filters, fuel lines and a new accelerator linkage. Exhaust pipes, a muffler exhaust stack, a heat shield, gaskets and other hardware make up the exhaust system and the new electrical system contains wiring harnesses, bulbs and dials, batteries and cables and a new battery box. The kit is complete with a new hood, fenders, instrument panel and other items.

All items in the kit have been fitted to US military vehicles and tested at the Aberdeen Proving Grounds, Maryland.

M41 (Cargo, Truck, 5 ton, 6 × 6)
This has single instead of dual rear wheels and can carry 6,804 kg of cargo and tow a 13,608 kg trailer on roads or 6,804 kg across country. It is powered by a Continental R6602 six-cylinder OHV petrol engine which develops 196 bhp at 2,800 rpm, coupled to a five-speed gearbox and a two-speed transfer box.

M54A2 (6 × 6) 5 ton cargo truck with bows stowed to rear of cab

M51 (6 × 6) 5 ton dump truck

M51 (Dump, Truck, 5 ton, 6 × 6)
This has a rear tipping all-steel dump body and is based on the M61 chassis, it can also be used to carry cargo. The engine is the same as in the M41 and is coupled to a Spicer Model 6352 gearbox, Timken Model T-138 two-speed transfer box and a Spicer WND-61 PTO mounted on the transmission, and powers the front-mounted winch (when fitted) and the hydraulic dump body mechanism. The equivalent vehicle in the M809 series is the M817. The M51A1 has a Mack Model ENDT-673 six-cylinder diesel engine which develops 211 bhp at 2,100 rpm and the M51A2 has an LD465 six-cylinder turbocharged multifuel engine which develops 230 hp.

M52 (Truck, Tractor, 5 ton, 6 × 6)
The chassis is designated the M61 and is powered by the same engine as the M41. Later models were the M52A1 and M52A2 with diesel and multifuel engines respectively. The M52 can tow a semi-trailer weighing up to 11,340 kg on roads or up to 6,804 kg across country. The equivalent vehicle in the M809 series is the M818.

M54 (Cargo, Truck, 5 ton, 6 × 6)
This is the basic cargo model and its chassis is designated the M40. It has the same engine as the M41 with later models being the M54A1 (diesel) and M54A2 (multifuel). The equivalent in the M809 series is the M813.

M55 (Cargo, Truck, 5 ton, 6 × 6)
This is the long wheelbase cargo model of the M54 and is based on the M63 chassis. It has the same engine as the M41 with later models being the M55A1 (diesel) and M55A2 (multifuel). The equivalent vehicle in the M809 series is the M814.

M62 (Truck, Wrecker, Medium, 5 ton, 6 × 6)
This is based on the M40C chassis with a walking beam rear suspension; details of this can be found in the *Recovery vehicles* section.

M64 (Cargo, Truck, Van, 5 ton, 6 × 6)
This has a fully enclosed rear body.

M246 (Truck, Tractor, Wrecker, 5 ton, 6 × 6)
This is based on the M63C chassis and can be used for recovery operations as well as towing trailers. Details of this can be found in the *Recovery vehicles* section.

M291 (Truck, Van, Expansible, 5 ton, 6 × 6)
This is fitted with a fully enclosed rear van body with sides which are extended when the vehicle is stationary. Later models were the M291A1 (diesel) and M291A2 (multifuel). The equivalent vehicle in the M809 series is the M820.

M328 (Truck, Stake, Bridging, 5 ton, 6 × 6)
This is used to carry bridge components. Later models were the M328A1 (diesel) and M328A2 (multifuel). The equivalent vehicle in the M809 series is the M821.

M543 (Truck, Wrecker, Medium, 5 ton, 6 × 6)
This is based on the M40C chassis with walking beam rear suspension. Later models are the M543A1 (diesel) and M543A2 (multifuel). The equivalent vehicle in the M809 series is the M816, details of which can be found in the *Recovery vehicles* section.

M748 (Bolster, Truck, 5 ton, 6 × 6)
Later models were designated the M748A1 (diesel) and M748A2 (multifuel). The equivalent vehicle in the M809 series is the M815.

M139 (Chassis, Truck, 5 ton, 6 × 6)
This chassis has been used for a variety of applications. The basic M139 is used for carrying bridging equipment and the M139C was designed to carry the Honest John surface-to-surface rocket (now withdrawn from service). The M139D has a different rear axle and jack bracket supports and both the M139C and M139D were equipped with high-reduction axles to increase tractive power and a modified front cross member.

The Italian CP 56 bridging truck used a chassis very similar to that of the M139.

Specifications

Designation	**M51**	**M52**	**M54**	**M55**
Type	dump	tractor	cargo	cargo (LWB)
Cab seating	1 + 2	1 + 2	1 + 2	1 + 2
Configuration	6 × 6	6 × 6	6 × 6	6 × 6
Weight (empty)	9,970 kg	8,616 kg	8,732 kg	10,915 kg
(loaded, road)	19,042 kg	19,956 kg	18,119 kg	20,146 kg
Weight on front axle (loaded, road)	4,112 kg	–	4,347 kg	–
Weight on rear axles (loaded, road)	15,088 kg	–	13,807 kg	–
Max load (road)	9,072 kg	n/a	9,072 kg	9,072 kg
(cross-country)	4,536 kg	n/a	4,536 kg	4,536 kg
Towed load (road)	13,608 kg	24,948 kg	13,608 kg	13,608 kg
(cross-country)	6,804 kg	13,608 kg	6,804 kg	6,804 kg
Load area	3.175 × 2.082 m	n/a	4.267 × 2.235 m	6.15 × 2.235 m
Length	7.146 m	6.933 m	7.974 m	9.797 m
Width	2.463 m	2.463 m	2.463 m	2.463 m
Height (overall)	2.809 m	2.638 m	2.946 m	2.946 m
Ground clearance	0.267 m	0.267 m	0.267 m	0.267 m
Track (front/rear)	1.869/1.828 m	1.869/1.828 m	1.869/1.828 m	1.869/1.828 m
Wheelbase	3.555 m + 1.371 m	3.555 m + 1.371 m	3.86 m + 1.371 m	4.775 m + 1.371 m
Angle of approach/departure	52.5°/69°	52.5°/69°	37°/38°	37°/23°
Max speed (road)	84 km/h	84 km/h	84 km/h	84 km/h
Range	785 km	483 km	344 km	344 km
Fuel capacity	416 litres	416 litres	295 litres	295 litres
Max gradient	70%	77%	50%	65%
Fording (without preparation)	0.762 m	0.762 m	0.762 m	0.762 m
(with preparation)	1.981 m	n/a	1.981 m	1.981 m
Gearbox	manual with 5 forward and 1 reverse gears			
Clutch	single dry plate	single dry plate	single dry plate	single dry plate
Transfer box	2-speed	2-speed	2-speed	2-speed
Steering	cam and lever	cam and lever	cam and lever	cam and lever
Tyres	11.00 × 20	11.00 × 20	11.00 × 20	11.00 × 20
Brakes	all have hydraulic brakes, air-actuated, and mechanical parking brakes			
Electrical system	24 V	24 V	24 V	24 V
Batteries	2	2	2	2

Status
Production complete. In service with the US Army and other armies that have received American aid, including South Korea, Spain (many of which have been re-engined) and Turkey.

Manufacturer
Last manufacturer of the M54 series was AM General Corporation, 105 North Niles Avenue, South Bend, Indiana 46617, USA.
Tel: +1 219 284 2942. Fax: +1 219 284 2959.

Similar vehicles were manufactured in Italy (as the CP 56) and Spain.

VERIFIED

M809 (6 × 6) 5 ton Cargo Truck Series

Development

The M809 series of (6 × 6) 5 ton cargo trucks is similar to the older M54 series but was fitted with a diesel engine in place of the multifuel engine by the AM General Corporation under a product engineering development and test programme.

AM General Corporation started production of the M809 series in 1970 and by the middle of 1980 had completed some 38,000 vehicles. In September 1988 it was announced that AM General would be withdrawing from the production of medium and heavy vehicles and their South Bend plant closed in late 1989.

Further development of the M809 series by AM General, under an engineering contract with the US Army, resulted in the much improved M939 series, for which there is a separate entry in this section.

Description

The chassis of the vehicle consists of two rail-type beams with six reinforced cross members. Both front and rear axles are of the hypoid, single-speed, double-reduction type. The layout is conventional, with the engine at the front; two-door cab in the centre with a windscreen which can be folded flat against the bonnet and a removable canvas top; and the cargo area at the rear. The basic cargo model has an all-steel rear cargo body with drop sides, removable bows, tarpaulin cover and troop seats down either side which enable 18 fully equipped troops to be carried.

There are three basic chassis in the series, the M809 (used for the M813, M813A1, M816 and M815), the M810 (used for the M817 and M818) and the M811/M812 (used for the M814, M819, M820 and M821). A variety of kits is available for the range including: A-frame, airbrake, closure hard top, deep water fording, level wind device, slave receptacle, thermal barrier, water personnel heater, winterisation personnel heater kit and a winterisation power plant kit. All models except the M820 can be fitted with a winch at the front.

M809 series vehicles can be fitted with the Enhanced Mobility System (EMS) for increased mobility over sand, mud and snow.

Variants

M813 (Cargo Truck, 5 ton, 6 × 6)
This is the basic cargo model and can carry 4,535 kg of cargo across country and 9,070 kg on roads. The foldable seats enable 26 fully equipped troops to be carried.

M813A1 (Drop side, Cargo Truck, 5 ton, 6 × 6)
This is similar to the M813 but has drop sides as well as a drop tailgate.

M814 (Long Cargo Truck, 5 ton, 6 × 6)
This is the long wheelbase model and can be delivered with removable bows and a tarpaulin cover.

M815 (Bolster Truck, 5 ton, 6 × 6)
This has the M809 chassis and can carry a maximum load of 4,536 kg across country or 9,072 kg on roads. Towed load is identical to the M813 cargo truck.

M816 (Wrecker, Truck, 5 ton, 6 × 6)
Details of this vehicle are given in the *Recovery Vehicles* section.

M817 (Dump Truck, 5 ton, 6 × 6)
This has an all-steel 3.8 m³ capacity rear dump body which can be fitted with bows and a tarpaulin cover.

M818 (Tractor, Truck, 5 ton, 6 × 6)
This has both a fifth wheel and a pintle tow hook and can tow semi-trailers weighing up to 24,970 kg on roads or 17,025 kg across country.

Designation	**M813**	**M813A1**	**M814**	**M817**	**M818**	**M820**
Type	cargo	cargo	cargo (LWB)	dump	tractor	van
Cab seating	1 + 2	1 + 2	1 + 2	1 + 2	1 + 2	1 + 2
Configuration	6 × 6	6 × 6	6 × 6	6 × 6	6 × 6	6 × 6
Weight (empty)	9,733 kg	10,043 kg	11,297 kg	10,776 kg	9,202 kg	12,474 kg
(loaded, road)	18,985 kg	18,993 kg	20,549 kg	20,028 kg	20,349 kg	29,459 kg
Weight on front axle (loaded)	5,015 kg	5,016 kg	5,326 kg	4,981 kg	4,950 kg	4,936 kg
Weight on rear axles (loaded)	13,969 kg	13,976 kg	15,223 kg	15,047 kg	15,399 kg	14,523 kg
Max load (road)	9,070 kg	9,070 kg	9,070 kg	9,070 kg	n/a	6,803 kg
(cross-country)	4,535 kg	4,535 kg	4,535 kg	4,535 kg	n/a	2,268 kg
Towed load (road)	13,620 kg	13,620 kg	13,620 kg	13,620 kg	24,943 kg	13,620 kg
(cross-country)	6,802 kg	6,802 kg	6,802 kg	6,802 kg	17,007 kg	6,802 kg
Load area	4.3 × 2.2 m	4.3 × 2.2 m	6.2 × 2.2 m	3.2 × 2.1 m	n/a	5.2 × 2.1 m
Length	7.652 m	7.652 m	9.609 m	6.86 m	6.71 m	9.144 m
Width	2.464 m	2.464 m	2.464 m	2.464 m	2.464 m	2.483 m
Height (overall)	2.946 m	2.946 m	2.946 m	2.946 m	2.946 m	3.467 m
(reduced)	2.172 m	2.172 m	2.172 m	2.832 m	2.172 m	3.467 m
(load area)	1.4 m	1.4 m	1.4 m	n/a	n/a	n/a
Ground clearance (axles)	0.295 m	0.295 m	0.295 m	0.295 m	0.295 m	0.295 m
Track (front/rear)	1.88/1.829 m	1.88/1.829 m	1.88/1.829 m	1.9/1.829 m	1.9/1.829 m	1.9/1.829 m
Wheelbase	3.86 m + 1.371 m	3.86 m + 1.371 m	4.775 m + 1.371 m	3.55 m + 1.371 m	3.55 m + 1.371 m	4.775 m + 1.371 m
Angle of approach/departure	46°/35°	46°/35°	46°/23.5°	34°/69°	45°/69°	46°/24°
Max speed (road)	84 km/h	84 km/h	84 km/h	84 km/h	84 km/h	84 km/h
Range	563 km	563 km	563 km	772 km	563 km	563 km
Fuel capacity	295 litres	295 litres	295 litres	295 litres	295 litres	295 litres
Max gradient (without towed load)	67%	67%	61%	61%	60%	65%
(with towed load)	42%	42%	38%	31%	42%	41%
Max side slope (cross-country)	20%	20%	20%	20%	20%	20%
Fording (without preparation)	0.76 m	0.76 m	0.76 m	0.76 m	0.76 m	0.76 m
(with preparation)	1.98 m	1.98 m	1.98 m	1.98 m	1.98 m	1.98 m
Engine	all powered by an NHC-250 6-cylinder in-line diesel developing 240 hp at 2,100 rpm					
Gearbox	all have manual gearbox with 5 forward and 1 reverse gears					
Clutch	single dry plate	single dry plate	single dry plate	single dry plate	single dry plate	single dry plate
Transfer box	2-speed	2-speed	2-speed	2-speed	2-speed	2-speed
Steering	power-assisted	power-assisted	power-assisted	power-assisted	power-assisted	power-assisted
Turning radius	12.75 m	12.75 m	14.3 m	12.75 m	12.75 m	14.53 m
Suspension (front/rear)	semi-elliptical springs/semi-elliptical springs, inverted					
Tyres	11.00 × 20	11.00 × 20	11.00 × 20	11.00 × 20	11.00 × 20	11.00 × 20
Brakes	main, air over hydraulic. Parking, dual grip on transfer case					
Electrical system	24 V	24 V	24 V	24 V	24 V	24 V
Batteries	4	4	4	4	4	4

M821 carrier for bridge erection boat (C R Zwart)

M813 5 ton (6 × 6) cargo truck (Michael Jerchel)

M819 (Tractor, Wrecker, 5 ton, 6 × 6)
Details of this vehicle are given in the *Recovery Vehicles* section.

M820 (Van, Expansible, Truck, 5 ton, 6 × 6)
This is provided with a fully enclosed body at the rear which is 2.1 m wide and 1.9 m high in its normal position. When static, the sides of the van are extended either side and the interior is then 4.2 m wide. Access to the body is via steps at the rear.

M821 (Bridge Transport, 5 ton, 6 × 6)
This is used to carry bridging components and is based on the M812 chassis.

Ribbon Bridge Carrier and Launching Vehicle
An M809 chassis is used to carry and launch the Ribbon Bridge system as well as carrying and launching a bridge erection boat.

Specifications
Specifications relate to a vehicle without winch, with the exception of the M817 which is fitted with a winch as standard equipment. All other vehicles, except M820, can be fitted with a front-mounted winch. When fitted with a front-mounted winch the vehicles have an angle of approach of 34°.

Status
Production complete. In service with the armed forces of the USA and other contries including El Salvador, Honduras, Jamaica, Jordan, South Korea, Lebanon, Liberia, Pakistan, Panama, Philippines, Saudi Arabia, Senegal, Somalia, Spain, Sudan, Taiwan, Thailand and Zaïre. An undisclosed number was supplied to China.

M817 (6 × 6) 5 ton dump truck (Michael Jerchel)

Manufacturer
AM General Corporation, 105 North Niles Avenue, South Bend, Indiana 46617, USA.
Tel: +1 219 284 2942. Fax: +1 219 284 2959.

VERIFIED

M939 (6 × 6) 5 ton cargo truck

Development
When the M809 was type classified it was intended that a product-improvement programme would be carried out in the areas of transmission, transfer case and brakes. Extensive testing of the M809 series in 1970 demonstrated that product improvements in these three areas would be most beneficial. The programme was shelved when the US Army withdrew from Vietnam but was reinstituted in 1975.

In October 1979 the M939 (6 × 6) 5 ton cargo truck was type classified for inclusion in US Army field units. In April 1981 AM General Corporation was awarded a contract for 11,394 M939 trucks, later increased to 22,789, by the US Army Tank Automotive Command. The total value of the five-year contract eventually reached $1.6 billion. Production began in the first half of 1982 at AM General's plant at South Bend, Indiana. The M939 series became available for export during late 1983.

During the mid-1980s, the basic M939 design was upgraded to the M939A1. This version was equipped with 14.00 R 20 'Super Single' tyres with a single tyre replacing the dual biased-ply tyres on the rear axles. Only a limited number of M939A1s was produced.

In May 1986 ARVECO, a joint venture between BMY Corporation and the General Automotive Corporation, won a contract to produce 15,218 M939A2 vehicles over a five-year period. The initial one-year contract was for 2,046 vehicles at a cost of $145 million, while the five-year contract was expected to be worth $1 billion. Following the contract award, BMY acquired the General Automotive Corporation's portion of ARVECO. HARSCO renamed this wholly owned organisation BMY-Wheeled Vehicles Division and began deliveries from its new truck plant in Marysville, Ohio, in early 1987. By late 1990 the base contract quantity was fixed at 17,092 vehicles, with production extending into 1992 following orders from the US government that included FMS quantities for Saudi Arabia.

AM General Corporation completed its five-year contract in September 1986 but production was extended until April 1987 by the award of a further contract. Production by AM General Corporation ceased in late 1989 when their South Bend plant closed.

Description
The M939 is essentially the existing M809 series but improved in three major areas: transmission, transfer case and brake system. The M809's transmission and transfer case were under capacity and mismatched to the engine/axle ratio and performance requirements; there was no way to prevent the engine from overspeeding (rpm too high) or labouring in many gear ratio selections. The M939 has a fully automatic transmission to eliminate these problems and it also requires less driver training, reduces fuel consumption, is more reliable/durable, lessens driver fatigue and improves safety.

The transfer case is pressure- rather than splash-lubricated. The driver controls the engagement of the front wheels for 6 × 6 drive with an air system, eliminating a mechanical sprag clutch which frequently failed. The transfer case on the M939 can be shifted from either range while the vehicle is moving.

Tests demonstrated that the full airbrakes on the M939 have four times the brake shoe life of the air-over-hydraulic brakes on the M809. The airbrakes are self-adjusting and backed by fail-safe mechanical spring brakes.

The M939 was the first tactical truck to meet Surgeon General standards for noise in the cab, even with the windows open. This was achieved by relocating intake/exhaust ports behind the cab.

The front-mounted winch is hydraulically driven and stops when overloaded and restarts when the overload is removed. The older mechanically driven winches used on the M809 require shear-pin replacement when similarly overloaded.

The M939 bonnet and bumpers tilt forward so maintenance can be carried out from the ground, whereas even opening the M809's bonnet necessitates climbing on to the bumpers.

Flat tyres are replaced using a boom just behind the cab. The cab holds three personnel, the M809 only two.

The M939 was the first truck built with special connectors for use with the US Army's diagnostic equipment (STE/ICE). This encourages the use of test equipment, reduces maintenance time and eliminates incorrect replacements based on poor diagnosis.

The upgraded M939A2, produced by BMY-Wheeled Vehicles Division (BMY-WVD) incorporates an Eaton Central Tyre Inflation System CTIS integrated with 14.00 R 20 'Super Single' radial tyres. The engine is changed to a Cummins 6CTA8.3 developing 240 hp at 2,100 rpm, providing an increase in range to 830 km. Due to the lighter engine the kerb weight is reduced. Modifications were also made to the cooling system.

Variants
BMY-WVD developed a 2,000 gallon (7,570 litre) hard-sided tanker with a GVW of 19,010 kg. The tanker, which is equipped with a CTIS, provides armed forces

M923A1 cargo (6 × 6) 5 ton truck (Michael Jerchel)

M932A1 (6 × 6) tractor trucks (Michael Jerchel)

with M939 fleets with a POL and water off-road transport capability not previously available.

BMY-WVD also developed an 8 ton variant of the M944A2 chassis. Using higher rated components from other M939A2 series vehicles, the 8 ton variant provides a basis for such tasks as Ribbon Bridge launcher, medium palletised loads and other operations not currently within the M939A2's capability.

M939 variants are as follows:

M923: drop side cargo
M924: cargo; equivalent to M814
M925: drop side cargo with winch
M926: cargo with winch
M927: long wheelbase cargo
M928: long wheelbase cargo with winch
M929: dump truck; equivalent to M817
M930: dump truck with winch
M931: tractor
M932: tractor with winch
M933: tractor wrecker with winch
M934: expansible body
M935: expansible van body
M936: wrecker with winch
M939: chassis with or without winch
M939: chassis with winch
M940: chassis with winch
M941: chassis without winch
M941: chassis with winch
M942: long wheelbase chassis with or without winch
M943: long wheelbase chassis with winch
M944: long wheelbase chassis
M945: long wheelbase chassis with winch.

M939A2 production models are as follows:

M923A2: drop side cargo without winch
M925A2: drop side cargo with winch
M927A2: long wheelbase cargo without winch
M928A2: long wheelbase cargo with winch
M929A2: dump truck without winch
M930A2: dump truck with winch
M931A2: tractor truck without winch
M932A2: tractor truck with winch

M939 series (6 × 6) truck carrying communications shelter (T J Gander) ***1997***

Specifications

Model	**M923A2**	**M925A2**	**M927A2**	**M928A2**	**M929A2**	**M930A2**
Type	cargo	cargo	LWB cargo	LWB cargo	dump	dump
Winch	no	yes	no	yes	no	yes
Cab seating	1 + 2	1 + 2	1 + 2	1 + 2	1 + 2	1 + 2
Configuration	6 × 6	6 × 6	6 × 6	6 × 6	6 × 6	6 × 6
Weight (kerb)	9,494 kg	9,993 kg	10,791 kg	11,290 kg	10,805 kg	11,304 kg
(GVW)	14,030 kg	14,529 kg	15,327 kg	15,826 kg	15,340 kg	15,840 kg
Payload	4,536 kg	4,536 kg	4,536 kg	4,536 kg	4,536 kg	4,536 kg
Towed load	6,804 kg	6,804 kg	6,804 kg	6,804 kg	6,804 kg	6,804 kg
Length	7.887 m	8.433 m	9.792 m	10.363 m	6.934 m	7.48 m
Width	2.474 m	2.474 m	2.474 m	2.474 m	2.474 m	2.474 m
Height (overall)	3.073 m	3.073 m	3.063 m	3.063 m	3.175 m	3.175 m
(reduced)	2.385 m	2.385 m	2.375 m	2.375 m	2.375 m	2.375 m
Track (front)	1.98 m	1.98 m	1.98 m	1.98 m	1.98 m	1.98 m
(rear)	2.065 m	2.065 m	2.065 m	2.065 m	2.065 m	2.065 m
Wheelbase	4.547 m	4.547 m	5.461 m	5.461 m	4.242 m	4.242 m
Ground clearance	0.333 m	0.333 m	0.333 m	0.333 m	0.333 m	0.333 m
Fording (normal)	0.76 m	0.76 m	0.76 m	0.76 m	0.76 m	0.76 m
(with kit)	1.98 m	1.98 m	1.98 m	1.98 m	1.98 m	1.98 m
Engine	Cummins 6CTA8.3 8.3 litre 6-cylinder in-line turbocharged after-cooled diesel developing 240 hp at 2,100 rpm					
Gearbox	Allison MT654CR 5-speed automatic					
Transfer box	2-speed with high ratio 0.732:1, low ratio 1.790:1					
Steering	Sheppard power-assisted					
Suspension	leaf springs front and rear					
Tyres	14.00 R 20 tread design, all-terrain					
Electrical system	24 V	24 V	24 V	24 V	24 V	24 V
Alternator	60 A	60 A	60 A	60 A	60 A	60 A

Model	**M931A2**	**M932A2**	**M934A2**	**M935A2**	**M936A2**
Type	tractor	tractor	exp van	exp van	wrecker
Winch	no	yes	no	no	yes
Cab seating	1 + 2	1 + 2	1 + 2	1 + 2	1 + 2
Configuration	6 × 6	6 × 6	6 × 6	6 × 6	6 × 6
Weight (kerb)	9,024 kg	9,523 kg	12,717 kg	13,624 kg	16,724 kg
(GVW)	26,034 kg	26,533 kg	14,985 kg	15,892 kg	19,918 kg
Payload	6,804 kg[1]	6,804 kg[1]	2,268 kg	2,268 kg	3,175 kg
Towed load	6,804 kg[2]	6,804 kg[2]	6,804 kg	6,804 kg	9,072 kg
Length	6.718 m	7.264 m	9.21 m	9.55 m	9.195 m
Width	2.474 m	2.474 m	2.489 m	2.489 m	2.474 m
Height (overall)	3.078 m	3.078 m	3.614 m	3.609 m	3.048 m
(reduced)	2.39 m	2.39 m	3.614 m	3.609 m	2.756 m
Track (front)	1.98 m	1.98 m	1.98 m	1.98 m	1.98 m
(rear)	2.065 m	2.065 m	2.065 m	2.065 m	2.065 m
Wheelbase	4.242 m	4.242 m	5.461 m	5.461 m	5.547 m
Ground clearance	0.333 m	0.333 m	0.333 m	0.333 m	0.333 m
Fording (normal)	0.76 m	0.76 m	0.76 m	0.76 m	0.76 m
(with kit)	1.98 m	1.98 m	1.98 m	1.98 m	1.98 m
Engine	Cummins 6CTA8.3 8.3 litre 6-cylinder in-line turbocharged after-cooled diesel developing 240 hp at 2,100 rpm				
Gearbox	Allison MT654CR 5-speed automatic				
Transfer box	2-speed with high ratio 0.732:1, low ratio 1.790:1				
Steering	Sheppard power-assisted				
Suspension	leaf springs front and rear				
Tyres	14.00 R 20 tread design, all-terrain				
Electrical system	24 V	24 V	24 V	24 V	24 V
Alternator	60 A	60 A	60 A	60 A	60 A

[1] on 5th wheel
[2] pintle

M934A2: expansible van without winch
M935A2: expansible van with hydraulic tailgate
M936A2: wrecker with winch
M942A2: chassis with or without winch
M944A2: chassis without winch
M945A2: chassis with winch.

Also produced are 7,570 litre water and fuel tankers and a cargo truck with a loading/unloading crane.

The following kits are available for the M939 and M939A2 series of trucks: airbrake, automatic chemical alarm, deep water fording, bow and tarpaulin cover, electric brake, engine coolant heater, front and rear lifting points, fuel burning heater, hardtop closure, radiator and hood cover and rifle racks. M939A2 series cargo and dump trucks fitted with a winch can also be fitted with a front-mounted A-frame kit. This can be fitted in the field and has a lifting capacity of up to 1,362 kg.

Status

Production complete. In service with the US Army, Egypt, Saudi Arabia and others.

Manufacturers

(M939 and M939A1 - production complete) AM General Corporation, 105 North Niles Avenue, South Bend, Indiana 46617, USA.
Tel: +1 219 284 2942. Fax: +1 219 284 2959.
(M939A2 - production complete) BMY-Wheeled Vehicles Division of HARSCO, 13311 Industrial Parkway, Marysville, Ohio 43040, USA.
Tel: +1 513 644 0041. Fax: +1 513 644 6802.

UPDATED

Oshkosh MK48 Series Logistics Vehicle System

Development

During the 1960s, Lockheed developed a vehicle known as the Twister which, following US Army trials, evolved into a commercial model known as the Dragon Wagon. After entering into a licence agreement with Lockheed, Oshkosh Truck Corporation developed an articulated high-mobility vehicle designated by Oshkosh as the DA series. The US Marine Corps were delivered four prototypes during 1981 and 1982.

As a result of extended user trials by the US Marine Corps, an order was placed in September 1983 for vehicles based on the DA Series known as the MK48 Series Logistic Vehicle System (LVS). This was a five-year contract with a production total of 1,600 vehicles.

Under the initial contract Oshkosh produced 4,136 LVS modules for the US Marine Corps. In August 1995 the Oshkosh Truck Corporation was awarded another contract for a base quantity of 149 LVS modules with an option for 64 additional units. The contract was valued at $21.7 million.

Description

The MK48 Series LVS consists of a front and rear body, connected by a centre articulation joint that provides 64° of powered steering motion co-ordinated with steering of the front axle. Articulated steering provides a 30 per cent shorter turning radius than a conventional truck of the same wheelbase. The articulated joint also provides 6° of roll capability between the front and rear modules. Roll capability and a new six-rod suspension with tapered leaf springs, provides superior wheel-to-ground contact for improved off-road capability. The rear body/module can be disconnected and another body/module of a different type quickly connected.

The two-man enclosed cab mounted at the front of the vehicle has an optional roof hatch. The engine is mounted behind the cab and is coupled to a four-speed automatic transmission and a two-speed transfer case. This provides full-time all-wheel drive and has driver-controlled lockout. The hydraulic boosted power steering for the front axle co-ordinates with twin hydraulic cylinders astride the articulated joint to induce rear-body yaw and ensure a tight turning circle.

Standard equipment includes: a voltmeter, diagnostic connector assembly for STE/ICE, blackout lights, suspension seats, winch, material handling crane, towing provision with air and electrical connections, traction control axle differentials, Jacobs engine brake, various axle and transfer ratios, spare tyre and wheel, spotlight, off-road onboard equipment, off-road cab and glass protection kit and various rear modules. Optional equipment includes 24.00 × 21 radial sand tyres, a strobe beacon and other special military options.

The MK48 series was produced as part of the US Marine Corps' LVS and includes the following models:
MK48/14: logistics platform truck
MK48/15: recovery vehicle
MK48/16: truck tractor
MK48/17: cargo truck with material handling crane (MHC)
MK48/18: load handling system vehicle.

The MK48/14 logistics platform truck uses a steel flat deck with standard container lashing points to carry either one 20 ft/6.096 m container, two 10 ft/3.048 m containers or six 6.66 ft/2.03 m containers. The platform length is 6.058 m and loading height is 1.549 m.

The MK48/15 recovery vehicle uses a steel cargo body. At the rear is a hydraulic crane with a capacity of 4,082 kg and a recovery winch with a pull capacity of 27,216 kg.

Oshkosh MK48/18A1 (8 × 8) vehicle using a container lift kit (CLK) and a PLS load handling system to carry an ISO container ***1997***

The MK48/16 truck tractor has a fully oscillating fifth wheel for a 3.5 in/88.9 mm kingpin mounted 1.6 m above ground level. Full trailer air and electrical connections are provided, as is a 27,216 kg recovery winch.

The MK48/17 cargo truck has a steel cargo body with fold-down sides and a 4,082 kg capacity hydraulic materials handling crane at the rear.

The MK48/18A1 utilises existing MK48/14 rear modules converted to carry a load handling system.

Specifications

Model	**MK48/14**	**MK48/15**	**MK48/16**	**MK48/17**
Type	logistics platform truck	recovery	truck tractor	cargo truck
Cab seating	1 + 1	1 + 1	1 + 1	1 + 1
Configuration	8 × 8	8 × 8	8 × 8	8 × 8
Weight (empty)	18,597 kg	23,133 kg	18,597 kg	21,769 kg
Gross vehicle combination weight				
(on road)	68,040 kg	69,840 kg	85,277 kg	69,840 kg
(off road)	47,628 kg	47,628 kg	47,628 kg	47,628 kg
Payload				
(on road)	20,412 kg	9,072 kg[1]	20,866 kg[2]	18,144 kg
(off road)	11,340 kg	n/a	n/a	9,072 kg
Length	11.582 m	11.278 m	10.09 m	11.582 m
Width	2.438 m	2.438 m	2.438 m	2.438 m
Height (cab)	2.591 m	2.591 m	2.591 m	2.591 m
Track	2.007 m	2.007 m	2.007 m	2.007 m
Wheelbase	1.524 m + 6.579 m +1.524 m	1.524 m + 6.579 m +1.524 m	1.524 m + 5.817 m +1.524 m	1.524 m + 6.579 m +1.524 m
Angle of approach/departure	45°/45°	45°/48°	45°/65°	45°/40°
Max speed (road)	84 km/h	84 km/h	84 km/h	84 km/h
Range (cruising)	483 km	483 km	483 km	483 km
Fuel capacity	2 × 284 litres	2 × 284 litres	2 × 284 litres	2 × 284 litres
Gradient	60%	60%	60%	60%
Side slope	30%	30%	30%	30%
Fording	1.524 m	1.524 m	1.524 m	1.524 m
Engine	Detroit Diesel 8V-92TA V-8 2-cycle water-cooled 12.06 litre diesel developing 445 hp at 2,100 rpm			
Gearbox	Allison HT740D 4-speed automatic with torque converter			
Transfer box	Oshkosh 2-speed			
Steering	hydraulic powered yaw steering, power-assisted on No 1 axle			
Turning radius	11.7 m	11.7 m	10.4 m	11.7 m
Suspension	leaf springs with steel saddle and 6 torque rods front and rear giving 356 mm wheel travel			
Tyres	16.00 × 21	16.00 × 21	16.00 × 21	16.00 × 21
Electrical system	24 V	24 V	24 V	24 V
Batteries	4 × 12 V	4 × 12 V	4 × 12 V	4 × 12 V
Alternator	24 V, 62 A	24 V, 62 A	24 V, 62 A	24 V, 62 A

[1] towed load 14,515 kg
[2] load on 5th wheel

Oshkosh incorporated the US Army's Palletised Load System (PLS) and added features such as a winch and an integral container lift kit (CLK). The MK48/18A1 can handle bridging, bridging boats, flatracks and ISO containers without them being flatrack mounted first. In May 1994 Oshkosh were awarded a contract to produce 164 MK48/18A1s for the US Marine Corps, with a 100 per cent option clause. The contract value was $9.7 million.

Status
In production for US Marine Corps (see under Development).

Manufacturer
Oshkosh Truck Corporation, PO Box 2566, Oshkosh, Wisconsin 54903-2566, USA.
Tel: +1 414 235 9150. Fax: +1 414 233 9540.

UPDATED

Oshkosh Heavy Expanded Mobility Tactical Truck (8 × 8) 9,979 kg

Development
After evaluating a number of proposals and testing 8 × 8 vehicles submitted by MAN and Pacific Car and Foundry in May 1981, the US Army Tank Automotive Command awarded a $251.13 million five-year contract to the Oshkosh Truck Corporation for production of the 10 ton (US) 8 × 8 Heavy Expanded Mobility Tactical Truck (HEMTT).

To date over 14,400 vehicles have been produced, the bulk of them under three different multiyear contracts. The US Army Tank Automotive and Armaments Command awarded the third contract in July 1994. This contract, worth $133 million, extended production until November 1996. A fourth contract award was issued in May 1996 for a base quantity of 84 vehicles. This contract covers customer requirements for three years at a planned rate of one vehicle per day.

To reduce procurement and life cycle costs, the HEMTT features extensive use of commercial automotive components including an Oshkosh truck cab, standard eight-cylinder diesel engine and a standard four-speed automatic transmission. Some of the components used in this vehicle are also found in the LVS (8 × 8) vehicles used by the US Marine Corps (see previous entry).

Description
The HEMTT chassis is of formed channel bolted construction with Grade 8 bolts, with heat-treated carbon manganese steel is used throughout. The chassis is provided with heavy-duty front bumper and skid plate, external hydraulic connection (not on tractor trucks), service and emergency airbrake connection, slave start connection and trailer electrical connector.

The two-man, two-door forward control cab is of heavy-duty welded steel construction with corrosion-resistant skins. Standard equipment includes: suspension seats for the driver and passenger, seat belts, heater and defroster, dual sun visors, interior lights, variable speed air windscreen wipers, windscreen washers and electric and air horns. A spare wheel is mounted to the rear of the cab and is provided with a davit to assist in lowering it to the ground.

All models have Oshkosh 46K front tandem axles that are used for steering, single reduction, 32° front turning angle, single cardan joint, closed-type steering ends with an inter-axle driver-controlled differential.

The rear axles, except for those on the M984A1, are Eaton DS-480 single reduction type with driver-controlled differential. The M984A1 has Eaton DS-650 axles with single reduction and driver-controlled differential.

Variants
M977 Cargo Truck
This is the basic member of the family with a light-duty materials handling crane at the very rear, produced with and without a self-recovery winch which can be used to the front or rear of the vehicle. The cargo area is 5.4 m long and is provided with drop sides.

M978 Fuel Servicing Truck
9,500 litre capacity tanker, produced with or without a self-recovery winch.

M983 Tractor Truck
Tractor truck with material handling crane and fifth wheel, produced with or without a self-recovery winch.

M984A1 Wrecker
See *Recovery vehicles* section.

M985 Cargo Truck
Similar to M977 but with a heavy-duty materials handling crane at rear, produced with or without a self-recovery winch. It is used, with a M989A1 trailer, to support the Multiple Launch Rocket System (MLRS).

10 × 10 PLS Truck
During 1990 it was announced that Oshkosh had been awarded a contract for 2,626 10 × 10 Palletized Load System (PLS) trucks based on the HEMTT.

8 × 8 PLS Truck
For details refer to separate PLS entry in this section.

HEMTT ESP
During 1994 a Extended Service Program (ESP) for the HEMTT series was announced by Oshkosh Truck Corporation. Designed to enhance reliability and performance, the ESP incorporates a modular independent suspension system to improve off-road mobility through increased wheel travel, greater axle clearance and consistent ground contact; the new suspension also reduces chassis wear and tear, improves payload ride and provides greater driver comfort. Also involved is a Central Tyre Inflation System (CTIS) to increase traction and tyre life. Other ESP features include lubed-for-life components and commonality with existing HEMTT and PLS models.

HEMTT Overhaul
The Oshkosh Truck Corporation is offering a service which involves rebuilding service-worn M977 series HEMTT into an as-new condition.

The programme involves a systematic strip-down to the bare frame which is steam cleaned, stripped of paint, rust and any corrosion. Individual components are disassembled and inspected, so that any wear or damage outside the Original Equipment Manufacturer (OEM) limits, results in replacement. Components are then reassembled using new gaskets and seals and repainted. Full reassembly is carried out on the Oshkosh production line, using new-build procedures, inspections and techniques and with components such

Specifications

Designation	**M977**	**M978**	**M983**	**M985**
Type	cargo	tanker	tractor	cargo
Cab seating	1 + 1	1 + 1	1 + 1	1 + 1
Configuration	8 × 8	8 × 8	8 × 8	8 × 8
Weight (empty)	17,600 kg	17,305 kg	14,583 kg	17,985 kg
(loaded)	27,080 kg	26,694 kg	27,125 kg	28,168 kg
Length	10.173 m	10.173 m	8.903 m	10.173 m
Width	2.438 m	2.438 m	2.438 m	2.438 m
Height (cab)	2.565 m	2.565 m	2.565 m	2.565 m
Wheelbase	5.334 m	5.334 m	4.597 m	5.334 m
Angle of approach/departure	43°/45°	43°/45°	43°/66°	43°/45°
Max speed	88 km/h	88 km/h	88 km/h	88 km/h
Gradient	60%	60%	60%	60%
Range	483 km	483 km	483 km	483 km
Fuel capacity	589 litres	589 litres	589 litres	589 litres
Engine	Detroit Diesel 8V-92TA, V-8 2-stroke 12.06 litre diesel developing 445 hp at 2,100 rpm			
Gearbox	Allison HT740D automatic with torque converter, 4 forward and 1 reverse gears			
Transfer box	Oshkosh 55000, 2-speed			
Steering	integral hydraulic, main and booster gears			
Suspension	Hendrickson RT340 spring with steel saddle and equalising beams, 250 mm vertical axle travel. M985 rear axle has Hendrickson RT380			
Tyres	16.00 × 20	16.00 × 20	16.00 × 20	16.00 × 20
Brakes (main)	air-operated, dual system, internal shoe			
(parking)	spring brakes mounted on Nos 3 and 4 axles			
Electrical system	24 V	24 V	24 V	24 V
Batteries	4 × 12 V	4 × 12 V	4 × 12 V	4 × 12 V

Oshkosh M985 (8 × 8) HEMTT cargo truck and trailer carrying MLRS rocket pods ***1997***

Oshkosh M983 (8 × 8) HEMTT tractor truck ***1997***

Oshkosh M985 (8 × 8) HEMTT cargo truck ***1997***

as wiring looms, hoses and fasteners being all-new items.

Final road tests are the same as for new vehicles, resulting in the provision of a new vehicle warranty but at a considerably lower cost than for a new vehicle. The end result will be able to meet the 1996 American federal heavy truck standards, including the installation of three-point seat belts.

In September 1995 it was announced that Oshkosh had been awarded a $10.1 million contract to carry out a complete overhaul of 123 HEMTTs. The contract included an option for a further 123 units. Further similar contracts are anticipated.

US Customs Service HEMTT

The US Customs Service operates a small number of HEMTTs which have been specially modified to act as mobile winch and anchor vehicles for aerosats (non-motorised blimps) equipped with General Electric radar. The vehicles are located along the US-Mexican border to counter drug-smuggling activities. Each vehicle carries a cable drum and winch gear to allow an aerosat to operate at heights up to 4,570 m.

Status

In production. In service with the US Army and in several Middle and Far East countries including Bahrain, Egypt, Israel, South Korea, Kuwait, Saudi Arabia and Taiwan. Commercial variants have also been sold.

Manufacturer

Oshkosh Truck Corporation, PO Box 2566, Oshkosh, Wisconsin 54903-2566, USA.
Tel: +1 414 235 9150. Fax: +1 414 233 9540.

UPDATED

Mack RM6866SX (6 × 6) 10,000 kg truck

Description

The Mack RM6866SX (6 × 6) 10,000 kg truck is one of the 'R' series of Mack 6 × 6 vehicles. It has an all-steel cab with an optional removable roof. Suspension seating is provided for a driver and one passenger. Left- or right-hand steering can be provided. A heavy-duty frame extending from the cab to the rearmost point of the vehicle is fitted. This is constructed from single channel frame rail with a 340 × 83 × 10 mm section. The engine is rated at 250 hp but other engines up to 400 hp can be fitted; all are turbocharged. The transmission is a Mack design and uses an aluminium case. The 'Maxitorque' transmission is equipped with three counter-shafts equally spaced around the main shaft to provide equal distribution of the torque load on the main shaft. An automatic transmission is available as an option. The Mack differential uses a 'top mount' with a high entry of drive-shaft to improve clearance. The bogies use camel-back springs and an automatic power divider.

The standard body on the RM6866SX is of the cargo type with drop sides and a tailgate. A hydraulic loading crane may be fitted if required. Bows and a tilt are provided as standard but other body configurations can be made available.

Mack RM6866SX (6 × 6) 10,000 kg truck

Specifications

Cab seating: 1 + 1
Configuration: 6 × 6
Weight:
(loaded chassis, front) 10,400 kg
(loaded chassis, rear) 17,200 kg
(GVW) 24,500 kg
Length: 7.6 m
Width: 2.463 m
Height: 2.994 m
Wheelbase: (front axle to centre of rear bogie) 4.677 m
Max speed: 84 km/h
Gradient: 60%
Fuel capacity: 359 litres
Engine: Mack 6-cylinder turbocharged diesel developing 300 hp
Transmission: Mack TRL1078 5-speed or optional automatic
Clutch: CL697
Transfer case: TC15, 2-speed
Tyres: 11.00 × 22
Brakes:
(front) air wedge type
(rear) air cam type
Electrical system: 12 or 24 V
Batteries: 4 × 12 V

Status

In production. In service with several nations - see also entry under Australia in this section.

Manufacturer

Mack Trucks Inc, Box M, Allentown, Pennsylvania 18105, USA.
Tel: +1 215 439 3011. Telex: 847429.
Fax: +1 215 439 3308.

VERIFIED

M915 Series of trucks

Development

One of the recommendations of the 'WHEELS' study was that the US Army should use commercial trucks wherever practicable. In January 1977 the US Army Automotive Material Readiness Command (TARCOM) issued a request for technical proposals for a series of heavy trucks ranging from 22,680 to 34,019 kg gross vehicle weights. The army subsequently received technical proposals from six truck manufacturers: AM General Corporation, FWD (Four Wheel Drive) Corporation, General Motors Truck and Coach, International Harvester Corporation, Kenworth Truck Corporation and White Autocar Corporation.

The six manufacturers that responded to the original request qualified to offer priced bids under the second step of the procurement. Bids were opened in June 1977 with AM General Corporation the lowest at US$252.8 million for the requirement of 5,507 trucks. In June 1977 a contract was awarded to AM General for 5,507 trucks and in addition the government had the right to exercise options on an additional 5,507 trucks, giving a potential of 11,014 units. The contract was for four years with the first test vehicles (M915 Line Haul Tractor) delivered early 1978.

The US government did not exercise its option on the additional 5,507 trucks but in September 1981 awarded AM General a contract for 2,511 M915A1 (6 × 4) tractor trucks, which was later completed.

The AM General series was based on the Centaur series of trucks built by the Crane Carrier Company of Tulsa, Oklahoma. A licence agreement between AM General and Crane Carriers gave rights to AM General to manufacture and sell vehicles to the US government.

The contract also required complete after-market support of the M915 including spare part provisioning and a full complement of maintenance and service publications, as well as the training of army instructors. The army received warranty covering defects in design, materials and workmanship for a period of 15 months or for 19,312 km. In addition, the army received 'free time' up to six months, for new vehicles stored in depots.

The latest production versions are the M915A2 and

M915A1 (6 × 4) 14 ton tractor truck (Michael Jerchel)

M917 (8 × 6) 20 ton dump truck

the M916A1. These versions are produced by the Freightliner Corporation which was awarded a contract to produce 797 trucks with a 200 per cent overbuy option.

In March 1995 it was announced that the Freightliner Corporation had formed an alliance with the Oshkosh Truck Corporation in which Freightliner would transfer the marketing, manufacturing and administrative functions of their specialised defence business to Oshkosh. In some regions existing Freightliner local arrangments will remain while in others Oshkosh will assume the Freightliner business. It is possible that in some regions a completely new arrangement may be formed.

Description

The series is based on commercial vehicles with the minimum of modifications to suit them for military use, such as forest green paint, rear-mounted pintle hooks, front and rear tow hooks and blackout lights. There are three basic chassis, 6 × 4 for road use and 6 × 6 and 8 × 6 for both road and cross-country use.

The layout of all vehicles is similar, with the engine at the front, two-door all-steel fully enclosed cab in the centre and the fifth wheel or body at the rear.

M915, (Tractor, Truck, 14 ton, 6 × 4)

This is powered by a Cummins Model NTC 400 six-cylinder diesel which develops 400 hp (gross) at 2,100 rpm, coupled to a Caterpillar semi-automatic gearbox with 16 forward and two reverse gears. The suspension consists of: front Rockwell Asymmetrical leaf pin and shackle; and rear Hendrickson RTE 380 walking beam.

The M915 is used primarily for the long distance movement of containers and normally tows the 34 ton (US) M872 semi-trailer. It replaced the 5 ton military series tractors and certain commercial tractors used in moving cargo from the port of embarkation to the division rear boundary. M52 and M818 truck tractors will continue to operate from within the division and brigade areas, performing the same types of mission as currently assigned.

M915A1, (Tractor, Truck, 14 ton, 6 × 4)

In September 1981 AM General Corporation was awarded a contract by the US Army Tank Automotive Command for 2,511 M915A1 (6 × 4) tractor trucks. This contract was worth US$130 million with a 100 per cent option for an additional 2,511 units. Production began at the company's South Bend plant in January 1983 and continued for 18 months.

The M915A1 is similar to the earlier M915 except that it has an Allison HT 754 CR automatic transmission with five forward and one reverse gears.

Specifications

Similar to the M915 except for the following details:
Length: 6.489 m
Width: 2.438 m
Height: 2.984 m
Fording: 0.616 m
Max road speed: 91.7 km/h
Speed on 3% slope: 45 km/h
Max gradient: 18.4%
Fuel capacity: 446 litres

M915A2, (Tractor, Truck, 14 ton, 6 × 4)

This Freightliner Corporation 6 × 4 tractor truck is powered by a Detroit Diesel DDE 12.7 litre unit developing 400 hp at 2,100 rpm, coupled to an Allison HT-740 automatic transmission. Weight empty is 8,481 kg and the vehicle is 7 m long, 2.49 m wide and 3.02 m high overall. Wheelbase is 4.12 m.

The M915A2 contract was awarded to Freightliner on 20 September, 1988, and was for 385 units. The first deliveries were made in May 1989 and the first production vehicles entered the US Army's inventory in November 1990 (see also following entry).

M916, (Tractor, Truck, 20 ton, 6 × 6)

This replaced the M123 (6 × 6) tractor and has the same engine and transmission as the M915, but also has an Oshkosh F-U29 transfer case. Suspension is the same as in the M915, with differential lock-ups on front axle and rear bogie. Mounted to the rear of the cab is a hydraulic winch with a capacity of 20,250 kg.

The M916 is used by units throughout the world as the prime mover for the M172A1 low bed trailer to transport heavy construction equipment within the theatre of operations. It also operates efficiently hauling the M870 40 ton low bed semi-trailer when fully loaded; however, a 6 in (152 mm) fifth-wheel spacer must be used for compatibility with the M870 kingpin height.

M916A1, (Tractor, Truck, 20 ton, 6 × 6)

This Freightliner Corporation 6 × 6 tractor truck is powered by a Detroit Diesel DDE 12.7 litre unit developing 400 hp at 2,100 rpm coupled to an Allison HT-740 automatic transmission. Weight empty is 12,594 kg. Length is 7.35 m, width 2.49 m and the height is 3.24 m overall. Wheelbase is 4.41 m.

The M916A1 contract was awarded to Freightliner, in September 1988, for 393 units. The first deliveries were made in May 1989 and the first production vehicles entered the US Army's inventory in November 1990.

M917, (Dump Truck, 20 ton, 8 × 6)

This has a rear tipping dump body by Fruehauf which has a heaped capacity of 10.7 m^3, and the same engine, gearbox, transfer box and suspension as the M916.

The M917 is used for earthmoving and construction projects, augmenting the current fleet of 20 ton dump

Specifications

Designation	**M915**	**M916**	**M917**	**M918**	**M919**	**M920**
Type	tractor	tractor	dump truck	bituminous	concrete	tractor
Cab seating	1 + 2	1 + 2	1 + 2	1 + 2	1 + 2	1 + 2
Configuration	6 × 4	6 × 6	8 × 6	6 × 6	8 × 6	8 × 6
Weight (empty)	8,446 kg	11,327 kg	14,768 kg	13,535 kg	15,791 kg	12,414 kg
(loaded)	22,213 kg	29,629 kg	33,070 kg	19,789 kg	32,056 kg	30,716 kg
Weight on front axle (loaded)	5,443 kg	6,543 kg	3,775 kg	5,600 kg	4,511 kg	3,864 kg
Weight on rear axles (loaded)	17,236 kg	23,036 kg	33,070 kg	14,210 kg	32,506 kg	30,716 kg
Gross combination weight	47,627 kg	57,153 kg	n/a	n/a	n/a	61,235 kg
Length	6.49 m	7.48 m	8.9 m	8.9 m	9.5 m	8.11 m
Width	2.49 m	2.49 m	2.49 m	2.49 m	2.49 m	2.49 m
Height (cab)	2.93 m	3.25 m	3.25 m	3.25 m	3.25 m	3.25 m
(overall)	3.31 m	3.61 m	3.61 m	3.61 m	3.61 m	3.61 m
Ground clearance	0.254 m	0.305 m	0.305 m	0.295 m	0.305 m	0.295 m
Track (front/rear)	1.99 m/1.82 m	1.98 m/1.85 m	1.98 m/1.85 m	1.98 m/1.85 m	1.98 m/1.85 m	1.98 m/1.85 m
Wheelbase (excluding pusher axle)	4.24 m + 1.32 m	4.72 m + 1.42 m	4.876 m + 1.42 m	4.72 m + 1.42 m	4.876 m + 1.42 m	4.876 m + 1.46 m
Angle of approach	38°	42°	41°	42°	42°	42°
Max speed (road)	107 km/h	103 km/h	107 km/h	103 km/h	107 km/h	103 km/h
(3.9% gradient)	40.2 km/h	40.2 km/h	40.2 km/h	40.2 km/h	40.2 km/h	40.2 km/h
Fuel capacity	416 litres	416 litres	416 litres	416 litres	416 litres	416 litres
Fording	0.51 m	0.73 m	0.61 m	0.61 m	0.61 m	0.61 m
Steering	all have Ross integral power system					
Turning radius	8.15 m	12.2 m	13.65 m	12.2 m	13.65 m	13.65 m
Tyres	10.00 × 20	11.00 × 24	11.00 × 24	11.00 × 24	11.00 × 24	11.00 × 24
Electrical system (24 V optional)	12 V	12 V	12 V	12 V	12 V	12 V
Batteries	4 × 12 V	4 × 12 V	4 × 12 V	4 × 12 V	4 × 12 V	4 × 12 V

Note: On the 8 × 6 model the second axle is of the pusher type and can be raised if required.

trucks and in some selected units replacing 5 ton dump trucks.

A M917A1 version is produced by the Freightliner Corporation.

M918, (Bituminous Distributor, 6 × 6)
This has a bituminous spreader on the rear manufactured by E D Etnyre and Company of Oregon. Engine, transmission, transfer box and suspension are the same as for the M916.

The M918 is used for distributing liquid bitumen for road and airfield construction and is provided with a hydrostatically driven bituminous pump.

M919, (Concrete Mobile, 8 cu yd, 8 × 6)
This has a concrete mixer to the rear manufactured by the National Concrete Machinery Company of Lancaster, Pennsylvania. The engine, transmission, transfer box and suspension are the same as in the M916. It can transport dry concrete ingredients and water, mix the ingredients in various increments and proportions, and pour the mixed concrete. The M919 is charged with dry ingredients by aggregate bins or scoop loaders and discharges into mixed concrete handling equipment. This vehicle can also be used as a central mix plant where large amounts of concrete are required at a single location.

A M919A1 version is produced by the Freightliner Corporation.

M920, (Tractor, Truck, 20 ton, 8 × 6)
This has a fifth wheel at the rear and is used to haul semi-trailers. Mounted to the rear of the cab is a hydraulic winch with a capacity of 20,250 kg. The engine, transmission, transfer box and suspension are the same as in the M916.

The M920 will, together with the M916 series, replace the M123, and is used to haul the M870 semi-trailer.

The Freightliner Corporation is producing a Freightliner Military Truck Family based on its M915A2 and M916A1 - see following entry for details.

Status
M915A2 and M916A1 in production. In service with the US Army.

Manufacturers
(M915 and M916 - production complete) AM General Corporation, 105 North Niles Avenue, South Bend, Indiana 46617, USA.
Tel: +1 219 284 2942. Fax: +1 219 284 2959.
(M915A2 and M916A1 - in production) Freightliner Corporation, 4747 North Channel Avenue, PO Box 3849, Portland, Oregon 97208, USA.
Tel: +1 503 735 8000.

UPDATED

Freightliner Military Truck Family

Description
The Freightliner Corporation is producing the Freightliner Military Truck Family based on its M915A2 and M916A1 tractor trucks already in service with the US Army (see previous entry). These two models are included in the family and, using them as a basis, Freightliner is also producing a 6 × 6 wrecker (see entry under *Recovery Vehicles* for details of this model), a 60 ton capacity Medium Equipment Transporter (MET) based on a tractor truck unit, a 6 × 6 dump truck (ordered by the US Army as the M917A1), and a mobile concrete mixer - this has been type standardised as the M919A2 and is in production.

All vehicles are based on the Detroit Diesel DDE Series 60 12.7 litre engine with microprocessor control and an Allison automatic transmission. All vehicles have a WABCO four-channel Anti-lock Brake System (ABS) and Ross integral power steering.

In March 1995 it was announced that the Freightliner Corporation had formed an alliance with the Oshkosh Truck Corporation in which Freightliner would transfer the marketing, manufacturing and administrative functions of its specialised defence business to Oshkosh. In some regions existing Freightliner local arrangments will remain while in others Oshkosh will assume the Freightliner business. It is possible that in some regions a completely new arrangement may be formed.

Specifications

Model	**M915A2**	**M916A1**	**M917A1**	**M919A2**	**—**
Type	tractor truck	tractor truck	dump	concrete mixer	MET
Cab seating	1 + 1	1 + 1	1 + 1	1 + 1	1 + 1
Configuration	6 × 4	6 × 6	6 × 6	6 × 6	6 × 6
Weight (kerb)	8,481 kg	12,594 kg	13,358 kg	15,442 kg	13,363 kg
Weight on front axle (loaded)	5,448 kg	7,258 kg	7,258 kg	7,258 kg	7,258 kg
Weight on rear axle (loaded)	18,160 kg	25,587 kg	25,587 kg	25,587 kg	25,587 kg
GCW	47,670 kg	59,020 kg	n/a	n/a	87,317 kg
Length	7 m	7.35 m	7.52 m	9.65 m	7.39 m
Width	2.49 m	2.49 m	2.49 m	2.49 m	2.49 m
Height (cab)	2.76 m	2.97 m	2.97 m	2.97 m	2.97 m
(overall)	3.02 m	3.24 m	3.24 m	3.24 m	3.24 m
Ground clearance	0.23 m	0.23 m	0.23 m	0.23 m	0.23 m
Wheelbase	4.12 m	4.41 m	4.41 m	4.83 m	4.75 m
Angle of approach	27°	37.5°	37.5°	37.5°	37.5°
Max speed (road)	90 km/h	85.3 km/h	85.3 km/h	85.3 km/h	78 km/h
(3% grade)	47 km/h	40.22 km/h	40.22 km/h	40.22 km/h	29 km/h
Fuel capacity	379 litres	379 litres	379 litres	379 litres	379 litres
Fording	0.51 m	0.51 m	0.51 m	0.51 m	0.51
Engine	Detroit Diesel Series 60 12.7 litres 6-cylinder diesel developing 400 hp at 2,100 rpm[1]				
Transmission	Allison HT-740 automatic with 4 forward and 1 reverse gears[2]				
Steering	Ross power	Ross power	Ross power	Ross power	Ross power
Turning radius	16.44 m	24.21 m	24.21 m	25.1 m	25.4 m
Brakes	air/mechanical	air/mechanical	air/mechanical	air/mechanical	air/mechanical
Tyres	315/80 R 22.5	315/80 R 22.5	315/80 R 22.5	315/80 R 22.5	315/80 R 22.5
Electrical system	12/24 V	12/24 V	12/24 V	12/24 V	12/24 V
Batteries	4 × 12 V	4 × 12 V	4 × 12 V	4 × 12 V	4 × 12 V

1 Engine on MET develops 450 hp at 2,100 rpm
2 Transmission on MET is Allison CLT-754 automatic with torque converter with 5 forward and 1 reverse gears

Freightliner M915A2, Tractor, Truck, 14 ton, 6 × 4

Freightliner M916A1, Tractor, Truck, 20 ton, 6 × 6

Freightliner 60 ton capacity (6 × 6) Medium Equipment Transporter

Freightliner M917A1 6 × 6 dump truck

Status
In production.

Manufacturer
Freightliner Corporation, 4747 North Channel Avenue, PO Box 3849, Portland, Oregon 97208-3849, USA.
Tel: +1 503 735 8000.

UPDATED

Oshkosh (10 × 10) Palletised Load System (PLS)

Development

The US Army Tank Automotive Command awarded prototype contracts to three contractors for prototype Palletised Load System (PLS) vehicles in January 1989. Each of the three firms involved delivered nine trucks, six trailers and 30 flatracks. Prototype hardware testing began in September 1989. The three firms involved in the programme were the Oshkosh Truck Corporation, the PACCAR Government Group, and General Motors, Military Vehicle Operations.

In September 1990 Oshkosh Truck Corporation was awarded a five-year US$860 million contract for 2,626 PLS trucks, 1,050 trailers (a LOHR design, see entry in *Materials handling equipment* section for details) and 11,030 flatracks. The contract included annual 100 per cent options raising the contract value to approximately US$1.1 billion. Production commenced in 1992 and will last until August 1997.

The US Army initially purchased the PLS for ammunition resupply. Other applications under development are the resupply of MLRS pods, fuel and water tanks, medical shelters, Ribbon Bridge components, portable kitchens, command/control shelters, construction plant, barrier material and tank ammunition. A PLS dump body with a payload capacity of 12,420 kg was produced for user evaluation commencing in March 1996. A PLS Patriot or THAAD lightweight launcher platform is undergoing testing.

Description

The Oshkosh PLS prime mover is a 10 × 10 full-time all-wheel drive truck. The vehicle has a front tandem unit with two powered steering axles and a rear tridem unit with two powered rigid axles plus one powered steering axle. The 10 × 10 configuration provides optimum turning capability and good weight distribution between all axles to provide mobility and capability above the desired level of the US Army specification. The vehicle is diesel powered with an automatic shift transmission, a two-speed transfer case, central tyre inflation, and full-time drive to all 10 wheels.

The PLS truck is capable of transporting 16.5 tons of flatrack-mounted cargo, and can tow a PLS trailer loaded with a 16.5 ton flatrack payload.

The PLS truck/trailer design includes an integral self-load/unload capability using the PLS flatrack. Oshkosh selected the Multilift Mark 5 load handling lift system and licence to produce the system for the PLS contract. A single operator can load or unload a 16.5 ton flatrack in less than one minute without leaving the cab.

The vehicles are air-transportable in C-5A and C-17 cargo aircraft and can be carried in C-141 aircraft with preparation.

One half of the trucks involved in the PLS contract are equipped with a Grove material handling crane with a capacity of 1,769 kg at 12.1 m and are designated the M1074. Trucks without the material handling crane are designated the M1075. The trailers and flatracks are the M1076 and M1077 respectively.

The General Purpose A-frame Flatrack Type M1077 is 6.058 m long, 2.438 m wide and 1.5915 m high over the A-frame. There is a further flatrack, the ISO-compatible Palletized Flatrack (IPF) Type M1. On this there are two end walls, one incorporating the A-frame. Both walls can fold down inwardly for stacking when empty. The dimensions are the same as for the M1077, apart from the end walls which have a height of 2.083 m.

Variants

XM1120 HEMTT-LHS (8 × 8 PLS)

The XM1120 HEMTT 8 × 8 chassis has been proposed for a number of PLS applications, ranging from 8 ton flatracks to compatible ISO containers. It is currently being used as the Common Bridge Transporter (CBT) to support engineer Ribbon Bridge units. Other proposed applications include the carrying of shelters (including the Standardized Integrated Command Post System (SICPS)), and THAAD missile launchers. Other applications could be included.

Forward Repair System Heavy

A Forward Repair System Heavy (FRS-H) prototype has been produced based on the Oshkosh PLS (10 × 10) chassis. For details see entry in the *Recovery vehicles* section.

Specifications

Cab seating: 1 + 1
Configuration: 10 × 10
Weights:
(kerb, with crane and flatrack) 24,948 kg
(kerb, without crane, with flatrack) 22,680 kg
(GVW) 39,916 kg
(GCW) 62,369 kg
Max PLS payload: 14,969 kg
Length:
(chassis) 10.671 m
(vehicle and trailer, max) 18.936 m
Width: 2.438 m
Height:
(overall) 3.284 m
(cab) 2.598 m
(flatbed) 1.801 m
Ground clearance: 0.4 m
Track: 1.925 m
Wheelbase: 5.682 m
Angle of approach/departure: 42°/61°
Max speed: (GCW, road) 91 km/h
Range: 541 km
Fuel capacity: 379 litres
Max gradient: 60%
Fording: 1.219 m
Engine: Detroit Diesel Model 8V-92TA 12.06 litre V-8 diesel equipped with electronic control system developing 500 hp at 2,100 rpm
Transmission: Allison CLT-755 5-speed automatic with

Oshkosh XM1120 (8 × 8) HEMTT-PLS truck

1997

Oshkosh XM1120 (8 × 8) HEMTT-PLS truck carrying a fuel rack (T J Gander)

1997

torque converter and equipped with Allison Transmission Electronic Controls (ATEC)
Transfer box: Oshkosh 55000 Series 2-speed with 30:70 differential
Steering: power-assisted front tandem co-ordinated with power-assisted rear axle steer
Turning radius: (wall to wall) 36.5 m
Suspension: Hendrickson walking beam front and rear with mid-axle equipped with Hendrickson-Turner air ride suspension
Tyres: 16.00 R 20 Michelin with central tyre inflation system
Brakes:
(main) Rockwell wedge drum, air-operated, dual system, internal shoe
(parking) spring brake mounted on rear tridem axles
Electrical system: 12/24 V
Batteries: 4 × 12 V
Alternator: 145 A

Status
Production commenced in early 1992 for the US Army (2,626 units).

Manufacturer
Oshkosh Truck Corporation, PO Box 2566, Oshkosh, Wisconsin 54903-2566, USA.
Tel: +1 414 235 9150. Fax: +1 414 233 9540.

UPDATED

Oshkosh (10 × 10) Palletised Load System (PLS) equipped with Multilift Mark 5 load handling system
1997

YUGOSLAVIA (SERBIA AND MONTENEGRO)

FAP 2220 BDS (6 × 4) and FAP 2026 BS (6 × 6) trucks

Description
The FAP 2220 BDS (6 × 4) truck was originally used for carrying and launching a truck-mounted scissors bridge. This vehicle can be distinguished from the FAP 2026 BS which has a central tyre pressure regulation system and single rear wheels.

The FAP 2026 BS/AV (6 × 6) was developed specifically for the former Yugoslav Army and features a central tyre pressure regulation system. The two-door all-steel cab is of the forward control type with the cargo area at the rear with a drop tailgate, removable bows and a tarpaulin cover. The engine for these vehicles was built under licence from the former British Leyland.

The FAP 2220 BDS and FAP 2026 BS (6 × 6) are used as the chassis for the 128 mm (32-round) M-77 Oganj multiple rocket system. Details of this equipment are given in *Jane's Armour and Artillery 1995-96,* pages 752 and 753.

The FAP 2832 BS/AV (8 × 8) 9,000 kg truck is described in the following entry in this section.

Specifications (FAP 2026 BS/AV (6 × 6))
Cab seating: 1 + 1 (20 in cargo area)
Configuration: 6 × 6
Weight:
(empty) 11,000 kg
(loaded) 21,000 kg
Weight on front axle: (loaded) 6,000 kg
Weight on rear axle: (loaded) 15,000 kg
Max load: 10,000 kg
Max towed load: 7,200 kg
Max winch capacity: 10,000 kg
Load area: 4.53 × 2.342 m
Length: 7.72 m
Width: 2.49 m
Height: (top of cab) 3.1 m
Wheelbase: 3.4 m + 1.4 m
Track: 2.02 m
Angle of approach/departure: 40°/40°
Max speed: 80 km/h
Fuel consumption: 33 l/100 km
Range: 600 km
Max gradient:
(less trailer) 60%
(with trailer) 40%
Fording: 0.3 m
Engine: V-8 direct injection water-cooled diesel developing 256 hp at 2,500 rpm
Gearbox: 6MS-80 with 6 forward and 1 reverse gears
Clutch: single dry plate
Transfer box: 2-speed
Steering: hydraulic ball, joint
Turning radius: 11 m
Suspension: leaf springs with additional rubber springs and telescopic shock-absorbers
Tyres: 15.00 × 21
Brakes:
(main) dual circuit air drum type
(parking) mechanical, air-actuated
Electrical system: 24 V
Number of batteries: 2 × 12 V, 143 Ah

Status
In service with the former Yugoslav Army and Saudi Arabia.

Manufacturer
Motor Vehicle Plant, Priboj.
Enquiries to: Holding Company YUGOIMPORT-SDPR, PO Box 89, Bulevar umetnosti 2, YU-11070 Novi Beograd, Yugoslavia.
Tel: +381 11 222 2743/222 3697. Telex: 71000 JU SDPR.
Fax: +381 11 324 8791.

UPDATED

FAP 2026 BS carrying 128 mm (32-round) M-77 Oganj multiple rocket system under canvas cover

FAP 2026 BS (6 × 6) truck

FAP 2832 BS/AV (8 × 8) 9,000 kg truck

Description

The FAP 2832 BS/AV (8 × 8) 9,000 kg truck is the 8 × 8 equivalent of the FAP 6 × 6 trucks (see previous entry). The vehicle has a forward control cab which is identical to that of the FAP 6 × 6 trucks and can be tipped forward for maintenance access. Behind the cab is a spare wheel carried in a handling device. The steel-sided cargo area has a tailgate and is covered by a tilt carried on curved bows. A central tyre pressure system is provided with control from the cab. A winch is a standard fitting and is located at the rear for forward or rear use.

The FAP 2832 BS/AV can be used to tow artillery weighing up to 11,000 kg and can carry a maximum load of 14,000 kg on roads.

FAP 2832 BS/AV (8 × 8) 9,000 kg truck

The FAP 3232 BDST/AV is a tractor truck version of this vehicle and is covered in an entry in the *Tank Transporters* section.

Specifications

Cab seating: 1 + 1
Configuration: 8 × 8
Weight:
(empty) 16,000 kg
(loaded, road) 30,000 kg
(loaded, cross-country) 25,000 kg
Max load:
(road) 14,000 kg
(cross-country) 9,000 kg
Towed load: 11,000 kg
Length: 9.02 m
Width: 2.5 m
Height: (top of tilt) 3.319 m
Ground clearance: 0.38 m
Track: 2.02 m
Wheelbase: 1.5 m + 3.2 m + 1.4 m
Angle of approach/departure: 41°/45°
Range: 600 km
Max gradient: 32%
Fording: 1.2 m
Engine: OM 403 15.95 litres V-10 4-stroke diesel developing 320 hp
Gearbox: 8 forward and 1 reverse gears
Clutch: single dry plate
Transfer box: 2-speed
Steering: hydraulic, ball, joint
Turning radius: 13 m
Suspension: parabolic leaf springs
Tyres: 15.00 × 21 T-101
Brakes:
(main) dual circuit air plus exhaust brake
(parking) mechanical with air actuation
Electrical system: 24 V
Batteries: 2 × 12 V, 210 Ah

Status

In service with the former Yugoslav Army.

Manufacturer

Motor Vehicle Plant, Priboj.
Enquiries to: FDSP - Federal Directorate of Supply and Procurement, PO Box 89, Bulevar umetnosti 2, YU-11070 Novi Beograd, Yugoslavia.
Tel: +381 11 222 2743. Telex: 71000 JU SPDR.
Fax: +381 11 324 8738.

UPDATED

TANK TRANSPORTERS

This section covers prime movers which are used principally for towing semi-trailers carrying MBTs. Many military trucks also exist as tractor trucks, these being used for towing semi-trailers carrying cargo or light armoured vehicles. For example the MAZ-535 is the tractor truck version of the MAZ-543 truck and is often used to tow a semi-trailer carrying a light tank. The reader is referred to the *Trucks* section for vehicles of this type. Details of the tank semi-trailers will be found in the *Trailers* section.

AUSTRIA

Steyr 1491.330.S34/6 × 6 and 3891/6 × 6 tractor trucks

Description

The Steyr 1491.330.S34/6 × 6 tractor truck is based on a heavy-duty all-wheel drive commercial vehicle adapted to meet military requirements. It was designed to pull semi-trailers on roads and prepared surfaces up to a weight of 85,000 kg. Most of the components used on this tractor truck are identical to those used on the Steyr 1491/6 × 6 M truck.

The forward control cab is all steel, has seating for the driver and two passengers and may be tilted forward 70° for engine access. An observation hatch in the roof is standard and a machine gun may be mounted on the roof. The chassis frame is of the parallel ladder type with an integrated subframe for the winch and the fifth wheel. There are two hydraulic winches, each with a pulling capacity of 20,000 kg and are mounted behind the cab for loading and offloading. The vehicle is fully rail-transportable and can be used with all conventional semi-trailers.

A version is also available with 14.00 R 20 MPT tyres, single at the front and dual on the rear axles. This is known as the Steyr 3891/6 × 6 and was produced for delivery to Saudi Arabia.

Specifications

Cab seating: 1 + 2
Configuration: 6 × 6
Weight:
(kerb) 13,000 kg
(5th wheel) 21,000 kg
(GCW) 85,000 kg
Length: 7.051 m
Width: 2.462 m
Height: 3.14 m
Ground clearance: 0.314 m
Wheelbase: 3.4 m + 1.35 m
Max speed: 60 km/h
Fuel capacity: 380 litres
Max gradient: (85 t GCW) 25%
Engine: Steyr Model WD 815.74 V-8 water-cooled direct injection turbocharged diesel developing 340 hp at 2,200 rpm
Gearbox: ZF 5 S 111 GP manual, synchronised with 8 speeds plus 1 crawler forward and 1 reverse
Transfer box: Steyr VG 1200 2-speed
Steering: ZF 8046 hydraulic power-assisted
Turning radius: 9 m
Suspension:
(front) semi-elliptical leaf springs with shock-absorbers
(rear) two swivel springs
Tyres: 12.00 × 20 (14.00 R 20 MPT optional)
Brakes:
(main) dual-circuit compressed air
(trailer) two-line compressed air
(parking) spring energy brake cylinder, air operated
Electrical system: 24 V
Batteries: 2 × 12 V, 135 Ah

Status

Steyr 3891/6 × 6 in service with Saudi Arabia.

Manufacturer

Steyr Nutzfahrzeuge AG, Schönauerstrasse 5, A-4400 Steyr, Austria.
Tel: +43 7252 585-0. Telex: 28200.
Fax: +43 7252 46746/48650.

VERIFIED

Steyr 3891/6 × 6 tractor truck as produced for Saudi Arabia

Steyr 1491.330.S34/6 × 6 tractor truck

Steyr 37M42/S38/6 × 6 tractor truck

Description

The Steyr 37M42/S38/6 × 6 tractor truck is a derivative of the Steyr 1491 series of trucks and is used to tow fifth wheel loads of up to 22,000 kg. It can be used as a heavy tractor truck for trailers and semi-trailers with payloads up to 60 tonnes. Two 25 tonne winches are provided to handle disabled vehicles.

Steyr 37M42/S38/6 × 6 tractor truck

Specifications

Cab seating: 1 + 2
Configuration: 6 × 6
Weight:
(empty) approx 15,000 kg
(loaded) 37,000 kg
(gross train weight) 90,000 kg
Weight on front axle: 8,000 kg
Weight on rear axle: 2 × 14,500 kg
Max load: 22,000 kg
Length: 7.764 m
Width: 2.072 m
Height: 3.246 m
Ground clearance: 0.38 m
Track: 2.072 m
Wheelbase: 3.8 m + 1.47 m
Angle of approach/departure: 31°/35°
Max speed: approx 80 km/h
Range: approx 500 km
Fuel capacity: 500 litres
Gradient: (fully loaded) 34.4%
Side slope: 30%
Fording: 0.8 m

Specifications

Engine: Steyr WD 815.76 12 litre V-8 direct injection, water-cooled, turbocharged, intercooled diesel developing 416 hp at 2,200 rpm
Gearbox: ZF 4 S 150 GP 8-speed synchromesh
Clutch: WSK 400 torque converter
Transfer box: Steyr VG 2000, 2-speed
Steering: ZF 8046 power-assisted
Turning radius: 10.8 m
Suspension: semi-elliptical springs with shock-absorbers at front
Tyres: 14.00 R 20
Brakes: dual-circuit, air
Electrical system: 24 V
Batteries: 2 × 12 V, 135 Ah each
Alternator: 28 V, 55 A

Status

In production.

Manufacturer

Steyr Nutzfahrzeuge AG, Schönauerstrasse 5, A-4400 Steyr, Austria.
Tel: +43 7252 585-0. Telex: 28200.
Fax: +43 7252 46746/48650.

VERIFIED

CANADA

Western Star M6966 (6 × 6) Heavy Equipment Transporter (HET)

Description

The Western Star M6966 (6 × 6) Heavy Equipment Transporter (HET) is a development of Western Star's heavy-duty off-road vehicle series. The layout is conventional with the engine in front of the galvanised all-steel cab. The standard cab provides seating for the driver and two passengers. Additional accommodation for four passengers is available if required.

Power is provided by a 515 hp Caterpillar 3406C diesel coupled to an Eaton/Fuller Ceemat nine-speed automatic transmission. Axles are planetary hub reduction and are equipped with central tyre inflation.

The comprehensive standard equipment includes two recovery winches rated at 25,000 kg each, a fifth wheel with a 3.5 in/88.9 mm SAE kingpin, tail roller, spare tyre and carrier, 12 and 24 V trailer connections and work lights. A full range of options is available to permit use in extreme climatic conditions, including an oversized radiator for operations in temperatures of over +60°C, air conditioning and a high-capacity heater.

Western Star M6966 (6 × 6) Heavy Equipment Transporter (HET) ***1997***

Specifications

Cab seating: 1 + 2 or 1 + 5
Configuration: 6 × 6
Weight:
(kerb) 17,241 kg
(GVW) 38,112 kg
(GCW) 110,000 kg
Load on 5th wheel: (nominal) 20,871 kg
Length: 9.12 m
Width: (with 24 R 21 tyres) 3.17 m
Height:
(overall) 3.28 m
(5th wheel) 1.7 m
Track: 2.54 m
Wheelbase: 6.1 m
Max speed: 80 km/h
Range: (cruising) 1,020 km
Fuel capacity: 925 litres
Gradient: (at max GCW) 30%
Fording: 1.05 m
Engine: Caterpillar 3406C 6-cylinder in-line diesel developing 515 hp at 2,100 rpm
Transmission: Eaton/Fuller Ceemat automatic with 9 forward speeds and 1 reverse gear
Transfer box: single speed
Steering: hydraulic assisted, dual gear
Turning radius: 12.3 m
Tyres: standard, 24 R 21; 16 R 20 or 14 R 24 optional
Brakes: dual-circuit, air
Electrical system: 24 V (12 V optional)
Alternator: 70 A

Status

In production.

Manufacturer

Western Star Trucks Inc, 2076 Enterprise Way, Kelowna, British Columbia, Canada V1Y 6H8.
Tel: +1 604 860 3319. Fax: +1 604 860 1252.

NEW ENTRY

CHINA, PEOPLE'S REPUBLIC

Hanyang HY 480 (8 × 8) tractor truck

Description

The HY 480 (8 × 8) heavy tractor truck is used for the moving of heavy loads carried on a semi-trailer and uses a standard fifth wheel (88.9 mm) carried between the third and fourth axles at the rear. The cab is of the forward control type with seating for a driver and one passenger. There is room behind the seats for a bunk or seating for four passengers. The cab can be tilted forwards for maintenance access to the engine and other components.

Specifications

Cab seating: 1 + 5
Configuration: 8 × 8
Weight:
(unladen) 13,900 kg
(max GVW) 34,000 kg
(fifth wheel loading) 20,000 kg
Length: 8.595 m
Width: 2.58 m
Height:
(top of cab) 2.95 m
(fifth wheel) 1.5 m
Ground clearance: 0.39 m
Wheelbase: 1.45 m + 3.5 m + 1.45 m
Track: 2.145 m
Max speed: 70 km/h
Max gradient: 35%
Engine: KHD BF 12 L 413 FC developing 500 hp at 2,500 rpm

Hanyang HY 480 (8 × 8) tractor truck

Hanyang HY 480 (8 × 8) tractor truck

Gearbox: 8TBF-150 8/8.14-1
Clutch: YB400
Turning radius: 12 m
Tyres: 14.00 × 20
Electrical system: 24 V

Status
In production.

Manufacturer
Hanyang Special Auto Works, Hanyang Bridge Jianghan Road, Wuhan, Hubei Province, People's Republic of China.

VERIFIED

Hanyang HY 473A 32.35 (6 × 6) tractor truck

Description
The Hanyang HY 473A 32.35 (6 × 6) tractor truck is used for the towing of semi-trailers carrying tanks up to approximately 50,000 kg in weight. It has a forward control cab that can be tilted forward for engine access and the cab is large enough to accommodate bunks behind the seats. Two winch drums are located behind the cab for loading and unloading. This vehicle is used to tow the HY 962J semi-trailer (see entry in the *Trailers* section).

The HY 472 and HY 473 are essentially similar vehicles intended mainly for commercial use and the HY 960 is a military version with an enlarged crew cab.

Specifications
Cab seating: 1 + 4 to 6
Configuration: 6 × 6
Weight:
(empty) 12,500 kg
(5th wheel loading) 19,500 kg
(GVW) 32,000 kg
(GCW) 62,000 kg
Length: 7.245 m
Width: 2.58 m
Height:
(cab) 2.9 m
(5th wheel) 1.45 m
Ground clearance: 0.34 m
Wheelbase: 3.5 m + 1.35 m
Track:
(front) 2.05 m
(rear) 1.92 m
Max speed: 64 km/h
Gradient: 24%
Engine: KHD F 12 L 413F developing 355 hp at 2,500 rpm
Gearbox: 9TBF-110 9/13.40-1
Turning radius: 10 m
Tyres: 12.00 × 20
Electrical system: 24 V

Status
In service with the People's Liberation Army.

Manufacturer
Hanyang Special Auto Works, Hanyang Bridge Jianghan Road, Wuhan, Hubei Province, People's Republic of China.

VERIFIED

Hanyang HY 960 (6 × 6) tractor truck with tank transporter semi-trailer (Christopher F Foss)

Hanyang HY 960 (6 × 6) tractor truck (Christopher F Foss)

FRANCE

Renault TRM 700-100 (6 × 6) tractor truck

Development
The Renault TRM 700-100 (6 × 6) tractor truck was first shown at the 1987 Satory exhibition and was designed as a tank transporter for the Leclerc MBT. In March 1994 the French Ministry of Defence placed an initial order for 179 of these vehicles with the first 35 units to be delivered to the French Army between January 1995 and February 1996. The remaining 144 will be delivered by the year 2000. A second order for 121 vehicles for delivery between the years 2000 and 2002 is anticipated, raising the total order to 300 units.

Description
The Renault TRM 700-100 (6 × 6) tractor truck uses a forward control four-door cab similar to that used on the Renault TRM 10 000 (6 × 6) 10,000 kg truck (see entry in the *Trucks* section) but enlarged to carry the driver and four passengers. The cab can be tilted forward hydraulically for maintenance purposes. The chassis has reinforced pressed steel side members with inner sections connected by cross members.

The engine is a Type E9 turbocharged diesel developing 700 hp coupled to an automatic six-speed gearbox (with integrated electronic control and a hydraulic retarder). The transfer box is integral with the gearbox. Hydraulic power-assisted steering is provided. Two optional hydraulic 15 tonne loading winches are located behind the cab.

The TRM 700-100 has been demonstrated towing a Nicolas SFD A6 semi-trailer with a capacity of 70,000 kg.

Specifications
Cab seating: 1 + 4
Configuration: 6 × 6
Weight:
(kerb) 16,800 kg
(GVW) 39,000 kg
(GTW) 106,000 kg
Towed load: 90,000 kg
Max load on front axle: 10,000 kg
Max load on rear axles: 30,000 kg
Max load on 5th wheel: 23,000 kg
Length: 8.045 m
Width: (overall) 2.68 m
Height: (unloaded) 3.283 m
Ground clearance: (rear axle, loaded) 0.341 m
Wheelbase: 4.325 m + 1.35 m

Renault TRM 700-100 (6 × 6) tractor truck (T J Gander) ***1997***

Renault TRM 700-100 (6 × 6) tractor truck (T J Gander) ***1997***

Track:
(front) 2.12 m
(rear) 2.03 m
Angle of approach: 42°
Max speed: 80 km/h
Fuel capacity: 1,000 litres (2 × 500 litres)
Range: 850 km
Max gradient: 20%
Engine: RVI Type E9 16.4 litre V-8 turbocharged intercooled diesel developing 700 hp at 2,500 rpm
Transmission: Type PS 226 automatic with 6 forward gears and 1 reverse; Type WR hydraulic torque converter with lock-up and hydraulic retarder
Transfer box: single-speed, integrated
Steering: Type 8098 hydraulic power-assisted
Turning radius: 14 m
Suspension:
(front) semi-elliptic leaf springs with auxiliary equaliser spring and hydraulic telescopic shock-absorbers
(rear) inverted semi-elliptic leaf springs, oscillating swivels and reaction rods
Tyres: 385/95 × T4 TL 1646
Brakes: dual-circuit, air
Electrical system: 24 V
Batteries: 4 × 12 V, 125 Ah
Alternator: 55 A

Renault TRM 700-100 (6 × 6) tractor truck towing semi-trailer carrying Leclerc MBT ***1997***

Status
In production for the French Army (initial order 179 units).

Manufacturer
Enquiries to: Renault Véhicules Industriels, 40 rue Pasteur, BP 302, F-92156 Suresnes Cedex, France.
Tel: +33 1 40 99 71 11. Telex: 620567 sdce f.
Fax: +33 1 40 99 71 08.

UPDATED

Renault R 390 (6 × 4) tractor truck

Description
The R 390 (6 × 4) tractor truck is a standard commercial vehicle adapted to meet military requirements. It was adopted by the French Army for road use and is used in conjunction with the Nicolas STA 43 semi-trailer. The vehicle can be used with other semi-trailers carrying MBTs up to a maximum weight of 55,000 kg.

The chassis consists of two U-shaped side members with five cross members welded and riveted into position. The vehicle is fitted with a model KB 2480 cab which is of the two-door forward control type and has seats for the driver, two passengers and two bunks at the rear of the cab. The cab is heated and soundproofed and can be tilted forward to an angle of 70° to allow maintenance work on the engine. Optional equipment includes a hydraulic winch with a capacity of 15,000 kg and 90 m of cable, dividing curtain, elbow rests, heating system, fog lamps, tachograph, radio equipment, blackout lights, searchlight, flashing light, combined air/electric cables and an anti-freeze device for the brake circuit.

The French Army uses a similar version of the Renault R390 (6 × 4) tractor truck for towing semi-trailers carrying fuel and other supplies. The French Army also uses the Berliet TR 260 (4 × 2) tractor truck for hauling a variety of semi-trailers.

Specifications
Cab seating: 1 + 4
Configuration: 6 × 4
Weight:
(empty) 10,300 kg
(loaded) 34,500 kg
Weight on front axle:
(empty) 5,300 kg
(loaded) 6,500 kg
Weight on rear bogie:
(empty) 5,000 kg
(loaded) 28,000 kg
Max weight on 5th wheel: 24,000 kg
Length: 6.53 m
Width: 2.49 m
Height: 3 m
Ground clearance: 0.303 m
Track:
(front) 2.028 m
(rear) 1.825 m
Wheelbase: 2.95 m + 1.35 m
Angle of approach/departure: 22°/45°
Max speed: 66 km/h
Range: 800 km
Fuel capacity: 650 litres
Max gradient: (laden with 70,600 kg gross weight) 15%
Fording: 0.7 m
Engine: Renault model MIVR 08-35-30 8-cylinder water-cooled supercharged diesel developing 390 hp at 2,100 rpm
Gearbox: Renault B 9 with 18 forward and 2 reverse gears
Steering: 8065 hydraulic power-assisted
Turning radius: 8.5 m
Suspension:
(front) leaf springs and shock-absorbers
(rear) torsion bar stabiliser; rear wheels have reduction gears in the wheel hubs, differential locking system between axles and wheels
Tyres: 13 × 22.5
Brakes: pneumatic
Electrical system: 24 V
Batteries: 4 × 12 V, 200 Ah

Renault R 390 (6 × 4) tractor truck towing Nicolas semi-trailer (M Bell)

Status
Production complete. In service with the French Army.

Manufacturer
Enquiries to: Renault Véhicules Industriels, 40 rue Pasteur, BP 302, F-92156 Suresnes Cedex, France.
Tel: +33 1 40 99 71 11. Telex: 620567 sdce f.
Fax: +33 1 40 99 71 08.

VERIFIED

Renault TRM 340.34 T (6 × 6) tractor truck

Description
The Renault TRM 340.34 T (6 × 6) tractor truck replaced the earlier TBH 280 (6 × 6) tractor truck in production and is a variant of the TRM 340.34 (6 × 6) 16,000 kg truck (see entry under *Trucks* for details).

The TRM 340.34 was designed for towing semi-trailers both on roads and across country up to a maximum weight of 65,000 kg. The chassis consists of U-shaped side members with the cross members bolted into position. The layout of the vehicle is conventional, with the engine at the front, cab in the centre and fifth wheel at the rear. The rear bogie has both inter-wheel and inter-axle differential locks.

Optional equipment includes a heating system and a sleeper cab with one bunk. The engine can have a preheating system, a heavy-duty air filter and a vertical exhaust. A power take-off can be provided on the transfer case. The chassis can be provided with a 3.5 in/89 mm fifth wheel coupling with single or double oscillation. Electrical equipment can include an interference suppressor and blackout lighting. Further options include twin 15 tonne hydraulic winches.

Specifications

Cab seating: 1 + 2
Configuration: 6 × 6
Weight:
(empty) 9,905 kg
(loaded) 34,000 kg
(max GCW, on road) 100,000 kg
Weight on front axle:
(empty) 5,049 kg
(loaded) 7,500 kg
Weight on rear bogie:
(empty) 4,856 kg
(loaded) 27,000 kg
Max weight on 5th wheel: 14,500 kg
Length: 7.78 m
Width: 2.49 m
Height:
(cab) 3.005 m
(5th wheel bracket) 1.365 m
Ground clearance: 0.312 m
Track:
(front) 1.91 m
(rear) 1.825 m
Wheelbase: 3.9 m + 1.35 m
Angle of approach: 28°
Max speed: 82 km/h
Range: 700 km
Fuel capacity: 250 litres
Gradient: (towing fully laden semi-trailer) 18%
Side slope: 30%
Fording: 0.85 m
Engine: Renault MIDR 06-35-40 6-cylinder turbocharged and intercooled diesel developing 336 hp at 1,900 rpm
Gearbox: Type B9 with 8 synchronised forward and 1 reverse gears
Transfer box: Type VG 1200, 2-speed
Steering: Type 8046 hydraulic integral power
Turning radius: 10.8 m
Suspension:
(front) semi-elliptical springs with rubber bump stops and double-acting hydraulic shock-absorbers
(rear) trunnion suspension using semi-elliptic leaf springs with axle torque rods
Tyres: 12.00 R 20
Brakes: pneumatic
Electrical system: 24 V
Batteries: 2 × 12 V, 143 Ah

Status

Production complete.

Manufacturer

Enquiries to: Renault Véhicules Industriels, 40 rue Pasteur, BP 302, F-92156 Suresnes Cedex, France.
Tel: +33 1 40 99 71 11. Telex: 620567 sdce f.
Fax: +33 1 40 99 71 08.

VERIFIED

LOHR DMC tank transporter system

Description

The LOHR DMC tank transporter system is described by the manufacturer as a combined tank transporter and recovery vehicle system. It consists of a hydraulically powered moving platform carried on the rear of a 6 × 6 open-framed truck chassis with a GVW of 26 to 32 tons and with a 300 to 400 hp engine. The DMC system can offload and recover the platform by the use of a hydraulic ram and for recovery a heavy-duty hydraulic winch (on the platform) is provided. The platform may be traversed 90° to one side or the other for offloading or recovery. The forward part of the platform has a raised guard frame to protect the carrier cab and this frame is also used as the carrier for the vehicle spare wheel. The LOHR DMC system has a payload of 16,000 to 18,000 kg. Locking devices at the front and back of the platform anchor the load when transported.

Status

In production.

LOHR DMC system being used to load AMX-10P APC onto IVECO 260 PM 35 (6 × 6) 10,000 kg truck

Manufacturer

LOHR SA, F-67980 Hangenbieten, France.
Tel: +33 3 88 38 98 00. Telex: 870082 f.
Fax: +33 3 88 96 06 36.

UPDATED

GERMANY

MAN (6 × 6) tractor trucks

Description

There are three MAN 6 × 6 tractor trucks which all use the same basic chassis, cab and equipment. They are the 32.365 DFAT with a maximum possible towed weight of 105,000 kg; the 40.440 DFAT with a maximum possible towed weight of 130,000 kg; and the 40.521 DFAT with a maximum possible towed weight of 150,000 kg. Differences are found mainly in engine capacity and power (see Specifications table).

All three models have forward control, all-steel cabs that can be tilted forward for engine access. The chassis frames have longitudinal members constructed from pressed U-sections with cross members both riveted and bolted in place. The front bumper is reinforced for winch operations. A sub-frame carries the fifth wheel.

All three models can be fitted with ZF gearboxes but the 32.365 may be fitted with optional Fuller RTO 11613, WSK 400, Allison or Renk automatic transmissions (the Specifications table provides the standard gearboxes). All gearboxes are provided with a power take-off.

The power take-off is used to power a 20,000 or 25,000 kg hydraulic winch. Both are equipped with 50 m of 24 mm diameter cable. A mechanical winch is also provided. This may be an Itag-Celle M 240 or M 180, the former with a 24,000 kg capacity and the latter with an 18,000 kg capacity. Both have 60 m of 26 mm cable. Power for both is taken from the vehicle's main engine.

Optional extra equipment for all models includes a variable speed governor for the fuel injection pump, tropical radiator (which raises the cab height by 80 mm), manual engine throttle, reinforced transmission, reinforced exhaust and tank brackets, roof air induction inlet, upswept exhaust stack, alternative Jost fifth-wheel coupling, raised axle bleed system, spare wheel on the winch guard, working platform behind the cab, towing coupling, battery charging socket and two 60 litre glass fibre water tanks.

MAN 40.440 DFAT (6 × 6) tractor truck in service with South African National Defence Force (T J Gander)
1997

Specifications

Model	**32.365 DFAT**	**40.440 DFAT**	**40.521 DFAT**
Cab seating	1 + 2	1 + 2	1 + 2
Configuration	6 × 6	6 × 6	6 × 6
Weight (loaded)	32,000 kg	40,000 kg	40,000 kg
Weight on front axle (loaded)	6,700 kg	8,500 kg	8,500 kg
Weight on rear axle (loaded)	13,000 kg	16,000 kg	16,000 kg
Permissible towed load	80,000 kg	100,000 kg	102,000 kg
Max towed load	105,000 kg	130,000 kg	150,000 kg
Length	7.57 m	7.74 m	8.81 m
Width (cab)	2.5 m	2.5 m	2.5 m
(rear wheels)	2.466 m	2.49 m	2.5 m
Height (cab, loaded)	3.122 m	3.115 m	3.189 m
(5th wheel base plate)	1.15 m	1.338 m	–
Track (front)	1.904 m	2.008 m	1.903 m
(rear)	1.804 m	1.816 m	1.804 m
Wheelbase	3.85 m + 1.35 m	3.825 m + 1.4 m	4 m + 1.4 m
Max speed (road)	65 km/h	78 km/h	79 km/h
(cross-country)	43 km/h	52 km/h	–
Fuel capacity (total)	630 litres	630 litres	400 litres
Max gradient (permissible load)	30%	20.2%	19.7%
Engine	MAN type D 2840 MF V-10 18.272 litre diesel developing 365 hp (DIN) at 2,500 rpm	MAN type D 2540 MTF V-10 15.953 litre diesel developing 440 hp (DIN) at 2,300 rpm	MAN type D 2840 LF V-10 18.272 litre diesel developing 520 hp (DIN) at 2,300 rpm
Gearbox	ZF Synchroma type, 4S-150, 8 forward and 1 reverse gears	ZF Synchroma type, 4S-150, 8 forward and 1 reverse gears	ZF Synchroma type, 16S-190A, 16 forward and 2 reverse gears
Transfer case	MAN type G 801 2-speed	MAN type G 801 2-speed	mounted on gearbox
Clutch	2-plate hydraulic	hydrodynamic torque converter	hydrodynamic torque converter
Steering	ZF hydro-steer type 8046	ZF hydro-steer type 8046	ZF hydro-steer type 8046
Suspension (front)	semi-elliptic leaf springs with hollow rubber springs and telescopic shock-absorbers	semi-elliptic leaf springs with hollow rubber springs and telescopic shock-absorbers	semi-elliptic leaf springs with hollow rubber springs and telescopic shock-absorbers
(rear)	semi-elliptic leaf springs	semi-elliptic leaf springs	semi-elliptic leaf springs
Tyres	12.00 × 24	12.00 × 24	12.00 × 24
Brakes	air, 2-circuit	air, 2-circuit	air, 2-circuit
Electrical system	24 V	24 V	24 V
Batteries	2 × 12 V, 170 Ah	2 × 12 V, 170 Ah	2 × 12 V, 170 Ah

Status
Production complete. In service with various countries, including South Africa.

Manufacturer
MAN Nutzfahrzeuge AG, Postfach 50 06 20, D-80976 Munich, Germany.
Tel: +49 89 1580-1. Telex: 523 211-0 ma d.
Fax: +49 89 150 3972.

UPDATED

Mercedes-Benz 2636 AS (6 × 6) tank transporter

Description
The Mercedes-Benz 2636 AS (6 × 6) tank transporter is used to tow tank loads up to 65,000 kg. It has a forward control cab complete with a frontal cab protector and has seating for the driver and two passengers with space for bunks at the rear. An air intake over the cab has a prefilter for use in dusty environments. The V-10 diesel engine powers a torque converter for a 16-speed transmission system and the axles use planetary gear hub reduction. There are differential locks in the front and rear axles as well as in the transfer case. For maintenance the cab can be tilted forward and a spare wheel is provided behind the cab, complete with a handling device.

The 2636 AS is supplied with a mechanically controlled ITAG Type 241 winch capable of pulling up to 65,000 kg using 60 m of 26 mm diameter cable. As an option a Rotzler Type 2 × 20000 H double-acting hydraulic winch may be supplied.

Specifications
Cab seating: 1 + 2
Configuration: 6 × 6
Weight:
(kerb) 11,900 kg
(GCW) 105,000 kg
(5th wheel load) 24,000 kg
Length: (tractor unit) 7.075 m
Width: 2.5 m
Height: (unloaded) 3.195 m
Ground clearance: (rear axle) 0.427 m
Track:
(front) 1.987 m
(rear) 1.804 m
Wheelbase: 3.5 m + 1.45 m
Angle of approach/departure: 35°/47°
Max speed: 79 km/h
Fuel capacity: 400 litres
Range: (approx) 1,000 km
Gradient:
(GCW 80,000 kg) 38%
(GCW 105,000 kg) 20%
Engine: Mercedes-Benz OM 423 V-10 18.273 litre water-cooled 4-stroke diesel developing 355 hp at 2,300 rpm
Clutch: torque converter
Transmission: ZF 16 S S-130 16-speed fully synchronised
Transfer case: VG 2000-3W/1.4 2-speed
Steering: power-assisted
Turning circle: 20.3 m
Tyres: 14.00 × 20

Mercedes-Benz 2636 AS (6 × 6) tank transporter

Brakes: two-circuit two-line with compressed air booster plus exhaust brake
Electrical system: 24 V
Batteries: 2 × 12 V, 110 Ah
Alternator: 28 V/55 A

Status
In service with Pakistan (100).

Manufacturer
Mercedes-Benz AG, Dept VN/RM - 2503, D-70322 Stuttgart, Germany.
Tel: +49 711 179 0326. Fax: +49 711 179 0934.

VERIFIED

Mercedes-Benz 3850 AS (6 × 6) tank transporter

Description
The Mercedes-Benz 3850 AS (6 × 6) tank transporter may be regarded as a development of the Mercedes-Benz 2636 AS (6 × 6) tank transporter and uses the same engine and many other components. It has a longer wheelbase and can operate at gross combination weights up to 110,000 kg.

The forward control cab has separate seating for a driver and two passengers, plus a folding bench at the rear for a further four passengers. For access to the roof hatch the centre front seat has a forward-folding backrest on which to stand. The cab is mounted at the front on two pivot bearings with flexible rubber sleeves and at the rear on four vibration-damped spring struts. An air conditioning unit or supplementary heater is optional. The entire cab can be tilted forward for access to the engine and transmission.

Mechanical or hydraulic single or dual drum winches can be fitted. Both can be operated from the cab or remotely. Various types of semi-trailers may be towed but for low-loader trailers a ballast platform on the tractor is necessary.

An automatic transmission is optional, as are single sand tyres and a tyre pressure regulation system.

Mercedes-Benz 3850 AS (6 × 6) tank transporter

Specifications
Cab seating: 1 + 6
Configuration: 6 × 6
Weight:
(kerb, without winch) 13,150 kg
(5th wheel load) 24,850 kg
(GCW, with semi-trailer) 110,000 kg
(GCW, with trailer) up to 220,000 kg
Front axle load: 8,000 kg
Rear axle load: 2 × 16,000 kg
Length: 7.6 m
Width: 2.765 m
Height: (unloaded) 3.46 m
Ground clearance: (rear axle) 0.39 m
Track:
(front) 2.07 m
(rear) 1.965 m
Wheelbase: 3.8 m + 1.45 m
Angle of approach/departure: 33°/55°
Max speed: 90 km/h
Fuel capacity: 600 litres
Range: (approx) 1,000 km
Gradient: (GCW 110,000 kg) 32%
Engine: Mercedes-Benz OM 423 LA 18.273 litre V-10 water-cooled diesel with intercooler developing 500 hp (DIN) at 2,300 rpm
Transmission: ZF 16 S 160 A 8-speed fully synchronised or
Allison CLBT 754 automatic with integrated retarder
Clutch: ZF-WSK 400 torque converter
Transfer box: 2-speed
Steering: power-assisted
Turning circle: (tractor) 19.4 m
Tyres: 14.00 × 20 XL (24 × 21 optional)
Brakes: dual-circuit two-line with compressed air booster plus exhaust retarder
Electrical system: 24 V
Batteries: 2 × 12 V, 136 Ah
Alternator: 28 V/55 A

Status
In production.

Manufacturer
Mercedes-Benz AG, Dept VN/RM - 2503, D-70322 Stuttgart, Germany.
Tel: +49 711 179 0326. Fax: +49 711 179 0934.

VERIFIED

Mercedes-Benz 4850 A (8 × 8) tank transporter

Description
The Mercedes-Benz 4850 A (8 × 8) tank transporter may be regarded as an 8 × 8 version of the Mercedes-Benz 3850 AS (6 × 6) tank transporter. There are two models in the range. The 4050 A has twin 14.00 × 20 R tyres on the rear axles while the 4850 A may have either twin 14.00 × 20 R tyres on the rear axles or single sand 24 × 21 R tyres on all axles. There are slight weight variations between all three models.

The Mercedes-Benz 4850 A shares many components with the smaller tank transport truck tractors in the Mercedes-Benz range, including the forward control cab which is tiltable for engine access. This has separate seating for the driver and two passengers with a folding bench seat for a further four passengers in the rear. Air conditioning or extra cabin heaters may be provided and a roof hatch is fitted. The area behind the cab may be occupied by mechanical or hydraulic drum winches as required and there is provision for carrying a spare wheel — more spare wheels may be carried on the semi-trailer or trailer. The fifth wheel is located directly over the third tractor axle and may be used to tow a variety of semi-trailers. If low-loader trailers are towed, a ballast platform is required.

The fuel tank capacity may be between 400 and 1,000 litres, as needed and another option is an Allison CLBT 754 automatic transmission with an integrated retarder. A tyre pressure regulation system is also optional.

Specifications
Data refers to Model 4850 A with dual rear tyres; data in square brackets refers to version with sand tyres where different
Cab seating: 1 + 6
Configuration: 8 × 8
Weight:
(kerb) 17,200 [18,100] kg
(5th wheel load) 31,000 [22,000] kg
(GCW, with semi-trailer) 110,000 kg
(GCW, with trailer) up to 220,000 kg
Front axle loads: 2 × 8,000 kg
Rear axle loads: 2 × 16,000 [2 × 12,000] kg
Length: 8.435 m
Width: 2.765 m
Height: 3.46 m
Ground clearance: (rear axle) 0.39 m
Track:
(front) 2.07 m
(rear) 1.965 m
Wheelbase: 1.6 m + 2.935 m + 1.48 m
Angle of approach/departure: 35°/60°
Max speed: 90 km/h
Fuel capacity: 400 to 1,000 litres
Range: variable according to fuel capacity
Gradient: (GCW 110,000 kg) 36%
Engine: Mercedes-Benz OM 423 LA 18.273 litre V-10 water-cooled diesel with intercooler and two turbochargers developing 500 hp at 2,300 rpm
Transmission: ZF 16 S - 190 8-speed fully synchromesh or
Allison CLBT 754 automatic with integrated retarder
Clutch: ZF-WSK 400 torque converter
Transfer box: VG 2000 - 3W, 2-speed
Steering: dual circuit power-assisted
Turning circle: (tractor) 22.5 [26.2] m
Tyres: 14.00 × 20 R [24 × 21 R]
Brakes: dual-circuit compressed air
Electrical system: 24 V
Batteries: 2 × 12 V, 135 Ah
Alternator: 28 V/55 A

Mercedes-Benz 4850 A (8 × 8) tank transporter fitted with single sand tyres on all axles

Status
In production.

Manufacturer
Mercedes-Benz AG, Dept VN/RM - 2503, D-70322 Stuttgart, Germany.
Tel: +49 711 179 0326. Fax: +49 711 179 0934.

VERIFIED

IVECO Magirus 330-40 ANWTM (6 × 6) tractor truck

Description
This 6 × 6 tractor truck is based on the IVECO Magirus range of cross-country vehicles. The 330-40 ANWTM was formerly the Magirus 400M33AS. (The IVECO Magirus 330-32 ANWTM is no longer offered.)

The IVECO Magirus 330-40 ANWTM has a turbocharged diesel engine and the same type of cab as that used on the IVECO Magirus cross-country trucks. There is a double articulated fifth wheel located over the rear axles.

Two hydraulic 20,000 kg winches are fitted. Other options are a cab roof hatch, a blackout lighting system and a twin-compartment crew cab. The 330-40 ANWTM may be fitted with special tyres for use over sandy terrain - single 14.00 R 20 at the front and single 24-20.5 at the rear.

IVECO Magirus 330-40 ANWTM (6 × 6) tractor truck

Specifications
Cab seating: 1 + 2
Configuration: 6 × 6
Weight:
(kerb) 14,200 kg
(on 5th wheel) 21,500 kg
(GVW) 35,700 kg
(GCW) up to 87,000 kg
Length: 7.86 m
Width:
(over cab) 2.5 m
(over rear wheels) 2.802 m
Height: (top of air intake) 2.965 m
Ground clearance: 0.372 m
Track:
(front) 2.002 m
(rear) 2.802 m
Wheelbase: 4 m + 1.45 m
Angle of approach/departure: 30°/54°
Max speed: 84 km/h
Range: 800 km
Fuel capacity: (standard) 400 + 300 litres
Max gradient: 63%
Side slope: 20%
Fording: 0.8 m
Engine: Deutz Diesel BF 10 L 513 15.953 litre 10-cylinder air-cooled diesel developing 415 hp at 2,300 rpm
Gearbox: ZF 4 S-150 GP manual, 8 forward and 1 reverse gears
Transfer box: 2-speed
Clutch: WSK 400 torque converter
Steering: ball and nut, power-assisted
Turning radius: 11.8 m
Suspension:
(front) semi-elliptic leaf springs, hydraulic shock-absorbers
(rear) cantilever leaf springs
Tyres: 14.00 R 20, twin at rear
Brakes:
(main) drum, air on all wheels plus exhaust retarder
(parking) mechanical on rear wheels
Electrical system: 24 V

Status
In production. In widespread service.

Manufacturer
IVECO Magirus AG, Defence Vehicles Division, D-89070 Ulm, Germany.
Tel: +49 731 729 6826. Fax: +49 731 76317.

VERIFIED

TADANO FAUN FS 42.75/42 (8 × 6) tractor truck

Description
The TADANO FAUN FS 42.75/42 (8 × 6) tractor truck was produced as a replacement for the Faun SLT 50-2 with the former West German Army.

The FS 42.75/42 follows the same overall layout as the SLT 50-2 and has a TADANO FAUN forward control cab constructed from steel and synthetic materials. Apart from the adjustable driver's seat there is seating for three passengers. There is a roof hatch with provision for a machine gun mounting and internally there is a stowage compartment. All walls are double-skinned and laminated glass is used for all windows. The windscreen is angled forward at an angle of 5°. An engine-driven heater is used to heat the cab interior and a Webasco heater, independent of the engine, can be used for heating and ventilation when the engine is switched off.

The chassis is of a die-pressed construction with U-shaped members. Lateral and longitudinal supports are bolted or riveted.

The area behind the cab is mainly occupied by a dual 18,600 kg capacity winch assembly provided with a 28 mm diameter cable. The winch speed is adjustable from zero to 24 m/min for loads up to 8,500 kg and up to 9 m/min for heavier loads. The fifth wheel is a Jost JSK 38 GL.

The two front axles are used for steering but only the front axle is driven. The rear-driven Starr planetary axles have a longitudinal differential lock between them and a transverse differential lock can be engaged.

An 8 × 8 version with 24 R 20.5 tyres is available.

Specifications

Cab seating: 1 + 3
Configuration: 8 × 6
Weight:
(empty) 19,700 kg
(loaded) 36,000 or 39,700 kg
(load on 5th wheel) 16,300 or 20,000 kg
(gross train weight) 95,000 kg
Length: 8.835 m
Width: 3.07 m
Height: 3.02 m
Track:
(front) 2.585 m
(rear) 2.612 m
Wheelbase: 1.5 m + 2.7 m + 1.5 m
Angle of approach/departure: 30°/45°
Max speed: 72 km/h
Fuel capacity: 800 litres
Gradient: 50%
Engine: KHD BF 12 L 513 C 19.144 litre 12-cylinder air-cooled diesel developing 525 hp at 2,300 rpm
Gearbox: ZF Transmatic with torque converter drive clutch ZF WSK 400 and Ecosplit-gearbox 16 S 190 A with 16 forward and 2 reverse gears
Steering: ZF semi-block hydraulic, dual circuit
Turning radius: 11.2 m
Suspension: (front and rear) semi-elliptical leaf springs and telescopic shock-absorbers
Tyres: 18.00 × 22.5 XS PR 20
Brakes:
(main) dual-circuit air
(parking) pneumatically operated exhaust valve
Electrical system: 24 V
Batteries: 4 × 12 V
Alternator: 95 A

TADANO FAUN FS 42.75/42 (8 × 6) tractor truck towing trailer carrying Leopard 2 MBT over Leguan bridge

Status

In service with the German Army.

Manufacturer

TADANO FAUN GmbH, Postfach 10 01 08, D-91205 Lauf a.d. Pegnitz, Germany.
Tel: +49 91 23185-0. Telex: 622990.
Fax: +49 91 233085.

VERIFIED

TADANO FAUN SLT 50-2 (8 × 8) tractor truck

Development

The development of the SLT 50-2, or Elefant as it is more commonly known, can be traced back to 1965 when the former West Germany and the USA decided to design a tank transporter known as the Heavy Equipment Transporter (HET) to carry the MBT-70 then under development by both countries. In the USA the Chrysler Corporation was the prime contractor for the tractor while in Germany Faun (now TADANO FAUN) was responsible for the tractor and Krupp for the trailer. In 1970 the MBT-70 and the HET were cancelled. Further development in the USA resulted in the XM746, which was standardised as the M746 and production was undertaken by Ward La France. Development continued in Germany and the German Army placed an order with Faun for 328 Elefants, the last of which were delivered in 1980. The SLT 50-2 and its semi-trailer can carry any of the AFVs in service with the Federal German Army, including the Leopard 2 MBT.

A more powerful model, known as the SLT 50-3, was available powered by a 750 hp engine coupled to a Renk automatic transmission with four forward gears and two reverse.

TADANO FAUN SLT 50-2 (8 × 8) Elefant tractor truck towing semi-trailer carrying M48 series MBT

Description

The SLT 50-2 is an 8 × 8 vehicle with powered steering on the front two axles. The fully enclosed cab is of steel and glass fibre construction and has seats for the driver and three passengers.

The drive train consists of a V-8 diesel engine coupled to a hot-shift transmission and a two-speed transfer case. The front and rear bogies are fitted with a lockable inter-axle differential and each axle has a lockable differential. The front and rear bogies are suspended by torque rods and taper leaf springs arranged in parallel with progressive springing.

Airbrakes are provided for both the tractor and semi-trailer and an automatic load-sensitive brake valve ensures equal braking under all load conditions. The parking brake consists of spring-loaded cylinders mounted to the rear wheel brakes. A retarder is connected directly to the gearbox and is also connected via an electric control line to the service brakes of the semi-trailer.

Mounted to the rear of the cab is a dual winch unit, each winch with a capacity of 18,600 kg. Both winches have a winch-up mechanism and the right winch can also be used to the front of the vehicle for self-recovery operations. Each winch is provided with 43 m of 28 mm diameter rope; maximum winding speed is 24 m/min at a capacity of 8,500 kg or 12 m/min at a capacity of 18,600 kg.

The spare wheel is mounted on the right side of the tractor and a small winch is provided to facilitate handling of the wheel.

Specifications

Cab seating: 1 + 3
Configuration: 8 × 8
Weight:
(empty) 23,030 kg
(on 5th wheel loaded) 18,300 kg
(GCW) 92,000 kg
Length: 8.83 m
(with trailer) 18.82 m
Width: 3.05 m
Height: 2.83 m
(to 5th wheel) 1.55 m
Ground clearance: 0.303 m
Track:
(front) 2.535 m
(rear) 2.593 m
Wheelbase: 1.5 m + 2.7 m + 1.5 m
Angle of approach/departure: 30°/50°
Max speed:
(road, without semi-trailer) 63.5 km/h
(road, with loaded semi-trailer) 40 km/h
(15% gradient with loaded semi-trailer) 9 km/h
Range: 500 km
Fuel capacity: 800 litres
Max gradient: (87,500 kg GCW) 30%
Fording: 0.8 m
Engine: MTU MB 837 Ea-500 V-8 diesel developing 730 hp at 2,100 rpm
Gearbox: ZF 4 PW 200H2 with 4 forward and 2 reverse gears
Torque converter: ZF 500-10
Transfer box: 2-speed
Steering: ZF semi-block hydraulic
Turning radius: (without trailer) 11.2 m
Tyres: 18.00 × 22.5
Brakes: air
Electrical system: 24 V

Status

Production complete. In service with the German Army.

Manufacturer

TADANO FAUN GmbH, Postfach 10 01 08, D-8560 Lauf a.d. Pegnitz, Germany.
Tel: +49 91 23185-0. Telex: 622990.
Fax: +49 91 233035.

VERIFIED

TADANO FAUN HZ 40.45/45 (6 × 6) tractor truck

Description

The TADANO FAUN HZ 40.45/45 (6 × 6) tractor truck is a militarised version of a commercial tractor truck in the TADANO FAUN range adapted for the tank transporter role. The TADANO FAUN Model Number is 2144.93 Version B.

The vehicle has a conventional layout, with the engine forward under a steel frame and reinforced glass fibre panel bonnet, the crew cab placed just to the rear of the front axle and the fifth wheel placed over the two rear axles. The steel cab has three doors, one on the left and two on the right, providing access to two rows of seating for the driver and six passengers. The front row has seating for the driver and two passengers, all on single seats. The rear seating consists of a single padded bench with the central portion tiltable to allow access to a centrally positioned 800 mm diameter roof hatch.

The chassis frame consists of pressed segments having a U-profile design and with cross members bolted or riveted to the longitudinal girders. Towing eyes are provided at the front and a NATO trailer coupling with a towing capacity up to 55 tonnes is located at the rear.

Optional equipment available for the vehicle includes a hydraulic twin-winch providing 2 × 18 tonnes drawbar pull on the first layer. The winches operate to the front or rear and each drum is provided with 55 m of 24 mm diameter cable. Transverse differential locks are available on both rear axles. Also available are: a set of six snow chains; up to four spare fuel cans; combined oil bath air filter; two 6 kg fire extinguishers; compressed air hose connection for slave filling of the brake system; slave starting device with 6 m of cable; and sliding rails to facilitate connecting a trailer to the fifth wheel.

TADANO FAUN HZ 40.45/45 (6 × 6) tractor truck (T J Gander)

Specifications

Cab seating: 1 + 6
Configuration: 6 × 6
Weight:
(empty) 19,000 kg
(loaded) 45,000 kg
Max fifth wheel load: 26,000 kg
Length: 9 m
Width: 2.75 m
Height: 3.4 m
Wheelbase: 5.4 m
Max speed: 63.6 km/h
Fuel capacity: 2 × 450 litres
Max gradient: 44.6%
Engine: Deutz Type BF 12 L 513 C 19.144 litre V-12 direct injection air-cooled diesel developing 525 hp at 2,300 rpm
Gearbox: Allison Type CL T 754 with 5 forward and 1 reverse gears
Torque converter: Allison Type TC 496
Transfer box: 2-speed
Steering: ZF semi-block steering with working cylinder, primary pump and secondary pump
Turning radius: approx 12.5 m
Suspension:
(front) semi-elliptic leaf springs, longitudinal guide struts for thrust transmission, telescopic shock-absorbers
(rear) semi-elliptic leaf springs in floating arrangement, longitudinal guide struts for thrust transmission
Tyres: 14.00 × 24
Brakes:
(main) dual-circuit air plus exhaust brake
(parking) operating on rear axles
Electrical system: 24 V
Batteries: 2 × 12 V, 170 Ah

Status

In service with Chad (38), Libya (69) and Turkey (40).

Manufacturer

TADANO FAUN GmbH, Postfach 10 01 08, D-8560 Lauf a.d. Pegnitz, Germany.
Tel: +49 91 23185-0. Telex: 622990.
Fax: +49 91 233085.

VERIFIED

GREECE

ELBO 26S31/S38/6X6 (6 × 6) tractor truck

Description

The ELBO 26S31/S38 (6 × 6) tractor truck is produced in Greece under licence from Steyr Trucks SA of Austria. The associated semi-trailer was developed and is produced in Greece by Kioleides SA and can carry MBTs, APCs, AIFVs and construction equipment.

The tractor cab has seating for the driver and two passengers. The semi-trailer has a a capacity of 35 tonnes, a steel frame and a timber load area. Options include a 9,000 kg capacity hydraulic front winch mounted behind the front bumper for self-recovery and a 20,000 kg capacity hydraulic rear winch mounted behind the cab for the loading of disabled tanks or construction equipment.

ELBO 26S31/S38/6X6 (6 × 6) tractor truck and semi-trailer carrying construction equipment **1996**

Specifications

Cab seating: 1 + 2
Configuration: 6 × 6
Weights:
(kerb, with winches) 14,000 kg
(GVW) 32,000 kg
Max front axle load: 7,500 kg
Max rear axle load: 2 × 13,000 kg
Towed load: 57,000 kg
Max load on 5th wheel: 18,000 kg
Length: (complete train) 18.5 m
Width: (overall) 3 m
Height: 3.5 m
Ground clearance: 0.33 m
Wheelbase: 3.8 m + 1.35 m
Track:
(front) 1.958 m
(rear) 1.8 m
Angle of approach/departure: 30°/40°
Max speed: 80 km/h
Fuel capacity: 400 litres
Range: 500 km
Max gradient: 35%
Side slope: 30°
Fording: 0.8 m
Engine: Steyr WD 615.78 9.73 litre direct injection turbocharged intercooled water-cooled diesel developing 306 hp at 2,200 rpm
Transmission: ZF Transmatic WSK400+4S150GP with torque converter, 8 forward and 1 reverse gears
Transfer box: Steyr VG 1200 2-speed
Steering: ZF 8046 power-assisted
Turning radius: 10.2 m
Suspension: (front) semi-elliptic leaf springs with telescopic shock-absorbers; (rear) inverted semi-elliptic leaf springs with torsion bars
Tyres: 12.00 R 20
Brakes: dual-circuit, air
Electrical system: 24 V
Batteries: 2 × 12 V, 143 Ah

Status

In service with the Greek armed forces.

Manufacturer

ELBO SA, Industrial Area Sindos, PO Box 10239, GR-541 10 Thessaloniki, Greece.
Tel: +30 31 798502. Telex: 412586 elbo gr.
Fax: +30 31 798426.

UPDATED

ITALY

IVECO 320-45 WTM (6 × 6) tractor truck

Development

The IVECO 320-45 WTM (6 × 6) tractor truck was designed to meet an Italian Army requirement for a vehicle capable of towing a semi-trailer carrying a Leopard 1 MBT both on- and off-road as well as having the capability to recover damaged and disabled vehicles. The first prototype was completed in 1978 and the vehicle is in service with the Italian Army as the ATC/81.

Description

The IVECO 320-45 WTM (6 × 6) tractor truck tows an OTO Melara-designed (now OTOBreda), Bartoletti-manufactured trailer, designated the Mod TCS 50 BO. The forward control four-passenger cab can be tilted forward to give access to the engine for maintenance. Two types of cab are available, one with a fully enclosed hardtop and the other with a windscreen that can be folded forward through 180°, removable door tops and side screens and a canvas roof that folds to the rear.

To the rear of the cab are two winches, each rated at 20,000 kg and with 50 m of cable. The standard 89 mm diameter king-pin is positioned over the two rear axles. The vehicle can be delivered with either single or dual rear wheels.

A variant was designed to operate over sandy terrain. It has a revised wheelbase (3.565 m + 1.45 m) and uses single tyres on all axles (14.00 × 24 at the front and 24 × 20.5 at the rear). Kerb weight is 14,700 kg.

IVECO 320-45 WTM (6 × 6) 'Sand Version' tractor truck towing semi-trailer carrying OF-40 ARV

Specifications

Cab seating: 1 + 3
Configuration: 6 × 6
Weight:
(kerb) 15,400 kg
(max load on 5th wheel) 21,000 kg
(GCW) 93,420 kg
Length: 7.52 m
Width: 2.775 m
Height:
(overall) 3.051 m
(5th wheel) 1.647 m
Ground clearance: 0.36 m
Track:
(front) 1.985 m
(rear) 1.97 m
Wheelbase: 3.6 m + 1.38 m
Angle of approach/departure: 41°/60°
Max speed: >65 km/h
Range: over 600 km
Fuel capacity: 500 litres
Gradient: over 30% with semi-trailer
Side slope: 20%
Fording: 0.85 m
Engine: Model 8280.22 17.174 litre V-8 water-cooled turbocharged diesel developing 450 hp (DIN) at 2,400 rpm
Transmission: 8 forward and 1 reverse gears, torque converter, transfer box with lockable torque divider, lockable differential at rear axles
Steering: powered
Turning radius: 9.5 m
Suspension: leaf springs
Tyres: 14.00 × 20
Brakes:
(main) drum, air operated
(parking) acting on rear axles
(exhaust) pneumatic
Electrical system: 24 V
Number of batteries: 2 × 12 V, 143 Ah
Generator: 650 W

Status

In production. In service with the Italian Army and several overseas countries.

Manufacturer

IVECO SpA, Defence Vehicle Division, Via Volta 6, I-39100 Bolzano, Italy.
Tel: +39 471 905111. Telex: 400541 ivedvd i.
Fax: +39 471 905444.

UPDATED

Astra HD 66.45 TIM (6 × 6) truck tractor

Description

The Astra HD 66.45 TIM (6 × 6) truck tractor is a development of the Astra HD series with the necessary modifications to upgrade it to a truck tractor. It is provided with a ZF torque converter between the engine and gearbox and wider axles have been installed to permit the fitting of dual 14.00 R 20 tyres. An oscillating fifth wheel is provided along with two 20 tonne winches (25 tonne winches are an option). The forward control cab has space for two crew beds and is air conditioned.

Specifications

Cab seating: 1 + 2
Configuration: 6 × 6
Weight:
(kerb) 15,500 kg
(GVW) 40,500 kg
(GCW) 100,000 kg
Load on 5th wheel: 25,000 kg
Length: 7.64 m
Width: 2.84 m
Height: 3.22 m
Track:
(front) 2.053 m
(rear) 2.017 m
Wheelbase: 3.8 m + 1.4 m
Max speed: 67 km/h
Fuel capacity: 400 + 400 litres
Engine: IVECO 8280 TCA turbo intercooled diesel developing 440 hp at 1,900 rpm
Transmission: ZF 16 S 221 gearbox with ZF WSK 400.90 torque converter providing 16 forward and 2 reverse gears
Steering: ZF Servocom 8099 dual circuit with 2 pumps (1 on engine, 1 on transfer box)
Turning diameter: 18.9 m
Suspension:
(front) semi-elliptic springs with shock-absorbers and torsion bar
(rear) reinforced oscillating springs with pivot system and torsion bar
Tyres: 14 R 20
Brakes:
(main) pneumatic
(parking) mechanical on rear axles
Electrical system: 24 V
Batteries: 2 × 12 V, 170 Ah
Alternator: 80 A

Status

In production.

Manufacturer

Astra Veicoli Industriali SpA, Via Caorsana 79, I-29100 Piacenza, Italy.
Tel: +39 523 543-. Telex: 530148 ASTRA I.
Fax: +39 523 593293.

NEW ENTRY

Astra HD 66.45 TIM (6 × 6) truck tractor **1997**

Astra HD 66.45 TIM (6 × 6) truck tractor **1997**

JAPAN

Mitsubishi Model FW455LRS2 (6 × 6) truck tractor

Description

The Mitsubishi Model FW455LRS2 (6 × 6) truck tractor is known to the Japanese Ground Self-Defence Force as the Type 84 truck tractor. It is designed to tow a three-axle semi-trailer carrying a tank with a maximum weight of 40,000 kg.

The Model FW455LRS2 uses a forward control cab of all-steel welded construction with seating for the driver and two passengers. The cab doors have extra vision panels and a ladder is provided for access to stowage space on the cab roof. A flat area behind the cab may be used for more stowage and carries the spare wheel and the fuel tank. The fifth wheel is located directly over the dual rear axles.

Mitsubishi Model FW455LRS2 (6 × 6) truck tractor

Specifications

Cab seating: 1 + 2
Configuration: 6 × 6
Weight:
(empty) 9,500 kg
(loaded) 26,240 kg
(on 5th wheel) 16,500 kg
Length: 7.16 m
Width: 2.49 m
Height:
(cab) 3.51 m
(5th wheel) 1.47 m
Track:
(front) 2.005 m
(rear) 1.845 m
Wheelbase: 3.73 m + 1.3 m
Max speed: 59 km/h
Engine: Mitsubishi 8DC9T2 turbocharged intercooled diesel developing 430 hp at 2,200 rpm
Turning radius: 9.6 m
Tyres: 10.00 × 20

Status

In service with Japanese Ground Self-Defence Force.

Manufacturer

Mitsubishi Motors Corporation, 33-8, Shiba 5-chome, Minato-ku, Tokyo, Japan.
Tel: +81 3 456 1111. Telex: BISIJIKO J26639/J26839.

VERIFIED

Mitsubishi Model NW204JR (6 × 6) truck tractor

Description

The Mitsubishi Model NW204JR (6 × 6) truck tractor was designed to tow a three-axle semi-trailer carrying an MBT with a maximum weight of 40,000 kg. The vehicle is designated the Type 73 truck tractor by the Japanese Ground Self-Defence Force. The layout of the vehicle is conventional, with the engine at the front, two-door fully enclosed cab in the centre and the fifth wheel at the rear.

Mitsubishi Model NW204JR (6 × 6) truck tractor towing semi-trailer carrying Type 61 MBT

Specifications

Cab seating: 1 + 2
Configuration: 6 × 6
Weight:
(empty) 9,500 kg
(loaded) 26,240 kg
(on 5th wheel) 16,500 kg
Length:
(tractor) 6.835 m
(tractor and trailer) 16.255 m
Width: 3.29 m
Height:
(cab) 2.92 m
(5th wheel) 1.47 m
Track:
(front) 1.915 m
(rear) 1.865 m
Wheelbase: 4.65 m
Max speed: 60 km/h
Engine: Model DED 10-cylinder diesel developing 375 hp at 2,500 rpm
Turning radius: 8.9 m
Tyres: 10.00 × 20

Status

Production complete. In service with the Japanese Ground Self-Defence Force.

Manufacturer

Mitsubishi Heavy Industries, 5-1, Marunouchi 2-chome, Chiyoda-ku, Tokyo, Japan.
Tel: +81 3 212 3111. Telex: 22282.
Fax: +81 3 201 6258.

VERIFIED

RUSSIAN FEDERATION AND ASSOCIATED STATES (CIS)

MAZ-535 and MAZ-537 (8 × 8) Series

Description

These 8 × 8 vehicles were seen for the first time in public during a parade held in Moscow in 1964 and are used for a wide variety of roles by the RFAS armed forces as well as civil use. They are closely related to the MAZ-543 (8 × 8) series of trucks.

They all have full 8 × 8 drive with powered steering on the front two pairs of wheels, a central tyre pressure regulation system, cab heater and an engine pre-heater. They are all powered by the same V-12 diesel

Cambodian Army MAZ-537 (8 × 8) tank transporter (T J Gander)
1996

Specifications

Designation	**MAZ-535A**	**MAZ-537A**	**MAZ-537**
Type	truck	truck	tractor truck
Configuration	8 × 8	8 × 8	8 × 8
Weight (empty)	18,975 kg	22,500 kg	21,600 kg
(loaded)	25,975 kg	37,500 kg	n/app
Weight on front axles (loaded)	n/a	14,890 kg	17,375 kg
Weight on rear axles (loaded)	n/a	22,610 kg	29,425 kg
Max load (road and cross-country)	6,000 kg	15,000 kg	n/app
Towed load (road)	50,000 kg	75,000 kg	65,000 kg
(dirt road)	15,000 kg	30,000 kg	25,000 kg
Load area	4.5 × 2.595 m	4.562 × 2.53 m	n/app
Length	8.78 m	9.13 m	8.96 m
Width	2.805 m	2.885 m	2.835 m
Height (cab)	2.915 m	2.8 m	3.1 m
(load area)	1.4 m	1.875 m	n/app
Ground clearance	0.475 m	0.5 m	0.5 m
Track	2.15 m	2.2 m	2.2 m
Wheelbase	1.7 m + 2.35 m + 1.7 m	1.7 m + 2.65 m + 1.7 m	1.8 m + 2.65 m + 1.7 m
Angle of approach/departure	38°/60°	38°/52°	38°/52°
Max speed (road)	60 km/h	60 km/h	55-60 km/h
Range	650 km	650 km	600 km
Fuel capacity	760 litres	840 litres	840 litres
Fuel consumption	110 litres/100 km	125 litres/100 km	125 litres/100 km
Max gradient	30°	8° (laden)	8° (laden)
Fording	1.3 m	1.3 m	1.3 m
Engine model	D12A-375	D12A-525	D12A-525
Engine type	V-12 water-cooled diesel developing 375 hp at 1,650 rpm	V-12 water-cooled diesel developing 525 hp at 2,100 rpm	V-12 water-cooled diesel developing 525 hp at 2,100 rpm
Gearbox	planetary, 3 speeds forward and 1 speed reverse, with smooth start device on low gear and reverse		
Transfer case	manual, 2-speed, with direct drive and reduction gears, pneumatic and manual back-up control		
Auxiliary reduction gear transmission	manual, with inter-axle self-locking differential, consisting of spur gear pair		
Torque converter	single stage	single stage	single stage
Overdrive	single-row 3-shaft reduction gear with spur skew gears		
Steering	hydraulic, screw with nut on moving balls and rack engaged with gear quadrant		
Suspension	MAZ-535A, independent, individual, lever torsion bar (rear suspension equaliser, springless on MAZ-537), with hydraulic shock-absorbers on all wheels. MAZ-537A, independent, individual, lever torsion bar, with hydraulic shock-absorbers on both sides of front axle; rear: springless equaliser		
Tyres	18.00 × 24	18.00 × 24	18.00 × 24
Brakes	air/hydraulic	air/hydraulic	air/hydraulic
Electrical system	24 V	24 V	24 V
Batteries	4 × 12-ST-70	4 × 12-ST-70	4 × 12-ST-70
Generator	1,500 W	1,500 W	1,500 W

used in some RFAS tanks. In the case of the MAZ-535 the engine has been derated to deliver 375 hp instead of 525 hp as in the MAZ-537.

The MAZ-535A, MAZ-537A and MAZ-537K (which has a small crane) are cargo trucks but are also used to tow trailers or heavy artillery. The other models are used to tow semi-trailers carrying missiles or armoured fighting vehicles. The tractor trucks are normally used with the ChMZAP-5247 and ChMZAP-5247G semi-trailers.

Variants

MAZ-535A: cargo truck
MAZ-537A: cargo truck
MAZ-537K: cargo truck
MAZ-537: tractor truck
MAZ-537D: tractor truck with additional generator to rear of engine compartment
MAZ-537E: tractor truck with generator, winch; can also be used with powered semi-trailer
MAZ-537G: tractor truck with winch
KET-T: heavy recovery version of MAZ-537G fitted with recovery crane, winches, stabilisers and towing gear.
A version of the MAZ-537 carrying a turntable crane has been observed but no details are available.

Status

In service with members of the former Warsaw Pact as well as Cambodia, Egypt, Finland, Iran, Syria and the former Yugoslavia.

Manufacturer

Minsk Wheel Tractor Plant, 150 pr-t Partizanski, 220021 Minsk, Belarus.
Tel: +7 172 469433. Telex: 252519 VIR BY.
Fax: +7 172 429861.

UPDATED

SOUTH AFRICA

Cavallo (8 × 8) truck tractor

Description

During 1994, the South African concern Armscor decided to procure the Spanish Kynos Aljaba (8 × 8) truck tractor (see following entry for details) for the South African National Defence Force (SANDF). It entered service under the name Cavallo. Chassis are obtained from Spain and once in South Africa they are fitted with mine-proof cabs.

With the SANDF the Cavallo is used more as a tank recovery vehicle than as a tank transporter. The cross-country performance and winches of the Cavallo are used to haul a stranded vehicle to a level surface where

Cavallo (8 × 8) truck tractor showing recovery crane erected (T J Gander)
1997

longer hauls are then made using MAN tractor trucks and semi-trailers. To assist in the recovery role each Cavallo has a folding crane support frame which can be erected behind the fifth wheel when the semi-trailer is not connected. Cables for the crane are routed from the main winches.

The Cavallo/Aljaba has been used as the basis for several special purpose vehicles, including a heavy recovery vehicle (see entry in *Trucks* section for details).

Status

In production. In service with the South African National Defence Force.

Manufacturer

FAF Engineering, 5 Berylium Road, Alrode 1450, South Africa.
Tel: +27 11 864 5033. Fax: +27 11 864 7510.

UPDATED

Cavallo (8 × 8) truck tractor towing semi-trailer carrying a G6 155 mm self-propelled howitzer
(T J Gander)
1997

SPAIN

Kynos Aljaba (8 × 8) truck tractor

Development

The Kynos Aljaba (8 × 8) truck tractor was originally developed as a private venture by Kynos SA, commencing in 1982 - Kynos SA is a subsidiary of the Agroman Group and normally specialises in construction equipment. The Spanish Ministry of Defence provided some funding and one prototype and three preseries vehicles were produced. The Spanish Army carried out field trials with one vehicle in May 1985, near Toledo. The first two production vehicles were delivered to the Spanish Marines during mid-1987.

This vehicle was selected by the South African National Defence Force (SANDF) as their tank transporter, the Cavallo. See previous entry for details.

Description

The Aljaba (Quiver) was designed to tow and retrieve MBTs up to a weight of 60,000 kg and is built on a flexible rectangular chassis that is resistant to the stresses imparted when travelling over uneven terrain. The chassis is reinforced by longitudinal members with special cross members attached. A swing-arm suspension with leaf springs and trailing arms is employed. The forward control cab is formed from a steel frame mounted on elastic shock-absorbers and has seating for the driver, crew member and three passengers.

The Aljaba is powered by a Deutz 19 litre V-12 turbocharged diesel and uses a ZF semi-automatic gearbox with torque converter. All four Rockwell 13 tonne axles are driven, the front two being steered with power-assistance. Each wheel hub has a planetary reduction gear and each pair of axles can be pneumatically locked. The dual-circuit pneumatic brakes act on all wheels and the brake circuits are also used to provide air for the de-icing, protection and alarm systems. An air take-off is provided each side of the vehicle for tyre inflation. The special 24-20.5XS-PR16 tyres may be fitted with chains for extra traction and the spare wheel is provided with handling gear.

For recovery the Aljaba is equipped with two 25,000 kg winches with 26 mm cable that normally operate to the rear, but the left-hand winch cable may be routed forward using a pulley fixed to the chassis. Operating speed is 8 m/min.

It has been proposed that the Aljaba may also be used to carry out a number of other roles including that of artillery tractor, logistics carrier, bridging equipment carrier, shelter/container for command and communications or hospitals and other such roles. It has also been suggested that the vehicle could be used as a missile system carrier. The semi-trailer demonstrated to date has two axles but it is possible that three- or four-axle semi-trailers or trailers could be developed.

Variant

Cavallo (8 × 8) Truck Tractor
See previous entry for details.

Specifications

Cab seating: 1 + 4
Configuration: 8 × 8
Weight:
(empty) 21,100 kg
(max load on chassis) 46,100 kg
Weight on front axles: 13,670 kg
Weight on rear axles: 25,680 kg
(GCW) 96,000 kg
Length:
(tractor) 9.875 m
(tractor and trailer) 22.269 m
Width:
(tractor) 3.3 m
(tractor and trailer) 4.025 m
Height: (overall) 3.278 m
Ground clearance: 0.365 m
Track: 2.6 m
Wheelbase: 1.524 m + 3.476 m + 1.524 m
Angle of approach/departure: 30°/75°
Max speed: 85 km/h

Kynos Aljaba (8 × 8) tank transporters
1996

Fuel capacity: 1,000 litres (500 + 500 litres)
Range:
(road) 1,000 km
(off road) 700-1,000 km
Gradient:
(tractor unit) 68%
(loaded) 31%
Vertical obstacle: 0.5 m
Trench: 2.1 m
Fording: 0.8 m
Engine: Deutz BF 12 L 513 C 19.144 litre V-12 turbocharged intercooled diesel developing 525 hp at 2,300 rpm
Transmission: ZF 16S-190A with ZF WSK 400 torque converter, 16 forward and 2 reverse gears
Transfer box: ZF, single-speed
Steering: power-assisted on front 2 axles
Turning radius: 13.407 m
Suspension: Hendrikson swing arm with leaf springs and trailing arms; rollbars on front axle
Tyres: 24-20.5XS-PR16
Brakes: dual-circuit air on all wheels
Electrical system: 24 V
Batteries: 2 × 12 V, 120 Ah

Status
In service with Spanish Marines.

Manufacturer
Kynos SA, Pol. Ind. AIMAYR, Parcela J-24, 28330 San Martin de la Vega, Madrid, Spain.
Tel: +34 1 691 5714.
Fax: +34 1 691 3870.

UPDATED

Pegaso 7345 (6 × 6) tractor truck

Description
The Pegaso 7345 (6 × 6) tractor truck was designed to meet a requirement from the Spanish Army for a tank transporter. In 1987 the prototype was approved by the Spanish Army and production and deliveries began in 1988.

The Pegaso 7345 has a forward control cab with seating for the driver and five passengers. The cab can be tilted forward for engine maintenance access. Power is provided by a Pegaso diesel engine developing 450 hp via a synchronised gearbox for a gross combination weight of 90,000 kg. Two winches behind the cab each have a capacity of 20,000 kg.

As well as being produced as a tank transporter the Pegaso 7345 could be produced as a cargo carrier, artillery tractor, multiple rocket launcher platform and as a bridging component transporter.

Specifications
Cab seating: 1 + 5
Configuration: 6 × 6
Weight:
(empty) 15,000 kg
(semi-trailer) 15,000 kg
(GCW) 90,000 kg
(towed load) 75,000 kg
Length: 7.5 m
Width: 2.48 m
Height: 3.4 m
Track: 2.062 m
Wheelbase: 3.815 m + 1.41 m
Max speed: 83 km/h
Range: 900 km
Fuel capacity: 560 litres
Max gradient: 30%
Side slope: 30%
Engine: Pegaso 12.8 litre turbo intercooling diesel with ZF torque converter developing 450 hp at 2,200 rpm
Gearbox: ZF synchronised with 16 forward and 2 reverse gears, easy shift
Transfer box: ZF single-speed with torque distributor
Steering: servo-hydraulic
Turning radius: 10.5 m
Suspension:
(front) semi-elliptical leaf springs and shock-absorbers
(rear) semi-elliptical leaf springs and Hendrikson type arms
Tyres: 12.00 × 24
Brakes: dual-circuit air
Electrical system: 24 V
Batteries: 2 × 12 V, 165 Ah

Status
In service with the Spanish Army.

Manufacturer
Pegaso, Empresa Nacional de Autocamiones SA, Defence Division, Avenida de Aragón 402, E-28022 Madrid, Spain.
Tel: +34 1 750 1000. Telex: 46972 maba e.
Fax: +34 1 747 1307.

UPDATED

Pegaso 7345 (6 × 6) tractor truck

Pegaso 7345 (6 × 6) tractor truck

SWEDEN

Volvo F16 (6 × 6) tank transporter

Description
The Volvo F16 (6 × 6) tank transporter is intended for the towing of heavy loads up to a gross combination weight of 150 tonnes under extreme climatic conditions. It has a forward control all-steel cab with seating for a driver and one or two passengers. A crew cab with space for up to five passengers is available. The cab may be air conditioned and provided with a roof hatch in various sizes and positions.

Volvo F16 (6 × 6) tank transporter
1997

Power is provided by a 500 hp Volvo six-cylinder turbocharged and intercooled diesel. This can be coupled to either a manual 14-speed gearbox or a Volvo Powertronic five- or six-speed automatic transmission. If required, the Powertronic transmission can be provided with an integrated primary retarder to assist the dual-circuit air brakes. The fuel tank capacity can be 410, 550 or 625 litres.

Two axle loadings are available. With 525/65 R 20 tyres the front axle has a 9 tonne loading while the rear pair has a 26 tonne loading. As an alternative for 12.00 R 24 tyres these loadings are 8 and 29 tonnes.

Two 25 tonne winches are provided, powered by hydraulic pumps; they can be operated separately. The winches work at two speeds, high speed with a 2 tonne traction force and low speed for 25 tonnes. High- and low-speed power take-offs are provided.

Volvo also produce the FL12 (6 × 6) tank transporter with a gross combination weight of 100 tonnes.

Status
In production.

Manufacturer
Volvo Truck Corporation, S-405 08 Göteborg, Sweden.
Tel: +46 31 666000. Telex: 27000 volvo s.
Fax: +46 31 510465.

NEW ENTRY

Volvo N12 (6 × 4) tractor truck

Description
The Volvo N12 (6 × 4) tractor truck was developed from a commercial vehicle to meet a Swedish Army requirement for a vehicle capable of carrying a Centurion or S-tank. It has a large enough cab to accommodate the tank crew of three or four in addition to the crew of the truck.

The prototype was delivered to the Swedish Army in October 1975 for field trials and first production tractor trucks were delivered in April 1977. With the Swedish Army the N12 tows the DAF YTS 10050 semi-trailer, although it can also tow other semi-trailers such as the Swedish HAFO H50-3-RLS or the H50-4-RLS, both of which have a maximum payload of 50,000 kg.

The layout of the N12 is conventional, with the engine at the front, a fully enclosed four-door all-steel cab in the centre and the fifth wheel at the rear. The vehicle is not fitted with a winch and so cannot recover disabled and damaged vehicles.

The chassis sides are of rolled U-profile steel and reinforced. The rear axle is a tandem bogie, single reduction with hub reduction, differential lock for wheels and shafts are pneumatically controlled from the cab. The fifth wheel is a Jost JSK 25 pivoting type.

Specifications
Configuration: 6 × 4
Max weight on front axle: 6,500 kg
Max weight on rear axle: 26,000 kg
Length: 7.56 m
Width: 2.5 m
Height: (overall) 3.1 m
Track:
(front) 1.945 m
(rear) 1.82 m
Wheelbase: 4.2 m + 1.37 m
Fuel capacity: 300 litres
Engine: 120 E 6-cylinder turbocharged OHV diesel developing 326 hp at 2,200 rpm
Gearbox: manual 8-speed range, supplemented with splinter section which provides 16 forward speeds
Clutch: twin dry plate
Steering: recirculating ball and nut with built-in servo
Tyres: 12.00 × 20
Brakes:
(main) air, dual-circuit
(parking) air-operated spring brakes operating directly on front and rear wheels
Suspension:
(front) semi-elliptical leaf springs with threaded spring bolt in front mounting and slipper-type anchorage at rear. Shock-absorbers and hollow rubber springs
(rear) multileaf springs with rubber springs at both ends

Volvo N12 (6 × 4) tractor truck towing DAF YTS 10050 semi-trailer carrying Centurion MBT

Status
In service with the Swedish Army.

Manufacturer
Volvo Truck Corporation, S-405 08 Göteborg, Sweden.
Tel: +46 31 666000. Telex: 27000 volvo s.
Fax: +46 31 510465.

VERIFIED

Scania R143EK and T143E (6 × 6) and (6 × 4) tractor trucks

Description
The Scania R143EK tractor truck has a forward control cab and the Scania T143E tractor truck has a bonnet. Both are designed for towing tank transporter semi-trailers and are similar in weight, capability and overall dimensions but differ in detail according to cab type and the engine fitted.

Both vehicles carry the driver and five passengers and there is provision for carrying the occupants' personal equipment. The forward control model has a four-door cab while the bonneted T model can be equipped with a shorter two-door cab or the four-door version.

Two Sepson MAS 30/30 recovery winches are fitted, each with a capacity of 20,000 kg and equipped with 40 m of 24 mm cable. Other equipment includes two snatch blocks, an electric engine heater, 12 pin trailer connector, 24 V emergency connection, fuel heater, battery master switch, reversing lights, 24 V sockets front and rear, 20 tonne hydraulic jack, two working lights and a rotating beacon.

The first of 19 Scania T143E tractors for the Swedish Army was handed over in March 1990; production of the rest of the batch continued until 1991. Production was completed during 1966. The T143Es are used with a four-axle semi-trailer with force-steered rear axles produced by HAFO of Göteborg.

It should be noted that while the gross train weight of the T143E is around 91 tonnes, the maximum peacetime gross train weight on Swedish roads is limited to 86 tonnes.

Specifications
Scania R143EK
Cab seating: 1 + 5
Configuration: 6 × 6
Weight:
(kerb) 14,800 kg
(GVW) 31,200 kg
(GTW) 100,000 kg
Length: 7.725 m
Width: 2.49 m
Height: 3.44 m
Ground clearance:
(front) 0.315 m
(bogie) 0.277 m
Track:
(front) 2.086 m
(rear) 1.83 m
Wheelbase: 3.93 m + 1.45 m
Fuel capacity: 400 litres
Engine: DSC 1403 diesel developing 450 hp at 1,900 rpm
Gearbox: GRH880 10-speed with torque converter

Scania R143EK (6 × 6) tractor truck towing semi-trailer carrying Leopard ARV

Scania T143E (6 × 4) tractor truck towing semi-trailer carrying Centurion MBT

Turning radius: 9.8 m
Tyres: 12.00 × 20 (options 12.00 × 24 or 315/80 × 22.5)
Brakes: dual-circuit air
Electrical system: 24 V
Batteries: 2 × 12 V, 160 Ah
Alternator: 55 A

Status

Production complete. T143E in service with the Swedish Army (19). R143EK in service with the Norwegian Army.

Manufacturer

Scania, S-151 87 Södertälje, Sweden.
Tel: +46 755 81000. Telex: 10200.

UPDATED

UNITED KINGDOM

Alvis Unipower M Series tank transporters

Development

The Alvis Unipower M Series is a new generation of heavy logistic vehicles developed to meet the needs of armed forces well into the 21st Century. The M Series tank transporters were designed to carry tanks weighing up to 75 tons at speeds up to 80 km/h. The two models available are the MH8875 8 × 8 750 hp tractor and the MH6660 6 × 6 600 hp tractor.

The MH6660 6 × 6 tractor entered service with the Royal Army of Oman in 1995 to carry the Challenger 2 main battle tank. A MH8875 8 × 8 tractor was built as a private venture for trials and demonstrations and was commissioned at the beginning of 1997. With the correctly configured trailer, produced in France by Nicolas, the MH8875 complies with European axle loading legislation and has exceptional mobility.

Description

Both models of M Series tank transporters are powered by a 19 litre six-cylinder Cummins diesel engine, rated at 750 hp for the 8 × 8 and 600 hp for the 6 × 6. The automatic transmission is a ZF 5HP 1500 five-speed gearbox with an integral two-speed transfer box and hydrodynamic torque converter with integral retarder.

The rear bogie comprises two Alvis Unipower axles while the front steer-drive axles are GKN. All axles have differential locks, both cross-axle and inter-axle. The rear suspension is a centrally pivoted spring pack providing equal axle loading over uneven ground and the front taper leaf springs provide a superior ride.

The engine-behind-cab configuration ensures there is no engine intrusion into the cab, maximising space for occupants and their equipment. The cab can be either a six seat crew cab or a twin sleeper cab. The cab is also low, facilitating rapid entry and exit and ease of camouflage. The ergonomically designed interior is comfortable and minimises driver fatigue.

Heavy tank casualties can be winched on to the transporter trailer with full lateral control by two Reynolds Boughton 25 tonne hydraulic winches mounted behind the engine. The cast steel fifth wheel is of the double oscillating type.

Status

In production. MH6660 in service with Oman (9).

Manufacturer

Unipower Vehicles Limited, 34 Greenhill Crescent, Watford Business Park, Watford, Hertfordshire WD1 8QU, UK.
Tel: +44 1923 816555.
Fax: +44 1923 228621.

UPDATED

Alvis Unipower MH8875 (8 × 8) 750 hp tractor unit with Nicolas multi-axle steer trailer **1997**

Specifications

Model	**MH8875**	**MH6660**
Cab seating	1 + 6	1 + 6
Configuration	8 × 8	6 × 6
Weights		
(1st front axle)	9,500 kg	10,000 kg
(2nd front axle)	9,500 kg	n/a
(1st rear axle)	14,200 kg	13,000 kg
(2nd rear axle)	14,200 kg	13,000 kg
(GVW)	47,400 kg	46,000 kg
(tractor, kerb)	21,000 kg	13,200 kg
(king-pin load)	26,000 kg	27,000 kg
Length	10.045 m	9.36 m
Width	3 m	2.89 m
Height	3.07 m	2.7 m
Wheelbase	6.2 m	5.5 m
Max speed	85 km/h	80 km/h
Fuel capacity	900 litres	900 litres
Range	1,000 km	1,000 km
Engine		
(make)	Cummins	Cummins
(model)	QSK-019-750	KTA-19-600
(capacity)	19 litre	19 litre
(power/revs)	750 bhp/2,100	600 bhp/2,100
Transmission	ZF HP1500 fully automatic, 5 forward and 1 reverse gears	
Transfer box	2-speed	2-speed
Steering	ZF 8098 power-assisted	
Brakes	air to all axles, dual circuit	
Tyres	24 R 21	395/85 R 20
Electrical system	24 V	24 V

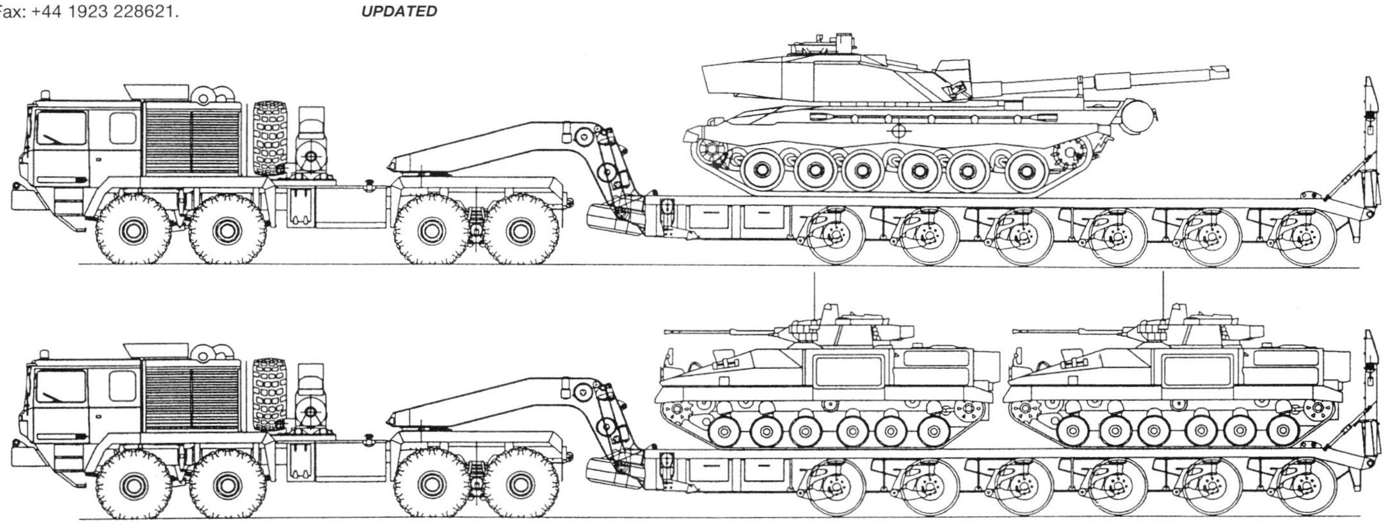

Alvis Unipower MH8875 (8 × 8) 750 hp unit with Nicolas multi-axle steer trailer suitable for carrying a tank up to 75 tons or two lighter vehicles **1997**

Alvis Unipower Contractor (6 × 4) tractor truck

Development

The Scammell (now Alvis Unipower Limited) Contractor (6 × 4) vehicle was introduced by Scammell in 1964 and has since been used for a variety of civil and military applications. It is used by the military for hauling semi-trailers carrying MBTs such as the M60 and large numbers were exported, especially to the Middle East and Africa.

Description

The layout of the Contractor is conventional, with the engine at the front, two-door fully enclosed all-steel cab in the centre and the fifth wheel at the rear. A wide range of different engines and transmissions is available, according to the role for which the vehicle is required.

Optional equipment includes: a larger four-door cab; three-line brake system (standard on British vehicles); engine brake; air conditioning system; canopy over roof; cab heater and demister; searchlights; seven pin plug at mid-chassis for semi-trailer; seven pin plug and socket at front and rear of frame for full trailer; tow hitches front and rear; winch with a capacity of 15,240 kg or a Darlington winch with a capacity of 22,680 kg; and a gearbox oil cooler. The winch has 131 m of cable and an interlock is provided to ensure that a winch brake and cut-out device operate continuously during loading. This prevents overloading and run-back cannot occur should the engine stop. A maximum pull of 50,800 kg is available by means of suitable sheaving; pull on first layer is 15,240 kg.

The series comprised seven basic models. Gross weights covered were 86,300 kg, 101,600 kg, 111,700 kg, 122,000 kg, 152,400 kg and 182,800 kg for articulated operation, or 111,700 kg, 162,500 kg, 193,000 kg, and 244,000 kg respectively as ballasted tractors for specialised heavy-duty haulage. The final production model had a maximum train weight of 244,000 kg. Standard models were 6 × 4, but 6 × 6 models were also available. Civilian applications include use as dump trucks and for logging.

Specifications

Cab seating: 1 + 2
Configuration: 6 × 4
Weight: (gross combination)
[CT85 C33F48] 86,360 kg
(loaded, tractor truck) 38,530 kg
(empty, tractor truck) 12,567 kg
Weight on front axle: (loaded) 8,130 kg
Weight on rear bogie: (loaded) 30,490 kg
Length: 7.773 m
Width: 2.489 m
Height: 2.955 m
Height: (5th wheel) 1.622 m
Track:
(front) 1.991 m
(rear) 1.845 m
Wheelbase: 4.748 m
Max speed:
(loaded) 77.14 km/h [CT85 C33F48] 86,360 kg GCW
Fuel capacity: 636 litres
Max gradient:
[GCW 116,840 kg] (loaded) 16.67%
Engine: Cummins NTC 6-cylinder 4-stroke in-line diesel developing 335 bhp (gross) at 2,100 rpm
Gearbox: RTO 12515 15-speed forward, 3 reverse, twin countershaft
Clutch: twin dry plate
Steering: power-assisted
Suspension:
(front) longitudinal semi-elliptical leaf springs with telescopic hydraulic dampers
(rear) fully articulated, inverted longitudinal springs, trunnion-mounted at centre
Brakes: air
Electrical system: 24 V
Batteries: 4 × 6 V, 195 Ah

Australian Army Contractor (6 × 4) tractor truck with semi-trailer (P Handel)

Status

Production complete. The Contractor is known to be in service in Australia, Jordan, Kenya and Libya.

Manufacturer

Alvis Unipower Limited, 34 Greenhill Crescent, Watford Business Park, Watford, Hertfordshire WD1 8QU, UK.
Tel: +44 1923 816555.
Fax: +44 1923 228621.

UPDATED

Alvis Unipower S24 Contractor Range (6 × 4) and (6 × 6) transporter tractors

Development

The S24 range was introduced in 1981 by Scammell, its design based on military and civilian experience gained with the earlier Contractor models.

In May 1988, Unipower Vehicles Limited (now Alvis Unipower Limited) acquired the design, manufacturing and support rights for the Scammell S24 range, their rear engined crash tenders, Commander, Contractor, Explorer, Super Constructor, Crusader, LDSS, Nubian Major and Thornycroft Antar models.

Description

The S24 Contractor range is available in 6 × 4 and 6 × 6 configuration with left- and right-hand normal control power-assisted steering.

Power outputs range from 400 to 600 bhp utilising Cummins N & K in-line series and Perkins RF CV8 power units, coupled to manual, semi- and fully automatic transmissions. Sustained performance figures up to 32 km/h on 1 in 40 gradients at 110 tonnes GTW are available with top speeds in the 74 to 80 km/h range.

King-pin imposed loads of 25 to 28 tonnes at unrestricted speeds are available with 12.00 R 24 and 14.00 R 20 tyre equipment on standard and wide track axles. All driving axles of the spiral bevel epicyclic hub reduction type are equipped with inter- and cross-lockable differentials actuated by air pressure from the cab. Front-wheel drive variants can be disengaged from the cab via the single-speed transposing box as dictated by terrain considerations.

Dual circuit primary air braking with secondary spring and parking brakes are provided, while manual transmission vehicles are equipped with engine brakes and the semi-automatic variants have integral hydraulic retarders.

Suspension is of the semi-elliptic leaf spring type with telescopic hydraulic damping at the front. High ground clearance is achieved by the fully articulating centre trunnion mounted rear springs.

An all-steel bonneted cab accommodates the driver and two passengers whilst the optional extended version can accommodate five seated passengers.

Dual hydraulically driven, independently operated 20 tonne winches are fitted behind the cab and are supplied with 50 m of 24 mm rope in an end drum.

A full set of tools and equipment associated with the transporter role are provided, together with up to two spare wheels and tyres.

Specifications

CA45
Cab seating: 1 + 2 or 1 + 5
Configuration: 6 × 6
Weight laden:
(front axle) 9,000 kg
(rear bogie) 36,000 kg
(GVW) 45,000 kg
(GCW) 100,000 kg
Length: 8.475 m
Width: 2.868 m
Height: 3.298 m
Wheelbase: 5.42 m
Max speed: 80 km/h
Fuel capacity: 900 litres
Engine: Cummins NTE 400 turbocharged and after-cooled developing 400 bhp at 2,100 rpm
Transmission: Eaton Fuller RTX 14615, 15 forward speeds
Clutch: twin plate (394 mm diam)
Transfer box: Unipower single speed
Tyres: 14.00 R 20
Brakes: dual circuit air
Electrical system: 24 V

Status

In service in the Middle East.

Manufacturer

Alvis Unipower Limited, 34 Greenhill Crescent, Watford Business Park, Watford, Hertfordshire WD1 8QU, UK.
Tel: +44 1923 816555.
Fax: +44 1923 228621.

UPDATED

Alvis Unipower S24TT8 tractor truck and trailer

Alvis Unipower Commander (6 × 4) tractor truck

Development

The Alvis Unipower (formerly Scammell) Commander tractor truck was developed to meet the requirements of the British Army for a new tractor truck to replace its ageing Thornycroft Antars and to carry the Challenger 1 MBT. Design work on the Commander began in 1976 with the first three prototypes being completed in 1978. Two of these were powered by the Rolls-Royce CV12 TCE diesel and the third by an American Cummins KTA 600 diesel.

In 1981 there was a General Staff Requirement (GSR) for a vehicle similar to the Commander but, as a result of defence spending cuts, the GSR was not endorsed at the time. However, in late 1982 an order for 125 Commanders for the British Army was placed. The first examples were delivered in late 1983, with production continuing into 1985.

In January 1992 Unipower (now Alvis Unipower) was awarded a contract for the in-depth repair of an undisclosed pilot quantity of Commander tractor trucks. The contract involved total strip-down and refurbishing, where necessary, with new parts.

Although designed for towing semi-trailers carrying AFVs weighing up to 65,000 kg, the Commander can also be used for high-speed haulage of heavy indivisible loads.

Description

The Alvis Unipower Commander tractor truck has the engine at the front, an all-steel cab in the centre, a winch located to the rear of the cab and the fifth wheel over the bogie at the rear.

The cab is arranged for left-hand drive and incorporates noise-insulating material. A two-piece flat glass windscreen with an electric heating element on the driver's side is provided. For crew access, large steps are fitted and full interior heating and ventilating equipment is standard. Air conditioning equipment is optional, as are various radio installations. The cab has individual seats for the commander and driver behind which is a bench seat for two personnel which can be positioned to form two individual bunks, one above the other.

The chassis is of steel channel side members with bolted-in, fabricated and tubular cross members. There are heavy-duty members at the front and rear for towing, lifting and recovery.

Alvis Unipower Commander (6 × 4) tractor truck (T J Gander) ***1997***

The front axle is a Scammell 12,200 kg capacity steer. Lock angles of 40° give high manoeuvrability. The rear axle is a Scammell 40,700 kg capability double drive bogie comprising two hub reduction axles linked by a lockable third differential for increased traction on poor surfaces. The fully articulated fifth wheel has an imposed load capacity of 34,600 kg.

A heavy-duty Rotzler winch is standard for the Commander for self-loading dead loads and the tractor is fully equipped with recovery fittings. The 20,300 kg line pull horizontal winch has 110 m of 26 mm diameter rope. The winch is fitted with an automatic pay-on gear and has a fail-safe brake. A high rope warning and an automatic overload cut-out is provided. The rope can be led out over the neck of the semi-trailer or through fairleads at the rear of the chassis. Winch controls are mounted behind the cab in a weatherproof enclosure.

Specifications

Cab seating: 1 + 3 or 4
Configuration: 6 × 4
Weight: (empty) 19,920 kg
Weight on front axle: (empty) 9,680 kg
Weight on rear bogie: (empty) 10,240 kg
Gross combination weight: 104,000 kg
Length: 9.01 m
Width: 3.25 m
Height: 3.5 m
Wheelbase: 5.03 m
Max speed: (with semi-trailer and 65 t MBT) 61 km/h
Fuel capacity: 817 litres
Gradient: 20%
Fording: 0.76 m
Engine: Rolls-Royce CV12 TCE turbocharged 60° V-12 diesel developing 625 bhp at 2,100 rpm
Gearbox: Allison CLBT 6061 6-speed epicyclic with torque converter permitting gear changes to be made under power. Hydraulic retarder for speed control on hills to supplement wheel brakes
Steering: hydraulic power-assisted
Suspension: leaf springs front and rear with telescopic dampers to front; 2-spring high-articulation rear bogie suspension ensures equal wheel loads
Tyres: 14.00 × 24
Brakes: twin leading show wedge brakes with automatic adjustment on all wheels, operated by 2 air circuits; 2-line couplings for trailer connections and additional couplings for double heading
Electrical system: 24 V
Batteries: 6 × 12 V, 100 Ah

Status

In service with the British Army (125).

Manufacturer

Alvis Unipower Limited, 34 Greenhill Crescent, Watford Business Park, Watford, Hertfordshire WD1 8QU, UK.
Tel: +44 1923 816555.
Fax: +44 1923 228621.

UPDATED

UNITED STATES OF AMERICA

Oshkosh M1070 (8 × 8) Heavy Equipment Transporter (HET)

Development

As the result of a US Army requirement to transport the M1A1 Abrams MBT, the Oshkosh Truck Corporation was awarded a contract to produce 1,044 M1070 Heavy Equipment Transporters (HET). This contract contained an option for a further 522 units. Production commenced in July 1992. To date 1,354 HETs have been produced, including 310 option vehicles. The full contract, with spares, was worth $275 million. Production for export customers began during 1993.

The Oshkosh Truck Corporation and Leyland Trucks Limited of Leyland, UK, have signed letters of intent to co-operate in promoting the M1070 HET for future UK HET requirements and to co-operate in the development of new military vehicles for combat support roles and missions in the UK and other countries.

M1070 (8 × 8) Heavy Equipment Transporter (HET) towing semi-trailer carrying an M1 Abrams MBT ***1997***

Description

The M1070 HET has full 8 × 8 drive and is powered by a Detroit Diesel 8V-92TA developing 500 hp coupled to an Allison five-speed automatic transmission. The single front axle is steered in co-ordination with the rear axle of the rear 'tridem' unit. All axles are Rockwell Model SVI 5MR units with planetary hub reductions and differential locks on the tridem.

The layout is conventional, with a large sloping bonnet in front of a fully enclosed cab which has seating for the driver and crewman and four passengers. Two 25 ton capacity loading winches are fitted as standard, as is a Central Tyre Inflation System (CTIS). The fifth wheel has a nominal load capacity of 20,866 kg and a diameter of 3.5 in/89 mm.

The M1070 HET tows the M1000 semi-trailer. The HET is transportable in C-5A and C-17 transport aircraft.

Specifications

Cab seating: 2 + 4
Configuration: 8 × 8
Weight:
(kerb) 18,598 kg
(GVW) 39,009 kg
(GCW) 104,961 kg
Load on 5th wheel: (nominal) 21,325 kg
Length: 9.093 m
Width: 2.591 m

Height: 3.962 m
(5th wheel) 1.616 m
Track: 2.083 m
Wheelbase: 5.461 m
Max speed:
(primary road at GCW) 72 km/h
(secondary road at GCW) 52 km/h
Range: (cruising) 724 km
Fuel capacity: 946 litres
Gradient: (at full GCW) 15%
Fording: 0.711 m
Engine: Detroit Diesel Model 8V-92TA 12.1 litre diesel developing 500 hp at 2,100 rpm with DDEC III electronic control system
Transmission: Allison CLT-754 automatic with 5 forward gears and torque converter
Transfer box: Oshkosh 55000 2-speed
Steering: power-assisted with front and rear axle co-ordinated steer
Turning radius: (wall to wall) 11.1 m
Suspension:
(front) springs and hydraulic shock-absorbers front
(rear) air ride
Tyres: 16.00 R 20 XLZT with central tyre inflation system
Brakes: drum type, air-actuated S-cam
Electrical system: 24 V start, 12 V lighting
Batteries: 4 × 12 V

Status
Production for US Army (1,044 plus 522 options) commenced July 1992. Current contracts extend to August 1998.

Manufacturer
Oshkosh Truck Corporation, PO Box 2566, Oshkosh, Wisconsin 54903-2566, USA.
Tel: +1 414 235 9150. Fax: +1 414 233 9540.

UPDATED

Oshkosh M911 (6 × 6) heavy equipment transporter

Development
In September 1976, Oshkosh was awarded an initial contract for 747 vehicles designated the M911, based on the commercial Oshkosh F2365 truck. A number of export customers purchased the M911 (some with Oshkosh-supplied trailers). Production commenced in 1976 and was completed with approximately 2,000 units produced.

The M911 is under consideration for a US Army Extended Service Life Program (ESLP) which will upgrade the vehicle to enable it to be modernised to extend its operational life while increasing its capabilities.

Description
The layout of the vehicle is conventional, with the engine at the front, all-steel two-door cab in the centre and the fifth wheel at the rear. Standard equipment for the cab includes: an adjustable seat for the driver, air-operated variable speed wipers, 22,000 BTU/h hot water heater and defroster.

The brakes have a dual air supply system, one for the front and one for the rear axles. If the air system that supplies the front brakes fails, the rear brakes can be operated normally. If the rear system fails, the front system as well as the rear spring chamber will still remain pressurised. A hydraulic retarder operated with a foot control in the transmission can absorb up to 80 per cent of the engine's horsepower.

The main transmission is fully automatic and will automatically upshift or downshift in all ranges above second gear, with a hold in each range. Built-in inhibitors prevent downshift or reverse shift at excessive speeds.

Standard equipment includes: two retrieval winches with a capacity of 20,412 kg each, rear decklights, spare tyre carrier with lift, rear pintle hook, trailer air and electrical connections, spotlights, fog lights, hydraulic jack, tyre inflation hose, oscillating fifth wheel, splash guards, radiator and headlight guard and towing eyes. Optional equipment includes air conditioning.

Specifications
Cab seating: 1 + 2
Configuration: 6 × 6
Weight:
(kerb) 18,144 kg
(GVW) 39,463 kg
(GCW) 102,514 kg
Load on 5th wheel: (nominal) 20,866 kg
Length: 9.373 m
Width:
(over bumpers) 2.438 m
(rear duals) 2.896 m
Height:
(overall) 3.404 m
(cab) 3.175 m
(5th wheel, no load) 1.626 m
Track:
(front) 2.083 m
(rear) 2.057 m
Wheelbase: 5.207 m + 1.524 m
Max speed:
(road, with GCW) 72 km/h
(3% gradient, fully loaded) 23 km/h
Range:
(no payload, maximum speed) 1,344 km
(at 86,183 kg GCW, economical speed) 990 km
Fuel capacity: 757 litres
Max gradient: (at full GCW) 20%
Fording: 0.711 m
Engine: Detroit Diesel model 8V-92TA-90 V-8 developing 450 hp at 2,100 rpm (500 hp optional)
Transmission: Allison CLBT-754 automatic with 5 forward and 1 reverse gears and hydraulic retarder
Transfer case: Oshkosh Model 7038 single-speed
Auxiliary transmission: Fuller AT 1202 2-speed
Steering: hydraulic, dual power-assisted at front
Turning radius: 13.72 m
Suspension:
(front) Hotchkiss type with semi-elliptical main springs with Berlin eye and semi-elliptical auxiliary spring
(rear) Hendrickson RT-650, steel spring, equalising beam design with 1.524 m spread
Tyres: 14 R 24
Brakes: drum, dual-system air S-type
Electrical system: 24 V
Batteries: 4 × 12 V, 100 Ah
Alternator: 65 A

Oshkosh M911 (6 × 6) truck tractor towing semi-trailer carrying M1 Abrams MBT ***1997***

Status
Production complete. In service with the US Army and countries in the Middle and Far East including Chad (16), Greece, Libya (18), Morocco, Oman, Saudi Arabia, Taiwan, Thailand, and Yemen.

Manufacturer
Oshkosh Truck Corporation, PO Box 2566, Oshkosh, Wisconsin 54903-2566, USA.
Tel: +1 414 235 9150. Fax: +1 414 233 9540.

UPDATED

Mack RD8226SX 52,000 kg (6 × 6) tank transporter

Description
The Mack RD8226SX is the tank transporter component of the Mack RM6006S range of vehicles. It is capable of towing semi-trailers carrying loads up to 46,700 kg and it may be equipped with a crew cab situated behind the normal driver's cab. The engine used is the Mack EE9 eight cylinder diesel developing 500 hp. This is located at the front of the vehicle with the driver's cab central and the fifth wheel to the rear. The Mack RD model has a full-time operating front drive axle. Driver comfort is catered for by an adjustable steering wheel, a convenient-to-view instrument layout and driving seat suspension. Many options are available to suit a wide range of requirements.

Mack RD8226SX 52,000 kg (6 × 6) tank transporter towing semi-trailer carrying M60 tank

Specifications
Cab seating: 1 + 2
Configuration: 6 × 6
Weight:
(loaded chassis, front) 10,400 kg
(loaded chassis, rear) 36,300 kg
(total) 46,700 kg
(GVW) 46,700 kg
(GCW) 92,254 kg
Length: 8.542 m
Width: 2.642 m
Height: 3.734 m
Wheelbase: 5.842 m
Max speed: 79 km/h
Fuel capacity: (total) 832 litres
Gradient: 20%
Engine: Mack EE9 8-cylinder turbocharged diesel developing 500 hp
Transmission: T2100 10-speed
Clutch: CLS795
Transfer case: TC152 2-speed with PTO
Suspension: spring type
Tyres: 12.00 × 24
Brakes:
(front) air/hydraulic disc
(rear) air cam
Batteries: 4 × 12 V

Status
In production. In service worldwide.

Manufacturer
Mack Trucks Inc, Box M, Allentown, Pennsylvania 18105, USA.
Tel: +1 215 439 3011. Telex: 84-7424.

VERIFIED

YUGOSLAVIA (SERBIA AND MONTENEGRO)

FAP 3232 BDST/AV (8 × 8) Tractor Truck

Description
First shown in 1985, the FAP 3232 BDST/AV (8 × 8) tractor truck is a variant of the FAP 2832 (8 × 8) 9,000 kg truck (see entry in *Trucks* section for details). The main change is that in place of the usual truck cargo body there is a fifth wheel coupling for towing semi-trailers weighing up to 65,000 kg and with a payload of 50,000 kg. There are two winches located on the platform behind the cab, one for forward use. In all other respects the FAP 3232 BDST/AV and FAP 2832 truck are identical.

Specifications
Cab seating: 1 + 1
Configuration: 8 × 8
Weight:
(unloaded) 16,000 kg
(loaded, on road) 34,000 kg
(load on 5th wheel) 18,000 kg
Towed load: 65,000 kg
Length: 8.612 m
Width:
(over tractor unit) 2.5 m
(over towed load) 3.5 m
Height:
(cab) 3.335 m
(5th wheel) 1.9 m
Ground clearance: 0.38 m
Track: 2.02 m
Wheelbase: 1.5 m + 3.2 m + 1.4 m
Angle of approach: 41°
Max speed: 60 km/h
Range: over 600 km
Max gradient: (tractor unit) 32%
Fording: 1.2 m
Engine: OM 403 15.95 litre V-10 diesel developing 320 hp
Gearbox: synchronised with auxiliary drive, 8 forward and 1 reverse gears
Clutch: single dry plate
Transfer box: 2-speed

FAP 3232 BDST/AV (8 × 8) tractor trucks towing semi-trailers carrying T-72 MBTs

Steering: hydraulic, ball joint
Turning radius: 13 m
Tyres: 15.00 × 21 T-101
Brakes:
(main) dual-circuit air plus exhaust brake
(parking) mechanical
Electrical system: 24 V
Batteries: 2 × 12 V, 210 Ah
Alternator: 85 A

Status
In service with the former Yugoslav Army.

Manufacturer
Sour FAP FAMOS 11 Oktomvri Taz, YU-11000 Beograd, Serbia.
Enquiries to: FDSP - Federal Directorate of Supply and Procurement, PO Box 89, Bulevar umetnosti 2, YU-11070 Novi Beograd, Yugoslavia.
Tel: +381 11 222 2743. Telex: 71000 JU SPDR.
Fax: +381 11 324 3738.

UPDATED

AMPHIBIANS

This section covers vehicles designed specifically for amphibious operations, such as crossing inland waterways or carrying cargo from ships offshore on to the beach, or inland. It does not include vehicles which require preparation to make them amphibious nor does it include amphibious over-snow vehicles or tracked prime movers, details of which will be found in their respective sections.

ITALY

IVECO Model 6640 G (4 × 4) 2,000 kg amphibious cargo carrier

Description

This vehicle, announced in 1980, is very similar to the IVECO Model 6640A (4 × 4) amphibious cargo carrier described in *Jane's Military Logistics 1988* page 573. The 6640 G has a much greater weight due to stronger construction and a slightly longer wheelbase; it also has a more powerful diesel engine coupled to automatic transmission and is propelled in the water by a water-jet rather than by a propeller as the earlier vehicle is.

The hull of the vehicle is of all-welded aluminium construction with a maximum thickness of 4 mm. The engine compartment is at the front of the hull and is separated from the crew compartment by a fireproof bulkhead. The two-man fully enclosed cab is heated and ventilated. The cargo area is to the rear of the cab and is provided with removable bows and a tarpaulin cover. Folding bench seats for 14 personnel run along both sides of the cargo area.

The vehicle is propelled in water by its wheels or the water-jet mounted under the hull at the rear; steering in the water is by a rear-mounted rudder linked to the steering wheel. In an emergency the rudder can be operated by hand. There are three bilge pumps, one in the bottom of the engine compartment and two below the cargo area.

As an option, a winch with a capacity of 4,500 kg on the first layer can be mounted at the front of the vehicle. A materials handling crane with a capacity of 700 kg can be fitted if required.

The latest version is the IVECO 6640 H which is powered by a 220 hp 5.861 litre engine. The automatic three-speed gearbox is fitted with a torque converter.

Specifications

Cab seating: 1 + 2
Configuration: 4 × 4
Weight:
(kerb) 6,700 kg
(loaded) 8,700 kg
(on front axle, loaded) 3,400 kg
(on rear axle, loaded) 5,300 kg
(payload) 2,000 kg
Load area: 3.21 × 1.95 m
Length: 8.2 m
Width: 2.5 m
Height:
(cab) approx 2.7 m
(tarpaulin) 3.16 m
Ground clearance: 0.35 m
Track: 1.96 m
Wheelbase: 3.1 m
Angle of approach/departure: 30°/25°
Max speed:
(road) 100 km/h
(water, propelled by water-jet) 11 km/h
Range:
(road) over 600 km
(water, propelled by water-jet) over 5 h
Max gradient: 60%
Max side slope: 30%
Fording: amphibious
Engine: Model 8062.24 6-cylinder 4-stroke liquid-cooled turbocharged diesel developing 195 hp at 3,200 rpm (6640 H, 220 hp at 3,000 rpm)
Gearbox: automatic, 3 forward and 1 reverse gears with power take-off for water-jet
Axle differential carriers: single reduction, spiral bevel drive. Air-operated differential lock at front and rear
Planetary drives: (front and rear) epicyclic gear train in wheel hubs
Steering: power-assisted
Turning radius: (land) 7.5 m
Suspension: (front and rear) independent strut and link type with helical spring and rubber bump stop plus hydraulic shock-absorbers
Tyres: 14.5 × 20 PS 12 (13.00 × 20 optional)
Brakes:
(main) disc, air over hydraulic, dual circuit
(parking) drum type, mounted on transfer rear output shaft
Electrical system: 24 V
Batteries: 2 × 12 V, 110 Ah

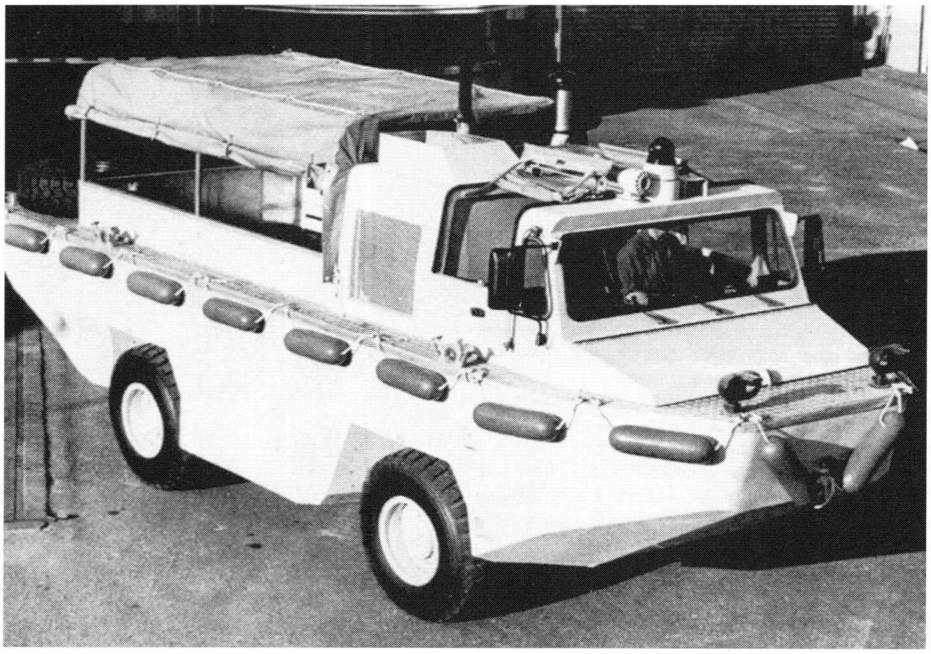

IVECO Model 6640 G (4 × 4) 2,000 kg amphibious cargo carrier with tarpaulin cover

Status

During 1984 a total of 17 sold to Italian Ministry of Interior (Direction of Civil Protection) plus a further 21 for Civil Defence.

Manufacturer

IVECO SpA, Defence Vehicle Division, Via Volta 6, I-39100 Bolzano, Italy.
Tel: +39 471 905111. Telex: 400541 ivedvd i.
Fax: +39 471 905444.

VERIFIED

RUSSIAN FEDERATION AND ASSOCIATED STATES (CIS)

PTS, PTS-M and PTS-2 tracked amphibious vehicles

Development

The PTS tracked amphibious vehicle entered service in the mid-1960s as the replacement for the earlier K-61 (GPT) vehicle. The main improvements over the K-61 were its higher water speed and its ability to transport 10,000 kg on land for 3 km in order to reach the water. The PTS-M, which first appeared in 1969, is a modified version while the later (1985) PTS-2 is the replacement for the PTS-M.

The Polish Army Engineeering Works at Deblin offers a repair, overhaul and rebuild service for the PTS-M. Also on offer are modifications to make the PTS-M more adaptable for operations in extremely cold regions.

The PTS-2 was offered for export sales, in 1994, at a unit price of US$130,000.

Description

PTS and PTS-M

These two vehicles are essentially similar. The crew compartment is at the front of the vehicle and is fully sealed against NBC attack. The crew enter the cab via two circular hatches in the roof. The cargo area is at the rear of the vehicle and vehicles are loaded via the hinged tailgate which also has integral loading ramps. Tiedown points are fitted for securing the vehicle. The engine is under the centre of the cargo compartment with the exhaust exits just above the top of the cargo compartment on each side. The vehicles are propelled in the water by two propellers in tunnels under the rear of the hull and steering is by two rudders at the rear of the hull.

The suspension is of the torsion bar type and consists of six roadwheels with the idler at the rear and the drive sprocket at the front.

A winch is mounted at the front of the vehicle and before entering the water a trim vane is erected at the front and the bilge pumps are switched on. The cargo area can be covered by bows and a tarpaulin cover and is sometimes used as an ambulance vehicle. Standard equipment includes infrared night vision equipment, intercom, radios and a searchlight mounted on the top of the crew compartment.

The vehicles were designed to carry 5,000 kg on land or 10,000 kg on water, or up to 70 personnel. In the ambulance role up to 12 stretchers can be carried in the

PTS-M tracked amphibious vehicle (Michael Jerchel) ***1996***

open cargo area. The PTS-M has also been used to lay portable trackway across beach areas.

The PKP trailer was designed specifically for use with the PTS-M. Basic specifications are: unladen weight 3,600 kg, overall length 10.3 m, width (travelling) 2.82 m, height 1.98 m, ground clearance (loaded) 0.4 m and track 1.89 m. The PKP is a boat-shaped trailer and has two small pontoons pivoted either side which rest on top of the trailer when it is travelling and are swung through 180° and locked in position before entering the water. Loading ramps are provided. The trailer is normally used to carry 122 mm howitzers, while the PTS carries the prime mover, for example the Ural-375D truck. The PKP can be towed at a speed of between 20 and 25 km/h laden or 25 to 30 km/h unladen.

PTS-2

Introduced in 1985, the PTS-2 is based on the chassis of the MT-T heavy tracked transporter which uses suspension components of the T-64 MBT. It is powered by a V-64-4 diesel derived from that used in the T-72 MBT, developing 710 hp. The PTS-2 is understood to be the eventual replacement for the PTS-M and has been in service for some time.

The PTS-2 has revised running gear consisting of seven roadwheels and four track-return rollers; the track links have rubber pads. The driving cab is enlarged and features an NBC protection system, while the trim vane at the bow is slightly curved. The sponson is higher than on earlier models and there is a rectangular exhaust outlet on both sides to accommodate the turbocharged engine mounted amidships. A ramp-like stern is provided.

The cargo space is enlarged to 8.3 × 2.6 m and, at 12,000 kg, the PTS-2 has a higher payload capacity than earlier models. Road speed is increased to 60 km/h and water speed is 12 km/h. Road range is 500 km and the vehicle can operate for up to 10 hours in the water. Dimensions are given as length 12 m, width 3.3 m and height 3.17 m.

Optional equipment available for the PTS-2 includes a multipurpose shovel, a dozer blade, a water-jet, air conditioning equipment and a self-recovery winch. It is possible to install radio location and navigation systems to ensure safe round-the-clock operations.

Variant

The Polish Army has used a number of PTS tracked amphibious vehicles fitted with rocket-propelled mineclearing equipment in the rear.

Specifications

PTS-M
Cab seating: 1 + 1 (70 troops in rear)
Weight:
(empty) 17,700 kg
(loaded, land) 22,700 kg
(loaded, water) 27,700 kg
Max load:
(land) 5,000 kg
(water) 10,000 kg
Load area: 7.9 × 2 6 m
Length: 11.426 m
Width: 3.3 m
Height: 2.6 m
Ground clearance: (loaded) 0.4 m
Track: 2.8 m
Track width: 480 mm
Length of track on ground: 5.63 m
Ground pressure:
(empty) 0.382 kg/cm^2
(with 5,000 kg load) 0.483 kg/cm^2
(with 10,000 kg load) 0.582 kg/cm^2
Max speed:
(land, with 5,000 kg load) 42 km/h
(water, with 10,000 kg load) 10.6 km/h
Range: up to 300 km
Fuel capacity: 705 litres
Max gradient:
(empty) 60%
(loaded) 20%
Vertical obstacle: 0.65 m
Trench: 2.5 m
Engine: V-54P diesel developing 350 hp at 1,800 rpm

Status

PTS-2 in production and offered for export sales. PTS and PTS-M in service with members of the former Warsaw Pact, including Bulgaria and Poland, as well as Algeria, Angola, Congo, Cuba, Egypt, India, Iran, Libya, Tanzania, Vietnam, Yemen and Yugoslavia.

Manufacturer

Coral Small State Enterprise, 195 Ligovskiy pr, PO Box 212, 192007 St Petersburg, Russia.
Tel: +7 812 166 4654. Telex: 614057 CORAL SU.
Fax: +7 812 167 0705.

UPDATED

SPAIN

VAP 3550/1 (4 × 4) 3,000 kg amphibious vehicle

Development

The VAP 3550/1 (4 × 4) 3,000 kg amphibious vehicle was developed by ENASA, to meet Spanish Navy requirements for a vehicle that can be launched from LSTs and other amphibious craft offshore, reach the coast under its own power and then travel inland over rough country. VAP is the export name of this vehicle, within Spain it is known as the Pegaso 3550, with the first production batch known as the 3550/1.

The vehicle uses many automotive components of the ENASA (Pegaso) range of 3045 (4 × 4) 3,000 kg and 3050 (6 × 6) 6,000 kg trucks, of which over 6,000 were supplied to the Spanish Army.

Description

The boat-shaped hull of the VAP 3550/1 is made of all-welded 6 mm thick steel plates and is divided into watertight compartments. The driver sits in the semi-enclosed cab, which has an open back towards the front, with two passengers seated on his right. A searchlight that can be operated from within the cab is mounted over the top of the cab. Immediately behind the cab is a hydraulic crane with a maximum lifting capacity of 350 kg.

The cargo area in the centre of the vehicle can be covered with removable bows and a tarpaulin cover. Removable benches for troops can be fitted down either side of the cargo compartment.

The engine compartment is at the rear with the air outlet/inlet louvres and exhaust pipe mounted in the top.

The VAP is fully amphibious, propelled in the water by two water-jets at the rear of the hull immediately behind the second axle. The two single water-jets are driven by a hydraulic system composed of a pump directly connected to the vehicle's engine and two hydraulic motors acting directly on the hydrojets. It features a pressurising system for the mechanical units in contact with the water which operates as soon as the VAP enters the water. When afloat, pivot turns can be accomplished.

The load compartment is equipped with two pumps which have a maximum capacity of 6,000 litres/h and there are two automatic bilge pumps in the hull with a maximum capacity of 3,600 litres/h. Mounted at the front is an optional winch with a maximum capacity of 4,500 kg. Other options include a pintle-mounted 7.62 mm machine gun and smoke grenade launchers.

VAP 3550/1 (4 × 4) 3,000 kg amphibious vehicle

Specifications

Cab seating: 1 + 2 (rear, 18 men)
Configuration: 4 × 4
Weight:
(empty) 9,500 kg
(loaded) 12,500 kg
Max load: 3,000 kg
Load area: 3.2 × 2.05 m
Length: 8.85 m
Width: 2.5 m
Height:
(cab) 2.5 m
(crane) 2.83 m
Ground clearance: 0.32 m
Track: 1.927 m

Wheelbase: 3.45 m
Angle of approach/departure: 33°/27°
Max speed:
(road) 87 km/h
(water) 5.5 kt
Range:
(road) 800 km
(water) 80 km
Fuel capacity: 250 litres
Max gradient: 60%
Max side slope: 30%
Fording: amphibious
Engine: Pegaso 9135/5 6.55 litre 6-cylinder in-line turbocharged diesel developing 170 hp at 2,600 rpm
Gearbox: Pegaso manual, 6 forward and 1 reverse gears
Clutch: single dry plate
Transfer box: Pegaso 2-speed
Steering: recirculating ball power-assisted
Turning radius: 9 m
Suspension: semi-elliptical springs and hydraulic double-acting shock-absorbers, both axles have a self-locking differential case
Tyres: 13.00 × 20
Brakes: drum, dual circuit, air
Electrical system: 24 V
Batteries: 2 × 12 V, 99 Ah
Alternator: 1,500 W

Status
Reported in service with Egypt. Seven delivered to Mexico in 1982. In service with the Spanish Marines.

Manufacturer
Pegaso SA, ENASA, Defence Division, Avenida de Aragón 402, E-28022 Madrid, Spain.
Tel: +34 1 750 1000. Telex: 42373 bara e.
Fax: +34 1 747 1307.

VERIFIED

UNITED KINGDOM

Glover Aquatrack tracked 5,000 kg amphibious vehicle

Description
The Glover Aquatrack tracked 5,000 kg amphibious vehicle, was developed to provide a vehicle capable of carrying up to 5 tonnes of payload or 40 seated personnel in a fully amphibious environment and over a wide variety of terrains. Power is provided by a single diesel engine driving twin variable-pitch propellers in Kort nozzles. Twin rudders assist in allowing the vehicle to manoeuvre and operate fully laden in Sea State 3 at water speeds of up to 10 km/h. The vehicle can operate in up to 3.05 m of plunging surf. It can also negotiate harsh terrain, mud, soft sand or frozen ground.

The Aquatrack hull incorporates five separate watertight compartments and has a low centre of gravity. The driver's cab is set well forward just behind the bow and the open load area behind the cab can carry vehicles the size of a Land Rover. The 'proven' suspension uses M113 pattern wheels and appears to be based upon the use of torsion bars.

It has been claimed that unit and running costs are low and that no highly specialised technical background or training are required for operations.

Glover Aquatrack tracked 5,000 kg amphibious vehicle ***1997***

Specifications
Cab seating: 1 + 3
Weight:
(empty) approx 13,000 kg
(loaded) approx 18,000 kg
(fuel and crew) 500 kg
Max load: 5,000 kg
Length: 9.2 m
Width: min 3.2 m
Height: approx 3.15 m
Deck height: 1.2 m
Load area: 4.3 × 2.6 m + 0.75 × 2.6 m
Ground clearance: 0.45-0.5 m
Angle of approach/departure: 40°/40°
Max speed:
(land) 40 km/h
(water) 10 km/h
Range: (road) 500 km
Fuel capacity: 400 litres
Max gradient: 50%
Vertical obstacle: 0.5 m
Trench: 2 m
Fording: amphibious
Engine: Deutz BF8 L513 13.382 litre air-cooled diesel developing 315 bhp at 2,300 rpm
Turning circle: pivot turn

Status
In service with the Philippines.

Manufacturers
GKN Defence, Hadley Castle Works, PO Box 106, Telford, Shropshire TF1 4QW, UK.
Tel: +44 1952 244321. Fax: +44 1952 243910.
Glover Webb Limited, Hamble Lane, Hamble, Hampshire SO31 4NY, UK.
Tel: +44 1703 452811. Fax: +44 1703 452964.

UPDATED

UNITED STATES OF AMERICA

LARC-5 (4 × 4) 4,545 kg amphibious cargo carrier

Development
The LARC-5 (Lighter, Amphibious, Resupply, Cargo, 5 ton) (4 × 4) was developed by the Borg Warner Corporation from 1958 under the direction of the US Transportation Engineering Command at Fort Eustis, Virginia. The first production contract was awarded to the Adams Division of Le Tourneau Westinghouse in June 1961 and between 1962 and 1968 950 LARC-5s were built. The last manufacturer was ConDiesel Mobile Equipment of Waterbury, Connecticut. The vehicle was designed to carry 4,545 kg of cargo, or 15 to 20 fully equipped troops from ships offshore to the beach or, if required, farther inland. It is issued on the scale of 34 per Army light amphibious company.

Description
The hull of the LARC-5 is of all-welded aluminium construction with reinforced aluminium frames. The cab is at the front of the vehicle and contains, in addition to the operating controls: a heater and windscreen defroster; portable lamp and cable; fire extinguisher; fabric cover for the back of the cab; radio; adjustable seat for the driver; two fixed seats for the other crew members and a magnetic compass. The cargo area is in the centre of the vehicle. Fabric curtains reinforced with stranded wire rope can be installed on each side of the cargo deck to protect the cargo.

The transfer transmission compartment is below the cargo deck and contains the transfer transmission, front wheel disconnects, drive shafts and service brakes.

The engine compartment is at the rear and is covered by two watertight hatches. Air is blown out of the compartment through a small grille between the two hatch covers. A fixed fire extinguisher is installed in the engine compartment and is controlled by a pull of a handle on the cargo deck rear bulkhead. Two manual bilge pumps are installed for use if the main hydraulic pump fails. The vehicle is propelled in the water by a three-bladed propeller under the rear of the hull.

Power is transmitted from the engine to a torque converter and hydraulic retarder installed on the flywheel end of the engine. The driver selects either forward or reverse by shifting the forward/reverse transmission lever. The main drive shaft connects the output of the forward/reverse transmission to the transfer transmission. This transmission has two gear ratios (high and low) for land operations and one gear ratio (marine) for water operations. The differential transmission transmits power to the four wheels. With the transmission in low or high range, power is always transmitted to the wheels. A mechanical disconnect can be used to apply power to the rear wheels only for two-wheel drive. Four drive shafts connect the differential transmission to the wheels. A right-angle drive assembly is installed at each wheel to apply the driving power to the wheels. The gearbox is used to apply the rotation of the four-drive shaft to the axle ends of the wheels.

Some vehicles were fitted with a hydraulically operated boom designed by the ConDiesel Mobile Equipment Division. The boom is 4.51 m long and can lift a maximum load of 2,500 kg.

Specifications
Cab seating: 1 + 2
Configuration: 4 × 4
Weight:
(empty) 9,507.5 kg
(loaded) 14,053 kg
Max load: 4,545 kg
Load area: 4.876 × 2.971 m

Length: 10.07 m
Width: 3.05 m
Height:
(overall) 3.1 m
(reduced) 2.41 m
Ground clearance: 0.406 m
Track: 2.565 m
Wheelbase: 4.876 m
Angle of approach/departure: 27°/20.7°
Max speed:
(road) 48.2 km/h
(water) 13.92 km/h
Range:
(land, empty) 400 km
(land, loaded) 322 km
(water, empty) 65 km
(water, loaded) 56 km
Fuel capacity: 547.2 litres
Max gradient: 60%
Max side slope: 25%
Fording: amphibious
Engine:
(early vehicles) 8-cylinder petrol developing 300 hp at 3,000 rpm
(late production vehicles) Cummins 4-cycle V-8 diesel developing 300 hp
Turning radius: 13.26 m
Suspension: rigid
Tyres: 18.00 × 25

LARC-5 (4 × 4) 4,545 kg amphibious cargo carrier on amphibious exercises

Status

Production complete. In service with Argentina, Australia (87), Portugal, Singapore, Thailand and the USA.

Manufacturer

ConDiesel Mobile Equipment, 84 Progress Lane, Waterbury, Connecticut 06705, USA.

UPDATED

TRACKED PRIME MOVERS, CARGO CARRIERS AND ARMOURED LOGISTIC VEHICLES

This section excludes tracked vehicles that have been specifically designed for amphibious use (such as the RFAS PTS-2) or for over-snow use (such as the Swedish Bv 202), details of which will be found in their respective sections.

BRAZIL

TECTRAN VBT-2028 (6 × 6) 10,000 kg armoured truck

Description

The TECTRAN VBT-2028 (6 × 6) 10,000 kg armoured truck was developed as a support and launch vehicle for the ASTROS II artillery rocket system. (For details of this system refer to *Jane's Armour and Artillery 1995-96* pages 695 to 697). With the ASTROS II system the VBT-2028 is used as a launch vehicle (AV-LMU); for ammunition supply (AV-RMD); as a battalion command and control vehicle (AV-VCC); for fire control (AV-UCF); and for mobile workshops. There is also a personnel carrier.

All versions use the same basic 6 × 6 drive configuration chassis with an armoured cab set forward. The cab has a wedge-shaped side profile with well-sloped armour and steel covers for the windows. Air conditioning is provided for the crew area. A roof hatch is provided with provision for a machine gun mounting and the cab roof can withstand the weight of at least two personnel. A hydraulic load-handling crane and winch are located behind the cab. Four stabiliser legs are lowered when the crane is in use.

The load area can take the form of a conventional cargo body complete with canvas tilt, drop sides and tailgate, but may also be of the flatbed type to carry shelters for the command and control or fire-control role. Rocket launch vehicles have an elevating and traverse system located over the rear axles.

Optional equipment includes a tyre pressure regulation system and a run-flat device.

Specifications

Configuration: 6 × 6
Gross vehicle weight: approx 20,000 kg
Payload: 10,000 kg

TECTRAN VBT-2028 (6 × 6) 10,000 kg armoured truck (left) loading an ASTROS II multiple rocket launcher (right)

Max speed: 90 km/h
Max gradient: 66%
Engine: Mercedes-Benz OM 422 14.618 litre V-8 diesel developing 280 hp at 2,300 rpm
Transmission: ZF AK 6S/90, 6-speed, with optional torque converter
Clutch: mechanical
Turning radius: 20.3 m
Tyres: 14.00 × 20
Brakes: dual circuit, air
Batteries: 2 × 12 V, 143 Ah

Status

In production. In service with Brazil, Iraq and Saudi Arabia.

Manufacturer

TECTRAN Engenharia, Indústria e Comércio SA, Rodovia Presidente Dutra, Km 155/156, Caixa Postal 165, São José dos Campos, São Paulo, Brazil.
Tel: +55 123 220142. Telex: 39431 ttte br.

UPDATED

CANADA

General Motors LAV (8 × 8) 2,470 kg armoured logistics carrier (LAV-L)

Development

The General Motors LAV (8 × 8) 2,470 kg armoured logistics carrier, or LAV-L, is one component in a series of vehicles known to the US Marine Corps as Mission-Role-Vehicles (MRVs). These are variants of the basic LAV light armoured vehicle and for details of this basic vehicle refer to *Jane's Armour and Artillery 1995-96*, pages 269 to 275. Originally the LAV-L was known as the General Motors (8 × 8) Logistics Vehicle.

The first LAV-L was handed over to the US Marine Corps in September 1985. The delivery of a batch of 94 vehicles began in November 1985 and was completed by August 1986.

Description

The LAV-L is intended to provide the US Marine Corps with mobile logistic support in the field. It uses many components of the basic LAV light armoured vehicle, including the suspension, drive train and hull. On the LAV-L the driver is seated forward to the left of the engine compartment under an armoured hatch. When this is in the closed down position the driver is provided with three wide-vision periscopes. Behind the driver is the raised commander's position which is provided with all-round vision devices when the armoured hatch is closed down. This position may be equipped with a pintle for a 7.62 mm M60 machine gun. There are also two four-barrel M257 smoke grenade launchers controlled from the commander's position. There is seating behind the commander for a third crew member.

The main cargo-carrying compartment has an unobstructed level floor measuring 2.59 × 1.65 m. Access to the area is via two outward-opening rear doors or through two roof hatches each with dimensions of 1.07 × 1.28 m. To the left rear of the roof hatch is a 500 kg telescopic crane operated from a folding platform over the left rear wheel. When the rear doors are open it is possible to load the cargo area using forklift trucks, internal net stowage and lashing points are also provided.

The LAV-L is amphibious and may be airlifted by a heavy helicopter. Extra internal seating can be provided, as can internal hoists and powered tailgates. A standard NATO towing hook is provided.

Specifications

Crew: 3
Configuration: 8 × 8
Weight: (combat) 12,730 kg
Max load: 2,470 kg
Length: 5.448 m
Width: 2.5 m
Height: 2.194 m
Max speed:
(road) 100 km/h
(water) 10 km/h
Range: 660 km
Gradient: 60%
Side slope: 30%
Vertical obstacle: 0.5 m
Trench: 2.057 m
Fording: amphibious
Engine: Detroit Diesel 6V-53T 6-cylinder diesel developing 275 hp at 2,800 rpm
Transmission: Allison MT-653 DR automatic, 5 forward and 1 reverse gears
Transfer case: Rockwell AG-VST (modified)
Steering: power-assisted on front 2 axles
Turning circle: 15.5 m
Suspension:
(front four wheels) independent coil springs and shock-absorbers
(rear four wheels) independent torsion bars and shock-absorbers
Tyres: 11.00 × 16 with Hutchinson run-flat inserts
Brakes:
(main) 8-wheel dual airbrake, transmission brake and transfer case lock
(parking) transmission brake and transfer case lock
Electrical system: 24 V
Batteries: 4 × 12 V
Alternator: 200 A
Armament: 1 × 7.62 mm M60 MG
2 × 4-barrel smoke grenade dischargers
Ammunition carried: 7.62 mm - 1,000 rounds
smoke grenades - 8 in launchers, 8 spare
Armour
hull front: 8 mm
hull sides: 10 mm

Status
Production complete. In service with US Marine Corps (94).

Manufacturer
Diesel Division, General Motors of Canada Limited, PO Box 5160, London, Ontario, Canada N6A 4N5.
Tel: +1 519 452 5184. Fax: +1 519 452 5688.

UPDATED

General Motors LAV (8 × 8) 2,470 kg armoured logistics carrier (LAV-L)

CHILE

FAMAE-MOWAG Piraña (8 × 8) armoured logistics carrier

Development

FAMAE of Chile obtained a licence to produce the 8 × 8 version of the Swiss MOWAG Piranha for the Chilean Army, where it is known as the Piraña. A prototype was built as an armoured personnel carrier and several other versions of this basic vehicle are under development, including a 120 mm mortar carrier, a 90 mm anti-tank gun carrier, a command and control vehicle and a repair and recovery vehicle. A riot control version has also been proposed. An armoured logistics carrier has also been developed and is in production for the Chilean Army.

For details of the armoured personnel carrier version of this vehicle refer to *Jane's Armour and Artillery 1995-96* pages 276 and 275. Details of the Swiss MOWAG Piranha will be found on pages 422 to 425.

Description

The FAMAE-MOWAG Piraña 8 × 8 armoured logistics carrier uses the main components of the basic MOWAG Piranha 8 × 8 including the drive train, suspension and hull configuration.

The vehicle has a basic crew of three: driver, commander and loader. The engine is located to the right of the driver and the commander is seated behind him. The commander is provided with all-round vision devices when the armoured hatch is closed down. This position may be provided with an external machine gun mounting. The loader is seated behind the commander.

The cargo compartment has an internal capacity of 5 m^3. Access to the cargo area is via a single downward-opening rear door or through two roof hatches. The compartment is provided with a 500 kg capacity telescopic crane used to load up to 2,500 kg of cargo. It is possible to use a forklift truck to load cargo via the rear door.

The Piraña 8 × 8 can be provided with two propeller blades at the rear for amphibious operations and the vehicle may be airlifted by a heavy helicopter.

FAMAE-MOWAG Piraña 8 × 8 armoured logistics carrier showing loading crane

Specifications

Crew: 3
Configuration: 8 × 8
Weight: (combat) 13,000 kg
Max load: 2,500 kg
Length: 6.373 m
Width: 2.5 m
Height: (top of hull) 1.85 m
Ground clearance:
(under differential) 0.39 m
(under hull) 0.5 m
Track:
(front) 2.18 m
(rear) 2.2 m
Wheelbase: 1.1 m + 1.335 m + 1.04 m
Angle of approach/departure: 40°/45°
Max speed:
(road) 100 km/h
(water) 10 km/h
Range: (road) 600 km
Fording: amphibious
Gradient: 60%
Side slope: 30%
Vertical obstacle: 0.5 m
Engine: Detroit Diesel 6V-53T turbocharged diesel developing 275 hp at 2,800 rpm
Transmission: Allison MT-653 automatic with 5 forward and 1 reverse gears
Steering: power-assisted on front two axles
Turning circle: ˉ5.5 m
Suspension:
(front) independent coil springs and shock-absorbers
(rear) independent torsion bars and shock-absorbers
Brakes: hydraulic/air, dual circuit
Tyres: 11.00 × 16 with run-flat inserts
Electrical system: 24 V

Status

In production for the Chilean Army.

Manufacturer

FAMAE, Fábricas y Maestranzas del Ejército, Avenida Pedro Montt 1606, PO Box 4100, Santiago, Chile.
Tel: +56 2 556 1011. Telex: 242346.
Fax: +56 2 555 0944.

UPDATED

CHINA, PEOPLE'S REPUBLIC

Type 59 artillery tractor

Description

This tractor was designed to tow artillery, such as the 122 mm Type 54 howitzer or the 122 mm Type 60 gun. The fully enclosed cab at the front of the vehicle has sufficient seats for most of the crew of the gun and has a circular observation hatch in the left side of the roof. The rear cargo area has a stake-type body and a drop tailgate, removable bows and a tarpaulin cover. The suspension is believed to be of the torsion bar type and consists of five roadwheels with the drive sprocket at the front and the idler at the rear. The Type 59 may be powered by a V-12 diesel truck engine.

Status

In service with China and Vietnam.

Type 59 artillery tractor towing 57 mm Type 59 anti-aircraft gun

Manufacturer

China North Industries Corporation (NORINCO), 7A Yue Tan Nan Jie, PO Box 2137 Beijing, Beijing, People's Republic of China.
Tel: +86 1 867570. Telex: 22339 cnic cn.
Fax: +86 1 867840.

VERIFIED

Type 60-1 tracked artillery tractor

Description

The Type 60-1 tracked artillery tractor appears to be a close development of the Type 59 artillery tractor, with differences occurring in details such as the cab shape.

The Type 60-1 can carry up to 5,000 kg of cargo and tow artillery or trailers weighing up to 15,000 kg. The forward control all-steel cab accommodates a driver and an artillery detachment. The roof is equipped with a hatch mounting a 12.7 mm Type 54 machine gun, the Chinese version of the Soviet DShK. There is stowage for the crew's personal equipment in lockers behind the cab and the main cargo area may be covered by bows and a canvas tarpaulin. A winch may be fitted, adding 1,000 kg to the vehicle's weight. This has a maximum capacity of 17,000 kg and is equipped with 75 m of cable. The engine starting time in winter with the ambient temperature at −35°C is stated to be 18 minutes. Rubber-covered track links are used.

The Type 60-1 is used as the carrier for the 273 mm Type 83 multiple rocket system. For details of this equipment refer to *Jane's Armour and Artillery 1995-96* pages 701 and 702.

Type 60-1 tracked artillery tractor

Specifications

Weight:
(with winch) 12,500 kg
(without winch) 11,500 kg
Payload: 5,000 kg
Towed load: 15,000 kg
Length: 6.117 m
Width: 2.6 m
Height:
(top of cab) 2.81 m
(top of tilt) 3.015 m
(to horizontal machine gun) 3.26 m
Ground clearance: 0.4 m
Track: 2.15 m
Length of track on ground: 3.248 m
Max speed: 48.8 km/h
Range: 400 km
Max gradient:
(with 5,000 kg load) 30°
(with 5,000 kg load and 15,000 kg trailer) 15°
Side slope: 26°
Fording: 1 m
Engine: 12150L-1 V-12 water-cooled 4-cycle direct injection diesel developing 300 hp at 1,600 rpm

Status

Offered for export sales.

Manufacturer

China North Industries Corporation (NORINCO), 7A Yue Tan Nan Jie, PO Box 2137 Beijing, Beijing, People's Republic of China.
Tel: +86 1 867570. Telex: 22339 cnic cn.
Fax: +86 1 867840.

UPDATED

FINLAND

Sisu NA-110 all-terrain vehicle

Development

The Sisu NA-110, originally known as the Sisu NA-140, was developed for military use in extreme terrain. Testing of the first prototype began in 1985. Preproduction of 11 units started in Autumn 1986 with full-scale production commencing during 1988. During 1988 the Finnish Defence Ministry placed an order worth US$35 million for 112 units, with first deliveries made during 1988. Since then 'several hundred' have been produced.

Description

The Sisu NA-110 is a two-compartment vehicle with articulated steering. Each compartment has two wide and flexible rubber tracks reinforced with composite cross members and longitudinal cords. The tracks have a guaranteed 7,500 km running life.

Standard versions manufactured before 1994 are powered by a General Motors V-8 diesel engine, coupled to a General Motors 3L89 HD 3+1 automatic transmission. Later examples are based on a L57 6.5 litre heavy-duty Detroit Diesel unit coupled to a Hydra-Matic 4L80-E 4+1 transmission. The engine is located in the front compartment and the compartments are connected by a flexible coupling which also accommodates the steering assembly through which the drive is transmitted to the rear unit. The axle system is based on Sisu BTN type drive axles. The main brake is a disc unit located in the coupling assembly and operating on the drive train. Track and roadwheel assemblies are interchangeable. Driving is stated to be easy, this is because of the powered steering, automatic transmission and the ground pressure is low. The vehicle is amphibious, capable of 4 km/h in the water.

The standard bodywork for both compartments is a single composite materials element on a light alloy frame. The front unit has four doors with space for the driver and five passengers or 700 kg of cargo; the rear unit can carry up to 12 personnel or 1,300 kg of cargo. A tow hook at the rear allows the towing of light weapons or a sledge. The rear unit can be configured to carry a 120 mm mortar.

The basic NA-110 chassis can be adapted to a wide variety of special bodies or optional equipment. Proposed variants have included: a logistics/cargo carrier with the rear unit configured as a load area; an ambulance container carrier; a fuel transport vehicle; a container carrier; and a mortar carrier. Civil versions have been delivered as firefighting and rescue vehicles.

Optional accessories include: a winch; a heavy machine gun mounting; add-on protection; a navigation system; various communications systems; and an AC electrical system.

Specifications

Cab seating: 1 + 5 (front unit - up to 12 in rear unit)
Weight:
(empty) 5,400 kg
(loaded) 7,400 kg
Max load: 2,200 kg
Towed load: 2,500 kg

Sisu NA-110 all-terrain vehicle ***1996***

Sisu NA-110 all-terrain vehicle ***1996***

Length:
(total, overall) 7.67 m
(front unit) 3.3 m
(rear unit) 3.45 m
Width: 1.91 m
Height: 2.35 m
Ground clearance: 0.3 m
Track width: 620 mm
Max speed:
(road) 55 km/h
(water) 3.6-6 km/h (depending on equipment)
Range: 500 km
Fuel capacity: 280 litres
Max gradient: friction limits
Side slope: 35°
Fording: amphibious when loaded
Engine: (until 1994) GM Powertrain 6.2 litre 4-stroke V-8 diesel
(current) Detroit Diesel 6.5 litre L57 four-stroke V-8 diesel developing 159 hp at 3,400 rpm
Transmission: (until 1994) GM Hydra-Matic 3L80 HD automatic with 3 forward and 1 reverse gears
(current) GM Hydra-Matic 4L80-E automatic with 4 forward and 1 reverse gears
Transfer box: Sisu L175 2-speed manual
Steering: articulated hydrostatic
Turning radius: 6 m
Suspension: transverse leaf springs
Brakes: single circuit, disc
Electrical system: 12/24 V
Batteries: 2 × 12 V

Status

In production. In service with the Finnish Army, China, Russia, Turkey and the UN.

Manufacturer

Sisu Defence Oy, PO Box 189, FIN-13101 Hämeenlinna, Finland.
Tel: +358 17 6451. Fax: +358 17 619 6710.

UPDATED

GERMANY

Kässbohrer Flexmobil FM 23.150K all-terrain vehicle

Description

The Kässbohrer Flexmobil FM 23.150K is based on the design of the Kässbohrer Pisten Bully, many thousands of which are in use in 40 countries. The Flexmobil is intended for use as a general load carrier and working vehicle and may be equipped with a load platform covered by a tilt or a passenger body accommodating up to 10 personnel including the driver. The passenger body has fold-up seats, a roof air-extractor, insulated walls and roof and a heater.

The Flexmobil uses rubber tracks fitted with bolt-on double steel web plates for difficult terrain such as rock, loose stones or gravel; one width of Summer track and two widths of Winter tracks, all steel, are available. The main roadwheels use a double track guide system and are self-cleaning. There are five roadwheels and three track-return rollers. A hydrostatic drive is used not only for the transmission but also for steering and braking. An 80 hp power take-off on the hydraulic system can be used to power tools and attachments. A wide range of attachments, including snow-clearing blades and blowers, cranes and a backhoe, may be fitted. The engine may be either a water-cooled Mercedes-Benz or an air-cooled Deutz. Both options are diesels. Also available as an option is a Rotzler 3,000 kg winch with 50 m of 11 mm cable. A special towing system for Winter accessories, such as snow-sweeping brushes, can be provided.

The Flexmobil can tow a special tracked trailer with a payload capacity of approximately 2,000 kg. Also available is a semi-trailer that is carried on a fifth wheel arrangement mounted on the Flexmobil's cargo bed. This has a payload of approximately 4 tonnes.

To transport the Flexmobil over long distances a special single-axle transport trailer, the ETP 9, is available. It is possible to lift the Flexmobil with heavy helicopters such as the CH-53E.

The latest version of the Flexmobil is the FM 23.150K Series 106. This can be fitted with either a KHD BF 6 L 913 152 hp diesel or a Mercedes-Benz OM 366A 150 hp diesel. A modified transmission reduces engine noise and fuel consumption. Other changes are made to the hydraulic drive and steering and the fuel capacity is increased to 165 litres. Various other detail design modifications have been made.

The Thyssen Henschel Breitkettenfahrzeug TH 439 is an armoured personnel carrier version of the Flexmobil FM 23.150K. For details refer to *Jane's Armour and Artillery 1995-96* page 363.

Specifications

Cab seating: (with personnel cab) 1 + 9
Weight:
(with load platform) 4,200 kg
(with personnel cab) 4,500 kg
(GVW) 5,400 kg
Payload: 1,200 kg
Load area: 2.2 × 2.2 m
Ground pressure:
(laden) 0.12 kg/cm²
(unladen) 0.09 kg/cm²
Length: 4.1 m
Width: 2.3 m
(over Summer tracks) 2.4 m
(over Winter tracks) 2.5 or 2.7 m
Height:
(to light on roof) 2.5 m
(top of cab) 2.4 m
(load area) 1.03 m
Ground clearance: 0.43 m
Track width:
(Summer steel track) 775 mm
(Winter steel track) 825 or 925 mm
Track: 1.55 m

Kässbohrer Flexmobil FM 23.150K all-terrain vehicle

Kässbohrer Flexmobil FM 23.150K all-terrain vehicle towing 4 tonne tracked semi-trailer

Max speed:
(road) 40 km/h
(off-road) 25 km/h
Fuel capacity: 116 litres
Gradient: 100%
Side slope: 40%
Fording: 0.8 m
Engine: Mercedes-Benz OM 352A water-cooled 6-cylinder turbocharged diesel, developing 150 hp or Deutz BF 6 L 913 air-cooled 6-cylinder turbocharged diesel developing 160 hp
Transmission: hydrostatic
Steering: hydrostatic
Turning radius: pivot turn

Status
In production. Has been used by UN personnel operating in Lebanon.

Manufacturer
Karl Kässbohrer Fahrzeugwerke GmbH, Postfach 2660, D-7900 Ulm/Donau, Germany.
Tel: +49 561 8011. Telex: 99750 thksr d.
Fax: +49 561 801 6733.

UPDATED

JAPAN

Type 73 tracked artillery tractor Type 87 tracked ammunition carrier

Description
The Type 73 tracked artillery tractor was developed from 1969 as the replacement for the M4 and M8 high-speed tractors used by the Japanese Ground Self-Defence Force. Trials with prototype vehicles were successfully completed in 1972 and the following year it was standardised as the Type 73 tracked artillery tractor, or tractor (prime mover). Production began in 1974 but only a small number were built owing to restrictions in the defence budget.

The Type 73 has a fully enclosed four-door cab at the front with the engine in the centre and the ammunition stowage area at the rear. The suspension is believed to be of the torsion bar type and consists of six roadwheels with the drive sprocket at the front, the sixth roadwheel acting as the idler. A 12.7 mm/0.50 M2 machine gun is mounted on the roof for anti-aircraft defence. Some Type 73s have been fitted with a hydraulically operated dozer blade at the front of the hull for clearing obstacles and preparing fire positions. The tractor was used to tow artillery pieces such as the 8 in/203 mm M115.

The Type 87 tracked ammunition carrier is based on the Type 73 tracked artillery tractor. Following the switch from towed to self-propelled artillery by the Japanese Ground Self-Defence Force, over 100 Type 73s will be converted to Type 87 carriers. The Type 87 can carry up to 50 rounds of 203 mm ammunition and is equipped with a crane.

Specifications
Type 73
Cab seating: 1 + 11
Weight: 19,800 kg
Max towed load: 16,000 kg
Length: 6.13 m
Width: 2.95 m
Height: 2.3 m
Max speed: (road) 45 km/h
Range: 300 km
Gradient: 60%
Engine: Mitsubishi ZF6 6-cylinder air-cooled diesel developing 400 hp at 2,200 rpm

Status
Production complete. In service with the Japanese Ground Self-Defence Force. Over 100 Type 73 tractors are being converted to Type 87 tracked ammunition carriers.

Manufacturer
Hitachi Manufacturing Company, Japan.

VERIFIED

Type 87 tracked ammunition carrier (Paul Beaver)

Type 73 tracked artillery tractor (Kensuke Ebata)

Special tracked prime mover

Description
This vehicle is based on the Type 73 armoured personnel carrier and is used to tow special equipment such as the Type 76 gun locating radar (JMPQ-P7). The vehicle has an enclosed crew cab and probably carries an electrical generator at the rear, but few details are available. For details of the Type 73 APC on which this vehicle is based refer to *Jane's Armour and Artillery 1995-96* pages 388 and 389.

Status
In service with the Japanese Ground Self-Defence Force.

Manufacturer
Mitsubishi Heavy Industries, 5-1 Marunouchi 2-chome, Chiyoda-ku, Tokyo, Japan.
Tel: +81 3 212 3111. Telex: 22282.
Fax: +81 3 201 6258.

VERIFIED

Special tracked prime mover towing Type 76 gun-locating radar (Kensuke Ebata)

KOREA, SOUTH

Daewoo Field Artillery Ammunition Support Vehicle

Description

The Daewoo Field Artillery Ammunition Support Vehicle (FAASV) is part of the proposed Daewoo Military Logistic Support Vehicle family and closely follows the overall design concept of the US M992 series FAASV (see separate entry in this section). It appears to be a follow-on design to an earlier Daewoo FAASV reconfigured to provide more internal space.

The FAASV is based on the same chassis as the Flying Tiger (Biho) twin 30 mm self-propelled anti-aircraft gun system (for details refer to *Jane's Land-based Air Defence 1995-96* pages 92 and 93). It differs from the earlier FAASV in having an armoured forward control cab-over-power pack cab for the driver and one passenger. The windows have armoured shutters and there is a roof hatch with a mounting for a 12.7 mm machine gun. The other two crew members are in an armoured superstructure to the rear, inside which are racks for 155 mm projectiles, charges and fuzes, a stacker mechanism, and a conveyor system which can be lowered through an open rear door. The FAASV has rack space for 139 155 mm projectiles, 148 charges and 167 fuzes. These can be stacked at a rate of 6 rds/min and passed along the conveyor at the same rate.

The Field Artillery Ammunition Support Vehicle uses a torsion bar suspension with six roadwheels each side. The track is of the steel double-pin type with detachable rubber pads. There are four shock-absorbers each side. Power is provided by a 520 hp Daewoo D2840L diesel.

Daewoo Field Artillery Ammunition Support Vehicle

Specifications

Crew: 4
Weight: (combat loaded) 28,600 kg
Length: 6.75 m
Width: 3.15 m
Height: 3.3 m
Track width: 392 mm
Max speed: 60 km/h
Range: (cruising) 400 km
Gradient: 60%
Side slope: 40%
Vertical obstacle: 0.7 m
Trench: 2 m
Fording: 1.2 m
Engine: Daewoo D2840L V-10 turbocharged diesel developing 520 hp
Transmission: Lockheed Martin HMPT 500-3 hydromechanical
Steering: hydrostatic
Suspension: torsion bar
Electrical system: 24 V

Status

Prototype.

Manufacturer

Daewoo Heavy Industries Limited, Special Products Division, 20th Floor, Daewoo Center, 541, 5ga, Namdaemun-ro, Jung-gu, Seoul, South Korea.
Tel: +82 2 726 3011. Telex: DHILTD K23301.
Fax: +82 2 756 2679.

UPDATED

POLAND

Bumar Labedy 668 high-speed tractor

Description

The Type 668 high-speed tractor, developed and produced by the Zaklady Mechaniczne 'Bumar Labedy' SA at Gliwice as a general purpose tractor unit, is capable of fulfilling numerous logistic and other support tasks. It can carry a maximum load of 3,000 kg, tow artillery pieces or trailers weighing up to 14,000 kg, and carry up to 12 troops in the cargo area. The Type 668 appears to be a close relative of the RFAS ATS-59G medium tracked artillery tractor.

The 668 has an all-steel forward control cab located over the engine. Seating is provided for the driver and up to five passengers. A roof hatch is provided on the right-hand side. The cabin is kept at an overpressure for operations in NBC-contaminated environments and extra defrosting equipment can be installed for use in cold environments.

The steel-sided load area located over the engine is used to carry supplies only. Only the rear part of the load area can be configured to carry passengers or supplies. The load area can be covered by a canvas tilt over bows. The towing position at the rear is equipped with shock-absorbers.

The tractor suspension consists of five pairs of torsion bars with shock-absorbers on the first and fifth roadwheel stations. Power is provided by a V-12 A-650-G diesel via a five-speed gearbox. An engine preheater system is provided and the engine may be started by electrical or compressed air systems.

The 668 is provided with a recovery winch with a capacity of 14,700 kg. A version of this vehicle without the winch is known as the 668 BL.

A version of the 668, known as the 668 S, is equipped with a front-mounted lattice crane with a lift capacity of 7,500 kg and an operating radius of 2.5 m. When fully raised the crane has a height of 7.8 m. The crane can be broken down into three sections for transport on the cargo area.

Other possible variants of the 668 include a firefighting vehicle and a cable-laying model capable of pulling a plough which enables cables to be laid to a depth of 0.7 m.

Specifications

Cab seating: 1 + 5 (up to 12 in rear)
Weight: (combat) 13,750 kg
Max load: 3,000 kg
Towed load: up to 14,000 kg
Length: 6.28 m
Width: 2.78 m
Height:
(top of tilt) 2.62 m
(cab) 2.58 m
(load area) 1.27 m
Ground clearance: (loaded) 0.42 m
Max speed:
(cruising) 39 km/h
(cross-country with towed load) 22-27 km/h
Range: (cross-country with towed load) 500 km
Fuel capacity: 580 litres (plus optional 240 litres)

Bumar Labedy 668 high-speed tractor

Fording:
(normal) 1.1 m
(with preparation) 1.5 m
Gradient:
(without towed load) 35°
(with towed load) 18°
Side slope: 25°
Engine: A-650-G V-12 diesel developing 295 hp
Transmission: synchromesh, 5 forward and 1 reverse gears
Suspension: independent torsion bar
Electrical system: 24 V
Batteries: 3 × 12 V, 210 Ah
Generator: 2,200 W

Status
Production as required.

Manufacturer
Zaklady Mechaniczne 'Bumar Labedy' SA, PL-44-109 Gliwice, ul. Mechanikow 9, Poland.
Tel: + 48 32 345111. Telex: 036237 zam pl.
Fax: + 48 32 342443.

VERIFIED

ROMANIA

ABAL armoured vehicle for combat supply

Description
The ABAL armoured vehicle for combat supply is a variant of the Romanian MLI-84 infantry combat vehicle which is based on the BMP-1 IFV. The ABAL also shares some components and features with the Romanian MLVM mountaineers' combat vehicle designed specifically for use in mountainous terrain. For details of both these vehicles refer to *Jane's Armour and Artillery 1995-96* pages 395 and 396.

The ABAL is a turretless armoured tracked carrier with the bulk of its load of ammunition (or other cargo) carried inside the hull, although ammunition may also be carried in ready-use lockers along each side of the upper hull. The maximum payload is 2,100 kg, which may be 64 rounds of 100 mm ammunition or 104 rounds of 76 mm ammunition. Other calibre/load combinations can be accommodated.

There are two crew and the only vehicle armament carried is a 7.62 mm machine gun mounted to the rear of the driver's position. Other equipment includes an NBC protection system, night driving equipment, firefighting equipment, a nuclear radiation detector and communications equipment.

Specifications
Crew: 2
Weight:
(vehicle) 8,400 kg
(combat) 10,500 kg
Payload: 2,100 kg
Towed load: 5,400 kg
Power-to-weight ratio: 15.47 hp/t
Ground pressure: 0.49 kg/cm^2
Length: 5.85 m
Width: 2.714 m
Height: 1.9 m
Ground clearance: 0.38 m
Track: 2.3 m
Angle of approach: 27°
Max speed: (road) 48 km/h
Range:
(roads) 666 km
(country roads) 500 km
(cross-country) 383 km
Fuel capacity: 450 litres
Fuel consumption:
(roads) 69 litres/100 km
(country roads) 92 litres/100 km
(cross-country) 120 litres/100 km
Fording: 0.6 m
Gradient: 60%
Side slope: 30%
Vertical obstacle: 0.7 m
Trench: 1.5 m
Engine: 4-stroke supercharged diesel
Armament: 1 × 7.62 mm MG
Elevation: –5 to +82°
Traverse: 360°

ABAL armoured vehicle for combat supply

Status
In service with the Romanian Army.

Marketing agency
Romtehnica, 9-11 Drumul Taberei Street, Bucharest, Romania.
Tel: +40 046 2087. Telex: 11608 TXKCO.
Fax: +40 046 0317.

UPDATED

RUSSIAN FEDERATION AND ASSOCIATED STATES (CIS)

ATS-59G medium-tracked artillery tractor

Description
The ATS-59G medium-tracked artillery tractor was first seen during 1972, towing a 130 mm M-46 field gun. Until its correct designation became known the vehicle was called the M1972. It is essentially an improved version of the earlier ATS-59 with a redesigned cab. The ATS-59G has a much larger forward control type cab with the cargo area at the rear. The suspension consists of five large roadwheels with the drive sprocket at the front and the idler at the rear. There are no track-return rollers.

In Egypt, ATS-59G tractors have been adapted as carriers for the SAKR 122 mm RL-21 30-round multiple rocket launcher role.

The Polish Bumar Labedy 668 high-speed tractor appears to be a close development of the ATS-59G - see separate entry in this section.

Specifications
Weight:
(empty) 13,750 kg
(loaded) 16,750 kg
Max load: 3,000 kg
Max towed load: 14,000 kg
Fuel capacity: 580 litres
Ground pressure: 0.55 kg/cm^2
Fording: 1.5 m

ATS-59G medium-tracked artillery tractor

Status
In service with Egypt, Yugoslavia, and members of the former Warsaw Pact.

Manufacturer
State factories.

UPDATED

MT-S medium-tracked transporter

Description

The MT-S medium-tracked transporter was developed around the same time as the MT-T heavy tracked transporter (see entry in this section), but it does not appear to have been produced in very large numbers in the transporter form. Few details regarding this vehicle have been released but it is believed to be powered by a diesel engine similar to that used by the MT-T. The basic chassis is derived from that used for the 152 mm self-propelled gun-howitzer 2S3 (S-152) Akatsiya and has been used for a number of other applications, including the GMZ minelayer (see entry in the *Minelaying equipment* section). It is also used as a carrier for components of the SA-11 'Gadfly' missile system including the associated radar vehicles.

Specifications

Provisional
Cab seating: 1 + 3
Weight: 23,500 kg
Max payload: 10,200 kg
Length: 7.8 m
Width: 3.34 m
Height: 1.846 m
Engine: diesel developing 710 hp

Status

In service with the RFAS.

Manufacturer

State factories.

VERIFIED

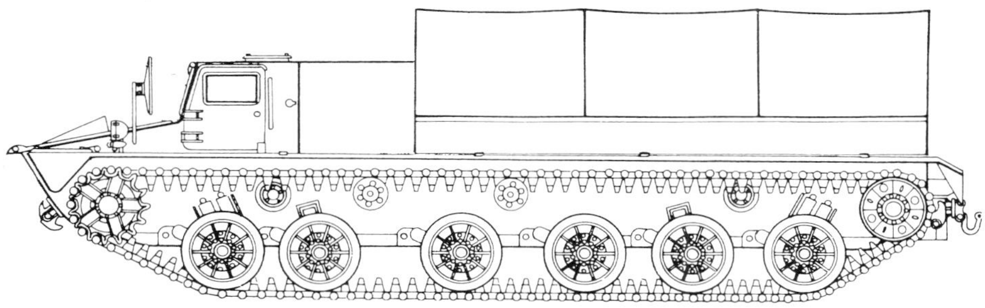

MT-S medium-tracked transporter (not to 1/76th scale) (Steve Zaloga)

MT-S medium-tracked transporter

AT-T heavy tracked artillery tractor

Description

The AT-T heavy tracked artillery tractor appeared in 1950. It was originally used to tow heavy artillery such as the 180 mm S-23 gun-howitzer, the 130 mm KS-30 anti-aircraft gun and the 130 mm SM-4-1 mobile coastal gun. It is now mainly relegated to the general purpose towing role for various systems and in specialised logistic roles towing heavy loads on trailers. It is also employed to carry specialised loads such as radar and other electronic shelters.

The engine is under the floor of the four-man cab, which is of all-steel construction and has a door either side. The cargo compartment is at the rear and is provided with a drop tailgate, removable bows and a tarpaulin cover. The fuel tanks are under the rear behind the engine and the winch is mounted at the very back.

The suspension is of the torsion bar type and consists of five large roadwheels with the drive sprocket at the front and the idler at the rear. There are no track-return rollers.

Variants

BAT and BAT-M Dozers

See entry in the *Field fortifications and related emplacements equipment* section.

Radar

The AT-T with a fully enclosed van-type body is used to mount the Track Dish (ARSOM-1). A lengthened version of the AT-T with seven roadwheels mounts the Long Track artillery surveillance radar.

BTM and MDK-2 Ditching Machines

For details see the *Field Fortifications and Related Emplacements Equipment* section.

AT-TA

This is a special version of the basic AT-T with extra wide tracks and other modifications for polar exploration.

Specifications

Cab seating: 1 + 3 (up to 16 in rear)
Weight:
(empty) 20,000 kg
(loaded) 25,000 kg
Max load: 5,000 kg
Towed load: 25,000 kg
Length: 7.043 m
Width: 3.154 m
Height: (cab) 2.845 m
Ground clearance: 0.425 m
Track: 2.64 m
Track width: 500 mm
Length of track on ground: 3.836 m
Ground pressure:
(empty) 0.52 kg/cm²
(loaded) 0.68 kg/cm²
Max speed: 38 km/h
Range: (with towed load) 1,100 km
Fuel: 1,415 litres
Fuel consumption:
(not towing) 140 litres/100 km
(towing) 190 litres/100 km
Max gradient: 60%
Vertical obstacle: 1 m
Trench: 2.1 m
Fording: 0.75 m (some can ford to 1 m)
Engine: V-401 V-12 water-cooled diesel developing 415 hp at 1,600 rpm

Status

Production complete. In service with members of the former Warsaw Pact, Yugoslavia and countries in the Middle East (including Iraq) and North Africa.

Manufacturer

Mzlyshev Transport Machine Plant, Khar'kov, Russia.

UPDATED

Lengthened version of AT-T heavy tracked artillery tractor with 'Long Hand' radar (T J Gander) ***1997***

MT-T heavy tracked transporter

Description

The MT-T heavy tracked transporter first appeared during the early 1980s and uses suspension components from the T-64 tank. The engine is the V-64-4 diesel derived from that used in the T-72 tank. On the MT-T a forward control cab is located in front of the main drive gear and the large cargo area to the rear can be covered by a canvas tilt.

The chassis of the MT-T is used as the basis for several other vehicles, including the PTS-2 amphibious tracked vehicle (see entry in the *Amphibians* section), the MDK-3 trenching machine and the BAT-2 tractor-mounted bulldozer (see entries in the *Field Fortifications and Related Emplacements Equipment* section), and the PMM-2 amphibious bridging and ferry system (see entry in the *Mechanised Bridges* section). Derivatives of the MT-T are used as the basis for the Antey S-300V SA-12 'Gladiator'/'Giant' missile system transporter and launcher vehicle. For details of this system refer to *Jane's Land-based Air Defence Systems 1995-96* pages 131 to 133.

Specifications

Provisional
Cab seating: 1 + 4 (up to 18 in rear)
Weight: approx 23,000 kg
Max load: 12,000 kg
Towed load: 12,000 kg

MT-T heavy tracked transporter

Length: 8.711 m
Width: 3.277 m
Height: 3.065 m
Engine: V-64-4 diesel developing 710 hp

Status
In production. In service with the RFAS and some members of the former Warsaw Pact.

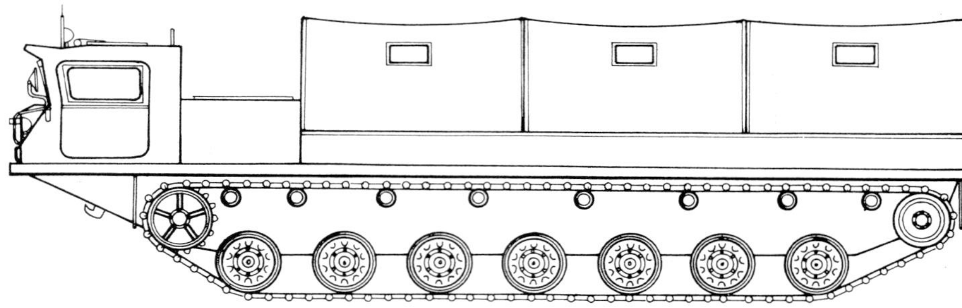

MT-T heavy tracked transporter (Steve Zaloga)

Manufacturer
State factories.

VERIFIED

GM-569 tracked tractor

Description
During 1990 the first reports of a GM-569 tracked tractor appeared in technical reports, mainly referring to the vehicle's novel hydromechanical transmission. The GM-569 appears to be a large vehicle, possibly utilising drive components of the T-64 MBT, with a low superstructure over the forward part of the upper hull and a built-up engine compartment at the rear. The driving and crew compartment extends the full width of the front hull. There are six roadwheels and four track-return rollers. No dimensions, weights or other technical details have yet been released.

The GM-569's hydromechanical transmission is considered to be a considerable improvement over mechanical transmissions and involves a reduction mechanism, a hydraulic torque converter, a planetary gearbox, an equalising block and a hydraulic steering mechanism imparting good manoeuvrability. Special driving techniques have to be employed to derive the best performance from the system and a column-mounted steering wheel is involved. Pivot turns are possible on hard surfaces. Hydraulically operated and self-adjusting dry plate disc brakes are used when on the move, although a mechanical brake is used for parking. To provide power supplies for the vehicle's various electrical and electronic control systems, two AC generators can be powered by a gas-turbine engine and there is a reserve system, used only when halted, powered by a reduction take-off from the main engine.

Status
Apparently in production.

Manufacturer
State factories.

VERIFIED

DT-10P, DT-30P and DT-30 tracked all-terrain transporters

Description
The DT-10P, DT-30P and DT-30 tracked all-terrain transporters have been in production since 1982. They are intended for carrying heavy loads over all manner of off-road terrain, including marsh and snow. All three vehicles have twin unit articulated configurations and are powered by a 710 hp multifuel diesel developing 710 hp. The engine has a preheater as the vehicles are intended for use over a temperature range of −50 to +40°C. All three vehicles have the same five-man forward control enclosed cab with the engine compartment behind.

All three vehicles have wide one-piece rubber tracks with steel welded cross-pieces. On the DT-10P the tracks are 960 mm wide. The DT-30P and DT-30 tracks are 1.1 m wide. An independent torsion bar suspension is employed. Steering is hydraulic using double-acting power cylinders acting on a double-hinged steering and coupler unit.

DT-10P
The DT-10P Vityaz is the smallest of the three vehicles, being a twin-unit vehicle carrying 10 tonnes or 10 passengers, including the crew of five. There are two load-carrying areas, one on the front section behind the cab and engine, and the other including the total platform area of the second unit. Both load areas may be covered by a tarpaulin. On the DT-10P the front load area measures 2.2 m long by 2.81 m wide. The rear area measures 6.2 × 2.81 m. Both load areas are waterproofed to allow the complete vehicle to be amphibious. The front unit has four roadwheels each side and the rear unit five roadwheels.

DT-30P tracked all-terrain transporter

DT-30P
The DT-30P, also named Vityaz, is an enlarged version of the DT-10P with the load capacity increased to 30 tonnes or 30 passengers, including the crew of five. The front load area measures 4.5 m long by 3.1 m wide. The rear area measures 6.6 × 3.1 m. Both load areas are waterproofed to allow the complete vehicle to be amphibious. Both the front and rear units have six roadwheels each side.

DT-30
The DT-30 also has a load capacity of 30 tonnes although it is not amphibious. On this vehicle, in place of the usual cargo bodies there is a single long cargo flatbed measuring 10.51 m long by 3.1 m wide supported by the front unit, which is structurally the same as that on the DT-30P apart from the cargo body. On the rear unit, again the same as that on the DT-30P, there is a gimbal arrangement on which the cargo area can pivot.

Specifications

Model	**DT-10P**	**DT-30P**	**DT-30**
Cab seating	5	5	5
Weight			
(empty)	22,500 kg	28,000 kg	29,000 kg
Max load	10,000 kg	30,000 kg	30,000 kg
Length			
(total, overall)	13.73 m	15.59 m	15.67 m
Width	2.81 m	3.1 m	3.1 m
Height	2.7 m	3.3 m	3.6 m
Track width	960 mm	1.1 m	1.1 m
Max speed			
(road)	36 km/h	36 km/h	36 km/h
(water)	5-6 km/h	4 km/h	n/a
Range	500 km	500 km	500 km
Max gradient	35°	30°	30°
Side slope	20°	15°	15°
Engine	multifuel liquid cooled diesel developing 710 hp		
Transmission	4 forward and 1 reverse gears		
Steering	articulated hydrostatic		
Suspension	torsion bars	torsion bars	torsion bars
Brakes	air	air	air
Electrical system	24 V	24 V	24 V

Status
In production. In service with the RFAS.

Manufacturer
Ishimbai Transport Machine Building Works, Industrial'noe Shosse 6, 453210 Ishimbai, Bashir Republic.
Tel: 2-46-22, 2-47-87. Telex: 350844 COCHA.

VERIFIED

SOUTH AFRICA

TFM Blesbok (4 × 4) 5,000 kg armoured freighter

Description

The Blesbok (4 × 4) 5,000 kg armoured freighter is a member of the Casspir Mark 11 armoured personnel carrier series. It was designed to carry supplies to military and paramilitary forces operating in remote areas. The main user was the Southwest African (now Namibian) Police which operated along mechanised infantry lines when on counter-insurgency operations. In total, 160 units were produced. Most vehicles are now operated by the South African National Defence Force (SANDF).

The Blesbok is built on the same chassis as the Casspir Mark 11 which was derived from that of a 15 ton commercial truck. The Blesbok also uses the same protected cab for the driver and passenger, and has many other components of the Casspir. The cab has seating for two personnel and houses the vehicle's 185 litre fuel tank. The cab has sloped armoured undersides for mine protection and the cab armour is proof against 5.56 and 7.62 mm NATO ball ammunition. The cab roof has a hatch and a small turret for a 7.62 mm machine gun. All windows use 52 mm thick laminated glass. The cab interior is cooled by two impeller fans and both adjustable seats have safety belts.

The open cargo area has drop sides and a tailgate and a tow hook is provided at the rear. Equipment carried includes a 200 litre drinking water tank and two 2.5 kg fire extinguishers. A towing pintle is provided at the front. An optional 1,000 litre fuel tank may be carried at the rear.

Variants

Duiker Armoured Tanker

The Duiker armoured tanker is a Blesbok with the cargo area replaced by a 5,000 litre diesel fuel tank fitted with a gravity feed system for dispensing fuel, with an electrical pump as an option. Loaded weight is 13,490 kg. A total of 30 units were produced.

Weapons Platforms

It was announced in 1996 that the Casspir range had been extended by the introduction of two armoured weapon platforms, one carrying an 81 mm mortar and the other a 106 mm recoilless rifle. The South African National Defence Force (SANDF) has ordered eight examples of each for trials.

Specifications

Blesbok
Cab seating: 1 + 1
Configuration: 4 × 4
Weight:
(empty) 9,400 kg
(loaded) 14,400 kg
Max load: 5,000 kg
Length: 7 m
Width: 2.5 m
Height: 3.12 m
Ground clearance: 0.355 m
Wheelbase: 4.3 m
Angle of approach: 46°
Max speed: 87 km/h
Range:
(road) 850 km
(cross-country) 560 km
Fuel capacity: 185 litres
Max gradient: 60%
Side slope: 30°
Vertical obstacle: 0.5 m
Trench crossing: 0.95 m
Fording: 1 m
Engine: ADE 352 ⁻ 5.675 litre 6-cylinder vertical in-line direct injection diesel developing 162.5 hp at 2,800 rpm
Gearbox: Mercedes-Benz DB 6/3/60 - 5/7.5 synchromesh with 5 forward and 1 reverse gears
Clutch: single dry p ate
Transfer box: Mercedes-Benz 2-speed
Steering: ball and nut, power-assisted
Turning radius: 9.15 m
Suspension: semi-elliptic leaf springs mounted on to needle roller bearings with telescopic shock-absorbers front and rear (double spring pack at rear)
Tyres: 14.00 × 20
Brakes:
(main) dual circuit pneumatic with engine-mounted exhaust brake
(parking) mechanical
Electrical system: 12 V
Battery: 1 × 12 V
Alternator: 55 A
Armament: 1 × 7.62 mm MG

Status

In service with the Namibian Police and South African National Defence Force (SANDF).

Manufacturer

TFM (Pty) Limited, FO Box 46131, Orange Grove 2119, South Africa.
Tel: +27 12 612106. Telex: 3-21041.
Fax: +27 12 611462.

UPDATED

Duiker armoured tanker with 5,000 litre diesel fuel tank

Blesbok (4 × 4) 5,000 kg armoured freighter

SWEDEN

Hägglunds Vehicle Bandvagn Bv 206 all-terrain carrier

Development

In 1974 the Swedish Defence Materiel Administration awarded Hägglund and Söner (now Hägglunds Vehicle AB) a contract worth SEK13 million, for the development of a new over-snow/all-terrain vehicle to succeed the Volvo Bv 202 in the Swedish Army. Three batches of Bv 206 vehicles were delivered to the Swedish Army for trials in 1976, 1977 and 1978. Hägglunds was awarded a first production contract in June 1979 from the Swedish Defence Administration. Under this SEK800 million contract the company delivered preproduction vehicles late in 1980 and began full-scale production in early 1981. Full rate deliveries were made from the middle of 1981 until 1988. In September 1981 the Norwegian Army ordered 220 units, followed in August 1985 by a further 2,004 Bv 206 vehicles in various configurations.

In September 1981 the British Army purchased four Bv 206s for trials in a variety of forms and a further number was purchased to replace Bv 202 vehicles then in service with the Army and Royal Marines; a further order (worth $40 million) for diesel-engined vehicles for the Royal Marines was placed in March 1990; these are known as the Bv 206D6. The UK Ministry of Defence plans to upgrade about 200 of its Bv 206s by replacing their petrol engines with diesel units to bring them up to Bv 206D6 standard. This will extend their service life by a further 12 years.

In 1983 the US Army ordered 318 Bv 206s for service in Alaska as the M973 Small Unit Support Vehicle. These entered service in 1983-84. During 1988 the US Army placed a new order for a further 390 M973 SUSV vehicles in various forms to enter service during 1989. By 1991 the US Army fleet numbered approximately 1,100 vehicles.

The first of a batch of 12 Bv 206Ds for the German Bundeswehr was delivered in mid-1985. A second order for 63 vehicles was awarded during 1989. The Royal Netherlands Marines ordered its first batch of Bv 206 vehicles in 1989 and continued to order vehicles with differing configurations during 1991 and 1992 (the total was approximately 200 units by 1992).

In October 1992 it was announced that the French Defence Ministry had awarded a contract worth SEK20 million for an initial batch of Bv 206 vehicles, mainly provided with load handling systems, for use by the French Army's 27th Mountain Division. It is anticipated that the French Army will eventually purchase as many as 300 to 400 Bv 206 vehicles in different configurations.

In early 1993 Singapore ordered a batch of 300 Bv 206s with first deliveries made during early 1994.

In November 1993 the 10,000th series produced Bv 206 was formally delivered to the Swedish Defence Materiel Administration (FMV).

Hägglunds Vehicle AB has produced over 10,000 Bv 206 vehicles, with military sales to more than 16 countries.

Description

The Bv 206 consists of two tracked units linked together with a steering unit, each unit consisting of a chassis with the body mounted on four rubber elements.

Each chassis consists of a central beam, a final drive assembly and two track assemblies. The chassis of the front and rear units are identical except that a two-step drop-down gearbox is mounted in the rear end of the front chassis. The track assemblies are mounted to the

Bv 206 all-terrain carrier armed with Bofors 90 mm recoilless rifle **1997**

Bv 206 all-terrain carrier configured for the RBS 90 air defence system

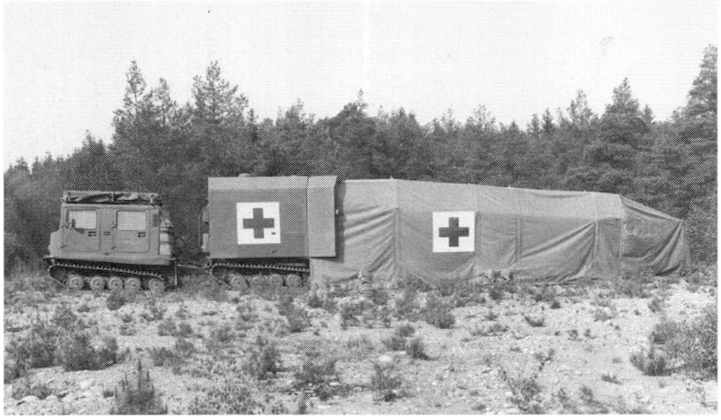

Bv 206 with rear unit configured to form a mobile first aid station

Bv 206 all-terrain carrier in amphibious mode towing Vammas BVK 191W all-terrain general purpose trailer

central beam by two transversal leaf springs. Each track assembly is built up around a tubular bar which carries the sprocket assembly, roadwheels and idler. In each track assembly there are four pairs of roadwheels on trailing arms sprung by rubber tension springs. The idler at the rear with a tensioning device is also supported by a rubber spring. All four track assemblies of the vehicle are identical and interchangeable. The tracks are rubber with longitudinal textile cord and integral steel profile reinforcements.

Steering is accomplished by changing the direction between the front and rear unit by two hydraulic cylinders, servo-controlled from a conventional steering wheel. The hydraulic system is built up of commercially available components. The steering unit is designed to permit a large freedom of movement between the two bodies.

The engine and transmission are mounted in the front unit. A shaft connects the gearbox with the drop-down gearbox. A disc brake is mounted on this shaft in front of the drop-down gearbox inside the body. Cardan shafts transmit power to the final drives on the front end of both chassis. Early units were fitted with petrol engines but current production examples are fitted with Mercedes-Benz 136 hp diesels.

The bodies are made of Glass fibre-Reinforced Plastic (GRP), which is fire resistant, with PVC foam insulation. Each body is built like a closed box with integrated roll-over protection. Heating of the units is by heat exchangers and the de-icing capacity in the front unit is sufficient to keep the windscreen clear down to a temperature of −40°C. The bodies have holders for lashing cargo in the rear unit and also in the rear part of the front unit. The rear unit is also provided with brackets for carrying four stretchers.

The Bv 206 is fully amphibious, being propelled in the water by its tracks.

Variants

Flatbed vehicle

This variant accommodates the driver and five passengers in the front unit. The load capacity is increased to 2,350 kg. This variant is available in amphibious and non-amphibious versions.

Anti-tank vehicle

Anti-tank vehicles may be armed with a Bofors 90 mm recoilless rifle, or a TOW, MILAN or Bofors RBS 56 BILL ATGW on a pivot mount which can be hydraulically raised to the required level. The open-front body is provided with roll bars which can be quickly lowered. The low profile rear body is designed to withstand the backblast of the weapon when it is fired and is used for storing ammunition.

RBS 90

The Bofors RBS 90 SAM system is carried in two Bv 206s, one carrying the disassembled missile stand and the other the fire-control equipment, including early warning radar; the crew is distributed between the two vehicles. For details of this system refer to *Jane's Land-based Air Defence 1995-96* pages 283 to 285.

Command Post/Radio vehicle

This can accommodate up to six operators in the rear body and driver plus four personnel in the front body. Door arrangements are identical to the standard carrier version. The vehicle has VHF transceivers in the rear body and all the radios can be operated from the front unit.

Radar and C³I vehicles

The high payload of the Bv 206 permits the installation of a very wide range of specialist equipment. For instance, a radar vehicle has a special generator and an Ericsson Giraffe 50 AT C-band search radar with an instrumented range of 50 km and a mast height of 7 m. A radio jamming vehicle is equipped with a separate jamming generator driven by the main engine, generating 15 kVA. The antenna is mounted on an 18 m rapidly erected mast.

A radio locating vehicle is equipped with radio intelligence and direction-finding systems. The antenna is mounted on an 18 m telescopic mast. Equipment in a radio relay vehicle consists of either VHF or UHF multichannel radios for automatic relay transmission. The rear vehicle carries spare batteries and a 3 kVA diesel-driven power unit. There is also a 20 m antenna mast for ground or vehicle mounting.

Mortar carriers

Special shock-absorbing flatbeds in various forms enable mortars to be fired from the rear unit of a Bv 206. Mortars tested with this system include TDA and Royal Ordnance 81 mm mortars, with each vehicle carrying approximately 100 rounds of ammunition. A 4.2 in mortar version carries 60 rounds of ammunition while a 120 mm mortar variant (in both smoothbore and rifled versions) can carry approximately 30 rounds.

Firefighting version

The Royal Air Force has three special fire and rescue Bv 206 variants based at RAF Mount Pleasant in the Falkland Islands. The rear units of these vehicles are occupied by fire and rescue equipment and are utilised as rapid intervention vehicles in the event of airfield accidents.

Bv 206D

Bv 206 all-terrain carriers used by the German Army are fitted with Mercedes-Benz OM 603.950 six-cylinder diesel engines developing 136 bhp at 4,600 rpm. This engine is the current standard installation for new vehicles.

Bv 208

This is the Swedish Army designation for vehicles fitted with diesel engines.

Bv 206 S

This variant appeared during mid-1990 and is an armoured personnel carrier version of the basic Bv 206. For details refer to entry in *Jane's Armour and Artillery 1995-96* pages 421 and 422.

Sea Crab

The Sea Crab D500 is an underwater remotely operated vehicle based on the hull, suspension and running gear of the Bv 206. A prototype has been built powered by a 40 kW electrohydraulic power unit with electrical power being delivered by an umbilical cord.

Skorpion-AB

This is an airborne open-topped version of the Bv 206 with the rear unit accommodating a Dynamit Nobel Skorpion mine launching system. It is in the final stages of development for the Swedish Army. For details see Skorpion entry under *Minelaying equipment*.

Ring Mount NM 165

When equipped with a Ring Mount NM 165 the Bv 206 can be armed with either a 0.50/12.7 mm machine gun or a 40 mm automatic grenade launcher.

Ambulance/first aid station vehicle
This variant carries a crew of five, four stretchers, a medical specialist and medical equipment. It is provided with a tent which folds onto the rear unit to allow the vehicle to be used as a first aid station. The front unit serves as a surgery while the tent acts as a treatment area for up to eight casualties.

Forward repair team vehicle
A typical example of this variant is provided with a hydraulic crane with a lifting capacity of 1,500 kg. A 3,600 kg capacity hydraulic winch is mounted at the front. The vehicle carries a repair crew and sufficient spares, tools and so on, to remain operational for several days. The example quoted is used by the British Army and Royal Marines.

SUSV
During 1983 the US Army Tank Automotive Command (TACOM) awarded a US$24.2 million contract for 268 Bv 206 all-terrain carriers (plus an option for a further 34, later exercised) known as the Small Unit Support Vehicle (SUSV). The contract followed a period of extensive trials carried out using a small batch of vehicles at the Cold Regions Test Center at Fort Greely, Alaska. Most of the vehicles involved in the contract, delivered from Sweden in 1983 and 1984, were issued to the US Army's 172nd Infantry Brigade in Alaska and the Alaskan National Guard. Early M973 SUSVs were powered by 3 litre four-stroke, five-cylinder in-line diesel engines developing 125 hp at 4,500 rpm. Vehicles delivered after 1987 are powered by a Mercedes-Benz six-cylinder turbocharged diesel engine developing 136 hp at 4,500 rpm.

There are four basic SUSV variants: M973/M973A1 cargo carrier, M1066 ambulance, M1067 flatbed, and M1067 command and control vehicle.

In 1988 the US Army ordered a further 390 M973 SUSVs from Hägglunds Vehicle AB. These vehicles were delivered during 1989. Continued purchases until 1992 resulted in a total fleet of approximately 1,100 vehicles.

Other variants
The Bv 206 has been used for a variety of other roles, not all of them military. Using a hydraulically operated load changer arm the Bv 206 can be used to carry a variety of containers, shelters and various load bodies. The trailer can also be used to carry a light crane or lifting platform and may be configured for a number of special purposes such as a remote area firefighting vehicle, as a mobile workshop or as a field ambulance.

Specifications
Cargo carrier
Cab seating:
(front unit) 5-6
(rear unit) 11
Weight:
(empty) 4,490 kg
(loaded) 6,740 kg
Weight of front unit: (empty) 2,740 kg
Weight of rear unit: (empty) 1,730 kg
Max load: 2,250 kg
(front unit) 610 kg
(rear unit) 1,640 kg
Towed load: 2,500 kg
Load area:
(front unit) 0.81 × 1.4 m
(rear unit) 2.5 × 1.4 m
Length: 6.9 m
Width: 1.87 or 2 m
Height: 2.4 m
Ground clearance: 0.35 m
Track width: 620 mm
Max speed:
(road) 52 km/h
(water) 3 km/h
Max gradient:
(hard surface) 100%
(snow) 30%
Max side slope: 90%
Fording: amphibious
Range: (roads) 300 km
Engine: Mercedes-Benz OM 603.950 2.996 litre 6-cylinder diesel developing 136 hp at 4,600 rpm (early production) Ford Model 2658 E V-6 water-cooled petrol developing 136 bhp at 5,200 rpm (could also be fitted with Mercedes-Benz 5-cylinder in-line turbocharged diesel engine developing 125 bhp at 4,500 rpm)
Gearbox: Daimler-Benz W4A-040 fully automatic with torque converter, with 4 forward and 1 reverse gears
Transfer box: Hägglunds 2-speed
Steering: articulated hydrostatic
Turning radius: 8 m (6 m with pitch control)
Electrical system: 24 V
Batteries: 2 × 12 V, 105 Ah
Alternator: 55 or 100 A

French Army Bv 206 with scraper blade and hydraulic load handling system on rear unit (R Stickland)
1997

Status
In production. More than 10,000 ordered or in military service with 16 countries. In service with Brazil, Canada (100), Chile, China, Finland (400), France, Germany, Italy, Netherlands (200), Norway (2,200), Pakistan, Singapore (300), Spain, Sweden (4,500), UK (600 for Army, Royal Marines and Royal Air Force) and US Army (approximately 1,100).

Manufacturer
Hägglunds Vehicle AB, S-891 82 Örnsköldsvik, Sweden.
Tel: +46 660 80000. Telex: 6051 haegg s.
Fax: +46 660 82612.

UPDATED

UNITED KINGDOM

Alvis Stormer High-Mobility Load Carrier (HMLC)

Description
Announced in June 1988, the Alvis Stormer high-mobility load carrier plays the same part in the Alvis Stormer family of vehicles as the earlier and lighter Streaker in the Scorpion/Spartan family (for full details of the Stormer armoured personnel carrier refer to *Jane's Armour and Artillery 1995-96* pages 435 to 438). It is a flatbed tracked cargo carrier which can be used in a variety of forms, including a platform for scatterable mine systems.

In general terms the Stormer High-Mobility Load Carrier (HMLC) follows the same lines as the smaller Streaker but there are six roadwheels each side in place of the five on the Streaker. The main load area and payload are thus larger. The commander has an armoured compartment with a cupola and an overpressure NBC system is provided for the crew.

The Stormer is powered by a Perkins T6.3544 six-cylinder diesel developing 250 bhp, coupled to an ATL T300 crossdrive transmission located at the front of the vehicle.

It has been proposed that the Stormer HMLC could be fitted with a hydraulic crane for use in the logistics

Alvis Stormer high-mobility load carrier carrying Minotaur Vehicle-Launched Scatterable Mine System (VLSMS)
1996

carrier role, or as a Barmine laying vehicle. It has also been proposed as the basis for a repair/recovery vehicle with a roof-mounted hydraulic crane. A variant has been sold to Indonesia as the carrier for a 15 m Light Assault Bridge (LAB).

Specifications

Crew: 1 + 1 or 2
Weight: (combat laden) 13,200 kg
Payload: 4,000 kg
Length: 6.223 m
Width: 2.7 m
Height: (overall) 2.236 m
Ground clearance: approx 0.406 m
Track: 2.05 m
Track width: 425 mm
Length of track on ground: 3.575 m
Max speed: 80 km/h
Range: (road) over 644 km
Fuel capacity: 322 litres
Gradient: 60%
Vertical obstacle: 0.6 m
Trench: 1.75 m
Fording: 1.15 m
Angle of approach/departure: 30°/28°
Engine: Perkins T6.3544 5.8 litre 6-cylinder in-line turbocharged diesel developing 250 hp at 2,600 rpm
Transmission: ATL T300 semi-automatic with 6 forward speeds
Suspension: transverse torsion bars with trailing arm geometry and lever type hydraulic dampers
Electrical system: 28 V
Batteries: 4 × 12 V 6TN, 100 Ah
Generator: 28 V, 220 A

Status

Initial production complete. Selected for the UK Ministry of Defence for Vehicle-Launched Scatterable Mine System (VLSMS), role to be known as Spider. Procured by Indonesia in flatbed form and as bridge carrier.

Manufacturer

Alvis Vehicles Limited, The Triangle, Walsgrave, Coventry, West Midlands CV2 2SP, UK.
Tel: +44 1203 535455. Fax: +44 1203 539287.

UPDATED

UNITED STATES OF AMERICA

United Defense M992 Series Field Artillery Ammunition Support Vehicles (FAASV)

Development

In 1979 prototypes of three armoured artillery resupply vehicles were tested during HELBAT Human Engineering Laboratory, Battalion Artillery Test) trials. The BMY Combat Systems (now United Defense LP) vehicle was based on the chassis of the proven M109 155 mm self-propelled howitzer chassis, as was the HEL/AAI vehicle, and was named the M109 Ammunition Delivery System by the company. The FMC Corporation provided a stretched and armoured M548 tracked vehicle with an additional roadwheel on either side.

Following completion of the initial trials and the Concept Evaluation Test, it was decided that the M109 chassis was most suitable for the Field Artillery Ammunition Support Vehicle. In August 1980 TACOM issued a request for proposals for the design, construction, test and integrated logistics support of the FAASV. In March 1981 BMY was awarded a contract, by TACOM, for the supply of five prototype FAASVs under the designation of XM992. These were delivered during November and December 1981 and underwent DT/OT II trials at Yuma Proving Ground and at Fort Sill until March 1982. Following these trials the XM992 was type classified as the M992 FAASV in the Autumn of 1982.

Research, development, test and evaluation costs for the FAASV up to Fiscal Year 1982 amounted to $5.6 million and in FY83 procurement costs totalled $29.7 million for 54 vehicles. FY84 procurement costs were $60 million for 120 vehicles, and FY85 costs were $78.2 million for 170 vehicles. The FY86 figures were $70.2 million for 142 vehicles and for FY87 they were $62.5 million for 141 vehicles. FY88 planning called for $28.7 million for 48 vehicles. Production ceased for a while until 1991 when it was restarted to meet an order for 60 FAASVs, believed to be for Saudi Arabia and worth more than $50 million. In September 1991 a contract for 125 US Army and 20 FMS FAASVs was announced; the contract was worth $42,340,041. In May 1993 an option for an additional 65 vehicles was awarded. Production of all these vehicles was completed in April 1994.

The following components are common to the M992 FAASV and the standard M109: hull structure; engine and auxiliary equipment; transmission and drive; complete suspension; commander's and driver's hatches and controls; heating and ventilation system; electrical components; internal communications equipment; and towing provisions.

The M992A1 and M109A2 are basically similar to the M992 but are configured for the support of the 155 mm M109A6 Paladin self-propelled howitzer. See entry under Variants for details.

Description

The hull is made of all-aluminium armour with the driver at the front of the hull on the left, the engine to his right and ammunition stowage at the rear.

The driver has a hinged hatch cover that opens to the rear, in front of which are three M45 periscopes which can be covered by small metal flaps to prevent damage.

The engine is coupled to the Lockheed Martin Allison Division XTG-411-4 crossdrive transmission which is at the front of the hull.

The torsion bar suspension consists of seven dual rubber-tyred roadwheels with the drive sprocket at the front and the idler at the rear. There are no track-return rollers. The tracks are of the single-pin, centre guide type with replaceable rubber pads.

In the forward part of the roof there is a three-part projectile rack removal hatch and behind it is the commander's cupola. This can be traversed through 360° and has a single-piece hatch cover. A 0.50/12.7 mm M2 machine gun can be mounted at this station.

Prototype FAASVs had a 626 kg capacity crane mounted at the front of the hull but this crane was not fitted to full production vehicles for the US Army (it was, however, fitted to vehicles supplied to Egypt). The 155 mm projectiles are handled in racks of 10 projectiles, for example two horizontal layers of five projectiles each. If required, projectiles can also be loaded individually using the powered conveyor to feed the empty racks.

The charges are located in containers either side of the FAASV at the rear with fuzes being stowed in the left side of the hull. Ammunition is transferred to the M109 through its lower rear door by a power-operated conveyor at a maximum rate of 6 rds/min, which is higher than the M109's rate of fire. Overhead protection is provided by the large upward-opening powered door of the vehicle. The conveyor has lateral protection from the M109 and M992 lower rear doors. When not in use the conveyor folds up and is stowed inside the vehicle.

The projectiles are transferred from their stowed position in the forward part of the vehicle to the conveyor belt by the X-Y stacker which is part of the ammunition handling system and can be moved vertically and horizontally. The tray of the stacker serves as a working platform for installing fuzes onto the projectiles.

The upward-swinging rear door is hydraulically powered, with the smaller lower door opening manually to the right. The commander's adjustable seat is over the rear part of the conveyor belt.

US Army FAASVs are configured to carry storage racks each holding 10 155 mm projectiles plus charges, fuzes and other cargo. For export the following maximum quantities can be carried:

Calibre	155 mm	203 mm
Projectiles	90	48
Propelling charges	99 or 110	53
Fuzes	104 or 128	56

In the US Army FAASV space has been allocated for special projectiles such as the M712 Copperhead Cannon-Launched Guided Projectile, the Remote Anti-Armor Mine System (M718/M741) and the Area Denial Artillery Munitions (M692/M731).

The auxiliary power unit is located in the forward part of the FAASV's superstructure and supplies hydraulic and electrical power to the vehicle systems as well as charging the batteries. In addition, it can also provide power to the self-propelled howitzer being serviced.

The FAASV also incorporates a number of improvements which are included in the M109A6 Paladin. These include: an automatic fire suppression system; NBC VFP protection system; simplified test equipment as fitted to more recent AFVs such as the M1 MBT, M2 IFV and M3 CFV; AN/VIC-1 intercom; and an AN/PRC-68 small unit radio. Also fitted are chemical detection and alarm units and chemical decontamination units.

US Army M992 FAASV in process of reloading an M109 series 155 mm self-propelled howitzer

US Army M992 FAASV

Variants

M992A1 and M992A2

A series of product improvements has been incorporated into the FAASV, leading to the type classification of the M992A1 and M992A2. These improvements include: electrical system enhancements; the integration of a low heat rejection engine and T-154 Track Assembly; removal of the X-Y stacker, plus container stowage and conveyor enhancements; and modification of the upward-swinging rear door adding two small doors to allow conveyor interface with the M109A6 with the rear door closed. These improvements were generated by the US Army's Initial Operational Test and Evaluation (IOTE) of the M109A6 HIP, later named Paladin. The improvements will also increase the operational capabilities of the FAASV with all versions of the M109 series.

M1050

The M1050 is a version of the FAASV for use with 8 in/203 mm howitzer ammunition and fitted with a stacker device. It has not yet been produced.

Fire Direction Centre Vehicle/Command Post Vehicle (FDCV/CPV)

BMY (now part of United Defense) delivered a Fire Direction Centre Vehicle/Command Post Vehicle (FDCV/CPV) based on the chassis of the FAASV to the US Army for trials. This model was fitted with the following equipment: complete hybrid NBC protection system, APU, Magnavox graphic display, Litton TACFIRE digital display terminal, PDP computer, Battery Computer System printer, power distribution system, pocket radio and a PRC-68 radio. Customers for this vehicle in specialised configurations included Egypt (72), Greece (41) and Taiwan (6).

Specifications

M992 FAASV
Cab seating: 2 (+6 passengers)
Weight:
(empty) 19,749 kg
(loaded) 26,105 kg
Max load: 6,109 kg
Load area: 3.38 × 3.05 × 2.21 m
Length: 6.67 m
Width: 3.15 m
Height: 3.24 m
Ground clearance: 0.368 m
Track width: 381 mm
Length of track on ground: 3.962 m
Max speed: (combat loaded, road) 56.3 km/h
Acceleration: (0-48 km/h) 19 s
Range: approx 354 km
Fuel capacity: 511 litres
Max gradient: 60%
Max side slope: 40%
Fording: 1.07 m
Vertical obstacle: 0.53 m
Trench: 1.83 m
Engine: Detroit Diesel Model 8V-71T turbocharged, 2-stroke, liquid-cooled 8-cylinder diesel developing 405 bhp at 2,350 rpm (440 hp on M992A1 and M992A2)
Transmission: Allison Transmission Division of Lockheed Martin XTG-411-4 crossdrive with 4 forward and 2 reverse gears
Suspension: torsion bar
Electrical system: 24 V
Batteries: 4 × 12 V 6TN
Alternator:
(M992) 100 A
(M992A1 and M992A2) 180 A

Status

In service with Egypt (51 FAASV, 72 FDCV/CPV), Greece (41 FDCV/CPV), Spain (6 FAASV), Saudi Arabia (60 FAASV), Taiwan (6 FDCV/CPV), Thailand (20 FAASV), and US Army (800 FAASV).

Manufacturer

United Defense LP, Combat Systems Division, PO Box 15512, York, Pennsylvania 17405-1512, USA.
Tel: +1 717 225 4781. Fax: +1 717 225 4615.

UPDATED

United Defense Fighting Vehicle Systems Carrier

Development

The Fighting Vehicle Systems Carrier (XM987) is part of the Bradley Fighting Vehicle Systems (BFVS) family, which also includes: the M2 Infantry Fighting Vehicle; the M3 Cavalry Fighting Vehicle; and the M993 Multiple Launch Rocket System (MLRS) carrier, all of which were developed by FMC Corporation (now part of United Defense) under contract to the US Army. Details of these vehicles are given in *Jane's Armour and Artillery 1995-96*, pages 452 to 457.

The vehicle is a highly mobile armoured carrier used for a wide range of requirements. Major features of the vehicle are its component commonality with the IFV and CFV, overpressure ventilation system, nuclear hardened electrical system, air transportability in the Lockheed C-141 Starlifter transport aircraft, tilt cab to facilitate maintenance and a current payload of 12,700 kg. For some applications the vehicles has been upgraded with the M2A2 power train and suspension which provides an increased payload of 15,812 kg while maintaining mobility characteristics.

Prototypes of the Fighting Vehicle Systems Carrier were originally built by FMC specifically for use with the General Support Rocket System (now known as the Multiple Launch Rocket System (MLRS)), and the first prototype vehicles, under the designation XM993, were delivered to the two competing contractors late in 1978. After trials, the Vought system was selected and in June 1980 Vought was awarded a $26.9 million contract for 1,374 missiles packed in launch containers and 16 self-propelled launcher loaders, to be built by FMC. This vehicle is designated the M993 and over 940 had been manufactured by mid-1996.

Description

The basic Fighting Vehicle Systems (FVS) Carrier has a cab-over-power pack arrangement with the cargo area at the rear.

The aluminium armour plate cab is fitted with noise attenuation materials and large ballistic windows to provide forward and side vision. The front windows are fitted with exterior louvres to provide protection during rocket firing and for nuclear survivability. The louvres can be opened or closed individually with levers inside the cab. In a tactical situation the louvres may be rotated to a stowed position on the cab roof. An 'Uparmor Kit' is available for vehicles fitted with the 600 hp engine.

The cab has accommodation for the three-man crew, an overpressure ventilation system, space for radios and provisions and an instrument panel for operating the vehicle. Sufficient space is provided to add a second control panel for weapon system operation.

An overhead hatch above the right-hand seat can be fully opened for use as an airguard, or fixed partially open for additional ventilation. As an option the cab roof can be fitted with a NATO mount for a 7.62 mm M240 machine gun.

The complete power pack, which is centered in the vehicle, is interchangeable with that of the IFV and CFV. It consists of a Cummins 500 or 600 hp diesel engine coupled to a Lockheed Martin crossdrive transmission. The power pack can be removed or installed in 30 minutes without breaking cooling or hydraulic lines and this arrangement enables the power pack to be operated on the ground outside the vehicle.

The power pack is wired to accommodate built-in test equipment for rapid fault isolation using STE/FVS (Simplified Test Equipment/Fighting Vehicle System), the same test equipment as used in the M1 Abrams MBT.

The power pack has a negative pressure system where cooling air is drawn through the radiator and is discharged through the exhaust grille above the right sponson. The fuel is carried in integral tanks under the floor plates at the rear of the vehicle.

The suspension is an elongated version of the IFV/CFV system and each side consists of six dual rubber-tyred roadwheels two dual support rollers, two single support rollers, front drive sprocket, raised rear idler and a high return track.

The single-pin track has forged steel blocks, rubber bushings and detachable rubber pads. Track tension is adjusted by a grease-filled cylinder between the hull and idler wheel. Vertical roadwheel travel is controlled by high-strength steel torsion bars splined to trailing road arms forming a fully independent suspension. Linear hydraulic shock-absorbers at the first, second and sixth roadwheels stabilise the vehicle on rough terrain.

XM4 Command and Control Vehicle mounted on Fighting Vehicle Systems Carrier

XM5 Electronic Fighting Vehicle System (EFVS) mounted on Fighting Vehicle Systems Carrier

A suspension lockout system is provided for the MLRS application of the carrier. The lockout is a hydraulically actuated, multidisc brake mounted concentric with the torsion bar. Lockout units can be installed at some or all the torsion bar stations depending on the degree of suspension stiffness required. The lockouts provide platform stability during both launching and loading operations.

In the left rear corner of the cab is the overpressure ventilation system. It consists of a 5 µm dust filter, 3 µm particulate filter, charcoal filter, bypass valve and an axial flow fan. Dual positioning of the bypass valve permits air to pass through all three filters or just through the dust filter, depending on the mission. As an option this system can be converted to a hybrid system. The present design has an M13A1 NBC unit in the cab. This secondary system provides for the crew's safety if the cabin air inadvertently becomes contaminated during the course of a mission.

Variants

MLRS carrier

This was the first application for the FVS Carrier, the chassis being designated the M993 for this purpose. Details of the MLRS are given in *Jane's Armour and Artillery 1995-96*, pages 725 to 729.

Cargo carrier

This is designated the XM987 but is not yet in production.

Armored, Forward-Area, Rearm Vehicle (AFARV)

The US Army tested the prototype of an Armoured, Forward-Area, Rearm Vehicle (AFARV) to resupply armoured and infantry units in the forward battlefield area.

That vehicle was fitted with a 1½ ton hydraulic crane to lift pallets of ammunition from trucks to the internal stowage compartment. The crane can also be used to unload pallets. A conveyor system enables the vehicle to offload single rounds of tank ammunition into the tank while the respective crews remain under armour protection. The AFARV can be used to carry 105 and 120 mm tank ammunition as well as TOW and Dragon ATGWs, LAWs and small arms ammunition.

Artillery Rearm Module (ARM)

The Artillery Rearm Module (ARM) is a Lockheed Martin (formerly General Electric Aerospace) project to resupply self-propelled artillery units operating in forward areas under adverse conditions. It is one component of the projected FARV-A (Fully Integrated/Armored Rearm) programme.

The ARM consists of an artillery ammunition module/container carried on the back of an FVSC with an automated transfer arm and conveyer moving ammunition (projectiles and charges) from the module and transferring them into the fighting compartment of a self-propelled artillery vehicle. Transfer rates can be up to 10 items/min. Operators use a remote handset to input and monitor the ARM while being able to remain under cover provided by the FVSC cab.

A prototype of the ARM was delivered for US Army trials commencing in October 1990.

Armored Maintenance Vehicle

A prototype Armored Maintenance Vehicle (AMV) fitted with a 10 ton crane was built and evaluated by the US Army in 1985. The vehicle was leased to the US Army. See entry in the *Armoured repair vehicles* section.

XM4 Command and Control Vehicle

Five prototypes of a Command and Control (C²) Vehicle have been built with an armoured module housing equipment and personnel for battle commander's staff. The enclosure is a self-contained unit which includes an NBC protection system, environmental control system (air conditioning and heating), a 43 kW 60 Hz primary power unit, and a 10 m telescopic mast, all remotely controlled from within the enclosure. The XM4 is completely self-contained and configured to accommodate both Command and Control (C²) and Joint STARS Ground Station Module mission equipment. With a change in software it may also be used as an artillery fire direction centre or an air defence battle management operations centre. The Command and Control (C²) Vehicle is equipped with the Cummins BFVA2 600 hp engine, Lockheed Martin transmission, final drives, steel roadwheels, improved track and torsion bars. The MLRS suspension lockout system has been removed. These changes are stated to provide significant growth capability for up-armoured alternatives.

In 1996, four preproduction prototypes underwent government testing. Low rate initial production was anticipated during early 1997.

XM5 Electronic Fighting Vehicle System (EFVS)

The XM5 Electronic Fighting Vehicle System (EFVS) is similar to the XM4 Command and Control Vehicle (see preceding text) and uses the same chassis and superstructure. However, it is configured to house the equipment and personnel for the Ground-Based Common Sensor - Heavy (GBCSH) programme. It provides an NBC overpressure system, an environmental control system, a 60 kW primary power unit and a 20 m telescopic mast controlled from inside the vehicle. A prototype was completed in 1986 and three engineering development models were completed by mid-1993.

Armored Treatment and Transport Vehicle (ATTV)

The Armored Treatment and Transport Vehicle (ATTV) is intended to be positioned well forward in battle areas to transport wounded soldiers to a place of care. It protects the casualties and medical attendants while carrying out close combat medical missions and is completely self-contained. Onboard support subsystems provide power for mission equipment and maintain an NBC contamination-free treatment environment. Casualties are carried inside an armoured shelter capsule on an FVS Carrier, the interior of which can be modified to suit requirements. Both seated and stretcher casualties can be accommodated. Up to 12 seated patients can be carried in addition to two medical attendants, three crew and the driver. In a stretcher configuration up to nine casualties can be carried, plus two medical attendants, three crew and the driver. A prototype has been produced and is undergoing government evaluation.

Gross vehicle weight is between 25,400 and 29,940 kg.

Proposed Variants

MLRS resupply vehicle

This vehicle can carry four pods of rockets and a crane for self-loading and unloading.

Fuel resupply vehicle

This vehicle can deliver 7,570 litres of fuel in one mission, sufficient to supply six M60A1 or five M1 Abrams MBTs. This unit can also provide oil and lubrication services during refuelling.

Firefinder vehicle

For trial purposes an FVS Carrier was fitted with the Hughes AN/TPQ-37 Firefinder artillery-locating radar. This trials vehicle is being reconfigured.

Specifications

Cab seating: 1 + 2 or 3
Weight:
(empty) 14,636 or 14,818 kg
(loaded) 25,545 or 30,630 kg
Load area:
(cargo bed between sponsons) 3.96 × 1.78 m
(above sponsons) 3.96 × 2.97 m
Length: 6.97 m
Width: 2.97 m
Height:
(cab) 2.59 m
(chassis) 1.206 m
(top of load area) 1.09 m
Ground clearance: 0.43 m
Track width: 533 mm
Length of track on ground: 4.33 m
Max speed:
(road) 65 km/h
(10% gradient) 26 km/h
Range: (at 40 km/h) 483 km
Fuel capacity: 726 litres
Max gradient: 60%
Max side slope: 40%
Fording: 1.02 m
Vertical obstacle: 0.91 m
Trench: 2.36 m
Engine: Cummins VTA-903 14.8 litre turbocharged 8-cylinder diesel developing 500 hp at 2,400 rpm or Cummins VTA-903T 14.8 litre turbocharged 8-cylinder diesel developing 600 hp at 2,400 rpm
Transmission: Lockheed Martin HMPT 500-3EC hydromechanical with electronic controller
Steering: hydrostatic
Suspension: steel torsion bar
Brakes: multidisc, oil-cooled
Electrical system: 28 V
Batteries: 4 × 12 V 6TN, 100 Ah
Alternator: 2 × 300 A

Status

In production for the US Army in MLRS role (Over 940 by mid-1996). Ordered for the British Army (71 for MLRS). Also ordered by Bahrain (9 for MLRS), France (56 for MLRS), Germany (202 for MLRS), Greece (9 for MLRS), Israel (48 for MLRS), Italy (24 for MLRS), Japan (48 for MLRS), Netherlands (22 for MLRS) and Turkey (15 for MLRS).

Manufacturer

United Defense LP, Ground Systems Division, 2830 De La Cruz Avenue, Santa Clara, California 95052, USA.
Tel: +1 408 289 0111. Telex: 6714210.
Fax: +1 408 289 2150.
Also being produced under licence in Europe for MLRS application.

UPDATED

United Defense M548 tracked cargo carrier

Development

The XM548 cargo carrier was designed for the US Army Signal Corps in 1960, using the basic automotive components of the M113 tracked armoured personnel carrier. At that time its primary purpose was to serve as a highly mobile transport carrier for the AN/MPQ-32 Hostile Artillery Radar System. This application did not reach production and a modified version with a diesel engine, designated the XM548E1, was designed using the power pack and automotive components common to the M113A1 APC. Engineer and service tests on three prototype vehicles were completed late in 1964 and the vehicle was type classified the following year. First production vehicles were completed early in 1966 by the FMC Corporation (now part of United Defense).

The vehicle is used in the US Army for a wide variety of roles including use as an ammunition resupply vehicle with self-propelled artillery units equipped with the M109 and M110 weapons.

The US Army did not procure any vehicles in FY76 to 1978 but 193 vehicles were bought with FY79 funding of $15.6 million and 242 with 1980 funding of $30 million. The FY82 request was for a further 160 vehicles at $20.9 million. In FY83 FMC produced 29 M548A1 carriers for the US Army plus six for other countries. As of March 1987, 3,683 M548s had been purchased by the US Army. Another 1,295 were produced for export. In total 254 M548A1s were purchased by the US Army with another 18 produced for export.

Total production of all versions of the M548 series was over 6,000 vehicles.

Description

The engine and crew compartment are at the front of the vehicle and the cab roof, sides, front and rear can be removed for air transport. The cargo compartment has a rear opening that can be secured by two watertight doors. Six hollow aluminium extruded plates bolted in place in either an upper or lower position form the cargo deck. The lower position allows more cargo to be carried or provides leg room for seated passengers. Tiedowns are available for both cargo deck positions. The cargo area can be enclosed using a standard vinyl-coated nylon cover supported by bows.

Canadian Forces M548 tracked cargo carriers (Michael Jerchel)

Australian Army M548 tracked cargo carrier (Barry Marriot)
1997

If required, an M66 ring mount for an air and local defence machine gun can be mounted over the top of the cab. If this is fitted with a 7.62 mm machine gun, 660 rounds of ammunition are carried, or if fitted with a 12.7 mm/0.50 heavy machine gun, 300 rounds of ammunition are carried.

The suspension is of the torsion bar type and consists of five roadwheels with the drive sprocket at the front and the idler at the rear. (The M548A1 has hydraulic shock-absorbers on the first, second and last roadwheel stations.) There are no track-return rollers. The first and last roadwheel stations are provided with a hydraulic shock-absorber and the tops of the tracks are covered by a rubber skirt.

The vehicle has the same wheels, sprocket, sprocket carrier, track adjuster, idler, idler wheel, shock-absorber and mount, wheels and track as the M113A1 but has a larger diameter torsion bar and the final drive assembly has a different gear ratio.

The M548 is fully amphibious and propelled in the water by its tracks. On US Army vehicles, however, the amphibious capability was removed. A winch with a 9,072 kg capacity is mounted at the front of the vehicle. Optional equipment includes a heater (personnel and cargo areas), heater (engine coolant and battery), an airbrake kit to actuate brakes on the towed trailer, and a materials handling hoist.

Variants

M1015

First shown in 1982, the XM1015 was a modified version of the M548 intended to carry equipment containers or containerised shelters for electronic systems. It was used to carry the first production units of the Emerson Electric Company's AN/MSQ-103A Teampack radar monitoring system developed for the US Army under a contract from the US Army Electronics Warfare Laboratories at Fort Monmouth, New Jersey. Teampack is used to monitor ground-based radar systems on the battlefield, which include mortar locating, surveillance and air defence radars. The equipment is contained in a shelter carried on the M1015. The M1015 has no provision for a tilt and when carrying the Teampack the vehicle has no provision for a machine gun mounting ring. A telescopic aerial is carried on the vehicle right front. The Teampack containerised shelter has ballistic protection for its contents and crew. The XM1015 was standardised as the M1015. In total, 125 were procured, they also carry the AN/TSQ-114A Trailblazer communications intercept and direction-finding system and the AN/MLQ-34 TACJAM system.

In August 1986 it was announced that the first of 127 Improved M1015 carriers had been ordered at a cost of $21 million. The Improved M1015 has a solid-state engine governor and revised shelter rails. It has a 60 kW generator for powering electronic systems.

By 1996 the M1015 was no longer in service with the US Army, its mission being absorbed by other systems.

Stretched M548A1E1

In 1977 the FMC Corporation (now United Defense) completed the prototype of a stretched M548A1E1, about 0.66 m longer than the basic model with an additional set of roadwheels. The standard 210 hp engine was replaced by a turbocharged model developing 300 hp. Other improvements included a transmission that featured hydrostatic steering and a modified cooling system.

M667

This is the basic vehicle for the Lance tactical missile system. The loader transporter is called the M688 and the launch vehicle is the M752. Net weight of the M667 is 6,455 kg. The suspension of the M667 can be locked to provide a more stable firing platform.

M730

This has four Chaparral SAMs on a launcher at the rear of the cab. Full details of this system are given in *Jane's Land-Based Air Defence 1995-96*, pages 177 to 180.

Recovery vehicles

A recovery version, known as the XM696, was developed to the prototype stage.

The Norwegian firm of Hägglunds Moelv AS fitted one of its NM84 wrecker cranes onto an M548 for the Norwegian Army. This crane has a maximum lift capacity of 5,000 kg at 3.5 m and a maximum winch pulling capacity of 18,000 kg. Maximum outreach is 5.3 m at which 3,400 kg can be lifted.

Radar vehicles

Many countries use the M548 for carrying radars: for example, Contraves of Switzerland has fitted its Skyguard anti-aircraft radar in the rear of the vehicle for trials.

M548A1

Introduced in 1982, this version has improved suspension and cooling systems. Both new production and depot rebuild versions were involved. In 1994 the US Army had approximately 2,000 M548A1s in service.

M548A3

In 1993 a prototype of a M548A3 was produced and tested. The vehicle is based on the M548A1 standard, fitted with a Detroit Diesel 6V-53T turbocharged diesel developing 275 hp, a new alternator, and hydrostatic steering, all part of the M113A3 Reliability Improved Selected Equipment (RISE) package. Basic specifications, where different from those of the M548A1, are:

Width: 2.69 m (2.54 m reduced)
Height: 2.68 m
Max road speed: 58.2 km/h
Range: 458 km
Engine: Detroit Diesel Model 6V-53T turbocharged diesel developing 275 hp
Transmission: Allison X200-4 hydrokinetic

The M548A3 also features an NBC crew protection system. Initial conversions were carried out in the USA (at the Red River Army Depot) and in Korea during 1994 and more were planned.

Future M548 options

Future upgrade and re-equipment options envisaged for the M548 series included: an armoured cab (which would add 1,360 kg to the vehicle weight); a material handling crane; drop side enhancements; a decontamination apparatus; a global positioning system; ammunition racks (possibly from the M992 Series FAASV); a winch; and trailer brakes. Not all of these options are expected come to fruition.

M548GA1

This is a low-sided vehicle that mounts a platform for the German Skorpion mine dispensing system. For details refer to the *Minelaying equipment* section.

LAR-160

AAI Inc of the USA acted as overall systems integrator to adapt the IMI LAR-160 light artillery rocket system to the M548. Also involved were Thiokol (Wasatch Division), Avco Systems Division and Westinghouse, with the intention of providing Rapid Deployment Force units with a long-range artillery rocket system. The LAR-160, has a body diameter of 160 mm, is 3.31 m long and weighs 110 kg at launch. As the body diameter is very similar, the LAR-160 can use 155 mm projectile warheads with HE, bomblet and other loads. On the M548 the LAR-160 is packed and fired from 18-round Launch Pod Containers (LPCs), two of which can be carried on one vehicle. The LPCs act as transport containers and are discarded once the rockets have been fired. Each LPC weighs 2,522 kg and the range is 30,000 m. The weight of an M548 loaded with two LAR-160 LPCs is approximately 12,800 kg.

Specifications M548A1

Cab seating: 1 + 3
Weight:
(empty) 7,439 kg
(loaded) 12,882 kg
Max load: 5,443 kg
Towed load: 6,350 kg
Load area: 3.32 × 2.45 m
Length: 5.892 m
Width: 2.69 m
Width: (over tracks) 2.54 m
Height:
(excl MG) 2.71 m
(reduced) 1.94 m
(load area) 1.21 m
Ground clearance: 0.43 m
Track: 2.159 m
Track width: 381 mm
Length of track on ground: 2.82 m
Ground pressure: 0.6 kg/cm²
Angle of approach/departure: 57°/35°
Max speed:
(road) 53.1 km/h
(10% gradient) 18.2 km/h
Range: 483 km
Fuel capacity: 397 litres
Max gradient: 60%
Max side slope: 30%
Fording: 1 m
Vertical obstacle: 0.609 m
Trench: 1.68 m
Engine: Detroit Diesel Model 6V-53 6-cylinder liquid-cooled diesel developing 215 hp at 2,800 rpm
Gearbox: Allison TX-100-1 3-speed; a torque converter gives 6 forward and 2 reverse speeds
Suspension: torsion bar
Electrical system: 24 V
Batteries: 2 × 12 V, 6TN, 100 Ah

Status

Production ceased 1994. In service with Australia, Canada, Egypt, Germany, Greece, Israel, Italy, New Zealand, Norway, Spain, Switzerland, Tunisia and the USA.

Manufacturer

United Defense LP, Ground Systems Division, 2830 De La Cruz Boulevard, Santa Clara, California 95052, USA.
Tel: +1 408 289 0111. Telex: 6714210.
Fax: +1 408 289 2150.

UPDATED

United Defense XM1108 Universal Carrier

Development

The XM1108 Universal Carrier, originally known as the M113A3 Universal Carrier was produced in early 1993 as a platform for the Chaparral/Hellfire Missile System. It is known as the Chaparral Chassis Service Life Extension Program (CCSLEP), originally termed Roadrunner. The Universal Carrier has also been proposed as a Standardised Integrated Command Post Shelter (SICPS) carrier, an anti-tank missile system carrier, a cargo carrier and as a platform for multiple launch rocket systems.

Description

The XM1108 Universal Carrier prototype incorporates a modified MLRS armoured cab on a stretched M113A3 APC lower chassis. Many major automotive components are shared with the M113A3 and many control cab components are common with the M993 MLRS vehicle. Refer to *Jane's Armour and Artillery 1995-96* page 463 for details of the M113A3 and page 725 for the MLRS.

The aluminium armour forward control cab can be tilted forward for maintenance and has seating for the driver and two passengers. A roof hatch can be used to mount a 7.62 mm M60 machine gun. The engine pack is located just behind the cab. There are six roadwheels each side.

XM1108 Universal Carrier

Specifications

Cab seating: 1 + 2
Weight: (max) 18,144 kg
Max load: 6,804 kg
Power to weight ratio:
(at 400 hp) 20 hp/ton
(at 350 hp) 17.5 hp/ton
Ground pressure: 0.6 kg/cm^2
Load area: (above sponsons) 4.04 × 2.54 m
Length: 6.37 m
Width: 2.84 m
Width: (over tracks) 2.54 m
Height: (overall) 2.58 m
Ground clearance: 0.42 m
Track width: 381 mm
Length of track on ground: 3.62 m
Max speed: 66 km/h
Range: 483 km
Fuel capacity: 492 litres
Max gradient: 60%
Max side slope: 40%
Vertical obstacle: 0.61 m
Trench: 2.18 m
Engine: Detroit Diesel 6V-53TA diesel developing 400 hp
Gearbox: Allison X-200-4A hydrokinetic
Suspension: torsion bar
Electrical system: 24 V
Batteries: 4 × 12 V, 6TL, 120 Ah
Generator: 200 or 300 A

Status

Prototype.

Manufacturer

United Defense LP, Ground Systems Division, 2830 De La Cruz Boulevard, Santa Clara, California 95052, USA.
Tel: +1 408 289 0111. Telex: 6714210.
Fax: +1 408 289 2150.

UPDATED

OVER-SNOW VEHICLES

CANADA

Bombardier BR-100+ over-snow vehicle

Description

The Bombardier BR-100+ vehicle was designed specifically for use in marginal terrain such as snow, swamplands and desert. It is an updated version of the Bombardier BR-100 (Bombi).

The chassis is the toboggan type with the fully enclosed three-man safety cab in the centre and the engine at the rear. Entry to the cab is by a forward-opening door in either side. The BR-100+ is delivered with a fully enclosed two-door cab.

The torsion bar suspension consists of four pneumatic tyres (4.60 × 10) either side with the drive sprocket at the front. There are no track-return rollers. Two types of track can be fitted: the Summer tracks are 711 mm wide and consist of rubber belts with steel crosslinks; and the wider (813 mm) Winter tracks consist of rubber belts with aluminium crosslinks.

Standard equipment includes: an automatic transmission; a pusher fan; full instrumentation; a differential oil cooler; shoulder and lap seat belts for all three occupants; a heater-defroster; and a lighting system.

Optional equipment includes: a spare wheel; tyre and inner tube mounted at the rear of the vehicle; a 600 W engine block heater; a brush guard; a front- or rear-mounted electric winch; pintle hook; hydraulic system; front blade; and an all-steel 453 kg payload trailer.

Bombardier BR-100 (Bombi) with fully enclosed cab

Specifications

Cab seating: 1 + 2
Weight:
(empty) 1,500 kg
(loaded) 1,950 kg
Max load: 450 kg
Length: 3.15 m
Width:
(with Summer tracks) 2 m
(with Winter tracks) 2.2 m
Height: 2.01 m
Ground clearance: 0.32 m
Track width:
(Winter tracks) 813 mm
(Summer tracks) 711 mm
Ground pressure:
(Summer tracks) 0.1 kg/cm^2
(Winter tracks) 0.08 kg/cm^2
Max speed: 22 km/h
Fuel capacity: 45 litres
Gradient: up to 75%
Side slope: up to 50%
Engine: Ford 2.3 litre 4-cylinder petrol developing 80 hp at 4,500 rpm
Gearbox: automatic, 3 forward and 1 reverse gears
Differential: planetary controlled
Steering: manual, through controlled differential
Turning radius: (inside) 1.68 m
Suspension: wheels independently mounted on rubber torsion bar
Brakes:
(main) disc brake on drive line
(parking) locking device on brake disc
Electrical system: 12 V
Battery: 1 × 12 V

Status

In production.

Manufacturer

Bombardier Inc, Industrial Equipment Division, Valcourt, Québec, Canada J0E 2L0.
Tel: +1 514 532 2211. Telex: 5832575.

VERIFIED

Bombardier Skidozer over-snow vehicle

Description

As with other members of the Bombardier range, the Skidozer was designed specifically for use in marginal terrain such as snow, swamplands and desert. There are two basic models in the series, 252G (petrol) and 252D (diesel).

The chassis of the Skidozer is formed of steel on a tubular subframe. The two-man fully enclosed cab at the front of the vehicle has two adjustable seats with safety belts, heater, two front windscreen wipers and thermal windscreens. Entry to the cab is through a large door in either side that opens to the rear.

The Skidozer can carry a maximum payload of 1,088 kg and is also available with a four-door 10-man fully enclosed cab that extends to the rear of the vehicle. On the latter model, in addition to the two normal doors, there are two large doors in the rear of the cargo/personnel compartment.

The torsion bar suspension consists of five pneumatic tyres each side and a drive sprocket at the rear. There are no return rollers. Two types of track can be fitted, both of the rubber belt type with steel or aluminium crosslinks and with a width of 740 or 1,040 mm.

Standard equipment includes: a dry air cleaner; suction-type fan; full instrumentation; differential oil cooler; and a lighting system. Optional equipment includes: six-way straight blade/six-way U-blade; power steering; two-speed gearbox; hydraulic systems; solid rubber tyres; rotating beacon; block heater; and mirrors.

Bombardier Skidozer over-snow vehicle with 10-man cab

Specifications
252G, military version
Cab seating: 1 + 9
Weight: 3,936 kg
Payload: 1,088 kg
Length: 4.09 m
Width: (overall) 2.946 m
Height:
(top of lamp) 2.546 m
(top of tilt) 2.35 m
Ground clearance: 0.305 m
Track width: 1.04 m
Max speed: 34.6 km/h
Fuel capacity: 170 litres
Gradient: up to 80% with front blade
Side slope: up to 60%
Engine: Ford 6-cylinder petrol developing 124 hp at 3,600 rpm
Gearbox: Ford C6 automatic with 3 forward and 1 reverse gears
Differential: planetary controlled
Steering: manual operating through controlled differential
Turning radius: (inside) 3.66 m
Suspension: wheels independently mounted on rubber torsion bar
Brakes:
(main) disc on drive line
(parking) mechanical on drive line
Electrical system: 12 V
Battery: 1 × 12 V, 85 Ah
Alternator: 60 A

Status
In service with Argentina, Canada, Greece, Italy, the Royal Air Force, Turkey (Army, Navy and Air Force) and the US Air Force.

Manufacturer
Bombardier Inc, Industrial Equipment Division, Valcourt, Québec, Canada J0E 2L0.
Tel: +1 514 532 2211. Telex: 5832575.

VERIFIED

JAPAN

Komatsu Type 60 over-snow vehicle

Description
In 1952 the Komatsu Manufacturing Company developed the KC-20 over-snow vehicle primarily for civilian use. Further development resulted in the KC-20-35 which entered production in 1960 and is widely used in Japan for civil applications.

The KC-20-35 was followed by the KC-20-3 which, with modifications, was adopted by the Japanese Ground Self-Defence Force in 1960 as the Type 60 Snow Mobile, or Medium Snow Mobile as it is sometimes called. It is widely used on the northern main Japanese island of Hokkaido.

The engine is at the front of the vehicle, with the crew and passenger/cargo area at the rear. The windscreen can be folded forwards onto the bonnet and the cargo area has bows, a tarpaulin-type cover with side windows and a drop tailgate at the rear.

The suspension is the bogie/torsion bar type with each side having eight dual roadwheels, drive sprocket, idler and two track-return rollers.

Type 60 over-snow vehicle (Kensuke Ebata)

Specifications
Cab seating: 1 + 9
Weight:
(empty) 2,870 kg
(loaded) 3,770 kg
Max load: 900 kg
Max towed load: 1,500 kg
Length: 4.07 m
Width: 1.98 m
Height: 2.05 m
Ground clearance: 0.3 m
Track: 1.34 m
Ground pressure: 0.105 kg/cm^2
Trench: 1.006 m
Max speed: 36 km/h
Range: 135 km
Fuel capacity: 90 litres
Max gradient: 60%
Engine: Toyota 6-cylinder water-cooled petrol developing 105 hp at 3,400 rpm
Gearbox: manual, 4 forward and 1 reverse gears
Suspension: bogie/torsion bar
Electrical system: 12 V

Status
Production complete. In service with the Japanese Ground Self-Defence Force.

Manufacturers
Komatsu Manufacturing Company and Ohara Ironworks, Japan.

VERIFIED

Komatsu Type 61 over-snow vehicle

Description
To meet the requirements of the Japanese Ground Self-Defence Force, Komatsu and Ohara Ironworks built a full-tracked over-snow vehicle designated the KC-50-2. The prototype was completed in 1955 and after modification was adopted as the Type 61 Large Snow Mobile in 1961. In addition to being used as a cargo/troop carrier it is also used to tow 105 and 155 mm artillery weapons mounted on skis.

The engine is at the front of the vehicle with the crew and passenger/cargo area at the rear. The engine is fitted with a turbocharger for high-altitude operations and aluminium is used wherever possible to save weight. The windscreen can be folded forwards onto the bonnet and the cargo area has a drop tailgate, bows and a tarpaulin cover with side windows.

The suspension is the bogie/torsion bar type with each side having eight dual roadwheels, drive sprocket, idler and three track-return rollers.

The Type 61 is distinguishable from the Type 60 by its larger size and much deeper engine compartment.

Specifications
Cab seating: 1 + 10
Weight:
(empty) 5,220 kg
(loaded) 6,500 kg
Max load: 1,280 kg
Max towed load: 3,200 kg
Length: 5.34 m
Width: 2.5 m
Ground clearance: 0.34 m
Track: 1.71 m
Track width: 790 mm
Length of track on ground: 3 m
Ground pressure: 0.135 kg/cm^2
Max speed: 35 km/h
Range: 166 km
Fuel capacity: 160 litres
Max gradient: 60%
Trench: 1.5 m
Engine: Isuzu DA-120T 6-cylinder water-cooled diesel developing 155 hp at 2,600 rpm
Gearbox: manual, 5 forward and 1 reverse gears
Suspension: bogie/torsion bar

Status
Production complete. In service with the Japanese Ground Self-Defence Force.

Manufacturer
Komatsu Manufacturing Company and Ohara Ironworks, Japan.

VERIFIED

RUSSIAN FEDERATION AND ASSOCIATED STATES (CIS)

GT-T tracked amphibious over-snow vehicle

Description

The GT-T is the largest in the GT range of tracked amphibious over-snow vehicles. It can carry a maximum of 2,000 kg of cargo and tow a trailer weighing up to 4,000 kg.

The engine is at the front of the hull and extends rearwards into the crew compartment. The driver is seated on the left side and the vehicle commander on the right, both with a side door, and the commander has a circular hatch in the roof. The load area is at the rear and is usually covered by a tarpaulin cover. Up to 10 personnel can be seated in the rear. An unusual feature of the GT-T is that the fuel tanks are positioned externally above the tracks on each side at the rear.

The torsion bar suspension consists of six roadwheels with the idler at the rear and the drive sprocket at the front. The roadwheels are similar to those used on the PT-76 light amphibious tank family, and are also used as the basis for the more recent MT-LB multipurpose tracked vehicle (refer to *Jane's Armour and Artillery 1995-96*, pages 307 to 309).

The GT-T is fully amphibious. Most sources state that it is propelled in the water by its tracks at a speed between 5 and 6 km/h.

Variants

A number of GT-Ts have had a fifth wheel mounted to the rear of the cab for towing semi-trailers carrying missiles such as the SA-2; they are designated the GT-TS.

A number of specialised civilian models have been developed, with a fully enclosed cabin at the rear. In 1965 the GT-T maintenance vehicle entered service with the designation MTO-SG. Mounted at the front of the vehicle is an A-frame for changing components and inside the hull rear is a workshop with a petrol-driven power generator, electric drill, electric grinder, arc welding equipment, compression pressure meter and a full set of tools. The A-frame can lift a maximum load of 1,500 kg. Radios are fitted as standard on all repair vehicles.

A field kitchen version has been produced and a version has been reported carrying a modified aircraft jet engine to carry out NBC decontamination of equipment and vehicles.

Troops boarding GT-T tracked amphibious over-snow vehicles during Winter exercises

Specifications

Cab seating: 1 + 2 (up to 10 in rear)
Weight:
(empty) 8,200 kg
(loaded) 10,200 kg
Max load: 2,000 kg
Towed load: 4,000 kg
Length: 6.34 m
Width: 3.14 m
Height: 2.16 m
Ground clearance: 0.45 m
Track width: 540 mm
Length of track on ground: 3.914 m
Ground pressure: (loaded) 0.24 kg/cm²
Max speed:
(road) 45.5 km/h
(water) 6 km/h
Range: 500 km
Max gradient: 60%
Engine: 1Z-6 6-cylinder water-cooled diesel developing 192 hp

Status

In service with the RFAS.

Manufacturer

State factories.

UPDATED

MT-L tracked amphibious over-snow vehicle

Development

The MT-L is thought to have been developed at the same time as the GT-T vehicle (which is specifically an over-snow vehicle) and is also used in swampy areas. The MT-L is not strictly an over-snow vehicle as its ground pressure is 0.42 kg/cm² compared with the 0.24 kg/cm² of the GT-T.

A special variant of the MT-L, the MT-LV, has a ground pressure of 0.27 kg/cm², which puts it in almost the same class as the GT-T. Its major difference from the MT-L is probably wider tracks and hence a lower ground pressure.

The MT-L is the basis for a complete family of full-tracked vehicles which replaces unarmoured over-snow vehicles, unarmoured artillery tractors, armoured artillery tractors and APCs.

The MT-LB multipurpose tracked vehicle, fully described and illustrated in *Jane's Armour and Artillery 1995-96*, pages 307 to 309, is a member of the family.

Description

The fully enclosed cab is at the front of the hull and can seat eight personnel including the vehicle commander and the driver. The engine of the MT-L is in the rear part of the cab. Normal means of entry is through the doors in either side of the hull, but there is also a roof hatch.

The load area is at the back and is provided with drop tailgate, bows and a tarpaulin cover. The torsion bar suspension consists of six roadwheels with the drive sprocket at the front and the idler at the rear. The MT-L can be fitted with extra wide 565 mm tracks and an aggressive grouser to facilitate over-snow and soft terrain operations. This model is designated the MT-LV and has a ground pressure of 0.27 kg/cm². There are no track-return rollers. The roadwheels are similar to those used on members of the PT-76 light amphibious tank family and are also used on the MT-LB. A hydraulic shock-absorber is mounted at the first and last roadwheel stations.

The MT-L is fully amphibious without preparation, being propelled in the water by its tracks at a maximum speed of 6 km/h. A bilge pump is fitted as standard.

MT-L tracked amphibious over-snow vehicle

Specifications

Cab seating: 1 + 7 (up to 10 in rear)
Weight:
(empty) 8,500 kg
(loaded, without trailer) 13,000 kg
(loaded, with trailer) 11,000 kg
Max load:
(without trailer) 4,500 kg
(loaded, with trailer) 2,500 kg
Towed load: 7,000 kg
Length: 6.364 m
Width: 2.85 m
Height: 2.013 m
Ground clearance: 0.4 m
Track width: 350 mm
Length of track on ground: 3.7 m
Track: 2.5 m
Ground pressure: 0.428 kg/cm²
Max speed:
(road, without trailer) 61.5 km/h
(road, with trailer) 46.8 km/h
(water) 5-6 km/h
Range: 500 km
Max gradient:
(without trailer) 60%
(with trailer) 40%
Side slope: 40%
Engine: YaMZ-238V V-8 diesel developing 230 hp at 2,100 rpm
Transmission: manual with 6 forward and 1 reverse gears
Steering: clutch and brake
Suspension: torsion bar
Electrical system: 24 V

Status

In service with the RFAS.

Manufacturer

State factories.

VERIFIED

SWEDEN

Volvo Bandvagn Bv 202 tracked over-snow vehicle

Development

After the Second World War the Swedish Army bought a number of war surplus Studebaker Weasel over-snow vehicles, however, by the early 1950s they were becoming expensive to maintain. Between 1954 and 1956 the Swedish Ordnance Department evaluated a number of vehicles as possible replacements for the Weasel but none of the vehicles offered met its requirements. In 1956 a number of Swedish companies were approached and asked if they could design an over-snow vehicle, but none showed any interest in the project. In 1957 the Swedish Army started its own project and, following trials with test rigs, two prototypes were completed in 1958. These were followed by another batch of 10 vehicles in 1960. In 1960 the completed design was sent out to Swedish companies for competitive bidding. Late in 1961 Volvo was awarded a contract to prepare the design for production. First production vehicles were completed by BolinderMunktell in 1962-63.

By early 1980 approximately 5,000 Bv 202s had been delivered, with final deliveries made in 1981.

Description

The Bv 202 consists of two tracked units joined by a universal coupling. The front and rear units have identical subframes and the bodies are of all-welded construction. The front unit contains the engine and transmission at the front and the fully enclosed cab at the rear. The driver is seated on the left with the vehicle commander to his right. The cab is insulated from the engine compartment by aluminium sheets covering a 25 mm compressed layer of mineral wool (rockwool). The cab has a heater as a standard fitting. The rear unit, which can be fitted with a heater as an optional extra, is provided with a drop tailgate and a tarpaulin cover. A maximum load of 800 to 1,000 kg can be carried in the rear unit.

Power from the engine is transmitted to a main gearbox which is combined with a transfer box. Power is transmitted from the transfer box to the drive axles at the front end of both units via propeller shafts.

The suspension system consists of torsion bars with pivoting and bogie arms and rubber helper springs. The torsion bars are attached to the bottom of the front and rear of each units, for example two torsion bars per unit. This means that each torsion bar carries four wheels, two on each side. The fifth and rear wheel on each side serves as a track tensioning wheel and therefore has an individual rubber suspension system. The five roadwheels each side all have pneumatic tyres. The drive sprocket is at the front of each unit and there is a single track-return roller. The tracks consist of endless rubber bands reinforced with stainless steel wire. They are moulded in one piece and have embedded steel plates which not only provide a meshing surface for the teeth of the drive sprocket, but also serve as retainers for the track guide tongues which are made of light alloy.

The steering system is mounted in the front unit and is of the hydrostatic type. It consists of a hydraulic pump, steering valve, steering wheel and steering cylinder. If the pump pressure should fail for any reason the vehicle can still be steered. The steering cylinder is double-acting and actuates the steering joint in such a way that the two vehicle units are deflected in relation to each other. Apart from lateral steering deflection, the two vehicle units are able to move horizontally in relation to each other by up to 40°. They can also move up to 34° vertically and if necessary can be lifted or lowered in relation to each other by a maximum of 180 mm. The steering joint has two damping units with strong rubber springs, one on the front unit and the other on the rear.

The basic vehicle is fully amphibious, being propelled in the water by its tracks. Standard equipment includes: two bilge pumps; jack, tools; towing cable; axe; crowbar; spade; and two jerricans. Optional equipment includes: a heater for the rear unit; tropical kit; fully enclosed rear unit; cold starting device; torque converter; and stowage bin on the roof of the rear unit.

The final model was the Bv 202 Mk 2 which has a more powerful engine and a different transmission. The Bv 203 has a 24 V electrical system, a hard top for the rear unit and is radio-suppressed.

Specifications

Bv 202 Mk 1; data in square brackets relate to Bv 202 Mk 2 where different

Seating:
(front) 1 + 1
(rear) 8-10

Weight:
(empty) 2,900 kg
(loaded) 4,200 [4,400] kg

Max load:
(road) 1,000 [1,200] kg
(cross-country) 800 [1,000] kg

Load area: (rear) 2.3 × 1.56 m
Length: 6.172 [6.175] m
Width: 1.759 m
Height: 2.21 m
Ground clearance: 0.3 [0.28] m
Track: 1.118 m
Ground pressure: 0.085 [0.09] kg/cm^2

Max speed:
(road) 39 km/h
(water) 3.3 [3] km/h

Range: 400 km
Fuel capacity: 156 litres
Max gradient: 60%
Vertical obstacle: 0.5 m
Engine: Volvo B18 [Volvo B20B]
Engine type: 4-cylinder petrol
Engine hp: 91 [97] bhp at 5,300 rpm
Gearbox: manual with 4 forward and 1 reverse gears
Clutch: single dry disc
Transfer box: 2-speed
Steering: power-assisted
Turning radius: 6.8 m
Brakes: hydraulic
Electrical system: 24 V
Batteries: 72 Ah

Status

Production complete. In service with Finland (including Mk 2 version), Netherlands (Marines, 60), Norway and Turkey.

Manufacturer

Volvo BM, S-631 85 Eskilstuna, Sweden.

VERIFIED

Bv 202 tracked over-snow vehicle in typical operating environment

Bv 202 tracked over-snow vehicle showing climbing capabilities (Simon Dunstan)

UNITED STATES OF AMERICA

LMC 1500 over-snow vehicle

Description

The LMC 1500 over-snow vehicle replaced the earlier LMC 1450 in production and is available in three versions. One is a two-man cab version with an uncovered cargo area measuring 1.7 × 1.778 m and another a five-man cab with a reduced cargo area measuring 1.122 × 1.7 m. A fully enclosed full cab version with a covered cargo area, or seating for the driver and up to seven passengers is also available.

In all forms the LMC 1500 has a maximum load capacity of 1,088 kg which is reduced to 454 kg when crossing deep unpacked snow. Two widths of track are available, 711 mm for Summer use and 914 mm to enable the vehicle to cross rough terrain as well as snow. The tracks are three-ply rubber with heat-treated steel cleats and steel grousers can be fitted to increase traction over rough ground. The suspension each side is three or four rubber-tyred roadwheels with trailing arms and there are no track-return rollers. The drive sprocket is at the rear.

The driver and passenger are provided with bucket seats and seat belts. Passengers in the five-seat version are also provided with bucket seats but rear passengers on the eight-man version use padded folding benches. The driver and forward passenger are provided with doors and on the cab models a rear door is fitted. All windows use safety glass.

The vehicle is powered by an AMC 258 petrol engine allied to a Chrysler A727 automatic transmission and a torque converter. Steering is provided by a planetary steering differential using two steering levers in front of the driver's seat. The levers actuate a master cylinder which in turn actuates slave cylinders connected by linkage to each steering band.

Rollbars may be fitted if required.

Specifications

Crew: 2 + up to 6

Weight: (base vehicle)
(empty, Summer tracks) 1,950 kg
(empty, wide tracks) 2,086 kg

Max load: 1,088 kg

Length:
(four axles) 2.972 m
(three axles) 2.667 m

Width:
(cab) 1.71 m
(overall, Summer tracks) 2.13 m
(overall, wide tracks) 2.489 m

Height: 2.11 m
Ground clearance: (with differential guard) 0.23 m
Tread width: 1.422 m
Track width: 711 or 914 mm
Max speed: (hard surface) 32 km/h
Fuel capacity: 110 litres
Fording: 0.355 m
Gradient: (hard snow) 80%
Vertical obstacle: 0.254 m
Engine: AMC 258 6-cylinder liquid-cooled petrol developing 120 hp at 3,500 rpm
Transmission: Chrysler A727 automatic with 3 forward and 1 reverse speeds
Turning radius: 3.048 m
Electrical system: 12 V
Battery: 1 × 12 V, 70 Ah
Alternator: 66 A

Status
In production. In service with US Army (6), Navy (6) and Marine Corps (2).

Manufacturer
Logan Manufacturing Company, PO Box 407, Logan, Utah 84321, USA.
Tel: +1 801 530220.

VERIFIED

LMC 1500 over-snow vehicle with rollbars, as delivered to US Navy

LMC 1200 over-snow vehicle

Description
Although designed primarily for over-snow operation, the LMC 1200 can be used over other types of marginal terrain such as swamps. The basic vehicle has a two-man fully enclosed cab at the front of the vehicle and optional rollbars can be fitted to enable it to withstand twice its weight. Entry to the cab is by a door in each side that opens to the front. There is a combined heater and windscreen defroster unit and an optional second heater can be fitted into the cab. In all models seat belts are standard for each occupant.

Fitted with the standard two-man cab the LMC 1200 has 4.08 m² of uncovered cargo area behind the cab. With a five-man cab it has 2.6 m² of uncovered cargo area and with the fully enclosed 10-man cab, with the bench seats folded, it has 3.9 m².

The engine is in the forward part of the cab, which is insulated to reduce noise. Steering is by a planetary steering differential providing drive to both track systems. The driver has two levers which actuate the hydraulic master cylinder, which in turn actuates the slave cylinders connected by linkage to each steering band.

The suspension each side consists of four rubber-tyred roadwheels with the drive sprocket at the rear. There are no track-return rollers and the track grousers are of steel alloy. The standard track is 0.914 m wide but the LMC 1200 is also available with 1.152 m tracks.

Specifications
LMC 1200 with 914 mm wide tracks
Cab seating: 2 + up to 10
Weight: (empty) 2,803 kg
Max load: 1,361 kg
Normal load: 453-907 kg
Length: 4.114 m
Width: 2.526 m

LMC 1200 over-snow vehicle with 10-man fully enclosed body

Height: 2.108 m
Ground clearance: 0.254 m
Track width: 914 mm
Ground pressure: 0.07 kg/cm²
Max speed: (with outboard reduction gearing) 27 km/h
Fuel capacity: 159 litres
Gradient:
(dirt) 80%
(snow) 60%
Engine: Ford 6-cylinder petrol developing 124 hp
Gearbox: Ford C-6 automatic, 3 forward and 1 reverse gears
Differential steering: planetary controlled
Turning radius: 4.572 m
Suspension: trailing arms in rubber
Brakes: (parking) double caliper disc
Electrical system: 12 V

Status
Production complete. In service with the US Army (41), Navy (25), Air Force (19), Marine Corps (7) and Coast Guard (13).

Manufacturer
Logan Manufactur ng Company, PO Box 407, Logan, Utah 84321, USA.
Tel: +1 801 753 0220.

VERIFIED

TRAILERS

Within each country, trailers appear in the following order:
general cargo trailers
tank-carrying trailers and flatbed type trailers used for carrying engineer plant and armoured vehicles
tanker trailers
specialised trailers including those used for carrying missiles, pontoons and radars

AUSTRALIA

Trailer, cargo, 1 tonne, GS 2-wheel

Description

The Australian Army 1 tonne cargo trailer was developed concurrently with the Trailer, Chassis, Light MC2 (see following entry), to replace existing equipment in service. It is used for the carriage of general cargo, packaged POL, ammunition or missiles and stores. The box-body, with a hinged tailgate, is manufactured from aluminium and is demountable from the trailer chassis.

Specifications

Weight:
(loaded) 1,510 kg
(empty) 510 kg
Max load: 1,000 kg
Length:
(overall) 3.33 m
(body, internal) 2.18 m
Width:
(overall) 2.06 m
(body, internal) 1.95 m
Height:
(overall, loaded) 1.2 m
(overall, empty) 1.6 m
(body, internal) 0.43 m
(lunette, loaded) 0.76 m
(lunette, empty) 0.87 m
(floor, loaded) 0.72 m
(floor, empty) 0.83 m
Track: 2 m
Suspension: semi-elliptic leaf springs and rubber bump stops
Brakes: hydraulic override
Tyres: 7.50 × 16
Electrical system: 12 V

Status

In service with the Australian Army.

Agency

Department of Defence (Army Office), Engineering Development Establishment, Private Bag 12, Ascot Vale PO, Victoria, Australia 3032.
Tel: +61 3 319 5511. Telex: Army AA30943.
Fax: +61 3 318 3421.

VERIFIED

Trailer, cargo, 1 tonne, GS 2-wheel

Trailer, cargo, 1 tonne, GS 2-wheel

Trailer, chassis, light MC2 (1 tonne)

Description

The Australian Army 1 tonne trailer chassis was developed to meet a requirement for the deployment of 10, 15 and 30 kVA generator sets in the field. Initially two variants were considered, a powered version and an unpowered version. After pilot model trials held in 1972-73, further development of the powered version was discontinued. Procurement detailing for the unpowered trailer began in 1976 and actual procurement was recommended in October 1980. Quantity production commenced in November 1980.

This trailer chassis is used for the carriage of twin-mounted 10 or 15 kVA generators, a single-mounted 30 kVA generator or a hydrogen generator. It complies with all NATO standards and may be towed behind all wheeled vehicles from 1,500 to 8,000 kg capacity. The braking system is a mechanical/electrical override hydraulic system modified to operate with either 12 or 24 V vehicle electrical systems. The trailer meets the requirements of Specification Army (Aust) 5798.

Specifications

Weight:
(loaded) 1,500 kg
(empty) 500 kg
Length: 3.33 m
Width: 2.06 m
Height:
(loaded) 0.72 m
(empty) 0.83 m
Track: 2 m
Tyres: 7.50 × 16
Electrical system: 12 V or 24 V

Status

In service with the Australian Army.

Agency

Department of Defence (Army Office), Engineering Development Establishment, Private Bag 12, Ascot Vale PO, Victoria, Australia 3032.
Tel: +61 3 319 5511. Telex: Army AA30943.
Fax: +61 3 318 3421.

VERIFIED

International Harvester semi-trailers

Description

In May 1987, it was announced that the Australian Army's existing semi-trailer and prime mover fleet would be replaced by a new fleet of 88 International Harvester SF2670 prime movers (fitted with Cummins NTC 365 engines) and 115 semi-trailers, plus support equipment. The contract was worth A$18 million, with over 70 per cent of the total spent in Australia.

The semi-trailers are used for carrying engineer plant, armoured vehicles, cargo and bulk water. They were produced by two subcontractors to International Harvester: Fruehauf at Somerton, Victoria and Walsh at Toowoomba, Queensland. The Walsh semi-trailers are used to carry tanks and large equipment and are capable of being closed up hydraulically to reduce road width when unloaded. Fruehauf produced the cargo and water tanker semi-trailers and converter dollies.

Status

In service with the Australian Army.

Main Contractor

International Harvester, Dandenong, Victoria, Australia.

VERIFIED

AUSTRIA

EMPL 3-axle 34 ton low bed semi-trailer

Description

The EMPL 3-axle 34 ton low bed tank transporter semi-trailer, has a main frame constructed of two I beams with heavy-duty cross members and with a goose neck construction for maximum strength and durability. Two loading ramps are located at the rear. The loading deck platform is covered with 8/10 mm chequered steel plates over cross members.

There are three 10,000 kg capacity heavy-duty axles provided, with two tyres each side forming a total axle capacity of 30,000 kg; two spare wheels are carried. The third axle is steerable with an additional electric-hydraulic steering system actuated by the driver from the cab at speeds of less than 20 km/h. A triple arrangement suspension is provided using three leaf springs. Dual circuit airbrakes are provided on all wheels.

Each semi-trailer is provided with a comprehensive set of accessories and tools.

EMPL 3-axle 34 ton low bed semi-trailer carrying two Steyr 4K 7FA G 127 APCs **1997**

Specifications

Weight: (empty) approx 12,000 kg
Max load: approx 34,000 kg
GCW: 46,000 kg
5th wheel load: approx 16,000 kg
Axle capacity: 30,000 kg
Length:
(overall) approx 12.2 m
(platform) approx 6.7 m
Width: (platform) 3.2 m
Height: (platform, loaded) approx 0.93 m
Brakes: dual circuit, air
Tyres: 235/75 (R) 17.5
Electrical system: 24 V

Status

In service with the Austrian Army plus Ethiopia, Germany, Ghana, Jordan, Poland and Turkey.

Manufacturer

EMPL Fahrzeugwerke GesmbH, PO Box 41, A-6272 Kaltenbach, Austria.
Tel: +43 5282 2248-0. Fax: +43 5283 224826.

NEW ENTRY

CHINA, PEOPLE'S REPUBLIC

Hanyang HY 962 50,000 kg semi-trailer

Description

This semi-trailer was designed for use with the Hanyang HY 473 tractor truck and has a maximum capacity of 50,000 kg. The semi-trailer has three axles with a total of 12 tyres. Two spare tyres are carried on supports on the top of the gooseneck. The two stabiliser legs at the forward end of the loading platform can be lowered to the ground for support when the tractor truck is removed. The two loading ramps are spring-loaded to assist in raising and lowering. The 88.9 mm towing pin is a standard component and has a loading capacity of 19,500 kg.

The Hanyang HY 960 is a simplified version of the HY 962 that may not always be fitted with spring-loaded loading ramps. It retains the 50,000 kg loading capacity but is 10.655 m long, 3.2 m wide and 1.994 m high. Its weight when empty is 11,500 kg. The normal tractor truck is the HY 417 F10.

HY 962 50,000 kg semi-trailer (Christopher F Foss)

Specifications

HY 962
Weight: (empty) 12,430 kg
Max payload: 50,000 kg
Axle load: 42,930 kg
Length:
(overall) 10.945 m
(platform) 6.965 m
Width: 3.22 m
Height:
(raised ramps) 2.92 m
(coupling face) 1.45 m
Wheelbase: 1.19 m + 1.19 m
Track: 2.358 m
Tyres: 10.00 × 20

Status

In service with the People's Liberation Army.

Manufacturer

Hanyang Special Auto Works, Hanyang Bridge Jianghan Road, Wuhan, Hubei Province, People's Republic of China.

VERIFIED

FINLAND

Vammas BVK 191W all-terrain general purpose trailer

Description

The Vammas BVK 191W all-terrain general purpose trailer was designed for use with the Swedish Bv 206 and Sisu NA-140 all-terrain vehicles (refer to entries in *the Tracked Prime Movers, Cargo Carriers and Armoured Logistic Vehicles* section for information on these vehicles). The 1,000 kg payload of the BVK 191W increases the potential payload of both vehicles by 50 per cent but the trailer does not affect the mobility of the tractor vehicle, to the point of being amphibious.

A preliminary series of BVK 191 trailers was trialled by the Finnish Defence Forces beginning in March 1987. Following minor changes to the design, series production of the resultant BVK 191W commenced during 1989.

The BVK 191W has a light but robust construction, having a steel frame with the body and deck made of aluminium. The trailer can be towed across country, marsh, sand, snow and in water as well as on roads. It can float when crossing water obstacles, even when fully loaded. The trailer is equipped with both wheels and skis, using an articulated suspension which allows the trailer to remain as level as possible in the most uneven terrain. When using the wide skis and when fully loaded, the ground pressure is as low as that of the tractor vehicle. A special filling is used for the four wheels to reduce damage from rough ground or tree stumps. The cargo area is watertight and can be covered with a tarpaulin. Cargo lashing points, a 76 mm towing eye and a tailgate are fitted.

Specifications

Weight: (empty) 725 kg
Payload: 1,000 kg
Length: 4.6 m
Width: 1.89 m
Height: 1.4 m
Cargo area dimensions: 2.8 × 1.7 × 0.7 m
Ground clearance:
(wheels) 0.6 m
(skis) 0.44 m
Tyres: 5.00 × 8/10 PR

Vammas BVK 191W all-terrain general purpose trailer

Vammas BVK 191W all-terrain general purpose trailer

Status
In production for the Finnish Defence Forces.

Manufacturer
Vammas Limited, PO Box 18, FIN-38201 Vammala, Finland.
Tel: +358 932 1971. Fax: +358 932 41148.

VERIFIED

FRANCE

LOHR RMTL 1.5 multipurpose trailer

Description
The LOHR RMTL 1.5 (*Remorque Militaire de Transport Logistique*) is a modular design intended to have a GVW of up to 3.1 tonnes, a payload of 1.5 tonnes and a cross-country capability. It is also intended to be transportable by aircraft or helicopter.

The RMTL 1.5 has a multileaf spring suspension (a pneumatic suspension is optional) with hydraulic shock-absorbers. A single rigid axle is provided with an air drum braking system (two or three lines) and a spare wheel is carried over the drawbar. Other equipment includes: a folding drawbar wheel, a tool box, two rear folding legs and a 24 V NATO lighting system.

The RMTL 1.5 can carry a series of modular bodies including standard cargo bodies (with optional covers), fuel tanker bodies, generators and other types. Optional equipment includes: two side boxes, a platform with four ISO corners, a tilt kit with a tarpaulin, a 1,800 litre water tank and other equipment to meet customer specifications.

Specifications
GVW: up to 3,100 kg
Payload: 1,500 kg
Length: 3.9 m
Width: 2.5 m
Height: (platform) 1 m
Ground clearance: 0.4 m
Load area: 2.1 × 2.6 m
Tyres: 9.00 R 16 XL tubeless
Departure angle: 45°
Electrical system: 24 V

Status
In service with the French Army.

Manufacturer
LOHR Industrie, F-67980 Hangenbieten, France.
Tel: +33 3 88 38 98 00. Telex: 870082 f.
Fax: +33 3 88 96 06 36.

UPDATED

Base chassis for LOHR RMTL 1.5 multipurpose trailer **1997**

LOHR RMTL 1.5 multipurpose trailer carrying a generator **1997**

Trailor 1,100 kg cross-country cargo trailer

Description
The Trailor 1,100 kg cross-country cargo trailer is sometimes referred to as the Africa-type trailer and is intended for off-road use in remote areas. The frame of the trailer is constructed from two channel-shaped variable-section high-tensile steel beams on which a steel-bodied cargo carrier is placed, supported on one-piece cross members. The body has steel sides 0.46 m high that can be swung downwards for loading or may be removed altogether. The floor is also steel. To extend the height of the sides, expanded metal extensions can be fitted to the sides and a tarpaulin cover may be fitted on bows. The axle is a 90 mm square beam and EU-approved dual airbrakes are fitted. Each trailer is equipped with a wheel nut wrench, a spare wheel carrier on the front panel, a tool box, a drinking water tank, a jerrican carrier and a rear stabilising strut. The trailer can be fitted with a fuel bowser or generating plant.

Specifications
Weight:
(empty) 1,400 kg
(loaded) 2,500 kg
Max load: 1,100 kg
Length: 4.05 m
Width: 2.2 m
Height:
(tarpaulin cover) 2.4 m
(load area) 1 m
(trail eye) 0.765 m
Load area: 2.71 × 2.06 m
Track: 1.953 m
Suspension: Trailor single-axle type SMU 7 SE with rubber-bushed radius rods, leaf springs and shock-absorbers
Brakes: dual circuit air
Tyres: 8.25 × 20 XL
Electrical system: 24 V

Status
In production. In service with several armed forces.

Manufacturer
Trailor SA, 5 route Nationale 10, Coigniegres, BP 49, F-78311 Maurepas Cedex, France.
Tel: +33 1 30 50 61 26. Telex: 698896 f.
Fax: +33 1 30 50 31 32.

VERIFIED

Trailor 1,100 kg cross-country cargo trailer

ACMAT RM 215S 1,500 kg cargo trailer

Description
The ACMAT RM 215S series of trailers was designed for maximum compatibility with the ACMAT range of vehicles. Each trailer has an integral 200 litre water tank and the standard specification includes eight jerricans and two sections of PSP-type sand channel. The RM 215S trailer is provided with canopy rails.

Specifications
Weight:
(loaded) 2,700 kg
(empty) 1,200 kg
Payload: 1,500 kg
Length: (overall) 4.125 m
Width: 2.07 m
Height:
(top of canopy rails) 2.4 m
(load area) 1 m
Lunette height: 0.765 m
Internal clearance: (inside canopy) 1.4 m
Track: 1.74 m
Load area: 2.69 × 1.94 m
Suspension: semi-elliptical springs
Tyres: 12.50 × 20

Status
In production.

Manufacturer
ACMAT, Ateliers de Construction Mécanique de l'Atlantique, Le Point du Jour, F-44600 Saint-Nazaire, France.
Tel: +33 2 40 22 33 71. Telex: 700913 f.
Fax: +33 2 40 66 30 96.

VERIFIED

ACMAT RM 215S 1,500 kg cargo trailer

LOHR RM16 16,000 kg trailer

Description
This two-axle trailer, known as the *Remorque Grande Capacité de 16 T PTC*, was designed to meet the requirements of the logistic transport squadrons of the French Army. The trailer has drop sides and a drop tailgate which can be quickly removed, bows that can be adjusted to any one of three heights and a quick *bâche* system for the rapid uncovering of the load.

Specifications
Weight:
(empty) 4,280 kg
(loaded) 16,000 kg
Max load: 11,720 kg
Length of platform: 6.3 m
Width of platform: 2.38 m
Tyres: 9.00 × 20

LOHR 16,000 kg trailer complete with bows and cover

Status
In production. In service with the French Army.

LOHR 16,000 kg trailer with sides and tailgate lowered

Manufacturer
LOHR Industrie, F-67980 Hangenbieten, France.
Tel: +33 3 88 38 98 00. Telex: 870082 f.
Fax: +33 3 88 96 06 36.

VERIFIED

Trailor 17,500 kg semi-trailer

Description

This semi-trailer was designed to meet the requirements of the French Army and also meets current TIR requirements. The cargo area has eight drop sides (four each side), doors at the rear, bows and a tarpaulin cover. The floor is of wood and has retractable rings for securing cargo. The trailer is fitted with dual airbrakes, one line acting on each of the rear axles, and a mechanical parking brake.

Specifications

Weight:
(empty) 5,500 kg
(loaded) 23,000 kg
Max load: 17,500 kg
Load area: 11.31 × 2.5 m

Status

In service with the French Army.

Manufacturer

Trailor SA, 5 route Nationale 10, Coigniegres, BP 49, F-78311 Maurepas Cedex, France.
Tel: +33 1 30 50 61 26. Telex: 698896 f.
Fax: +33 1 30 50 31 32.

VERIFIED

Decauville SRPC 36/46 tonne tank transporter semi-trailer

Description

This semi-trailer was designed for transporting MBTs such as the AMX-30, Leopard 1, T-54/T-55, T-62 and similar tanks. Its chassis consists of two longitudinal girders connected by crossbars and gussets, with a serrated steel sheet covering. Two loading ramps are provided at the rear of the trailer.

The suspension consists of four axles on oscillating balance bars. Airbrakes are fitted on all wheels and a mechanical parking brake is provided. Shoe-type stabilisers are provided at the front of the trailer.

Standard equipment includes stowage boxes, front roller, adjustable blocks and a jack. Optional equipment includes hydraulically operated rear-loading ramps, rear rollers, shrouded pulley wheels, drinking water tank, jerrican support racks and a camouflage net.

Decauville SRPC 36/46 tonne tank transporter semi-trailer carrying AMX-30 MBT

Specifications

Weight:
(empty) 12,900 kg
(loaded) 58,900 kg
Max load: 46,000 kg
Length:
(overall) 12.05 m
(of platform) 7.35 m
Width:
(overall) 3.65 m
(of loading ramps) 0.62 m
Height: (over loading ramps in vertical position) 2.6 m
Ground clearance:
(under axles) 0.35 m
(under front of semi-trailer) 1.65 m
Wheelbase: (from king-pin to centre of rear bogie) 9 m
Max towing speed: 60 km/h
Tyres: 9.00 × 20 or 12.00 × 20

Status

Has been exported to a number of countries in the Middle East.

Manufacturer

Decauville SA, BP 38, F-91102 Corbeil-Essonnes, France.

VERIFIED

Fruehauf medium tank transporter semi-trailer

Description

This semi-trailer was designed for the long distance transport of tanks weighing up to 45,000 kg. The trailer has manually operated 0.6 m wide rear-loading ramps, removable track guides and heavy-duty manually operated stabilisers mounted towards the front. It can be towed on roads at a maximum speed of 80 km/h.

Specifications

Max load: 45,000 kg
Length:
(overall) 11.75 m
(of loading platform) 7.92 m
Width: 3.1 m
King-pin size: 89 mm
Tyres: 8.25 × 15

Status

In production. In service with at least one army outside France.

Manufacturer

Fruehauf France, 2 avenue de l'Aunette, F-91130 Ris-Orangis, France.
Tel: +33 4 69 43 30 00. Telex: 601381 f.
Fax: +33 4 69 43 33 66.

VERIFIED

Fruehauf medium tank transporter semi-trailer

Fruehauf medium tank transporter semi-trailer

Fruehauf heavy tank transporter semi-trailer

Description

This semi-trailer was designed for carrying tanks weighing up to 60,000 kg. The trailer has manually operated 0.6 m wide rear-loading ramps, removable track guides and heavy-duty manually operated stabilisers mounted towards the front of the trailer.

Fruehauf heavy tank transporter semi-trailer

Specifications
Max load: 60,000 kg
Length:
(overall) 12.5 m
(of loading platform) 7 m
Width: 3.6 m
King-pin size: 89 mm
Tyres: 12.00 × 20

Status
In production. In service with at least one country outside France.

Manufacturer
Fruehauf France, 2 avenue de l'Aunette, F-91130 Ris-Orangis, France.
Tel: +33 4 69 43 30 00. Telex: 601381 f.
Fax: +33 4 69 43 33 66.

VERIFIED

LOHR SMC tank transporter semi-trailers

Description
Three semi-trailers in the LOHR SMC range are the SMC 60 DT (DT - desert), the SMC 60 RD (RD - *route et desert)* and the SMC 40 PL. The SMC 60 DT has a maximum load of 60,000 kg and uses a hydraulically operated rear ramp. The SMC 60 RD also has a maximum load of 60,000 kg and uses manually lifted ramps. Other differences between the two are the types of tyres fitted and the chassis construction. The SMC 40 PL has a maximum load of 44,000 kg and uses manually lifted ramps.

On all three models the chassis and deck are made of all-welded steel and the suspension consists of twin unsprung walking beams on which are mounted the eight tyres. Winch cable guides and rollers are fitted and there are also guide blocks for the tracks of the vehicles being loaded. Spare wheels are carried under the loading deck.

Specifications

Model	**SMC 60 DT**	**SMC 60 RD**	**SMC 40 PL**
Weight (loaded)	78,500 kg	77,000 kg	60,000 kg
Max load	60,000 kg	60,000 kg	44,000 kg
Length	13.05 m	11.55 m	12.93 m
Length of load on deck	7.215 m	6.6 m	7.76 m
Length of overhang at rear	1.9 m	1.75 m	1.52 m
Height of load deck	1.54 m	1.38 m	1.35 m
Width	4.13 m	3.4 m	3.3 m
Tyres	24.00 × 20.5 sand	18.00 × 22.5 sand	13.00 × 20 P
Electrical system	24 V	24 V	24 V

A special version of the SMC semi-trailers, the SMC 60 EM 5 LT, is described in the following entry.

A cargo version of the SMC 60 DT, which remains capable of carrying an AMX-30 MBT, is available.

Status
In production. SMC 60 DT exported to the Middle East and Africa. SMC 60 RD in service with the French Army and several other countries.

Manufacturer
LOHR Industrie, F-67980 Hangenbieten, France.
Tel: +33 3 88 38 98 00. Telex: 870082 f.
Fax: +33 3 88 96 06 36.

VERIFIED

LOHR SMC 60 RD tank transporter semi-trailer loaded with 155 mm GCT self-propelled gun; the towing vehicle is a Renault R400 (6 × 6) tractor truck

LOHR SMC 60 series tank transporter semi-trailer

LOHR SMC 60 EM 5 LT tank transporter semi-trailer

Description
The LOHR SMC 60 EM 5 LT tank transporter semi-trailer was designed to act as a road and rough-terrain transporter semi-trailer for the latest generation of MBTs. It has been tested in the USA as a possible carrier for the M1 Abrams series, a similar one has been tested by the French Army for the Leclerc.

The chassis consists of a steel heavy-duty low bed with two longitudinal girders braced by cross members. The all-steel platform is welded to the chassis and the rear part is inclined with square bars for track grips. There are perforations for track guides and blocks, with a roller and pulley located at the rear for the winch cable. There are two two-piece loading ramps provided at the rear, with spring assistance for handling. The running gear is based on 10 pendular hydraulic axles arranged in five rows, four steerable and one fixed (the forward row). Dual circuit airbrakes are provided. Two support legs are provided at the base of the gooseneck. A standard 3.5 in/89 mm king-pin coupling is used.

Standard equipment provided with the semi-trailer includes four track guides and blocks, three lashing rings each side plus two lashing eyes on the gooseneck, a wheel wrench and a 25 tonne capacity hydraulic jack. Optional equipment includes: four adjustable chains for the load, three stowage boxes each side of the platform plus another on the gooseneck, a 300 litre water tank, a 300 litre fuel tank in the gooseneck and heavy-duty hydraulic legs on each side. Further optional equipment includes: hydraulic loading ramps, an emergency pump, a hydraulic loading/unloading mechanism for a spare wheel and blackout electrical equipment.

Specifications
Weight: (kerb) 20,000 kg
Max load: 65,000 kg
Length:
(overall, ramps raised) 15.8 m
(platform, useful) 10.08 m
Width:
(overall) 3.6 m
(ramps) 0.96 m
Height: (platform) 1.17 m
Loading angle: 18°
Tyres: 215/75 R 17 5XZA
Electrical system: 24 V

Status
In production. Tested by US Army. A similar trailer has been tested by the French Army for the Leclerc MBT.

Manufacturer
LOHR Industrie, F-67980 Hangenbieten, France.
Tel: +33 3 88 38 98 00. Telex: 870082 f.
Fax: +33 3 88 96 06 36.

UPDATED

LOHR SMC 60 EM 5 LT tank transporter semi-trailer on tow by Oshkosh M983 (8 × 8) tractor truck

LOHR SMC tank transporter semi-trailers

Description

LOHR Industries has developed a new range of SMC tank transporter semi-trailers. Basic data for the four models announced to date is provided in the Specifications table.

The SMC 65-3 semi-trailer is equipped with 395/85 R 20 XZL tyres.

The SMC 70-2+1 is equipped with 24 × 20.5 ribbed tyres. Of the five axles, four are transversal and longitudinal oscillating units with one central rigid axle.

The SMC 70-30+1 has 14.00 R 20 XZL tyres, three rigid front axles and one self-turning axle at the rear.

The SMC 70-5.2 has 24 × 20.5 XS tyres, three rigid axles and two steering axles driven from the fifth wheel.

Specifications

Model	SMC 65-3	SMC 70-2+1	SMC 70-30+1	SMC 70-5.2
Max load	65,000 kg	70,000 kg	70,000 kg	70,000 kg
Length	16.38 m	16.38 m	16.38 m	16.38 m
Length of platform	11 m	11 m	11 m	11 m
Width	3.6 m	4 m	3.6 m	3.6 m
Height	1.36 m	1.52 m	1.36 m	1.52 m
Tyres	12	10	14	10
Axles	3	5	4	5

Status

Available.

Manufacturer

LOHR Industrie, F-67980 Hangenbieten, France.
Tel: +33 3 88 38 98 00. Telex: 870082 f.
Fax: +33 3 88 96 06 36.

VERIFIED

LOHR SRPB 60 tank transporter semi-trailer

Description

The LOHR SRPB 60 (Semi-Remorque Porte Blindés 60 tonnes) tank transporter semi-trailer is described by the manufacturer as a combined tank transporter and recovery system. The semi-trailer has a maximum on-road payload of 61,500 kg and was designed, according to French Army requirements, for on-road transport of the Leclerc MBT and other heavy vehicles over long distances at speeds up to 85 km/h.

The semi-trailer has six axles. The front three axles are fixed while the rear three are steerable being automatically controlled by the tractor-trailer angle. For the suspension, on each axle two hydraulic cylinders are fitted with accumulators with, on each side, a hydraulic circuit connecting the six cylinders to equalise the load. A central lubrication system is provided. Dual-circuit airbrakes meet EC regulations. In all there are 24 roadwheels and two spares.

The two loading ramps are hydraulically operated for one-man operation. Standard equipment includes pulleys and cable guide rollers for loading and unloading. A diesel power pack is mounted on the goose neck to provide power for all the hydraulic operations on the semi-trailer.

Specifications

Weights: (loaded/unloaded)
(total) 79,900/19,900 kg
(on axles) 61,900/15,900 kg
(on 5th wheel) 18,000/4,000 kg
Max load: 65,000 kg
Length:
(overall) 15.22 m
(useful) 10.94 m
Width: (overall) 3.35 m
Height:
(overall, empty, ramps lifted) 3.26 m
(coupling, loaded) 1.55 m
(upper deck, loaded) 1.1 m
Access ramp slope: 30%
Access ramp width: 0.8 m
Access ramp track: 2.42 m
Wheelbase: 9.81 m
Ground clearance: (loaded) 0.45 m
Rear overhang: 1.39 m
Tyres: 265/70 R 19.5 143/141 J
Electrical system: 24 V

Status

Available. In service with the French Army.

Manufacturer

LOHR Industrie, F-67980 Hangenbieten, France.
Tel: +33 3 88 38 98 00. Telex: 870082 f.
Fax: +33 3 88 96 06 36.

NEW ENTRY

LOHR SRPB 60 tank transporter semi-trailer **1997**

LOHR SRPB 60 tank transporter semi-trailer carrying Leclerc MBT **1997**

Nicolas tank transporter semi-trailers

Description

Nicolas Industrie SA manufactures a wide range of semi-trailers for military and commercial use. It has recently expanded its military range by the introduction of a wide range of tank transporter trailers. The range includes 14 separate models varying in payload from 16,500 up to 95,000 kg. The various semi-trailers differ not only in size but also in the number of short axles used, the tyre sizes and the number of tanks that can be carried. Two of the smaller models (the STA C 30.15 and the STA L 30.20) can each carry two light tanks of the AMX-13 type, while the STA 100 can carry two AMX-30 tanks using eight axle units and 32 tyres. The full range of models and their main characteristics can be seen on the Model table.

Renault TRH 350 (6 × 4) tractor towing Nicolas STA 43 semi-trailer

A typical example from the range is the STA S 60.20.5. This model can be used to carry most models of MBT in service and has been mainly constructed from high-tensile steel. The loading deck has two central main beams and two side rails connected by welded beams of 'I' section high-tensile steel. The steel decking is 2 mm thick, apart from the main channels which are 2 to 8 mm thick. Panels over the twin axles can be removed for access. A 250 mm adjustment for width is possible on the loading ramps, which fold upwards for travelling, and an optional hydraulic lift mechanism is available. The axles have a 10° oscillation and airbrakes are used. One spare wheel is provided with an optional rack for another and a small crane is provided for handling the spare wheel. Optional equipment includes: a tool box, cable guides on the gooseneck, lashing chains, blocks, track guides, blackout lighting and fuel and water tanks.

Variants

Nicolas Industrie SA produces a tank transporter semi-trailer with powered rear axles for use with its Tractomas tractor trucks. By varying the type of Tractomas tractor and the type of powered semi-trailer, loads of 40,000, 50,000 or 60,000 kg can be carried. The hydraulically powered axles provide extra traction and power when travelling with loads across country and across soft terrain such as sand. All types can travel at 7 km/h with the power-assist facility selected. All the semi-trailers use 24 × 20.5 tyres and the number of axles used may be two or three per side.

The hydraulic system used to power the wheels of the semi-trailer is based on the use of a three-part motor produced by Poclain. One part is the cylinder block which supports the hydraulic cylinders and the associated rollers. The rollers controlled by the hydraulic cylinders press on cams on the second part which is the motor casing fastened onto the hub.

List of models

Model	Weight* (empty)	Length*	Height (platform)	Payload*	Load* (5th wheel)	Load* (rear axle)	Axles (pairs)	Tyres (number)
STA 18.15	6,000 kg	9.815 m	0.89 m	17,500 kg	7,800 kg	15,700 kg	2	8.25 × 15 (8)
STA 18.20	6,300 kg	9.815 m	1.1 m	17,500 kg	7,800 kg	16,000 kg	2	9.00 × 20 (8)
STA D 30.15	12,000 kg	14.515 m	1.17 m	30,000 kg	15,000 kg	30,000 kg	3	10.00 × 15 (12)
STA L 30.20	10,500 kg	14.615 m	1.29 m	33,000 kg	15,500 kg	28,000 kg	2	12.00 × 20 (8)
STA 45.15	10,600 kg	10.815 m	0.998 m	45,000 kg	19,600 kg	36,000 kg	2	8.25 × 15 (16)
STA 40.20	10,800 kg	10.815 m	1.29 m	38,000 kg	15,800 kg	33,000 kg	2	12.00 × 20 (8)
STA 3 35.15	10,400 kg	11.115 m	0.83 m	40,400 kg	14,000 kg	31,500 kg	3	8.25 × 15 (12)
STA 3 45.22.5	11,200 kg	11.115 m	1.27 m	43,000 kg	17,200 kg	37,000 kg	3	18 × 22.5 (6)
STA 40.24	11,600 kg	11.915 m	1.38 m	40,000 kg	19,600 kg	32,000 kg	2	12.00 × 24 (8)
STA 50.20	12,600 kg	11.915 m	1.39 m	50,600 kg	21,600 kg	42,000 kg	3	12.00 × 20 (12)
STA 3 45.15	11,500 kg	11.115 m	1.17 m	44,500 kg	17,000 kg	39,000 kg	3	10.00 × 15 (12)
STA S 60.20.5	17,000 kg	12.62 m	1.6 m	60,000 kg	21,000 kg	56,000 kg	2	20.5 × 24 (8)
STA 60.20	16,500 kg	11.5 m	1.36 m	58,000 kg	22,500 kg	52,000 kg	2	12.00 × 20 (16)
STA 100	20,000 kg	18.215 m	1.15 m	95,000 kg	27,000 kg	88,000 kg	4	8.25 × 15 (32)

* approximate figures

Pressure on the cam makes the hub and wheel turn. When not in use the rollers are retracted and held away from the cams by springs. The third section of the motor is the timing section. The entire motor can be removed without removal of the wheel or any part of the axle. Power for the motor comes from a multicylinder pump on a declutchable power take-off on the gearbox. Each cylinder of the pump feeds one axle motor. The system is operated from a remote-control box carried in the vehicle cab.

Specifications
STA S 60.20.5 only
Weight:
(empty) 17,000 kg
(on 5th wheel) 21,000 kg
(on axles) 56,000 kg
Max load: 60,000 kg
Max gross weight: 77,000 kg
Length:
(ramp folded) 13.1 m
(loading deck) 7.2 m
(loading ramp) 2.875 m
Width:
(overall) 4.195 m
(gooseneck) 2.5 m
(loading ramps) 0.8 m
Height:
(loading deck) 1.6 m
(fifth wheel) 1.8 m
Wheelbase: (5th wheel to centre of axle unit) 9.12 m
Tyres: 20.5 × 24
Electrical system: 24 V

Status
The company is currently building several hundred STA 45.15 semi-trailers for the French Army.

Manufacturer
Nicolas Industrie SA, PO Box 3, F-89290 Champs-sur-Yonne, France.
Tel: +33 3 86 53 30 09. Telex: 800938 f.
Fax: +33 3 86 53 79 74.

UPDATED

Trailor tank transporter semi-trailers

Description
There are three Trailor tank transporters and all are similar, differing only in load capacities and tyre sizes. All are of conventional two-axle configuration, with all-steel frames and decks. Three-line airbraking and 89 mm king-pins are standard and each trailer also has track guides, chocks and winch cable rollers. The rear-loading ramps are spring-balanced and mechanically operated.

Specifications

Model	**54T**	**62T**	**74T**
Length	12.96 m	11.95 m	12.97 m
Width	3.1 m	—	—
Deck length	6.6 m	5.18 m	6.15 m
Capacity	38,000 kg	45,000 kg	55,000 kg
Tyres	7.50 × 15	E20 size	12.00 × 20

Status
In production.

Manufacturer
Trailor SA, 5 route Nationale 10, Coigniegres, BP 49, F-78311 Maurepas Cedex, France.
Tel: +33 1 30 50 61 26. Telex: 698896 f.
Fax: +33 1 30 50 31 32.

VERIFIED

Trailor 54T tank transporter semi-trailer

Trailor 62T tank transporter semi-trailer

Fruehauf flatbed semi-trailers

Description
Fruehauf France flatbed semi-trailers are produced with varying load-carrying capacities for the transport of general cargoes over long distances. Versions with capacities of between 20 and 60 tonnes are available. Tilt-bodies, cargo vans and container carriers with associated locking and other systems may be adapted to these semi-trailers.

The data provided in the specifications table are for a typical 40 tonne capacity flatbed semi-trailer producing a GVW of 50 tonnes.

Fruehauf 40 tonne flatbed semi-trailer

Specifications
Max load: 40,000 kg
Length: 12.2 m
Width: 2.5 m
King-pin: 2 or 3.5 in
Tyres: 12.00 × 20

Status
In service with at least one army outside France.

Manufacturer
Fruehauf France, 2 avenue de l'Aunette, F-91130 Ris-Orangis, France.
Tel: +33 4 69 43 30 00. Telex: 601381 f.
Fax: +33 4 69 43 33 66.

VERIFIED

ACMAT ALM Saharian trailer Model RM 215 SC 1,250 litre tanker

Description
This trailer may be used to carry water, petrol or oil in a 1,250 litre steel-plated tank. The tank, together with its associated stowage boxes, is carried in a 2 mm thick steel body on the standard ACMAT ALM Saharian trailer chassis. A Japy manual pump is fitted as standard but for the petrol version a reel with piping and connection pipes are carried. For the water tanker version a filter system is supplied. A 4 litre fire extinguisher is also carried in the petrol tanker version.

Standard fittings carried on all types of trailer include: one spare wheel, stowage for eight jerricans, a tool set (in two boxes) and sand channels. Oil and airbrakes are standard and a mechanical handbrake is also fitted for parking. A jockey wheel with a diameter of 0.4 m is provided.

ACMAT ALM Saharian trailer Model RM 215 SC

Specifications
Weight:
(loaded) 2,850 kg
(empty) 1,600 kg
Max load: 1,250 kg
Capacity of tank: 1,250 litres
Length: 4.125 m
Width: 2.07 m
Height:
(overall) 1.915 m
(lunette) 0.765 m
Track: 1.71 m
Tyres: 12.5 × 20 XL

Status
In production.

Manufacturer
ACMAT, Ateliers de Construction Mécanique de l'Atlantique, Le Point du Jour, F-44600 Saint-Nazaire, France.
Tel: +33 2 40 22 33 71. Telex: 700913 f.
Fax: +33 2 40 66 30 96.

VERIFIED

Fruehauf 20,000 litre tanker semi-trailer

Description
This semi-trailer was designed for transporting various fuels, including petrol. It has remote-controlled drainage pipes with safety valves, a pump operated by an auxiliary engine, a shielded electrical system, manually operated stabilisers mounted towards the front of the trailer and external fully enclosed containers on either side of the tank for stowage of fuel pipes.

A generally similar semi-trailer with an aluminium tank having a capacity of 30,000 litres is also produced. This has a gross weight of 32,000 kg.

Specifications
Capacity: 20,000 litres
Length: 9.92 m
Width: 2.5 m
King-pin size: 89 mm
Tyres: 12.00 × 20

Status
In production. In service with the French Army.

Manufacturer
Fruehauf France, 2 avenue de l'Aunette, F-91130 Ris-Orangis, France.
Tel: +33 4 69 43 30 00. Telex: 601381 f.
Fax: +33 4 69 43 33 66.

VERIFIED

Fruehauf 20,000 litre tanker semi-trailer

LOHR SMT 50 D special tanker semi-trailer

Description
This all-terrain tanker semi-trailer, was developed specifically for use in desert areas and can carry 50,000 litres of dispensable fuel. The fuel is carried in five 10,000 litre compartments, with an additional 2,000 litre compartment used to supply fuel for an air-cooled 36 hp auxiliary motor to power a compressor or to provide services for the tractor vehicle. There are two 670 litres/min fuel pumps provided and four fuel-dispensing nozzles can each deliver 250 litres/min; there are manual pumping facilities. There is also a 300 litre water tank. A 12 or 24 V electrical system can be provided.

Specifications
Capacity: 50,000 litres
Length:
(overall) 12.8 m
(tank) 11.4 m
Width: 4.28 m
Height: 4.8 m

Tank diameter: 2.5 m
Tyres: 24 × 20.5 S

Status
In production. In service with at least one nation.

Manufacturer
LOHR Industrie, F-67980 Hangenbieten, France.
Tel: +33 3 88 38 98 00. Telex: 870082 f.
Fax: +33 3 88 96 06 36.

UPDATED

LOHR SMT 50 D special tanker semi-trailer for desert use

Trailor tanker semi-trailers

Description
The 30,000 litre tank is of all-aluminium construction and has two compartments, upper and lower. The top of the tank is provided with a non-skid walkway and handrails. The trailer has two-speed manually operated telescopic stabilisers, hoses, hose racks and winder, airbrakes (one line acting on each axle) and a shielded electrical system.

The 37,500 litre tank is of all-aluminium construction and has five compartments. Other equipment is similar to that fitted to the 30,000 litre tanker.

Specifications

Capacity	30,000 litres	37,500 litres
Length	10.36 m	12.36 m
Width	2.5 m	2.5 m
Tyres	18.00 × 22.5	18.00 × 22.5

Status
In production. In service with the French Army.

Manufacturer
Trailor SA, 5 route Nationale 10, Coigniegres, BP 49, F-78311 Maurepas Cedex, France.
Tel: +33 1 30 50 61 26. Telex: 698896 f.
Fax: +33 1 30 50 31 32.

VERIFIED

ACMAT RM 215 SA workshop trailer

Description
The ACMAT RM 215 SA is built onto the standard ACMAT trailer, as described earlier in this section. The workshop trailer was designed to carry out vehicle field repairs up to fourth echelon level, even in desert areas. Equipment provided with the trailer includes: a 13 kVA generator, two 180 litre oil tanks for engine and transmission oils, an air compressor, injection calibrating equipment, a bench-mounted press, engine and transmission oil and grease lines, a work bench with a 125 mm vice, an electric grinder and a vertical drill. Portable equipment includes: hydraulic circuit maintenance equipment, battery maintenance equipment, electric and gas welding equipment, two large tool chests and two spare parts cabinets. A mast with floodlights is also supplied.

ACMAT RM 215 SA workshop trailer

Specifications
Weight: (operating) 3,950 kg
Length: 4.155 m
Width: (travelling) 2.07 m
Height:
(travelling) 2.034 m
(towing eye) 0.765 m
Track: 1.71 m

Status
In production.

Manufacturer
ACMAT, Ateliers de Construction Mécanique de l'Atlantique, Le Point du Jour, F-44600 Saint-Nazaire, France.
Tel: +33 2 40 22 33 71. Telex: 700913 f.
Fax: +33 2 40 66 30 96.

VERIFIED

ACMAT RM 215 GCR trailer-mounted field kitchen

Description
The RM 215 GCR is an air-portable, trailer-mounted field kitchen capable of serving 350 meals in 2 hours. The kitchen equipment is mounted on a standard ACMAT trailer chassis, as described previously.

The kitchen is equipped with four 100 litre cooking containers and may be fired by wood or diesel fuel, depending upon availability. Ovens or heating plates can be substituted for cooking containers. The equipment can also be butane-fired.

In the working position, the kitchen is surrounded by folding walkways and protected by a canvas/nylon canopy. A 400 litre water tank is incorporated in the trailer chassis and four jerricans are also carried. A fuel tank is incorporated and this allows over 3 hours' continuous cooking on all burners. There are two utensil cupboards and two working surfaces also included, and a jib and block are provided for handling cooking pots.

Specifications
On tow
Weight: (laden) 2,870 kg
Length: (overall) 4.225 m
Body length: 2.3 m
Height: (overall) 2.4 m
Width: (overall) 2.07 m
Track: 1.71 m
Lunette height: 0.765 m

Working position
Cooking area: 1.7 × 1.25 m
Walking area: 4.12 × 3.45 m
Height above ground: 1 m
Height: (overall) 4.3 m

ACMAT RM 215 GCR trailer-mounted field kitchen in travelling order

ACMAT RM 215 GCR trailer-mounted field kitchen in working configuration

Status
In production.

Manufacturer
ACMAT, Ateliers de Construction Mécanique de l'Atlantique, Le Point du Jour, F-44600 Saint-Nazaire, France.
Tel: +33 2 40 22 33 71. Telex: 700913 f.
Fax: +33 2 40 66 30 96.

VERIFIED

ACMAT semi-trailers

Description
ACMAT produces a range of semi-trailers for use with its TPK 6.35 TSR tractor truck. These semi-trailers are used to carry standard military shelter/containers and three types have been produced with capacities ranging from 4,300 to 6,000 kg. A 9,000 litre fuel or water tank can be installed if necessary. The semi-trailers maintain a high level of commonality with existing ACMAT trucks and trailers and the suspension is the same as that used on the rear suspension of the TPK 6.35 TSR. A 200 litre water tank and a spare wheel are fitted to each semi-trailer. The full NATO specification is respected for the 24 V suppressed electrical system and also for the air transport hold-down points.

ACMAT semi-trailer carrying communications shelter and towed by TPK 6.35 TSR tractor truck

Specifications
SR 490
Weight: (empty) 10,000 kg
Length: 8.53 m
Width: 2.4 m
Height:
(load platform) 1.12 m
(gooseneck) 1.97 m
Track: 1.8 m
Wheelbase:
(fifth wheel to centre of rear bogie) 6.25 m
(between rear axles) 1.22 m

Status
In production.

Manufacturer
ACMAT, Ateliers de Construction Mécanique de l'Atlantique, Le Point du Jour, F-44600 Saint-Nazaire, France.
Tel: +33 2 40 22 33 71. Telex: 700913 f.
Fax: +33 2 40 66 30 96.

VERIFIED

ARE 2 tonne F2 ammunition trailer

Description
The F2 trailer was designed to fill the need for additional ammunition to accompany the 155 mm self-propelled gun Mk F3. It was specifically designed to be towed by the AMX-13 VCA which normally transports the gun detachment and some ammunition. The body is steel and is open at the top, although the sides provide a measure of ballistic protection to the contents. Removable racks and cases provide stowage for 30 155 mm projectiles and charges and six hinged cases for storing warhead fuze containers. The rear door is hinged along its base, forming a working platform when open and a watertight seal when closed, allowing the trailer to ford to a depth of about 1 m. The wheels are fitted with bulletproof tyres and are mounted on torsion bars. After removing the ammunition racks the trailer may be used for general cargo.

ARE 2 tonne F2 ammunition trailer on tow behind AMX VCI infantry combat vehicle

Specifications
Length: 4.8 m
Width: 2.5 m
Height: 1.465 m
Weight:
(empty) 2,500 kg
(loaded) 4,400 kg
Capacity: 2.9 m^3
Ground clearance: 0.48 m

Status
In service with the French Army, Qatar and Venezuela.

Manufacturer
Giat Industries, 13 route de la Miniére, Satory, F-78022 Versailles Cedex, France.
Tel: +33 1 39 49 35 30. Fax: +33 1 39 49 82 79.

UPDATED

LOHR RF 3.5 tonne refrigerated container trailer

Description

This special trailer was designed to carry refrigerated containers for food and other perishable products. It consists of an open side-walled trailer onto which the refrigerated containers can be loaded. The towing bar has a dolly wheel and a towing bar link which can accommodate towing hitch heights from 0.8 to 1.5 m. Two 50 litre tanks are provided to service a carried container. One is located on each side, the left-hand one for fuel and the other for water. Removable steps are provided for access to the container interior. Telescopic stabiliser arms are provided on the towing arm (one) and at the rear (two). A spare wheel is provided.

Specifications

Weight:
(empty) 1,090 kg
(loaded) 3,870 kg
Length: 4.236 m
Width: 2.23 m
Height: 1.25 m
Track: 1.9 m
Tyres: 11.00 R 16 X type S
Electrical system: 24 V

Status

In production.

Manufacturer

LOHR Industrie, F-67980 Hangenbieten, France.
Tel: +33 3 88 38 98 00. Telex: 870082 f.
Fax: +33 3 88 96 06 36.

VERIFIED

LOHR RF 3.5 tonne refrigerated container trailer

LOHR RM 19 load handling trailer

Description

The LOHR RM 19 load handling trailer is part of the French Army's *Véhicule de Transport Logistique* (VTL) programme. LOHR Industrie was awarded a contract to produce 2,500 RM 19 trailers and 12,000 of the associated 20 ft/6.096 m ISO flatracks for the programme. The RM 19 trailer has two axles with four double-tyred wheels and is so arranged that the VTL prime mover's load handling system can load and off-load flatracks from the trailer. It is possible to erect covers for the complete trailer.

Specifications

Weight: (loaded) 19,000 kg
Max load: 13,400 kg
Length: (with flatrack) 8.37 m
Width: 2.5 m
Height: (with loaded flatrack) 1.475 m

Status

In production for the French Army.

LOHR RM 19 load handling trailer

Manufacturer

LOHR Industrie, F-67980 Hangenbieten, France.
Tel: +33 3 88 38 98 00. Telex: 870082 f.
Fax: +33 3 88 96 06 36.

VERIFIED

LOHR RM 22 PLM load handling trailer

Description

The LOHR RM 22 PLM load handling trailer, was selected as the M1076 trailer component of the US Army's Palletised Loading System (PLS) which involves the Oshkosh M1074/M1075 (10 × 10) truck. The US Army ordered 1,050 RM 22 trailers in October 1990 and it was anticipated that more would be ordered.

The RM 22 uses a mainframe assembled using electric welding, with a secondary frame supporting the front axle on a turning table and the drawbar; a steering lock allows the steering to be fixed when required. Two axles are provided at the rear and braking is on all three axles. Flatracks can be on- and off-loaded from the trailer by the prime mover's load handling system.

Specifications

Weight:
(kerb) 7,484 kg
(gross) 22,463 kg
Max payload: (with flatrack) 14,969 kg
Length: 8.316 m

LOHR RM 22 PLM load handling trailer on tow by Oshkosh (10 × 10) PLS tractor

Width: 2.431 m
Height:
(overall) 2.941 m
(flatbed) 1.801 m
Ground clearance: 0.46 m
Track: 2.04 m
Wheelbase: 3.793 m
Max road speed: 90 km/h
Side slope: 30%
Tyres: 13.00 R 20 XL

Status
In production for the US Army.

Manufacturer
LOHR Industrie, F-67980 Hangenbieten, France.
Tel: +33 3 88 38 98 00. Telex: 870082 f.
Fax: +33 3 88 96 06 36.

UPDATED

LOHR MTS trailer

Description
The LOHR MTS trailer was originally designed to carry a water treatment unit for off-road deployment. It uses a Caterpillar Mobil-trac Tracked Suspension (MTS - see entry under *United States of America* for details) and is able to carry a 10 ft/3.048 m ISO container, a cargo version with or without a tilt, or a generator. This trailer has also been referred to as the RTC 6000.

In addition to the MTS suspension the trailer has a welded steel chassis with an adjustable jockey wheel, and six welded points for mounting a water treatment unit. The triple circuit airbrakes are operated from the towing vehicle and a mechanical parking brake is also provided.

The LOHR MTS trailer is air-transportable in C-130 and C160 aircraft and may be slung under a helicopter.

Cargo version of LOHR MTS trailer
1997

Specifications
Weights:
(empty) 1,900 kg
(loaded) 6,000 kg
Length: (overall) 5.3 m ±15 mm
Width: (overall) 2.49 m
Height:
(overall, loaded) 1.18 m ±50 mm
(coupling ring, loaded) 0.9 m ±50 mm
Angle of approach/departure: 16°30°
Track width: 0.558 m
Fording: 0.9 m
Max towing speed: 80 km/h

Status
In production. In service with the French Army.

Manufacturer
LOHR Industrie, F-67980 Hangenbieten, France.
Tel: +33 3 88 38 98 00. Telex: 870082 f.
Fax: +33 3 88 96 06 36.

NEW ENTRY

LOHR CRC 150 trailer-mounted field kitchen

Description
The LOHR CRC 150 trailer-mounted field kitchen is mounted on a single-axle trailer. It can cook or reheat meals for up to 150 personnel at one time. The CRC 150 uses a central kitchen unit arranged around a heater powered by a 2.5 kWA power unit and supplying a cooking pot, a pressure cooker, a grill pan and a bain-marie - a folding chimney is provided. A folding work platform is provided all around the central unit and weather protection is provided by an awning over supports which can be folded down for transport.

There is independent pneumatic suspension on each wheel.

Specifications
Weight: 1,450 kg
Length: 3.92 m
Width: 2.06 m
Height: (max) 1.9 m

Status
Available.

Manufacturer
LOHR Industrie, F-67980 Hangenbieten, France.
Tel: +33 3 88 38 98 00. Telex: 870082 f.
Fax: +33 3 88 96 06 36.

VERIFIED

LOHR CRC 150 trailer-mounted field kitchen

GERMANY

SLT 50-2 tank transporter semi-trailer

Development
In 1965 West Germany and the USA agreed on the joint development of a tank transporter for the MBT-70. The firm of Krupp was responsible for the design of the trailer. Even before the cancellation of the MBT-70 project each country had decided to go its own way on the design of the tank transporter, although in major design features the versions of both looked very similar. The SLT 50-2 semi-trailer (heavy load carrier, *Schwerlasttransporter*) was adopted by the German Army and 328 were produced.

Description
The semi-trailer has four axles, each with two single-tyred steerable wheels. Steering of the wheels is hydraulically controlled and depends on the angle between the tractor and semi-trailer. Each axle is independently suspended on semi-leaf springs which also locate the axles. Swing arms and a connecting linkage provide for load equalisation and an additional mechanism, allows the load platform to be lowered by about 0.1 m to the required underpass height of 3.9 m when loaded. All four axles are equipped with airbrakes and parking brakes are fitted to two axles.

The load platform has a frame of box construction to resist longitudinal flexing and a bearing surface of sandwich construction. Folding, spring-balanced ramps are fitted and make an angle of only 14° for loading, yet may be lifted by only two personnel. Cable guides are provided on the trailer gooseneck to assist in winching loads on to and off the trailer.

Specifications
Weight: 16,600 kg
Max load: 52,000 kg
Axle loadings:
(loaded) 13,000 kg
(unloaded) 3,478 kg
Length: (overall) 13.1 m
Width: 3.15 m
Track: 2.65 m
Height: (to top of raised ramps) 3.15 m
Fifth wheel height: (loaded) 1.55 m
Loading platform
Length: 8 m
Height:
(unloaded) 1.37 m
(loaded) 1.28 m
(loaded, lowered) 1.16 m
Ground clearance: 0.31 m
Departure angle: 35°
Turning circle radius: (with SLT 50-2 tractor) 9 m
Tyres: 18.00 × 22.5, 20 PR
Ramps:
(length) 4.18 m
(width) 0.75 m each
King-pin: 89 mm

Status
Production complete. In service with the German Army.

Manufacturer
Karl Kässbohrer Fahrzeugwerke GmbH, Postfach 2660, D-7900 Ulm/Donau, Germany.
Tel: +49 731 1811. Telex: 712766.

VERIFIED

Kärcher TFK 250 mobile kitchen trailer

Description

This tactical field kitchen is a mobile catering unit mounted on a single-axle 3.5 tonne trainer. It uses stainless steel (chrome-nickel 18/10) for the cooking appliances as well as other high-quality materials.

The trailer is fitted with the following equipment: 150 litre double-walled GASTRONOM pressure cookers; a 55 litre GASTRONOM pressure frier; an integrated 75 litre GASTRONOM baking area in which two working levels are heated by one burner; a 28 litre hot water generator and tea maker; and a four-unit pressure vaporisation burner that can use diesel oil, light fuel, petrol or kerosene. Using the TFK 250 two cooks can provide meals for 250 personnel or a single-course meal for 500 personnel in 2 hours.

The trailer has a cross-country capability and in use is provided with four support legs. The trailer has an integral weather shelter and four tarpaulins can be opened semi-automatically within seconds.

Specifications

Weight: (unloaded) approx 2,075 kg
Load capacity: approx 425 kg
Length: 4.17 m
Width: 2.16 m
Height:
(overall) 2.62 m
(working table) 0.96 m
Ground clearance: 0.32 m
Track: 1.908 m

Kärcher TFK 250 mobile kitchen trailer

Status

In production. In service with Belgium (air force), China, Germany (1,283 on order between 1990 and 1997), Ireland, Luxembourg, NATO headquarters, New Zealand, Norway, Qatar, Saudi Arabia, UK and USA, plus some nations in the Middle East and Africa.

Manufacturer

Alfred Kärcher GmbH & Co, Alfred Kärcher-Strasse 28-40, D-71364 Winnenden, Germany.
Tel: +49 7195 142797. Telex: 724432.
Fax: +49 7195 142720.

VERIFIED

Kärcher Modular Field Kitchen (MFK)

Description

The Kärcher modular field kitchen MFK can be assembled to meet a number of field catering needs and can be mounted on a single axle-trailer. The design is based on that of the TFK 250 mobile kitchen (see previous entry) but the various cooking modules can be used separately on the ground or as a complete kitchen unit on the trailer.

Equipment carried includes the following: two 125 litre boiling kettles, one single-walled, one double-walled; a 25 litre frying/baking module; a frying pan module; a grill module; a stowage compartment module; lateral stowage compartments equipped with stainless steel working plates; working tables; integral weather protection; two pressure vaporisation burners; a propane burner; and a solid fuel burner. All cooking equipment is to GASTRONOM/DIN 66075 regulations.

Specifications

Weight:
(empty) 520 kg
(loaded) 2,000 kg
Width: 2 m
Height:
(with tilted chimney) 1.38 m
(with weather shelter) 2.6 m
Electrical system: 24 V

Status

In production. In service with Austrian and German civil defence and emergency rescue organisations and Spain (approximately 250).

Manufacturer

Alfred Kärcher GmbH & Co, Alfred Kärcher-Strasse 28-40, D-71364 Winnenden, Germany.
Tel: +49 7195 142797. Telex: 724432.
Fax: +49 7195 142720.

UPDATED

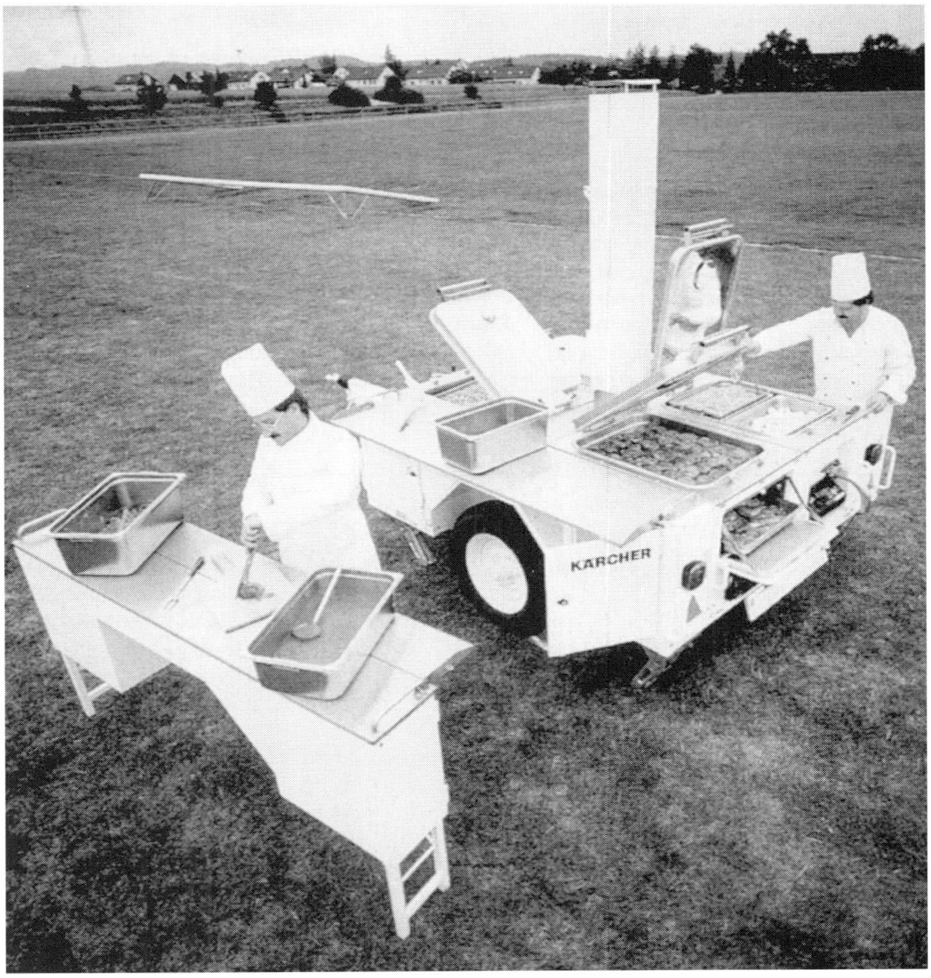

Kärcher modular field kitchen MFK

GREECE

NK AP-130/60 T tank transporter semi-trailer

Description

This tank transporter semi-trailer was designed to carry tanks up to the weight of a Chieftain or Leopard 2, and heavy engineering plant. The chassis frame is formed from two longitudinal 'I' section beams connected by cross members and bridges through stress diffusers. The deck is hardwood, with the sides, rear area and foredeck covered with 10 mm steel sheet. The suspension is formed from a pair of Model MZP 2/14000 walking beam units each side.

The king-pin is a bolted 89 mm Jost unit mounted on a heavy-duty upper coupler plate 16 mm thick. Jost also supplies the two Model HDK 240T side-operated two-speed loading winches, providing a static capacity of 60,000 kg and a lift capacity of 30,000 kg. Two hydraulically operated loading ramps, each 2.2 m long by 0.7 m wide, are provided. Power for the ramps is provided by a hydraulic pump located on the right-hand side of the trailer.

Equipment carried on the trailer includes two spare wheels mounted on the gooseneck and provided with handling winches. Also carried are two pulley assemblies, cable guides, a set of wheel chocks and two storage boxes under the front deck.

A generally similar semi-trailer, the NK AP-140/58T has a payload of 58,000 kg and a gross vehicle weight of 73,000 kg.

NK AP-130/60 T tank transporter semi-trailer carrying M32 armoured recovery vehicle

Specifications

Weight: (unloaded) 19,000 kg
Max payload: 60,000 kg
Length:
(overall, ramps lowered) 14.8 m
(deck) 7.6 m
Width: 3.6 m
Height:
(overall) 2.75 m
(king-pin) 1.67 m
Ground clearance: 0.63 m
Tyres: 12.00 × 18 PR
Electrical system: 12/24 V

Status

In production.

Manufacturer

N Kioleides SA, PO Box 51156, GR-145 10 Kifissia, Athens, Greece.
Tel: +30 1 816 1902/3. Telex: 214500 nk gr.
Fax: +30 1 816 1739.

UPDATED

INDIA

Mahindra Owen 500 kg general service trailer

Description

The Mahindra Owen 500 kg general service trailer, is a single-axle trailer with leaf spring suspension and mechanical overrun brakes. Welded steel construction is used for the chassis and the platform is sheet steel. The trailer is fitted with an adjustable draught eye to match pintle hook heights of 500 or 700 mm. The body panels are detachable. Two folding spring-loaded handles are provided for manual handling.

Mahindra Owen 500 kg general service trailer

Specifications

Weight:
(empty) 350 kg
(loaded) 850 kg
Max load: 500 kg
Length: (incl towbar, overall) 2.93 m
Width: (overall) 1.416 m
Height:
(platform) 540 mm
(panel) 455 mm
Track: 1.23 m
Tyres: 6.00 × 15-6 ply
Max towing speed:
(road) 80 km/h
(cross-country) 48 km/h

Status

In production. In service with the Indian Army.

Manufacturer

Mahindra Engineering & Chemical Products Limited, Mahindra Owen Division, 148 Bombay Pune Road, Pimpri, Pune-411 018, India.
Tel: +91 212 775045-6. Telex: 0146-233 MOL-IN.
Fax: +91 212 771286.

VERIFIED

Mahindra Owen 1,000 kg general service trailer

Description

The Mahindra Owen 1,000 kg general service trailer, is a single-axle trailer with leaf spring suspension and mechanical overrun brakes. Welded steel construction is used for the chassis and the platform is sheet steel. The trailer is fitted with an adjustable draught eye to match pintle hook heights of 575 or 775 mm. The body panels are detachable. Two folding spring-loaded handles are provided for manual handling.

Specifications

Weight:
(empty) 780 kg
(loaded) 1,780 kg
Max load: 1,000 kg
Length: (incl towbar, overall) 3.555 m
Width: (overall) 1.935 m
Height:
(platform) 700 mm
(panel) 575 mm
Track: 1.65 m
Tyres: 8.25 × 20-10 ply
Max towing speed:
(road) 80 km/h
(cross-country) 48 km/h

Status
In production. In service with the Indian Army.

Manufacturer
Mahindra Engineering & Chemical Products Limited, Mahindra Owen Division, 148 Bombay Pune Road, Pimpri, Pune-411 018, India.
Tel: +91 212 775045-6. Telex: 0146-233 MOL-IN.
Fax: +91 212 771286.

VERIFIED

Mahindra Owen 1,000 kg general service trailer

Mahindra Owen 500 litre water tanker trailer

Description
The Mahindra Owen 500 litre water tanker trailer, is a single-axle trailer with leaf spring suspension and mechanical overrun brakes. Welded steel construction is used for the chassis. The trailer is fitted with an adjustable draught eye to match pintle hook heights of 500 or 700 mm. Two folding spring-loaded handles are provided for manual handling. The water tank is detachable and is provided with one manhole cover at the top and three dispensing taps at the rear.

A 1,000 litre water tanker trailer is also produced.

Specifications
Weight:
(empty) 360 kg
(loaded) 860 kg
Length: (incl towbar, overall) 2.835 m
Width: (overall) 1.41 m
Height: (overall) 1.32 m
Track: 1.225 m
Tyres: 6.00 × 15-6 ply
Max towing speed:
(road) 80 km/h
(cross-country) 48 km/h
Water tank inside length: 1.075 m
Water tank inside width: 965 mm
Water tank inside depth: 510 mm

Status
In production. In service with the Indian Army.

Mahindra Owen 500 litre water tanker trailer

Manufacturer
Mahindra Engineering & Chemical Products Limited, Mahindra Owen Division, 148 Bombay Pune Road, Pimpri, Pune-411 018, India.
Tel: +91 212 775045-6. Telex: 0146-233 MOL-IN.
Fax: + 91 212 771286.

VERIFIED

Mahindra Owen 1,000 kg generator set trailer

Description
The Mahindra Owen 1,000 kg generator set trailer, is a single-axle trailer with leaf spring suspension and mechanical overrun brakes. Welded steel construction is used for the chassis and the platform is sheet steel. The trailer is fitted with an adjustable draught eye to match pintle hook heights of 575 or 775 mm. Two folding spring-loaded handles are provided for manual handling.

Mahindra Owen 1,000 kg generator set trailer

Specifications

Weight:
(empty) 700 kg
(loaded) 1,700 kg
Max load: 1,000 kg
Length: (incl towbar, overall) 3.481 m
Width: (overall) 1.935 m
Height: (platform) 700 mm
Track: 1.65 m
Tyres: 8.25 × 20-10 ply
Max towing speed:
(road) 80 km/h
(cross-country) 48 km/h

Status

In production. In service with the Indian Army.

Manufacturer

Mahindra Engineering & Chemical Products Limited, Mahindra Owen Division, 148 Bombay Pune Road, Pimpri, Pune-411 018, India.
Tel: +91 212 775045-6. Telex: 0146-233 MOL-IN.
Fax: + 91 212 771286.

VERIFIED

Mahindra Owen 2,000 kg generator set trailer

Description

The Mahindra Owen 2,000 kg generator set trailer, is a twin-axle trailer with semi-elliptical heavy-duty leaf spring suspension and a dual line airbrake system. The front axle uses an Ackermann steering arrangement. Welded steel construction is used for the chassis and the platform is sheet steel.

Specifications

Weight:
(empty) 1,440 kg
(loaded) 3,440 kg
Max load: 2,000 kg
Length: (incl towbar, overall) 6.935 m
Width: (overall) 2.1 m
Height: (platform) 750 mm
Track: 1.7 m
Tyres: 8.25 × 20-10 ply
Max towing speed:
(road) 72 km/h
(cross-country) 24 km/h

Mahindra Owen 2,000 kg generator set trailer

Status

In production. In service with the Indian Army.

Manufacturer

Mahindra Engineering & Chemical Products Limited, Mahindra Owen Division, 148 Bombay Pune Road, Pimpri, Pune-411 018, India.
Tel: +91 212 775045-6. Telex: 0146-233 MOL-IN.
Fax: +91 212 771286.

VERIFIED

Mahindra Owen 7,500 kg trailer for mobile radar

Description

The Mahindra Owen 7,500 kg trailer for mobile radar, is a single-axle semi-low-bed trailer for carrying containerised radar equipment. The trailer has semi-elliptical leaf spring suspension with twin shock-absorbers. The braking system consists of a dual line air system plus a manual parking brake. Low alloy welded steel construction is used for the chassis. Removable loading ramps are provided. The loading ramps and platform are fitted with roller tracks for loading and unloading equipment containers. The trailer has a 50 mm king-pin and a 24 V electrical system.

A 10,000 kg version of this trailer is produced.

Specifications

Weight:
(empty) 3,500 kg
(loaded) 11,050 kg
Length: (overall) 10.175 m
Width: (overall) 2.3 m
Height: (platform) 1.165 m
Track: 1.73 m
Tyres: 9.00 × 20-14 ply
Electrical system: 24 V

Status

In production. In service with the Indian Air Force.

Manufacturer

Mahindra Engineering & Chemical Products Limited, Mahindra Owen Division, 148 Bombay Pune Road, Pimpri, Pune-411 018, India.
Tel: +91 212 775045-6. Telex: 0146-233 MOL-IN.
Fax: +91 212 771286.

VERIFIED

20 tonne low deck full trailer

Description

Bharat Earth Movers Limited and Mahindra & Mahindra produce a number of trailers and semi-trailers, among which is a 20 tonne low deck full trailer. Designed for on- and off-road use it has a front and rear suspension that provides a lateral articulation of 200 mm for the equal distribution of the load during off-road operations. The front suspension has two longitudinally mounted beams pivoted in the front and spring supported in the rear, with two single wheels on each beam. The rear suspension has four rear wheels in one row fitted to half axles supported by longitudinally mounted suspension beams pivoted front and rear.

Standard accessories include: two spare wheels, lashing shackles, a tool box, plus hinged and extension ramps.

Specifications

Weight: (unladen) 12,050 kg
Payload: 20,000 kg
Bed area: 6 × 3.1 m
Length: (overall) 12.8 m
Width: 3.1 m
Height:
(overall) 2.259 m
(platform) 1.1 m
(drawbar) 0.87 m
Wheelbase: 8.4 m
Tyres:
(front) 14.00 × 20, 18 PR (4)
(rear) 9.00 × 20, 14 PR (8)
Electrical system: 24 V

Mahindra Owen-produced 20 tonne low deck full trailer

Status

In production.

Manufacturers

Bharat Earth Movers Limited, BEML Soudha, Mission Road, Bangalore-560 027, India.
Tel: +91 80 222 4141. Telex: 0845-2278.
Fax: +91 80 222 6883.

Mahindra Engineering & Chemical Products Limited, Mahindra Owen Division, 148 Bombay Pune Road, Pimpri, Pune-411 018, India.
Tel: +91 212 775045-6. Telex: 0146-233 MOL-IN.
Fax: +91 212 771286.

VERIFIED

Mahindra Owen 50 tonne heavy-duty tank transporter trailer

Description

The Mahindra Owen 50 tonne heavy-duty tank transporter trailer, is a low deck independent trailer intended for on- and off-road use. The lateral articulation provided on the front axle and the lateral and longitudinal articulation provided on the rear axles, give equal load distribution during off-road operations. The trailer is provided with two rows of two in-line axles at the rear and a single axle at the front. There are 24 tyres in all. The chassis is of welded steel construction while the platform is chequered steel plate. Dual line airbrakes and manual parking brakes are provided.

The trailer is equipped with rear-loading ramps, adjustable track guides, wheel chocks and a roller arrangement for use when winching disabled vehicles onto the platform. Spares wheels and tyres are provided. Standard equipment includes a tool box, a lashing chain and hydraulic jacks.

Mahindra Owen 50 tonne heavy-duty tank transporter trailer carrying T-54/55 series MBT

Specifications

Weight:
(empty) 18,000 kg
(loaded) 68,000 kg
Max load: 50,000 kg
Length: (overall) 12.705 m
Width: (overall) 3.42 m
Height: (platform) 1.295 m
Wheelbase: 7.1 m
Tyres: 11.00 × 20-16 ply
Electrical system: 24 V

Status

In production. In service with the Indian Army.

Manufacturer

Mahindra Engineering & Chemical Products Limited, Mahindra Owen Division, 148 Bombay Pune Road, Pimpri, Pune-411 018, India.
Tel: +91 212 775045-6. Telex: 0146-233 MOL-IN.
Fax: +91 212 771286.

VERIFIED

ISRAEL

AAI MT-100 1,000 kg general purpose trailer

Description

The MT-100 1,000 kg general purpose trailer was designed for use with the AAI M-325 Commandcar vehicles (see entry in *Trucks* section), but can be used with other vehicles. It has a welded steel chassis with trailing arm suspension using rubber element springs. The MT-100 can be towed on roads at speeds up to 80 km/h and the wheels and tyres are the same as those used on the M-325. Service braking is by means of an overrun unit operating mechanical brakes through a spring damper. A safety chain will operate the brakes if the trailer is released from the towing vehicle and a spring damper is fitted to the towing eye. The electrical system is 24 V connected through a plug for the stop lights, turn signals and side lights.

The trailer may be produced in three forms: a flatbed trailer for carrying special equipment and generators; a water tank with a capacity of 770 litres constructed from stainless steel; or for general cargo use a welded steel cargo body may be fitted. The last is supplied with lashing points for the load and a PVC-coated tarpaulin cover if required.

MT-100 trailer with general cargo body and cover in place

Specifications

Weight: (chassis, unladen) 290 kg
Max load: 1,000 kg
Length: 3.2 m
Width: 1.95 m
Track: 1.66 m
Fording: 0.76 m
Tyres: 9.00 × 16-10 ply
Electrical system: 24 V

Status

Production complete. In service with the Israeli armed forces.

Manufacturer

Automotive Industries Limited, PO Box 535, Nazareth Illit, Israel.
Tel: +972 6 558111. Telex: 46217 ail il.
Fax: +972 6 558103.

VERIFIED

Urdan ARTRAIL ammunition resupply trailer

Description

The Urdan ARTRAIL ammunition resupply trailer was designed and developed to overcome some of the ammunition resupply problems associated with self-propelled artillery in the field. In essence, the ARTRAIL is a trailer that can be towed into action by a self-propelled artillery system as a mobile auxiliary ammunition supply source. The ARTRAIL carries 44 projectiles, propellant charges and fuzes. It has been tested with the M109A2 155 mm self-propelled howitzer system.

The ARTRAIL is a two-wheeled trailer consisting of three units: a stowage compartment for the ammunition, a two-wheeled assembly with a connecting shaft, and a conveyor assembly with two conveyors and one pick-up device.

When required for action the front-mounted doors on the ammunition compartment are swung open. The

ARTRAIL ammunition resupply trailer in action with an M109A2 155 mm self-propelled howitzer

doors contain the propellant charges and when open reveal the projectiles stowed in horizontal and slightly inclined stowage cradles. Up to eight smoke projectiles are stored vertically in side-mounted compartments. All projectiles may be carried fuzed or unfuzed and there is capacity to carry more fuzes than projectiles to accommodate variations in fire plans.

The conveyor assembly takes the projectile into the self-propelled artillery system fighting compartment by gravity. The conveyor assembly is balanced and easy to raise or lower. Each projectile is released from its stowage cradle in the trailer ammunition compartment by a pick-up device described as a forklift, and moves down the conveyor into the fighting compartment to a height of 400 mm above the fighting compartment floor, that is at a height convenient for loading. Propellant charges can be taken direct from their locations in the trailer doors and the vertically stowed projectiles can be moved into the fighting compartment by hand. Up to 10 projectiles per minute can be delivered.

After use, reloading the ARTRAIL takes up to 15 minutes. If required, an empty ARTRAIL can be unhitched and directly replaced by a loaded unit. ARTRAIL can be located by any 5 ton truck or armoured personnel carrier. The rear door of the self-propelled artillery system involved can be modified to accommodate the ARTRAIL in 30 minutes under field conditions. The transition time for ARTRAIL from road transport to field conditions takes 5 minutes. In action ARTRAIL provides the same degree of protection to its contents as the self-propelled artillery system involved but, as an option, composite armour can be added to the entire stowage area or to selected sensitive areas.

The ARTRAIL is mounted on a two-wheel assembly fitted with combat tyres, hydropneumatic suspension and a braking system.

Specifications
Weight:
(empty) approx 3,990 kg
(loaded) approx 6,105 kg
Length: 4.2 m
Width: 3.2 m
Height:
(empty) 2.5 m
(loaded) 2.4 m

Status
Development complete. Ready for production.

Manufacturer
Urdan Industries Limited, Industrial Zone, IL-42378 Netanya, Israel.
Tel: +972 9 338074. Telex: 341822 usaf il.
Fax: +972 9 610246.

VERIFIED

ITALY

Bartoletti single-axle 1,000 kg cargo trailer Type B.10

Description
The Type B.10 single-axle trailer has: an all-steel body with drop tailgate; canvas supports and canvas cover; leaf spring suspension; 510 mm disc wheels; hydraulic overrun brakes; 12-pole 24 V electric system; standard and blackout lights; and a retractable dolly wheel.

Specifications
Weight:
(empty) 750 kg
(loaded) 1,750 kg
Max load: 1,000 kg
Inside length of body: 2.445 m
Inside width of body: 1.165 m
Height of side boards: 0.46 m
Length: 3.72 m
Width: 1.81 m
Height: 1.8 m
Track: 1.51 m
Tyres: 8.25 × 20

Status
In production.

Bartoletti single-axle 1,000 kg cargo trailer Type B.10 with canvas cover

Manufacturer
E Bartoletti, Via Leonardo Da Vinci 4, I-47100 Forli, Italy.
Tel: +39 543 734111. Telex: 550064 barto 1.
Fax: +39 543 739029.

VERIFIED

Bartoletti single-axle 2,000 kg cargo trailer

Description
This single-axle trailer can be fitted with a variety of different bodies and used for specialised applications such as carrying radar. The trailer has: semi-elliptic simple flexible leaf spring suspension with hydraulic dampers; 510 mm disc wheels; compressed air two-line (automatically adjustable) brake booster; 12 pole 24 V electric system; standard and blackout lights; removable screw-type dolly wheel; and two stabilisers that can be lowered at the rear.

Bartoletti single-axle 2,000 kg cargo trailer

Specifications
Weight:
(empty) 1,000 kg
(loaded) 3,000 kg
Payload: 2,000 kg
Length:
(incl towbar) 4.19 m
(excl towbar) 2.3 m
Width: 2.07 m
Chassis width: 1.35 m
Height to upper edge of chassis: 1 m
Track: 1.77 m
Tyres: 11.00 × 20

Status
In production.

Manufacturer
E Bartoletti, Via Leonardo Da Vinci 4, I-47100 Forli, Italy.
Tel: +39 543 734111. Telex: 550064 barto 1.
Fax: +39 543 739029.

VERIFIED

Bartoletti 3,000 kg light cargo trailer Type 2L-30 M

Description

This all-steel trailer has: drop sides and a drop tailgate with the front axle mounted on a ball-type turntable; simple flexible semi-elliptic leaf spring suspension; 380 mm spoke wheels; 12-pole 24 V electric system; standard and blackout lights; and a compressed air two-line (automatically adjustable) brake booster.

Specifications

Weight:
(empty) 1,300 kg
(loaded) 4,300 kg
Payload: 3,000 kg
Length:
(incl towbar) 4.325 m
(excl towbar) 3 m
Width: 1.56 m
Height of deck from ground: (loaded) 0.92 m
Height of sideboards: 0.4 m
Track: 1.295 m
Tyres: 7.50 × 15

Bartoletti 3,000 kg light cargo trailer Type 2L-30 M

Status

In production.

Manufacturer

E Bartoletti, Via Leonardo Da Vinci 4, I-47100 Forli, Italy.
Tel: +39 543 734111. Telex: 550064 barto 1.
Fax: +39 543 739029.

VERIFIED

Bartoletti single-axle flatbed trailer Type B.40 MGB

Description

The Type B.40 MGB single-axle trailer was designed for the transport of components of the British Williams Fairey Medium Girder Bridge (MGB). The trailer has: semi-elliptic flexible leaf spring suspension; 510 mm disc wheels; compressed air two-line (automatically adjustable) brake booster; 12 pole 24 V electric system; standard and blackout lights; and a removable screw-type loading gear. Stabilisers are mounted at the rear of the trailer.

Bartoletti single-axle flatbed trailer Type B.40 MGB

Specifications

Weight:
(empty) 1,600 kg
(loaded) 5,700 kg
Max load: 4,100 kg
Length:
(incl towbar) 5.5 m
(excl towbar) 3.6 m
Width: 2.5 m
Deck width: 2.31 m
Deck height from ground: (loaded) 1.145 m
Track: 2.05 m
Tyres: 11.00 × R20

Status

In production. In service with the Italian Army.

Manufacturer

E Bartoletti, Via Leonardo Da Vinci 4, I-47100 Forli, Italy.
Tel: +39 543 734111. Telex: 550064 barto 1.
Fax: +39 543 739029.

VERIFIED

Bartoletti 24,000 kg container transport trailer Type 24 RD

Description

This two-axle trailer has a flat deck designed for the transportation of standard 20 ft ISO containers or general cargo. The trailer has: semi-elliptic double flexible leaf springs on silent blocks with stabilisers; 510 mm disc wheels; 12 pole 24 V electric system; standard and blackout lights; and compressed air two-line (automatically adjustable) brake booster.

Specifications

Weight:
(empty) 4,750 kg
(loaded) 24,000 kg
Payload: 19,250 kg
Length:
(deck) 6.5 m
(incl towbar) 8.505 m
Width:
(overall) 2.5 m
(deck) 2.4 m
Height of deck from ground level: 1.29 m
Tyres: 12.00 × 20 PR 18

Status

In production.

Bartoletti 24,000 kg container transport trailer Type 24 RD

Manufacturer

E Bartoletti, Via Leonardo Da Vinci 4, I-47100 Forli, Italy.
Tel: +39 543 734111. Telex: 550064 barto 1.
Fax: +39 543 739029.

VERIFIED

Bartoletti 1,800 litre water tanker Type B.10

Description

This single-axle trailer has: an 1,800 litre capacity drinking water tank; 500 mm diameter safety manhole; gravity outlet through a 65 mm gate valve and 26 mm taps; leaf spring suspension; 510 mm disc wheels; hydraulic overrun brake; 12 pole 24 V electric system with standard blackout lights; and a retractable dolly wheel.

Specifications

Weight:
(empty) 1,000 kg
(loaded) 2,800 kg
Max load: 1,800 kg
Length: (incl towbar) 3.72 m
Width: 1.81 m
Track: 1.51 m
Tyres: 8.25 × 20
Max diameter of tank: 1.46 m
Min diameter of tank: 0.86 m
Height of tank: 2 m

Status

In production.

Bartoletti 1,800 litre water tanker Type B.10

Manufacturer

E Bartoletti, Via Leonardo Da Vinci 4, I-47100 Forli, Italy.
Tel: +39 543 734111. Telex: 550064 barto 1.
Fax: +39 543 739029.

VERIFIED

Bartoletti 8,000 litre fuel tanker trailer Type 2M-55

Description

The front axle of this two-axle trailer is mounted on a turntable. The 8,000 litre oval section tank is of all-steel construction and is provided with a single 255 mm loading mouth, metric rod and fuel level indicator, remote-controlled foot valve, 80 mm gate valves with quick coupling mouth and standard accessories.

Specifications

Weight:
(empty) 4,200 kg
(loaded) 11,000 kg
Max load: 6,800 kg
Length:
(incl towbar) 7.32 m
(trailer only) 5.75 m
Width: 2.47 m
Height: 2.84 m
Track: 1.727 m
Tyres: 12.00 × 20 PR 18

Status

In production.

Bartoletti 8,000 litre fuel tanker trailer Type 2M-55

Manufacturer

E Bartoletti, Via Leonardo Da Vinci 4, I-47100 Forli, Italy.
Tel: +39 543 734111. Telex: 550064 barto 1.
Fax: +39 543 739029.

VERIFIED

Bartoletti 2PX trailer

Description

The trailer was designed for the transport of tracked vehicles such as armoured fighting vehicles and engineering equipment. The trailer has: aligned axles on independent dollies; low bed; steering by means of equaliser and guide arms; sliding rear dollies with jacks; simple flexible semi-elliptical oscillating leaf spring suspension; 510 mm spoke wheels; 12 pole 24 V electric system; standard and blackout lights; compressed air two-line (automatically adjustable) brake booster; and rear-loading ramps.

Bartoletti 2PX trailer in travelling configuration

Specifications

Weight:
(empty) 7,350 kg
(loaded) 27,350 kg
Payload: 20,000 kg
Length:
(incl towbar) 9.56 m
(trailer only) 8.3 m
(bed) 4.7 m
Width:
(overall) 2.5 m
(bed) 2.5 m
Height of low bed: (loaded) 0.63 m
Tyres: 11.00 × 20 PR 16

Status

In production.

Manufacturer

E Bartoletti, Via Leonardo Da Vinci 4, I-47100 Forli, Italy.
Tel: +39 543 734111. Telex: 550064 barto 1.
Fax: +39 543 739029.

VERIFIED

Bartoletti tank semi-trailer Type TCS 60 BO

Description

The TCS 60 BO semi-trailer was designed for carrying MBTs weighing up to 60,000 kg. It can be towed by tractor trucks such as the IVECO 320-45 WTM (6 × 6).

The trailer has: three axles each with four tyres; lowered chassis with double hydraulically retractable loading ramps at the rear; balanced suspension with semi-elliptical leaf springs and reaction arms; pneumatic retractable forward landing gear; and adjustable track guides to suit different types of tracked vehicles. The semi-trailer also has a system of rollers and pulleys for easy loading and unloading of vehicles.

Standard equipment includes 12 pole 24 V electric system with standard and blackout lights, compressed air (two-line) automatically adjustable service brake and parking brake.

Bartoletti tank semi-trailer Type TCS 60 BO towed by IVECO 320-45 WTM (6 × 6) tractor truck

Specifications

Weight:
(empty) 20,000 kg
(loaded) 80,000 kg
Payload: 60,000 kg
Length: 11.95 m
Width: 4 m
Platform height from ground: (loaded) 1.42 m
Length of platform: 7.08 + 0.95 m
Track: 2.6 m
Tyres: 16.00 × 20 XS

Status

In production.

Manufacturer

E Bartoletti, Via Leonardo Da Vinci 4, I-47100 Forli, Italy.
Tel: +39 543 734111. Telex: 550064 barto 1.
Fax: +39 543 739029.

VERIFIED

Bartoletti TCS 50 BO tank transporter semi-trailer

Description

In the same way as the German SLT 50-2 and US M747 semi-trailers, the Bartoletti TCS 50 BO semi-trailer owes its origins to the US/FRG HET-70 programme and is based on the design of the SLT 50-2. It is intended for the transportation of MBTs up to 50,000 kg, such as the M60 series, without exceeding a height of 4.5 m from ground level to permit passage through road tunnels. The Bartoletti TCS 50 BO is a four-axle semi-trailer with a lowered chassis and retractable loading ramps. The suspension is a combination of semi-elliptical springs and air springs, the outer axles can be lifted to reduce tyre wear during unloaded running and to improve manoeuvrability at low speed. Rollers and pulleys are provided to facilitate the loading and unloading of damaged vehicles.

Bartoletti TCS 50 BO tank transporter semi-trailer

Specifications

Weight: (unladen) 14,000 kg
Payload: 50,000 kg
Max fifth wheel load: 20,000 kg
Length: 12.42 m
Width: 3.15 m
Track: 2.7 m
Load deck length: 7.67 m
Platform height: (loaded) 1.23 m
Height:
(over folded ramps) 2.93 m
(over spare wheel) 3.12 m
King-pin:
(height) 1.64 m
(diameter) 89 mm
Distance between axle centres: 1.25 m
Tyres: 18 × 22.5 XS
Brakes: dual air
Electrical system: 24 V

Status

In production, under licence from OTOBreda.

Manufacturer

E Bartoletti, Via Leonardo Da Vinci 4, I-47100 Forli, Italy.
Tel: +39 543 734111. Telex: 550064 barto 1.
Fax: +39 543 739029.

VERIFIED

Cometto MPS50 tank transporter semi-trailer

Description

The twin-axle rows of the MPS50 Cometto tank transporter semi-trailers may be fitted with either four or eight tyres to assist in weight distribution to suit the payload. An extra pair of tyres may be carried under the loading platform for use as spares; these two spare tyre carriers have lifting devices to assist stowage. The MPS50 is fitted with a five-speed roller and pulley system for loading disabled vehicles and when not on the tractor truck, a folding stand is situated to the front of the trailer. The two loading ramps are spring-loaded to provide quick and easy loading and for extra stability two foldable rear stabiliser legs on the ramp hinges may be used. The loading ramps measure 1.7 × 0.75 m. Sliding track guides are fitted as standard as are load-fastening hooks and chains. Accessories include a tool kit and box and a 150 litre water tank.

Cometto MPS50 tank transporter semi-trailer carrying a M60A1 tank

Specifications

Number of tyres per axle row	**4**	**8**
Weight (empty)	16,000 kg	16,600 kg
(loaded)	66,000 kg	77,000 kg
Max load	50,000 kg	60,000 kg
Length (overall)	12.34 m	12.34 m
(loading platform)	7.925 m	7.925 m
Width	3.505 m	3.505 m
Height (loading platform)	1.25-1.3 m	1.25-1.3 m
(5th wheel, approx)	1.5 m	1.5 m
Axle wheelbase	1.5 m	1.5 m
King-pin	89 mm	89 mm
Tyres	18.00 × 22.5	12.00 × 20

Status

In production.

Manufacturer

Cometto Industriale Srl, Via Cuneo 38, I-12011 Borgo S Dalmazzo (Cuneo), Italy.
Tel: +39 171 76331. Telex: 210375.

VERIFIED

Cometto S53 tank transporter semi-trailer

Description

The S53 is a heavy-duty semi-trailer for the transport of tanks and heavy equipment. Its running gear has 20 wheels, suspended independently in five pairs on each side. It is thought that there is provision for raising and steering at least some of the wheels hydraulically, to assist manoeuvrability as an auxiliary power unit is mounted on the gooseneck. The operation and stabilisation of the rear-loading ramps are also hydraulic.

Cometto S53 tank transporter semi-trailer

Specifications

Weight: 27,000 kg
Payload: 51,000 kg
Load on fifth wheel: 21,500 kg
Length: (overall) 14.77 m
Width: 3.4 m
Load deck height: 1.4 m (can be adjusted by 0.28 m up or down)
Tyres: 18-19.5 × 5

Status

In production.

Manufacturer

Cometto Industriale Srl, Via Cuneo 38, I-12011 Borgo S Dalmazzo (Cuneo), Italy.
Tel: +39 171 76331. Telex: 210375.

VERIFIED

De Filippi Trailers

De Filippi manufactures a wide range of trailers and semi-trailers, many of which are suited for military use. Brief specifications are given below:

Status

In production.

Manufacturer

Costruzioni Meccaniche A De Filippi SAS, Via Garibaldi 124, I-12061 Carrù (CN), Italy.
Tel: +39 173 75101. Fax: +39 173 75112.

VERIFIED

Model	Type	Payload	Empty weight	Length	Length of platform	Width	Tyres	Number of tyres	Axles	Notes
R-6	full trailer	4,300 kg	1,700 kg	5.6 m	4 m	2.38 m	7.50 × 15	4	1	
R-10	full trailer	9,500 kg	3,000 kg	5.8 m	5.6 m	2.5 m	7.50 × 15	8	2	Can be fitted with side panels
R-20	full trailer	13,400 kg	4,500 kg	6.9 m	5.2 m	2.5 m	8.25 × 15	8	2	Fitted with hydraulic loading ramps
R-25A	full trailer	19,800 kg	6,000 kg	7.93 m	5.2 m	2.5 m	8.25 × 15	12	3	Fitted with hydraulic loading ramps
R-28NA	full trailer	25,800 kg	6,200 kg	7.93 m	6 m	2.5 m	8.25 × 15	12	3	
S-36C	semi-trailer	28,100 kg	7,900 kg	10 m	6.95 m	2.5 m	8.25 × 15	12	3	Can be fitted with side panels
S-36L	semi-trailer	27,500 kg	8,500 kg	13.6 m	9.4 m	2.5 m	265/70 R 19.5	12	3	Spring-assisted loading ramps
R-42	full trailer	33,100 kg	8,900 kg	8 m	5.8 m	2.5 m	8.25 × 15	16	4	Fitted with hydraulic loading ramps
S-40	semi-trailer	40,000 kg	10,000 kg	11 m	6.5 m	2.5 m	12.00 × 20	8	2	
S-48	semi-trailer	40,000 kg	9,000 kg	10.5 m	7 m	2.5 m	8.25 × 15	12	3	Spring-assisted loading ramps
SR-60	semi-trailer	50,000 kg	15,000 kg	12.5 m	9 m	3 m	12.00 × 20	12	3	
S-56	semi-trailer	60,000 kg	15,000 kg	11.3 m	7.4 m	2.5 m	265/70 R 19.5	16	2	Fitted with spring-assisted loading ramps

De Filippi S-48 semi-trailer

De Filippi R-42 trailer

KOREA, SOUTH

SsangYong light trailers

Description

The SsangYong Motor Company produces five types of light trailer in the ¼ to 1½ ton payload range, including a water tank trailer. All five types are licence-produced versions of US service trailers and the US designation is retained in the Korean designations, prefixed by the letters DA. In nearly every respect the Korean trailers are the same as the US originals but the data quoted by the Korean manufacturer differ from the US specifications and are as shown here.

Status

In production. In service with the South Korean armed forces.

Specifications

Model	**DA-M100**	**DA-M101**	**DA-M105**	**DA-M332**	**DA-M106**
Type	cargo	cargo	cargo	cargo	water tank
Max load	250 kg	750 kg	1,500 kg	1,500 kg	1,500 litres
Length	2.735 m	3.78 m	4.23 m	3.76 m	4.23 m
Width	1.545 m	1.87 m	2.105 m	2.413 m	2.06 m
Height	1.31 m	2.17 m	2.413 m	1.532 m	2.26 m
Ground clearance	0.263 m	0.315 m	0.45 m	0.27 m	0.35 m
Track	1.27 m	1.48 m	1.714 m	2.032 m	1.714 m
Tyres	7.00 × 16	7.50 × 20	9.00 × 20	9.00 × 20	9.00 × 20
Electrical system	24 V	24 V	24 V	24 V	24 V

Manufacturer

SsangYong Motor Company Limited, SsangYong Building, 24-1, 2-ka, Jeo-dong, Chung-ku, Seoul, South Korea 100-748.
Tel: +82 2 273 4181. Telex: 27596 ssymc k.
Fax: +82 2 274 5062.

VERIFIED

SsangYong low bed semi-trailers

Description

The SsangYong Motor Company produces four types of low bed semi-trailer with capacities of 25,000 kg, 32,000 kg, 40,000 kg and 60,000 kg. All may be used to carry a wide range of military vehicles and equipment ranging from construction equipment to self-propelled artillery and armoured fighting vehicles. All four types have rear-mounted loading ramps, with the largest size at least using hydraulic power to lift the ramps after use.

Status

In production.

Specifications

Capacity	**25,000 kg**	**32,000 kg**	**40,000 kg**	**60,000 kg**
Weight (empty)	7,200 kg	9,900 kg	11,590 kg	21,000 kg
Length	10.75 m	11.68 m	11.68 m	12.885 m
Width	2.98 m	2.98 m	2.98 m	3.64 m
Height	1.65 m	1.65 m	1.71 m	2.256 m
Offset	3.63 m	4.25 m	4.25 m	4.75 m
Wheelbase	8.73 m	9.75 m	9.67 m	10.49 m
Tyres	9.00 × 20	10.00 × 20	11.00 × 20	11.00 × 20

Manufacturer

SsangYong Motor Company Limited, SsangYong Building, 24-1, 2-ka, Jeo-dong, Chung-ku, Seoul, South Korea 100-748.
Tel: +82 2 273 4187. Telex: 27596 ssymc k.
Fax: +82 2 274 5062.

VERIFIED

NETHERLANDS

GEMCO GPT 400 General Cargo Military Trailer

Description

The GEMCO GPT 400 General Cargo Military Trailer has a payload capacity of 400 kg and was designed to match the Mercedes-Benz G-Wagen and Peugeot P4 (4 × 4) light vehicles, having the same wheels, tyres and track. When attached to the Mercedes-Benz vehicle the trailer does not interfere with the full opening of the vehicle rear door.

The reinforced steel chassis is manufactured using 80 × 40 × 4 mm beams and is fitted with a NATO 40 mm towing eye. Overrun brakes are provided and there is a manually operated parking brake and a reversing brake lock. A 200 kg capacity jockey wheel with a pneumatic tyre is provided, along with two adjustable support legs at the rear. A complete set of military and civilian lighting equipment is fitted, along with blackout lighting.

Optional equipment includes: flat tarpaulin cover; superstructure and tarpaulin hood; steel hinged cover; opening rear panel; fire extinguisher; equipment brackets and belts; decontamination equipment; brackets for fuel and water cans; camouflage finish; and a camouflage net and rack. Special options include racks for Stinger missiles and there is an increased payload variant.

Specifications

Weight: 750 kg
Payload: 400 kg
Length:
(overall) 4 m
(internal) 2.23 m
Width:
(overall) 1.65 m
(internal) 1.07 m
Height:
(without hood) 1.06 m
(with hood) 1.52 m
(panel) 0.455 m
Tyres: as towing vehicle

Status

In production.

Manufacturer

GEMCO Mobile Systems BV, Sciencepark Eindhoven 5053, NL-5692 EB SON, Netherlands.
Tel: +31 499 462525. Fax: +31 499 462500.

UPDATED

GEMCO GPT 400 general cargo military trailer on tow behind Mercedes-Benz (4 × 4) light vehicle

GEMCO GPT 400 general cargo military trailer

DAF tank transporter semi-trailer Type YTS 10050

Description

The DAF tank transporter semi-trailer Type YTS 10050 was designed to carry all MBTs currently in service. Provision was made in the design of the semi-trailer to winch disabled tracked vehicles onto and off the trailer with the aid of the winch on the tractor.

The chassis frame consists of two steel box-section girders which form a single welded unit together with the cross-pulley members, outriggers, outer rails, floor plates, fifth-wheel support frame, tool boxes, pulley housing and other components. The king-pin is a standard 89 mm size.

The platform bed is provided with track guides to prevent sideways movement of the tank and these track guides can be adjusted laterally to suit different types of tracked vehicle.

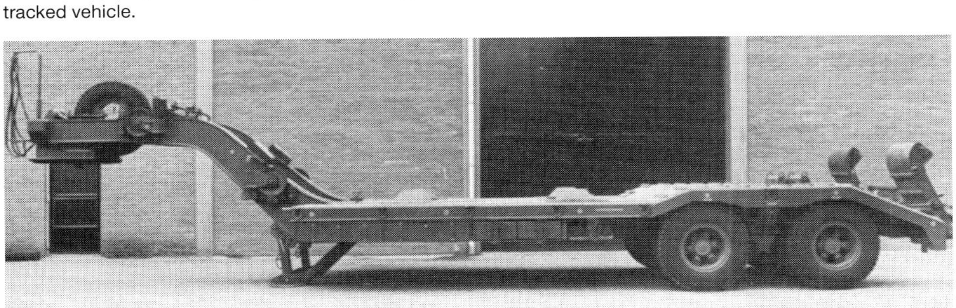

DAF tank transporter semi-trailer Type YTS 10050

The running gear comprises two separate and identical independently oscillating tandem axle units next to each other, each axle with four sets of twin wheels and tyres. The front and rear axles can oscillate 7° longitudinally in relation to the rocker beam brackets. The transverse oscillation of the wheel axles in relation to the rocker beam is also 7°.

The two ramps at the rear of the trailer can be raised and lowered by two personnel. Two mechanically operated, pivot-mounted front supports are raised and locked to the main girder at the gooseneck when travelling. As an optional extra, the semi-trailer can be equipped with two hydraulic front supports.

The trailer is fitted with a two-line brake system. Four of the eight diaphragm-type brake chambers are safety actuators with a locking mechanism functioning as a parking brake as well as a service brake. Standard equipment includes chains and two hydraulic jacks each with a capacity of 25,000 kg. The YTS 10060 semi-trailer is similar to the YTS 10050.

Specifications

Weight:
(empty) 15,000 kg
(loaded) 70,000 kg
Max load: 55,000 kg
Length: (overall) 11.72 m
Width:
(overall) 3.4 m
(of platform) 3.04 m
Height:
(top of platform) 1.27 m
(over loading ramps) 1.81 m
Ground clearance: (under chassis) 0.61 m
Track: 1.78 m
Wheelbase: (king-pin to centre of rear bogie) 8.7 m
Max towing speed: 50 km/h
Tyres: 11.00 × 20

Status

Production complete. In service with Belgium (29), Denmark (14), Netherlands (36), Spain (made under licence) and Sweden (91).

Manufacturer

DAF Military Sales, Hugo van der Goeslaan 1, NL-5643 TW Eindhoven, Netherlands.
Tel: +31 40 143440. Telex: 51085 daf nl.
Fax: +31 40 144378.

VERIFIED

DAF tanker trailer, 2-wheel, 900 litre Type YEW 400

Description

This single-axle two-wheeled trailer was designed for transporting up to 900 litres of drinking water and has a double-acting hand pump with suction hose, one 26 mm and four 12.7 mm taps and four supporting legs.

Specifications

Weight:
(empty) 670 kg
(loaded) 1,620 kg
Capacity: 900 litres
Length:
(overall) 3.61 m
(of tank) 1.55 m
Width:
(overall) 1.55 m
(of tank) 1.65 m
Height:
(overall) 1.6 m
(to towing eye) 0.89 m
Ground clearance: 0.4 m
Track: 1.5 m
Tyres: 9.00 × 16

Status

Production as required. In service with the Netherlands Army.

Manufacturer

DAF Military Sales, Hugo van der Goeslaan 1, NL-5643 TW Eindhoven, Netherlands.
Tel: +31 40 143440. Telex: 51085 daf nl.
Fax: +31 40 144318.

DAF tanker trailer, two-wheel, 900 litre Type YEW 400 (C R Zwart)

VERIFIED

DAF 2,500 kg trailer Type M53 (YEP 600)

Description

This trailer was designed for carrying pontoons for the Engineer Corps of the Netherlands Army. It has mechanical brakes and a single front support mounted to the rear of the towing lunette. The DAF trailer types YRE 600 and YRI 600 are similar to the YEP 600 but have torsion bar suspension and are used to transport radar and other electronic equipment.

Specifications

Weight:
(empty) 1,280 kg
(loaded) 3,780 kg
Max load: 2,500 kg
Length:
(overall) 4.59 m
(of mounting) 2.2 m
Width:
(over wheels) 2.23 m
(over mounting) 2.08 m
Height: (to top of mounting) 1.2 m
Ground clearance: 0.45 m
Track: 1.97 m
Wheelbase: (towing eye to centre of axle) 3 34 m
Tyres: 9.00 × 20

Status

Production complete. In service with the Netherlands Army.

DAF 2,500 kg trailer Type YEP 600

Manufacturer

DAF Military Sales, Hugo van der Goeslaan 1, NL-5643 TW Eindhoven, Netherlands.
Tel: +31 40 143440. Telex: 51085 daf nl.
Fax: +31 40 144318.

VERIFIED

NORWAY

Hägglunds Moelv MB 58 Off-road Trailer

Description

The Hägglunds Moelv MB 58 is a general purpose cargo trailer designed for extreme cross-country operations. The special Hägglunds Moelv pendulum bogie suspension provides excellent cross-country and road performance. The cargo platform is constructed of impregnated pine boards and the drop sides, tailboard and frontboard are plywood. The platform has an area of 9.36 m^2.

This trailer may be reinforced for total weights up to 12,000 kg on request.

Specifications

Weight: (chassis) 3,060 kg
Max load: 6,140 kg
Length: 5.47 m
Width: 2.56 m
Height: 2.35 m
Track: 2.03 m
Tyres: 315/80 × 22.5
Electrical system: 24 V

Status
Production on request.

Manufacturer
Hägglunds Moelv AS, Postboks 244, N-2391 Moelv, Norway.
Tel: +47 6234 6000. Fax: +47 6234 6001.

VERIFIED

Hägglunds Moelv MB 58 off-road trailer showing pendulum bogie suspension

Hägglunds Moelv MB 59 all-terrain trailer

Description
The Hägglunds Moelv MB 59 is a general purpose cargo trailer, developed to meet tough environmental and terrain conditions. The system is based on the Hägglunds Moelv pendulum bogie system providing good cross-country and road travelling capabilities. The materials used are a combination of high-tensile steel and aluminium.

The MB 59 can be used on- and off-road, over rough terrain, over swamps and in deep snow. The trailer is amphibious carrying a full load and for operations in deep snow the bogie system is fitted with skis. For the latter the lower wheel sections extend below the skis to allow the trailer to be towed from snow to dry road conditions.

The cargo platform may be covered by a canvas or hard top. The trailer system can be reinforced for total weights of up to 2,500 kg.

Hägglunds Moelv MB 59 all-terrain trailer

Specifications
Weight:
(total) 2,000 kg
(chassis, winter with skis) 900 kg
Max load: 1,100 kg
Load area: 1.67 × 2.38 m
Length: 4.2 m
Width: 1.84 m
Height: 1.89 m
Track: 1.546 m
Tyres: 18.5 × 14
Electrical system: 24 V

Status
Production as required.

Manufacturer
Hägglunds Moelv AS, Postboks 244, N-2391 Moelv, Norway.
Tel: +47 6234 6000. Fax: +47 6234 6001.

VERIFIED

Hägglunds Moelv MB 63 Multi trailer

Description
The Hägglunds Moelv MB 63 Multi trailer was designed and built specifically for transporting ISO containers and light tracked vehicles. The trailer features a unique system for loading and unloading standard ISO containers and the complete loading platform can be tilted to the rear to enable vehicles to be driven directly onto the platform.

The standard Multi trailer can carry loads up to 12,500 kg. To meet special requirements the trailer system can be reinforced for higher total weights.

At the front of the cargo platform are loading ramps to enable tracked vehicles to be driven directly onto the cargo platform. Anchoring fixtures are provided to secure vehicles and general cargo.

The design of the MB 63 Multi is based on the Hägglunds Moelv pendulum bogie system to provide enhanced cross-country and road mobility. The wide interspace between the wheels and the low loading height add to trailer stability, even if the cargo has a high centre of gravity.

Hägglunds Moelv MB 63 Multi trailer

Specifications
Weight:
(chassis) 4,500 kg
(total) 17,000 kg
Max load: 12,500 kg
Length: 8.45 m
Width: 2.5 m
Height: (top of platform) 1.15 m

Status
In production.

Manufacturer
Hägglunds Moelv AS, Postboks 244, N-2391 Moelv, Norway.
Tel: +47 6234 6000. Fax: +47 6234 6001.

VERIFIED

PAKISTAN

60 ton tank transporter semi-trailer

Description

The Military Vehicles Research and Development Establishment (MVRDE) at Rawalpindi, has developed a 60 ton tank transporter semi-trailer for the Pakistan armed forces. The semi-trailer is suited to the transport of MBTs, APCs and heavy construction plant up to a payload of 60 tons.

The semi-trailer uses a built-in gooseneck construction crimped upwards in its frontal part, thus providing an optimum angle of lock to the prime mover. The low level bed platform is covered with chequered steel plates while the chassis frame is constructed using high-quality steel, employing the latest welding techniques. The rear part of the platform slopes downwards, presenting a loading angle of 20°. Two Roker beam axles are provided, each with a load capacity of 1,400 kg.

The loading ramps can be adjusted laterally by 400 mm. The distance between the loading ramps can also be adjusted according to the track width of various vehicle types.

60 ton tank transporter semi-trailer carrying M88A1 ARV **1996**

Specifications

Weight: 77 tons
Max load: 60 tons
Length: (overall) 12.6 m
Width: (overall) 3.65 m
Height: (top of platform) 1.3 m
Ground clearance: 350 mm
King-pin diameter: 89 mm

Status

Ready for production.

Development Agency

Military Vehicles Research and Development Establishment, (MVRDE), Ministry of Defence, PO Box 240, Abid Majeed Road, Rawalpindi, Pakistan.
Telex: 54102 MVRDR PK.
Fax: +92 51 584786.

VERIFIED

POLAND

Polish heavy transport trailers and semi-trailers P-40, P-50, NP-40 and PN-600

Description

The P-40 and P-50 are conventional three-axle trailers with folding ramps at the rear of the trailer bed. The NP-40 is a semi-trailer and is towed by the single-axle Zg-201A tractor. The PN-600 is a four-axle full trailer with a single-axle dolly and three axles under the loading platform. There are folding ramps at the rear of the platform.

Specifications

Model	**P-40**	**NP-40**	**PN-600**
Weight			
(empty)	13,575 kg	14,000 kg	13,500 kg
(loaded)	53,575 kg	54,000 kg	73,500 kg
Max load	40,000 kg	40,000 kg	60,000 kg
Length			
(overall)	11.2 m	n/a	11.75 m
(trailer only)	n/a	n/a	9.88 m
Length of platform	5.1 m	8.5 m	6.5 m
Width	2.9 m	3.15 m	3.1 m
Height	2 m	n/a	n/a
(to platform)	1 m	n/a	1.1 m
Towing speed			
(empty)	15 km/h	40 km/h	80 km/h
(loaded)	–	–	60 km/h
Tyres	–	–	8.25 × 15 (32)

Status

The P-40 and P-50 are known to be in service with the Polish Army. The NP-40 and the PN-600 both have military applications.

Manufacturer

(PN-600) ZREMB Factory, Wroclaw, Poland.

VERIFIED

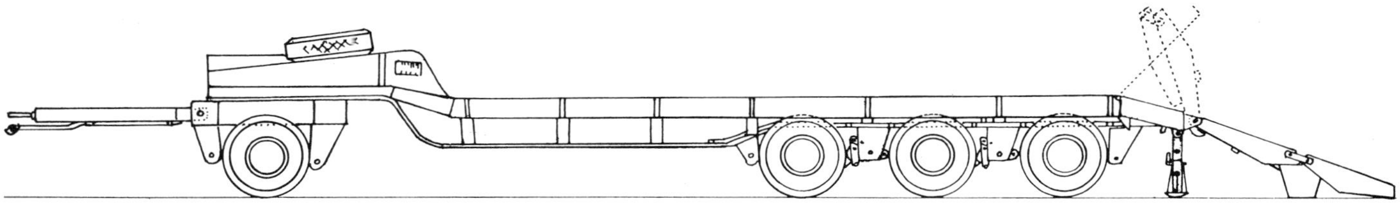

Polish PN-600 60,000 kg trailer

ROMANIA

200-man field kitchen trailer

Description

This single-axled kitchen trailer was designed to prepare hot food for up to 200 personnel at one time. The heater for the system may be either diesel oil or petrol and is used to heat up to four boilers, one 120 litre, one 70 litre and two 60 litre. If required, one of the 60 litre boilers may be replaced by a cooking range which will then be an addition to a further two cooking ranges already provided as fixtures.

Along each side of the trailer are two containers, the sides of which can be opened up to form working surfaces. The containers contain various kitchen utensils to assist food preparation and serving.

The trailer weighs 1,400 kg and is 3.565 m long, 2.08 m wide and 2.28 m high.

Status

In production. In service with the Romanian armed forces. Offered for export sales.

Marketing Agency

ROMTEHNICA, 9-11 Drumul Taberei Street, Bucharest, Romania.
Tel: +40 046 2087. Telex: 11608 TXKCO.
Fax: +40 046 0317.

VERIFIED

200-man field kitchen trailer

RUSSIAN FEDERATION AND ASSOCIATED STATES (CIS)

ChMZAP-2303V 20 tonne heavy load trailer

Description

This three-axle heavy load trailer is normally towed by a KrAZ-255B (6 × 6) 7,500 kg truck. If required, the four-wheeled dolly can be removed and the trailer is then used as a semi-trailer and can be towed by a KrAZ-221 or similar vehicle. The suspension of the dolly consists of semi-elliptical springs with check springs, while the rear suspension consists of semi-elliptical springs. The trailer is provided with drum-type air-operated brakes on the rear axle, with a mechanical parking brake operating on one rear axle. Standard equipment includes loading ramps, guide bars and guide rollers for the tractor winch cable.

Specifications

Weight: (empty) 9,750 kg
Max load:
(as trailer) 20,500 kg
(as semi-trailer) 22,000 kg
Length:
(overall) 12.95 m
(of platform) 6.59 m
Width: 3 m
Height:
(overall) 2.09 m
(loading) 1.35 m
Ground clearance: 0.28 m
Track: 1.92 m
Max towing speed:
(road) 50 km/h
(dirt road) 25 km/h
Tyres: 12.00 × 20

Status

In service with the RFAS and other armed forces.

Manufacturer

Chelyabinsk Machine Building Plant for Motor Vehicle and Tractor Trailers, Chelyabinsk, RFAS.

VERIFIED

ChMZAP-5523 heavy flatbed trailer

Description

This trailer was designed for hauling heavy equipment on both hard surface and dirt roads. The front dolly can be removed enabling the trailer to be used as a semi-trailer. Two hydraulically operated folding ramps are provided at the rear of the hull.

The suspension consists of longitudinal semi-elliptical springs with check springs while the rear bogie is provided with semi-elliptical springs. Drum-type brakes are provided on all eight wheels of the rear axle and a parking brake operates on one of the rear axles.

Specifications

With front dolly
Weight:
(loaded) 31,000 kg
(empty) 9,750 kg
(on front dolly unladen) 3,750 kg
(on rear bogie unladen) 6,000 kg
Max load: 21,250 kg
Length:
(overall) 12.95 m
(of platform) 6.43 m
Width:
(overall) 3 m
(of platform) 3 m
Height:
(overall) 2.08 m
(of platform) 1.35 m
Ground clearance: (under axle of rear suspension equaliser arm) 0.28 m
Max towing speed: 50 km/h
Tyres: 12.00 × 20

Status

In service with the RFAS and other armed forces.

Manufacturer

Chelyabinsk Machine Building Plant for Motor Vehicle and Tractor Trailers, Chelyabinsk, RFAS.

VERIFIED

ChMZAP-5524 23 tonne heavy load trailer

Description

The ChMZAP-5524 can be used either as a conventional trailer, or as a semi-trailer with its single-axle four-wheeled bogie removed. Its front suspension consists of semi-elliptical springs with check springs, while its rear suspension consists of a bogie with semi-elliptical springs. The trailer is normally towed by a KrAZ-255B (6 × 6) 7,500 kg truck or a tractor truck such as the KrAZ-255V (6 × 6) when being used as a semi-trailer.

The ChMZAP-5524A is similar but is extended by 1 m.

Specifications

Weight:
(loaded) 30,000 kg
(empty) 6,300 kg
Max load:
(as trailer) 23,700 kg
(as semi-trailer) 25,500 kg
Length: (overall) 11.36 m
Width: (overall) 2.64 m
Height:
(overall) 1.6 m
(loading) 1.09 m
Ground clearance: 0.28 m
Track: 1.92 m
Max towing speed:
(road) 50 km/h
(dirt road) 20 km/h
Tyres: 12.00 × 20

Status

In service with the RFAS and other armed forces.

Manufacturer

Chelyabinsk Machine Building Plant for Motor Vehicle and Tractor Trailers, Chelyabinsk, RFAS.

VERIFIED

ChMZAP-5208 40 tonne heavy flatbed trailer

Description

This trailer entered production at the Chelyabinsk plant in 1957 and was designed for hauling heavy cargo on hard-surfaced roads. The trailer has drum-type air-operated brakes on all wheels, with a hydraulically operated parking brake on the rear axle. Loading ramps, guide bars and load-securing gear are also fitted. The ChMZAP-5208 is normally towed by the MAZ-537A (8 × 8) tractor truck.

Specifications

Weight:
(loaded) 51,000 kg
(empty) 11,000 kg
(on front dolly, unloaded) 3,900 kg
(on rear bogie, unloaded) 7,100 kg
Max load: 40,000 kg
Length:
(overall) 9.33 m
(of platform) 4.88 m
Width:
(overall) 3.2 m
(of platform) 3.2 m
Height:
(overall) 1.94 m
(of platform) 1.44 m
Ground clearance: (loaded under front suspension cross member) 0.35 m
Track: (centre of outer wheels) 2.41 m
Wheelbase:
(centre of front dolly to centre of rear bogie) 4.75 m
(rear bogie) 1.19 m
Max towing speed: 40 km/h
Tyres: 8.25 × 20

Status

In service with the RFAS and other armed forces.

Manufacturer

Chelyabinsk Machine Building Plant for Motor Vehicle and Tractor Trailers, Chelyabinsk, RFAS.

VERIFIED

ChMZAP-5247G 50 tonne heavy load semi-trailer

Description

The ChMZAP-5247G 50 tonne semi-trailer is widely used for transporting MBTs such as the T-62 and is towed by the MAZ-537 (8 × 8) tractor truck. All eight wheels have air-operated drum brakes with a mechanical parking brake. Standard equipment includes powered loading ramps, guide bars and load-securing gear.

The ChMZAP-5247B is a 45 tonne semi-trailer 14.68 m long and 2.64 m wide. The rear section is not provided with loading ramps and has no incline.

Specifications

Weight:
(loaded) 68,000 kg
(empty) 18,000 kg
Max load: 50,000 kg
Length:
(overall) 15.23 m
(of platform) 5.69 m
Width:
(overall) 3.38 m
(of platform) 3.23 m
Height:
(overall) 2.78 m
(of platform) 1.16 m
Ground clearance: 0.35 m
Track: 2.09 m
Max towing speed: 50 km/h

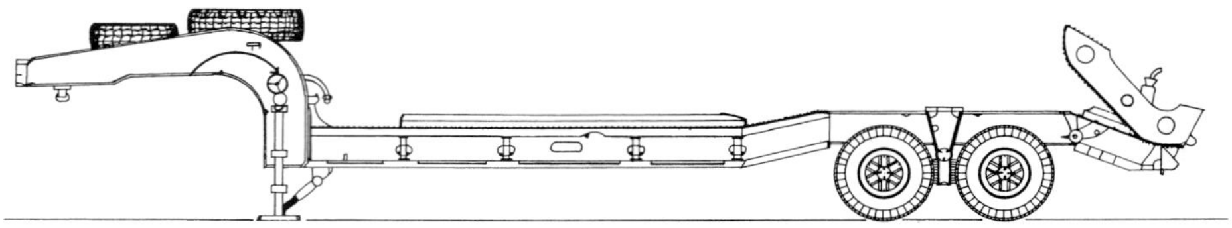

ChMZAP-5247G 50 tonne heavy load semi-trailer

Tyres: 15.00 × 20
Electrical system: 24 V

Status

In service with members of the former Warsaw Pact and with countries in Africa and the Middle East.

Manufacturer

Chelyabinsk Machine Building Plant for Motor Vehicle and Tractor Trailers, Chelyabinsk, RFAS.

VERIFIED

ChMZAP-5212 60 tonne heavy flatbed trailer

Description

This four-axle heavy flatbed trailer was designed for carrying heavy loads on highways and is normally towed by the MAZ-537A (8 × 8) tractor truck. The trailer has drum brakes, air-operated on all wheels, loading ramps, guide bars and load-securing gear.

Specifications

Weight:
(loaded) 74,500 kg
(empty) 14,500 kg
(on front dolly empty) 7,340 kg
(on rear bogie empty) 7,160 kg
Max load: 60,000 kg
Length:
(overall) 11.37 m
(of load platform) 5.5 m
Width: (overall) 3.3 m
Height:
(overall) 2.07 m
(of load platform) 1.14 m
Ground clearance: (at full load under suspension equalisers) 0.19 m
Track: (on centres of external dual wheels) 2.48 m
Wheelbase: 1.2 m + 4 m + 1.2 m
Max towing speed: 32 km/h
Tyres: 9.00 × 15

Status

In service with the RFAS and other armed forces.

Manufacturer

Chelyabinsk Machine Building Plant for Motor Vehicle and Tractor Trailers, Chelyabinsk, RFAS.

VERIFIED

RFAS Trailers

Designation	Type	Unladen weight	Payload	Max tow speed	Length overall	Height overall	Width	Wheel-base	Track	Ground clearance	Load area (l × w)	No of tyres	Electrical system	Towing vehicle(s)
GAZ-704	cargo	340 kg	500 kg	75 km/h	2.56 m	1.15 m	1.645 m	–	1.44 m	0.28 m	1.66 × 1.07 m	2	12 V	GAZ-69, GAZ-69A
UAZ-8109	cargo	350 kg	500 kg	75 km/h	2.785 m	1.19 m	1.72 m	–	1.44 m	0.29 m	1.66 × 1.07 m	2	12 V	UAZ-469, UAZ-469B
TAPZ-755, TAPZ-755A	special	470 kg	1,500 kg	80 km/h	3.025 m	0.8 m	2.07 m	–	1.77 m	0.375 m	1.73 × 0.936 m	2 + 1 spare	12 V	GAZ-66
GKB-8302	cargo	950 kg	1,200 kg	90 km/h	3.82 m	2.52 m	2.24 m	–	1.82 m	0.35 m	2.153 × 1.82 m	2 + 1 spare	12 V	GAZ-66
GKB-83021	special	650 kg	1,500 kg	90 km/h	3.25 m	1.115 m	2.2 m	–	1.82 m	0.35 m	1.685 × 1.56 m	2 + 1 spare	12 V	GAZ-66
IAZP-738	special	570 kg	1,500/ 1,800 kg	80 km/h	3.85 m	1.65 m	2.1 m	–	1.77 m	0.378 m	2.285 × 1.87 m	2 + 1 spare	12 V	GAZ-66
IAZP-739	special	470 kg	1,000 kg	80 km/h	3.85 m	1.65 m	2.1 m	–	1.77 m	0.355 m	2.285 × 1.87 m	2 + 1 spare	12 V	GAZ-66
GKB-8301	cargo	1,610 kg	2,500 kg	80 km/h	4.825 m	2.9 m	2.5 m	–	1.89 m	0.36 m	3.225 × 2.325 m	2 + 1 spare	12 V	ZIL-131
GKB-83011	special	1,110 kg	3,000 kg	80 km/h	4.055 m	1.16 m	2.73 m	–	1.88 m	0.36 m	no platform	2 + 1 spare	12 V	ZIL-131
SMZ-710V	cargo	1,500 kg	2,000 kg	60 km/h	5.64 m	2.715 m	2.32 m	3.7 m	1.59 m	0.305 m	5.3 × 2.11 m	4 + 1 spare	12 V	ZIL-157K, ZIL-131
SMZ-710B	special	1,250 kg	2,500 kg	60 km/h	5.64 m	1.13 m	1.89 m	2.4 m	1.59 m	0.305 m	3.95 × 1.89 m	4 + 1 spare	12 V	ZIL-157K, ZIL-131
SMZ-8325	cargo	2,030 kg	2,100 kg	80 km/h	6.22 m	2.78 m	2.48 m	2.5 m	1.82 m	0.42 m	4.22 × 2.25 m	4 + 1 spare	12 V	ZIL-131
SMZ-8326	special	1,650 kg	2,500 kg	80 km/h	6.2 m	1.2 m	2.23 m	2.5 m	1.82 m	0.42 m	4.535 × 0.925 m	4 + 1 spare	12 V	ZIL-131
SMZ-810	cargo	2,400 kg	4,000 kg	50 km/h	6.24 m	2.315 m	2.35 m	2.95 m	1.97 m	0.3 m	4.21 × 2.185 m	4 + 1 spare	12 V	Ural-375D
SMZ-810A	special	1,900 kg	4,500 kg	50 km/h	6.34 m	1.275 m	2.35 m	2.95 m	1.97 m	0.3 m	4.5 × 1.8 m	4 + 1 spare	12 V	Ural-375D
782V	cargo	3,030 kg	4,000 kg	75 km/h	6.875 m	2.9 m	2.455 m	2.95 m	2 m	0.35 m	4.87 × 2.3 m	4 + 1 spare	12 V	Ural-4320, Ural-375N
782B	special	2,320 kg	4,680 kg	75 km/h	6.8 m	1.34 m	2.34 m	2.95 m	2 m	0.35 m	4.93 × 1.8 m	4 + 1 spare	12 V	Ural-4320, Ural-375N
GKB-817	cargo	2,540 kg	5,500 kg	80 km/h	6.68 m	1.87 m	2.5 m	3 m	1.8 m	0.37 m	4.95 × 2.35 m	4 + 1 spare	12 V	ZIL-130, ZIL-130G
GKB-817A	cargo	2,450 kg	5,500 kg	80 km/h	6.68 m	1.87 m	2.5 m	3 m	1.8 m	0.37 m	4.95 × 2.35 m	4 + 1 spare	12 V	ZIL-130, ZIL-130G
GKB-817V	special	2,640 kg	5,400 kg	80 km/h	n/a	n/a	n/a	n/a	1.8 m	0.37 m	n/a	4 + 1 spare	12 V	ZIL-130, ZIL-130G
MAZ-5207VSh	special	2,450 kg	6,750 kg	50 km/h	6.55 m	1.14 m	2.405 m	3 m	1.97 m	0.31 m	4.91 × 2.405 m	4 + 1 spare	24 V	KrAZ-255B
MAZ-8925	special	3,000 kg	7,000 kg	85 km/h	7.785 m	1.14 m	2.5 m	3.7 m	1.97 m	0.43 m	5.56 × 1.7 m	4 + 1 spare	24 V	KrAZ-255B
MAZ-8926	cargo	4,000 kg	6,000/ 8,000 kg	85 km/h	7.71 m	2.79 m	2.5 m	3.7 m	1.97 m	0.43 m	5.5 × 2.365 m	4 + 1 spare	24 V	KrAZ-255B
GKB-8350	cargo	3,500 kg	8,000 kg	80 km/h	8.29 m	1.81 m	2.5 m	4.34 m	2.315 m	0.38 m	6.1 × 2.315 m	8 + 1 spare	24 V	KamAZ-5320
MAZ-5224V	special	4,000 kg	10,500 kg	50 km/h	8.355 m	1.06 m	2.87 m	3.7 m	2.15 m	0.38 m	6.46 × 2.8 m	8 + 2 spare	24 V	KrAZ-255B
MAZ-8950	special	4,500 kg	10,500 kg	70 km/h	8.92 m	1.2 m	2.82 m	4.1 m	2.16 m	0.4 m	6.78 × 2.8 m	4 + 1 spare	24 V	KrAZ-255B
MAZ-8378	cargo	5,500 kg	14,500 kg	85 km/h	11.57 m	3.735 m	2.5 m	7.375 m	2.365 m	0.43 m	9.67 × 2.365 m	8 + 1 spare	24 V	MAZ-5335, MAZ-5336
ChMZAP-5208	transporter	10,900 kg	40,000 kg	40 km/h	9.33 m	1.74 m	3.2 m	4.13 m	2.55 m	0.26 m	7.43 × 3.2 m	24 + 2 spare	24 V	MAZ-537P
ChMZAP-8386	transporter	13,170 kg	40,000 kg	60 km/h	11.23 m	1.63 m	3.15 m	3.85 + 1.3 m	–	0.25 m	8.2 × 3.15 m	12 + 2 spare	24 V	MAZ-537P

VERIFIED

RFAS Semi-trailers

Designation	Type	Weight	Payload	Max tow speed	Length overall	Height overall	Width	Coupling to wheel-base	Track	Ground clearance	Load area (l × w)	No of tyres	Electrical system	Towing vehicle(s)
OdAZ-885	cargo	2,850 kg	7,500 kg	80 km/h	6.385 m	2.03 m	2.455 m	4.43 m	1.79 m	0.315 m	6.08 × 2.22 m	4 + 1 spare	12 V	ZIL-130V1, ZIL-MMZ-164N, KAZ-608, KAZ-606, KAZ-606A
MAZ-5245	cargo	3,800 kg	13,500 kg	75 km/h	8.12 m	2.355 m	2.5 m	5.18 m	1.866 m	0.38 m	7.875 × 2.32 m	4	24 V	MAZ-504A
MAZ-93801	cargo	4,100 kg	13,500 kg	85 km/h	8.745 m	2.035 m	2.5 m	5.75 m	1.86 m	0.43 m	8.535 × 2.365 m	4 + 1 spare	24 V	MAZ-504A, MAZ-5429
OdAZ-9350	cargo	5,500 kg	9,500 kg	75 km/h	10.57 m	3.49 m	2.5 m	4.545 + 1.42 m	2 m	0.4 m	10 × 2.335 m	4 + 1 spare	12 V	Ural-4420, Ural-375S
MAZ-938	special	7,500 kg	13,000 kg	70 km/h	13.05 m	2.23 m	2.5 m	8.46 + 1.4 m	2.16 m	0.36 m	9.16 × 2.8 m	4	24 V	KrAZ-255V
MAZ-938B	special	7,000 kg	13,500 kg	70 km/h	13.05 m	2.23 m	2.5 m	8.46 + 1.4 m	2.16 m	0.36 m	9.16 × 2.8 m	4	24 V	KrAZ-255V
OdAZ-9370	cargo	3,700 kg	15,400 kg	80 km/h	9.64 m	2.03 m	2.5 m	6.14 + 1.32 m	1.85 m	0.26 m	9.18 × 2.32 m	8 + 1 spare	24 V	KamAZ-5410
MAZ-5205A	cargo	5,700 kg	20,000 kg	80 km/h	10.18 m	3.615 m	2.5 m	5.53 + 1.54 m	1.9 m	0.34 m	9.965 × 2.32 m	8 + 1 spare	24 V	MAZ-504V
OdAZ-9385	cargo	4,100 kg	21,000 kg	80 km/h	10.67 m	1.98 m	2.5 m	6.19 + 1.32 m	1.85 m	0.285 m	10.17 × 2.32 m	8 + 1 spare	24 V	KamAZ-54112
MAZ-9397	cargo	6,000 kg	21,000 kg	85 km/h	11.47 m	3.715 m	2.5 m	6.5 + 1.54 m	1.79 m	0.35 m	11.24 × 2.4 m	8 + 1 spare	24 V	MAZ-5432
MAZ-941	cargo	6,700 kg	25,000 kg	80 km/h	13.22 m	3.755 m	2.5 m	8.85 + 1.54 m	1.86 m	0.42 m	12.795 × 2.365 m	8 + 1 spare	24 V	MAZ-515B
ChMZAP-5524P	special	4,400 kg	25,600 kg	68 km/h	9.8 m	1.6 m	2.638 m	4.52 + 1.4 m	1.92 m	0.24 m	n/a	8 + 2 spare	24 V	KrAZ-258
MAZ-9398	cargo	6,500 kg	26,500 kg	85 km/h	12.33 m	3.715 m	2.5 m	6.485 + 1.5 + 1.5 m	1.97 m	0.42 m	12.12 × 2.4 m	6 + 1 spare	24 V	MAZ-6422
ChMZAP-5523A	transporter	7,000 kg	25,000 kg	70 km/h	12.83 m	n/a	3 m	6.83 + 1.4 m	n/a	0.24 m	6.765 × 3 m	8 + 1 spare	24 V	KrAZ-258
MAZ-5247/ ChMZAP-5247G	transporter	18,000 kg	50,000 kg	50 km/h	15.23 m	2.78 m	3.38 m	10.84 + 1.58 m	2.09 m	0.35 m	n/a	8 + 1 spare	24 V	MAZ-537
ChMZAP-9990	transporter	18,000 kg	52,000 kg	60 km/h	14.42 m	3.19 m	3.15 m	8.4 + 1.3 + 1.3 m	n/a	0.25 m	8.96 × 3.15 m	12 + 2 spare	24 V	MAZ-537

VERIFIED

Other RFAS trailers

Designation	Type	Manufacturer	Capacity	Unladen weight	Length	Width	Height	Notes
2-ASP-4.5	dump	Nal'chik	4,500 kg	2,190 kg	5.02 m	2.26 m	2.25 m	Side dump trailer
2-PN-2 (SMZ-710V)	low bed, 2-axle		2,000 kg	1,500 kg	5.75 m	2.32 m	1.75 m	Towed by ZIL-157K or ZIL-131[1]
2-PN-4 (SMZ-810)	low bed, 2-axle		4,000 kg	2,400 kg	6.24 m	2.35 m	1.45 m	Towed by Ural-375D[2]
2-PN-6 (MAZ-520)	low bed, 2-axle		6,000 kg	3,200 kg	6.58 m	2.5 m	1.52 m	Towed by MAZ or KrAZ[3]
GKB-817 (2-P-5)	2-axle	Lugansk, 1967 Petropavlovsk, 1968	5,000 kg	2,540 kg	6.6 m	2.5 m	2.22 m	Wooden platform with hinged sides towed by Ural-375
GKB-816	2-axle		3,000 kg	1,500 kg	5.73 m	2.25 m	1.73 m	Towed by ZIL-131
KAZ-717	semi-trailer	Kutaisi, 1965	11,500 kg	4,000 kg	7.69 m	2.48 m	1.98 m	Wooden platform, 3 hinged sides
MAZ-524Z	2-axle	Minsk, 1960	6,800 kg	3,200 kg	6.97 m	2.5 m	2.05 m	MAZ-524ZB has canopy
MAZ-584B	semi-trailer	Mytishchi, 1956 Saransk, 1961	7,000 kg	2,525 kg	6.3 m	2.46 m	2 m	Hinged side boards and tailgate, towed by ZIL-164A
MAZ-886 (2-P-8)	2-axle	Minsk, 1968	8,500 kg	3,500 kg	7.12 m	2.5 m	3.31 m	Hinged on 3 sides, fitted canopy and bows, towed by MAZ-500 or KrAZ-255B
MAZ-938	semi-trailer	Minsk	15,000 kg	7,500 kg	12.2 m	2.8 m	2.33 m	Towed by KrAZ-255
MAZ-941	semi-trailer	Minsk	24,700 kg	13,700 kg	10.67 m	2.5 m	2.3 m	Towed by MAZ-525
MAZ-5206	flatbed, 2-axle	Minsk	10,000 kg	5,400 kg	9.23 m	2.53 m	1.53 m	Towed by wheeled or tracked vehicle
MAZ-5224	low bed, 2-axle	Minsk	10,000 kg	5,000 kg	8.04 m	2.68 m	1.77 m	Towed by KrAZ-255B[4]
MAZ-5243	2-axle	Minsk	6,800 kg	3,200 kg	6.97 m	2.5 m	2.05 m	Towed by MAZ-500 truck
MAZ-5245	semi-trailer	Minsk, 1961	14,000 kg	3,800 kg	8.12 m	2.5 m	2.33 m	Towed by MAZ-504[5]
OdAZ-740	semi-trailer	Odessa	23,000 kg	7,000 kg	11.33 m	2.5 m	3.45 m	Van type, towed by KrAZ-221B
OdAZ-760	semi-trailer	Odessa	14,000 kg	4,800 kg	8.84 m	2.5 m	2.27 m	Towed by Ural-377
OdAZ-784	semi-trailer	Odessa, 1959	7,000 kg	2,950 kg	8 m	2.57 m	3.3 m	Van type, towed by Kaz-606
OdAZ-794	semi-trailer	Odessa, 1966	7,500 kg	3,000 kg	6.93 m	2.5 m	3.22 m	Van type
OdAZ-795	semi-trailer	Odessa, 1965	13,250 kg	4,200 kg	9.38 m	2.5 m	3.32 m	Van type, towed by MAZ-504
OdAZ-832	semi-trailer	Odessa, 1965	12,000 kg	4,000 kg	9.5 m	2.5 m	3.5 m	Van type, towed by MAZ-200V
OdAZ-828	semi-trailer	Odessa	5,500 kg	5,800 kg	8.46 m	2.37 m	1.97 m	Van type, towed by ZIL-157KV
OdAZ-935	semi-trailer	Odessa, 1965	13,500 kg	4,800 kg	9.38 m	2.5 m	3.32 m	Van type, towed by Ural-377S
PAZS-3137	mobile fuelling unit trailer		4,200 litres	2,806 kg	5.99 m	2.31 m	2.84 m	Fitted with pumps and hoses
TAPZ-738 (1-P-1.5)	trailer chassis	Irbitsk	1,800 kg	600 kg	3.94 m	2.07 m	1.77 m	Towed by GAZ-53A, GAZ-63 and GAZ-66
TAPZ-739 (1-P-1)	trailer chassis	Irbitsk	1,000 kg	500 kg	3.55 m	1.81 m	0.92 m	Towed by GAZ-53A, GAZ-63[6] and GAZ-66
TAPZ-754V	2-axle	Irbitsk, 1958	4,000 kg	1,900 kg	6.05 m	2.39 m	2.13 m	Wooden platform, 3 hinged sides, towed by ZIL-164
TAPZ-755A	trailer chassis	Irbitsk	1,500 kg	470 kg	3.03 m	1.82 m	0.93 m	Towed by GAZ-51A, GAZ-53A and GAZ-66[7]
TsV-50	single-axle trailer	Biysk	1,000 litres	910 kg	3.92 m	1.66 m	2.08 m	Drinking water trailer

Designation	Type	Manufacturer	Capacity	Unladen	Length	Width	Height	Notes
PTs-4.2-754V	tanker-trailer	Grabovskiy, 1966	4,200 litres	2,268 kg	6 m	2.28 m	2.75 m	Fitted with dispensing equipment
PTs-6	tanker-trailer	Grabovskiy	6,700 litres	3,700 kg	5.06 m	2.4 m	2.55 m	Fitted with dispensing equipment
VMG-40-51	tanker-trailer		950 litres (water) 300 litres (oil)		4.2 m	2.13 m	2.28 m	Water and oil, heater fitted
PKhZ	semi-trailer, 2-axle	–	–		–	–		Field bakery
SP-32	4-axle	–	–		–	–		Dining hall

1 SMZ-710B trailer chassis is a modified version of 2-PN-2 with 2,500 kg load capacity. The 2-PN-2M (SMZ-8325) trailer is used for the PPK material preservation trailer
2 Modified versions of 2-PN-4 are SMZ-810A trailer chassis with 4,500 kg load capacity and SMZ-810PA trailer chassis for special bodies
3 Modified version of 2-PN-6 is MAZ-5207VSh trailer chassis with 6,750 kg capacity, which is fitted with various types of body, and MAZ-847 dump trailer with 6,000 kg capacity
4 Modified version of MAZ-5224 is MAZ-5224V chassis for van bodies
5 MAZ-5245B is provided with canopy. Between 1956 and 1961 plant produced MAZ-5215B semi-trailer with 12,500 kg capacity
6 This chassis is used for 1-P-1 1,000 kg cargo trailer. TAPZ-739K chassis is used for KP-125M field kitchen
7 MNUM-50M trailer-mounted pump comprises 3 V 40/25 pump on TAPZ-755A chassis

Manufacturers

Biysk Molmashstroy Plant.
Grabovskiy Specialised Automobile Plant.
Irbitsk Motor Vehicle Trailer Plant, Irbit.
Kutaisi Motor Vehicle Plant, Kutaisi, Georgia.
Lugansk Motor Vehicle Assembly Plant.
Minsk Wheel Tractor Plant, Minsk, Belarus.
Mytishchi Machine Building Plant, Mytishchi.
Nal'chik Machinery Construction Plant.
Odessa Motor Vehicle Assembly Vehicle Plant, Odessa, Ukraine.
Petropavlovsk Machine Construction Plant.
Saransk Dump Truck Plant, Saranski.
Zelenokumsk Plant.

VERIFIED

SOUTH AFRICA

10 tonne cargo trailer

Description

This general purpose cargo trailer is the standard South African Defence Forces load-carrying trailer and can be towed by SAMIL 100 trucks. The trailer is a twin-axled design with 650 kPa airbrakes and an all-steel body. There are four downward-opening side panels, two each side, and steel wire mesh panel extensions are permanently fitted for oversize loads. A spare wheel is carried under the cargo body, and a full set of rear lights is provided.

Specifications

Weight:
(empty) 4,200 kg
(fully loaded) 14,200 kg
Payload: 10,000 kg
Length:
(incl towbar) 6.66 m
(body internal) 4.85 m
Height:
(incl cant rails) 2.85 m
(towing eye, coupling) 1 m
(body internal) 0.46 m
Width: (body internal) 2.335 m
Wheelbase: 3.2 m
Track: 2.03 m
Tyres: 12.00 × 20 × 14 ply

10 tonne cargo trailer

Status

In service with the South African National Defence Forces.

Manufacturer

Enquiries to: Armscor, Private Bag X337, Pretoria 0001, South Africa.
Tel: +27 12 428 1911. Telex: 30217.
Fax: +27 12 428 5635.

VERIFIED

3 tonne 4-wheel workshop trailers

Description

There are a number of these workshop trailers, all basically the same in layout and appearance but differing in equipment and internal layout. A general repair trailer is one of the more numerous types but there are also armourers' workshop, welding workshop and vehicle repair trailers. Each trailer has a fully enclosed body when being towed but, once in position, side and rear-mounted flaps are opened upwards to reveal the interior. Once emplaced, each trailer is usually covered by a canvas tent with the trailer roof and side flaps supporting the tent roof. The tent, when erected, is 8.5 m long, 5.7 m wide and 2.5 m high. Each trailer has its own 10 kVA 380 V three-phase 50 Hz generator with a convertor supplying 265 V at 200 Hz. A 220 V supply is used for internal lighting and two fans in the trailer roof. The general chassis voltage supply is 24 V from batteries. Airbrakes are fitted. The trailers are supplied complete with all the necessary machine and other tools.

Specifications

Weight:
(empty) 3,780 kg
(equipped) 5,500 kg
Length:
(incl drawbar) 7.765 m
(interior) 5.6 m
Height:
(total) 2.49 m
(interior) 1.28 m
(interior, over gooseneck) 1.06 m
(tow eye) 0.9 m

3 tonne 4-wheel workshop trailer

Width: (overall) 2.3 m
Track: 2.03 m
Tyres: 9.00 × 16 × 12 ply
Electrical system: 24 V

Status
In service with the South African National Defence Forces.

Manufacturer
AFRIT (Pty) Ltd, PO Box 911374, Rosslyn 0200, South Africa.
Tel: +27 12 541 2123. Telex: 32-0439 SA.

VERIFIED

2 tonne 4-wheel shower unit trailer

Description
This trailer body is basically the same as that used on an office trailer but is fitted with shower facilities for up to eight personnel at one session. Each shower head can be operated individually but normally an operator controls the shower cycle. Water is pumped from an outside source through a filter unit to three liquid petroleum gas heaters, although any one or two of these heaters can be eliminated if required. Internal lighting is supplied from a 12 V battery and a tunnel tent can be provided for connection from the rear access door to a standard military tent nearby. Airbrakes are fitted. The water pump is powered by a petrol/gas-operated engine carried behind an access flap.

2 tonne 4-wheel shower unit trailer

Specifications
Weight:
(empty) 3,000 kg
(equipped) 4,000 kg
Length:
(incl drawbar) 6.65 m
(internal) 4.79 m
Width: (internal) 2.21 m
Height:
(total) 3.075 m
(internal) 2.1 m
Track: 2.03 m
Tyres: 7.50 × 16 × 10 ply
Electrical system: 24 V

Status
In service with the South African National Defence Forces.

Manufacturer
Enquiries to: Armscor, Private Bag X337, Pretoria 0001, South Africa.
Tel: +27 12 428 1911. Telex: 30217.
Fax: +27 12 428 5635.

VERIFIED

250-man mobile field kitchen trailer

Description
This mobile kitchen unit is mounted on a heavy-duty two-wheeled trailer fitted with airbrakes. The cooking range, three pressure cookers, griddle plates, oven and hot water tank are enclosed by side flaps which can open upwards to the equipment and provide shelter for the operator. The cooking equipment is heated by gas supplied by low-pressure gas cylinders through a vaporiser unit. All the control valves are fitted with flame failure devices and the oven is thermostatically controlled.

The full equipment carried on the trailer includes: three pressure cookers; three pans; three gas lights; a poker lighter; four 14 kg liquid withdrawal gas cylinders; two fire extinguishers; a spare wheel and heavy-duty jockey wheel; canvas sides and roof cover; a handbrake; four adjustable corner jacks with levellers; and a removable stainless steel worktop with a chopping board on the reverse side. The canvas sides are provided for blackout use at night and for protection in bad weather.

250-man mobile field kitchen trailer

Status
In service with the South African National Defence Forces.

Manufacturer
Enquiries to: Armscor, Private Bag X337, Pretoria 0001, South Africa.
Tel: +27 12 428 1911. Telex: 30217.
Fax: +27 12 428 5635.

VERIFIED

Other South African trailers

Description
South Africa has the ability to produce about 80 types of trailer and semi-trailer. Trailers are produced with capacities from 1,000 kg to 16,000 kg and semi-trailers (of which there are nearly 40 types) with capacities from 7,000 kg to 60,000 kg. Most of these are produced to match the standard South African National Defence Forces truck capacities of 1,000, 2,000, 5,000 and 10,000 kg but many are specially produced for specific roles, some of which have been described in the preceding entries. Apart from specialist roles such as training media and office trailers, trailers are also produced for such roles as mobile food pantries and corpse embalming.

Status
Most types are available. In service with the South African National Defence Forces.

Manufacturer
Enquiries to: Armscor, Private Bag X337, Pretoria 0001, South Africa.
Tel: +27 12 428 1911. Telex: 30217.
Fax: +27 12 428 5635.

VERIFIED

TURKEY

Ibrahim Örs military trailers

Description

Ibrahim Örs AS produces a wide range of military and commercial trailers and associated equipment at its Kirikkale facility, including the following military trailers:

Model ORS AR-1, 1,000 kg payload single-axle cargo trailer similar to the US M105 series

Model ORS AR-05, 500 kg payload single-axle trailer intended for towing by Jeep-type vehicles

Model ORS AT-2, double-axle fuel tank trailer carrying a 2,000 litre tank

Model ORS AR-2, double-axle trailer intended for the carriage of up to 2,000 kg of ammunition and similar loads

Model ORS SM-250, mobile kitchen with three cookers and the capacity to cook meals for up to 250 personnel

Model ORS MV-1, twin-axle mobile van-bodied trailer for use as a mobile darkroom or electronic equipment carrier

Model ORS AS, single-axle water tank trailer carrying up to 1,250 litres of drinking water

Model ORS A-7, 750 kg capacity tipping body trailer

Model ORS M3-MTK, carrier for M3 machine guns for Turkish Navy

Model ORS TTR, twin-axle carrier for torpedos

Model ORS PTR, glider container

Model ORS MTR, mine carrier trailer for NATO

Model ORS STR, single-axle water purification trailer

Model ORS BYT, bomb loading trailer for aircraft

Model ORS IS, 3,000 to 4,000 litre fire tender trailer

Model ORS AND, asphalt distribution trailer.

In addition to the above there is also a special electrical generator-carrying single-axle trailer and a stabilised trailer with an elevating servicing platform that can be raised to a height of 8 m.

Status

All the above are in production. In service with the Turkish armed forces and exported to some other countries.

Manufacturer

Ibrahim Örs AS, GMK Bulvari, Onur Isçhani Kat:6, No 146, T-06440 Kizilay, Ankara, Turkey.
Tel: +90 4 418 3946. Fax: +90 4 425 3925.

UPDATED

Model ORS AS 1,250 litre water tank trailer produced for Turkish Army and Gendarmerie

ORS generator trailer

UNITED KINGDOM

Scottorn 750 kg trailers

Description

Reynolds Boughton Limited, Scottorn Division, produces many military pattern trailers in various weight capacities. There are two 750 kg trailers: the Bushranger trailer is lighter and simpler than the Military L model, and has a hinged removable tailboard.

Specifications

Model	**Bushranger**	**Military L (510/760 kg)**
Weight	279 kg	355 kg
Length		
(overall)	2.695 m	2.795 m
(of cargo area)	1.754 m	1.83 m
Width		
(overall)	1.587 m	1.475 m
(of cargo area)	1.015 m	1.475 m
Height		
(without canopy)	1.015 m	1.015 m
(of towing eye)	0.56 m	0.535 m
Track	1.333 m	1.335 m
Tyres	6.00 × 16	6.00 × 16 (Land Rover wheels)
Suspension	6-leaf semi-elliptic leaf springs	

Status

In production.

Manufacturer

Reynolds Boughton Limited, Scottorn Division, Bell Lane, Amersham, Buckinghamshire HP6 6PE, UK.
Tel: +44 1494 764411. Telex: 83132.
Fax: +44 1494 765218.

VERIFIED

Scottorn Military L (510/760 kg) cargo trailer

750 kg cargo trailer FV2381

Description

The FV2381 cargo trailer has a capacity of 750 kg and is of all-welded steel construction. Two stabilising jacks are provided at the rear and an adjustable jockey wheel is fitted in front. The trailer is equipped with lashing points for the load and for a canvas canopy. Service braking is by means of an overrun unit which operates hydraulic disc brakes and there is also a mechanical parking brake. A hydraulic damper isolates the trailers from shock transmitted from the prime mover, prolonging the operational life of the trailers.

As cargo carriers these trailers meet the NATO requirement of fording to a depth of 0.76 m in fresh or seawater without preparation. The limit for the protection of cargo while wading is 0.61 m. When fully loaded the trailers will normally be towed by any suitable vehicle up to 1 tonne capacity, and a NATO standard intervehicle connector is fitted.

The FV2380 0.75 tonne trailer chassis is identical to the above but without the cargo body. This chassis can be adapted to take a variety of equipment such as water tanks and generators.

The FV2381 trailer is the standard trailer to MVEE Specification 679 and uses the FV2380 0.75 tonne trailer chassis.

Specifications

Weight: 408 kg
Length: 2.85 m
Width: 1.68 m
Track: 1.42 m

FV2381 750 kg Mk III cargo trailer

Max towing speed:
(road) 72 km/h
(cross-country) 24 km/h
Tyres: 6.50 × 16

Status

In production. This and other similar trailers are in service with the British Army.

Manufacturer

Reynolds Boughton Limited, Scottorn Division, Bell Lane, Amersham, Buckinghamshire HP6 6PE, UK.
Tel: +44 1494 764411. Telex: 83132.
Fax: +44 1494 765218.

VERIFIED

Scottorn Military 1,000/1,250 kg general purpose trailer

Description

This trailer is designed for towing by vehicles of the Land Rover class and, as with many other Scottorn trailers, the wheels are interchangeable with the Land Rover. The trailer body is welded and has a removable hinged tailboard. Stop lights, tail lights, direction indicators and an electrical plug are provided to suit the towing vehicle. The square section beam axle is fitted with 9.00 × 16 tyres on Land Rover wheels and it is suspended on semi-elliptical leaf springs with shock-absorbers and Aeon rubber helper springs. Mechanical overrun and parking brakes are fitted.

Status

In production.

Manufacturer

Reynolds Boughton Limited, Scottorn Division, Bell Lane, Amersham, Buckinghamshire HP6 6PE, UK.
Tel: +44 1494 764411. Telex: 83132.
Fax: +44 1494 765218.

VERIFIED

Scottorn Military 1,000/1,250 kg general purpose cargo trailer with optional canopy supports

Scottorn Military 1,750 kg cargo trailer

Description

This trailer was produced to meet the MVEE/LV98 specification for a trailer with a gross weight of 2.8 tonnes. It is conventional in design, with a drawbar mounting a standard NATO towing eye and with connections for the hydraulic overrun service brakes and the full military lighting system. The single axle uses a torsion bar and Aeon rubber suspension. A mechanical parking brake is provided. The cargo body has a tailgate.

A light repair workshop unit can be mounted on this trailer. The unit consists of a steel frame with lockers, a fold-out bench, a davit, enclosed working lights and electrical controls. An all-round canvas awning is provided for use in all weathers. This workshop body is in service with the British Army.

Scottorn 1,750 kg cargo trailer carrying a light repair workshop

Specifications
Basic trailer
Weight:
(unladen) 1,040 kg
(gross) 2,800 kg
(payload) 1,750 kg
Length:
(overall) 3.8 m
(cargo area) 2.425 m
Width:
(overall) 2.425 m
(cargo area) 1.72 m
Height:
(overall) 1.225 m
(cargo area) 0.45 m
Track: 2.09 m
Tyres: 8.25 × 16, 14 ply

Status
In service with the UK Ministry of Defence.

Manufacturer
Reynolds Boughton Limited, Scottorn Division, Bell Lane, Amersham, Buckinghamshire HP6 6PE, UK.
Tel: +44 1494 764411. Telex: 83132.
Fax: +44 1494 765218.

UPDATED

1,750 kg cargo trailer FV2408

Description
The FV2408 cargo trailer was designed for the carriage of general cargo. The trailer is suitable for a wide range of equipment including water tanks, weapon systems, air conditioning units and so on.

Welded steel construction is used for the chassis, with trailer arm torsion bar suspension and Aeon helpers. The steel cargo body has removable sides and a hinged tailgate. The platform is constructed of exterior grade plywood and incorporates load lashing loops and wear strips. Service braking is by means of an overrun unit operating hydraulic brakes through a hydraulic damper linkage to the master cylinder. A mechanical handbrake is used for parking. A spare wheel and carrier are fitted to the underside at the rear.

The usual tractor is a 4 × 4, 4 tonne truck. Electrical equipment conforms to EC regulations and a NATO intervehicle connector is fitted.

1,750 kg cargo trailer FV2408 ***1996***

Specifications
Weight: (unladen) 1,046 kg
Max load: 1,750 kg
Length: 4.2 m
Width: 2.489 m
Height: (top of headboard) 1.213 m
Track: 2.242 m
Max towing speed:
(roads) 72 km/h
(rough roads) 24 km/h

Status
In service with the British Army.

Manufacturer
Reynolds Boughton Limited, Scottorn Division, Bell Lane, Amersham, Buckinghamshire HP6 6PE, UK.
Tel: +44 1494 764411. Telex: 83132.
Fax: +44 1494 765218.

VERIFIED

2,500 kg trailers

Description
There are three trailers in the 2,500 kg special purpose trailer range. The base element is the FV2405 Mark 2 Trailer Chassis (2.5 tonne). Then comes the FV2406 Mark 2 Flat Platform Trailer (2.5 tonne) and the FV2406 Mark 2 Trailer Cargo (2.5 tonne).

The FV2406 Mark 2 Flat Platform Trailer (2.5 tonne) was designed primarily for the carriage of a specialist 24/16 kW generator but the trailer is suitable for a wide range of equipment including water tanks, weapon systems, air conditioning units and so on. Welded steel construction is used for the chassis with uprated trailing arm suspension using torsion bars and Aeon spring assisters. The platform is constructed of exterior grade plywood and incorporates load lashing loops and wear strips. Three stabilising jacks are fitted as is a jockey wheel. Braking is by means of a dual line airbrake system; a mechanical handbrake is fitted for parking. A spare wheel and carrier are fitted to the underside at the rear.

The normal tractor is a 4 × 4, 4 tonne truck. Electrical equipment conforms to EC regulations and a NATO intervehicle connector is fitted.

FV2406 Mark 2 flat platform trailer (2.5 tonne)

Specifications
FV2406
Weight: (empty) 1,089 kg
Max load: 2,500 kg
Length: 4.25 m
Width: 2.489 m
Height: (top of mudguard) 1.002 m
Track: 2.242 m
Max towing speed:
(roads) 72 km/h
(rough roads) 24 km/h

Status
In service with the British Army.

Manufacturer
Reynolds Boughton Limited, Scottorn Division, Bell Lane, Amersham, Buckinghamshire HP6 6PE, UK.
Tel: +44 1494 764411. Telex: 83132.
Fax: +44 1494 765218.

UPDATED

Scottorn 1,000 to 2,500 kg flatbed trailers

Description
There are three Scottorn flatbed trailers, each of which consists of a heavy gauge sheet steel decking mounted on a rolled steel channel chassis, the drawbar is provided with a 76 mm NATO towing eye. The trailers are designed to be towed by four-wheel drive vehicles ranging from Land Rovers to 4 tonne trucks. Typical loads could include generators, radar systems or ammunition packs, with the larger trailer carrying either NATO standard pallets or two 682 litre (150 gallon) water tanks. General cargo can be carried when the optional side panels are fitted.

Status
In production.

Manufacturer
Reynolds Boughton Limited, Scottorn Division, Bell Lane, Amersham, Buckinghamshire HP6 6PE, UK.
Tel: +44 1494 764411. Telex: 83132.
Fax: +44 1494 765218.

VERIFIED

Specifications

Payload	**1,000 kg**	**1,750 kg**	**2,500 kg**
Weight	400 kg	900 kg	1,050 kg
Length (overall)	2.85 m	3.8 m	4.25 m
(deck)	1.8 m	2.25 m	2.25 m
Width (overall)	1.67 m	2.46 m	2.46 m
(deck)	1 m	1.72 m	1.72 m
Track	1.4 m	2.2 m	2.2 m
Tyres	7.50 × 16	8.25 × 16	9.00 × 20

Scottorn flatbed trailer

Crane Fruehauf 35 tonne engineer plant semi-trailer (FV3541A)

Description
This is a cranked framed trailer where the vehicle is loaded over the rear suspension. The foredeck is narrow and fitted with tool lockers between the beams. The main frame is of high-tensile steel universal beam 406 × 178 mm on the main deck with 356 × 171 mm on the foredeck. Frame cross members are 152 × 89 mm rolled steel joists with a steel box-section member of 203 × 152 mm at the extreme rear. The suspension is of the unsprung beam type with oscillating axles mounted at the ends of each beam to give full compensation between all wheels. Axles are constructed in solid form and oscillate on the rocker beams on plain bushes. The rear-loading ramps are made from steel channel section with 34.9 mm finished hardwood boards, plus auxiliary ramps. The landing gear is screw-operated with feet fitted at the crank. Rear support jacks are mounted under the tail of the frame to give support while loading.

Crane Fruehauf 35 tonne commercial type semi-trailer being towed by Scammell Constructor (6 × 4) tractor

Specifications
Length:
(overall) 13.98 m
(main deck from base of crank to start of suspension area) 5.51 m
(drop deck, incl area over suspension and beavertail) 9.7 m
Width: (overall) 3.2 m
Height: (of platform) 1 m
Ground clearance: (min empty) 0.51 m
Slope of ramp: 1 in 5
Bogie wheeltrack: 0.84 m
Wheelbase: (king-pin to centreline of rear bogie) 10.08 m

Status
In service with the British Army.

Manufacturer
Crane Fruehauf, Toftwood, Dereham, Norfolk NR19 1JF, UK.
Tel: +44 1362 695353. Telex: 97251.

VERIFIED

Crane Fruehauf tank transporter semi-trailer CF 2000

Description
This semi-trailer was designed to carry MBTs weighing up to 65,000 kg. The chassis is of all-welded steel with a mild steel decking welded into position. The suspension consists of heavy-duty unsprung rocker beam design with phosphor-bronze bushes on the pivots. The suspension is fitted with bump stops to limit fore and aft oscillation. The semi-trailer has four short

Crane Fruehauf tank transporter semi-trailer CF 2000

axles mounted in two lines of two axles. Each axle is fitted with a taper roller bearing. Brakes are air operated and a screw-type handbrake operates on one row of axles only. Standard fittings include manually operated rear-loading gear, track guides, stowage boxes and tiedown points.

Status
Production complete. In service with several armies including those of Jordan, Kenya, Kuwait and Libya.

Manufacturer
Crane Fruehauf, Toftwood, Dereham, Norfolk NR19 1JF, UK.
Tel: +44 1362 695353. Telex: 97251.

VERIFIED

King TT72/3WB-20.5 tank transporter semi-trailer

Description
The King TT72/3WB-20.5 is a 72 tonne capacity six-axle tank transporter semi-trailer, one of a range designed for the carrying of MBTs at speeds up to 80 km/h. The frame is of heavy-duty construction with a chequer plate deck and is fitted with adjustable track guides, stops and shackling points. Two light armoured vehicles can be carried on the main deck. All winch rope guides and pulleys are fitted to allow for the shortest possible loading time of both mobile and disabled tanks. The tyres can be filled with either heavy-duty foam or self-sealing fluids to allow continued operations in hostile conditions.

King TT72/3WB-20.5 tank transporter semi-trailer

The heavy-duty, high-articulation, six-axle pendle walking beam suspension offers good load equalisation. A high-articulation, narrow type neck is suitable for on/off-road conditions and is fitted with spare wheels and tyres, a loading davit, a water tank, a tool box and camouflage net stowage.

Loading is over the rear beavertail via two heavy-duty ramps. These are manually operated via springs with hydraulic assistance and are adjustable across the width of the trailer.

Specifications
Weight: approx 24,000 kg
Max load: approx 72,000 kg
Length: (overall) 11.5 m
Width: 4.15 m
Tyres: 525/65 R 20.5 XS

Status
In service worldwide.

Manufacturer
King Trailers Limited, Riverside, Market Harborough, Leicestershire LE16 7PX, UK.
Tel: +44 1858 467361. Telex: 341835 king g.
Fax: +44 1858 467161.

VERIFIED

King 35 tonne tank transporter stepframe semi-trailer

Description
The King 35 tonne tank transporter stepframe semi-trailer was designed mainly to transport lightweight armoured vehicles and armoured personnel carriers. It can also carry general military plant and machinery and 20 ft/6.096 m ISO containers.

The main load platform is 9.6 m long and 3 m wide, although the latter can be extended to 3.3 m using outriggers. The deck is covered in 8 mm chequerplate and fitted with twistlocks for ISO containers, and attachments for track guides for tracked vehicles.

Machinery is loaded over the rear via a beavertail and 2.5 × 1 m spring-assisted ramps which are wide enough to accommodate various track widths and wheeled vehicles, including Land Rovers.

SAF Triaxle three-leaf parabolic spring suspension is fitted, incorporating one fixed and two king pin power-steered axles which substantially reduce the turning circle and tyre wear. The semi-trailer is approved for operations in Germany.

King 35 tonne tank transporter stepframe semi-trailer

Specifications
Weight: approx 13,000 kg
Max load: approx 35,000 kg
Length: (overall) 13.885 m
Width: 3 m
Tyres: 9.5 R 17.5

Status
In service with the British Army in Germany (30).

Manufacturer
King Trailers Limited, Riverside, Market Harborough, Leicestershire LE16 7PX, UK.
Tel: +44 1858 467361. Telex: 341835 king g.
Fax: +44 1858 467161.

VERIFIED

King TT72/6-19.5 tank transporter stepframe semi-trailer

Description
The King TT72/6-19.5 is one of a range of stepframe tank transporters offered by King Trailers, specifically designed to comply with EC legislation applicable to military operations in Europe. The six-axle TT72/6-19.5 has a 72 tonne capacity for carrying MBTs at highway speeds over metalled roads; the deck length design also allows the alternative transport of two lighter armoured vehicles. The trailer can be designed to be compatible with all main truck/tractors, and can be ordered with a high-articulation narrow neck if on/off-road conditions dictate.

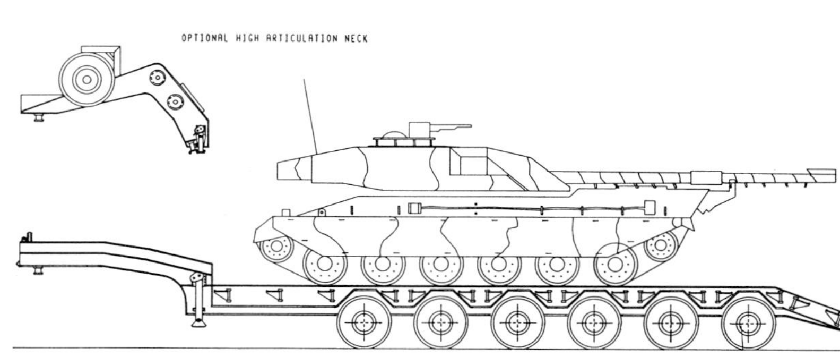

Drawing of King TT72/6-19.5 tank transporter stepframe semi-trailer

The heavy-duty frame construction incorporates a full width gooseneck which, like the main deck, is covered in steel chequerplate. To ease load equalisation and positioning, location adjustment holes are provided on the main deck to suit the removable track guides and stops. The all-steel beavertail is fitted with grip strips and adjustable inner track guides to further assist loading. Loading is over the rear via two all-steel heavy-duty ramps, also fitted with grip strips. A total of 16 pairs of swing-out type outriggers are provided for the main deck and beavertail. The axles are designed at up to 12.5 tonnes capacity and are fitted with fully equalising air suspension, with a raise and lower facility provided as standard. A combination of fixed and forced steer axles reduce tyre wear and increase manoeuvrability. The tyres are fitted in twin formations; two spares are provided.

An alternative hydraulic suspension can be specified for higher axle loads.

Specifications

Weight: approx 22,000 kg
Max load: approx 72,000 kg
Length: (overall) 13.95 m
Width: 3 m
Tyres: 285/70 R 19.5

Status

Available for production.

Manufacturer

King Trailers Limited, Riverside, Market Harborough, Leicestershire LE16 7PX, UK.
Tel: +44 1858 467361. Telex: 341835 king g.
Fax: +44 1858 467161.

VERIFIED

Scottorn tanker trailers

Description

The Scottorn Division of Reynolds Boughton Limited produces a range of tanker trailers of various capacities, of which three are described here. The smallest is a 682 litre (150 gallon) tanker which can be towed by any 750 kg vehicle such as a Land Rover. The other trailers differ in dimensions, capacities and fittings. All models have 3.2 mm welded steel plate tanks with 4.8 mm welded steel plate ends. Transverse baffles are incorporated in the tanks to prevent the liquid from surging. The tanks are filled through a 406 mm manhole with a 254 mm filling aperture, which is fitted with a lockable quick release cover with a pressure/vacuum valve. On the 680 and 1,140 litre tankers a drainage pump, fitted with a 40 mm bsp gate valve, is provided at the rear of the tank. The 1,350 litre tanker trailer is fitted with a water purification filter. The filtered liquid can be discharged either by gravity feed to four 12.5 mm bibcocks and one 25 mm bibcock, or pumped out by a 25 mm semi-rotary pump with 15 m of 25 mm delivery hose. Similar filtering and pumping arrangements are also available as optional extras for the other trailers.

Specifications

Type	**680 litres**	**1,140 litres**	**1,350 litres**
Weight (empty)	350 kg	582 kg	860 kg
Length (overall)	2.49 m	2.794 m	3.404 m
(of tank)	–	1.524 m	1.829 m
Width (overall)	1.65 m	1.85 m	2.11 m
(of tank)	–	1.22 m	1.219 m
Height (overall)	1.12 m	1.473 m	1.524 m
(of towing eye)	–	0.66 m	0.99 m
Track	1.33 m	1.6 m	1.79 m
Tyres	6.00 × 16	7.50 × 16	10.00, 11.00 or 12.00 × 20 or 10.50 × 16
Wheels	5-stud	5-stud	6- or 8-stud to suit towing vehicle
Brakes	mechanical overrun and parking		service brake: mechanical overrun or single-line airbrake
Suspension	6-leaf springs semi-elliptical	8-leaf springs semi-elliptical	semi-elliptical springs with Aeon rubber assists

Status

In production.

Manufacturer

Reynolds Boughton Limited, Scottorn Division, Bell Lane, Amersham, Buckinghamshire HP6 6PE, UK.
Tel: +44 1494 764411. Telex: 83132.
Fax: +44 1494 765218.

VERIFIED

Scottorn 680 litre tanker trailer

Scottorn Military 1,350 litre tanker trailer

Scottorn 'Flat Top' 2,280 to 10,000 litre tanker trailers

Description

The Scottorn 'Flat Top' all-steel tanker trailer has been developed over the years to form a comprehensive range of four (and in some cases eight) wheel tankers with a capacity of 2,280 to 10,000 litres (500 to 2,200 gallons). The design features an integral chassis and a 'D' shaped tank, allowing the tankers to be used for a variety of purposes. The trailers can be equipped as fuel bowsers, fire tenders, drinking water and foam compound tanks; carrying liquids such as diesel, oil, petrol, water and chemicals. As there is a flat top to the tank there is a capability for storage and for carrying equipment.

Specifications

4,560 litre (1,000 gallon) version
Weight: 2,250 kg
Length: (over drawbar) 5.185 m
Width: 1.83 m
Height: 1.75 m
Track: 1.664 m

Status

In service with the UK Ministry of Defence.

Scottorn 'Flat Top' tanker trailer

Manufacturer

Reynolds Boughton Limited, Scottorn Division, Bell Lane, Amersham, Buckinghamshire HP6 6PE, UK.
Tel: +44 1494 764411. Telex: 83132.
Fax: +44 1494 765218.

VERIFIED

Tecalemit Model SM 6001 lubricating and service unit trailer

Description

Following extensive tests, Tecalemit Garage Equipment Company Limited was awarded a contract to produce 100 mobile lubrication and service unit trailers for the British Army during late 1986. This order was followed by another for 110 units during July 1987 and a further 240 units in 1990.

The self-contained trailers are known as the Model SM 6001 and are intended for towing behind 4,000 and 8,000 kg trucks. They consist of a single-axle trailer base complete with overrun brakes, lights and a NATO 76 mm towing eye. The base frame is constructed from 100 × 50 × 3 mm folded steel channel-section, welded and decked with 3 mm steel plate. Under-deck bracing gives additional support to equipment carried and also permits lifting by forklift trucks. There are four lifting eyes provided. A 300 kg capacity jockey wheel is positioned on the towing eye A-frame. There are two adjustable prop standards provided for support and stability at the rear end of the trailer.

Carried on the trailer is a diesel-driven air compressor for operating the lubrication pumps, tyre inflation and tools. Also carried are a high-pressure grease-dispensing system, three low-pressure oil-dispensing systems, a low-pressure transfer pump for filling oil drums, a hand-held penetrating oil sprayer, a tyre inflator, a storage locker, a spare 50 litre oil drum, a 150 litre coolant tank, a 2.5 kg fire extinguisher and a tool box. A bank of four 7.6 m long hoses is provided. The equipment is protected by a canvas canopy with zipped side panels.

Tecalemit Model SM 6001 lubricating and service unit trailer

Specifications

Weight:
(with full lubricant and coolant tank) 1,350 kg
(less liquids) 950 kg
(on towing eye, fully laden) 70 kg
Length: (with towing eye) 3.5 m
Width: 1.77 m
Height:
(over canopy) 1.8 m
(towing eye, nominal) 0.8 m
Max towing speed:
(roads) 80.5 km/h
(rough tracks) 32 km/h
Tyres: 6.00 × 16 × 6 PR

Status

In service with the British Army (210).

Manufacturer

Tecalemit Garage Equipment Company Limited, Belliver Industrial Estate, Roborough, Plymouth, Devon PL6 7BW, UK.
Tel: +44 1752 701212. Telex: 45124.
Fax: +44 1752 796329.

VERIFIED

Scottorn mobile servicing 510 kg lubrication trailer and Scottorn LAD 510 kg workshop trailer

Description

Both trailers are constructed on a chassis fabricated from rolled steel channel. There are four retractable snapjack supports fitted, one at each corner of the body, for added stability when the unit is in use. A tubular support frame is provided for a canvas tilt, which is raised when the unit is in use and lowered for travelling.

The lubrication trailer is fitted with a petrol engined compressor, grease pump, two oil pumps, grease reel, two oil reels, air reel, tyre inflator, penetrating oil reel, static sprayer and a spark plug cleaner. Optional extras include an all-steel canopy, hubs and brakes fully interchangeable with the Land Rover Defender 110.

Scottorn mobile servicing 510 kg lubrication trailer

The LAD workshop trailer is provided with: a 6 kVA, single-phase, 250 V, 50 Hz alternator set with a direct coupled air-cooled diesel engine complete with starter; starter battery and starter battery dynamo; totally enclosed sheet steel switchboard mounted on the alternator; and a distribution panel incorporating one 30 A fused isolator, one 25 A Nipham output socket and seven output sockets. Equipment includes: heavy-duty drill; heavy-duty grinder; bench grinder and polisher; pillar drill stand; portable air-cooled arc-welding set; portable air compressor; battery charger; two vices; and a 9.14 m extension cable.

Specifications

Type	**Mobile servicing**	**LAD workshop**
Weight	825 kg	1,029 kg
Length (overall)	2.794 m	2.794 m
(of body)	1.829 m	1.829 m
Width	1.525 m	1.624 m
Height (overall)	1.955 m	2.134 m
(over equipment)	1.65 m	n/a
(to towing eye)	0.56 m	0.584 m
Track	1.333 m	1.333 m

Status

In production.

Manufacturer

Reynolds Boughton Limited, Scottorn Division, Bell Lane, Amersham, Buckinghamshire HP6 6PE, UK.
Tel: +44 1494 764411. Telex: 83132.
Fax: +44 1494 765218.

VERIFIED

1 tonne special purpose trailers

Description

This range of trailers shares the same basic suspension. Welded steel construction is used for the chassis, while the suspension is of the trailing arm type using rubber springs. Service braking is by means of an overrun unit operating hydraulic brakes through a hydraulic damper linkage to the master cylinder, and a mechanical handbrake is fitted for parking.

The trailers are normally towed by vehicles of 250 to 1,000 kg capacity, except in the case of the FV2415, which is described below.

The electrical equipment and intervehicle connector comply with NATO requirements. Optional fitments for the FV2411, FV2412, FV2413 and FV2415 include suspension retraction for air-portability, front jockey wheels and support legs.

The Trailer, Missile Resupply, 1 tonne FV2411, was designed to carry containerised spare missiles for the Rapier SAM system; while the Trailer, Cargo, 1 tonne FV2412, was designed for the carriage of general cargo and specialist equipment for the Rapier system.

The 1 tonne flat platform trailers FV2413 and FV2415 were designed for the carriage of specialist equipment such as generators for use in conjunction with weapon systems. The two trailers differ from each other in that the FV2415 has a raised and lengthened drawbar to permit towing by a 3 or 4 tonne vehicle, while the FV2413 is normally towed by vehicles of 250 to 1,000 kg capacity.

The 1 tonne air-portable bridge trailer FV2420 was designed to carry sections of the Laird Class 16 air-portable bridge. Each trailing suspension arm is retractable to assist in loading and offloading the bridge sections and to reduce the overall height when the trailer is transported by air. The electrical system is mounted on a quickly detachable bar and extension cable, enabling it to be mounted either on the rear cross

Australian Army GKN Defence 1 tonne cargo trailer (specialist) (FV2412) (P Handel)

1 tonne missile resupply trailer (FV2411) (T J Gander)

member of the trailer or on the bottom of the box on the bridge load. The side lamps are mounted on hinged bars to facilitate stacking of trailers. Turnbuckle-type lashing screws are provided for securing the bottom box of the bridge to the chassis. These are also used for lashing stacked trailers together for transport by air.

The flat platform generator trailer FV2421, rated at 1.5 tonnes, was designed primarily for the carriage of a specialist 24/16 kW generator used in conjunction with Electronic Repair Vehicles (ERVs), but is also suitable for a wide range of equipment such as air conditioning systems and water tanks. Its chassis is fitted with uprated suspension units, and three stabilising jacks and a jockey wheel.

The FV2425 trailer was designed for the carriage of the THORN EMI Cymbeline mortar locating radar (Radar FA No 15). Four steadying legs are fitted and the suspension may be retracted hydraulically to assist the functioning of the radar.

Specifications

Model	**FV2411**	**FV2412**	**FV2413**	**FV2415**	**FV2420**	**FV2421**	**FV2425**
Weight	534 kg	534 kg	533 kg	533 kg	340 kg	533 kg	447 kg
Length	3.63 m	3.63 m	3.63 m	3.95 m	3.81 m	4.15 m	3.35 m
Width	1.75 m	1.75 m	1.75 m	1.75 m	2.01 m	2.02 m	1.78 m
Height	1.75 m	1.35 m	1.07 m	1.07 m	0.71 m	0.83 m	0.71 m
Track	1.5 m	1.5 m	1.5 m	1.5 m	1.75 m	1.77 m	1.5 m
Towing speed (good roads)	88 km/h	88 km/h	88 km/h	88 km/h	72 km/h	72 km/h	72 km/h
(rough roads)	40 km/h	40 km/h	40 km/h	40 km/h	24 km/h	24 km/h	24 km/h

Status

FV2411/2412: In service with countries using the Rapier SAM system: Australia, Brunei, Indonesia, Iran, Oman, Singapore, Switzerland, Turkey, United Arab Emirates, UK, USA and Zambia.
FV2413/2415: In service with the British Army and other armed forces.
FV2420: In service with Australia, Canada, Nigeria and the UK.
FV2421: In service with the British Army.
FV2425: In service with the British Army and other armed forces.

Manufacturer

GKN Defence, PO Box 106, Hadley Castle Works, Telford, Shropshire TF1 4QW, UK.
Tel: +44 1952 244321. Telex: 35248 sanhad g.
Fax: +44 1952 243910.

VERIFIED

GKN Defence High-Mobility Trailer

Description

The GKN Defence High-Mobility Trailer was designed to be towed by the Warrior MCRV (see entry in the *Armoured recovery vehicles* section) but can be towed by other vehicles, including B vehicles, MBTs and self-propelled artillery; 45 were ordered for the Warrior recovery and repair role. The trailer has a gross weight of 10,500 kg and is used to carry major assemblies such as two Warrior IFVs or one Challenger MBT power pack. It is also being evaluated for carrying ammunition and bridging components.

The High-Mobility Trailer is fitted with an airbrake system and is provided with a four-damper jacking system to maintain a safe level platform when not connected for towing. It can also traverse 40° side slopes. All controls for the system are contained in a compartment at the front end of the load area. The detachable towing eye is fitted with an umbilical winch to aid one-man coupling to towing vehicles.

This trailer has a high volume and weight capability and is being assessed for other roles such as armoured unit, ammunition and missile resupply and for carrying radars. It can be towed behind tanks, self-propelled artillery, mortar vehicles and other vehicles.

GKN Defence High-Mobility Trailer (T J Gander)

Specifications

Weight:
(kerb) 5,500 kg
(gross) 12,000 kg
Length: 4.94 m
Width: 2.89 m
Height: 1.97 m
Ground clearance: (laden) 0.5 m
Track: 2.49 m
Wheelbase: 1.37 m
Tyres: 14.00 × 20

Status

In service with the British Army (45) and Kuwait.

Manufacturer

GKN Defence, PO Box 106, Hadley Castle Works, Telford, Shropshire TF1 4QW, UK.
Tel: +44 1952 244321. Telex: 35248 sanhad g.
Fax: +44 1952 243910.

VERIFIED

Rubery Owen Medium Girder Bridge (MGB) Trailer (FV2842)

Description

This trailer was designed to carry palletised loads of the Williams Fairey Medium Girder Bridge (MGB), with a secondary role as a general purpose flat trailer. The trailer is the single-axle type fitted with semi-elliptical suspension springs with rubber bump stops with wire rope rebound slings. The chassis is constructed of rectangular tubular members on to which an aluminium corrugated decking is fitted. There is a two-line air pressure braking system and a mechanical parking brake. The trailer was designed to be towed on roads at a maximum speed of 56 km/h and cross-country at a maximum speed of 24 km/h.

Specifications

Weight:
(empty) 1,461 kg
(loaded) 4,939 kg
Length: (overall) 5.867 m
Width: (overall) 2.565 m
Height:
(overall, loaded) 2.62 m
(overall, empty) 1.02 m
Track: 2.261 m
Tyres: 10.00 × 15, 14-ply rating

Status

In service with most countries using the Williams Fairey Medium Girder Bridge.

Manufacturer

Defence Equipment Division, Rubery Owen Group Services Limited, Darlaston PO Box 10, Wednesbury, West Midlands WS10 8JD, UK.
Tel: +44 121 526 3131. Fax: +44 121 526 2869.

VERIFIED

Medium Girder Bridge (MGB) trailer with load of MGB sections

Rubery Owen trailer, four-wheeled close-coupled 1,000 kg, medium mobility

Description

The chassis of this trailer is of welded steel, hollow section and is fitted with two torsion bar axle units. The trailing/leading arm suspension is integral with the axle tubes and comprises laminated square section torsion bars for wheel articulation. A hydraulically damped towing eye minimises towing shock loads. Hydraulic brakes on all four wheels are actuated by the towing eye through an overrun device. For parking, a mechanical handbrake operates on the leading axle.

Variants dedicated to the carriage of radar equipment and generator sets have undergone development. Available options include a spare wheel, rugged mudwings, steady legs, levelling jacks and robust longitudinal mounting rails for generator sets.

Specifications

Weight: (unladen) 738 kg
Length: 3.98 m
Width: 1.88 m
Height: (chassis) 0.6 m
Track: 1.67 m

Status

In service with the British Army.

Rubery Owen trailer, four-wheeled close-coupled 1,000 kg, medium mobility

Manufacturer

Defence Equipment Division, Rubery Owen Group Services Limited, Darlaston PO Box 10, Wednesbury, West Midlands WS10 8JD, UK.
Tel: +44 121 526 3131. Fax: +44 121 526 2869.

VERIFIED

Lolode 8 tonne recovery transporter ground loading trailer

Description

These Lolode trailers were designed primarily for the recovery of forklift trucks. They can be used for the recovery of disabled equipment or plant and machinery with poor gradability or low ground clearance.

The trailer is fitted with the Lolode ground loading suspension system. A hydraulic device allows the trailer bed to be lowered to the ground to enable equipment to be loaded and unloaded at ground level. The hydraulic system, operated via an electric start, petrol engined power pack with a hand pump backup, then raises the trailer bed and the load up to the travelling position. Locking pins are designed to withstand all imposed loads when towing.

The main load platform measures 4.26 × 1.76 m. The timber deck is overlaid with three strips of 3 mm aluminium tread plate to distribute concentrated loadings. An adjustable height drawbar is fitted so that the trailer does not require uncoupling from the tractive unit when the platform is lowered for loading and unloading. The loading ramp is hydraulically operated.

Provided with the trailer is a 2 tonne capacity manual winch, a tool box and a spare wheel and tyre.

Lolode 8 tonne recovery transporter ground loading trailer

Specifications

Weight: 2,360 kg
Max load: 5,640 kg
Length: (overall) 5.86 m
Width: 2.5 m
Tyres: 215 × 75R 17.5

Status

In service with the British Army (20).

Manufacturer

King Trailers Limited, Riverside, Market Harborough, Leicestershire LE16 7PX, UK.
Tel: +44 1858 467361. Telex: 341835 king g.
Fax: +44 1858 467161.

VERIFIED

Stead and Wilkins Ammunition Mover SWAM II

Description

The Stead and Wilkins Ammunition Mover, SWAM II, was originally developed to be towed by such vehicles as self-propelled artillery (or similar) and to carry two Unit Load Containers (ULCs) cross-country. A typical ULC could contain 17 155 mm projectiles, 17 charges and would weigh 1,320 kg.

The SWAM II has a structural steel chassis made up mainly from rectangular hollow sections. It is towed using a drawbar with a patented device to enhance cross-country mobility and a damped coupling with a capacity of 3,500 kg. The coupling is fitted with a handbrake and an emergency breakaway cable. The twin-axle sprung undergear uses a tandem set of leaf springs with a centre rocker beam. There is overrun braking on all four wheels which are fitted with auto-reverse hubs.

The two ULCs are secured in position using four webbing straps with a ratchet breaking load of 3,000 kg/strap. ULCs can be carried back-to-back or side-to-side. Other options include mudguards, an adjustable height drawbar and a lighting system.

In addition to use as a ULC carrier, the SWAM II can be configured as an emergency water carrier, a heavy-duty general purpose trailer with a timber or sheet metal body, a generator or air conditioning unit carrier, or as a personnel accommodation shelter carrier.

In 1991 the UK Ministry of Defence ordered 75 SWAM IIs for use in the air conditioning unit transport role; these were supplied for use in the Gulf region.

During 1996 the SWAM II was being redesigned to be attached to British Army 155 mm AS90 self-propelled howitzers. This will result in a change of designation to SWAM 90.

A Stead and Wilkins Ammunition Mover (SWAM II) on tow behind an M109A2 155 mm self-propelled howitzer

Specifications

Load carrying capacity: 3,300 kg
Length overall: 3.95 m
Width: 1.83 m
Height: (with ULC) 1.9 m
Tyres: 7.5 × 15 Avon Ranger 11
Towing speed: (max load, cross-country) 50 km/h
Side slope: (max, with 2 ULCs) 40°

Status

In service with the British Army.

Manufacturer

Stead and Wilkins Fabrications Limited, Jolly Farmers Wharf, Thames Road, Crayford, Dartford, Kent DA1 4HQ, UK.
Tel: +44 1322 529134. Fax: +44 1322 550314.

UPDATED

AVRE Trailer

Description

The AVRE Trailer was designed to be towed by the Chieftain AVRE (Armoured Vehicle Royal Engineers) and may also be towed by 14 tonne trucks and DROPS vehicles. It is produced by Reynolds Boughton Limited and completed Reliability Qualification Trials, ready for an in-service date during 1995.

The AVRE Trailer has an 8,250 kg payload capacity and can carry either eight Barmine pallets, two Class 60 trackways, a fascine, a Universal Dozer blade, or general cargo. The trailer has a walking beam suspension with close coupled axles and 14.00 × 20 tyres fitted with run-flat inserts. A dual line anti-lock airbrake system is provided with an air supply from an onboard compressor. All axle hubs and drums are fully sealed for operations at temperatures down to −40°C.

AVRE Trailer on tow behind Chieftain AVRE

Specifications

Payload: 8,250 kg
Length: 7.3 m
Width: 3.03 m
Height: (overall) 2.07 m
Tyres: 14.00 × 20

Status

Entered service with the British Army during 1995.

Manufacturer

Reynolds Boughton Limited, Scottorn Division, Bell Lane, Amersham, Buckinghamshire HP6 6PE, UK.
Tel: +44 1494 764411. Telex: 83132.
Fax: +44 1494 765218.

VERIFIED

Trailer, mineclearing equipment, Giant Viper, No 2 Mk 3 (FV3705)

Description

This trailer was developed specifically as the transport and launch platform for the Giant Viper and Improved Giant Viper (Python) mine clearance device. It provides a cross-country capability for a payload of some 3,000 kg comprising explosive hose and launching rocket(s). It can be towed by wheeled or tracked vehicles. The chassis is of all-steel welded construction, supported by semi-elliptical leaf springs on the single beam axle.

Rubery Owen FV3705 Giant Viper trailer

Specifications
Length: (overall) 5.9 m
Width: (overall) 2.476 m
Height:
(overall) 2.217 m
(platform) 1.089 m
Track: 2.134 m
Wheelbase: (axle to towing eye) 3.569 m
Gross vehicle weight: 4,460 kg
Brakes: 3-line air system operation 394 mm diameter × 76 mm wide 'S' cam brakes
Tyres: 12.00 × 20 with run-flat capability

Status
In service with the British Army.

Manufacturer
Defence Equipment Division, Rubery Owen Group Services Limited, Darlaston PO Box 10, Wednesbury, West Midlands WS10 8JD, UK.
Tel: +44 121 526 3131. Telex: 338236/7.
Fax: +44 121 526 2869.

UPDATED

King rapid runway repair mat carrier drawbar trailer

Description
The King rapid runway repair mat carrier drawbar trailer is used for the transport, loading and unloading of rapid runway repair mats. For details of the Rapid Runway Repair System involving the mats refer to the entry in the *Rapid Runway Repair Equipment/Portable Runways* section.

The 20 tonne gross unit consists of a front and rear bogie, connected by a long single spine fitted with hydraulic lifting gear by which the runway repair mat is handled. The front bogie features a turntable steered front axle mounted on two-spring, seven-leaf suspension connected to a 1.83 m drawbar towing facility. The rear bogie has hydraulic override steering complete with a wanderlead control mounted on a ball-bearing turntable. The tyres are filled with urethane foam for puncture protection and to provide improved heat resistance.

The 13.41 m central spine has a detachable centre portion to reduce overall length and facilitate aircraft loading. Two hydraulic boosters for unloading and loading operations are fitted either end of the spine and are operated off a fully independent power pack. Two hydraulically operated jack legs are fitted beneath each booster for stability during operation.

King rapid runway repair mat carrier drawbar trailer

Specifications
Weight: 5,000 kg
Capacity: 15,340 kg
Length: (overall) 16.37 m
Width: 2.5 m
Tyres: 8.25 × 15

Status
In service with the British Army (20).

Manufacturer
King Trailers Limited, Riverside, Market Harborough, Leicestershire LE16 7PX, UK.
Tel: +44 1858 467361. Telex: 341835 king g.
Fax: +44 1858 467161.

VERIFIED

Reynolds Boughton DROPS/PLS pallet trailer

Description
This trailer is a drawbar two-axle close-coupled type of skeletal construction designed specifically for the transportation of ISO-configured flatracks and their loads. The trailer is towed behind a tractive unit, fitted with a demountable swapbody system such as the Boughton (6 × 6) DROPS/PLS truck.

The design incorporates a patented telescopic pivoting drawbar. This enables the trailer to be coupled to and uncoupled from the prime mover by one person (the vehicle driver) without assistance. The flatrack can be transferred to and from the trailer with the trailer coupled to the vehicle.

Pallet location sockets and stops are fitted to the rear with standard ISO locks to the front and rear. A 76 mm diameter towing eye is fitted.

The unit is mounted on a Ridewell Dynaflex suspension. Two air line brakes are fitted, together with a full 24 V lighting system.

Reynolds Boughton DROPS/PLS pallet trailer with a flatrack carrying a Scorpion tracked reconnaissance vehicle
1996

Specifications
Length:
(drawbar extended) 7.68 m
(drawbar retracted) 6.48 m
Width: 2.438 m
Height:
(laden) 1.2 m
(unladen) 1.25 m
Track: 2.97 m
Departure angle: 30°
Tyres: 16.5 × 22.5.385-65 R 22.5
Electrical system: 24 V

Status
In production. In service with the UK Ministry of Defence, the US Army (21) and some other armies.

Manufacturer
Reynolds Boughton Limited, Bell Lane, Amersham, Buckinghamshire HP6 6PE, UK.
Tel: +44 1494 764411. Telex: 83132.
Fax: +44 1494 765218.

VERIFIED

King DROPS DB20/2-19.5 tandem-axle drawbar trailer

Description
The King DROPS DB20/2-19.5 tandem-axle drawbar trailer can combine with either Foden or Leyland Trucks 8 × 6 vehicles to make up a DROPS combination. The trailer was designed specifically to carry ammunition flatracks or 20 ft/6.096 m ISO containers and have a payload capacity of 16,500 kg.

King Trailers designed and developed the trailer to UK Ministry of Defence requirements. The design consists of a skeletal construction with built-in guides having a gathering-in facility to allow easy loading of flatracks by DROPS vehicles.

Although the initial production run was five units, the total order, spread over five years, was for 290 trailers.

Specifications
Weight:
(axle) 10,170 kg
(GTW) 20,340 kg
Payload: 16,940 kg

Length: (overall) 7.73 m
Width: 2.49 m
Wheelbase: 4.714 m
Tyres: 265 - 70R - 19.5
Electrical system: 24 V

Status
In service with the British Army (290).

Manufacturer
King Trailers Limited, Riverside, Market Harborough, Leicestershire LE16 7PX, UK.
Tel: +44 1858 467361. Telex: 341835 king g.
Fax: +44 1858 467161.

VERIFIED

King DROPS DB20/2-19.5 tandem-axle drawbar trailer (T J Gander)

SUPACAT Fork-Lift Pallet Trailer (FLPT)

Description
Designed and developed by SUPACAT Limited, in response to a foreseen military requirement, the Fork-Lift Pallet Trailer (FLPT) was developed principally for the SUPACAT (6 × 6) 1,000 kg light vehicle (see entry in the *Light Vehicles* section for details) but can be used with other vehicles. The FLPT can extend the load moving capacity of the towing vehicle by one or two times. It was developed primarily for military airborne and air mobile operations and can be delivered by helicopter or parachute.

The FLPT has a tilting load platform on which are mounted folding forklift forks. The platform and forks are rotated by an electrically or manually operated hydraulic system and the FLPT can self-load and transport a standard pallet weighing up to 1,400 kg. Once loaded, the pallet lies on its side and is retained by the lifting forks, but a tiedown can be used when traversing rough terrain.

The FLPT is light and can be stowed to take up minimum space. The drawbar can be retracted and the forks folded. If required, the FLPT can be stowed vertically, with or without a load, and can be self-loaded onto a SUPACAT The trailer has a rear tow hook to enable one FLPT to be towed behind another in a 'train' when conditions are favourable. As an alternative the FLPT tow hook may be used to tow a mortar or 105 mm Light Gun.

The trailer can be quickly converted to carry loose loads such as ammunition boxes or personal kit by fitting four corner posts and ratchet straps which are carried on the trailer. Fittings are included to carry a water carriage pack and special brackets are provided to enable three stretchers to be transported.

Another trailer designed specifically for the SUPACAT is a 500 kg payload capacity unit. This is a general purpose box trailer with a drop tailboard and a retractable towbar. Weight of this trailer is 280 kg.

SUPACAT Fork-Lift Pallet Trailer (FLPT) loaded with ammunition and on tow behind a SUPACAT (6 × 6) 1,000 kg light vehicle, also towing a 105 mm Pack Howitzer

Specifications
Weight: (unloaded) 457 kg
Load capacity: 1,400 kg
Length overall:
(drawbar extended) 2.75 m
(drawbar retracted) 1.9 m
Width: (overall) 2.01 m
Height: (unloaded) 0.95 m
Tyres: 31 × 15.5 × 15 low ground pressure

Status
Available. In service with the British Army. A replacement for this trailer has been requested.

Manufacturer
SUPACAT Limited, The Airfield, Dunkeswell, Honiton, Devon EX14 0RA, UK.
Tel: +44 1404 891777. Fax: +44 1404 891776.

UPDATED

UNITED STATES OF AMERICA

Trailer, High-Mobility, 2-wheel

Description
The US Army High Mobility Trailer (HMT) is a ruggedised trailer designed for use with all variants of the US Army's High-Mobility Multipurpose Wheeled Vehicle (HMMWV) and with commercial utility vehicles. It replaces the M101 series trailer on an attrition basis.

There are three configurations available: Light Cargo (M1101, HMT-L, 907 kg payload); Heavy Cargo (M1102, HMT-H, 1,270 kg payload); and Heavy Chassis (HMT-HC, 1,393 kg payload). The trailer has a riveted aluminium construction for light weight, high strength, high payload, anti-corrosion and durability. The HMT meets MIL-STD requirements for rail, sea and air transport operations.

The HMT meets the mission profile of the HMMWV for operating cross-country or over unimproved roads, meeting the maximum gradient (60 per cent) and side

Base chassis for High-Mobility Trailer (HMT) for use with HMMWV ***1997***

slope (40 per cent) of the HMMWV. Other HMMWV commonality features include the same tyres, 24 V electrical connectors, hydraulic brake connections and towhook/lunette compatibility.

An independent trailing arm suspension is provided along with hydraulic surge brakes with automatic breakaway actuation. An adjustable and retractable jack is located on the left-hand side of the drawbar and two adjustable support legs are located at the rear.

Cargo versions can be provided with a soft top. Several end-user configured systems are under qualification testing or development, including tactical quiet power generator units ranging from 3 to 15 kW, and a lighting tower.

Specifications

Model	HMT-H	HMT-L	HMT-HC
Designation	M1102	M1101	n/av
Weight	635 kg	635 kg	511.7 kg
Payload	1,270 kg	907 kg	1,393 kg
Length overall	3.35 m	3.35 m	3.35 m
Width	2.18 m	2.18 m	2.18 m
Height (deck)	0.84 m	0.84 m	n/av
Track	1.84 m	1.84 m	1.84 m
Ground clearance	0.406 m	0.406 m	0.406 m

Status

Production for US Army commenced March 1995.

Manufacturer

Electrospace Systems Inc, PO Box 831359, Richardson, Texas 75083-1359, USA.
Tel: +1 214 470 2899. Fax: +1 214 470 2466.

UPDATED

Trailer, cargo: ¼ ton, 2-wheel, M416 and M416A1

Description

The M416 single-axle trailer was the replacement for the earlier ¼ ton M100 trailer. The all-welded body, bolted to the M569 chassis, is watertight and will float with a load of 226.8 kg. Two drain holes are provided in the trailer floor, which has a support leg and a tarpaulin cover. Its suspension consists of semi-elliptical springs. A mechanical parking brake is fitted as standard.

The M416A1 is the replacement for the earlier M100 and M416 and has an inertia braking system added. The M367 and later M716 are variants used to install and repair electrical transmission lines.

Trailer, cargo: ¼ ton, 2-wheel, M416 with non-standard top (Michael Ledford)

Specifications

M416

Weight:
- (empty) 258 kg
- (loaded road) 599 kg
- (loaded cross-country) 485 kg

Max load:
- (road) 340 kg
- (cross-country) 227 kg

Length:
- (overall) 2.75 m
- (inside body) 2.44 m

Width: 1.54 m
- (inside body) 1.04 m

Height: 1.07 m
Depth: (inside body) 0.28 m
Coupling height: 0.58 or 0.66 m
Track: 1.44 m
Wheelbase: 1.87 m
Tyres: 7.00 × 16

Status

In service with the US Army and the armed forces of other countries.

UPDATED

Trailer, cargo: ¾ ton, 2-wheel, M101 series

Description

This trailer was designed to transport cargo both on-road and cross-country. Its body is all steel and is provided with a hinged tailgate, bows and a tarpaulin cover. The drawbar assembly is attached to the front of the chassis and a retractable pivoted front support is attached to the drawbar bracket. The chassis of the M101 trailer is designated the M116.

The latest version is the M101A2.

Specifications

Weight:
- (empty) 608 kg
- (loaded road) 1,588 kg
- (loaded cross-country) 1,288 kg

Max load:
- (road) 1,020 kg
- (cross-country) 680 kg

Length:
- (overall) 3.73 m
- (inside body) 2.44 m

Width:
- (overall) 1.87 m
- (over wheels) 1.83 m

Height: (over tarpaulin) 2.11 m
Ground clearance: 0.36 m

Trailer, cargo: ¾ ton, 2-wheel, M101

Track: 1.57 m
Wheelbase: (towing eye to centre of axle) 2.54 m
Tyres: 9.00 × 16

Status

In service with the US Army and the armed forces of other countries.

Manufacturers

(M101A2) Kasel Manufacturing Company, Ebensburg, Pennsylvania, USA.
UMC Electronics Company, North Haven, Connecticut, USA.

UPDATED

Trailer, cargo: 1½ ton, 2-wheel, M105 series

Description

The M105 single-axle cargo trailer was designed to transport cargo both on roads and cross-country and is normally towed by a 2½ ton truck. The body of the trailer is of the box type and is provided with lattice-type side extensions, tailgate, bows and a tarpaulin cover (optional). The brakes of the trailer are air over hydraulic with a mechanical parking brake.

The M105A1 and M105A2 are slightly smaller than the M105. The body of the M105A2 is wood and steel where the others are all steel, and the M105A2 has an optional support leg.

Systems and Electronics Inc produces a mobile clothing repair shop based on the M105A2. Weight complete is 3,289 kg.

Specifications

Weight: (empty) 1,202 kg
Max load: 1,361 kg

Length:
- (overall) 4.2 m
- (inside body) 2.79 m

Width:
- (overall) 2.11 m
- (inside body) 1.88 m

Height:
(overall) 2.49 m
(to towing eye) 0.87 m
Ground clearance: 0.42 m
Track: 1.71 m
Wheelbase: (towing eye to centre of axle) 2.79 m
Tyres: 9.00 × 20

Status
In service with the US Army and the armed forces of other countries.

UPDATED

Trailer, cargo: 1½ ton, 2-wheel, M105, complete with side extensions (Ray Young)

Semi-trailer, van: cargo, 6 ton, 2-wheel, M119 and M119A1

Description
The M119 semi-trailer van is used for carrying cargo both by road and cross-country and is normally towed by a 2½ ton truck. The semi-trailer consists of a van body mounted on an M117 chassis. The body of the trailer consists of an angle iron framework, covered on the outside with sheet metal, and on the inside with plywood panels. The chassis consists of the frame, semi-elliptical springs mounted on an axle which is supported by dual wheels. The M119 has air-operated brakes while the later M119A1 has air over hydraulic brakes. Dual mechanically operated landing gear is mounted under the forward part of the chassis.

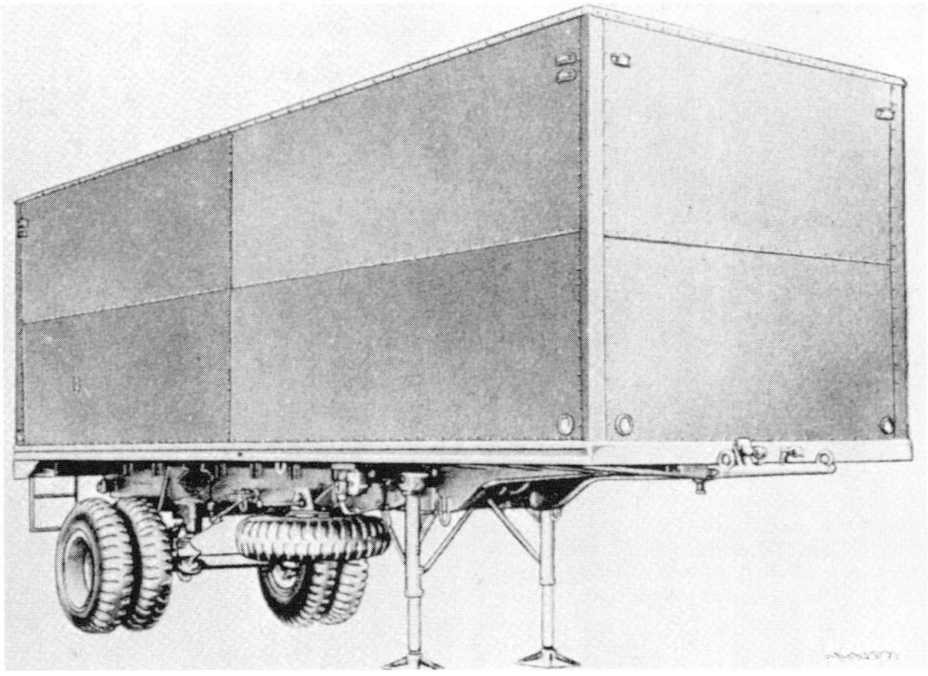

Semi-trailer, van: cargo, 6 ton, 2-wheel, M119 (US Army)

Specifications
Weight:
(empty) 3,257 kg
(loaded road) 10,605 kg
(loaded cross-country) 8,700 kg
Max load:
(road) 7,348 kg
(cross-country) 5,443 kg
Length:
(overall) 6.97 m
(inside body) 6.78 m
Width:
(overall) 2.53 m
(inside body) 2.23 m
Height:
(overall) 3.37 m
(coupling) 1.19 m
Ground clearance: 0.49 m
Track: 1.78 m
Wheelbase: (king-pin to centre of rear axle) 5.18 m
Tyres: 9.00 × 20

Status
In service with the US Army and the armed forces of other countries.

VERIFIED

Semi-trailer, stake: 12 ton, 4-wheel, M127, M127A1, M127A1C and M127A2C

Description
The M127 series of semi-trailers are used to transport general cargo and are towed by 5 ton tractor trucks. The body frame consists of pressed steel side rails, a cross member, and short cross members welded together forming one integral unit. The chassis consists of two drop-frame I-section longitudinal frame rails and intermediate cross members with an upper fifth wheel plate, king-pin, two axles mounted on a leaf spring suspension and two foot-type landing legs.

The basic M127 has airbrakes but all later models have air over hydraulic brakes. The M127A1 is provided with chains which support the panels and also lifting rings for hoisting the semi-trailer. The M127A1C has a

Semi-trailer, stake: 12 ton, 4-wheel, M127 (US Army)

voltage control box mounted on the underside of the body and the M127A2C has improved landing legs.

Specifications

Weight:
(empty) 6,123 kg [M127]
6,531 kg [M127A1, M127A1C]
Max load:
(road) 16,329 kg
(cross-country) 10,886 kg
Length:
(overall) 8.75 m
(inside body) 8.53 m
Width:
(overall) 2.46 m
(inside body) 2.26 m
Height: (overall) 2.76 m
Ground clearance: 0.3 m
Track: 1.83 m
Wheelbase: (king-pin to centre of rear bogie) 6.12 m
Max towing speed:
(road) 80 km/h
(cross-country) 48 km/h
Tyres: 11.00 × 20

Status

In service with the US Army and the armed forces of other countries.

VERIFIED

Trailer, flatbed: 10 ton, 4-wheel, M345

Description

The M345 flatbed trailer is used for carrying equipment, both by road and cross-country and can be towed by a 5 ton truck provided it has an air supply.

The trailer is constructed of structural and pressed steel with the two axles forming a bogie unit. The trailer has an air over hydraulic brake system that is controlled from the towing vehicle. As the trailer has no parking brakes, chocks are provided to stop the trailer rolling when parked. Swing-mounted landing gear is provided under the front and rear of the trailer.

Trailer, flatbed: 10 ton, 4-wheel, M345 (Larry Provo)

Specifications

Weight:
(empty) 5,107 kg
(loaded road) 16,901 kg
(loaded cross-country) 14,179 kg
Max load:
(road) 11,793 kg
(cross-country) 9,072 kg
Length:
(overall) 8.38 m
(platform) 7.01 m
Width: 1.93 m
Height: 1.38 m
Track: 1.83 m
Tyres: 11.00 × 20

Status

In service with the US Army. A 12 ton flatbed trailer is under development.

VERIFIED

FMTV cargo trailers

Description

There are two cargo trailers being manufactured as components of the Family of Medium Tactical Vehicles (FMTV—for full details of these vehicles refer to entry in *Trucks* section). The lightest is the M1082 for use with the LMTV cargo truck and has a matching payload of 2,268 kg, thereby doubling the carrying capacity of the LMTV cargo truck. The second trailer is the M1095 for use with the MTV cargo truck, which has a matching payload of 4,536 kg, thereby doubling the carrying capacity of the MTV cargo truck.

Both trailers share many components, including the Rockwell axles (one on the M1082 and two on the M1095), an anti-lock airbrake system, a 12/24 V electrical system with blackout lights, tiedown points, and integral sling bars. Both trailers also use the same 395 R 20 tyres. A cargo cover with bows is optional, as is a tyre inflation system.

The main variable between the two trailers is the size of the steel cargo bodies which duplicate those of the appropriate towing vehicle. Both have removable drop sides with storage rails for these once removed. An access ladder is provided at the rear of both trailers.

Specifications

Model	**M1082**	**M1095**
Used with	LMTV	MTV
Weight (kerb)	2,953 kg	4,173 kg
(GVW)	5,221 kg	8,709 kg
Payload	2,268 kg	4,536 kg
Length	5.321 m	5.829 m
Width	2.438 m	2.438 m
Deck height, empty	1.532 m	1.582 m
Deck height, loaded	1.448 m	1.524 m
Cargo bed length	3.78 m	4.288 m
Cargo bed width	2.306 m	2.306 m
Ground clearance, loaded	343 mm	312 mm
Gradient	60%	60%
Side slope	30%	30%
Fording	1.524 m	1.524 m
Tyres	395/80 R 20	395/80 R 20
Electrical system	12/24 V	12/24 V

Status

In production. In service with the US Army.

Manufacturer

Stewart & Stevenson Services Inc, Tactical Vehicle Systems Division, 5000 I-10 West Sealy, Texas 77474, USA.
Tel: +1 713 856 7040. Fax: +1 409 885 7910.

UPDATED

M1082 cargo trailer for LMTV **1996**

M1095 cargo trailer for MTV **1996**

Semi-trailer, Heavy Equipment Transporter, M1000

Development

The Semi-trailer, Heavy Equipment Transporter (HET), M1000, with automatically steered axles, was originally developed as a private venture by Southwest Mobile Systems (now Systems and Electronics Inc (SEI)) as a response to a possible US Army requirement for transporting M1 and M1A1 MBTs. This requirement materialised as the M1000 and the HET was involved in US Army trials, completed during 1986. A production order for 1,066 M1000 units was placed by the US Army in 1989 with Southwest Mobile Systems (SEI). An additional 111 units were ordered in 1992. As of 2 June 1994 a total of 1,376 units had been ordered by the US government (plus an option for 101 additional units). Total requirement is for 2,452 units. First article approval for the US Army was given in February 1994.

The M1000 has been investigated by the US Army Engineer Center as a transporter of heavy engineer equipment. In this role the towing vehicle is the M916 tractor truck series.

Description

The M1000 semi-trailer uses hydraulic/mechanical steering on four of the five axles; steering is automatically controlled by the tractor-trailer angle. The semi-trailer has a maximum on- and off-road payload of 80,000 kg and is compatible with a variety of prime movers such as the M911, MK48/16, M1070, MAN, Mercedes-Benz, Volvo and German SLT 50-2 tractors. The M1000 is air-transportable in a C-5 Galaxy complete with tractor, and the C-17 (semi-trailer only).

When coupled to an M911 the HET can negotiate 90° intersections of 9.144 m roads in a single pass. There are five axle lines with two half-width axles per line. Each axle is mounted on a hydraulic pendular suspension providing a ±254 mm stroke, fully equalised. There is lateral oscillation provided to accommodate crowned roads and rough terrain. A hydraulic suspension system is also provided on the pivoting gooseneck to equalise fifth wheel loads. Hydraulics are used for the control of the deck height, gooseneck angle and axle jacking. Spring-assisted loading ramps can be manually lowered or raised by one person. An auxiliary diesel power pack is mounted on the gooseneck for loading, unloading and maintenance. Any axle may be raised by one person without having to remove the payload and the axles can be rotated to allow access to the inboard tyres for ease of changing flat tyres.

Specifications

Weight: 22,860 kg
Payload: 80,000 kg
Length:
(overall) 15.8 m
(deck) 10.24 m
Width:
(overall) 3.66 m
(deck) 3.05 m (3.66 m optional)
Height:
(deck) 1.074 m ± 0.254 m
(5th wheel) 1.6 m - variable
King-pin: 88.9 mm (3½ in)
Max speed:
(road) 72.4 km/h
(secondary road) 64 km/h
(cross-country) 24 km/h
Tyres: 215/75 R 17.5
Loading ramp angle: 18° (can be reduced to 14°)

Status

In production for US Army, Army Reserves, National Guard and Marine Corps (1,376 units).

Manufacturer

Systems and Electronics Inc, 201 Evans Lane, PO Box 5639, St Louis, Missouri 63121-1126, USA.
Tel: +1 314 553 4950. Fax: +1 314 553 4909.

UPDATED

Overview of Semi-trailer, Heavy Equipment Transporter (HET) M1000

Semi-trailer, Heavy Equipment Transporter (HET) M1000 showing steering feature

Talbert 64 tonne tank transporter semi-trailer

Description

For use during operations associated with operations Desert Shield and Desert Storm the US Army procured 150 Talbert Model T4LW-64M-FG 64 tonne capacity tank transporter semi-trailers. These commercially designed semi-trailers were used for the road transport of either a single M1 Abrams MBT or two M2/M3 Bradley fighting vehicles.

The Talbert 64 tonne semi-trailer has a usable cargo bed length of 12.041 m and a width of 3.609 m. Maximum towing speed is 88 km/h. The main spring beam suspension consists of four axles, carrying two roadwheels each side. Spring-loaded and balanced loading ramps are provided. The gooseneck has provision for a spare wheel and carrier and a water tank.

A shorter and lighter version of this semi-trailer is available, as is a version with an off-road suspension.

M1 Abrams MBT being loaded on to Talbert 64 tonne tank transporter semi-trailer

Specifications
Weight: (empty) 18,642 kg
Max load: 64,012 kg
Length:
(overall) 15.773 m
(cargo bed) 12.041 m
Width: 3.609 m
Height:
(5th wheel, loaded) 1.689 m
(overall) 3.359 m
Tyres: 315/80 R 22.5

Status
Available. In service with US Army.

Manufacturer
Talbert Manufacturing Inc, R.R.5, Box 195, Rensselaer, Indiana 47978, USA.
Tel: +1 219 866 7141. Fax: +1 219 866 5437.

VERIFIED

Semi-trailer, flatbed, 22½ ton, break bulk/container transporter M871 and semi-trailer, flatbed, 34 ton, break bulk/ container transporter M872/ M872A3

Description

The M871 and M872 semi-trailers were procured to carry containers and other military freight in rear areas, and are intended for operation at sustained high speeds on main roads. The M872 is normally operated from a port area as far forward as the Corps General Support Supply Activity (GSSA), while the M871 is used for the transport of containers, shelters and cargo forward from the Corps GSSA to the Division Support Command (DISCOM) or even to user units.

From 1979 Southwest Mobile Systems (now Systems and Electronics Inc (SEI)) delivered 3,367 M871 units and a further 246 M871A1s were delivered by Shoals America; the total requirement for the M871A2 was understood to be 3,420 units. From Fiscal Year 1986 8,877 M872 units were purchased, plus a further 346 M872A3 units from Southwest Mobile Systems (SEI).

The M871 semi-trailer was designed to carry containers and shelters up to 6.1 m in length, while the M872 was designed for the carriage of containers up to 12.2 m long. Both can also carry break bulk (loose) items of cargo. The two semi-trailers use many identical components and differ only in length and running gear. Each has a mainframe on which the load bed and a headboard are fitted. The M871 is a two-axle trailer while the M872 has a three-axle bogie. Two landing jacks are provided and there are two storage lockers below the load deck. A yard kit is supplied with removable guide assemblies to aid positioning of containers before being secured with twistlock devices. Side racks are also provided with removable side and rear panels and cover plates. The side panels are hinged to allow for the carriage of loose break bulk cargo without spoilage. The M872A3 is the latest production version of the M872. It retains all the original capabilities plus specific features to facilitate the transport of two M113 APCs (there are tiedowns on the platform and lower platform heights to allow a 4 m overhead clearance).

Specifications

Type	M871	M872/ M872A3
Weight	7,090 kg	8,300/ 8,700 kg
Payload	20,412 kg	30,481 kg
Length	9.1 m	12.45 m
Width	2.44 m	2.44 m

Type	M871	M872/ M872A3
Deck (length)	9 m	12.3 m
(width)	2.44 m	2.44 m
(height)	1.4 m	1.52/1.4 m
King-pin height	1.27 m	1.27 m
Track	1.79 m	1.8 m
Ground clearance	0.38 m	0.36/0.27 m
Tyres (commercial highway tread)	11.20 × 20, 12 PR	10.00 × 20, 14 PR
Number of tyres	8 + 1 spare	12 + 1 spare
Towing vehicle		
(primary)	M915 line haul tractor	
(secondary)	M818 5 ton tractor	
(terminal)	M878 yard tractor	

Status

In production for the US Army and Air Force. The US Army has a requirement for a total of 3,982 M871 and 11,488 M872 semi-trailers.

Manufacturers

M871: Systems and Electronics Inc, St Louis, Missouri, USA.
M871A2: Dynaweld Inc, Naperville, Illinois, USA.
M872: Heller Truck Body System, Systems and Electronics Inc and Theurer-Greenville Corporation.

UPDATED

Semi-trailer, flatbed, 34 ton, break bulk/container transporter M872A3 carrying two M113 APCs

Semi-trailer, flatbed, 34 ton, break bulk/container transporter M872

Trailer, tank: water, 1½ ton, 2-wheel, 400 gallon, M106 series

Description

The M106 single-axle trailer is used to transport, store and distribute drinking water. The aluminium tank has a capacity of 1,514 litres and has an elliptical cross-section. The trailer is provided with a hand water pump and a 7.62 m suction hose for filling.

The trailer was designed for fording where the trailer may be completely submerged and is normally towed by a 2½ ton truck, provided the towing vehicle has an air supply.

The M106A1 has a cover plate over the piping between the two welded metal tap boxes at the front of the tank. The M106A2 employs two hydraulic wheel cylinders for each wheel service brake. The M106A1 and M106A2 do not have a support leg, which is fitted only on the M106 trailer.

Trailer, tank: water, 1½ ton, 2-wheel, 400 gallon, M106 (Ray Young)

Specifications
M106; data in square brackets relate to M106A1 where different
Weight:
(empty) 1,034 [1,070] kg
(loaded) 2,547 [2,583] kg
Max load: 1,514 litres
Length: 4.23 m
Width: 2.36 m
(over tyres) 2.06 m
Height: 2.03 m
Ground clearance: 0.5 m
Track: 1.77 m
Wheelbase: 2.88 m
Tyres: 11.00 × 20
Max towing speed:
(road) 80 km/h
(cross-country) 48 km/h
Fording: complete submergence

Status
In service with the US Army and Marine Corps.

VERIFIED

Trailer, tank: water, 1½ ton, 2-wheel, 400 gallon, M107 series

Description
The M107 tanker trailer was designed to transport, store and dispense drinking water, and is normally towed by a 2½ ton truck, provided the truck has an air supply.

The trailer has a 1,514 litre aluminium water tank with an elliptical cross-section. It is fitted with a manual water pump and a 7.62 m suction hose for filling. Brakes on the trailer are of the air over hydraulic type, with a mechanical parking brake. The basic M107 is provided with a support leg, which is not fitted to the M107A1 and M107A2.

On the M107A1 and M107A2 there is a cover plate over the piping between the two welded metal tap boxes at the front of the tank, and the castor is raised and locked in a mounting bracket welded to the nose of the chassis frame. The M107A2 also employs two hydraulic wheel cylinders for each wheel service brake.

Trailer, tank: water, 1½ ton, 2-wheel, 400 gallon, M107A1 of Spanish Army

Specifications
M107; data in square brackets relate to M107A1 where different
Weight:
(empty) 1,034 kg
(loaded) 2,546 kg
Max load: 1,514 litres
Length: (overall) 4.32 [4.14] m
Width: (overall) 2.27 [2.08] m
Height: 1.91 [1.96] m
Ground clearance: 0.41 m
Track: 1.72 m
Max towing speed:
(road) 88 km/h
(cross-country) 24 km/h
Tyres: 9.00 × 20
Fording: complete submergence

Status
In service with the US Army and the armed forces of other countries.

VERIFIED

Trailer, tank: water, 1½ ton, 2-wheel, 400 gallon M149 series

Description
This single-axle two-wheeled trailer is used to transport, store and dispense drinking water, and is normally towed by a 2½ ton truck. The trailer has a filling cap on top of the tank and two dispensing taps in boxes in front of the tank. The main brakes are air over hydraulic with a mechanical parking brake also provided. The M149A1 has an improved lighting system. Glass fibre tanks are fitted in all versions except the M149A2, where the tank is stainless steel.

Trailer, tank: water, 1½ ton, 2-wheel, 400 gallon, M149

Specifications
Weight: 1,288 kg
Max load: 1,514 litres
Length: 4.09 m
Width: 2.09 m
Height: 1.94 m
(towing eye) 0.76 to 1.04 m
Ground clearance: 0.43 m
Track: 1.71 m
Wheelbase: (towing eye to centre of axle) 2.87 m
Tyres: 9.00 × 20

Status
In production. In service with the US Army and some other countries including Kenya (M149A2).

Manufacturer
(M149A2) Kogen Industries, Patton, Pennsylvania, USA.

VERIFIED

Semi-trailer, tank, fuel, 5,000 gallon, 4-wheel, M900 series, M967, M969, M970

Description
The M900 series of fuel semi-trailers is the replacement for the earlier M131 series. It consists of the M967 bulk haul refueller, the M969 automotive refueller and the M970 aircraft refueller.

Status
In service with the US Army and (M970) US Marine Corps. The initial order was for approximately 850 semi-trailers of the M900 series. The FY82 budget included a request for funds for 703 fuel semi-trailers. M900 series semi-trailers have been supplied to Portugal.

VERIFIED

Trailer, ammunition: 1½ ton, 2-wheel, M332

Description

The M332 ammunition trailer consists of a frame supported by an axle assembly with leaf spring suspension. A retractable support is provided at the front end of the trailer. Brakes are air over hydraulic and there is also a mechanical parking brake. If required, the top of the trailer can be covered by a tarpaulin which is stowed in a box on the forward part of the trailer when not in use.

Specifications

Weight:
(empty) 1,270 kg
(loaded) 2,630 kg
Max load: 1,360 kg
Length: (overall) 3.76 m
Width: (overall) 2.48 m
Height: 1.35 m
(to towing eye) 0.85 m
Track: 2.03 m
Towing speed:
(road) 80 km/h
(cross-country) 40 km/h
Tyres: 9.00 × 20

Trailer, ammunition: 1½ ton, 2-wheel, M332

Status

In service with the US Army.

VERIFIED

Chassis, trailer: 2 ton, 2-wheel, M390 and M390C

Description

These chassis trailers are used with the HAWK SAM system. The M390 is used to transport electronic equipment while the M390C carries three missiles. Each trailer consists of a chassis, deck plate (not on the M390), suspension system, two wheels air over hydraulic brakes, individual handbrakes for each wheel, retractable landing gear, levelling support jacks and a towing eye.

These trailers were also built in the Netherlands by DAF under the designation M390-17 and M390-17C.

Specifications

M390; data in square brackets relate to M390C where different
Weight:
(empty) 1,711 [1,656] kg
(loaded) 3,525 [3,470] kg
Length: 4.77 m
Width:
(travelling) 2.47 m
(emplaced) 2.82 m
Height: (without equipment) 0.93 m
Ground clearance: 0.43 m
Track: 2.03 m
Wheelbase: (towing eye to centre of axle) 3.23 m
Tyres: 9.00 × 20

M390C trailer of Japanese Self-Defence Force with three-round pallet of HAWK SAMs

Towing speed:
(road) 80 km/h
(cross-country) 40 km/h

Status

In service with countries using the HAWK SAM system including Bahrain, Belgium, Denmark, Egypt, France, Germany, Greece, Iran, Israel, Italy, Japan, Jordan, South Korea, Kuwait, Netherlands, Saudi Arabia, Spain, Sweden, Taiwan and the USA. Most trailers for European countries were supplied by DAF.

VERIFIED

Chassis, trailer: 1 ton, 2-wheel, M514

Description

The M514 trailer consists of a heavy-duty frame supported by an independent trailing arm suspension. Mounted at the front of the chassis is an A-shaped extension on which the retractable swivel castor and towing eye are mounted. The chassis has three levelling jacks, one at the rear and one either side. The brakes are the air over hydraulic type with two mechanical handbrakes for parking. This trailer is used with the HAWK SAM system and is used to mount the Radar Set, CW, Acquisition, AN/MPQ-34 and the Radar Set, CW, Illuminator, AN/MPQ-33.

Radar set, CW, illuminator, AN/MPQ-33 mounted on Spanish Army M514 1 ton trailer chassis (J I Taibo)

Specifications
Weight:
(empty) 1,034 kg
(loaded) 2,150 kg
Max load: 1,116 kg
Length: (without radar) 4.35 m
Width: (without radar) 2.39 m
Height: (without radar) 1.67 m
Ground clearance: 0.36 m
Towing eye height: adjustable from 0.56-0.86 m
Track: 2.03 m
Wheelbase: 2.69 m
Tyres: 9.00 × 20
Towing speed:
(road) 80 km/h
(cross-country) 40 km/h

Status
In service with countries using the HAWK SAM system including Bahrain, Belgium, Denmark, Egypt, France, Germany, Greece, Iran, Israel, Italy, Japan, Jordan, South Korea, Kuwait, Netherlands, Saudi Arabia, Spain, Sweden, Taiwan and the USA.

VERIFIED

Chassis, Trailer, Generator, 2½ ton, 2-wheel, M200A1

Description
The M200A1 trailer was originally a single-axle trailer with two roadwheels each side. There were 1,704 basic chassis procured following the Fiscal Year 1982 budget.

While the M200A1 was originally intended to be a mobile platform for electrical generators and similar equipment, the trailer was also used for carrying the M58 MICLIC mineclearing system. However, following operational experience in the Gulf region, it was appreciated that the M200A1 could not carry out its MICLIC tasks due to a general lack of mobility over soft terrain. As a result an initial total of 250 kits was procured to convert the M200A1 trailers to accommodate the Caterpillar Mobil-trac tracked suspension system which considerably reduces the ground pressure of the trailer and enhances its mobility over mud, sand and snow. The first kit conversions were carried out in March 1992 at Army depots.

See separate entry in this section for full details of the Mobil-trac system.

Status
In service with the US Army. Mobil-trac system conversions being carried out by US Army depots.

Manufacturer (Mobil-trac system)
Caterpillar Inc, Defense & Federal Products Department, Peoria, Illinois 61629, USA.
Tel: +1 309 675 6938. Telex: 404435.

VERIFIED

Chassis, Trailer, Generator, 2½ ton, 2-wheel, M200A1 carrying a 60 kW generator and with the Caterpillar Mobil-trac system installed

Chassis, trailer: 3½ ton, 2-wheel, M353

Description
The M353 chassis was designed to mount various types of bodies and other equipment such as generators. It consists of a heavy-duty frame supported by an offset axle assembly with leaf spring suspension and has two retractable swivel castors mounted at the front. The trailer has air over hydraulic brakes operated from the towing vehicle and a mechanical parking brake. The castors are secured in the raised or lowered positions by gravity pin and chain assembly. The fuel tank for the Patriot system generator is mounted on the M353 trailer.

Chassis, trailer: 3½ ton, 2-wheel, M353 fitted with generator (Michael Ledford)

Specifications
Weight: (empty) 1,202 kg
Max load:
(road) 3,629 kg
(cross-country) 3,175 kg
Length: 4.58 m
Width: 2.43 m
Height:
(overall) 1.23 m
(to platform) 0.69 m
(to towing eye) 0.86 or 0.96 m
Track: 2.07 m
Wheelbase: 3.36 m
Tyres: 11.00 × 20
Max towing speed:
(road) 80 km/h
(cross-country) 40 km/h

Status
In service with the US Army.

VERIFIED

Patriot missile semi-trailer M860A1

Description
Under contract to TACOM, Southwest Mobile Systems Corporation (now Systems and Electronics Inc (SEI)) was the sole producer of 419 M860A1 semi-trailers for the Patriot air defence guided missile system. In early 1987 Urdan Industries of Israel was awarded a multiyear contract to produce 72 M860A1 semi-trailers per year, 70 per cent of the work being carried out in Netanya, Israel, and the rest in Lebanon, Pennsylvania. Southwest (SEI) completed a technology transfer programme with the TOKYU Car Corporation of Yokohama, Japan, where M860A1s have been produced since 1987.

The M860A1 semi-trailer is designed to carry either the Patriot missile-launching stations or associated radar equipment, and is constructed to a high degree of precision so that the deck mounting pads for the equipment do not vary more than 0.015 in (0.381 mm) in height over the deck length. A stowage box is situated under the sides of the deck and houses lashing equipment and spares. Two triple-legged joists support the semi-trailer when it is not attached to a tractor. In use at a firing location, the semi-trailer is removed from the tractor, raised off the ground and levelled by four heavy-duty outriggers on the ends of folding stays. These operate under electric power supplied by the tractor truck's 100 A alternator. When on the move, the outriggers are raised vertically and lie alongside the carried load which is in shelter/container form. A panel at the front of the gooseneck connects the semi-trailer electrical system to that of the tractor truck.

Specifications

Weight: (kerb) 10,182 kg
Max load: 12,273 kg
Length: 10.08 m
Width:
(outriggers up) 2.87 m
(outriggers down) 6.55 m
Height: (outriggers up) 3 m
Max towing speed:
(road) 80 km/h
(cross-country) 32 km/h

Status

In production. In service with the US Army. The Patriot missile system is used by Israel, Netherlands, Germany, Saudi Arabia and Japan.

Manufacturers

Systems and Electronics Inc, 201 Evans Lane, PO Box 5639, St Louis, Missouri 63121-1126, USA.
Urdan Industries Limited, 30161 Southfield Road, Southfield, Michigan 48076, USA.

M860A1 semi-trailer for Patriot air defence missile system

TOKYU Car Corporation, 1 Kamajiya-cho, Kanazawa-ku, Yokohama 236, Japan.

VERIFIED

M989A1 Heavy Expanded Mobility Ammunition Trailer (HEMAT)

Development

The M989A1 Heavy Expanded Mobility Ammunition Trailer (HEMAT - originally the XM989) is intended for the carrying of Multiple Launch Rocket System (MLRS) rocket pods. Other applications include transporting ammunition pallets as well as fuel bladders for aviation refuelling.

A contract was awarded in Fiscal Year 1982 for 400 trailers but this contract was suspended in 1985 following the discovery of a stability problem after 353 trailers had been delivered. In 1986 the US Army solicited bids for a new multiyear production phase. Bids were submitted by the Southwest Mobile Systems Corporation (now Systems and Electronics Inc (SEI)), Teledyne Continental Motors, Landoll, and Prototype Development. In 1987 the Southwest Mobile Systems Corporation (now SEI) was awarded a four year contract for 1,046 M989A1 HEMAT trailers with a 50 per cent option. Testing of the first article units was carried out during 1989. Contract deliveries commenced during 1990.

Description

The HEMAT is a two-axle trailer with the front axle featuring an Ackerman steering design; the rear axle is stationary. The brakes are air-actuated while the air suspension is designed for on- and off-road mobility. The trailer meets all the requirements for a 0.55 *g* lateral stability.

The treated hardwood deck is surrounded by 22 tiedown rings evenly spaced for securing MLRS pods and other payloads up to 10,000 kg. The HEMAT is normally towed by the M985 HEMTT but can be towed by other military or commercial tractors as well as the MLRS launch vehicle. The HEMAT can be air-transported, unloaded, in a C-130 transport aircraft, or in a loaded state by C-141 and C-5 aircraft. It can also be carried, loaded, by a CH-47D helicopter.

Specifications

Weight:
(gross) 14,982 kg
(kerb) 4,994 kg
Max payload: 10,000 kg
Length: 7.85 m
Width: 2.44 m
Height: 1.45 m
Cargo area: 4.47 × 2.44 m
Ground clearance:
(bottom of axles) 0.37 m
(trailer structure) 0.55 m
Wheelbase: 5.94 m
Gradient: 40%
Side slope: 30%
Brakes: air
Suspension: air ride

Status

In production for US Army.

Manufacturer

Systems and Electronics Inc, 201 Evans Lane, PO Box 5639, St Louis, Missouri 63121-1126, USA.
Tel: +1 314 553 4950. Fax: +1 314 553 4909.

VERIFIED

M989A1 Heavy Expanded Mobility Ammunition Trailers (HEMAT)

Caterpillar Combat Support Trailer (CST)

Description

The Caterpillar Combat Support Trailer (CST) is intended for off-road employment in forward areas carrying a variety of combat engineer and other supply loads. The trailer uses a Mobil-trac (qv) rubber-belted two-wheeled undercarriage arrangement that provides up to 24° of vertical articulation over rough terrain while maintaining the level of the flat cargo bed. Air over hydraulic disc brakes are provided on all four wheels.

The trailer has been demonstrated carrying bridging equipment, mineclearing systems and up to 6,804 kg of payload. It can be towed by a 5 ton truck or a tracked vehicle such as an M113 APC and may be air-transported in a C-130 transport aircraft.

Specifications

Weight:
(unloaded) 2,721 kg
(fully loaded) 9,525 kg
Payload: 6,804 kg
Length: 4.267 m
Width: 2.438 m
Height: 1.09 m

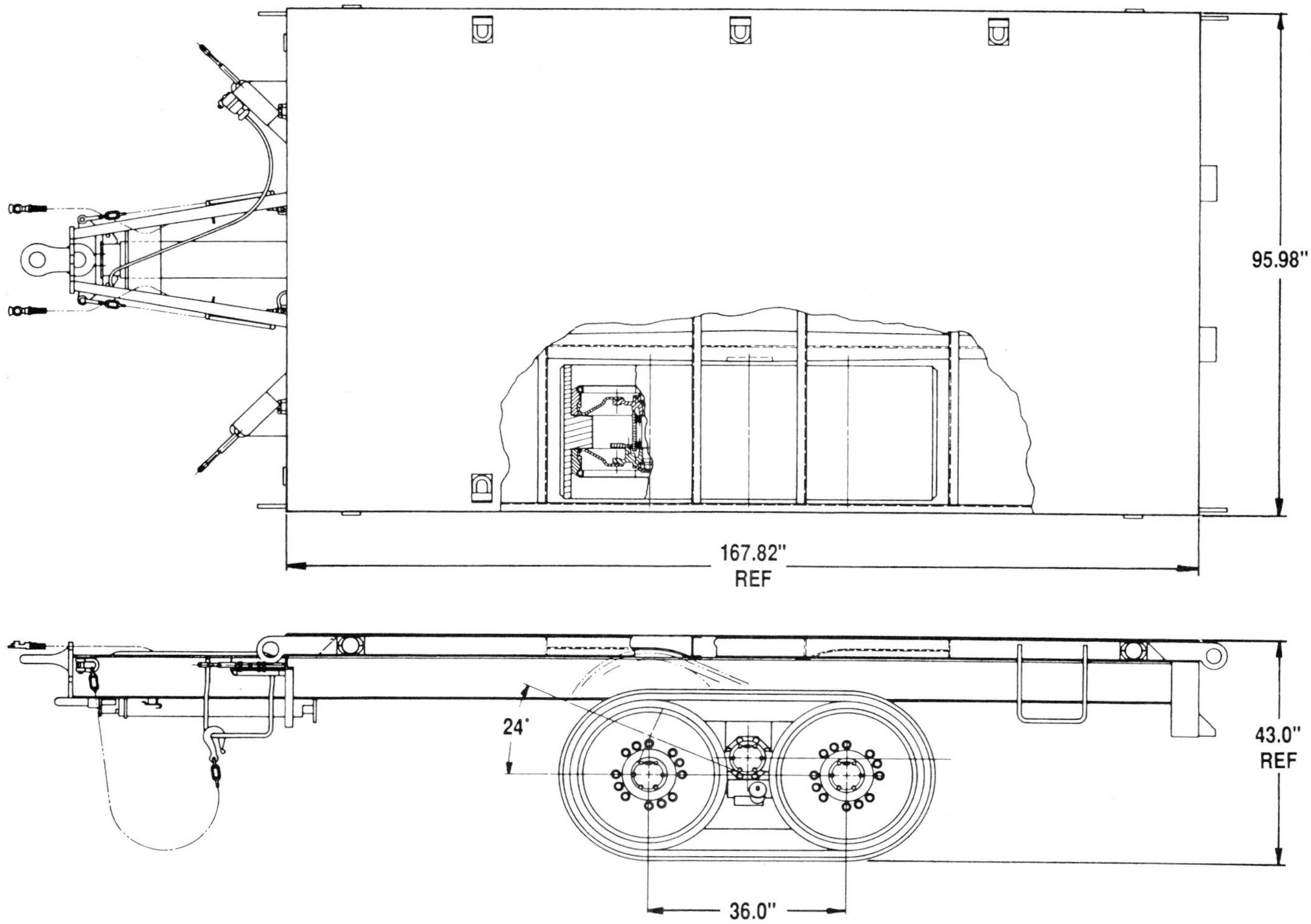

General outline drawings of an early version of the Caterpillar Combat Support Trailer (CST)

Flatbed dimensions: 4.263 × 2.438 m
Ground clearance: 0.419 m
Wheelbase: 0.914 m
Gradient: (loaded) 60%
Side slope: (loaded) 40%
Vertical obstacle: 0.38 m
Trench: 1.524 m

Towing speed:
(5 ton truck, paved road) 80 km/h
(5 ton truck, secondary road) 48 km/h
(tracked vehicle, cross-country) 32 km/h

Status
Ready for production.

Manufacturer
Caterpillar Inc, Defense & Federal Products Department, Mossville, Illinois 61552-0470, USA.
Tel: +1 309 578 6950.
Fax: +1 309 578 3739.

UPDATED

Caterpillar Mobil-trac Tracked Suspension (MTS) for trailers

Development
The Caterpillar Mobil-trac Tracked Suspension (MTS) has been used commercially on agricultural tractors since 1986, to lower ground pressure and increase mobility. Following experience during the 1992 Operation Desert Storm, the US Army Tank Automotive Command (TACOM) requested Caterpillar to design, build and install MTS kits on four M200 series trailers. Within four months government testing was completed and TACOM awarded Caterpillar a contract for the production of 250 MTS kits for M200 trailers (see separate entry for details of M200 trailer with MTS). MTS kits have also been produced for use with heavier trailers.

In March 1993 an agreement was formalised between Reynolds Boughton of the United Kingdom and Caterpillar to manufacture and market trailers equipped with MTS. Following this agreement, an MTS equipped trailer was used for trials with the Giant Viper mineclearing system.

The MTS is also used on the French LOHR MTS trailer - see separate entry in this section.

Description
The Caterpillar MTS lowers the ground pressure of the M200 series trailers by 78 per cent and increases the ground surface contact area by 450 per cent. The system encapsulates trailer tyres with a heavy-duty steel-reinforced rubber belt. Designed for sand and snow use, the MTS is also suitable for any areas where an increase in mobility is required.

A complete MTS kit intended for trailers in the 3 to 5 ton payload range features a simple walking beam design built around a two-piece telescoping mainframe. Commercial spindles, brakes and tyres mount to the frame and the belt wraps around the tyres. Guideblocks run between the tyres and provide lateral stability. A tension bolt and alignment screws enable proper adjustment to ensure the track runs true.

Apart for their use with M200 series trailers, MTS kits have also been installed on M149 400 gallon water tankers, M1048 and M1073 medium-capacity trailers carrying power packs and shelter truck trailers. One special MTS application is for a 22,680 kg capacity trailer for use in Antarctica; this trailer has four MTS units.

A further research and development contract awarded to Caterpillar involves the use of MTS technology on a US Marine Corps MK48/14 truck.

Caterpillar Mobil-trac System (MTS) fitted to medium-capacity trailer

Specifications
MTS undercarriage for light trailer, typical
Weights:
(complete kit) 862 kg
(belt, each) 136 kg
Length: 1.785 m
Width: 0.559 m
Height: 0.82 m
Ground contact length: (per track) 0.94 m
Ground contact area: (per track) 5.25 m^2

Status
In production - see text.

Manufacturer
Caterpillar Inc, Defense & Federal Products Department, Peoria, Illinois 61629, USA.
Tel: +1 309 675 6938. Telex: 404435.

UPDATED

Other US trailers

Many other trailers are in service, or have been in service, with the US Army. Those listed below include some that have been declared obsolete, but may still be found in other armies' inventories.

Designation	Notes
Trailer, mount, 20 mm automatic gun, 2-wheel, M42	For M167 towed Vulcan Air Defense System
Chassis, trailer: 1½ ton, 2-wheel, M102	For M104 and M106 trailers
Chassis, trailer: 1½ ton, 2-wheel, M103	For M105 and M107 trailers
Trailer, flatbed, 5 ton, 4-wheel, M106A1	997 units ordered from Utility Tool and Body Company for use as fuel unit carriers to replace M105 series
Trailer, low bed, 3½ ton, 4-wheel, M114	M113 chassis
Chassis, trailer: ¼ ton, 2-wheel, M115	For M100 and M367 trailers
Chassis, trailer: ¾ ton, 2-wheel, M116	For M101 trailer
Chassis, semi-trailer: 6 ton, 2-wheel, M117	For M118, M119 and M508 semi-trailers
Chassis, semi-trailer: 12 ton, 4-wheel, M126	For M127, M128 and M129 semi-trailers
Semi-trailer, van, cargo, 12 ton, 4-wheel, M128 series	Production of 44 M128A2C semi-trailers at a cost of $1.4 m was included in FY82 budget request
Semi-trailer, van, supply, 12 ton, 4-wheel, M129 series	Similar to M128 series but with interior fittings
Trailer, bomb, 2 ton, 4-wheel, M143	M143A1 is shorter
Semi-trailer, van, shop, 6 ton, 2-wheel, M146	M146C has windows
Semi-trailer, low bed, 60 ton, 8-wheel, M162	
Semi-trailer, van, office, 6 ton, 2-wheel, M164	
Dolly, trailer converter: 4 ton, 2-wheel, M196	Converts 3 and 4 ton semi-trailers to full trailers
Dolly, trailer converter: 6 ton, 2-wheel, M197	For 6 ton semi-trailers
Dolly, trailer converter: 8 ton, 4-wheel, M198	For 12 ton semi-trailers
Dolly, trailer converter: 18 ton, 4-wheel, M199	For larger semi-trailers
Trailer, van, fire-control mount, 3 ton, 4-wheel, M242	
Trailer, flatbed: fire-control/acquisition radar, 2 ton, 4-wheel, M243	
Trailer, van, fire control, 2 ton, 4-wheel, M244	
Tracking station, guided missile, trailer-mounted, M248	
Trailer, van, radar tracking control, M258	Used with Nike-Hercules system
Trailer, van, director station, M259	
Trailer, low bed, antenna mount, M260	
Trailer, flatbed, guided missile, M261	
Trailer, van, launching control, M262	
Chassis, trailer, generator, 2½ ton, 2-wheel, M267	
Trailer, bolster, pole handing, 3½ ton, 2-wheel, M271	M271A1 is later model
Semi-trailer, low bed, 12 ton, wrecker, 4-wheel, M269 and M269A1	
Chassis, semi-trailers, 6 ton, 4-wheel, M295	
Trailer, basic utility, 2½ ton, 2-wheel, M296	
Electrical shop, trailer-mounted, M304	
Semi-trailer, tank, water, 4,000 gallon, 4-wheel, M308	Similar to M131 series of fuel semi-trailers
Trailer, cable reel, 3½ ton, 2-wheel, M310	Signal Corps designation was K37-B
Semi-trailer, van, expansible, 6 ton, 4-wheel M313	
Semi-trailer, refrigerator, 15 ton, 4-wheel, M347	
Semi-trailer van, electronic, 3 ton, 2-wheel, M348	M348A1 and M348A2 are also 3 ton but M348A2C, -D, -F and -G models are 6 ton
Semi-trailer, van, refrigerator, 7½ ton, 2-wheel, M349 series	M349A1 is later model
Dolly, trailer converter, 18 ton, 4-wheel, M354	
Trailer, van, electronic shop, 2½ ton, 4-wheel, M359	
Dolly, trailer converter, 3 ton, 2-wheel, M363	
Dolly, trailer converter, 6 ton, 2-wheel, M364	
Dolly, trailer converter, 10 ton, 2-wheel, M365	
Trailer, maintenance, cable splicer, ¼ ton, 2-wheel, M367	Based on M100 trailer
Semi-trailer, van, electronic, 3 ton, 2-wheel, M373	Contract for 330 M373A2 worth $8,120,373 awarded to Miller Trailers Inc, Bradenton, Florida, August 1989
Trailer, van, electronic shop, 6 ton, 4-wheel, M382	
Trailer, van, electronic shop, 5 ton, 4-wheel, M383	
Semi-trailer, van, dental clinic, 3 ton, 2-wheel, M393	M393A1 is electronic shop
Semi-trailer, van, medical, 3 ton, 2-wheel, M394	
Semi-trailer, van, medical, 6 ton, 2-wheel, M395	
Trailer, antenna, 2 ton, 4-wheel, M406	For Nike-Hercules system
Trailer, guided missile director, M424	For Nike-Hercules system
Trailer, guided missile, tracking station, M428	Used on M242, M244, M248, M258, M259, M260, M261, M262, and M359 trailers
Dolly, trailer, front, M429 series	
Dolly, trailer, rear, M430 series	
Dolly, trailer, front, M431	Single-wheel dollies with electric brakes
Dolly, trailer, rear, M432	
Semi-trailer, van, shop, folding sides, 6 ton, 4-wheel, M447 and M447C	
Trailer, van, shop, folding side, 1½ ton, 2-wheel, M448	
Chassis, trailer, 2½ ton, 2-wheel, M454	Mounts AN/MPQ-4 mortar locating radar
Semi-trailer, maintenance, weapon, mechanical unit, M457	Based on M508 semi-trailer
Semi-trailer, maintenance, weapon, electric unit, M458	Based on M508 semi-trailer
Semi-trailer, maintenance, weapon, connecting unit, M459	Based on M508C semi-trailer
Trailer, air conditioner, 1½ ton, 2-wheel, M463	
Semi-trailer, van, shop, 6 ton, 2-wheel, M508	On M119 chassis; basis of M457 and M458
Semi-trailer, van, electrical, 6 ton, 2-wheel, M513	

Designation	Notes
Laundry unit, trailer-mounted, 80 lb capacity, M532	On M536 chassis
Bakery oven, trailer-mounted, M533	On M537 chassis
Chassis, trailer, laundry, 3½ ton, 2-wheel, M536	
Chassis, trailer, bakery oven, 2½ ton, 2-wheel, M537	
Chassis, trailer, bakery, dough mixing, 1½ ton, 2-wheel, M538	
Trailer, van, electronic, 10 ton, 4-wheel, M564	
Dolly, trailer, front, electronic shop, M565	
Dolly, trailer, rear, electronic shop, M566	
Chassis, trailer, ¼ ton, 2-wheel, M569	For M416 and M716 trailers
Semi-trailer, van, 10 ton, electronic, 4-wheel, XM574	
Chassis, trailer, 1½ ton, 2-wheel, M580	
Trailer, van, electronic, 1½ ton, 2-wheel, M581	
Trailer, van, electronic, 2 ton, 4-wheel, M582	
Trailer, van, electronic, 2 ton, 4-wheel, M583	
Dolly, trailer, front, M584	
Dolly, trailer, front, electronic shop, M589	
Dolly, trailer, rear, electronic shop, M590	
Semi-trailer, tank transporter, jointed, 52½ ton, 16-wheel, M627 ('Trackporter')	Previously designated XM793; believed to be for foreign military sales only
Semi-trailer, low bed, 15 ton, 4-wheel, M674	For transport of Nike-Hercules missile
Semi-trailer, van, 15 ton, 4-wheel, M681	
Semi-trailer, van, 15 ton, 4-wheel, M682	
Semi-trailer, van, 15 ton, 4-wheel, M683	
Dolly set, lifting, transportable shelter, M689	Comprises M690 and M691 dollies
Dolly, front, M690	Together form M689 dolly set
Dolly, rear, M691	
Dolly set, lift, M707	
Trailer, stake and platform, utility, ¾ ton, M709	
Trailer, cable splicer, ¼ ton, 2-wheel, M716	Based on M416 trailer
Dolly set, lifting, transportable shelter, M720	Comprises M721 and M722 dollies; for shelters S-141/G, S-280/G, S-285/G
Dolly, front, M721	
Dolly, rear, M722	
Semi-trailer, van, repair parts storage, 6 ton, 4-wheel, M749	
Semi-trailer, van, repair, shop, 6 ton, 4-wheel, M750	
Trailer, platform, ¾ ton, 2-wheel, M762	
Dolly set, lift, MUST, M840	Used with MUST hospital
Semi-trailer, low bed, construction equipment transporter, 40 ton, M870	Manufactured by Load King

Recent development model trailers

Designation	Notes
Trailer, aircraft cargo loading, 3½ ton, 4-wheel, XM712	
Trailer, flatbed, tilt loading, 6 ton, 4-wheel, XM714	
Semi-trailer, van, telephone, 10 ton, 4-wheel, XM738	
Semi-trailer, van, switchboard, 10 ton, 4-wheel, XM739	
Trailer, flatbed, tilt loading, ¾ ton, 2-wheel, XM789	
Semi-trailer, van, M822E1	
Semi-trailer, van, electronic, 10 ton, 4-wheel, XM823	
Semi-trailer, van, electronic, 10 ton, 4-wheel, XM824	
Dolly set, lift, transportable shelter, general support, XM832	
Chassis, semi-trailer, guided missile launching station, 4-wheel, M869 (for M901)	
Semi-trailer, line haul, 22½ ton, M871A2	
Launching station, guided missile, semi-trailer mounted, M901 (Patriot)	
Semi-trailer, tank, 12 ton, bulk haul, 5,000 gallon, self load/unload, M967A1	
Semi-trailer, tank, 12 ton, fuel dispensing, 5,000 gallon automotive, M969A1	
Semi-trailer, tank, 12 ton, fuel dispensing, 5,000 gallon under/over wing aircraft, M970A1	
Semi-trailer, van, 10 ton, 4-wheel, XM971	For electrical equipment
Semi-trailer, tank, potable water, 5,000 gallon, 4-wheel, XM972	
Semi-trailer, flatbed, 4-wheel, XM974	
Semi-trailer, guided missile transport, 4-wheel, M976 (Patriot)	
Trailer, flatbed, 5 ton, M979	For GEMSS
Trailer, ammunition, 10 ton, M989A1	
Semi-trailer, van, 6 ton, 2-wheel, XM990	
Semi-trailer, van, 10 ton, 4-wheel, XM991	For mobile test equipment
Semi-trailer, van, 10 ton, 4-wheel, XM995	For mobile test equipment
Semi-trailer, low bed, 70 ton	
Semi-trailer, van, repair parts storage, 12 ton, 4-wheel	
Dolly set, 7.5 ton, M1022	
Trailer, 5 ton, XM1034	For generator used with AN/TSQ-114A tactical intercept system
Trailer, 6 ton, XM1048	For AN/TPQ-37 Firefinder antenna-receiver unit
Trailer, flatbed, 5 ton, 4-wheel, M1061A1	For laundry units, generators and fuel pods
Semi-trailer, tank, 7,500 gallon, M1062	Replacement for M967, 727 units ordered from Fruehauf
Semi-trailer, van, 12 ton, 40 ft, XM1063	For electronic equipment
Semi-trailer, 35 ft, XM1065	For ATE
Trailer, flatbed, 7½ ton, XM1073	65 ordered from Utility Tool and Body Company, 1989

VERIFIED

YUGOSLAVIA (SERBIA AND MONTENEGRO)

FNNPP 50/3U 50,000 kg low bed semi-trailer

Description

The FNNPP 50/3U low bed semi-trailer is normally towed by the FAP 3232 BDST/AV tractor truck (see entry in *Tank transporters* section) but reports have indicated that it has also been used by MAZ-537G (8 × 8) tractor trucks, in which combination it is known as the M-60.

The semi-trailer frame is welded and comprises two steel longitudinal 'I' beams with cross members and side rails. Fixed to the frame are the axle bogies, loading ramps, spare wheel carriers, air and hydraulic systems and other equipment. The front part of the frame is in the form of a gooseneck while the frame platform is flat with a 20° tapered rear end. The loading ramps are 600 mm wide and are raised and lowered with the assistance of hydraulic pumps. The ramps are laterally movable within limits of 2.5 to 3 m and their upper surface is ribbed steel. The platform is hardwood and the side rails are provided with load securing hooks.

The four axles have a loading of 12,000 kg each. The front axle is fixed and the remainder are steerable. The suspension uses flexible leaf springs supported at their ends by spring hangers. Axle steering is hydromechanical and airbrakes are used. Load distribution between axles is performed by equalisers. Each axle has four tyres.

Loading can be assisted by using a winch system driven by a power take-off on the tractor. Two pulleys are provided on the gooseneck for this system.

FNNPP 50/3U 50,000 kg low bed semi-trailer

Specifications

Weight:
(empty) 16,000 kg
(loaded) 66,000 kg
Max load: 50,000 kg
Max 5th wheel load: 18,000 kg
Length:
(overall) 12.287 m
(platform) 7.697 m
Width: 3.5 m
Height:
(5th wheel, loaded) 1.83 m
(5th wheel, unloaded) 1.025 m
(platform, loaded) 0.972 m
(platform, unloaded) 1.025 m
Wheelbase:
(axles) 1.245 m + 1.245 m + 1.245 m
(from 5th wheel to bogie centre) 8.7 m
Track: 2.55 m
Tyres: 8.25 × 15
Electrical system: 12 or 24 V

Status

In service with the former Yugoslav Army.

Contractor

FDSP - Federal Directorate of Supply and Procurement, PO Box 89, Bulevar umetnosti 2, YU-11070 Novi Beograd, Yugoslavia.
Tel: +381 11 222 2743. Telex: 71000 JU SPDR.
Fax: +381 11 324 8738.

UPDATED

MATERIALS HANDLING EQUIPMENT

AUSTRIA

Voest-Alpine TROSS 130 military multipurpose vehicle

Description

The TROSS 130 (Technical and Rear Operation Support System) was the result of a joint development programme between the Austrian Army and Voest-Alpine. This military multipurpose equipment is a four-wheeled, all-wheel steering vehicle with a swivel mount on which the various hydraulically operated attachments, such as a front-loading bucket, lifting forks and a crane, can be fitted using a rapid-change device.

The TROSS 130 was designed for earthmoving and recovery operations as well as for handling military loads under difficult conditions. All the necessary auxiliary equipment, including a spare wheel and tools, can be carried with the TROSS 130 and with a high cruising speed of 62 km/h, it is able to travel in convoy with other military vehicles.

The TROSS 130 has an automatic differential lock and a dual-circuit brake system. Standard equipment includes an emergency steering system that allows towing of the vehicle after breakdown, a cold starting device, three-phase dynamo and special military-proof indicator and control instruments. The driver's steel safety cabin is sound insulated and has two adjustable seats, a heater, defroster and stowage for weapons and a full pack. Optional equipment includes air conditioning, cab ventilation, tropical roof, tyre inflation system, blackout lights and an infrared suppressed colour scheme.

Specifications

Cab seating: 1 + 1
Weight: 15,950 kg
Length: 8.3 m
Width: 2.65 m
Height: 3.13 m
Ground clearance: 0.435 m
Fording depth: 1 m
Turning radius: 5.35 m
Fuel capacity: 250 litres
Max speed: 62 km/h
Max gradient: 70%
Engine: MAN Type D 2565 ME water-cooled diesel developing 160 hp at 2,100 rpm
Tyres: 20.5 × 25 EM 16 PR
Electrical system: 24 V
Max load:
(front) 6,000 kg
(turning) 4,000 kg
Bucket capacity: 1.8 m^3
Winch capacity: 8,000 kg
Winch cable length: 55 m

Status

In service with the Austrian Army.

Manufacturer

Voest-Alpine AG, A-4010 Linz, Austria.

UPDATED

TROSS 130 with crane attachment

TROSS 130 with forklift attachment

Steyr LBW 1500 electrohydraulic tailgate

Description

The Steyr LBW 1500 electrohydraulic tailgate lift is used for the loading and unloading of bulky and heavy loads or equipment on to a platform truck. The hydraulic system pump is driven by an electro-motor from the vehicle batteries. Control may be by a control box in the cab or using a remote-control unit on a cable. The basic carrier vehicle is a standard troop carrier without bench seats or a tailgate. The structure is bolted to the rear section of the chassis frame. Warning lamps are provided in the platform for safety.

The lifting capacity is 1,500 kg at 0.5 m.

Status

In production. In service with Austria.

Manufacturer

Steyr Nutzfahrzeuge AG, Schönauerstrasse 5, A-4400 Steyr, Austria.
Tel: +43 7252 25351. Telex: 28200.
Fax: +43 7252 46746/48650.

VERIFIED

The Steyr LBW 1500 electrohydraulic tailgate fitted to a 12 M 18 (4 × 4) truck

FINLAND

Sisu TD1612HS high-speed forklift truck

Development

The Sisu TD1612HS high-speed forklift truck was developed in response to a French Army requirement for a palletised load handling system. In total, there are 57 in operation with the French Army in various locations.

Description

The 4 × 2 TD1612HS has a top speed on roads of 70 km/h and can handle general cargo or containers weighing up to 16,000 kg at a load distance of 1.2 m. While working, the forklift mechanism is at the front of the vehicle with steering on the rear axle. For high-speed travelling the seat and steering wheel are rotated 180°. For forklift operation the suspension is lowered and placed out of use although, as an option, it is possible to operate loads up to 8,000 kg with the suspension raised providing more ground clearance over rough terrain. For normal travelling without a load, the suspension is raised to an intermediate position and, when travelling over rough terrain, the suspension can be fully raised to provide a ground clearance of nearly 0.5 m.

The standard lift mast is duplex with a 3.2 m lift height. Also available are 2 or 2.5 m single section simplex masts with lift carriages an approximate 1 m front face distance from the drive axle. The standard lift carriage has hydraulic side shift and fork side shifting, or hydraulic side shift, fork side shifting and tilting.

The engine is a Renault MIDS 0620458 9.5 litre or a Cummins CTA8.3C 8.3 litre turbocharged diesel although other engines may be fitted. The hydraulic power shift transmission is equipped with torque converter lock-up.

Sisu TD1612HS high-speed forklift truck

Specifications

Cab seating: 1
Configuration: 4 × 4
Weight: (unladen) 24,680 kg
Length: (forks lowered) 8.12 m
Width: 2.5 m
Height: 3.1 m
Ground clearance: (max) 0.49 m
Turning radius: 5.2 m
Fuel capacity: 560 litres
Road range: 800 km
Max speed: 70 km/h
Max gradient: 33%
Engine: Renault MIDS 0620458 9.48 litre 6-cylinder direct injection turbocharged diesel developing 270 hp at 2,200 rpm
Transmission: Clark 34000 hydraulic power shift with torque converter lock-up providing 4 forward speeds and 2 reverse
Tyres:
(front) 12.00 R 20
(rear) 15.5/80 R 20
Max load: 16,000 kg

Status

In production. In service with the French Army (57).

Manufacturer

Sisu Terminal Systems Inc, Lentokonetehtaantie 10, PO Box 387, FIN-33101 Tampere, Finland.
Tel: +358 31 658 111. Telex: 22140 valma sf.
Fax: +358 31 658 223.

UPDATED

FRANCE

Marrel Ampliroll system

Description

The Marrel Ampliroll system is a load handling system designed to fit any type of truck with a load capacity from 3,000 to 25,000 kg. To meet military requirements, the systems selected for 6 × 4 6 × 6 and 8 × 8 trucks generally have a load capacity of between 12 and 20 tonnes.

The Ampliroll system consists of: a chassis reinforcing subframe; central beam to which the tipping rams are hinged; short hinged connecting arm framework which controls the extent of rearward movement; telescopic jib sliding in front of the central beams and carrying at its upper end the unit front lifting hook and rear hinges and incorporating flanged rollers on bronze bushes to centralise the unit. The hydraulic system contains: a pump complete with a safety valve and transmission; two main arms providing the normal tipping action and for lowering the platform; double-acting ram to actuate the sliding jib beam; and a direction and braking valve block actuated from the vehicle cab by a remote servo control.

Military applications include: ammunition handling, either on a flatrack or in an ISO container; fuel handling using the Marrel 9,000 or 12,000 litre roll-off fuel tank and filling unit, complete with a hydraulic pump serving five discharge points; mobile shelter handling for field hospitals; and general cargo handling using flatracks with suitable cargo bodies and covers. The system may also be used for vehicle recovery using a multipurpose flatrack with adjustable track guides allowing for the recovery of light armoured vehicles such as the M113 APC. With some additional components the Ampliroll system can be used to launch floating bridge components (such as the Ribbon Bridge) and their associated bridging boats. Flatracks may also be used

Marrel Ampliroll system in use loading an ammunition flatrack

Marrel Ampliroll system in use on French Army Renault TRM 10 000 (6 × 6) truck

1997

Marrel Ampliroll system in use with 9,000 litre fuel tank and filling unit

Marrel Ampliroll system in use with French Army Renault G 290 (6 × 4) truck loading a 20 ft/6.096 m ISO container with an integrated skid **1997**

to carry engineer equipment or may be converted to carry weapon systems such as air or coastal defence weapons. Other flatrack loads could include electrical generators, desalination plant and communication centres. There is also an Ampliroll platform configured to carry personnel on fitted benches.

If required, the lifting hook can be used as a crane both to lift loads and to haul disabled vehicles on to a flatrack ready for loading.

Status
In production. Quantities have been sold worldwide for military applications in 11 countries. In service with the French Army (2,800 units).

Manufacturer
Marrel SA, BP 56, F-42160 Andrezieux-Bouthéon, France.
Tel: +33 4 77 36 28 28. Fax: +33 4 77 36 28 00.

UPDATED

LOHR PLM 17 materials handling system

Description
The LOHR PLM 17 materials handling system was designed for the transportation of vehicles (wheeled or tracked) and for the logistic transport of ammunition and other combat supplies on 20 ft/6.096 m ISO flatracks. It can be mounted on all types of 6 × 4 and 6 × 6 trucks with a GVW of 26,000 kg and a 300 hp engine. It has been mounted on Renault TRM 10 000 (6 × 6) 10,000 kg trucks in conjunction with a RM 19 trailer to which flatrack loads can be transferred and unloaded by the PLM 17 system.

The system uses a hydraulic system to either load a platform on to a truck or unload it on to the ground. It utilises a two-part subframe which bolts on to the vehicle chassis. Two conical rollers, located at the rear, centre the platform during loading operations. The loading arm is constructed using two rectangular and longitudinal beams braced by cross members and is articulated round the rear part of the frame. A bracket assembly on the loading platform completes the structural system.

The hydraulic system is based on a Leduc L 65 pump operating at a maximum pressure of 400 bar. The pump is mounted on the vehicle gearbox and is directly driven. The hydraulic circuit has three parts for blocking the vehicle suspension, controlling the lifting bracket and the lifting arm. Two cylinders are used, one for the bracket, one for the arm. A tank associated with the hydraulic system contains 100 litres of oil.

The platform is a standard 20 ft/6.096 m DIN unit and is made of beams braced by cross members. A total of 12 lashing rings is provided. Load covers are available.

LOHR PLM 17 materials handling system in use on Renault G 290.26 VTL truck towing LOHR RM 19 trailer **1997**

Specifications
System on TRM 10 000 truck
Weight on front axle: (loaded) 7,480 kg
Weight on rear axle: (loaded) 18,520 kg
Weight, total: 26,000 kg
Loading capacity: 17,000 kg
Payload: 15,000 kg
Flatrack weight: 2,000 kg
Length: (overall) 9.2 m
Width: 2.5 m
Height:
(overall) 3.7 m
(platform, loaded) 1.7 m

RM 19 trailer
GVW: 19,000 kg
Payload: 13,400 kg
Length: (with flatrack) 8.37 m
Width: 2.5 m
Height: (with loaded flatrack) 1.475 m

Status
Available.

Manufacturer
LOHR SA, F-67980 Hangenbieten, France.
Tel: +33 3 88 38 98 00. Telex: 870082 f.
Fax: +33 3 88 96 06 36.

UPDATED

FDI Sambron J24 series wheeled materials handlers

Description
The FDI Sambron J24 series encompasses a number of rough-terrain materials handlers designed to handle bulk materials, palletised loads or stores such as munitions.

The basic model is the J24 2004S loader with a lift capacity of 2,000 kg to a height of 3.6 m. It is powered by a Deutz F 3 L 912 diesel developing 52 hp at 2,500 rpm. All other models (L and T ranges) are powered by a Perkins 4.236 diesel and are available in a choice of two capacities (2,000 or 2,500 kg) with two- or four-wheel drive. Four-wheel drive models are available with straight axles or dropped hub axles for higher ground clearance, with four-wheel steering as an option. The L range has a rigid arm, giving lift heights up to 4.6 m. The T range has a telescopic arm which provides a maximum forward reach of 3.81 m ahead of the front wheels and a maximum lift height of 6.2 m.

All the J24 series vehicles have a dual range hydrostatic transmission. The arm can be fitted with a variety of earthmoving or handling attachments such as buckets, diggers, pallet forks, jibs, clamps and aerial platforms, all secured by quick-fit systems.

Specifications
J24 2504T
Cab seating: 1
Configuration: 4 × 4
Weight without attachment: 6,980 kg
Length:
(overall, with forks, straight axles) 5.64 m
(overall, with forks, dropped hub axles) 5.53 m
Width:
(over body) 2.03 m
(over wheels, straight axles) 2.05 m
(over wheels, dropped hub axles) 2.18 m
Height:
(straight axles) 2.34 m
(dropped hub axles) 2.51 m
Ground clearance:
(straight axles) 0.31 m
(dropped hub axles) 0.46 m
Wheelbase: 1.92 m
Max speed:
(low range) 10 km/h
(high range) 25 km/h
Fuel capacity: 100 litres
Engine: Perkins 4.236 water-cooled diesel developing 72 hp at 2,250 rpm
Transmission: dual range hydrostatic
Steering: hydrostatic
Turning radius:
(outer, straight axles) 5.36 m
(outer, dropped hub axles) 5.65 m

FDI Sambron J24 2504T wheeled materials handlers awaiting delivery

Tyres:
(straight axles) 12.5 × 18
(dropped hub axles) 16/70 × 20
Brakes: hydrostatic plus inboard wet disc brakes
Max lift height:
(straight axles) 6.03 m
(dropped hub axles) 6.2 m
Capacity at max lift height: 2,500 kg
Max forward reach:
(straight axles) 3.81 m
(dropped hub axles) 3.74 m
Capacity at max forward reach: 1,200 kg

Status
In production.

Manufacturer
FDI Sambron, Route de Nantes, BP 71, F-44160 Pontchateau, France.
Tel: +33 2 40 00 73 50. Telex: 710788 f.
Fax: +33 2 40 00 73 60.

VERIFIED

GERMANY

ADK 160 16 tonne mobile crane

Description
AD 160 or ADK 160 (for *Autodrehkran* 160) was the German designation of a 16 tonne hydraulic crane mounted on the Czech/Slovak TATRA 138 or 148 (6 × 6) chassis. The chassis differ from the standard truck chassis only in the addition of four stabilising jacks. The crane, made by CKD, is separately powered and has a two-section telescopic jib.

Specifications
Chassis: see under Czech Republic and Slovakia in the *Trucks* section *except* max speed: 71 km/h
Weight: (travelling) 24,600 kg
Length: 9.85 m
Width: 2.5 m
(over extended stabilisers) 4.5 m
Height: 3.4 m
Crane capacity: max 16,000 kg at 3.8 m radius
Traverse: 360°

Status
Believed to be in service with several former Warsaw Pact armies.

Manufacturers
Chassis: TATRA, Národni Podnik, Kohprivinice, Czech Republic.
Crane: CKD, Germany.

VERIFIED

ADK 160 mobile crane on TATRA 148 chassis

Klaus Kranmobil mobile container handling system

Description
The Klaus Kranmobil mobile container handling system has been in production for over 20 years and is now produced in two basic forms: double-side operating (KM 32, KM 26) or single-side operating (KM 32 E, KM 24 E, KM 13 E). Both types have a lifting capacity from 13 to 36 tonnes and can handle containers or flatracks.

The Kranmobil system can be used in conjunction with vehicles or railway trucks and enables the carrier vehicle to load and offload loaded containers and flatracks as required. The larger systems are usually carried on semi-trailers while the smaller systems can be carried on large trucks. Containers can be loaded on to the carrier vehicle or directly on to another vehicle or railway truck. The double-sided loaders can lift and load from one side to the other in one sweep if required. Containers can be stacked two high.

When operating, the lifter-loader is stabilised by hydraulically operated stabiliser legs. This enables the system to be used on unprepared surfaces. The system can be used with toplift container spreaders, rope slings or chains. Control is usually carried out remotely by means of a portable control box.

When not in use, the system arms can be folded down or to the side to lower the overall height for air transport.

The Klaus RTE is a towed semi-trailer equipment

Klaus Kranmobil KM 32 container handling system on Mercedes-Benz tractor truck

Klaus Kranmobil KM 26-298 container handling system as used by German forces during the UN mission to Somalia, 1993

1996

Specifications

Model	**KM 32**	**KM 26**	**KM 32 E**	**KM 24 E**	**KM 13 E**
Loading sides	both	both	single	single	single
Container size	20-45 ft	20 ft	20-45 ft	20 ft	20 ft
Lifting capacity	32,000 kg	26,000 kg	36,000 kg	24,000 kg	13,000 kg
Carrier (typical)	3-axle semi-trailer	2-axle semi-trailer	3-axle semi-trailer	2-axle semi-trailer	8 × 8 truck
Length (approx, with tractor)	17.4 m	12 m	16 m	11 m	9.8 m
Width	2.5 m	2.5 m	2.5 m	2.5 m	2.5 m
Height (approx)	4 m	4 m	4 m	4 m	4 m
Hydraulic circuit (dual circuit)	300 bar	200 bar	270 bar	250 bar	230 bar
Electrical system	24 V	24 V	24 V	24 V	24 V

used to unload flatracks and ISO containers weighing up to 16.5 tonnes from rail cars. The RTE moves along either side of a train, lifts off a flatrack using a special chain-lift device which can be safely used under overhead power cables and places it on the ground nearby. The RTE can handle up to 12 flatracks/hour. The British Army has 30 of these KM 20-298 (RTE) units in service for its DROPS system; they are carried on Leyland Trucks MMLC vehicles. In 1991 (6) and 1995 (20) units were equipped with a top spreader attachment for easier and faster handling of containers.

Status

In production. In widespread use, including with the British Army (over 70 units, with 30 units employed with DROPS system), US Army (14 KM 26-298 for 20 ft ISO containers) and US Air Force in Europe (30 KM 32-298 and 6 KM 26-298).

Manufacturer

Klaus Transportsysteme Vertriebs GmbH, PO Box 11 41, D-87681 Memmingen, Germany; Schlachthofstrasse 46, D-87700 Memmingen, Germany.
Tel: +49 8331 160. Fax: +49 8331 16212.

UPDATED

INDIA

Mahindra Manitou rough terrain forklift trucks

Description

The Mahindra Manitou MC and MCE ranges of forklift trucks are manufactured in collaboration with Manitou BF of France. The MC range is intended for outdoor use and is available with 2,000, 2,500, 3,000 and 4,000 kg capacities, while the MCE has both indoor and outdoor functions and is available with 2,000, 2,500 and 3,000 kg capacities.

All trucks are fitted with a four-cylinder 45 hp diesel engine and a mechanical transmission. A duplex mast with a 3.6 m lift height is standard. Free lift and triplex masts are also available, as is a wide range of specialised attachments.

The Indian Ordnance Services have standardised on the 2,000 and 3,000 kg MC models.

Status

In production. In service with the Indian Ordnance Services.

Manufacturer

Mahindra Engineering & Chemical Products Limited, Mahindra Manitou Division, 148 Bombay Pune Road, Pimpri, Pune-411 018, India.
Tel: +91 212 775045-6. Telex: 0146-233 MOL-IN.
Fax: +91 212 771286.

VERIFIED

Mahindra Manitou rough terrain forklift trucks awaiting delivery

NETHERLANDS

KOOI-AAP portable forklift truck

Description

The KOOI-AAP military portable forklift truck is a derivative of the commercial KOOI-AAP portable forklift truck which has been sold worldwide since 1968. The KOOI-AAP truck can be mounted or dismounted within 30 seconds on or from the rear of a truck or trailer by a special mounting system and can also be mounted on flatracks. It can lift itself to the transport position by inserting its forks in the rear of the load platform of the carrying vehicle, then lowering the forks.

Hydraulic power for driving and lifting is provided by a diesel-powered hydraulic pump. The constant all-wheel-drive KOOI-AAP has a good cross-country capability provided by a Trac-lock system and puncture-proof low pressure tyres.

The KOOI-AAP is available in a wide range of types and specifications with lift capacities from 1 to 4 tonnes and lift heights up to 3.9 m. Special models to meet specific customer requirements are also produced.

The KOOI-AAP is air transportable by the majority of NATO helicopters and transport aircraft.

Multidirectional KOOI-AAPs are in service with the RAF and RNAF for the handling of large missile containers and/or missiles in and around airfield ammunition storage bunkers. It is also available for the

KOOI-AAP portable forklift truck with scissor reachmast ***1997***

KOOI-AAP portable forklift truck in Dutch IFOR service loading a container on a DAF YAC 2300 (6 × 6) 10,000 kg truck **1997**

KOOI-AAP portable forklift truck carried by Bedford MK (4 × 4) 4,000 kg truck used to support RAF Harrier field operations **1997**

distribution and handling of missiles in the field. Carried by the Bedford MK (4 ×4) 4,000 kg truck, the KOOI-AAP is also used to resupply bombs and missiles to RAF Harrier strike aircraft and AH-64D attack helicopters deployed in the field. Since 1996 the RNAF has been using the KOOI-AAP for IFOR logistic duties in Bosnia, carried by the DAF YAC 2300 (6 × 6) 10,000 kg truck. In service with the Royal Netherlands Navy the vehicle is used for the distribution and handling of gas bottle racks. The Swedish Army has been using these vehicles for tactical services since 1983. It is also used to tow mortars and supply ammunition from carrier trucks to the mortar.

The military KOOI-AAP can also be used to drill field emplacements. See separate entry under *Field fortifications* for details and an illustration.

Kooi bv, is an Allegheny Teledyne Company with sales and service offices worldwide.

Status
In production. In service with the Royal Netherlands Navy, Army and Air Force, Royal Air Force and Swedish Navy.

Manufacturer
Kooi bv, Molen 25, NL-9077 SW Vrouwenparochie, Netherlands.
Tel: +31 518 1741. Fax: +31 518 1024.

NEW ENTRY

RUSSIAN FEDERATION AND ASSOCIATED STATES (CIS)

Truck-mounted cranes

Description
Some of the more recent RFAS truck-mounted cranes are described briefly here. Many of these are in service with the RFAS and other former Warsaw Pact armed forces, but the exact status of many is uncertain.

9T31M1 crane on Ural-375D (6 × 6) truck chassis

Model	Operation	Traverse	Chassis*	Capacity Max load/ radius	Load at max radius	Length	Width	Height	Speed	Remarks
LAZ-690	mechanical	360°	ZIL-164A (ZIL-130) (ZIL-150)	3 t/2.5 m	1 t/5.5 m	8.88 m	2.4 m	3.45 m	45 km/h	Obsolescent
KS-1571	hydraulic	360°	GAZ-53A	4 t/3.3 m	—/9.35 m	—	—	—	—	Military use not confirmed
K-67	electric	360°	MAZ-500 (MAZ-200)	6.3 t/3.5 m	2 t/7.5 m	8.2 m	2.6 m	3.35 m	40 km/h	180 hp engine
K-68	electric	360°	MAZ-200 (MAZ-500)	6.3 t/5.5 m	2 t/7.5 m	8.9 m	2.71 m	3.3 m	40 km/h	Also K-68A
KS-2571	hydraulic	360°	ZIL-130	6.3 t/3.3 m	—/9.8 m	—	—	—	—	Military use not confirmed
8T210	hydraulic	360°	Ural-375D	6.3 t/3.5 m	1.8 t/7.5 m	8.3 m	2.45 m	—	70 km/h	Special military crane originally for missile units
9T31M1	—	360°	Ural-375D	—	—	—	—	—	—	Special military crane for missile units
AK-75	mechanical	360°	ZIL-164A (ZIL-130)	7.5 t/2.8 m	1.65 t/7 m	10.1 m	2.5 m	3.56 m	40 km/h	97 hp engine
KS-3562A	electric	360°	MAZ-500A	10 t/4 m	—	—	—	—	60 km/h	
KS-3571	hydraulic	360°	MAZ-500A	10 t/4 m	—/13.2 m	—	—	5 m	75 km/h	Military use not confirmed
KS-3572	hydraulic	360°	KrAZ-255B	10 t/4 m	—/9.1 m	—	—	—	70 km/h	
K-162	electric	360°	KrAZ-219 KrAZ-257	16 t/3.9 m	2.35 t/10 m	14 m	2.75 m	3.96 m	30 km/h	Also K-162M
KS-4571	hydraulic	360°	KrAZ-257	16 t/3.8 m	—/20.25 m	—	—	—	70 km/h	Military use not confirmed

AK-75 7.5 tonne crane

MPK-30 crane mounted on MAZ-529V chassis

Model	Operation	Traverse	Chassis*	Capacity Max load/ radius	Load at max radius	Length	Width	Height	Speed	Remarks
MPK-30	electric	360°	MAZ-529V	30 t/3.7 m	6 t/27 m	12.8 m	4 m	4.2 m	25 km/h	Articulated, with 2-wheel tractor
8T26	—	—	—	10 t/4.5 m	2 t/9.5 m	—	—	—	40 km/h	Special military crane for missile units
KM-61	mechanical	80°	KrAZ-214 KrAZ-255B	3.2 t/2 m	2 t/2.8 m	—	—	—	—	Used with KMT-5 mineclearing apparatus
KS-2561D	—	360°	—	6.3 t/—	1.5 t/—	—	—	—	—	150 hp engine

* Chassis designations in brackets are older or less common variants.

VERIFIED

SOUTH AFRICA

Bond Miskruier rough terrain forklift truck

Description

The Bond Miskruier forklift truck is a compact three-wheel rough terrain vehicle intended for use in forward areas to handle stores and munitions. The vehicle utilises two front wheels with a single wheel at the rear. It turns on a turntable and yoke steering assembly operated via a hydrostatic steering valve and a hydraulic rack and pinion. Drive to all three wheels is hydrostatic with power provided by a 2.5 litre ADE 152 diesel.

The driver is seated on the right-hand side of the vehicle on a fully adjustable seat; a tubular rollbar assembly is provided over the driver's position. The driver uses his left hand to operate the forklift controls.

The mast assembly has a lift height of 2 m and a side shift of 200 mm. The mast assembly can be tilted 12° back and 6° forward. Lift capacity is 2,500 kg at 600 mm load centre.

Optional attachments include a 0.5 m^3 bucket, and a 200 to 500 kg capacity rotating grab. Other attachments can be produced to meet customer requirements.

Bond Miskruier rough terrain forklift truck

Specifications

Seating: 1
Configuration: 3 × 3
Weight:
(kerb) 3,800 kg
(loaded) approx 6,300 kg
Max load: 2,500 kg at 600 mm load centre
Length: 2.73 m
Width: 1.6 m
Height: (overall) 2.[illegible]84 m
Ground clearance: 0.25 m
Wheelbase: 1.7 m
Max speed: 12 km/h
Engine: ADE 152 2.5 litre 3-cylinder in-line diesel developing 46 hp at 2,500 rpm
Transmission: hydrostatic
Steering: hydrostatic
Turning radius: 2 m
Tyres: 400 × 15.5 10 ply

Status

In production.

Manufacturer

Bond Industries, 5 Teak Avenue/30 Vulcan Street Klerkindustria, PO Box 914 Klerksdorp 2570, South Africa.
Tel: +27 18 462 9737. Fax: +27 18 462 5910.

VERIFIED

UNITED KINGDOM

Boughton MK18 load handling system

Development

The Reynolds Boughton (Vehicle Division) MK18 load handling system (LHS) was designed for the loading/unloading and transport of 20 ft/6.096 m ISO containers without the assistance of other mechanical handling equipment.

Description

The Boughton MK10 load handling system is mounted on a modified MK14 trailer coupled to an Oshkosh MK48/18 (8 × 8) truck. Due to its unique pick-up system the MK18 LHS can handle any body which has a 20 ft/6.096 m ISO compatible baseframe. Bodies of varying height can therefore be handled, such as half-height containers, shelters, fuel bowsers, flat beds, etc. The system can also handle ISO configuration bases of less than 20 ft, 10 ft long containers or combinations of these.

Conversion of the equipment to a flat bed body requires the insertion of metal floor panels which, when placed in position, make some of the hydraulic functions inoperable.

The operation of all the hydraulic functions of the Boughton MK18 load handling system is carried out either by manual lever operation at the side of the vehicle or by a remote-control lead up to 12.2 m in length, together with a proportional control chest-mounted valve pack.

The system is also designed for the loading, launching and transport of Ribbon Bridge sections. The unit can handle the interior bay and ramp sections which are 7 m long, 3.3 m wide and weigh a minimum of 7,000 kg. Launching can be undertaken in shallow and high bank modes and utilises a Boughton hydraulic winch.

US Marine Corps MK48/18 truck fitted with Boughton MK18 load handling system

Specifications

Weight: (system only) 5,000 kg
Length: 6.5 m
Width:
(with ISO container) 2.44 m
(with Ribbon Bridge) 3.3 m
Height: (overall) 1.86 m

Status

Available.

Manufacturer

Reynolds Boughton Limited, Vehicle Division, Bell Lane, Amersham, Buckinghamshire HP6 6PE, UK.
Tel: +44 1494 764411. Telex: 83132.
Fax: +44 1494 765218.

UPDATED

US Marine Corps MK48/18 truck fitted with Boughton MK18 load handling system launching a Ribbon Bridge erection boat

US Marine Corps MK48/18 truck fitted with Boughton MK18 load handling system launching a Ribbon Bridge section

Boughton DROPS/PLS hydraulic loading system

Development

The Boughton DROPS/PLS hydraulic loading system evolved from the original Ampliroll system developed by Bennes Marrel of France. Vehicles modified by Boughton were used by the British Army for trials and development over a period of six years and all equipment was supplied on a single tender basis. The British Army term used for the loading system is DROPS (Demountable Rack Off-loading and Pick-up System). The US Army originally required only the hydraulic loading system but procured the complete Boughton system (see following entries) including trucks and trailers. The US Army uses the term PLS (Palletised Load System).

Development of the system concentrated on improvements in operation on rough terrain, on height and weight reductions and generally militarising the basic commercial concept. 'Retrofit DROPS/PLS' or the fitting of the 'load handling system only' to an existing vehicle is an option for any armed service not requiring maximum performance in terms of mobility, payload or overall cost-effectiveness. The system can be fitted to almost any chassis with varying degrees of effectiveness, but even in the worst cases, it may be expected to provide some of the basic benefits of the DROPS/PLS concept.

Boughton DROPS/PLS hydraulic loading system fitted to Irish Army Hino truck

Description

The Boughton DROPS/PLS features a hydraulically operated self-loading cargo system designed to improve the productivity of a chassis by reducing turnround times and freeing the chassis from dedicated roles. The system enables a conventional chassis to carry more in payload terms and allows an armed service to increase its total logistic capability by switching chassis from secondary roles to meet peak requirements.

Roles envisaged for the Boughton DROPS/PLS include: a flatrack cargo vehicle (especially for carrying pallets of ammunition), recovery vehicle, fuel or water tanker, box-bodied vehicle (command post, ambulance, medical centre or workshop) and specialist engineer uses such as carrying, launching and recovering bridging pontoons, carrying palletised loads of bridging components and for carrying and laying portable roadways.

Status

In service with the armed forces of Ireland, Tanzania, Thailand and the USA.

Manufacturer

Hearncrest Boughton Engineering Limited, Bell Lane, Amersham, Buckinghamshire HP6 6PE, UK.
Tel: +44 1494 764411. Telex: 83132.
Fax: +44 1494 765218.

VERIFIED

Boughton (6 × 6) DROPS/PLS truck (Heavy PLS)

Development

Experience gained during the British Army trials of the Boughton DROPS system indicated to Boughton that the only satisfactory DROPS/PLS system would be based on a fundamentally new design of truck and trailer. The inherent compromises involved in retrofit solutions and modifications to existing vehicles all involved loss of mobility, loss of payload, some lateral instability, greater complexity, excessive size and less value for money. Accordingly a new design was initiated and the result is a purpose-built Boughton chassis with an integral hydraulic load handling system coupled with a special Boughton DROPS/PLS close-coupled trailer (for details see entry in this section). All the bodies are Boughton designed and built numerous features which are patented or including patent applied for.

Three sets of prime movers, trailers and bodies were supplied to the US Army in mid-1983 for concept evaluation at Fort Lewis in Washington and were joined by three more complete sets. These vehicles, known as 'Heavy PLS', led to support for the concept and its expansion to incorporate a breakbulk/distribution capability. This gave rise to a need for a smaller truck known as 'Medium PLS' (see following entry).

Boughton (6 × 6) DROPS/PLS truck loading a loaded flatrack on to a DROPS/PLS trailer **1996**

Description

The Boughton truck is a 6 × 6 vehicle with a payload of 14,000 kg. The truck was designed to meet the Medium Mobility criteria of the British Army and, when used with the special DROPS/PLS trailer, enables 28,000 kg of payload to be moved into forward areas. A forward control cab is used with seating for the driver and one or two passengers. Twin hydraulic cylinders under the flatrack are used to power the DROPS/PLS system. The engine is a Perkins diesel developing 450 hp and coupled to an Allison five-speed automatic transmission. During US trials the truck was used as an artillery tractor for the 155 mm M198 howitzer. During mobility trials the flatrack was used to free the vehicle from muddy locations by unloading the flatrack using the DROPS/PLS system and using the loading system to 'lift' the vehicle free. It is air-portable in the C-141 Starlifter with preparation.

The latest Boughton DROPS/PLS system is integrated with a Bedford 6 × 6 chassis and is capable of carrying a 16,500 kg payload, including fuel tank flatracks. When used with the special DROPS/PLS trailer a 40,000 kg gross train weight can be achieved.

Status

In production. In service with some overseas armies.

Manufacturer

The Boughton Group (Hearncrest Boughton/Reynolds Boughton/Scottorn/TTB Fabrication), Bell Lane, Amersham, Buckinghamshire HP6 6PE, UK.
Tel: +44 1494 764411. Telex: 83132.
Fax: +44 1494 765218.

UPDATED

Boughton (4 × 4) DROPS/PLS truck (Medium PLS)

Development

Following the performance success of the Boughton (6 × 6) Heavy PLS (see previous entry) and trailers used by the US Army Development and Employment Agency (ADEA) for concept trials since mid-1983, the US Army sought offers for a new 'Medium PLS'. The Medium PLS was to have roughly half the payload of the Heavy PLS. A clear preference was stated for achieving this on a 10 ft ISO flatrack, so that two Medium PLS flatracks could be carried 'piggyback' on a Heavy PLS 20 ft flatrack.

The Boughton Group won the competitive tender by producing a new vehicle where the 10 ft Medium PLS body can be handled with 6,350 kg of payload on a Medium PLS vehicle weighing only 6,350 kg. This preserved the aim of the US Army Light Division logistic planners for a one-to-one payload-to-weight ratio and achieving the objective of a two-to-one load split with the Heavy PLS. The first of 15 leased Medium PLS sets was delivered to the US Army 9th Infantry Division in March 1985 and the remainder soon after, seven with three-person cabs and eight with six-person crew cabs.

Boughton (4 × 4) DROPS/PLS truck (Medium PLS) with six-person crew cab and towing a Medium PLS trailer

Description

The Boughton (4 × 4) Medium PLS was designed to carry a payload of 6,350 kg. It is powered by a GM 205 hp engine allied to an Allison automatic transmission. It can tow a 3.05 m long Medium PLS trailer with a 6,350 kg useful payload. As the vehicle is 5.893 m long it can fit within 20 ft ISO flatrack dimensions and can be carried on a Heavy PLS flatrack. With preparation and a reduced weight of 5,216 kg, the vehicle can be carried in a C-130 transport aircraft or slung under a CH-47D Chinook helicopter.

Status

The US Army has 15 vehicles in service.

Manufacturer

The Boughton Group (Hearncrest Boughton/Reynolds Boughton/Scottorn/TTB Fabrication), Bell Lane, Amersham, Buckinghamshire HP6 6PE, UK.
Tel: +44 1494 764411. Telex: 83132.
Fax: +44 1494 765218.

VERIFIED

Cargotec load handling systems

Development

Multilift load handling systems for defence applications are designed, manufactured and marketed by Cargotec (UK) Limited, a division of the Scandinavian multinational Partek Corporation. This organisation is dedicated to supplying chassis-mounted materials handling equipment according to specific customer requirements from the Cargotec range of equipment, including Multilift and LeeBur hooklifts and HIAB cranes. The range includes sliding and pivoting hooks, with both tipping and non-tipping variants.

Description

DROPS load handling system

The Demountable Rack Off-load and Pick-up System (DROPS - Multilift Mark 4) Load Handling System (LHS) was designed and developed from the outset for

Cargotec HL8M LHS installed on DRA (6 × 6) high mobility technical demonstrator vehicle **1997**

Multilift Mark 4 LHS carried by a Foden (8 × 6) DROPS IMMLC logistic support vehicle (C F Foss) **1997**

Multilift and Bridge Adaptor Pallet (BAP) retrieving a Ribbon Bridge module

military applications. Mounted in a light and flexible 'compression frame', the LHS structures (the middle frame and hook arm) are pivoted on large plain bearings, while the operating cylinders utilise spherical bearings. This arrangement, combined with a special 'transit' circuit within the hydraulic system, allows flexibility whilst guaranteeing load security and ensures that no stresses are passed into the carrier truck chassis when manoeuvring over rough terrain. The system can lift loads at its rated lift capacity from 500 mm below ground level.

The patented rear roller assembly design is unique to Multilift. Its four rollers allow the alignment of flatracks that are picked up from a misalignment of 15° left or right. The rear rollers also support the rear end of the flatrack during transit.

The control system uses automatic single lever operation with two modes of back-up for emergencies. A driver can easily operate the one-touch sequence control to load or unload in less than one minute, without leaving the cab.

The DROPS LHS is a conventional DIN conforming pick-up system used in conjunction with 20 ft/6.096 m flatracks embodying a simple one-piece hook A-frame. A high power-to-weight ratio maximises payload within the vehicle maximum weight.

DROPS LHS are in service with the British Army. Leyland Trucks MMLC and Foden Trucks IMMLC are in use (1,500 and 400 units respectively). See entries in *Trucks* section for details of these vehicles.

In response to US Army requirements, a Multilift LHS was fitted to Oshkosh 10 × 10 trucks. The system, a modified version of the DROPS equipment, was designed with an air-portability feature. The system is in production under licence to Oshkosh to meet an order for 2,800 M1074 series trucks. (For details of the vehicle involved refer to entry in *Trucks* section).

Cargotec have installed DROPS LHS for the following countries:

Canada, on Kenworth (6 × 6), Western Star (6 × 6) M48665 and Steyr (6 × 6) HLWV chassis

Italy, on IVECO Astra 8 × 4 and 8 × 8 chassis

Malaysia on ROMAN (6 × 6) chassis

Norway, on Scania (8 × 6) P113 and Mercedes-Benz 2628 (6 × 6) chassis

Netherlands, on DAF 6 × 6 chassis

Sweden, on Scania (8 × 6) P113 chassis.

Many other nations worldwide have systems on trial.

HL8M

Cargotec have developed a prototype 8 tonne LHS currently under test and evaluation on a UK Defence Research Agency (DRA) 6 × 6 high-mobility load technical demonstrator vehicle. The system allows a 20 ft/6.096 m load to be halved and carried on 10 ft/3.048 m flatracks.

A Multilift 2 tonne hooklift, the HL2M, has been installed on the Hagglunds Bv 206 all-terrain vehicle for concept evaluation. The vehicle can carry one rack, one third the length of a 20 ft/6.096 m load.

Container handling unit

The Container Handling Unit (CHU), developed by Cargotec, Government Business Operations, allows unmodified ISO containers to be picked up and transported by vehicles equipped with DROPS/PLS load handling systems. See separate entry under *Shelters and Containers* for details.

Bridging systems

Cargotec have supplied a number of LHS mounted on Steyr/UTDC Percheron (6 × 6) vehicles and Ribbon Bridge launchers for the Canadian Forces Heavy Logistic Vehicle Wheeled (HLVW) programme. This HLVW system is modified from the standard DROPS equipment to conform to Canadian requirements. The Ribbon Bridge launching equipment is based on a standard DROPS unit to which a Bridge Adaptor Pallet (BAP) is added with all the necessary guides, rollers, locks and winches for Ribbon Bridge unit handling; the weight of a BAP is 1,600 kg.

Mounted on the Oshkosh HEMTT, the system has been ordered for the US Army's Improved Ribbon Bridge Programme. Using the BAP full bridge and Combat Support Boat, launching and retrieval capability is possible. NATO flatracks can also be carried.

Tipping load handling system

There are 75 Multilift tipping LHS units in service with the UK Royal Engineers. This enhanced product, combines the full DROPS/PLS performance characteristics and interoperability with an additional 20 tonne tipping capability. The double-acting rams provide speed and performance comparable with specialised earthmoving vehicles, tipping the engineer skip body to angles in excess of 60° in less than 15 seconds. A wide range of engineer and logistic equipment can be accommodated.

HL and LBS load handling systems

Manufactured in Finland and the Netherlands by Cargotec, the Multilift HL and LeeBur LBS patented range of LHS are suitable for 2-, 3- and 4-axle rigid chassis, having lifting capacities from 3 to 40 tons. The range includes sliding and pivoting hooks, with both tipping and non-tipping variants. The entire systems are hydraulically operated from within the cab, with full locking, warning and fail-safe facilities.

HIAB Military Cranes

Cargotec HIAB cranes are in service with the armed forces of many nations. Typical uses are missile handling, armoured vehicle powerpack replacement, and the precision handling of ammunition. Naval applications for the Cargotec HIAB range of marinised cranes include boat launch and retrieval as well as a full capability for materials handling tasks.

Cargotec HIAB cranes incorporate the use of multifunction remote control as well as computer monitored safety systems. Known as the HIAB SPACE and COMBIDRIVE, these systems offer precision and sensitivity of operation with enhanced safety.

A recent application of this technology is the introduction into service of the HIAB 8801 DINO crane. This equipment, rear mounted on a VME frame steer articulated off-road vehicle, was selected to perform handling and maintenance work within the Swedish Army's self-propelled artillery batteries.

Specifications

DROPS LHS

Installed weight: 1,910 kg

Rated lift capacity: (ground level) 16,500 kg

Overload capacity: (ground level) 20,000 kg

Multilift Mark 4T 'tipping' load handling system installed on a Terex 3066 LC high mobility frame steer truck **1997**

Multilift Mark 4 LHS installed on an Astra 8 × 4 chassis for the Italian STL logistic vehicle programme **1996**

Multilift HL2M LHS installed on Bv 206 **1997**

The Multilift Mark 4 DROPS system as selected by the Norwegian Army for UN support **1996**

Flatrack or body length: 6.055 m
Flatrack or body width: 2.435 m
Misaligned tolerance: 15° either side
Load/unload cycle time: 58 s

Status

Mark 4 LHS - in production. See text for users.
Mark 5 LHS - in production (under licence) for the US Army's PLS.
Mark 6 LHS - ordered for the US Army.
Mark 4T LHS - in service with British Army.
Multilift LHS are undergoing trials with a number of armed forces worldwide.

Manufacturer

Cargotec (UK) Limited, Government Business Operations, Harlescott Lane, Shrewsbury, Shropshire SY1 3AG, UK.
Tel: +44 1743 442 232. Fax: +44 1743 469 935.

UPDATED

Marshall DROPS flatracks

Description

In 1983, Marshall SPV Limited was selected by the UK Ministry of Defence to produce for evaluation a range of DROPS flatracks to be used with the Leyland Trucks and Foden DROPS logistic supply vehicles. Each flatrack can carry a payload of 15,000 kg.

The test programme was completed in 1984 and in 1986 interoperability trials were carried out with UK and German vehicles. In late 1986 it was announced that Marshall SPV was to be a subcontractor to Foden for the supply of 9,500 flatracks for the British Army's DROPS programme. Production of standard and other flatracks is carried out in a purpose-built semi-automated plant with a capacity of more than 2,000 units per year.

Marshall SPV have also produced 3,000 gallon/11,356 litre DROPS fuel and water racks and a 'sided' DROPS cargo flatrack. There are also enclosed shelter systems.

Status

In production for the British Army's DROPS programme.

Manufacturer

Marshall SPV Limited, The Airport, Cambridge, Cambridgeshire CB5 8RX, UK.
Tel: +44 1223 373475. Fax: +44 1223 373046.

VERIFIED

Marshall DROPS flatrack suitable for ammunition or other special purpose loads

---

Reynolds Boughton 16 tonne DROPS/PLS Flatrack

Description

This flatrack, designed to carry up to 15,000 kg of supplies, is capable of interoperability with the complete range of British Army and NATO DROPS/PLS equipment. Being of ISO configuration it can also be transported on road and rail flatbeds, is transportable by ship and air, and can be carried underslung from helicopters.

Features of the Boughton flatrack are its patented retractable rollers and its extendable rear ISO locks. The retractable roller enables the flatrack to be easily manoeuvred on the ground and in vehicle-to-vehicle transfer operations, while the extendable rear locks enable standard 20 ft/6.096 m ISO containers to be accommodated.

Status

In production.

Manufacturer

Reynolds Boughton Limited, Bell Lane, Amersham, Buckinghamshire HP6 6PE, UK.
Tel: +44 1494 764411. Telex: 83132.
Fax: +44 1494 765218.

VERIFIED

Reynolds Boughton 16 tonne DROPS/PLS flatrack

Reynolds Boughton MLRS/ DROPS flatrack

Description

This skeletal flatrack transports up to four Multiple Launch Rocket System (MLRS) rocket pods and is capable of interoperability with the complete range of British Army DROPS equipment. Being of ISO configuration, it can also be transported on road and rail flat beds, is ship and air transportable and can be carried underslung by helicopters.

A unique feature of the Boughton flatrack is a patented retractable roller which enables the flatrack to be easily manoeuvred on the ground and assists in vehicle-to-vehicle transfer operations.

Reynolds Boughton MLRS/DROPS flatrack

Status

In production. In service with the UK Ministry of Defence.

Manufacturer

Reynolds Boughton Limited, Bell Lane, Amersham, Buckinghamshire HP6 6PE, UK.
Tel: +44 1494 764411. Telex: 83132.
Fax: +44 1494 765218.

VERIFIED

Hyster forklift trucks

Description

Hyster materials handling equipment comprises over 80 models of forklift truck, reach stackers for container and swap-body handling. The lift trucks have capacities from 1 to 52 tonnes, using electric and internal combustion engine power. The trucks are manufactured at three plants in Europe: Irvine and Craigavon in the UK and Nijmegen in the Netherlands.

The Challenger range of trucks comprises LPG or diesel engine powered pneumatic-tyred trucks. The range extends from the H1.50XM, which can handle loads up to 1.5 tonnes at 500 mm, to the H7.00XL. Above this is the diesel-powered Big Trucks range which can handle anything from 8 tonnes up to fully laden containers, swap-bodies and semi-trailers. Included in the range are trucks with dedicated laden and empty containers handling attachments. The ReachStacker series also provides the option of top and bottom lift capability, with an extended reach of up to 6.4 m.

The Electric Range has capacities of from 1 to 5.5 tonnes. Trucks in the 1.5 to 3.2 tonnes capacity range have the option of cushion or pneumatic shaped tyres. For extra manoeuvrability, 1 to 2 tonne models are available in three-wheel configuration. The SpaceSaver range of LPG powered trucks, running on cushion tyres, is specially designed to operate in confined spaces. Capacities are from 1.5 to 7 tonnes.

Status

In production. Hyster has provided the British armed forces with some 1,200 trucks from its range during the last 15 years. Hyster trucks are in service with the US Army and other defence forces.

Manufacturer

Enquiries to: Hyster Europe Limited, Berk House, Basing View, Basingstoke, Hampshire RG21 4HQ, UK.
Tel: +44 1256 461171. Telex: 858384.
Fax: +44 1256 56733.

UPDATED

Hyster H2.00J lift truck used by the British Army

Hyster H2.00J lift truck stacking ammunition boxes

Kalmar forklift trucks

Description

Kalmar Limited supplies a wide range of both engine-powered and electric front-loading forklift trucks from 1.5 to 90 tonne lifting capacity. Included in this range is the Climax range of trucks (up to 3 tonnes) which can also be supplied in explosion-proof versions under the PROTECTION name. This incorporates over 30 years' experience of supplying trucks to operate in potentially explosive or hazardous atmospheres (for example ammunition handling). Kalmar also supplies a range of electric moving mast reach trucks from 1.2 to 2.4 tonnes and sideloaders from 2 to 52 tonnes.

Status

In production. The company is registered on the UK Defence Contractor's list to supply forklift trucks to the UK armed forces. Models are in service with the defence forces of Belgium, Canada, Denmark, Egypt, Norway, UK and the USA.

Manufacturer

Kalmar Limited, Siskin Drive, Coventry, West Midlands CV3 4FJ, UK.
Tel: +44 1203 555355. Telex: 312414.
Fax: +44 1203 229417.

VERIFIED

Kalmar Climax Series 2000 'PROTECTION 2' forklift truck

Boughton forklifts

Description

Reynolds Boughton Limited manufactures a range of forklift attachments for wheeled front end loaders with lift capacities from 1,500 to 6,000 kg at 0.5 m load centre. Optional features include hydraulic operation carriage side shift, carriage rotation, fork load levelling and fork spread. Fitment to the loader is either by direct pin mounting to the loader arms or via a quick coupler attachment.

Status

In production. In service with the British Army and French Army.

Manufacturer

Reynolds Boughton Limited, Engineering Division, Bell Lane, Amersham, Buckinghamshire HP6 6PE, UK.
Tel: +44 1494 764411. Telex: 83132.
Fax: +44 1494 765218.

VERIFIED

Boughton 1,850 kg forklift on Muir Hill A5000 front loader

Boughton 4,200 kg forklift on Case 721 wheeled tractor in service with the British Army

JCB 410M rough terrain forklift truck

Description

The JCB 410M is a military rough terrain forklift designed specifically to handle NATO pallets. Its lift capacity is 1,815 kg. Developed for military use, the 410M has a low profile cab for air-portability and a higher road speed to keep pace with military convoys; it can also be towed at convoy speeds. It has a forward reach capability to allow loading and unloading from one side of a cargo truck or rail wagon. The operator's cab is on the front module to provide good visibility with minimum engine noise and vibration. It has a full three-speed power-shift transmission. Steering is by power-operated centre pivot articulation for increased manoeuvrability, giving a 9.26 m maximum turning circle. The 410M can be fitted with a range of attachments such as buckets, crane hooks, sweepers and so on, to increase the machine's versatility. A NATO tow hitch is fitted as standard.

Fully winterised and waterproofed versions have been supplied.

British Army JCB 410M rough terrain forklift truck (R Stickland) ***1997***

Specifications

Cab seating: 1
Configuration: 4 × 4
Weight: 6,850 kg
Operating load: 1,815 kg at 0.61 m
Length:
(overall) 5.625 m
(without forks) 4.5 m
Height: (top of cab) 2.67 m
Ground clearance: 0.4 m
Track: 1.8 m
Wheelbase: 2.7 m
Angle of approach/departure: 45°/45°
Max speed: 57.02 km/h
Fuel capacity: 95 litres
Max gradient: 100 %
Fording: 0.75 m
Engine: Perkins T4.236 3.86 litre 4-cylinder direct injection diesel developing 102 hp at 2,600 rpm
Transmission: Clark 18000 series full powershift with 3 forward and 3 reverse gears
Steering: power-assisted
Turning radius: (outside tyres) 9.26 m
Tyres: 15.5/80 × 20R XL
Brakes:
(main) hydraulic multidisc
(parking) axle-mounted disc on front differential shaft
Electrical system: 24 V
Batteries: 2 × 12 V, 96 Ah
Alternator: 40 A

Status

In service with the British Army, Royal Navy and Royal Air Force and other armed forces.

Manufacturer

JCB Military Products Division, Rocester, Staffordshire ST14 5JP, UK.
Tel: +44 1889 590312. Telex: 36372.
Fax: +44 1889 590583.

VERIFIED

JCB 926 and 930 rough terrain forklift trucks

Description

The JCB 926 and 930 rough terrain forklift trucks are basically similar, with the JCB 926 having a lift capacity of 2,600 kg and the JCB 930 3,000 kg. Both are built around a specially designed chassis and can be fitted with duplex, triplex (heights of 3.6, 5.5 and 6.55 m available) or free lift (4.5 m) masts and both can be produced in 4 × 2 and 4 × 4 configurations. Access to the one-person cab is from either side.

The engine used is a Perkins 4.40 diesel driving through a JCB Syncro-Shuttle gearbox via a torque converter. The JCB epicyclic hub reduction front axle incorporates inboard-mounted oil-immersed multidisc brakes.

Specifications

Data relates to JCB 926; data for JCB 930 in square brackets where different
Cab seating: 1
Configuration: 4 × 2 or 4 × 4

JCB 926 rough terrain forklift truck; JCB 930 is visually similar

Weight:
(4 × 2) 5,955 [6,685] kg
(4 × 4) 6,025 [6,755] kg
Length: (less forks) 3.58 [3.66] m
Width: 2.22 m
Height: (cab) 2.56 m
Track: (front) 1.8 m
Wheelbase:
(4 × 2) 2.06 m
(4 × 4) 2.12 m
Engine: Perkins 4.40 water-cooled 4-cylinder diesel developing 76 hp
Transmission: torque converter with powershuttle reverser and full synchromesh 4-speed gearbox

Status
In production. In service with the defence forces of several countries.

Manufacturer
JCB Military Products Division, Rocester, Staffordshire ST14 5JP, UK.
Tel: +44 1889 590312. Telex: 36372.
Fax: +44 1889 590588.

VERIFIED

JCB Loadalls

Description
JCB is the largest producer of variable reach handling machines in the world. Production of the first model, the JCB 520, began in 1977. Machines in the range have lift capacities from 2,000 to 3,700 kg and lift heights from 5.5 to 13 m. The range includes both two- and three-stage telescopic arms.

The Loadalls have the features of conventional rough terrain trucks with the advantage of forward reach, which permits 'one side' loading and unloading of road and rail trucks. Telescopic machines can stand back from hazardous areas or soft ground and still lift or place a load accurately.

In the latest range all machines now have four equal-sized wheels which enable them to steer in three modes: all-wheel steer, front-wheel steer and crab steer.

All Loadalls are powered by four-cylinder Perkins diesel engines and drive via a torque converter through a JCB synchromesh gearbox and JCB drive axles, which incorporate oil-immersed multidisc brakes. The single operator all-weather cab meets all ROPS/FOPS requirements.

JCB Loadall 525B in service with the Royal Navy

Specifications
JCB 525-67
Cab seating: 1
Configuration: 4 × 4
Weight: 6,420 kg
Length: (fork carriage) 4.78 m
Width: 2.34 m
Height: (to top of cab) 2.59 m
Ground clearance: (minimum) 0.42 m
Track: (front) 1.9 m
Wheelbase: 2.48 m
Max road speed: 29 km/h
Engine: Perkins 4.40 water-cooled 4-cylinder diesel developing 76, 95 or 106 hp
Transmission: full powershift torque converter with 4 forward and reverse gear ranges
Steering: twin ram hydrostatic power on each axle with 3 steer modes: all-wheel, front-wheel and crab
Tyres: (front and rear) 15.5 × 24 traction
Capacity at 0.5 m load centres: 2,700 kg at radius of 1.63 m to maximum height of 6.7 m; 800 kg at maximum radius of 3.55 m to maximum height of 6.7 m

Status
In production. In service with the Royal Navy and the defence forces of several countries; some models are remotely controlled 'specials'.

Manufacturer
JCB Military Products Division, Rocester, Staffordshire ST14 5JP, UK.
Tel: +44 1889 590312. Telex: 36372.
Fax: +44 1889 590588.

VERIFIED

Lansing Linde forklift trucks

Description
Lansing Linde Limited manufactures an extensive range of materials handling equipment ranging from 1 to 42 tonnes capacity in both battery electric and internal combustion engine forms. Products cater for virtually every form of materials handling from hand pallet trucks to the stacking of heavy ISO containers.

Several models have been the subject of UK Ministry of Defence 'Best Buy' agreements, the reach truck, sea-going counterbalanced forklift truck and the shipboard pallet truck being the prime examples. Lansing Linde remains an approved supplier to the UK Ministry of Defence under the terms of the current Enabling Arrangement for engine-powered trucks.

The company has production plants in the UK as well as in Germany and France. In the UK the factories are registered to British Standard (BS) 5750 Pt 1 and have been assessed by the Ministry of Defence to AQAP 1 in quality control systems. Maintenance and spares support services are available, as are training courses for operators, engineers and instructors.

Status
Numerous models in production and in service worldwide, including with the British armed forces and the US Army, Navy and Air Force.

Manufacturer
Lansing Linde Limited, Kingsclere Road, Basingstoke, Hampshire RG21 6XJ, UK.
Tel: +44 1256 342000. Telex: 858120 lbba g.
Fax: +44 1256 342923.

UPDATED

Three Lansing Linde models: R 16 reach truck (left); H 16 engine-powered counterbalanced truck (right); and T 20 pallet truck (below)

Lansing Linde C4130 reach stacker

1997

BOSS 'G' Series Rough Terrain container handling front-lift trucks

Description

BOSS 'G' Series container handling Rough Terrain (RT) front-lifts were specifically designed to handle and transport laden containers up to 25,000 kg on rough surface conditions ranging from relatively poor uneven surfaces to rough cross-country terrain.

A choice of two- or four-wheel drive models is available (designated RT and AT respectively, also known as the Bear) sharing many design features. Many major components are used in the orthodox 'G' Series container handling front-lift Series for port operations, with the advantage of spare parts commonality and servicing familiarity for operators.

High-flotation tyres, high-under-clearances, long wheelbase and optimum weight distribution combine with the high-power turbocharged diesel and power-shift transmission to give good traction and stability, laden or unladen. Panoramic visibility from the operator's offset control centre and the BOSS 'VisionMast' ensure that the driver can see clearly in all weather conditions. A forklift may be attached.

A BOSS 20 ft/6.096 m fork-mounted top-lift attachment is fitted to the inverted fork frame and the all-hydraulic operation allows for effective transportation and stacking of laden containers up to 25,000 kg.

BOSS 'G' Series ATCH with container

Specifications

Cab seating: 1
Configuration: 4 × 2 (RT) or 4 × 4 (AT)
Weight: 46,620 kg
Length: (including top-lift attachment) 10.125 m
Height: 4.99 m
Width: (excluding top-lift attachment) 4.48 m
Wheelbase: 5.5 m
Track: (front axle) 4.05 m
Max road speed: (unladen) 34 km/h
Engine: Volvo TD.120G 11.98 litre turbocharged diesel developing 350 bhp at 2,200 rpm
Transmission: Allison CRT5643 (torque converter and powershift: 3 forward/3 reverse speeds)
Tyres: front, 26.4R × 25; rear, 23.5 × 25.
Capacity: (under top-lift attachment) 25,000 kg

Status

In production. In service with the British Army (4, AT).

Manufacturer

BOSS Group Limited, Grovebury Road, Leighton Buzzard, Bedfordshire LU7 8SR, UK.
Tel: +44 1525 372031. Telex: 825781.
Fax: +44 1525 383071.

UPDATED

Grove Coles mobile cranes

Description

Grove Coles is a well established supplier of cranes and lifting equipment for military purposes and over 40 types of Grove Coles cranes are at present in military use throughout the world. Grove Coles produces the widest range of mobile cranes in the world including industrial, port, all-terrain, rough terrain and truck-mounted, all of which can be used for military purposes, plus specialised military cranes with lifting capacities from 6 to 250 tonnes.

Hydra Truck, Supertruck and Octag

These ranges of truck-mounted telescopic boom cranes have lifting capacities of 12 to 135 tonnes. They have limited off-road capabilities and are therefore more suited to use in rear areas.

The Hydra Truck 12/14T general duties crane can be made as a special variant for use aboard aircraft carriers.

Hydra Speedcranes

The Speedcrane range consists of conventional truck-mounted telescopic hydraulically operated cranes with lifting capacities of 6 to 9 tonnes. The 4 × 2 chassis makes the Speedcrane range suitable for use in permanent installations.

The Speedcrane 8/9T Mk II is the standard Royal Air Force 7 tonne crane and is in use for general stores handling and aircraft maintenance work in the Royal Air Force and with the Fleet Air Arm, both ashore and aboard.

Hydra Husky, Transit and Ranger

The Husky series of cranes is specifically designed for use on rough terrain and therefore has clear military applications. The Hydra Husky 150T has been the standard Medium Field Support Crane in the British Army for over a decade and is ideally suited to military applications.

The 6/8TCC Airportable Husky, suitable for transportation in the C-130 Hercules transport aircraft, is used in quantity by the former Yugoslav Army.

The Model 315M is the current 'Crane, Field, Medium' used by the Royal Engineers for bridge building and general stores handling from base to forward areas, by the REME for major unit changes on AFVs in the field and for workshop use and by the RLC in their general stores and overall support role along the British Army's lines of communication.

The Transit 517 Lightweight is a specially developed version of the 517 for use both on flight decks and in hangar operations on aircraft carriers.

The Ranger series of all-terrain cranes has the on-road performance of a truck crane and the off-road capabilities of a rough terrain crane with lifting capacities from 15 to 30 tonnes and again is suited to military purposes.

Starlift

The Starlift range of self-propelled, telescopic aerial

Grove Coles Hydra Husky 150T used by the British Army (T J Gander)

Grove Coles Hydra Husky 6/8TCC being loaded on to an RAF Hercules transport aircraft

Grove Coles 315M, the British Army's Crane. Field, Medium (T J Gander)

Specifications

Model	Hydra Speedcrane 8/9T Mk II	Husky 6/8TCC	Model 315M	Hydra Truck 12/14T	Ranger (Transit) 517	Model 335M	Ranger 520	Ranger 530	Husky 680
Configuration	4 × 2	4 × 4	4 × 4	4 × 2	4 × 4	4 × 2	4 × 4	4 × 4	4 × 4
Traverse	360°	360°	360°	360°	360°	360°	360°	360°	360°
Lifting capacity									
(min radius)	8,130 kg/2.45 m	7,320 kg/2.8 m	15,000 kg/3 m	12,200 kg/3 m	18,000 kg/3 m	40,800 kg/3.5 m	20,000 kg/3 m	30,000 kg/3 m	75,000 kg/3 m
(max radius)	710 kg/11 m	1,650 kg/25 m	2,050 kg/10 m	820 kg/15 m	1,400 kg/16.5 m	1,150 kg/36 m	900 kg/24 m	600 kg/28 m	2,560 kg/30 m
Length	6.55 m	7.27 m	9.14 m	8.6 m	8.71 m	15 m	10.7 m	11.08 m	14.63 m
Width	2.44 m	2.5 m	2.5 m	2.5 m	2.5 m	2.9 m	2.5 m	2.75 m	3.65 m
Height	2.82 m	2.67 m	3.6 m	3.26 m	3.62 m	3.9 m	3.4 m	3.66 m	4 m
Weight	13,590 kg	14,860 kg	21,130 kg	16,642 kg	18,430 kg	36,240 kg	21,250 kg	23,970 kg	50,460 kg
Engine	100 hp	158 hp	196 hp	146 hp	160 hp	165 hp	150 hp	240 hp	210 hp
Gears	5 forward, 1 reverse	4 forward, 4 reverse	6 forward, 3 reverse	6 forward, 1 reverse	6 forward, 1 reverse	4 forward, 4 reverse	6 forward, 3 reverse	5 forward, 1 reverse	6 forward, 6 reverse
Max speed	—	—	75 km/h	—	65 km/h	61 km/h	70 km/h	61 km/h	30 km/h

work platforms can place men and their equipment to working heights from 17 to 27.5 m and at maximum outreach from 13.5 to 24 m. Typical military applications include the maintenance and repair of buildings and aircraft. The standard versions are self-propelled, but truck-mounted variants are available on such chassis as the military version of the Bedford TK 4 × 2 truck or on the specialised Grove Coles 315M crane chassis.

Status

Various models of Grove Coles Cranes are in service with the British armed forces, Belgian Air Force, former Yugoslav Army and numerous other armed forces.

Manufacturer

Grove Coles, Harefield, Uxbridge, Middlesex UB9 6QG, UK.
Tel: +44 1895 656281. Telex: 53484 cranes g.
Fax: +44 1895 109242.

VERIFIED

UNITED STATES OF AMERICA

Case rough terrain forklift trucks

Description

The Case M4K-B rough terrain forklift truck was designed to load and unload material from ISO containers. The fork free lift, side shift and vehicle dimensions allow the forklift to enter a container and pick or place loads. Adequate side shift is provided to place loads against either wall of a container. The M4K-B is towable with standard drawbar and safety equipment. Canopies or cabs are provided with ROPS and FOPS meeting the latest international requirements. Lift and tie-down brackets are standard equipment and are designed for parachute delivery or helicopter transport. The vehicle is fitted with high flotation tyres and standard driving and blackout lights. Other features include hydraulic fork rotation and front and rear axle disconnects for towing.

The M6K-B articulated rough terrain forklift has reach for loading trucks from one side. Features include power rotation, side shifting and fork spacing. The unit has been tested and is in service in saltwater applications, both onshore and shipboard.

The M13K articulated rough terrain forklift is designed with the primary mission of loading and unloading cargo aircraft such as the C-130. The forks can be equipped with conveyor ramps which match aircraft cargo systems. The cab, load guard and counterweight may be removed so that the unit can be transported by C-130 aircraft.

The M6K-B and M13K have a parallel linkage system which ensures that the forks remain in a level position throughout the lift arc. The M13K is also equipped with the necessary safety equipment for handling nuclear systems.

Specifications

Model	M4K-B	M6K-B*	M13K*
Cab seating	1	1	1
Configuration	4 × 4	4 × 4	4 × 4
Weight (without load)	4,536 kg	9,100 kg	10,400 kg
Length	5.35 m	6.25 m	7.35 m
Height (cab/canopy)	2.03 m	3.05 m	3.12 m
Width	2.01 m	2.16 m	2.31 m
Ground clearance	334 mm	406 mm	432 mm
Wheelbase	2.34 m	2.54 m	2.74 m
Engine (diesel)	Case 4-390	Case 6-590	Case A504BD
Hp/rpm (SAE net)	66/2,200	90/2,200	110/2,200
Gearbox	3/3	3/3	4/2
Transmission	full powershift	full powershift	full powershift
Max road speed	35.9 km/h	34.76 km/h	37.49 km/h
Turning radius	n/a	10 m	9.93 m
Brakes (type)	hydraulic disc	air/hydraulic disc	air/hydraulic disc
Tyres	15 × 19.5	15.5 × 25	17.5 × 25
Electrical system	24 V	24 V	24 V
Forklift (type)	mast	boom	boom
Lift capacity rated	1,814 kg at 610 mm	2,722 kg at 610 mm	5,897 kg at 1.22 m
Max lift height	2.54 m	3.05 m	1.98 m
Tilt (forward)	11°	30°	10°
(back)	22°	33°	16.5°
Side shift (left/right)	559 mm/559 mm	1.27 m/1.27 m	none
Fork rotation (total)	22°	21°	none

* Weight and height can be reduced for aircraft transport in C-130.

Status

M4K-B, M6K-B and M13K, production complete. Over 3,000 Case M-series military rough terrain forklifts are in service with the US armed forces. The M4K and its predecessor (the MC4000) are no longer in production but remain in service. The M6K and M6K-B are in service with the US Navy. The M13K and M10K are in service with the US Air Force.

Manufacturer

J I Case Company, Government Marketing, 700 State Street, Racine, Wisconsin 53404, USA.

VERIFIED

Case M6K-B rough terrain forklift truck for US Navy and US Air Force

US Air Force Case M13K loading C-130 transport aircraft

TEREX 72-31 Series rough terrain forklift trucks

Description
TEREX produced two main types of military forklift truck for the US armed forces, both derived from the commercial TEREX 72-31 articulated front end loader. One was the TEREX 72-31MP2U/R for the US Marine Corps which was first produced in 1967; production continued until 1972. These were produced as scoop loaders but, once in service, trials demonstrated that the vehicle was capable of acting as a 4,545 kg forklift vehicle after it was fitted with rear-mounted counterweights and a modified US Air Force 463L fork carriage. The vehicle was then reclassified as a Tractor, Wheeled, Industrial with the capability of use as a forklift truck or a loader with a bucket attachment. The US Marine Corps had 418 units. During 1984 a further 92 72-31MP2U units were delivered.

In 1977 it was decided to rebuild these machines to new specifications at the TEREX factory. During the rebuild a Rollover Protective Structure (ROPS), water compatible disc brakes, larger tyres, heavier counterweights and a side shift fork carriage were fitted as part of a general upgrade package which took place from 1978 to 1981. A total of 369 vehicles was involved.

Another military version of the TEREX 72-31 was the TEREX 72-31F Adverse Terrain Forklift Truck (US Air Force Type A/S32 H-15) produced for the US Air Force to use as part of its 463L Material Handling System. As such, it has an aluminium half-cab that can be easily removed by two personnel without tools and a segmented counterweight system to allow the machine to be airlifted in either a C-130 or C-141 aircraft. The TEREX 72-31F is widely used as a loader for the C-130 transport aircraft and the US Air Force alone has over 300. As with the 72-31M, the vehicle is rated as a 4,545 kg payload vehicle but the US Air Force uses the 72-31F at a rating of 6,136 kg for short periods.

Specifications

Model	**72-31M**	**72-31F**
Used by	USMC	USAF
Cab seating	1	1
Configuration	4 × 4	4 × 4
Weight	15,727 kg	11,766 kg
Operating load	4,545 kg	4,545 kg
Length	7.2 m	7.2 m
Width	2.624 m	2.489 m
Height	3.086 m	3.01 m
Wheelbase	2.515 m	2.5 m
Ground clearance	0.343 m	0.343 m
Max speed	41.9 km/h	32.9 km/h
Fording	1.524 m	n/a
Engine	Detroit Diesel Model 4-71N 2-cycle diesel developing 140 hp at 2,300 rpm	Detroit Diesel Model 4-53N 2-cycle diesel developing 129 hp at 2,800 rpm
Transmission	Allison powershift CRT 3331-1 with 3 forward and 3 reverse gears	Allison twin turbine powershift Model TRT2220-1 with 4 forward and 4 reverse gears
Brakes	air over hydraulic	air over hydraulic
Tyres	20.5 × 25 16PR	17.5 × 25 12PR
Electrical system	24 V	24 V

Status
72-31M in service with US Marine Corps. 72-31F in service with the Argentine, Belgian, Israeli and US air forces.

Manufacturer
TEREX Corporation, 5405 Darrow Road, Hudson, Ohio 44236, USA.
All TEREX production is now concentrated at their facility in Scotland.

VERIFIED

TEREX 72-31MP2U/R industrial wheeled tractor as used by US Marine Corps

TEREX 72-31F rough terrain forklift truck as used by US Air Force

TRAK All-Terrain Lifter Army System (ATLAS)

Development
TRAK International Inc announced in May 1995 that it had been awarded a contract by the Department of the Army (TACOM) to supply the next generation of military, variable reach, rough terrain forklift vehicles.

The new forklift is capable of lifting 10,000 lb/4,536 kg and has a maximum fork reach of 22 ft/6.7 m. The US Army defines the forklift as an 'All-Terrain Lifter Army System' or ATLAS. The primary military mission requirement is the unloading, transporting and loading containers, boxes and palletised ammunition loads. The ATLAS is capable of mobility over rough terrain and is transportable in C-130 aircraft.

The ATLAS forklift is specifically designed to handle the US Air Force 10,000 Pallet (463L Pallet). It also has the capability of accepting dual carriages, 2,724 kg at a 609 mm load centre and 4,540 kg at 1.219 m load centre, utilising a quick attach feature. In 1993, TRAK International Inc provided the New Zealand Army with a 6,000 lb/2,721 kg forklift modified to accept a 4.536 kg load centre. This dual 6K/10K application is similar to the ATLAS design.

Total requirements under this contract, if all options are exercised, could exceed 1,500 ATLAS vehicles (the initial contract called for 1,198 vehicles at an estimated cost of US$125 million). TRAK will manufacture the forklifts at its Port Washington, Wisconsin, facility. Full scale production will be undertaken by mid-1997.

TRAK All-Terrain Lifter Army System (ATLAS) ***1997***

Description
The ATLAS vehicle has a one-person enclosed cab. There is a full-time planetary four-wheel drive with dry disc brakes and the front axle is equipped with a 'No-Spin' differential and parking brake. The hydraulically actuated forks are located at the articulating end of a telescopic three-stage welded box-section boom constructed from high strength alloy steel. The maximum reach from the front tyres to the 609 mm load centre is 7.25 m and maximum lift capacity is 4,536 kg. Maximum lift height is 8.5 m.

Specifications
ATLAS 10,000M SKY TRAK
Cab seating: 1
Configuration: 4 × 4
Weight: (operating with driver and max fuel)
(with 6K carriage) 13,920 kg
(with 10K carriage) 14.520 kg

Max lift capacity: 4,536 kg
Length: (with forks) 8.96 m
Width: 2.6 m
Height: 2.57 m
Ground clearance: (axle) 0.363 m
Track: 2.6 m
Wheelbase: 3.2 m
Speed: 37 km/h
Fuel capacity: 166.6 litres
Engine: Cummins 6BT 5.9-C165 diesel developing 165 hp at 2,500 rpm
Transmission: powershift with 3 forward and 3 reverse speeds
Turning radius: 8.2 m

Max lift capacity: 4,536 kg
Max lift height: 8.5 m
Max lift below grade: 0.8 m
Max reach:
(from front tyres to 609 mm load centre) 7.2 m
(at max height to 609 mm load centre) 4.2 m
Frame tilt: (left or right) 9°

Status
The contract requires preproduction qualification testing and first article testing. Full production anticipated mid-1997.

Manufacturer
TRAK International Inc, 369 W. Western Avenue, Port Washington, Wisconsin 53074, USA.
Tel: +1 414 284 5571. Fax: +1 414 284 4955.

UPDATED

Caterpillar 988 50,000 lb (22,680 kg) Rough Terrain Container Handler (RTCH)

Description
The Rough Terrain Container Handler (RTCH) is a Military Adapted Commercial Item (MACI), consisting of an AH-60 lift mast and carriage mounted on the basic chassis of the Caterpillar 988B Wheeled Loader; the latest models are the Caterpillar 988F Series II standard (see Specifications). It was designed to lift, carry and load containers 20 ft/6.096 m, 35 ft/10.67 m and 40 ft/12.19 m long, up to 9 ft/2.74 m high and up to 50,000 lb/22,680 kg in weight. It can stack them two high or place them on railway wagons or flat bed trucks. The RTCH is equipped with carriage side shift as well as mast tilt to allow easy positioning of the container and lifting pins are locked in place when the container is being lifted.

The RTCH is capable of wading in sea water up to 1.52 m deep as it may have to operate at beach heads. Special radial tyres provide flotation on sand and traction in mud and the rear axle oscillates to maintain ground contact on all types of terrain.

In competition with three other major manufacturers, Caterpillar was awarded an US$89 million contract from the US Army Mobility Equipment Research and Development Command (MERADCOM, now the Belvoir Research, Development and Engineering Center at Fort Belvoir) in September 1978 for the production of 344 units; this contract was completed. A further contract was awarded by the US Marine Corps at the end of July 1986. This was for 42 units with a contract value of approximately US$9.5 million.

RTCH stacking composite container made from two standard US Army 20 ft MILVANs

Specifications
Caterpillar 988F Series II
Cab seating: 1
Configuration: 4 × 4
Weight: (with 40 ft top handler) 53,050 kg
Length: 10.731 m
Width: (over tyres) 3.555 m
Height: (top of cab) 4.115 m
Ground clearance: 0.403 m
Fuel capacity: 624.5 litres
Engine: Caterpillar 3408E HEUI diesel developing 430 hp at 2,000 rpm
Steering: centre-point frame articulation, hydraulic, steering angle 30°
Tyres: 35/65-30PR
Brakes: hydraulic on all wheels

Status
In service with the US Marine Corps and US Army (344).

Manufacturer
Caterpillar Inc, Defense & Federal Products, Mossville, Illinois 61552-0470, USA.
Tel: +1 309 578 6941. Fax: +1 309 578 3739.

UPDATED

Truck, Tractor, Yard Type, 4 × 2, M878 and M878A1

Description
The original M878 was a commercially available vehicle primarily used to shuttle semi-trailers within permanent port installations or rail transfer areas. In 1976, 28 were ordered for the US Army and an additional 16 in 1977. In 1981, 175 examples of a revised model, the M878A1, were procured, bringing the number of Yard Tractors in service up to 219. The M878A1 differs from the original version by having a turbocharged 6V-53T Detroit Diesel engine in place of the original model 6V-53.

The M878 and M878A1 tractors are highly manoeuvrable vehicles with an automatic locking, hydraulic-lift fifth wheel which facilitates semi-trailer coupling and disengagement and allows movement of the semi-trailer or chassis without the need to retract landing legs.

Truck, Tractor, Yard Type, 4 × 2, M878

Specifications
M878A1
Cab seating: 1
Configuration: 4 × 2
Weight: 7,384 kg
Length: 4.64 m
Width: 2.49 m
Height: 2.89 m
Wheelbase: 2.94 m
5th wheel height: 1.21 to 1.63 m
Engine: Detroit Diesel turbocharged 6V-53T

Status
Production complete. In service with the US Army.

Manufacturer
Ottawa Truck Division, Gulf and Western.

VERIFIED

Crane, truck-mounted, hydraulic, 25 ton (CCE) (Grove TMS 300-5)

Description

The Grove TMS 300-5 25 ton crane is a commercial truck-mounted crane. It consists of a hydraulically operated telescopic crane with a full 360° traverse mounted on an eight-wheeled carrier. The operator controls the crane from an electric control panel in the superstructure cab. Four outriggers are used to stabilise the crane in operation. The crane is used by engineer units in the construction and repair of roads, airfields, pipelines and bridges and can also be used for port, marine and beach facilities. In addition to lifting, it can also be used for pile-driving and clamshell operations.

Specifications

Weight: 28,250 kg
Length: (travelling) 12.8 m
Width: (travelling) 2.44 m

Crane
Boom length: 2-section hydraulically extended boom with third lattice section, giving total length of 24.4 m
Counterweight: 4,310 kg
Capacity: (max) 25,400 kg

Carrier
Configuration: 8 × 4, front 2 axles steering
Engine: GM 6-71N diesel, developing 203 hp (net) at 2,100 rpm
Transmission: Fuller Roadranger RTO 613, providing 13 forward and 3 reverse gears
Tyres:
(front) 11.00 × 14-ply
(rear) 11.00 × 12-ply

Status

Production comp ete. There are 133 in service with the US Army.

Manufacturer

Grove Worldwide Box 21, Shady Grove, Pennsylvania 17256, USA.
Tel: +1 717 597 8121. Telex: 1842308.
Fax: +1 717 597 4062.

VERIFIED

Harnischfeger MT-250 25 ton truck-mounted hydraulic crane

Model M315T 15 ton capacity truck-mounted crane

Grove mobile hydraulic cranes

Description

Grove Worldwide produces a wide range of mobile hydraulic cranes for military and commercial purposes. Most of the military models are basically commercial ones converted for military use. Five main types of hydraulic crane are produced:

1. Truck-mounted cranes from 30 to 140 tons
2. Rough terrain cranes from 20 to 85 tons
3. All-purpose warehouse cranes from 18 to 159 tons
4. All-terrain cranes from 20 to 125 tons
5. Hydraulic lattice boom cranes of 134 tons capacity

Grove Worldwide also produces a wide range of aerial work platforms, both scissors and boom type, up to a working height of 35.35 m.

Status

In production and service with the US defence forces and with other armed forces.

Manufacturer

Grove Worldwide, Box 21, Shady Grove, Pennsylvania 17256, USA.
Tel: +1 717 597 8121. Telex: 1842308.
Fax: +1 717 597 4062.

VERIFIED

Grove TMS760 60 ton truck crane used by the US Navy

Grove RT875CC rough terrain container crane, 256 of which were ordered for the US Army with deliveries from 1989 to 1991

Rough terrain crane (4 × 4) 5 ton Hanson H-446A

Description

This rough terrain crane was developed to meet the requirements of the US Army by the Hanson Company of Tiffin, Ohio. The telescopic jib has a maximum reach of 7.62 m and can be rotated through a full 360°. A folding stabiliser leg is provided at each corner of the vehicle and most models have a bulldozer blade mounted at the front of the hull. The fully enclosed cab is at the front of the vehicle with the engine at the rear. Gross vehicle weight is 16,647 kg.

Status

In service with the US Army.

Manufacturer

Hanson Machinery Company, Tiffin, Ohio, USA.

VERIFIED

Hanson rough terrain crane (4 × 4) 5 ton (Larry Provo)

US dual role cargo pallet

Description

A cargo pallet, designed to fit the US Army's Ribbon Bridge transporter with a dual role as a general cargo carrier, was developed by the US Army Mobility Equipment Research and Development Command (MERADCOM) (now the Belvoir Research, Development and Engineering Center at Fort Belvoir).

An M812 5 ton (6 × 6) transporter equipped with a hydraulic boom is used to carry, launch and retrieve complete bays of the Ribbon Bridge. When not performing its bridging mission, it can launch and retrieve the specially designed 5.791 × 3.048 m pallet loaded with up to 5 tons (US) of equipment or supplies to double as a cargo hauler.

Cargo pallet in use with Ribbon Bridge transporter showing units being winched back on to truck (US Army)

The second capability to self-load, transport and off-load cargo was added late in the Ribbon Bridge transporter development, when cutbacks in military vehicles made maximum utilisation imperative. R & D pallets were procured, tested and type classified in only eight months. An initial production quantity of 120 pallets was delivered to the US Army in 1977.

Status

In service with the US Army.

VERIFIED

MISCELLANEOUS EQUIPMENT

FIELD FORTIFICATIONS AND RELATED EMPLACEMENTS EQUIPMENT

BULGARIA

OZ-82 foxhole excavation set

Description

The OZ-82 foxhole excavation set consists of two parts, a shaped charge initiation charge assembly and a high-explosive charge. In use the shaped charge is pressed into the ground using a plastic prong spacer to hold the charge steady. After the shaped charge has been fired the high-explosive charge is lowered into the resultant hole, which will be from 80 to 120 mm in diameter and detonated. The finished crater will measure from 1 to 1.5 m in diameter and will be from 0.6 to 1.2 m deep, all dimensions being dependent on the type of soil involved. The entire process takes from 4 to 8 minutes. Multiples of OZ-82 sets can be used to create trenches.

Specifications

Weight of complete set: 3.1 kg
Weight of HE charge: 1.56 kg
Finished hole diameter: 1 to 1.5 m
Finished hole depth: 0.6 to 1.2 m

Status

In production.

Manufacturer

KINTEX, PO Box 209, 66 James Baucher Street, 1407 Sofia, Bulgaria.
Tel: +359 2 662311. Telex: 23 243.
Fax: +359 2 658191.

VERIFIED

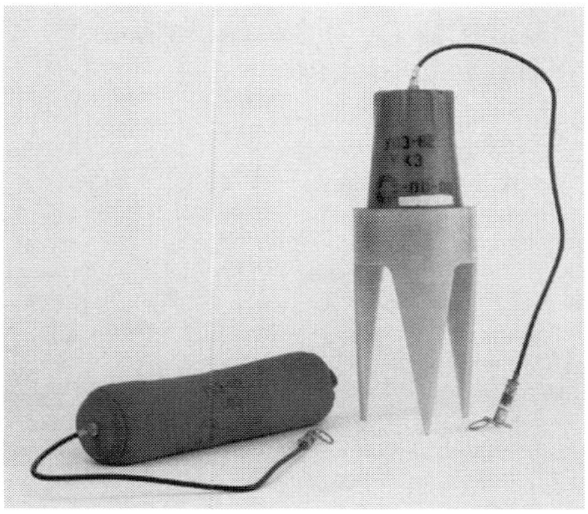

OZ-82 foxhole excavation set

DUGA light framework shelter

Description

The DUGA light framework shelter is issued as a kit and is intended to be installed underground in prepared excavations. It is then covered with spoil to create protected command posts or personnel accommodation. The basis of the shelter is a light tubular semi-circular framework which is covered by reinforced fabric once the framework has been assembled inside an excavated area. Entrance to the shelter is via a hatch which leads down a short ladder to an entrance compartment which could be made gas-tight. Internal fittings will vary with the application.

A complete set of DUGA components weighs 1,445 kg and is assembled into a shelter 7.175 m long, 2.083 m wide at floor level and 2.083 m high. A single shelter could be assembled in 75 minutes. Three or four sets can be carried on a 5 tonne truck. Each set can be used a minimum of 10 times.

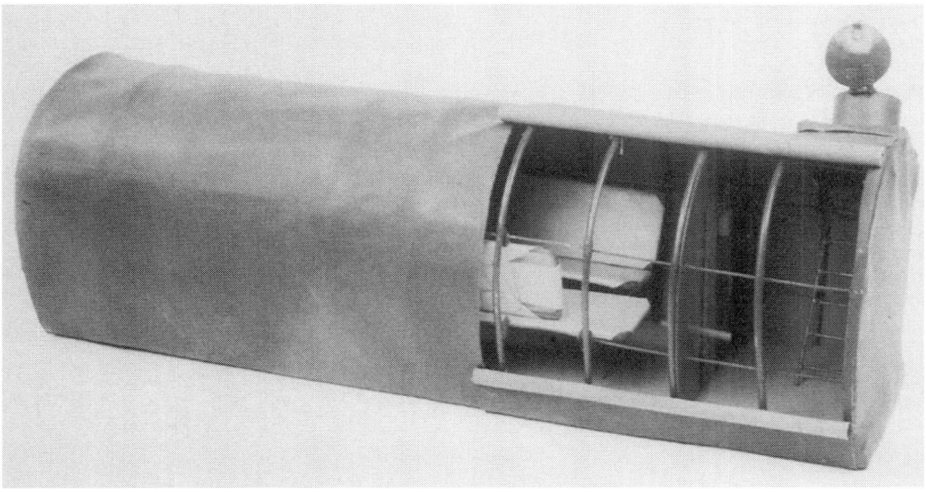

Status

In production.

Manufacturer

KINTEX, PO Box 209, 66 James Baucher Street, 1407 Sofia, Bulgaria.
Tel: +359 2 662311. Telex: 23 243.
Fax: +359 2 658191.

VERIFIED

Model of a DUGA light framework shelter

CHINA, PEOPLE'S REPUBLIC

NORINCO trench digger

Description

The NORINCO trench digger is based on the chassis and hull of the Type 83 152 mm self-propelled gun-howitzer (for details refer to *Jane's Armour and Artillery 1995-96* pages 539 and 540). The vehicle carries a circular trench-digging machine similar in appearance to that used on the RFAS MDK-2 series of trench-digging machines. This digging machine is electro-hydraulically lifted upwards and over the hull rear for transport and lowered to behind the vehicle for digging. The top of the hull carries an armoured enclosed cab and a straight dozer blade is carried on the front of the hull.

Power for the digging operation is taken by a multipurpose gear system from the vehicle's main engine, which is mounted in the forward part of the hull together with the transmission. Mechanical and hydraulic safety devices are incorporated.

The superstructure cab is air conditioned and the driver is provided with an infrared night vision device and an adjustable shock-absorber seat. No armament appears to be carried other than the crew's personal weapons.

NORINCO trench digger

The digging machine can remove up to 300 m^3/h of soft soil and up to 150 m^3/h of harder ground. The maximum digging depth for each digging cup is 0.45 m and the maximum digging depth 3.5 m. A typical finished trench can be 4.1 to 4.5 m wide at the top and 3.5 m wide at the bottom.

Specifications

Crew: 3
Weight: (combat) 32,000 kg
Length:
(travelling) 8.074 m
(operating) 10 m

Width: 3.6 m
Height:
(top of digging equipment) 4.195 m
(cab roof) 2.96 m
Ground clearance: 0.45 m
Track: 2.62 m
Track width: 480 mm
Length of track on ground: 4.601 m
Max speed: (road) 52 km/h
Gradient: 20°
Side slope: 10°
Engine: Type 12150L diesel developing 520 hp

Status
In service with the People's Liberation Army.

Manufacturer
China North Industries Corporation (NORINCO), 7A Yue Tan Nan Jie, PO Box 2137 Beijing, Beijing, People's Republic of China.
Tel: +86 1 7570. Telex: 22339 CNIN CN.
Fax: +86 1 7840.

UPDATED

NORINCO Type 82 tracked military bulldozer

Description
Designed from the outset for combat engineer use, the Type 82 tracked military bulldozer has a four-person cab. It is fitted with a dozer blade that can be angled about a central point from the concave to the convex. The blade is raised and lowered about a point between the second and third roadwheels on each side and is actuated vertically by a large hydraulic cylinder in front of the cab. The blade can be removed if required.

The cab is provided with a roof hatch equipped with a mounting for a 12.7 mm machine gun. The cab is equipped with two infrared vision systems: one for the driver with a 30° field of view; and another for the commander, a Type 85 device with a 76° horizontal field of vision and a 20° vertical field of view. A roof periscope is also provided.

The area behind the cab and over the engine and transmission covers is used as a general cargo or equipment-carrying area and is provided with steel drop sides. There is no tailgate other than a load restraint frame.

The Type 82 can travel at a maximum speed of 47.4 km/h to enable it to maintain station in convoys — average convoy speed is 32 km/h. When operating, the minimum bulldozing speed is 2.4 km/h and when operating within a distance of 30 m the earthmoving capacity is 262 m^3/h. At an operating distance of 50 m this is reduced to 232 m^3/h. The Type 82 can operate for periods of between 20 and 25 hours. Maximum push is 16 tonnes.

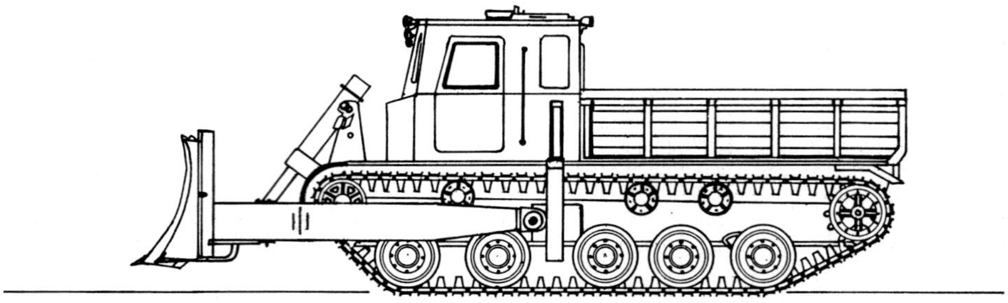

Type 82 tracked military bulldozer

Specifications
Crew: 4
Weight: (combat) 20,500 kg
Length:
(with blade) 7.17 m
(without blade) 5.81 m
Width:
(operating) 4.33 m
(travelling) 3.36 m
(without blade) 2.92 m
Height:
(to AA MG) 3.21 m
(top of cab) 2.76 m
Ground clearance: 0.45 m
Track: 2.49 m
Length of track on ground: 3.385 m
Ground pressure: 0.7 kg/cm^2
Max speed:
(road) 47.4 km/h
(cruising) 32 km/h
(reverse) 21.5 km/h
Range:
(road) 400-450 km
(cross-country) 350-400 km
Fording: 1.3 m
Gradient: 30°
Vertical obstacle: 0.7 m
Trench: 2.6 m
Engine: Type 12150 L-3 diesel developing 430 hp
Batteries: 4 × Type 65
Armament: 1 × 12.7 mm MG
Ammunition: 200 rounds

Dozer blade
Width: 4.33 to 4.4 m
Height: 1.1 m
Lift height: 1 m
Depth of cut: 370 mm

Status
In service with the Chinese armed forces and offered for export.

Manufacturer
China North Industries Corporation (NORINCO), 7A Yue Tan Nan Jie, PO Box 2137 Beijing, Beijing, People's Republic of China.
Tel: +86 1 7570. Telex: 22339 CNIN CN.
Fax: +86 1 7840.

VERIFIED

CZECH REPUBLIC AND SLOVAKIA

DOK wheeled engineer tractor

Description
The DOK (dozer on wheels) is a wheeled engineer tractor designed specifically for military applications. The DOK has an articulated chassis and a rear-mounted diesel engine; drive is electric. An electrically driven winch is located at the rear of the cab. In addition to the normal multipurpose bucket, a snow plough attachment is also available. The cab is hermetically sealed and has a filtered ventilation system which enables the vehicle to operate in an NBC environment.

There are three variants of the DOK: the DOK-L, DOK-M and DOK-R. The DOK-L has a universal shovel and the DOK-M, a modified DOK-L, has the universal shovel with the addition of a sawtooth edge and a central ridge. The DOK-M also has hydraulic steering, improved brakes and an improved hydraulic system. The DOK-R has a V-shaped blade that can be adjusted to form a straight blade for dozing.

DOK-L wheeled engineer tractor

Specifications
Cab seating: 1
Configuration: 4 × 4
Weight: 28,000 kg
Towed load: 65,000 kg
Length: 10.53 m
Width: 3.15 m
Height: 3.15 m
Ground clearance: 0.45 m
Track: 2.45 m
Wheelbase: 5 m
Max speed: (road) 50 km/h
Range: (cross-country) 250 km
Fuel capacity: 500 litres
Engine: T-930-42 V-12 diesel developing 255 hp at 1,800 rpm
Tyres: 21 × 28

Status
In service with the Czech and Slovak armed forces. The DOK-M appears to have been used only by the now disbanded East German Army.

Manufacturer
Czech and Slovak state factories.

UPDATED

DENMARK

Hydrema multipurpose machines

Description

Hydrema multipurpose machines can be used for combat engineering and logistics tasks, combining the attributes of a 1.2 m³ front-end loader, an 8 tonne hydraulic excavator and a conventional backhoe loader. All machines in the range feature high cross-country mobility due to a patented pivot steer system with oscillation in the centre pivot providing automatic load distribution to the four large wheels. Various types of buckets, shovels and other ancillaries can be fitted, including a forklift in place of the front shovel. The machines are air-portable in C-130 or similar aircraft and can be carried slung under a CH-47D Chinook helicopter. A road speed of 40 km/h can be achieved with a maximum load.

The product range consists of two models: a hydraulic side shift model, the Hydrema 906, offering 280° of slew with the excavator (purchased by the British Army and deployed in Saudi Arabia); and a fixed kingpost model offering a 200° slew with the excavator (purchased by the Danish Army). Winterised versions capable of operating in temperatures down to −40°C are available.

The Hydrema 910 10 tonne articulated dump truck, built on the same major components as the multipurpose machines, has been purchased by the British Army. It is powered by a Perkins 1000 series 96 hp turbo-diesel and has constant four-wheel drive on four equal sized wheels. Cross-country capabilities are enhanced by oscillation about a central pivot on the articulated chassis.

Status

In service with the British Army and Danish Army. Tested by the French, German and Canadian Forces.

Manufacturer

A/S Hydrema, GL. Kirkevej 16, DK-9530 Stovring, Denmark.
Tel: +45 98 37 13 33. Telex: 9381805.
Fax: +45 98 37 19 96.

UPDATED

Hydrema 906 military loader/excavator **1997**

Hydrema 910 10 tonne articulated dump truck **1997**

EGYPT

Trench fortification system

Description

This Egyptian trench fortification system is intended to convert open communication or combat trenches into fully protected shelters and firing positions. It consists of a series of U-section supports placed every 0.6 m along a trench and connected by tie units. Corrugated steel sheets are used to line each side of the trench. The roof may be covered by either flat or curved roof panels which are covered over with spoil to provide maximum protection against enemy action and the effects of the weather. The roof panels may be corrugated steel sheets, double panels with added insulation material or strengthened fibre sheets. At intervals, provision can be made to add pre-fabricated open firing pits, each with a diameter of 0.75 m.

The layout of a completed fortified trench system can be either all-in-line or in a zig-zag pattern. The main trench requirements are an initial depth of 1.45 m which becomes 1.25 m when the reinforcements are added. Trench width is 0.85 m.

Model of Egyptian trench fortification system **1997**

Status

In production. In service with the Egyptian armed forces.

Manufacturer

Helwan Company for Metallic Appliances (Factory 360), Helwan, Egypt.
Tel: +20 781688. Fax: +20 782983.

NEW ENTRY

FRANCE

Matenin trench digger

Description

The Matenin trench digger was designed for the rapid excavation of trenches and defences in forward areas and can dig straight or curved trenches or one-man foxholes.

The equipment consists of a cross-country 4 × 4 chassis with a forward control cab. The trench-digging equipment, mounted on the rear of the chassis, operates in the vertical position and can be swung into the horizontal position for travelling. Digging can be controlled from within the cab of the vehicle or by means of a remote-control system.

The excavator itself consists of chain-driven buckets made of moulded steel with tilting bases, mounted on a boom. A conveyor belt is provided to discharge earth to the right or the left and the boom is fitted with a scraper to clean the bottom of the trench.

The power-shift hydromechanical transmission has six ranges, allowing for road travel, cross-country, digging and winch drive.

There are two models of the Matenin trench digger: the MXD, which digs a trench 0.85 m wide and up to 2 m deep; and the Model NX 7B3, which digs a trench 0.6 m wide and 1.8 m deep. Both models can dig a vertical trench on a 15 per cent side slope and have a typical capacity of 250 m³/h.

Specifications

Cab seating: 1 + 2
Configuration: 4 × 4, with locking differentials
Weight:
(NX 7B3) 14,500 kg
(MXD) 15,500 kg
Length: 7.76 m
Width: 2.5 m
Height:
(over cab) 2.8 m
(travelling) 3.89 m
Ground clearance: 0.51 m
Wheelbase: 3.3 m
Track: 1.98 m
Angle of approach/departure: 38°/34°
Max speed: (road) 70 km/h
Max gradient: 50%
Max side slope: 15%
Fording: 1.2 m

Matenin trench digger (T J Gander) **1997**

Engine: 6-cylinder water-cooled diesel developing 225 hp at 2,300 rpm
Transmission: Powershift hydromechanical, with torque converter and two 4-gear ranges
Steering: power-assisted
Tyres: 16.00 × 25, low pressure
Brakes: air
Electrical system: 24 V
Winch: driven by hydraulic motor; 5,000 kg capacity line pull

Status

In service with the French Army and several other armies.

Manufacturer

Matenin SA, 34 avenue des Champs Elysées, F-75008 Paris, France.
Tel: +33 1 43 59 22 00. Telex: 801 853 f.
Fax: +33 1 42 25 44 19.

UPDATED

MFRD/F1 mobile drilling machine

Description

The MFRD/F1 is a highly mobile drilling machine mounted on a Pinguely rough terrain carrier known as the RTC10. The drilling equipment can be used for a variety of field engineering purposes, but the primary role is that of drilling holes into which demolition charges can be placed (MFRD - *moyen de forage rapide de destruction*). The French Army ordered 122 examples of this system. Production is complete.

The drilling machine can be used at any angle from the vertical to the horizontal at 7.5° intervals. The drill uses an auger in soft soils but in rock a bottom hammer is employed. Both have a working diameter of 220 mm. Maximum drilling depth is 6 m. When drilling through rock the following drilling speeds can be attained:

Limestone	12 m/h
Granite	17 m/h
Schist	32 m/h

The RCT10 rough terrain carrier has a 4 × 4 configuration with a forward drive cab and seating for a driver/operator and one passenger. The drilling rig is carried on the area behind the cab with the gantry pivot overhanging the rear of the vehicle. An air compressor driven from a power take-off on the gearbox delivers 19 m^3/min at 7 bar. A 6,000 to 8,000 kg winch can be fitted.

MFRD/F1 mobile drilling machines

Specifications

Cab seating: 1 + 1
Configuration: 4 × 4
Weight:
(total) 16,800 kg
(front axle) 8,300 kg
(rear axle) 8,500 kg
Length:
(overall, travelling) 9.75 m
(chassis) 7.565 m
Width: 2.49 m
Height:
(overall) 2.384 m
(cab) 2.68 m
Ground clearance: 0.52 m
Wheelbase: 4.1 m
Track:
(front) 1.971 m
(rear) 2.038 m
Angle of approach/departure: 42°/35°
Fording: 1 m
Vertical obstacle: 0.5 m
Gradient: 50%
Max speed: (road) 70-75 km/h
Range: (road) 600 km
Engine: turbocharged diesel developing 260 hp at 2,200 rpm
Transmission: torque converter semi-automatic hydromechanical gearbox with 3 forward and 3 reverse gears
Suspension: double-action leaf springs and hydraulic shock-absorbers
Turning radius: 11 m
Tyres: 16.00 × 25

Status

In service with the French Army (122).

Manufacturers

Carrier: HAULOTTE - c/o SOLEM SA, 32 avenue Marcel-Cachin, F-69120 Vaulx en Velin, France.
Drilling equipment: Constructions Industrielles d'Anjou, ZI Ecouflant, F-49000 Angers, France.

VERIFIED

Mle F1 foxhole charge

Description

Intended for the loosening of soil to allow the rapid manual excavation of an infantry foxhole, the Mle F1 foxhole charge consists of three explosive charges, a 50 second delay exploder and detonating fuze.

The three explosive charges are in rod form and are sealed in a plastic tube 1.06 m long and 25 mm in diameter. Three of these charges are driven vertically into the ground 0.6 m apart and connected by a detonating fuze to a 50 second delay exploder. After initiating the exploder and the 50 second delay all three charges will detonate together and loosen and partially remove topsoil over a length of 1.8 m, width of 0.6 m and to a depth of 1.2 m (all foxhole dimensions approximate).

Each main charge weighs 0.5 kg of which 0.222 kg is explosive.

Status

In production.

Manufacturer

Société d'Armement et d'Etudes Alsetex, 4 rue Tronchet, F-75009 Paris, France.
Tel: +33 1 42 65 50 16. Telex: 640961 f.
Fax: +33 1 42 65 24 87.

VERIFIED

CNIM Soldier FIghting Cover (SFC)

Development

During the 1990-91 Gulf War the US Army encountered problems involving foxholes without overhead protection. In December 1992 the US Army issued a Request For Quotation relating to existing foxhole covers which could meet the overhead cover requirement. Constructions Industrielles de la Méditerranée (CNIM) responded to the Request For Quotation and its cover, the Soldier Fighting Cover (SFC), was successfully tested by the French and US armies using 155 mm artillery fire. The CNIM cover was scheduled to undergo additional testing by the US Army from the end of 1994 onwards.

Description

Once assembled, the CNIM Soldier Fighting Cover (SFC) forms an arch 1.35 m long, 0.61 m wide and 1.25 m high. The arch supports soil cover with the two walls forming sides to prevent soil from entering. Two half arches form the full arch with a joining strip at the apex and six struts holding the base apart. The cover is formed using very thin glass fibre sheets which can be retrieved after installation for reuse.

A complete SFC weighs 15.87 kg and can be carried by one person. Each SFC is intended to be used for two soldiers as the complete installation involves a trench 1.83 m long and 0.61 m wide; the depth is sufficient to reach a soldier's armpit. Installation of the cover takes only a few minutes. Once emplaced, an SFC can support a saturated soil load up to 0.61 m thick, sufficient to provide protection against a 155 mm projectile overhead or near-miss blast and fragments.

A completed CNIM Soldier Fighting Cover (SFC) in use

The SFC can be supplied stacked on pallets.

The French term for the SFC is Couverture d'Emplacement de Combat (CEC).

Status

Prototypes.

Manufacturer

Constructions Industrielles de la Méditerranée (CNIM), Zone Industrielle de Brégaillon, BP 208, F-83507 La-Seyne-sur-Mer Cedex, France.
Tel: +33 3 44 31 11 00. Fax: +33 3 44 31 11 30.
Enquiries to: Constructions Industrielles de la Méditerranée (CNIM), 35 rue de Bassano, F-75008 Paris, France.
Tel: +33 3 44 31 11 00. Fax: +33 3 44 31 11 30.

UPDATED

A CNIM Soldier Fighting Cover (SFC) in its unassembled state

Joining the two halves of a CNIM Soldier Fighting Cover (SFC)

Placing sandbags over an installed CNIM Soldier Fighting Cover (SFC), showing the depth of protection which can be provided

GERMANY

Schaeff FL 412 frontloader and HT 12 backhoe on Mercedes-Benz Unimog U 110

Description

The Mercedes-Benz Unimog with short wheelbase, power take-offs and hydraulic connections at the front and rear for attached implements, is used for a variety of applications as a mobile workhorse.

The Unimog U 110 (wheelbase 2.83 m) replaced the U 900 as a carrier for implements. It has a cut-away

Schaeff FL 412 frontloader and HT 12 backhoe on Mercedes-Benz Unimog U 110

section in the engine hood to provide the driver with an excellent view for the control of front-mounted implements. A typical attachment configuration consists of the Schaeff FL 412 frontloader and an HT 12 backhoe for digging command posts, combat positions and many other combat engineer and construction tasks.

Status
In production.

Manufacturers
Attachments: Karl Schaeff GmbH & Co Maschinenfabrik, Seestrasse 17, D-74595 Langenburg/Württ, Germany.
Unimog: Mercedes-Benz AG, Dept VN/RM - 2503, D-70322 Stuttgart, Germany.
Tel: +49 711 179 0326. Fax: +49 711 179 0934.

VERIFIED

Schaeff HT 11A backacter on Unimog U84/406 (Erdarbeitsgerät)

Description
The *Erdarbeitsgerät* (earthwork equipment) is used by non-mechanised units of the German armed forces to enable field defences to be quickly prepared. It consists of a standard commercial backacter mounted on the rear of a Mercedes-Benz Unimog cross-country vehicle, which can also be equipped with a bulldozer blade at the front. In medium soil, up to 18 two-person slit trenches (1.6 m long × 0.6 m wide × 1.2 m deep) can be dug in one hour. The backacter is operated from a 30 hp hydraulic pump driven by the vehicle engine and can be speedily disconnected and removed from the vehicle. The entire assembly can be pivoted by the two-person crew through 90° into its travelling position.

The overall mobility of the Unimog vehicle is scarcely affected by the addition of the backacter. It can maintain a road speed of 74 km/h and can be transported in both the Transall C-160 transport aircraft and the CH-53G helicopter.

Status
There are 377 examples of the system in service with the German armed forces. A similar system mounted on the Unimog U 900 is in service with the US Army under the designation SEE (qv).

Manufacturers
Backacter: Karl Schaeff GmbH & Co Maschinenfabrik, Seestrasse 17, D-74595 Langenburg/Württ, Germany.
Tel: +49 7905 58-0. Telex: 749 811.
Fax: +49 7905 58114.
Unimog: Mercedes-Benz AG, Dept VN/RM - 2503, D-70322 Stuttgart, Germany.
Tel: +49 711 179 0326. Fax: +49 711 179 0934.

VERIFIED

Unimog Erdarbeitsgerät in travelling order and in operation (Federal German Ministry of Defence)

Defendress Type S and Type T battlefield shelters

Description
The Defendress battlefield shelters are steel drums 1.25 m high and 1.4 m in diameter fitted with a steel floor and an armoured roof. Each shelter can accommodate two personnel or act as a mortar or light weapon pit. A communication tunnel 1 m in diameter and at least 2 m long can connect each shelter to others nearby. The armoured hatch can be lifted by a hydraulic pump for observation and once raised it can be turned aside for weapon firing. The hatch has a raised rim to allow soil and camouflage to be inserted. To close the hatch it is turned into the lowering position and a position lock is released; the shelter is then secure. With soil and other top cover the armoured hatch can weigh up to 1,400 kg, so a manually operated hydraulic pump is provided with a maximum lift capacity of 15 tonnes. One stroke of the pump handle raises the hatch 15 mm and the maximum lift height is 250 mm.

The Defendress Type S is a welded and prefabricated steel structure with a total weight of approximately 850 kg. It is totally sealed, consisting of an upright cylinder with a connection for the communication tunnel pipe and a pivoting shaft for the armoured hatch, the removable armoured hatch and the hydraulic system. The Type T shelter uses a prefabricated kit weighing 550 kg that can be assembled by two personnel using a simple wedge system. The heaviest component weighs 83 kg. Parts include: eight steel shell segments (one of them with a welded pivoting shaft for the hatch); ground plate with a welded supporting skirt; top ring; ring with a pinion connection and support plate for the armoured hatch; the hydraulic system; joining wedges and connection elements for the communication tunnel pipe.

Status
In production.

Manufacturer
Protection and Defence Systems (PDS), Bäckerstrasse 53, D-8000 Munich 60, Germany.
Tel: +49 89 260 3841. Telex: 522072 is g.
Fax: +49 89 260 3395.

VERIFIED

NETHERLANDS

SPIT Prefab Foxhole Twin (PFT)

Description
The SPIT Prefab Foxhole Twin or PFT is a two-person shelter or foxhole, prefabricated from steel components and delivered complete for installation. The kit consists of two identical sections forming the walls of the foxhole and all other necessary accessories including sandbags, a weather cover and camouflage net. The complete kit weighs 65 kg and uses steel plates with aluminium profiles.

To install the kit two personnel can carry it to the selected site. A hole approximately 1.9 × 0.7 × 1.3 m is dug and, once the foxhole assembly has been constructed, it can be placed in the hole and spoil used to infill the sides. If required, insulating material such as dry hay, leaves and so on can be used for the infill preventing condensation from forming on the interior walls. It is possible to construct the foxhole in such a manner that there are two foxholes at each end with a sheltered centre section covering a stowage area in the form of a shelf. This can then be used for stowing ammunition and provisions. Sandbags can be stowed around the periphery for increased protection. When the PFT is no longer required in one location it can be levered out of the ground using pioneer tools inserted into holes in the sides.

If required, the PFT can be constructed to stand on the ground surface and then be filled with soil to provide increased protection for tactical positions; earth banks may also be formed against the PFT. A special roof container can be used to provide overhead protection for a normal PFT and will be held in place by six struts. It is possible to arrange the struts and roof container to allow weapons to be fired through gaps between the roof and the ground.

Specifications
PFT
Weight: (kit) approx 65 kg
Length: (assembled) 1.83 m
Width: 0.62 m
Height: 1.4 m

Roof container
Weight: (empty) approx 90 kg
Length: 2.5 m
Width: 1.23 m
Height: 0.3, 0.6 or 0.9 m
Adjustable height: 0.3 m
Max load: 4,000 kg

Status
In production.

Manufacturer
Special Projects and International Trading, PO Box 300, NL-6236 ZN Maastricht, Netherlands.
Tel: +31 4402 70182. Telex: 56235 spit nl.

VERIFIED

Complete PFT ready for use and showing centre section of roof in place

RUSSIAN FEDERATION AND ASSOCIATED STATES (CIS)

BTM series of high-speed ditching machines

Description
The BTM trenching machine consists of the ETR-409 ditching machine mounted on the rear of the AT-T heavy tracked artillery tractor. The equipment can dig a trench 0.8 m wide to a maximum depth of 1.5 m at a rate of 1,120 m/h. The BTM-TMG is a variant of the basic BTM designed to dig trenches in frozen ground. Once a special cable and anchor assembly has been deployed it can dig a trench 0.6 m wide at 100 m/h in frozen soil. One further model, the BTM-3 weighing 27,300 kg, is unable to dig trenches in frozen ground. There is also a BTM-TMG2S model, which has a larger ditching machine.

From 1992 onwards, the above models were joined by the BTM-4 based on the chassis of the MT-T general purpose tracked carrier. This model weighs 41 tonnes, has a two-person crew and is powered by a 710 hp diesel. For transport, the BTM-4 can carry its trench-digging machine lowered on to the cargo bed behind the forward control armoured cab. When digging, spoil is dispensed to either side of the trench. The BTM-4 can dig trenches in frozen soil at increased speeds without any need for prior preparation. In unfrozen ground the BTM-4 can dig a 1.1 m wide trench at a rate of 1,200 m/h; in frozen ground the rate is 400 m/h. When a 1.5 m wide trench is dug in unfrozen ground the rate is 1,000 m/h; in frozen ground the rate is 300 m/h.

The latest machine in the series is the BTM-4M Tundra based on the tracked chassis of the 203 mm 2S7M self-propelled gun (for details of this gun refer to *Jane's Armour and Artillery 1995-96* pages 543 and 544). Development was carried out by the Special Engineering JSC (Spetsmash JSC) in St Petersburg. The general layout of the BTM-4M follows that of the earlier models, but the digger at the rear is of a new type. It is driven by two sets of the GST-90 hydrostatic transmission which enable it to operate in frozen ground. Using the digger the BTM-4M can cut 250 m of a trench 1.5 m deep and 0.6 m wide in one hour. If the trench is 1.1 m deep the length per hour is increased to 300 m. In unfrozen ground up to 1,000 m of 1.5 m trench can be dug in one hour. A dozer blade is provided at the front of the vehicle.

The BTM-4M Tundra has a crew of two, weighs 43.9 tonnes and is powered by a V-94T diesel powerpack providing a maximum output of 840 hp. Cruising speed is up to 50 km/h. 2,200 litres of fuel are carried.

Specifications

Model	BTM	BTM-TMG	BTM-TMG2S
Crew	2	2	2
Weight	26,500 kg	30,000 kg	32,000 kg
Length			
(travelling)	7.35 m	7.6 m	11.5 m
(operating)	10.85 m	—	11.5 m
Width	3.2 m	3.2 m	4.6 m
Height			
(travelling)	4.3 m	4.3 m	4.6 m
(operating)	3.5 m	—	3.2 m
Speed	35 km/h	36 km/h	36 km/h
Cruising range	500 km	400 km	400 km
Fuel capacity	810 litres	—	—
Trench crossing ability	2.1 m	2.1 m	2.1 m
Max vertical obstacle	1 m	1 m	1 m
Gradient	33°	36°	36°
Side slope	17°	17°	17°
Fording depth	0.75 m	0.75 m	0.75 m
Working capacity			
(0.8 m trench)	1,120 m/h	1,120 m/h (Summer)	—
Depth of trench	1.5 m	—	3 m
Width of trench			
(top)	1.1 m	0.6 m	1.1 m
(bottom)	0.6 m	0.6 m	—
Height of parapet	0.4 m	—	—

Note: automotive details are as for the basic AT-T tractor in the *Tracked prime movers, cargo carriers and armoured logistic vehicles* section.

Status
BTM-4M is in production and offered for export sales. Most of the other models are in service with members of the former Warsaw Pact and some Middle Eastern and North African countries.

Manufacturer
State factories.

UPDATED

BTM digging machine with digger stowed (Michael Jerchel)

BTM digging machine with digger stowed

1996

MDK-2 series of trench-digging machines

Description

The MDK-2 trench-digging machine *(maschina dorozhnoy kopatelnoy)* is based on the chassis of the AT-T heavy tracked artillery tractor. The circular digging machine is carried horizontally on the rear of the chassis and is swung 90° into the vertical for ditching operations. The MDK-2 is used for digging weapon trenches and pits for vehicles, guns and other equipment. Depending on the soil conditions, the MDK-2 can dig a maximum of 300 m³/h. The dug ditch has a maximum depth of 4.5 m and is 3.5 m wide at the bottom and 4 m wide at the top. An OTT hydraulically operated dozer blade is mounted at the front of the vehicle. There is also a MDK-2M model.

Specifications

Data in square brackets relates to MDK-2M where different from MDK-2

Crew: 2
Weight: 27,000 [28,000] kg
Length:
(travelling) 8 m
(operating) 10.23 m
Width:
(travelling) 4 [3.4] m
(operating) n/avail [4.05] m
Height:
(travelling) 3.95 m
(operating) 3.48 m
Speed: 35 km/h
Max vertical obstacle: 1 [0.65] m
Max gradient: 36°
Side slope: 10°
Working speed: (class 1 and 2 soil) 300 [387] m³/h
Depth of ditch: 4.5 m
Width of ditch:
(top) 4 m
(bottom) 3.5 m

Status

In service with members of the former Warsaw Pact and some countries in North Africa and the Middle East.

Manufacturer

State factories.

VERIFIED

MDK-2M trench-digging machine (Michael Jerchel)

MDK-2M trench-digging machine (Michael Jerchel)

MDK-3 trench-digging machine

Description

The MDK-3 trench-digging machine *(maschina dorozhnoy kopatelnoy)* is intended to be the replacement for the MDK-2 and MDK-2M (see previous entry) and is based on the chassis of the MT-T tracked carrier (see entry in the *Tracked prime movers, cargo carriers and armoured logistic vehicles* section).

The MDK-3 chassis, known as the Kharkov engineer chassis, uses suspension elements from the T-64 tank and is powered by a V-64-4 V-12 diesel engine. The cab is armoured, fitted with a radio and may be fitted with an NBC protection system. At the front of the vehicle is a hydraulically operated dozer blade which can be tilted. Mounted at the rear of the vehicle is a six-blade digging wheel capable of excavating a trench just under 3 m in depth in a single pass. Hydraulic rams allow the digging wheel to be positioned for excavating; when travelling, the digging wheel is positioned horizontally over the chassis rear. When digging, the vehicle is driven slowly backwards and spoil is pushed out to one side to form a loose berm alongside the ditch. Digging blades on the wheel are bolted into position and can be changed.

MDK-3 trench-digging machine (Jane's Intelligence Review)

Specifications

Provisional

Crew: 2
Weight: 40,000 kg
Length:
(digger raised) 10 m
(digger lowered to ground level) 12 m
Width:
(over hull) 3.2 m
(over digger) 3.5 m
Height: (digger raised or lowered) 4 m
Max speed: 50 km/h
Range: approx 500 km
Engine: V-64-4 V-12 water-cooled diesel developing 737 hp
Trench digging speed: 200 m/h
Trench width: 3.5 m
Trench depth: 2.9 m

Status

In service with the RFAS.

Manufacturer

State factories.

VERIFIED

TMK-2 trench-digging machine

Description

The TMK-2 trench-digging machine is based on the chassis of the MAZ-538 wheeled tractor (using a longer wheelbase) and shares many components with the PKT road-building machine. It has a crew of two: a driver and a digger assembly operator. Both are seated in a cab equipped with radiation warning and navigation equipment plus an air filtration system. External equipment stowage space is provided on the cab roof.

As well as its trench-digging function the TMK-2 can be used to clear bushes and scrub, scrape topsoil from contaminated areas and cut approaches and exits for water crossings, using a hydraulically operated dozer blade at the front. The main trench-digging assembly is located to the rear of the vehicle and consists of a circular cutter that is raised and lowered hydraulically. The assembly has no digging shovels but relies upon cutting teeth that produce a specific pressure on their cutting edges. Excavated earth is lifted upwards and adheres to the wheel assembly by friction before being discharged to both sides once clear of the trench. The fastest trench-digging speed possible in soft soil is 1.3 km/h, controlled by a hand wheel in the cab. Maximum digging depth is 1.5 m and the width of the trench can vary from 0.6 to 1.1 m.

The engine has a preheating facility enabling it to start at temperatures below −15°C.

Specifications

Crew: 2
Weight: 27,200 kg
Length: 9.75 m
Width: 3.15 m
Height: 4.18 m

Road speed: between 30 and 60 km/h
Operating endurance with full tank: 21 h
Engine: D 12 A 375 A 12-cylinder diesel developing 362 hp
Transmission: hydraulic with 3 forward and 1 reverse gears
Electrical system: 24 V
Batteries: 4 × 24 V, 280 Ah

Status
In service with the RFAS.

Manufacturer
Dmitrov Excavator Plant, 1 Pushkin Street, Dmitrov 141800, Moscow Region, RFAS.
Tel: +7 95 587 3120.

UPDATED

TMK-2 trench-digging machine in travelling configuration
1996

PZM and PZM-2 regimental trench-digging machines

Description
The PZM regimental trench-digging machine is built on the basis of the T-150K wheeled tractor. The more recent PZM-2 is based on the T-155 wheeled tractor powered by a 165 hp diesel engine. The PZM-2 has a roller chain with digging buckets driven mechanically from the main tractor engine. A front-mounted winch can be used to propel the machine when digging, as the transmission can be disconnected when the excavator is in use. The winch is driven hydromechanically and exerts a 5,000 kg pull at a rate of 60 m/h. The digger is raised and lowered by hand. A front-mounted dozer blade is provided.

Variant
ETTs-200
The ETTs-200 is based on the PZM-2 and is powered by the same 165 hp diesel engine but lacks the front-mounted dozer blade and is apparently intended for civilian construction purposes.

PZM-2 regimental trench-digging machine in travelling order (T J Gander) ***1996***

Specifications
Crew: 2
Weight: 13,200 kg
Length:
(travelling) 7 m
(operating) 9.75 m
Width: 2.52 m
Height:
(travelling) 3.75 m
(operating) 2.82 m
Speed: 45 km/h
Range: 500 km
Engine: diesel deve oping 165 hp

Trench digging speed:
(normal soils) 180 m/h
(frozen soils) 35 m/h
Trench dimensions, normal soils:
(width, top) 0.9 m
(width, bottom) 0.65 m
(depth) up to 1.2 m
Trench dimensions, frozen soils:
(width, top and bottom) 0.65 m
(depth) up to 1.2 m
Winch pull capacity: 5,000 kg

Status
Both the PZM and the PZM-2 are in service with the RFAS. Offered for export sales.

Manufacturer
Kiev Plant for Engineering Machines, 21 Degtyarevskaya Str, Kiev 254119, Ukraine, RFAS.

VERIFIED

High-speed tractor-mounted bulldozers BAT-1, BAT-M, OLT, OST, OTT

Description
Most of the RFAS full-tracked artillery tractors described in the *Tracked prime movers, cargo carriers and armoured logistic vehicles* section can be fitted with bulldozer blades. One vehicle, the BAT, has been specially modified as a bulldozer.

The BAT tractor dozer consists of the AT-T heavy tractor with a large dozer blade mounted at the front of the hull. It is designed for the hasty preparation of roads and approaches to bridges and crossing sites also for filling in ditches and similar obstacles. It can also fell trees, uproot stumps and boulders and dig emplacements. The basic demolition blade can be fitted with attachments to form a V-blade, bulldozer and angledozer and is provided with routers and a float.

The more recent BAT-M is an improved model and is electrohydraulically operated, whereas the BAT (also known as the BAT-1) is electropneumatically operated. The BAT-M also has a hydraulic crane and the dozer blade can be swung rearwards to improve the vehicle's load distribution when travelling.

The BAT tractor dozers should not be confused with the OTT dozer blade mounted on the AT-T tractor, for which data is also given here. Details of the basic tractor chassis are given in the *Tracked prime movers, cargo carriers and armoured logistic vehicles* section.

Specifications

Model	BAT (BAT-1)	BAT-M	OLT	OST	OTT
Tractor	(integral)	(integral)	AT-L, AT-LM	AT-S*, ATS-59	AT-T
Weight (with blade)	25,300 kg	27,500 kg	7,000 kg	13,392 kg	22,000 kg
Length (with blade)	10 m	7 m (travelling)	n/avail	n/avail	n/avail
Width (with blade)	4.73 m	4.85 m	2.5 m	2.8 m	3.5 m
Height (travelling)	2.95 m	–	2.18 m	2.54 m	2.58 m
Max speed	35 km/h	35 km/h	42 km/h	35 km/h	35 km/h
Cruising range	700 km	550 km	300 km	380 km	700 km
Fuel consumption	140-190 litres/100 km	140-190 litres/100 km	140-190 litres/100 km	140-190 litres/100 km	140-190 litres/100 km
Trench	1.53 m	2.1 m	1 m	1.45 m	2.1 m
Max vertical obstacle	1 m	1 m	0.6 m	0.6 m	1 m
Max gradient	45%	n/avail	60°	53°	60°
Fording depth	0.75 m	0.75 m	0.6 m	1 m	0.75 m
Ground pressure	0.65 kg/cm²	0.71 kg/cm²	0.45 kg/cm²	0.58 kg/cm²	0.68 kg/cm²
Working speed	1.5-10 km/h (moderate terrain)	1.5-10 km/h	4-6 km/h (moderate terrain)	4 km/h	n/avail
Working capacity (depends on soil conditions)	120-140 m³/h	max 150 m³/h (excavation) max 200 m³/h (dozing)	40 m³/h	80-90 m³/h (light soil) 40-50 m³/h (medium soil)	100 m³/h
Crane capacity	n/app	2,000 kg	n/app	n/app	n/app
Winch capacity	n/app	25,000 kg (100 m cable)	n/app	n/app	n/app
Blade control	electropneumatic	electrohydraulic	n/avail	cable operated	n/avail
Blade angle	n/avail	straight (5 m wide) 55° (4 m wide) 110° (4.5 m wide)	n/avail	n/avail	n/avail
Tractive effort	n/avail	16,300 kg	n/avail	n/avail	n/avail

* Data refers to OST dozer on AT-S tractor.

Status
All the tractor-mounted dozers are in service with the former Warsaw Pact nations. The BAT tractor dozer is also in service with some North African and Middle Eastern countries.

Manufacturers
State factories.

VERIFIED

SWEDEN

Kockums SMAC shelter

Description
Developed by Kockums Industrier, the SMAC low-weight shelter is intended for both military and civilian protection purposes. It is designed in such a manner that, when installed below ground level and surrounded by soil, its strength factor, as well as the protection factor provided, is increased by the depth at which it is buried. All components are already in use by the Swedish building industry and construction involves steel over, or combined with, glass reinforced plastic or other composite materials.

The construction of the SMAC is modular so it can be installed in numerous configurations. The normal soil covering is up to about 4.5 m deep, resulting in a weight on the shelter of about 9 tonnes/m². Once assembled, the shelter is gastight.

The weight of an SMAC entrance unit, measuring 9.6 × 1.6 m, is approximately 2.4 tonnes. A fully fitted accommodation unit weighs approximately 3.9 tonnes, measures 8.6 × 2.4 m and could accommodate about 15 people. Both units could be carried ready assembled by trucks and could be installed in prepared positions in about 30 minutes. Due to the modular construction, extensions of the basic units could be carried out rapidly.

A full-scale prototype of the SMAC shelter was successfully tested by the Research Department of The Royal Swedish Fortifications Administration, using bombs and explosive charges.

Installing a Kockums SMAC low-weight shelter

Status
Available.

Manufacturer
Kockums Industrier, S-205 55 Malmö, Sweden.
Tel: +46 40 34 80 00. Fax: +46 40 97 32 81.

VERIFIED

Atlas fast foxholes

Description

The Atlas Copco Ground Entrenchment-Foxhole Explosive System (GE-FES) was developed to produce a foxhole rapidly on the battlefield. Using Atlas Copco Berema Rock drill-breakers, a two-person team can rapidly drive two charge pipes, insert charges and blast a foxhole crater measuring 1.8 × 0.7 × 1.8 m. Loose material can then be shovelled out.

The GE-FES system can also be used for the preparation of hull-down positions for MBTs, APCs, self-propelled artillery and so on.

Atlas Copco Berema Rock drills and breakers are in service with defence forces in over 30 countries.

Status

In production. In service with Swedish infantry, artillery and tank battalions. Evaluated by Belgian Engineer Corps, British Army, Danish Army and NBC Defence School, Netherlands Engineer Corps and Miners and Sappers School and the US Army.

Manufacturer

Atlas Copco Berema AB, Uddvägen 7, Box 767, S-131 24 Nacka, Sweden.
Tel: +46 8 743 9641. Fax: +46 8 743 9650

UPDATED

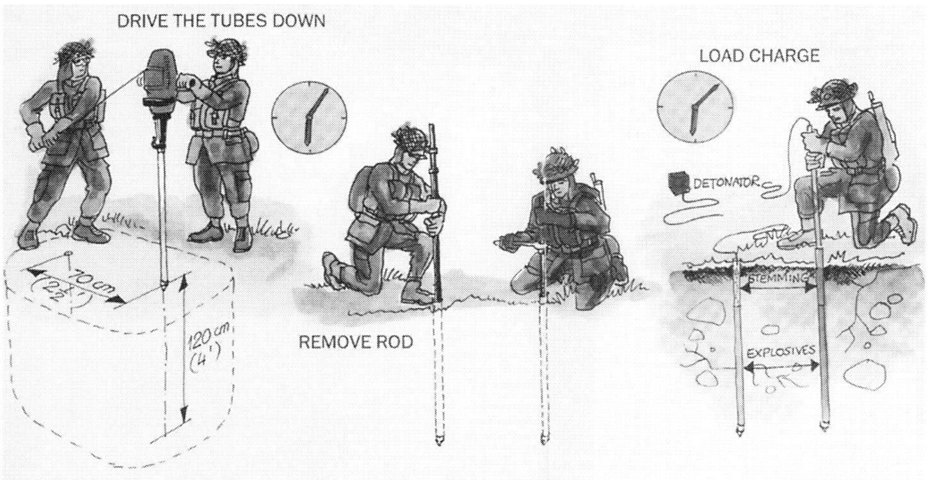

Outline drawings of the sequence of preparing the Atlas Copco Ground Entrenchment-Foxhole Explosive System (GE-FES)
1997

SWITZERLAND

Type 88 field shelter

Description

Developed for use by the Swiss infantry, the Type 88 field shelter can be assembled in the field without special tools or training. The trench for the shelter can be dug using manual or mechanical methods. The cylindrical shelters are 6 m long and have a diameter of 2.5 m. Each shelter consists of 12 2.5 mm thick corrugated steel sections fixed by screws to 500 mm wide hoops. When assembled, each shelter has space for 18 seats or 12 bunks and a simple ventilation system is supplied. The shelters can also be used as command or first-aid posts.

Status

In production for the Swiss Army (4,500 ordered).

Manufacturer

Dr König Company, Dietikon, Switzerland.

VERIFIED

Type 88 field shelter assembled in above ground position for demonstration purposes (T J Gander)

UNITED KINGDOM

Hesco Bastion Concertainer Defence Wall

Description

The Hesco Bastion Concertainer Defence Wall is a rapidly built solid blast barrier wall constructed from heavy-duty corrosion- and rot-proof materials, supplied in a collapsed state for easy transport.

A standard NATO pallet holds five Concertainer walls, each of which, when erected and filled, is 10 m long, 1.37 m high and 1.067 m wide. As supplied, the walls have no top or bottom and are constructed from heavy-duty wire mesh galvanised for corrosion protection. A standard wall has nine compartments formed by panels joined together with spiral spring hinges, also in galvanised wire. The wire sides of each panel are lined with a geo-membrane approximately 0.5 mm thick. Linings are available in camouflage colours to suit arctic, desert, jungle or European environments. A bag of accessories is also supplied.

Once removed from a pallet, collapsed walls are laid horizontally on the ground. Using handles at one end, two personnel can then pull the wall in the desired direction. As the wall is pulled it will concertina from the horizontal to the vertical and become self-supporting. The wall can then be pulled to its maximum length of 10 m for alignment and levelling. Longer walls can be produced by joining erected units together using wire from the accessory bag. Each nine-compartment wall can be shortened or split along the panel lines, before infill, into two or more smaller walls. It is possible to produce walls higher than the standard 1.37 m. Once erected, a 10 m Concertainer can be filled with 35 tonnes of infill material, such as sand, mud, gravel, soil or snow, in approximately 17 minutes, using a four-in-one mechanical digger.

Hesco Bastion Concertainer Defence Wall in use during Operation Granby

The protection provided by the Concertainer has been tested using small arms, machine guns, mortars and 30 mm RARDEN cannon. During operations in the Gulf during 1991 they were used extensively by the Royal Air Force for the protection of ammunition, fuel, stores and aircraft on four airfields. Concertainers can also be used for the rapid construction of defence or logistic shelters and reservoirs and have been employed as emergency flood barriers.

Status
In production. In service with the Royal Air Force, Denmark, Netherlands, UN Protection Forces (UNPROFOR) in the former Yugoslavia and with armed forces throughout the Middle East. Purchased by the US Army Waterways Experimental Station, Vicksburg.

Manufacturer
Hesco Bastion Limited, Unit 37, Knowsthorpe Gate, Cross Green Industrial Estate, Leeds LS9 0NP, UK.
Tel: +44 113 2486633. Telex: 94016703 hesi g.
Fax: +44 113 2483501.
Marketed in North America by: IDEEA Inc, 6233 Nelway Drive, McLean, Virginia 22101, USA.
Tel: +1 703 760 0762. Fax: +1 703 760 0764.

VERIFIED

JCB backhoe loaders

Description
JCB backhoe loaders are in use worldwide for a number of military and commercial purposes as they have proved to be compact, manoeuvrable and ruggedly built. They can combine the capabilities of a loader and excavator but both the loader arms and excavator arm can be fitted with special attachments to perform the functions of a forklift, crane, grader, bulldozer, rock breaker, grab and other special functions. All can be remotely attached from the cab using the JCB Hydraulic Quick Hitch.

Armoured cabs can be provided for front-line operations such as digging vehicle scrapes, combat trenches, command and first-aid posts. All controls can be operated from within the cab. With attachments fitted, JCB backhoe loaders can assist with other combat duties such as clearing debris, repairing roads and runways, mine clearance, bomb disposal and vehicle recovery.

Military models have been supplied to meet air-portability requirements and some have been specially prepared for air-drops with fold-down cabs. Remote-control facilities can be fitted.

Naturally aspirated or turbocharged engines developing up to 98 hp are available to provide road speeds of up to 32 km/h.

Equal size wheels with two-wheel, four-wheel and crab steering are available.

JCB backhoe loader

Specifications
Backhoe
Dig depth: up to 6 m
Reach at ground level: (to rear wheel centre) up to 8 m
Bucket width: up to 1.1 m
Lift capacity: (excluding bucket) up to 1,243 kg

Loader
Bucket capacity: up to 1.2 m^3
Loadover height: up to 3.1 m
Forward reach at ground level: up to 1.5 m

Forklift
Fork capacity: up to 2,000 kg
Lift height: up to 2.9 m
Forward reach at ground level: up to 2.7 m

Status
In production. In service with the British Army, the Royal Air Force and the defence forces of more than 20 countries.

Manufacturer
JCB Military Products Division, Rocester, Staffordshire ST14 5JP, UK.
Tel: +44 1889 590312. Telex: 36372.
Fax: +44 1889 590588.

VERIFIED

Case 721 wheeled loader

Description
In March 1992, the UK Ministry of Defence placed a £5.6 million order for 63 Model 721 wheeled loaders with J I Case Europe Limited. In service, the machines' primary functions are earthmoving and excavations, but they have also been used for general material handling duties. They have been deployed in Bosnia, Germany and the UK.

The 721s involved in the order were essentially standard production machines customised by Case to British Army specifications. These included the fitment of a Boughton 2.1 m^3 multipurpose four-in-one bucket, front loader mounted ripper and pallet fork attachments and a 10 tonne capacity hydraulic winch on the rear of the chassis. The hydraulic systems were adapted to supply power on demand for hand operated tools, including a jack and bead breaker used with a loader mounted tyre handling system.

Case 721 wheeled loader with accessories ***1996***

Specifications
Seating: 1
Configuration: 4 × 4
Weight: (operating) 15,140 kg
Length: (bucket down) 8.52 m
Width: (over tyres) 2.76 m
Height: 3.25 m
Ground clearance: 0.37 m
Wheelbase: 1.93 m
Max speed: 36 km/h
Fuel capacity: 235 litres
Engine: Case 6T-830 turbocharged diesel developing 157 hp at 2,100 rpm
Transmission: ZF 4WG 180 power-shift with 4 forward and 3 reverse gears
Steering: hydrostatic
Brakes: disc, dual circuit
Tyres: 20.5 × 25
Electrical system: 24 V
Batteries: 2 × 12 V
Alternator: 65 A

Status
In production. In service with the British Army.

Manufacturer
Case Corporation. Customised and supplied by Case UK Limited, PO Box 121, Wheatley Hall Road, Doncaster, South Yorkshire DN2 4PN, UK.
Tel: +44 1302 733393. Fax: +44 1302 760314.

VERIFIED

Case 721BXT wheeled loader

Description

To meet the demanding specifications of the Royal Marines, Case customised 17 of its 721BXT wheeled loaders to operate in water up to 1.5 m deep. Eight were also winterised for operations in Arctic climates. Extensive tests and trials under simulated operational conditions were necessary as the machines are transported by seagoing vessels and then by landing craft to carry out applications such as the preparation of beachheads and laying trackway. In most sea to shore landings the machines will have to wade through varying depths of water.

When the latest order for a further 19 of these units was completed in 1996, the British Army became the largest single user of Case wheeled loaders (99, including the Case 721 wheeled loaders detailed in the previous entry) in Europe.

Case 721BXT wheeled loader with forklift carrying trackway ***1996***

Specifications

Seating: 1
Configuration: 4 × 4
Weight: (operating) 14,860 kg
Length: (bucket down) 8.215 m
Width: (over tyres) 2.5 m
Height: 3.25 m
Ground clearance: 0.37 m
Wheelbase: 2.9 m
Max speed: 36 km/h
Fuel capacity: 235 litres
Engine: Case 6T-830 turbocharged diesel developing 157 hp at 2,100 rpm
Transmission: ZF 4WG 180 power-shift with 4 forward and 3 reverse gears
Steering: hydrostatic
Brakes: disc, dual circuit
Tyres: 20.5 × 25
Electrical system: 24 V
Batteries: 2 × 12 V
Alternator: 65 A

Status

In production. In service with the British Army and Royal Marines.

Manufacturer

Case Corporation. Customised and supplied by Case UK Limited, PO Box 121, Wheatley Hall Road, Doncaster, South Yorkshire DN2 4PN, UK.
Tel: +44 1302 733393. Fax: +44 1302 760314.

UPDATED

Case 660 trencher

Description

Eight Case 660 trenchers were supplied to the British Army for use by the Royal Corps of Signals primarily for the laying of communication cables.

The machines are capable of accepting a combination of up to four different attachments at the same time. These include a dozer blade and backhoe on the front, plus a trencher and mole plough to the rear. Other attachments included in the Army specification include a rockwheel, a cable reel carrier, and compactors.

The machines are carried on a dedicated trailer towed by a 4 tonne truck.

Case 660 trencher ***1996***

Specifications

Seating: 1
Configuration: 4 × 4
Weight: (base unit including backhoe, blade and trencher) 5,710 kg
Length: (with backfill blade and trencher attachment) 4.738 m
Width: (over backfill blade) 1.778 m
Height: 2.323 m
Ground clearance: 0.188 m
Wheelbase: 1.65 m
Max speed: 11.9 km/h
Fuel capacity: 64.3 litres
Engine: Case 4-390 diesel developing 56 hp at 2,100 rpm
Transmission: hydrostatic via variable displacement axial piston pump
Steering: hydrostatic 4-wheel
Brakes: neutral position of hydrostatic drive with engine running. Parking via foot-operated pedal with disc and caliper on transmission
Tyres: 31 × 15.5 or 35 × 19.00 - 16.1
Electrical system: 12 V
Batteries: 1 × 12 V
Alternator: 65 A

Status

In production. In service with the British Army.

Manufacturer

Case Corporation. Customised and supplied by Case UK Limited, PO Box 121, Wheatley Hall Road, Doncaster, South Yorkshire DN2 4PN, UK.
Tel: +44 1302 733393. Fax: +44 1302 760314.

VERIFIED

JCB Mini Excavators

Description

JCB Mini Excavators in a range of sizes are effective, easily transported digging and dozing equipments with many field engineering applications, including Explosive Ordnance Disposal (EOD) duties.

Optional items available include a range of bucket sizes, hydraulic breaker and pipework, fully glazed or canopy cab, steel or rubber tracks and standard or military trailers.

Status

In production. In service with the British Army.

Manufacturer

JCB Military Products Division, Rocester, Staffordshire ST14 5JP, UK.
Tel: +44 1889 590312. Telex: 36372.
Fax: +44 1889 590588.

VERIFIED

JCB 801 Mini Excavator on its transporter trailer

Boughton Ulrich Simplex multipurpose bucket

Description

The Boughton Ulrich Simplex multipurpose bucket is available for most makes of wheeled or crawler loader in capacities from 0.6 to 5.5 m³ and can be mounted either direct on the loader arms or via a quicker-coupler attachment.

Every bucket is designed to match the requirement of each loader model to dig and load, forward dump, doze, back grade, grab, scrape and bottom dump a wide range of materials with maximum efficiency. The buckets are manufactured entirely from high-tensile steel with high abrasion resistant cutting edges. The bucket back is fitted with a bolt-on reversible dozer edge and bolt-on dozing skid plates.

Status

In production. In service with the British Army, US Marine Corps, US Air Force, and other armed forces.

Manufacturer

Reynolds Boughton Limited, Engineering Division, Bell Lane, Amersham, Buckinghamshire HP6 6PE, UK.
Tel: +44 1494 764411. Telex: 83132.
Fax: +44 1494 765218.

VERIFIED

Boughton Ulrich Simplex 1.72 m³ multipurpose bucket fitted to Volvo 4400 wheeled loader

Boughton Ulrich Simplex 2.1 m³ multipurpose bucket fitted to Case 721 medium wheeled tractor in service with the British Army

---

Boughton ripper scarifiers and quick couplers

Description

Reynolds Boughton Limited manufactures ripper scarifiers for direct or quick coupler mounting to most makes of wheeled loaders. They can be used to break up asphalt, unreinforced concrete or frozen ground for route denial or during fortification preparations.

Reynolds Boughton also produces hydraulic and manual quick couplers which allow the ripper scarifiers and other front end attachments to be changed within minutes on most makes of wheeled loader.

Status

In production. In service with the British Army.

Manufacturer

Reynolds Boughton Limited, Engineering Division, Bell Lane, Amersham, Buckinghamshire HP6 6PE, UK.
Tel: +44 1494 764411. Telex: 83132.
Fax: +44 1494 765218.

VERIFIED

Boughton ripper scarifiers fitted to Case 721 medium wheeled tractor in service with the British Army

Typical Boughton quick couplers which allow various front end attachments to be fitted to most makes of wheeled loader

Heywood Williams Field Shelter MEXE Mk III

Description

The Field Shelter MEXE Mk III (or Mexeshelter) was developed by the Military Engineering Experimental Establishment at Christchurch (now the RARDE Christchurch), with production and marketing being undertaken by Heywood Williams Limited. The shelter provides protection against conventional attack, nearmiss nuclear attack and protection against chemical and bacteriological attack. Protection against the first of these two categories is given by the standard shelter with protection against chemical and bacteriological attack being provided by a kit described below.

The Mk III was originally conceived as an underground shelter for Command Posts (CPs), Observation Posts (OPs) and Regimental Aid Posts (RAPs), for use in forward and rear area tactical situations, and for the underground protection of personnel on air stations and in areas not directly related with the tactical battlefield itself. It is also used as a casualty clearing and first aid station, storage for medical supplies, including plasma, drugs, dressings; fuel, food and other combat support items; and as temporary magazines for ordnance items. In the CP role it is used at battalion and divisional level.

The Mk III consists of four components plus accessories: three metal structural members, pickets, spacers, arches; plus flexible revetting material to clothe the metal structural members which is held in place by earth backfill. All components are packaged in manportable kits and can be installed by unskilled personnel with no special tools; all work can be accomplished with a pick and shovel. With the use of mechanical equipment such as a bulldozer, scraper or shovel, approximately 2 hours are required to fully install the Mk III. In addition to the mechanical equipment, approximately 15 hours of troop labour are required.

The flexible revetting material used to line the walls and roof of the shelter is a PVC-coated jute fabric reinforced with wire. The PVC layer is impermeable and provides an obstacle to chemical agent penetration. When combined with the additional barrier of 450 to 600 mm of soil over the shelter roof, it is likely that there will be no hazard in the shelter even with heavy chemical agent contamination on the ground outside. The installation of an NBC filtration unit and liner as outlined below, provides the modifications necessary to give chemical protection without the use of respirators and NBC clothing.

The liner, including the airlock, is constructed of translucent reinforced plastic and folds down in similar fashion to a tent. It is self-supporting, even when there is no inflation from the AFU and takes approximately 6 minutes to inflate. The airlock has easy access 'fold-away' doors and adjustable ports to vary the liner/airlock pressure. The liner was developed by Driclad Limited of Sittingbourne, Kent, and liners for other configurations of shelter can be made available.

Status

Production complete. In service with British, Canadian, Kuwaiti and Saudi Arabian armed forces.

Manufacturer

Heywood Williams Limited, Military Engineering Division, Bayhall Works, Huddersfield, West Yorkshire HD1 5EJ, UK.
Tel: +44 1484 710111. Telex: 51544.

VERIFIED

MEXE shelter covering prior to earth backfill

Assembling MEXE shelter and showing general construction method

UNITED STATES OF AMERICA

Caterpillar Deployable Universal Combat Earthmover (DEUCE)

Development

The Caterpillar Deployable Universal Combat Earthmover (DEUCE - originally known as the Caterpillar 30/30 Engineer Support Tractor) was developed as a private venture to act as a high-speed combat support machine intended for use in several forward area roles. Following an initial US$3.3 million development contract issued during 1995, two test machines were delivered to the US Army in June 1996. An initial order for 15 units with a total value of US$8.7 million was placed during late 1996, with the first examples delivered during May 1997. US Army initial requirements were for 45 units with a total requirement understood to be 184 units until 2003.

Caterpillar Deployable Universal Combat Earthmover (DEUCE)

Description

The Caterpillar DEUCE is a self-deployable unit capable of traversing rough terrain and is fitted with a hydraulically operated angled dozer blade at the front. The driver/operator is seated in an air conditioned forward control cab fitted with bulletproof glass windows. An automatic fire detection and suppression system is provided. About 70 per cent of DEUCE components are standard commercial items.

The tracks use the Caterpillar Mobil-trac System with a rubber track hydropneumatic undercarriage that allows on- and off-road travel at speeds of up to 53 km/h. The sound signature is stated to be low.

Power is provided by a Caterpillar 3126 7.2 litre diesel coupled to a Caterpillar planetary power shift. The engine has two outputs, 265 hp for the self-deployable mode for travelling on roads or cross-country and 185 hp for earthmoving. Sufficient fuel is carried for 10 hours of earthmoving or 322 km of road travel.

For drive-on, drive-off transport in a C-130 aircraft, the overall height can be lowered by tilting forward the spotlamp array over the cab. For low velocity air drops the entire cab shell can be removed to lower the overall height to 2.286 m.

Accessories include a 6,804 kg capacity winch, pintle hook, towing lugs and a hydraulic tool connection.

Specifications

Cab seating: 1
Weight: 15,513 kg

Length:
(dozer blade angled) 6.426 m
(dozer blade straight) 5.842 m
Width:
(dozer blade angled) 2.718 m
(dozer blade straight) 2.946 m
Height:
(working) 2.718 m
(air transport) 2.565 m
(air drop) 2.286 m
Ground clearance: 0.406 m
Track: 1.829 m
Length of track on ground: 2.438 m
Max speed:
(self-deployable) 53 km/h
(earthmoving) 11.7 km/h
(reverse, both modes) 20.3 km/h
Range: (roads) 322 km
Gradient: 60%
Side slope: 30%
Vertical obstacle: 0.305 m
Fording: 0.914 m
Engine: Caterpillar Model 3126 7.2 litre diesel developing 265 hp (self-deployable) or 185 hp (earthmoving)
Transmission: Caterpillar planetary power shift automatic providing (self-deployable) 6 forward and 2 reverse gears or (earthmoving) 3 forward and 2 reverse gears
Suspension: (self-deployable) hydropneumatic with 254 mm vertical travel; (earthmoving) rigid
Electrical system: 24 V

Status
In production for US Army.

Manufacturer
Caterpillar Inc, Defense & Federal Products Department, PO Box 470, Mossville, Illinois 61552-0470, USA.
Tel: +1 309 578 0470. Fax: +1 309 578 3739.

UPDATED

Freightliner Small Emplacement Excavator (SEE)

Development
The Small Emplacement Excavator (SEE) programme was initially undertaken by Euclid Inc, a subsidiary of Mercedes-Benz AG; this concern was taken over by the Freightliner Corporation. Based on a modified Unimog U 900, the SEE underwent four years of testing and evaluation with the US Army and in 1985 Freightliner was awarded a five-year contract for 922 vehicles with a 200 per cent overbuy option. The first four vehicles were handed over to the US Army in July 1985. Some 170 SEEs were issued to the US Marine Corps.

With the US Army the SEE is issued to all active and reserve forces and to the National Guard, replacing John Deere 410 tractors on a one-to-one basis from Fiscal Year 1987 onwards. Production is complete.

Description
The SEE is based on a modified Mercedes-Benz Unimog U 900 chassis with a 2.38 m wheelbase. This chassis can be provided with a variety of equipment for digging, loading, lifting, trenching, grading and powering hydraulic tools. The standard attachments are a front end loader and backhoe. Typical power tools that can be powered by the SEE include a chainsaw, a hammer drill and a concrete breaker. The tractor can be equipped with quickly interchangeable attachments such as a backhoe-mounted breaker and tamper for rapid runway repair, front-mounted sweeper, snow-plough and a winch. In addition the tractor provides a platform for shelters and other functions.

Data for the SEE is provided in the Specifications table but data for the standard attachments is as follows.

Side view of Small Emplacement Excavator (SEE) (Michael Jerchel)

Front End Loader
Bucket width: 2.07 m
Lift height: 2.5 m
Breakout force: 2,722 kg
Lift capacity: 1,497 kg
Bucket capacity: 0.57 m³

Backhoe
Bucket capacity: 0.2 m³
Digging depth: 4.26 m
Digging radius: 5.39 m
Loading height: 3.35 m

Swing arc: 180°
Digging force: 4,536 kg

Specifications
Cab seating: 1 + 1
Configuration: 4 × 4
Max weight: 7,250 kg
Length: 6.35 m
Width: 2.44 m
Height: 2.58 m
Ground clearance: 0.43 m
Track: 1.63 m
Wheelbase: 2.38 m
Angle of approach/departure: 40°/34°
Max speed: (road) 80 km/h
Fuel capacity: 114 litres
Max gradient: 60%
Side slope: 30%
Fording: 0.76 m
Engine: OM 352 4-stroke direct injection diesel developing 110 hp at 2,800 rpm
Transmission: fully synchromesh, 16 forward and 8 reverse gears
Clutch: single dry plate
Steering: hydraulic power-assisted
Turning circle: 10.9 m
Suspension:
(front) coil springs, telescopic shock-absorbers and stabiliser
(rear) coil springs with helper springs, telescopic shock-absorbers and stabiliser
Tyres: 12.5 × 20 12PR MPT
Brakes: hydraulic, dual-circuit disc
Electrical system: 24 V with blackout drive
Batteries: 2 × 12 V, 100 Ah

Status
In service with US Army and Marine Corps (2,206 delivered).

Manufacturer
Freightliner Corporation, 4747 North Channel Avenue, PO Box 3849, Portland, Oregon 97208-3849, USA.
Tel: +1 503 735 8000.

UPDATED

Overhead Foxhole Cover (OFC)

Description
The Overhead Foxhole Cover (OFC) is a US Army standard issue item and is carried and used by an individual soldier. A mylar-coated polyester fabric sheet, it serves to protect a soldier using a foxhole from the effects of the weather. It provides no other form of protection. Along each side of the cover, which is 2.34 m long and 1.625 m wide, are long pockets which are filled with spoil as the foxhole is excavated. The cover is then placed over the foxhole and layered with spoil to camouflage the emplacement. Setting-up takes 5 minutes.

Status
In service with the US Army.

VERIFIED

Demolition Kit, Blasting: Lightweight Fighting Position Excavator (FPE) M300 (Badger)

Development
The M300, or 'Badger' Lightweight Fighting Position Excavator (FPE) System, was developed by Alliant Techsystems Inc in response to a US Army need for a lightweight, simple and reliable system to rapidly create a two-person fighting position. Technical and operation tests were conducted by the US Army during the second half of 1995 in various locations, climates and soil types.

Description
The M300 involves a hand auger and a soil-loosening charge. The steel bucket auger is a lightweight military version of a design used in forestry. The soil-loosening charge is a high energy binary explosive widely used in the mining industry and is completely inert during storage and transport.

The auger is used to bore two 76 mm diameter holes about 1 m apart and 1 m deep; each hole takes from 1 to 5 minutes to bore, depending on soil conditions. The binary explosive is then mixed in its container and one container is placed into each hole. The explosive is detonated from a safe position more than 20 m away using a Primaline connected to a standard detonator. All components of the initiation system are either type classified by the US military or are widely used commercial explosives.

The detonation of the explosives fractures and loosens the soil, which can then be cleared using a standard entrenching tool, resulting in a two-person combat position measuring approximately 1.8 m long, 0.6 m wide and 1.2 m deep. A complete position can be created in less than one hour.

A complete M300 kit can be carried in a soldier's pack and weighs approximately 3.8 kg. M301 reload kits are available, carried in identical packs. Each reload kit contains all consumable items needed to replenish the M300, either in the field or in a supply depot, allowing multiple use of the M300 kit.

Status

In production for US Army.

Manufacturer

Alliant Techsystems Inc, 600 Second Street Northeast, Hopkins, Minnesota 55343, USA.
Tel: +1 612 931 5295. Fax: +1 612 931 6654.

UPDATED

Preparing a combat position using the Demolition Kit, Blasting: Fighting Position Excavator M300 (Badger)
1996

BULK FUEL STORAGE AND DISTRIBUTION SYSTEMS

AUSTRALIA

Tank and pump unit 4.5 m³ capacity

Description

The Australian Army tank and pump unit is based on two in-service 1,892.5 litre tanks and a 38 mm pump. The Australian requirement was for fitment to a 5,000 kg truck. An additional Australian provision was for the use of the equipment in the static role, for example with tanks mounted on a stand.

Procurement detailing began in 1970 and actual procurement was recommended in August 1975. Trials and testing were conducted during 1976 and, as a result, modified units were tested early in 1978.

Tank and pump unit 4.5 m³ capacity mounted on rear of 5,000 kg (6 × 6) truck (Australian Army)

Manufacture of production models began in June 1978 and was completed by December 1978.

The equipment consists of two 2.25 m³ aluminium tanks and a 30 mm transfer pump. The two tanks and pump unit are fitted on a common base suitable for mounting on a 5,000 kg truck. The pump is driven by a diesel engine identical to the 38 mm general purpose fuel pump. Three types of pump unit are based on functional needs:

Aviation units: the pump is fitted with a filter, separator, air eliminator, flow meter and one hose complete with nozzle and hose reel, all transportable on a 5,000 kg truck.

RAAC: the pump unit is fitted with two hoses complete with nozzles and reels. Transportability as for aviation units.

RAE units: the pump is fitted with two hoses complete with nozzles and reels. The equipment is carried on a 5,000 kg truck, or two tanks can be fitted on separate tank stands. Gravity flow fills the vehicles whereas the pump unit is used to fill the tanks on the stands.

Status

In service with the Australian Army.

VERIFIED

AUSTRIA

Steyr fuel tankers

Description

Steyr produces a wide range of fuel tankers based on 4 × 4 and 6 × 6 vehicles. The usual capacity for 4 × 4 medium trucks is 6,000 litres (12M and 14M series), 11,000 litres for the 4 × 4 heavy truck (19S series) and 20,000 litres for the 6 × 6.

Vehicles are available with power take-off driven centrifugal pumps for unloading, metering and self-loading operations. Water tankers with a special plastic interior coating are also available.

Specifications (typical)

Chassis type	4 × 4	4 × 4	6 × 6
Length overall	7 m	7.5 m	9 m
Length of tank	4 m	5 m	6 m
Width	2.4 m	2.4 m	2.4 m
Height	2.9 m	3 m	3 m
Capacity	6,000 litres	11,000 litres	20,000 litres

Status

In production. In service with Canadian Forces, Ghana, Nigeria and Saudi Arabia.

Manufacturer

Steyr Nutzfahrzeuge AG, Schönauerstrasse 5, A-4400 Steyr, Austria.
Tel: +43 7252 25351-0. Telex: 28200.
Fax: +43 7252 26746, 28650.

VERIFIED

Steyr 32S29/6 × 6 20,000 litre fuel tanker showing pumping and metering unit

Steyr 19S25/4 × 4 11,000 litre fuel tanker

CANADA

SEI Industries tank and pump systems

Description

SEI Industries is a single source for the engineering, manufacturing and on-site installation of fuel and water distribution systems anywhere in the world. The company has completed tank farm designs and installations for the UN, the Canadian Department of National Defense and oil and gas operations in Burma (Myanmar), Ecuador, Nigeria and other countries. Applications have included fuel storage and supply in remote areas and support of drilling operations and fuel logistics in Arctic areas.

Status

In production – see text.

Manufacturer

SEI Industries, 7400 Wilson Avenue, Delta, British Columbia, Canada V4G 1E5.
Tel: +1 604 946 3131. Fax: +1 604 940 9566.

VERIFIED

SEI Industries filtration and metering skid (foreground) and pump skid and tank (background) ***1996***

SEI Industries Onion Tank with a foam flotation collar ***1996***

SEI Industries Forward Area Refuelling Equipment (FARE) pump unit ***1996***

SEI Industries Terra Tanks

Description

The SEI Industries Terra Tank system covers a range of collapsible pillow tanks ranging in capacity from 100 US gallons (380 litres) up to 50,000 gallons (189,300 litres); larger sizes can be produced. Versions are produced for liquids such as petrol (Petro Guard), water (Aqua Guard) and chemicals (Chem Guard), all with coatings suitable for the liquid involved.

Petro Guard tanks are constructed using urethane-coated nylon fabric which meets and exceeds US Military specifications. Corrosion-proof fittings are available for tanks to be used with volatile solutions.

Each tank is provided with a fill/drain fitting plus a second fitting at the centre of the tank for vent/overfill protection. A wide range of other valves, flanges and fittings in several sizes can be supplied. Optional items include camouflage covers, sunshades, berm liners, ground sheets, flow meters and pumping systems.

A typical SEI Industries Terra Tank installation supporting helicopter operations

Specifications

Capacity	Approximate dimensions filled:	Width	Length	Weight empty
100 US gal	380 litres	1.1 m	1.4 m	34 kg
250 US gal	945 litres	1.3 m	2.5 m	42 kg
500 US gal	1,890 litres	1.9 m	2.5 m	48 kg
750 US gal	2,840 litres	2.4 m	2.4 m	52 kg
1,000 US gal	3,785 litres	2.2 m	3.9 m	62 kg
1,500 US gal	5,670 litres	2.7 m	3.8 m	68 kg
2,000 US gal	7,570 litres	3.1 m	3.7 m	77 kg
2,500 US gal	9,460 litres	3.5 m	3.7 m	83 kg
3,000 US gal	11,355 litres	3.2 m	5.2 m	97 kg
4,000 US gal	15,140 litres	3.8 m	5 m	102 kg
5,000 US gal	18,295 litres	4.3 m	4.9 m	117 kg
7,500 US gal	28,380 litres	4.7 m	6.3 m	151 kg
10,000 US gal	37,850 litres	5.5 m	6.2 m	169 kg
15,000 US gal	56,775 litres	7.2 m	7.5 m	197 kg
20,000 US gal	75,710 litres	6.8 m	8.9 m	273 kg
50,000 US gal	189,300 litres	8 m	17.5 m	564 kg

Status
In production.

Manufacturer
SEI Industries, 7400 Wilson Avenue, Delta, British Columbia, Canada V4G 1E5.
Tel: +1 604 946 3131. Fax: +1 604 940 9566.

VERIFIED

SEI Industries Fuel-Easy helicopter-transportable tank system

Description
The SEI Industries Fuel-Easy helicopter-transportable tank system is a fully collapsible, modular external load fuel carrier intended to support helicopter operations. The system involves a flexible fuel container carried inside a frame which remains aerodynamic during flight, even when partially filled and is stable on the ground. Filling is via the top of the tank and emptying through the bottom, so utilising all available volume.

Fuel-Easy tanks are available in seven sizes, from 120 US gallons (454 litres) to 1,260 US gallons (4,775 litres). Each tank is cradled within a collapsible high-strength aluminium frame (steel and bronze frames are available) which can be assembled without tools in about 5 minutes. The tank fabric is puncture-resistant but can be repaired easily using a field repair kit.

Special Fuel-Easy tanks are available for chemicals or potable water.

A partially filled SEI Industries Fuel-Easy tank slung under a helicopter

Specifications

Model	Capacity US gallons	Litres	Weight (approx) Gross	Empty
F-100	120	454	415 kg	35 kg
F-150	180	680	610 kg	41 kg
F-250	300	1,140	1,010 kg	68 kg
F-430	515	1,955	1,710 kg	85 kg
F-775	930	3,525	3,055 kg	100 kg
F-950	1,140	4,320	3,730 kg	110 kg
F-1050	1,260	4,775	4,135 kg	140 kg

Status
In production.

Manufacturer
SEI Industries, 7400 Wilson Avenue, Delta, British Columbia, Canada V4G 1E5.
Tel: +1 604 946 3131. Fax: +1 604 940 9566.

VERIFIED

FRANCE

Trailor 5,000 litre bowser equipment for AFVs

Description
The Trailor 5,000 litre bowser equipment was developed specifically for refuelling tanks and other AFVs in the field and can refuel up to four vehicles at one time. The tank used has an elliptical outline and is manufactured from steel with a Rhomeyte interior lining. The tank has two compartments, one with a capacity of 2,000 litres and the other of 3,000 litres and may be filled from either top or bottom. Each compartment has a manhole cover on top and is fitted with safety vents for internal vacuum or pressure. Access to the top of the tank is via ladders and there is a working platform covered with anti-skid and anti-spark coatings. The distribution panel at the rear is under a cover which hinges upwards and locks in place. Inside the cover is a pump driven from the vehicle gearbox power take-off and this embodies a bypass set for an output of 20,000 litres/h. Refuelling can be carried out with or without metering.

The full control panel contains a meter, two valves, manifold linked to four hose reels each with a hose 20 m long (each hose has a diameter of 30 mm and is fitted with a nozzle), two-way valve for self-filling or bottom filling, three-way valve, three cabinets for 80 mm diameter hoses, accessories and tools.

The complete equipment may be transported by rail or by C-160 Transall transport aircraft. Cross-country tyres are fitted and the entire equipment may be covered by a full tarpaulin camouflage cover. It is normally carried on a 6 × 6 truck chassis.

Status
In service with the French Army.

Manufacturer
Trailor SA, 5 route Nationale 10, BP 49, F-78311 Coignières, Maurepas Cedex, France.
Tel: +33 1 30 50 61 26. Telex: 698896 f.
Fax: +33 1 30 50 31 32.

VERIFIED

Distribution panel for Trailor 5,000 litre AFV bowser

Superflexit flexible storage and transport tanks

Description

The Superflexit Division d'AERAZUR manufactures a range of flexible tanks for transport (capacity 500 to 20,000 litres) and for storage (capacity 500 to 300,000 litres). They are made of high-performance synthetic fabrics coated with elastomer to store fuel and other liquids such as water as well as chemicals. The tanks are provided with standard equipment according to French and US specifications but can be equipped to suit customer requirements. All the tanks in the range can be folded when empty and are easy to set up on site. They are suitable for a wide range of climatic conditions.

Status

In production. In service with the French and other armed forces.

Manufacturer

Superflexit Division d'AERAZUR, 58 boulevard Galliéni, F-92137 Issy-les-Moulineaux Cedex, France.
Tel: +33 5 45 54 92 80. Telex: 270887 f.
Fax: +33 5 45 54 23 55.

VERIFIED

Superflexit fuel transport tank carried by French Army Berliet GBC 8 KT (6 × 6) 4,000 kg truck

Specifications

Capacity	25,000 litres	80,000 litres	150,000 litres	300,000 litres
Weight (empty)	124 kg	287 kg	480 kg	878 kg
Volume (folded)	0.43 m³	0.76 m³	1.25 m³	2.81 m³
Length (empty)	8.1 m	9.1 m	12 m	17.65 m
(full)	7.75 m	9.05 m	11.15 m	17.25 m
Width (empty)	4.25 m	8.2 m	11.5 m	13.7 m
(full)	3.7 m	7.85 m	11.15 m	13.5 m
Height (full)	1.1 m	1.3 m	1.35 m	1.4 m

Superflexit helicopter-transportable tanks

Description

These tanks are used to transport fuel or drinking water and can be slung under helicopters. They are manufactured from synthetic fabric coated on each side with elastomer and have capacities of 500, 1,000 and 1,500 litres. A standard fitting is a 76 mm fill/drain flange on the top but other equipment can be fitted to suit customer specifications.

Status

In production. In service with the French and some other armed forces.

Manufacturer

Superflexit Division d'AERAZUR, 58 boulevard Galliéni, F-92137 Issy-les-Moulineaux Cedex, France.
Tel: +33 5 45 54 92 80. Telex: 270887 f.
Fax: +33 5 45 54 23 55.

VERIFIED

Superflexit helicopter-transportable tank

Superflexit 20 litre air-droppable fuel containers

Description

Superflexit manufactures a 20 litre flexible fuel container that can be dropped in free fall from helicopters or aircraft flying at altitudes up to 2,000 m. The containers may be dropped individually or in clusters of 20 from a tilting platform.

They may also be used to carry drinking water.

Status

In production. In service with the French and other armed forces.

Manufacturer

Superflexit Division d'AERAZUR, 58 boulevard Galliéni, F-92137 Issy-les-Moulineaux Cedex, France.
Tel: +33 5 45 54 92 80. Telex: 270887 f.
Fax: +33 5 45 54 23 55.

VERIFIED

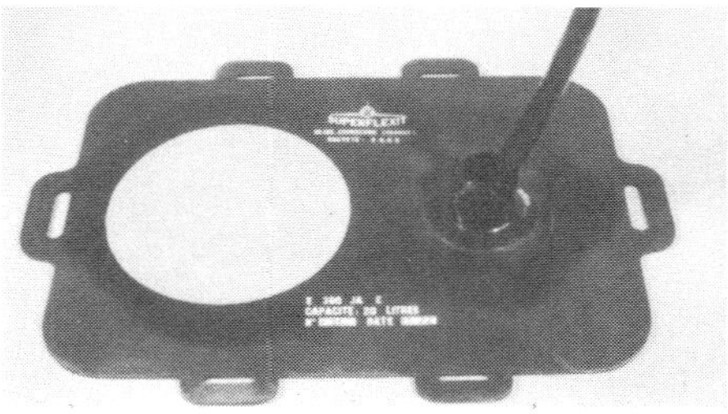

Superflexit 20 litre air-droppable fuel container

Superflexit 20 litre flexible jerrican

Description

Superflexit 20 litre flexible jerricans are containers for fuel or water that require minimum space when empty and are self-supporting when full. The containers are manufactured from rot-proof materials using flexible coated fabric resistant to hydrocarbons. They have been designed for long storage periods and when full can withstand being dropped from heights of up to 10 m. The filler hole allows the use of flexible metallic pouring adaptors used by the French Army. The carrying handles can be arranged to ensure the container remains fully closed up when folded.

Status

In production. In service with the French Army.

Manufacturer

Superflexit Division d'AERAZUR, 58 boulevard Galliéni, F-92137 Issy-les-Moulineaux Cedex, France.
Tel: +33 5 45 54 92 80. Telex: 270887 f.
Fax: +33 5 45 54 23 55.

VERIFIED

Full Superflexit 20 litre flexible jerrican

Superflexit Supessence 100 litre fuel tank

Description

This self-contained 100 litre fuel tank is intended to increase the range of cargo- or troop-carrying vehicles by a factor of about 3.5. The principle is that each vehicle can carry up to four of these tanks hooked on the vehicle side rails. Each tank can be placed in position by two personnel using the integral carrying handles provided and the tank can be filled through a filling port on the top of the tank. The tank is emptied by using a drain hose placed directly into the vehicle main tank.

The tanks are manufactured using a coated fabric resistant to hydrocarbons with an interchangeable abrasion-resistant cover on the side next to the carrier vehicle. The tank has a self-bracing system that allows it to hold its position when on the ground, even at decreased capacity. The drain hoses can be connected and disconnected to provide the correct refuelling lengths.

Status

In production.

Manufacturer

Superflexit Division d'AERAZUR, 58 boulevard Galliéni, F-92137 Issy-les-Moulineaux Cedex, France.
Tel: +33 5 45 54 92 80. Telex: 270887 f.
Fax: +33 5 45 54 23 55.

VERIFIED

Superflexit Supessence 100 litre fuel tank being carried by Renault TRM 2000 (4 × 4) 2,000 kg truck

PRONAL flexible storage and transport tanks

Description

PRONAL produces a wide range of flexible tanks for both storage and transport purposes and can produce these tanks to suit any customer requirement. For fuel storage purposes the tanks are constructed from a nitrile-coated nylon fabric. Storage tanks have a 2 in/51 mm inlet/outlet valve on one side and a centrally located vent valve. Transport tanks are similar but have a series of lashing points along each side to allow them to be carried on almost any type of truck.

The figures provided in the specifications table are not comprehensive, as the tanks can be produced to suit almost any customer requirement. Tanks up to 200,000 litres capacity can be produced.

Status

In production. In service with the Belgian, French and Swiss armed forces and some other nations.

Manufacturer

PRONAL SA, Z. I. Roubaix Est, BP 18, F-59115 Leers, France.
Tel: +33 3 20 99 75 00. Fax: +33 3 20 99 75 20.

VERIFIED

Typical PRONAL flexible fuel storage tank

Specifications

Flexible storage tanks

Capacity	Length (empty)	Width (empty)	Height (full)	Weight	Volume (folded)
1,000 litres	2.44 m	1.37 m	0.65 m	25 kg	0.15 m³
2,000 litres	2.4 m	2.12 m	1.1 m	35 kg	0.2 m³
5,000 litres	4.6 m	2.12 m	1.1 m	55 kg	0.25 m³
10,000 litres	8.2 m	2.12 m	1.1 m	85 kg	0.4 m³
15,000 litres	4.97 m	4.13 m	1.2 m	90 kg	0.45 m³
20,000 litres	6.29 m	4.13 m	1.2 m	115 kg	0.55 m³
25,000 litres	6.01 m	4.96 m	1.3 m	135 kg	0.65 m³
30,000 litres	6.99 m	4.96 m	1.3 m	155 kg	0.75 m³
35,000 litres	7.98 m	4.96 m	1.3 m	175 kg	0.85 m³
40,000 litres	8.96 m	4.96 m	1.3 m	190 kg	0.95 m³
45,000 litres	8.65 m	5.63 m	1.3 m	210 kg	1.05 m³
50,000 litres	9.49 m	5.63 m	1.3 m	230 kg	1.15 m³
60,000 litres	8.88 m	6.97 m	1.3 m	265 kg	1.3 m³
70,000 litres	8.47 m	8.14 m	1.35 m	300 kg	1.5 m³
80,000 litres	9.14 m	8.47 m	1.35 m	330 kg	1.65 m³
100,000 litres	11.15 m	8.47 m	1.35 m	420 kg	2.1 m³
125,000 litres	13.66 m	8.47 m	1.35 m	510 kg	2.55 m³
150,000 litres	16.17 m	8.47 m	1.35 m	595 kg	2.95 m³

Transport tanks

Capacity	Length (empty)	Width (empty)	Height (full)	Weight	Volume (folded)
1,000 litres	1.8 m	1.8 m	0.7 m	30 kg	0.15 m³
2,000 litres	2.4 m	2.12 m	1.1 m	40 kg	0.2 m³
3,000 litres	3.15 m	2.12 m	1.1 m	50 kg	0.25 m³
4,000 litres	3.85 m	2.12 m	1.1 m	60 kg	0.3 m³
5,000 litres	4.6 m	2.12 m	1.1 m	70 kg	0.35 m³
6,000 litres	5.3 m	2.12 m	1.1 m	80 kg	0.4 m³
7,000 litres	6 m	2.12 m	1.1 m	90 kg	0.45 m³
8,000 litres	6.75 m	2.12 m	1.1 m	100 kg	0.5 m³
9,000 litres	7.45 m	2.12 m	1.1 m	110 kg	0.55 m³
10,000 litres	8.2 m	2.12 m	1.1 m	120 kg	0.6 m³
11,000 litres	8.9 m	2.12 m	1.1 m	130 kg	0.65 m³
12,000 litres	9.65 m	2.12 m	1.1 m	140 kg	0.7 m³

PRONAL helicopter-transportable and towable flexible tanks

Description

PRONAL flexible tanks are drum-shaped tanks used to transport hydrocarbon fuels, oils or water. They are made of nylon fabric coated with synthetic nitrile rubber which is anti-ultraviolet radiation treated. There are three capacities - 200, 1,000 and 1,900 litres - with the material thickness of the two smaller tanks being 5 mm and the largest 9 mm.

The 200 litre tank can be dropped from a height of 12 m and, once on the ground, is rollable. The two largest tanks can be towed. All three can be carried slung under helicopters. The 200 litre tank has a 0.75 in/19 mm adaptor and vent, and a filling pipe with a 0.75 in/19 mm coupling and semi-symmetrical DN 40 connection or Kamlock coupling. The larger tanks have a self-plugging DN 50 adaptor and there is also a filling and emptying pipe 1.6 m in length - this is fitted with DN 40 Kamlock connections and caps. Also available for the larger tanks are towing, pushing or lifting bars and turning plates with lift rings.

Status

In production. In service with the French Army and some other nations.

Manufacturer

PRONAL SA, Z. I. Roubaix Est, BP 18, F-59115 Leers, France.
Tel: +33 3 20 99 75 00. Fax: +33 3 20 99 75 20.

VERIFIED

Typical PRONAL helicopter-transportable and towable flexible tank on tow behind a Land Rover (T J Gander)

1996

Specifications

Capacity	**200 litres**	**1,000 litres**	**1,900 litres**
Diameter	0.6 m	1.1 m	1.35 m
Length	1 m	1.3 m	1.6 m
Weight (empty)	20 kg	85 kg	135 kg
Dimensions folded	0.85 × 0.95 × 0.25 m	1.2 × 1.5 × 0.25 m	1.5 × 1.8 × 0.25 m
Working pressure	0.3 bar	0.3 bar	0.3 bar

PRONAL RSH helicopter-transportable tanks

Description

PRONAL RSH (RSH - *reservoirs souples heliportables*) tanks are produced in three sizes: 500, 1,000 and 1,500 litres. A patented design is used in which the tank is supported in a Maltese cross-shaped skirt that immediately spreads the load of the vehicle once it is lifted from its single clip-on ring. Once in flight, the skirt passes air through its panels and acts as a drogue to prevent the tank from revolving beneath the carrying helicopter. Each tank has one slinging ring, an outlet with a 2 in/51 mm valve coupling and plug, and a symmetrical Guillemin system. The tanks may be used to carry water.

Puma helicopter carrying PRONAL RSH tank

Specifications

Model	**0.5 RSH**	**1.0 RSH**	**1.5 RSH**
Capacity	500 litres	1,000 litres	1,500 litres
Weight (tank and fittings)	25 kg	32 kg	40 kg
Dimensions empty	1.4 × 1.4 × 0.15 m	1.8 × 1.8 × 0.15 m	2.2 × 2.2 × 0.15 m
Dimensions loaded (hanging)	1.01 × 1.01 × 1.2 m	1.4 × 1.4 × 1.55 m	1.7 × 1.7 × 1.9 m

Status

In production. Approved for use by the French Army Air Arm.

Manufacturer

PRONAL SA, Z. I. Roubaix Est, BP 18, F-59115 Leers, France.
Tel: +33 3 20 99 75 00. Fax: +33 3 20 99 75 20.

VERIFIED

GERMANY

Aluminium storage tank

Description

This storage tank was designed for bulk storage of liquids such as oil, petrol or water. Its main advantages are its ease of transportation, short assembly time and its capability of being erected on various types of surface, including concrete, earth and sand. The tank consists of 16 sections with tongue-and-groove sealing and an adjustable mast which is used as a roof support when the tank is being erected, as the upper sections are assembled first and the lower sections last. The shell and supporting parts are aluminium with the bottom and roof being a fuel-resistant rubber tarpaulin.

Specifications

Capacity: 450 m^3
Diameter: 12 m
Height:
(to eaves) 4.2 m
(filling) 4 m
(overall) 5.3 m

Aluminium storage tank

Close-up of discharge outlet on storage tank

Weight of tank: 8,300 kg
Total weight: (with transport frames) 10,780 kg

Status
In service with Argentina, Egypt, Libya and the German Army.

Manufacturer
Kurt Steinborn, Neustrasse 17, D-7830 Emmendingen, Germany.

VERIFIED

ISRAEL

Achidatex collapsible fuel containers

Description
Achidatex produces a range of collapsible containers suitable for the transport and storage of fuels, chemicals, drinking water and other potable liquids.

One type of container is the pillow tank, produced in sizes ranging from 250 to 100,000 gallons (1,136.5 to 454,600 litres). Similar tanks are also produced for drinking water but the largest of these holds 50,000 gallons (300,227 litres). Also produced are collapsible tanks for container transport (a standard 20 ft ISO container collapsible tank can carry a payload weighing 18,000 kg). The smallest collapsible containers produced are described as jerry-bags with dimensions (full) of 650 × 450 × 180 mm. They are suitable for fuel or water.

Achidatex collapsible containers are produced in three material types, as follows:

ACH-5, with a base fabric of tyre cord strength polyester yarn. This has a high temperature resistant, tough, resilient coating with a non-extractable plasticiser.

Urethane-coated polyester ACH-7714, a specially formulated and produced membrane fabric for containing leaded or unleaded petrol, jet fuels, methanol and ethanol.

Polyester membrane fabrics ACH-8028 and 8424, approved by the US Food and Drug Administration for the containment of drinking water and other liquids.

Status
All the above are in production and in service with the Israeli defence forces.

Manufacturer
Achidatex, PO Box 2156, IL-52151 Ramat Gan, Israel.
Tel: +972 3 647 6956. Fax: +972 3 497307.

VERIFIED

ITALY

Bartoletti system

Description
The Bartoletti system is used by the Italian Army for the transport and distribution of petrol, diesel and kerosene fuels and in service, is mounted on three different chassis: the IVECO CM 52, CP 62 and the CP 70. A special model of the CP 70 is also in service fitted with a micro-filter for refuelling light aircraft and helicopters.

The tanker is provided with both left- and right-side service points and a drum filling unit.

The left-side service point contains a self-priming centrifugal pump driven by the vehicle engine which can deliver 500 litres/min, a 77 mm filter, intake for suction from outside, visual indicator and three operating gate valves for circuit formation. The right-hand service point contains a control board with rev counter, pressure gauge, vacuum gauge, two remote controls, engine accelerator remote control, volumetric measuring unit, discharge mouth, an intake for closed-cycle connection, set of gate valves, reel with hose, and a 51 mm distributor nozzle. The drum filling unit has four 38 mm gate valves with rapid couplings, four rubber hoses and four distributor nozzles.

Bartoletti system on IVECO CP 70 (4 × 4) chassis (Italian Army)

Specifications
Mounted on CP 70 chassis
Configuration: 4 × 4
Weight:
(empty) 7,960 kg
(loaded) 12,500 kg
Tank capacity:
(petrol) 5,800 litres
(diesel) 5,380 litres
(kerosene) 5,750 litres
Useful load: 4,540 kg
Length: 6.554 m
Width: 2.5 m
Height: 2.97 m
Range: 650 km
Fuel capacity: 230 litres
Gradient: 60%

Status
In production. In service with the Italian Army.

Manufacturer
E Bartoletti, Via Leonardo da Vinci 4, I-47100 Forli, Italy.
Tel: +39 6 1122. Telex: 550064 bartfo 1.

VERIFIED

RUSSIAN FEDERATION AND ASSOCIATED STATES (CIS)

Fabric-reinforced rubber fuel tanks

Description

This type of fuel tank was introduced into service in 1961 and since then many variants have been produced. The first family of tanks comprised the MR-2.5, MR-4, MR-10 and MR-12. These were all orthodox reinforced rubber fuel tanks and all, except the MR-10, were used to convert flatbed trucks into mobile fuel carriers. The MR-10 was used primarily as a storage tank, not having the internal baffles fitted to the other types.

The latest family of fabric-reinforced rubber fuel tanks comprises the MR-4, MR-6, MR-25, MR-50, MR-150 and MR-250. Only the MR-4 and MR-6 are now used for the mobile carrying of fuel and other liquids, and are fitted with metal loading brackets on all corners and diagonal partitions internally. All the tanks have an inspection manhole, filling/draining and air connections, T-pipes and air pipes. Handles are provided for folding, unfolding and handling of each and these are situated around the periphery of each tank. The MR-150 and MR-250 have underside connectors with elbows and flexible pipes for flushing residual liquids from the tank. When folded, each tank has an external pocket into which the various fittings can be stowed. When transported, each tank is carried inside a cover fitted with loading loops. For storage, the tanks may be kept either folded or unfolded under cover, or outdoors under canvas.

Fabric-reinforced rubber fuel tanks

Specifications

Tank type	**MR-4**	**MR-6**	**MR-25**	**MR-50**	**MR-150**	**MR-250**
Rated capacity	4,000 litres	6,000 litres	25,000 litres	50,000 litres	150,000 litres	250,000 litres
Length (filled)	3.6 m	3.8 m	9 m	17.6 m	18 m	18 m
Width (filled)	2.6 m	2.5 m	3.7 m	3.7 m	7 m	10 m
Height (filled)	0.65 m	0.9 m	1 m	1 m	1.4 m	1.4 m
Weight (incl cover)	125 kg	135 kg	290 kg	580 kg	1,050 kg	1,450 kg
Number of filling/ draining connectors	1	1	2	2	2	2
Number of air connectors	1	1	1	2	2	2

Status

The original MR-2.5, MR-4, MR-10 and MR-12 are no longer in production but may be held in reserve. The later MR-4, MR-6, MR-25, MR-50, MR-150 and MR-250 are in production and service.

Manufacturer

State factories.

VERIFIED

PMTP-100 and PTMP-150 tactical pipelines

Description

PMTP-100 and PTMP-150 tactical POL pipelines are used by the pipeline regiments assigned to each Russian Army Front. The pipe is laid in 6 m lengths, each with quick-connect junctions, from tractor-towed TUM-150B machines (see following entry) which automatically lay the pipes at the rate of 2 to 3 km/h. The pipes can also be connected manually at a much slower rate. Once connected, the pipes are designed to be easily removed and replaced if damaged and the entire pipeline can be retrieved for later use.

Both pipelines are issued in sets sufficient to lay 150 km of pipeline, each set containing 27,000 pipes.

Section of PMTP-100 tactical pipeline

The PTMP-100 set contains pipes with a diameter of 100 mm. Pipe diameter for the PTMP-150 is 150 mm. An assembled PTMP-100 set has a capacity of 1,200 tonnes of fuel per day while the PTMP-150 can handle 2,600 tonnes per day.

Also included with both sets of pipelines are 32 PNU-100/200M mobile pumping stations, six PSG-160 fuel pumping stations and four ZIF-55 compressor stations.

A complete PTMP-100 set weighs approximately 1,100 tonnes. Weight of a PTMP-150 set is approximately 2,500 kg.

Status

In service with former Warsaw Pact forces. Offered for export sales.

Manufacturer

63 Boiler-Welding Enterprise, Khriplinskaya st.11, Ivano-Frankivsk 284002, Ukraine, RFAS.
Tel: +38 7 3422 25 95. Telex: 292302 Bereza.
Fax: +38 7 3422 22148.
Marketed by: Progress, 15 Piririchna st, Kiev 252216, Ukraine, RFAS.
Tel: +38 7 44 411 0589. Fax: +38 7 44 411 0848.

VERIFIED

TUM-150 pipelaying machine

Description

The TUM-150 pipelaying machine is used by pipelaying brigades assigned to each Russian Army Front to lay PTMP-100 and PTMP-150 tactical pipelines during offensive operations (see previous entry); it may also be used at divisional level in static situations. The TUM-150 is a trailer-mounted system and lays pipes 6 m long while travelling at a speed of 3 km/h. A complete load of 70 pipes is placed on the machine by an onboard hydraulic crane mounted on the forward part of the trailer. The pipes are loaded from the left-hand side in layers of 10 and are unloaded into a shallow trench to the rear via a chute located at the lower right-hand side and extending to the rear. A full load of 70 pipes with a total length of 427 m can be laid in approximately 8.5 minutes.

With its towing vehicle, the TUM-150 has an overall length of approximately 18 m and a width of 3.5 m. The towing vehicle may be a crawler tractor or a 4 × 4 wheeled tractor. Pipes may be loaded into the machine while it is in operation and on tow.

Status

In service with former Warsaw Pact forces.

Manufacturer

63 Boiler-Welding Enterprise, Khriplinskaya St.11, Ivano-Frankivsk 284002, Ukraine, RFAS.
Tel: +38 3422 25195. Telex: 292302 Bereza.
Fax: +38 3422 22148.
Marketed by: Progress, 15 Piririchna St, Kiev 252216, Ukraine, RFAS.
Tel: +38 44 411 0589. Fax: +38 44 411 0848.

VERIFIED

SOUTH AFRICA

SARMCOL static storage tanks

Description

Intended for the storage of fuel, oil products and water, these static storage tanks are for use in forward areas. They are made from nylon-reinforced nitrile rubber with abrasion-, ozone- and sunlight-resistant external surfaces. Each tank is fitted with an air vent, quick connection dry-break couplings, an inspection manhole and handles for lifting when empty. Each tank is delivered packed in a wooden crate that also contains a repair kit for on-the-spot repairs. Five sizes are produced (see specifications table).

Typical SARMCOL static storage tank

Specifications

Capacity	Length*	Width
4,500 litres	3.64 m	2.64 m
13,000 litres	5.08 m	4.11 m
22,000 litres	5.38 m	5.44 m
45,000 litres	7.06 m	6.91 m
50,000 litres	7.49 m	6.91 m

* empty

Status
In service with the South African National Defence Forces.

Manufacturer
SARMCOL SA (Pty) Limited.
Enquiries to: Armscor, Private Bag X337, Pretoria 0001, South Africa.
Tel: +27 12 428 1911. Telex: 30217.
Fax: +27 12 428 5635.

VERIFIED

SARMCOL transportable fuel tanks

Description
Produced in sizes up to 10,000 litres, these tanks are intended to convert flatbed or box-body trucks into fuel tankers. They are made of high-strength and durable synthetic fibre and rubber materials. The ends of each tank are sealed with special reinforcing metal clamps with a fuelling/defuelling coupling at one end only. A simple manually operated air vent is situated at the top centre of the tank to allow the escape of air which might be present in the tank before filling. A special retaining harness is used made from high-tensile nylon webbing to provide a safe and sure anchorage for the tank on standard flatbed trucks. Each tank is delivered in a special wooden crate that contains not only the tank but all its accessories including a repair kit. When not in use the tank may be stored in the crate or rolled up.

A 10,000 litre tank laid flat is 6.7 m long and 2.3 m wide.

Status
In service with the South African National Defence Forces.

Manufacturer
SARMCOL SA (Pty) Limited.
Enquiries to: Armscor, Private Bag X337, Pretoria 0001, South Africa.
Tel: +27 12 428 1911. Telex: 30217.
Fax: +27 12 428 5635.

VERIFIED

Transportable fuel tank on a truck

Container, vehicle fuel pump unit, 2 tonne, 1DX

Description
This fuel pump unit is based on a steel construction 1DX ISO container with canvas flaps at the sides and back. These give access to a diesel engine-driven pump, filters, suction hoses, meters and delivery hoses. The unit can be used to refuel up to 10 vehicles simultaneously from an outside source or for transferring fuel in bulk. Capacity is 27,000 litres/h.

When refuelling, fuel is pumped from an outside source by the engine and a self-priming pump, through the filter system and meter to delivery hoses. The hoses are fitted with automatic shut-off delivery nozzles and when transferring fuel in bulk it is pumped directly to a 64 mm delivery hose.

The container may be carried on a container carrier trailer, a SAMIL 20 (4×4) 2,000 kg truck fitted with twistlocks, or inside a SAMIL 50 (4×4) 4,800 kg truck drop-side body.

The unit weighs 1,800 kg, is 3 m long, 2.44 m wide and 1.45 m high.

Status
In service with the South African National Defence Forces.

Manufacturer
Enquiries to: Armscor, Private Bag X337, Pretoria 0001, South Africa.
Tel: +27 12 428 1911. Telex: 30217.
Fax: +27 12 428 5635.

VERIFIED

Container, Vehicle Fuel Pump Unit, 2 tonne, 1DX

TFM tanktainer

Description
The TFM Tanktainer consists of a steel or stainless steel tank held in a rigid angle frame assembly that conforms to ISO standards and can therefore be handled by standard ISO material handling systems; ISO twistlocks are provided. The tank can hold a total of 21,000 litres in three equal compartments, each compartment having a lockable manhole and a sump. Associated piping is 80 mm welded steel and each compartment has a butterfly valve with a locking mechanism and seal. Fire extinguisher brackets are provided close to the discharge valves.

Specifications
Weight: (empty) 4,250 kg
Length: 6.058 m
Width: 2.438 m
Height: (steel tank) 2.42 m

Status
Available.

Manufacturer
TFM (Pty) Limited, PO Box 46131, Orange Grove, Transvaal, South Africa.
Tel: +27 12 316 2106. Fax: +27 12 316 1462.

VERIFIED

TFM Tanktainer ready for delivery

UNITED KINGDOM

Thompson (8 × 6) 20,000 litre bulk fuel tanker

Development
The original Thompson Multidrive Tankers were produced for the British Army in 1990 and served in the Gulf. These consisted of simple petroleum tanks on a steel chassis.

In 1995, following a competition, Thompson Carmichael was awarded a contract by the UK Ministry of Defence for the supply of an undisclosed number of (8 × 6) 20,000 litre off-road bulk fuel transporter systems. These were delivered to the British Army late in 1995 and were deployed to Bosnia and the UK. A second order was placed for an additional quantity of systems and these were delivered in the first half of 1996.

On the late production version the semi-trailer chassis was replaced by an aluminium alloy monocoque tank. To allow the vehicles to be loaded at civilian petroleum terminals, vapour recovery and overspill protection equipment was added. Pumping and metering equipment was also included adding to the versatility of the system.

Description
The vehicles are designed to operate on all roads, from motorways to heavily damaged roads and cross-country. They comprise a Foden 4 × 2 tractor unit connected via a turntable to a 20,000 litre four compartment tank trailer. The trailer rear bogie is also turntable-mounted and linked to the tractor via a Multidrive system to provide rear wheel drive and steering.

While Thompson Carmichael was the prime contractor for the (8 × 6) 20,000 litre off-road bulk fuel transporter, there were two other key subcontractors, with Foden being responsible for the 4 × 2 tractor truck and Multidrive for the rear power elements which provide the combination with its 8 × 6 drive and off-road capability. All driven axles have cab operated cross-differential locks for maximum traction on soft ground.

The Foden 4 × 2 4380 tractor has a fully enclosed forward control GRP cab that tilts forward for maintenance purposes using a manually operated hydraulic system with a safety lock. Both the driver and passenger seats have air suspension and to the rear is a lower bunk with stowage space underneath. A detachable independent night heater is fitted as standard.

Provision is made for mounting two sets of 5.56 mm SA80 weapon clips. The tractor has steel spring suspension.

The electrical system is 24 V with standard civilian pattern lights front and rear, including reversing lights, halogen headlights and front fog lights. A NATO/MoD blackout lighting system is fitted as standard.

The rear trailer has a patented powered and steered trailer bogie and rubber walking beam rear suspension which gives high articulation and requires no maintenance.

The double ball race and heavy-duty turntables of the fifth wheel and bogie have capacities of 16 and 26 tonnes respectively and give a safety factor of two.

Thompson Carmichael (8 × 6) 20,000 litre bulk fuel tanker ***1996***

Two spare wheels and tyres are carried with a special quick dismount system which was designed for one-person operation. Large toolboxes are provided for clean and dirty kit with cab stowage.

Safety equipment carried includes fire extinguishers, earthing reels and spill kit, automatic slack adjusters to the trailer and automatic grease lubrication system for the complete vehicle.

The 20,000 litre tank has four compartments which allows it to carry four different types of fuel to be discharged left and right. Fuel can be loaded through the top or through either side. All compartments are fitted with bolted manways and each compartment has two full-length anti-surge baffles.

A pneumatic fuel control system, fully interlocked against operator error, is fitted as standard. A power take-off driven pump allows self-loading or pump discharge at 1,000 litres/min.

The full flowmeter allows metered delivery and all tank couplings are 'dry break' for maximum safety in the event of an accident. A lockable plastic undercover prevents road and dirt contamination of the fuel system and inhibits tampering.

The petroleum handling equipment is designed to meet Oil Industry Standards in addition to military ones and incorporates the necessary safety and environmental measures required to obtain safe loading passes for Civilian Oil Terminals. The discharge of fuel is by gravity or through an integral pump and metering system, either into bulk storage or via a manifold and trigger nozzles into drums or cans.

Double 5 m hose trays are provided and there is a full-length walkway with a pneumatically operated collapsible handrail on the tank top for safe top loading; this is quickly removable for maintenance.

Optional extras include hoses to suit local requirements, load sensing system on truck and trailer suspension, ABS system, NATO towing jaw to rear, NATO recovery eyes to front and rear, snow chains, ferry lashing rings, wire mesh windscreen protection, alternative 24,000 litre capacity and European side under-run guards.

Specifications
Crew: 1 + 1
Configuration: 8 × 6
Weight
(empty) 18,380 kg
(loaded) 35,000 kg
Length: 12.8 m
Width: 2.9 m
Height:
(cab, laden) 3.508 m
(cab, unladen) 3.508 m

Coupling height:
(laden) 0.951 m
(unladen) 1.051 m
Wheelbase:
(tractor) 3.6 m
(trailer) 6 m
(system) 10.16 m
Angle of approach/departure: 21°/23°
Max road speed: 96 km/h
Fuel capacity: 400 litres
Engine: Cummins L325 6-cylinder turbocharged and air-to-air after-cooled 4-stroke direct injection diesel developing 325 bhp at 1,900 rpm
Transmission: Eaton RTO 14613A 12-speed overdrive, constant mesh, twin countershaft
Steering: ZF, power-assisted booster ram
Turning radius: 7.3 m, kerb to kerb
Axles:
(front) GKN, forged beam, rated at 7,100 kg
(rear) Rockwell S1600E hypoid single reduction, rated at 11,500 kg with differential lock
(drive axles) SISU heavy duty
Tyres:
(front) Michelin XZA 295/80R 22.5 tubeless, single
(rear) Michelin XDA 295/80R 22.5 tubeless, twins
Brakes: air
Electrical system: 24 V
Batteries: 2 × 12 V

Status
In production. In service with the British Army.

Manufacturer
Thompson Carmichael Limited, Great Bridge Road, Bilston, West Midlands WV14 8NP.
Tel: +44 1902 353141. Fax: +44 1902 405509.

UPDATED

Demountable bulk fuel dispensing unit

Description
The demountable bulk fuel dispensing unit comprises two tanks, one pumping and dispensing pack, a two-tier rack and mounting beams.

The unit was designed for mounting on a Bedford MK, or Leyland Trucks (4 × 4) flatbed truck, but can be adapted to fit other chassis with suitable capacity. The unit can be removed from the vehicle by a forklift truck, crane or, in an emergency, by hand.

The tanks are manufactured from AA5454 aluminium sheet and each tank has a capacity of 2,100 litres. Access to the tank interior is through a 406 × 356 mm collar in the top skin. The filler cap assembly incorporates a 254 mm diameter EMCO open-fill quick-release manhole and a single pressure/vacuum vent. A dip point containing a captive dipstick is provided and the filler cap and dip point are contained within a spillage band. An automatic bottom-loading facility fitted to each tank enables loading rates up to 682 litres/min and off-loading rates of 455 litres/min to be achieved. A drain plug is provided in the tank bottom skin and each tank is pressure tested to 0.703 kg/cm^2.

The dispensing equipment may be operated independently of the tanks if desired. The pump set is resiliently mounted inside the dispensing pack frame and is adequately protected against spillage by a firescreen and boxed-in exhaust which meets Ministry of Defence (Army) safety requirements.

A two-tier rack fitted with a drip tray and drain plug is provided. The lower tier of the rack is designed as a roller carriage which can be withdrawn from either side of the vehicle. An Alan Cobham filter/water separator with differential pressure gauge is incorporated in the design.

Two 12.192 m long 32 mm diameter smoothbore delivery hoses terminating in automatic shut-off nozzles are stowed on a self-rewind Dean hose reel. Also 10 m of 63.5 mm layflat hose is provided. A cross pumping off-take complete with an Avery Hardoll 63.5 mm self-sealing coupling for cross pumping to a tanker vehicle is connected to a secondary discharge outlet.

A rigid suction pipe connects at one end to the pump and at the other to bifurcated suction hoses, which connect to the two tanks via the CC41 coupling units. The suction hoses have a bore of 63.5 mm.

Demountable bulk fuel dispensing unit on rear of Bedford MK (4 × 4) 4,000 kg truck (T J Gander) ***1997***

The whole system is retained on two longitudinal beams mounted on the vehicle platform. The equipment is fully bonded to the vehicle and is capable of being earthed. Two Dean bonding reels are provided, each with 12.2 m of bonding cable terminating with crocodile clip-type clamps. Two fire extinguishers are mounted on the nearside and offside of the two-tier rack.

Optional equipment includes a manifold assembly box for can filling, comprising 63.5 mm Avery Hardoll hose unit type CC40, 10 m of 38 mm bore hose to BS 3158, five-way manifold, 5 m of 19 mm bore hose to BS 3395 (five off) and 19 mm automatic shut-off nozzles (five off).

The lid of the stowage box acts as holder for the cans during filling.

Specifications
Weight: 1,798 kg
Payload: 2,967 kg
Length: 3.66 m
Width: 2.44 m
Height: 1.22 m

Status
In service with the British Army and other armed forces.

Manufacturer
Simon Gloster Saro Limited, Gloucester Trading Estate, Hucclecote, Gloucestershire GL3 4AD, UK.
Tel: +44 1452 371321. Telex: 43134 glosro g.
Fax: +44 1452 371024.

UPDATED

Marshall DROPS Fuel Rack

Description
The Marshall DROPS Fuel Rack was developed for the UK Ministry of Defence to provide a self-contained, stand-alone dispenser of a wide range of vehicle and aviation fuels. The 12,000 litre fuel tank is divided into compartments, having internal baffling and a low centre of gravity to enhance handling characteristics during transport and deployment.

The Fuel Rack tank and pumping system (with an integral diesel engine) is enclosed within a framework incorporating welded steel ISO castings allowing interface with vehicles and toplift handling. There are four separate dispensing hoses capable of filling up to four vehicles at a distance of 18 m.

Within the British Army, these racks are carried by Leyland Trucks (8 × 6) DROPS vehicles. Transport is

Marshall DROPS Fuel Rack being loaded on to a Bedford MT26-30 (6 × 6) chassis fitted with a Cargotec load handling system
1997

possible by vehicle, underslung from a helicopter (such as a CH-47 Chinook), or within a C-130 Hercules transport aircraft. The Fuel Rack is designed to be stowed in a three-high stack or transported by sea in a two-high stack.

The Marshall DROPS Fuel Rack is compliant to UK and European Petrochemical legislation and compatible with NATO DROPS and PLS handling systems.

Status
In service with the British Army.

Manufacturer
Marshall SPV Limited, The Airport, Cambridge, Cambridgeshire CB5 8RX, UK.
Tel: +44 1223 373045. Fax: +44 1223 373046.

NEW ENTRY

Reynolds Boughton DROPS fuel flatracks

Description
The Reynolds Boughton DROPS fuel flatrack has a capacity of 12,000 litres of diesel fuel. The tank is divided into three internal compartments which are lined. Pumping is by means of a diesel engine-driven pump rated at 450 litres/min. The piping system incorporates a water/filter separator and is complete with four hose reels for multiple refuelling. The tank may be filled from either the top or bottom.

The system is built, where possible, to Institute of Petroleum Code of Practice standards and other current UK regulations. The equipment has been situated so that all fuelling and defuelling can take place with the flatrack on or off a vehicle.

The Reynolds Boughton DROPS fuel flatrack is fully interoperable with NATO DROPS and PLS systems.

Status
Available.

Reynolds Boughton DROPS fuel flatrack

Manufacturer
Reynolds Boughton (Devon) Limited, Winkleigh Airfield, Winkleigh, Devon EX19 8DR, UK.
Tel: +44 1837 83555. Telex: 83132.
Fax: +44 1837 83768.

VERIFIED

Reynolds Boughton tankers

Description
Reynolds Boughton produce a range of fuel and water tankers which can be mounted on commercially available 4 × 4 chassis. The tanks are manufactured from 3 mm/4 mm mild steel and are elliptical in section. Units are fitted with pumping equipment suitable for their respective loads. The tanks can be lined or unlined depending on the type of fuel or water carried. All tankers are fitted with commercially available equipment to ensure efficient and reliable back-up in the field.

Reynolds Boughton also produce specialised cesspit/sewage tankers.

Status
Available.

Reynolds Boughton tankers ***1997***

Manufacturer
Reynolds Boughton (Devon) Limited, Winkleigh Airfield, Winkleigh, Devon EX19 8DR, UK.
Tel: +44 1837 83555. Telex: 83132.
Fax: +44 1837 83768.

NEW ENTRY

Dunlop Dracone barges

Description
The idea of the Dracone barge dates from 1956, with the concept of transporting liquids having lower specific gravity than the surrounding water through which the cargo is towed, in specially designed flexible containers. Dunlop became involved in the programme during 1963. Since 1965 it has developed the Dracone barge to the stage where it is now widely used by both military and civil concerns.

The Dracone barge combines strength of construction with good towing capabilities in open waters. Each Dracone barge has a tapered nose and rounded tail section and is moulded in individual seamless units joined to the parallel centre section by circumferential seams. The nose section will withstand a pull of 30 tonnes without damage. Body seams are both stitched and hot vulcanised. The high-tensile nylon fabrics used are coated with neoprene for the outer proofing giving resistance to abrasion, sunlight, oils and seawater. On the inside, nitrile is normally used for proofing.

Dracone barges can be used to carry a wide range of mineral and vegetable oils and can also be used to transport water. Being flexible, the barges can slide over obstructions without friction when wet and the barge acts as its own shock-absorber. For open sea towing a length of tow of approximately 100 m is usual, although this can be shortened for towing in sheltered waters or moving in confined anchorages. Nylon tow ropes are used. The towing hose, through which the barge is loaded and discharged, is designed to accept the maximum towing loads with ample safety margins. The smaller sizes can be towed at speeds in excess of 12 kt and the largest can travel at 7 kt. From size D5 and upwards all Dracone barges are fitted with a rear stabiliser to prevent snaking. The power required to tow a Dracone is considerably less than that required for a conventional barge of similar capacity.

Dracones can be launched in a number of ways, including dropping a rolled unit from a cargo net, pulling a rolled unit from a slipway by a small launch, rolling bodily from a beach or launching over the stern of a vessel. Built-in buoyancy panels keep the barge afloat even when it is completely empty.

Cargo is loaded and discharged through a length of flexible hose fitted to the nose as an integral part of the tow to make handling easier. Small barges (up to and including the D5) are fitted with a 50 mm bore hose, while the larger sizes have a 100 mm hose as standard. Larger bore hoses are available on request. All hoses are supplied in standard lengths of 4.57 m that are bolted together to suit. At the end of the hose is a lightweight corrosion-resistant float to which the loading/discharge connection is fitted. A quick-acting coupling can be fitted. Loading can be by gravity or pump.

Empty Dracone barges may be carried as deck cargo but are normally left afloat once launched. A special gripper bar has been developed for recovering empty barges. Little or no maintenance other than cleaning is normally required.

Dracone barges can also be used as alternatives to ship-to-shore pipelines and can be employed as floating fuel reservoirs.

Specifications

Model	A1	A2	D5	D10	E	F	J	L	O
100% capacity	4.55 m^3	9.1 m^3	22.75 m^3	45.5 m^3	100 m^3	191 m^3	385 m^3	519 m^3	1,100 m^3
85% capacity	3.9 m^3	7.8 m^3	19.3 m^3	38.6 m^3	85 m^3	162 m^3	327.25 m^3	441 m^3	935 m^3
Length	7.5 m	14.1 m	15.07 m	31 m	37.64 m	49.2 m	66 m	66 m	91.45 m
Diameter	0.94 m	0.94 m	1.42 m	1.42 m	1.87 m	2.347 m	2.82 m	3.28 m	4.225 m
Weight empty	270 kg	310 kg	430 kg	780 kg	1,000 kg	2,275 kg	3,540 kg	4,060 kg	6,500 kg
Towing hose bore	50 mm	50 mm	100 mm	100 mm	100 mm	100 mm	100 mm	100 mm	100 mm
Stabiliser	no	no	yes	yes	yes	yes	yes	yes	yes

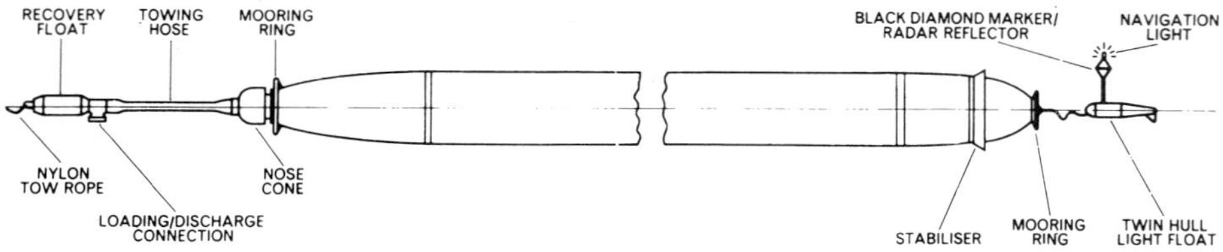

Main components of a typical Dracone barge

Status
In production and widespread service.

Manufacturer
Dunlop Limited, GRG Division, PO Box 151, Cambridge Street, Manchester M60 1PD, UK.
Tel: +44 161 236 2131. Telex: 667225 dunman g.
Fax: +44 161 236 1599.

VERIFIED

Dunlop collapsible containers

Description
Dunlop collapsible containers can be used to store a wide variety of liquids ranging from petrol, oil and lubricants to water and chemicals for almost any purpose. The range produced is wide, but all tanks are constructed from a flexible two-ply textile and polymer laminate. Both sides of the laminate are treated to ensure long life and resistance to abrasions and hard knocks. For extra strength each corner is reinforced by metal clamping bars. Joint reinforcing straps fabricated from panels of vulcanised polymer-coated textiles add extra strength to the main structure and all the ports and fittings are constructed from either steel or brass. The filled tanks may be either placed on prepared surfaces or laid on earth or sand foundations. The smaller tanks may be carried on flatbed trucks or the decks of ships. When not in use the collapsible tanks may be cleaned and rolled up for storage ready for reuse.

45,000 litre tank intended primarily for petroleum products undergoing final examination before delivery to British Army

Status
In production. In service with several armed forces.

Manufacturer
Dunlop Limited, GRG Division, PO Box 151, Cambridge Street, Manchester M60 1PD, UK.
Tel: +44 161 236 2131. Telex: 667225 dunman g.
Fax: +44 161 236 1599.

Specifications

Tank capacity	225 litres	1,125 litres	2,250 litres	4,500 litres	9,000 litres	11,250 litres	22,500 litres	45,000 litres	90,000 litres	135,000 litres	225,000 litres
Length	2.06 m	1.52 m	2.59 m	4.72 m	4.88 m	5.49 m	9.76 m	10.21 m	11.58 m	17.15 m	18.59 m
Width	0.69 m	2.06 m	2.06 m	2.06 m	3.43 m	3.43 m	3.43 m	4.8 m	7.45 m	8.92 m	11.66 m
Height (filled)	0.38 m	0.79 m	0.79 m	0.79 m	0.91 m	0.91 m	0.91 m	1.22 m	1.22 m	1.07 m	1.22 m

VERIFIED

UNITED STATES OF AMERICA

SIXCON fuel storage and dispensing system

Description
The SIXCON fuel storage and dispensing system is intended for both stationary and mobile use as a source of fuel or water – for the SIXCON water system refer to entry under *Water supplies*. The system can be used for fuelling, defuelling or recirculation.

Each SIXCON fuel pump or fuel tank module uses a specially configured shipping frame ensuring that it can be interconnected with five other modules and transported in standard 20 ft/6.096 m ISO configuration. Each fuel pump module incorporates a SIXCON frame, 125 gal/min (473 litres/min) pump, diesel engine, filter separator, monitor and meter, static discharge reel, hose reel and hoses and couplings and fittings, including an earth rod. Each fuel tank module also has a SIXCON frame, 934 gallon 3,535 litre) stainless steel tank, fuel level gauge, inlet, outlet and vent valves, covered manhole and a drain pipe.

A SIXCON fuel storage and dispensing system being unloaded

Specifications
Fuel pump module
Weight: approx 1,043 kg
Length, external: 1.969 m
Width, external: 2.438 m
Height, external: 1.204 m
Flow rate: 378.5 litres/min

Fuel tank module
Weight:
(empty) approx 1,179 kg
(full) approx 4,082 kg
Capacity: approx 3,535 litres
Tank length: 1.79 m
Tank width: 2.21 m
Tank height: 0.965 m

Status
In production. In service with the US Navy and Marine Corps.

Marketing Agency
Tradeways Limited, 184 Duke of Gloucester Street, Annapolis, Maryland 21401, USA.
Tel: +1 410 295 0813. Fax: +1 410 295 0821.

UPDATED

Flexcel liquid containers

Description

The basis of the Flexcel container is a cylindrical storage tank manufactured from acrylonitrile-butadiene synthetic rubber reinforced with nylon cords for the interior bladder and polychloroprene for the strong outer cover. The exterior is fitted with straps enabling the container to be carried in or on almost any type of vehicle or slung under a helicopter. One end of the container has a shroud containing a hose and a dispensing nozzle plus junctions and refuelling fittings.

There are various sizes of Flexcel container, with the most usual being 2.6 m long, 0.356 m wide and weighing 56.7 kg empty. The storage capacity of this version is 55 gallons/208 litres. Also available is a 30 gallons/113.5 litres tank. An 80 gallons/303 litres tank is available on request.

The Flexcel container is strong enough to be free-dropped from a helicopter even when full. It has been demonstrated being dropped from a helicopter flying at a height of 12.2 m at a speed of 80 knots. When the Flexcel has been dropped from a helicopter or vehicle it is ready for immediate use. The dispenser hose and nozzle are removed from within the shroud and are attached to the Flexcel by a quick-connect device. The vehicle or AFV to be refuelled is then driven on to the Flexcel to provide the pressure needed to force the liquid from the Flexcel as the nozzle is placed in the vehicle fuel tank. The complete process can take as little as 5 minutes. After use the Flexcel container can be retrieved and used again.

The Flexcel can be used for liquids other than fuel such as water, decontaminants, oil and other fluids.

Status

In production. In service with US Marine Corps and US Army.

Manufacturer

General Dynamics Land Systems, 38500 Mound Road, Sterling Heights, Michigan 48310, USA.
Tel: +1 313 825 4000.

UPDATED

Flexcel fuel container being carried on side of M60 MBT

Refuelling nozzle of Flexcel fuel container

GTA fuel drums and superdrums

Description

GTA manufactures drums for the storage and transportation of fuel, drinking water, fluid food products and so on. Two types of fuel drum are available, a standard drum constructed using rubber-coated fabric and the GTA Superdrum constructed of elastomeric coated fabric. The GTA Superdrum is significantly lighter but is more rugged and durable.

Three standard sizes are available; 55 gallons (208 litres), 250 gallons (946 litres), and 500 gallons (1,893 litres). The fabric used in all types of construction is puncture- and tear-resistant and each drum is subjected to a hydropressure test following assembly. Empty drums collapse to 10 per cent of their original volume for transport or storage, while filled drums can be rolled, floated, towed or lifted by helicopter. Drums can also be dropped free-fall (the 250 gallon drum can be dropped free-fall from 15 feet/4.5 m) or by parachute.

GTA fuel drums are coloured black and can be temporarily or permanently repaired in the field when necessary. Hoses, nozzles and valves have quick-connect/disconnect fittings.

These drums have numerous alternative applications including flotation devices (surface or submerged, for example for the construction of temporary bridges), as pneumatic lifting devices or ship fenders and for the underground storage of fuel supplies.

GTA Superdrum on tow

Specifications

Capacity	55 US gal	250 US gal	500 US gal
	208 litres	946 litres	1,893 litres
Weight (drum)	22.7 kg	81.8 kg	104.5 kg
(Superdrum)	13.6 kg	63.6 kg	72.7 kg
Length	0.876 m	1.524 m	1.524 m
Diameter	0.597 m	1.016 m	1.27 m

Status

In production.

Manufacturer

GTA Containers Inc, 1410 W. Napier Street, South Bend, Indiana 4660-, USA.
Tel: +1 219 288 3459. Fax: +1 219 289 6060.

VERIFIED

GTA static fuel storage tanks

Description

GTA Containers Inc manufactures a range of static fuel storage pillow tanks that are available in sizes from 3,000 gallons (11,361 litres) to 225,000 gallons

Specifications

Capacity	10,000 US gal	20,000 US gal	50,000 US gal
	37,869 litres	75,738 litres	189,345 litres
Length	6.71 m	7.32 m	7.62 m
Width	6.71 m	8.53 m	19.81 m
Weight (empty)	140.9 kg	187.2 kg	415.9 kg

A typical GTA 50,000 gallon (189,345 litre) static fuel storage tank

(851,715 litres). All tanks are manufactured in accordance with US military requirements and are thus lightweight, rugged and durable with high abrasion and puncture resistance.

Each tank is constructed using tear-resistant fabric with an elastomeric coating on both sides. Chafing patches are included as part of the standard design and are located on the interior surface directly opposite each fitting providing a double-wall thickness supporting every piece of hardware. Handles, each capable of lifting 454 kg, are moulded on to the tank.

All tanks are equipped with standard 4 in/101.6 mm quick-disconnect filler/discharge fittings. Bottom discharge fittings are provided for gravity unloading. A 4 in/101.6 mm reinforced hose with quick-disconnects is also provided.

Each tank is fully collapsible and folds down into a package that occupies 2 per cent of the volume of the filled tank.

Status
In production for the US Army and others.

Manufacturer
GTA Containers Inc, 1410 W. Napier Street, South Bend, Indiana 46601, USA.
Tel: +1 219 288 3459. Fax: +1 219 289 6060.

VERIFIED

Bell Avon flexitanks for fuel

Description
Bell Avon Inc produces a range of flexible tanks for transporting and storing fuel. They are made from high strength, durable polymer-coated fabric and are available in a wide range of sizes from 5 to 500,000 gallons. There is also a 210,000 gallon capacity bulk fuel tank assembly (see following entry).

When not in use, or for ease of shipping, the tanks can be folded, transported and packed into a convenient crate. They are suitable for operation throughout a wide temperature range.

Status
In production. Available from inventory or short lead time.

Manufacturer
Bell Avon Inc, 1200 Martin Luther King Jr. Boulevard, Picayune, Mississippi 39466, USA.
Tel: +1 601 799 1217. Fax: +1 601 799 1360.
European sales: Avon Technical Products, PO Box 2, Melksham, Wiltshire SN12 8AA, UK.
Tel: +44 1225 791823. Fax: +44 1225 705585.

VERIFIED

A typical Bell Avon flexible fuel tank in operation

210,000 gallon collapsible fabric tank for fuel

Description
This 210,000 gallon (794,934 litre) collapsible fabric tank for fuel is known to the US Army as a Bulk Fuel Tank Assembly (BFTA). It was developed during the early 1980s as part of the Inland Petroleum Distribution System (IPDS — see separate entry in this section) and is used by the US Army as a component of the Tactical Petroleum Terminal.

The tank is made from a woven nylon cloth coated on both sides with an elastomer (usually polyester polyurethane). A typical installation would entail a berm with a liner.

Bulk Fuel Tank Assembly (US Army)

Specifications
Capacity: 210,000 gallons, 794,934 litres
Weight: (crated) 1,361 kg
Length: 20.73 m
Width: 20.73 m
Height: 2.03 m

Status
In service with the US Army.

Development Agency
US Army Tank Automotive Command, Mobility Technology Center - Belvoir, Fort Belvoir, Virginia 22060-5843, USA.

Manufacturer
Bell Avon Inc, 1200 Martin Luther King Jr. Boulevard, Picayune, Mississippi 39466, USA.
Tel: +1 601 799 1217. Fax: +1 601 799 1360.
European sales: Avon Technical Products, PO Box 2, Melksham, Wiltshire SN12 8AA, UK.
Tel: +44 1225 791823. Fax: +44 1225 705585.

VERIFIED

Amfuel Sealdrums

Description

Amfuel Sealdrums are portable, collapsible rubber containers for storing and transporting POL products, water, liquid chemicals and other fluids. They are circular drums with metal 'hubs' with the bulk of the container being constructed of elastomeric-coated rayon cord. The outer cover is neoprene with the inner lining being a nitrile material. This casing is puncture-resistant and can withstand a minimum of three successive free-fall drops from a height of 3.8 m. Once filled, the Sealdrums are non-vented and hermetically sealed, with extremely low permeability. When empty, Sealdrums collapse to about 15 per cent of their filled size. When filled, they can be transported by truck, aircraft or helicopter and dropped off the rear of a truck or para-dropped. They are the main item in the FARE (see entry in this section), Arctic fuelling and other fuel systems.

There is an Amfuel range of similar containers known as Sealdbins which are used for the storing and carrying of dry materials.

Specifications

Capacity	55 US gal	250 US gal	500 US gal	515 US gal
	208.2 litres	946.25 litres	1,892.5 litres	1,949.3 litres
Length	0.876 m	1.524 m	1.575 m	2.032 m
Diameter	0.597 m	1.016 m	1.349 m	1.168 m
Weight (empty)	22.68 kg	113.4 kg	129.3 kg	129.3 kg

Amfuel Sealdrum undergoing trials in South-east Asia

Status

In production. In service with the US Army and many other nations.

Manufacturer

American Fuel Cel and Coated Fabrics Company (Amfuel), 124 South Jackson, Magnolia, Arkansas 71753, USA.
Tel: +1 501 234 3381. Fax: +1 501 235 7270.

VERIFIED

Amfuel static storage tanks

Description

Amfuel produces a wide range of static storage tanks to store an equally wide range of liquids, including fuel and water (for details of the water tanks see entry in the *Water supplies* section). The tanks intended for fuel use are made from tough polymer-coated nylon fabric. Chafing patches at all fitting and hard-wearing locations provide double-wall thickness and protection. Handles are moulded on to each tank to assist positioning. All the tanks are free-standing. When not in use, the tanks can be stored and carried in wooden crates.

Specifications

Capacity		**Approximate dimensions filled:**		
		Width	**Length**	**Height**
1,000 US gal	3,785 litres	2.388 m	3.073 m	0.61 m
2,000 US gal	7,570 litres	3.353 m	3.353 m	0.813 m
3,000 US gal	11,355 litres	3.581 m	4.267 m	0.914 m
5,000 US gal	18,925 litres	4.47 m	4.623 m	1.118 m
10,000 US gal	37,850 litres	6.096 m	6.096 m	1.219 m
15,000 US gal	56,775 litres	7.468 m	7.315 m	1.219 m
20,000 US gal	75,700 litres	6.7 m	8.23 m	1.727 m
25,000 US gal	94,625 litres	6.7 m	10.363 m	1.727 m
50,000 US gal	189,250 litres	6.7 m	19.2 m	1.727 m
100,000 US gal	378,500 litres	18.136 m	18.212 m	1.219 m

Amfuel static storage tanks in use by German Army

Status

In production. In service with US forces and many other nations.

Manufacturer

American Fuel Cel and Coated Fabrics Company (Amfuel), 124 South Jackson, Magnolia, Arkansas 71753, USA.
Tel: +1 501 234 3381. Fax: +1 501 235 7270.

VERIFIED

Inland Petroleum Distribution System (IPDS)

Development

Starting in 1977, a study established that a rapidly deployed bulk fuel storage, distribution and dispensing system using pipelines could be a viable proposition. In 1983 the Joint Chiefs of Staff tasked the US Army to construct and operate major bulk petroleum distribution and storage systems in Southeast Asia using, where possible, an accelerated non-developmental item procurement strategy. Type classification of the Inland Petroleum Distribution System (IPDS) was made in July 1984. Equipment costing 'several hundred million dollars' was procured and used during Operations Desert Shield and Restore Hope. Recent developments concentrated on reconstituting used equipment, procuring missing equipment and packaging the system into 20 ft/6.096 m ISO containers.

Description

The IPDS consists of Tactical Petroleum Terminals (TPT), pipeline pump stations, 5 mile (8.05 km) pipeline sets, 1,250 gallon/min (4,732 litre/min) pumps, pressure reducing stations, critical gap crossings, suspension bridges, IPDS recovery equipment and pipeline support equipment. The system is capable of supplying bulk fuel in an undeveloped theatre of operations and begins at a port of disembarkation. There, fuel is unloaded into a TPT and pumped through a pipeline for dispersing.

Each TPT has a bulk storage capacity of 3.6 million gallons. An average rate of fuel transfer is 960,000 gallons/day (3,633,984 litres/day). Pipeline can be emplaced at the rate of 4.8 km/day.

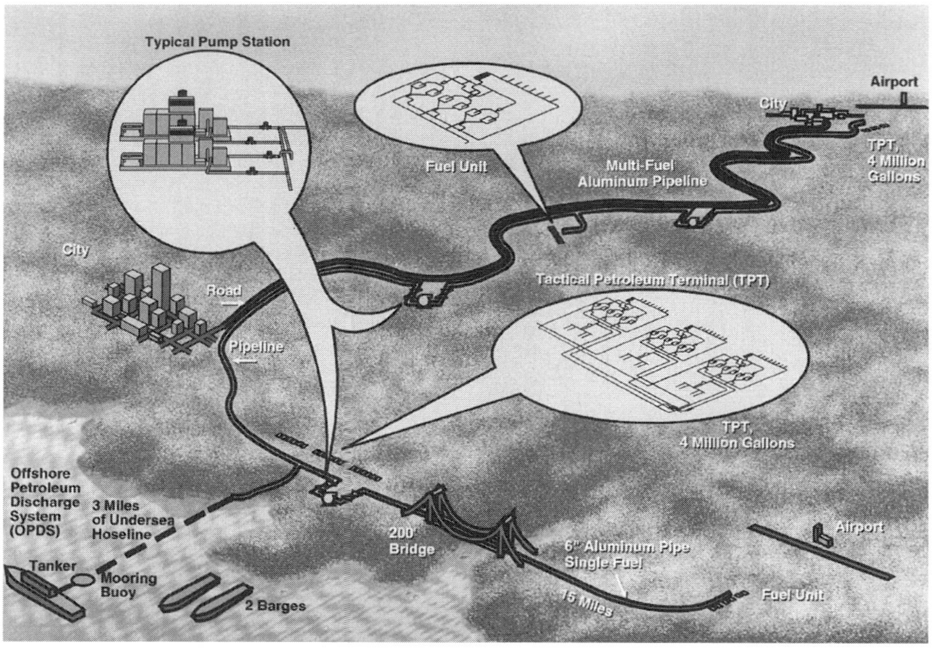

The original IPDS procurement exercise involved 17 TPTs, 860 miles (1,384 km) of 76.2 mm assault pipeline, 172 mainline pumps and associated support equipment.

Status

In service with the US Army.

Development Agency

US Army Tank Automotive Command, Mobility Technology Center - Belvoir, Fort Belvoir, Virginia 22060-5843, USA.

VERIFIED

Schematic outline of Inland Petroleum Distribution System (IPDS) **1996**

Forward Area Refuelling Equipment (FARE)

Description

FARE is a lightweight, air-transportable refuelling system intended for refuelling helicopters in forward area combat operations. As a secondary function FARE provides a means for safe rapid refuelling of all US Army aircraft, ground vehicles and other equipment. FARE was type classified as Standard A and replaced the Pumping Assembly, Flammable Liquid, Bulk Transfer, and Pump Centrifugal, Gas Driven, Base Mounted, 38 mm, 378.5 litres/min.

FARE includes: a petrol engine-driven centrifugal pump with a rated capacity of 378.5 litres/min at 30.48 m total dynamic head; lightweight military design filter-separator with a rated capacity of 378.5 litres/min; two closed-circuit refuelling nozzles with adaptors for conventional gravity fuelling; 60.96 m of lightweight discharge hose with 50.8 mm nominal inside diameter; 18.28 m of suction hose with 50.8 mm nominal inside diameter; and miscellaneous fittings, valves and accessories.

FARE being used in conjunction with standard 1,892.5 litre collapsible drums to refuel Bell UH-1 Iroquois helicopter (US Army)

FARE is capable of dispensing fuel through two nozzles simultaneously at flow rates up to 189.25 litres/min through each nozzle. The maximum flow rate possible when dispensing fuel through a single nozzle is approximately 302.8 litres/min.

Any available bulk fuel storage tank may be used to supply fuel to the FARE — the standard 1,892.5 litre (500 US gallon) collapsible drum is the primary source of bulk fuel supply.

Status

In service with the US Army.

VERIFIED

Arctic Fuels Dispensing Equipment (AFDE)

Description

Arctic weather can have many severe effects on orthodox fuelling systems, such as making collapsible fuel storage tanks brittle and liable to split. An increase in fuel viscosity, brought on by cold weather, can make fuels harder to pump and cause engine operating difficulties. To overcome these problems, the US Army Research, Development and Engineering Center began development work on a new system of Arctic Fuels Dispensing Equipment (AFDE). The programme introduces a new generation of fuel equipment which is simple, lightweight, air-transportable and collapsible. The fuel tanks and hoses involved are designed for use between −60 and +95°F. They are fabricated from materials based on low-temperature elastomers such as polyurethane and polypropylene. Another AFDE feature is easy starting and improved pumping power made possible by the use of gas-turbine engines. Two types of basic installation are envisaged: an AFDE Supply Point with collapsible tanks to store up to 120,000 US gallons of fuel and a pumping capability of 600 US gal/min and a Forward Area Refuelling Equipment made up of 500 US gallon collapsible fabric drums with a fuel pumping capability of 200 US gal/min.

Status

Development complete. Production planned.

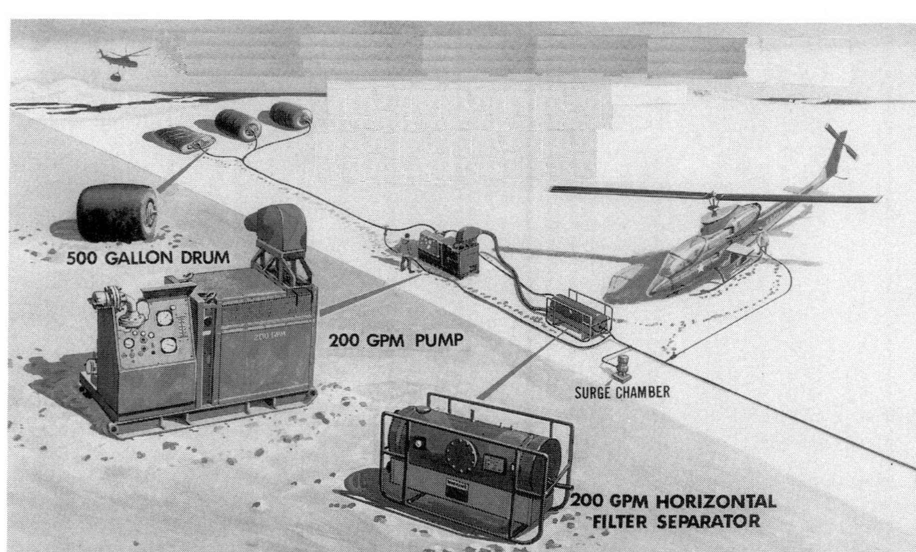

Artist's impression of Arctic Fuels Dispensing Equipment (AFDE) in use (US Army)

Development Agency

US Army Tank Automotive Command, Mobility Technology Center - Belvoir, Fort Belvoir, Virginia 22060-5843, USA.

VERIFIED

Fuel System, Supply Point Portable, 60,000 gallon (US) capacity

Description

The Fuel System, Supply Point Portable is the US Army's primary means of distributing and issuing bulk petroleum to combat units under tactical conditions. The system can handle a total, at any one time, up to 60,000 US gallons (227,100 litres) in three 3,000 or 20,000 US gallon (11,356 or 75,708 litre) collapsible bulk fuel static storage tanks. The system consists of: two pumping assemblies; two filter/separators; three 3,000 or 20,000 US gallon (11,356 or 75,708 litre) collapsible tanks; three bottom-loading points for loading tank trucks and semi-trailers; two 500 US gallon (1,892.5 litre) Sealdrum filling points and three refuelling points for filling vehicles. The combination and types of filling points can be altered according to local requirements.

Status

In service with the US Army and some Far Eastern and European armed forces.

Part of Fuel System, Supply Point Portable showing hoses and tanks

Manufacturer

La Barge Products Inc, 2900 Brannon Avenue, St Louis, Missouri, USA.

VERIFIED

Force Provider Fuel Supply System

Description

Commencing during Fiscal Year 1995, the US Army Tank Automotive Command, Mobility Technology Centre - Belvoir, started to procure the Force Provider Fuel Supply System. This consists of four subsystems; a bulk diesel fuel storage and distribution system to supply fuel to a tank and pump unit and material handling equipment, a bulk petroleum fuel storage and distribution system to supply fuel to kitchen stoves and water chillers, a 500 gallon/1,893 litre drum diesel fuel storage system to support tactical generator power groups and a bulk diesel fuel storage system to support 750 kW power generators.

The bulk fuel storage and distribution system can store up to 20,000 gallons (75,708 litres) of diesel fuel or JP-8. It is capable of receiving fuel from tankers and dispensing fuel to all tanks and containers within the full system. The bulk petroleum fuel storage and distribution system can store up to 3,000 gallons (11,356 litres) and can receive petroleum from tankers for dispensing to 5 gallon/18.9 litre fuel cans. The 500 gallon/1,893 litre drum diesel fuel storage system can supply fuel to three tactical generators and is refilled by a tank and pump unit. The generator bulk diesel fuel storage systems can supply 10,000 gallons (37,854 litres) of fuel to each 750 kW generator.

Status

Procurement started during FY95.

Development Agency

US Army Tank Automotive Command, Mobility Technology Center - Belvoir, Fort Belvoir, Virginia 22060-5843, USA.

UPDATED

Low profile tank and pump unit

Description

This tank and pump unit is mounted on a 5 ton cargo carrier and consists of two 500 gallon aluminium tanks and a 125 gal/min electric motor-driven pump, together with the related fuel-dispensing equipment mounted on an A-frame. The unit will convert a general purpose military vehicle into an air-transportable fuel-dispensing unit for refuelling military vehicles, ground equipment, aircraft, storage drums and 5 gallon (18.9 litre) cans. Production of these units began during FY89.

Status

Entered US Army service during FY90.

Development Agency

US Army Tank Automotive Command, Mobility Technology Center - Belvoir, Fort Belvoir, Virginia 22060-5843, USA.

VERIFIED

Angus Chemicoil System

Description

The Angus Chemicoil System was designed for the delivery of multipurpose fuels and chemicals overland in bulk from fixed pipelines and storage points. The system may be used over long distances and may be laid by helicopters or vehicles using a minimum of manpower. It may also be used for ship-to-shore deliveries, aircraft refuelling and marine fuel bunkering. The system can be used with JP-4, JP-5 and JP-8 aircraft fuels, AVGAS, COMBATGAS and DF-2 and AM-2 diesel fuels.

The pipes used with the Chemicoil system have a high-quality, polyurethane inner lining and a circular woven and rot-proof reinforcement which may use copper-coated textile strands for static protection. The cover uses high quality polyurethane that is abrasion-, ozone- and weather-resistant. The covers may be coloured camouflage sand, black, or NATO green. Pipes up to 200 m long may be supplied, with diameters of 25, 38, 45, 51, 63.5, 76, 102 and 152 mm. Pipe with a diameter of 25 mm, weighs 0.138 kg/m and 152 mm pipe weighs 1.558 kg/m.

Status

In production. In service with US armed forces and other nations around the world.

Manufacturer

Angus Fire Armour Corporation, 1000 Junny Road, Angier, North Carolina 27501, USA.
Tel: +1 919 639 6151. Fax: +1 919 639 6131.

VERIFIED

Chemicoil system in use by NATO forces

Fuel pumping assembly

Description

The fuel pumping assembly has a capacity of 800 gallons/min (3,028 litres/min) and is a component within the Inland Petroleum Distribution System (IPDS - see separate entry in this section). It was type classified in 1985 and 172 were procured to support 86 IPDS mainline pump stations.

The pump is skid mounted and is a multistage, centrifugal noise suppressed unit configured to be permanently mounted on to a 20 ft/6.096 m ISO flatrack.

Another fuel pumping assembly employed by the IPDS has a capacity of 350 gallons/min (1,325 litres/min). These wheel-mounted units were type classified in 1960. A further 122 noise-suppressed units are scheduled to be procured by the year 2000.

Status

In service with the US Army.

Development Agency

US Army Tank Automotive Command, Mobility Technology Center - Belvoir, Fort Belvoir, Virginia 22060-5843, USA.

UPDATED

800 gallons/min Fuel Pumping Unit ***1996***

WATER SUPPLIES

AUSTRALIA

Permutit water purification units

Description

Permutit Australia manufactured 59 CPC 7.5 and CPC 20 trailer-mounted water purification equipments for the Australian Defence Forces and later supplied 16 air-transportable MRO 10 units, used to supply up to 10,000 litres/h.

The CPC units use a process of chemical coagulation, settling and filtration to clarify water using a high rate settling technique and operate on the concept of a low head clarifier using corrugated interceptor plates. Filtration after settling polishes the water while chlorine sterilises it. The units will make any naturally occurring dirty, muddy, unsafe and undrinkable water potable.

The units are self-contained and on arrival on site require levelling using the jacks mounted on the equipment. After hose connection to the water source, potable water is available within 1 hour.

The CPC 20 unit is mounted on a four-wheeled trailer and can produce 20,000 litres/h. The CPC 7.5 unit is mounted on a two-wheeled trailer and can produce 7,500 litres/h.

Status

MRO, CPC 7.5 and CPC 20 in service with the Australian Army and in Papua New Guinea.

Water purification trailer, Permutit-Boby CPC 20

Manufacturer

Permutit Australia, PO Box 117, Brookvale, New South Wales 2100, Australia.
Tel: +61 2 938 4666. Fax: +61 2 938 6569.

VERIFIED

CANADA

Zenon Advanced Double Pass Reverse Osmosis Water Purification Unit (ADROWPU)

Description

The Zenon Advanced Double Pass Reverse Osmosis Water Purification Unit (ADROWPU) was designed for field use and can be used to treat water that has been contaminated by NBC agents, as well as fresh, brackish and seawater. The ADROWPU employs a double pass reverse osmosis process, that is, a pressure-driven membrane separation process that separates dissolved solutes and suspended substances from water.

The ADROWPU is a fully integrated self-contained system with its own 40 kW diesel power generator and a semi-automatic control system, digital controls and instrumentation. Automatic self-cleaning and pretreatment features are built-in and a pillow tank for storing processed water is provided with each system. Installation takes 20 minutes.

The system is arranged in a self-contained palletised enclosure which enables it to be transported via all modes of military transport, including NATO standard palletised loading systems.

The ADROWPU can also fit inside a standard ISO container. In this form it is known as the Containerized Reverse Osmosis Water Purification System, or CROWPU.

The ADROWPU weighs 8,200 kg when fully loaded and measures 5.4 × 2 × 1.7 m. It can be operated at temperatures ranging from −40 to +50°C.

Typical ADROWPU outputs are as follows:

Typical outputs (litres/day)

Input water	Fresh	Brackish	Sea
Basic unit:			
Without NBC	84,700	81,520	52,390
With NBC	59,250	58,950	52,390
Expanded unit:			
Without NBC	112,650	108,450	69,680
With NBC	78,830	78,380	69,680

Status

In production. Has been deployed operationally to Cambodia, Qatar, Rwanda, Saudi Arabia, Somalia and the former Yugoslavia.

Manufacturer

Zenon Environmental Systems Inc, 845 Harrington Court, Burlington, Ontario, Canada L7N 3P3.
Tel: +1 416 639 6320. Fax: +1 416 639 1812.

UPDATED

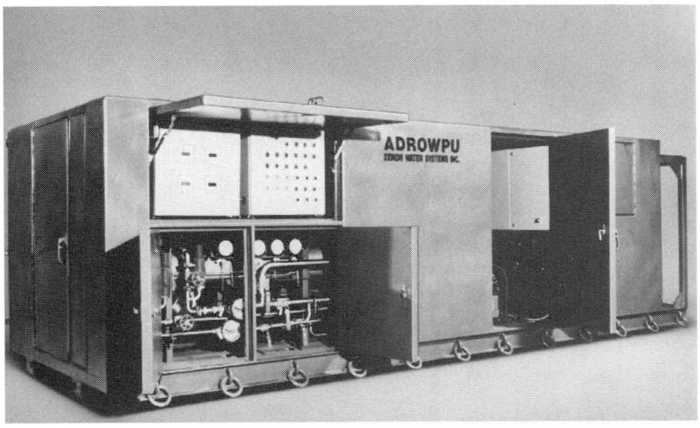

Stand-alone version of Zenon Advanced Double Pass Reverse Osmosis Water Purification Unit (ADROWPU)

Truck-mounted version of Zenon Advanced Double Pass Reverse Osmosis Water Purification Unit (ADROWPU) in travelling configuration

Zenon Seagold Mini-ROWPU water purification modules

Description

The Zenon Seagold Mini-ROWPU family of water purification modules was designed to meet the water purification requirements of mobile rapid-deployment forces. The modules are compact and lightweight and can be used in any terrain to treat fresh, brackish and seawater. The modular system can be configured for double pass operation to treat water contaminated with NBC agents.

There are four main types of Seagold Mini-ROWPU module. They are the Pretreatment - Screen Filter (PTSF); Pretreatment - Nanofilter (PTNF); Reverse Osmosis Single Pass treatment (ROSP); and Power Supply Unit (PSU). The latter may be a self-contained petrol unit (PSU/G) or electrical (PSU/E) using a generator. Typical dimensions for each module are 0.6 × 0.6 × 1.5 m.

The PTSF has a typical output of 500 litres/h (PTSF-500) while the PTNF output is 500 or 1,000 litres/h (PTNF-500 or PTNF-1000). A typical ROSP output is 250 litres/h (ROSP-250) but ROSP modules connected in parallel can boost output, for example, two ROSP-250 modules placed in parallel and supplied from a PTSF-500 can have an output of 500 litres/h. For NBC-contaminated water treatment a typical configuration could be a PTNF-500 feeding two ROSP-250 modules in parallel, with their outputs combining to supply a further ROSP-250, thus providing 250 litres/h of water free from NBC contamination.

Status

In production.

Manufacturer

Zenon Environmental Systems Inc, 845 Harrington Court, Burlington, Ontario, Canada L7N 3P3.
Tel: +1 416 639 6320. Fax: +1 416 639 1812.

VERIFIED

Stand-alone example of Zenon Seagold Mini-ROWPU water purification modules

FRANCE

ACMAT UMTE 48 mobile water treatment unit

Description

The ACMAT UMTE 48 mobile water treatment unit is carried on a standard ACMAT RM 215 S single-axle trailer and can process raw water containing 120 mg/litre of suspended matter at the rate of 2 m³/h. The unit is self-contained and automatic under the control of an electronic controller. The system can be used either for plain filtration or filtration with flocculants. If required, the unit can be removed from the trailer and placed on a skid mount. Power for the system can be obtained from a 220 V mains supply or from a generator.

Status

In production.

Manufacturer

ACMAT (Ateliers de Construction Mécanique de l'Atlantique), Le Point du Jour, F-44600 Saint-Nazaire, France.
Tel: +33 2 40 22 33 71. Telex: 700913 f.
Fax: +33 2 40 66 30 96.

VERIFIED

ACMAT UMTE 48 mobile water treatment unit

ACMAT UMTE 48 mobile water treatment unit in use

MEP 200 water treatment unit

Description

Developed by Mengin for the French Army, the MEP 200 water treatment unit is portable, with a transit case transport weight of 25 kg. The unit has a flow rate of 200 litres/h using a patented filtration process including filtration-adsorption with activated carbon and anti-bacteriological sterilisation. Powered by a hand pump, the unit can treat water from any kind of surface water. It eliminates organic, mineral and bacterial pollutants up to a maximum of 1,200 litres, depending on the quality of the raw water. Changing filtration cartridges takes about 10 minutes, while setting up the unit for initial use takes about 5 minutes.

MEP 200 water treatment unit **1996**

MEP 200 water treatment unit transport case with consumables pack **1996**

Specifications
Flow rate: 200 litres/h
Capacity: up to 1,200 litres
Weight:
(dry) 18 kg
(transport) 25 kg
Dimensions: (transport) 580 × 260 × 370 mm

Status
In production.

Manufacturer
Compagnie Mengin - Group GMC, ZI Amilly, BP 901, F-45209 Montargis Cedex, France.
Tel: +33 2 38 89 83 00. Fax: +33 2 38 89 83 09.

Marketed by Giat Industries, 13 route de la Minière, F-78034 Versailles Cedex, France.
Tel: +33 2 39 49 81 02. Fax: +33 2 39 49 81 06.

VERIFIED

PRONAL small capacity water storage tanks

Description
These small capacity tanks are manufactured from polyester fabric coated with non-toxic PVC and are resistant to ultraviolet radiation and all climatic conditions. They are produced in capacities from 20 to 500 litres; the 20 litre size is fitted with carrying handles for use when full. The larger sizes are used laid flat upon the ground. All are fitted with a 19 mm filling and emptying hose and a symmetrical coupling and plug. When not in use they can be rolled up for storage or carrying.

Specifications

Capacity	Length (empty)	Width (empty)	Height (full)	Weight (empty)	Volume (folded)
20 litres	0.62 m	0.4 m	0.14 m	1.2 kg	5 dm³
50 litres	0.75 m	0.62 m	0.17 m	1.65 kg	6.5 dm³
100 litres	1.24 m	0.6 m	0.22 m	2.45 kg	10 dm³
200 litres	1.24 m	0.95 m	0.26 m	3.4 kg	12.5 dm³
500 litres	1.7 m	1.24 m	0.39 m	5.5 kg	22 dm³

Status
In production. The 20 litre tanks are in service with the Swiss Army.

Manufacturer
PRONAL SA, Z. I. Roubaix Est, BP 18, F-59115 Leers, France.
Tel: +33 3 20 99 75 00. Fax: +33 3 20 99 75 20.

VERIFIED

PRONAL flexible open vat water storage tanks

Description
These open vat tanks are used for the large scale storage of water and other liquids such as decontamination fluids. They have an advantage over conventional storage tanks in that the tank may be filled and emptied very quickly and more than one user point may be employed at any one time. The tanks are placed on circular groundsheets and are filled from an over stand pipe or hose which is usually a 2 in/51 mm diameter component. The sides of the tank rise with the volume contained and, if required, a tarpaulin may be used to cover the contents. The tanks are constructed of polyester fabric coated with PVC.

Specifications

Capacity	Diameter	Height	Weight	Volume (folded)
5,000 litres	3 m	1 m	40 kg	0.45 m³
10,000 litres	4.4 m	1 m	62 kg	0.7 m³

Status
In production.

Manufacturer
PRONAL SA, Z. I. Roubaix Est, BP 18, F-59115 Leers, France.
Tel: +33 3 20 99 75 00. Fax: +33 3 20 99 75 20.

VERIFIED

PRONAL flexible water storage tanks

Description
PRONAL produces a wide range of flexible water storage tanks to suit specific customer requirements, therefore examples given in the Specifications table should be regarded as typical rather than applicable to individual cases. The material used for the tanks is a PVC-coated polyester fabric with a weight of 1.1 kg/m². The usual fittings are 2 in/51 mm filling/discharging valves and caps which can be connected by adaptors to 51 or 76 mm hoses.

Status
In production. In service with the French armed forces and other nations.

Manufacturer
PRONAL SA, Z. I. Roubaix Est, BP 18, F-59115 Leers, France.
Tel: +33 3 20 99 75 00. Fax: +33 3 20 99 75 20.

VERIFIED

Specifications

Capacity	Length (empty)	Width (empty)	Height (full)	Weight (empty)	Volume (folded)
1,000 litres	2.46 m	1.35 m	0.6 m	20 kg	0.09 m³
2,000 litres	2.46 m	2 m	0.9 m	25 kg	0.11 m³
5,000 litres	3.69 m	2.5 m	1.1 m	35 kg	0.17 m³
10,000 litres	4 m	3.69 m	1.2 m	50 kg	0.25 m³
15,000 litres	4.92 m	4.1 m	1.2 m	65 kg	0.3 m³
20,000 litres	5 m	4.92 m	1.3 m	75 kg	0.4 m³
25,000 litres	6 m	4.92 m	1.3 m	85 kg	0.45 m³
30,000 litres	7 m	4.92 m	1.3 m	100 kg	0.5 m³
35,000 litres	6.15 m	6.1 m	1.4 m	110 kg	0.55 m³
40,000 litres	6.8 m	6.15 m	1.4 m	120 kg	0.6 m³
45,000 litres	7.5 m	6.15 m	1.4 m	130 kg	0.65 m³
50,000 litres	8.2 m	6.15 m	1.4 m	140 kg	0.7 m³
60,000 litres	7.8 m	7.38 m	1.45 m	160 kg	0.8 m³
70,000 litres	8.9 m	7.38 m	1.45 m	180 kg	0.9 m³
80,000 litres	10 m	7.38 m	1.45 m	200 kg	1 m³
100,000 litres	10.4 m	8.61 m	1.45 m	245 kg	1.2 m³
125,000 litres	12.7 m	8.61 m	1.45 m	295 kg	1.5 m³
150,000 litres	13.05 m	9.84 m	1.45 m	345 kg	1.75 m³
200,000 litres	17 m	9.84 m	1.45 m	445 kg	2.25 m³

Individual water treatment equipment

Description

The hand-operated Individual Water Treatment Equipment is intended for use by individual soldiers to remove suspended particles and bacterial, mineral and organic pollution from untreated water sources. Water to be treated is sucked into the system through a flexible tube fitted with a filter; the maximum pumping height is 1 m. The water is then passed through a processing device comprising a disposable cartridge which remains efficient for a maximum of 50 litres having a feed flow rate of approximately 6 litres/h. Processed water is recovered directly into a standard water bottle.

The equipment is intended for military use and has an external volume of 1.5 litres and weighs 1.5 kg. Start-up time is less than 1 minute and reversion to the transport mode takes less than 2 minutes.

Status

In production for the French Army.

Manufacturer

SRTI-SODETEG, 9 avenue Réaumur, BP 22, F-92352 Le Plessis Robinson, France.
Tel: +33 5 45 37 86 00. Telex: 631096 tcsf f.
Fax: +33 5 45 37 88 30.

VERIFIED

SRTI-SODETEG individual water treatment equipment

GERMANY

PBU 50M well-boring equipment

Description

The PBU 50M well-boring plant is used to sink bored, tube and dug wells. The bored and dug wells are sunk by the rotary boring method and tube wells by the boring and cable drilling methods.

The complete equipment comprises two ZIL-131 (6 × 6) 3,500 kg trucks and two twin-axled trailers. One ZIL-131 serves as the drilling vehicle and the rigs, spares, tools and accessories are carried on the other truck and the two trailers. The normal crew is four persons.

The boring equipment is hinged on the rear frame of the drilling vehicle and during transportation it lies horizontally on the vehicle. Hydraulics are used to raise the rig and carry out the drilling, with the power for the two hydraulic pumps and gears coming from the vehicle engine. Two support jacks support the bore frame in the operating position and serve at the same time to align the bore frame.

The drilling vehicle is equipped with a cable winch coupled to a ram. The cable winch is used to raise and lower the drilling tools, the casing for tube wells and the concrete or timber collars for dug wells. The ram enables bores to be sunk by the percussion drill method, the travel of the drilling tool being 1 m at a maximum of 48 actions/minute. Torque transmission to the earth auger is carried out by a rotation gear pivoted at the bore frame which has an interchangeable liner inserted to form the connection with the bore rod. The liner can be disengaged and can be swung into two diametrically opposed directions to allow drilling tools to be connected to the rotation gear. Below the upper mitre gear there is a safety clutch with a shear pin as an overload protection device — normally only one shear pin is used.

Operating controls are located on the left-hand side of the vehicle with an auxiliary drive lever behind the driver's cab. The controls are for the hydraulics, drilling transmission, vehicle clutch, speed regulator, cable drum, ram and the cable drum brake. There are also gauges and monitoring instruments behind the driver's cab for the hydraulic fluid pressure, oil and coolant temperatures, the running time and the drill speed. Some of these indicators are duplicated in the driver's cab.

Using the PBU 50M it is possible to sink two bored wells up to 50 m deep and run them simultaneously, with each well delivering an average of 2.8 to 4.5 m^3/h of water. Dug wells can be made to a depth of 15 m.

Among the drilling tools used by the PBU 50M are augers with an external diameter of 200 mm and a useful length of 2.5 m. Using the PBU 50M it is possible to dig well shafts up to 1 m in diameter.

Status

In service with former Warsaw Pact armed forces.

Manufacturer

(Drilling equipment only): state factories.

VERIFIED

Kärcher WATERCLEAN 1000 water treatment system

Description

The Kärcher WATERCLEAN 1000 water treatment system can continually process water contaminated by NBC agents into water of drinking quality. The unit meets World Health Organisation (WHO) requirements.

The operational capacity of the WATERCLEAN 1000 system is a minimum of 1,000 litres/h. Various chemicals are added to destroy chemical and biological agents, viruses and bacteria for operating periods up to 8 hours, after which filters have to be changed and chemicals replaced. The treatment process is fully automatic using a memory-programmable control system with minimal operator intervention. Radioactive ions can be removed using reverse osmosis; this feature allows the desalination of seawater. Each unit has its own integral diesel-powered generator but the unit may be run from an external power supply.

During a 24 hour working day the WATERCLEAN 1000 can purify a minimum of 24,000 litres of chemically or biologically contaminated water, approximately 15,000 litres of radioactively contaminated water, or 15,000 litres of sea or brackish water, including groundwater containing an excess of nitrates.

Kärcher WATERCLEAN 1000 water treatment system in use ***1997***

The system, complete with all optional modules, weighs approximately 1,445 kg when empty. It is 2.405 m long, 1.755 m wide and 1.494 m high. The unit can be mounted on any truck or trailer with a 1.5 tonne carrying capacity.

A wide range of accessories is available. The specially designed double axle trailer provides enough space to carry a WATERCLEAN 1000 with chemicals sufficient for treating one million litres of drinking water and a water distribution kit complete with a 1,500 litre storage tank.

In situations where processed water is to be packaged in bottles or sachets, specially designed packaging machine is available.

Status

In production.

Manufacturer

Alfred Kärcher GmbH & Co, Alfred-Kärcher-Strasse 28-40, D-71364 Winnenden, Germany.
Tel: +49 7195 142797. Telex: 724432.
Fax: +49 7195 14212.

UPDATED

Kärcher WATERCLEAN 500 series water treatment system

Description

The key features of the Kärcher WATERCLEAN 500 series water treatment system are a modular design to suit the system according to customer requirements, simple and fully automatic operation.

The WATERCLEAN 500 series consists of two models, the WATERCLEAN 500 RO and the WATERCLEAN 500 BC. Both systems require no chemicals during operation and are stackable up to three units. The WATERCLEAN 500 RO is designed to desalinate water up to 55,000 ppm at +35°C and produces 300 to 700 litres/h of drinking water depending on the raw water salt content; it has an integrated UV-disinfection system and integrated post-chlorination. The WATERCLEAN 500 BC is designed to decontaminate water containing chemical and/or biological agents using a combination of ozone, UV and activated carbon. The system can produce 500 litres/h without any water loss.

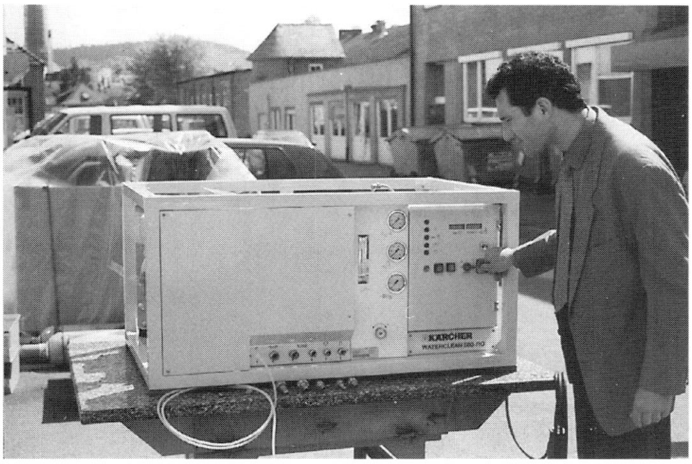

Kärcher WATERCLEAN 500 RO water treatment system ***1997***

Kärcher WATERCLEAN 500 BC water treatment system ***1997***

Specifications
WATERCLEAN 500 RO
Weight: (total) 175 kg
Length: 1.2 m
Width: 800 mm
Height: 600 mm
Capacity:
(at 55,000 ppm) 250 litres/h
(at 35,000 ppm) 350 litres/h
(at 10,000 ppm) 500 litres/h
Power consumption: (max) 3.2 kW

Status
In production.

Manufacturer
Alfred Kärcher GmbH & Co, Alfred-Kärcher-Strasse 28-40, D-71364 Winnenden, Germany.
Tel: +49 7195 142797. Telex: 724432.
Fax: +49 7195 14212.

UPDATED

ISRAEL

Aquaport desalination plants

Description
Aquaport desalination plants are an advanced type of low-temperature evaporator designed to convert sea or brackish water into pure (distilled) water for drinking or industrial purposes. The systems may be used in remote or isolated areas and are designed for simple operation, easy maintenance and low-energy consumption. The plants are rugged and of simple mechanical design. They are produced in various standard models with varying capacities. Mobile units can produce up to 150 m^3 a day. The units are all electrical, fully automatic self-contained units and require only push-button operation.

Status
In production.

Manufacturer
IDE Technologies Limited, PO Box 591, IL-43104 Ra'anana, Israel.
Tel: +972 9 909777. Fax: +972 9 909715.

VERIFIED

ITALY

EUROVINIL foldable container for drinking water

Description
The EUROVINIL foldable container for drinking water is produced in various sizes and capacities using polyester fabrics covered with non-toxic special resins. All containers are provided with filling and dispensing accessories as well as care and repair kits for emergency repairs. The smaller portable containers are also provided with a special strap for stowage on vehicles.

Versions for use with fuels and other liquids are available. These can be provided with internal separation walls to allow the transport and storage of more than one type of liquid at any one time.

Status
In production.

Manufacturer
EUROVINIL Industries SpA, Via Genova 3, I-58100 Grosseto, Italy.
Tel: +39 564 487300. Fax: +39 564 487222.

VERIFIED

A typical EUROVINIL foldable container for drinking water

NETHERLANDS

Promac Aquaset MLT mobile drinking water units

Description
The Promac Aquaset MLT mobile drinking water units use a specially developed and patented combination of membrane separation and other purification techniques to produce high-quality water from any source of contaminated surface water. Two sets are available.

The Aquaset MLT Mark III was designed to produce clean water from sources polluted by NBC agents and systems are also available for brackish and seawater feed sources. The first unit was designed and built in co-operation with the TNO PML Royal Netherlands Army Research Laboratory and underwent an initial two-year period of laboratory and field testing.

Several units are operational. The unit is contained in a military container measuring 4.25 × 2.2 × 2.09 m and weighing approximately 4,800 kg. The unit is operated by two personnel and can commence clean water production within 30 minutes of arrival on site. Output is a continuous 3,500 litres/h. The unit can remain in operation for 30 days without support materials.

The Aquaset MLT Mark I was developed for the Royal Netherlands Marines. The Aquaset Mark I is derived from the Aquaset MLT Mark III but produces 500 litres/h and measures 2.5 × 1.32 × 1.36 m. The unit is supplied in an insulated container for operations in cold environments.

Status
In service.

Manufacturer
Promac Watertreatment Division of Promac BV, Schimminck 1, PO Box 22, NL-5300 AA Zaltbommel, Netherlands.
Tel: +31 4180 13855. Fax: +31 4180 12400.

VERIFIED

RUSSIAN FEDERATION AND ASSOCIATED STATES (CIS)

Mobile water desalination plants OPS and POU

Description
These mobile water desalination plants were produced mainly for use in desert regions and both are mounted on truck chassis. The smallest unit is the POU, typically mounted on a ZIL-157 (6 × 6) 2,500 kg truck chassis, which has an output capacity of 320 litres/h. The unit is powered by a petrol engine. The OPS has a capacity of approximately 2,000 litres/h and is carried on a KrAZ-214 (6 × 6) 7,000 kg truck chassis. The process used with the OPS not only distils the water but also purifies it biologically. Power for the OPS is provided by a towed generator.

Status
In service with the RFAS.

Manufacturer
State factories.

VERIFIED

Mobile water desalination plant OPS

LBU-200 mobile well-drilling equipment

Description
The LBU-200 mobile well-drilling equipment is capable of drilling for water, testing its purity and providing limited storage capacity. The equipment is mounted on three KrAZ-255B (6 × 6) 7,500 kg trucks and three 2-PN-6M trailers. One truck is fitted with the drilling rig, while a second carries the LGR-3 laboratory. The third truck, fitted with a hydraulic loading crane, carries hose, purification equipment and other ancillaries. RBD-5000 water tank components are carried on the trailers.

Status
In service with the RFAS.

Manufacturer
State factories.

VERIFIED

SOUTH AFRICA

Waterbuffel water treatment system

Description
The Waterbuffel (Water Buffalo) water treatment system is carried in a standard 3 m container and can be placed on level ground next to a source of raw water to generate up to 3,500 litres/h of fresh water. The system includes pretreatment, filtering and chlorination stages. Processed water can be stored within storage tanks until it is required.

For transportation the system can be carried on a SAMIL 20 (4 × 4) 2,000 kg truck or any 3 tonne flatbed truck. It can also be towed using a set of transporter wheels. The container is equipped with lifting eyes for handling.

Status
In service with the South African National Defence Force.

Manufacturer
Enquiries to: Armscor, Private Bag X337, Pretoria 0001, South Africa.
Tel: +27 12 428 1911. Telex: 30217.
Fax: +27 12 428 5635.

VERIFIED

Waterbuffel water treatment system in use

Waterbuffel water treatment system container on tow behind a SAMIL 50 (4 × 4) 4,800 kg truck

SWITZERLAND

Katadyn Pocket Filter (PF) individual water treatment equipment

Description

The Katadyn Pocket Filter (PF) produces bacteria free water instantly from sources of dubious quality. It can purify even heavily sedimented turbid water with high bacteriological content or parasitic contamination levels. It has been in service with various armies for over 30 years.

Purification is carried out by placing the intake tube, with strainer, into the water source and operating the integral hand pump. The water is then filtered mechanically through a microporous deep bed ceramic element, with the mineral contents of the water remaining unchanged. Output capacity is approximately 1 litre/min.

The PF has an unlimited shelf life, as silver within the filter prevents contamination. Maintenance is carried out by brushing the surface of the ceramic filter. An enclosed measuring gauge displays the state of exhaustion of the filter element. Construction overall uses aluminium alloy

Each PF is issued with a carrying case, measuring gauge, cleaning brush, spare O-ring and a water- and tear-resistant user manual.

Specifications

Capacity: approx 10,000 litres
Flow rate: 1 litres/min
Weight: 535 g
Length: 245 mm
Diameter: 50 mm
Intake length: 1 m

Status

In production. In service with the US Marine Corps.

Manufacturer

Katadyn Products Inc, Water Treatment, Birkenweg 4, CH-8304 Wallisellen, Switzerland.
Tel: +41 1 839 2111. Fax: +41 1 830 7942.

VERIFIED

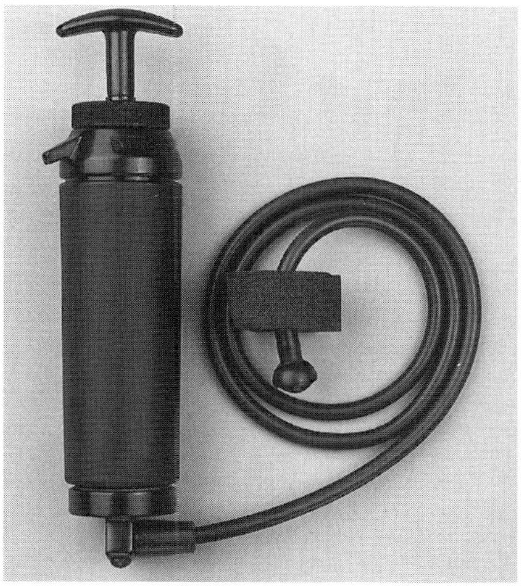

Katadyn Pocket Filter (PF) individual water treatment equipment for US Marine Corps ***1997***

Katadyn Expedition Filter (KFT) portable squad water treatment equipment

Description

The Katadyn Expedition Filter (KFT) portable squad water treatment equipment features the same Katadyn ceramic filter technology as the PF (see previous entry) but is provided with a metal reinforced industrial filter element having a higher output and capacity. It is self-contained, with an integral double action piston pump actuated by a large handle; suction height is up to 4 m. The flexible intake tube is reinforced against wear and tear. Coarse and fine metal screens in the strainer section filter particles from incoming water. A safety device relieves internal pressures when silt-laden water threatens to clog the unit.

The unit construction overall is to heavy duty standard. The footrest, which can be adjusted in height, is made of steel. The unit is packed in a canvas carrying bag complete with maintenance tools and accessories.

Specifications

Capacity: approx 25,000 litres
Flow rate: 3.5 litres/min
Weight: 4.4 kg
Length: 580 mm
Diameter: 140 mm
Width overall: 200 mm
Intake length: 2 m

Status

In production. In service with the German Army.

Manufacturer

Katadyn Products Inc, Water Treatment, Birkenweg 4, CH-8304 Wallisellen, Switzerland.
Tel: +41 1 839 2111. Fax: +41 1 830 7942.

VERIFIED

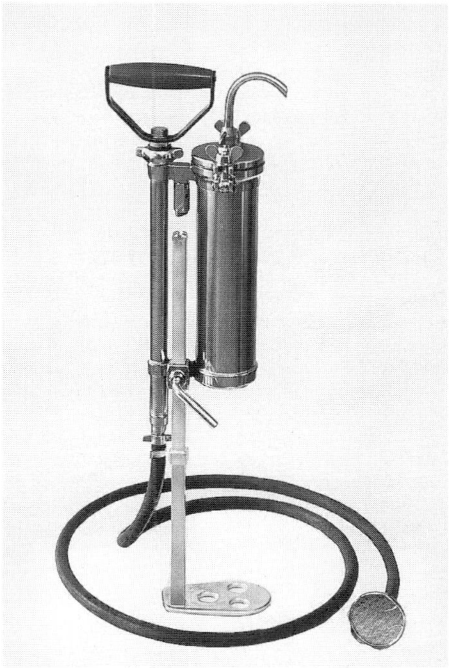

Katadyn Expedition Filter (KFT) portable squad water treatment equipment ***1996***

UNITED KINGDOM

Acrokool Nomad water filter and cooling trailer

Description

The Acrokool Nomad Mark III is a mobile water filter and cooling unit carried on a single-axle trailer. The filters used are Acrokool BF5 filters fitted with five Super Sterasyl silver-impregnated ceramic elements with activated carbon cores to remove unpleasant tastes and odours. The water is passed through the filter system by an internal pump driven from the internal petrol/diesel engine, which also drives the cooling unit. The single-cylinder, air-cooled four-stroke engine also drives a generator for battery charging. Fuel for the engine comes from a 24.6 litre (nominal) tank. The main water tank has a 378.5 litre capacity and the filter/cooling system can supply 94.6 litres/h of treated water. The fuel tank will provide the engine with an average non-stop running time of 15 hours.

The trailer unit has two jacks at the rear and a jockey wheel near the towing point. The simple control panel is on the right-hand side facing forward and at the rear are

Acrokool Nomad water filter and cooling trailer

four outlet taps, three for glass-filling and one for filling larger containers. Spare filters are carried in a dustproof box inside the unit housing.

Specifications

Shipping dimensions
Weight: 747 kg
Length: 2.921 m
Width: 1.727 m
Height: 1.27 m

Status

In production.

Manufacturer

Acrokool Limited, Unit 1, The Shires, Shirehill Industrial Estate, Thaxted Road, Saffron Walden, Essex CB11 3AN, UK.
Tel: +44 1799 513631. Fax: +44 1799 513635.

VERIFIED

Acro-Well water filtration systems

Description

The basic Acro-Well is a mobile, self-contained unit for the provision of germ-free drinking water, clean water for domestic uses and the transfer of natural water for irrigation, firefighting and so on. The power of the self-priming centrifugal pump is provided by a petrol engine, although diesel or electric units may be used as alternatives.

The Acro-Well will provide germ-free drinking water at the rate up to 4,000 litres/h or 5,000 litres/h of clean water or 50,000 litres/h of natural water. It can also be used as a water pump. Included with the equipment are a reinforced suction hose and foot strainer and a 7.62 m minimum length of hose. The pump can lift water from a height of 7.5 m and is fitted with a non-return valve. There is an SF2/NP2 prefilter and three BF5/NP5 Super Sterasyl purifiers together with the associated pipework and control valves. The entire unit is mounted on a steel frame with two 250 mm diameter wheels for mobility and two levelling jacks for stability.

Each of the Super Sterasyl filters can be cleaned independently (with supplied brushes) while the unit is operating. Each of these filters uses silver-impregnated ceramic elements with activated carbon cores to remove unpleasant tastes and odours and each filter is contained in a nylon-coated aluminium housing.

The net weight of an Acro-Well unit is 200 kg. Each unit is 1.67 m long, 1.06 m wide and 0.91 m high.

Associated Acro-Well products are as follows:

Mini Acro-Well
The Mini Acro-Well is electrically driven and is completely automatic. It can supply up to 1,000 litres/h of germ-free drinking water or 4,500 litres/h of unfiltered water for domestic purposes. The Mini Acro-Well weighs 35 kg and dimensions are: length 0.67 m, width 0.45 m and height 0.49 m.

Mini Acro-Well with prefilter
This unit is similar to the Mini Acro-Well but has an additional prefilter to remove suspended solids when dirty water has to be used. It has the same dimensions and weight as the Mini Acro-Well.

Hand-operated Mini Acro-Well
This is basically a Mini Acro-Well with prefilter with the electric motor replaced by a hand-operated pump for use in remote areas. The maximum flow rate of this unit is 350 litres/h of drinking water and the unit is mounted in a heavy steel frame with roll-over bars and carrying handles. Shipping weight is 80 kg.

Acro-Well storage tank
This temporary storage tank can hold up to 9,000 litres of water but can also be used to hold other liquids such as fuel oils, or dry materials. It is transported in a robust carrying case and can be erected by one man within minutes. Aluminium alloy extrusions are used to support the chemically resistant PVC-coated fabric liner. Options for the tank include a cover and disposable liners. When erected, the tank has a diameter of 3.26 m and is 1.5 m high. When packed in its case the tank weighs 80 kg.

Status

All the above items are in production. The Acro-Well is in service with the UK Ministry of Defence in the Falkland Islands and is also used by several other armed forces.

Manufacturer

Acrokool Limited, Unit 1, The Shires, Shirehill Industrial Estate, Thaxted Road, Saffron Walden, Essex CB11 3AN, UK.
Tel: +44 1799 513631. Fax: +44 1799 513635.

VERIFIED

Airborne inflatable water tanks

Description

Airborne inflatable water tanks are widely used for both civil and military applications where the ability to store water or set up an open water relay is required at short notice. The tanks are man-portable and each is packed in a valise.

The tank is deployed as follows: the tank is removed from the valise and laid on a flat surface; the ring capping the top of the tank wall, which is formed from a 350 mm diameter tube, is inflated and floats on the water as the tank is filled, supporting the wall. A webbing strap and buckle keep the filling hose in position while the tank is being filled. The tank has a cover supported on an inflatable float, which protects the contents from contamination and growth of algae. A water outlet connection is provided at the base of the wall. The tank takes a maximum of 3 minutes to inflate and can be used on inclines up to 1 in 7.

Other sizes to a maximum capacity of 113,650 litres are available to special order.

Specifications

Capacity	**22,730 litres**	**11,365 litres**	**1,820 litres**
Weight	77.18 kg	44 kg	19 kg
Dimension filled (top)	5.02 m	3 m	1.5 m
(base)	5.58 m	3.6 m	1.8 m
(height)	1.4 m	1.4 m	1.1 m
Dimensions packed	1.68 × 0.76 × 0.46 m	1.25 × 0.76 × 0.46 m	0.76 × 0.48 × 0.38 m

Status

In production. In service with the British Army.

Manufacturer

Airborne Industries Limited, Airborne Industrial Estate, Arterial Road, Leigh-on-Sea, Essex SS9 4EF, UK.
Tel: +44 1702 525265. Telex: 99412 airbrn g.
Fax: +44 1702 510454.

VERIFIED

Airborne inflatable water tank being filled

Filled Airborne inflatable water tank

Reynolds Boughton water tanker

Description

Reynolds Boughton (Devon) Limited produces a mobile de-ionising water unit mounted on a 6 × 4 chassis. The unit can produce de-ionised water from any potable water source and has a tank capacity of 11,750 litres. Two outlets deliver or dispose de-ionised water at a rate of 500 litres/min. The quality of the de-ionised water is constantly monitored by an automatic onboard system.

Reynolds Boughton also produces a comprehensive

range of water (and fuel) tankers of the rigidly-mounted, trailer-mounted and demountable DROPS type.

Status

In production. In service with the UK Ministry of Defence.

Manufacturer

Reynolds Boughton (Devon) Limited, Winkleigh Airfield, Winkleigh, Devon EX19 8DR, UK.
Tel: +44 1837 83555. Telex: 83132.
Fax: +44 1837 83768.

VERIFIED

Reynolds Boughton water tanker

Stellar water purification equipment

Description

Stella-Meta manufactures mobile water purification equipment to treat fresh, saline and NBC-contaminated water. Capacities range from the hand-portable ST1 producing 6.8 m^3/h through to the Type 10/A5 trailer-mounted unit producing 23 m^3/h.

Each Stellar unit typically comprises a pump set module, a precoat filtration module and a sterilising module. The individual units can be used separately when required to provide greater mobility in use.

In addition to the water purification equipment, Stella-Meta is able to supply a complete system of equipment to store and distribute treated water, using the Stellaflex range of flexible tanks (see separate entry in this section).

Equipment is either tubular frame-mounted hand-portable with separate filtration, power and sterilisation modules, or two-wheel trailer-mounted as an option. Sterilisation is carried out with either a stand-alone hypochlorite powder dosing module, or is incorporated into the power source in the form of an electro-chlorinator.

Status

In production. In service with armed forces throughout the world.

Manufacturer

TM Products Limited, Winnal Valley Road, Winchester, Hampshire SO23 0LL, UK.
Tel: +44 1962 705200. Fax: +44 1962 841344.

VERIFIED

Stellar Type ST1 transportable water purification set

Specifications

Type	AB1A	AB3	ST1
Filter			
Max rated capacity	2.7 m^3/h	4.5 m^3/h	6.8 m^3/h
Type	precoat	precoat	precoat
Weight	47 kg	81 kg	78 kg
Volume	0.23 m^3	0.27 m^3	0.49 m^3
Power Module			
Pump type	positive displacement	centrifugal	centrifugal
Engine type	petrol	diesel	diesel
Weight	73 kg	146 kg	98 kg
Volume	0.25 m^3	0.45 m^3	0.21 m^3
Sterilisation			
Type	sterilant doser	electro-chlorination	sterilant doser
Weight	33 kg	in power module	33 kg
Volume	0.18 m^3	in power module	0.22 m^3
Frame	tubular	tubular	tubular
Option	trailer	trailer*	trailer

* can be dropped by parachute

Stellar AB3 air-portable water purification equipment

Stellar Type AB1A transportable water purification set

Stellar Type 10A/5 trailer-mounted water purification equipment

Description

The Stellar Type 10A/5 trailer-mounted water purification set has a nominal rated capacity of 23 m^3/h of drinking water. It is a self-contained unit with its own diesel engine/pump/alternator set, incorporating a filter powder pretreatment plant, hypochlorite steriliser and a Stellar pressure filter with its own cleaning system.

Accessories, tools and spares are stowed within the trailer and bins are provided for storing the filter powder and hypochlorite, an efficient sterilising agent which takes immediate effect in the water.

The unit is so arranged, that the pump can be used alone, the pump and filter can be used without the steriliser, or the pump and steriliser can be used without the filter. This saves filter powder and sterilant where filtration and/or sterilising is unnecessary, for example for washing water and similar domestic purposes.

All the equipment is mounted on a four-wheel trailer with suitable bodywork, side and end screens. The side and end screens are mounted on metal frames which are hinged to give protection to the operator.

The Type 10A/5 weighs 2,700 kg, with dimensions of 4.5 × 2.3 × 2.2 m.

Status

In production.

Manufacturer

TM Products Limited, Winnal Valley Road, Winchester, Hampshire SO23 0LL, UK.
Tel: +44 1962 705200. Fax: +44 1962 841344.

VERIFIED

Stellar Type 10A/5 trailer-mounted water purification equipment

Stellar saline water treatment units

Description

When brackish or seawater has to be used as a water source Stella-Meta Filters offers two types of equipment to treat both these fluids. Stellar units will convert saline waters to potable fluid that is in accordance with World Health Organisation standards. The complete equipment comprises a low-pressure pump, pretreatment filtration, a high-pressure pump, reverse osmosis modules and post sterilisation. All the equipment is mounted on a single trailer chassis and a diesel unit provides the motive power.

Specifications

Type	DS1/A (/B)	DS2/A (/B)	DS3/A (/C)
Sea and brackish water flow rate	0.45 m^3/h (1.1)	0.9 m^3/h (2.4)	2.6 m^3/h (5)
Weight	1,364 kg	1,364 kg	5,000 kg
Length	4.2 m	4.2 m	7.07 m
Width	1.88 m	1.88 m	2.41 m
Height	1.83 m	1.83 m	2.78 m

Status

In production. In service in several Middle Eastern countries.

Manufacturer

TM Products Limited, Winnal Valley Road, Winchester, Hampshire SO23 0LL, UK.
Tel: +44 1962 705200. Fax: +44 1962 841344.

VERIFIED

Stellar NBC decontamination water purification equipments

Description

The Stellar range of Military Transportable Water Purification Equipment has been augmented to cope with Nuclear, Biological and Chemical (NBC) contamination removal. The treatment process, designed to UK Ministry of Defence specifications, includes precoat filtration, reverse osmosis, activated carbon treatment and post-disinfection capable of removing ionic, organic and bacterial contaminants.

The range of equipments is capable of achieving output capacities of 8.2 m^3/h from fresh water or 2.3 m^3/h of NBC-contaminated water. Variants are available that will also treat seawater. The units can either be fully integral self-contained modules or 'add-on' modules compatible with non-NBC Stellar water purification sets.

The latest equipment in service with the British armed forces is the Type NBC6 with a capacity up to 2.3 m^3/h. The Type NBC6 employs a modular construction incorporating a demountable transfer pump, Stellasep crossflow microfilter, slurry feed pump, diesel-driven high-pressure pump, reverse osmosis module, final trap activated carbon unit and a post-chlorination system. A demountable pump and a pillow tank complete the equipment. The Type NBC6 is mounted in a self-contained space frame unit that can be carried on a truck, trailer, skid base or pallet.

Stellar Type NBC6 NBC decontamination water purification equipment for British Army mounted on a two-wheeled trailer

Stellar Type NBC6 NBC decontamination water purification equipment in its modular space frame

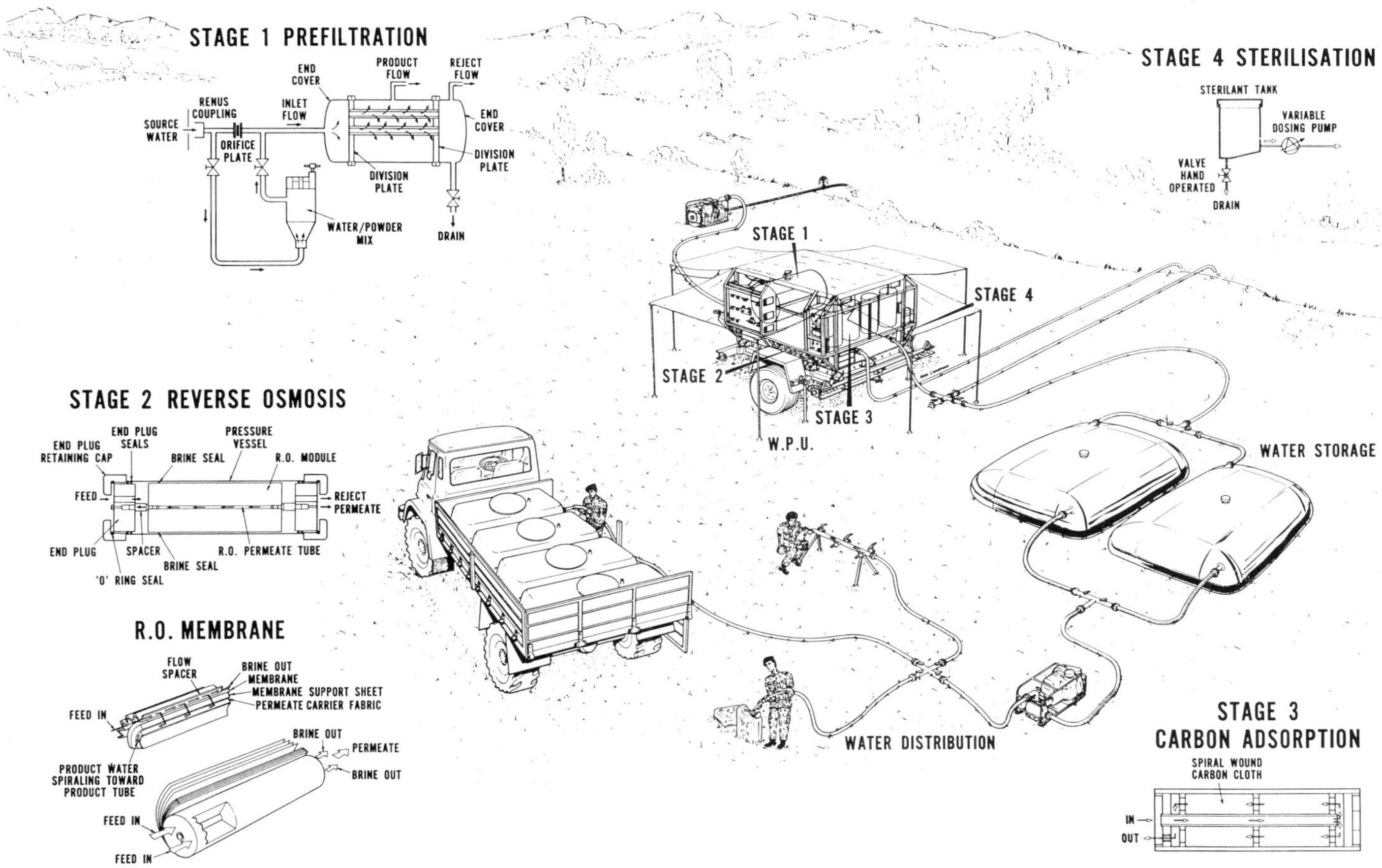

Artist's impression of a typical water processing unit in an NBC environment

Specifications

Type	**NBC6(F)**	**NBC6(SW)**
Flow rate	2.3 m^3/h	1.8 m^3/h
Weight	3,300 kg	3,300 kg
Length, trailer, on tow	4 m	4 m
Width	2.5 m	2.5 m
Height	2.4 m	2.4 m

Status
In production for the British Army.

Manufacturer
TM Products Limited, Winnal Valley Road, Winchester, Hampshire SO23 0LL, UK.
Tel: +44 1962 705200. Fax: +44 1962 841344.

VERIFIED

Stellaflex flexible water tank systems

Description

Stella-Meta has developed a complete flexible water tank system enabling the treatment, storage and distribution of drinking water on a totally mobile basis. The Stellaflex flexible tank range is manufactured in materials approved to international standards for drinking water and has been extensively field proven and climatically tested to UK MoD standard 001/issue 2 A1 and B3.

Available sizes range from 2 to 20 m^3 in the following types:

Onion tanks or open top tanks where access to stored water is required

Pillow tanks, sealed to prevent contamination of stored water

Flexible tanks for instant transport of stored water on flatbed vehicles.

The tanks are lightweight, foldable and fully transportable when not in use and are manufactured from chemically resistant material.

Status
In production. In service with armed forces worldwide.

Manufacturer
TM Products Limited, Winnal Valley Road, Winchester, Hampshire SO23 0LL, UK.
Tel: +44 1962 705200. Fax: +44 1962 841344.

VERIFIED

Stellaflex onion tank

Stellaflex pillow tank in use

Stellaflex flexible water tank in use

Pre-Mac Model PWP personal water purifier

Description

The Model PWP personal water purifier was designed for military use in consultation with UK MoD (Army) medical authorities. It produces safe drinking water from sources in which high levels of bacteria, viral and parasitic contamination are present. It will also remove the organic matter existing in most natural water sources, improving the taste, smell and appearance of the water. The Model PWP can treat at least 500 litres of raw water before replacement and is active for five years.

The Model PWP weighs 500 g, is 140 mm high and 95 mm in diameter. It is made of impact-resistant plastics approved for food usage. The unit is available to fit UK MoD standard water bottles and by means of a simple modification, can fit any water bottle. After screwing the unit on to a bottle, raw water is poured into the top chamber. A 1 litre water bottle can be filled in 8 minutes, including a 2 minute resting time for final purification.

In the Model PWP, a coarse reticulated foam retains large particulate matter and can be removed for cleaning. This is followed by a silver-impregnated activated carbon prefilter to remove further particulate matter and organic contamination. A primary disinfection chamber contains a resin-iodine complex acting as a contact microbiocide to release iodine at a concentration of 4 ppm into the water passing through to carry out secondary disinfection in the water bottle. If required, an optional outlet spout can be supplied to remove residual iodine prior to consumption.

The PWP is marketed in the Far East by TM Products Limited of Winchester, Hampshire, UK.

Status

In production. Supplied to a number of overseas armed forces. Undergoing trials with the British Army and US Marine Corps.

Manufacturer

Pre-Mac (Kent) Limited, 40 Holden Park Road, Southborough, Tunbridge Wells, Kent TN4 0ER, UK.
Tel: +44 1892 534361. Fax: +44 1892 515770.

VERIFIED

The Pre-Mac Model PWP personal water purifier in use

Pre-Mac Model MWP and SWP personal water purifiers

Description

Following the technology developed in the PWP and FWP personal water purifiers (see previous entry), Pre-Mac developed the Model MWP to meet operational requirements where size and weight are critical. The Model MWP is a small hand-operated pump unit in which source water is initially filtered to clean the water and remove organic matter as well as chemical contamination. This is followed by disinfection using an iodine-resin complex which acts as a contact microbiocide releasing a low level of iodine into the water for secondary disinfection. Secondary disinfection occurs during the recommended holding period (3 minutes). The Model MWP will produce 1 litre of water in 5 minutes and its rated capacity is 100 litres. The filtration and disinfection tubes can be changed quickly without any tools.

The Pre-Mac Model MWP in use

The Pre-Mac Model SWP in use

The Model SWP was designed to meet water treatment requirements in emergency or survival conditions and has a capacity of 25 litres, after which it should be discarded. The water treatment process follows the same lines as the Model MWP.

Specifications

Model	MWP	SWP
Capacity	100 litres	25 litres
Flow rate	400 ml/min	200 ml/min
Weight	180 g	60 g
Length	135 mm	135 mm
Diameter	45 mm	20 mm

Status

In production. Supplied to UN, UK Special Forces and overseas armed forces. Both models are NATO coded.

Manufacturer

Pre-Mac (Kent) Limited, 40 Holden Park Road, Southborough, Tunbridge Wells, Kent TN4 0ER, UK.
Tel: +44 1892 534361. Fax: +44 1892 515770.

VERIFIED

Pre-Mac Model JWP portable water purification sets

Description

The Model JWP portable water purification sets were designed to supply volumes of potable water from any groundwater source to groups up to 100 personnel. Model JWP units employ progressive stages of filtration to remove particulate matter to below 3 microns, bonded activated charcoal to remove chemical contaminants and an iodine-resin complex for sterilisation by elimination of pathogenic micro-organisms.

The filtration and disinfection cartridges are mounted within a lightweight, corrosion-proof frame. Cartridge housings are manufactured using tough polypropylene and connections are made using PVC fittings approved for potable water. Model JWP purifiers are fitted with a totalising water meter to monitor throughput.

In standard form, Model JWP purifiers are fitted with a high performance hand pump and deliver 4 litres of

Pre-Mac Model JWP portable water purifier in operation

clean, safe drinking water per minute. As options, an electrical pump can be fitted in place of the hand pump or, in a dual form, both electric and hand pumps are fitted. Electric pumps are available to operate from both AC and DC supplies, enabling power supplies from field generators or vehicles to be utilised.

Model JWP units are available with rated capacities of 2,500 or 5,000 litres. A simple and quick operation can fit a set of replacement cartridges at the point of exhaustion.

Specifications

Model	JWP 2500	JWP 5000
Capacity	2,500 litres	5,000 litres
Flow rate	4 l/min	4 l/min
Weight	11 kg	14 kg
Dimensions	500 × 300 × 420 mm	620 × 300 × 420 mm

Status

In production. Supplied to the UK and overseas armed forces as well as emergency/relief organisations. Model JWP units are NATO coded.

Manufacturer

Pre-Mac (Kent) Limited, 40 Holden Park Road, Southborough, Tunbridge Wells, Kent TN4 0ER, UK.
Tel: +44 1892 534361. Fax: +44 1892 515770.

VERIFIED

Pre-Mac solar powered water purifier Solarwater 5 (SW5)

Description

The Pre-Mac solar powered water purifier Solarwater 5 (SW5) is a stand alone system intended to provide safe, clean potable water from contaminated water sources for up to 100 personnel. It is quick and easy to set up by an individual and is then ready for immediate operation.

Pre-Mac solar powered water purifier Solarwater 5 (SW5) ***1996***

The SW5 employs three stages of filtration to remove particulate matter to below 3 µm, including a carbon treatment to remove organic matter and an iodine resin treatment to eliminate harmful micro-organisms.

The SW5 is powered by four 2.6 A solar panels that unfold from a tubular frame housing a six-stage patented water purification system. A hand pump is incorporated in the unit if a power source is not available, or the unit can be connected to a 12 V external power supply. An optional lead is available to connect the unit to 240 V field generators and vehicle batteries.

The SW5 has a purifying rate of 4 to 6 litres/min with a capacity to treat 5,000 litres. Replacement cartridges are available. The filtration and disinfection cartridges are mounted within polypropylene housings in a transport frame.

The four solar panels utilise single crystal silicon cells which are laminated to tempered glass and are anti-reflective coated. The SW5 is provided with a water totalising meter to monitor the throughput and a 12 V pump is powered by the solar cells; a hand pump is provided as a back-up facility. All switches and connectors are suitable for outdoor use.

An OF3 outlet filter is available as an option. This is used to remove residual iodine from the treated water and operates at a rate of 4 litres/min, with a capacity of 5,000 litres between cartridge changes.

Specifications

Weight: 30 kg
Length: 950 mm
Width: 420 mm
Height: 550 mm
Flow rate: 4 to 6 litres/min
Capacity: 5,000 litres

Status

In production.

Manufacturer

Pre-Mac (Kent) Limited, 40 Holden Park Road, Southborough, Tunbridge Wells, Kent TN4 0ER, UK.
Tel: +44 1892 534361. Fax: +44 1892 515770.

VERIFIED

Airborne solar still

Description

The Airborne solar still is an inflatable unit that can produce drinking water from seawater or impure water sources. It is primarily intended for use in emergency situations and is widely used as an integral part of military survival packs.

The solar still consists of an inflatable buoyancy ring that allows the unit to float on water if required. Inside the ring is a black solar collector with a reservoir for impure water underneath. Above the solar collector is a clear cone. Impure water is fed into the solar collector where it evaporates, the water vapour condensing on the inside of the clear cone. The purified water runs down the inside of the cone into a collecting gutter from where it runs into an external collecting bag for use.

The performance of the solar still will depend on the level of solar radiation received which, in turn, is dependent on the geographic location and the local climate. Typical examples are: in the UK on an average February day the still will produce 0.5 litre; on an average March day in the Indian Ocean it will produce 1.5 litres; and on an October day in North Africa, 1.685 litres will be produced.

A typical Airborne solar still has an inflated diameter of 775 mm and a height of 540 mm. Packed weight is 0.85 kg and dimensions (when packed) are 225 × 325 × 35 mm. nstructions for use are printed on the outer waterproof pack.

Status

In production and in widespread use.

Manufacturer

Airborne Industries Limited, Airborne Industrial Estate, Arterial Road, Leigh-on-Sea, Essex SS9 4EF, UK.
Tel: +44 1702 525265. Telex: 99412 airbrn g.
Fax: +44 1702 510454.

VERIFIED

PDTM water carriage packs

Description

PDTM Water Carriage Packs (WCP) were developed to carry drinking water to British Army units in the field, replacing all previous tanks in service. The basic unit of a WCP is a 680 litre polyethylene tank which can be provided with a variety of delivery hoses, a restraint frame, dispenser manifolds and lashing arrangements. The tank includes a manhole for cleaning.

The WCP can be used on standard 750 kg and 1.75 tonne trailers, Land Rovers and 4 tonne trucks as well as other vehicles. An empty WCP can be lifted by two personnel.

Status

In production.

Manufacturer

PD Technical Mouldings plc, Rotation House, 20 Mayday Road, Thornton Heath, Surrey CR7 7HL, UK.
Tel: +44 181 689 4336. Fax: +44 181 683 2491.

VERIFIED

UNITED STATES OF AMERICA

SIXCON water storage and dispensing system

Description

The SIXCON water storage and dispensing system is intended for both stationary and mobile use as a source of potable water — for the SIXCON fuel system and an illustration refer to entry under *Fuel storage and distribution systems.*

Each SIXCON water pump or stainless steel water tank module uses a specially configured shipping frame ensuring that it can be interconnected with five other modules and transported in standard 20 ft/6.096 m ISO configuration. Each water pump module incorporates a SIXCON frame, 125 gal/min

A SIXCON water storage and dispensing system being delivered by CH-53 helicopter

(473 litres/min) pump, diesel engine, nine hoses and fire nozzles. Each water tank module also has a SIXCON frame, 934 gallon (3,535 litre) stainless steel tank, water level gauge, drain valve, hose attachment, covered manhole, connectors and hose storage.

Specifications

Water pump module
Weight:
(empty) approx 794 kg
(with equipment) approx 1,034 kg
Length, external: 1.969 m
Width, external: 2.438 m
Height, external: 1.204 m

Water tank module
Weight:
(empty) approx 1,193 kg
(full) approx 4,536 kg
Capacity: approx 3,535 litres
Tank length: 1.79 m
Tank width: 2.21 m
Tank height: 0.965 m

Status

In production. In service with the US Navy and Marine Corps.

Marketing Agency

Tradeways Limited, 184 Duke of Gloucester Street, Annapolis, Maryland 21401, USA.
Tel: +1 (410) 295 0813. Fax: +1 (410) 295 0821

UPDATED

600 ft water well drilling system

Description

The US Army's 600 ft (182.9 m) well drilling system supports rapid deployment forces during operations in arid areas of the world where no surface water sources exist. The system includes a drilling machine, support vehicle, well completion kit and accessories. The system can be loaded for long-range transport into three C-130 transport aircraft. Navistar F-1954 (6 × 6) trucks are used to transport the system by road.

The system is used by US Army active, reserve and National Guard drilling units and by US Navy Construction Battalions (Seabees).

Status

In service with the US Army and Navy.

Development Agency

US Army Tank Automotive Command, Mobility Technology Center - Belvoir, Fort Belvoir, Virginia 22060-5843, USA.

Manufacturer

George E. Failing Company, Enid, Oklahoma, USA.

600 ft water well drilling system drilling machine being carried on a Navistar F-1954 (6 × 6) support vehicle

VERIFIED

Tactical Water Distribution System (TWDS)

Description

In service since 1981 and intended for use in a rapid deployment force-type situation, the Tactical Water Distribution System (TWDS) is a flexible distribution, storage and receiving system. It can issue a maximum of 720,000 US gallons (2,200,725 litres) of water during the course of a 20 hour working day.

A complete TWDS set includes 10 miles (16.1 km) of hoseline, six pump stations, two collapsible tanks, two distribution points plus the connecting hoses, valves and hardware. Each TWDS is manned by 30 personnel.

The TWDS is designed to distribute water up to 112.4 km. Working pressure is 150 to 200 psi. The basic hardware, apart from the various smaller accessories, consists of 152.4 mm lightweight layflat hose, 20,000 to 50,000 US gallon (75,700 to 189,250 litre) collapsible tanks and 2,271 litres/min centrifugal, diesel-driven pumps.

Tactical Water Distribution System undergoing 30 day trials in California

Status

Over 100 10 mile sets delivered.

Manufacturers

Tanks: Bell Avon Inc, 1200 Martin Luther King Jr. Boulevard, Picayune, Mississipi 39466, USA.
American Fuel Cell and Coated Fabrics Company (Amfuel), 124 South Jackson, PO Box 887, Magnolia, Arkansas 71753, USA.
Hoses: Angus Fire Armour Corporation, PO Box 879, Angier, North Carolina 27501, USA.

VERIFIED

Forward Area Water Point Supply System (FAWPSS)

Description

The Forward Area Water Point Supply System (FAWPSS) is intended to provide large volumes of drinking water to troops in remote areas where no water supplies of any type are available. Water is delivered to a central point by air, truck or any other local method and from there is distributed as required through a network of tanks, pumps and hose-lines. The complete system consists of six 2,271 litre collapsible water drums, one 189 litres/min water pump assembly, three suction hose assemblies, six discharge hose assemblies, two valve assemblies, four 'Y' junctions, four dispensing nozzles, one towing and lifting yoke and the associated hoses. The complete system is air-transportable and can be set up by two personnel. Once in use only one person is required for operation.

Status

In service with US Army.

Manufacturer

American Fuel Cell and Coated Fabrics Company (Amfuel), 124 South Jackson, Magnolia, Arkansas 71753, USA.
Tel: +1 501 234 3381. Fax: +1 501 235 7270.

VERIFIED

GTA drums and superdrums

Description

GTA manufactures drums for the storage and transportation of fuel, drinking water, fluid food products and so on. Two types of water drum are available, a standard drum constructed using rubber-coated fabric and the GTA Superdrum constructed of elastomeric coated fabric. The GTA Superdrum is significantly lighter, but is more rugged and durable.

Three standard sizes are available; 55 gallons (208 litres), 250 gallons (946 litres) and 500 gallons (1,893 litres). The fabric used in all types of construction is puncture- and tear-resistant and each drum is subjected to a hydropressure test following assembly. The inside surface of the potable water drum meets US Food and Drug Administration requirements and imparts neither taste nor odour to the water.

Empty drums collapse to 10 per cent of their original volume for transport or storage, while filled drums can be rolled, floated, towed and lifted by helicopter. Drums can also be dropped free-fall (the 250 gallon drum can be dropped free-fall from 15 ft/4.5 m) or by parachute.

GTA fuel drums are sand coloured and they can be temporarily or permanently repaired in the field when necessary. Hoses, nozzles and valves have quick-connect/disconnect fittings.

These drums have numerous alternative applications including flotation devices (surface or submerged, for example for the construction of temporary bridges), as pneumatic lifting devices or ship fenders and for the underground storage of water supplies.

GTA 55 gallon water drum (left) and 250 gallon Superdrum (right)

Specifications

Capacity	55 US gal 208 litres	250 US gal 946 litres	500 US gal 1,893 litres
Weight (drum)	22.7 kg	81.8 kg	104.5 kg
Weight (Superdrum)	13.6 kg	63.6 kg	72.7 kg
Length	0.876 m	1.524 m	1.524 m
Diameter	0.597 m	1.016 m	1.27 m

Status

In production for the US Army and others.

Manufacturer

GTA Containers Inc, 1410 W. Napier Street, South Bend, Indiana 46601, USA.
Tel: +1 219 288 3459. Fax: +1 219 289 6060.

VERIFIED

GTA static water storage tanks

Description

GTA Containers Inc manufactures a range of static water storage pillow tanks which are available in sizes from 3,000 gallons (11,361 litres) to 225,000 gallons (851,715 litres). All tanks are manufactured in accordance with US military requirements and are thus lightweight, rugged and durable with high abrasion and puncture resistance.

Each tank is constructed using tear-resistant fabric with an elastomeric coating on both sides. The inside surface of the potable water drum meets US Food and Drug Administration requirements and imparts neither taste nor odour to the water. Chafing patches are included as part of the standard design and are located on the interior surface directly opposite each fitting to provide a double-wall thickness supporting every piece of hardware. Handles, each capable of lifting 454 kg, are moulded on to the tank.

All tanks are equipped with standard 4 in/101.6 mm quick-disconnect filler/discharge fittings. Bottom discharge fittings are provided for gravity unloading. A 4 in/101.6 mm reinforced hose with quick-disconnects is also provided.

Each tank is fully collapsible and folds down into a package that occupies 2 per cent of the volume of the filled tank.

Specifications

Capacity	3,000 US gal 11,361 litres	10,000 US gal 37,869 litres	20,000 US gal 75,738 litres	50,000 US gal 189,345 litres
Length	4.27 m	6.71 m	7.32 m	7.62 m
Width	4.27 m	6.71 m	8.53 m	19.81 m
Weight (empty)	61.8 kg	140.9 kg	187.2 kg	415.9 kg

Status

In production.

Manufacturer

GTA Containers Inc, 1410 W. Napier Street, South Bend, Indiana 46601, USA.
Tel: +1 219 288 3459. Fax: +1 219 289 6060.

VERIFIED

GTA Mini Tanks

Description

The GTA Mini Tank is a flexible pillow tank which may be carried by a pick-up truck or van. Three standard sizes are available; 120 gallons (454 litres), 250 gallons (946 litres) and 500 gallons (1,893 litres). The fabric used is puncture- and tear-resistant and each tank is subjected to a hydropressure test following assembly. Empty tanks collapse to 10 per cent of their original volume for transport or storage.

The US Army employs a 140 gallon (530 litre) Mini Tank which is transported on a HMMWV or a towed ¾ ton trailer.

Each tank is constructed using puncture-resistant fabric with an elastomeric coating on both sides. The inside surface of the potable water tanks meets US Food and Drug Administration requirements and imparts neither taste nor odour to the water. Chafing patches are part of the standard design and handles are bonded on to the tank. A tie-down webbing assembly consists of a nylon/polyester harness with four galvanised D-rings attached. Tanks are equipped with a manual pressure relief valve and one inlet/outlet

GTA Mini Tank with hose assembly installed in pick-up truck

fitting and a male quick connect adaptor with a dust cap.

Optional equipment includes a 51 mm × 2.44 m hose assembly and a field repair kit.

Specifications

Capacity	120 US gal	250 US gal	500 US gal
	454 litres	946 litres	1,893 litres
Weight (empty)	9.1 kg	12.7 kg	17.7 kg
Length	1.22 m	1.83 m	3.35 m
Width	1.52 m	1.52 m	1.52 m

Status

In production.

Manufacturer

GTA Containers Inc, 1410 W. Napier Street, South Bend, Indiana 46601, USA.
Tel: +1 219 288 3459. Fax: +1 219 289 6060.

VERIFIED

GTA Trailer Tanks

Description

The GTA Trailer Tank converts an open or closed trailer, truck or van into a tanker unit; the tanks may also be installed in barges or ships. Two personnel can unroll a flatbed tank and convert a trailer or vehicle into a tanker in approximately 15 minutes. Empty tanks collapse to 2 per cent of their original volume for transport or storage.

5,000 gallon GTA Trailer Tank in use

Each tank is constructed using abrasion- and puncture-resistant fabric. The inside surface of the potable water tank meets US Food and Drug Administration requirements and imparts neither taste nor odour to the water.

These tanks may be carried on highways at speeds up to 105 km/h; braking at that speed will not cause shifting or leaking, even when full. The tanks can be transported off road over unimproved terrain up to gradients of 30 per cent without shifting or leaking while maintaining trailer stability.

GTA Trailer Tanks can be unloaded using a gravity drain system to empty the tank in 30 minutes or less.

Specifications

Capacity	3,000 gallons	5,000 gallons
	11,355 litres	18,925 litres
Weight empty	200 kg	250 kg
Length	8.53 m	12.19 m
Width	2.44 m	2.44 m

Status

In production.

Manufacturer

GTA Containers Inc, 1410 W. Napier Street, South Bend, Indiana 46601, USA.
Tel: +1 219 288 3459. Fax: +1 219 289 6060.

VERIFIED

Amfuel static water tanks

Description

Amfuel produces a wide range of static flexible water tanks for both military and commercial purposes. The tanks produced for water and water-based products are free-standing and made from a chlorobutyl rubber-based material that is fully vulcanised and reinforced with tough nylon basket-weave fabric. These tanks can be placed at almost any location and have double-thickness panels at handling points and positions where hardware is attached. Positioning handles are provided. For transport and storage the tanks can be placed into wooden crates.

The US armed forces currently employ three main types of these tanks, with capacities of 10,000 gallons (37,850 litres), 20,000 gallons (75,700 litres) and 50,000 gallons (189,250 litres). They are used as part of the Tactical Water Distribution System (TWDS – qv).

Unpacking an Amfuel static flexible tank

Typical Amfuel static flexible water tank

Specifications

Capacity		Approximate dimensions filled: Width	Length	Height
1,000 US gal	3,785 litres	2.388 m	3.073 m	0.61 m
2,000 US gal	7,570 litres	3.353 m	3.353 m	0.813 m
3,000 US gal	11,355 litres	3.581 m	4.267 m	0.914 m
5,000 US gal	18,295 litres	4.47 m	4.623 m	1.118 m
10,000 US gal	37,850 litres	6.096 m	6.096 m	1.219 m
15,000 US gal	56,775 litres	7.468 m	7.315 m	1.219 m
20,000 US gal	75,700 litres	6.7 m	8.23 m	1.727 m
25,000 US gal	94,625 litres	6.7 m	10.363 m	1.727 m
50,000 US gal	189,250 litres	6.7 m	19.2 m	1.727 m
100,000 US gal	378,500 litres	18.136 m	18.212 m	1.219 m

Status

In production. In service with the US armed forces.

Manufacturer

American Fuel Ce l and Coated Fabrics Company (Amfuel), 124 South Jackson, Magnolia, Arkansas 71753, USA.
Tel: +1 501 234 3381. Fax: +1 501 235 7270.

VERIFIED

Amfuel collapsible water drums

Description

The US Army uses two main types of Amfuel collapsible water drum that are similar in form to the Uniroyal Sealdrum tanks (see separate entry in *Bulk fuel storage and distribution systems* section). The two models in service have a 55 US gallon (208 litre) capacity and a 250 US gallon (946 litre) capacity. Both are cylindrical tanks constructed of water-resistant synthetic rubber-coated fabric and can be towed at slow speeds over smooth surfaces for short distances. Both can be airlifted by helicopter.

Specifications

Capacity	55 US gal	250 US gal
	208 litres	946 litres
Length	0.876 m	1.524 m
Diameter	0.597 m	1.016 m
Weight (full)	211.4 kg	1,043 kg
(empty)	22.68 kg	93 kg

Status

In service with the US Army.

Manufacturer

American Fuel Cell and Coated Fabrics Company (Amfuel), 124 South Jackson, Magnolia, Arkansas 71753, USA.
Tel: +1 501 234 3381. Fax: +1 501 235 7270.

VERIFIED

Bell Avon Flexitanks for water

Description

Bell Avon Inc produces a range of flexible tanks for transporting and storing water. They are made from high-strength, durable polymer-coated fabric and are available in a wide range of sizes from 5 gallons to 500,000 gallons. Bell Avon Inc also produces open onion tanks.

When not in use, or for ease of shipping, the tanks can be folded, transported and packed into a convenient crate. They are suitable for operations throughout a wide temperature range.

A typical Bell Avon 50,000 gallon (189,250 litre) flexible water tank installation

Status

In production for the US Army and others. Available from inventory or short lead time.

Manufacturer

Bell Avon Inc, 1200 Martin Luther King Jr. Boulevard, Picayune, Mississippi 39466, USA.
Tel: +1 601 799 1217. Fax: +1 601 799 1360.
European sales: Avon Technical Products, PO Box 2, Melksham, Wiltshire SN12 8AA, UK.
Tel: +44 1225 791823. Fax: +44 1225 705585.

VERIFIED

Aqua-Chem 11,355 litres/h Reverse Osmosis Water Purification Unit (ROWPU)

Description

In August 1984, the Belvoir Research, Development and Engineering Center (now the US Army Tank Automotive Command, Mobility Technology Center - Belvoir) awarded two contracts for the design, fabrication, assembly and testing of prototype Reverse Osmosis Water Purification Units (ROWPUs) capable of purifying 11,355 litres (3,000 US gallons) of water per hour. Brunswick Corporation Defense Division of Deland, Florida, received a $2,060,680 contract while a $3,309,099 contract was awarded to Aqua-Chem's Water Technologies Division. Each company produced three prototypes by December 1984 for testing at Aberdeen Proving Ground (Developmental) and Forts Story and Eustis (Operational) during 1985. Testing was completed by late 1986.

In November 1987 it was announced that Aqua-Chem Inc had been awarded a contract for 98 units with deliveries commencing during 1990.

The specifications called for the purification of 11,355 litres/h (3,000 gallons/h) at +25°C using fresh and brackish water sources and 7,570 litres/h at 25°C using seawater sources, both including NBC contamination. The units are mounted in standard ISO containers with the NBC treatment unit outside and may be transported by truck, rail or aircraft. The units are designed to be installed or packed up in 90 minutes.

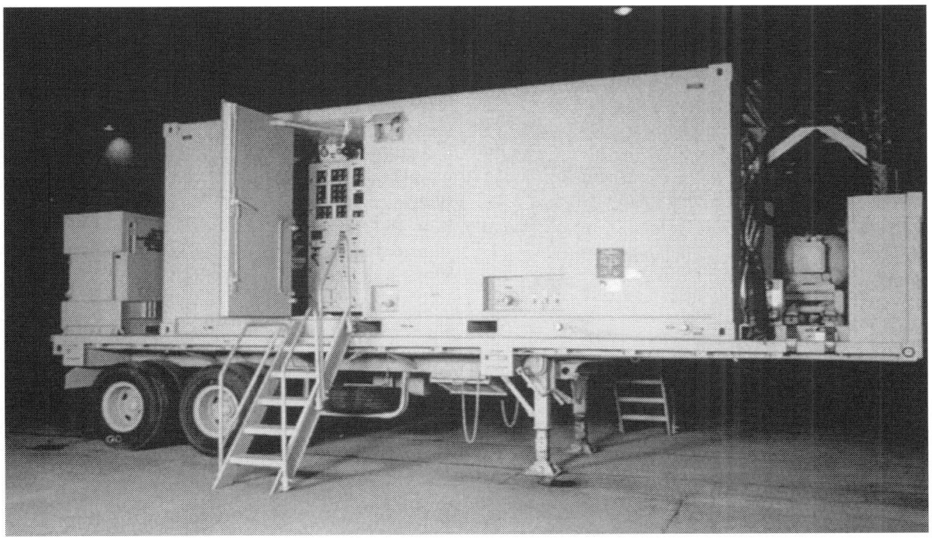

Aqua-Chem 11,355 litres/h Reverse Osmosis Water Purification Unit (ROWPU)

Status

In service with the US Army (98).

Manufacturer

Aqua-Chem Inc, Water Technologies Division, 7800 North 113th Street, PO Box 421, Milwaukee, Wisconsin 53201, USA.
Tel: +1 414 577 2944. Telex: 26679 aqm mil.
Fax: +1 414 577 2723.

VERIFIED

Univox 2,271 litres/h Reverse Osmosis Water Purification System (ROWPU)

Description

This equipment was developed by the US Army Mobility Equipment Research and Development Command (MERADCOM) (now the US Army Tank Automotive Command, Mobility Technology Center - Belvoir) and Univox California Inc, to replace four existing water purification units. A contract was placed in January 1981 with Univox California for 11 units. Total production orders were 455 units at a cost of $55 million.

The 2,271 litres/h reverse osmosis water purification unit was the first and smallest in a family of multipurpose units developed for field support. This system replaced the ERDLator, 568 litres/h distillation unit, CW-BW decontamination kit and the mobile ion exchange unit.

The Reverse Osmosis (RO) system is capable of producing drinking water from polluted fresh water, seawater and brackish water and removing chemical and radiological contaminants from water. Reverse osmosis is the process of pressurising raw water above its osmotic pressure which forces it through a cellulose acetate or polyamide membrane fabricated into a spiral wound element. Eight elements are used in the new unit.

The development of a dry membrane represents a significant advancement in reverse osmosis membrane technology. Unlike commercially available wet membranes, the dry membranes can be stored dry and are wet/dry reversible. Incoming water is pretreated with a polymer to aid filtration before final passage through the RO elements. The unit is mobile and can be air-dropped to forward tactical units, or towed on a trailer. In both cases, it will supply enough water to support at least 2,000 personnel per day.

2,271 litres/h water purification system (US Army)

Specifications

Output: 2,271 litres/h
Weight:
(on trailer) 7,639 kg
(on skid mount) 5,121 kg
Feed requirement: 132.4 l/min at 30 psi
Length:
(trailer) 5.8 m
(skid mount) 3 m
Height:
(trailer) 2.4 m
(skid mount) 2.4 m
Width:
(trailer) 2.4 m
(skid mount) 2.4 m
Electrical power consumption: 22 kW
Generator power required: 30 kW

Status

In service with the US Army, Air Force and Marine Corps.

Manufacturer

Univox California Inc, 6551 Loisdale Court, Springfield, Virginia 22150, USA.

VERIFIED

Univox Reverse Osmosis Water Purification Units (ROWPUs)

Description

Univox California Inc produces a range of Reverse Osmosis Water Purification Units (ROWPUs), one of which, the 2,271 litres/h system, is covered in a separate entry but is included here to provide a complete survey of the range. All Univox ROWPUs use the same basic system in which water contaminants are removed by pressurising untreated water through a cellulose acetate or polyamide membrane, fabricated into a spiral wound element on a drum. A series of these filters completes the process. The units are highly reliable, require only relatively low power sources and are available in different outputs to meet various military and civil requirements. They may be skid-mounted for use from trailers or semi-trailers, or for para-dropping and air transport, or they may be mounted on their own

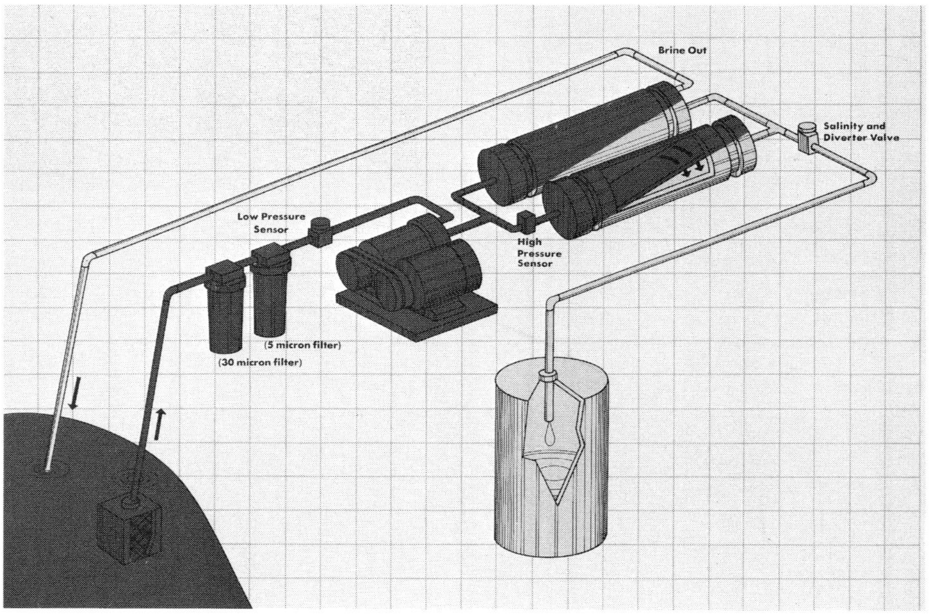

Diagrammatic representation of reverse osmosis process

Specifications

Model	150S	300S	600S	1000S	ROWPU 600[1]	ROWPU 1000[1]
Output	567 litres/h	1,135 litres/h	2,271 litres/h	3,785 litres/h	2,271 litres/h	3,785 litres/h
Input (at 30 psi)	37.8 litres/min	94.6 litres/min	132.5 litres/min	170.5 litres/min	132.5 litres/min	170.5 litres/min
Weight on skid mount	2,115 kg	2,610 kg	2,880 kg	3,015 kg	3,240 kg	3,600 kg
Electrical power consumption	6 kW	12 kW	15 kW	19 kW	20 kW	22 kW
Primary motor supplies	7.5 hp	15 hp	20 hp	25 hp	20 hp	25 hp

[1] Includes full NBC filtration

special trailers. If required, a special prefiltration package may be added to each model to provide full NBC treatment.

Skid mount dimensions (all models)
Length: 2.895 m
Height: 1.702 m
Width: 2.108 m

Trailer dimensions (trailer is optional for all models)
Weight: 4,500 kg
Length: 5.842 m
Height: 0.762 m
Width: 2.438 m

Status
In production.

Manufacturer
Univox California Inc, 6551 Loisdale Court, Springfield, Virginia 22150, USA.

VERIFIED

Angus Wellmaster drop pipe system

Description
The Angus Wellmaster rising main drop pipe for submersible pumps is in use in many parts of the world. It was developed following experiences in the Middle East, where it was found that conventional steel mains have a limited life due to corrosion problems. The Wellmaster pipe uses a composite construction of high strength synthetic polyester and thermoplastic elastomer and can be used at well depths up to 182.9 m (600 ft). The pipe is corrosion resistant, fast and easy to install or remove, flexible and layflat and available in lengths up to 182.9 m. The pipe is also easy to transport, store and handle. It is available with internal nominal diameters of 2 in (51 mm), 3 in (76.2 mm) and 4 in (101.6 mm).

Accessories for use with the Wellmaster pipe include stainless steel couplings and adaptors, assembly tools for the couplings, a hose installation set comprising a rotating drum and support frame and two hose clamps, rolls of rubber strapping and studs for power cable retention and self-lock nylon ties to attach the strapping to the hose ridge.

Status
In production. In service with the US Army.

Manufacturer
Angus Fire Armour Corporation, 1000 Junny Road, Angier, North Carolina 27501, USA.
Tel: +1 919 639 6151. Fax: +1 919 639 6131.

UPDATED

Wellmaster rising main drop pipe being lowered into well while undergoing evaluation trials held in New Mexico (US Army)

Recovery Engineering hand-operated watermakers

Description
The Recovery Engineering range of hand-operated watermakers, known as Survivors, combines reverse osmosis and energy recovery technology to make pure drinkable water from salt, brackish or contaminated fresh water. Designed for use as an emergency water supply for naval lifeboats and military units separated from central water supplies, the Survivors are in use with the US Army, Navy and Coast Guard. Three models are available: the Survivor 06, Survivor 35 and the Survivor 35CS.

To operate any of the Survivors, the user pumps the unit handle back and forth; on the Survivor 06 the unit is operated by lever action with the unit resting on the operator's thigh. This pressurises the feedwater and forces it against a reverse osmosis membrane which allows only pure water molecules to pass through. The units will supply purified water as long as the user pumps.

The Explorer is a pump-type unit that can purify 1 litre/min of water without clogging. Weight is 603 g.

Status
In production. In service with US Army, Navy and Coast Guard. Survivor 06 and Survivor 35 are in service with the UK armed forces.

Manufacturer
Recovery Engineering Inc, 2229 Edgewood Avenue South, Minneapolis, Minnesota 55426, USA.
Tel: +1 612 541 1313. Telex: 290828.
Fax: +1 612 541 1230.

VERIFIED

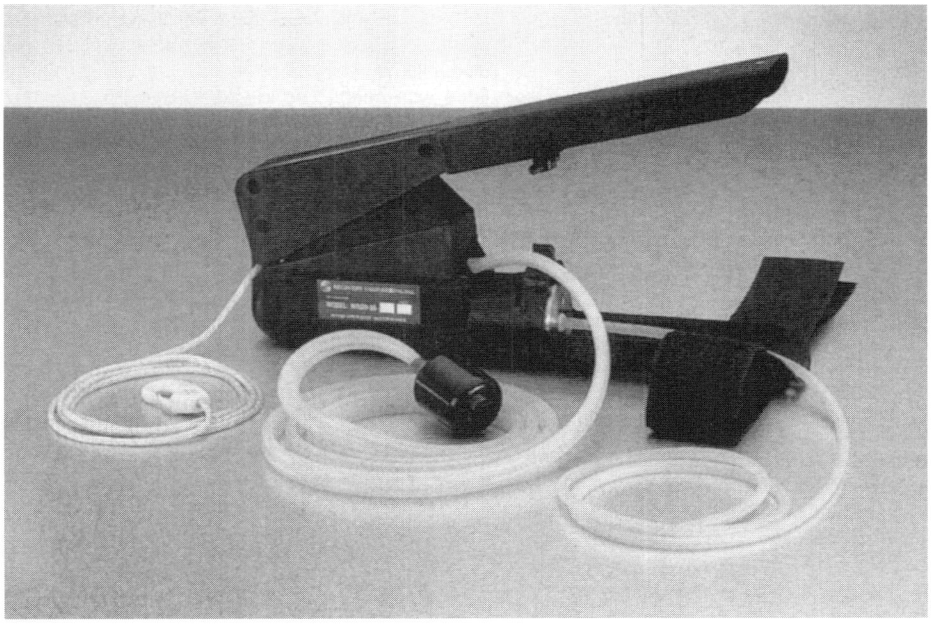

Recovery Engineering Survivor 06 hand-operated watermaker

Specifications

Model	**Survivor 06**	**Survivor 35**	**Survivor 35CS**
Dry weight	1.6 kg	3.2 kg	3.7 kg
Purification rate	1.1 litres/h	5.4 litres/h	6.41 litres/h
Length	200 mm	560 mm	560 mm
Height	130 mm	1.4 m	1.4 m
Width	60 mm	90 mm	90 mm

YUGOSLAVIA (SERBIA AND MONTENEGRO)

Hand-operated water filters

Description

There are three hand-operated water filters in this range, all employing the same basic principle. Polluted water is hand- or gravity-pumped through a filter element in which particles and bacteria are mechanically separated or absorbed, along with an element of ion exchange, by a multilayer paper filter insert. The paper insert is saturated/impregnated with active charcoal and other constituents and has to be changed when clogged.

The hand-operated filter FR-2 has a capacity of 1 to 2 litres/h. Two filter elements can provide up to 2 litres of pure water from NBC-polluted water and up to 10 litres from ordinary pollution. Operated by one person, the FR-2 has two filter elements, a pump, storage vessel, plus hoses and fittings. After filtration the pure water is chlorinated with a tablet. The FR-2 kit is carried in a bag weighing 0.76 kg and measuring 190 × 155 × 88 mm.

The hand-operated filter FR-30 has a capacity of 30 litres/h and uses three filter elements to provide up to 150 litres of pure water. For ordinary water purification the capacity is 45 litres/h, providing up to 1,000 litres. The FR-30 is contained in two metal boxes and is operated by one person. A hand-operated reciprocating pump is employed and after purification, the water is chlorinated. The complete kit is carried in a haversack and weighs 14.5 kg.

The hand-operated filter FR-300/150 has a capacity of 300 litres/h for ordinary water purification and 150 litres/h for more involved processing. This system uses a process of preliminary treatment (coagulation, hyper-chlorination and settling tanks) before filtration through two or four filter columns. The kit can filter NBC agents. The complete kit weighs 132 kg and breaks down into three manpack loads.

Status

In service with the former Yugoslav Army.

Contractor

FDSP - Federal Directorate of Supply and Procurement, PO Box 89, Bulevar umetnosti 2, YU-11070 Novi Beograd, Yugoslavia.
Tel: +381 11 222 2743. Telex: 71000 JU SPDR.
Fax: +381 11 324 8738.

UPDATED

Hand-operated filter FR-2 in use

CAMOUFLAGE EQUIPMENT AND DECOYS

AUSTRIA

Blaschke camouflage systems

Description

J Blaschke Maschinen- und Pumpenbau produces a variety of camouflage systems.

One of its base products is a camouflage net and support system which utilises a series of rapid erection and flexible 'umbrella' modules which can be rapidly erected and placed around any vehicle or item of military equipment to break up the outline. The modules are reversible so that they can carry two types of camouflage finish, and may be used in conjunction with covering camouflage nets. When not in use the modules fold up to occupy a small volume.

Blaschke also produces a series of special weapon camouflage sets which can be rapidly erected around weapons such as anti-aircraft guns or artillery pieces. Once erected the sets present differing appearances from varying angles and do not need to be removed before the weapon can be operated.

One special camouflage product is a patented windscreen camouflage system for vehicles. With this system camouflage panels are placed over all windscreens to completely remove reflections while allowing unimpaired visibility from within the vehicle, thus allowing normal driving. The panels may be one colour or multicolour. Camouflage paints are available for headlights.

Status

In production.

Manufacturer

J Blaschke Wehrtechnik GmbH, Wienerbergstrasse 42-44, A-1120 Wien, Austria.
Tel: +43 1 810 0909. Fax: +43 1 8100 90933.

UPDATED

A Steyr-Daimler-Puch Pinzgauer (6 × 6) 1,500 kg vehicle with windscreens covered by Blaschke windscreen camouflage panels **1997**

A light air defence gun under Blaschke special weapon camouflage

A military vehicle partially concealed beneath Blaschke camouflage netting and support systems **1996**

A military vehicle partially concealed beneath Blaschke camouflage netting and support systems **1997**

BELGIUM

Seyntex camouflage nets

Description

Seyntex produces a wide range of camouflage nets manufactured from modern synthetic materials which have been treated to avoid visual detection, even when using reconnaissance equipment such as infrared optical devices and infrared photography. For camouflage purposes various shades of net can be produced and the size of net varies making it suitable both for personnel use and for large items of equipment. Seyntex also produces supporting poles and mushrooms for the nets. A quick attach and release system that can raise and lower the nets within 5 seconds is also available.

Seyntex camouflage net being erected over light vehicle

Status
In production.

Manufacturer
Seyntex nv/sa, Seyntexlaan 1, Industriepark Zuid, B-8880 Tielt, Belgium.
Tel: +32 51 402424. Telex: 81 947.

VERIFIED

BULGARIA

Camouflage nets and materials

Description
The Bulgarian defence industry manufactures a wide variety of camouflage products ranging from personal camouflage to the concealment of major sites against sensor surveillance. Those aspects listed below are only the main headings.

Personal camouflage
Three types of helmet camouflage are available. The Model A is a two-sided fabric cover with a leaf outline while the Model B is a fabric. Both are effective in the visual and near infrared regions. The Model C is a fabric extending concealment to the radar wave region.

Also available is a range of camouflage clothing that can be supplied in a variety of forms and colours to suit customer requirements. Camouflage creams are available for personal use, in 'lipstick' or cream form. Colours available include light or dark green, brown and sandy yellow.

Camouflage nets
A camouflage net system is available, capable of providing protection against the visible, near infrared and S, X and K radar bands. Made from PVC film elements this net is produced in 3 × 6 m panels. A standard set comprises of six panels 20 fixing elements, a packing bag and six supporting rods.

Camouflage paints
A range of camouflage paints is produced to provide protection of equipment in the visible, near infrared and radar regions.

The CM-1 series is based on pentaphthalic resins (CM-1A) or epoxy-polyester resins (CM-1B) to provide visual and near infrared protection.

The CM-2 series provides anti-corrosion and waterproofing as well as concealment against S-, X- and K-band radars. CM-2A is a multilayer coating based on synthetic rubbers while CM-2B can be used with textiles.

The CM-3 series is a more complex preparation combining the assets of the CM-1 and CM-2 series.

CM-1A, CM-2A and CM-3A paints are not proof against decontamination solutions of the DS2 type.

Anti-radar coatings
Intended to provide protection of tanks and other armoured equipment against S-, X- and K-band radars, anti-radar coatings are available in two forms. Based on the use of rubber blended with various fillers, Type A is a two-layer coating 3 to 4 mm thick and providing 14 to 20 dB attenuation in the X band. Type B is 10 to 14 mm thick and provides radar attenuation of 25 to 30 dB.

A further preparation extends protection to the thermal region. A standard set of this material is provided in sheets measuring 3 × 4 m.

Other products include sets of multispectral materials, including the above items, various forms of equipment decoy, and radar-reflecting systems.

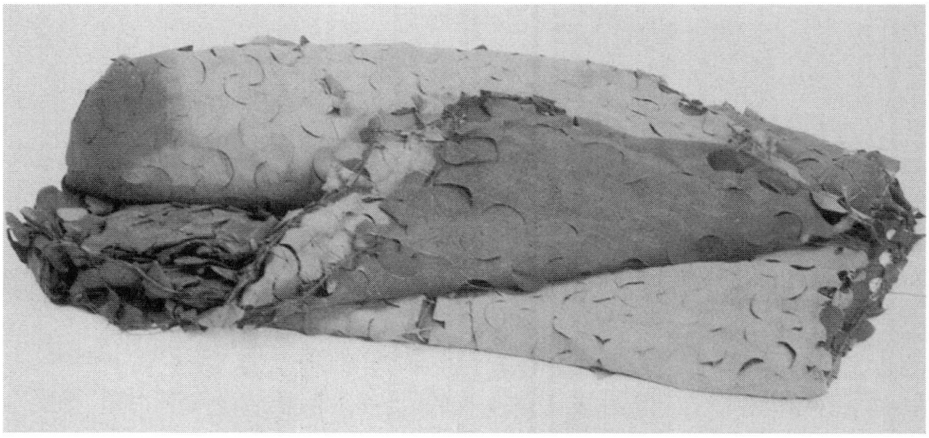

A typical KINTEX camouflage net module showing Autumn-Winter and Spring-Summer aspects

Status
In production.

Agency
KINTEX, 66 James Baucher Street, PO Box 209, 1407 Sofia, Bulgaria.
Tel: +359 2 662311. Fax: +359 2 658191/8101.

VERIFIED

CZECH REPUBLIC

B.O.I.S. anti-radar camouflage nets

Description
These anti-radiolocation camouflage nets are of unusual design and were developed by the Vojensky Techniky Ustav Ochrany (VTUO) of Brno. As well as providing protection against detection by numerous types of radar, they also provide protection in the visual and near infrared spectrum as well as in the microwave region.

The nets are based on a polyamide mesh with dimensions of approximately 34 × 34 mm. The mesh is filled with a self-extinguishing polyamide foam of

Example of the structure of a B.O.I.S. anti-radar camouflage net
1997

varying thicknesses to provide an optical density of from 5 to 100 per cent. The borders of each net module are strengthened with sewn ribbon. Colours can include white, grey, grey-brown, light green, dark green and sand, all in various patterns according to requirements. The nets have a weight of up to 350 g/m^2 and can be used over a temperature range of from −40 to +80°C. Service life is up to one year if used continuously. The basic net module dimensions are 3 × 6 m, although other sizes up to 12 × 12 m can be produced.

The nets provide camouflage in the 8 to 12 GHz band and the visible and infrared spectrum from 400 to 1,200 nm. Positive results can also be obtained in the 3 to 14 μm spectral region.

Status

In production. In service with the Czech armed forces.

Manufacturer

B.O.I.S. Engineering Praha sro, Holečkova 31, 150 95 Praha 5, Czech Republic.
Tel: +42 2 535288. Fax: +42 2 535259.

NEW ENTRY

DENMARK

CAMTEX texture mats

Description

CAMTEX texture mats are part of a new generation of Camouflage, Concealment and Deception (CCD) systems developed by the CAMTEX Camouflage Division of GCA APS. The mats are produced using a multispectral material which can be fitted permanently or temporarily on armoured and other vehicles, installations, ships, and helicopters and other aircraft on the ground. When fitted permanently on vehicles they do not impede mobility and minimise time spent on fitting and removing normal camouflage systems.

The texture mats are made of a structured material which has diffuse reflective and absorption properties, including visual spectrum properties such as gloss and shine. The material, which is supplied in mats or rolls and cut to shape for permanent installation, has built-in capabilities against radar, ultraviolet, and near and full infrared sensors.

Other CCD systems offered by CAMTEX include a Chemical Tone Down Method used to camouflage large fixed installations such as airfields. The company also offers a camouflage consultancy service.

Status

In production. In service with NATO nations.

Manufacturer

CAMTEX Camouflage Division, A Division of GCA APS, Store Kongensgade 110A, DK-1264 Copenhagen K, Denmark.
Tel: +45 33 911525. Telex: 52628 gccab dk.
Fax: +45 33 322535.

VERIFIED

FINLAND

Scantarp camouflage products

Description

Scantarp produces a range of camouflage products that includes camouflage nets, umbrellas, trees and vehicle covers. A typical product is the Scantarp Camouflage Net Summer 25 K. This is made by gluing pieces of PVC-coated fabric of various colours onto a polyamide/polyester support net. There is an edge cord sewn around the net and there are cord loops that enable several nets to be joined together. Standard net sizes are 5 × 5 m, 6 × 8 m and 8 × 12 m, although net sizes can be produced to customer requirements. The standard colours are light green, dark green and black. Other colour reflection requirements in visible and near infrared ranges are available.

The Summer 25 K camouflage net has a weight of 290 g/m^2 and a thickness of 0.25 mm. Mesh width is 85 × 85 mm ±7 mm.

Status

In production.

Manufacturer

OY Scantarp Ab, PO Box 1766, FIN-70421 Kuopio, Finland.
Tel: +358 71 288188. Fax: +358 71 465 1762.

VERIFIED

GREECE

VALPAK camouflage nets

Description

VALPAK SA is the only camouflage netting manufacturer in Greece and has an exclusive five-year contract with the Hellenic Army. It produces camouflage netting for use in the visual, infrared and radar spectra, the last for use in the X and Ku bands.

Nets produced to NATO standards utilise a polyamide material coated with PVC and polyurethane. The material is flame and water-resistant. Garnishing materials are nylon fabric. Colour shades can vary according to requirements with nets being shaded on one or both sides.

Standard net sizes are 3.4 × 3.4 m, 3.4 × 6.8 m, 6.8 × 6.8 m, 6.8 × 10.2 m and 10.2 × 10.2 m, although other sizes can be produced on request. The nets can be connected using high strength plastic clips although rapid connect/disconnect loops are provided around the periphery of each net.

Status

In production for the Greek Army and Cyprus.

Manufacturer

VALPAK SA, 7 P P Germanou, GR-10561 Athens, Greece.
Tel: +30 1 323 2538/325 2126.
Telex: 226631 VALP GR. Fax: +30 1 325 2153.

VERIFIED

HUNGARY

The HUNCAM Hungarian Camouflage System

Description

The HUNCAM Hungarian Camouflage System is made up of a number of camouflage materials with special multifrequency absorption characteristics. These extend into the radar and infrared spectra and the materials involved may be used for personal camouflage clothing and concealment nets. The main components of the range are as follows:

Camouflage net RABSORB CND

Made from bulked polyester yarn with varying sizes of mesh, this net has radar absorption properties and is made in standard sizes of 5 × 5 m, 6 × 6 m and 8 × 12 m. (These sizes apply to all HUNCAM nets.) It is produced in colours to suit customer requirements and has a weight of 160 g/m^2.

Camouflage net RABSORB CPV

This net uses pieces of PVC-coated fabric of varying colours and sizes glued onto a polyamide/polyester support net and with an edge cord around the net. Produced in various colours the weight varies between 200 and 300 g/m^2 according to the exact type of net required.

Camouflage net RABSORB CNS

This is a lightweight net with a colour-printed surface which may be applied on one or both sides. Weight is between 110 and 180 g/m^2.

Camouflage net RABSORB RTN

This net provides both radar and near infrared thermal as well as visual camouflage and uses pieces of incised PVC foil interlayered with special additions to produce a three-dimensional texture. The main wavelength efficiency is in the X and Ku bands. Weight is 250 to 400 g/m^2 according to type.

Camouflage net RABSORB TCN

This net is made from 2 m wide panels with the garnish consisting of incised matt PVC film cemented to the net material, thus providing a three-dimensional effect with infrared reflection properties. Weight is 200 to 300 g/m^2, according to type.

Individual net RABSORB PCN

These nets are intended for use by individual soldiers and are made from incised matt PVC film cemented to a supporting net. This net is normally used together with the Helmet Cover CC made from the same materials.

Camouflage uniform RABSORB OCS

These uniforms are made from 100 per cent cotton or polyester/cotton fabric. They provide protection against ultraviolet, visual, infrared and multispectrum observation and can be produced in various colours as required.

Camouflage Leaves CV

This is a series of artificial camouflage leaves for attenuating multispectral radiation by absorption, reflection and/or scattering. The incised leaves have a thin electrically conductive intermediate layer with particles distributed at intervals and with the two outer layers reflecting visible and near infrared light. When placed onto supporting cords the leaves use a self-locking system that allows the leaves to be distributed as required. Weight is between 200 and 300 g/m².

Camouflage Trees CT

These are a series of artificial trees using a framework onto which camouflage leaves are distributed. Objects to be concealed are placed behind the trees where they are concealed by the multispectral absorption and reflection properties of the materials involved.

Camouflage Paints CP

There are two types of camouflage paint available. The CP paints are based on a water-soluble polymer and can be removed by water; they are infrared reflective and seven colours are available. The PCP paints are medium oil alkyd, epoxy or polyurethane based. Both paints provide a matt infrared reflective finish.

Status

In service with the Hungarian armed forces.

Marketing Agency

IDEX Foreign Trading, Contracting and Engineering Company Limited, PO Box 24, H-1251 Budapest, Hungary.
Tel: +36 1 115 0090. Telex: 22 4541 idex h.
Fax: +36 1 135 1393.

VERIFIED

ISRAEL

Fibrotex camouflage nets

Description

Fibrotex Limited is the sole producer and supplier of camouflage nets to the Israel Defence Forces. They have developed a new concept involving a two dimensional polyester warp knitted camouflage net providing protection against visual infrared detection, good thermal attenuation and radar scattering properties. It is very light in weight and can be adapted to any topographic area. Sizes and colours can be varied to suit customer requirements. Several types of net are available for applications including the concealment of helicopters, aircraft, tanks, light vehicles and personnel. The net is claimed to answer threats in both the visual and thermal field without using an additional thermal blanket.

Fibrotex Limited have a research and development department with a staff having over 20 years of experience in camouflage development.

Status

In production. In service with the Israeli defence forces and several other armed forces.

Manufacturer

Fibrotex Limited, PO Box 18064, IL-61180 Tel Aviv, Israel.
Tel: +972 3 922 3585. Fax: +972 3 922 3311.

UPDATED

A Fibrotex camouflage net being used to cover a helicopter ***1997***

ITALY

Pirelli camouflage nets

Description

Pirelli produces a range of conventional camouflage netting which is delivered in relatively small sections each measuring 4 × 4 m or 6 × 6 m. The net material support is made up of rubber-coated fabric with the addition of a special synthetic and vulcanised elastomer to which various pigments have been added. This enables the outer surface of the net to be in three or four colours according to requirements, while the inner surface is in two shades only in a 60:40 ratio. The nets are not visible to infrared or ultraviolet photographic reconnaissance and are flame resistant. If required, both the support and the rubber-coated net can be provided with further support from a conventional camouflage net. Weight of the supporting net is 500 g/m² and the rubber-coated material weighs 300 g/m².

Status

In service with the Italian armed forces.

Manufacturer

Moldip SpA - Pirelli Group, Via Milano 8, I-20038 Seregno (Milano), Italy.
Tel: +39 362 237711. Telex: 330191 pirese i.
Fax: +39 362 220412.

VERIFIED

SOUTH AFRICA

Alnet camouflage nets

Description

The Alnet garnished camouflage net was developed to meet the stringent requirements of the South African defence forces and to withstand the extreme climatic conditions of South Africa. In the development stage of producing a net to meet these requirements, a technique of obtaining spectral data of the environment within a region was evolved to the extent that specific colours and disruptive patterns can be produced to resemble the local foliage.

The nets use a basic support net with a square 50 × 50 mm mesh. Low mass, high-strength garnishing material is clipped onto the support net to resemble the foliage in the area in which the net is to be deployed. The garnishing material is painted on both sides, one side being for Winter use and the other for Summer

Ratel MICV about to be covered by Alnet camouflage nets

Ratel MICV covered by Alnet camouflage nets

use. Joining rings on the perimeter of each net and the included joining cord and clips allow for the rapid joining and separation of two or more nets - the net sizes can be produced to suit. The nets are flame retardant and provide effective protection in the visual and near infrared spectra.

Weight of a complete net is 350 g/m^2. Alnet also produces helmet camouflage nets and shade cloth.

Status
In service with the South African National Defence Forces.

Manufacturer
Alnet, PO Box 4995, Cape Town 8000, South Africa.
Tel: +27 21 54232⁻. Fax: +27 21 544003.

UPDATED

SWEDEN

BARRACUDA Technologies signature management systems

Description
BARRACUDA Technologies AB approach the subject of camouflage by adopting a signature management approach. They consider that static camouflage is only one part of the protection needed in scenarios where numerous sensors may be operating so new techniques are required to protect forces against detection, identification and attack, even when moving. Signature management includes all possible passive measures in applicable parts of the electromagnetic spectrum. It includes structural design, application of materials and the necessary verification methods, from modelling to full scale measurements.

BARRACUDA signature management begins at the design stage of an object. Significant signatures are then adapted and camouflaged to suit the operative environment and are customised to meet specific needs. In this manner advanced camouflage, deception techniques, mechanical modifications, adaptation of signatures, plus education of personnel, all combine in a customised and integrated programme for signature management.

In addition to the usual visual detection methods, near infrared photography can capture reflected light just beyond visual light and is thus used for night reconnaissance. All BARRACUDA equipment protects in the near infrared as well as visual parts of the spectrum. Signature management also includes protection against all types of thermal sensor, plus protection against radars, including millimetric radar used with final-phase targeting systems. Systems also include ultraviolet (UV) light where snow observed through a UV filter remains white whereas paint stands out as black. BARRACUDA winter camouflage with high UV reflection includes protection against UV reconnaissance and has the same reflection as snow. All systems can be provided with multispectral properties.

BMS-ULCAS
The BARRACUDA Multispectral Ultra Lightweight Camouflage Screen (BMS-ULCAS) has camouflage properties covering the entire spectrum. It is much lighter than conventional camouflage with similar properties and requires less handling time due to a base scrim that eliminates snagging. Only one customised screen is required per object.

MCS
The BARRACUDA Mobile Camouflage System (MCS) is an individually designed camouflage kit for combat vehicles. MCS provides object adapted signature reduction that provides full protection against smart munitions together with a considerable reduction of the detection range even for moving objects. MCS comprises a combination of camouflage materials with signature adaption properties in four different versions: visual and near infrared; visual, near infrared and thermal; visual, near infrared and radar; visual, near infrared, thermal and radar.

Each of these combinations can be supplemented with a Heat Transfer Reduction (HTR) system to reduce the heat penetration through the hull into vehicles operating in extremely hot climates. All parts are built into and firmly fixed to the host vehicle and can be prepared before use and attached for combat within one hour by the vehicle crew. The MCS can be washed by high pressure c eaning.

HMBS
To counter systems with detection and attack in real time BARRACUDA designs and produces object adapted, multispectral camouflage systems for rapid deployment. These are combined under the term High Mobility on Board System, or HMBS.

The time required to deploy HMBS featuring the Rapid Camouflage System (RAPCAM®) to complete the signature adaptation of an object can be reduced by up to 90 per cent compared to conventional camouflage. The HMBS has optimum camouflage properties as each system is customised for each object. All supporting systems are integrated and both firing and other actions are possible with the camouflage in the deployed state. Thermal infrared properties permit the use of idling power packs and generators. All components can be carried in containers fixed to the object while TOPCAM® umbrellas can be added. These are reversible and can be provided with double-sided finishes for use in varying terrain.

The BARRACUDA COMCAM® system features easy to handle, add-on modules individually configured to eliminate the regularity of an object while adapting the signature of the object to the local environment. All HMBS components can remain in place during moves.

A 40 mm/L/70 bofors air defence gun covered with the BARRACUDA High Mobility on Board System (HMBS) combined with the Rapid Camouflage (RAPCAM®) system ***1997***

A Leopard 2 MBT covered by the BARRACUDA Mobile Camouflage System (MCS) about to be covered by a BARRACUDA Multispectral Ultra Lightweight Camouflage Screen (BMS-ULCAS) ***1997***

A Leopard 2 MBT covered by the BARRACUDA Mobile Camouflage System (MCS) ***1997***

A Leclerc MBT covered with the BARRACUDA Mobile Camouflage System (MCS) combined with the Heat Transfer Reduction (HTR) system ***1997***

Deception

Deception involving decoys remains highly profitable for tactical and strategic purposes. BARRACUDA produces durable, easily transportable and easily mountable decoys of all kinds of military objects, complete with the required visual, near infrared, thermal and radar signatures. The decoys have a soft cover over a structural element and the signatures are integrated. Their construction can withstand bad weather and can absorb multiple hits and fragments without collapsing.

Status

In production. In service with Australia, Canada, France, India, Japan, Sweden and many other armed forces.

Manufacturer

BARRACUDA Technologies AB, Box 160, S-594 23 Gamleby, Sweden.
Tel: +46 493 14800. Fax: +46 493 12064.

UPDATED

UNITED KINGDOM

Courtaulds Aerospace Temporary Camouflage Coating (TCC)

Description

Courtaulds Aerospace Temporary Camouflage Coating (TCC) was developed as a 'quick change' finish in response to the growing need for mission-specific camouflage coatings for vehicles and equipment which can be applied and removed with ease as required. TCC may be brushed, sprayed or applied using aerosol dispensers and, after softening with an alkaline gel solution, it may be hosed away with water. The cycle may be repeated indefinitely. Once applied in a specific shade to suit a mission or region, the 'skin' finish provides protection in the visual, near infrared and ultraviolet fields. TCC may be applied over any type of existing paint finish.

TCC meets the UK Ministry of Defence's DEF STAN 80-128 specifications.

Status

In production.

Manufacturer

Courtaulds Aerospace, Sealants, Adhesives and Coatings Division, Darlington Road, Shildon, Co Durham DL4 2QP, UK.
Tel: +44 1388 772541. Fax: +44 1388 774373.

VERIFIED

Bridport Aviation Products camouflage and concealment equipment and decoys

Description

Bridport Aviation Products is a division of Bridport-Gundry (UK) Limited — Europe's largest manufacturer of nets and netting — which has been a contractor to the British government for over 200 years. The company specialises in the design, manufacture, supply and installation of modern camouflage and concealment systems for both mobile equipment (vehicles, weapons, fixed-wing aircraft, helicopters and personnel) and fixed installation targets in all types of terrain.

For more than 50 years Bridport Aviation Products has been involved with the innovation, design and manufacture of camouflage material and techniques for the British armed forces, the US Air Force and other NATO countries, the armed forces of the member countries of the Gulf Co-operation Council, ASEAN Pact members, various armed forces in Latin America, as well as many other countries worldwide.

Major features of the company's camouflage systems are custom design, flexibility of size, low weight and ease of repair. The company's camouflage nets are manufactured from synthetic materials which are rotproof and waterproof, and are lighter than comparable products. The company offers an associated range of support equipment which blends the contours of the camouflage net to the surrounding terrain, and has also developed concealment systems for use with artillery, radar equipment, self-propelled artillery, aircraft and helicopters.

Radar scattering camouflage

To complete the multispectral range of camouflage, nets can be manufactured to give a broad/band attenuation to radar.

Thermal camouflage

Developed by the UK Ministry of Defence, thermal camouflage material extends protection into the infrared (thermal) region of the spectrum. The material used has a special composite construction which reduces thermal emissions from warm or hot objects to a level comparable with their surroundings. Far infrared emission characteristics can be combined with near infrared reflectivity and visual colouring in a single continuous membrane system to enable concealment over a wide range of wavelengths. This camouflage component was designed for low weight and bulk and will not snag in use. It can be incorporated with an incised camouflage set which can be designed and tailored for all vehicles and equipment of any size and shape.

Camouflage net systems

Camouflage net systems degrade hostile surveillance and acquisition system performance primarily by giving the target the optical and textural appearance of its environment. In reducing target recognition and identification probability, immediate attack is less likely and an enemy is less likely to gain a correct assessment of military functions and intentions.

Simple repair procedures involving the attachment of new/replacement garnishing patches with plastic tie-wraps can extend the life of the company's camouflage nets.

Camouflage concealment sets

Purpose-designed as complete systems, concealment

Military post under Bridport desert camouflage from distance of 50 m

Bridport Aviation Products two-dimensional decoy of Chieftain tank in hull-down position

sets are primarily for use with artillery, aircraft and helicopters, and generally for equipment difficult to camouflage due to bulk or manpower limitations, or for protecting equipment of high tactical value. Specific designs allow artillery to be serviced or fired, radar to operate and aircraft or helicopters to be refuelled and re-armed from concealed positions. Sets are available for 105 and 155 mm artillery (FH-70 and M198 howitzers), 81 and 120 mm mortars, self-propelled guns, the Rapier SAM system and fixed-wing aircraft (F-16, Harrier and Tornado). The company can design and quote for any custom-made set to meet client needs and to match local environmental conditions.

Helicopter sets

The company offers concealment sets for use in forward or rear areas to protect various types of helicopter. A specialist design team can develop sets for use with any helicopter.

Camouflage screens

Modern practices for the camouflage of parked aircraft have resulted in the development of a new generation of products including camouflage screens stemming from the desert raschel net produced by Bridport Aviation Products for over 20 years. The screen is based on a two-dimensional lightweight mesh screen which can be produced with a large repeat pattern and may be reversible to suit different areas or climatic changes. Multispectral properties are achieved. By the addition of garnish, three-dimensional properties can be enhanced to achieve lower detection ranges.

The screens are snag free and Foreign Object Damage (FOD) free. Weights are 120 g/m² for the reversible mesh and 170 g/m² for the garnished mesh. The family also includes a version of the screen for use as a False Operating Surface (FOS).

Ancillary equipment

To achieve effective camouflage it is important for the external net shape to be contoured to match the surrounding terrain. The company's equipment range includes support poles, mushroom caps and ground pegs to help produce natural formations. A quick release system is also available to allow rapid opening for engaging targets or breaking from cover.

Personnel camouflage

The company offers a range of helmet and individual camouflage nets for any type of terrain. The helmet nets are manufactured from synthetic materials to fit most standard helmets and can be supplied with or without garnishing materials.

Fixed installation camouflage

The key to camouflaging airbases and other fixed installations is to deceive oncoming attacking aircraft into targeting systems on non-vital areas. Aircraft shelters are particularly vulnerable to attack due to the easily detected large semi-elliptical shadows created by door areas. Camouflage packages for such installations include the deployment of camouflage nets and screens, decoys and paint schemes for runways, equipment and buildings. A company design team can advise on the camouflage of any fixed installation.

Decoys

Bridport Aviation Products designs and manufactures to client's specific requirements a full range of decoy systems which can authentically duplicate the visual, near infrared, far infrared (thermal) and radar signatures of the original equipment. Any MBT, artillery piece or other land vehicle or aircraft (both fixed- and rotary-wing) can be produced as a full-size realistic decoy which can convince surveillance and guidance systems that the decoy is the real thing.

Status

In production. In service with British and other armed forces.

Manufacturer

Bridport Aviation Products, The Court, Bridport, Dorset DT6 3QU, UK.
Tel: +44 1308 456666. Telex: 41132 netsbt g.
Fax: +44 1308 456605.

VERIFIED

Airborne inflatable dummy targets

Description

These inflatable dummy targets were developed to provide realistic recognition training to Forward Air Controllers (FACs) directing ground attack aircraft against AFVs and other battle formations. They are used in this role by the Royal Air Force.

At ranges of between 2,000 and 3,000 m, through binoculars, they are immediately identifiable as armoured vehicles, at 900 to 1,000 m they are identifiable as to type. Only at a distance of less than 300 m are they identifiable as dummies.

The targets are manufactured in synthetic rubber-coated nylon and consist of a framework of low-pressure inflatable tubes which are covered to simulate the vehicle outline. Paint is used to highlight prominent features such as wheels and hatches. Support poles and guys ensure stability. Pressure relief valves are incorporated to limit the pressure to 0.07 kg/cm² during inflation or temperature increases.

Inflation is by a small battery-operated blower and can be accomplished in 5 to 8 minutes. The support poles and guys take a further 2 to 3 minutes to position. Several outlet tubes give fast deflation and normally two people can completely deflate, pack and stow a target in 10 minutes. Each target is packed in a valise for ease of handling and storage.

The current range includes the T-72 and T-62 MBT, BMP-1 MICV, BTR-50PK and BTR-60PB APCs, ZSU-23-4 air defence system, and a Rapier SAM system. A 155 mm GCT SPG has been delivered to the French Army. Other AFVs can be designed and the company has designed a T-12 100 mm anti-tank gun dummy target.

Specifications

Model	**T-72**	**T-62**	**BMP-1**	**BTR-60PB**
Weight	90 kg	88 kg	76 kg	94 kg
Length (inflated)	9.5 m	9.45 m	6.71 m	7.16 m
Width (inflated)	3.5 m	3.35 m	3.02 m	2.74 m
Height (inflated)	2.45 m	2.29 m	2.44 m	2.44 m
Pack size	all are approximately 1.07 × 0.6 × 0.4 m			

Status

Available. In service with the Royal Air Force and French Army.

Manufacturer

Airborne Industries Limited, Airborne Industrial Estate, Arterial Road, Leigh-on-Sea, Essex SS9 4EF, UK.
Tel: +44 1702 525265. Telex: 99412 airbrn g.
Fax: +44 1702 510454.

VERIFIED

T-62 MBT dummy target manufactured by Airborne Industries (UK Ministry of Defence)

BMP-1 MICV dummy target manufactured by Airborne Industries (UK Ministry of Defence)

UNITED STATES OF AMERICA

Small Area Camouflage Cover (SACC)

Description

The US Army Belvoir Research, Development and Engineering Center at Fort Belvoir (now the US Army Tank Automotive Command, Mobility Technology Center - Belvoir) developed a Small Area Camouflage Cover (SACC) designed to conceal individual troops or to be attached together for use over weapon emplacements, combat positions and supply caches. The SACC comes in woodland, desert, arctic and tropical colour versions and provides concealment against visual, near infrared, radar and ultraviolet light detection (in the case of the arctic SACC). The

Soldier in firing position without camouflage cover

Soldier in firing position with camouflage cover provided by Small Area Camouflage Cover (SACC)

woodland and tropical SACCs are patterned on both sides with a seasonal colour change on the reverse. Desert and arctic are monotone-coloured on one side and patterned on the reverse. SACCs are made of incised vinyl-coated scrim cloth, with the exception of the arctic version which is made of unincised spun bonded cloth. Each unit weighs less than 518 g and is small enough (2.13 × 1.37 m) to be folded to fit into a uniform pocket.

Status
Development complete.

Development Agency
US Army Tank Automotive Command, Mobility Technology Center - Belvoir, Fort Belvoir, Virginia 22060-5843, USA.

VERIFIED

Camouflage paints and pattern painting

Description
Fast-drying, dull, high quality alkyd paints are applied by spray gun or brush in patterns to disrupt signature characteristics of vehicles, and to reduce contrast with soil and vegetation in the background. Light green, dark green, forest green, field drab, sand, earth yellow, earth brown, olive drab, black and white colours were provided for blends that counter visual and near infrared surveillance and target acquisition. Solar and heat-reflecting paints and easily removable white coating were also provided for special applications. By January 1977 all US Army tactical equipment was pattern painted.

As the result of a programme that began in August 1978, the US Army turned from alkyd paints to a polyurethane paint. The new finish has a much improved resistance to chemical agents which are unable to penetrate the surface. A surface film is formed which is resistant to the ingress of moisture in any form and thus NBC decontamination can be carried out in the field by washing the affected areas. Once applied, the new paint is stated to be far more durable than the alkyd paints. These coatings prevent surfaces from absorbing chemical agents and enable soldiers to decontaminate their equipment without breaking down and dissolving the paint.

Three-colour camouflage patterns developed by the Belvoir Research, Development and Engineering Center (now the US Army Tank Automotive Command, Mobility Technology Center - Belvoir) replaced the four-colour pattern formerly used on tactical equipment. The new pattern is composed of brown, green and black pattern shapes. In theory, the broad patches of colour used will break up the vehicle's silhouette making it harder to identify at a distance as well as blending better with its background at close range.

For Europe the US Army decided to adopt the new pattern developed by Germany. By standardising the camouflage used by US and German armies, enemy forces cannot ascertain a vehicle's country of origin by its pattern. After a series of tests, the three-colour German pattern was shown to provide better protection than the four-colour American design. Conversion to the three-colour pattern was carried out in conjunction with the introduction of the new chemical agent resistant coatings also developed by the Belvoir Center.

Status
See text.

Development Agency
US Army Tank Automotive Command, Mobility Technology Center - Belvoir, Fort Belvoir, Virginia 22060-5843, USA.

VERIFIED

Model showing three-colour Winter camouflage pattern applied to M1 Abrams MBT

Model showing three-colour Winter camouflage pattern applied to M2 Bradley IFV

Brunswick camouflage and other defensive systems

Description
Brunswick camouflage screens comprise a minimal number of parts and are manufactured from durable synthetics to withstand use and reuse in all environments. Metal fibres distributed throughout the camouflage effectively conceal military equipment from radar systems as described below. Research and development have resulted in colour pigmentations and textures which, in addition to achieving good visual and radar camouflage, provide effective concealment from spectrozal and false colour film combinations, including colour infrared film.

Three standard colour/texture combinations are available to meet woodland, desert and snow conditions. Each screen has a different colour-texture pattern on each side for greater versatility of application. One side of the woodland screen simulates Spring and Summer while the reverse simulates Autumn. Desert screens simulate both tan and grey desert terrain and the Arctic screen provides concealment in total as well as partial snow cover with optimum reflectance in the ultraviolet necessary to preclude detection in this spectral region. Special colour/texture formulations can be developed by Brunswick to meet specific environmental requirements. The screens meet US Military Specifications MIL-C-52771 (ME) and MIL-C-52765 (ME).

Camouflage screens produced by Brunswick have been designed to scatter some of the impinging radar energy, absorb some and allow only enough to be transmitted back to the radar detector to give the military object the same general return as that of the terrain and thus prevent detection. This effect is achieved in the screens by the random dispersion of short, thin steel fibres throughout the garnishing material in combination with selected incising patterns. This combination in the woodland screen results in a proper balance of reflection, absorption, transmission and scatter to achieve the desired simulation of natural woodland surroundings.

Camouflage kits manufactured by Brunswick consist of two subsystems: a screen system and a support system. Each is packed separately in its own vinyl-coated nylon transport case. One screen system consists of a simple hexagon-shaped screen and a rhombic-shaped screen, carrying case and a repair kit. Screen edges are a uniform 4.9 m for joining to make larger screens. Quick disconnect brackets are permanently attached to all screen edges for quick joining. The screens can be quickly disconnected by pulling a lanyard. The support system contains support poles, spikes, spreader assemblies and a carrying case.

The modular concept was designed to meet the diverse field needs encountered. The rhombic or diamond screen can be used for smaller requirements such as personnel or weapons emplacements, while hexagonal nets meet the need for shielding equipment in the ¼ to ½ ton truck range, two- and three-module combinations provide cover for 2½ ton trucks, medium tanks and mobile artillery, regular or irregular configuration module combinations can be assembled to meet almost any installation need.

In October 1990 the US Army ordered 20,000 units of standard lightweight camouflage worth US$6.3 million in a new 'Saudi Tan' colour. The order brought the number of lightweight camouflage units produced by Brunswick to over one million since 1974.

In 1990 Brunswick introduced an 'ultra-lightweight' camouflage with improved radar-scattering and anti-thermal imaging properties. This new camouflage weighs about half as much as standard netting and uses a lightweight garnish cut in a fine pattern and quilted to a fine mesh scrim that is virtually 'snagproof'. The US Army ordered an initial batch of 500 units in 'Saudi Tan' in October 1990, for use with AH-64A Apache helicopters.

Status

In production for the US Army.

Manufacturer

Brunswick Corporation, Defense Division, One Brunswick Plaza, Skokie, Illinois 60076, USA.
Tel: +1 312 470 4827/8. Telex: 190127.
Fax: +1 312 470 4938.

VERIFIED

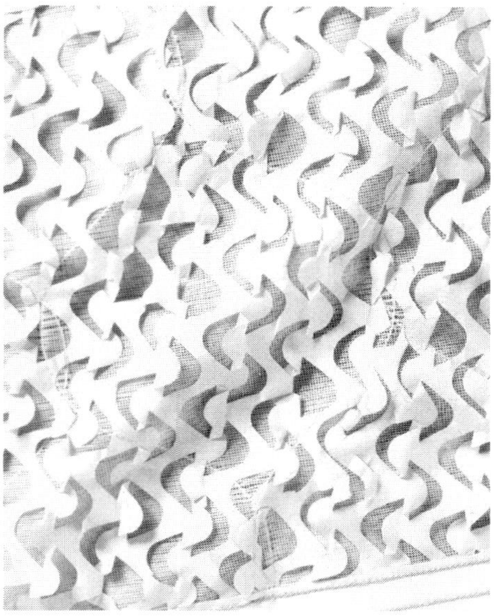

Close-up of Brunswick ultra-lightweight camouflage produced for US Army in a 'Saudi Tan' colour

PORTABLE ROADWAYS

CZECH REPUBLIC AND SLOVAKIA

Truck-mounted roadway laying system

Description

This system consists of a TATRA 815 (8 × 8) truck chassis (originally TATRA 813 8 × 8 trucks were employed) with two cassettes, each loaded with 40 linked steel plates, behind the cab. The roadway is laid with the vehicle travelling in reverse. As the truck travels backwards, the trackways unfold into the horizontal, pass over rollers and are laid under each of the rear wheels. This system is used to prepare the exit points of bridges such as the PMP pontoon system. For this latter application the truck may be fitted with a front-mounted dozer blade.

The roadway, once emplaced can carry wheeled vehicles weighing up to 25 tonnes or tracked vehicles weighing up to 40 tonnes. Fully laden the truck carrying the roadway system weighs 26,400 kg. Overall length carrying a dozer blade is 10.42 m.

Status

In service with the Czech and Slovak armies.

Manufacturer

State factories.

UPDATED

Truck-mounted roadway laying system on TATRA 813 (8 × 8) truck chassis showing method of operation

FRANCE

TEMP TRAIL tactical deployable mobility matting

Description

TEMP TRAIL tactical deployable mobility matting is constructed using an advanced woven nylon-based monofilament method which is light enough to be placed in position by hand but can withstand the weight of a tank or an aircraft. It can be used to rapidly create vehicle or aircraft operating surfaces over any terrain and without the need for ground preparation. The matting is flexible, provides a non-slip surface for vehicles, can follow ground contours and is resistant to a temperature range from −40 to +160°C It is laid in rolls and can be recovered for further use.

There are three types of TEMP TRAIL matting. Type LX is described as standard duty and can withstand axial loads of more than 7 tonnes, yet its weight is only 0.7 kg/m². Type LX is available in standard widths of 2.2, 3.4 or 4.2 m.

Type A2X is described as medium duty. It can withstand axial loads of more than 15 tonnes and has a weight of 1.65 kg/m². The standard width is 4.2 m. Type A2X can be used to form a 64 m² helipad emplaced by hand by two persons in approximately 30 minutes.

Type TX matting is described as extra heavy duty and is primarily intended for aircraft applications, being suitable for bearing the loads of transport aircraft and for use during crash recovery operations. The axial load capacity is over 40 tonnes and weight is 7.5 kg/cm². Standard widths are 2.1 and 4.2 m.

All types of TEMP TRAIL matting are available in widths up to 6.8 m on request. Standard roll lengths are 5, 10, 25, 50 and 100 m.

It is claimed that the matting has a life of from 8 to 10 years. It is UV-resistant, petrol, oils and fuel spill resistant, rotproof and tear-resistant.

Status

Available.

Manufacturer

Deschamps, Usine de Bourisson, BP 20, F-16440 La Couronne, France.
Tel: +33 5 45 67 70 30. Fax: +33 5 45 67 81 60.

UPDATED

TEMP TRAIL tactical deployable mobility matting in use during an amphibious operation **1997**

TEMP TRAIL tactical deployable mobility matting forming a temporary roadway for trucks **1997**

TEMP TRAIL tactical deployable mobility matting bearing the weight of an aircraft **1997**

GERMANY

Krauss-Maffei folding roadway system

Description

During the late 1960s personnel from France, Germany and Italy undertook a study to improve the crossing of rough terrain by heavy vehicles. These studies led in 1974 to a concept and definition phase conducted by France and Germany. This was completed in late 1982. By then 24 proposals had been submitted, of which four were selected for further study. The decision ultimately favoured a system of loosely interlocked hexagonal plates forming a temporary roadway capable of compensating for longitudinal and lateral inclinations without suffering damage. Special joints interconnect the 4.2 m wide and 2.77 m long hexagonal plate roadway sections. A 50 m standard logistic roadway is made up of 18 sections which can be rapidly disconnected or connected to form shorter or longer roadways.

The laying mechanism developed to lay and recover the roadway can be mounted on any type of truck with a payload of greater than 10 tonnes. The latest development is a laying mechanism designed to operate as a multilift load handling system.

During 1989 Krauss-Maffei, the prime contractor, commenced series production. Subcomponents are produced in France by LOHR SA and Fonderie Brea.

The equipment is carried on German Army MAN (8 × 8) 10,000 kg high mobility trucks while the French Army uses the Renault TRM 10 000 (6 × 6) 10,000 kg truck. The MAN truck carries approximately 50 m of roadway while the Renault vehicle originally carried 40 m and used shorter (2.2 m) sections. However, a revised version of the TRM 10 000 can carry the same 50 m length as the MAN vehicle, and is known within France as MATS (Moyen d'Aide a la Traficabilite des Sols). The carrying and laying equipment can be mounted on other existing truck chassis.

The laying mechanism, which weighs 3,600 kg, is constructed of welded steel and is supported at three points. A swivel arm is erected hydraulically to guide the belt during the laying and/or recovery process. The laying mechanism can be lowered for rail transport. The main frame is the base for the roadway carrier. It is constructed of torsional elastic steel and consists of individual cross members which are hinged to the vehicle chassis and support rails which are bolted to the cross members. The roadway sits directly on the roadway carrier, which has an upper and lower cradle. To position the roadway the lower cradle can be moved longitudinally, and the upper cradle can rotate 90° in either direction. Both assemblies are of welded steel construction and the lower cradle has a jointed cross-shaft axle to compensate for chassis distortion. The launching guide is a tubular steel lattice framework hydraulically pivoted over the top of an upright support. Rollers are mounted on the framework to reduce friction. The upright support is a welded aluminium two-part unit.

Folding Roadway System on MAN (8 × 8) 10,000 kg high mobility truck ***1997***

In use, the carrying truck is backed up to the site to be covered by the roadway. The roadway carrier is then rotated and moved into the laying position before a belt is attached to the beginning of the folding roadway. The laying operation starts as the vehicle moves onto the roll-on lanes attached to the first roadway section, which have first been passed to the rear wheels over the launching guide. The roadway is launched by continuously backing up the equipment. Unfolding of the roadway is stabilised and braked by the guide belt running off the winch.

The hexagonal roadway plates are plated chilled aluminium castings while joints and locks are aluminium.

The laying process takes approximately three minutes for every 50 m of roadway and the roadway can be laid on longitudinal slopes of up to 30 per cent and 15 per cent transverse slopes, as well as under water. The recovery process, using the central belt and rear winch is the reverse of the laying procedure with the carrying truck moving forward one section of plates at a time. The recovery process takes 20 minutes. Laying and recovery involves a vehicle driver and one other man.

Introduced in 1996, a variant known as the MATS/DEP-96 became available. This involves the same folding roadway system as described above but the entire system is mounted on a flatrack and is thus able to be carried on, and deployed from, any truck with a 13-tonne capacity (or upwards) fitted with a hydraulic load handling system.

Specifications

Total length: 50 m
Length of each section: 2.77 m
Width: (nominal) 4.2 m
Weight:
(each plate) 10 kg
(each section) 510 kg
(50 m roadway) 9,200 kg
Continuous load capacity:
(wheeled) MLC 30
(tracked) MLC 70

Status

In production for French, German and Netherlands armies.

Manufacturer

Krauss-Maffei Wehrtechnik GmbH, Krauss-Maffei-Strasse 2, D-80997 Munich 50, Germany.
Tel: +49 89 8899-0. Fax: +49 89 8899 2399.

UPDATED

Portable Roadway - Thyssen system

Description

The basis of this heavy portable roadway system is a hexagonal steel plate which is connected to adjoining plates by angled iron brackets. Each plate weighs 23 kg and a standard truck width can be formed by joining nine plates which form a roadway 4.2 m wide. The roadway is constructed by hand and up to eight men can lay 1 m of roadway per minute. To support a 5 tonne truck a mat of 200 plates is necessary, and for a 7 tonne truck, 288 plates. To enable the system to be used for water crossings a method has been evolved in which a completed mat is towed across the water obstacle by a vehicle with a winch. One of the advantages of this system is that it can be assembled under cover and then dragged by winch across the ground to be covered, enabling heavy concentrations to be built up rapidly in forward areas.

Status

In service with German forces and the Danish Army.

VERIFIED

Thyssen portable roadway system being positioned at river crossing point

Close-up of Thyssen portable roadway system

KRUPP Rollmat matting

Description

KRUPP Rollmat matting may be used in conditions where normal road traffic would be unable to move, such as sandy ground or marshy meadowland. Under such conditions the use of this matting enables vehicles up to MLC Class 60 to cross. Rollmat matting may be laid by hand or from special laying units carried on 10,000 kg trucks. The basic element of the matting is a multichamber aluminium alloy strip which is 250 mm wide and 57 mm deep. Each element is foam-filled and can be hinge-connected to other elements in varying widths. These widths may be as little as 1 m for walkways to 16 m for the assembly of temporary aircraft landing strips. For the rapid laying of roads from the vehicle-carried laying unit, widths of 4.6 to 5 m are more usual. The weight per square metre of the basic elements is 33 kg. When laid from rolls, lengths may be up to 50 m which will produce a roll diameter of 2.2 m. If required, elements may be joined together to form helicopter landing mats.

Apart from the vehicle-carried laying unit, accessories available for use with KRUPP Rollmat include wedge-shaped panels for access ramps, safety dowels, anchors and tools.

Status

In production.

Manufacturer

KRUPP Fördertechnik GmbH, Alterdorfer Strasse 120, D-45143 Essen, Germany.
Tel: +49 201 82801. Fax: +49 201 8282566.

UPDATED

KRUPP Rollmat matting in use carrying MLC 60 Leopard 2 MBT

KRUPP Rollmat matting being rolled out from its laying equipment

GREECE

MICROTECH Class 60 trackway

Description

The MICROTECH Class 60 Trackway is used to support armoured and other vehicles over soft ground. It uses the same components and general construction methods as the MICROTECH Runway Repair Decking described in the following section. The trackway is designed for both tracked and wheeled vehicles and is constructed from extruded sections made of heat-treated aluminium alloy. The trackway consists of a number of panels with a tongue on one side and a groove in the other. Each panel has a corrugated shape which, together with additional grooves on the section faces, provides an anti-slip surface. The panels are provided in 2.3 and 4.6 m lengths which can be easily handled by one man or, more conveniently, by two men using special handling tools. Two end ramp panel sections are supplied to enable vehicles to get onto the trackway easily.

The tongue-and-groove connecting method joins panels together. The panels are locked in place with special aluminium alloy bolts which may be removed for disassembly, cleaning and storage. If the trackway is to be laid on a slope it should be anchored in place using a special chain and shackle assembly. Eight section-handling tools are supplied for ease of carrying and assembly of the panels as well as placement of the locking bolts. Optional accessories include towing adaptors.

To produce a standard trackway 40.4 m long and 4.6 m wide a full set as supplied will consist of the following: 176 4.6 m long panels, eight 2.3 m long panels, two ramp panels, eight handling tools, four towing adaptors, 400 locking bolts, four strap assemblies, four 18-link welded chains, 24 steel guy stakes and 12 shackles. Recovery of the mat is facilitated by the use of a strong 50 m long recovery strap.

The trackway can also be used to construct a mat to cover an area of soft ground for depot, maintenance area and other uses. A standard set can be produced to construct a mat 18.4 m wide and 11.7 m long. The tools and accessories supplied are the same as those used for a standard trackway set but the mat requires 185 4.6 m panels, 54 2.3 m panels and eight ramp panels.

MICROTECH Class 60 trackway in use

Specifications

Main section width:
(overall) 242 mm
(effective) 225 mm
Length:
(long panel) 4.6 m
(short panel) 2.3 m
Weight:
(long panel) 32 kg
(short panel) 16 kg
Ramp section width:
(overall) 184 mm
(effective) 152 mm
Ramp section length: (long panel) 4.6 m
Ramp section weight: (long panel) 37 kg
Breaking load: 150 kg/cm of length of standard panel

Status

In production. In service with the Greek armed forces.

Manufacturer

MICROTECH Limited, 14 Adonidos Street, GR-11253 Athens, Greece.
Tel: +30 1 865 5030. Fax: +30 1 867 8541.

VERIFIED

INDIA

Mat Fording

Description

The Mat Fording was developed by the Research and Development Establishment (Engineers) at Dighi, Pune. It is intended for deployment at the approaches to water obstacle fording points and may be used at water depths of up to 1 m. It can support vehicles up to MLC 60 and provides a firm roadway 5 m wide.

Mat Fording is formed by laterally joining a series of extruded aluminium planks which are interlocked to form a standard length of 50 m which is rolled onto a spool. A swivelling platform supports the spool and is mounted on a Bharat TATRA 815 VVN 8 × 8 truck. Other similar vehicles could be employed.

The system is operated by a hydraulic unit powered by a power take-off on the vehicle. In the transport mode the spool is locked in position along the chassis. In operation the spool is turned through 90° to position the spool across the chassis.

For laying the vehicle is positioned with the rear wheels just ahead of the position where laying is to commence. The mat is laid over the rear of the vehicle and placed under the rear wheels. As the vehicle reverses the mat rolls off the spool until the full 50 m length is in position. More 50 m lengths can be laid if required.

The mat is recovered in the reverse sequence.

Specifications

Weight of mat: 34 kg/m^2
Weight of 50 m length: 8,500 kg
Length: 50 m
Width 5 m

Status

In production. In service with the Indian Army and offered for export sales.

Development Agency

Research and Development Establishment (Engineers), Defence Research and Development Organisation, Ministry of Defence, Government of India, Dighi, Pune 411 015, India.
Tel: +91 11 301 3209. Fax: +91 11 379 3008.

UPDATED

Mat Fording in transport mode on a TATRA 815 VVN 8 × 8 truck (T J Gander) ***1997***

Mat Fording in the ready for launch position ***1996***

PAKISTAN

Assault Trackway Class 30 and Class 60

Description

The Military Vehicles Research and Development Establishment at Rawalpindi has developed two forms of assault trackway, Class 30 and Class 60.

Assault Trackway Class 30

Assault Trackway Class 30 is assembled from a number of high tensile extruded aluminium alloy planks, 3.352 m long and 257 mm wide, with interlocking tongue and groove joints, forming a continuous non-skid surface 3.352 m wide and 50 m long. The trackway can be used by wheeled vehicles and tracked vehicles with rubber tyres. Heavier vehicles can use the trackway when ground conditions are favourable. Tanks up to Class 50 may cross at an angle provided they do not slew when negotiating the trackway.

The Class 30 trackway can be launched and recovered by its transport vehicle. A spool carrying 50 m of trackway is mounted on a 5 tonne forward control cab truck. The carriage assembly comprises a swivel type turntable on which a steel stand carrying a three-flanged spool and steel wire rope spool drum is fitted. The trackway is wrapped around the spool and, using the turntable, can be turned to the appropriate position for launching the trackway. The spool can be rotated mechanically in either direction while laying or retrieving the trackway using a 7 tonne pull capacity winch powered through an auxiliary drive from the carrier vehicle's transmission. A manual handbrake on one side of the spool is used to control the speed during the uncoiling of the spool.

The launching assembly consists of a structural steel frame to carry the trackway over the carrier vehicle cab during the launching operation. The assembly can be removed for carrying on the vehicle body during moves and can be replaced within minutes.

The leading edge of the Class 30 trackway is first laid on the ground by hand with the trackway passing above the vehicle cab and with the leading edge placed under the front wheels. The vehicle then moves forward so that the trackway unwinds and passes under the truck wheels. Recovery is over the rear of the vehicle. The preparation and launching operation can be carried out in about 10 minutes by a crew of three. Recovery, takes about 10 to 15 minutes.

Class 30 trackway ready to be launched ***1996***

Tank Access Trackway Class 60

This heavy duty trackway is intended for both wheeled and tracked vehicles and can support MBTs up to 60 tonnes over very soft ground. Its construction is similar to that of the Class 30 trackway except that each panel in 4.572 m long and 228 mm wide. Each panel weighs 32.12 kg.

This trackway is launched and recovered hydraulically by its transporter vehicles. A complete 45 m length set is carried on two Yasoob 6 × 6 trucks, each carrying a 22.5 m length.

For this trackway the launching assembly consists of a steel framework which carries the trackway over the vehicle platform during a launching operation. The assembly can be removed and carried on the vehicle body during moves and can be replaced within minutes.

For launch, a spool carrying the two joined halves of

the trackway is carried towards the rear of the carrier vehicle. The swivelling stand carrying the spool is hydraulically powered to the lateral operational position and a short length of trackway is fed off the spool and tucked under the rear wheels. The vehicle then reverses and launches the trackway as it proceeds. Recovery is carried out in the opposite direction using a winch to position the trackway back on the spool.

Specifications

Type	**Class 30**	**Class 60**
Length	50 m	45 m
Weight	3,500 kg	7,350 kg
Width	3.352 m	4.572 m
Launch system	vehicle	hydraulic

Status
In production.

Development Agency
Military Vehicles Research and Development Establishment (Ministry of Defence), PO Box 240, Abid Majeed Road, Rawalpindi, Pakistan.
Telex: 54102 MCRDR PK.
Fax: +92 51 584786.

VERIFIED

RUSSIAN FEDERATION AND ASSOCIATED STATES (CIS)

Heavy portable roadway sections

Description
There are three basic forms of portable heavy roadway used by the RFAS, two of which make use of the great timber resources of the RFAS with timber girders and timber planks on timber ribs. As such there are many different forms and sizes and many are made either by local state factories or by army workshops. Although the sizes may vary, for supply purposes each section using wood planks is estimated to weigh between 160 and 220 kg less crosspieces, and the girders are estimated to weigh between 250 and 300 kg. Roadways using such timber sections are laid either by hand or by cranes and the rate of laying can vary between 40 and 60 m/h.

The third form of roadway section is manufactured from corrugated steel sheets, each weighing between 100 and 110 kg. Each section measures 2 × 1.05 × 0.08 m, although this may vary. Roadways using these steel sheets can be assembled either by hand or by crane.

Portable roadway sections with KrAZ-255B (6 × 6) truck acting as carrying and laying vehicle

Specifications
Metal sections only
Dimensions: 2 × 1.05 × 0.08 m
Wearing quality: 50,000 to 70,000 vehicles
Weight:
(single section) 100-110 kg
(section for 1 km road) 100,000 to 120,000 kg
Number of trucks: (to transport 1 km of road) 30 to 35
Rate of laying: up to 100 m/h

Status
In service with the former Warsaw Pact forces.

Manufacturer
All types: state factories. Can be assembled in army workshops.

VERIFIED

Glued plywood roadway sections - SRDP

Description
The lightest of the RFAS types of portable roadway is the SRDP (*Sborno-Razbornoye-Dorozhnoye Pokrytiye*). The basic element of this system is a glued plywood panel measuring 2.5 × 1 × 0.07 m and weighing between 100 and 120 kg. Each panel is held in place by wooden inserts placed into butt-brackets at each end. The panels can be laid to form either strips or wider roadways. They are normally laid by hand with teams of 10 to 12 men, but a mechanical laying method has been developed which enables strips to be laid from amphibious PTS vehicles. This method replaces the wooden end inserts with steel cables and the panels are preprepared as they are loaded onto the PTS.

The panels are prepared using methods which have been specially developed to take advantage of the large timber resources of the RFAS. The wooden sheets that make up each panel are joined together by phenol-formaldehyde glues and are then rapidly passed to a further process which pressure-coats the panels with a layer of bakelite material. The ends are covered by strips of similarly treated plywood and surface gripping strips are added. The panels then require no further maintenance and can be stored in the open if necessary.

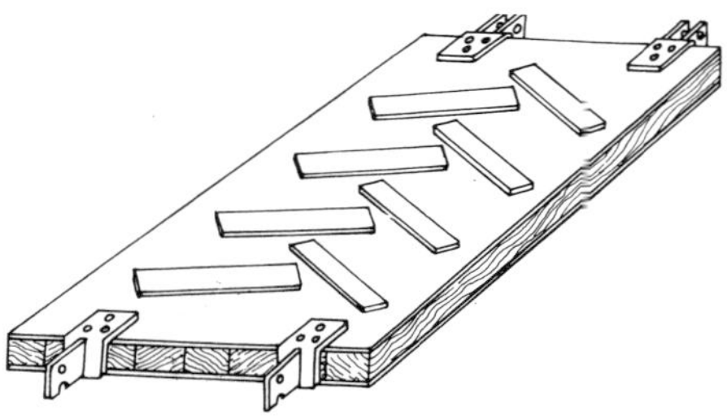

SRDP glued plywood roadway section

Specifications
Panel dimensions: 2.5 × 1 × 0.07 m
Wearing quality: 40,000 to 50,000 vehicles
Weight:
(single panel) 100 to 120 kg
(panels for 1 km road) 80,000 to 100,000 kg
Number of ZIL-164 trucks: (to transport 1 km of road) 25 to 30
Rate of laying: 100 to 150 running m/h

SRDP glued plywood roadway in position

Status
In service with the RFAS.

Manufacturer
State factories. Simplified versions may be assembled in army workshops.

VERIFIED

SWEDEN

Columbus Mat

Description

The Columbus Mat was designed by Curt F Lundin and is used for a wide variety of civilian and military applications including reinforcement of soft ground to enable tracked or wheeled vehicles to cross difficult country, at river exit points, and as military bridge access mats. It is also used as recovery tracks for aircraft recovery after abortive take-offs and landing overruns. (The Royal Thai Air Force and Scandinavian airports have purchased sets for this purpose.) The standard 5 m mat consists of polythene tubing held together by 15 steel cables. At the ends of the mats there are three tubes 65 mm in diameter and 15 mm thick. The other tubes (71 in all) have an external diameter of 65 mm and are 5.8 mm thick. The cables have 171 threads and a diameter of 8 mm, and the cable ends are drawn in a loop back through the heavy end tube. Coupling links at the cable ends enable mats to be joined end-to-end to form a roadway of any length.

Laying and recovery require no special tools or equipment and the mat can be transported by truck either flat, rolled or folded.

For use by heavy vehicles at water crossings a special system called the Vehicle Mat 2MT was developed. The complete system consists of: four mats each with 140 mm outside diameter end tubes and heavy-duty cables with coupling links, two steel anchoring beams, four climbing frames, 12 chains, 10 soil anchor frames, 20 soil rod sets plus cables and other accessories. Weight of the complete equipment is 5,600 kg. The system was developed in close co-operation with the Swedish Army.

Method of joining two sections of Columbus Vehicle Mat 2MT and showing polythene tube construction

The basic idea is that the mats and the climbing frames, which consist of four Bailey bridge panels, are assembled on the bank on one side of the river and then winched down the river bank by a winch on the far bank. Once in place, the mats are held firmly in position by the soil anchor frames which are staked to the ground and connected to the anchoring beam by chains.

Specifications

Basic mat
Width: 4.5 m (other widths available in 300 mm multiples)
Length: max 15 m (normal standard 5 m)
Weight: 20 kg/m^2

Status

In service with the armed forces of Germany, Norway, Sweden, Switzerland, Thailand, UK and the USA. Used with aircraft recovery kits by many countries.

Manufacturer

Vårgårda Plast AB, Columbus Mat Division, S-447 00 Vårgårda, Sweden.
Information and sales: Curt F Lundin, PO Box 1037, S-144 01 Rönninge 1, Sweden.
Tel: +46 753 54677. Telex: 13362 grevtur s.

VERIFIED

UNITED KINGDOM

Faun portable trackways

Description

The Military Load Class (MLC) 30 and 60 Portable Trackways made by Faun MV Limited were originally designed for the British Army by the Military Engineering Experimental Establishment. Later development by Laird (Anglesey) Limited (now subsumed into Faun MV Limited) in conjunction with the British Aluminium Company, the Portable Trackway systems continue to be supplied to armies worldwide. The Trackways consist of interlocking aluminium planks able to be unwound from a spool to provide non-skid surfaces which are firm enough for going over ground that is, or would quickly become, impassable to vehicles. After use Trackway can be recovered, rewound, relocated and reused many times.

Class 30 Trackway

Class 30 Trackway has an almost unlimited life when used by military vehicles up to MLC 30. This includes most wheeled logistic vehicles and tracked vehicles with rubber tracks. It can also be used by heavier wheeled and tracked vehicles, up to MLC 50, when ground conditions are favourable although some damage to the Trackway may occur.

Class 30 Trackway is assembled from a number of interlocking, captive tongue-and-groove jointed, extruded aluminium alloy planks wound onto a spool. When unwound these planks form a continuous non-skid surface 3.35 m wide. A typical 4 tonne flatbed military truck can carry the Laying and Recovery Equipment (LRE) which contains all the stores necessary for laying or recovery of Trackway, and a spool with some 32 m of Trackway. Carried along the length of the truck during road delivery, the spool can be rotated through 90° on its turntable assembly so that Trackway can be laid either over the front of the truck or over the rear. Preparation on site and launching of 32 m spools can be completed by a team of five in 10 minutes. Trackway is recovered over the rear of the truck using ratchet spanners by a team of five in about 15 minutes. When a Trackway longer than 32 m is required a vehicle-mounted LRE can be reloaded with further spools delivered by general purpose transport and, once laid, the second and subsequent lengths of Trackway can be clamped to those already laid. Anchorage stores are provided for use on steeply sloping ground.

An important application of Class 30 Trackway is beach landing. The Trackway on its carrier frame, matched to a waterproofed Rough Terrain Fork Lift Truck, is unrolled onto the beach as the truck moves through the shallows from a landing craft onto the beach.

Specifications

Class 30 trackway
Length: typically 32 m
Width: 3.35 m
Weight:
(per metre run) 68 kg
(LRE) 1,141 kg
(total, Trackway and LRE) 3,325-4,280 kg
Individual plank dimensions: 3.35 × 0.23 m

Class 60 Trackway

Heavy-duty Class 60 Trackway, originally designed to carry tracked and wheeled vehicles up to MLC 60 has also been approved for use by MLC 70 vehicles. When used to reinforce any heavily used route, such as defiles at a bridge approach and exit, Class 60 Trackway is usually delivered in a roll and unwound at the site. The Trackway can be laid as a roll up to 15 m long from under armour when carried by tanks or other combat vehicles, power-laid from a roll up to 50 m long from 10 to 16 tonne payload logistic vehicles, or laid by hand either as a 15 m roll or as an area mat. Class 60 Trackway has other applications for the rapid repair of battle damage to runways for fast jet aircraft or for the rapid construction of extra hardstandings and dispersals for aircraft. Typical military vehicles can carry rolls up to 30 m on 8 tonne, 45 m on 10 tonne, and 50 m on 16 tonne trucks. Both rolls and mats of

Class 60 Trackway being carried by a Mercedes-Benz (6 × 6) truck **1997**

British Army Leyland Trucks (4 × 4) 4,000 kg truck carrying Class 30 Trackway **1996**

Class 60 Trackway are built from interlocking aluminium alloy planks secured by captive bolts. A standard plank weighs 33.1 kg, is 4.6 m long and provides an effective width of 200 mm. Half planks are used to stagger the joints during mat assembly and to allow easy sub-division of long (50 m) rolls.

When transported on suitably equipped logistic vehicles, 50 m of Class 60 Trackway can be laid or recovered by a team of three in 5 to 15 minutes. The Trackway is carried on a detachable spool supported by a swivelling stand with a hydraulic power unit. The LRE can be mounted directly on the vehicle chassis or clamped to the vehicle deck. Trackway is laid or recovered over the rear of the vehicle.

Specifications

Class 60 Trackway
Length: typically in 15 m or up to 50 m rolls
Width: 4.6 m
Weight:
(standard plank) 33.1 kg
(laid as a trackway) 156 kg/m run
(laid as a mat) 34.2 kg/m²
(laid as a 16.46 × 18.3 m area mat) 10,300 kg

Status

In production. In service with 25 countries worldwide including Canada, Germany, Greece, Iraq, Japan, Netherlands, Saudi Arabia, Singapore, UK and the USA.

Manufacturer

Faun MV Limited (formerly Laird (Anglesey) Limited), Beaumaris, Anglesey, Gwynedd LL58 8HY, UK.
Tel: +44 1248 810431. Fax: +44 1248 810300.

UPDATED

UNITED STATES OF AMERICA

MO-MAT roadway system

Description

MO-MAT was designed for laying over terrain such as mud, sand and snow to allow the passage of wheeled vehicles. It is also used for a variety of other roles including use as a helicopter pad.

MO-MAT is fabricated from a glassfibre-reinforced plastic called Stratoglas, developed by the Stratoglas Division of the Air Logistics Corporation. MO-MAT is moulded into a structural shape resembling that of a waffle with an overall thickness of 16 mm. It is also available in thicknesses of 2.2 and 3.2 mm forming a lattice cross-section 16 mm thick overall and has a durable non-skid material bound to the top surface.

The standard duty MO-MAT (2.2 mm) weighs 4.9 kg/m² and the heavy-duty MO-MAT (3.2 mm) 7.3 kg/m². MO-MAT is supplied in standard panels 3.709 m wide and 14.782 m long. Utility panels 3.709 m wide and 6.6 m long and sheets 3.6 m long and 1.8 m wide are also available. Precision holes are provided around the periphery for interconnecting panels or sheets to any desired length or width. Panels or sheets may also be attached to frames to form a variety of structures.

The MO-MAT panels reduce the ground pressure by spreading the wheel loads over a wider surface area. Excessive crowning of the roadway caused by heavy traffic can be mitigated by doubling the thickness of the roadway — laying one panel on top of the other — or increasing the width of the roadway and spreading the traffic pattern.

The sheets are supplied rolled up on a pallet for ease of transport and can be manually deployed, re-rolled or assembled without the use of special tools or equipment.

MO-MAT roll being held ready for use by US Marine Corps bulldozer (Ray Young)

Status

Production on receipt of large order. In service with US and other armed forces.

Manufacturer

Air Logistics Corporation, 3600 E Foothill Boulevard, Pasadena, California 81109, USA.
Tel: +1 818 795 9971. Fax: +1 818 795 2528.

VERIFIED

RAPID RUNWAY REPAIR EQUIPMENT/PORTABLE RUNWAYS

FRANCE

DALLEXPRESS emergency runway repair system for runways and taxiways

Description

The DALLEXPRESS emergency runway repair system for runways and taxiways was developed by Lafarge Fondu International, a subsidiary of the Lafarge Coppee Group; it is distributed by Automatismes & Techniques Nouvelles (ATN).

The DALLEXPRESS system employs new concrete technologies, producing a concrete slab in a period as short as 30 minutes. With this process a binder in the form of a liquid grout is pumped onto a course of aggregates through which it percolates to occupy all voids and spaces. At the end of the operation the liquid grout entirely covers the aggregate course. The thickness of the finished slab (corresponding to the thickness of the aggregate course) can be adjusted according to traffic requirements. The resultant surface is flat and even and can be matched to the surrounding material by using a broom. Traffic can be resumed 30 to 40 minutes after the application of the grout.

The process can be used over a temperature range of −20 to +50°C. As an example, a transport aircraft of the C-130 Hercules type will require a slab thickness of 370 mm on an aggregate course.

Status

Available.

Manufacturer

ATN SA, Le Clos de Villarceaux, F-78770 Thoiry, France.
Tel: +33 5 34 87 55 88. Telex: 696819 f.
Fax: +33 5 34 87 40 77.

VERIFIED

GREECE

MICROTECH runway repair decking

Description

MICROTECH runway repair decking provides a rapid repair system to create a firm running surface over bomb damaged runways. The decking is constructed from extruded sections of heat-treated aluminium alloy and consists of panels and accessories that fit together to form a mat. The main sections have been designed to join with a tongue at one end and a groove at the other. Once in position, the connections are locked in place using special aluminium alloy bolts. The sections are supplied in two sizes, each size being man-portable. Each section is corrugated in cross-section and has an anti-skid upper surface. Ramp sections can be fitted to the sides or ends of the mat. Special handling tools are supplied for carrying, assembly and disassembly of the decking. Other accessories include towing adaptors and expansion bolts to anchor the decking.

To form a standard runway repair decking 18.4 m wide and 11.7 m long the following items are required: 185 long panels, 54 short panels, eight long ramp panels, eight handling tools, four towing adaptors, 500 locking devices and 40 expansion bolts. The resultant mat may be rolled for stowage if required.

The MICROTECH runway repair decking can be used, without anchoring, to form a forward area helicopter landing pad that can be rolled up and moved to a new location as required.

Moving rolled repair decking mat using mobile crane

The components used for the MICROTECH runway repair decking are the same as those used for the MICROTECH Class 60 Trackway described in the previous section.

Specifications

Main section width:
(overall) 242 mm
(effective) 225 mm
Length:
(long panel) 4.6 m
(short panel) 2.3 m
Weight:
(long panel) 32 kg
(short panel) 16 kg
Ramp section width:
(overall) 184 mm
(effective) 152 mm
Ramp section length: (long panel) 4.6 m
Ramp section weight: (long panel) 37 kg
Breaking load: 150 kg/cm of length of standard panel

Status

In production. In service with the Greek Air Force.

Manufacturer

MICROTECH Limited, 14 Adonidos Street, GR-11253 Athens, Greece.
Tel: +30 1 865 5030. Fax: +30 1 867 8541.

VERIFIED

UNITED KINGDOM

Croker rapid runway repair system

Description

Croker Limited developed its mobile flood grouter mixer within a flush capping system for rapid runway repairs to incorporate an independently hydraulically driven demountable system. This diesel-driven modular system is also suitable for mounting on flatbeds and trailers. The system can be mounted on flatracks compatible with DROPS vehicles and trailers. The system incorporates a hydraulic folding crane mounted on a flatrack, with the mixer driven by a single diesel engine through a hydraulic transmission system.

For operating flexibility the materials to be mixed are palletised for mounting on standard flatracks so that all the materials and equipment used with the system are DROPS compatible. The flush capping system incorporates a mixer with an integral folding crane on a flatrack, with the grout and water mounted on a separate flatrack with a similar folding crane. The

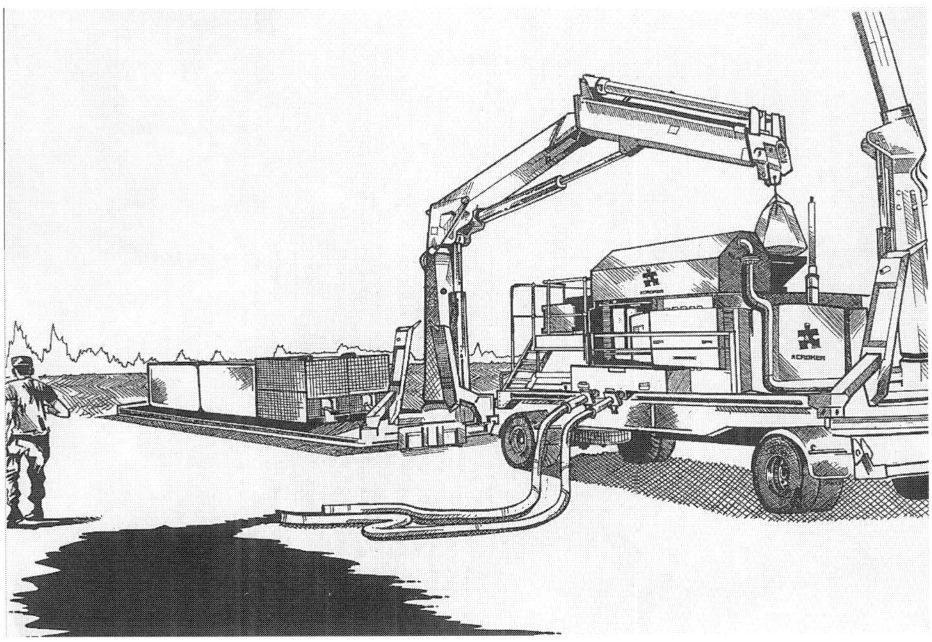

Drawing of Croker rapid runway repair system in use

1997

second flatrack is also provided with a diesel engine and incorporates a suction pump for filling storage tanks also mounted on the flatrack, and also for dispensing water to the mixer.

The Croker system can mix ASR1 grout powder with a measured amount of water in the minimum cycle time. Water is measured by volume from water tanks above a Croker Rotating Pan RP1500 mixer. A large bagsplitter with static knives enables powder to be dispensed from large storage tote bags into the mixer at the required rate. The mixer uses the Cumflow mixing technique and achieves an efficient mix approximately 1 minute after all ingredients have been fully charged. The mixed product is then discharged into a grout holding tank situated below the mixer, before being dispensed into the damaged crater via the grout hose distribution system. The system can produce the mixed product at a rate of 20 m³/h.

Status
Available.

Manufacturer
Croker Limited, Bournemouth Road, Chandler's Ford, Eastleigh, Hampshire SO53 3ZL, UK.
Tel: +44 1703 262311. Fax: +44 1703 271012.

NEW ENTRY

Faun rapid runway repair system

Rapid runway repair mat being laid (UK Ministry of Defence)

Description
The Faun rapid runway repair system is NATO approved and underwent extensive trials with both the Royal and US air forces. It was accepted for service by the UK in 1970 and entered service in 1972.

It is based on the rapid runway repair mat which consists of an area of interlocking Class 60 Trackway aluminium alloy panels, a full description of which is given in the entry for Faun Portable Trackways in the *Portable Roadways* section. The mat can be built to any desired size although the 22 × 16 m mat is the accepted size for repairing of bomb craters caused by 340 to 450 kg bombs.

The mat is stowed rolled on special chocks and when required is lifted and transported on a runway repair mat trolley, enabling the mat to be positioned at the crater site without additional lifting equipment. The laden trolley can be towed by a medium tractor.

Following a strike on a runway, the first task of the repair team is to locate the damage and transfer the information to a runway plan. A minimum operating strip, possibly 1,500 m long and 15 m wide, is then plotted to establish the least amount of initial repair work necessary to restart operations.

The area around the bomb crater is first cleared of debris, and fallback material is taken from inside the crater. The crater is then filled with selected aggregate, stockpiled within the perimeter of the airfield. If aggregate is not available crater debris may be used and pushed back into the crater.

Once the fill has been compacted, the damaged area is levelled with a purpose-built screed beam and the repair mat is moved to the site, lowered to the ground and unrolled over the levelled fill. The mat is then tensioned and fastened to the undamaged section of the runway by expanding foundation bolts which pass through the fairing panels. Special fairing panels which enable aircraft to run onto and off the repair mat are connected to each end.

The final stage is to clear the area of all loose material with a motorised roller. The complete operation can take as little as 1 hour 30 minutes from the time tractors start clearing debris.

To aid concealment, the trackway and fairing panels are treated with pylumin, a chemical dip producing a grey/green finish. The Class 60 mat can also be used for aircraft dispersal pads and portable runways.

Status
In service with Greece, Jordan, Saudi Arabia, Switzerland, Turkey, UK and other armed forces.

Manufacturer
Faun MV Limited (formerly Laird (Anglesey) Limited), Beaumaris, Anglesey, Gwynedd LL58 8HY, UK.
Tel: +44 1248 810431. Fax: +44 1248 810300.

UPDATED

Faun airfield landing mat

Description
Designed and developed by the MVEE (Christchurch) in conjunction with the British Aluminium Company Limited, the Airfield Landing Mat is usually referred to as the PSA (Prefabricated Surface Aluminium). The standard mat, known as the PSA1 and which weighs 15 kg/m², is suitable for most freight and passenger aircraft, and for combat aircraft with low-pressure tyres; the maximum tyre pressure is 5.62 kg/cm². The mats are made from aluminium panels and, apart from their general use for aircraft landing strip assembly, can be used for operating pads for aircraft such as the Harrier and also for rapid runway repair.

The Airfield Landing Mat uses six basic components, as follows:

Basic panel — This is 2.74 m long with an effective width of 250 mm. It is light enough to be carried by one man as the PSA1 panel weighs 9.5 kg. Each panel has slots and lugs to connect it with its neighbouring panels in a brickwork pattern at an angle of 45° to the line of the runway.

Double female panel — This panel is 2.7 m long and 44.5 mm wide. It forms the centreline of the runway from which panels can be laid on both sides.

End anchor panel — This is the same as the basic panel apart from six 25 mm picket holes which provide anchorage.

Picket — The picket is 1.2 m long and 22 mm in diameter. It is made from galvanised steel rod and is used to anchor the end anchor panel. It has a T-shaped head that fits flush into the panel corrugation.

Repair panel — This is a two-part basic panel with a longitudinal interlocking joint. The two halves can be held in place by nine countersunk screws.

Edge restraint device — To prevent the ripple effect that forms when aircraft land and brake on airfield surface mats, the edge restraint device is fitted to the edge of all airfield landing mats. It consists of a fabric strip filled with earth or sand, and with the fabric tied around the filling to form a long flexible cylinder. The fabric is laid on a connected plywood strip 305 mm wide.

To build a runway, some site preparation is needed to ensure the area is reasonably flat and drained. In wet

Laying PSA1 mat (UK Ministry of Defence)

Fitting end anchor panel to PSA1 mat (UK Ministry of Defence)

areas a layer of neoprene-coated nylon fabric (PSN) may be laid. The airfield landing mat can be used by unskilled labour under engineer supervision, and it is possible to lay a complete airfield runway in one or two days. Once laid, the mat requires only a minimum of maintenance, but a well-laid and maintained runway can be recovered and re-used with only a minimum of panels needing replacement.

Status
In service with the Royal Air Force and in some other countries.

Manufacturer
Faun MV Limited, (formerly Laird (Anglesey) Limited), Beaumaris, Anglesey, Gwynedd LL58 8HY, UK.
Tel: +44 1248 810431. Fax: +44 1248 810300.

UPDATED

UNITED STATES OF AMERICA

Lockheed Martin aluminium Rapid Runway Repair (RRR) kit

Description
The Rapid Runway Repair kit (RRR) was developed by Martin Marietta Aluminium (now part of Lockheed Martin) from its AM-2 airfield matting system and is standard equipment with the US Air Force and is approved for use by NATO countries. Selected forward operational bases are issued with prepositioned RRR kits, each base being provided with equipment, tools and portable airfield matting to repair three 340 kg bomb craters within four hours.

Each kit provides a patch of AM-2 matting 16.459 m wide and 24.621 m long, containing 35 half mats (0.61 × 1.829 m × 38 mm), 15 full mats (0.61 × 3.65 m × 38 mm) and 18 ramps each 1.829 m long. All components have an anti-skid coating and are packed for shipment and storage in the same manner as the standard AM-2 matting.

The mats have been designed to withstand single wheel loads of up to 13,608 kg each. Half mats are contained in each pallet assembly so that the mats can be laid in a brick-type pattern. This staggered joint arrangement provides the required stability across the mat and the necessary flexibility in the direction of travel. The sides of the mat panels are constructed to interlock with a rotating motion. The end connectors are arranged with the prongs up on one end and down on the other and by properly placing the end connector on one mat over the end connector of another, a continuous layer of matting is formed. Locking couplers are then inserted into the common slot to form a bond between the plates.

There are six main steps in the repair cycle: identifying the craters for repair and establishing a temporary runway centreline; repairing the crater; delivering, stockpiling and placing the selected fill material; assembling the AM-2 matting patch; cleaning, sweeping and painting the new runway; and finally placing and anchoring the ramps.

A temporary runway may be a portion of the runway itself, a combination of the parking areas and the taxiway, or other adjacent surfaced area. A temporary runway is normally 15.24 m wide and 1,524 m long. All material which has been blown out of the crater is pushed into the crater void and compacted with a dozer. While the crater is being refilled, a grader clears areas on the runway surface for the AM-2 matting assembly. The matting assembly is positioned to allow for a parallel pull over the crater but a sufficient distance away to allow crater repair to continue uninterrupted. When the crater is filled to the final 0.305 m the selected fill is pushed in, compacted and levelled.

The standard patch can be assembled by 17 trained personnel in less than one hour, and when assembled is pulled into place over the filled crater. When the final positioning is made, ramps are attached to each end to alleviate the step or contour change between the mat and the runway. Should it become practicable to rebuild the runway, the patch sections may be easily removed, dismantled and stored for future use.

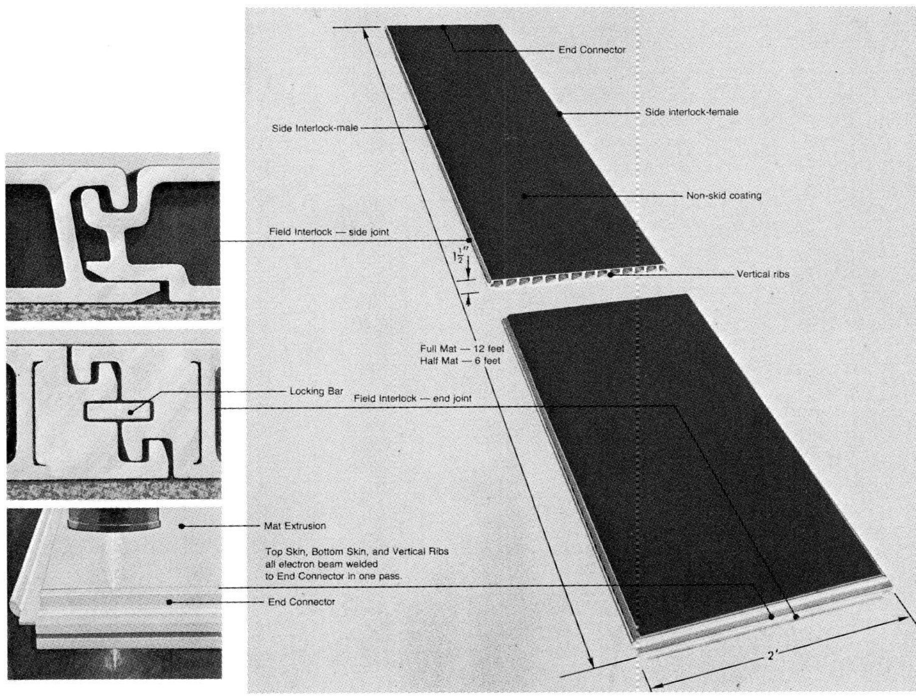

Main components of Lockheed Martin aluminium rapid runway repair kit mat

Status
In service with the US Air Force.

Manufacturer
Lockheed Martin, 19200 South Western Avenue, Torrance, California 90509, USA.
Sales handled by Transaero Inc, 80 Crossways Park Drive, Woodbury, New York 11797, USA.
Tel: +1 516 248 0400.

VERIFIED

SHELTERS AND CONTAINERS

AUSTRALIA

BONDOR Australian Army medical shelter

Description

Under contract to the Australian Army BONDOR Pty Limited designed an air-transportable medical shelter, one of many types for use in a mobile field hospital. The shelter has the same dimensions as an ISO 1C container. It is rated at 8,000 kg and can be stacked two high. The shelter has aluminium-polyurethane wall panels inserted into a structural steel frame. The integral high lift jacks enable the shelter to be dismounted from its carrying vehicle using a hydraulic system or an air line hose from the vehicle. The jacks are normally pushed into the wall housing for transportation. The double doors at both ends provide access for wide loads and the side door is used for normal entry and exit. The shelters are delivered complete, apart from air conditioning, furniture and other such specialised items.

Status

In production.

Manufacturer

BONDOR Pty Limited, 260 Musgrave Road, Coopers Plains, Brisbane, Queensland 4108, Australia.
Tel: +61 7 277 4044. Telex: AA41214.
Fax: +61 7 275 1760.

VERIFIED

Australian Army medical shelter raised on outrigger legs ready for loading on to truck

Australian Army medical shelter emplaced ready for use

BELGIUM

Baeten shelters

Description

Often referred to as the Plasti-Baeten Shelter, the Type PB 7601 is made from lightweight non-metallic materials. The basic panel material consists of a 40 mm layer of non-hygroscopic PVC foam covered with a 2.5 mm skin of polyester-reinforced glass fibre. The foam has a density of 60 kg/m^3. The interior floor has a strong non-slip surface and the underside of the shelter is fitted with two longitudinal aluminium skids 4.4 m long, 32 mm high and 180 mm wide. There is a single door 1.78 m high and 0.8 m wide with blocking devices incorporated. The interior sides can be fitted with up to three windows and there is provision for blackout screens. There is an aluminium ladder for access to the shelter when it is carried on a truck and folding handholds to gain access to the shelter roof. Lashing and lifting points are integral. Mains electrical points are provided.

In addition to the above, Baeten also produces a multipurpose 4 tonne shelter and a 2 or 4 tonne shelter with EMI shielding for the protection of electrical and electronic equipment against high altitude nuclear (EMP) explosions. Tests have shown that the EMI shelter has an attenuation of over 140 dB and may be built with sandwich panels having an inner and outer skin in polyester or aluminium or a combination of both. Several types have been produced.

Specifications

Type PB 7601
Length:
(internal) 4.31 m
(external) 4.4 m
Width:
(internal) 2.11 m
(external) 2.2 m
Height:
(internal) 1.9 m
(external) 2 m
Weight: 940 kg
Payload: 3,500 kg

Status

In production. In service with the Belgian Army.

Manufacturer

nv Baeten, Autostradweg 1, B-9230 Melle, Belgium.
Tel: +32 91 52 18 52. Telex: 11087 baeten b.
Fax: +32 91 52 37 93.

VERIFIED

Baeten Shelter Type PB 7601

Baeten non-shielded 4 tonne shelter carried on MAN (4 × 4) truck

BRAZIL

TECTRAN straddle carriers

Description

TECTRAN straddle carriers are intended to move bulky loads such as ISO containers across reasonably flat and paved open areas. The exact form of the carrier can be modified to suit customer requirements and the data provided is only a guide.

There are six basic models, as follows:

TT-2000, maximum load, 20,000 kg
TT-3000, maximum load, 30,000 kg
TT-4000, maximum load, 40,000 kg
TT-5000, maximum load, 55,000 kg

The TTC-2020 and TTC-3040 are special ISO container carriers for 20 ft/6.096 m and 40 ft/12.19 m containers respectively.

The TECTRAN straddle carrier is based around a welded steel frame. It uses hydrostatic transmission connected by planetary gears installed in the two rear driving wheel hubs and is highly manoeuvrable, being able to make 360° pivot turns. The driver/operator is seated in a heated, air conditioned cab over the main engine on the left-hand side of the vehicle and controls two hoists, which are also driven by a hydrostatic transmission connected to planetary gears installed in the hoist cable drum. The hoists are located in the front and rear arches and are two-speed. Loads can be traversed on a trolley operated by a steel cable and pulley arrangement driven by a hydraulic motor and a worm gear reduction box. Steering is hydraulic, using the 'Orbitrol' system actuating the two rear driving wheels with double-action hydraulic cylinders and with mechanical articulation around each front wheel.

Specifications

TT-2000
Cab seating: 1
Max length: 5.8 m
Max width: 5.3 to 9.3 m
Inside clear width: 3.5 to 7.5 m
Max height: 5.45 to 8.45 m
Height to hood saddle: 4.75 to 7.5 m
Wheelbase: 4.5 m
Max speed: 8 km/h
Gradient:
(unloaded) 10%
(loaded) 4%
Engine: Mercedes-Benz OM-314 developing 76 hp at 2,600 rpm
Transmission: hydrostatic
Steering: 'Orbitrol' system
Turning radius: 360° pivot turns
Brakes:
(service) disc on front wheels, vacuum-assisted
(parking) drum on rear wheels
Tyres: 14.00 × 24
Hoist speed, full load: 10 m/min
Trolley traverse speed: 45 m/min

Status

In production.

Manufacturer

TECTRAN Engenharia, Indústria e Comércio SA, Rodovia Presidente Dutra, Km 155/156, Caixa Postal 165, CEP 12200 São José dos Campos - SP, Brazil.
Tel: +55 123 220142. Telex: 39431 ttte br.

VERIFIED

TECTRAN TT-2000 straddle carrier

TECTRAN TTC-2020 straddle carrier for 20 ft/6.096 m ISO containers

DENMARK

Container Load Trailer (CLT)

Description

The Container Load Trailer (CLT), developed in co-operation with the Danish Army, is a highly mobile container handling system that can be used in various configurations to handle and carry 20 and 40 ft ISO containers, shelters and associated loads. It is particularly suitable for use in remote areas where normal ISO handling equipment is often not available. Containers can be handled as end or side loads and the CLT can also be used as a trailer, carrying a container or, with a flatbed, other loads. Towing vehicles may be trucks or tracked armoured vehicles such as APCs.

The CLT consists of two units: a front section with a fixed steerable axle and drawbar and a rear unit with a non-steerable pendular axle. Each unit has two hydraulic primary cylinders and two secondary cylinders. The primary cylinders are used for lifting and the secondary cylinders for positioning when loading or unloading. Each unit is fitted with chains used for sideways lifting, a hydraulic unit and tank, a diesel engine, valves, lock fittings and wheels with all-terrain or rough terrain balloon tyres.

When lifting, the CLT is completely self-contained using its own hydraulic system. The system has a maximum lifting height of 1.8 m and a lifting capacity of 25 tonnes both at end and side lifting. Acting as a trailer the system has a capacity of 21.5 tonnes. Total weight of the CLT is 6,000 kg.

The CLT may also be used with the special CLT/C-130 kit which enables it to load and unload two ISO containers from a C-130 in 20 to 25 minutes.

When not in use the CLT can be stored or transported inside a standard 20 ft ISO container, towed as a short four-wheel trailer or air-transported in a C-130.

A 7.5 ton CLT system is also available from ESCO Datron; for details see under United States in this section. ESCO Datron also markets the CLT in the USA.

Container Load Trailer (CLT) carrying a container as a side load

Specifications

Payload:
(lift) 25,000 kg
(transport) 21,300 kg
Max lift height: 1.8 m
Max towed speed: (loaded) 80 km/h
Weight, empty: 6,000 kg
Length: (with 20 ft container) 12 m
Width: 2.45 m
Height: adjustable
Engine: 5.6 hp diesel

Tyres:
(standard) 16.5 R 22.5
(rough terrain) 23.6 R 22.5

Status
In production. In service with the Danish Army and Air Force, Swiss Army, Royal Air Force and UN peacekeeping forces.

Manufacturer
Danish Camp Supply ApS, Virkelyst 11, DK-9400 Norresundby, Denmark.
Tel: +45 98191300. Telex: 69 925 camp dk.
Fax: +45 98190700.

VERIFIED

FRANCE

Giat Industries shelters

Description
Giat Industries mass-produces the types of shelters listed below for the French and other armies. They may be delivered bare or fully equipped. The different installations available allow each of these units to respond to the various demands of environment, logistic army, air force and marine employment.

Cadre JVH
This is a monoshell cabin with PVC, polyester resin and aluminium sandwich panels plus metal reinforcements. It is used by French Army RITA communications system units and also for the Sirocco meteorological station.

Cadre JVA
This shelter uses a metal frame with welded steel components plus plywood facing and thermal insulation. It is used for second and third line mobile mechanical and electronic workshops.

Cadre JVB
Using PVC, resin polyester and aluminium panels, this shelter has an aluminium frame and a door that employs a microwave braided cable surround. It is used in support of the French Army DIADEME programme.

Cadre JVF
This is a specially hardened shelter with sandwich panels continuously welded to an aluminium ISO construction frame. The door uses folded hinges. It is used for the main station of the HADES data transmission system.

Giat Industries JMV shelter

Cadre JVG
This is another hardened shelter with high mechanical and thermal performance sandwich panels and continuous welding to provide integrated ballistic and thermal protection. ISO corners are provided and there is a folded hinge door. This shelter is used with various command, control and communication systems.

ATSF
The ATSF (*Abri Technique Semi Fixé*) is a TEMPEST-hardened shelter proof against the EMP produced by a nuclear attack. It has an airlock and is intended to be covered by earth or concrete for complete protection. It has its own integral fire extinguishing system, air conditioning and NBC system. A series of 60 was produced for the French Army for transport by rail.

ATM
The ATM (*Abri Technique Mobile*) is a 10 ft ISO shelter intended to contain the utilities required by a field hospital, including electrical power generation, the housing of medical gases, air conditioning and filtering and other services. A total of 100 were ordered for the French Army. The ATN 20 DD is a variation with extending side walls and ends for use as an operating theatre and is a 20 ft/6.096 m ISO shelter; 200 were ordered for the French Army.

A variant of the basic ATM shelter is constructed with an Ampliroll compatible pick-up system at one end. This enables the container to be handled by the Ampliroll system without recourse to interface attachments. With the ATM bottom guide rails folded this pick-up system allows the ATM shelter to be loaded directly into transport aircraft. For details of the Marrel Ampliroll system refer to the entry in the *Materials handling equipment* section.

Status
In production. In service with the French Army and some other armies.

Manufacturer
Giat Industries, 13 route de la Minière, F-78034 Versailles Cedex, France.
Tel: +33 1 30 97 36 48. Fax: +33 1 30 97 39 58.

UPDATED

A typical Giat Industries shelter carried on a Renault TRM 10 000 (6 × 6) truck

LOHR UMC multipurpose shelter

Description
The LOHR UMC (*Unité Mobile de Campagne*) multipurpose 20 ft shelter was designed to withstand a variety of environments while providing an effective facility for various systems and/or equipment. Applications could include command post, mobile office, communications, workshop, bakery, medical post and so on.

The UMC shelter is constructed using insulated sandwich panels and stands on four retractable legs, one at each corner and extending 500 mm. The interior can be extended by lowering the side walls to extend the floor area. Tarpaulins can be used to protect the interior and access platforms are available. The UMC can be handled by forklift or various load handling systems.

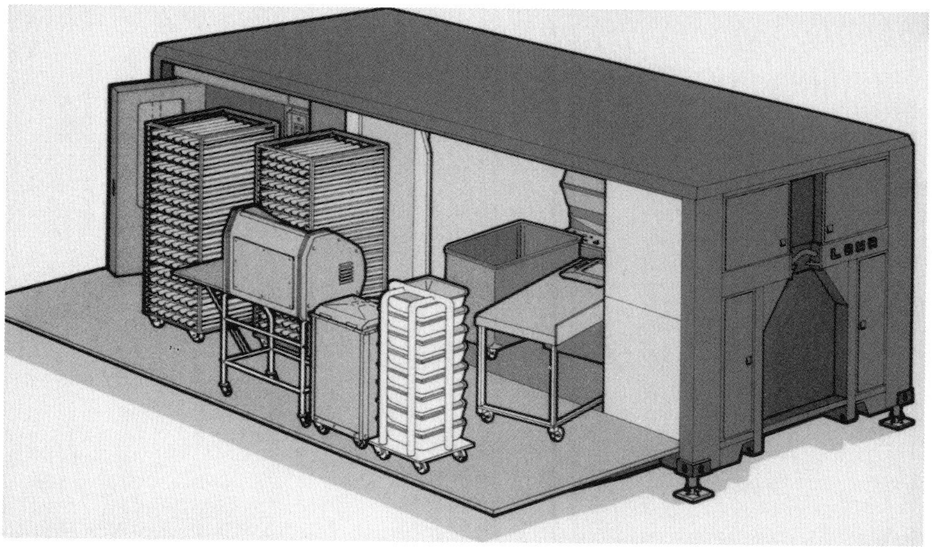

Drawing of LOHR UMC shelter configured as field bakery and with side wall lowered

Specifications
Weight empty: 2,900 kg
Length: 6.055 m
Width:
(closed) 2.435 m
(open) 4.38 m
Height: 2.435 m

Status
In production. In service with the French Army and several other armies.

Manufacturer
LOHR Industrie, F-67980 Hangenbieten, France.
Tel: +33 3 88 38 98 00. Telex: 870 082 f.
Fax: +33 3 88 96 06 36.

VERIFIED

LOHR RMTS 7.5 container handling and carrying trailer

Description
The LOHR RMTS 7.5 container handling and carrying trailer was designed and developed for the logistic handling and transport of shelter/containers on roads and over rough terrain. The trailer can handle 10 and 20 ft ISO shelters and can be used to load and unload containers into and from C-130 or C-160 transport aircraft. The French Army uses this trailer with mobile field hospital shelters.

There is also an RMTS 10 with a payload of 10 tonnes.

Specifications
Weight:
(empty) 3,000 kg
(GVW) 11,000 kg
Payload: 8,000 kg
Length:
(overall) 8.67 m
(folded) 6.535 m
Width: 2.47 m
Max towing speed: (road) 70 km/h
Tyres: 9.5 R 17.5 XTA

Status
In production. In service with French Army.

Manufacturer
LOHR Industrie, F-67980 Hangenbieten, France.
Tel: +33 3 88 38 98 00. Telex: 870 082 f.
Fax: +33 3 88 96 06 36.

VERIFIED

LOHR RMTS 7.5 container handling and carrying trailer

LOHR SLCT 8 container handling and carrying trailer

Description
The LOHR SLCT 8 container handling and carrying trailer follows the same general lines as the RMTS 7.5 (see previous entry) but uses two carrying axles. It is intended primarily for the transport and handling of 20 ft/6.096 m ISO containers only. For loading and unloading, the trailer remains attached to the towing vehicle while the towing bar is hydraulically driven downwards to lower the rear end of the trailer to the ground at an angle of 14°. Containers are loaded on to the trailer using an electrical winch with a pull force of 5,440 kg.

Two or three trailers can be stacked for air transport.

LOHR SLCT 8 container handling and carrying trailer

Specifications
Weight:
(empty) 3,500 kg
(GVW) 11,500 kg
Payload: 8,000 kg
Length: (overall) 8.68 m
Width: 2.42 m
Loading height: (unloaded) 0.92 m
Tyres: 305/70 R 19.5 XZA
Electrical system: 12 and 24 V

Status
In production. In service with the French Army and several other armies. On trial with the US Army.

Manufacturer
LOHR Industrie, F-67980 Hangenbieten, France.
Tel: +33 3 88 38 98 00. Telex: 870 082 f.
Fax: +33 3 88 96 06 36.

VERIFIED

Marrel ISO 20 ft container with integrated skid for load handling systems

Description
The Marrel ISO 20 ft/6.096 m container with an integrated skid is designed to be handled and transported by a truck equipped with an Ampliroll system, or any other Load Handling System (LHS) complying with the STANAG 2413 specification, without the need for additional interfaces.

Marrel ISO 20 ft container with integrated skid for load handling systems
1997

The ISO 20 ft/6.096 m container has a closed top, with rails, an A-frame and a lifting bar for handling by hydraulic LHS. The lift bar, A-frame and rails are built within the structure and outside dimensions as defined by ISO-668 for sea, rail, air and road transportation. The rear double door opens fully and is foldable to the sides. One option is detachable rear rollers for towing on hard surfaces and for handling on trailers.

Status

In production. In service with the French Army.

Manufacturer

Marrel SA, BP 56, F-42161 Andrezieux-Bouthéon Cedex, France.
Tel: +33 4 77 36 28 28. Fax: +33 4 77 36 28 00.

NEW ENTRY

Marrel Amplitainer

Description

The Marrel Amplitainer is a handling unit which operates as a complement to the Marrel Ampliroll Load Handling System (LHS - for details see entry in *Materials handling equipment* section). It enables autonomous loading and transport of standard ISO 20 ft/6.096 m containers without using any interface flatrack, skid, or specialised container handling equipment.

The Amplitainer has three parts. A front frame holds the container by its four front corner fittings. Two telescopic rear supports centre, tilt and lock the container. A storage frame on the truck, behind the cab, is used to store the front frame when the vehicle is used to carry a flatrack.

The Amplitainer is adaptable to any kind or make of truck and is compatible with all types of 20 ft/6.096 m container. No special tools are required during use.

Status

In production.

Manufacturer

Marrel SA, BP 56, F-42161 Andrezieux-Bouthéon Cedex, France.
Tel: +33 4 77 36 28 28. Fax: +33 4 77 36 28 00.

NEW ENTRY

Marrel Amplitainer in handling position ***1997***

GERMANY

MAN/Doll containers and container carriers

Description

MAN Nutzfahrzeuge GmbH and Doll Fahrzeugbau GmbH combined to produce a series of special shelter/containers and carrier vehicles to meet a variety of requirements. The containers have been developed for use on the full range of MAN high mobility military vehicles.

One of the Doll shelters has external dimensions of 6.1 × 2.5 × 2.4 m and a weight of 8,000 kg, although other Doll shelters can be accommodated on the MAN chassis. The shelter/container has a steel framework with steel covering panels insulated by 50 mm thick sandwich-type foam. A material handling crane may be installed in front of the shelter operating electrically or hydraulically. The shelter roof may be folded backwards for loading and unloading and the entire shelter and crane may be lifted by an overhead crane or gantry. The shelter/container may be fitted to suit customer requirements. Doors, roof walkways and so on can be fitted to suit customer requirements and various forms of internal and external lighting can be produced.

MAN Type 27.365 VFAE carrying opened Doll workshop shelter/container

A special workshop vehicle using the shelter/container concept has been produced using the MAN (8 × 8) carrier vehicle.

Doll also produces two smaller trailer-borne aluminium shelter/containers. One has dimensions of 2.5 × 2.2 × 2 m and can be carried on a single-axle

MAN Type 27.365 VFAE carrying special Doll container and material handling crane

Typical Doll 20 ft/6.096 m ISO container converted for workshop use

trailer. It can be produced fully or partially insulated, with RFI shielding and with air conditioning. It is also possible to fit a generator. A shelter/container produced to similar specifications has dimensions of 3.5 × 2.2 × 2 m and is carried on a twin-axle trailer. Both units can be carried slung from a helicopter. The larger unit can be fitted with stabilising jacks at each corner.

Status
In production. In service with an unspecified nation.

Manufacturers
Doll Fahrzeugbau GmbH, Industriestrasse 13, D-7603 Oppenau, Germany.

MAN Nutzfahrzeuge GmbH, Postfach 500620, D-8000 München 50, Germany.
Tel: +49 89 150 3869. Telex: 523211.

VERIFIED

Zeppelin shelters

Description
Zeppelin Mobile Systeme produces a wide range of containerised shelters for a wide variety of purposes ranging from mobile offices to radar cabins. The entire range is constructed from basic sandwich monocoque with the side walls, roof and floor units all having aluminium skins. The core material is polyurethane foam. The joints of the exterior cover sheets at the edges of the door are welded to the door profiles all round. An insulating layer is interposed between the framing sections. All external riveting is air and watertight and the corner fittings are replaceable. Skids are mounted lengthwise under the floor unit with each skid being 51 mm high and approximately 100 mm wide. They are riveted or bolted to the integral stiffening members of the floor unit. The shelter door is located in one end wall and mounted on the right-hand side by three heavy-duty hinge units. All four corners are rounded and there is a three-point locking device fitted. The flooring is constructed from 20 mm thick marine plywood.

Various optional doors, hatches and windows can be fitted to either end and to the side walls. An air conditioning unit is available. The shelters can be lifted for transport by four heavy-duty jacks that are attached to each corner of the shelter and for some vehicle-carried shelters a special hydraulic jack is used. Most of the Zeppelin shelters are air-transportable.

Zeppelin Mobile system shelters configured as a field hospital ***1997***

Status
In production.

Manufacturer
Zeppelin Mobile Systeme GmbH, Albert-Einstein-Strasse 2, D-77656 Offenburg, Germany.
Tel: +49 781 6105-0. Telex: 752900 zepso.
Fax: +49 781 6105-60.

UPDATED

Kärcher container field kitchen CFK 250

Description
The Kärcher container field kitchen CFK 250 uses a special water- and sandstorm-proof 20 ft/6.096 m container. It is the only product of its kind in a range that includes container kitchens which cater for 250 up to 1,500 people. Units can be provided with or without NBC protection and may be towed as a trailer using a twin-axle trailer arrangement.

The CFK 250 is fully equipped for its role and contains two baking and frying units and two 150 litre pressure cookers, all covered by two 2,800 m³/h ventilation hoods. Working tables and shelves are provided, along with tables and cupboards, food preparation machines, sinks, water pumps and a hinged table for meal distribution. Also provided is a warm air generator and a 6,000 m³/h air conditioning inlet. A canopy can be erected over the container.

The CFK 250 weighs 6,000 kg, is 6.058 m long, 2.438 m wide and 2.438 m high (with canopy).

Kärcher container field kitchen CFK 250

Status
In production. In service with a NATO headquarters, German border and state police and the Red Cross.

Manufacturer
Alfred Kärcher GmbH & Co, Alfred-Kärcher-Strasse 28-40, D-71364 Winnenden, Germany.
Tel: +49 7195 142797. Telex: 724432.
Fax: +49 7195 142720.

VERIFIED

Kärcher High Mobil Containerised Kitchen (HMCK)

Description
Immediate readiness for action of the Kärcher High Mobil Containerised Kitchen HMCK is ensured by a generator with large power reserves to provide energy around the clock. The kitchen is contained inside an ISO 20 ft/6.096 m container. Fixing and transport are easy, using the standardised lock system on ships, trucks and trains as well as inside aircraft.

A Kärcher 150 litre Gastronorm pressure cooker allows fast pressure cooking while the maintenance-free stainless steel double wall avoids burning of food. In addition, a high performance 70 litre pressure frying kettle is provided, as is a diesel-heated Kärcher Combi-oven.

Due to the variable controls the burner technology for all cooking and grilling modules offers the same user-friendliness as an electric stove. Four or five burners are available to provide up to 100 kW of

immediate heat per container. Depending on the equipment, three cooks are able to prepare a maximum of 300 complete meals within two hours. A refrigerator module with a 1,200 litre capacity and high insulation levels permits the storage of up to 600 rations.

Status
In production.

Interior of Kärcher High Mobil Containerised Kitchen HMCK
1997

Manufacturer
Alfred Kärcher GmbH & Co, Alfred-Kärcher-Strasse 28-40, D-71364 Winnenden, Germany.
Tel: +49 7195 142797. Telex: 724432.
Fax: +49 7195 142720.

NEW ENTRY

Kärcher CFL 60 mobile field laundry

Description
The Kärcher CFL 60 mobile field laundry was developed in close co-operation with the German armed forces. It is used for the washing and disinfection of all kinds of laundry coming from all arms of the services (army, air force and navy) but it also has applications in the hospital and disaster relief roles. It is possible to disinfect or decontaminate infected or contaminated material in the CFL 60.

Areas inside the container body are defined into clean and dirty sections. A 65°C disinfection programme uses 300 litres/h by involving a water recycling plant installed in the CFL 60 which reuses rinsing water in the following washing cycle.

The CFL 60 container is equipped with a lifting device which raises it to a height of 1.65 m; the operation is electrical with automatic levelling. Inside the container are two washer units, each with a capacity of 25 kg of dry laundry and one drier. The container also has its own 125 kVA generator powered by a turbocharged diesel engine.

The CFL 60 weighs approximately 10,000 kg. Ready for use the length is 6.5 m, height 2.5 m and width 7.5 m (all dimensions approximate). Minimum space required, apart from peripherals, is 10 × 8 m.

Status
In production.

Manufacturer
Alfred Kärcher GmbH & Co, Alfred-Kärcher-Strasse 28-40, D-71364 Winnenden, Germany.
Tel: +49 7195 142797. Telex: 724432.
Fax: +49 7195 142720.

UPDATED

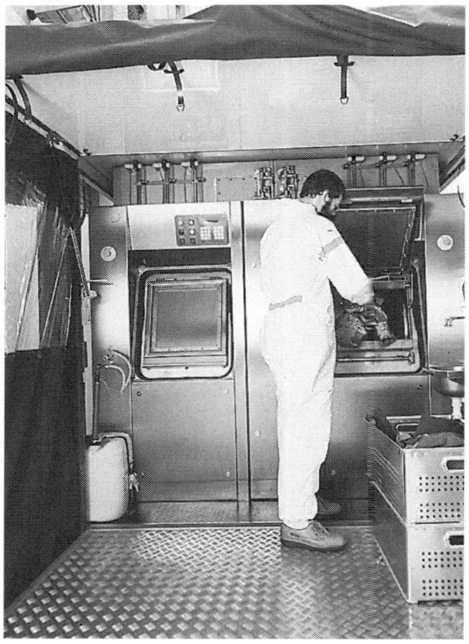

Interior of Kärcher CFL 60 mobile field laundry
1997

Kärcher CFL 60 mobile field laundry

ITALY

Improved Boneschi shelters

Description
The Boneschi range of shelters uses a basic container body which is made from light-alloy sheets with polyurethane fillings. The framework is of aluminium sections. The overall construction is orthodox and the shelters can be adapted to carry out a wide range of military duties. There are two optional methods of moving the Boneschi shelters, both of them incorporated into the shelter system when required. One is the 'Rolbon' system which has four hand-operated worm screws and wheels fitted to each corner of the shelter by rolling arms. The other system is the 'Uniborn' lifting system which has four hand-operated worm screws fitted to each corner by pivoting arms. This latter system enables the shelter to be raised and lowered on to flatbed trucks.

One specialised form of Boneschi shelter is the 'Lifeshelter'. This is a specialised light alloy container which is insulated and encloses a complete medical

Boneschi shelter loaded on to IVECO 6601 truck using Uniborn system

Boneschi shelter fitted out as fire-control shelter on IVECO 75 PM (4 × 4) 2,000 kg truck

unit ready for use. Equipment in the container includes a simple stretcher-type operating table, racks for medical equipment and instruments (all included in the standard kit), oxygen and other such equipment, lighting and air conditioning. The 'Lifeshelter' is air-transportable and is fitted with the 'Uniborn' lifting system for handling. If required, the 'Lifeshelter' can be carried slung under a helicopter.

Status
In production. In service with the Italian Army.

Manufacturer
Carrozzeria Boneschi srl, Via G Boneschi 1, I-20040 Cambiago (Milan), Italy.
Tel: +39 2 9530 8045. Fax: +39 2 9530 8204.

Piaggio shelters

Description
Piaggio started building military shelters in 1964 and since then has produced shelters for both military and civil purposes. There are now 10 basic Piaggio shelters with a far larger number being produced for special purposes, but the overall construction remains the same.

Each cabin has eight horizontal angles all in light alloy and four vertical angles in galvanised steel or light alloy. The corner blocks may be standard ISO container types and all the wall and floor or ceiling panels are constructed from a sandwich formed from lightalloy sheets 0.8 to 1.6 mm thick with a core of 50 mm thick polyurethane foam. Internal strengthening ribs are fitted and there are three underside skids. The floor has a drain plug, and the end access door is built in several dimensions according to customer requirements. Other standard fittings are access steps, roof reinforcement in the centre and an electrical earthing point.

A wide range of accessories is available and numerous fits can be produced to suit customer requirements. Piaggio was the first manufacturer to respond to NATO's requirements for EMP shelters (such as shelters protected against electromagnetic pulse from nuclear blast) and is one of the most important suppliers for this type of shelter to NATO and NATO affiliated nations.

Piaggio EMP shelters differ from conventional shelters in that the outer skin of the cabin is fusion welded to guarantee a flawless electrical continuity.

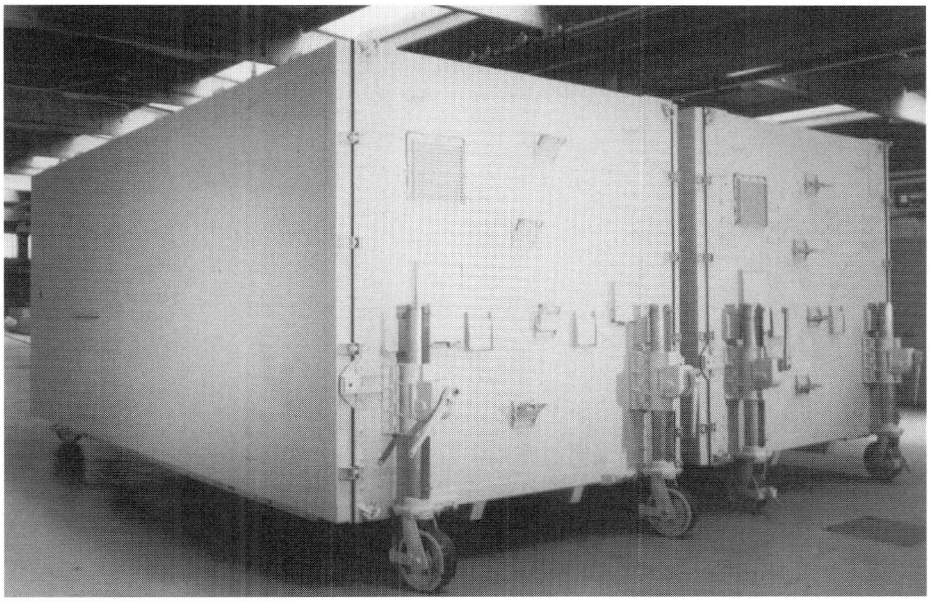

Piaggio modular shelters awaiting delivery

Besides the above types Piaggio has developed and manufactured, on its own or in co-operation with other companies, a full range of special shelters including shelters with bulletproof protection, nuclear-hardened shelters, and modular shelters.

Specifications

Type	ACE1	ACE2	ACE3	UEO1	UEO2
Length	2.9 m	4.25 m	5 m	2.71 m	3.983 m
Width	2.05 m	2.2 m	2.2 m	1.74 m	1.915 m
Height	1.881 m	2.131 m	2.131 m	1.9 m	2.1 m
Weight	700 kg	900 kg	1,200 kg	600 kg	750 kg
Capacity (useful load)	1,300 kg	3,600 kg	4,300 kg	1,400 kg	3,250 kg

Status
In production. In service with the Italian armed forces and several other nations.

Manufacturer
I.A.M. Rinaldo Piaggio SpA, Via Cibrario 4, I-16154 Genoa, Italy.
Tel: +39 10 64811. Telex: 270695 aerpia i.
Fax: +39 10 6520160/6503378.

UPDATED

SAI Ambrosini containers and shelters

Description
SAI Ambrosini has been manufacturing containers and shelters since 1970 and produces a wide range of specialist containers for military purposes.

The containers are constructed from a sandwich material made up of light aluminium alloy heat bonded to a polyurethane core. Watertight rivets and stainless steel bolts and nuts are used to hold the panels

Examples of SAI Ambrosini logistic shelters and containers

together. All panels have a special lattice of light aluminium struts to absorb stresses from internal and external loads and to provide lashing anchors for internal loads.

SAI Ambrosini also produces a range of air conditioned containers for special purposes. There are three main models in this range, the smallest with a 4 kW conditioner, then a 9 kW conditioner and the largest has a 20 kW conditioner unit. These units can be used to provide refrigeration if required

Both standard and cooled containers can be supplied with a range of accessories that includes jacks for lifting and loading, ladders, special or emergency exits and hatches, special electrical looms or lighting and RF suppression. Numerous changes to suit customer requirements can be introduced.

SAI Ambrosini shelters are built to standard dimensions or to special requirements. Air conditioning units are usually installed and the units can be equipped with mechanical systems for loading and unloading from trucks, including over the sides of drop side trucks. Lifting points are provided and the shelters can be carried on special SAI Ambrosini trailers and trailer systems.

Shelters can be supplied in armoured, RF screened, EMP- and NBC-protected forms. SAI Ambrosini shelters, containers and handling systems can be combined to form the modular SAI Ambrosini Mobile Flight Support System for helicopter unit support.

Specifications

Containers			
Capacity	3,000 kg	5,000 kg	8,000 kg
Weight	1,600 kg	2,700 kg	2,800 kg
Length	3.25 m	4.55 m	4.55 m
Width	2.1 m	2.2 m	2.3 m

Status

In production. In service with the Italian Army.

Manufacturer

SAI Ambrosini - Società Aeronautica Italiana SpA, Viale Roma 25, I-06065 Passignano sul Trasimeno (PG), Italy.
Tel: +39 6 475 6774. Fax: +39 6 474 2909.

VERIFIED

SAI Ambrosini wheel-coupled shelter and platform transporters

Description

To complete its shelter and container systems SAI Ambrosini designed and constructed a wheel-coupled trailer for the transport of shelters and platforms. The wheel-coupling system meets Mil-M-8090F with reference to Mobility Type III, Group C.

The system can be towed by a number of types of military vehicle. The two rolling trains can be coupled independently at the front or rear of the shelter or platform and lift the load for towing. If necessary, the load can be lifted to a height of 700 mm for towing through fords or over rough terrain.

When not carrying a load, the transporter sections can be connected to form a compact unit for towing.

SAI Ambrosini wheel-coupled shelter and platform transporters

Specifications

Model	**TP-3**	**TP-5**	**TP-8**
Length	2.45 m	3.4 m	3.4 m
Width	2.25 m	2.3 m	2.3 m
Height	1.43 m	1.47 m	1.47 m
Track	1.93 m	2.01 m	2.01 m
Weight	1,300 kg	2,700 kg	2,800 kg
Max capacity	3,500 kg	7,500 kg	7,500 kg

Status

In production. In service with the Italian Army.

Manufacturer

SAI Ambrosini - Società Aeronautica Italiana SpA, Viale Roma 25, I-06065 Passignano sul Trasimeno (PG), Italy.
Tel:+39 6 475 6774. Fax: +39 6 474 2909.

VERIFIED

EUROVINIL (EV) Inflatable shelters

Description

EUROVINIL (EV) inflatable shelters are modular structures and can thus be constructed in many different configurations as field hospitals, personnel accommodation, storage shelters, vehicle garages, conference areas, and so on. The shelters are manufactured using electronic high-frequency welding involving an EV patented fabric known as EMC. This entails a high-tenacity textile support coated by immersing the support in liquid PVC.

Completed EV shelters can be erected in less than 10 minutes and once erected have a high resistance to winds of up to 100 km/h. They can withstand snow loads up to 120 mm thick and can be used over a temperature range of from −20 to +60°C.

There are a number of standard shelters in the EV range but several other options are available on request.

EV inflatable shelters configured as a field hospital ***1996***

Specifications

3 arches, 2 doors
Dimensions: 5.25 × 5.15 × 2.55 m
Weight: 115 kg

4 arches, 2,3 or 4 doors
Dimensions: 5.25 × 7.25 × 2.55 m
Weight: 175 kg

5 arches, 2 doors
Dimensions: 5.25 × 9.55 × 2.55 m
Weight: 215 kg

Medium-sized tent
Dimensions: 8.25 × 12.5 × 4.1 m
Weight: 566 kg

Status

In production.

Manufacturer

EUROVINIL INDUSTRIES SpA, Via Genova 3, I-58100 Grosseto, Italy.
Tel:+39 564 487300. Fax: +39 564 487222.

VERIFIED

SINGAPORE

Mobile field hospital

Description

This mobile field hospital was designed to provide a full range of medical facilities in any field or emergency environment. It is based on a specially designed container that can be configured as an operating theatre, X-ray unit, laboratory, blood bank, dental clinic, nursing ward, casualty clearing unit, kitchen, laundry or for other specialised medical purposes.

Each container is fully equipped for its purpose and can be carried into position using a platform truck or a tractor truck with a flat bed semi-trailer. On-site the containers are placed on the ground in a suitable formation but each unit can be used independently if required. Power is supplied by towed generators. When not in use, the containers can be stacked up to five high.

Each container has external dimensions of 6.058 × 2.438 × 2.438 m.

Status

In service with the Singapore armed forces.

Manufacturer

Singapore Technologies Automotive Limited (A Singapore Technologies Company), 5 Portsdown Road, Singapore 139296.
Tel: +65 473 6311. Telex: 25755 SINENG RS.
Fax: +65 471 0662.

VERIFIED

Mobile field hospital container towed by Hino tractor truck

SOUTH AFRICA

Modular operating theatre

Description

This modular operating theatre, based on ICX ISO container dimensions, is constructed from glass fibre panels supported on a steel framework. It has two pairs of access doors, while internally it is fully equipped as an operating theatre for all types of medical operations. It uses a 380 V three-phase electrical system input and a 24 V battery-operated emergency power supply. A penthouse is provided which fits over the theatre to provide accommodation for personnel on either side. The module is transported by means of a specialised semi-trailer.

The module is 6.058 m long, 2.438 m wide and 2.54 m high.

Status

In service with the Surgeon General's arm of the South African National Defence Forces.

Manufacturer

Enquiries to: Armscor, Private Bag X337, Pretoria 0001, South Africa.
Tel: +27 12 428 1911. Telex: 30217.
Fax: +27 12 428 5635.

VERIFIED

Modular operating theatre in position

SPAIN

SPA NATO shelters

Description

SPA shelters are built to military standards and have been certified to comply with the applicable MIL specifications (such as MIL-STD-907B) and their NATO equivalent (SHCPE 6516).

The shelters feature a foam and beam structure together with several innovative features such as ISO type corner fittings and a door locking device that provides them with a distinctive appearance.

Applications include telecommunications, radar control, command post, medical centre, air traffic control and field laboratory. Versions include RFI shielding to MIL-STD-285, nuclear hardening and various types of other protection, including EMP, NBC TEMPEST and bal istic. Available models include the most widely used, both in the military field (S-250, S-280, NATO ACE types I, II and III) and in the commercial (ISO 10 ft, 15 ft and 20 ft). The two types mainly used by the Spanish Army are the NATO II with

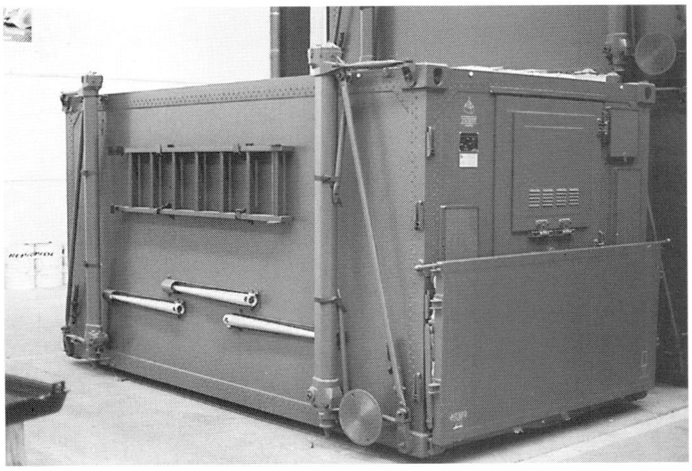

SPA NATO II shelter ***1997***

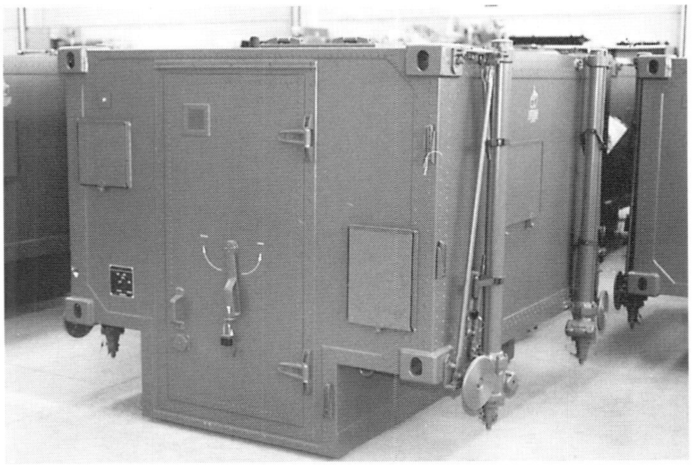

SPA S-250E shelter ***1997***

internal dimensions of 3.645 × 1.915 × 1.9 m, and the S-250E with internal dimensions of 1.864 × 1.834/0.954 × 1.627 m.

Accessories include shelter lifting and fastening slings, lifting and levelling devices, a platform for the entry door, air conditioners of varying capacities, and NBC filters and output valves.

Status

In production. In service with the Spanish armed forces.

Manufacturer

Servicios y Proyectos Avanzados SA (SPA), Torres Quevedo 24, Naves 2-3, E-28022 Madrid, Spain.
Tel: +34 1 329 4455. Fax: +34 1 329 4805.

NEW ENTRY

EINSA tactical operations centre Type M/PMU-01

Description

This tactical operations centre consists of a number of special containerised shelters, normally four, carried by 4 × 4 all-terrain vehicles that can be configured on-site in a number of formations. The Tactical Operations Centre (TOC) shelters are connected to the vehicles by means of a special kit in such a way that each will protrude 1.2 m over the end of the vehicle, allowing the shelter doors to remain free from obstruction and to allow variations in the method of interconnecting the shelters. Each shelter has three sliding doors, one at each side rear and the third at the right rear side. Each shelter is supplied with an access ladder that can fit under any door but the variations in positioning the TOC shelters may allow direct access from one shelter to another through the doors. The shelters may also be interconnected to other services including 220 V AC and 24 V DC power lines, intercom telephones and air conditioning. All doors are equipped for blackout fittings and each shelter has eight telephone connection points and may be fitted with aerial connection points.

The shelters are constructed from tubular profiles with hooks on the upper surface for crane emplacement. The walls have 40 mm polyester boards for insulation and the roof insulation is 60 mm thick. Each unit has a roof window and some units have side windows as well. Provision is made for air conditioning and each unit is equipped with an inter-unit closure device. Fittings are supplied for securing camouflage netting.

Extras include quartz wall clocks, fire extinguishers, first aid kits, axes and clothing hooks.

Specifications

Weight: 1,200 kg
Length:
(internal) 5.3 m
(external) 5.4 m
Width:
(internal) 2.2 m
(external) 2.3 m
Minimum height:
(internal) 2.04 m
(external) 2.2 m
Maximum height:
(internal) 2.09 m
(external) 2.25 m
Volume:
(internal) 24 m^3
(external) 27.6 m^3
Floor surface:
(internal) 11.8 m^2
(external) 12.4 m^2

Status

In service with the Spanish armed forces.

Manufacturer

Equipos Industriales de Manutencion SA (EINSA), Antigua Ctra. de Barcelona, km 27.300, E-28812 Alcala de Henares, Madrid, Spain.
Tel: +34 1 880 9000. Fax: +34 1 880 9086.

VERIFIED

Typical EINSA tactical operations centre shelter with side windows

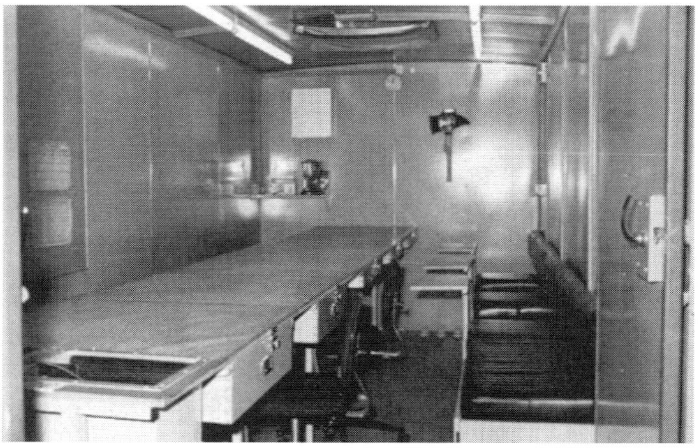

Interior of typical tactical operations centre shelter

UNITED KINGDOM

Marconi ISOLATOR electronic equipment shelters

Description

The Marconi ISOLATOR shelter is a high-strength, low-weight transportable shelter for both military and commercial applications. It protects electronic equipment from the adverse effects of the environment including the elements, EMI, RFI and NEMP and provides TEMPEST shielding if required.

It is constructed using advanced sandwich panel technology. Finite Element Analysis of the structural design ensures optimum strength, weight and payload ratios at minimum cost. The end product provides the customer with total flexibility in the location of doors, apertures, escape hatches and interface panels.

ISOLATOR shelters have been supplied for many applications including telecommunications, airfield

Marconi ISOLATOR sandwich panel technology adapted for a vehicle hard top application

A semi-trailer mounted ISOLATOR shelter with its associated air conditioning unit for a NATO application

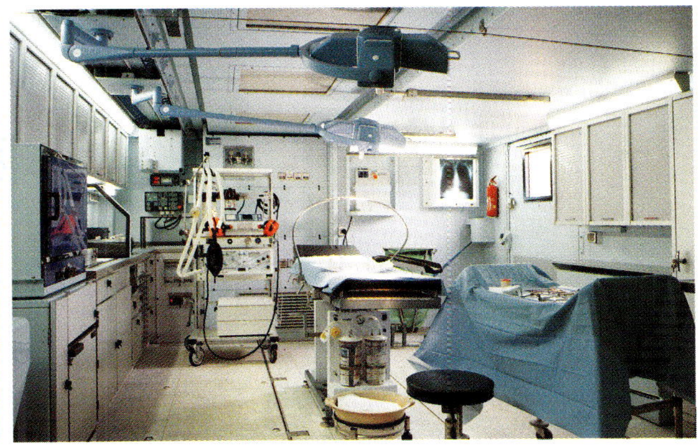

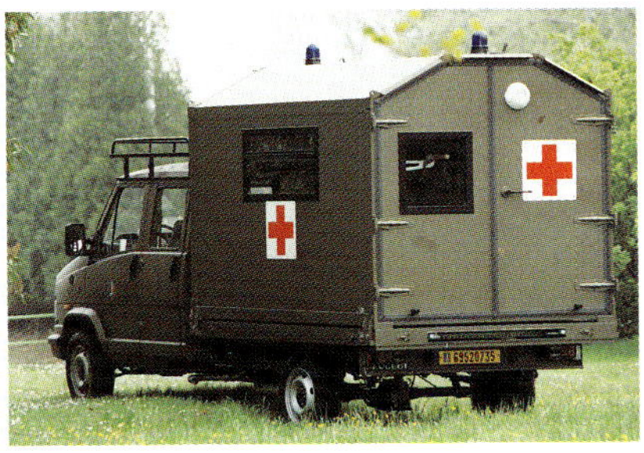

TransHospital®.
Medical Service like at Home.

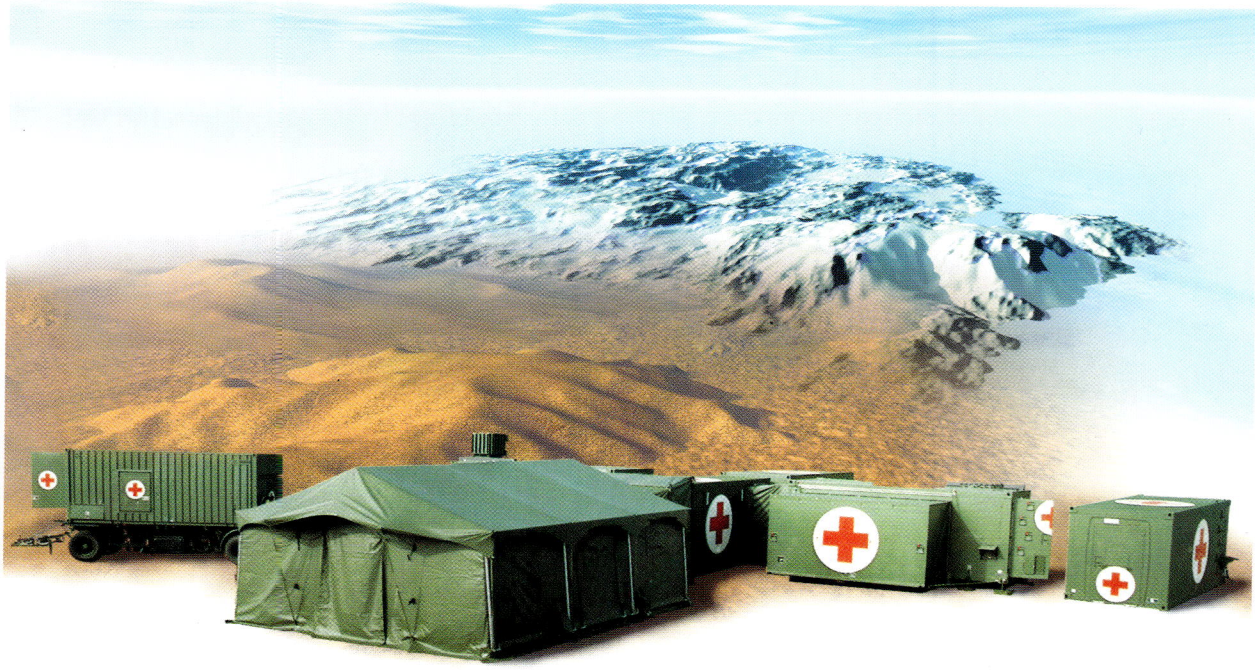

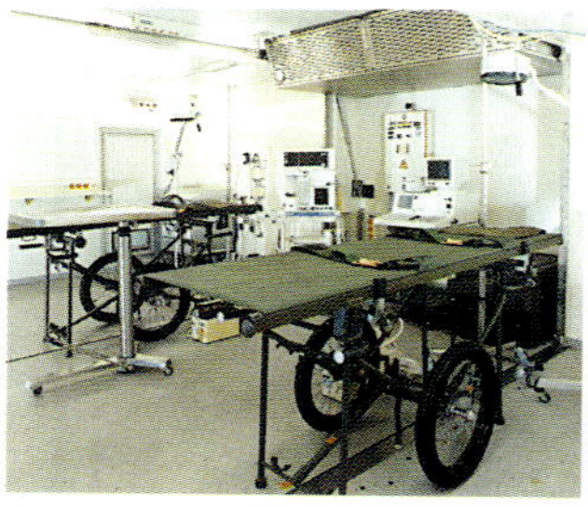

The best possible medical care is essential for accomplishing the crisis missions. With the help of *TransHospital,* the responsible persons are now in a position to apply the same medical standard as at home. Worldwide.

TransHospital is based on a modular containerized concept. From a single mobile emergency station, via a basic medical service hospital up to the field hospital for enhanced medical treatment, it allows almost any combination of the containers and adaptation to various tasks.

The containers, equipped with state-of-the-art medical technology, are easily transported and made fully operational within a few hours.

The central areas are provided with an air-conditioning system and operate even under extreme environmental conditions. In addition, an NBC-protection unit is incorporated for personnel protection. National components can be easily integrated in many cases.

Dornier GmbH
VIM 41
88039 Friedrichshafen/Germany
Phone +49 75 45 8-91 38
Fax +49 75 45 8-20 34

Daimler-Benz Aerospace
Dornier

Defense and Civil Systems

radar and command and control systems. Among these are the Phoenix UAV system for the British Army, the Home Defence Radio System, UKADGE and the British Aerospace electro-optical tracking system.

A lightweight version incorporates many features of the standard shelter but is much lighter and was specifically designed for truck or trailer use.

Status
In production for the British Army, NATO and some overseas defence organisations.

Manufacturer
Marconi Defence Systems (Gateshead), Bill Quay, Gateshead, Tyne and Wear NE10 0UR, UK.
Tel: +44 191 495 0111. Telex: 53312.
Fax: +44 191 495 0144.

VERIFIED

CB 300 and CB 100 series of transportable shelters

Description
The CB 300 series of transportable shelters was designed by the UK Ministry of Defence and developed by Marshall SPV from 1962. The UK MoD requirement was for a lightweight, demountable body which would provide the best possible working conditions under extremes of temperature, would have an extended service life under minimal maintenance requirements and would have maximum utilisation by both the British Army and the Royal Air Force in mobile and static roles. The first production units were completed in 1964.

The transportable shelter was designed for the operational installation of specialist pneumatic, hydraulic, signals, electronic and similar equipment, for use as specialist repair modules and for use as tactical offices or command posts.

The shelter is constructed of aluminium alloy, is lightweight and fully insulated to provide protected working conditions for personnel and equipment under extremes of temperature. It is weather-sealed and may also be air conditioned. The basic design concept proved sound but, whereas NBC was covered initially, with only minor changes the shelter can be modified to meet EMP and RFI requirements and supplied in armoured or standard form. It was designed to be carried on any flat platform truck or trailer and can be operated dismounted in a static role. It is air-transportable in a C-130 transport aircraft. Its modular construction allows doors, windows and access panels to be positioned to meet specific requirements. Strongpoints in the main framework facilitate the attachment of accessories and equipment, while heavily reinforced corners simplify the fitting of running and handling gear. Typical roles include use as a fuel injection repair shop or machine shop/automotive repair shop, when it has a wide range of installed equipment, including a combined lathe/milling machine, bench drill, battery charger and a range of specialist items and hand tools.

In addition to the CB 300 series, Marshall SPV manufactured the CB 100 range of shelters.

CB 300 series transportable shelter mounted on Bedford TM 4-4 (4 × 4) 8,000 kg truck

Specifications

Model	CB 100	CB 300
Weight (empty)	685 kg	from 925 kg[1]
(loaded)	1,703 kg	up to 4,500 kg[1]
Length (external)	1.173 m	4.483 m
(internal)	0.958 m	4.318 m
Width (external)	0.88 m	2.489 m
(internal)	0.835 m	2.324 m
Height (external)	0.73 m	2.13 m
(internal)	0.684 m	1.93 m

[1] depending on role

Status
Production complete. In service with the UK armed forces and Saudi Arabia.

Manufacturer
Marshall SPV Limited, Airport Works, Cambridge CB5 8RX, UK.
Tel: +44 1223 373475. Fax: +44 1223 373046.

VERIFIED

Marshall Matrix expandable shelter

Description
The Marshall Matrix expandable shelter was developed to provide a rigid Matrix shelter capable of transportation and deployment as a DROPS compatible container with expandable sides to provide an internal floor area of almost three times the closed shelter area. When deployed, the shelter is capable of operation either on the ground or as a vehicle-based unit.

The shelter can be expanded and deployed within 30 minutes and used in varied roles such as mechanical/electronic repair, simulator housing and complex together, mobile field headquarters and hospital centres.

Status
Production and specifications as required.

Manufacturer
Marshall SPV Limited, The Airport, Cambridge CB5 8RX, UK.
Tel: +44 1223 373045. Fax: +44 1223 373046.

NEW ENTRY

Marshall Matrix expandable shelter **1997**

Marshall Matrix shelters

Description
The Matrix family of military shelters produced by Marshall SPV is constructed of sandwich panels formed from a variable range of materials to suit specific applications. The structural sandwich assembly, having no intermediate structure, is capable of accepting attachments and apertures in any position to suit the required installation.

The panels are contained in a peripheral metal frame. The vertical corners can incorporate locations for the attachment of transfer gear, jacks, ISO corner fittings, running gear, roof access steps and crane or helicopter lift points. The fitment of any of these features does not

Marshall Matrix shelter, trailer-mounted, for use with live fire monitoring equipment in support of Challenger MBT

degrade the electrical integrity of the shelter. Aluminium skids can be fitted longitudinally below the shelter floor panel and equipment mounting rails can be fitted to suit individual requirements.

Basic versions of the Matrix shelter are fitted with a single entrance door located in the shelter rear wall but alternative door positions can be offered in sides or ends to suit special installations. Hinged flaps can be fitted to act as escape hatches, power input panels, air conditioner flaps, and so on. All doors and external flaps can incorporate security fittings.

The standard sandwich panels provide good thermal insulation and various panels are available to provide thermal transmittance figures. Sealing of the shelter enables an efficient overpressure to be created within the container by the fitment of a suitable NBC filtration kit. The shelter, because of its double-skinned insulated structure can provide a high level of RFI and EMP attenuation. Doors and windows can be fitted with metallic gaskets to ensure continuity of the outside skin screening. Shelters required to operate in an EMP/RFI environment can be fitted with input vaults for power and signal cables.

Matrix shelters can be produced in virtually any size as the modular construction system employed, means that alternative sizes can be readily accommodated. Optional features include air conditioning, equipment lifting gantries, roof antenna mountings, single or double doors, detachable access panels, fixed, sliding or double-glazed windows, window blackout panels, emergency exits, access steps, entrance platforms, underfloor skids, equipment mounting rails and penthouses. Protection can be provided against NBC, EMP, RFI, TEMPEST and ballistic projectiles.

Systems installations developed for the CB 100 and CB 300 shelters are capable of being installed into the Matrix series shelters when manufactured to compatible dimensions utilising the same attachment features.

Status

In production for the UK Ministry of Defence.

Manufacturer

Marshall SPV Limited, The Airport, Cambridge CB5 8RX, UK.
Tel: +44 1223 373045. Fax: +44 1223 373046.

VERIFIED

Marshall Matrix 'Drop-In' shelter

Description

The Marshall Matrix 'Drop-In' transportable shelter was developed to meet the requirements of the electronics and aerospace industry for a small shelter capable of being carried in pick-up truck bodies. It was designed to enable dimensional variations to be made to suit specific requirements.

Construction is based on the use of aluminium-skinned sandwich panels and provides structural strength for the installation of floor- and wall-mounted equipment. It can be fitted with ISO corners and may be carried as external cargo beneath helicopters. The basic shelter provides a high level of attenuation for RFI and EMP shielding because of its double-skinned structure, and conductivity seals are provided around aperture areas.

The shelter permits a 'drop-in' fit to ¾ and 2 tonne trucks and can be produced to fit any suitable pick-up truck.

Status

Production as required.

Manufacturer

Marshall SPV Limited, The Airport, Cambridge CB5 8RX, UK.
Tel: +44 1223 373045. Fax: +44 1223 373046.

UPDATED

Marshall Matrix 'Drop-In' shelter in use on Reynolds Boughton RB-44 (4 × 4) truck

Penman mobile shelters and custom-built containers

Description

Penman Engineering Limited designs and manufactures a range of shelters and cabins for a variety of uses. Whilst able to build standard NATO tactical shelters, Penman specialises in tailoring specifications to meet the customer's precise needs. The materials and techniques used are fully approved for military applications and are closely controlled by the Quality Assurance department.

Typical cabin construction comprises sides, roof, floor and end panels manufactured from one-piece aluminium-faced RTM Styrofoam core, and sandwich panels fully bonded together with an extruded aluminium peripheral framework. Doors, windows and escape hatches are located to meet specifications, together with integral stiffening members to suit internal fixtures.

A range of internal layouts can be supplied to meet most military requirements, including command posts, field workshops, communications centres and mobile training units.

Optional equipment for the cabins includes lifting and securing slings, levelling jacks, demount jack legs, power input panels, interior electrical systems, interior partitions and furniture, and RFI screening. Cabins can be supplied on shock-absorbing subframes, demountable frames compatible with DROPS, or rigidly

Typical Penman demountable communications cabin on Bedford MK (4 × 4) 4,000 kg truck

Penman self-contained mobile workshop

mounted on vehicle chassis or trailers. Cabins can be constructed to meet most military specifications, including armour to various levels.

Status
In production.

Manufacturer
Penman Engineering Limited, Heathhall, Dumfries DG1 3NY, UK.
Tel: +44 1387 52784. Telex: 779771.
Fax: +44 1387 67332.

VERIFIED

CSC shelters and containers

Description
CSC Specialised Vehicles Limited design and construct a wide variety of both land- and air-transportable shelters for multirole purposes. Typical are a series of shelters equipped as the cross-connect interface for telecommunications designed to replace damaged or destroyed static installations. A feature of the ISO shelter is the provision of cable glands at low level to allow the connection of upwards of 800 signal cables. Insulated panel construction, heat pump air conditioning and de-humidifying equipment allow the unit to function in all environments from sub-arctic to tropical.

Interior of CSC shelter configured as a cross-connect interface for telecommunications ***1997***

Other units supplied by CSC include medical facilities, workshops, command posts, and field kitchens.

Status
In production.

Manufacturer
CSC (Specialised Vehicles) Limited, Somerden Road, Kingston Upon Hull, Humberside HU9 5PE, UK.
Tel: +44 1482 798400. Fax: +44 1482 787238.

UPDATED

CSC shelter configured as a cross-connect interface for telecommunications ***1997***

CSC demountable shelter bodies

Description
CSC Specialised Vehicles have designed and manufactured many variations of demountable shelters for over 30 years. One project carried out in conjunction with Dytecna Limited was to design and manufacture two operationally compatible shelter bodies which would fit the Steyr-Daimler-Puch Pinzgauer and the Reynolds Boughton TUH.

Both shelters are demountable and air transportable, utilising aluminium-faced insulated panels by Paneltex to achieve the RFI shielding specification for Ptarmigan

CSC demountable shelter body fitted on Steyr-Daimler-Puch Pinzgauer (6 × 6) 1,500 kg vehicles of the Omani armed forces

communications equipment. The TUH vehicle posed interface problems which were overcome by a specially designed mounting frame retaining the inherent flexibility of the chassis but providing a stable base for the shelter. This frame is being developed to accept other configurations of demountable loads.

The shelters have been fitted with BATES and BMets equipment by Dytecna, with communications electronics by Marconi to form part of an area communications system deployed in Bosnia.

Status

In production. In service with the British Army.

Manufacturer

CSC (Specialised Vehicles) Limited, Somerden Road, Kingston Upon Hull, Humberside HU9 5PE, UK.
Tel: +44 1482 798400. Fax: +44 1482 787238.

UPDATED

CSC demountable shelter body fitted on Reynolds-Boughton TUH **1997**

UBK Kabin system

Description

Developed specifically for the military to meet current rapid deployment demands in all parts of the world, the UBK Kabin system can be quickly and easily deployed and recovered by a team of six time after time. The system is designed to convert from a flat-pack to a rigid, weatherproof structure in about 30 minutes without the need for tools or lifting gear.

Up to five complete UBK Kabins measuring 5.2 × 2.4 m can be carried on a DROPS flatrack. Whether packed or deployed the standard UBK Kabin is DROPS compatible and can be transported using most military vehicles, medium helicopters and transport aircraft.

The UBK Kabin can be finished to suit operations in arctic, European, tropical or desert regions. The modularity of the system allows for easy extension and expansion to meet most operational requirements as side walls can be easily removed to create larger areas. Modules can be added to create longer buildings or removed for smaller applications. Rapid repairs can be easily effected without recourse to a workshop by replacing damaged panels. All components are either interchangeable or replaceable.

Additional units can be stacked where space is at a premium. Stacking is facilitated by means of a simple stacking kit which can be supplied as an optional extra. When not required the UBK Kabin can revert back to its original form.

UBK Kabins are supplied as specified or with a basic 'fitted for role' facility, from field accommodation to rebroadcast station, security post or medical facility.

Basic UBK Kabins are supplied with a built-in electrical system complete with lighting, light switches, power points and consumer unit, one lockable door and two windows. Alternative materials, fittings and colours can be utilised to meet customer specifications. A wide range of folding equipment, including folding bunk beds, lockers, cages and furniture, is available to support the system.

An assembled UBK Kabin in the process of being protected by sandbags and a defence wall **1997**

UBK Kabin panels are formed from a nominal 56 mm thick rigid closed cell polyurethane foam core with two 0.5 mm zintec coated mild steel sheets finished with a 200 µm high-performance coloured plastisol. The central frame and all pressed steel sections are fabricated from zintec coated mild steel sheet grade 43. Floor panels have an overall thickness of 66 mm with the polyurethane core treated against insects and damp and a non-slip waterproof floor covering is applied. Double glazed windows are provided and may be fitted with anti-glare, anti-blast film as an option. The roof is manufactured using 75 mm glass reinforced plastic panels.

A basic UBK Kabin weighs between 1,000 and 1,200 kg.

Status

In production. Supplied to Bosnia, Cyprus, Germany, South Korea, Russian Federation and West Africa.

Manufacturer

Kudos 2000 Limited, Duckett's Wharf, 107-109 South Street, Bishops Stortford, Hertfordshire CM23 3AR, UK.
Tel: +44 1279 656565. Fax: +44 1279 656566.

UPDATED

A standard UBK Kabin fully erected **1997**

A standard UBK Kabin fully erected **1997**

MSI-Defence Systems tactical shelters

Description

MSI-Defence Systems design and produce a wide variety of tactical shelters for military equipment. The shelters are suitable for rapid reaction situations and can be configured to provide protection for both personnel and equipment from EMI/RFI, N-EMP, TEMPEST, lightning, thermal flash, NBC and projectile attack. The shelter units may be complexed or butt connected to form integrated centres while maintaining full protection.

The shelters are constructed from steel or aluminium panels and are equipped and customised to customer specifications.

Status

In production. In service with the British armed forces.

Manufacturer

MSI-Defence Systems Limited, Salhouse Road, Norwich, Norfolk NR7 9AY, UK.
Tel: +44 1603 484065. Fax: +44 1603 415649.

NEW ENTRY

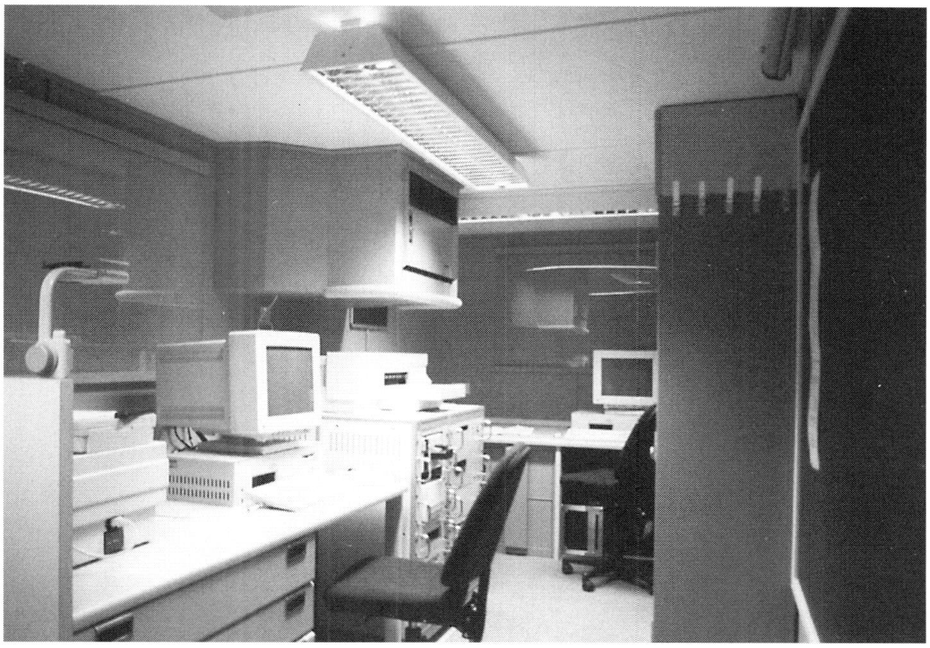

Interior of a tactical shelter designed and supplied by MSI-Defence Systems for the Royal Navy **1997**

Multilift Container Handling Unit (CHU)

Description

The Container Handling Unit (CHU) was developed by Multilift, Government Business Operations, and allows unmodified ISO containers to be picked up and transported by vehicles equipped with load handling systems such as the British Army DROPS or the US Army PLS (see entries under *Materials handling equipment* for details of these systems). This enables all types of 20 ft/6.906 m ISO shelter/containers to be efficiently transported without using specialised ISO handling equipment. As no intermediate flatrack is required, the payload of the load handling system may also be increased. Reconfiguration between the flatrack carrying mode and the container handling mode is quick and simple, requiring no special tools. Incorporation of the CHU does not degrade the flatrack handling performance of the load handling system.

The CHU consists of an H-frame which attaches to the end of a container and provides a hook-lift attachment point. The frame attaches to the container using self-aligning locking pins, enabling easy deployment by one man. The height of the H-frame is automatically adjustable in use to cater for 8 to 8.5 ft (2.438 to 2.59 m) containers. The rear roller assembly guides, supports and protects the container during loading and unloading.

A pair of twist locks integral with the rear support section secure the rear end of the container during transit and facilitate loading and unloading. The CHU can be carried on a load handling system when not in use; a stand is provided for storage.

Container Handling Unit (CHU) and ISO container being handled by a Mercedes-Benz truck **1996**

Status

In service with the British, Dutch and Norwegian armies. Under evaluation in the USA, Sweden, Italy and Canada and other armies worldwide.

Manufacturer

Cargotec (UK) Limited, Government Business Operations, Harlescott Lane, Shrewsbury, Shropshire SY1 3AG, UK.
Tel: +44 1743 442232. Fax: +44 1743 469935.

VERIFIED

Blatchford container handling equipment

Description

Blatchford Transport Equipment, a member of the Herbert Pool Group, produces a range of trailer and rail mounted container handling and transport systems.

Sideloader

The Sideloader trailer-mounted container handling system is a high-capacity (up to 34,000 kg) low-tare weight design suitable for 44,000 kg (or more) GCW on roads, off-road or in depot, depending on the trailer design. One person remote operation, low cycle times and a simple and stable layout, combine to ensure no risk, safe and rapid ground level loading and unloading or transfer to road or rail. The system's hydraulic cranes operate in tandem or individually and move longitudinally to accommodate 20, 30 or 40 ft ISO containers. The system is also fitted to rigid chassis for handling 20 ft ISO containers.

T-lift

The T-Lift system is suitable for both trailer and rail operation. It can handle 20, 30 and 40 ft ISO containers and swapbodies of up to 13.6 m. Maximum capacity is 34,000 kg.

The T-lift system has two sets of side-folding arms which are raised to form an end-on 'T' configuration with the top of the 'T' forming the crane arms along which an ISO container, under a toplift frame with four powered twist locks, travels. Power for the system is provided by an integral diesel power pack. Operations are controlled by a remote control box on a wanderlead.

The RTL-30 is mounted on a custom-built self-

Blatchford Sideloader container handling system mounted on a semi-trailer

propelled railway wagon which can travel at speeds up to 11 km/h during operations. For long distance moves the wagon can be towed at speeds up to 120 km/h.

STAG

The STAG is intended for trailer-mounted use with 20 to 40 ft ISO containers with capacities of up to 32,000 kg. Models are available from single side-loaders to all-purpose dual-sided loaders. The general principle is that articulated side legs can be hydraulically unfolded to form horizontal sliding arms along which containers may be side-shifted for loading on to another vehicle or grounding. The all-purpose dual-sided machines can transfer containers from either own vehicle or another vehicle and transfer them to another or on to the ground. Single-sided machines may have lift systems for container stacking.

Status

In production.

Manufacturer

Blatchford Transport Equipment, a Division of Herbert Pool Limited, 95 Fleet Road, Fleet, Hampshire GU13 8PJ, UK.
Tel: +44 1252 620444. Fax: +44 1252 622292.

1996 ***VERIFIED***

Blatchford 30 ton single-sided STAG for 40 ft ISO containers in operation

Blatchford T-Lift container handling system in operation

UNITED STATES OF AMERICA

Gichner shelter, electrical equipment S-250

Description

The S-250 is a T-shape, non-expansible shelter of adhesive-bonded construction with seamless aluminium interior and exterior skins, extruded aluminium framework structure and a polyurethane foam core. It is used to carry communications, radar, teletype and maintenance electronic equipment and is mounted on the rear of a 1¼ or 1½ ton truck. It can be transported by fixed-wing aircraft, helicopter, rail and ship.

The shelter meets military specifications MIL-S-55541 and is available with RFI integrity.

Specifications

Weight:
(without RFI kit) 349±9 kg
(with RFI kit) 355±9 kg
Payload: 862 kg

Gichner shelter, electrical equipment S-250

Interior dimensions:
(length) 1.628 m
(width, upper section) 1.908 m
(width, lower section) 1.333 m
(height, centre section) 1.79 m
(height) 1.191 m
Exterior dimensions:
(length) 1.953 m
(width, upper section) 2.234 m
(width, lower section) 1.857 m
(height) 1.78 m

Status
In production. In service with the US and other armed forces.

Manufacturer
Gichner Systems Group, 490 Locust Street, Dallastown, Pennsylvania 17313, USA.
Tel: +1 717 246 5479. Fax: +1 717 246 5496.

UPDATED

Gichner shelter, electrical equipment S-280

Description
The S-280 is a rigid rectangular, non-expansible shelter of adhesive-bonded construction consisting of seamless aluminium interior and exterior skins, extruded aluminium framework structure and polyurethane foam core. It is used to carry communications, radar, maintenance equipment and electronic equipment, and is mounted on the rear of a standard 2½ ton truck (for example the M35). It can also be transported by a transporter/mobiliser dolly set (models M720, M832 or M840), fixed-wing aircraft, helicopter, rail or ship.

The shelter meets military specification MIL-S-55286E and is available with RFI integrity.

Specifications
Weight: 612 kg
Payload: 2,268 kg
Interior dimensions:
(length) 3.504 m
(width) 2.07 m
(height) 1.959 m
Exterior dimensions:
(length) 3.733 m
(width) 2.209 m
(height) 2.193 m

Status
In production. In service with the US and other armed forces.

Manufacturer
Gichner Systems Group, 490 Locust Street, Dallastown, Pennsylvania 17313, USA.
Tel: +1 717 246 5479. Fax: +1 717 246 5496.

UPDATED

Gichner shelter, electrical equipment S-280

Gichner HMMWV compatible shelters

Description
The Gichner Systems Group produces a range of shelters which is compatible with the M998 HMMWV and M1097 'Heavy' HMMWV. The models are as follows:

S-787/G and S-788/G

These shelters are known as the Standard Integrated Command Post Shelters (SICPS) and are produced in SICPS and SICPS-LMS forms. Both are lightweight, high-strength enclosures specifically for use on the HMMWV. Construction involves aluminium-faced, honeycomb core, hot-bonded panels. The electromagnetic shielding is a minimum of 60 dB attenuation for electric, magnetic fields and plane wave in the 150 kHz to 10 GHz frequency range. Both shelters can be driven on to and off a C-130 Hercules transport aircraft.

The SICPS was selected by the US Army as its primary electronics platform for wheeled tactical vehicles. A CUCV version is available.

Weight:
(SICPS, empty) 286 kg
(SICPS-LMS, empty) 268 kg
Exterior dimensions:
(length) 2.59 m
(width) 2.13 m
(height) 1.74 m
Interior dimensions:
(length) 2.53 m
(width) 2.07 m
(height) 1.64 m
Usable interior volume: 8.35 m^3

S-710/M

The S-710/M is a lightweight shelter designed specifically for use with the M1037 HMMWV and M1028 CUCV vehicles. The construction techniques and materials are the same as those used for the S-250 and S-280 shelters. RFI integrity in excess of 60 dB per MIL-STD-285 is a standard feature, making this shelter suitable for C^{3}I applications. The dimensions of the shelter allow C-130 transport aircraft drive-on capability when used with either vehicle.

Weight: 308 kg
Payload: 862 kg

Gichner S-787/G SICPS shelter on HMMWV

Gichner S-805/G shelter on HMMWV

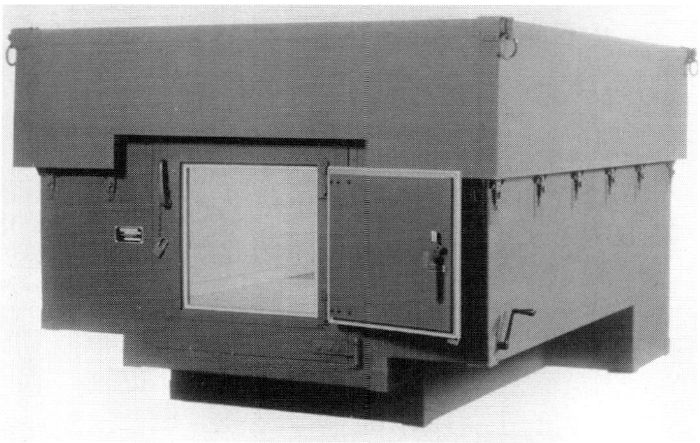

Gichner NOMAD 1 shelter in transport configuration

Gichner NOMAD 2 shelter in open configuration

Exterior dimensions:
(length) 2.695 m
(width, upper section) 2.148 m
(width, lower section) 1.268 m
(height) 1.753 m
Interior dimensions:
(length) 2.604 m
(width, upper section) 2.057 m
(width, lower section) 1.194 m
(height, centre section) 1.651 m
(height) 1.297 m

S-805/G

The S-805/G is a lightweight, high-strength foam-and-beam enclosure designed for use on the HMMWV. The foam and beam construction facilitates ease of shelter wall repair. The electromagnetic shielding provides over 60 dB of attenuation for electric, magnetic and plane wave fields in the 150 kHz to 10 GHz frequency range. This shelter can be driven on to and off C-130 transport aircraft.

Weight: approx 249 kg
Payload: 1,533 kg
Exterior dimensions:
(length) 2.59 m
(width) 2.13 m
(height) 1.74 m
Interior dimensions:
(length) 2.47 m
(width) 2.05 m
(height) 1.65 m

NOMAD 1 and 2

The NOMAD is the newest shelter design for the Heavy HMMWV variant. When mounted on the HMMWV or CUCV it does not exceed an overall height of 2.59 m and can be driven onto a C-130 or C-141 aircraft. When deployed the shelter roof can be raised to allow a full 2.03 m of interior height. In the extended mode the NOMAD can provide EMI/RFI shielding to 60 dB. The interior dimensions are the same as those for the S-710/M.

Status

All the above are in production.

Manufacturer

Gichner Systems Group, 490 Locust Street, Dallastown, Pennsylvania 17313, USA.
Tel: +1 717 246 5479. Fax: +1 717 246 5496.

UPDATED

Gichner HMMWV satellite terminal pallet, electronic equipment, GSS-4501

Description

The GSS-4501 is a satellite communications (SATCOM) system housed in a custom foam- and beam- adhesive-bonded construction pallet with seamless aluminium interior and exterior skins. The pallet is available with RFI integrity and can be obtained in several configurations with a list of modular options. The primary configurations include an open architecture with slide-out shock-mounted electronic equipment racks, power distribution and climate control systems, and an onboard generator version with a tunnel accommodating the generator, power control and distribution equipment. Options include, but are not limited to, dish antennas, environmental control systems, telescopic antenna masts and rack systems.

Although the system was designed as a tri-band SATCOM platform for the M1097 HMMWV it can be adapted to other prime movers and can accommodate many electronic systems, including direction finding, radio relay, emergency broadcasting, signal intelligence, cellular phone hub, and many others. The system can be transported by fixed-wing aircraft, helicopters, rail and ship, both on and off the HMMWV.

Gichner HMMWV Satellite Terminal Pallet Electronic Equipment, GSS-4501

Specifications

Weight: 454 kg
Payload: 1,996 kg
Exterior dimensions:
(length) 2.63 m
(width) 2.24 m
(height) 1.1 m
Interior dimensions:
(length) 2.52 m
(width) 2.06 m
(height) 1.02 m

Status

Prototype complete. Second-generation enhanced design imminent.

Manufacturer

Gichner Systems Group, 490 Locust Street, Dallastown, Pennsylvania 17313, USA.
Tel: +1 717 246 5479. Fax: +1 717 246 5496.

UPDATED

Gichner 20 ft ISO shelter/container

Description

This is a rigid, rectangular, non-expansible shelter/container of adhesive-bonded sandwich panelling consisting of seamless aluminium exterior skins, extruded aluminium framework structure and a polyurethane foam core. It has integral ISO corner fittings which meet ANSI MH 5.1 requirements. It is used as a multipurpose electronic equipment enclosure (radar, communication, maintenance, housing, laboratory and so on), with a special butting kit feature and slide-in/slide-out air conditioner unit optional. Large doors are provided for the installation and removal of equipment.

The shelter/container meets military specification MIL-M-81957 and can be transported by truck, transporter (M1022 end mount or van undercarriage type), fixed-wing aircraft, rail and ship. It is stackable via ISO couplers.

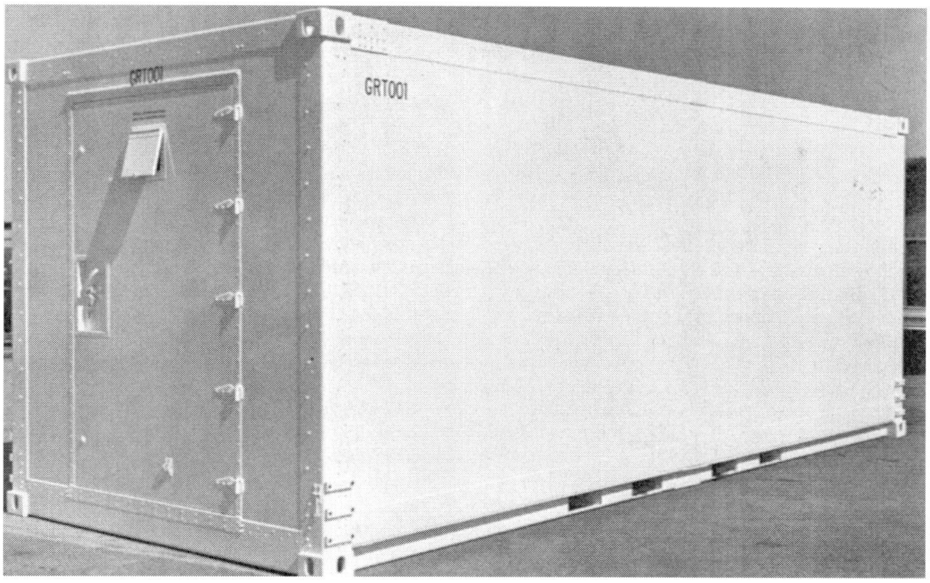

Gichner 20 ft ISO shelter/container

Specifications

Weight: 1,996 kg
Payload: 4,536 kg
Exterior dimensions:
(length) 6.057 m
(width) 2.438 m
(height, less skids) 2.438 m
(height, with skid for van mobiliser interface) 2.514 m
Interior dimensions:
(length) 5.897 m
(width) 2.23 m
(height) 2.141 m

Status

In production. In service with the US and other armed forces.

Manufacturer

Gichner Systems Group, 490 Locust Street, Dallastown, Pennsylvania 17313, USA.
Tel: +1 717 246 5479. Fax: +1 717 246 5496.

UPDATED

Gichner modular electronic equipment shelter GMS-451 series

Description

The GMS-451 is a rigid rectangular modular side-to-side butting shelter and is of adhesive-bonded sandwich panel construction consisting of seamless interior and exterior skins, extruded aluminium framework structure and a polyurethane foam core. It has removable side panels with GMS shear-loc design.

The shelter, which meets military specification MIL-S-55286 Modified (RFI optional) can be used for a variety of roles including communications, radar, teletype, laboratory, dormitory and hospital. It can be transported by truck, transporter, aircraft, rail or ship.

Specifications

Model	**GMS-451**	**GMS-451-479**	**GMS-451-537**	**GMS-451-571**
Weight (basic unit)	907 kg	1,089 kg	1,724 kg	1,134 kg
Payload	2,268 kg	2,722 kg	4,536 kg	2,722 kg
Exterior (length)	3.733 m	4.242 m	6.096 m	4.241 m
(width)	2.209 m	2.209 m	2.438 m	2.438 m
(height)	2.336 m	2.336 m	2.54 m	2.387 m
Interior (length)	3.505 m	4.019 m	5.893 m	4.019 m
(width)	2.032 m	2.07 m	2.286 m	2.286 m
(height)	2.082 m	2.063 m	2.26 m	2.063 m

Status

In production. In service with the US and other armed forces.

Manufacturer

Gichner Systems Group, 490 Locust Street, Dallastown, Pennsylvania 17313, USA.
Tel: +1 717 246 5479. Fax: +1 717 246 5496.

UPDATED

Gichner GMS-451-571 seven-shelter complex built for Melpar Division of E-Systems

Interior view of GMS-451 series shelters coupled together

Gichner transporter/mobiliser dolly sets

Description

Tactical shelters require the use of rugged wheeled devices (end-mounted and van undercarriage type) capable of transporting the shelters via a towing vehicle over rough terrain as well as airfields and roads. This is accomplished via militarised tactical mobilisers of various capacities and types. There are two of the standard units available from Gichner listed below.

Specifications

	Applicable specification	**Gross load capacity**	**Type**
M1022	M1022	6,804 kg	end-mount
A/M32U-17	CP600530	7,711 kg	end-mount

Status

In production. In service with the US and other armed forces.

Manufacturer

Gichner Systems Group, 490 Locust Street, Dallastown, Pennsylvania 17313, USA.
Tel: +1 717 246 5479. Fax: +1 717 246 5496.

UPDATED

Gichner transportable material handling pallet SPA-5000

Description

This pallet provides tactical transport/handling capability for miscellaneous accessory equipment (jacks, air conditioners, cable reels and generators and so on) used in conjunction with the transportable shelter system.

Its adhesive-bonded sandwich panel constructed base consists of aluminium interior and exterior skins, extruded aluminium framework structure and foam core material.

The pallet, which meets military specification MIL-D-27925, is transportable by truck, transporter/mobiliser dolly set (M720, M832 and M689), fixed-wing aircraft, helicopter, rail and ship.

Specifications

Weight: (with sling) 226.8 kg
Payload: 2,268 kg
Payload area: (inside stanchions) 1.831 × 2.352 m
Overall exterior: (including mobiliser brackets and handles) 2.057 × 2.659 m
Standard stanchion height: 0.695 m

Status

In service with the US armed forces.

Manufacturer

Gichner Systems Group, 490 Locust Street, Dallastown, Pennsylvania 17313, USA.
Tel: +1 717 246 5479. Fax: +1 717 246 5496.

UPDATED

Gichner transportable material handling pallet SPA-5000, complete with jacks and air conditioning equipment

Gichner levelling, alignment and lifting jack systems

Description

The type of jacking system depends on the basic requirements of the shelter: whether it is fixed or modular/expansible, its gross load and size, jacking height required; and its purpose; levelling, truck-loading, or both. Gichner has two basic jack designs which within themselves have been modified to incorporate features which will satisfy all tactical shelter jacking requirements for all tactical military shelters. The two basic types are the fixed and the roller base types, brief specifications for which are given here.

Status

In production.

Manufacturer

Gichner Systems Group, 490 Locust Street, Dallastown, Pennsylvania 17313, USA.
Tel: +1 717 246 5479. Fax: +1 717 246 5496.

UPDATED

Gichner 20 ft shelter with lifting jacks in position

Specifications

Reference	**LWJ 2500**	**RBJ 2500**	**AWJ5000**	**AWJ5000-AE**
Type base	fixed	roller	fixed	fixed
Lift capacity	1,134 kg	1,134 kg	2,268 kg	2,268 kg
Travel height	0.457-0.736 m	0.457-0.763 m	1.524 m	1.524 m
General application	S-280	expansible	8 ft × 8 ft × 20 ft modular	truck loading shelter
Approx weight (each)	13.6 kg	14.5 kg	40.9 kg	54.43 kg

Gichner transportable material handling pallet SPA-10003

Description

This pallet provides tactical transport/handling capability for miscellaneous equipment (jacks, air conditioners, cable reels and generators and so on) used in conjunction with the transportable shelter system.

Its adhesive sandwich panel constructed base consists of aluminium interior and exterior skins, extruded aluminium framework structure and foam core material.

The pallet, which meets military specification MIL-D-27925 (modified), is transportable by truck, transporter/mobiliser dolly set (M832 or M689), fixed-wing aircraft, helicopter, rail or ship.

Specifications

Weight: (with sling) 363 kg
Payload: 4,538 kg

Payload area: (inside stanchions) 1.872 × 3.342 m
Overall exterior: (including mobiliser brackets and handles, less 463-L rails) 2.085 × 3.768 m
Standard stanchion height: 0.695 m

Status
In service with the US armed forces.

Manufacturer
Gichner Systems Group, 490 Locust Street, Dallastown, Pennsylvania 17313, USA.
Tel: +1 717 246 5479. Fax: +1 717 246 5496.

UPDATED

Gichner transportable material handling pallet SPA-10003

Craig standard and ISO shelters

Description
Craig Systems is one of the largest manufacturers of mobile shelters in the USA and produces a very wide range of standard mobile shelters, shelter vans and accessories. The ISO range of shelters is produced with floors, walls and roofs of single-piece inner and outer aluminium skins with foamed-in-place polyurethane filling. Aluminium extrusions are used for structural strength and for any internal fittings. As with all Craig Systems shelters, the ISO range can be adapted for a wide range of purposes with the internal and external fittings and accessories altered accordingly.

Status
In production. In widespread service.

Manufacturer
Craig Systems, 10 Industrial Way, Amesbury, Massachusetts 01913, USA.
Tel: +1 617 688 6961. Fax: +1 508 388 0845.

VERIFIED

Specifications

Model	**S-138**	**S-250**	**S-280**	**H-376**	**H-395**	**H-454**	**H-556**	**H-581**	**H-584**	**H-586**	**H-587**
Length	2.438 m	1.981 m	3.505 m	3.556 m	3.607 m	6.705 m	5.944 m	1.93 m	2.591 m	3.861 m	3.404 m
Width	1.93 m	1.905 m	2.07 m	2.134 m	1.93 m	2.286 m	2.286 m	1.93 m	1.93 m	1.93 m	1.93 m
Height	1.867 m	1.625 m	1.892 m	2.007 m	1.867 m	2.184 m	2.184 m	1.93 m	1.93 m	1.867 m	1.93 m
Weight	454 kg	347 kg	626.5 kg	726 kg	567.5 kg	1,816 kg	1,589 kg	390 kg	463 kg	590 kg	545 kg

Craig Model H-833 electronic equipment shelter on Craig Model D-74 1B transporter

Craig S-280 shelter with external air conditioner mounted on M548 tracked cargo carrier

Craig N-1080 and N-1050 nuclear hardened tactical shelters

Description
The Craig N-1080 nuclear hardened shelter was developed by Harry Diamond Laboratories together with the Italian concern Miki SpA to house and shelter complex and expensive equipment that modern armed forces use in the front line. Such equipment includes electronics and various forms of communications equipment, and in some cases the contents of a mobile military shelter can cost millions, whatever the currency involved. The N-1080 was designed to protect such equipment against the effects of tactical nuclear weapons. It is a conventional shelter in appearance but

Craig N-1080 nuclear hardened shelter

Craig N-1050 nuclear hardened shelter

the walls, floors and ceiling are Kevlar laminates and many of the fittings are Kevlar extrusions. The door has special blastproof hinges and a Craig V-846 anti-blast valve is fitted above the door. The N-1080 has a specially strengthened air conditioning and filter unit at the front end of the shelter. The door has an emergency escape hatch, and special corner and lifting castings are provided.

The N-1080 is 3.734 m long, 2.21 m wide and 2.12 m high.

The N-1050 is a smaller nuclear hardened shelter suitable for carrying on small trucks. It is 2.073 m long, 1.994 m wide and 1.717 m high. Weight is 544 kg.

Status
In production.

Manufacturer
Craig Systems, 10 Industrial Way, Amesbury, Massachusetts 01913, USA.
Tel: 1 617 688 6961. Fax: +1 508 388 0845.

VERIFIED

Craig shelter transporters

Description
There are four main types of transporter in the Craig range, the D-741B, D-741Z, D-741 (ISO) and the M-832.

The D-741B transporter consists of front and rear half sections which are permanently attached to the shelter. The payload is up to 7,711 kg. The front half is steerable, primarily to facilitate loading large shelters onto cargo aircraft. The D-741B is in service with the US Marine Corps.

The D-741 (ISO) has the same capabilities as the D-741B and is designed for use with ISO shelters.

The D-741Z provides type 'V' mobility without hydraulic cylinders to raise and lower the shelter on to the transporter. A third wheel is provided and the manual system of loading and unloading shelters can be performed with minimal personnel and effort.

The M-832 provides mobility for medium-range payloads up to 4,763 kg. The front and rear dolly assemblies are attached to the payload Hydraulic cylinders permit lowering and raising of the shelter suspension bars and strut assemblies lock the trailer in the raised position for transport. Air over hydraulic brakes are provided on all four wheels of the dolly set. Tail, stop and blackout lights are mounted on the rear trailer. The M-832 is used by the US Marine Corps.

Specifications

Model	**D-741Z**	**M-832**	**D-741B**	**D-741 (ISO)**
Weight	1,361 kg	1,664.7 kg	1,927.75 kg	2,565 kg
Payload (max)	6,804 kg	4,763 kg	7,711 kg	7,711 kg
Length	front 1.435 m rear 1.384 m	3.038 m	front 1.435 m rear 1.384 m	front 1.448 m rear 1.422 m
Width	2.438 m	2.438 m	2.438 m	2.438 m
Height	1.168 m	1.321 m	n/a	2.337 m
Track width	2.438 m	2.134 m	2.134 m	2.134 m
Road clearance	0.432 m	0.432 m	0.432 m	0.432 m
Shelter levelling capability	–	0 to 0.305 m	0 to 0.102 m	–
Shelter lift capability	–	0 to 0.432 m	0 to 0.143 m	0 to 0.635 m
Hydraulic lifting	no	standard	standard	standard
Mechanical lifting	yes	no	no	no
Tyres	11.00 × 20	9.00 × 20	11.00 × 20	11.00 × 20

Status
In production. In service with the US Army, US Marine Corps and many other armed forces.

Manufacturer
Craig Systems, 10 Industrial Way, Amesbury, Massachusetts 01913, USA.
Tel: 1 617 688 6961. Fax: +1 508 388 0845.

VERIFIED

Craig Model M-832 transporter

Craig Model D-741 (ISO) being used to load shelter into US Air Force C-130 Hercules transport aircraft

Marion Composites Army Standard Family (ASF) shelters

Description
Developed by the US Army, the Army Standard Family (ASF) of relocatable/expandable shelters is used by all branches of the US armed forces and military services worldwide. ASF shelters are available in three configurations: non-expandable; one-side expandable; and two-side expandable. All three configurations measure 2.4 × 2.4 × 6 m when in the shipping mode. When deployed the one-side and two-side expandable units expand to 4.43 × 5.82 m and 6.55 × 5.82 m respectively to provide up to 36.6 m² of floor space.

The shelters, constructed of honeycomb core with aluminium-faced panels, are highly mobile and can be set up on ground slopes of up to 7.5 per cent. The shelters conform to ISO shipping container standards and can be transported by ship, rail, truck, cargo air, sea and helicopter lift.

Primary uses for ASF shelters include deployable medical units, aviation maintenance shops and command centres.

Marion Composites Army Standard Family (ASF) non-expandable shelter
1997

Marion Composites Army Standard Family (ASF) one-side expandable shelter *1997*

Marion Composites Army Standard Family (ASF) two-side expandable shelter *1997*

Specifications

Non-expandable
Military specification: MIL-S-44196
Weight: (empty) 1,751 kg
Payload: 5,053 kg
Dimensions:
(exterior, shipping) 6.06 × 2.44 × 2.44 m
(interior, shipping) 5.82 × 2.17 × 2.31 m
(interior, deployed) 5.82 × 2.17 × 2.31 m
Volume:
(shipping) 28.3 m^3
(deployed) 28.3 m^3
Floor area:
(shipping) 13.5 m^2
(deployed) 13.5 m^2

One-side expandable
Military specification: MIL-S-44197
Weight: (empty) 2,504 kg
Payload: 4,300 kg
Dimensions:
(exterior, shipping) 6.06 × 2.49 × 2.49 m
(interior, shipping) 5.82 × 2.14 × 2.17 m
(interior, deployed) 5.82 × 4.43 × 2.17 m
Volume:
(shipping) 27.2 m^3
(deployed) 53.8 m^3
Floor area:
(shipping) 12.4 m^2
(deployed) 25.3 m^2

Two-side expandable
Military specification: MIL-S-44195
Weight: (empty) 3,153 kg
Payload: 3,652 kg
Dimensions:
(exterior, shipping) 6.06 × 2.49 × 2.49 m
(interior, shipping) 5.82 × 1.98 × 2.17 m
(interior, deployed) 5.82 × 6.55 × 2.17 m
Volume:
(shipping) 24.9 m^3
(deployed) 79.2 m^3
Floor area:
(shipping) 11.5 m^2
(deployed) 37.2 m^2

Status

In production. In service with the US armed forces and other military services worldwide.

Marketing Office

Marion Composites, 150 Johnston Road, Marion, Virginia 24354, USA.
Tel: +1 540 783 3121. Fax: +1 540 783 9665.

NEW ENTRY

Marion Composites general purpose rigid ISO shelter

Description

Designed and developed for the US Marine Corps, this lightweight honeycomb-constructed 2.4 and 6 m general purpose rigid ISO shelters with their 'mount anywhere capability' can be complexed into large suites by removing the sidewalls and connecting the units with a complexing kit. ISO fittings and forklift tunnels assure handling and transport under maximum ISO load conditions.

Construction involves aluminium-faced, honeycomb sandwich panels hot bonded with epoxy adhesives. There are electrical service entry and distribution panels for 120/208 V, 60 Hz, 60 A, three-phase, five-wire power supplies. Also provided are 15 A utility outlets while ceiling mounted fluorescent fixtures are standard.

Accessories include blackout curtains, levelling jacks, complexing kit, exhaust fan and a joining corridor.

The shelters can be mounted at any location using high-strength threaded inserts providing up to 726 kg pullout and 43 Nm torque. The shelters are US Coast Guard certified (ISP 1496/1) for sea transport including six high stacking. The units can also be transported by rail, truck and air.

Marion Composites general purpose rigid ISO shelter *1997*

Specifications

Rigid 10 ft ISO
Military specification: MIL-S-29434(MC)
Weight: (empty) 1,080 kg
Payload: 2,322 kg
Dimensions:
(exterior) 2.99 × 2.44 × 2.44 m
(interior) 2.72 × 2.25 × 2.14 m

Rigid 20 ft ISO
Military specification: MIL-S-29410(MC)
Weight: (empty) 1,724 kg
Payload: 5,080 kg
Dimensions:
(exterior) 6.06 × 2.44 × 2.44 m
(interior) 5.78 × 2.25 × 2.14 m

Status

In service with the US armed forces.

Marketing Office

Marion Composites, 150 Johnston Road, Marion, Virginia 24354, USA.
Tel: +1 540 783 3121. Fax: +1 540 783 9665.

NEW ENTRY

Marion Composites EMI ISO shelter

Description

These Electromagnetic Interference (EMI) attenuated 10 ft and 20 ft ISO shelters are elements within the US Marine Corps Expeditionary Shelter System (MCESS). Honeycomb construction provides a mounting platform for heavy equipment and allows maximum flexibility for the location of equipment. Aluminium-faced sandwich construction provides shielding properties meeting or exceeding 60 dB RFI/EMI attenuation through the entire 150 kHz to 10 GHz range.

Construction involves aluminium-faced, honeycomb sandwich panels hot bonded with epoxy adhesives. There are electrical service entry and distribution panels for 120/208 V, 60 Hz, 60 A, three-phase, five-wire power supplies. Also provided are 15 A utility outlets while ceiling mounted RFI/EMI suppressed fluorescent fixtures are standard.

Accessories include blackout curtains, levelling jacks and a joining corridor.

Both sizes of shelter provide magnetic field attenuation from 150 kHz to 1.5 MHz, electric field attenuation from 150 to 400 kHz, and plane wave attenuation from 400 MHz to 10 GHz.

Specifications

EMI 10 ft ISO
Military specification: MIL-S-29411(MC)
Weight: (empty) 1,338 kg
Payload: 2,064 kg
Dimensions:
(exterior) 2.99 × 2.44 × 2.44 m
(interior) 2.74 × 2.3 × 2.17 m

EMI 20 ft ISO
Military specification: MIL-S-29411(MC)
Weight: (empty) 2,109 kg
Payload: 4,695 kg
Dimensions:
(exterior) 6.06 × 2.44 × 2.44 m
(interior) 5.8 × 2.3 × 2.17 m

Status

In service with the US armed forces.

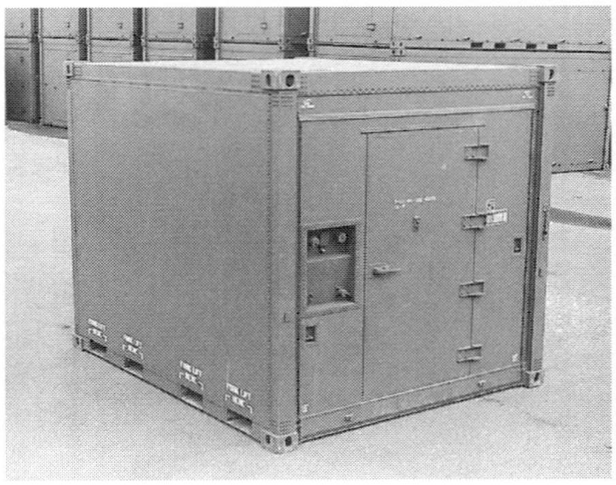

Marion Composites EMI ISO shelter **1997**

Marketing Office

Marion Composites, 150 Johnston Road, Marion, Virginia 24354, USA.
Tel: +1 540 783 3121. Fax: +1 540 783 9665.

NEW ENTRY

Marion Composites Knockdown ISO shelter

Description

The Knockdown ISO shelter, developed for the US Marine Corps, is a lightweight 20 ft ISO shelter which easily knocks down into a shipping configuration only 530 mm high. This allows four shelters to be shipped in the same space and for the same cost as one assembled shelter. Two ISO stacks (eight shelters) constitute one C-130 load of Knockdown shelters. Assembly of individual shelters can take place within minutes.

Construction involves aluminium-faced, honeycomb sandwich panels hot bonded with epoxy adhesives. There are electrical service entry and distribution panels for 120/208 V, 60 Hz, 60 A, three-phase, five-wire power supplies. Also provided are 15 A utility outlets while ceiling mounted fluorescent fixtures are standard.

Accessories include blackout curtains, levelling jacks, complexing kit, exhaust fan and a joining corridor.

The shelters can be mounted at any location using high-strength threaded inserts providing up to 726 kg pullout and 43 Nm torque. The shelters are US Coast Guard certified (ISP 1496/1) for sea transport including six high stacking. The units can also be transported by rail, truck and air.

Specifications

Military specification: MIL-S-29409(MC)
Weight: (empty) 1,660 kg
Payload: 6,641 kg
Dimensions:
(exterior, 4-high stack) 6.06 × 2.44 × 2.44 m
(exterior, individual, set up) 6.06 × 2.44 × 2.44 m
(interior, individual, set up) 5.87 × 2.25 × 2.15 m

Status

In service with the US armed forces.

Marketing Office

Marion Composites, 150 Johnston Road, Marion, Virginia 24354, USA.
Tel: +1 540 783 3121. Fax: +1 540 783 9665.

NEW ENTRY

Marion Composites Knockdown ISO shelter in partially assembled state **1997**

Assembling a side panel of a Marion Composites Knockdown ISO shelter **1997**

Marion Composites relocatable hospitals

Description

Developed in conjunction with the US armed forces, this modular hospital system was designed for use in outlying or remote locations and has a full spectrum of uses, from primary care and emergency/disaster relief to a semi-permanent hospital installation capable of offering a wide range of medical services. Modules may be configured in various combinations and layouts to meet the needs of a particular site or hospital service requirements. Configurations ranging from mobile medical clinics to 200-bed hospital systems are available.

Basic modules include an operating room, an emergency receiving area, an eight-bed ward, laboratory, pharmacy and X-ray options, a central material supply room, a dental surgery, post-operative or intensive care units, and a power and water supply module. Optional units include a kitchen, a dining area and a laundry.

All units are constructed using aluminium faced honeycomb sandwich panels.

Specifications

Basic unit
Weight: (maximum, equipped) 6,804 kg
Dimensions:
(shipping mode, external) 6.06 × 2.44 × 2.44 m
(deployed, internal) 5.59 × 6.55 × 2.16 m

A five-unit Marion Composites relocatable hospital **1997**

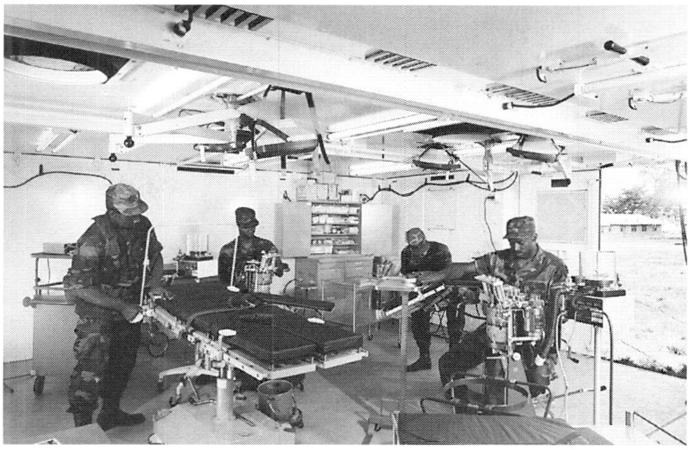

A typical operating room in a Marion Composites relocatable hospital **1997**

Status
In service with the US armed forces. Deployed worldwide.

Marketing Office
Marion Composites, 150 Johnston Road, Marion, Virginia 24354, USA.
Tel: +1 540 783 3121. Fax: +1 540 783 9665.

NEW ENTRY

Marion Composites HMMWV shelters

Description
Marion Composite's High Medium Mobility Wheeled Vehicle (HMMWV - qv) shelters are deployed worldwide in applications including communications, command, surveillance, electronics, maintenance and medical systems.

The Lightweight Multipurpose Shelter (LMS) offers lightweight, high-strength environmental and EMI protection. An advanced version of the LMS is the Standardized Integrated Command Post Shelter (SICPS). One application of the SICPS is mounted on a Heavy HMMWV to provide chemical and biological protection employing a gas particulate filtration and overpressure system.

Construction of these shelters involves aluminium facing sheets bonded to a honeycomb core. EMI shielding involves a minimum of 60 dB attenuation for electrical, magnetic and plane waves over a frequency range of 150 kHz to 10 GHz.

Standard equipment includes an access ladder, roof access steps, a mounting kit and a pintle extension.

Specifications
Military specification: MIL-S-44408
Weight: (empty) 276 kg
(mounting kit) 36 kg
Payload: 1,497 kg
Dimensions:
(external) 2.59 × 2.13 × 1.7 m
(internal) 2.53 × 2.07 × 1.64 m
Volume: 8.07 m^3
Floor area: 5.2 m^2
Overall height on prime mover: 2.59 m

Status
In production. In service with the US armed forces.

Marketing Office
Marion Composites, 150 Johnston Road, Marion, Virginia 24354, USA.
Tel: +1 540 783 3121. Fax: +1 540 783 9665.

NEW ENTRY

Marion Composites HMMWV lightweight multipurpose Shelter (LMS) **1997**

Marion Composites HMMWV Standardized Integrated Command Post Shelter (SICPS) **1997**

Mobile Frame Structures M720 Mobiliser Dolly Set

Description
Mobile Frame Structures manufactures specially engineered trailers and transport equipment including the M720 Mobiliser Dolly Set. This is used by the US Army Tank Automotive Command (TACOM) as the dedicated transport vehicle for the S-280 shelter. The M720 meets the requirements of MIL-M-8090F with reference to Mobility Type III Group C.

Used for the deployment of containerised battlefield communications and radar equipment, the M720 enables the user to easily place containers in field

Mobile Frame Structures M720 mobiliser dolly set

locations. Dolly/shelter combinations can then be moved from location to location and decoupled within minutes. The unloaded dolly can then be used to transport other shelters, eliminating the need for special materials handling equipment.

Generally towed by the M924 or similar 5 ton trucks, the M720 can be loaded and mounted on to a shelter by two persons. Braking is air/hydraulic on all wheels and automotive-type steering is used. The hydraulic lift mechanism can be operated independently at each end to mount or level the shelter.

The M720 is rated for a 3 ton payload. Larger shelters than the S-280 require the use of a higher capacity mobiliser which is under development by Mobile Frame Structures.

The M720 weighs 1,000 kg unladen and has a capacity of 2,800 kg. Length is 2.35 m, width 2.43 m and height 2.43 m (0.79 m reduced).

Status
In production and widespread use.

Manufacturer
Mobile Frame Structures Inc, 1408 Calcon Hook Road, Sharon Hill, Pennsylvania 19079, USA.

VERIFIED

ESCO Datron Container Load Trailer and CLT Dolly Set

Description
The ESCO Datron Container Load Trailer (CLT) and CLT Dolly Set are combined container handling and transport devices designed to fill three military-related mission essential functions:

Container Handling — Side lifting permits containers to be loaded, off-loaded or cross-loaded to and from flat bed trucks and trailers at origin, destination or intermediate points. Special CLT adaptor dollies also facilitate close-in nesting or complexing of military shelters.

Container Transport — End lifting permits transport on roads at 80 km/h or over rough terrain at 30 km/h.

Transport Aircraft Loading — When end-mounted to an 8 or 8.5 ft tall (2.44 or 2.59 m) container the CLT rolls on and off all military transport aircraft, including the C-130. With optional loading dollies, two 20 ft/6.096 m containers can be loaded onto a single C-130, either consecutively or tied together in tandem.

The CLT is available in two basic capacity variants; the CLT Dolly Set for military shelters and lightweight containers and the 25 ton CLT for maximum weight ISO containers.

Each CLT half includes a diesel-powered hydraulic system to perform all final alignment, connection and lifting functions within minutes. With the optional self-powered wheels, one operator can perform these functions. The CLT includes a unique centre-of-gravity shift capability which facilitates handling of unloaded dollies in a three wheel mode.

The CLT is compatible with all ISO shelters and containers ranging in size from 8 to 8.5 ft tall (2.44 to 2.59 m) and from 4 to 40 ft (1.219 to 12.19 m) in length.

Specifications
CLT Dolly Set
Payload: (lift and transport) 9,091 kg
Max lift height: 1.5 m
Max towed speed: (loaded) 80 km/h
Weight, empty: 2,700 kg
Length: (with 20 ft container) 11 m
Width: 2.44 m
Height: adjustable
Engine: 5.6 hp diesel
Tyres: 11.00 R 20

Status
CLT Dolly Set in production (500+ units) for the US Army, Navy and Air Force. More than 100 systems are fielded with the Japan Self Defence Force, Canadian Armed Forces, Royal Australian Air Force and numerous defence related prime contractors. Special adaptors are also in production for compatibility with S-280 shelters.

The 25 ton CLT is in service with other armed forces (see separate entry under Denmark for details).

Manufacturer
Engineered Systems (ESCO Datron), 2550 Market Street, Aston, Pennsylvania 19014, USA.
Tel: +1 610 494 8000. Fax: +1 610 494 8989.

UPDATED

CLT Dolly Set in use with US Army tactical shelter

Side-shift loading of a shelter onto a truck

Deployable Rapid Assembly Shelter (DRASH)

Description
The Deployable Rapid Assembly Shelter (DRASH) was developed as a private venture and is in use with the US Air Force and Marine Corps, Austrian, British and Israeli armed forces, and others. DRASH meets the requirement for a rugged multiclimate man-portable shelter system and can be deployed by four persons in less than 5 minutes under practically any weather conditions.

A DRASH Model 6 stowed in its transport bag complete with ground cover and flooring

DRASH shelters are manufactured using aluminium alloy and titanite tubing connected to a network of patented articulating hubs. They can fold and pack down to about 2 per cent of their deployed size. Interior and exterior covers are made from state-of-the-art MIL-SPEC fabrics and are pre-attached to the frames. Tests have shown that repairs to frames and covers can be made in the field by unskilled personnel in less than 5 minutes.

Each DRASH shelter has four entrances and all have universal connectors to join two or more shelters together. Special 'boots' are available to provide a weathertight seal with other types of hard- or soft-wall

A Land Rover booted to a DRASH Model 2T to create a mobile first aid post

Two Model 6 DRASH shelters deployed

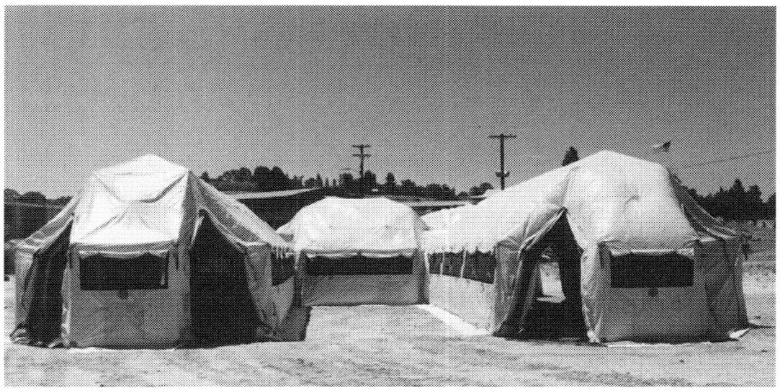

Two Model 6 DRASH units connected by a Model 2 to form a DRASH System 2 providing 92 m² of floor space

shelters and with a variety of wheeled and tracked vehicles.

The T-Series of DRASH shelters is exactly the same as the standard shelters except that one end is truncated to allow it to be attached to a vehicle; in this form the internal volume is slightly reduced. Standard shelters can be configured to form command posts or tactical operations centres.

Each DRASH shelter has the following equipment as standard: aluminium/titanite frame; exterior cover; interior cover; ground cover; flooring; integrated duct interface with plenum; doorways; universal connectors for networking shelters; screened windows with flaps; dual staking/securing system; transport bag; and a repair kit. Accessories include vestibule curtains for ingress/egress during blackout periods, plastic windows, an insulated/cushion subfloor and a Utilities Transport Trailer. The trailer offers an integrated utilities package including a 15 kW generator set, a 30 kW HVAC unit and an electrical distribution panel with four single-phase outlets.

DRASH shelters are available in six sizes. Basic features are as follows (the weight and dimensions stowed include the shelter, ground cover and flooring):

Model 1
Area: 10.5 m²
Dimensions stowed: 1.68 × 0.7 × 0.56 m
Weight: 86 kg

Model 2
Area: 15.9 m²
Dimensions stowed: 1.68 × 0.76 × 0.56 m
Weight: 108 kg

Model 3
Area: 21.35 m²
Dimensions stowed: 1.68 × 0.83 × 0.56 m
Weight: 131 kg

Model 4
Area: 26.7 m²
Dimensions stowed: 1.68 × 0.89 × 0.56 m
Weight: 156 kg

Model 5
Area: 32.1 m²
Dimensions stowed: 1.68 × 0.95 × 0.56 m
Weight: 176 kg

Model 6
Area: 37.5 m²
Dimensions stowed: 1.68 × 1.02 × 0.56 m
Weight: 199 kg

Status

In production. In service with the Austrian, British, Chinese (PRC) Israeli and Singapore armed forces and the US Marine Corps, US Air Force, US Navy SEALs, the US Army Medical Department and US Army Rapid Deployment Forces.

Manufacturer

DHS Systems LLC, 33 Kings Highway, Orangeburg, New York 10962-1302, USA.
Tel: +1 914 359 6066. Fax: +1 914 365 2114.

UPDATED

VEHICLE BODIES

FRANCE

Trailor rolling tilt drop side cargo body

Description

The Trailor rolling tilt drop side cargo body combines the attributes of a drop side cargo body with a rapidly and easily moved tilt cover. It was specifically designed for use with trucks carrying NATO cargo pallets or 20 ft/6.096 m ISO containers and has been fitted to Renault G 260 (6 × 4) trucks. It can also be used on a Trailor 14,400 kg drawbar trailer or with a Trailor 32,000 kg semi-trailer; the latter is in service as a logistics load carrier with the French Army, as is the Renault G 260 truck.

The body has a fixed forward wall in one or two sections. For the latter the lower part is a steel panel with an upper canvas wall. The side and rear drop panels can all be removed, as can the side posts. Along the top of the panels are recessed rails in which the rolling tilt can move. Fixed bows are covered by a polyamide or polyester cover and the tilt is so arranged that one man can rapidly open or close the tilt using only one hand by pulling a handle. When open, the bows and cover can be stowed either in the forward or rear position. The cargo floor is manufactured from reinforced timber planks.

Trailor rolling tilt drop side body fitted to Renault G260 (6 × 4) truck

Specifications

Length: (internal) 7.42 m
Width: (internal) 2.42 m
Height: (under tarpaulin, minimum) 2.1 m
Floor height: (unladen) 1.47 m

Status

In service with the French Army.

Manufacturer

Trailor SA, 5 route Nationale 10, BP 49, Coignières, F-78311 Maurepas Cedex, France.
Tel: +33 1 30 50 61 26. Telex: 698896 f.
Fax: +33 1 30 50 31 32.

VERIFIED

GERMANY

Marrel torsion-free military platform

Description

The Marrel torsion-free military platform is intended for use on vehicles that have to carry electronic equipment and cargo such as palletised loads over rough terrain. The all-steel body is usually mounted on two swivel joints at the front and rear which allow an optimum turning motion between the mountings and the chassis. This allows the platform to maintain a straight and level position even under extreme off-road conditions.

A typical installation is that used on the Mercedes-Benz 2628 A (6 × 6) 12,000 kg truck. The platform used allows a maximum on-road payload of 13,800 kg with the platform being 5.5 m long and 2.14 m wide. Loading height is 1.7 m. The platform has folding bench seating for 20 personnel, and is equipped with a storage box for chains and a camouflage net, a spare wheel, a fire extinguisher box, two 20 litre jerrican racks, a toolbox, two wheel chocks, a drinking water container, racks for entrenching tools and provision for a winch cable.

Another vehicle that can use a Marrel torsion-free military platform is the Mercedes-Benz L 911B/42 (4 × 4) truck. On this vehicle the platform is 4.5 m long, and the loading height is approximately 1.215 m unladen. Folding seating is provided for 18 personnel.

Status

In production.

Manufacturer

Marrel GmbH, Kruppstrasse 44, D-5603 Wülfrath, Germany.

VERIFIED

Marrel torsion-free military platform on Mercedes-Benz 2628 A (6 × 6) 12,000 kg truck

NETHERLANDS

GEMCO stretcher carrier set (GMS 2058)

Description

GEMCO Mobile Systems specialises in the development and production of custom-made solutions for mobile applications. Their stretcher carrier set (GMS 2058) was developed for use in M113 APCs but is equally applicable for many wheeled vehicles. The GMS 2058 was developed to facilitate the loading, unloading and transport of up to seven casualties as well as a medical assistant and the vehicle crew.

The GMS 2058 was designed to meet the latest health and safety regulations to reduce the risk of back injuries to orderlies. Due to a rail loading system it is not necessary to lift stretchers into the vehicle or even lift the stretcher while bending over. All types of NATO stretcher can be used without the need for special tools or prior preparation of the set. The GMS 2058 is sturdy, simple and will operate under the harshest environmental conditions, including in dust and sand.

The GMS 2058 can be used in four different configurations:

Seven seated casualties
Three stretcher casualties
Three seated plus two stretcher casualties
Four seated p us one stretcher casualty.

Status

In production.

Manufacturer

GEMCO Mobile Systems BV, Science Park Eindhoven 5053, NL-5692 EB SON, Netherlands.
Tel: +31 499 462525. Telex: 59388 Gemco nl.
Fax: +31 499 462500.

NEW ENTRY

GEMCO stretcher carrier set (GMS 2058) configured for three stretcher casualties in a M113 APC **1997**

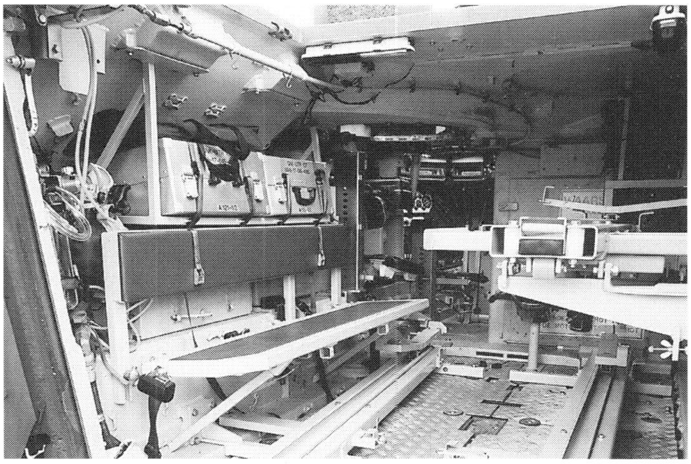

GEMCO stretcher carrier set (GMS 2058) configured for seated casualties **1997**

SINGAPORE

Tactical Ambulance Mark II

Description

This tactical ambulance body is built on a Land Rover Defender 110 (4 × 4) chassis. The ambulance body can accommodate a crew of three (driver, co-driver and one medical attendant) and either four stretcher casualties or six seated casualties.

In this ambulance the stretcher berths can be folded away to form personnel seating. A wash basin with water tap also folds away to become a bench rest. These features allow the vehicle to be used for general purposes as well as an ambulance.

The body has fresh and waste water tanks, an air extractor, a medical stores stowage area, four hook rails, recessed high/low lighting, a siren, a rotating beacon and ambulance signs.

Status

In service with the Singapore armed forces.

Manufacturer

Singapore Technologies Automotive Limited (A Singapore Technologies Company), 5 Portsdown Road, Singapore 139296.
Tel: +65 473 6311. Telex: 25755 SINENG RS.
Fax: +65 471 0662.

UPDATED

Tactical Ambulance Mark II

UNITED KINGDOM

Marshall SPV vehicle bodies

Description

Marshall SPV is the largest manufacturer of military vehicle bodies in the UK. The following are typical of the range of military products available.

Ambulances

The range of Marshall 4 × 4 ambulances has been developed over more than 25 years. Over 4,000 vehicles have been produced for both the UK Ministry of Defence and many overseas customers. The bodywork was designed to suit either the Land Rover Defender 110 or 127 4 × 4 chassis integral body or box body. Designs are available to meet customers' specific requirements.

Interior arrangements offer a capability of accepting up to four stretcher patients or two stretcher patients and four sitting patients, or eight sitting patients. An attendant's seat is fitted and locker space for storage of medical equipment is provided. A wide range of optional equipment can be offered, including air conditioning, oxygen systems, portable resuscitation equipment and emergency first aid equipment.

Land Rover conversions

A wide range of conversions and bodies has been supplied for the Land Rover series. The vehicle has been the base for ambulances (see above), signal vehicles, weapon platforms, radar stations, command

Ambulance built by Marshall SPV using Land Rover Defender 110 (4 × 4) ***1997***

posts, workshops, refuelling and missile support vehicles. A cargo-bodied version has been supplied through Land Rover to several overseas territories.

Body General Service/Cargo Drop Side 4 tonne
This was developed from 1967 for the Bedford MK (4 × 4) 4,000 kg truck and entered production the following year. Its FVRDE (MVEE) specification is 9722. The body is capable of being used as a flat platform, fitted with drop sides, as a troop carrier, or for the carriage of a distributed load of 4,000 kg The steel body is of welded construction, with the front bulkhead bolted to forward pillars mounted to the underframe. The hardwood floor is fitted directly to the cross member. It is protected by steel wearing strips which cover the joints. Rotating flush-fitting lashing rings mounted directly to the cross member are positioned to provide the most efficient lashing points for lashing down bulky loads. Additional heavy-duty lashing points are positioned along the main sides. The four drop sides and tailboards can be removed, and the rear and intermediate pillars unbolted to provide a flat platform. A tubular steel superstructure is fitted with a tailored tilt cover of cotton-polyester material. Secured to the underframes are two lockers and a carrier for two jerricans. Provision is made on the forward face of the front bulkhead for stowing a pickaxe head and handle, a shovel and a crowbar.

Body General Service 8 tonne
Development of this GS cargo body began during 1974. The body can be used as a flat platform or fitted with drop sides and a tailboard for carrying a distributed load of 8,000 kg. Typical loads can be up to six standard pallets, shelters up to 4.5 m long, loose stores, and ISO containers. The body is of welded steel with a hardwood floor, both designed with a view to mass production and providing the maximum protection against corrosion. Rotating flush-fitting lashing rings are secured directly to the body underframe and, with additional lashing points secured to the sides, are positioned to provide an efficient lashing pattern for securing all types of load. The conversion to a flat platform mode is easy and convenient. A cotton-polyester tilt cover for a tubular steel superstructure and tailboard is supplied. The overall length of the body is 6.637 m and it is 2.5 m wide. The load area measures 4.53 × 2.145 m.

Status
Production as required.

Manufacturer
Marshall SPV Limited, The Airport, Cambridge CB5 8RX, UK.
Tel: +44 1223 373045. Fax: +44 1223 373046.

UPDATED

Land Rover Defender 90 anti-tank conversion by Marshall SPV carrying 106 mm RCL ***1997***

Marshall SPV Land Rover Defender 110 covert patrol vehicle with retractable 0.50/12.7 mm M2 Browning machine gun ***1997***

Pilcher-Greene vehicles

Description
Pilcher-Greene as a long-standing company has developed and produced a wide range of vehicles based on all major manufacturers' chassis. These include emergency vehicles encompassing conventional 4 × 2 ambulances together with ambulances based on 4 × 4 chassis, with a range of light fire appliances and rapid intervention vehicles.

Included in the standard range are patient transfer vehicles and many specialised vehicles such as laboratories, clinics, radio command units and other bespoke vehicles are available tailored to clients' specific individual requirements.

The 4 × 4 range of vehicles utilises chassis from around the world and in particular the Land Rover Defender in all its forms. The vehicles are produced using a variety of composite construction techniques from steel, aluminium and GRP and all are corrosion protected and thermally insulated.

The ambulance can be supplied from the basic to full paramedic level and in addition to the medical equipment that is available, numerous non-medical

Pilcher-Greene 8303 Type 'S' interior 15 Land Rover ambulance with air conditioning

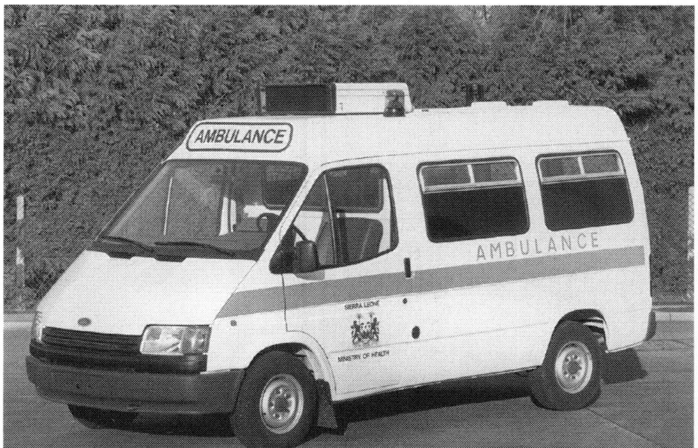

Pilcher-Greene Ford Transit 120 ambulance with air conditioning

options may be specified, from heating for arctic conditions, to air conditioning and rescue equipment.

The light fire appliances and particularly those on 4 × 4 chassis have application in airfield, off-road and forestry situations and where space or access is limited.

The accessible vehicles may be fitted with access ramps and passenger lifts or lowering suspensions dependant upon their suitability for the given application. Variations in layouts are available and the ability to change the function of the vehicle is also possible.

Status

In production. Pilcher-Greene vehicles are in service in South America, Africa, Middle East, Indian sub-continent, Asia and Eastern Europe, with Brunei, Qatar, Dubai, Bangladesh and the Jordanian armed forces as well as many other authorities throughout the world.

Manufacturer

Pilcher-Greene Limited, Consort Way, Burgess Hill, West Sussex RH15 9NA, UK.
Tel: +44 1444 235707. Fax: +44 1444 241730.

VERIFIED

Edbro Vehicle Bodies and Hydraulic Systems

Description

Approved to ISO 9001 accredited quality assurance standard, Edbro plc is involved in the design and manufacture of vehicle bodywork for all types of material handling to UK Ministry of Defence specifications. The manufacturing facility incorporates CAD/CAM and extensive use is made of FE stress analysis in body design. CNC machine tools, friction welding, shot blast, hard chrome plating and low bake ovens for IRR paint schemes feature in the production plant.

Typical products include flat, cargo, winch and crane bodies for the British Army's Leyland Trucks 4 × 4 4 tonne trucks. Edbro also completed an order for military dump trucks based on the IVECO-Ford 330-25 (6 × 6) chassis. These vehicles, intended for high-speed airfield runway repair work, have special features for low temperature operation, including exhaust heated bodywork.

Edbro has produced dump bodies and hoists for the British Army's Foden 8 × 4 Low Mobility tipper trucks and Bedford TM 4-4 tippers. The flat cargo and winch bodies fitted to the Bedford TM 6-6 14 tonne vehicles were manufactured by Edbro.

Types of bodywork include flat platform, truck, cargo, winch, crane, tipper, dump truck and dust cart. For over 75 years, since 1916 when the company was founded, Edbro hydraulic systems have featured in automotive applications throughout the world.

Status

In production. In service with the British Army.

Manufacturer

Edbro plc, Nelson Street, Bolton, Lancashire BL3 2JJ, UK.
Tel: +44 1204 528888. Telex: 63196 edbroh g.
Fax: +44 1204 531957.

VERIFIED

IVECO-Ford 330-25 (6 × 6) chassis with Edbro DK14 double-acting hoist and 9 m^3 steel rock dump body

Edbro cargo body with canvas cover fitted to Leyland Trucks (4 × 4) 4 tonne truck

Reynolds Boughton general service cargo body

Description

The Reynolds Boughton general service cargo body was designed to carry general cargo and is mounted on the Renault G300.18D 4 × 4 8000 kg truck. The body is constructed from welded steel and uses an underframe of pressed steel, longitudinal members and cross-bearers. The sides, headboard and tailgate are constructed from aluminium. The floor is hardwood, treated with a preservative after machining, with wearing irons over all the joints.

Optional equipment includes a mid-mounted Boughton H7500 hydraulic winch, enabling self-recovery and recovery in both a forward and rearward direction. A hydraulic loader may be fitted between the cab and body to permit the self-loading and unloading of cargo.

Specifications

Interior length: 4.3 m
Interior width: 2.33 m
Height of sides: 0.61 m

Status

In production. Similar bodies are in service with many overseas armed forces, especially in Asia and the Middle East.

Manufacturer

Reynolds Boughton Limited, Vehicle Division, Bell Lane, Amersham, Buckinghamshire HP6 6PE, UK.
Tel: +44 1494 764411. Telex: 83132.
Fax: +44 1494 765218.

VERIFIED

Reynolds Boughton general service cargo body
1996

Reynolds Boughton general service tipper body

Description

The Reynolds Boughton general service tipper body is of welded construction and is manufactured from high-tensile steel. Both under-body and front-end tipping gears are available and bodies are produced to suit a wide variety of general service cargo vehicles.

Status

In production. In service with a number of overseas armed forces, especially in Asia.

Manufacturer

Reynolds Boughton Limited, Engineering Division, Bell Lane, Amersham, Buckinghamshire HP6 6PE, UK.
Tel: +44 1494 764411. Telex: 83132.
Fax: +44 1494 765218.

VERIFIED

Reynolds Boughton general service tipper body

UNITED STATES OF AMERICA

Marion Composites Multipurpose Maintenance and Shop Van (MMSV)

Description

Marion Composites' Multipurpose Maintenance and Shop Van (MMSV) was designed for field maintenance personnel. The MMSV, mounted on the US Army's M1079 Light Medium Tactical Vehicle (LMTV) 2½ ton truck, is fabricated using an aluminium skin, foam and beam core sandwich panels. Standard equipment includes power entry and distribution, lights, blackout provisions, double rear doors, windows, removable steps, and provisions for an optional heater and air conditioner. The MMSV is supplied with a base frame for quick mounting and dismounting from the LMTV. Interfaces can be developed to adapt the MMSV for other applications.

Marion Composites Multipurpose Maintenance and Shop Van (MMSV) on M1079 Light Medium Tactical Vehicle (LMTV)
1997

Specifications

Weight:
(MMSV only) 1,297 kg
(with mounting frame) 1,887 kg
Dimensions:
(exterior) 3.78 × 2.4 × 2.17 m
(interior) 3.67 × 2.29 × 2.02 m
Volume: 17 m^3
Floor area: 8.4 m^2
Overall height on prime mover: 3.51 m

Status

In production.

Manufacturer

Marion Composites, 150 Johnston Road, Marion, Virginia 24354, USA.
Tel: +1 540 783 3121. Fax: +1 540 783 9665.

NEW ENTRY

Marion Composites cargo bed covers

Description

Marion Composites has developed a family of general purpose truck and trailer covers that provide environmental protection for personnel and cargo, plus security, for mission equipment. The bonded aluminium skin/honeycomb enclosures are ground, rail and air transportable when mounted on the HMMWV, M105A2 cargo trailer, and M35A2 and LMTV 2½ ton trucks. Standard equipment includes mounting provisions, hoist fittings, a personnel access door, and air vents. Enhancement kits are optional for electrical and signal entry panels, antenna mountings, internal lighting, windows, and a cab access port for the HMMWV cargo bed cover.

Specifications

HMMWV
Weight: 136 kg
Dimensions:
(exterior) 2.07 × 2.14 × 1.62 m
(interior) 2.02 × 2.09 × 1.6 m
Volume: 6.2 m^3
Floor area: 4.2 m^2
Overall height on prime mover: 2.59 m

Marion Composites cargo cover on M35A2 2½ ton truck ***1997***

Marion Composites cargo cover on HMMWV ***1997***

Marion Composites cargo cover on M105A2 trailer ***1997***

M105A2 trailer
Weight: 277 kg
Dimensions:
(exterior) 2.78 × 1.93 × 1.61 m
(interior) 2.73 × 1.88 × 1.59 m
Volume: 7.6 m³
Floor area: 5.1 m²
Overall height on prime mover: 2.54 m

2½ ton trucks
Weight: 136 kg
Dimensions:
(exterior, operational) 3.4 × 2.21 × 1.91 m
(exterior, collapsed) 3.4 × 2.21 × 0.98 m
(interior, operational) 3.28 × 2.08 × 1.87 m
(interior, collapsed) 3.28 × 2.08 × 0.97 m
Volume:
(operational) 12.4 m³
(collapsed) 6.2 m³
Floor area: 6.8 m²
Overall height on prime mover:
(M35A2, operational) 3.28 m
(M35A2, collapsed) 2.26 m
(LMTV, operational) 3.505 m
(LMTV, collapsed) 2.591 m

Status
In production.

Manufacturer
Marion Composites, 150 Johnston Road, Marion, Virginia 24354, USA.
Tel: +1 540 783 3121. Fax: +1 540 783 9665.

NEW ENTRY

SEI Special Workshop Bodies

Description
Over the last 20 years Systems and Electronics Inc (SEI, formerly Southwest Mobile Systems Corporation) has produced some 3,000 special shop bodies of several types and produces five workshop body 'sets' as follows:

Set 3 - Contact maintenance shop
This is a self-propelled, self-contained, maintenance workshop for wheeled and tracked vehicle maintenance in the field. Early production of this set was mounted on a Dodge chassis with later production (105 units for the US Marine Corps) mounted on the M1031 CUCV, on which the usual onboard generator is replaced by a power take-off driven 12 kW alternator from the vehicle engine. Equipment carried in the body includes the following: a 200 A welder, an air compressor, electric motor-driven portable drills, a 3 ton chain hoist, a hydraulic hand jack, ignition timing light, a multimeter, a tachometer, a bench and pipe vice, six electrical test engine adaptors, a comprehensive tool kit and chest, soldering and welding kits, support materials, equipment and supplies, floodlights and masts.

On the CUCV the Set 3 (with vehicle) is 5.41 m long, 2.06 m wide and 1.94 m high when loaded. Weight is 2,381 kg.

A lightweight body has been prototyped for fitting to the HMMWV M998 series. A US Army Concept Evaluation Program (CEP) was sponsored by the Ordnance Corps from Aberdeen, Maryland.

Set 4 - Electrical repair shop
This is an electrical repair shop mounted in an M447C Van Semi-trailer ordered by the US Army (158 units). The repair shop is carried in a four-wheel semi-trailer with the steel and aluminium body fitted with benches, diagnostic equipment, tools, five load banks for electrical power testing of generators, a hydraulic test stand, electrical and other tools and the necessary support supplies. Major equipment carried in the repair shop includes the following: a power supply amplifier, a battery charger, frequency test meters, a petrol engine generating set, a grinding machine, a 1 ton chain hoist, various oscilloscopes and multimeters, a rotary actuator test stand, a variable power transformer, a 2 ton portable hoist trestle, heaters and an air conditioner. A comprehensive tool kit, containers, measurement devices and various support materials and supplies are all carried inside the repair shop body.

The Set 4 semi-trailer is 8.204 m long, 2.4 m wide and 3.343 m high when loaded. Weight is 10,433 kg.

Set 11 - General purpose machine shop
This is a semi-trailer-mounted general purpose machine shop intended for use in the repair of heavy mechanical equipment such as construction equipment; 128 units were produced for the US Army, US Marine Corps and US Navy Construction Battalions. The unit houses a 300 A welder, an engine lathe with milling, drilling, grinding and slotting attachments, and over 400 tools, test equipments and accessories, plus supplies. The shop contains its own diesel engine-driven 25 kW alternator power generating system, fuel supply and space heater. Large upper side doors provide easy access and open with hydraulic assists. Canvas covers, secured from the raised side and rear doors, can be used to form a spacious heated field repair centre.

Length of the complete semi-trailer is 8.928 m, width 2.502 m and height 3.188 m. Weight is 13,675 kg.

Set 12 - Organisational repair shop
This is a truck-mounted organisational level repair set equipped for the inspection, testing, maintenance and repair of vehicles, automotive and other equipment under on- or off-road conditions. The M809 (6 × 6) 5 ton cargo truck was originally used as the chassis for this set while later contracts (240 units for the US Army) utilised the M944 version of the M939 5 ton cargo truck.

The unit contains over 350 separate types of tools, equipment, accessories and supplies. Canvas covers, secured from the raised side and rear doors, can be used to form a spacious heated repair centre. The set's power tools, including a lathe and 300 A welding kit, can be powered either by an external source or from the truck engine through a power take-off driven alternator system. Other major items of equipment

Set 4 electrical repair shop mounted in an M447C van semi-trailer

Set 12 organisational repair shop in travelling configuration on M944 chassis

carried include a space heater with blower, an electric motor-driven air compressor, floodlights and masts, and other lighting accessories. There is also a small reference library.

Overall length of the set on an M944 chassis is 9.042 m, width is 2.5 m and height 3.175 m. Weight complete is 14,800 kg.

Clothing Repair Shop

This mobile clothing repair shop is used by the US Army (36 units). The shop is carried in a two-wheel M105A2 trailer and is equipped with six clothing machines, one button machine and one darner machine as well as the associated accessories. The set was designed for on- and off-road transport of the equipment, and to allow field set-up of the equipment. Power is provided by a 3 kW power source stored in the shop. Length of the complete trailer is 4.3 m, width 2.11 m and height 2.46 m. Weight is 3,289 kg.

Status

In service with the US armed forces.

Manufacturer

Systems and Elecronics Inc, 201 Evans Lane, PO Box 5639, St Louis, Missouri 63121-1126, USA.
Tel: +1 314 553 4900. Fax: +1 314 553 4950.

VERIFIED

OTHER EQUIPMENT

GERMANY

Kärcher hot air generator FB 60E

Description

The hot air generator FB 60E was designed to meet the requirements of the German armed forces and is also employed by the medical corps. It is also used for the heating of depots and storage shelters. Other uses include the heating of diesel engines in low temperatures and the preflight heating of helicopters and aircraft on the ground.

Kärcher hot air generator FB 60E

With the help of a special medical-use extension set clean air and constant temperatures can be achieved in hospital tents without creating turbulence. This is achieved by the use of specially treated and perforated air outlets and distribution ductings. The FB 60E is equipped with a remote thermostat with a built-in sensor accurate to within 0.2°C. In Summer, the unit can be used to cool tents, shelters, aircraft cockpits and cabins.

All the main components of the FB 60E are installed in a torsionally stiff tubular aluminium frame with pick-up points for fork-lifts or air transport. It can be easily moved by one person using the integral wheels and handles.

The main fan and fuel pump are driven by a central 220 V motor. Fresh air is sucked in via a radial fan and heated by the burner and heat exchanger. The rated thermal output is 60 kW by a rated air flow of 3,000 m^3. Fuel consumption is 7.5 litres/h.

Optimised combustion is achieved by a specially shaped burner sphere. This and the heat exchanger are made of high-temperature resistant material, both of which are gas-tight welded to ensure that no combustion gases are mixed with the heated fresh air.

Due to its two integrated fuel preheaters, the FB 60E is able to work in temperatures down to −30°C. The fuel used can be diesel, diesel-petrol mixtures or kerosene.

Specifications

Weight: 200 kg
Length: 1.72 m
Width: 0.74 m
Height: 0.91 m

Status

In production. In service with the Australian, German and US armed forces.

Manufacturer

Alfred Kärcher GmbH and Co, Alfred-Kärcher-Strasse 28-40, D-71364 Winnenden, Germany.
Tel: +49 7195 142797. Telex: 724432.
Fax: +49 7195 142720.

VERIFIED

INTERNATIONAL

Munters dry air method

Description

The Munters dry air method was first used by the Swedish Army during the late 1950s. Since then its use has become widespread for the protection of equipment against depredations caused by moisture in the atmosphere.

The Munters dry air method uses dehumidifiers that will operate efficiently in any climatic area of the world. The essence of the system is to envelop weapons systems, subsystems, system components or logistical supplies in dry air. This can be done by placing the systems or supplies in a dehumidified building or room or by blowing a current of dry air into the area, even if the systems involved are still operational. Corrosion is thus halted and supplies are protected against mould and fungi.

Status

In production for over 30 years. In service with all NATO countries plus Australia, Austria, Finland, Israel, Japan, Sweden and Switzerland.

Defence Contact Point

Munters Defence Business Co-ordination, PO Box 229, NL-2400 AE Alphen a/d Rijn, Netherlands.
Tel: +31 1720 33231. Fax: +31 1720 42960.

VERIFIED

The Munters dry air method in use to protect British Army Tracked Rapier fire units

The Munters dry air method in use with a Royal Swedish Air Force AJ-37 Viggen

ISRAEL

Achidatex protective equipment covers

Description

The Achidatex dry storage system maintains a 'climatic bubble' around a stored item, isolating it from the effects of humidity, rain, frost, sand erosion and the chemical influence of bacteria, fungi, insects and acid rain. Being flame retardant Achidatex covers also limit the danger from fire. Stored items are thus protected against rust and corrosion, thereby eliminating the need for storage sheds and reducing equipment life-cycle costs.

Achidatex coverings are made from PVC, urethane or other material suitable for local climatic conditions. A dehumidifier maintains dry air inside the cover at corrosion-prevention levels by absorbing and expelling excess water vapours through a regulated cycle of operations. Units can be connected to a standard electrical supply or a small generator.

Achidatex produces dry storage systems to protect outdoor equipment such as aircraft and helicopters of all sizes. In addition it provides common shell dry storage systems for vehicles, tanks and artillery as well as standard dry storage tunnels for equipment such as engines and spare parts. Storage systems can be placed in large warehouses to provide extra protection and humidity control for sensitive systems and parts.

The Achidatex Common Structure Dry Storage Cell provides cover for outsized equipments such as armoured vehicles. They may be used for temporary storage or as cover for extended periods.

In addition to the above Achidatex also produces multipurpose dry storage shelters, hanging liners and special containers.

Status

In production. In service with the Israeli armed forces, the German Army and the US Army.

Manufacturer

Achidatex, PO Box 2156, IL-52151 Ramat Gan, Israel.
Tel: +972 3 647 6946. Fax: +972 3 497307.

VERIFIED

Driving an M109AL 155 mm self-propelled howitzer into an Achidatex common structure dry storage cell ***1996***

Achidatex covers protecting aircraft engines

Chemoplast dry storage systems

Description

The Chemoplast dry storage systems use a high frequency welded cover system in both one-piece cover and tunnel forms.

The one-piece envelope covers are tailor-made to suit specific requirements and can be used in extremes of climate. With these covers the seams generally overlap 20 mm and are formed by continuous heat welding for maximum strength. Extruded closures ensure perfect seals and maximum separation strength. The material has very low water permeability and provides protection against corrosion. Dehumidification by chemical or mechanical means is incorporated. Also provided around high stress areas are padded blankets and reinforced PVC.

The Chemoplast dry storage tunnel system is intended for the bulk storage of high-value products for extended periods. On the tunnel the roof and sides form a single unit supported on a plastic frame system of assorted GRP elements which can be easily assembled and dismantled. An extruded all-around closure ensures top and bottom sealing. The tunnel measures 15 × 2.8 × 2 m and provides a covered volume of 80 m^3, although other sizes can be produced to meet customer requirements.

Status

In production. In service with the Israeli armed forces.

Manufacturer

Chemoplast Limited, PO Box 2110, IL-18301 Afula, Israel.
Tel: +972 6 523044/5. Telex: 451881 attn chemoplast aran il.
Fax: +972 6 597422.

VERIFIED

Chemoplast dry storage covers in use

An assembled Chemoplast dry storage tunnel

ITALY

Pirelli protective covers

Description

Pirelli produces a rubber-coated fabric which is widely used for the protection of equipment in storage and in transit. This fabric can be produced in a wide range of shapes and sizes and has a weight of between 650 and 800 g/m^2. The fabric, usually coloured olive brown on one side and black on the other, has a minimum tensile strength of 306 kg for every 50 mm and is fire-resistant. It is also resistant to the effects of extremes of temperature and protection is provided, not only against atmospheric agents, but particularly against NBC attacks and infrared photographic observations. Contaminated covers can be reclaimed.

One of the various articles produced using this rubber-coated fabric, is a cover for use as a weather-defeating barrier on equipment in transit or in storage, but used specifically for Leopard MBTs. It is 8 m long and 4.5 m wide and weighs approximately 25 kg. When packed this cover occupies about 0.15 m^3 and in use is lashed around the equipment to be protected.

Status

In service with the Italian armed forces.

Manufacturer

MOLDIP SpA - Pirelli Group, Via Milano 8, I-20038 Seregno, Italy.
Tel: +39 362 237711. Telex: 330191 pirese i.
Fax: +39 362 220412.

VERIFIED

UNITED KINGDOM

Bog-Cogs

Description
Bog-Cogs are a bolt-on device that can be added to vehicle wheels to improve performance when crossing soft terrain. When fitted to a vehicle they increase flotation and reduce the likelihood of sinking in soft ground, and increase traction and stability.

Bog-Cogs are moulded in high-performance resilient plastic and are attached to vehicle wheels using a special wheel nut and retaining assembly system. Once in place, the hollow units add to the width of wheel and the serrated edges provide extra traction, flotation and stability as the tyres sink into soft ground. There is no need to remove the Bog-Cogs once soft conditions have been left behind and a fitted vehicle can continue to travel at speeds of up to 64 km/h.

On each vehicle wheel a Bog-Cog retaining assembly is fitted onto permanently fitted special wheel nuts and retained by linch pins. The Bog-Cogs are then offered

Cutaway example of Bog-Cog showing method of attachment

up to the retaining assembly and a single nut holds them in place. Four Bog-Cogs can be fitted in 5 minutes. A complete kit for a Land Rover-type vehicle consists of four Bog-Cogs (one for each wheel), four flange interfaces with seals, four retaining assemblies with linch pins, 12 special wheel nuts and one socket spanner and bar. The shipping weight of a complete set for a Toyota Land Cruiser is 70 kg and a set can be carried with the units nested together to reduce storage space.

Bog-Cogs can be fitted to any type of wheeled vehicle and there is no limit, large or small, to the sizes that can be produced. As well as being used on all wheel drive or all-terrain vehicles, Bog-Cogs can be fitted to non-driven axles, trailers and artillery pieces.

Status
Available.

Manufacturer
Wessex (UK) plc, Unit 4, Porsham Close, Belliver Industrial Estate, Plymouth, Devon PL6 7DB, UK.
Tel: +44 1752 766833. Fax: +44 1752 766811.

VERIFIED

Airflex system

Description
Airflex environmental shelters and containers provide efficient and cost-effective long-term protection for military vehicles, weapon systems or support hardware of any configuration, even in the most hostile environment. Equipment protected by the Airflex system may be left in battle-ready condition in forward areas. The system is easy to install, requires no permanent electricity, such as dehumidifiers, and is reusable indefinitely.

Status
In production. In service with armed forces worldwide.

Manufacturer
Air-Log Limited, North Lane, Aldershot, Hampshire GU12 4QH, UK.
Tel: +44 1252 24411. Fax: +44 1252 333790.

VERIFIED

Scorpion reconnaissance vehicle with Airflex kit in crate

Airflex flexible barrier shelter on a Scorpion reconnaissance vehicle ***1996***

EPS Driclad protection and storage system

Description
The EPS Driclad system is the result of applied research into the causes of corrosion and deterioration of equipment during storage and in transit. It is based on the enclosure of stores or equipment in a dry and clean atmosphere within a flexible plastic cover.

The EPS Driclad system was accepted into British Army service in 1965 after extensive trials in Europe and Asia lasting over two years. The system is now in service with many countries all over the world.

The reusable covers are an effective barrier to water vapour in extreme conditions of climate, temperature and humidity. They also provide protection against damage by chemical gases, salt-laden air, sand, dust, fungi and insects. A system can be designed by EPS Driclad for any shape.

The system comprises a dust- and moisture-proof cover with a patented EPS 'Drilok' closure, mechanical dehumidification or a desiccant and a humidity or metering device. Equipment protected by EPS Driclad requires no special building or packing material.

Maintenance requirements are greatly reduced and, in some instances, maintenance procedures become redundant. At certain locations used by the British Army only two technicians were required to keep up to 80 tanks, protected by EPS Driclad, in a state of permanent operational readiness.

Typical uses of the EPS Driclad system include the protection and preservation of tanks, artillery, helicopters and missiles. It is also used by the Royal Navy in the laying-up of ships and the temporary shore-based protection of ships' stores. Many other items of equipment, large and small, are currently being given short-term and long-term protection against environmental attack.

Status
In production. In service with approximately 40 countries.

Manufacturer
EPS Logistics Technology Limited, Staplehurst Road, Sittingbourne, Kent ME10 1XS, UK.
Tel: +44 1795 424433. Fax: +44 1795 436035.

VERIFIED

EPS Driclad covers being used for long-term protection of tanks

MOWAG Piranha 8 × 8 APC protected in long term storage by the EPS Driclad system
1996

Armour Barpack Limited protective packaging

Description

One of the base materials used in the Armour Barpack range of military engineered packaging is the Texikoon system. Texikoon covers and shelters protect components and equipment, including complete tanks and aircraft, against chemical gases, sand, dust, salt-laden air, fungi and ultraviolet light. They are designed to allow easy access to the stored equipment and enable simple resealing to restore the controlled environment. They are repeatedly reusable and cost-effective. Texikoon shelters and covers can be manufactured to meet various defence specifications.

An addition to the product range is Storlina, a system which utilises Texikoon with dehumidifiers for the conversion of existing buildings into environmentally controlled warehouses. Storlina is in use by the UK Ministry of Defence.

Status

In production. In service with the British armed forces.

Manufacturer

Armour Barpack Limited, Alma Park, Grantham, Lincolnshire NG31 9SN, UK.
Tel: +44 1476 62323. Telex: 37551 barpak g.
Fax: +44 1476 64858.

VERIFIED

Texikoon covering Sea King helicopter

Pipe fascines

Description

Pipe fascines are a modern development of an ancient combat engineering technique used to provide emergency crossings of gap-type obstacles and were developed to their current standard by the then Royal Armament Research and Development Establishment, Christchurch (RARDE(C)).

The pipe fascine comprises bundles of lightweight plastic pipes chained together in such a way that they take up the shape of the gap, trench or hole into which they are dropped. Vehicles can then cross the gap involved. In contrast to the earlier wood fascines used until the introduction of the pipe fascine, the modern unit is much lighter, more durable, does not create dams in wet gaps and can be recovered and reused many times. It is also easier to store for long periods, easier to handle and can withstand heavier loads.

The pipe fascine used with heavy armoured vehicles is the MAXI. This weighs 2,500 kg and can fill a gap 5 m wide for tracked vehicles; more can be added for wider gaps. Vehicles such as the British Army's Chieftain AVRE can carry up to three pipe fascines and launch two simultaneously, but many other vehicles, such as bulldozers or dump trucks, can carry and launch them.

There is also a MINI pipe fascine which weighs 210 kg. These can be carried on the sides of APCs such as the M113, and can be dropped to fill ditches or small gaps for wheeled vehicles, or used in conjunction with MAXI pipe fascines.

MAXI pipe facines stacked on Chieftain AVRE ***1997***

Status

In production. In service with the Canadian, Dutch British and US armies.

Manufacturer

PD Technical Mouldings plc, Rotation House, 20 Mayday Road, Thornton Heath, Surrey CR7 7HL, UK.
Tel: +44 181 689 4336. Fax: +44 181 683 2491.

UPDATED

Inventory

For ease of reference, the inventory is divided up into five key sections: armoured engineer vehicles, recovery vehicles (tracked and wheeled), mechanised bridges (tracked and wheeled), light vehicles and trucks.

In addition to the vehicle type, the following additional information is provided where possible: model number, cross-country payload, configuration, number delivered and the approximate delivery date. It should be noted that the total number normally includes many variants.

Abbreviations used in this section:

ABFS	amphibious bridge and ferry system
AEV	armoured engineer vehicle
ALV	air-portable light vehicle
ARV	armoured recovery vehicle
AVLB	armoured vehicle-launched bridge
CEV	combat engineer vehicle
ES	entering service
LAV	Light Armored Vehicle
MB	Mercedes-Benz
PO	Phasing Out of Service
RV	recovery vehicle
RVI	Renault Véhicules Industriels
SDP	Steyr-Daimler-Puch
W	wrecker

AFGHANISTAN

Recovery vehicles
T-54/T-55 ARV

Mechanised bridges
MTU-20 AVLB
MTU AVLB

Light vehicles
UAZ-469B 600 kg (4 × 4)

Trucks
ZIL-131 3,500 kg (6 × 6)
Ural-375D 4,000 kg (6 × 6)

ALBANIA

Recovery vehicles
T-54/T-55 ARV

Light vehicles
UAZ-469B 600 kg (4 × 4)

Trucks
ZIL-131 3,500 kg (6 × 6)

ALGERIA

Recovery vehicles
T-54/T-55 ARV (?)

Light vehicles
UAZ-469B 600 kg (4 × 4)
MB 750 kg (4 × 4) (90)

Trucks
RVI TRM 1200 1,200 kg (4 × 4)
Berliet GBC 8 KT 4,000 kg (6 × 6)
MB Unimog (4 × 4)
MAN 6000/10,000 kg (280)
RVI TRM 9000 9,000 kg (6 × 6) (500)

ANGOLA

Recovery vehicles
T-54/T-55

Light vehicles
ENGESA EE-12 (4 × 4)
Land Rover (4 × 4)
UAZ-469B 600 kg (4 × 4)
UMM (4 × 4) light vehicle

Trucks
ENGESA EE-15 1,500 kg (4 × 4)
ENGESA EE-25 2,500 kg (6 × 6)
ENGESA EE-50 5,000 kg (6 × 6) (255)
MB Unimog (4 × 4)
GAZ-66 2,000 kg (4 × 4)
RVI TRM 2000 2,000 kg (4 × 4) (85)
ZIL-131 3,500 kg (6 × 6)
Ural-375D 4,000 kg (6 × 6)

ARGENTINA

Recovery vehicles
SDP 4KH 7FA SB 20 Greif ARV (10)
VCRT ARV (trials)
AAVR7 ARV (1) (Marines)

Mechanised bridges
AMX-13 AVLB

Light vehicles
LOHR FL 500 (4 × 4)
MB 750 kg (4 × 4) (1,200)

Trucks
MB (4 × 4)

AUSTRALIA

Recovery vehicles
Leopard 1 ARV (8)
LAV recovery (8 × 8)
ASLAV-F (fitters) (10)
M806A1 light ARV
Mack MC3 heavy wrecker (6 × 6)
M543/M543A1 W (6 × 6)
M816 W (6 × 6)
UNIMOG 2450l (6 × 6) (55)
(1997-98)

Mechanised bridges
Biber AVLB (5)

Light vehicles
Land Rover Defender 110 (4 × 4)
Land Rover LWB (4 × 4)
Land Rover 1,000 kg (4 × 4) (PO)

Trucks
Land Rover 110 heavy duty (6 × 6) (400) (1989-91)
Ford O913 (4 × 2) (372)
Mack Model RM6866RS 8,000 kg (6 × 6) (906)
MB UL 1700 L 4,000 kg (4 × 4) (1,295) (1982-87)
Volvo NL10 (4 × 4)

AUSTRIA

Armoured engineer vehicles
SDP 4KH 7FA AVE (18)

Recovery vehicles
SDP 4KH 7FA SB 20 Greif ARV (50)
M60/M108 W (6 × 6)

Light vehicles
SDP 700 AP (4 × 4)
VW 181 400 kg (4 × 2)
MB 750 kg (4 × 4) (1,200)

Trucks
SDP Pinzgauer (4 × 4 and 6 × 6)
SDP 680 M 4,500 kg (4 × 4)
SDP 680 M3 3,500 kg (6 × 6)
Steyr 480 4,000 kg (4 × 2)
SDP 12 M 18 5,000 kg (4 × 4) (2,000+)
SDP 12 M 21 5,000 kg (4 × 4)
G & S LAVI–9F/2H 6,000 kg (4 × 4)
G & S ZAVT–9F/1 6,000 kg (4 × 4)
G & S ZA-200/1 A 10,000 kg (6 × 6)
OAF 20.320 10,000 kg (6 × 6) (410)
Berliet GBC 8 KT 4,000 kg (6 × 6)

BAHRAIN

Recovery vehicles
M88A1 ARV (4)

Light vehicles
Land Rover (4 × 4)

Trucks
HMMWV (4 × 4) (277)
Bedford TM 4-4 8,000 kg (4 × 4)
Oshkosh HEMTT 9,979 kg (8 × 8)

BANGLADESH

Recovery vehicles
Type 653 ARV
T-54/T-55 ARV

Light vehicles
Land Rover (4 × 4)
Mahindra (4 × 4)

Trucks
Bedford MK 4,000 kg (4 × 4)

BELGIUM

Armoured engineer vehicles
Leopard 1 AEV (6)

Recovery vehicles
Leopard 1 ARV (28)
M113A1-B-REC ARV
Samson ARV
Berliet TBU 15 CLD W (6 × 6)

Mechanised bridges
Mobile Assault Bridge/Ferry (4 × 4)
M48 AVLB (14)

Light vehicles
Iltis 500 kg (4 × 4) (2,673)
Land Rover ALV 564 kg (4 × 4)
Land Rover Defender 110 (4 × 4)

Trucks
Berliet GBU 15 6,000 kg (6 × 6)
MB Unimog (4 × 4)
Bedford MK 4,000 kg (4 × 4)
MAN 11.136 HA 5,000 kg (4 × 4) (3,000)
MAN 630 5,000 kg (4 × 4) (PO)
Volvo N10/N12 (6 × 6) (1,228)
MAN (8 × 8) (82)

BELIZE

Light vehicles
Land Rover (4 × 4)

Trucks
Bedford MK 4,000 kg (4 × 4)

BENIN

Trucks
ACMAT (4 × 4)

BOLIVIA

Recovery vehicles
SDP 4KH 7FA SB 20 Greif ARV (2)
M578 ARV

Trucks
HMMWV (4 × 4) (30)
ENGESA EE-25 2,500 kg (6 × 6)
M35/M44 2,268 kg (6 × 6)

BOTSWANA

Light vehicles
Land Rover (4 × 4)

Trucks
TJ 10,000 kg (82) (1990-91)

BRAZIL

Recovery vehicles
M548 ARV
AAVR7 ARV (2) (Marines)

Bridging systems
XLP-10 AVLB

Light vehicles
X-12 Gurgel series (4 × 2)
ENGESA EE-34 (4 × 4)

Trucks
ENGESA EE-15 1,500 kg (4 × 4)
ENGESA EE-25 2,500 kg (6 × 6)
ENGESA EE-50 5,000 kg (6 × 6)
TEREX UAI M1-50 5,000 kg (6 × 6)
MB (4 × 4 and 6 × 6)
M34/M44 2,268 kg (6 × 6)
Scania T113 HK (6 × 6)
Volvo NL10 (4 × 4)

BRUNEI

Recovery vehicles
Samson ARV (1)
Reynolds Boughton RV (4 × 4) (6+)

Light vehicles
Land Rover ALV 564 kg (4 × 4)

Trucks
Bedford MK 4,000 kg (4 × 4)
Leyland 4,000 kg (4 × 4) (41) (1995-96)

BULGARIA

Recovery vehicles
T-54/T-55 ARV
MTP-1 ARV

Mechanised bridges
BLG-67 AVLB
TMM (6 × 6)
TMM modified (6 × 6)

Light vehicles
UAZ-469B 600 kg (4 × 4)

Trucks
ZIL-157 2,500 kg (6 × 6)
ZIL-131 2,500 kg (6 × 6)

BURKINA FASO

Light vehicles
Auverland/SAMO (4 × 4)

Trucks
ACMAT (4 × 4)
Pegaso 3045 3,000 kg (4 × 4) (33)

BURMA

Light vehicles
Mitsubishi (4 × 4)

BURUNDI

Light vehicles
Auverland/SAMO (4 × 4)
Citröen A FAF (4 × 4)

CAMBODIA

Recovery vehicles
T-54/T-55 ARV

Light vehicles
UAZ-469B 600 kg (4 × 4)

Trucks
ZIL-157 2,500 kg (6 × 6)
ZIL-151 2,500 kg (6 × 6)

CAMEROON

Light vehicles
Iltis 500 kg (4 × 4)
Auverland/SAMO (4 × 4)

Trucks
ACMAT (4 × 4)

CANADA

Armoured engineer vehicles
Leopard AEV (8)
M113 Engineering Specially Equipped Vehicle (70)

Recovery vehicles
Leopard 1 ARV Taurus (8)
M578 ARV
Husky repair vehicle (6 × 6) (27)
Bison (8 × 8) MRTV (16)
MAN (6 × 6)
Percheron RV (6 × 6) (124)

Mechanised bridges
Biber AVLB (9)

Light vehicles
Iltis 500 kg (4 × 4) (2,500) (1984-85)

Trucks
Chevrolet 1,250 kg (4 × 4) (3,484)
Bombardier 2,500 kg (6 × 6) (2,765)
Steyr 8,000 kg (4 × 4)
MAN 10,000 kg (8 × 8) (42)
Percheron 10,000 kg (6 × 6) (1,200) (1988-91)
IVECO 40-10 WM 1,500 kg (4 × 4) (2,250)

CAPE VERDE

Light vehicles
UMM (4 × 4) light vehicle

CENTRAL AFRICAN REPUBLIC

Light vehicles
Auverland/SAMO (4 × 4)

CHAD

Light vehicles
Beijing BJ-212/212A (4 × 4)
Auverland/SAMO (4 × 4)

Trucks
HMMWV (4 × 4) (20)
ACMAT (4 × 4)
M35/M44 2,268 kg (6 × 6)

CHILE

Recovery vehicles
AMX-13 ARV (?)

Trucks
ENGESA EE-15 1,500 kg (4 × 4)
ENGESA EE-25 2,500 kg (6 × 6)
MB Unimog (4 × 4)
Pegaso 3045 3,000 kg (4 × 4)

CHINA, PEOPLE'S REPUBLIC

Recovery vehicles
Type 653/653A ARV
T-54/T-55 ARV

Mechanised bridges
Type 84 AVLB
TMM (6 × 6)
KMM (6 × 6)
Type 84A heavy mechanised bridge

Light vehicles
Beijing BJ-212/212A (4 × 4)

Trucks
NJ-230 (4 × 4)
Berliet GBC 8 KT 4,000 kg (6 × 6)
Berliet GBU 15 6,000 kg (6 × 6)
SDP 1291.280.4 × 4 M 6,000 kg (4 × 4)
SDP 1491.6 × 6 M 107,000 kg (6 × 6)
CA-30 2,500 kg (6 × 6)
Dong Feng EQ2080E4DY 2,500 kg (6 × 6)
Dong Feng EQ2100E 3,500 kg (6 × 6)
Jiefang CA-10 3,540 kg (4 × 2)
Jiefang CA-141 5,000 kg (4 × 2)
Jiefang CA-1091 5,000 kg (4 × 4)
EQD142 6,400 kg (4 × 2)
Hongyan CQ 261 8,250 kg (6 × 6)
ZIL-157 2,500 kg (6 × 6)
ZIL-151 2,500 kg (6 × 6)

COLOMBIA

Trucks
HMMWV (4 × 4) (86)
ENGESA EE-15 1,500 kg (4 × 4)
ENGESA EE-25 2,500 kg (6 × 6)
CUCV (4 × 4)

RUSSIAN FEDERATION AND ASSOCIATED STATES (CIS)

Armoured engineer vehicles
IMR-2 CEV (T-72 chassis)
IMR CEV (T-55 chassis)
BAT-2 CEV
IRM engineer reconnaissance vehicle
MT-LB engineer vehicle

Recovery vehicles
BREM-1 ARV (T-72 chassis)
M1977 ARV
T-54/T-55 ARV
BTR-50PK (B) amphibious ARV
BMP-1 light ARV
MTP technical support vehicle (BTR-50P)
MTP-LB technical support vehicle
Ural-375D RV (6 × 6)
Ural-4320 RV (6 × 6)
KET-L RV (6 × 6)

Mechanised bridges
MTU-72 AVLB
MT-55 AVLB
MTU-20 AVLB
MTU AVLB
PMM-2 ABFS

Light vehicles
LuAZ-967M (4 × 4)
UAZ-469B 600 kg (4 × 4)
UAZ-452D 800 kg (4 × 4)
UAZ-3251 800 kg (4 × 4)

Trucks
GAZ-51 2,000 kg (4 × 2)
GAZ-63 2,000 kg (4 × 4)
GAZ-66 2,000 kg (4 × 4)
ZIL-157 2,500 kg (4 × 4)
ZIL-151 2,500 kg (4 × 4)
GAZ-53 3,000 kg (4 × 2)
ZIL-131 3,500 kg (6 × 6)
Ural-375 4,000 kg (6 × 6)
Ural-4320 4,500 kg (6 × 6)
Ural-377 7,500 kg (6 × 6)
YaAZ-214 7,000 kg (6 × 6)
KrAZ-214 7,000 kg (6 × 6)
KrAZ-255B 7,500 kg (6 × 6)
KamAZ-5320 8,000 kg (6 × 4)
KrAZ-260 9,000 kg (6 × 6)
KrAZ-257 12,000 kg (6 × 6)
ZIL-135 (8 × 8)
MAZ-543 (8 × 8)

CONGO

Light vehicles
Auverland/SAMO (4 × 4)
UMM (4 × 4) light vehicle

CUBA

Recovery vehicles
T-54/T-55 ARV

Mechanised bridges
MTU-20 AVLB
TMM (6 × 6)
KMM (6 × 6)

Light vehicles
UAZ-469B 600 kg (4 × 4)
GAZ-69/69A 500 kg (4 × 4)

Trucks
GAZ-51 2,000 kg (4 × 2)
GAZ-63 2,000 kg (4 × 4)
GAZ-66 2,000 kg (4 × 4)
ZIL-157 2,500 kg (6 × 6)
ZIL-151 2,500 kg (6 × 6)

CYPRUS

Recovery vehicles
AMX-30D ARV (2) (1988/1990)
Steyr Type 19S25 (4 × 4) RV
Steyr Type 32S29 (6 × 6) RV

Trucks
MB Unimog
SDP 14 M 14 8,000 kg (4 × 4)
Steyr 24 M 12,000 kg (6 × 6)

CZECH REPUBLIC

Recovery vehicles
VT-55A ARV
MT-55 ARV
WPT-TOPAS ARV (OT-62A chassis)
VPV ARV (BMP-1 chassis)
AD-90 (6 × 6)

Mechanised bridges
MT-55A AVLB
AM-50 (6 × 6)

Light vehicles
UAZ-469B 600 kg (4 × 4)

Trucks
Praga V3S 3,000 kg (6 × 6)
BAZ SNA 4,000 kg (6 × 6)
ZIL-131 3,500 kg (6 × 6)
Ural-375D 4,000 kg (6 × 6)
Tatra 813 7,900 kg (8 × 8)
Tatra 815 8,000 kg (8 × 8)
Tatra 148 14,580 kg (6 × 6)
Tatra 138 11,850 kg (6 × 6)

DENMARK

Recovery vehicles
M578 ARV

Mechanised bridges
Biber AVLB (10)

Light vehicles
VW 181 400 kg (4 × 2)
MB 750 kg (4 × 4) (1,300)

Trucks
Bedford RL 4,000 kg (4 × 4)
MB Unimog (4 × 4)
MAN 8.136 FAE 3,000 kg (4 × 4) (725) (1985-90)
IVECO Magirus 110-16 AWM 5,000 kg (4 × 4)

DJIBOUTI

Trucks
ACMAT (4 × 4)
HMMWV 1,542 kg (4 × 4) (18)
M34/M44 2,268 kg (6 × 6)
M929A1 4,536 kg (6 × 6) (89)

DOMINICAN REPUBLIC

Trucks
M34/M44 2,268 kg (6 × 6)

ECUADOR

Light vehicles
M151 362 kg (4 × 4)

Trucks
M34/M44 2,268 kg (6 × 6)
HMMWV (4 × 4) (95)

EGYPT

Recovery vehicles
T-54/T-55 ARV
M88A1 ARV (221)
M113 ARV
M578 ARV (43)
BMR 3560.55 ARV (6 × 6)

Mechanised bridges
MTU-20 AVLB
MTU AVLB
KMM (6 × 6)

Light vehicles
M151 (4 × 4)
Jeep (CJ & YJ) (4 × 4)
Santana Model 88 500 kg (4 × 4)
Santana Model 109 1,000 kg (4 × 4)
UAZ-469B 600 kg (4 × 4)
GAZ-69/69A 500 kg (4 × 4)
UAZ-452D 800 kg (4 × 4)
Land Rover 1,000 kg (4 × 4)
Toyota Land Cruiser (4 × 4)

Trucks
HMMWV (4 × 4) (681)
Jeep AM 720 1,135 kg (4 × 4)
Jeep J-20 (4 × 4)
MB Unimog (4 × 4)
GAZ-66 2,000 kg (4 × 4)
ZIL-131 3,500 kg (6 × 6)
Ural-375D 4,000 kg (6 × 6)
NASR 4,000 kg (4 × 4)
RVI TRM 9000 9,000 kg (6 × 6) (50)
Pegaso 3046 3,000 kg (4 × 4) (9,000+)
M923A2 4,540 kg (6 × 6)
Steyr 17 M 29 7,000 kg (4 × 4)
Oshkosh HEMTT 9,979 kg (8 × 8)
Steyr 24 M 12,000 kg (6 × 6)

EL SALVADOR

Light vehicles
M37 680 kg (4 × 4)

Trucks
M35/M44 2,268 kg (6 × 6)
M809 4,536 kg (6 × 6)

EQUATORIAL GUINEA

Recovery vehicles
T-54/T-55 ARV (?)

ETHIOPIA

Trucks
ZIL-157 2,500 kg (6 × 6)
ZIL-151 2,500 kg (6 × 6)

FINLAND

Recovery vehicles
T-54/T-55 ARV

Mechanised bridges
MTU-20 AVLB
MTU AVLB

Light vehicles
GAZ-69/GAZ-69A 500 kg (4 × 4)

Trucks
GAZ-66 2,000 kg (4 × 4)
ZIL-131 3,500 kg (6 × 6)
SISU A-45 3,000 kg (4 × 4)
SISU SA-130 VK 6,500 kg (4 × 4)
SISU SA-241 10,000 kg (6 × 6)
Scania SBA 111 4,500 kg (4 × 4) (33)

FRANCE

Armoured engineer vehicles
AMX-30 EBG (71)

Recovery vehicles
AMX-30D ARV (134)
TRM 10,000 CLD RV (6 × 6)
Berliet TBC 8 KT W (6 × 6)
Berliet TBU 15 CLD W (6 × 6)
Brimot TASE RV (4 × 4)

Mechanised bridges
AMX-13 AVLB
PAA (4 × 4) (55) (1974-78)
Gillois Series 2 (60)
Gillois Series 1 (4 × 4) (PO)
EFA (4 × 4) (80)

Light vehicles
LOHR FL 500 500 kg (4 × 4) (300)
Citroën Méhari (4 × 2)
Peugeot P4 750 kg (4 × 4) (12,500) (1983-92)
VW 181 400 kg (4 × 2)
Auverland Type A3 (4 × 4)
UMM (4 × 4) light vehicle

Trucks
Peugeot 504 Dangel 1,110 kg (4 × 4)
ACMAT (4 × 4)
RVI TRM 1200 1,200 kg (4 × 4) (4,000+)
SUMB MH 600 BS 1,500 kg (4 × 4)
SUMB MH 600 BS 3,000 kg (4 × 4)
Simca-Unic F 594 WML 3,000 kg (4 × 4)
RVI TRM 2000 2,000 kg (4 × 4) (12,000, total requirement)
RVI TRM 4000 4,000 kg (4 × 4) (7,000)
Berliet GBC 8 KT 4,000 kg (6 × 6)
Berliet GBU 15 6,000 kg (6 × 6)
RVI TRM 10,000 10,000 kg (5,000, total requirement)
MB Unimog
ACMAT 6,000 kg (8 × 8)

GABON

Light vehicles
Peugeot P4 750 kg (4 × 4)

Trucks
ACMAT (4 × 4)
ENGESA EE-15 1,500 kg (4 × 4)
ENGESA EE-25 2,500 kg (6 × 6)

GAMBIA, THE

Light vehicles
M151 (4 × 4)

Trucks
ACMAT (4 × 4)

GERMANY

Armoured engineer vehicles
Pionierpanzer 2 Dachs (137)
Leopard 1 AEV (36)

Recovery vehicles
Büffel ARV (75)
Leopard 1 ARV (440)
M88A1 ARV (2)
Faun RTF 15 M RV (4 × 4)
Faun RTF 30-3 M RV (6 × 6)

Mechanised bridges
Biber AVLB (104)
M48 AVLB
M3 (4 × 4) (430) (ES)
M2 (4 × 4)

Light vehicles
VW 181 400 kg (4 × 2)
VW Iltis 500 kg (4 × 4) (8,470)
MB 750 kg (4 × 4) (12,000)

Trucks
MB Unimog 2,000 kg (4 × 4) (17,000)
MB L 508 PG MA 2 150 kg (4 × 2)
Magirus 5,000 kg (7,000)
MB Model 1017 5,450 kg (4 × 2)
MB Model 1017 5,450 kg (4 × 4)
MAN 5,000 kg (4 × 4) }
MAN 7,000 kg (6 × 6) } (8,615)
MAN 10,000 kg (8 × 8) }
MAN 7,000 kg (4 × 2) (1,500)
MAN 10,000 kg (6 × 4) (2,500)
MAN 6300 5,000 kg (4 × 4) (PO)

GHANA

Recovery vehicles
Steyr Type 19S25 (4 × 4) RV
Steyr Type 32S29 (6 × 6) RV

Trucks
SDP Pinzgauer
Steyr 17 M 29 7,000 kg (4 × 4)

GREECE

Recovery vehicles
Leopard 1 ARV (5)
M88A1 ARV (66)
AMX-30D ARV (14)

Light vehicles
MB 750 kg (4 × 4) (5,000)
Namco Pony (4 × 2)
M151 362 kg (4 × 4)
M37 680 kg (4 × 4)

Trucks
M35/M44 2,268 kg (6 × 6)
SDP 680 M 4,500 kg (4 × 4)
SDP 680 M3 3,500 kg (6 × 6)
SDP 14 M 14 8,000 kg (4 × 4)
Scania P113 HK (6 × 6)

GRENADA

Trucks
CUCV (4 × 4)

GUATEMALA

Light vehicles
M151 362 kg (4 × 4)
MB 750 kg (4 × 4) (50)

Trucks
M35/M44 2,268 kg (6 × 6)

GUYANA

Light vehicles
Land Rover ALV 564 kg (4 × 4)

HAITI

Recovery vehicles
V-150 ARV (4 × 4)

Trucks
M715 1,360 kg (4 × 4)
M35/M44 2,268 kg (6 × 6)

HONDURAS

Trucks
CUCV (4 × 4)
M35/M44 2,268 kg (4 × 4)
M809 4,536 kg (6 × 6)
HMMWV (4 × 4) (125)

HUNGARY

Recovery vehicles
T-54/T-55 ARV
BMP-1VPV

Mechanised bridges
BLG-60 AVLB
MTU AVLB
TMM (6 × 6)

Light vehicles
UAZ-469B 600 kg (4 × 4)

Trucks
Robur 1800 A 1,800 kg (4 × 4)
Csepel D-566 5,000 kg (6 × 6)
Csepel D-564 4,000 kg (4 × 4)
IFA L60 3,000 kg (4 × 4)
IFA W50 3,000 kg (4 × 4)
Ural-375 4,000 kg (6 × 6)
Ural-4320 4,500 kg (6 × 6)
Tatra 148 14,580 kg (6 × 6)
DAC 665T 5,000 kg (6 × 6)

INDIA

Armoured engineer vehicles
Combat Engineer Tractor (15)

Recovery vehicles
T-54/T-55 ARV
VT-72B ARV
Vijayanta

Mechanised bridges
MT-55 AVLB
BLG-60 AVLB
MTU-20 AVLB
Kartik AVLB
AM-50 (6 × 6)

Light vehicles
Mahindra (4 × 4)
Jonga (4 × 4)
Carrier 750 kg (4 × 4)

Trucks
MAN 630 5,000 kg (4 × 4)
Shaktiman 4,000 kg (4 × 4)
Scania SBAT 111S 6,000 kg (6 × 6) (600)
Tatra 813 7,900 kg (8 × 8)
Tatra 815 8,000 kg (8 × 8)

INDONESIA

Recovery vehicles
AMX-13 ARV (2)
T-54/T-55 ARV
Stormer ARV

Mechanised bridges
AMX-13 AVLB
Alvis Stormer AVLB

Light vehicles
SDP 700 AP (4 × 4) (1,000)
Nissan Q4W73 750 kg (4 × 4) (600)
DAF 66 YA 400 kg (4 × 2) (150)
Land Rover ALV 564 kg (4 × 4) (812)
M151 362 kg (4 × 4)

Trucks
Leyland 4,000 kg (4 × 4) (14)
MB Unimog (4 × 4)
SDP 680 M 4,500 kg (4 × 4) (750)
Bedford MK 4,000 kg (4 × 4)
Steyr 17 M 29 7,000 kg (4 × 4) (200+)

IRAN

Recovery vehicles
Chieftain ARV
M578 ARV

Mechanised bridges
Chieftain AVLB (14)

Light vehicles
UAZ-469B 600 kg (4 × 4)
Land Rover (4 × 4)
Land Rover 1,000 kg (4 × 4)

Trucks
GAZ-66 2,000 kg (4 × 4)
M34/M44 2,268 kg (6 × 6)
Volvo NL10 (4 × 4)
MB (4 × 4)

IRAQ

Recovery vehicles
Type 653 ARV
T-54/T-55 ARV
Chieftain ARV
ENGESA EE-11 repair and recovery (6 × 6)

Mechanised bridges
MT-55 AVLB
BLG-60 AVLB

Light vehicles
Land Rover (4 × 4)
UAZ-469B 600 kg (4 × 4)

Trucks
GAZ-66 2,000 kg (4 × 4)
Ural-375D 4,000 kg (6 × 6)
MB Unimog (4 × 4)
Scania (4 × 4)
Berliet GBC 8KT 4,000 kg (6 × 6)
W 50 LA/A 3,000 kg (4 × 4)
L 60 LA/PVB 6,200 kg (4 × 4)

IRELAND

Recovery vehicles
Berliet TBU 15 CLD W (6 × 6)

Light vehicles
Land Rover (4 × 4)
Land Rover Defender 110 (4 × 4)

Trucks
ACMAT (4 × 4)
Leyland 4,000 kg (4 × 4) (2)
Bedford MK 4,000 kg (4 × 4)
Bedford RL 4,000 kg (4 × 4)
MAN 6,000 kg (4 × 4)

ISRAEL

Recovery vehicles
T-54/T-55 ARV
M88A1 ARV (25)
Centurion Mk 2 ARV

Mechanised bridges
MTU-20 AVLB
MTU AVLB
MAB/F (4 × 4)
M48 AVLB

Light vehicles
M151 362 kg (4 × 4)
M-242 540 kg (6 × 6)

Trucks
HMMWV (4 × 4) (16)
M-462 ABIR 1,800 kg (4 × 4)
M-325 Commandcar 1,800 kg (4 × 4)
CUCV (4 × 4)
M715 1,360 kg (4 × 4)
M35/M44 2,268 kg (6 × 6)
M54 4,536 kg (6 × 6)
Oshkosh HEMTT 9,979 kg (8 × 8)

ITALY

Armoured engineer vehicles
Leopard 1 AEV (40)
M113 Combat Engineer Support Vehicle

Recovery vehicles
Leopard 1 ARV (69)
IVECO 6605 RV (6 × 6)

Mechanised bridges
Biber AVLB (64)

Light vehicles
FIAT AR-59 500 kg (4 × 4)
SDP 700 AP (4 × 4)
Fresia F18 400 kg (4 × 4)
FIAT Campagnola 1107AD 750 kg (4 × 4)

Trucks
Leoncino 1,000 kg (4 × 4)
CL 51 1,800 kg (4 × 4)
IVECO 40-10 WM 1,500 kg (4 × 4) (1,300+)
IVECO 75-14 WM 2,500 kg (4 × 4)
IVECO 90-17 WM 4,000 kg (4 × 4)
IVECO 6601 4,000 kg (4 × 4)
IVECO 6602 6,000 kg (4 × 4)
IVECO 6607 CM 6,000 kg (6 × 6)
IVECO 6605 (6 × 6)

JAMAICA

Light vehicles
Land Rover ALV 564 kg (4 × 4)

Trucks
CUCV (4 × 4)
M809 4,536 kg (6 × 6)

JAPAN

Armoured engineer vehicles
Type 67 AEV
Type 75 armoured dozer

Recovery vehicles
Type 90 ARV
Type 78 ARV
Type 70 ARV
Type 73 RV (6 × 6)
Model FW415M1 RV (6 × 6)
Type 73 light wrecker (6 × 6)

Mechanised bridges
Type 91 AVLB
Type 67 AVLB
Type 70 pontoon bridge (4 × 4)
Type 81 (6 × 6)

Light vehicles
Mitsubishi (4 × 4)
Nissan Q4W73 750 kg (4 × 4)
Toyota 2FQ15L 750 kg (4 × 4)

Trucks
Type 73 2,000 kg (4 × 4)
Isuzu 2,500 kg (4 × 4)
Isuzu 2,500 kg (6 × 6)
Toyota 2,500 kg (6 × 6)
Type 73 3,500 kg (6 × 6)
Hino 4,000 kg (6 × 6)
Mitsubishi W121P 6,000 kg (6 × 6)
Mitsubishi FW415 7,000 kg (6 × 6)
Mitsubishi FW115 10,000 kg (6 × 6)
Type 74 10,000 kg (6 × 6)

JORDAN

Recovery vehicles
Chieftain ARV
Centurion Mk 2 ARV
M578 ARV
M88A1 ARV (32)

Light vehicles
Land Rover (4 × 4)

Trucks
M35/M44 2,268 kg (6 × 6)
M809 4,536 kg (6 × 6)
HMMWV (4 × 4) (129)

KENYA

Recovery vehicles
Vickers ARV (7)

Light vehicles
Land Rover (4 × 4)

Trucks
MB Unimog (4 × 4)
Bedford MK 4,000 kg (4 × 4)

KOREA, NORTH

Recovery vehicles
T-54/T-55 ARV

Light vehicles
UAZ-469A 600 kg (4 × 4)

Trucks
Victory-58 (4 × 4)
Isuzu (6 × 6)

KOREA, SOUTH

Armoured engineer vehicles
M9 armoured combat earthmover

Recovery vehicles
M47 ARV
K-1 ARV (200 on order)
M88A1 ARV (38)
K288A1 ARV
AAVR7 ARV (3) (Marines)
KM502 W (6 × 6)
M816 W (6 × 6)

Mechanised bridges
K-1 AVLB (56) (ordered 1993)

Light vehicles
M151 362 kg (4 × 4)
Nissan Q4W73 750 kg (4 × 4)
Toyota 2FQ15L 750 kg (4 × 4)
KM41 series (4 × 4)

Trucks
KM45 family 1,530 kg (4 × 4)
KM25 family 2,500 kg (6 × 6)
KM50 family 5,000 kg (6 × 6)
M35/M44 2,268 kg (6 × 6)
M54 4,536 kg (6 × 6)
M809 4,536 kg (6 × 6)
Oshkosh HEMTT 9,979 kg (8 × 8)

KUWAIT

Exact status of Kuwaiti vehicles is uncertain.

Recovery vehicles
M88A1 (10)
M88 IRV (14)
Warrior mechanised recovery vehicle
Warrior mechanised combat repair vehicle

Light vehicles
Jeep (YJ) (4 × 4)
MB 750 kg (4 × 4) (400)
Land Rover Defender (4 × 4)

Trucks
MB Unimog (4 × 4)
Steyr 5,000 kg (4 × 4)
Volvo FL6 5,000 kg (4 × 4)
HMMWV (4 × 4) (770)

LEBANON

Recovery vehicles
AMX-13 ARV (18)
M113 ARV

Light vehicles
Land Rover (4 × 4)

Trucks
M35/M44 2,268 kg (6 × 6)
M809 4,536 kg (6 × 6)

LIBERIA

Light vehicles
M151 362 kg (4 × 4)

Trucks
CUCV (4 × 4)
M35/M44 2,268 kg (6 × 6)
M809 4,536 kg (6 × 6)

LIBYA

Recovery vehicles
T-54/T-55 ARV

Light vehicles
UAZ-469B 600 kg (4 × 4)
Land Rover (4 × 4)

Trucks
GAZ-66 2,000 kg (4 × 4)
ENGESA EE-25 2,500 kg (6 × 6)
Ural-375D 4,000 kg (6 × 6)
MB Unimog
IVECO 6605 (6 × 6)

LUXEMBOURG

Light vehicles
MB 750 kg (4 × 4) (38)
Land Rover 1,000 kg (4 × 4) (57)

Trucks
HMMWV 1,542 kg (4 × 4) (69)
MAN (4 × 4 and 6 × 6)

MADAGASCAR

Light vehicles
Auverland/SAMO (4 × 4)

MALAYSIA

Recovery vehicles
SIBMAS ARV (6 × 6) (24)
Volvo F10 RV (4 × 4) (30)

Light vehicles
Land Rover (4 × 4)

Trucks
Volvo 4140 (4 × 4)
Volvo 4140 (6 × 6)
Stonefield (4 × 4) (1983-86)
Bedford RL 4,000 kg (4 × 4)
MB (4 × 4)
Leyland 4,000 kg (4 × 4) (40)

MALTA

Light vehicles
Land Rover (4 × 4)

Trucks
Bedford RL 4,000 kg (4 × 4)

MEXICO

Trucks
HMMWV (4 × 4) (1838)
M561 (6 × 6)

MONGOLIA

Recovery vehicles
T-54/T-55 ARV

Light vehicles
UAZ-469 600 kg (4 × 4)

Trucks
Ural-375D 4,000 kg (6 × 6)

MOROCCO

Recovery vehicles
SDP 4KH 7FA SB 20 Greif ARV (11)
M578 ARV
M88A1 ARV (18)

Light vehicles
Santana Model 88 500 kg (4 × 4)
Santana Model 109 1,000 kg (4 × 4)

Trucks
ACMAT (4 × 4)
RVI TRM 1200 1,200 kg (4 × 4)
RVI TRM 2000 2,000 kg (4 × 4) (190)
RVI TRM 9000 9,000 kg (6 × 6) (1,500)
Berliet GBC 8 KT 4,000 kg (6 × 6)
MB Unimog (4 × 4)
Pegaso 3046 3,000 kg (4 × 4)
Pegaso 7323 6,000 kg (6 × 6)
M35/M44 2,268 kg (6 × 6)
ACMAT 6,000 kg (8 × 8)
Volvo NL10 (4 × 4)

MOZAMBIQUE

Light vehicles
UAZ-469 600 kg (4 × 4)
Jonga (4 × 4) (200)

Trucks
MB Unimog (4 × 4)
Ural-375D 4,000 kg (6 × 6)
SAMIL 20 2,000 kg (4 × 4) (4)
SAMIL 50 5,000 kg (4 × 4) (30)

NAMIBIA

Trucks
TRM 2000 2,000 kg (4 × 4) (50)

NETHERLANDS

Armoured engineer vehicles
Leopard 1 AEV (25)

Recovery vehicles
Leopard 1 ARV (37)
Büffel ARV (25)
M113 ARV
M578 ARV
DAF YBZ 3300 DKX (6 × 6) HRV (255) (1991-93)

Mechanised bridges
Biber AVLB (17)

Light vehicles
VW 181 400 kg (4 × 2)
MB 750 kg (4 × 4) (3,195)
DAF 66 YA 400 kg (4 × 2) (PO)
Land Rover ALV 564 kg (4 × 4)
Land Rover Defender 110 (4 × 4)
M151 362 kg (4 × 4)
Lohr VLA (180)

Trucks
DAF YA 314 3,000 kg (4 × 4) (PO)
DAF YA 4442 DNT 4,000 kg (4 × 4) (5,125) (1986-93)
DAF YA 4440 4,000 kg (4 × 4) (7,250) (1977-83)
DAF YA 5444 DNT 5,000 kg (4 × 4)
DAF YA 5441/5,442 5,000 kg (4 × 4)
DAF YA 2300 DHTD 7,000 kg (6 × 6) (65) (NATO)
DAF YAZ 2300 10,000 kg (6 × 6) (1,400+)
Bedford MK 4,000 kg (4 × 4)

NEW ZEALAND

Light vehicles
Land Rover V8 750 kg (4 × 4) (567)

Trucks
MB U 1300 L 1,500 kg (4 × 4) (210)
MB UL 1700 L 4,000 kg (4 × 4) (412)
MB 2228/4 8,000 kg (6 × 4) (118)

NICARAGUA

Recovery vehicles
T-54/T-55 ARV

Light vehicles
UAZ-469 600 kg (4 × 4)

Trucks
Pegaso 3045 3,000 kg (4 × 4)
Ural-375D 4,000 kg (4 × 4)

NIGERIA

Recovery vehicles
SDP 4KH 7FA SB 20 Greif ARV (15)
T-54/T-55 ARV (?)
Vickers ARV (15)
Steyr Type 19S25 (4 × 4) RV
Steyr Type 32S29 (6 × 6) RV

Mechanised bridges
MTU-20 AVLB
Vickers AVLB (18)

Light vehicles
Land Rover (4 × 4)
SDP 700 AP (4 × 4)

Trucks
SDP Pinzgauer
SDP 680 M 4,500 kg (4 × 4)
MB Unimog (4 × 4)
Steyr 14 M 14 8,000 kg (4 × 4)

NORWAY

Armoured engineer vehicles
Leopard AEV (22) (on order)

Recovery vehicles
Leopard 1 ARV (14)
M88A1 ARV (3)
M578 ARV
Scania P113 HK (6 × 6)

Mechanised bridges
Leguan (8 × 8) (26)
Leopard Leguan (9) (on order)

Light vehicles
Land Rover (4 × 4)
MB 750 kg (4 × 4) (3,500)

Trucks
Scania (4 × 4/6 × 6) (1,700+) (1986-95)

OMAN

Recovery vehicles
Challenger ARV (4)
Samson (3)
M88A1 (2)
MOWAG Piranha ARV (8 × 8)
Reynolds Boughton 6 t (4 × 4)

Light vehicles
Jeep (CJ) (4 × 4)
Iltis 500 kg (4 × 4)
MB 750 kg (4 × 4)

Trucks
SDP Pinzgauer (6 × 6) (95)
MAN 6000/10,000 kg (95)
Bedford TM 4-4 8,000 kg (4 × 4)

PAKISTAN

Recovery vehicles
Type 653 ARV
T-54/T-55 ARV
M88A1 ARV (52)

Mechanised bridges
M48 AVLB
M47M AVLB (trials)

Light vehicles
Beijing BJ-212/212A (4 × 4)
Nispak 400 kg (4 × 4)
M151 362 kg (4 × 4) (1,000+)

Trucks
MB Unimog (4 × 4)
IVECO 40-10 WM 1,500 kg (4 × 4) (2,200)
M880 series 1,133 kg (4 × 4 and 4 × 2)
M35/M44 2,268 kg (6 × 6)
RL 4,000 kg (4 × 4)
M809 4,536 kg (6 × 6)
MB 2028A 10,000 kg (6 × 6)
Steyr 5,000 kg (4 × 4)

PANAMA

Trucks
CUCV (4 × 4)
M35/M44 2,268 kg (6 × 6)
M809 4,536 kg (6 × 6)

PAPUA NEW GUINEA

Light vehicles
Land Rover (4 × 4)

PERU

Recovery vehicles
T-54/T-55 ARV (?)

Light vehicles
M151 362 kg (4 × 4)

Trucks
MB Unimog (4 × 4)
Pegaso 3046 3,000 kg (4 × 4)
MAN 6000/10,000 kg (165)
Volvo NL10 (4 × 4)

PHILIPPINES

Recovery vehicles
M578 ARV
M816 W (6 × 6) (36)

Light vehicles
Delta Explorer (4 × 4)
M151 362 kg (4 × 4)

Trucks
CM-125 1,250 kg (4 × 4)
HMMWV 1,542 kg (4 × 4) (440)
M35/M44 2,268 kg (6 × 6)
M809 4,536 kg (6 × 6)

POLAND

Armoured engineer vehicles
IWT CEV (T-55 chassis)
MT-LB engineer reconnaissance vehicle

Recovery vehicles
T-54/T-55 ARV
WZT-3 ARV
WPT-TOPAS ARV
MT-LB technical support vehicle

Mechanised bridges
BLG-67M2 AVLB
SMT-1 (6 × 6)

Light vehicles
UAZ-468B 600 kg (4 × 4)

Trucks
Lubin-51 (4 × 2)
Robur LO 1800 A 1,800 kg (4 × 4)
Star 66 2,500 kg (6 × 6)
Star 266 3,500 kg (6 × 6)
Star 244 5,000 kg (4 × 4)

PORTUGAL

Recovery vehicles
M88A1 (6)

Light vehicles
Citroën A FAF (4 × 4)
Bravia Comando Mk II 605 kg (4 × 4)
UMM Alter (4 × 4)
UMM 490 1,000 kg (4 × 4)
M151 362 kg (4 × 4)

Trucks
Bravia Gazela 1,134 kg (4 × 4)
MWG 1,500 kg (6 × 6)
Bravia Leopardo 3,000 kg (6 × 6)
IVECO 90-17 WM 4,000 kg (4 × 4)
Bravia Pantera 6,000 kg (6 × 6)
Bravia Elephant 10,000 kg (6 × 6)
DAF YA 4440 4,000 kg (4 × 4) (300) (1984-85)
Berliet GBC 8 KT 4,000 kg (4 × 4)
MB Unimog (4 × 4)

QATAR

Recovery vehicles
AMX-30D ARV (1)
Piranha repair and recovery vehicles (2) (on order)

Light vehicles
Land Rover

Trucks
RVI TRM 2000 2,000 kg (4 × 4) (10)

ROMANIA

Recovery vehicles
T-54/T-55 ARV

Mechanised bridges
MT-55A AVLB
BLG-67M AVLB
BLG-67M2 AVLB

Light vehicles
ARO (4 × 4)
UAZ-469B 600 kg (4 × 4)

Trucks
SR-101 3,500 kg (4 × 2)
SR-132 2,000 kg (6 × 6)
Ural-375D 4,000 kg (6 × 6)
DAC 665 T 5,000 kg (6 × 6)
ROMAN 12135 DFAE 10,000 kg (6 × 6)

SAUDI ARABIA

Armoured engineer vehicles
M728 CEV (15)

Recovery vehicles
AMX-30D ARV (57)
M88A1 ARV (139)
M578 ARV (88)
AMX-10 ECH RV
LAV (8 × 8) (67)
LAV-150 ARV (4 × 4)
SANTA BARBARA BMR 3560.55 (6 × 6) ARV
Steyr Type 19S25 (4 × 4) RV
Steyr Type 32S29 (6 × 6) RV
M984A1 (8 × 8) RV

Light vehicles
Land Rover APV 564 kg (4 × 4)
M151 362 kg (4 × 4)

Mechanised bridges
AMX-30 AVLB (12)

Trucks
MB Unimog (4 × 4)
M35/M44 2,268 kg (6 × 6)
M809 4,536 kg (6 × 6)
HMMWV (4 × 4) (3,829)
Oshkosh HEMTT 9,979 kg (8 × 8)
M939A2 4,536 kg (6 × 6) (2,633) (1991/1992)
RVI TRM 9000 9,000 kg (6 × 6) (38)
RVI TRM 10,000 10,000 kg (6 × 6) (67)
TAM 110 T7 BV 1,500/2,500 kg (4 × 4)
TAM 140 T11 BV 3,000/5,000 kg (6 × 6)
FAP 2026 BS (6 × 6)

SENEGAL

Light vehicles
M151 362 kg (4 × 4)

Trucks
ACMAT (4 × 4)
M35/M44 2,268 kg (6 × 6)
M809 4,536 kg (6 × 6)

SINGAPORE

Armoured engineer vehicles
M728 CEV (8)
Combat Engineer Tractor (18 + 18)

Recovery vehicles
V-200 ARV (4 × 4)
M806A1 ARV
M113A1 fitters

Mechanised bridges
M60 AVLB (12)
M2 system (4 × 4) (36)

Light vehicles
MB 750 kg (4 × 4) (311)
Land Rover 110 (4 × 4)
M151 (4 × 4)

Trucks
MB Unimog (4 × 4)
MB (4 × 4)
MAN 6000/10,000 kg (165)
IVECO 90-17 4,000 kg (4 × 4) (1990/1991)

SLOVAKIA

Recovery vehicles
VT-55A ARV
MT-55 ARV
WPT-TOPAS ARV (OT-62A chassis)
VPV ARV (BMP-1 chassis)
AD-90 (6 × 6)

Mechanised bridges
MT-55A AVLB
AM-50 (6 × 6)

Light vehicles
UAZ-469B 600 kg (4 × 4)

Trucks
Praga V3S 3,000 kg (6 × 6)
BAZ SNA 4,000 kg (6 × 6)
ZIL-131 3,500 kg (6 × 6)
Ural-375D 4,000 kg (6 × 6)
Tatra 813 7,900 kg (8 × 8)
Tatra 815 8,000 kg (8 × 8)
Tatra 148 14,580 kg (6 × 6)
Tatra 138 11,850 kg (6 × 6)

SOMALIA

Light vehicles
M151 362 kg (4 × 4)

Trucks
ACMAT (4 × 4)
M35/M44 2,268 kg (6 × 6)
Pegaso 3046 3,000 kg (4 × 4)
M809 4,536 kg (6 × 6)
IVECO 90-17 WM 4,000 kg (4 × 4) (100)
IVECO 75-14 WM 2,500 kg (4 × 4)
IVECO 6605 (6 × 6) (291)

Owing to the current situation in this country, exact status of equipment is uncertain.

SOUTH AFRICA

Recovery vehicles
Olifant ARV (Centurion chassis)
Centurion ARV
Ratel RV (6 × 6)
SAMIL 50 RV (4 × 4)
SAMIL 100 RV (6 × 6)

Mechanical bridges
Olifant AVLB (28)

Light vehicles
Land Rover (6 × 6)
Mitsubishi (4 × 4)
Toyota Land Cruiser (4 × 4)
Jakkals 350 kg (4 × 4)

Trucks
MB Unimog (4 × 4)
Bedford MK 4,000 kg (4 × 4)
SAKOM 5,000 kg (4 × 2)
SAMIL 20 2,000 kg (4 × 4)
SAMIL 50 5,000 kg (4 × 4)
SAMIL 100 10,000 kg (6 × 6)

SPAIN

Recovery vehicles
AMX-30D ARV (10)
M88A1 ARV (1)
GEMESA M47VR ARV (22) (1996-97)
M578 ARV
AAVR7 (1)
SANTA BARBARA BMR 3560.55 (6 × 6) ARV
M543 W (6 × 6)

Mechanised bridges
M60 AVLB

Light vehicles
LOHR Fardier (4 × 4)
Citroën Méhari Armée (4 × 2)
Santana Model 88 500 kg (4 × 4)
Santana Model 109 1,000 kg (4 × 4)
M151 362 kg (4 × 4)
M37 680 kg (4 × 4)

Trucks
Santana S-2000 2,000 kg (4 × 4)
URO 115 PM 2,000 kg (4 × 4)
Pegaso 3045 3,000 kg (4 × 4)
Pegaso 3046 3,000 kg (4 × 4)
Pegaso 3050 6,000 kg (6 × 6)
Pegaso 3055 6,000 kg (6 × 6)
Pegaso 7323 6,000 kg (6 × 6)
M35/M44 2,268 kg (6 × 6)
M54 4,356 kg (6 × 6)
M809 4,536 kg (6 × 6)

SRI LANKA

Light vehicles
Land Rover (4 × 4)
Mahindra

SUDAN

Recovery vehicles
T-54/T-55 ARV
M113 ARV
M88A1 ARV (2)
M3 repair vehicle (4 × 4)

Light vehicles
Land Rover ALV 564 kg (4 × 4)

Trucks
M35/M44 2,268 kg (6 × 6)
M809 4,536 kg (6 × 6)
SDP Pinzgauer

SWEDEN

Recovery vehicles
Bgbv 82 ARV (24)
Bgbv 81 ARV (Centurion Mk 2)
CV 90 ARV

Mechanised bridges
Brobv AVLB (17)

Light vehicles
Volvo L3304 (4 × 4)
Volvo L3314 Laplander (4 × 4)

Trucks
MB L 508 DG MA 2,105 kg (4 × 2)
Volvo 4140 (4 × 4)
Volvo 4140 (6 × 6)
Volvo 4151 (4 × 4)
Volvo L2204 1,500 kg (6 × 6)
Volvo L3154 3,000 kg (6 × 6)
Scania SBA 111 4,500 kg (4 × 4) } (2,500)
Scania SBAT 111S 6,000 kg (6 × 6) }
Volvo L4854 3,000 kg (6 × 6)
Scania-Vabis L-36A/L-50 5,920 kg (4 × 2) (800+)
Volvo NL10 (4 × 4)
Volvo FL6 5,000 kg (4 × 4)

SWITZERLAND

Recovery vehicles
Entp Pz 65/88 ARV (Pz 68 chassis)

Mechanised bridges
Bru Pz 68/88 AVLB

Light vehicles
SDP 700 AP (4 × 4)
MB 750 kg (4 × 4) (4,100)

Trucks
SDP Pinzgauer (4 × 4 and 6 × 6)
SDP 680 M 4,500 kg (4 × 4)
MAN (4 × 4 and 6 × 6)
Saurer 2DM 4,500 kg (4 × 4)
Berna 2VM 4,500 kg (4 × 4)
Saurer 6 DM 6,000 kg (4 × 4)
Saurer 10 DM 10,000 kg (6 × 6)
DURO (4 × 4 and 6 × 6) (2,000 ordered in 1993 for delivery from 1994 to 1999)

SYRIA

Recovery vehicles
T-54/T-55 ARV

Mechanised bridges
MTU-20 AVLB
MTU AVLB

Light vehicles
UAZ-469B 600 kg (4 × 4)
GAZ-69/69A 500 kg (4 × 4)

Trucks
MB Unimog (4 × 4)
Ural-375D (4 × 4)
GAZ-66 2,000 kg (4 × 4)

TAIWAN

Recovery vehicles
M88A1 ARV (33)

Mechanised bridges
M48 AVLB

Light vehicles
M151 362 kg (4 × 4)

Trucks
CUCV (4 × 4)
HMMWV (4 × 4) (5,429)
M35/M44 2,268 kg (6 × 6)
M809 4,536 kg (6 × 6)
Oshkosh HEMTT 9,979 kg (8 × 8)

TANZANIA

Recovery vehicles
Vickers ARV (3) (1990)

Light vehicles
Land Rover (4 × 4)

Trucks
MB Unimog (4 × 4)

THAILAND

Recovery vehicles
Samson ARV
M88A1 ARV (16)
AAVR7 ARV (1) (Marines)
Type 653 ARV
M113A3 maintenance/recovery (10)

Mechanised bridges
Type 84 AVLB

Light vehicles
Mitsubishi (4 × 4)
M151 362 kg (4 × 4)
M37 680 kg (4 × 4)

Trucks
HMMWV 1,542 kg (4 × 4) (275)
M35/M44 2,268 kg (6 × 6)
M809 4,536 kg (6 × 6)
Dong Feng EQ2080E4DY 2,500 kg (6 × 6) (235)
MB (6 × 4)
Steyr 5,000 kg (4 × 4)
Steyr 14 M 14 8,000 kg (4 × 4)

TOGO

Light vehicles
Peugeot P4 750 kg (4 × 4)

Trucks
ACMAT (4 × 4)

TUNISIA

Recovery vehicles
M88A1 ARV (6)

Light vehicles
FIAT Campagnola 1107 AD 750 kg (4 × 4) (200)
LOHR Fardier (4 × 4)

Trucks
SDP Pinzgauer
MB Unimog (4 × 4)
HMMWV (4 × 4) (341)

TURKEY

Recovery vehicles
Leopard 1 ARV (12)

Light vehicles
Land Rover Defender 110 (4 × 4)
Land Rover Defender 130 (4 × 4)
M151 362 kg (4 × 4)
M37 680 kg (4 × 4)

Trucks
M35/M44 2,268 kg (6 × 6)
M54 4,536 kg (6 × 6)
MB Unimog (4 × 4)
Bedford MK 4,000 kg (4 × 4) (300)
MANAS 16.210 FAE 7,500 kg (4 × 4)
MANAS 26.281 DFAE 13,800 kg (6 × 6)
SISU NA-140

UGANDA

Light vehicles
Land Rover (4 × 4)

Trucks
Bedford MK 4,000 kg (4 × 4)
Steyr 14 M 14 8,000 kg (4 × 4)

UNITED ARAB EMIRATES

Recovery vehicles
OF-40 ARV (4)
AMX-30D ARV (4)
Leclerc ARV (46) (1994-98)
Berliet TBU CLD (6 × 6)

Light vehicles
Land Rover (4 × 4)
MB 750 kg (4 × 4) (25)

Trucks
HMMWV (4 × 4) (4)
Bedford MK 4,000 kg (4 × 4)
Bedford TM 4-4 8,000 kg (4 × 4)
Bedford TM 6-6 14,000 kg (6 × 6)
Bedford TM 30-30 18,000 kg (6 × 6)
Berliet GBU 15 6,000 kg (6 × 6)
MB 2028 A 10,000 kg (6 × 6)
MB Unimog (4 × 4)
IVECO 230-35 WM 10,000 kg (6 × 6)
Tatra T815 (8 × 8) (1997 onwards)

UNITED KINGDOM

Armoured engineer vehicles
Combat Engineer Tractor (141)
Chieftain AVRE (48) (1992-94)

Recovery vehicles
Challenger ARV (80) (1990-93)
Chieftain ARV
M578 ARV
Samson ARV
Saxon ARV (4 × 4) (38)
FV432 ARV
FV434 armoured repair vehicle
Warrior mechanised combat repair vehicle (101)
Warrior mechanised recovery vehicle (39)
Reynolds Boughton 6 t RV (4 × 4)
Foden RV (6 × 6) (333)
Scammell Crusader RV (6 × 4) (130)

Mechanised bridges
Chieftain AVLB (48)
M3 ABFS (4 × 4) (38) (ES)
M2 ABFS (4 × 4)
BR90 (ES)(1995-98)

Light vehicles
SUPACAT Mk 2 1,000 kg (4 × 4) (50)
Land Rover Defender 90 (4 × 4)
Land Rover ALV 564 kg (4 × 4)
Land Rover Defender 110 (4 × 4)
Land Rover (4 × 4) (8,000) (1997-2001)

Trucks
MB Unimog (4 × 4)
Reynolds Boughton RB-44 2,000 kg (4 × 4) (840)
SDP Pinzgauer (4 × 4) (400 +) (1994-96)
Leyland 4,000 kg (4 × 4) (4,500) (1990-95)
Bedford MK 4,000 kg (4 × 4)
Foden low mobility (6 × 4 and 8 × 4) (1,275)
Foden medium mobility (6 × 6) (116)
Bedford TM 6-6 14,000 kg (1,045) (1986-90)
Bedford TM 4-4 8,000 kg (4 × 4) (2,099)
Leyland DROPS (8 × 6) (1,648)
(1989-96)
Foden DROPS (8 × 6) (400)
1994-98

UNITED STATES OF AMERICA

Armoured engineer vehicles
M728 CEV
M9 armored combat earthmover (434)

Recovery vehicles
M578 ARV
AAVR7 ARV (54) (Marines)
M88A1 ARV (81) (Marines)
M88A1 ARV (2,470) (Army)
M88 IRV (52) (on order)
LAV repair and recovery (45) (Marines)
M60/M108 W (6 × 6)
M62/M246/M543 W (6 × 6)
M816/M819 W (6 × 6)
M984A1 RV (8 × 8)

Mechanised bridges
M48/M60 AVLB
M1 based HAB (2 + 6) (trials)

Light vehicles
M151 362 kg (4 × 4)
Fast attack vehicle (4 × 2)
MB 750 kg (4 × 4) (480)

Trucks
CUCV (4 × 4) (60,000)
M715 1,360 kg (4 × 4) (PO)
M880 1,133 kg (4 × 4, 4 × 2) (PO)
HMMWV (4 × 4) (120,000)
LMTV 2,268 kg (4 × 4)
MTV 4,536 kg (6 × 6)
M35/M44 2,268 kg (6 × 6)
M54 4,536 kg (6 × 6)
M809 4,536 kg (6 × 6)
M939 4,536 kg (6 × 6) (17,092)
Oshkosh MK48 9,072 kg (8 × 8) (1,700+) (Marines)
Oshkosh HEMTT 9,979 kg (8 × 8) (13,000)

VENEZUELA

Recovery vehicles
AMX-30D ARV (4)
Samson ARV
AAVR7 ARV (1) (Marines)
IVECO 90 PM 16 W (4 × 4) (60)

Light vehicles
M151 362 kg (4 × 4)

Trucks
M35/M44 2,268 kg (6 × 6)
MAN 6000/10,000 kg (405)

VIETNAM

Recovery vehicles
T-54/T-55 ARV

Light vehicles
UAZ-469B 600 kg (4 × 4)
UAZ-69/69A 500 kg (4 × 4)

Trucks
ZIL-157 2,500 kg (6 × 6)
ZIL-151 2,500 kg (6 × 6)

YEMEN

Recovery vehicles
T-54/T-55 ARV (?)

Light vehicles
Land Rover (4 × 4)
UAZ-469B 600 kg (4 × 4)
UAZ-69/69A 500 kg (4 × 4)

Owing to the current situation in this country, exact status of equipment is uncertain.

ZAÏRE

Recovery vehicles
M816 W (6 × 6)

Light vehicles
Auverland/SAMO (4 × 4)
M151 362 kg (6 × 6)

Trucks
ACMAT
M35/M44 2,268 kg (6 × 6)
M809 4,536 kg (6 × 6)

YUGOSLAVIA (SERBIA AND MONTENEGRO)

Armoured engineer vehicles
IWT CEV (equivalent to Russian IMR)

Recovery vehicles
T-54/T-55 ARV

Mechanised bridges
MT-55 AVLB
TMM (6 × 6)

Light vehicles
MB 750 kg (4 × 4) (500)
Zastava AR-51 500 kg (4 × 4)
FIAT Campagnola 1,107 AD 750 kg (4 × 4)
IMR A-0.75 750/1,000 kg (4 × 4)

Trucks
SDP Pinzgauer
IMR TATA 1,200 kg (4 × 4)
TAM 110 T7 BV 1,500/2,500 kg (4 × 4)
TAM 150 T11 BV 3,000/5,000 kg (6 × 6)
TAM 4500 4,500 kg (4 × 4)
TAM 5000 5,000 kg (4 × 4)
TAM 5500 5,500 kg (4 × 4)
TAM 6500 6,500 kg (4 × 4)
FAP 2220 BDS (6 × 4)
FAP 2026 BS/AV 10,000 kg (6 × 6)
FAP 2832 BS/AV 9,000 kg (6 × 6)

Owing to the current situation in this country, exact status of equipment is uncertain.

ZAMBIA

Recovery vehicles
T-54/T-55 ARV (?)

Light vehicles
MB 750 kg (4 × 4) (50)
Land Rover (4 × 4)

ZIMBABWE

Recovery vehicles
T-54/T-55 ARV (?)

Light vehicles
Land Rover (4 × 4)

Trucks
Bedford MK 4,000 kg (4 × 4)

ADDENDA

Dornier TransHospital mobile field hospital

Description

Dornier GmbH (Friedrichshafen), a subsidiary of Daimler-Benz Aerospace AG (Dasa/Munich), is the prime contractor for the TransHospital mobile field hospital. It is being developed for the German Armed Forces in support of various international organisations involved in humanitarian disaster and other relief operations. The first complete TransHospital was planned for 1997.

The TransHospital is based on a modular system consisting of laterally expandable containers. The complete hospital is equipped to the same standard as a general hospital with a capacity for 200 patients. It contains modules dedicated to specific medical tasks such as emergency ward, diagnostics, specialist sections, surgery, intensive care, as well as sterilisation, laboratory and pharmacy modules. Also provided are all necessary services, including electrical power supplies, water supply and disposal, laundry, kitchen, workshop and satellite communication services.

The smallest unit is an emergency ward consisting of 22 containers. A complete modular field hospital is planned to include 62 containers. All containers can be easily combined with tents.

Associated with Dornier GmbH in the production of the TransHospital are BINZ Ambulance- und Umweittechnik GmbH of Ilmenau/Thue (equipment integration), Robert Bosch GmbH of Möglingen (electrical power supplies), Drägerwerk AG of Essen (medical equipments), M Schall & Co KG of Merzenich Girbelsrath (containers and tents) and Weiss Umwelttechnik of Reiskirchen-Lindensruth (hygiene air conditioning).

Elements of a Dornier TransHospital mobile field hospital ***1997***

Status

In production for the German Armed Forces.

Manufacturer

Daimler-Benz Aerospace AG, Defense and Civil Systems Group, D-81663 München, Germany.
Tel: +89 6 07-3 42 58. Fax: +89 07-3 42 87.
Daimler-Benz Aerospace AG, Dornier Mobile Systems, An der Bundesstrasse 31, D-88039 Friedrichshafen, Germany.
Tel: +49 7545 8-9138. Fax: +49 7545 8-2034.

NEW ENTRY

Elements of a Dornier TransHospital mobile field hospital ***1997***

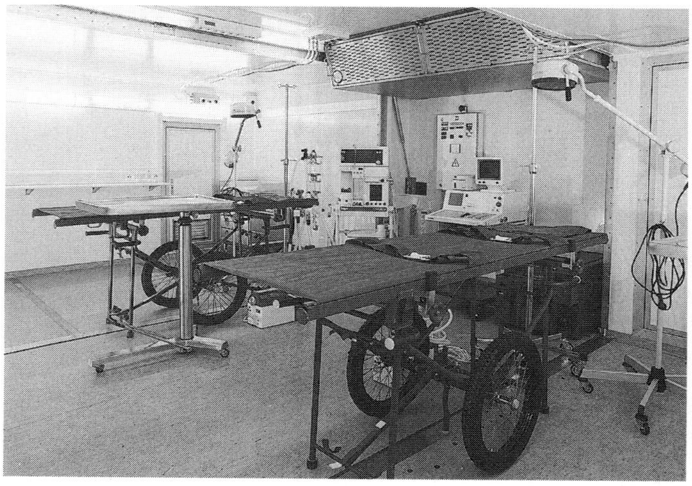

Interior of a typical Dornier TransHospital mobile field hospital ***1997***

Manufacturers Index

C

D

Alphabetical Index

A

B

G

H

I

J